FORMULAS FROM GEOMETRY

Area A; circumference C; volume V; curved surface area S; altitude h; radius r

RIGHT TRIANGLE

Pythagorean Theorem: $c^2 = a^2 + b^2$

TRIANGLE

$$A = \tfrac{1}{2}bh \qquad C = a + b + c$$

EQUILATERAL TRIANGLE

$$h = \frac{\sqrt{3}}{2}s \qquad A = \frac{\sqrt{3}}{4}s^2$$

RECTANGLE

$$A = lw \qquad C = 2l + 2w$$

PARALLELOGRAM

$$A = bh$$

TRAPEZOID

$$A = \tfrac{1}{2}(a + b)h$$

CIRCLE

$$A = \pi r^2 \qquad C = 2\pi r$$

CIRCULAR SECTOR

$$A = \tfrac{1}{2}r^2\theta \qquad s = r\theta$$

CIRCULAR RING

$$A = \pi(R^2 - r^2)$$

RECTANGULAR BOX

$$V = lwh \qquad S = 2(hl + lw + hw)$$

SPHERE

$$V = \tfrac{4}{3}\pi r^3 \qquad S = 4\pi r^2$$

RIGHT CIRCULAR CYLINDER

$$V = \pi r^2 h \qquad S = 2\pi rh$$

RIGHT CIRCULAR CONE

$$V = \tfrac{1}{3}\pi r^2 h \qquad S = \pi r \sqrt{r^2 + h^2}$$

FRUSTUM OF A CONE

$$V = \tfrac{1}{3}\pi h(r^2 + rR + R^2)$$

PRISM

$V = Bh$ with B the area of the base

CALCULUS

DEDICATION

This Sixth Edition of *Calculus* is dedicated to the memory of Earl W. Swokowski, whose high standards of mathematical exposition guided our efforts.

A NOTE FROM THE PUBLISHER

Earl W. Swokowski died on June 2, 1992. We at PWS are grateful to have been associated with one of the premier mathematics authors of the last twenty-five years.

Calculus

EARL W. SWOKOWSKI

MICHAEL OLINICK
Middlebury College

DENNIS PENCE
Western Michigan University

With the Assistance of
JEFFERY A. COLE
Anoka-Ramsey Community College

PWS PUBLISHING COMPANY · BOSTON

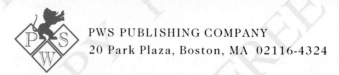

PWS PUBLISHING COMPANY
20 Park Plaza, Boston, MA 02116-4324

PWS Publishing Company is a division of Wadsworth, Inc.

I⟨T⟩P™

International Thomson Publishing
The trademark ITP is used under license.

Library of Congress Cataloging-in-Publication Data

Swokowski, Earl William
  Calculus / Earl W. Swokowski, Michael Olinick, Dennis Pence ; with the assistance of Jeffery A. Cole.—6th ed.
    p. cm.—(The Prindle, Weber & Schmidt series in calculus and upper-division mathematics)
  Includes bibliographical references and index.
  ISBN 0-534-93624-5
  1. Calculus. I. Olinick, Michael. II. Pence, Dennis. III. Title. IV. Series.
QA303.S94 1994   93-35389
515'.15—dc20   CIP

Chapter Opening Photos: *Precalculus Review* (p. 3) © 1988 Larry Dale Gordon / The Image Bank; *Chapter 1* (p. 83) © Stephen R. Swinburne / Stock Boston; *Chapter 2* (p. 143) © Mitchell Funk / The Image Bank; *Chapter 3* (p. 247) © 1989 Harald Sund / The Image Bank; *Chapter 4* (p. 343) © Steve Dunwell / The Image Bank; *Chapter 5* (p. 431) © 1987 Sobel / Klonsky / The Image Bank; *Chapter 6* (p. 515) © 1988 Jake Rajs / The Image Bank; *Chapter 7* (p. 629) © Nancie Battaglia; *Chapter 8* (p. 693) © Kay Chernush / The Image Bank; *Chapter 9* (p. 785) © 1988 Kaz Mori / The Image Bank; *Chapter 10* (p. 845) © Alan Becker / The Image Bank; *Chapter 11* (p. 921) © 1991 K. Iwasaki / The Stock Market; *Chapter 12* (p. 973) © 1992 Eric Meola / The Image Bank; *Chapter 13* (p. 1075) © Guido Alberto Rossi / The Image Bank; *Chapter 14* (p. 1167) © 1992 A. & J. Verkaik / The Stock Market; *Chapter 15* (p. 1235) © 1989 Garry Gay / The Image Bank. **Cover Photos:** *Dunes* © 1990 S. Achernarl / The Image Bank and *Sand* © 1989 Weinberg-Clark / The Image Bank.

Sponsoring Editor: *Steve Quigley*
Managing Developmental Editor: *David Dietz*
Production Coordinator: *Robine Andrau*
Marketing Manager: *Marianne Rutter*
Interior Designer: *Elise S. Kaiser*
Manufacturing Coordinator: *Marcia A. Locke*
Editorial Assistant: *John Ward*
Developmental Editor: *Sylvia Dovner/Technical Texts*
Production: *Geri Davis and Martha Morong/Quadrata, Inc.*
Interior Illustrator: *Scientific Illustrators*
Cover Designer: *Julia D. Gecha*
Typesetter: *Interactive Composition Corporation, Inc.*
Cover Printer: *New England Book Components*
Text Printer and Binder: *Arcata Graphics/Hawkins*

Printed and bound in the United States of America
93 94 95 96 97 98 — 10 9 8 7 6 5 4 3 2 1

CONTENTS

CHAPTER·1

Contents

ADVANCES IN COMPUTING TECHNOLOGY and changing emphases and goals in mathematics education are transforming the teaching of calculus in the 1990s. These changes offer great opportunities for improvement of textbooks for the standard three-semester calculus course, including Swokowski's *Calculus*. Our aim in this edition has been to preserve the strengths that characterized earlier editions, while continuing the evolution of the book toward fuller incorporation of modern applications. The approach remains a student-oriented one, with comprehensive discussions of the concepts of calculus and an impressively large collection of worked examples and illustrative figures. We have upheld the tradition of thoroughly checking the exposition, examples, and exercises to ensure a text that is as error-free as possible. In making the changes described here, we strove to maintain the integrity of Earl Swokowski's now-classic treatment.

ORGANIZATIONAL CHANGES AND ADDITIONS IN THE SIXTH EDITION

In response to suggestions for improvements from students, instructors, and reviewers, we have streamlined the structure of the book by some reorganization of topics. We have also made some significant additions. The principal changes are highlighted in the following list:

Precalculus Review   We have added a section on exponentials and logarithms to this preliminary chapter and included the analytic geometry of the conic sections. Examples and exercises on the conic sections that require derivatives and integrals now appear in the chapters in which these topics are introduced.

Limits and Continuity   In Chapter 1, we have retained the approach to limits that first explores the concept intuitively before presenting a rigorous definition. We have added new examples that show how graphing calculators can provide numerical and visual evidence for the existence and value of limits. A new example shows how the Michaelis–Menten law in biology leads to a real-world application of limits involving infinity.

Derivative and Its Applications   The derivatives of the trigonometric functions introduced in Section 2.4 are now used throughout the early chapters on differentiation and integration. Chapter 2 also presents the chain rule earlier without reference to differentials. We have rewritten the treatment of differentials in Section 2.8 to emphasize linear approximations. In this edition, we have strengthened the treatment of numerical methods by introducing Newton's method earlier (in Chapter 2) and carrying it forward throughout the book in examples and exercises where analytic solutions cannot be obtained. In Chapter 3, a new section briefly explores mathematical models in economics and the social and life sciences and shows how elementary properties of the derivative may provide qualitative and quantitative predictions about the behavior of complex systems.

Integral and Its Applications   We begin our treatment of the integral in Chapter 4 by establishing the connection between antiderivatives and solutions of differential equations. Section 4.7, which highlights this edition's more modern and comprehensive introduction to numerical integration, shows the common threads connecting the classic approaches of the midpoint rule, the trapezoidal rule, and Simpson's rule. We also present examples that explore estimating the errors associated with these methods. Several new applications of the definite integral to problems in economics have been added to Chapter 5.

Transcendental Functions   In Chapter 6, we begin with a discussion of the derivative of an inverse function and then apply the results to the natural logarithm and exponential functions, logarithms and exponentials to arbitrary bases, the inverse trigonometric functions, and the hyperbolic functions and their inverses. This chapter also discusses separable differential equations and shows their application to problems of growth and decay in Section 6.6. We conclude the chapter with l'Hôpital's rule for differentiating indeterminate forms that often occur in applications using transcendental functions.

Techniques of Integration   In Chapter 7, we have rewritten the section on tables of integrals to include a discussion of computer algebra systems and illustrations of their use in finding indefinite integrals. A new example shows how several different techniques of integration are combined to solve the brachistochrone problem in Section 7.5.

Infinite Series   We have expanded the treatment of infinite sequences in Chapter 8 to include a discussion of exploring recursively defined sequences on a calculator. A new example in Section 8.2 shows how infinite series may be used to determine the long-term impact of a chemical plant discharging pesticide into a river once a week. We have also added examples in Sections 8.7 and 8.8 that illustrate graphically how Taylor polynomials of increasing degree approximate a given function.

Parametric Equations and Polar Coordinates   One of the goals of Chapter 9 is to emphasize the benefit of examining plane curves using several different representations, including alternative coordinate systems, so we have added a section presenting translation and rotation of axes. As a new application of parametric equations in Section 9.1, we briefly introduce Bezier curves and their applications to computer-aided design and the mathematical representation of type fonts.

Multivariable Calculus   In Chapters 10–14, we maintain the same overall approach to vectors and surfaces, vector-valued functions, partial differentiation, multiple integrals, and vector calculus as in the fifth edition. Some new applications include the use of paraboloids of revolution for work with radio telescopes and for determining the surface area and volume of lakes. In these chapters, we discuss the numerical evaluation of iterated integrals and line integrals as well as the use of computer software to view surfaces. We also show how vector representation of lines, calculus methods for finding slopes of tangent lines, and a graphics calculator may be combined to approximate the points of intersection of two curves and the angles between the tangent lines at these points.

Differential Equations   In Chapter 15, we have provided an extended treatment of the numerical methods of solving differential equations. Emphasis is placed on the Euler, improved Euler, and Runge–Kutta methods. We discuss the connection between these methods and analogous methods for numerical integration. The treatment of vibrations in Section 15.5 has been expanded to include the resonance case of undamped vibration subjected to a forcing function, and we show how this resonance can lead to the collapse of physical structures.

NEW FEATURES IN THE SIXTH EDITION

Technology Examples and Exercises   We have expanded the use of graphics calculators and computer software. We use these tools in examples to explore the concepts of calculus and to solve problems numerically where analytic, closed-form solutions are not possible. As a reflection of the high degree of precision available on these tools, numerical answers presented in the text are sometimes given to several decimal places. Since a wide variety of calculators or computers is used by instructors and students, we have made our treatment as generic as possible. (Results computed on different devices may differ slightly.) The text does not attempt to teach the reader how to use particular tools or software packages, but rather concentrates on how the features available on almost all of these devices help illuminate the behavior of functions, their derivatives, and their integrals. We also discuss in several settings the inherent limitations of computer/calculator technologies to develop in the reader a critical attitude toward the output they produce. The technology examples and exercises, which are clearly identified, may be skipped by instructors who do not use calculators or computers.

Extended Problems and Group Projects   Beginning with Chapter 1, we have added a small set of Extended Problems and Group Projects at the end of each chapter for instructors and students interested in exploring topics in greater depth than is possible in homework assignments. These open-ended discovery problems are designed to stimulate investigative thinking on the part of students and to sharpen their writing and reporting skills. Many of the problems are suitable for use by small cooperative learning groups. Several of them ask the reader to attack aspects of a problem whose full solution may be presented later in the text when more sophisticated calculus concepts have been introduced.

Mathematicians and Their Times   We have added short biographical/ historical sketches called Mathematicians and Their Times to each chapter. These sketches attempt to show the more human side of mathematics and the manner in which social, political, and intellectual structures of particular periods influenced the development of calculus. They discuss not only the achievements of mathematicians but also the prejudices and obstacles many of them overcame in order to pursue their studies. We hope that these brief introductions will whet the appetites of readers to explore the fascinating history of mathematics on their own.

FEATURES OF THE TEXT

Applications   As in previous editions, applied problems are drawn from a broad range of the natural and social sciences, engineering, and technology. Applications from physics, chemistry, biology, economics, physiology, sociology, psychology, ecology, oceanography, meteorology, radiotherapy, astronautics, and transportation are included. Among the many applications new to this edition are the population of an endangered species, the spread of the AIDS virus, the concentration of glucose in an enzyme-glucose complex, predator–prey interaction, consumers' surplus, computer-aided design, the breakdown of pesticides, the flow of discharge at an industrial plant, and the surface area and volume of a lake. Each chapter opens with a photograph illustrating a specific field of application of the material introduced in the chapter; and within each chapter, we revisit the particular application setting, usually in an example, to give students a fuller understanding of how the calculus is applied.

Examples   Well-structured and graded examples provide detailed solutions of problems similar to those that appear in exercise sets. The examples frequently include applications that indicate the usefulness of a topic; and many contain graphs, charts, or tables to help readers understand procedures and solutions. New to this edition are about 70 examples showing the power and limitations of graphics calculators and computer algebra systems to explore calculus topics. These examples are identified by a technology icon in the margin. In keeping with the enhanced numerical emphasis in this edition, other examples include both an analytic solution and a computational method solution, which is also marked with the technology icon. Brief examples, labeled Illustrations, appear throughout the book and demonstrate the use of definitions, laws, and theorems.

Exercises   The exercise sets at the end of each section usually begin with routine drill problems and progress gradually to more difficult types. Applied problems generally appear near the end of a set to help students gain confidence in manipulations and new ideas before attempting questions that require analyses of practical situations. Approximately 400 new exercises are designed to be carried out with a graphics calculator or a computer algebra system. These are designated by a boxed "**c**." Review exercises, which may be used to prepare for examinations, appear at the end of every numbered chapter, followed by Extended Problems and Group Projects. As noted earlier, these problems offer open-ended topics for exploration and investigation by students. Working alone or in small groups, students will need a week or two to complete one of these assignments and prepare a detailed written report of their conclusions.

Answers   The answer section at the end of the text provides answers for most of the odd-numbered exercises. Considerable thought and effort have been devoted to making this section a learning device rather than merely a place to check answers. If an answer is a region's area or a solid's volume, for example, a suitable definite integral is often given along with its value. Graphs, proofs, and hints are included when appropriate. Numerical answers for many exercises appear in both an exact and an approximate form.

Text Design and Figures A new design for the text that emphasizes important concepts makes discussions easier to follow. Chapter introductions have been expanded to highlight an important application of the chapter topic. Cautionary notes have been added, where appropriate, to alert students to common pitfalls in applying formulas and drawing conclusions from theorems. Many of the graphics have been redone for this edition, and more than 200 new figures have been added. All graphs of functions are plotted to a high degree of accuracy, and two or more colors have been used in all figures. To distinguish between different parts of a graph, for example, blue may be used for a function's graph and red for a tangent line to the graph. Accompanying labels are in the same color as the parts of the graph they identify. The use of several colors also makes surfaces and solids easier to visualize. For this edition, realistic screen images have been reproduced from calculator and computer screens for the examples focusing on the use of these devices.

Flexibility Syllabi from schools that have used previous editions attest to the flexibility of the text. Instructors can rearrange sections and chapters in different ways, depending on the objectives and length of the course.

Michael Olinick
Dennis Pence

ACKNOWLEDGMENTS

AUTHORS' ACKNOWLEDGMENTS

Our primary debt of gratitude, of course, goes to the late Earl Swokowski. In developing the excellent calculus text that forms the basis of this revision, he was committed to keeping the student's needs in the forefront and to presenting calculus in a clear, careful, and well-motivated approach, with interesting applications and ample examples and exercises. We hope that he would have found this revision a natural extension of his successful previous editions.

We are also profoundly grateful to Jeffery Cole of Anoka-Ramsey Community College. Jeff carefully read the entire manuscript, gently pointing out errors and enthusiastically offering many useful suggestions for improvement. Jeff's close association with Earl Swokowski in the development of earlier editions of *Calculus* and other texts provided us with valuable insight.

We wish to thank Gary Rockswold of Mankato State University for his extensive work in helping to prepare answers and solutions to the exercises.

We are indebted to Sylvia Dovner of Technical Texts for her heroic efforts to transform sometimes imperfect prose and syntax into understandable exposition in a coherent and consistent manner. Tirelessly and with good humor, she devoted herself to helping the authors and the manuscript.

We are also thankful for the excellent cooperation of the staff of PWS Publishing Company, particularly Senior Editor Steve Quigley and Managing Developmental Editor David Dietz, who initiated the process that led to our participation in the preparation of this edition and continued to provide encouragement, information, and advice at critical points along the way. Elise Kaiser designed the book, Robine Andrau coordinated the production, and John Ward rendered invaluable editorial assistance.

We also acknowledge the care and attention to detail provided by Geri Davis and Martha Morong of Quadrata, Inc., who were responsible for the production process, and George Morris of Scientific Illustrators, who rendered the figures.

We are both deeply grateful to our wives, Judy and Becky, and to our children—Eli, Abby, Sasha, and Anne Olinick and Adam Pence—for their support, understanding, patience, and love.

In addition to the persons named here, we express our sincere appreciation to the many students and teachers who have helped shape our views on the teaching and learning of calculus. Last, we thank the reviewers of the manuscript for this edition of *Calculus,* and we share Earl Swokowski's gratitude to those who reviewed previous editions. Please feel free to write us about any aspect of this text. We value your opinions and suggestions.

PUBLISHER'S ACKNOWLEDGMENTS

We at PWS Publishing Company would like to thank the following instructors who have taught from the fifth edition of Swokowski's *Calculus* and who responded to our calculus survey, the results of which were used to help plan the direction of this new edition:

John E. Atkinson, *Missouri Western State College*

Richard Bisk, *Fitchburg State College*

Eddy J. Brackin, *University of North Alabama*

Edith Cook, *Suffolk University*

Roy W. Daughdrill, *Copiah-Lincoln Community College*

John W. Davenport, *Georgia Southern University*

Stephen H. Fast, *Bluefield College*

J. William Friel, *University of Dayton*

Anthony M. Gaglione, *United States Naval Academy*

Robert Garfunkel, *Montclair State College*

Herbert A. Gindler, *San Diego State University*

Charles R. Hampton, *The College of Wooster*

James L. Hartman, *The College of Wooster*

Jean Harvey, *Jones County Junior College*

Bruce M. Landman, *University of North Carolina—Greensboro*

Elizabeth M. Markham, *St. Joseph College*

John R. Martin, *Tarrant County Junior College—Northeast Campus*

Evan L. Parker, *Valley Forge College*

Mike Penna, *Indiana University—Purdue University at Indianapolis*

John R. Ramsay, *The College of Wooster*

Sybil Rogert, *San Diego Mesa College*

Leon Sagan, *Anne Arundel Community College*

Mansour Samimi, *Winston-Salem State University*

Teresa B. Sink, *Davidson County Community College*

Patricia P. Taylor, *Thomas Nelson Community College*

William P. Wardlaw, *United States Naval Academy*

We at PWS are also greatly indebted to the following people who reviewed the previous edition of this text and provided us with helpful suggestions for the current edition:

David F. Anderson, *University of Tennessee*

Robert Beezer, *University of Puget Sound*

Paul Wayne Britt, *Louisiana State University*

Andrew Demetropoulos, *Montclair State College*

Susanna S. Epp, *DePaul University*

C. J. Knickerbocker, *St. Lawrence University*

Jim Lewis, *University of Nebraska*

James E. Moran, *Diablo Valley College*

David Price, *Tarrant County Junior College*

Ron Smit, *University of Portland*

David Sprows, *Villanova University*

Juan Tolosa, *Stockton State College*

Carroll G. Wells, *Western Kentucky University*

Joyce W. Williams, *University of Massachusetts—Lowell*

John Wolfskill, *Northern Illinois University*

The following instructors, whose invaluable insights aided our selection of co-authors for the sixth edition, have our sincerest gratitude:

Jorge Cossio, *Miami-Dade Community College*

Tom Davis, *Sam Houston State University*

Toni Kasper, *Borough of Manhattan Community College*

James R. McKinney, *California State Polytechnic University*

Walter Martens, *University of Alabama—Birmingham*

Eldon L. Miller, *University of Mississippi*

Gerald Francis Morrell, *City College of San Francisco*

A. N. V. Rao, *University of South Florida*

K. Salkauskas, *The University of Calgary*

PWS and the authors would like to thank the following individuals who lent their mathematical expertise in reviewing portions of the sixth edition manuscript:

David F. Anderson, *University of Tennessee*

Ali Hajjafar, *The University of Akron*

Carlton A. Lane, *Hillsborough Community College*

James E. Moran, *Diablo Valley College*

David Price, *Tarrant County Junior College*

Robert P. Roe, *University of Missouri—Rolla*

Joyce W. Williams, *University of Massachusetts—Lowell*

Finally, PWS and the authors thank the numerous people who reviewed previous editions of this text and therefore contributed to the evolution of this book:

Alfred D. Andrew, *Georgia Institute of Technology;* Stephen Andrilli, *LaSalle University;* Jan Frederick Andrus, *University of New Orleans;* Richard D. Armstrong, *St. Louis Community College at Florissant Valley;* Jacqueline E. Barab, *University of Nevada, Las Vegas;* Phillip W. Bean, *Mercer University;* Robert Beezer, *University of Puget Sound;* Daniel D. Benice, *Montgomery College;* Delmar L. Boyar, *University of Texas—El Paso;* Christian C. Braunschweiger, *Marquette University;* Paul W. Britt, *Louisiana State University;* Robert M. Brooks, *University of Utah;* Ronald E. Bruck, *University of Southern California;* David C. Buchthal, *University of Akron;* Dawson Carr, *Sandhills Community College;* Clifton Clarridge, *Santa Monica College;* James L. Cornette, *Iowa State University;* Andrew Demetropoulos, *Montclair State College;* John E. Derwent, *University of Notre Dame;* Daniel Drucker, *Wayne State University;* Joseph M. Egar, *Cleveland State University;* Ronald D. Ferguson, *San Antonio State College;* William R. Fuller, *Purdue University;* August J. Garver, *University of Missouri;* Stuart Goldenberg, *California State Polytechnic University;* Richard Grassl, *University of Northern Colorado;* Joe A. Guthrie, *University of Texas—El Paso;* Gary Haggard, *Bucknell University;* Mark P. Hale, Jr., *University of Florida;* Clemens B. Hanneken, *Marquette University;* Alan Heckenbach, *Iowa State University;* Simon Hellerstein, *University of Wisconsin;* John C. Higgins, *Brigham Young University;* Arthur M. Hobbs, *Texas A & M University;* David Hoff, *Indiana University;* Dale T. Hoffman, *Bellevue Community College;* Adam J. Hulin, *University of New Orleans;* Michael Iannone, *Trenton State College;* George Johnson, *University of South Carolina;* Elgin H. Johnston, *Iowa State University;* Herbert M. Kamowitz, *University of Massachusetts—Boston;* Andrew Karantinos, *University of South Dakota;* Harvey B. Keynes, *University of Minnesota;* Eleanor Killam,

University of Massachusetts; Karl R. Klose, *Susquehanna University;* William James Lewis, *University of Nebraska—Lincoln;* Margaret Lial, *American River College;* James T. Loats, *Metropolitan State College;* Phil Locke, *University of Maine—Orono;* Robert H. Lohman, *Kent State University;* Stanley M. Lukawecki, *Clemson University;* Wayne McDaniel, *University of Missouri—St. Louis;* James E. McKenna, *SUNY College at Fredonia;* James R. McKinney, *California State Polytechnic University—Pomona;* Judith R. McKinney, *California State Polytechnic University—Pomona;* Francis E. Masat, *Glassboro State College;* Burnett Meyer, *University of Colorado at Boulder;* Joseph Miles, *University of Illinois;* David Minda, *University of Cincinnati;* Chester L. Miracle, *University of Minnesota;* John A. Nohel, *University of Wisconsin;* Norman K. Nystrom, *American River College;* James Osterburg, *University of Cincinnati;* Richard R. Patterson, *Indiana University—Purdue University at Indianapolis;* Charles V. Peele, *Marshall University;* Ada Peluso, *CUNY—Hunter College;* Michael A. Penna, *Indiana University—Purdue University at Indianapolis;* David A. Petrie, *Cypress College;* Richard A. Quint, *Ventura College;* Neal C. Raber, *University of Akron;* William H. Robinson, *Ventura College;* Jean E. Rubin, *Purdue University;* Leon F. Sagan, *Anne Arundel Community College;* John T. Scheick, *Ohio State University;* Eugene P. Schlereth, *University of Tennessee;* Jon W. Scott, *Montgomery College;* Richard D. Semmeler, *Northern Virginia Community College—Annandale;* Leonard Shapiro, *University of Minnesota;* Donald Sherbert, *University of Illinois;* Eugene R. Speer, *Rutgers University;* David J. Sprows, *Villanova University;* Ronald Stoltenberg, *Sam Houston State University;* Monty J. Strauss, *Texas Tech University;* John Tung, *Miami University;* Jan Vandever, *South Dakota State University;* Charles Van Gordon, *Millersville State College;* Richard G. Vinson, *University of South Alabama;* Roman W. Voronka, *New Jersey Institute of Technology;* Dale W. Walston, *University of Texas at Austin;* Frederick R. Ward, *Boise State University;* Alan Wiederhold, *San Jacinto College;* Kenneth L. Wiggins, *Walla Walla College;* Loyd V. Wilcox, *Golden West College;* T. J. Worosz, *Metropolitan State College;* Dennis Wortman, *University of Massachusetts—Boston*

ANCILLARIES

An array of ancillaries are available with *Calculus, Sixth Edition:*

Student's Solutions Manual, Volume 1 (Chapters 1–9) and *Volume 2 (Chapters 10–15)* contain solutions, worked out in detail, to a subset of the odd-numbered exercises from the text.

Student Study Guide, Volume 1 (Chapters 1–9) and *Volume 2 (Chapters 10–15)* lead students through all major topics in the text and reinforce their understanding through review sections, drills, and self-tests.

Calculus Laboratories are text-specific manuals for use with *Derive, Maple,* or *Mathematica* computer algebra systems, which include over 25 laboratories with corresponding problem sets. Emphasis is on simple, direct development of concepts and mastery of important key strokes.

Instructor's Solutions Manual, Volume 1 (Chapters 1–9) and *Volume 2 (Chapters 10–15)* give solutions or answers for all exercises in the text.

Test Bank is available to all instructors; contains sample tests for each chapter and a listing of all test items in printed form.

Transparencies are available to all instructors; consist of four-color acetates of selected figures from the text.

SOFTWARE

Student Edition of Theorist is an easy-to-use student version of a popular symbolic algebra and graphing package developed for the Macintosh by Prescience Corporation. This software program provides the visual, interactive problem-solving capabilities that will find immediate utility in calculus problems. *Theorist* boasts powerful mathematical manipulation features such as two- and three-dimensional graphing, animation graphing, accurate display, and WYSIWYG printing of mathematical notation. The package contains a 512-page manual and two disks and requires a Macintosh system with 1 MB RAM (System 6 or 7).

The PWS Notebook Series™ are electronic problem sets keyed to the text with additional problems, practical problem-solving suggestions, and other pedagogical enhancements. *Notebooks* are compatible for use with four different commercially available graphing and symbolic algebra software programs: *Theorist* (Macintosh platform), *Derive* (IBM PC–compatible), *Maple* (Mac and IBM PC), and *Mathematica* (Mac and IBM PC).

Grapher, Version 2.0 is a flexible graphing software program for the Macintosh. This versatile collection of graphing utilities can plot rectangular curves, polar curves, and interpolating polynomials, as well as graph parametric equations, systems of two first-order differential equations, series, direction fields, and more.

TrueBASIC Calculus is a disk/manual package for the IBM PC platform that enables the student to select from among a wide range of calculus topics for self-study, computational drill, and problem solving. The record and playback feature of this program make *TrueBASIC Calculus* ideal for classroom demonstrations.

COMPUTERIZED TESTING

EXPTest is a testing program for IBM PCs and compatibles that allows users to view and edit all tests, adding to, deleting from, and modifying existing questions. Any number of student tests can be created, including multiple forms for larger sections or single tests for individual use. Users can create a question bank using mathematical symbols and notation. A graphics importation feature permits display and printing of graphs, diagrams, and maps provided with the test banks. The package includes easy-to-follow documentation, with a quick-start guide. A demonstration disk is available.

ExamBuilder is a Macintosh testing program that allows users to create, view, and edit tests. Questions can be stored by objectives, and the user can create questions using multiple-choice, true/false, fill-in-the-blank, essay, and matching formats. The order of alternatives and the order of questions can be scrambled. A demonstration disk is available (the user must have HyperCard to run the demo disk).

GRAPHIC CALCULATOR SUPPLEMENTS

Calculus Activities for Graphic Calculators (*Pence*) is available for students and instructors. It offers exercises and examples utilizing the graphics calculator (Sharp, Casio, and HP-28S). Classroom tested, these varied activities will enhance understanding of calculus topics for those using graphics calculators.

Calculus Activities for the TI Graphic Calculator (*Pence*) is available for students and instructors using the TI-family of graphic calculators and includes a multitude of examples and exercises designed specifically for calculus.

ALTERNATE VERSION OF THE TEXT

An abbreviated version of *Calculus, Sixth Edition* is available. *Calculus of a Single Variable, Second Edition* comprises the Precalculus Review and Chapters 1–9 of the parent text and is appropriate for use in a full-year course that emphasizes single-variable calculus. The material in this brief volume is identical to that of the corresponding chapters in *Calculus, Sixth Edition.*

ANCILLARIES FOR STUDENTS

PRODUCT NAME	TEXT-SPECIFIC	PLATFORM/PROGRAM	CAPSULE DESCRIPTION	FORMAT
Student's Solutions Manual Vol. 1 *(Chs. 1–9)* Vol. 2 *(Chs. 10–15)*	yes	N/A	Detailed solutions for selected odd-numbered exercises	Print only
Student Study Guide Vol. 1 *(Chs. 1–9)* Vol. 2 *(Chs. 10–15)*	yes	N/A	Review, drill, and practice, plus self-tests	Print only
Calculus Laboratories	yes	IBM/Derive Mac, IBM/Maple Mac, IBM/Mathematica	Supplemental lab exercises	Print only
Student Edition of Theorist	no	Macintosh	Powerful graphing and symbolic algebra software	Software + manual
The PWS Notebook Series™	yes	Mac/Theorist IBM/Derive Mac, IBM/Maple Mac, IBM/Mathematica	Electronic problem sets keyed to the text	Software only (data disks)
Grapher, Version 2.0	no	Macintosh	Flexible graphing software program for (1) calculus (2) differential equations (3) advanced engineering math	Software only (data disks)
TrueBASIC Calculus	no	IBM	Software for self-study, solving problems, and classroom demos for (1) precalculus (2) calculus	Print + software (data disks)
Pence, Calculus Activities for Graphic Calculators	no	Sharp Casio HP-28S	Exercises and examples using graphic calculators for calculus	Print only
Pence, Calculus Activities for the TI Graphic Calculator	no	TI-81 TI-85	Graphing exercises and examples for the TI family of calculators for calculus	Print only

ANCILLARIES FOR INSTRUCTORS

PRODUCT NAME	CAPSULE DESCRIPTION	FORMAT
Instructor's Solutions Manual Vol. 1 *(Chs. 1–9)* Vol. 2 *(Chs. 10–15)*	Fully worked solutions to all the problems in the text	Print only
Test Bank	Printed test items	Print only
Transparencies	Selected figures from the text	Four-color acetates
EXPTest	IBM-based software to view, edit, and create tests, with manual; demo available	Software test generator
ExamBuilder	Macintosh software to view, edit, and create tests; requires HyperCard; demo available	Software test generator

ALCULUS IS ONE OF THE SUPREME creations of human thought. It combines analytical and geometric ideas both to forge powerful tools for the solution of important problems and to develop concepts of central importance in mathematics.

Calculus was invented in the seventeenth century to investigate problems that involve motion. We use algebra and trigonometry to study objects moving at constant speeds along linear or circular paths, but we need calculus if the speed varies or if the path is irregular. An accurate description of motion requires precise definitions of *velocity* (the rate at which distance changes per unit time) and *acceleration* (the rate at which velocity changes). We obtain these definitions by using one of the fundamental concepts of calculus—the *derivative*.

Although calculus was developed to solve physical problems, many diverse fields of study use its power and versatility. Modern-day applications of the derivative include investigating the rate of growth of populations, predicting the outcome of chemical reactions, measuring instantaneous changes in electrical current, describing the behavior of atomic particles, estimating tumor shrinkage in radiation therapy, forecasting economic profits and losses, determining the spread of epidemics, examining the impact of automobile emissions on ozone depletion, and analyzing vibrations in mechanical systems.

We also use the derivative in solving optimization problems such as manufacturing the least expensive rectangular box that has a given volume, calculating the greatest distance a rocket will travel, obtaining the maximum safe flow of traffic across a long bridge, determining the number of wells to drill in an oil field for the most efficient production, finding the point between two light sources at which illumination will be greatest, and maximizing corporate revenue for a particular product. Mathematicians often use derivatives to find tangent lines to curves and to analyze graphs of complicated functions.

Another fundamental concept of calculus—the *definite integral*—arose from the problem of finding areas of regions with curved boundaries. Scientists use definite integrals as extensively as derivatives and in as many different fields. Some applications are finding the center of mass or moment of inertia of a solid, determining the work required to send a space probe to another planet, calculating the blood flow through an arteriole, estimating depreciation of equipment in a manufacturing plant, and interpreting the amount of dye dilution in physiological tests that involve tracer methods. We also use integrals to investigate mathematical concepts such as area of a curved surface, volume of a geometric solid, or length of a curve.

We define *derivative* and *definite integral* by limiting processes. The notion of *limit* is the basic idea that separates calculus from elementary mathematics. Sir Isaac Newton (1642–1727) and Gottfried Wilhelm Leibniz (1646–1716), who independently discovered the connection between derivatives and integrals, are credited with the invention of calculus. Many other mathematicians have added greatly to the development in the last 300 years.

The applications of calculus mentioned here represent just a few of the many we consider in this book. We cannot possibly discuss all uses of calculus, and more are being developed with every advance in technology. Whatever your field of interest, calculus is probably used in some pure or applied investigations. Perhaps *you* will discover a new application for this branch of science.

CALCULUS

INTRODUCTION

I **N THE FIRST MOMENTS** after the launch of a space shuttle, many changes occur rapidly. The rocket gains altitude as it accelerates to higher speeds. Its mass decreases as fuel burns up. Inside the shuttle, an astronaut feels increasing force due to the acceleration. As the distance from the earth gets larger, the astronaut's weight decreases. Indeed the values of many variables change dramatically during this time period.

Calculus is the mathematics of change. Wherever there is motion or growth or where variations in one quantity produce alterations in another, calculus helps us understand the changes that occur. We can use calculus, for example, to predict the height and speed of the rocket at each instant of time after launch. We also use calculus to study important geometric properties of curves, such as their tangent lines and the amount of area they enclose.

In this preliminary chapter, we briefly review topics from precalculus mathematics that are essential for the study of calculus, beginning with inequalities, equations, absolute values, and graphs of lines and circles. We then turn our attention to functions and their graphs. We also discuss some very important functions that occur frequently in applications of calculus: trigonometric functions, exponential functions, and logarithmic functions. We conclude the precalculus review with an examination of the elementary geometry of conic sections: parabolas, ellipses, and hyperbolas.

To say that the concept of function is important in calculus is an understatement. It is literally the foundation of calculus and the backbone of the entire subject. You will find the word *function* and the symbol f or $f(x)$ used frequently on many pages of this text.

In precalculus courses, we study properties of functions by using algebra and graphical methods that include plotting points and determining symmetry. These techniques are adequate for obtaining a rough sketch of a graph. Calculus is required, however, to find precisely where graphs of functions rise or fall, exact coordinates of high or low points, slopes of tangent lines, and many other useful facts. We can often successfully attack applied problems in science, engineering, economics, and the social sciences that cannot be solved by means of algebra, geometry, or trigonometry if we represent physical quantities in terms of functions and then apply the tools of calculus.

With the preceding remarks in mind, carefully read Section B on functions and their graphs. You should have a good understanding of this material before beginning your study of calculus in the next chapter.

Planning a complex mission like the launch of a space shuttle requires extensive use of calculus and precalculus mathematics.

Precalculus Review

A ALGEBRA

This section reviews topics from algebra that are prerequisites for calculus. We shall state important facts and work examples without supplying detailed reasons to justify our work. Texts on precalculus mathematics provide more extensive coverage of this material.

All concepts in calculus are based on properties of the set $\mathbb{R}$ of real numbers. There is a one-to-one correspondence between $\mathbb{R}$ and points on a *coordinate line* (or *real line*) l, as Figure 1 illustrates. The point O is called the *origin* and corresponds to the number 0 (*zero*), which is neither positive nor negative. The real number associated with a point on the line is called a **coordinate** of the point.

Figure 1

If a and b are real numbers, then $a > b$ (**a is greater than b**) if $a - b$ is positive. An equivalent statement is $b < a$ (**b is less than a**). Referring to the coordinate line in Figure 1, we see that $a > b$ if and only if the point A corresponding to a lies to the right of the point B corresponding to b. Other types of inequality symbols include $a \leq b$, which means $a < b$ or $a = b$, and $a < b \leq c$, which means $a < b$ and $b \leq c$.

ILLUSTRATION

- $5 > 3$
- $-7 < -2$
- $(-3)^2 > 0$
- $a^2 \geq 0$ for every real number a

The following properties can be proved for real numbers a, b, and c.

Properties of Inequalities 1

(i)   If $a > b$ and $b > c$, then $a > c$.

(ii) If $a > b$, then $a + c > b + c$.

(iii) If $a > b$, then $a - c > b - c$.

(iv) If $a > b$ and c is positive, then $ac > bc$.

(v) If $a > b$ and c is negative, then $ac < bc$.

Analogous properties are true if the inequality signs are reversed. Thus, if $a < b$ and $b < c$, then $a < c$; if $a < b$, then $a + c < b + c$; and so on.

The **absolute value** $|a|$ of a real number a is defined as follows:

$$|a| = \begin{cases} a & \text{if } a \geq 0 \\ -a & \text{if } a < 0 \end{cases}$$

If a is the coordinate of the point A on the coordinate line in Figure 1, then $|a|$ is the number of units (that is, the distance) between A and the origin O. If a and b are real numbers, then $|a - b|$ represents the distance between a and b.

ILLUSTRATION

- $|3| = 3$
- $|-3| = -(-3) = 3$
- $|0| = 0$
- $|3 - \pi| = -(3 - \pi) = \pi - 3$

The following properties can be proved.

Properties of Absolute Values ($b > 0$) 2

(i) $|a| < b$ if and only if $-b < a < b$
(ii) $|a| > b$ if and only if either $a > b$ or $a < -b$
(iii) $|a| = b$ if and only if $a = b$ or $a = -b$

An **equation** (in x) is a statement such as

$$x^2 = 3x - 4 \quad \text{or} \quad 5x^3 + 2\sin x - \sqrt{x} = 0.$$

A **solution** (or **root**) is a number b that produces a true statement when b is substituted for x. To *solve an equation* means to find all the solutions.

EXAMPLE ■ 1   Solve each equation:

(a) $x^3 + 3x^2 - 10x = 0$
(b) $2x^2 + 5x - 6 = 0$
(c) $7.3x^2 - 31.7x + 15.2 = 0$

SOLUTION
(a) Factoring the left-hand side yields

$$x(x^2 + 3x - 10) = 0, \quad \text{or} \quad x(x - 2)(x + 5) = 0.$$

Setting each factor equal to zero gives us the solutions 0, 2, and -5.
(b) Using the *quadratic formula*,

$$x = \frac{-b \pm \sqrt{b^2 - 4ac}}{2a},$$

with $a = 2$, $b = 5$, and $c = -6$, we obtain

$$x = \frac{-5 \pm \sqrt{(5)^2 - (4)(2)(-6)}}{(2)(2)} = \frac{-5 \pm \sqrt{73}}{4}.$$

Thus, the solutions are $\dfrac{-5 + \sqrt{73}}{4}$ and $\dfrac{-5 - \sqrt{73}}{4}$.

(c) Again using the quadratic formula, we obtain

$$x = \frac{31.7 \pm \sqrt{(-31.7)^2 - (4)(7.3)(15.2)}}{(2)(7.3)}$$

$$= \frac{31.7 \pm \sqrt{1004.89 - 443.84}}{14.6}$$

$$= \frac{31.7 \pm \sqrt{561.05}}{14.6}.$$

In this case, the solutions are $\dfrac{31.7 + \sqrt{561.05}}{14.6}$ and $\dfrac{31.7 - \sqrt{561.05}}{14.6}$.

 COMPUTATIONAL METHOD We may want to use a calculator or a computer to obtain numerical answers in decimal form for results similar to the two algebraic solutions in Example 1(c). In such cases, the ability to estimate roots is a useful skill for checking whether the formula has been keyed in correctly. We can quickly estimate the two roots in (c) by rounding the coefficients to convenient integers or fractions in order to approximate the solutions as follows:

$$x \approx \frac{30 \pm \sqrt{(-30)^2 - 4(7.5)(15)}}{2(7.5)}$$

$$= \frac{30 \pm \sqrt{900 - 30(15)}}{15}$$

$$= \frac{30 \pm \sqrt{450}}{15}$$

$$\approx \frac{30 \pm 20}{15} = \frac{50}{15} \text{ and } \frac{10}{15} \approx 3.3 \text{ and } 0.7$$

To obtain more exact numerical answers with a calculator or a computer, we use the quadratic formula and key in the values in a format similar to the following:

```
(31.7+√(-31.7^2-4*7.3*15.2))/(2*7.3)
```

We then evaluate the expression, obtaining 3.79359548233 as an approximate answer, which is close to the estimated value of 3.3. A similar calculation (replacing the first + sign with a − sign when keying in the formula values) yields 0.548870271098 for the second root in (c), which is also close to the estimated value.

CAUTION When entering complex numerical formulas on an algebraic calculator, it is essential to use parentheses properly. See the reference manual for your calculator for details.

An **inequality** (in x) is a statement that contains at least one of the symbols $<, >, \le$, or $\ge$, such as

$$5x - 4 > x^2 \quad \text{or} \quad -3 < 4x + 2 \le 5.$$

The notions of **solution** of an inequality and *solving* an inequality are similar to the analogous concepts for equations.

In calculus, we often use *intervals*. In the definitions that follow, we employ the *set notation* $\{x : \quad \}$, where the space after the colon is used to specify restrictions on the variable x. The notation $\{x : a < x \le b\}$, for example, denotes the set of all real numbers greater than a and less than or equal to b — the equivalent interval notation for this set is $(a, b]$. In the following chart, we call (a, b) an **open interval**, $[a, b]$ a **closed interval**, $[a, b)$ and $(a, b]$ **half-open intervals**, and intervals defined in terms of ∞ (*infinity*) or $-\infty$ (*minus* or *negative infinity*) **infinite intervals** or **rays**.

Intervals 3

Notation	Definition	Graph
(a, b)	$\{x : a < x < b\}$	
$[a, b]$	$\{x : a \le x \le b\}$	
$[a, b)$	$\{x : a \le x < b\}$	
$(a, b]$	$\{x : a < x \le b\}$	
(a, ∞)	$\{x : x > a\}$	
$[a, \infty)$	$\{x : x \ge a\}$	
$(-\infty, b)$	$\{x : x < b\}$	
$(-\infty, b]$	$\{x : x \le b\}$	
$(-\infty, \infty)$	$\mathbb{R}$	

E X A M P L E ■ 2 Solve each inequality, and then sketch the graph of its solution:

(a) $-5 \le \dfrac{4 - 3x}{2} < 1$

(b) $x^2 - 10 > 3x$

SOLUTION

(a) $-5 \le \dfrac{4 - 3x}{2} < 1$ given

$-10 \le 4 - 3x < 2$ multiply by 2

$-14 \le -3x < -2$ subtract 4

$\dfrac{14}{3} \ge x > \dfrac{2}{3}$ divide by -3, reverse the inequality signs

$\dfrac{2}{3} < x \le \dfrac{14}{3}$ equivalent inequality

Hence the solutions are the numbers in the half-open interval $(\frac{2}{3}, \frac{14}{3}]$. The graph is sketched in Figure 2.

(b) $\quad x^2 - 10 > 3x$ given

$x^2 - 3x - 10 > 0$ subtract $3x$

$(x - 5)(x + 2) > 0$ factor

We next examine the signs of the factors $x - 5$ and $x + 2$, as shown in Figure 3. Since $(x - 5)(x + 2) > 0$ if both factors have the same sign, the solutions are the real numbers in the union $(-\infty, -2) \cup (5, \infty)$, as illustrated in Figure 3.

Figure 2

Figure 3

Sign of factor

Inequalities involving absolute value occur frequently in calculus.

EXAMPLE ▪ 3   Solve each inequality, and then sketch the graph of its solution:

(a) $|x - 3| < 0.5$

(b) $|2x - 7| > 3$

SOLUTION

(a) $|x - 3| < 0.5$ given

$-0.5 < x - 3 < 0.5$ property (i) of absolute values

$2.5 < x < 3.5$ add 3

Figure 4

The solutions are the real numbers in the open interval $(2.5, 3.5)$, as shown in Figure 4.

(b) $|2x - 7| > 3$ given

$\qquad 2x - 7 < -3 \quad$ or $\quad 2x - 7 > 3 \qquad$ property (ii) of absolute values

$\qquad\qquad 2x < 4 \quad$ or $\qquad 2x > 10 \quad$ add 7

$\qquad\qquad\quad x < 2 \quad$ or $\qquad\quad x > 5 \quad$ divide by 2

Figure 5

The solutions are given by $(-\infty, 2) \cup (5, \infty)$. The graph is sketched in Figure 5.

NOTE   We can also solve the inequalities in Example 3 graphically (that is, in terms of distance) by observing that $|x - 3| < 0.5$ means that x is less than 0.5 unit from 3. Hence, x must lie between $3 - 0.5$ and $3 + 0.5$, or, equivalently, $2.5 < x < 3.5$. Similarly, for $|2x - 7| > 3$, we note that $2x$ is more than 3 units away from 7. Thus, if $2x < 7 - 3$ or $2x > 7 + 3$, we obtain the same inequalities: $2x < 4$ or $2x > 10$, or, equivalently, $x < 2$ or $x > 5$.

Inequalities often occur in applications to physical problems, as the next example demonstrates.

EXAMPLE ▪ 4   As the altitude of a space shuttle increases, an astronaut's weight decreases until a state of weightlessness is achieved. The weight of a 125-lb astronaut at an altitude of x kilometers above sea level is given by

$$W = 125 \left(\frac{6400}{6400 + x} \right)^2.$$

At what altitudes is the astronaut's weight less than 5 lb?

SOLUTION   We need to find the values of x for which $W < 5$ — that is,

$$125 \left(\frac{6400}{6400 + x} \right)^2 < 5.$$

Dividing each side of the inequality by 125 gives us

$$\left(\frac{6400}{6400 + x} \right)^2 < \frac{1}{25}.$$

Taking the square root of each side yields

$$\frac{6400}{6400 + x} < \frac{1}{5}.$$

(Since x is positive, the fraction $6400/(6400 + x)$ will also be positive. Thus, we can ignore the negative square root.)

Now we can multiply both sides of the last inequality by the positive expression $5(6400 + x)$ to obtain

$$(5)(6400) < (1)(6400 + x)$$

or

$$x > (5)(6400) - 6400 = (4)(6400) = 25{,}600.$$

The astronaut's weight will be less than 5 lb at altitudes greater than 25,600 km.

A **rectangular coordinate system** is an assignment of *ordered pairs* (a, b) to points in a plane, as illustrated in Figure 6. The plane is called a **coordinate plane**, or an **xy-plane**. Note that in this context (a, b) is not an open interval. It should always be clear from our discussion whether (a, b) represents a point or an interval.

Figure 6

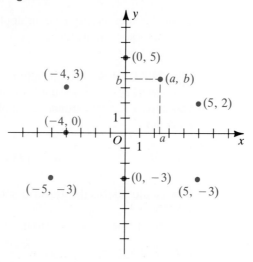

Two important formulas that show how geometric properties of line segments can be obtained from the coordinate-plane representation of points are the *distance formula* and the *midpoint formula*. Both of these formulas can be proved.

Distance Formula 4

The distance between P_1 and P_2 is

$$d(P_1, P_2) = \sqrt{(x_2 - x_1)^2 + (y_2 - y_1)^2}.$$

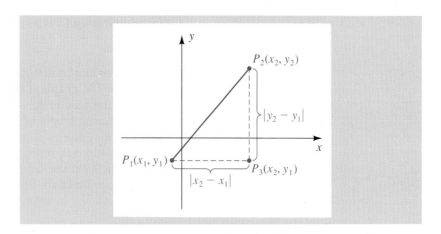

Midpoint Formula **5**

The midpoint M of segment $P_1 P_2$ is

$$M(\overline{P_1 P_2}) = M\left(\frac{x_1 + x_2}{2}, \frac{y_1 + y_2}{2}\right).$$

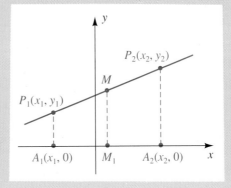

EXAMPLE ▪ 5   Given $A(-2, 3)$ and $B(4, -2)$, find:

(a) the distance between A and B

(b) the midpoint M of segment AB

SOLUTION   Using the formulas in (4) and (5), we obtain the following:

(a) $d(A, B) = \sqrt{(4 + 2)^2 + (-2 - 3)^2} = \sqrt{36 + 25} = \sqrt{61}$

(b) $M(\overline{AB}) = M\left(\frac{-2 + 4}{2}, \frac{3 + (-2)}{2}\right) = M\left(1, \frac{1}{2}\right)$

The points are plotted in Figure 7.

Figure 7

An **equation in x and y** is an equality such as

$$2x + 3y = 5, \qquad y = x^2 - 5x + 2, \quad \text{or} \quad y^2 + \sin x = 8.$$

A **solution** is an ordered pair (a, b) that produces a true statement when $x = a$ and $y = b$. The **graph** of the equation consists of all points (a, b) in a plane that correspond to the solutions. We shall assume that you have experience in sketching graphs of basic equations in x and y.

The concept of *symmetry* is useful in calculus. It enables us to sketch only half a graph and then reflect that half through an axis or the origin. Some graphs in the xy-plane are symmetric with respect to the y-axis, the x-axis, or the origin. There are simple tests, given in (6), that we can apply to an equation in x and y in order to determine symmetry.

Symmetries of Graphs 6

(i) y-axis	(ii) x-axis	(iii) Origin
Test (i): Substitution of $-x$ for x leads to the same equation.	*Test* (ii): Substitution of $-y$ for y leads to the same equation.	*Test* (iii): Substitution of *both* $-x$ for x *and* $-y$ for y leads to the same equation.

In the next example, we shall plot several points on each graph to illustrate solutions of the equations. However, *a principal objective in graphing is to obtain an accurate sketch without plotting many (or any) points.*

EXAMPLE ■ 6 Sketch the graph of each of the following equations:

(a) $y = \frac{1}{2}x^2$ **(b)** $y^2 = x$ **(c)** $4y = x^3$

SOLUTION

(a) By symmetry test (i), the graph of $y = \frac{1}{2}x^2$ is symmetric with respect to the y-axis. Some points (x, y) on the graph are listed in the following table.

x	0	1	2	3	4
y	0	$\frac{1}{2}$	2	$\frac{9}{2}$	8

Plotting, drawing a smooth curve through the points, and then using symmetry gives us the sketch in Figure 8. The graph is a *parabola,* with *vertex* (0, 0) and *axis* along the y-axis. We discuss parabolas in more detail in Section E of this chapter.

(b) By symmetry test (ii), the graph of $y^2 = x$ is symmetric with respect to the x-axis. Points above the x-axis are given by $y = \sqrt{x}$. Several such points are (0, 0), (1, 1), (4, 2), and (9, 3). Plotting and using symmetry gives us Figure 9. The graph is a parabola with vertex (0, 0) and axis along the x-axis.

(c) By symmetry test (iii), the graph of $4y = x^3$ is symmetric with respect to the origin. Several points on the graph are (0, 0), $(1, \frac{1}{4})$, and (2, 2). Plotting and using symmetry gives us the sketch in Figure 10.

Figure 8

Figure 9

Figure 10

Figure 11

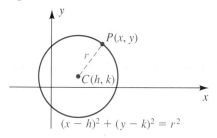

A very symmetric geometric figure in the plane is a circle with its center at the origin, since it is symmetric with respect to both coordinate axes *and* with respect to the origin. A circle with center $C(h, k)$ and radius r is illustrated in Figure 11. If $P(x, y)$ is any point on the circle, then by the distance formula (4), $d(P, C) = r$, or $[d(P, C)]^2 = r^2$, which in turn yields the following equation.

Equation of a Circle **7**

$$(x - h)^2 + (y - k)^2 = r^2$$

If the radius r is 1, then the circle is called a **unit circle**. A unit circle U with center at the origin has the equation $x^2 + y^2 = 1$.

EXAMPLE ▪ 7 Find an equation of the circle that has center $C(-2, 3)$ and contains the point $D(4, 5)$.

SOLUTION The circle is illustrated in Figure 12. Since D is on the circle, the radius r is $d(C, D)$. By the distance formula,

$$r = \sqrt{(4+2)^2 + (5-3)^2} = \sqrt{36+4} = \sqrt{40}.$$

Figure 12

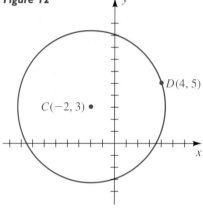

Using the equation of a circle (7) with $h = -2$, $k = 3$, and $r = \sqrt{40}$ gives us

$$(x+2)^2 + (y-3)^2 = 40.$$

In calculus we often consider lines in a coordinate plane. The following formulas are used for finding their equations.

Lines **8**

(i) Slope m:

$$m = \frac{y_2 - y_1}{x_2 - x_1}$$

(ii) Point–Slope Form:

$$y - y_1 = m(x - x_1)$$

(iii) Slope–Intercept Form:

$$y = mx + b$$

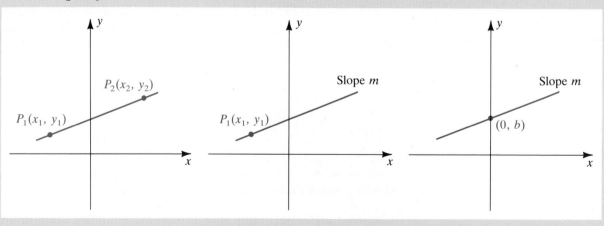

Some special types of lines and properties of their slopes are given in (9).

Special Lines | **9**

(i) Vertical: m undefined
    Horizontal: $m = 0$

(ii) Parallel: $m_1 = m_2$

(iii) Perpendicular: $m_1 m_2 = -1$

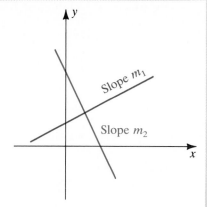

EXAMPLE ■ 8   Sketch the line through each pair of points, and find its slope:

(a) $A(-1, 4)$ and $B(3, 2)$ **(b)** $A(2, 5)$ and $B(-2, -1)$

(c) $A(4, 3)$ and $B(-2, 3)$ **(d)** $A(4, -1)$ and $B(4, 4)$

SOLUTION   The lines are sketched in Figure 13.

Figure 13
(a) $m = -\frac{1}{2}$ **(b)** $m = \frac{3}{2}$ **(c)** $m = 0$ **(d)** m undefined

From the slope formula (8)(i),

(a) $m = \dfrac{2 - 4}{3 - (-1)} = \dfrac{-2}{4} = -\dfrac{1}{2}$ **(b)** $m = \dfrac{5 - (-1)}{2 - (-2)} = \dfrac{6}{4} = \dfrac{3}{2}$

(c) $m = \dfrac{3 - 3}{4 - (-2)} = \dfrac{0}{6} = 0,$ which indicates that the line is horizontal.

(d) $m = \dfrac{4 - (-1)}{4 - 4} = \dfrac{5}{0},$ which is undefined. Note that the line is vertical.

A **linear equation** in x and y is an equation of the form $ax + by = c$ (or $ax + by + d = 0$) with a and b not both zero. The graph of a linear equation is a line.

EXAMPLE ▪ 9 Find a linear equation for the line through $A(1, 7)$ and $B(-3, 2)$.

SOLUTION The slope m of the line is

$$m = \frac{7 - 2}{1 - (-3)} = \frac{5}{4}.$$

We may use the coordinates of either A or B for (x_1, y_1) in the point–slope form (8)(ii). Using $A(1, 7)$ gives us

$$y - 7 = \tfrac{5}{4}(x - 1),$$

which is equivalent to

$$4y - 28 = 5x - 5, \quad \text{or} \quad 5x - 4y = -23.$$

EXAMPLE ▪ 10

(a) Find the slope of the line l with equation $2x - 5y = 9$.

(b) Find linear equations for the lines through $P(3, -4)$ that are parallel to l and perpendicular to l.

SOLUTION

(a) If we rewrite the equation as $5y = 2x - 9$ and divide both sides by 5, we obtain

$$y = \tfrac{2}{5}x - \tfrac{9}{5}.$$

Comparing this equation with the slope–intercept form $y = mx + b$, we see that the slope is $m = \tfrac{2}{5}$.

(b) By (ii) and (iii) of (9), the line through $P(3, -4)$ parallel to l has slope $\tfrac{2}{5}$ and the line perpendicular to l has slope $-\tfrac{5}{2}$. The corresponding equations are

$$y + 4 = \tfrac{2}{5}(x - 3), \quad \text{or} \quad 2x - 5y = 26,$$

and
$$y + 4 = -\tfrac{5}{2}(x - 3), \quad \text{or} \quad 5x + 2y = 7.$$

EXERCISES A

Exer. 1–8: Rewrite the expression without using the absolute value symbol.

1 (a) $(-5)|3 - 6|$ (b) $|-6|/(-2)$ (c) $|-7| + |4|$

2 (a) $(4)|6 - 7|$ (b) $5/|-2|$ (c) $|-1| + |-9|$

3 (a) $|4 - \pi|$ (b) $|\pi - 4|$ (c) $|\sqrt{2} - 1.5|$

4 (a) $|\sqrt{3} - 1.7|$ (b) $|1.7 - \sqrt{3}|$ (c) $|\tfrac{1}{5} - \tfrac{1}{3}|$

5 $|3 + x|$ if $x < -3$

6 $|5 - x|$ if $x > 5$

7 $|2 - x|$ if $x < 2$

8 $|7 + x|$ if $x \geq -7$

Exer. 9–12: Solve the equation by factoring.

9 $15x^2 - 12 = -8x$ 10 $15x^2 - 14 = 29x$

11 $2x(4x + 15) = 27$ 12 $x(3x + 10) = 77$

Exer. 13–16: Solve the equation by using the quadratic formula.

13 $x^2 + 4x + 2 = 0$ 14 $x^2 - 6x - 3 = 0$

15 $2x^2 - 3x - 4 = 0$ 16 $3x^2 + 5x + 1 = 0$

Exer. 17–38: Solve the inequality and express the solution in terms of intervals whenever possible.

17 $2x + 5 < 3x - 7$ 18 $x - 8 > 5x + 3$

19 $3 \le \dfrac{2x - 3}{5} < 7$ 20 $-2 < \dfrac{4x + 1}{3} \le 0$

21 $x^2 - x - 6 < 0$

22 $x^2 + 4x + 3 \ge 0$

23 $x^2 - 2x - 5 > 3$

24 $x^2 - 4x - 17 \le 4$

25 $x(2x + 3) \ge 5$ 26 $x(3x - 1) \le 4$

27 $\dfrac{x + 1}{2x - 3} > 2$ 28 $\dfrac{x - 2}{3x + 5} \le 4$

29 $\dfrac{1}{x - 2} \ge \dfrac{3}{x + 1}$ 30 $\dfrac{2}{2x + 3} \le \dfrac{2}{x - 5}$

31 $|x + 3| < 0.01$ 32 $|x - 4| \le 0.03$

33 $|x + 2| \ge 0.001$ 34 $|x - 3| > 0.002$

35 $|2x + 5| < 4$ 36 $|3x - 7| \ge 5$

37 $|6 - 5x| \le 3$ 38 $|-11 - 7x| > 6$

Exer. 39–40: Describe the set of all points $P(x, y)$ in a coordinate plane that satisfy the given condition.

39 (a) $x = -2$ (b) $y = 3$ (c) $x \ge 0$ (d) $xy > 0$
     (e) $y < 0$ (f) $|x| \le 2$ and $|y| \le 1$

40 (a) $y = -2$ (b) $x = -4$ (c) $x/y < 0$ (d) $xy = 0$
     (e) $y > 1$ (f) $|x| \ge 2$ and $|y| \ge 3$

Exer. 41–42: Find (a) $d(A, B)$ and (b) the midpoint of AB.

41 $A(4, -3)$, $B(6, 2)$ 42 $A(-2, -5)$, $B(4, 6)$

43 Show that the triangle with vertices $A(8, 5)$, $B(1, -2)$, and $C(-3, 2)$ is a right triangle, and find its area.

44 Show that the points $A(-4, 2)$, $B(1, 4)$, $C(3, -1)$, and $D(-2, -3)$ are vertices of a square.

Exer. 45–56: Sketch the graph of the equation.

45 $y = 2x^2 - 1$ 46 $y = -x^2 + 2$

47 $x = \frac{1}{4}y^2$ 48 $x = -2y^2$

49 $y = x^3 - 8$ 50 $y = -x^3 + 1$

51 $y = \sqrt{x} - 4$ 52 $y = \sqrt{x - 4}$

53 $(x + 3)^2 + (y - 2)^2 = 9$

54 $x^2 + (y - 2)^2 = 25$

55 $y = -\sqrt{16 - x^2}$ 56 $y = \sqrt{4 - x^2}$

Exer. 57–60: Find an equation of the circle that satisfies the given conditions.

57 Center $C(2, -3)$; radius 5

58 Center $C(-4, 6)$; passing through $P(1, 2)$

59 Tangent to both axes; center in the second quadrant; radius 4

60 Endpoints of a diameter $A(4, -3)$ and $B(-2, 7)$

Exer. 61–66: Find an equation of the line that satisfies the given conditions.

61 Through $A(5, -3)$; slope -4

62 Through $A(-1, 4)$; slope $\frac{2}{3}$

63 x-intercept 4; y-intercept -3

64 Through $A(5, 2)$ and $B(-1, 4)$

65 Through $A(2, -4)$; parallel to the line $5x - 2y = 4$

66 Through $A(7, -3)$; perpendicular to the line
$$2x - 5y = 8$$

c **Exer. 67–70: Use the quadratic formula to solve the equation. Give approximations to two decimal places.**

67 $0.7x^2 + 3.2x + 1.5 = 0$

68 $\sqrt{3}x^2 + \frac{3}{7}x - \frac{5}{13} = 0$

69 $375x^2 - 921x + 47 = 0$

70 $x^4 - 8x^2 + 5 = 0$

71 The cost C (in dollars) of renting a luxury car for one week is given by $C = 0.25m + 150$, where m is the number of miles driven. What range of miles will result in a rental charge that is between $200 and $300?

72 A coin is considered fair if it has an equal probability of landing with heads up or tails up when tossed. An experimenter tosses a coin 100 times and counts the number of heads H. From statistical theory, the coin will be considered fair if

$$\left| \frac{H - 50}{5} \right| \le 1.645.$$

For what range of values of H will the experimenter declare the coin fair?

73 Shown in the figure is a simple magnifier consisting of a convex lens. The object to be magnified is positioned so that its distance p from the lens is less than the focal length f. The linear magnification M is the ratio of the image size to the object size. It is shown in physics that $M = f/(f - p)$. If $f = 6$ cm, how far should the object be placed from the lens so that its image appears at least three times as large?

Exercise 73

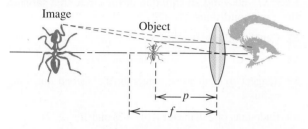

c 74 The *escape velocity* is the initial velocity v_0 with which a rocket must leave the surface of a planet so that it can eventually rise as far as desired. The escape velocity satisfies the inequality

$$\frac{1}{2} v_0^2 \geq \frac{km}{R},$$

where m and R are the mass and the radius of the planet, respectively, and k is a constant. If mass is given in kilograms and radius in meters, then $k = 6.67 \times 10^{-11}$, with units chosen so that v_0 is measured in meters per second. For which initial velocities can a rocket escape the earth, Mars, and the moon? Use the data in the following table.

	m (kg)	R (m)
Earth	6.0×10^{24}	6.2×10^6
Mars	6.4×10^{23}	3.3×10^6
Moon	7.3×10^{22}	1.7×10^6

75 The rate at which a tablet of vitamin C begins to dissolve depends on the surface area of the tablet. One brand of tablet is 2 cm long and is in the shape of a cylinder with hemispheres of diameter 0.5 cm attached to both ends (see figure). A second brand of tablet is to be manufactured in the shape of a right circular cylinder of altitude 0.5 cm.

Exercise 75

(a) Find the diameter of the second tablet so that its surface area is equal to that of the first tablet.

(b) Find the volume of each tablet.

76 A manufacturer of tin cans wishes to construct a right circular cylindrical can of height 20 cm and of capacity 3000 cm^3 (see figure). Find the inner radius r of the can.

Exercise 76

77 The braking distance d (in feet) of a car traveling v mi/hr is approximated by $d = v + (v^2/20)$. Determine velocities that result in braking distances of less than 75 ft.

78 In order for a drug to have a beneficial effect, its concentration in the bloodstream must exceed a certain value, the *minimum therapeutic level*. Suppose that the concentration c of a drug t hours after it has been taken orally is given by $c = 20t/(t^2 + 4)$ mg/L. If the minimum therapeutic level is 4 mg/L, determine when this level is exceeded.

79 The electrical resistance R (in ohms) for a pure metal wire is related to its temperature T (in °C) by the formula $R = R_0(1 + aT)$ for positive constants a and R_0.

(a) For what temperature is $R = R_0$?

(b) Assuming that the resistance is 0 if $T = -273$ °C (absolute zero), find a.

(c) Silver wire has a resistance of 1.25 ohms at 0 °C. At what temperature is the resistance 2 ohms?

80 Pharmacological products must specify recommended dosages for adults and children. Two formulas for modification of adult dosage levels for young children are

$$\text{Cowling's rule:} \quad y = \tfrac{1}{24}(t + 1)a$$
$$\text{Friend's rule:} \quad y = \tfrac{2}{25}ta,$$

where a denotes the adult dose (in milligrams) and t denotes the age of the child (in years).

(a) If $a = 100$, graph the two linear equations on the same axes for $0 \leq t \leq 12$.

(b) For what age do the two formulas specify the same dosage?

Mathematicians and Their Times

HYPATIA

HYPATIA, THE FIRST WOMAN MATHEMATICIAN whose achievements we know, was a brilliant scholar and gifted teacher who suffered a horrible death at the hands of a mob blinded by religious hatred.

Hypatia was born around A.D. 370. Her father, Theon, was a mathematician at the Alexandrian Museum, a university that Egypt's rulers had founded 700 years earlier. "In an era in which the domains of intellect and politics were almost exclusively male," notes one biographer,* "Theon was an unusually liberated person who taught an unusually gifted daughter and encouraged her to achieve things that, as far as we know, no woman before her did or perhaps even dreamed of doing."

Theon supervised his daughter's education, immersing her in an environment of learning and exploration and passing on his own great love of mathematics. Hypatia's remarkable intellectual skills, combined with great eloquence, modesty, and beauty, attracted many enthusiastic students from Europe, Africa, and Asia. She lectured on philosophy as well as mathematics and was recognized as the leader of the Neoplatonic philosophers. Students gathered in her home or followed her in the streets to hear more of her brilliant philosophical discussions or expositions on mathematics. Hypatia authored commentaries on the *Conics* of Appolonius, the *Arithmetica* of Diophantus, and the astronomical work of Ptolemy. These expositions were designed to help students understand difficult classic texts.

The scientific rationalism of the Neoplatonists challenged the more doctrinaire beliefs of the early Christian Church, whose leaders condemned the Greeks as "pagans". When Cyril became Alexandria's Christian patriarch in 412, he began a systematic plan of oppression aimed at all he saw as heretics. He led an attack against the Jews, destroying their synagogue, looting their homes, and finally expelling them from the city. When Orestes, the head of the civil government, complained,

*Ian Mueller, "Hypatia," in Louise S. Grinstein and Paul J. Campbell, eds., *Women of Mathematics: A Bibliographic Sourcebook.* New York: Greenwood Press, 1987.

a band of Cyril's supporters attacked him with stones. Rescued from the mob, Orestes tortured and executed the monk who had wounded him. Cyril, in turn, demanded the sacrifice of a virgin who followed the Greek religion and advised Orestes. Rumors spread that Hypatia was a major force inciting Orestes against Cyril. Cyril's supporters responded swiftly. In March 415, a fanatical mob barbarously murdered Hypatia.

Hypatia's tragic death also brought an eclipse to significant scientific and mathematical thought in the West, one that unfortunately lasted nearly 1000 years.

B   FUNCTIONS AND THEIR GRAPHS

The notion of *function* is basic for much of our work in calculus. We often study the effect that a change in one variable has on the values of a second variable when the second variable is a function of the first.

Definition 10

> A **function** f from a set D to a set E is a correspondence that assigns to each element x of the set D exactly one element y of the set E.

Figure 14

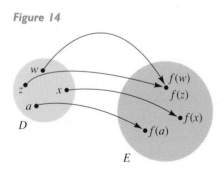

The element y of E is the **value** of f at x and is denoted by $f(x)$, read "f of x." The set D is the **domain** of the function f, and the set E is the **codomain** of f. The **range** of f is the subset of the codomain E consisting of all possible function values $f(x)$ for x in D.

We sometimes depict functions as shown in Figure 14, where the sets D and E are represented by points within regions in a plane. The curved arrows indicate that the elements $f(x)$, $f(w)$, $f(z)$, and $f(a)$ of E correspond to the elements x, w, z, and a, respectively, of D. It is important to remember that *to each x in D, there is assigned exactly one function value $f(x)$ in E.* Different elements of D, such as w and z in Figure 14, may yield the same function value in E. Until we reach Chapter 11, the phrase *f is a function* will mean that the domain and the range of f are sets of real numbers. We say that f is a **one-to-one** function if $f(x) \neq f(y)$ whenever $x \neq y$.

We usually define a function f by stating a formula or rule for finding $f(x)$, such as $f(x) = \sqrt{x - 2}$. The domain is then assumed to be the set of all real numbers such that $f(x)$ is real. Thus, for $f(x) = \sqrt{x - 2}$, the domain is the infinite interval $[2, \infty)$. If x is in the domain, we say that f **is defined at** x, or that $f(x)$ **exists**. If S is a subset of the domain, then f **is defined on** S. The terminology f **is undefined at** x means that x is not in the domain of f.

EXAMPLE ■ 1 Let $f(x) = \dfrac{\sqrt{4+x}}{1-x}$.

(a) Find the domain of f. **(b)** Find $f(5)$, $f(-2)$, $f(-a)$, and $-f(a)$.

SOLUTION

(a) Note that $f(x)$ is a real number if and only if the radicand $4+x$ is nonnegative and the denominator $1-x$ is not equal to 0. Thus, $f(x)$ exists if and only if

$$4 + x \geq 0 \quad \text{and} \quad 1 - x \neq 0$$

or, equivalently, $x \geq -4 \quad \text{and} \quad x \neq 1.$

Hence, the domain is $[-4, 1) \cup (1, \infty)$.

(b) To find values of f, we substitute for x:

$$f(5) = \frac{\sqrt{4+5}}{1-5} = \frac{\sqrt{9}}{-4} = -\frac{3}{4} \qquad f(-2) = \frac{\sqrt{4+(-2)}}{1-(-2)} = \frac{\sqrt{2}}{3}$$

$$f(-a) = \frac{\sqrt{4+(-a)}}{1-(-a)} = \frac{\sqrt{4-a}}{1+a} \qquad -f(a) = -\frac{\sqrt{4+a}}{1-a} = \frac{\sqrt{4+a}}{a-1}$$

In calculus, we often work with the **difference quotient** of a function. If f is a function, then its difference quotient is an expression of the form

$$\frac{f(x+h)-f(x)}{h}, \qquad \text{where } h \neq 0.$$

EXAMPLE ■ 2 Simplify the difference quotient

$$\frac{f(x+h)-f(x)}{h}$$

using the function $f(x) = x^2 + 6x - 4$.

SOLUTION We have

$$\frac{f(x+h)-f(x)}{h}$$

$$= \frac{[(x+h)^2 + 6(x+h) - 4] - [x^2 + 6x - 4]}{h} \qquad \text{definition of } f$$

$$= \frac{(x^2 + 2xh + h^2 + 6x + 6h - 4) - (x^2 + 6x - 4)}{h} \qquad \text{expand}$$

$$= \frac{2xh + h^2 + 6h}{h} \qquad \text{collect like terms}$$

$$= \frac{h(2x + h + 6)}{h} \qquad \text{factor out } h$$

$$= 2x + h + 6. \qquad \text{cancel } h \neq 0$$

Thus, the difference quotient simplifies to $2x + h + 6$.

Many formulas that occur in mathematics and the sciences determine functions. For instance, the formula $A = \pi r^2$ for the area A of a circle of radius r assigns to each positive real number r exactly one value of A. The letter r, which represents an arbitrary number from the domain, is an **independent variable**. The letter A, which represents a number from the range, is a **dependent variable,** since its value *depends* on the number assigned to r. If two variables r and A are related in this manner, we say that A is *a function of r*. As another example, if an automobile travels at a uniform rate of 50 mi/hr, then the distance d (in miles) traveled in time t (in hours) is given by $d = 50t$, and hence the distance *d is a function of time t*.

EXAMPLE ■ 3   A steel storage tank for propane gas is to be constructed in the shape of a right circular cylinder of altitude 10 ft with a hemisphere attached to each end. The radius r is yet to be determined. Express the volume V of the tank as a function of r.

SOLUTION   The tank is sketched in Figure 15. We may find the volume of the cylindrical part of the tank by multiplying the altitude 10 by the area πr^2 of the base of the cylinder:

$$\text{volume of cylinder} = 10(\pi r^2) = 10\pi r^2$$

The two hemispherical ends, taken together, form a sphere of radius r.

Figure 15

Using the formula for the volume of a sphere, we obtain

$$\text{volume of the two ends} = \tfrac{4}{3}\pi r^3.$$

Thus, the volume V of the tank is

$$V = \tfrac{4}{3}\pi r^3 + 10\pi r^2 = \tfrac{2}{3}\pi r^2(2r + 15).$$

This formula expresses V as a function of r.

If f is a function, we may use a graph to illustrate the change in the function value $f(x)$ as x varies through the domain of f. The **graph of a function** f with domain D is the graph of the equation $y = f(x)$ for x in D. The graph is the set of all points $(x, f(x))$, where x is in D. If a point $P(a, b)$ is on the graph, then the y-coordinate b is the function value $f(a)$. Figure 16 shows the graph of f and indicates the domain and the range. In this figure, the domain and the range are shown as closed intervals. In other examples, they may be infinite intervals or other sets of real numbers.

Figure 16

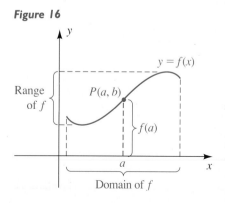

Since there is exactly one value $f(a)$ for each a in the domain, only *one* point on the graph has x-coordinate a. Thus, *every vertical line intersects the graph of a function in at most one point.* Consequently, the graph of a function cannot be a figure such as a circle, which can be intersected by a vertical line in more than one point.

The x-intercepts of the graph of a function f are the solutions of the equation $f(x) = 0$. These numbers are the **zeros** of the function. The y-intercept of the graph is $f(0)$, if it exists.

If f is an **even function**—that is, if $f(-x) = f(x)$ for every x in the domain of f—then the graph of f is symmetric with respect to the y-axis, by symmetry test (i) of (6). If f is an **odd function**—that is, if $f(-x) = -f(x)$ for every x in the domain of f—then the graph of f is symmetric with respect to the origin, by symmetry test (iii). Most functions in calculus are neither even nor odd.

The next illustration contains sketches of graphs of some common functions and indicates the symmetry, the domain, and the range for each.

ILLUSTRATION

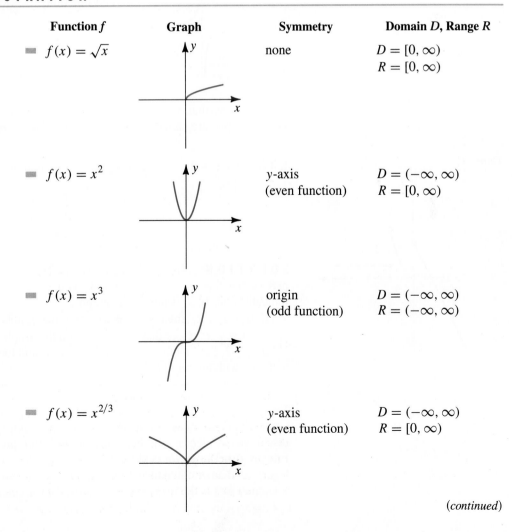

Function f	Graph	Symmetry	Domain D, Range R
$f(x) = \sqrt{x}$		none	$D = [0, \infty)$ $R = [0, \infty)$
$f(x) = x^2$		y-axis (even function)	$D = (-\infty, \infty)$ $R = [0, \infty)$
$f(x) = x^3$		origin (odd function)	$D = (-\infty, \infty)$ $R = (-\infty, \infty)$
$f(x) = x^{2/3}$		y-axis (even function)	$D = (-\infty, \infty)$ $R = [0, \infty)$

(continued)

Function f		Symmetry	Domain D, Range R
$f(x) = x^{1/3}$		origin (odd function)	$D = (-\infty, \infty)$ $R = (-\infty, \infty)$
$f(x) = \lvert x \rvert$		y-axis (even function)	$D = (-\infty, \infty)$ $R = [0, \infty)$
$f(x) = \dfrac{1}{x}$		origin (odd function)	$D = (-\infty, 0) \cup (0, \infty)$ $R = (-\infty, 0) \cup (0, \infty)$

Functions that are described by more than one expression, as in the next example, are called **piecewise-defined functions**.

Figure 17

EXAMPLE ■ 4 Sketch the graph of the function f defined as follows:

$$f(x) = \begin{cases} 2x + 3 & \text{if } x < 0 \\ x^2 & \text{if } 0 \le x < 2 \\ 1 & \text{if } x \ge 2 \end{cases}$$

SOLUTION If $x < 0$, then $f(x) = 2x + 3$, and the graph of f is part of the line $y = 2x + 3$, as indicated in Figure 17. The open circle indicates that $(0, 3)$ is not on the graph.

If $0 \le x < 2$, then $f(x) = x^2$, and the graph of f is part of the parabola $y = x^2$. Note that $(2, 4)$ is not on the graph.

If $x \ge 2$, the function values are always 1, and the graph is a horizontal half-line with endpoint $(2, 1)$.

In Example 4, we see a function whose graph is made up of several disconnected pieces. Another function with this property is the **greatest integer function** f defined by $f(x) = [\![x]\!]$, where $[\![x]\!]$ is the greatest integer less than or equal to x. If we identify $\mathbb{R}$ with points on the coordinate line, then $[\![x]\!]$ is the first integer *to the left of* (or *equal to*) x.

The following illustration gives some specific values for the greatest integer function, and Figure 18 graphically illustrates the location of x and $[\![x]\!]$ for each of these values.

ILLUSTRATION

- $[\![0.5]\!] = 0$
- $[\![1.8]\!] = 1$
- $[\![\sqrt{5}]\!] = 2$
- $[\![3]\!] = 3$
- $[\![-3]\!] = -3$
- $[\![-2.7]\!] = -3$
- $[\![-\sqrt{3}]\!] = -2$
- $[\![-0.5]\!] = -1$

Figure 18

EXAMPLE ■ 5 Sketch the graph of the greatest integer function.

SOLUTION The x- and y-coordinates of some points on the graph may be listed as follows:

Figure 19

Values of x	$f(x) = [\![x]\!]$
⋮	⋮
$-2 \le x < -1$	-2
$-1 \le x < 0$	-1
$0 \le x < 1$	0
$1 \le x < 2$	1
$2 \le x < 3$	2
⋮	⋮

Whenever x is between successive integers, the corresponding part of the graph is a segment of a horizontal line. Part of the graph is sketched in Figure 19. The graph continues indefinitely to the right and to the left.

If we know the graph of $y = f(x)$, then it is easy to sketch the graphs of functions obtained from f by *transformations* involving shifts, stretching, compressing, or reflecting. Adding or subtracting a positive constant c to each function value $f(x)$ produces a **vertical shift**. Adding c shifts the graph of f upward a distance of c units, and subtracting c shifts the graph downward, as illustrated in Figure 20. The graphs of $y = f(x + c)$

Figure 20 Vertical shifts, $c > 0$

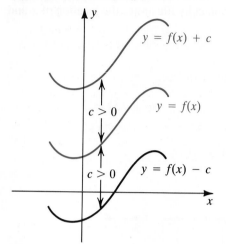

Figure 21 Horizontal shifts, $c > 0$

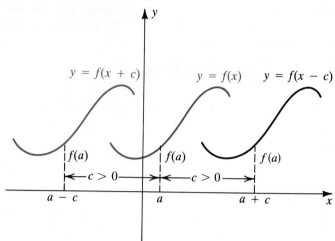

Figure 22 Vertical stretch, $c > 1$

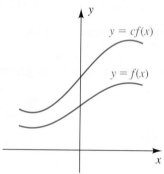

Figure 23 Vertical compression, $0 < c < 1$

Figure 24 Reflection

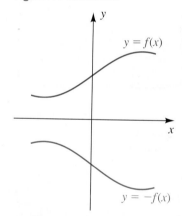

and $y = f(x - c)$ are **horizontal shifts** of the graph of $y = f(x)$, c units to the left and c units to the right, respectively, as shown in Figure 21.

If we multiply each function value $f(x)$ by a positive constant c to obtain $y = cf(x)$, then we have a **vertical stretch** if $c > 1$ (Figure 22) and a **vertical compression** if $0 < c < 1$ (Figure 23). The graphs of $y = f(x)$ and $y = -f(x)$ are **reflections** of each other across the x-axis, as shown in Figure 24. It should be noted that the x-intercepts of the graph of $y = cf(x)$ are the same as those of $y = f(x)$.

In the graphs of functions we have seen thus far, the two coordinate axes have had equal scales: One unit along the x-axis represents the same length as one unit along the y-axis. We assume equal scales on all coordinate graphs that have no scale markings or numbered "tics."

It is often desirable, however, to use graphs with unequal scales. For some functions f, a relatively small x-value may give a relatively large value for $f(x)$. For example, if $f(x) = x^4 + 100$, as x increases from 0 to 5, $f(x)$ has values between 100 and 725. If we were to use equal scales to graph this function, we would have a large amount of wasted space in which no part of the graph appears.

Figure 25
$-10 \le x \le 10, x_{scl} = 2$
$-62 \le y \le 17,529, y_{scl} = 1000$

Figure 26
$-10 \le x \le 20, x_{scl} = 2$
$-62 \le y \le 17,529, y_{scl} = 1000$

Figure 27
$-2 \le x \le 6, x_{scl} = 1$
$-62 \le y \le 89, y_{scl} = 10$

Figure 28
$-15.3 \le x \le 20.8, x_{scl} = 10$
$-4.3 \le y \le 19.3, y_{scl} = 10$

To minimize such wasted space on the screen, most computer and calculator graphs use unequal scaling. While many graphing utilities set the scales automatically, some also permit the user to set the scales x_{scl} and y_{scl} and thus designate specific units between the tic marks on each axis.

For many functions, the domain or the range of the function may be all real numbers. The computer or calculator, however, can display only a finite rectangle called the *viewing window*. The user must specify the x-interval for the viewing window by giving the left-endpoint x_{min} and the right-endpoint x_{max}. The user may also specify the y-interval, or the graphing utility may automatically calculate y_{min} and y_{max} so that the graph fits within the viewing window.

EXAMPLE ■ 6 Let $f(x) = x^4 - 7x^3 + 6x^2 + 8x + 9$. Use a graphing utility to

(a) view f with x-interval $[-10, 10]$ and y-interval $[y_{min}, y_{max}]$, where y_{min} and y_{max} represent the smallest and largest values of f, respectively, on the given x-interval

(b) estimate, without changing the y-interval from part (a), the number b such that the graph of f on the x-interval $[-10, b]$ stays within the viewing window

(c) investigate the behavior of the function near the origin

(d) view the graph with equal scales near the origin

SOLUTION

(a) To view the graph with $[-10, 10]$ as the x-interval, we set the x-range at $-10 \le x \le 10$. If your graphing utility has an automatic scaling feature, utilize it to determine that the smallest y-value is approximately -62 (at $x \approx 4.5$) and that the largest y-value is 17,529 (at $x = -10$). If this feature is not available, find these values by examining several viewing windows and then tracing to the low point (the high point at $x = -10$ should be easy to detect). Now set the y-range to $-62 \le y \le 17,529$ and graph f to obtain a figure similar to Figure 25.

(b) To obtain the same view as shown in Figure 26, we change the x-range to $-10 \le x \le 20$ while leaving the y-range alone. We then use the tracing feature to follow along on the curve until the graph of f leaves the viewing window in the first quadrant at $x = b \approx 13.5$.

(c) To place the origin in the viewing window, as in Figure 27, we change the x- and y-ranges and both scales and then trace the curve for x between -2 and 6. We determine that the function is negative for the (approximate) x-interval $[2.37, 5.63]$.

(d) Figure 28 shows the origin in the viewing window when the scales are set equal to 10. Note that much of this viewing window contains no part of the curve while part of the graph of the function, including its lowest point, falls outside this viewing window.

A function f is a **polynomial function** if $f(x)$ is a polynomial — that is, if

$$f(x) = a_n x^n + a_{n-1} x^{n-1} + \cdots + a_1 x + a_0,$$

where the coefficients $a_0, a_1, \ldots, a_n$ are real numbers and the exponents are nonnegative integers. If $a_n \neq 0$, then f has **degree** n. The following are special cases, where $a \neq 0$:

degree 0: $f(x) = a$ constant function

degree 1: $f(x) = ax + b$ linear function

degree 2: $f(x) = ax^2 + bx + c$ quadratic function

A **rational function** is a quotient of two polynomial functions. Later in the text, we shall use methods of calculus to investigate graphs of polynomial and rational functions.

An **algebraic function** is a function that can be expressed in terms of sums, differences, products, quotients, or rational powers of polynomials. For example, if

$$f(x) = 5x^4 - 2\sqrt[3]{x} + \frac{x(x^2 + 5)}{\sqrt{x^3 + \sqrt{x}}},$$

then f is an algebraic function. Functions that are not algebraic are termed **transcendental**. The trigonometric, exponential, and logarithmic functions considered later are examples of transcendental functions.

In calculus, we often build complicated functions from simpler functions by combining them in various ways, using arithmetic operations and composition. If f and g are functions, we define the **sum** $f + g$, the **difference** $f - g$, the **product** fg, and the **quotient** f/g as follows:

$$(f + g)(x) = f(x) + g(x)$$

$$(f - g)(x) = f(x) - g(x)$$

$$(fg)(x) = f(x)g(x)$$

$$\left(\frac{f}{g}\right)(x) = \frac{f(x)}{g(x)}$$

The domain of $f + g$, $f - g$, and fg is the *intersection* of the domains of f and g — that is, the numbers that are *common* to both domains. The domain of f/g consists of all numbers x in the intersection such that $g(x) \neq 0$.

EXAMPLE ■ 7   Let $f(x) = \sqrt{4 - x^2}$ and $g(x) = 3x + 1$. Find the sum, difference, product, and quotient of f and g, and specify the domain of each.

SOLUTION The domain of f is the closed interval $[-2, 2]$, and the domain of g is $\mathbb{R}$. Consequently, the intersection of their domains is $[-2, 2]$, and we obtain the following:

$$(f + g)(x) = \sqrt{4 - x^2} + (3x + 1) \qquad -2 \leq x \leq 2$$
$$(f - g)(x) = \sqrt{4 - x^2} - (3x + 1) \qquad -2 \leq x \leq 2$$
$$(fg)(x) = \sqrt{4 - x^2}(3x + 1) \qquad -2 \leq x \leq 2$$
$$\left(\frac{f}{g}\right)(x) = \frac{\sqrt{4 - x^2}}{3x + 1} \qquad -2 \leq x \leq 2 \text{ and } x \neq -\tfrac{1}{3}$$

We can also combine two functions to form a new function by the process of composition — that is, by applying one function to the result obtained from the other. Starting with functions f and g, we obtain *composite functions* $f \circ g$ and $g \circ f$ (read "f circle g" and "g circle f," respectively). The function $f \circ g$ is defined as follows.

Definition **11**

The **composite function** $f \circ g$ of f and g is defined by
$$(f \circ g)(x) = f(g(x)).$$
The domain of $f \circ g$ is the set of all x in the domain of g such that $g(x)$ is in the domain of f.

Figure 29

Figure 29 illustrates relationships between f, g, and $f \circ g$. Note that for x in the domain of g, we *first find* $g(x)$ (which must be in the domain of f) and then, *second, find* $f(g(x))$.

For the composite function $g \circ f$, we reverse this order, first finding $f(x)$ and then finding $g(f(x))$. The domain of $g \circ f$ is the set of all x in the domain of f such that $f(x)$ is in the domain of g.

EXAMPLE ■ 8 If $f(x) = x^2 - 1$ and $g(x) = 3x + 5$, find

(a) $(f \circ g)(x)$ and the domain of $f \circ g$
(b) $(g \circ f)(x)$ and the domain of $g \circ f$

SOLUTION

(a) $(f \circ g)(x) = f(g(x))$ definition of $f \circ g$
$\qquad\qquad = f(3x + 5)$ definition of g
$\qquad\qquad = (3x + 5)^2 - 1$ definition of f
$\qquad\qquad = 9x^2 + 30x + 24$ simplifying

The domain of both f and g is $\mathbb{R}$. Since for each x in $\mathbb{R}$ (the domain of g) the function value $g(x)$ is in $\mathbb{R}$ (the domain of f), the domain of $f \circ g$ is also $\mathbb{R}$.

(b) $\begin{aligned}(g \circ f)(x) &= g(f(x)) && \text{definition of } g \circ f \\ &= g(x^2 - 1) && \text{definition of } f \\ &= 3(x^2 - 1) + 5 && \text{definition of } g \\ &= 3x^2 + 2 && \text{simplifying}\end{aligned}$

Since for each x in $\mathbb{R}$ (the domain of f) the function value $f(x)$ is in $\mathbb{R}$ (the domain of g), the domain of $g \circ f$ is $\mathbb{R}$.

Note that in Example 8, $f(g(x))$ and $g(f(x))$ are not always the same; that is, $f \circ g \neq g \circ f$.

If two functions f and g both have domain $\mathbb{R}$, then the domain of $f \circ g$ and $g \circ f$ is also $\mathbb{R}$, as was illustrated in Example 8. The next example shows that the domain of a composite function may differ from those of the two given functions.

EXAMPLE ■ 9   If $f(x) = x^2 - 16$ and $g(x) = \sqrt{x}$, find

(a) $(f \circ g)(x)$ and the domain of $f \circ g$

(b) $(g \circ f)(x)$ and the domain of $g \circ f$

SOLUTION   We first note that the domain of f is $\mathbb{R}$ and the domain of g is the set of all nonnegative real numbers — that is, the interval $[0, \infty)$. We may proceed as follows.

(a) $\begin{aligned}(f \circ g)(x) &= f(g(x)) && \text{definition of } f \circ g \\ &= f(\sqrt{x}) && \text{definition of } g \\ &= (\sqrt{x})^2 - 16 && \text{definition of } f \\ &= x - 16 && \text{simplifying}\end{aligned}$

If we consider only the final expression $x - 16$, we might be led to believe that the domain of $f \circ g$ is $\mathbb{R}$, since $x - 16$ is defined for every real number x. However, this is not the case. By definition, the domain of $f \circ g$ is the set of all x in $[0, \infty)$ (the domain of g) such that $g(x)$ is in $\mathbb{R}$ (the domain of f). Since $g(x) = \sqrt{x}$ is in $\mathbb{R}$ for every x in $[0, \infty)$, it follows that the domain of $f \circ g$ is $[0, \infty)$.

(b) $\begin{aligned}(g \circ f)(x) &= g(f(x)) && \text{definition of } g \circ f \\ &= g(x^2 - 16) && \text{definition of } f \\ &= \sqrt{x^2 - 16} && \text{definition of } g\end{aligned}$

By definition, the domain of $g \circ f$ is the set of all x in $\mathbb{R}$ (the domain of f) such that $f(x) = x^2 - 16$ is in $[0, \infty)$ (the domain of g). The statement

$x^2 - 16$ is in $[0, \infty)$ is equivalent to each of the inequalities

$$x^2 - 16 \geq 0, \qquad x^2 \geq 16, \quad \text{and} \quad |x| \geq 4.$$

Thus, the domain of $g \circ f$ is $(-\infty, -4] \cup [4, \infty)$. Note that this domain is different from the domains of both f and g.

If f and g are functions such that

$$y = f(u) \quad \text{and} \quad u = g(x),$$

then substituting for u in $y = f(u)$ yields

$$y = f(g(x)).$$

For certain problems in calculus, we *reverse* this procedure; that is, given $y = h(x)$ for some function h, we find a *composite function form* $y = f(u)$ and $u = g(x)$ such that $h(x) = f(g(x))$.

EXAMPLE ■ 10   Express $y = (2x + 5)^8$ in composite function form.

SOLUTION   A simple method for solving this problem is to assume that we want to evaluate the expression $(2x + 5)^8$ by using a calculator. We might first calculate $2x + 5$ and then raise the result to the eighth power. This procedure suggests that we let

$$u = 2x + 5 \quad \text{and} \quad y = u^8,$$

which is a composite function form for $y = (2x + 5)^8$.

The method of the preceding example can be extended to other functions. In general, suppose we are given $y = h(x)$. To choose the *inside* expression $u = g(x)$ in a composite function form, ask the following question: If you were using a calculator, which part of the expression $h(x)$ would you evaluate first? The answer often leads to a suitable choice for $u = g(x)$. After choosing u, refer to $h(x)$ to determine $y = f(u)$. The following illustration provides some typical examples.

ILLUSTRATION

Function value	Choice for $u = g(x)$	Choice for $y = f(u)$
$y = (x^3 - 5x + 1)^4$	$u = x^3 - 5x + 1$	$y = u^4$
$y = \sqrt{x^2 - 4}$	$u = x^2 - 4$	$y = \sqrt{u}$
$y = \dfrac{2}{3x + 7}$	$u = 3x + 7$	$y = \dfrac{2}{u}$

The composite function form is never unique. For example, consider the first expression in the preceding illustration:

$$y = (x^3 - 5x + 1)^4$$

If n is any nonzero integer, we could choose

$$u = (x^3 - 5x + 1)^n \quad \text{and} \quad y = u^{4/n}.$$

Thus, there are an *unlimited* number of composite function forms. Generally, our goal is to choose a form such that the expression for y is simple, as we did in the illustration.

As a general rule, the composition $f \circ g$ of two functions will be more complex than either f or g. In some instances, however, the composition may turn out to be particularly simple, as the following example illustrates.

EXAMPLE ■ 11   If $f(x) = x^3 + 1$ and $g(x) = \sqrt[3]{x - 1}$, find

(a) $(f \circ g)(x)$ and the domain of $f \circ g$

(b) $(g \circ f)(x)$ and the domain of $g \circ f$

SOLUTION

(a) $(f \circ g)(x) = f(g(x))$ definition of $f \circ g$

$\qquad\qquad\quad = f(\sqrt[3]{x - 1})$ definition of g

$\qquad\qquad\quad = (\sqrt[3]{x - 1})^3 + 1$ definition of f

$\qquad\qquad\quad = x - 1 + 1 = x$ simplifying

Since for each x in $\mathbb{R}$ (the domain of g) the function value $g(x)$ is in $\mathbb{R}$ (the domain of f), the domain of $f \circ g$ is $\mathbb{R}$.

(b) A similar computation shows that $(g \circ f)(x) = x$ for all real numbers x and the domain of $g \circ f$ is also $\mathbb{R}$.

An **identity function** is a function h with the property that $h(x) = x$ for all x in the domain of h. The graph of an identity function lies along the line $y = x$. For the functions f and g of Example 11, both $f \circ g$ and $g \circ f$ are identity functions.

If the composition of two functions f and g is an identity function, then the functions are **inverses** of each other; that is, applying g to $f(x)$ returns x and applying f to $g(x)$ returns x. For inverse functions, it follows that if the point (a, b) lies on the graph of one of the functions, then the point (b, a) lies on the graph of the other. Thus, for inverse functions, the graph of either function is the reflection of the graph of the other across the line $y = x$.

EXERCISES B

1 If $f(x) = \sqrt{x-4} - 3x$, find $f(4)$, $f(8)$, and $f(13)$.

2 If $f(x) = \dfrac{x}{x-3}$, find $f(-2)$, $f(0)$, and $f(3.01)$.

Exer. 3–6: If a and h are real numbers, find and simplify
(a) $f(a)$, (b) $f(-a)$, (c) $-f(a)$, (d) $f(a+h)$, (e) $f(a) + f(h)$,
and (f) $\dfrac{f(a+h) - f(a)}{h}$, provided $h \neq 0$.

3 $f(x) = 5x - 2$ **4** $f(x) = 3 - 4x$

5 $f(x) = x^2 - x + 3$ **6** $f(x) = 2x^2 + 3x - 7$

Exer. 7–10: Find the domain of f.

7 $f(x) = \dfrac{x+1}{x^3 - 4x}$ **8** $f(x) = \dfrac{4x}{6x^2 + 13x - 5}$

9 $f(x) = \dfrac{\sqrt{2x - 3}}{x^2 - 5x + 4}$ **10** $f(x) = \dfrac{\sqrt{4x - 3}}{x^2 - 4}$

Exer. 11–12: Determine whether f is even, odd, or neither even nor odd.

11 (a) $f(x) = 5x^3 + 2x$ (b) $f(x) = |x| - 3$
(c) $f(x) = (8x^3 - 3x^2)^3$

12 (a) $f(x) = \sqrt{3x^4 + 2x^2 - 5}$
(b) $f(x) = 6x^5 - 4x^3 + 2x$
(c) $f(x) = x(x - 5)$

Exer. 13–18: Sketch the graph of f.

13 $f(x) = \begin{cases} x+2 & \text{if } x \leq -1 \\ x^3 & \text{if } |x| < 1 \\ -x+3 & \text{if } x \geq 1 \end{cases}$

14 $f(x) = \begin{cases} x-3 & \text{if } x \leq -2 \\ -x^2 & \text{if } -2 < x < 1 \\ -x+4 & \text{if } x \geq 1 \end{cases}$

15 $f(x) = \begin{cases} \dfrac{x^2 - 1}{x + 1} & \text{if } x \neq -1 \\ 2 & \text{if } x = -1 \end{cases}$

16 $f(x) = \begin{cases} \dfrac{x^2 - 4}{2 - x} & \text{if } x \neq 2 \\ 1 & \text{if } x = 2 \end{cases}$

17 (a) $f(x) = [\![x - 3]\!]$ (b) $f(x) = [\![x]\!] - 3$
(c) $f(x) = 2[\![x]\!]$ (d) $f(x) = [\![2x]\!]$

18 (a) $f(x) = [\![x + 2]\!]$ (b) $f(x) = [\![x]\!] + 2$
(c) $f(x) = \frac{1}{2}[\![x]\!]$ (d) $f(x) = [\![\frac{1}{2}x]\!]$

Exer. 19–28: Sketch, on the same coordinate plane, the graphs of f for the given values of c. (Make use of symmetry, vertical shifts, horizontal shifts, stretching, or reflecting.)

19 $f(x) = |x| + c$; $c = 0, 1, -3$

20 $f(x) = |x - c|$; $c = 0, 2, -3$

21 $f(x) = 2\sqrt{x} + c$; $c = 0, 3, -2$

22 $f(x) = \sqrt{9 - x^2} + c$; $c = 0, 1, -3$

23 $f(x) = 2\sqrt{x - c}$; $c = 0, 1, -2$

24 $f(x) = -2(x - c)^2$; $c = 0, 1, -2$

25 $f(x) = c\sqrt{4 - x^2}$; $c = 1, 3, -2$

26 $f(x) = (x + c)^3$; $c = 0, 1, -2$

27 $f(x) = (x - c)^{2/3} + 2$; $c = 0, 4, -3$

28 $f(x) = (x - 1)^{1/3} - c$; $c = 0, 2, -1$

Exer. 29–30: The graph of a function f with domain $0 \leq x \leq 4$ is shown in the figure. Sketch the graph of the given equation.

29

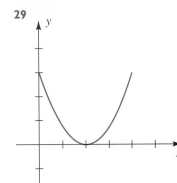

(a) $y = f(x + 3)$
(b) $y = f(x - 3)$
(c) $y = f(x) + 3$
(d) $y = f(x) - 3$
(e) $y = -3f(x)$
(f) $y = -\frac{1}{3}f(x)$
(g) $y = -f(x + 2) - 3$
(h) $y = f(x - 2) + 3$

30

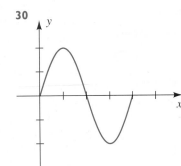

(a) $y = f(x - 2)$
(b) $y = f(x + 2)$
(c) $y = f(x) - 2$
(d) $y = f(x) + 2$
(e) $y = -2f(x)$
(f) $y = -\frac{1}{2}f(x)$
(g) $y = -f(x + 4) - 2$
(h) $y = f(x - 4) + 2$

Exer. 31–40: Use a graphing utility to examine several different views of the graph of the function f. Copy one that displays the important features of the function. Clearly indicate the scaling or the range of the viewing window selected.

31 $f(x) = \dfrac{1}{\sqrt{x^4 + 1}}$

32 $f(x) = x|x^2 - 7|$

33 $f(x) = \sqrt[3]{x^3 - 2}$

34 $f(x) = \sqrt[3]{x^3 - 8x}$

35 $f(x) = \dfrac{\sqrt{1 + x} - 1}{x}$

36 $f(x) = \dfrac{\sqrt{2 + x} - \sqrt{5}}{x - 3}$

37 $f(x) = |x^3 - x + 1|$

38 $f(x) = \frac{1}{5}x^4 + \frac{2}{3}x^3 + 1$

39 $f(x) = x^4 + 5x^3 - 6x^2 - 7x - 8$

40 $f(x) = x^5 - 7x^3 + 8x + 5$

Exer. 41–44: **(a)** Find $(f + g)(x), (f - g)(x), (fg)(x),$ and $(f/g)(x)$. **(b)** Find the domain of $f + g, f - g,$ and fg; and find the domain of f/g.

41 $f(x) = \sqrt{x + 5}$; $g(x) = \sqrt{x + 5}$

42 $f(x) = \sqrt{3 - 2x}$; $g(x) = \sqrt{x + 4}$

43 $f(x) = \dfrac{2x}{x - 4}$; $g(x) = \dfrac{x}{x + 5}$

44 $f(x) = \dfrac{x}{x - 2}$; $g(x) = \dfrac{3x}{x + 4}$

Exer. 45–52: **(a)** Find $(f \circ g)(x)$ and the domain of $f \circ g$. **(b)** Find $(g \circ f)(x)$ and the domain of $g \circ f$.

45 $f(x) = x^2 - 3x$; $g(x) = \sqrt{x + 2}$

46 $f(x) = \sqrt{x - 15}$; $g(x) = x^2 + 2x$

47 $f(x) = \sqrt{x - 2}$; $g(x) = \sqrt{x + 5}$

48 $f(x) = \sqrt{3 - x}$; $g(x) = \sqrt{x + 2}$

49 $f(x) = \sqrt{25 - x^2}$; $g(x) = \sqrt{x - 3}$

50 $f(x) = \sqrt{3 - x}$; $g(x) = \sqrt{x^2 - 16}$

51 $f(x) = \dfrac{x}{3x + 2}$; $g(x) = \dfrac{2}{x}$

52 $f(x) = \dfrac{x}{x - 2}$; $g(x) = \dfrac{3}{x}$

Exer. 53–60: Find a composite function form for y.

53 $y = (x^2 + 3x)^{1/3}$

54 $y = \sqrt[4]{x^4 - 16}$

55 $y = \dfrac{1}{(x - 3)^4}$

56 $y = 4 + \sqrt{x^2 + 1}$

57 $y = (x^4 - 2x^2 + 5)^5$

58 $y = \dfrac{1}{(x^2 + 3x - 5)^3}$

59 $y = \dfrac{\sqrt{x + 4} - 2}{\sqrt{x + 4} + 2}$

60 $y = \dfrac{\sqrt[3]{x}}{1 + \sqrt[3]{x}}$

61 If

$$f(x) = \sqrt{x^2 - 1.7} \quad \text{and} \quad g(x) = \dfrac{x^3 - x + 1}{\sqrt{x}},$$

approximate $(f \circ g)(2.4)$ and $(g \circ f)(2.4)$.

62 If $f(x) = \sqrt{x^3 + 1} - 1$, approximate $f(0.0001)$. In order to avoid calculating a zero value for $f(0.0001)$, rewrite the formula for f as

$$f(x) = \dfrac{x^3}{\sqrt{x^3 + 1} + 1}.$$

63 An open box is to be made from a rectangular piece of cardboard 20 in. × 30 in. by cutting out identical squares of area x^2 from each corner and turning up the sides (see figure). Express the volume V of the box as a function of x.

Exercise 63

64 An open-top aquarium of height 1.5 ft is to have a volume of 6 ft^3. Let x denote the length of the base, and let y denote the width (see figure on the following page).

(a) Express y as a function of x.

(b) Express the total number of square feet S of glass needed as a function of x.

Exercise 64

65 A hot-air balloon is released at 1:00 P.M. and rises vertically at a rate of 2 m/sec. An observation point is situated 100 m from a point on the ground directly below the balloon (see figure). If t denotes the time (in seconds) after 1:00 P.M., express the distance d between the balloon and the observation point as a function of t.

Exercise 65

66 Refer to Example 3. A steel storage tank for propane gas is to be constructed in the shape of a right circular cylinder of altitude 10 ft with a hemisphere attached to each end. The radius r is yet to be determined. Express the surface area S of the tank as a function of r.

67 From an exterior point P that is h units from a circle of radius r, a tangent line is drawn to the circle (see figure). Let y denote the distance from the point P to the point of tangency T.

(a) Express y as a function of h. (*Hint:* If C is the center of the circle, then PT is perpendicular to CT.)

(b) If r is the radius of the earth and h is the altitude of a space shuttle, then we can derive a formula for the maximum distance (to the earth) that an astronaut

can see from the shuttle. In particular, if $h = 200$ mi and $r \approx 4000$ mi, approximate y.

Exercise 67

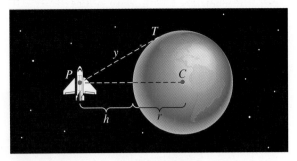

68 Triangle ABC is inscribed in a semicircle of diameter 15 (see figure).

(a) If x denotes the length of side AC, express the length y of side BC as a function of x, and state its domain. (*Hint:* Angle ACB is a right angle.)

(b) Express the area of triangle ABC as a function of x.

Exercise 68

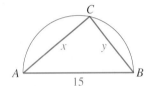

69 The relative positions of an airport runway and a 20-ft-tall control tower are shown in the figure. The beginning of the runway is at a perpendicular distance of 300 ft from the base of the tower. If x denotes the distance that an airplane has moved down the runway, express the distance d between the airplane and the control booth as a function of x.

Exercise 69

70 An open rectangular storage shelter consisting of two vertical sides, 4 ft wide, and a flat roof is to be attached to an existing structure as illustrated in the figure on the following page. The flat roof is made of tin that costs $5 per square foot, and the other two sides are made of plywood that costs $2 per square foot.

(a) If $400 is to be spent on construction, express the length y as a function of the height x.

(b) Express the volume V inside the shelter as a function of x.

Exercise 70

71 The shape of the first spacecraft in the Apollo program was a frustum of a right circular cone, a solid formed by truncating a cone by a plane parallel to its base. For the frustum shown in the figure, the radii a and b have already been determined.

Exercise 71

(a) Use similar triangles to express y as a function of h.

(b) Express the volume of the frustum as a function of h.

(c) If $a = 6$ ft and $b = 3$ ft, for what value of h is the volume of the frustum 600 ft³?

72 Suppose 5 in³ of water is poured into a conical filter and subsequently drips into a cup, as shown in the figure. Let x denote the height of the water in the filter, and let y denote the height of the water in the cup.

(a) Express the radius r shown in the figure as a function of x. (*Hint:* Use similar triangles.)

(b) Express the height y of the water in the cup as a function of x. (*Hint:* What is the sum of the two volumes shown in the figure?)

Exercise 72

C   TRIGONOMETRY

Trigonometry helps us understand angles, triangles, and circles through the use of six special *trigonometric functions*. In this section, we review some of the basic ideas and formulas of trigonometry that are especially important for calculus.

Figure 30

Figure 31

Figure 33

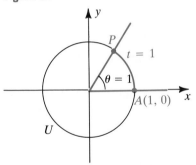

Figure 34

$\theta = t$ radians

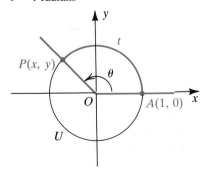

ANGLES

An angle is determined by two rays, or line segments, having the same initial point O (the **vertex** of the angle). If A and B are points on the rays l_1 and l_2 in Figure 30, we refer to **angle** *AOB*, or $\angle AOB$.

We may also interpret $\angle AOB$ as a rotation about O of the ray l_1 (the **initial side** of the angle) to a position specified by l_2 (the **terminal side**). There is no restriction on the amount or direction of rotation. We can let l_1 make several full revolutions in either direction about O before stopping at l_2, as shown by the curved arrows in Figure 31. Thus, many different angles have the same initial and terminal sides.

In a rectangular coordinate system, the **standard position** of an angle has the vertex at the origin and the initial side along the positive x-axis (see Figure 32). A counterclockwise rotation of the initial side produces a **positive angle**, whereas a clockwise rotation gives a **negative angle**. Lower-case Greek letters such as α, β, and θ are often used to denote angles.

Figure 32

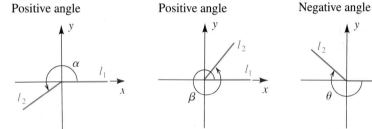

The magnitude of an angle is expressed in either degrees or radians. An angle of **degree measure** $1°$ corresponds to $1/360$ of a complete counterclockwise revolution. An angle of **radian measure** 1 corresponds to $1/(2\pi)$ of a complete counterclockwise revolution. In calculus, the radian is a more important unit of angular measure. To visualize radian measure, consider a circle of radius 1 with center at the vertex of the angle. The radian measure of an angle is the length of the arc on the circle that lies between the initial and the terminal sides. If the length of arc AP (sometimes denoted $\overset{\frown}{AP}$) is 1 unit, as in Figure 33, then θ is an angle of 1 radian. Figure 34 shows a more general case in which the radian measure of angle θ is the length t of arc AP. For convenience, we show the angle θ in Figures 33 and 34 in standard position.

Since the circumference of the unit circle is 2π, it follows that

$$2\pi \text{ radians} = 360°.$$

From this relationship between degrees and radians, we find that

$$1 \text{ radian} = \left(\frac{180}{\pi}\right)° \approx 57.29578° \quad \text{and} \quad 1° \approx 0.01745 \text{ radian}.$$

The following rules are a more general consequence of these relationships.

Conversion Rules for Radians and Degrees | **12**

(i) To change radian measure to degrees, multiply by $180/\pi$.

(ii) To change degree measure to radians, multiply by $\pi/180$.

This table displays the relationship between the radian and the degree measures of several common angles.

Radians	0	$\dfrac{\pi}{6}$	$\dfrac{\pi}{4}$	$\dfrac{\pi}{3}$	$\dfrac{\pi}{2}$	$\dfrac{2\pi}{3}$	$\dfrac{3\pi}{4}$	$\dfrac{5\pi}{6}$	π	$\dfrac{7\pi}{6}$	$\dfrac{5\pi}{4}$	$\dfrac{4\pi}{3}$	$\dfrac{3\pi}{2}$	$\dfrac{5\pi}{3}$	$\dfrac{7\pi}{4}$	$\dfrac{11\pi}{6}$	2π
Degrees	0°	30°	45°	60°	90°	120°	135°	150°	180°	210°	225°	240°	270°	300°	315°	330°	360°

CAUTION   *When radian measure of an angle is used, no units are indicated.* Thus, if an angle θ has radian measure 5, we write $\theta = 5$ instead of $\theta = 5$ radians. There should be no confusion as to whether radian or degree measure is intended, since if θ has degree measure 5°, we write $\theta = 5°$, *not* $\theta = 5$.

EXAMPLE ■ 1

(a) Express $7\pi/9$ radians in degrees.

(b) Express 105° in radians.

SOLUTION

(a) By (12)(i), to convert radians to degrees, we multiply $7\pi/9$ by $180/\pi$ to obtain 140°.

(b) By (12)(ii), to convert degrees to radians, we multiply 105° by $\pi/180$ to obtain $7\pi/12$ radians.

Figure 35

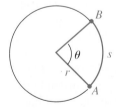

A **central angle** of a circle is an angle θ whose vertex is at the center of the circle, as illustrated in Figure 35. We say that arc *AB subtends* θ or that θ *is subtended by arc AB*. A central angle also determines a circular sector *AOB*. The length of the arc *AB* and the area of the sector *AOB* are functions of the radian measure of the angle and the radius of the circle.

Length of a Circular Arc and Area of a Circular Sector | **13**

If an arc of length s on a circle of radius r subtends a central angle of radian measure θ, and if A is the area of the circular sector determined by θ, then

(i) $$s = r\theta$$

and

(ii) $$A = \tfrac{1}{2}r^2\theta.$$

EXAMPLE ■ 2   An arc of length 6 cm on a circle of radius 3 cm subtends a central angle θ.

(a) Find the radian measure of θ.

(b) Find the area of the circular sector determined by θ.

SOLUTION

(a) From (13)(i), $s = r\theta$, so

$$\theta = \frac{s}{r} = \frac{6}{3} = 2 \text{ radians.}$$

(b) From (13)(ii), the area A of the circular sector is

$$\tfrac{1}{2}r^2\theta = \tfrac{1}{2}(3^2)(2) = 9 \text{ cm}^2.$$

TRIGONOMETRIC FUNCTIONS

The six trigonometric functions are the **sine, cosine, tangent, cosecant, secant,** and **cotangent**. We denote them by **sin, cos, tan, csc, sec,** and **cot,** respectively.

We may define the trigonometric functions in terms of either an angle θ or a real number x. We begin with the angle approach. Let θ be any angle in standard position, and let C be a circle of radius r with center at the origin. Let P be a point on the circle that lies on the terminal side of the angle and has coordinates (x, y). The trigonometric functions are defined as ratios involving the values x, y, and r.

Trigonometric Functions
of Any Angle 14

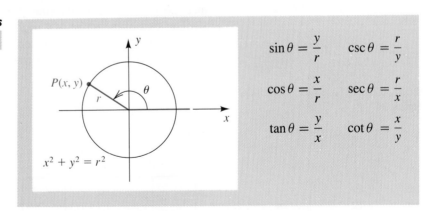

$$\sin \theta = \frac{y}{r} \qquad \csc \theta = \frac{r}{y}$$

$$\cos \theta = \frac{x}{r} \qquad \sec \theta = \frac{r}{x}$$

$$\tan \theta = \frac{y}{x} \qquad \cot \theta = \frac{x}{y}$$

In the special case where θ is an acute angle (between 0 and $\pi/2$), the vertical segment PQ from P to a point Q on the x-axis determines a right triangle POQ. The value of the x-coordinate of P is equal to the length of the segment OQ, and the value of the y-coordinate of P is equal to the length of the segment PQ.

The trigonometric functions of θ can also be expressed as ratios involving the hypotenuse c, the adjacent side a, and the opposite side b of a right triangle.

Trigonometric Functions
of an Acute Angle | 15

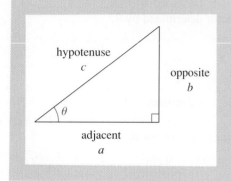

$$\sin \theta = \frac{\text{opposite}}{\text{hypotenuse}} = \frac{b}{c} \qquad \csc \theta = \frac{\text{hypotenuse}}{\text{opposite}} = \frac{c}{b}$$

$$\cos \theta = \frac{\text{adjacent}}{\text{hypotenuse}} = \frac{a}{c} \qquad \sec \theta = \frac{\text{hypotenuse}}{\text{adjacent}} = \frac{c}{a}$$

$$\tan \theta = \frac{\text{opposite}}{\text{adjacent}} = \frac{b}{a} \qquad \cot \theta = \frac{\text{adjacent}}{\text{opposite}} = \frac{a}{b}$$

Now that we have a definition of the trigonometric functions using the angle approach, it is easy to define these functions for an arbitrary real number x.

Trigonometric Functions
of a Real Number | 16

The **value of a trigonometric function at a real number x** is its value at an angle of x radians.

From this definition, we see that there is no difference between trigonometric functions of angles measured in radians and trigonometric functions of real numbers. We can interpret sin 2, for example, as *either* the sine of an angle of 2 radians *or* the sine of the real number 2.

The sign of the value of a trigonometric function of an angle depends on the quadrant containing the terminal side of θ. For example, if θ is in quadrant IV (as in Figure 36), then the point $P(x, y)$ has $x > 0$ and $y < 0$,

Figure 36

Figure 37 Positive functions

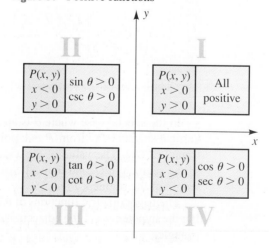

so $\cos\theta = x/r$ and $\sec\theta = r/x$ are positive while the other four functions are negative. Figure 37 indicates the *positive* trigonometric functions for each quadrant.

EXAMPLE ■ 3 Find the values of the trigonometric functions for $\theta = 3\pi/4$.

SOLUTION For $\theta = 3\pi/4$, the point $P(x, y)$ is in quadrant II on the unit circle U and on the line $y = -x$, as illustrated in Figure 38. Since $x^2 + y^2 = 1$ and $y = -x$, we have $x^2 + (-x)^2 = 1$, so $2x^2 = 1$. Thus,

$$\cos\frac{3\pi}{4} = x = -\frac{\sqrt{2}}{2} \quad \text{and} \quad \sin\frac{3\pi}{4} = y = \frac{\sqrt{2}}{2}.$$

The other trigonometric function values are

$$\tan\frac{3\pi}{4} = -1, \qquad \cot\frac{3\pi}{4} = -1,$$

$$\sec\frac{3\pi}{4} = -\sqrt{2}, \quad \text{and} \quad \csc\frac{3\pi}{4} = \sqrt{2}.$$

Figure 38

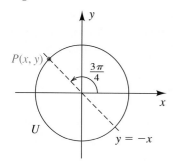

Let us now consider the domain of the trigonometric functions. Since these functions are ratios, there is the possibility of an undefined fraction occurring because a denominator is 0. Since r, the radius of a circle, is always positive, $\sin\theta = y/r$ and $\cos\theta = x/r$ are defined for all angles. Hence, the domain of the sine and the cosine functions consists of all real numbers. The cosecant and the cotangent functions are undefined when $y = 0$, which occurs when the terminal side of the angle lies along the x-axis; that is, when θ is an integer multiple of π. Similarly, the secant and the tangent functions are undefined when $x = 0$, which occurs when the terminal side lies along the y-axis; that is, when $\theta = \pi/2$ plus an integer multiple of π.

From the definition of the trigonometric functions of any angle, $|x| \le r$ and $|y| \le r$ or, equivalently, $|x/r| \le 1$ and $|y/r| \le 1$. Thus,

$$|\sin\theta| \le 1, \qquad |\cos\theta| \le 1, \qquad |\csc\theta| \ge 1, \quad \text{and} \quad |\sec\theta| \ge 1$$

for every θ in the domains of these functions.

TRIGONOMETRIC IDENTITIES

We next examine some important relationships or identities that exist among the trigonometric functions. Trigonometric identities provide us with ways in which to rewrite expressions in forms that may be simpler to work with.

Several **fundamental identities** follow directly from the definition of the trigonometric functions of any angle.

Reciprocal and Ratio Identities **17**

$$\csc \theta = \frac{1}{\sin \theta} \qquad \tan \theta = \frac{\sin \theta}{\cos \theta}$$

$$\sec \theta = \frac{1}{\cos \theta} \qquad \cot \theta = \frac{\cos \theta}{\sin \theta}$$

$$\cot \theta = \frac{1}{\tan \theta}$$

A second set of identities, called the **Pythagorean identities,** can be formulated from the observation that if $P(x, y)$ is a point on the unit circle centered at the origin, then $\sin \theta = y$ and $\cos \theta = x$. Thus, the equation of the circle, $x^2 + y^2 = 1$, or

$$y^2 + x^2 = 1, \quad \text{is equivalent to} \quad \sin^2 \theta + \cos^2 \theta = 1.$$

CAUTION   The notation $\sin^2 \theta$ represents the square of the sine of θ; that is, $\sin^2 \theta = (\sin \theta)(\sin \theta)$. To indicate the sine of the square of θ, we write $\sin(\theta^2)$.

Dividing both sides of the identity $\sin^2 \theta + \cos^2 \theta = 1$ by $\sin^2 \theta$ or $\cos^2 \theta$ yields two more useful identities.

Pythagorean Identities **18**

$$\sin^2 \theta + \cos^2 \theta = 1$$

$$1 + \tan^2 \theta = \sec^2 \theta$$

$$1 + \cot^2 \theta = \csc^2 \theta$$

EXAMPLE ▪ 4   Express $\sqrt{16 - x^2}$ in terms of a trigonometric function of θ without radicals by making the trigonometric substitution

$$x = 4 \sin \theta \quad \text{for} \quad -\frac{\pi}{2} \le \theta \le \frac{\pi}{2}.$$

SOLUTION   We let $x = 4 \sin \theta$. Then

$$\sqrt{16 - x^2} = \sqrt{16 - (4 \sin \theta)^2}$$

$$= \sqrt{16 - 16 \sin^2 \theta}$$

$$= \sqrt{16(1 - \sin^2 \theta)}$$

$$= \sqrt{16 \cos^2 \theta}$$

$$= 4 \cos \theta.$$

The last equality is true because if $-\pi/2 \le \theta \le \pi/2$, then $\cos \theta \ge 0$ and so $\sqrt{\cos^2 \theta} = \cos \theta$.

Some trigonometric identities state relationships that hold among the lengths of sides of a triangle and the sine and cosine of the angles of the triangle. Of particular usefulness in applications are the *law of sines* and the *law of cosines*.

Law of Sines and
Law of Cosines   19

If *ABC* is a triangle labeled as shown, then the following relationships are true.

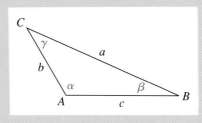

The law of sines:
$$\frac{\sin \alpha}{a} = \frac{\sin \beta}{b} = \frac{\sin \gamma}{c}$$

The law of cosines:
$$a^2 = b^2 + c^2 - 2bc \cos \alpha$$
$$b^2 = a^2 + c^2 - 2ac \cos \beta$$
$$c^2 = a^2 + b^2 - 2ab \cos \gamma$$

Note that if *ABC* is a right triangle with $\gamma = \pi/2$, then the third equation in the law of cosines becomes the familiar Pythagorean theorem, $c^2 = a^2 + b^2$, since $\cos \pi/2 = 0$. Thus, we may regard the law of cosines as a generalization of the Pythagorean theorem.

Many other important relationships exist among the trigonometric functions.

Additional Trigonometric
Identities   20

Formulas for negatives:

$$\sin(-\theta) = -\sin \theta \qquad \cos(-\theta) = \cos \theta \qquad \tan(-\theta) = -\tan \theta$$
$$\csc(-\theta) = -\csc \theta \qquad \sec(-\theta) = \sec \theta \qquad \cot(-\theta) = -\cot \theta$$

for any real number θ.

Addition and subtraction formulas for the sine and cosine:

$$\sin(\alpha + \beta) = \sin \alpha \cos \beta + \sin \beta \cos \alpha$$
$$\sin(\alpha - \beta) = \sin \alpha \cos \beta - \sin \beta \cos \alpha$$
$$\cos(\alpha + \beta) = \cos \alpha \cos \beta - \sin \alpha \sin \beta$$
$$\cos(\alpha - \beta) = \cos \alpha \cos \beta + \sin \alpha \sin \beta$$

for any real numbers α and β.

Double-angle formulas for the sine and cosine:

$$\sin 2\theta = 2 \sin \theta \cos \theta$$
$$\cos 2\theta = \cos^2 \theta - \sin^2 \theta = 1 - 2 \sin^2 \theta = 2 \cos^2 \theta - 1$$

(continued)

> *Half-angle formulas* for the sine and cosine:
>
> $$\sin^2 \theta = \frac{1 - \cos 2\theta}{2}$$
>
> $$\cos^2 \theta = \frac{1 + \cos 2\theta}{2}$$
>
> for any real number θ.

The negative formulas show that the sine, tangent, cosecant, and cotangent functions are odd and the cosine and secant functions are even. Other trigonometric identities useful in calculus are listed on the inside back cover of this text.

EXAMPLE ■ 5 Verify the following addition formula for the tangent function.

$$\tan(\alpha + \beta) = \frac{\tan \alpha + \tan \beta}{1 - \tan \alpha \tan \beta}$$

SOLUTION

$$\tan(\alpha + \beta) = \frac{\sin(\alpha + \beta)}{\cos(\alpha + \beta)} \qquad \text{tangent identity}$$

$$= \frac{\sin \alpha \cos \beta + \sin \beta \cos \alpha}{\cos \alpha \cos \beta - \sin \alpha \sin \beta} \qquad \begin{array}{l}\text{addition formulas for} \\ \text{sine and cosine}\end{array}$$

If $\cos \alpha \cos \beta \neq 0$, then we may divide the numerator and the denominator by $\cos \alpha \cos \beta$, thereby obtaining 1 as the first term in the denominator.

$$\tan(\alpha + \beta) = \frac{\dfrac{\sin \alpha \cos \beta}{\cos \alpha \cos \beta} + \dfrac{\sin \beta \cos \alpha}{\cos \alpha \cos \beta}}{\dfrac{\cos \alpha \cos \beta}{\cos \alpha \cos \beta} - \dfrac{\sin \alpha \sin \beta}{\cos \alpha \cos \beta}} \qquad \text{divide by } \cos \alpha \cos \beta$$

$$= \frac{\tan \alpha + \tan \beta}{1 - \tan \alpha \tan \beta} \qquad \text{simplify}$$

EVALUATING TRIGONOMETRIC FUNCTIONS

There are a variety of ways to find the values of a trigonometric function, including the use of scientific calculators. For certain important special cases, we can obtain them from familiar right triangles. Figure 39 shows a right triangle with acute angles of $\pi/6$ and $\pi/3$ and an isosceles right triangle with acute angles of $\pi/4$. From these triangles, the following values can be determined.

Special Values of the Trigonometric Functions **21**

θ (Radians)	θ (Degrees)	$\sin\theta$	$\cos\theta$	$\tan\theta$	$\cot\theta$	$\sec\theta$	$\csc\theta$
$\dfrac{\pi}{6}$	$30°$	$\dfrac{1}{2}$	$\dfrac{\sqrt{3}}{2}$	$\dfrac{\sqrt{3}}{3}$	$\sqrt{3}$	$\dfrac{2\sqrt{3}}{3}$	2
$\dfrac{\pi}{4}$	$45°$	$\dfrac{\sqrt{2}}{2}$	$\dfrac{\sqrt{2}}{2}$	1	1	$\sqrt{2}$	$\sqrt{2}$
$\dfrac{\pi}{3}$	$60°$	$\dfrac{\sqrt{3}}{2}$	$\dfrac{1}{2}$	$\sqrt{3}$	$\dfrac{\sqrt{3}}{3}$	2	$\dfrac{2\sqrt{3}}{3}$

Figure 39

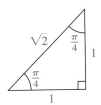

Since these particular values occur frequently in work involving trigonometry, it is a good idea either to memorize the table or to be able to find the values quickly by using the triangles in Figure 39.

Another method for finding values of trigonometric functions for an angle θ uses the **reference angle** of θ, which is the acute angle θ_R that the terminal side of θ makes with the x-axis when θ is in standard position. Figure 40 illustrates the reference angle θ_R for an angle in each of the four quadrants. To find the value of a trigonometric function at angle θ, we first determine the value for the reference angle θ_R of θ and then prefix with the appropriate sign.

Figure 40 Reference angles

(a) Quadrant I

(b) Quadrant II

(c) Quadrant III

(d) Quadrant IV

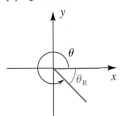

E X A M P L E ▪ 6 Find $\sin\theta$ and $\cos\theta$ for the following:

(a) $\theta = \dfrac{5\pi}{6}$ **(b)** $\theta = \dfrac{7\pi}{4}$

Figure 41

(a)

(b)

Figure 42

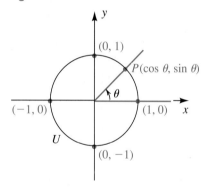

SOLUTION   We sketch the angles and their reference angles in Figure 41. Using the table of special values (20) gives the following:

(a) $\sin \dfrac{5\pi}{6} = \sin \dfrac{\pi}{6} = \dfrac{1}{2}$

$\cos \dfrac{5\pi}{6} = -\cos \dfrac{\pi}{6} = -\dfrac{\sqrt{3}}{2}$

(b) $\sin \dfrac{7\pi}{4} = -\sin \dfrac{\pi}{4} = -\dfrac{\sqrt{2}}{2}$

$\cos \dfrac{7\pi}{4} = \cos \dfrac{\pi}{4} = \dfrac{\sqrt{2}}{2}$

GRAPHS OF THE TRIGONOMETRIC FUNCTIONS

To graph the sine function, we first study the variation of $\sin \theta$ as θ increases. For convenience, consider arcs along the unit circle U in Figure 42. Since $r = 1$, the formulas $\sin \theta = y/r$ and $\cos \theta = x/r$ take on the simpler forms $\sin \theta = y$ and $\cos \theta = x$. Thus, the coordinates (x, y) of the point P corresponding to θ can be written as $(\cos \theta, \sin \theta)$. At $\theta = 0$, P is the point $(1, 0)$. As θ increases from 0 to 2π, the point $P(\cos \theta, \sin \theta)$ travels around the unit circle once in a counterclockwise direction. Observation of the y-coordinate leads to the following facts, where arrows are used to indicate the variations of θ and $\sin \theta$. (For example, $0 \to \pi/2$ means that θ increases from 0 to $\pi/2$, and $0 \to 1$ means that $\sin \theta$ increases from 0 to 1.)

$$\theta: \quad 0 \quad \to \quad \frac{\pi}{2} \quad \to \quad \pi \quad \to \quad \frac{3\pi}{2} \quad \to \quad 2\pi$$

$$\sin \theta: \quad 0 \quad \to \quad 1 \quad \to \quad 0 \quad \to \quad -1 \quad \to \quad 0$$

If we let P continue to travel around U, the same pattern repeats in θ-intervals $[2\pi, 4\pi]$ and $[4\pi, 6\pi]$. In general, the values of $\sin \theta$ repeat in all successive intervals of length 2π. A function f with domain D is **periodic** if there is a positive real number k such that $x + k$ is in D and $f(x + k) = f(x)$ for every x in D. If a smallest such positive number k exists, it is called the **period** of f. We have seen that the sine function is periodic with period 2π. Using these facts and plotting several points corresponding to the special values of θ gives the graph of the sine function, shown in Figure 43, where we have used $\theta = x$ for the independent variable (measured in radians or real numbers).

The graph of the cosine function can be found in a similar fashion by studying the behavior of the horizontal component of P as θ increases. The graphs of all the trigonometric functions are given in Figure 43. Note that the period of the tangent and the cotangent functions is π.

Figure 43

$y = \sin x$

$y = \cos x$

$y = \tan x$

$y = \csc x$

$y = \sec x$

$y = \cot x$

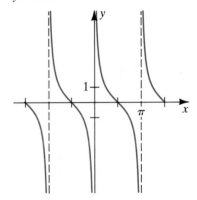

EXAMPLE ■ 7   Sketch the graph of the function $f(x) = 2 \sin x$.

SOLUTION   We begin by sketching the graph of $\sin x$, as in Figure 43. We can then stretch this graph by multiplying each of the y-coordinates by a factor of 2 to obtain the graph of $y = 2 \sin x$, shown in Figure 44.

Figure 44

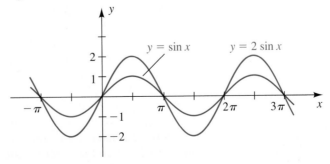

TRIGONOMETRIC EQUATIONS

A **trigonometric equation** is an equation that contains trigonometric expressions. Each fundamental identity is an example of a trigonometric equation where every number (or angle) in the domain of the variable is a solution of the equation. If a trigonometric equation is not an identity, we often find solutions by using techniques similar to those used for algebraic equations. The main difference is that we first solve the trigonometric equation for sin x, cos θ, and so on, and then find values of x or θ that satisfy the equation. *If degree measure is not specified, then solutions of a trigonometric equation should be expressed in radian measure (or as real numbers).*

E X A M P L E ■ 8   Find the solutions of the equation $\sin \theta = \frac{1}{2}$ if

(a) θ is in the interval $[0, 2\pi)$

(b) θ is any real number

S O L U T I O N

(a) If $\sin \theta = \frac{1}{2}$, then the reference angle for θ is $\theta_R = \pi/6$. If we regard θ as an angle in standard position, then, since $\sin \theta > 0$, the terminal side is in either quadrant I or quadrant II, as illustrated in Figure 45. Thus there are two solutions for $0 \le \theta < 2\pi$:

$$\theta = \frac{\pi}{6} \quad \text{and} \quad \theta = \pi - \frac{\pi}{6} = \frac{5\pi}{6}$$

(b) Since the sine function has period 2π, we may obtain all solutions by adding multiples of 2π to $\pi/6$ and $5\pi/6$. This procedure gives us

$$\theta = \frac{\pi}{6} + 2\pi n \quad \text{and} \quad \theta = \frac{5\pi}{6} + 2\pi n \quad \text{for every integer } n.$$

Figure 45

NOTE   An alternative (graphical) solution involves determining where the graph of $y = \sin \theta$ intersects the horizontal line $y = \frac{1}{2}$, as shown in Figure 46.

Figure 46

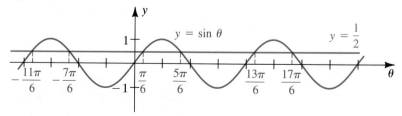

Most calculators have ⎡SIN⎤, ⎡COS⎤, and ⎡TAN⎤ keys to approximate values for these trigonometric functions. The values of the cosecant, secant, and cotangent functions can also be found using a calculator (by using the ⎡1/x⎤ key) and the formulas in (17).

CAUTION   Calculators have both a radian and a degree mode. Choosing the wrong mode on a calculator is a very common error made when evaluating trigonometric functions. For example, in radian mode, the calculator gives the approximate value

$$\sin 1.3 \approx 0.963558185417,$$

but in degree mode, it yields a different value:

$$\sin 1.3 \approx 0.022687333573$$

The next example illustrates the use of a calculator in solving a trigonometric equation.

    EXAMPLE ▪ 9   Approximate, to the accuracy of your calculator (in radian mode), the solutions of the following equation in the interval $[0, 2\pi)$:

$$5 \sin\theta \tan\theta - 10 \tan\theta + 3 \sin\theta - 6 = 0$$

SOLUTION

$5 \sin\theta \tan\theta - 10 \tan\theta + 3 \sin\theta - 6 = 0$	given
$5 \tan\theta (\sin\theta - 2) + 3(\sin\theta - 2) = 0$	factor groups
$(5 \tan\theta + 3)(\sin\theta - 2) = 0$	factor out $\sin\theta - 2$
$5 \tan\theta + 3 = 0, \quad \sin\theta - 2 = 0$	set each factor equal to 0
$\tan\theta = -\frac{3}{5}, \qquad \sin\theta = 2$	solve for $\tan\theta$ and $\sin\theta$

The equation $\sin\theta = 2$ has no solution, since $\sin\theta \le 1$ for every θ. To solve $\tan\theta = -\frac{3}{5}$, we need to find the number θ whose tangent is $-\frac{3}{5}$.

Many scientific calculators have a key labeled $\boxed{\text{TAN}^{-1}}$ that can be used to find such a number. In this case, the calculator gives

$$\theta = \tan^{-1} \approx -0.540419500271.$$

(We will discuss inverse trigonometric functions more in Chapter 6.) Hence, the reference angle is $\theta_R \approx 0.540419500271$, which we store temporarily in a calculator memory. Then, without re-entering any numbers, we obtain the following solutions in quadrants II and IV:

$$\theta = \pi - \theta_R \approx 2.60117315332,$$

$$\theta = 2\pi - \theta_R \approx 5.74276580691$$

We may not always report all the digits shown on the final calculator screen, but we try not to round intermediate results and not to re-enter numbers.

The next example illustrates how a graphing utility can aid in solving trigonometric equations.

EXAMPLE ■ 10   Find the solutions of the following equation that are in the interval $[0, 2\pi)$:

$$\sin x + \sin 2x + \sin 3x = 0$$

SOLUTION   Since $2\pi \approx 6.3$ and $|\sin \theta| \leq 1$ for $\theta = x, 2x$, and $3x$, we choose the viewing window $[0, 6.3]$ by $[-3, 3]$ and obtain a sketch similar to Figure 47. Using the zoom and tracing features, we obtain the following approximations for the x-intercepts — that is, the *approximate* solutions of the given equation in $[0, 2\pi)$:

$$0, \quad 1.57, \quad 2.09, \quad 3.14, \quad 4.19, \quad 4.71$$

Figure 47

$0 \leq x \leq 6.3, -3 \leq y \leq 3$

The approximate solution 3.14 might lead us to guess that π is a solution. Checking $x = \pi$ in the given equation confirms that π is an exact solution.

We will now apply algebraic methods to find the *exact* solutions, *knowing that there should be six solutions in this interval.* We use the addition and double-angle formulas (19) to change the form of the given equation:

$$\sin x + \sin 2x + \sin 3x = 2 \sin x \cos x (2 \cos x + 1) = 0$$

Setting the factors equal to 0 gives us

$$\sin x = 0 \qquad \text{or} \quad x = 0, \pi$$

$$\cos x = 0 \qquad \text{or} \quad x = \frac{\pi}{2}, \frac{3\pi}{2}$$

$$\cos x = -\frac{1}{2} \quad \text{or} \quad x = \frac{2\pi}{3}, \frac{4\pi}{3}.$$

Frequently, exact solutions are "lost" when careless algebraic work is performed. By comparing the exact solutions

$$0, \quad \frac{\pi}{2}, \quad \frac{2\pi}{3}, \quad \pi, \quad \frac{4\pi}{3}, \quad \frac{3\pi}{2}$$

with the numerical estimates obtained from the graph, we confirm that the number of solutions and their approximate values agree.

In the preceding example, we were able to use a graphing utility to help us find the exact solutions of the equation. For many equations that occur in applications, however, it is possible only to approximate the solutions.

EXAMPLE ■ 11   In Boston, the number of hours of daylight $D(t)$ at a particular time of the year may be approximated by

$$D(t) = 3 \sin \left[\frac{2\pi}{365}(t - 79) \right] + 12,$$

with t in days and $t = 0$ corresponding to January 1. How many days of the year have more than 10.5 hr of daylight?

SOLUTION The graph of D is shown in Figure 48. If we can find two numbers a and b with $D(a) = 10.5$, $D(b) = 10.5$, and $0 < a < b < 365$, then there will be more than 10.5 hr of daylight in the tth day of the year if $a < t < b$.

Let us solve the equation $D(t) = 10.5$ as follows:

$$3 \sin\left[\frac{2\pi}{365}(t - 79)\right] + 12 = 10.5 \qquad \text{let } D(t) = 10.5$$

$$3 \sin\left[\frac{2\pi}{365}(t - 79)\right] = -1.5 \qquad \text{subtract 12}$$

$$\sin\left[\frac{2\pi}{365}(t - 79)\right] = -0.5 = -\frac{1}{2} \qquad \text{divide by 3}$$

If $\sin \theta = -\frac{1}{2}$, then the reference angle is $\pi/6$ and the angle θ is in either quadrant III or quadrant IV. Thus, we can find the numbers a and b by solving the equations

$$\frac{2\pi}{365}(t - 79) = \frac{7\pi}{6} \quad \text{and} \quad \frac{2\pi}{365}(t - 79) = \frac{11\pi}{6}.$$

From the first of these equations, we obtain

$$t - 79 = \frac{7\pi}{6} \cdot \frac{365}{2\pi} = \frac{2555}{12} \approx 213,$$

and hence, $\qquad t \approx 213 + 79, \quad \text{or} \quad t \approx 292.$

Similarly, the second equation gives us $t \approx 414$. Since the period of the function D is 365 days (see Figure 48), we obtain

$$t \approx 414 - 365, \quad \text{or} \quad t \approx 49.$$

Thus, there will be at least 10.5 hr of daylight from $t = 49$ to $t = 292$ — that is, for 242 days of the year.

Figure 48

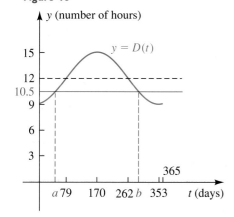

EXERCISES C

Exer. 1–2: Find the exact radian measure of the angle.

1 (a) $150°$ (b) $120°$ (c) $450°$ (d) $-60°$

2 (a) $225°$ (b) $210°$ (c) $630°$ (d) $-135°$

Exer. 3–4: Find the exact degree measure of the angle.

3 (a) $\dfrac{2\pi}{3}$ (b) $\dfrac{5\pi}{6}$ (c) $\dfrac{3\pi}{4}$ (d) $-\dfrac{7\pi}{2}$

4 (a) $\dfrac{11\pi}{6}$ (b) $\dfrac{4\pi}{3}$ (c) $\dfrac{11\pi}{4}$ (d) $-\dfrac{5\pi}{2}$

Exer. 5–6: Find the length of arc that subtends a central angle θ on a circle of diameter d and the area of the circular sector that θ determines.

5 $\theta = 50°$; $d = 16$

6 $\theta = 2.2$; $d = 120$

Exer. 7–8: Find the values of x and y in the figure.

7

8

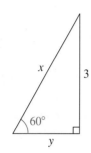

Exer. 9–12: Find the values of the trigonometric functions if θ is an acute angle.

9 $\sin\theta = \frac{3}{5}$ **10** $\cos\theta = \frac{8}{17}$

11 $\tan\theta = \frac{5}{12}$ **12** $\cot\theta = 1$

Exer. 13–14: If θ is in standard position and Q is on the terminal side of θ, find the values of the trigonometric functions of θ.

13 $Q(4, -3)$ **14** $Q(-8, -15)$

Exer. 15–20: Refer to Example 4. Make the indicated trigonometric substitution and use fundamental identities to obtain a simplified trigonometric expression that contains no radicals.

15 $\sqrt{16 - x^2}$; $x = 4\sin\theta$ for $-\dfrac{\pi}{2} \le \theta \le \dfrac{\pi}{2}$

16 $\dfrac{x^2}{\sqrt{9 - x^2}}$; $x = 3\sin\theta$ for $-\dfrac{\pi}{2} < \theta < \dfrac{\pi}{2}$

17 $\dfrac{x}{\sqrt{25 + x^2}}$; $x = 5\tan\theta$ for $-\dfrac{\pi}{2} < \theta < \dfrac{\pi}{2}$

18 $\dfrac{\sqrt{x^2 + 4}}{x^2}$; $x = 2\tan\theta$ for $-\dfrac{\pi}{2} < \theta < \dfrac{\pi}{2}$

19 $\dfrac{\sqrt{x^2 - 9}}{x}$; $x = 3\sec\theta$ for $0 < \theta < \dfrac{\pi}{2}$

20 $x^3\sqrt{x^2 - 25}$; $x = 5\sec\theta$ for $0 < \theta < \dfrac{\pi}{2}$

Exer. 21–26: Find the exact value.

21 (a) $\sin(2\pi/3)$ **(b)** $\sin(-5\pi/4)$

22 (a) $\cos 150°$ **(b)** $\cos(-60°)$

23 (a) $\tan(5\pi/6)$ **(b)** $\tan(-\pi/3)$

24 (a) $\cot 120°$ **(b)** $\cot(-150°)$

25 (a) $\sec(2\pi/3)$ **(b)** $\sec(-\pi/6)$

26 (a) $\csc 240°$ **(b)** $\csc(-330°)$

c **Exer. 27–32:** Find approximate values using a calculator.

27 (a) $\sin 67°$ **(b)** $\csc 25°$

28 (a) $\sin(-2.743)$ **(b)** $\csc 51.314$

29 (a) $\cos(-12°)$ **(b)** $\sec 39°$

30 (a) $\cos(-4.2)$ **(b)** $\sec 15.9$

31 (a) $\tan 15$ **(b)** $\cot 5$

32 (a) $\tan 1.8$ **(b)** $\cot(-3)$

Exer. 33–38: Sketch the graph of f, making use of stretching, reflecting, or shifting.

33 (a) $f(x) = \frac{1}{4}\sin x$ **(b)** $f(x) = -4\sin x$

34 (a) $f(x) = \sin(x - \pi/2)$ **(b)** $f(x) = \sin x - \pi/2$

35 (a) $f(x) = 2\cos(x + \pi)$ **(b)** $f(x) = 2\cos x + \pi$

36 (a) $f(x) = \frac{1}{3}\cos x$ **(b)** $f(x) = -3\cos x$

37 (a) $f(x) = 4\tan x$ **(b)** $f(x) = \tan(x - \pi/4)$

38 (a) $f(x) = \frac{1}{4}\tan x$ **(b)** $f(x) = \tan(x + 3\pi/4)$

Exer. 39–42: Find a composite function form for y.

39 $y = \sqrt{\tan^2 x + 4}$ **40** $y = \cot^3(2x)$

41 $y = \sec(x + \pi/4)$ **42** $y = \csc\sqrt{x - \pi}$

43 If $f(x) = \cos x$, show that

$$\frac{f(x + h) - f(x)}{h} = \cos x\left(\frac{\cos h - 1}{h}\right) - \sin x\left(\frac{\sin h}{h}\right).$$

44 If $f(x) = \sin x$, show that

$$\frac{f(x + h) - f(x)}{h} = \sin x\left(\frac{\cos h - 1}{h}\right) + \cos x\left(\frac{\sin h}{h}\right).$$

Exer. 45–54: Verify the identity.

45 $(1 - \sin^2 t)(1 + \tan^2 t) = 1$

46 $\sec\beta - \cos\beta = \tan\beta\sin\beta$

47 $\dfrac{\csc^2\theta}{1 + \tan^2\theta} = \cot^2\theta$

48 $\cot t + \tan t = \csc t\sec t$

49 $\dfrac{1 + \csc\beta}{\sec\beta} - \cot\beta = \cos\beta$

50 $\dfrac{1}{\csc z - \cot z} = \csc z + \cot z$

51 $\sin 3u = \sin u(3 - 4\sin^2 u)$

52 $2\sin^2 2t + \cos 4t = 1$

53 $\cos^4(\theta/2) = \frac{3}{8} + \frac{1}{2}\cos\theta + \frac{1}{8}\cos 2\theta$

54 $\sin^4 2x = \frac{3}{8} - \frac{1}{2}\cos 4x + \frac{1}{8}\cos 8x$

Exer. 55–56: Find all solutions of the equation.

55 $2\cos 2\theta - \sqrt{3} = 0$ **56** $2\sin 3\theta + \sqrt{2} = 0$

Exer. 57–64: Find the solutions of the equation in $[0, 2\pi)$.

57 $2\sin^2 u = 1 - \sin u$ **58** $\cos\theta - \sin\theta = 1$

59 $2\tan t - \sec^2 t = 0$

60 $\sin x + \cos x \cot x = \csc x$

61 $\sin 2t + \sin t = 0$ **62** $\cos u + \cos 2u = 0$

63 $\tan 2x = \tan x$ **64** $\sin\frac{1}{2}u + \cos u = 1$

c **Exer. 65–70: Approximate, to the accuracy of your calculator or computer in radians, the solutions of the equation that are in the interval $[0, 2\pi)$.**

65 $\sin\theta = -0.5640$ **66** $\cos\theta = 0.7490$

67 $\tan\theta = 2.798$ **68** $\cot\theta = -0.9601$

69 $\sec\theta = -1.116$ **70** $\csc\theta = 1.485$

c **71** Use a graphing utility to graph $f(x) = (\sin x)/(x - \pi)$. Zoom in several times near $x = \pi$ and investigate the behavior of f.

c **72** Approximate the solution of the equation $x = \frac{1}{2}\cos x$ by using the following procedure.

(1) Graph $y = x$ and $y = \frac{1}{2}\cos x$ on the same coordinate axes.

(2) Use the graphs in (1) to find a first approximation x_1 to the solution.

(3) Find successive approximations $x_2, x_3, \ldots$ by using the formulas $x_2 = \frac{1}{2}\cos x_1$, $x_3 = \frac{1}{2}\cos x_2, \ldots$ until accuracy to six decimal places is obtained.

c **73** Graph $y = (\sin x - \cos \pi x)/\cos x$ for $-1 \le x \le 1$ and estimate the x-intercepts.

Exer. 74–75: The *angle of elevation* of an object is the angle between a horizontal line at an observer's position and the line of sight from the observer to the object. Use the angle of elevation to estimate the heights specified.

74 From a point on level ground 135 ft from the base of a tower, the angle of elevation θ of the top of the tower is 1 radian. Approximate the height of the tower.

Exercise 74

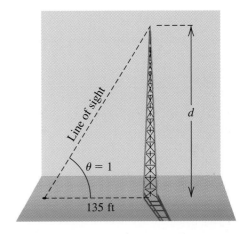

75 A motorist, traveling along a level highway at a speed of 60 km/hr directly toward a mountain, observes that between 1:00 P.M. and 1:10 P.M. the angle of elevation changes from $\theta = 0.17$ to $\theta = 1.2$. Approximate the height of the mountain.

D EXPONENTIALS AND LOGARITHMS

Exponential and logarithmic functions play a major role in calculus. They are examples of *transcendental functions*. We defer a complete rigorous definition of exponential and logarithmic functions until we have developed the necessary tools of calculus. We review some of their properties in this section.

EXPONENTIAL FUNCTIONS

Exponential functions involve raising a constant base to a variable exponent. Two simple examples are $f(x) = 10^x$ and $g(x) = (\frac{1}{3})^x$.

Exponential Functions 22

> The **exponential function with base** *a* is defined by
>
> $$f(x) = a^x,$$
>
> where $a > 0$, $a \neq 1$, and x is any real number.

From algebra, we know how to evaluate a^x if x is a positive or a negative integer or if x is a rational number. If x is a positive integer, then a^x is the product of x factors of a.

$$a^x = \underbrace{(a)(a) \cdots (a)}_{x \text{ times}} \quad \text{if } x \text{ is a positive integer}$$

If x is a negative integer, then $x = -n$ for some positive integer n and

$$a^x = a^{-n} = \frac{1}{a^n}.$$

If x is a rational number of the form $x = m/n$, where m and n are integers with $n > 0$, then a^x is well-defined as

$$a^x = a^{m/n} = \sqrt[n]{a^m} = (\sqrt[n]{a})^m.$$

ILLUSTRATION

Exponential notation a^x

- $2^4 = (2)(2)(2)(2) = 16$

- $\left(\frac{1}{5}\right)^3 = \left(\frac{1}{5}\right)\left(\frac{1}{5}\right)\left(\frac{1}{5}\right) = \frac{1}{125}$

Exponential notation a^{-n}

- $3^{-5} = \frac{1}{3^5} = \frac{1}{243}$

- $\frac{1}{4^{-3}} = \frac{1}{1/4^3} = 4^3 = 64$

Exponential notation $a^{m/n}$

- $2^{3/5} = \sqrt[5]{2^3} = \sqrt[5]{8} \approx 1.5157$

- $\left(\frac{1}{9}\right)^{-5/2} = \frac{1}{(1/9)^{5/2}} = \frac{1}{\left(\sqrt{1/9}\right)^5} = \frac{1}{(1/3)^5} = 3^5 = 243$

We can make use of the $\boxed{x^y}$ or $\boxed{\wedge}$ key on a calculator for computation of a positive number raised to a rational power.

It can be shown algebraically that if x_1 and x_2 are any two rational numbers with $x_1 < x_2$, then $a^{x_1} < a^{x_2}$ if $a > 1$ and $a^{x_1} > a^{x_2}$ if $0 < a < 1$. Thus, if $a > 1$, then $f(x) = a^x$ is an increasing function, sometimes called an **exponential growth function,** whose graph rises. If $0 < a < 1$, then $f(x) = a^x$ is a decreasing function, sometimes called an **exponential decay function,** whose graph falls. In the graphs of $y = a^x$ shown in Figure 49, the dots indicate that only the points with *rational x*-coordinates are on the graphs. There is a *hole* in the graph whenever the *x*-coordinate of a point is irrational.

Figure 49

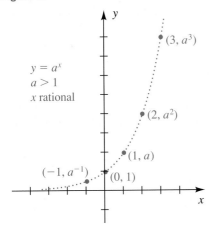

$y = a^x$
$a > 1$
x rational

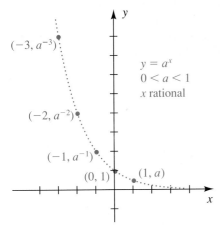

$y = a^x$
$0 < a < 1$
x rational

To extend the domain of the exponential function a^x to all real numbers, we must define a^x for *irrational* values of the exponent x. For example, in order to rigorously define 2^π, we need some knowledge of *limits* (the subject of Chapter 1), but for now, we use the nonterminating decimal representing 3.1415926... for π and consider the following *rational* powers of 2:

$$2^3, \quad 2^{3.1}, \quad 2^{3.14}, \quad 2^{3.141}, \quad 2^{3.1415}, \quad 2^{3.14159}, \ldots$$

We will show, in Chapter 6, that each successive power gets closer to a unique real number, which is designated as 2^π. The numerical value of 2^π is the nonterminating decimal 8.824977827.... We use the same technique for any other irrational value of x; that is, we find a sequence of rational numbers $x_1, x_2, x_3, \ldots$ that approaches x and let a^x be the unique real number approached by the numbers $a^{x_1}, a^{x_2}, a^{x_3}, \ldots$. Note that each of the values $a^{x_1}, a^{x_2}, a^{x_3}, \ldots$ is well-defined and has a numerical value that is easily approximated on a scientific calculator.

To sketch the graph of $y = a^x$ with x a real number, we replace any hole in the graph in Figure 49 with a point. The following chart summarizes this discussion and shows typical graphs.

Definition	Graph of f for $a > 1$	Graph of f for $0 < a < 1$
$f(x) = a^x$ for every x in $\mathbb{R}$, where $a > 0$ and $a \neq 1$		

The graphs merely indicate the *general* appearance; the *exact* shape of each depends on the value of a. Since a^x is either strictly increasing or strictly decreasing, it never takes on the same value twice. Thus, exponential functions are one-to-one functions.

One-to-One Property of
Exponential Functions 23

The exponential function f given by

$$f(x) = a^x \quad \text{for } 0 < a < 1 \text{ or } a > 1$$

is one-to-one; that is, for any real numbers x_1 and x_2:

(i) If $x_1 \neq x_2$, then $a^{x_1} \neq a^{x_2}$.
(ii) If $a^{x_1} = a^{x_2}$, then $x_1 = x_2$.

CAUTION We do not define exponential functions for $a = 1$, $a = 0$, or negative values of a. For such choices of a, the values of a^x do not give a one-to-one function whose domain is the set of all real numbers. If $a = 1$, then $a^x = 1$ for all values of x and we have a constant function. If $a = 0$, then a^x is undefined if $x < 0$ and has the constant value 0 for $x > 0$. If $a < 0$, then a^x is undefined for many values of x. For example, $(-2)^{1/2} = \sqrt{-2}$ is not a real number.

Exponential functions also satisfy the familiar laws of exponents.

Laws of Exponents 24

If u and v are any two real numbers, then

    (i) $a^u a^v = a^{u+v}$

    (ii) $\dfrac{a^u}{a^v} = a^{u-v}$

    (iii) $(a^u)^v = a^{uv}$

Note too that the domain of an exponential function is the set of all real numbers and the range is the set of positive numbers ($a^x > 0$ for all x).

We frequently use the base 10 for exponential functions because of our familiarity with the decimal representation of numbers. In Chapter 6, we will study another important base, the irrational number e, which has a nonterminating decimal expansion that begins $2.7182818284\ldots$. Computer scientists often use exponentials with base 2 because computers store numbers internally in base 2 format. Most calculators provide special keys to compute 10^x and e^x.

An **exponential equation** is an equation involving exponential functions. We can often solve exponential equations by using the one-to-one property of exponential functions.

EXAMPLE ▪ I   Solve the exponential equation $5^{5x} = 5^{4x+7}$.

SOLUTION   From (23)(ii),

$$5^{5x} = 5^{4x+7} \quad \text{implies} \quad 5x = 4x + 7.$$

Subtracting $4x$ from each side of the equation gives the solution $x = 7$. Checking the answer by substituting 7 for x in the original equation, we obtain the identity $5^{35} = 5^{35}$.

EXAMPLE ▪ 2   Solve the exponential function $2^{5x-8} = 4^{x+2}$.

SOLUTION   We first express 4^{x+2} with the base 2:

$$4^{x+2} = (2^2)^{x+2} = 2^{2(x+2)} = 2^{2x+4}$$

By (23)(ii),

$$2^{5x-8} = 2^{2x+4} \quad \text{implies} \quad 5x - 8 = 2x + 4,$$

which simplifies to $3x = 12$, so $x = 4$. Checking the answer by substituting 4 for x in the original equation yields the identity $2^{12} = 4^6$.

EXAMPLE ▪ 3 Suppose it is observed experimentally that the number of bacteria in a given culture doubles every day. If 1000 bacteria are present at the start, then we obtain the following table, where t is the time in days and $f(t)$ is the bacteria count at time t.

t (time in days)	0	1	2	3	4
$f(t)$ (bacteria count)	1000	2000	4000	8000	16,000

(a) Determine a function of the form $f(t) = ba^t$ that can be used to predict the number of bacteria present at any time $t \geq 0$.

(b) Sketch the graph of f from part (a) and approximate the number of bacteria present after $1\frac{1}{2}$ days.

Figure 50

SOLUTION

(a) Since $f(t) = 1000$ when $t = 0$, we have $1000 = ba^0$ or, equivalently, $b = 1000$. Because the number of bacteria are doubling every day, $a = 2$. Hence,

$$f(t) = (1000)2^t.$$

(b) The graph of f is sketched in Figure 50. The number of bacteria present after $1\frac{1}{2}$ days is

$$f(\tfrac{3}{2}) = (1000)2^{3/2} \approx 2828.$$

LOGARITHMIC FUNCTIONS

If a is a positive number (other than 1), then the exponential function with base a is a one-to-one function whose range is the set of positive real numbers. Thus, given a positive number x, there will be a unique number y such that $x = a^y$. The number y is called the *logarithm of x with base a*. We denote this number as $\log_a(x)$ or as $\log_a x$ (read "the logarithm of x with base a").

Logarithmic Function 25

> If a is a positive real number other than 1, then the **logarithm of x with base a** is defined by
>
> $$y = \log_a x \quad \text{if and only if} \quad x = a^y$$
>
> for every $x > 0$ and every real number y.

Note that the domain of a logarithmic function is the set of positive real numbers ($\log_a x$ is defined only if $x > 0$), and the range is the set of all real numbers. The two equations in (25) are equivalent; they assert the same relationship between the variables x and y. We call the first equation the **logarithmic form** and the second the **exponential form**. Consider the following equivalent forms.

ILLUSTRATION

Logarithmic form	Exponential form
$\log_5 u = 2$	$5^2 = u$
$\log_b 8 = 3$	$b^3 = 8$
$r = \log_p q$	$p^r = q$
$w = \log_4(2t + 3)$	$4^w = 2t + 3$
$\log_3 x = 5 + 2z$	$3^{5+2z} = x$

We can use equivalent forms to verify a number of general properties of logarithmic functions.

Properties of Logarithms and Equivalent Exponential Forms 26

Property of $\log_a x$	Exponential form
(i) $\log_a 1 = 0$	$a^0 = 1$
(ii) $\log_a a = 1$	$a^1 = a$
(iii) $\log_a a^x = x$	$a^x = a^x$
(iv) $\log_a x = \log_a x$	$a^{\log_a x} = x$

Figure 51

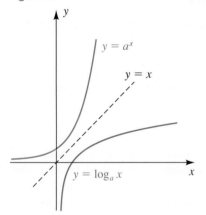

To obtain graphs of logarithmic functions, we first show that $\log_a x$ and a^x are inverses of each other. If $f(x) = \log_a x$ and $g(x) = a^x$, then the composite function $f \circ g$ is computed by

$$
\begin{aligned}
(f \circ g)(x) &= f(g(x)) & &\text{definition of } f \circ g \\
&= f(a^x) & &\text{definition of } g \\
&= \log_a a^x & &\text{definition of } f \\
&= x. & &\text{by (25)(iii)}
\end{aligned}
$$

Figure 52

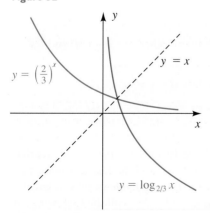

A similar computation using the exponential form of (26)(iv) shows that $(g \circ f)(x) = x$ for all positive real numbers x. The functions $\log_a x$ and a^x are inverses since both $f \circ g$ and $g \circ f$ are identity functions. Because $\log_a x$ and a^x are inverses of each other, the graph of either function is the reflection of the graph of the other across the line $y = x$. Figure 51 shows typical graphs of these functions for $a > 1$.

EXAMPLE ■ 4   Sketch the graphs of $y = (2/3)^x$ and $y = \log_{2/3} x$.

SOLUTION   We begin with the graph of $y = (2/3)^x$. Since we have $0 < 2/3 < 1$, the graph will decrease as x increases, with positive values for all real numbers x. Reflecting this graph across the line $y = x$ yields the graph of $y = \log_{2/3} x$. Figure 52 shows both graphs.

Logarithmic functions either strictly increase or strictly decrease in their domains and hence are one-to-one functions.

One-to-One Property of Logarithmic Functions 27

The logarithmic function f given by

$$f(x) = \log_a x \quad \text{for } 0 < a < 1 \text{ or } a > 1$$

is one-to-one; that is, for any two positive real numbers x_1 and x_2:

(i) If $x_1 \neq x_2$, then $\log_a x_1 \neq \log_a x_2$.

(ii) If $\log_a x_1 = \log_a x_2$, then $x_1 = x_2$.

Other properties of logarithms may be stated as laws, which correspond to the laws of exponents.

Laws of Logarithms 28

If u and v are any two positive real numbers, then

(i) $\log_a (uv) = \log_a u + \log_a v$

(ii) $\log_a \left(\dfrac{u}{v}\right) = \log_a u - \log_a v$

(iii) $\log_a (u^c) = c \log_a u$ for every real number c

We will prove (28)(i) here; the others have similar proofs.

PROOF Let $x = \log_a u$ and $y = \log_a v$.

Then $\qquad a^x = u$ and $a^y = v$. by definition of the logarithm

Now $\qquad uv = a^x a^y = a^{x+y}$. by the properties of exponents

The exponential equation

$$uv = a^{x+y}$$

has the equivalent logarithmic form

$$\log_a uv = x + y.$$

But since $x = \log_a u$ and $y = \log_a v$, the last equation can be written as

$$\log_a uv = \log_a u + \log_a v. \quad \blacksquare$$

Logarithms with base 10 are called **common logarithms,** and the symbol **log x** is an abbreviation for $\log_{10} x$. A second widely used logarithm is the **natural logarithm,** denoted **ln x**, which has the irrational number e for its base.

Most calculators have a $\boxed{\text{LOG}}$ key for the calculation of common logarithms and an $\boxed{\text{LN}}$ key for natural logarithms. To numerically calculate logarithms with bases other than 10 and e, we need to use the following change-of-base formula.

Change-of-Base Formula for Logarithms **29**

If $x > 0$ and if a and b are positive real numbers other than 1, then

$$\log_b x = \frac{\log_a x}{\log_a b}.$$

PROOF Let $u = \log_b x$.

Then $$b^u = x.$$

If we take the logarithm with base a of both sides of this equation, we obtain

$$\log_a x = \log_a(b^u) = u \log_a b,$$

which we can write as

$$\log_a x = (\log_b x)(\log_a b).$$

Dividing each side by $\log_a b$ gives the formula. ■

EXAMPLE ■ 5 Approximate $\log_7 32$ using common logarithms.

SOLUTION Using the change-of-base formula with $a = 10$, we have

$$\log_7 32 = \frac{\log_{10} 32}{\log_{10} 7} = \frac{\log 32}{\log 7}.$$

We can now use the ⃞LOG key to obtain

$$\frac{\log 32}{\log 7} \approx \frac{1.5051}{0.8451} \approx 1.7810.$$

Note that we could have also used $(\ln 32)/(\ln 7)$ to obtain the approximation. We can check our approximation by using the ⃞x^y key to evaluate $7^{1.7810}$.

A **logarithmic equation** is an equation involving logarithmic functions. We can often solve logarithmic equations by using the one-to-one property of logarithmic functions.

EXAMPLE ■ 6 Solve the logarithmic equation

$$\log_3(4x - 5) = \log_3(2x + 1).$$

SOLUTION Since logarithmic functions are one-to-one, if there is a solution, then $4x - 5$ must equal $2x + 1$ or, equivalently, $x = 3$. We must check that $x = 3$ does not make $4x - 5$ or $2x + 1$ zero or negative, because then the logarithms in the given equation would be undefined. In this case, $4x - 5$ and $2x + 1$ both equal 7, so $x = 3$ is a valid solution.

EXAMPLE ■ 7 If the number N of bacteria in a culture after t days is given by $N = (1000)2^t$,

(a) express t as a logarithmic function of N with base 2

(b) determine the time when the number of bacteria is 8000

SOLUTION

(a) From $N = (1000)2^t$, we have

$$2^t = \frac{N}{1000} \quad \text{or, equivalently,} \quad t = \log_2 \frac{N}{1000}.$$

(b) Using the result of part (a) with $N = 8000$,

$$t = \log_2 \frac{8000}{1000} = \log_2 8 = \log_2 2^3 = 3.$$

EXAMPLE ■ 8 Use a graphing utility to estimate the x-intercepts of $f(x) = \cos(\ln x)$ for $0.1 \leq x \leq 6$.

Figure 53

$0.1 \leq x \leq 6, -1 \leq y \leq 1$

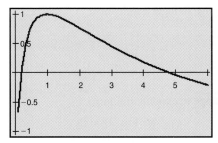

SOLUTION Since the range of the cosine function is the interval $[-1, 1]$, we set the viewing window so that $-1 \leq y \leq 1$. Using a graphing utility, we obtain the graph of the function, shown in Figure 53. We estimate the x-intercepts to be 0.21 and 4.81. An interesting problem arises if you investigate the x-intercepts on the interval $0 < x \leq 0.1$.

The next example is a good illustration of the power of a graphing utility, since it is impossible to find the exact solution using only algebraic methods.

EXAMPLE ■ 9 Estimate the point of intersection of the graphs of

$$f(x) = \log_3 x \quad \text{and} \quad g(x) = \log_6(x + 2).$$

Figure 54

$-2 \leq x \leq 4, -2 \leq y \leq 2$

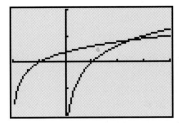

SOLUTION Most graphing utilities work directly only with common and natural logarithmic functions. Thus, we first use the change-of-base formula to rewrite f and g as

$$f(x) = \frac{\ln x}{\ln 3} \quad \text{and} \quad g(x) = \frac{\ln(x + 2)}{\ln 6}.$$

We then use a graphing utility with a viewing window of $-2 \leq x \leq 4$ and $-2 \leq y \leq 2$ to obtain graphs like those in Figure 54. We see that there is a point of intersection in the first quadrant with $2 < x < 3$. Using the tracing and zoom features, we find that the point of intersection is approximately $(2.52, 0.84)$.

EXERCISES D

Exer. 1–6: Sketch the graph of f.

1 $f(x) = 2^x$

2 $f(x) = -3^x$

3 $f(x) = 2(5)^x$

4 $f(x) = 7^x + 3$

5 $f(x) = 4^{-x}$

6 $f(x) = (\frac{1}{2})^x$

Exer. 7–14: Solve the equation.

7 $5^{x+8} = 5^{3x-2}$

8 $8^{7-x} = 8^{2x+1}$

9 $5^{(x^2)} = 5^{2x+3}$

10 $25^{(x^2)} = 5^{3x+2}$

11 $(\frac{1}{2})^{5-x} = 2$

12 $2^{-100x} = (0.5)^{x-4}$

13 $27^{4-x} = 9^{x-3}$

14 $8^{x-1} = 4^{2x-3}$

15 A colony of an endangered species originally numbering 1000 was predicted to have a population N after t years given by the equation $N(t) = 1000(0.9)^t$. Estimate the population after

(a) 1 year

(b) 5 years

(c) 10 years

16 The number of bacteria in a certain culture increased from 600 to 1800 between 8 A.M. and 10 A.M. Assuming the growth is exponential, the number $f(t)$ of bacteria t hours after 8 A.M. is given by $f(t) = 600(3)^{t/2}$.

(a) Estimate the number of bacteria at 9 A.M., 11 A.M., and noon.

(b) Sketch the graph of f.

17 Prescription drugs that enter the body are eventually eliminated through excretion. For an initial dose of 20 mg, suppose that the amount $A(t)$ remaining in the body t hours later is given by $A(t) = 20(0.7)^t$.

(a) Estimate the amount of the drug in the body 8 hr after the initial dose.

(b) What percentage of the drug still in the body is eliminated each hour?

18 An important problem in oceanography is to determine the amount of light that can penetrate to various ocean depths. The Beer–Lambert law asserts that the exponential function given by $I(x) = I_0 c^x$ is a model for this phenomenon (see figure). For a certain location, $I(x) = 10(0.4)^x$ is the amount of light (in calories/cm²/sec) reaching a depth of x meters.

(a) Find the amount of light at a depth of 2 m.

(b) Sketch the graph of I for $0 \le x \le 5$.

Exercise 18

Exer. 19–24: Change to logarithmic form.

19 $5^3 = 125$

20 $5^{-3} = \frac{1}{125}$

21 $3^x = 7 + t$

22 $m^n = p$

23 $(0.7)^t = \frac{2}{3}$

24 $3^{-2x} = P/F$

Exer. 25–30: Change to exponential form.

25 $\log_2 32 = 5$

26 $\log_3 27 = 3$

27 $\log_{10} 1000 = 3$

28 $\log_2 \frac{1}{64} = -6$

29 $\log_7 m = 5x + 3$

30 $\log_a 1994 = 7$

Exer. 31–34: Solve for t using logarithms with base a.

31 $2a^{t/5} = 5$

32 $5a^{3t} = 63$

33 $A = Ba^{Ct} + D$

34 $C = Ba^{t/D} - Q$

Exer. 35–40: Find the number, if possible.

35 $\log_9 1$

36 $\log_6 6$

37 $\log_9(-3)$

38 $\log_3 3^2$

39 $17^{\log_{17} 8}$

40 $\log_2 1024$

Exer. 41–44: Solve the logarithmic equation.

41 $\log_4 x = \log_4(8 - x)$

42 $\log_3(x + 4) = \log_3(1 - x)$

43 $\log x^2 = \log(-3x - 2)$

44 $\log x^2 = -4$

Exer. 45–50: Sketch the graph of f.

45 $f(x) = \log_6 x$

46 $f(x) = -\log_6 x$

47 $f(x) = 2\log_6 x$

48 $f(x) = 3 + \log_6 x$

49 $f(x) = \log_6(x - 2)$

50 $f(x) = \log_6|x|$

51 The loudness of a sound, as experienced by the human ear, is based on its intensity level. The intensity level α (in decibels) that corresponds to a sound intensity I is $\alpha = 10\log(I/I_0)$, where I_0 is a special value of I agreed to be the weakest sound that can be detected by the human ear under certain conditions. Find α if

   (a) I is 10 times as great as I_0

   (b) I is 1000 times as great as I_0

52 A sound intensity level of 140 decibels produces pain in the average human ear (refer to Exercise 51). Approximately how many times greater than I_0 must I be in order for α to reach this level?

c **53** The population $N(t)$ of the United States (in millions) t years after 1990 may be approximated by the formula $N(t) = 253(2.72)^{0.007t}$.

   (a) Estimate the population in 1990 and 2000.

   (b) Approximately when will the population be twice what it was in 1990?

54 Find the error in the following "solution" to the problem: Solve

$$\log_3(5x - 17) = \log_3(4x - 14).$$

"*Solution:* Since logarithmic functions are one-to-one, we must have $5x - 17 = 4x - 14$, which implies that $5x - 4x = 17 - 14$, so $x = 3$."

c **Exer. 55–62: Use a graphing utility to obtain a graph of f on the indicated x-interval. Adjust the y-interval so that the viewing window contains the entire graph.**

55 $f(x) = 2^x + 2^{-x};$ $-2 \le x \le 2$

56 $f(x) = \dfrac{2}{\sqrt{\pi}}\, 2^{-x^2};$ $-3 \le x \le 3$

57 $f(x) = \log(2^x + 5);$ $-2 \le x \le 10$

58 $f(x) = \dfrac{5(3)^x}{3^x + 1};$ $-8 \le x \le 15$

59 $f(x) = \log(\sin x);$ $0.1 \le x \le 3$

60 $f(x) = \sin(\log x);$ $0.1 \le x \le 800$

61 $f(x) = \log(x(1.2 + \sin x));$ $0.1 \le x \le 20$

62 $f(x) = 2.56(3)^{-0.22x}\cos 4.9x;$ $0 \le x \le 12$

E   CONIC SECTIONS

We now review some of the elementary geometric properties of the conic sections. In later chapters, we will use the methods of calculus to solve problems that include finding equations of tangent lines to conics and calculating areas and volumes of regions determined by conics.

Intersecting a double-napped right circular cone with a plane produces curves known as the *conic sections*. By varying the position of the plane, we can obtain a *circle,* an *ellipse,* a *parabola,* or a *hyperbola,* as Figure 55 illustrates on the following page. *Degenerate conics* occur if the plane intersects the cone in a single point or along either one or two lines that lie on the cone.

PARABOLAS

We first define a *parabola* and present equations for parabolas that have a vertical or a horizontal axis.

Figure 55

(a) Circle   (b) Ellipse   (c) Parabola   (d) Hyperbola

Definition 30

A **parabola** is the set of all points in a plane equidistant from a fixed point F (the **focus**) and a fixed line l (the **directrix**) that lie in the plane.

Figure 56

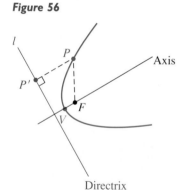

If the focus F lies on the directrix l, then we have a degenerate parabola, the line through F perpendicular to l. Thus, we assume that F does not lie on l. If P is a point in the plane, then the distance from P to the line l is the distance $d(P, P')$, where P' is the point determined by the line through P that is perpendicular to l (see Figure 56). The point P is on the parabola if and only if $d(P, P') = d(P, F)$. The **axis** of the parabola is the line through F that is perpendicular to the directrix. The **vertex** of the parabola is the point V on the axis halfway from F to l.

If the axis of the parabola is the y-axis and the vertex is at the origin with focus F at $(0, p)$, then the equation of the parabola is

$$x^2 = 4py.$$

If $p > 0$, the parabola opens upward, as in Figure 57. If $p < 0$, the parabola opens downward. The graph is symmetric with respect to the y-axis; substitution of $-x$ for x does not change the equation $x^2 = 4py$.

Interchanging the roles of x and y yields the similar equations

$$x^2 = 4py \quad \text{and} \quad y^2 = 4px,$$

which are the equations of a parabola with vertex at the origin and focus $F(0, p)$ and $F(p, 0)$, respectively. If $p > 0$, the parabola opens upward or to the right, and if $p < 0$, it opens downward or to the left.

We see from (31) that for any nonzero real number a, the graph of $y = ax^2$ or $x = ay^2$ is a parabola with vertex $V(0, 0)$. Moreover, $a = 1/(4p)$,

Figure 57

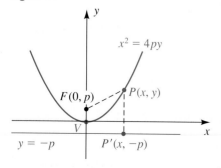

Parabolas with Vertex $V(0,0)$ **31**

Equation	Graph for $p > 0$	Graph for $p < 0$
$x^2 = 4py,$ or $y = \dfrac{1}{4p}x^2$		
$y^2 = 4px,$ or $x = \dfrac{1}{4p}y^2$		

or, equivalently, $p = 1/(4a)$, where $|p|$ is the distance between the focus F and the vertex V. To find the directrix l, recall that l is also a distance $|p|$ from V.

EXAMPLE ■ 1 Find the focus and the directrix of the parabola having equation $y = -\frac{1}{6}x^2$, and sketch its graph.

SOLUTION The equation has the form $y = ax^2$ with $a = -\frac{1}{6}$. As in (31), $a = 1/(4p)$, or

$$p = \frac{1}{4a} = \frac{1}{4(-\frac{1}{6})} = -\frac{3}{2}.$$

The parabola opens downward and has focus $F(0, -\frac{3}{2})$, as illustrated in Figure 58. The directrix is the horizontal line $y = \frac{3}{2}$, which is a distance $\frac{3}{2}$ above V, as shown in the figure.

Figure 58

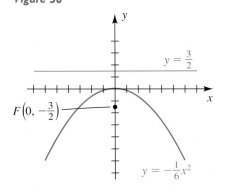

EXAMPLE ▪ 2

(a) Find an equation of a parabola that has vertex at the origin, opens right, and passes through the point $P(7, -3)$.

(b) Find the focus.

SOLUTION

Figure 59

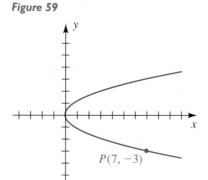

$P(7, -3)$

(a) The parabola is sketched in Figure 59. By (31), the equation of the parabola has the form $x = ay^2$ for some number a. If $P(7, -3)$ is on the graph, then

$$7 = a(-3)^2, \quad \text{or} \quad a = \tfrac{7}{9}.$$

Hence an equation of the parabola is $x = \tfrac{7}{9}y^2$.

(b) The focus is a distance p to the right of the vertex, where

$$p = \frac{1}{4a} = \frac{1}{4(\tfrac{7}{9})} = \frac{9}{28}.$$

Thus, the focus has coordinates $(\tfrac{9}{28}, 0)$.

If the vertex V of a parabola lies at any point (h, k) in the xy-plane, then we can also find equations for the parabola if the directrix is horizontal or vertical. If the focus is $F(h, k + p)$ and the directrix is the horizontal line $y = k - p$, an equation of the parabola is

$$(x - h)^2 = 4p(y - k).$$

Expanding the left-hand side of $(x - h)^2 = 4p(y - k)$ and simplifying leads to an equation of the form

$$y = ax^2 + bx + c,$$

where a, b, and c are real numbers. Conversely, if $a \neq 0$, then the graph of $y = ax^2 + bx + c$ is a parabola with a vertical axis. As with $y = ax^2$, we can show that $a = 1/(4p)$. The parabola opens upward if $p > 0$ and downward if $p < 0$.

Similarly, if the directrix is a vertical line $x = h - p$ and the vertex is at $V(h, k)$ with focus $F(h + p, k)$, then an equation is

$$(y - k)^2 = 4p(x - h),$$

with the parabola opening to the right if $p > 0$ and to the left if $p < 0$. We may write this equation in the form $x = ay^2 + by + c$, where $a = 1/(4p)$.

For convenience, in the following summary of this discussion, we have taken $V(h, k)$ in the first quadrant of the figures.

Parabolas with Vertex $V(h, k)$ **32**

Equation	Graph for $p > 0$	Graph for $p < 0$
$(x - h)^2 = 4p(y - k)$, or $y = ax^2 + bx + c$, where $p = \dfrac{1}{4a}$		
$(y - k)^2 = 4p(x - h)$, or $x = ay^2 + by + c$, where $p = \dfrac{1}{4a}$		

If the equation of a parabola is in the form $y = ax^2 + bx + c$, then we can find the vertex $V(h, k)$ algebraically by completing the square in x and changing the equation to the form $(x - h)^2 = 4p(y - k)$.

EXAMPLE ▪ 3   Discuss and sketch the graph of $y = 2x^2 - 6x + 4$.

SOLUTION   By (32), the graph is a parabola with vertical axis. We rewrite the equation as $y/2 = x^2 - 3x + 2$ and complete the square in x:

$$x^2 - 3x + 2 = x^2 - 2(\tfrac{3}{2})x + \tfrac{9}{4} - \tfrac{9}{4} + 2 = (x - \tfrac{3}{2})^2 - \tfrac{1}{4}$$

so we have

$$\left(x - \frac{3}{2}\right)^2 = \frac{y}{2} + \frac{1}{4} = 4\left(\frac{1}{8}\right)\left(y - \left(-\frac{1}{2}\right)\right).$$

Hence, the vertex is $V(\tfrac{3}{2}, -\tfrac{1}{2})$ and $p = \tfrac{1}{8}$.

Figure 60

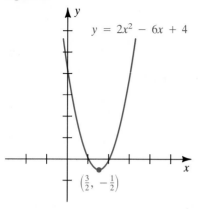

$y = 2x^2 - 6x + 4$

$\left(\frac{3}{2}, -\frac{1}{2}\right)$

Since the parabola opens upward, the focus F is a distance $p = \frac{1}{8}$ above V, which gives us

$$F\left(\tfrac{3}{2}, -\tfrac{1}{2} + \tfrac{1}{8}\right) = F\left(\tfrac{3}{2}, -\tfrac{3}{8}\right).$$

The directrix is the horizontal line l that is a distance $p = \frac{1}{8}$ below V. Therefore, an equation for l is

$$y = -\tfrac{1}{2} - \tfrac{1}{8}, \quad \text{or, equivalently,} \quad y = -\tfrac{5}{8}.$$

The graph is sketched in Figure 60. Note that the y-intercept is 4 and the x-intercepts (the solutions of $2x^2 - 6x + 4 = 0$) are 1 and 2.

For equations of the form $x = ay^2 + by + c$, we complete the square in y and write the equation in the form $(y - k)^2 = 4p(x - h)$.

EXAMPLE ■ 4   Discuss and sketch the graph of $2x = y^2 + 8y + 22$.

SOLUTION   By (32), the graph is a parabola with horizontal axis. Note that

$$x = \tfrac{1}{2}(y^2 + 8y + 22) = \tfrac{1}{2}(y^2 + 8y + 16 + 6) = \tfrac{1}{2}(y + 4)^2 + 3,$$

which we may rewrite as

$$\tfrac{1}{2}(y + 4)^2 = x - 3$$

so that

$$(y + 4)^2 = 2x - 6 = 4(\tfrac{1}{2})(x - 3).$$

Hence, the vertex is $V(3, -4)$ and $p = \frac{1}{2}$. Since the parabola opens to the right, the focus F is a distance $p = \frac{1}{2}$ to the right of V, which gives us

$$F\left(3 + \tfrac{1}{2}, -4\right) = F\left(\tfrac{7}{2}, -4\right).$$

The directrix is the vertical line l that is a distance $p = \frac{1}{2}$ to the left of V. Therefore, an equation for l is

$$x = 3 - \tfrac{1}{2}, \quad \text{or} \quad x = \tfrac{5}{2}.$$

The parabola is sketched in Figure 61.

Figure 61

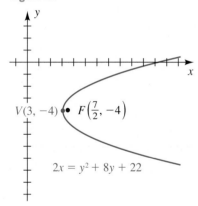

$V(3, -4)$ $F\left(\tfrac{7}{2}, -4\right)$

$2x = y^2 + 8y + 22$

EXAMPLE ■ 5   Find an equation of the parabola with vertex $V(-4, 2)$ and directrix $y = 5$.

SOLUTION   The vertex and the directrix are shown in Figure 62. The dashes indicate a possible shape for the parabola. From (32), we have the following equation of the parabola:

$$(x - h)^2 = 4p(y - k),$$

Figure 62

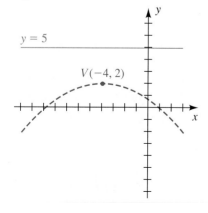

$y = 5$

$V(-4, 2)$

with $h = -4$, $k = 2$, and $p = -3$, since V is 3 units *below* the directrix. Substituting gives us

$$(x + 4)^2 = -12(y - 2).$$

This equation can be expressed in the form $y = ax^2 + bx + c$, as follows:

$$x^2 + 8x + 16 = -12y + 24$$
$$12y = -x^2 - 8x + 8$$
$$y = -\tfrac{1}{12}x^2 - \tfrac{2}{3}x + \tfrac{2}{3}$$

ELLIPSES

We may define an ellipse as follows. (Note that *foci* is the plural of *focus*.)

Definition 33

> An **ellipse** is the set of all points in a plane, the sum of whose distances from two fixed points (the **foci**) in the plane is constant.

Figure 63

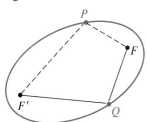

Figure 63 shows an ellipse with foci F and F'. If P and Q are any two points on the ellipse, then $d(P, F) + d(P, F') = d(Q, F) + d(Q, F')$. When F and F' are close to each other, the ellipse is almost circular. If $F = F'$, then we obtain a circle with center F. The midpoint of the segment FF' is the **center** of the ellipse.

If the foci lie along the x-axis and the center is at the origin, then the ellipse has a simple equation. Suppose F has coordinates $(c, 0)$ so that F' has coordinates $(-c, 0)$. Let $2a$ denote the constant sum of the distances of P from F and F', and let $b = \sqrt{a^2 - c^2}$. Note that $b < a$. It can be shown that the coordinates (x, y) of every point on the ellipse satisfy

$$\frac{x^2}{a^2} + \frac{y^2}{b^2} = 1.$$

Figure 64 illustrates this case. We may find the x-intercepts of the ellipse by letting $y = 0$ in the equation. Doing so gives $x^2 = a^2$, so there are two

Figure 64

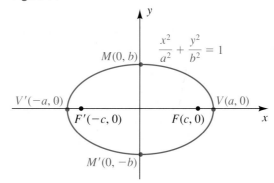

x-intercepts, a and $-a$. The corresponding points $V(a, 0)$ and $V'(-a, 0)$ on the graph are the **vertices** of the ellipse. The line segment $V'V$ is the **major axis**. Similarly, letting $x = 0$ in the equation, we obtain $y^2/b^2 = 1$, or $y^2 = b^2$. Hence, the *y*-intercepts are b and $-b$. The segment between $M'(0, -b)$ and $M(0, b)$ is the **minor axis** of the ellipse. The major axis is always longer than the minor axis, since $a > b$. The foci are always on the major axis.

Applying tests for symmetry, we see that the ellipse is symmetric with respect to the *x*-axis, the *y*-axis, and the origin.

The preceding discussion may be summarized as follows.

Theorem 34

The graph of the equation

$$\frac{x^2}{a^2} + \frac{y^2}{b^2} = 1$$

for $a^2 > b^2$ is an ellipse with vertices $(\pm a, 0)$. The endpoints of the minor axis are $(0, \pm b)$. The foci are $(\pm c, 0)$, where $c^2 = a^2 - b^2$.

EXAMPLE ■ 6 Sketch the graph of $2x^2 + 9y^2 = 18$, and find the foci.

SOLUTION To obtain the form in Theorem (34), we divide both sides of the equation by 18 and simplify to get

$$\frac{x^2}{9} + \frac{y^2}{2} = 1,$$

which is in the proper form, with $a^2 = 9$ and $b^2 = 2$. Thus, $a = 3$ and $b = \sqrt{2}$; hence the endpoints of the major axis are $(\pm 3, 0)$, and the endpoints of the minor axis are $(0, \pm\sqrt{2})$. Since

$$c^2 = a^2 - b^2 = 9 - 2 = 7, \quad \text{or} \quad c = \sqrt{7},$$

the foci are $(\pm\sqrt{7}, 0)$. The graph is sketched in Figure 65.

Figure 65

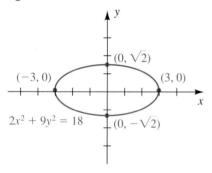

EXAMPLE ■ 7 Find an equation of the ellipse with vertices $(\pm 4, 0)$ and foci $(\pm 2, 0)$.

SOLUTION Using the notation of Theorem (34), we conclude that $a = 4$ and $c = 2$. Since $c^2 = a^2 - b^2$, we have $b^2 = a^2 - c^2 = 16 - 4 = 12$. Hence, an equation of the ellipse is

$$\frac{x^2}{16} + \frac{y^2}{12} = 1.$$

We sometimes choose the major axis of the ellipse along the y-axis. If the foci are $(0, \pm c)$, then by the same type of argument used previously, we obtain the following.

Theorem 35

The graph of the equation

$$\frac{x^2}{b^2} + \frac{y^2}{a^2} = 1$$

for $a^2 > b^2$ is an ellipse with vertices $(0, \pm a)$. The endpoints of the minor axis are $(\pm b, 0)$. The foci are $(0, \pm c)$, where $c^2 = a^2 - b^2$.

A typical graph is sketched in Figure 66.

The preceding discussion shows that an equation of an ellipse with center at the origin and foci on a coordinate axis can always be written in the form

$$\frac{x^2}{p} + \frac{y^2}{q} = 1, \quad \text{or} \quad qx^2 + py^2 = pq,$$

with p and q positive and $p \neq q$. If $p > q$, the major axis is on the x-axis, and if $q > p$, the major axis is on the y-axis. It is unnecessary to memorize these facts, because in any given problem the major axis can be determined by examining the x- and y-intercepts.

Figure 66

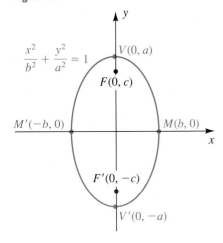

EXAMPLE ■ 8   Sketch the graph of $9x^2 + 4y^2 = 25$, and find the foci.

SOLUTION   The graph is an ellipse with center at the origin and foci on one of the coordinate axes. To find x-intercepts, we let $y = 0$, obtaining

$$9x^2 = 25, \quad \text{or} \quad x = \pm\tfrac{5}{3}.$$

Similarly, to find the y-intercepts, we let $x = 0$, obtaining

$$4y^2 = 25, \quad \text{or} \quad y = \pm\tfrac{5}{2}.$$

These results enable us to sketch the ellipse (see Figure 67). Since $\tfrac{5}{3} < \tfrac{5}{2}$, the major axis is on the y-axis.

To find the foci, we first calculate

$$c^2 = a^2 - b^2 = (\tfrac{5}{2})^2 - (\tfrac{5}{3})^2 = \tfrac{125}{36}.$$

Thus, $c = 5\sqrt{5}/6$ and the foci are $(0, \pm 5\sqrt{5}/6)$.

Figure 67

Figure 68

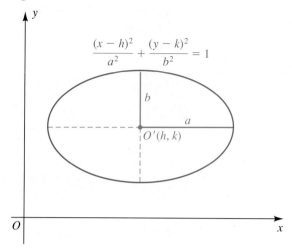

For an ellipse with center (h, k) at any point in the xy-plane and with major and minor axes that are horizontal or vertical (see Figure 68), the equation takes the form

$$\frac{(x - h)^2}{a^2} + \frac{(y - k)^2}{b^2} = 1.$$

Squaring the indicated terms and simplifying gives us an equation of the form

$$Ax^2 + Cy^2 + Dx + Ey + F = 0,$$

where the coefficients are real numbers and both A and C are positive. Conversely, if we start with such an equation, then by completing squares, we can obtain a form that displays the center of the ellipse and the lengths of the major and minor axes. This technique is illustrated in the next example.

EXAMPLE ■ 9   Discuss and sketch the graph of the equation

$$16x^2 + 9y^2 + 64x - 18y - 71 = 0.$$

SOLUTION   We begin by writing the equation in the form

$$16(x^2 + 4x \quad) + 9(y^2 - 2y \quad) = 71.$$

Completing the squares gives us

$$16(x^2 + 4x + 4) + 9(y^2 - 2y + 1) = 71 + 64 + 9,$$

which may be written as

$$16(x + 2)^2 + 9(y - 1)^2 = 144.$$

Dividing by 144, we obtain

$$\frac{(x + 2)^2}{9} + \frac{(y - 1)^2}{16} = 1,$$

Figure 69

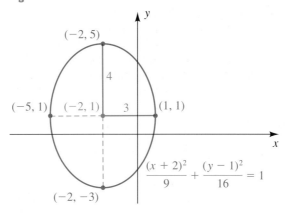

which is of the form

$$\frac{(x-h)^2}{a^2} + \frac{(y-k)^2}{b^2} = 1$$

with $h = -2$ and $k = 1$. The graph of the equation is an ellipse with center $(-2, 1)$ (see Figure 69). Since $16 > 9$, the major axis is on the vertical line $x = -2$.

To find the foci, note that $c^2 = 16 - 9 = 7$. Thus, the distance from the center of the ellipse to either focus is $c = \sqrt{7}$. Since the center is $(-2, 1)$, the foci are $(-2, 1 \pm \sqrt{7})$.

Ellipses can be very flat or almost circular. To obtain information about the *roundness* of an ellipse, we sometimes use the term *eccentricity*.

Definition 36

The **eccentricity** e of an ellipse is

$$e = \frac{c}{a} = \frac{\sqrt{a^2 - b^2}}{a}.$$

Consider the ellipse $(x^2/a^2) + (y^2/b^2) = 1$, and suppose that the length $2a$ of the major axis is fixed and the length $2b$ of the minor axis is variable. Since $\sqrt{a^2 - b^2} < a$, we see that $0 < e < 1$. If $e \approx 1$, then $\sqrt{a^2 - b^2} \approx a$ and $b \approx 0$. Thus, the ellipse is very flat. If $e \approx 0$, then $\sqrt{a^2 - b^2} \approx 0$ and $a \approx b$. Thus, the ellipse is almost circular.

Many comets have elliptical orbits with the sun at a focus. In this case the eccentricity e is close to 1, and the ellipse is very flat. In the next example, we use the **astronomical unit** (AU)—that is, the average distance from the earth to the sun—to specify large distances (1 AU $\approx$ 93,000,000 miles).

E X A M P L E ■ 10   Halley's comet has an elliptical orbit with eccentricity $e = 0.967$. The closest that Halley's comet comes to the sun is

0.587 AU. Approximate the maximum distance of the comet from the sun, to the nearest 0.1 AU.

SOLUTION   Figure 70 illustrates the orbit of the comet, where c is the distance from the center of the ellipse to a focus (the sun) and $2a$ is the length of the major axis.

Since $a - c$ is the minimum distance between the sun and the comet, we have (in AU)

$$a - c = 0.587, \quad \text{or} \quad a = c + 0.587.$$

Since $e = c/a = 0.967$,

$$c = 0.967a = 0.967(c + 0.587)$$
$$c \approx 0.967c + 0.568.$$

Thus,

$$0.033c \approx 0.568 \quad \text{and} \quad c \approx \frac{0.568}{0.033} \approx 17.2.$$

Consequently,

$$a = c + 0.587$$
$$a \approx 17.2 + 0.587 \approx 17.8,$$

and the maximum distance between the sun and the comet is

$$a + c \approx 17.8 + 17.2, \quad \text{or} \quad a + c \approx 35.0 \text{ AU}.$$

Figure 70

Halley's comet

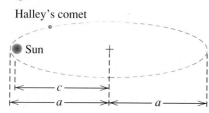

HYPERBOLAS

The definition of a hyperbola is similar to that of an ellipse. The only change is that instead of using the *sum* of distances from two fixed points, we use the *difference*.

Definition 37

A **hyperbola** is the set of all points in the plane, the difference of whose distances from two fixed points (the **foci**) in the plane is a positive constant.

Figure 71

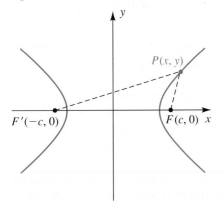

The **center** of a hyperbola is the midpoint of the segment FF'. If the foci lie along the x-axis and the center is at the origin, then the hyperbola has a simple equation. Let P be a point on the hyperbola. Suppose F has coordinates $(c, 0)$ so that F' has coordinates $(-c, 0)$ (see Figure 71). Let $2a$ denote the constant difference of the distances of P from F and F', and let $b^2 = c^2 - a^2$. It can be shown that P is on the hyperbola if and only if its coordinates (x, y) satisfy the equation

$$\frac{x^2}{a^2} - \frac{y^2}{b^2} = 1.$$

By tests for symmetry, the hyperbola is symmetric with respect to both axes and the origin. The x-intercepts are a and $-a$. The corresponding points $V(a, 0)$ and $V'(-a, 0)$ are the **vertices,** and the line segment $V'V$ is the **transverse axis** of the hyperbola.

The preceding discussion may be summarized as follows.

Theorem 38

> The graph of the equation
>
> $$\frac{x^2}{a^2} - \frac{y^2}{b^2} = 1$$
>
> is a hyperbola with vertices $(\pm a, 0)$. The foci are $(\pm c, 0)$, where $c^2 = a^2 + b^2$.

If we solve the equation $(x^2/a^2) - (y^2/b^2) = 1$ for y, we obtain

$$y = \pm \frac{b}{a}\sqrt{x^2 - a^2}.$$

If a graph approaches a line as the absolute value of x gets increasingly large, then the line is called an **asymptote** for the graph. It can be shown that the lines $y = (b/a)x$ and $y = -(b/a)x$ are asymptotes for the hyperbola. These asymptotes serve as excellent guides for sketching the graph. A convenient way to sketch the asymptotes is to first plot the vertices $V(a, 0)$, $V'(-a, 0)$ and the points $W(0, b)$, $W'(0, -b)$ (see Figure 72). The line segment $W'W$ of length $2b$ is the **conjugate axis** of the hyperbola. If horizontal and vertical lines are drawn through the endpoints of the conjugate and transverse axes, respectively, then the diagonals of the resulting rectangle have slopes b/a and $-b/a$. Hence, by extending these diagonals, we obtain lines with equations $y = (\pm b/a)x$, which are

Figure 72 $\dfrac{x^2}{a^2} - \dfrac{y^2}{b^2} = 1$

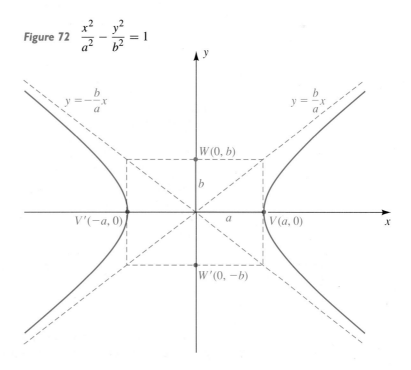

the asymptotes. The hyperbola is then sketched as in Figure 72, using the asymptotes as guides. The two curves that make up the hyperbola are the **branches** of the hyperbola.

EXAMPLE ▪ 11   Discuss and sketch the graph of $9x^2 - 4y^2 = 36$. Then find the foci and the equations of the asymptotes.

SOLUTION   The graph is a hyperbola with the center at the origin. Dividing both sides of the given equation by 36 and simplifying gives

$$\frac{x^2}{4} - \frac{y^2}{9} = 1,$$

which is of the form stated in Theorem (38), with $a^2 = 4$ and $b^2 = 9$. Hence, $a = 2$ and $b = 3$. The vertices $(\pm 2, 0)$ and the endpoints $(0, \pm 3)$ of the conjugate axis determine a rectangle whose diagonals (extended) give us the asymptotes. The graph of the equation is sketched in Figure 73.

To find the foci, we calculate

$$c^2 = a^2 + b^2 = 4 + 9 = 13.$$

Thus, $c = \sqrt{13}$ and the foci are $(\pm\sqrt{13}, 0)$.

The equations of the asymptotes, $y = \pm\frac{3}{2}x$, can be found by referring to the graph or to the equations $y = \pm(b/a)x$.

Figure 73
$9x^2 - 4y^2 = 36$

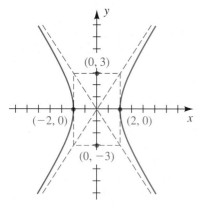

The preceding example indicates that for hyperbolas it is not always true that $a > b$, as is the case for ellipses. Indeed, we may have $a < b$, $a > b$, or $a = b$.

EXAMPLE ▪ 12   A hyperbola has vertices $(\pm 3, 0)$ and passes through the point $P(5, 2)$. Find its equation, the foci, and the equations of the asymptotes.

SOLUTION   We begin by sketching a hyperbola with vertices $(\pm 3, 0)$ that passes through the point $P(5, 2)$, as in Figure 74.

An equation of the hyperbola has the form

$$\frac{x^2}{3^2} - \frac{y^2}{b^2} = 1.$$

Since $P(5, 2)$ is on the hyperbola, the x- and y-coordinates satisfy the last equation; that is,

$$\frac{25}{9} - \frac{4}{b^2} = 1.$$

Solving for b^2 gives us $b^2 = \frac{9}{4}$, so the desired equation is

$$\frac{x^2}{9} - \frac{y^2}{\frac{9}{4}} = 1,$$

Figure 74

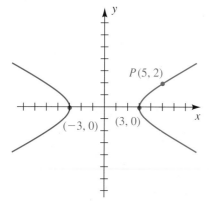

or, equivalently,

$$x^2 - 4y^2 = 9.$$

To find the foci, we first calculate

$$c^2 = a^2 + b^2 = 9 + \tfrac{9}{4} = \tfrac{45}{4}.$$

Hence, $c = \sqrt{\tfrac{45}{4}} = \tfrac{3}{2}\sqrt{5}$, and the foci are $(\pm\tfrac{3}{2}\sqrt{5}, 0)$.

The general equations of the asymptotes are $y = \pm(b/a)x$. Substituting $a = 3$ and $b = \tfrac{3}{2}$ gives us $y = \pm\tfrac{1}{2}x$.

If the foci of a hyperbola are the points $(0, \pm c)$ on the y-axis, then by the same type of argument used previously, we obtain the following theorem.

Theorem   39

> The graph of the equation
>
> $$\frac{y^2}{a^2} - \frac{x^2}{b^2} = 1$$
>
> is a hyperbola with vertices $(0, \pm a)$. The foci are $(0, \pm c)$, where $c^2 = a^2 + b^2$.

For the hyperbola in the preceding theorem, the endpoints of the conjugate axis are $W(b, 0)$ and $W'(-b, 0)$. We find the asymptotes as before, by using the diagonals of the rectangle determined by these points, the vertices, and lines parallel to the coordinate axes. The graph is sketched in Figure 75. The equations of the asymptotes are $y = \pm(a/b)x$. Note the difference between these equations and the equations $y = \pm(b/a)x$ for the asymptotes of the hyperbola considered first in this section.

Figure 75 $\dfrac{y^2}{a^2} - \dfrac{x^2}{b^2} = 1$

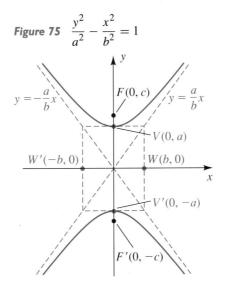

EXAMPLE ▪ 13   Discuss and sketch the graph of $4y^2 - 2x^2 = 1$. Then find the foci and the equations of the asymptotes.

SOLUTION   We may obtain the form in Theorem (39) by writing the equation as

$$\frac{y^2}{\frac{1}{4}} - \frac{x^2}{\frac{1}{2}} = 1.$$

Thus, $a^2 = \frac{1}{4}, \qquad b^2 = \frac{1}{2}, \qquad c^2 = a^2 + b^2 = \frac{3}{4},$

and hence, $a = \frac{1}{2}, \qquad b = \frac{\sqrt{2}}{2}, \qquad c = \frac{\sqrt{3}}{2}.$

The vertices are $(0, \pm\frac{1}{2})$, the foci are $(0, \pm\sqrt{3}/2)$, and the endpoints of the conjugate axes are $(\pm\sqrt{2}/2, 0)$. The graph is sketched in Figure 76.

Figure 76 $4y^2 - 2x^2 = 1$

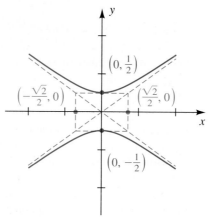

To find equations of the asymptotes, we can use $y = \pm(a/b)x$, obtaining $y = \pm(\sqrt{2}/2)x$.

If the center of a hyperbola is at any point (h, k) in the xy-plane, then it has the equation

$$\frac{(x - h)^2}{a^2} - \frac{(y - k)^2}{b^2} = 1$$

if the foci lie on a horizontal line, or the form of the equation is

$$\frac{(y - k)^2}{a^2} - \frac{(x - h)^2}{b^2} = 1$$

if the foci lie on a vertical line.

Squaring the indicated terms in these equations and simplifying allows us to write the equation for the hyperbola in the form

$$Ax^2 + Cy^2 + Dx + Ey + F = 0,$$

where the coefficients are real numbers and A and C have opposite signs (one is positive and the other is negative). Conversely, if we begin with such an equation, then by completing squares, we can obtain a form that displays the center of the hyperbola and the transverse and conjugate axes. The next example illustrates this technique.

EXAMPLE ▪ 14   Discuss and sketch the graph of the equation

$$9x^2 - 4y^2 - 54x - 16y + 29 = 0.$$

SOLUTION   We arrange our work as follows:

$$9(x^2 - 6x \quad) - 4(y^2 + 4y \quad) = -29$$

$$9(x^2 - 6x + 9) - 4(y^2 + 4y + 4) = -29 + 81 - 16$$

$$9(x - 3)^2 - 4(y + 2)^2 = 36$$

$$\frac{(x - 3)^2}{4} - \frac{(y + 2)^2}{9} = 1$$

Note that $h = 3$ and $k = -2$. The graph of the equation is a hyperbola with center $(3, -2)$. The foci lie on the horizontal line $y = -2$ through the center. We see that

$$a^2 = 4, \qquad b^2 = 9, \qquad c^2 = a^2 + b^2 = 13.$$

Hence, $\qquad a = 2, \qquad b = 3, \qquad c = \sqrt{13}.$

As illustrated in Figure 77, the vertices are $(3 \pm 2, -2)$—that is, $(5, -2)$ and $(1, -2)$. The endpoints of the conjugate axis are $(3, -2 \pm 3)$—that is, $(3, 1)$ and $(3, -5)$. The foci are $(3 \pm \sqrt{13}, -2)$, and the equations of the asymptotes are

$$y + 2 = \pm \tfrac{3}{2}(x - 3).$$

Figure 77

$$\frac{(x - 3)^2}{4} - \frac{(y + 2)^2}{9} = 1$$

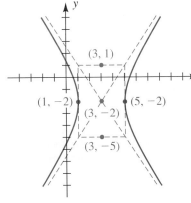

The results of the last three sections indicate that the graph of every equation of the form

$$Ax^2 + Cy^2 + Dx + Ey + F = 0$$

is a conic, except for certain degenerate cases in which a point, one or two lines, or no graph is obtained. Although we have considered only special examples, our methods are perfectly general. If A and C are equal and not 0, then the graph, when it exists, is a circle or, in exceptional cases, a point. If A and C are unequal but have the same sign, then by completing squares and properly translating axes, we obtain an equation whose graph, when it exists, is an ellipse (or a point). If A and C have opposite signs, an equation of a hyperbola is obtained or possibly, in the degenerate case, two intersecting straight lines. If either A or C (but not both) is 0, the graph is a parabola or, in certain cases, a pair of parallel lines.

EXERCISES E

Exer. 1–6: Find the vertex, the focus, and the directrix of the parabola. Sketch its graph, showing the focus and the directrix.

1. $y = -\frac{1}{12}x^2$

2. $x = 2y^2$

3. $2y^2 = -3x$

4. $x^2 = -3y$

5. $y = 8x^2$

6. $y^2 = -100x$

Exer. 7–16: Find the vertex and the focus of the parabola. Sketch its graph, showing the focus.

7. $y = x^2 - 4x + 2$

8. $y = 8x^2 + 16x + 10$

9. $y^2 - 12 = 12x$

10. $y^2 - 20y + 100 = 6x$

11. $y^2 - 4y - 2x - 4 = 0$

12. $y^2 + 14y + 4x + 45 = 0$

13. $4x^2 + 40x + y + 106 = 0$

14. $y = 40x - 97 - 4x^2$

15. $x^2 + 20y = 10$

16. $4x^2 + 4x + 4y + 1 = 0$

Exer. 17–24: Find an equation of the parabola that satisfies the given conditions.

17. focus $F(2, 0)$;   directrix $x = -2$

18. focus $F(0, -4)$;   directrix $y = 4$

19. vertex $V(3, -5)$;   directrix $x = 2$

20. vertex $V(-2, 3)$;   directrix $y = 5$

21. vertex $V(-1, 0)$;   focus $F(-4, 0)$

22. vertex $V(1, -2)$;   focus $F(1, 0)$

23. vertex at the origin; symmetric to the y-axis; and passing through the point $(2, -3)$

24. vertex $V(-3, 5)$; axis parallel to the x-axis; and passing through the point $(5, 9)$

Exer. 25–38: Find the vertices and the foci of the ellipse. Sketch its graph, showing the foci.

25. $\dfrac{x^2}{9} + \dfrac{y^2}{4} = 1$

26. $\dfrac{x^2}{25} + \dfrac{y^2}{16} = 1$

27. $4x^2 + y^2 = 16$

28. $y^2 + 9x^2 = 9$

29. $5x^2 + 2y^2 = 10$

30. $\frac{1}{2}x^2 + 2y^2 = 8$

31. $4x^2 + 25y^2 = 1$

32. $10y^2 + x^2 = 5$

33. $4x^2 + 9y^2 - 32x - 36y + 64 = 0$

34. $x^2 + 2y^2 + 2x - 20y + 43 = 0$

35. $9x^2 + 16y^2 + 54x - 32y - 47 = 0$

36. $4x^2 + 9y^2 + 24x + 18y + 9 = 0$

37. $25x^2 + 4y^2 - 250x - 16y + 541 = 0$

38. $4x^2 + y^2 = 2y$

Exer. 39–48: Find an equation for the ellipse that has its center at the origin and satisfies the given conditions.

39. vertices $V(\pm 8, 0)$;   foci $F(\pm 5, 0)$

40. vertices $V(0, \pm 7)$;   foci $F(0, \pm 2)$

41. vertices $V(0, \pm 5)$;   minor axis of length 3

42. foci $F(\pm 3, 0)$;   minor axis of length 2

43. vertices $V(0, \pm 6)$;   passing through $(3, 2)$

44. passing through $(2, 3)$ and $(6, 1)$

45. eccentricity $\frac{3}{4}$;   vertices $V(0, \pm 4)$

46. eccentricity $\frac{1}{2}$;   vertices on the x-axis; passing through $(1, 3)$

47. x-intercepts ± 2; y-intercepts $\pm\frac{1}{3}$

48. x-intercepts $\pm\frac{1}{2}$; y-intercepts ± 4

Exer. 49–66: Find the vertices and the foci of the hyperbola. Sketch its graph, showing the asymptotes and the foci.

49. $\dfrac{x^2}{9} - \dfrac{y^2}{4} = 1$

50. $\dfrac{y^2}{49} - \dfrac{x^2}{16} = 1$

51. $\dfrac{y^2}{9} - \dfrac{x^2}{4} = 1$

52. $\dfrac{x^2}{49} - \dfrac{y^2}{16} = 1$

53. $y^2 - 4x^2 = 16$

54. $x^2 - 2y^2 = 8$

55. $x^2 - y^2 = 1$

56. $y^2 - 16x^2 = 1$

57. $x^2 - 5y^2 = 25$

58. $4y^2 - 4x^2 = 1$

59. $3x^2 - y^2 = -3$

60. $16x^2 - 36y^2 = 1$

61. $25x^2 - 16y^2 + 250x + 32y + 109 = 0$

62. $y^2 - 4x^2 - 12y - 16x + 16 = 0$

63. $4y^2 - x^2 + 40y - 4x + 60 = 0$

64. $25x^2 - 9y^2 + 100x - 54y + 10 = 0$

65. $9y^2 - x^2 - 36y + 12x - 36 = 0$

66. $4x^2 - y^2 + 32x - 8y + 49 = 0$

Exer. 67–76: Find an equation for the hyperbola that has its center at the origin and satisfies the given conditions.

67 foci $F(0, \pm 4)$;   vertices $V(0, \pm 1)$

68 foci $F(\pm 8, 0)$;   vertices $V(\pm 5, 0)$

69 foci $F(\pm 5, 0)$;   vertices $V(\pm 3, 0)$

70 foci $F(0, \pm 3)$;   vertices $V(0, \pm 2)$

71 foci $F(0, \pm 5)$;   conjugate axis of length 4

72 vertices $V(\pm 4, 0)$;   passing through $(8, 2)$

73 vertices $V(\pm 3, 0)$;   asymptotes $y = \pm 2x$

74 foci $F(0, \pm 10)$;   asymptotes $y = \pm \frac{1}{3}x$

75 x-intercepts ± 5;   asymptotes $y = \pm 2x$

76 y-intercepts ± 2;   asymptotes $y = \pm \frac{1}{4}x$

77 The graphs of the equations

$$\frac{x^2}{a^2} - \frac{y^2}{b^2} = 1 \quad \text{and} \quad \frac{x^2}{a^2} - \frac{y^2}{b^2} = -1$$

are called *conjugate hyperbolas*. Sketch the graphs of both equations on the same coordinate plane, with $a = 5$ and $b = 3$. Describe the relationship between the two graphs.

78 The parabola $y^2 = 4p(x + p)$ has its focus at the origin and its axis along the x-axis. By assigning different values to p, we obtain a family of *confocal parabolas*, as shown in the figure. Families of this type occur in the study of electricity and magnetism. Show that there are exactly two parabolas in the family that pass through a given point $P(x_1, y_1)$ if $y_1 \neq 0$.

Exercise 78

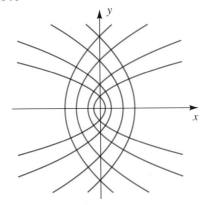

79 The arch of a bridge is semielliptical, with major axis horizontal. The base of the arch is 30 ft across, and the highest part is 10 ft above the horizontal roadway, as shown in the figure. Find the height of the arch 6 ft from the center of the base.

Exercise 79

80 Assume that the length of the major axis of the earth's orbit is 186,000,000 mi and the eccentricity is 0.017. Find, to the nearest 1000 mi, the maximum and minimum distances between the earth and the sun.

81 In 1911, the physicist Ernest Rutherford (1871–1937) discovered that when alpha particles are shot toward the nucleus of an atom, they are eventually repulsed away from the nucleus along hyperbolic paths. The figure illustrates the path of a particle that starts toward the origin along the line $y = \frac{1}{2}x$ and comes within 3 units of the nucleus. Find an equation of the path.

Exercise 81

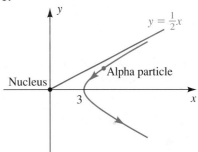

82 A cruise ship is traveling a course that is 100 mi from, and parallel to, a straight shoreline. The ship sends out a distress signal, which is received by two Coast Guard stations A and B, located 200 mi apart, as shown in the figure. By measuring the difference in signal reception times, officials determine that the ship is 160 mi closer to B than to A. Where is the ship?

Exercise 82

INTRODUCTION

IN THE GRACEFUL FLOW of a river, the smooth passage of time, and the majestic ascent of a hot air balloon, we perceive *continuity* of motion. There are no abrupt changes, no jumping over intermediate points in the movement. Through the idea of limits, calculus provides the means to study continuity rigorously.

The concept of *limit* is the central idea of calculus. All the fundamental notions — continuous functions, derivatives, integrals, and convergent series — are limits in one sense or another. Limits are also essential for understanding the geometric properties of curves and surfaces, such as length, slope, area, and volume. The principal features of motion — velocity, speed, acceleration — are best comprehended as limits as well.

Paradoxes about limits play a vital role in the history of thought. Zeno (495 – 435 B.C.) posed many vexing paradoxes. In *Achilles and the Tortoise,* Zeno argues that a swift runner (Achilles) cannot overtake a slow opponent (the tortoise) if the latter has a head start. While Achilles runs to the tortoise's initial position, the tortoise moves forward to a new spot. While Achilles races to that spot, the tortoise moves to a further position. This process continues indefinitely, and Achilles always remains behind the tortoise! The paradox challenges our common sense that given enough time, a fast runner will pass a slower one. At the heart of the paradox lie sequences of numbers (the positions of the two racers) and the limit of these sequences.

As calculus developed in the eighteenth century, mathematicians treated the limit concept intuitively: Limits exist if a function's outputs get close to some value as its inputs get close to another value. In Section 1.1, we explore this intuitive definition, flawed by its use of the imprecise word *close*. A scientist may consider a measurement as being close to a value L if it is within 10^{-6} cm of L. A marathon runner is close to the finish line when 100 yd are left in the race. An astronomer may measure closeness in terms of light-years.

To avoid ambiguity, we require a definition of limit not containing the word *close*. In Section 1.2, we present the traditional $\epsilon - \delta$ *definition of limit of a function*. This definition is precise and applicable to every situation we consider. The formal definition leads to theorems we use to determine the values of limits or to verify that a limit does not exist. The theorems give us techniques to find many limits, without applying the $\epsilon - \delta$ definition directly. Section 1.3 examines some of these techniques.

We consider in Section 1.4 limits of a function $f(x)$ where either $|x|$ or $|f(x)|$ grows unboundedly large. Finally, we use limits to define *continuous functions*, a concept used extensively throughout calculus.

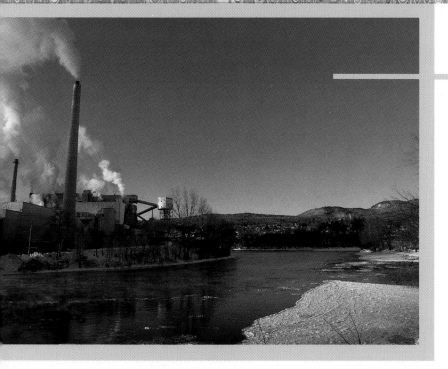

Limits provide a powerful tool for the mathematical analysis of objects, such as a river, that flow in a continuous manner.

Limits and Continuity

1.1   INTRODUCTION TO LIMITS

In calculus and its applications, we are often interested in function values $f(x)$ of a function f when x is *close* to a number a, *but not necessarily equal to a*. In fact, there are many instances where a is not in the domain of f; that is, $f(a)$ is undefined. For example, consider

$$f(x) = \frac{x^3 - 2x^2}{3x - 6}$$

with $a = 2$. Note that 2 is not in the domain of f, since substituting $x = 2$ gives us the undefined expression 0/0. The following table, obtained with a calculator, lists some function values (to eight-decimal-place accuracy) for x close to 2.

x	$f(x)$	x	$f(x)$	x	$f(x)$
1.97	1.293633333	1.9997	1.332933363	1.999997	1.333329333
1.98	1.306800000	1.9998	1.333066680	1.999998	1.333330667
1.99	1.320033333	1.9999	1.333200003	1.999999	1.333332000
2.01	1.346700000	2.0001	1.333466670	2.000001	1.333334667
2.02	1.360133333	2.0002	1.333600013	2.000002	1.333336000
2.03	1.373633333	2.0003	1.333733363	2.000003	1.333337333

It *appears* from the table that the closer x is to 2, the closer $f(x)$ is to $\frac{4}{3}$; however, we cannot be certain of this because we have merely calculated several function values for x near 2. To give a more convincing argument, let us factor the numerator and the denominator of $f(x)$ as follows:

$$f(x) = \frac{x^2(x - 2)}{3(x - 2)}$$

Figure 1.1

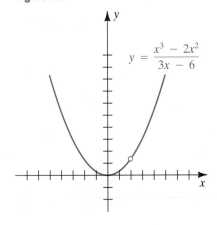

If $x \neq 2$, then we may cancel the common factor $x - 2$ and observe that $f(x)$ is given by $\frac{1}{3}x^2$. Thus the graph of f is the parabola $y = \frac{1}{3}x^2$ with the point $(2, \frac{4}{3})$ deleted, as shown in Figure 1.1. It is geometrically evident that as x gets closer to 2, $f(x)$ gets closer to $\frac{4}{3}$, as indicated in the preceding table.

In general, if a function f is defined throughout an open interval containing a real number a, except possibly at a itself, we may ask the following questions:

1. As x gets closer to a (but $x \neq a$), does the function value $f(x)$ get closer to some real number L?

2. Can we make the function value $f(x)$ *as close to L as desired* by choosing x sufficiently close to a (but $x \neq a$)?

If the answers to these questions are *yes,* we use the notation

$$\lim_{x \to a} f(x) = L$$

and say that *the limit of $f(x)$, as x approaches a, is L,* or that *$f(x)$ approaches L as x approaches a.* We may also write

$$f(x) \to L \quad \text{as} \quad x \to a.$$

Thus the point $(x, f(x))$ on the graph of f approaches the point (a, L) as x approaches a. Using the limit notation, we denote the result in our example as follows:

$$\lim_{x \to 2} \frac{x^3 - 2x^2}{3x - 6} = \frac{4}{3}$$

Note that in this section we define *limit* using the phrases *close to* and *approaches* in an intuitive manner; the next section contains a formal definition of limit that avoids this terminology. This discussion of the intuitive meaning of limit may be summarized as follows.

Limit of a Function   1.1

Notation	Intuitive meaning	Graphical interpretation
$\lim\limits_{x \to a} f(x) = L$	We can make $f(x)$ as close to L as desired by choosing x sufficiently close to a, and $x \neq a$.	

If $f(x)$ approaches some number as x approaches a, but we do not know what that number is, we use the phrase $\lim_{x \to a} f(x)$ *exists.*

The graph of f shown in (1.1) illustrates only one way in which $f(x)$ might approach L as x approaches a. We have used an open circle, or "hole," in the graph rather than a point with x-coordinate a because, when using the limit concept given in (1.1), we always assume that $x \neq a$; that is, *the function value $f(a)$ is completely irrelevant.* As we shall see, $f(a)$ may be different from L, may equal L, or may not exist, depending on the nature of the function f.

In our discussion of $f(x) = (x^3 - 2x^2)/(3x - 6)$, it was possible to simplify $f(x)$ by factoring the numerator and the denominator. In many cases, such algebraic simplifications are impossible. In particular, when we consider derivatives of trigonometric functions later in the text, it will be necessary to answer the following question.

$$\text{Question:} \quad \text{Does } \lim_{x \to 0} \frac{\sin x}{x} \text{ exist?}$$

Note that substituting 0 for x gives us the undefined expression 0/0. The following table lists some approximations of $f(x) = (\sin x)/x$ for x near

x	$f(x) = \dfrac{\sin x}{x}$
± 2.0	0.454648713
± 1.0	0.841470985
± 0.5	0.958851077
± 0.4	0.973545856
± 0.3	0.985067356
± 0.2	0.993346654
± 0.1	0.998334166
± 0.01	0.999983333
± 0.001	0.999999833
± 0.0001	0.999999998

Figure 1.2

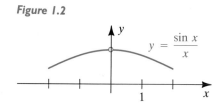

0, *where x is a real number or the radian measure of an angle.* The graph of f is sketched in Figure 1.2 beside the table.

Referring to the table or the graph, we arrive at the following conjecture.

$$\text{Educated guess:} \quad \lim_{x \to 0} \frac{\sin x}{x} = 1$$

As indicated, *we have merely guessed at the answer.* The table indicates that $(\sin x)/x$ gets closer to 1 as x gets closer to 0; however, we cannot be absolutely sure of this fact. The function values could conceivably deviate from 1 if x were closer to 0 than are those x-values listed in the table. Although a calculator may help us guess if a limit exists, it cannot be used in proofs. We will return to this limit in Section 2.4, where we will *prove* that our guess is correct.

It is easy to find $\lim_{x \to a} f(x)$ if $f(x)$ is a simple algebraic expression. For example, if $f(x) = 2x - 3$ and $a = 4$, it is evident that the closer x is to 4, the closer $f(x)$ is to $2(4) - 3$, or 5. This example gives us the first limit in the following illustration. The remaining two limits may be obtained in the same intuitive manner.

ILLUSTRATION

- $\displaystyle \lim_{x \to 4} (2x - 3) = 2(4) - 3 = 8 - 3 = 5$

- $\displaystyle \lim_{x \to -3} (x^2 + 1) = (-3)^2 + 1 = 9 + 1 = 10$

- $\displaystyle \lim_{x \to 7} \sqrt{x + 2} = \sqrt{7 + 2} = \sqrt{9} = 3$

In the preceding illustrations, the limits as $x \to a$ can be found by merely substituting the number a for x. For a special class of functions called *continuous functions,* which are discussed in Section 1.5, limits can always be found by such a substitution. The next illustration shows that we cannot use this substitution technique for *every* algebraic function f. In the next illustration, it is important to note that

$$\frac{x^2 + x - 2}{x - 1} = \frac{(x - 1)(x + 2)}{x - 1} = x + 2, \quad provided \ x \neq 1.$$

(If $x \neq 1$, then $x - 1 \neq 0$, and it is permissible to cancel the common factor $x - 1$ in the numerator and the denominator.) It follows that the graphs of the equations $y = (x^2 + x - 2)/(x - 1)$ and $y = x + 2$ are the same *except* for $x = 1$. Thus, the point $(1, 3)$ is on the graph of $y = x + 2$, but is not on the graph of $y = (x^2 + x - 2)/(x - 1)$, as indicated in the illustration.

ILLUSTRATION

Function value	Graph	Limit as $x \to 1$
$f(x) = x + 2$	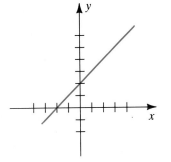	$\displaystyle\lim_{x \to 1} f(x) = 3$
$g(x) = \dfrac{x^2 + x - 2}{x - 1}$	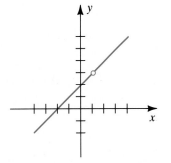	$\displaystyle\lim_{x \to 1} g(x) = 3$
$h(x) = \begin{cases} \dfrac{x^2 + x - 2}{x - 1} & \text{if } x \neq 1 \\ 2 & \text{if } x = 1 \end{cases}$	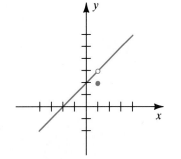	$\displaystyle\lim_{x \to 1} h(x) = 3$

In the preceding illustration, the limit of each function as x approaches 1 is 3; but in the first case, $f(1) = 3$; in the second, $g(1)$ does not exist; and in the third, $h(1) = 2 \neq 3$.

The following two examples illustrate how algebraic manipulations may be used to help find certain limits.

EXAMPLE ▪ 1   If $f(x) = \dfrac{2x^2 - 5x + 2}{5x^2 - 7x - 6}$, find $\lim\limits_{x \to 2} f(x)$.

SOLUTION   The number 2 is not in the domain of f since the meaningless expression $0/0$ is obtained if 2 is substituted for x. Factoring the numerator and the denominator gives us

$$f(x) = \frac{(x - 2)(2x - 1)}{(x - 2)(5x + 3)}.$$

We cannot cancel the factor $x - 2$ at this stage; however, if we take the *limit* of $f(x)$ as $x \to 2$, this cancellation *is* allowed, because by (1.1), $x \neq 2$ and hence $x - 2 \neq 0$. Thus,

$$\lim_{x \to 2} f(x) = \lim_{x \to 2} \frac{2x^2 - 5x + 2}{5x^2 - 7x - 6}$$

$$= \lim_{x \to 2} \frac{(x - 2)(2x - 1)}{(x - 2)(5x + 3)}$$

$$= \lim_{x \to 2} \frac{2x - 1}{5x + 3} = \frac{3}{13}.$$

EXAMPLE ▪ 2   Let $f(x) = \dfrac{x - 9}{\sqrt{x} - 3}$.

(a) Find $\lim\limits_{x \to 9} f(x)$.

(b) Sketch the graph of f and illustrate the limit in part (a) graphically.

SOLUTION

(a) Note that the number 9 is not in the domain of f. To find the limit, we shall change the form of $f(x)$ by rationalizing the denominator as follows:

$$\lim_{x \to 9} f(x) = \lim_{x \to 9} \frac{x - 9}{\sqrt{x} - 3}$$

$$= \lim_{x \to 9} \left(\frac{x - 9}{\sqrt{x} - 3} \cdot \frac{\sqrt{x} + 3}{\sqrt{x} + 3} \right)$$

$$= \lim_{x \to 9} \frac{(x - 9)(\sqrt{x} + 3)}{x - 9}$$

By (1.1), when investigating the limit as $x \to 9$, we assume that $x \neq 9$. Hence $x - 9 \neq 0$, and we can divide the numerator and the denominator

Figure 1.3

$$f(x)\frac{x-9}{\sqrt{x}-3}$$

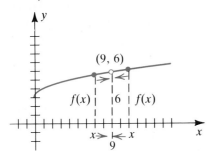

by $x - 9$ (that is, we can *cancel* the expression $x - 9$) and thus obtain

$$\lim_{x \to 9} f(x) = \lim_{x \to 9} (\sqrt{x} + 3) = \sqrt{9} + 3 = 6.$$

(b) If we rationalize the denominator of $f(x)$ as in part (a), we see that the graph of f is the same as the graph of the equation $y = \sqrt{x} + 3$, *except for the point* (9, 6), as illustrated in Figure 1.3. As x gets closer to 9, the point $(x, f(x))$ on the graph of f gets closer to the point (9, 6). Note that $f(x)$ never actually attains the value 6; however, $f(x)$ can be made as close to 6 as desired by choosing x sufficiently close to 9.

The first two examples show that we can often use algebraic manipulations to find limits. In many cases, however, there is no apparent algebraic simplification to try. In such instances, we may obtain numerical evidence for the value of a limit by constructing a table of values or by graphing the function. The next example illustrates both of these approaches.

EXAMPLE ■ 3   Lend numerical support for the claim that

$$\lim_{x \to 1} f(x) \approx 0.4343, \quad \text{where} \quad f(x) = \frac{\log_{10} x}{x - 1}$$

(a) by creating a table of function values for x close to 1

(b) by using a graphing utility to repeatedly zoom in on the graph of f near $x = 1$

Figure 1.4

$0.5 \le x \le 1.5, 0.3 \le y \le 0.6$

SOLUTION

(a) We construct the following table of function values for x very close to but not equal to 1.

x	$f(x)$	x	$f(x)$	x	$f(x)$
0.97	0.440942191	0.997	0.434947229	0.9997	0.434359639
0.98	0.438696215	0.998	0.434729356	0.9998	0.434337917
0.99	0.436480540	0.999	0.434511774	0.9999	0.434316198
1.01	0.432137378	1.001	0.434077479	1.0001	0.434272769
1.02	0.430008588	1.002	0.433860766	1.0002	0.434251058
1.03	0.427907490	1.003	0.433644340	1.0003	0.434229351

Figure 1.5

$0.999 \le x \le 1.001,$
$0.43408 \le y \le 0.43451$

(b) We graph the function with x-interval $0.5 \le x \le 1.5$ to view the graph near $x = 1$, as in Figure 1.4, where we see that y is close to 0.4. Repeatedly zooming in to obtain the graph in Figure 1.5 allows us to estimate the value of y to be 0.4343 in the x-interval [0.999, 1.001].

Both procedures provide strong circumstantial evidence that the limit is approximately 0.4343. Neither procedure, however, gives an exact answer.

The finite limitations of calculators and graphing utilities allow us to get close to $x = 1$, but not *arbitrarily* close. For example, if b is the smallest positive number available on our calculating device, then the function f could behave very differently inside the interval $(1 - b, 1 + b)$ (where we cannot evaluate f) than it does outside that interval. In Chapter 6, we will introduce a new idea that will permit us to determine the *exact* value of limits like the one in this example.

Figure 1.6
$$f(x) = \frac{1}{x}$$

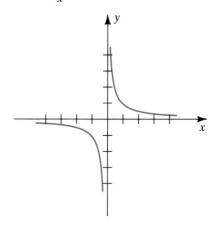

The next two examples involve functions that have no limit as x approaches 0. The solutions are intuitive in nature. Rigorous proofs require the formal definition of limit discussed in the next section.

EXAMPLE ■ 4 Show that $\lim\limits_{x \to 0} \dfrac{1}{x}$ does not exist.

SOLUTION The graph of $f(x) = 1/x$ is sketched in Figure 1.6. Note that we can make $|f(x)|$ as large as desired by choosing x sufficiently close to 0 (but $x \neq 0$). For example, if we want $f(x) = -1,000,000$, we choose $x = -0.000001$. For $f(x) = 10^9$, we choose $x = 10^{-9}$. Since $f(x)$ does not approach a specific number L as x approaches 0, the limit does not exist.

EXAMPLE ■ 5 Show that $\lim\limits_{x \to 0} \sin \dfrac{1}{x}$ does not exist.

SOLUTION Let us first determine some of the characteristics of the graph of $y = \sin(1/x)$. To find the x-intercepts, we note that the following statement is true for every integer n:

$$\sin \frac{1}{x} = 0 \quad \text{if and only if} \quad \frac{1}{x} = \pi n, \quad \text{or} \quad x = \frac{1}{\pi n}$$

Some specific x-intercepts (with $n = \pm 1, \pm 2, \pm 3, \ldots, \pm 100$) are

$$\pm \frac{1}{\pi}, \pm \frac{1}{2\pi}, \pm \frac{1}{3\pi}, \ldots, \pm \frac{1}{100\pi}.$$

If we let x approach 0, then the distance between successive x-intercepts decreases and, in fact, approaches 0. Similarly,

$$\sin \frac{1}{x} = 1 \quad \text{if and only if} \quad x = \frac{1}{(\pi/2) + 2\pi n}$$

and $\qquad \sin \dfrac{1}{x} = -1 \quad \text{if and only if} \quad x = \dfrac{1}{(3\pi/2) + 2\pi n}$,

where n is any integer. Thus, as x approaches 0, the function values $\sin(1/x)$ oscillate between -1 and 1, and the corresponding waves on the graph become very compressed horizontally, as illustrated in Figure

Figure 1.7 $f(x) = \sin \dfrac{1}{x}$

1.7. Hence $\lim_{x \to 0} \sin(1/x)$ does not exist, because the function values do not approach a specific number L as x approaches 0.

We sometimes use *one-sided* limits of the following types.

One-Sided Limits **1.2**

Notation	Intuitive meaning	Graphical interpretation
$\lim\limits_{x \to a^-} f(x) = L$ (left-hand limit)	We can make $f(x)$ as close to L as desired by choosing x sufficiently close to a, and $x < a$.	
$\lim\limits_{x \to a^+} f(x) = L$ (right-hand limit)	We can make $f(x)$ as close to L as desired by choosing x sufficiently close to a, and $x > a$.	

For a left-hand limit, the function f must be defined in (at least) an open interval of the form (c, a) for some real number c. For a right-hand limit, f must be defined in (a, c) for some c. The notation $x \to a^-$ is read *x approaches a from the left,* and $x \to a^+$ is read *x approaches a from the right.*

EXAMPLE ■ 6 If $f(x) = \sqrt{x - 2}$, sketch the graph of f and find, if possible,

(a) $\lim\limits_{x \to 2^+} f(x)$ **(b)** $\lim\limits_{x \to 2^-} f(x)$ **(c)** $\lim\limits_{x \to 2} f(x)$

SOLUTION The graph of f is sketched in Figure 1.8.
(a) If $x > 2$, then $x - 2 > 0$ and hence $f(x) = \sqrt{x - 2}$ *is* a real number;

Figure 1.8

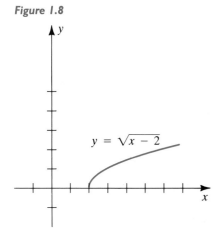

$y = \sqrt{x - 2}$

that is, $f(x)$ is defined. Thus,

$$\lim_{x \to 2^+} \sqrt{x - 2} = \sqrt{2 - 2} = 0.$$

(b) If $x < 2$, then $x - 2 < 0$ and hence $f(x) = \sqrt{x - 2}$ is *not* a real number. Thus, the left-hand limit does not exist.

(c) The limit of f as x approaches 2 does not exist because $f(x) = \sqrt{x - 2}$ is not defined throughout an open interval containing 2 — that is, an interval containing numbers that are less than 2 *and* numbers that are greater than 2.

The relationship between one-sided limits and limits is described in the next theorem.

Theorem   1.3

$$\lim_{x \to a} f(x) = L \quad \text{if and only if} \quad \lim_{x \to a^-} f(x) = L = \lim_{x \to a^+} f(x)$$

Theorem (1.3), which can be proved using definitions in Section 1.2, tells us that *the limit of $f(x)$ as x approaches a exists if and only if both the right-hand and left-hand limits exist and are equal to some real number L.*

Figure 1.9

$$f(x) = \frac{|x|}{x}$$

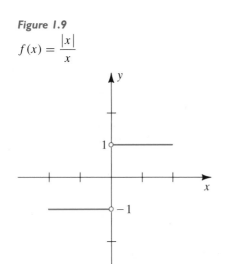

EXAMPLE ∎ 7   If $f(x) = \dfrac{|x|}{x}$, sketch the graph of f and find, if possible,

(a) $\lim\limits_{x \to 0^-} f(x)$ **(b)** $\lim\limits_{x \to 0^+} f(x)$ **(c)** $\lim\limits_{x \to 0} f(x)$

SOLUTION   The function f is undefined at $x = 0$. If $x > 0$, then $|x| = x$ and $f(x) = x/x = 1$. Hence for $x > 0$, the graph of f is the horizontal line $y = 1$. If $x < 0$, then $|x| = -x$ and $f(x) = -x/x = -1$. These results give us the sketch in Figure 1.9. Referring to the graph, we see that

(a) $\lim\limits_{x \to 0^-} f(x) = -1$ the left-hand limit is -1

(b) $\lim\limits_{x \to 0^+} f(x) = 1$ the right-hand limit is 1

(c) Since the left-hand and right-hand limits are not equal, it follows from Theorem (1.3) that $\lim_{x \to 0} f(x)$ does not exist.

In the next example, we consider a piecewise-defined function.

EXAMPLE ∎ 8   Sketch the graph of the function f defined as follows:

$$f(x) = \begin{cases} 3 - x & \text{if } x < 1 \\ 4 & \text{if } x = 1 \\ x^2 + 1 & \text{if } x > 1 \end{cases}$$

Find $\lim\limits_{x \to 1^-} f(x)$, $\lim\limits_{x \to 1^+} f(x)$, and $\lim\limits_{x \to 1} f(x)$.

Figure 1.10

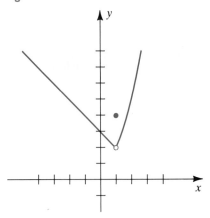

SOLUTION   The graph is sketched in Figure 1.10. The one-sided limits are

$$\lim_{x \to 1^-} f(x) = \lim_{x \to 1^-} (3 - x) = 2$$

and

$$\lim_{x \to 1^+} f(x) = \lim_{x \to 1^+} (x^2 + 1) = 2.$$

Since the left-hand and right-hand limits both equal 2, it follows from Theorem (1.3) that

$$\lim_{x \to 1} f(x) = 2.$$

Note that the function value $f(1) = 4$ is irrelevant in finding the limit.

The following application involves one-sided limits.

Figure 1.11

EXAMPLE ▪ 9   A gas (such as water vapor or oxygen) is held at a constant temperature in the piston shown in Figure 1.11. As the gas is compressed, the volume V decreases until a certain critical pressure is reached. Beyond this pressure, the gas assumes liquid form. Use the graph in Figure 1.11 to find and interpret

(a) $\lim_{P \to 100^-} V$ **(b)** $\lim_{P \to 100^+} V$ **(c)** $\lim_{P \to 100} V$

SOLUTION

(a) We see from Figure 1.11 that when the pressure P (in torrs) is low, the substance is a gas and the volume V (in liters) is large. (The definition of the unit of pressure, the *torr*, may be found in textbooks on physics.) If P approaches 100 through values less than 100, V decreases and approaches 0.8; that is,

$$\lim_{P \to 100^-} V = 0.8.$$

The limit 0.8 represents the volume at which the substance begins to change from a gas to a liquid.

(b) If $P > 100$, the substance is a liquid. If P approaches 100 through values greater than 100, the volume V increases very slowly (since liquids are nearly incompressible), and

$$\lim_{P \to 100^+} V = 0.3.$$

The limit 0.3 represents the volume at which the substance begins to change from a liquid to a gas.

(c) $\lim_{P \to 100} V$ does not exist since the left-hand and right-hand limits in parts (a) and (b) are not equal. (At $P = 100$, the gas and liquid forms exist together in equilibrium, and the substance cannot be classified as either a gas or a liquid.)

EXERCISES 1.1

Exer. 1–10: Find the limit.

1. $\lim\limits_{x \to -2} (3x - 1)$

2. $\lim\limits_{x \to 3} (x^2 + 2)$

3. $\lim\limits_{x \to 4} x$

4. $\lim\limits_{x \to -3} (-x)$

5. $\lim\limits_{x \to 100} 7$

6. $\lim\limits_{x \to 7} 100$

7. $\lim\limits_{x \to -1} \pi$

8. $\lim\limits_{x \to \pi} (-1)$

9. $\lim\limits_{x \to -1} \dfrac{x + 4}{2x + 1}$

10. $\lim\limits_{x \to 5} \dfrac{x + 2}{x - 4}$

Exer. 11–24: Use an algebraic simplification to help find the limit, if it exists.

11. $\lim\limits_{x \to -3} \dfrac{(x + 3)(x - 4)}{(x + 3)(x + 1)}$

12. $\lim\limits_{x \to -1} \dfrac{(x + 1)(x^2 + 3)}{x + 1}$

13. $\lim\limits_{x \to 2} \dfrac{x^2 - 4}{x - 2}$

14. $\lim\limits_{x \to 3} \dfrac{2x^3 - 6x^2 + x - 3}{x - 3}$

15. $\lim\limits_{r \to 1} \dfrac{r^2 - r}{2r^2 + 5r - 7}$

16. $\lim\limits_{r \to -3} \dfrac{r^2 + 2r - 3}{r^2 + 7r + 12}$

17. $\lim\limits_{k \to 4} \dfrac{k^2 - 16}{\sqrt{k} - 2}$

18. $\lim\limits_{x \to 25} \dfrac{\sqrt{x} - 5}{x - 25}$

19. $\lim\limits_{h \to 0} \dfrac{(x + h)^2 - x^2}{h}$

20. $\lim\limits_{h \to 0} \dfrac{(x + h)^3 - x^3}{h}$

21. $\lim\limits_{h \to -2} \dfrac{h^3 + 8}{h + 2}$

22. $\lim\limits_{h \to 2} \dfrac{h^3 - 8}{h^2 - 4}$

23. $\lim\limits_{z \to -2} \dfrac{z - 4}{z^2 - 2z - 8}$

24. $\lim\limits_{z \to 5} \dfrac{z - 5}{z^2 - 10z + 25}$

Exer. 25–30: Find each limit, if it exists:

(a) $\lim\limits_{x \to a^-} f(x)$ (b) $\lim\limits_{x \to a^+} f(x)$ (c) $\lim\limits_{x \to a} f(x)$

25. $f(x) = \dfrac{|x - 4|}{x - 4}$; $a = 4$

26. $f(x) = \dfrac{x + 5}{|x + 5|}$; $a = -5$

27. $f(x) = \sqrt{x + 6} + x$; $a = -6$

28. $f(x) = \sqrt{5 - 2x} - x^2$; $a = \frac{5}{2}$

29. $f(x) = \dfrac{1}{x^3}$; $a = 0$

30. $f(x) = \dfrac{1}{x - 8}$; $a = 8$

Exer. 31–40: Refer to the graph to find each limit, if it exists:

(a) $\lim\limits_{x \to 2^-} f(x)$ (b) $\lim\limits_{x \to 2^+} f(x)$ (c) $\lim\limits_{x \to 2} f(x)$

(d) $\lim\limits_{x \to 0^-} f(x)$ (e) $\lim\limits_{x \to 0^+} f(x)$ (f) $\lim\limits_{x \to 0} f(x)$

31

32

33

Exercises 1.1

34

35

36

37

38

39

40

Exer. 41–46: Sketch the graph of f and find each limit, if it exists:

(a) $\lim\limits_{x \to 1^-} f(x)$ (b) $\lim\limits_{x \to 1^+} f(x)$ (c) $\lim\limits_{x \to 1} f(x)$

41   $f(x) = \begin{cases} x^2 - 1 & \text{if } x < 1 \\ 4 - x & \text{if } x \geq 1 \end{cases}$

42   $f(x) = \begin{cases} x^3 & \text{if } x \leq 1 \\ 3 - x & \text{if } x > 1 \end{cases}$

43   $f(x) = \begin{cases} 3x - 1 & \text{if } x \leq 1 \\ 3 - x & \text{if } x > 1 \end{cases}$

44   $f(x) = \begin{cases} |x - 1| & \text{if } x \neq 1 \\ 1 & \text{if } x = 1 \end{cases}$

45 $f(x) = \begin{cases} x^2 + 1 & \text{if } x < 1 \\ 1 & \text{if } x = 1 \\ x + 1 & \text{if } x > 1 \end{cases}$

46 $f(x) = \begin{cases} -x^2 & \text{if } x < 1 \\ 2 & \text{if } x = 1 \\ x - 2 & \text{if } x > 1 \end{cases}$

47 A country taxes the first $20,000 of an individual's income at a rate of 15%, and all income over $20,000 is taxed at 20%.

(a) Find a piecewise-defined function T for the total tax on an income of x dollars.

(b) Find

$$\lim_{x \to 20,000^-} T(x) \quad \text{and} \quad \lim_{x \to 20,000^+} T(x).$$

48 A telephone company charges 25 cents for the first minute of a long-distance call and 15 cents for each additional minute.

(a) Find a piecewise-defined function C for the total cost of a long-distance call of x minutes.

(b) If n is an integer greater than 1, find

$$\lim_{x \to n^-} C(x) \quad \text{and} \quad \lim_{x \to n^+} C(x).$$

49 A mail-order company adds a shipping and handling fee of $4 for any order that weighs up to 10 lb with an additional 40 cents for each pound over 10 lb.

(a) Find a piecewise-defined function S for the shipping and handling fee on an order of x pounds.

(b) If a is an integer greater than 10, find

$$\lim_{x \to a^-} S(x) \quad \text{and} \quad \lim_{x \to a^+} S(x).$$

50 The Campus Cinema charges $3 admission for children (under age 12), $6 for adults, and $4.50 for senior citizens (over age 60).

(a) Find a piecewise-defined function T for the ticket price for a person x years old.

(b) For which values of a are $\lim_{x \to a^-} T(x)$ and $\lim_{x \to a^+} T(x)$ equal and for which values of a are they unequal?

51 The figure shows a graph of the g-forces experienced by astronauts during the takeoff of a spacecraft with two rocket boosters. (A force of $2g$'s is twice that of gravity, $3g$'s is three times that of gravity, etc.) If $F(t)$ denotes the g-force t minutes into the flight, find and interpret

(a) $\lim_{t \to 0^+} F(t)$

(b) $\lim_{t \to 3.5^-} F(t)$ and $\lim_{t \to 3.5^+} F(t)$

(c) $\lim_{t \to 5^-} F(t)$ and $\lim_{t \to 5^+} F(t)$

Exercise 51

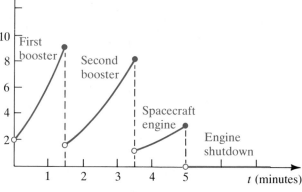

52 A hospital patient receives an initial 200-mg dose of a drug. Additional doses of 100 mg each are then administered every 4 hr. The amount $f(t)$ of the drug present in the bloodstream after t hours is shown in the figure. Find and interpret $\lim_{t \to 8^-} f(t)$ and $\lim_{t \to 8^+} f(t)$.

Exercise 52

c Exer. 53–60: The stated limit (of the form $\lim_{x \to a} f(x) = L$) may be verified by methods developed later in the text. Lend *numerical* support for the stated result by (a) creating a table of function values for x close to a and (b) using a graphing utility to repeatedly zoom in on the graph of f near $x = a$. Give additional digits if the stated limit is an approximation.

53 $\lim_{x \to 0} (1 + x)^{1/x} \approx 2.72$

54 $\lim_{x \to 0} (1 + 2x)^{3/x} \approx 403.4$

55 $\lim_{x \to 2} \dfrac{3^x - 9}{x - 2} \approx 9.89$

56 $\lim_{x \to 1} \dfrac{2^x - 2}{x - 1} \approx 1.39$

57 $\lim_{x \to 0} \left(\dfrac{4^{|x|} + 9^{|x|}}{2} \right)^{1/|x|} = 6$

58 $\lim_{x \to 0} |x|^x = 1$

59 $\lim_{x \to 0} \dfrac{\sin x - 7x}{x \cos x} = -6$

60 $\lim_{x \to 3} \dfrac{\sin(\pi x)}{x - 3} = -\pi$

c **61** (a) Given that $f(x) = \cos(1/x) - \sin(1/x)$, investigate $\lim_{x \to 0} f(x)$ by first letting $x = 3.1830989 \times 10^{-n}$ for $n = 2, 3,$ and 4 and then letting $x = 3 \times 10^{-n}$ for $n = 2, 3,$ and 4.

(b) What is the limit in part (a)?

c **62** (a) Given that $f(x) = x^{1/100} - 0.933$, investigate $\lim_{x \to 0^+} f(x)$ by letting $x = 10^{-n}$ for $n = 20, 40, 60,$ and 80. (If your calculator allows, use $n = 200, 400, 600,$ and 800.)

(b) What *appears* to be the limit in part (a)?

1.2 DEFINITION OF LIMIT

We state the precise meaning of a limit of a function in Definition (1.4) of this section. Because the definition is rather abstract, let us begin with a physical illustration that may make it easier to understand.

Scientists often investigate the manner in which quantities vary and whether they approach specific values under certain conditions. Suppose that an industrial plant discharges waste water into a nearby river. The discharge contains a chemical that in large concentrations is toxic to humans. Several miles downstream from the plant, the river flows through a small town (Figure 1.12a on the following page). Since the townspeople use the water from the river for drinking, washing, and cooking, they are concerned about the possible dangers.

The operators of the plant have promised to keep the concentration of the chemical at the point of discharge small enough so that by the time the water reaches the town, its concentration will be low enough not to cause any harm. Meters are installed at the plant and at the water station in the town to measure the concentration of the chemical in the water at both points. We let x denote the concentration indicated by the meter at the plant and y represent the concentration indicated by the meter in town. The plant operators agree to regulate the discharge so that the concentration y at the town's meter will be near a desired level of L, low enough so that the chemical poses no hazard to the town.

Workers at the plant and officials of the town observe that when the concentration x of the chemical is near the level a on the plant's meter, the measurement y on the town's meter is close to L. They also note that the closer x is to a, the closer y is to L. Because of random fluctuations in the operation of the plant, it is not possible to maintain the concentration of the discharge exactly at a for extended periods; there will always be fluctuations in the concentration.

We use these meters to give a precise meaning to the statement y *approaches L as x approaches a*, or, symbolically,

$$\lim_{x \to a} y = L.$$

Figure 1.12

(a)

Industrial plant

Water
station
with town's
meter

(b)

Concentration at plant's meter

Concentration at town's meter

When monitoring these meters, we would not expect the concentration *y* in town to remain *exactly* at *L* over a long period of time. Instead, our goal might be to force *y* to remain very close to *L* by restricting *x* to values *near a*. In particular, if ϵ (epsilon) denotes a small positive real number, let us suppose it is sufficient that

$$L - \epsilon < y < L + \epsilon,$$

as indicated on the town's meter in Figure 1.12(b). An equivalent statement using absolute values is

$$|y - L| < \epsilon.$$

If these inequalities are true, we say that *y* **has** ϵ**-tolerance at L.** Thus, the statement *y has 0.01-tolerance at L* means that $|y - L| < 0.01$; that is, *y* is within 0.01 unit of *L*. This tolerance may be sufficiently accurate for our purposes.

Similarly, we consider a small positive number δ (delta) and define δ-tolerance at *a* on the plant's meter in Figure 1.12(b). In our later work with functions, it will be important that $x \neq a$. Anticipating this restriction, we say that *x* **has** δ**-tolerance at a** if

$$0 < |x - a| < \delta$$

or, equivalently, if

$$a - \delta < x < a + \delta \quad \text{and} \quad x \neq a.$$

Let us now consider the following question.

Question: Given any $\epsilon > 0$, is there a $\delta > 0$ such that
if *x* has δ-tolerance at *a*, then *y* has ϵ-tolerance at *L*?

If the answer to this question is *yes*, we write

$$\lim_{x \to a} y = L.$$

It is important to note that if $\lim_{x \to a} y = L$, then *no matter how small the number ϵ, we can always find a $\delta > 0$ such that if x is restricted to the interval $(a - \delta, a + \delta)$ on the plant's meter* (and $x \neq a$), then *y* will lie in the interval $(L - \epsilon, L + \epsilon)$ on the town's meter.

This example has provided a more precise interpretation of the limit concept than that given in Section 1.1, where we used words such as *close to* and *approaches*. If we rephrase the last question and its answer in terms of inequalities, we obtain the following statement:

$$\lim_{x \to a} y = L$$

means that for every $\epsilon > 0$, there is a $\delta > 0$ such that

$$\text{if} \quad 0 < |x - a| < \delta, \quad \text{then} \quad |y - L| < \epsilon$$

It is now a small step to formulate the definition of a limit of a function *f*. Letting $y = f(x)$ in the preceding discussion gives us the following definition, which also states the conditions required for the function *f*.

**Definition of Limit
of a Function   1.4**

Let a function f be defined on an open interval containing a, except possibly at a itself, and let L be a real number. The statement

$$\lim_{x \to a} f(x) = L$$

means that for every $\epsilon > 0$, there is a $\delta > 0$ such that

$$\text{if} \quad 0 < |x - a| < \delta, \quad \text{then} \quad |f(x) - L| < \epsilon.$$

We sometimes call the inequality $0 < |x - a| < \delta$ a δ-*tolerance state-ment* and the inequality $|f(x) - L| < \epsilon$ an ϵ-*tolerance statement*.

If we wish to use a form of Definition (1.4) that does not contain absolute value symbols, we note that

(i) $0 < |x - a| < \delta$ is equivalent to $a - \delta < x < a + \delta$ and $x \neq a$

(ii) $|f(x) - L| < \epsilon$ is equivalent to $L - \epsilon < f(x) < L + \epsilon$

The inequalities in (i) and (ii) are represented graphically on real lines in Figure 1.13. We may restate Definition (1.4) as follows.

**Alternative Definition of
Limit   1.5**

$$\lim_{x \to a} f(x) = L$$

means that for every $\epsilon > 0$, there is a $\delta > 0$ such that if x is in the open interval $(a - \delta, a + \delta)$ and $x \neq a$, then $f(x)$ is in the open interval $(L - \epsilon, L + \epsilon)$.

Figure 1.13
(i) $0 < |x - a| < \delta$

(ii) $|f(x) - L| < \epsilon$

If $f(x)$ *has a limit as x approaches a, then that limit is unique.* A proof of this fact is given at the beginning of Appendix I.

In using either Definition (1.4) or (1.5) to show that $\lim_{x \to a} f(x) = L$, *it is very important to remember the order in which we consider the numbers ϵ and δ.* Always use the following steps:

Step 1   Consider *any* $\epsilon > 0$.

Step 2   Show that there is a $\delta > 0$ such that if x has δ-tolerance at a, then $f(x)$ has ϵ-tolerance at L.

The number δ in the limit definitions is not unique, for if a specific δ can be found, then any *smaller* positive number δ_1 will also satisfy the requirements.

Before considering examples, let us rephrase the preceding discussion in terms of the graph of the function f. In particular, for $\epsilon > 0$ and $\delta > 0$, we have the following graphical interpretations of tolerances, where $P(x, f(x))$ denotes a point on the graph of f.

Tolerance statement	Graphical interpretation
$f(x)$ has ϵ-tolerance at L.	$P(x, f(x))$ lies between the horizontal lines $y = L \pm \epsilon$.
x has δ-tolerance at a.	x is in the interval $(a - \delta, a + \delta)$ on the x-axis, and $x \neq a$.

The two steps in showing that $\lim_{x \to a} f(x) = L$ may now be interpreted graphically as follows:

Step 1 For any $\epsilon > 0$, consider the horizontal lines $y = L \pm \epsilon$ (shown in Figure 1.14).

Step 2 Show that there is a $\delta > 0$ such that if x is in the open interval $(a - \delta, a + \delta)$ and $x \neq a$, then $P(x, f(x))$ lies between the horizontal lines $y = L \pm \epsilon$ (that is, inside the shaded rectangular region shown in Figure 1.15).

Figure 1.14

Figure 1.15

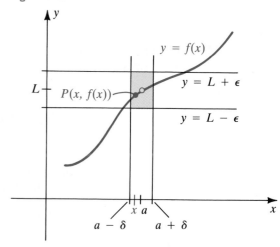

EXAMPLE 1 Use Definition (1.4) to prove that

$$\lim_{x \to 4} (3x - 5) = 7.$$

SOLUTION If, in Definition (1.4), we let $f(x) = 3x - 5$, $a = 4$, and $L = 7$, then we must show that given any $\epsilon > 0$, we can find a $\delta > 0$ such that

$(*)$ if $\quad 0 < |x - 4| < \delta, \quad$ then $\quad |(3x - 5) - 7| < \epsilon.$

To solve an inequality problem of this type, we can often obtain a clue to a proper choice for δ by first examining the ϵ-tolerance statement. Doing so leads to the following list of equivalent inequalities:

$$|(3x - 5) - 7| < \epsilon \qquad \epsilon\text{-tolerance statement}$$
$$|3x - 12| < \epsilon \qquad \text{simplifying}$$
$$|3(x - 4)| < \epsilon \qquad \text{common factor 3}$$
$$3\,|x - 4| < \epsilon \qquad \text{properties of absolute value}$$
$$|x - 4| < \tfrac{1}{3}\epsilon \qquad \text{multiply by } \tfrac{1}{3}$$

The final inequality in the list gives us the needed clue. Specifically, we choose δ such that $\delta \leq \tfrac{1}{3}\epsilon$ and obtain the following equivalent inequalities:

$$0 < |x - 4| < \delta \qquad \delta\text{-tolerance statement}$$
$$0 < |x - 4| < \tfrac{1}{3}\epsilon \qquad \text{choice of } \delta \leq \tfrac{1}{3}\epsilon$$
$$0 < 3\,|x - 4| < \epsilon \qquad \text{multiply by 3}$$
$$0 < |3x - 12| < \epsilon \qquad \text{properties of absolute value}$$
$$0 < |(3x - 5) - 7| < \epsilon \qquad \text{equivalent form}$$

These equivalent inequalities verify $(*)$ and hence complete the proof.

The next example illustrates how the geometric process shown in Figures 1.14 and 1.15 may be applied to a specific function.

EXAMPLE ■ 2 Prove that $\lim\limits_{x \to a} x^2 = a^2$.

SOLUTION Let us consider the case $a > 0$. We shall apply the alternative definition (1.5) with $f(x) = x^2$ and $L = a^2$. Thus, given any $\epsilon > 0$, we must find a $\delta > 0$ such that

$(*)$ if x is in $(a - \delta,\, a + \delta)$ and $x \neq a$, then x^2 is in $(a^2 - \epsilon,\, a^2 + \epsilon)$.

We can obtain a clue to a proper choice for δ by examining graphical interpretations of tolerance statements. Thus, as in step (1) on page 100, consider the horizontal lines $y = a^2 \pm \epsilon$. As shown in Figure 1.16, these lines intersect the graph of $y = x^2$ at points with x-coordinates $\sqrt{a^2 - \epsilon}$ and $\sqrt{a^2 + \epsilon}$. Note that if x is in the open interval $(\sqrt{a^2 - \epsilon}, \sqrt{a^2 + \epsilon})$, then the point (x, x^2) on the graph of f lies between the horizontal lines. If we choose a positive number δ *smaller* than both $\sqrt{a^2 + \epsilon} - a$ and $a - \sqrt{a^2 - \epsilon}$, with $a^2 - \epsilon > 0$, as illustrated in Figure 1.17, then when x has δ-tolerance at a, the point (x, x^2) lies between the horizontal lines $y = a^2 \pm \epsilon$ (that is, x^2 has ϵ-tolerance at a^2). This geometric demonstration proves $(*)$. Although we have considered only $a > 0$, a similar argument applies if $a \leq 0$.

Figure 1.16

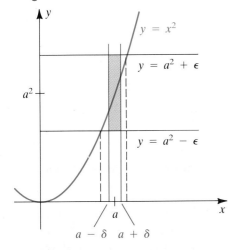

Figure 1.17

The next two examples, which were also discussed in Section 1.1, indicate how the geometric process illustrated in Figure 1.15 may be used to show that certain limits do *not* exist.

EXAMPLE ■ 3   Show that $\lim\limits_{x \to 0} \dfrac{1}{x}$ does not exist.

SOLUTION   Let us proceed in an indirect manner. Thus, *suppose* that

$$\lim_{x \to 0} \frac{1}{x} = L$$

for some number L. Consider any pair of horizontal lines $y = L \pm \epsilon$, as illustrated in Figure 1.18. Since we are assuming that the limit exists, it should be possible to find an open interval $(0 - \delta, 0 + \delta)$, or, equivalently, $(-\delta, \delta)$, such that if $-\delta < x < \delta$ and $x \neq 0$, then the point $(x, 1/x)$ on the graph lies between the horizontal lines. However, since $|1/x|$ can be made as large as desired by choosing x close to 0, some points on the graph will lie either above or below the lines. Hence our supposition is false; that is, $\lim_{x \to 0}(1/x) \neq L$ for any real number L. Thus, the limit does not exist.

Figure 1.18 $f(x) = \dfrac{1}{x}$

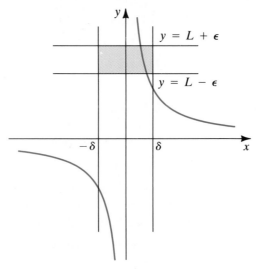

Figure 1.19

$$f(x) = \frac{|x|}{x}$$

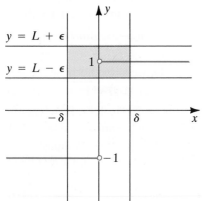

EXAMPLE ■ 4   If $f(x) = \dfrac{|x|}{x}$, show that $\lim\limits_{x \to 0} f(x)$ does not exist.

SOLUTION   The graph of f is sketched in Figure 1.19. If we consider any pair of horizontal lines $y = L \pm \epsilon$, with $0 < \epsilon < 1$, then there are always some points on the graph that do not lie between these lines. In the figure, we have illustrated a case for $L = 1$; however, our proof is valid for every L. Since we cannot find a $\delta > 0$ such that step (2) on page 100 is true, the limit does not exist.

The following theorem states that *if a function f has a positive limit as x approaches a, then $f(x)$ is positive throughout some open interval containing a, with the possible exception of a.*

Theorem 1.6

If $\lim_{x \to a} f(x) = L$ and $L > 0$, then there is an open interval $(a - \delta, a + \delta)$ containing a such that $f(x) > 0$ for every x in $(a - \delta, a + \delta)$, except possibly $x = a$.

PROOF If $L > 0$ and we let $\epsilon = \frac{1}{2}L$, then the horizontal lines $y = L \pm \epsilon$ are above the x-axis, as illustrated in Figure 1.20. By Definition (1.5), there is a $\delta > 0$ such that if $a - \delta < x < a + \delta$ and $x \neq a$, then $L - \epsilon < f(x) < L + \epsilon$. Since $f(x) > L - \epsilon$ and $L - \epsilon > 0$, it follows that $f(x) > 0$ for these values of x. ∎

Figure 1.20

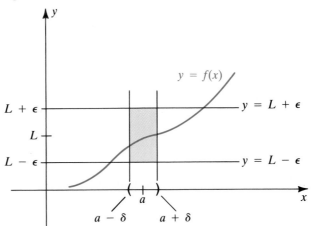

We can also prove that *if f has a negative limit as x approaches a, then there is an open interval containing a such that $f(x) < 0$ for every x in the interval, with the possible exception of $x = a$.*

Formal definitions can be given for one-sided limits. For the right-hand limit $x \to a^+$, we replace the condition $0 < |x - a| < \delta$ in Definition (1.4) by $a < x < a + \delta$. In terms of the alternative definition (1.5), we restrict x to the *right* half $(a, a + \delta)$ of the interval $(a - \delta, a + \delta)$. Similarly, for the left-hand limit $x \to a^-$, we replace $0 < |x - a| < \delta$ in (1.4) by $a - \delta < x < a$. This is equivalent to restricting x to the *left* half $(a - \delta, a)$ of the interval $(a - \delta, a + \delta)$ in (1.5).

EXERCISES 1.2

Exer. 1–2: Express the limit statement in the form of (a) Definition (1.4) and (b) Alternative Definition (1.5).

1 $\lim_{t\to c} v(t) = K$ **2** $\lim_{t\to b} f(t) = M$

Exer. 3–6: Express the one-sided limit statement in a form similar to (a) Definition (1.4) and (b) Alternative Definition (1.5).

3 $\lim_{x\to p^-} g(x) = C$ **4** $\lim_{z\to a^-} h(z) = L$

5 $\lim_{z\to t^+} f(z) = N$ **6** $\lim_{x\to c^+} s(x) = D$

Exer. 7–14: For the given $\lim_{x\to a} f(x) = L$ and ϵ, use the graph of the function f to find the largest value of δ such that if $0 < |x - a| < \delta$, then $|f(x) - L| < \epsilon$.

7 $\lim_{x\to 3/2} \dfrac{4x^2 - 9}{2x - 3} = 6;$ $\epsilon = 0.01$

8 $\lim_{x\to -2/3} \dfrac{9x^2 - 4}{3x + 2} = -4;$ $\epsilon = 0.1$

9 $\lim_{x\to 4} x^2 = 16;$ $\epsilon = 0.1$

10 $\lim_{x\to 3} x^3 = 27;$ $\epsilon = 0.01$

11 $\lim_{x\to 16} \sqrt{x} = 4;$ $\epsilon = 0.1$

12 $\lim_{x\to 27} \sqrt[3]{x} = 3;$ $\epsilon = 0.1$

13 $\lim_{x\to \pi/3} \tan x = \sqrt{3};$ $\epsilon = 0.1$

14 $\lim_{x\to \pi/3} \cos x = \tfrac{1}{2};$ $\epsilon = 0.1$

Exer. 15–24: Use Definition (1.4) to prove that the limit exists.

15 $\lim_{x\to 3} 5x = 15$ **16** $\lim_{x\to 5} (-4x) = -20$

17 $\lim_{x\to -3} (2x + 1) = -5$ **18** $\lim_{x\to 2} (5x - 3) = 7$

19 $\lim_{x\to -6} (10 - 9x) = 64$ **20** $\lim_{x\to 4} (15 - 8x) = -17$

21 $\lim_{x\to 3} 5 = 5$ **22** $\lim_{x\to 5} 3 = 3$

23 $\lim_{x\to a} c = c$ for all real numbers a and c

24 $\lim_{x\to a} (mx + b) = ma + b$ for all real numbers m, b, and a

Exer. 25–30: Use the graphical method illustrated in Example 2 to verify the limit for $a > 0$.

25 $\lim_{x\to -a} x^2 = a^2$ **26** $\lim_{x\to a} (x^2 + 1) = a^2 + 1$

27 $\lim_{x\to a} x^3 = a^3$ **28** $\lim_{x\to a} x^4 = a^4$

29 $\lim_{x\to a} \sqrt{x} = \sqrt{a}$ **30** $\lim_{x\to a} \sqrt[3]{x} = \sqrt[3]{a}$

Exer. 31–38: Use the method illustrated in Examples 3 and 4 to show that the limit does not exist.

31 $\lim_{x\to 3} \dfrac{|x - 3|}{x - 3}$ **32** $\lim_{x\to -2} \dfrac{x + 2}{|x + 2|}$

33 $\lim_{x\to -1} \dfrac{3x + 3}{|x + 1|}$ **34** $\lim_{x\to 5} \dfrac{2x - 10}{|x - 5|}$

35 $\lim_{x\to 0} \dfrac{1}{x^2}$ **36** $\lim_{x\to 4} \dfrac{7}{x - 4}$

37 $\lim_{x\to -5} \dfrac{1}{x + 5}$ **38** $\lim_{x\to 1} \dfrac{1}{(x - 1)^2}$

39 A country taxes the first $20,000 of an individual's income at a rate of 15%, and all income over $20,000 is taxed at 20%. For an income of x dollars, let $T(x)$ be the total tax and $P(x)$ the percentage owed in taxes on the next dollar earned. Explain why $\lim_{x\to 20,000} T(x)$ exists but $\lim_{x\to 20,000} P(x)$ does not exist.

40 Prove that if f has a negative limit as x approaches a, then there is an open interval containing a such that $f(x) < 0$ for every x in the interval, with the possible exception of $x = a$.

41 Give an example of a function f that is defined at a such that $\lim_{x\to a} f(x)$ exists and $\lim_{x\to a} f(x) \neq f(a)$.

42 If f is the greatest integer function (see Example 5 of Section B in the Precalculus Review) and a is any integer, show that $\lim_{x\to a} f(x)$ does not exist.

43 Let f be defined as follows:

$$f(x) = \begin{cases} 0 & \text{if } x \text{ is rational} \\ 1 & \text{if } x \text{ is irrational} \end{cases}$$

Prove that for every real number a, $\lim_{x\to a} f(x)$ does not exist.

Mathematicians and Their Times

RENÉ DESCARTES

DURING THE NIGHT of November 10, 1619, a twenty-three-year-old Frenchman, René Descartes, had three vivid dreams that changed the course of intellectual history. In these dreams, filled with whirlwinds, terrifying phantoms, thunderclaps, and sparks, Descartes felt a supernatural force pointing to the unification and illumination of all knowledge by a single method, the method of *reason*. These nightmares and revelations marked, in the words of mathematicians Philip Davis and Reuben Hersh, the beginning of "the modern world, our world of triumphant rationality."*

The middle of the seventeenth century was one of the most critical periods in the history of mathematics. In this period, France was the undisputed mathematical center of the world, boasting such great thinkers as Pierre de Fermat, Blaise Pascal, Gilles Persone de Roberval, and Girard Desargues. But few men before or since have achieved as much distinction in philosophy and mathematics as Descartes (1596–1650). Regarded by many historians as the "father of modern philosophy," Descartes is perhaps best known to us through his dictum "Cogito, ergo sum" – that is, "I think; therefore I am."

Descartes' most important contribution to mathematics was the creation of *analytic geometry,* the linking of algebra and geometry. Beginning with the representation of a point in the plane by a pair of numbers (now called *Cartesian coordinates* in his honor), Descartes set out to show, in his own words, "how the calculations of arithmetic are related to the operations of geometry." He demonstrated how the familiar figures of Euclidean geometry — lines, polygons, circles, ellipses, and the other conic sections — corresponded to algebraic equations. Geometric questions could be translated into algebraic equations and algebraic operations could be represented in the language of geometry. Thus two of the main branches of mathematics, which had previously

*Philip J. Davis and Reuben Hersh, *Descartes' Dream: The World According to Mathematics.* New York: Harcourt Brace Jovanovich, 1986.

been treated as distinct and independent fields, were now seen to be unified. If the classic geometric techniques failed to solve a problem in geometry, there was now hope of recasting it into algebra where a new set of tools was available. Similarly, one could often gain great insight into difficult algebra problems by interpreting them geometrically.

At age 8, Descartes went to a Jesuit school, where he spent the mornings in bed because of his delicate health. He used these periods productively for study and contemplation, following the custom of remaining in bed until late morning for the rest of his life. In 1649, Sweden's Queen Christina invited him to become her tutor. Perhaps tempted by the glamour of royalty, Descartes accepted, but the cruel winter proved too much for him; he died of pneumonia in 1650.

1.3   TECHNIQUES FOR FINDING LIMITS

Figure 1.21 $f(x) = c$

It would be an excruciating task to verify every limit by means of Definition (1.4) or (1.5). The purpose of this section is to introduce theorems that may be used to simplify problems involving limits. Before stating the first theorem, let us consider the limits of two very simple functions:

(i) the constant function f given by $f(x) = c$
(ii) the linear function g given by $g(x) = x$

The graph of f is the horizontal line $y = c$ shown in Figure 1.21 for the case $c > 0$. Since

$$|f(x) - c| = |c - c| = 0 \quad \text{for every } x$$

and since 0 is less than any $\epsilon > 0$, it follows from Definition (1.4) that $f(x)$ has the limit c as x approaches a. Thus,

$$\lim_{x \to a} f(x) = \lim_{x \to a} c = c.$$

This limit is often described by the phrase *the limit of a constant is the constant.*

The graph of the linear function g given in (ii) is shown in Figure 1.22, and the limit can also be established by means of Definition (1.4). As x approaches a, $g(x)$ approaches a; that is,

$$\lim_{x \to a} g(x) = \lim_{x \to a} x = a.$$

Figure 1.22 $g(x) = x$

The facts demonstrated by these examples are given for reference in the following theorem.

Theorem 1.7

(i) $\lim\limits_{x \to a} c = c$

(ii) $\lim\limits_{x \to a} x = a$

ILLUSTRATION

$\lim\limits_{x \to 3} 8 = 8$ $\lim\limits_{x \to 8} 3 = 3$

$\lim\limits_{x \to \sqrt{2}} x = \sqrt{2}$ $\lim\limits_{x \to -4} x = -4$

The preceding illustration gives simple examples of the limits in Theorem (1.7), but as we shall see, the limits in (1.7) can be used as building blocks for finding limits of very complicated expressions.

Many functions can be expressed as sums, differences, products, and quotients of other functions. Suppose f and g are functions and L and M are real numbers. If

$$f(x) \to L \quad \text{and} \quad g(x) \to M \quad \text{as} \quad x \to a,$$

we would expect that

$$f(x) + g(x) \to L + M \quad \text{as} \quad x \to a.$$

The next theorem states that this expectation is true and gives analogous results for products and quotients.

Theorem 1.8

If $\lim\limits_{x \to a} f(x)$ and $\lim\limits_{x \to a} g(x)$ both exist, then

(i) $\lim\limits_{x \to a} [f(x) + g(x)] = \lim\limits_{x \to a} f(x) + \lim\limits_{x \to a} g(x)$

(ii) $\lim\limits_{x \to a} [f(x) \cdot g(x)] = \lim\limits_{x \to a} f(x) \cdot \lim\limits_{x \to a} g(x)$

(iii) $\lim\limits_{x \to a} \left[\dfrac{f(x)}{g(x)} \right] = \dfrac{\lim\limits_{x \to a} f(x)}{\lim\limits_{x \to a} g(x)}, \quad$ provided $\lim\limits_{x \to a} g(x) \neq 0$

(iv) $\lim\limits_{x \to a} [cf(x)] = c \left[\lim\limits_{x \to a} f(x) \right], \quad$ for any number c

(v) $\lim\limits_{x \to a} [f(x) - g(x)] = \lim\limits_{x \to a} f(x) - \lim\limits_{x \to a} g(x)$

We may state the properties in Theorem (1.8) as follows:

(i) The limit of a sum is the sum of the limits.

(ii) The limit of a product is the product of the limits.

(iii) The limit of a quotient is the quotient of the limits, provided the denominator has a nonzero limit.

(iv) The limit of a constant times a function is the constant times the limit of the function.

(v) The limit of a difference is the difference of the limits.

Proofs for (i) – (iii), based on Definition (1.4), are given in Appendix I. Part (iv) of the theorem follows readily from part (ii) and from Theorem (1.7)(i):

$$\lim_{x \to a} [cf(x)] = \left[\lim_{x \to a} c \right] \left[\lim_{x \to a} f(x) \right]$$
$$= c \left[\lim_{x \to a} f(x) \right]$$

To prove (v), we may write

$$f(x) - g(x) = f(x) + (-1)g(x)$$

and then use parts (i) and (iv) (with $c = -1$).

We now use the preceding theorems to establish the following.

Theorem 1.9

If m, b, and a are real numbers, then

$$\lim_{x \to a} (mx + b) = ma + b.$$

PROOF   By Theorem (1.7),

$$\lim_{x \to a} x = a \quad \text{and} \quad \lim_{x \to a} b = b.$$

We next use (i) and (iv) of Theorem (1.8) to obtain

$$\lim_{x \to a} (mx + b) = \lim_{x \to a} (mx) + \lim_{x \to a} b$$
$$= m \left(\lim_{x \to a} x \right) + b$$
$$= ma + b. \quad \blacksquare$$

This result can also be proved directly from Definition (1.4).

ILLUSTRATION

▪ $\lim_{x \to -2} (5x + 2) = 5(-2) + 2 = -10 + 2 = -8$

▪ $\lim_{x \to 6} (4x - 11) = 4(6) - 11 = 24 - 11 = 13$

It is easy to find the limit in the next two examples by means of Theorems (1.8) and (1.9). To obtain a better appreciation of the power of these theorems, you could try to verify the limits in each by using only Definition (1.4).

EXAMPLE ▪ 1 Find $\lim\limits_{x \to 2} \dfrac{3x + 4}{5x + 7}$.

SOLUTION From Theorem (1.9), we know that the limits of the numerator and the denominator exist. Moreover, the limit of the denominator is not 0. Hence, by Theorem (1.8)(iii) and Theorem (1.9),

$$\lim_{x \to 2} \frac{3x + 4}{5x + 7} = \frac{\lim\limits_{x \to 2} (3x + 4)}{\lim\limits_{x \to 2} (5x + 7)} = \frac{3(2) + 4}{5(2) + 7} = \frac{10}{17}.$$

EXAMPLE ▪ 2 If Achilles runs at a rate of 600 ft/min while the tortoise pokes along at 100 ft/min,

(a) find an expression for the distance between them as a function of time x if the tortoise is given a head start of 2000 ft

(b) determine the limit of this distance as $x \to 4$

SOLUTION Let $A(x)$ and $T(x)$ denote the position along the race course (in feet) for Achilles and the tortoise, respectively, where x is the time (in minutes).

(a) Since each racer runs at a constant speed, we have $A(x) = 600x$ and $T(x) = 2000 + 100x$. The distance between them is given by

$$T(x) - A(x) = (2000 + 100x) - 600x = 2000 - 500x.$$

(b) By Theorem (1.9), with $m = -500$ and $b = 2000$, we have

$$\lim_{x \to 4} [T(x) - A(x)] = \lim_{x \to 4} [2000 - 500x] = 0,$$

so it appears that Achilles will catch up to the tortoise at time $x = 4$ min.

Theorem (1.8) can be extended to limits of sums, differences, products, and quotients that involve any number of functions. In the next example, we use part (ii) for a product of three (equal) functions.

EXAMPLE ▪ 3 Prove that $\lim\limits_{x \to a} x^3 = a^3$.

SOLUTION Since $\lim\limits_{x \to a} x = a$,

$$\lim_{x \to a} x^3 = \lim_{x \to a} (x \cdot x \cdot x)$$

$$= \left(\lim_{x \to a} x \right) \cdot \left(\lim_{x \to a} x \right) \cdot \left(\lim_{x \to a} x \right)$$

$$= a \cdot a \cdot a = a^3.$$

The method used in Example 3 can be extended to x^n for any positive integer n. We merely write x^n as a product $x \cdot x \cdot x \cdot \cdots \cdot x$ of n factors and then take the limit of each factor. Thus, we obtain (i) of the next theorem. Part (ii) may be proved in similar fashion by using Theorem (1.8)(ii). Another method of proof is to use mathematical induction.

Theorem 1.10

If n is a positive integer, then

(i) $\lim\limits_{x \to a} x^n = a^n$

(ii) $\lim\limits_{x \to a} [f(x)]^n = \left[\lim\limits_{x \to a} f(x) \right]^n$,   provided $\lim\limits_{x \to a} f(x)$ exists

EXAMPLE ▪ 4   Find $\lim\limits_{x \to 2} (3x + 4)^5$.

SOLUTION   Applying Theorems (1.10)(ii) and (1.9), we have

$$\lim_{x \to 2} (3x + 4)^5 = \left[\lim_{x \to 2} (3x + 4) \right]^5$$
$$= [3(2) + 4]^5$$
$$= 10^5 = 100,000.$$

EXAMPLE ▪ 5   Find $\lim\limits_{x \to -2} (5x^3 + 3x^2 - 6)$.

SOLUTION   We may proceed as follows, with the reasons justifying each step as indicated:

$$\lim_{x \to -2} (5x^3 + 3x^2 - 6)$$

$$= \lim_{x \to -2} (5x^3) + \lim_{x \to -2} (3x^2) + \lim_{x \to -2} (-6) \quad \text{Theorem (1.8)(i)}$$

$$= \lim_{x \to -2} (5x^3) + \lim_{x \to -2} (3x^2) - 6 \quad \text{Theorem (1.7)(i)}$$

$$= 5 \lim_{x \to -2} (x^3) + 3 \lim_{x \to -2} (x^2) - 6 \quad \text{Theorem (1.8)(iv)}$$

$$= 5(-2)^3 + 3(-2)^2 - 6 \quad \text{Theorem (1.10)(i)}$$

$$= 5(-8) + 3(4) - 6 = -34 \quad \text{simplify}$$

The limit in Example 5 is the number obtained by substituting -2 for x in $5x^3 + 3x^2 - 6$. The next theorem states that the same is true for the limit of *every* polynomial.

Theorem 1.11

If f is a polynomial function and a is a real number, then

$$\lim_{x \to a} f(x) = f(a).$$

PROOF   Since f is a polynomial function,

$$f(x) = b_n x^n + b_{n-1} x^{n-1} + \cdots + b_0$$

for real numbers $b_n, b_{n-1}, \ldots, b_0$. As in Example 5,

$$\lim_{x \to a} f(x) = \lim_{x \to a} (b_n x^n) + \lim_{x \to a} (b_{n-1} x^{n-1}) + \cdots + \lim_{x \to a} b_0$$

$$= b_n \lim_{x \to a} (x^n) + b_{n-1} \lim_{x \to a} (x^{n-1}) + \cdots + \lim_{x \to a} b_0$$

$$= b_n a^n + b_{n-1} a^{n-1} + \cdots + b_0 = f(a). \quad \blacksquare$$

Corollary 1.12

If q is a rational function and a is in the domain of q, then

$$\lim_{x \to a} q(x) = q(a).$$

PROOF Since q is a rational function, $q(x) = f(x)/h(x)$, where f and h are polynomial functions. If a is in the domain of q, then $h(a) \neq 0$. Using Theorems (1.8)(iii) and (1.11) gives us

$$\lim_{x \to a} q(x) = \frac{\lim_{x \to a} f(x)}{\lim_{x \to a} h(x)} = \frac{f(a)}{h(a)} = q(a). \quad \blacksquare$$

NOTE Corollary (1.12) also remains true if q is a trigonometric, exponential, or logarithmic function. We will examine proofs and examples for limits of such transcendental functions in Chapters 2 and 6.

EXAMPLE ▪ 6 Find $\lim\limits_{x \to 3} \dfrac{5x^2 - 2x + 1}{4x^3 - 7}$.

SOLUTION Applying Corollary (1.12) yields

$$\lim_{x \to 3} \frac{5x^2 - 2x + 1}{4x^3 - 7} = \frac{5(3)^2 - 2(3) + 1}{4(3)^3 - 7} = \frac{45 - 6 + 1}{108 - 7} = \frac{40}{101}.$$

The next theorem states that for positive integral roots of x, we may determine a limit by substitution. A proof, using Definition (1.4), may be found in Appendix I.

Theorem 1.13

If $a > 0$ and n is a positive integer, or if $a \leq 0$ and n is an odd positive integer, then

$$\lim_{x \to a} \sqrt[n]{x} = \sqrt[n]{a}.$$

If m and n are positive integers and $a > 0$, then using Theorems (1.10)(ii) and (1.13) gives us

$$\lim_{x \to a} (\sqrt[n]{x})^m = \left(\lim_{x \to a} \sqrt[n]{x} \right)^m = (\sqrt[n]{a})^m.$$

In terms of rational exponents,

$$\lim_{x \to a} x^{m/n} = a^{m/n}.$$

This limit formula may be extended to negative exponents by writing $x^{-r} = 1/x^r$ and then using Theorem (1.8)(iii).

EXAMPLE ▪ 7   Find $\displaystyle\lim_{x \to 8} \frac{x^{2/3} + 3\sqrt{x}}{4 - (16/x)}$.

SOLUTION   We may proceed as follows (supply reasons):

$$\lim_{x \to 8} \frac{x^{2/3} + 3\sqrt{x}}{4 - (16/x)} = \frac{\displaystyle\lim_{x \to 8}(x^{2/3} + 3\sqrt{x})}{\displaystyle\lim_{x \to 8}[4 - (16/x)]}$$

$$= \frac{\displaystyle\lim_{x \to 8} x^{2/3} + \lim_{x \to 8} 3\sqrt{x}}{\displaystyle\lim_{x \to 8} 4 - \lim_{x \to 8}(16/x)}$$

$$= \frac{8^{2/3} + 3\sqrt{8}}{4 - (16/8)} = \frac{4 + 6\sqrt{2}}{4 - 2} = 2 + 3\sqrt{2}$$

Theorem 1.14

If a function f has a limit as x approaches a, then

$$\lim_{x \to a} \sqrt[n]{f(x)} = \sqrt[n]{\lim_{x \to a} f(x)},$$

provided either n is an odd positive integer or n is an even positive integer and $\displaystyle\lim_{x \to a} f(x) > 0$.

The preceding theorem will be proved in Section 1.5. In the meantime, we shall use it whenever applicable to gain experience in finding limits that involve roots of algebraic expressions.

EXAMPLE ▪ 8   Find $\displaystyle\lim_{x \to 5} \sqrt[3]{3x^2 - 4x + 9}$.

SOLUTION   Using Theorems (1.14) and (1.11), we obtain

$$\lim_{x \to 5} \sqrt[3]{3x^2 - 4x + 9} = \sqrt[3]{\lim_{x \to 5}(3x^2 - 4x + 9)}$$

$$= \sqrt[3]{75 - 20 + 9} = \sqrt[3]{64} = 4.$$

The next theorem concerns three functions f, h, and g such that $h(x)$ is "sandwiched" between $f(x)$ and $g(x)$. If f and g have a common limit L as x approaches a, then, as stated in the theorem, h must have the same limit.

Sandwich Theorem **1.15**

Suppose $f(x) \le h(x) \le g(x)$ for every x in an open interval containing a, except possibly at a.

If $\displaystyle\lim_{x \to a} f(x) = L = \lim_{x \to a} g(x)$, then $\displaystyle\lim_{x \to a} h(x) = L$.

Figure 1.23

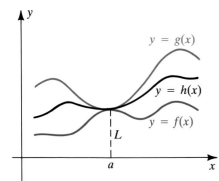

If $f(x) \le h(x) \le g(x)$ for every x in an open interval containing x, then the graph of h lies between the graphs of f and g in that interval, as illustrated in Figure 1.23. If f and g have the same limit L as x approaches a, then it appears from the graphs that h also has the limit L. A proof of the sandwich theorem based on the definition of limit may be found in Appendix I.

EXAMPLE▪9 Use the sandwich theorem (1.15) to prove that

$$\lim_{x \to 0} x^2 \sin \frac{1}{x^2} = 0.$$

SOLUTION Since $-1 \le \sin t \le 1$ for every real number t,

$$-1 \le \sin \frac{1}{x^2} \le 1$$

Figure 1.24

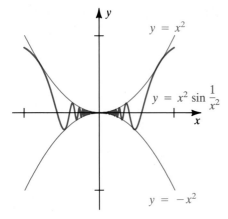

for every $x \ne 0$. Multiplying by x^2 (which is positive if $x \ne 0$), we obtain

$$-x^2 \le x^2 \sin \frac{1}{x^2} \le x^2.$$

This inequality implies that the graph of $y = x^2 \sin(1/x^2)$ lies between the parabolas $y = -x^2$ and $y = x^2$ (see Figure 1.24). Since

$$\lim_{x \to 0} (-x^2) = 0 \quad \text{and} \quad \lim_{x \to 0} x^2 = 0,$$

it follows from the sandwich theorem, with $f(x) = -x^2$ and $g(x) = x^2$, that

$$\lim_{x \to 0} x^2 \sin \frac{1}{x^2} = 0.$$

Theorems similar to the limit theorems given in this section can be proved for one-sided limits. For example,

$$\lim_{x \to a^+} [f(x) + g(x)] = \lim_{x \to a^+} f(x) + \lim_{x \to a^+} g(x)$$

and

$$\lim_{x \to a^+} \sqrt[n]{f(x)} = \sqrt[n]{\lim_{x \to a^+} f(x)}$$

with the usual restrictions on the existence of limits and nth roots. Analogous results are true for left-hand limits.

Figure 1.25

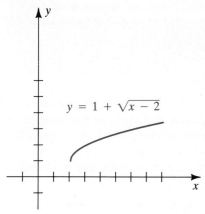

$$y = 1 + \sqrt{x - 2}$$

EXAMPLE■10 Find $\lim_{x \to 2^+} (1 + \sqrt{x - 2})$.

SOLUTION The graph of $f(x) = 1 + \sqrt{x - 2}$ is sketched in Figure 1.25. Using (one-sided) limit theorems, we obtain

$$\lim_{x \to 2^+} (1 + \sqrt{x - 2}) = \lim_{x \to 2^+} 1 + \lim_{x \to 2^+} \sqrt{x - 2}$$

$$= 1 + \sqrt{\lim_{x \to 2^+} (x - 2)}$$

$$= 1 + 0 = 1.$$

Note that since $\sqrt{x - 2}$ is not a real number if $x < 2$, there is no left-hand limit, nor is there a limit of f as x approaches 2.

Figure 1.26

EXAMPLE■11 Let c denote the speed of light (approximately 3.0×10^8 m/sec, or 186,000 mi/sec). In Einstein's theory of relativity, the *Lorentz contraction formula*

$$L = L_0 \sqrt{1 - \frac{v^2}{c^2}}$$

specifies the relationship between (1) the length L of an object that is moving at a velocity v with respect to an observer and (2) its length L_0 at rest (see Figure 1.26). The formula implies that the length of the object measured by the observer is shorter when it is moving than when it is at rest. Find and interpret $\lim_{v \to c^-} L$, and explain why a left-hand limit is necessary.

SOLUTION Using (one-sided) limit theorems yields

$$\lim_{v \to c^-} L = \lim_{v \to c^-} L_0 \sqrt{1 - \frac{v^2}{c^2}}$$

$$= L_0 \lim_{v \to c^-} \sqrt{1 - \frac{v^2}{c^2}}$$

$$= L_0 \sqrt{\lim_{v \to c^-} \left(1 - \frac{v^2}{c^2}\right)}$$

$$= L_0 \sqrt{0} = 0.$$

Thus, if the velocity of an object could approach the speed of light, then its length, as measured by an observer at rest, would approach 0. This result is sometimes used to help justify the theory that the speed of light is the ultimate speed in the universe; that is, no object can have a velocity that is greater than or equal to c.

A left-hand limit is necessary because if $v > c$, then $\sqrt{1 - (v^2/c^2)}$ is not a real number.

EXERCISES 1.3

Exer. 1–48: Use theorems on limits to find the limit, if it exists.

1. $\lim\limits_{x \to \sqrt{2}} 15$

2. $\lim\limits_{x \to 15} \sqrt{2}$

3. $\lim\limits_{x \to -2} x$

4. $\lim\limits_{x \to 3} x$

5. $\lim\limits_{x \to 4} (3x - 4)$

6. $\lim\limits_{x \to -2} (-3x + 1)$

7. $\lim\limits_{x \to -2} \dfrac{x - 5}{4x + 3}$

8. $\lim\limits_{x \to 4} \dfrac{2x - 1}{3x + 1}$

9. $\lim\limits_{x \to 1} (-2x + 5)^4$

10. $\lim\limits_{x \to -2} (3x - 1)^5$

11. $\lim\limits_{x \to 3} (3x - 9)^{100}$

12. $\lim\limits_{x \to 1/2} (4x - 1)^{50}$

13. $\lim\limits_{x \to -2} (3x^3 - 2x + 7)$

14. $\lim\limits_{x \to 4} (5x^2 - 9x - 8)$

15. $\lim\limits_{x \to \sqrt{2}} (x^2 + 3)(x - 4)$

16. $\lim\limits_{t \to -3} (3t + 4)(7t - 9)$

17. $\lim\limits_{x \to \pi} (x - 3.1416)$

18. $\lim\limits_{x \to \pi} (\tfrac{1}{2}x - \tfrac{11}{7})$

19. $\lim\limits_{s \to 4} \dfrac{6s - 1}{2s - 9}$

20. $\lim\limits_{x \to 1/2} \dfrac{4x^2 - 6x + 3}{16x^3 + 8x - 7}$

21. $\lim\limits_{x \to 1/2} \dfrac{2x^2 + 5x - 3}{6x^2 - 7x + 2}$

22. $\lim\limits_{x \to 2} \dfrac{x - 2}{x^3 - 8}$

23. $\lim\limits_{x \to 2} \dfrac{x^2 - x - 2}{(x - 2)^2}$

24. $\lim\limits_{x \to -2} \dfrac{x^2 + 2x - 3}{x^2 + 5x + 6}$

25. $\lim\limits_{x \to -2} \dfrac{x^3 + 8}{x^4 - 16}$

26. $\lim\limits_{x \to 16} \dfrac{x - 16}{\sqrt{x} - 4}$

27. $\lim\limits_{x \to 2} \dfrac{(1/x) - (1/2)}{x - 2}$

28. $\lim\limits_{x \to -3} \dfrac{x + 3}{(1/x) + (1/3)}$

29. $\lim\limits_{x \to 1} \left(\dfrac{x^2}{x - 1} - \dfrac{1}{x - 1} \right)$

30. $\lim\limits_{x \to 1} \left(\sqrt{x} + \dfrac{1}{\sqrt{x}} \right)^6$

31. $\lim\limits_{x \to 16} \dfrac{2\sqrt{x} + x^{3/2}}{\sqrt[4]{x} + 5}$

32. $\lim\limits_{x \to -8} \dfrac{16x^{2/3}}{4 - x^{4/3}}$

33. $\lim\limits_{x \to 4} \sqrt[3]{x^2 - 5x - 4}$

34. $\lim\limits_{x \to -2} \sqrt{x^4 - 4x + 1}$

35. $\lim\limits_{x \to 3} \sqrt[3]{\dfrac{2 + 5x - 3x^3}{x^2 - 1}}$

36. $\lim\limits_{x \to \pi} \sqrt[5]{\dfrac{x - \pi}{x + \pi}}$

37. $\lim\limits_{h \to 0} \dfrac{4 - \sqrt{16 + h}}{h}$

38. $\lim\limits_{h \to 0} \left(\dfrac{1}{h} \right) \left(\dfrac{1}{\sqrt{1 + h}} - 1 \right)$

39. $\lim\limits_{x \to 1} \dfrac{x^2 + x - 2}{x^5 - 1}$

40. $\lim\limits_{x \to 2} \dfrac{x^2 - 7x + 10}{x^6 - 64}$

41. $\lim\limits_{v \to 3} v^2(3v - 4)(9 - v^3)$

42. $\lim\limits_{k \to 2} \sqrt{3k^2 + 4} \sqrt[3]{3k + 2}$

43. $\lim\limits_{x \to 5^+} \left(\sqrt{x^2 - 25} + 3 \right)$

44. $\lim\limits_{x \to 3^-} x\sqrt{9 - x^2}$

45. $\lim\limits_{x \to 3^+} \dfrac{\sqrt{(x - 3)^2}}{x - 3}$

46. $\lim\limits_{x \to -10^-} \dfrac{x + 10}{\sqrt{(x + 10)^2}}$

47. $\lim\limits_{x \to 5^+} \dfrac{1 + \sqrt{2x - 10}}{x + 3}$

48. $\lim\limits_{x \to 4^+} \dfrac{\sqrt[4]{x^2 - 16}}{x + 4}$

Exer. 49–52: Find each limit, if it exists:

(a) $\lim\limits_{x \to a^-} f(x)$ (b) $\lim\limits_{x \to a^+} f(x)$ (c) $\lim\limits_{x \to a} f(x)$

49. $f(x) = \sqrt{5 - x}; \quad a = 5$

50. $f(x) = \sqrt{8 - x^3}; \quad a = 2$

51. $f(x) = \sqrt[3]{x^3 - 1}; \quad a = 1$

52. $f(x) = x^{2/3}; \qquad a = -8$

Exer. 53–56: Let n denote an arbitrary integer. Sketch the graph of f and find $\lim_{x \to n^-} f(x)$ and $\lim_{x \to n^+} f(x)$.

53. $f(x) = (-1)^n \quad$ if $n \le x < n + 1$

54. $f(x) = n \quad$ if $n \le x < n + 1$

55. $f(x) = \begin{cases} x & \text{if } x = n \\ 0 & \text{if } x \ne n \end{cases}$

56. $f(x) = \begin{cases} 0 & \text{if } x = n \\ 1 & \text{if } x \ne n \end{cases}$

Exer. 57–60: Let $[\![\ ]\!]$ denote the greatest integer function and n an arbitrary integer. Find

(a) $\lim\limits_{x \to n^-} f(x)$ (b) $\lim\limits_{x \to n^+} f(x)$

57. $f(x) = [\![x]\!]$

58. $f(x) = x - [\![x]\!]$

59. $f(x) = -[\![-x]\!]$

60. $f(x) = [\![x]\!] - x^2$

Exer. 61–64: Use the sandwich theorem to verify the limit.

61. $\lim\limits_{x \to 0} (x^2 + 1) = 1$ (*Hint:* Use $\lim\limits_{x \to 0} (|x| + 1) = 1$.)

62. $\lim\limits_{x \to 0} \dfrac{|x|}{\sqrt{x^4 + 4x^2 + 7}} = 0$

(*Hint:* Use $f(x) = 0$ and $g(x) = |x|$.)

63. $\lim\limits_{x \to 0} x \sin(1/x) = 0$

(*Hint:* Use $f(x) = -|x|$ and $g(x) = |x|$.)

64. $\lim\limits_{x \to 0} x^4 \sin(1/\sqrt[3]{x}) = 0$ (*Hint:* See Example 9.)

65. If $0 \le f(x) \le c$ for some real number c, prove that $\lim_{x \to 0} x^2 f(x) = 0$.

66 If $\lim_{x \to a} f(x) = L \neq 0$ and $\lim_{x \to a} g(x) = 0$, prove that $\lim_{x \to a}[f(x)/g(x)]$ does not exist. (*Hint:* Assume there is a number M such that $\lim_{x \to a}[f(x)/g(x)] = M$ and consider $\lim_{x \to a} f(x) = \lim_{x \to a}[g(x) \cdot f(x)/g(x)]$.)

67 Explain why $\lim_{x \to 0} \left(x \sin \dfrac{1}{x} \right) \neq \left(\lim_{x \to 0} x \right) \left(\lim_{x \to 0} \sin \dfrac{1}{x} \right)$.

68 Explain why $\lim_{x \to 0} \left(\dfrac{1}{x} + x \right) \neq \lim_{x \to 0} \dfrac{1}{x} + \lim_{x \to 0} x$.

69 Charles's law for gases states that if the pressure remains constant, then the relationship between the volume V that a gas occupies and its temperature T (in $^\circ$C) is given by $V = V_0(1 + \frac{1}{273}T)$. The temperature $T = -273\,^\circ$C is *absolute zero*.

(a) Find $\lim\limits_{T \to -273^+} V$.

(b) Why is a right-hand limit necessary?

70 According to the theory of relativity, the length of an object depends on its velocity v (see Example 11). Einstein also proved that the mass m of an object is related to v by the formula

$$m = \frac{m_0}{\sqrt{1 - (v^2/c^2)}},$$

where m_0 is the mass of the object at rest.

(a) Investigate $\lim_{v \to c^-} m$.

(b) Why is a left-hand limit necessary?

71 A convex lens has focal length f centimeters. If an object is placed a distance p centimeters from the lens, then the distance q centimeters of the image from the lens is related to p and f by the *lens equation*

$$\frac{1}{p} + \frac{1}{q} = \frac{1}{f}.$$

As shown in the figure, p must be greater than f for the rays to converge.

(a) Investigate $\lim_{p \to f^+} q$.

Exercise 71

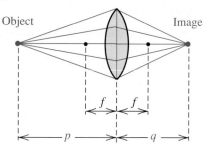

Exercise 71

(b) What is happening to the image as $p \to f^+$?

72 Shown in the figure is a simple magnifier consisting of a convex lens. The object to be magnified is positioned so that its distance p from the lens is less than the focal length f. The linear magnification M is the ratio of the image size to the object size. Using similar triangles, we obtain $M = q/p$, where q is the distance between the image and the lens.

(a) Find $\lim_{p \to 0^+} M$ and explain why a right-hand limit is necessary.

(b) Investigate $\lim_{p \to f^-} M$ and explain what is happening to the image size as $p \to f^-$.

Exercise 72

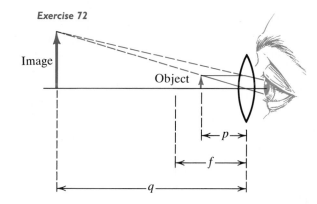

1.4 LIMITS INVOLVING INFINITY

When investigating $\lim_{x \to a^-} f(x)$ or $\lim_{x \to a^+} f(x)$, we may find that as x approaches a, the function value $f(x)$ either increases without bound or decreases without bound. To illustrate, let us consider

$$f(x) = \frac{1}{x - 2}.$$

Figure 1.27

$$f(x) = \frac{1}{x - 2}$$

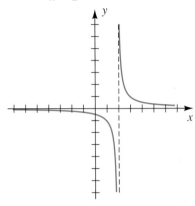

The graph of f is sketched in Figure 1.27. We can show, as in Example 4 of Section 1.1 and Example 3 of Section 1.2, that

$$\lim_{x \to 2} \frac{1}{x - 2} \quad \text{does not exist.}$$

Some function values for x near 2, with $x > 2$, are listed in the following table.

x	2.1	2.01	2.001	2.0001	2.00001	2.000001
$f(x)$	10	100	1000	10,000	100,000	1,000,000

As x approaches 2 from the right, $f(x)$ _increases without bound_ in the sense that _we can make $f(x)$ as large as desired by choosing x sufficiently close to 2 and $x > 2$._ We denote this by writing

$$\lim_{x \to 2^+} \frac{1}{x - 2} = \infty, \quad \text{or} \quad \frac{1}{x - 2} \to \infty \quad \text{as} \quad x \to 2^+.$$

The symbol ∞ (**infinity**) _does not represent a real number._ It is a notation we use to denote how certain functions behave. Thus, although we may state that as _x approaches 2 from the right,_ $1/(x - 2)$ _approaches_ ∞ (_or tends to_ ∞), or that _the limit of_ $1/(x - 2)$ _equals_ ∞, we do _not_ mean that $1/(x - 2)$ gets closer to some specific real number nor do we mean that $\lim_{x \to 2^+}[1/(x - 2)]$ _exists._

The symbol $-\infty$ (**minus infinity**) is used in similar fashion to denote that $f(x)$ _decreases without bound_ (takes on very large _negative_ values) as x approaches a real number. Thus, for $f(x) = 1/(x - 2)$ (see Figure 1.27), we write

$$\lim_{x \to 2^-} \frac{1}{x - 2} = -\infty, \quad \text{or} \quad \frac{1}{x - 2} \to -\infty \quad \text{as} \quad x \to 2^-.$$

Figure 1.28 shows typical (partial) graphs of arbitrary functions that approach ∞ or $-\infty$ in various ways. We have pictured a as positive; however, we can also have $a \le 0$.

Figure 1.28

$\lim\limits_{x \to a^-} f(x) = \infty$, or

$f(x) \to \infty$ as $x \to a^-$

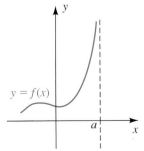

$\lim\limits_{x \to a^+} f(x) = \infty$, or

$f(x) \to \infty$ as $x \to a^+$

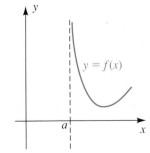

$\lim\limits_{x \to a^-} f(x) = -\infty$, or

$f(x) \to -\infty$ as $x \to a^-$

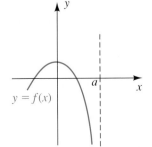

$\lim\limits_{x \to a^+} f(x) = -\infty$, or

$f(x) \to -\infty$ as $x \to a^+$

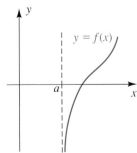

Figure 1.29

$$\lim_{x \to a} f(x) = \infty, \text{ or}$$
$$f(x) \to \infty \text{ as } x \to a$$

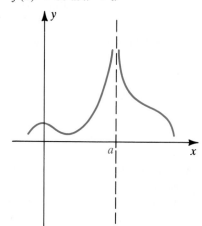

$$\lim_{x \to a} f(x) = -\infty, \text{ or}$$
$$f(x) \to -\infty \text{ as } x \to a$$

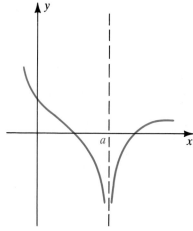

We now consider the two-sided limits illustrated in Figure 1.29. The line $x = a$ in Figures 1.28 and 1.29 is called a **vertical asymptote** for the graph of f.

Note that for $f(x)$ to approach ∞ as x approaches a, *both* the right-hand and left-hand limits must be ∞. For $f(x)$ to approach $-\infty$, *both* one-sided limits must be $-\infty$. If the limit of $f(x)$ from one side of a is ∞ and from the other side of a is $-\infty$, as in Figure 1.27, we say that $\lim_{x \to a} f(x)$ *does not exist*.

It is possible to investigate many algebraic functions that approach ∞ or $-\infty$ by reasoning intuitively, as in the following examples. A formal definition that can be used for rigorous proofs is stated at the end of this section.

Figure 1.30

$$f(x) = \frac{1}{(x - 2)^2}$$

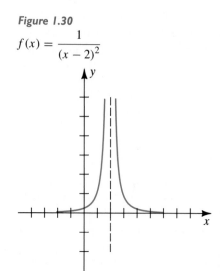

EXAMPLE ▪ 1 Find $\lim_{x \to 2} \dfrac{1}{(x - 2)^2}$, if it exists.

SOLUTION If x is close to 2 and $x \neq 2$, then $(x - 2)^2$ is positive and close to 0. Hence, the reciprocal of $(x - 2)^2$, $1/(x - 2)^2$, is positive and large. There is no real number L that is the limit of $1/(x - 2)^2$ as x approaches 0. The limit does not exist, because we can make $1/(x - 2)^2$ as large as desired by choosing x sufficiently close to 2. Since $1/(x - 2)^2$ increases without bound, we may write

$$\lim_{x \to 2} \frac{1}{(x - 2)^2} = \infty.$$

The graph of $y = 1/(x - 2)^2$ is sketched in Figure 1.30. The line $x = 2$ is a vertical asymptote for the graph.

EXAMPLE ■ 2 Find each limit, if it exists.

(a) $\lim\limits_{x \to 4^-} \dfrac{1}{(x-4)^3}$ **(b)** $\lim\limits_{x \to 4^+} \dfrac{1}{(x-4)^3}$ **(c)** $\lim\limits_{x \to 4} \dfrac{1}{(x-4)^3}$

SOLUTION In all three cases, the limit does not exist, because the denominator approaches 0 as x approaches 4 and hence the fraction has an unbounded absolute value.

Figure 1.31

$$f(x) = \frac{1}{(x-4)^3}$$

(a) If x is close to 4 and $x < 4$, then $x - 4$ is close to 0 and *negative,* and

$$\lim_{x \to 4^-} \frac{1}{(x-4)^3} = -\infty.$$

(b) If x is close to 4 and $x > 4$, then $x - 4$ is close to 0 and *positive,* and

$$\lim_{x \to 4^+} \frac{1}{(x-4)^3} = \infty.$$

(c) Since the one-sided limits are not both ∞ or both $-\infty$, we can only conclude that

$$\lim_{x \to 4} \frac{1}{(x-4)^3} \quad \text{does not exist.}$$

The graph of $y = 1/(x-4)^3$ is sketched in Figure 1.31. The line $x = 4$ is a vertical asymptote for the graph.

Formulas that represent physical quantities may lead to limits involving infinity. Obviously, a physical quantity cannot approach infinity, but an analysis of a hypothetical situation in which that *could* occur may suggest uses for other related quantities. For example, consider Ohm's law in electrical theory, which states that $I = V/R$, where R is the resistance (in ohms) of a conductor, V is the potential difference (in volts) across the conductor, and I is the current (in amperes) that flows through the conductor (see Figure 1.32). The resistance of certain alloys approaches zero as the temperature approaches absolute zero (approximately $-273\ °\mathrm{C}$), and the alloy becomes a *superconductor* of electricity. If the voltage V is fixed, then, for such a superconductor,

Figure 1.32

$$\lim_{R \to 0^+} I = \lim_{R \to 0^+} \frac{V}{R} = \infty;$$

that is, the current increases without bound. Superconductors allow very large currents to be used in generating plants or motors. They also have applications in experimental high-speed ground transportation, where the strong magnetic fields produced by superconducting magnets enable trains to levitate so that there is essentially no friction between the wheels and the track. Perhaps the most important use for superconductors is in circuits for computers, because such circuits produce very little heat.

$$f(x) = 2 + \frac{1}{x}$$

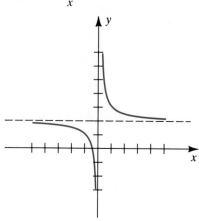

Let us next discuss functions whose values approach some number L as $|x|$ becomes very large. Consider

$$f(x) = 2 + \frac{1}{x},$$

the graph of which is sketched in Figure 1.33. Some values of $f(x)$ if x is large are listed in the following table.

x	100	1000	10,000	100,000	1,000,000
$f(x)$	2.01	2.001	2.0001	2.00001	2.000001

We can make $f(x)$ as close to 2 as desired by choosing x sufficiently large. We denote this fact by

$$\lim_{x \to \infty} \left(2 + \frac{1}{x}\right) = 2,$$

which may be read *the limit of $2 + (1/x)$ as x approaches ∞ is 2*.

Once again, remember that ∞ is not a real number, and hence ∞ *should never be substituted for the variable x*. Note that the terminology *x approaches ∞* does not mean that x gets close to some real number. Intuitively, we think of x as increasing without bound or being assigned arbitrarily large values.

If we let x *decrease* without bound — that is, if we let x take on very large *negative* values — then, as indicated by the second-quadrant portion of the graph shown in Figure 1.33, $2 + (1/x)$ again approaches 2, and we write

$$\lim_{x \to -\infty} \left(2 + \frac{1}{x}\right) = 2.$$

Before considering additional examples, let us state definitions for such limits involving infinity, using ϵ-tolerances for $f(x)$ at L. When we considered $\lim_{x \to a} f(x) = L$ in Section 1.2, we wanted $|f(x) - L| < \epsilon$ whenever x was close to a and $x \neq a$. In the present situation, we want $|f(x) - L| < \epsilon$ whenever x is *sufficiently large* — say, larger than any given positive number M. The precise definition for the limit of a function as x increases without bound follows next.

Definition 1.16

Let a function f be defined on an infinite interval (c, ∞) for a real number c, and let L be a real number. The statement

$$\lim_{x \to \infty} f(x) = L$$

means that for every $\epsilon > 0$, there is a number $M > 0$ such that

$$\text{if } x > M, \quad \text{then} \quad |f(x) - L| < \epsilon.$$

If $\lim_{x \to \infty} f(x) = L$, we say that *the limit of $f(x)$ as x approaches ∞ is L*, or that *$f(x)$ approaches L as x approaches ∞* . We sometimes write

$$f(x) \to L \quad \text{as} \quad x \to \infty.$$

We may give a graphical interpretation of $\lim_{x \to \infty} f(x) = L$ as follows. Consider any horizontal lines $y = L \pm \epsilon$, as in Figure 1.34. According to Definition (1.16), if x is larger than some positive number M, the point $P(x, f(x))$ on the graph lies between these horizontal lines. Intuitively, we know that the graph of f gets closer to the line $y = L$ as x gets larger. We call the line $y = L$ a **horizontal asymptote** for the graph of f. As illustrated in Figure 1.34, *a graph may cross a horizontal asymptote*. The line $y = 2$ in Figure 1.33 is a horizontal asymptote for the graph of $f(x) = 2 + (1/x)$.

Figure 1.34 $\lim_{x \to \infty} f(x) = L$

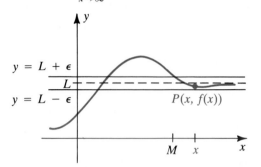

Figure 1.35 $\lim_{x \to -\infty} f(x) = L$

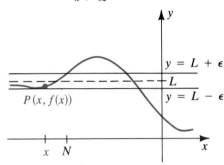

In Figure 1.34, the graph of f approaches the asymptote $y = L$ from below — that is, with $f(x) < L$. A graph can also approach $y = L$ from above — that is, with $f(x) > L$ — or in other ways, such as with $f(x)$ alternately greater than and less than L as $x \to \infty$.

The next definition covers the case in which x is a *large negative number*.

Definition 1.17

Let a function f be defined on an infinite interval $(-\infty, c)$ for a real number c, and let L be a real number. The statement

$$\lim_{x \to -\infty} f(x) = L$$

means that for every $\epsilon > 0$, there is a number $N < 0$ such that

$$\text{if} \quad x < N, \quad \text{then} \quad |f(x) - L| < \epsilon.$$

If $\lim_{x \to -\infty} f(x) = L$, we say *the limit of $f(x)$ as x approaches $-\infty$ is L*, or that *$f(x)$ approaches L as x approaches $-\infty$.*

Definition (1.17) is illustrated in Figure 1.35. If we consider any horizontal lines $y = L \pm \epsilon$, then every point $P(x, f(x))$ on the graph lies between these lines if x is *less* than some negative number N. The line $y = L$ is a horizontal asymptote for the graph of f.

Limit theorems that are analogous to those in Section 1.3 may be established for limits involving infinity. In particular, Theorem (1.8) concerning limits of sums, products, and quotients is true for $x \to \infty$ or $x \to -\infty$. Similarly, Theorem (1.14) on the limit of $\sqrt[n]{f(x)}$ holds if $x \to \infty$ or $x \to -\infty$. We can also show that

$$\lim_{x \to \infty} c = c \quad \text{and} \quad \lim_{x \to -\infty} c = c.$$

A proof of the next theorem, using Definition (1.16), is given in Appendix I.

Theorem   1.18

If k is a positive rational number and c is any real number, then

$$\lim_{x \to \infty} \frac{c}{x^k} = 0 \quad \text{and} \quad \lim_{x \to -\infty} \frac{c}{x^k} = 0,$$

provided x^k is always defined.

Theorem 1.18 is useful for investigating limits of rational functions. Specifically, *to find* $\lim_{x \to \infty} f(x)$ *or* $\lim_{x \to -\infty} f(x)$ *for a rational function* f, *first divide the numerator and the denominator of* $f(x)$ *by* x^n, *where n is the highest power of x that appears in the denominator, and then use limit theorems.* This technique is illustrated in the next examples.

EXAMPLE ▪ 3   Find $\displaystyle \lim_{x \to -\infty} \frac{2x^2 - 5}{3x^2 + x + 2}$.

SOLUTION   The highest power of x in the denominator is 2. Hence, by the rule stated in the preceding paragraph, we divide the numerator and the denominator by x^2 and then use limit theorems. Thus,

$$\lim_{x \to -\infty} \frac{2x^2 - 5}{3x^2 + x + 2} = \lim_{x \to -\infty} \frac{\dfrac{2x^2 - 5}{x^2}}{\dfrac{3x^2 + x + 2}{x^2}} = \lim_{x \to -\infty} \frac{2 - \dfrac{5}{x^2}}{3 + \dfrac{1}{x} + \dfrac{2}{x^2}}$$

$$= \frac{\displaystyle \lim_{x \to -\infty} \left(2 - \frac{5}{x^2}\right)}{\displaystyle \lim_{x \to -\infty} \left(3 + \frac{1}{x} + \frac{2}{x^2}\right)}$$

$$= \frac{\displaystyle \lim_{x \to -\infty} 2 - \lim_{x \to -\infty} \frac{5}{x^2}}{\displaystyle \lim_{x \to -\infty} 3 + \lim_{x \to -\infty} \frac{1}{x} + \lim_{x \to -\infty} \frac{2}{x^2}}$$

$$= \frac{2 - 0}{3 + 0 + 0} = \frac{2}{3}.$$

It follows that the line $y = \frac{2}{3}$ is a horizontal asymptote for the graph of f.

EXAMPLE 4 Find $\displaystyle\lim_{x \to \infty} \frac{2x^3 - 5}{3x^2 + x + 2}$.

SOLUTION The highest power of x in the denominator is 2, so we first divide the numerator and the denominator by x^2, obtaining

$$\lim_{x \to \infty} \frac{2x^3 - 5}{3x^2 + x + 2} = \lim_{x \to \infty} \frac{2x - \dfrac{5}{x^2}}{3 + \dfrac{1}{x} + \dfrac{2}{x^2}}.$$

Since each term of the form c/x^k approaches 0 as $x \to \infty$, we see that

$$\lim_{x \to \infty} \left(2x - \frac{5}{x^2} \right) = \infty$$

and

$$\lim_{x \to \infty} \left(3 + \frac{1}{x} + \frac{2}{x^2} \right) = 3.$$

It follows that

$$\lim_{x \to \infty} \frac{2x^3 - 5}{3x^2 + x + 2} = \infty.$$

EXAMPLE 5 If $f(x) = \dfrac{\sqrt{9x^2 + 2}}{4x + 3}$, find $\displaystyle\lim_{x \to \infty} f(x)$.

SOLUTION If x is large and positive, then

$$\sqrt{9x^2 + 2} \approx \sqrt{9x^2} = 3x$$

and

$$4x + 3 \approx 4x$$

and hence

$$f(x) = \frac{\sqrt{9x^2 + 2}}{4x + 3} \approx \frac{3x}{4x} = \frac{3}{4}.$$

This approximation *suggests* that $\lim_{x \to \infty} f(x) = \frac{3}{4}$. To give a rigorous proof, we may write

$$\lim_{x \to \infty} \frac{\sqrt{9x^2 + 2}}{4x + 3} = \lim_{x \to \infty} \frac{\sqrt{x^2 \left(9 + \dfrac{2}{x^2} \right)}}{4x + 3}$$

$$= \lim_{x \to \infty} \frac{\sqrt{x^2}\sqrt{9 + \dfrac{2}{x^2}}}{4x + 3}.$$

If x is positive, then $\sqrt{x^2} = x$, and dividing the numerator and the denominator of the last fraction by x gives us

$$\lim_{x \to \infty} \frac{\sqrt{9x^2 + 2}}{4x + 3} = \lim_{x \to \infty} \frac{\sqrt{9 + \dfrac{2}{x^2}}}{4 + \dfrac{3}{x}}$$

$$= \frac{\sqrt{9 + 0}}{4 + 0} = \frac{3}{4}.$$

EXAMPLE ∎ 6   Glucose is transported within an enzyme-glucose complex through the placenta from the mother to the fetus. The enzyme acts as a catalyst, accelerating the transportation process. The *Michaelis–Menten law,*

$$C(x) = \frac{ax}{x + b},$$

approximates the relationship between the concentration C of the enzyme-glucose complex and the concentration x of glucose (a and b are positive constants).

(a) Determine $\lim_{x \to \infty} C(x)$.

(b) Sketch the graph of C for $a = 8$ and $b = 3$.

SOLUTION

(a) Since the highest power of x in the denominator is 1, we first divide the numerator and the denominator by x, obtaining

$$\lim_{x \to \infty} C(x) = \lim_{x \to \infty} \frac{ax}{x + b} = \lim_{x \to \infty} \frac{a}{1 + \dfrac{b}{x}} = \frac{a}{1 + 0} = a.$$

(b) From $C(x) = 8x/(x + 3)$, we have $C(0) = 0$ and $C(x) > 0$ when $x > 0$. Writing $C(x)$ as $8/[1 + (3/x)]$, we see that $C(x)$ increases as x increases (the numerator is constant and the denominator decreases in value). In the following table, we show values of $C(x)$ for several values of x.

x	0	1	3	9	21	50	100	200
$C(x)$	0	2	4	6	7	7.55	7.77	7.88

Plotting these points and using the fact that the limit of $C(x)$ as $x \to \infty$ is $a = 8$, we obtain the graph shown in Figure 1.36.

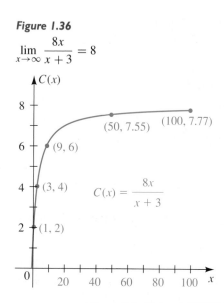

Figure 1.36

$$\lim_{x \to \infty} \frac{8x}{x + 3} = 8$$

$C(x)$

$(50, 7.55)$ $(100, 7.77)$

$(9, 6)$

$(3, 4)$ $C(x) = \dfrac{8x}{x + 3}$

$(1, 2)$

We may also consider cases in which *both* x and $f(x)$ approach ∞ or $-\infty$. For example, the limit statement

$$\lim_{x \to -\infty} f(x) = \infty$$

means that $f(x)$ increases without bound as x decreases without bound, as would be the case for $f(x) = x^2$.

The preceding types of limits involving ∞ occur in applications. To illustrate, Newton's law of universal gravitation may be stated: *Every particle in the universe attracts every other particle with a force that is proportional to the product of their masses and inversely proportional to the square of the distance between them.* In symbols, this statement may be represented by

$$F = G\,\frac{m_1 m_2}{r^2},$$

where F is the force on each particle, m_1 and m_2 are their masses, r is the distance between them, and G is a gravitational constant. Assuming that m_1 and m_2 are constant, we obtain

$$\lim_{r \to \infty} F = \lim_{r \to \infty} G\,\frac{m_1 m_2}{r^2} = 0,$$

which tells us that as the distance between the particles increases without bound, the force of attraction approaches 0. Theoretically, there is always *some* attraction; however, if r is very large, the attraction cannot be measured with conventional laboratory equipment.

We shall conclude this section by stating a formal definition of $\lim_{x \to a} f(x) = \infty$. The main difference from our work in Section 1.2 is that instead of showing that $|f(x) - L| < \epsilon$ whenever x is near a, we consider any (large) positive number M and show that $f(x) > M$ whenever x is near a.

Definition 1.19

Let a function f be defined on an open interval containing a, except possibly at a itself. The statement

$$\lim_{x \to a} f(x) = \infty$$

means that for every $M > 0$, there is a $\delta > 0$ such that

$$\text{if}\quad 0 < |x - a| < \delta, \quad \text{then}\quad f(x) > M.$$

Figure 1.37
$\lim_{x \to a} f(x) = \infty$

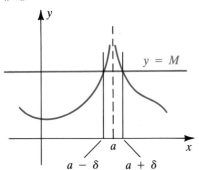

For a graphical interpretation of Definition (1.19), consider any horizontal line $y = M$, as in Figure 1.37. If $\lim_{x \to a} f(x) = \infty$, then whenever x is in a suitable interval $(a - \delta, a + \delta)$ and $x \neq a$, the points on the graph of f lie *above* the horizontal line.

To define $\lim_{x \to a} f(x) = -\infty$, we may alter Definition (1.19), replacing $M > 0$ by $N < 0$ and $f(x) > M$ by $f(x) < N$. Then if we consider any horizontal line $y = N$ (with N negative), the graph of f lies *below* this line whenever x is in a suitable interval $(a - \delta, a + \delta)$ and $x \neq a$.

EXERCISES 1.4

Exer. 1–10: For the given $f(x)$, express each of the following limits as ∞, $-\infty$, or DNE (Does Not Exist):

(a) $\lim\limits_{x \to a^-} f(x)$ (b) $\lim\limits_{x \to a^+} f(x)$ (c) $\lim\limits_{x \to a} f(x)$

1 $f(x) = \dfrac{5}{x-4}$; $a = 4$

2 $f(x) = \dfrac{5}{4-x}$; $a = 4$

3 $f(x) = \dfrac{8}{(2x+5)^3}$; $a = -\dfrac{5}{2}$

4 $f(x) = \dfrac{-4}{7x+3}$; $a = -\dfrac{3}{7}$

5 $f(x) = \dfrac{3x}{(x+8)^2}$; $a = -8$

6 $f(x) = \dfrac{3x^2}{(2x-9)^2}$; $a = \dfrac{9}{2}$

7 $f(x) = \dfrac{2x^2}{x^2-x-2}$; $a = -1$

8 $f(x) = \dfrac{4x}{x^2-4x+3}$; $a = 1$

9 $f(x) = \dfrac{1}{x(x-3)^2}$; $a = 3$

10 $f(x) = \dfrac{-1}{(x+1)^2}$; $a = -1$

Exer. 11–24: Find the limit, if it exists.

11 $\lim\limits_{x \to \infty} \dfrac{5x^2-3x+1}{2x^2+4x-7}$

12 $\lim\limits_{x \to \infty} \dfrac{3x^3-x+1}{6x^3+2x^2-7}$

13 $\lim\limits_{x \to -\infty} \dfrac{4-7x}{2+3x}$

14 $\lim\limits_{x \to -\infty} \dfrac{(3x+4)(x-1)}{(2x+7)(x+2)}$

15 $\lim\limits_{x \to -\infty} \dfrac{2x^2-3}{4x^3+5x}$

16 $\lim\limits_{x \to \infty} \dfrac{2x^2-x+3}{x^3+1}$

17 $\lim\limits_{x \to \infty} \dfrac{-x^3+2x}{2x^2-3}$

18 $\lim\limits_{x \to -\infty} \dfrac{x^2+2}{x-1}$

19 $\lim\limits_{x \to -\infty} \dfrac{2-x^2}{x+3}$

20 $\lim\limits_{x \to \infty} \dfrac{3x^4+x+1}{x^2-5}$

21 $\lim\limits_{x \to \infty} \sqrt[3]{\dfrac{8+x^2}{x(x+1)}}$

22 $\lim\limits_{x \to -\infty} \dfrac{4x-3}{\sqrt{x^2+1}}$

23 $\lim\limits_{x \to \infty} \sin x$

24 $\lim\limits_{x \to \infty} \cos x$

c **Exer. 25–26:** Investigate the limit by letting $x = 10^n$ for $n = 1, 2, 3,$ and 4.

25 $\lim\limits_{x \to \infty} \dfrac{1}{x} \tan\left(\dfrac{\pi}{2} - \dfrac{1}{x}\right)$ 26 $\lim\limits_{x \to \infty} \sqrt{x} \sin\dfrac{1}{x}$

Exer. 27–36: Find the vertical and horizontal asymptotes for the graph of f.

27 $f(x) = \dfrac{1}{x^2-4}$ 28 $f(x) = \dfrac{5x}{4-x^2}$

29 $f(x) = \dfrac{2x^2}{x^2+1}$ 30 $f(x) = \dfrac{3x}{x^2+1}$

31 $f(x) = \dfrac{1}{x^3+x^2-6x}$ 32 $f(x) = \dfrac{x^2-x}{16-x^2}$

33 $f(x) = \dfrac{x^2+3x+2}{x^2+2x-3}$ 34 $f(x) = \dfrac{x^2-5x}{x^2-25}$

35 $f(x) = \dfrac{x+4}{x^2-16}$ 36 $f(x) = \dfrac{\sqrt[3]{16-x^2}}{4-x}$

Exer. 37–40: A function f satisfies the given conditions. Sketch a possible graph for f, assuming that it does not cross a horizontal asymptote.

37 $\lim\limits_{x \to -\infty} f(x) = 1$; $\lim\limits_{x \to \infty} f(x) = 1$;
$\lim\limits_{x \to 3^-} f(x) = -\infty$; $\lim\limits_{x \to 3^+} f(x) = \infty$

38 $\lim\limits_{x \to -\infty} f(x) = -1$; $\lim\limits_{x \to \infty} f(x) = -1$;
$\lim\limits_{x \to 2^-} f(x) = \infty$; $\lim\limits_{x \to 2^+} f(x) = -\infty$

39 $\lim\limits_{x \to -\infty} f(x) = -2$; $\lim\limits_{x \to \infty} f(x) = -2$;
$\lim\limits_{x \to 3^-} f(x) = \infty$; $\lim\limits_{x \to 3^+} f(x) = -\infty$;
$\lim\limits_{x \to -1^-} f(x) = -\infty$; $\lim\limits_{x \to -1^+} f(x) = \infty$

40 $\lim\limits_{x \to -\infty} f(x) = 3$; $\lim\limits_{x \to \infty} f(x) = 3$;
$\lim\limits_{x \to 1^-} f(x) = \infty$; $\lim\limits_{x \to 1^+} f(x) = -\infty$;
$\lim\limits_{x \to -2^-} f(x) = -\infty$; $\lim\limits_{x \to -2^+} f(x) = \infty$

41 Salt water of concentration 0.1 lb of salt per gallon flows into a large tank that initially contains 50 gal of pure water.

(a) If the flow rate of salt water into the tank is 5 gal/min, find the volume $V(t)$ of water and the amount $A(t)$ of salt in the tank after t minutes.

(b) Find a formula for the salt concentration $c(t)$ (in pounds per gallon) after t minutes.

(c) What happens to $c(t)$ over a long period of time?

42 An important problem in fishery science is predicting next year's adult breeding population R (the recruits) from the number S that are presently spawning. For some species (such as North Sea herring), the relationship between R and S is given by $R = aS/(S+b)$, where a and b are positive constants. What happens as the number of spawners increases?

1.5 CONTINUOUS FUNCTIONS

In everyday usage, we say that time is continuous, since it proceeds in an uninterrupted manner. On any given day, time does not jump from 1:00 P.M. to 1:01 P.M., leaving a gap of one minute. If an object is dropped from a hot air balloon, we regard its subsequent motion as continuous. If the initial altitude is 500 ft above ground, the object passes through every altitude between 500 and 0 ft before it hits the ground. The concentration of a chemical at a particular spot along a river may vary continuously, increasing at some times and decreasing at others. In this section, we will use our knowledge of limits to define *continuous functions* and examine their behavior.

Intuitively, we regard a continuous function as a function whose graph has no breaks, holes, or vertical asymptotes. To illustrate, the graph of each function in Figure 1.38 is *not continuous* at the number c.

Figure 1.38

(a)

(b)

(c)

(d)

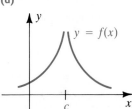

Note that in part (a) of the figure, $f(c)$ is not defined. In part (b), $f(c)$ is defined; however, $\lim_{x \to c} f(x) \neq f(c)$. In part (c), $\lim_{x \to c} f(x)$ does not exist. In part (d), $f(c)$ is undefined and, in addition, $\lim_{x \to c} f(x) = \infty$. The graph of a function f is *not* one of these types if f satisfies the three conditions listed in the next definition.

Definition 1.20

A function f is **continuous** at a number c if the following conditions are satisfied:

(i) $f(c)$ is defined

(ii) $\lim_{x \to c} f(x)$ exists

(iii) $\lim_{x \to c} f(x) = f(c)$

Whenever this definition is used to show that a function f is continuous at c, it is sufficient to verify only the third condition, because if $\lim_{x \to c} f(x) = f(c)$, then $f(c)$ must be defined and also $\lim_{x \to c} f(x)$ must exist; that is, the first two conditions are satisfied automatically.

Intuitively, we know that condition (iii) implies that as x gets closer to c, the function value $f(x)$ gets closer to $f(c)$. More precisely, we can make $f(x)$ *as close to $f(c)$ as desired* by choosing x sufficiently close to c.

If one (or more) of the three conditions in Definition (1.20) is not satisfied, we say that f is **discontinuous** at c, or that f has a **discontinuity** at c. Certain types of discontinuities are given special names. The discontinuities in parts (a) and (b) of Figure 1.38 are **removable discontinuities,** because we could remove each discontinuity by defining the function value $f(c)$ appropriately. The discontinuity in part (c) is a **jump discontinuity,** so named because of the appearance of the graph. If $f(x)$ approaches ∞ or $-\infty$ as x approaches c from either side, as, for example, in part (d), we say that f has an **infinite discontinuity** at c.

In general, if a function f is not continuous at c, then it has a removable discontinuity at c if the right-hand and left-hand limits exist at c and are equal; a jump discontinuity at c if they are not equal; and an infinite discontinuity if $|f(x)|$ can be made arbitrarily large near c.

In the following illustration, we reconsider some specific functions that were discussed in Sections 1.1 and 1.2.

ILLUSTRATION

Function value	Graph	Discontinuity
$f(x) = x + 2$		None, since for every c, $\lim_{x \to c} f(x) = c + 2 = f(c)$.
$g(x) = \dfrac{x^2 + x - 2}{x - 1}$		Removable discontinuity at $c = 1$ since $\lim_{x \to 1^-} g(x) = 3 = \lim_{x \to 1^+} g(x)$

(continued)

Function value	Graph	Discontinuity
$h(x) = \begin{cases} \dfrac{x^2 + x - 2}{x - 1} & \text{if } x \neq 1 \\ 2 & \text{if } x = 1 \end{cases}$ 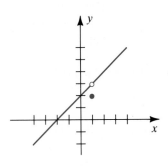		Removable discontinuity at $c = 1$ since $\displaystyle\lim_{x \to 1^-} h(x) = 3 = \lim_{x \to 1^+} h(x)$
$h(x) = \dfrac{1}{x}$ 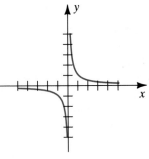		Infinite discontinuity at $c = 0$ since $\lvert h(x) \rvert$ can be arbitrarily large if x is arbitrarily close to 0.
$p(x) = \dfrac{\lvert x \rvert}{x}$ 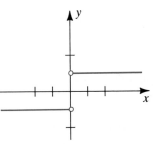		Jump discontinuity at $c = 0$ since $\displaystyle\lim_{x \to 0^-} p(x) = -1,$ $\displaystyle\lim_{x \to 0^+} p(x) = 1,$ but $-1 \neq 1.$

The next theorem states that polynomial functions and rational functions (quotients of polynomial functions) are continuous at every number in their domains.

Theorem 1.21

  (i) A polynomial function f is continuous at every real number c.

  (ii) A rational function $q = f/g$ is continuous at every number except the numbers c such that $g(c) = 0$.

PROOF

  (i) If f is a polynomial function and c is a real number, then, by Theorem (1.11), $\lim_{x \to c} f(x) = f(c)$. Hence, f is continuous at every real number c.

  (ii) If $g(c) \neq 0$, then c is in the domain of $q = f/g$ and, by Theorem (1.12), $\lim_{x \to c} q(x) = q(c)$; that is, q is continuous at c. ∎

A similar version of this theorem is true for the trigonometric, exponential, and logarithmic functions: If q is one of these functions and a is in the domain of q, then q is continuous at a. We will present proofs in Chapters 2 and 6.

EXAMPLE ▪ I   If $f(x) = |x|$, show that f is continuous at every real number c.

SOLUTION   The graph of f is sketched in Figure 1.39. If $x > 0$, then $f(x) = x$. If $x < 0$, then $f(x) = -x$. Since x and $-x$ are polynomials, it follows from Theorem (1.21)(i) that f is continuous at every nonzero real number. It remains to be shown that f is continuous at 0. The one-sided limits of $f(x)$ at 0 are

$$\lim_{x \to 0^+} |x| = \lim_{x \to 0^+} x = 0$$

and

$$\lim_{x \to 0^-} |x| = \lim_{x \to 0^-} (-x) = 0.$$

Since the right-hand and left-hand limits exist and are equal, it follows from Theorem (1.3) that

$$\lim_{x \to 0} f(x) = \lim_{x \to 0} |x| = 0 = |0| = f(0).$$

Hence, f is continuous at 0, and therefore continuous at every real number.

Figure 1.39
$f(x) = |x|$

EXAMPLE ▪ 2   If $f(x) = \dfrac{x^2 - 1}{x^3 + x^2 - 2x}$, find the discontinuities of f.

SOLUTION   Since f is a rational function, it follows from Theorem (1.21) that the only discontinuities occur at the zeros of the denominator, $x^3 + x^2 - 2x$. By factoring, we obtain

$$x^3 + x^2 - 2x = x(x^2 + x - 2) = x(x + 2)(x - 1).$$

Setting each factor equal to zero, we see that the discontinuities of f are at $0, -2,$ and 1.

If a function f is continuous at every number in an open interval (a, b), we say that f **is continuous on the interval** (a, b). Similarly, a function is continuous on an infinite interval of the form (a, ∞) or $(-\infty, b)$ if it is continuous at every number in the interval. The next definition covers the case of a closed interval.

Definition   1.22

Let a function f be defined on a closed interval $[a, b]$. The **function f is continuous on $[a, b]$** if it is continuous on (a, b) and if, in addition,

$$\lim_{x \to a^+} f(x) = f(a) \quad \text{and} \quad \lim_{x \to b^-} f(x) = f(b).$$

If a function f has either a right-hand or a left-hand limit of the type indicated in Definition (1.22), we say that f **is continuous from the right at a** or that f **is continuous from the left at b,** respectively.

EXAMPLE ▪ 3   If $f(x) = \sqrt{9 - x^2}$, sketch the graph of f and prove that f is continuous on the closed interval $[-3, 3]$.

SOLUTION   The graph of $x^2 + y^2 = 9$ is a circle with center at the origin and radius 3. Solving for y gives us $y = \pm\sqrt{9 - x^2}$, and hence the graph of $y = \sqrt{9 - x^2}$ is the upper half of that circle (see Figure 1.40).

If $-3 < c < 3$, then, using Theorem (1.14), we obtain

$$\lim_{x \to c} f(x) = \lim_{x \to c} \sqrt{9 - x^2} = \sqrt{9 - c^2} = f(c).$$

Hence f is continuous at c by Definition (1.20). All that remains is to check the endpoints of the interval $[-3, 3]$ using one-sided limits as follows:

$$\lim_{x \to -3^+} f(x) = \lim_{x \to -3^+} \sqrt{9 - x^2} = \sqrt{9 - 9} = 0 = f(-3)$$

$$\lim_{x \to 3^-} f(x) = \lim_{x \to 3^-} \sqrt{9 - x^2} = \sqrt{9 - 9} = 0 = f(3)$$

Thus, f is continuous from the right at -3 and from the left at 3. By Definition (1.22), f is continuous on $[-3, 3]$.

Figure 1.40

$y = \sqrt{9 - x^2}$

Strictly speaking, the function f in Example 3 is discontinuous at every number c *outside* of the interval $[-3, 3]$, because $f(c)$ is not a real number if $x < -3$ or $x > 3$. However, it is *not* customary to use the phrase *discontinuous at c* if c is in an open interval throughout which f is undefined.

We may also define continuity on other types of intervals. For example, a function f is continuous on $[a, b)$ or $[a, \infty)$ if it is continuous at every number greater than a in the interval and if, in addition, f is continuous from the right at a. For intervals of the form $(a, b]$ or $(-\infty, b]$, we require continuity at every number less than b in the interval and also continuity from the left at b.

Using facts stated in Theorem (1.8), we can prove the following.

Theorem 1.23

If two functions f and g are continuous at a real number c, then the following are also continuous at c:

(i) the sum $f + g$
(ii) the difference $f - g$
(iii) the product fg
(iv) the quotient f/g,   provided $g(c) \neq 0$

PROOF   If f and g are continuous at c, then

$$\lim_{x \to c} f(x) = f(c) \quad \text{and} \quad \lim_{x \to c} g(x) = g(c).$$

By definition of the sum of two functions,

$$(f + g)(x) = f(x) + g(x).$$

Consequently,

$$\lim_{x \to c}(f + g)(x) = \lim_{x \to c}[f(x) + g(x)]$$

$$= \lim_{x \to c} f(x) + \lim_{x \to c} g(x)$$

$$= f(c) + g(c)$$

$$= (f + g)(c).$$

We have thus proved that $f + g$ is continuous at c. Parts (ii)–(iv) are proved in similar fashion.   ■

If f and g are continuous on an interval, then $f + g$, $f - g$, and fg are continuous on the interval. If, in addition, $g(c) \neq 0$ for every c in the interval, then f/g is continuous on the interval. These results may be extended to more than two functions; that is, sums, differences, products, or quotients involving any number of continuous functions are continuous (provided zero denominators do not occur).

EXAMPLE ■ 4   If $k(x) = \dfrac{\sqrt{9 - x^2}}{3x^4 + 5x^2 + 1}$, prove that k is continuous on the closed interval $[-3, 3]$.

SOLUTION   Let $f(x) = \sqrt{9 - x^2}$ and $g(x) = 3x^4 + 5x^2 + 1$. From Example 3, f is continuous on $[-3, \ 3]$, and from Theorem (1.21), g is continuous at every real number. Moreover, $g(c) \neq 0$ for every number c in $[-3, 3]$. Hence, by Theorem (1.23)(iv), the quotient $k = f/g$ is continuous on $[-3, 3]$.

A proof of the next result on the limit of a composite function $f \circ g$ is given in Appendix I.

Theorem   1.24

If $\lim_{x \to c} g(x) = b$ and if f is continuous at b, then

$$\lim_{x \to c} f(g(x)) = f(b) = f\left(\lim_{x \to c} g(x)\right).$$

The principal use of Theorem (1.24) is to prove other theorems. To illustrate, let us use Theorem (1.24) to prove Theorem (1.14) from Section 1.3, in which we assumed that $\lim_{x \to c} g(x)$ and the indicated nth roots exist.

Conclusion of Theorem 1.14

$$\lim_{x \to c} \sqrt[n]{g(x)} = \sqrt[n]{\lim_{x \to c} g(x)}$$

PROOF Let $f(x) = \sqrt[n]{x}$. Applying Theorem (1.24), which states that

$$\lim_{x \to c} f(g(x)) = f\left(\lim_{x \to c} g(x)\right),$$

we obtain
$$\lim_{x \to c} \sqrt[n]{g(x)} = \sqrt[n]{\lim_{x \to c} g(x)}. \quad \blacksquare$$

The next theorem follows from Theorem (1.24) and the definitions of a continuous function and of the composite function $f \circ g$.

Theorem 1.25

If g is continuous at c and if f is continuous at $g(c)$, then the composite function $f \circ g$ is continuous at c; that is,

$$\lim_{x \to c} f(g(x)) = f\left(\lim_{x \to c} g(x)\right) = f(g(c)).$$

EXAMPLE 5 If $k(x) = \left|3x^2 - 7x - 12\right|$, show that k is continuous at every real number.

SOLUTION If we let

$$f(x) = |x| \quad \text{and} \quad g(x) = 3x^2 - 7x - 12,$$

then $k(x) = f(g(x)) = (f \circ g)(x)$. Since both f and g are continuous functions (see Example 1 and (i) of Theorem (1.21)), it follows from Theorem (1.25) that the composite function $k = f \circ g$ is continuous at c.

A proof of the following property of continuous functions may be found in more advanced texts on calculus.

Intermediate Value Theorem 1.26

If f is continuous on a closed interval $[a, b]$ and if w is any number between $f(a)$ and $f(b)$, then there is at least one number c in $[a, b]$ such that $f(c) = w$.

Figure 1.41

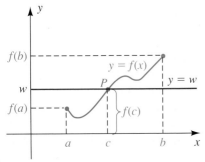

The intermediate value theorem states that *as x varies from a to b, the continuous function f takes on every value between f(a) and f(b).* If the graph of the continuous function f is regarded as extending in an unbroken manner from the point $(a, f(a))$ to the point $(b, f(b))$, as illustrated in Figure 1.41, then for any number w between $f(a)$ and $f(b)$, the horizontal line with y-intercept w intersects the graph in at least one point P. The x-coordinate c of P is a number such that $f(c) = w$.

A consequence of the intermediate value theorem is that *if $f(a)$ and $f(b)$ have opposite signs, then there is a number c between a and b such that $f(c) = 0$; that is, f has a zero at c.* Thus, if the point $(a, f(a))$ on the graph of a continuous function lies below the x-axis and the point $(b, f(b))$ lies above the x-axis, or vice versa, then the graph crosses the x-axis at some point $(c, 0)$ for $a < c < b$.

We can use this consequence of the intermediate value theorem to help locate zeros of a function, as in the next example.

EXAMPLE ■ 6   Let $f(x) = x^5 + 2x^4 - 6x^3 + 2x - 3$.

(a) Use the intermediate value theorem (1.26) to show that f has three zeros in the interval $[-4, 2]$.

(b) Use a graphing utility to approximate these zeros to two decimal places.

SOLUTION

(a) We compute the value of f at the integers from -4 to 2, as shown.

x	-4	-3	-2	-1	0	1	2
$f(x)$	-139	72	41	2	-3	-4	17

Figure 1.42
$-4 \le x \le 2, -50 \le y \le 50$

Since f is a polynomial, it is continuous at all values of x. By the intermediate value theorem, f has a zero between -4 and -3 since $f(-4)$ and $f(-3)$ are of opposite sign. Similarly, f has a zero between -1 and 0 and another zero between 1 and 2.

(b) With the aid of a graphing utility, we can look for other possible zeros. Figure 1.42 shows the graph of f, which indicates that only these three zeros exist on the x-interval $[-4, 2]$. Using the *trace* and *zoom* features, or a *solve* feature, we determine that the zeros are approximately -3.60, -0.88, and 1.63.

Analytic and algebraic techniques can be combined with the effective use of a graphing utility to locate the discontinuities of a given function, as demonstrated in the next example.

EXAMPLE ■ 7   Approximate the discontinuities of the function

$$f(x) = 1 + \frac{\sqrt{x + 6}}{x^2 + 2x - 5}.$$

Figure 1.43
$-6 \le x \le 5, -2 \le y \le 4.5$

SOLUTION   The term $\sqrt{x + 6}$ restricts the domain of f to those values for which $x + 6 \ge 0$; that is, $x \ge -6$. We select the viewing window $-6 \le x \le 5$ and $-2 \le y \le 4.5$ to obtain Figure 1.43. Note that the graph of f begins at $x = -6$.

You may obtain a graph slightly different from Figure 1.43, depending on the size and the resolution of your screen and the graphing utility you use. In Figure 1.43, the exact behavior of the graph of f may not be entirely

Figure 1.44

$$g(x) = x^2 + 2x - 5$$
$$-5 \leq x \leq 5, \ -6 \leq y \leq 30$$

Figure 1.45

$$f(x) = 1 + \frac{\sqrt{x+6}}{x^2 + 2x - 5}$$
$$-3.48 \leq x \leq -3.42, \ -400 \leq y \leq 400$$

Figure 1.46

(a) Graph of f
(normal screen view)

(b) DOT or POINT mode
(magnified screen view)

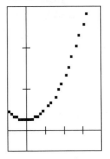

(c) LINE or CONNECTED mode
(magnified screen view)

clear, but we *can* see that something unusual occurs at two different values of x. In such cases, before zooming in on a segment of the graph, it is useful to do some preliminary analysis of the function.

Because the quantity $x^2 + 2x - 5$ appears in the denominator of $f(x)$, we cannot have values of x where $x^2 + 2x - 5 = 0$. To find these values, we either solve this quadratic equation or estimate the values from the graph of the function $g(x) = x^2 + 2x - 5$. Figure 1.44 shows the graph of g, where it appears that there is a zero of g between -4 and -3 and another zero between 1 and 2.

Repeatedly zooming in gives zeros for g at $x_1 \approx -3.449$ and $x_2 \approx 1.449$. By Theorem (1.23)(iv) on the continuity of the quotient of continuous functions, the function f will be continuous except possibly at the zeros of g. We can examine the behavior of f near x_1 by viewing the graph of f in a window with $-3.48 \leq x \leq -3.42$, as shown in Figure 1.45.

From Figure 1.45, we can easily infer what is occurring near x_1: As $x \to x_1^-$, we have $f(x) \to \infty$, but as $x \to x_1^+$, we have $f(x) \to -\infty$. Since neither the right-hand nor the left-hand limit of f exists as $x \to x_1$, we have a discontinuity of f at x_1. A similar analysis at x_2 shows that $\lim_{x \to x_2^-} f(x) = -\infty$ and $\lim_{x \to x_2^+} f(x) = \infty$, so there is a discontinuity at x_2 as well.

In Example 7, the graph of f near x_1 has two branches, one for values less than x_1 and a second for values greater than x_1. The graphing utility, however, shows a nearly vertical line connecting these two branches. Although this line is not really part of the graph of f, it is displayed because the underlying software of the graphing utility assumes that functions are continuous.

A graphing utility can evaluate a function only at a finite number of points. For each of these computed values, it darkens a picture element (called a *pixel*) on the screen. Assuming continuity, it draws a line segment between adjacent plotted pixels to represent the intermediate values that a continuous function assumes. If sufficiently small segments are used, the human eye perceives a smooth curve.

Some graphing utilities have a DOT or POINT mode which displays only the pixels for computed function values. In LINE or CONNECTED mode, the default mode for graphing utilities, the connecting line segments are also displayed. Figure 1.46 shows magnified views of the graph of a function in both DOT mode and CONNECTED mode. In generating Figure 1.46, the graphing utility calculated a large positive value to the left of x_1 and a large negative value to the right of x_1, and then darkened the corresponding pixels as well as the line segment between them, thereby producing the unusual view of the function.

EXERCISES 1.5

Exer. 1–10: The graph of a function f is given. Classify the discontinuities of f as removable, jump, or infinite.

1

2

3

4

5

6

7

8

9

10

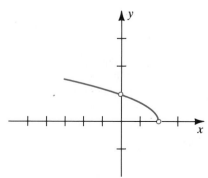

Exer. 11–18: Classify the discontinuities of f as removable, jump, or infinite.

11 $f(x) = \begin{cases} x^2 - 1 & \text{if } x < 1 \\ 4 - x & \text{if } x \geq 1 \end{cases}$

12 $f(x) = \begin{cases} x^3 & \text{if } x \leq 1 \\ 3 - x & \text{if } x > 1 \end{cases}$

13 $f(x) = \begin{cases} |x + 3| & \text{if } x \neq -2 \\ 2 & \text{if } x = -2 \end{cases}$

14 $f(x) = \begin{cases} |x - 1| & \text{if } x \neq 1 \\ 1 & \text{if } x = 1 \end{cases}$

15 $f(x) = \begin{cases} x^2 + 1 & \text{if } x < 1 \\ 1 & \text{if } x = 1 \\ x + 1 & \text{if } x > 1 \end{cases}$

16 $f(x) = \begin{cases} -x^2 & \text{if } x < 1 \\ 2 & \text{if } x = 1 \\ x - 2 & \text{if } x > 1 \end{cases}$

c 17 $f(x) = x^{-1/3} \sin\left[\cos\left(\dfrac{\pi}{2} - x^2\right)\right]$

c 18 $f(x) = \dfrac{\sin(x^2 - 1)}{(x - 1)^2}$

Exer. 19–22: Show that f is continuous at a.

19 $f(x) = \sqrt{2x - 5} + 3x; \qquad a = 4$

20 $f(x) = \sqrt[3]{x^2 + 2}; \qquad a = -5$

21 $f(x) = 3x^2 + 7 - \dfrac{1}{\sqrt{-x}}; \quad a = -2$

22 $f(x) = \dfrac{\sqrt[3]{x}}{2x + 1}; \qquad a = 8$

Exer. 23–30: Explain why f is not continuous at a.

23 $f(x) = \dfrac{3}{x + 2}; \qquad a = -2$

24 $f(x) = \dfrac{1}{x - 1}; \qquad a = 1$

25 $f(x) = \begin{cases} \dfrac{x^2 - 9}{x - 3} & \text{if } x \neq 3 \\ 4 & \text{if } x = 3 \end{cases} \quad a = 3$

26 $f(x) = \begin{cases} \dfrac{x^2 - 9}{x + 3} & \text{if } x \neq -3 \\ 2 & \text{if } x = -3 \end{cases} \quad a = -3$

27 $f(x) = \begin{cases} 1 & \text{if } x \neq 3 \\ 0 & \text{if } x = 3 \end{cases} \quad a = 3$

28 $f(x) = \begin{cases} \dfrac{|x - 3|}{x - 3} & \text{if } x \neq 3 \\ 1 & \text{if } x = 3 \end{cases} \quad a = 3$

c 29 $f(x) = \begin{cases} \dfrac{\sin x}{x} & \text{if } x \neq 0 \\ 0 & \text{if } x = 0 \end{cases}$ $a = 0$

c 30 $f(x) = \begin{cases} \dfrac{1 - \cos x}{x} & \text{if } x \neq 0 \\ 1 & \text{if } x = 0 \end{cases}$ $a = 0$

Exer. 31–34: Find all numbers at which f is discontinuous.

31 $f(x) = \dfrac{3}{x^2 + x - 6}$ 32 $f(x) = \dfrac{5}{x^2 - 4x - 12}$

33 $f(x) = \dfrac{x - 1}{x^2 + x - 2}$ 34 $f(x) = \dfrac{x - 4}{x^2 - x - 12}$

Exer. 35–38: Show that f is continuous on the given interval.

35 $f(x) = \sqrt{x - 4}$; $[4, 8]$

36 $f(x) = \sqrt{16 - x}$; $(-\infty, 16]$

37 $f(x) = \dfrac{1}{x^2}$; $(0, \infty)$

38 $f(x) = \dfrac{1}{x - 1}$; $(1, 3)$

Exer. 39–54: Find all numbers at which f is continuous.

39 $f(x) = \dfrac{3x - 5}{2x^2 - x - 3}$

40 $f(x) = \dfrac{x^2 - 9}{x - 3}$

41 $f(x) = \sqrt{2x - 3} + x^2$

42 $f(x) = \dfrac{x}{\sqrt[3]{x - 4}}$

43 $f(x) = \dfrac{x - 1}{\sqrt{x^2 - 1}}$ 44 $f(x) = \dfrac{x}{\sqrt{1 - x^2}}$

45 $f(x) = \dfrac{|x + 9|}{x + 9}$ 46 $f(x) = \dfrac{x}{x^2 + 1}$

47 $f(x) = \dfrac{5}{x^3 - x^2}$

48 $f(x) = \dfrac{4x - 7}{(x + 3)(x^2 + 2x - 8)}$

49 $f(x) = \dfrac{\sqrt{x^2 - 9}\,\sqrt{25 - x^2}}{x - 4}$

50 $f(x) = \dfrac{\sqrt{9 - x}}{\sqrt{x - 6}}$

51 $f(x) = \tan 2x$ 52 $f(x) = \cot \frac{1}{3}x$

53 $f(x) = \csc \frac{1}{2}x$ 54 $f(x) = \sec 3x$

55 Suppose that
$$f(x) = \begin{cases} cx^2 - 3 & \text{if } x \leq 2 \\ cx + 2 & \text{if } x > 2 \end{cases}$$
Find a value of c such that f is continuous on $\mathbb{R}$.

56 Suppose that
$$f(x) = \begin{cases} c^2x & \text{if } x < 1 \\ 3cx - 2 & \text{if } x \geq 1 \end{cases}$$
Determine all values of c such that f is continuous on $\mathbb{R}$.

57 Suppose that
$$f(x) = \begin{cases} c & \text{if } x = -3 \\ \dfrac{9 - x^2}{4 - \sqrt{x^2 + 7}} & \text{if } |x| < 3 \\ d & \text{if } x = 3 \end{cases}$$
Find values of c and d such that f is continuous on $[-3, 3]$.

58 Suppose that
$$f(x) = \begin{cases} 4x & \text{if } x \leq -1 \\ cx + d & \text{if } -1 < x < 2 \\ -5x & \text{if } x \geq 2 \end{cases}$$
Find values of c and d such that f is continuous on $\mathbb{R}$.

Exer. 59–62: Verify the intermediate value theorem (1.26) for f on the stated interval $[a, b]$ by showing that if $f(a) \leq w \leq f(b)$, then $f(c) = w$ for some c in $[a, b]$.

59 $f(x) = x^3 + 1$; $[-1, 2]$

60 $f(x) = -x^3$; $[0, 2]$

61 $f(x) = x^2 - x$; $[1, 3]$

62 $f(x) = 2x - x^2$; $[-2, -1]$

63 If $f(x) = x^3 - 5x^2 + 7x - 9$, use the intermediate value theorem (1.26) to prove that there is a real number a such that $f(a) = 100$.

64 Prove that the equation $x^5 - 3x^4 - 2x^3 - x + 1 = 0$ has a solution between 0 and 1.

65 Use the intermediate value theorem (1.26) to show that the graphs of the functions $f(x) = x^4 - 5x^2$ and $g(x) = 2x^3 - 4x + 6$ intersect between $x = 3$ and $x = 4$. (*Hint:* Consider $h = f - g$.)

66 Show that if a function f is continuous and has no zeros on an interval, then either $f(x) > 0$ or $f(x) < 0$ for every x in the interval.

67 In models for free fall, it is generally assumed that the gravitational acceleration g is the constant 9.8 m/sec^2 (or 32 ft/sec^2). Actually, g varies with latitude. If θ is the latitude (in degrees), then a formula that approximates g is

$$g(\theta) = 9.78049(1 + 0.005264\sin^2\theta + 0.000024\sin^4\theta).$$

Use the intermediate value theorem (1.26) to show that $g = 9.8$ somewhere between latitudes $35°$ and $40°$.

68 The temperature T (in °C) at which water boils may be approximated by the formula

$$T(h) = 100.862 - 0.0415\sqrt{h + 431.03},$$

where h is the elevation (in meters above sea level). Use the intermediate value theorem (1.26) to show that water boils at 98 °C at an elevation somewhere between 4000 and 4500 m.

Exer. 69–74: Approximate the zeros of f to three decimal places.

69 $f(x) = x^5 - x + 3$

70 $f(x) = x^3 - \sin x + 0.5$

71 $f(x) = 8x^4 - 14x^3 - 9x^2 + 12x + 2$

72 $f(x) = 3x^5 - 10x^4 + 10x^3 + 3x + 7$

73 $f(x) = \ln(1 + x^2) - 5$

74 $f(x) = 3^x - x^3 - 3x - 3$

Exer. 75–77: Approximate the discontinuities of f to three decimal places.

75 $f(x) = \dfrac{\sqrt{x + 4}}{x^2 - 14x + 47}$

76 $f(x) = \dfrac{x + 3}{|2\cos x - 1|}$

77 $f(x) = \dfrac{1}{x^3 - x + 2}$

CHAPTER I REVIEW EXERCISES

Exer. 1–26: Find the limit, if it exists.

1 $\displaystyle\lim_{x \to 3} \frac{5x + 11}{\sqrt{x + 1}}$

2 $\displaystyle\lim_{x \to -2} \frac{6 - 7x}{(3 + 2x)^4}$

3 $\displaystyle\lim_{x \to -2} \left(2x - \sqrt{4x^2 + x}\right)$

4 $\displaystyle\lim_{x \to 4^-} \left(x - \sqrt{16 - x^2}\right)$

5 $\displaystyle\lim_{x \to 3/2} \frac{2x^2 + x - 6}{4x^2 - 4x - 3}$

6 $\displaystyle\lim_{x \to 2} \frac{3x^2 - x - 10}{x^2 - x - 2}$

7 $\displaystyle\lim_{x \to 2} \frac{x^4 - 16}{x^2 - x - 2}$

8 $\displaystyle\lim_{x \to 3^+} \frac{1}{x - 3}$

9 $\displaystyle\lim_{x \to 0^+} \frac{1}{\sqrt{x}}$

10 $\displaystyle\lim_{x \to 5} \frac{(1/x) - (1/5)}{x - 5}$

11 $\displaystyle\lim_{x \to 1/2} \frac{8x^3 - 1}{2x - 1}$

12 $\displaystyle\lim_{x \to 2} 5$

13 $\displaystyle\lim_{x \to 3^+} \frac{3 - x}{|3 - x|}$

14 $\displaystyle\lim_{x \to 2} \frac{\sqrt{x} - \sqrt{2}}{x - 2}$

15 $\displaystyle\lim_{h \to 0} \frac{(a + h)^4 - a^4}{h}$

16 $\displaystyle\lim_{h \to 0} \frac{(2 + h)^{-3} - 2^{-3}}{h}$

17 $\displaystyle\lim_{x \to -3} \sqrt[3]{\frac{x + 3}{x^3 + 27}}$

18 $\displaystyle\lim_{x \to 5/2^-} \left(\sqrt{5 - 2x} - x^2\right)$

19 $\displaystyle\lim_{x \to -\infty} \frac{(2x - 5)(3x + 1)}{(x + 7)(4x - 9)}$

20 $\displaystyle\lim_{x \to \infty} \frac{2x + 11}{\sqrt{x + 1}}$

21 $\displaystyle\lim_{x \to -\infty} \frac{6 - 7x}{(3 + 2x)^4}$

22 $\displaystyle\lim_{x \to \infty} \frac{x - 100}{\sqrt{x^2 + 100}}$

23 $\displaystyle\lim_{x \to 2/3^+} \frac{x^2}{4 - 9x^2}$

24 $\displaystyle\lim_{x \to 3/5^-} \frac{1}{5x - 3}$

25 $\displaystyle\lim_{x \to 0^+} \left(\sqrt{x} - \frac{1}{\sqrt{x}}\right)$

26 $\displaystyle\lim_{x \to 1} \frac{x - 1}{\sqrt{(x - 1)^2}}$

Exer. 27–32: Sketch the graph of the piecewise-defined function f and, for the indicated value of a, find each limit, if it exists:

(a) $\displaystyle\lim_{x \to a^-} f(x)$ (b) $\displaystyle\lim_{x \to a^+} f(x)$ (c) $\displaystyle\lim_{x \to a} f(x)$

27 $f(x) = \begin{cases} 3x & \text{if } x \le 2 \\ x^2 & \text{if } x > 2 \end{cases}$ $a = 2$

28 $f(x) = \begin{cases} x^3 & \text{if } x \le 2 \\ 4 - 2x & \text{if } x > 2 \end{cases}$ $a = 2$

29 $f(x) = \begin{cases} \dfrac{1}{2 - 3x} & \text{if } x < -3 \\ \sqrt[3]{x + 2} & \text{if } x \geq -3 \end{cases}$ $a = -3$

30 $f(x) = \begin{cases} \dfrac{9}{x^2} & \text{if } x \leq -3 \\ 4 + x & \text{if } x > -3 \end{cases}$ $a = -3$

31 $f(x) = \begin{cases} x^2 & \text{if } x < 1 \\ 2 & \text{if } x = 1 \\ 4 - x^2 & \text{if } x > 1 \end{cases}$ $a = 1$

32 $f(x) = \begin{cases} \dfrac{x^4 + x}{x} & \text{if } x \neq 0 \\ 2 & \text{if } x = 0 \end{cases}$ $a = 0$

33 Use Definition (1.4) to prove that $\lim_{x \to 6}(5x - 21) = 9$.

34 Let f be defined as follows:

$$f(x) = \begin{cases} 1 & \text{if } x \text{ is rational} \\ -1 & \text{if } x \text{ is irrational} \end{cases}$$

Prove that for every real number a, $\lim_{x \to a} f(x)$ does not exist.

Exer. 35–38: Find all numbers at which f is discontinuous.

35 $f(x) = \dfrac{|x^2 - 16|}{x^2 - 16}$ **36** $f(x) = \dfrac{1}{x^2 - 16}$

37 $f(x) = \dfrac{x^2 - x - 2}{x^2 - 2x}$ **38** $f(x) = \dfrac{x + 2}{x^3 - 8}$

Exer. 39–42: Find all numbers at which f is continuous.

39 $f(x) = 2x^4 - \sqrt[3]{x} + 1$ **40** $f(x) = \sqrt{(2 + x)(3 - x)}$

41 $f(x) = \dfrac{\sqrt{9 - x^2}}{x^4 - 16}$ **42** $f(x) = \dfrac{\sqrt{x}}{x^2 - 1}$

Exer. 43–44: Show that f is continuous at the number a.

43 $f(x) = \sqrt{5x + 9}; \ a = 8$

44 $f(x) = \sqrt[3]{x^2} - 4; \ a = 27$

<u>c</u> **Exer. 45–48:** Lend numerical support for the stated result (of the form $\lim_{x \to a} f(x) = L$) by **(a)** creating a table of function values for x close to a and **(b)** using a graphing utility to repeatedly zoom in on the graph of f near $x = a$.

45 $\lim_{x \to 3} \dfrac{x^3 + 2x^2 - 9x - 18}{x - 3} = 30$

46 $\lim_{x \to 0} \dfrac{x}{\tan x} = 1$

47 $\lim_{x \to 3/2} \dfrac{\cos(\pi x)}{x - (3/2)} = \pi$

48 $\lim_{x \to 0^+} (\sin x)^x = 1$

<u>c</u> **Exer. 49–50:** Approximate the zeros of f to three decimal places.

49 $f(x) = x^4 - x^3 - 2x - 3$

50 $f(x) = 2 \sin x - x^2 + 1$

<u>c</u> **Exer. 51–52:** Approximate the discontinuities of f to three decimal places.

51 $f(x) = \dfrac{\sqrt{x + 3}}{x^2 + x - 1}$ **52** $f(x) = \dfrac{x + 5}{|2 \sin x - x|}$

EXTENDED PROBLEMS AND GROUP PROJECTS

1 If f is continuous on the closed interval $[0, 1]$ with $0 \leq f(x) \leq 1$ for all x, then

   (a) Show that the graph of f lies inside the square with vertices at $(0, 0)$, $(0, 1)$, $(1, 1)$, and $(1, 0)$,

   (b) Prove that f has a *fixed point*; that is, there is a number c in $[0, 1]$ with $f(c) = c$. (*Hint:* Apply the intermediate value theorem (1.26) to the function $g(x) = f(x) - x$, examining the signs of $g(0)$ and $g(1)$.)

   (c) Part (b) shows that the graph of f must hit the diagonal that runs from $(0, 0)$ to $(1, 1)$. Must the graph of f also hit the diagonal between $(0, 1)$ and $(1, 0)$?

   (d) Suppose that f is a continuous function on a closed interval I and that for each x in I, $f(x)$ also belongs to I. Does the function f necessarily have at least one fixed point?

   (e) Investigate the question posed in part (d) if the interval I is not closed.

2 Show that a function f has a removable discontinuity at a if $\lim_{x \to a} f(x)$ exists but the limit does not equal $f(a)$. We say that f has an **essential discontinuity** at a if $\lim_{x \to a} f(x)$ does not exist. Investigate the distinction between these types of discontinuity. Construct an example of a function that has an essential discontinuity at every real number. Construct a function that has an infinite number of removable discontinuities on an interval I. Can you construct a function that has a removable discontinuity at every point of an interval I? Discuss your attempts to build such an example or to prove that it cannot be done.

3 The intermediate value theorem (1.26) suggests a procedure for obtaining approximate solutions to equations of the form $f(x) = 0$, where f is a given continuous function.

Step 1 Find numbers a and b with $a < b$ so that $f(a)$ and $f(b)$ have opposite signs (one is positive and the other is negative). Let I be the closed interval $[a, b]$.

Repeat Step 2 until the length of I is less than 10^{-3}.

Step 2 Let m be the midpoint of the interval I and determine the sign of $f(m)$. If $f(a)$ and $f(m)$ have opposite signs, then let I^* be the closed left-hand half of I; otherwise, let I^* be the closed right-hand half of I. Let $I = I^*$.

(a) Show that at each step, there is a solution to $f(x) = 0$, which lies in I.

(b) Show that each new interval I is half the length of the preceding interval.

(c) If the length of $[a, b]$ is less than 1, then show that the procedure will stop in 10 or fewer repetitions of step (2).

(d) Show that if we repeat step (2) 20 times, then we will have located a solution of $f(x) = 0$ inside an interval of length less than $(b - a)/10^6$.

(e) Use this procedure with $f(x) = x^2 - 2$ to approximate $\sqrt{2}$, starting with $a = 1$ and $b = 2$.

(f) Use this procedure to determine the positive cube root of 100 to three decimal places.

(g) Use this procedure to determine a positive-number solution of the equation $\sin x = x$ to two decimal places.

INTRODUCTION

A **S WE WATCH A FORMATION** of swans or geese fly across the late afternoon sky, our focus may shift back and forth between the individual and the group. When we consider the individual bird, we may wonder about its motion: How fast does it fly? How quickly can it adjust its speed? When our perspective moves to the formation, we have questions about the smooth curve that we imagine connecting the birds: How does the curve change shape? Does the same curve occur in other natural phenomena? What geometric properties does this curve have? In this chapter, we begin to study the *derivative*, a principal tool of calculus designed to help answer some of these questions.

We begin the chapter by considering two applied problems in Section 2.1. The first is to find the slope of the tangent line at a point on the graph of a function, and the second is to define the velocity of an object moving along a line. Remarkably, these seemingly diverse applications lead to the same concept: the derivative.

Our discussion provides insight into the power and generality of mathematics. Specifically, we eliminate the geometric and physical aspects of the two problems and define the derivative in Section 2.2 as the limit of an expression involving a function f. This allows us to apply the derivative concept to any quantity that can be represented by a function. Since quantities of this type occur in nearly every field of knowledge, applications of the derivative are numerous and varied, but each concerns a *rate of change*. Thus, returning to the two problems that started it all, we see that the slope of the tangent line may be used to describe the rate at which a graph rises (or falls) and velocity is the rate at which distance changes with respect to time.

Our main objective in Section 2.2 is to define derivatives and develop rules to find them without using limits. Section 2.3 presents the basic techniques for differentiation. We examine ways to determine derivatives for polynomials and trigonometric functions (Section 2.4) and for more complicated functions that can be built up from them by addition, subtraction, multiplication, division, and composition. We consider the chain rule, which is fundamental for the differentiation of composite functions, in Section 2.5. We then turn to derivatives where functions are described either explicitly or implicitly in Section 2.6, which presents implicit differentiation techniques.

In the final sections of the chapter, we consider two important applications of the derivative that involve estimation and approximation: linear approximations and differentials in Section 2.8 and Newton's method in Section 2.9. We shall discuss many more applications in subsequent chapters.

*The flight of a flock of birds suggests
questions about motion and curves
that the derivative helps answer.*

The Derivative

2.1　TANGENT LINES AND RATES OF CHANGE

In this section, we examine two general problems whose solutions use limits of the same form. First, we consider how to define the tangent line to the graph of a function. Then, we turn to the problem of measuring rates of change, with particular emphasis on velocity as the rate of change of position of a moving object.

TANGENT LINES

Tangent lines to graphs are useful in many applications of calculus. In geometry, the tangent line l at a point P on a circle may be interpreted as the line that intersects the circle only at P, as illustrated in Figure 2.1. We cannot extend this interpretation to the graph of a function f, since a line may "touch" the graph of f at some isolated point P and then intersect it again at another point, as illustrated in Figure 2.2. Our plan is to define the *slope* of the tangent line at P, for if the slope is known, we can find an equation for l by using the point–slope form of the equation for lines (p. 14).

To define the slope of the tangent line l at $P(a, f(a))$ on the graph of f, we first choose another point $Q(x, f(x))$ (see Figure 2.3a) and consider the line through P and Q. This line is called a **secant line** for the graph.

We shall use the following notation:

$$l_{PQ}: \quad \text{the secant line through } P \text{ and } Q$$
$$m_{PQ}: \quad \text{the slope of } l_{PQ}$$
$$m_a: \quad \text{the slope of the tangent line } l \text{ at } P(a, f(a))$$

If Q is close to P, it appears that m_{PQ} is an approximation to m_a. Moreover, we would expect this approximation to improve if we take Q closer to P. With this in mind, we let Q *approach* P— that is, we (intuitively) let Q get closer to P— but $Q \neq P$. If Q approaches P from the right, we have the situation illustrated in Figure 2.3(b), where dashed lines indicate possible positions for l_{PQ}. In Figure 2.3(c), Q approaches P from

Figure 2.1

Figure 2.2

Figure 2.3

(a)

(b)

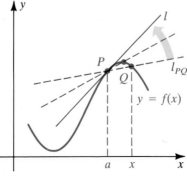

(c)

the left. We could also let Q approach P in other ways, such as by taking points on the graph that are alternately to the left and to the right of P. If m_{PQ} has a limiting value — that is, if m_{PQ} gets closer to some number as Q approaches P — then that number is the slope m_a of the tangent line l.

Let us rephrase this discussion in terms of the function f. Referring to Figure 2.3 and using the coordinates of $P(a, f(a))$ and $Q(x, f(x))$, we see that the slope of the secant line l_{PQ} is

$$m_{PQ} = \frac{f(x) - f(a)}{x - a}.$$

Note that in order to have a secant line, P and Q must be distinct points, so we must have $x \neq a$. If f is continuous at a, we can make $Q(x, f(x))$ approach $P(a, f(a))$ by letting x approach a. This leads to the following definition for the slope m_a of l at $P(a, f(a))$:

$$m_a = \lim_{x \to a} \frac{f(x) - f(a)}{x - a},$$

provided the limit exists.

It is often desirable to use an alternative form for m_a, which can be obtained by changing from the variable x to a variable h as follows.

$$\text{Let} \quad h = x - a, \quad \text{or, equivalently,} \quad x = a + h.$$

If h is small, then x is close to a and the secant line through P and Q will be close to the tangent line at P. Referring to Figure 2.4 and using the coordinates $P(a, f(a))$ and $Q(a + h, f(a + h))$, we see that the slope m_{PQ} of the secant line is

$$m_{PQ} = \frac{f(a + h) - f(a)}{h}.$$

Since $x \to a$ is equivalent to $h \to 0$, our definition of the slope m_a of the tangent line l may be stated as follows.

Figure 2.4

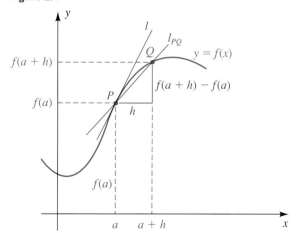

Definition 2.1

The **slope m_a of the tangent line** to the graph of a function f at $P(a, f(a))$ is

$$m_a = \lim_{h \to 0} \frac{f(a+h) - f(a)}{h},$$

provided the limit exists.

If the limit in Definition (2.1) does not exist, then the slope of the tangent line at $P(a, f(a))$ is undefined.

EXAMPLE ▪ I   Let $f(x) = x^2$, and let a be any real number.

(a) Find the slope of the tangent line to the graph of f at $P(a, a^2)$.

(b) Find an equation for the tangent line at $R(-2, 4)$.

SOLUTION

(a) The graph of $y = x^2$ and a typical point $P(a, a^2)$ are shown in Figure 2.5. Applying Definition (2.1), we see that the slope of the tangent line at P is

$$m_a = \lim_{h \to 0} \frac{f(a+h) - f(a)}{h}$$

$$= \lim_{h \to 0} \frac{(a+h)^2 - a^2}{h}$$

$$= \lim_{h \to 0} \frac{a^2 + 2ah + h^2 - a^2}{h}$$

$$= \lim_{h \to 0} \frac{2ah + h^2}{h}$$

$$= \lim_{h \to 0} (2a + h) = 2a.$$

Figure 2.5

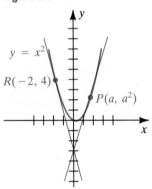

(b) The slope of the tangent line at the point $R(-2, 4)$ is the special case of the formula $m_a = 2a$ with $a = -2$; that is,

$$m_{-2} = 2(-2) = -4.$$

Using the point–slope form, we can express an equation for the tangent line as

$$y - 4 = -4(x + 2), \quad \text{or} \quad y = -4x - 4.$$

RATES OF CHANGE

Limits of the form given in Definition (2.1) occur in many applied problems where we wish to measure the rate of change of one variable with respect to another. Let us begin with the familiar problem of determining the velocity of a moving object. We consider **rectilinear motion,** in which an object travels along a line. Here the *average velocity* during a time interval is the ratio of the net distance traveled to the time elapsed.

Definition 2.2

The **average velocity** v_{av} of an object that travels a net distance d in time t is

$$v_{av} = \frac{d}{t}.$$

Figure 2.6

To illustrate, if an automobile leaves city A at 1:00 P.M. and travels along a straight highway, arriving at city B, 150 mi from A, at 4:00 P.M. (see Figure 2.6), then using Definition (2.2) with $d = 150$ and $t = 3$ (hours) yields the average velocity during the time interval [1, 4]:

$$v_{av} = \frac{150}{3} = 50 \text{ mi/hr}$$

This result is the velocity that, if maintained for 3 hr, would enable the automobile to travel the 150 mi from A to B.

The average velocity gives no information whatsoever about the velocity at any instant. At 2:30 P.M., for example, the automobile may be standing still or its speedometer may register 40 or 60 mi/hr. We can estimate the velocity at 2:30 P.M. if we know the position *near* this time. For example, suppose that at 2:30 P.M. the automobile is 80 mi from A and 5 min later, at 2:35 P.M., it is 84 mi from A, as Figure 2.7 illustrates. The net distance traveled in this 5 min, or $\frac{1}{12}$ hr, is 4 mi, and the average velocity during this time interval is

Figure 2.7

$$v_{av} = \frac{4}{\frac{1}{12}} = 48 \text{ mi/hr}.$$

Note that this result is not necessarily an accurate indication of the velocity at 2:30 P.M., since, for example, the automobile may have been traveling very slowly at 2:30 P.M. and then increased speed considerably to arrive at the point 84 mi from A at 2:35 P.M. Evidently, we obtain a better approximation by using the average velocity during a smaller time interval, say from 2:30 P.M. to 2:31 P.M. The best procedure seems to require taking smaller and smaller time intervals near 2:30 P.M. and studying the average velocity in each time interval. The approach leads us into a limiting process similar to that discussed for tangent lines.

To make our discussion more precise, let us represent the position of an object moving rectilinearly by a point P on a coordinate line l. We sometimes refer to the motion of the *point P on l*, or the motion of an object whose position is specified by P. We shall assume that we know the position of P at every instant in a given interval of time. If $s(t)$ denotes the coordinate of P at time t, then the function s is called the **position function** for P. If we keep track of time by means of a clock, then, as illustrated in Figure 2.8, for each t the point P is $s(t)$ units from the origin.

To define the velocity of P at time a, we first determine the average velocity in a (small) time interval near a. Thus, we consider times a and

Figure 2.8

Time   Position of P

Figure 2.9

Change in   Change in
time   position of P

$a + h$, where h is a (small) nonzero real number. The corresponding positions of P are $s(a)$ and $s(a + h)$, as illustrated in Figure 2.9. The amount of change in the position of P is $s(a + h) - s(a)$. This number may be positive, negative, or zero. Note that $s(a + h) - s(a)$ is not necessarily the distance traveled by P between times a and $a + h$ since, for example, P may have moved beyond the point corresponding to $s(a + h)$ and then returned to that point at time a.

By Definition (2.2), the average velocity of P between times a and $a + h$ is

$$v_{av} = \frac{\text{change in distance}}{\text{change in time}} = \frac{s(a + h) - s(a)}{h}.$$

As in our previous discussion, we assume that the closer h is to 0, the closer v_{av} is to the velocity of P at time a. Thus, we *define* the velocity as the limit, as h approaches 0, of v_{av}, as in the following definition.

Definition 2.3

Suppose a point P moves on a coordinate line l such that its coordinate at time t is $s(t)$. The **velocity** v_a of P at time a is

$$v_a = \lim_{h \to 0} \frac{s(a + h) - s(a)}{h},$$

provided the limit exists.

The limit in Definition (2.3) is also called the **instantaneous velocity** of P at time a.

If $s(t)$ is measured in centimeters and t in seconds, then the unit of velocity is centimeters per second (cm/sec). If $s(t)$ is in miles and t in hours, then the unit of velocity is miles per hour. Other units of measurement may, of course, be used.

We shall return to the velocity concept in Chapter 3, where we will show that if the velocity is positive in a given time interval, then the point is moving in the positive direction on l. If the velocity is negative, the point is moving in the negative direction. Although these facts have not been proved, we shall use them in the following example.

Figure 2.10

512 ft $s(t)$

EXAMPLE ▪2   A sandbag is dropped from a hot-air balloon that is hovering at a height of 512 ft above the ground. If air resistance is disregarded, then the distance $s(t)$ from the ground to the sandbag after t seconds is given by

$$s(t) = -16t^2 + 512.$$

Find the velocity of the sandbag at

(a) $t = a$ sec   **(b)** $t = 2$ sec   **(c)** the instant it strikes the ground

SOLUTION

(a) As shown in Figure 2.10, we consider the sandbag to be moving along a vertical coordinate line l with origin at ground level. Note that at the instant it is dropped, $t = 0$ and

$$s(0) = -16(0) + 512 = 512 \text{ ft.}$$

To find the velocity of the sandbag at $t = a$, we use Definition (2.3), obtaining

$$
\begin{aligned}
v_a &= \lim_{h \to 0} \frac{s(a+h) - s(a)}{h} \\
&= \lim_{h \to 0} \frac{[-16(a+h)^2 + 512] - (-16a^2 + 512)}{h} \\
&= \lim_{h \to 0} \frac{-16(a^2 + 2ah + h^2) + 512 + 16a^2 - 512}{h} \\
&= \lim_{h \to 0} \frac{-32ah - 16h^2}{h} \\
&= \lim_{h \to 0} (-32a - 16h) = -32a \text{ ft/sec.}
\end{aligned}
$$

The negative sign indicates that the motion of the sandbag is in the negative direction (downward) on l.

(b) To find the velocity at $t = 2$, we substitute 2 for a in the formula $v_a = -32a$, obtaining

$$v_2 = -32(2) = -64 \text{ ft/sec.}$$

(c) The sandbag strikes the ground when the distance above the ground is zero — that is, when

$$s(t) = -16t^2 + 512 = 0, \quad \text{or} \quad t^2 = \tfrac{512}{16} = 32.$$

This result gives us $t = \sqrt{32} = 4\sqrt{2} \approx 5.7$ sec. If we use the formula $v_a = -32a$ from part (a) with $a = 4\sqrt{2}$, we obtain the following impact velocity:

$$-32(4\sqrt{2}) = -128\sqrt{2} \approx -181 \text{ ft/sec}$$

There are many other applications that require limits similar to those in (2.1) and (2.3). In some, the independent variable is time t, as in the definition of velocity. For example, over a period of time, a chemist may be interested in the rate at which a certain substance dissolves in water; an electrical engineer may wish to know the rate of change of current in part of an electrical circuit; a biologist may be concerned with the rate at which the bacteria in a culture increase or decrease. In the social sciences, an economist may wish to determine the rate at which the gross national product is growing; a demographer or geographer may wish to analyze the rate of urbanization of a population; a sociologist may study the rate at which measures of alienation fluctuate; a political scientist may be concerned with the rate at which the public's approval of a national leader changes.

We can also consider rates of change with respect to quantities other than time. To illustrate, Boyle's law for a confined gas states that if the temperature remains constant, then the volume v and pressure p are related by the formula $v = c/p$ for some constant c. If the pressure is changing, a typical problem is to find the rate at which the volume is changing per unit change in pressure. This rate is known as *the instantaneous rate of change of v with respect to p*. To develop general methods that can be applied to different problems of this type, let us use x and y for variables and suppose that $y = f(x)$ for some function f. (In the preceding illustration, $y = v$, $x = p$, and $f(x) = c/x$.) We define rates of change of a variable y with respect to a variable x as follows.

Definition 2.4

Let $y = f(x)$, where f is defined on an open interval containing a.

(i) The **average rate of change** of $y = f(x)$ with respect to x on the interval $[a, a + h]$ is

$$y_{av} = \frac{f(a + h) - f(a)}{h}.$$

(ii) The **instantaneous rate of change** of y with respect to x at a is

$$y_a = \lim_{h \to 0} \frac{f(a + h) - f(a)}{h},$$

provided the limit exists.

We shall use the phrase *rate of change* interchangeably with *instantaneous rate of change*.

If, in Definition (2.4), we consider the special case $x = t$ (time) and $y = s(t)$ (position on a coordinate line), we obtain the following interpretations for rectilinear motion:

average velocity (v_{av}): the average rate of change of s with respect to t in some interval of time

velocity v_a: the instantaneous rate of change of s with respect to t at time a.

To interpret Definition (2.4)(ii) geometrically, imagine a point P traveling from left to right along the graph of $y = f(x)$ in Figure 2.11. The

Figure 2.11

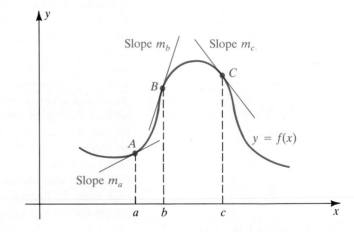

instantaneous rate of change of y with respect to x gives us information about the rate at which the graph rises or falls per unit change in x. In Figure 2.11, m_a (the slope of the tangent line at A) is less than m_b (the slope of the tangent line at B), and the rate y_a at which y changes with respect to x at a is less than the rate y_b at which y changes with respect to x at b. Also note that since $m_c < 0$, the slope of the tangent line at C is negative, and y *decreases* as x increases.

The next two examples are physical and social science applications of Definition (2.4).

EXAMPLE ■ 3   The voltage in a certain electrical circuit is 100 volts. If the current (in amperes) is I and the resistance (in ohms) is R, then by Ohm's law, $I = 100/R$. If R is increasing, find the instantaneous rate of change of I with respect to R at

(a) any resistance R **(b)** a resistance of 20 ohms

SOLUTION

(a) Using Definition (2.4)(ii) with $y = I$, $x = R$, and $f(R) = 100/R$ yields the instantaneous rate of change of I with respect to R at a resistance of R ohms:

$$I_R = \lim_{h \to 0} \frac{f(R+h) - f(R)}{h}$$

$$= \lim_{h \to 0} \frac{\dfrac{100}{R+h} - \dfrac{100}{R}}{h}$$

$$= \lim_{h \to 0} \frac{100R - 100(R+h)}{h(R+h)R}$$

$$= \lim_{h \to 0} \frac{-100h}{h(R+h)R}$$

$$= \lim_{h \to 0} \frac{-100}{(R+h)R}$$

$$= -\frac{100}{R^2}$$

The negative sign indicates that the current is decreasing as the resistance is increasing.

(b) Using the formula $I_R = -100/R^2$ from part (a), we find the instantaneous rate of change of I with respect to R at $R = 20$ to be

$$I_{20} = -\frac{100}{20^2} = -\frac{1}{4}.$$

Thus, when $R = 20$, the current is *decreasing* at a rate of $\frac{1}{4}$ ampere per ohm.

EXAMPLE ▪ 4   The expression $P = \sqrt{at + b}$, with $a = 920$ and $b = 151.3^2$, gives a good approximation for the population P (in millions) of the United States during the period 1950–1990, where $t = 0$ corresponds to the year 1950. Find the instantaneous rate of change of P with respect to t at

(a) any time t **(b)** $t = 39$ (the year 1989)

SOLUTION

(a) Using Definition (2.4)(ii) with $y = P$, $x = t$, and $f(t) = \sqrt{at + b}$ gives the instantaneous rate of change of P with respect to t at time t years:

$$P_t = \lim_{h \to 0} \frac{f(t + h) - f(t)}{h}$$

$$= \lim_{h \to 0} \frac{\sqrt{a(t + h) + b} - \sqrt{at + b}}{h}$$

Evaluating this limit by rationalizing the numerator and then simplifying yields

$$P_t = \lim_{h \to 0} \frac{\sqrt{a(t + h) + b} - \sqrt{at + b}}{h} \cdot \frac{\sqrt{a(t + h) + b} + \sqrt{at + b}}{\sqrt{a(t + h) + b} + \sqrt{at + b}}$$

$$= \lim_{h \to 0} \frac{[a(t + h) + b] - (at + b)}{h(\sqrt{a(t + h) + b} + \sqrt{at + b})}$$

$$= \lim_{h \to 0} \frac{at + ah + b - at - b}{h(\sqrt{a(t + h) + b} + \sqrt{at + b})}$$

$$= \lim_{h \to 0} \frac{ah}{h(\sqrt{a(t + h) + b} + \sqrt{at + b})}$$

$$= \lim_{h \to 0} \frac{a}{\sqrt{a(t + h) + b} + \sqrt{at + b}} \qquad (h \neq 0)$$

$$= \frac{a}{\sqrt{at + b} + \sqrt{at + b}}$$

$$= \frac{a}{2\sqrt{at + b}}.$$

(b) Using the formula for P_t from part (a), with $a = 920$ and $b = 151.3^2$, we find the instantaneous rate of change of P with respect to t at $t = 39$ to be

$$P_{39} = \frac{920}{2\sqrt{920(39) + 151.3^2}} = \frac{460}{\sqrt{58,771.69}} \approx 1.897.$$

Thus, in 1989, the United States population was growing at an approximate rate of 1.9 million people per year.

To find slopes of tangent lines and instantaneous rates of change, we need to determine limits of the form given in Definition (2.1). In Examples 1–4, we have determined the limits algebraically. In Section 2.3, we will discuss other algebraic techniques for evaluating similar limits. Calculators and computers can help us *approximate* values for such limits by evaluating the difference quotient $[f(a + h) - f(a)]/h$ for values of h close to 0, but there are some difficulties to be considered. For example, difficulty in evaluating calculations can occur for values of h extremely close to 0. Most computing devices perform arithmetic with a finite number of significant digits, usually between 7 and 14 digits. When h is close to 0, the numerator $f(a + h) - f(a)$ may be difficult to evaluate because the terms $f(a + h)$ and $f(a)$ differ by an amount too small for the calculator to distinguish from zero. Hence, the approximations of the difference quotient may get worse rather than better, as illustrated in the next example.

EXAMPLE ▪ 5   Use Definition (2.1) to approximate the slope m_a of the tangent line to the graph of $f(x) = \sin(\sin x)$ at $P(a, f(a))$ when $a = 1$ by letting $h = 100^{-n}$ for $n = 1, 2, \ldots, 7$.

SOLUTION   We use a calculator to generate the results in the following table:

n	h	$f(a + h)$	$f(a)$	$f(a+h) - f(a)$	$\dfrac{f(a+h) - f(a)}{h}$
1	0.01	0.749185709107	0.745624141665	0.003561567441	0.356156744109
2	0.0001	0.745660141723	0.745624141665	3.600005706E-5	0.3600005706
3	0.000001	0.745624501705	0.745624141665	3.6003911E-7	0.36003911
4	0.00000001	0.745624145266	0.745624141665	3.60045E-9	0.360045
5	0.0000000001	0.745624141702	0.745624141665	3.605E-11	0.3605
6	0.000000000001	0.745624141666	0.745624141665	4E-13	0.4
7	0.00000000000001	0.745624141665	0.745624141665	0	0

It appears from the table's first four lines that the limit is approximately 0.36004. Examining a graph of the function indicates that this approximate value appears to be correct for the slope of the tangent line. Figure 2.12 shows a graph of f and a graph of the tangent line through $(1, \sin(\sin 1))$ with slope 0.36004. In Section 2.5, we will see that this approximate value is correct to five significant digits.

Note that below the fourth line of the table, where the values of h are extremely small, the estimates of the difference quotient are not getting closer to 0.36004, but rather, farther away. Because of finite precision arithmetic, the numbers in the third and fourth columns become so close that there are few significant digits in the fifth column. When h gets so small that both $f(a + h)$ and $f(a)$ round to exactly the same value in the calculator, the value 0 is obtained in the last two columns.

Figure 2.12

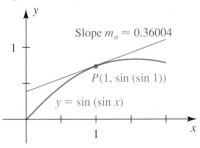

EXERCISES 2.1

Exer. 1–6: (a) Use Definition (2.1) to find the slope of the tangent line to the graph of f at $P(a, f(a))$. (b) Find an equation of the tangent line at $P(2, f(2))$.

1 $f(x) = 5x^2 - 4x$

2 $f(x) = 3 - 2x^2$

3 $f(x) = x^3$

4 $f(x) = x^4$

5 $f(x) = 3x + 2$

6 $f(x) = 4 - 2x$

Exer. 7–10: (a) Use Definition (2.1) to find the slope of the tangent line to the graph of the equation at the point with x-coordinate a. (b) Find an equation of the tangent line at P. (c) Sketch the graph and the tangent line at P.

7 $y = \sqrt{x}$; $P(4, 2)$

8 $y = \sqrt[3]{x}$; $P(-8, -2)$

9 $y = 1/x$; $P(2, \frac{1}{2})$

10 $y = 1/x^2$; $P(2, \frac{1}{4})$

Exer. 11–12: (a) Sketch the graph of the equation and the tangent lines at the points with x-coordinates $-2, -1, 1$, and 2. (b) Find the point on the graph at which the slope of the tangent line is the given number m.

11 $y = x^2$; $m = 6$

12 $y = x^3$; $m = 9$

Exer. 13–14: The position function s of a point P moving on a coordinate line l is given, with t in seconds and $s(t)$ in centimeters. (a) Find the average velocity of P in the following time intervals: [1, 1.2], [1, 1.1], and [1, 1.01]. (b) Find the velocity of P at $t = 1$.

13 $s(t) = 4t^2 + 3t$

14 $s(t) = 2t - 3t^2$

15 A rescue helicopter drops a crate of supplies from a height of 160 ft. After t seconds, the crate is $160 - 16t^2$ feet above the ground.

   (a) Find the velocity of the crate at $t = 1$.

   (b) With what velocity does the crate strike the ground?

16 A projectile is fired directly upward from the ground with an initial velocity of 112 ft/sec. Its distance above the ground after t seconds is $112t - 16t^2$ feet.

   (a) Find the velocity of the projectile at $t = 2$, $t = 3$, and $t = 4$.

   (b) When does the projectile strike the ground?

   (c) Find the velocity at the instant it strikes the ground.

17 In the video game shown in the figure, airplanes fly from left to right along the path $y = 1 + (1/x)$ and can shoot their bullets in the tangent direction at creatures placed along the x-axis at $x = 1, 2, 3, 4$, and 5. Determine whether a creature will be hit if the player shoots when the plane is at

(a) $P(1, 2)$ (b) $Q(\frac{3}{2}, \frac{5}{3})$

Exercise 17

18 An athlete runs the hundred-meter dash in such a way that the distance $s(t)$ run after t seconds is given by $s(t) = \frac{1}{5}t^2 + 8t$ meters (see figure). Find the athlete's velocity at

(a) the start of the dash   (b) $t = 5$ sec

(c) the finish line

Exercise 18

Exer. 19–20: (a) Find the average rate of change of y with respect to x on the given interval. (b) Find the instantaneous rate of change of y with respect to x at the left endpoint of the interval.

19 $y = x^2 + 2$;   [3, 3.5]   20 $y = 3 - 2x^2$;   [2, 2.4]

21 Boyle's law states that if the temperature remains constant, the pressure p and volume v of a confined gas are related by $p = c/v$ for some constant c. If, for a certain gas, $c = 200$ and v is increasing, find the instantaneous rate of change of p with respect to v at

(a) any volume v (b) a volume of 10

Exercises 2.1

22 Using the Lorentz contraction formula,

$$L = L_0\sqrt{1 - v^2/c^2},$$

find the instantaneous rate of change of the length L of an object with respect to the velocity v at

(a) any velocity v

(b) $v = 0.9c$

c 23 Graph $f(x) = \sin(\pi x)$ on the interval $[0, 2]$.

(a) Use the graph to estimate the slope of the tangent line at $P(1.4, f(1.4))$.

(b) Use Definition (2.1) with $h = \pm 0.0001$ to approximate the slope in part (a).

c 24 Graph $f(x) = \dfrac{10\cos x}{x^2 + 4}$ on the interval $[-2, 2]$.

(a) Use the graph to estimate the slope of the tangent line at $P(-0.5, f(-0.5))$.

(b) Use Definition (2.1) with $h = \pm 0.0001$ to approximate the slope in part (a).

(c) Find an (approximate) equation of the tangent line to the graph at P.

c 25 An object's position on a coordinate line is given by

$$s(t) = \frac{\cos^2 t + t^2 \sin t}{t^2 + 1},$$

where $s(t)$ is in feet and t is in seconds. Approximate its velocity at $t = 2$ by using Definition (2.3) with $h = 0.01, 0.001$, and 0.0001.

c 26 The position function s of an object moving on a coordinate line is given by

$$s(t) = \frac{t - t^2 \sin t}{t^2 + 1},$$

where $s(t)$ is in meters and t is in minutes.

(a) Graph s for $0 \le t \le 10$.

(b) Approximate the time intervals in which its velocity is positive.

c Exer. 27–30: (a) Use Definition (2.1) with the given value of h to approximate the slope of the tangent line at the indicated points. (b) Graph the function and the three approximated tangent lines over the given interval.

27 $f(x) = \sin(\pi x)$ on $[0, 2]$; $\quad P(0.7, f(0.7))$, $P(1.1, f(1.1))$, $P(1.4, f(1.4))$; $h = 0.001$

28 $f(x) = 3^{-x^2}$ on $[-2, 2]$; $\quad P(-0.67, f(-0.67))$, $P(0.3, f(0.3))$, $P(1.14, f(1.14))$; $h = 0.0002$

29 $f(x) = 0.625\sqrt{64 - x^2}$ on $[-8, 8]$; $\quad P(-7, f(-7))$, $P(3, f(3))$, $P(5, f(5))$; $h = 0.0005$

30 $f(x) = 0.4\sqrt{25 + x^2}$ on $[-10, 10]$; $\quad P(-5, f(-5))$, $P(-1, f(-1))$, $P(9, f(9))$; $h = 0.0001$

c 31 An object's position on a coordinate line is given by

$$s(t) = \sin(\sin t),$$

where $s(t)$ is in miles and t is in hours. Use Definition (2.3), with $h = -0.1, -0.05$, and -0.001 to approximate its velocity at

(a) $t = -1$ **(b)** $t = 2$

Mathematicians and Their Times

PIERRE DE FERMAT

PIERRE DE FERMAT WAS PERHAPS the greatest mathematician of the seventeenth century. He made fundamental contributions to analytic geometry, calculus, probability, and number theory. Most astounding to us in this age of specialized knowledge and major research centers is that Fermat was not a professional mathematician. He did not even have a degree in mathematics.

To others, it appeared that Fermat's life was quiet and uneventful. Born in Beaumont-de-Lomagne, France, in August 1601, Fermat was a shy and retiring person. His father was a leather merchant. His mother's family boasted a number of public service lawyers; Fermat followed this occupation. He rose to the rank of King's Councilor in the Parlement of Toulouse and discharged this position with great skill and integrity for 17 years until his death on January 12, 1665.

Fermat's vocation may have been the law and public service, but his passionate avocation was mathematics. Although Fermat and Descartes were independent inventors of analytic geometry, Fermat went considerably further than Descartes, introducing perpendicular axes and finding equations for straight lines, circles, ellipses, parabolas, and hyperbolas.

While Newton and Leibniz share credit for the invention of calculus, Fermat had made critically important discoveries in this field more than a decade before either of them was born. He found the equations of tangent lines, located the maximum and minimum points, and computed the area beneath many different curves. He was also able to solve these three principal problems of calculus for a wide variety of functions.

Fermat's favorite branch of mathematics was number theory: the study of integers and relations between them. For 350 years, the most famous unsolved problem in mathematics was *Fermat's Last Theorem*. Next to a passage on integer solutions to the equation $x^2 + y^2 = z^2$ (examples: $3^2 + 4^2 = 5^2$, and $5^2 + 12^2 = 13^2$), Fermat penned the most memorable margin note in the history of science: "I have discovered a truly wonderful proof which this margin is too narrow to contain." He meant that if $n > 2$, then there are no integer solutions to $x^n + y^n = z^n$. In June 1993, Andrew Wiles, a Princeton University mathematician, announced that he had proved the truth of Fermat's claim.

Although primarily interested in "pure" mathematics, Fermat also made profound discoveries in applications as well. He formulated the idea that the path along which a light ray travels as it moves from one medium to another — say, from air to water — is the path that minimizes the total travel time. *Fermat's principle of least time,* as it is called today, led to the calculus of variations and provided the basis for Hamilton's principle of least action, a powerful unifying idea in physics.

2.2 DEFINITION OF DERIVATIVE

In the preceding section, we examined several different problems whose solutions all involved similar limits. Whether in determining the slope of the tangent to a curve, or finding the velocity of an object moving along a line, or discovering the instantaneous rate of change of an electrical current with respect to voltage, we ultimately use limits of the form

$$\lim_{h \to 0} \frac{f(a+h) - f(a)}{h},$$

or, equivalently,

$$\lim_{x \to a} \frac{f(x) - f(a)}{x - a}.$$

This limit is the basis for one of the fundamental concepts of calculus, the *derivative*. The derivative occurs throughout calculus in problems concerned with rates of change and thus has applications in many fields of study.

In this section, we begin with equivalent definitions of the derivative, given in terms of limits, that can be applied to any function. We then look at some simple rules that allow us to find derivatives without directly evaluating limits. We also consider some basic properties of the derivative and its notation.

DEFINING THE DERIVATIVE

Definition **2.5**

The **derivative** of a function f is the function f' whose value at x is given by

$$f'(x) = \lim_{h \to 0} \frac{f(x+h) - f(x)}{h},$$

provided the limit exists.

The symbol f' in Definition (2.5) is read f *prime*. It is important to note that in determining $f'(x)$, we regard x as an arbitrary real number and consider the limit as h approaches 0. Once we have obtained $f'(x)$, we can find $f'(a)$ for a specific real number a by substituting a for x.

The statement $f'(x)$ *exists* means that the limit in Definition (2.5) exists. If $f'(x)$ exists, we say that **f is differentiable at x**, or that **f has a derivative at x**. If the limit does not exist, then f is not differentiable at x. The terminology *differentiate* $f(x)$ or *find the derivative of* $f(x)$ means to find $f'(x)$.

Occasionally we will find it convenient to use the following alternative form of Definition (2.5) to find $f'(a)$.

Alternative Definition of Derivative **2.6**

$$f'(a) = \lim_{x \to a} \frac{f(x) - f(a)}{x - a}$$

This formula was first used to define m_a (see p. 145).

The following applications are restatements of Definitions (2.1) and (2.4)(ii) using $f'(x)$. These interpretations of the derivative are very important and will be used in many examples and exercises throughout the text.

Applications of the Derivative **2.7**

(i) Tangent line: The slope of the tangent line to the graph of the function $y = f(x)$ at the point $(a, f(a))$ is $f'(a)$.

(ii) Rate of change: If $y = f(x)$, the instantaneous rate of change of y with respect to x at a is $f'(a)$.

As a special case of (2.7)(ii), recall from Definition (2.3) that if $x = t$ denotes time and $y = s(t)$ is the position of a point P on a coordinate line, then $s'(a)$ is the velocity of P at time a.

EXAMPLE ▪ I   If $f(x) = 3x^2 - 12x + 8$, find

(a) $f'(x)$ **(b)** $f'(4)$, $f'(-2)$, and $f'(a)$

SOLUTION

(a) By Definition (2.5),

$$f'(x) = \lim_{h \to 0} \frac{f(x+h) - f(x)}{h}$$

$$= \lim_{h \to 0} \frac{[3(x+h)^2 - 12(x+h) + 8] - (3x^2 - 12x + 8)}{h}$$

$$= \lim_{h \to 0} \frac{(3x^2 + 6xh + 3h^2 - 12x - 12h + 8) - (3x^2 - 12x + 8)}{h}$$

$$= \lim_{h \to 0} \frac{6xh + 3h^2 - 12h}{h} = \lim_{h \to 0} (6x + 3h - 12) = 6x - 12.$$

(b) Substituting for x in $f'(x) = 6x - 12$, we obtain

$$f'(4) = 6(4) - 12 = 12,$$
$$f'(-2) = 6(-2) - 12 = -24,$$

and
$$f'(a) = 6a - 12.$$

EXAMPLE ■ 2   If $y = 3x^2 - 12x + 8$, use the results of Example 1 to find

(a) the slope of the tangent line to the graph of this equation at the point $P(3, -1)$

(b) the point on the graph at which the tangent line is horizontal

Figure 2.13

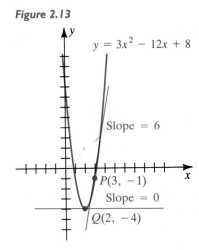

SOLUTION

(a) If we let $f(x) = 3x^2 - 12x + 8$, then, by (2.7)(i) and Example 1, the slope of the tangent line at $(x, f(x))$ is $f'(x) = 6x - 12$. In particular, the slope at $P(3, -1)$ is

$$f'(3) = 6(3) - 12 = 6.$$

(b) Since the tangent line is horizontal if the slope $f'(x)$ is 0, we solve $6x - 12 = 0$, obtaining $x = 2$. The corresponding value of y is -4. Hence the tangent line is horizontal at the point $Q(2, -4)$.

The graph of f (a parabola) and the tangent lines at P and Q are sketched in Figure 2.13. Note that the vertex of the parabola is the point $Q(2, -4)$.

EXAMPLE ■ 3   As a swan begins its flight, its height above the ground (in meters) after x seconds is observed to be given by the function

$$f(x) = x + \sin 2x.$$

Use a graphing utility to obtain a graph of f on the interval $[0, 3]$, estimate the times when the tangent line is horizontal, and interpret the results in terms of rates of change of the swan's distance from the ground.

Figure 2.14
$0 \le x \le 3, 0 \le y \le 3$

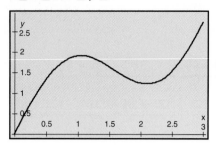

SOLUTION Using a graphing utility and the viewing window shown, we obtain the graph of $f(x) = x + \sin 2x$ in Figure 2.14. If we sketch or visualize tangent lines at various points on the graph, we can determine that they have positive slopes on the open interval extending from 0 to approximately 1.05 and again on the open interval from approximately 2.10 to 3. On these intervals, the rate of change of the swan's height above the ground is positive, and the distance between the swan and the ground is increasing. On the open interval extending from 1.05 to 2.10, the tangent lines have negative slopes. During this time interval, the swan's height above the ground is decreasing. At approximate times of 1.05 and 2.10, the tangent lines are horizontal. At these two times, the swan's distance above the ground is neither decreasing nor increasing.

BASIC RULES OF DIFFERENTIATION

The process of finding a derivative by means of Definition (2.5) can be very tedious if $f(x)$ is a complicated expression. Fortunately, we can establish general formulas and rules that enable us to find $f'(x)$ without using limits.

If f is a linear function, then $f(x) = mx + b$ for real numbers m and b. The graph of f is the line with slope m and y-intercept b (see Figure 2.15).

Figure 2.15

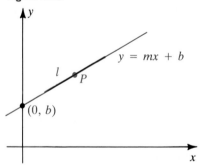

As indicated in the figure, the tangent line l at any point P coincides with the graph of f and hence has slope m. From Definition (2.7)(i), we conclude that $f'(x) = m$ for every x. Thus, we obtain the following rule, which we can also prove directly from Definition (2.5).

Derivative of a Linear Function 2.8

$$\text{If} \quad f(x) = mx + b, \quad \text{then} \quad f'(x) = m.$$

The following result is the special case of (2.8) with $m = 0$.

Derivative of a Constant Function 2.9

$$\text{If} \quad f(x) = b, \quad \text{then} \quad f'(x) = 0.$$

The preceding result is also graphically evident, because the graph of a constant function is a horizontal line and hence has slope 0.

Some special cases of (2.8) and (2.9) are given in the following illustration.

ILLUSTRATION

$f(x)$	$3x - 7$	$-4x + 2$	$7x$	x	13	π^2	$\sqrt[3]{10}$
$f'(x)$	3	-4	7	1	0	0	0

Many algebraic expressions contain a variable x raised to some power n. The next result, appropriately called the *power rule,* provides a simple formula for finding the derivative if n is an integer.

Power Rule 2.10

Let n be an integer.

$$\text{If} \quad f(x) = x^n, \quad \text{then} \quad f'(x) = nx^{n-1},$$

provided $x \neq 0$ when $n \leq 0$.

PROOF By Definition (2.5),

$$f'(x) = \lim_{h \to 0} \frac{f(x + h) - f(x)}{h}$$

$$= \lim_{h \to 0} \frac{(x + h)^n - x^n}{h}.$$

If n is a positive integer, then we can expand $(x + h)^n$ by using the binomial theorem, obtaining

$$f'(x) = \lim_{h \to 0} \frac{\left[x^n + nx^{n-1}h + \dfrac{n(n - 1)}{2!}x^{n-2}h^2 + \cdots + nxh^{n-1} + h^n \right] - x^n}{h}$$

$$= \lim_{h \to 0} \left[nx^{n-1} + \frac{n(n - 2)}{2!}x^{n-2}h + \cdots + nxh^{n-2} + h^{n-1} \right].$$

Since each term within the brackets, except the first, contains a power of h, we see that $f'(x) = nx^{n-1}$.

If n is negative and $x \neq 0$, then we can write $n = -k$ with k positive. Thus,

$$f'(x) = \lim_{h \to 0} \frac{(x + h)^{-k} - x^{-k}}{h}$$

$$= \lim_{h \to 0} \frac{x^k - (x + h)^k}{h(x + h)^k x^k}.$$

As before, if we use the binomial theorem to expand $(x + h)^k$ and then simplify and take the limit, we obtain

$$f'(x) = -kx^{-k-1} = nx^{n-1}.$$

If $n = 0$ and $x \neq 0$, then the power rule is also true, for in this case $f(x) = x^0 = 1$ and, by (2.9), $f'(x) = 0 = 0 \cdot x^{0-1}$. ■

Some special cases of the power rule are listed in the next illustration.

ILLUSTRATION

$f(x)$	x^2	x^3	x^4	x^{100}	$x^{-1} = \dfrac{1}{x}$	$x^{-2} = \dfrac{1}{x^2}$	$x^{-10} = \dfrac{1}{x^{10}}$
$f'(x)$	$2x$	$3x^2$	$4x^3$	$100x^{99}$	$(-1)x^{-2} = -\dfrac{1}{x^2}$	$-2x^{-3} = -\dfrac{2}{x^3}$	$-10x^{-11} = -\dfrac{10}{x^{11}}$

We can extend the power rule to rational exponents. In particular, in Appendix I we show that for every positive integer n,

$$\text{if} \quad f(x) = x^{1/n}, \quad \text{then} \quad f'(x) = \frac{1}{n}x^{(1/n)-1},$$

provided these expressions are defined. By using the power rule for functions (2.27), proved in Section 2.5, we can then show that for any rational number m/n,

$$\text{if} \quad f(x) = x^{m/n}, \quad \text{then} \quad f'(x) = \frac{m}{n}x^{(m/n)-1}.$$

In Chapter 6, we will prove that the power rule holds for every *real* number n. Some special cases of the power rule for rational exponents are given in the next illustration.

ILLUSTRATION

$f(x)$	$f'(x)$
$\sqrt{x} = x^{1/2}$	$\dfrac{1}{2}x^{-1/2} = \dfrac{1}{2\sqrt{x}}$
$\sqrt[3]{x^2} = x^{2/3}$	$\dfrac{2}{3}x^{-1/3} = \dfrac{2}{3\sqrt[3]{x}}$
$\sqrt[4]{x^5} = x^{5/4}$	$\dfrac{5}{4}x^{1/4} = \dfrac{5}{4}\sqrt[4]{x}$
$\dfrac{1}{x^{1/3}} = x^{-1/3}$	$-\dfrac{1}{3}x^{-4/3} = -\dfrac{1}{3x^{4/3}}$
$\dfrac{1}{x^{3/2}} = x^{-3/2}$	$-\dfrac{3}{2}x^{-5/2} = -\dfrac{3}{2x^{5/2}}$

By using the same type of proof that was used for the power rule (2.10), we can prove the following for any real number c.

Theorem 2.11

$$\text{If} \quad f(x) = cx^n, \quad \text{then} \quad f'(x) = (cn)x^{n-1}.$$

In words, *to differentiate cx^n, multiply the coefficient c by the exponent n and reduce the exponent by* 1.

EXAMPLE ■ 4 A spherical balloon is being inflated. Find the instantaneous rate of change of the surface area S of the balloon with respect to the radius x.

SOLUTION Using the formula for surface area S of a sphere, $S = f(x) = 4\pi x^2$, and Theorem (2.11), we can readily find the instantaneous rate of change of the surface area with respect to the radius:

$$f'(x) = (4\pi \cdot 2)x = 8\pi x$$

CONTINUITY AND DIFFERENTIABILITY

Not every function $f(x)$ is differentiable at every value of x in its domain. As we shall see from Theorem (2.12), if f is not continuous at a, then f is not differentiable at a. Moreover, many continuous functions can also fail to be differentiable.

To formally determine whether a function is differentiable or not, we must examine the limit in (2.5) or (2.6) and decide whether the limit exists or not. Informally, however, we can say that a function is continuous at a point if its graph has no breaks or jumps at the point, and if it is also differentiable, it passes through the point in a "smooth" fashion with no corners or vertical tangents. Our next example is perhaps the simplest familiar function that fails to be differentiable at a point where it is continuous.

EXAMPLE ■ 5 If $f(x) = |x|$, show that f is not differentiable at 0.

SOLUTION The graph of f is sketched in Figure 2.16. We can prove that $f'(0)$ does not exist by showing that the limit in Alternative Definition (2.6) does not exist. With $a = 0$ and $f(x) = |x|$, we have

$$\frac{f(x) - f(a)}{x - a} = \frac{|x| - |0|}{x - 0} = \frac{|x|}{x}.$$

But, by the definition of the absolute value function,

$$|x| = \begin{cases} x & \text{if } x \geq 0 \\ -x & \text{if } x < 0 \end{cases}$$

Figure 2.16

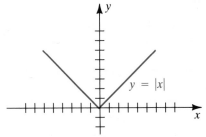

$y = |x|$

we have

$$\frac{|x|}{x} = \begin{cases} 1 & \text{if } x > 0 \\ -1 & \text{if } x < 0 \end{cases}$$

and this expression has no limit as x approaches 0 (see Example 7 of Section 1.1). Since $\lim_{x \to a}[f(x) - f(a)]/(x - a)$ does not exist at $a = 0$, the function $f(x) = |x|$ is not differentiable at 0.

We observe from Example 5 that the absolute value function is continuous at 0 (see Example 1 of Section 1.5), but it is not differentiable at 0. The graph of $f(x) = |x|$ shown in Figure 2.16 displays a geometric distinction between continuous and differentiable functions. The graph of f is unbroken and has no jumps as it passes through $x = 0$, so f is continuous. However, since it has a corner at $(0, 0)$ and therefore no tangent line at that point, it is not differentiable. (A formal definition of *corner* is given later in this section.)

CAUTION   Not every continuous function is differentiable. In contrast, as the next theorem states, *every differentiable function is continuous.*

Theorem 2.12

If a function f is differentiable at a, then f is continuous at a.

PROOF   We shall use Alternative Definition (2.6):

$$f'(a) = \lim_{x \to a} \frac{f(x) - f(a)}{x - a}.$$

We may write $f(x)$ in a form that contains $[f(x) - f(a)]/(x - a)$ as follows, provided $x \neq a$:

$$f(x) = \frac{f(x) - f(a)}{x - a}(x - a) + f(a).$$

Using limit theorems, we find that

$$\lim_{x \to a} f(x) = \lim_{x \to a} \frac{f(x) - f(a)}{x - a} \cdot \lim_{x \to a}(x - a) + \lim_{x \to a} f(a)$$

$$= f'(a) \cdot 0 + f(a) = f(a).$$

Thus, by Definition (1.20), f is continuous at a.   ■

DIFFERENTIABILITY ON AN INTERVAL

Thus far, we have considered the differentiability of a function $f(x)$ at a particular single value of x. We now extend the concept to *differentiability on an interval*.

A function f **is differentiable on an open interval** if $f'(x)$ exists for every x in that interval. For closed intervals, we use the following convention, which is analogous to the definition of continuity on a closed interval given in Definition (1.22).

Definition **2.13**

> A function f **is differentiable on a closed interval** $[a, b]$ if f is differentiable on the open interval (a, b) and if the following limits exist:
>
> $$\lim_{h \to 0^+} \frac{f(a + h) - f(a)}{h} \quad \text{and} \quad \lim_{h \to 0^-} \frac{f(b + h) - f(b)}{h}.$$

The one-sided limits in Definition (2.13) are sometimes referred to as the **right-hand derivative** and the **left-hand derivative** of f at a and b, respectively.

Note that for the right-hand derivative, $h \to 0^+$, and $a + h$ approaches a from the right. In this case, the point $Q(a + h, f(a + h))$ in Figure 2.17 lies to the right of $P(a, f(a))$. The quotient $[f(a + h) - f(a)]/h$ is the slope of the secant line through P and Q on the graph of f to the right of P. The right-hand derivative is the limiting value of the slope of the secant lines through P and Q as Q approaches P from the right.

Figure 2.17

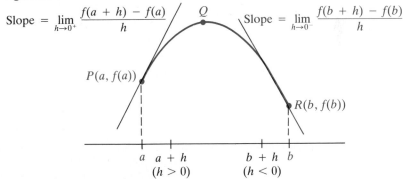

$$\text{Slope} = \lim_{h \to 0^+} \frac{f(a + h) - f(a)}{h} \qquad \text{Slope} = \lim_{h \to 0^-} \frac{f(b + h) - f(b)}{h}$$

For the left-hand derivative, $h \to 0^-$ and $b + h$ approaches b from the left. The left-hand derivative is the limiting value of the slope of the secant lines through P and Q as Q approaches P from the left.

If a function f is defined only on a closed interval $[a, b]$, then the right-hand and left-hand derivatives define the slopes of the tangent lines at the points $P(a, f(a))$ and $R(b, f(b))$. Figure 2.17 shows the graph of such a function f with the tangent lines at P and R. By using one-sided limits, we can extend Theorem (2.12) to functions that are differentiable on a closed interval.

Differentiability on an interval of the form $[a, b)$, $[a, \infty)$, $(a, b]$, or $(-\infty, b]$ is defined in similar fashion, using a one-sided limit at an endpoint.

A number c is in the **domain of the derivative** f' if f is differentiable at c or if c is an endpoint of the domain of f at which the appropriate one-sided limit exists.

If f is defined on an open interval containing a, then $f'(a)$ exists if *and only if both the right-hand and left-hand derivatives at a exist and are equal.* Thus, a function f may fail to be differentiable at a if either one or both one-sided derivatives fail to exist, as we will see in Example 7. The function may also fail to be differentiable at a if both one-sided derivatives exist at a but are not equal to each other. In Example 5, we saw that the absolute value function $f(x) = |x|$ is not differentiable at 0. This function has a right-hand derivative of 1 and a left-hand derivative of -1 at $x = 0$.

The functions whose graphs are sketched in Figure 2.18 have right-hand and left-hand derivatives at a that give the slopes of the lines l_1 and l_2, respectively. Since the slopes of l_1 and l_2 are unequal, $f'(a)$ does not exist. The graph of f has a **corner** at $P(a, f(a))$ if f is continuous at a and if the right-hand and left-hand derivatives at a exist and are unequal or if *one* of those derivatives exists at a and $|f'(x)| \to \infty$ as $x \to a^-$ or $x \to a^+$.

Figure 2.18

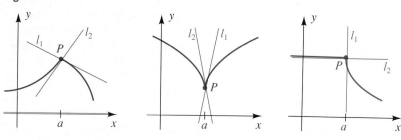

As indicated in the next definition, a *vertical tangent line* may occur at $P(a, f(a))$ if $f'(a)$ does not exist. If P is an endpoint of the domain of f, we can state a similar definition using a right-hand or a left-hand derivative, as appropriate.

Definition 2.14

> The graph of a function f has a **vertical tangent line** $x = a$ at the point $P(a, f(a))$ if f is continuous at a and if
>
> $$\lim_{x \to a} |f'(x)| = \infty.$$

The next example shows a function f with a vertical tangent line at an endpoint of the domain of f.

EXAMPLE ■ 6

(a) sketch the graph of f **(b)** find $f'(x)$ and the domain of f'

SOLUTION

(a) The graph of f is sketched in Figure 2.19. Note that the domain of f consists of all nonnegative numbers.

(b) Since $x = 0$ is an endpoint of the domain of f, we shall examine the cases $x > 0$ and $x = 0$ separately.

If $x > 0$, then, by Definition (2.5),

$$f'(x) = \lim_{h \to 0} \frac{\sqrt{x+h} - \sqrt{x}}{h}.$$

Figure 2.19

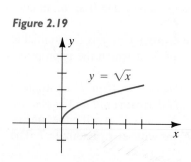

$y = \sqrt{x}$

To find the limit, we first rationalize the numerator of the quotient and then simplify:

$$f'(x) = \lim_{h \to 0} \frac{\sqrt{x+h} - \sqrt{x}}{h} \cdot \frac{\sqrt{x+h} + \sqrt{x}}{\sqrt{x+h} + \sqrt{x}}$$

$$= \lim_{h \to 0} \frac{(x+h) - x}{h(\sqrt{x+h} + \sqrt{x})}$$

$$= \lim_{h \to 0} \frac{1}{\sqrt{x+h} + \sqrt{x}}$$

$$= \frac{1}{\sqrt{x} + \sqrt{x}} = \frac{1}{2\sqrt{x}}$$

Since $x = 0$ is an endpoint of the domain of f, we must use a one-sided limit to determine if $f'(0)$ exists. Using Definition (2.13) with $x = 0$, we obtain

$$\lim_{h \to 0^+} \frac{f(0+h) - f(0)}{h} = \lim_{h \to 0^+} \frac{\sqrt{0+h} - \sqrt{0}}{h}$$

$$= \lim_{h \to 0^+} \frac{\sqrt{h}}{h} = \lim_{h \to 0^+} \frac{1}{\sqrt{h}} = \infty.$$

Since the limit does not exist, the domain of f' is the set of positive real numbers. The last limit shows that the graph of f has a vertical tangent line (the y-axis) at the point $(0, 0)$.

Figure 2.20 illustrates some typical cases of vertical tangent lines. In Figure 2.20(b), $f'(x) \to \infty$ as x approaches a from either side. In contrast, in Figure 2.20(c), $f'(x) \to \infty$ as x approaches a from the left, but $f'(x) \to -\infty$ as x approaches a from the right. The resulting sharp peak (or spike) at P is called a *cusp*, formally defined as follows.

Definition 2.15

The graph of f has a **cusp** at $P(a, f(a))$ if f is continuous at a and if the following two conditions hold:

    (i) $f'(x) \to \infty$ as x approaches a from one side
    (ii) $f'(x) \to -\infty$ as x approaches a from the other side

Figure 2.20
(a)

(b)

(c)

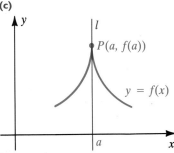

EXAMPLE ■ 7   Determine the nature of the graph of the function $f(x) = 1 + x^{2/3}$ near the point (0, 1).

SOLUTION   Since $\lim_{x \to 0} f(x) = f(0) = 1$, the function f is continuous at (0, 1). If $x \neq 0$, then we have

$$f'(x) = \frac{2}{3}x^{-1/3} = \frac{2}{3x^{1/3}}.$$

Since $\lim_{x \to 0^+} f'(x) = \infty$ and $\lim_{x \to 0^-} f'(x) = -\infty$, there is a cusp at the point (0, 1). The function $f(x) = 1 + x^{2/3}$ is not differentiable at 0 because the right-hand and left-hand derivatives do not exist. Figure 2.21 shows a graph of the function f, where the y-axis is a vertical tangent line.

Figure 2.21

If f is a complicated continuous function, we may have difficulty finding the values of x at which f is not differentiable using only algebraic techniques. We may be able to make good approximations of such values by looking for cusps and corners on the graph of f.

EXAMPLE ■ 8   Graph the function

$$f(x) = 2 + \left| x^4 - 5x^2 + 4 \right|$$

on the interval [0, 2.5], and estimate the values of x at which the function is not differentiable.

SOLUTION   We have graphed $f(x) = 2 + \left| x^4 - 5x^2 + 4 \right|$ in the viewing window shown in Figure 2.22. Note that the graph appears smooth at all values of x on the interval [0, 2.5] except at approximately $x = 1$ and $x = 2$, where there appear to be corners. By zooming in near the point $(1, f(1))$, we are led to believe that the right-hand and left-hand derivatives exist at $x = 1$, but are not equal. A similar situation occurs at $x = 2$, and we conclude that f is not differentiable at $x = 1$ and $x = 2$.

Figure 2.22
$0 \le x \le 2.5, 0 \le y \le 12.5$

DERIVATIVE NOTATIONS

We conclude this section by considering the various notations for derivatives.

Notations for the Derivative
of $y = f(x)$ **2.16**

$$f'(x) = y' = \frac{dy}{dx} = \frac{d}{dx}f(x) = D_x f(x) = D_x y$$

All of these notations are used in mathematics and applications, and you should become familiar with the different forms. For example, we can now write

$$D_x f(x) = \frac{d}{dx}f(x) = \lim_{h \to 0} \frac{f(x+h) - f(x)}{h}.$$

The letter x in D_x and d/dx denotes the independent variable. If we use a different independent variable, say t, then we write

$$f'(t) = D_t f(t) = \frac{d}{dt} f(t).$$

Each of the symbols D_x and d/dx is called a **differential operator**. Standing alone, D_x or d/dx has no practical significance; however, when either symbol has an expression to its right, it denotes a derivative. We say that D_x or d/dx *operates* on the expression, and we call $D_x y$ or dy/dx **the derivative of y with respect to x**. We shall justify the notation dy/dx in Section 2.8, where the concept of a *differential* is defined.

The next illustration contains some examples of the use of (2.16) and Theorem (2.11).

ILLUSTRATION

$$\blacksquare \quad \frac{d}{dx}(3x^7) = (3 \cdot 7)x^6 = 21x^6$$

$$\blacksquare \quad \frac{d}{dt}(\tfrac{1}{2}t^{12}) = (\tfrac{1}{2} \cdot 12)t^{11} = 6t^{11}$$

$$\blacksquare \quad \frac{d}{dx}(9x^{4/3}) = (9 \cdot \tfrac{4}{3})x^{1/3} = 12x^{1/3}$$

$$\blacksquare \quad \frac{d}{dr}(2r^{-4}) = 2(-4)r^{-5} = -\frac{8}{r^5}$$

Note that in (2.16), $D_x y$, y', and dy/dx are used for the derivative of y with respect to x. If we wish to denote the *value* of the derivative $D_x y$, y', or dy/dx at some number $x = a$, we often use a single or double bracket and write

$$D_x y]_{x=a} \qquad \frac{dy}{dx}\bigg]_{x=a} \qquad [D_x y]_{x=a} \quad \text{or} \quad \left[\frac{dy}{dx}\right]_{x=a}$$

as in the following examples:

$$\left[\frac{d}{dx}(x^3)\right]_{x=5} = [3x^2]_{x=5} = 3(5^2) = 75$$

$$\left[\frac{d}{dx}(9x^{4/3})\right]_{x=8} = [12x^{1/3}]_{x=8} = 12(8^{1/3}) = 24$$

In calculus, we sometimes consider *derivatives of derivatives*. As we have seen, if we differentiate a function f, we obtain another function denoted f'. If f' has a derivative, it is denoted f'' (read f *double prime*) and is called the **second derivative** of f. Thus,

$$f''(x) = [f'(x)]' = \frac{d}{dx}(f'(x)) = \frac{d}{dx}\left(\frac{d}{dx}(f(x))\right) = \frac{d^2}{dx^2}(f(x)),$$

where we use the operator symbol d^2/dx^2 for second derivatives. The **third derivative** of f, denoted f''', is the derivative of the second derivative. Thus,

$$f'''(x) = [f''(x)]' = \frac{d}{dx}(f''(x)) = \frac{d}{dx}\left(\frac{d^2}{dx^2}(f(x))\right) = \frac{d^3}{dx^3}(f(x)).$$

In general, if n is a positive integer, then $f^{(n)}$ denotes the **nth derivative** of f and is found by starting with f and differentiating, successively, n times. In operator notation, $f^{(n)}(x) = \frac{d^n}{dx^n}(f(x))$, where the integer n is the **order** of the derivative $f^{(n)}(x)$. The following summarizes various notations used for these **higher derivatives,** with $y = f(x)$.

Notations for Higher Derivatives   2.17

$f'(x),$	$f''(x),$	$f'''(x),$	$f^{(4)}(x),$	$\dots,$	$f^{(n)}(x)$
$y',$	$y'',$	$y''',$	$y^{(4)},$	$\dots,$	y^n
$\dfrac{dy}{dx},$	$\dfrac{d^2y}{dx^2},$	$\dfrac{d^3y}{dx^3},$	$\dfrac{d^4y}{dx^4},$	$\dots,$	$\dfrac{d^ny}{dx^n}$
$D_x y,$	$D_x^2 y,$	$D_x^3 y,$	$D_x^4 y,$	$\dots,$	$D_x^n y$

EXAMPLE 9   Find the first four derivatives of $f(x) = 4x^{3/2}$.

SOLUTION   We use (2.17) and Theorem (2.11) four times:

$$f'(x) = (4 \cdot \tfrac{3}{2})x^{1/2} = 6x^{1/2}$$

$$f''(x) = (6 \cdot \tfrac{1}{2})x^{-1/2} = 3x^{-1/2}$$

$$f'''(x) = 3(-\tfrac{1}{2})x^{-3/2} = -\tfrac{3}{2}x^{-3/2}$$

$$f^{(4)}(x) = -\tfrac{3}{2}(-\tfrac{3}{2})x^{-5/2} = \tfrac{9}{4}x^{-5/2}$$

EXERCISES 2.2

Exer. 1–4: (a) Use Definition (2.5) to find $f'(x)$. (b) Find the domain of f'. (c) Find an equation of the tangent line to the graph of f at P. (d) Find the points on the graph at which the tangent line is horizontal.

1   $f(x) = -5x^2 + 8x + 2$; $P(-1, -11)$

2   $f(x) = 3x^2 - 2x - 4$; $P(2, 4)$

3   $f(x) = x^3 + x$; $P(1, 2)$

4   $f(x) = x^3 - 4x$; $P(2, 0)$

Exer. 5–12: (a) Use (2.8)–(2.11) to find $f'(x)$. (b) Find the domain of f'. (c) Find an equation of the tangent line to the graph of f at P. (d) Find the points on the graph at which the tangent line is horizontal.

5   $f(x) = 9x - 2$; $P(3, 25)$

6   $f(x) = -4x + 3$; $P(-2, 11)$

7   $f(x) = 37$; $P(0, 37)$

8   $f(x) = \pi^2$; $P(5, \pi^2)$

9 $f(x) = 1/x^3$; $P(2, \frac{1}{8})$

10 $f(x) = 1/x^4$; $P(1, 1)$

11 $f(x) = 4x^{1/4}$; $P(81, 12)$

12 $f(x) = 12x^{1/3}$; $P(-27, -36)$

Exer. 13 – 16: Find the first three derivatives.

13 $f(x) = 3x^6$

14 $f(x) = 6x^4$

15 $f(x) = 9\sqrt[3]{x^2}$

16 $f(x) = 3x^{7/3}$

17 If $z = 25t^{9/5}$, find $D_t^2 z$.

18 If $y = 3x + 5$, find $D_x^3 y$.

19 If $y = -4x + 7$, find $\dfrac{d^3 y}{dx^3}$.

20 If $z = 64\sqrt[4]{t^3}$, find $\dfrac{d^2 z}{dt^2}$.

Exer. 21 – 22: Is f differentiable on the given interval? Explain.

21 $f(x) = 1/x$ **(a)** $[0, 2]$ **(b)** $[1, 3)$

22 $f(x) = \sqrt[3]{x}$ **(a)** $[-1, 1]$ **(b)** $[-2, -1)$

Exer. 23 – 24: Use the graph of f to determine if f is differentiable on the given interval.

23 $f(x) = \sqrt{4 - x}$ **(a)** $[0, 4]$ **(b)** $[-5, 0]$

24 $f(x) = \sqrt{4 - x^2}$ **(a)** $[-2, 2]$ **(b)** $[-1, 1]$

Exer. 25 – 30: Determine whether f has (a) a vertical tangent line at $(0, 0)$ and (b) a cusp at $(0, 0)$.

25 $f(x) = x^{1/3}$ **26** $f(x) = x^{5/3}$ **27** $f(x) = x^{2/5}$

28 $f(x) = x^{1/4}$ **29** $f(x) = 5x^{3/2}$ **30** $f(x) = 7x^{4/3}$

Exer. 31 – 32: Estimate $f'(-1)$, $f'(1)$, $f'(2)$, and $f'(3)$, whenever they exist.

31

32

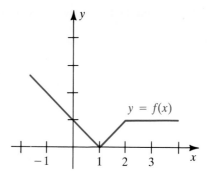

Exer. 33 – 36: Use right-hand and left-hand derivatives to prove that f is not differentiable at a.

33 $f(x) = |x - 5|$; $a = 5$

34 $f(x) = |x + 2|$; $a = -2$

35 $f(x) = [\![x - 2]\!]$; $a = 2$

36 $f(x) = [\![x]\!] - 2$; $a = 2$

Exer. 37 – 40: Use the graph of f to find the domain of f'.

37 $f(x) = \begin{cases} 2x & \text{if } x \le 0 \\ x^2 & \text{if } x > 0 \end{cases}$

38 $f(x) = \begin{cases} 2x - 1 & \text{if } x \le 1 \\ x^2 & \text{if } x > 1 \end{cases}$

39 $f(x) = \begin{cases} -x^2 & \text{if } x < -1 \\ 2x + 3 & \text{if } x \ge -1 \end{cases}$

40 $f(x) = \begin{cases} x^2 - 2 & \text{if } x < 0 \\ -3 & \text{if } x \ge 0 \end{cases}$

Exer. 41 – 42: Each figure is the graph of a function f. Sketch the graph of f' and determine where f is not differentiable.

41

42

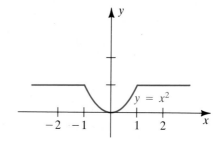

Exer. 43–44: (a) Find the first derivative and the second derivative on each of the three subintervals of the domain of the function. (b) Use the right-hand and left-hand derivatives to determine whether f' exists at a and b.

43 $a = 3; b = 5$

$$f(x) = \begin{cases} 3x^2 - 6x + 3 & \text{if } 1 \le x < a \\ -7x^2 + 54x - 87 & \text{if } a \le x < b \\ 8x^2 - 96x + 288 & \text{if } b \le x \le 6 \end{cases}$$

44 $a = -1; b = 1$

$$f(x) = \begin{cases} 3x^2 + 3x + 1 & \text{if } x < a \\ -x^3 & \text{if } a \le x < b \\ x^3 - 6x^2 + 6x - 2 & \text{if } x \ge b \end{cases}$$

Exer. 45–46: Given the position function s of a point P moving on a coordinate line l, find the times at which the velocity is the given value k.

45 $s(t) = 3t^{2/3}; \quad k = 4$

46 $s(t) = 4t^3; \quad k = 300$

47 The relationship between the temperature F on the Fahrenheit scale and the temperature C on the Celsius scale is given by $C = \frac{5}{9}(F - 32)$. Find the rate of change of F with respect to C.

48 Charles's law for gases states that if the pressure remains constant, then the relationship between the volume V that a gas occupies and its temperature T (in $^\circ$C) is given by $V = V_0(1 + \frac{1}{273}T)$. Find the rate of change of T with respect to V.

49 Show that the rate of change of the volume of a sphere with respect to its radius is numerically equal to the surface area of the sphere.

50 Poiseuille's law states that the velocity v of blood in a small artery with a circular cross section of radius R is given by

$$v(r) = A(R^2 - r^2),$$

where r is the distance from the center of the artery and A is a constant. Find $v'(r)$.

51 An oil spill is increasing such that the surface covered by the spill is always circular. Find the rate at which the area A of the surface is changing with respect to the radius r of the circle at

(a) any value of r (b) $r = 500$ ft

52 A spherical balloon is being inflated. Find the rate at which its volume V is changing with respect to the radius r of the balloon at

(a) any value of r (b) $r = 10$ ft

53 In some applications, function values $f(x)$ may be known only for several values of x near a. In these situations, $f'(a)$ is frequently approximated by the formula

$$f'(a) \approx \frac{f(a+h) - f(a-h)}{2h}.$$

(a) Interpret this formula graphically.

(b) Show that $\lim\limits_{h \to 0} \dfrac{f(a+h) - f(a-h)}{2h} = f'(a)$.

(c) If $f(x) = 1/x^2$, use the approximation formula to estimate $f'(1)$ with $h = 0.1, 0.01$, and 0.001.

(d) Find the exact value of $f'(1)$.

54 (a) Use the approximation formula in Exercise 53 to show that if $h \approx 0$, then

$$f''(a) \approx \frac{f(a+h) - 2f(a) + f(a-h)}{h^2}.$$

(b) If $f(x) = 1/x^2$, use part (a) to estimate $f''(1)$ with $h = 0.1, 0.01$, and 0.001.

(c) Find the exact value of $f''(1)$.

Exer. 55–56: Use the following table, which lists the approximate number of feet $s(t)$ that a car travels in t seconds to reach a velocity of 60 mi/hr in 6 sec.

t	0	1	2	3	4	5	6	7
$s(t)$	0	11.7	42.6	89.1	149.0	220.1	303.7	396.7

55 Use Exercise 53 to approximate the velocity of the car at

(a) $t = 3$ (b) $t = 6$

56 Use Exercise 54 to approximate the rate of change of the velocity of the car with respect to t at

(a) $t = 3$ (b) $t = 6$

57 Graph $f(x) = |x^5 - 2x^4 + 3x^3 - x + 1|$ on the interval $[-1, 1]$ and estimate where f is not differentiable.

58 Graph $f(x) = x^4 - 3x^3 + 2x - 1$ on the interval $[-1, 3]$ and estimate the x-coordinates of points at which the tangent line is horizontal.

Exer. 59–64: (a) Graph the function on the interval. (b) From the graph, estimate the x-coordinates of points where the function is not differentiable or where it has a horizontal tangent line. (c) Many graphing utilities can display the function and its first and second derivatives without explicitly requiring the formulas for the derivatives. If you have access to such a utility, graph the first and second derivatives on the same viewing window as the function. Discuss the behavior of the derivatives at points where the function has a horizontal tangent or near points where it is not differentiable.

59 $f(x) = \dfrac{10 \cos x}{x^2 + 4}$ on $-2 \le x \le 2$

60 $f(x) = \dfrac{(x - x^2) \sin x}{x^2 + 1}$ on $-1 \le x \le 4$

61 $f(x) = x \sin x$ on $-\pi \le x \le \pi$

62 $f(x) = \sin(\frac{1}{2}x + 1)$ on $0 \le x \le 4\pi$

63 $f(x) = |x^4 - 2x^3 - x^2 + 2x|$ on $-1.5 \le x \le 2.5$

64 $f(x) = \sqrt{4 - x^2}$ on $-2 \le x \le 2$

65 If f is differentiable at a and g is defined by $g(x) = x(f(x))$ for all x in the domain of f, then use Definition (2.5) to show that

$$g'(x) = x(f'(x)) + f(x).$$

66 Use the result of Exercise 65 and mathematical induction to give an alternative proof to the power rule (2.10) for $n \ge 1$. (*Hint:* $x^{n+1} = xx^n$.)

2.3 TECHNIQUES OF DIFFERENTIATION

This section contains some general rules that simplify the task of finding derivatives. The rules are stated in terms of the differential operator d/dx, where $(d/dx)(f(x)) = f'(x)$. In the rules, f and g denote differentiable functions, c, m, and b are real numbers, and n is a rational number. The first three parts of the following theorem were proved in Section 2.2 and are restated here for completeness.

Theorem 2.18

(i) $\dfrac{d}{dx}(c) = 0$

(ii) $\dfrac{d}{dx}(mx + b) = m$

(iii) $\dfrac{d}{dx}(x^n) = nx^{n-1}$

(iv) $\dfrac{d}{dx}(cf(x)) = c\dfrac{d}{dx}(f(x))$

(v) $\dfrac{d}{dx}(f(x) + g(x)) = \dfrac{d}{dx}(f(x)) + \dfrac{d}{dx}(g(x))$

(vi) $\dfrac{d}{dx}(f(x) - g(x)) = \dfrac{d}{dx}(f(x)) - \dfrac{d}{dx}(g(x))$

PROOF

(iv) Applying the definition of the derivative to $cf(x)$, we have

$$\frac{d}{dx}(cf(x)) = \lim_{h \to 0} \frac{cf(x + h) - cf(x)}{h} = \lim_{h \to 0} c\frac{f(x + h) - f(x)}{h}$$

$$= c \lim_{h \to 0} \frac{f(x + h) - f(x)}{h} = c\frac{d}{dx}(f(x)).$$

(v) Applying the definition of the derivative to $f(x) + g(x)$ yields

$$\frac{d}{dx}(f(x) + g(x)) = \lim_{h \to 0} \frac{[f(x+h) + g(x+h)] - [f(x) + g(x)]}{h}$$

$$= \lim_{h \to 0} \left[\frac{f(x+h) - f(x)}{h} + \frac{g(x+h) - g(x)}{h} \right]$$

$$= \lim_{h \to 0} \frac{f(x+h) - f(x)}{h} + \lim_{h \to 0} \frac{g(x+h) - g(x)}{h}$$

$$= \frac{d}{dx}(f(x)) + \frac{d}{dx}(g(x)).$$

We can prove (vi) in similar fashion, or we can write

$$f(x) - g(x) = f(x) + (-1)g(x)$$

and then use (v) and (iv).   ■

If we use the differential operator D_x in place of d/dx, the rules in Theorem (2.18) take on the following forms:

$$D_x(c) = 0,$$

$$D_x(mx + b) = m,$$

$$D_x(x^n) = nx^{n-1},$$

$$D_x(cf(x)) = cD_x(f(x)),$$

$$D_x(f(x) \pm g(x)) = D_x(f(x)) \pm D_x(g(x)).$$

Parts (v) and (vi) of Theorem (2.18) may be stated as follows:

(v) *The derivative of a sum is the sum of the derivatives.*

(vi) *The derivative of a difference is the difference of the derivatives.*

These results can be extended to sums or differences involving any finite number of functions. We may use these results to obtain easily the derivative of a polynomial, since a polynomial is a sum of terms of the form cx^n, where c is a real number and n is a nonnegative integer. The next example illustrates this process.

EXAMPLE ■ 1   If $f(x) = 2x^4 - 5x^3 + x^2 - 4x + 1$, find $f'(x)$.

SOLUTION   By Theorem (2.18), we have

$$f'(x) = \frac{d}{dx}(2x^4 - 5x^3 + x^2 - 4x + 1)$$

$$= \frac{d}{dx}(2x^4) - \frac{d}{dx}(5x^3) + \frac{d}{dx}(x^2) - \frac{d}{dx}(4x) + \frac{d}{dx}(1)$$

$$= 8x^3 - 15x^2 + 2x - 4.$$

To find the equation of a tangent line to the graph of a function f at a particular point $P(a, f(a))$, we need to find the slope of the line, which we know is given by $f'(a)$. We may use the differentiation rules to first determine the general form for the function $f'(x)$ and then substitute a for x to compute $f'(a)$, as in the next example.

EXAMPLE ■ 2

(a) Find an equation of the tangent line to the graph of the function $y = 1.2\sqrt[3]{x^2} - (0.8/\sqrt{x})$ at $P(1, 0.4)$.

(b) Use a graphing utility to graph the function and the tangent line.

SOLUTION

(a) We first express y in terms of rational exponents and then use Theorem (2.18) to find dy/dx:

$$y = 1.2x^{2/3} - 0.8x^{-1/2}$$

$$\frac{dy}{dx} = \frac{d}{dx}(1.2x^{2/3}) - \frac{d}{dx}(0.8x^{-1/2})$$
$$= 0.8x^{-1/3} - (-0.4x^{-3/2})$$
$$= \frac{0.8}{x^{1/3}} + \frac{0.4}{x^{3/2}}$$

To find the slope of the tangent line at $P(1, 0.4)$, we evaluate dy/dx at $x = 1$:

$$\frac{dy}{dx}\bigg]_{x=1} = \frac{0.8}{1} + \frac{0.4}{1} = 1.2.$$

Using the point–slope form, we can express an equation of the tangent line as

$$y - 0.4 = 1.2(x - 1),$$

or

$$y = 1.2x - 0.8.$$

(b) We use equal scaling and the viewing window shown in Figure 2.23 to graph the function $f(x) = 1.2\sqrt[3]{x^2} - (0.8/\sqrt{x})$ and the tangent line $L(x) = 1.2x - 0.8$. We observe that both the tangent line and the curve are rising as x increases through values near 1. We also note that the tangent line remains close to the curve near the point of tangency.

Figure 2.23
$-4 \le x \le 6, -3 \le y \le 3$

Formulas for derivatives of products or quotients are more complicated than those for sums and differences. In particular, *the derivative of a product generally is not equal to the product of the derivatives*. We may illustrate this fact by using the product $x^2 \cdot x^5$ as follows:

$$\frac{d}{dx}(x^2 \cdot x^5) = \frac{d}{dx}(x^7) = 7x^6$$

$$\frac{d}{dx}(x^2) \cdot \frac{d}{dx}(x^5) = (2x) \cdot (5x^4) = 10x^5$$

Hence
$$\frac{d}{dx}(x^2 \cdot x^5) \neq \frac{d}{dx}(x^2) \cdot \frac{d}{dx}(x^5).$$

The derivative of any product $f(x)g(x)$ may be expressed in terms of derivatives of $f(x)$ and $g(x)$ as in the following rule.

Product Rule 2.19

$$\frac{d}{dx}(f(x)g(x)) = f(x)\frac{d}{dx}(g(x)) + g(x)\frac{d}{dx}(f(x))$$

PROOF Let $y = f(x)g(x)$. Using the definition of the derivative, we write

$$\frac{dy}{dx} = \lim_{h \to 0} \frac{f(x+h)g(x+h) - f(x)g(x)}{h}.$$

To change the form of the quotient so that the limit may be evaluated, we subtract and add the expression $f(x+h)g(x)$ in the numerator. Thus,

$$\frac{dy}{dx} = \lim_{h \to 0} \frac{f(x+h)g(x+h) - f(x+h)g(x) + f(x+h)g(x) - f(x)g(x)}{h}$$

$$= \lim_{h \to 0} \left[f(x+h) \cdot \frac{g(x+h) - g(x)}{h} + g(x) \cdot \frac{f(x+h) - f(x)}{h} \right]$$

$$= \lim_{h \to 0} f(x+h) \cdot \lim_{h \to 0} \frac{g(x+h) - g(x)}{h} + \lim_{h \to 0} g(x) \cdot \lim_{h \to 0} \frac{f(x+h) - f(x)}{h}.$$

Since f is differentiable at x, it is continuous at x (see Theorem (2.12)). Hence, $\lim_{h \to 0} f(x+h) = f(x)$. Also, $\lim_{h \to 0} g(x) = g(x)$, since x is fixed in this limiting process. Finally, applying the definition of derivative to $f(x)$ and $g(x)$, we obtain

$$\frac{dy}{dx} = f(x)g'(x) + g(x)f'(x). \quad \blacksquare$$

The product rule may be phrased as follows: *The derivative of a product equals the first factor times the derivative of the second factor, plus the second times the derivative of the first.*

EXAMPLE ▪ 3   If $y = (x^3 + 1)(2x^2 + 8x - 5)$, find dy/dx.

SOLUTION   Using the product rule (2.19), we have

$$\frac{dy}{dx} = (x^3 + 1)\frac{dy}{dx}(2x^2 + 8x - 5) + (2x^2 + 8x - 5)\frac{dy}{dx}(x^3 + 1)$$

$$= (x^3 + 1)(4x + 8) + (2x^2 + 8x - 5)(3x^2)$$

$$= (4x^4 + 8x^3 + 4x + 8) + (6x^4 + 24x^3 - 15x^2)$$

$$= 10x^4 + 32x^3 - 15x^2 + 4x + 8.$$

EXAMPLE ▪ 4   If $f(x) = x^{1/3}(x^2 - 3x + 2)$, find

(a) $f'(x)$

(b) the x-coordinate of the points on the graph of f at which the tangent line is either horizontal or vertical

SOLUTION

(a) By the product rule (2.19),

$$f'(x) = x^{1/3}\frac{d}{dx}(x^2 - 3x + 2) + (x^2 - 3x + 2)\frac{d}{dx}(x^{1/3})$$

$$= x^{1/3}(2x - 3) + (x^2 - 3x + 2)(\tfrac{1}{3}x^{-2/3})$$

$$= \frac{3x(2x - 3) + (x^2 - 3x + 2)}{3x^{2/3}}$$

$$= \frac{7x^2 - 12x + 2}{3x^{2/3}}.$$

(b) The tangent line to the graph of f is horizontal if its slope is zero. Setting $f'(x) = 0$ and using the quadratic formula, we obtain

$$x = \frac{12 \pm \sqrt{144 - 56}}{2(7)} = \frac{12 \pm \sqrt{88}}{14} = \frac{6 \pm \sqrt{22}}{7}.$$

Referring to $f'(x)$, we see that the denominator $3x^{2/3}$ is zero at $x = 0$. Since f is continuous at 0 and $\lim_{x \to 0} |f'(x)| = \infty$, it follows from Definition (2.14) that the graph of f has a vertical tangent line at $x = 0$—that is, the point $(0, 0)$ (the origin).

We shall next obtain a formula for the derivative of a quotient. Note that *the derivative of a quotient generally is not equal to the quotient of the derivatives.* We may illustrate this with the quotient x^5/x^2 as follows:

$$\frac{d}{dx}\left(\frac{x^5}{x^2}\right) = \frac{d}{dx}(x^3) = 3x^2$$

$$\frac{\frac{d}{dx}(x^5)}{\frac{d}{dx}(x^2)} = \frac{5x^4}{2x} = \frac{5}{2}x^3$$

Hence
$$\frac{d}{dx}\left(\frac{x^5}{x^2}\right) \neq \frac{\dfrac{d}{dx}(x^5)}{\dfrac{d}{dx}(x^2)}.$$

The derivative of any quotient $f(x)/g(x)$ may be expressed in terms of the derivatives of $f(x)$ and $g(x)$ as in the following rule.

Quotient Rule   2.20

$$\frac{d}{dx}\left(\frac{f(x)}{g(x)}\right) = \frac{g(x)\dfrac{d}{dx}(f(x)) - f(x)\dfrac{d}{dx}(g(x))}{(g(x))^2}$$

P R O O F   Let $y = f(x)/g(x)$. From the definition of derivative,

$$\frac{dy}{dx} = \lim_{h \to 0} \frac{\dfrac{f(x+h)}{g(x+h)} - \dfrac{f(x)}{g(x)}}{h}$$

$$= \lim_{h \to 0} \frac{g(x)f(x+h) - f(x)g(x+h)}{hg(x+h)g(x)}.$$

Subtracting and adding $g(x)f(x)$ in the numerator of the last quotient, we obtain

$$\frac{dy}{dx} = \lim_{h \to 0} \frac{g(x)f(x+h) - g(x)f(x) + g(x)f(x) - f(x)g(x+h)}{hg(x+h)g(x)}$$

$$= \lim_{h \to 0} \frac{g(x)[f(x+h) - f(x)] - f(x)[g(x+h) - g(x)]}{hg(x+h)g(x)}$$

$$= \lim_{h \to 0} \frac{g(x)\left[\dfrac{f(x+h) - f(x)}{h}\right] - f(x)\left[\dfrac{g(x+h) - g(x)}{h}\right]}{g(x+h)g(x)}.$$

Taking the limit of the numerator and the denominator gives us the quotient rule.   ■

The quotient rule may be stated as follows: *The derivative of a quotient is equal to the denominator times the derivative of the numerator minus the numerator times the derivative of the denominator, divided by the square of the denominator.*

E X A M P L E ▪ 5   Find $\dfrac{dy}{dx}$ if $y = \dfrac{3x^2 - x + 2}{4x^2 + 5}$.

SOLUTION By the quotient rule (2.20),

$$\frac{dy}{dx} = \frac{(4x^2 + 5)\dfrac{d}{dx}(3x^2 - x + 2) - (3x^2 - x + 2)\dfrac{d}{dx}(4x^2 + 5)}{(4x^2 + 5)^2}$$

$$= \frac{(4x^2 + 5)(6x - 1) - (3x^2 - x + 2)(8x)}{(4x^2 + 5)^2}$$

$$= \frac{(24x^3 - 4x^2 + 30x - 5) - (24x^3 - 8x^2 + 16x)}{(4x^2 + 5)^2}$$

$$= \frac{4x^2 + 14x - 5}{(4x^2 + 5)^2}.$$

If we let $f(x) = 1$ in the quotient rule (2.20), then, since $(d/dx)(1) = 0$, we obtain the following.

Reciprocal Rule 2.21

$$\frac{d}{dx}\left(\frac{1}{g(x)}\right) = -\frac{\dfrac{d}{dx}(g(x))}{(g(x))^2}$$

ILLUSTRATION

■ $\dfrac{d}{dx}\left(\dfrac{1}{x}\right) = -\dfrac{\dfrac{d}{dx}(x)}{(x)^2} = -\dfrac{1}{x^2}$

■ $\dfrac{d}{dx}\left(\dfrac{1}{3x^2 - 5x + 4}\right) = -\dfrac{\dfrac{d}{dx}(3x^2 - 5x + 4)}{(3x^2 - 5x + 4)^2} = -\dfrac{6x - 5}{(3x^2 - 5x + 4)^2}$

The differentiation formulas in Theorem (2.18) are stated in terms of the function values $f(x)$ and $g(x)$. If we wish to state such rules without referring to the variable x, we may write

$$(cf)' = cf', \quad (f + g)' = f' + g', \quad \text{and} \quad (f - g)' = f' - g'.$$

Using this notation for the product, quotient, and reciprocal rules and at the same time commuting some of the factors that appear in (2.19) and (2.20), we obtain

$$(fg)' = f'g + fg', \quad \left(\frac{f}{g}\right)' = \frac{f'g - fg'}{g^2}, \quad \text{and} \quad \left(\frac{1}{g}\right) = \frac{-g'}{g^2}.$$

You may find it helpful to memorize these formulas. To obtain the quotient rule, change the $+$ sign in the formula for $(fg)'$ to $-$ and divide by g^2.

The next example gives an application that makes use of the quotient rule for differentiation.

EXAMPLE ▪ 6 A convex lens of focal length f is shown in Figure 2.24. If an object is a distance p from the lens as shown, then the distance q from the lens to the image is related to p and f by the *lens equation*

$$\frac{1}{f} = \frac{1}{p} + \frac{1}{q}.$$

Figure 2.24

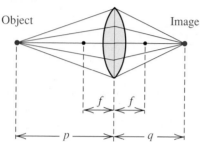

If, for a particular lens, $f = 2$ cm and p is increasing, find

(a) a general formula for the rate of change of q with respect to p

(b) the rate of change of q with respect to p if $p = 22$ cm

SOLUTION

(a) By (2.7)(ii), the rate of change of q with respect to p is given by the derivative dq/dp. If $f = 2$, then the lens equation gives us

$$\frac{1}{2} = \frac{1}{p} + \frac{1}{q}, \quad \text{or} \quad \frac{1}{q} = \frac{1}{2} - \frac{1}{p} = \frac{p-2}{2p}.$$

Hence

$$q = \frac{2p}{p-2}.$$

Applying the quotient rule (with $x = p$) yields

$$\frac{dq}{dp} = \frac{(p-2)\dfrac{d}{dp}(2p) - (2p)\dfrac{d}{dp}(p-2)}{(p-2)^2}$$

$$= \frac{(p-2)(2) - (2p)(1)}{(p-2)^2}$$

$$= \frac{-4}{(p-2)^2}.$$

(b) Substituting $p = 22$ in the formula obtained in part (a), we get

$$\frac{dq}{dp}\bigg]_{p=22} = \frac{-4}{(22-2)^2} = \frac{-4}{400} = -\frac{1}{100}.$$

Thus, if $p = 22$ cm, the image distance q is decreasing at a rate of $\frac{1}{100}$ centimeter per centimeter change in p.

We have introduced several rules in this section that ease the task of finding derivatives of functions. These rules permit us to differentiate a complicated function that has been built up from simpler functions in particular ways (addition, subtraction, multiplication, and division) by combining the derivatives of the simpler functions in the correct manner. These formulas and others to be studied later make differentiation a relatively straightforward process of applying the various rules.

In recent years, the possibility of carrying out these rules with the assistance of computational devices has become a reality. We discussed earlier how calculators and computers can provide *numerical* estimates for derivatives. Computer scientists have also created sophisticated programs that accept and operate on algebraic expressions and functions. Such programs can perform a variety of algebraic operations on command, including symbolic differentiation. These programs, called **computer algebra systems (CAS)** or **computer mathematics systems (CMS),** combine the capability for algebraic manipulation with provisions for graphing and numerical evaluations. In effect, a CAS has stored the rules for differentiation. It performs pattern matching on a given symbolic expression to determine which rules to apply, much as you are learning to do.

Since electronic devices can perform symbolic differentiation, you may well ask whether we still need to learn the rules for differentiation. The answer is *yes,* for much the same reasons that the ability of calculators to add and multiply numbers quickly and accurately has not ended the need to learn the rules of arithmetic. In both instances, the emphasis shifts away from developing great skill in applying the rules to very complicated expressions to a greater need for understanding and interpretation. The properties of differentiation reflected in the rules are used throughout calculus and other branches of mathematics to gain new insights.

If you were going to add a few small integers, you would probably do so mentally, instead of searching for a calculator. Similarly, you will find it easier to differentiate a polynomial or a rational function on paper than to type these expressions into the CAS. When you must perform a very important, complicated differentiation, the speed and accuracy of a CAS provides a valuable tool. In summary, the availability of a CAS no more replaces the need to perform basic symbolic manipulation than the availability of a calculator replaces the need to perform basic arithmetic operations.

EXERCISES 2.3

Exer. 1–34: Find the derivative.

1 $g(t) = 6t^{5/3}$

2 $h(z) = 8z^{3/2}$

3 $f(s) = 15 - s + 4s^2 - 5s^4$

4 $f(t) = 12 - 3t^4 + 4t^6$

5 $f(x) = 3x^2 + \sqrt[3]{x^4}$

6 $g(x) = x^4 - \sqrt[4]{x^3}$

7 $g(x) = (x^3 - 7)(2x^2 + 3)$

8 $k(x) = (2x^2 - 4x + 1)(6x - 5)$

9 $f(x) = x^{1/2}(x^2 + x - 4)$

10 $h(x) = x^{2/3}(3x^2 - 2x + 5)$

11 $h(r) = r^2(3r^4 - 7r + 2)$

12 $k(v) = v^3(-2v^3 + v - 3)$

13 $g(x) = (8x^2 - 5x)(13x^2 + 4)$

14 $H(z) = (z^5 - 2z^3)(7z^2 + z - 8)$

15 $f(x) = \dfrac{4x - 5}{3x + 2}$

16 $h(x) = \dfrac{8x^2 - 6x + 11}{x - 1}$

17 $h(z) = \dfrac{8 - z + 3z^2}{2 - 9z}$

18 $f(w) = \dfrac{2w}{w^3 - 7}$

19 $G(v) = \dfrac{v^3 - 1}{v^3 + 1}$

20 $f(t) = \dfrac{8t + 15}{t^2 - 2t + 3}$

21 $g(t) = \dfrac{\sqrt[3]{t^2}}{3t - 5}$

22 $f(x) = \dfrac{\sqrt{x}}{2x^2 - 4x + 8}$

23 $f(x) = \dfrac{1}{1 + x + x^2 + x^3}$

24 $p(x) = 1 + \dfrac{1}{x} + \dfrac{1}{x^2} + \dfrac{1}{x^3}$

25 $h(x) = \dfrac{7}{x^2 + 5}$

26 $k(z) = \dfrac{6}{z^2 + z - 1}$

27 $F(t) = t^2 + \dfrac{1}{t^2}$

28 $s(x) = 2x + \dfrac{1}{2x}$

29 $K(s) = (3s)^{-4}$

30 $W(s) = (3s)^4$

31 $h(x) = (5x - 4)^2$

32 $S(w) = (2w + 1)^3$

33 $g(r) = (5r - 4)^{-2}$

34 $S(x) = (3x + 1)^{-2}$

c | **Exer. 35–40:** (a) Find the derivative. (b) Plot the function and its derivative in the given viewing window.

35 $f(x) = x^3 - 5x^2 + 8x - 25$,
$-3.5 \le x \le 7, -150 \le y \le 150$

36 $f(x) = x^4 - x^3 - 13x^2 + x + 12$,
$-4 \le x \le 4, -70 \le y \le 110$

37 $f(x) = \left(4 + \dfrac{1}{x}\right)\left(6x - \dfrac{1}{x^2}\right)$,
$-3 \le x \le 3, -50 \le y \le 50$

38 $f(x) = \left(x^2 - \dfrac{1}{x^2}\right)\left(x^2 + \dfrac{1}{x^2}\right)$,
$-3 \le x \le 3, -50 \le y \le 50$

39 $f(x) = \dfrac{(4x + 1)(x - 3)}{3x + 2}$,
$-6 \le x \le 6, -15 \le y \le 10$

40 $f(x) = \left(\dfrac{x + 3}{x + 1}\right)(x^2 - 2x - 1)$,
$-11 \le x \le 5, -25 \le y \le 35$

Exer. 41–44: Solve the equation $dy/dx = 0$.

41 $y = 2x^3 - 3x^2 - 36x + 4$

42 $y = 4x^3 + 21x^2 - 24x + 11$

43 $y = \dfrac{2x^2 + 3x - 6}{x - 2}$

44 $y = \dfrac{x^2 + 2x + 5}{x + 1}$

Exer. 45–46: Solve the equation $d^2y/dx^2 = 0$.

45 $y = 6x^4 + 24x^3 - 540x^2 + 7$

46 $y = 6x^5 - 5x^4 - 30x^3 + 11x$

Exer. 47–50: Find dy/dx by (a) using the quotient rule, (b) using the product rule, and (c) simplifying algebraically and using Theorem (2.18).

47 $y = \dfrac{3x - 1}{x^2}$

48 $y = \dfrac{x^2 + 1}{x^4}$

49 $y = \dfrac{x^2 - 3x}{\sqrt[3]{x^2}}$

50 $y = \dfrac{2x + 3}{\sqrt{x^3}}$

Exer. 51–52: Find d^2y/dx^2.

51 $y = \dfrac{3x + 4}{x + 1}$

52 $y = \dfrac{x + 3}{2x + 3}$

Exer. 53–54: Find an equation of the tangent line to the graph of f at P.

53 $f(x) = \dfrac{5}{1 + x^2}$; $P(-2, 1)$

54 $f(x) = 3x^2 - 2\sqrt{x}$; $P(4, 44)$

55 Find the x-coordinates of all points on the graph of $y = x^3 + 2x^2 - 4x + 5$ at which the tangent line is
(a) horizontal (b) parallel to the line $2y + 8x = 5$

56 Find the point P on the graph of $y = x^3$ such that the tangent line at P has x-intercept 4.

57 Find the points on the graph of $y = x^{3/2} - x^{1/2}$ at which the tangent line is parallel to the line $y - x = 3$.

58 Find the points on the graph of $y = x^{5/3} + x^{1/3}$ at which the tangent line is perpendicular to the line $2y + x = 7$.

Exer. 59–60: Sketch the graph of the equation and find the vertical tangent lines.

59 $y = \sqrt{x} - 4$

60 $y = x^{1/3} + 2$

61 A weather balloon is released and rises vertically such that its distance $s(t)$ above the ground during the first 10 sec of flight is given by $s(t) = 6 + 2t + t^2$, where $s(t)$ is in feet and t is in seconds. Find the velocity of the balloon at
(a) $t = 1, t = 4$, and $t = 8$
(b) the instant the balloon is 50 ft above the ground

62 A ball rolls down an inclined plane such that the distance (in centimeters) that it rolls in t seconds is given by $s(t) = 2t^3 + 3t^2 + 4$ for $0 \le t \le 3$ (see figure).
(a) Find the velocity of the ball at $t = 2$.
(b) At what time is the velocity 30 cm/sec?

Exercise 62

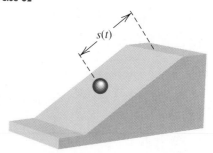

Exer. 63–64: An equation of a classical curve and its graph are given for positive constants a and b. (Consult books on analytic geometry for further information.) Find the slope of the tangent line at the point P.

63 Witch of Agnesi: $y = \dfrac{a^3}{a^2 + x^2}$; $P(a, a/2)$

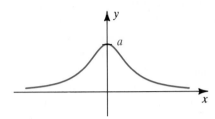

64 Serpentine curve: $y = \dfrac{abx}{a^2 + x^2}$; $P(a, b/2)$

Exer. 65–66: Find equations of the lines through P that are tangent to the graph of the equation.

65 $P(5, 9)$; $y = x^2$

66 $P(3, 1)$; $xy = 4$

Exer. 67–70: If f and g are functions such that $f(2) = 3$, $f'(2) = -1$, $g(2) = -5$, and $g'(2) = 2$, evaluate the expression.

67 (a) $(f + g)'(2)$ (b) $(f - g)'(2)$ (c) $(4f)'(2)$

     (d) $(fg)'(2)$ (e) $(f/g)'(2)$ (f) $(1/f)'(2)$

68 (a) $(g - f)'(2)$ (b) $(g/f)'(2)$

     (c) $(4g)'(2)$ (d) $(ff)'(2)$

69 (a) $(2f - g)'(2)$ (b) $(5f + 3g)'(2)$

     (c) $(gg)'(2)$ (d) $\left(\dfrac{1}{f + g}\right)'(2)$

70 (a) $(3f - 2g)'(2)$ (b) $(5/g)'(2)$

     (c) $(6f)'(2)$ (d) $\left(\dfrac{f}{f + g}\right)'(2)$

71 If f, g, and h are differentiable functions of x, use the product rule to prove that

$$\frac{d}{dx}(fgh) = f'gh + fg'h + fgh'.$$

As a corollary, let $f = g = h$ to prove that

$$\frac{d}{dx}(f(x))^3 = 3[f(x)]^2 f'(x).$$

72 Extend Exercise 71 to the derivative of a product of four functions, and then find a formula for $(d/dx)(f(x))^4$.

Exer. 73–76: Use Exercise 71 to find dy/dx.

73 $y = (8x - 1)(x^2 + 4x + 7)(x^3 - 5)$

74 $y = (3x^4 - 10x^2 + 8)(2x^2 - 10)(6x + 7)$

75 $y = x(2x^3 - 5x - 1)(6x^2 + 7)$

76 $y = 4x(x - 1)(2x - 3)$

77 As a spherical balloon is being inflated, its radius r (in centimeters) after t minutes is given by $r = 3\sqrt[3]{t}$ for $0 \le t \le 10$. Find the rate of change for each of the following with respect to t at $t = 8$:

(a) the radius r (b) the volume V of the balloon

(c) the surface area S of the balloon

78 The volume V (in cubic feet) of water in a small reservoir during spring runoff is given by the formula $V = 5000(t + 1)^2$ for t in months and $0 \le t \le 3$. The rate of change of volume with respect to time is the instantaneous *flow rate* into the reservoir. Find the flow rate at times $t = 0$ and $t = 2$. What is the flow rate when the volume is 11,250 ft^3?

79 A stone is dropped into a pond, causing water waves that form concentric circles. If, after t seconds, the radius of one of the waves is $40t$ centimeters, find the rate of change, with respect to t, of the area of the circle caused by the wave at

(a) $t = 1$ (b) $t = 2$ (c) $t = 3$

80 Boyle's law for confined gases states that if the temperature remains constant, then $pv = c$, where p is the pressure, v is the volume, and c is a constant. Suppose that at time t (in minutes) the pressure is $20 + 2t$ centimeters of mercury for $0 \le t \le 10$. If the volume is 60 cm^3 at $t = 0$, find the rate at which the volume is changing with respect to t at $t = 5$.

81 When a bright light is directed toward the eye, the pupil contracts. Suppose that the relationship between R, the area of the pupil (in square millimeters), and x, the brightness of the light source (in lumens), is given by

$$R = \frac{40 + 23.7x^4}{1 + 3.95x^4}.$$

The rate of change dR/dx is called the *sensitivity at stimulus level x*.

(a) Show that R decreases from 40 to 6 as x increases without bound.

(b) Find a formula for the sensitivity as a function of x.

c **(c)** Using the result of part (b), plot the graph of the sensitivity as a function of x for $x \geq 0$. Approximate the value of x for which the absolute value of the sensitivity is largest.

82 To win a point in racquetball, a player must have the serve and then win a rally. If p is the probability of winning a rally, then the probability S of achieving a shutout (winning 21–0) is given by

$$S = \frac{1+p}{2}\left(\frac{p}{1-p+p^2}\right)^{21},$$

provided each player is equally likely to have the initial serve. Note that a probability is always a number between 0 and 1.

(a) Find the rate of change of S with respect to p.

c **(b)** Estimate this rate of change when $p = \frac{1}{2}$.

c **83 (a)** If $f(x) = x^3 - 2x + 2$, approximate $f'(1)$ using Exercise 53 of Section 2.2 with $h = 0.1$.

(b) Graph the following on the same coordinate axes: $y = f(x)$, the secant line l_1 through $(1, f(1))$ and $(1.1, f(1.1))$, and the secant line l_2 through $(0.9, f(0.9))$ and $(1.1, f(1.1))$.

(c) Find $f'(1)$ and explain why the slope of l_2 is a better approximation to $f'(1)$ than is the slope of l_1.

c **84 (a)** If $f(x) = x^{2/3} + 1$, approximate $f'(0)$ using Exercise 53 of Section 2.2 with $h = 0.1$.

(b) Graph the following on the same coordinate axes: $y = f(x)$, the secant line through $(0, f(0))$ and $(0.1, f(0.1))$, and the secant line through $(-0.1, f(-0.1))$ and $(0.1, f(0.1))$.

(c) Why don't the slopes of the secant lines in part (b) approximate $f'(0)$?

2.4 DERIVATIVES OF THE TRIGONOMETRIC FUNCTIONS

In this section, we examine limits and derivatives involving the trigonometric functions. To obtain formulas for the derivatives of these functions, we must first prove several results about limits. Whenever we discuss limits of trigonometric expressions involving $\sin \theta$, $\cos t$, $\tan x$, and so on, *we shall assume that each variable represents the radian measure of an angle or a real number.*

Let θ denote an angle in standard position on a rectangular coordinate system, and consider the unit circle U in Figure 2.25. According to the definition of the sine and cosine functions, the coordinates of the indicated point P are $(\cos \theta, \sin \theta)$. It appears that if $\theta \to 0$, then $\sin \theta \to 0$ and $\cos \theta \to 1$. This suggests the theorem on the following page.

Figure 2.25

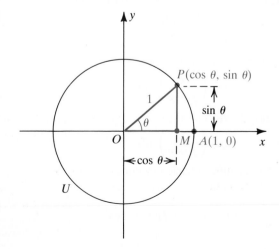

Theorem 2.22

(i) $\lim\limits_{\theta \to 0} \sin \theta = 0$

(ii) $\lim\limits_{\theta \to 0} \cos \theta = 1$

PROOF

(i) Let us first show that $\lim_{\theta \to 0^+} \sin \theta = 0$. If $0 < \theta < \pi/2$, then, referring to Figure 2.25, we see that

$$0 < MP < \overset{\frown}{AP},$$

where MP denotes the length of the line segment joining M to P and $\overset{\frown}{AP}$ denotes the length of the circular arc between A and P. By the definition of radian measure of an angle, $\overset{\frown}{AP} = \theta$, and therefore the preceding inequality can be written

$$0 < \sin \theta < \theta.$$

Since $\lim_{\theta \to 0^+} \theta = 0$ and $\lim_{\theta \to 0^+} 0 = 0$, it follows from the sandwich theorem (1.15) that $\lim_{\theta \to 0^+} \sin \theta = 0$.

To complete the proof of (i), it suffices to show that $\lim_{\theta \to 0^-} \sin \theta = 0$. If $-\pi/2 < \theta < 0$, then $0 < -\theta < \pi/2$ and hence, from the first part of the proof,

$$0 < \sin(-\theta) < -\theta.$$

Using the trigonometric identity $\sin(-\theta) = -\sin \theta$ and then multiplying by -1 gives us

$$\theta < \sin \theta < 0.$$

Since $\lim_{\theta \to 0^-} \theta = 0$ and $\lim_{\theta \to 0^-} 0 = 0$, it follows from the sandwich theorem that $\lim_{\theta \to 0^-} \sin \theta = 0$.

(ii) Using $\sin^2 \theta + \cos^2 \theta = 1$, we obtain $\cos \theta = \pm\sqrt{1 - \sin^2 \theta}$. If $-\pi/2 < \theta < \pi/2$, then $\cos \theta$ is positive, and hence $\cos \theta = \sqrt{1 - \sin^2 \theta}$. Consequently,

$$\begin{aligned} \lim_{\theta \to 0} \cos \theta &= \lim_{\theta \to 0} \sqrt{1 - \sin^2 \theta} \\ &= \sqrt{\lim_{\theta \to 0} (1 - \sin^2 \theta)} \\ &= \sqrt{1 - 0} = 1. \quad \blacksquare \end{aligned}$$

In Section 1.1, we used a calculator and a graph to guess the limit stated in the next theorem (see page 86), which we can now prove.

Theorem 2.23

$$\lim_{\theta \to 0} \frac{\sin \theta}{\theta} = 1$$

Figure 2.26

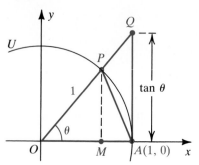

PROOF If $0 < \theta < \pi/2$, we have the situation illustrated in Figure 2.26, where U is a unit circle. Note that

$$MP = \sin\theta \quad \text{and} \quad AQ = \tan\theta.$$

From the figure, we see that

$$\text{Area of } \triangle AOP < \text{area of sector } AOP < \text{area of } \triangle AOQ.$$

From geometry and trigonometry,

$$\text{Area of } \triangle AOP = \tfrac{1}{2}bh = \tfrac{1}{2}(1)(MP) = \tfrac{1}{2}\sin\theta,$$
$$\text{Area of sector} AOP = \tfrac{1}{2}r^2\theta = \tfrac{1}{2}(1)^2\theta = \tfrac{1}{2}\theta,$$
$$\text{Area of } \triangle AOQ = \tfrac{1}{2}bh = \tfrac{1}{2}(1)(AQ) = \tfrac{1}{2}\tan\theta.$$

Hence the preceding inequality may be written

$$\tfrac{1}{2}\sin\theta < \tfrac{1}{2}\theta < \tfrac{1}{2}\tan\theta.$$

Using the identity $\tan\theta = (\sin\theta)/(\cos\theta)$ and then dividing by $\tfrac{1}{2}\sin\theta$ leads to the following equivalent inequalities:

$$1 < \frac{\theta}{\sin\theta} < \frac{1}{\cos\theta}$$

$$1 > \frac{\sin\theta}{\theta} > \cos\theta$$

$$\cos\theta < \frac{\sin\theta}{\theta} < 1$$

The last inequality is also true if $-\pi/2 < \theta < 0$, for in this case we have $0 < -\theta < \pi/2$ and hence

$$\cos(-\theta) < \frac{\sin(-\theta)}{-\theta} < 1.$$

Using the identities $\cos(-\theta) = \cos\theta$ and $\sin(-\theta) = -\sin\theta$, we again obtain

$$\cos\theta < \frac{\sin\theta}{\theta} < 1.$$

Since $\lim_{\theta\to 0}\cos\theta = 1$ and $\lim_{\theta\to 0}1 = 1$, the statement of the theorem follows from the sandwich theorem. ■

We shall also make use of the following result.

Theorem 2.24

$$\lim_{\theta\to 0}\frac{1-\cos\theta}{\theta} = 0$$

PROOF If we let $\theta = 0$ in the expression $(1-\cos\theta)/\theta$, we obtain $0/0$. Hence we must change the form of the quotient. Remembering from trigonometry that $1 - \cos^2\theta = \sin^2\theta$, we multiply the numerator and the denominator of the expression by $1 + \cos\theta$ and then simplify as follows:

$$\frac{1 - \cos\theta}{\theta} = \frac{1 - \cos\theta}{\theta} \cdot \frac{1 + \cos\theta}{1 + \cos\theta}$$

$$= \frac{1 - \cos^2\theta}{\theta(1 + \cos\theta)} = \frac{\sin^2\theta}{\theta(1 + \cos\theta)} = \frac{\sin\theta}{\theta} \cdot \frac{\sin\theta}{1 + \cos\theta}$$

Consequently,

$$\lim_{\theta \to 0} \frac{1 - \cos\theta}{\theta} = \lim_{\theta \to 0} \left(\frac{\sin\theta}{\theta} \cdot \frac{\sin\theta}{1 + \cos\theta} \right)$$

$$= \lim_{\theta \to 0} \frac{\sin\theta}{\theta} \cdot \lim_{\theta \to 0} \frac{\sin\theta}{1 + \cos\theta}$$

$$= 1 \cdot \frac{0}{1 + 1} = 1 \cdot 0 = 0. \quad \blacksquare$$

The next three examples illustrate the use of Theorems (2.22), (2.23), and (2.24) when finding limits of trigonometric expressions.

EXAMPLE ▪ 1 Find $\displaystyle\lim_{x \to 0} \frac{\sin 5x}{2x}$.

SOLUTION We cannot apply Theorem (2.23) directly, since the given expression is not in the form $(\sin t)/t$. However, we may introduce this form (with $t = 5x$) by using the following algebraic manipulation:

$$\lim_{x \to 0} \frac{\sin 5x}{2x} = \lim_{x \to 0} \frac{1}{2} \frac{\sin 5x}{x}$$

$$= \lim_{x \to 0} \frac{5}{2} \frac{\sin 5x}{5x}$$

$$= \frac{5}{2} \lim_{x \to 0} \frac{\sin 5x}{5x}$$

It follows from the definition of limit that $x \to 0$ may be replaced by $5x \to 0$. Hence, by Theorem (2.23), with $t = 5x$, we see that

$$\frac{5}{2} \lim_{x \to 0} \frac{\sin 5x}{5x} = \frac{5}{2}(1) = \frac{5}{2}.$$

EXAMPLE ▪ 2 Find $\displaystyle\lim_{t \to 0} \frac{\tan t}{2t}$.

SOLUTION Using the fact that $\tan t = (\sin t)/(\cos t)$ yields

$$\lim_{t \to 0} \frac{\tan t}{2t} = \lim_{t \to 0} \left(\frac{1}{2} \cdot \frac{\sin t}{t} \cdot \frac{1}{\cos t} \right)$$

$$= \tfrac{1}{2} \cdot 1 \cdot \tfrac{1}{1} = \tfrac{1}{2}.$$

EXAMPLE ■ 3   Find $\displaystyle\lim_{x \to 0} \frac{2x + 1 - \cos x}{3x}$.

SOLUTION   We plan to use Theorem (2.24). With this in mind, we begin by isolating the part of the quotient that involves $(1 - \cos x)/x$ and then proceed as follows:

$$\lim_{x \to 0} \frac{2x + 1 - \cos x}{3x} = \lim_{x \to 0} \left(\frac{2x}{3x} + \frac{1 - \cos x}{3x} \right)$$

$$= \lim_{x \to 0} \left(\frac{2x}{3x} \right) + \lim_{x \to 0} \frac{1}{3} \left(\frac{1 - \cos x}{x} \right)$$

$$= \lim_{x \to 0} \frac{2}{3} + \frac{1}{3} \lim_{x \to 0} \frac{1 - \cos x}{x}$$

$$= \tfrac{2}{3} + \tfrac{1}{3} \cdot 0 = \tfrac{2}{3}$$

We may now establish the formulas listed in the following theorem, where x *denotes a real number or the radian measure of an angle.*

Derivatives of the Trigonometric Functions **2.25**

$$\frac{d}{dx}(\sin x) = \cos x \qquad \frac{d}{dx}(\cos x) = -\sin x$$

$$\frac{d}{dx}(\tan x) = \sec^2 x \qquad \frac{d}{dx}(\cot x) = -\csc^2 x$$

$$\frac{d}{dx}(\sec x) = \sec x \tan x \qquad \frac{d}{dx}(\csc x) = -\csc x \cot x$$

PROOF   Applying Definition (2.5) with $f(x) = \sin x$ and then using the addition formula for the sine function, we obtain

$$\frac{d}{dx}(\sin x) = \lim_{h \to 0} \frac{\sin(x + h) - \sin x}{h}$$

$$= \lim_{h \to 0} \frac{\sin x \cos h + \cos x \sin h - \sin x}{h}$$

$$= \lim_{h \to 0} \frac{\sin x (\cos h - 1) + \cos x \sin h}{h}$$

$$= \lim_{h \to 0} \left[\sin x \left(\frac{\cos h - 1}{h} \right) + \cos x \left(\frac{\sin h}{h} \right) \right].$$

By Theorems (2.24) and (2.23),

$$\lim_{h \to 0} \frac{\cos h - 1}{h} = 0 \quad \text{and} \quad \lim_{h \to 0} \frac{\sin h}{h} = 1,$$

and hence $\displaystyle\frac{d}{dx}(\sin x) = (\sin x)(0) + (\cos x)(1) = \cos x.$

We have shown that the derivative of the sine function is the cosine function. We may obtain the derivative of the cosine function in similar fashion:

$$\frac{d}{dx}(\cos x) = \lim_{h \to 0} \frac{\cos(x + h) - \cos x}{h}$$

$$= \lim_{h \to 0} \frac{\cos x \cos h - \sin x \sin h - \cos x}{h}$$

$$= \lim_{h \to 0} \frac{\cos x(\cos h - 1) - \sin x \sin h}{h}$$

$$= \lim_{h \to 0} \left[\cos x \left(\frac{\cos h - 1}{h} \right) - \sin x \left(\frac{\sin h}{h} \right) \right]$$

$$= (\cos x)(0) - (\sin x)(1) = -\sin x$$

Thus the derivative of the cosine function is the *negative* of the sine function.

To find the derivative of the tangent function, we begin with the fundamental identity $\tan x = (\sin x)/(\cos x)$ and then apply the quotient rule as follows:

$$\frac{d}{dx}(\tan x) = \frac{d}{dx} \left(\frac{\sin x}{\cos x} \right)$$

$$= \frac{\cos x \dfrac{d}{dx}(\sin x) - \sin x \dfrac{d}{dx}(\cos x)}{\cos^2 x}$$

$$= \frac{\cos x \cos x - \sin x(-\sin x)}{\cos^2 x}$$

$$= \frac{\cos^2 x + \sin^2 x}{\cos^2 x} = \frac{1}{\cos^2 x} = \sec^2 x$$

For the secant function, we first write $\sec x = 1/\cos x$ and then use the reciprocal rule (2.21):

$$\frac{d}{dx}(\sec x) = \frac{d}{dx} \left(\frac{1}{\cos x} \right)$$

$$= -\frac{\dfrac{d}{dx}(\cos x)}{\cos^2 x}$$

$$= -\frac{-\sin x}{\cos^2 x}$$

$$= \frac{\sin x}{\cos^2 x}$$

$$= \frac{1}{\cos x} \frac{\sin x}{\cos x} = \sec x \tan x$$

Proofs of the formulas for $(d/dx)(\cot x)$ and $(d/dx)(\csc x)$ are left as exercises. ∎

We can use (2.25) to obtain information about the continuity of the trigonometric functions. For example, since the sine and cosine functions are differentiable at every real number, it follows from Theorem (2.12) that these functions are continuous throughout $\mathbb{R}$. Similarly, the tangent function is continuous on the open intervals $(-\pi/2, \pi/2), (\pi/2, 3\pi/2)$, and so on, since it is differentiable at each number in these intervals.

EXAMPLE■4   Find y' if $y = \dfrac{\sin x}{1 + \cos x}$.

SOLUTION   By the quotient rule and (2.25),

$$y' = \frac{(1 + \cos x)\dfrac{d}{dx}(\sin x) - \sin x \dfrac{d}{dx}(1 + \cos x)}{(1 + \cos x)^2}$$

$$= \frac{(1 + \cos x)\cos x - \sin x(0 - \sin x)}{(1 + \cos x)^2}$$

$$= \frac{\cos x + \cos^2 x + \sin^2 x}{(1 + \cos x)^2}$$

$$= \frac{\cos x + 1}{(1 + \cos x)^2}$$

$$= \frac{1}{1 + \cos x}.$$

In the solution to Example 4, we used the fundamental identity $\cos^2 x + \sin^2 x = 1$. This and other trigonometric identities are often useful in simplifying problems that involve derivatives of trigonometric functions.

EXAMPLE■5   Find $g'(x)$ if $g(x) = \sec x \tan x$.

SOLUTION   By the product rule and (2.25),

$$g'(x) = \sec x \frac{d}{dx}(\tan x) + \tan x \frac{d}{dx}(\sec x)$$

$$= \sec x \sec^2 x + \tan x(\sec x \tan x)$$

$$= \sec^3 x + \sec x \tan^2 x$$

$$= \sec x(\sec^2 x + \tan^2 x).$$

The formula for $g'(x)$ can be written in many other ways. For example, because

$$\sec^2 x = \tan^2 x + 1, \quad \text{or} \quad \tan^2 x = \sec^2 x - 1,$$

we can write

$$g'(x) = \sec x(2 \tan^2 x + 1), \quad \text{or} \quad g'(x) = \sec x(2 \sec^2 x - 1).$$

E X A M P L E ▪ 6   Find $dy/d\theta$ if $y = \sec\theta\cot\theta$.

S O L U T I O N   We could use the product rule as in Example 5; however, it is simpler to first change the form of y by using fundamental identities as follows:

$$y = \sec\theta\cot\theta = \frac{1}{\cos\theta}\frac{\cos\theta}{\sin\theta} = \frac{1}{\sin\theta} = \csc\theta$$

Applying (2.25) yields

$$\frac{dy}{d\theta} = \frac{d}{d\theta}(\csc\theta) = -\csc\theta\cot\theta.$$

E X A M P L E ▪ 7

(a) Find the slopes of the tangent lines to the graph of $y = \sin x$ at the points with x-coordinates 0, $\pi/3$, $\pi/2$, $2\pi/3$, and π.

(b) Sketch the graph of $y = \sin x$ and the tangent lines of part (a).

(c) For what values of x is the tangent line horizontal?

S O L U T I O N

(a) The slope of the tangent line at the point (x, y) on the graph of the equation $y = \sin x$ is given by the derivative $y' = \cos x$. The slopes at the desired points are listed in the following table.

x	0	$\dfrac{\pi}{3}$	$\dfrac{\pi}{2}$	$\dfrac{2\pi}{3}$	π
$y' = \cos x$	1	$\dfrac{1}{2}$	0	$-\dfrac{1}{2}$	-1

(b) A portion of the graph of $y = \sin x$ and the tangent lines of part (a) are sketched in Figure 2.27.

Figure 2.27

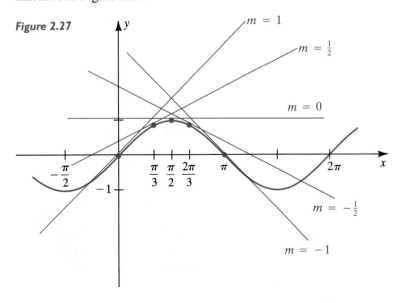

(c) A tangent line is horizontal if its slope is zero. Since the slope of the tangent line at the point (x, y) is y', we must solve the equation

$$y' = 0; \quad \text{that is,} \quad \cos x = 0.$$

Thus the tangent line is horizontal if $x = \pm\pi/2$, $x = \pm 3\pi/2$, and, in general, if $x = (\pi/2) + \pi n$ for any integer n.

Figure 2.28

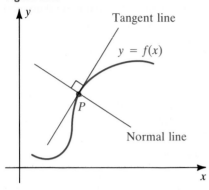

If f is a differentiable function, then the **normal line** at a point $P(a, f(a))$ on the graph of f is the line through P that is perpendicular to the tangent line, as illustrated in Figure 2.28. If $f'(a) \neq 0$, then, by (9)(iii) on page 15, the slope of the normal line is $-1/f'(a)$. If $f'(a) = 0$, then the tangent line is horizontal, and in this case the normal line is vertical and has the equation $x = a$.

EXAMPLE■8 Find an equation of the normal line to the graph of $y = \tan x$ at the point $P(\pi/4, 1)$, and illustrate it graphically.

Figure 2.29

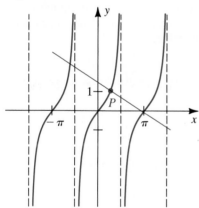

SOLUTION Since $y' = \sec^2 x$, the slope m of the tangent line at P is

$$m = \sec^2 \frac{\pi}{4} = (\sqrt{2})^2 = 2$$

and hence the slope of the normal line is $-1/m = -1/2$.

Using the point–slope form, we can express an equation for the normal line as

$$y - 1 = -\frac{1}{2}\left(x - \frac{\pi}{4}\right),$$

or

$$y = -\frac{1}{2}x + \frac{\pi}{8} + 1.$$

The graph of $y = \tan x$ for $-3\pi/2 < x < 3\pi/2$ and the normal line at P are sketched in Figure 2.29.

The next example illustrates an application involving the derivatives of the trigonometric functions.

EXAMPLE■9 If a projectile is fired from ground level with an initial velocity of v ft/sec and at an angle of θ, then the range R of the projectile is given by

$$R = \frac{v^2}{16} \sin\theta \cos\theta \quad \text{for } 0 < \theta < \frac{\pi}{2}.$$

(a) If $v = 80$ ft/sec, find the rate of change of R with respect to θ.

(b) Determine the values of θ for which the rate of change in part (a) is positive.

SOLUTION

(a) The rate of change of R with respect to θ is given by $dR/d\theta$. With $v = 80$, we have $R = 400 \sin \theta \cos \theta$. Using the product rule (2.19) gives

$$\frac{dR}{d\theta} = 400\left[\sin \theta \frac{d}{d\theta}(\cos \theta) + \cos \theta \frac{d}{d\theta}(\sin \theta)\right]$$

$$= 400[\sin \theta(-\sin \theta) + \cos \theta \cos \theta]$$

$$= 400[-\sin^2 \theta + \cos^2 \theta],$$

which we can rewrite, using a double-angle formula, as $400 \cos 2\theta$. Thus the rate of change of R with respect to θ is $400 \cos 2\theta$ feet per radian. For a particular value of θ, $dR/d\theta$ is an estimate of the number of feet the range of the projectile changes for each radian change in θ.

(b) From part (a), $dR/d\theta = 400 \cos 2\theta$, which is positive if $\cos 2\theta > 0$; that is, if $0 \le 2\theta < \pi/2$, which corresponds to $0 \le \theta < \pi/4$. If θ is a positive angle less than $\pi/4$, then a small increase in θ means a positive value for the rate of change of the projectile's range. Figure 2.30 shows a graph of the function $R = 400 \sin \theta \cos \theta$. We see that R is increasing for values of θ between 0 and $\pi/4$ and R is decreasing for values of θ between $\pi/4$ and $\pi/2$. Do not mistake the graph in Figure 2.30 with the graph of the path of the projectile.

Figure 2.30

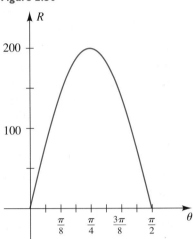

EXERCISES 2.4

Exer. 1–26: Find the limit, if it exists.

1 $\displaystyle\lim_{x \to 0} \frac{x}{\sin x}$

2 $\displaystyle\lim_{x \to 0} \frac{\sin x}{\sqrt[3]{x}}$

3 $\displaystyle\lim_{t \to 0} \frac{\sin^3 t}{(2t)^3}$

4 $\displaystyle\lim_{\theta \to 0} \frac{3\theta + \sin \theta}{\theta}$

5 $\displaystyle\lim_{\to 0} \frac{2 + \sin x}{3 + x}$

6 $\displaystyle\lim_{t \to 0} \frac{1 - \cos 3t}{t}$

7 $\displaystyle\lim_{\theta \to 0} \frac{2 \cos \theta - 2}{3\theta}$

8 $\displaystyle\lim_{x \to 0} \frac{x^2 + 1}{x + \cos x}$

9 $\displaystyle\lim_{x \to 0} \frac{\sin(-3x)}{4x}$

10 $\displaystyle\lim_{x \to 0} \frac{x \sin x}{x^2 + 1}$

11 $\displaystyle\lim_{x \to 0} \frac{1 - \cos x}{x^{2/3}}$

12 $\displaystyle\lim_{x \to 0} \frac{1 - 2x^2 - 2\cos x + \cos^2 x}{x^2}$

13 $\displaystyle\lim_{t \to 0} \frac{4t^2 + 3t \sin t}{t^2}$

14 $\displaystyle\lim_{x \to 0} \frac{x \cos x - x^2}{2x}$

15 $\displaystyle\lim_{t \to 0} \frac{\cos t}{1 - \sin t}$

16 $\displaystyle\lim_{t \to 0} \frac{\sin t}{1 + \cos t}$

17 $\displaystyle\lim_{t \to 0} \frac{1 - \cos t}{\sin t}$

18 $\displaystyle\lim_{x \to 0} \frac{\sin \frac{1}{2}x}{x}$

19 $\displaystyle\lim_{x \to 0} \frac{x + \tan x}{\sin x}$

20 $\displaystyle\lim_{t \to 0} \frac{\sin^2 2t}{t^2}$

21 $\displaystyle\lim_{x \to 0} x \cot x$

22 $\displaystyle\lim_{x \to 0} \frac{\csc 2x}{\cot x}$

23 $\displaystyle\lim_{\alpha \to 0} \alpha^2 \csc^2 \alpha$

24 $\displaystyle\lim_{x \to 0} \frac{\sin 3x}{\sin 5x}$

25 $\displaystyle\lim_{v \to 0} \frac{\cos(v + \frac{1}{2}\pi)}{v}$

26 $\displaystyle\lim_{x \to 0} \frac{\sin^2 \frac{1}{2}x}{\sin x}$

Exer. 27–30: Establish the limit for all nonzero real numbers a and b.

27 $\displaystyle\lim_{x \to 0} \frac{\sin ax}{bx} = \frac{a}{b}$

28 $\displaystyle\lim_{x \to 0} \frac{1 - \cos ax}{bx} = 0$

29 $\displaystyle\lim_{x \to 0} \frac{\sin ax}{\sin bx} = \frac{a}{b}$

30 $\displaystyle\lim_{x \to 0} \frac{\cos ax}{\cos bx} = 1$

Exer. 31–58: Find the derivative.

31 $f(x) = 4 \cos x$

32 $H(z) = 7 \tan z$

33 $G(v) = 5v \csc v$

34 $f(x) = 3x \sin x$

35 $k(t) = t - t^2 \cos t$

36 $p(w) = w^2 + w \sin w$

37 $f(\theta) = \dfrac{\sin \theta}{\theta}$

38 $g(\alpha) = \dfrac{1 - \cos \alpha}{\alpha}$

39 $g(t) = t^3 \sin t$

40 $T(r) = r^2 \sec r$

41 $f(x) = 2x \cot x + x^2 \tan x$

42 $f(x) = 3x^2 \sec x - x^3 \tan x$

43 $h(z) = \dfrac{1 - \cos z}{1 + \cos z}$

44 $R(w) = \dfrac{\cos w}{1 - \sin w}$

45 $g(x) = \dfrac{1}{\sin x \tan x}$

46 $k(x) = \dfrac{1}{\cos x \cot x}$

47 $g(x) = (x + \csc x) \cot x$

48 $K(\theta) = (\sin \theta + \cos \theta)^2$

49 $p(x) = \sin x \cot x$

50 $g(t) = \csc t \sin t$

51 $f(x) = \dfrac{\tan x}{1 + x^2}$

52 $h(\theta) = \dfrac{1 + \sec \theta}{1 - \sec \theta}$

53 $k(v) = \dfrac{\csc v}{\sec v}$

54 $q(t) = \sin t \sec t$

55 $g(x) = \sin(-x) + \cos(-x)$

56 $s(z) = \tan(-z) + \sec(-z)$

57 $H(\phi) = (\cot \phi + \csc \phi)(\tan \phi - \sin \phi)$

58 $f(x) = \dfrac{1 + \sec x}{\tan x + \sin x}$

Exer. 59–60: Find equations of the tangent line and the normal line to the graph of f at the point $(\pi/4, f(\pi/4))$.

59 $f(x) = \sec x$

60 $f(x) = \csc x + \cot x$

Exer. 61–64: Shown is a graph of the function f with restricted domain. Find the points at which the tangent line is horizontal.

61 $f(x) = \cos x + \sin x, \quad 0 \le x \le 2\pi$

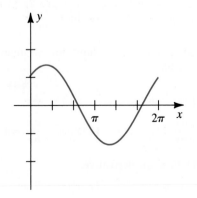

62 $f(x) = \cos x - \sin x, \quad 0 \le x \le 2\pi$

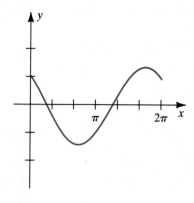

63 $f(x) = \csc x + \sec x, \quad 0 < x < \pi/2$

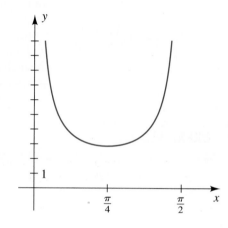

64 $f(x) = 2 \sec x - \tan x, \quad -\pi/2 < x < \pi/2$

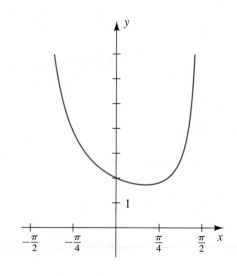

Exer. 65–66: (a) Find the x-coordinates of all points on the graph of f at which the tangent line is horizontal. (b) Find an equation of the tangent line to the graph of f at P.

65 $f(x) = x + 2\cos x$; $P(0, f(0))$

66 $f(x) = x + \sin x$; $P(\pi/2, f(\pi/2))$

67 If $y = 3 + 2\sin x$, find

    (a) the x-coordinates of all points on the graph at which the tangent line is parallel to the line $y = \sqrt{2}x - 5$

    (b) an equation of the tangent line to the graph at the point on the graph with x-coordinate $\pi/6$

68 If $y = 1 + 2\cos x$, find

    (a) the x-coordinates of all points on the graph at which the tangent line is perpendicular to the line

$$y = \frac{1}{\sqrt{3}}x + 4$$

    (b) an equation of the tangent line to the graph at the point where the graph crosses the y-axis

69 Graph $f(x) = |\sin^2 x - \cos x \sin(\frac{1}{2}\pi x)|$ on the interval $[0, 5]$ and estimate where f is not differentiable.

70 Graph $f(x) = 4/(16\sin 2x - x)$ on the interval $[0, 4]$ and estimate the x-coordinates of points at which the tangent line is horizontal.

Exer. 71–72: A point P moving on a coordinate line l has the given position function s. When is its velocity 0?

71 $s(t) = t + 2\cos t$ **72** $s(t) = t - \sqrt{2}\sin t$

Exer. 73–74: A point $P(x, y)$ is moving from left to right along the graph of the equation. Where is the rate of change of y with respect to x equal to the given number a?

73 $y = x^{3/2} + 2x$; $a = 8$

74 $y = x^{5/3} - 10x$; $a = 5$

75 (a) Find the first four derivatives of $f(x) = \cos x$.

    (b) Find $f^{(99)}(x)$.

76 Find $f'''(x)$ if $f(x) = \cot x$.

77 Find $\dfrac{d^3y}{dx^3}$ if $y = \tan x$.

78 Find $\dfrac{d^3y}{dx^3}$ if $y = \sec x$.

Exer. 79–82: Prove each formula.

79 $\dfrac{d}{dx}(\cot x) = -\csc^2 x$

80 $\dfrac{d}{dx}(\csc x) = -\csc x \cot x$

81 $\dfrac{d}{dx}(\sin 2x) = 2\cos 2x$ (*Hint:* $\sin 2x = 2\sin x \cos x$.)

82 $\dfrac{d}{dx}(\cos 2x) = -2\sin 2x$ (*Hint:* $\cos 2x = 1 - 2\sin^2 x$.)

83 Use Theorem (2.22) and the addition formula for the sine to show that the sine function is continuous at $x = a$. (*Hint:* Show $\lim_{h \to 0} \sin(a + h) = \sin a$.)

84 Work Exercise 83 for the cosine rather than the sine.

2.5 THE CHAIN RULE

In this section, we will study perhaps the single most powerful tool for differentiation: the chain rule for differentiating composite functions. The rules for derivatives obtained in previous sections are limited in scope because they can be used only for sums, differences, products, and quotients that involve x^n, $\sin x$, $\cos x$, $\tan x$, and so on. There is no rule that can be applied *directly* to expressions such as $\sin 2x$ or $\sqrt{x^2 + 1}$. Note that

$$\frac{d}{dx}(\sin 2x) \neq \cos 2x,$$

for if we use the identity $\sin 2x = 2\sin x \cos x$ and apply the product rule, as in Example 9 of Section 2.4, we obtain

$$\frac{d}{dx}(\sin 2x) = 2\cos 2x.$$

Since these manipulations are rather cumbersome, we seek a more direct method of finding the derivative of $y = \sin 2x$. The key is to regard y as a composite function of x. Thus, for functions f and g,

$$\text{if} \quad y = f(u) \quad \text{and} \quad u = g(x), \quad \text{then} \quad y = f(g(x)),$$

provided $g(x)$ is in the domain of f. The function given by $y = f(g(x))$ is the composite function $f \circ g$ defined on p. 29. The functions f and g are called the *components* of the composition. Note that $y = \sin 2x$ may be expressed in this way because

$$\text{if} \quad y = \sin u \quad \text{and} \quad u = 2x, \quad \text{then} \quad y = \sin 2x.$$

If we can find a general rule for differentiating $y = f(g(x))$, then, as a special case, we may apply it to $y = \sin 2x$ and, in fact, to $y = \sin g(x)$ for any differentiable function g.

To get an idea of the type of rule to expect, let us consider a composite function that we can easily differentiate by changing its form. If $y = (x^3 - 1)^2$, then we can represent y in a composite function form by letting $u = g(x) = x^3 - 1$ and $f(u) = u^2$ so that $y = f(g(x))$. We can find the derivative of y with respect to x by expanding the original expression for y to obtain

$$y = u^2 = (x^3 - 1)^2 = x^6 - 2x^3 + 1,$$

and then differentiating, term by term, to obtain

$$y' = [f(g(x))]' = \frac{dy}{dx} = 6x^5 - 6x^2,$$

which can be written as

$$6x^5 - 6x^2 = 6x^2(x^3 - 1) = 2(x^3 - 1)(3x^2).$$

Note here that

$$2(x^3 - 1) = 2u = \frac{dy}{du} = f'(g(x)) \quad \text{and} \quad 3x^2 = \frac{du}{dx} = g'(x),$$

so we can write the derivative of $f(g(x))$ in the following equivalent forms:

$$f'(u) = f'(u)g'(x) \quad \text{or} \quad [f(g(x))]' = f'(g(x))g'(x) \quad \text{or} \quad \frac{dy}{dx} = \frac{dy}{du}\frac{du}{dx}.$$

These results suggest a rule indicating that the derivative of the composite function is a product of derivatives of the component functions.

Note too that this rule also holds true for the derivative $y = \sin 2x$, for if we write

$$y = \sin u \quad \text{and} \quad u = 2x$$

and use the suggested rule, we obtain

$$\frac{dy}{dx} = \frac{dy}{du}\frac{du}{dx} = (\cos u)(2) = 2\cos u = 2\cos 2x.$$

The fact that the same rule gives the correct answer for both of these examples of composite functions is no accident. The *chain rule* for differentiation states that the derivative of a composite function is always the product of the derivatives of the component functions, provided they exist.

Chain Rule 2.26

> If $y = f(u)$, $u = g(x)$, and the derivatives dy/du and du/dx both exist, then the composite function defined by $y = f(g(x))$ has a derivative given by
>
> $$\frac{dy}{dx} = \frac{dy}{du}\frac{du}{dx} = f'(u)g'(x) = f'(g(x))g'(x).$$

PARTIAL PROOF Using Definition (2.5), we must show that

$$\lim_{h \to 0} \frac{f(g(x+h)) - f(g(x))}{h} = f'(g(x))g'(x).$$

If h is close to 0 and $g(x+h) \neq g(x)$, we can write the left-hand side of the equation as

$$\lim_{h \to 0} \left[\frac{f(g(x+h)) - f(g(x))}{g(x+h) - g(x)} \cdot \frac{g(x+h) - g(x)}{h} \right].$$

If each factor has a limit, then by Theorem (1.8)(ii), the above limit can be written as

$$\lim_{h \to 0} \frac{f(g(x+h)) - f(g(x))}{g(x+h) - g(x)} \lim_{h \to 0} \frac{g(x+h) - g(x)}{h}.$$

By Definition (2.5),

$$\lim_{h \to 0} \frac{g(x+h) - g(x)}{h} = g'(x),$$

so it remains to be shown that

$$\lim_{h \to 0} \frac{f(g(x+h)) - f(g(x))}{g(x+h) - g(x)} = f'(g(x)).$$

To see why this last result is true, first note that since g is differentiable, it is continuous. Thus, as $h \to 0$, $g(x+h) \to g(x)$. Note too that Alternative Definition (2.6) can be written in the form

$$f'(a) = \lim_{t \to a} \frac{f(t) - f(a)}{t - a}.$$

If we now let $a = g(x)$ and $t = g(x+h)$,

$$\lim_{h \to 0} \frac{f(g(x+h)) - f(g(x))}{g(x+h) - g(x)} = \lim_{h \to 0} \frac{f(t) - f(a)}{t - a}$$

$$= \lim_{g(x+h) \to g(x)} \frac{f(t) - f(a)}{t - a}$$

$$= \lim_{t \to a} \frac{f(t) - f(a)}{t - a}$$

$$= f'(a) = f'(g(x)),$$

which is what we needed to prove. ■

In many applications of the chain rule, the function $u = g(x)$ has the property that $g(x + h) \neq g(x)$ for values of h sufficiently close to 0, a property that we assumed at the beginning of the proof. If g does not satisfy this property, then every open interval containing x contains a number $x + h$ for which $g(x + h) = g(x)$, so that $g(x + h) - g(x) = 0$. In such cases, our proof is invalid, since the expression $g(x + h) - g(x)$ occurs in the denominator. To construct a proof that takes functions of this type into account requires additional techniques. A complete proof of the chain rule is given in Appendix I.

EXAMPLE 1 Find $\dfrac{dy}{dx}$ if $y = \sqrt{u}$ and $u = x^2 + 1$.

SOLUTION If we substitute $x^2 + 1$ for u in $y = \sqrt{u} = u^{1/2}$, we obtain

$$y = \sqrt{x^2 + 1} = (x^2 + 1)^{1/2}.$$

We cannot find dy/dx by using previous differentiation formulas; however, using the chain rule (2.26), we have

$$\frac{dy}{dx} = \frac{dy}{du}\frac{du}{dx} = \left(\frac{1}{2}u^{-1/2}\right)2x = \frac{x}{\sqrt{u}}$$

and hence

$$\frac{dy}{dx} = \frac{x}{\sqrt{x^2 + 1}}.$$

In Example 1, the composite function was given by a power of $x^2 + 1$. Since powers of functions occur frequently in calculus, it will save us time to state a general differentiation rule that can be applied to such special cases. In the following, we assume that n is any rational number, g is a differentiable function, and zero denominators do not occur. We shall see later that the rule can be used for any *real* number n.

Power Rule for Functions 2.27

If $y = u^n$ and $u = g(x)$, then

$$\frac{d}{dx}(u^n) = nu^{n-1}\frac{du}{dx},$$

or, equivalently, $\quad \dfrac{d}{dx}(g(x))^n = n[g(x)]^{n-1}\dfrac{d}{dx}(g(x)).$

PROOF By the chain rule,

$$\frac{dy}{dx} = \frac{dy}{du}\frac{du}{dx} = nu^{n-1}\frac{du}{dx} = n[g(x)]^{n-1}\frac{d}{dx}(g(x)). \quad \blacksquare$$

Note that if $u = x$, then $du/dx = 1$ and (2.27) reduces to (2.10).

EXAMPLE ■ 2   Find $f'(x)$ if $f(x) = (x^5 - 4x + 8)^7$.

SOLUTION   Using the power rule (2.27) with $u = x^5 - 4x + 8$ and $n = 7$ yields

$$f'(x) = \frac{d}{dx}(x^5 - 4x + 8)^7$$

$$= 7(x^5 - 4x + 8)^6 \frac{d}{dx}(x^5 - 4x + 8)$$

$$= 7(x^5 - 4x + 8)^6(5x^4 - 4).$$

EXAMPLE ■ 3   Find $\dfrac{dy}{dx}$ if $y = \dfrac{1}{(4x^2 + 6x - 7)^3}$.

SOLUTION   Writing $y = (4x^2 + 6x - 7)^{-3}$ and using the power rule with $u = 4x^2 + 6x - 7$ and $n = -3$, we have

$$\frac{dy}{dx} = \frac{d}{dx}(4x^2 + 6x - 7)^{-3}$$

$$= -3(4x^2 + 6x - 7)^{-4}\frac{d}{dx}(4x^2 + 6x - 7)$$

$$= -3(4x^2 + 6x - 7)^{-4}(8x + 6)$$

$$= \frac{-6(4x + 3)}{(4x^2 + 6x - 7)^4}.$$

EXAMPLE ■ 4   Find $f'(x)$ if $f(x) = \sqrt[3]{5x^2 - x + 4}$.

SOLUTION   Writing $f(x) = (5x^2 - x + 4)^{1/3}$ and using the power rule with $u = 5x^2 - x + 4$ and $n = \frac{1}{3}$, we obtain

$$f'(x) = \frac{1}{3}(5x^2 - x + 4)^{-2/3}\frac{d}{dx}(5x^2 - x + 4)$$

$$= \left(\frac{1}{3}\right)\frac{1}{(5x^2 - x + 4)^{2/3}}(10x - 1) = \frac{10x - 1}{3\sqrt[3]{(5x^2 - x + 4)^2}}.$$

EXAMPLE ■ 5   Find $F'(z)$ if $F(z) = (2z + 5)^3(3z - 1)^4$.

SOLUTION   Using first the product rule, second the power rule, and then factoring the result gives us

$$F'(z) = (2z + 5)^3\frac{d}{dz}(3z - 1)^4 + (3z - 1)^4\frac{d}{dz}(2z + 5)^3$$

$$= (2z + 5)^3 \cdot 4(3z - 1)^3(3) + (3z - 1)^4 \cdot 3(2z + 5)^2(2)$$

$$= 6(2z + 5)^2(3z - 1)^3[2(2z + 5) + (3z - 1)]$$

$$= 6(2z + 5)^2(3z - 1)^3(7z + 9).$$

EXAMPLE ■ 6   Find y' if $y = (3x + 1)^6\sqrt{2x - 5}$.

SOLUTION   Since $y = (3x + 1)^6(2x - 5)^{1/2}$, we have, by the product and power rules,

$$y' = (3x + 1)^6 \tfrac{1}{2}(2x - 5)^{-1/2}(2) + (2x - 5)^{1/2}6(3x + 1)^5(3)$$

$$= \frac{(3x + 1)^6}{\sqrt{2x - 5}} + 18(3x + 1)^5\sqrt{2x - 5}$$

$$= \frac{(3x + 1)^6 + 18(3x + 1)^5(2x - 5)}{\sqrt{2x - 5}} = \frac{(3x + 1)^5(39x - 89)}{\sqrt{2x - 5}}.$$

The next example is of interest because it illustrates the fact that after the power rule is applied to $[g(x)]^r$, it may be necessary to apply it again in order to find $g'(x)$.

EXAMPLE ■ 7   Find $f'(x)$ if $f(x) = (7x + \sqrt{x^2 + 6})^4$.

SOLUTION   Applying the power rule yields

$$f'(x) = 4(7x + \sqrt{x^2 + 6})^3 \frac{d}{dx}(7x + \sqrt{x^2 + 6})$$

$$= 4(7x + \sqrt{x^2 + 6})^3 \left[\frac{d}{dx}(7x) + \frac{d}{dx}(\sqrt{x^2 + 6})\right].$$

Again applying the power rule, we have

$$\frac{d}{dx}(\sqrt{x^2 + 6}) = \frac{d}{dx}(x^2 + 6)^{1/2} = \frac{1}{2}(x^2 + 6)^{-1/2}\frac{d}{dx}(x^2 + 6)$$

$$= \frac{1}{2\sqrt{x^2 + 6}}(2x) = \frac{x}{\sqrt{x^2 + 6}}.$$

Therefore,

$$f'(x) = 4(7x + \sqrt{x^2 + 6})^3\left(7 + \frac{x}{\sqrt{x^2 + 6}}\right).$$

As another application of the chain rule, we can prove the following.

Theorem   2.28

If $u = g(x)$ and g is differentiable, then

$$\frac{d}{dx}(\sin u) = \cos u \frac{du}{dx} \qquad \frac{d}{dx}(\cos u) = -\sin u \frac{du}{dx}$$

$$\frac{d}{dx}(\tan u) = \sec^2 u \frac{du}{dx} \qquad \frac{d}{dx}(\cot u) = -\csc^2 u \frac{du}{dx}$$

$$\frac{d}{dx}(\sec u) = \sec u \tan u \frac{du}{dx} \qquad \frac{d}{dx}(\csc u) = -\csc u \cot u \frac{du}{dx}$$

PROOF If we let $y = \sin u$, then, by (2.25),

$$\frac{dy}{du} = \cos u.$$

Applying the chain rule (2.26) yields

$$\frac{dy}{dx} = \frac{dy}{du}\frac{du}{dx} = \cos u \frac{du}{dx}.$$

The remaining formulas may be obtained in similar fashion. ■

Note that Theorem (2.25) is the special case of Theorem (2.28) in which $u = x$.

EXAMPLE ■ 8 If $y = \cos(5x^3)$, find dy/dx and d^2y/dx^2.

SOLUTION Using the formula for $(d/dx)(\cos u)$ in Theorem (2.28) with $u = 5x^3$, we have

$$\frac{dy}{dx} = \frac{d}{dx}(\cos(5x^3))$$

$$= -\sin(5x^3)\frac{d}{dx}(5x^3)$$

$$= -\sin(5x^3)(15x^2)$$

$$= -15x^2 \sin(5x^3).$$

To find d^2y/dx^2, we differentiate $dy/dx = -15x^2 \sin(5x^3)$. Using the product rule and Theorem (2.28) gives us

$$\frac{d^2y}{dx^2} = -15x^2\frac{d}{dx}(\sin(5x^3)) + \sin(5x^3)\frac{d}{dx}(-15x^2)$$

$$= -15x^2 \cos(5x^3)\frac{d}{dx}(5x^3) + \sin(5x^3)(-30x)$$

$$= -15x^2 \cos(5x^3)(15x^2) - 30x \sin(5x^3)$$

$$= -225x^4 \cos(5x^3) - 30x \sin(5x^3).$$

EXAMPLE ■ 9 Find $f'(x)$ if $f(x) = \tan^3 4x$.

SOLUTION First note that $f(x) = \tan^3 4x = (\tan 4x)^3$. Applying the power rule with $u = \tan 4x$ and $n = 3$ yields

$$f'(x) = 3(\tan 4x)^2\frac{d}{dx}(\tan 4x) = (3 \tan^2 4x)\frac{d}{dx}(\tan 4x).$$

Next, by Theorem (2.28),

$$\frac{d}{dx}(\tan 4x) = \sec^2 4x\frac{d}{dx}(4x) = (\sec^2 4x)(4) = 4 \sec^2 4x.$$

Thus

$$f'(x) = (3 \tan^2 4x)(4 \sec^2 4x) = 12 \tan^2 4x \sec^2 4x.$$

EXAMPLE ■ 10   If $f(x) = \sin(\sin x)$, find

(a) $f'(x)$

(b) the slope of the tangent line to the graph of f at $P(1, f(1))$.

SOLUTION

(a) Let $u = \sin x$ and apply Theorem (2.28) to obtain

$$f'(x) = \frac{d}{dx}(\sin u) = \cos u \frac{du}{dx} = \cos(\sin x) \cos x.$$

(b) By Definition (2.1), the slope of the tangent line at P is $f'(1)$, which by part (a) is

$$f'(1) = \cos(\sin 1) \cos(1)$$
$$\approx \cos(0.84147) \cos(1)$$
$$\approx (0.66637)(0.54030) \approx 0.36004.$$

Compare this result with the numerical approximation discussed in Example 5 of Section 2.1.

EXAMPLE ■ 11   A graph of $y = \cos 2x + 2\cos x$ for $0 \le x \le 2\pi$ is shown in Figure 2.31. Find the points at which the tangent line is horizontal.

Figure 2.31

SOLUTION   Differentiating, we obtain

$$\frac{dy}{dx} = -\sin 2x \frac{d}{dx}(2x) + 2(-\sin x)$$
$$= -2\sin 2x - 2\sin x.$$

The tangent line is horizontal if its slope dy/dx is 0 — that is, if

$$-2\sin 2x - 2\sin x = 0, \quad \text{or} \quad \sin 2x + \sin x = 0.$$

Using the double-angle formula, $\sin 2x = 2\sin x \cos x$, gives us

$$2\sin x \cos x + \sin x = 0,$$

or, equivalently,

$$\sin x(2\cos x + 1) = 0.$$

Thus, either

$$\sin x = 0 \quad \text{or} \quad 2\cos x + 1 = 0;$$

that is,

$$\sin x = 0 \quad \text{or} \quad \cos x = -\tfrac{1}{2}.$$

The solutions of these equations for $0 \le x \le 2\pi$ are

$$0, \quad \pi, \quad 2\pi, \quad 2\pi/3, \quad 4\pi/3.$$

The two solutions $x = 0$ and $x = 2\pi$ tell us that there are horizontal tangent lines at the endpoints of the interval $[0, 2\pi]$. The remaining solutions $2\pi/3$, π, and $4\pi/3$ are the x-coordinates of the points P, Q, and R shown in Figure 2.31. Using $y = \cos 2x + 2\cos x$, we see that horizontal tangent lines occur at the points

$$(0, 3), \quad (2\pi/3, -1.5), \quad (\pi, -1), \quad (4\pi/3, -1.5), \quad (2\pi, 3).$$

If only approximate solutions are desired, then, to the nearest tenth, we obtain

$$(0, 3), \quad (2.1, -1.5), \quad (3.1, -1), \quad (4.2, -1.5), \quad (6.3, 3).$$

CAUTION Failure to use the chain rule properly is a common error that can be avoided if you write the function to be differentiated as a composition of simpler functions.

The chain rule is vitally important in calculus. It provides the power to differentiate complicated expressions involving many layers of composition. In the vast majority of calculus exercises or in applications of calculus to real-world problems, differentiation plays a critical role. In virtually every instance, the functions you will encounter will be compositions of simpler functions, and you will need to use the chain rule to complete the differentiation correctly.

EXERCISES 2.5

Exer. 1–6: Use the chain rule to find dy/dx, and express the answer in terms of x.

1 $y = u^2;$ $\quad u = x^3 - 4$

2 $y = \sqrt[3]{u};$ $\quad u = x^2 + 5x$

3 $y = 1/u;$ $\quad u = \sqrt{3x - 2}$

4 $y = 3u^2 + 2u;$ $\quad u = 4x$

5 $y = \tan 3u;$ $\quad u = x^2$

6 $y = u \sin u;$ $\quad u = x^3$

Exer. 7–62: Find the derivative.

7 $f(x) = (x^2 - 3x + 8)^3$

8 $f(x) = (4x^3 + 2x^2 - x - 3)^2$

9 $g(x) = (8x - 7)^{-5}$

10 $k(x) = (5x^2 - 2x + 1)^{-3}$

11 $f(x) = \dfrac{x}{(x^2 - 1)^4}$

12 $g(x) = \dfrac{x^4 - 3x^2 + 1}{(2x + 3)^4}$

13 $f(x) = (8x^3 - 2x^2 + x - 7)^5$

14 $g(w) = (w^4 - 8w^2 + 15)^4$

15 $F(v) = (17v - 5)^{1000}$

16 $s(t) = (4t^5 - 3t^3 + 2t)^{-2}$

17 $N(x) = (6x - 7)^3(8x^2 + 9)^2$

18 $f(w) = (2w^2 - 3w + 1)(3w + 2)^4$

19 $g(z) = \left(z^2 - \dfrac{1}{z^2}\right)^6$

20 $S(t) = \left(\dfrac{3t + 4}{6t - 7}\right)^3$

21 $k(r) = \sqrt[3]{8r^3 + 27}$

22 $h(z) = (2z^2 - 9z + 8)^{-2/3}$

23 $F(v) = \dfrac{5}{\sqrt[5]{v^5 - 32}}$

24 $k(s) = \dfrac{1}{\sqrt{3s - 4}}$

25 $g(w) = \dfrac{w^2 - 4w + 3}{w^{3/2}}$

26 $K(x) = \sqrt{4x^2 + 2x + 3}$

27 $H(x) = \dfrac{2x + 3}{\sqrt{4x^2 + 9}}$

28 $f(x) = (7x + \sqrt{x^2 + 3})^6$

29 $k(x) = \sin(x^2 + 2)$

30 $f(t) = \cos(4 - 3t)$

31 $H(\theta) = \cos^5 3\theta$

32 $g(x) = \sin^4(x^3)$

33 $g(z) = \sec(2z + 1)^2$

34 $k(z) = \csc(z^2 + 4)$

35 $H(s) = \cot(s^3 - 2s)$

36 $f(x) = \tan(2x^2 + 3)$

37 $f(x) = \cos(3x^2) + \cos^2 3x$

38 $g(w) = \tan^3 6w$

39 $F(\phi) = \csc^2 2\phi$

40 $M(x) = \sec(1/x^2)$

41 $K(z) = z^2 \cot 5z$

42 $G(s) = s \csc(s^2)$

43 $h(\theta) = \tan^2 \theta \sec^3 \theta$

44 $H(u) = u^2 \sec^3 4u$

45 $N(x) = (\sin 5x - \cos 5x)^5$

46 $p(v) = \sin 4v \csc 4v$

47 $T(w) = \cot^3(3w + 1)$

48 $g(r) = \sin(2r + 3)^4$

49 $h(w) = \dfrac{\cos 4w}{1 - \sin 4w}$

50 $f(x) = \dfrac{\sec 2x}{1 + \tan 2x}$

51 $f(x) = \tan^3 2x - \sec^3 2x$

52 $h(\phi) = (\tan 2\phi - \sec 2\phi)^3$

53 $f(x) = \sin \sqrt{x} + \sqrt{\sin x}$

54 $f(x) = \tan \sqrt[3]{5 - 6x}$

55 $k(\theta) = \cos^2 \sqrt{3 - 8\theta}$

56 $r(t) = \sqrt{\sin 2t - \cos 2t}$

57 $g(x) = \sqrt{x^2 + 1} \tan \sqrt{x^2 + 1}$

58 $h(\phi) = \dfrac{\cot 4\phi}{\sqrt{\phi^2 + 4}}$

59 $M(x) = \sec \sqrt{4x + 1}$

60 $F(s) = \sqrt{\csc 2s}$

61 $h(x) = \sqrt{4 + \csc^2 3x}$

62 $f(t) = \sin^2 2t \sqrt{\cos 2t}$

Exer. 63–68: (a) Find equations of the tangent line and the normal line to the graph of the equation at P. (b) Find the x-coordinates on the graph at which the tangent line is horizontal.

63 $y = (4x^2 - 8x + 3)^4$;　$P(2, 81)$

64 $y = (2x - 1)^{10}$;　　　$P(1, 1)$

65 $y = \left(x + \dfrac{1}{x}\right)^5$;　　$P(1, 32)$

66 $y = \sqrt{2x^2 + 1}$;　　$P(-1, \sqrt{3})$

67 $y = 3x + \sin 3x$;　　$P(0, 0)$

68 $y = x + \cos 2x$;　　$P(0, 1)$

Exer. 69–74: Find the first and second derivatives.

69 $g(z) = \sqrt{3z + 1}$

70 $k(s) = (s^2 + 4)^{2/3}$

71 $k(r) = (4r + 7)^5$

72 $f(x) = \sqrt[5]{10x + 7}$

73 $f(x) = \sin^3 x$

74 $G(t) = \sec^2 4t$

75 If an object of mass m has velocity v, then its *kinetic energy* K is given by $K = \frac{1}{2}mv^2$. If v is a function of time t, use the chain rule to find a formula for dK/dt.

76 As a spherical weather balloon is being inflated, its radius r is a function of time t. If V is the volume of the balloon, use the chain rule to find a formula for dV/dt.

77 When a space shuttle is launched into space, an astronaut's body weight decreases until a state of weightlessness is achieved. The weight W of a 150-lb astronaut at an altitude of x kilometers above sea level is given by

$$W = 150 \left(\frac{6400}{6400 + x}\right)^2.$$

If the space shuttle is moving away from the earth's surface at the rate of 6 km/sec, at what rate is W decreasing when $x = 1000$ km?

78 The length–weight relationship for Pacific halibut is well described by the formula $W = 10.375L^3$, where L is the length (in meters) and W is the weight (in kilograms). The rate of growth in length dL/dt is given by $0.18(2 - L)$, where t is time (in years).

(a) Find a formula for the rate of growth in weight dW/dt in terms of L.

(b) Use the formula in part (a) to estimate the rate of growth in weight of a halibut weighing 20 kg.

79 If $k(x) = f(g(x))$ and if $f(2) = -4$, $g(2) = 2$, $f'(2) = 3$, and $g'(2) = 5$, find $k(2)$ and $k'(2)$.

80 Let p, q, and r be functions such that $p(z) = q(r(z))$. If $r(3) = 3$, $q(3) = -2$, $r'(3) = 4$, and $q'(3) = 6$, find $p(3)$ and $p'(3)$.

81 If $f(t) = g(h(t))$ and if $f(4) = 3$, $g(4) = 3$, $h(4) = 4$, $f'(4) = 2$, and $g'(4) = -5$, find $h'(4)$.

82 If $u(x) = v(w(x))$ and if $v(0) = -1$, $w(0) = 0$, $u(0) = -1$, $v'(0) = -3$, and $u'(0) = 2$, find $w'(0)$.

c **83** Let $h = f \circ g$ be a differentiable function. The following tables list some values of f and g. Use Exercise 53 of Section 2.2 to approximate $h'(1.12)$.

x	2.2210	2.2320	2.2430
$f(x)$	4.9328	4.9818	5.0310

x	1.1100	1.1200	1.1300
$g(x)$	2.2210	2.2320	2.2430

c **84** Let $h = f \circ g$ be a differentiable function. The following tables list some values of f and g. Use Exercise 53 of Section 2.2 to approximate $h'(-2)$.

x	−8.48092	−8.46000	−8.43908
$f(x)$	−2.03930	−2.03762	−2.03594

x	−2.00400	−2.00000	−1.99600
$g(x)$	−8.48092	−8.46000	−8.43908

85 Let f be differentiable. Use the chain rule to prove that

(a) if f is even, then f' is odd

(b) if f is odd, then f' is even

Use polynomial functions to give examples of parts (a) and (b).

86 Use the chain rule, the derivative formula for $(d/dx)(\sin u)$, together with the identities

$$\cos x = \sin\left(\frac{\pi}{2} - x\right)$$

and

$$\sin x = \cos\left(\frac{\pi}{2} - x\right)$$

to obtain the formula $(d/dx)(\cos x)$.

87 Pinnipeds are a suborder of aquatic carnivorous mammals, such as seals and walruses, whose limbs are modified into flippers. The length–weight relationship during fetal growth is well described by the formula $W = (6 \times 10^{-5})L^{2.74}$, where L is the length (in centimeters) and W is the weight (in kilograms).

(a) Use the chain rule to find a formula for the rate of growth in weight with respect to time t.

(b) If the weight of a seal is 0.5 kg and is changing at a rate of 0.4 kg per month, how fast is the length changing?

88 The formula for the adiabatic expansion of air is $pv^{1.4} = c$, where p is the pressure, v is the volume, and c is a constant. Find a formula for the rate of change of pressure with respect to volume.

89 The deflection d of a diving board at a position s feet from the stationary end is given by

$$d = cs^2(3L - s) \quad \text{for} \quad 0 \le s \le L,$$

where L is the length of the board and c is a positive constant that depends on the weight of the diver and on the physical properties of the board. If the board is 10 ft long, find the rate of change of d with respect to s.

Exercise 89

90 When an individual is walking, the magnitude F of the vertical force of one foot on the ground (see figure) can be described by

$$F = A(\cos bt - a\cos 3bt),$$

where t is the time (in seconds), $A > 0, b > 0$, and $0 < a < 1$. Use the chain rule to find the rate of change of F with respect to time t.

Exercise 90

91 A common form of cardiovascular branching is bifurcation, in which an artery splits into two smaller blood vessels. The bifurcation angle θ is the angle formed by the two smaller arteries. In the figure, the line through A and D bisects θ and is perpendicular to the line through B and C.

(a) Show that the length L of the artery from A to B is given by

$$L = a + \frac{b}{2}\tan\frac{\theta}{4}.$$

(b) Use the chain rule to find the rate of change of L with respect to θ.

Exercise 91

2.6 IMPLICIT DIFFERENTIATION

Our objective in this section is to find derivatives of functions that are given in implicit form. If we have an equation such as

$$y = 2x^2 - 3,$$

we sometimes say that y is an **explicit function** of x, since we can write

$$y = f(x) \quad \text{with} \quad f(x) = 2x^2 - 3.$$

The equation

$$4x^2 - 2y = 6$$

determines the same function f, since solving for y gives us

$$-2y = -4x^2 + 6, \quad \text{or} \quad y = 2x^2 - 3.$$

For the case $4x^2 - 2y = 6$, we say that y (or f) is an **implicit function** of x, or that f is determined *implicitly* by the equation. If we substitute $f(x)$ for y in $4x^2 - 2y = 6$, we obtain

$$4x^2 - 2f(x) = 6$$
$$4x^2 - 2(2x^2 - 3) = 6$$
$$4x^2 - 4x^2 + 6 = 6.$$

The last equation is an identity, since it is true for every x in the domain of f. This is a characteristic of every function f determined implicitly by an equation in x and y; that is, *f is implicit if and only if substitution of $f(x)$ for y leads to an identity*. Since $(x, f(x))$ is a point on the graph of f, the last statement implies that *the graph of the implicit function f coincides with a portion (or all) of the graph of the equation*.

In the next example, we show that an equation in x and y may determine more than one implicit function.

EXAMPLE ■ I   How many different functions are determined implicitly by the equation $x^2 + y^2 = 1$?

SOLUTION   The graph of $x^2 + y^2 = 1$ is the unit circle with center at the origin. Solving the equation for y in terms of x, we obtain

$$y = \pm\sqrt{1 - x^2}.$$

Two functions f and g determined implicitly by the equation are given by

$$f(x) = \sqrt{1 - x^2} \quad \text{and} \quad g(x) = -\sqrt{1 - x^2}.$$

Figure 2.32

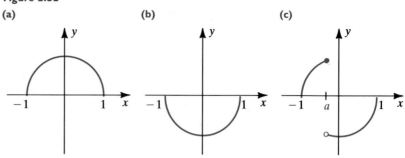

The graphs of f and g are the upper and lower halves, respectively, of the unit circle (see Figure 2.32a and b). To find other implicit functions, we may let a be any number between -1 and 1 and then define the function k by

$$k(x) = \begin{cases} \sqrt{1 - x^2} & \text{if } -1 \le x \le a \\ -\sqrt{1 - x^2} & \text{if } a < x \le 1 \end{cases}$$

The graph of k is sketched in Figure 2.32(c). Note that there is a jump discontinuity at $x = a$. The function k is determined implicitly by the equation $x^2 + y^2 = 1$, since

$$x^2 + (k(x))^2 = 1$$

for every x in the domain of k. By letting a take on different values, we can obtain as many implicit functions as desired. Many other functions are determined implicitly by $x^2 + y^2 = 1$, and the graph of each is a portion of the graph of the equation.

If the equation

$$y^4 + 3y - 4x^3 = 5x + 1$$

determines an implicit function f, then

$$(f(x))^4 + 3(f(x)) - 4x^3 = 5x + 1$$

for every x in the domain of f; however, there is no obvious way to solve for y in terms of x to obtain $f(x)$. It is possible to state conditions under which an implicit function exists and is differentiable at numbers in its domain; however, the proof requires advanced methods and hence is omitted. In the examples that follow, we will assume that a given equation in x and y determines a differentiable function f such that if $f(x)$ is substituted for y, the equation is an identity for every x in the domain of f. The derivative of f may then be found by the method of **implicit differentiation,** in which we differentiate each term of the equation with respect to x. In using implicit differentiation, it is often necessary to consider $(d/dx)(y^n)$ for some unknown function y of x, say, $y = f(x)$. By the power rule (2.27) with $y = u$, we can write $(d/dx)(y^n)$ in any of the following forms:

$$\frac{d}{dx}(y^n) = ny^{n-1}\frac{dy}{dx} = ny^{n-1}y'$$

Since the dependent variable y represents the expression $f(x)$, it is *essential* to multiply ny^{n-1} by the derivative y' when we differentiate y with respect to x. Thus,

$$\frac{d}{dx}(y^n) \neq ny^{n-1}, \quad \text{unless} \quad y = x.$$

EXAMPLE • 2   Assuming that the equation $y^4 + 3y - 4x^3 = 5x + 1$ determines, implicitly, a differentiable function f such that $y = f(x)$, find its derivative.

SOLUTION   We regard y as a symbol that denotes $f(x)$ and consider the equation as an identity for every x in the domain of f. Since derivatives of both sides are equal, we obtain the following:

$$\frac{d}{dx}(y^4 + 3y - 4x^3) = \frac{d}{dx}(5x + 1)$$

$$\frac{d}{dx}(y^4) + \frac{d}{dx}(3y) - \frac{d}{dx}(4x^3) = \frac{d}{dx}(5x) + \frac{d}{dx}(1)$$

$$4y^3y' + 3y' - 12x^2 = 5 + 0$$

We now solve for y', obtaining

$$(4y^3 + 3)y' = 12x^2 + 5,$$

or

$$y' = \frac{12x^2 + 5}{4y^3 + 3},$$

provided $4y^3 + 3 \neq 0$. Thus, if $y = f(x)$, then

$$f'(x) = \frac{12x^2 + 5}{4(f(x))^3 + 3}.$$

The last two equations in the solution of Example 2 bring out a disadvantage of using the method of implicit differentiation: The formula for y' (or $f'(x)$) may contain the expression y (or $f(x)$). However, these formulas can still be very useful in analyzing f and its graph.

In the next example, we use implicit differentiation to find the slope of the tangent line at a point $P(a, b)$ on the graph of an equation. In problems of this type, we shall assume that the equation determines an implicit function f whose graph coincides with the graph of the equation for every x in some open interval containing a. Note that since $P(a, b)$ is a point on the graph, the ordered pair (a, b) must be a solution of the equation.

EXAMPLE ■ 3 Find the slope of the tangent line to the graph of

$$y^4 + 3y - 4x^3 = 5x + 1$$

at the point $P(1, -2)$.

SOLUTION The point $P(1, -2)$ is on the graph, since substituting $x = 1$ and $y = -2$ gives us

$$(-2)^4 + 3(-2) - 4(1)^3 = 5(1) + 1, \quad \text{or} \quad 6 = 6.$$

The slope of the tangent line at $P(1, -2)$ is the value of the derivative y' when $x = 1$ and $y = -2$. The given equation is the same as that in Example 2, where we found that $y' = (12x^2 + 5)/(4y^3 + 3)$. Substituting 1 for x and -2 for y gives us the following, where $y'|_{(1,-2)}$ denotes the value of y' when $x = 1$ and $y = -2$:

$$y'|_{(1,-2)} = \frac{12(1)^2 + 5}{4(-2)^3 + 3} = -\frac{17}{29}.$$

EXAMPLE ■ 4 If $y = f(x)$, where f is determined implicitly by the equation $x^2 + y^2 = 1$, find y'.

SOLUTION In Example 1, we showed that there is an unlimited number of implicit functions determined by $x^2 + y^2 = 1$. As in Example 2, we differentiate both sides of the equation with respect to x, obtaining

$$\frac{d}{dx}(x^2) + \frac{d}{dx}(y^2) = \frac{d}{dx}(1)$$
$$2x + 2yy' = 0$$
$$yy' = -x$$
$$y' = -\frac{x}{y} \quad \text{if} \quad y \neq 0.$$

The method of implicit differentiation provides the derivative of *any* differentiable function determined by an equation in two variables. For example, the equation $x^2 + y^2 = 1$ determines many implicit functions (see Example 1). From Example 4, the slope of the tangent line at the point (x, y) on any of the graphs in Figure 2.32 is given by $y' = -x/y$, provided the derivative exists.

EXAMPLE ■ 5 Find y' if $4xy^3 - x^2y + x^3 - 5x + 6 = 0$.

SOLUTION Differentiating both sides of the equation with respect to x yields

$$\frac{d}{dx}(4xy^3) - \frac{d}{dx}(x^2y) + \frac{d}{dx}(x^3) - \frac{d}{dx}(5x) + \frac{d}{dx}(6) = \frac{d}{dx}(0).$$

Since y denotes $f(x)$ for some function f, the product rule must be applied to $(d/dx)(4xy^3)$ and $(d/dx)(x^2y)$. Thus,

$$\frac{d}{dx}(4xy^3) = 4x\frac{d}{dx}(y^3) + y^3\frac{d}{dx}(4x)$$
$$= 4x(3y^2y') + y^3(4)$$
$$= 12xy^2y' + 4y^3$$

and $\qquad \dfrac{d}{dx}(x^2y) = x^2\dfrac{dy}{dx} + y\dfrac{d}{dx}(x^2) = x^2y' + y(2x).$

Substituting these expressions in the first equation of the solution and differentiating the other terms leads to

$$(12xy^2y' + 4y^3) - (x^2y' + 2xy) + 3x^2 - 5 = 0.$$

Collecting the terms containing y' and transposing the remaining terms to the right-hand side of the equation gives us

$$(12xy^2 - x^2)y' = 5 - 3x^2 + 2xy - 4y^3.$$

Consequently, $\qquad y' = \dfrac{5 - 3x^2 + 2xy - 4y^3}{12xy^2 - x^2},$

provided $12xy^2 - x^2 \neq 0$.

EXAMPLE ■ 6   Find y' if $y = x^2 \sin y$.

SOLUTION   Differentiating both sides of the equation with respect to x and using the product rule, we obtain

$$\frac{dy}{dx} = (x^2)\frac{d}{dx}(\sin y) + \sin y \frac{d}{dx}(x^2).$$

Since $y = f(x)$ for some (implicit) function f, we have, by Theorem (2.28),

$$\frac{d}{dx}(\sin y) = \cos y \frac{dy}{dx}.$$

Using this equation and the fact that $(d/dx)(x^2) = 2x$, we may rewrite the first equation of our solution as

$$\frac{dy}{dx} = (x^2 \cos y)\frac{dy}{dx} + \sin y(2x),$$

or $\qquad y' = (x^2 \cos y)y' + 2x \sin y.$

Finally, we solve for y' as follows:

$$y' - (x^2 \cos y)y' = 2x \sin y$$
$$(1 - x^2 \cos y)y' = 2x \sin y$$
$$y' = \frac{2x \sin y}{1 - x^2 \cos y},$$

provided $1 - x^2 \cos y \neq 0$.

In the next example, we find the second derivative of an implicit function.

EXAMPLE ■ 7 Find y'' if $y^4 + 3y - 4x^3 = 5x + 1$.

SOLUTION The equation was considered in Example 2, where we found that

$$y' = \frac{12x^2 + 5}{4y^3 + 3}.$$

Hence

$$y'' = \frac{d}{dx}(y') = \frac{d}{dx}\left(\frac{12x^2 + 5}{4y^3 + 3}\right).$$

We now use the quotient rule, differentiating implicitly as follows:

$$y'' = \frac{(4y^3 + 3)\dfrac{d}{dx}(12x^2 + 5) - (12x^2 + 5)\dfrac{d}{dx}(4y^3 + 3)}{(4y^3 + 3)^2}$$

$$= \frac{(4y^3 + 3)(24x) - (12x^2 + 5)(12y^2 y')}{(4y^3 + 3)^2}$$

Substituting for y' yields

$$y'' = \frac{(4y^3 + 3)(24x) - (12x^2 + 5) \cdot 12y^2\left(\dfrac{12x^2 + 5}{4y^3 + 3}\right)}{(4y^3 + 3)^2}$$

$$= \frac{(4y^3 + 3)^2(24x) - 12y^2(12x^2 + 5)^2}{(4y^3 + 3)^3}.$$

EXAMPLE ■ 8 Use implicit differentiation to find an equation for the tangent line to the ellipse $9x^2 + 4y^2 = 40$ at the point $P(2, 1)$, as shown in Figure 2.33.

Figure 2.33

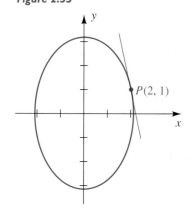

SOLUTION We verify first that the point $P(2, 1)$ is on the ellipse by showing that its coordinates satisfy the equation $9x^2 + 4y^2 = 40$:

$$9(2^2) + 4(1^2) = (9)(4) + (4)(1) = 36 + 4 = 40.$$

The slope of the tangent line will be the value of the derivative y' evaluated at $P(2, 1)$. We find y' by differentiating both sides of the equation of the ellipse with respect to x:

$$\frac{d}{dx}(9x^2) + \frac{d}{dx}(4y^2) = \frac{d}{dx}(40)$$
$$18x + 8yy' = 0$$
$$8yy' = -18x$$
$$y' = \frac{-18x}{8y} = \frac{-9x}{4y}$$

Thus, the slope of the tangent line at $P(2, 1)$ is

$$\frac{(-9)(2)}{(4)(1)} = \frac{-9}{2}.$$

We can now find an equation of the tangent line by using the point–slope formula:

$$(y - 1) = \frac{-9}{2}(x - 2), \quad \text{or, equivalently,} \quad 9x + 2y = 20$$

EXERCISES 2.6

Exer. 1–18: Assuming that the equation determines a differentiable function f such that $y = f(x)$, find y'.

1 $8x^2 + y^2 = 10$

2 $4x^3 - 2y^3 = x$

3 $2x^3 + x^2y + y^3 = 1$

4 $5x^2 + 2x^2y + y^2 = 8$

5 $5x^2 - xy - 4y^2 = 0$

6 $x^4 + 4x^2y^2 - 3xy^3 + 2x = 0$

7 $\sqrt{x} + \sqrt{y} = 100$

8 $x^{2/3} + y^{2/3} = 4$

9 $x^2 + \sqrt{xy} = 7$

10 $2x - \sqrt{xy} + y^3 = 16$

11 $\sin^2 3y = x + y - 1$

12 $x = \sin(xy)$

13 $y = \csc(xy)$

14 $y^2 + 1 = x^2 \sec y$

15 $y^2 = x \cos y$

16 $xy = \tan y$

17 $x^2 + \sqrt{\sin y} - y^2 = 1$

18 $\sin \sqrt{y} - 3x = 2$

Exer. 19–22: The equation of a classical curve and its graph are given for positive constants a and b. (Consult books on analytic geometry for further information.) Find the slope of the tangent line at the point P for the stated values of a and b.

19 *Ovals of Cassini:* $(x^2 + y^2 + a^2)^2 - 4a^2x^2 = b^4$; $a = 2, \quad b = \sqrt{6}, \quad P(2, \sqrt{2})$

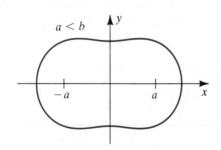

20 *Folium of Descartes:* $x^3 + y^3 - 3axy = 0$; $a = 4, \quad P(6, 6)$

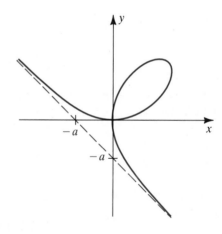

21 *Lemniscate of Bernoulli:* $(x^2 + y^2)^2 = 2a^2xy$; $a = \sqrt{2}, \quad P(1, 1)$

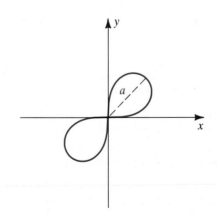

22 *Conchoid of Nicomedes:* $(y-a)^2(x^2+y^2) = b^2 y^2$; $a = 2$, $b = 4$, $P(\sqrt{15}, 1)$

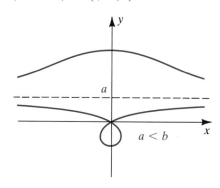

$a < b$

Exer. 23–28: Find the slope of the tangent line to the graph of the equation at P.

23 $xy + 16 = 0$; $P(-2, 8)$

24 $y^2 - 4x^2 = 5$; $P(-1, 3)$

25 $2x^3 - x^2 y + y^3 - 1 = 0$; $P(2, -3)$

26 $3y^4 + 4x - x^2 \sin y - 4 = 0$; $P(1, 0)$

27 $x^2 y + \sin y = 2\pi$; $P(1, 2\pi)$

28 $xy^2 + 3y = 27$; $P(2, 3)$

Exer. 29–34: Assuming that the equation determines a function f such that $y = f(x)$, find y'', if it exists.

29 $3x^2 + 4y^2 = 4$ **30** $5x^2 - 2y^2 = 4$

31 $x^3 - y^3 = 1$ **32** $x^2 y^3 = 1$

33 $\sin y + y = x$ **34** $\cos y = x$

Exer. 35–38: How many implicit functions are determined by the equation?

35 $x^4 + y^4 - 1 = 0$ **36** $x^4 + y^4 = 0$

37 $x^2 + y^2 + 1 = 0$ **38** $\cos x + \sin y = 3$

39 Show that the equation $y^2 = x$ determines an infinite number of implicit functions.

40 Use implicit differentiation to show that if P is any point on the circle $x^2 + y^2 = a^2$, then the tangent line at P is perpendicular to OP.

41 If tangent lines to the ellipse $9x^2 + 4y^2 = 36$ intersect the y-axis at the point $(0, 6)$, find the points of tangency.

42 If tangent lines to the hyperbola $9x^2 - y^2 = 36$ intersect the y-axis at the point $(0, 6)$, find the points of tangency.

43 Find an equation of a line through $P(-2, 3)$ that is tangent to the ellipse $5x^2 + 4y^2 = 56$.

44 Find an equation of a line through $P(2, -1)$ that is tangent to the hyperbola $x^2 - 4y^2 = 16$.

Exer. 45–46: Find equations of the tangent line and the normal line to the ellipse at the point P.

45 $5x^2 + 4y^2 = 56$; $P(-2, 3)$

46 $9x^2 + 4y^2 = 72$; $P(2, 3)$

Exer. 47–48: Find equations of the tangent line and the normal line to the hyperbola at the point P.

47 $2x^2 - 5y^2 = 3$; $P(-2, 1)$

48 $3y^2 - 2x^2 = 40$; $P(2, -4)$

49 For the ellipse $(x^2/a^2) + (y^2/b^2) = 1$, where we have $a > b > 0$:

 (a) Use implicit differentiation to find a formula for the slope of the tangent line at the point $P(x_1, y_1)$.

 (b) Determine at which points the tangent line is horizontal or vertical.

 (c) Show that an equation of the tangent line at $P(x_1, y_1)$ is

$$\frac{x_1 x}{a^2} + \frac{y_1 y}{b^2} = 1.$$

50 Prove that an equation of the tangent line to the graph of the hyperbola $(x^2/a^2) - (y^2/b^2) = 1$ at the point $P(x_1, y_1)$ is

$$\frac{x_1 x}{a^2} - \frac{y_1 y}{b^2} = 1.$$

51 Prove that if a normal line to each point on an ellipse passes through the center of the ellipse, then the ellipse is a circle.

52 Let l denote the tangent line at a point P on a hyperbola (see figure). If l intersects the asymptotes at Q and R, prove that P is the midpoint of QR.

Exercise 52

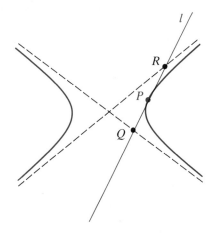

2.7 RELATED RATES

In many applications of calculus, we encounter situations in which there are variables — say, x and y — that satisfy some relationship over some interval of time t. We usually have some knowledge of the rate of change of one of these variables with respect to time, and we wish to find the rate of change with respect to time of the other variables. These problems are called *related rate problems* and are the focus of study in this section.

Suppose that two variables x and y are functions of another variable t, say,

$$x = f(t) \quad \text{and} \quad y = g(t).$$

By (2.7)(ii), we may interpret the derivatives dx/dt and dy/dt as the rates of change of x and y with respect to t. As a special case, if f and g are position functions for points moving on coordinate lines, then dx/dt and dy/dt are the velocities of these points (see (2.2)). In other situations, these derivatives may represent rates of change of physical quantities.

In certain applications, x and y may be related by means of an equation, such as

$$x^2 - y^3 - 2x + 7y^2 - 2 = 0.$$

If we differentiate this equation implicitly *with respect to t*, we obtain

$$\frac{d}{dt}(x^2) - \frac{d}{dt}(y^3) - \frac{d}{dt}(2x) + \frac{d}{dt}(7y^2) - \frac{d}{dt}(2) = \frac{d}{dt}(0).$$

Using the power rule (2.27) with t as the independent variable gives us

$$2x\frac{dx}{dt} - 3y^2\frac{dy}{dt} - 2\frac{dx}{dt} + 14y\frac{dy}{dt} = 0.$$

The derivatives dx/dt and dy/dt are called **related rates**, since they are related by means of an equation. This equation can be used to find one of the rates when the other is known. The following examples give several illustrations.

EXAMPLE ▪ I Two variables x and y are functions of a variable t and are related by the equation

$$x^3 - 2y^2 + 5x = 16.$$

If $dx/dt = 4$ when $x = 2$ and $y = -1$, find the corresponding value of dy/dt.

SOLUTION We differentiate the given equation implicitly with respect to t as follows:

$$\frac{d}{dt}(x^3) - \frac{d}{dt}(2y^2) + \frac{d}{dt}(5x) = \frac{d}{dt}(16)$$

$$3x^2\frac{dx}{dt} - 4y\frac{dy}{dt} + 5\frac{dx}{dt} = 0$$

$$(3x^2 + 5)\frac{dx}{dt} = 4y\frac{dy}{dt}$$

$$\frac{dy}{dt} = \frac{3x^2 + 5}{4y}\frac{dx}{dt}$$

The last equation is a *general* formula relating dy/dt and dx/dt. For the special case $dx/dt = 4$, $x = 2$, and $y = -1$, we obtain

$$\frac{dy}{dt} = \frac{3(2)^2 + 5}{4(-1)} \cdot 4 = -17.$$

EXAMPLE ▪ 2 A ladder 20 ft long leans against the wall of a vertical building. If the bottom of the ladder slides away from the building horizontally at a rate of 2 ft/sec, how fast is the ladder sliding down the building when the top of the ladder is 12 ft above the ground?

Figure 2.34

SOLUTION We begin by sketching a general position of the ladder as in Figure 2.34, where x denotes the distance from the base of the building to the bottom of the ladder and y denotes the distance from the ground to the top of the ladder.

We next consider the following problem involving the rates of change of x and y with respect to t:

$$\text{Given:} \quad \frac{dx}{dt} = 2 \text{ ft/sec}$$

$$\text{Find:} \quad \frac{dy}{dt} \text{ when } y = 12 \text{ ft}$$

An equation that relates the variables x and y can be obtained by applying the Pythagorean theorem to the right triangle formed by the building, the ground, and the ladder (see Figure 2.34):

$$x^2 + y^2 = 400$$

Differentiating both sides of this equation implicitly with respect to t, we obtain

$$\frac{d}{dt}(x^2) + \frac{d}{dt}(y^2) = \frac{d}{dt}(400)$$

$$2x\frac{dx}{dt} + 2y\frac{dy}{dt} = 0$$

$$\frac{dy}{dt} = -\frac{x}{y}\frac{dx}{dt},$$

provided $y \neq 0$.

The last equation is a general formula relating the two rates of change dx/dt and dy/dt. Let us now consider the special case $y = 12$. The corresponding value of x may be determined from

$$x^2 + 12^2 = 400, \quad \text{or} \quad x^2 = 256.$$

Thus, $x = \sqrt{256} = 16$ when $y = 12$. Substituting these values into the

general formula for dy/dt, we obtain

$$\frac{dy}{dt} = -\frac{16}{12}(2) = -\frac{8}{3} \text{ ft/sec.}$$

The following guidelines may be helpful for solving related rate problems of the type illustrated in Example 2.

Guidelines for Solving Related Rate Problems 2.29

1   Read the problem carefully several times, and think about the given facts and the unknown quantities that are to be found.

2   Sketch a picture or diagram and label it appropriately, introducing variables for unknown quantities.

3   Write down all the known facts, expressing the given and unknown rates as derivatives of the variables introduced in guideline (2).

4   Formulate a *general* equation that relates the variables.

5   Differentiate the equation formulated in guideline (4) implicitly with respect to t, obtaining a *general* relationship between the rates.

6   Substitute the *known* values and rates, and then find the unknown rate of change.

CAUTION   A common error is introducing specific values for the rates and variable quantities *too early* in the solution. Always remember to obtain a *general* formula that involves the rates of change at *any* time t. *Specific values should not be substituted for variables until the final steps of the solution.*

EXAMPLE • 3   At 1:00 P.M., ship A is 25 mi due south of ship B. If ship A is sailing west at a rate of 16 mi/hr and ship B is sailing south at a rate of 20 mi/hr, find the rate at which the distance between the ships is changing at 1:30 P.M.

Figure 2.35

SOLUTION   Let t denote the number of hours after 1:00 P.M. In Figure 2.35, P and Q are the positions of the ships at 1:00 P.M., x and y are the number of miles they have traveled in t hours, and z is the distance between the ships after t hours. Our problem may be stated as follows:

$$Given: \quad \frac{dx}{dt} = 16 \text{ mi/hr} \quad \text{and} \quad \frac{dy}{dt} = 20 \text{ mi/hr}$$

$$Find: \quad \frac{dz}{dt} \text{ when } t = \frac{1}{2} \text{ hr}$$

Applying the Pythagorean theorem to the triangle in Figure 2.35 gives us the following general equation relating the variables x, y, and z:

$$z^2 = x^2 + (25 - y)^2$$

Differentiating implicitly with respect to t and using the power rule and the chain rule, we obtain

$$\frac{d}{dt}(z^2) = \frac{d}{dt}(x^2) + \frac{d}{dt}(25 - y)^2$$

$$2z\frac{dz}{dt} = 2x\frac{dx}{dt} + 2(25 - y)\left(0 - \frac{dy}{dt}\right)$$

$$z\frac{dz}{dt} = x\frac{dx}{dt} + (y - 25)\frac{dy}{dt}.$$

At 1:30 P.M., the ships have traveled for half an hour and

$$x = \tfrac{1}{2}(16) = 8, \qquad y = \tfrac{1}{2}(20) = 10, \quad \text{and} \quad 25 - y = 15.$$

Consequently,

$$z^2 = 64 + 225 = 289, \quad \text{or} \quad z = \sqrt{289} = 17.$$

Substituting into the last equation involving dz/dt, we have

$$17\frac{dz}{dt} = 8(16) + (-15)(20), \quad \text{or} \quad \frac{dz}{dt} = -\frac{172}{17} \approx -10.12 \text{ mi/hr}.$$

The negative sign indicates that the distance between the ships is decreasing at 1:30 P.M.

Another method of solution is to write $x = 16t$, $y = 20t$, and

$$z = [x^2 + (25 - y)^2]^{1/2} = [256t^2 + (25 - 20t)^2]^{1/2}.$$

The derivative dz/dt may then be found, and substitution of $\tfrac{1}{2}$ for t produces the desired rate of change.

EXAMPLE ■ 4 A water tank has the shape of an inverted right circular cone of altitude 12 ft and base radius 6 ft. If water is being pumped into the tank at a rate of 10 gal/min, approximate the rate at which the water level is rising when the water is 3 ft deep (1 gal ≈ 0.1337 ft^3).

SOLUTION We begin by sketching the tank as in Figure 2.36, letting r denote the radius of the surface of the water when the depth is h. Note that r and h are functions of time t.

The problem can now be stated as follows:

$$\text{Given:} \quad \frac{dV}{dt} = 10 \text{ gal/min}$$

$$\text{Find:} \quad \frac{dh}{dt} \text{ when } h = 3 \text{ ft}$$

The volume V of water in the tank corresponding to depth h is

$$V = \tfrac{1}{3}\pi r^2 h.$$

This formula for V relates V, r, and h. Before differentiating implicitly with respect to t, let us express V in terms of one variable. Referring to Figure 2.36 and using similar triangles, we obtain

$$\frac{r}{h} = \frac{6}{12}, \quad \text{or} \quad r = \frac{h}{2}.$$

Figure 2.36

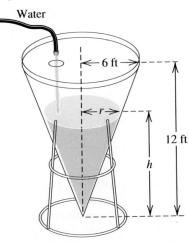

Water

6 ft

r

12 ft

h

Consequently, at depth h,

$$V = \frac{1}{3}\pi \left(\frac{h}{2}\right)^2 h = \frac{1}{12}\pi h^3.$$

Differentiating the last equation implicitly with respect to t gives us the following general relationship between the rates of change of V and h at any time t:

$$\frac{dV}{dt} = \frac{1}{4}\pi h^2 \frac{dh}{dt}$$

If $h \neq 0$, an equivalent formula is

$$\frac{dh}{dt} = \frac{4}{\pi h^2}\frac{dV}{dt}.$$

Finally, we let $h = 3$ and $dV/dt = 10$ gal/min ≈ 1.337 ft^3/min, obtaining

$$\frac{dh}{dt} \approx \frac{4}{\pi(9)}(1.337) \approx 0.189 \text{ ft/min}.$$

EXAMPLE • 5 A revolving beacon in a lighthouse makes one revolution every 15 sec. The beacon is 200 ft from the nearest point P on a straight shoreline. Find the rate at which a ray from the light moves along the shore at a point 400 ft from P.

Figure 2.37

SOLUTION The problem is diagrammed in Figure 2.37, where B denotes the position of the beacon and ϕ is the angle between BP and a light ray to a point S on the shore x units from P.

Since the light revolves four times per minute, the angle ϕ changes at a rate of $4 \cdot 2\pi$ radians per minute; that is, $d\phi/dt = 8\pi$. Using triangle PBS, we see that

$$\tan \phi = \frac{x}{200},$$

or

$$x = 200 \tan \phi.$$

The rate at which the ray of light moves along the shore is

$$\frac{dx}{dt} = 200 \sec^2 \phi \frac{d\phi}{dt} = (200 \sec^2 \phi)(8\pi) = 1600\pi \sec^2 \phi.$$

If $x = 400$, then $BS = \sqrt{200^2 + 400^2} = 200\sqrt{5}$, and

$$\sec \phi = \frac{200\sqrt{5}}{200} = \sqrt{5}.$$

Hence

$$\frac{dx}{dt} = 1600\pi(\sqrt{5})^2 = 8000\pi \approx 25,133 \text{ ft/min}.$$

EXAMPLE • 6 Figure 2.38, on the following page, shows a solar panel that is 10 ft in width and is equipped with a hydraulic lift. As the sun rises, the panel is adjusted so that the sun's rays are perpendicular to the panel's surface.

Figure 2.38

10 ft

Hydraulic lift

B

Panel

A

ϕ

C

y

θ

(a) Find the relationship between the rate dy/dt at which the panel should be lowered and the rate $d\theta/dt$ at which the angle of inclination of the sun increases.

(b) If $d\theta/dt = \pi/12$ radian/hr when $\theta = \pi/6$, find dy/dt.

SOLUTION

(a) If we let ϕ denote angle BAC in Figure 2.38, then, from plane geometry, $\phi = \frac{1}{2}\pi - \theta$. Since $d\phi/dt = -d\theta/dt$, ϕ decreases at the rate that θ increases.

Referring to right triangle BAC, we see that

$$\sin\phi = \frac{y}{10},$$

or

$$y = 10\sin\phi = 10\sin(\tfrac{1}{2}\pi - \theta).$$

Differentiating implicitly with respect to t and using the cofunction identity $\cos(\frac{1}{2}\pi - \theta) = \sin\theta$ yields

$$\frac{dy}{dt} = 10\cos\left(\frac{1}{2}\pi - \theta\right)\left(0 - \frac{d\theta}{dt}\right) = -10\sin\theta\frac{d\theta}{dt}.$$

(b) We substitute $d\theta/dt = \pi/12$ radian/hr and $\theta = \pi/6$ in the formula for dy/dt from part (a), obtaining

$$\frac{dy}{dt} = -10\left(\frac{1}{2}\right)\left(\frac{\pi}{12}\right) = -\frac{5\pi}{12} \approx -1.3 \text{ ft/hr.}$$

EXERCISES 2.7

Exer. 1–8: Assume that all variables are functions of t.

1 If $A = x^2$ and $dx/dt = 3$ when $x = 10$, find dA/dt.

2 If $S = z^3$ and $dz/dt = -2$ when $z = 3$, find dS/dt.

3 If $V = -5p^{3/2}$ and $dV/dt = -4$ when $V = -40$, find dp/dt.

4 If $P = 3/w$ and $dP/dt = 5$ when $P = 9$, find dw/dt.

5 If $x^2 + 3y^2 + 2y = 10$ and $dx/dt = 2$ when $x = 3$ and $y = -1$, find dy/dt.

6 If $2y^3 - x^2 + 4x = -10$ and $dy/dt = -3$ when $x = -2$ and $y = 1$, find dx/dt.

7 If $3x^2y + 2x = -32$ and $dy/dt = -4$ when $x = 2$ and $y = -3$, find dx/dt.

8 If $-x^2y^2 - 4y = -44$ and $dx/dt = 5$ when $x = -3$ and $y = 2$, find dy/dt.

9 As a circular metal griddle is being heated, its diameter changes at a rate of 0.01 cm/min. Find the rate at which the area of one side is changing when the diameter is 30 cm.

10 A fire has started in a dry, open field and spreads in the form of a circle. The radius of the circle increases at a rate of 6 ft/min. Find the rate at which the fire area is increasing when the radius is 150 ft.

11 Gas is being pumped into a spherical balloon at a rate of 5 ft³/min. Find the rate at which the radius is changing when the diameter is 18 in.

12 Suppose a spherical snowball is melting and the radius is decreasing at a constant rate, changing from 12 in. to 8 in. in 45 min. How fast was the volume changing when the radius was 10 in.?

13 A ladder 20 ft long leans against a vertical building. If the bottom of the ladder slides away from the building horizontally at a rate of 3 ft/sec, how fast is the ladder sliding down the building when the top of the ladder is 8 ft from the ground?

14 A girl starts at a point A and runs east at a rate of 10 ft/sec. One minute later, another girl starts at A and runs north at a rate of 8 ft/sec. At what rate is the distance between them changing 1 min after the second girl starts?

15 A light is at the top of a 16-ft pole. A boy 5 ft tall walks away from the pole at a rate of 4 ft/sec (see figure). At what rate is the tip of his shadow moving when he is 18 ft from the pole? At what rate is the length of his shadow increasing?

Exercise 15

16 A man on a dock is pulling in a boat using a rope attached to the bow of the boat 1 ft above water level and passing through a simple pulley located on the dock 8 ft above water level (see figure). If he pulls in the rope at a rate of 2 ft/sec, how fast is the boat approaching the dock when the bow of the boat is 25 ft from a point that is directly below the pulley?

Exercise 16

17 The top of a silo has the shape of a hemisphere of diameter 20 ft. If it is coated uniformly with a layer of ice and if the thickness is decreasing at a rate of $\frac{1}{4}$ in./hr, how fast is the volume of the ice changing when the ice is 2 in. thick?

18 As sand leaks out of a hole in a container, it forms a conical pile whose altitude is always the same as its radius. If the height of the pile is increasing at a rate of 6 in./min, find the rate at which the sand is leaking out when the altitude is 10 in.

19 A person flying a kite holds the string 5 ft above ground level, and the string is payed out at a rate of 2 ft/sec as the kite moves horizontally at an altitude of 105 ft (see figure). Assuming there is no sag in the string, find the rate at which the kite is moving when 125 ft of string has been payed out.

Exercise 19

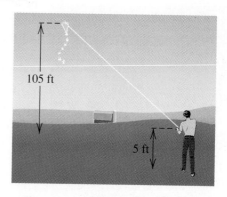

20 A hot-air balloon rises vertically as a rope attached to the base of the balloon is released at a rate of 5 ft/sec. The pulley that releases the rope is 20 ft from the platform where passengers board (see figure). At what rate is the balloon rising when 500 ft of rope has been payed out?

Exercise 20

21 Boyle's law for confined gases states that if the temperature is constant, $pv = c$, where p is pressure, v is volume, and c is a constant. At a certain instant, the volume is 75 in^3, the pressure is 30 lb/in^2, and the pressure is decreasing at a rate of 2 lb/in^2 every minute. At what rate is the volume changing at this instant?

22 A 100-ft-long cable of diameter 4 in. is submerged in seawater. Because of corrosion, the surface area of the cable decreases at a rate of 750 in^2/yr. Ignoring the corrosion at the ends of the cable, find the rate at which the diameter is decreasing.

23 The ends of a water trough 8 ft long are equilateral triangles whose sides are 2 ft long (see figure on the following page). If water is being pumped into the trough at a rate of 5 ft^3/min, find the rate at which the water level is rising when the depth of the water is 8 in.

Exercise 23

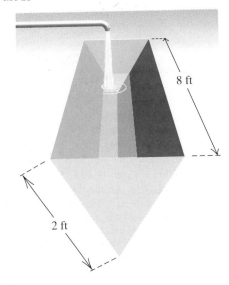

24 Work Exercise 23 if the ends of the trough have the shape of the graph of $y = 2|x|$ between the points $(-1, 2)$ and $(1, 2)$.

25 The area of an equilateral triangle is decreasing at a rate of 4 cm²/min. Find the rate at which the length of a side is changing when the area of the triangle is 200 cm².

26 Gas is escaping from a spherical balloon at a rate of 10 ft³/hr. At what rate is the radius changing when the volume is 400 ft³?

27 A stone is dropped into a lake, causing circular waves whose radii increase at a constant rate of 0.5 m/sec. At what rate is the circumference of a wave changing when its radius is 4 m?

28 A softball diamond has the shape of a square with sides 60 ft long. If a player is running from second base to third base at a speed of 24 ft/sec, at what rate is her distance from home plate changing when she is 20 ft from third base?

29 When two resistors R_1 and R_2 are connected in parallel (see figure), the total resistance R is given by the equation $1/R = (1/R_1) + (1/R_2)$. If R_1 and R_2 are increasing at rates of 0.01 ohm/sec and 0.02 ohm/sec, respectively, at what rate is R changing at the instant that $R_1 = 30$ ohms and $R_2 = 90$ ohms?

Exercise 29

30 The formula for the adiabatic expansion of air is $pv^{1.4} = c$, where p is the pressure, v is the volume, and c is a constant. At a certain instant, the pressure is 40 dyn/cm² and is increasing at a rate of 3 dyn/cm² per second. If, at that same instant, the volume is 60 cm³, find the rate at which the volume is changing.

31 If a spherical tank of radius a contains water that has a maximum depth h, then the volume V of water in the tank is given by $V = \frac{1}{3}\pi h^2(3a - h)$. Suppose a spherical tank of radius 16 ft is being filled at a rate of 100 gal/min. Approximate the rate at which the water level is rising when $h = 4$ ft (1 gal ≈ 0.1337 ft³).

32 A spherical water storage tank for a small community is coated uniformly with a 2-in. layer of ice. As the ice melts, the rate at which the volume of the ice decreases is directly proportional to its surface area. Show that the outside diameter is decreasing at a constant rate.

33 From the edge of a cliff that overlooks a lake 200 ft below, a boy drops a stone and then, 2 sec later, drops another stone from exactly the same position. Discuss the rate at which the distance between the two stones is changing during the next second. (Assume that the distance an object falls in t seconds is $16t^2$ feet.)

34 A metal rod has the shape of a right circular cylinder. As it is being heated, its length is increasing at a rate of 0.005 cm/min and its diameter is increasing at 0.002 cm/min. At what rate is the volume changing when the rod has length 40 cm and diameter 3 cm?

35 An airplane is flying at a constant speed of 360 mi/hr and climbing at an angle of 45°. At the moment the plane's altitude is 10,560 ft, it passes directly over an air traffic control tower on the ground. Find the rate at which the airplane's distance from the tower is changing 1 min later (neglect the height of the tower).

36 A north–south highway A and an east–west highway B intersect at a point P. At 10:00 A.M., an automobile crosses P traveling north on highway A at a speed of 50 mi/hr. At that same instant, an airplane flying east at a speed of 200 mi/hr and an altitude of 26,400 ft is directly above the point on highway B that is 100 mi west of P. If the airplane and the automobile maintain the same speed and direction, at what rate is the distance between them changing at 10:15 A.M.?

37 A paper cup containing water has the shape of a frustum of a right circular cone of altitude 6 in. and lower and upper base radii 1 in. and 2 in., respectively. If water is leaking out of the cup at a rate of 3 in³/hr, at what rate is the water level decreasing when the depth of the water is 4 in.? (*Note:* The volume V of a frustum of a right circular cone of altitude h and base radii a and b is given by $V = \frac{1}{3}\pi h(a^2 + b^2 + ab)$.)

38 The top part of a swimming pool is a rectangle of length 60 ft and width 30 ft. The depth of the pool varies uniformly from 4 ft to 9 ft through a horizontal distance of 40 ft and then is level for the remaining 20 ft, as illustrated by the cross-sectional view in the figure. If the pool is being filled with water at a rate of 500 gal/min, approximate the rate at which the water level is rising when the depth of the water at the deep end of the pool is 4 ft (1 gal ≈ 0.1337 ft³).

Exercise 38

39 An airplane at an altitude of 10,000 ft is flying at a constant speed on a line that will take it directly over an observer on the ground. If, at a given instant, the observer notes that the angle of elevation of the airplane is 60° and is increasing at a rate of 1° per second, find the speed of the airplane.

40 In Exercise 16, let θ be the angle that the rope makes with the horizontal. Find the rate at which θ is changing at the instant that $\theta = 30°$.

41 An isosceles triangle has equal sides 6 in. long. If the angle θ between the equal sides is changing at a rate of 2° per minute, how fast is the area of the triangle changing when $\theta = 30°$?

42 A ladder 20 ft long leans against a vertical building. If the bottom of the ladder slides away from the building horizontally at a rate of 2 ft/sec, at what rate is the angle between the ladder and the ground changing when the top of the ladder is 12 ft above the ground?

43 The relative positions of an airport runway and a 20-ft-tall control tower are shown in the figure. The beginning of the runway is at a perpendicular distance of 300 ft from the base of the tower. If an airplane reaches a speed of 100 mi/hr after having traveled 300 ft down the runway, at approximately what rate is the distance between the airplane and the top of the control tower increasing at this time?

Exercise 43

44 The speed of sound in air at 0 °C (or 273 °K) is 1087 ft/sec, but this speed increases as the temperature rises. If T is temperature in °K, the speed of sound v at this temperature is given by $v = 1087\sqrt{T/273}$. If the temperature increases at the rate of 3 °C per hour, approximate the rate at which the speed of sound is increasing when $T = 30$ °C (or 303 °K).

45 An airplane is flying at a constant speed and altitude on a line that will take it directly over a radar station located on the ground. At the instant that the airplane is 60,000 ft from the station, an observer in the station notes that the airplane's angle of elevation is 30° and is increasing at a rate of 0.5° per second. Find the speed of the airplane.

46 A missile is fired vertically from a point that is 5 mi from a tracking station and at the same elevation (see figure). For the first 20 sec of flight, its angle of elevation θ changes at a constant rate of 2° per second. Find the velocity of the missile when the angle of elevation is 30°.

Exercise 46

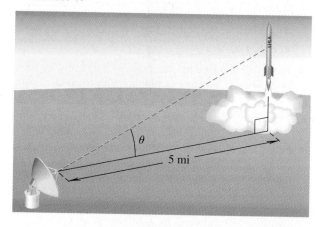

47 The sprocket assembly for a 28-in. bicycle is shown in the figure on the following page. Find the relationship between the angular velocity $d\theta/dt$ (in radians per second) of the pedal assembly and the ground speed of the bicycle (in miles per hour).

Exercises 2.7

14 in.

5 in.

θ

2 in.

48 A 100-candlepower spotlight is located 20 ft above a stage (see figure). The illuminance E (in footcandles) in the small lighted area of the stage is given by $E = (I \cos \theta)/s^2$, where I is the intensity of the light, s is the distance the light must travel, and θ is the indicated angle. As the spotlight is rotated through ϕ degrees, find the relationship between the rate of change in illumination dE/dt and the rate of rotation $d\phi/dt$.

Exercise 48

ϕ

20 ft

θ

49 A *conical pendulum* consists of a mass m, attached to a string of fixed length l, that travels around a circle of radius r at a fixed velocity v (see figure). As the velocity of the mass is increased, both the radius and the angle θ increase. Given that $v^2 = rg \tan \theta$, where g is a gravitational constant, find a relationship between the related rates

(a) dv/dt and $d\theta/dt$ (b) dv/dt and dr/dt

50 Water in a paper conical filter drips into a cup as shown in the figure. Let x denote the height of the water in the filter and y the height of the water in the cup. If 10 in³ of water is poured into the filter, find the relationship between dy/dt and dx/dt.

Exercise 49

θ

l

r

Exercise 50

2 in.

4 in.

r

x

y

4 in.

c **51** Ship A is sailing north, and ship B is sailing east. Using an xy-plane, radar records the coordinates (in miles) of each ship at intervals of 1.25 min as shown in the following tables. Approximate the rate (in miles per hour) at which the distance between the ships is changing at $t = 5$.

Ship A:

t (min)	1.25	2.50	3.75	5.00
x (mi)	1.77	1.77	1.77	1.77
y (mi)	2.71	3.03	3.35	3.67

Ship B:

t (min)	1.25	2.50	3.75	5.00
x (mi)	5.24	5.52	5.80	6.08
y (mi)	1.24	1.24	1.24	1.24

c **52** Two variables x and y are functions of a variable t and are related by the formula

$$1.31 \sin(2.56x) + \sqrt{y} = (x - 1)^2.$$

If $dy/dt \approx 3.68$ when $x \approx 1.71$ and $y \approx 3.03$, approximate the corresponding value of dx/dt.

2.8   LINEAR APPROXIMATIONS AND DIFFERENTIALS

We now examine a fundamental geometric property of the tangent line to the graph of a function: The tangent line stays close to the graph near the point of tangency. We will also introduce some additional notation and terminology that scientists and engineers commonly use in approximations involving derivatives.

Let us consider a differentiable function $y = f(x)$ where we know the behavior of the function at a point $P(x_0, f(x_0))$; in particular, assume that we know the values of $f(x_0)$ and $f'(x_0)$. Suppose we wish to approximate the value of f at x_1, where x_1 is close to x_0. Figure 2.39 illustrates this situation with Q being the point $(x_1, f(x_1))$. In some instances, we may not be able to determine $f(x_1)$ exactly because we do not have an explicit formula for f. In other cases, the numbers x_0 and x_1 may be two slightly different measurements of a physical quantity, and we want to estimate quickly the difference between the values of $f(x_0)$ and $f(x_1)$. In such cases, our needs might be met by finding a good approximation for $f(x_1)$. We will consider in this section a simple way to approximate $f(x_1)$ using the known values $f(x_0)$ and $f'(x_0)$.

Figure 2.39

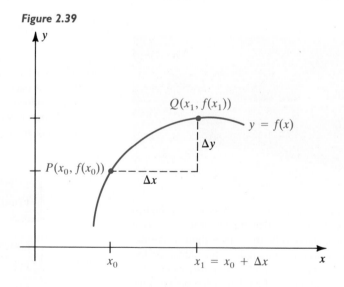

If the variable x has an initial value x_0 and is then assigned a different value x_1, the change or difference $x_1 - x_0$ is called an **increment of x**. In calculus, it is traditional to denote an increment of x by the symbol Δx (read *delta x*). The corresponding change to the value of $y = f(x)$, namely, $f(x_1) - f(x_0)$, is called an **increment of y** and is traditionally labeled Δy. We summarize this notation in the following definition.

Definition 2.30

> If $y = f(x)$ and the variable x has an initial value of x_0 that is changed to x_1, then the **increment Δx of x** is
>
> $$\Delta x = x_1 - x_0$$
>
> and the corresponding **increment Δy of y** is
>
> $$\Delta y = f(x_0 + \Delta x) - f(x_0).$$

Definition (2.30) implies that Δy is the *change in y* corresponding to the *change in x* of Δx. Since x_1 is not equal to x_0, Δx must be either positive or negative. The resulting Δy may be positive, negative, or zero. In Figure 2.39, both changes, Δx and Δy, are positive.

From Definition (2.30), we have $f(x_0) + \Delta y = f(x_0 + \Delta x) = f(x_1)$. Thus we can obtain an accurate approximation for the value $f(x_1)$ if we can accurately estimate Δy. The ratio $\Delta y / \Delta x$ is the slope m_{PQ} of the secant line through P and Q (see Figure 2.40). Since $\Delta y / \Delta x = m_{PQ}$, we have $\Delta y = m_{PQ} \Delta x$. We already know that the value of Δx is $x_1 - x_0$, so if we can obtain an estimate for m_{PQ}, we can then estimate Δy and $f(x_1)$. In Section 2.1, we defined the slope of the tangent line to $y = f(x)$ at P to be the limit of slopes of secant lines through P and Q. In Section 2.2, we defined $f'(x_0)$ as the notation for this limit. Thus m_{PQ} is approximately equal to $f'(x_0)$ if x_1 is approximately equal to x_0. Thus, we have

$$f(x_1) = f(x_0) + \Delta y = f(x_0) + m_{PQ} \Delta x \approx f(x_0) + f'(x_0)\Delta x.$$

This approximation allows us to estimate $f(x_1)$ using the known values $f(x_0)$ and $f'(x_0)$. The approximation $f(x_1) \approx f(x) + f'(x_0)\Delta x$ is particularly useful when x_1 is close to a value x_0 where it is easier to compute $f(x_0)$ and $f'(x_0)$ than to calculate $f(x_1)$ directly.

Figure 2.40

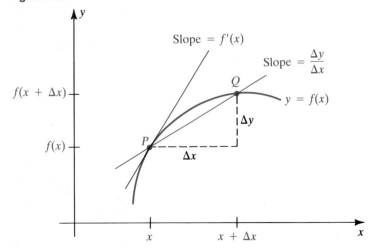

To make this discussion more precise, we use the increment notation to rewrite the definition of the derivative of a function, substituting Δx for h in Definition (2.5) as follows.

Increment Definition of Derivative 2.31

$$f'(x_0) = \lim_{\Delta x \to 0} \frac{f(x_0 + \Delta x) - f(x_0)}{\Delta x} = \lim_{\Delta x \to 0} \frac{\Delta y}{\Delta x}$$

If f is differentiable, then as Δx approaches 0, the ratio $\Delta y / \Delta x$ approaches $f'(x_0)$, as illustrated in Figure 2.40.

In earlier sections, we used quotients

$$\frac{f(x_1) - f(x_0)}{x_1 - x_0} = \frac{f(x_0 + \Delta x) - f(x_0)}{\Delta x} = \frac{\Delta y}{\Delta x}$$

to approximate numerically the derivative $f'(x_0)$. Here we reverse the process and use $f'(x_0)$ to approximate $\Delta y / \Delta x$.

Approximation Formula for Δy 2.32

$$\Delta y \approx f'(x_0) \Delta x \quad \text{if } \Delta x \approx 0$$

It is helpful to consider the *graphical* interpretation of this approximation formula (see Figure 2.41). The slope $f'(x_0)$ is the slope of the line tangent to the graph at $P(x_0, f(x_0))$. An equation for this line is

$$y = f(x_0) + f'(x_0)(x - x_0).$$

This *tangent line l* is the graph of the function L, where

$$L(x) = f(x_0) + f'(x_0)(x - x_0).$$

Evaluating L at x_0 and at $x_1 = x_0 + \Delta x$ gives $L(x_0) = f(x_0)$ and $L(x_0 + \Delta x) = f(x_0) + f'(x_0)\Delta x$, respectively. Thus,

$$\Delta L = L(x_0 + \Delta x) - L(x_0) = f'(x_0)\Delta x.$$

We can interpret (2.32) as stating that the *change in y* is approximately the *change in the tangent function L*. We have replaced the graph of the function f by the line representing L to obtain this estimate, and this function L is described as a **linear approximation** for f at x_0. Figure 2.41 illustrates this approximation.

Figure 2.41

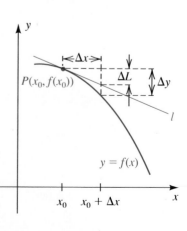

**Linear Approximation
Formula   2.33**

If $y = f(x)$, with f differentiable at x_0, then

$$f(x) \approx L(x) = f(x_0) + f'(x_0)(x - x_0) \quad \text{for } x \text{ near } x_0.$$

There are many practical uses for linear approximations to functions. We will see one such application in Section 2.9 when we develop Newton's method for approximating zeros of functions. The next example shows how we can use linear approximations to estimate square roots.

 EXAMPLE ▪ I   For the function $y = f(x) = \sqrt{3 + x}$:

(a) Find the linear approximation at $x_0 = 6$.
(b) Use this linear approximation to estimate $\sqrt{8}$, $\sqrt{8.9}$, and $\sqrt{9.3}$.
(c) Compare these approximations to values obtained with a calculator.

SOLUTION

(a) We use the linear approximation formula (2.33). For the function $f(x) = \sqrt{3 + x} = (3 + x)^{1/2}$, we have

$$f'(x) = \frac{1}{2}(3 + x)^{-1/2} = \frac{1}{2\sqrt{3 + x}}.$$

Evaluating f and f' at $x_0 = 6$ gives

$$f(6) = \sqrt{3 + 6} = 3 \quad \text{and} \quad f'(6) = \frac{1}{2\sqrt{3 + 6}} = \frac{1}{6}.$$

Thus the linear approximation to f at $x_0 = 6$ is

$$L(x) = f(x_0) + f'(x_0)(x - x_0)$$
$$= 3 + \tfrac{1}{6}(x - 6).$$

(b) For values of x close to 6, $3 + 6$ is close to 9, and so we can use the approximation $f(x) \approx L(x)$ to estimate square roots of numbers close to 9. For example,

$$\sqrt{8} = \sqrt{3 + 5} = f(5) \approx L(5) = 3 + \tfrac{1}{6}(-1) \approx 2.83333333333.$$

Similarly, we have

$$\sqrt{8.9} = \sqrt{3 + 5.9} = f(5.9) \approx L(5.9) = 3 + \tfrac{1}{6}(-0.1) \approx 2.98333333333,$$
$$\sqrt{9.3} = \sqrt{3 + 6.3} = f(6.3) \approx L(6.3) = 3 + \tfrac{1}{6}(0.3) = 3.05.$$

(c) The following table lists the approximated square roots obtained with a linear approximation and with a calculator:

Square root	$\sqrt{8}$	$\sqrt{8.9}$	$\sqrt{9.3}$
Linear approximation	2.83333333333	2.98333333333	3.05
Calculator	2.82842712475	2.98328677804	3.04959013640

Note that the linear approximations are close to the calculator values and do not require the computation of a square root.

EXAMPLE ■ 2   Use a linear approximation to estimate $\sin 0.05$ and compare the result to that obtained with a calculator.

SOLUTION   If $f(x) = \sin x$, then $f'(x) = \cos x$. Evaluating f and f' at $x_0 = 0$ gives

$$f(0) = 0 \quad \text{and} \quad f'(0) = 1.$$

By Theorem (2.23), the linear approximation to f at $x_0 = 0$ is given by $L(x) = 0 + 1(x - 0)$, or

$$\sin x \approx x \quad \text{for } x \text{ near } 0,$$

and the linear approximation to $\sin 0.05$ is 0.05. Using a calculator, we get

$$\sin 0.05 \approx 0.049979169271.$$

You may be asking yourself: Why use linear approximations for square roots or trigonometric functions when a scientific calculator can do the job more efficiently and accurately? There are several answers. First, the *process of linear approximation* is widely used throughout mathematics and in applications. It is important then to consider the geometric reasoning behind linear approximation. It is also easier to examine this process in elementary problems in which we can check the approximations against more precise answers. Second, computers and calculators themselves use algorithms, such as linear approximation, to produce approximate values of elementary functions. To gain a better understanding of the powers and limitations of calculating devices, we need to study approximation techniques, beginning with linear ones. We will consider other approximation techniques in later chapters.

Linear approximations are closely connected with the idea of *differentials*. It is traditional to use the differential expression dy to denote the actual change in the tangent line corresponding to a change Δx in x. The expression dy is another notation for the quantity we have previously labeled as ΔL.

Definition 2.34

Let $y = f(x)$, where f is a differentiable function. The **differential** dy is defined by the expression

$$dy = f'(x)\,\Delta x.$$

The vertical change in the graph of the function over an interval is the change in the value of the function over that interval. Note that dy measures the vertical change in the *tangent line* and Δy measures the actual change in the value of the *function* for the same change in x. The next example illustrates this distinction.

E X A M P L E ▪ 3   Let $y = 3x^2 - 5$ and let Δx be an increment of x.

(a) Find general formulas for Δy and dy.

(b) If x changes from 2 to 2.1, find the values of Δy and dy.

S O L U T I O N

(a) If $y = f(x) = 3x^2 - 5$, then, by Definition (2.30) with $x = x_0$,

$$\begin{aligned}
\Delta y &= f(x + \Delta x) - f(x) \\
&= [3(x + \Delta x)^2 - 5] - (3x^2 - 5) \\
&= [3(x^2 + 2x(\Delta x) + (\Delta x)^2) - 5] - (3x^2 - 5) \\
&= 3x^2 + 6x(\Delta x) + 3(\Delta x)^2 - 5 - 3x^2 + 5 \\
&= 6x(\Delta x) + 3(\Delta x)^2.
\end{aligned}$$

To find dy, we use Definition (2.34):

$$dy = f'(x)\Delta x = 6x\,\Delta x$$

(b) We wish to find Δy and dy if $x = 2$ and $\Delta x = 0.1$. Substituting in the formula for Δy obtained in part (a) gives us

$$\Delta y = 6(2)(0.1) + 3(0.1)^2 = 1.23.$$

Thus, y changes by 1.23 if x changes from 2 to 2.1. We could also find Δy directly as follows:

$$\begin{aligned}
\Delta y &= f(2.1) - f(2) \\
&= [3(2.1)^2 - 5] - [3(2)^2 - 5] = 1.23
\end{aligned}$$

Similarly, using the formula $dy = 6x\,\Delta x$, with $x = 2$ and $\Delta x = 0.1$, yields

$$dy = (6)(2)(0.1) = 1.2.$$

Note that the approximation $dy = 1.2$ is correct to the nearest tenth.

Using differential notation, we can state an alternative form of the linear approximation formula (2.33).

Alternative Linear Approximation Formula **2.35**

> If $y = f(x)$ is a differentiable function, then
>
> $$f(x + \Delta x) \approx f(x) + dy,$$
>
> where $dy = f'(x)\Delta x$.

The next example illustrates the use of this alternative linear approximation formula.

E X A M P L E ▪ 4

(a) Use differentials to approximate the change in $\sin\theta$ if θ changes from $\pi/3 = 60\pi/180$ to $61\pi/180$.

(b) Use a linear approximation to estimate $\sin(61\pi/180)$.

SOLUTION

(a) If $y = \sin\theta = f(\theta)$, then

$$dy = f'(\theta)\Delta\theta = \cos\theta\,\Delta\theta.$$

The change $\Delta\theta$ in θ is $61\pi/180 - 60\pi/180 = \pi/180$. Thus if we let $\theta = \pi/3$ and $\Delta\theta = \pi/180$ in formula (2.34) for dy, we obtain

$$dy = \left(\cos\frac{\pi}{3}\right)\left(\frac{\pi}{180}\right) = \left(\frac{1}{2}\right)\left(\frac{\pi}{180}\right) = \frac{\pi}{360} \approx 0.0087.$$

(b) If we use the linear approximation formula (2.35) with $x = \theta$ and $y = \sin\theta$, we have

$$\sin(\theta + \Delta\theta) \approx \sin\theta + dy.$$

Letting $\theta = \pi/3$, $\Delta\theta = \pi/180$, and $dy = 0.0087$ (from part a), we obtain

$$\sin\left(\frac{61\pi}{180}\right) \approx \sin\frac{\pi}{3} + dy$$

$$\approx \frac{\sqrt{3}}{2} + 0.0087$$

$$\approx 0.8660 + 0.0087 = 0.8747.$$

Using a calculator and rounding to four decimal places, we get $\sin(61\pi/180) \approx 0.8746$. Thus the error involved in using the linear approximation is roughly 0.0001.

The next example illustrates the use of differentials in estimating errors that may occur because of approximate measurements. As indicated in the solution, *it is important to first consider general formulas involving the variables* that are being considered. Specific values should *not* be substituted for variables until the final steps of the solution.

EXAMPLE ▪ 5   The radius of a spherical balloon is measured as 12 in., with a maximum error in measurement of ± 0.06 in. Approximate the maximum error in the calculated volume of the sphere.

Figure 2.42

SOLUTION   We begin by considering *general* formulas involving the radius and the volume. Thus we let

$$x = \text{the measured value of the radius}$$

and $\Delta x = \text{the maximum error in } x.$

Assuming that Δx is positive, we have

$$x - \Delta x \le \text{the exact radius} \le x + \Delta x.$$

If Δx is negative, we may use $|\Delta x|$ in place of Δx. A cross-sectional view of the balloon, indicating the possible error Δx, is shown in Figure 2.42. If the volume V of the balloon is calculated using the measured value x, then $V = \frac{4}{3}\pi x^3$.

Let ΔV be the change in V that corresponds to Δx. We may interpret ΔV as *the error in the calculated volume* caused by the error Δx. We approximate ΔV by means of dV as follows:

$$\Delta V \approx dV = \frac{dV}{dx}(\Delta x) = 4\pi x^2 \Delta x$$

Finally, we substitute specific values for x and Δx. If $x = 12$ and if $\Delta x = \pm 0.06$, then

$$dV = 4\pi(12^2)(\pm 0.06) = \pm(34.56)\pi \approx \pm 109.$$

Thus the maximum error in the calculated volume due to the error in measurement of the radius is approximately ± 109 in^3.

The radius of the balloon in Example 5 was measured as 12 in., with a maximum error of ± 0.06 in. The maximum error is also referred to as the **absolute change**. The ratio of ± 0.06 to 12 is called the *relative error* in the measurement of the radius. We may also refer to the relative error as the **average error** or the **relative change**. Thus, for Example 5,

$$\text{relative error} = \frac{\pm 0.06}{12} = \pm 0.005.$$

The significance of this number is that the error in measurement of the radius is, *on average,* ± 0.005 inch per inch. The *percentage error* (*change*) is defined as the average error (change) multiplied by 100%. In this example,

$$\text{percentage error} = (\pm 0.005)(100\%) = \pm 0.5\%.$$

The general definition of these concepts follows.

Definition 2.36

Let $y = f(x)$ and suppose that y changes from y_0 to y_1 as x changes from x_0 to x_1. We can describe the change in y as follows:

	Exact value	**Approximate value**
Absolute change (error)	$\Delta y = y_1 - y_0$	$dy = f'(x_0)\,\Delta x$
Relative change (error)	$\dfrac{\Delta y}{y_0}$	$\dfrac{dy}{y_0}$
Percentage change (error)	$\dfrac{\Delta y}{y_0} \times 100\%$	$\dfrac{dy}{y_0} \times 100\%$

EXAMPLE 6   The radius of a spherical balloon is measured as 12 in., with a maximum error in measurement of ± 0.06 in. Approximate the relative error and the percentage error for the calculated value of the volume.

SOLUTION   As in Figure 2.42, let x denote the measured radius of the balloon and Δx the maximum error in x. Let V denote the calculated volume and ΔV the error in V caused by Δx. Applying Definition (2.36) to the volume $V = \frac{4}{3}\pi x^3$ yields

$$\text{relative error} = \frac{\Delta V}{V} \approx \frac{dV}{V} = \frac{4\pi x^2 \Delta x}{\frac{4}{3}\pi x^3} = \frac{3\Delta x}{x}.$$

For the special case $x = 12$ and $\Delta x = \pm 0.06$, we obtain

$$\text{relative error} \approx \frac{3(\pm 0.06)}{12} = \pm 0.015.$$

From Definition (2.36),

$$\text{percentage error} \approx (\pm 0.015) \times (100\%) = \pm 1.5\%.$$

Thus, *on average,* there is an error of ± 0.015 in^3 per in^3 of calculated volume. Note that this leads to a percentage error of $\pm 1.5\%$ for the volume.

EXAMPLE 7   A sperm whale is spotted by a merchant ship, and crew members estimate its length L to be 32 ft, with a possible error of ± 2 ft. Whale research has shown that the weight W (in metric tons) is related to L by the formula $W = 0.000137L^{3.18}$. Use differentials to approximate

(a) the error in estimating the weight of the whale (to the nearest tenth of a metric ton)

(b) the relative error and the percentage error

SOLUTION   Let ΔL denote the error in the estimation of L, and let ΔW be the corresponding error in the calculated value of W. This error may be approximated by dW.

(a) Applying Definition (2.34) yields

$$\Delta W \approx dW = (0.000137)(3.18)L^{2.18}\,\Delta L.$$

Substituting $L = 32$ and $\Delta L = \pm 2$, we obtain

$$\Delta W \approx (0.000137)(3.18)(32)^{2.18}(\pm 2) \approx \pm 1.7 \text{ metric tons.}$$

(b) By Definition (2.36),

$$\text{relative error} = \frac{\Delta W}{W} \approx \frac{dW}{W} = \frac{(0.000137)(3.18)L^{2.18}\,\Delta L}{(0.000137)L^{3.18}} = \frac{3.18\,\Delta L}{L}.$$

Substituting $\Delta L = \pm 2$ and $L = 32$, we have

$$\text{relative error} \approx \frac{3.18(\pm 2)}{32} \approx \pm 0.20.$$

By Definition (2.36),

$$\text{percentage error} \approx (\pm 0.20) \times (100\%) = \pm 20\%.$$

Estimates of vertical wind shear are of great importance to pilots during take-offs and landings. If we assume that the wind speed v at a height h above the ground is given by $v = f(h)$, where f is a differentiable

Figure 2.43

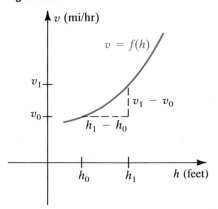

function, then **vertical (scalar) wind shear** is defined as dv/dh (the instantaneous rate of change of v with respect to h). Since it is impossible to know the wind speed v at every height h, the wind shear must be estimated by using only a finite number of function values. Consider the situation illustrated in Figure 2.43, where we know only the wind speeds v_0 and v_1 at heights h_0 and h_1, respectively. An estimate of wind shear at height h_1 may be obtained by using the approximation formula

$$\frac{dv}{dh}\bigg]_{h=h_1} \approx \frac{v_1 - v_0}{h_1 - h_0}.$$

The empirical relation

$$\left(\frac{v_0}{v_1}\right) = \left(\frac{h_0}{h_1}\right)^P$$

may also be employed, where the exponent P is determined by observation and depends on many factors. For strong winds, the value $P = \frac{1}{7}$ is sometimes used.

EXAMPLE 8 Suppose that at a height of 20 ft above the ground the wind speed is 28 mi/hr. On the basis of the preceding discussion (with $P = \frac{1}{7}$), estimate the vertical wind shear 200 ft above the ground.

SOLUTION Using the notation of the preceding discussion, we let

$$h_0 = 20, \quad v_0 = 28, \quad \text{and} \quad h_1 = 200.$$

Solving $(v_0/v_1) = (h_0/h_1)^P$ for v_1 and then substituting values, we obtain

$$v_1 = v_0 \left(\frac{h_1}{h_0}\right)^P = 28 \left(\frac{200}{20}\right)^{1/7} \approx 39 \text{ mi/hr.}$$

At $h_1 = 200$,

$$\frac{dv}{dh}\bigg]_{h=h_1} \approx \frac{v_1 - v_0}{h_1 - h_0} \approx \frac{39 - 28}{200 - 20} \approx 0.06.$$

Thus, at a height of 200 ft, the vertical wind shear is approximately 0.06 (mi/hr)/ft, which is a common value. Wind shear values greater than 0.1 are considered high.

EXERCISES 2.8

Exer. 1–8: Use a linear approximation to estimate $f(b)$ if the independent variable changes from a to b.

1 $f(x) = 4x^5 - 6x^4 + 3x^2 - 5$; $\quad a = 1, \quad b = 1.03$

2 $f(x) = -3x^3 + 8x - 7$; $\qquad a = 4, \quad b = 3.96$

3 $f(x) = x^4$; $\qquad\qquad\qquad\quad a = 1, \quad b = 0.98$

4 $f(x) = x^4 - 3x^3 + 4x^2 - 5$; $\quad a = 2, \quad b = 2.01$

5 $f(\theta) = 2\sin\theta + \cos\theta$; $\qquad a = 30°, \quad b = 27°$

6 $f(\phi) = \csc\phi + \cot\phi$; $\qquad a = 45°, \quad b = 46°$

7 $f(\alpha) = \sec\alpha$; $\qquad\qquad a = 60°, \quad b = 62°$

8 $f(\beta) = \tan\beta$; $\qquad\qquad a = 30°, \quad b = 28°$

Exer. 9–12: (a) Find general formulas for Δy and dy. (b) If, for the given values of a and Δx, x changes from a to $a + \Delta x$, find the values of Δy and dy.

9 $y = 2x^2 - 4x + 5$; $a = 2$, $\Delta x = -0.2$

10 $y = x^3 - 4$; $a = -1$, $\Delta x = 0.1$

11 $y = 1/x^2$; $a = 3$, $\Delta x = 0.3$

12 $y = \dfrac{1}{2+x}$; $a = 0$, $\Delta x = -0.03$

Exer. 13–18: Find (a) Δy, (b) dy, and (c) $dy - \Delta y$.

13 $y = 4 - 9x$

14 $y = 7x + 12$

15 $y = 3x^2 + 5x - 2$

16 $y = 4 - 7x - 2x^2$

17 $y = 1/x$

18 $y = 1/x^2$

c 19 (a) If $f(x) = \sin(\tan x - 1)$, find an (approximate) equation of the tangent line to the graph of f at $(2.5, f(2.5))$, using Exercise 53 in Section 2.2.

(b) Use the equation found in part (a) to approximate $f(2.6)$.

(c) Use (2.35) with $x = 2.5$ to approximate $f(2.6)$.

(d) Compare the two approximations in parts (b) and (c).

c 20 (a) If $f(x) = x^3 + 3x^2 - 2x + 5$, find an (approximate) equation of the tangent line to the graph of f at $(0.4, f(0.4))$.

(b) Use the equation found in part (a) to approximate $f(0.43)$.

(c) Use (2.35) with $x = 0.4$ to approximate $f(0.43)$.

(d) Compare the two approximations in parts (b) and (c).

Exer. 21–26: (a) Use differentials to approximate the value.

c (b) Compare the approximation in part (a) with the result obtained from evaluating the number with a calculator.

21 $\sqrt[3]{65}$
(*Hint:* Let $y = \sqrt[3]{x}$.)

22 $\sqrt[5]{35}$

23 $7^{2/3}$

24 $1/\sqrt{50}$

25 $\cos 59°$

26 $\tan(\pi/4 + 0.05)$

Exer. 27–30: Let x denote a measurement with a maximum error of Δx. Use differentials to approximate the relative error and the percentage error for the calculated value of y.

27 $y = 3x^4$; $x = 2$, $\Delta x = \pm 0.01$

28 $y = x^3 + 5x$; $x = 1$, $\Delta x = \pm 0.1$

29 $y = 4\sqrt{x} + 3x$; $x = 4$, $\Delta x = \pm 0.2$

30 $y = 6\sqrt[3]{x}$; $x = 8$, $\Delta x = \pm 0.03$

31 If $A = 3x^2 - x$, find dA for $x = 2$ and $dx = 0.1$.

32 If $P = 6t^{2/3} + t^2$, find dP for $t = 8$ and $dt = 0.2$.

33 If $y = 4x^3$ and the maximum percentage error in x is $\pm 15\%$, approximate the maximum percentage error in y.

34 If $z = 40\sqrt[5]{w^2}$ and the maximum relative error in w is ± 0.08, approximate the maximum relative error in z.

35 If $A = 15\sqrt[3]{s^2}$ and the allowable maximum relative error in A is to be ± 0.04, determine the allowable maximum relative error in s.

36 If $S = 10\pi x^2$ and the allowable maximum percentage error in S is to be $\pm 10\%$, determine the allowable maximum percentage error in x.

37 The radius of a circular manhole cover is estimated to be 16 in., with a maximum error in measurement of ± 0.06 in. Use differentials to estimate the maximum error in the calculated area of one side of the cover. Approximate the relative error and the percentage error.

38 The length of a side of a square floor tile is estimated as 1 ft, with a maximum error in measurement of $\pm \frac{1}{16}$ in. Use differentials to estimate the maximum error in the calculated area. Approximate the relative error and the percentage error.

39 Use differentials to approximate the increase in volume of a cube if the length of each edge changes from 10 in. to 10.1 in. What is the absolute change in volume?

40 A spherical balloon is being inflated with gas. Use differentials to approximate the increase in surface area of the balloon if the diameter changes from 2 ft to 2.02 ft.

41 One side of a house has the shape of a square surmounted by an equilateral triangle. If the length of the base is measured as 48 ft, with a maximum error in measurement of ± 1 in., calculate the area of the side. Use differentials to estimate the maximum error in the calculation. Approximate the relative error and the percentage error.

42 Small errors in measurements of dimensions of large containers can have a marked effect on calculated volumes. A silo has the shape of a right circular cylinder surmounted by a hemisphere (see figure on the following page). The altitude of the cylinder is exactly 50 ft. The circumference of the base is measured as 30 ft, with a maximum error in measurement of ± 6 in. Calculate the volume of the silo from these measurements, and use differentials to estimate the maximum error in the calculation. Approximate the relative error and the percentage error.

Exercise 42

$C = 30$ ft

50 ft

43 As sand leaks out of a container, it forms a conical pile whose altitude is always the same as the radius. If, at a certain instant, the radius is 10 cm, use differentials to approximate the change in radius that will increase the volume of the pile by 2 cm^3.

44 An isosceles triangle has equal sides of length 12 in. If the angle θ between these sides is increased from $30°$ to $33°$, use differentials to approximate the change in the area of the triangle.

45 Newton's law of gravitation states that the force F of attraction between two particles having masses m_1 and m_2 is given by $F = Gm_1 m_2/s^2$, where G is a constant and s is the distance between the particles. If $s = 20$ cm, use differentials to approximate the change in s that will increase F by 10%.

46 The formula $T = 2\pi \sqrt{l/g}$ relates the length l of a pendulum to its period T, where g is a gravitational constant. What percentage change in the length corresponds to a 30% increase in the period?

47 Constriction of arterioles is a cause of high blood pressure. It has been verified experimentally that as blood flows through an arteriole of fixed length, the pressure difference between the two ends of the arteriole is inversely proportional to the fourth power of the radius. If the radius of an arteriole decreases by 10%, use differentials to find the percentage change in the pressure difference.

48 The electrical resistance R of a wire is directly proportional to its length and inversely proportional to the square of its diameter. If the length is fixed, how accurately must the diameter be measured (in terms of percentage error) to keep the percentage error in R between -3% and 3%?

49 If an object of weight W pounds is pulled along a horizontal plane by a force applied to a rope that is attached to the object and if the rope makes an angle θ with the horizontal, then the magnitude of the force is given by

$$F(\theta) = \frac{\mu W}{\mu \sin\theta + \cos\theta},$$

where μ is a constant called the *coefficient of friction*. Suppose that a 100-lb box is being pulled along a floor and that $\mu = 0.2$ (see figure). If θ is changed from $45°$ to $46°$, use differentials to approximate the change in the force that must be applied.

Exercise 49

50 It will be shown in Chapter 11 that if a projectile is fired from a cannon with an initial velocity v_0 and at an angle α to the horizontal, then its maximum height h and range R are given by

$$h = \frac{v_0^2 \sin^2 \alpha}{2g} \quad \text{and} \quad R = \frac{2v_0^2 \sin\alpha \cos\alpha}{g}.$$

Suppose $v_0 = 100$ ft/sec and $g = 32$ ft/sec^2. If α is increased from $30°$ to $30.5°$, use differentials to estimate the changes in h and R.

51 At a point 20 ft from the base of a flagpole, the angle of elevation of the top of the pole is measured as $60°$, with a possible error of $\pm 0.25°$. Use differentials to approximate the error in the calculated height of the pole.

52 A spacelab circles the earth at an altitude of 380 mi. When an astronaut views the horizon, the angle θ shown in the figure is $65.8°$, with a possible maximum error of $\pm 0.5°$. Use differentials to approximate the error in the astronaut's calculation of the radius of the earth.

Exercise 52

380 mi

θ

r

53 The Great Pyramid of Egypt has a square base of 230 m (see figure). To estimate the height h of this massive structure, an observer stands at the midpoint of one of the sides and views the apex of the pyramid. The angle of elevation ϕ is found to be 52°. How accurate must this angle measurement be to keep the error in h between -1 m and 1 m?

Exercise 53

54 As a point light source moves on a semicircular track, as shown in the figure, the illuminance E on the surface is inversely proportional to the square of the distance s from the source and is directly proportional to the cosine of the angle θ between the direction of light flow and the normal to the surface. If θ is decreased from 21° to 20° and s is constant, use differentials to approximate the percentage increase in illuminance.

Exercise 54

55 Boyle's law states that if the temperature is constant, the pressure p and the volume v of a confined gas are related by the formula $pv = c$, where c is a constant or, equivalently, by $p = c/v$ with $v \neq 0$. Show that dp and Δv are related by the formula $p \, \Delta v + v \, dp = 0$.

56 In electrical theory, Ohm's law states that $I = V/R$, where I is the current (in amperes), V is the electromotive force (in volts), and R is the resistance (in ohms). Show that dI and ΔR are related by the formula $R \, dI + I \, \Delta R = 0$.

57 The area A of a square of side s is given by $A = s^2$. If s increases by an amount Δs, illustrate dA and $\Delta A - dA$ geometrically.

58 The volume V of a cube of edge s is given by $V = s^3$. If s increases by an amount Δs, illustrate dV and $\Delta V - dV$ geometrically.

59 The curved surface area S of a right circular cone having altitude h and base radius r is given by $S = \pi r \sqrt{r^2 + h^2}$. For a certain cone, $r = 6$ cm. The altitude is measured as 8 cm, with a maximum error in measurement of ± 0.1 cm.

    (a) Calculate S from the measurements and use differentials to estimate the maximum error in the calculation.

    (b) Approximate the percentage error.

60 The period T of a simple pendulum of length l may be calculated by means of the formula $T = 2\pi \sqrt{l/g}$, where g is a gravitational constant. Use differentials to approximate the change in l that will increase T by 1%.

61 Suppose that $3x^2 - x^2 y^3 + 4y = 12$ determines a differentiable function f such that $y = f(x)$. If $f(2) = 0$, use differentials to approximate the change in $f(x)$ if x changes from 2 to 1.97.

62 Suppose that $x^3 + xy + y^4 = 19$ determines a differentiable function f such that $y = f(x)$. If $P(1, 2)$ is a point on the graph of f, use differentials to approximate the y-coordinate b of the point $Q(1.1, b)$ on the graph.

  c **63** Suppose that $x^2 + xy^3 = 4.0764$ determines a differentiable function f such that $y = f(x)$.

    (a) If $P(1.2, 1.3)$ and $Q(1.23, b)$ are on the graph of f, use (2.35) to approximate b.

    (b) Apply the method in part (a), using $Q(1.23, b)$ to approximate the y-coordinate of $R(1.26, c)$. (This process, called *Euler's method*, can be repeated to approximate additional points on the graph.)

  c **64** Suppose that $\sin x + y \cos y = -2.395$ determines a differentiable function f such that $y = f(x)$.

    (a) If $P(2.1, 3.3)$ is an approximation to a point on the graph of f, use (2.35) to approximate the y-coordinate of $Q(2.12, b)$.

    (b) Apply the method in part (a), using $Q(2.12, b)$ to approximate the y-coordinate of $R(2.14, c)$.

2.9 NEWTON'S METHOD

Many problems and applications in mathematics require solving equations. We can write equations involving one variable in the form $f(x) = 0$ for some function f. The cubic equation $2x^3 + x^2 - 16x - 15 = 0$, for example, has this form, where $f(x) = 2x^3 + x^2 - 16x - 15$. Recall that a *zero* of a function f is a value r such that $f(r) = 0$. The zeros of the function are also called the *roots* of the equation.

In this section, we explore methods for finding a real zero of a function f. One such method for a polynomial or a rational function is to *factor* the numerator. Since the cubic polynomial $f(x) = 2x^3 + x^2 - 16x - 15$ factors as $f(x) = (2x + 5)(x + 1)(x - 3)$, the only real numbers r satisfying $f(r) = 0$ are $r = -2.5$, -1, and 3. When f is a quadratic polynomial, the quadratic formula gives the zeros (real and complex). For many functions, it is not possible to use factoring or an exact formula to find the zeros. In such cases, we may be able only to approximate values of the zeros. In this section, we focus on *Newton's method* for carrying out such approximations

Newton's method for approximating a zero r of a differentiable function f is based on the idea that the tangent line stays close to the curve near the point of tangency. With this method, we begin with some approximation x_1 for the zero and consider the tangent line l to the graph of $y = f(x)$ at $(x_1, f(x_1))$ (see Figure 2.44). The tangent line and the graph of f should intersect the x-axis near each other since the tangent line remains close to the graph of f. Thus, we can approximate a zero for f by finding a zero for the tangent line. Because the equation of the tangent line is linear, it is easy to determine where it has a zero.

A graph of the function is often very helpful for suggesting a "good" first approximation x_1. Lacking a graphing utility, we might evaluate the function several times. The intermediate value theorem (1.26) guarantees a zero in any interval where the values of a continuous function at the endpoints have opposite signs.

Next, we consider the tangent line l to the graph of f at the point $(x_1, f(x_1))$. If x_1 is sufficiently close to r, then, as illustrated in Figure 2.44, the x-intercept x_2 of l should be a better approximation to r. Since the slope of l is $f'(x_1)$, an equation of the tangent line is

$$y - f(x_1) = f'(x_1)(x - x_1).$$

The x-intercept x_2 of l corresponds to the point $(x_2, 0)$ on l, so

$$0 - f(x_1) = f'(x_1)(x_2 - x_1).$$

If $f'(x_1) \neq 0$, the preceding equation is equivalent to

$$x_2 = x_1 - \frac{f(x_1)}{f'(x_1)}.$$

Figure 2.44

Taking x_2 as a second approximation to r, we may repeat the process by using the tangent line at $(x_2, f(x_2))$. If $f'(x_2) \neq 0$, a third approximation x_3 is given by

$$x_3 = x_2 - \frac{f(x_2)}{f'(x_2)}.$$

We continue the process until the desired degree of accuracy is obtained. This technique of successive approximations of real zeros is called **Newton's method**.

Newton's Method 2.37

Let f be a differentiable function, and suppose r is a real zero of f. If x_n is an approximation to r, then the next approximation x_{n+1} is given by

$$x_{n+1} = x_n - \frac{f(x_n)}{f'(x_n)},$$

provided $f'(x_n) \neq 0$.

Figure 2.45

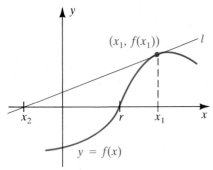

Newton's method does not guarantee that x_{n+1} is a better approximation to r than x_n for every n. In particular, we must be careful in choosing the first approximation x_1. If x_1 is not sufficiently close to r, it is possible for the second approximation x_2 to be worse than x_1. Figure 2.45 shows such a case. It is clear that we should not choose a number x_n such that $f'(x_n)$ is close to 0, for then the tangent line l is almost horizontal.

If $x_1 \approx r$, f'' is continuous near r, and $f'(r) \neq 0$, then we can show that the approximations $x_2, x_3, \ldots$ approach r rapidly, with the number of decimal places of accuracy nearly doubling with each successive approximation. If $f(x)$ has a factor $(x - r)^k$ with $k > 1$ and if $x_n \neq r$ for each n, then the approximations approach r more slowly, because $f'(r) = 0$.

We will use the following rule when applying Newton's method: *If an approximation to k decimal places is required, we continue the process until two consecutive approximations give exactly the same k decimal places.* If we use a computer or calculator, we cannot go beyond the precision of the machine. Working by hand, we can calculate each approximation x_2, $x_3, \ldots$ to at least k decimal places. The following examples illustrate the process.

Figure 2.46

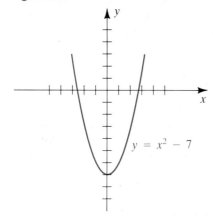

$y = x^2 - 7$

EXAMPLE ■ I Use Newton's method to approximate $\sqrt{7}$ to five decimal places.

SOLUTION This problem is equivalent to that of approximating the positive real zero r of $f(x) = x^2 - 7$. Figure 2.46 shows the graph of f. Since $f(2) = -3$ and $f(3) = 2$, it follows from the continuity of f that $2 < r < 3$. From the graph, we believe that f has only one zero in the open interval $(2, 3)$. If x_n is any approximation to r, then Newton's method

(2.37) gives the next approximation x_{n+1} as

$$x_{n+1} = x_n - \frac{f(x_n)}{f'(x_n)} = x_n - \frac{x_n^2 - 7}{2x_n}.$$

Let us choose $x_1 = 2.5$ as a first approximation. Using the formula for x_{n+1} with $n = 1$ gives us

$$x_2 = 2.5 - \frac{(2.5)^2 - 7}{2(2.5)} = 2.65.$$

Again using the formula (with $n = 2$), we obtain the next approximation,

$$x_3 = 2.65 - \frac{(2.65)^2 - 7}{2(2.65)} \approx 2.64575.$$

Repeating the procedure (with $n = 3$) yields

$$x_4 = 2.64575 - \frac{(2.65475)^2 - 7}{2(2.64575)} \approx 2.64575.$$

Since two consecutive values of x_n are the same (to the desired degree of accuracy), we have $\sqrt{7} \approx 2.64575$. Note that $(2.64575)^2 = 6.9999930625$.

NOTE   A 13-digit calculator gives $\sqrt{7} \approx 2.645751311065$. Some early computers used a procedure very similar to this one to calculate square roots, but even faster algorithms are now used.

   EXAMPLE ▪ 2   Find the largest positive real root of $x^3 - 3x + 1 = 0$ to ten decimal places.

Figure 2.47

$y = x^3 - 3x + 1$

SOLUTION   If we let $f(x) = x^3 - 3x + 1$, then the problem is equivalent to finding the largest real zero of f. Figure 2.47 shows the graph of f. Note that f has three real zeros. We wish to find the zero that lies between 1 and 2. Since $f'(x) = 3x^2 - 3$, the formula for x_{n+1} in Newton's method is

$$x_{n+1} = x_n - \frac{x_n^3 - 3x_n + 1}{3x_n^2 - 3}.$$

Referring to the graph, we take $x_1 = 1.5$ as a first approximation. On a calculator with which we can store a calculation in a variable memory labeled $\boxed{\text{X}}$ using an operation symbolized by $\boxed{\rightarrow}$, we first store 1.5 in $\boxed{\text{X}}$. Then we create a command line of the form

$$\text{x} - (\text{x}^\wedge 3 - 3\text{x} + 1) / (3\text{x}^\wedge 2 - 3) \rightarrow \text{x}$$

Repeated execution of this line produces the successive approximations of Newton's method:

$$x_2 \approx 1.5333333333$$
$$x_3 \approx 1.5320906433$$
$$x_4 \approx 1.5320888862$$
$$x_5 \approx 1.5320888862$$

Thus the desired approximation is 1.5320888862. The remaining two real roots can be approximated in similar fashion (see Exercise 19).

EXAMPLE ▪ 3 Approximate the real root of $x - \cos x = 0$.

SOLUTION We wish to find a value of x such that $\cos x = x$. This value coincides with the x-coordinate of the point of intersection of the graphs of $y = \cos x$ and $y = x$. It appears from Figure 2.48 that $x_1 = 0.7$ is a reasonable first approximation. (Note that the figure also indicated that there is only one real root of the given equation.) If we let $f(x) = x - \cos x$, then $f'(x) = 1 + \sin x$ and the formula in Newton's method is

$$x_{n+1} = x_n - \frac{x_n - \cos x_n}{1 + \sin x_n}.$$

Although we can compute several successive approximations on a calculator, if we use a computer with greater precision, we can see the approximate doubling of the correct digits with each additional step of Newton's method. The following calculations were performed on a computer with 33-digit precision:

$$x_2 \approx \underline{0.73943649784805819542871591}1443437$$
$$x_3 \approx \underline{0.7390851604651073986023420}91722227$$
$$x_4 \approx \underline{0.73908513321516080561647}3437040986$$
$$x_5 \approx \underline{0.739085133215160641655312087673879}$$
$$x_6 \approx \underline{0.739085133215160641655312087673873}$$
$$x_7 \approx \underline{0.739085133215160641655312087673873}$$

If the results here are *rounded* to the underscored number of digits, then each result will be correct to that number of digits.

Figure 2.48

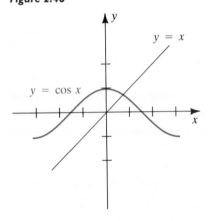

There is seldom a need for 33 decimal places in an answer to a problem in the physical world. Still it is comforting to know that Newton's method often works this well. A final warning about implementing this procedure on a computer is appropriate here. Although we *hope* to continue the process until two successive results agree, it may not happen. To avoid continuing the process indefinitely, we can write code asking for a finite number of approximations. For instance, the code used for Example 3 computed $x_2, x_3, \ldots, x_{10}$. Since the last seven approximations were exactly the same, only two of them are given in the list.

There are many other ways to approximate zeros of functions. Some of these are explored in the exercises of this section and in later chapters. Since approximating zeros of a function is equivalent to the important problem of solving equations in one variable, many calculators and most computer algebra systems provide a simple command to carry out this task. Explore the manual for the device you are using to understand the exact syntax for the "solver" routine. You can be confident that an approximation very much like Newton's method is being used.

EXERCISES 2.9

c | For Exercises 1–26, use Newton's method to at least the accuracy indicated.

Exer. 1–4: (a) Approximate to four decimal places. (b) Compare the approximation to one obtained from a calculator.

1 $\sqrt{11}$

2 $\sqrt{57}$

3 $\sqrt[3]{2}$

4 $\sqrt[5]{3}$

Exer. 5–8: Approximate, to four decimal places, the root of the equation that lies in the interval.

5 $x^4 + 2x^3 - 5x^2 + 1 = 0$; [1, 2]

6 $x^4 - 5x^2 + 2x - 5 = 0$; [2, 3]

7 $x^5 + x^2 - 9x - 3 = 0$; [-2, -1]

8 $\sin x + x \cos x = \cos x$; [0, 1]

Exer. 9–10: Approximate the largest zero of $f(x)$ to four decimal places.

9 $f(x) = x^4 - 11x^2 - 44x - 24$

10 $f(x) = x^3 - 36x - 84$

Exer. 11–14: Approximate the real root to two decimal places.

11 The root of $x^3 + 5x - 3 = 0$

12 The largest root of $2x^3 - 4x^2 - 3x + 1 = 0$

13 The positive root of $2x - 3 \sin x = 0$

14 The root of $\cos x + x = 2$

Exer. 15–22: Approximate all real roots of the equation to two decimal places.

15 $x^4 = 125$

16 $10x^2 - 1 = 0$

17 $x^4 - x - 2 = 0$

18 $x^5 - 2x^2 + 4 = 0$

19 $x^3 - 3x + 1 = 0$

20 $x^3 + 2x^2 - 8x - 3 = 0$

21 $2x - 5 - \sin x = 0$

22 $x^2 - \cos 2x = 0$

Exer. 23–26: Approximate, to two decimal places, the x-coordinates of the points of intersection of the graphs of the equations.

23 $y = x^2$; $y = \sqrt{x + 3}$

24 $y = x^3$; $y = 7 - x^2$

25 $y = \cos \frac{1}{2}x$; $y = 9 - x^2$

26 $y = \sin 2x$; $y = 6x - 6$

27 Approximations to π may be obtained by applying Newton's method to $f(x) = \sin x$ and letting $x_1 = 3$.
   (a) Find the first five approximations to π.
   (b) What happens to the approximations if $x_1 = 6$?

28 A dramatic example of the phenomenon of *resonance* occurs when a singer adjusts the pitch of her voice to shatter a wine glass. Functions having the form $f(x) = ax \cos bx$ occur in the mathematical analysis of such vibrations. Shown in the figure is a graph of $f(x) = x \cos 2x$. Use Newton's method to approximate, to three decimal places, the value of x that lies between 1 and 2 such that $f'(x) = 0$.

Exercise 28

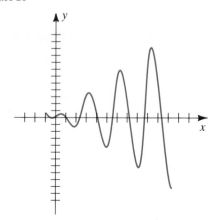

29 The graph of a function f is shown. Explain why Newton's method fails to approximate the zero of f if $x_1 = 0.5$.

Exercise 29

$$f(x) = 16x^3 - 24x^2 + 12x - 1$$

30 If $f(x) = x^{1/3}$, show that Newton's method fails for any first approximation $x_1 \neq 0$.

c | Exer. 31–32: The functions f and g have a zero at $x = 1$. (a) Let $x_1 = 1.1$ in formula (2.37), and find x_2, x_3, and x_4 for each function. (b) Why are the approximations for the zero of g more accurate than those for the zero of f?

31 $f(x) = (x - 1)^3(x^2 - 3x + 7)$;
$g(x) = (x - 1)(x^2 - 3x + 7)$

32 $f(x) = (x - 1)^2\sqrt{x + 7}$;
$g(x) = (x - 1)\sqrt{x + 7}$

c | Exer. 33–34: If it is difficult to calculate $f'(x)$, the formula in (2.37) may be replaced by

$$x_{n+1} = x_n - \frac{f(x_n)}{m},$$

where

$$m = \frac{f(x_n) - f(x_{n-1})}{x_n - x_{n-1}} \approx f'(x_n).$$

Two initial values, x_1 and x_2, are required to use this method (called the *secant method*). Use the secant method to approximate, to three decimal places, the zero of f that is in $[0, 1]$.

33 $f(x) = \tan^2(\cos^2 x - x + 0.25) - 0.5x$;
$$x_1 = 0.5, \quad x_2 = 0.55$$

34 $f(x) = \dfrac{1}{\cos^2 x - x} - 5\sqrt{x}$;
$$x_1 = 0.4, \quad x_2 = 0.5$$

c | Exer. 35–36: Graph f and g on the same coordinate axes. (a) Estimate, to one decimal place, the x-coordinate x_1 of the point of intersection of the graphs. (b) Use Newton's method to approximate the x-coordinate in part (a) to two decimal places.

35 $f(x) = \frac{1}{4}x^3 + x - 1$;
$g(x) = \sin^2 x$

36 $f(x) = x^3 - x^2 + x - 1$;
$g(x) = -x^3 - 1.1x^2 - x - 1.9$

c | **37** Many of the equations we solved can be rewritten in the form $g(x) = x$. A simple way to try to approximate the solution is to find a close initial x_1 as we did for Newton's method. Then proceed to find successive approximations by using $x_{n+1} = g(x_n)$. Use this method (called a *Picard iteration*) to approximate, to five decimal places, the desired solutions.

(a) $\sin x + 1 = x$ (b) $\frac{1}{2}\cos x = x$

CHAPTER 2 REVIEW EXERCISES

Exer. 1–2: Find $f'(x)$ directly from the definition of the derivative.

1 $f(x) = \dfrac{4}{3x^2 + 2}$ **2** $f(x) = \sqrt{5 - 7x}$

Exer. 3–24: Find the first derivative.

3 $f(x) = 2x^3 - 7x + 2$ **4** $k(x) = \dfrac{1}{x^4 - x^2 + 1}$

5 $g(t) = \sqrt{6t + 5}$ **6** $h(t) = \dfrac{1}{\sqrt{6t + 5}}$

7 $F(z) = \sqrt[3]{7z^2 - 4z + 3}$ **8** $f(w) = \sqrt[5]{3w^2}$

9 $G(x) = \dfrac{6}{(3x^2 - 1)^4}$ **10** $H(x) = \dfrac{(3x^2 - 1)^4}{6}$

11 $F(r) = (r^2 - r^{-2})^{-2}$

12 $h(z) = \left[(z^2 - 1)^5 - 1\right]^5$

13 $g(x) = \sqrt[5]{(3x + 2)^4}$

14 $P(x) = (x + x^{-1})^2$

15 $r(s) = \left(\dfrac{8s^2 - 4}{1 - 9s^3}\right)^4$

16 $g(w) = \dfrac{(w - 1)(w - 3)}{(w + 1)(w + 3)}$

17 $F(x) = (x^6 + 1)^5 (3x + 2)^3$

18 $k(z) = [z^2 + (z^2 + 9)^{1/2}]^{1/2}$

19 $k(s) = (2s^2 - 3s + 1)(9s - 1)^4$

20 $p(x) = \dfrac{2x^4 + 3x^2 - 1}{x^2}$

21 $f(x) = 6x^2 - \dfrac{5}{x} + \dfrac{2}{\sqrt[3]{x^2}}$

22 $F(t) = \dfrac{5t^2 - 7}{t^2 + 2}$ **23** $f(w) = \sqrt{\dfrac{2w + 5}{7w - 9}}$

24 $S(t) = \sqrt{t^2 + t + 1}\sqrt[3]{4t - 9}$

Exer. 25–32: Find the limit, if it exists.

25 $\displaystyle\lim_{x \to 0} \dfrac{x^2}{\sin x}$ **26** $\displaystyle\lim_{x \to 0} \dfrac{x^2 + \sin^2 x}{4x^2}$

27 $\displaystyle\lim_{x \to 0} \dfrac{\sin^2 x + \sin 2x}{3x}$ **28** $\displaystyle\lim_{x \to 0} \dfrac{2 - \cos x}{1 + \sin x}$

29 $\displaystyle\lim_{x \to 0} \dfrac{2\cos x + 3x - 2}{5x}$ **30** $\displaystyle\lim_{x \to 0} \dfrac{3x + 1 - \cos^2 x}{\sin x}$

31 $\displaystyle\lim_{x \to 0} \dfrac{x \sin x}{1 - \cos x}$ **32** $\displaystyle\lim_{x \to 0} \dfrac{\cos x - 1}{2x}$

Exer. 33–50: Find the first derivative.

33 $g(r) = \sqrt{1 + \cos 2r}$ **34** $g(z) = \csc\left(\dfrac{1}{z}\right) + \dfrac{1}{\sec z}$

35 $f(x) = \sin^2(4x^3)$ **36** $H(t) = (1 + \sin 3t)^3$

37 $h(x) = (\sec x + \tan x)^5$ **38** $K(r) = \sqrt[3]{r^3 + \csc 6r}$

39 $f(x) = x^2 \cot 2x$ **40** $P(\theta) = \theta^2 \tan^2(\theta^2)$

41 $K(\theta) = \dfrac{\sin 2\theta}{1 + \cos 2\theta}$ **42** $g(v) = \dfrac{1}{1 + \cos^2 2v}$

43 $g(x) = (\cos \sqrt[3]{x} - \sin \sqrt[3]{x})^3$

44 $f(x) = \dfrac{x}{2x + \sec^2 x}$ **45** $G(u) = \dfrac{\csc u + 1}{\cot u + 1}$

46 $k(\phi) = \dfrac{\sin \phi}{\cos \phi - \sin \phi}$

47 $F(x) = \sec 5x \tan 5x \sin 5x$

48 $H(z) = \sqrt{\sin \sqrt{z}}$ **49** $g(\theta) = \tan^4(\sqrt[4]{\theta})$

50 $f(x) = \csc^3 3x \cot^2 3x$

Exer. 51–56: Assuming that the equation determines a differentiable function f such that $y = f(x)$, find y'.

51 $5x^3 - 2x^2y^2 + 4y^3 - 7 = 0$

52 $3x^2 - xy^2 + y^{-1} = 1$

53 $\dfrac{\sqrt{x} + 1}{\sqrt{y} + 1} = y$ **54** $y^2 - \sqrt{xy} + 3x = 2$

55 $xy^2 = \sin(x + 2y)$ **56** $y = \cot(xy)$

Exer. 57–58: Find equations of the tangent line and the normal line to the graph of f at P.

57 $y = 2x - \dfrac{4}{\sqrt{x}}$; $P(4, 6)$

58 $x^2y - y^3 = 8$; $P(-3, 1)$

59 Find the x-coordinates of all points on the graph of the equation $y = 3x - \cos 2x$ at which the tangent line is perpendicular to the line $2x + 4y = 5$.

60 If $f(x) = \sin 2x - \cos 2x$ for $0 \le x \le 2\pi$, find the x-coordinates of all points on the graph of f at which the tangent line is horizontal.

Exer. 61–62: Find y', y'', and y'''.

61 $y = 5x^3 + 4\sqrt{x}$

62 $y = 2x^2 - 3x - \cos 5x$

63 If $x^2 + 4xy - y^2 = 8$, find y'' by implicit differentiation.

64 If $f(x) = x^3 - x^2 - 5x + 2$, find

  (a) the x-coordinates of all points on the graph of f at which the tangent line is parallel to the line through $A(-3, 2)$ and $B(1, 14)$

  (b) the value of f'' at each zero of f'

65 If $y = 3x^2 - 7$, find

  (a) Δy **(b)** dy **(c)** $dy - \Delta y$

66 If $y = 5x/(x^2 + 1)$, find dy and use it to approximate the change in y if x changes from 2 to 1.98. What is the exact change in y?

67 The side of an equilateral triangle is estimated to be 4 in., with a maximum error of ± 0.03 in. Use differentials to estimate the maximum error in the calculated area of the triangle. Approximate the percentage error.

68 If $s = 3r^2 - 2\sqrt{r + 1}$ and $r = t^3 + t^2 + 1$, use the chain rule to find the value of ds/dt at $t = 1$.

69 If $f(x) = 2x^3 + x^2 - x + 1$ and $g(x) = x^5 + 4x^3 + 2x$, use differentials to approximate the change in $g(f(x))$ if x changes from -1 to -1.01.

70 Use differentials to find a linear approximation of $\sqrt[3]{64.2}$.

71 Suppose f and g are functions such that $f(2) = -1$, $f'(2) = 4$, $f''(2) = -2$, $g(2) = -3$, $g'(2) = 2$, and $g''(2) = 1$. Find the value of each of the following at $x = 2$.

(a) $(2f - 3g)'$ (b) $(2f - 3g)''$ (c) $(fg)'$

(d) $(fg)''$ (e) $(f/g)'$ (f) $(f/g)''$

72 Refer to Exercise 85 in Section 2.5. Let f be an odd function and g an even function such that $f(3) = -3$, $f'(3) = 7$, $g(3) = -3$, and $g'(3) = -5$. Find $(f \circ g)'(3)$ and $(g \circ f)'(3)$.

73 Determine where the graph of f has a vertical tangent line or a cusp.

(a) $f(x) = 3(x + 1)^{1/3} - 4$

(b) $f(x) = 2(x - 8)^{2/3} - 1$

74 Let $f(x) = \begin{cases} (2x - 1)^3 & \text{if } x \geq 2 \\ 5x^2 + 34x - 61 & \text{if } x < 2 \end{cases}$

Determine if f is differentiable at 2.

75 The Stefan–Boltzmann law states that the radiant energy emitted from a unit area of a black surface is given by $R = kT^4$, where R is the rate of emission per unit area, T is the temperature (in $°K$), and k is a constant. If the error in the measurement of T is 0.5%, find the resulting percentage error in the calculated value of R.

76 Let V and S denote the volume and surface area, respectively, of a spherical balloon. If the diameter is 8 cm and the volume increases by 12 cm^3, use differentials to approximate the change in S.

77 A right circular cone has height $h = 8$ ft, and the base radius r is increasing. Find the rate of change of its surface area S with respect to r when $r = 6$ ft.

78 The intensity of illumination from a source of light is inversely proportional to the square of the distance from the source. If a student works at a desk that is a certain distance from a lamp, use differentials to find the percentage change in distance that will increase the intensity by 10%.

79 The ends of a horizontal water trough 10 ft long are isosceles trapezoids with lower base 3 ft, upper base 5 ft, and altitude 2 ft. If the water level is rising at a rate of $\frac{1}{4}$ in./min when the depth of the water is 1 ft, how fast is water entering the trough?

80 Two cars are approaching the same intersection along roads that run at right angles to each other. Car A is traveling at 20 mi/hr, and car B is traveling at 40 mi/hr. If, at a certain instant, A is $\frac{1}{4}$ mi from the intersection and B is $\frac{1}{2}$ mi from the intersection, find the rate at which they are approaching each other at that instant.

81 Boyle's law states that $pv = c$, where p is pressure, v is volume, and c is a constant. Find a formula for the rate of change of p with respect to v.

82 A railroad bridge is 20 ft above, and at right angles to, a river. A man in a train traveling 60 mi/hr passes over the center of the bridge at the same instant that a man in a motorboat traveling 20 mi/hr passes under the center of the bridge (see figure). How fast are the two men moving away from each other 10 sec later?

Exercise 82

83 A large ferris wheel is 100 ft in diameter and rises 110 ft off the ground, as illustrated in the figure. Each revolution of the wheel takes 30 sec.

(a) Express the distance h of a seat from the ground as a function of time t (in seconds) if $t = 0$ corresponds to a time when the seat is at the bottom.

(b) If a seat is rising, how fast is the distance changing when $h = 55$ ft?

Exercise 83

84 A piston is attached to a crankshaft, as shown in the figure. The connecting rod *AB* has length 6 in., and the radius of the crankshaft is 2 in.

(a) If the crankshaft rotates counterclockwise 2 times per second, find formulas for the position of point *A* at *t* seconds after *A* has coordinates $(2, 0)$.

(b) Find a formula for the position of point *B* at time *t*.

(c) How fast is *B* moving when *A* has coordinates $(0, 2)$?

c **85** Use Newton's method to approximate, to three decimal places, the root of the equation $\sin x - x \cos x = 0$ between π and $3\pi/2$.

c **86** Use Newton's method to approximate $\sqrt[4]{5}$ to three decimal places.

Exercise 84

EXTENDED PROBLEMS AND GROUP PROJECTS

1 How far can you proceed in determining the derivative of the function $f(x) = 2^x$? Some suggestions follow.

(a) Assuming that $f(x) = 2^x$ is defined and continuous for all real values of x, show that $f'(0) = \lim_{h \to 0}(2^h - 1)/h$ if this limit exists.

c (b) Evaluate the difference quotient $(2^h - 1)/h$ for a number of different values of h close to 0. Does the difference quotient appear to reach a limit as h approaches 0? What is that limit?

c (c) If you have a graphing calculator or access to graphing software, examine the graph of $(2^h - 1)/h$ for $h \neq 0$.

(d) Suppose $\lim_{h \to 0}(2^h - 1)/h$ does exist and has value M. Use the definition of the derivative to show that $f'(x)$ would equal $M(2^x)$ for all real values of x. Thus, if we can show that 2^x is differentiable at $x = 0$, then we know it is differentiable at all x.

(e) Can you determine analytically what

$$\lim_{h \to 0} \frac{2^h - 1}{h}$$

really is?

(f) What can you say about the differentiability of other exponential functions such as 3^x, 10^x, and $(\frac{1}{2})^x$?

(g) Write up your methods of investigation and conclusions.

2 Suppose that the function f is differentiable at every value x in an interval I, and let f' be its derivative.

(a) Give several examples of functions f for which f' is itself differentiable at every x in I.

(b) Give an example in which f' is continuous at every x in I, but not differentiable everywhere.

(c) Can you construct an example of a function f for which f' is continuous at every x in I, but has a derivative at no point in I?

(d) Find a function f such that $f'(x) = |x|$.

(e) Prove that it is impossible to have a differentiable function f with the property that

$$f'(x) = \begin{cases} 0 & \text{if } x < 0 \\ 1 & \text{if } x \geq 0 \end{cases}$$

(f) If f' is the derivative of f, can the function f' have any simple discontinuities?

3 (a) Show that for each rational number $r \neq -1$, there is a rational number q such that the derivative of $f(x) = (1/q)x^q$ is x^r.

(b) Show that there is no rational number q such that the function $f(x) = (1/q)x^q$ has $f'(x) = 1/x$.

(c) Try to construct a function f such that $f'(x) = 1/x$ for all positive values of x. Can you build such a function f out of polynomials or rational functions or trigonometric functions? What properties must such a function f have?

INTRODUCTION

WIND TURBINES EXEMPLIFY EFFORTS to find energy sources that maximize electrical power while keeping costs and pollution at acceptably low levels. In many other real-world situations, there is a similar challenge to optimize the behavior of a system while meeting prescribed constraints. Knowledge about derivatives helps solve such optimization problems.

In this chapter, we use derivatives to obtain important details about function values $f(x)$ as x varies through an interval. These facts enable us to sketch the graph of a function accurately and to describe precisely where it rises or falls—something that generally cannot be accomplished with precalculus methods.

Of major importance is finding the high and low points on a graph, since they provide the largest and smallest function values. The determination of these *extreme values* plays a significant role in applications that involve time, temperature, volume, pressure, gasoline consumption, air pollution, business profits, corporate expenses, and, in general, *any* quantity that can be represented by a function. In Section 3.1, we show that local extreme points of a function correspond to points where the derivative has value 0 or the function is not differentiable.

In Sections 3.2–3.4, we develop further the theory that relates derivatives to extreme values. We obtain this relationship by using one of the most important results in calculus: the *mean value theorem*, which we prove in Section 3.2. This theorem also plays a crucial role in Chapter 4, where we will examine another fundamental concept of calculus: the *definite integral*. We extend the theory in Section 3.3 to show how the sign of the first derivative provides information about where the function is increasing and decreasing. In Section 3.4, we illustrate how the sign of the second derivative is related to the concavity of the graph of the function. We consider applications of the theory to *graphing functions* in Section 3.5 and to *optimization problems* in Section 3.6.

In the final sections of the chapter, we examine applications of the derivative to a variety of problems in the physical and social sciences. In Section 3.7, particular emphasis is given to motion along a straight line, examining the relationship among position, velocity, speed, and acceleration for objects exhibiting simple harmonic motion or objects subjected to constant acceleration. We then discuss, in Section 3.8, some recent applications of differentiation to problems in a range of the social and life sciences and economics.

The derivative is a key tool in finding optimal solutions to problems such as maximizing electric power or minimizing costs and pollution.

Applications of the Derivative

3.1   EXTREMA OF FUNCTIONS

We begin this section by looking informally at the property that a curve and its tangent line exhibit similar behavior near the point of tangency. We then make our observations more precise and turn to the problem of finding extreme values for a function f.

In Figure 3.1, the line l is tangent to the graph of the function f at the point $P(a, f(a))$. Note that l has a positive slope: It *rises* vertically as it moves from left to right. The graph must exhibit this same rising property near $x = a$. In terms of the function f, we see that as x increases through a, the values of $f(x)$ also increase. Since the slope of the tangent line at P is the derivative $f'(a)$ of the function f at $x = a$, we conclude: *If $f'(a) > 0$, then f is increasing at a.*

Examining the situation at the point $Q = (b, f(b))$ on the graph of f, we see that the tangent line l^*, which has a negative slope, *falls* vertically as it moves from left to right. The graph of f near Q also drops as it passes through $x = b$. We conclude: *If $f'(b) < 0$, then f is decreasing at b.*

These conclusions are clearly valid only near the point of tangency. The similarity between the behavior of the tangent line l and the behavior of the function f holds only when the line and the curve are close to each other, and the tangent line is guaranteed to be close to the curve only near the point of tangency. The line and the curve can drift far apart as we move away from the point of tangency. The tangent line l at point P in Figure 3.1 continues to rise forever, but this behavior is not true for the graph of f. Indeed, we have seen that f is decreasing at Q.

At the point $R(c, f(c))$ on the graph of f, in Figure 3.1, the tangent line l' is horizontal: It is neither rising nor falling. The graph in Figure 3.1 shows the behavior of the function near this point: The function f is decreasing as it approaches $x = c$ and is increasing afterward. We can say that the function f has a local minimum value at $x = c$—that is, near $x = c$, all the values of $f(x)$ are greater than $f(c)$.

Our reasoning here has been geometric and informal. To make our conclusions more precise, let us now consider a graph that represents variations in some physical quantity such as temperature, resistance in an elec-

Figure 3.1

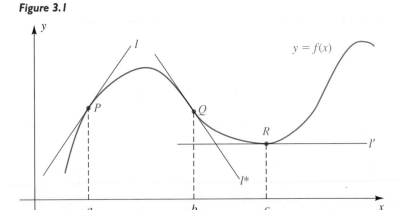

trical circuit, blood pressure of an individual, the amount of chemical in a solution, or the bacteria count in a culture. Suppose that the graph in Figure 3.2 was made by an instrument that measures one of these physical quantities. The x-axis represents time and the y-axis represents measurements of the quantity made at particular times.

Figure 3.2

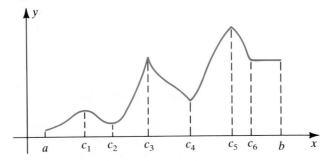

The graph indicates that the quantity increased in the time interval $[a, c_1]$, decreased in $[c_1, c_2]$, increased in $[c_2, c_3]$, decreased in $[c_3, c_4]$, and so on. We can extend these changes in the physical quantity at different time intervals to definitions of the behavior of any function.

Definition 3.1

Let a function f be defined on an interval I, and let x_1, x_2 denote numbers in I.

(i) f is **increasing** on I if $f(x_1) < f(x_2)$ whenever $x_1 < x_2$.
(ii) f is **decreasing** on I if $f(x_1) > f(x_2)$ whenever $x_1 < x_2$.
(iii) f is **constant** on I if $f(x_1) = f(x_2)$ for every x_1 and x_2.

Figure 3.3 contains graphical illustrations of Definition (3.1). We see that if a function is increasing, then its graph rises as x increases. If a function is decreasing, then its graph falls as x increases. Note that we use

Figure 3.3

(i) Increasing function

(ii) Decreasing function

(iii) Constant function

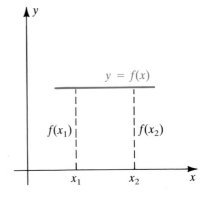

the phrases *f is increasing* and *f(x) is increasing* interchangeably, as we also do for *decreasing*.

If Figure 3.2 is the graph of a function f, we see that f is increasing on the intervals $[a, c_1]$, $[c_2, c_3]$, and $[c_4, c_5]$, decreasing on $[c_1, c_2]$, $[c_3, c_4]$, and $[c_5, c_6]$, and constant on the interval $[c_6, b]$. If we restrict our attention to the interval $[c_1, c_4]$, we note that the quantity has its largest (or maximum) value at c_3 and its smallest (or minimum) value at c_2. In other intervals, the largest and smallest values are different. Over the *entire* interval $[a, b]$, we see that the maximum value occurs at c_5 and the minimum value at a. We define the maximum and minimum values of a function f as follows.

Definition 3.2

Let a function f be defined on a set S of real numbers, and let c be a number in S.

(i) $f(c)$ is the **maximum value** of f on S if $f(x) \leq f(c)$ for every x in S.

(ii) $f(c)$ is the **minimum value** of f on S if $f(x) \geq f(c)$ for every x in S.

Maximum and minimum values are illustrated in Figures 3.4 and 3.5. Here S is shown as a closed interval, although Definition (3.2) may be applied to other types of intervals or sets of real numbers. Each graph in these figures has a maximum or minimum value $f(c)$ with $a < c < b$, but such values may also occur at endpoints of intervals or at endpoints of domains of functions.

Figure 3.4 Maximum value $f(c)$

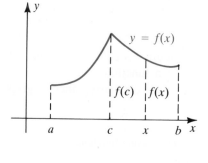

Figure 3.5 Minimum value $f(c)$

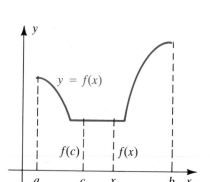

Figure 3.6 $f(x) = 4 - x^2$

(a) $[-2, 1]$

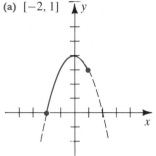

Max: $f(0) = 4$
Min: $f(-2) = 0$

(b) $(-2, 1)$

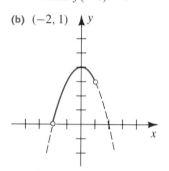

Max: $f(0) = 4$
Min: none

(c) $(1, 2]$

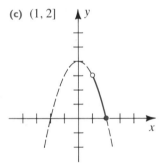

Max: none
Min: $f(2) = 0$

(d) $(1, 2)$

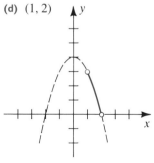

Max: none
Min: none

If $f(c)$ is the maximum value of f on S, we say that f *takes on* its maximum value at c. The point $(c, f(c))$ is a highest point on the graph of f. If $f(c)$ is the minimum value of f, we say that f *takes on* its minimum value at c, and $(c, f(c))$ is a lowest point on the graph of f. Maximum and minimum values are sometimes called **extreme values**, or **extrema**, of f. A function can take on a maximum or a minimum value more than once. If f is a constant function, then $f(c)$ is both a maximum and a minimum value of f for *every* real number c.

If D is the domain of f, then the maximum and minimum values of f on D, if they exist, are called the **global** (or **absolute**) **maximum** and **global** (or **absolute**) **minimum**.

EXAMPLE ■ 1 Let $f(x) = 4 - x^2$. Find the extrema of f on the following intervals:

(a) $[-2, 1]$ (b) $(-2, 1)$ (c) $(1, 2]$ (d) $(1, 2)$

SOLUTION The graph of f (a parabola) is sketched with dashes in Figure 3.6, where the solid portions correspond to the intervals (a)–(d). The extrema in each interval (denoted by Max and Min) are listed under each graph.

There is no minimum value of f in (b); if c is any number in $(-2, 1)$, we can find a number a in $(-2, 1)$ such that $f(a) < f(c)$. Similarly, there is no maximum value in (c) and neither a maximum nor a minimum value in (d).

EXAMPLE ■ 2 Let $f(x) = 1/x^2$. Find the extrema of f on

(a) $[-1, 2]$ (b) $[-1, 2)$

SOLUTION Portions of the graph of f and the extrema on the given intervals are shown in Figure 3.7. Note that f is not continuous at 0.

Figure 3.7 $f(x) = 1/x^2$

(a) $[-1, 2]$ (b) $[-1, 2)$

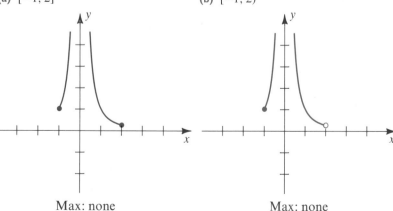

Max: none   Max: none
Min: $f(2) = \frac{1}{4}$ Min: none

The preceding examples indicate that the existence of maximum or minimum values may depend on the type of interval and on the continuity of the function. The next theorem states conditions under which a function takes on a maximum value and a minimum value on an interval. More advanced texts on calculus may be consulted for the proof.

Extreme Value Theorem 3.3

> If a function f is continuous on a closed interval $[a, b]$, then f takes on a minimum value and a maximum value at least once in $[a, b]$.

The importance of this theorem is that it *guarantees* the existence of extrema if f is continuous on a closed interval. However, as indicated by our examples, extrema *may* occur on intervals that are not closed and for functions that are not continuous.

We shall also be interested in *local extrema* of f, defined as follows.

Definition 3.4

> Let c be a number in the domain of a function f.
>
> **(i)** $f(c)$ is a **local maximum** of f if there exists an open interval (a, b) containing c such that $f(x) \leq f(c)$ for every x in (a, b) that is in the domain of f.
> **(ii)** $f(c)$ is a **local minimum** of f if there exists an open interval (a, b) containing c such that $f(x) \geq f(c)$ for every x in (a, b) that is in the domain of f.

The term *local* is used because we *localize* our attention to a sufficiently small *open* interval containing c such that f takes on a largest (or smallest) value at c. Outside of that open interval, f may take on larger (or smaller) values. The word *relative* is sometimes used in place of *local*. Each local maximum or minimum is called a **local extremum** of f, and the totality of such numbers are the **local extrema** of f. Several examples of local extrema are illustrated in Figure 3.8. As indicated in the figure,

Figure 3.8

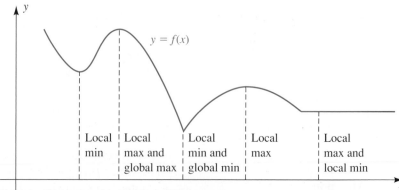

it is possible for a local minimum to be *larger* than a local maximum. It is also possible for a local minimum to be a global minimum and a local maximum to be a global maximum.

NOTE Whenever we use the phrases *maximum value* and *minimum value*, without including the adjective *local*, we mean the function values in Definition (3.2)—that is, we are referring to the *global*, or *absolute*, values. When referring to the extrema on a small open interval in Definition (3.4), we always use the word *local*.

The local extrema on an interval may not include the maximum or minimum values of f. For example, in Figure 3.2, $f(a)$ is the minimum value of f on $[a, b]$; however, it is not a local minimum, since there is no *open* interval I contained in $[a, b]$ such that $f(a)$ is the least value of f on I.

At a point corresponding to a local extremum of the function graphed in Figure 3.8, the tangent line is horizontal or the graph has a corner. The x-coordinate of such a point is a number at which the derivative either is zero or does not exist. The next theorem specifies that these facts are generally true. A proof is given at the end of this section.

Theorem 3.5

> If a function f has a local extremum at a number c in an open interval, then either $f'(c) = 0$ or $f'(c)$ does not exist.

The following is an immediate consequence of Theorem (3.5).

Corollary 3.6

> If $f'(c)$ exists and $f'(c) \neq 0$, then $f(c)$ is not a local extremum of the function f.

A result similar to Theorem (3.5) is true for the maximum and minimum values of a function that is continuous on a closed interval $[a, b]$, provided the extrema occur on the *open* interval (a, b). The theorem may be stated as follows.

Theorem 3.7

> If a function f is continuous on a closed interval $[a, b]$ and has its maximum or minimum value at a number c in the open interval (a, b), then either $f'(c) = 0$ or $f'(c)$ does not exist.

The proof of Theorem (3.7) is similar to that of Theorem (3.5) given at the end of this section.

It follows from Theorems (3.5) and (3.7) that the numbers at which the derivative either is zero or does not exist play a crucial role in the search for extrema of a function. We give these numbers and the associated points on the graph of the function special names.

Definition 3.8

> A number c in the domain of a function f is a **critical number** of f if either $f'(c) = 0$ or $f'(c)$ does not exist.

If c is a critical number of f such that $f'(c) = 0$, then c must be in an open interval of the domain of f, because the limit in Definition (2.6) exists, with $a = c$. A critical number c such that $f'(c)$ does not exist may occur at an endpoint of the domain of f.

Referring to Theorem (3.7), we see that if f is continuous on a closed interval $[a, b]$, then the maximum and minimum values of f occur either at a critical number of f in (a, b) or at the endpoints a or b of the interval. It is also possible for a critical number c to be one of the endpoints of the interval $[a, b]$. If either $f(a)$ or $f(b)$ is an extremum of f on $[a, b]$, it is called an **endpoint extremum**. The sketches in Figure 3.9 illustrate this concept.

Figure 3.9   Endpoint extrema of f on $[a, b]$

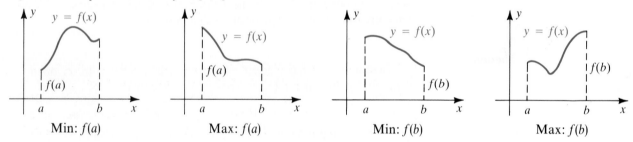

Min: $f(a)$ Max: $f(a)$ Min: $f(b)$ Max: $f(b)$

The preceding discussion gives us the following guidelines.

Guidelines for Finding the Extrema of a Continuous Function f on $[a, b]$ 3.9

> **1** Find all the critical numbers of f in (a, b).
> **2** Calculate $f(c)$ for each critical number c found in guideline (1).
> **3** Calculate the endpoint values $f(a)$ and $f(b)$.
> **4** The maximum and minimum values of f on $[a, b]$ are the largest and smallest function values calculated in guidelines (2) and (3).

EXAMPLE • 3   If $f(x) = x^3 - 12x$, find the maximum and minimum values of f on the closed interval $[-3, 5]$ and sketch the graph of f.

SOLUTION   Using Guidelines (3.9), we begin by finding the critical numbers of f. Differentiating yields

$$f'(x) = 3x^2 - 12 = 3(x^2 - 4) = 3(x + 2)(x - 2).$$

Since the derivative exists for every x, the only critical numbers are those for which the derivative is zero — that is, -2 and 2. Since f is continuous on $[-3, 5]$, it follows from our discussion that the maximum and minimum values are among the numbers $f(-2)$, $f(2)$, $f(-3)$, and $f(5)$. Calculating these values (see guidelines 2 and 3), we obtain the following table.

Value of x	Classification of x	Function value $f(x)$
-2	critical number of f	$f(-2) = 16$
2	critical number of f	$f(2) = -16$
-3	endpoint of $[-3, 5]$	$f(-3) = 9$
5	endpoint of $[-3, 5]$	$f(5) = 65$

By guideline (4), the minimum value of f on $[-3, 5]$ is the smallest function value $f(2) = -16$, and the maximum value is the endpoint extremum $f(5) = 65$.

The graph of f is sketched in Figure 3.10, with different scales on the x- and y-axes. The tangent line is horizontal at the point corresponding to each of the critical numbers, -2 and 2. It will follow from our work in Section 3.4 that $f(-2) = 16$ is a *local* maximum for f, as indicated by the graph.

Figure 3.10

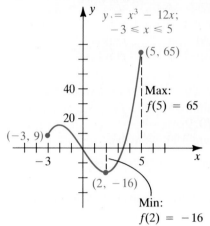

CAUTION   It is important to distinguish between an extreme value of a function and the point or points in the domain at which it occurs. Note that in Example 3 the minimum point occurs at $x = 2$, and the minimum value is $f(2) = -16$. The maximum point occurs at $x = 5$, and the maximum value is $f(5) = 65$. When asked to find a minimum or a maximum, do not merely find the value of x at which it occurs: *Be sure to complete the problem by calculating the function value.* In an economics application, for example, a manufacturer would want to know *both* the maximum profit $f(x)$ as well as the optimal price x that produces that largest profit.

In determining the extreme values of a function, we can often make use of algebraic tools as well as the ideas of calculus, as the next example shows.

EXAMPLE ■ 4   If $f(x) = (x - 1)^{2/3} + 2$, find the maximum and minimum values of f on the interval $[0, 9]$, and sketch the graph of f.

SOLUTION   We note first that $(x - 1)^{2/3} = [(x - 1)^2]^{1/3}$ is defined for all real values of x. In fact, since this term is the cube root of the nonnegative number $(x - 1)^2$, it always has a nonnegative value. The value will be 0 when $x = 1$; otherwise, it will be positive. Hence, we see that $f(x) = 2$ when $x = 1$ and $f(x) > 2$ for all other values of x. Our function has a minimum value of 2 at $x = 1$.

The function f has derivative

$$f'(x) = \frac{2}{3}(x - 1)^{-1/3} = \frac{2}{3(x - 1)^{1/3}},$$

provided $x \neq 1$. To find the critical numbers, we note that $f'(x)$ never equals 0 and that $f'(x)$ does not exist at $x = 1$. Thus, 1 is the only critical number in $(0, 9)$.

Let us tabulate our work as in Example 3.

Value of x	Classification of x	Function value $f(x)$
1	critical number of f	$f(1) = 2$
0	endpoint of $[0, 9]$	$f(0) = 3$
9	endpoint of $[0, 9]$	$f(9) = 6$

Thus, by Guidelines (3.9), f has a minimum value $f(1) = 2$ and a maximum value $f(9) = 6$ on the interval $[0, 9]$.

The graph of f is sketched in Figure 3.11. Note that

$$\lim_{x \to 1^-} f'(x) = -\infty \quad \text{and} \quad \lim_{x \to 1^+} f'(x) = \infty.$$

Since f is continuous at $x = 1$, the graph has a cusp at $(1, 2)$ by Definition (2.15).

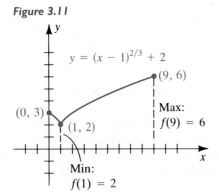

Figure 3.11

We see from Theorem (3.5) that if a function has a *local* extremum, then it *must* occur at a critical number; however, not every critical number leads to a local extremum, as illustrated by the next example.

EXAMPLE ■ 5 If $f(x) = x^3$, prove that f has no local extremum.

SOLUTION The graph of f is sketched in Figure 3.12. The derivative is $f'(x) = 3x^2$, which exists for every x and is zero only if $x = 0$. Consequently, 0 is the only critical number. However, if $x < 0$, then $f(x)$ is negative, and if $x > 0$, then $f(x)$ is positive. Thus, $f(0)$ is neither a local maximum nor a local minimum. Since a local extremum *must* occur at a critical number (see Theorem 3.5), it follows that f has no local extrema. Note that the tangent line is horizontal and crosses the graph at the point $(0, 0)$.

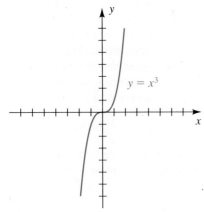

Figure 3.12

It sometimes requires considerable effort to find the critical numbers of a function. In the next two examples, we shall find these numbers for functions that are more complicated than those considered in the preceding examples. We could again use Guidelines (3.9) to determine extrema on a closed interval, but to do so would merely be repetitive of our previous work. In Sections 3.3 and 3.4, we discuss methods for determining the local extrema of a function by using critical numbers and derivatives.

EXAMPLE ■ 6 Find the critical numbers of the function f if $f(x) = (x + 5)^2 \sqrt[3]{x - 4}$.

SOLUTION Differentiating $f(x) = (x + 5)^2(x - 4)^{1/3}$, we obtain

$$f'(x) = (x + 5)^2 \tfrac{1}{3}(x - 4)^{-2/3} + 2(x + 5)(x - 4)^{1/3}.$$

To find the critical numbers, we simplify $f'(x)$ as follows:

$$f'(x) = \frac{(x + 5)^2}{3(x - 4)^{2/3}} + 2(x + 5)(x - 4)^{1/3}$$

$$= \frac{(x + 5)^2 + 6(x + 5)(x - 4)}{3(x - 4)^{2/3}}$$

$$= \frac{(x + 5)\,[(x + 5) + 6(x - 4)]}{3(x - 4)^{2/3}}$$

$$= \frac{(x + 5)(7x - 19)}{3(x - 4)^{2/3}}$$

Hence, $f'(x) = 0$ if $x = -5$ or $x = \frac{19}{7}$. The derivative $f'(x)$ does not exist at $x = 4$. Thus, f has the three critical numbers -5, $\frac{19}{7}$, and 4.

EXAMPLE ■ 7 If $f(x) = x^2 + 3x + 7\cos x$,

(a) approximate the critical numbers of f that are in the interval $[-10, 10]$ using a graphing utility and Newton's method

(b) estimate the extrema of f

SOLUTION

(a) By calculating a number of values of the function in this interval, by zooming as needed, or by using the automatic y-range selection procedure of the graphing utility, we obtain the graph of f in Figure 3.13.

To locate the critical numbers of f, we examine the graph for points at which the tangent line should be horizontal. Apparently, there are three such points where the derivative is zero; they correspond to the approximate x-values $-3, 0$, and 2. We zoom in once to obtain Figure 3.14 with a viewing window that has $-4.5 \leq x \leq 3$. Tracing, we find the critical numbers u_1, v_1, and w_1, where $u_1 \approx -2.78$, $v_1 \approx 0.68$, and $w_1 \approx 1.87$. Evaluation of f at these points gives points with approximate coordinates:

$$(u_1, f(u_1)) \approx (-2.78, -7.16) \qquad \text{global minimum}$$
$$(v_1, f(v_1)) \approx (0.68, 7.95) \qquad \text{local maximum}$$
$$(w_1, f(w_1)) \approx (1.87, 7.04) \qquad \text{local minimum}$$

To obtain more exact values for the critical numbers, we use Newton's method. We first search for the zeros of f' by letting g denote the function f'—that is, $g(x) = f'(x) = 2x + 3 - 7\sin x$. Then $g'(x) = f''(x) = 2 - 7\cos x$. Since the critical numbers of f are the zeros of g, we apply Newton's method to the function g:

$$x_{n+1} = x_n - \frac{2x_n + 3 - 7\sin x_n}{2 - 7\cos x_n}$$

Figure 3.13
$-10 \leq x \leq 10, -10 \leq y \leq 125$

Figure 3.14
$-4.5 \leq x \leq 3, -9 \leq y \leq 11$

We choose the values of u_1, v_1, and w_1 obtained by tracing as the first approximation and repeat the procedure to obtain the results in the following table.

n	u_n	v_n	w_n
1	−2.78	0.68	1.87
2	−2.77021338073	0.667931728395	1.85744882538
3	−2.77019934640	0.668022962424	1.85731625131
4	−2.77019934637	0.668022967587	1.85731623647
5	−2.77019934637	0.668022967587	1.85731623647

(b) To estimate the extrema, we evaluate the function f using the values in the last row of the table obtained by Newton's method and find the following extreme values:

$f(u_5)$	$f(v_5)$	$f(w_5)$
−7.15935181636	7.94565848661	7.04326223660

Summarizing our results (to four decimal places), we have the following estimates for extreme values on the interval $[-10, 10]$:

global minimum of -7.1594 at $x = -2.7702$

local minimum of 7.9457 at $x = 0.6680$

local minimum of 7.0433 at $x = 1.8573$

Since it is easy to make a mistake keying in a formula or implementing Newton's method, we always compare our numerical results to our estimates from the graph. In this case, there is very good agreement.

We conclude this section with a restatement and proof of Theorem (3.5). (Note that the proof for Theorem (3.7) is exactly the same if we change the phrase *local maximum* to *global maximum*.)

Theorem 3.5

If a function f has a local extremum at a number c in an open interval, then either $f'(c) = 0$ or $f'(c)$ does not exist.

PROOF We give a proof for the case that f has a local maximum at c. (You should provide a parallel proof for the case that f has a local minimum at c.)

If $f'(c)$ does not exist, then there is nothing more to prove. If $f'(c)$ does exist, then we need to show that it has value 0. We begin by reexamining Definition (2.6), which gives the limit of the difference quotient:

$$f'(c) = \lim_{x \to c} \frac{f(x) - f(c)}{x - c}$$

Since f has a local maximum at c, $f(x) \le f(c)$ for all x in the interval sufficiently close to c. Thus, the numerator of the difference quotient is nonpositive for all x close to c (since $f(x) - f(c) \le 0$). However, the denominator $(x - c)$ is negative when $x < c$ and positive when $x > c$. Hence, the difference quotient will be nonnegative when $x < c$ and nonpositive when $x > c$—that is,

$$\frac{f(x) - f(c)}{x - c} \ge 0 \text{ for } x < c \quad \text{and} \quad \frac{f(x) - f(c)}{x - c} \le 0 \text{ for } x > c.$$

Consequently, we must have

$$f'(c) \ge 0 \quad \text{and} \quad f'(c) \le 0.$$

We conclude that $f'(c)$ must equal 0. ■

EXERCISES 3.1

Exer. 1–2: Use the graph to estimate the absolute maximum, the absolute minimum, and the local extrema of f.

1

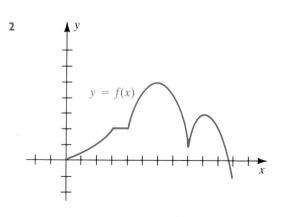

2

Exer. 3–4: Use the graph to estimate the extrema of f on each interval.

3 (a) $[-3, 3)$ (b) $(-3, \sqrt{3})$ (c) $[-\sqrt{3}, 1)$ (d) $[0, 3]$

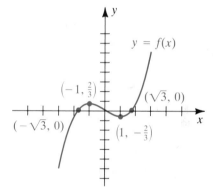

4 (a) $[-2, 2]$ (b) $(0, 2)$ (c) $(-1, 1]$ (d) $[-2, -1)$

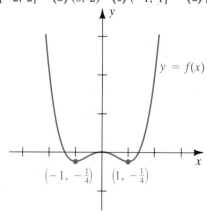

Exer. 5–6: Sketch the graph of f, and find the extrema on each interval.

5 $f(x) = \frac{1}{2}x^2 - 2x$

   (a) [0, 5)   **(b)** (0, 2)   **(c)** (0, 4)   **(d)** [2, 5]

6 $f(x) = (x - 1)^{2/3} - 4$

   (a) [0, 9]   **(b)** (1, 2]   **(c)** (−7, 2)   **(d)** [0, 1)

Exer. 7–10: Find the extrema of f on the given interval.

7 $f(x) = 5 - 6x^2 - 2x^3$; [−3, 1]

8 $f(x) = 3x^2 - 10x + 7$; [−1, 3]

9 $f(x) = 1 - x^{2/3}$;   [−1, 8]

10 $f(x) = x^4 - 5x^2 + 4$; [0, 2]

Exer. 11–36: Find the critical numbers of the function.

11 $f(x) = 4x^2 - 3x + 2$

12 $g(x) = 2x + 5$

13 $s(t) = 2t^3 + t^2 - 20t + 4$

14 $K(z) = 4z^3 + 5z^2 - 42z + 7$

15 $F(w) = w^4 - 32w$

16 $k(r) = r^5 - 2r^3 + r - 12$

17 $f(z) = \sqrt{z^2 - 16}$

18 $M(x) = \sqrt[3]{x^2 - x - 2}$

19 $h(x) = (2x - 5)\sqrt{x^2 - 4}$

20 $T(v) = (4v + 1)\sqrt{v^2 - 16}$

21 $g(t) = t^2\sqrt[3]{2t - 5}$ **22** $g(x) = (x - 3)\sqrt{9 - x^2}$

23 $G(x) = \dfrac{2x - 3}{x^2 - 9}$ **24** $f(s) = \dfrac{s^2}{5s + 4}$

25 $f(t) = \sin^2 t - \cos t$

26 $g(t) = 4\sin^3 t + 3\sqrt{2}\cos^2 t$

27 $K(\theta) = \sin 2\theta + 2\cos \theta$

28 $f(x) = 8\cos^3 x - 3\sin 2x - 6x$

29 $f(x) = \dfrac{1 + \sin x}{1 - \sin x}$

30 $g(\theta) = 2\sqrt{3}\,\theta + \sin 4\theta$

31 $k(u) = u - \tan u$ **32** $p(z) = 3\tan z - 4z$

33 $H(\phi) = \cot \phi + \csc \phi$ **34** $g(x) = 2x + \cot x$

35 $f(x) = \sec(x^2 + 1)$ **36** $s(t) = \dfrac{\sec t + 1}{\sec t - 1}$

37 (a) If $f(x) = x^{1/3}$, prove that 0 is the only critical number of f and that $f(0)$ is not a local extremum.

(b) If $f(x) = x^{2/3}$, prove that 0 is the only critical number of f and that $f(0)$ is a local minimum of f.

38 If $f(x) = |x|$, prove that 0 is the only critical number of f, that $f(0)$ is a local minimum of f, and that the graph of f has no tangent line at the point (0, 0).

Exer. 39–40: (a) Prove that f has no local extrema. **(b)** Sketch the graph of f. **(c)** Prove that f is continuous on the interval (0, 1) but has neither a maximum nor a minimum value on (0, 1). **(d)** Explain why part (c) does not contradict Theorem (3.3).

39 $f(x) = x^3 + 1$ **40** $f(x) = 1/x^2$

41 (a) Prove that a polynomial function f of degree 1 has no extrema on the interval $(-\infty, \infty)$.

(b) Discuss the extrema of f on a closed interval $[a, b]$.

42 If f is a constant function and (a, b) is any open interval, prove that $f(c)$ is both a local and an absolute extremum of f for every number c in (a, b).

43 If f is the greatest integer function, prove that every number is a critical number of f.

44 Let f be defined by the following conditions: $f(x) = 0$ if x is rational and $f(x) = 1$ if x is irrational. Prove that every number is a critical number of f.

45 Prove that a quadratic function has exactly one critical number on $(-\infty, \infty)$.

46 Prove that a polynomial function of degree 3 has either two, one, or no critical numbers on $(-\infty, \infty)$, and sketch graphs that illustrate how each of these possibilities can occur.

47 Let $f(x) = x^n$ for a positive integer n. Prove that f has either one or no local extremum on $(-\infty, \infty)$ depending on whether n is even or odd, respectively, and sketch a typical graph illustrating each case.

48 Prove that a polynomial function of degree n can have at most $n - 1$ local extrema on $(-\infty, \infty)$.

  c **Exer. 49–50:** Graph f on the given interval, and use the graph to estimate the extrema of f.

49 $f(x) = x^6 - x^5 + 3x^3 - 2x + 1$; [−1, 1]

50 $f(x) = \dfrac{x^2 - \cos 3x}{\sin 2x - x + 2}$;   [−π, 1]

  c **Exer. 51–52:** Graph f on the given interval, and use the graph to estimate the critical numbers of f.

51 $f(x) = \left|\frac{1}{3}x^3 + x^2 - x - 1.5\right|$; [−3, 2]

52 $f(x) = x\sin^2 x\cos^2 x$;   [−2, 2]

| c | Exer. 53–58: Approximate the critical numbers of f using a graphing utility and Newton's method (or a solving routine) on a calculator or computer.

53 $f(x) = \dfrac{x^4 - x^3 - 10x^2 + 2}{x^2 + 4}$

54 $f(x) = \dfrac{x^4 - 3x^3 - 5x + 27}{x^2 + x + 3}$

55 $f(x) = \begin{cases} -x^2 + x + 5 & \text{if } x < 0 \\ x^3 - 4x^2 + x + 5 & \text{if } 0 \leq x < 3 \\ x^2 - 8x + 14 & \text{if } x \geq 3 \end{cases}$

56 $f(x) = \begin{cases} x^2 & \text{if } 0 \leq x < 1 \\ -3(x-1)^2 + 2(x-1) + 1 & \text{if } 1 \leq x < 2 \\ 3(x-2)^2 - 4(x-2) & \text{if } 2 \leq x < 3 \\ -(x-4)^2 & \text{if } 3 \leq x \leq 4 \end{cases}$

57 $f(x) = x^5 - 20x^4 + 15x^3 + 7x^2 + 6$

58 $f(x) = x^5 - x^4 - \sqrt{3x + 1} - 2$

3.2 THE MEAN VALUE THEOREM

We now discuss one of the most influential theorems in calculus, the *mean value theorem*. First stated by Joseph Louis Lagrange (1736–1813) in 1797, the mean value theorem is used to establish many fundamental results in calculus, as we shall see later in this text. In order to lay the groundwork for our study of this significant theorem, we begin with a theorem credited to the French mathematician Michel Rolle (1652–1719).

The theorem discovered by Rolle provides sufficient conditions for the existence of a critical number. It is stated for a function f that is continuous on a closed interval $[a, b]$, is differentiable on the open interval (a, b), and satisfies the condition $f(a) = f(b)$. Some typical graphs of functions of this type are sketched in Figure 3.15.

Figure 3.15

  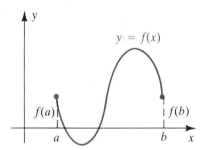

From the sketches in Figure 3.15 we expect that there is at least one number c between a and b such that the tangent line at the point $(c, f(c))$ is horizontal—that is, $f'(c) = 0$, which is precisely the conclusion of Rolle's theorem.

Rolle's Theorem 3.10

If f is continuous on a closed interval $[a, b]$ and differentiable on the open interval (a, b) and if $f(a) = f(b)$, then $f'(c) = 0$ for at least one number c in (a, b).

PROOF The function f must fall into at least one of the following three categories:

(i) $f(x) = f(a)$ *for every* x *in* (a, b). In this case, f is a constant function, and hence $f'(x) = 0$ for every x. Consequently, *every* number c in (a, b) is a critical number.

(ii) $f(x) > f(a)$ *for some* x *in* (a, b). In this case, the maximum value of f in $[a, b]$ is greater than $f(a)$ or $f(b)$ and, therefore, must occur at some number c in the *open* interval (a, b). Since the derivative exists throughout (a, b), we conclude from Theorem (3.7) that $f'(c) = 0$.

(iii) $f(x) < f(a)$ *for some* x *in* (a, b). In this case, the minimum value of f in $[a, b]$ is less than $f(a)$ or $f(b)$ and must occur at some number c in (a, b). As in (ii), $f'(c) = 0$. ∎

Corollary 3.11

If f is continuous on a closed interval $[a, b]$ and if $f(a) = f(b)$, then f has at least one critical number in the open interval (a, b).

PROOF If f' does not exist at some number c in (a, b), then, by Definition (3.8), c is a critical number. Alternatively, if f' exists throughout (a, b), then, by Rolle's theorem, a critical number exists. ∎

EXAMPLE ■ 1 Let $f(x) = 4x^2 - 20x + 29$. Show that f satisfies the hypotheses of Rolle's theorem on the interval $[1, 4]$, and find all real numbers c in the open interval $(1, 4)$ such that $f'(c) = 0$. Illustrate the results graphically.

SOLUTION Since f is a polynomial function, it is continuous and differentiable for every x. In particular, it is continuous on $[1, 4]$ and differentiable on $(1, 4)$. Moreover,

$$f(1) = 4 - 20 + 29 = 13,$$
$$f(4) = 64 - 80 + 29 = 13,$$

and hence $f(1) = f(4)$. Thus, f satisfies the three hypotheses of Rolle's theorem on the interval $[1, 4]$.

Differentiating $f(x)$, we have

$$f'(x) = 8x - 20.$$

Setting $f'(x) = 0$ gives us $8x = 20$, or $x = \frac{5}{2}$. Hence,

$$f'(\tfrac{5}{2}) = 0 \quad \text{and} \quad 1 < \tfrac{5}{2} < 4.$$

The graph of f (a parabola) is sketched in Figure 3.16. Since $f'(\frac{5}{2}) = 0$, the tangent line is horizontal at the vertex $(\frac{5}{2}, 4)$.

Figure 3.16

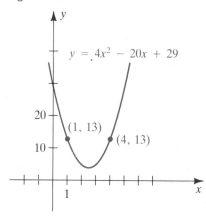

Although Rolle's theorem is of interest in itself, our principal use for it is in the proof of one of the most important tools in calculus: the **mean**

Figure 3.17

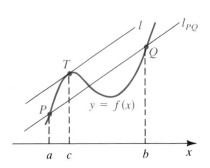

value theorem. In order to illustrate graphically the mean value theorem, let f be any function that is continuous on $[a, b]$ and differentiable on (a, b) (possibly $f(a) \neq f(b)$). Consider the points $P(a, f(a))$ and $Q(b, f(b))$ on the graph of f, as illustrated by any of the sketches in Figure 3.17.

If f' exists throughout the open interval (a, b), then there appears to be at least one point $T(c, f(c))$ on the graph at which the tangent line l is parallel to the secant line l_{PQ} through P and Q. In terms of slopes,

$$\text{slope of } l = \text{slope of } l_{PQ},$$

or

$$f'(c) = \frac{f(b) - f(a)}{b - a}.$$

If we multiply both sides of the last equation by $b - a$, we obtain the second formula stated in the mean value theorem.

Mean Value Theorem 3.12

If f is continuous on a closed interval $[a, b]$ and differentiable on the open interval (a, b), then there exists a number c in (a, b) such that

$$f'(c) = \frac{f(b) - f(a)}{b - a},$$

or, equivalently,

$$f(b) - f(a) = f'(c)(b - a).$$

Figure 3.18

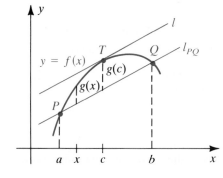

PROOF For any x in the interval $[a, b]$, let $g(x)$ be the vertical (signed) distance from the secant line l_{PQ} to the graph of f, as illustrated in Figure 3.18. (The term *signed distance* means that $g(x)$ is positive or negative if the graph of f lies above or below l_{PQ}, respectively.)

It appears that if $T(c, f(c))$ is a point at which the tangent line l is parallel to l_{PQ}, then the distance $g(c)$ is a local extremum of g. This conclusion suggests that we find the critical numbers of the function g.

We may obtain a formula for $g(x)$ as follows. First, by the point–slope form, an equation of the secant line l_{PQ} is

$$y - f(a) = \frac{f(b) - f(a)}{b - a}(x - a),$$

or, equivalently,

$$y = f(a) + \frac{f(b) - f(a)}{b - a}(x - a).$$

As illustrated in Figure 3.18, $g(x)$ is the difference of the distances from the x-axis to the graph of f and to the line l_{PQ}. That is,

$$g(x) = f(x) - \left[f(a) + \frac{f(b) - f(a)}{b - a}(x - a) \right].$$

We shall use Rolle's theorem to find a critical number of g. We first observe that since f is continuous on $[a, b]$ and differentiable on (a, b), the same is true for the function g. Differentiating, we obtain

$$g'(x) = f'(x) - \frac{f(b) - f(a)}{b - a}.$$

Moreover, by direct substitution, we see that $g(a) = g(b) = 0$, and hence the function g satisfies the hypotheses of Rolle's theorem. Consequently, there exists a number c in the open interval (a, b) such that $g'(c) = 0$, or, equivalently,

$$f'(c) - \frac{f(b) - f(a)}{b - a} = 0.$$

The last equation may be written in the forms stated in the conclusion of the theorem. ▬

EXAMPLE■2   If $f(x) = \frac{1}{4}x^2 + 1$, show that f satisfies the hypotheses of the mean value theorem on the interval $[-1, 4]$, and find a number c in $(-1, 4)$ that satisfies the conclusion of the theorem. Illustrate the results graphically.

SOLUTION   The quadratic function f is continuous on $[-1, 4]$ and differentiable on $(-1, 4)$; hence, by the mean value theorem, there is a number c in $(-1, 4)$ such that

$$\frac{f(4) - f(-1)}{4 - (-1)} = f'(c).$$

Using $f(4) = 5$, $f(-1) = \frac{5}{4}$, and $f'(x) = \frac{1}{2}x$ gives us

$$\frac{5 - \frac{5}{4}}{5} = \frac{1}{2}c, \quad \text{or} \quad \frac{3}{4} = \frac{1}{2}c.$$

Thus, $c = \frac{3}{2}$.

The graph of f (a parabola) is sketched in Figure 3.19. The points $P(-1, \frac{5}{4})$ and $Q(4, 5)$ correspond to the endpoints of the interval $[-1, 4]$. The point $T(\frac{3}{2}, \frac{25}{16})$ is obtained by using $c = \frac{3}{2}$ and is the point at which the tangent line l is parallel to the secant line l_{PQ}.

Figure 3.19

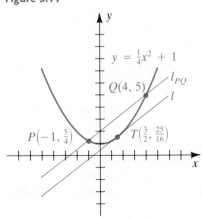

EXAMPLE■3   If $f(x) = x^4 - 5x^3 - 33x^2 + 113x + 50$,

(a) show that f satisfies the hypotheses of the mean value theorem on the interval $[0, 7]$

(b) approximate all numbers c in the open interval $(0, 7)$ that satisfy the conclusion of the theorem

SOLUTION

(a) Since f is a polynomial function, it is continuous and differentiable for all real numbers. In particular, it is continuous on $[0, 7]$ and differentiable on the open interval $(0, 7)$. Thus, the function does satisfy all the hypotheses of the mean value theorem.

Figure 3.20

$0 \leq x \leq 7, -250 \leq y \leq 150$

(b) By the mean value theorem, there is at least one number c in $(0, 7)$ such that

$$\frac{f(7) - f(0)}{7 - 0} = \frac{-90 - 50}{7} = f'(c).$$

Since $f'(x) = 4x^3 - 15x^2 - 66x + 113$, we must solve

$$4c^3 - 15c^2 - 66c + 113 = -20,$$

or

$$4c^3 - 15c^2 - 66c + 133 = 0.$$

Using Newton's method or a solving routine on a calculator, we find that the approximate solutions are -3.54638161, 1.66487475, and 5.63150686. However, the first solution is not in the interval $(0, 7)$. The desired numbers are $c \approx 1.66487475$ and $c \approx 5.63150686$.

Figure 3.21

$0 \leq x \leq 7, -150 \leq y \leq 150$

The values of c can also be obtained from a graphing utility. Figure 3.20 shows the graph of f on the interval $[0, 7]$ along with the line segment of slope -20 connecting the endpoints $(0, 50)$ and $(7, -90)$. Figure 3.21 displays a plot of the graph of f' and the horizontal line $y = -20$ for $0 \leq x \leq 7$. The desired values of c are the x-coordinates of the points of intersection of the graph of f' and this horizontal line.

The mean value theorem is also called the **theorem of the mean**. In the study of statistics, the term *mean* is used for the *average* of a collection of numbers. The word has a similar connotation in *mean value theorem*. As an illustration, if a point P is moving on a coordinate line and if $s(t)$ denotes the coordinate of P at time t, then, by Definition (2.2), the average velocity of P during the time interval $[a, b]$ is

$$v_{av} = \frac{s(b) - s(a)}{b - a}.$$

According to the mean value theorem, this average (mean) velocity is equal to the velocity $s'(c)$ at some time c between a and b. The next example illustrates this application of the mean value theorem.

EXAMPLE ■ 4   The speedometer of an automobile registers 50 mi/hr as it passes a mileage marker along a highway. Four minutes later, as the automobile passes a marker 5 mi from the first, the speedometer registers 55 mi/hr. Use the mean value theorem to prove that the velocity exceeded 70 mi/hr at some time while the automobile was traveling between the two markers.

Figure 3.22

$t = 0$ $t = \frac{1}{15}$

SOLUTION   We may assume that the automobile is moving along a straight highway with the two mileage markers at points A and B, as illustrated in Figure 3.22. Let t denote the elapsed time (in hours) after the automobile passes A, and let $s(t)$ denote its distance from A (in miles) at time t. Since the automobile passes B at $t = \frac{4}{60} = \frac{1}{15}$ hr, the average velocity during the trip from A to B is

$$v_{\mathrm{av}} = \frac{s(\frac{1}{15}) - s(0)}{\frac{1}{15} - 0} = \frac{5}{\frac{1}{15}} = 75 \text{ mi/hr.}$$

If we assume that s is a differentiable function, then applying the mean value theorem (3.12) to s, with $a = 0$ and $b = \frac{1}{15}$, proves that there is a time c in the time interval $[0, \frac{1}{15}]$ at which the velocity $s'(c)$ of the automobile is 75 mi/hr. Note that the velocity at A or B is irrelevant.

We now briefly examine two corollaries of the mean value theorem.

Corollary 3.13

If $f'(x) = 0$ for all x in some interval I, then there is a constant C such that $f(x) = C$ for all x in I. That is, if the derivative of a function is identically zero, then the function must be a constant.

PROOF   Let a and b be any two points in the interval I. By the mean value theorem,

$$f(b) = f(a) + f'(c)(b - a)$$

for some c between a and b. But $f'(c)$ is 0. Thus, $f(b) = f(a)$, and we see that the function has the same value at any two numbers.   ■

Corollary 3.14

If $f'(x) = g'(x)$ for all x in an interval I, then there is a constant C such that $f(x) = g(x) + C$ for all x in I. That is, two functions with identical derivatives differ by some constant value.

PROOF   The function $h(x) = f(x) - g(x)$ has derivative $h'(x) = f'(x) - g'(x) = 0$ for all x in the interval I. By Corollary (3.13), there is a constant C such that $h(x) = C$ for all x in I. Thus, $C = h(x) = f(x) - g(x)$.   ■

Corollary (3.14) has important implications in the study of integration, which we take up in Chapter 4. One of the major concerns in integration is to discover the explicit form of a function if we are given its derivative. In other words, if we are given f', what function f gives rise to it? Corollary (3.14) tells us that there is essentially only one answer to this question since any two functions that have the same derivative can differ only by a constant. Stated geometrically, Corollary (3.14) asserts that if two graphs have parallel tangent lines at every point, then the graph of one is a simple vertical translation of the other.

This result may be surprising to you. For example, differentiation of the function $f(x) = (1 - x)/(1 + x)$ gives $f'(x) = -2/(x + 1)^2$. It is far from obvious that we could not determine some other fundamentally different function that also had $-2/(x + 1)^2$ as its derivative. Corollary (3.14) tells us, however, that the only successes we will ever have will be of the form $(1 - x)/(1 + x) + C$ for some constant C.

EXAMPLE■5 Find all functions f that have the derivative given by $f'(x) = 7x^6 + \sec x \tan x - 9$.

SOLUTION Since $g(x) = x^7 + \sec x - 9x$ has $g'(x) = f'(x)$, the function f must have the form $f(x) = x^7 + \sec x - 9x + C$, where C is some constant.

EXERCISES 3.2

Exer. 1–2: Use the graph of f to estimate the numbers in [1, 10] that satisfy the conclusion of the mean value theorem.

1

2

Exer. 3–8: Show that f satisfies the hypotheses of Rolle's theorem on $[a, b]$, and find all numbers c in (a, b) such that $f'(c) = 0$.

3 $f(x) = 3x^2 - 12x + 11$;   [0, 4]

4 $f(x) = 5 - 12x - 2x^2$;   [−7, 1]

5 $f(x) = x^4 + 4x^2 + 1$;   [−3, 3]

6 $f(x) = x^3 - x$;   [−1, 1]

7 $f(x) = \sin 2x$;   [0, π]

8 $f(x) = \cos 2x + 2 \cos x$;   [0, 2π]

Exer. 9–24: Determine whether f satisfies the hypotheses of the mean value theorem on $[a, b]$, and, if so, find all numbers c in (a, b) such that

$$f(b) - f(a) = f'(c)(b - a).$$

9 $f(x) = 5x^2 - 3x + 1$;   [1, 3]

10 $f(x) = 3x^2 + x - 4$;   [1, 5]

11 $f(x) = \dfrac{1}{(x - 1)^2}$;   [0, 2]

12 $f(x) = \dfrac{x + 3}{x - 2}$;   [−2, 3]

13 $f(x) = x^{2/3}$;   [−8, 8]

14 $f(x) = |x - 3|$;   [−1, 4]

15 $f(x) = x + (4/x)$;   [1, 4]

16 $f(x) = 3x^5 + 5x^3 + 15x$;   [−1, 1]

17 $f(x) = x^3 - 2x^2 + x + 3$;   [−1, 1]

18 $f(x) = 1 - 3x^{1/3}$;   [−8, −1]

19 $f(x) = 4 + \sqrt{x-1}$; [1, 5]

20 $f(x) = (x+2)^{2/3}$; [-1, 6]

21 $f(x) = x^3 + 1$; [-2, 4]

22 $f(x) = x^3 + 4x$; [-3, 6]

23 $f(x) = \sin x$; [0, $\pi/2$]

24 $f(x) = \tan x$; [0, $\pi/4$]

c | **Exer. 25–32:** Determine whether f satisfies the hypotheses of the mean value theorem on $[a, b]$, and, if so, approximate the numbers c satisfying the conclusion of the theorem.

25 $f(x) = 0.2x^2 + \sin x$; [-1, 2]

26 $f(x) = 0.1x^3 + \cos x$; [-1, 2]

27 $f(x) = x^4 + 2x^3 - 39x^2 - 72x + 108$; [-2, 1]

28 $f(x) = x^5 + 5x^3 - 10x^2 + 15$; [-4, 5]

29 $f(x) = \dfrac{x^3 - 8x^2 + 5}{x+2}$; [1, 8]

30 $f(x) = \dfrac{x^3 - 2x^2 + 1}{x^2 + 1}$; [-0.5, 1.5]

31 $f(x) = |x^4 - 8x^2 - 7x + 1|$; [1, 4]

32 $f(x) = \begin{cases} -x^2 + x + 5 & \text{if } x < 0 \\ x^3 - 4x^2 + x + 5 & \text{if } x \geq 0 \end{cases}$ [-1, 1]

33 If $f(x) = |x|$, show that $f(-1) = f(1)$ but $f'(c) \neq 0$ for every number c in the open interval $(-1, 1)$. Why doesn't this contradict Rolle's theorem?

34 If $f(x) = 5 + 3(x-1)^{2/3}$, show that $f(0) = f(2)$ but $f'(c) \neq 0$ for every number c in the open interval $(0, 2)$. Why doesn't this contradict Rolle's theorem?

35 If $f(x) = 4/x$, prove that there is no real number c such that $f(4) - f(-1) = f'(c)[4 - (-1)]$. Why doesn't this contradict the mean value theorem applied to the interval $[-1, 4]$?

36 If f is the greatest integer function and if a and b are real numbers such that $b - a \geq 1$, prove that there is no real number c such that $f(b) - f(a) = f'(c)(b - a)$. Why doesn't this contradict the mean value theorem?

37 If f is a linear function, prove that f satisfies the hypotheses of the mean value theorem on every closed interval $[a, b]$ and that *every* number c satisfies the conclusion of the theorem.

38 If f is a quadratic function and $[a, b]$ is any closed interval, prove that precisely one number c in the interval (a, b) satisfies the conclusion of the mean value theorem.

39 If f is a polynomial function of degree 3 and $[a, b]$ is any closed interval, prove that at most two numbers in

(a, b) satisfy the conclusion of the mean value theorem. Generalize to polynomial functions of degree n for any positive integer n.

40 If $f(x)$ is a polynomial of degree 3, use Rolle's theorem to prove that f has at most three real zeros. Extend this result to polynomials of degree n.

Exer. 41–49: Use the mean value theorem.

41 If f is continuous on $[a, b]$ and if $f'(x) > 0$ for every x in (a, b), prove that $f(b) > f(a)$.

42 If f is continuous on $[a, b]$ and if $f'(x) = c$ for every x in (a, b), prove that $f(x) = cx + d$ for some real number d.

43 A straight highway 50 mi long connects two cities A and B. Prove that it is impossible to travel from A to B by automobile in exactly 1 hr without having the speedometer register 50 mi/hr at least once.

44 The electrical charge Q on a capacitor increases from 2 millicoulombs to 10 millicoulombs in 15 milliseconds. Show that the current $I = dQ/dt$ exceeded $\frac{1}{2}$ ampere at some instant during this short time interval. (*Note:* 1 ampere = 1 coulomb/second.)

45 If W denotes the weight (in pounds) of an individual and t denotes time (in months), then dW/dt is the rate of weight gain or loss (in pounds per month). The current speed record for weight loss is a drop in weight from 487 lb to 130 lb over an eight-month period. Show that the rate of weight loss exceeded 44 lb/mo at some time during the eight-month period.

46 Let T denote the temperature (in °F) at time t (in hours). If the temperature is decreasing, then dT/dt is the *rate of cooling*. The greatest temperature variation during a twelve-hour period occurred in Montana in 1916, when the temperature dropped from 44 °F to a chilling −56 °F. Show that the rate of cooling exceeded −8 °F/hr at some time during the period of change.

47 Prove that if u and v are any real numbers, then

$$|\sin u - \sin v| \leq |u - v|.$$

(*Hint:* Apply the mean value theorem with $f(x) = \sin x$.)

48 Prove that $\sqrt{1+h} < 1 + \frac{1}{2}h$ for $h > 0$. (*Hint:* Apply the mean value theorem with $f(x) = \sqrt{1+x}$.)

49 If I denotes the number of people infected with a particular disease at time t (in months), then dI/dt is the rate of spread of the disease. The Centers for Disease Control reported 82,764 cases of AIDS in the United States as of December 31, 1988, and 242,146 cases as of September 30, 1992. Show that at some time in this period, AIDS was spreading at a rate exceeding 3500 cases/month.

Exer. 50–53: For each of the given functions g, find all functions f such that f' = g.

50 $g(x) = 3x^2$

51 $g(x) = 1/x^2$

52 $g(x) = 5x^4 - x^2 + 6x + 11$

53 $g(x) = \cos x$

54 (a) Show that the functions

$$f(x) = \tan x \quad \text{and} \quad g(x) = \sin x \sec x$$

have the same derivative.

(b) Does this fact contradict Corollary (3.14)? Explain.

55 (a) Show that the functions

$$f(x) = \sin^2 x \quad \text{and} \quad g(x) = -\cos^2 x$$

have the same derivative.

(b) Does this fact contradict Corollary (3.14)? Explain.

56 Suppose that $f'(x) = g'(x) + x$ for every x in some interval I. How different can the functions f and g be?

 Exer. 57–58: Graph, on the same coordinate axes, $y = f(x)$ for $0 \le x \le 1$ and the line through $(0, f(0))$ and $(1, f(1))$. Use the graph to estimate the numbers in [0, 1] that satisfy the conclusion of the mean value theorem.

57 $f(x) = 2.1x^4 - 1.4x^3 + 0.8x^2 - 1$

58 $f(x) = \dfrac{\sin 2x + \cos x}{2 + \cos \pi x}$

3.3 THE FIRST DERIVATIVE TEST

In this section, we see how the sign of the derivative f' may be used to determine where a function f is increasing and where it is decreasing. This information will enable us to classify the local extrema of the function.

The graph of $y = f(x)$ in Figure 3.23 indicates that if the slope of the tangent line is positive in an open interval I (that is, if $f'(x) > 0$ for every x in I), then f is increasing on I. Similarly, it appears that if the slope is negative (that is, if $f'(x) < 0$), then f is decreasing.

Figure 3.23

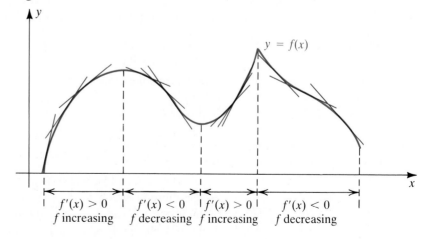

That these intuitive observations are actually true is a consequence of the following theorem.

Theorem 3.15

Let f be continuous on $[a, b]$ and differentiable on (a, b).

(i) If $f'(x) > 0$ for every x in (a, b), then f is increasing on $[a, b]$.

(ii) If $f'(x) < 0$ for every x in (a, b), then f is decreasing on $[a, b]$.

P R O O F   To prove (i), suppose that $f'(x) > 0$ for every x in (a, b), and consider any numbers x_1, x_2 in $[a, b]$ such that $x_1 < x_2$. We wish to show that $f(x_1) < f(x_2)$. Applying the mean value theorem (3.12) to the interval $[x_1, x_2]$ gives us

$$f(x_2) - f(x_1) = f'(c)(x_2 - x_1)$$

for some number c in the open interval (x_1, x_2). Since $x_2 - x_1 > 0$ and since, by hypothesis, $f'(c) > 0$, the right-hand side of the preceding equation is positive; that is, $f(x_2) - f(x_1) > 0$. Hence, $f(x_2) > f(x_1)$, which is what we wished to show. The proof of (ii) is similar. ■

We may also show that if $f'(x) > 0$ throughout an infinite interval $(-\infty, a)$ or (b, ∞), then f is increasing on $(-\infty, a]$ or $[b, \infty)$, respectively, provided f is continuous on those intervals. An analogous result holds for decreasing functions if $f'(x) < 0$.

To apply Theorem (3.15), we must determine intervals in which $f'(x)$ is either always positive or always negative. Theorem (1.26) is useful in this respect; see Exercise 66 of Section 1.5. Specifically, if the *derivative* f' is continuous and has no zeros on an interval, then either $f'(x) > 0$ or $f'(x) < 0$ for every x in the interval. Thus, if we choose *any* number k in the interval and if $f'(k) > 0$, then $f'(x) > 0$ for *every* x in the interval. Similarly, if $f'(k) < 0$, then we have $f'(x) < 0$ throughout the interval. We shall call $f'(k)$ a **test value** of $f'(x)$ for the interval.

E X A M P L E ▪ 1   If $f(x) = x^3 + x^2 - 5x - 5$,

(a) find the intervals on which f is increasing and the intervals on which f is decreasing

(b) sketch the graph of f

S O L U T I O N

(a) First we differentiate $f(x)$:

$$f'(x) = 3x^2 + 2x - 5 = (3x + 5)(x - 1)$$

By Theorem (3.15), it is sufficient to find the intervals in which $f'(x) > 0$ and those in which $f'(x) < 0$. The factored form of $f'(x)$ and the critical numbers $-\frac{5}{3}$ and 1 suggest the open intervals $(-\infty, -\frac{5}{3})$, $(-\frac{5}{3}, 1)$, and $(1, \infty)$. On each of these intervals, f' is continuous and has no zeros, and therefore $f'(x)$ has the same sign throughout the interval. We can determine the sign in two different ways.

First, we can determine the sign by choosing a suitable test value for the interval. The following table displays our work. The values of k were chosen for convenience. We chose $k = -2$ in the interval $(-\infty, -\frac{5}{3})$, but *any* number, such as -3 or -10, could have been used.

Interval	$(-\infty, -\frac{5}{3})$	$(-\frac{5}{3}, 1)$	$(1, \infty)$
k	-2	0	2
Test value $f'(k)$	$f'(-2) = 3 > 0$	$f'(0) = -5 < 0$	$f'(2) = 11 > 0$
Sign of $f'(k)$	$+$	$-$	$+$
Conclusion	f is increasing on $(-\infty, -\frac{5}{3}]$	f is decreasing on $[-\frac{5}{3}, 1]$	f is increasing on $[1, \infty)$

We can also determine the sign of f' by using the fact that a product of numbers is negative if it has an odd number of negative factors and positive if it has an even number of negative factors. In this case, we have

$$f'(x) = (3x + 5)(x - 1).$$

We draw a number line for each factor and mark the regions where it is positive and where it is negative. Figure 3.24 illustrates how the sign of the product can then be determined from the pattern of signs of the factors. From the last line of the figure, we see that the sign of $f'(x)$ is positive on $(-\infty, -\frac{5}{3})$, negative on $(-\frac{5}{3}, 1)$, and positive on $(1, \infty)$.

Figure 3.24

Sign of $(3x + 5)$

Sign of $(x - 1)$

Sign of $(3x + 5)(x - 1)$

Figure 3.25

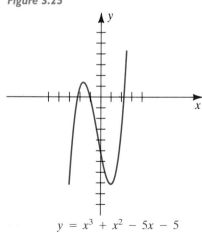

$y = x^3 + x^2 - 5x - 5$

(b) As an aid to sketching the graph of f, we shall find the x-intercepts by solving the equation $f(x) = 0$. Since

$$\begin{aligned} f(x) &= x^3 + x^2 - 5x - 5 \\ &= x^2(x + 1) - 5(x + 1) \\ &= (x^2 - 5)(x + 1), \end{aligned}$$

we see that the x-intercepts are $\sqrt{5}$, $-\sqrt{5}$, and -1. The y-intercept is $f(0) = -5$. The points corresponding to the critical numbers are $(-\frac{5}{3}, \frac{40}{27})$ and $(1, -8)$. Plotting these six points and using the information in the table gives us the sketch in Figure 3.25.

We saw in Section 3.1 that if a function has a local extremum, then it must occur at a critical number; however, not every critical number leads to a local extremum (see Example 5 of Section 3.1). To find the local extrema, we begin by locating all the critical numbers of the function. Next, we test each critical number to determine whether or not a local extremum occurs. There are several methods for conducting this test. The following theorem is based on the sign of the first derivative of f. In the statement of the theorem, the terminology f' *changes from positive to negative at c* means that there is an open interval (a, b) containing c such that $f'(x) > 0$ if $a < x < c$ and $f'(x) < 0$ if $c < x < b$. Analogous meanings are applied to the phrases f' *changes from negative to positive at c* and f' *does not change sign at c*.

First Derivative Test   3.16

Let c be a critical number for f, and suppose that f is continuous at c and differentiable on an open interval I containing c, except possibly at c itself.

   (i) If f' changes from positive to negative at c, then $f(c)$ is a local maximum of f.

  (ii) If f' changes from negative to positive at c, then $f(c)$ is a local minimum of f.

 (iii) If $f'(x) > 0$ or if $f'(x) < 0$ for every x in I except $x = c$, then $f(c)$ is not a local extremum of f.

Figure 3.26   Local maximum $f(c)$

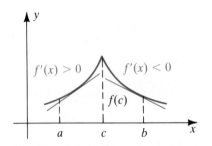

PROOF   If f' changes from positive to negative at c, then there is an open interval (a, b) containing c such that $f'(x) > 0$ for $a < x < c$ and $f'(x) < 0$ for $c < x < b$. Furthermore, we may choose (a, b) such that f is continuous on $[a, b]$. Hence, by Theorem (3.15), f is increasing on $[a, c]$ and decreasing on $[c, b]$. Thus, $f(x) < f(c)$ for every x in (a, b) different from c; that is, $f(c)$ is a local maximum for f. This proves (i). Similar proofs may be given for (ii) and (iii).   ■

To remember the first derivative test, think of the graphs in Figures 3.26 and 3.27. For a local maximum, the slope $f'(x)$ of the tangent line at $P(x, f(x))$ is positive if $x < c$ and negative if $x > c$. For a local minimum, the opposite situation occurs. Figure 3.28 illustrates (iii) of (3.16).

Figure 3.27   Local minimum $f(c)$

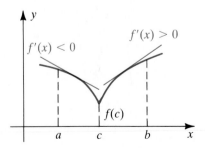

Figure 3.28 $f(c)$ is not a local extremum

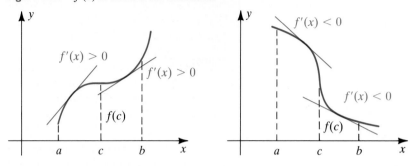

EXAMPLE▪2 If $f(x) = x^3 + x^2 - 5x - 5$, find the local extrema of f.

SOLUTION We considered this function in Example 1. The critical numbers are $-\frac{5}{3}$ and 1. We see from the table in Example 1 that the sign of $f'(x)$ changes from positive to negative as x increases through $-\frac{5}{3}$. Hence, by the first derivative test, f has a local maximum at $-\frac{5}{3}$. This maximum value is $f(-\frac{5}{3}) = \frac{40}{27}$ (see Figure 3.25).

A local minimum occurs at 1, since the sign of $f'(x)$ changes from negative to positive as x increases through 1. This minimum value is $f(1) = -8$.

EXAMPLE▪3 If $f(x) = x^{1/3}(8 - x)$, find the local extrema of f, and sketch the graph of f.

SOLUTION By the product rule,

$$f'(x) = x^{1/3}(-1) + (8 - x)\tfrac{1}{3}x^{-2/3}$$
$$= \frac{-3x + (8 - x)}{3x^{2/3}}$$
$$= \frac{4(2 - x)}{3x^{2/3}}.$$

Hence the critical numbers of f are 0 and 2. As in Example 1, these results suggest that we consider the sign of $f'(x)$ in each of the intervals $(-\infty, 0)$, $(0, 2)$, and $(2, \infty)$. Since f' is continuous and has no zeros on each interval, we may determine the sign of $f'(x)$ by using a suitable test value $f'(k)$. It is unnecessary to actually evaluate $f'(k)$; all we need to know is its sign. Thus, if we choose $k = 3$ in $(2, \infty)$, then

$$f'(3) = \frac{4(2 - 3)}{3(3^{2/3})},$$

and we can tell, *without evaluating*, that the numerator is negative and the denominator is positive. Hence $f'(3) < 0$, as shown in the following table.

Interval	$(-\infty, 0)$	$(0, 2)$	$(2, \infty)$
k	-1	1	3
Test value $f'(k)$	$f'(-1) > 0$	$f'(1) > 0$	$f'(3) < 0$
Sign of $f'(x)$	$+$	$+$	$-$
Conclusion	f is increasing on $(-\infty, 0]$	f is increasing on $[0, 2]$	f is decreasing on $[2, \infty)$

By the first derivative test, f has a local maximum at 2, since f' changes from positive to negative at 2. Thus we have

$$\text{Local max:} \quad f(2) = 2^{1/3}(8 - 2)$$
$$= 6\sqrt[3]{2} \approx 7.6.$$

The function does not have an extremum at 0, since f' does not change sign at 0.

To sketch the graph, we first plot points corresponding to the critical numbers. From $f(x) = x^{1/3}(8 - x)$, we see that the x-intercepts of the graph are 0 and 8. The graph is sketched in Figure 3.29. Note that

$$\lim_{x \to 0} f'(x) = \infty.$$

Since f is continuous at $x = 0$, the graph has a vertical tangent line at $(0, 0)$ by Definition (2.14).

Figure 3.29

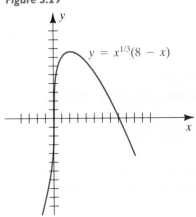

$y = x^{1/3}(8 - x)$

EXAMPLE ■ 4 If $f(x) = x^{2/3}(x^2 - 8)$, find the local extrema, and sketch the graph of f.

SOLUTION Applying the product rule, we obtain

$$f'(x) = x^{2/3}(2x) + (x^2 - 8)(\tfrac{2}{3}x^{-1/3})$$
$$= \frac{6x^2 + 2(x^2 - 8)}{3x^{1/3}} = \frac{8(x^2 - 2)}{3x^{1/3}}.$$

The critical numbers are the solutions of $x^2 - 2 = 0$ and $x^{1/3} = 0$—that is, $-\sqrt{2}, 0$, and $\sqrt{2}$. We find the sign of $f'(x)$ in each of the intervals $(-\infty, -\sqrt{2})$, $(-\sqrt{2}, 0)$, $(0, \sqrt{2})$, and $(\sqrt{2}, \infty)$. Arranging our work in tabular form as in previous examples, we obtain the following.

Interval	$(-\infty, -\sqrt{2})$	$(-\sqrt{2}, 0)$	$(0, \sqrt{2})$	$(\sqrt{2}, \infty)$
k	-2	-1	1	2
Test value $f'(k)$	$f'(-2) < 0$	$f'(-1) > 0$	$f'(1) < 0$	$f'(2) > 0$
Sign of $f'(x)$	$-$	$+$	$-$	$+$
Conclusion	f is decreasing on $(-\infty, -\sqrt{2}]$	f is increasing on $[-\sqrt{2}, 0]$	f is decreasing on $[0, \sqrt{2}]$	f is increasing on $[\sqrt{2}, \infty)$

Figure 3.30

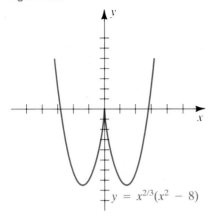

$y = x^{2/3}(x^2 - 8)$

By the first derivative test, f has local minima at $-\sqrt{2}$ and $\sqrt{2}$ and a local maximum at 0. The corresponding function values give us the following results:

Local max: $f(0) = 0$

Local min: $f(\sqrt{2}) = f(-\sqrt{2}) = -6\sqrt[3]{2} \approx -7.6$

Note that $f'(0)$ does not exist. Since

$$\lim_{x \to 0^-} f'(x) = \infty \quad \text{and} \quad \lim_{x \to 0^+} f'(x) = -\infty$$

and since f is continuous at $x = 0$, the graph has a cusp at $(0, 0)$, by Definition (2.15).

The graph of f is sketched in Figure 3.30.

Recall from (3.9) that if a function f is continuous on a closed interval $[a, b]$ and we wish to determine the maximum and minimum values, then in addition to finding all the local extrema, we should calculate the values $f(a)$ and $f(b)$ of f at the endpoints a and b of the interval $[a, b]$. The largest number among the local extrema and the values $f(a)$ and $f(b)$ is the maximum of f on $[a, b]$. The smallest of these numbers is the minimum of f on $[a, b]$. To illustrate these remarks, we shall consider the function discussed in the previous example and restrict our attention to certain intervals.

EXAMPLE ▪ 5   If $f(x) = x^{2/3}(x^2 - 8)$, find the maximum and minimum values of f on each of the following intervals:

(a) $[-1, \frac{1}{2}]$ **(b)** $[-1, 3]$ **(c)** $[-3, -2]$

SOLUTION   The graph in Figure 3.30 indicates the local extrema and the intervals on which f is increasing or decreasing. Figure 3.31 illustrates the part of the graph of f that corresponds to each of the intervals (a), (b), and (c).

Figure 3.31

(a) $[-1, \frac{1}{2}]$

(b) $[-1, 3]$

(c) $[-3, -2]$

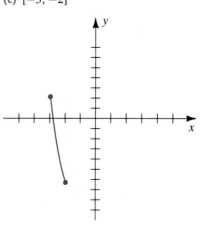

Referring to these sketches, we obtain the following table (check each entry).

Interval	Minimum value	Maximum value
$[-1, \frac{1}{2}]$	$f(-1) = -7$	$f(0) = 0$
$[-1, 3]$	$f(\sqrt{2}) = -6\sqrt[3]{2}$	$f(3) = \sqrt[3]{9}$
$[-3, -2]$	$f(-2) = -4\sqrt[3]{4}$	$f(-3) = \sqrt[3]{9}$

Note that on some intervals the maximum or minimum value of f is also a local extremum; on other intervals, this is not the case.

We often use the first derivative to determine the qualitative behavior of a function f at the endpoints of an interval. If f is differentiable on a closed interval $[a, b]$ and $f'(a)$ is positive, then f is increasing at $x = a$, so f cannot have a maximum at $x = a$ but does have a possible minimum at $x = a$. If $f'(a)$ is negative, then analogously, f has a possible maximum at $x = a$. The results are reversed at the right-hand endpoint: If $f'(b) > 0$, then there is a possible maximum at $x = b$, whereas if $f'(b) < 0$, then there is a possible minimum at $x = b$.

Looking again at Example 5(c), we see that

$$f'(x) = \frac{8(x^2 - 2)}{3x^{1/3}},$$

so that

$$f'(-3) = \frac{8(9 - 2)}{3(-3)^{1/3}} < 0 \quad \text{and} \quad f'(-2) = \frac{8(4 - 2)}{3(-2)^{1/3}} < 0.$$

Thus, f has a possible maximum at $x = -3$ and a possible minimum at $x = -2$. Since f is continuous on $[-3, -2]$, it has a maximum value and a minimum value on the interval. The function f has no critical numbers on this interval, so the maximum value must occur at the only possible maximum ($x = -3$) and the minimum value must occur at the only possible minimum ($x = -2$).

If c is in the domain of a differentiable function f and it is not an endpoint, then it is a critical number if $f'(c) = 0$. We may be able to obtain exact values for the critical numbers for a differentiable function by explicitly solving the equation $f'(x) = 0$. Quite often, however, it is not possible to solve the equation exactly; we could then use a graphing utility and numerical approximation techniques to locate the critical points. The next example shows such an approach.

EXAMPLE ▪ 6   Graph the function

$$f(x) = (x + 2)(x + 1)(x - 3)(x - 5),$$

and approximate the local extrema of f.

7 $f(x) = 10x^3(x-1)^2$ **8** $f(x) = (x^2 - 10x)^4$

9 $f(x) = x^{4/3} + 4x^{1/3}$ **10** $f(x) = x(x-5)^{1/3}$

11 $f(x) = x^{2/3}(x-7)^2 + 2$

12 $f(x) = x^{2/3}(8-x)$

13 $f(x) = x^2\sqrt[3]{x^2 - 4}$

14 $f(x) = 8 - \sqrt[3]{x^2 - 2x + 1}$

15 $f(x) = x\sqrt{x^2 - 9}$ **16** $f(x) = x\sqrt{4 - x^2}$

Exer. 17–22: Find the local extrema of f on $[0, 2\pi]$ and the subintervals on which f is increasing or is decreasing. Sketch the graph of f.

17 $f(x) = \cos x + \sin x$ **18** $f(x) = \cos x - \sin x$

19 $f(x) = \frac{1}{2}x - \sin x$ **20** $f(x) = x + 2\cos x$

21 $f(x) = 2\cos x + \sin 2x$ **22** $f(x) = 2\cos x + \cos 2x$

Exer. 23–28: Find the local extrema of f.

23 $f(x) = \sqrt[3]{x^3 - 9x}$ **24** $f(x) = \sqrt{x^2 + 4}$

25 $f(x) = (x-2)^3(x+1)^4$

26 $f(x) = x^2(x-5)^4$

27 $f(x) = \dfrac{\sqrt{x-3}}{x^2}$ **28** $f(x) = \dfrac{x^2}{\sqrt{x+7}}$

Exer. 29–32: Find the local extrema of f on the given interval.

29 $f(x) = \sec \frac{1}{2}x;$ $[-\pi/2, \pi/2]$

30 $f(x) = \cot^2 x + 2\cot x;$ $[\pi/6, 5\pi/6]$

31 $f(x) = 2\tan x - \tan^2 x;$ $[-\pi/3, \pi/3]$

32 $f(x) = \tan x - 2\sec x;$ $[-\pi/4, \pi/4]$

Exer. 33–34: Shown in the figure is a graph of the equation for $-2\pi \le x \le 2\pi$. Determine the x-coordinates of the points that correspond to local extrema.

33 $y = \frac{1}{2}x + \cos x$

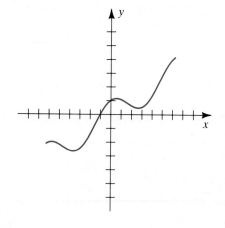

34 $y = \dfrac{\sqrt{3}}{2}x - \sin x$

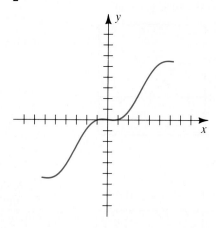

Exer. 35–40: Sketch the graph of a differentiable function f that satisfies the given conditions.

35 $f(0) = 3;$ $f(-2) = f(2) = -4;$ $f'(0)$ is undefined; $f'(-2) = f'(2) = 0;$ $f'(x) > 0$ if $-2 < x < 0$ or $x > 2;$ $f'(x) < 0$ if $x < -2$ or $0 < x < 2$

36 $f(3) = 5;$ $f(5) = 0;$ $f'(5)$ is undefined; $f'(3) = 0;$ $f'(x) > 0$ if $x < 3$ or $x > 5;$ $f'(x) < 0$ if $3 < x < 5$

37 $f(3) = 5;$ $f(5) = 0;$ $f'(3) = f'(5) = 0;$ $f'(x) > 0$ if $x < 3$ or $x > 5;$ $f'(x) < 0$ if $3 < x < 5$

38 $f(0) = 3;$ $f(-2) = f(2) = -4;$ $f'(-2) = f'(0) = f'(2) = 0;$ $f'(x) > 0$ if $-2 < x < 0$ or $x > 2;$ $f'(x) < 0$ if $x < -2$ or $0 < x < 2$

39 $f(-5) = 4;$ $f(0) = 0;$ $f(5) = -4;$ $f'(-5) = f'(0) = f'(5) = 0;$ $f'(x) > 0$ if $|x| > 5;$ $f'(x) < 0$ if $0 < |x| < 5$

40 $f(a) = a$ and $f'(a) = 0$ for $a = 0, \pm 1, \pm 2, \pm 3;$ $f'(x) > 0$ for all other values of x

c **Exer. 41–44:** Graph f on the indicated interval. **(a)** Use the graph to estimate the local extrema of f. **(b)** Estimate where f is increasing or is decreasing.

41 $f(x) = \dfrac{x^2 - 1.5x + 2.1}{0.3x^4 + 2.3x + 2.7}$ on $[-2, 2]$

42 $f(x) = \dfrac{10\cos 2x}{x^2 + 4}$ on $[-2, 2]$

43 $f(x) = |x^3 - 5x^2 + 8x - 105| + 2x$ on $[-4, 10]$

44 $f(x) = |x^4 - 4x^2 - x - 8|$ on $[-5, 1]$

| c | **Exer. 45–48:** Graph f' on the indicated interval. Estimate the x-coordinates of the local extrema of f and classify each local extremum.

45 $f'(x) = x - \cos \pi x - \sin x$ on $[-1, 1]$

46 $f'(x) = 6x^3 - 3x^2 - 1.3x + 0.5$ on $[-1, 1]$

47 $f'(x) = \sin(x^2 - 7x + 3)$ on $[0, 6]$

48 $f'(x) = x^4 - x^3 - 13x^2 + x + 12$ on $[-4, 6]$

3.4 CONCAVITY AND THE SECOND DERIVATIVE TEST

In the preceding section, we used the sign of the derivative f' to determine where a function f was increasing and where it was decreasing. In this section, we use the sign of the *second* derivative f'' to determine where the derivative f' is increasing and where it is decreasing. This approach utilizes the geometric property of the graph of the function f known as *concavity*.

CONCAVITY

If $f''(x) > 0$ on an open interval I, then $f'(x)$ is increasing on I, and thus the slope of the tangent line to the graph of f increases as x increases (see Figure 3.34a). Furthermore, if c is a point in I at which $f'(c) = 0$, then f' is increasing from negative to positive values, so the original function f must have a local minimum at $x = c$. As the next definition states, the graph is *concave upward* on I. If $f''(x) < 0$ on an open interval I, then $f'(x)$ is decreasing on I. In this case, the slope of the tangent line to the graph of f decreases as x increases (see Figure 3.34b), and the definition tells us that the graph is *concave downward* on I.

Figure 3.34

(a) Concave upward graph

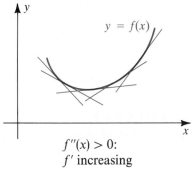

$f''(x) > 0$:
f' increasing

(b) Concave downward graph

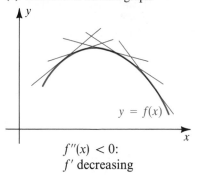

$f''(x) < 0$:
f' decreasing

Definition 3.17

Let f be differentiable on an open interval I. The graph of f is

    (i) concave upward on I if f' is increasing on I

    (ii) concave downward on I if f' is decreasing on I

It can be proved that for upward concavity on an interval, the graph lies *above* every tangent line to the graph (see Figure 3.34a), and for downward concavity, the graph lies *below* every tangent line (see Figure 3.34b).

The next theorem is a restatement, in terms of concavity, of our discussion about the sign of $f''(x)$.

Test for Concavity **3.18**

If the second derivative f'' of f exists on an open interval I, then the graph of f is

(i) **concave upward** on I if $f''(x) > 0$ on I

(ii) **concave downward** on I if $f''(x) < 0$ on I

EXAMPLE ■ 1 If $f(x) = x^3 + x^2 - 5x - 5$, determine intervals on which the graph of f is concave upward or is concave downward, and illustrate the results graphically.

SOLUTION The function f was considered in Examples 1 and 2 of the preceding section and is resketched in Figure 3.35. Since $f'(x) = 3x^2 + 2x - 5$,

$$f''(x) = 6x + 2 = 2(3x + 1).$$

Hence, $f''(x) < 0$ if $3x + 1 < 0$ — that is, if $x < -\frac{1}{3}$. Similarly, we have $f''(x) > 0$ if $x > -\frac{1}{3}$.

Applying the test for concavity (3.18) gives us the following.

Interval	$(-\infty, -\frac{1}{3})$	$(-\frac{1}{3}, \infty)$
Sign of $f''(x)$	−	+
Concavity	downward	upward

These facts are illustrated in Figure 3.35, where P is the point with x-coordinate $-\frac{1}{3}$.

Figure 3.35

Concave downward   Concave upward

$y = x^3 + x^2 - 5x - 5$

EXAMPLE ■ 2 If $f(x) = \sin x$, determine where the graph of f is concave upward and where it is concave downward, and illustrate the results graphically.

SOLUTION Differentiating $f(x)$ twice, we obtain

$$f'(x) = \cos x, \quad f''(x) = -\sin x.$$

Since $f''(x) = -f(x)$, we see that $f''(x) < 0$ whenever $f(x) > 0$; hence, by the test for concavity (3.18), the graph is concave downward whenever it lies above the x-axis. Similarly, $f''(x) > 0$ whenever $f(x) < 0$, so the

Figure 3.36

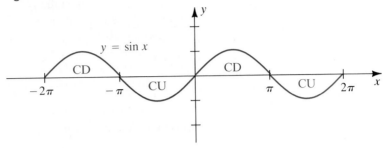

graph is concave upward whenever it lies below the x-axis. These facts are partially illustrated in Figure 3.36, in which the abbreviations CD and CU are used for concave downward and concave upward, respectively.

Each point on the graph of f at which the concavity changes from upward to downward, or vice versa, is called a *point of inflection* (P.I.). The precise definition may be stated as follows.

Definition 3.19

A point $(c, f(c))$ on the graph of f is a **point of inflection** if the following two conditions are satisfied:

(i) f is continuous at c.

(ii) There is an open interval (a, b) containing c such that the graph is concave upward on (a, c) and concave downward on (c, b), or vice versa.

Statement (ii) of Definition (3.19) is sometimes abbreviated by stating that *the concavity changes* at $P(c, f(c))$.

The point P in Figure 3.35 is a point of inflection. In Figure 3.36, every point $P(\pi n, 0)$, where n is an integer, is a point of inflection. The sketch in Figure 3.37 displays typical points of inflection on the graph of a function. Observe that a corner or cusp may or may not be a point of inflection.

Figure 3.37

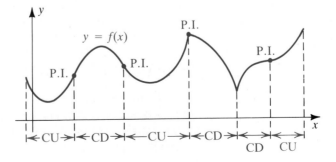

By the test for concavity (3.18), if the second derivative f'' changes sign at a number c, then the point $P(c, f(c))$ is a point of inflection. If, in addition, f'' is continuous at c, then we must have $f''(c) = 0$. However, it is also possible that $f''(c)$ may not exist at a point of inflection. Thus, *to find points of inflection, we begin by finding the zeros of f'' and the numbers at which f'' does not exist.* Each of these numbers is then tested to determine if it is an x-coordinate of a point of inflection.

SECOND DERIVATIVE TEST

The following test for local extrema is based on the second derivative.

Second Derivative Test **3.20**

> Suppose that f is differentiable on an open interval containing c and that $f'(c) = 0$.
>
> **(i)** If $f''(c) < 0$, then f has a local maximum at c.
> **(ii)** If $f''(c) > 0$, then f has a local minimum at c.

Figure 3.38
Local maximum $f(c)$

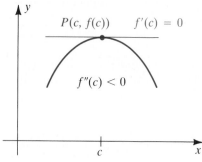

PROOF   If $f'(c) = 0$, then the tangent line to the graph at $P(c, f(c))$ is horizontal. If, in addition, $f''(c) < 0$, then the graph is concave downward at c, and hence there is an interval (a, b) containing c such that the graph lies below the tangent lines. It follows that $f(c)$ is a local maximum for f, as illustrated in Figure 3.38. We thus prove (i). A similar proof may be given for (ii), as illustrated in Figure 3.39. ▪

If $f''(c) = 0$, the second derivative test is not applicable. In such cases, the first derivative test should be used.

Figure 3.39
Local minimum $f(c)$

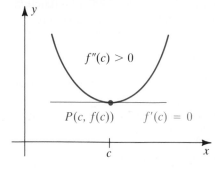

EXAMPLE ■ 3   If $f(x) = 12 + 2x^2 - x^4$, use the second derivative test to find the local extrema of f. Discuss concavity, find the points of inflection, and sketch the graph of f.

SOLUTION   Differentiating $f(x)$ twice yields

$$f'(x) = 4x - 4x^3 = 4x(1 - x^2),$$
$$f''(x) = 4 - 12x^2 = 4(1 - 3x^2).$$

The expression for $f'(x)$ is used to find the critical numbers 0, 1, and -1. The values of f'' at these numbers are

$$f''(0) = 4 > 0, \qquad f''(1) = -8 < 0, \quad \text{and} \quad f''(-1) = -8 < 0.$$

Hence, by the second derivative test, the function has a local minimum at 0 and local maxima at 1 and -1. The corresponding function values are $f(0) = 12$ and $f(1) = 13 = f(-1)$. The following table summarizes our discussion.

Critical number c	$f''(c)$	Sign of $f''(c)$	Conclusion
-1	-8	$-$	local max: $f(-1) = 13$
0	4	$+$	local min: $f(0) = 12$
1	-8	$-$	local max: $f(1) = 13$

To locate the possible points of inflection, we solve the equation $f''(x) = 0$ (that is, $4(1 - 3x^2) = 0$), obtaining the solutions $-\sqrt{3}/3$ and $\sqrt{3}/3$. We next examine the sign of $f''(x)$ in each of the intervals

$$(-\infty, -\sqrt{3}/3), \quad (-\sqrt{3}/3, \sqrt{3}/3), \quad \text{and} \quad (\sqrt{3}/3, \infty).$$

Since f'' is continuous and has no zeros on each interval, we may use test values to determine the sign of $f''(x)$. Let us arrange our work in tabular form as follows. The last row is a consequence of (3.18).

Figure 3.40

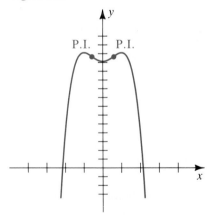

Interval	$(-\infty, -\sqrt{3}/3)$	$(-\sqrt{3}/3, \sqrt{3}/3)$	$(\sqrt{3}/3, \infty)$
k	-1	0	1
Test value $f''(k)$	$f''(-1) = -8$	$f''(0) = 4$	$f''(1) = -8$
Sign of $f''(x)$	$-$	$+$	$-$
Concavity	downward	upward	downward

Since $f''(x)$ changes sign at $-\sqrt{3}/3$ and $\sqrt{3}/3$, the corresponding points $(\pm\sqrt{3}/3, 113/9)$ on the graph are points of inflection. These are the points at which the concavity changes. As shown in the table, the graph is concave upward on the open interval $(-\sqrt{3}/3, \sqrt{3}/3)$ and concave downward outside of $[-\sqrt{3}/3, \sqrt{3}/3]$. The graph is sketched in Figure 3.40.

EXAMPLE ■ 4 If $f(x) = x^5 - 5x^3 + x^2 \sin x$, use a graphing utility to graph the function on the interval $[-2.5, 2.5]$, and approximate the local extrema of f. Find approximate points of inflection and discuss concavity.

SOLUTION We begin by differentiating $f(x)$ twice:

$$f'(x) = 5x^4 - 15x^2 + x^2 \cos x + 2x \sin x$$

$$f''(x) = 20x^3 - 30x - x^2 \sin x + 4x \cos x + 2 \sin x$$

Figure 3.41
$-2.5 \le x \le 2.5, -11.5 \le y \le 11.5$
(a) $y = f(x) = x^5 - 5x^3 + x^2 \sin x$

(b) $y = f'(x)$
$= 5x^4 - 15x^2$
$+ x^2 \cos x + 2x \sin x$

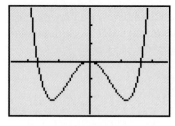

(c) $y = f''(x)$
$= 20x^3 - 30x - x^2 \sin x$
$+ 4x \cos x + 2 \sin x$

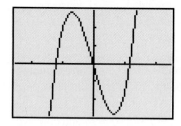

Next, we graph f, f', and f'', shown in Figures 3.41(a), (b), and (c), respectively.

To approximate the local extrema, we trace the graph of the first derivative (Figure 3.41b), obtaining estimates for the critical numbers: $-1.7, 0$, and 1.7. The critical number 0 is exact, as we can easily verify by substituting 0 into the derivative formula. Since $f(-x) = -f(x)$, the graph of f is symmetric about the origin, and the graph of f' is symmetric about the y-axis since $f'(-x) = f'(x)$. Thus, we need to approximate more accurately only one of the other two critical numbers. Newton's method or a solving routine yields the approximate value 1.66749833368. Rounding off to 1.6675 and evaluating, we find $f(1.6675) \approx -7.5231$. Thus, we have the following critical points on the graph:

$$(-1.6675, 7.5231), \quad (0, 0), \quad \text{and} \quad (1.6675, -7.5231).$$

The first point represents a local maximum for f, and the third point represents a local minimum. The point $(0, 0)$, yielded by the critical number 0, represents neither a local maximum nor a local minimum.

To find the approximate points of inflection, we estimate the zeros on the graph of the second derivative (see Figure 3.41c)—that is, the numbers where the second derivative changes sign. We find that the approximate values are $-1.2, 0$, and 1.2. Using a solving routine to obtain more accurate estimates for the zeros of the second derivative, we obtain 1.18366627376. Rounding off to 1.1837 and evaluating gives us $f(1.1837) \approx -4.6711$. We now estimate the points of inflection:

$$(-1.1837, 4.6711), \quad (0, 0), \quad \text{and} \quad (1.1837, -4.6711).$$

Viewing these points on the graph of the function f (see Figure 3.41a), we can easily determine concavity:

concave down on $(-2.5, -1.1837)$

concave up on $(-1.1837, 0)$

concave down on $(0, 1.1837)$

concave up on $(1.1837, 2.5)$

In Examples 1–4 , f'' is continuous. It is also possible for $(c, f(c))$ to be a point of inflection if either $f'(c)$ or $f''(c)$ does not exist, as illustrated in the next example.

EXAMPLE ■ 5 If $f(x) = 1 - x^{1/3}$, find the local extrema, discuss concavity, find the points of inflection, and sketch the graph of f.

SOLUTION Differentiating $f(x)$ twice yields

$$f'(x) = -\frac{1}{3}x^{-2/3} = -\frac{1}{3x^{2/3}},$$

$$f''(x) = \frac{2}{9}x^{-5/3} = \frac{2}{9x^{5/3}}.$$

The first derivative does not exist at $x = 0$, and 0 is the only critical number for f. Since $f''(0)$ is undefined, the second derivative test is not

Figure 3.42

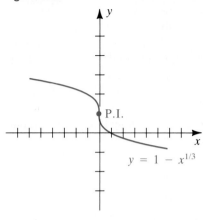

$$y = 1 - x^{1/3}$$

useful. However, if $x \neq 0$, then $x^{2/3} > 0$ and $f'(x) = -1/(3x^{2/3}) < 0$, which means that f is decreasing throughout its domain. Consequently, $f(0)$ is not a local extremum.

Using test values gives us the following table.

Interval	$(-\infty, 0)$	$(0, \infty)$
Sign of $f''(x)$	$-$	$+$
Concavity	downward	upward

The concavity changes at $x = 0$ and f is continuous at 0, so the point $(0, 1)$ is a point of inflection.

The graph is sketched in Figure 3.42. Note that there is a vertical tangent line at $(0, 1)$, since f is continuous at $x = 0$ and $\lim_{x \to 0} |f'(x)| = \infty$.

EXAMPLE ▪ 6 If $f(x) = x^{2/3}(5 + x)$, find the local extrema, discuss concavity, find points of inflection, and sketch the graph of f.

SOLUTION Writing $f(x) = 5x^{2/3} + x^{5/3}$ and differentiating twice gives us the following:

$$f'(x) = \frac{10}{3}x^{-1/3} + \frac{5}{3}x^{2/3} = \frac{5}{3}\left(\frac{2+x}{x^{1/3}}\right),$$

$$f''(x) = -\frac{10}{9}x^{-4/3} + \frac{10}{9}x^{-1/3} = \frac{10}{9}\left(\frac{x-1}{x^{4/3}}\right).$$

Referring to $f'(x)$, we see that the critical numbers for f are -2 and 0. We apply the second derivative test, as indicated in the following table.

Critical number c	-2	0
Sign of $f''(c)$	$-$	none
Conclusion	local max: $f(-2) = (-2)^{2/3}(3) \approx 4.8$	no conclusion

Since the second derivative test is not applicable at $c = 0$, let us apply the first derivative test. Using test values, we see that the sign of $f'(x)$ changes from $-$ to $+$ at $c = 0$. Hence, f has a local minimum at $(0, 0)$.

To determine concavity, first we note that $f''(x) = 0$ at $x = 1$ and $f''(x)$ does not exist at $x = 0$. Next, we examine the sign of $f''(x)$ for the cases $x < 0$, $0 < x < 1$, and $x > 1$. Using test values for f'' leads to the following table.

Interval	$(-\infty, 0)$	$(0, 1)$	$(1, \infty)$
Sign of $f''(x)$	$-$	$-$	$+$
Concavity	downward	downward	upward

Figure 3.43

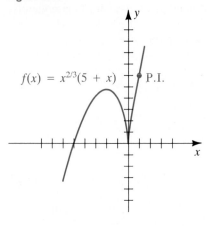

$f(x) = x^{2/3}(5 + x)$ P.I.

We see from the table that the graph of f has a point of inflection at $(1, 6)$, but not at $(0, 0)$. The graph is sketched in Figure 3.43. Note that

$$\lim_{x \to 0^-} f'(x) = -\infty \quad \text{and} \quad \lim_{x \to 0^+} f'(x) = \infty.$$

Since f is continuous at $x = 0$, the graph has a cusp at $(0, 0)$ by Definition (2.15).

EXAMPLE■7 If $f(x) = 2 \sin x + \cos 2x$, find the local extrema and sketch the graph of f on the interval $[0, 2\pi]$.

SOLUTION We differentiate $f(x)$ twice:

$$f'(x) = 2 \cos x - 2 \sin 2x$$
$$f''(x) = -2 \sin x - 4 \cos 2x$$

The critical numbers for f are the values of x for which $f'(x) = 0$. Solving this equation gives us

$$
\begin{array}{ll}
2 \cos x - 2 \sin 2x = 0 & f'(x) = 0 \\
\cos x - 2 \sin x \cos x = 0 & \text{divide by 2; double-angle formula} \\
(1 - 2 \sin x) \cos x = 0 & \text{factor out } \cos x \\
1 - 2 \sin x = 0, \quad \cos x = 0 & \text{set each factor equal to 0} \\
\sin x = \tfrac{1}{2}, \quad \cos x = 0 & \text{solve for } \sin x
\end{array}
$$

In the interval $[0, 2\pi]$, this result gives us values of $x = \pi/6, 5\pi/6, \pi/2,$ and $3\pi/2$.

Substituting these critical numbers for x in $f''(x)$, we obtain

$$f''(\pi/6) = -3, \quad f''(5\pi/6) = -3, \quad f''(\pi/2) = 2, \quad \text{and} \quad f''(3\pi/2) = 6.$$

Applying the second derivative test, we see that there are local maxima at $\pi/6$ and $5\pi/6$ and local minima at $\pi/2$ and $3\pi/2$. Thus we have

$$
\begin{array}{lll}
\text{Local max:} & f(\pi/6) = 3/2 & \text{and} \quad f(5\pi/6) = 3/2 \\
\text{Local min:} & f(\pi/2) = 1 & \text{and} \quad f(3\pi/2) = -3
\end{array}
$$

Using this information and plotting several more points gives us the sketch in Figure 3.44.

Figure 3.44

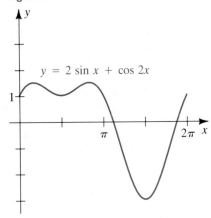

$y = 2 \sin x + \cos 2x$

CONCAVITY AND THE LAW OF DIMINISHING RETURNS

Thus far our study of concavity has yielded two applications: (1) more accurate sketches of the qualitative behavior of the graph of a function and (2) the second derivative test, which provides another tool in identifying extreme values for a function. Concavity is also an important concept in many applications of mathematics to the social sciences.

Figure 3.45 shows the graph of a function that is strictly increasing, but concave down. As x increases, so does $f(x)$, but the slopes of the tangent lines decrease: The *rate* of increase is slowing down. We can also

Figure 3.45

Increasing function, concave down

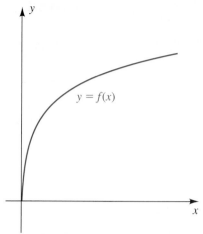

Figure 3.45

Increasing function, concave down

characterize this behavior by noting that although every unit increase in x corresponds to an increase in $f(x)$, each additional unit increase in x results in a *smaller* increase in $f(x)$. Thus, by the *law of diminishing returns*, if x represents a measure of effort and $y = f(x)$ represents the reward, then the same amount of additional effort at higher levels of x will yield smaller gains in y than it will at lower levels of x. For example, the extra knowledge gained in the seventh hour of studying for a test is much smaller than the additional wisdom accumulated in the third hour.

In the social sciences, psychologists, sociologists, economists, and others have observed that even in situations with monetary rewards, theories built on assumptions that people act to maximize wealth often do not yield correct predictions. Instead, many people appear to attach a measure of *utility*, or level of happiness, to different amounts of wealth. Utility increases with increasing amounts of money, but at a decreasing *rate*—that is, if $u(x)$ is the utility of x dollars, then $u'(x)$ is positive but $u''(x)$ is negative. In simple terms, an additional dollar means more to a poor person (for whom x is small) than to a rich one (who operates at a large value of x). Many current theories of behavior are predicated on the idea that people strive to maximize utility rather than wealth.

Conversely, some real-life situations are modeled well by functions whose graphs are concave up. In the area of investment opportunities, it is frequently the case that the larger the amount of money one has to invest, the greater is the possible rate of return. Banks, for example, may offer certificates of deposit with a 4.5% interest rate for amounts less than $1000, a 5% interest rate for amounts between $1000 and $10,000, and a 5.5% interest rate for investments above $10,000. In this case, a graph of interest rate as a function of amount invested will be concave up.

COMPARING THE FIRST AND THE SECOND DERIVATIVE TESTS

When asked to determine the nature of a function f where $f'(c) = 0$, we are faced with the question of whether to use the first derivative test or the second derivative test. There is no hard and fast answer to this question. The use of test values and the analysis of signs with the first derivative test can often be tedious. The second derivative test may seem preferable because it often gives an immediate answer: If $f''(c) > 0$, then $f(c)$ is a local maximum; if $f''(c) < 0$, then $f(c)$ is a local minimum.

The second derivative test also has its drawbacks, however. The function f may not be twice differentiable—that is, $f''(c)$ may not exist. If it does exist, it may be difficult to compute. (Recall those cases in which you had to find the second derivative of a rational function.) Even when it is easy to find the second derivative, it may turn out that $f''(c) = 0$. In this case, no conclusion could be made about the nature of f at $x = c$.

As a very simple example of this last point, consider the function $f(x) = x^4$. Here, $f'(x) = 4x^3$ and $f''(x) = 12x^2$. The only critical point is at $x = 0$, where both $f'(0)$ and $f''(0)$ are 0. The second derivative test does not help us. On the other hand, examination of the first derivative easily shows that $f'(x) < 0$ for $x < 0$ and $f'(x) > 0$ for $x > 0$, so there is a local minimum at $x = 0$.

EXERCISES 3.4

Exer. 1–18: Find the local extrema of f, using the second derivative test whenever applicable. Find the intervals on which the graph of f is concave upward or is concave downward, and find the x-coordinates of the points of inflection. Sketch the graph of f.

1 $f(x) = x^3 - 2x^2 + x + 1$

2 $f(x) = x^3 + 10x^2 + 25x - 50$

3 $f(x) = 3x^4 - 4x^3 + 6$ 4 $f(x) = 8x^2 - 2x^4$

5 $f(x) = 2x^6 - 6x^4$ 6 $f(x) = 3x^5 - 5x^3$

7 $f(x) = (x^2 - 1)^2$

8 $f(x) = x^4 - 4x^3 + 10$

9 $f(x) = \sqrt[5]{x} - 1$

10 $f(x) = 2 - \sqrt[3]{x^2}$

11 $f(x) = \sqrt[3]{x^2}(3x + 10)$

12 $f(x) = x^{2/3}(1 - x)$

13 $f(x) = x^2(3x - 5)^{1/3}$

14 $f(x) = x\sqrt[3]{3x + 2}$

15 $f(x) = 8x^{1/3} + x^{4/3}$

16 $f(x) = 6x^{1/2} + x^{3/2}$

17 $f(x) = x^2\sqrt{9 - x^2}$

18 $f(x) = x\sqrt{4 - x^2}$

Exer. 19–24: Use the second derivative test to find the local extrema of f on the interval $[0, 2\pi]$. (These exercises are the same as Exercises 17–22 in Section 3.3, for which the method of solution involved the first derivative test.)

19 $f(x) = \cos x + \sin x$ 20 $f(x) = \cos x - \sin x$

21 $f(x) = \frac{1}{2}x - \sin x$ 22 $f(x) = x + 2\cos x$

23 $f(x) = 2\cos x + \sin 2x$ 24 $f(x) = 2\cos x + \cos 2x$

Exer. 25–28: Use the second derivative test to find the local extrema of f on the given interval. (See Exercises 29–32 of Section 3.3.)

25 $f(x) = \sec \frac{1}{2}x;$ $[-\pi/2, \pi/2]$

26 $f(x) = \cot^2 x + 2\cot x;$ $[\pi/6, 5\pi/6]$

27 $f(x) = 2\tan x - \tan^2 x;$ $[-\pi/3, \pi/3]$

28 $f(x) = \tan x - 2\sec x;$ $[-\pi/4, \pi/4]$

Exer. 29–30: Shown in the figure is a graph of the equation for $-2\pi \le x \le 2\pi$. (See Exercises 33–34 of Section 3.3.) Use the second derivative test to find the local extrema of f.

29 $y = \frac{1}{2}x + \cos x$

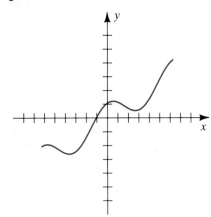

30 $y = \frac{\sqrt{3}}{2}x - \sin x$

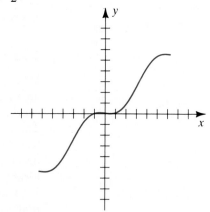

Exer. 31–38: Sketch the graph of a continuous function f that satisfies all of the stated conditions.

31 $f(0) = 1$; $f(2) = 3$; $f'(0) = f'(2) = 0$;
$f'(x) < 0$ if $|x - 1| > 1$; $f'(x) > 0$ if $|x - 1| < 1$;
$f''(x) > 0$ if $x < 1$; $f''(x) < 0$ if $x > 1$

32 $f(0) = 4$; $f(2) = 2$; $f(5) = 6$; $f'(0) = f'(2) = 0$;
$f'(x) > 0$ if $|x - 1| > 1$; $f'(x) < 0$ if $|x - 1| < 1$;
$f''(x) < 0$ if $x < 1$ or if $|x - 4| < 1$;
$f''(x) > 0$ if $|x - 2| < 1$ or if $x > 5$

33 $f(0) = 2; f(2) = f(-2) = 1; f'(0) = 0;$
$f'(x) > 0$ if $x < 0; f'(x) < 0$ if $x > 0;$
$f''(x) < 0$ if $|x| < 2; f''(x) > 0$ if $|x| > 2$

34 $f(1) = 4; f'(x) > 0$ if $x < 1; f'(x) < 0$ if $x > 1;$
$f''(x) > 0$ for every $x \neq 1$

35 $f(-2) = f(6) = -2; f(0) = f(4) = 0;$
$f(2) = f(8) = 3;$
f' is undefined at 2 and 6; $f'(0) = 1;$
$f'(x) > 0$ throughout $(-\infty, 2)$ and $(6, \infty);$
$f'(x) < 0$ if $|x - 4| < 2;$
$f''(x) < 0$ throughout $(-\infty, 0), (4, 6),$ and $(6, \infty);$
$f''(x) > 0$ throughout $(0, 2)$ and $(2, 4)$

36 $f(0) = 2; f(2) = 1; f(4) = f(10) = 0; f(6) = -4;$
$f'(2) = f'(6) = 0;$
$f'(x) < 0$ throughout $(-\infty, 2), (2, 4),$
    $(4, 6),$ and $(10, \infty);$
$f'(x) > 0$ throughout $(6, 10);$
$f'(4)$ and $f'(10)$ do not exist;
$f''(x) > 0$ throughout $(-\infty, 2), (4, 10),$ and $(10, \infty);$
$f''(x) < 0$ throughout $(2, 4)$

37 If n is an odd integer, then $f(n) = 1$ and $f'(n) = 0;$ if
n is an even integer, then $f(n) = 0$ and $f'(n)$ does not
exist; if n is any integer, then

(1) $f'(x) > 0$ whenever $2n < x < 2n + 1$

(2) $f'(x) < 0$ whenever $2n - 1 < x < 2n$

(3) $f''(x) < 0$ whenever $2n < x < 2n + 2$

38 $f(x) = x$ if $x = -1, 2, 4,$ or $8;$
$f'(x) = 0$ if $x = -1, 4, 6,$ or $8;$
$f'(x) < 0$ throughout $(-\infty, -1), (4, 6),$ and $(8, \infty);$
$f'(x) > 0$ throughout $(-1, 4)$ and $(6, 8);$
$f''(x) > 0$ throughout $(-\infty, 0), (2, 3),$ and $(5, 7);$
$f''(x) < 0$ throughout $(0, 2), (3, 5),$ and $(7, \infty)$

39 Prove that the graph of a quadratic function has no point
of inflection. State conditions for which the graph is
always

(a) concave upward   **(b)** concave downward

40 Prove that the graph of a polynomial function of degree
3 has exactly one point of inflection.

C **Exer. 41–42: Graph f on $[-1, 1]$. (a) Estimate where the
graph of f is concave upward or is concave downward.
(b) Estimate the x-coordinate of each point of inflection.**

41 $f(x) = 4x^5 + x^4 + 3x^2 - 2x + 1$

42 $f(x) = (x - 0.1)^2\sqrt{1.08 - 0.9x^2}$

C **Exer. 43–44: Graph f'' on $[0, 3]$. (a) Estimate where the
graph of f is concave upward or is concave downward.
(b) Estimate the x-coordinate of each point of inflection.**

43 $f''(x) = x^4 - 5x^3 + 7.57x^2 - 3.3x + 0.4356$

44 $f''(x) = 2.1 \sin \pi x + 1.4 \cos x - 0.6$

C **Exer. 45–50: Graph f on the indicated interval. (a) Estimate the local extrema of f. (b) Find approximate points
of inflection and discuss the concavity.**

45 $f(x) = x^4 - 3x^3 - 4x + 17$ on $[-10, 10]$

46 $f(x) = x^5 - 2x^4 - \sqrt{4x + 1}$ on $[0, 1.3]$

47 $f(x) = \sin(x^2 - 7x + 3)$ on $[0, 6]$

48 $f(x) = |x^4 - 4x^2 - x - 8|$ on $[-5, 1]$

49 $f(x) = x^2(x + 2)(x - 1)(x - 4)$ on $[-4, 6]$

50 $f(x) = x(x - 1)(x - 2)(x - 5)(x - 6)$ on $[-2, 8]$

Mathematicians and Their Times

ISAAC NEWTON

"Nature, and Nature's Laws lay hid in Night.
God said, Let Newton be! and All was Light."

SO WROTE POET ALEXANDER POPE of Isaac Newton. Called by many "the noblest intellect the human race has produced," Newton fundamentally altered our view of the universe.

In his youth, Newton faced many obstacles. His illiterate father died three months before Newton's birth on Christmas day in 1642 to a young widow only eight months married. The premature Isaac was small and weak at birth, with little hope of survival. When he was three, his mother remarried, moved away, and left him in his grandmother's care. She returned eight years later, after his stepfather died, and soon sent Newton away to boarding school, only to demand that at age 14 he quit school to become a farmer. Fortunately, others recognized his potential and persuaded her to let Newton continue his education.

When he entered Cambridge University, Newton had not yet read a single mathematics book. He graduated in 1665, showing no exceptional ability or particular distinction. In the next two years, however, he made spectacular discoveries in mathematics, physics, and optics. Between 1665 and 1667, as historian I. Bernard Cohen put it, Newton "invented a new kind of mathematics, discovered the secret of light and color, and showed how the universe is held together."

These accomplishments led to his appointment as Lucasian Professor of Mathematics at Cambridge in 1669. His major work, *Philosophiae Naturalis Principia Mathematica*, presents a unified system of scientific principles linking, via the theory of gravitation, the motion we see every day on the earth and the movements of heavenly bodies. Newton made many contributions to mathematics, including fundamental work in calculus on the relation between differentiation and integration.

Newton's England suffered much political conflict and rapid social change: civil war, revolution, the beheading of a king, Cromwell's rule, and the monarchy's restoration. The country began to change from a traditional hierarchical society to a more modern one. Newton played a part in the political life of his times, serving as a member of Parliament. He left Cambridge in 1696 to head the Royal Mint in London. He died in March 1727.

In assessing Newton's significance, physicist E. N. da Costa Andrade observed, "From time to time in the history of mankind a man arises who is of universal significance, whose work changes the current of human thought or of human experience, so that all that comes after him bears evidence of his spirit. Such a man was Shakespeare, such a man was Beethoven, such a man was Newton, and, of the three, his kingdom is the most widespread."

3.5 SUMMARY OF GRAPHICAL METHODS

Many applications of calculus involve analyzing the behavior of the function values $f(x)$ as x varies through a set of real numbers. An excellent way to exhibit this behavior is to sketch the graph of f. Not only does a graph display properties of f that may otherwise be hidden or unclear, it also provides an easy way to see the qualitative behavior of the function: concavity, local extrema, and intervals of increasing (or decreasing) values.

Since throughout the past chapters, we have discussed a variety of ideas useful for sketching graphs, our objective in this section is to summarize the ideas and consider several additional points. Our summary consists of a set of guidelines that are helpful whether we are sketching a graph by hand or using a graphing utility.

Guidelines for Sketching the Graph of $y = f(x)$ **3.21**

1 **Domain of f** Find the domain of f—that is, all real numbers x such that $f(x)$ is defined.

2 **Range of f** Determine the regions where the graph of f lies above the x-axis and where it lies below the x-axis—that is, find the values of x for which $f(x)$ is positive and those for which $f(x)$ is negative.

3 **Continuity of f** Determine whether f is continuous on its domain and, if not, find and classify the discontinuities.

4 **x- and y-intercepts** The x-intercepts are the solutions of the equation $f(x) = 0$, if any. The y-intercept is the function value $f(0)$, if it exists.

(continued)

5 Symmetry If f is an even function, the graph is symmetric with respect to the y-axis. If f is an odd function, the graph is symmetric with respect to the origin.

6 Critical numbers and local extrema Find $f'(x)$ and determine the critical numbers—that is, the values of x such that $f'(x) = 0$ or $f'(x)$ does not exist. Use the first derivative test to classify these local extrema. Determine whether there are corners or cusps on the graph. Note where the function is increasing ($f'(x) > 0$) and where it is decreasing ($f'(x) < 0$). Evaluate f at each critical number, whenever possible.

7 Concavity and points of inflection Find $f''(x)$ and use the second derivative test whenever appropriate. If $f''(x) > 0$ on an open interval, the graph is concave upward. If $f''(x) < 0$, the graph is concave downward. If f is continuous at c and if $f''(x)$ changes sign at c, then $P(c, f(c))$ is a point of inflection.

8 Asymptotes If $\lim_{x \to \infty} f(x) = L$ or $\lim_{x \to -\infty} f(x) = L$, then the line $y = L$ is a *horizontal asymptote*. If $\lim_{x \to a^+} f(x)$ or $\lim_{x \to a^-} f(x)$ is either ∞ or $-\infty$, then the line $x = a$ is a *vertical asymptote*.

Figure 3.46

Among the simplest functions to graph is a polynomial function f; however, if the degree of $f(x)$ is large, it may take considerable effort to find the zeros of f, f', and f''. Every derivative of a polynomial function f is continuous; therefore, the graph has a smooth appearance and possibly many high and low points, as illustrated in Figure 3.46. Each critical number c determines a local extremum $f(c)$ or a point of inflection, as we have seen in numerous examples and exercises.

Recall that a function f is a *rational function* if $f(x) = g(x)/h(x)$, where g and h are polynomial functions. If g and h have no common factors, then for every zero c of the denominator $h(x)$, the line $x = c$ is a vertical asymptote. In the next example, we shall use Guidelines (3.21) to sketch the graph of a rational function.

EXAMPLE ▪ 1 If $f(x) = \dfrac{2x^2}{9 - x^2}$, discuss and sketch the graph of f.

SOLUTION We shall follow Guidelines (3.21).

Guideline 1 The domain of f consists of all real numbers except -3 and 3.

Guideline 2 In determining the range of the values of x, we see that the numerator, $2x^2$, is nonnegative for all x, which means that the sign of $f(x)$ depends on the sign of the denominator, $9 - x^2$. The quantity $9 - x^2$ is positive for $-3 < x < 3$ and negative otherwise. In this interval, the values for $f(x)$ are positive between -3 and 3, except at 0, where $f(0) = 0$. For all other values of x, $f(x)$ is negative. We summarize our work in the following table.

Interval	$(-\infty, -3)$	$(-3, 0)$	$(0, 3)$	$(3, \infty)$
Sign of $f(x)$	$-$	$+$	$+$	$-$

Guideline 3 The function f has infinite discontinuities at -3 and 3 and is continuous at all other real numbers.

Guideline 4 To find the x-intercepts, we solve the equation $f(x) = 0$, obtaining $x = 0$. The y-intercept is $f(0) = 0$. Therefore, the graph intersects both the x-axis and the y-axis at the origin.

Guideline 5 Since $f(-x) = f(x)$, f is an even function and the graph is symmetric with respect to the y-axis.

Guideline 6 We differentiate $f(x)$:

$$f'(x) = \frac{(9 - x^2)(4x) - 2x^2(-2x)}{(9 - x^2)^2}$$

$$= \frac{36x}{(9 - x^2)^2}$$

Since $f'(x) = 0$ if $x = 0$, 0 is a critical number. The numbers -3 and 3 are not critical numbers because they are not in the domain of f.

Using test values gives us the following table.

Interval	$(-\infty, -3)$	$(-3, 0)$	$(0, 3)$	$(3, \infty)$
Sign of $f'(x)$	$-$	$-$	$+$	$+$
Conclusion	f is decreasing	f is decreasing	f is increasing	f is increasing

The sign of the derivative $f'(x)$ changes from negative to positive at $x = 0$, so, by the first derivative test, $f(0) = 0$ is a local minimum for f.

Guideline 7 We differentiate $f'(x)$:

$$f''(x) = \frac{(9 - x^2)^2(36) - (36x)(2)(9 - x^2)(-2x)}{(9 - x^2)^4}$$

$$= \frac{108(x^2 + 3)}{(9 - x^2)^3}$$

The numerator is always positive, so the sign of $f''(x)$ is determined by $(9 - x^2)^3$. Using test values gives us the following.

Interval	$(-\infty, -3)$	$(-3, 3)$	$(3, \infty)$
Sign of $f''(x)$	$-$	$+$	$-$
Concavity	downward	upward	downward

Figure 3.47

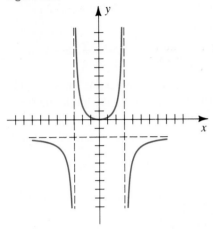

Since f is not continuous at -3 or 3, there are no points of inflection. As a check on local extrema (see guideline 6), we note that $f''(0) > 0$, and hence, by the second derivative test, $f(0) = 0$ is a local minimum.

Guideline 8 To find horizontal asymptotes, we use the methods in Section 1.4, obtaining

$$\lim_{x \to \infty} \frac{2x^2}{9 - x^2} = -2 \quad \text{and} \quad \lim_{x \to -\infty} \frac{2x^2}{9 - x^2} = -2.$$

Thus the line $y = -2$ is a horizontal asymptote.

The vertical asymptotes correspond to the zeros of the denominator $9 - x^2$ and hence are $x = -3$ and $x = 3$.

Using the results of the guidelines and referring to the table developed from guideline (6) to obtain the behavior of f near the vertical asymptotes ($x = \pm 3$) gives us the sketch in Figure 3.47.

The graph of $y = f(x)$ in the preceding example had no points of inflection. It is often difficult to find the points of inflection, whenever they exist, for the graph of a rational function. The reason is that the quotient rule is used to find $f'(x)$ and again to find $f''(x)$. Hence finding the solutions of $f''(x) = 0$ may require solving a polynomial equation of degree greater than 2. Solutions of such equations are often approximated by using Newton's method or a computer program. Because of this difficulty, in the next example and in most of the exercises, we shall not find the points of inflection or discuss concavity for graphs of rational functions.

EXAMPLE■2 If $f(x) = \dfrac{x^2}{x^2 - x - 2}$, discuss and sketch the graph of f.

SOLUTION We again follow the guidelines.

Guideline 1 The denominator equals $(x - 2)(x + 1)$, so the domain of f consists of all numbers except -1 and 2.

Guideline 2 The numerator is positive for all values of x except $x = 0$. The denominator $(x - 2)(x + 1)$ is negative on the interval $(-1, 2)$ and positive outside this interval. Thus, $f(x)$ is positive on $(-\infty, -1)$, negative on $(-1, 0)$ and $(0, 2)$, and positive on $(2, \infty)$.

Guideline 3 The function has infinite discontinuities at -1 and 2 and is continuous at all other numbers.

Guideline 4 To find the x-intercepts, we solve the equation $f(x) = 0$, obtaining $x = 0$. The y-intercept is $f(0) = 0$. Hence, as in Example 1, the graph intersects the x-axis and the y-axis at the origin.

Guideline 5 Since f is neither even nor odd, the graph is not symmetric to the y-axis or to the origin.

Guideline 6 We differentiate $f(x)$:

$$f'(x) = \frac{(x^2 - x - 2)(2x) - x^2(2x - 1)}{(x^2 - x - 2)^2} = -\frac{x(x + 4)}{(x^2 - x - 2)^2}$$

Solving $f'(x) = 0$ gives us the critical numbers 0 and -4. The zeros of the denominator, -1 and 2, are not critical numbers since $f(-1)$ and $f(2)$ do not exist. Using test values, we obtain the following table.

Interval	$(-\infty, -4)$	$(-4, -1)$	$(-1, 0)$	$(0, 2)$	$(2, \infty)$
Sign of $f'(x)$	$-$	$+$	$+$	$-$	$-$
Conclusion	f is decreasing	f is increasing	f is increasing	f is decreasing	f is decreasing

By the first derivative test, f has a local minimum $f(-4) = \frac{8}{9}$ and a local maximum $f(0) = 0$.

Guideline 7 We will not discuss concavity or find points of inflection. You may wish to verify that

$$f''(x) = \frac{2(x^3 + 6x^2 + 4)}{(x^2 - x - 2)^3}.$$

To solve the equation $f''(x) = 0$, we must solve the cubic equation

$$x^3 + 6x^2 + 4 = 0.$$

It can be shown that this equation has one real root r. To the nearest tenth, $r \approx -6.1$. The point of inflection is *approximately* $(-6.1, 0.9)$, slightly higher than the low (minimum) point $(-4, \frac{8}{9})$.

Guideline 8 To find horizontal asymptotes, we consider

$$\lim_{x \to \infty} \frac{x^2}{x^2 - x - 2} = 1 \quad \text{and} \quad \lim_{x \to -\infty} \frac{x^2}{x^2 - x - 2} = 1.$$

Thus, the graph has a horizontal asymptote $y = 1$.

The vertical asymptotes are obtained from the solutions of the equation $x^2 - x - 2 = 0$, so they are $x = -1$ and $x = 2$.

Using the preceding results and referring to the table developed from guideline (6) to obtain the behavior of f near the vertical asymptotes leads to the sketch in Figure 3.48.

The graph intersects the horizontal asymptote $y = 1$. To find the x-coordinate of the point of intersection, we solve the equation $f(x) = 1$ as follows:

$$\frac{x^2}{x^2 - x - 2} = 1$$
$$x^2 = x^2 - x - 2$$
$$x = -2$$

Thus the point of intersection is $(-2, 1)$.

Figure 3.48

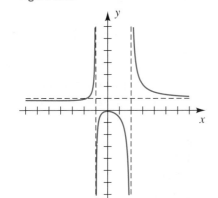

If $f(x) = g(x)/h(x)$ for polynomials $g(x)$ and $h(x)$ and *if the degree of $g(x)$ is 1 greater than the degree of $h(x)$*, then the graph of f has an **oblique asymptote** $y = ax + b$; that is, the vertical distance between the graph of f and this line approaches 0 as $x \to \infty$ or $x \to -\infty$. To establish this fact, we may use long division to express $f(x)$ in the form

$$f(x) = \frac{g(x)}{h(x)} = (ax + b) + \frac{r(x)}{h(x)},$$

where either $r(x) = 0$ or the degree of $r(x)$ is less than the degree of $h(x)$. It follows that $\lim_{x \to \infty} r(x)/h(x) = 0$ and $\lim_{x \to -\infty} r(x)/h(x) = 0$. Consequently, $f(x)$ gets closer to $ax + b$ as x approaches either ∞ or $-\infty$. If the graph of f has an oblique asymptote, then it has no horizontal asymptote.

EXAMPLE ■ 3   If $f(x) = \dfrac{x^2 - 9}{2x - 4}$, discuss and sketch the graph of f.

SOLUTION

Guidelines I and 3 The domain of f consists of all real numbers except $x = 2$, where there is an infinite discontinuity.

Guideline 2 Analysis of the signs of $x^2 - 9$ and $2x - 4$ shows that $f(x)$ is negative on $(-\infty, -3)$ and $(2, 3)$, whereas $f(x)$ is positive on $(-3, 2)$ and $(3, \infty)$.

Guideline 4 The x-intercepts are -3 and 3, and the y-intercept is $f(0) = \frac{9}{4}$.

Guideline 5 The graph is symmetric with respect to neither the y-axis nor the origin.

Guidelines 6 and 7 You should verify that

$$f'(x) = \frac{x^2 - 4x + 9}{2(x - 2)^2} \quad \text{and} \quad f''(x) = -\frac{5}{(x - 2)^3}.$$

Since $f'(x) \neq 0$ for every $x \neq 2$, there are no local extrema (see Corollary (3.6)).

Since $f''(x) > 0$ if $x < 2$ and $f''(x) < 0$ if $x > 2$, the graph of f is concave upward on $(-\infty, 2)$ and concave downward on $(2, \infty)$. There is no point of inflection.

Guideline 8 The degree of the numerator $x^2 - 9$ is 1 greater than that of the denominator $2x - 4$, so there is an oblique asymptote, and we use long division to express $f(x)$ as follows:

$$\frac{x^2 - 9}{2x - 4} = \left(\frac{1}{2}x + 1\right) - \frac{5}{2x - 4}$$

From the discussion preceding this example, the line $y = \frac{1}{2}x + 1$ is an oblique asymptote.

Figure 3.49

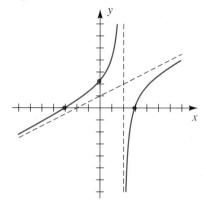

Since the graph has an oblique asymptote, there is no horizontal asymptote. There is a vertical asymptote $x = 2$ corresponding to the zero of the denominator $2x - 4$.

Representing the asymptotes by dashed lines, plotting the intercepts, and using the other information obtained by following the guidelines gives us the sketch in Figure 3.49.

We may well ask why, in this age of computers and graphing calculators, it is necessary to go through all the analysis suggested in the guidelines to get a picture of the graph of a function. As a first answer, when we use a graphing utility to plot the graph of a function, we must decide on a viewing window. We want to select an x-range that includes the points where the function displays particularly interesting properties; for example, where it takes on a local extreme value or where its graph switches concavity. Finding the critical numbers for f and f' helps us choose appropriate viewing windows. Checking at each step of the guidelines helps us ensure that the important features do not lie outside the current viewing window.

If the calculator- or computer-generated graph lacks sufficient detail, we may also miss some important features of the graph. The guidelines are useful in identifying intervals where we might want to use a "zoom" feature to study the graph more closely.

We must also be aware that graphing utilities have limitations that can produce misleading pictures. Many graphing utilities are programmed to plot a fixed number of points and then to join them by tiny line segments so that the resulting graph appears to the human eye to be a smooth curve. If g is any function, then the function f defined by

$$f(x) = x(x - 1)(x - 2)(x - 3) \cdots (x - 1000)(\sin x) + g(x)$$

has the same values as $g(x)$ for each of the 1001 values $x = 0, 1, 2, \ldots,$ 1000. A graphing utility that proceeds by plotting 1001 points may produce the same picture for the graphs of f and g even though in reality, f has a much more complicated graph. Careful use of the Guidelines (3.21) will reveal where the graphs of f and g differ.

If we wish to sketch a graph by hand, the guidelines also help us determine which points to plot. We first find the values of x at which the first or second derivative is either zero or does not exist. We then use these values to break up the domain of the function into a finite number of intervals over each of which the function is qualitatively the same: increasing and concave downward, for example, or decreasing and concave upward. Plotting the points associated with the endpoints and one or two test values inside each of these intervals, we should obtain an accurate sketch of the graph.

Whenever the graph and your hand calculations do not agree, check all your work. You may have entered the function incorrectly into the graphing utility, or you may have made an error differentiating or simplifying an expression. In the next example, we illustrate the use of the guidelines in conjunction with a graphing calculator.

EXAMPLE ■ 4   If

$$f(x) = \frac{5(x^2 - 8x + 6)}{\sqrt{x^4 + 2x^2 + 10}},$$

discuss the graph of f and examine several views of the graph of f.

SOLUTION

Guideline 1 Since

$$x^4 + 2x^2 + 10 \geq 10 \quad \text{for all } x,$$

the denominator $\sqrt{x^4 + 2x^2 + 10}$ is never 0, so the domain of f is $\mathbb{R}$.

Guideline 2 The sign of $f(x)$ is the same as the sign of $x^2 - 8x + 6$. By the quadratic formula, $x^2 - 8x + 6 = 0$ has solutions $x = 4 - \sqrt{10} \approx 0.84$ and $x = 4 + \sqrt{10} \approx 7.16$. The function $f(x)$ is negative on the interval $(4 - \sqrt{10}) < x < (4 + \sqrt{10})$, or approximately $(0.84, 7.16)$. The values of $f(x)$ are positive for $x < 0.84$ and for $x > 7.16$.

Guidelines 3-5 The function f is continuous at every real number with x-intercepts at approximately 0.84 and 7.16. The y-intercept is $f(0) = 30/\sqrt{10} \approx 9.49$. There is no symmetry.

We now have enough information to obtain the first view on a graphing utility. Note that the function values of f change sign at 0.84 and 7.16 so we will want the viewing window to include the interval $(0.84, 7.16)$. We begin with an x-interval of $-10 \leq x \leq 10$, and generate a graph of f using autoscaling of the y-interval, as shown in Figure 3.50.

Guideline 6 Applying the quotient rule yields

$$f'(x) = 10 \left[\frac{4x^4 - 5x^3 + 4x - 40}{(x^4 + 2x^2 + 10)^{3/2}} \right].$$

The critical numbers for f are the zeros of f'. Since the denominator of f' is positive, the critical numbers are the zeros of the numerator. One zero for the numerator of f' is -1.60, which leads to the displayed maximum of 22.94. The only other real zero is 2.10, leading to the displayed minimum of -5.16. These features are currently visible in Figure 3.50. If they were not, we could adjust the viewing window.

Guideline 7 Applying the quotient rule again, we find that

$f''(x)$

$$= 10 \left[\frac{-8x^7 + 15x^6 + 8x^5 - 20x^4 + 400x^3 - 166x^2 + 240x + 40}{(x^4 + 2x^2 + 10)^{5/2}} \right].$$

The three roots of the numerator for f'' lead to three points of inflection, at approximately $(-2.57, 20.29)$, $(-0.15, 11.35)$, and $(3.27, -3.91)$. Zooming in on Figure 3.50 or examining the sign of f'' on a graph of f'' (see Figure 3.51) indicates that the graph of f is concave upward for $x < -2.57$, concave downward on the interval $(-2.57, -0.15)$, con-

Figure 3.50
$y = f(x)$
$-10 \leq x \leq 10, -5.2 \leq y \leq 23$

Figure 3.51
$y = f''(x)$
$-3 < x < 4, -20 \leq y < 10$

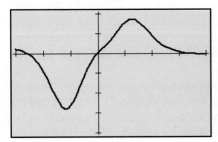

cave upward on the interval $(-0.15, 3.27)$, and concave downward for $x > 3.27$.

Guideline 8 To find any horizontal asymptotes, we rewrite f by dividing each term by x^2:

$$f(x) = \frac{5(x^2 - 8x + 6)}{\sqrt{x^4 + 2x^2 + 10}} = \frac{5\left(1 - 8\dfrac{1}{x} + 6\dfrac{1}{x^2}\right)}{\sqrt{1 + 2\dfrac{1}{x^2} + 10\dfrac{1}{x^4}}}$$

Since $\lim_{x \to \infty} f(x) = 5$ and $\lim_{x \to -\infty} f(x) = 5$, the graph has a horizontal asymptote at $y = 5$, which we can visualize more easily if we use the viewing window shown in Figure 3.52.

Figure 3.52
$y = f(x)$
$-70 \le x \le 70, -5 \le y \le 15$

EXAMPLE ▪ 5 Sketch the graph of $y = |4 - x^2|$.

SOLUTION We could use the fact that $|a| = \sqrt{a^2}$ to write $y = |4 - x^2|$ as $y = \sqrt{(4 - x^2)^2}$, and then use the derivatives to find local extrema and points of inflection, but this method would be fairly difficult. A simpler method is to sketch the graph of $y = 4 - x^2$, as shown in Figure 3.53(a). This graph is the same as that of $y = |4 - x^2|$ if $-2 \le x \le 2$. If $x > 2$ or $x < -2$, then $4 - x^2 < 0$, and the graph of $y = |4 - x^2|$ is the reflection of the graph of $y = 4 - x^2$ through the x-axis, as shown in Figure 3.53(b). The extrema, concavity, and points of inflection should be self-evident.

Figure 3.53

(a)   (b)

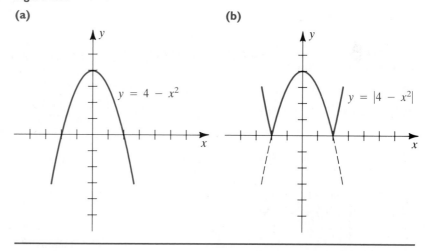

$y = 4 - x^2$ $y = |4 - x^2|$

In general, the best approach to sketching the graph of $y = |f(x)|$ is to first sketch the graph of $f(x)$ and then reflect across the x-axis any portion of the curve lying below the axis.

EXERCISES 3.5

Exer. 1–18: Find the extrema and sketch the graph of f.

1 $f(x) = \dfrac{2x - 5}{x + 3}$

2 $f(x) = \dfrac{3 - x}{x + 2}$

3 $f(x) = \dfrac{x^2 + x - 6}{x^2 - 1}$

4 $f(x) = \dfrac{4x}{x^2 - 4x + 3}$

5 $f(x) = \dfrac{3x^2 - 6x}{x^2 - x - 12}$

6 $f(x) = \dfrac{-2x^2 + 14x - 24}{x^2 + 2x}$

7 $f(x) = \dfrac{2x^2}{x^2 + 1}$

8 $f(x) = \dfrac{x - 1}{x^2 + 1}$

9 $f(x) = \dfrac{x + 4}{\sqrt{x}}$

10 $f(x) = \dfrac{x - 4}{\sqrt[3]{x^2}}$

11 $f(x) = \dfrac{-3x}{\sqrt{x^2 + 4}}$

12 $f(x) = \dfrac{2x}{\sqrt{x^2 + x + 2}}$

13 $f(x) = \dfrac{x^2 - x - 6}{x + 1}$

14 $f(x) = \dfrac{2x^2 - x - 3}{x - 2}$

15 $f(x) = \dfrac{x^2}{x + 1}$

16 $f(x) = \dfrac{-x^2 - 4x - 4}{x + 1}$

17 $f(x) = \dfrac{4 - x^2}{x + 3}$

18 $f(x) = \dfrac{8 - x^3}{2x^2}$

$\boxed{\text{c}}$ **Exer. 19–26: Find the extrema and graph the function using a graphing utility.**

19 $f(x) = x(x - 1)(x - 2)(x - 3)(x - 4)$

20 $f(x) = 64x^7 - 12x^5 + 56x^3 - 7x$

21 $f(x) = 1.5x^8 - 8.4x^4 + 3.2x^2 - 4.8$

22 $f(x) = 32x^6 - 16x^5 - 48x^4 + 20x^3 + 18x^2 - 5x - 1$

23 $f(x) = \begin{cases} 0 & \text{if } x < 1 \\ 3(x - 1)^2 & \text{if } 1 \le x < 3 \\ -7x^2 + 54x - 87 & \text{if } 3 \le x < 5 \\ 8(6 - x)^2 & \text{if } 5 \le x < 6 \\ 0 & \text{if } x \ge 6 \end{cases}$

24 $f(x) = \begin{cases} -1.33x^2 - 4.67x + 7 & \text{if } -3 \le x \le 0 \\ 1.2x^2 - 6.6x + 7 & \text{if } 0 < x < 5 \\ 0.25x^2 - 4.5x + 20.25 & \text{if } 5 \le x \le 7 \end{cases}$

25 $f(x) = |x^3 - 10x^2 - 2x + 65|$

26 $f(x) = |x^4 - 4x^2 - x - 8|$

Exer. 27–32: Find the extrema and the points of inflection, and sketch the graph of f.

27 $f(x) = \dfrac{3x}{(x + 8)^2}$

28 $f(x) = \dfrac{3x^2}{(2x - 9)^2}$

29 $f(x) = \dfrac{3x}{x^2 + 1}$

30 $f(x) = \dfrac{-4}{x^2 + 1}$

31 $f(x) = x^2 - \dfrac{27}{x^2}$

32 $f(x) = x^3 + \dfrac{3}{x}$

Exer. 33–36: Simplify $f(x)$ and sketch the graph of f.

33 $f(x) = \dfrac{2x^2 + x - 6}{x^2 + 3x + 2}$

34 $f(x) = \dfrac{x^2 - x - 6}{x^2 - 2x - 3}$

35 $f(x) = \dfrac{x - 1}{1 - x^2}$

36 $f(x) = \dfrac{x + 2}{x^2 - 4}$

Exer. 37–42: Sketch the graph of f.

37 $f(x) = |x^2 - 6x + 5|$

38 $f(x) = |8 + 2x - x^2|$

39 $f(x) = |x^3 + 1|$

40 $f(x) = |x^3 - x|$

41 $f(x) = -|\sin x|$

42 $f(x) = |\cos x| + 2$

Exer. 43–46: Use the information given to sketch the graph of f.

43 $f(x) = \dfrac{x - 3}{x^2 - 1} = \dfrac{x - 3}{(x + 1)(x - 1)}$;

$f'(x) = \dfrac{-x^2 + 6x - 1}{(x^2 - 1)^2}$;

$f'(x) = 0$ if $x = 3 \pm 2\sqrt{2}$ (approximately $5.83, 0.17$);

$f''(x) = \dfrac{2x^3 - 18x^2 + 6x - 6}{(x^2 - 1)^3}$;

$f''(x) = 0$ if $x = 2\sqrt[3]{2} + 2\sqrt[3]{4} + 3 \approx 8.69$;

$f(5.83) \approx 0.09$; $f(0.17) \approx 2.91$; $f(8.69) \approx 0.08$

44 $f(x) = \dfrac{x + 4}{x^2 - 4} = \dfrac{x + 4}{(x + 2)(x - 2)}$;

$f'(x) = \dfrac{-x^2 - 8x - 4}{(x^2 - 4)^2}$;

$f'(x) = 0$ if $x = -4 \pm 2\sqrt{3}$
(approximately $-0.54, -7.46$);

$f''(x) = \dfrac{2x^3 + 24x^2 + 24x + 32}{(x^2 - 4)^3}$;

$f''(x) = 0$ if $x = -2\sqrt[3]{3} - 2\sqrt[3]{9} - 4 \approx -11.04$;

$f(-0.54) \approx -0.93$; $f(-7.46) \approx -0.07$;

$f(-11.04) \approx -0.06$

45 $f(x) = \dfrac{2x^2 - 2x - 4}{x^2 + x - 12} = \dfrac{2(x+1)(x-2)}{(x+4)(x-3)}$;

$f'(x) = \dfrac{4x^2 - 40x + 28}{(x^2 + x - 12)^2}$;

$f'(x) = 0$ if $x = 5 \pm 3\sqrt{2}$ (approximately 9.24, 0.76);

$f''(x) = \dfrac{-8x^3 + 120x^2 - 168x + 424}{(x^2 + x - 12)^3}$;

$f''(x) = 0$ if $x = \sqrt[3]{36} + 3\sqrt[3]{6} + 5 \approx 13.75$;

$f(9.24) \approx 1.79$; $f(0.76) \approx 0.41$; $f(13.75) \approx 1.82$

46 $f(x) = \dfrac{-3x^2 - 3x + 6}{x^2 - 9} = \dfrac{-3(x+2)(x-1)}{(x+3)(x-3)}$;

$f'(x) = \dfrac{3x^2 + 42x + 27}{(x^2 - 9)^2}$;

$f'(x) = 0$ if $x = -7 \pm 2\sqrt{10}$
(approximately $-0.68, -13.32$);

$f''(x) = \dfrac{-6x^3 - 126x^2 - 162x - 378}{(x^2 - 9)^3}$;

$f''(x) = 0$ if $x = -2\sqrt[3]{20} - 2\sqrt[3]{50} - 7 \approx -19.80$;

$f(-0.68) \approx -0.78$; $f(-13.32) \approx -2.89$;

$f(-19.80) \approx -2.90$

47 Coulomb's law in electricity asserts that the force of attraction F between two charged particles is directly proportional to the product of the charges and inversely proportional to the square of the distance between the particles. Suppose that a particle of charge $+1$ is placed on a coordinate line between two particles of charge -1 (see figure).

(a) Show that the net force acting on the particle of charge $+1$ is given by

$$F(x) = -\dfrac{k}{x^2} + \dfrac{k}{(x-2)^2},$$

where k is a positive constant.

(b) Let $k = 1$ and sketch the graph of F for $0 < x < 2$.

Exercise 47

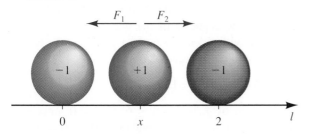

48 Biomathematicians have proposed many different functions for describing the effect of light on the rate at which photosynthesis can take place. If the function is to be realistic, then it must exhibit the *photoinhibition effect*; that is, the rate of production P of photosynthesis must decrease to 0 as the light intensity I reaches high levels (see figure). Which of the following formulas for P, where a and b are constants, may be used? Give reasons for your answer.

(a) $P = \dfrac{aI}{b + I}$ 　　　 **(b)** $P = \dfrac{aI}{b + I^2}$

Exercise 48

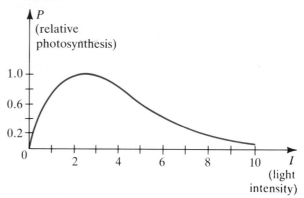

c Exer. 49–50: Graph f on the given interval. **(a)** Estimate where the graph of f is concave upward or is concave downward. **(b)** Estimate the x-coordinate of each point of inflection.

49 $f(x) = \dfrac{0.5x^3 + 3x + 7.1}{6 - x^2}$; $\quad [-2, 2]$

50 $f(x) = \dfrac{x^2 + 2x + 1}{0.7x^2 - x + 1}$; $\quad [-6, 6]$

51 We can often gain insight into the graph of a rational function by carrying out the long division of polynomials, so that $f(x)$ can be expressed as a polynomial plus a fraction, where the degree of the numerator of the fraction is smaller than the degree of the denominator.

(a) Show that the function in Example 1 can be written as

$$f(x) = -2 + \dfrac{18}{9 - x^2}.$$

(b) Using the form derived in part (a), show that if $|x| > 3$, then $f(x) < -2$.

(c) Using this form, show also that the graph of f has vertical asymptotes at $x = 3$ and $x = -3$.

(d) Show that there is a horizontal asymptote of $y = -2$.

52 Apply the approach discussed in Exercise 51 to the function of Example 2.

53 Suppose that $f(x) = g(x)/h(x)$, where g and h are polynomials of degree m and n, respectively.

(a) Show that the numerator of $f'(x)$ (before any simplification) is a polynomial of degree $m + n - 1$.

(b) What is the degree of the numerator of $f''(x)$?

3.6   OPTIMIZATION PROBLEMS

In this section, we will examine applications in which we need to find the maximum or minimum values of a function. For example, a physical or geometric quantity Q is often described by means of some formula $Q = f(x)$, where f is a function. Thus, Q might represent the temperature of a substance at time x, the current in an electrical circuit when the resistance is x, or the volume of gas in a spherical balloon of radius x. Of course, we often use other symbols for variables, such as T for temperature, t for time, I for current, R for resistance, V for volume, and r for radius. If $Q = f(x)$ and f is differentiable, then the derivative $dQ/dx = f'(x)$ can be used to help find the maximum or minimum values of Q. In applications, these extreme values are sometimes called **optimal values**, because they are, in a sense, the best or most favorable values of the quantity Q. The task of finding these values is called an **optimization problem**.

If an optimization problem is stated in words, then it is often necessary to convert the statement into an appropriate formula, such as $Q = f(x)$, in order to find critical numbers. In most cases, there will be only one critical number c. If, in addition, f is continuous on a closed interval $[a, b]$ containing c, then, by Guidelines (3.9), the extrema of f are the largest and smallest of the values $f(a)$, $f(b)$, and $f(c)$. Hence, it is often unnecessary to apply a derivative test. However, if it is easy to calculate $f''(x)$, we sometimes apply the second derivative test to verify an extremum, as illustrated in the next example.

EXAMPLE ▪ I   A long rectangular sheet of metal, 12 in. wide, is to be made into a rain gutter by turning up two sides so that they are perpendicular to the sheet. How many inches should be turned up in order to give the gutter its greatest capacity?

SOLUTION   The gutter is illustrated in Figure 3.54, where x denotes the number of inches turned up on each side. The width of the base of the gutter is $12 - 2x$ inches. The capacity of the gutter will be greatest when the area of the rectangle with sides of lengths x and $12 - 2x$ has its greatest value. Letting $f(x)$ denote this area, we obtain

$$f(x) = x(12 - 2x) = 12x - 2x^2.$$

Since $0 \leq 2x \leq 12$, the domain of f is $0 \leq x \leq 6$. If $x = 0$, or $x = 6$, no gutter is formed (the area of the rectangle would be $f(0) = 0 = f(6)$).

Differentiating yields

$$f'(x) = 12 - 4x = 4(3 - x);$$

Figure 3.54

$12 - 2x$

thus, the only critical number of f is 3. Since $f''(x) = -4 < 0$, $f(3)$ is a local maximum for f. It follows that 3 in. should be turned up to achieve maximum capacity.

COMPUTATIONAL METHOD   Once we have represented the area as a function with a prescribed domain, we can use a graphing utility to obtain a graph of the function and trace it to the maximum value. In this example, when we graph $f(x) = 12x - 2x^2$ over the interval $[0, 6]$, we obtain Figure 3.55. Using the trace option, we find that the maximum occurs at the point $(3, 18)$.

Because the types of optimization problems are unlimited, it is difficult to state specific rules for finding solutions. However, we can develop a general strategy for attacking such problems. The following guidelines are often helpful. When using the guidelines, don't become discouraged if you are unable to solve a given problem quickly. It takes a great deal of effort and practice to become proficient in solving optimization problems. Keep trying!

Figure 3.55
$0 \le x \le 6, 0 \le y \le 20$

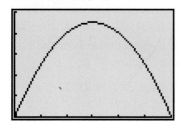

Guidelines for Solving Optimization Problems 3.22

1 Read the problem carefully several times, and think about the given facts as well as the unknown quantities that are to be found.

2 If possible, sketch a picture or diagram and label it appropriately, introducing variables for unknown quantities. Words such as *what, find, how much, how far,* or *when* should alert you to the unknown quantities.

3 Write down the known facts together with any relationships involving the variables.

4 Determine which variable is to be maximized or minimized, and express this variable as a function of *one* of the other variables.

5 Find the critical numbers of the function obtained in guideline (4).

6 Determine the extrema by using Guidelines (3.9) or the first or second derivative test. Check for endpoint extrema whenever appropriate.

The use of Guidelines (3.22) is illustrated in the next example.

EXAMPLE ▪ 2   An open box with a rectangular base is to be constructed from a rectangular piece of cardboard 16 in. wide and 21 in. long by cutting a square from each corner and then bending up the resulting sides. Find the size of the corner square that will produce a box having the largest possible volume. (Disregard the thickness of the cardboard.)

SOLUTION

Guideline 1   Read the problem at least one more time.

Guideline 2   Sketch the cardboard, as in Figure 3.56(a), introducing a variable x for the length of the side of the square to be cut from each corner.

Guideline 3   If the cardboard is folded along the dashed lines in Figure 3.56(a), the base of the resulting box has dimensions $21 - 2x$ and $16 - 2x$.

Guideline 4   The quantity to be maximized is the volume V of the box. Referring to Figure 3.56(b), we express V as a function of x:

$$V = x(16 - 2x)(21 - 2x) = 2(168x - 37x^2 + 2x^3)$$

Since $0 \le 2x \le 16$, the domain of V is $0 \le x \le 8$.

Figure 3.56

(a)

(b)

Guideline 5   To find the critical numbers for the function in guideline (4), differentiate V with respect to x:

$$\frac{dV}{dx} = 2(168 - 74x + 6x^2)$$
$$= 4(3x^2 - 37x + 84)$$
$$= 4(3x - 28)(x - 3)$$

Thus the possible critical numbers are $\frac{28}{3}$ and 3. Since $\frac{28}{3}$ is outside the domain of V, the only critical number is 3.

Guideline 6   Since V is continuous on $[0, 8]$, we shall use Guidelines (3.9) to determine the extrema. The endpoints $x = 0$ and $x = 8$ of the domain yield the minimum value $V = 0$. For the critical number $x = 3$, we obtain $V = 450$, which is a maximum value. Consequently, a 3-in. square should be cut from each corner of the cardboard in order to maximize the volume of the resulting box.

Figure 3.57
$0 \le x \le 8, 0 \le y \le 475$

 COMPUTATIONAL MÉTHOD   Using a graphing utility for

$$V(x) = 2(168x - 37x^2 + 2x^3)$$

over the interval $[0, 8]$ yields the graph shown in Figure 3.57, from which we see that the maximum value of V occurs at $x = 3$.

In the remaining examples, we shall not always point out the guidelines used. You should be able to determine specific guidelines by studying the solutions.

E X A M P L E ▪ 3 A circular cylindrical metal container, open at the top, is to have a capacity of 24π in³. The cost of the material used for the bottom of the container is 15 cents per in², and that of the material used for the curved part is 5 cents per in². If there is no waste of material, find the dimensions that will minimize the cost of the material.

S O L U T I O N We begin by sketching a typical container, as in Figure 3.58, letting r denote the radius of the base and h the altitude (both in inches). *The quantity we wish to minimize is the cost C of the material.* Since the costs per square inch for the base and the curved part are 15 cents and 5 cents, respectively, we have, in terms of cents,

$$\text{cost of container} = 15(\text{area of base}) + 5(\text{area of curved part}).$$

Thus,

$$C = 15(\pi r^2) + 5(2\pi rh),$$

or

$$C = 5\pi(3r^2 + 2rh).$$

We can express C as a function of one variable r by expressing h in terms of r. Since the volume of the container is 24π in³, we see that

$$\pi r^2 h = 24\pi, \quad \text{or} \quad h = \frac{24}{r^2}.$$

Substituting $24/r^2$ for h in the latter formula for C gives us

$$C = 5\pi\left(3r^2 + 2r \cdot \frac{24}{r^2}\right) = 5\pi\left(3r^2 + \frac{48}{r}\right).$$

The domain of C is $(0, \infty)$.

Next, we find critical numbers by differentiating C with respect to r:

$$\frac{dC}{dr} = 5\pi\left(6r - \frac{48}{r^2}\right) = 30\pi\left(\frac{r^3 - 8}{r^2}\right)$$

Since $dC/dr = 0$ if $r = 2$, we see that 2 is the only critical number. Since $dC/dr < 0$ if $r < 2$, and $dC/dr > 0$ if $r > 2$, it follows from the first derivative test that C has its minimum value if the radius of the cylinder is 2 in. The corresponding value for the altitude (obtained from $h = 24/r^2$) is $\frac{24}{4}$, or 6 in.

E X A M P L E ▪ 4 Find the maximum volume of a right circular cylinder that can be inscribed in a cone of altitude 12 cm and base radius 4 cm, if the axes of the cylinder and cone coincide.

Figure 3.59

(a)

(b)

SOLUTION The problem is sketched in Figure 3.59, where (b) represents a cross section through the axes of the cone and the cylinder. *The quantity we wish to maximize is the volume V of the cylinder.* From geometry,

$$V = \pi r^2 h.$$

Next, we express V in terms of one variable by finding a relationship between r and h. Referring to Figure 3.59(b) and using similar triangles, we see that

$$\frac{h}{4-r} = \frac{12}{4} = 3, \quad \text{or} \quad h = 3(4-r).$$

Consequently,

$$V = \pi r^2 h = \pi r^2 \cdot 3(4-r) = 3\pi r^2 (4-r).$$

The domain of V is $0 \le r \le 4$.

If either $r = 0$ or $r = 4$, we see that $V = 0$, and hence the maximum volume is not an endpoint extremum. It is sufficient, therefore, to search for local maxima. Since $V = 3\pi(4r^2 - r^3)$,

$$\frac{dV}{dr} = 3\pi(8r - 3r^2) = 3\pi r(8 - 3r).$$

Thus the critical numbers for V are $r = 0$ and $r = \frac{8}{3}$. At $r = \frac{8}{3}$, we have

$$V = \pi \left(\frac{8}{3}\right)^2 (4) = \frac{256\pi}{9} \approx 89.4 \text{ cm}^3,$$

which, by Guidelines (3.9), is a maximum value for the volume of the inscribed cylinder.

EXAMPLE ■ 5 A north–south highway intersects an east–west highway at a point P. An automobile crosses P at 10:00 A.M., traveling east at a constant speed of 20 mi/hr. At that same instant, another automobile is 2 mi north of P, traveling south at 50 mi/hr. Find the time at which they are closest to each other, and approximate the minimum distance between the automobiles.

Figure 3.60

SOLUTION Typical positions of the automobiles are illustrated in Figure 3.60. If t denotes the number of hours after 10:00 A.M., then the slower automobile is $20t$ miles east of P. The faster automobile is $50t$ miles south of its position at 10:00 A.M., and hence its distance from P is $2 - 50t$. By the Pythagorean theorem, the distance d between the automobiles is

$$d = \sqrt{(2-50t)^2 + (20t)^2}$$
$$= \sqrt{4 - 200t + 2500t^2 + 400t^2} = \sqrt{4 - 200t + 2900t^2}.$$

We wish to find the time t at which d has its smallest value, which will occur when the expression under the radical is minimal because d increases if and only if $4 - 200t + 2900t^2$ increases. Thus, we may simplify our work by letting

$$f(t) = 4 - 200t + 2900t^2$$

and finding the value of t for which f has a minimum. Since
$$f'(t) = -200 + 5800t,$$
the only critical number for f is
$$t = \frac{200}{5800} = \frac{1}{29}.$$

Moreover, $f''(t) = 5800$, so the second derivative is always positive. Therefore, f has a local minimum at $t = \frac{1}{29}$, and $f(\frac{1}{29}) = \frac{16}{29}$. Since the domain of t is $[0, \infty)$ and since $f(0) = 4$, there is no endpoint extremum. Consequently, the automobiles will be closest at $\frac{1}{29}$ hour (or approximately 2.07 min) after 10:00 A.M. The minimum distance is

$$\sqrt{f(\tfrac{1}{29})} = \sqrt{\tfrac{16}{29}} \approx 0.74 \text{ mi.}$$

EXAMPLE • 6   A person in a rowboat 2 mi from the nearest point on a straight shoreline wishes to reach a house 6 mi farther down the shore. If the person can row at a rate of 3 mi/hr and walk at a rate of 5 mi/hr, find the least amount of time required to reach the house.

SOLUTION   Figure 3.61 illustrates the problem: A denotes the position of the boat, B the nearest point on shore, C the house, D the point at which the boat reaches shore, and x the distance between B and D. By the Pythagorean theorem, the distance between A and D is $\sqrt{x^2 + 4}$, where $0 \le x \le 6$. Using the formula

$$\text{time} = \frac{\text{distance}}{\text{rate}},$$

we obtain

$$\text{time to row from } A \text{ to } D = \frac{\text{distance from } A \text{ to } D}{\text{rowing rate}} = \frac{\sqrt{x^2 + 4}}{3},$$

$$\text{time to walk from } D \text{ to } C = \frac{\text{distance from } D \text{ to } C}{\text{walking rate}} = \frac{6 - x}{5}.$$

Hence the total time T for the trip is

$$T = \frac{\sqrt{x^2 + 4}}{3} + \frac{6 - x}{5},$$

Figure 3.61

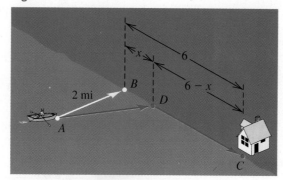

or, equivalently, $\quad T = \frac{1}{3}(x^2 + 4)^{1/2} + \frac{6}{5} - \frac{1}{5}x.$

We wish to find the minimum value for T. Note that $x = 0$ corresponds to the extreme situation in which the person rows directly to B and then walks the entire distance from B to C. If $x = 6$, then the person rows directly from A to C. These numbers may be considered as endpoints of the domain of T. If $x = 0$, then, from the formula for T,

$$T = \frac{\sqrt{4}}{3} + \frac{6}{5} - 0 = \frac{28}{15},$$

which is 1 hr 52 min. If $x = 6$, then

$$T = \frac{\sqrt{40}}{3} + \frac{6}{5} - \frac{6}{5} = \frac{2\sqrt{10}}{3} \approx 2.11,$$

or approximately 2 hr 7 min.

Differentiating the general formula for T, we see that

$$\frac{dT}{dx} = \frac{1}{3} \cdot \frac{1}{2}(x^2 + 4)^{-1/2}(2x) - \frac{1}{5},$$

or

$$\frac{dT}{dx} = \frac{x}{3(x^2 + 4)^{1/2}} - \frac{1}{5}.$$

In order to find the critical numbers, we let $dT/dx = 0$, obtaining the following equations:

$$\frac{x}{3(x^2 + 4)^{1/2}} = \frac{1}{5}$$

$$5x = 3(x^2 + 4)^{1/2}$$

$$25x^2 = 9(x^2 + 4)$$

$$x^2 = \frac{36}{16}$$

$$x = \frac{6}{4} = \frac{3}{2}$$

Thus, $\frac{3}{2}$ is the only critical number. The time T that corresponds to $x = \frac{3}{2}$ is

$$T = \frac{1}{3}(\frac{9}{4} + 4)^{1/2} + \frac{6}{5} - \frac{3}{10} = \frac{26}{15},$$

or, equivalently, 1 hr 44 min.

We have already examined the values of T at the endpoints of the domain, obtaining 1 hr 52 min and approximately 2 hr 7 min, respectively. Hence the minimum time of 1 hr 44 min occurs at $x = \frac{3}{2}$. Therefore, the boat should land at D, $1\frac{1}{2}$ mi from B, in order to minimize T. For a similar problem, but one in which the endpoints of the domain lead to minimum time, see Exercise 12.

Figure 3.62

$0 \le x \le 6, \ 1.7 \le y \le 2.1$

COMPUTATIONAL METHOD   Examining the function

$$T(x) = \frac{\sqrt{x^2 + 4}}{3} + \frac{6 - x}{5}$$

on $[0, 6]$ with a graphing utility produces the graph shown in Figure 3.62. We can use the tracing operation to find that the minimum time occurs at $x = \frac{3}{2}$.

CAUTION   We can err in solving optimization problems by formulating an appropriate function $f(x)$, finding where $f'(x) = 0$, and then declaring that we have located the extreme point. The critical points at which the derivative is zero may not be the extreme we are seeking. A critical point could turn out, for example, to be the minimum of the function when we needed to find the maximum. We may also have values c where $f'(c) = 0$, but the real-world constraints on the variables may put c outside the acceptable domain of f. The next two examples illustrate what can happen in such cases.

EXAMPLE ■ 7   A farmer has 2040 ft of fencing and wishes to fence off two separate fields. As Figure 3.63 shows, one of the fields is to be a rectangle with the length twice as long as the width, while the other field is to be square. Determine the dimensions of the fields if the farmer wishes to maximize the total area of the two fields.

Figure 3.63

$s = (2040 - 6x)/4$

SOLUTION   The total area A is the sum of the areas of the rectangle and the square. Let x represent the width of the rectangular field. We will write A as a function of x.

If x is the width, then the length is $2x$. The area of this rectangle is $2x^2$ square feet and its perimeter is $6x$ feet. The amount of fencing left after the rectangle has been built is $2040 - 6x$ feet. If the $2040 - 6x$ feet of fencing is used for the square, then each of its sides is $(2040 - 6x)/4 = 510 - 1.5x$ feet. Thus, the area of the square is $(510 - 1.5x)^2$ square feet.

The total area of both fields is

$$A(x) = 2x^2 + (510 - 1.5x)^2 \text{ square feet.}$$

We find the derivative $A'(x)$ to be

$$A'(x) = 4x + 2(510 - 1.5x)(-1.5) = 8.5x - 1530.$$

Solving $A'(x) = 0$ yields a unique value $x = c = 180$.

We may be tempted to advise the farmer to construct one rectangular field of dimensions 180 ft by 360 ft and one square field of sides 240 ft, which will achieve a "maximum" total area of $(180)(360) + 240^2 = 122{,}400 \text{ ft}^2$.

We need to check, however, whether the critical value $c = 180$ really is a maximum for the function. We apply the second derivative test. Since

$A''(x) = 8.5$, we have $A''(c) = A''(180) = 8.5$, which is positive. Thus, there is a local *minimum* at c, not a maximum.

To determine where the maximum of $A(x)$ actually occurs, we note that A is a differentiable function, so the only other possible candidates for the local extrema are at the endpoints of the interval of the domain of A. We need to determine this interval.

Since x represents a length, we must have $x \geq 0$. The smallest possible value for x is 0, which corresponds to having no rectangle and putting all the fencing into building the square. The corresponding total area is

$$A(0) = 510^2 = 260,100 \text{ ft}^2.$$

On the other hand, since the amount of fencing available for the square must be nonnegative, we have $2040 - 6x \geq 0$ or, equivalently, $x \leq 340$. The largest possible value for x is 340, which corresponds to using all the fencing to construct the rectangle. The associated area is

$$A(340) = 2(340)^2 = 231,200 \text{ ft}^2.$$

The interval is [0, 340], and the maximum value of A occurs at $x = 0$. The best advice to the farmer is to build a single square 510 ft on a side. If the farmer insists on having two separate fields, each of positive area, then the rectangular field should be made as small as possible.

The graph of $A(x)$ on the interval [0, 340] is shown in Figure 3.64. If we were to fail to analyze the nature of the critical point, then we would incorrectly find the minimum value, rather than the maximum value, and give the farmer the worst possible advice.

Figure 3.64

Figure 3.65

EXAMPLE ▪ 8   A recycling company transports recyclable paper and cardboard from city A to a processing plant in city B along a highway (see Figure 3.65). Materials are carried in trucks that travel at a speed of x miles per hour. Legal speeds on the highway are between 35 and 55 mi/hr. Assume that diesel fuel costs $1.25 per gallon and is consumed at the rate of $4 + (x^2/500)$ gallons per hour. The recycler pays its drivers $11 per hour and reimburses them for the cost of fuel as well. At what speed should the trucks be driven in order to minimize the recycler's total cost?

SOLUTION   The total cost C is the sum of the cost of wages and of fuel:

$$\text{total cost } C = \text{cost of wages } + \text{ cost of fuel}$$

We note that

$$\text{cost of wages} = (\text{wages per hour})(\text{number of hours})$$

and

$$\text{cost of fuel} = (\text{cost per gallon})(\text{number of gallons})$$
$$= (\text{cost per gallon})(\text{gallons per hour})(\text{number of hours}).$$

Since the trucks travel at x miles per hour, we have

$$\text{number of hours} = \frac{\text{distance}}{x}.$$

Hence, total cost $C(x)$ is given by

$$C(x) = 11\left(\frac{\text{distance}}{x}\right) + 1.25\left(4 + \frac{x^2}{500}\right)\left(\frac{\text{distance}}{x}\right),$$

which we can write as

$$C(x) = \left(\frac{\text{distance}}{x}\right)\left[11 + 1.25\left(4 + \frac{x^2}{500}\right)\right].$$

The distance traveled is not specified in the statement of the problem. We note, however, that since the distance is a positive number, the quantity $C(x)$ is minimized when the function

$$f(x) = \left(\frac{1}{x}\right)\left[11 + 1.25\left(4 + \frac{x^2}{500}\right)\right] = \left(\frac{1}{x}\right)\left(16 + \frac{1.25x^2}{500}\right)$$

is minimized.

Differentiating f and simplifying yields

$$f'(x) = \frac{1.25}{500} - \frac{16}{x^2} \quad \text{and} \quad f''(x) = \frac{32}{x^3}.$$

We then solve $f'(x) = 0$ to find two roots, $x = 80$ and $x = -80$. The negative root can be ignored since x represents the speed of a truck, which must be positive. Thus we have a unique critical point at $x = 80$. We check the second derivative, $f''(80) = 32/80^3$, which is positive, so we indeed have a local minimum at $x = 80$. Another critical point at $x = 0$, where the first derivative fails to exist, can also be ignored because the trucks must move at a positive speed. (For the purposes of this problem, the domain of $C(x)$ consists only of positive numbers.)

We cannot advise the company to instruct its truckers to drive at 80 mi/hr in violation of the legal speed limits, which are on the interval [33, 55]. We thus conclude that the global minimum is not available. Examination of the derivative $f'(x)$ shows that it is negative for all positive x less than 80, so cost is lowest when the speed is the highest legal one, 55 mi/hr. Note that for $x = 55$, $f(x) \approx 0.4284$ (about 43 cents per mile), so the minimum total cost is about 0.4284(distance). Figure 3.66 is a graph of f, which shows the location of the endpoint extrema.

Figure 3.66

$$y = f(x) = \left(\frac{1}{x}\right)\left(16 + \frac{1.25x^2}{500}\right)$$

Figure 3.67

(a)

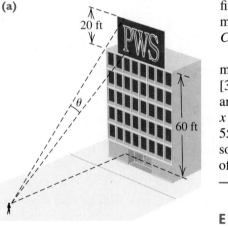

(b)

EXAMPLE ■ 9 A billboard 20 ft tall is located on top of a building, with its lower edge 60 ft above the level of a viewer's eye, as shown in Figure 3.67(a). How far from a point directly below the sign should a viewer stand to maximize the angle θ between the lines of sight of the top and bottom of the billboard? (This angle should result in the best view of the billboard.)

SOLUTION The problem is sketched in Figure 3.67(b), using right triangles AOC and BOC having common side OC of (variable) length x. We see that

$$\tan\alpha = \frac{80}{x} \quad \text{and} \quad \tan\beta = \frac{60}{x}.$$

The angle $\theta = \alpha - \beta$ is a function of x and

$$\tan \theta = \tan(\alpha - \beta) = \frac{\tan \alpha - \tan \beta}{1 + \tan \alpha \tan \beta}.$$

Substituting for $\tan \alpha$ and $\tan \beta$ and simplifying, we obtain

$$\tan \theta = \frac{(80/x) - (60/x)}{1 + (80/x)(60/x)} = \frac{20x}{x^2 + 4800}.$$

The extrema of θ occur if $d\theta/dx = 0$. Differentiating implicitly with respect to x and using the quotient rule gives us

$$\sec^2 \theta \frac{d\theta}{dx} = \frac{(x^2 + 4800)(20) - 20x(2x)}{(x^2 + 4800)^2} = \frac{96{,}000 - 20x^2}{(x^2 + 4800)^2}.$$

Since $\sec^2 \theta > 0$, it follows that $d\theta/dx = 0$ if and only if

$$96{,}000 - 20x^2 = 0, \quad \text{or} \quad x^2 = 4800.$$

Thus the only critical number of θ is

$$x = \sqrt{4800} = 40\sqrt{3}.$$

We may verify that the sign of $d\theta/dx$ changes from positive to negative at $\sqrt{4800}$, and hence a maximum value of θ occurs at $x = 40\sqrt{3}$ ft ≈ 69.3 ft.

EXERCISES 3.6

Exer. 1–6: You will need to formulate a function and find its extreme values on some interval, which you must also determine. In addition to Guidelines (3.9) and the first and second derivative tests, you may wish to use a graphing utility to examine the graph of the function on the chosen interval.

1 Find the maximum value of Z if $Z = xw$, where $x + w = 30$.

2 Find the maximum value of B if $B = st$, where $4s + 3t = 48$.

3 Find the minimum value of A if $A = 4y + x^2$, where $(x^2 + 1)y = 324$.

4 Find the maximum value of S if $S = 8x - 512y^2$, where $x(y^2 + 1) = 64$.

5 Find the minimum value of P if $P = x^2 + y^2$, where $x - y = 40$.

6 Find the minimum value of C if $C = \sqrt{x^2 + y^2}$, where $xy = 9$.

7 If a box with a square base and an open top is to have a volume of 4 ft³, find the dimensions that require the least material. (Disregard the thickness of the material and waste in construction.)

8 Work Exercise 7 if the box has a closed top.

9 A metal cylindrical container with an open top is to hold 1 ft³. If there is no waste in construction, find the dimensions that require the least amount of material. (Compare with Example 3.)

10 If the circular base of the container in Exercise 9 is cut from a square sheet and the remaining metal is discarded, find the dimensions that require the least amount of material.

11 One thousand feet of chain link fence will be used to construct six cages for a zoo exhibit, as shown in the figure on the following page. Find the dimensions that maximize the enclosed area A. (*Hint:* First express y as a function of x, and then express A as a function of x.)

Exercise 11

12 Refer to Example 6. If the person is in a motorboat that can travel at an average rate of 15 mi/hr, what route should be taken to arrive at the house in the least amount of time?

13 At 1:00 P.M., ship A is 30 mi due south of ship B and is sailing north at a rate of 15 mi/hr. If ship B is sailing west at a rate of 10 mi/hr, find the time at which the distance d between the ships is minimal (see figure).

Exercise 13

14 A window has the shape of a rectangle surmounted by a semicircle. If the perimeter of the window is 15 ft, find the dimensions that will allow the maximum amount of light to enter.

15 A fence 8 ft tall stands on level ground and runs parallel to a tall building (see figure). If the fence is 1 ft from the

Exercise 15

building, find the length of the shortest ladder that will extend from the ground over the fence to the wall of the building. (*Hint*: Use similar triangles.)

16 A page of a book is to have an area of 90 in^2, with 1-in. margins at the bottom and sides and a $\frac{1}{2}$-in. margin at the top. Find the dimensions of the page that will allow the largest printed area.

17 A builder intends to construct a storage shed having a volume of 900 ft^3, a flat roof, and a rectangular base whose width is three-fourths the length. The cost per square foot of the materials is $4 for the floor, $6 for the sides, and $3 for the roof. What dimensions will minimize the cost?

18 A water cup in the shape of a right circular cone is to be constructed by removing a circular sector from a circular sheet of paper of radius a and then joining the two straight edges of the remaining paper (see figure). Find the volume of the largest cup that can be constructed.

Exercise 18

19 A farmer has 500 ft of fencing with which to enclose a rectangular field. A straight riverbank will be used as part of the fencing on one side of the field (see figure). Prove that the area of the field is greatest when the rectangle is square.

Exercise 19

20 Refer to Exercise 19. Suppose the farmer wants the area of the rectangular field to be A ft^2. Prove that the least amount of fencing is required when the rectangle is a square.

21 A hotel that charges $80 per day for a room gives special rates to organizations that reserve between 30 and 60 rooms. If more than 30 rooms are reserved, the charge per room is decreased by $1 times the number of rooms over 30. Under these conditions, how many rooms must be rented if the hotel is to receive the maximum income per day?

22 Refer to Exercise 21. Suppose that for each room rented it costs the hotel $6 per day for cleaning and maintenance. In this case, how many rooms must be rented to obtain the greatest net income?

23 A steel storage tank for propane gas is to be constructed in the shape of a right circular cylinder with a hemisphere at each end (see figure). The construction cost per square foot for the end pieces is twice that for the cylindrical piece. If the desired capacity is 10π ft^3, what dimensions will minimize the cost of construction?

Exercise 23

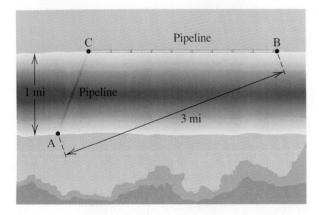

24 A pipeline for transporting oil will connect two points A and B that are 3 mi apart and on opposite banks of a straight river 1 mi wide (see figure). Part of the pipeline will run under water from A to a point C on the opposite bank, and then above ground from C to B. If the cost per mile of running the pipeline under water is four times the cost per mile of running it above ground, find the location of C that will minimize the cost (disregard the slope of the river bed).

Exercise 24

25 Find the dimensions of the rectangle of maximum area that can be inscribed in a semicircle of radius a, if two vertices lie on the diameter (see figure).

Exercise 25

26 Find the dimensions of the rectangle of maximum area that can be inscribed in an equilateral triangle of side a, if two vertices of the rectangle lie on one of the sides of the triangle.

27 Of all possible right circular cones that can be inscribed in a sphere of radius a, find the volume of the one that has maximum volume.

28 Find the dimensions of the right circular cylinder of maximum volume that can be inscribed in a sphere of radius a.

29 Find the point on the graph of $y = x^2 + 1$ that is closest to the point $(3, 1)$.

30 Find the point on the graph of $y = x^3$ that is closest to the point $(4, 0)$.

31 The strength of a rectangular beam is directly proportional to the product of its width and the square of the depth of a cross section. Find the dimensions of the strongest beam that can be cut from a cylindrical log of radius a (see figure).

Exercise 31

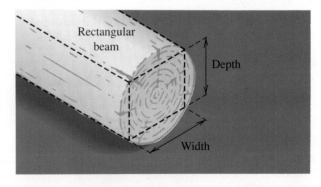

32 The illumination from a light source is directly proportional to the strength of the source and inversely proportional to the square of the distance from the source. If two light sources of strengths S_1 and S_2 are d units apart, at what point on the line segment joining the two sources is the illumination minimal?

33 A wholesaler sells running shoes at $20 per pair if fewer than 50 pairs are ordered. If 50 or more pairs are ordered (up to 600), the price per pair is reduced by 2 cents times the number ordered. What size order will produce the maximum amount of money for the wholesaler?

34 A paper cup is to be constructed in the shape of a right circular cone. If the volume desired is 36π in³, find the dimensions that require the least amount of paper. (Disregard any waste that may occur in the construction.)

35 A wire 36 cm long is to be cut into two pieces. One of the pieces will be bent into the shape of an equilateral triangle and the other into the shape of a rectangle whose length is twice its width. Where should the wire be cut if the combined area of the triangle and rectangle is to be **(a)** minimized? **(b)** maximized?

36 An isosceles triangle has base b and equal sides of length a. Find the dimensions of the rectangle of maximum area that can be inscribed in the triangle if one side of the rectangle lies on the base of the triangle.

37 A window has the shape of a rectangle surmounted by an equilateral triangle. If the perimeter of the window is 12 ft, find the dimensions of the rectangle that will produce the largest area for the window.

38 Two vertical poles of lengths 6 ft and 8 ft stand on level ground, with their bases 10 ft apart. Approximate the minimal length of cable that can reach from the top of one pole to some point on the ground between the poles and then to the top of the other pole.

39 Prove that the rectangle of largest area having a given perimeter p is a square.

40 A right circular cylinder is generated by rotating a rectangle of perimeter p about one of its sides. What dimensions of the rectangle will generate the cylinder of maximum volume?

41 The owner of an apple orchard estimates that if 24 trees are planted per acre, then each mature tree will yield 600 apples per year. For each additional tree planted per acre, the number of apples produced by each tree decreases by 12 per year. How many trees should be planted per acre to obtain the most apples per year?

42 A real estate company owns 180 efficiency apartments, which are fully occupied when the rent is $300 per month. The company estimates that for each $10 increase in rent, 5 apartments will become unoccupied. What rent should be charged in order to obtain the largest gross income?

43 A package can be sent by parcel post only if the sum of its length and girth (the perimeter of the base) is not more than 108 in. Find the dimensions of the box of

maximum volume that can be sent, if the base of the box is a square.

44 A north–south highway A and an east–west highway B intersect at a point P. At 10:00 A.M., an automobile crosses P traveling north on highway A at a speed of 50 mi/hr. At that same instant, an airplane flying east at a speed of 200 mi/hr and an altitude of 26,400 ft is directly above the point on highway B that is 100 mi west of P. If the automobile and the airplane maintain the same speeds and directions, at what time will they be closest to each other?

45 Two factories A and B that are 4 mi apart emit particles in smoke that pollute the area between the factories. Suppose that the number of particles emitted from each factory is directly proportional to the amount of smoke and inversely proportional to the cube of the distance from the other factory. If factory A emits twice as much smoke as factory B, at what point between A and B is the pollution minimal?

46 An oil field contains 8 wells, which produce a total of 1600 barrels of oil per day. For each additional well that is drilled, the average production per well decreases by 10 barrels per day. How many additional wells should be drilled to obtain the maximum amount of oil per day?

47 A canvas tent is to be constructed in the shape of a pyramid with a square base. A steel pole, placed in the center of the tent, will form the support (see figure). If S ft² of canvas is available for the four sides and x is the length of the base, show that

(a) the volume V of the tent is $V = \frac{1}{6}x\sqrt{S^2 - x^4}$

(b) V has its maximum value when x equals $\sqrt{2}$ times the length of the pole

Exercise 47

48 A boat must travel 100 mi upstream against a 10-mi/hr current. When the velocity of the boat relative to the water is v mi/hr, the number of gallons of gasoline consumed each hour is directly proportional to v^2.

(a) If a constant velocity of v mi/hr is maintained, show that the total number y of gallons of gasoline consumed is given by $y = 100kv^2/(v - 10)$ for $v > 10$ and for some positive constant k.

(b) Find the speed that minimizes the number of gallons of gasoline consumed during the trip.

49 Cars are crossing a bridge that is 1 mi long. Each car is 12 ft long and is required to stay a distance of at least d ft from the car in front of it (see figure).

(a) Show that the greatest number of cars that can be on the bridge at one time is $[\![5280/(12 + d)]\!]$, where $[\![\ ]\!]$ denotes the greatest integer function.

(b) If the velocity of each car is v mi/hr, show that the maximum traffic flow rate F (in cars per hour) is given by $F = [\![5280v/(12 + d)]\!]$.

(c) The stopping distance (in feet) of a car traveling v mi/hr is approximately $0.05v^2$. If $d = 0.025v^2$, find the speed that maximizes the traffic flow across the bridge.

Exercise 49

12 ft d

50 Prove that the shortest distance from a point (x_1, y_1) to the graph of a differentiable function f is measured along a normal line to the graph — that is, a line perpendicular to the tangent line.

51 A railroad route is to be constructed from town A to town C, branching out from a point B toward C at an angle of θ degrees (see figure). Because of the mountains between A and C, the branching point B must be at least 20 mi east of A. If the construction costs are $50,000 per mile between A and B and $100,000 per mile between B and C, find the branching angle θ that minimizes the total construction cost.

Exercise 51

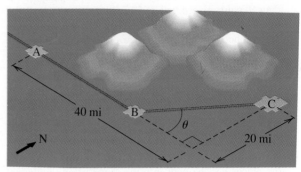

A

40 mi   B   C

θ

N   20 mi

52 A long rectangular sheet of metal, 12 in. wide, is to be made into a rain gutter by turning up two sides at angles of 120° to the sheet. How many inches should be turned up to give the gutter its greatest capacity?

53 Refer to Exercise 18. Find the central angle of the sector that will maximize the volume of the cup.

54 A square picture having sides 2 ft long is hung on a wall such that the base is 6 ft above the floor. If a person whose eye level is 5 ft above the floor looks at the picture and if θ is the angle between the line of sight and the top and bottom of the picture, find the person's distance from the wall at which θ has its maximum value.

55 A rectangle made of elastic material will be made into a cylinder by joining edges AD and BC (see figure). To support the structure, a wire of fixed length L is placed along the diagonal of the rectangle. Find the angle θ that will result in the cylinder of maximum volume.

Exercise 55

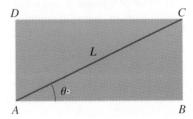

D   C

L

θ

A   B

56 When a person is walking, the magnitude F of the vertical force of one foot on the ground (see figure) can be approximated by $F = A(\cos bt - a\cos 3bt)$ for time t (in seconds), with $A > 0$, $b > 0$, and $0 < a < 1$.

(a) Show that $F = 0$ when $t = -\pi/(2b)$ and $t = \pi/(2b)$. (The time $t = -\pi/(2b)$ corresponds to the instant when the foot first touches the ground and the weight of the body is being supported by the other foot.)

(b) Show that the maximum force occurs when $t = 0$ or when $\sin^2 bt = (9a - 1)/(12a)$.

(c) If $a = \frac{1}{3}$, express the maximum force in terms of A.

(d) If $0 < a \le \frac{1}{9}$, express the maximum force in terms of A.

Exercise 56

57 A battery having fixed voltage V and fixed internal resistance r is connected to a circuit that has variable resistance R (see figure). By Ohm's law, the current I in the circuit is $I = V/(R + r)$. If the power output P is given by $P = I^2 R$, show that the maximum power occurs if $R = r$.

Exercise 57

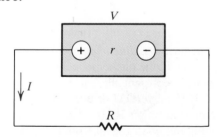

58 The power output P of an automobile battery is given by $P = VI - I^2 r$ for voltage V, current I, and internal resistance r of the battery. What current corresponds to the maximum power?

59 Two corridors 3 ft and 4 ft wide, respectively, meet at a right angle. Find the length of the longest nonbendable rod that can be carried horizontally around the corner, as shown in the figure. (Disregard the thickness of the rod.)

60 Light travels from one point to another along the path that requires the least amount of time. Suppose that light has velocity v_1 in air and v_2 in water, where $v_1 > v_2$. If light travels from a point P in air to a point Q in water

Exercise 59

Exercise 60

(see figure), show that the path requires the least amount of time if

$$\frac{\sin \theta_1}{\sin \theta_2} = \frac{v_1}{v_2}.$$

(This is an example of *Snell's law of refraction*.)

61 A circular cylinder of fixed radius R is surmounted by a cone (see figure). The ends of the cylinder are open, and the total volume is to be a specified constant V.

(a) Show that the total surface area S is given by

$$S = \frac{2V}{R} + \pi R^2 \left(\csc \theta - \frac{2}{3} \cot \theta \right).$$

(b) Show that S is minimized when $\theta \approx 48.2°$.

Exercise 61

62 In the classic honeycomb-structure problem, a hexagonal prism of fixed radius (and side) R is surmounted by adding three identical rhombuses that meet in a common vertex (see figure). The bottom of the prism is open, and the total volume is to be a specified constant V. A more elaborate geometric argument than that in Exercise 61 establishes that the total surface area S is given by

$$S = \frac{4}{3}\sqrt{3}\frac{V}{R} - \frac{3}{2}R^2\cot\theta + \frac{3\sqrt{3}}{2}R^2\csc\theta.$$

Show that S is minimized when $\theta \approx 54.7°$. (Remarkably, bees construct their honeycombs so that the amount of wax S is minimized.)

Exercise 62

3.7 VELOCITY AND ACCELERATION

In this section, we use derivatives to describe and analyze several important types of motion that occur in physical situations. One of the greatest early achievements in the history of calculus was Newton's derivation in the seventeenth century of Kepler's laws of planetary motions (see Section 11.6 for more details). In the succeeding three hundred years, calculus has repeatedly helped scientists study moving objects. Our focus will be objects moving along a straight line.

RECTILINEAR MOTION

As we saw in Section 2.1, the term *rectilinear motion* is used to describe movement of a point along a line. In the mathematical model of rectilinear motion in this section, we will represent the car as a point P and the highway as a straight line l. If l is a vertical or horizontal coordinate line and if the coordinate of point P at time t is $s(t)$, then s is considered the *position function* of P (see Figure 3.68). Recall from Definition (2.3) that the *velocity* of P at time t, the rate of change of P with respect to t, is the derivative $s'(t)$. The velocity is also denoted as $v(t)$.

The *acceleration* $a(t)$ of P at time t is defined as the rate of change of velocity with respect to time: $a(t) = v'(t)$. Thus the acceleration is the second derivative $(d/dt)(s'(t)) = s''(t)$. The next definition summarizes this discussion and also introduces the notion of the *speed* of P.

Figure 3.68

Time   Position of P

O P

0 $s(t)$ l

Definition 3.23

Let $s(t)$ be the coordinate of a point P on a coordinate line l at time t.

(i) The **velocity** of P is $v(t) = s'(t)$.

(ii) The **speed** of P is $|v(t)|$.

(iii) The **acceleration** of P is $a(t) = v'(t) = s''(t)$.

We shall call v the **velocity function** of P and a the **acceleration function** of P. We sometimes use the notation

$$v = \frac{ds}{dt} \quad \text{and} \quad a = \frac{dv}{dt}.$$

If t is in seconds and $s(t)$ is in centimeters, then $v(t)$ is in cm/sec and $a(t)$ is in cm/sec^2 (centimeters per second per second). If t is in hours and $s(t)$ is in miles, then $v(t)$ is in mi/hr and $a(t)$ is in mi/hr^2 (miles per hour per hour).

If $v(t)$ is positive in a time interval, then $s'(t) > 0$, and, by Theorem (3.15), $s(t)$ is increasing—that is, the point P is moving in the positive direction on l. If $v(t)$ is negative, the motion is in the negative direction. The velocity is zero at a point where P changes direction. If the acceleration $a(t) = v'(t)$ is positive, the velocity is increasing. If $a(t)$ is negative, the velocity is decreasing.

NOTE We make a distinction between the *velocity* $v(t)$ and the *speed* $|v(t)|$ of a moving object. The speed conveys only how fast the object is moving; it contains no information about the direction of motion. We will make use of the speed in Chapter 5 in determining the total distance that an object moves. The velocity conveys not only the speed of motion but also whether the object is moving in a positive or a negative direction along the coordinate line.

EXAMPLE ■ 1 The position function s of a point P on a coordinate line is given by

$$s(t) = t^3 - 12t^2 + 36t - 20,$$

with t in seconds and $s(t)$ in centimeters. Describe the motion of P during the time interval $[-1, 9]$.

SOLUTION Differentiating, we obtain

$$v(t) = s'(t) = 3t^2 - 24t + 36 = 3(t - 2)(t - 6),$$
$$a(t) = v'(t) = 6t - 24 = 6(t - 4).$$

Let us determine when $v(t) > 0$ and when $v(t) < 0$, since this will tell us when P is moving to the right and to the left, respectively. Since $v(t) = 0$ at $t = 2$ and $t = 6$, we examine the following time subintervals of $[-1, 9]$:

$$(-1, 2), \quad (2, 6), \quad \text{and} \quad (6, 9)$$

We may determine the sign of $v(t)$ by using test values, as indicated in the table (check each entry):

Time interval	$(-1, 2)$	$(2, 6)$	$(6, 9)$
k	0	3	7
Test value $v(k)$	36	−9	15
Sign of $v(t)$	+	−	+
Direction of motion	right	left	right

The next table lists the values of the position, velocity, and acceleration functions at the endpoints of the time interval $[-1, 9]$ and the times at which the velocity or acceleration is zero.

t	-1	2	4	6	9
$s(t)$	-69	12	-4	-20	61
$v(t)$	63	0	-12	0	63
$a(t)$	-30	-12	0	12	30

It is convenient to represent the motion of P schematically, as in Figure 3.69. *The curve above the coordinate line is not the path of the point,* but rather a scheme for showing the manner in which P moves on the line l.

Figure 3.69

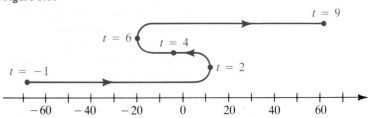

As indicated by the tables and Figure 3.69, at $t = -1$ the point is 69 cm to the left of the origin and is moving to the right with a velocity of 63 cm/sec. The negative acceleration -30 cm/sec^2 indicates that the velocity is decreasing at a rate of 30 cm/sec, each second. The point continues to move to the right, slowing down until it has zero velocity at $t = 2$, 12 cm to the right of the origin. The point P then reverses direction and moves until, at $t = 6$, it is 20 cm to the left of the origin. It then again reverses direction and moves to the right for the remainder of the time interval, with increasing velocity. The direction of motion is indicated by the arrows on the curve in Figure 3.69.

Since the point began at position -69 and ended at position 61, the net change in its position is $61 - (-69) = 130$ units. The total distance traveled by the point, however, is more. In the time interval $[-1, 2]$, it moved $12 - (-69) = 81$ units. In the time interval $[2, 6]$, it moved a distance of 32 units, and in the interval $[6, 9]$, it moved an additional 81 units. The total distance traveled was $81 + 32 + 81 = 194$ units.

EXAMPLE ▪ 2   A projectile is fired straight upward with a velocity of 400 ft/sec. From physics, its distance above the ground after t seconds is $s(t) = -16t^2 + 400t$.

(a) Find the time and the velocity at which the projectile hits the ground.

(b) Find the maximum altitude achieved by the projectile.

(c) Find the acceleration at any time t.

Figure 3.70

$s(t) = -16t^2 + 400$

SOLUTION

(a) Let us represent the path of the projectile on a vertical coordinate line *l* with origin at ground level and positive direction upward, as illustrated in Figure 3.70. The projectile is on the ground when $-16t^2 + 400t = 0$— that is, when $-16t(t - 25) = 0$. This gives us $t = 0$ and $t = 25$. Hence the projectile hits the ground after 25 sec. The velocity at time *t* is

$$v(t) = s'(t) = -32t + 400.$$

In particular, at $t = 25$, we obtain the *impact velocity*:

$$v(25) = -32(25) + 400 = -400 \text{ ft/sec.}$$

The negative velocity indicates that the projectile is moving in the negative direction on *l* (downward) at the instant that it strikes the ground. Note that the *speed* at this time is

$$|v(25)| = |-400| = 400 \text{ ft/sec.}$$

(b) The maximum altitude occurs when the velocity is zero —that is, when $s'(t) = -32t + 400 = 0$. Solving for *t* gives us $t = \frac{400}{32} = \frac{25}{2}$, and hence the maximum altitude is

$$s(\tfrac{25}{2}) = -16(\tfrac{25}{2})^2 + 400(\tfrac{25}{2}) = 2500 \text{ ft.}$$

(c) The acceleration at any time *t* is

$$a(t) = v'(t) = -32 \text{ ft/sec}^2.$$

This constant acceleration is caused by the force of gravity.

SIMPLE HARMONIC MOTION

Simple harmonic motion takes place in waves. It involves trigonometric functions and is defined as follows.

Definition 3.24

> A point *P* moving on a coordinate line *l* is in **simple harmonic motion** if its distance $s(t)$ from the origin at time *t* is given by either
>
> $$s(t) = k \sin(\omega t + b) \quad \text{or} \quad s(t) = k \cos(\omega t + b),$$
>
> where k, ω, and b are constants, with $\omega > 0$.

Simple harmonic motion may also be defined by requiring that the acceleration $a(t)$ satisfy the condition

$$a(t) = -\omega^2 s(t)$$

for every *t*. It can be shown that this condition is equivalent to Definition (3.24).

In simple harmonic motion, the point *P* oscillates between the points on *l* with coordinates $-k$ and k. The **amplitude** of the motion is the

Figure 3.71

maximum displacement $|k|$ of the point from the origin. The **period** is the time $2\pi/\omega$ required for one complete oscillation. The **frequency** $\omega/2\pi$ is the number of oscillations per unit of time.

Simple harmonic motion takes place in many different types of waves, such as water waves, sound waves, radio waves, light waves, and distortional waves present in vibrating bodies. Functions of the type defined in (3.24) also occur in the analysis of electrical circuits that contain an alternating electromotive force and current.

As another example of simple harmonic motion, consider a spring with an attached weight that is oscillating vertically relative to a coordinate line, as illustrated in Figure 3.71. The number $s(t)$ represents the coordinate of a fixed point P in the weight, and we assume that the amplitude $|k|$ of the motion is constant. In this case, there is no frictional force retarding the motion. If friction is present, then the amplitude decreases with time, and the motion is *damped*.

EXAMPLE • 3 Suppose that the weight shown in Figure 3.71 is oscillating and

$$s(t) = 10\cos\frac{\pi}{6}t,$$

where t is in seconds and $s(t)$ is in centimeters. Discuss the motion of the weight.

SOLUTION Comparing the given equation with the general form $s(t) = k\cos(\omega t + b)$ in Definition (3.24), we obtain $k = 10$, $\omega = \pi/6$, and $b = 0$, which gives us the following:

amplitude: $\quad k = 10$ cm

period: $\quad \dfrac{2\pi}{\omega} = \dfrac{2\pi}{\pi/6} = 12$ sec

frequency: $\quad \dfrac{\omega}{2\pi} = \dfrac{1}{12}$ oscillation/sec

Let us examine the motion during the time interval $[0, 12]$. The velocity and acceleration functions are given by the following:

$$v(t) = s'(t) = 10\left(-\sin\frac{\pi}{6}t\right)\cdot\frac{\pi}{6} = -\frac{5\pi}{3}\sin\frac{\pi}{6}t,$$

$$a(t) = v'(t) = -\frac{5\pi}{3}\left(\cos\frac{\pi}{6}t\right)\cdot\frac{\pi}{6} = -\frac{5\pi^2}{18}\cos\frac{\pi}{6}t.$$

The velocity is 0 at $t = 0$, $t = 6$, and $t = 12$, since $\sin[(\pi/6)t] = 0$ for these values of t. The acceleration is 0 at $t = 3$ and $t = 9$, since in these cases, $\cos[(\pi/6)t] = 0$. The times at which the velocity and acceleration are 0 lead us to examine the time intervals $(0, 3)$, $(3, 6)$, $(6, 9)$, and $(9, 12)$. The following table displays the main characteristics of the motion. The signs of $v(t)$ and $a(t)$ in the intervals can be determined using test values (verify each entry).

Time interval	Sign of $v(t)$	Direction of motion	Sign of $a(t)$	Variation of $v(t)$	Speed $|v(t)|$
(0, 3)	−	downward	−	decreasing	increasing
(3, 6)	−	downward	+	increasing	decreasing
(6, 9)	+	upward	+	increasing	increasing
(9, 12)	+	upward	−	decreasing	decreasing

Note that if $0 < t < 3$, the velocity $v(t)$ is negative and decreasing; that is, $v(t)$ becomes *more* negative. Hence the absolute value $|v(t)|$, the speed, is *increasing*. If $3 < t < 6$, the velocity is negative and increasing ($v(t)$ becomes *less* negative); that is, the speed of P is *decreasing* in the time interval $(3, 6)$. Similar remarks can be made for the intervals $(6, 9)$ and $(9, 12)$.

We may summarize the motion of P as follows: At $t = 0$, $s(0) = 10$ and the point P is 10 cm above the origin O. It then moves downward, gaining speed until it reaches the origin O at $t = 3$. It then slows down until it reaches a point 10 cm below O at the end of 6 sec. The direction of motion is then reversed, and the weight moves upward, gaining speed until it reaches O at $t = 9$, after which it slows down until it returns to its original position at the end of 12 sec. The direction of motion is then reversed again, and the same pattern is repeated indefinitely.

Figure 3.72

FREE FALL

According to Newton's second law of motion, the product of the mass and acceleration of an object is equal to the sum of the forces acting on it. Using this as a first model for the fall of an object toward the surface of a planet from a starting position not far from the surface, we ignore all forces except the gravitational attraction of the planet, which we take to be constant.

With this model, Newton's second law becomes

$$m\, a(t) = mg,$$

where m is the mass of the object and g is the gravitational constant. This equation simplifies to

$$a(t) = g.$$

If we set up our coordinate line as in Figure 3.72 with 0 at the surface of the planet and the positive side above the surface, then the force of gravity is in the negative direction; thus g will have a negative value. A useful convention is to assign time t to be 0 at the instant the object begins to move.

Since $a(t) = v'(t)$ is the constant function whose value is g, the velocity function $v(t)$ must be of the form

$$v(t) = gt + C$$

for some constant C. If we evaluate this equation at the initial time $t = 0$, we have

$$v(0) = g \cdot 0 + C = C,$$

so we see that C is the *initial velocity* $v(0)$, which we will denote as v_0.

Now, using the relationship that velocity is the derivative of the position function, we have

$$v(t) = s'(t) = gt + v_0.$$

One possible candidate for the position function is an expression of the form $(g/2)t^2 + v_0 t$, since this expression has derivative $gt + v_0$. By Corollary (3.14), any other function with the same derivative, $gt + v_0$, must differ from $(g/2)t^2 + v_0 t$ by a constant. Hence, the position function must have the form

$$s(t) = \frac{g}{2}(t^2) + v_0 t + C$$

for some constant C. Evaluation again at the initial time $t = 0$ gives $s(0) = 0 + 0 + C$, so the constant C is just the *initial position* $s(0)$, which we will denote by s_0.

Putting these results together, we find that the rectilinear motion of an object falling near the surface of a planet is given by the position function

$$s(t) = \frac{gt^2}{2} + v_0 t + s_0.$$

For an object near the earth's surface, the value of g is approximately -32 ft/sec^2, or -9.8 m/sec^2.

EXAMPLE ▪ 4   A student accidentally drops her calculus book from an upper-story window of her dormitory 144 ft above the ground. How fast is the book moving when it strikes the ground?

SOLUTION   Taking the instant the book is dropped to be $t = 0$, we have an initial position $s_0 = 144$ and an initial velocity $v_0 = 0$ (since the book is dropped rather than thrown downward or hurled upward). Since the distance is measured in feet, we use $g = -32$. The equations for position and velocity thus take the form

$$s(t) = -16t^2 + 144 \quad \text{and} \quad v(t) = -32t.$$

The book hits the ground when $s(t) = 0$, so the value of t at the instant of impact satisfies

$$-16t^2 + 144 = 0$$

or

$$t^2 = \tfrac{144}{16},$$

which yields

$$t = \pm 3.$$

We can ignore the negative value because all the action takes place after $t = 0$. We conclude that the book strikes the ground in 3 sec. At that moment, it is traveling with a velocity of $v(3) = -32(3) = -96$ ft/sec.

EXAMPLE ■ 5 To practice his skill catching fly balls, a baseball player hurls the ball straight up in the air, releasing it from his outstretched arm at a point 8 ft above the ground. How high does the ball go if he can impart an initial velocity of 60 mi/hr?

SOLUTION Since 60 mi/hr translates into 88 ft/sec, we have $v_0 = 88$ and $s_0 = 8$. The position and velocity functions become

$$s(t) = -16t^2 + 88t + 8 \quad \text{and} \quad v(t) = -32t + 88.$$

The ball reaches its maximum height when the velocity is 0, which occurs when

$$-32t + 88 = 0; \quad \text{that is, } t = \tfrac{11}{4} \text{ sec.}$$

The height of the ball at this time is

$$s(\tfrac{11}{4}) = -16(\tfrac{11}{4})^2 + 88(\tfrac{11}{4}) + 8 = 129 \text{ ft.}$$

The model we have discussed in this section, one of constant acceleration, is a relatively simple one in that it ignores other forces that are often quite important. Air resistance, for example, often plays a crucial role, especially for a relatively light object with a relatively large size. The model also treats the gravitational force as constant; a more realistic one would take into account that the force due to gravity varies with the distance between the object and the center of attracting planetary mass. Still, this simple model gives reasonably good predictions. You may wish to explore some generalizations of our model.

CONSTANT ACCELERATION

In many situations, the motion of an object is governed by constant acceleration or deceleration, as the next examples illustrate.

EXAMPLE ■ 6 An automobile manufacturer claims that its new model can accelerate "from 0 to 60 miles per hour in 11 seconds." Find the constant acceleration that makes this rate of speed possible.

SOLUTION Let the unknown constant acceleration be a ft/sec^2. At time $t = 0$, the velocity v_0 is 0, so the velocity at time t is given by

$$v(t) = gt + v_0 = at.$$

We are given that $v(11) = 88$ ft/sec. Hence,

$$88 = 11a, \quad \text{or, equivalently,} \quad a = 8 \text{ ft/sec}^2.$$

EXAMPLE ■ 7 Suppose that the automobile of Example 6 is involved in an accident on a quiet residential street. Police officers investigating the incident conclude from physical evidence and reports of witnesses that the car traveled 196 ft after the driver slammed on the brakes

Figure 3.73

before it stopped (see Figure 3.73). Assuming that the car decelerated at a rate of 8 ft/sec², determine how fast the car was going when the brakes were applied.

SOLUTION   Let $t = 0$ be the moment when the brakes are applied. When the car comes to rest at unknown time t^*, its velocity $v(t^*) = 0$ and its position $s(t^*) = 196$. We wish to find v_0.

If we use $g = -8$, the velocity and position functions become

$$v(t) = -8t + v_0 \quad \text{and} \quad s(t) = -4t^2 + v_0 t.$$

Evaluating these functions at time t^*, we find

$$0 = -8t^* + v_0 \quad \text{and} \quad 196 = -4(t^*)^2 + v_0 t^*.$$

From the first equation, we have the relationship $t^* = v_0/8$, which we substitute into the second equation to obtain

$$196 = -4 \left(\frac{v_0}{8} \right)^2 + v_0 \left(\frac{v_0}{8} \right).$$

Simplifying gives

$$\frac{v_0^2}{16} = 196, \quad \text{so} \quad v_0 = 56 \text{ ft/sec}.$$

Converting to miles per hour, we find that the car was traveling at a speed slightly above 38 mi/hr.

EXERCISES 3.7

Exer. 1–8: A point moving on a coordinate line has position function s. Find the velocity and the acceleration at time t, and describe the motion of the point during the indicated time interval. Illustrate the motion by means of a diagram of the type shown in Figure 3.69.

1 $s(t) = 3t^2 - 12t + 1$; $[0, 5]$

2 $s(t) = t^2 + 3t - 6$; $[-2, 2]$

3 $s(t) = t^3 - 9t + 1$; $[-3, 3]$

4 $s(t) = 24 + 6t - t^3$; $[-2, 3]$

5 $s(t) = -2t^3 + 15t^2 - 24t - 6$; $[0, 5]$

6 $s(t) = 2t^3 - 12t^2 + 6$; $[-1, 6]$

7 $s(t) = 2t^4 - 6t^2$; $[-2, 2]$

8 $s(t) = 2t^3 - 6t^5$; $[-1, 1]$

Exer. 9–10: An automobile rolls down an incline, traveling $s(t)$ feet in t seconds. (a) Find its velocity at $t = 3$. (b) After how many seconds will the velocity be k ft/sec?

9 $s(t) = 5t^2 + 2$; $k = 28$

10 $s(t) = 3t^2 + 7$; $k = 88$

Exer. 11–12: A projectile is fired directly upward with an initial velocity of v_0 ft/sec, and its height (in feet) above the ground after t seconds is given by $s(t)$. Find (a) the velocity and acceleration after t seconds, (b) the maximum height, and (c) the duration of the flight.

11 $v_0 = 144;$ $s(t) = 144t - 16t^2$

12 $v_0 = 192;$ $s(t) = 100 + 192t - 16t^2$

Exer. 13–16: A particle in simple harmonic motion has position function s, and t is the time in seconds. Find the amplitude, the period, and the frequency.

13 $s(t) = 5 \cos \dfrac{\pi}{4} t$ 　　　　14 $s(t) = 4 \sin \pi t$

15 $s(t) = 6 \sin \dfrac{2\pi}{3} t$ 　　　16 $s(t) = 3 \cos 2t$

17 The electromotive force V and current I in an alternating-current circuit are given by

$$V = 220 \sin 360\pi t,$$

$$I = 20 \sin \left(360\pi t - \dfrac{\pi}{4}\right).$$

Find the rates of change of V and I with respect to time at $t = 1$.

18 The annual variation in temperature T (in °C) in Vancouver, B.C., may be approximated by the formula

$$T = 14.8 \sin \left[\dfrac{\pi}{6}(t - 3)\right] + 10,$$

where t is in months, with $t = 0$ corresponding to January 1. Approximate the rate at which the temperature is changing at time $t = 3$ (April 1) and at time $t = 10$ (November 1). At what time of the year is the temperature changing most rapidly?

19 The graph in the figure shows the rise and fall of the water level in Boston Harbor during a particular 24-hr period.

Exercise 19

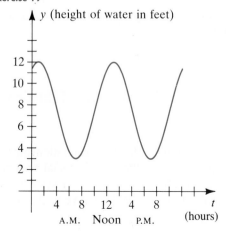

y (height of water in feet)

12

10

8

6

4

2

| 4 | 8 | 12 | 4 | 8 | t
A.M. 　 Noon 　 P.M. 　(hours)

(a) Approximate the water level y by means of an expression of the form

$$y = a \sin(bt + c) + d,$$

with $t = 0$ corresponding to midnight.

(b) Approximately how fast is the water level rising at 12 noon?

20 A *tsunami* is a tidal wave caused by an earthquake beneath the sea. These waves can be more than 100 ft in height and can travel at great speeds. Engineers sometimes represent tsunamis by an equation of the form $y = a \cos bt$. Suppose that a wave has a height $h = 25$ ft and period 30 min and is traveling at the rate of 180 ft/sec.

(a) Let (x, y) be a point on the wave represented in the figure. Express y as a function of t if $y = 25$ ft when $t = 0$.

(b) How fast is the wave rising (or falling) when $y = 10$ ft?

Exercise 20

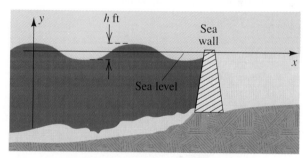

21 A cork bobs up and down in a lake. The distance from the bottom of the lake to the center of the cork at time $t \geq 0$ is given by $s(t) = \cos \pi t + 12$, where $s(t)$ is in inches and t is in seconds (see figure).

(a) Find the velocity of the cork at $t = 0, \frac{1}{2}, 1, \frac{3}{2},$ and 2.

(b) During what time intervals is the cork rising?

Exercise 21

22 A particle in a vibrating spring is moving vertically such that its distance $s(t)$ from a fixed point on the line of vibration is given by $s(t) = 4 + \frac{1}{25} \sin 100\pi t$, where $s(t)$ is in centimeters and t is in seconds.

(a) How long does it take the particle to make one complete vibration?

(b) Find the velocity of the particle at $t = 1, 1.005,$ 1.01, and 1.015.

Exer. 23–24: Show that $s''(t) = -\omega^2 s(t)$.

23 $s(t) = k\cos(\omega t + b)$ 24 $s(t) = k\sin(\omega t + b)$

25 A point $P(x, y)$ is moving at a constant rate around the circle $x^2 + y^2 = a^2$. Prove that the projection $Q(x, 0)$ of P onto the x-axis is in simple harmonic motion.

26 If a point P moves on a coordinate line such that
$$s(t) = a\cos\omega t + b\sin\omega t,$$
show that P is in simple harmonic motion by

(a) using the remark following Definition (3.24)

(b) using only trigonometric methods (*Hint:* Show that $s(t) = A\cos(\omega t - c)$ for some constants A and c.)

c **Exer. 27–28: A point moving on a coordinate line has position $s(t)$. (a) Graph $y = s(t)$ for $0 \le t \le 5$. (b) Approximate the point's position, velocity, and acceleration at $t = 0, 1, 2, 3, 4,$ and 5.**

27 $s(t) = \dfrac{10\sin t}{t^2 + 1}$ 28 $s(t) = \dfrac{5\tan(\frac{1}{4}t)}{2t + 1}$

29 Emergency food supplies are dropped from a helicopter and hit the ground 10 sec later.

(a) What is the height h of the helicopter?

Exercise 29

(b) The box in which the supplies are packed is strong enough to withstand a speed of 180 mi/hr on impact. Will the supplies be intact?

(c) What is the maximum height at which the helicopter can be positioned to guarantee that the box will not break up when it hits the ground?

30 A golf ball projected vertically from the ground lands back on the surface in 8 sec. What was the initial velocity?

31 Judy's roommate drops Judy's car keys from the dormitory window, which is 144 ft above the ground, to Judy, who is standing on the ground below the window.

(a) How long must Judy wait for the keys?

(b) If Judy needs the keys in 2 sec, with what initial speed should her roommate throw the keys?

32 If a stone is hurled vertically upward, show that it takes the same amount of time for the stone to achieve its maximum height as it takes to drop from that spot back to the ground.

33 A truck driver speeding down a narrow street suddenly sees a bicyclist s feet ahead. The driver slams on the brakes, imparting a constant deceleration a to the truck. If the bicycle is moving at a rate of v_1 mi/hr in the same direction as the truck, find the maximum value for the initial speed v_0 of the truck so that the vehicles will not collide.

34 In Vermont, a straight stretch of U.S. highway 7 connects Burlington to Vergennes. A car begins in Burlington at $t = 0$ and heads toward Vergennes with a velocity given by $v(t) = 60t - 12t^2$, measured in miles per hour. When the automobile arrives at the Vergennes city limit, it is clocked at a speed of 48 mi/hr and it is speeding up.

(a) How far apart are Burlington and Vergennes?

(b) What are the minimum and maximum velocities experienced during the trip? When do they occur?

(c) How would the answer be affected if the car were observed to be slowing down rather than speeding up when it reached Vergennes?

3.8   APPLICATIONS TO ECONOMICS, SOCIAL SCIENCES, AND LIFE SCIENCES

A primary focus of calculus is **change** in functional relationships. Since change is a characteristic property of most natural and social systems, calculus provides powerful techniques to understand these systems. In this section, we examine some applications of the derivative to the social and life sciences. We concentrate first on some applications of the derivative in economics. We then examine how the calculus we have developed so far can help us make some qualitative conclusions about complex mathematical systems that we cannot quantitatively solve.

ECONOMICS

Calculus has become an important tool for solving problems that occur in economics because of its power to analyze functional relationships. Revenues and profits, for example, are functions of fluctuating costs and prices, and these in turn depend on varying supply and demand. Economists often face optimizing problems that involve making the most efficient use of scarce resources to achieve societal goals.

If x is the number of units of a commodity, economists often use the functions C, c, R, and P, defined as follows:

Cost function:	$C(x) = $ cost of producing x units
Average cost function:	$c(x) = C(x)/x$
	$= $ average cost of producing one unit
Revenue function:	$R(x) = $ revenue received for selling x units
Profit function:	$P(x) = R(x) - C(x)$
	$= $ profit in selling x units

To use the techniques of calculus, we regard x as a real number, even though this variable may take on only integer values. We always assume that $x \geq 0$, since the production of a negative number of units has no practical significance.

EXAMPLE ■ 1   A manufacturer of miniature tape decks has a monthly fixed cost of $10,000, a production cost of $12 per unit, and a selling price of $20 per unit.

(a) Find $C(x)$, $c(x)$, $R(x)$, and $P(x)$.

(b) Find the function values in part (a) if $x = 1000$.

(c) How many units must be manufactured in order to break even?

SOLUTION

(a) The production cost of manufacturing x units is $12x$. Since there is also a fixed monthly cost of $10,000, the total monthly cost of manufacturing x units is

$$C(x) = 12x + 10,000.$$

The remaining functions are given by

$$c(x) = \frac{C(x)}{x} = 12 + \frac{10,000}{x},$$
$$R(x) = 20x,$$
$$P(x) = R(x) - C(x) = 8x - 10,000.$$

(b) Substituting $x = 1000$ in part (a) gives us the following values:

$$C(1000) = 22,000 \quad \text{cost of manufacturing 1000 units}$$
$$c(1000) = 22 \quad\quad \text{average cost of manufacturing one unit}$$
$$R(1000) = 20,000 \quad \text{total revenue received for 1000 units}$$
$$P(1000) = -2000 \quad \text{profit in manufacturing 1000 units}$$

Note that the manufacturer incurs a loss of $2000 per month if only 1000 units are produced and sold.

(c) The break-even point corresponds to zero profit—that is, when we have $8x - 10,000 = 0$. This result gives us

$$8x = 10,000, \quad \text{or} \quad x = 1250.$$

Thus to break even, it is necessary to produce and sell 1250 units per month.

If a function f is used to describe some economic entity, the adjective *marginal* is used to specify the derivative f'. The derivatives C', c', R', and P' are called the **marginal cost function,** the **marginal average cost function**, the **marginal revenue function**, and the **marginal profit function,** respectively. The number $C'(x)$ is referred to as the **marginal cost** associated with the production of x units. If we interpret the derivative as a rate of change, then $C'(x)$ is the rate at which the cost changes with respect to the number x of units produced. Similar statements can be made for $c'(x)$, $R'(x)$, and $P'(x)$.

If C is a cost function and n is a positive integer, then, by Definition (2.5),

$$C'(n) = \lim_{h \to 0} \frac{C(n + h) - C(n)}{h}.$$

Hence, if h is small, then

$$C'(n) \approx \frac{C(n + h) - C(n)}{h}.$$

If the number n of units produced is large, economists often let $h = 1$ in the preceding formula to approximate the marginal cost, obtaining

$$C'(n) \approx C(n + 1) - C(n).$$

In this context, *the marginal cost associated with the production of n units is (approximately) the cost of producing one more unit.*

Some companies find that the cost $C(x)$ of producing x units of a commodity is given by a formula such as

$$C(x) = a + bx + dx^2 + kx^3.$$

The constant a represents a fixed overhead charge for items like rent, heat, and light that are independent of the number of units produced. If the cost of producing one unit were b dollars and no other factors were involved, then the second term bx in the formula would represent the cost of producing x units. If x becomes very large, then the terms dx^2 and kx^3 may significantly affect production costs.

EXAMPLE ▪ 2 An electronics company estimates that the cost (in dollars) of producing x components used in electronic toys is given by

$$C(x) = 200 + 0.05x + 0.0001x^2.$$

(a) Find the cost, the average cost, and the marginal cost of producing 500 units, 1000 units, and 5000 units.

(b) Compare the marginal cost of producing 1000 units with the cost of producing the 1001st unit.

SOLUTION

(a) The average cost of producing x components is

$$c(x) = \frac{C(x)}{x} = \frac{200}{x} + 0.05 + 0.0001x.$$

The marginal cost is

$$C'(x) = 0.05 + 0.0002x.$$

You should verify the entries in the following table, where numbers in the last three columns represent dollars.

Units	Cost	Average cost	Marginal cost
x	$C(x)$	$c(x) = \dfrac{C(x)}{x}$	$C'(x)$
500	250.00	0.50	0.15
1000	350.00	0.35	0.25
5000	2950.00	0.59	1.05

(b) Using the cost function yields

$$C(1001) = 200 + 0.05(1001) + (0.0001)(1001)^2$$
$$\approx 350.25.$$

Hence the cost of producing the 1001st unit is

$$C(1001) - C(1000) \approx 350.25 - 350.00$$
$$= 0.25,$$

which is approximately the same as the marginal cost $C'(1000)$.

A company must consider many factors in order to determine a selling price for each product. In addition to the cost of production and the profit desired, the company should be aware of the manner in which consumer demand will vary if the price increases. For some products, there is a constant demand, and changes in price have little effect on sales. For items that are not necessities of life, a price increase will probably lead to a decrease in the number of units sold. Suppose a company knows from past experience that it can sell x units when the price per unit is given by $p(x)$ for some function p. We sometimes say that $p(x)$ is the price per unit when there is a **demand** for x units, and we refer to p as the **demand function** for the commodity. The total income, or revenue, is the number of units sold times the price per unit—that is, $x \cdot p(x)$. Thus,

$$R(x) = xp(x).$$

The derivative p' is called the **marginal demand function**.

If $S = p(x)$, then S is the selling price per unit associated with a demand of x units. Since a decrease in S would ordinarily be associated with an increase in x, a demand function p is usually decreasing; that is, $p'(x) < 0$ for every x. Demand functions are sometimes defined implicitly by an equation involving S and x, as in the next example.

EXAMPLE ▪ 3   The demand for x units of a product is related to a selling price of S dollars per unit by the equation $2x + S^2 - 12{,}000 = 0$.

(a) Find the demand function, the marginal demand function, the revenue function, and the marginal revenue function.

(b) Find the number of units and the price per unit that yield the maximum revenue.

(c) Find the maximum revenue.

SOLUTION

(a) Since $S^2 = 12{,}000 - 2x$ and S is positive, we see that the demand function p is given by

$$S = p(x) = \sqrt{12{,}000 - 2x}.$$

The domain of p consists of every x such that $12{,}000 - 2x > 0$, or, equivalently, $2x < 12{,}000$. Thus, $0 \le x < 6000$. The graph of p is sketched in Figure 3.74. In theory, there are no sales if the selling price is $\sqrt{12{,}000}$, or approximately 109.54, and when the selling price is close to 0, the demand is close to 6000.

The marginal demand function p' is given by

$$p'(x) = \frac{-1}{\sqrt{12{,}000 - 2x}}.$$

The negative sign indicates that a decrease in price is associated with an increase in demand.

The revenue function R is given by

$$R(x) = xp(x) = x\sqrt{12{,}000 - 2x}.$$

Differentiating and simplifying gives us the marginal revenue function R':

$$R'(x) = \frac{12{,}000 - 3x}{\sqrt{12{,}000 - 2x}}$$

Figure 3.74

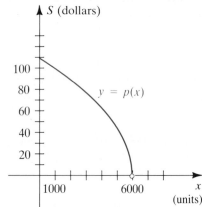

(b) A critical number for the revenue function R is $x = 12{,}000/3 = 4000$. Since $R'(x)$ is positive if $0 \leq x < 4000$ and negative if $4000 < x < 6000$, the maximum revenue occurs when 4000 units are produced and sold. This corresponds to a selling price per unit of

$$p(4000) = \sqrt{12{,}000 - 2(4000)} \approx \$63.25.$$

(c) The maximum revenue, obtained from selling 4000 units at \$63.25 per unit, is

$$4000(63.25) = \$253{,}000.$$

A MODEL FROM SOCIOLOGY AND GEOGRAPHY

Sociologists and geographers often study a phenomenon called **social diffusion**; that is, the spreading of a piece of information, technological innovation, or cultural fad among a population. The individuals in the population can be divided into those who have the information and those who do not.

In a fixed population, it is reasonable to assume that the rate of diffusion is proportional to the number who have the information and the number yet to receive it. The rate of diffusion should be proportional to the number of encounters between individuals of the two groups. If both populations are large, then there will be a relatively large number of such contacts, while if either population is small, there will be relatively few such meetings.

If x represents the number of people in a population of N individuals who have the information, then the rate of diffusion r is the rate of change of x—that is, $r(x) = dx/dt = x'(t)$. A mathematical model for the rate $r(x)$ at which the information is spreading is the equation

(∗) $$r(x) = kx(N - x) = kNx - kx^2,$$

where k is a positive proportionality constant.

If the rate of diffusion in a population of N people is given by (∗), then we may want to be able to find when the rate is zero and interpret the result. We are also interested in determining when the information is spreading most rapidly.

The rate of information spread is zero when $r(x) = 0$—that is, when $kx(N - x) = 0$. The possible values of x are $x = 0$ and $x = N$. We can draw the following conclusion: *If no one has the information, it cannot diffuse; if everyone has it, it cannot spread any further. So long as there are some people who have the information and some who do not, the information will continue to spread.*

To determine when the information is spreading most rapidly, we compute the derivative of r as

$$r'(x) = kN - 2kx.$$

Thus, $r'(x) > 0$ when $0 < x < N/2$ and $r'(x) < 0$ when $N/2 < x < N$. Accordingly, the rate of information spread increases until half the population is informed, and then it begins to decrease. Information is spreading most rapidly when $x = N/2$.

A MODEL FROM EPIDEMIOLOGY

Scientists in different fields often use essentially the same mathematical model to represent the dynamics of what appear to be widely different situations. As an example, the diffusion-of-information model that we have just examined assumes that the rate of change of the informed population is jointly proportional to the numbers of informed and uninformed individuals. Precisely the same concept lies at the heart of many models for the transmission of a communicable disease.

In such a model, two important subgroups of the population are the *infectives* (those who are currently infected with the disease and are capable of spreading it) and the *susceptibles* (those who are currently uninfected but could contract the illness). One of the important equations in this model is

$$S'(t) = -\beta S(t)\, I(t),$$

where β is a positive constant and $S = S(t)$ and $I = I(t)$ are the number of susceptibles and infectives at time t, respectively. Note that since the populations of susceptibles and infectives must remain nonnegative and β is positive, we have $S'(t) \leq 0$, so the susceptible population is nonincreasing.

We want to use the model to show that if the population is constant and is made up entirely of susceptibles S and infectives I, then the rate of change of the susceptible population has the same form as the rate of information spread.

We let N represent the constant population size. Then

$$S + I = N \quad \text{or, equivalently,} \quad I = N - S.$$

The equation for the rate of change of susceptibles has the form

$$S' = -\beta SI = -\beta S(N - S),$$

which has the same form as the social diffusion equation

$$x' = kx(N - x).$$

A MODEL FROM ECOLOGY

We now examine a model from ecology that uses the same central concept. Imagine a simple ecosystem with two animal species, one of which preys on the other. To enliven the model, think of rabbits as the prey species and foxes as the predators. We assume that the rabbits live on clover, which is in abundant supply, but the foxes have only a single source of food, the rabbits. The classic predator–prey model represents the growth rate of both the rabbit and the fox populations over time by a linked system of equations:

$$R'(t) = a\, R(t) - b\, R(t)\, F(t),$$
$$F'(t) = m\, R(t)\, F(t) - n\, F(t),$$

where $R(t)$ and $F(t)$ are the rabbit and the fox populations at time t, respectively, and a, b, m, and n are positive constants.

We write this system in a slightly condensed notation, suppressing the explicit mention of the variable t:

$$R' = aR - bRF \quad \text{and} \quad F' = mRF - nF.$$

Although the full solution of this pair of equations lies beyond our current understanding of calculus, we can still do quite a bit of fruitful analysis of the model. First, let us note that in the absence of foxes ($F = 0$), the growth function for the rabbits becomes

$$R' = aR.$$

Since $R'/R = a$ in this situation, the rabbits would grow at a constant percentage rate. We will study the consequences of such growth systematically in Chapter 6. As you might imagine, the rabbits will experience rapid increases in numbers.

Second, if there are no rabbits ($R = 0$), then the dynamics of the fox population reduces to

$$F' = -nF,$$

so the foxes would experience a constant percentage *decline* in numbers. It is not surprising that the foxes would face extinction. (Chapter 6 provides the tools for the quantitative analysis.)

Let us turn then to the more interesting case in which both rabbits and foxes are running around. Each of the growth equations involves the product RF. We are making the not unreasonable assumption that the number of kills of rabbits by foxes is proportional to the frequency of encounters between the two species, which, in turn, is proportional to the product of the two populations. There will be few kills if there are few rabbits or few foxes, and many kills only when both populations are large. Each kill diminishes the rabbit population and enhances the likelihood for growth in the number of foxes.

If we rewrite the equations as

$$R' = R(a - bF) \quad \text{and} \quad F' = F(mR - n),$$

then we see (since $R > 0$ and $F > 0$) that the sign of R' is the same as the sign of $(a - bF)$ and the sign of F' is the sign of $(mR - n)$. Since positive signs for the derivative correspond to increases in the function and negative signs to decreases, we have

$$\begin{cases} R \text{ increases if } F < a/b \\ R \text{ decreases if } F > a/b \end{cases} \quad \text{and} \quad \begin{cases} F \text{ increases if } R > n/m \\ F \text{ decreases if } R < n/m \end{cases}$$

We can gain further insight into the dynamics of the predator–prey relationship by constructing a Cartesian coordinate system with the rabbit population graphed along the horizontal axis and the fox population along the vertical axis, as in Figure 3.75 on the following page.

Figure 3.75

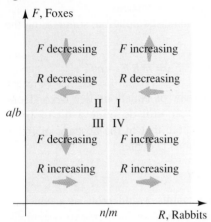

We have pictured the first quadrant only since the variables R and F represent nonnegative numbers. Along the horizontal line $F = a/b$, we have $R' = 0$, and along the vertical line $R = n/m$, we have $F' = 0$. The point of intersection of these two lines $((n/m), (a/b))$ represents a **stable point**. If the populations were to reach this point, both rates of change would be zero and the populations would remain at this level. At every other point in the first quadrant, at least one of the populations would be changing.

What happens if the initial population levels are at some other point? Suppose we begin in region I. Here both fox and rabbit populations are relatively large, which is at first good for the foxes since there will be lots of prey. The fox population will increase as the rabbit population decreases. As time goes by, we will find that the population level has moved to a point to the northwest of the initial point. This northwest movement continues until we reach the critical vertical line $R = n/m$. As we cross this line, the rabbits become scarcer and the fox population also begins to decrease. For a while, there will be dwindling numbers of both foxes and rabbits. The population level moves in a southwesterly direction until it ultimately hits the critical horizontal line $F = a/b$. Now the number of foxes has dropped and there is less danger for the rabbits. The rabbit population begins to increase, but since it is relatively small, the fox population will continue to decrease. The population level moves in a southeasterly direction, continuing this path until the rabbit population passes the vertical line $R = n/m$. Now there are sufficiently many rabbits to support a growing fox population. Both species continue to prosper and the population level moves in a northeasterly direction. Eventually, however, it passes the horizontal line at height a/b and we are back in region I. The process then proceeds as before.

We can see from this analysis that the population point moves in a counterclockwise direction, visiting all four regions in turn. We wish to determine what will happen in the long run—that is, whether the population point will spiral in toward the stable point, spiral out, or form some sort of elliptical closed orbit. Answering these questions must be deferred until we have developed sufficiently powerful tools of calculus.

EXERCISES 3.8

Exer. 1–4: If C is the cost function for a particular product, find (a) the cost of producing 100 units and (b) the average and the marginal cost functions and their values at $x = 100$.

1 $C(x) = 800 + 0.04x + 0.0002x^2$

2 $C(x) = 6400 + 6.5x + 0.003x^2$

3 $C(x) = 250 + 100x + 0.001x^3$

4 $C(x) = 200 + 0.01x + (100/x)$

5 A manufacturer of small motors estimates that the cost (in dollars) of producing x motors per day is given by $C(x) = 100 + 50x + (100/x)$. Compare the marginal cost of producing five motors with the cost of producing the sixth motor.

6 A company conducts a pilot test for production of a new industrial solvent and finds that the cost of producing x liters of each pilot run is given by the formula $C(x) = 3 + x + (10/x)$. Compare the marginal cost of

producing 10 liters with the cost of producing the 11th liter.

Exer. 7–8: For the given demand and cost functions, find (a) the marginal demand function, (b) the revenue function, (c) the profit function, (d) the marginal profit function, (e) the maximum profit, and (f) the marginal cost when the demand is 10 units.

7 $p(x) = 50 - 0.1x;$ $\quad C(x) = 10 + 2x$

8 $p(x) = 80 - \sqrt{x-1};$ $\; C(x) = 75x + 2\sqrt{x-1}$

9 A travel agency estimates that, in order to sell x package-deal vacations, it must charge a price per vacation of $1800 - 2x$ dollars for $1 \le x \le 100$. If the cost to the agency for x vacations is $1000 + x + 0.01x^2$ dollars, find

(a) the revenue function

(b) the profit function

(c) the number of vacations that will maximize the profit

(d) the maximum profit

10 A manufacturer determines that x units of a product will be sold if the selling price is $400 - 0.05x$ dollars for each unit. If the production cost for x units is $500 + 10x$, find

(a) the revenue function

(b) the profit function

(c) the number of units that will maximize the profit

(d) the price per unit when the marginal revenue is 300

11 A kitchen specialty company determines that the cost of manufacturing and packaging x pepper mills per day is $500 + 0.02x + 0.001x^2$. If each mill is sold for $8.00, find

(a) the rate of production that will maximize the profit

(b) the maximum daily profit

12 A company that conducts bus tours found that when the price was $9.00 per person, the average number of customers was 1000 per week. When the company reduced the price to $7.00 per person, the average number of customers increased to 1500 per week. Assuming that the demand function is linear, what price should be charged to obtain the greatest weekly revenue?

Exer. 13–14: Analyze each model using the technique developed in the investigation of the predator–prey model.

13 The competitive-hunters model represents an ecosystem with two species, each of which requires the same resource for survival. If x and y are the populations of two species at time t, then the model has the form

$$x'(t) = ax - bxy,$$
$$y'(t) = my - nxy,$$

where $a, b, m,$ and n are positive constants.

14 The Richardson arms race model (Lewis F. Richardson, *Arms and Insecurity*, Pittsburgh: Boxwood Press, 1960) represents the arms expenditures x and y of two nations as the system of equations

$$x'(t) = ay - mx + r,$$
$$y'(t) = bx - ny + s,$$

where $a, b, m, n, r,$ and s are constants, the first four of which are positive.

CHAPTER 3 REVIEW EXERCISES

Exer. 1–2: Find the extrema of f on the given interval.

1 $f(x) = -x^2 + 6x - 8;$ $\; [1, 6]$

2 $f(x) = 3x^3 + x^2;$ $\qquad (-1, 0]$

Exer. 3–4: Find the critical numbers of f.

3 $f(x) = (x+2)^3(3x-1)^4$

4 $f(x) = \sqrt{x-1}(x-2)^3$

Exer. 5–8: Use the first derivative test to find the local extrema of f. Find the intervals on which f is increasing or is decreasing, and sketch the graph of f.

5 $f(x) = -4x^3 + 9x^2 + 12x$

6 $f(x) = \dfrac{1}{x^2 + 1}$

7 $f(x) = (4-x)x^{1/3}$ $\qquad$ **8** $f(x) = \sqrt[3]{x^2 - 9}$

Exer. 9–12: Use the second derivative test (whenever applicable) to find the local extrema of f. Find the intervals on which the graph of f is concave upward or is concave downward, and find the x-coordinates of the points of inflection. Sketch the graph of f.

9 $f(x) = \sqrt[3]{8 - x^3}$

10 $f(x) = -x^3 + 4x^2 - 3x$

11 $f(x) = \dfrac{1}{x^2 + 1}$

12 $f(x) = 40x^3 - x^6$

13 If $f(x) = 2\sin x - \cos 2x$, find the local extrema, and sketch the graph of f for $0 \le x \le 2\pi$.

14 If $f(x) = 2\sin x - \cos 2x$, find equations of the tangent and normal lines to the graph of f at the point $(\pi/6, 1/2)$.

Exer. 15–16: Sketch the graph of a continuous function f that satisfies all the stated conditions.

15 $f(0) = 2;\ f(-2) = f(2) = 0;$
$f'(-2) = f'(0) = f'(2) = 0;$
$f'(x) > 0$ if $-2 < x < 0;$
$f'(x) < 0$ if $x < -2$ or $x > 0;$
$f''(x) > 0$ if $x < -1$ or $1 < x < 2;$
$f''(x) < 0$ if $-1 < x < 1$ or $x > 2$

16 $f(0) = 4;\ f(-3) = f(3) = 0;$
$f'(-3) = 0;\ f'(0)$ is undefined;
$f'(x) > 0$ if $-3 < x < 0;$
$f'(x) < 0$ if $x < -3$ or $x > 0;$
$f''(x) > 0$ if $x < 0$ or $0 < x < 2;$
$f''(x) < 0$ if $x > 2$

Exer. 17–22: Find the extrema and sketch the graph of f.

17 $f(x) = \dfrac{3x^2}{9x^2 - 25}$

18 $f(x) = \dfrac{x^2}{(x - 1)^2}$

19 $f(x) = \dfrac{x^2 + 2x - 8}{x + 3}$

20 $f(x) = \dfrac{x^4 - 16}{x^3}$

21 $f(x) = \dfrac{x - 3}{x^2 + 2x - 8}$

22 $f(x) = \dfrac{x}{\sqrt{x + 4}}$

23 If $f(x) = x^3 + x^2 + x + 1$, find a number c that satisfies the conclusion of the mean value theorem on the interval $[0, 4]$.

24 The posted speed limit on a 125-mi toll highway is 65 mi/hr. When an automobile enters the toll road, the driver is issued a ticket on which is printed the exact time. If the driver completes the trip in 1 hr 40 min or less, a speeding citation is issued when the toll is paid. Use the mean value theorem to explain why this citation is justified.

25 A man wishes to put a fence around a rectangular field and then subdivide this field into three smaller rectangular plots by placing two fences parallel to one of the sides. If he can afford only 1000 yd of fencing, what dimensions will give him the maximum area?

26 An open rectangular storage shelter 4 ft deep, consisting of two vertical sides and a flat roof, is to be attached to an existing structure, as illustrated in the figure. The flat roof is made of tin and costs $5 per ft^2. The two sides are made of plywood costing $2 per ft^2. If $400 is available for construction, what dimensions will maximize the volume of the shelter?

Exercise 26

27 A V-shaped water gutter is to be constructed from two rectangular sheets of metal 10 in. wide. Find the angle between the sheets that will maximize the carrying capacity of the gutter.

28 Find the altitude of the right circular cylinder of maximum curved surface area that can be inscribed in a sphere of radius a.

29 The interior of a half-mile race track consists of a rectangle with semicircles at two opposite ends. Find the dimensions that will maximize the area of the rectangle.

30 A cable television firm presently serves 5000 households and charges $20 per month. A marketing survey indicates that each decrease of $1 in monthly charge will result in 500 new customers. Find the monthly charge that will result in the maximum monthly revenue.

31 A wire 5 ft long is to be cut into two pieces. One piece is to be bent into the shape of a circle and the other into the shape of a square. Where should the wire be cut so that the sum of the areas of the circle and square is

(a) a maximum   **(b)** a minimum

32 In biochemistry, the general threshold-response curve is given by $R = kS^n/(S^n + a^n)$, where R is the chemical response that corresponds to a concentration S of a substance for positive constants k, n, and a. An example is the rate R at which the liver removes alcohol from the bloodstream when the concentration of alcohol is S. Show that R is an increasing function of S and that $R = k$ is a horizontal asymptote for the curve.

33 The position function of a point moving on a coordinate line is given by $s(t) = (t^2 + 3t + 1)/(t^2 + 1)$. Find the velocity and the acceleration at time t, and describe the motion of the point during the time interval $[-2, 2]$.

34 The position of a moving point on a coordinate line is given by

$$s(t) = a \sin(kt + m) + b \cos(kt + m)$$

for constants a, b, k, and m. Prove that the magnitude of the acceleration is directly proportional to the distance from the origin.

35 A manufacturer of microwave ovens determines that the cost of producing x units is given by

$$C(x) = 4000 + 100x + 0.05x^2 + 0.0002x^3.$$

Compare the marginal cost of producing 100 ovens with the cost of producing the 101st oven.

36 The cost function for producing a microprocessor component is given by $C(x) = 1000 + 2x + 0.005x^2$. If 2000 units are produced, find the cost, the average cost, the marginal cost, and the marginal average cost.

37 An electronics company estimates that the cost of producing x calculators per day is

$$C(x) = 500 + 6x + 0.02x^2.$$

If each calculator is sold for $18, find

(a) the revenue function

(b) the profit function

(c) the daily production that will maximize the profit

(d) the maximum daily profit

38 A small office building is to contain 500 ft^2 of floor space. Simplified floor plans are shown in the figure. If the walls cost $100 per running foot and if the wall space above the doors is disregarded,

(a) show that the cost $C(x)$ of the walls is

$$C(x) = 100[3x - 6 + (1000/x)]$$

(b) find the vertical and oblique asymptotes, and sketch the graph of $C(x)$ for $x > 0$

(c) find the design that minimizes the cost

Exercise 38

39 A NASA rocket is propelled straight upward from the ground. If the acceleration is constant and the rocket achieves a height of 49 ft in 1 sec, what is the rate of acceleration?

40 The "Alpine Slide" is a way to enjoy sledding when there is no snow. The slide itself is a long concrete trough that runs downhill parallel to a ski trail. The rider takes a chair lift to the top of the slide and then races down the slide on a small sled. The sled has metal rollers on the bottom and a control stick that permits the rider to slow down while moving along the track.

A sled is initially moving at a rate of 44 ft/sec. It decelerates to 32 ft/sec over a distance of 114 ft at an unknown constant rate. It continues to decelerate at that same rate until it comes to a full stop.

(a) How long does it take to reduce the speed to 32 ft/sec?

(b) What is the acceleration of the sled?

(c) How long does it take before the sled comes to a complete stop?

(d) How many feet does the sled travel before it comes to a stop?

Exercise 40

$\boxed{c}$ Exer. 41–44: Approximate a number c that satisfies the conclusion of the mean value theorem on the interval given.

41 $f(x) = x^3 + x^2 + x + 1$; [0, 4]

42 $f(x) = x^3 + 2x^2 - 3x - 4$; [-3, 2]

43 $f(x) = \sin(\sin x)$; $[-\pi/2, \pi/2]$

44 $f(x) = \sin(\cos x)$; $[0, \pi]$

$\boxed{c}$ Exer. 45–50: Graph f on the given interval and approximate the extrema and the points of inflection to four decimal places.

45 $f(x) = x^2 - 3\sqrt{x} - \sqrt{x^3} - 7$; [0, 7]

46 $f(x) = \sin x - x \cos x$; [-5, 10]

47 $f(x) = |x^3 + 24x^2 - 18x + 3|$; [-7, 5]

48 $f(x) = x^5 - 5x^3 + 3x + 4$; [-25, 2.5]

49 $f(x) = 5 \cos(\cos x) - 2x$; [-2, 2]

50 $f(x) = 2 \sin x + 3 \sin \pi x$; [-2, 2]

EXTENDED PROBLEMS AND GROUP PROJECTS

1 Here is an important generalization of the mean value theorem: *If f and g are continuous on $[a, b]$ and differentiable on (a, b), then there is a number c in (a, b) such that $[f(b) - f(a)]/[g(b) - g(a)] = f'(c)/g'(c)$.*

(a) Show that the mean value theorem is a special case of this generalization. (*Hint:* Use $g(x) = x$.)

(b) Prove this theorem by applying the mean value theorem to the function $h(x)$ defined by $h(x) = [f(b) - f(a)]g(x) - [g(b) - g(a)]f(x)$.

$\boxed{c}$ (c) Suppose that f and g are differentiable on an open interval (a, b) containing c, except possibly at c itself. If $\lim_{x \to c} f(x) = 0 = \lim_{x \to c} g(x)$, then show that

$$\lim_{x \to c} \frac{f(x)}{g(x)} = \lim_{x \to c} \frac{f'(x)}{g'(x)},$$

provided $\lim_{x \to c} \dfrac{f'(x)}{g'(x)}$ exists.

Obtain first some numerical and graphical evidence that this result is true by examining various examples of pairs of functions f and g. Then use the generalization of the mean value theorem to prove the result.

2 Suppose that f is a function with the property that $f'(x) = 1/x$ for all $x > 0$ and $f(1) = 0$. (We have not yet seen this function, but in this problem, we will investigate what the mean value theorem and its consequences imply about such a function.)

(a) Show that f is a strictly increasing function.

(b) Find $f''(x)$ and show that the graph of f is concave downward.

(c) Let c be any positive number. Define the function $g(x)$ by $g(x) = f(cx)$. By the chain rule, show that $g'(x) = 1/x$. Why does this result imply that $g(x) = f(x) + C$ for some constant C? Use the fact that $f(1) = 0$ to find C.

(d) Show that f satisfies one of the properties of logarithmic functions—namely, $f(cx) = f(c) + f(x)$ for all positive numbers c and x.

(e) Let n be a nonzero rational number. Use the chain rule to show that the functions $g(x) = f(x^n)$ and $h(x) = nf(x)$ have the same derivative. Thus, $g(x) = h(x) + C$ for some constant C. Use the fact that $f(1) = 0$ to determine C.

(f) Show that f satisfies another property of logarithmic functions: $f(x^n) = nf(x)$.

(g) Show that if $1 < c < 2$, then $f'(c) > 1/2$. Use the mean value theorem to show that $f(2) > 1/2$. By part (f), show that this implies that $f(2^n) > n/2$ if n is a positive integer. Conclude that

$$\lim_{x \to \infty} f(x) = \infty.$$

(h) Determine $\lim_{x \to 0^+} f(x)$.

(i) Using the properties of f derived above, sketch a graph of f.

3 Suppose that f is a function with the property that $f'(x) = f(x)$ for all x and $f(0) = 1$. (We have not yet seen this function, but in this problem, we will investigate what the mean value theorem and its consequences imply about such a function.)

(a) Show that f is a strictly increasing function.

(b) Find $f''(x)$ and show that the graph of f is concave upward.

(c) Use the chain rule to show that if $h(x) = f(x + c)$, then $h'(x) = h(x)$. Apply the quotient rule to the function $g(x) = f(x + c)/f(x)$ to show that $g'(x) = 0$ for all x. Conclude that f has an important property of exponential functions: $f(x + c) = f(x)f(c)$, for all numbers x and c.

(d) Show that f also has another important property of exponential functions: $(f(x))^n = f(xn)$.

(e) Use the mean value theorem to conclude that $f(n) > 1 + n$ if $n > 0$ and hence show that $\lim_{x \to \infty} f(x) = \infty$.

(f) Determine $\lim_{x \to -\infty} f(x)$.

(g) Using the properties of f derived above, sketch a graph of f.

INTRODUCTION

HUMAN CIVILIZATIONS AROSE in the fertile river valleys of China, Egypt, Africa, India, and Mesopotamia. As societies grew highly complex and interdependent, governmental units were created to provide services. To fund these efforts, people paid taxes that were often based then, as now, on the amount and value of their land. Since the annual flooding of the rivers swept away land masses or affected their agricultural value, there occurred early in human history a need to measure accurately land regions, a need we continue to have today.

Land masses and their boundaries are highly irregular. They are subject to significant changes due to such forces as oceans, rivers, and earthquakes. We require accurate ways to estimate precisely lengths, areas, and volumes. As technology will increasingly have an impact on society, there will be an expanding set of situations in which complex quantities must be determined accurately. Calculus provides a powerful means of making such measurements.

This chapter begins with the seemingly unrelated problem of reversing the procedure for finding derivatives: Given a function f, find a function F such that $F' = f$. This problem leads to the definition in Section 4.1 of the closely related ideas of antiderivatives and indefinite integrals. We also consider some elementary differential equations, an important modeling tool for applications. In Section 4.2, we study the technique of change of variables for finding indefinite integrals.

We then turn to a more careful examination of the problem of finding the area of an irregular region, beginning with the area under the graph of a function. In Section 4.3, such an area is defined as a limit of areas of inscribed or circumscribed rectangles. This approach is generalized in Section 4.4, where we give a careful definition of the definite integral of a function as a limit of Riemann sums. We state and show the application of basic properties of the definite integral in Section 4.5.

The principal result in this chapter is the *fundamental theorem of calculus*, proved in Section 4.6. This important theorem enables us to find exact values of definite integrals by using an *antiderivative* or indefinite integral. In addition to providing an important evaluation process, the fundamental theorem shows that there is a relationship between derivatives and integrals—a key result in calculus.

The chapter closes in Section 4.7 with a discussion of methods of *numerical integration*, which are used to approximate definite integrals that we cannot evaluate by the fundamental theorem. These methods are readily programmable for use with calculators and computers and are employed in a wide variety of applied fields.

The definite integral provides a
powerful tool for computing lengths of
curves, areas of regions, and volumes
of solids.

Integrals

4.1 ANTIDERIVATIVES, INDEFINITE INTEGRALS, AND SIMPLE DIFFERENTIAL EQUATIONS

We begin this section with the problem of reversing the process of differentiation and examine two closely related concepts: *antiderivative* and *indefinite integral*. Then we take a first look at using indefinite integrals to solve simple differential equations where we seek to obtain explicit information about a function from given information about its derivative.

ANTIDERIVATIVES

In our previous work, we solved problems of the following type: *Given a function f, find the derivative f'.* We now consider the reverse process: *Given a function f, find a function F such that $F' = f$.* In the next definition, we give F a special name.

Definition 4.1

> A function F is an **antiderivative** of the function f on an interval I if $F'(x) = f(x)$ for every x in I.

We shall also call $F(x)$ an antiderivative of $f(x)$. The process of finding F, or $F(x)$, is called **antidifferentiation**.

To illustrate, $F(x) = x^2$ is an antiderivative of $f(x) = 2x$, because

$$F'(x) = \frac{d}{dx}(x^2) = 2x = f(x).$$

There are many other antiderivatives of $2x$, such as $x^2 + 2$, $x^2 - \frac{5}{3}$, and $x^2 + \sqrt{3}$. In general, if C is *any* constant, then $x^2 + C$ is an antiderivative of $2x$, because

$$\frac{d}{dx}(x^2 + C) = 2x + 0 = 2x.$$

Thus there is a *family of antiderivatives* of $2x$ of the form $F(x) = x^2 + C$, where C is any constant. Graphs of several members of this family are sketched in Figure 4.1.

The next illustration contains other examples of antiderivatives, where C is a constant.

Figure 4.1

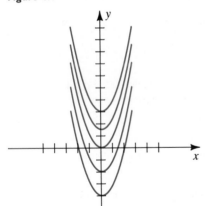

ILLUSTRATION

$f(x)$	Antiderivatives of $f(x)$
x^2	$\frac{1}{3}x^3, \quad \frac{1}{3}x^3 + 8, \quad \frac{1}{3}x^3 + C$
$8x^3$	$2x^4, \quad 2x^4 - \sqrt[3]{7}, \quad 2x^4 + C$
$\cos x$	$\sin x \quad \sin x + \frac{4}{9}, \quad \sin x + C$

As in the preceding illustration, if $F(x)$ is an antiderivative of $f(x)$, then so too is $F(x) + C$ for any constant C. It is a consequence of the mean value theorem (3.12) that *every* antiderivative is of this form. In Corollary (3.14), we proved that two functions with identical derivatives can differ only by some constant. The next theorem restates this result in the language of antiderivatives.

Theorem 4.2

Let F be an antiderivative of f on an interval I. If G is any antiderivative of f on I, then

$$G(x) = F(x) + C$$

for some constant C and every x in I.

We refer to the constant C in Theorem (4.2) as an **arbitrary constant**. If $F(x)$ is an antiderivative of $f(x)$, then *all* antiderivatives of $f(x)$ can be obtained from $F(x) + C$ by letting C range through the set of real numbers.

INDEFINITE INTEGRALS

We shall use the following notation for a family of antiderivatives of the type given in Theorem (4.2).

Definition 4.3

The notation

$$\int f(x)\, dx = F(x) + C,$$

where $F'(x) = f(x)$ and C is an arbitrary constant, denotes the family of all antiderivatives of $f(x)$ on an interval I.

CAUTION Theorem (4.2) may be false if the interval I is replaced by some other set of real numbers. For example, if A is the set of nonzero real numbers, then the function $F(x) = 1/x$ has derivative $-1/x^2$ for all x in A, as does the function G defined by

$$G(x) = \begin{cases} 1/x & \text{if } x < 0 \\ (1/x) + 1 & \text{if } x > 0 \end{cases}$$

but there is no constant C such that $G(x) = F(x) + C$ for every x in A. Thus, in problems involving antiderivatives for a function f, we always assume, even if not explicitly stated, that the domain of f is an interval.

The symbol $\int$ used in Definition (4.3) is an **integral sign**. We call $\int f(x)\, dx$ the **indefinite integral** of $f(x)$. The expression $f(x)$ is the **integrand**, and C is the **constant of integration**. The process of finding $F(x) + C$, when given $\int f(x)\, dx$, is referred to as **indefinite integration, evaluating the integral,** or **integrating** $f(x)$. The adjective *indefinite* is used because $\int f(x)\, dx$ represents a *family* of antiderivatives, not any *specific* function. Later in the chapter, when we discuss definite integrals, we shall see the reasons for using the integral sign and the differential expression dx that appears to the right of the integrand $f(x)$. For now, we regard dx merely as a symbol that specifies the independent variable x, which we refer to as the **variable of integration**. If we use a different variable of integration, such as t, we write

$$\int f(t)\,dt = F(t) + C,$$

where $F'(t) = f(t)$.

ILLUSTRATION

- $\displaystyle\int x^4\,dx = \frac{1}{5}x^5 + C$ because $\displaystyle\frac{d}{dx}\left(\frac{1}{5}x^5\right) = x^4.$

- $\displaystyle\int t^{-3}\,dt = -\frac{1}{2}t^{-2} + C$ because $\displaystyle\frac{d}{dt}\left(-\frac{1}{2}t^{-2}\right) = t^{-3}.$

- $\displaystyle\int \cos u\,du = \sin u + C$ because $\displaystyle\frac{d}{du}(\sin u) = \cos u.$

- $\displaystyle\int x \cos x\,dx = x \sin x + \cos x + C$ because $\displaystyle\frac{d}{dx}(x \sin x + \cos x) = x \cos x.$

Note that, in general,

$$\int \frac{d}{dx}(f(x))\,dx = f(x) + C$$

because $f'(x) = (d/dx)(f(x))$. This result allows us to use any derivative formula to obtain a corresponding formula for an indefinite integral, as illustrated in the next table. As shown in Formula (1), it is customary to abbreviate $\int 1\,dx$ by $\int dx$.

Brief Table of Indefinite Integrals 4.4

Derivative $\dfrac{d}{dx}(f(x))$	Indefinite integral $\displaystyle\int \dfrac{d}{dx}(f(x))\,dx = f(x) + C$
$\dfrac{d}{dx}(x) = 1$	(1) $\displaystyle\int 1\,dx = \int dx = x + C$
$\dfrac{d}{dx}\left(\dfrac{x^{r+1}}{r+1}\right) = x^r\,(r \neq -1)$	(2) $\displaystyle\int x^r\,dx = \dfrac{x^{r+1}}{r+1} + C\ (r \neq -1)$
$\dfrac{d}{dx}(\sin x) = \cos x$	(3) $\displaystyle\int \cos x\,dx = \sin x + C$
$\dfrac{d}{dx}(-\cos x) = \sin x$	(4) $\displaystyle\int \sin x\,dx = -\cos x + C$
$\dfrac{d}{dx}(\tan x) = \sec^2 x$	(5) $\displaystyle\int \sec^2 x\,dx = \tan x + C$
$\dfrac{d}{dx}(-\cot x) = \csc^2 x$	(6) $\displaystyle\int \csc^2 x\,dx = -\cot x + C$
$\dfrac{d}{dx}(\sec x) = \sec x \tan x$	(7) $\displaystyle\int \sec x \tan x\,dx = \sec x + C$
$\dfrac{d}{dx}(-\csc x) = \csc x \cot x$	(8) $\displaystyle\int \csc x \cot x\,dx = -\csc x + C$

Formula (2) is called the *power rule for indefinite integration.* As in the following illustration, it is often necessary to rewrite an integrand before applying the power rule or one of the trigonometric formulas.

ILLUSTRATION

$$\blacksquare \quad \int x^3 \cdot x^5 \, dx = \int x^8 \, dx = \frac{x^{8+1}}{8+1} + C = \frac{x^9}{9} + C$$

$$\blacksquare \quad \int \frac{1}{x^3} \, dx = \int x^{-3} \, dx = \frac{x^{-3+1}}{-3+1} + C = -\frac{1}{2x^2} + C$$

$$\blacksquare \quad \int \sqrt[3]{x^2} \, dx = \int x^{2/3} \, dx = \frac{x^{2/3+1}}{\frac{2}{3}+1} + C = \frac{3}{5} x^{5/3} + C$$

$$\blacksquare \quad \int \frac{\tan x}{\sec x} \, dx = \int \cos x \, \frac{\sin x}{\cos x} \, dx = \int \sin x \, dx = -\cos x + C$$

It is a good idea to check indefinite integrations (such as those in the preceding illustration) by differentiating the final expression to see if either the integrand or an equivalent form of the integrand is obtained.

The next theorem indicates that differentiation and indefinite integration are inverse processes, because each, in a sense, undoes the other. In (i), we assume that f is differentiable, and in (ii), that f has an antiderivative on some interval.

Theorem 4.5

(i) $\displaystyle \int \frac{d}{dx}(f(x)) \, dx = f(x) + C$

(ii) $\displaystyle \frac{d}{dx}\left(\int f(x) \, dx \right) = f(x)$

PROOF We have already proved (i). To prove (ii), let F be an antiderivative of f and write

$$\frac{d}{dx}\left(\int f(x) \, dx \right) = \frac{d}{dx}(F(x) + C) = F'(x) + 0 = f(x). \quad \blacksquare$$

EXAMPLE 1 Verify Theorem (4.5) for the special case $f(x) = x^2$.

SOLUTION

(i) If we first differentiate x^2 and then integrate,

$$\int \frac{d}{dx}(x^2) \, dx = \int 2x \, dx = x^2 + C.$$

(ii) If we first integrate x^2 and then differentiate,

$$\frac{d}{dx}\left(\int x^2 \, dx \right) = \frac{d}{dx}\left(\frac{x^3}{3} + C \right) = x^2.$$

The next theorem is useful for evaluating many types of indefinite integrals. In the statements, we assume that $f(x)$ and $g(x)$ have antiderivatives on an interval I.

Theorem 4.6

(i) $\displaystyle\int cf(x)\,dx = c\int f(x)\,dx$ for any nonzero constant c

(ii) $\displaystyle\int [f(x) + g(x)]\,dx = \int f(x)\,dx + \int g(x)\,dx$

(iii) $\displaystyle\int [f(x) - g(x)]\,dx = \int f(x)\,dx - \int g(x)\,dx$

PROOF We shall prove (ii). The proofs of (i) and (iii) are similar. If F and G are antiderivatives of f and g, respectively,

$$\frac{d}{dx}(F(x) + G(x)) = F'(x) + G'(x) = f(x) + g(x).$$

Hence, by Definition (4.3),

$$\int [f(x) + g(x)]\,dx = F(x) + G(x) + C,$$

where C is an arbitrary constant. Similarly,

$$\int f(x)\,dx + \int g(x)\,dx = F(x) + C_1 + G(x) + C_2$$

for arbitrary constants C_1 and C_2. These give us the same family of antiderivatives, since for any special case, we can choose values of the constants such that $C = C_1 + C_2$. We thus prove (ii). ■

Theorem (4.6)(i) is sometimes stated as follows: *A constant factor in the integrand may be taken outside the integral sign.*

CAUTION It is *not* permissible to take expressions involving variables outside the integral sign in this manner. Note, for example, that

$$\int x \cos x\,dx = x \sin x + \cos x + C,$$

whereas

$$x\int \cos x\,dx = x(\sin x + C) = x \sin x + Cx.$$

These two expressions do not differ by a constant. Hence,

$$\int x \cos x\,dx \neq x\int \cos x\,dx.$$

EXAMPLE ■ 2   Evaluate $\int (5x^3 + 2\cos x)\, dx$.

SOLUTION   We first use (ii) and (i) of Theorem (4.6) and then formulas from (4.4):

$$\int (5x^3 + 2\cos x)\, dx = \int 5x^3\, dx + \int 2\cos x\, dx$$

$$= 5\int x^3\, dx + 2\int \cos x\, dx$$

$$= 5\left(\frac{x^4}{4} + C_1\right) + 2(\sin x + C_2)$$

$$= \tfrac{5}{4}x^4 + 5C_1 + 2\sin x + 2C_2$$

$$= \tfrac{5}{4}x^4 + 2\sin x + C,$$

where $C = 5C_1 + 2C_2$.

In Example 2, we added the two constants $5C_1$ and $2C_2$ to obtain one arbitrary constant C. We can always manipulate arbitrary constants in this way, so it is not necessary to introduce a constant for each indefinite integration as we did in Example 2. Instead, if an integrand is a sum, *we integrate each term of the sum without introducing constants and then add one arbitrary constant C after the last integration*. We also often bypass the step $\int cf(x)\, dx = c \int f(x)\, dx$, as in the next example.

EXAMPLE ■ 3   Evaluate $\int \left(8t^3 - 6\sqrt{t} + \dfrac{1}{t^3}\right) dt$.

SOLUTION   First we find an antiderivative for each of the three terms in the integrand and then add an arbitrary constant C. We rewrite $\sqrt{t}$ as $t^{1/2}$ and $1/t^3$ as t^{-3} and then use the power rule for integration:

$$\int \left(8t^3 - 6\sqrt{t} + \frac{1}{t^3}\right) dt = \int (8t^3 - 6t^{1/2} + t^{-3})\, dt$$

$$= 8 \cdot \frac{t^4}{4} - 6 \cdot \frac{t^{3/2}}{\frac{3}{2}} + \frac{t^{-2}}{-2} + C$$

$$= 2t^4 - 4t^{3/2} - \frac{1}{2t^2} + C$$

EXAMPLE ■ 4   Evaluate $\int \dfrac{(x^2 - 1)^2}{x^2}\, dx$.

SOLUTION   First we change the form of the integrand, because the degree of the numerator is greater than or equal to the degree of the denominator. We then find an antiderivative for each term, adding an arbitrary constant C after the last integration:

$$\int \frac{(x^2 - 1)^2}{x^2}\, dx = \int \frac{x^4 - 2x^2 + 1}{x^2}\, dx$$

$$= \int (x^2 - 2 + x^{-2})\, dx$$

$$= \frac{x^3}{3} - 2x + \frac{x^{-1}}{-1} + C$$

$$= \frac{1}{3}x^3 - 2x - \frac{1}{x} + C$$

EXAMPLE ▪ 5   Evaluate $\displaystyle\int \frac{1}{\cos u \cot u}\, du$.

SOLUTION   We use trigonometric identities to change the integrand and then apply Formula (7) from Table (4.4):

$$\int \frac{1}{\cos u \cot u}\, du = \int \sec u \tan u \, du$$

$$= \sec u + C$$

NOTE   While most work on computers involves numerical calculations, computer algebra systems (CAS) can perform operations on symbolic formulas. Using CAS software, you may be able to enter the expression for a function and then request the indefinite integral. If an antiderivative can be found using the techniques discussed in this text, there is an excellent chance that the CAS will be successful in finding one. If you have access to a CAS, you should investigate the rules for entering functions symbolically and requesting derivatives and antiderivatives.

SIMPLE DIFFERENTIAL EQUATIONS

An applied problem may be stated in terms of a **differential equation**— that is, an equation that involves derivatives or differentials of an unknown function. A function f is a **solution** of a differential equation if it satisfies the equation — that is, if substitution of f for the unknown function produces a true statement. To **solve** a differential equation means to find all solutions. Sometimes, in addition to the differential equation, we may know certain values of f or f', called **initial conditions**.

Indefinite integrals are useful for solving certain differential equations, because if we are given a derivative $f'(x)$, we can integrate and use Theorem (4.5)(i) to obtain an equation involving the unknown function f:

$$\int f'(x)\, dx = f(x) + C$$

If we are also given an initial condition for f, it may be possible to find $f(x)$ explicitly, as in the next example.

EXAMPLE ▪ 6 Solve the differential equation

$$f'(x) = 6x^2 + x - 5$$

subject to the initial condition $f(0) = 2$.

SOLUTION We proceed as follows:

$$f'(x) = 6x^2 + x - 5$$

$$\int f'(x)\, dx = \int (6x^2 + x - 5)\, dx$$

$$f(x) = 2x^3 + \tfrac{1}{2}x^2 - 5x + C$$

for some number C. (It is unnecessary to add a constant of integration to *each* side of the equation.) Letting $x = 0$ and using the given initial condition $f(0) = 2$ gives us

$$f(0) = 0 + 0 - 0 + C, \quad \text{or} \quad 2 = C.$$

Hence the solution f of the differential equation with the initial condition $f(0) = 2$ is

$$f(x) = 2x^3 + \tfrac{1}{2}x^2 - 5x + 2.$$

If we are given a *second* derivative $f''(x)$, then we must employ two successive indefinite integrals to find $f(x)$. First we use Theorem (4.5)(i) as follows:

$$\int f''(x)\, dx = \int \frac{d}{dx}(f'(x))\, dx = f'(x) + C$$

After finding $f'(x)$, we proceed as in Example 6.

EXAMPLE ▪ 7 Solve the differential equation

$$f''(x) = 5\cos x + 2\sin x$$

subject to the initial conditions $f(0) = 3$ and $f'(0) = 4$.

SOLUTION We proceed as follows:

$$f''(x) = 5\cos x + 2\sin x$$

$$\int f''(x)\, dx = \int (5\cos x + 2\sin x)\, dx$$

$$f'(x) = 5\sin x - 2\cos x + C$$

Letting $x = 0$ and using the initial condition $f'(0) = 4$ gives us

$$f'(0) = 5\sin 0 - 2\cos 0 + C$$
$$4 = 0 - 2\cdot 1 + C, \quad \text{or} \quad C = 6.$$

Thus,

$$f'(x) = 5\sin x - 2\cos x + 6.$$

We integrate a second time:

$$\int f'(x)\,dx = \int (5\sin x - 2\cos x + 6)\,dx$$
$$f(x) = -5\cos x - 2\sin x + 6x + D$$

Letting $x = 0$ and using the initial condition $f(0) = 3$, we find that

$$f(0) = -5\cos 0 - 2\sin 0 + 6 \cdot 0 + D$$
$$3 = -5 - 0 + 0 + D, \quad \text{or} \quad D = 8.$$

Therefore, the solution of the differential equation with the given initial condition is

$$f(x) = -5\cos x - 2\sin x + 6x + 8.$$

Suppose that a point P is moving on a coordinate line with an acceleration $a(t)$ at time t, and the corresponding velocity is $v(t)$. By Definition (3.23), $a(t) = v'(t)$ and hence

$$\int a(t)\,dt = \int v'(t)\,dt = v(t) + C$$

for some constant C.

Similarly, if we know $v(t)$, then since $v(t) = s'(t)$, where s is the position function of P, we can find a formula that involves $s(t)$ by indefinite integration:

$$\int v(t)\,dt = \int s'(t)\,dt = s(t) + D$$

for some constant D. In the next example, we shall use this technique to find the position function for an object that is moving with a given acceleration function $a(t)$.

EXAMPLE ▪ 8   A particle moving along a coordinate line at time $t = 0$ is at a position 3 cm from the origin and traveling at a velocity of 7 cm/sec. If the acceleration of the particle is given by

$$a(t) = 2 - 2(t + 1)^{-3},$$

find the velocity and the position of the particle as functions of t.

SOLUTION   Since the velocity $v(t) = \int v'(t)\,dt = \int a(t)\,dt$, we have

$$v(t) = \int [2 - 2(t + 1)^{-3}]\,dt$$
$$= 2t + (t + 1)^{-2} + C$$

for some number C. Substituting 0 for t and using the fact that $v(0) = 7$ gives us $7 = 0 + 1 + C$, or $C = 6$. Consequently,

$$v(t) = 2t + (t + 1)^{-2} + 6.$$

Since $s'(t) = v(t)$, we obtain

$$s'(t) = 2t + (t + 1)^{-2} + 6$$

$$\int s'(t) \, dt = \int [2t + (t + 1)^{-2} + 6] \, dt$$

$$s(t) = t^2 - (t + 1)^{-1} + 6t + D$$

for some number D. Using the fact that $s(0) = 3$ gives $3 = 0 - 1 + 0 + D$, or $D = 4$. Thus, the position of the particle from the origin at time t is given by

$$s(t) = t^2 - (t + 1)^{-1} + 6t + 4 \text{ cm},$$

and the particle travels at a velocity of

$$v(t) = 2t + (t + 1)^{-2} + 6 \text{ cm/sec}.$$

In economics applications, if a marginal function is known (see page 330), then we can use indefinite integration to find the function, as illustrated in the next example.

EXAMPLE ▪ 9 A manufacturer finds that the marginal cost (in dollars) associated with the production of x units of a photocopier component is given by $30 - 0.02x$. If the cost of producing one unit is \$35, find the cost function and the cost of producing 100 units.

SOLUTION If C is the cost function, then the marginal cost is the rate of change of C with respect to x—that is,

$$C'(x) = 30 - 0.02x.$$

Hence

$$\int C'(x) \, dx = \int (30 - 0.02x) \, dx$$

and

$$C(x) = 30x - 0.01x^2 + K$$

for some K. Letting $x = 1$ and using $C(1) = 35$, we obtain

$$35 = 30 - 0.01 + K, \quad \text{or} \quad K = 5.01.$$

Consequently,

$$C(x) = 30x - 0.01x^2 + 5.01.$$

In particular, the cost of producing 100 units is

$$C(100) = 3000 - 100 + 5.01$$
$$= \$2905.01.$$

EXERCISES 4.1

Exer. 1–40: Evaluate.

1 $\displaystyle\int (4x + 3)\,dx$

2 $\displaystyle\int (4x^2 - 8x + 1)\,dx$

3 $\displaystyle\int (9t^2 - 4t + 3)\,dt$

4 $\displaystyle\int (2t^3 - t^2 + 3t - 7)\,dt$

5 $\displaystyle\int \left(\frac{1}{z^3} - \frac{3}{z^2}\right) dz$

6 $\displaystyle\int \left(\frac{4}{z^7} - \frac{7}{z^4} + z\right) dz$

7 $\displaystyle\int \left(3\sqrt{u} + \frac{1}{\sqrt{u}}\right) du$

8 $\displaystyle\int \left(\sqrt{u^3} - \tfrac{1}{2}u^{-2} + 5\right) du$

9 $\displaystyle\int (2v^{5/4} + 6v^{1/4} + 3v^{-4})\,dv$

10 $\displaystyle\int (3v^5 - v^{5/3})\,dv$

11 $\displaystyle\int (3x - 1)^2\,dx$

12 $\displaystyle\int \left(x - \frac{1}{x}\right)^2 dx$

13 $\displaystyle\int x(2x + 3)\,dx$

14 $\displaystyle\int (2x - 5)(3x + 1)\,dx$

15 $\displaystyle\int \frac{8x - 5}{\sqrt[3]{x}}\,dx$

16 $\displaystyle\int \frac{2x^2 - x + 3}{\sqrt{x}}\,dx$

17 $\displaystyle\int \frac{x^3 - 1}{x - 1}\,dx$

18 $\displaystyle\int \frac{x^3 + 3x^2 - 9x - 2}{x - 2}\,dx$

19 $\displaystyle\int \frac{(t^2 + 3)^2}{t^6}\,dt$

20 $\displaystyle\int \frac{(\sqrt{t} + 2)^2}{t^3}\,dt$

21 $\displaystyle\int \tfrac{3}{4}\cos u\,du$

22 $\displaystyle\int -\tfrac{1}{5}\sin u\,du$

23 $\displaystyle\int \frac{7}{\csc x}\,dx$

24 $\displaystyle\int \frac{1}{4\sec x}\,dx$

25 $\displaystyle\int (\sqrt{t} + \cos t)\,dt$

26 $\displaystyle\int \left(\sqrt[3]{t^2} - \sin t\right) dt$

27 $\displaystyle\int \frac{\sec t}{\cos t}\,dt$

28 $\displaystyle\int \frac{1}{\sin^2 t}\,dt$

29 $\displaystyle\int (\csc v \cot v \sec v)\,dv$

30 $\displaystyle\int (4 + 4\tan^2 v)\,dv$

31 $\displaystyle\int \frac{\sec w \sin w}{\cos w}\,dw$

32 $\displaystyle\int \frac{\csc w \cos w}{\sin w}\,dw$

33 $\displaystyle\int \frac{(1 + \cot^2 z)\cot z}{\csc z}\,dz$

34 $\displaystyle\int \frac{\tan z}{\cos z}\,dz$

35 $\displaystyle\int \frac{d}{dx}\left(\sqrt{x^2 + 4}\right) dx$

36 $\displaystyle\int \frac{d}{dx}\left(\sqrt[3]{x^3 - 8}\right) dx$

37 $\displaystyle\int \frac{d}{dx}(\sin\sqrt[3]{x})\,dx$

38 $\displaystyle\int \frac{d}{dx}(\sqrt{\tan x})\,dx$

39 $\displaystyle\frac{d}{dx}\int (x^3\sqrt{x - 4})\,dx$

40 $\displaystyle\frac{d}{dx}\int \left(x^4\sqrt[3]{x^2 + 9}\right) dx$

41 Show that $\displaystyle\int x^2\,dx \neq x\int x\,dx$.

42 Show that $\displaystyle\int (1 + x)\,dx \neq 1 + \int x\,dx$.

Exer. 43–48: Evaluate the integral if a and b are constants.

43 $\displaystyle\int a^2\,dx$

44 $\displaystyle\int ab\,dx$

45 $\displaystyle\int (at + b)\,dt$

46 $\displaystyle\int \left(\frac{a}{b^2}t\right) dt$

47 $\displaystyle\int (a + b)\,du$

48 $\displaystyle\int (b - a^2)\,du$

Exer. 49–56: Solve the differential equation subject to the given conditions.

49 $f'(x) = 12x^2 - 6x + 1$; $f(1) = 5$

50 $f'(x) = 9x^2 + x - 8$; $f(-1) = 1$

51 $\dfrac{dy}{dx} = 4x^{1/2}$; $y = 21$ if $x = 4$

52 $\dfrac{dy}{dx} = 5x^{-1/3}$; $y = 70$ if $x = 27$

53 $f''(x) = 4x - 1$; $f'(2) = -2$; $f(1) = 3$

54 $f''(x) = 6x - 4$; $f'(2) = 5$; $f(2) = 4$

55 $\dfrac{d^2y}{dx^2} = 3\sin x - 4\cos x$; $y = 7$ and $y' = 2$ if $x = 0$

56 $\dfrac{d^2y}{dx^2} = 2\cos x - 5\sin x$;

   $y = 2 + 6\pi$ and $y' = 3$ if $x = \pi$

Exer. 57–58: If a point is moving on a coordinate line with the given acceleration $a(t)$ and initial conditions, find $s(t)$.

57 $a(t) = 2 - 6t$; $v(0) = -5$; $s(0) = 4$

58 $a(t) = 3t^2$; $v(0) = 20$; $s(0) = 5$

59 A projectile is fired vertically upward from ground level with a velocity of 1600 ft/sec. Disregarding air resistance, find

   (a) its distance $s(t)$ above the ground at time t

   (b) its maximum height

60 An object is dropped from a height of 1000 ft. Disregarding air resistance, find

   (a) the distance it falls in t seconds

   (b) its velocity at the end of 3 sec

   (c) when it strikes the ground

61 A stone is thrown directly downward from a height of 96 ft with an initial velocity of 16 ft/sec. Find

   (a) its distance above the ground after t seconds

   (b) when it strikes the ground

   (c) the velocity at which it strikes the ground

62 A gravitational constant for objects near the surface of the moon is 5.3 ft/sec^2.

   (a) If an astronaut on the moon throws a stone directly upward with an initial velocity of 60 ft/sec, find the maximum altitude of the stone.

   (b) If, after returning to earth, the astronaut throws the same stone directly upward with the same initial velocity, find the maximum altitude of the stone.

63 If a projectile is fired vertically upward from a height of s_0 feet above the ground with a velocity of v_0 ft/sec, prove that if air resistance is disregarded, its distance $s(t)$ above the ground after t seconds is given by $s(t) = -\frac{1}{2}gt^2 + v_0 t + s_0$, where g is a gravitational constant.

64 A ball rolls down an inclined plane with an acceleration of 2 ft/sec^2.

   (a) If the ball is given no initial velocity, how far will it roll in t seconds?

   (b) What initial velocity must be given for the ball to roll 100 ft in 5 sec?

65 If an automobile starts from rest, what constant acceleration will enable it to travel 500 ft in 10 sec?

66 If a car is traveling at a speed of 60 mi/hr, what constant (negative) acceleration will enable it to stop in 9 sec?

67 A small country has natural gas reserves of 100 billion ft^3. If $A(t)$ denotes the total amount of natural gas consumed after t years, then dA/dt is the *rate of consumption*. If the rate of consumption is predicted to be $5 + 0.01t$ billion ft^3/yr, in approximately how many years will the country's natural gas reserves be depleted?

68 Refer to Exercise 67. Based on U.S. Department of Energy statistics, the rate of consumption of gasoline in the United States (in billions of gallons per year) is approximated by $dA/dt = 2.74 - 0.11t - 0.01t^2$, with $t = 0$ corresponding to the year 1980. Estimate the number of gallons of gasoline consumed in the United States between 1980 and 1984.

69 A sportswear manufacturer determines that the marginal cost in dollars of producing x warmup suits is given by $20 - 0.015x$. If the cost of producing one suit is \$25, find the cost function and the cost of producing 50 suits.

70 If the marginal cost function of a product is given by $2/x^{1/3}$ and if the cost of producing 8 units is \$20, find the cost function and the cost of producing 64 units.

c **Exer. 71–76: Use the commands of a computer algebra system (CAS) to find the derivative and the indefinite integral of the following functions.**

71 $f(x) = 2x^5 + x^4 + 9x^3 - 5x^2 + 4x + 10$

72 $f(x) = 3x^4 + 6x^2 + 8\sqrt{x}$

73 $g(x) = x^2 e^{3x} \cos 4x$

74 $g(x) = x^4 e^x \sin 2x$

75 $s(t) = \dfrac{t - 5}{2t^3 + 3t^2 - 5t - 6}$

76 $s(t) = \dfrac{t^2 + 1}{2t^3 + 3t^2 - 5t - 6}$

77 (a) Show that each of the functions $\sin^2 x$, $-\cos^2 x$, and $-\frac{1}{2}\cos 2x$ are each antiderivatives of $2\sin x \cos x$.

   (b) Reconcile the results of part (a) with the conclusion of Theorem (4.2).

Mathematicians and Their Times

GOTTFRIED WILHELM LEIBNIZ

THE THIRTY YEARS' WAR began as a religious struggle between German Protestants and Roman Catholics in 1618 and spread to a general European struggle for territory and political power. At the war's end, Germany was in ruins. Thousands of people had been killed and entire cities and towns had disappeared.

In this unhappy period of European history were also sown the seeds of modern rationalist thought. Such profound thinkers as Galileo, Newton, Descartes, Pascal, Bacon, Spinoza, Locke, and Leibniz, "the great teachers of the seventeenth century," as one historian called them, "disciplined the minds of men for impartial inquiry and . . . produced a passionate love of truth which has revolutionized all departments of knowledge."* Gottfried Wilhelm Leibniz was the most versatile genius of all, making notable contributions to logic, philosophy, law, history, geology, theology, physics, and mathematics while carrying out an active diplomatic career.

Born in Leipzig, Germany, on July 1, 1646, Leibniz showed an early interest in his studies. He mastered Latin and Greek as an essentially self-taught youth. At age 15, he entered the University of Leipzig, where he earned a philosophy degree at age 17. By age 20, he had completed a brilliant doctoral thesis. Leibniz entered the diplomatic service, first for the Elector of Mainz and later, for 40 years, for the Elector of Hanover. He died on November 14, 1716.

Leibniz formulated many plans to avoid a recurrence of the bloodshed prompted by earlier religious and political rivalries, seeking, unsuccessfully, to reconcile Catholicism and Protestantism. Leibniz went to Paris to try to persuade Louis XIV, the king of France, to turn his attention from attacks against the German states to seizing Egypt. Although Louis XIV chose not to attack Egypt, Leibniz spent four fruitful years

---

*W. E. Lecky, *History of the Rise and Influence of the Spirit of Rationalism in Europe*. London: Longmans, 1866.

in Paris, absorbing the latest scientific advances and training himself in mathematics. During this period, he conceived the principal features of calculus, developing a general method for the calculation of derivatives and integrals and discovering the fundamental theorem of calculus.

A bitter controversy arose concerning the "discovery" of calculus. Newton apparently made his own discoveries in 1666, but did not publish them until 1692. Leibniz independently reached the same results in 1676, publishing them in 1686. Some British mathematicians unfairly charged Leibniz with plagiarizing Newton's ideas; the resulting furor drove a wedge between the British and Continental intellectual communities that hampered the development of calculus.

4.2 CHANGE OF VARIABLES IN INDEFINITE INTEGRALS

The formulas for indefinite integrals in Table (4.4) are limited in scope, because we cannot use them directly to evaluate integrals such as

$$\int \sqrt{5x + 7}\, dx \quad \text{or} \quad \int \cos 4x\, dx.$$

In this section, we shall develop a simple but powerful method for changing the variable of integration so that these integrals (and many others) can be evaluated by using the formulas in Table (4.4).

To justify this method, we shall apply Formula (i) of Theorem (4.5) to a *composite* function. We intend to consider several functions f, g, and F, so it will simplify our work if we state the formula in terms of a function h as follows:

$$\int \frac{d}{dx}(h(x))\, dx = h(x) + C$$

Suppose that F is an antiderivative of a function f and that g is a differentiable function such that $g(x)$ is in the domain of F for every x in some interval. If we let h denote the composite function $F \circ g$, then $h(x) = F(g(x))$ and hence

$$\int \frac{d}{dx}(F(g(x)))\, dx = F(g(x)) + C.$$

Applying the chain rule (2.26) to the integrand $(d/dx)(F(g(x)))$ and using the fact that $F' = f$, we obtain

$$\frac{d}{dx}(F(g(x))) = F'(g(x))g'(x) = f(g(x))g'(x).$$

Substitution in the preceding indefinite integral gives us

$$(*) \qquad \int f(g(x))g'(x)\,dx = F(g(x)) + C.$$

We can use differential notation to help remember this formula. We formally identify the expression dx with the increment Δx—that is, $dx = \Delta x$. We then introduce the variable $u = g(x)$ and note that Definition (2.34) gives us the statement:

$$\text{If} \quad u = g(x), \quad \text{then} \quad du = g'(x)\,dx$$

If we *formally substitute* u and du into $(*)$, we obtain

$$\int f(u)\,du = F(u) + C.$$

This equation has the same *form* as the integral in Definition (4.3); however, u represents a *function*, not an independent variable x, as before. The equation indicates that $g'(x)\,dx$ in $(*)$ may be regarded as the product of $g'(x)$ and dx. Since the variable x has been replaced by a new variable u, finding indefinite integrals in this way is referred to as a **change of variable**, or as the **method of substitution**. We may summarize our discussion as follows, where we assume that f and g have the properties described previously.

Method of Substitution 4.7

If F is an antiderivative of f, then

$$\int f(g(x))g'(x)\,dx = F(g(x)) + C.$$

If $u = g(x)$ and $du = g'(x)\,dx$, then

$$\int f(u)\,du = F(u) + C.$$

After we have made the substitution $u = g(x)$ as indicated in (4.7), it may be necessary to insert a constant factor k into the integrand in order to arrive at the proper form $\int f(u)\,du$. We must then also multiply by $1/k$ to maintain equality, as illustrated in the next examples.

EXAMPLE■I Evaluate $\int \sqrt{5x + 7}\,dx$.

SOLUTION We let $u = 5x + 7$ and calculate du:

$$u = 5x + 7, \quad du = 5\,dx$$

Since du contains the factor 5, the integral is not in the proper form $\int f(u)\,du$ required by (4.7). However, we can *introduce* the factor 5 into the integrand, provided we also multiply by $\frac{1}{5}$. Doing so and using Theorem (4.6)(i) gives us

$$\int \sqrt{5x+7}\, dx = \int \sqrt{5x+7}\left(\tfrac{1}{5}\right)5\, dx$$

$$= \tfrac{1}{5}\int \sqrt{5x+7}\, 5\, dx.$$

We now substitute and use the power rule for integration:

$$\int \sqrt{5x+7}\, dx = \tfrac{1}{5}\int \sqrt{u}\, du$$

$$= \tfrac{1}{5}\int u^{1/2}\, du$$

$$= \tfrac{1}{5}\frac{u^{3/2}}{\frac{3}{2}} + C$$

$$= \tfrac{2}{15}u^{3/2} + C$$

$$= \tfrac{2}{15}(5x+7)^{3/2} + C$$

In the future, after inserting a factor k into an integrand, as in Example 1, we shall simply multiply the integral by $1/k$, skipping the intermediate steps of first writing $(1/k)k$ and then bringing $1/k$ *outside* — that is, to the left of — the integral sign.

E X A M P L E ▪ 2   Evaluate $\displaystyle\int \cos 4x\, dx$.

S O L U T I O N   We make the substitution

$$u = 4x, \quad du = 4\, dx.$$

Since du contains the factor 4, we adjust the integrand by multiplying by 4 and compensate by multiplying the integral by $\tfrac{1}{4}$ before substituting:

$$\int \cos 4x\, dx = \tfrac{1}{4}\int (\cos 4x)4\, dx$$

$$= \tfrac{1}{4}\int \cos u\, du$$

$$= \tfrac{1}{4}\sin u + C$$

$$= \tfrac{1}{4}\sin 4x + C$$

It is not always easy to decide what substitution $u = g(x)$ is needed to transform an indefinite integral into a form that can be readily evaluated. It may be necessary to try several different possibilities before finding a suitable substitution. In most cases, *no* substitution will simplify the integrand properly. The following guidelines may be helpful.

Guidelines for Changing Variables in Indefinite Integrals **4.8**

1 Decide on a reasonable substitution $u = g(x)$.

2 Calculate $du = g'(x)\,dx$.

3 Using guidelines (1) and (2), try to transform the integral into a form that involves only the variable u. If necessary, introduce a *constant* factor k into the integrand and compensate by multiplying the integral by $1/k$. If any part of the resulting integrand contains the variable x, use a different substitution in guideline (1).

4 Evaluate the integral obtained in guideline (3), obtaining an antiderivative involving u.

5 Replace u in the antiderivative obtained in guideline (4) by $g(x)$. The final result should contain only the variable x.

The next examples illustrate the use of the guidelines.

EXAMPLE ▪ 3   Evaluate $\int (2x^3 + 1)^7 x^2 \, dx$.

SOLUTION   If an integrand involves an expression raised to a power, such as $(2x^3 + 1)^7$, we substitute u for the expression. Thus we let

$$u = 2x^3 + 1, \quad du = 6x^2 \, dx.$$

Comparing $du = 6x^2 \, dx$ with $x^2 \, dx$ in the integral suggests that we introduce the factor 6 into the integrand. Doing so and compensating by multiplying the integral by $\frac{1}{6}$, we obtain the following:

$$\int (2x^3 + 1)^7 x^2 \, dx = \frac{1}{6} \int (2x^3 + 1)^7 6x^2 \, dx$$

$$= \frac{1}{6} \int u^7 \, du$$

$$= \frac{1}{6} \left(\frac{u^8}{8} \right) + C$$

$$= \frac{1}{48} (2x^3 + 1)^8 + C$$

A substitution in an indefinite integral can sometimes be made in several different ways. To illustrate, another method for evaluating the integral in Example 3 is to consider

$$u = 2x^3 + 1, \quad du = 6x^2 \, dx, \quad \frac{1}{6} du = x^2 \, dx.$$

We then substitute $\frac{1}{6} du$ for $x^2 \, dx$,

$$\int (2x^3 + 1)^7 x^2 \, dx = \int u^7 \frac{1}{6} \, du = \frac{1}{6} \int u^7 \, du,$$

and integrate as before.

EXAMPLE ▪ 4   Evaluate $\int x\sqrt[3]{7 - 6x^2}\, dx$.

SOLUTION   Note that the integrand contains the term $x\, dx$. If the factor x were missing or if x were raised to a higher power, the problem would be more complicated. For integrands that involve a radical, we often substitute for the expression under the radical sign. Thus we let

$$u = 7 - 6x^2, \qquad du = -12x\, dx.$$

Next, we introduce the factor -12 into the integrand, compensate by multiplying the integral by $-\frac{1}{12}$, and proceed as follows:

$$\int x\sqrt[3]{7 - 6x^2}\, dx = -\tfrac{1}{12}\int \sqrt[3]{7 - 6x^2}(-12)x\, dx$$

$$= -\tfrac{1}{12}\int \sqrt[3]{u}\, du = -\tfrac{1}{12}\int u^{1/3}\, du$$

$$= -\frac{1}{12}\left(\frac{u^{4/3}}{4/3}\right) + C = -\frac{1}{16}u^{4/3} + C$$

$$= -\tfrac{1}{16}(7 - 6x^2)^{4/3} + C$$

We could also have written

$$u = 7 - 6x^2, \qquad du = -12x\, dx, \qquad -\tfrac{1}{12}\, du = x\, dx$$

and substituted directly for $x\, dx$. Thus,

$$\int \sqrt[3]{7 - 6x^2}\, x\, dx = \int \sqrt[3]{u}\left(-\tfrac{1}{12}\right) du = -\tfrac{1}{12}\int \sqrt[3]{u}\, du.$$

The remainder of the solution would proceed exactly as before.

EXAMPLE ▪ 5   Evaluate $\int \dfrac{x^2 - 1}{(x^3 - 3x + 1)^6}\, dx$.

SOLUTION   Let

$$u = x^3 - 3x + 1, \qquad du = (3x^2 - 3)\, dx = 3(x^2 - 1)\, dx$$

and proceed as follows:

$$\int \frac{x^2 - 1}{(x^3 - 3x + 1)^6}\, dx = \frac{1}{3}\int \frac{3(x^2 - 1)}{(x^3 - 3x + 1)^6}\, dx$$

$$= \frac{1}{3}\int \frac{1}{u^6}\, du = \frac{1}{3}\int u^{-6}\, du$$

$$= \frac{1}{3}\left(\frac{u^{-5}}{-5}\right) + C = -\frac{1}{15}\left(\frac{1}{u^5}\right) + C$$

$$= -\frac{1}{15}\frac{1}{(x^3 - 3x + 1)^5} + C$$

If the variable of integration is different from x, we can make use of Guidelines (4.8), with an appropriate change of notation. In the next example, t is the original variable of integration; our substitution takes the form $u = g(t)$ for an appropriately chosen function g.

EXAMPLE • 6 Evaluate $\int \cos^3 5t \sin 5t \, dt$.

SOLUTION The form of the integrand suggests that we use the power rule (2) in (4.4) with $\int u^3 \, du = \frac{1}{4}u^4 + C$. Thus we let

$$u = g(t) = \cos 5t, \qquad du = -5 \sin 5t \, dt.$$

The form of du indicates that we should introduce the factor -5 into the integrand, multiply the integral by $-\frac{1}{5}$, and then integrate as follows:

$$\int \cos^3 5t \sin 5t \, dt = -\frac{1}{5} \int \cos^3 5t (-5 \sin 5t) \, dt$$

$$= -\frac{1}{5} \int u^3 \, du$$

$$= -\frac{1}{5} \left(\frac{u^4}{4} \right) + C$$

$$= -\frac{1}{20} \cos^4 5t + C$$

The method of substitution or change of variable is also quite useful in solving differential equations. The next example illustrates an application of differential equations in which we make use of a change of variable.

EXAMPLE • 7 Studies have shown that the rate at which students learn new vocabulary words in a foreign language decreases as the size of the known vocabulary increases. If $W(t)$ is the number of words known after t days and the rate of change of W is modeled by the differential equation

$$W'(t) = \frac{8200}{W(t)} \quad \text{for} \quad 0 \le t \le 365,$$

(a) find W as an explicit function of t, if the student knows 400 words at time $t = 0$

(b) find the number of words known after 1 day and 2 days

SOLUTION

(a) From the differential equation, we have

$$W(t) W'(t) = 8200 \quad \text{for} \quad 0 \le t \le 365.$$

Since the expressions on each side of the equation are identical on an interval $[0, 365]$, the indefinite integrals of the functions they represent will be identical. Thus, we have

$$\int W(t) W'(t) \, dt = \int 8200 \, dt.$$

We make the substitution

$$u = W(t), \qquad du = W'(t)\, dt$$

on the left-hand side of this equation, obtaining

$$\int u\, du = \int 8200\, dt$$

so that

$$\frac{u^2}{2} = 8200t + C.$$

Changing back to our original variables, we have

$$\frac{[W(t)]^2}{2} = 8200t + C.$$

Evaluating at $t = 0$ (with $W(0) = 400$) yields

$$\frac{400^2}{2} = 8200(0) + C = 0 + C = C,$$

so $C = 80{,}000$. Hence,

$$\frac{[W(t)]^2}{2} = 8200t + 80{,}000.$$

This result gives us

$$[W(t)]^2 = 16{,}400t + 160{,}000$$

and

$$W(t) = \sqrt{16{,}400t + 160{,}000} = 20\sqrt{41t + 400}.$$

Thus the number of words $W(t)$ known after t days is $20\sqrt{41t + 400}$.

(b) After 1 day of additional study, the number of words the student knows is given by

$$W(1) = 20\sqrt{41 + 400} = 20\sqrt{441} = (20)(21) = 420.$$

After 2 days, the number of words is

$$W(2) = 20\sqrt{482} \approx 439.$$

In this model of learning, the student masters an additional 20 words on the first day, but only 19 more words on the second day.

EXERCISES 4.2

Exer. 1–8: Evaluate the integral using the given substitution, and express the answer in terms of x.

1 $\displaystyle \int x(2x^2 + 3)^{10}\, dx; \quad u = 2x^2 + 3$

2 $\displaystyle \int \frac{x}{(x^2 + 5)^3}\, dx; \quad u = x^2 + 5$

3 $\displaystyle \int x^2 \sqrt[3]{3x^3 + 7}\, dx; \quad u = 3x^3 + 7$

4 $\displaystyle \int \frac{5x}{\sqrt{x^2 - 3}}\, dx; \quad u = x^2 - 3$

5 $\displaystyle \int \frac{(1 + \sqrt{x})^3}{\sqrt{x}}\, dx; \quad u = 1 + \sqrt{x}$

6 $\int \dfrac{1}{(5x-4)^{10}}\,dx;\quad u = 5x-4$

7 $\int \sqrt{x}\cos\sqrt{x^3}\,dx;\quad u = x^{3/2}$

8 $\int \tan x\,\sec^2 x\,dx;\quad u = \tan x$

Exer. 9–48: Evaluate the integral.

9 $\int \sqrt{3x-2}\,dx$

10 $\int \sqrt[4]{2x+5}\,dx$

11 $\int \sqrt[3]{8t+5}\,dt$

12 $\int \dfrac{1}{\sqrt{4-5t}}\,dt$

13 $\int (3z+1)^4\,dz$

14 $\int (2z^2-3)^5 z\,dz$

15 $\int v^2\sqrt{v^3-1}\,dv$

16 $\int v\sqrt{9-v^2}\,dv$

17 $\int \dfrac{x}{\sqrt[3]{1-2x^2}}\,dx$

18 $\int (3-x^4)^3 x^3\,dx$

19 $\int (s^2+1)^2\,ds$

20 $\int (3-s^3)^2 s\,ds$

21 $\int \dfrac{(\sqrt{x}+3)^4}{\sqrt{x}}\,dx$

22 $\int \left(1+\dfrac{1}{x}\right)^{-3}\left(\dfrac{1}{x^2}\right)dx$

23 $\int \dfrac{t-2}{(t^2-4t+3)^3}\,dt$

24 $\int \dfrac{t^2+t}{(4-3t^2-2t^3)^4}\,dt$

25 $\int 3\sin 4x\,dx$

26 $\int 4\cos\tfrac{1}{2}x\,dx$

27 $\int \cos(4x-3)\,dx$

28 $\int \sin(1+6x)\,dx$

29 $\int v\sin(v^2)\,dv$

30 $\int \dfrac{\cos\sqrt[3]{v}}{\sqrt[3]{v^2}}\,dv$

31 $\int \cos 3x\sqrt[3]{\sin 3x}\,dx$

32 $\int \dfrac{\sin 2x}{\sqrt{1-\cos 2x}}\,dx$

33 $\int (\sin x+\cos x)^2\,dx$ (*Hint:* $\sin 2\theta = 2\sin\theta\cos\theta$.)

34 $\int \dfrac{\sin 4x}{\cos 2x}\,dx$ (*Hint:* $\sin 2\theta = 2\sin\theta\cos\theta$.)

35 $\int \sin x(1+\cos x)^2\,dx$

36 $\int \sin^3 x\cos x\,dx$

37 $\int \dfrac{\sin x}{\cos^4 x}\,dx$

38 $\int \sin 2x\sec^5 2x\,dx$

39 $\int \dfrac{\cos t}{(1-\sin t)^2}\,dt$

40 $\int (2+5\cos t)^3\sin t\,dt$

41 $\int \sec^2(3x-4)\,dx$

42 $\int \dfrac{\csc 2x}{\sin 2x}\,dx$

43 $\int \sec^2 3x\tan 3x\,dx$

44 $\int \dfrac{1}{\tan 4x\sin 4x}\,dx$

45 $\int \dfrac{1}{\sin^2 5x}\,dx$

46 $\int \dfrac{x}{\cos^2(x^2)}\,dx$

47 $\int x\cot(x^2)\csc(x^2)\,dx$

48 $\int \sec\left(\dfrac{x}{3}\right)\tan\left(\dfrac{x}{3}\right)dx$

Exer. 49–52: Solve the differential equation subject to the given conditions.

49 $f'(x)=\sqrt[3]{3x+2};\quad f(2)=9$

50 $\dfrac{dy}{dx}=x\sqrt{x^2+5};\quad y=12$ if $x=2$

51 $f''(x)=16\cos 2x-3\sin x;\quad f(0)=-2;\ f'(0)=4$

52 $f''(x)=4\sin 2x+16\cos 4x;\quad f(0)=6;\quad f'(0)=1$

Exer. 53–56: Evaluate the integral by (a) the method of substitution and (b) expanding the integrand. In what way do the constants of integration differ?

53 $\int (x+4)^2\,dx$

54 $\int (x^2+4)^2 x\,dx$

55 $\int \dfrac{(\sqrt{x}+3)^2}{\sqrt{x}}\,dx$

56 $\int \left(1+\dfrac{1}{x}\right)^2\dfrac{1}{x^2}\,dx$

57 A charged particle is moving on a coordinate line in a magnetic field such that its velocity (in centimeters per second) at time t is given by $v(t)=\tfrac{1}{2}\sin(3t-\tfrac{1}{4}\pi)$. Show that the motion is simple harmonic (see page 321).

58 The acceleration of a particle that is moving on a coordinate line is given by $a(t)=k\cos(\omega t+\phi)$ for constants k,ω, and ϕ and time t (in seconds). Show that the motion is simple harmonic (see page 321).

59 A reservoir supplies water to a community. In summer, the demand A for water (in cubic feet per day) changes according to the formula
$$dA/dt = 4000 + 2000\sin(\tfrac{1}{90}\pi t)$$
for time t (in days), with $t=0$ corresponding to the beginning of summer. Estimate the total amount of water consumption during 90 days of summer.

60 The pumping action of the heart consists of the systolic phase, in which blood rushes from the left ventricle into the aorta, and the diastolic phase, during which the heart muscle relaxes. The graph shown in the figure on the following page is sometimes used to model one complete cycle of the process. For a particular individual, the systolic phase lasts $\tfrac{1}{4}$ sec and has a maximum flow rate dV/dt of 8 L/min, where V is the volume of blood in the heart at time t.

(a) Show that $dV/dt = 8\sin(240\pi t)$ L/min.

(b) Estimate the total amount of blood pumped into the aorta during a systolic phase.

Exercise 60

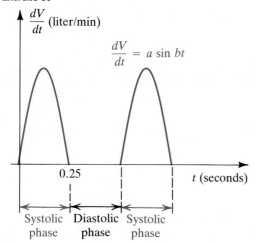

$\frac{dV}{dt}$ (liter/min)

$\frac{dV}{dt} = a \sin bt$

0.25

t (seconds)

Systolic   Diastolic   Systolic
phase   phase   phase

61 The rhythmic process of breathing consists of alternating periods of inhaling and exhaling. For an adult, one complete cycle normally takes place every 5 sec. If V denotes the volume of air in the lungs at time t, then dV/dt is the flow rate.

(a) If the maximum flow rate is 0.6 L/sec, find a formula $dV/dt = a \sin bt$ that fits the given information.

(b) Use part (a) to estimate the amount of air inhaled during one cycle.

62 Many animal populations fluctuate over 10-yr cycles. Suppose that the rate of growth of a rabbit population is given by $dN/dt = 1000 \cos(\frac{1}{5}\pi t)$ rabbits/yr, where N denotes the number in the population at time t (in years) and $t = 0$ corresponds to the beginning of a cycle. If the population after 5 yr is estimated to be 3000 rabbits, find a formula for N at time t and estimate the maximum population.

63 Show, by evaluating in three different ways, that

$$\int \sin x \cos x \, dx = \tfrac{1}{2}\sin^2 x + C$$
$$= -\tfrac{1}{2}\cos^2 x + D$$
$$= -\tfrac{1}{4}\cos 2x + E.$$

How can all three answers be correct?

4.3 SUMMATION NOTATION AND AREA

In this section, we lay the foundation for the definition of the *definite integral*. At the outset, it is virtually impossible to see any connection between definite integrals and indefinite integrals. In Section 4.6, however, we show that there is a very close relationship: *Indefinite integrals can be used to evaluate definite integrals.*

In our development of the definite integral, we shall employ sums of many numbers. To express such sums compactly, it is convenient to use **summation notation**. Given a collection of numbers $\{a_1, a_2, \ldots, a_n\}$, the symbol $\sum_{k=1}^{n} a_k$ represents their sum as follows.

Summation Notation 4.9

$$\sum_{k=1}^{n} a_k = a_1 + a_2 + a_3 + \cdots + a_n$$

The Greek capital letter Σ (sigma) indicates a sum, and a_k represents the kth term of the sum. The letter k is the **index of summation**, or the **summation variable**, and assumes successive integer values. The integers 1 and n indicate the extreme values of the summation variable.

EXAMPLE ▪ 1   Evaluate $\displaystyle\sum_{k=1}^{4} k^2(k-3)$.

SOLUTION   Comparing the sum with (4.9), we have $a_k = k^2(k-3)$ and $n = 4$. To find the sum, we substitute 1, 2, 3, and 4 for k and add the resulting terms. Thus,

$$\sum_{k=1}^{4} k^2(k-3) = 1^2(1-3) + 2^2(2-3) + 3^2(3-3) + 4^2(4-3)$$

$$= (-2) + (-4) + 0 + 16 = 10.$$

Letters other than k can be used for the summation variable. To illustrate,

$$\sum_{k=1}^{4} k^2(k-3) = \sum_{i=1}^{4} i^2(i-3) = \sum_{j=1}^{4} j^2(j-3) = 10.$$

If $a_k = c$ for every k, then

$$\sum_{k=1}^{2} a_k = a_1 + a_2 = c + c = 2c = \sum_{k=1}^{2} c,$$

$$\sum_{k=1}^{3} a_k = a_1 + a_2 + a_3 = c + c + c = 3c = \sum_{k=1}^{3} c.$$

In general, the following result is true for every positive integer n.

Theorem 4.10

$$\sum_{k=1}^{n} c = nc$$

The domain of the summation variable does not have to begin at 1. For example,

$$\sum_{k=4}^{8} a_k = a_4 + a_5 + a_6 + a_7 + a_8.$$

EXAMPLE ▪ 2   Evaluate $\displaystyle\sum_{k=0}^{3} \frac{2^k}{(k+1)}$.

SOLUTION

$$\sum_{k=0}^{3} \frac{2^k}{(k+1)} = \frac{2^0}{(0+1)} + \frac{2^1}{(1+1)} + \frac{2^2}{(2+1)} + \frac{2^3}{(3+1)}$$

$$= 1 + 1 + \tfrac{4}{3} + 2 = \tfrac{16}{3}$$

The next theorem states some elementary properties of summation.

Theorem 4.11

If n is any positive integer and $\{a_1, a_2, \ldots, a_n\}$ and $\{b_1, b_2, \ldots, b_n\}$ are sets of real numbers, then

(i) $\displaystyle\sum_{k=1}^{n} (a_k + b_k) = \sum_{k=1}^{n} a_k + \sum_{k=1}^{n} b_k$

(ii) $\displaystyle\sum_{k=1}^{n} ca_k = c\left(\sum_{k=1}^{n} a_k\right),$ for every real number c

(iii) $\displaystyle\sum_{k=1}^{n} (a_k - b_k) = \sum_{k=1}^{n} a_k - \sum_{k=1}^{n} b_k$

PROOF To prove (i), we begin with

$$\sum_{k=1}^{n} (a_k + b_k) = (a_1 + b_1) + (a_2 + b_2) + (a_3 + b_3) + \cdots + (a_n + b_n).$$

Rearranging terms on the right, we obtain

$$\sum_{k=1}^{n} (a_k + b_k) = (a_1 + a_2 + a_3 + \cdots + a_n) + (b_1 + b_2 + b_3 + \cdots + b_n)$$

$$= \sum_{k=1}^{n} a_k + \sum_{k=1}^{n} b_k.$$

For (ii),

$$\sum_{k=1}^{n} (ca_k) = ca_1 + ca_2 + ca_3 + \cdots + ca_n$$

$$= c(a_1 + a_2 + a_3 + \cdots + a_n) = c\left(\sum_{k=1}^{n} a_k\right).$$

To prove (iii), we write $a_k - b_k = a_k + (-1)b_k$ and use (i) and (ii). ∎

The formulas in the following theorem will be useful later in this section. They may be proved by mathematical induction.

Theorem 4.12

(i) $\displaystyle\sum_{k=1}^{n} k = 1 + 2 + \cdots + n = \frac{n(n+1)}{2}$

(ii) $\displaystyle\sum_{k=1}^{n} k^2 = 1^2 + 2^2 + \cdots + n^2 = \frac{n(n+1)(2n+1)}{6}$

(iii) $\displaystyle\sum_{k=1}^{n} k^3 = 1^3 + 2^3 + \cdots + n^3 = \left[\frac{n(n+1)}{2}\right]^2$

EXAMPLE • 3   Evaluate $\displaystyle\sum_{k=1}^{100} k$ and $\displaystyle\sum_{k=1}^{20} k^2$.

SOLUTION   Using (i) and (ii) of Theorem (4.12), we obtain

$$\sum_{k=1}^{100} k = 1 + 2 + \cdots + 100 = \frac{100(101)}{2} = 5050$$

and

$$\sum_{k=1}^{20} k^2 = 1^2 + 2^2 + \cdots + 20^2 = \frac{20(21)(41)}{6} = 2870.$$

EXAMPLE • 4   Express $\displaystyle\sum_{k=1}^{n}(k^2 - 4k + 3)$ in terms of n.

SOLUTION   We use Theorems (4.11), (4.12), and (4.10):

$$\sum_{k=1}^{n}(k^2 - 4k + 3) = \sum_{k=1}^{n} k^2 - 4\sum_{k=1}^{n} k + \sum_{k=1}^{n} 3$$

$$= \frac{n(n + 1)(2n + 1)}{6} - 4\frac{n(n + 1)}{2} + 3n$$

$$= \tfrac{1}{3}n^3 - \tfrac{3}{2}n^2 + \tfrac{7}{6}n$$

We will be working quite extensively with sums of the form $\sum_{k=1}^{n} f(k)$ in our examination of areas of planar regions and in our study of the definite integral in this chapter. The definition of the definite integral (to be given in Section 4.4) is closely related to the areas of certain regions in a coordinate plane. We can easily calculate the area if the region is bounded by lines. For example, the area of a rectangle is the product of its length and width. The area of a triangle is one-half the product of an altitude and the corresponding base. The area of any polygon can be found by subdividing it into triangles.

In order to find areas of regions whose boundaries involve graphs of functions, however, we utilize a limiting process and then use methods of calculus. In particular, let us consider a region R in a coordinate plane, bounded by the vertical lines $x = a$ and $x = b$, by the x-axis, and by the graph of a function f that is continuous and nonnegative on the closed interval $[a, b]$. A region of this type is illustrated in Figure 4.2. Since $f(x) \geq 0$ for every x in $[a, b]$, no part of the graph lies below the x-axis. For convenience, we shall refer to R as **the region under the graph of f from a to b**. We wish to define the area A of R.

To arrive at a satisfactory definition of A, we shall consider many rectangles of equal width such that each rectangle lies completely under the graph of f and intersects the graph in at least one point, as illustrated in Figure 4.3. The boundary of the region formed by the totality of these

Figure 4.2 Region under the graph of f

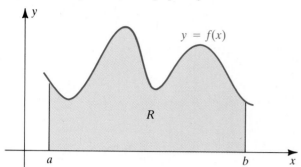

Figure 4.3 An inscribed rectangular polygon

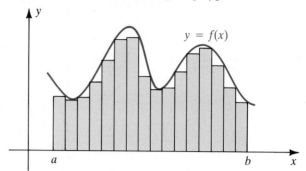

rectangles is called an **inscribed rectangular polygon**. We shall use the following notation:

$$A_{IP} = \text{area of an inscribed rectangular polygon}$$

If the width of the rectangles in Figure 4.3 is small, then it appears that

$$A_{IP} \approx A.$$

This result suggests that we let the width of the rectangles approach zero and define A as a limiting value of the areas A_{IP} of the corresponding inscribed rectangular polygons. The notation discussed next will allow us to carry out this procedure rigorously.

If n is any positive integer, we divide the interval $[a, b]$ into n subintervals, all having the same length $\Delta x = (b - a)/n$. We choose the numbers $x_0, x_1, x_2, \ldots, x_n$, and let $a = x_0$, $b = x_n$, and

$$x_k - x_{k-1} = \frac{b - a}{n} = \Delta x$$

for $k = 1, 2, \ldots, n$, as indicated in Figure 4.4 on the following page. Note that

$$x_0 = a, \quad x_1 = a + \Delta x, \quad x_2 = a + 2\Delta x, \quad x_3 = a + 3\Delta x, \ldots$$
$$x_k = a + k\Delta x, \quad \ldots, \quad x_n = a + n\Delta x = b.$$

The function f is continuous on each subinterval $[x_{k-1}, x_k]$, and hence, by the extreme value theorem (3.3), f takes on a minimum value at some number u_k in $[x_{k-1}, x_k]$. For each k, let us construct a rectangle of width $\Delta x = x_k - x_{k-1}$ and height equal to the minimum distance $f(u_k)$ from the x-axis to the graph of f (see Figure 4.4). The area of the kth rectangle is $f(u_k)\Delta x$. The area A_{IP} of the resulting inscribed rectangular polygon is the sum of the areas of the n rectangles — that is,

$$A_{IP} = f(u_1)\Delta x + f(u_2)\Delta x + \cdots + f(u_n)\Delta x.$$

Using summation notation, we may write

$$A_{IP} = \sum_{k=1}^{n} f(u_k)\Delta x,$$

where $f(u_k)$ is the minimum value of f on $[x_{k-1}, x_k]$.

Figure 4.4

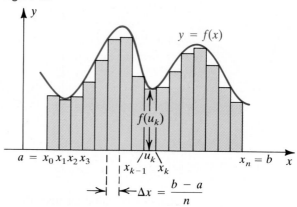

If n is very large, or, equivalently, if Δx is very small, then the sum A_{IP} of the rectangular areas should approximate the area of the region R. Intuitively, we know that if there exists a number A such that $\sum_{k=1}^{n} f(u_k)\Delta x$ gets closer to A as Δx gets closer to 0 (but $\Delta x \neq 0$), we can call A the **area** of R and write

$$A = \lim_{\Delta x \to 0} A_{\text{IP}} = \lim_{\Delta x \to 0} \sum_{k=1}^{n} f(u_k)\Delta x.$$

The meaning of this *limit of sums* is not the same as that of the limit of a function, introduced in Chapter 1. To eliminate the word *closer* and arrive at a satisfactory definition of A, let us take a slightly different point of view. If A denotes the area of the region R, then the difference

$$A - \sum_{k=1}^{n} f(u_k)\Delta x$$

is the area of the portion in Figure 4.4 that lies *under* the graph of f and *over* the inscribed rectangular polygon. This number may be regarded as the error in using the area of the inscribed rectangular polygon to approximate A. We should be able to make this error as small as desired by choosing the width Δx of the rectangles sufficiently small. This procedure is the motivation for the following definition of the area A of R. The notation is the same as that used in the preceding discussion.

Definition 4.13

Let f be continuous and nonnegative on $[a, b]$. Let A be a real number, and let $f(u_k)$ be the minimum value of f on $[x_{k-1}, x_k]$. The notation

$$A = \lim_{\Delta x \to 0} \sum_{k=1}^{n} f(u_k)\Delta x$$

means that for every $\epsilon > 0$, there is a $\delta > 0$ such that if $0 < \Delta x < \delta$, then

$$A - \sum_{k=1}^{n} f(u_k)\Delta x < \epsilon.$$

If A is the indicated limit and we let $\epsilon = 10^{-9}$, then Definition (4.13) states that by using rectangles of sufficiently small width Δx, we can make the difference between A and the area of the inscribed polygon less than one-billionth of a square unit. Similarly, if $\epsilon = 10^{-12}$, we can make this difference less than one-trillionth of a square unit. In general, the difference can be made less than *any* preassigned ϵ.

If f is continuous on $[a, b]$, it is shown in more advanced texts that a number A satisfying Definition (4.13) actually exists. We shall call A **the area under the graph of f from a to b.**

The area A may also be obtained by means of **circumscribed rectangular polygons** of the type illustrated in Figure 4.5. In this case, we select the number v_k in each interval $[x_{k-1}, x_k]$ such that $f(v_k)$ is the *maximum* value of f on $[x_{k-1}, x_k]$.

Figure 4.5   A circumscribed rectangular polygon

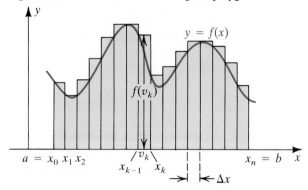

Let

$$A_{\text{CP}} = \text{area of a circumscribed rectangular polygon.}$$

Using summation notation, we have

$$A_{\text{CP}} = \sum_{k=1}^{n} f(v_k)\Delta x,$$

where $f(v_k)$ is the maximum value of f on $[x_{k-1}, x_k]$. Note that

$$\sum_{k=1}^{n} f(u_k)\Delta x \le A \le \sum_{k=1}^{n} f(v_k)\Delta x.$$

The limit of A_{CP} as $\Delta x \to 0$ is defined as in (4.13). The only change is that we use

$$\sum_{k=1}^{n} f(v_k)\Delta x - A < \epsilon,$$

since we want this difference to be nonnegative. It can be proved that the same number A is obtained using either inscribed or circumscribed rectangles.

The next example illustrates how close the areas of the inscribed and circumscribed rectangles become if we use a small value for Δx.

EXAMPLE ■ 5 Let $f(x) = \sqrt{x}$, and let R be the region under the graph of f from 1 to 5. Approximate the area A of R using

(a) an inscribed rectangular polygon with $\Delta x = 0.1$

(b) a circumscribed rectangular polygon with $\Delta x = 0.1$

SOLUTION With $\Delta x = 0.1 = 1/10$, the interval $[1, 5]$ is divided into 40 subintervals. Since $x_0 = 1$ and $x_n = 5$ with $n = 40$, we have $x_k = 1 + k\,\Delta x = 1 + k/10$.

(a) Since f is increasing on the interval $[1, 5]$, we obtain inscribed rectangles by selecting $u_k = x_{k-1}$, the left-hand endpoint of each subinterval $[x_{k-1}, x_k]$. Thus,

$$u_k = 1 + \frac{k-1}{10} \quad \text{and} \quad f(u_k) = \sqrt{1 + \frac{k-1}{10}}.$$

Using a computational device to sum the 40 terms, we find that the inscribed rectangular polygon has area

$$A_{\text{IP}} = \sum_{k=1}^{40} \sqrt{1 + \frac{k-1}{10}} \left(\frac{1}{10}\right) \approx 6.72485958283.$$

(b) We obtain circumscribed rectangles over $[1, 5]$ by selecting $u_k = x_k$, the right-hand endpoint of each subinterval $[x_{k-1}, x_k]$. Hence, the area of the circumscribed rectangular polygon is

$$A_{\text{CP}} = \sum_{k=1}^{40} \sqrt{1 + \frac{k}{10}} \left(\frac{1}{10}\right) \approx 6.84846638058.$$

Thus, we can conclude that the area A satisfies

$$6.72485958283 < A < 6.84846638058.$$

For larger values of Δx, there are fewer rectangles, and we may be able to compute the areas of the inscribed and circumscribed polygons by hand. The next example illustrates this approach and also shows that there may be a considerable gap between the numbers A_{IP} and A_{CP}.

EXAMPLE ■ 6 One side of a farmer's field is bordered by a straight stretch of highway. The opposite side is bordered by a river whose path traces a curve that is modeled by the function $f(x) = 16 - x^2$. The farmer measures the side of the field along the highway, placing markers at every $\frac{1}{2}$ km for a total of 3 km, as shown in Figure 4.6. Approximate the area A of the field using

(a) an inscribed rectangular polygon with $\Delta x = \frac{1}{2}$

(b) a circumscribed rectangular polygon with $\Delta x = \frac{1}{2}$

Figure 4.6

$\longleftarrow$ 3 km $\longrightarrow$

Figure 4.7

(a)

(b)

Figure 4.8

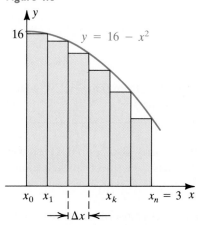

SOLUTION

(a) We model the farmer's field by graphing the function $f(x) = 16 - x^2$ and considering the area of the region under the graph from 0 to 3 on the x-axis, which is the side of the field measured in units of $\frac{1}{2}$ km. The graph of f and the inscribed rectangular polygon with $\Delta x = \frac{1}{2}$ are sketched in Figure 4.7(a) (with different scales on the x- and y-axes). Note that f is decreasing on $[0, 3]$, and hence the minimum value $f(u_k)$ on the kth subinterval occurs at the right-hand endpoint of the subinterval. Since there are six rectangles to consider, the formula for A_{IP} is

$$A_{\text{IP}} = \sum_{k=1}^{6} f(u_k)\Delta x$$

$$= f(\tfrac{1}{2}) \cdot \tfrac{1}{2} + f(1) \cdot \tfrac{1}{2} + f(\tfrac{3}{2}) \cdot \tfrac{1}{2} + f(2) \cdot \tfrac{1}{2} + f(\tfrac{5}{2}) \cdot \tfrac{1}{2} + f(3) \cdot \tfrac{1}{2}$$

$$= \tfrac{63}{4} \cdot \tfrac{1}{2} + 15 \cdot \tfrac{1}{2} + \tfrac{55}{4} \cdot \tfrac{1}{2} + 12 \cdot \tfrac{1}{2} + \tfrac{39}{4} \cdot \tfrac{1}{2} + 7 \cdot \tfrac{1}{2}$$

$$= \tfrac{293}{8} = 36.625.$$

Using inscribed rectangles, we find that the area of the field is approximately 36.6 km^2.

(b) The graph of f and the circumscribed rectangular polygon are sketched in Figure 4.7(b). Since f is decreasing on $[0, 3]$, the maximum value $f(v_k)$ occurs at the left-hand endpoint of the kth subinterval. Hence,

$$A_{\text{CP}} = \sum_{k=1}^{6} f(v_k)\Delta x$$

$$= f(0) \cdot \tfrac{1}{2} + f(\tfrac{1}{2}) \cdot \tfrac{1}{2} + f(1) \cdot \tfrac{1}{2} + f(\tfrac{3}{2}) \cdot \tfrac{1}{2} + f(2) \cdot \tfrac{1}{2} + f(\tfrac{5}{2}) \cdot \tfrac{1}{2}$$

$$= 16 \cdot \tfrac{1}{2} + \tfrac{63}{4} \cdot \tfrac{1}{2} + 15 \cdot \tfrac{1}{2} + \tfrac{55}{4} \cdot \tfrac{1}{2} + 12 \cdot \tfrac{1}{2} + \tfrac{39}{4} \cdot \tfrac{1}{2}$$

$$= \tfrac{329}{8} = 41.125.$$

Using circumscribed rectangles, we find that the area of the field is approximately 41.1 km^2.

It follows that $36.625 < A < 41.125$. In the next example, we prove that $A = 39$.

EXAMPLE ■ 7 Referring to Example 6, determine the area of the farmer's field, which is the area of the region under the graph of f from 0 to 3.

SOLUTION The graph of $f(x) = 16 - x^2$ and the inscribed rectangular polygon in Figure 4.7(a) is resketched in Figure 4.8. If the interval $[0, 3]$ is divided into n equal subintervals, then the length Δx of each subinterval is $3/n$. Employing the notation used in Figure 4.4, with $a = 0$ and $b = 3$, we have

$$x_0 = 0, \quad x_1 = \Delta x, \quad x_2 = 2\Delta x, \quad \ldots, \quad x_k = k\Delta x, \quad \ldots, \quad x_n = n\Delta x = 3.$$

Since $\Delta x = 3/n$,

$$x_k = k\Delta x = k\frac{3}{n} = \frac{3k}{n}.$$

Since f is decreasing on $[0, 3]$, the number u_k in $[x_{k-1}, x_k]$ at which f takes on its minimum value is always the right-hand endpoint x_k of the subinterval; that is, $u_k = x_k = 3k/n$. Thus,

$$f(u_k) = f\left(\frac{3k}{n}\right) = 16 - \left(\frac{3k}{n}\right)^2 = 16 - \frac{9k^2}{n^2},$$

and the summation in Definition (4.13) is

$$\sum_{k=1}^{n} f(u_k)\Delta x = \sum_{k=1}^{n}\left[\left(16 - \frac{9k^2}{n^2}\right)\cdot\frac{3}{n}\right]$$

$$= \frac{3}{n}\sum_{k=1}^{n}\left(16 - \frac{9k^2}{n^2}\right),$$

where the last equality follows from (ii) of Theorem (4.11). (Note that $3/n$ does not contain the summation variable k.) We next use Theorems (4.11), (4.10), and (4.12) to obtain

$$\sum_{k=1}^{n} f(u_k)\Delta x = \frac{3}{n}\left(\sum_{k=1}^{n} 16 - \frac{9}{n^2}\sum_{k=1}^{n} k^2\right)$$

$$= \frac{3}{n}\left[n\cdot 16 - \frac{9}{n^2}\frac{n(n+1)(2n+1)}{6}\right]$$

$$= 48 - \frac{9}{2}\frac{(n+1)(2n+1)}{n^2}.$$

To find the area of the region, we let Δx approach 0. Since $\Delta x = 3/n$, we can accomplish this by letting n increase without bound. Although our discussion of limits involving infinity in Section 1.4 was concerned with a real variable x, a similar discussion can be given if the variable is an integer n. Assuming that it is and that we can replace $\Delta x \to 0$ by $n \to \infty$, we obtain

$$\lim_{\Delta x \to 0}\sum_{k=1}^{n} f(u_k)\Delta x = \lim_{n \to \infty}\left[48 - \frac{9}{2}\frac{(n+1)(2n+1)}{n^2}\right]$$

$$= 48 - \tfrac{9}{2}\cdot 2 = 39.$$

Thus the area of the region under the graph of f from 0 to 3 is 39, which means that the area of the farmer's field is 39 km^2.

The area of a region under the graph of f may also be found by using circumscribed rectangular polygons. In this case, we select, in each subinterval $[x_{k-1}, x_k]$, the number $v_k = (k - 1)(3/n)$ at which f takes on its maximum value.

The next example illustrates the use of circumscribed rectangles in finding an area.

EXAMPLE • 8 If $f(x) = x^3$, find the area under the graph of f from 0 to b for any $b > 0$.

SOLUTION Subdividing the interval $[0, b]$ into n equal parts (see Figure 4.9), we obtain a circumscribed rectangular polygon such that $\Delta x = b/n$ and $x_k = k\,\Delta x$.

Figure 4.9

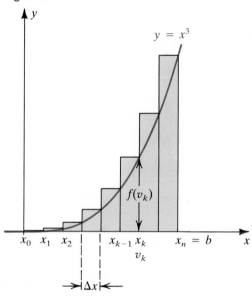

Since f is an increasing function, the maximum value $f(v_k)$ in the interval $[x_{k-1}, x_k]$ occurs at the right-hand endpoint—that is,

$$v_k = x_k = k\Delta x = k\frac{b}{n} = \frac{bk}{n}.$$

The sum of the areas of the circumscribed rectangles is

$$\sum_{k=1}^{n} f(v_k)\Delta x = \sum_{k=1}^{n}\left[\left(\frac{bk}{n}\right)^3 \cdot \frac{b}{n}\right] = \sum_{k=1}^{n}\frac{b^4}{n^4}k^3$$

$$= \frac{b^4}{n^4}\sum_{k=1}^{n}k^3 = \frac{b^4}{n^4}\left[\frac{n(n+1)}{2}\right]^2$$

$$= \frac{b^4}{4}\cdot\frac{n^2(n+1)^2}{n^4},$$

where we have used Theorem (4.12)(iii). If we let Δx approach 0, then n increases without bound and the expression involving n approaches 1. It follows that the area under the graph is

$$\lim_{\Delta x \to 0}\sum_{k=1}^{n} f(v_k)\Delta x = \frac{b^4}{4}.$$

The last example shows how we can use knowledge of the area under the graph of f on the interval $[0, b]$ for any $b > 0$ to find the area under the graph for a subinterval of $[0, b]$.

EXAMPLE ■ 9 If $f(x) = x^3$, find the area A of the region under the graph of f from $\frac{1}{2}$ to 2.

SOLUTION The region is sketched in Figure 4.10. If we let

$$A_1 = \text{ area under the graph of } f \text{ from 0 to } \tfrac{1}{2}$$

and

$$A_2 = \text{ area under the graph of } f \text{ from 0 to 2,}$$

the area A can be found by subtracting A_1 from A_2:

$$A = A_2 - A_1$$

In Example 8, we found that the area under the graph of $y = x^3$ from 0 to b is $\frac{1}{4}b^4$. Hence, using $b = \frac{1}{2}$ for A_1 and $b = 2$ for A_2 yields

$$A = \frac{2^4}{4} - \frac{(\frac{1}{2})^4}{4} = 4 - \frac{1}{64} \approx 3.98.$$

Figure 4.10

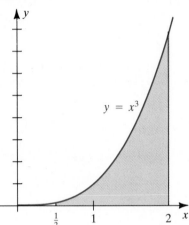

$y = x^3$

EXERCISES 4.3

Exer. 1–8: Evaluate the sum.

1 $\displaystyle\sum_{j=1}^{4} (j^2 + 1)$

2 $\displaystyle\sum_{j=1}^{4} (2^j + 1)$

3 $\displaystyle\sum_{k=0}^{5} k(k-1)$

4 $\displaystyle\sum_{k=0}^{4} (k-2)(k-3)$

5 $\displaystyle\sum_{n=1}^{10} [1 + (-1)^n]$

6 $\displaystyle\sum_{n=1}^{4} (-1)^n \left(\frac{1}{n}\right)$

7 $\displaystyle\sum_{i=1}^{50} 10$

8 $\displaystyle\sum_{k=1}^{1000} 2$

Exer. 9–12: Express the sum in terms of n (see Example 4).

9 $\displaystyle\sum_{k=1}^{n} (k^2 + 3k + 5)$

10 $\displaystyle\sum_{k=1}^{n} (3k^2 - 2k + 1)$

11 $\displaystyle\sum_{k=1}^{n} (k^3 + 2k^2 - k + 4)$

12 $\displaystyle\sum_{k=1}^{n} (3k^3 + k)$

Exer. 13–18: Express in summation notation.

13 $1 + 5 + 9 + 13 + 17$

14 $2 + 5 + 8 + 11 + 14$

15 $\frac{1}{2} + \frac{2}{5} + \frac{3}{8} + \frac{4}{11}$

16 $\frac{1}{4} + \frac{2}{9} + \frac{3}{14} + \frac{4}{19}$

17 $1 - \dfrac{x^2}{2} + \dfrac{x^4}{4} - \dfrac{x^6}{6} + \cdots + (-1)^n \dfrac{x^{2n}}{2n}$

18 $1 + x + \dfrac{x^2}{2} + \dfrac{x^3}{3} + \cdots + \dfrac{x^n}{n}$

c **Exer. 19–26: Approximate the sum using a calculator or a computer. Write a short program or use a built-in summation procedure in your calculator or CAS.**

19 $\displaystyle\sum_{k=1}^{20} \frac{2^k}{k}$

20 $\displaystyle\sum_{k=1}^{15} \frac{3^k}{k}$

21 $\displaystyle\sum_{k=1}^{1000} \frac{1}{k}$

22 $\displaystyle\sum_{k=1}^{50} \frac{1}{k^2}$

23 $\displaystyle\sum_{k=1}^{40} \frac{\sin(k/40)}{k}$

24 $\displaystyle\sum_{k=1}^{50} \frac{\cos[-1 + (k/25)]}{k}$

25 $\displaystyle\sum_{k=1}^{40} k^4$

26 $\displaystyle\sum_{k=1}^{30} k^5$

Exer. 27–32: Let A be the area under the graph of the given function f from a to b. Approximate A by dividing $[a, b]$ into subintervals of equal length Δx and using (a) A_{IP} and (b) A_{CP}.

27 $f(x) = 3 - x$; $a = -2$, $b = 2$; $\Delta x = 1$

28 $f(x) = x + 2$; $a = -1$, $b = 4$; $\Delta x = 1$

29 $f(x) = x^2 + 1$; $a = 1$, $b = 3$; $\Delta x = \frac{1}{2}$

30 $f(x) = 4 - x^2$; $a = 0$, $b = 2$; $\Delta x = \frac{1}{2}$

$\boxed{\text{c}}$ **31** $f(x) = \sqrt{\sin x}$; $a = 0$, $b = 1.5$; $\Delta x = 0.15$

$\boxed{\text{c}}$ **32** $f(x) = \dfrac{1}{\sqrt{x^3 + 1}}$; $a = 0$, $b = 3$; $\Delta x = 0.3$

Exer. 33–38: Refer to Examples 7 and 8. Find the area under the graph of the given function f from 0 to b using (a) inscribed rectangles and (b) circumscribed rectangles.

33 $f(x) = 2x + 3$; $b = 4$

34 $f(x) = 8 - 3x$; $b = 2$

35 $f(x) = 9 - x^2$; $b = 3$

36 $f(x) = x^2$; $b = 5$

37 $f(x) = x^3 + 1$; $b = 2$

38 $f(x) = 4x + x^3$; $b = 2$

Exer. 39–40: Refer to Example 8. Find the area under the graph of f corresponding to the interval (a) [1, 3] and (b) $[a, b]$.

39 $f(x) = x^3$

40 $f(x) = x^3 + 2$

4.4 THE DEFINITE INTEGRAL

Our objective in this section is a careful definition, using Riemann sums, of the definite integral of a function on a closed interval. We examine the concept of an *integrable function* and discuss the relationship between continuous functions and integrable functions.

In Section 4.3, we defined the area under the graph of a function f from a to b as a limit of the form

$$\lim_{\Delta x \to 0} \sum_{k=1}^{n} f(w_k)\Delta x.$$

In our discussion, we restricted f and Δx as follows:

1. The function f is continuous on the closed interval $[a, b]$.

2. $f(x)$ is nonnegative for every x in $[a, b]$.

3. All the subintervals $[x_{k-1}, x_k]$ have the same length Δx.

4. The number w_k is chosen such that $f(w_k)$ is always the minimum (or maximum) value of f on $[x_{k-1}, x_k]$.

There are many applications involving this type of limit in which one or more of these conditions is not satisfied. Thus it is desirable to allow the following changes in (1)–(4):

1′ The function f may be discontinuous at some number in $[a, b]$.

2′ $f(x)$ may be negative for some x in $[a, b]$.

3′ The lengths of the subintervals $[x_{k-1}, x_k]$ may be different.

4′ The number w_k may be *any* number in $[x_{k-1}, x_k]$.

Note that if $(2')$ is true, part of the graph lies under the x-axis, and therefore the limit is no longer the area under the graph of f.

Let us introduce some new terminology and notation. A **partition** P of a closed interval $[a, b]$ is any decomposition of $[a, b]$ into subintervals of the form

$$[x_0, x_1], \quad [x_1, x_2], \quad [x_2, x_3], \quad \ldots, \quad [x_{n-1}, x_n]$$

for a positive integer n and numbers x_k such that

$$a = x_0 < x_1 < x_2 < x_3 < \cdots < x_{n-1} < x_n = b.$$

The length of the kth subinterval $[x_{k-1}, x_k]$ will be denoted by Δx_k—that is,

$$\Delta x_k = x_k - x_{k-1}.$$

A typical partition of $[a, b]$ is illustrated in Figure 4.11. The largest of the numbers $\Delta x_1, \Delta x_2, \ldots, \Delta x_n$ is the **norm** of the partition P and is denoted by $\|P\|$.

Figure 4.11   A partition of $[a, b]$

EXAMPLE ▪ I   The numbers $\{1, 1.7, 2.2, 3.3, 4.1, 4.5, 5, 6\}$ determine a partition P of the interval $[1, 6]$. Find the lengths $\Delta x_1, \Delta x_2, \ldots, \Delta x_n$ of the subintervals in P and the norm of the partition.

SOLUTION   The lengths Δx_k of the subintervals are found by subtracting successive numbers in P. Thus,

$$\Delta x_1 = 0.7, \ \Delta x_2 = 0.5, \ \Delta x_3 = 1.1, \ \Delta x_4 = 0.8,$$
$$\Delta x_5 = 0.4, \ \Delta x_6 = 0.5, \ \Delta x_7 = 1.0.$$

The norm of P is the largest of these numbers. Hence,

$$\|P\| = \Delta x_3 = 1.1.$$

The following concept, named after the nineteenth-century mathematician G. F. B. Riemann (see *Mathematicians and Their Times*, Chapter 11), is fundamental to the definition of the definite integral.

Definition   4.14

Let f be defined on a closed interval $[a, b]$, and let P be a partition of $[a, b]$. A **Riemann sum** of f (or $f(x)$) for P is any expression R_P of the form

$$R_P = \sum_{k=1}^{n} f(w_k) \Delta x_k,$$

where w_k is in $[x_{k-1}, x_k]$ and $k = 1, 2, \ldots, n$.

Figure 4.12

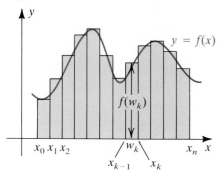

In Definition (4.14), $f(w_k)$ is not necessarily a maximum or minimum value of f on $[x_{k-1}, x_k]$. If we construct a rectangle of length $|f(w_k)|$ and width Δx_k, as illustrated in Figure 4.12, the rectangle may be neither inscribed nor circumscribed. Moreover, since $f(x)$ may be negative, certain terms of the Riemann sum R_P may be negative. Consequently, R_P does not always represent a sum of areas of rectangles.

We may interpret the Riemann sum R_P in (4.14) geometrically, as follows. For each subinterval $[x_{k-1}, x_k]$, we construct a vertical line segment through the point $(w_k, f(w_k))$, thereby obtaining a collection of rectangles. If $f(w_k)$ is positive, the rectangle lies above the x-axis, as illustrated by the lighter rectangles in Figure 4.13, and the product $f(w_k)\Delta x_k$ is the area of this rectangle. If $f(w_k)$ is negative, then the rectangle lies below the x-axis, as illustrated by the darker rectangles in Figure 4.13. In this case, the product $f(w_k)\Delta x_k$ is the *negative* of the area of a rectangle. It follows that R_P is the sum of the areas of the rectangles that lie above the x-axis and the *negatives* of the areas of the rectangles that lie below the x-axis.

Figure 4.13

Figure 4.14

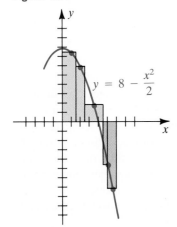

EXAMPLE ▪ 2 Let $f(x) = 8 - \frac{1}{2}x^2$, and let P be the partition of $[0, 6]$ into the five subintervals determined by

$$x_0 = 0, \quad x_1 = 1.5, \quad x_2 = 2.5, \quad x_3 = 4.5, \quad x_4 = 5, \quad x_5 = 6.$$

Find the norm of the partition and the Riemann sum R_P if

$$w_1 = 1, \quad w_2 = 2, \quad w_3 = 3.5, \quad w_4 = 5, \quad w_5 = 5.5.$$

SOLUTION The graph of f is sketched in Figure 4.14, where we have also shown the points that correspond to w_k and the rectangles of lengths $|f(w_k)|$ for $k = 1, 2, 3, 4,$ and 5. Thus,

$$\Delta x_1 = 1.5, \quad \Delta x_2 = 1, \quad \Delta x_3 = 2, \quad \Delta x_4 = 0.5, \quad \Delta x_5 = 1.$$

The norm $\|P\|$ of the partition is Δx_3, or 2.

Using Definition (4.14) with $n = 5$, we have

$$R_P = \sum_{k=1}^{5} f(w_k) \Delta x_k$$
$$= f(w_1) \Delta x_1 + f(w_2) \Delta x_2 + f(w_3) \Delta x_3 + f(w_4) \Delta x_4 + f(w_5) \Delta x_5$$
$$= f(1)(1.5) + f(2)(1) + f(3.5)(2) + f(5)(0.5) + f(5.5)(1)$$
$$= (7.5)(1.5) + (6)(1) + (1.875)(2) + (-4.5)(0.5) + (-7.125)(1)$$
$$= 11.625.$$

We shall not always specify the number n of subintervals in a partition P of $[a, b]$. A Riemann sum (4.14) will then be written

$$R_P = \sum_k f(w_k) \Delta x_k,$$

and we will assume that terms of the form $f(w_k) \Delta x_k$ are to be summed over all subintervals $[x_{k-1}, x_k]$ of the partition P.

Using the same approach as in Definition (4.13), we next define

$$\lim_{\|P\| \to 0} \sum_k f(w_k) \Delta x_k = L$$

for a real number L.

Definition 4.15

Let f be defined on a closed interval $[a, b]$, and let L be a real number. The statement

$$\lim_{\|P\| \to 0} \sum_k f(w_k) \Delta x_k = L$$

means that for every $\epsilon > 0$, there is a $\delta > 0$ such that if P is a partition of $[a, b]$ with $\|P\| < \delta$, then

$$\left| \sum_k f(w_k) \Delta x_k - L \right| < \epsilon$$

for any choice of numbers w_k in the subintervals $[x_{k-1}, x_k]$ of P. The number L is a **limit of (Riemann) sums**.

For every $\delta > 0$, there are infinitely many partitions P of $[a, b]$ with $\|P\| < \delta$. Moreover, for each such partition P, there are infinitely many ways of choosing the number w_k in $[x_{k-1}, x_k]$. Consequently, an infinite number of different Riemann sums may be associated with *each* partition P. However, if the limit L exists, then for any $\epsilon > 0$, every Riemann sum is within ϵ units of L, provided a small enough norm is chosen. Although Definition (4.15) differs from the definition of the limit of a function, we may use a proof similar to that given for the uniqueness theorem in Appendix I to show that if the limit L exists, then it is unique.

We next define the definite integral as a limit of a sum, where w_k and Δx_k have the same meanings as in Definition (4.15).

Definition 4.16

Let f be defined on a closed interval $[a, b]$. The **definite integral of f from a to b,** denoted by $\int_a^b f(x)\,dx$, is

$$\int_a^b f(x)\,dx = \lim_{\|P\|\to 0} \sum_k f(w_k)\Delta x_k,$$

provided the limit exists.

If the limit in Definition (4.16) exists, then f is **integrable** on $[a, b]$, and we say that the definite integral $\int_a^b f(x)\,dx$ **exists**. The process of finding the limit is called **evaluating the integral**. Note that the value of a definite integral is a *real number*, not a family of antiderivatives, as was the case for indefinite integrals.

The integral sign in Definition (4.16), which may be thought of as an elongated letter S (the first letter of the word *sum*), is used to indicate the connection between definite integrals and Riemann sums. The numbers a and b are the **limits of integration**, a being the **lower limit** and b the **upper limit**. In this context, *limit* refers to the smallest or largest number in the interval $[a, b]$ and is not related to definitions of limits given earlier in the text. The expression $f(x)$, which appears to the right of the integral sign, is the *integrand*, as it is with indefinite integrals. The differential symbol dx that follows $f(x)$ may be associated with the increment Δx_k of a Riemann sum of f. This association will be useful in later applications.

EXAMPLE ▪ 3   Express the following limit of sums as a definite integral on the interval $[3, 8]$:

$$\lim_{\|P\|\to 0} \sum_{k=1}^n (5w_k^3 + \sqrt{w_k} - 4\sin w_k)\Delta x_k,$$

where w_k and Δx_k are as in Definition (4.15).

SOLUTION   The given limit of sums has the form stated in Definition (4.16), with

$$f(x) = 5x^3 + \sqrt{x} - 4\sin x.$$

Hence the limit can be expressed as the definite integral

$$\int_3^8 (5x^3 + \sqrt{x} - 4\sin x)\,dx.$$

Letters other than x may be used in the notation for the definite integral. If f is integrable on $[a, b]$, then

$$\int_a^b f(x)\,dx = \int_a^b f(s)\,ds = \int_a^b f(t)\,dt$$

and so on. For this reason, the letter x in Definition (4.16) is called a **dummy variable.**

Whenever an interval $[a, b]$ is used, we assume that $a < b$. Consequently, Definition (4.16) does not take into account the cases in which the lower limit of integration is greater than or equal to the upper limit. The definition may be extended to include the case where the lower limit is greater than the upper limit, as follows.

Definition 4.17

$$\text{If } c > d, \text{ then } \int_c^d f(x)\,dx = -\int_d^c f(x)\,dx.$$

Definition (4.17) may be phrased as follows: *Interchanging the limits of integration changes the sign of the integral.*

The case in which the lower and upper limits of integration are equal is covered by the next definition.

Definition 4.18

$$\text{If } f(a) \text{ exists, then } \int_a^a f(x)\,dx = 0.$$

If f is integrable, then the limit in Definition (4.16) exists no matter how w_k is chosen in the subinterval $[x_{k-1}, x_k]$ and no matter what type of partition is used (provided the norm of the partition gets small). Thus we may choose the partition and the w_k's in a computationally or theoretically convenient manner.

For numerical approximations, it is convenient to select w_k as either the left-hand endpoint x_{k-1}, the right-hand endpoint x_k, or the midpoint of every subinterval. For theoretical considerations, we may want to select each w_k so that $f(w_k)$ is the minimum value for f on $[x_{k-1}, x_k]$ as with inscribed rectangles for area, or so that $f(w_k)$ is the maximum value for f on $[x_{k-1}, x_k]$ as with circumscribed rectangles for area. For both numerical and theoretical work, it is convenient to select partitions in which all the subintervals $[x_{k-1}, x_k]$ are of equal length. Such a partition is called a **regular partition**.

If a regular partition of $[a, b]$ contains n subintervals, then $\Delta x = (b - a)/n$. The requirement that $\|P\| \to 0$ is equivalent to $\Delta x \to 0$ or $n \to \infty$. For a regular partition, Definition (4.16) takes the form

$$\int_a^b f(x)\,dx = \lim_{n \to \infty} \sum_{k=1}^{n} f(w_k)\Delta x.$$

If we specialize Definition (4.16) to **left-hand endpoints for a regular partition,** for example, we have

$$\int_a^b f(x)\,dx = \lim_{n \to \infty} \sum_{k=1}^{n} f(x_{k-1})\Delta x.$$

We can use this result to obtain a numerical approximation to the definite integral. The **left endpoint approximation** L_n, where $w_k = x_{k-1}$ for

all k, has the form

$$\int_a^b f(x)\, dx \approx L_n = \sum_{k=1}^n f(x_{k-1}) \Delta x.$$

Similarly, if we let $w_k = x_k$ for all k, then we obtain a **right endpoint approximation R_n,**

$$\int_a^b f(x)\, dx \approx R_n = \sum_{k=1}^n f(x_k) \Delta x.$$

Often one of these endpoint approximations tends to be a lower estimate (as with inscribed rectangles) for the area under the graph, while the other endpoint approximation gives a higher estimate (as with circumscribed rectangles).

The **midpoint approximation**, which uses the midpoint $(x_{k-1} + x_k)/2$ as w_k for all k, often gives more accurate approximations. It has the form

$$\int_a^b f(x)\, dx \approx M_n = \sum_{k=1}^n f\left(\frac{x_{k-1} + x_k}{2}\right) \Delta x.$$

EXAMPLE ▪ 4   Use midpoint approximations with $n = 10, 20, 40,$ and 80 to estimate $\int_0^\pi \cos(x^2)\, dx$.

SOLUTION   For a regular partition, $\Delta x = (\pi - 0)/n = \pi/n$, so that $x_k = k\pi/n$. Thus,

$$\frac{x_{k-1} + x_k}{2} = \frac{\dfrac{(k-1)\pi}{n} + \dfrac{k\pi}{n}}{2} = \frac{k\pi + k\pi - \pi}{2n} = \frac{2k\pi - \pi}{2n} = \frac{k\pi}{n} - \frac{\pi}{2n}.$$

With $f(x) = \cos(x^2)$, we have

$$f\left(\frac{x_{k-1} + x_k}{2}\right) = \cos\left(\frac{k\pi}{n} - \frac{\pi}{2n}\right)^2.$$

We use a calculator to compute the desired summations for the midpoint approximations:

$$\int_0^\pi \cos(x^2)\, dx \approx M_n = \sum_{k=1}^n \left[\cos\left(\frac{k\pi}{n} - \frac{\pi}{2n}\right)^2\right] \frac{\pi}{n}$$

$$M_{10} = 0.553751825506$$
$$M_{20} = 0.562860413669$$
$$M_{40} = 0.564995257214$$
$$M_{80} = 0.565519579069$$

For an arbitrary partition of the interval $[a, b]$, evaluation of the sum $\sum_{k=1}^n f(w_k) \Delta x_k$ involves the addition of n terms, each of which is the product of two numbers $f(w_k)$ and Δx_k. Thus we have n additions and

n multiplications to perform. For a regular partition, $\Delta x = (b-a)/n$ is independent of k, and we have

$$\sum_{k=1}^{n} f(w_k)\Delta x = \sum_{k=1}^{n}\left[f(w_k)\left(\frac{b-a}{n}\right)\right] = \frac{b-a}{n}\sum_{k=1}^{n} f(w_k).$$

In this case, since evaluation of the sum involves n additions but only 1 multiplication, it is more efficient (computationally) to work with a regular partition than with an arbitrary partition. We shall give more careful consideration to numerical approximation for definite integrals in Section 4.7.

The following theorem is a first application of the special Riemann sums. Many other applications will be discussed in Chapter 5.

Theorem 4.19

If f is integrable and $f(x) \geq 0$ for every x in $[a, b]$, then the area A of the region under the graph of f from a to b is

$$A = \int_{a}^{b} f(x)\,dx.$$

Figure 4.15

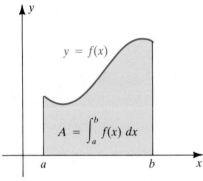

PROOF   From the preceding section, we know that the area A is a limit of sums $\sum_k f(u_k)\Delta x$, where $f(u_k)$ is the minimum value of f on $[x_{k-1}, x_k]$. Since these are Riemann sums, the conclusion follows from Definition (4.16). ∎

Theorem (4.19) is illustrated in Figure 4.15. It is important to keep in mind that area is merely our first application of the definite integral. *There are many instances where $\int_a^b f(x)\,dx$ does not represent the area of a region.* In fact, if $f(x) < 0$ for some x in $[a, b]$, then the definite integral may be negative or zero.

If f is continuous and $f(x) \geq 0$ on $[a, b]$, then Theorem (4.19) can be used to evaluate $\int_a^b f(x)\,dx$, provided we can find the area of the region under the graph of f from a to b. This will be true if the graph is a line or part of a circle, as in the following examples. (We consider more complicated definite integrals later in this chapter.) When evaluating a definite integral using this empirical technique, remember that the area of the region and the value of the integral are *numerically equal*; that is, if the area is A square units, the value of the integral is the real number A.

Figure 4.16

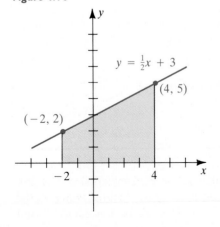

EXAMPLE ▪ 5   Evaluate $\int_{-2}^{4} (\tfrac{1}{2}x + 3)\,dx$.

SOLUTION   If $f(x) = \tfrac{1}{2}x + 3$, then the graph of f is the line sketched in Figure 4.16. By Theorem (4.19), the value of the integral is numerically equal to the area of the region under this line from $x = -2$ to $x = 4$. The region is a trapezoid with bases parallel to the y-axis of lengths

2 and 5 and altitude on the x-axis of length 6. Using the formula for the area of a trapezoid, we obtain

$$\int_{-2}^{4} (\tfrac{1}{2}x + 3)\, dx = \tfrac{1}{2}(2 + 5)6 = 21.$$

Figure 4.17

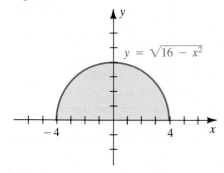

$y = \sqrt{16 - x^2}$

EXAMPLE ▪ 6 Evaluate $\int_{-4}^{4} \sqrt{16 - x^2}\, dx$.

SOLUTION If $f(x) = \sqrt{16 - x^2}$, then the graph of f is the semi-circle shown in Figure 4.17. By Theorem (4.19), the value of the integral is numerically equal to the area of the region under this semicircle from $x = -4$ to $x = 4$. Hence,

$$\int_{-4}^{4} \sqrt{16 - x^2}\, dx = \tfrac{1}{2} \cdot \pi(4)^2 = 8\pi.$$

EXAMPLE ▪ 7 Evaluate

(a) $\displaystyle\int_{4}^{-4} \sqrt{16 - x^2}\, dx$ **(b)** $\displaystyle\int_{4}^{4} \sqrt{16 - x^2}\, dx$

SOLUTION
(a) Using Definition (4.17) and Example 6, we have

$$\int_{4}^{-4} \sqrt{16 - x^2}\, dx = -\int_{-4}^{4} \sqrt{16 - x^2}\, dx = -8\pi.$$

(b) By Definition (4.18),

$$\int_{4}^{4} \sqrt{16 - x^2}\, dx = 0.$$

The next theorem states that functions that are continuous on closed intervals are integrable. This fact will play a crucial role in the proof of the fundamental theorem of calculus in Section 4.6.

Theorem 4.20

If f is continuous on $[a, b]$, then f is integrable on $[a, b]$.

A proof of Theorem (4.20) may be found in texts on advanced calculus.
Definite integrals of discontinuous functions may or may not exist, depending on the types of discontinuities. In particular, *functions that have infinite discontinuities on a closed interval are not integrable on that in-*

Figure 4.18

Nonintegrable discontinuous function

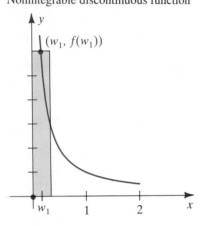

Figure 4.19

Integrable discontinuous function

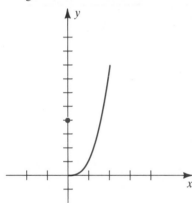

terval. To illustrate, we consider the piecewise-defined function f with domain $[0, 2]$ such that

$$f(x) = \begin{cases} 0 & \text{if } x = 0 \\ \dfrac{1}{x} & \text{if } 0 < x \le 2 \end{cases}$$

The graph of f is sketched in Figure 4.18. Note that $\lim_{x \to 0^+} f(x) = \infty$. If M is any (large) positive number, then in the first subinterval $[x_0, x_1]$ of any partition P of $[a, b]$, we can find that there exists a number w_1 such that $f(w_1) > M/\Delta x_1$, or, equivalently, $f(w_1)\Delta x_1 > M$. It follows that there are Riemann sums $\sum_k f(w_k)\Delta x_k$ that are arbitrarily large, and hence the limit in Definition (4.16) cannot exist. Thus, f is not integrable. A similar argument can be given for *any* function that has an infinite discontinuity in $[a, b]$. Consequently, *if a function f is integrable on $[a, b]$, then it is bounded on $[a, b]$—that is, there is a real number M such that $|f(x)| \le M$ for every x in $[a, b]$.*

As an illustration of a discontinuous function that *is* integrable, we consider the piecewise-defined function f with domain $[0, 2]$ such that

$$f(x) = \begin{cases} 4 & \text{if } x = 0 \\ x^3 & \text{if } 0 < x \le 2 \end{cases}$$

The graph of f is sketched in Figure 4.19. Note that f has a jump discontinuity at $x = 0$. From Example 8 of Section 4.3, the area under the graph of $y = x^3$ from 0 to 2 is $2^4/4 = 4$. Thus, by Theorem (4.19), $\int_0^2 x^3 \, dx = 4$. We can also show that $\int_0^2 f(x) \, dx = 4$. Hence, f is integrable.

We have shown that a function that is discontinuous on a closed interval may or may not be integrable. However, by Theorem (4.20), functions that are *continuous* on a closed interval are *always* integrable.

EXERCISES 4.4

Exer. 1–4: The given numbers determine a partition P of an interval. (a) Find the length of each subinterval of P. (b) Find the norm $\|P\|$ of the partition.

1 $\{0, 1.1, 2.6, 3.7, 4.1, 5\}$ 2 $\{2, 3, 3.7, 4, 5.2, 6\}$

3 $\{-3, -2.7, -1, 0.4, 0.9, 1\}$

4 $\{1, 1.6, 2, 3.5, 4\}$

Exer. 5–10: Find the Riemann sum R_P for the given function f on the indicated partition P by choosing on each subinterval of P (a) the left-hand endpoint, (b) the right-hand endpoint, and (c) the midpoint.

5 $f(x) = 2x + 3$; $P = \{1, 3, 4, 5\}$, $n = 3$

6 $f(x) = 3 - 4x$; $P = \{-1, 0, 2, 4, 6\}$, $n = 4$

7 $f(x) = 8 - x^2$; $P = \{-1, -0.5, 0.3, 0.8, 1\}$, $n = 4$

8 $f(x) = 8 - \frac{1}{2}x^2$; $P = \{0, 1.5, 3, 4.5, 6\}$, $n = 4$

9 $f(x) = x^3$; $P = \{-2, 0, 1, 3, 4, 6\}$, $n = 5$

10 $f(x) = \sqrt{x}$; $P = \{1, 3, 4, 9, 12, 16\}$, $n = 5$

c **Exer. 11–14: Find the Riemann sum R_P for the given function f on the indicated interval with a regular partition P of size n by choosing on each subinterval of P (a) the left-hand endpoint, (b) the right-hand endpoint, and (c) the midpoint.**

11 $f(x) = x^3$; $[-2, 6]$, $n = 32$

12 $f(x) = \sqrt{x}$; $[1, 16]$, $n = 30$

13 $f(x) = x^2\sqrt{\cos x}$; [0, 1], $n = 25$

14 $f(x) = \sin(\cos x)$; [−1, 1], $n = 40$

Exer. 15–18: Use Definition (4.16) to express each limit as a definite integral on the given interval $[a, b]$.

15 $\displaystyle\lim_{\|P\|\to 0} \sum_{k=1}^{n} (3w_k^2 - 2w_k + 5)\Delta x_k$; $[-1, 2]$

16 $\displaystyle\lim_{\|P\|\to 0} \sum_{k=1}^{n} \pi(w_k^2 - 4)\Delta x_k$; $[2, 3]$

17 $\displaystyle\lim_{\|P\|\to 0} \sum_{k=1}^{n} 2\pi w_k(1 + w_k^3)\Delta x_k$; $[0, 4]$

18 $\displaystyle\lim_{\|P\|\to 0} \sum_{k=1}^{n} (\sqrt[3]{w_k} + 4w_k)\Delta x_k$; $[-4, -3]$

Exer. 19–24: Given $\displaystyle\int_1^4 \sqrt{x}\,dx = \frac{14}{3}$, evaluate the integral.

19 $\displaystyle\int_4^1 \sqrt{x}\,dx$

20 $\displaystyle\int_1^4 \sqrt{s}\,ds$

21 $\displaystyle\int_1^4 \sqrt{t}\,dt$

22 $\displaystyle\int_1^4 \sqrt{x}\,dx + \int_4^1 \sqrt{x}\,dx$

23 $\displaystyle\int_4^4 \sqrt{x}\,dx + \int_4^1 \sqrt{x}\,dx$

24 $\displaystyle\int_4^4 \sqrt{x}\,dx$

Exer. 25–28: Express the area of the region in the figure as a definite integral.

25

$5x + 4y = 20$

26

$3x - 4y = -11$

27

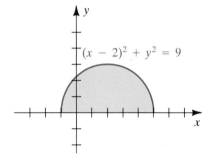

$(x - 2)^2 + y^2 = 9$

28

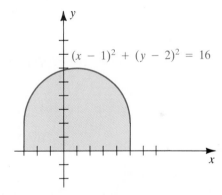

$(x - 1)^2 + (y - 2)^2 = 16$

Exer. 29–38: Evaluate the definite integral by regarding it as the area under the graph of a function.

29 $\displaystyle\int_{-1}^{5} 6\,dx$

30 $\displaystyle\int_{-2}^{3} 4\,dx$

31 $\displaystyle\int_{-3}^{2} (2x + 6)\,dx$

32 $\displaystyle\int_{-1}^{2} (7 - 3x)\,dx$

33 $\displaystyle\int_{0}^{3} |x - 1|\,dx$

34 $\displaystyle\int_{-1}^{4} |x|\,dx$

35 $\displaystyle\int_{0}^{3} \sqrt{9 - x^2}\,dx$

36 $\displaystyle\int_{0}^{a} \sqrt{a^2 - x^2}\,dx,\ a > 0$

37 $\displaystyle\int_{-2}^{2} (3 + \sqrt{4 - x^2})\,dx$

38 $\displaystyle\int_{-2}^{2} (3 - \sqrt{4 - x^2})\,dx$

4.5 PROPERTIES OF THE DEFINITE INTEGRAL

This section contains some fundamental properties of the definite integral. Most of the proofs are difficult and have been placed in Appendix I.

Theorem 4.21

If c is a real number, then

$$\int_a^b c \, dx = c(b - a).$$

PROOF   Let f be the constant function defined by $f(x) = c$ for every x in $[a, b]$. If P is a partition of $[a, b]$, then for every Riemann sum of f,

$$\sum_k f(w_k) \Delta x_k = \sum_k c \Delta x_k = c \sum_k \Delta x_k = c(b - a).$$

(The last equality is true because the sum $\sum_k \Delta x_k$ is the length of the interval $[a, b]$.) Consequently,

$$\left| \sum_k f(w_k) \Delta x_k - c(b - a) \right| = |c(b - a) - c(b - a)| = 0,$$

which is less than any positive number ϵ *regardless* of the size of $\|P\|$. Thus, by Definition (4.15), with $L = c(b - a)$,

$$\lim_{\|P\| \to 0} \sum_k f(w_k) \Delta x_k = \lim_{\|P\| \to 0} \sum_k c \Delta x_k = c(b - a).$$

By Definition (4.16), we thus obtain

$$\int_a^b f(x) \, dx = \int_a^b c \, dx = c(b - a). \quad \blacksquare$$

Note that if $c > 0$, then Theorem (4.21) agrees with Theorem (4.19): As illustrated in Figure 4.20, the graph of f is the horizontal line $y = c$, and the region under the graph from a to b is a rectangle with sides of lengths c and $b - a$. Hence the area $\int_a^b f(x) \, dx$ of the rectangle is $c(b - a)$.

Figure 4.20

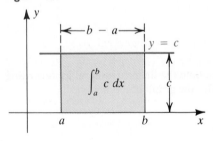

EXAMPLE I   Evaluate $\int_{-2}^{3} 7 \, dx$.

SOLUTION   Using Theorem (4.21) yields

$$\int_{-2}^{3} 7 \, dx = 7[3 - (-2)] = 7(5) = 35.$$

If $c = 1$ in Theorem (4.21), we shall abbreviate the integrand as follows:

$$\int_a^b dx = b - a$$

If a function f is integrable on $[a, b]$ and c is a real number, then, by Theorem (4.11)(ii), a Riemann sum of the function cf may be written

$$\sum_k cf(w_k)\Delta x_k = c \sum_k f(w_k)\Delta x_k.$$

We can prove that the limit of the sums on the left of the last equation is equal to c times the limit of the sums on the right. This gives us the next theorem. A proof may be found in Appendix I.

Theorem 4.22

> If f is integrable on $[a, b]$ and c is any real number, then cf is integrable on $[a, b]$ and
>
> $$\int_a^b cf(x)\, dx = c \int_a^b f(x)\, dx.$$

Theorem (4.22) is sometimes stated as follows: *A constant factor in the integrand may be taken outside the integral sign.* It is *not* permissible to take expressions involving variables outside the integral sign in this manner.

If two functions f and g are defined on $[a, b]$, then, by Theorem (4.11)(i), a Riemann sum of $f + g$ may be written

$$\sum_k [f(w_k) + g(w_k)]\Delta x_k = \sum_k f(w_k)\Delta x_k + \sum_k g(w_k)\Delta x_k$$

We can show that if f and g are integrable, then the limit of the sums on the left may be found by adding the limits of the two sums on the right. This fact is stated in integral form in (i) of the next theorem. A proof of (i) may be found in Appendix I. The analogous result for differences is stated in (ii) of the theorem.

Theorem 4.23

> If f and g are integrable on $[a, b]$, then $f + g$ and $f - g$ are integrable on $[a, b]$ and
>
> (i) $\displaystyle \int_a^b [f(x) + g(x)]\, dx = \int_a^b f(x)\, dx + \int_a^b g(x)\, dx$
>
> (ii) $\displaystyle \int_a^b [f(x) - g(x)]\, dx = \int_a^b f(x)\, dx - \int_a^b g(x)\, dx$

Theorem (4.23)(i) may be extended to any finite number of functions. Thus, if $f_1, f_2, \ldots, f_n$ are integrable on $[a, b]$, then so is their sum and

$$\int_a^b [f_1(x) + f_2(x) + \cdots + f_n(x)] \, dx$$

$$= \int_a^b f_1(x) \, dx + \int_a^b f_2(x) \, dx + \cdots + \int_a^b f_n(x) \, dx.$$

EXAMPLE • 2 It will follow from the results in Section 4.6 that

$$\int_0^2 x^3 \, dx = 4 \quad \text{and} \quad \int_0^2 x \, dx = 2.$$

Use these facts to evaluate $\int_0^2 (5x^3 - 3x + 6) \, dx$.

SOLUTION We may proceed as follows:

$$\int_0^2 (5x^3 - 3x + 6) \, dx = \int_0^2 5x^3 \, dx - \int_0^2 3x \, dx + \int_0^2 6 \, dx$$

$$= 5 \int_0^2 x^3 \, dx - 3 \int_0^2 x \, dx + 6(2 - 0)$$

$$= 5(4) - 3(2) + 12 = 26$$

If f is continuous on $[a, b]$ and $f(x) \geq 0$ for every x in $[a, b]$, then, by Theorem (4.19), the integral $\int_a^b f(x) \, dx$ is the area under the graph of f from a to b. Similarly, if $a < c < b$, then the integrals $\int_a^c f(x) \, dx$ and $\int_c^b f(x) \, dx$ are the areas under the graph of f from a to c and from c to b, respectively, as illustrated in Figure 4.21. Since the area from a to b is the sum of the two smaller areas, we have

$$\int_a^b f(x) \, dx = \int_a^c f(x) \, dx + \int_c^b f(x) \, dx.$$

The next theorem shows that the preceding equality is true under a more general hypothesis. The proof is given in Appendix I.

Figure 4.21

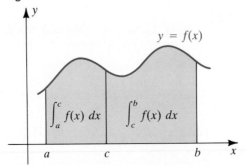

Theorem 4.24

> If $a < c < b$ and if f is integrable on both $[a, c]$ and $[c, b]$, then f is integrable on $[a, b]$ and
>
> $$\int_a^b f(x)\, dx = \int_a^c f(x)\, dx + \int_c^b f(x)\, dx.$$

The following result is a generalization of Theorem (4.24) to the case where c is not necessarily between a and b.

Theorem 4.25

> If f is integrable on a closed interval and if a, b, and c are any three numbers in the interval, then
>
> $$\int_a^b f(x)\, dx = \int_a^c f(x)\, dx + \int_c^b f(x)\, dx.$$

PROOF If a, b, and c are all different, then there are six possible ways of arranging these three numbers. The theorem should be verified for each of these cases and also for the cases in which two or all three of the numbers are equal. We shall verify one case. Suppose that $c < a < b$. By Theorem (4.24),

$$\int_c^b f(x)\, dx = \int_c^a f(x)\, dx + \int_a^b f(x)\, dx,$$

which, in turn, may be written

$$\int_a^b f(x)\, dx = -\int_c^a f(x)\, dx + \int_c^b f(x)\, dx.$$

The conclusion of the theorem now follows from the fact that interchanging the limits of integration changes the sign of the integral (see Definition 4.17). ■

EXAMPLE 3 Express as one integral:

$$\int_2^7 f(x)\, dx - \int_5^7 f(x)\, dx$$

SOLUTION First we interchange the limits of the second integral using Definition (4.17) and then use Theorem (4.25) with $a = 2$, $b = 5$, and $c = 7$:

$$\int_2^7 f(x)\, dx - \int_5^7 f(x)\, dx = \int_2^7 f(x)\, dx + \int_7^5 f(x)\, dx$$

$$= \int_2^5 f(x)\, dx$$

As an alternative solution, by recognizing that

$$\int_2^7 f(x)\,dx = \int_2^5 f(x)\,dx + \int_5^7 f(x)\,dx,$$

the previous result immediately follows.

If f and g are continuous on $[a, b]$ and $f(x) \geq g(x) \geq 0$ for every x in $[a, b]$, then the area under the graph of f from a to b is greater than or equal to the area under the graph of g from a to b. The corollary to the next theorem is a generalization of this fact to arbitrary integrable functions. The proof of the theorem is given in Appendix I.

Theorem 4.26

If f is integrable on $[a, b]$ and $f(x) \geq 0$ for every x in $[a, b]$, then

$$\int_a^b f(x)\,dx \geq 0.$$

Corollary 4.27

If f and g are integrable on $[a, b]$ and $f(x) \geq g(x)$ for every x in $[a, b]$, then

$$\int_a^b f(x)\,dx \geq \int_a^b g(x)\,dx.$$

PROOF By Theorem (4.23), $f - g$ is integrable. Moreover, since $f(x) \geq g(x)$, $f(x) - g(x) \geq 0$ for every x in $[a, b]$. Hence, by Theorem (4.26),

$$\int_a^b [f(x) - g(x)]\,dx \geq 0.$$

Applying Theorem (4.23)(ii) leads to the desired conclusion. ■

Figure 4.22

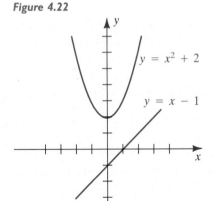

$y = x^2 + 2$

$y = x - 1$

EXAMPLE 4 Show that $\int_{-1}^{2} (x^2 + 2)\,dx \geq \int_{-1}^{2} (x - 1)\,dx$.

SOLUTION The graphs of $y = x^2 + 2$ and $y = x - 1$ are sketched in Figure 4.22. Since

$$x^2 + 2 \geq x - 1$$

for every x in $[-1, 2]$, the conclusion follows from Corollary (4.27).

Suppose, in Theorem (4.26), that f is continuous and that, in addition to the condition $f(x) \geq 0$, we have $f(c) > 0$ for some c in $[a, b]$. In this case, $\lim_{x \to c} f(x) > 0$, and, by Theorem (1.6), there is a subinterval

Figure 4.23

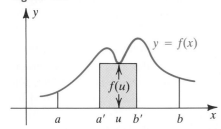

$[a', b']$ of $[a, b]$ throughout which $f(x)$ is positive. If $f(u)$ is the minimum value of f on $[a', b']$ (see Figure 4.23), then the area under the graph of f from a to b is at least as large as the area $f(u)(b' - a')$ of the pictured rectangle. Consequently, $\int_a^b f(x)\, dx > 0$. It now follows, as in the proof of Corollary (4.27), that if f and g are continuous on $[a, b]$, if $f(x) \geq g(x)$ throughout $[a, b]$, and if $f(x) > g(x)$ for some x in $[a, b]$, then $\int_a^b f(x)\, dx > \int_a^b g(x)\, dx$. This fact will be used in the proof of the next theorem.

Mean Value Theorem for Definite Integrals 4.28

> If f is continuous on a closed interval $[a, b]$, then there is a number z in the open interval (a, b) such that
> $$\int_a^b f(x)\, dx = f(z)(b - a).$$

Figure 4.24

PROOF   If f is a constant function, then $f(x) = c$ for some number c, and by Theorem (4.21),

$$\int_a^b f(x)\, dx = \int_a^b c\, dx = c(b - a) = f(z)(b - a)$$

for every number z in (a, b).

Next, assume that f is not a constant function and suppose that m and M are the minimum and maximum values of f, respectively, on $[a, b]$. Let $f(u) = m$ and $f(v) = M$ for some u and v in $[a, b]$, as illustrated in Figure 4.24 for the case in which $f(x)$ is positive throughout $[a, b]$. Since f is not a constant function, $m < f(x) < M$ for some x in $[a, b]$. Hence, by the remark immediately preceding this theorem,

$$\int_a^b m\, dx < \int_a^b f(x)\, dx < \int_a^b M\, dx.$$

Applying Theorem (4.21) yields

$$m(b - a) < \int_a^b f(x)\, dx < M(b - a).$$

Dividing by $b - a$ and recalling that $m = f(u)$ and $M = f(v)$ gives us

$$f(u) < \frac{1}{b - a} \int_a^b f(x)\, dx < f(v).$$

Since $[1/(b - a)] \int_a^b f(x)\, dx$ is a number between $f(u)$ and $f(v)$, it follows from the intermediate value theorem (1.26) that there is a number z, with $u < z < v$, such that

$$f(z) = \frac{1}{b - a} \int_a^b f(x)\, dx.$$

Multiplying both sides by $b - a$ gives us the conclusion of the theorem. ■

Figure 4.25

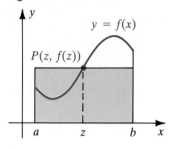

The number z of Theorem (4.28) is not necessarily unique; however, the theorem guarantees that *at least* one number z will produce the desired result.

The mean value theorem has an interesting geometric interpretation if $f(x) \geq 0$ on $[a, b]$. In this case, $\int_a^b f(x)\,dx$ is the area under the graph of f from a to b. If, as in Figure 4.25, a horizontal line is drawn through the point $P(z, f(z))$, then the area of the rectangular region bounded by this line, the x-axis, and the lines $x = a$ and $x = b$ is $f(z)(b - a)$, which, according to Theorem (4.28), is the same as the area under the graph of f from a to b.

EXAMPLE ▪ 5   It will follow from the results of Section 4.6 that $\int_0^3 x^2\,dx = 9$. Find a number z that satisfies the conclusion of the mean value theorem (4.28) for this definite integral.

SOLUTION   The graph of $f(x) = x^2$ for $0 \leq x \leq 3$ is sketched in Figure 4.26. By the mean value theorem, there is a number z between 0 and 3 such that

Figure 4.26

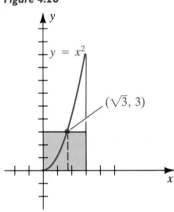

$$\int_0^3 x^2\,dx = f(z)(3 - 0) = z^2(3).$$

This result implies that

$$9 = 3z^2, \quad \text{or} \quad z^2 = 3.$$

The solutions of the last equation are $z = \pm\sqrt{3}$; however, $-\sqrt{3}$ is not in $[0, 3]$. The number $z = \sqrt{3}$ satisfies the conclusion of the theorem.

If we consider the horizontal line through $P(\sqrt{3}, 3)$, then the area of the rectangle bounded by this line, the x-axis, and the lines $x = 0$ and $x = 3$ is equal to the area under the graph of f from $x = 0$ to $x = 3$ (see Figure 4.26).

In statistics, the term **arithmetic mean** is used for the **average** of a set of numbers. Therefore, the arithmetic mean of two numbers a and b is $(a + b)/2$, the arithmetic mean of three numbers a, b, and c is $(a + b + c)/3$, and so on. To see the relationship between arithmetic means and the word *mean* used in *mean value theorem*, let us rewrite the conclusion of (4.28) as

$$f(z) = \frac{1}{b - a} \int_a^b f(x)\,dx$$

and express the definite integral as a limit of sums. If we specialize Definition (4.16) by using a regular partition P with n subintervals, then

$$f(z) = \frac{1}{b - a} \lim_{n \to \infty} \sum_{k=1}^n f(w_k)\Delta x = \lim_{n \to \infty} \sum_{k=1}^n \left[f(w_k) \frac{\Delta x}{b - a} \right]$$

for any number w_k in the kth subinterval of P and $\Delta x = (b-a)/n$. Since $\Delta x/(b-a) = 1/n$, we obtain

$$f(z) = \lim_{n \to \infty} \sum_{k=1}^{n} \left[f(w_k)\frac{\Delta x}{b-a} \right] = \lim_{n \to \infty} \sum_{k=1}^{n} \left[f(w_k)\frac{1}{n} \right],$$

or

$$f(z) = \lim_{n \to \infty} \left[\frac{f(w_1) + f(w_2) + \cdots + f(w_n)}{n} \right].$$

This result shows that we may regard the number $f(z)$ in the mean value theorem (4.28) as a limit of the arithmetic means (averages) of the function values $f(w_1), f(w_2), \ldots, f(w_n)$ as n increases without bound. This fact is the motivation for the next definition.

Definition 4.29

Let f be continuous on $[a, b]$. The **average value** f_{av} of f on $[a, b]$ is

$$f_{av} = \frac{1}{b-a} \int_a^b f(x)\,dx.$$

Note that, by the mean value theorem for definite integrals, if f is continuous on $[a, b]$, then

$$f_{av} = f(z) \quad \text{for some } z \text{ in } [a, b].$$

EXAMPLE ■ 6   Given $\int_0^3 x^2\,dx = 9$, find the average value of f on $[0, 3]$.

SOLUTION   By Definition (4.29), with $a = 0, b = 3$, and $f(x) = x^2$,

$$f_{av} = \frac{1}{3-0} \int_0^3 x^2\,dx = \frac{1}{3} \cdot 9 = 3.$$

In the interval $[0, 3]$, the function values $f(x) = x^2$ range from $f(0) = 0$ to $f(3) = 9$. Note that the function f *takes on* its average value 3 at the number $z = \sqrt{3}$.

EXERCISES 4.5

Exer. 1–6: Evaluate the integral.

1 $\int_{-2}^{4} 5\,dx$ 2 $\int_1^{10} \sqrt{2}\,dx$ 3 $\int_6^2 3\,dx$ 4 $\int_4^{-3} dx$ 5 $\int_{-1}^{1} dx$ 6 $\int_2^2 100\,dx$

Exer. 7–10: It will follow from the results in Section 4.6 that

$$\int_1^4 x^2 \, dx = 21 \quad \text{and} \quad \int_1^4 x \, dx = \tfrac{15}{2}.$$

Use these facts to evaluate the integral.

7 $\displaystyle\int_1^4 (3x^2 + 5) \, dx$

8 $\displaystyle\int_1^4 (6x - 1) \, dx$

9 $\displaystyle\int_1^4 (2 - 9x - 4x^2) \, dx$

10 $\displaystyle\int_1^4 (3x + 2)^2 \, dx$

Exer. 11–16: Verify the inequality without evaluating the integrals.

11 $\displaystyle\int_1^2 (3x^2 + 4) \, dx \geq \int_1^2 (2x^2 + 5) \, dx$

12 $\displaystyle\int_1^4 (2x + 2) \, dx \leq \int_1^4 (3x + 1) \, dx$

13 $\displaystyle\int_2^4 (x^2 - 6x + 8) \, dx \leq 0$

14 $\displaystyle\int_2^4 (5x^2 - x + 1) \, dx \geq 0$

15 $\displaystyle\int_0^{2\pi} (1 + \sin x) \, dx \geq 0$

16 $\displaystyle\int_{-\pi/3}^{\pi/3} (\sec x - 2) \, dx \leq 0$

Exer. 17–22: Express as one integral.

17 $\displaystyle\int_5^1 f(x) \, dx + \int_{-3}^5 f(x) \, dx$

18 $\displaystyle\int_4^1 f(x) \, dx + \int_6^4 f(x) \, dx$

19 $\displaystyle\int_c^d f(x) \, dx + \int_e^c f(x) \, dx$

20 $\displaystyle\int_{-2}^6 f(x) \, dx - \int_{-2}^2 f(x) \, dx$

21 $\displaystyle\int_c^{c+h} f(x) \, dx - \int_c^h f(x) \, dx$

22 $\displaystyle\int_c^m f(x) \, dx - \int_d^m f(x) \, dx$

Exer. 23–30: The given integral $\int_a^b f(x) \, dx$ may be verified using the results in Section 4.6. (a) Find a number z that satisfies the conclusion of the mean value theorem (4.28). (b) Find the average value of f on $[a, b]$.

23 $\displaystyle\int_0^3 3x^2 \, dx = 27$

24 $\displaystyle\int_{-4}^{-1} \frac{3}{x^2} \, dx = \frac{9}{4}$

25 $\displaystyle\int_{-2}^1 (x^2 + 1) \, dx = 6$

26 $\displaystyle\int_{-1}^3 (3x^2 - 2x + 3) \, dx = 32$

27 $\displaystyle\int_{-1}^8 3\sqrt{x + 1} \, dx = 54$

28 $\displaystyle\int_{-2}^{-1} \frac{8}{x^3} \, dx = -3$

29 $\displaystyle\int_1^2 (4x^3 - 1) \, dx = 14$

30 $\displaystyle\int_1^4 (2 + 3\sqrt{x}) \, dx = 20$

[c] Exer. 31–32: The given integral may be verified using results in Section 4.6. Use Newton's method to approximate, to three decimal places, a number z that satisfies the conclusion of the mean value theorem (4.28).

31 $\displaystyle\int_{-2}^3 (8x^3 + 3x - 1) \, dx = 132.5$

32 $\displaystyle\int_{\pi/6}^{\pi/4} (1 - \cos 4x) \, dx = \frac{\pi}{12} + \frac{\sqrt{3}}{8}$

33 Let f and g be integrable on $[a, b]$. If c and d are any real numbers, prove that

$$\int_a^b [cf(x) + dg(x)] \, dx = c \int_a^b f(x) \, dx + d \int_a^b g(x) \, dx.$$

34 If f is continuous on $[a, b]$, prove that

$$\left| \int_a^b f(x) \, dx \right| \leq \int_a^b |f(x)| \, dx.$$

(Hint: $-|f(x)| \leq f(x) \leq |f(x)|$.)

4.6   THE FUNDAMENTAL THEOREM OF CALCULUS

This section contains one of the most important theorems in calculus. In addition to being useful in evaluating definite integrals, the theorem also exhibits the relationship between derivatives and definite integrals. This theorem, aptly called *the fundamental theorem of calculus*, was discovered independently by Sir Isaac Newton and by Gottfried Wilhelm Leibniz. It

Figure 4.27

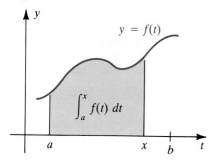

$$\int_a^x f(t)\,dt$$

is primarily because of this discovery that both men are credited with the invention of calculus.

To avoid confusion in the following discussion, we shall use t as the independent variable and denote the definite integral of f from a to b by $\int_a^b f(t)\,dt$. If f is continuous on $[a, b]$ and $a \leq x \leq b$, then f is continuous on $[a, x]$; therefore, by Theorem (4.20), f is integrable on $[a, x]$. Consequently, the formula

$$G(x) = \int_a^x f(t)\,dt$$

determines a function G with domain $[a, b]$, since for each x in $[a, b]$, there corresponds a unique number $G(x)$.

To obtain a geometric interpretation of $G(x)$, suppose that $f(t) \geq 0$ for every t in $[a, b]$. In this case, we see from Theorem (4.19) that $G(x)$ is the area of the region under the graph of f from a to x (see Figure 4.27).

As a specific illustration, consider $f(t) = t^3$ with $a = 0$ and $b > 0$ (see Figure 4.28). In Example 8 of Section 4.3, we proved that the area under the graph of f from 0 to b is $\frac{1}{4}b^4$. Hence the area from 0 to x is

Figure 4.28

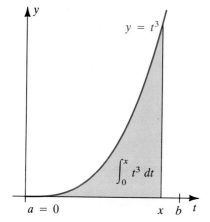

$$\int_0^x t^3\,dt$$

$$G(x) = \int_0^x t^3\,dt = \tfrac{1}{4}x^4.$$

This gives us an explicit form for the function G if $f(t) = t^3$. Note that in this illustration,

$$G'(x) = \frac{d}{dx}(\tfrac{1}{4}x^4) = x^3 = f(x).$$

Thus, by Definition (4.1), G is an antiderivative of f. This result is not an accident. Part I of the next theorem brings out the remarkable fact that if f is *any* continuous function and $G(x) = \int_a^x f(t)\,dt$, then G is an antiderivative of f. Part II of the theorem shows how *any* antiderivative may be used to find the value of $\int_a^b f(x)\,dx$.

Fundamental Theorem of Calculus   4.30

Suppose f is continuous on a closed interval $[a, b]$.

Part I If the function G is defined by

$$G(x) = \int_a^x f(t)\,dt$$

for every x in $[a, b]$, then G is an antiderivative of f on $[a, b]$.

Part II If F is any antiderivative of f on $[a, b]$, then

$$\int_a^b f(x)\,dx = F(b) - F(a).$$

P R O O F   To establish Part I, we must show that if x is in $[a, b]$, then $G'(x) = f(x)$—that is,

$$\lim_{h \to 0} \frac{G(x + h) - G(x)}{h} = f(x).$$

Figure 4.29

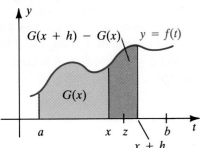

Before giving a formal proof, let us consider some geometric aspects of this limit. If $f(x) \geq 0$ throughout $[a, b]$, then $G(x)$ is the area under the graph of f from a to x, as illustrated in Figure 4.29. If $h > 0$, then the difference $G(x + h) - G(x)$ is the area under the graph of f from x to $x + h$, and the number h is the length of the interval $[x, x + h]$. We shall show that

$$\frac{G(x + h) - G(x)}{h} = f(z)$$

for some value z between x and $x + h$. Apparently, if $h \to 0$, then $z \to x$ and $f(z) \to f(x)$, which is what we wish to prove.

Let us now give a rigorous proof that $G'(x) = f(x)$. If x and $x + h$ are in $[a, b]$, then using the definition of G together with Definition (4.17) and Theorem (4.24) yields

$$G(x + h) - G(x) = \int_{a}^{x+h} f(t)\,dt - \int_{a}^{x} f(t)\,dt$$

$$= \int_{a}^{x+h} f(t)\,dt + \int_{x}^{a} f(t)\,dt$$

$$= \int_{x}^{x+h} f(t)\,dt.$$

Consequently, if $h \neq 0$, then

$$\frac{G(x + h) - G(x)}{h} = \frac{1}{h} \int_{x}^{x+h} f(t)\,dt.$$

If $h > 0$, then, by the mean value theorem (4.28), there is a number z in the open interval $(x, x + h)$ such that

$$\int_{x}^{x+h} f(t)\,dt = f(z)h$$

and, therefore,

$$\frac{G(x + h) - G(x)}{h} = f(z).$$

Since $x < z < x + h$, it follows from the continuity of f that

$$\lim_{h \to 0^+} f(z) = \lim_{z \to x^+} f(z) = f(x)$$

and hence

$$\lim_{h \to 0^+} \frac{G(x + h) - G(x)}{h} = \lim_{h \to 0^+} f(z) = f(x).$$

If $h < 0$, then we may prove in similar fashion that

$$\lim_{h \to 0^-} \frac{G(x + h) - G(x)}{h} = f(x).$$

The two preceding one-sided limits imply that

$$G'(x) = \lim_{h \to 0} \frac{G(x + h) - G(x)}{h} = f(x).$$

This completes the proof of Part I.

To prove Part II, let F be any antiderivative of f and let G be the special antiderivative defined in Part I. From Theorem (4.2), we know that there is a constant C such that

$$G(x) = F(x) + C$$

for every x in $[a, b]$. Hence, from the definition of G,

$$\int_a^x f(t)\, dt = F(x) + C$$

for every x in $[a, b]$. If we let $x = a$ and use the fact that $\int_a^a f(t)\, dt = 0$, we obtain $0 = F(a) + C$, or $C = -F(a)$. Consequently,

$$\int_a^x f(t)\, dt = F(x) - F(a).$$

This is an identity for every x in $[a, b]$, so we may substitute b for x, obtaining

$$\int_a^b f(t)\, dt = F(b) - F(a).$$

Replacing the variable t by x gives us the conclusion of Part II. ■

We often denote the difference $F(b) - F(a)$ either by $F(x)]_a^b$ or by $[F(x)]_a^b$. Part II of the fundamental theorem may then be expressed as follows.

Corollary 4.31

If f is continuous on $[a, b]$ and F is any antiderivative of f, then

$$\int_a^b f(x)\, dx = F(x)\Bigg]_a^b = F(b) - F(a).$$

The formula in Corollary (4.31) is also valid if $a \geq b$. If $a > b$, then, by Definition (4.17),

$$\int_a^b f(x)\, dx = -\int_b^a f(x)\, dx$$

$$= -[F(a) - F(b)]$$
$$= F(b) - F(a).$$

If $a = b$, then by Definition (4.18),

$$\int_a^a f(x)\, dx = 0 = F(a) - F(a).$$

Corollary (4.31) allows us to evaluate a definite integral very easily if we can find an antiderivative of the integrand. For example, since an antiderivative of x^3 is $\frac{1}{4}x^4$, we have

$$\int_0^b x^3 \, dx = \frac{1}{4}x^4 \Big]_0^b = \frac{1}{4}b^4 - \frac{1}{4}(0)^4 = \frac{1}{4}b^4.$$

Those who doubt the importance of the fundamental theorem should compare this simple computation with the limit of sums calculation discussed in Example 8 of Section 4.3.

EXAMPLE ▪ 1   Evaluate $\displaystyle\int_{-2}^3 (6x^2 - 5)\, dx$.

SOLUTION   An antiderivative of $6x^2 - 5$ is $F(x) = 2x^3 - 5x$. Applying Corollary (4.31), we get

$$\int_{-2}^3 (6x^2 - 5)\, dx = \left[2x^3 - 5x\right]_{-2}^3$$
$$= [2(3)^3 - 5(3)] - [2(-2)^3 - 5(-2)]$$
$$= [54 - 15] - [-16 + 10] = 45.$$

Note that if $F(x) + C$ is used in place of $F(x)$ in Corollary (4.31), the same result is obtained, since

$$\left[F(x) + C\right]_a^b = [F(b) + C] - [F(a) + C]$$
$$= F(b) - F(a)$$
$$= \left[F(x)\right]_a^b.$$

In particular, since

$$\int f(x)\, dx = F(x) + C,$$

where $F'(x) = f(x)$, we obtain the following theorem.

Theorem 4.32

$$\int_a^b f(x)\, dx = \left[\int f(x)\, dx\right]_a^b$$

Theorem (4.32) states that *a definite integral can be evaluated by evaluating the corresponding indefinite integral.* As with previous cases, when using Theorem (4.32), it is unnecessary to include the constant of integration C for the indefinite integral.

EXAMPLE ▪ 2   Find the area A of the region between the graph of $y = \sin x$ and the x-axis from $x = 0$ to $x = \pi$.

Figure 4.30

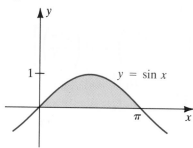

$y = \sin x$

S O L U T I O N The region is sketched in Figure 4.30. Applying Theorems (4.19) and (4.32) gives us the following:

$$A = \int_0^\pi \sin x \, dx = \left[\int \sin x \, dx \right]_0^\pi$$

$$= \left[-\cos x \right]_0^\pi$$

$$= -\cos \pi - (-\cos 0)$$

$$= -(-1) + 1 = 2$$

By Theorem (4.32), we can use any formula for indefinite integration to obtain a formula for definite integrals. To illustrate, using Table (4.4), we obtain

$$\int_a^b x^r \, dx = \left[\frac{x^{r+1}}{r+1} \right]_a^b \quad \text{if } r \neq -1$$

$$\int_a^b \sin x \, dx = \left[-\cos x \right]_a^b$$

$$\int_a^b \sec^2 x \, dx = \left[\tan x \right]_a^b.$$

E X A M P L E ■ 3 Evaluate $\int_{-1}^2 (x^3 + 1)^2 \, dx$.

S O L U T I O N We first square the integrand and then apply the power rule to each term as follows:

$$\int_{-1}^2 (x^3 + 1)^2 \, dx = \int_{-1}^2 (x^6 + 2x^3 + 1) \, dx$$

$$= \left[\frac{x^7}{7} + 2 \cdot \frac{x^4}{4} + x \right]_{-1}^2$$

$$= \left[\frac{2^7}{7} + 2 \cdot \frac{2^4}{4} + 2 \right] - \left[\frac{(-1)^7}{7} + 2 \frac{(-1)^4}{4} + (-1) \right]$$

$$= \frac{405}{14}$$

E X A M P L E ■ 4 Evaluate $\int_1^4 \left(5x - 2\sqrt{x} + \frac{32}{x^3} \right) dx$.

SOLUTION   We begin by changing the form of the integrand so that the power rule may be applied to each term. Thus,

$$\int_1^4 (5x - 2x^{1/2} + 32x^{-3})\, dx = \left[5\left(\frac{x^2}{2}\right) - 2\left(\frac{x^{3/2}}{3/2}\right) + 32\left(\frac{x^{-2}}{-2}\right)\right]_1^4$$

$$= \left[\frac{5}{2}x^2 - \frac{4}{3}x^{3/2} - \frac{16}{x^2}\right]_1^4$$

$$= \left[\frac{5}{2}(4)^2 - \frac{4}{3}(4)^{3/2} - \frac{16}{4^2}\right] - \left[\frac{5}{2} - \frac{4}{3} - 16\right]$$

$$= \frac{259}{6}.$$

CAUTION   A common *misuse* of the fundamental theorem of calculus is to make the false interpretation that Corollary (4.31) asserts that if F is a function such that $F'(x) = f(x)$ for some function f, then $\int_a^b f(x)\, dx = F(b) - F(a)$. Guided by this false interpretation, we might make the following fallacious argument:

"If $F(x) = \dfrac{-1}{x},$ then $F'(x) = \dfrac{1}{x^2}$

and so $\displaystyle\int_{-1}^1 \frac{1}{x^2}\, dx = F(1) - F(-1) = -2.$"

This reasoning is incorrect because it tries to make use of the *conclusion* of Corollary (4.31) in a situation in which the *hypothesis* is not true. Theorems in mathematics are of the form: If a certain set of conditions holds (the hypothesis), then certain conclusions must be true. In Corollary (4.31), the hypothesis is that the function f is continuous on the interval $[a, b]$. In this instance, the function f is not continuous on the interval $[-1, 1]$: Not only is f undefined at $x = 0$, but $\lim_{x \to 0} f(x)$ does not even exist.

Before we apply any theorem in a particular situation, we must check that all of the conditions in the hypothesis are true.

The method of substitution developed for indefinite integrals may also be used to evaluate a definite integral. We could use (4.7) to find an indefinite integral (that is, an antiderivative) and then apply the fundamental theorem of calculus. Another method, which is sometimes shorter, is to change the limits of integration. Using (4.7) together with the fundamental theorem gives us the following formula, with $F' = f$:

$$\int_a^b f(g(x))g'(x)\, dx = F(g(x))\Big]_a^b$$

The number on the right may be written

$$F(g(b)) - F(g(a)) = F(u)\Big]_{g(a)}^{g(b)} = \int_{g(a)}^{g(b)} f(u)\, du.$$

This result gives us the following theorem, provided f and g' are integrable.

Theorem 4.33

If $u = g(x)$, then $\displaystyle\int_a^b f(g(x))g'(x)\,dx = \int_{g(a)}^{g(b)} f(u)\,du$.

Theorem (4.33) states that after making the substitution $u = g(x)$ and $du = g'(x)\,dx$, we may use the values of g that correspond to $x = a$ and $x = b$, respectively, as the limits of the integral involving u. It is then unnecessary to return to the variable x after integrating. This technique is illustrated in the next example.

EXAMPLE ▪ 5 Evaluate $\displaystyle\int_2^{10} \frac{3}{\sqrt{5x-1}}\,dx$.

SOLUTION Let us begin by writing the integral as

$$3 \int_2^{10} \frac{1}{\sqrt{5x-1}}\,dx.$$

The expression $\sqrt{5x-1}$ in the integrand suggests the following substitution:

$$u = 5x - 1, \qquad du = 5\,dx$$

The form of du indicates that we should introduce the factor 5 into the integrand and then compensate by multiplying the integral by $\frac{1}{5}$, as follows:

$$3 \int_2^{10} \frac{1}{\sqrt{5x-1}}\,dx = \frac{3}{5} \int_2^{10} \frac{1}{\sqrt{5x-1}}\,5\,dx$$

We next calculate the values of $u = 5x - 1$ that correspond to the limits of integration $x = 2$ and $x = 10$:

(i) If $x = 2$, then $u = 5(2) - 1 = 9$.

(ii) If $x = 10$, then $u = 5(10) - 1 = 49$.

Substituting in the integrand and changing the limits of integration as in Theorem (4.33) gives us

$$3 \int_2^{10} \frac{1}{\sqrt{5x-1}}\,dx = \frac{3}{5} \int_2^{10} \frac{1}{\sqrt{5x-1}}\,5\,dx$$

$$= \frac{3}{5} \int_9^{49} \frac{1}{\sqrt{u}}\,du = \frac{3}{5} \int_9^{49} u^{-1/2}\,du$$

$$= \left[\left(\frac{3}{5}\right) \frac{u^{1/2}}{1/2} \right]_9^{49} = \frac{6}{5}[49^{1/2} - 9^{1/2}] = \frac{24}{5}.$$

E X A M P L E ■ 6   Evaluate $\displaystyle\int_0^{\pi/4} (1 + \sin 2x)^3 \cos 2x \, dx$.

S O L U T I O N   The integrand suggests the power rule $\int_a^b u^3 \, du = \left[\frac{1}{4}u^4\right]_a^b$. Thus, we let

$$u = 1 + \sin 2x, \qquad du = 2 \cos 2x \, dx.$$

The form of du indicates that we should introduce the factor 2 into the integrand and multiply the integral by $\frac{1}{2}$, as follows:

$$\int_0^{\pi/4} (1 + \sin 2x)^3 \cos 2x \, dx = \frac{1}{2} \int_0^{\pi/4} (1 + \sin 2x)^3 \, 2 \cos 2x \, dx.$$

We next calculate the values of $u = 1 + \sin 2x$ that correspond to the limits of integration $x = 0$ and $x = \pi/4$:

(i) If $x = 0$, then $u = 1 + \sin 0 = 1 + 0 = 1$.

(ii) If $x = \dfrac{\pi}{4}$, then $u = 1 + \sin \dfrac{\pi}{2} = 1 + 1 = 2$.

Substituting in the integrand and changing the limits of integration gives us

$$\int_0^{\pi/4} (1 + \sin 2x)^3 \cos 2x \, dx = \frac{1}{2} \int_1^2 u^3 \, du$$

$$= \frac{1}{2} \left[\frac{u^4}{4}\right]_1^2 = \frac{1}{8}[16 - 1] = \frac{15}{8}.$$

The following theorem illustrates a useful technique for evaluating certain definite integrals.

Theorem 4.34

Let f be continuous on $[-a, a]$.

(i) If f is an even function,

$$\int_{-a}^{a} f(x) \, dx = 2 \int_0^a f(x) \, dx.$$

(ii) If f is an odd function,

$$\int_{-a}^{a} f(x) \, dx = 0.$$

Figure 4.31

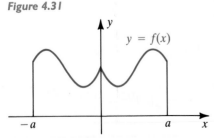

P R O O F   We shall prove (i). If f is an even function, then the graph of f is symmetric with respect to the y-axis. As a special case, if $f(x) \geq 0$ for every x in $[0, a]$, we have a situation similar to that in Figure 4.31, and hence the area under the graph of f from $x = -a$ to $x = a$ is twice that from $x = 0$ to $x = a$. This gives us the formula in (i).

To show that the formula is true if $f(x) < 0$ for some x, we may proceed as follows. Using, successively, Theorem (4.24), Definition (4.17), and Theorem (4.22), we have

$$\int_{-a}^{a} f(x)\,dx = \int_{-a}^{0} f(x)\,dx + \int_{0}^{a} f(x)\,dx$$

$$= -\int_{0}^{-a} f(x)\,dx + \int_{0}^{a} f(x)\,dx$$

$$= \int_{0}^{-a} f(x)(-dx) + \int_{0}^{a} f(x)\,dx.$$

Since f is even, $f(-x) = f(x)$, and the last equality may be written

$$\int_{-a}^{a} f(x)\,dx = \int_{0}^{-a} f(-x)(-dx) + \int_{0}^{a} f(x)\,dx.$$

If, in the first integral on the right, we substitute $u = -x$, $du = -dx$ and observe that $u = a$ when $x = -a$, we obtain

$$\int_{-a}^{a} f(x)\,dx = \int_{0}^{a} f(u)\,du + \int_{0}^{a} f(x)\,dx.$$

The last two integrals on the right are equal, since the variables are *dummy variables*, and, therefore,

$$\int_{-a}^{a} f(x)\,dx = 2\int_{0}^{a} f(x)\,dx. \quad \blacksquare$$

EXAMPLE 7   Evaluate

(a) $\displaystyle\int_{-1}^{1} (x^4 + 3x^2 + 1)\,dx$

(b) $\displaystyle\int_{-1}^{1} (x^5 + 3x^3 + x)\,dx$

(c) $\displaystyle\int_{-5}^{5} (2x^3 + 3x^2 + 7x)\,dx$

SOLUTION

(a) Since the integrand determines an even function, we may apply Theorem (4.34)(i):

$$\int_{-1}^{1} (x^4 + 3x^2 + 1)\,dx = 2\int_{0}^{1} (x^4 + 3x^2 + 1)\,dx$$

$$= 2\left[\frac{x^5}{5} + x^3 + x\right]_{0}^{1} = \frac{22}{5}$$

(b) The integrand is odd, so we apply Theorem (4.34)(ii):

$$\int_{-1}^{1} (x^5 + 3x^3 + x)\,dx = 0$$

(c) The function given by $2x^3 + 7x$ is odd but the function given by $3x^2$ is even, so we apply Theorem (4.34)(ii) *and* (i):

$$\int_{-5}^{5} (2x^3 + 3x^2 + 7x)\,dx = \int_{-5}^{5} (2x^3 + 7x)\,dx + \int_{-5}^{5} 3x^2\,dx$$

$$= 0 + 2\int_{0}^{5} 3x^2\,dx$$

$$= 2\left[x^3\right]_0^5 = 250$$

The technique of defining a function by means of a definite integral, as in Part I of the fundamental theorem of calculus (4.30), will have a very important application in Chapter 6, when we consider logarithmic functions. Recall, from (4.30), that if f is continuous on $[a, b]$ and $G(x) = \int_a^x f(t)\,dt$ for $a \leq x \leq b$, then G is an antiderivative of f—that is, $(d/dx)(G(x)) = f(x)$. This result may be stated in integral form as follows:

$$\frac{d}{dx}\int_a^x f(t)\,dt = f(x)$$

The preceding formula is generalized in the next theorem.

Theorem 4.35

Let f be continuous on $[a, b]$. If $a \leq c \leq b$, then for every x in $[a, b]$,

$$\frac{d}{dx}\int_c^x f(t)\,dt = f(x).$$

PROOF   If F is an antiderivative of f, then

$$\frac{d}{dx}\int_c^x f(t)\,dt = \frac{d}{dx}(F(x) - F(c))$$

$$= \frac{d}{dx}(F(x)) - \frac{d}{dx}(F(c))$$

$$= f(x) - 0 = f(x). \quad \blacksquare$$

EXAMPLE 8   If $G(x) = \int_1^x \frac{1}{t}\,dt$ and $x > 0$, find $G'(x)$.

SOLUTION   We apply Theorem (4.35) with $c = 1$ and $f(x) = 1/x$. If we choose a and b such that $0 < a \leq 1 \leq b$, then f is continuous on $[a, b]$. Hence, by Theorem (4.35), for every x in $[a, b]$,

$$G'(x) = \frac{d}{dx}\int_1^x \frac{1}{t}\,dt = \frac{1}{x}.$$

In (4.29), we defined the *average value* f_{av} of a function f on $[a, b]$ as follows:

$$f_{av} = \frac{1}{b-a} \int_a^b f(x)\, dx$$

The next example indicates why this terminology is appropriate in applications.

EXAMPLE ▪ 9   Suppose that a point P moving on a coordinate line has a continuous velocity function v. Show that the average value of v on $[a, b]$ equals the average velocity during the time interval $[a, b]$.

SOLUTION   By Definition (4.29) with $f = v$,

$$v_{av} = \frac{1}{b-a} \int_a^b v(t)\, dt.$$

If s is the position function of P, then $s'(t) = v(t)$ — that is, $s(t)$ is an antiderivative of $v(t)$. Hence, by the fundamental theorem of calculus,

$$\int_a^b v(t)\, dt = \int_a^b s'(t)\, dt = s(t)\Big]_a^b = s(b) - s(a).$$

Substituting in the formula for v_{av} give us

$$v_{av} = \frac{s(b) - s(a)}{b - a},$$

which is the average velocity of P on $[a, b]$ (see Definition 2.2).

Results similar to that in Example 9 occur in discussions of average acceleration, average marginal cost, average marginal revenue, and many other applications of the derivative (see Exercises 49–54).

EXERCISES 4.6

Exer. 1–36: Evaluate the integral.

1 $\displaystyle\int_1^4 (x^2 - 4x - 3)\, dx$

2 $\displaystyle\int_{-2}^3 (5 + x - 6x^2)\, dx$

3 $\displaystyle\int_{-2}^3 (8z^3 + 3z - 1)\, dz$

4 $\displaystyle\int_0^2 (z^4 - 2z^3)\, dz$

5 $\displaystyle\int_7^{12} dx$

6 $\displaystyle\int_{-6}^{-1} 8\, dx$

7 $\displaystyle\int_1^2 \frac{5}{x^6}\, dx$

8 $\displaystyle\int_1^4 \sqrt{16x^5}\, dx$

9 $\displaystyle\int_4^9 \frac{t - 3}{\sqrt{t}}\, dt$

10 $\displaystyle\int_{-1}^{-2} \frac{2t - 7}{t^3}\, dt$

11 $\displaystyle\int_{-8}^8 (\sqrt[3]{s^2} + 2)\, ds$

12 $\displaystyle\int_1^0 s^2(\sqrt[3]{s} - \sqrt{s})\, ds$

13 $\displaystyle\int_{-1}^0 (2x + 3)^2\, dx$

14 $\displaystyle\int_1^2 (4x^{-5} - 5x^4)\, dx$

15 $\displaystyle\int_3^2 \frac{x^2 - 1}{x - 1}\, dx$

16 $\displaystyle\int_0^{-1} \frac{x^3 + 8}{x + 2}\, dx$

17 $\int_1^1 (4x^2 - 5)^{100}\, dx$

18 $\int_5^5 \sqrt[3]{x^2 + \sqrt{x^5 + 1}}\, dx$

19 $\int_1^3 \dfrac{2x^3 - 4x^2 + 5}{x^2}\, dx$

20 $\int_{-2}^{-1} \left(x - \dfrac{1}{x}\right)^2 dx$

21 $\int_{-3}^6 |x - 4|\, dx$

22 $\int_{-1}^5 |2x - 3|\, dx$

23 $\int_1^4 \sqrt{5 - x}\, dx$

24 $\int_1^5 \sqrt[3]{2x - 1}\, dx$

25 $\int_{-1}^1 (v^2 - 1)^3 v\, dv$

26 $\int_{-2}^0 \dfrac{v^2}{(v^3 - 2)^2}\, dv$

27 $\int_0^1 \dfrac{1}{(3 - 2x)^2}\, dx$

28 $\int_0^4 \dfrac{x}{\sqrt{x^2 + 9}}\, dx$

29 $\int_1^4 \dfrac{1}{\sqrt{x}(\sqrt{x} + 1)^3}\, dx$

30 $\int_0^1 (3 - x^4)^3 x^3\, dx$

31 $\int_{\pi/2}^{\pi} \cos(\tfrac{1}{3}x)\, dx$

32 $\int_0^{\pi/2} 3\sin(\tfrac{1}{2}x)\, dx$

33 $\int_{\pi/4}^{\pi/3} (4\sin 2\theta + 6\cos 3\theta)\, d\theta$

34 $\int_{\pi/6}^{\pi/4} (1 - \cos 4\theta)\, d\theta$

35 $\int_{-\pi/6}^{\pi/6} (x + \sin 5x)\, dx$

36 $\int_0^{\pi/3} \dfrac{\sin x}{\cos^2 x}\, dx$

Exer. 37–40: Is the calculation or argument valid? Explain.

37 $\int_0^{\pi} \sec^2 x\, dx = \big[\tan x\big]_0^{\pi} = \tan \pi - \tan 0 = 0 - 0 = 0$

38 $\int_0^{\pi} \cos^2 x\, dx = \left[\dfrac{x}{2} + \dfrac{\sin 2x}{4}\right]_0^{\pi}$

$= \left(\dfrac{\pi}{2} + 0\right) - (0 + 0) = \dfrac{\pi}{2}$

39 If $f(x) = x^3$, then since $f(-x) = -f(x)$, we have $\int_{-1}^0 f(x)\, dx = -\int_0^1 f(x)\, dx$ and hence $\int_{-1}^1 f(x)\, dx = 0$.

40 If $f(x) = 1/x^3$, then since $f(-x) = -f(x)$, we have $\int_{-1}^0 f(x)\, dx = -\int_0^1 f(x)\, dx$ and hence $\int_{-1}^1 f(x)\, dx = 0$.

Exer. 41–44: (a) Find a number z that satisfies the conclusion of the mean value theorem (4.28) for the given integral $\int_a^b f(x)\, dx$. **(b)** Find the average value of f on $[a, b]$.

41 $\int_0^4 \dfrac{x}{\sqrt{x^2 + 9}}\, dx$

42 $\int_{-2}^0 \sqrt[3]{x + 1}\, dx$

43 $\int_0^5 \sqrt{x + 4}\, dx$

44 $\int_{-3}^2 \sqrt{6 - x}\, dx$

Exer. 45–48: Find the derivative without integrating.

45 $\dfrac{d}{dx} \int_0^3 \sqrt{x^2 + 16}\, dx$

46 $\dfrac{d}{dx} \int_0^1 x\sqrt{x^2 + 4}\, dx$

47 $\dfrac{d}{dx} \int_0^x \dfrac{1}{t + 1}\, dt$

48 $\dfrac{d}{dx} \int_0^x \dfrac{1}{\sqrt{1 - t^2}}\, dt, \quad |x| < 1$

49 A point P is moving on a coordinate line with a continuous acceleration function a. If v is the velocity function, then the *average acceleration* on a time interval $[t_1, t_2]$ is

$$\dfrac{v(t_2) - v(t_1)}{t_2 - t_1}.$$

Show that the average acceleration is equal to the average value of a on $[t_1, t_2]$.

50 If a function f has a continuous derivative on $[a, b]$, show that the average rate of change of $f(x)$ with respect to x on $[a, b]$ (see Definition 2.4) is equal to the average value of f' on $[a, b]$.

51 The vertical distribution of velocity of the water in a river may be approximated by $v = c(d - y)^{1/6}$, where v is the velocity (in meters per second) at a depth of y meters below the water surface, d is the depth of the river, and c is a positive constant.

 (a) Find a formula for the average velocity v_{av} in terms of d and c.

 (b) If v_0 is the velocity at the surface, show that $v_{av} = \tfrac{6}{7} v_0$.

52 In the electrical circuit shown in the figure, the alternating current I is given by $I = I_M \sin \omega t$, where t is the time and I_M is the maximum current. The rate P at which heat is being produced in the resistor of R ohms is given by $P = I^2 R$. Compute the *average rate* of production of heat over one complete cycle (from $t = 0$ to $t = 2\pi/\omega$). (*Hint:* Use the half-angle formula for the sine.)

Exercise 52

53 If a ball is dropped from a height of s_0 feet above the ground and air resistance is negligible, then the distance that it falls in t seconds is $16t^2$ feet. Use Definition (4.29) to show that the average velocity for the ball's journey to the ground is $4\sqrt{s_0}$ ft/sec.

54 A meteorologist determines that the temperature T (in °F) on a cold winter day is given by

$$T = \tfrac{1}{20}t(t - 12)(t - 24),$$

where t is time (in hours) and $t = 0$ corresponds to midnight. Find the average temperature between 6 A.M. and 12 noon.

55 If g is differentiable and f is continuous for every x, prove that

$$\frac{d}{dx}\int_a^{g(x)} f(t)\, dt = f(g(x))g'(x).$$

56 Extend the formula in Exercise 55 to

$$\frac{d}{dx}\int_{k(x)}^{g(x)} f(t)\, dt = f(g(x))g'(x) - f(k(x))k'(x).$$

Exer. 57–60: Use Exercises 55 and 56 to find the derivative.

57 $\dfrac{d}{dx}\displaystyle\int_2^{x^4} \dfrac{t}{\sqrt{t^3 + 2}}\, dt$

58 $\dfrac{d}{dx}\displaystyle\int_0^{x^2} \sqrt[3]{t^4 + 1}\, dt$

59 $\dfrac{d}{dx}\displaystyle\int_{3x}^{x^3} (t^3 + 1)^{10}\, dt$

60 $\dfrac{d}{dx}\displaystyle\int_{1/x}^{\sqrt{x}} \sqrt{t^4 + t^2 + 4}\, dt$

4.7 NUMERICAL INTEGRATION

In this section, we will study several techniques of numerical integration that help us approximate definite integrals to any desired degree of accuracy. Evaluating a definite integral $\int_a^b f(x)\, dx$ by the fundamental theorem of calculus requires having an antiderivative for f. If we cannot obtain an antiderivative, we may use these numerical methods to obtain very accurate approximations. To emphasize their geometric nature, we illustrate these methods for functions with $f(x) \geq 0$ on $[a, b]$.

RECTANGLE RULES

Recalling Definition (4.16) and assuming that the definite integral $\int_a^b f(x)\, dx$ exists, we approximate its value, as a sum of areas of rectangles, using any Riemann sum of f. In particular, if we use a regular partition with $\Delta x = (b - a)/n$, then $x_k = a + k\Delta x$ for $k = 0, 1, 2, \ldots, n$, and

$$\int_a^b f(x)\, dx \approx \sum_{k=1}^n f(w_k)\Delta x,$$

where w_k is any number in the kth subinterval $[x_{x-1}, x_k]$ of the partition. (Refer to Figure 4.12 on page 379.) Each term $f(w_k)\Delta x$ in the sum is the area of a rectangle of width Δx and height $f(w_k)$. The accuracy of such an approximation to $\int_a^b f(x)\, dx$ by rectangles is affected by both the location of w_k within each subinterval and the width Δx of the rectangles.

As we saw in Section 4.4, by locating each w_k at a left-hand endpoint x_{k-1}, we obtain a left endpoint approximation. Alternatively, by locating each w_k at a right-hand endpoint x_k, we obtain a right endpoint approximation. A third possibility is to let w_k be the midpoint of each subinterval;

then $w_k = (x_{k-1} + x_k)/2$. This choice of location for w_k gives a midpoint approximation. Using the notation $x_{k-1/2}$ to indicate this midpoint, $(x_{k-1} + x_k)/2$, we formalize the three choices for the location of w_k in the following rules.

Rectangle Rules   4.36

For a regular partition of an interval $[a, b]$ with n subintervals, each of width $\Delta x = (b - a)/n$, the definite integral $\int_a^b f(x)\, dx$ is approximated by

(i) the **left rectangle rule:**

$$L_n = \sum_{k=1}^{n} f(x_{k-1}) \Delta x$$

(ii) the **right rectangle rule:**

$$R_n = \sum_{k=1}^{n} f(x_k) \Delta x$$

(iii) the **midpoint rule:**

$$M_n = \sum_{k=1}^{n} f(x_{k-1/2}) \Delta x$$

If a function is strictly increasing or strictly decreasing over the interval, then the endpoint rules in (4.36) give the areas of the inscribed and circumscribed rectangles. Figure 4.32 shows a function f that is decreasing over the interval $[a, b]$. In Figure 4.32(a), the left rectangle rule L_n gives the sum of the areas of the circumscribed rectangles; it overestimates the definite integral. The gray-shaded area represents the error resulting from the left rectangle rule. That is, the gray-shaded regions are contained within the circumscribed rectangles but are not under the graph of f.

Figure 4.32

(a) $L_n = \sum_{k=1}^{n} f(x_{k-1}) \Delta x$

(b) $R_n = \sum_{k=1}^{n} f(x_k) \Delta x$

(c) $M_n = \sum_{k=1}^{n} f(x_{k-1/2}) \Delta x$

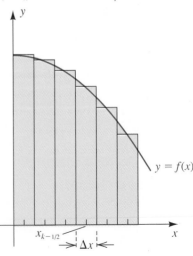

Similarly, for a decreasing function, as shown in Figure 4.32(b), the right rectangle rule gives the sum of the areas of the inscribed rectangles, and it underestimates the definite integral. The gray-shaded area shows the resulting error, which is made up of the regions under the graph that are not included within the inscribed rectangles. Finally, we see in Figure 4.32(c) that the midpoint rule appears to give a better approximation of the definite integral. As indicated by the gray-shaded area, the resulting error includes regions under the graph that are not within the rectangles as well as portions of the rectangles that are not under the graph. These areas of error may partially offset each other and yield a more accurate estimate of the original definite integral. Thus, the midpoint rule M_n often gives a number that lies between the left rectangle rule L_n and the right rectangle rule R_n.

In the next example, we apply the left rectangle, the right rectangle, and the midpoint rules to determine approximations for the definite integral of a specific function on a prescribed interval.

EXAMPLE ■ 1 Approximate $\int_1^2 1/x\, dx$ using a regular partition with $n = 4$, using

(a) the midpoint rule M_n　　　　**(b)** the left rectangle rule L_n
(c) the right rectangle rule R_n

SOLUTION With $n = 4$, we have $\Delta x = (b - a)/n = (2 - 1)/4 = 1/4$, and the function f is given by $f(x) = 1/x$. The endpoints of the subintervals are $x_0 = 1, x_1 = \frac{5}{4}, x_2 = \frac{3}{2}, x_3 = \frac{7}{4}$, and $x_4 = 2$.

(a) The midpoints are $x_{1/2} = \frac{9}{8}, x_{3/2} = \frac{11}{8}, x_{5/2} = \frac{13}{8}$, and $x_{7/2} = \frac{15}{8}$ (see Figure 4.33). By (4.36)(iii), we obtain

$$\int_a^b f(x)\, dx = \int_1^2 \frac{1}{x}\, dx \approx M_4 = \sum_{k=1}^4 f(x_{k-1/2})\Delta x$$

$$= \sum_{k=1}^4 \left(\frac{1}{x_{k-1/2}}\right)\left(\frac{1}{4}\right)$$

$$= \frac{1}{4}\sum_{k=1}^4 \left(\frac{1}{x_{k-1/2}}\right)$$

$$= \tfrac{1}{4}(\tfrac{8}{9} + \tfrac{8}{11} + \tfrac{8}{13} + \tfrac{8}{15}) = \frac{4448}{6435}$$

$$\approx 0.6912198912.$$

(b) The left-hand endpoints are $1, \frac{5}{4}, \frac{3}{2}$, and $\frac{7}{4}$. By (4.36)(i),

$$\int_1^2 \frac{1}{x}\, dx \approx L_4 = \frac{1}{4}\left(1 + \frac{4}{5} + \frac{2}{3} + \frac{4}{7}\right) = \frac{319}{420}$$

$$\approx 0.7595238095.$$

Figure 4.33
$f(x) = 1/x$

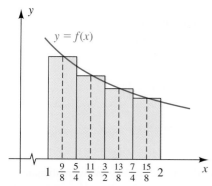

(c) The right-hand endpoints are $\frac{5}{4}, \frac{3}{2}, \frac{7}{4}$, and 2. By (4.36)(ii),

$$\int_1^2 \frac{1}{x}\, dx \approx R_4 = \frac{1}{4}\left(\frac{4}{5} + \frac{2}{3} + \frac{4}{7} + \frac{1}{2}\right)$$

$$= \frac{533}{840} \approx 0.6345238095.$$

In Chapter 6, we will see that the correct value to ten decimal places for $\int_1^2 1/x\, dx$ is 0.6931471806. We note that the midpoint rule, which yields a number between those given by the left and the right rectangle rules, gives a better approximation than either endpoint rule.

Note that in each of the rectangle rules (4.36), Δx is a constant factor so that we can also write these rules as

$$L_n = \Delta x \sum_{k=1}^{n} f(x_{k-1}), \quad R_n = \Delta x \sum_{k=1}^{n} f(x_k), \quad M_n = \Delta x \sum_{k=1}^{n} f(x_{k-1/2}).$$

Once we compute the left endpoint or the right endpoint approximation, it is easy to determine the other endpoint approximation since the right-hand endpoint of one subinterval is the left-hand endpoint of the next interval:

$$L_n = \Delta x\left[f(x_0) + \sum_{k=1}^{n-1} f(x_k)\right] \quad \text{and} \quad R_n = \Delta x\left[\sum_{k=1}^{n-1} f(x_k) + f(x_n)\right]$$

If we let $C = \sum_{k=1}^{n-1} f(x_k)$ and note that $x_0 = a$ and $x_n = b$, we have

$$R_n - L_n = \Delta x[C + f(b)] - \Delta x[f(a) + C] = \Delta x[f(b) - f(a)]$$

or, equivalently,

$$L_n = R_n + \Delta x[f(a) - f(b)].$$

Thus, for the case of Example 1, we can find the left endpoint approximation from the right endpoint approximation, as follows:

$$L_4 = R_4 + (\tfrac{1}{4})[f(1) - f(2)]$$
$$\approx 0.6345238095 + (\tfrac{1}{4})(1 - \tfrac{1}{2})$$
$$= 0.7595238095$$

TRAPEZOIDAL RULES

Since the left and right rectangle rules often yield under- or overestimates, it is natural to consider a numerical integration rule based on their average, T_n:

$$T_n = \tfrac{1}{2}(L_n + R_n) = (\tfrac{1}{2})\left(\sum_{k=1}^{n} f(x_{k-1})\Delta x + \sum_{k=1}^{n} f(x_k)\Delta x\right)$$

$$= \sum_{k=1}^{n} (\tfrac{1}{2})\left[f(x_{k-1}) + f(x_k)\right]\Delta x.$$

Trapezoidal Rule **4.37**

For a regular partition of an interval $[a, b]$ with n subintervals, each of width $\Delta x = (b - a)/n$, the definite integral $\int_a^b f(x)\, dx$ is approximated by the **trapezoidal rule**:

$$T_n = \tfrac{1}{2}(L_n + R_n) = \sum_{k=1}^{n} \tfrac{1}{2}\left[f(x_{k-1}) + f(x_k)\right]\Delta x$$

$$= \frac{b - a}{2n}\left[f(x_0) + 2f(x_1) + 2f(x_2) + \cdots + 2f(x_{n-1}) + f(x_n)\right]$$

Note that since each term in the sum for T_n has a constant factor $(\tfrac{1}{2})\Delta x = (b - a)/2n$, we can also write the sum as

$$T_n = \frac{b - a}{2n}\sum_{k=1}^{n}\left[f(x_{k-1}) + f(x_k)\right].$$

The last equality in (4.37) follows from the relation between the left rectangle and right rectangle rules,

$$L_n = R_n + \Delta x[f(a) - f(b)] = R_n + \Delta x[f(x_0) - f(x_n)],$$

and the definition of the trapezoidal rule as the average of L_n and R_n:

$$T_n = \tfrac{1}{2}(L_n + R_n) = \tfrac{1}{2}(R_n + \Delta x[f(x_0) - f(x_n)] + R_n)$$

$$= \tfrac{1}{2}\left(2R_n + \Delta x[f(x_0) - f(x_n)]\right)$$

$$= \tfrac{1}{2}\left(2\Delta x \sum_{k=1}^{n} f(x_k) + \Delta x\left[f(x_0) - f(x_n)\right]\right)$$

$$= \frac{\Delta x}{2}\left(2\left[f(x_1) + f(x_2) + \cdots + f(x_{n-1}) + f(x_n)\right] + f(x_0) - f(x_n)\right)$$

$$= \frac{b - a}{2n}\left[f(x_0) + 2f(x_1) + 2f(x_2) + \cdots + 2f(x_{n-1}) + f(x_n)\right]$$

Figure 4.34 provides a graphical interpretation of the trapezoidal rule. Each term $\tfrac{1}{2}[f(x_{k-1}) + f(x_k)]\Delta x$ in the sum is the area of a trapezoid

Figure 4.34

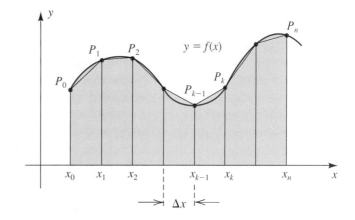

formed by the sec   line joining the endpoints of the graph over the kth subinterval $[x_{k-1}, x_k]$, the interval itself, and the vertical segments above x_{k-1} and x_k. The gray-shaded regions in the figure show the error when we approximate the area under the graph of f by the area of the trapezoid.

EXAMPLE ■ 2   Approximate $\int_1^2 1/x \, dx$ using a regular partition with $n = 4$, using the trapezoidal rule T_n.

SOLUTION   With the results of Example 1, we have

$$T_4 = \tfrac{1}{2}(L_4 + R_4)$$
$$= \tfrac{1}{2}(0.7595238095 + 0.6345238095)$$
$$= 0.6970238095,$$

which is closer to the correct value (to ten decimal places) of 0.6931471806 than either L_4 or R_4.

Alternatively, we can compute T_4 directly from the last form of the trapezoidal rule in (4.37):

$$T_n = \frac{2-1}{2(4)} \left[f(1) + 2f\left(\frac{5}{4}\right) + 2f\left(\frac{3}{2}\right) + 2f\left(\frac{7}{4}\right) + f(2) \right]$$
$$= \tfrac{1}{8}\left[1 + 2(\tfrac{4}{5}) + 2(\tfrac{2}{3}) + 2(\tfrac{4}{7}) + (\tfrac{1}{2}) \right]$$
$$= \tfrac{1}{8}(1 + \tfrac{8}{5} + \tfrac{4}{3} + \tfrac{8}{7} + \tfrac{1}{2})$$
$$= \tfrac{1}{8}(\tfrac{1171}{210}) = \tfrac{1171}{1680} \approx 0.6970238095$$

Figure 4.35

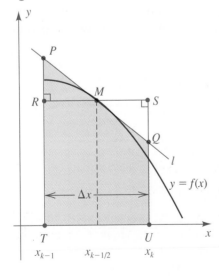

We can obtain other trapezoidal approximations for the area under the graph of f over a subinterval. For example, as illustrated in Figure 4.35, we can construct a nonvertical line l through the point M, which lies on the graph over the midpoint of the interval. Extending this line until it meets the vertical lines over x_{k-1} and x_k at points P and Q, respectively, forms a trapezoid $TPQU$. Adding a horizontal line through M forms a rectangle $TRSU$, whose area is one of the terms of the midpoint rule. Using elementary geometry, it can be shown that the area of the trapezoid $TPQU$ is equal to the area of the rectangle $TRSU$.

Note that this result is independent of the shape of the graph of f. If we take any nonvertical line through M, *the resulting trapezoid has the same area as the midpoint rectangle.* Thus, in addition to having a second trapezoidal approximation, we also have an alternative geometric way of viewing the midpoint rule. By an appropriate choice of the line l, we may be able to see if the midpoint rule gives an underestimate or an overestimate of the definite integral. We may be able to choose the line l, as in Figure 4.36, so that the entire area under the curve is below the line; thus the midpoint rule will overestimate the definite integral on this subinterval.

Figure 4.36

 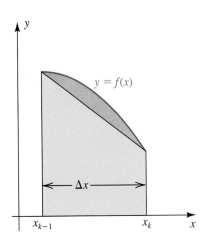

If we examine the difference between the midpoint rule and the trapezoidal rule on one subinterval (Figure 4.36), we see that the midpoint rule is an overestimate and the trapezoidal rule is an underestimate. The midpoint rule has about half the error of the trapezoidal rule.

SIMPSON'S RULE

Recall that the trapezoidal rule, which averages the results of the left rectangle rule and the right rectangle rule, is an improvement over each of them. We may do even better by combining the midpoint rule and the trapezoidal rule. The British mathematician Thomas Simpson (1710–1761) suggested a combination, using a "weighted average," where M_n is counted twice as heavily as T_n.

Simpson's Rule 4.38

For a regular partition of an interval $[a, b]$ with n subintervals, each of width $\Delta x = (b - a)/n$, the definite integral $\int_a^b f(x)\,dx$ is approximated by **Simpson's rule**:

$$S_n = \tfrac{1}{3}(2M_n + T_n)$$

$$= \frac{b - a}{6n}[f(x_0) + 4f(x_{1/2}) + 2f(x_1) + 4f(x_{3/2})$$
$$+ 2f(x_2) + \cdots + 2f(x_{n-1}) + 4f(x_{n-1/2}) + f(x_n)]$$

The last equality in (4.38) follows from the fact that

$$M_n = [f(x_{1/2}) + f(x_{3/2}) + f(x_{5/2}) + \cdots + f(x_{n-1/2})]\Delta x$$

can be combined with the final expression for the trapezoidal rule in (4.37).

The next example shows the result of applying Simpson's rule to the definite integral given in Examples 1 and 2.

EXAMPLE ▪ 3   Approximate $\int_1^2 1/x \, dx$ using a regular partition with $n = 4$, using Simpson's rule S_n.

SOLUTION   Using the results of Examples 1 and 2, we have

$$S_4 = \tfrac{1}{3}(2M_4 + T_4)$$

$$\approx \tfrac{1}{3}[2(0.6912198912) + 0.6970238095] \approx 0.6931545306.$$

Comparing this result to the correct value (to ten decimal places) of 0.6931471806, we see that Simpson's rule gives the best approximation, followed by the midpoint rule and then the trapezoidal rule. Alternatively, we can compute S_4 directly from the last form of Simpson's rule in (4.38):

$$S_4 = \frac{2-1}{6(4)}\left[f(1) + 4f\left(\frac{9}{8}\right) + 2f\left(\frac{5}{4}\right) + 4f\left(\frac{11}{8}\right) + 2f\left(\frac{3}{2}\right)\right.$$

$$\left. + 4f\left(\frac{13}{8}\right) + 2f\left(\frac{7}{4}\right) + 4f\left(\frac{15}{8}\right) + f(2)\right]$$

$$= \tfrac{1}{24}(1 + \tfrac{32}{9} + \tfrac{8}{5} + \tfrac{32}{11} + \tfrac{4}{3} + \tfrac{32}{13} + \tfrac{8}{7} + \tfrac{32}{15} + \tfrac{1}{2})$$

$$= \tfrac{1}{24}\left(\frac{35,969,064}{2,162,160}\right) = \left(\frac{1,498,711}{2,162,160}\right) \approx 0.6931545307$$

The numerical integration techniques we have considered up to now approximate the *region* under the graph lying over a small subinterval by a simpler region (a rectangle or a trapezoid) whose area is found by simple geometric formulas. Another conceptual approach to numerical integration also leads to Simpson's rule: We replace the *function f* by a simpler function g whose graph closely approximates the graph of f on each subinterval. We then integrate the simpler function by finding its antiderivative and approximate $\int_{x_{k-1}}^{x_k} f(x)\, dx$ by $\int_{x_{k-1}}^{x_k} g(x)\, dx$. In this perspective, note that a rectangle rule replaces f by a constant function. The trapezoidal rule replaces f by a linear function that matches the values of f at the endpoints of the subinterval.

For Simpson's rule, on each subinterval, we replace the function f by a *quadratic function* g that matches the value of f at the endpoints and the midpoint—that is, on each subinterval, the graph of g is a parabola with three points in common with the graph of f, as shown in Figure 4.37.

A quadratic function can be written in the form $g(x) = c + bx + ax^2$ for constants c, b, and a. It will be easier to use an equivalent form

$$g(x) = c_0 + c_1(x - x_{k-1/2}) + c_2(x - x_{k-1/2})^2$$

on the subinterval $[x_{k-1}, x_k]$.

We must first determine the values of the coefficients $c_0, c_1,$ and c_2 so that the values of f and g are equal at the endpoints and at the midpoint of the subinterval. To do so, we need to satisfy the conditions

$$g(x_{k-1}) = f(x_{k-1}), \quad g(x_{k-1/2}) = f(x_{k-1/2}), \quad \text{and} \quad g(x_k) = f(x_k),$$

and then use these equations to determine $c_0, c_1,$ and c_2.

Figure 4.37

1. At the midpoint $x = x_{k-1/2}$, we have

$$g(x_{k-1/2}) = c_0 + c_1(0) + c_2(0)^2 = c_0.$$

For agreement at the midpoint, we need $g(x_{k-1/2}) = f(x_{k-1/2})$, so we have $c_0 = f(x_{k-1/2})$.

2. At the left endpoint $x = x_{k-1}$, we have

$$x - x_{k-1/2} = x_{k-1} - x_{k-1/2} = -\tfrac{1}{2}\Delta x,$$

so $\quad g(x_{k-1}) = f(x_{k-1/2}) - \tfrac{1}{2}c_1\Delta x + \tfrac{1}{4}c_2(\Delta x)^2.$

To obtain agreement with f at the left endpoint, we need $g(x_{k-1}) = f(x_{k-1})$—that is,

(I) $\qquad f(x_{k-1/2}) - \tfrac{1}{2}c_1\Delta x + \tfrac{1}{4}c_2(\Delta x)^2 = f(x_{k-1}).$

3. At the right endpoint $x = x_k$, we have $x - x_{k-1/2} = \tfrac{1}{2}\Delta x$, so

$$g(x_k) = f(x_{k-1/2}) + \tfrac{1}{2}c_1\Delta x + \tfrac{1}{4}c_2(\Delta x)^2.$$

To achieve agreement with f at the right endpoint, we must have $g(x_k) = f(x_k)$ or, equivalently,

(II) $\qquad f(x_{k-1/2}) + \tfrac{1}{2}c_1\Delta x + \tfrac{1}{4}c_2(\Delta x)^2 = f(x_k).$

Thus, in order for g and f to agree at the three points, we must solve the equations (I) and (II) for c_1 and c_2. Doing so yields

$$c_1 = \left\{ \frac{f(x_k) - f(x_{k-1})}{\Delta x} \right\}, \quad c_2 = \left\{ \frac{2[f(x_k) - 2f(x_{k-1/2}) + f(x_{k-1})]}{(\Delta x)^2} \right\}.$$

Thus, we have determined the coefficients (c_0, c_1, and c_2) of the quadratic function g.

Once we have explicitly found the function g, we approximate $\int_{x_{k-1}}^{x_k} f(x)\,dx$ by $\int_{x_{k-1}}^{x_k} g(x)\,dx$. The integral of the quadratic function is

$$\int_{x_{k-1}}^{x_k} \left[c_0 + c_1(x - x_{k-1/2}) + c_2(x - x_{k-1/2})^2 \right] dx$$

$$= \left[c_0 x + \tfrac{1}{2}c_1(x - x_{k-1/2})^2 + \tfrac{1}{3}c_2(x - x_{k-1/2})^3 \right]_{x_{k-1}}^{x_k}$$

$$= \left[c_0 x_k + \tfrac{1}{2}c_1(x_k - x_{k-1/2})^2 + \tfrac{1}{3}c_2(x_k - x_{k-1/2})^3 \right]$$

$$\quad - \left[c_0 x_{k-1} + \tfrac{1}{2}c_1(x_{k-1} - x_{k-1/2})^2 + \tfrac{1}{3}c_2(x_{k-1} - x_{k-1/2})^3 \right]$$

$$= c_0(x_k - x_{k-1}) + \tfrac{1}{2}c_1(\tfrac{1}{2}\Delta x)^2 + \tfrac{1}{3}c_2(\tfrac{1}{2}\Delta x)^3$$

$$\quad - \tfrac{1}{2}c_1(-\tfrac{1}{2}\Delta x)^2 - \tfrac{1}{3}c_2(-\tfrac{1}{2}\Delta x)^3$$

$$= c_0(\Delta x) + \tfrac{1}{12}c_2(\Delta x)^3,$$

which becomes, after substituting the values for c_0 and c_2 we found above,

$$\tfrac{1}{6}\left[f(x_{k-1}) + 4f(x_{k-1/2}) + f(x_k) \right]\Delta x,$$

which are exactly the terms we have for Simpson's rule. This result gives an alternative justification for Simpson's rule: We can derive Simpson's rule either by beginning with a weighted average of the midpoint and trapezoidal rules or by approximating the graph of f with parabolas.

In some treatments of numerical integration, a different form of Simpson's rule is used. Instead of using quadratics whose values match those of the function f at the endpoints and the midpoint of each subinterval $[x_{k-1}, x_k]$, this other form uses an *even* value for n. It then divides the interval $[a, b]$ into $n/2$ subintervals and uses a quadratic for each of these $n/2$ subintervals. In this approach, the first quadratic matches f at x_0, x_1, x_2, the next quadratic matches f at x_2, x_3, x_4, and so forth. Note that Simpson's rule (4.38) can be used for an odd or an even value of n. If you use a software package on a computer or a built-in function on a calculator for numerical integration, consult the reference manual to determine which form of Simpson's rule is being used.

In the next example of numerical integration, we approximate the definite integral of a function known only by a table of function values with equally spaced x-coordinates. In applications, results obtained from an experiment frequently provide only function *values*, rather than a formula for the function.

Figure 4.38

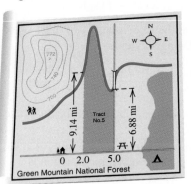

Green Mountain National Forest

EXAMPLE 4 Aerial surveys of a tract of the Green Mountain National Forest shown in Figure 4.38 measured the width of the forest (in miles) at regularly spaced intervals, $\frac{3}{10}$ mi apart. The gathered data are shown in the following table.

x	2.0	2.3	2.6	2.9	3.2	3.5	3.8	4.1	4.4	4.7	5.0
y	9.14	11.82	13.41	13.72	12.87	11.27	9.42	7.81	6.78	6.49	6.88

The Forest Service estimates that, on average, there are 125 mature trees per acre. Approximate the total number of mature trees in this tract of the forest using the rules of numerical integration.

SOLUTION We must first obtain estimates for the forest's land area in square miles. To do so, we use the data in the table and consider the forest area as the definite integral of the function $y = f(x)$ over the interval $[2, 5]$.

For the left and right rectangle rules and for the trapezoidal rule, we can choose $n = 10$ and $\Delta x = (5.0 - 2.0)/10 = 0.3$. Using our earlier observation that $R_n = \Delta x \left[\sum_{k=1}^{n-1} f(x_k) + f(x_n) \right]$, we have

$$R_{10} = (0.3) \left[\sum_{k=1}^{9} f(x_k) + f(x_{10}) \right]$$

$$= (0.3)[(11.82 + 13.41 + 13.72 + 12.87 + 11.27$$
$$+ 9.42 + 7.81 + 6.78 + 6.49) + 6.88]$$

$$= (0.3)(93.59 + 6.88) = 30.141$$

and $$L_{10} = R_{10} + \Delta x[f(a) - f(b)]$$
$$= 30.141 + (0.3)(9.14 - 6.88) = 30.819,$$

so that $$T_{10} = \tfrac{1}{2}(L_{10} + R_{10}) = 30.48.$$

For the midpoint rule and Simpson's rule, we consider every other x-value as a midpoint so that $n = 5$ and $\Delta x = 0.6$. Then we compute

$$L_5 = (9.14 + 13.41 + 12.87 + 9.42 + 6.78)(0.6) = 30.972,$$
$$R_5 = (13.41 + 12.87 + 9.42 + 6.78 + 6.88)(0.6) = 29.616,$$
$$M_5 = (11.82 + 13.72 + 11.27 + 7.81 + 6.49)(0.6) = 30.666,$$
$$T_5 = (L_5 + R_5)/2 = 30.294,$$

and $$S_5 = (2M_5 + T_5)/3 = 30.542.$$

The computed results are summarized in the following table, where we have rounded figures to two decimal places because the given data on the width of the forest can be assumed accurate to only two places.

n	L_n	R_n	M_n	T_n	S_n
5	30.97	29.62	30.67	30.29	30.54
10	30.82	30.14	—	30.48	—

From these figures, we estimate the tract of forest to be about 30.5 mi². Since there are 640 acres in a square mile, the forest is about 19,520 acres in extent. With an average of 125 trees per acre, the forest contains approximately $(19,520)(125) = 2,440,000$ mature trees.

DEPENDENCE ON Δx

We noted earlier in this section that both the location for w_k and the size Δx affect the accuracy of $\sum_{k=1}^{n} f(w_k)\Delta x$ as an approximation for $\int_a^b f(x)\,dx$. We have considered several different choices for locating w_k and now examine the size of Δx. Since the width $\Delta x = (b - a)/n$ depends on the number n of rectangles, the discussion will focus on n. Increasing n may increase the accuracy but it introduces more terms in the sum to calculate. We can find the approximations using a calculator or a computer program for different choices for n to see how increasing n improves the accuracy. The next example shows the numerical results for a particular definite integral using a program on a calculator that displays 12 significant digits and works internally with 14 digits.

 EXAMPLE ▪ 5 Use the numerical integration rules to approximate the definite integral $\int_0^1 [4/(1 + x^2)]\,dx$ for $n = 2, 6, 18, 54,$ and 162.

SOLUTION   The following table displays the results of running the computer program:

n	L_n	R_n	M_n	T_n	S_n
2	3.6	2.6	3.16235294118	3.1	3.14156862745
6	3.30362973314	2.9702963998	3.14390742722	3.13696306647	3.14159264031
18	3.19663380591	3.08552269480	3.14184985518	3.14107825036	3.14159265357
54	3.16005401619	3.12301697915	3.14162123155	3.14153549767	3.14159265359
162	3.14775914244	3.13541346343	3.14159582892	3.14158630293	3.14159265359

We can make several observations on the basis of an examination of the table.

As we increase n, the values given by the midpoint, the trapezoidal, and Simpson's rules all seem to approach a number whose first six significant digits are 3.14159. For the left rectangle rule, increases in n produce decreases in the values of L_n. These values also get closer to 3.14159. For the right rectangle rule, increases in n produce increases in the values of R_n, which also get closer to 3.14159.

In going from one value of n to the next value, we see that the change in the approximated values is greater for the rectangle rules than for either the trapezoidal or Simpson's rule. For example, when n increases from 6 to 18, the value for L_n changes by 0.10699592723, whereas the value for S_n changes by only -0.00000001326.

To gain a better understanding of the effect on the approximations of increases in n, we can compare the results in the table of Example 5 with the exact value of the definite integral. The next example discusses such a comparison.

EXAMPLE ▪ 6   The definite integral $\int_0^1 [4/(1 + x^2)]\,dx$ has a value of π. (We will prove this fact in Chapter 6.)

(a) Using the results of Example 5, compute the errors for each approximation by finding the difference between π and the approximation.

(b) Investigate the ratios of error for each successive pair of values for n.

SOLUTION

(a) Given the known value π for the result, we use a calculator to compute each error, that is, the correct value π minus the approximated value. The following table lists the results.

n	L_n	R_n	M_n	T_n	S_n
2	−4.584 E −1	5.416 E −1	−2.076 E −2	4.159 E−2	2.403 E −5
6	−1.620 E −1	1.713 E −1	−2.315 E −3	4.630 E −3	1.328 E −8
18	−5.504 E −2	5.607 E −2	−2.572 E −4	5.144 E −4	1.82 E−11
54	−1.846 E −2	1.858 E −2	−2.858 E −5	5.716 E −5	3 E −13
162	−6.166 E −3	6.179 E −3	−3.175 E −6	6.351 E −6	1 E −13

The data in the table indicate that the error decreases as n increases. We note too that when Simpson's rule is used, the error is extremely small even when $n = 2$.

(b) The next table shows the ratio of a column entry and the entry below it.

$n, n + 1$	L_n/L_{n+1}	R_n/R_{n+1}	M_n/M_{n+1}	T_n/T_{n+1}	S_n/S_{n+1}
2, 6	2.83	3.16	8.97	8.98	1809
6, 18	2.94	3.06	9.00	9.00	729.7
18, 54	2.98	3.02	9.00	9.00	60.7
54, 162	2.99	3.01	9.00	9.00	3

Note that each successive value of n is 3 times the preceding value. The errors for the left and right rectangle rules were approximately divided by 3 for each tripling of n, and the errors for the midpoint and trapezoidal rules were approximately divided by $9 = 3^2$. We see no pattern in the errors for Simpson's rule, which may be due to round-off errors that occur because one very small number is being divided by another.

The patterns we observed in the table of Example 6(b) are not coincidental. They follow from more general results about error estimates.

ERROR ESTIMATES

For the five numerical integration rules that we have considered in this section, we can find *error estimates*, or *bounds*, on the size of the error even if we do not know the exact value of the definite integral. If I is the actual value of the definite integral $\int_a^b f(x)\, dx$ and A_n is an approximated value using n rectangles, then the size of the error is $|I - A_n|$. By a **bound** on the size of the error, we mean a number B such that $|I - A_n| \leq B$. We can obtain bounds that depend on the number n of subintervals and the maximum value of derivatives of the function f. We state without proof the theorem describing these error estimates.

Theorem **4.39**

Let $I = \int_a^b f(x)\, dx$ be the definite integral being approximated. If f' is continuous and if K_1 is a positive number such that $|f'(x)| \le K_1$ for every x in $[a, b]$, then the **error estimates for the left rectangle rule** L_n and the **right rectangle rule** R_n are given by

$$|I - L_n| \le K_1 \frac{(b-a)^2}{2n} \quad \text{and} \quad |I - R_n| \le K_1 \frac{(b-a)^2}{2n}.$$

If f'' is continuous and if K_2 is a positive real number such that $|f''(x)| \le K_2$ for every x in $[a, b]$, then the **error estimates for the midpoint rule** M_n and the **trapezoidal rule** T_n are given by

$$|I - M_n| \le K_2 \frac{(b-a)^3}{24n^2} \quad \text{and} \quad |I - T_n| \le K_2 \frac{(b-a)^3}{12n^2}.$$

If $f^{(4)}$ is continuous and if K_4 is a positive real number such that $|f^{(4)}(x)| \le K_4$ for every x in $[a, b]$, then the **error estimate for Simpson's rule** S_n is given by

$$|I - S_n| \le K_4 \frac{(b-a)^5}{2880n^4}.$$

The next example illustrates how the error estimates in (4.39) can be used. If we can find values for K_1, K_2, or K_4, then we may use the estimates in (4.39) to determine how large n should be in order to ensure that a particular approximation is within a given margin of error.

EXAMPLE ▪ 7   Determine how large n must be in order to use the trapezoidal rule to approximate $I = \int_1^3 1/x\, dx$ with an error less than 10^{-3}.

SOLUTION   From Theorem (4.39), we have

$$|I - T_n| \le K_2 \frac{(3-1)^3}{12n^2} = \frac{2K_2}{3n^2},$$

where K_2 is a bound on the absolute value of the second derivative of $f(x) = 1/x$ on the interval $[1, 3]$. Since $f''(x) = 2/x^3$ is positive and decreasing on $[1, 3]$, its maximum value is $f''(1) = 2$. Therefore, we have $|I - T_n| \le 4/(3n^2)$. To ensure that the error is less than 10^{-3}, we must choose n so that

$$\frac{4}{3n^2} < 10^{-3},$$

which is equivalent to

$$n^2 > \tfrac{4000}{3}, \quad \text{or} \quad n > \sqrt{\tfrac{4000}{3}}$$
$$\approx 36.5.$$

Hence we should choose n to be at least 37 in order to guarantee an error less than 10^{-3}.

Before the availability of electronic computing devices, great efforts were made to estimate the constants K_1, K_2, and K_4. Once these numbers were known, the inequalities in (4.39) could be solved for n, as in Example 7, to determine how many subintervals n to use in order to obtain an approximation within a prescribed error. Today with inexpensive computing power (including hand-held programmable calculators), there is an alternative approach. We can obtain an approximation that is within a given margin of error by repeatedly computing a numerical integration rule for increasing values of n and observing the convergence of the estimates as n grows larger. We will illustrate this approach in Example 8.

This alternative approach is based on the fact that the error estimates (4.39) give the expected decrease in the error when we *change n* by a certain multiple. To illustrate, if we compare the error estimates for the trapezoidal rule with n and $5n$ subintervals, respectively, on the same definite integral, we have, by (4.39),

$$E_n = |I - T_n| \le K_2 \frac{(b-a)^3}{12n^2}$$

and

$$E_{5n} = |I - T_{5n}| \le K_2 \frac{(b-a)^3}{12(5n)^2}.$$

The ratio of these two error estimates is

$$\frac{E_n}{E_{5n}} = \frac{K_2[(b-a)^3/12n^2]}{K_2[(b-a)^3/12(5n)^2]} = 5^2 = 25.$$

Since $E_{5n} = \frac{1}{25} E_n$, we expect the error to decrease by a factor of 25 when we increase the number of subintervals from n to $5n$. We obtain similar expected decreases for the other numerical integration rules by examining the power of n in the denominator of the error estimates. For example, if we multiply n by 3, then, by (4.39), we expect the error in the left and right rectangle rules to be divided by 3, the error in the midpoint and trapezoidal rules to be divided by $3^2 = 9$, and the error in Simpson's rule to be divided by $3^4 = 81$.

EXAMPLE ■ 8

(a) Use Simpson's rule to approximate the definite integral $\int_1^{12} 1/x \, dx$ for $n = 5, 10, 20, 40,$ and 80.

(b) Discuss the expected accuracy of the final result.

SOLUTION

(a) We use Simpson's rule (4.38) for the integrand $f(x) = 1/x$, and display the results in a table:

n	S_n
5	2.50179046384
10	2.48685897261
20	2.48507069664
40	2.48491806727
80	2.48490738622

(b) Each time we double n, the error estimation in (4.39) predicts that the error in Simpson's rule will be divided by $2^4 = 16$, which means that we will add at least one correct decimal digit each time we double n. Thus, the estimate

$$\int_1^{12} \frac{1}{x}\, dx \approx 2.4849$$

is correct to at least four decimal places.

For the function $f(x) = 1/x$, we can easily find $f^{(4)}(x) = 24x^{-5}$. The largest value for this positive decreasing function on the interval $[1, 12]$ is $f^{(4)}(1) = 24$. Using the formal error estimate in (4.39) yields

$$|I - S_{80}| \le 24\frac{(12-1)^5}{2880(80)^4}$$
$$\approx 3.28 \times 10^{-5}.$$

CAUTION In some instances, increasing the size of n does not necessarily lead to a more accurate approximation to the value of a definite integral. When a calculator or a computer is used to implement one of the numerical integration rules, round-off errors can occur when the size of the numbers becomes so small that they cannot be stored precisely in the machine. When n is very large, the numerical integration rules add a very large number of terms. If there are sufficiently many terms (a large value of n), the sum of the round-off errors can be large enough to produce a less accurate estimate for the value of the definite integral than does a smaller value of n. Courses in numerical analysis explore such issues in greater depth.

EXERCISES 4.7

 Exer. 1–4: Use all five numerical integration rules with an appropriate n to approximate the definite integral of the function $y = f(x)$ over the interval $[2, 5]$ when the function values are as given in the table.

1

x	2.0	2.5	3.0	3.5	4.0	4.5	5.0
y	3.2	2.7	4.1	3.8	3.5	4.6	5.2

2

x	2.0	2.75	3.5	4.25	5.0
y	15.2	17.1	18.6	19.2	20.4

3

x	y
2.00	4.12
2.375	3.76
2.75	3.21
3.125	3.58
3.50	3.94
3.875	4.15
4.25	4.69
4.625	5.44
5.00	7.52

4

x	y
2.0	12.1
2.3	10.4
2.6	8.4
2.9	6.2
3.2	5.8
3.5	5.3
3.8	5.9
4.1	6.4
4.4	7.6
4.7	9.0
5.0	12.1

 Exer. 5–6: **(a)** Use Riemann sums with both left-hand endpoints and right-hand endpoints to approximate the definite integral of the function $y = f(x)$ over the interval $[4, 6]$ when the unequally spaced function values are as given in the table. **(b)** Find a trapezoidal rule estimate for the given partition of $[4, 6]$ by averaging the two estimates from part **(a)**.

5

x	y
4.00	0.386
4.15	0.423
4.35	0.470
4.50	0.504
4.75	0.558
5.10	0.629
5.30	0.668
5.65	0.732
6.00	0.792

6

x	y
4.000	3.812
4.587	1.392
4.954	-2.250
5.223	-3.128
5.434	-2.435
5.608	-1.225
5.756	-0.029
5.886	0.950
6.000	1.637

Exer. 7–14: **(a)** Approximate the definite integral using the indicated rule for the given values of n. **(b)** Evaluate the definite integral exactly, and compute the errors for each approximation. **(c)** Determine how the error changes for successive computations.

7 $\displaystyle\int_{1}^{1.6} (2x - 1)\, dx$;   left rectangle rule for $n = 3, 6$, and 12

8 $\displaystyle\int_{1}^{3} (x^2 + 1)\, dx$;   right rectangle rule for $n = 4, 8$. and 16

9 $\displaystyle\int_{1}^{5} x^3\, dx$;   midpoint rule for $n = 2, 4$, and 8

10 $\displaystyle\int_{-1}^{1} (x^2 + 5x + 1)\, dx$;   midpoint rule for $n = 1, 4$, and 16

11 $\displaystyle\int_{1}^{5} x^3\, dx$;   trapezoidal rule for $n = 2, 4$, and 8

12 $\displaystyle\int_{-1}^{1} (x^2 + 5x + 1)\, dx$;   trapezoidal rule for $n = 1, 4$, and 16

13 $\displaystyle\int_{1}^{5} x^3\, dx$;   Simpson's rule for $n = 2, 4$, and 8

14 $\displaystyle\int_{-1}^{1} (x^2 + 5x + 1)\, dx$;   Simpson's rule for $n = 1, 4$, and 16

 Exer. 15–18: **(a)** Approximate the definite integral using the indicated rule for the given values of n. **(b)** On the basis of the pattern of values, determine the expected accuracy for the approximation corresponding to the largest n.

15 $\displaystyle\int_{1}^{3} \sqrt{1 + x^3}\, dx$;   trapezoidal rule for $n = 5, 10, 20$, and 40

16 $\displaystyle\int_{0}^{5} 2^{-x^2}\, dx$;   trapezoidal rule for $n = 2, 8, 32$, and 128

17 $\displaystyle\int_{0}^{\pi} \cos(\sin x)\, dx$;   Simpson's rule for $n = 2, 6, 18$, and 54

18 $\displaystyle\int_{0}^{4} \frac{1}{1 + x^3}\, dx$;   Simpson's rule for $n = 8, 16, 32$, and 64

Exer. 19–22: Use Theorem (4.39) to estimate the maximum error in approximating the definite integral for the given value of *n*, using (a) the trapezoidal rule and (b) Simpson's rule.

19 $\int_{-2}^{3} (\frac{1}{360}x^6 + \frac{1}{60}x^5)\, dx; \quad n = 25$

20 $\int_{0}^{3} (-\frac{1}{12}x^4 + \frac{2}{3}x^3)\, dx; \quad n = 24$

21 $\int_{1}^{5} \frac{1}{x^2}\, dx; \quad\quad\quad\quad n = 16$

22 $\int_{1}^{4} \frac{1}{35}x^{7/2}\, dx; \quad\quad n = 15$

Exer. 23–26: Using Theorem (4.39), find the least integer *n* such that the error estimate in approximating the definite integral is less than the given *E* when using (a) the left rectangle rule, (b) the midpoint rule, and (c) Simpson's rule.

23 $\int_{1}^{8} 81x^{8/3}\, dx; \quad E = 0.05$

24 $\int_{1}^{2} \frac{1}{120x^2}\, dx; \quad E = 0.1$

25 $\int_{1/2}^{1} \frac{1}{x}\, dx; \quad\quad E = 0.02$

26 $\int_{0}^{3} \frac{1}{x+1}\, dx; \quad E = 0.005$

27 If $f(x)$ is a polynomial of degree less than 4, prove that Simpson's rule gives the exact value of $\int_{a}^{b} f(x)\, dx$.

28 Suppose that f is continuous and that both f and f'' are nonnegative throughout $[a, b]$. Prove that $\int_{a}^{b} f(x)\, dx$ is less than the approximation given by the trapezoidal rule.

29 The graph in the figure was recorded by an instrument used to measure a physical quantity. Estimate *y*-coordinates of points on the graph, and approximate the area of the shaded region by using (a) the trapezoidal rule, with $n = 6$, and (b) Simpson's rule, with $n = 3$.

30 An artificially created lake has the shape illustrated in the figure, with adjacent measurements 20 ft apart. Use the trapezoidal rule to estimate the surface area of the lake.

Exercise 30

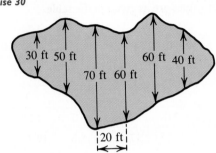

c 31 An important aspect of water management is the production of reliable data on *streamflow*, the number of cubic meters of water passing through a cross section of a stream or river per second. A first step in this computation is the determination of the average velocity $\bar{v}_x$ at a distance *x* meters from the river bank. If *k* is the depth of the stream at a point *x* meters from the bank and $v(y)$ is the velocity (in meters per second) at a depth of *y* meters (see figure), then

$$\bar{v}_x = \frac{1}{k} \int_{0}^{k} v(y)\, dy$$

(see Definition (4.29)). With the *six-point method*, velocity readings are taken at the surface; at depths $0.2k, 0.4k, 0.6k,$ and $0.8k$; and near the river bottom. The trapezoidal rule is then used to estimate $\bar{v}_x$. Given the data in the following table, estimate $\bar{v}_x$.

y (m)	0	0.2*k*	0.4*k*	0.6*k*	0.8*k*	*k*
v(*y*) (m/sec)	0.28	0.23	0.19	0.17	0.13	0.02

Exercise 29

Exercise 31

32 Refer to Exercise 31. The streamflow F (in cubic meters per second) can be approximated using the formula

$$F = \int_0^L \bar{v}_x h(x)\, dx,$$

where $h(x)$ is the depth of the stream at a distance x meters from the bank and L is the length of the cross section. Given the data in the following table, use Simpson's rule to estimate F.

x (m)	0	3	6	9	12
$h(x)$ (m)	0	0.51	0.73	1.61	2.11
$\bar{v}_x$ (m/sec)	0	0.09	0.18	0.21	0.36

x (m)	15	18	21	24
$h(x)$ (m)	2.02	1.53	0.64	0
$\bar{v}_x$ (m/sec)	0.32	0.19	0.11	0

Exer. 33–34: Use Simpson's rule, with $n = 4$, to approximate the average value of f on the given interval.

33 $f(x) = \dfrac{1}{x^4 + 1}$; $[0, 4]$

34 $f(x) = \sqrt{\cos x}$; $[-1, 1]$

Exer. 35–36: If f is determined by the given differential equation and initial condition $f(0)$, approximate $f(1)$ using the trapezoidal rule with $n = 10$.

35 $f'(x) = \dfrac{\sqrt{x}}{x^2 + 1}$; $f(0) = 1$

36 $f'(x) = \sqrt{\tan x}$; $f(0) = 2$

Exer. 37–38: Let a regular partition of $[a, b]$ be determined by $a = x_0, x_1, \ldots, x_{m-1}, x_m = b$.

37 Show that Simpson's rule can be expressed as

$$S_n = \sum_{k=1}^{n} \frac{1}{6}\left[f(x_{k-1}) + 4f(x_{k-1/2}) + f(x_k) \right] \Delta x.$$

38 If m is an even integer, show that Simpson's rule can be expressed as

$$S_{m/2} = \frac{b-a}{3m}\left[f(x_0) + 4f(x_1) + 2f(x_2) + 4f(x_3) + \cdots \right.$$
$$\left. + 2f(x_{m-2}) + 4f(x_{m-1}) + f(x_m) \right].$$

CHAPTER 4 REVIEW EXERCISES

Exer. 1–42: Evaluate.

1 $\displaystyle\int \frac{8x^2 - 4x + 5}{x^4}\, dx$

2 $\displaystyle\int (3x^5 + 2x^3 - x)\, dx$

3 $\displaystyle\int 100\, dx$

4 $\displaystyle\int x^{3/5}(2x - \sqrt{x})\, dx$

5 $\displaystyle\int (2x + 1)^7\, dx$

6 $\displaystyle\int \sqrt[3]{5x + 1}\, dx$

7 $\displaystyle\int (1 - 2x^2)^3 x\, dx$

8 $\displaystyle\int \frac{(1 + \sqrt{x})^2}{\sqrt[3]{x}}\, dx$

9 $\displaystyle\int \frac{1}{\sqrt{x}(1 + \sqrt{x})^2}\, dx$

10 $\displaystyle\int (x^2 + 4)^2\, dx$

11 $\displaystyle\int (3 - 2x - 5x^3)\, dx$

12 $\displaystyle\int (x + x^{-1})^2\, dx$

13 $\displaystyle\int (4x + 1)(4x^2 + 2x - 7)^2\, dx$

14 $\displaystyle\int \frac{\sqrt[4]{1 - (1/x)}}{x^2}\, dx$

15 $\displaystyle\int (2x^{-3} - 3x^2)\, dx$

16 $\displaystyle\int (x^{3/2} + x^{-3/2})\, dx$

17 $\displaystyle\int_0^1 \sqrt[3]{8x^7}\, dx$

18 $\displaystyle\int_1^2 \frac{x^2 - x - 6}{x + 2}\, dx$

19 $\displaystyle\int_0^1 \frac{x^2}{(1 + x^3)^2}\, dx$

20 $\displaystyle\int_1^9 \sqrt{2x + 7}\, dx$

21 $\displaystyle\int_1^2 \frac{x + 1}{\sqrt{x^2 + 2x}}\, dx$

22 $\displaystyle\int_1^2 \frac{x^2 + 2}{x^2}\, dx$

23 $\displaystyle\int_0^2 x^2 \sqrt{x^3 + 1}\, dx$

24 $\displaystyle\int_1^1 3x^2 \sqrt{x^3 + x}\, dx$

25 $\displaystyle\int_0^1 (2x - 3)(5x + 1)\, dx$

26 $\displaystyle\int_{-1}^1 (x^2 + 1)^2\, dx$

27 $\displaystyle\int_0^4 \sqrt{3x}(\sqrt{x} + \sqrt{3})\, dx$

28 $\int_{-1}^{1} (x+1)(x+2)(x+3)\, dx$

29 $\int \sin(3-5x)\, dx$

30 $\int x^2 \cos(2x^3)\, dx$

31 $\int \cos 3x \sin^4 3x\, dx$

32 $\int \dfrac{\sin(1/x)}{x^2}\, dx$

33 $\int \dfrac{\cos 3x}{\sin^3 3x}\, dx$

34 $\int (3\cos 2\pi t - 5\sin 4\pi t)\, dt$

35 $\int_{0}^{\pi/2} \cos x\sqrt{3+5\sin x}\, dx$

36 $\int_{-\pi/4}^{0} (\sin x + \cos x)^2\, dx$

37 $\int_{0}^{\pi/4} \sin 2x \cos^2 2x\, dx$

38 $\int_{\pi/6}^{\pi/4} (\sec x + \tan x)(1-\sin x)\, dx$

39 $\int \dfrac{d}{dx}\sqrt[5]{x^4+2x^2+1}\, dx$

40 $\int_{0}^{\pi/2} \dfrac{d}{dx}(x\sin^3 x)\, dx$

41 $\dfrac{d}{dx}\int_{0}^{1}(x^3+x^2-7)^5\, dx$

42 $\dfrac{d}{dx}\int_{0}^{x}(t^2+1)^{10}\, dt$

Exer. 43–44: Solve the differential equation subject to the given conditions.

43 $\dfrac{d^2y}{dx^2}=6x-4;\quad y=4$ and $y'=5$ if $x=2$

44 $f''(x)=x^{1/3}-5;\quad f'(1)=2; f(1)=-8$

Exer. 45–46: Let $f(x)=9-x^2$ for $-2 \le x \le 3$, and let P be the regular partition of $[-2,3]$ into five subintervals.

45 Find the Riemann sum R_P if f is evaluated at the midpoint of each subinterval of P.

46 Find **(a)** A_{IP} and **(b)** A_{CP}.

Exer. 47–48: Verify the inequality without evaluating the integrals.

47 $\int_{0}^{1} x^2\, dx \ge \int_{0}^{1} x^3\, dx$

48 $\int_{1}^{2} x^2\, dx \le \int_{1}^{2} x^3\, dx$

Exer. 49–50: Express as one integral.

49 $\int_{c}^{e} f(x)\, dx + \int_{a}^{b} f(x)\, dx - \int_{c}^{b} f(x)\, dx - \int_{d}^{a} f(x)\, dx$

50 $\int_{a}^{d} f(x)\, dx - \int_{t}^{b} f(x)\, dx - \int_{g}^{b} f(x)\, dx$
$+ \int_{m}^{b} f(x)\, dx + \int_{t}^{a} f(x)\, dx$

51 A stone is thrown directly downward from a height of 900 ft with an initial velocity of 30 ft/sec.

(a) Determine the stone's distance above the ground after t seconds.

(b) Find its velocity after 5 sec.

(c) Determine when it strikes the ground.

52 Is the following argument valid? Explain. The function f defined by

$$f(x) = \begin{cases} 1 & \text{if } x \text{ is rational and } x > 0 \\ -1 & \text{if } x \text{ is rational and } x < 0 \\ 0 & \text{if } x \text{ is irrational} \end{cases}$$

is defined for all numbers in $[-1, 1]$ and has the property that $f(-x) = -f(x)$ for all x in $[-1, 1]$. Thus, $\int_{-1}^{1} f(x)\, dx = 0$.

53 Find a definite integral for which

$$\sum_{k=1}^{50} \sqrt{1+3\left(-2+\dfrac{k}{10}\right)^2}\left(\dfrac{1}{10}\right)$$

is a right rectangle rule approximation.

54 Given $\int_{1}^{4}(x^2+2x-5)\, dx$, find

(a) a number z that satisfies the conclusion of the mean value theorem for integrals (4.28)

(b) the average value of x^2+2x-5 on $[1,4]$

Exer. 55–58: Approximate the definite integral using the indicated rule for the stated values of n.

55 $\int_{0}^{2} \sin(x^2)\, dx;\quad$ midpoint rule for $n=5$ and 10

56 $\int_{0}^{1} \cos\sqrt{x}\, dx;\quad$ trapezoidal rule for $n=10$ and 20

57 $\int_{2}^{4} \sqrt{x^3+x}\, dx;\quad$ Simpson's rule for $n=4$ and 8

58 $\int_{0}^{5} \dfrac{1}{x+2}\, dx;\quad$ Simpson's rule for $n=5$ and 20

59 To monitor the thermal pollution of a river, a biologist takes hourly temperature readings (in °F) from 9 A.M. to 5 P.M. The results are shown in the following table.

Time of day	9	10	11	12	1
Temperature	75.3	77.0	83.1	84.8	86.5

Time of day	2	3	4	5
Temperature	86.4	81.1	78.6	75.1

Use Simpson's rule and Definition (4.29) to estimate the average water temperature between 9 A.M. and 5 P.M.

EXTENDED PROBLEMS AND GROUP PROJECTS

1 Let $f(t) = 1/(1+t^2)$.

(a) Sketch the graph of f and discuss its symmetries.

(b) Show that f is continuous for all real numbers t.

(c) Prove that the function $F(x) = \int_0^x f(t)\,dt$ exists and is differentiable for all real numbers x.

(d) Find $F(0)$.

(e) Show that $F'(x) = 1/(1+x^2)$.

(f) Show that F is a strictly increasing function.

(g) Find $F''(x)$ and determine the intervals over which F is concave upward and concave downward. Find all points of inflection.

(h) Use the information obtained so far to sketch a graph of F.

(i) Show that F must have an inverse function T. Assuming that T is differentiable, use the identity $F(T(x)) = x$, differentiation, and the chain rule to conclude that $T'(x) = 1 + [T(x)]^2$.

(j) Show that the tangent function satisfies

$$(\tan x)' = 1 + [\tan x]^2.$$

(k) Discuss the similarity between the results of parts (i) and (j).

2 Let $f(t)$ be a continuous function and define

$$F(x) = \int_0^x x\, f(t)\,dt.$$

(a) Find $F'(x)$ if $f(t) = t$.

(b) Find $F'(x)$ if $f(t) = t^2$.

(c) Find $F'(x)$ if $f(t) = \cos t$.

(d) Formulate a general result about $F'(x)$.

(e) Prove that

$$\int_0^x \left(\int_0^u f(t)\,dt \right) du = \int_0^x f(u)(x-u)\,du.$$

(*Hint:* Differentiate both sides, using the result of part (d).)

(f) Prove that

$$\int_0^x f(u)(x-u)^2\,du = 2 \int_0^x \left[\int_0^w \left(\int_0^v f(t)\,dt \right) dv \right] dw.$$

3 One way to measure the effectiveness of a numerical integration method is to test the method on polynomials. A method is **exact for polynomials of degree n** if it produces zero error for any polynomial of degree at most n, but does produce error for some polynomial of degree $n+1$.

(a) Show that a method is exact for polynomials of degree n if and only if it produces zero error for monomials, $f(x) = x^j$ for $j = 0, 1, \ldots, n$, but has some error for monomials x^{n+1}.

(b) Show that the midpoint rule and the trapezoidal rule are exact for polynomials of degree 1.

(c) Show that Simpson's rule is exact for polynomials of degree 3.

(d) Show that the following numerical integration rule (called a *Gaussian rule*) is exact for polynomials of degree 3:

$$\int_a^b f(x)\,dx \approx \sum_{k=1}^n \left\{ f\left[x_{k-1} + \left(1 - \sqrt{\tfrac{1}{3}} \right) \frac{\Delta x}{2} \right] \right.$$

$$\left. + f\left[x_{k-1} + \left(1 + \sqrt{\tfrac{1}{3}} \right) \frac{\Delta x}{2} \right] \right\} \frac{\Delta x}{2}$$

(e) Discuss the advantages and disadvantages of implementing the Gaussian rule in part (d) over implementing Simpson's rule.

(f) Test the following numerical integration rule (called *Simpson's 3/8 rule*) to determine its polynomial exactness:

$$\int_a^b f(x)\,dx \approx \sum_{k=1}^n \frac{3}{8} \left[f\left(x_{k-1} \right) + 3f\left(x_{k-1} + \frac{\Delta x}{3} \right) \right.$$

$$\left. + 3f\left(x_{k-1} + 2\frac{\Delta x}{3} \right) + f\left(x_k \right) \right] \frac{\Delta x}{3}$$

(g) Can a numerical integration rule that is exact for polynomials of degree 2 be designed?

INTRODUCTION

I **N DESIGNING A DAM** and projecting its cost, engineers must determine how much concrete is needed for construction. This amount depends on the volume of the dam. The volume, in turn, is a function of the shape of the dam and its thickness at various levels. The dam must be thick enough to withstand the force of the enormous amount of water held in the dam's reservoir. To compute the force of the water, the required thickness at each height, and the resulting volume of concrete, engineers set up and evaluate numerous definite integrals.

Hoover Dam, for example, which supplies much electrical power and water for a large region of the American Southwest, is one of the world's largest concrete dams. The dam is 726 ft high, the equivalent of a 50-story building, and 1244 ft wide, and its reservoir, Lake Mead, can store approximately 1.3 trillion ft^3 of water. To withstand the resulting pressure, the dam's base is 660 ft thick, and the total amount of concrete is over 118 million ft^3, enough to pave a two-lane highway from San Francisco to New York.

In this chapter, we discuss some of the many uses for the definite integral. We begin by reconsidering in Section 5.1 the application that motivated the definition of this mathematical concept: determining the area of a region in the xy-plane. Then, in turn, we use definite integrals to find volumes (Sections 5.2–5.4), lengths of graphs and surface areas of solids (Section 5.5), work done by a variable force (Section 5.6), and moments and the center of mass (the balance point) of a flat plate (Section 5.7). Definite integrals are applicable because each of these quantities can be expressed as a limit of sums.

Because of the multitude of other quantities that can be similarly expressed, the definite integral is useful in a wide variety of applications, some of which are considered in Section 5.8: finding the force exerted by a liquid against a wall (water on a dam, gasoline on one end of a storage tank, oil on the walls of an ocean tanker), measuring cardiac output and blood flow in arteries, estimating the future wealth of a corporation, calculating the thickness of the ozone layer, determining the amount of radon gas in a home, and finding the number of calories burned during a workout on an exercise bicycle.

As you proceed through this chapter and whenever you encounter definite integrals in applications, keep the following words in mind: *limit of sums, limit of sums, limit of sums.*

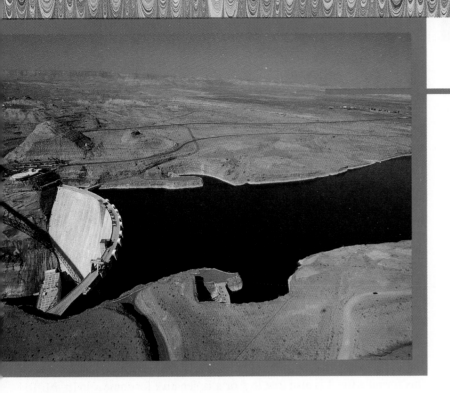

The design of large engineering projects such as a dam requires the calculation of many physical quantities that are most accurately described by definite integrals.

Applications of the Definite Integral

5.1 AREA

If a function f is continuous and $f(x) \geq 0$ on $[a, b]$, then, by Theorem (4.19), the area of the region under the graph of f from a to b is given by the definite integral $\int_a^b f(x)\, dx$. In this section, we consider the region that lies *between* the graphs of two functions.

If f and g are continuous and $f(x) \geq g(x) \geq 0$ for every x in $[a, b]$, then the area A of the region R bounded by the graphs of f, g, $x = a$, and $x = b$ (see Figure 5.1) can be found by subtracting the area of the region under the graph of g (the **lower boundary** of R) from the area of the region under the graph of f (the **upper boundary** of R), as follows:

$$A = \int_a^b f(x)\, dx - \int_a^b g(x)\, dx$$

$$= \int_a^b [f(x) - g(x)]\, dx.$$

This formula for A is also true if f or g is negative for some x in $[a, b]$. To verify this fact, choose a *negative* number d that is less than the minimum value of g on $[a, b]$, as illustrated in Figure 5.2(a). Next, consider the functions f_1 and g_1, defined as follows:

$$f_1(x) = f(x) - d = f(x) + |d|$$
$$g_1(x) = g(x) - d = g(x) + |d|$$

The graphs of f_1 and g_1 can be obtained by vertically shifting the graphs of f and g a distance $|d|$. If A is the area of the region in Figure 5.2(b),

Figure 5.1

Figure 5.2

(a)

(b)

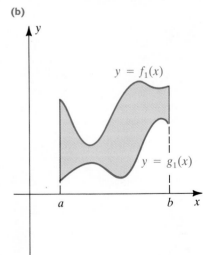

then

$$A = \int_a^b [f_1(x) - g_1(x)]\,dx$$

$$= \int_a^b [(f(x) - d) - (g(x) - d)]\,dx$$

$$= \int_a^b [f(x) - g(x)]\,dx.$$

We may summarize our discussion as follows.

Theorem 5.1

If f and g are continuous and $f(x) \geq g(x)$ for every x in $[a, b]$, then the area A of the region bounded by the graphs of $f, g, x = a$, and $x = b$ is

$$A = \int_a^b [f(x) - g(x)]\,dx.$$

Figure 5.3

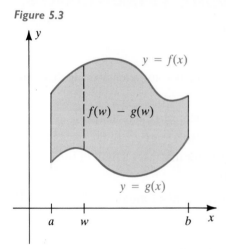

We may interpret the formula for A in Theorem (5.1) as a limit of sums. If we let $h(x) = f(x) - g(x)$ and if w is in $[a, b]$, then $h(w)$ is the vertical distance between the graphs of f and g for $x = w$ (see Figure 5.3). As in our discussion of Riemann sums in Chapter 4, let P denote a partition of $[a, b]$ determined by $a = x_0, x_1, \ldots, x_n = b$. For each k, let $\Delta x_k = x_k - x_{k-1}$, and let w_k be any number in the kth subinterval $[x_{k-1}, x_k]$ of P. By the definition of h,

$$h(w_k)\Delta x_k = [f(w_k) - g(w_k)]\Delta x_k,$$

which is the area of the rectangle of length $f(w_k) - g(w_k)$ and width Δx_k shown in Figure 5.4.

Figure 5.4

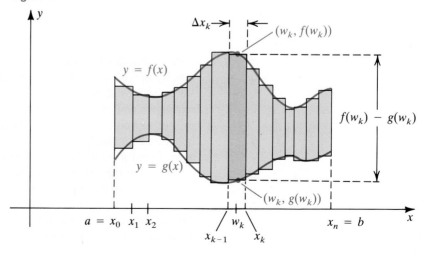

The Riemann sum

$$\sum_k h(w_k)\Delta x_k = \sum_k [f(w_k) - g(w_k)]\Delta x_k$$

is the sum of the areas of the rectangles in Figure 5.4 and is therefore an approximation to the area of the region between the graphs of f and g from a to b. By the definition of the definite integral,

$$\lim_{\|P\|\to 0} \sum_k h(w_k)\Delta x_k = \int_a^b h(x)\,dx.$$

Since $h(x) = f(x) - g(x)$, we obtain the following corollary of Theorem (5.1).

Corollary 5.2

$$A = \lim_{\|P\|\to 0} \sum_k [f(w_k) - g(w_k)]\Delta x_k = \int_a^b [f(x) - g(x)]\,dx$$

Figure 5.5

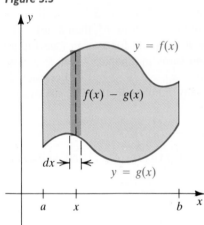

We may use the following intuitive method for remembering this limit of sums formula (see Figure 5.5):

1. Use dx for the width Δx_k of a typical vertical rectangle.
2. Use $f(x) - g(x)$ for the length $f(w_k) - g(w_k)$ of the rectangle.
3. Regard the symbol $\int_a^b$ as an operator that takes a limit of sums of the rectangular areas $[f(x) - g(x)]\,dx$.

This method allows us to interpret the area formula in Theorem (5.1) as follows:

$$A = \int_a^b [f(x) - g(x)]\,dx$$

limit of   length of a   width of a
sums   rectangle   rectangle

When using this technique, we visualize summing areas of vertical rectangles by moving through the region from left to right. Later in this section, we consider different types of regions, finding areas by using *horizontal* rectangles and integrating with respect to y.

Let us call a region an R_x **region** (for integration with respect to x) if it lies between the graphs of two equations $y = f(x)$ and $y = g(x)$, with f and g continuous, and $f(x) \geq g(x)$ for every x in $[a, b]$, where a and b are the smallest and largest x-coordinates, respectively, of the points (x, y) in the region. The regions in Figures 5.1–5.5 are R_x regions. Several others are sketched in Figure 5.6 on the following page. Note that the graphs of $y = f(x)$ and $y = g(x)$ may intersect one or more times; however, $f(x) \geq g(x)$ throughout the interval.

Figure 5.6 R_x regions

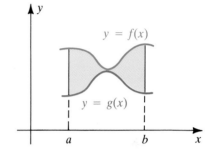

The following guidelines may be helpful when working problems.

Guidelines for Finding the Area of an R_x Region 5.3

1 Sketch the region, labeling the upper boundary $y = f(x)$ and the lower boundary $y = g(x)$. Find the smallest value $x = a$ and the largest value $x = b$ for points (x, y) in the region.

2 Sketch a typical vertical rectangle and label its width dx.

3 Express the area of the rectangle in guideline (2) as

$$[f(x) - g(x)] \, dx.$$

4 Apply the limit of sums operator $\int_a^b$ to the expression in guideline (3) and evaluate the integral.

Figure 5.7

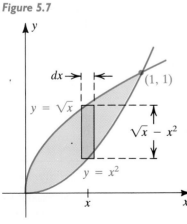

EXAMPLE ■ 1   Find the area of the region bounded by the graphs of the equations $y = x^2$ and $y = \sqrt{x}$.

SOLUTION   Following guidelines (1)–(3), we sketch and label the region and show a typical vertical rectangle (see Figure 5.7). The points $(0, 0)$ and $(1, 1)$ at which the graphs intersect can be found by solving the equations $y = x^2$ and $y = \sqrt{x}$ simultaneously. Referring to the figure, we obtain the following facts:

upper boundary: $y = \sqrt{x}$

lower boundary: $y = x^2$

width of rectangle: dx

length of rectangle: $\sqrt{x} - x^2$

area of rectangle: $(\sqrt{x} - x^2)\,dx$

Next, we follow guideline (4) with $a = 0$ and $b = 1$, remembering that applying $\int_0^1$ to the expression $(\sqrt{x} - x^2)\,dx$ represents taking a limit of sums of areas of vertical rectangles. We thus obtain

$$A = \int_0^1 (\sqrt{x} - x^2)\,dx = \int_0^1 (x^{1/2} - x^2)\,dx$$

$$= \left[\frac{x^{3/2}}{\frac{3}{2}} - \frac{x^3}{3} \right]_0^1 = \frac{2}{3} - \frac{1}{3} = \frac{1}{3}.$$

EXAMPLE■2   Find the area of the region bounded by the graphs of $y + x^2 = 6$ and $y + 2x - 3 = 0$.

SOLUTION   The region and a typical rectangle are sketched in Figure 5.8. The points of intersection $(-1, 5)$ and $(3, -3)$ of the two graphs may be found by solving the two given equations simultaneously. To apply guideline (1), we must label the upper and lower boundaries $y = f(x)$ and $y = g(x)$, respectively, and hence we solve each of the given equations for y in terms of x, as shown in Figure 5.8. Here we obtain

Figure 5.8

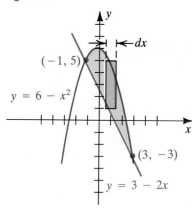

upper boundary: $y = 6 - x^2$

lower boundary: $y = 3 - 2x$

width of rectangle: dx

length of rectangle: $(6 - x^2) - (3 - 2x)$

area of rectangle: $[(6 - x^2) - (3 - 2x)]\,dx$

Next, we use guideline (4), with $a = -1$ and $b = 3$, regarding $\int_{-1}^3$ as an operator that takes a limit of sums of areas of rectangles. Thus,

$$A = \int_{-1}^3 [(6 - x^2) - (3 - 2x)]\,dx = \int_{-1}^3 (3 - x^2 + 2x)\,dx$$

$$= \left[3x - \frac{x^3}{3} + x^2 \right]_{-1}^3$$

$$= [9 - \tfrac{27}{3} + 9] - [-3 - (-\tfrac{1}{3}) + 1] = \tfrac{32}{3}.$$

The following example illustrates that it is sometimes necessary to subdivide a region into several R_x regions and then use more than one definite integral to find the area.

Figure 5.9

Figure 5.10

Figure 5.11

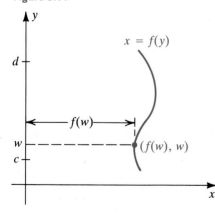

EXAMPLE ▪ 3 Find the area of the region R bounded by the graphs of $y - x = 6$, $y - x^3 = 0$, and $2y + x = 0$.

SOLUTION The graphs and the region are sketched in Figure 5.9. Each equation has been solved for y in terms of x, and the boundaries have been labeled as in guideline (1). Typical vertical rectangles are shown extending from the lower boundary to the upper boundary of R. Since the lower boundary consists of portions of two different graphs, the area cannot be found by using only one definite integral. However, if R is divided into two R_x regions, R_1 and R_2, as shown in Figure 5.10, then we can determine the area of each and add them together. Let us arrange our work as follows.

	Region R_1	Region R_2
upper boundary:	$y = x + 6$	$y = x + 6$
lower boundary:	$y = -\frac{1}{2}x$	$y = x^3$
width of rectangle:	dx	dx
length of rectangle:	$(x + 6) - (-\frac{1}{2}x)$	$(x + 6) - x^3$
area of rectangle:	$[(x + 6) - (-\frac{1}{2}x)]\,dx$	$[(x + 6) - x^3]\,dx$

Applying guideline (4), we find the areas A_1 and A_2 of R_1 and R_2:

$$A_1 = \int_{-4}^{0} [(x + 6) - (-\tfrac{1}{2}x)]\,dx$$

$$= \int_{-4}^{0} \left(\frac{3}{2}x + 6\right) dx = \left[\frac{3}{2}\left(\frac{x^2}{2}\right) + 6x\right]_{-4}^{0}$$

$$= 0 - (12 - 24) = 12$$

$$A_2 = \int_{0}^{2} [(x + 6) - x^3]\,dx$$

$$= \left[\frac{x^2}{2} + 6x - \frac{x^4}{4}\right]_{0}^{2}$$

$$= (2 + 12 - 4) - 0 = 10$$

The area A of the entire region R is

$$A = A_1 + A_2 = 12 + 10 = 22.$$

We have now evaluated many integrals similar to those in Example 3. For this reason, we sometimes merely *set up* an integral—that is, we express it in the proper form but do not find its numerical value.

If we consider an equation of the form $x = f(y)$, where f is continuous for $c \le y \le d$, then we *reverse the roles of x and y in the previous discussion, treating y as the independent variable and x as the dependent variable.* A typical graph of $x = f(y)$ is sketched in Figure 5.11. Note that if a value w is assigned to y, then $f(w)$ *is an x-coordinate of the* corresponding point on the graph.

Figure 5.12

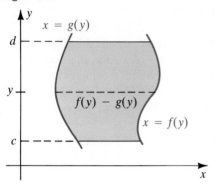

An R_y **region** is a region that lies between the graphs of two equations of the form $x = f(y)$ and $x = g(y)$, with f and g continuous, and with $f(y) \geq g(y)$ for every y in $[c, d]$, where c and d are the smallest and largest y-coordinates, respectively, of points in the region. One such region is illustrated in Figure 5.12. We call the graph of f the **right boundary** of the region and the graph of g the **left boundary**. For any y, the number $f(y) - g(y)$ is the horizontal distance between these boundaries, as shown in Figure 5.12.

We can use limits of sums to find the area A of an R_y region. We begin by selecting points on the y-axis with y-coordinates $c = y_0, y_1, \ldots, y_n = d$, obtaining a partition of the interval $[c, d]$ into subintervals of width $\Delta y_k = y_k - y_{k-1}$. For each k, we choose a number w_k in $[y_{k-1}, y_k]$ and consider horizontal rectangles that have areas $[f(w_k) - g(w_k)]\Delta y_k$, as illustrated in Figure 5.13. This procedure leads to

$$A = \lim_{\|P\| \to 0} \sum_k [f(w_k) - g(w_k)]\Delta y_k = \int_c^d [f(y) - g(y)]\, dy.$$

The last equality follows from the definition of the definite integral.

Figure 5.13

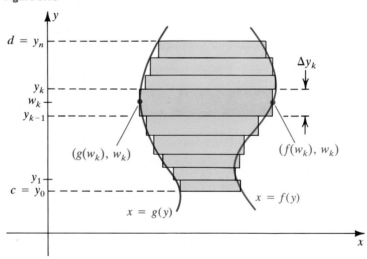

Using notation similar to that for R_x regions, we represent the width Δy_k of a horizontal rectangle by dy and the length $f(w_k) - g(w_k)$ of the rectangle by $f(y) - g(y)$ in the following guidelines.

Guidelines for Finding the Area of an R_y Region   5.4

1   Sketch the region, labeling the right boundary $x = f(y)$ and the left boundary $x = g(y)$. Find the smallest value $y = c$ and the largest value $y = d$ for points (x, y) in the region.

2   Sketch a typical horizontal rectangle and label its width dy.

3 Express the area of the rectangle in guideline (2) as

$$[f(y) - g(y)]\,dy.$$

4 Apply the limit of sums operator $\int_c^d$ to the expression in guideline (3) and evaluate the integral.

In guideline (4), we visualize summing areas of horizontal rectangles by moving from the lowest point of the region to the highest point.

Figure 5.14

Figure 5.15

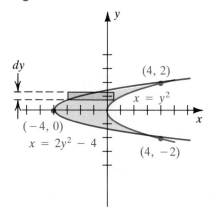

EXAMPLE■4 Find the area of the region bounded by the graphs of the equations $2y^2 = x + 4$ and $y^2 = x$.

SOLUTION The region is sketched in Figures 5.14 and 5.15. Figure 5.14 illustrates the use of vertical rectangles (integration with respect to x), and Figure 5.15 illustrates the use of horizontal rectangles (integration with respect to y). Referring to Figure 5.14, we see that several integrations with respect to x are required to find the area. However, for Figure 5.15, we need only one integration with respect to y. Thus we apply Guidelines (5.4), solving each equation for x in terms of y. Referring to Figure 5.15, we obtain the following:

$$\begin{aligned}
\text{right boundary:} &\quad x = y^2 \\
\text{left boundary:} &\quad x = 2y^2 - 4 \\
\text{width of rectangle:} &\quad dy \\
\text{length of rectangle:} &\quad y^2 - (2y^2 - 4) \\
\text{area of rectangle:} &\quad [y^2 - (2y^2 - 4)]\,dy
\end{aligned}$$

We could now use guideline (4) with $c = -2$ and $d = 2$, finding A by applying the operator $\int_{-2}^{2}$ to $[y^2 - (2y^2 - 4)]\,dy$. Another method is to use the symmetry of the region with respect to the x-axis and find A by doubling the area of the part that lies above the x-axis. Thus,

$$\begin{aligned}
A &= \int_{-2}^{2} [y^2 - (2y^2 - 4)]\,dy \\
&= 2 \int_{0}^{2} (4 - y^2)\,dy \\
&= 2\left[4y - \frac{y^3}{3}\right]_0^2 = 2\left(8 - \frac{8}{3}\right) = \frac{32}{3}.
\end{aligned}$$

In following Guidelines (5.3) or (5.4) for finding the area of a region, we may need to use a graphing utility and numerical methods to obtain an accurate sketch of the region, find the smallest and largest x- or y-values in the region, and approximate the area. Our next example illustrates such a case.

EXAMPLE ▪ 5   For the region of the plane bounded by the curves $y = \cos(0.3x^2)$ and $y = x^2 + 0.6x - 2$,

(a) use a graphing utility to sketch the curves and determine the region

(b) find numerical approximations for the intersection points of the bounding curves

(c) set up a definite integral representing the area of the region

(d) approximate the area

SOLUTION

(a) After examining several different viewing windows, we obtain the view of the desired area shown in Figure 5.16.

(b) By tracing on the graph, we obtain first approximations for the intersection points as $(-1.9, 0.5)$ and $(1.4, 0.8)$. At the intersection points, the bounding curves have equal y-values. Thus, $\cos(0.3x^2) = x^2 + 0.6x - 2$, so $\cos(0.3x^2) - [x^2 + 0.6x - 2] = 0$. We can use a solving routine or apply Newton's method to the function

$$f(x) = \cos(0.3x^2) - x^2 - 0.6x + 2$$

with starting values of $x = -1.9$ and then $x = 1.4$ to obtain values $a \approx -1.89968629228$ and $b \approx 1.40826496779$. Substituting these values into the equation for either bounding curve gives the other coordinates for the points of intersection, $y_a \approx 0.468996233702$ and $y_b \approx 0.828169200187$.

(c) In Figure 5.17, we have labeled the bounding curves and points of intersection and shaded the region. We see there that the graph of $y = \cos(0.3x^2)$ is above the graph of $y = x^2 + 0.6x - 2$ on the interval $[a, b]$. The integral representing the area of the region is

$$A = \int_a^b \left[\cos(0.3x^2) - (x^2 + 0.6x - 2)\right] dx$$

$$\approx \int_{-1.89968629228}^{1.40826496779} \left[\cos(0.3x^2) - x^2 - 0.6x + 2\right] dx.$$

(d) We compute numerical approximations for the definite integral in part (c) using Simpson's rule for several different values of n to approximate the area. For example, when $n = 64$, $A \approx 6.93542681443$ and when $n = 128$, $A \approx 6.93542681577$. Thus, we have confidence in the approximation that, to seven decimal places, $A \approx 6.9354268$.

As an application of the area between two curves, let us consider what economists call **capital formation**—that is, the process of increasing or decreasing a given holding of capital over time. If $K(t)$ is the amount of capital at time t, then dK/dt denotes the **rate of capital formation**. Economists consider the rate of capital formation to be identical to the **net investment flow**, which we will denote by $I(t)$. We can look at the

Figure 5.16
$-3.8 \le x \le 3.8,\ -2.5 \le y \le 2$

Figure 5.17

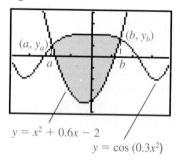

$y = x^2 + 0.6x - 2$

$y = \cos(0.3x^2)$

relationship between capital formation and net investment flow in two ways: in a derivative formulation,

$$\frac{dK}{dt} = I(t)$$

and in an integral form,

$$K(t) = \int I(t)\, dt.$$

Note that with $I(t) \geq 0$ for $a \leq t \leq b$, the amount of capital accumulation in this time interval is $\int_a^b I(t)\, dt$, the area under the graph of the function $I(t)$.

If we know the amount of capital $K(t)$ accumulated at time t, we may differentiate with respect to t to find the investment flow. Alternatively, if we are given the investment flow $I(t)$, we may integrate with respect to t to find the amount of capital—that is, $K(t)$ represents the total change in capital or the capital accumulation. As a derivative, the investment flow $I(t)$ is a rate of change of capital. That is, the value of $I(t)$ at a particular time t is the rate at which investment is flowing in or out of the given holding of capital, measured in units of capital per unit of time. For example, if $I(t) = 4 - t^2 + 2t$, where capital is measured in millions of dollars and time is measured in years, at time $t = 1$ year, we have $I(1) = 4 - 1 + 2 = 5$ million dollars per year. Hence, capital is increasing at an annual rate of \$5 million. At $t = 4$, $I(4) = 4 - 16 + 8 = -4$, so at time $t = 4$ years, capital is decreasing at an annual rate of \$4 million.

EXAMPLE ■ 6   If the net investment flow is $I(t) = 4 - t^2 + 2t$ millions of dollars per time unit, find the capital formation during the time interval $[1, 2]$.

SOLUTION   The capital formation is given by

$$\int_1^2 I(t)\, dt = \int_1^2 (4 - t^2 + 2t)\, dt.$$

We can evaluate the definite integral by finding an antiderivative for $I(t)$:

$$\int_1^2 (4 - t^2 + 2t)\, dt = \left[4t - \frac{t^3}{3} + t^2 \right]_1^2$$

$$= \left[8 - \frac{8}{3} + 4 \right] - \left[4 - \frac{1}{3} + 1 \right] = 4\frac{2}{3}$$

Thus, the capital accumulation is about \$4.67 million.

EXAMPLE ■ 7   Consider two different net investment flows given by $I_1(t) = 4 - t^2 + 2t$ and $I_2(t) = 4 - t$ (both in millions of dollars per year at year t).

(a) Find the time interval during which the first investment flow I_1 is at least as great as the second investment flow I_2.

(b) For the time interval found in part (a), determine how much more capital accumulates under the first investment flow than under the second investment flow.

SOLUTION

(a) We need to find the interval $[a, b]$ during which $I_1(t) \geq I_2(t)$. We first sketch the graphs of the two functions (Figure 5.18) and then find the points of intersection by solving the equations $y = 4 - t^2 + 2t$ and $y = 4 - t$ simultaneously:

$$4 - t^2 + 2t = 4 - t$$
$$3t - t^2 = 0$$
$$t(3 - t) = 0$$
$$t = 0 \quad \text{and} \quad t = 3$$

Figure 5.18

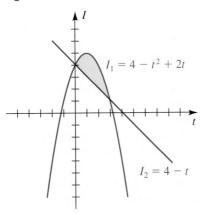

Thus, we see from the graph that $I_1(t) \geq I_2(t)$ on the interval $[0, 3]$. If $t = 0$ corresponds to the present time, then the investment flow I_1 will exceed the investment flow I_2 for the next three years.

(b) The difference in capital accumulation between the two investment flows is the area of the region between the two curves I_1 and I_2 over the interval $[0, 3]$. That is,

$$\int_0^3 [I_1(t) - I_2(t)] \, dt = \int_0^3 [4 - t^2 + 2t - (4 - t)] \, dt$$
$$= \int_0^3 [3t - t^2] \, dt$$
$$= \left[\frac{3t^2}{2} - \frac{t^3}{3} \right]_0^3 = \left[\frac{27}{2} - \frac{27}{3} \right] - [0 - 0] = \frac{9}{2}.$$

Thus, the first investment flow will generate $4.5 million more in accumulated capital than the second investment flow during the next three years.

Figure 5.19

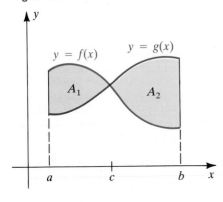

Throughout this section, we have assumed that the graphs of the functions (or equations) do not cross one another in the interval under discussion. If the graphs of f and g cross at one point $P(c, d)$, with $a < c < b$, and we wish to find the area bounded by the graphs from $x = a$ to $x = b$, then the methods developed in this section may still be used; however, *two* integrations are required, one corresponding to the interval $[a, c]$ and the other to $[c, b]$, as is illustrated in Figure 5.19, with $f(x) \geq g(x)$ on $[a, c]$ and $g(x) \geq f(x)$ on $[c, b]$. The area A is given by

$$A = A_1 + A_2 = \int_a^c [f(x) - g(x)] \, dx + \int_c^b [g(x) - f(x)] \, dx.$$

If the graphs cross *several* times, then several integrals may be necessary. Problems in which graphs cross one or more times appear in Exercises 31–36.

EXERCISES 5.1

Exer. 1–4: Set up an integral that can be used to find the area of the shaded region.

1

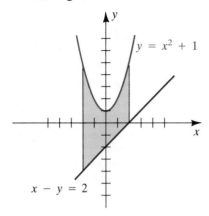

$y = x^2 + 1$

$x - y = 2$

2

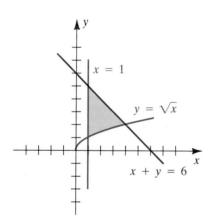

$x = 1$

$y = \sqrt{x}$

$x + y = 6$

3

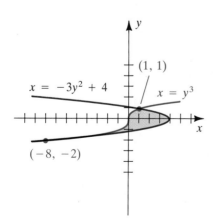

$(1, 1)$

$x = -3y^2 + 4$

$x = y^3$

$(-8, -2)$

4

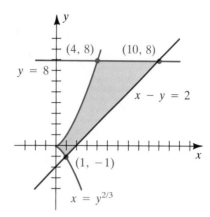

$(4, 8)$ $(10, 8)$

$y = 8$

$x - y = 2$

$(1, -1)$

$x = y^{2/3}$

Exer. 5–22: Sketch the region bounded by the graphs of the equations and find its area.

5 $y = x^2;$	$y = 4x$	
6 $x + y = 3;$	$y + x^2 = 3$	
7 $y = x^2 + 1;$	$y = 5$	
8 $y = 4 - x^2;$	$y = -4$	
9 $y = 1/x^2;$	$y = -x^2;$ $x = 1;$ $x = 2$	
10 $y = x^3;$	$y = x^2$	
11 $y^2 = -x;$	$x - y = 4;$ $y = -1;$ $y = 2$	
12 $x = y^2;$	$y - x = 2;$ $y = -2;$ $y = 3$	
13 $y^2 = 4 + x;$	$y^2 + x = 2$	
14 $x = y^2;$	$x - y = 2$	
15 $x = 4y - y^3;$	$x = 0$	
16 $x = y^{2/3};$	$x = y^2$	
17 $y = x^3 - x;$	$y = 0$	
18 $y = x^3 - x^2 - 6x;$	$y = 0$	
19 $x = y^3 + 2y^2 - 3y;$	$x = 0$	
20 $x = 9y - y^3;$	$x = 0$	
21 $y = x\sqrt{4 - x^2};$	$y = 0$	
22 $y = x\sqrt{x^2 - 9};$	$y = 0;$ $x = 5$	

Exer. 23–24: Find the area of the region between the graphs of the two equations from $x = 0$ to $x = \pi$.

23 $y = \sin 4x;$ $y = 1 + \cos \frac{1}{3}x$

24 $y = 4 + \cos 2x;$ $y = 3 \sin \frac{1}{2}x$

Exer. 25–26: Set up sums of integrals that can be used to find the area of the shaded region by integrating with respect to **(a)** *x* and **(b)** *y*.

25

26

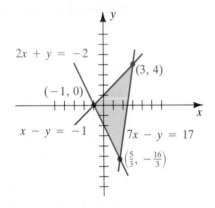

Exer. 27–30: Set up sums of integrals that can be used to find the area of the region bounded by the graphs of the equations by integrating with respect to **(a)** *x* and **(b)** *y*.

27 $y = \sqrt{x}$; $y = -x$; $x = 1$; $x = 4$

28 $y = 1 - x^2$; $y = x - 1$

29 $y = x + 3$; $x = -y^2 + 3$

30 $x = y^2$; $x = 2y^2 - 4$

Exer. 31–36: Find the area of the region between the graphs of *f* and *g* if *x* is restricted to the given interval.

31 $f(x) = 6 - 3x^2$; $g(x) = 3x$; $[0, 2]$

32 $f(x) = x^2 - 4$; $g(x) = x + 2$; $[1, 4]$

33 $f(x) = x^3 - 4x + 2$; $g(x) = 2$; $[-1, 3]$

34 $f(x) = x^2$; $g(x) = x^3$; $[-1, 2]$

35 $f(x) = \sin x$; $g(x) = \cos x$; $[0, 2\pi]$

36 $f(x) = \sin x$; $g(x) = \frac{1}{2}$; $[0, \pi/2]$

Exer. 37–38: Let *R* be the region bounded by the graph of *f* and the *x*-axis, from $x = a$ to $x = b$. Set up a sum of integrals, not containing the absolute value symbol, that can be used to find the area of *R*.

37 $f(x) = |x^2 - 6x + 5|$; $a = 0$, $b = 7$

38 $f(x) = |-x^2 + 2x + 3|$; $a = -3$, $b = 4$

39 Show that the area of the region bounded by an ellipse whose major and minor axes have lengths $2a$ and $2b$, respectively, is πab. (*Hint:* Use an equation of the ellipse to show first that the area is given by $2(b/a) \int_{-a}^{a} \sqrt{a^2 - x^2}\, dx$, and then interpret the definite integral as the area of a semicircle of radius *a*.)

40 Suppose that the function values of *f* and *g* in the following table were obtained empirically. Assuming that *f* and *g* are continuous, approximate the area between their graphs from $x = 1$ to $x = 5$ using **(a)** the trapezoidal rule, with $n = 8$, and **(b)** Simpson's rule, with $n = 4$.

x	1	1.5	2	2.5	3	3.5	4	4.5	5
$f(x)$	3.5	2.5	3	4	3.5	2.5	2	2	3
$g(x)$	1.5	2	2	1.5	1	0.5	1	1.5	1

C 41 Graph $f(x) = |x^3 - 0.7x^2 - 0.8x + 1.3|$ on $[-1.5, 1.5]$. Set up a sum of integrals, not containing the absolute value symbol, that can be used to approximate the area of the region bounded by the graph of *f*, the *x*-axis, and the lines $x = -1.5$ and $x = 1.5$.

C 42 Graph, on the same coordinate axes, $f(x) = \sin x$ and $g(x) = x^3 - x + 0.2$ for $-2 \le x \le 2$. Set up a sum of integrals that can be used to approximate the area of the region bounded by the graphs.

C **Exer. 43–46:** Plot the graphs of the equations. **(a)** Find numerical approximations for the intersection points of the different bounding curves. **(b)** Set up a definite integral representing the area of the bounded region. **(c)** Approximate this area to four-decimal-place accuracy using Simpson's rule.

43 $y = x^3 - 2x^2 - x + 1$; $y = \sqrt{10x}$

44 $y = 4x^4 - 8x^2 + x - 1$; $y = -2x^2 - x + 4$

45 $y = 50\cos(0.5x)$; $y = x^2 - 20$

46 $y = 0.2x^4 - x^3 + 0.4x^2 - 2$; $y = \cos(0.7x)$

C **Exer. 47–50:** Plot the graphs of the equations. **(a)** Set up a definite integral representing the area of the bounded region. **(b)** Approximate this area to four-decimal-place accuracy using Simpson's rule.

47 $y = \sqrt{25 - x^2}$; $y = \sqrt{29 - x^2} - 2$

48 $y = \sin[\pi(x^2 - 1)];$ $\quad y = 1 - x^2$

49 $y = \sin x;$ $\qquad y = \sin(\sin x);$
$\qquad\qquad\qquad\qquad x = 0, \quad x = \pi$

50 $y = 1 + 1.6x - 0.3x^2;$ $\quad y = \sqrt{1 + x^3}$

Exer. 51–54: For each pair of net investment flows $I_1(t)$ and $I_2(t)$, **(a)** find the time interval during which I_1 is at least as great as I_2, and **(b)** for the time interval found in part (a), determine how much more capital accumulates under the first investment flow than the second investment flow.

51 $I_1(t) = t;$ $\qquad I_2(t) = t^2$

52 $I_1(t) = 4(1 - t^2);$ $\quad I_2(t) = 1 - t^2$

53 $I_1(t) = 2(1 - t^2);$ $\quad I_2(t) = t^2 - 1$

54 $I_1(t) = -t^2 + 4t;$ $\quad I_2(t) = 3t/2$

[c] **Exer. 55–56:** Graph, on the same coordinate axes, the given ellipses. **(a)** Estimate their points of intersection.

(b) Set up an integral that can be used to approximate the area of the region bounded by and inside both ellipses.

55 $\dfrac{x^2}{2.9} + \dfrac{y^2}{2.1} = 1;$ $\quad \dfrac{x^2}{4.3} + \dfrac{(y - 2.1)^2}{4.9} = 1$

56 $\dfrac{x^2}{3.9} + \dfrac{y^2}{2.4} = 1;$ $\quad \dfrac{(x + 1.9)^2}{4.1} + \dfrac{y^2}{2.5} = 1$

[c] **Exer. 57–58:** Graph, on the same coordinate axes, the given hyperbolas. **(a)** Estimate their first-quadrant point of intersection. **(b)** Set up an integral that can be used to approximate the area of the region in the first quadrant bounded by the hyperbolas and a coordinate axis.

57 $\dfrac{(y - 0.1)^2}{1.6} - \dfrac{(x + 0.2)^2}{0.5} = 1;$

$\dfrac{(y - 0.5)^2}{2.7} - \dfrac{(x - 0.1)^2}{5.3} = 1$

58 $\dfrac{(x - 0.1)^2}{0.12} - \dfrac{y^2}{0.1} = 1;$ $\quad \dfrac{x^2}{0.9} - \dfrac{(y - 0.3)^2}{2.1} = 1$

5.2 SOLIDS OF REVOLUTION

The volume of an object plays an important role in many problems in the physical sciences. In this section and the next two sections, we consider several methods for computing volumes. Since it is difficult to determine the volume of an irregularly shaped object, we begin with objects that have simple shapes, including the solids of revolution.

If a region in a plane is revolved about a line in the plane, the resulting solid is a **solid of revolution**, and we say that the solid is **generated** by the region. The line is an **axis of revolution**. In particular, if the R_x region shown in Figure 5.20(a) is revolved about the x-axis, we obtain the solid illustrated in Figure 5.20(b). As a special case, if f is a constant function,

Figure 5.20

(a)

(b)

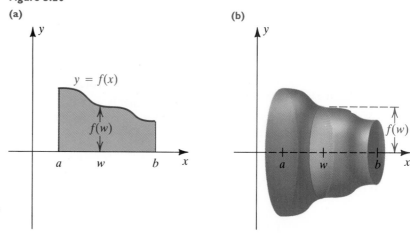

say $f(x) = k$, then the region is rectangular and the solid generated is a right circular cylinder. If the graph of f is a semicircle with endpoints of a diameter at the points $(a, 0)$ and $(b, 0)$, then the solid of revolution is a sphere. If the region is a right triangle with base on the x-axis and two vertices at the points $(a, 0)$ and $(b, 0)$ with the right angle at one of these points, then the solid generated is a right circular cone.

If a plane perpendicular to the x-axis intersects the solid shown in Figure 5.20(b), a circular cross section is obtained. If, as indicated in the figure, the plane passes through the point on the axis with x-coordinate w, then the radius of the circle is $f(w)$, and hence its area is $\pi[f(w)]^2$. We shall arrive at a definition for the volume of such a solid of revolution by using Riemann sums.

Let us partition the interval $[a, b]$, as we did for areas in Section 5.1, and consider the rectangles in Figure 5.21(a). The solid of revolution generated by these rectangles has the shape shown in Figure 5.21(b). Beginning with Figure 5.25, we shall remove, or cut out, parts of solids of revolution to help us visualize portions generated by typical rectangles. When referring to such figures, remember that the entire solid is obtained by one *complete* revolution about an axis, not a partial one.

Figure 5.21

(a)   (b)

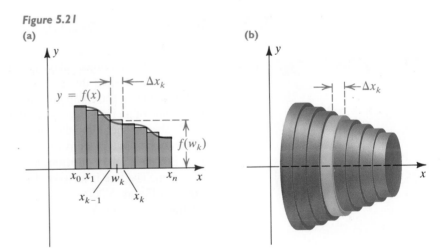

Observe that the kth rectangle generates a **circular disk** (a flat right circular cylinder) of base radius $f(w_k)$ and altitude (thickness) $\Delta x_k = x_k - x_{k-1}$. The volume of this disk is the area of the base times the altitude—that is, $\pi[f(w_k)]^2 \Delta x_k$. The volume of the solid shown in Figure 5.21(b) is the sum of the volumes of all such disks:

$$\sum_k \pi[f(w_k)]^2 \Delta x_k$$

This sum may be regarded as a Riemann sum for $\pi[f(x)]^2$. If the norm $\|P\|$ of the partition is close to zero, then the sum should be close to the volume of the solid. Hence we define the volume of the solid of revolution as a limit of these sums.

Definition 5.5

Let f be continuous on $[a, b]$, and let R be the region bounded by the graph of f, the x-axis, and the vertical lines $x = a$ and $x = b$. The **volume** V of the solid of revolution generated by revolving R about the x-axis is

$$V = \lim_{\|P\| \to 0} \sum_{k} \pi [f(w_k)]^2 \Delta x_k = \int_a^b \pi [f(x)]^2 \, dx.$$

Figure 5.22
(a)

$y = f(x)$

(b)

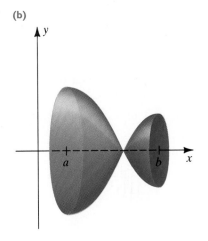

The fact that the limit of sums in this definition equals $\int_a^b \pi [f(x)]^2 \, dx$ follows from the definition of the definite integral. We shall not ordinarily specify the units of measure for volume. If the linear measurement is inches, the volume is in cubic inches (in^3). If x is measured in centimeters, then V is in cubic centimeters (cm^3), and so on.

The requirement that $f(x) \geq 0$ was omitted intentionally in Definition (5.5). If f is negative for some x, as in Figure 5.22(a), and if the region bounded by the graphs of f, $x = a$, $x = b$, and the x-axis is revolved about the x-axis, we obtain the solid shown in Figure 5.22(b). This solid is the same as that generated by revolving the region under the graph of $y = |f(x)|$ from a to b about the x-axis. Since $|f(x)|^2 = [f(x)]^2$, the limit in Definition (5.5) gives us the volume.

Let us interchange the roles of x and y and revolve the R_y region in Figure 5.23(a) about the y-axis, obtaining the solid illustrated in Figure 5.23(b). If we partition the y-interval $[c, d]$ and use *horizontal* rectangles of width Δy_k and length $g(w_k)$, the same type of reasoning that gave us (5.5) leads to Definition (5.6) on the following page.

Figure 5.23
(a)
(b)

Definition 5.6

Figure 5.24

$$V = \lim_{\|P\| \to 0} \sum_k \pi[g(w_k)]^2 \Delta y_k = \int_c^d \pi[g(y)]^2 \, dy$$

Since we may revolve a region about the x-axis, the y-axis, or some other line, *it is not advisable to merely memorize the formulas in (5.5) and (5.6)*. It is better to remember the following general rule for finding the volume of a circular disk (see Figure 5.24).

Volume V of a Circular Disk   5.7

$$V = \pi(\text{radius})^2 \cdot (\text{thickness})$$

When working problems, we shall use the intuitive method developed in Section 5.1, replacing Δx_k or Δy_k by dx or dy, and so on. The following guidelines may be helpful.

Guidelines for Finding the Volume of a Solid of Revolution Using Disks   5.8

1 Sketch the region R to be revolved, and label the boundaries. Show a typical vertical rectangle of width dx or a horizontal rectangle of width dy.

2 Sketch the solid generated by R and the disk generated by the rectangle in guideline (1).

3 Express the radius of the disk in terms of x or y, depending on whether its thickness is dx or dy.

4 Use (5.7) to find a formula for the volume of the disk.

5 Apply the limit of sums operator $\int_a^b$ or $\int_c^d$ to the expression in guideline (4) and evaluate the integral.

EXAMPLE ▪ 1   The region bounded by the x-axis, the graph of the equation $y = x^2 + 1$, and the lines $x = -1$ and $x = 1$ is revolved about the x-axis. Find the volume of the resulting solid.

SOLUTION   As specified in guideline (1), we sketch the region and show a vertical rectangle of width dx (see Figure 5.25a). Following guideline (2), we sketch the solid generated by R and the disk generated by the rectangle (see Figure 5.25b). As specified in guidelines (3) and (4), we note the following:

thickness of disk: dx

radius of disk: $x^2 + 1$

volume of disk: $\pi(x^2 + 1)^2 \, dx$

Figure 5.25

(a)   **(b)**

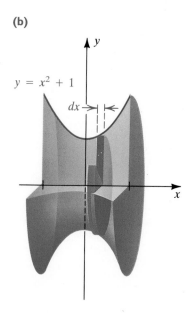

We could next apply guideline (5) with $a = -1$ and $b = 1$, finding the volume V by regarding $\int_{-1}^{1}$ as an operator that takes a limit of sums of volumes of disks. Another method is to use the symmetry of the region with respect to the y-axis and find V by applying $\int_{0}^{1}$ to $\pi(x^2 + 1)^2\, dx$ and doubling the result. Thus,

$$V = \int_{-1}^{1} \pi(x^2 + 1)^2\, dx$$

$$= 2\int_{0}^{1} \pi(x^4 + 2x^2 + 1)\, dx$$

$$= 2\pi \left[\frac{x^5}{5} + 2\left(\frac{x^3}{3}\right) + x\right]_{0}^{1}$$

$$= 2\pi(\tfrac{1}{5} + \tfrac{2}{3} + 1) = \tfrac{56}{15}\pi \approx 11.7.$$

Figure 5.26

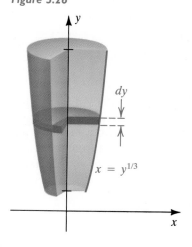

EXAMPLE = 2   The region bounded by the y-axis and the graphs of $y = x^3$, $y = 1$, and $y = 8$ is revolved about the y-axis. Find the volume of the resulting solid.

SOLUTION   The region and the solid are sketched in Figure 5.26, together with a disk generated by a typical horizontal rectangle. Since we plan to integrate with respect to y, we solve the equation $y = x^3$ for x in terms of y, obtaining $x = y^{1/3}$. We note the following facts (see guidelines 3 and 4):

$$\text{thickness of disk:}\quad dy$$
$$\text{radius of disk:}\quad y^{1/3}$$
$$\text{volume of disk:}\quad \pi(y^{1/3})^2\, dy$$

Finally, we apply guideline (5), with $c = 1$ and $d = 8$, regarding $\int_1^8$ as an operator that takes a limit of sums of disks:

$$V = \int_1^8 \pi (y^{1/3})^2 \, dy = \pi \int_1^8 y^{2/3} \, dy = \pi \left[\frac{y^{5/3}}{\frac{5}{3}} \right]_1^8$$

$$= \tfrac{3}{5}\pi \left[y^{5/3} \right]_1^8 = \tfrac{3}{5}\pi \, [32 - 1] = \tfrac{93}{5}\pi \approx 58.4$$

Let us next consider an R_x region of the type illustrated in Figure 5.27(a). If this region is revolved about the x-axis, we obtain the solid illustrated in Figure 5.27(b). Note that if $g(x) > 0$ for every x in $[a, b]$, there is a hole through the solid.

Figure 5.27

(a)

(b)

(c)

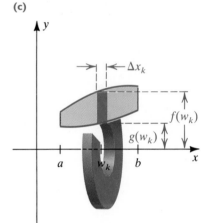

The volume V of the solid may be found by subtracting the volume of the solid generated by the smaller region from the volume of the solid generated by the larger region. Using Definition (5.5) gives us

$$V = \int_a^b \pi [f(x)]^2 \, dx - \int_a^b \pi [g(x)]^2 \, dx$$

$$= \int_a^b \pi \{[f(x)]^2 - [g(x)]^2\} \, dx.$$

The last integral has an interesting interpretation as a limit of sums. As illustrated in Figure 5.27(c), a vertical rectangle extending from the graph of g to the graph of f, through the points with x-coordinate w_k, generates a washer-shaped solid whose volume is

$$\pi [f(w_k)]^2 \Delta x_k - \pi [g(w_k)]^2 \Delta x_k = \pi \{[f(w_k)]^2 - [g(w_k)]^2\} \Delta x_k.$$

Summing the volumes of all such washers and taking the limit gives us the desired definite integral. When working problems of this type, it is convenient to use the following general rule.

Volume V of a Washer | 5.9

$$V = \pi[(\text{outer radius})^2 - (\text{inner radius})^2] \cdot (\text{thickness})$$

In applying (5.9), a common error is to use the square of the difference of the radii instead of the difference of the squares. Note that

volume of a washer $\neq \pi[(\text{outer radius}) - (\text{inner radius})]^2 \cdot (\text{thickness})$.

Guidelines similar to (5.8) can be stated for problems involving washers. The principal differences are that in guideline (3), we find expressions for the outer radius and inner radius of a typical washer, and in guideline (4), we use (5.9) to find a formula for the volume of the washer.

EXAMPLE 3   The region bounded by the graphs of the equations $x^2 = y - 2$ and $2y - x - 2 = 0$ and the vertical lines $x = 0$ and $x = 1$ is revolved about the x-axis. Find the volume of the resulting solid.

SOLUTION   The region and a typical vertical rectangle are sketched in Figure 5.28(a). Since we wish to integrate with respect to x, we solve the first two equations for y in terms of x, obtaining $y = x^2 + 2$ and $y = \frac{1}{2}x + 1$. The solid and a washer generated by the rectangle are illustrated in Figure 5.28(b). Using (5.9) we obtain the following:

thickness of washer: dx

outer radius: $x^2 + 2$

inner radius: $\frac{1}{2}x + 1$

volume: $\pi[(x^2 + 2)^2 - (\frac{1}{2}x + 1)^2] \, dx$

We take a limit of sums of volumes of washers by applying $\int_0^1$:

$$V = \int_0^1 \pi[(x^2 + 2)^2 - (\tfrac{1}{2}x + 1)^2] \, dx$$

$$= \pi \int_0^1 (x^4 + \tfrac{15}{4}x^2 - x + 3) \, dx$$

$$= \pi \left[\frac{x^5}{5} + \frac{15}{4}\left(\frac{x^3}{3}\right) - \frac{x^2}{2} + 3x \right]_0^1 = \frac{79\pi}{20} \approx 12.4$$

Figure 5.28

(a)

(b)

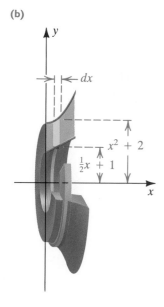

EXAMPLE ▪ 4   Find the volume of the solid generated by revolving the region described in Example 3 about the line $y = 3$.

SOLUTION   The region and a typical vertical rectangle are re-sketched in Figure 5.29(a), together with the axis of revolution $y = 3$. The solid and a washer generated by the rectangle are illustrated in Figure 5.29(b). We note the following:

$$\text{thickness of washer:} \quad dx$$
$$\text{outer radius:} \quad 3 - (\tfrac{1}{2}x + 1) = 2 - \tfrac{1}{2}x$$
$$\text{inner radius:} \quad 3 - (x^2 + 2) = 1 - x^2$$
$$\text{volume:} \quad \pi[(2 - \tfrac{1}{2}x)^2 - (1 - x^2)^2]\,dx$$

Figure 5.29

(a)   (b)

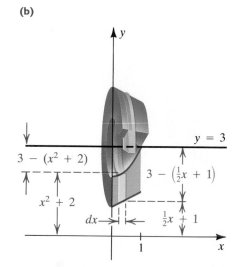

Applying the limit of sums operator $\int_0^1$ gives us the volume:

$$V = \int_0^1 \pi[(2 - \tfrac{1}{2}x)^2 - (1 - x^2)^2]\,dx$$

$$= \pi \int_0^1 (3 - 2x + \tfrac{9}{4}x^2 - x^4)\,dx$$

$$= \pi \left[3x - x^2 + \frac{9}{4}\left(\frac{x^3}{3}\right) - \frac{x^5}{5} \right]_0^1$$

$$= \frac{51\pi}{20} \approx 8.01$$

Figure 5.30

(a)

(b)

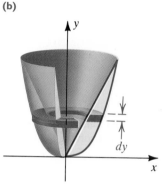

EXAMPLE ■ 5 The region in the first quadrant bounded by the graphs of $y = \frac{1}{8}x^3$ and $y = 2x$ is revolved about the y-axis. Find the volume of the resulting solid.

SOLUTION The region and a typical horizontal rectangle are shown in Figure 5.30(a). We wish to integrate with respect to y, so we solve the given equations for x in terms of y, obtaining

$$x = \tfrac{1}{2}y \quad \text{and} \quad x = 2y^{1/3}.$$

Figure 5.30(b) illustrates the volume generated by the region and the washer generated by the rectangle. We note the following:

thickness of washer: dy

outer radius: $2y^{1/3}$

inner radius: $\frac{1}{2}y$

volume: $\pi[(2y^{1/3})^2 - (\tfrac{1}{2}y)^2]\,dy = \pi(4y^{2/3} - \tfrac{1}{4}y^2)\,dy$

Applying the limit of sums operator $\int_0^8$ gives us the volume:

$$V = \int_0^8 \pi(4y^{2/3} - \tfrac{1}{4}y^2)\,dy$$

$$= \pi\left[\tfrac{12}{5}y^{5/3} - \tfrac{1}{12}y^3\right]_0^8 = \tfrac{512}{15}\pi \approx 107.2$$

EXERCISES 5.2

Exer. 1–4: Set up an integral that can be used to find the volume of the solid obtained by revolving the shaded region about the indicated axis.

2

3

4

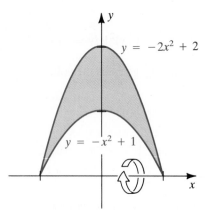

Exer. 5–24: Sketch the region R bounded by the graphs of the equations, and find the volume of the solid generated if R is revolved about the indicated axis.

5 $y = 1/x$, $x = 1$, $x = 3$, $y = 0$; x-axis

6 $y = \sqrt{x}$, $x = 4$, $y = 0$; x-axis

7 $y = x^2 - 4x$, $y = 0$; x-axis

8 $y = x^3$, $x = -2$, $y = 0$; x-axis

9 $y = x^2$, $y = 2$; y-axis

10 $y = 1/x$, $y = 1$, $y = 3$, $x = 0$; y-axis

11 $x = 4y - y^2$, $x = 0$; y-axis

12 $y = x$, $y = 3$, $x = 0$; y-axis

13 $y = x^2$, $y = 4 - x^2$; x-axis

14 $x = y^3$, $x^2 + y = 0$; x-axis

15 $y = x$, $x + y = 4$, $x = 0$; x-axis

16 $y = (x-1)^2 + 1$, $y = -(x-1)^2 + 3$; x-axis

17 $y^2 = x$, $2y = x$; y-axis

18 $y = 2x$, $y = 4x^2$; y-axis

19 $x = y^2$, $x - y = 2$; y-axis

20 $x + y = 1$, $x - y = -1$, $x = 2$; y-axis

21 $y = \sin 2x$, $x = 0$, $x = \pi$, $y = 0$; x-axis
(*Hint:* Use a half-angle formula.)

22 $y = 1 + \cos 3x$, $x = 0$, $x = 2\pi$, $y = 0$; x-axis
(*Hint:* Use a half-angle formula.)

23 $y = \sin x$, $y = \cos x$, $x = 0$, $x = \pi/4$; x-axis
(*Hint:* Use a double angle formula.)

24 $y = \sec x$, $y = \sin x$, $x = 0$, $x = \pi/4$; x-axis

Exer. 25–26: Sketch the region R bounded by the graphs of the equations, and find the volume of the solid generated if R is revolved about the given line.

25 $y = x^2$, $y = 4$
    (a) $y = 4$ (b) $y = 5$
    (c) $x = 2$ (d) $x = 3$

26 $y = \sqrt{x}$, $y = 0$, $x = 4$
    (a) $x = 4$ (b) $x = 6$
    (c) $y = 2$ (d) $y = 4$

Exer. 27–28: Set up an integral that can be used to find the volume of the solid generated by revolving the shaded region about the line (a) $y = -2$, (b) $y = 5$, (c) $x = 7$, and (d) $x = -4$.

27

28

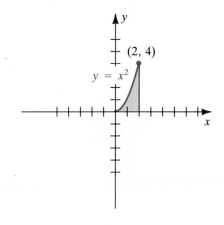

Exer. 29–34: Sketch the region R bounded by the graphs of the equations, and set up integrals that can be used to find the volume of the solid generated if R is revolved about the given line.

29 $y = x^3$, $y = 4x$; $y = 8$

30 $y = x^3$, $y = 4x$; $x = 4$

31 $x + y = 3$, $y + x^2 = 3$; $x = 2$

32 $y = 1 - x^2$, $x - y = 1$; $y = 3$

33 $x^2 + y^2 = 1$; $x = 5$

34 $y = x^{2/3}$, $y = x^2$; $y = -1$

Exer. 35–40: Use a definite integral to derive a formula for the volume of the indicated solid.

35 A right circular cylinder of altitude h and radius r

36 A cylindrical shell of altitude h, outer radius R, and inner radius r

37 A right circular cone of altitude h and base radius r

38 A sphere of radius r

39 A frustum of a right circular cone of altitude h, lower base radius R, and upper base radius r

40 A spherical segment of altitude h in a sphere of radius r

41 If the region shown in the figure is revolved about the x-axis, use the trapezoidal rule with $n = 6$ to approximate the volume of the resulting solid.

Exercise 41

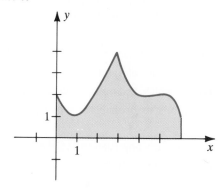

42 If the region shown in the figure is revolved about the x-axis, use Simpson's rule with $n = 4$ to approximate the volume of the resulting solid.

Exercise 42

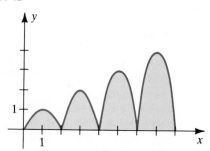

c   **Exer. 43–44:** Graph f and g on the same coordinate axes for $0 \le x \le \pi$. **(a)** Estimate the x-coordinates of the points of intersection of the graphs. **(b)** If the region bounded by the graphs of f and g is revolved about the x-axis, use Simpson's rule with $n = 2$ to approximate the volume of the resulting solid.

43 $f(x) = \dfrac{\sin x}{1 + x}$; $g(x) = 0.3$

44 $f(x) = \sqrt[4]{|\sin x|}$; $g(x) = 0.2x + 0.7$

45 Find the volume of the solid obtained by revolving the region bounded by the ellipse $b^2 x^2 + a^2 y^2 = a^2 b^2$ about the x-axis.

46 Work Exercise 45 with the region revolved about the y-axis.

47 A *paraboloid of revolution* is formed by revolving a parabola about its axis. Paraboloids are the basic shape for a wide variety of collectors and reflectors. Shown in the figure is a (finite) paraboloid of altitude h and radius of base r.

    (a) The *focal length* of the paraboloid is the distance p between the vertex and the focus of the parabola. Express p in terms of r and h.

    (b) Find the volume of the paraboloid.

Exercise 47

Mathematicians and Their Times

JOHN BERNOULLI

WE OFTEN HEAR OF SCIENTIFIC or artistic genius emerging in an individual of modest circumstances whose forebears gave little evidence of greatness. Such was the case with Newton. There have been notable instances, however, of enormous talent displayed by several generations of the same family. In music, the Bach family included a score of eminent artists. In mathematics, the premier example is the Bernoulli family, with eight mathematicians in three generations and dozens of other distinguished descendants who played a leading part in developing calculus and making it accessible to a wider audience.

The first generation of Bernoulli mathematicians were the three sons of a druggist who fled from Antwerp to Switzerland to escape religious persecution. Two of these brothers, James (1654–1705) and John (1667–1748), are the most eminent mathematicians among the Bernoullis. Their brother Nicholas (1662–1716), also a gifted mathematician, first earned a degree in philosophy at age 16 and then turned to law before joining the mathematics faculty at St. Petersburg Academy.

James Bernoulli, a mathematics professor at the University of Basel, introduced the term *integral* into the field and developed the calculus significantly beyond the state in which Leibniz and Newton left it. He also made important contributions to probability and the calculus of variations.

John Bernoulli began his career as a physician and studied mathematics under his older brother James, eventually replacing him as mathematics professor at the University of Basel. He became deeply interested in calculus and was indirectly responsible for the first calculus textbook (1696), published by the French marquis G. F. A. de l'Hôpital. Bernoulli tutored l'Hôpital and sold the marquis the rights to a number of his own mathematical discoveries. Later Bernoulli virtually accused l'Hôpital of plagiarism.

John also became locked in a bitter quarrel over mathematics with his brother James, exchanging words that later writers characterized as more in keeping with horse thieves or street brawlers than well-known scientists. When the French Academy of Sciences awarded his son Daniel a prize, John was so jealous that he expelled Daniel from his home.

Most notable of the second generation of Bernoullis were John's sons: Nicholas, Daniel, and John II. Daniel's discoveries in science were so extensive that he is considered the founder of mathematical physics. John II began his career in law, later became a professor of eloquence, and eventually succeeded his father as Basel's mathematics professor.

In the third generation, the sons of John II, John III and Jacob, also did significant work in the sciences. John III became the royal astronomer at Berlin, but his brother's promising career was tragically cut short by drowning at age 30.

5.3 VOLUMES BY CYLINDRICAL SHELLS

In the preceding section, we found volumes of solids of revolution by using circular disks or washers. In this section, we shall see that for certain types of solids, it is convenient to use hollow circular cylinders—that is, thin **cylindrical shells** of the type illustrated in Figure 5.31, where r_1 is the *outer radius*, r_2 is the *inner radius*, h is the *altitude*, and $\Delta r = r_1 - r_2$ is the *thickness* of the shell. The **average radius** of the shell is $r = \frac{1}{2}(r_1 + r_2)$. We can find the volume of the shell by subtracting the volume $\pi r_2^2 h$ of the inner cylinder from the volume $\pi r_1^2 h$ of the outer cylinder. If we do so and change the form of the resulting expression, we obtain

$$\pi r_1^2 h - \pi r_2^2 h = \pi (r_1^2 - r_2^2) h$$
$$= \pi (r_1 + r_2)(r_1 - r_2) h$$
$$= 2\pi \cdot \tfrac{1}{2}(r_1 + r_2) h (r_1 - r_2)$$
$$= 2\pi r h \Delta r,$$

which gives us the following general rule.

Figure 5.31

Volume V of a
Cylindrical Shell   5.10

$$V = 2\pi(\text{average radius})(\text{altitude})(\text{thickness})$$

If the R_x region in Figure 5.32(a) is revolved about the y-axis, we obtain the solid illustrated in Figure 5.32(b).

Let P be a partition of $[a, b]$, and consider the typical vertical rectangle in Figure 5.32(c), where w_k is the midpoint of $[x_{k-1}, x_k]$. If we revolve this rectangle about the y-axis, we obtain a cylindrical shell of average radius w_k, altitude $f(w_k)$, and thickness Δx_k. Hence, by (5.10), the volume of the shell is

$$2\pi w_k f(w_k) \Delta x_k.$$

Revolving the rectangular polygon formed by *all* the rectangles determined by P gives us the solid illustrated in Figure 5.32(d). The volume of this solid is a Reimann sum:

$$\sum_k 2\pi w_k f(w_k) \Delta x_k$$

The smaller the norm $\|P\|$ of the partition, the better the sum approximates the volume V of the solid shown in Figure 5.32(b). This discussion provides the motivation for Definition (5.11) on the following page.

Figure 5.32

(a)   (b)

(c)   (d)

Definition 5.11

Let f be continuous and suppose $f(x) \geq 0$ on the interval $[a, b]$, where $0 \leq a \leq b$. Let R be the region under the graph of f from a to b. The volume V of the solid of revolution generated by revolving R about the y-axis is

$$V = \lim_{\|P\| \to 0} \sum_k 2\pi w_k f(w_k) \Delta x_k = \int_a^b 2\pi x f(x)\, dx.$$

It can be proved that if the methods of Section 5.2 are also applicable, then both methods lead to the same answer.

We may also consider solids obtained by revolving a region about the x-axis or some other line. The following guidelines may be useful.

***Guidelines for Finding the Volume
of a Solid of Revolution Using
Cylindrical Shells 5.12***

I Sketch the region R to be revolved, and label the boundaries. Show a typical vertical rectangle of width dx or a horizontal rectangle of width dy.

2 Sketch the cylindrical shell generated by the rectangle in guideline (1).

3 Express the average radius of the shell in terms of x or y, depending on whether its thickness is dx or dy. *Remember that x represents a distance from the y-axis to a vertical rectangle, and y represents a distance from the x-axis to a horizontal rectangle.*

4 Express the altitude of the shell in terms of x or y, depending on whether its thickness is dx or dy.

5 Use (5.10) to find a formula for the volume of the shell.

6 Apply the limit of sums operator $\int_a^b$ or $\int_c^d$ to the expression in guideline (5) and evaluate the integral.

Figure 5.33
(a)

(b)

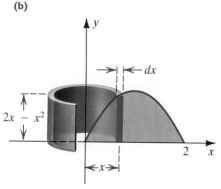

EXAMPLE ▪ I The region bounded by the graph of $y = 2x - x^2$ and the x-axis is revolved about the y-axis. Find the volume of the resulting solid.

SOLUTION The region to be revolved is sketched in Figure 5.33(a), together with a typical vertical rectangle of width dx. Figure 5.33(b) shows the cylindrical shell generated by revolving the rectangle about the y-axis. Note that x represents the distance from the y-axis to the midpoint of the rectangle (the average radius of the shell). Referring to the figure and using (5.10) gives us the following:

thickness of shell: dx

average radius: x

altitude: $2x - x^2$

volume: $2\pi x(2x - x^2)\, dx$

To sum all such shells, we move from left to right through the region from $a = 0$ to $b = 2$ (do *not* sum from -2 to 2). Hence, the limit of sums is

$$V = \int_0^2 2\pi x(2x - x^2)\, dx = 2\pi \int_0^2 (2x^2 - x^3)\, dx$$

$$= 2\pi \left[2\left(\frac{x^3}{3}\right) - \frac{x^4}{4} \right]_0^2 = \frac{8\pi}{3} \approx 8.4.$$

The volume V can also be found using washers; however, the calculations would be much more involved, since the equation $y = 2x - x^2$ would have to be solved for x in terms of y.

Figure 5.34

(a)

(b)

EXAMPLE ▪2   The region bounded by the graphs of $y = x^2$ and $y = x + 2$ is revolved about the line $x = 3$. Set up the integral for the volume of the resulting solid.

SOLUTION   The region is sketched in Figure 5.34(a), together with a typical vertical rectangle extending from the lower boundary $y = x^2$ to the upper boundary $y = x + 2$. Also shown is the axis of revolution $x = 3$. In Figure 5.34(b), we have illustrated both the cylindrical shell and the solid that are generated by revolving the rectangle and the region about the line $x = 3$. It is important to note that since x is the distance from the y-axis to the rectangle, the radius of the shell is $3 - x$. Referring to Figure 5.34 and using (5.10) gives us the following:

thickness of shell: dx

average radius: $3 - x$

altitude: $(x + 2) - x^2$

volume: $2\pi(3 - x)(x + 2 - x^2)\, dx$

To sum all such shells, we move from left to right through the region from $a = -1$ to $b = 2$. Hence, the limit of sums is

$$V = \int_{-1}^2 2\pi(3 - x)(x + 2 - x^2)\, dx.$$

EXAMPLE ▪3   The region in the first quadrant bounded by the graph of the equation $x = 2y^3 - y^4$ and the y-axis is revolved about the x-axis. Set up the integral for the volume of the resulting solid.

SOLUTION   The region is sketched in Figure 5.35(a), together with a typical horizontal rectangle. Figure 5.35(b) shows the cylindrical shell and the solid that are generated by the revolution about the x-axis. Referring to the figure and using (5.10) gives the following:

thickness of shell: dy

average radius: y

altitude: $2y^3 - y^4$

volume: $2\pi y(2y^3 - y^4)\, dy$

Figure 5.35

(a)

(b)

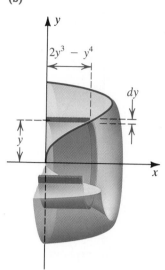

To sum all such shells, we move upward through the region from $c = 0$ to $d = 2$. Hence, the limit of sums is

$$V = \int_0^2 2\pi y(2y^3 - y^4)\, dy.$$

It is worth noting that in the preceding example we were forced to use shells and to integrate with respect to y, since the use of washers and integration with respect to x would require that we solve the equation $x = 2y^3 - y^4$ for y in terms of x, a rather formidable task.

EXERCISES 5.3

Use cylindrical shells for each exercise.

Exer. 1–4: Set up an integral that can be used to find the volume of the solid obtained by revolving the shaded region about the indicated axis.

1

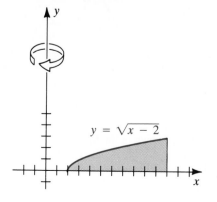

$y = \sqrt{x - 2}$

2

$y = x^2 + 1$

3

2x + y = 6

4

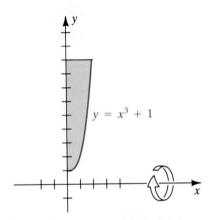

y = x³ + 1

Exer. 5–18: Sketch the region R bounded by the graphs of the equations, and find the volume of the solid generated if R is revolved about the indicated axis.

5 $y = \sqrt{x}$,	$x = 4$,	$y = 0$;	y-axis
6 $y = 1/x$;	$x = 1$,	$x = 2, y = 0$;	y-axis
7 $y = x^2$,	$y^2 = 8x$;		y-axis
8 $16y = x^2$,	$y^2 = 2x$;		y-axis
9 $2x - y = 12$,	$x - 2y = 3$,	$x = 4$;	y-axis
10 $y = x^3 + 1$,	$x + 2y = 2$,	$x = 1$;	y-axis
11 $2x - y = 4$,	$x = 0$,	$y = 0$;	y-axis
12 $y = x^2 - 5x$,	$y = 0$;		y-axis
13 $x^2 = 4y$,	$y = 4$;		x-axis
14 $y^3 = x$,	$y = 3$,	$x = 0$;	x-axis
15 $y = 2x$,	$y = 6$,	$x = 0$;	x-axis
16 $2y = x$,	$y = 4$,	$x = 1$;	x-axis
17 $y = \sqrt{x + 4}$,	$y = 0$,	$x = 0$;	x-axis
18 $y = -x$,	$x - y = -4$,	$y = 0$;	x-axis

Exer. 19–26: Let R be the region bounded by the graphs of the equations. Set up an integral that can be used to find the volume of the solid generated if R is revolved about the given line.

19 $y = x^2 + 1$, $x = 0$, $x = 2$, $y = 0$
   (a) $x = 3$ (b) $x = -1$

20 $y = 4 - x^2$, $y = 0$
   (a) $x = 2$ (b) $x = -3$

21 $y = x^2$, $y = 4$
   (a) $y = 4$ (b) $y = 5$ (c) $x = 2$ (d) $x = -3$

22 $y = \sqrt{x}$, $y = 0$, $x = 4$
   (a) $x = 4$ (b) $x = 6$ (c) $y = 2$ (d) $y = -4$

23 $x + y = 3$, $y + x^2 = 3$; $x = 2$

24 $y = 1 - x^2$, $x - y = 1$; $y = 3$

25 $x^2 + y^2 = 1$; $x = 5$

26 $y = x^{2/3}$, $y = x^2$; $y = -1$

Exer. 27–30: Let R be the region bounded by the graphs of the equations. Set up integrals that can be used to find the volume of the solid generated if R is revolved about the given axis using (a) cylindrical shells and (b) disks or washers.

27 $y = 1/\sqrt{x}$, $x = 1$, $x = 4$, $y = 0$; x-axis

28 $y = 9 - x^2$, $x = 0$, $x = 2$, $y = 0$; x-axis

29 $y = x^2 + 2$, $x = 0$, $x = 1$, $y = 0$; y-axis

30 $y = x + 1$, $x = 0$, $x = 1$, $y = 0$; y-axis

31 If the region shown in the figure is revolved about the y-axis, use the trapezoidal rule, with $n = 6$, to approximate the volume of the resulting solid.

Exercise 31

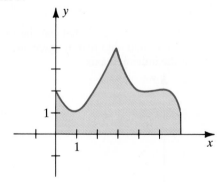

32 If the region shown in the figure on the following page is revolved about the y-axis, use Simpson's rule, with $n = 4$, to approximate the volume of the resulting solid.

Exercise 32

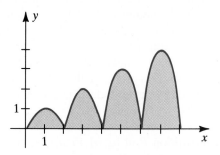

c 33 Graph $f(x) = -x^4 + 2.21x^3 - 3.21x^2 + 4.42x - 2$.

(a) Estimate the x-intercepts of the graph.

(b) If the region bounded by the graph of f and the x-axis is revolved about the y-axis, set up an integral that can be used to approximate the volume of the resulting solid.

c 34 Graph, on the same coordinate axes, $f(x) = \csc x$ and $g(x) = x + 1$ for $0 < x < \pi$.

(a) Use Newton's method to approximate, to four decimal places, the x-coordinates of the points of intersection of the graphs.

(b) If the region bounded by the graphs is revolved about the y-axis, use the trapezoidal rule, with $n = 6$, to approximate the volume of the resulting solid.

35 Let R be the region bounded by the parabola $x^2 = 4y$ and the line l through the focus that is perpendicular to the axis of the parabola.

(a) Find the area of R.

(b) If R is revolved about the y-axis, find the volume of the resulting solid.

(c) If R is revolved about the x-axis, find the volume of the resulting solid.

36 Work (a)–(c) of Exercise 35 if R is the region bounded by the graphs of $y^2 = 2x - 6$ and $x = 5$.

Exer. 37–38: Let R be the region bounded by the hyperbola with equation $b^2x^2 - a^2y^2 = a^2b^2$ and a vertical line through a focus.

37 Show that the area of the region R is given by

$$\frac{2b}{a} \int_a^c \sqrt{x^2 - a^2}\, dx, \quad \text{where } c = \sqrt{a^2 + b^2}.$$

38 Find the volume of the solid obtained by revolving R about the y-axis.

5.4	**VOLUMES BY CROSS SECTIONS**

If a plane intersects a solid, then the region common to the plane and the solid is a **cross section** of the solid. In Section 5.2, we used circular and washer-shaped cross sections to find volumes of solids of revolution. In this section, we shall study solids that have the following property (see Figure 5.36): For every x in $[a, b]$, the plane perpendicular to the x-axis at x intersects the solid in a cross section whose area is $A(x)$, where A is a continuous function on $[a, b]$.

Figure 5.36

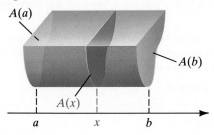

Figure 5.37

$A(a)$

$A(b)$

$A(x)$

a x b

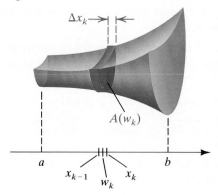

Figure 5.38

Δx_k

$A(w_k)$

a x_{k-1} w_k x_k b

Volumes by Cross Sections 5.13

Figure 5.39

(a)

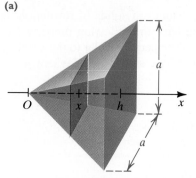

a

O x h x

a

(b)

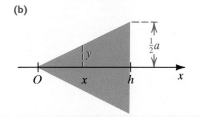

$\frac{1}{2}a$

y

O x h x

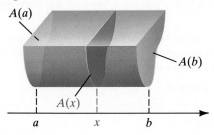

The solid is called a **cylinder** if, as illustrated in Figure 5.37 a line parallel to the x-axis that traces the boundary of the cross section corresponding to a also traces the boundary of the cross section corresponding to every x in $[a, b]$. The cross sections determined by the planes through $x = a$ and $x = b$ are the **bases** of the cylinder. The distance between the bases is the **altitude** of the cylinder. By definition, the volume of the cylinder is the area of a base multiplied by the altitude. Thus, the volume of the solid in Figure 5.37 is $A(a) \cdot (b - a)$.

To find the volume of a noncylindrical solid of the type illustrated in Figure 5.38, we begin with a partition P of $[a, b]$. Planes perpendicular to the x-axis at each x_k in the partition slice the solid into smaller pieces. If we choose any number w_k in $[x_{k-1}, x_k]$, the volume of a typical slice can be approximated by the volume $A(w_k)\Delta x_k$ of the red cylinder shown in Figure 5.38. If V is the volume of the solid and if the norm $\|P\|$ is small, then

$$V \approx \sum_k A(w_k)\Delta x_k.$$

Since this approximation improves as $\|P\|$ gets smaller, we define the volume of the solid by

$$V = \lim_{\|P\| \to 0} \sum_k A(w_k)\Delta x_k = \int_a^b A(x)\, dx,$$

where the last equality follows from the definition of the definite integral. We may summarize our discussion as follows.

Let S be a solid bounded by planes that are perpendicular to the x-axis at a and b. If, for every x in $[a, b]$, the cross-sectional area of S is given by $A(x)$, where A is continuous on $[a, b]$, then the volume of S is

$$V = \int_a^b A(x)\, dx.$$

An analogous result can be stated for a y-interval $[c, d]$ and a cross-sectional area $A(y)$.

EXAMPLE ■ I   Find the volume of a right pyramid with a square base of side a and altitude h.

SOLUTION   As in Figure 5.39(a), let us take the vertex of the pyramid at the origin, with the x-axis passing through the center of the square base, a distance h from O. Cross sections by planes perpendicular to the x-axis are squares. Figure 5.39(b) is a side view of the pyramid. Since $2y$ is the length of the side of the square cross section corresponding to x, the cross-sectional area $A(x)$ is

$$A(x) = (2y)^2 = 4y^2.$$

Using similar triangles in Figure 5.39(b), we have

$$\frac{y}{x} = \frac{\frac{1}{2}a}{h}, \quad \text{or} \quad y = \frac{ax}{2h}.$$

Hence,

$$A(x) = 4y^2 = \frac{4a^2x^2}{4h^2} = \frac{a^2}{h^2}x^2.$$

Applying (5.13) yields

$$V = \int_0^h A(x)\,dx = \int_0^h \left(\frac{a^2}{h^2}\right)x^2\,dx$$

$$= \left(\frac{a^2}{h^2}\right)\left[\frac{x^3}{3}\right]_0^h = \frac{a^2}{h^2}\frac{h^3}{3} = \frac{1}{3}a^2h.$$

EXAMPLE ▪ 2 A solid has, as its base, the circular region in the xy-plane bounded by the graph of $x^2 + y^2 = a^2$ with $a > 0$. Find the volume of the solid if every cross section by a plane perpendicular to the x-axis is an equilateral triangle with one side in the base.

Figure 5.40

(a)

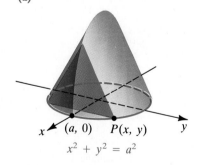

$$(a, 0) \quad P(x, y)$$
$$x^2 + y^2 = a^2$$

(b)

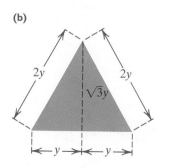

SOLUTION A triangular cross section by a plane x units from the origin is illustrated in Figure 5.40(a). If the point $P(x, y)$ is on the circle and $y > 0$, then the lengths of the sides of this equilateral triangle are $2y$. Referring to Figure 5.40(b), we see, by the Pythagorean theorem, that the altitude of the triangle is

$$\sqrt{(2y)^2 - y^2} = \sqrt{3y^2} = \sqrt{3}y.$$

Hence, the area $A(x)$ of the cross section is

$$A(x) = \tfrac{1}{2}(2y)(\sqrt{3}y) = \sqrt{3}y^2 = \sqrt{3}(a^2 - x^2).$$

Applying (5.13) gives us

$$V = \int_{-a}^a A(x)\,dx = \int_{-a}^a \sqrt{3}(a^2 - x^2)\,dx$$

$$= \sqrt{3}\left[a^2x - \frac{x^3}{3}\right]_{-a}^a = \frac{4\sqrt{3}}{3}a^3.$$

EXERCISES 5.4

Exer. 1–8: Let R be the region bounded by the graphs of $x = y^2$ and $x = 9$. Find the volume of the solid that has R as its base if every cross section by a plane perpendicular to the x-axis has the given shape.

1 A square

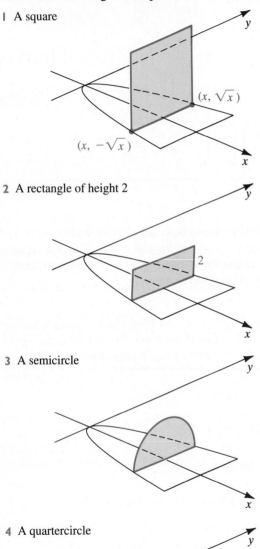

$(x, \sqrt{x})$

$(x, -\sqrt{x})$

2 A rectangle of height 2

2

3 A semicircle

4 A quartercircle

5 An equilateral triangle

6 A triangle with height equal to $\frac{1}{4}$ the length of the base

7 A trapezoid with lower base in the xy-plane, upper base equal to $\frac{1}{2}$ the length of the lower base, and height equal to $\frac{1}{4}$ the length of the lower base

8 A parallelogram with base in the xy-plane and height equal to twice the length of the base

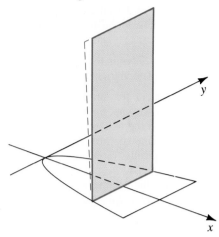

9 A solid has as its base the circular region in the xy-plane bounded by the graph of $x^2 + y^2 = a^2$ with $a > 0$. Find the volume of the solid if every cross section by a plane perpendicular to the x-axis is a square.

10 Work Exercise 9 if every cross section is an isosceles triangle with base on the xy-plane and altitude equal to the length of the base.

11 A solid has as its base the region in the xy-plane bounded by the graphs of $y = 4$ and $y = x^2$. Find the volume of the solid if every cross section by a plane perpendicular to the x-axis is an isosceles right triangle with hypotenuse on the xy-plane.

12 Work Exercise 11 if every cross section is a square.

13 Find the volume of a pyramid of the type illustrated in Figure 5.39 if the altitude is h and the base is a rectangle of dimensions a and $2a$.

14 A solid has as its base the region in the xy-plane bounded by the graphs of $y = x$ and $y^2 = x$. Find the volume of the solid if every cross section by a plane perpendicular to the x-axis is a semicircle with diameter in the xy-plane.

15 A solid has as its base the region in the xy-plane bounded by the graphs of $y^2 = 4x$ and $x = 4$. If every cross section by a plane perpendicular to the y-axis is a semicircle, find the volume of the solid.

16 A solid has as its base the region in the xy-plane bounded by the graphs of $x^2 = 16y$ and $y = 2$. Every cross section by a plane perpendicular to the y-axis is a rectangle whose height is twice that of the side in the xy-plane. Find the volume of the solid.

17 A log having the shape of a right circular cylinder of radius a is lying on its side. A wedge is removed from the log by making a vertical cut and another cut at an angle of $45°$, both cuts intersecting at the center of the log (see figure). Find the volume of the wedge.

Exercise 17

18 The axes of two right circular cylinders of radius a intersect at right angles. Find the volume of the solid bounded by the cylinders.

19 The base of a solid is the circular region in the xy-plane bounded by the graph of $x^2 + y^2 = a^2$ with $a > 0$. Find the volume of the solid if every cross section by a plane perpendicular to the x-axis is an isosceles triangle of constant altitude h. (*Hint:* Interpret $\int_{-a}^{a} \sqrt{a^2 - x^2}\, dx$ as an area.)

20 Cross sections of a horn-shaped solid by planes perpendicular to its axis are circles. If a cross section that is s inches from the smaller end of the solid has diameter $6 + \frac{1}{36}s^2$ inches and if the length of the solid is 2 ft, find its volume.

21 A tetrahedron has three mutually perpendicular faces and three mutually perpendicular edges of lengths 2, 3, and 4 cm, respectively. Find its volume.

22 *Cavalieri's theorem* states that if two solids have equal altitudes and if all cross sections by planes parallel to their bases and at the same distances from their bases have equal areas, then the solids have the same volume (see figure). Prove Cavalieri's theorem.

Exercise 22

23 The base of a solid is an isosceles right triangle whose equal sides have length a. Find the volume if cross sections that are perpendicular to the base and to one of the equal sides are semicircular.

24 Work Exercise 23 if the cross sections are ʌegular hexagons with one side in the base.

25 Show that the disk and washer methods discussed in Section 5.2 are special cases of (5.13).

26 A circular swimming pool has diameter 28 ft. The depth of the water changes slowly from 3 ft at a point A on one side of the pool to 9 ft at a point B diametrically opposite A (see figure). Depth readings $h(x)$ (in feet) taken along the diameter AB are given in the following table, where x is the distance (in feet) from A.

x	0	4	8	12	16	20	24	28
$h(x)$	3	3.5	4	5	6.5	8	8.5	9

Use the trapezoidal rule, with $n = 7$, to estimate the volume of water in the pool. Approximate the number of gallons of water contained in the pool (1 gal $\approx$ 0.134 ft^3).

Exercise 26

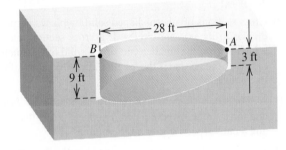

27 The base of a solid is a region bounded by an ellipse with major and minor axes of lengths 16 and 9, respectively. Find the volume of the solid if every cross section by a plane perpendicular to the major axis has the shape of a square.

28 Work Exercise 27 with the cross section having the shape of an equilateral triangle.

29 A common model for human limbs is the *elliptical frustum* shown in the figure, where cross sections perpendicular to the axis of the frustum are elliptical and have the same eccentricity. For human limbs, the eccentricity typically varies from 0.6 to values near 1. If $k = a_1/b_1 = a_2/b_2$ and if L is the length of the limb, show that the volume V is given by the equation $V = (\frac{1}{3}\pi L/k)(a_1^2 + a_1 a_2 + a_2^2)$. (*Hint*: Use Exercise 39 in Section 5.1.)

Exercise 29

30 The base of a right elliptic cone has major and minor axes of lengths $2a$ and $2b$, respectively. Find the volume if the altitude of the cone is h. (*Hint*: Use Exercise 39 in Section 5.1.)

5.5 ARC LENGTH AND SURFACES OF REVOLUTION

In earlier sections, we considered the volume of the solid created when the graph of a function is revolved about an axis. In this section, we will determine the *length* of the graph and the *surface area* of the solid.

For some applications, we must determine the *length* of the graph of a function. To obtain a suitable formula, we shall employ a process similar to one that could be used to approximate the length of a bent wire. Let us imagine dividing the wire into many small pieces by placing dots at

Figure 5.41
Bent wire

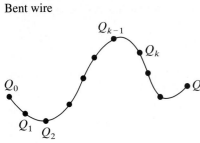

$Q_0, Q_1, Q_2, \ldots, Q_n$, as illustrated in Figure 5.41. We may then approximate the length of the wire between Q_{k-1} and Q_k (for each k) by measuring the distance $d(Q_{k-1}, Q_k)$ with a ruler. The sum of all these distances is an approximation for the total length of the wire. Evidently, the closer together we place the dots, the better the approximation. The process we shall use for the graph of a function is similar; however, we shall find the *exact* length by taking a *limit of sums* of lengths of line segments. This process leads to a definite integral. To guarantee that the integral exists, we must place restrictions on the function, as indicated in the following discussion.

A function f is **smooth** on an interval if it has a derivative f' that is continuous throughout the interval. Intuitively, this means that a small change in x produces a small change in the slope $f'(x)$ of the tangent line to the graph of f. Thus, the graph has no corners or cusps. We shall define the **length of arc** between two points A and B on the graph of a smooth function.

If f is smooth on a closed interval $[a, b]$, the points $A(a, f(a))$ and $B(b, f(b))$ are called the **endpoints** of the graph of f. Let P be the partition of $[a, b]$ determined by $a = x_0, x_1, x_2, \ldots, x_n = b$, and let Q_k denote the point with coordinates $(x_k, f(x_k))$ on the graph of f, as illustrated in Figure 5.42. If we connect each Q_{k-1} to Q_k by a line segment of length $d(Q_{k-1}, Q_k)$, the length L_P of the resulting broken line is

$$L_P = \sum_{k=1}^{n} d(Q_{k-1}, Q_k).$$

Using the distance formula, we get

$$d(Q_{k-1}, Q_k) = \sqrt{(x_k - x_{k-1})^2 + [f(x_k) - f(x_{k-1})]^2}.$$

By the mean value theorem (3.12),

$$f(x_k) - f(x_{k-1}) = f'(w_k)(x_k - x_{k-1})$$

Figure 5.42

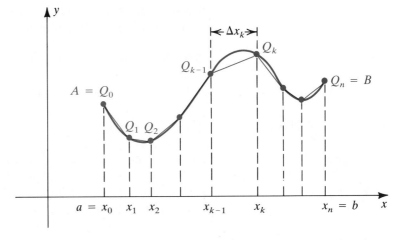

for some number w_k in the open interval (x_{k-1}, x_k). Substituting for $f(x_k) - f(x_{k-1})$ in the preceding formula and letting $\Delta x_k = x_k - x_{k-1}$, we obtain

$$d(Q_{k-1}, Q_k) = \sqrt{(\Delta x_k)^2 + [f'(w_k)\Delta x_k]^2}$$
$$= \sqrt{1 + [f'(w_k)]^2}\, \Delta x_k.$$

Consequently,

$$L_P = \sum_{k=1}^{n} \sqrt{1 + [f'(w_k)]^2}\, \Delta x_k.$$

Observe that L_P is a Riemann sum for $g(x) = \sqrt{1 + [f'(x)]^2}$. Moreover, g is continuous on $[a, b]$, since f' is continuous. If the norm $\|P\|$ is small, then the length L_P of the broken line approximates the length of the graph of f from A to B. This approximation should improve as $\|P\|$ decreases, so we define the *length* (also called the *arc length*) of the graph of f from A to B as the limit of sums L_P. Since $g = \sqrt{1 + (f')^2}$ is a continuous function, the limit exists and equals the definite integral $\int_a^b \sqrt{1 + [f'(x)]^2}\, dx$. This arc length will be denoted by L_a^b.

Definition 5.14

> Let f be smooth on $[a, b]$. The **arc length of the graph** of f from $A(a, f(a))$ to $B(b, f(b))$ is
>
> $$L_a^b = \int_a^b \sqrt{1 + [f'(x)]^2}\, dx.$$

Definition (5.14) will be extended to more general graphs in Chapter 9. If a function f is defined implicitly by an equation in x and y, then we shall also refer to the *arc length of the graph of the equation*.

EXAMPLE ■ 1 If $f(x) = 3x^{2/3} - 10$, find the arc length of the graph of f from the point $A(8, 2)$ to $B(27, 17)$.

SOLUTION The graph of f is sketched in Figure 5.43. Since

$$f'(x) = 2x^{-1/3} = \frac{2}{x^{1/3}},$$

we have, by Definition (5.14),

$$L_8^{27} = \int_8^{27} \sqrt{1 + \left(\frac{2}{x^{1/3}}\right)^2}\, dx = \int_8^{27} \sqrt{1 + \frac{4}{x^{2/3}}}\, dx$$

$$= \int_8^{27} \sqrt{\frac{x^{2/3} + 4}{x^{2/3}}}\, dx = \int_8^{27} \sqrt{x^{2/3} + 4}\, \frac{1}{x^{1/3}}\, dx.$$

Figure 5.43

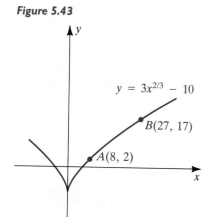

$y = 3x^{2/3} - 10$

$B(27, 17)$

$A(8, 2)$

To evaluate this integral, we make the substitution

$$u = x^{2/3} + 4, \quad du = \frac{2}{3}x^{-1/3}\, dx = \frac{2}{3}\frac{1}{x^{1/3}}\, dx.$$

The integral can be expressed in a suitable form for integration by introducing the factor $\frac{2}{3}$ in the integrand and compensating by multiplying the integral by $\frac{3}{2}$:

$$L_8^{27} = \frac{3}{2}\int_8^{27} \sqrt{x^{2/3} + 4}\,\left(\frac{2}{3}\frac{1}{x^{1/3}}\right) dx$$

We next calculate the values of $u = x^{2/3} + 4$ that correspond to the limits of integration $x = 8$ and $x = 27$:

   (i) If $x = 8$, then $u = 8^{2/3} + 4 = 8$.

   (ii) If $x = 27$, then $u = 27^{2/3} + 4 = 13$.

Substituting in the integrand and changing the limits of integration gives us the arc length:

$$L_8^{27} = \frac{3}{2}\int_8^{13}\sqrt{u}\,du = \left[u^{3/2}\right]_8^{13} = 13^{3/2} - 8^{3/2} \approx 24.2$$

Interchanging the roles of x and y in Definition (5.14) gives us the following formula for integration with respect to y.

Definition 5.15

Figure 5.44

Let $x = g(y)$ with g smooth on the interval $[c, d]$. The arc length of the graph of g from $(g(c), c)$ to $(g(d), d)$ (see Figure 5.44) is

$$L_c^d = \int_c^d \sqrt{1 + [g'(y)]^2}\, dy.$$

The integrands $\sqrt{1 + [f'(x)]^2}$ and $\sqrt{1 + [g'(y)]^2}$ in formulas (5.14) and (5.15) often result in expressions that have no obvious antiderivatives. In such cases, numerical integration may be used to approximate arc length, as illustrated in the next example.

EXAMPLE 2

(a) Set up an integral for finding the arc length of the graph of the equation $y^3 - y - x = 0$ from $A(0, -1)$ to $B(6, 2)$.

(b) Approximate the integral in part (a) to at least four-decimal-place accuracy by using Simpson's rule (4.38).

Figure 5.45

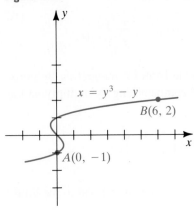

SOLUTION

(a) Since the equation is not of the form $y = f(x)$, we cannot apply Definition (5.14) directly. However, if we write $x = y^3 - y$, then we can use (5.15) with $g(y) = y^3 - y$. The graph of the equation is sketched in Figure 5.45. Using (5.15) with $c = -1$ and $d = 2$ yields

$$L_{-1}^2 = \int_{-1}^{2} \sqrt{1 + (3y^2 - 1)^2} \, dy$$

$$= \int_{-1}^{2} \sqrt{9y^4 - 6y^2 + 2} \, dy.$$

(b) We compute Simpson's rule repeatedly, beginning with $n = 1$ and then successively doubling the number of subintervals—that is, $n = 1, 2, 4, 8, \ldots$. The following table shows the results of our calculations.

n	S_n
1	8.70226731015
2	8.94490388877
4	8.70891806925
8	8.72498046484
16	8.72499726385
32	8.72500017224

From the table, we see that to four decimal places we obtain the approximation 8.7250 for $n = 8$, 16, and 32. We know, furthermore, from Theorem (4.39) that another doubling of n would divide the error by 16. Thus, subsequent changes in the approximation based on larger values of n would not affect the first four decimal places. Our conclusion is

$$\int_{-1}^{2} \sqrt{9y^4 - 6y^2 + 2} \, dy \approx 8.7250.$$

A function f is **piecewise smooth** on its domain if the graph of f can be decomposed into a finite number of parts, each of which is the graph of a smooth function. We define the arc length of the graph as the sum of the arc lengths of the individual graphs.

To avoid any misunderstanding in the following discussion, we shall denote the variable of integration by t. In this case, the arc length formula in Definition (5.14) is written

$$L_a^b = \int_a^b \sqrt{1 + [f'(t)]^2} \, dt.$$

If f is smooth on $[a, b]$, then f is smooth on $[a, x]$ for every number x in $[a, b]$, and the length of the graph from the point $A(a, f(a))$ to the point $Q(x, f(x))$ is

$$L_a^x = \int_a^x \sqrt{1 + [f'(t)]^2} \, dt.$$

Figure 5.46

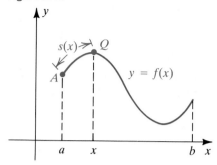

If we change the notation and use the symbol $s(x)$ in place of L_a^x, then s may be regarded as a function with domain $[a, b]$, since to each x in $[a, b]$ there corresponds a unique number $s(x)$. As shown in Figure 5.46, $s(x)$ is the length of arc of the graph of f from $A(a, f(a))$ to $Q(x, f(x))$. We shall call s the *arc length function* for the graph of f, as in the next definition.

Definition 5.16

Let f be smooth on $[a, b]$. The **arc length function** s for the graph of f on $[a, b]$ is defined by

$$s(x) = \int_a^x \sqrt{1 + [f'(t)]^2}\, dt$$

for $a \le x \le b$.

If s is the arc length function, the differential ds is called the **differential of arc length**. The next theorem specifies formulas for finding ds.

Theorem 5.17

Let f be smooth on $[a, b]$, and let s be the arc length function for the graph of $y = f(x)$ on $[a, b]$. If Δx is an increment in the variable x, then

(i) $\dfrac{ds}{dx} = \sqrt{1 + [f'(x)]^2}$

(ii) $ds = \sqrt{1 + [f'(x)]^2}\, \Delta x$

PROOF   By Definition (5.16) and Theorem (4.35),

$$\frac{ds}{dx} = \frac{d}{dx}\left[\int_a^x \sqrt{1 + [f'(t)]^2}\, dt\right] = \sqrt{1 + [f'(x)]^2}.$$

Applying Definition (2.34) yields Theorem (5.17)(ii). ∎

EXAMPLE ■ 3 Approximate the arc length of the graph of the equation $y = x^3 + 2x$ from $A(1, 3)$ to $B(1.2, 4.128)$ by

(a) differentials at $x_0 = 1$

(b) the line segment from A to B

(c) the trapezoidal rule applied to the definite integral from Definition (5.15) with $n = 10$ and $n = 20$

SOLUTION

(a) If we let $g(x) = x^3 + 2x$, then $g'(x) = 3x^2 + 2$, so by Theorem (5.17)(ii),

$$ds = \sqrt{1 + (3x^2 + 2)^2}\,\Delta x.$$

We obtain an approximation by letting $x = 1$ and $\Delta x = 0.2$:

$$\Delta s \approx ds = \sqrt{1 + 5^2}(0.2) = \sqrt{26}(0.2) \approx 1.01980$$

(b) By the distance formula (4) on page 10, the length of the line segment from $A(1, 3)$ to $B(1.2, 4.128)$ is

$$d(A, B) = \sqrt{(1.2 - 1)^2 + (4.128 - 3)^2} = \sqrt{(0.2)^2 + (1.128)^2} \approx 1.14559.$$

(c) For the trapezoidal rule, with $n = 10$, we have

$$\Delta x = \frac{b - a}{n} = \frac{1.2 - 1}{10} = 0.02 \quad \text{so that} \quad x_k = 1 + 0.02k.$$

Figure 5.47

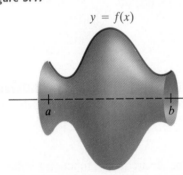

$y = f(x)$

From (4.37), we have

$$T_{10} = \frac{\Delta x}{2}[f(x_0) + 2f(x_1) + 2f(x_2) + \cdots + 2f(x_9) + f(x_{10})]$$
$$= (0.01)[f(1) + 2f(1.02) + 2f(1.04) + \cdots + 2f(1.18) + f(1.2)]$$

with $f(x) = \sqrt{1 + (3x^2 + 2)^2}$. Calculating each term in this sum and then adding them together is a task best left to a computing device with a built-in program for the trapezoidal rule. Using such a program, which allows the user to enter the values for a, b, and n and an expression for the function f, we obtain $T_{10} \approx 1.456709$ and $T_{20} \approx 1.456811$.

Let f be a function that is nonnegative throughout a closed interval $[a, b]$. If the graph of f is revolved about the x-axis, a **surface of revolution** is generated (see Figure 5.47). For example, if $f(x) = \sqrt{r^2 - x^2}$ for a positive constant r, then the graph of f on $[-r, r]$ is the upper half of the circle $x^2 + y^2 = r^2$, and a revolution about the x-axis produces a sphere of radius r having surface area $4\pi r^2$.

If the graph of f is the line segment shown in Figure 5.48, then the surface generated is a frustum of a cone having base radii r_1 and r_2 and slant height s. It can be shown that the surface area is

Figure 5.48

$$\pi(r_1 + r_2)s = 2\pi\left(\frac{r_1 + r_2}{2}\right)s.$$

You may remember this formula as follows.

Surface Area S of a Frustum of a Cone 5.18

$$S = 2\pi \text{ (average radius)(slant height)}$$

We shall use this fact in the following discussion.

Let f be a smooth function that is nonnegative on $[a, b]$, and consider the surface generated by revolving the graph of f about the x-axis (see Figure 5.47). We wish to find a formula for the area S of this surface. Let P be a partition of $[a, b]$ determined by $a = x_0, x_1, \ldots, x_n = b$, and for each k, let Q_k denote the point $(x_k, f(x_k))$ on the graph of f (see Figure 5.49). If the norm $\|P\|$ is close to zero, then the broken line l_P obtained by connecting Q_{k-1} to Q_k for each k is an approximation to the graph of f, and hence the area of the surface generated by revolving l_P about the x-axis should approximate S. The line segment $Q_{k-1}Q_k$ generates a frustum of a cone having base radii $f(x_{k-1})$ and $f(x_k)$ and slant height $d(Q_{k-1}, Q_k)$. By (5.18), its surface area is

$$2\pi \frac{f(x_{k-1}) + f(x_k)}{2} d(Q_{k-1}, Q_k).$$

Figure 5.49

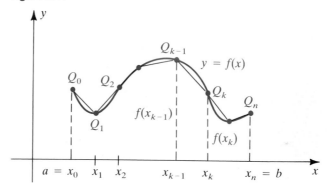

Summing terms of this form from $k = 1$ to $k = n$ gives us the area S_P of the surface generated by the broken line l_P. If we use the expression for $d(Q_{k-1}, Q_k)$ on page 474, then

$$S_P = \sum_{k=1}^{n} 2\pi \frac{f(x_{k-1}) + f(x_k)}{2} \sqrt{1 + [f'(w_k)]^2} \, \Delta x_k,$$

where $x_{k-1} < w_k < x_k$. We define the area S of the surface of revolution as

$$S = \lim_{\|P\| \to 0} S_P.$$

From the form of S_P, it is reasonable to expect that the limit is given by

$$\int_a^b 2\pi \frac{f(x) + f(x)}{2} \sqrt{1 + [f'(x)]^2} \, dx = \int_a^b 2\pi f(x) \sqrt{1 + [f'(x)]^2} \, dx.$$

The proof of this fact requires results from advanced calculus and is omitted. The following definition summarizes our discussion.

Definition 5.19

If f is smooth and $f(x) \geq 0$ on $[a, b]$, then the **area** S of the surface generated by revolving the graph of f about the x-axis is

$$S = \int_a^b 2\pi f(x)\sqrt{1 + [f'(x)]^2}\, dx.$$

If f is negative for some x in $[a, b]$, then the following extension of Definition (5.19) can be used to find the surface area S:

$$S = \int_a^b 2\pi\, |f(x)|\, \sqrt{1 + [f'(x)]^2}\, dx$$

We can use (5.18) to remember the formula for S in Definition (5.19). As in Figure 5.50, let (x, y) denote an arbitrary point on the graph of f and, as in Theorem (5.17)(ii), consider the differential of arc length

$$ds = \sqrt{1 + [f'(x)]^2}\,\Delta x.$$

Next, regard ds as the slant height of the frustum of a cone that has average radius $y = f(x)$ (see Figure 5.50). Applying (5.18), the surface area of this frustum is given by

$$2\pi f(x)\, ds = 2\pi y\, ds.$$

As with our work in Sections 5.1–5.3, applying $\int_b^a$ may be regarded as taking a limit of sums of these areas of frustums. Thus,

$$S = \int_a^b 2\pi f(x)\, ds = \int_a^b 2\pi y\, ds.$$

Figure 5.50

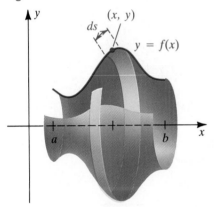

EXAMPLE 4 The graph of $y = \sqrt{x}$ from $(1, 1)$ to $(4, 2)$ is revolved about the x-axis. Find the area of the resulting surface.

SOLUTION The surface is illustrated in Figure 5.51. Using Definition (5.19) or the previous discussion, we have

$$S = \int_1^4 2\pi y\, ds$$

$$= \int_1^4 2\pi x^{1/2}\sqrt{1 + \left(\frac{1}{2x^{1/2}}\right)^2}\, dx$$

$$= \int_1^4 2\pi x^{1/2}\sqrt{\frac{4x + 1}{4x}}\, dx = \pi \int_1^4 \sqrt{4x + 1}\, dx$$

$$= \frac{\pi}{6}\left[(4x + 1)^{3/2}\right]_1^4 = \frac{\pi}{6}(17^{3/2} - 5^{3/2}) \approx 30.85 \text{ square units.}$$

Figure 5.51

Figure 5.52

If we interchange the roles of x and y in the preceding discussion, then a formula analogous to (5.19) can be obtained for integration with respect to y. Thus, if $x = g(y)$ and g is smooth and nonnegative on $[c, d]$, then the area S of the surface generated by revolving the graph of g about the y-axis (see Figure 5.52) is

$$S = \int_c^d 2\pi g(y)\sqrt{1 + [g'(y)]^2}\, dy.$$

EXERCISES 5.5

Exer. 1–4: Set up an integral that can be used to find the arc length of the graph from A to B by integrating with respect to **(a)** x and **(b)** y.

1

3

2

4

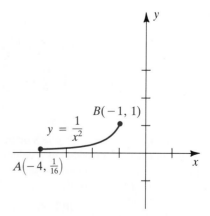

Exer. 5–12: Find the arc length of the graph of the equation from A to B.

5 $y = \frac{2}{3}x^{2/3}$; $A(1, \frac{2}{3})$, $B(8, \frac{8}{3})$

6 $(y + 1)^2 = (x - 4)^3$; $A(5, 0)$, $B(8, 7)$

7 $y = 5 - \sqrt{x^3};$ $A(1, 4),$ $B(4, -3)$

8 $y = 6\sqrt[3]{x^2} + 1;$ $A(-1, 7),$ $B(-8, 25)$

9 $y = \dfrac{x^3}{12} + \dfrac{1}{x};$ $A(1, \tfrac{13}{12}),$ $B(2, \tfrac{7}{6})$

10 $y + \dfrac{1}{4x} + \dfrac{x^3}{3} = 0;$ $A(2, \tfrac{67}{24}),$ $B(3, \tfrac{109}{12})$

11 $30xy^3 - y^8 = 15;$ $A(\tfrac{8}{15}, 1),$ $B(\tfrac{271}{240}, 2)$

12 $x = \dfrac{y^4}{16} + \dfrac{1}{2y^2};$ $A(\tfrac{9}{8}, -2),$ $B(\tfrac{9}{16}, -1)$

Exer. 13–14: Set up an integral for finding the arc length of the graph of the equation from A to B.

13 $2y^3 - 7y + 2x = 8;$ $A(3, 2),$ $B(4, 0)$

14 $11x - 4x^3 - 7y = -7;$ $A(1, 2),$ $B(0, 1)$

15 Find the arc length of the graph of $x^{2/3} + y^{2/3} = 1$.
(*Hint:* Use symmetry with respect to the line $y = x$.)

16 Find the arc length of the graph of $y = \dfrac{3x^8 + 5}{30x^3}$
from $(1, \tfrac{4}{5})$ to $(2, \tfrac{773}{240})$.

c **Exer. 17–22:** (a) Set up an integral for finding the arc length of the graph of f between A and B, and approximate the integral by using Simpson's rule with $n = 4$. (b) Approximate the arc length from A to B by differentials. (c) Approximate the arc length from A to B by computing the length of the line segment from A to B.

17 $f(x) = \sqrt[3]{x^2};$ $A(1, 1),$ $B\left(1.1, \sqrt[3]{1.1^2}\right)$

18 $f(x) = \sqrt{x^3};$ $A(1, 1),$ $B\left(1.1, \sqrt{1.1^3}\right)$

19 $f(x) = x^2;$ $A(2, 4),$ $B(2.1, 4.41)$

20 $f(x) = -x^3;$ $A(1, -1),$ $B(1.1, -1.331)$

21 $f(x) = \cos x;$ $A\left(\dfrac{\pi}{6}, \dfrac{\sqrt{3}}{2}\right),$ $B\left(\dfrac{31\pi}{180}, \cos\dfrac{31\pi}{180}\right)$

22 $f(x) = \sin x;$ $A(0, 0),$ $B\left(\dfrac{\pi}{90}, \sin\dfrac{\pi}{90}\right)$

c **Exer. 23–26:** Use Simpson's rule, with $n = 8$ or larger, or use the numerical integration provided by a calculator or computer application to approximate the arc length of the graph of the equation from A to B.

23 $y = x^2 + x + 3;$ $A(-2, 5),$ $B(2, 9)$

24 $y = x^3;$ $A(0, 0),$ $B(2, 8)$

25 $y = \csc x;$ $A\left(\dfrac{\pi}{4}, \sqrt{2}\right),$ $B\left(\dfrac{3\pi}{4}, \sqrt{2}\right)$

26 $y = \tan x;$ $A(0, 0),$ $B\left(\dfrac{\pi}{4}, 1\right)$

c **Exer. 27–28:** Consider the arc length L_a^b of the graph of f from the point $A(a, f(a))$ to the point $B(b, f(b))$. Let $a = x_0, x_1, \ldots, x_n = b$ be a regular partition of $[a, b]$ with $\Delta x = (b - a)/n$. (a) Approximate L_a^b by $\sum_{k=1}^{n} d(Q_{k-1}, Q_k)$, where Q_k is the point $(x_k, f(x_k))$ on the graph, for $n = 4$ and $n = 8$. In general, for any n, how does this approximation compare to L_a^b? (b) Set up a definite integral for L_a^b and approximate this integral using the trapezoidal rule for $n = 4$ and $n = 8$.

27 $f(x) = \sin x;$ $A(0, 0),$ $B(\pi, 0)$

28 $f(x) = \sin(\sin x);$ $A(0, 0),$ $B(\pi, 0)$

Exer. 29–32: The graph of the equation from A to B is revolved about the x-axis. Find the area of the resulting surface.

29 $4x = y^2;$ $A(0, 0),$ $B(1, 2)$

30 $y = x^3;$ $A(1, 1),$ $B(2, 8)$

31 $8y = 2x^4 + x^{-2};$ $A(1, \tfrac{3}{8}),$ $B(2, \tfrac{129}{32})$

32 $y = 2\sqrt{x + 1};$ $A(0, 2),$ $B(3, 4)$

Exer. 33–34: The graph of the equation from A to B is revolved about the y-axis. Find the area of the resulting surface.

33 $y = 2\sqrt[3]{x};$ $A(1, 2),$ $B(8, 4)$

34 $x = 4\sqrt{y};$ $A(4, 1),$ $B(12, 9)$

Exer. 35–36: If the smaller arc of the circle $x^2 + y^2 = 25$ between the points $(-3, 4)$ and $(3, 4)$ is revolved about the given axis, find the area of the resulting surface.

35 The y-axis   36 The x-axis

Exer. 37–39: Use a definite integral to derive a formula for the surface area of the indicated solid.

37 A right circular cone of altitude h and base radius r

38 A spherical segment of altitude h in a sphere of radius r

39 A sphere of radius r

40 Show that the area of the surface of a sphere of radius a between two parallel planes depends only on the distance between the planes. (*Hint:* Use Exercise 38.)

41 If the graph in Figure 5.50 is revolved about the y-axis, show that the area of the resulting surface is given by

$$\int_a^b 2\pi x \sqrt{1 + [f'(x)]^2}\, dx.$$

42 Use Exercise 41 to find the area of the surface generated by revolving the graph of $y = 3\sqrt[3]{x}$ from $A(1, 3)$ to $B(8, 6)$ about the y-axis.

c Exer. 43–44: Let S be the area of the surface generated by revolving the graph of f from $A(a, f(a))$ to $B(b, f(b))$ about the x-axis. Let $a = x_0, x_1, \ldots, x_n = b$ be a regular partition of $[a, b]$ with $\Delta x = (b - a)/n$. (a) Approximate S by

$$\sum_{k=1}^{n} 2\pi \left[\frac{f(x_{k-1}) + f(x_k)}{2} \right] d(Q_{k-1}, Q_k),$$

where Q_k is the point $(x_k, f(x_k))$ on the graph for $n = 4$ and $n = 8$. In general, for any n, how does this approximation compare to S? (b) Set up a definite integral for S and approximate this integral using the trapezoidal rule for $n = 4$ and $n = 8$.

43 $f(x) = \sin x;$ $\quad A(0, 0), \quad B(\pi, 0)$

44 $f(x) = 1 - x^3;$ $\quad A(0, 1), \quad B(1, 0)$

c **45** An American football has the approximate shape of the solid generated by revolving the arc of a circle, $x^2 + (y + k)^2 = r^2$, where $y \geq 0$ and $0 \leq k < r$. For a full-sized football, the arc from point to point measures about 14 in. along a seam. Around the widest part, the circumference measures about 22 in. Approximate the surface area of the football.

c **46** For a junior-sized football, the arc from point to point measures about 13 in. along a seam. (Assume the same model for a football used in Exercise 45.) Around the widest part, the circumference measures about 18 in. Approximate the surface area for a junior-sized football.

47 One section of a suspension bridge has its weight uniformly distributed between twin towers that are 400 ft apart and that rise 90 ft above the horizontal roadway. A cable strung between the tops of the towers has the shape of a parabola, with center point 10 ft above the roadway. Suppose coordinate axes are introduced, as shown in the figure.

Exercise 47

(a) Find an equation for the parabola.

(b) Set up an integral whose value is the length of the cable.

(c) If nine equispaced vertical cables are used to support the parabolic cable, find the total length of these supports.

48 Let R be the region bounded by the parabola $x = ay^2$ and the line l through the focus that is perpendicular to the axis of the parabola. Find the area of the curved surface obtained by revolving R about the x-axis.

49 A radio telescope has the shape of a paraboloid of revolution (see Exercise 47 of Section 5.2) with focal length p and diameter of base $2a$.

(a) Show that the surface area S available for collecting radio waves is

$$S = \frac{8\pi p^2}{3} \left[\left(1 + \frac{a^2}{4p^2} \right)^{3/2} - 1 \right].$$

c (b) One of the largest radio telescopes, located in Jodrell Bank, Cheshire, England, has diameter 250 ft and focal length 50 ft. Approximate S to the nearest square foot.

50 (a) Show that the circumference C of the ellipse with the equation

$$(x^2/a^2) + (y^2/b^2) = 1$$

is given by

$$C = 4a \int_0^{\pi/2} \sqrt{1 - e^2 \sin^2 \theta} \, d\theta,$$

where e is the eccentricity. (This is called an *elliptic integral*, and it cannot be evaluated exactly using methods we have presented to this point.)

c (b) The planet Mercury travels in an elliptical orbit with $e = 0.206$ and $a = 0.387$ AU. Use part (a) and Simpson's rule, with $n = 10$, to approximate the length of the orbit.

(c) Find the maximum and minimum distances between Mercury and the sun.

5.6 WORK

We may consider the concept of *force* as a push or pull on an object. For example, we need a force to push or pull furniture across a floor, to lift an object off the ground, to stretch or compress a spring, or to move a charged particle through an electromagnetic field. In this section, we discuss the *work* done by a continuously varying force.

If an object weighs 10 lb, then by definition the force required to lift it (or hold it off the ground) is 10 lb. A force of this type is a **constant force,** since its magnitude does not change while it is being applied to the object.

The concept of *work* is used when a force acts through a distance. The following definition covers the simplest case, in which the object moves along a line in the same direction as the applied force.

Definition 5.20

> If a constant force F acts on an object, moving it a distance d in the direction of the force, the **work** W done is
>
> $$W = Fd.$$

The following table lists units of force and work in the British system and the International System (abbreviated SI, for the French *Système International*). In SI units, 1 Newton is the force required to impart an acceleration of 1 m/sec^2 to a mass of 1 kilogram.

System	Unit of force	Unit of distance	Unit of work
British	pound (lb)	foot (ft)	foot-pound (ft-lb)
		inch (in.)	inch-pound (in.-lb)
International (SI)	Newton (N)	meter (m)	Newton-meter (N-m)

A Newton-meter is also called a *joule* (J). It can be shown that

$$1 \text{ N} \approx 0.225 \text{ lb} \quad \text{and} \quad 1 \text{ N-m} \approx 0.74 \text{ ft-lb.}$$

For simplicity, in examples and most exercises we will use the British system, in which the magnitude of the force is the same as the weight, in pounds, of the object. In using SI units, it is often necessary to consider a gravitational constant a (9.81 m/sec^2) and use Newton's second law of motion, $F = ma$, to change a mass m (in kilograms) to a force F (in Newtons).

EXAMPLE ▪ 1   Find the work done in pushing an automobile a distance of 20 ft along a level road while exerting a constant force of 90 lb.

SOLUTION   The problem is illustrated in Figure 5.53. Since the constant force is $F = 90$ lb and the distance that the automobile moves is $d = 20$ ft, it follows from Definition (5.20) that the work done is

$$W = (90)(20) = 1800 \text{ ft-lb.}$$

Figure 5.53

Anyone who has pushed an automobile (or some other object) is aware that the force applied is seldom constant. Thus, if an automobile is stalled, a larger force may be required to get it moving than to keep it in motion. The force may also vary because of friction, since part of the road may be smooth and another part rough. A force that is not constant is a **variable force.** We next develop a method for determining the work done by a variable force in moving an object rectilinearly in the same direction as the force.

Suppose that a force moves an object along the x-axis from $x = a$ to $x = b$ and that the force at x is given by $f(x)$, where f is continuous on $[a, b]$. (The phrase *force at x* means the force acting at the point with coordinate x.) As shown in Figure 5.54, we begin by considering a partition P of $[a, b]$ determined by

$$a = x_0, x_1, x_2, \ldots, x_n = b \quad \text{with} \quad \Delta x_k = x_k - x_{k-1}.$$

Figure 5.54

If ΔW_k is the **increment of work**—that is, the amount of work done from x_{k-1} to x_k—then the work W done from a to b is the sum

$$W = \Delta W_1 + \Delta W_2 + \cdots + \Delta W_n = \sum_{k=1}^{n} \Delta W_k.$$

To approximate ΔW_k, we choose any number z_k in $[x_{k-1}, x_k]$ and consider the force $f(z_k)$ at z_k. If the norm $\|P\|$ is small, then intuitively we know that the function values change very little on $[x_{k-1}, x_k]$—that is, f is *almost constant* on this interval. Applying Definition (5.20) gives us

$$\Delta W_k \approx f(z_k)\Delta x_k$$

and hence

$$W = \sum_{k=1}^{n} \Delta W_k \approx \sum_{k=1}^{n} f(z_k)\Delta x_k.$$

Since this approximation should improve as $\|P\| \to 0$, we define W as a limit of such sums. This limit leads to a definite integral.

Definition 5.21

If $f(x)$ is the force at x and if f is continuous on $[a, b]$, then the **work** W done in moving an object along the x-axis from $x = a$ to $x = b$ is

$$W = \lim_{\|P\|\to 0} \sum_k f(z_k)\Delta x_k$$

$$= \int_a^b f(x)\, dx.$$

An analogous definition can be stated for an interval on the y-axis by replacing x with y throughout our discussion.

Definition (5.21) can be used to find the work done in stretching or compressing a spring. To solve problems of this type, it is necessary to use the following law from physics.

Hooke's Law: The force $f(x)$ required to stretch a spring x units beyond its natural length is given by $f(x) = kx$, where k is a constant called the **spring constant.**

EXAMPLE 2   A force of 9 lb is required to stretch a spring from its natural length of 6 in. to a length of 8 in. Find the work done in stretching the spring

(a) from its natural length to a length of 10 in.

(b) from a length of 7 in. to a length of 9 in.

SOLUTION

(a) Let us introduce an x-axis as shown in Figure 5.55, with one end of the spring attached to a point to the left of the origin and the end to be pulled located at the origin. According to Hooke's law, the force $f(x)$ required to stretch the spring x units beyond its natural length is $f(x) = kx$ for some constant k. Since a 9-lb force is required to stretch the spring 2 in. beyond its natural length, we have $f(2) = 9$. We let $x = 2$ in $f(x) = kx$:

$$9 = k \cdot 2, \quad \text{or} \quad k = \tfrac{9}{2}$$

Figure 5.55

Consequently, for this spring, Hooke's law has the form

$$f(x) = \tfrac{9}{2}x.$$

Applying Definition (5.21) with $a = 0$ and $b = 4$, we can determine the work done in stretching the spring 4 in.:

$$W = \int_0^4 \frac{9}{2}x\, dx = \frac{9}{2}\left[\frac{x^2}{2}\right]_0^4 = 36 \text{ in.-lb}$$

(b) We again use the force $f(x) = \tfrac{9}{2}x$ obtained in part (a). By Definition (5.21), the work done in stretching the spring from $x = 1$ to $x = 3$ is

$$W = \int_1^3 \frac{9}{2}x\, dx = \frac{9}{2}\left[\frac{x^2}{2}\right]_1^3 = 18 \text{ in.-lb.}$$

Figure 5.56

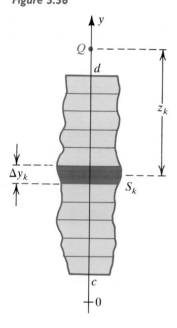

In some applications, we wish to determine the work done in pumping out a tank containing a fluid or in lifting an object, such as a chain or a cable, that extends vertically between two points. A general situation is illustrated in Figure 5.56, which shows a solid that extends along the y-axis from $y = c$ to $y = d$. We wish to vertically lift all particles contained in the solid to the level of point Q. Let us consider a partition P of $[c, d]$ and imagine slicing the solid by means of planes perpendicular to the y-axis at each number y_k in the partition. As shown in the figure, $\Delta y_k = y_k - y_{k-1}$, and S_k represents the kth slice. We next introduce the following notation:

$$z_k = \text{the (approximate) distance } S_k \text{ is lifted}$$
$$\Delta F_k = \text{the (approximate) force required to lift } S_k$$

If ΔW_k is the work done in lifting S_k, then, by Definition (5.20),

$$\Delta W_k \approx \Delta F_k \cdot z_k = z_k \cdot \Delta F_k.$$

We define the work W done in lifting the entire solid as a limit of sums.

Definition 5.22

$$W = \lim_{\|P\| \to 0} \sum_k z_k \cdot \Delta F_k$$

The limit leads to a definite integral. Note the difference between this type of problem and that in our earlier discussion. To obtain (5.21), we considered *distance increments* Δx_k and the force $f(z_k)$ that acts through Δx_k. In the present situation, we consider *force increments* ΔF_k and the distance z_k through which ΔF_k acts. The next two examples illustrate this technique. As in preceding sections, we shall use dy to represent a typical increment Δy_k and y to denote a number in $[c, d]$.

Figure 5.57

30 ft

EXAMPLE ▪ 3   A uniform cable 30 ft long and weighing 60 lb hangs vertically from a pulley system at the top of a building, as shown in Figure 5.57. A steel beam weighing 500 lb is attached to the end of the cable. Find the work required to pull it to the top.

SOLUTION   Let W_B denote the work required to pull the beam to the top, and let W_C denote the work required for the cable. Since the beam weighs 500 lb and must move through a distance of 30 ft, we have, by Definition (5.20),

$$W_B = 500 \cdot 30 = 15{,}000 \text{ ft-lb.}$$

The work required to pull the cable to the top may be found by the method used to obtain (5.22). Consider a y-axis with the lower end of the cable at the origin and the upper end at $y = 30$, as in Figure 5.58. Let dy denote an increment of length of the cable. Since each foot of cable weighs $60/30 = 2$ lb, the weight of the increment (and hence the force required to lift it) is $2\,dy$. If y denotes the distance from O to a point in the increment, then we have the following:

increment of force:	$2\,dy$
distance lifted:	$30 - y$
increment of work:	$(30 - y)2\,dy$

Applying $\int_0^{30}$ takes a limit of sums of the increments of work. Hence,

$$W_C = \int_0^{30} (30 - y)2\,dy$$

$$= 2\left[30y - \tfrac{1}{2}y^2\right]_0^{30} = 900 \text{ ft-lb.}$$

The total work required is

$$W = W_B + W_C = 15{,}000 + 900 = 15{,}900 \text{ ft-lb.}$$

Figure 5.58

EXAMPLE ▪ 4   A right circular conical tank of altitude 20 ft and radius of base 5 ft has its vertex at ground level and axis vertical. If the tank is full of water weighing 62.5 lb/ft³, find the work done in pumping all the water over the top of the tank.

SOLUTION   We begin by introducing a coordinate system, as shown in Figure 5.59. The cone intersects the xy-plane along lines of slope -4 and 4 through the origin. An equation for the line with slope 4 is

$$y = 4x, \quad \text{or} \quad x = \tfrac{1}{4}y.$$

Let us imagine subdividing the water into slices, using planes perpendicular to the y-axis, from $y = 0$ to $y = 20$. If dy represents the width of a typical slice, then its volume may be approximated by the circular disk

Figure 5.59

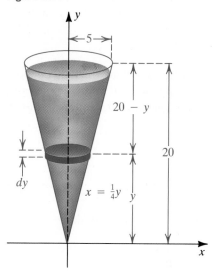

shown in Figure 5.59. As we did in our work with volumes of revolution in Section 5.2, we obtain

$$\text{volume of disk} = \pi x^2 \, dy = \pi(\tfrac{1}{4}y)^2 \, dy.$$

Since water weighs 62.5 lb/ft^3, the weight of the disk, and hence the force required to lift it, is $62.5\pi(\tfrac{1}{4}y)^2 \, dy$. Thus, we have

increment of force: $\quad 62.5\pi(\tfrac{1}{16}y^2)\, dy$

distance lifted: $\quad 20 - y$

increment of work: $\quad (20 - y)62.5\pi(\tfrac{1}{16}y^2)\, dy$

Applying $\int_0^{20}$ takes a limit of sums of the increments of work. Hence,

$$W = \int_0^{20} (20 - y)62.5\pi(\tfrac{1}{16}y^2)\, dy \;=\; \frac{62.5}{16}\pi \int_0^{20} (20y^2 - y^3)\, dy$$

$$= \frac{62.5}{16}\pi \left[20\left(\frac{y^3}{3}\right) - \frac{y^4}{4} \right]_0^{20} = \frac{62.5}{16}\pi \left(\frac{40{,}000}{3}\right) \approx 163{,}625 \text{ ft-lb.}$$

The next example is another illustration of how work may be calculated by means of a limit of sums—that is, by a definite integral.

EXAMPLE ■5 A confined gas has pressure p (lb/in^2) and volume v (in^3). If the gas expands from $v = a$ to $v = b$, show that the work done (in.-lb) is given by

$$W = \int_a^b p \, dv.$$

SOLUTION Since the work done is independent of the shape of the container, we may assume that the gas is enclosed in a right circular cylinder of radius r and that the expansion takes place against a piston head, as illustrated in Figure 5.60. As in the figure, let dv denote the

Figure 5.60

$dv = $ change in volume

$h = $ change in position of piston head

change in volume that corresponds to a change of h inches in the position of the piston head. Thus,

$$dv = \pi r^2 h, \quad \text{or} \quad h = \frac{1}{\pi r^2} \, dv.$$

If p denotes the pressure at some point in the volume increment shown in Figure 5.60, then the force against the piston head is the product of p and the area πr^2 of the piston head. Thus, we have the following for the indicated volume increment:

$$\begin{aligned}
\text{force against piston head:} &\quad p(\pi r^2) \\
\text{distance piston head moves:} &\quad h \\
\text{increment of work:} &\quad (p\pi r^2)h = (p\pi r^2)\frac{1}{\pi r^2} \, dv = p \, dv
\end{aligned}$$

Applying $\int_a^b$ to the increments of work gives us the work done as the gas expands from $v = a$ to $v = b$:

$$W = \int_a^b p \, dv$$

EXERCISES 5.6

1 A 400-lb gorilla climbs a vertical tree 15 ft high. Find the work done if the gorilla reaches the top in

(a) 10 sec   **(b)** 5 sec

2 Find the work done in lifting an 80-lb sandbag a height of 4 ft.

3 A spring of natural length 10 in. stretches 1.5 in. under a weight of 8 lb. Find the work done in stretching the spring

(a) from its natural length to a length of 14 in.

(b) from a length of 11 in. to a length of 13 in.

4 A force of 25 lb is required to compress a spring of natural length 0.80 ft to a length of 0.75 ft. Find the work done in compressing the spring from its natural length to a length of 0.70 ft.

5 If a spring is 12 in. long, compare the work W_1 done in stretching it from 12 in. to 13 in. with the work W_2 done in stretching it from 13 in. to 14 in.

6 It requires 60 in.-lb of work to stretch a certain spring from a length of 6 in. to 7 in. and another 120 in.-lb of work to stretch it from 7 in. to 8 in. Find the spring constant and the natural length of the spring.

7 A freight elevator weighing 3000 lb is supported by a 12-ft-long cable that weighs 14 lb per linear foot. Approximate the work required to lift the elevator 9 ft by winding the cable onto a winch.

8 A construction worker pulls a 50-lb motor from ground level to the top of a 60-ft-high building using a rope that weighs $\frac{1}{4}$ lb/ft. Find the work done.

9 A bucket containing water is lifted vertically at a constant rate of 1.5 ft/sec by means of a rope of negligible weight. As the bucket rises, water leaks out at a rate of 0.25 lb/sec. If the bucket weighs 4 lb when empty and if it contained 20 lb of water at the instant that the lifting began, determine the work done in raising the bucket 12 ft.

10 In Exercise 9, find the work required to raise the bucket until half the water has leaked out.

11 A fishtank has a rectangular base of width 2 ft and length 4 ft, and rectangular sides of height 3 ft. If the tank is filled with water weighing 62.5 lb/ft^3, find the work required to pump all the water over the top of the tank.

12 Generalize Example 4 to the case of a conical tank of altitude h feet and radius of base a feet that is filled with a liquid weighing ρ lb/ft^3.

13 A vertical cylindrical tank of diameter 3 ft and height 6 ft is full of water. Find the work required to pump all the water

  (a) over the top of the tank

  (b) through a pipe that rises to a height of 4 ft above the top of the tank

14 Work Exercise 13 if the tank is only half full of water.

15 The ends of an 8-ft-long water trough are equilateral triangles having sides of length 2 ft. If the trough is full of water, find the work required to pump all of it over the top.

16 A cistern has the shape of the lower half of a sphere of radius 5 ft. If the cistern is full of water, find the work required to pump all the water to a point 4 ft above the top of the cistern.

17 Refer to Example 5. The volume and the pressure of a certain gas vary in accordance with the law $pv^{1.2} = 115$, where the units of measurement are inches and pounds. Find the work done if the gas expands from 32 in^3 to 40 in^3.

18 Refer to Example 5. The pressure and the volume of a quantity of enclosed steam are related by the formula $pv^{1.14} = c$, where c is a constant. If the initial pressure and volume are p_0 and v_0, respectively, find a formula for the work done if the steam expands to twice its volume.

19 Newton's law of gravitation states that the force F of attraction between two particles having masses m_1 and m_2 is given by $F = Gm_1m_2/s^2$, where G is a gravitational constant and s is the distance between the particles. If the mass m_1 of the earth is regarded as concentrated at the center of the earth and a rocket of mass m_2 is on the surface (a distance 4000 mi from the center), find a general formula for the work done in firing the rocket vertically upward to an altitude h (see figure).

Exercise 19

20 In the study of electricity, the formula $F = kq/r^2$, where k is a constant, is used to find the force (in Newtons) with which a positive charge Q of strength q units repels a unit positive charge located r meters from Q. Find the work done in moving a unit charge from a point d centimeters from Q to a point $\frac{1}{2}d$ centimeters from Q.

C | **Exer. 21–22: Suppose the table was obtained experimentally for a force $f(x)$ acting at the point with coordinate x on a coordinate line. Use the trapezoidal rule to approximate the work done on the interval $[a, b]$, where a and b are the smallest and largest values of x, respectively.**

21

x (ft)	0	0.5	1.0	1.5	2.0	2.5
$f(x)$ (lb)	7.4	8.1	8.4	7.8	6.3	7.1

x (ft)	3.0	3.5	4.0	4.5	5.0
$f(x)$ (lb)	5.9	6.8	7.0	8.0	9.2

22

x (m)	1	2	3	4	5
$f(x)$ (N)	125	120	130	146	165

x (m)	6	7	8	9
$f(x)$ (N)	157	150	143	140

23 The force (in Newtons) with which two electrons repel each other is inversely proportional to the square of the distance (in meters) between them.

  (a) If one electron is held fixed at the point $(5, 0)$, find the work done in moving a second electron along the x-axis from the origin to the point $(3, 0)$.

  (b) If two electrons are held fixed at the points $(5, 0)$ and $(-5, 0)$, respectively, find the work done in moving a third electron from the origin to $(3, 0)$.

24 If the force function is constant, show that Definition (5.21) reduces to Definition (5.20).

5.7   MOMENTS AND CENTERS OF MASS

In this section, we consider some topics involving the mass of an object. The terms *mass* and *weight* are sometimes confused with each other. Weight is determined by the force of gravity. For example, the weight of an object on the moon is approximately one-sixth its weight on earth, because the force of gravity is weaker. However, the mass is the same. Newton used the term *mass* synonymously with *quantity of matter* and related it to force by his *second law of motion*, $F = ma$, where F denotes the force acting on an object of mass m that has acceleration a. In the British system, we often approximate a by 32 ft/sec^2 and use the **slug** as the unit of mass. In SI units, $a \approx 9.81$ m/sec^2, and the kilogram is the unit of mass. It can be shown that

$$1 \text{ slug } \approx 14.6 \text{ kg} \quad \text{and} \quad 1 \text{ kg} \approx 0.07 \text{ slug.}$$

In applications, we generally assume that the mass of an object is concentrated at a point, and we refer to the object as a **point-mass**, regardless of its size. For example, using the earth as a frame of reference, we may regard a human being, an automobile, or a building as a point-mass.

In an elementary physics experiment, we consider two point-masses m_1 and m_2 attached to the ends of a thin rod, as illustrated in Figure 5.61, and then locate the point P at which a fulcrum should be placed so that the rod balances. (This situation is similar to balancing a seesaw with a person sitting at each end.) If the distances from m_1 and m_2 to P are d_1 and d_2, respectively, then it can be shown experimentally that P is the balance point if

$$m_1 d_1 = m_2 d_2.$$

Figure 5.61

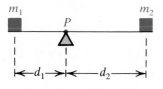

In order to generalize this concept, let us introduce an x-axis, as illustrated in Figure 5.62, with m_1 and m_2 located at points with coordinates x_1 and x_2. If the coordinate of the balance point P is $\bar{x}$, then using the formula $m_1 d_1 = m_2 d_2$ yields

$$m_1(\bar{x} - x_1) = m_2(x_2 - \bar{x})$$
$$m_1 \bar{x} + m_2 \bar{x} = m_1 x_1 + m_2 x_2$$
$$\bar{x} = \frac{m_1 x_1 + m_2 x_2}{m_1 + m_2}.$$

Figure 5.62

This gives us a formula for locating the balance point P.

If a mass m is located at a point on the axis with coordinate x, then the product mx is called the *moment* M_0 *of the mass about the origin.* Our formula for $\bar{x}$ states that to find the coordinate of the balance point, we may divide the sum of the moments about the origin by the total mass. The point with coordinate $\bar{x}$ is called the *center of mass* (or *center of gravity*) of the two point-masses. The next definition extends this discussion to many point-masses located on an axis, as shown in Figure 5.63.

Figure 5.63

Definition 5.23

Let S denote a system of point-masses $m_1, m_2, \ldots, m_n$ located at $x_1, x_2, \ldots, x_n$ on a coordinate line, and let $m = \sum_{k=1}^{n} m_k$ denote the total mass.

(i) The **moment of S about the origin** is $M_0 = \sum_{k=1}^{n} m_k x_k$.

(ii) The **center of mass** of S is given by $\bar{x} = M_0/m$.

The point with coordinate $\bar{x}$ is the balance point of the system S in the same sense as in our seesaw illustration.

EXAMPLE ▪ 1   Three point-masses of 40, 60, and 100 kg are located at -2, 3, and 7, respectively, on an x-axis. Find the center of mass.

Figure 5.64

SOLUTION   If we denote the three masses by m_1, m_2, and m_3, we have the situation illustrated in Figure 5.64, with $x_1 = -2$, $x_2 = 3$, and $x_3 = 7$. Applying Definition (5.23) gives us the coordinate $\bar{x}$ of the center of mass:

$$\bar{x} = \frac{40(-2) + 60(3) + 100(7)}{40 + 60 + 100} = \frac{800}{200} = 4$$

Figure 5.65

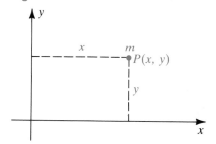

Let us next consider a point-mass m located at $P(x, y)$ in a coordinate plane (see Figure 5.65). We define the moments M_x and M_y of m about the coordinate axes as follows:

moment about the x-axis: $M_x = my$

moment about the y-axis: $M_y = mx$

In words, to find M_x we multiply m by the y-coordinate of P, and to find M_y we multiply m by the x-coordinate. To find M_x and M_y for a *system* of point-masses, we add the individual moments, as in (i) and (ii) of the next definition.

Definition 5.24

Let S denote a system of point-masses $m_1, m_2, \ldots, m_n$ located at $(x_1, y_1), (x_2, y_2), \ldots, (x_n, y_n)$ in a coordinate plane, and let $m = \sum_{k=1}^{n} m_k$ denote the total mass.

(i) The **moment of S about the x-axis** is $M_x = \sum_{k=1}^{n} m_k y_k$.

(continued)

(ii) The **moment of S about the y-axis** is $M_y = \displaystyle\sum_{k=1}^{n} m_k x_k$.

(iii) The **center of mass** of S is the point $(\bar{x}, \bar{y})$ such that

$$\bar{x} = \frac{M_y}{m}, \quad \bar{y} = \frac{M_x}{m}.$$

Figure 5.66

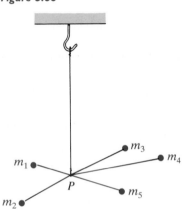

From (iii) of this definition,

$$m\bar{x} = M_y \quad \text{and} \quad m\bar{y} = M_x.$$

Since $m\bar{x}$ and $m\bar{y}$ are the moments about the y-axis and x-axis, respectively, of a single point-mass m located at $(\bar{x}, \bar{y})$, we may interpret the center of mass as the point at which the total mass can be concentrated to obtain the moments M_y and M_x of S.

We might think of the n point-masses in (5.24) as being fastened to the center of mass P by weightless rods, as spokes of a wheel are attached to the center of the wheel. The system S would balance if supported by a cord attached to P, as illustrated in Figure 5.66. The appearance would be similar to that of a mobile having all its objects in the same horizontal plane.

EXAMPLE ■2 Point-masses of 4, 8, 3, and 2 kg are located at $(-2, 3)$, $(2, -6)$, $(7, -3)$, and $(5, 1)$, respectively. Find M_x, M_y, and the center of mass of the system.

Figure 5.67

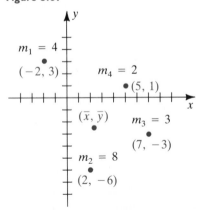

SOLUTION The masses are illustrated in Figure 5.67, in which we have also anticipated the position of $(\bar{x}, \bar{y})$. Applying Definition (5.24) gives us

$$M_x = (4)(3) + (8)(-6) + (3)(-3) + (2)(1) = -43$$
$$M_y = (4)(-2) + (8)(2) + (3)(7) + (2)(5) = 39.$$

Since $m = 4 + 8 + 3 + 2 = 17$,

$$\bar{x} = \frac{M_y}{m} = \frac{39}{17} \approx 2.3 \quad \text{and} \quad \bar{y} = \frac{M_x}{m} = -\frac{43}{17} \approx -2.5.$$

Thus, the center of mass is $(\frac{39}{17}, -\frac{43}{17})$.

Later in the text we shall consider solid objects that are **homogeneous** in the sense that the mass is uniformly distributed throughout the solid. In physics, the **density** ρ (rho) of a homogeneous solid of mass m and volume V is defined by $\rho = m/V$. Thus, *density is mass per unit volume.* The SI unit for density is kg/m³; however, g/cm³ is also used. The British unit is lb/ft³ or lb/in³.

Figure 5.68

(a)   (b)

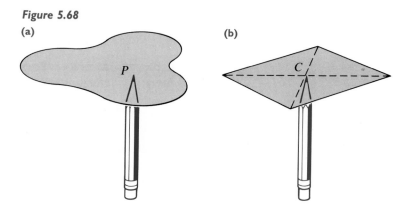

In this section, we restrict our discussion to homogeneous **laminas** (thin flat plates) that have **area density** (mass per unit area) ρ. Area density is measured in kg/m^2, lb/ft^2, and so on. If the area of one face of a lamina is A and the area density is ρ, then its mass m is given by $m = \rho A$. We wish to define the center of mass P such that if the tip of a sharp pencil were placed at P, as illustrated in Figure 5.68, the lamina would balance in a horizontal position. As in Figure 5.68(b), we shall assume that *the center of mass of a rectangular lamina is the point C at which the diagonals intersect.* We call C the *center* of the rectangle. Thus, for problems involving mass, we may assume that a rectangular lamina is a point-mass located at the center of the rectangle. This assumption is the key to our definition of the center of mass of a lamina.

Consider a lamina that has area density ρ and the shape of the R_x region in Figure 5.69. Since we have had ample experience using limits of Riemann sums for definitions in Sections 5.1–5.6, let us proceed directly to the method of representing the width of the rectangle in the figure by dx (instead of Δx_k), obtaining

$$\text{area of rectangle:} \quad [f(x) - g(x)]\, dx.$$

Figure 5.69

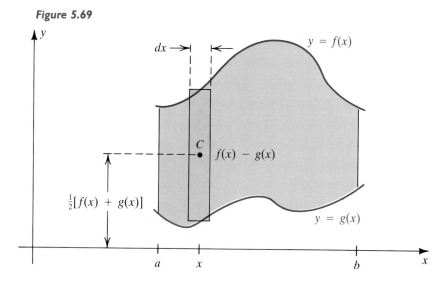

Since the area density of the lamina is ρ, we may write

mass of rectangular lamina: $\rho[f(x) - g(x)] \, dx$.

If, as in previous sections, we regard $\int_a^b$ as an operator that takes limits of sums, we arrive at the following definition for the mass m of the lamina:

$$m = \int_a^b \rho[f(x) - g(x)] \, dx$$

We next assume that the rectangular lamina in Figure 5.69 is a point-mass located at the center C of the rectangle. Since, by the midpoint formula (5), on page 11, the distance from the x-axis to C is $\frac{1}{2}[f(x) + g(x)]$, we obtain the following result for the rectangular lamina:

moment about the x-axis: $\frac{1}{2}[f(x) + g(x)] \cdot \rho[f(x) - g(x)] \, dx$

Similarly, since the distance from the y-axis to C is x,

moment about the y-axis: $x \cdot \rho[f(x) - g(x)] \, dx$.

Taking limits of sums by applying $\int_a^b$ leads to the next definition.

Definition 5.25

Let a lamina L of area density ρ have the shape of the R_x region in Figure 5.69.

(i) The **mass** of L is $m = \int_a^b \rho[f(x) - g(x)] \, dx$.

(ii) The **moments** of L **about the x-axis and y-axis** are

$$M_x = \int_a^b \frac{1}{2}[f(x) + g(x)] \cdot \rho[f(x) - g(x)] \, dx$$

and $$M_y = \int_a^b x \cdot \rho[f(x) - g(x)] \, dx.$$

(iii) The **center of mass** of L is the point $(\bar{x}, \bar{y})$ such that

$$\bar{x} = \frac{M_y}{m} \quad \text{and} \quad \bar{y} = \frac{M_x}{m}.$$

An analogous definition can be stated if L has the shape of an R_y region and the integrations are with respect to y. We could also obtain formulas for moments with respect to lines other than the x-axis or y-axis; however, it is advisable to remember the *technique* for finding moments—multiplying a mass by a distance from an axis—instead of memorizing formulas that cover all possible cases.

EXAMPLE ▪ 3   A lamina of area density ρ has the shape of the region bounded by the graphs of $y = x^2 + 1$, $x = 0$, $x = 1$, and $y = 0$. Find the center of mass.

Figure 5.70

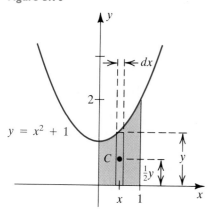

$y = x^2 + 1$

SOLUTION   The region and a typical rectangle of width dx and height y are sketched in Figure 5.70. As indicated in the figure, the distance from the x-axis to the center C of the rectangle is $\frac{1}{2}y$, and the distance from the y-axis to C is x. Hence, *for the rectangular lamina*, we have the following:

$$\text{mass:} \quad \rho y\, dx = \rho(x^2 + 1)\, dx$$

$$\text{moment about } x\text{-axis:} \quad \tfrac{1}{2}y \cdot \rho y\, dx = \tfrac{1}{2}\rho(x^2 + 1)^2\, dx$$

$$\text{moment about } y\text{-axis:} \quad x \cdot \rho y\, dx = \rho x(x^2 + 1)\, dx$$

We now take a limit of sums of these expressions by applying the operator $\int_0^1$:

$$m = \int_0^1 \rho(x^2 + 1)\, dx = \rho\left[\tfrac{1}{3}x^3 + x\right]_0^1 = \tfrac{4}{3}\rho$$

$$M_x = \int_0^1 \tfrac{1}{2}\rho(x^2 + 1)^2\, dx = \tfrac{1}{2}\rho \int_0^1 (x^4 + 2x^2 + 1)\, dx$$

$$= \tfrac{1}{2}\rho\left[\tfrac{1}{5}x^5 + \tfrac{2}{3}x^3 + x\right]_0^1 = \tfrac{14}{15}\rho$$

$$M_y = \int_0^1 \rho x(x^2 + 1)\, dx = \rho \int_0^1 (x^3 + x)\, dx$$

$$= \rho\left[\tfrac{1}{4}x^4 + \tfrac{1}{2}x^2\right]_0^1 = \tfrac{3}{4}\rho$$

To find the center of mass $(\bar{x}, \bar{y})$, we use Definition (5.25)(iii):

$$\bar{x} = \frac{M_y}{m} = \frac{\tfrac{3}{4}\rho}{\tfrac{4}{3}\rho} = \frac{9}{16} \quad \text{and} \quad \bar{y} = \frac{M_x}{m} = \frac{\tfrac{14}{15}\rho}{\tfrac{4}{3}\rho} = \frac{7}{10}$$

When we found $(\bar{x}, \bar{y})$ in Example 3, the constant ρ in the numerator and the denominator canceled. This will always be the case for a homogeneous lamina. Hence, the center of mass is independent of the area density ρ; that is, $\bar{x}$ and $\bar{y}$ depend only on the shape of the lamina. For this reason, the point $(\bar{x}, \bar{y})$ is sometimes referred to as the center of mass of a *region* in the plane, or as the **centroid** of the region. We can obtain formulas for moments of centroids by letting $\rho = 1$ and $m = A$ (the area of the region) in our previous work.

EXAMPLE ▪ 4   Find the centroid of the region bounded by the graphs of $y = 6 - x^2$ and $y = 3 - 2x$.

SOLUTION   The region is the same as that considered in Example 2 of Section 5.1 and is resketched in Figure 5.71 on the following page. To find the moments and the centroid, we take $\rho = 1$ and $m = A$. Referring

Figure 5.71

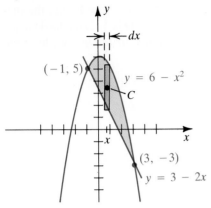

to the typical rectangle with center C shown in Figure 5.71, we obtain the following:

area of rectangle: $[(6 - x^2) - (3 - 2x)]\,dx$

distance from x-axis to C: $\frac{1}{2}[(6 - x^2) + (3 - 2x)]$

moment about x-axis: $\frac{1}{2}[(6 - x^2) + (3 - 2x)] \times$
$$[(6 - x^2) - (3 - 2x)]\,dx$$

distance from y-axis to C: x

moment about y-axis: $x[(6 - x^2) - (3 - 2x)]\,dx$

We now take a limit of sums by applying the operator $\int_{-1}^{3}$:

$$M_x = \int_{-1}^{3} \tfrac{1}{2}[(6 - x^2) + (3 - 2x)] \cdot [(6 - x^2) - (3 - 2x)]\,dx$$

$$= \tfrac{1}{2} \int_{-1}^{3} [(6 - x^2)^2 - (3 - 2x)^2]\,dx$$

$$= \tfrac{1}{2} \int_{-1}^{3} (x^4 - 16x^2 + 12x + 27)\,dx = \tfrac{416}{15}$$

$$M_y = \int_{-1}^{3} x[(6 - x^2) - (3 - 2x)]\,dx$$

$$= \int_{-1}^{3} (3x + 2x^2 - x^3)\,dx = \tfrac{32}{3}$$

Using $A = \frac{32}{3}$ and Definition (5.25)(iii), we determine the centroid:

$$\bar{x} = \frac{M_y}{m} = \frac{32/3}{32/3} = 1 \quad \text{and} \quad \bar{y} = \frac{M_x}{m} = \frac{416/15}{32/3} = \frac{13}{5}$$

We could have found the centroid by using Definition (5.25) with $f(x) = 6 - x^2$, $g(x) = 3 - 2x$, $a = -1$, and $b = 3$, but that would merely teach you how to substitute and not how to think.

If a homogeneous lamina has the shape of a region that has an axis of symmetry, then the center of mass must lie on that axis. This fact is used in the next example.

Figure 5.72

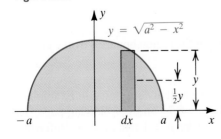

EXAMPLE ▪ 5   Find the centroid of the semicircular region bounded by the x-axis and the graph of $y = \sqrt{a^2 - x^2}$ with $a > 0$.

SOLUTION   The region is sketched in Figure 5.72. By symmetry, the centroid is on the y-axis; that is, $\bar{x} = 0$. Hence, we need find only $\bar{y}$. Referring to the rectangle in Figure 5.72 and using $\rho = 1$ gives us the following result:

moment about x-axis: $\frac{1}{2}y \cdot y \, dx = \frac{1}{2}y^2 \, dx = \frac{1}{2}(a^2 - x^2) \, dx$

We now take a limit of sums by applying the operator $\int_{-a}^{a}$:

$$M_x = \int_{-a}^{a} \tfrac{1}{2}(a^2 - x^2) \, dx = 2 \int_{0}^{a} \tfrac{1}{2}(a^2 - x^2) \, dx$$

$$= \left[a^2 x - \tfrac{1}{3}x^3 \right]_0^a = \tfrac{2}{3}a^3$$

Using $m = A = \frac{1}{2}\pi a^2$ gives us

$$\bar{y} = \frac{M_x}{m} = \frac{\frac{2}{3}a^3}{\frac{1}{2}\pi a^2} = \frac{4a}{3\pi} \approx 0.42a.$$

Thus, the centroid is the point $\left(0, \dfrac{4}{3\pi}a \right)$.

We conclude this section by stating a useful theorem about solids of revolution. To illustrate a special case of the theorem, consider an R_x region R of the type shown in Figure 5.69. Using $\rho = 1$ and $m = A$ (the area of R), we find that the moment of R about the y-axis is given by

$$M_y = \int_a^b x[f(x) - g(x)] \, dx.$$

If R is revolved about the y-axis, then using cylindrical shells, we find that the volume V of the resulting solid is given by

$$V = \int_a^b 2\pi x[f(x) - g(x)] \, dx.$$

Comparing these two equations, we see that

$$M_y = \frac{V}{2\pi}.$$

If $(\bar{x}, \bar{y})$ is the centroid of R, then, by Definition (5.25)(iii),

$$\bar{x} = \frac{M_y}{m} = \frac{(V/2\pi)}{A} = \frac{V}{2\pi A}$$

and hence, $V = 2\pi \bar{x} A.$

Since $\bar{x}$ is the distance from the y-axis to the centroid of R, the last formula states that the volume V of the solid of revolution may be found by multiplying the area A of R by the distance $2\pi \bar{x}$ that the centroid travels when R is revolved once about the y-axis. A similar statement is true if R is revolved about the x-axis. In Chapter 13, we shall prove the following more general theorem, named after the mathematician Pappus of Alexandria (ca. A.D. 300).

Theorem of Pappus **5.26**

Let R be a region in a plane that lies entirely on one side of a line l in the plane. If R is revolved once about l, the volume of the resulting solid is the product of the area of R and the distance traveled by the centroid of R.

Figure 5.73

E X A M P L E ▪ 6   The region bounded by a circle of radius a is revolved about a line l, in the plane of the circle, that is a distance b from the center of the circle, where $b > a$ (see Figure 5.73). Find the volume V of the resulting solid. (The surface of this doughnut-shaped solid is called a **torus**.)

S O L U T I O N   The region bounded by the circle has area πa^2, and the distance traveled by the centroid is $2\pi b$. Hence, by the theorem of Pappus,

$$V = (2\pi b)(\pi a^2) = 2\pi^2 a^2 b.$$

EXERCISES 5.7

Exer. 1–2: The table lists point-masses (in kilograms) and their coordinates (in meters) on an *x*-axis. Find *m*, M_0, and the center of mass.

1

Mass	100	80	70
Coordinate	−3	2	4

2

Mass	50	100	50
Coordinate	−10	2	3

Exer. 3–4: The table lists point-masses (in kilograms) and their locations (in meters) in an *xy*-plane. Find *m*, M_x, M_y, and the center of mass of the system.

3

Mass	2	7	5
Location	(4, −1)	(−2, 0)	(−8, −5)

4

Mass	10	3	4	1	8
Location	(−5, −2)	(3, 7)	(0, −3)	(−8, −3)	(0, 0)

Exer. 5–14: Sketch the region bounded by the graphs of the equations, and find *m*, M_x, M_y, and the centroid.

5 $y = x^3$, $y = 0$, $x = 1$

6 $y = \sqrt{x}$, $y = 0$, $x = 9$

7 $y = 4 - x^2$, $y = 0$

8 $2x + 3y = 6$, $y = 0$, $x = 0$

9 $y^2 = x$, $2y = x$

10 $y = x^2$, $y = x^3$

11 $y = 1 - x^2$, $x - y = 1$

12 $y = x^2$, $x + y = 2$

13 $x = y^2$, $x - y = 2$

14 $x = 9 - y^2$, $x + y = 3$

15 Find the centroid of the region in the first quadrant bounded by the circle $x^2 + y^2 = a^2$ and the coordinate axes.

16 Let R be the region in the first quadrant bounded by part of the parabola $y^2 = cx$ with $c > 0$, the x-axis, and the vertical line through the point (a, b) on the parabola, as shown in the figure on the following page. Find the centroid of R.

Exercise 16

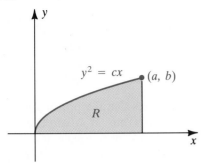

$y^2 = cx$ (a, b)

R

17 A region has the shape of a square of side $2a$ surmounted by a semicircle of radius a. Find the centroid. (*Hint:* Use Example 5 and the fact that the moment of the region is the sum of the moments of the square and the semicircle.)

18 Let the points P, Q, R, and S have coordinates $(-b, 0)$, $(-a, 0)$, $(a, 0)$, and $(b, 0)$, respectively, with $0 < a < b$. Find the centroid of the region bounded by the graphs of $y = \sqrt{b^2 - x^2}$, $y = \sqrt{a^2 - x^2}$, and the line segments PQ and RS. (*Hint:* Use Example 5.)

19 Prove that the centroid of a triangle coincides with the intersection of the medians. (*Hint:* Take the vertices at the points $(0, 0)$, (a, b), and $(0, c)$, with a, b, and c positive.)

20 A region has the shape of a square of side a surmounted by an equilateral triangle of side a. Find the centroid. (*Hint:* See Exercise 19 and the hint given for Exercise 17.)

Exer. 21–24: Use the theorem of Pappus.

21 Let R be the rectangular region with vertices $(1, 2)$, $(2, 1)$, $(5, 4)$, and $(4, 5)$. Find the volume of the solid generated by revolving R about the y-axis.

22 Let R be the triangular region with vertices $(1, 1)$, $(2, 2)$, and $(3, 1)$. Find the volume of the solid generated by revolving R about the y-axis.

23 Find the centroid of the region in the first quadrant bounded by the graph of $y = \sqrt{a^2 - x^2}$ and the coordinate axes.

24 Find the centroid of the triangular region with vertices $O(0, 0)$, $A(0, a)$, and $B(b, 0)$ for positive numbers a and b.

c 25 A lamina of area density ρ has the shape of the region bounded by the graphs of $f(x) = \sqrt{|\cos x|}$ and $g(x) = x^2$. Graph f and g on the same coordinate axes.

(a) Set up an integral that can be used to approximate the mass of the lamina.

(b) Use Simpson's rule, with $n = 2$, to approximate the integral in part (a).

c 26 Use Simpson's rule, with $n = 2$, to approximate the centroid of the region bounded by the graphs of $y = 0$, $y = (\sin x)/x$, $x = 1$, and $x = 2$.

5.8   OTHER APPLICATIONS

It should be evident from our work in this chapter that if a quantity can be approximated by a sum of many terms, then it is a candidate for representation as a definite integral. The main requirement is that as the number of terms increases, the sums approach a limit. In this section, we consider several miscellaneous applications of the definite integral. Let us begin with the force exerted by a liquid on a submerged object.

In physics, the **pressure** p at a depth h in a fluid is defined as the weight of fluid contained in a column that has a cross-sectional area of one square unit and an altitude h. Pressure may also be regarded as the force per unit area exerted by the fluid. If a fluid has density ρ, then the pressure p at depth h is given by

$$p = \rho h.$$

The following illustration is for water, with $\rho = 62.5$ lb/ft^3.

ILLUSTRATION

Density ρ (lb/ft^3)	Depth h (ft)	Pressure $p = \rho h$ (lb/ft^2)
62.5	2	125
62.5	4	250
62.5	6	375

Figure 5.74

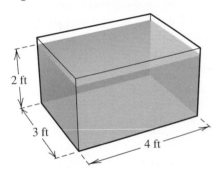

Pascal's principle in physics states that the pressure at a depth h in a fluid is the same in every direction. Thus, if a flat plate is submerged in a fluid, then the pressure on one side of the plate at a point that is h units below the surface is ρh, regardless of whether the plate is submerged vertically, horizontally, or obliquely (see Figure 5.74, where the pressure at points A, B, and C is ρh).

If a rectangular tank, such as a fish aquarium, is filled with water (see Figure 5.75), the total force exerted by the water on the base may be calculated as follows:

$$\text{force on base} = (\text{pressure at base}) \cdot (\text{area of base})$$

For the tank in Figure 5.75, we use $\rho = 62.5$ lb/ft^3 and $h = 2$ ft to obtain

$$\text{force on base} = (125 \text{ lb/ft}^2) \cdot (12 \text{ ft}^2) = 1500 \text{ lb.}$$

This corresponds to 12 columns of water, each having cross-sectional area 1 ft^2 and each weighing 125 lb.

It is more complicated to find the force exerted on one of the sides of the aquarium, since the pressure is not constant there but increases as the depth increases. Instead of investigating this particular problem, let us consider the following more general situation.

Figure 5.75

Suppose a flat plate is submerged in a fluid of density ρ such that the face of the plate is perpendicular to the surface of the fluid. Let us introduce a coordinate system as shown in Figure 5.76, where the width of the plate extends over the interval $[c, d]$ on the y-axis. Assume that for each y in $[c, d]$, the corresponding depth of the fluid is $h(y)$ and the length of the plate is $L(y)$, where h and L are continuous functions.

We shall use our standard technique of considering a typical horizontal rectangle of width dy and length $L(y)$, as illustrated in Figure 5.76. If dy is small, then the pressure at any point in the rectangle is approximately $\rho h(y)$. Thus, the force on one side of the rectangle can be approximated by

$$\text{force on rectangle} \approx (\text{pressure}) \cdot (\text{area of rectangle}),$$

or

$$\text{force on rectangle} \approx \rho h(y) \cdot L(y)\, dy.$$

Taking a limit of sums of these forces by applying the operator $\int_c^d$ leads to the following definition.

Figure 5.76

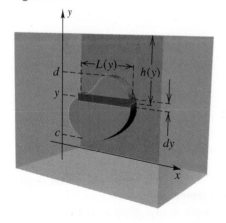

Definition **5.27**

> The **force F exerted by a fluid** of constant density ρ on one side of a submerged region of the type illustrated in Figure 5.76 is
>
> $$F = \int_c^d \rho h(y) \cdot L(y)\, dy.$$

If a more complicated region is divided into subregions of the type illustrated in Figure 5.76, we apply Definition (5.27) to each subregion and add the resulting forces.

The coordinate system may be introduced in various ways, as the next two examples illustrate. In Example 1, we choose the x-axis at the base of the liquid and the positive direction of the y-axis upward. In Example 2, we choose the x-axis along the surface of the liquid and the positive direction of the y-axis downward.

EXAMPLE ■ I One end of a reservoir presses against the wall of a small dam. The wall follows the depth contours of the reservoir and is generally in the shape of a parabola. If the wall of the dam is 60 ft deep at its center and 90 ft across at the water level, find the total force of the water in the reservoir against this wall of the dam.

Figure 5.77
$x = 45\sqrt{y/60}$

SOLUTION Figure 5.77 illustrates the end of the dam superimposed on a rectangular coordinate system. An equation for the parabola is $y = (60/45^2)x^2$, or, equivalently, $x = \pm 45\sqrt{y/60}$. Referring to Figure 5.77 gives us the following, for a horizontal rectangle of width dy:

$$\begin{aligned}
\text{length:} \quad & 2x = 2 \cdot 45\sqrt{y/60} = 90\sqrt{y/60} \\
\text{area:} \quad & 90\sqrt{y/60}\, dy \\
\text{depth:} \quad & 60 - y \\
\text{pressure:} \quad & 62.5(60 - y) \\
\text{force:} \quad & 62.5(60 - y)\, 90\sqrt{y/60}\, dy
\end{aligned}$$

Taking a limit of sums by applying the operator $\int_0^{60}$, we obtain, as in Definition (5.27),

$$\begin{aligned}
F &= \int_0^{60} 62.5(60 - y)\, 90\sqrt{\frac{y}{60}}\, dy \\
&= \frac{(62.5)(90)}{\sqrt{60}} \int_0^{60} [(60 - y)y^{1/2}]\, dy \\
&= \frac{(62.5)(90)}{\sqrt{60}} \int_0^{60} [60y^{1/2} - y^{3/2}]\, dy \\
&= \frac{(62.5)(90)}{\sqrt{60}} \left[\frac{2}{3}(60y^{3/2}) - \frac{2}{5}y^{5/2} \right]_0^{60} \\
&= \frac{(62.5)(90)}{\sqrt{60}} \left[y^{1/2}\left(40y - \frac{2}{5}y^2 \right) \right]_0^{60} \\
&= \frac{(62.5)(90)}{\sqrt{60}} \left\{ \sqrt{60}\left[(40)(60) - \frac{2}{5}(60)^2 \right] \right\} \\
&= (62.5)(90)[(40)(60) - (24)(60)] \\
&= (62.5)(90)(16)(60) = 5,400,000 \text{ lb.}
\end{aligned}$$

In the preceding example, the *length* of the reservoir is irrelevant when we consider the force on the dam. The same is true for the length of the oil tank in the next example.

EXAMPLE ▪ 2   A cylindrical oil storage tank 6 ft in diameter and 10 ft long is lying on its side. If the tank is half full of oil that weighs 58 lb/ft^3, set up an integral for the force exerted by the oil on one end of the tank.

SOLUTION   Let us introduce a coordinate system such that the end of the tank is a circle of radius 3 ft with the center at the origin. The equation of the circle is $x^2 + y^2 = 9$. If we choose the positive direction of the y-axis *downward*, then referring to the horizontal rectangle in Figure 5.78 gives us the following:

Figure 5.78

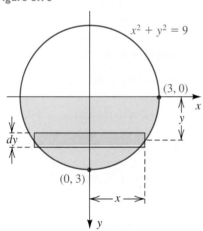

$$\text{length:}\quad 2x = 2\sqrt{9 - y^2}$$

$$\text{area:}\quad 2\sqrt{9 - y^2}\, dy$$

$$\text{depth:}\quad y$$

$$\text{pressure:}\quad 58y$$

$$\text{force:}\quad 58y \cdot 2\sqrt{9 - y^2}\, dy$$

Taking a limit of sums by applying $\int_0^3$, we obtain

$$F = \int_0^3 116y\sqrt{9 - y^2}\, dy.$$

Evaluating the integral by using the method of substitution gives us

$$F = 1044 \text{ lb.}$$

Definite integrals can be applied to dye-dilution or tracer methods used in physiological tests and elsewhere. One example involves the measurement of cardiac output—that is, the rate at which blood flows through the aorta. A simple model for tracer experiments is sketched in Figure 5.79, where a liquid (or gas) flows into a tank at A and exits at B, with a constant flow rate F (in liters per second). Suppose that at time $t = 0$, Q_0 grams of tracer (or dye) are introduced into the tank at A and that a stirring mechanism thoroughly mixes the solution at all times. The concentration $c(t)$ (in grams per liter) of tracer at time t is monitored at B. Thus, the amount of tracer passing B at time t is given by

$$\text{(flow rate)} \cdot \text{(concentration)} = F \cdot c(t) \text{ g/sec.}$$

If the amount of tracer in the tank at time t is $Q(t)$, where Q is a differentiable function, then the rate of change $Q'(t)$ of Q is given by

$$Q'(t) = -F \cdot c(t)$$

(the negative sign indicates that Q is decreasing).

Figure 5.79

Stirring mechanism

If T is a time at which all the tracer has left the tank, then $Q(T) = 0$ and, by the fundamental theorem of calculus,

$$\int_0^T Q'(t)\,dt = Q(t)\Big]_0^T = Q(T) - Q(0)$$
$$= 0 - Q_0 = -Q_0.$$

We may also write

$$\int_0^T Q'(t)\,dt = \int_0^T [-F \cdot c(t)]\,dt = -F\int_0^T c(t)\,dt.$$

Equating the two forms for the integral gives us the following formula.

Flow Concentration Formula 5.28

$$Q_0 = F\int_0^T c(t)\,dt$$

Usually an explicit form for $c(t)$ will not be known, but, instead, a table of function values will be given. By using numerical integration, we may find an approximation to the flow rate F (see Exercises 11 and 12).

Let us next consider another aspect of the flow of liquids. If a liquid flows through a cylindrical tube and if the velocity is a constant v_0, then the volume of liquid passing a fixed point per unit time is given by $v_0 A$, where A is the area of a cross section of the tube (see Figure 5.80).

A more complicated formula is required to study the flow of blood in an arteriole. In this case, the flow is in layers, as illustrated in Figure 5.81. In the layer closest to the wall of the arteriole, the blood tends to stick to the wall, and its velocity may be considered zero. The velocity increases as the layers approach the center of the arteriole.

For computational purposes, we may regard the blood flow as consisting of thin cylindrical shells that slide along, with the outer shell fixed and the velocity of the shells increasing as the radii of the shells decrease (see Figure 5.81). If the velocity in each shell is considered constant, then from the theory of liquids in motion, the velocity $v(r)$ in a shell having average radius r is

$$v(r) = \frac{P}{4vl}(R^2 - r^2),$$

where R is the radius of the arteriole (in centimeters), l is the length of the arteriole (in centimeters), P is the pressure difference between the two ends of the arteriole (in dyn/cm^2), and v is the viscosity of the blood (in dyn-sec/cm^2). Note that the formula gives zero velocity if $r = R$ and maximum velocity $PR^2/(4vl)$ as r approaches 0. If the radius of the kth shell is r_k and the thickness of the shell is Δr_k, then, by (5.10), the volume of blood in this shell is

$$2\pi r_k v(r_k)\Delta r_k = \frac{2\pi r_k P}{4vl}(R^2 - r_k^2)\Delta r_k.$$

Figure 5.80

Figure 5.81

If there are n shells, then the total flow in the arteriole per unit time may be approximated by

$$\sum_{k=1}^{n} \frac{2\pi r_k P}{4vl}(R^2 - r_k^2)\Delta r_k.$$

To estimate the total flow F (the volume of blood per unit time), we consider the limit of these sums as n increases without bound. This leads to the following definite integral:

$$F = \int_0^R \frac{2\pi r P}{4vl}(R^2 - r^2)\,dr$$

$$= \frac{2\pi P}{4vl}\int_0^R (R^2 r - r^3)\,dr$$

$$= \frac{\pi P}{2vl}\left[\frac{1}{2}R^2 r^2 - \frac{1}{4}r^4\right]_0^R$$

$$= \frac{\pi P R^4}{8vl}\ \text{cm}^3$$

This formula for F is not exact, because the thickness of the shells cannot be made arbitrarily small. The lower limit is the width of a red blood cell, or approximately 2×10^{-4} cm. We may assume, however, that the formula gives a reasonable estimate. It is interesting to observe that a small change in the radius of an arteriole produces a large change in the flow, since F is directly proportional to the fourth power of R. A small change in pressure difference has a lesser effect, since P appears to the first power.

In many types of employment, a worker must perform the same assignment repeatedly. For example, a bicycle shop employee may be asked to assemble new bicycles. As more and more bicycles are assembled, the time required for each assembly should decrease until a certain minimum assembly time is reached. Another example of this process of learning by repetition is that of a data processor who must keyboard information from written forms into a computer system. The time required to process each entry should decrease as the number of entries increases. As a final illustration, the time required for a person to trace a path through a maze should improve with practice.

Let us consider a general situation in which a certain task is to be repeated many times. Suppose experience has shown that the time required to perform the task for the kth time can be approximated by $f(k)$ for a continuous decreasing function f on a suitable interval. The total time required to perform the task n times is given by the sum

$$\sum_{k=1}^{n} f(k) = f(1) + f(2) + \cdots + f(n).$$

If we consider the graph of f, then, as illustrated in Figure 5.82, the preceding sum equals the area of the pictured inscribed rectangular polygon and, therefore, may be approximated by the definite integral $\int_0^n f(x)\,dx$. Evidently, the approximation will be close to the actual sum if f decreases slowly on $[0, n]$. If f changes rapidly per unit change in x, then an integral should not be used as an approximation.

Figure 5.82

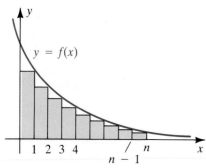

EXAMPLE ■ 3 A company that conducts polls via telephone interviews finds that the time required by an employee to complete one interview depends on the number of interviews that the employee has completed previously. Suppose it is estimated that, for a certain survey, the number of minutes required to complete the kth interview is given by $f(k) = 6(1 + k)^{-1/5}$ for $0 \le k \le 500$. Use a definite integral to approximate the time required for an employee to complete 100 interviews and 200 interviews. If an interviewer receives \$4.80 per hour, estimate how much more expensive it is to have two employees each conduct 100 interviews than to have one employee conduct 200 interviews.

SOLUTION From the preceding discussion, the time required for 100 interviews is approximately

$$\int_0^{100} 6(1 + x)^{-1/5}\, dx = 6 \cdot \tfrac{5}{4}(1 + x)^{4/5} \Big]_0^{100} \approx 293.5 \text{ min.}$$

The time required for 200 interviews is approximately

$$\int_0^{200} 6(1 + x)^{-1/5}\, dx \approx 514.4 \text{ min.}$$

Since an interviewer receives \$0.08 per minute, the cost for one employee to conduct 200 interviews is roughly (\$0.08)(514.4), or \$41.15. If two employees each conduct 100 interviews, the cost is about 2(\$0.08)(293.5), or \$46.96, which is \$5.81 more than the cost of one employee. Note, however, that the time saved in using two people is approximately 221 min.

Using a computer, we have

$$\sum_{k=1}^{100} 6(1 + k)^{-1/5} \approx 291.75$$

and

$$\sum_{k=1}^{200} 6(1 + k)^{-1/5} \approx 512.57.$$

Hence, the results obtained by integration (the area under the graph of f) are roughly 2 min more than the value of the corresponding sum (the area of the inscribed rectangular polygon).

In economics, the price p at which there is a demand for x units of a particular product may be given by a function, $p = f(x)$. Figure 5.83 illustrates the graph of such a function, which is called the *price–demand curve*. It reflects the assumption that decreases in price correspond to increases in demand. Point $P(x_c, p_c)$ represents the current price p_c (in dollars) at any point in time and the corresponding current demand of x_c units. Point $Q(x_1, p_1)$ is the higher price ($p_1 > p_c$) consumers are hypothetically willing to pay for the same product when the demand is smaller ($x_1 < x_c$).

We can use a definite integral to determine the *consumer's surplus*, which is the savings or total difference between what they are willing

Figure 5.83

to pay at higher prices and what they actually pay at the current price. We need to consider all possible prices greater than p_c dollars. From the price–demand curve, we see that the prices exceeding p_c dollars correspond to demands for fewer than x_c units. We partition the interval $[0, x_c]$ into n equal subintervals of width Δx and choose a point w_k in each subinterval. The corresponding price is $f(w_k)$, so the savings per unit is $[f(w_k) - p_c]$. If the price remained constant on the kth subinterval, then the savings to consumers over this subinterval would be

$$\text{(savings per unit)} \cdot \text{(number of units)} = [f(w_k) - p_c]\Delta x.$$

Thus, we can approximate the total savings by

$$\sum_{k=1}^{n} [f(w_k) - p_c]\Delta x.$$

This approximation improves as Δx approaches zero. But this sum is also a Riemann sum for $\int_0^{x_c}[f(x) - p_c]\,dx$ and so its limit as Δx approaches zero is the definite integral. We summarize our discussion in the next definition.

Definition   5.29

If (x_c, p_c) is a point representing current demand of x_c units of a particular good or service and current price p_c on the graph of a continuous price–demand function $p = f(x)$, then the **consumers' surplus** is given by

$$\int_0^{x_c}[f(x) - p_c]\,dx,$$

which represents the consumers' savings or total difference between what they are hypothetically willing to pay and what they actually pay.

EXAMPLE ▪ 4   The price–demand function for a particular product is given by $p = f(x) = 50 - \frac{1}{10}x$. Determine the consumers' surplus for this price–demand function at a price level of $10.

SOLUTION   For the price–demand function $p = 50 - (x/10)$, we note that when $x = 0$, $p = 50$. Thus, at a price of $50, there is no demand for the product. When $x = 200$, $p = f(200) = 50 - (200/10) = 30$. At $30 per unit, there is a demand for 200 units. To find the consumers' surplus, we first determine the demand x_c at the current price $p_c = 10$: Solving

$$10 = 50 - \tfrac{1}{10}x_c$$

for x_c yields $x_c = 400$. Thus, the consumers' surplus is given by the definite integral

$$\int_0^{400} [(50 - \tfrac{1}{10}x) - 10]\, dx = \int_0^{400} (40 - \tfrac{1}{10}x)\, dx$$

$$= \left[40x - \frac{x^2}{20} \right]_0^{400} = 8000,$$

and the consumers' surplus is \$8000.

Note that by Theorem (4.23)(ii), the consumers' surplus is equal to

$$\int_0^{x_c} [f(x) - p_c]\, dx = \int_0^{x_c} f(x)\, dx - \int_0^{x_c} p_c\, dx.$$

By Theorem (4.21),

$$\int_0^{x_c} p_c\, dx = p_c x_c.$$

The product $p_c x_c$ is the total amount paid by consumers for x_c units at the current price of p_c. Since $f(x) > 0$ for $0 < x < x_c$, the definite integral

$$\int_0^{x_c} f(x)\, dx$$

is the area under the price–demand curve between 0 and x_c, and we can interpret the consumers' surplus at a price level of p_c to be the amount by which the area under the price–demand curve exceeds the total amount paid for demanded goods at the current price level. The area of the shaded region in Figure 5.84 represents the consumers' surplus.

Any quantity that can be interpreted as an area of a region in a plane may be investigated by means of a definite integral. Conversely, definite integrals allow us to represent physical quantities as areas. In the following illustrations, a quantity is *numerically equal* to an area of a region; that is, we *disregard units of measurement*, such as centimeter, foot-pound, and so on.

Suppose $v(t)$ is the velocity, at time t, of an object that is moving on a coordinate line. If s is the position function, then $s'(t) = v(t)$ and

$$\int_a^b v(t)\, dt = \int_a^b s'(t)\, dt = s(t) \Big]_a^b = s(b) - s(a).$$

If $v(t) > 0$ throughout the time interval $[a, b]$, this tells us that the area under the graph of the function v from a to b represents the distance that the object travels, as illustrated in Figure 5.85. This observation is useful to an engineer or physicist, who may not have an explicit form for $v(t)$ but merely a graph (or table) indicating the velocity at various times. The distance traveled may then be estimated by approximating the area under the graph.

If $v(t) < 0$ at certain times in $[a, b]$, the graph of v may resemble that in Figure 5.86. The figure indicates that the object moved in the negative direction from $t = c$ to $t = d$. The distance that it traveled during that time

Figure 5.84

Figure 5.85

Figure 5.86

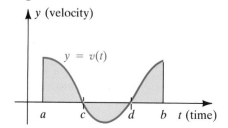

is given by $\int_c^d |v(t)|\,dt$. It follows that $\int_a^b |v(t)|\,dt$ is the *total* distance traveled in $[a, b]$, whether $v(t)$ is positive or negative.

EXAMPLE ▪ 5   As an object moves along a straight path, its velocity $v(t)$ (in feet per second) at time t is recorded each second for 6 sec. The results are given in the following table.

t	0	1	2	3	4	5	6
$v(t)$	1	3	4	6	5	5	3

Approximate the distance traveled by the object.

SOLUTION   The points $(t, v(t))$ are plotted in Figure 5.87. If we assume that v is a continuous function, then, as in the preceding discussion, the distance traveled during the time interval $[0, 6]$ is $\int_0^6 v(t)\,dt$. Let us approximate this definite integral by means of Simpson's rule, with $n = 3$:

$$\int_0^6 v(t)\,dt \approx \frac{6-0}{6 \cdot 3}[v(0) + 4v(1) + 2v(2) + 4v(3)$$
$$+ 2v(4) + 4v(5) + v(6)]$$
$$= \tfrac{1}{3}[1 + 4 \cdot 3 + 2 \cdot 4 + 4 \cdot 6 + 2 \cdot 5 + 4 \cdot 5 + 3] = 26 \text{ ft}$$

Figure 5.87

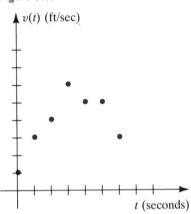

$v(t)$ (ft/sec)

t (seconds)

In (5.21), we defined the work W done by a variable force $f(x)$ that acts along a coordinate line from $x = a$ to $x = b$ by $W = \int_a^b f(x)\,dx$. Suppose that $f(x) \geq 0$ throughout $[a, b]$. If we sketch the graph of f as illustrated in Figure 5.88, then the work W is numerically equal to the area under the graph from a to b.

Figure 5.88

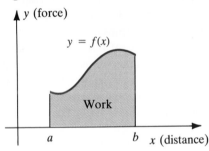

y (force)

$y = f(x)$

Work

a b x (distance)

EXAMPLE ▪ 6   An engineer obtains the graph in Figure 5.89, which shows the force (in pounds) acting on a small cart as it moves 25 ft along horizontal ground. Estimate the work done.

SOLUTION   If we assume that the force is a continuous function f for $0 \leq x \leq 25$, then the work done is

$$W = \int_0^{25} f(x)\,dx.$$

We do not have an explicit form for $f(x)$; however, we may estimate function values from the graph and approximate W by means of numerical integration.

Let us apply the trapezoidal rule with $a = 0$, $b = 25$, and $n = 5$. Referring to the graph to estimate function values gives us the table on the following page.

Figure 5.89

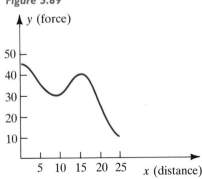

y (force)

50

40

30

20

10

5   10 15 20 25 x (distance)

k	x_k	$f(x_k)$
0	0	45
1	5	35
2	10	30
3	15	40
4	20	25
5	25	10

Since $(b - a)/(2n) = (25 - 0)/10 = 2.5$, the trapezoidal rule (4.37) gives

$$T_5 = (2.5)[f(x_0) + 2f(x_1) + 2f(x_2) + 2f(x_3) + 2f(x_4) + f(x_5)]$$
$$= (2.5)[45 + 70 + 60 + 80 + 50 + 10] = (2.5)(315) = 790.$$

It follows that

$$W = \int_0^{25} f(x)\, dx \approx 790 \text{ ft-lb.}$$

Suppose that the amount of a physical entity, such as oil, water, electric power, money supply, bacteria count, or blood flow, is increasing or decreasing in some manner, and that $R(t)$ is the rate at which it is changing at time t. If $Q(t)$ is the amount of the entity present at time t and if Q is differentiable, then $Q'(t) = R(t)$. If $R(t) > 0$ (or $R(t) < 0$) in a time interval $[a, b]$, then the amount that the entity increases (or decreases) between $t = a$ and $t = b$ is

$$Q(b) - Q(a) = \int_a^b Q'(t)\, dt = \int_a^b R(t)\, dt.$$

This number may be represented as the area of the region in a ty-plane bounded by the graphs of R, $t = a$, $t = b$, and $y = 0$.

EXAMPLE ▪ 7 Starting at 9:00 A.M., oil is pumped into a storage tank at a rate of $(150t^{1/2} + 25)$ gal/hr, for time t (in hours) after 9:00 A.M. How many gallons will have been pumped into the tank at 1:00 P.M.?

SOLUTION Letting $R(t) = 150t^{1/2} + 25$ in the preceding discussion, we obtain the following:

$$\int_0^4 (150t^{1/2} + 25)\, dt = \left[100t^{3/2} + 25t \right]_0^4$$

$$= 900 \text{ gal}$$

We have given only a few illustrations of the use of definite integrals. The interested reader may find many more in books on the physical and biological sciences, economics, and business, and even such areas as political science and sociology.

EXERCISES 5.8

1 A glass aquarium tank is 3 ft long and has square ends of width 1 ft. If the tank is filled with water, find the force exerted by the water on

(a) one end   **(b)** one side

2 If one of the square ends of the tank in Exercise 1 is divided into two parts by means of a diagonal, find the force exerted on each part.

3 The ends of a water trough 6 ft long have the shape of isosceles triangles with equal sides of length 2 ft and the third side of length $2\sqrt{3}$ ft at the top of the trough. Find the force exerted by the water on one end of the t if the trough is

(a) full of water   **(b)** half full of water

4 The ends of a water trough have the shape of the region bounded by the graphs of $y = x^2$ and $y = 4$, with x and y measured in feet. If the trough is full of water, find the force on one end.

5 A cylindrical oil storage tank 4 ft in diameter and 5 ft long is lying on its side. If the tank is half full of oil weighing 60 lb/ft^3, find the force exerted by the oil on one end of the tank.

6 A rectangular gate in a dam is 5 ft long and 3 ft high. If the gate is vertical, with the top of the gate parallel to the surface of the water and 6 ft below it, find the force of the water against the gate.

7 A plate having the shape of an isosceles trapezoid with upper base 4 ft long and lower base 8 ft long is submerged vertically in water such that the bases are parallel to the surface. If the distances from the surface of the water to the lower and upper bases are 10 ft and 6 ft, respectively, find the force exerted by the water on one side of the plate.

8 A circular plate of radius 2 ft is submerged vertically in water. If the distance from the surface of the water to the center of the plate is 6 ft, find the force exerted by the water on one side of the plate.

9 A rectangular plate 3 ft wide and 6 ft long is submerged vertically in oil weighing 50 lb/ft^3, with its short side parallel to, and 2 ft below, the surface.

(a) Find the total force exerted on one side of the plate.

(b) If the plate is divided into two parts by means of a diagonal, find the force exerted on each part.

c **10** A flat, irregularly shaped plate is submerged vertically in water (see figure). Measurements of its width, taken at successive depths at intervals of 0.5 ft, are compiled in the following table.

Water depth (ft)	1	1.5	2	2.5	3	3.5	4
Width of plate (ft)	0	2	3	5.5	4.5	3.5	0

Estimate the force of the water on one side of the plate by using **(a)** the trapezoidal rule, with $n = 6$, and **(b)** Simpson's rule, with $n = 3$.

Exercise 10

c **11** Refer to (5.28). To estimate cardiac output F (the number of liters of blood per minute that the heart pumps through the aorta), a 5-mg dose of the tracer indocyanine-green is injected into a pulmonary artery, and dye concentration measurements $c(t)$ are taken every minute from a peripheral artery near the aorta. The results are given in the following table. Use Simpson's rule, with $n = 6$, to estimate the cardiac output.

t (min)	$c(t)$ (mg/L)
0	0
1	0
2	0.15
3	0.48
4	0.86
5	0.72
6	0.48
7	0.26
8	0.15
9	0.09
10	0.05
11	0.01
12	0

c 12 Refer to (5.28). Suppose that 1200 kg of sodium dichromate is mixed into a river at point A, and sodium dichromate samples are taken every 30 sec at a point B downstream. The concentration $c(t)$ at time t is recorded in the following table. Use the trapezoidal rule, with $n = 12$, to estimate the river flow rate F.

t (sec)	$c(t)$ (mg/L or g/m^3)
0	0
30	2.14
60	3.89
90	5.81
120	8.95
150	7.31
180	6.15
210	4.89
240	2.98
270	1.42
300	0.89
330	0.29
360	0

13 A manufacturer estimates that the time required for a worker to assemble a certain item depends on the number of this item the worker has previously

assembled. If the time (in minutes) required to assemble the kth item is given by $f(k) = 20(k + 1)^{-0.4} + 3$, use a definite integral to approximate the time, to the nearest minute, required to assemble

(a) 1 item   (b) 4 items

(c) 8 items   (d) 16 items

14 A data processor keyboards registration data for college students from written forms to electronic files. The number of minutes required to process the kth registration is estimated to be approximately $f(k) = 6(1 + k)^{-1/3}$. Use a definite integral to estimate the time required for

(a) one person to keyboard 600 registrations

(b) two people to keyboard 300 registrations each

15 The number of minutes needed for a person to trace a path through a certain maze without error is estimated to be $f(k) = 5(1 + k)^{-1/2}$, where k is the number of trials previously completed. Use a definite integral to approximate the time required to complete 10 trials.

16 Anne has found that if she is making string necklaces, it takes her $7(2 + k)^{-2/3}$ minutes to complete the kth necklace. Use a definite integral to estimate the time that she needs to finish 10 necklaces.

Exer. 17–18: Use a definite integral to approximate the sum, and round the answer to the nearest integer.

17 $\displaystyle\sum_{k=1}^{100} k(k^2 + 1)^{-1/4}$

18 $\displaystyle\sum_{k=1}^{200} 5k(k^2 + 10)^{-1/3}$

c 19 The velocity (in miles per hour) of an automobile as it traveled along a freeway over a 12-min interval is indicated in the figure. Use the trapezoidal rule to approximate the distance traveled to the nearest mile.

Exercise 19

c 20 The acceleration (in feet per second per second) of an automobile over a period of 8 sec is indicated in the figure. Use the trapezoidal rule to approximate the net change in velocity in this time period.

Exercise 20

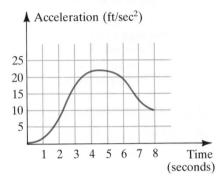

21 The following table was obtained by recording the force $f(x)$ (in Newtons) acting on a particle as it moved 6 m along a coordinate line from $x = 1$ to $x = 7$. Estimate the work done using

(a) the trapezoidal rule, with $n = 6$

(b) Simpson's rule, with $n = 3$

x	1	2	3	4	5	6	7
f(x)	20	23	25	22	26	30	28

22 A bicyclist pedals directly up a hill, recording the velocity $v(t)$ (in feet per second) at the end of every 2 sec. Referring to the results recorded in the following table, use the trapezoidal rule to approximate the distance traveled.

t	0	2	4	6	8	10
v(t)	24	22	16	10	2	0

23 A motorboat uses gasoline at the rate of $t\sqrt{9 - t^2}$ gal/hr. If the motor is started at $t = 0$, how much gasoline is used in 2 hr?

24 The population of a city has increased since 1985 at a rate of $1.5 + 0.3\sqrt{t} + 0.006t^2$ thousand people per year, where t is the number of years after 1985. Assuming that this rate continues and that the population was 50,000 in 1985, estimate the population in 1994.

25 A simple thermocouple, in which heat is transformed into electrical energy, is shown in the figure. To determine the total charge Q (in coulombs) transferred to the copper wire, current readings (in amperes) are recorded every $\frac{1}{2}$ sec, and the results are shown in the following table.

t (sec)	0	0.5	1.0	1.5	2.0	2.5	3.0
I (amp)	0	0.2	0.6	0.7	0.8	0.5	0.2

Use the fact that $I = dQ/dt$ and the trapezoidal rule, with $n = 6$, to estimate the total charge transferred to the copper wire during the first 3 sec.

Exercise 25

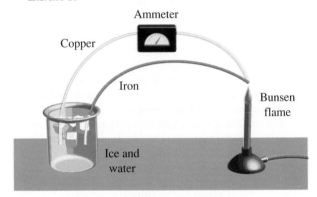

c 26 Suppose that $\rho(x)$ is the density (in centimeters per kilometer) of ozone in the atmosphere at a height of x kilometers above the ground. For example, if $\rho(6) = 0.0052$, then at a height of 6 km there is effectively a thickness of 0.0052 cm of ozone for each kilometer of atmosphere. If ρ is a continuous function, the thickness of the ozone layer between heights a and b can be found by evaluating $\int_a^b \rho(x)\,dx$. Values for $\rho(x)$ found experimentally are shown in the following table.

x (km)	ρ(x) (spring)	ρ(x) (autumn)
0	0.0034	0.0038
6	0.0052	0.0043
12	0.0124	0.0076
18	0.0132	0.0104
24	0.0136	0.0109
30	0.0084	0.0072
36	0.0034	0.0034
42	0.0017	0.0016

(a) Use the trapezoidal rule to estimate the thickness of the ozone layer between the altitudes of 6 and 42 km during both spring and autumn.

(b) Work part (a) using Simpson's rule.

c **27** Radon gas can pose a serious health hazard if inhaled. If $V(t)$ is the volume of air (in cubic centimeters) in an adult's lungs at time t (in minutes), then the rate of change of V can often be approximated by

$$V'(t) = 12{,}450\pi \sin(30\pi t).$$

Inhaling and exhaling correspond to $V'(t) > 0$ and $V'(t) < 0$, respectively. Suppose an adult lives in a home that has a radioactive energy concentration due to radon of 4.1×10^{-12} joule/cm^3.

(a) Approximate the volume of air inhaled by the adult with each breath.

(b) If inhaling more than 0.02 joule of radioactive energy in one year is considered hazardous, is it safe for the adult to remain at home?

28 A stationary exercise bicycle is programmed so that it can be set for different intensity levels L and workout times T. It displays the elapsed time t (in minutes), for $0 \le t \le T$, and the number of calories $C(t)$ that are being burned per minute at time t, where

$$C(t) = 5 + 3L - 6\frac{L}{T}\left|t - \frac{1}{2}T\right|.$$

Suppose that an individual exercises for 16 min, with $L = 3$ for $0 \le t \le 8$ and with $L = 2$ for $8 \le t \le 16$. Find the total number of calories burned during the workout.

29 The rate of growth R (in centimeters per year) of an average boy who is t years old is shown in the following table for $10 \le t \le 15$.

t (yr)	10	11	12	13	14	15
R (cm/yr)	5.3	5.2	4.9	6.5	9.3	7.0

Use the trapezoidal rule, with $n = 5$, to approximate the number of centimeters the boy grows between his tenth and fifteenth birthdays.

30 To determine the number of zooplankton in a portion of an ocean 80 m deep, marine biologists take samples at successive depths of 10 m, obtaining the following table, where $\rho(x)$ is the density (in number per cubic meter) of zooplankton at a depth of x meters.

x	0	10	20	30	40	50	60	70	80
$\rho(x)$	0	10	25	30	20	15	10	5	0

Use Simpson's rule, with $n = 4$, to estimate the total number of zooplankton in a *water column* (a column of water) having a cross section 1 m square extending from the surface to the ocean floor.

Exer. 31–34: Find the consumers' surplus for the given demand function $f(x)$ and the given price level p_c.

31 $f(x) = 20 - \frac{1}{20}x$; $p_c = 4$

32 $f(x) = 30 - \frac{1}{5}x$; $p_c = 10$

33 $f(x) = 400 - \frac{3}{8}x$; $p_c = 100$

34 $f(x) = 60 - \frac{2}{7}x$; $p_c = 40$

CHAPTER 5 REVIEW EXERCISES

Exer. 1–2: Sketch the region bounded by the graphs of the equations, and find the area by integrating with respect to (a) x and (b) y.

1 $y = -x^2$, $y = x^2 - 8$

2 $y^2 = 4 - x$, $x + 2y = 1$

Exer. 3–4: Find the area of the region bounded by the graphs of the equations.

3 $x = y^2$, $x + y = 1$

4 $y = -x^3$, $y = \sqrt{x}$, $7x + 3y = 10$

5 Find the area of the region between the graphs of the equations $y = \cos\frac{1}{2}x$ and $y = \sin x$, from $x = \pi/3$ to $x = \pi$.

6 The region bounded by the graph of $y = \sqrt{1 + \cos 2x}$ and the x-axis, from $x = 0$ to $x = \pi/2$, is revolved about the x-axis. Find the volume of the resulting solid.

Exer. 7–10: Sketch the region R bounded by the graphs of the equations, and find the volume of the solid generated by revolving R about the indicated axis.

7 $y = \sqrt{4x + 1}$, $y = 0$, $x = 0$, $x = 2$; x-axis

8 $y = x^4$, $y = 0$, $x = 1$; y-axis

9 $y = x^3 + 1$, $x = 0$, $y = 2$; y-axis

10 $y = \sqrt[3]{x}$, $y = \sqrt{x}$; x-axis

Exer. 11–12: The region bounded by the x-axis and the graph of the given equation, from $x = 0$ to $x = b$, is revolved about the y-axis. Find the volume of the resulting solid.

11 $y = \cos x^2$; $b = \sqrt{\pi/2}$

12 $y = x \sin x^3$; $b = 1$

13 Find the volume of the solid generated by revolving the region bounded by the graphs of $y = 4x^2$ and $4x + y = 8$ about

(a) the x-axis   (b) $x = 1$ (c) $y = 16$

14 Find the volume of the solid generated by revolving the region bounded by the graphs of $y = x^3$, $x = 2$, and $y = 0$ about

(a) the x-axis   (b) the y-axis   (c) $x = 2$

(d) $x = 3$ (e) $y = 8$ (f) $y = -1$

15 Find the arc length of the graph of $(x + 3)^2 = 8(y - 1)^3$ from $A(-2, \frac{3}{2})$ to $B(5, 3)$.

16 A solid has for its base the region in the xy-plane bounded by the graphs of $y^2 = 4x$ and $x = 4$. Find the volume of the solid if every cross section by a plane perpendicular to the x-axis is an isosceles right triangle with one of its equal sides on the base of the solid.

17 An above-ground swimming pool has the shape of a right circular cylinder of diameter 12 ft and height 5 ft. If the depth of the water in the pool is 4 ft, find the work required to empty the pool by pumping the water out over the top.

18 As a bucket is raised a distance of 30 ft from the bottom of a well, water leaks out at a uniform rate. Find the work done if the bucket originally contains 24 lb of water and one-third leaks out. Assume that the weight of the empty bucket is 4 lb, and disregard the weight of the rope.

19 A square plate of side 4 ft is submerged vertically in water such that one of the diagonals is parallel to the surface. If the distance from the surface of the water to the center of the plate is 6 ft, find the force exerted by the water on one side of the plate.

20 Use differentials to approximate the arc length of the graph of $y = 2 \sin \frac{1}{3}x$ between the points with x-coordinates π and $91\pi/90$.

Exer. 21–22: Sketch the region bounded by the graphs of the equations, and find m, M_x, M_y, and the centroid.

21 $y = x^3 + 1$, $x + y = -1$, $x = 1$

22 $y = x^2 + 1$, $y = x$, $x = -1$, $x = 2$

23 The graph of the equation $12y = 4x^3 + (3/x)$ from $A(1, \frac{7}{12})$ to $B(2, \frac{67}{24})$ is revolved about the x-axis. Find the area of the resulting surface.

24 The shape of a reflector in a searchlight is obtained by revolving a parabola about its axis. If, as shown in the figure, the reflector is 4 ft across at the opening and 1 ft deep, find its surface area.

Exercise 24

25 The velocity $v(t)$ of a rocket that is traveling directly upward is given in the following table. Use the trapezoidal rule to approximate the distance that the rocket travels from $t = 0$ to $t = 5$.

t (sec)	0	1	2	3	4	5
$v(t)$ (ft/sec)	100	120	150	190	240	300

c 26 An electrician suspects that a meter showing the total consumption Q in kilowatt hours (kWh) of electricity is not functioning properly. To check the accuracy, the electrician measures the consumption rate R directly every 10 min, obtaining the results in the following table.

t (min)	0	10	20	30
R (kWh/min)	1.31	1.43	1.45	1.39

t (min)	40	50	60
R (kWh/min)	1.36	1.47	1.29

(a) Use Simpson's rule to estimate the total consumption during this 1-hr period.

(b) If the meter read 48,792 kWh at the beginning of the experiment and 48,953 kWh at the end, what should the electrician conclude?

27 Interpret $\int_0^1 2\pi x^4 \, dx$ in the following ways:

(a) as the area of a region in the xy-plane

(b) as the volume of a solid obtained by revolving a region in the xy-plane about

  (i) the x-axis

  (ii) the y-axis

(c) as the work done by a force

28 Let R be the semicircular region in the xy-plane with endpoints of its diameter at $(4, 0)$ and $(10, 0)$. Use the theorem of Pappus to find the volume of the solid obtained by revolving R about the y-axis.

c Exer. 29–32: Plot the graphs of the equations. (a) Approximate the points of intersection. (b) Approximate the area bounded by the graphs of the equations.

29 $y = \sqrt{1 + x^3}$; $\qquad y = x^2$

30 $y = 5e^{-x^2}$; $\qquad y = \ln(x + 4)$

31 $y = x^3 - 4x^2 - x + 3$; $y = \sqrt{20x}$

32 $y = \sin(\sin x)$; $\qquad y = \sin(\cos x)$;
$x = -\dfrac{3\pi}{4}$; $\qquad x = \dfrac{\pi}{4}$

EXTENDED PROBLEMS AND GROUP PROJECTS

1 Explore an alternative approach to determining the length of the graph of a function between points A and B using the notation that we developed in Section 5.5. In particular, Q_k is the point with coordinates $(x_k, f(x_k))$. Now let P_k be the point $(x_k, f(x_{k+1}))$, and let

$$T_P = \sum_{k=1}^{n} [d(Q_{k-1}, P_k) + d(P_k, Q_k)].$$

(a) Show that P_k lies on a vertical line through Q_{k-1} and a horizontal line through Q_k (see figure).

(b) Discuss why T_P appears to be a good approximation for the length of the graph between A and B. In particular, show that as $\|P\|$ decreases, the distance between each P_k and the curve approaches zero.

(c) Discuss why T_P is not a good approximation for the length of the graph.

Problem 1

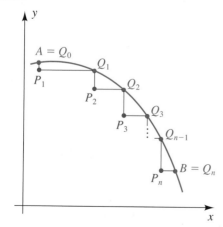

c 2 The model for a football using the approximate shape of the solid generated by revolving the arc of a circle $x^2 + (y + k)^2 = r^2$, where $y \geq 0$ and $0 < k < r$, does not quite match measurable dimensions. For a full-sized football, the distance from endpoint to endpoint along the axis of revolution is about 11 in. and the arc on the surface from endpoint to endpoint along a seam is about 14 in. long. Around the widest part, the circumference measures about 22 in. Explore using the shape of the solid generated by revolving the arc of an ellipse $a^2 x^2 + (y + k)^2 = r^2$, where $y \geq 0$ and $0 < k < r$. Approximate

(a) the volume

(b) the surface area for this new model for a football

3 Let f be a smooth function with $f(x) \geq 0$ on $[a, b]$. Partition the interval $[a, b]$ into n subintervals of equal width, and inscribe a rectangle under the graph of f over each subinterval. Then revolve each rectangle about the x-axis. Determine the surface area of the resulting solid, and let R_n be the sum of these surface areas. Let S be the area of the surface generated by revolving the graph of f about the x-axis. In what sense is R_n a good approximation to S? Will the limit of R_n as $n \to \infty$ be equal to S? Will we do better by taking *circumscribed* rather than inscribed rectangles?

INTRODUCTION

IN 1948, THE FINNISH-BORN AMERICAN ARCHITECT Eero Saarinen (1910–1961) submitted the winning design for a new national park, the Thomas Jefferson Westward Expansion Memorial in St. Louis. The center of his design was a great gleaming stainless-steel arch. Saarinen wanted "to create a monument which would have lasting significance and would be a landmark of our time. An absolutely simple shape . . . seemed to be the basis of the great memorials that have kept their significance and dignity over time." Saarinen designed his arch to be the purest expression of the forces within. This arch . . . is a catenary curve—the curve of a hanging chain—a curve in which the forces of thrust are continuously kept within the center of the legs of the arch. The mathematical precision seemed to enhance the timelessness of the form, but at the same time its dynamic quality seemed to link it to our own time.

To understand the mathematics of Saarinen's Gateway Arch to the West, we need to examine the natural exponential function. This function and its inverse, the natural logarithm, are perhaps the most important functions in applications of calculus to the natural world. They are examples of *transcendental functions*, the main topic of this chapter. We begin in Section 6.1 with a brief review of inverse functions and develop a formula for the derivative of an inverse function that will be useful throughout the entire chapter. Next, we employ a definite integral to introduce in Section 6.2 the *natural logarithm function*, which is then used to define in Section 6.3 the *natural exponential function* as the inverse of the natural logarithm. The natural logarithmic and exponential functions occur in many indefinite integral problems, a number of which are studied in Section 6.4. There are many other pairs of exponential and logarithmic functions; we analyze the general case in Section 6.5. After developing the theory of logarithms and exponentials, we explore in Section 6.6 a number of applications that involve these functions as solutions to first-order separable differential equations, an important modeling tool.

In Sections 6.7 and 6.8, we introduce other important transcendental functions: the inverse trigonometric functions and the hyperbolic functions and their inverses. We derive the equation for the catenary curve as an application of the hyperbolic functions. The chapter concludes with l'Hôpital's rule, which provides a direct way to evaluate limits of quotients in which both the numerator and the denominator approach 0 or both approach ∞ or $-\infty$. Such limits often occur when dealing with transcendental functions.

Transcendental functions frequently occur in the descriptions of curves that possess both aesthetic appeal and important structural properties of stability.

Transcendental Functions

515

6.1   THE DERIVATIVE OF THE INVERSE FUNCTION

A function f may have the same value for different numbers in its domain. For example, if $f(x) = x^2$, then $f(2) = 4 = f(-2)$ but $2 \neq -2$. In order to define the *inverse of a function*, it is essential that different numbers in the domain *always* give different values of f. Such functions are called *one-to-one functions*.

Definition   6.1

A function f with domain D and range R is a **one-to-one function** if whenever $a \neq b$ in D, then $f(a) \neq f(b)$ in R.

Figure 6.1

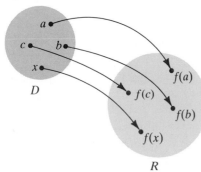

The diagram in Figure 6.1 illustrates a one-to-one function, because each function value in the range R corresponds to *exactly one* element in the domain D. The function whose graph is illustrated in Figure 6.2 is *not* one-to-one, because $a \neq b$ but $f(a) = f(b)$. Note that the horizontal line $y = f(a)$ (or $y = f(b)$) intersects the graph in more than one point. Thus, *if any horizontal line intersects the graph of a function f in more than one point, then f is not one-to-one*. Every increasing function is one-to-one, because if $a < b$, then $f(a) < f(b)$, and if $b < a$, then $f(b) < f(a)$. Thus, if $a \neq b$, then $f(a) \neq f(b)$. Similarly, every decreasing function is one-to-one.

If f is a one-to-one function with domain D and range R, then for each number y in R, there is *exactly one* number x in D such that $y = f(x)$, as illustrated by the arrow in Figure 6.3(a). Since x is *unique*, we may define a function g from R to D by means of the rule $x = g(y)$. As in Figure 6.3(b), g *reverses the correspondence given by f*. We call g the *inverse function of f*, as in the following definition.

Figure 6.2

Figure 6.3

(a)

D

R

(b)

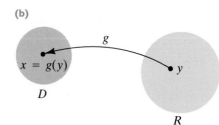

D

R

Definition   6.2

Let f be a one-to-one function with domain D and range R. A function g with domain R and range D is the **inverse function** of f, provided the following condition is true for every x in D and every y in R:

$$y = f(x) \quad \text{if and only if} \quad x = g(y)$$

The following theorem can be used to verify that a function g is the inverse of f.

Theorem 6.3

Let f be a one-to-one function with domain D and range R. If g is a function with domain R and range D, then g is the inverse function of f if and only if both of the following conditions are true:

(i) $g(f(x)) = x$ for every x in D

(ii) $f(g(y)) = y$ for every y in R

PROOF   Let us first prove that if g is the inverse function of f, then conditions (i) and (ii) are true. By the definition of an inverse function,

$$y = f(x) \quad \text{if and only if} \quad x = g(y)$$

for every x in D and every y in R. If we substitute $f(x)$ for y in the equation $x = g(y)$, we obtain condition (i): $x = g(f(x))$. Similarly, if we substitute $g(y)$ for x in the equation $y = f(x)$, we obtain condition (ii): $y = f(g(y))$. Thus, if g is the inverse function of f, then conditions (i) and (ii) are true.

Conversely, let g be a function with domain R and range D, and suppose that conditions (i) and (ii) are true. To show that g is the inverse function of f, we must prove that

$$y = f(x) \quad \text{if and only if} \quad x = g(y)$$

for every x in D and every y in R.

First suppose that $y = f(x)$. Since (i) is true, $g(f(x)) = x$—that is, $g(y) = x$. Thus, if $y = f(x)$, then $x = g(y)$.

Next suppose that $x = g(y)$. Since (ii) is true, $f(g(y)) = y$—that is, $f(x) = y$. Thus, if $x = g(y)$, then $y = f(x)$, which completes the proof. ∎

A one-to-one function f can have only one inverse function. Conditions (i) and (ii) of Theorem (6.3) imply that if g is the inverse function of f, then f is the inverse function of g. We say that f *and g are inverse functions of each other*.

If a function f has an inverse function g, we often denote g by f^{-1}. The -1 used in this notation should not be mistaken for an exponent—that is, $f^{-1}(y)$ *does not mean* $1/[f(y)]$. The reciprocal $1/[f(y)]$ may be denoted by $[f(y)]^{-1}$. It is important to remember the following relationships.

Domains and Ranges of
f and f^{-1} 6.4

$$\text{domain of } f^{-1} = \text{range of } f$$
$$\text{range of } f^{-1} = \text{domain of } f$$

When we discuss functions, we often let x denote an arbitrary number in the domain. Thus, for the inverse function f^{-1}, we may consider

$f^{-1}(x)$, *where x is in the domain of f^{-1}*. In this case, the two conditions in Theorem (6.3) are written as follows:

(i) $f^{-1}(f(x)) = x$ for every x in the domain of f
(ii) $f(f^{-1}(x)) = x$ for every x in the domain of f^{-1}

In some cases, we can find the inverse of a one-to-one function by solving the equation $y = f(x)$ for x in terms of y, obtaining an equation of the form $x = g(y)$. If the two conditions $g(f(x)) = x$ and $f(g(x)) = x$ are true for every x in the domains of f and g, respectively, then g is the required inverse function f^{-1}. The following guidelines summarize this procedure. In guideline (2), in anticipation of finding f^{-1}, we shall write $x = f^{-1}(y)$ instead of $x = g(y)$.

Guidelines for Finding
f^{-1} in Simple Cases   6.5

1   Verify that f is a one-to-one function (or that f is increasing or is decreasing) throughout its domain.
2   Solve the equation $y = f(x)$ for x in terms of y, obtaining an equation of the form $x = f^{-1}(y)$.
3   Verify the two conditions

$$f^{-1}(f(x)) = x \quad \text{and} \quad f(f^{-1}(x)) = x$$

for every x in the domains of f and f^{-1}, respectively.

The success of this method depends on the nature of the equation $y = f(x)$, since we must be able to solve for x in terms of y. For this reason, we include *simple cases* in the title of the guidelines.

EXAMPLE ▪ 1   Let $f(x) = 3x - 5$. Find the inverse function of f.

SOLUTION   We shall follow the three guidelines. First, we note that the graph of the linear function f is a line of slope 3. Since f is increasing throughout $\mathbb{R}$, f is one-to-one, and hence the inverse function f^{-1} exists. Moreover, since the domain and the range of f are $\mathbb{R}$, the same is true for f^{-1}.

As in guideline (2), we consider the equation

$$y = 3x - 5$$

and then solve for x in terms of y, obtaining

$$x = \frac{y + 5}{3}.$$

We now let

$$f^{-1}(y) = \frac{y + 5}{3}.$$

Since the symbol used for the variable is immaterial, we may also write

$$f^{-1}(x) = \frac{x + 5}{3}.$$

We next verify the conditions (i) $f^{-1}(f(x)) = x$ and (ii) $f(f^{-1}(x)) = x$:

(i) $\begin{aligned} f^{-1}(f(x)) &= f^{-1}(3x - 5) && \text{definition of } f \\ &= \frac{(3x - 5) + 5}{3} && \text{definition of } f^{-1} \\ &= x && \text{simplifying} \end{aligned}$

(ii) $\begin{aligned} f(f^{-1}(x)) &= f\left(\frac{x + 5}{3}\right) && \text{definition of } f^{-1} \\ \\ &= 3\left(\frac{x + 5}{3}\right) - 5 && \text{definition of } f \\ &= x && \text{simplifying} \end{aligned}$

Thus, by Theorem (6.3), the inverse function of f is given by $f^{-1}(x) = (x + 5)/3$.

EXAMPLE ■ 2   Let $f(x) = x^2 - 3$ for $x \geq 0$. Find the inverse function of f.

SOLUTION   The graph of f is sketched in Figure 6.4. The domain of f is $[0, \infty)$, and the range is $[-3, \infty)$. Since f is increasing, it is one-to-one and hence has an inverse function f^{-1} that has domain $[-3, \infty)$ and range $[0, \infty)$.

As in guideline (2), we consider the equation

$$y = x^2 - 3$$

and solve for x, obtaining

$$x = \pm\sqrt{y + 3}.$$

Since x is nonnegative, we reject $x = -\sqrt{y + 3}$ and let

$$f^{-1}(y) = \sqrt{y + 3}, \quad \text{or, equivalently,} \quad f^{-1}(x) = \sqrt{x + 3}.$$

Finally, we verify that (i) $f^{-1}(f(x)) = x$ for x in $[0, \infty)$ and that (ii) $f(f^{-1}(x)) = x$ for x in $[-3, \infty)$:

(i) $\begin{aligned} f^{-1}(f(x)) &= f^{-1}(x^2 - 3) \\ &= \sqrt{(x^2 - 3) + 3} = \sqrt{x^2} = |x| = x \quad \text{if } x \geq 0 \end{aligned}$

(ii) $\begin{aligned} f(f^{-1}(x)) &= f(\sqrt{x + 3}) \\ &= (\sqrt{x + 3})^2 - 3 = (x + 3) - 3 = x \quad \text{if } x \geq -3 \end{aligned}$

Thus, the inverse function is given by $f^{-1}(x) = \sqrt{x + 3}$ for $x \geq -3$.

There is an interesting relationship between the graph of a function f and the graph of its inverse function f^{-1}. We first note that $b = f(a)$ is equivalent to $a = f^{-1}(b)$. These equations imply that *the point (a, b) is on the graph of f if and only if the point (b, a) is on the graph of f^{-1}.*

Figure 6.4

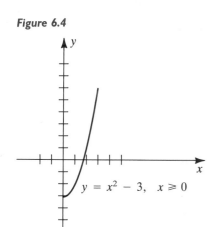

$y = x^2 - 3, \quad x \geqslant 0$

Figure 6.5

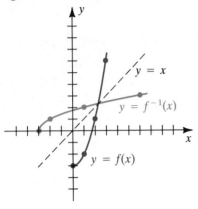

As an illustration, in Example 2 we found that the functions f and f^{-1} given by

$$f(x) = x^2 - 3 \quad \text{and} \quad f^{-1}(x) = \sqrt{x + 3}$$

are inverse functions of each other, provided that x is suitably restricted. Some points on the graph of f are $(0, -3)$, $(1, -2)$, $(2, 1)$, and $(3, 6)$. Corresponding points on the graph of f^{-1} are $(-3, 0)$, $(-2, 1)$, $(1, 2)$, and $(6, 3)$. The graphs of f and f^{-1} are sketched on the same coordinate plane in Figure 6.5. If the page is folded along the line $y = x$ that bisects quadrants I and III (as indicated by the dashed line in the figure), then the graphs of f and f^{-1} coincide. The two graphs are *reflections* of each other through the line $y = x$. This reflective property is typical of the graph of every function f that has an inverse function f^{-1}(see Exercise 14).

Figure 6.6 illustrates the graphs of an arbitrary one-to-one function f and its inverse function f^{-1}. As indicated in the figure, (c, d) is on the graph of f if and only if (d, c) is on the graph of f^{-1}. Thus, if we restrict the domain of f to the interval $[a, b]$, then the domain of f^{-1} is restricted to $[f(a), f(b)]$. If f is continuous, then the graph of f has no breaks or holes, and hence the same is true for the (reflected) graph of f^{-1}. Thus, we see intuitively that if f is continuous on $[a, b]$, then f^{-1} is continuous on $[f(a), f(b)]$. We can also show that if f is increasing, then so is f^{-1}. The next theorem states these facts, and Appendix I contains a proof.

Figure 6.6

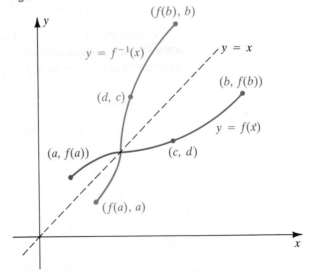

Theorem 6.6

If f is continuous and increasing on $[a, b]$, then f has an inverse function f^{-1} that is continuous and increasing on $[f(a), f(b)]$.

We can also prove the analogous result obtained by replacing the word *increasing* in Theorem (6.6) with the word *decreasing*.

The next theorem provides a method for finding the derivative of an inverse function.

Theorem 6.7

If a differentiable function f has an inverse function $g = f^{-1}$ and if $f'(g(c)) \neq 0$, then g is differentiable at c and

$$g'(c) = \frac{1}{f'(g(c))}.$$

PROOF By Definition (2.6),

$$g'(c) = \lim_{x \to c} \frac{g(x) - g(c)}{x - c}.$$

Let $y = g(x)$ and $a = g(c)$. Since f and g are inverse functions of each other,

$$g(x) = y \quad \text{if and only if} \quad f(y) = x$$

and $\qquad\qquad g(c) = a \quad \text{if and only if} \quad f(a) = c.$

Because f is differentiable, it is continuous and hence, by Theorem (6.6), so is the inverse function $g = f^{-1}$. Thus, if $x \to c$, then $g(x) \to g(c)$; that is, $y \to a$. If $y \to a$, then $f(y) \to f(a)$. Thus, we may write

$$g'(c) = \lim_{x \to c} \frac{g(x) - g(c)}{x - c}$$

$$= \lim_{y \to a} \frac{y - a}{f(y) - f(a)}$$

$$= \lim_{y \to a} \frac{1}{\dfrac{f(y) - f(a)}{y - a}}$$

$$= \frac{1}{\displaystyle\lim_{y \to a} \frac{f(y) - f(a)}{y - a}}$$

$$= \frac{1}{f'(a)} = \frac{1}{f'(g(c))}. \quad \blacksquare$$

It is convenient to restate Theorem (6.7) as follows.

Corollary 6.8

If g is the inverse function of a differentiable function f and if $f'(g(x)) \neq 0$, then

$$g'(x) = \frac{1}{f'(g(x))}.$$

Theorem (6.7) and Corollary (6.8) are useful because they enable us to compute the derivative of the inverse of a function without having an explicit formula for the inverse function. In the next example, we need the derivative of an inverse function so that we can find the slope of the tangent line to a point on its graph.

EXAMPLE ▪ 3   If $f(x) = x^3 + 2x - 1$, prove that f has an inverse function g, and find the slope of the tangent line to the graph of g at the point $P(2, 1)$.

SOLUTION   Since $f'(x) = 3x^2 + 2 > 0$ for every x, f is increasing and hence is one-to-one. Thus, f has an inverse function g. Since $f(1) = 2$, it follows that $g(2) = 1$, and consequently the point $P(2, 1)$ is on the graph of g. It would be difficult to find g using Guidelines (6.5), because we would have to solve the equation $y = x^3 + 2x - 1$ for x in terms of y. However, even if we cannot find g explicitly, we *can* find the slope $g'(2)$ of the tangent line to the graph of g at $P(2, 1)$. Thus, by Theorem (6.7),

$$g'(2) = \frac{1}{f'(g(2))} = \frac{1}{f'(1)} = \frac{1}{5}.$$

An easy way to remember Corollary (6.8) is to let $y = f(x)$. If g is the inverse function of f, then $g(y) = g(f(x)) = x$. From (6.8),

$$g'(y) = \frac{1}{f'(g(y))} = \frac{1}{f'(x)}.$$

This shows that, in a sense, the derivative of the inverse function g is the reciprocal of the derivative of f. A disadvantage of this formula is that it is not stated in terms of the independent variable for the inverse function. To illustrate, in Example 3, let $y = x^3 + 2x - 1$ and $x = g(y)$. Then

$$g'(y) = \frac{1}{3x^2 + 2} = \frac{1}{3(g(y))^2 + 2}.$$

We may also write this in the form

$$g'(x) = \frac{1}{3(g(x))^2 + 2}.$$

Consequently, to find $g'(x)$, it is necessary to know $g(x)$, just as in Corollary (6.8).

We may use a graphing utility to obtain the graphs of a function and its inverse simultaneously by exploiting the result that the point (a, b) is on the graph of f if and only if the point (b, a) is on the graph of f^{-1}. Once the utility has plotted a point (a, b), we ask it to plot the point (b, a) as well. The formal mechanism for achieving this result utilizes *parametric equations*, which we will study in more depth in Chapter 9. For now, we represent the graph of $y = f(x)$ for some domain $a \le x \le b$ by the points $(t, f(t))$ for $a \le t \le b$. The variable t is called a *parameter*. We actually have a pair of parametric equations:

$$x = t, \qquad y = f(t) \quad \text{for} \quad a \le t \le b$$

The graph of the inverse function (if it exists) is obtained by reversing the roles of x and y. This reversal is easily accomplished by plotting a second pair of parametric equations:

$$x = f(t), \qquad y = t, \quad \text{for} \quad a \le t \le b$$

If f is not one-to-one on the interval $[a, b]$, then the plot of this second pair of equations is not the graph of a function. To see that the plot of the first pair of equations is a reflection of the second pair about the line $y = x$, we also plot the line by using a third pair of parametric equations:

$$x = t, \qquad y = t, \quad \text{for} \quad a \le t \le b$$

In using a graphing utility, we must indicate by special notation or command that what is being requested is the plotting of parametric equations.

EXAMPLE ■ 4

(a) Use a graphing utility and parametric equations to view the graphs of the function f given by $f(x) = x^3 + 0.3x - 2$, the inverse of f, and the line $y = x$ for $-1 \le x \le 2$.

(b) Verify that the function f is one-to-one on this interval.

SOLUTION

(a) We set the graphing utility to plot the following parametric equations:

$X_{1T} = T$,	$Y_{1T} = T \wedge 3 + 0.3T - 2$	equations for f
$X_{2T} = Y_{1T}$,	$Y_{2T} = T$	equations for its inverse
$X_{3T} = T$,	$Y_{3T} = T$	equations for the line $y = x$

Note that we use an uppercase T instead of the lower-case t since most graphing calculators use uppercase letters. The graphing utility plots each of the points $(t, t^3 + 0.3t - 2)$, $(t^3 + 0.3t - 2, t)$, and (t, t) for each t in the interval $-1 \le t \le 2$ to produce the graphs shown in Figure 6.7.

(b) Both a visual inspection of the graph and observation of the fact that $f'(x) = 3x^2 + 0.3 > 0$ confirm that f is strictly increasing on $[-1, 2]$, and hence f is one-to-one on this interval.

Figure 6.7
$-9 \le x \le 9, -4 \le y \le 8$

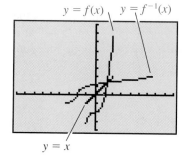

EXAMPLE ■ 5

(a) Graph $f(x) = \cos[\cos(0.9x)]$ on the interval $[0, 3]$.

(b) Estimate the largest interval $[a, b]$ with $0 \le a < b \le 3$ on which f is one-to-one.

(c) If g is the function with domain interval $[a, b]$ such that $g(x) = f(x)$ for $a \le x \le b$, estimate the domain and the range of g^{-1}.

(d) Make use of parametric equations to view the graphs of g and g^{-1} on the same coordinate axes.

SOLUTION

(a) Using a graphing utility gives the graph shown in Figure 6.8.

Figure 6.8
$0 \le x \le 3, 0 \le y \le 1$

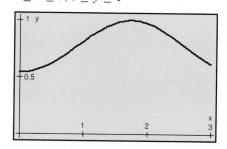

(b) From the graph in part (a), we see that as x increases from 0, the function f initially increases from a value of $\cos(\cos 0) \approx 0.5403023$ until it reaches a maximum value and then begins to decrease. Using the trace feature on the graphing utility, we find that the maximum occurs at approximately $x = 1.7453293$ where the value of the function is $\cos[\cos(0.9)(1.7453293)] \approx 1$. Thus, f is one-to-one on the interval $[0, 1.7453293]$ and also on the interval $[1.7453293, 3]$. Since the first interval is longer, we select $[a, b] = [0, 1.7453293]$.

(c) From the analysis in part (b), we have $g(x) = \cos[\cos(0.9x)]$ with domain $[0, 1.7453293]$ and range $[0.5403023, 1]$. Hence, the inverse g^{-1} has domain $[0.5403023, 1]$ and range $[0, 1.7453293]$.

(d) We use parametric equations to generate the points $(t, \cos[\cos(0.9t)])$ and $(\cos[\cos(0.9t)], t)$ for $0 \le t \le 1.7453293$ to obtain the graphs of g and g^{-1} shown in Figure 6.9.

Figure 6.9
$0 \le x \le 2.7, 0 \le y \le 1.8$

$y = g^{-1}(x)$

$y = g(x)$

EXERCISES 6.1

Exer. 1–12: Find $f^{-1}(x)$.

1 $f(x) = 3x + 5$

2 $f(x) = 7 - 2x$

3 $f(x) = \dfrac{1}{3x - 2}$

4 $f(x) = \dfrac{1}{x + 3}$

5 $f(x) = \dfrac{3x + 2}{2x - 5}$

6 $f(x) = \dfrac{4x}{x - 2}$

7 $f(x) = 2 - 3x^2, \quad x \le 0$

8 $f(x) = 5x^2 + 2, \quad x \ge 0$

9 $f(x) = \sqrt{3 - x}$

10 $f(x) = \sqrt{4 - x^2}, \ 0 \le x \le 2$

11 $f(x) = \sqrt[3]{x} + 1$

12 $f(x) = (x^3 + 1)^5$

13 (a) Prove that the linear function defined by $f(x) = ax + b$ with $a \ne 0$ has an inverse function, and find $f^{-1}(x)$.

(b) Does a constant function have an inverse? Explain.

14 Show that the graph of f^{-1} is the reflection of the graph of f through the line $y = x$ by verifying the following conditions:

(i) If $P(a, b)$ is on the graph of f, then $Q(b, a)$ is on the graph of f^{-1}.

(ii) The midpoint of line segment PQ is on the line $y = x$.

(iii) The line PQ is perpendicular to the line $y = x$.

Exer. 15–18: The graph of a one-to-one function f is shown in the figure. (a) Use a reflection to sketch the graph of f^{-1}. **(b)** Find the domain and the range of f. **(c)** Find the domain and the range of f^{-1}.

15

16

17

18

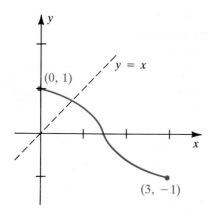

$\boxed{c}$ Exer. 19–24: Graph f on the given interval. (a) Estimate the largest interval $[a, b]$ with $a < b$ on which f is one-to-one. (b) If g is the function with domain $[a, b]$ such that $g(x) = f(x)$ for $a \le x \le b$, estimate the domain and the range of g^{-1}. (c) Use parametric equations to view the graphs of g and g^{-1} on the same coordinate axes.

19 $f(x) = 2.1x^3 - 2.98x^2 - 2.11x + 3$; $[-1, 2]$

20 $f(x) = 16x^5 + 8x^4 - 20x^3 - 8x^2 + 5x + 1$; $[-1, 1]$

21 $f(x) = \sin[\sin(1.1x)]$; $[-2, 2]$

22 $f(x) = \sin(x^3 + 2x^2 - 0.3)$; $[-1, 1]$

23 $f(x) = 2^{\cos(x-1)}$; $[-3, 2]$

24 $f(x) = 3^{(2x^2 - 3x - 1)}$; $[-1, 2]$

Exer. 25–30: (a) Prove that f has an inverse function g. (b) State the domain of g. (c) Use Corollary (6.8) to find $g'(x)$.

25 $f(x) = \sqrt{2x + 3}$

26 $f(x) = \sqrt[3]{5x + 2}$

27 $f(x) = 4 - x^2$, $x \ge 0$

28 $f(x) = x^2 - 4x + 5$, $x \ge 2$

29 $f(x) = 1/x$, $x \ne 0$

30 $f(x) = \sqrt{9 - x^2}$, $0 \le x \le 3$

Exer. 31–36: (a) Use f' to prove that f has an inverse function. (b) Find the slope of the tangent line at the point P on the graph of f^{-1}.

31 $f(x) = x^5 + 3x^3 + 2x - 1$; $P(5, 1)$

32 $f(x) = 2 - x - x^3$; $P(-8, 2)$

33 $f(x) = -2x + (8/x^3)$, $x > 0$; $P(-3, 2)$

34 $f(x) = 4x^5 - (1/x^3)$, $x > 0$; $P(3, 1)$

35 $f(x) = x^3 + 4x - 1$; $P(15, 2)$

36 $f(x) = x^5 + x$; $P(2, 1)$

Mathematicians and Their Times

LEONHARD EULER

IN THE EIGHTEENTH CENTURY, Europe witnessed a growing conflict between religion and science and the first of the political revolutions that overturned monarchies and founded new democratic forms of government. In his own life, Leonhard Euler (1707–1783), the leading mathematician and theoretical physicist of his time, balanced traditional religious beliefs with the demanding rationalist logic of deductive thought.

Euler was the most prolific mathematician in history, perhaps the most prolific author in any field. His writings fill 100 large books and contain contributions to mechanics, optics, acoustics, hydrodynamics, astronomy, chemistry, and medicine, as well as especially profound work in every branch of pure and applied mathematics. Born in Switzerland, Euler first studied to become a Calvinist minister as his father was, but mathematics led him down another path.

As a young man, Euler lost the sight in one eye, from an illness brought on by prolonged scientific work. Later in life, cataracts took the vision in his other eye. Although he was completely blind for 17 years, Euler's research never slackened. Blessed with a phenomenal memory and the ability to concentrate on difficult problems while surrounded by playing children (he had 13!), Euler could accurately complete complex problems mentally. As the physicist Arago noted, "He calculated without apparent effort, as men breathe, or as eagles sustain themselves in the wind."

In Euler's time, the major centers of scientific research were often academies funded by royalty. As a young student of John Bernoulli, Euler became friends with Bernoulli's sons Daniel and Nicolaus, who helped him secure a position at the Russian Academy in St. Petersburg. Euler remained there from 1727 until 1741 when Frederick the Great invited him to join the Berlin Academy. Euler did most of his best work during the quarter century he spent in Berlin. Since relations with Frederick were never very cordial (the emperor derided Euler as a "mathematical cyclops"), Euler gladly accepted the invitation of Catherine the Great to return to St. Petersburg in 1766.

Besides hundreds of research monographs, Euler also wrote influential mathematical textbooks at all levels. He introduced or established the use of many common mathematical notations: $f(x)$ for a function of x, π, e, $\sum$, $\log x$, $\sin x$, $\cos x$, and i. He also discovered the relationship

$$e^{\pi i} + 1 = 0,$$

the "mystical formula" linking the five most significant numbers in mathematics. George F. Simmons accurately describes Euler as "the Shakespeare of mathematics—universal, richly detailed, and inexhaustible."

6.2 THE NATURAL LOGARITHM FUNCTION

In this section, we define the natural logarithm function as a definite integral. At first you may think it strange to do so; later, however, you will see that the function we obtain obeys the familiar laws of logarithms considered in precalculus courses.

Let f be a function that is continuous on a closed interval $[a, b]$. As in the proof of Part I of the fundamental theorem of calculus (4.30), we can define a function F by

$$F(x) = \int_a^x f(t)\, dt$$

for x in $[a, b]$. If $f(t) \geq 0$ throughout $[a, b]$, then $F(x)$ is the area under the graph of f from a to x, as illustrated in Figure 6.10. For the special case $f(t) = t^n$, where n is a rational number and $n \neq -1$, we can find an explicit form for F. Thus, by the power rule for integrals,

$$F(x) = \int_a^x t^n\, dt = \left[\frac{t^{n+1}}{n+1} \right]_a^x$$

$$= \frac{1}{n+1} (x^{n+1} - a^{n+1}) \quad \text{if} \quad n \neq -1.$$

As indicated, we cannot use $t^{-1} = 1/t$ for the integrand, since $1/(n+1)$ is undefined if $n = -1$. Up to this point in our work, we have been unable to determine an antiderivative of $1/x$—that is, a function F such that $F'(x) = 1/x$. The next definition will remedy this situation.

Figure 6.10

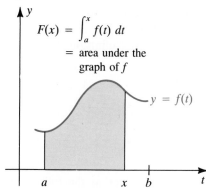

$F(x) = \int_a^x f(t)\, dt$
= area under the graph of f

$y = f(t)$

Definition **6.9**

The **natural logarithm function**, denoted by **ln**, is defined by

$$\ln x = \int_1^x \frac{1}{t}\, dt$$

for every $x > 0$.

The expression $\ln x$ (read *ell-en of x*) is called the **natural logarithm of x**. We use this terminology because, as we shall see, ln has the same properties as the logarithmic functions studied in precalculus courses. The restriction $x > 0$ is necessary because if $x \le 0$, the integrand $1/t$ has an infinite discontinuity between x and 1 and hence $\int_1^x (1/t)\, dt$ does not exist.

If $x > 1$, the definite integral $\int_1^x (1/t)\, dt$ may be interpreted as the area of the region under the graph of $y = 1/t$ from $t = 1$ to $t = x$ (see Figure 6.11a).

Figure 6.11

(a)

(b)

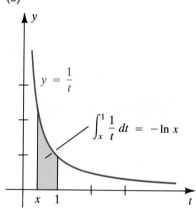

If $0 < x < 1$, then, since

$$\int_1^x \frac{1}{t}\, dt = -\int_x^1 \frac{1}{t}\, dt,$$

the integral is the *negative* of the area of the region under the graph of $y = 1/t$ from $t = x$ to $t = 1$ (see Figure 6.11b). Thus, $\ln x$ *is negative for $0 < x < 1$ and positive for $x > 1$*. Also note that, by Definition (4.18),

$$\ln 1 = \int_1^1 \frac{1}{t}\, dt = 0.$$

Applying Theorem (4.35) yields

$$\frac{d}{dx} \int_1^x \frac{1}{t}\, dt = \frac{1}{x}$$

for every $x > 0$. Substituting $\ln x$ for $\int_1^x (1/t)\, dt$ gives us the following theorem.

Theorem 6.10

$$\frac{d}{dx}(\ln x) = \frac{1}{x}$$

By Theorem (6.10), $\ln x$ *is an antiderivative of* $1/x$. Since $\ln x$ is differentiable and its derivative $1/x$ is positive for every $x > 0$, it follows from Theorems (2.12) and (3.15) that *the natural logarithmic function is continuous and increasing throughout its domain.* Also note that

$$\frac{d^2}{dx^2}(\ln x) = \frac{d}{dx}\left(\frac{d}{dx}(\ln x)\right) = \frac{d}{dx}\left(\frac{1}{x}\right) = -\frac{1}{x^2},$$

which is negative for every $x > 0$. Hence, by (3.18), the graph of the natural logarithmic function is concave downward on $(0, \infty)$.

Let us sketch the graph of $y = \ln x$. If $0 < x < 1$, then $\ln x < 0$ and the graph is below the x-axis. If $x > 1$, the graph is above the x-axis. Since $\ln 1 = 0$, the x-intercept is 1. We may approximate y-coordinates of points on the graph by applying the trapezoidal rule or Simpson's rule. If $x = 2$, then, by Example 3 in Section 4.7,

$$\ln 2 = \int_1^2 \frac{1}{t}\, dt \approx 0.693.$$

We will show in Theorem (6.12) that if $a > 0$, then $\ln a^r = r \ln a$ for every rational number r. Using this result yields the following:

$$\ln 4 = \ln 2^2 = 2 \ln 2 \approx 2(0.693) \approx 1.386$$
$$\ln 8 = \ln 2^3 = 3 \ln 2 \approx 2.079$$
$$\ln \tfrac{1}{2} = \ln 2^{-1} = -\ln 2 \approx -0.693$$
$$\ln \tfrac{1}{4} = \ln 2^{-2} = -2 \ln 2 \approx -1.386$$
$$\ln \tfrac{1}{8} = \ln 2^{-3} = -3 \ln 2 \approx -2.079$$

Plotting the points that correspond to the y-coordinates we have calculated and using the fact that $\ln$ is continuous and increasing gives us the sketch in Figure 6.12.

At the end of this section, we prove that

$$\lim_{x \to \infty} \ln x = \infty \quad \text{and} \quad \lim_{x \to 0^+} \ln x = -\infty.$$

The first of these results tells us that $y = \ln x$ increases without bound as $x \to \infty$. Note, however, that the *rate of change* of y with respect to x is very small if x is large. For example, if $x = 10^6$, then

$$\frac{dy}{dx}\Bigg]_{10^6} = \frac{1}{x}\Bigg]_{10^6} = \frac{1}{10^6} = 0.000001.$$

Thus, the tangent line is *almost* horizontal at the point on the graph with x-coordinate 10^6, and hence the graph is very flat near that point. The fact that $\lim_{x \to 0^+} \ln x = -\infty$ tells us that the line $x = 0$ (the y-axis) is a vertical asymptote for the graph (see Figure 6.12).

The next result generalizes Theorem (6.10).

Figure 6.12

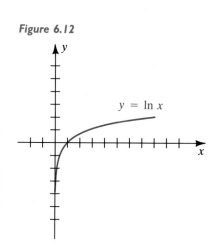

$y = \ln x$

Theorem **6.11**

If $u = g(x)$ and g is differentiable, then

(i) $\dfrac{d}{dx}(\ln u) = \dfrac{1}{u}\dfrac{du}{dx}$ if $u > 0$

(ii) $\dfrac{d}{dx}(\ln |u|) = \dfrac{1}{u}\dfrac{du}{dx}$ if $u \neq 0$

PROOF

(i) If we let $y = \ln u$ and $u = g(x)$, then, by the chain rule and Theorem (6.10),

$$\frac{d}{dx}(\ln u) = \frac{dy}{dx} = \frac{dy}{du}\frac{du}{dx} = \frac{1}{u}\frac{du}{dx}.$$

(ii) If $u > 0$, then $|u| = u$ and, by part (i),

$$\frac{d}{dx}(\ln |u|) = \frac{d}{dx}(\ln u) = \frac{1}{u}\frac{du}{dx}.$$

If $u < 0$, then $|u| = -u > 0$ and, by part (i),

$$\frac{d}{dx}(\ln |u|) = \frac{d}{dx}(\ln(-u)) = \frac{1}{-u}\frac{d}{dx}(-u)$$

$$= -\frac{1}{u}(-1)\frac{du}{dx} = \frac{1}{u}\frac{du}{dx}. \quad \blacksquare$$

In examples and exercises, if a function is defined in terms of the natural logarithm function, its domain will not usually be stated explicitly. Instead *we shall tacitly assume that x is restricted to values for which the logarithmic expression has meaning.* Thus, in Example 1, we assume $x^2 - 6 > 0$; that is, $|x| > \sqrt{6}$. In Example 2, we assume $x + 1 > 0$.

EXAMPLE ■ 1   If $f(x) = \ln(x^2 - 6)$, find $f'(x)$.

SOLUTION   Letting $u = x^2 - 6$ in Theorem (6.11)(i) yields

$$f'(x) = \frac{d}{dx}\left(\ln(x^2 - 6)\right) = \frac{1}{x^2 - 6}\frac{d}{dx}(x^2 - 6) = \frac{2x}{x^2 - 6}.$$

EXAMPLE ■ 2   If $y = \ln\sqrt{x + 1}$, find dy/dx.

SOLUTION   Letting $u = \sqrt{x + 1}$ in Theorem (6.11)(i) gives us

$$\frac{dy}{dx} = \frac{d}{dx}(\ln\sqrt{x + 1})$$

$$= \frac{1}{\sqrt{x + 1}}\frac{d}{dx}(\sqrt{x + 1}) = \frac{1}{\sqrt{x + 1}} \cdot \frac{1}{2}(x + 1)^{-1/2}$$

$$= \frac{1}{\sqrt{x + 1}} \cdot \frac{1}{2}\frac{1}{\sqrt{x + 1}} = \frac{1}{2(x + 1)}.$$

EXAMPLE ■ 3   If $f(x) = \ln |4 + 5x - 2x^3|$, find $f'(x)$.

SOLUTION   Using Theorem (6.11)(ii) with $u = 4 + 5x - 2x^3$ yields

$$f'(x) = \frac{d}{dx} \left(\ln |4 + 5x - 2x^3| \right)$$

$$= \frac{1}{4 + 5x - 2x^3} \frac{d}{dx} (4 + 5x - 2x^3) = \frac{5 - 6x^2}{4 + 5x - 2x^3}.$$

The next result states that natural logarithms satisfy the laws of logarithms studied in precalculus mathematics courses.

Laws of Natural Logarithms   6.12

If $p > 0$ and $q > 0$, then

    (i) $\ln pq = \ln p + \ln q$

    (ii) $\ln \dfrac{p}{q} = \ln p - \ln q$

    (iii) $\ln p^r = r \ln p$ for every rational number r

PROOF

    (i) If $p > 0$, then using Theorem (6.11) with $u = px$ gives us

$$\frac{d}{dx} (\ln px) = \frac{1}{px} \frac{d}{dx} (px) = \frac{1}{px} p = \frac{1}{x}.$$

Thus, $\ln px$ and $\ln x$ are both antiderivatives of $1/x$, and hence, by Theorem (4.2),

$$\ln px = \ln x + C$$

for some constant C. Letting $x = 1$, we obtain

$$\ln p = \ln 1 + C.$$

Since $\ln 1 = 0$, we see that $C = \ln p$, and therefore

$$\ln px = \ln x + \ln p.$$

Substituting q for x in the last equation gives us

$$\ln pq = \ln q + \ln p,$$

which is what we wished to prove.

    (ii) Using the formula $\ln p + \ln q = \ln pq$ with $p = 1/q$, we see that

$$\ln \frac{1}{q} + \ln q = \ln \left(\frac{1}{q} \cdot q \right) = \ln 1 = 0$$

and hence

$$\ln \frac{1}{q} = - \ln q.$$

Consequently,

$$\ln \frac{p}{q} = \ln \left(p \cdot \frac{1}{q} \right) = \ln p + \ln \frac{1}{q} = \ln p - \ln q.$$

(iii) If r is a rational number and $x > 0$, then, by Theorem (6.11) with $u = x^r$,

$$\frac{d}{dx}(\ln x^r) = \frac{1}{x^r}\frac{d}{dx}(x^r) = \frac{1}{x^r}rx^{r-1} = r\left(\frac{1}{x}\right) = \frac{r}{x}.$$

By Theorems (2.18)(iv) and (6.7), we may also write

$$\frac{d}{dx}(r \ln x) = r\frac{d}{dx}(\ln x) = r\left(\frac{1}{x}\right) = \frac{r}{x}.$$

Since $\ln x^r$ and $r \ln x$ are both antiderivatives of r/x, it follows from Theorem (4.2) that

$$\ln x^r = r \ln x + C$$

for some constant C. If we let $x = 1$ in the last formula, we obtain

$$\ln 1 = r \ln 1 + C.$$

Since $\ln 1 = 0$, this implies that $C = 0$ and, therefore,

$$\ln x^r = r \ln x.$$

In Section 6.5, we shall extend this law to irrational exponents.   ■

As shown in the following illustration, sometimes it is convenient to use laws of natural logarithms *before* differentiating.

ILLUSTRATION

$f(x)$	$f(x)$ after using laws of logarithms	$f'(x)$
▬ $\ln[(x + 2)(3x - 5)]$	$\ln(x + 2) + \ln(3x - 5)$	$\dfrac{1}{x + 2} + \dfrac{1}{3x - 5} \cdot 3 = \dfrac{6x + 1}{(x + 2)(3x - 5)}$
▬ $\ln \dfrac{x + 2}{3x - 5}$	$\ln(x + 2) - \ln(3x - 5)$	$\dfrac{1}{x + 2} - \dfrac{1}{3x - 5} \cdot 3 = \dfrac{-11}{(x + 2)(3x - 5)}$
▬ $\ln(x^2 + 1)^5$	$5 \ln(x^2 + 1)$	$5 \cdot \dfrac{1}{x^2 + 1} \cdot 2x = \dfrac{10x}{x^2 + 1}$
▬ $\ln \sqrt{x + 1}$	$\dfrac{1}{2} \ln(x + 1)$	$\dfrac{1}{2} \cdot \dfrac{1}{x + 1} = \dfrac{1}{2(x + 1)}$

An advantage of using laws of logarithms before differentiating may be seen by comparing the method of finding $(d/dx) \ln \sqrt{x + 1}$ in the preceding illustration with the solution of Example 2.

In the next two examples, we apply laws of logarithms to complicated expressions before differentiating.

EXAMPLE ▪ 4 If $f(x) = \ln[\sqrt{6x-1}(4x+5)^3]$, find $f'(x)$.

SOLUTION We first write $\sqrt{6x-1} = (6x-1)^{1/2}$ and then use laws of logarithms (i) and (iii):

$$f(x) = \ln[(6x-1)^{1/2}(4x+5)^3]$$
$$= \ln(6x-1)^{1/2} + \ln(4x+5)^3$$
$$= \tfrac{1}{2}\ln(6x-1) + 3\ln(4x+5)$$

By Theorem (6.11),

$$f'(x) = \left(\frac{1}{2} \cdot \frac{1}{6x-1} \cdot 6\right) + \left(3 \cdot \frac{1}{4x+5} \cdot 4\right)$$
$$= \frac{3}{6x-1} + \frac{12}{4x+5}$$
$$= \frac{84x+3}{(6x-1)(4x+5)}.$$

EXAMPLE ▪ 5 If $y = \ln\sqrt[3]{\dfrac{x^2-1}{x^2+1}}$, find $\dfrac{dy}{dx}$.

SOLUTION We first use laws of logarithms to change the form of y as follows:

$$y = \ln\left(\frac{x^2-1}{x^2+1}\right)^{1/3} = \frac{1}{3}\ln\left(\frac{x^2-1}{x^2+1}\right)$$
$$= \tfrac{1}{3}[\ln(x^2-1) - \ln(x^2+1)]$$

Next we use Theorem (6.11) to obtain

$$\frac{dy}{dx} = \frac{1}{3}\left(\frac{1}{x^2-1} \cdot 2x - \frac{1}{x^2+1} \cdot 2x\right)$$
$$= \frac{2x}{3}\left(\frac{1}{x^2-1} - \frac{1}{x^2+1}\right)$$
$$= \frac{2x}{3}\left[\frac{2}{(x^2-1)(x^2+1)}\right] = \frac{4x}{3(x^2-1)(x^2+1)}.$$

Given $y = f(x)$, we may sometimes find dy/dx by **logarithmic differentiation**. This method is especially useful if $f(x)$ involves complicated products, quotients, or powers. In the following guidelines, it is assumed that $f(x) > 0$; however, we shall show that the same steps can be used if $f(x) < 0$.

Guidelines for Logarithmic
Differentiation **6.13**

1	$y = f(x)$	given
2	$\ln y = \ln f(x)$	take natural logarithms and simplify
3	$\dfrac{d}{dx}(\ln y) = \dfrac{d}{dx}(\ln f(x))$	differentiate implicitly
4	$\dfrac{1}{y}\dfrac{dy}{dx} = \dfrac{d}{dx}(\ln f(x))$	by Theorem (6.11)
5	$\dfrac{dy}{dx} = f(x)\dfrac{d}{dx}(\ln f(x))$	multiply by $y = f(x)$

Of course, to complete the solution we must differentiate $\ln f(x)$ at some stage after guideline (3). If $f(x) < 0$ for some x, then guideline (2) is invalid, since $\ln f(x)$ is undefined. In this event, we can replace guideline (1) with $|y| = |f(x)|$ and take natural logarithms, obtaining $\ln|y| = \ln|f(x)|$. If we now differentiate implicitly and use Theorem (6.11)(ii), we again arrive at guideline (4). Thus, negative values of $f(x)$ do not change the outcome, and we are not concerned whether $f(x)$ is positive or negative. The method should not be used to find $f'(a)$ if $f(a) = 0$, since $\ln 0$ is undefined.

EXAMPLE■6 If

$$y = \frac{(5x - 4)^3}{\sqrt{2x + 1}},$$

use logarithmic differentiation to find dy/dx.

SOLUTION As in guideline (2), we begin by taking the natural logarithm of each side, obtaining

$$\ln y = \ln(5x - 4)^3 - \ln \sqrt{2x + 1}$$
$$= 3 \ln(5x - 4) - \tfrac{1}{2} \ln(2x + 1).$$

Applying guidelines (3) and (4), we differentiate implicitly with respect to x and then use Theorem (6.8) to obtain

$$\frac{1}{y}\frac{dy}{dx} = \left(3 \cdot \frac{1}{5x - 4} \cdot 5\right) - \left(\frac{1}{2} \cdot \frac{1}{2x + 1} \cdot 2\right)$$
$$= \frac{25x + 19}{(5x - 4)(2x + 1)}.$$

Finally, as in guideline (5), we multiply both sides of the last equation by y (that is, by $(5x-4)^3/\sqrt{2x+1}$) to get

$$\frac{dy}{dx} = \frac{25x+19}{(5x-4)(2x+1)} \cdot \frac{(5x-4)^3}{\sqrt{2x+1}}$$

$$= \frac{(25x+19)(5x-4)^2}{(2x+1)^{3/2}}.$$

We could check this result by applying the quotient rule to y.

An application of natural logarithms to growth processes is given in the next example. Many additional applied problems involving ln appear in other examples and exercises of this chapter.

EXAMPLE ■ 7 The *Count model* is an empirically based formula that can be used to predict the height of a preschooler. If $h(x)$ denotes the height (in centimeters) at age x (in years) for $\frac{1}{4} \le x \le 6$, then $h(x)$ can be approximated by

$$h(x) = 70.228 + 5.104x + 9.222 \ln x.$$

(a) Predict the height and rate of growth when a child reaches age 2.

(b) When is the rate of growth largest?

SOLUTION

(a) The height at age 2 is approximately

$$h(2) = 70.228 + 5.104(2) + 9.222 \ln 2 \approx 86.8 \text{ cm.}$$

The rate of change of h with respect to x is

$$h'(x) = 5.104 + 9.222\left(\frac{1}{x}\right).$$

Letting $x = 2$ gives us

$$h'(2) = 5.104 + 9.222(\tfrac{1}{2}) = 9.715.$$

Hence the rate of growth on the child's second birthday is about 9.7 cm/yr.

(b) To determine the maximum value of the rate of growth $h'(x)$, we first find the critical numbers of h'. Differentiating $h'(x)$, we obtain

$$h''(x) = 9.222\left(-\frac{1}{x^2}\right) = -\frac{9.222}{x^2}.$$

Since $h''(x)$ is negative for every x in $[\frac{1}{4}, 6]$, h' has no critical numbers in $(\frac{1}{4}, 6)$. It follows from Theorem (3.15) that h' is decreasing on $[\frac{1}{4}, 6]$. Consequently, the maximum value of $h'(x)$ occurs at $x = \frac{1}{4}$; that is, the rate of growth is largest at the age of 3 months.

Figure 6.13

$0.25 \le x \le 6, 0 \le y \le 120$

COMPUTATIONAL METHOD   The graphs of the functions h and h' on the interval [0.25, 6] are shown in Figure 6.13. Using the trace operation, we find that $h(2) \approx 86.8$. We also see that h' decreases on the interval so that its maximum is at the left endpoint, $x = 0.25$.

We conclude this section by examining $\ln x$ as $x \to \infty$ and as $x \to 0^+$. If $x > 1$, we may interpret the integral $\int_1^x (1/t)\, dt = \ln x$ as the area of the region shown in Figure 6.14. The sum of the areas of the three rectangles shown in Figure 6.15 is

$$\frac{1}{2} + \frac{1}{3} + \frac{1}{4} = \frac{13}{12}.$$

Since the area under the graph of $y = 1/t$ from $t = 1$ to $t = 4$ is $\ln 4$, we see that

$$\ln 4 > \frac{13}{12} > 1.$$

It follows that if M is any positive rational number, then

$$M \ln 4 > M, \quad \text{or} \quad \ln 4^M > M.$$

If $x > 4^M$, then since $\ln$ is an increasing function,

$$\ln x > \ln 4^M > M.$$

This proves that $\ln x$ can be made as large as desired by choosing x sufficiently large—that is,

$$\lim_{x \to \infty} \ln x = \infty.$$

To investigate the case $x \to 0^+$, we first note that

$$\ln \frac{1}{x} = \ln 1 - \ln x = 0 - \ln x = -\ln x.$$

Hence,

$$\lim_{x \to 0^+} \ln x = \lim_{x \to 0^+} \left(-\ln \frac{1}{x} \right).$$

Figure 6.14

Figure 6.15

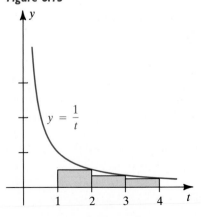

As x approaches zero through positive values, $1/x$ increases without bound and, therefore, so does $\ln(1/x)$. Consequently, $-\ln(1/x)$ *decreases* without bound and

$$\lim_{x \to 0^+} \ln x = -\infty.$$

EXERCISES 6.2

Exer. 1–34: Find $f'(x)$ if $f(x)$ is the given expression.

1 $\ln(9x + 4)$

2 $\ln(x^4 + 1)$

3 $\ln(3x^2 - 2x + 1)$

4 $\ln(4x^3 - x^2 + 2)$

5 $\ln|3 - 2x|$

6 $\ln|4 - 3x|$

7 $\ln|2 - 3x|^5$

8 $\ln|5x^2 - 1|^3$

9 $\ln\sqrt{7 - 2x^3}$

10 $\ln\sqrt[3]{6x + 7}$

11 $x\ln x$

12 $\ln(\ln x)$

13 $\ln\sqrt{x} + \sqrt{\ln x}$

14 $\ln x^3 + (\ln x)^3$

15 $\dfrac{1}{\ln x} + \ln\dfrac{1}{x}$

16 $\dfrac{x^2}{\ln x}$

17 $\ln[(5x - 7)^4(2x + 3)^3]$

18 $\ln[\sqrt[3]{4x - 5}(3x + 8)^2]$

19 $\ln\dfrac{\sqrt{x^2 + 1}}{(9x - 4)^2}$

20 $\ln\dfrac{x^2(2x - 1)^3}{(x + 5)^2}$

21 $\ln\sqrt{\dfrac{x^2 - 1}{x^2 + 1}}$

22 $\ln\sqrt{\dfrac{4 + x^2}{4 - x^2}}$

23 $\ln(x + \sqrt{x^2 - 1})$

24 $\ln(x + \sqrt{x^2 + 1})$

25 $\ln\cos 2x$

26 $\cos(\ln 2x)$

27 $\ln\tan^3 3x$

28 $\ln\cot(x^2)$

29 $\ln\ln\sec 2x$

30 $\ln\csc^2 4x$

31 $\ln|\sec x|$

32 $\ln|\sin x|$

33 $\ln|\sec x + \tan x|$

34 $\ln|\csc x - \cot x|$

Exer. 35–38: Use implicit differentiation to find y'.

35 $3y - x^2 + \ln xy = 2$

36 $y^2 + \ln(x/y) - 4x = -3$

37 $x\ln y - y\ln x = 1$

38 $y^3 + x^2\ln y = 5x + 3$

Exer. 39–44: Use logarithmic differentiation to find dy/dx.

39 $y = (5x + 2)^3(6x + 1)^2$

40 $y = (x + 1)^2(x + 2)^3(x + 3)^4$

41 $y = \sqrt{4x + 7}(x - 5)^3$

42 $y = \sqrt{(3x^2 + 2)\sqrt{6x - 7}}$

43 $y = \dfrac{(x^2 + 3)^5}{\sqrt{x + 1}}$

44 $y = \dfrac{(x^2 + 3)^{2/3}(3x - 4)^4}{\sqrt{x}}$

45 Find an equation of the tangent line to the graph of $y = x^2 + \ln(2x - 5)$ at the point $P(3, 9)$.

46 Find an equation of the tangent line to the graph of $y = x + \ln x$ that is perpendicular to the line whose equation is $2x + 6y = 5$.

47 Shown in the figure is a graph of $y = 5\ln x - \frac{1}{2}x$. Find the coordinates of the highest point, and show that the graph is concave downward for $x > 0$.

Exercise 47

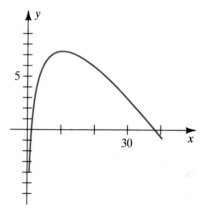

48 Shown in the figure is a graph of $y = \ln(x^2 + 1)$. Find the points of inflection.

Exercise 48

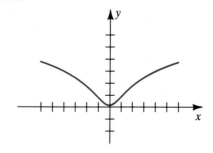

49 An approximation to the age T (in years) of a female blue whale can be obtained from a length measurement L (in feet) using $T = -2.57 \ln[(87 - L)/63]$. A blue whale has been spotted by a research vessel, and her length is estimated to be 80 ft. If the maximum error in estimating L is ± 2 ft, use differentials to approximate the maximum error in T.

50 The *Ehrenberg relation*, $\ln W = \ln 2.4 + 0.0184h$, is an empirically based formula relating the height h (in centimeters) to the weight W (in kilograms) for children aged 5–13. The formula, with minor changes in constants, has been verified in many different countries. Find the relationship between the rates of change dW/dt and dh/dt, for time t (in years).

51 A rocket of mass m_1 is filled with fuel of mass m_2, which will be burned at a constant rate of b kg/sec. If the fuel is expelled from the rocket at a constant rate, the distance $s(t)$ (in meters) that the rocket has traveled after t seconds is

$$s(t) = ct + \frac{c}{b}(m_1 + m_2 - bt) \ln\left(\frac{m_1 + m_2 - bt}{m_1 + m_2}\right)$$

for some constant $c > 0$.

(a) Find the initial velocity and the initial acceleration of the rocket.

(b) Burnout occurs when $t = m_2/b$. Find the velocity and the acceleration at burnout.

52 One method of estimating the thickness of the ozone layer is to use the formula $\ln(I/I_0) = -\beta T$, where I_0 is the intensity of a particular wavelength of light from the sun before it reaches the atmosphere, I is the intensity of the same wavelength after passing through a layer of ozone T centimeters thick, and β is the absorption coefficient for that wavelength. Suppose that for a wavelength of 3055×10^{-8} cm with $\beta \approx 2.7$, I_0/I is measured as 2.3.

(a) Approximate the thickness of the ozone layer to the nearest 0.01 cm.

(b) If the maximum error in the measured value of I_0/I is ± 0.1, use differentials to approximate the maximum error in the approximation obtained in part (a).

53 Describe the difference between the graphs of $y = \ln(x^2)$ and $y = 2 \ln x$.

54 Sketch the graphs of

(a) $y = \ln |x|$ **(b)** $y = |\ln x|$

c **Exer. 55–60:** Graph f on the given interval. **(a)** Approximate the range of the function defined on this interval so that the graph just fits in the viewing window. **(b)** Identify x-intercepts, y-intercepts, and relative extrema in this viewing window, if any. Estimate these features to two decimal places.

55 $f(x) = \ln(\sin x)$;   [0.1, 3.1]

56 $f(x) = \sin(\ln x)$;   [0.1, 800]

57 $f(x) = \dfrac{\ln[4(x + 1)]}{\ln[2(x + 1)]}$;   [−0.1, 3]

58 $f(x) = x^2 \ln x$;   [0, 2]

59 $f(x) = \ln[x(1.3 + \sin x)]$;   [0.1, 20]

60 $f(x) = \ln(x^4 - 4x^2 - 0.8x + 5.4)$;   [−3, 3]

c **Exer. 61–64:** Use Newton's method or a solving routine to approximate the real root(s) of the equation to four decimal places. Use a graphing utility to ensure that all roots are found.

61 $\ln x + x = 0$

62 $\ln(x^2 - 1.8x + 1) = 3$

63 $2 - 0.3x - 0.2x^2 - \ln[\ln(x^2 + 1.5)] = 0$

64 $\ln[x(1 - 0.9 \cos x)] = 0$

c **Exer. 65–68:** Use a numerical integration method or routine to approximate the definite integral to four decimal places.

65 $\displaystyle\int_1^4 \ln(\sin x - x \cos x) \, dx$

66 $\displaystyle\int_1^{50} \ln(1 + \ln x) \, dx$

67 $\displaystyle\int_{0.5}^2 \frac{\ln 4x}{\ln 3x} \, dx$

68 $\displaystyle\int_2^5 \frac{x^2}{\ln x} \, dx$

c **69** Approximate the area bounded by the graphs of $y = \ln x$, $y = 1/x$, and $x = 10$.

c **70** Approximate the volume of the solid generated by revolving the graph of $y = (\ln x)^2/x$, $1 \le x \le 5$, about the x-axis.

c **71** Approximate the arc length of the part of the curve $y = \ln x$ that lies inside the circle $x^2 + y^2 = 25$.

6.3 THE EXPONENTIAL FUNCTION

In Section 6.2, we saw that

$$\lim_{x \to \infty} \ln x = \infty \quad \text{and} \quad \lim_{x \to 0^+} \ln x = -\infty.$$

These facts are used in the proof of the following result.

Theorem 6.14

> To every real number x there corresponds exactly one positive real number y such that $\ln y = x$.

PROOF First note that if $x = 0$, then $y = 1$. Moreover, since ln is an increasing function, 1 is the only value of y such that $\ln y = 0$.

If x is positive, then we may choose a number b such that

$$\ln 1 < x < \ln b.$$

Since ln is continuous, it takes on every value between $\ln 1$ and $\ln b$ (see the intermediate value theorem (1.26)). Thus, there is a number y between 1 and b such that $\ln y = x$. Since ln is an increasing function, there is only one such number.

Finally, if x is negative, then there is a number $b > 0$ such that

$$\ln b < x < \ln 1,$$

and, as before, there is exactly one number y between b and 1 such that $\ln y = x$. ■

It follows from Theorem (6.14) that the range of the natural logarithms is $\mathbb{R}$. Since ln is an increasing function, it is one-to-one and therefore has an inverse function, to which we give the following special name.

Definition 6.15

> The **natural exponential function**, denoted by **exp**, is the inverse of the natural logarithm function.

The reason for the term *exponential* in this definition will become clear shortly. Since exp is the inverse of ln, its domain is $\mathbb{R}$ and its range is $(0, \infty)$. Moreover, as in (6.2),

$$y = \exp x \quad \text{if and only if} \quad x = \ln y,$$

where x is any real number and $y > 0$. By Theorem (6.3), we may also write

$$\ln(\exp x) = x \quad \text{and} \quad \exp(\ln y) = y.$$

Figure 6.16

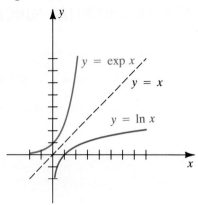

As we observed in Section 6.1, if two functions are inverses of each other, then their graphs are reflections through the line $y = x$. Hence the graph of $y = \exp x$ can be obtained by reflecting the graph of $y = \ln x$ through this line, as illustrated in Figure 6.16. Note that

$$\lim_{x \to \infty} \exp x = \infty \quad \text{and} \quad \lim_{x \to -\infty} \exp x = 0.$$

By Theorem (6.14), there exists exactly one positive real number whose natural logarithm is 1. This number is denoted by e. The great Swiss mathematician Leonhard Euler was among the first to study its properties extensively. (See *Mathematicians and Their Times*.)

Definition of e 6.16

> The letter e denotes the positive real number such that $\ln e = 1$.

Several values of ln were calculated in Section 6.2. We can show, by means of the trapezoidal rule, that

$$\int_1^{2.7} \frac{1}{t} \, dt < 1 < \int_1^{2.8} \frac{1}{t} \, dt.$$

Applying Definitions (6.9) and (6.16) yields

$$\ln 2.7 < \ln e < \ln 2.8$$

and hence

$$2.7 < e < 2.8.$$

Later, in Theorem (6.32), we show that

$$e = \lim_{h \to 0} (1 + h)^{1/h}.$$

This limit formula can be used to approximate e to any degree of accuracy. In Section 6.5, the first five decimal places in the following 32-decimal-place approximation for e will be justified in this way.

Approximation to e   6.17

$$e \approx 2.718281828459045235360287471352266$$

It can be shown that e is an irrational number.
If r is any *rational* number, then

$$\ln e^r = r \ln e = r(1) = r.$$

The formula $\ln e^r = r$ may be used to motivate a definition of e^x for every *real* number x. Specifically, we shall *define* e^x as the real number y such that $\ln y = x$. The following statement is a convenient way to remember this definition.

Definition of e^x 6.18

If x is any real number, then

$$e^x = y \quad \text{if and only if} \quad \ln y = x.$$

Since exp is the inverse function of ln,

$$\exp x = y \quad \text{if and only if} \quad \ln y = x.$$

Comparing this relationship with Definition (6.18), we see that

$$e^x = \exp x \quad \text{for every } x.$$

This result shows the reason for calling exp an *exponential* function and referring to it as the **exponential function with base e**. The graph of $y = e^x$ is the same as that of $y = \exp x$, illustrated in Figure 6.16. Hereafter *we shall use e^x instead of $\exp x$ to denote values of the natural exponential function.* The most commonly used exponential function in mathematics and its applications is the natural exponential function exp. For this reason, the function exp is often called "the exponential function."

The fact that $\ln(\exp x) = x$ for every x and $\exp(\ln x) = x$ for every $x > 0$ may now be written as follows:

Theorem   6.19

(i) $\ln e^x = x$ for every x
(ii) $e^{\ln x} = x$ for every $x > 0$

Some special cases of this theorem are given in the following illustration.

ILLUSTRATION

- $\ln e^5 = 5$
- $e^{\ln 5} = 5$
- $e^{3 \ln x} = e^{\ln(x^3)} = x^3$

- $\ln e^{\sqrt{x+1}} = \sqrt{x+1}$
- $e^{\ln \sqrt{x+1}} = \sqrt{x+1}$
- $e^{k \ln x} = e^{\ln(x^k)} = x^k$

NOTE   To obtain numerical approximations for e^x, use the natural exponential function provided by your calculator or computer. It gives a better approximation than entering an approximation for the number e and then raising this approximation to a power.

The next theorem states that the laws of exponents are true for powers of e.

Theorem 6.20

> If p and q are real numbers and r is a rational number, then
>
> **(i)** $e^p e^q = e^{p+q}$
>
> **(ii)** $\dfrac{e^p}{e^q} = e^{p-q}$
>
> **(iii)** $(e^p)^r = e^{pr}$

PROOF   Using Theorems (6.12) and (6.19), we obtain

$$\ln e^p e^q = \ln e^p + \ln e^q = p + q = \ln e^{p+q}.$$

Since the natural logarithm function is one-to-one,

$$e^p e^q = e^{p+q}.$$

Thus, we have proved (i). The proofs for (ii) and (iii) are similar. We show in Section 6.5 that (iii) is also true if r is irrational.   ■

By Theorem (6.7), the inverse function of a differentiable function is differentiable, and hence $(d/dx)(e^x)$ exists. The next theorem states that e^x *is its own derivative.*

Theorem 6.21

> $$\frac{d}{dx}(e^x) = e^x$$

PROOF   By (i) of Theorem (6.19),

$$\ln e^x = x.$$

Differentiating each side of this equation and using Theorem (6.11)(i) with $u = e^x$ gives us the following:

$$\frac{d}{dx}(\ln e^x) = \frac{d}{dx}(x)$$

$$\frac{1}{e^x}\frac{d}{dx}(e^x) = 1$$

$$\frac{d}{dx}(e^x) = e^x \qquad ■$$

EXAMPLE 1 If $f(x) = x^2 e^x$, find $f'(x)$.

SOLUTION By the product rule and Theorem (6.21),

$$f'(x) = x^2 \frac{d}{dx}(e^x) + e^x \frac{d}{dx}(x^2)$$
$$= x^2 e^x + e^x(2x) = xe^x(x+2).$$

The next result is a generalization of Theorem (6.21).

Theorem 6.22

If $u = g(x)$ and g is differentiable, then

$$\frac{d}{dx}(e^u) = e^u \frac{du}{dx}.$$

PROOF Letting $y = e^u$ with $u = g(x)$, and using the chain rule and Theorem (6.21), we have

$$\frac{d}{dx}(e^u) = \frac{dy}{dx} = \frac{dy}{du}\frac{du}{dx} = e^u \frac{du}{dx}. \quad \blacksquare$$

If $u = x$, then Theorem (6.22) reduces to (6.21).

EXAMPLE 2 If $y = e^{\sqrt{x^2+1}}$, find dy/dx.

SOLUTION By Theorem (6.22),

$$\frac{dy}{dx} = \frac{d}{dx}\left(e^{\sqrt{x^2+1}}\right) = e^{\sqrt{x^2+1}}\frac{d}{dx}\left(\sqrt{x^2+1}\right)$$
$$= e^{\sqrt{x^2+1}}\frac{d}{dx}((x^2+1)^{1/2})$$
$$= e^{\sqrt{x^2+1}}(\tfrac{1}{2})(x^2+1)^{-1/2}(2x)$$
$$= e^{\sqrt{x^2+1}} \cdot \frac{x}{\sqrt{x^2+1}}$$
$$= \frac{xe^{\sqrt{x^2+1}}}{\sqrt{x^2+1}}.$$

EXAMPLE 3 The function f defined by $f(x) = e^{-x^2/2}$ occurs in the branch of mathematics called *probability*. Find the local extrema of f, discuss concavity, find the points of inflection, and sketch the graph of f.

SOLUTION   By Theorem (6.22),

$$f'(x) = e^{-x^2/2} \frac{d}{dx} \left(-\frac{x^2}{2} \right) = e^{-x^2/2} \left(-\frac{2x}{2} \right) = -xe^{-x^2/2}.$$

Since $e^{-x^2/2}$ is always positive, the only critical number of f is 0. If $x < 0$, then $f'(x) > 0$, and if $x > 0$, then $f'(x) < 0$. It follows from the first derivative test that f has a local maximum at 0. The maximum value is $f(0) = e^{-0} = 1$.

Applying the product rule to $f'(x)$ yields

$$\begin{aligned} f''(x) &= -x \frac{d}{dx} \left(e^{-x^2/2} \right) + e^{-x^2/2} \frac{d}{dx}(-x) \\ &= -xe^{-x^2/2}(-2x/2) - e^{-x^2/2} \\ &= e^{-x^2/2}(x^2 - 1), \end{aligned}$$

and hence the second derivative is zero at -1 and 1. If $-1 < x < 1$, then $f''(x) < 0$ and, by (3.18), the graph of f is concave downward in the open interval $(-1, 1)$. If $x < -1$ or $x > 1$, then $f''(x) > 0$ and, therefore, the graph is concave upward throughout the infinite intervals $(-\infty, -1)$ and $(1, \infty)$. Consequently, $P(-1, e^{-1/2})$ and $Q(1, e^{-1/2})$ are points of inflection. From the expression

$$f(x) = \frac{1}{e^{x^2/2}}$$

it is evident that as x increases numerically, $f(x)$ approaches 0. We can prove that $\lim_{x \to \infty} f(x) = 0$ and $\lim_{x \to -\infty} f(x) = 0$—that is, the x-axis is a horizontal asymptote. The graph of f is sketched in Figure 6.17.

Figure 6.17

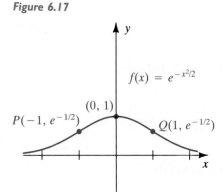

Exponential functions play an important role in the field of *radiotherapy*, the treatment of tumors by radiation. The fraction of cells in a tumor that survive a treatment, called the *surviving fraction*, depends not only on the energy and nature of the radiation, but also on the depth, size, and characteristics of the tumor itself. The exposure to radiation may be thought of as a number of potentially damaging events, where only one *hit* is required to kill a tumor cell. Suppose that each cell has exactly one *target* that must be hit. If k denotes the average target size of a tumor cell and if x is the number of damaging events (the *dose*), then the surviving fraction $f(x)$ is given by

$$f(x) = e^{-kx}$$

and is called the *one-target–one-hit surviving fraction*.

Suppose next that each cell has n targets and that hitting each target once results in the death of a cell. In this case, the *n-target–one-hit surviving fraction* is given by

$$f(x) = 1 - (1 - e^{-kx})^n.$$

In the next example, we examine the case where $n = 2$.

EXAMPLE 4 If each cell of a tumor has two targets, then the two-target–one-hit surviving fraction is given by

$$f(x) = 1 - (1 - e^{-kx})^2,$$

where k is the average size of a cell. Analyze the graph of f to determine what effect increasing the dosage x has on decreasing the surviving fraction of tumor cells.

SOLUTION First note that if $x = 0$, then $f(0) = 1$; that is, if there is no dose, then all cells survive. Differentiating, we obtain

$$f'(x) = 0 - 2(1 - e^{-kx})\frac{d}{dx}(1 - e^{-kx})$$

$$= -2(1 - e^{-kx})(ke^{-kx})$$

$$= -2ke^{-kx}(1 - e^{-kx}).$$

Since $f'(x) < 0$ for every $x > 0$ and $f'(0) = 0$, the function f is decreasing and the graph has a horizontal tangent line at the point $(0, 1)$. We may verify that the second derivative is

$$f''(x) = 2k^2 e^{-kx}(1 - 2e^{-kx}).$$

We see that $f''(x) = 0$ if $1 - 2e^{-kx} = 0$—that is, if $e^{-kx} = \frac{1}{2}$, or, equivalently, $-kx = \ln\frac{1}{2} = -\ln 2$. We thus obtain

$$x = \frac{1}{k}\ln 2.$$

It can be verified that if $0 \le x < (1/k)\ln 2$, then $f''(x) < 0$, and hence the graph is concave downward. If $x > (1/k)\ln 2$, then $f''(x) > 0$, and the graph is concave upward. The implication is that a point of inflection exists at x-coordinate $(1/k)\ln 2$. The y-coordinate of this point is

$$f\left(\frac{1}{k}\ln 2\right) = 1 - (1 - e^{-\ln 2})^2$$

$$= 1 - (1 - \tfrac{1}{2})^2 = \tfrac{3}{4}.$$

The graph is sketched in Figure 6.18 for the case $k = 1$. The *shoulder* on the curve near the point $(0, 1)$ represents the threshold nature of the treatment—that is, a small dose results in very little tumor elimination. Note that if x is large, then an increase in dosage has little effect on the surviving fraction. To determine the ideal dose that should be administered to a given patient, specialists in radiation therapy must also take into account the number of healthy cells that are killed during a treatment.

Figure 6.18

Surviving fraction of tumor cells after a radiation treatment

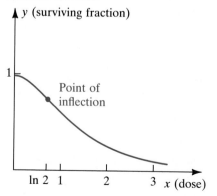

EXERCISES 6.3

Exer. 1–30: Find $f'(x)$ if $f(x)$ equals the given expression.

1 e^{-5x}

2 e^{3x}

3 e^{3x^2}

4 e^{1-x^3}

5 $\sqrt{1 + e^{2x}}$

6 $1/(e^x + 1)$

7 $e^{\sqrt{x+1}}$

8 xe^{-x}

9 $x^2 e^{-2x}$

10 $\sqrt{e^{2x} + 2x}$

11 $e^x/(x^2 + 1)$

12 $x/e^{(x^2)}$

13 $(e^{4x} - 5)^3$

14 $(e^{3x} - e^{-3x})^4$

15 $e^{1/x} + (1/e^x)$

16 $e^{\sqrt{x}} + \sqrt{e^x}$

17 $\dfrac{e^x - e^{-x}}{e^x + e^{-x}}$

18 $e^{x \ln x}$

19 $e^{-2x} \ln x$

20 $\ln e^x$

21 $\sin e^{5x}$

22 $e^{\sin 5x}$

23 $\ln \cos e^{-x}$

24 $e^{-3x} \cos 3x$

25 $e^{3x} \tan \sqrt{x}$

26 $\sec e^{-2x}$

27 $\sec^2(e^{-4x})$

28 $e^{-x} \tan^2 x$

29 $xe^{\cot x}$

30 $\ln(\csc e^{3x})$

Exer. 31–34: Use implicit differentiation to find y'.

31 $e^{xy} - x^3 + 3y^2 = 11$

32 $xe^y + 2x - \ln(y + 1) = 3$

33 $e^x \cot y = xe^{2y}$

34 $e^x \cos y = xe^y$

35 Find an equation of the tangent line to the graph of $y = (x - 1)e^x + 3 \ln x + 2$ at the point $P(1, 2)$.

36 Find an equation of the tangent line to the graph of $y = x - e^{-x}$ that is parallel to the line $6x - 2y = 7$.

Exer. 37–42: Find the local extrema of f. Determine where f is increasing or is decreasing, discuss concavity, find the points of inflection, and sketch the graph of f.

37 $f(x) = xe^x$

38 $f(x) = x^2 e^{-2x}$

39 $f(x) = e^{1/x}$

40 $f(x) = xe^{-x}$

41 $f(x) = x \ln x$

42 $f(x) = (1 - \ln x)^2$

43 A radioactive substance decays according to the formula $q(t) = q_0 e^{-ct}$, where q_0 is the initial amount of the substance, c is a positive constant, and $q(t)$ is the amount remaining after time t. Show that the rate at which the substance decays is proportional to $q(t)$.

44 The current $I(t)$ at time t in an electrical circuit is given by $I(t) = I_0 e^{-Rt/L}$, where R is the resistance, L is the inductance, and I_0 is the current at time $t = 0$. Show that the rate of change of the current at any time t is proportional to $I(t)$.

45 If a drug is injected into the bloodstream, then its concentration t minutes later is given by

$$C(t) = \frac{k}{a - b}(e^{-bt} - e^{-at})$$

for positive constants a, b, and k.

(a) At what time does the maximum concentration occur?

(b) What can be said about the concentration after a long period of time?

46 If a beam of light that has intensity k is projected vertically downward into water, then its intensity $I(x)$ at a depth of x meters is $I(x) = ke^{-1.4x}$.

(a) At what rate is the intensity changing with respect to depth at 1 m? 5 m? 10 m?

(b) At what depth is the intensity one-half its value at the surface? one-tenth its value?

47 The *Jenss model* is generally regarded as the most accurate formula for predicting the height of a preschooler. If $h(x)$ denotes the height (in centimeters) at age x (in years) for $\frac{1}{4} \le x \le 6$, then $h(x)$ can be approximated by

$$h(x) = 79.041 + 6.39x - e^{3.261 - 0.993x}.$$

(Compare with Example 7 of Section 6.2.)

(a) Predict the height and the rate of growth when a child reaches the age of 1.

(b) When is the rate of growth largest, and when is it smallest?

48 For a population of female African elephants, the weight $W(t)$ (in kilograms) at age t (in years) may be approximated by a *von Bertanlanffy growth function W* such that

$$W(t) = 2600(1 - 0.51e^{-0.075t})^3.$$

(a) Approximate the weight and the rate of growth of a newborn.

(b) Assuming that an adult female weighs 1800 kg, estimate her age and her rate of growth at present.

(c) Find and interpret $\lim_{t \to \infty} W(t)$.

(d) Show that the rate of growth is largest between the ages of 5 and 6.

49 Gamma distributions, which are important in traffic control studies and probability theory, are determined by $f(x) = cx^n e^{-ax}$ for $x > 0$, a positive integer n, a positive constant a, and $c = a^{n+1}/n!$. Shown in the figure are graphs corresponding to $a = 1$ for $n = 2, 3$, and 4.

(a) Show that f has exactly one local maximum.

(b) If $n = 4$, determine where $f(x)$ is increasing most rapidly.

Exercise 49

50 The relative number of gas molecules in a container that travel at a velocity of v cm/sec can be computed by means of the *Maxwell–Boltzmann speed distribution*, $F(v) = cv^2 e^{-mv^2/(2kT)}$, where T is the temperature (in °K), m is the mass of a molecule, and c and k are positive constants. Show that the maximum value of F occurs when $v = \sqrt{2kT/m}$.

51 An *urban density model* is a formula that relates the population density (in number per square mile) to the distance r (in miles) from the center of the city. The formula $D = ae^{-br+cr^2}$, where a, b, and c are positive constants, has been found to be appropriate for certain cities. Determine the shape of the graph for $r \geq 0$.

52 The effect of light on the rate of photosynthesis can be described by

$$f(x) = x^a e^{(a/b)(1-x^b)}$$

for $x > 0$ and positive constants a and b.

(a) Show that f has a maximum at $x = 1$.

(b) Conclude that if $x_0 > 0$ and $y_0 > 0$, then $g(x) = y_0 f(x/x_0)$ has a maximum $g(x_0) = y_0$.

53 The rate R at which a tumor grows is related to its size x by the equation $R = rx \ln(K/x)$, where r and K are positive constants. Show that the tumor is growing most rapidly when $x = e^{-1}K$.

54 If p denotes the selling price (in dollars) of a commodity and x is the corresponding demand (in number sold per day), then the relationship between p and x may be given by $p = p_0 e^{-ax}$ for positive constants p_0 and a. Suppose $p = 300e^{-0.02x}$. Find the selling price that will maximize daily revenues (see page 332).

55 In statistics, the probability density function for the normal distribution is defined by

$$f(x) = \frac{1}{\sigma\sqrt{2\pi}} e^{-z^2/2} \quad \text{with} \quad z = \frac{x - \mu}{\sigma}$$

for real numbers μ and $\sigma > 0$ (μ is the *mean* and σ^2 is the *variance* of the distribution). Find the local extrema of f, and determine where f is increasing or is decreasing. Discuss concavity, find points of inflection, find $\lim_{x \to \infty} f(x)$ and $\lim_{x \to -\infty} f(x)$, and sketch the graph of f (see Example 3).

C 56 The integral $\int_a^b e^{-x^2}\, dx$ has applications in statistics. Use the trapezoidal rule, with $n = 10$, to approximate this integral if $a = 0$ and $b = 1$.

C Exer. 57–60: Graph f on the given interval. (a) Approximate the range of f on this interval. (b) Identify x-intercepts, y-intercepts, and relative extrema in this viewing window, if any. Estimate these features to two decimal places.

57 $f(x) = 1 - (1 - e^{-x})^3$;   [0, 4]
(This function is the *three-target–one-hit surviving fraction*, with $k = 1$; see Example 4.)

58 $f(x) = \dfrac{150}{1 + 5e^{-0.4x}}$;   [−5, 25]

59 $f(x) = 6e^{-0.39x} \cos(1.87x)$;   [−1, 8]

60 $f(x) = \ln(8 + e^x)$;   [−4, 15]

C Exer. 61–64: Use Newton's method or a solving routine to approximate the real root(s) of the equation to four decimal places. Use a graphing utility to ensure that all roots are found.

61 $e^{-x} = x$

62 $e^{3x} - 5e^{2x} + 7e^x = 2$

63 $xe^x = 4$

64 $0.2e^{0.6x} + 1.3e^{-0.1x} = 4 - 2.8x^2$

65 Nerve impulses in the human body travel along nerve fibers that consist of an *axon*, which transports the impulse, and an insulating coating surrounding the axon, called the *myelin sheath* (see figure). The nerve fiber is similar to an insulated cylindrical cable, for which the velocity v of an impulse is given by $v = -k(r/R)^2 \ln(r/R)$, where r is the radius of the cable and R is the insulation radius. Find the value of r/R that maximizes v. (In most nerve fibers, $r/R \approx 0.6$.)

Exercise 65

Myelin sheath

R

r

Axon

6.4 INTEGRATION USING NATURAL LOGARITHM AND EXPONENTIAL FUNCTIONS

We may use differentiation formulas for ln to obtain formulas for integration. In particular, by Theorem (6.11),

$$\frac{d}{dx}\left(\ln|g(x)|\right) = \frac{1}{g(x)}g'(x),$$

which gives us the integration formula

$$\int \frac{1}{g(x)}g'(x)\,dx = \ln|g(x)| + C.$$

This result is restated in the next theorem in terms of the variable u.

Theorem 6.23

If $u = g(x) \neq 0$ and g is differentiable, then

$$\int \frac{1}{u}\,du = \ln|u| + C.$$

Of course, if $u > 0$, then the absolute value sign may be deleted. A special case of Theorem (6.23) is

$$\int \frac{1}{x}\,dx = \ln|x| + C.$$

EXAMPLE 1 Evaluate $\displaystyle\int \frac{x}{3x^2 - 5}\,dx.$

SOLUTION Rewriting the integral as

$$\int \frac{x}{3x^2 - 5}\,dx = \int \frac{1}{3x^2 - 5}x\,dx$$

suggests that we use Theorem (6.23) with $u = 3x^2 - 5$. Thus, we make the substitution

$$u = 3x^2 - 5, \qquad du = 6x\,dx.$$

Introducing a factor 6 in the integrand and using Theorem (6.23) yields

$$\int \frac{x}{3x^2 - 5}\, dx = \frac{1}{6}\int \frac{1}{3x^2 - 5} 6x\, dx = \frac{1}{6}\int \frac{1}{u}\, du$$

$$= \frac{1}{6}\ln|u| + C = \frac{1}{6}\ln|3x^2 - 5| + C.$$

Another technique is to replace the expression $x\, dx$ in the integral by $\frac{1}{6}\, du$ and then integrate.

EXAMPLE ■ 2   Evaluate $\displaystyle\int_2^4 \frac{1}{9 - 2x}\, dx$.

SOLUTION   Since $1/(9 - 2x)$ is continuous on $[2, 4]$, the definite integral exists. One method of evaluation consists of using an indefinite integral to find an antiderivative of $1/(9 - 2x)$. We let

$$u = 9 - 2x, \qquad du = -2\, dx$$

and proceed as follows:

$$\int \frac{1}{9 - 2x}\, dx = -\frac{1}{2}\int \frac{1}{9 - 2x}(-2)\, dx$$

$$= -\frac{1}{2}\int \frac{1}{u}\, du = -\frac{1}{2}\ln|u| + C$$

$$= -\frac{1}{2}\ln|9 - 2x| + C$$

Applying the fundamental theorem of calculus yields

$$\int_2^4 \frac{1}{9 - 2x}\, dx = -\frac{1}{2}\Big[\ln|9 - 2x|\Big]_2^4$$

$$= -\frac{1}{2}(\ln 1 - \ln 5) = \frac{1}{2}\ln 5.$$

Another method is to use the same substitution in the *definite* integral and change the limits of integration. Since $u = 9 - 2x$, we obtain the following:

(i) If $x = 2$,   then $u = 5$.
(ii) If $x = 4$,   then $u = 1$.

Thus,

$$\int_2^4 \frac{1}{9 - 2x}\, dx = -\frac{1}{2}\int_2^4 \frac{1}{9 - 2x}(-2)\, dx$$

$$= -\frac{1}{2}\int_5^1 \frac{1}{u}\, du = -\frac{1}{2}\Big[\ln|u|\Big]_5^1$$

$$= -\frac{1}{2}(\ln 1 - \ln 5) = \frac{1}{2}\ln 5.$$

EXAMPLE ■ 3   Evaluate $\int \dfrac{\sqrt{\ln x}}{x}\, dx$.

SOLUTION   Two possible substitutions are $u = \sqrt{\ln x}$ and $u = \ln x$. If we use

$$u = \ln x, \qquad du = \frac{1}{x}\, dx,$$

then

$$\int \frac{\sqrt{\ln x}}{x}\, dx = \int \sqrt{\ln x} \cdot \frac{1}{x}\, dx = \int u^{1/2}\, du = \frac{u^{3/2}}{3/2} + C$$

$$= \tfrac{2}{3}(\ln x)^{3/2} + C.$$

The substitution $u = \sqrt{\ln x}$ could also be used; however, the algebraic manipulations would be somewhat more involved.

The derivative formula $(d/dx)(e^{g(x)}) = e^{g(x)}g'(x)$ gives us the following integration formula for the natural exponential function:

$$\int e^{g(x)} g'(x)\, dx = e^{g(x)} + C$$

This result is restated in the next theorem in terms of the variable u.

Theorem   6.24

If $u = g(x)$ and g is differentiable, then

$$\int e^u\, du = e^u + C.$$

As a special case of Theorem (6.24), if $u = x$, then

$$\int e^x\, dx = e^x + C.$$

EXAMPLE ■ 4   Evaluate:

(a) $\displaystyle\int \frac{e^{3/x}}{x^2}\, dx$ **(b)** $\displaystyle\int_1^2 \frac{e^{3/x}}{x^2}\, dx$

SOLUTION

(a) Rewriting the integral as

$$\int \frac{e^{3/x}}{x^2}\, dx = \int e^{3/x} \frac{1}{x^2}\, dx$$

suggests that we use Theorem (6.24) with $u = 3/x$. Thus, we make the substitution

$$u = \frac{3}{x}, \qquad du = -\frac{3}{x^2}\, dx.$$

The integrand may be written in the form of Theorem (6.24) by introducing the factor -3. Doing this and compensating by multiplying the integral by $-\frac{1}{3}$, we obtain

$$\int \frac{e^{3/x}}{x^2}\, dx = -\frac{1}{3} \int e^{3/x} \left(-\frac{3}{x^2} \right) dx$$

$$= -\frac{1}{3} \int e^u\, du$$

$$= -\frac{1}{3} e^u + C$$

$$= -\frac{1}{3} e^{3/x} + C.$$

(b) Using the antiderivative found in part (a) and applying the fundamental theorem of calculus yields

$$\int_1^2 \frac{e^{3/x}}{x^2}\, dx = -\frac{1}{3} \left[e^{3/x} \right]_1^2$$

$$= -\frac{1}{3}(e^{3/2} - e^3) \approx 5.2.$$

We can also evaluate the integral by using the method of substitution. As in part (a), we let $u = 3/x$, $du = (-3/x^2)\, dx$, and we note that if $x = 1$, then $u = 3$, and if $x = 2$, then $u = \frac{3}{2}$. Consequently,

$$\int_1^2 \frac{e^{3/x}}{x^2}\, dx = -\frac{1}{3} \int_1^2 e^{3/x} \left(-\frac{3}{x^2} \right) dx$$

$$= -\frac{1}{3} \int_3^{3/2} e^u\, du$$

$$= -\frac{1}{3} \left[e^u \right]_3^{3/2} = -\frac{1}{3}(e^{3/2} - e^3) \approx 5.2.$$

The integral $\int e^{ax}\, dx$, with $a \neq 0$, occurs frequently. We can show that

$$\int e^{ax}\, dx = \frac{1}{a} e^{ax} + C$$

either by using Theorem (6.24) or by showing that $(1/a)e^{ax}$ is an antiderivative of e^{ax}.

ILLUSTRATION

- $\displaystyle\int e^{3x}\, dx = \frac{1}{3}e^{3x} + C$ ∎ $\displaystyle\int e^{-5x}\, dx = -\frac{1}{5}e^{-5x} + C$

- $\displaystyle\int e^{-x}\, dx = -e^{-x} + C$

In the next example, we solve a differential equation that contains exponential expressions.

EXAMPLE ■ 5   Solve the differential equation

$$\frac{dy}{dx} = 3e^{2x} + 6e^{-3x}$$

subject to the initial condition $y = 4$ if $x = 0$.

SOLUTION   As in Example 6 of Section 4.1, we may multiply both sides of the equation by dx and then integrate as follows:

$$dy = (3e^{2x} + 6e^{-3x})\, dx$$

$$\int dy = \int (3e^{2x} + 6e^{-3x})\, dx = 3\int e^{2x}\, dx + 6\int e^{-3x}\, dx$$

$$y = 3(\tfrac{1}{2})e^{2x} + 6(-\tfrac{1}{3})e^{-3x} + C$$

$$= \tfrac{3}{2}e^{2x} - 2e^{-3x} + C$$

Using the initial condition $y = 4$ if $x = 0$ gives us

$$4 = \tfrac{3}{2}e^{0} - 2e^{0} + C = \tfrac{3}{2} - 2 + C.$$

Hence, $C = 4 - \tfrac{3}{2} + 2 = \tfrac{9}{2}$, and the solution of the differential equation is

$$y = \tfrac{3}{2}e^{2x} - 2e^{-3x} + \tfrac{9}{2}.$$

EXAMPLE ■ 6   Find the area of the region bounded by the graphs of the equations $y = e^{x}$, $y = \sqrt{x}$, $x = 0$, and $x = 1$.

SOLUTION   The region and a typical rectangle of the type considered in Chapter 5 are shown in Figure 6.19. As usual, we list the following:

$$\text{width of rectangle:}\quad dx$$
$$\text{length of rectangle:}\quad e^{x} - \sqrt{x}$$
$$\text{area of rectangle:}\quad (e^{x} - \sqrt{x})\, dx$$

We next take a limit of sums of these rectangular areas by applying the operator $\int_{0}^{1}$:

$$\int_{0}^{1} (e^{x} - \sqrt{x})\, dx = \int_{0}^{1} (e^{x} - x^{1/2})\, dx$$

$$= \left[e^{x} - \tfrac{2}{3}x^{3/2} \right]_{0}^{1} = e - \tfrac{5}{3} \approx 1.05$$

Figure 6.19

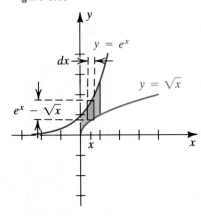

In Chapter 4, we obtained integration formulas for the sine and cosine functions. We were unable to consider the remaining four trigonometric functions at that time because, as indicated in the next theorem, their integrals are logarithmic functions. In the theorem, we assume that $u = g(x)$, with g differentiable whenever the function is defined.

Theorem 6.25

(i) $\displaystyle\int \tan u \ du = -\ln |\cos u| + C$

(ii) $\displaystyle\int \cot u \ du = \ln |\sin u| + C$

(iii) $\displaystyle\int \sec u \ du = \ln |\sec u + \tan u| + C$

(iv) $\displaystyle\int \csc u \ du = \ln |\csc u - \cot u| + C$

PROOF   It is sufficient to consider the case $u = x$, since the formulas for $u = g(x)$ then follow from the Chain Rule, Theorem (4.7).

To find $\int \tan x \ dx$, we first use a trigonometric identity to express $\tan x$ in terms of $\sin x$ and $\cos x$ as follows:

$$\int \tan x \ dx = \int \frac{\sin x}{\cos x} \ dx = \int \frac{1}{\cos x} \sin x \ dx$$

The form of the integrand on the right suggests that we make the substitution

$$v = \cos x, \qquad dv = -\sin x \ dx.$$

This gives us

$$\int \tan x \ dx = -\int \frac{1}{v} \ dv.$$

If $\cos x \neq 0$, then by Theorem (6.11)(ii),

$$\int \tan x \ dx = -\ln |v| + C = -\ln |\cos x| + C.$$

A formula for $\int \cot x \ dx$ may be obtained in similar fashion by first writing $\cot x = (\cos x)/(\sin x)$.

To find a formula for $\int \sec x \ dx$, we begin as follows:

$$\int \sec x \ dx = \int \sec x \frac{\sec x + \tan x}{\sec x + \tan x} \ dx$$

$$= \int \frac{\sec^2 x + \sec x \tan x}{\sec x + \tan x} \ dx$$

$$= \int \frac{1}{\sec x + \tan x} (\sec x \tan x + \sec^2 x) \ dx$$

Using the substitution

$$v = \sec x + \tan x, \qquad dv = (\sec x \tan x + \sec^2 x) \ dx$$

gives us

$$\int \sec x \, dx = \int \frac{1}{v} \, dv$$

$$= \ln |v| + C$$

$$= \ln |\sec x + \tan x| + C.$$

A similar proof can be given for (iv). ∎

If we use $\cos u = 1/\sec u$, $\sin u = 1/\csc u$, and $\ln(1/v) = -\ln v$, then formulas (i) and (ii) of Theorem (6.25) can be written as follows:

$$\int \tan u \, du = \ln |\sec u| + C$$

$$\int \cot u \, du = -\ln |\csc u| + C$$

EXAMPLE ■ 7   Evaluate $\int x \cot x^2 \, dx$.

SOLUTION   To obtain the form $\int \cot u \; du$, we make the substitution

$$u = x^2, \qquad du = 2x \, dx.$$

We next introduce the factor 2 in the integrand as follows:

$$\int x \cot x^2 \, dx = \tfrac{1}{2} \int (\cot x^2) 2x \, dx$$

Since $u = x^2$ and $du = 2x \, dx$,

$$\int x \cot x^2 \, dx = \tfrac{1}{2} \int \cot u \, du = \tfrac{1}{2} \ln |\sin u| + C$$

$$= \tfrac{1}{2} \ln |\sin x^2| + C.$$

EXAMPLE ■ 8   Evaluate $\int_0^{\pi/2} \tan \frac{x}{2} \, dx$.

SOLUTION   We make the substitution

$$u = \frac{x}{2}, \qquad du = \frac{1}{2} \, dx$$

and note that $u = 0$ if $x = 0$, and $u = \pi/4$ if $x = \pi/2$. Thus,

$$\int_0^{\pi/2} \tan \frac{x}{2} \, dx = 2 \int_0^{\pi/2} \tan \frac{x}{2} \cdot \frac{1}{2} \, dx$$

$$= 2 \int_0^{\pi/4} \tan u \, du = 2 \left[\ln \sec u \right]_0^{\pi/4} .$$

In this case, we may drop the absolute value sign given in Theorem (6.25)(iii), because $\sec u$ is positive if u is between 0 and $\pi/4$. Since $\ln \sec(\pi/4) = \ln \sqrt{2} = \frac{1}{2} \ln 2$ and $\ln \sec 0 = \ln 1 = 0$, it follows that

$$\int_0^{\pi/2} \tan \frac{x}{2} \, dx = 2 \cdot \frac{1}{2} \ln 2 = \ln 2 \approx 0.69.$$

EXAMPLE ■ 9 Evaluate $\int e^{2x} \sec e^{2x} \, dx$.

SOLUTION We let

$$u = e^{2x}, \qquad du = 2e^{2x} \, dx$$

and proceed as follows:

$$\int e^{2x} \sec e^{2x} \, dx = \frac{1}{2} \int (\sec e^{2x}) 2 e^{2x} \, dx$$

$$= \frac{1}{2} \int \sec u \, du$$

$$= \frac{1}{2} \ln |\sec u + \tan u| + C$$

$$= \frac{1}{2} \ln |\sec e^{2x} + \tan e^{2x}| + C$$

EXAMPLE ■ 10 Evaluate $\int (\csc x - 1)^2 \, dx$.

SOLUTION

$$\int (\csc x - 1)^2 \, dx = \int (\csc^2 x - 2 \csc x + 1) \, dx$$

$$= \int \csc^2 x \, dx - 2 \int \csc x \, dx + \int dx$$

$$= -\cot x - 2 \ln |\csc x - \cot x| + x + C.$$

We shall discuss additional methods for integrating trigonometric expressions in Chapter 7.

EXERCISES 6.4

Exer. 1–36: Evaluate the integral.

1 (a) $\displaystyle \int \frac{1}{2x+7} \, dx$
  (b) $\displaystyle \int_{-2}^{1} \frac{1}{2x+7} \, dx$

3 (a) $\displaystyle \int \frac{4x}{x^2-9} \, dx$
  (b) $\displaystyle \int_{1}^{2} \frac{4x}{x^2-9} \, dx$

2 (a) $\displaystyle \int \frac{1}{4-5x} \, dx$
  (b) $\displaystyle \int_{-1}^{0} \frac{1}{4-5x} \, dx$

4 (a) $\displaystyle \int \frac{3x}{x^2+4} \, dx$
  (b) $\displaystyle \int_{1}^{2} \frac{3x}{x^2+4} \, dx$

5 (a) $\displaystyle\int e^{-4x}\,dx$ (b) $\displaystyle\int_1^3 e^{-4x}\,dx$

6 (a) $\displaystyle\int x^2 e^{3x^3}\,dx$ (b) $\displaystyle\int_1^2 x^2 e^{3x^3}\,dx$

7 (a) $\displaystyle\int \tan 2x\,dx$ (b) $\displaystyle\int_0^{\pi/8} \tan 2x\,dx$

8 (a) $\displaystyle\int \cot \tfrac{1}{3}x\,dx$ (b) $\displaystyle\int_{3\pi/2}^{9\pi/4} \cot \tfrac{1}{3}x\,dx$

9 (a) $\displaystyle\int \csc \tfrac{1}{2}x\,dx$ (b) $\displaystyle\int_{\pi}^{5\pi/3} \csc \tfrac{1}{2}x\,dx$

10 (a) $\displaystyle\int \sec 3x\,dx$ (b) $\displaystyle\int_0^{\pi/12} \sec 3x\,dx$

11 $\displaystyle\int \frac{x-2}{x^2-4x+9}\,dx$ 12 $\displaystyle\int \frac{x^3}{x^4-5}\,dx$

13 $\displaystyle\int \frac{(x+2)^2}{x}\,dx$ 14 $\displaystyle\int \frac{(2+\ln x)^{10}}{x}\,dx$

15 $\displaystyle\int \frac{\ln x}{x}\,dx$ 16 $\displaystyle\int \frac{1}{x(\ln x)^2}\,dx$

17 $\displaystyle\int (x+e^{5x})\,dx$ 18 $\displaystyle\int \frac{e^{\sqrt{x}}}{\sqrt{x}}\,dx$

19 $\displaystyle\int \frac{3\sin x}{1+2\cos x}\,dx$ 20 $\displaystyle\int \frac{\sec^2 x}{1+\tan x}\,dx$

21 $\displaystyle\int \frac{(e^x+1)^2}{e^x}\,dx$ 22 $\displaystyle\int \frac{e^x}{(e^x+1)^2}\,dx$

23 $\displaystyle\int \frac{e^x-e^{-x}}{e^x+e^{-x}}\,dx$ 24 $\displaystyle\int \frac{e^x}{e^x+1}\,dx$

25 $\displaystyle\int \frac{\cot \sqrt[3]{x}}{\sqrt[3]{x^2}}\,dx$ 26 $\displaystyle\int e^x(1+\tan e^x)\,dx$

27 $\displaystyle\int \frac{1}{\cos 2x}\,dx$ 28 $\displaystyle\int (x+\csc 8x)\,dx$

29 $\displaystyle\int \frac{\tan e^{-3x}}{e^{3x}}\,dx$ 30 $\displaystyle\int e^{\cos x}\sin x\,dx$

31 $\displaystyle\int \frac{\cos^2 x}{\sin x}\,dx$ 32 $\displaystyle\int \frac{\tan^2 2x}{\sec 2x}\,dx$

33 $\displaystyle\int \frac{\cos x \sin x}{\cos^2 x - 1}\,dx$ 34 $\displaystyle\int (\tan 3x + \sec 3x)\,dx$

35 $\displaystyle\int (1+\sec x)^2\,dx$ 36 $\displaystyle\int \csc x(1-\csc x)\,dx$

Exer. 37–38: Find the area of the region bounded by the graphs of the given equations.

37 $y = e^{2x}$, $y = 0$, $x = 0$, $x = \ln 3$

38 $y = 2\tan x$, $y = 0$, $x = 0$, $x = \pi/4$

Exer. 39–40: Find the volume of the solid generated if the region bounded by the graphs of the equations is revolved about the indicated axis.

39 $y = e^{-x^2}$, $x = 0$, $x = 1$, $y = 0$;   y-axis

40 $y = \sec x$, $x = -\pi/3$, $x = \pi/3$, $y = 0$;   x-axis

Exer. 41–44: Solve the differential equation subject to the given conditions.

41 $y' = 4e^{2x} + 3e^{-2x}$; $y = 4$ if $x = 0$

42 $y' = 3e^{4x} - 8e^{-2x}$; $y = -2$ if $x = 0$

43 $y'' = 3e^{-x}$; $y = -1$ and $y' = 1$ if $x = 0$

44 $y'' = 6e^{2x}$; $y = -3$ and $y' = 2$ if $x = 0$

Exer. 45–46: A nonnegative function f defined on a closed interval $[a,b]$ is called a *probability density function* if $\int_a^b f(x)\,dx = 1$. Determine c so that the resulting function is a probability density function.

45 $f(x) = \dfrac{cx}{x^2+4}$ for $0 \le x \le 3$

46 $f(x) = cxe^{-x^2}$ for $0 \le x \le 10$

47 A culture of bacteria is growing at a rate of $3e^{0.2t}$ per hour, with t in hours and $0 \le t \le 20$.

(a) How many new bacteria will be in the culture after the first five hours?

(b) How many new bacteria are introduced in the sixth through the fourteenth hours?

(c) For approximately what value of t will the culture contain 150 new bacteria?

48 If a savings bond is purchased for $500 with interest compounded continuously at 7% per year, then after t years the bond will be worth $500e^{0.07t}$ dollars.

(a) Approximately when will the bond be worth $1000?

(b) Approximately when will the value of the bond be growing at a rate of $50 per year?

49 The specific heat c of a metal such as silver is constant at temperatures T above $200\,°\text{K}$. If the temperature of the metal increases from T_1 to T_2, the area under the curve $y = c/T$ from T_1 to T_2 is called the *change in entropy* ΔS, a measurement of the increased molecular disorder of the system. Express ΔS in terms of T_1 and T_2.

50 The 1952 earthquake in Assam had a magnitude of 8.7 on the Richter scale—the largest ever recorded. (The October 1989 San Francisco earthquake had a magnitude of 7.1.) Seismologists have determined that if the largest earthquake in a given year has magnitude R, then the energy E (in joules) released by all earthquakes

in that year can be estimated by using the formula

$$E = 9.13 \times 10^{12} \int_0^R e^{1.25x}\, dx.$$

Find E if $R = 8$.

51 In a circuit containing a 12-volt battery, a resistor, and a capacitor, the current $I(t)$ at time t is predicted to be $I(t) = 10e^{-4t}$ amperes. If $Q(t)$ is the charge (in coulombs) on the capacitor, then $I = dQ/dt$.

   (a) If $Q(0) = 0$, find $Q(t)$.

   (b) Find the charge on the capacitor after a long period of time.

52 A country that presently has coal reserves of 50 million tons used 6.5 million tons last year. On the basis of population projections, the rate of consumption R (in million tons/year) is expected to increase according to the formula $R = 6.5e^{0.02t}$, where t is the time in years. If the country uses only its own resources, estimate how many years the coal reserves will last.

53 A very small spherical particle (on the order of 5 microns in diameter) is projected into still air with an initial velocity of v_0 m/sec, but its velocity decreases because of drag forces. Its velocity after t seconds is given by $v(t) = v_0 e^{-t/k}$ for some positive constant k.

   (a) Express the distance that the particle travels as a function of t.

   (b) The *stopping distance* is the distance traveled by the particle before it comes to rest. Express the stopping distance in terms of v_0 and k.

54 If the temperature remains constant, the pressure p and the volume v of an expanding gas are related by the equation $pv = k$ for some constant k. Show that the work done if the gas expands from v_0 to v_1 is $k \ln(v_1/v_0)$. (*Hint:* See Example 5 of Section 5.6.)

c Exer. 55–58: Use a numerical integration method or routine to approximate the definite integral to four decimal places.

55 $\displaystyle\int_0^1 e^{-x^2}\, dx$

56 $\displaystyle\int_{-4}^8 e^{-x^2}\, dx$

57 $\displaystyle\int_{0.5}^{6.5} \frac{e^x}{x}\, dx$

58 $\displaystyle\int_0^3 \sqrt{x+1}\, e^x\, dx$

c **59** Approximate the area bounded by the graphs of $y = e^x$ and $y = 4 - x^2$.

c **60** Approximate the volume of the solid generated by revolving the graph of $y = e^x$, $-10 \le x \le 1$, about the x-axis.

c **61** Approximate the arc length of the part of the curve $y = e^x$ that lies inside the circle $x^2 + y^2 = 25$.

6.5 GENERAL EXPONENTIAL AND LOGARITHMIC FUNCTIONS

Throughout this section, a will denote a positive real number. Let us begin by defining a^x for every real number x. If the exponent is a *rational* number r, then applying Theorems (6.19)(ii) and (6.12)(iii) yields

$$a^r = e^{\ln a^r} = e^{r \ln a}.$$

This formula is the motivation for the following definition of a^x.

Definition of a^x 6.26

$$a^x = e^{x \ln a}$$

for every $a > 0$ and every real number x.

- $2^\pi = e^{\pi \ln 2} \approx e^{2.1775860903} \approx 8.82497782708$
- $(\frac{1}{2})^{\sqrt{3}} = e^{\sqrt{3}\ln(1/2)} \approx e^{-1.20056613385} \approx 0.301023743931$

If $f(x) = a^x$, then f is the **exponential function with base** a. Since e^x is positive for every x, so is a^x. To approximate values of a^x, we may use a calculator or refer to standard tables of logarithmic and exponential functions.

It is now possible to prove that the law of logarithms stated in Theorem (6.12)(iii) is also true for irrational exponents. Thus, if u is any *real* number, then, by Definition (6.26) and Theorem (6.19)(i),

$$\ln a^u = \ln e^{u \ln a} = u \ln a.$$

The next theorem states that properties of rational exponents from elementary algebra are also true for real exponents.

Laws of Exponents 6.27

Let $a > 0$ and $b > 0$. If u and v are any real numbers, then
$$a^u a^v = a^{u+v} \qquad (a^u)^v = a^{uv} \qquad (ab)^u = a^u b^u$$

$$\frac{a^u}{a^v} = a^{u-v} \qquad \left(\frac{a}{b}\right)^u = \frac{a^u}{b^u}$$

PROOF To show that $a^u a^v = a^{u+v}$, we use Definition (6.26) and Theorem (6.20)(i) as follows:

$$a^u a^v = e^{u \ln a} e^{v \ln a}$$
$$= e^{u \ln a + v \ln a}$$
$$= e^{(u+v) \ln a}$$
$$= a^{u+v}$$

To prove that $(a^u)^v = a^{uv}$, we first use Definition (6.26) with a^u in place of a and $v = x$ to write

$$(a^u)^v = e^{v \ln a^u}.$$

Using the fact that $\ln a^u = u \ln a$ and then applying Definition (6.26), we obtain

$$(a^u)^v = e^{vu \ln a} = a^{vu} = a^{uv}.$$

The proofs of the remaining laws are similar. ■

As usual, in part (ii) of the next theorem, $u = g(x)$, where g is differentiable.

Theorem   6.28

(i) $\dfrac{d}{dx}(a^x) = a^x \ln a$

(ii) $\dfrac{d}{dx}(a^u) = (a^u \ln a)\dfrac{du}{dx}$

PROOF   Applying Definition (6.26) and Theorem (6.22), we obtain

$$\frac{d}{dx}(a^x) = \frac{d}{dx}(e^{x \ln a})$$

$$= e^{x \ln a}\frac{d}{dx}(x \ln a)$$

$$= e^{x \ln a}(\ln a).$$

Since $e^{x \ln a} = a^x$, this gives us formula (i):

$$\frac{d}{dx}(a^x) = a^x \ln a$$

Formula (ii) follows from the chain rule.   ■

Note that if $a = e$, then Theorem (6.28)(i) reduces to Theorem (6.21), since $\ln e = 1$.

ILLUSTRATION

■ $\dfrac{d}{dx}(3^x) = 3^x \ln 3$

■ $\dfrac{d}{dx}(10^x) = 10^x \ln 10$

■ $\dfrac{d}{dx}(3^{\sqrt{x}}) = (3^{\sqrt{x}} \ln 3)\dfrac{d}{dx}(\sqrt{x}) = (3^{\sqrt{x}} \ln 3)\left(\dfrac{1}{2\sqrt{x}}\right) = \dfrac{3^{\sqrt{x}} \ln 3}{2\sqrt{x}}$

■ $\dfrac{d}{dx}(10^{\sin x}) = (10^{\sin x} \ln 10)\dfrac{d}{dx}(\sin x) = (10^{\sin x} \ln 10)\cos x$

If $a > 1$, then $\ln a > 0$ and, therefore, $(d/dx)(a^x) = a^x \ln a > 0$. Hence a^x is increasing on the interval $(-\infty, \infty)$ if $a > 1$.

If $0 < a < 1$, then $\ln a < 0$ and $(d/dx)(a^x) = a^x \ln a < 0$. Thus, a^x is decreasing for every x if $0 < a < 1$.

Figure 6.20   *Figure 6.21*

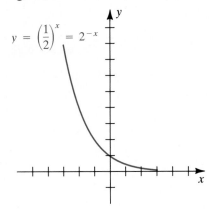

The graphs of $y = 2^x$ and $y = \left(\frac{1}{2}\right)^x = 2^{-x}$ are sketched in Figures 6.20 and 6.21. The graph of $y = a^x$ has the general shape illustrated in Figure 6.20 or 6.21 if $a > 1$ or $0 < a < 1$, respectively.

If $u = g(x)$, *it is important to distinguish between expressions of the form a^u and u^a.* To differentiate a^u, we use Theorem (6.28); for u^a, the power rule must be employed, as illustrated in the next example.

EXAMPLE ▪ 1   Find y' if $y = (x^2 + 1)^{10} + 10^{x^2+1}$.

SOLUTION   Using the power rule for functions and Theorem (6.28), we obtain

$$y' = 10(x^2 + 1)^9(2x) + (10^{x^2+1} \ln 10)(2x)$$
$$= 20x[(x^2 + 1)^9 + 10^{x^2} \ln 10].$$

The integration formula in (i) of the next theorem may be verified by showing that the integrand is the derivative of the expression on the right side of the equation. Formula (ii) follows from Theorem (6.28)(ii), where $u = g(x)$.

Theorem 6.29

(i) $\displaystyle\int a^x \, dx = \left(\frac{1}{\ln a}\right) a^x + C$

(ii) $\displaystyle\int a^u \, du = \left(\frac{1}{\ln a}\right) a^u + C$

EXAMPLE ▪ 2   Evaluate:

(a) $\displaystyle\int 3^x \, dx$ **(b)** $\displaystyle\int x 3^{(x^2)} \, dx$

SOLUTION

(a) Using (i) of Theorem (6.29) yields

$$\int 3^x \, dx = \left(\frac{1}{\ln 3}\right) 3^x + C.$$

(b) To use (ii) of Theorem (6.29), we make the substitution

$$u = x^2, \qquad du = 2x \, dx$$

and proceed as follows:

$$\int x 3^{(x^2)} \, dx = \frac{1}{2} \int 3^{(x^2)} (2x) \, dx = \frac{1}{2} \int 3^u \, du$$

$$= \frac{1}{2}\left(\frac{1}{\ln 3}\right) 3^u + C = \left(\frac{1}{2\ln 3}\right) 3^{(x^2)} + C$$

EXAMPLE ∎ 3   An important problem in oceanography is determining the light intensity at different ocean depths. The *Beer–Lambert law* states that at a depth x (in meters), the light intensity $I(x)$ (in calories/cm^2/sec) is given by $I(x) = I_0 a^x$, where I_0 and a are positive constants.

(a) What is the light intensity at the surface?

(b) Find the rate of change of the light intensity with respect to depth at a depth x.

(c) If $a = 0.4$ and $I_0 = 10$, find the average light intensity between the surface and a depth of x meters.

(d) Show that $I(x) = I_0 e^{kx}$ for some constant k.

SOLUTION

(a) At the surface, $x = 0$ and

$$I(0) = I_0 a^0 = I_0.$$

Hence the light intensity at the surface is I_0.

(b) The rate of change of $I(x)$ with respect to x is $I'(x)$. Thus,

$$I'(x) = I_0(a^x \ln a) = (\ln a)(I_0 a^x) = (\ln a)I(x).$$

Hence the rate of change $I'(x)$ at depth x is directly proportional to $I(x)$, and the constant of proportionality is $\ln a$.

(c) If $I(x) = 10(0.4)^x$, then, by Definition (4.29), the average value of I on the interval $[0, 5]$ is

$$I_{av} = \frac{1}{5 - 0} \int_0^5 10(0.4)^x \, dx = 2 \int_0^5 (0.4)^x \, dx$$

$$= 2\left[\frac{1}{\ln(0.4)}(0.4)^x\right]_0^5 = \frac{2}{\ln(0.4)}[(0.4)^5 - (0.4)^0]$$

$$= \frac{-1.97952}{\ln(0.4)} \approx 2.16.$$

(d) Using Definition (6.26) yields

$$I(x) = I_0 a^x = I_0 e^{x \ln a} = I_0 e^{kx},$$

where $k = \ln a$.

If $a \neq 1$ and $f(x) = a^x$, then f is a one-to-one function. Its inverse function is denoted by $\log_a$ and is called the **logarithmic function with base a**. Another way of stating this relationship is as follows.

Definition of $\log_a x$ 6.30

$$y = \log_a x \quad \text{if and only if} \quad x = a^y$$

The expression $\log_a x$ is called **the logarithm of x with base a**. In this terminology, natural logarithms are logarithms with base e—that is,

$$\ln x = \log_e x.$$

Laws of logarithms similar to Theorem (6.12) are true for logarithms with base a.

To obtain the relationship between $\log_a$ and ln, consider $y = \log_a x$, or, equivalently, $x = a^y$. Taking the natural logarithm of both sides of the last equation gives us $\ln x = y \ln a$, or $y = (\ln x)/(\ln a)$ and thus proves that

$$\log_a x = \frac{\ln x}{\ln a}.$$

Differentiating both sides of the last equation leads to (i) of the next theorem. Using the chain rule and generalizing to absolute values as in Theorem (6.11) gives us (ii), where $u = g(x)$.

Theorem   6.31

(i) $\dfrac{d}{dx}(\log_a x) = \dfrac{d}{dx}\left(\dfrac{\ln x}{\ln a}\right) = \dfrac{1}{\ln a} \cdot \dfrac{1}{x}$

(ii) $\dfrac{d}{dx}(\log_a |u|) = \dfrac{d}{dx}\left(\dfrac{\ln |u|}{\ln a}\right) = \dfrac{1}{\ln a} \cdot \dfrac{1}{u}\dfrac{du}{dx}$

ILLUSTRATION

- $\dfrac{d}{dx}(\log_2 x) = \dfrac{d}{dx}\left(\dfrac{\ln x}{\ln 2}\right) = \dfrac{1}{\ln 2} \cdot \dfrac{1}{x} = \dfrac{1}{(\ln 2)x}$

- $\dfrac{d}{dx}(\log_2 |x^2 - 9|) = \dfrac{d}{dx}\left(\dfrac{\ln |x^2 - 9|}{\ln 2}\right) = \dfrac{1}{\ln 2} \cdot \dfrac{1}{x^2 - 9} \cdot 2x = \dfrac{2x}{(\ln 2)(x^2 - 9)}$

Logarithms with base 10 are useful for certain applications (see Exercises 50–54). We refer to such logarithms as **common logarithms** and use the symbol **log x** as an abbreviation for $\log_{10} x$. This notation is used in the next example where we perform, in a new context, the familiar task of determining the tangent line to a graph.

EXAMPLE ▪ 4 If $f(x) = \log \sqrt[3]{(2x + 5)^2}$,

(a) find $f'(x)$

(b) graph both the function f and the line tangent to its graph at $x = -0.6$.

SOLUTION

(a) Although most graphing utilities can work with both common and natural logarithms, we express the function in terms of natural logarithms to make differentiating easier. We first write $f(x) = \log(2x + 5)^{2/3}$. The law $\log u^r = r \log u$ is true only if $u > 0$; however, since $(2x + 5)^{2/3} = |2x + 5|^{2/3}$, we may proceed as follows:

$$
\begin{aligned}
f(x) &= \log(2x + 5)^{2/3} \\
&= \log |2x + 5|^{2/3} \\
&= \tfrac{2}{3} \log |2x + 5| \\
&= \frac{2}{3} \frac{\ln |2x + 5|}{\ln 10}
\end{aligned}
$$

Differentiating yields

$$
f'(x) = \frac{2}{3} \cdot \frac{1}{\ln 10} \cdot \frac{1}{2x + 5}(2) = \frac{4}{3(2x + 5) \ln 10}.
$$

(b) We begin by using a graphing utility to plot the function $f(x) = \log(2x + 5)^{2/3}$. To graph the tangent line, we must first find an equation for it using the point–slope formula. Since

$$
f(-0.6) = \tfrac{2}{3} \log 3.8 \approx 0.386522,
$$

Figure 6.22
$-6 \le x \le 1.5, -0.8 \le y \le 0.8$

the point of tangency is $(-0.6, 0.386522)$. The slope of the tangent line is the value of the derivative f' at $x = -0.6$. From part (a), we have

$$
f'(-0.6) = \frac{4}{11.4 \ln 10} \approx 0.152384.
$$

Thus, an equation for the line tangent to the graph at $x = -0.6$ is approximately $y = 0.386522 + 0.152384(x + 0.6)$. We plot the tangent line with the graph of the function on the same coordinate axes, obtaining the results shown in Figure 6.22.

Now that we have defined irrational exponents, we may consider the **general power function** f given by $f(x) = x^c$ for any real number c. If c is irrational, then, by definition, the domain of f is the set of positive real numbers. Using Definition (6.26) and Theorems (6.22) and (6.11)(i), we have

$$\frac{d}{dx}(x^c) = \frac{d}{dx}(e^{c \ln x}) = e^{c \ln x} \frac{d}{dx}(c \ln x)$$

$$= e^{c \ln x} \left(\frac{c}{x}\right) = x^c \left(\frac{c}{x}\right) = cx^{c-1}.$$

This result proves that the power rule is true for irrational as well as rational exponents. The power rule for functions may also be extended to irrational exponents.

ILLUSTRATION

∎ $\dfrac{d}{dx}(x^{\sqrt{2}}) = \sqrt{2}x^{\sqrt{2}-1}$

∎ $\dfrac{d}{dx}(1 + e^{2x})^{\pi} = \pi(1 + e^{2x})^{\pi-1}\dfrac{d}{dx}(1 + e^{2x})$

$$= \pi(1 + e^{2x})^{\pi-1}(2e^{2x}) = 2\pi e^{2x}(1 + e^{2x})^{\pi-1}$$

EXAMPLE ▪ 5   If $y = x^x$ and $x > 0$,

(a) find dy/dx

(b) graph both the function and its derivative

SOLUTION

(a) Since the exponent in x^x is a variable, the power rule may not be used. Similarly, Theorem (6.28) is not applicable, since the base a is not a fixed real number. However, by Definition (6.26), $x^x = e^{x \ln x}$ for every $x > 0$, and hence

$$\frac{d}{dx}(x^x) = \frac{d}{dx}(e^{x \ln x})$$

$$= e^{x \ln x}\frac{d}{dx}(x \ln x)$$

$$= e^{x \ln x}\left[x\left(\frac{1}{x}\right) + (1)\ln x\right] = x^x(1 + \ln x).$$

Another way of solving this problem is to use the method of logarithmic differentiation introduced in the preceding section. In this case, we take the natural logarithm of both sides of the equation $y = x^x$ and then differentiate implicitly as follows:

$$\ln y = \ln x^x = x \ln x$$

$$\frac{dy}{dx}(\ln y) = \frac{d}{dx}(x \ln x)$$

$$\frac{1}{y}\frac{dy}{dx}(y) = 1 + \ln x$$

$$\frac{dy}{dx}(y) = y(1 + \ln x) = x^x(1 + \ln x)$$

Figure 6.23
$$f(x) = x^x$$
$$f'(x) = x^x(1 + \ln x)$$
$$0 \le x \le 2.5, \ -1.5 \le y \le 6$$

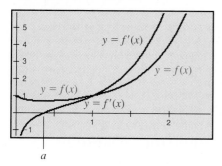

(b) In part (a), we saw that the expression x^x is unusual because both the base and the exponent are variable. The expression for the derivative f' is even more complicated, since it involves both x^x and $(1 + \ln x)$. To gain a better understanding of the function $f(x) = x^x$, we use a graphing utility to plot its graph and the graph of its derivative, as shown in Figure 6.23. Several features of the function are evident from these graphs. First, when $x = 0$, the expression x^x becomes the undefined algebraic expression 0^0, but from the graph it appears that the function $f(x) = x^x$ approaches 1 as x approaches 0 from the right. We shall prove this result in Section 6.9. Second, if we examine the graph of the derivative f', we see that it is negative for $0 < x < a$ and positive for $x > a$. We can determine the value of a by examining the sign of $f'(x)$. Since $x^x > 0$, $f'(x) < 0$ if $1 + \ln x < 0$ or, equivalently, $x < e^{-1}$. Hence, $a = e^{-1} \approx 0.37$. Thus, the function $f(x) = x^x$ decreases to an absolute minimum at $x = a$ and then increases for $x > a$. It also appears from the graph of the derivative that it is unbounded in the negative direction as $x \to 0^+$, so the function $f(x) = x^x$ is not differentiable at $x = 0$.

We conclude this section by expressing the number e as a limit.

Theorem 6.32

$$\text{(i)} \ \lim_{h \to 0} (1 + h)^{1/h} = e \qquad \text{(ii)} \ \lim_{n \to \infty} \left(1 + \frac{1}{n}\right)^n = e$$

PROOF Applying the definition of derivative (2.5) to $f(x) = \ln x$ and using laws of logarithms yields

$$f'(x) = \lim_{h \to 0} \frac{\ln(x + h) - \ln x}{h} = \lim_{h \to 0} \frac{1}{h} \ln \frac{x + h}{x}$$

$$= \lim_{h \to 0} \frac{1}{h} \ln \left(1 + \frac{h}{x}\right) = \lim_{h \to 0} \ln \left(1 + \frac{h}{x}\right)^{1/h}.$$

Since $f'(x) = 1/x$, we have, for $x = 1$,

$$1 = \lim_{h \to 0} \ln(1 + h)^{1/h}.$$

We next observe, from Theorem (6.19), that

$$(1 + h)^{1/h} = e^{\ln(1+h)^{1/h}}.$$

Since the natural exponential function is continuous at 1, it follows from Theorem (1.25) that

$$\lim_{h \to 0} (1 + h)^{1/h} = \lim_{h \to 0} [e^{\ln(1+h)^{1/h}}]$$

$$= e^{[\lim_{h \to 0} \ln(1+h)^{1/h}]} = e^1 = e.$$

This establishes part (i) of the theorem. The limit in part (ii) may be obtained by introducing the change of variable $n = 1/h$ with $h > 0$. ∎

The formulas in Theorem (6.32) are sometimes used to *define* the number e. You may find it instructive to calculate $(1 + h)^{1/h}$ for numerically small values of h. Some approximate values are given in the following table.

h	$(1+h)^{1/h}$	h	$(1+h)^{1/h}$
-0.01	2.73199902643	0.01	2.70481382942
-0.001	2.71964221644	0.001	2.71692393224
-0.0001	2.71841775501	0.0001	2.71814592683
-0.00001	2.71829541999	0.00001	2.71826823717
-0.000001	2.71828318760	0.000001	2.71828046932
-0.0000001	2.71828196437	0.0000001	2.71828169255

To five decimal places, $e \approx 2.71828$.

EXERCISES 6.5

Exer. 1–24: Find $f'(x)$ if $f(x)$ is the given expression.

1 7^x

2 5^{-x}

3 8^{x^2+1}

4 $9^{\sqrt{x}}$

5 $\log(x^4 + 3x^2 + 1)$

6 $\log_3 |6x - 7|$

7 5^{3x-4}

8 3^{2-x^2}

9 $(x^2 + 1)10^{1/x}$

10 $(10^x + 10^{-x})^{10}$

11 $\log(3x^2 + 2)^5$

12 $\log \sqrt{x^2 + 1}$

13 $\log_5 \left| \dfrac{6x + 4}{2x - 3} \right|$

14 $\log \left| \dfrac{1 - x^2}{2 - 5x^3} \right|$

15 $\log \ln x$

16 $\ln \log x$

17 $x^e + e^x$

18 $x^\pi \pi^x$

19 $(x + 1)^x$

20 x^{4+x^2}

21 $2^{\sin^2 x}$

22 $4^{\sec 3x}$

23 (a) e^e (b) x^5 (c) $x^{\sqrt{5}}$ (d) $(\sqrt{5})^x$ (e) $x^{(x^2)}$

24 (a) π^π (b) x^4 (c) x^π (d) π^x (e) x^{2x}

[c] **Exer. 25–28: Plot the graph of the function and the line tangent to the graph at the point $(a, f(a))$.**

25 $f(x) = 5^{3x-4}$; $a = 1$

26 $f(x) = 3^{2-x^2}$; $a = -1.5$

27 $f(x) = \log(3x^2 + 2)^5$; $a = 5$

28 $f(x) = \log \sqrt{x^2 + 1}$; $a = 10$

Exer. 29–44: Evaluate the integral.

29 (a) $\displaystyle\int 7^x \, dx$ (b) $\displaystyle\int_{-2}^{1} 7^x \, dx$

30 (a) $\displaystyle\int 3^x \, dx$ (b) $\displaystyle\int_{-1}^{0} 3^x \, dx$

31 (a) $\displaystyle\int 5^{-2x} \, dx$ (b) $\displaystyle\int_{1}^{2} 5^{-2x} \, dx$

32 (a) $\displaystyle\int 2^{3x-1} \, dx$ (b) $\displaystyle\int_{-1}^{1} 2^{3x-1} \, dx$

33 $\displaystyle\int 10^{3x} \, dx$

34 $\displaystyle\int 5^{-5x} \, dx$

35 $\displaystyle\int x(3^{-x^2}) \, dx$

36 $\displaystyle\int \dfrac{(2^x + 1)^2}{2^x} \, dx$

37 $\displaystyle\int \dfrac{2^x}{2^x + 1} \, dx$

38 $\displaystyle\int \dfrac{3^x}{\sqrt{3^x + 4}} \, dx$

39 $\displaystyle\int \dfrac{1}{x \log x} \, dx$

40 $\displaystyle\int \dfrac{10^{\sqrt{x}}}{\sqrt{x}} \, dx$

41 $\displaystyle\int 3^{\cos x} \sin x \, dx$

42 $\displaystyle\int \dfrac{5^{\tan x}}{\cos^2 x} \, dx$

43 (a) $\int \pi^{\pi} \, dx$ (b) $\int x^4 \, dx$

(c) $\int x^{\pi} \, dx$ (d) $\int \pi^x \, dx$

44 (a) $\int e^e \, dx$ (b) $\int x^5 \, dx$

(c) $\int x^{\sqrt{5}} \, dx$ (d) $\int (\sqrt{5})^x \, dx$

45 Find the area of the region bounded by the graphs of $y = 2^x$, $x + y = 1$, and $x = 1$.

46 The region under the graph of $y = 3^{-x}$ from $x = 1$ to $x = 2$ is revolved about the x-axis. Find the volume of the resulting solid.

47 An economist predicts that the buying power $B(t)$ of a dollar t years from now will decrease according to the formula $B(t) = (0.95)^t$.

(a) At approximately what rate will the buying power be decreasing two years from now?

(b) Estimate the *average buying power* of the dollar over the next two years.

48 When a person takes a 100-mg tablet of an asthma drug orally, the rate R at which the drug enters the bloodstream is predicted to be $R = 5(0.95)^t$ mg/min. If the bloodstream does not contain any trace of the drug when the tablet is taken, determine the number of minutes needed for 50 mg to enter the bloodstream.

49 One thousand trout, each one year old, are introduced into a large pond. The number still alive after t years is predicted to be $N(t) = 1000(0.9)^t$.

(a) Approximate the death rate dN/dt at times $t = 1$ and $t = 5$. At what rate is the population decreasing when $N = 500$?

(b) The weight $W(t)$ (in pounds) of an individual trout is expected to increase according to the formula $W(t) = 0.2 + 1.5t$. After approximately how many years is the total number of pounds of trout in the pond a maximum?

50 The vapor pressure P (in psi), a measure of the volatility of a liquid, is related to its temperature T (in °F) by the *Antoine equation*: $\log P = a + [b/(c + T)]$, for constants a, b, and c. Vapor pressure increases rapidly with an increase in temperature. Find conditions on a, b, and c that guarantee that P is an increasing function of T.

51 Chemists use a number denoted by pH to describe quantitatively the acidity or basicity of solutions. By definition, $pH = -\log [H^+]$, where $[H^+]$ is the hydrogen ion concentration in moles per liter. For

a certain brand of vinegar, it is estimated (with a maximum percentage error of $\pm 0.5\%$) that $[H^+] \approx 6.3 \times 10^{-3}$. Calculate the pH and use differentials to estimate the maximum percentage error in the calculation.

52 The magnitude R (on the Richter scale) of an earthquake of intensity I may be found by means of the formula $R = \log(I/I_0)$, where I_0 is a certain minimum intensity. Suppose the intensity of an earthquake is estimated to be 100 times I_0. If the maximum percentage error in the estimate is $\pm 1\%$, use differentials to approximate the maximum percentage error in the calculated value of R.

53 Let $R(x)$ be the reaction of a subject to a stimulus of strength x. For example, if the stimulus x is *saltiness* (in grams of salt per liter), $R(x)$ may be the subject's estimate of how salty the solution tasted on a scale from 0 to 10. A function that has been proposed to relate R to x is given by the *Weber–Fechner formula*: $R = a \log(x/x_0)$, where a is a positive constant.

(a) Show that $R = 0$ for the *threshold stimulus* $x = x_0$.

(b) The derivative $S = dR/dx$ is the *sensitivity* at stimulus level x and measures the ability to detect small changes in stimulus level. Show that S is inversely proportional to x, and compare $S(x)$ to $S(2x)$.

54 The loudness of sound, as experienced by the human ear, is based on intensity level. A formula used for finding the intensity level α that corresponds to a sound intensity I is $\alpha = 10 \log(I/I_0)$ decibels, where I_0 is a special value of I agreed to be the weakest sound that can be detected by the ear under certain conditions. Find the rate of change of α with respect to I if

(a) I is 10 times as great as I_0

(b) I is 1000 times as great as I_0

(c) I is 10,000 times as great as I_0 (This is the intensity level of the average voice.)

55 If a principal of P dollars is invested in a savings account for t years and the yearly interest rate r (expressed as a decimal) is compounded n times per year, then the amount A in the account after t years is given by the *compound interest formula*:

$$A = P[1 + (r/n)]^{nt}.$$

(a) Let $h = r/n$ and show that

$$\ln A = \ln P + rt \ln(1 + h)^{1/h}.$$

(b) Let $n \to \infty$ and use the expression in part (a) to establish the formula $A = Pe^{rt}$ for interest *compounded continuously*.

56 Establish Theorem (6.32)(ii) by using the limit in part (i) and the change of variable $n = 1/h$.

57 Prove that $\lim_{n \to \infty}[1 + (x/n)]^n = e^x$ by letting $h = x/n$ and using Theorem (6.32)(i).

$\boxed{c}$ **58** Graph, on the same coordinate axes, $y = 2^{-x}$ and $y = \log_2 x$.

    (a) Estimate the x-coordinate of the point of intersection of the graphs.

(b) If the region R bounded by the graphs and the line $x = 1$ is revolved about the x-axis, set up an integral that can be used to approximate the volume of the resulting solid.

(c) Use Simpson's rule, with $n = 2$, to approximate the integral in part (b).

6.6   SEPARABLE DIFFERENTIAL EQUATIONS AND LAWS OF GROWTH AND DECAY

Suppose that a physical quantity varies with time and that the magnitude of the quantity at time t is given by $q(t)$, where q is differentiable and $q(t) > 0$ for every t. The derivative $q'(t)$ is the rate of change of $q(t)$ with respect to time. In many applications, this rate of change is directly proportional to the magnitude of the quantity at time t—that is,

$$q'(t) = cq(t)$$

for some constant c. The number of bacteria in certain cultures behaves in this way. If the number of bacteria $q(t)$ is small, then the rate of increase $q'(t)$ is small; however, as the number of bacteria increases, the *rate of increase* also increases. The decay of a radioactive substance obeys a similar law: As the amount of matter decreases, the rate of decay—that is, the amount of radiation—also decreases. As a final illustration, suppose an electrical condenser is allowed to discharge. If the charge on the condenser is large at the outset, the rate of discharge is also large, but as the charge weakens, the condenser discharges less rapidly.

In applied problems, the equation $q'(t) = cq(t)$ is often expressed in terms of differentials. Thus, if $y = q(t)$, we may write

$$\frac{dy}{dt} = cy, \quad \text{or} \quad dy = cy\,dt.$$

Dividing both sides of the last equation by y, we obtain

$$\frac{1}{y}\,dy = c\,dt.$$

Since it is possible to **separate the variables** y and t—in the sense that they can be placed on opposite sides of the equals sign—the differential equation $dy/dt = cy$ is a **separable differential equation**. We will study such equations in more detail later in the text and will show that solutions can be found by integrating both sides of the "separated" equation $(1/y)\,dy = c\,dt$. Thus,

$$\int \frac{1}{y}\,dy = \int c\,dt$$

and, assuming $y > 0$,

$$\ln y = ct + d$$

for some constant d. It follows that

$$y = e^{ct+d} = e^d e^{ct}.$$

If y_0 denotes the initial value of y (that is, the value corresponding to $t = 0$), then letting $t = 0$ in the last equation gives us

$$y_0 = e^d e^0 = e^d,$$

and hence the solution $y = e^d e^{ct}$ may be written

$$y = y_0 e^{ct}.$$

We have proved the following theorem.

Theorem 6.33

> Let y be a differentiable function of t such that $y > 0$ for every t, and let y_0 be the value of y at $t = 0$. If $dy/dt = cy$ for some constant c, then
>
> $$y = y_0 e^{ct}.$$

The preceding theorem states that *if the rate of change of $y = q(t)$ with respect to t is directly proportional to y, then y may be expressed in terms of an exponential function.* If y increases with t, the formula $y = y_0 e^{ct}$ is a **law of growth** ($c > 0$), and if y decreases, it is a **law of decay** ($c < 0$).

EXAMPLE ■ 1   The number of bacteria in a culture increases from 600 to 1800 in 2 hr. Assuming that the rate of increase is directly proportional to the number of bacteria present, find

(a) a formula for the number of bacteria at time t

(b) the number of bacteria at the end of 4 hr

SOLUTION

(a) Let $y = q(t)$ denote the number of bacteria after t hours. Thus, $y_0 = q(0) = 600$ and $q(2) = 1800$. By hypothesis,

$$\frac{dy}{dt} = cy.$$

Following exactly the same steps used in the proof of Theorem (6.33), we obtain

$$y = y_0 e^{ct} = 600 e^{ct}.$$

Since $y = 1800$ when $t = 2$, we obtain the following equivalent equations:

$$1800 = 600 e^{2c}, \qquad 3 = e^{2c}, \qquad e^c = 3^{1/2}.$$

Substituting for e^c in $y = 600 e^{ct}$ gives us

$$y = 600(3^{1/2})^t, \quad \text{or} \quad y = 600(3)^{t/2}.$$

(b) Letting $t = 4$ in $y = 600(3)^{t/2}$ yields

$$y = 600(3)^{4/2} = 600(9) = 5400.$$

EXAMPLE ■ 2   Radium decays exponentially and has a half-life of approximately 1600 yr—that is, given any quantity, one half of it will disintegrate in 1600 yr.

(a) Find a formula for the amount y remaining from 50 mg of pure radium after t years.

(b) When will the amount remaining be 20 mg?

SOLUTION

(a) If we let $y = q(t)$, then

$$y_0 = q(0) = 50 \quad \text{and} \quad q(1600) = \tfrac{1}{2}(50) = 25.$$

Since $dy/dt = cy$ for some c, it follows from Theorem (6.33) that

$$y = 50e^{ct}.$$

Since $y = 25$ when $t = 1600$,

$$25 = 50e^{1600c}, \quad \text{or} \quad e^{1600c} = \tfrac{1}{2}.$$

Hence,

$$e^c = (\tfrac{1}{2})^{1/1600} = 2^{-1/1600}.$$

Substituting for e^c in $y = 50e^{ct}$ gives us

$$y = 50(2^{-1/1600})^t, \quad \text{or} \quad y = 50(2)^{-t/1600}.$$

(b) Using $y = 50(2)^{-t/1600}$, we see that the value of t at which $y = 20$ is a solution of the equation

$$20 = 50(2)^{-t/1600}, \quad \text{or} \quad 2^{t/1600} = \tfrac{5}{2}.$$

Taking the natural logarithm of each side, we obtain

$$\frac{t}{1600} \ln 2 = \ln \frac{5}{2},$$

or

$$t = \frac{1600 \ln \frac{5}{2}}{\ln 2} \approx 2115 \text{ yr.}$$

EXAMPLE ■ 3   According to Newton's law of cooling, the rate at which an object cools is directly proportional to the difference in temperature between the object and the surrounding medium. If an object cools from 125 °F to 100 °F in half an hour when surrounded by air at a temperature of 75 °F, find its temperature at the end of the next half hour.

SOLUTION   Let y denote the temperature of the object after t hours of cooling. Since the temperature of the surrounding medium is 75°, the difference in temperature is $y - 75$, and therefore, by Newton's law of cooling,

$$\frac{dy}{dt} = c(y - 75)$$

for some constant c. We separate variables and integrate as follows:

$$\frac{1}{y - 75}\, dy = c\, dt$$

$$\int \frac{1}{y - 75}\, dy = \int c\, dt$$

$$\ln(y - 75) = ct + b$$

for some constant b. The last equation is equivalent to

$$y - 75 = e^{ct+b} = e^b e^{ct}.$$

Since $y = 125$ when $t = 0$,

$$125 - 75 = e^b e^0 = e^b, \quad \text{or} \quad e^b = 50.$$

Hence,

$$y - 75 = 50 e^{ct}, \quad \text{or} \quad y = 50 e^{ct} + 75.$$

Using the fact that $y = 100$ when $t = \frac{1}{2}$ leads to the following equivalent equations:

$$100 = 50 e^{c/2} + 75, \qquad e^{c/2} = \tfrac{25}{50} = \tfrac{1}{2}, \qquad e^c = \tfrac{1}{4}$$

Substituting $\frac{1}{4}$ for e^c in $y = 50 e^{ct} + 75$ gives us a formula for the temperature after t hours:

$$y = 50(\tfrac{1}{4})^t + 75$$

In particular, if $t = 1$,

$$y = 50(\tfrac{1}{4}) + 75 = 87.5\,^\circ\text{F}.$$

In biology, a function G is sometimes used, as follows, to estimate the size of a quantity at time t:

$$G(t) = k e^{(-A e^{-Bt})}$$

for positive constants k, A, and B. The function G is called a **Gompertz growth function**. It is always positive and increasing, but has a limit as t increases without bound. The graph of G is called a **Gompertz growth curve**.

EXAMPLE ■ 4 Discuss and sketch the graph of the Gompertz growth function G.

SOLUTION We first observe that the y-intercept is $G(0) = k e^{-A}$ and that $G(t) > 0$ for every t. Differentiating twice, we obtain

$$G'(t) = k e^{(-A e^{-Bt})} (-A e^{-Bt})'$$

$$= A B k e^{(-Bt - A e^{-Bt})}$$

$$G''(t) = A B k e^{(-Bt - A e^{-Bt})} (-Bt - A e^{-Bt})'$$

$$= A B k (-B + A B e^{-Bt}) e^{-Bt - A e^{-Bt}}.$$

Figure 6.24

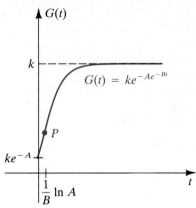

$$G(t) = ke^{-Ae^{-Bt}}$$

Since $G'(t) > 0$ for every t, the function G is increasing on $[0, \infty)$. The second derivative $G''(t)$ is zero if

$$-B + ABe^{-Bt} = 0, \quad \text{or} \quad e^{Bt} = A.$$

Solving the last equation for t gives us $t = (1/B) \ln A$, which is a critical number for the function G'. We leave it as an exercise to show that at this time the rate of growth G' has a maximum value Bk/e. We can also show that

$$\lim_{t \to \infty} G'(t) = 0 \quad \text{and} \quad \lim_{t \to \infty} G(t) = k.$$

Hence, as t increases without bound, the rate of growth approaches 0 and the graph of G has a horizontal asymptote $y = k$. A typical graph is sketched in Figure 6.24. The point P on the graph, corresponding to $t = (1/B) \ln A$, is a point of inflection, and the concavity changes from upward to downward at P.

In the next example, we consider a physical quantity that increases to a maximum value and then decreases asymptotically to 0.

EXAMPLE ▪ 5   When uranium disintegrates into lead, one step in the process is the radioactive decay of radium into radon gas. Radon gas enters homes by diffusing through the soil into basements, where it presents a health hazard if inhaled. If a quantity Q of radium is present initially, then the amount of radon gas present after t years is given by

$$A(t) = \frac{c_1 Q}{c_2 - c_1}(e^{-c_1 t} - e^{-c_2 t}),$$

where $c_1 = \frac{1}{1600} \ln 2$ and $c_2 = \frac{1}{0.0105} \ln 2$ are the *decay constants* for radium and radon gas, respectively.

(a) Find the amount of radon gas present initially and after an extended period of time.

(b) Use a graphing utility to graph $A(t)$.

(c) Determine the maximum amount A_M of radon gas and when that amount is reached.

(d) After the maximum amount A_M has been reached, estimate how long it would take the radon gas to decrease to 90% of the maximum.

SOLUTION

(a) The initial amount of radon gas is

$$A(0) = \frac{c_1 Q}{c_2 - c_1}(e^0 - e^0) = 0.$$

If we let t increase without bound, then

$$\lim_{t \to \infty} A(t) = \frac{c_1 Q}{c_2 - c_1} \lim_{t \to \infty} (e^{-c_1 t} - e^{-c_2 t})$$

$$= \frac{c_1 Q}{c_2 - c_1}(0 - 0) = 0.$$

Figure 6.25

$$A(t) = \frac{c_1 Q}{c_2 - c_1}(e^{-c_1 t} - e^{-c_2 t})$$

$0 \le t \le 1, 0 \le A \le 10^{-5} Q$

Hence, over a long period of time, the amount of radon gas decreases to 0.

(b) We use a graphing utility to obtain Figure 6.25, which illustrates the graph of $A(t)$. In the viewing window, $0 \le t \le 1$, it appears that the radon gas rises fairly quickly to its maximum level and then levels off or perhaps decreases very, very slowly. In part (a), we concluded that the amount of radon gas would eventually decrease to 0 but that is not evident from the graph in Figure 6.25. In parts (c) and (d), we do further analysis to determine how quickly the maximum is reached and how slowly the gas disappears.

(c) To find the critical numbers of A, we differentiate, obtaining

$$A'(t) = \frac{c_1 Q}{c_2 - c_1}\left[-\left(c_1 e^{-c_1 t} + c_2 e^{-c_2 t}\right)\right].$$

Thus, $A'(t) = 0$ if

$$c_1 e^{-c_1 t} = c_2 e^{-c_2 t}, \quad \text{or} \quad e^{(c_2 - c_1)t} = \frac{c_2}{c_1}.$$

It follows that

$$(c_2 - c_1)t = \ln \frac{c_2}{c_1},$$

or

$$t = \frac{\ln(c_2/c_1)}{c_2 - c_1}.$$

This value of t yields the maximum value of A. Substituting this value for t into the function, we find (after a fair amount of algebraic manipulation) that the maximum value is

$$A_{\text{M}} = A\left(\frac{\ln(c_2/c_1)}{c_2 - c_1}\right) = \left(\frac{c_1}{c_2}\right)^{c_2/(c_2 - c_1)} Q.$$

For the given values of the constants c_1 and c_2, these two numbers are approximately

$$t_{\text{M}} \approx 0.181 \text{ years} \approx 66 \text{ days} \quad \text{and} \quad A_{\text{M}} \approx (6.562)10^{-6} Q.$$

(d) To find the value $t_1 > t_{\text{M}}$ that yields $A(t_1) = 0.90 A_{\text{M}}$, we first divide both sides by Q so that we can work numerically. Using the solving routine on a computational device (or Newton's method), we find that the solution to

$$\frac{c_1}{c_2 - c_1}\left(e^{-c_1 t_1} - e^{-c_2 t_1}\right) = (0.9)(6.562)10^{-6} Q$$

for the given values of c_1 and c_2 is $t_1 \approx 243$ years. (See Figure 6.26 for a viewing window displaying this part of the graph of the function.) The fact that the radon decreases to 0 over a long period of time may not be very comforting to a homeowner since the decrease takes place so slowly.

Figure 6.26

$$A(t) = \frac{c_1 Q}{c_2 - c_1}(e^{-c_1 t} - e^{-c_2 t})$$

$0 \le t \le 400, 0 \le A \le 10^{-5} Q$

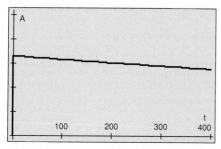

EXERCISES 6.6

1 The number of bacteria in a culture increases from 5000 to 15,000 in 10 hr. Assuming that the rate of increase is proportional to the number of bacteria present, find a formula for the number of bacteria in the culture at any time t. Estimate the number at the end of 20 hr. When will the number be 50,000?

2 The polonium isotope ^{210}Po has a half-life of approximately 140 days. If a sample weighs 20 mg initially, how much remains after t days? Approximately how much will be left after two weeks?

3 If the temperature is constant, then the rate of change of barometric pressure p with respect to altitude h is proportional to p. If $p = 30$ in. at sea level and $p = 29$ in. when $h = 1000$ ft, find the pressure at an altitude of 5000 ft.

4 The population of a city is increasing at the rate of 5% per year. If the present population is 500,000 and the rate of increase is proportional to the number of people, what will the population be in 10 yr?

5 Agronomists use the assumption that a quarter acre of land is required to provide food for one person and estimate that there are 10 billion acres of tillable land in the world. Hence a maximum population of 40 billion people can be sustained if no other food source is available. The world population at the beginning of 1993 was approximately 5.5 billion. Assuming that the population increases at a rate of 2% per year and the rate of increase is proportional to the number of people, when will the maximum population be reached?

6 A metal plate that has been heated cools from 180°F to 150°F in 20 min when surrounded by air at a temperature of 60°F. Use Newton's law of cooling (see Example 3) to approximate its temperature at the end of 1 hr of cooling. When will the temperature be 100°F?

7 An outdoor thermometer registers a temperature of 40°F. Five minutes after it is brought into a room where the temperature is 70°F, the thermometer registers 60°F. When will it register 65°F?

8 The rate at which salt dissolves in water is directly proportional to the amount that remains undissolved. If 10 lb of salt is placed in a container of water and 4 lb dissolves in 20 min, how long will it take for two more pounds to dissolve?

9 According to Kirchhoff's first law for electrical circuits, $V = RI + L(dI/dt)$, where the constants V, R, and L denote the electromotive force, the resistance, and the inductance, respectively, and I denotes the current at time t. If the electromotive force is terminated at time $t = 0$ and if the current is I_0 at the instant of removal, prove that $I = I_0 e^{-Rt/L}$.

10 A physicist finds that an unknown radioactive substance registers 2000 counts per minute on a Geiger counter. Ten days later, the substance registers 1500 counts per minute. Approximate its half-life.

11 The air pressure P (in atmospheres) at an elevation of z meters above sea level is a solution of the differential equation $dP/dz = -9.81\rho(z)$, where $\rho(z)$ is the density of air at elevation z. Assuming that air obeys the ideal gas law, this differential equation can be rewritten as $dP/dz = -0.0342P/T$, where T is the temperature (in °K) at elevation z. If $T = 288 - 0.01z$ and if the pressure is 1 atmosphere at sea level, express P as a function of z.

12 During the first month of growth for crops such as maize, cotton, and soybeans, the rate of growth (in grams per day) is proportional to the present weight W. For a species of cotton, $dW/dt = 0.21W$. Predict the weight of a plant at the end of the month ($t = 30$) if the plant weighs 70 mg at the beginning of the month.

13 Radioactive strontium-90, ^{90}Sr, with a half-life of 29 yr, can cause bone cancer in humans. The substance is carried by acid rain, soaks into the ground, and is passed through the food chain. The radioactivity level in a particular field is estimated to be 2.5 times the safe level S. For approximately how many years will this field be contaminated?

14 The radioactive tracer ^{51}Cr, with a half-life of 27.8 days, can be used in medical testing to locate the position of a placenta in a pregnant woman. Often the tracer must be ordered from a medical supply lab. If 35 units are needed for a test and delivery from the lab requires 2 days, estimate the minimum number of units that should be ordered.

15 Veterinarians use sodium pentobarbital to anesthetize animals. Suppose that to anesthetize a dog, 30 mg is required for each kilogram of body weight. If sodium pentobarbital is eliminated exponentially from the bloodstream and half is eliminated in 4 hr, approximate the single dose that will anesthetize a 20-kg dog for 45 min.

16 In the study of lung physiology, the following differential equation is used to describe the transport of a substance across a capillary wall:

$$\frac{dh}{dt} = -\frac{V}{Q}\left(\frac{h}{k+h}\right),$$

where h is the hormone concentration in the blood-stream, t is time, V is the maximum transport rate, Q is the volume of the capillary, and k is a constant that measures the affinity between the hormones and enzymes that assist with the transport process. Find the general solution of the differential equation.

17 A space probe is shot upward from the earth. If air resistance is disregarded, a differential equation for the velocity after burnout is $v(dv/dy) = -ky^{-2}$, where y is the distance from the center of the earth and k is a positive constant. If y_0 is the distance from the center of the earth at burnout and v_0 is the corresponding velocity, express v as a function of y.

18 At high temperatures, nitrogen dioxide, NO_2, decomposes into NO and O_2. If $y(t)$ is the concentration of NO_2 (in moles per liter), then, at $600\,^\circ K$, $y(t)$ changes according to the *second-order reaction law* $dy/dt = -0.05y^2$ for time t in seconds. Express y in terms of t and the initial concentration y_0.

19 The technique of carbon-14 dating is used to determine the age of archeological or geological specimens. This method is based on the fact that the unstable isotope carbon-14 (^{14}C) is present in the CO_2 in the atmosphere. Plants take in carbon from the atmosphere; when they die, the ^{14}C that has accumulated begins to decay, with a half-life of approximately 5700 yr. By measuring the amount of ^{14}C that remains in a specimen, it is possible to approximate when the organism died. Suppose that a bone fossil contains 20% as much ^{14}C as an equal amount of carbon in present-day bone. Approximate the age of the bone.

20 Refer to Exercise 19. The hydrogen isotope ^3_1H, which has a half-life of 12.3 yr, is produced in the atmosphere by cosmic rays and is brought to earth by rain. If the wood siding of an old house contains 10% as much ^3_1H as the siding on a similar new house, approximate the age of the old house.

21 The earth's atmosphere absorbs approximately 32% of the sun's incoming radiation. The earth also emits radiation (mostly in the form of heat), and the atmosphere absorbs approximately 93% of this outgoing radiation. This difference in absorption of incoming and outgoing radiation by the atmosphere is called the *greenhouse effect*. Changes in this balance will affect the earth's climate. Suppose I_0 is the intensity of the sun's radiation and I is the intensity of the radiation after traveling a distance x through the atmosphere. If $\rho(h)$ is the density of the atmosphere at height h, then the *optical thickness* is $f(x) = k \int_0^x \rho(h)\,dh$, where k is an absorption constant, and I is given by $I = I_0 e^{-f(x)}$. Show that $dI/dx = -k\rho(x)I$.

22 Certain learning processes may be illustrated by the graph of $f(x) = a + b(1 - e^{-cx})$ for positive constants a, b, and c. Suppose a manufacturer estimates that a new employee can produce 5 items the first day on the job. As the employee becomes more proficient, the daily production increases until a certain maximum production is reached. Suppose that on the nth day on the job, the number $f(n)$ of items produced is approximated by $f(n) = 3 + 20(1 - e^{-0.1n})$.

(a) Estimate the number of items produced on the fifth day, the ninth day, the twenty-fourth day, and the thirtieth day.

(b) Sketch the graph of f from $n = 0$ to $n = 30$. (Graphs of this type are called *learning curves* and are used frequently in education and psychology.)

(c) What happens as n increases without bound?

23 A spherical cell has volume V and surface area S. A simple model for cell growth before mitosis assumes that the rate of growth dV/dt is proportional to the surface area of the cell. Show that $dV/dt = kV^{2/3}$ for some $k > 0$, and express V as a function of t.

24 In Theorem (6.33), we assumed that the rate of change of a quantity $q(t)$ at time t is directly proportional to $q(t)$. Find $q(t)$ if its rate of change is directly proportional to $[q(t)]^2$.

25 Refer to Example 4.

(a) Verify that Bk/e is a maximum value for G'.

(b) Show that $\lim_{t\to\infty} G'(t) = 0$ and $\lim_{t\to\infty} G(t) = k$.

(c) Sketch the graph of G if $k = 10$, $A = 2$, and $B = 1$.

26 Graph the Gompertz growth function G on the interval $[0, 5]$ for $k = 1.1$, $A = 3.2$, and $B = 1.1$.

Exer. 27–31: Each function contains a constant term for which four values are given. Plot the four versions of the function in the same viewing window.

27 $y = 10e^{ct}$ on $-5 \le t \le 8$
for $c = 0.05, 0.1, 0.2,$ and 0.4

28 $y = 10e^{-Ae^{-1.1t}}$ on $-3 \le t \le 8$
for $A = 0.8, 2.0, 3.2,$ and 4.4

29 $y = 10e^{-3.2e^{-Bt}}$ on $-3 \le t \le 8$
for $B = 0.2, 0.8, 1.1,$ and 2.0

30 $y = 10(e^{-c_1 t} - e^{-c_2 t})$ on $0 \le t \le 5$
for $c_1 = 1$
and $c_2 = 3, 5, 10,$ and 25

31 $y = 10(e^{-c_1 t} - e^{-c_2 t})$ on $0 \le t \le 5$
for $c_1 = 0.5, 1, 2,$ and 4
and $c_2 = 10$

6.7 INVERSE TRIGONOMETRIC FUNCTIONS

In this section, we discuss the inverse trigonometric functions, their derivatives, and their integrals. Since we may regard the values of the inverse trigonometric functions as angles, these functions have a broad range of applications, such as rates of change in the angle of elevation as an observer tracks a moving object, speed of rotation of a searchlight, and optimal angles to minimize energy loss in blood flows.

DEFINING THE INVERSE TRIGONOMETRIC FUNCTIONS

Since the trigonometric functions are not one-to-one, they do not have inverse functions (see Section 6.1). By restricting their domains, however, we may obtain one-to-one functions that have the same values as the trigonometric functions and that *do* have inverses over these restricted domains.

We consider first the graph of the sine function, whose domain is $\mathbb{R}$ and whose range is the closed interval $[-1, 1]$ (see Figure 6.27). The sine function is not one-to-one, since a horizontal line such as $y = \frac{1}{2}$ intersects the graph at more than one point. If we restrict the domain to $[-\pi/2, \pi/2]$, then, as the solid portion of the graph in Figure 6.27 illustrates, we obtain an increasing function that takes on every value of the sine function exactly once. This new function, with domain $[-\pi/2, \pi/2]$ and range $[-1, 1]$, is continuous and increasing and hence, by Theorem (6.6), has an inverse function that is continuous and increasing. The inverse function has domain $[-1, 1]$ and range $[-\pi/2, \pi/2]$.

If we restrict the domain of the cosine function to the interval $[0, \pi]$, as shown in the solid portion of the graph in Figure 6.28, we obtain a one-to-one continuous decreasing function that has a continuous decreasing inverse function. The inverse cosine function has domain $[-1, 1]$ and range $[0, \pi]$.

We formalize this discussion in the following definition.

Figure 6.27

Figure 6.28

Definition 6.34

The **inverse sine function**, denoted $\sin^{-1}$, is defined by

$$y = \sin^{-1} x \quad \text{if and only if} \quad x = \sin y$$

for $-1 \le x \le 1$ and $-\pi/2 \le y \le \pi/2$.

The **inverse cosine function**, denoted $\cos^{-1}$, is defined by

$$y = \cos^{-1} x \quad \text{if and only if} \quad x = \cos y$$

for $-1 \le x \le 1$ and $0 \le y \le \pi$.

The inverse sine and inverse cosine functions are also called the **arcsine function** (denoted $\arcsin x$) and **arccosine function** (denoted $\arccos x$), respectively. The -1 in $\sin^{-1}$ and $\cos^{-1}$ is not regarded as an exponent, but rather as a means of denoting an inverse function. We may read the notation $y = \sin^{-1} x$ as *y is the inverse sine of x* and the notation $y = \cos^{-1} x$ as *y* is the inverse cosine of *x*. The equations $x = \sin y$ and $x = \cos y$ in the definition allow us to regard *y* as an angle. Thus, we often read the inverse functions as *y is the angle whose sine is x* or *y is the angle whose cosine is x*. Note that

$$-\pi/2 \le \sin^{-1} x \le \pi/2 \quad \text{and} \quad 0 \le \cos^{-1} x \le \pi.$$

ILLUSTRATION

- If $y = \sin^{-1} \dfrac{1}{2}$, then $\sin y = \dfrac{1}{2}$ and $-\dfrac{\pi}{2} \le y \le \dfrac{\pi}{2}$. Hence, $y = \dfrac{\pi}{6}$.

- If $y = \arcsin\left(-\dfrac{1}{2}\right)$, then $\sin y = -\dfrac{1}{2}$ and $-\dfrac{\pi}{2} \le y \le \dfrac{\pi}{2}$. Hence, $y = -\dfrac{\pi}{6}$.

- If $y = \cos^{-1} \dfrac{1}{2}$, then $\cos y = \dfrac{1}{2}$ and $0 \le y \le \pi$. Hence, $y = \dfrac{\pi}{3}$.

- If $y = \arccos\left(-\dfrac{1}{2}\right)$, then $\cos y = -\dfrac{1}{2}$ and $0 \le y \le \pi$. Hence, $y = \dfrac{2\pi}{3}$.

Since the graphs of a function f and its inverse f^{-1} are reflections of each other through the line $y = x$, we can sketch the graphs of $y = \sin^{-1} x$ and $y = \cos^{-1} x$ by reflecting the solid portions of the graphs in Figures 6.27 and 6.28. The graphs of the inverse sine and inverse cosine functions are shown in Figures 6.29 and 6.30. We can also use the equations $x = \sin y$ with $-\pi/2 \le y \le \pi/2$ and $x = \cos y$ with $0 \le y \le \pi$ to find points on the graphs of the inverse functions.

On a calculator, the inverse sine function may be approximated by using a single $\boxed{\text{SIN}^{-1}}$ or $\boxed{\text{ASIN}}$ key, if available, or a two-stroke combination $\boxed{\text{INV}}\,\boxed{\text{SIN}}$. The inverse cosine function is implemented in an analogous fashion. Be sure to set the calculator to *radian mode*. For example, $\boxed{\text{SIN}^{-1}}$ 0.85 yields the approximate result 1.01598529 radians, but if the calculator is set in degrees, the result is 58.21°.

We can proceed in a similar manner to find an inverse for the tangent function. If we restrict the domain of the tangent to the open interval $(-\pi/2, \pi/2)$, we obtain a continuous increasing function (see Figure 6.31 on the following page). We use this *new* function to define the *inverse tangent function*.

Figure 6.29   **Figure 6.30**

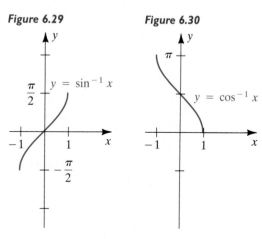

Definition   6.35

The **inverse tangent function**, or **arctangent function**, denoted by $\tan^{-1}$, or arctan, is defined by

$$y = \tan^{-1} x = \arctan x \quad \text{if and only if} \quad x = \tan y$$

for every x and $-\pi/2 < y < \pi/2$.

The domain of the arctangent function is $\mathbb{R}$ and the range is the open interval $(-\pi/2, \pi/2)$. We can obtain the graph of $y = \arctan x$ shown in Figure 6.32 by reflecting the graph in Figure 6.31 through the line $y = x$.

Figure 6.31

Figure 6.32

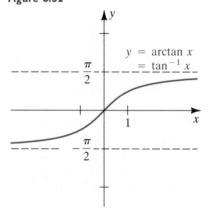

ILLUSTRATION

■   If $y = \arctan(-1)$, then $\tan y = -1$ and $-\dfrac{\pi}{2} < y < \dfrac{\pi}{2}$. Hence, $y = -\dfrac{\pi}{4}$.

■   If $y = \arctan(\sqrt{3})$, then $\tan y = \sqrt{3}$ and $-\dfrac{\pi}{2} < y < \dfrac{\pi}{2}$. Hence, $y = \dfrac{\pi}{3}$.

The relationships $f(f^{-1}(x)) = x$ and $f^{-1}(f(x)) = x$ that hold for any inverse function f^{-1} give us the following properties.

Properties of Inverse
Trigonometric Functions   6.36

(i) $\sin(\sin^{-1} x) = \sin(\arcsin x) = x$ if $-1 \leq x \leq 1$

(ii) $\sin^{-1}(\sin x) = \arcsin(\sin x) = x$ if $-\pi/2 \leq x \leq \pi/2$

(iii) $\cos(\cos^{-1} x) = \cos(\arccos x) = x$ if $-1 \leq x \leq 1$

(iv) $\cos^{-1}(\cos x) = \arccos(\cos x) = x$ if $0 \leq x \leq \pi$

(v) $\tan(\tan^{-1} x) = \tan(\arctan x) = x$ for every x

(vi) $\tan^{-1}(\tan x) = \arctan(\tan x) = x$ if $-\pi/2 < x < \pi/2$

ILLUSTRATION

- $\sin(\sin^{-1}\frac{1}{2}) = \frac{1}{2}$ since $-1 \le \frac{1}{2} \le 1$

- $\arcsin\left(\sin\dfrac{\pi}{4}\right) = \dfrac{\pi}{4}$ since $-\dfrac{\pi}{2} \le \dfrac{\pi}{4} \le \dfrac{\pi}{2}$

- $\cos\left[\cos^{-1}\left(-\frac{1}{2}\right)\right] = -\frac{1}{2}$ since $-1 \le -\frac{1}{2} \le 1$

- $\arccos\left(\cos\dfrac{2\pi}{3}\right) = \dfrac{2\pi}{3}$ since $0 \le \dfrac{2\pi}{3} \le \pi$

- $\tan(\tan^{-1} 1000) = 1000$ by (6.36)(v)

- $\tan^{-1}\left(\tan\dfrac{\pi}{4}\right) = \dfrac{\pi}{4}$ since $-\dfrac{\pi}{2} < \dfrac{\pi}{4} < \dfrac{\pi}{2}$

- $\cos^{-1}\left[\cos\left(-\dfrac{\pi}{4}\right)\right] = \cos^{-1}\left(\dfrac{\sqrt{2}}{2}\right) = \dfrac{\pi}{4}$

- $\arctan(\tan \pi) = \arctan 0 = 0$

- $\sin^{-1}\left(\sin\dfrac{2\pi}{3}\right) = \sin^{-1}\left(\dfrac{\sqrt{3}}{2}\right) = \dfrac{\pi}{3}$

Figure 6.33
$f(x) = \sin^{-1}(\sin x)$

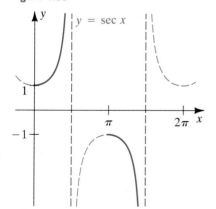

Figure 6.34

Figure 6.35

Be careful when using (6.36). In the final part of the preceding illustration, for example, $2\pi/3$ is *not* between $-\pi/2$ and $\pi/2$, and hence we cannot use (6.36)(ii). Instead, we use properties of reference angles (page 45) to first evaluate $\sin(2\pi/3)$ and then find $\sin^{-1}(\sqrt{3}/2)$. As we see in this illustration, in general, $\sin^{-1}(\sin x) \ne x$. The function $f(x) = \sin^{-1}(\sin x)$ is defined and continuous for all real numbers x and has an interesting graph, part of which is shown in Figure 6.33.

EXAMPLE ■ 1   Find the exact value of $\sec(\arctan\frac{2}{3})$.

SOLUTION   If we let $y = \arctan\frac{2}{3}$, then $\tan y = \frac{2}{3}$. We wish to find $\sec y$. Since $-\pi/2 < \arctan x < \pi/2$ for every x and $\tan y > 0$, it follows that $0 < y < \pi/2$. Thus, we may regard y as the radian measure of an angle of a right triangle such that $\tan y = \frac{2}{3}$, as illustrated in Figure 6.34. By the Pythagorean theorem, the hypotenuse is $\sqrt{3^2 + 2^2} = \sqrt{13}$. Referring to the triangle, we obtain

$$\sec\left(\arctan\frac{2}{3}\right) = \sec y = \frac{\sqrt{13}}{3}.$$

If we consider the graph of $y = \sec x$, there are many ways to restrict x so that we obtain a one-to-one function that takes on every value of the secant function. There is no universal agreement on how this should be done. It is convenient to restrict x to the intervals $[0, \pi/2)$ and $[\pi, 3\pi/2)$, as indicated by the solid portion of the graph of $y = \sec x$ in Figure 6.35, rather than to the "more natural" intervals $[0, \pi/2)$ and $(\pi/2, \pi]$, because the differentiation formula for the inverse secant is simpler. We show in the next section that $(d/dx)(\sec^{-1} x) = 1/(x\sqrt{x^2 - 1})$. Thus, the slope of the tangent line to the graph of $y = \sec^{-1} x$ is negative if $x < -1$ or positive if $x > 1$. For the more natural intervals, the slope is always positive, and we would have $(d/dx)(\sec^{-1} x) = 1/(|x| \sqrt{x^2 - 1})$.

Definition 6.37

> The **inverse secant function**, or **arcsecant function**, denoted by $\sec^{-1}$, or arcsec, is defined by
>
> $$y = \sec^{-1} x = \operatorname{arcsec} x \quad \text{if and only if} \quad x = \sec y$$
>
> for $|x| \geq 1$ and y in $[0, \pi/2)$ or in $[\pi, 3\pi/2)$.

Figure 6.36

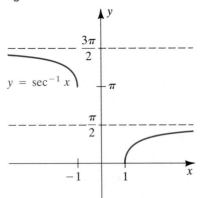

The graph of $y = \sec^{-1} x$ is sketched in Figure 6.36.

The inverse cotangent function, $\cot^{-1}$, and the inverse cosecant function, $\csc^{-1}$, can be defined in similar fashion (see Exercises 25–26).

Most calculators and many computer applications do not provide for the direct evaluation of the secant function or the inverse secant function. We evaluate $\sec x$ by computing the reciprocal of $\cos x$, but there is no simple way to evaluate the inverse secant function. The next example suggests a procedure.

EXAMPLE 2 Use a calculator to approximate $\operatorname{arcsec}(-14.3)$.

SOLUTION From the graph of the inverse secant function in Figure 6.36, we see that $\operatorname{arcsec}(-14.3)$ will lie between π and $3\pi/2$. If we let $y = \operatorname{arcsec}(-14.3)$, then $\sec y = -14.3$ and $\cos y = -(1/14.3)$. Since the range of the inverse cosine function is $[0, \pi]$ with $\pi/2 < \arccos x \leq \pi$ when $x < 0$, a calculator provides the approximation

$$\tilde{y} = \cos^{-1}(-1/14.3) \approx 1.64078352.$$

To find the desired answer in the interval $[\pi, 3\pi/2)$, we treat $\tilde{y}$ as a reference angle and compute the answer as $y = \pi + (\pi - \tilde{y}) \approx 4.64240179$.

EXAMPLE 3 If $-1 \leq x \leq 1$, rewrite $\cos(\sin^{-1} x)$ as an algebraic expression in x.

SOLUTION Let

$$y = \sin^{-1} x, \quad \text{or, equivalently,} \quad \sin y = x.$$

We wish to express $\cos y$ in terms of x. Since $-\pi/2 \leq y \leq \pi/2$, it follows that $\cos y \geq 0$, and hence

$$\cos y = \sqrt{1 - \sin^2 y} = \sqrt{1 - x^2}.$$

Figure 6.37

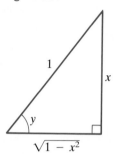

Consequently, $\qquad \cos(\sin^{-1} x) = \sqrt{1 - x^2}.$

The last identity can also be seen geometrically if $0 < x < 1$. In this case, $0 < y < \pi/2$, and we may regard y as the radian measure of an angle of a right triangle such that $\sin y = x$, as illustrated in Figure 6.37. (The side of length $\sqrt{1 - x^2}$ is found by using the Pythagorean theorem.) Referring to the triangle, we have

$$\cos(\sin^{-1} x) = \cos y = \frac{\sqrt{1 - x^2}}{1} = \sqrt{1 - x^2}.$$

DIFFERENTIATING AND INTEGRATING INVERSE TRIGONOMETRIC FUNCTIONS

We consider next the derivatives and integrals of the inverse trigonometric functions and integrals that result in inverse trigonometric functions. We concentrate on the inverse sine, cosine, tangent, and secant functions. The next two theorems provide formulas with $u = g(x)$ differentiable and x restricted to values for which the indicated expressions have meaning. You may find it surprising to learn that although we use trigonometric functions to define inverse trigonometric functions, their derivatives are *algebraic* functions.

Theorem 6.38

(i) $\dfrac{d}{dx}(\sin^{-1} u) = \dfrac{1}{\sqrt{1 - u^2}} \dfrac{du}{dx}$

(ii) $\dfrac{d}{dx}(\cos^{-1} u) = -\dfrac{1}{\sqrt{1 - u^2}} \dfrac{du}{dx}$

(iii) $\dfrac{d}{dx}(\tan^{-1} u) = \dfrac{1}{1 + u^2} \dfrac{du}{dx}$

(iv) $\dfrac{d}{dx}(\sec^{-1} u) = \dfrac{1}{u\sqrt{u^2 - 1}} \dfrac{du}{dx}$

PROOF We shall consider only the special case $u = x$, since the formulas for $u = g(x)$ may then be obtained by applying the chain rule.

If we let $f(x) = \sin x$ and $g(x) = \sin^{-1} x$ in Theorem (6.7), then it follows that the inverse sine function g is differentiable if $|x| < 1$. We shall use implicit differentiation to find $g'(x)$. First note that the equations

$$y = \sin^{-1} x \quad \text{and} \quad \sin y = x$$

are equivalent if $-1 < x < 1$ and $-\pi/2 < y < \pi/2$. Differentiating $\sin y = x$ implicitly, we have

$$\cos y \frac{dy}{dx} = 1$$

and hence
$$\frac{d}{dx}(\sin^{-1} x) = \frac{dy}{dx} = \frac{1}{\cos y}.$$

Since $-\pi/2 < y < \pi/2$, $\cos y$ is positive and, therefore,
$$\cos y = \sqrt{1 - \sin^2 y} = \sqrt{1 - x^2}.$$

Thus,
$$\frac{d}{dx}(\sin^{-1} x) = \frac{1}{\sqrt{1 - x^2}}$$

for $|x| < 1$. The inverse sine function is not differentiable at ± 1. This fact is evident from Figure 6.29, since vertical tangent lines occur at the endpoints of the graph.

The formula for $(d/dx)(\cos^{-1} x)$ can be obtained in similar fashion.

It follows from Theorem (6.7) that the inverse tangent function is differentiable at every real number. Let us consider the equivalent equations
$$y = \tan^{-1} x \quad \text{and} \quad \tan y = x$$

for $-\pi/2 < y < \pi/2$. Differentiating $\tan y = x$ implicitly, we have
$$\sec^2 y \frac{dy}{dx} = 1.$$

Consequently,
$$\frac{d}{dx}(\tan^{-1} x) = \frac{dy}{dx} = \frac{1}{\sec^2 y}.$$

Using the fact that $\sec^2 y = 1 + \tan^2 y = 1 + x^2$ gives us
$$\frac{d}{dx}(\tan^{-1} x) = \frac{1}{1 + x^2}.$$

Finally, consider the equivalent equations
$$y = \sec^{-1} x \quad \text{and} \quad \sec y = x$$

for y in either $(0, \pi/2)$ or $(\pi, 3\pi/2)$. Differentiating $\sec y = x$ implicitly yields
$$\sec y \tan y \frac{dy}{dx} = 1.$$

Since $0 < y < \pi/2$ or $\pi < y < 3\pi/2$, it follows that $\sec y \tan y \neq 0$ and, hence,
$$\frac{d}{dx}(\sec^{-1} x) = \frac{dy}{dx} = \frac{1}{\sec y \tan y}.$$

Using the fact that $\tan y = \sqrt{\sec^2 y - 1} = \sqrt{x^2 - 1}$, we obtain
$$\frac{d}{dx}(\sec^{-1} x) = \frac{1}{x\sqrt{x^2 - 1}}$$

for $|x| > 1$. The inverse secant function is not differentiable at $x = \pm 1$. Note that the graph has vertical tangent lines at the points with these x-coordinates (see Figure 6.36). ■

ILLUSTRATION

$f(x)$	$f'(x)$
$\sin^{-1} 3x$	$\dfrac{1}{\sqrt{1-(3x)^2}}\dfrac{d}{dx}(3x) = \dfrac{3}{\sqrt{1-9x^2}}$
$\arccos(\ln x)$	$-\dfrac{1}{\sqrt{1-(\ln x)^2}}\dfrac{d}{dx}(\ln x) = -\dfrac{1}{x\sqrt{1-(\ln x)^2}}$
$\tan^{-1} e^{2x}$	$\dfrac{1}{1+(e^{2x})^2}\dfrac{d}{dx}(e^{2x}) = \dfrac{2e^{2x}}{1+e^{4x}}$
$\operatorname{arcsec}(x^2)$	$\dfrac{1}{x^2\sqrt{(x^2)^2-1}}\dfrac{d}{dx}(x^2) = \dfrac{2}{x\sqrt{x^4-1}}$

The next example illustrates an application involving derivatives of the inverse trigonometric functions. Exercises 66 and 70 demonstrate other important applications.

EXAMPLE ■ 4   A rocket is fired directly upward with initial velocity 0 and burns fuel at a rate that produces a constant acceleration of 50 ft/sec^2 for $0 \le t \le 5$, with time t in seconds. As illustrated in Figure 6.38, an observer 400 ft from the launching pad visually follows the flight of the rocket.

(a) Express the angle of elevation θ of the rocket as a function of t.

(b) The observer perceives the rocket to be rising fastest when $d\theta/dt$ is largest. (Of course, this is an illusion, since the velocity is steadily increasing.) Determine the height of the rocket at the moment of perceived maximum velocity.

Figure 6.38

SOLUTION

(a) Let $s(t)$ denote the height of the rocket at time t (see Figure 6.38). The fact that the acceleration is always 50 gives us the differential equation

$$s''(t) = 50,$$

subject to the initial conditions $s'(0) = 0$ and $s(0) = 0$. Integrating with respect to t, we obtain

$$\int s''(t)\,dt = \int 50\,dt$$
$$s'(t) = 50t + C$$

for some constant C. Substituting $t = 0$ and using $s'(0) = 0$ gives us $0 = 50(0) + C$, or $C = 0$. Hence,

$$s'(t) = 50t.$$

Integrating again, we have

$$\int s'(t)\,dt = \int 50t\,dt$$
$$s(t) = 25t^2 + D$$

for some constant D. If we substitute $t = 0$ and use $s(0) = 0$, we obtain $0 = 25(0) + D$, or $D = 0$. Hence,

$$s(t) = 25t^2.$$

Referring to Figure 6.38, with $s(t) = 25t^2$, we find

$$\tan\theta = \frac{25t^2}{400} = \frac{t^2}{16}, \quad \text{or} \quad \theta = \arctan\frac{t^2}{16}.$$

(b) By Theorem (6.38), the rate of change of θ with respect to t is

$$\frac{d\theta}{dt} = \frac{1}{1 + (t^2/16)^2}\left(\frac{2t}{16}\right) = \frac{32t}{256 + t^4}.$$

Since we wish to find the maximum value of $d\theta/dt$, we begin by finding the critical numbers of $d\theta/dt$. Using the quotient rule, we obtain

$$\frac{d}{dt}\left(\frac{d\theta}{dt}\right) = \frac{d^2\theta}{dt^2} = \frac{(256 + t^4)(32) - 32t(4t^3)}{(256 + t^4)^2} = \frac{32(256 - 3t^4)}{(256 + t^4)^2}.$$

Considering $d^2\theta/dt^2 = 0$ gives us the critical number $t = \sqrt[4]{256/3}$. It follows from the first (or second) derivative test that $d\theta/dt$ has a maximum value at $t = \sqrt[4]{256/3} \approx 3.04$ sec. The height of the rocket at this time is

$$s(\sqrt[4]{256/3}) = 25(\sqrt[4]{256/3})^2 = 25\sqrt{256/3} \approx 230.9 \text{ ft.}$$

We may use differentiation formulas (i), (ii), and (iv) of Theorem (6.38) to obtain the following integration formulas:

(1) $\displaystyle\int \frac{1}{\sqrt{1 - u^2}}\,du = \sin^{-1}u + C$

(2) $\displaystyle\int \frac{1}{1 + u^2}\,du = \tan^{-1}u + C$

(3) $\displaystyle\int \frac{1}{u\sqrt{u^2 - 1}}\,du = \sec^{-1}u + C$

These formulas can be generalized for $a > 0$ as follows.

Theorem 6.39

(i) $\displaystyle\int \frac{1}{\sqrt{a^2 - u^2}}\,du = \sin^{-1}\frac{u}{a} + C$

(ii) $\displaystyle\int \frac{1}{a^2 + u^2}\,du = \frac{1}{a}\tan^{-1}\frac{u}{a} + C$

(iii) $\displaystyle\int \frac{1}{u\sqrt{u^2 - a^2}}\,du = \frac{1}{a}\sec^{-1}\frac{u}{a} + C$

PROOF   Let us prove (ii). As usual, it is sufficient to consider the case $u = x$. We begin by writing

$$\int \frac{1}{a^2 + x^2}\, dx = \frac{1}{a^2}\int \frac{1}{1 + (x/a)^2}\, dx.$$

Next we make the substitution $v = x/a$, $dv = (1/a)\, dx$. Introducing the factor $1/a$ in the integrand, compensating by multiplying the integral by a, and using formula (2), preceding this theorem, gives us the following:

$$\int \frac{1}{a^2 + x^2}\, dx = \frac{1}{a}\int \frac{1}{1 + (x/a)^2} \cdot \frac{1}{a}\, dx$$

$$= \frac{1}{a}\int \frac{1}{1 + v^2}\, dv$$

$$= \frac{1}{a}\tan^{-1} v + C$$

$$= \frac{1}{a}\tan^{-1} \frac{x}{a} + C$$

The remaining formulas may be proved in similar fashion.   ■

In Example 5 of Section 4.7, we obtained numerical approximations for the value of the definite integral $\int_0^1 [4/(1 + x^2)]\, dx$. We can now use our knowledge of the inverse tangent function to show that the exact value of this integral is π.

EXAMPLE ■ 5   Evaluate $\displaystyle\int_0^1 \frac{4}{1 + x^2}\, dx$.

SOLUTION   Using (6.39)(ii), we have

$$\int_0^1 \frac{4}{1 + x^2}\, dx = 4\int_0^1 \frac{1}{1 + x^2}\, dx$$

$$= 4\big[\arctan x\big]_0^1$$

$$= 4(\arctan 1 - \arctan 0)$$

$$= 4\left(\frac{\pi}{4} - 0\right) = \pi.$$

EXAMPLE ■ 6   Evaluate $\displaystyle\int \frac{e^{2x}}{\sqrt{1 - e^{4x}}}\, dx$.

SOLUTION   The integral may be written as in Theorem (6.39)(i) by letting $a = 1$ and using the substitution

$$u = e^{2x}, \qquad du = 2e^{2x}\, dx.$$

We introduce a factor 2 in the integrand and proceed as follows:

$$\int \frac{e^{2x}}{\sqrt{1 - e^{4x}}}\, dx = \frac{1}{2} \int \frac{1}{\sqrt{1 - (e^{2x})^2}} 2e^{2x}\, dx$$

$$= \frac{1}{2} \int \frac{1}{\sqrt{1 - u^2}}\, du$$

$$= \tfrac{1}{2} \sin^{-1} u + C$$

$$= \tfrac{1}{2} \sin^{-1} e^{2x} + C$$

EXAMPLE ▪ 7   Evaluate $\displaystyle\int \frac{x^2}{5 + x^6}\, dx$.

SOLUTION   The integral may be written as in Theorem (6.39)(ii) by letting $a^2 = 5$ and using the substitution

$$u = x^3, \qquad du = 3x^2\, dx.$$

We introduce a factor 3 in the integrand and proceed as follows:

$$\int \frac{x^2}{5 + x^6}\, dx = \frac{1}{3} \int \frac{1}{5 + (x^3)^2} 3x^2\, dx$$

$$= \frac{1}{3} \int \frac{1}{(\sqrt{5})^2 + u^2}\, du$$

$$= \frac{1}{3} \cdot \frac{1}{\sqrt{5}} \tan^{-1} \frac{u}{\sqrt{5}} + C$$

$$= \frac{\sqrt{5}}{15} \tan^{-1} \frac{x^3}{\sqrt{5}} + C$$

EXAMPLE ▪ 8   Evaluate $\displaystyle\int \frac{1}{x\sqrt{x^4 - 9}}\, dx$.

SOLUTION   The integral may be written as in Theorem (6.39)(iii) by letting $a^2 = 9$ and using the substitution

$$u = x^2, \qquad du = 2x\, dx.$$

We introduce $2x$ in the integrand by multiplying the numerator and the denominator by $2x$ and then proceed as follows:

$$\int \frac{1}{x\sqrt{x^4 - 9}}\, dx = \int \frac{1}{2x \cdot x\sqrt{(x^2)^2 - 3^2}} 2x\, dx$$

$$= \frac{1}{2} \int \frac{1}{u\sqrt{u^2 - 3^2}}\, du$$

$$= \frac{1}{2} \cdot \frac{1}{3} \sec^{-1} \frac{u}{3} + C$$

$$= \frac{1}{6} \sec^{-1} \frac{x^2}{3} + C$$

The formulas we developed in previous chapters for such quantities as areas, volumes of solids of revolution, arc length, and surface area may also be applied to transcendental functions. The resulting definite integrals may be evaluated by finding antiderivatives where possible or may be approximated by the numerical methods discussed in Section 4.7. In the next example, we approximate the arc length of a piece of the inverse secant function using Simpson's rule.

 EXAMPLE ▪ 9 Approximate the arc length of the graph of the function $f(x) = \text{arcsec } x$ from $x = 2$ to $x = 3$ to four decimal places.

SOLUTION From the definition (5.14) of arc length, we know that the arc length of this graph is

$$\int_2^3 \sqrt{1 + [f'(x)]^2} \, dx.$$

If $f(x) = \text{arcsec } x$, then by Theorem (6.38)(iv), $f'(x) = 1/(x\sqrt{x^2 - 1})$, so

$$1 + [f'(x)]^2 = 1 + \frac{1}{x^2(x^2 - 1)},$$

and the arc length is

$$\int_2^3 \sqrt{1 + \frac{1}{x^2(x^2 - 1)}} \, dx.$$

We evaluate this definite integral numerically by using Simpson's rule with $n = 10$, 20, and 40. For each of these values of n, we obtain 1.01783 as an approximation. Thus to four decimal places, the arc length of the graph of $y = \text{arcsec } x$ from $x = 2$ to $x = 3$ is 1.0178.

We conclude this section with a brief look at indefinite integrals of the inverse trigonometric integrals.

Integrals of Inverse Trigonometric Functions 6.40

(i) $\displaystyle\int \sin^{-1} u \, du = u \sin^{-1} u + \sqrt{1 - u^2} + C$

(ii) $\displaystyle\int \cos^{-1} u \, du = u \cos^{-1} u - \sqrt{1 - u^2} + C$

(iii) $\displaystyle\int \tan^{-1} u \, du = u \tan^{-1} u - \frac{1}{2} \ln(1 + u^2) + C$

(iv) $\displaystyle\int \sec^{-1} u \, du = u \sec^{-1} u - \ln \left| u + \sqrt{u^2 - 1} \right| + C$

We derive these forms of the indefinite integrals using the approach demonstrated in the next example.

EXAMPLE ■ 10

(a) Find $f'(x)$ if $f(x) = x \arcsin x$.

(b) Use the result of part (a) to find $\int \arcsin x \, dx$.

SOLUTION

(a) By the product rule (2.19),

$$f'(x) = (x \arcsin x)'$$
$$= (x)'(\arcsin x) + (x)(\arcsin x)'$$
$$= 1 \arcsin x + x \frac{1}{\sqrt{1 - x^2}}$$
$$= \arcsin x + \frac{x}{\sqrt{1 - x^2}}.$$

Thus, $\qquad (x \arcsin x)' = \arcsin x + \frac{x}{\sqrt{1 - x^2}}.$

(b) From part (a), we have

$$\arcsin x = (x \arcsin x)' - \frac{x}{\sqrt{1 - x^2}}.$$

Integrating each side of this equation with respect to x gives

$$\int \arcsin x \, dx = \int \left[(x \arcsin x)' - \frac{x}{\sqrt{1 - x^2}} \right] dx$$
$$= \int (x \arcsin x)' \, dx - \int \frac{x}{\sqrt{1 - x^2}} \, dx$$
$$= x \arcsin x + \sqrt{1 - x^2} + C.$$

EXERCISES 6.7

Exer. 1–11: Find the exact value of the expression, whenever it is defined.

1 (a) $\sin^{-1}\left(-\frac{\sqrt{2}}{2}\right)$ (b) $\cos^{-1}\left(-\frac{1}{2}\right)$

(c) $\tan^{-1}(-\sqrt{3})$

2 (a) $\arcsin \frac{\sqrt{3}}{2}$ (b) $\arccos \frac{\sqrt{2}}{2}$

(c) $\arctan \frac{1}{\sqrt{3}}$

3 (a) $\sin^{-1} \frac{\pi}{3}$ (b) $\cos^{-1} \frac{\pi}{2}$

(c) $\tan^{-1} 1$

4 (a) $\sin[\arcsin(-\frac{3}{10})]$ (b) $\cos(\arccos \frac{1}{2})$

(c) $\tan(\arctan 14)$

5 (a) $\sin^{-1}\left(\sin \frac{\pi}{3}\right)$ (b) $\cos^{-1}\left(\cos \frac{5\pi}{6}\right)$

(c) $\tan^{-1}\left[\tan\left(-\frac{\pi}{6}\right)\right]$

6 (a) $\arcsin\left(\sin \frac{5\pi}{4}\right)$ (b) $\arccos\left(\cos \frac{5\pi}{4}\right)$

(c) $\arctan\left(\tan \frac{7\pi}{4}\right)$

7 (a) $\sin^{-1}\left(\sin\dfrac{2\pi}{3}\right)$ (b) $\cos^{-1}\left(\cos\dfrac{4\pi}{3}\right)$

  (c) $\tan^{-1}\left(\tan\dfrac{7\pi}{6}\right)$

8 (a) $\sin[\cos^{-1}(-\tfrac{1}{2})]$ (b) $\cos(\tan^{-1}1)$

  (c) $\tan[\sin^{-1}(-1)]$

9 (a) $\sin(\tan^{-1}\sqrt{3})$ (b) $\cos(\sin^{-1}1)$

  (c) $\tan(\cos^{-1}0)$

10 (a) $\cot(\sin^{-1}\tfrac{2}{3})$ (b) $\sec[\tan^{-1}(-\tfrac{3}{5})]$

  (c) $\csc[\cos^{-1}(-\tfrac{1}{4})]$

11 (a) $\cot[\sin^{-1}(-\tfrac{2}{5})]$ (b) $\sec(\tan^{-1}\tfrac{7}{4})$

  (c) $\csc(\cos^{-1}\tfrac{1}{5})$

12 If $-1 \le x \le 1$, is it always possible to find the value of $\sin^{-1}(\sin^{-1}x)$ by pressing the calculator key sequence $\boxed{\text{INV}}\ \boxed{\text{SIN}}$ twice? If not, determine the permissible values of x.

---

c   **Exer. 13–16: Find a four-decimal-place approximation of the expression, whenever it is defined.**

13 (a) $\sin^{-1}(-0.931)$ (b) $\tan^{-1}(0.278)$

14 (a) $\cos^{-1}(-0.265)$ (b) $\sec^{-1}(15.4)$

15 (a) $\sec[\sin^{-1}(-0.582)+\tan^{-1}(0.304)]$

  (b) $\cos[\sin^{-1}(0.179)+\tan^{-1}(-1.89)]$

16 (a) $\tan[\sin^{-1}(0.783)+\sec^{-1}(8.54)]$

  (b) $\sin[\cos^{-1}(0.496)+\tan^{-1}(6.12)]$

Exer. 17–20: Rewrite as an algebraic expression in x for $x > 0$.

17 $\sin(\tan^{-1}x)$ **18** $\tan(\arccos x)$

19 $\sec\left(\sin^{-1}\dfrac{x}{3}\right)$ **20** $\cot\left(\sin^{-1}\dfrac{1}{x}\right)$

Exer. 21–26: Sketch the graph of the equation.

21 $y = \sin^{-1}2x$ **22** $y = \tfrac{1}{2}\sin^{-1}x$

23 $y = \cos^{-1}\tfrac{1}{2}x$ **24** $y = 2\cos^{-1}x$

25 $y = \cos(2\arccos x)$ **26** $y = \cos(3\cos^{-1}x)$

27 (a) Define $\cot^{-1}$ by restricting the domain of the cotangent function to the interval $(0, \pi)$.

  (b) Sketch the graph of $y = \cot^{-1}x$.

  (c) Show that

$$\frac{d}{dx}(\cot^{-1}u) = -\frac{1}{1+u^2}\frac{du}{dx} = -\frac{d}{dx}(\tan^{-1}u).$$

28 (a) Define $\csc^{-1}$ by restricting the domain of the cosecant function to $[-\pi/2, 0) \cup (0, \pi/2]$.

  (b) Sketch the graph of $y = \csc^{-1}x$.

  (c) Show that

$$\frac{d}{dx}(\csc^{-1}u) = -\frac{1}{u\sqrt{u^2-1}}\frac{du}{dx}.$$

29 As shown in the figure, a sailboat is following a straight-line course l. The shortest distance from a tracking station T to the course is d miles. As the boat sails, the tracking station records its distance k from T and its direction θ with respect to T. Angle α specifies the direction of the sailboat.

  (a) Express α in terms of d, k, and θ.

  (b) Estimate α to the nearest degree if $d = 50$ mi, $k = 210$ mi, and $\theta = 53.4°$.

Exercise 29

30 An art critic whose eye level is 6 ft above the floor views a painting that is 10 ft in height and is mounted 4 ft above the floor, as shown in the figure.

  (a) If the critic is standing x feet from the wall, express the viewing angle θ in terms of x.

Exercise 30

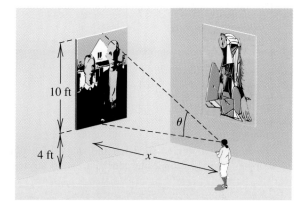

(b) Use the addition formula for the tangent to show that

$$\theta = \tan^{-1}\left(\frac{10x}{x^2 - 16}\right).$$

(c) For what value of x is $\theta = 45°$?

Exer. 31–48: Find $f'(x)$ if $f(x)$ is the given expression.

31 $\sin^{-1}\sqrt{x}$

32 $\sin^{-1}\frac{1}{3}x$

33 $\tan^{-1}(3x - 5)$

34 $\tan^{-1}(x^2)$

35 $e^{-x}\operatorname{arcsec} e^{-x}$

36 $\sqrt{\operatorname{arcsec} 3x}$

37 $\ln\arctan(x^2)$

38 $\arcsin \ln x$

39 $(1 + \cos^{-1} 3x)^3$

40 $\cos^{-1}\cos e^x$

41 $\cos(x^{-1}) + (\cos x)^{-1} + \cos^{-1} x$

42 $x\arccos\sqrt{4x + 1}$

43 $3^{\arcsin(x^3)}$

44 $\left(\dfrac{1}{x} - \arcsin\dfrac{1}{x}\right)^4$

45 $\dfrac{\arctan x}{x^2 + 1}$

46 $(\sin 2x)(\sin^{-1} 2x)$

47 $\sqrt{x}\sec^{-1}\sqrt{x}$

48 $(\tan x)^{\arctan x}$

Exer. 49–50: Find y'.

49 $x^2 + x\sin^{-1} y = ye^x$

50 $\ln(x + y) = \tan^{-1} xy$

Exer. 51–62: Evaluate the integral.

51 (a) $\displaystyle\int \frac{1}{x^2 + 16}\,dx$

(b) $\displaystyle\int_0^4 \frac{1}{x^2 + 16}\,dx$

52 (a) $\displaystyle\int \frac{e^x}{1 + e^{2x}}\,dx$

(b) $\displaystyle\int_0^1 \frac{e^x}{1 + e^{2x}}\,dx$

53 (a) $\displaystyle\int \frac{x}{\sqrt{1 - x^4}}\,dx$

(b) $\displaystyle\int_0^{\sqrt{2}/2} \frac{x}{\sqrt{1 - x^4}}\,dx$

54 $\displaystyle\int \frac{\sin x}{\cos^2 x + 1}\,dx$

55 $\displaystyle\int \frac{1}{\sqrt{x}(1 + x)}\,dx$

56 $\displaystyle\int \frac{\cos x}{\sqrt{9 - \sin^2 x}}\,dx$

57 $\displaystyle\int \frac{e^x}{\sqrt{16 - e^{2x}}}\,dx$

58 $\displaystyle\int \frac{\sec x\tan x}{1 + \sec^2 x}\,dx$

59 $\displaystyle\int \frac{x}{x^2 + 9}\,dx$

60 $\displaystyle\int \frac{1}{x\sqrt{x^6 - 4}}\,dx$

61 $\displaystyle\int \frac{1}{\sqrt{e^{2x} - 25}}\,dx$

62 $\displaystyle\int \frac{1}{x\sqrt{x - 1}}\,dx$

63 The floor of a storage shed has the shape of a right triangle. The sides opposite and adjacent to an acute angle θ of the triangle are measured as 10 ft and 7 ft, respectively, with a possible error of ± 0.5 in. in the 10-ft measurement. Use the differential of an inverse trigonometric function to approximate the error in the calculated value of θ.

64 Use differentials to approximate the arc length of the graph of $y = \tan^{-1} x$ from $A(0, 0)$ to $B(0.1, \tan^{-1} 0.1)$.

65 An airplane at a constant altitude of 5 mi and a speed of 500 mi/hr is flying in a direction away from an observer on the ground. Use inverse trigonometric functions to find the rate at which the angle of elevation is changing when the airplane flies over a point 2 mi from the observer.

66 A searchlight located $\frac{1}{8}$ mi from the nearest point P on a straight road is trained on an automobile traveling on the road at a rate of 50 mi/hr. Use inverse trigonometric functions to find the rate at which the searchlight is rotating when the car is $\frac{1}{4}$ mi from P.

67 A billboard 20 ft high is located on top of a building, with its lower edge 60 ft above the level of a viewer's eye. Use inverse trigonometric functions to find how far from a point directly below the sign a viewer should stand to maximize the angle between the lines of sight of the top and bottom of the billboard (see Example 9 of Section 3.6).

68 The velocity, at time t, of a point moving on a coordinate line is $(1 + t^2)^{-1}$ ft/sec. If the point is at the origin at $t = 0$, find its position at the instant that the acceleration and the velocity have the same absolute value.

69 A missile is fired vertically from a point that is 5 mi from a tracking station and at the same elevation. For the first 20 sec of flight, its angle of elevation changes at a constant rate of $2°$ per second. Use inverse trigonometric functions to find the velocity of the missile when the angle of elevation is $30°$.

70 Blood flowing through a blood vessel causes a loss of energy due to friction. According to *Poiseuille's law*, this energy loss E is given by $E = kl/r^4$, where r is the radius of the blood vessel, l is the length, and k is a constant. Suppose that a blood vessel of radius r_2 and length l_2 branches off, at an angle θ, from a blood vessel of radius r_1 and length l_1, as illustrated in the figure on the following page, where the white arrows indicate the direction of blood flow. The energy loss is then the sum of the individual energy losses—that is,

$$E = \frac{kl_1}{r_1^4} + \frac{kl_2}{r_2^4}.$$

Express l_1 and l_2 in terms of a, b, and θ, and find the angle that minimizes the energy loss.

Exercise 70

Exer. 71–74: (a) Verify the correctness by differentiation. (b) Derive the formulas using the approach illustrated in Example 10.

71 Formula (i) of (6.40)   **72** Formula (ii) of (6.40)

73 Formula (iii) of (6.40)

74 Formula (iv) of (6.40) (Hint for part (b): Verify first that if $g(x) = \ln|x + \sqrt{x^2 - 1}|$, then $g'(x) = 1/\sqrt{x^2 - 1}$.)

Exer. 75–78: Evaluate the integral.

75 $\displaystyle\int \sin^{-1} 2x \, dx$ **76** $\displaystyle\int \cos^{-1} \tfrac{1}{3}x \, dx$

77 $\displaystyle\int x \tan^{-1}(x^2) \, dx$ **78** $\displaystyle\int \frac{\sec^{-1} \sqrt{x}}{\sqrt{x}} \, dx$

c **Exer. 79–82: Approximate the arc length of the graph of the function between A and B. Use Simpson's rule or numerical integration provided on a calculator or a computer to ensure at least four correct decimal places.**

79 $y = \arcsin x$; $A(0, 0)$, $B\left(\dfrac{1}{2}, \dfrac{\pi}{6}\right)$

80 $y = \arccos x$; $A\left(-\dfrac{\sqrt{2}}{2}, \dfrac{3\pi}{4}\right)$, $B\left(\dfrac{\sqrt{2}}{2}, \dfrac{\pi}{4}\right)$

81 $y = \arctan x$; $A(0, 0)$, $B\left(\sqrt{3}, \dfrac{\pi}{3}\right)$

82 $y = \arctan x$; $A(-5, -\arctan 5)$, $B(5, \arctan 5)$

c **Exer. 83–84: Approximate the surface area generated if the graph of the function between A and B is revolved about the x-axis. Use Simpson's rule or numerical integration provided on a calculator or a computer to ensure at least four correct decimal places.**

83 $y = 4\arctan(x^2)$; $A(0, 0)$, $B(1, \pi)$

84 $y = \text{arcsec } x$; $A(2, \text{arcsec } 2)$, $B(10, \text{arcsec } 10)$

6.8 HYPERBOLIC AND INVERSE HYPERBOLIC FUNCTIONS

The hyperbolic functions and their inverses, which we investigate in this section, are used to solve a variety of problems in the physical sciences and engineering.

HYPERBOLIC FUNCTIONS

Many of the advanced applications of calculus involve the exponential expressions

$$\frac{e^x - e^{-x}}{2} \quad \text{and} \quad \frac{e^x + e^{-x}}{2},$$

which define the hyperbolic functions. The properties of these expressions are similar in many ways to those of $\sin x$ and $\cos x$. Later in our discussion, we shall see why they are called the *hyperbolic sine* and the *hyperbolic cosine* of x.

Definition 6.41

The **hyperbolic sine function**, denoted by **sinh**, and the **hyperbolic cosine function**, denoted by **cosh**, are defined by

$$\sinh x = \frac{e^x - e^{-x}}{2} \quad \text{and} \quad \cosh x = \frac{e^x + e^{-x}}{2}$$

for every real number x.

We pronounce $\sinh x$ and $\cosh x$ as *sinch x* and *kosh x*, respectively.

The graph of $y = \cosh x$ may be found by **addition of y-coordinates**. Noting that $\cosh x = \frac{1}{2}e^x + \frac{1}{2}e^{-x}$, we first sketch the graphs of $y = \frac{1}{2}e^x$ and $y = \frac{1}{2}e^{-x}$ on the same coordinate plane, as shown with dashes in Figure 6.39. We then add the y-coordinates of points on these graphs to obtain the graph of $y = \cosh x$. Note that the range of cosh is $[1, \infty)$.

We may find the graph of $y = \sinh x$ by adding y-coordinates of the graphs of $y = \frac{1}{2}e^x$ and $y = -\frac{1}{2}e^{-x}$, as shown in Figure 6.40.

Figure 6.39

Figure 6.40

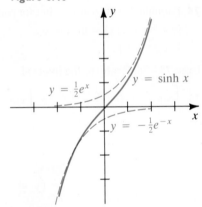

Some scientific calculators have keys that can be used to find values of sinh and cosh directly. We can also substitute numbers for x in Definition (6.41), as in the following illustration.

ILLUSTRATION

- $\sinh 3 = \dfrac{e^3 - e^{-3}}{2} \approx 10.0179$ ■ $\cosh 0.5 = \dfrac{e^{0.5} + e^{-0.5}}{2} \approx 1.1276$

The hyperbolic cosine function can be used to describe the shape of a uniform flexible cable, or chain, whose ends are supported from the same height. As illustrated in Figure 6.41, telephone or power lines may be strung between poles in this manner. The shape of the cable appears to be

Figure 6.41

a parabola, but is actually a **catenary** (after the Latin word for *chain*). If we introduce a coordinate system, as in Figure 6.41, we will later show that an equation corresponding to the shape of the cable is $y = a \cosh(x/a)$ for some real number a.

The hyperbolic cosine function also occurs in the analysis of motion in a resisting medium. If an object is dropped from a given height and if air resistance is disregarded, then the distance y that it falls in t seconds is $y = \frac{1}{2}gt^2$, where g is a gravitational constant. However, air resistance cannot always be disregarded. As the velocity of the object increases, air resistance may significantly affect its motion. For example, if the air resistance is directly proportional to the square of the velocity, then the distance y that the object falls in t seconds is given by

$$y = A \ln(\cosh Bt)$$

for constants A and B (see Exercise 42). Another application is given in Example 2 of this section.

Many identities similar to those for trigonometric functions hold for the hyperbolic sine and cosine functions. For example, if $\cosh^2 x$ and $\sinh^2 x$ denote $(\cosh x)^2$ and $(\sinh x)^2$, respectively, we have the following identity.

Theorem 6.42

$$\cosh^2 x - \sinh^2 x = 1$$

PROOF By Definition (6.41),

$$\cosh^2 x - \sinh^2 x = \left(\frac{e^x + e^{-x}}{2}\right)^2 - \left(\frac{e^x - e^{-x}}{2}\right)^2$$

$$= \frac{e^{2x} + 2 + e^{-2x}}{4} - \frac{e^{2x} - 2 + e^{-2x}}{4}$$

$$= \frac{e^{2x} + 2 + e^{-2x} - e^{2x} + 2 - e^{-2x}}{4}$$

$$= \frac{4}{4} = 1. \quad \blacksquare$$

Theorem (6.42) is analogous to the identity $\cos^2 x + \sin^2 x = 1$. Other hyperbolic identities are stated in the exercises. To verify an identity, it is sufficient to express the hyperbolic functions in terms of exponential functions and show that one side of the equation can be transformed into the other, as illustrated in the proof of Theorem (6.42). The hyperbolic identities are similar to (but not always the same as) certain trigonometric identities—differences usually involve signs of terms.

If t is a real number, there is an interesting geometric relationship between the points $P(\cos t, \sin t)$ and $Q(\cosh t, \sinh t)$ in a coordinate plane. Let us consider the graphs of $x^2 + y^2 = 1$ and $x^2 - y^2 = 1$, sketched in

Figure 6.42 $x^2 + y^2 = 1$ **Figure 6.43** $x^2 - y^2 = 1$

Figures 6.42 and 6.43. The graph in Figure 6.42 is the unit circle with center at the origin. The graph in Figure 6.43 is a *hyperbola*. (Hyperbolas and their properties are discussed in the Precalculus Review Chapter.) Note first that since $\cos^2 t + \sin^2 t = 1$, the point $P(\cos t, \sin t)$ is on the circle $x^2 + y^2 = 1$. Next, by Theorem (6.42), $\cosh^2 t - \sinh^2 t = 1$, and hence the point $Q(\cosh t, \sinh t)$ is on the hyperbola $x^2 - y^2 = 1$. These are the reasons for referring to cos and sin as *circular* functions and to cosh and sinh as *hyperbolic* functions.

The graphs in Figures 6.42 and 6.43 are related in another way. If $0 < t < \pi/2$, then t is the radian measure of angle POB, shown in Figure 6.42. The area A of the shaded circular sector is $A = \frac{1}{2}(1)^2 t = \frac{1}{2}t$, and hence $t = 2A$. Similarly, if $Q(\cosh t, \sinh t)$ is the point in Figure 6.43, then $t = 2A$ for the area A of the shaded hyperbolic sector (see Exercise 33).

The impressive analogies between the trigonometric and the hyperbolic sine and cosine functions motivate us to define hyperbolic functions that correspond to the four remaining trigonometric functions. The **hyperbolic tangent, hyperbolic cotangent, hyperbolic secant,** and **hyperbolic cosecant functions,** denoted by tanh, coth, sech, and csch, respectively, are defined as follows.

Definition 6.43

(i) $\tanh x = \dfrac{\sinh x}{\cosh x} = \dfrac{e^x - e^{-x}}{e^x + e^{-x}}$

(ii) $\coth x = \dfrac{\cosh x}{\sinh x} = \dfrac{e^x + e^{-x}}{e^x - e^{-x}}, \quad x \neq 0$

(iii) $\operatorname{sech} x = \dfrac{1}{\cosh x} = \dfrac{2}{e^x + e^{-x}}$

(iv) $\operatorname{csch} x = \dfrac{1}{\sinh x} = \dfrac{2}{e^x - e^{-x}}, \quad x \neq 0$

Figure 6.44

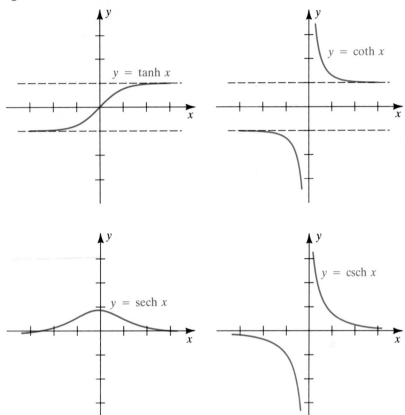

We pronounce the four function values in the preceding definition as *tansh x, cotansh x, setch x,* and *cosetch x.* Their graphs are sketched in Figure 6.44.

If we divide both sides of the identity $\cosh^2 x - \sinh^2 x = 1$ (see (6.42)) by $\cosh^2 x$, we obtain

$$\frac{\cosh^2 x}{\cosh^2 x} - \frac{\sinh^2 x}{\cosh^2 x} = \frac{1}{\cosh^2 x}.$$

Using the definitions of $\tanh x$ and $\operatorname{sech} x$ gives us (i) of the next theorem. Formula (ii) may be obtained by dividing both sides of (6.42) by $\sinh^2 x$.

Theorem 6.44

> **(i)** $1 - \tanh^2 x = \operatorname{sech}^2 x$ **(ii)** $\coth^2 x - 1 = \operatorname{csch}^2 x$

Note the similarities and differences between (6.44) and the analogous trigonometric identities.

Derivative formulas for the hyperbolic functions are listed in the next theorem, where $u = g(x)$ and g is differentiable.

Theorem 6.45

$$(i)\ \ \frac{d}{dx}(\sinh u) = \cosh u\, \frac{du}{dx}$$

$$(ii)\ \ \frac{d}{dx}(\cosh u) = \sinh u\, \frac{du}{dx}$$

$$(iii)\ \ \frac{d}{dx}(\tanh u) = \operatorname{sech}^2 u\, \frac{du}{dx}$$

$$(iv)\ \ \frac{d}{dx}(\coth u) = -\operatorname{csch}^2 u\, \frac{du}{dx}$$

$$(v)\ \ \frac{d}{dx}(\operatorname{sech} u) = -\operatorname{sech} u \tanh u\, \frac{du}{dx}$$

$$(vi)\ \ \frac{d}{dx}(\operatorname{csch} u) = -\operatorname{csch} u \coth u\, \frac{du}{dx}$$

PROOF   As usual, we consider only the case $u = x$. Since $(d/dx)(e^x) = e^x$ and $(d/dx)(e^{-x}) = -e^{-x}$,

$$\frac{d}{dx}(\sinh x) = \frac{d}{dx}\left(\frac{e^x - e^{-x}}{2}\right) = \frac{e^x + e^{-x}}{2} = \cosh x$$

and

$$\frac{d}{dx}(\cosh x) = \frac{d}{dx}\left(\frac{e^x + e^{-x}}{2}\right) = \frac{e^x - e^{-x}}{2} = \sinh x.$$

To differentiate $\tanh x$, we apply the quotient rule as follows:

$$\frac{d}{dx}(\tanh x) = \frac{d}{dx}\left(\frac{\sinh x}{\cosh x}\right)$$

$$= \frac{\cosh x (d/dx)(\sinh x) - \sinh x (d/dx)(\cosh x)}{\cosh^2 x}$$

$$= \frac{\cosh^2 x - \sinh^2 x}{\cosh^2 x}$$

$$= \frac{1}{\cosh^2 x}$$

$$= \operatorname{sech}^2 x$$

The remaining formulas can be proved in similar fashion. ■

EXAMPLE ■ 1   If $f(x) = \cosh(x^2 + 1)$, find $f'(x)$.

SOLUTION Applying Theorem (6.45)(i), with $u = x^2 + 1$, we obtain

$$f'(x) = \sinh(x^2 + 1) \cdot \frac{d}{dx}(x^2 + 1)$$

$$= 2x \sinh(x^2 + 1).$$

Figure 6.45

Basement wall

Radon gas

Basement

Soil

x

EXAMPLE ■ 2 Radon gas can readily diffuse through solid materials such as brick and cement. If the direction of diffusion in a basement wall is perpendicular to the surface, as illustrated in Figure 6.45, then the radon concentration $f(x)$ (in joules/cm^3) in the air-filled pores within the wall at a distance x from the outside surface can be approximated by

$$f(x) = A \sinh(qx) + B \cosh(qx) + k,$$

where the constant q depends on the porosity of the wall, the half-life of radon, and a diffusion coefficient; the constant k is the maximum radon concentration in the air-filled pores; and A and B are constants that depend on initial conditions. Show that $y = f(x)$ is a solution of the *diffusion equation*

$$\frac{d^2y}{dx^2} - q^2y + q^2k = 0.$$

SOLUTION Differentiating $y = f(x)$ twice gives us

$$\frac{dy}{dx} = qA \cosh(qx) + qB \sinh(qx)$$

and

$$\frac{d^2y}{dx^2} = q^2A \sinh(qx) + q^2B \cosh(qx).$$

Since $y = A \sinh(qx) + B \cosh(qx) + k$, we have

$$q^2y = q^2A \sinh(qx) + q^2B \cosh(qx) + q^2k.$$

Subtracting the expressions for d^2y/dx^2 and q^2y yields

$$\frac{d^2y}{dx^2} - q^2y = -q^2k$$

and hence

$$\frac{d^2y}{dx^2} - q^2y + q^2k = 0.$$

The integration formulas that correspond to the derivative formulas in Theorem (6.45) are as follows.

Theorem   6.46

(i) $\displaystyle\int \sinh u \, du = \cosh u + C$

(ii) $\displaystyle\int \cosh u \, du = \sinh u + C$

(iii) $\displaystyle\int \text{sech}^2 u \, du = \tanh u + C$

(iv) $\displaystyle\int \text{csch}^2 u \, du = -\coth u + C$

(v) $\displaystyle\int \text{sech } u \tanh u \, du = -\text{sech } u + C$

(vi) $\displaystyle\int \text{csch } u \coth u \, du = -\text{csch } u + C$

EXAMPLE ▪ 3   Evaluate $\displaystyle\int x^2 \sinh x^3 \, dx$.

SOLUTION   If we let $u = x^3$, then $du = 3x^2 \, dx$ and

$$\int x^2 \sinh x^3 \, dx = \tfrac{1}{3} \int (\sinh x^3) 3x^2 \, dx$$

$$= \tfrac{1}{3} \int \sinh u \, du = \tfrac{1}{3} \cosh u + C = \tfrac{1}{3} \cosh x^3 + C.$$

INVERSE HYPERBOLIC FUNCTIONS

We now investigate the inverses of the hyperbolic functions, which frequently occur in evaluating certain types of integrals. We will also see how an inverse hyperbolic function is used in the derivation of the equation for a hanging cable.

The hyperbolic sine function is continuous and increasing for every x and hence, by Theorem (6.6), has a continuous, increasing inverse function, denoted by $\sinh^{-1}$. Since $\sinh x$ is defined in terms of e^x, we might expect that $\sinh^{-1}$ can be expressed in terms of the inverse, ln, of the natural exponential function. The first formula of the next theorem shows that this is the case.

Theorem   6.47

(i) $\sinh^{-1} x = \ln(x + \sqrt{x^2 + 1})$

(ii) $\cosh^{-1} x = \ln(x + \sqrt{x^2 - 1}), \quad x \geq 1$

(iii) $\tanh^{-1} x = \dfrac{1}{2} \ln \dfrac{1 + x}{1 - x}, \quad |x| < 1$

(iv) $\text{sech}^{-1} x = \ln \dfrac{1 + \sqrt{1 - x^2}}{x}, \quad 0 < x \leq 1$

PROOF   To prove (i), we begin by noting that

$$y = \sinh^{-1} x \quad \text{if and only if} \quad x = \sinh y.$$

The equation $x = \sinh y$ can be used to find an explicit form for $\sinh^{-1} x$. Thus, if

$$x = \sinh y = \frac{e^y - e^{-y}}{2},$$

then

$$e^y - 2x - e^{-y} = 0.$$

Multiplying both sides by e^y, we obtain

$$e^{2y} - 2xe^y - 1 = 0.$$

Applying the quadratic formula yields

$$e^y = \frac{2x \pm \sqrt{4x^2 + 4}}{2}, \quad \text{or} \quad e^y = x \pm \sqrt{x^2 + 1}.$$

Since $x - \sqrt{x^2 + 1} < 0$ and e^y is never negative, we must have

$$e^y = x + \sqrt{x^2 + 1}.$$

The equivalent logarithmic form is

$$y = \ln(x + \sqrt{x^2 + 1});$$

that is,

$$\sinh^{-1} x = \ln(x + \sqrt{x^2 + 1}).$$

Formulas (ii)–(iv) are obtained in similar fashion. As with trigonometric functions, some inverse functions exist only if the domain is restricted. For example, if the domain of cosh is restricted to the set of nonnegative real numbers, then the resulting function is continuous and increasing, and its inverse function $\cosh^{-1}$ is defined by

$$y = \cosh^{-1} x \quad \text{if and only if} \quad \cosh y = x, \quad y \geq 0.$$

Employing the process used for $\sinh^{-1} x$ leads us to (ii). Similarly,

$$y = \tanh^{-1} x \quad \text{if and only if} \quad \tanh y = x \quad \text{for} \quad |x| < 1.$$

Using Definition (6.43), we may write $\tanh y = x$ as

$$\frac{e^y - e^{-y}}{e^y + e^{-y}} = x.$$

Solving for y gives us (iii).

Finally, if we restrict the domain of sech to nonnegative numbers, the result is a one-to-one function, and we define

$$y = \operatorname{sech}^{-1} x \quad \text{if and only if} \quad \operatorname{sech} y = x, \quad y \geq 0.$$

Again, introducing the exponential form leads to (iv). ■

In the next theorem, $u = g(x)$, where g is differentiable and x is suitably restricted.

Theorem 6.48

(i) $\dfrac{d}{dx}(\sinh^{-1} u) = \dfrac{1}{\sqrt{u^2 + 1}}\dfrac{du}{dx}$

(ii) $\dfrac{d}{dx}(\cosh^{-1} u) = \dfrac{1}{\sqrt{u^2 - 1}}\dfrac{du}{dx}, \quad u > 1$

(iii) $\dfrac{d}{dx}(\tanh^{-1} u) = \dfrac{1}{1 - u^2}\dfrac{du}{dx}, \quad |u| < 1$

(iv) $\dfrac{d}{dx}(\operatorname{sech}^{-1} u) = \dfrac{-1}{u\sqrt{1 - u^2}}\dfrac{du}{dx}, \quad 0 < u < 1$

PROOF   By Theorem (6.47)(i),

$$\frac{d}{dx}(\sinh^{-1} x) = \frac{d}{dx}(\ln(x + \sqrt{x^2 + 1}))$$

$$= \frac{1}{x + \sqrt{x^2 + 1}}\left(1 + \frac{x}{\sqrt{x^2 + 1}}\right)$$

$$= \frac{\sqrt{x^2 + 1} + x}{(x + \sqrt{x^2 + 1})\sqrt{x^2 + 1}}$$

$$= \frac{1}{\sqrt{x^2 + 1}}.$$

This formula can be extended to $(d/dx)(\sinh^{-1} u)$ by applying the chain rule. The remaining formulas can be proved in similar fashion. ■

EXAMPLE 4   If $y = \sinh^{-1}(\tan x)$, find dy/dx.

SOLUTION   Using Theorem (6.48)(i) with $u = \tan x$, we have

$$\frac{dy}{dx} = \frac{1}{\sqrt{\tan^2 x + 1}}\frac{d}{dx}\tan x = \frac{1}{\sqrt{\sec^2 x}}\sec^2 x$$

$$= \frac{1}{|\sec x|}|\sec x|^2 = |\sec x|.$$

The following theorem may be verified by differentiating the right-hand side of each formula.

Theorem 6.49

(i) $\displaystyle\int \frac{1}{\sqrt{a^2 + u^2}}\, du = \sinh^{-1}\frac{u}{a} + C, \quad a > 0$

(ii) $\displaystyle\int \frac{1}{\sqrt{u^2 - a^2}}\, du = \cosh^{-1}\frac{u}{a} + C, \quad 0 < a < u$

(iii) $\displaystyle\int \frac{1}{a^2 - u^2}\, du = \frac{1}{a}\tanh^{-1}\frac{u}{a} + C, \quad |u| < a$

(iv) $\displaystyle\int \frac{1}{u\sqrt{a^2 - u^2}}\, du = -\frac{1}{a}\operatorname{sech}^{-1}\frac{|u|}{a} + C, \quad 0 < |u| < a$

If we use Theorem (6.47), then each of the integration formulas in the preceding theorem can be expressed in terms of the natural logarithm function. To illustrate,

$$\int \frac{1}{\sqrt{a^2 + u^2}}\, du = \sinh^{-1}\frac{u}{a} + C$$

$$= \ln\left(\frac{u}{a} + \sqrt{\left(\frac{u}{a}\right)^2 + 1}\right) + C.$$

We can show that if $a > 0$, then the last formula can be written as

$$\int \frac{1}{\sqrt{a^2 + u^2}}\, du = \ln(u + \sqrt{a^2 + u^2}) + D,$$

where D is a constant. In Section 7.3, we shall discuss another method for evaluating the integrals in Theorem (6.49).

EXAMPLE 5   Evaluate $\displaystyle\int \frac{1}{\sqrt{25 + 9x^2}}\, dx$.

SOLUTION   We may express the integral as in Theorem (6.49)(i), by using the substitution

$$u = 3x, \quad du = 3\, dx.$$

Since du contains the factor 3, we adjust the integrand by multiplying by 3 and then compensate by multiplying the integral by $\frac{1}{3}$ before substituting:

$$\int \frac{1}{\sqrt{25 + 9x^2}}\, dx = \frac{1}{3}\int \frac{1}{\sqrt{5^2 + (3x)^2}}\, 3\, dx$$

$$= \frac{1}{3}\int \frac{1}{\sqrt{5^2 + u^2}}\, du$$

$$= \frac{1}{3}\sinh^{-1}\frac{u}{5} + C$$

$$= \frac{1}{3}\sinh^{-1}\frac{3x}{5} + C$$

EXAMPLE ▪ 6   Evaluate $\int \dfrac{e^x}{16 - e^{2x}}\, dx$.

SOLUTION   Substituting $u = e^x$, $du = e^x\, dx$ and applying Theorem (6.49)(iii) with $a = 4$, we have

$$\int \frac{e^x}{16 - e^{2x}}\, dx = \int \frac{1}{4^2 - (e^x)^2}\, e^x\, dx$$

$$= \int \frac{1}{4^2 - u^2}\, du$$

$$= \frac{1}{4} \tanh^{-1} \frac{u}{4} + C$$

$$= \frac{1}{4} \tanh^{-1} \frac{e^x}{4} + C$$

for $|u| < a$ (that is, $e^x < 4$).

We now consider how the hyperbolic cosine and the inverse hyperbolic sine functions are used in describing the shape of the curve along which a hanging cable lies. We first derive a differential equation for the function whose graph is the curve, and then we solve the differential equation.

Figure 6.41 shows a hanging cable in the form of a power line strung between two towers. A section of the cable is shown in Figure 6.46, where we have set up a coordinate system with the vertical y-axis running through the lowest point $A(0, y_0)$ of the cable. Consider a section of the cable running upward from A to a point P. Figure 6.46 also shows the forces acting on the cable: There is a horizontal tension H at the point A, a tangential tension T at the point P, and a downward gravitational force ws.

The tangential tension can be resolved in a horizontal component $T\cos\theta$ and a vertical component $T\sin\theta$, where θ is the angle that the tangent line to the cable at P makes with the horizontal. (This angle is also the angle of inclination of the tangent line.) Thus, the derivative dy/dx at P is equal to $\tan\theta$. The force due to gravity is equal to the weight of the section of the cable, expressed as ws, where w is the weight per unit length and s is the length of the section.

Since the cable is not moving, the forces acting on any section of it must cancel out. Since the cable is not moving to the right or the left, the magnitude of the horizontal tension at point A equals the magnitude of the horizontal tension at point P:

$$T\cos\theta = H$$

But the cable is also stationary in the vertical direction, so the magnitude of the gravitational force equals the vertical tension at P:

$$T\sin\theta = ws$$

We can now write

$$\frac{ws}{H} = \frac{T\sin\theta}{T\cos\theta} = \tan\theta = \frac{dy}{dx},$$

Figure 6.46
Hanging cable

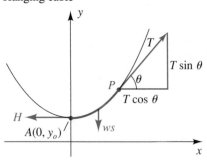

or simply,
$$\frac{dy}{dx} = \frac{ws}{H}.$$

If we differentiate this equation with respect to x and use Theorem (5.17), we obtain

$$\frac{d^2y}{dx^2} = \frac{w}{H}\frac{ds}{dx} = \frac{w}{H}\sqrt{1 + \left(\frac{dy}{dx}\right)^2}.$$

Letting $a = w/H$ gives

$$\frac{d^2y}{dx^2} = a\sqrt{1 + \left(\frac{dy}{dx}\right)^2},$$

for some constant a, as the differential equation satisfied by the equation $y = f(x)$ of the curve formed by a hanging cable. We can now solve the differential equation to find an explicit expression for the function f.

EXAMPLE ■ 7   Solve the differential equation

$$\frac{d^2y}{dx^2} = a\sqrt{1 + \left(\frac{dy}{dx}\right)^2}$$

to find an explicit formula for the curve of a hanging cable.

SOLUTION   The differential equation

$$\frac{d^2y}{dx^2} = a\sqrt{1 + \left(\frac{dy}{dx}\right)^2}$$

is a second-order differential equation, because it involves the second derivative of y with respect to x. We first reduce it to a first-order differential equation by the substitution

$$z = \frac{dy}{dx}$$

so that

$$\frac{d^2y}{dx^2} = \frac{d}{dx}\frac{dy}{dx} = \frac{dz}{dx}.$$

This result converts the original differential equation to a first-order equation

$$\frac{dz}{dx} = a\sqrt{1 + z^2}$$

in which the variables separate. Dividing each side by $\sqrt{1 + z^2}$ and integrating, we obtain

$$\int \frac{1}{\sqrt{1 + z^2}}\,dz = \int a\,dx.$$

The integrand on the left-hand side of the equation, $1/\sqrt{1 + z^2}$, is the derivative of the inverse hyperbolic sine of z. By Theorem (6.49)(i), we have

$$\sinh^{-1} z = ax + C \quad \text{and hence} \quad z = \sinh(ax + C).$$

Since $z = dy/dx$, the last equation becomes

$$\frac{dy}{dx} = \sinh(ax + C).$$

Because $(0, y_0)$ is the minimum point on the curve, the tangent line to the curve at $(0, y_0)$ is horizontal. Thus, at $x = 0$, $dy/dx = 0$, and

$$0 = \sinh(a \cdot 0 + C) = \sinh C.$$

Therefore, $\qquad\qquad C = 0 \quad$ and $\quad \dfrac{dy}{dx} = \sinh ax.$

Thus, $\qquad\qquad y = \displaystyle\int \sinh ax \, dx = \frac{1}{a}(\cosh ax) + C.$

To find the constant of integration, we first use the fact that $y = y_0$ at $x = 0$:

$$y_0 = \frac{1}{a}(\cosh 0) + C = \frac{1}{a}(1) + C = \frac{1}{a} + C$$

Hence, if we choose our coordinate system so that $y_0 = 1/a$, we have $C = 0$ and the equation for the hanging cable is

$$y = \frac{1}{a}(\cosh ax).$$

Note that if the coordinate system has already been established in such a way that $y_0 \neq 1/a$, then the equation for the catenary has the more general form

$$y = \left(y_0 - \frac{1}{a}\right) + \frac{1}{a} \cosh ax.$$

Another commonly used form for the equation of the catenary is

$$y = b + a \cosh\left(\frac{x}{a}\right),$$

where a and b are constants and the lowest point on the curve occurs at $x = 0$.

 EXAMPLE ∎ 8   A cable television line hangs between two 30-ft poles that are 36 ft apart. At its lowest point, the cable is 16 ft above the level ground. Determine the height of the cable above a point on the ground that is 6 ft from the poles.

SOLUTION   We use the form

$$y = b + a \cosh\left(\frac{x}{a}\right)$$

and determine first the values of the constants a and b. Since the lowest point occurs at $x = 0$, we have

$$16 = b + a \cosh\left(\frac{0}{a}\right) = b + a \cosh 0 = b + a,$$

so that
$$b = 16 - a.$$

We also have $y = 30$ when $x = 18$ since the poles are 36 ft apart. Thus,

$$30 = b + a \cosh\left(\frac{18}{a}\right),$$

or
$$b = 30 - a \cosh\left(\frac{18}{a}\right)$$

Equating the two expressions for b, we have

$$16 - a = 30 - a \cosh\left(\frac{18}{a}\right)$$

or, equivalently,

$$a - a \cosh\left(\frac{18}{a}\right) + 14 = 0.$$

We use Newton's method to solve for a, obtaining $a \approx 13.42$. So $b = 16 - a \approx 2.58$, and the equation for the catenary becomes

$$y = 2.58 + 13.42 \cosh\left(\frac{x}{13.42}\right).$$

At a point 6 ft from one of the poles, we have $x = \pm 12$. When $x = \pm 12$, $y = 2.58 + 13.42 \cosh(\pm 12/13.42) \approx 21.73$. Thus, at a point 12 ft from the lowest point on the cable television line, the height of the cable is approximately 21.73 ft.

The analysis we have seen for hanging cables also applies to the Gateway Arch to the West in St. Louis. All the internal forces are in equilibrium when a cable hangs freely. There are no transverse forces pushing the cable out of shape. Constructing an arch in the shape of an inverted hyperbolic cosine creates a structure for which there are also no transverse forces that might cause the arch to collapse. This inherent stability of the inverted catenary, along with its beauty, led Saarinen to choose it for his design of the Gateway Arch.

As with other functions that we have studied, we can gain an understanding of compositions of functions that use inverse hyperbolic functions as components by combining the techniques of calculus with the graphs that a graphing utility can display. The next example illustrates this process.

EXAMPLE ■ 9   For the function $f(x) = \ln[\sinh^{-1}(x^2 + 1)]$,

(a) determine the domain of the function f

(b) find the derivative f'

(c) use a graphing utility to plot both the function and its derivative in the viewing window $-5 \le x \le 5$, $-1 \le y \le 1.5$

SOLUTION

(a) The function f is a composition of functions, requiring that we first add 1 to the square of x, then compute an inverse hyperbolic sine, and finally determine the natural logarithm of the resulting number. Since $x^2 + 1$ and the inverse hyperbolic sine are defined for all real numbers, the only step that may cause difficulty in computing $f(x)$ is that the natural logarithm is defined only for positive values.

We note first that by Theorem 6.47(i),

$$\sinh^{-1}(x^2 + 1) = \ln\left[(x^2 + 1) + \sqrt{(x^2 + 1)^2 + 1}\right] = \ln u,$$

where $u = x^2 + 1 + \sqrt{(x^2 + 1)^2 + 1}$. Now u is strictly positive and has its minimum value $1 + \sqrt{2}$ when $x = 0$. Hence,

$$\sinh^{-1}(x^2 + 1) \geq \ln(1 + \sqrt{2}) \approx \ln 2.4142136 \approx 0.8813736.$$

Since $\sinh^{-1}(x^2 + 1)$ is always positive, $f(x) = \ln(\sinh^{-1}(x^2 + 1))$ is defined for all real numbers x. Thus, the domain of f consists of all real numbers.

(b) We use the chain rule twice to find the derivative:

$$f'(x) = \frac{[\sinh^{-1}(x^2 + 1)]'}{\sinh^{-1}(x^2 + 1)} = \frac{\dfrac{1}{\sqrt{(x^2 + 1)^2 + 1}}(x^2 + 1)'}{\sinh^{-1}(x^2 + 1)}$$

$$= \frac{2x}{\sqrt{(x^2 + 1)^2 + 1}\,\sinh^{-1}(x^2 + 1)}$$

Figure 6.47
$-5 \leq x \leq 5, -1 \leq y \leq 1.5$

(c) We use a graphing utility to plot f and f' in the specified viewing window, as shown in Figure 6.47. From the figure, it appears that the graph of f is symmetric about the y-axis and the graph of f' is symmetric about the origin. We can confirm these observations by substituting into the expressions for the function and its derivative to find that $f(-x) = f(x)$ and that $f'(-x) = -f'(x)$. Using the trace feature, we find that f' has a maximum of 0.6580 at approximately $x = 0.7477$, where the graph of f has a point of inflection. By symmetry, there is also a point of inflection for f at $x = -0.7477$, where f' has a minimum.

EXERCISES 6.8

Exer. 1–2: Approximate to four decimal places.

1 (a) $\sinh 4$ (b) $\cosh \ln 4$ (c) $\tanh(-3)$

    (d) $\coth 10$ (e) $\operatorname{sech} 2$ (f) $\operatorname{csch}(-1)$

2 (a) $\sinh \ln 4$ (b) $\cosh 4$ (c) $\tanh 3$

    (d) $\coth(-10)$ (e) $\operatorname{sech}(-2)$ (f) $\operatorname{csch} 1$

Exer. 3–14: Find $f'(x)$ if $f(x)$ is the given expression.

3 $\sinh 5x$

4 $\sinh(x^2 + 1)$

5 $\cosh(x^3)$

6 $\cosh^3 x$

7 $\sqrt{x}\,\tanh\sqrt{x}$

8 $\arctan \tanh x$

9 $\coth(1/x)$

10 $\coth x / \cot x$

11 $\dfrac{\text{sech}(x^2)}{x^2+1}$

12 $\sqrt{\text{sech }5x}$

13 $\text{csch}^2 6x$

14 $x \text{ csch } e^{4x}$

Exer. 15–18: (a) Find the domain of the function. (b) Find $f'(x)$. (c) Plot f and f' in the indicated viewing window.

15 $f(x) = \cosh\sqrt{4x^2+3};\quad -3 \le x \le 3,\ -25 \le y \le 50$

16 $f(x) = \dfrac{1+\cosh x}{1-\cosh x};\quad -8 \le x \le 8,\ -5 \le y \le 2$

17 $f(x) = \dfrac{1}{\tanh x + 1};\quad -3 \le x \le 3,\ -25 \le y \le 50$

18 $f(x) = \ln|\tanh x|;\quad -2 \le x \le 2,\ -10 \le y \le 10$

Exer. 19–30: Evaluate the integral.

19 $\displaystyle\int x^2 \cosh(x^3)\,dx$

20 $\displaystyle\int \dfrac{1}{\text{sech }7x}\,dx$

21 $\displaystyle\int \dfrac{\sinh\sqrt{x}}{\sqrt{x}}\,dx$

22 $\displaystyle\int x \sinh(2x^2)\,dx$

23 $\displaystyle\int \dfrac{1}{\cosh^2 3x}\,dx$

24 $\displaystyle\int \text{sech}^2(5x)\,dx$

25 $\displaystyle\int \text{csch}^2(\tfrac{1}{2}x)\,dx$

26 $\displaystyle\int (\sinh 4x)^{-2}\,dx$

27 $\displaystyle\int \tanh 3x \text{ sech } 3x\,dx$

28 $\displaystyle\int \sinh x \text{ sech}^2 x\,dx$

29 $\displaystyle\int \cosh x \text{ csch}^2 x\,dx$

30 $\displaystyle\int \coth 6x \text{ csch } 6x\,dx$

31 Find the points on the graph of $y = \sinh x$ at which the tangent line has slope 2.

32 Find the arc length of the graph of $y = \cosh x$ from $(0, 1)$ to $(1, \cosh 1)$.

33 If A is the region shown in Figure 6.43, prove that $t = 2A$.

34 The region bounded by the graphs of $y = \cosh x$, $x = -1$, $x = 1$, and $y = 0$ is revolved about the x-axis. Find the volume of the resulting solid.

35 The Gateway Arch to the West in St. Louis has the shape of an inverted catenary (see figure). Rising 630 ft at its center and stretching 630 ft across its base, the shape of the arch can be approximated by

$$y = -127.7 \cosh(x/127.7) + 757.7$$

for $-315 \le x \le 315$.

(a) Approximate the total open area under the arch.

(b) Approximate the total length of the arch.

Exercise 35

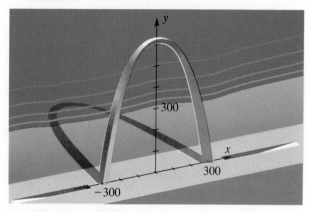

36 A uniform flexible cable supported by poles at $x = -c$ and $x = c$ takes the shape of the graph of the equation $y = b + a \cosh(x/a)$ for $-c \le x \le c$ (see figure).

(a) Find the height of the cable on the poles at each end.

(b) Find the height of the cable at its lowest point.

(c) Find the arc length of the cable hanging between the two poles.

Exercise 36

Exer. 37–40: Refer to Exercise 36.

37 A power line is strung between two 21-ft poles that are 33 ft apart. At its lowest point, the cable is 16 ft above level ground. Find the arc length of the cable hanging between the two poles.

38 A rope 12 ft long is hung between two 5-ft high poles that are 10 ft apart. How high will the rope be off the level ground at its lowest point?

39 Two children pick up a 15-ft rope to play jump rope. Each child grasps the rope 6 in. from an end and holds the rope 3.5 ft above level ground. The two move together until the rope just touches the ground hanging

between their hands before they start to swing the rope. How far apart will they be?

40 A telephone line is to be strung across a city street between two 25-ft poles that are 30 ft apart. To allow large trucks to pass under the line, the lowest point should be at least 19 ft high. Find the arc length of the line between the two poles if it has a lowest point of exactly 19 ft.

41 If an object falls through the air toward the ground in such a way that the air resistance is proportional to the square of the velocity,

(a) show that position y of the object satisfies the differential equation

$$y'' = g - \alpha(y')^2$$

(b) make the change of variable

$$z = y'$$

as in Example 7 and solve the differential equation in part (a)

42 If a steel ball of mass m is released into water and the force of resistance is directly proportional to the square of the velocity, then the distance y that the ball travels in t seconds is given by

$$y = km \ln \cosh\left(\sqrt{\frac{g}{km}}\, t\right),$$

where g is a gravitational constant and $k > 0$. Show that y is a solution of the differential equation

$$m\frac{d^2y}{dt^2} + \frac{1}{k}\left(\frac{dy}{dt}\right)^2 = mg.$$

43 If a wave of length L is traveling across water of depth h (see figure), the velocity v, or *celerity*, of the wave is related to L and h by the formula

$$v^2 = \frac{gL}{2\pi}\tanh\frac{2\pi h}{L},$$

where g is a gravitational constant.

(a) Find $\lim_{h\to\infty} v^2$ and conclude that $v \approx \sqrt{gL/(2\pi)}$ in deep water.

Exercise 43

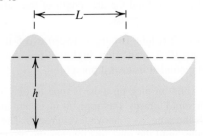

(b) If $x \approx 0$ and f is a continuous function, then, by the mean value theorem (3.12), $f(x) - f(0) \approx f'(0)\, x$. Use this fact to show that $v \approx \sqrt{gh}$ if $h/L \approx 0$. Conclude that wave velocity is independent of wave length in shallow water.

44 A soap bubble formed by two parallel concentric rings is shown in the figure. If the rings are not too far apart, it can be shown that the function f whose graph generates this surface of revolution is a solution of the differential equation $yy'' = 1 + (y')^2$, where $y = f(x)$. If A and B are positive constants, show that $y = A \cosh Bx$ is a solution if and only if $AB = 1$. Conclude that the graph is a catenary.

Exercise 44

c 45 Graph, on the same coordinate axes, $y = \tanh x$ and $y = \operatorname{sech}^2 x$ for $0 \le x \le 2$.

(a) Estimate the x-coordinate a of the point of intersection of the graphs.

(b) Use Newton's method to approximate a to three decimal places.

c 46 Graph, on the same coordinate axes, $y = \cosh^2 x$ and $y = 2$.

(a) Set up integrals for estimating the centroid of the region R bounded by the graphs.

(b) Use Simpson's rule, with $n = 2$, to approximate the coordinates of the centroid of R.

Exer. 47 – 58: Verify the identity.

47 $\cosh x + \sinh x = e^x$

48 $\sinh(-x) = -\sinh x$

49 $\sinh(x + y) = \sinh x \cosh y + \cosh x \sinh y$

50 $\cosh(x + y) = \cosh x \cosh y + \sinh x \sinh y$

51 $\sinh(x - y) = \sinh x \cosh y - \cosh x \sinh y$

52 $\tanh(x + y) = \dfrac{\tanh x + \tanh y}{1 + \tanh x \tanh y}$

53 $\sinh 2x = 2 \sinh x \cosh x$

54 $\cosh 2x = \cosh^2 x + \sinh^2 x$

55 $\sinh^2 \dfrac{x}{2} = \dfrac{\cosh x - 1}{2}$

56 $\cosh^2 \dfrac{x}{2} = \dfrac{\cosh x + 1}{2}$

57 $(\cosh x + \sinh x)^n = \cosh nx + \sinh nx$ for every positive integer n (*Hint:* Use Exercise 47.)

58 $(\cosh x - \sinh x)^n = \cosh nx - \sinh nx$ for every positive integer n

⟨c⟩ **Exer. 59–60: Approximate to four decimal places.**

59 (a) $\sinh^{-1} 1$ (b) $\cosh^{-1} 2$
  (c) $\tanh^{-1}\left(-\tfrac{1}{2}\right)$ (d) $\operatorname{sech}^{-1}\tfrac{1}{2}$

60 (a) $\sinh^{-1}(-2)$ (b) $\cosh^{-1} 5$
  (c) $\tanh^{-1}\tfrac{1}{3}$ (d) $\operatorname{sech}^{-1}\tfrac{3}{5}$

Exer. 61–68: Find $f'(x)$ if $f(x)$ is the given expression.

61 $\sinh^{-1} 5x$ **62** $\sinh^{-1} e^x$

63 $\cosh^{-1} \sqrt{x}$ **64** $\sqrt{\cosh^{-1} x}$

65 $\tanh^{-1}(-4x)$ **66** $\tanh^{-1} \sin 3x$

67 $\operatorname{sech}^{-1} x^2$ **68** $\operatorname{sech}^{-1} \sqrt{1-x}$

⟨c⟩ **Exer. 69–72: (a) Find the domain of the function. (b) Find $f'(x)$. (c) Plot f and f' in the indicated viewing window.**

69 $f(x) = \ln \cosh^{-1} 4x$; $0 \le x \le 10,\ 0 \le y \le 2$

70 $f(x) = \cosh^{-1} \ln 4x$; $0 \le x \le 10,\ 0 \le y \le 2$

71 $f(x) = \tanh^{-1}(x+1)$; $-2 \le x \le 0,\ -3 \le y \le 5$

72 $f(x) = \tanh^{-1} x^3$; $-1 \le x \le 1,\ -3 \le y \le 5$

Exer. 73–80: Evaluate the integral.

73 $\displaystyle\int \dfrac{1}{\sqrt{81 + 16x^2}}\, dx$ **74** $\displaystyle\int \dfrac{1}{\sqrt{16x^2 - 9}}\, dx$

75 $\displaystyle\int \dfrac{1}{49 - 4x^2}\, dx$ **76** $\displaystyle\int \dfrac{\sin x}{\sqrt{1 + \cos^2 x}}\, dx$

77 $\displaystyle\int \dfrac{e^x}{\sqrt{e^{2x} - 16}}\, dx$ **78** $\displaystyle\int \dfrac{2}{5 - 3x^2}\, dx$

79 $\displaystyle\int \dfrac{1}{x\sqrt{9 - x^4}}\, dx$ **80** $\displaystyle\int \dfrac{1}{\sqrt{5 - e^{2x}}}\, dx$

81 A point moves along the line $x = 1$ in a coordinate plane with a velocity that is directly proportional to its distance from the origin. If the initial position of the point is $(1, 0)$ and the initial velocity is 3 ft/sec, express the y-coordinate of the point as a function of time t (in seconds).

82 The rectangular coordinate system shown in the figure illustrates the problem of a dog seeking its master. The dog, initially at the point $(1, 0)$, sees its master at the point $(0, 0)$. The master proceeds

up the y-axis at a constant speed, and the dog runs directly toward its master at all times. If the speed of the dog is twice that of the master, it can be shown that the path of the dog is given by $y = f(x)$, where y is a solution of the differential equation $2xy'' = \sqrt{1 + (y')^2}$. Solve this equation by first letting $z = dy/dx$ and solving $2xz' = \sqrt{1 + z^2}$, obtaining $z = \tfrac{1}{2}[\sqrt{x} - (1/\sqrt{x})]$. Finally, solve the equation $y' = \tfrac{1}{2}[\sqrt{x} - (1/\sqrt{x})]$.

Exercise 82

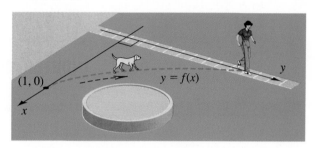

Exer. 83–86: Sketch the graph of the equation.

83 $y = \sinh^{-1} x$ **84** $y = \cosh^{-1} x$

85 $y = \tanh^{-1} x$ **86** $y = \operatorname{sech}^{-1} x$

Exer. 87–91: (a) Derive the formula. (b) and (c) Verify the formula.

87 (a) $\cosh^{-1} x = \ln(x + \sqrt{x^2 - 1}),$ $x \ge 1$

  (b) $\dfrac{d}{dx}(\cosh^{-1} u) = \dfrac{1}{\sqrt{u^2 - 1}}\dfrac{du}{dx},$ $u > 1$

  (c) $\displaystyle\int \dfrac{1}{\sqrt{u^2 - a^2}}\, du = \cosh^{-1} \dfrac{u}{a} + C,$ $0 < a < u$

88 (a) $\tanh^{-1} x = \dfrac{1}{2}\ln\dfrac{1 + x}{1 - x},$ $|x| < 1$

  (b) $\dfrac{d}{dx}(\tanh^{-1} u) = \dfrac{1}{1 - u^2}\dfrac{du}{dx},$ $|u| < 1$

  (c) $\displaystyle\int \dfrac{1}{a^2 - u^2}\, du = \tanh^{-1}\dfrac{u}{a} + C,$ $|u| < a$

89 (a) $\operatorname{sech}^{-1} x = \ln \dfrac{1 + \sqrt{1 - x^2}}{x},$ $0 < x \le 1$

  (b) $\dfrac{d}{dx}(\operatorname{sech}^{-1} u) = -\dfrac{1}{u\sqrt{1 - u^2}}\dfrac{du}{dx},$ $0 < u < 1$

  (c) $\displaystyle\int \dfrac{1}{u\sqrt{a^2 - u^2}}\, du = -\dfrac{1}{a}\operatorname{sech}^{-1}\dfrac{|u|}{a} + C,$
     $0 < |u| < a$

90 (a) $\coth^{-1} x = \dfrac{1}{2} \ln \dfrac{x+1}{x-1} = \tanh^{-1}\left(\dfrac{1}{x}\right), \quad |x| > 1$

(b) $\dfrac{d}{dx}(\coth^{-1} u) = \dfrac{1}{1-u^2}\dfrac{du}{dx}, \qquad |u| > 1$

(c) $\displaystyle\int \dfrac{1}{a^2-u^2}\, du = \dfrac{1}{a}\coth^{-1}\dfrac{u}{a} + C, \qquad |u| > a$

91 (a) $\operatorname{csch}^{-1} x = \ln\left(\dfrac{1}{x} + \dfrac{\sqrt{1+x^2}}{|x|}\right) = \sinh^{-1}\left(\dfrac{1}{x}\right), \quad x \neq 0$

(b) $\dfrac{d}{dx}(\operatorname{csch}^{-1} u) = \dfrac{-1}{|u|\sqrt{1+u^2}}\dfrac{du}{dx}, \qquad |u| \neq 0$

(c) $\displaystyle\int \dfrac{1}{u\sqrt{a^2+u^2}}\, du = -\dfrac{1}{a}\operatorname{csch}^{-1}\dfrac{|u|}{a} + C, \quad u \neq 0$

6.9   INDETERMINATE FORMS AND l'HÔPITAL'S RULE

In Chapter 1, we considered limits of quotients such as

$$\lim_{x\to 3}\frac{x^2-9}{x-3} \quad \text{and} \quad \lim_{x\to 0}\frac{\sin x}{x}.$$

In each case, taking the limits of the numerator and the denominator separately gives the undefined expression 0/0. In a limit of the form

$$\lim_{x\to 0}(\cos x - 1)^{(x^2)},$$

we obtain an undefined expression of the form 0^0 if we take the limits of the base and the exponent separately. For such *indeterminate forms*, we have used algebraic, geometric, and trigonometric methods accompanied by an ingenious manipulation to calculate limits. In this section, we develop other techniques that allow us to proceed in a more direct manner to evaluate several different types of indeterminate forms that occur in both theoretical settings and applications such as electric circuits and insulated cables.

THE FORMS 0/0 AND ∞/∞

We first consider the **indeterminate form 0/0** for limits of quotients where both the numerator and the denominator have limit 0 and the **indeterminate form ∞/∞** where both the numerator and the denominator approach ∞ or −∞. The following table displays general definitions of these forms.

Indeterminate form	Limit form: $\displaystyle\lim_{x\to c}\frac{f(x)}{g(x)}$
$\dfrac{0}{0}$	$\displaystyle\lim_{x\to c} f(x) = 0 \quad \text{and} \quad \lim_{x\to c} g(x) = 0$
$\dfrac{\infty}{\infty}$	$\displaystyle\lim_{x\to c} f(x) = \infty \text{ or } -\infty \quad \text{and} \quad \lim_{x\to c} g(x) = \infty \text{ or } -\infty$

The main tool for investigating these indeterminate forms is *l'Hôpital's rule*. The proof of this rule makes use of the following formula, which bears the name of the French mathematician Augustin Cauchy (1789–1857). (See *Mathematicians and Their Times*, Chapter 9.)

Cauchy's Formula 6.50

> If f and g are continuous on $[a, b]$ and differentiable on (a, b) and if $g'(x) \neq 0$ for every x in (a, b), then there is a number w in (a, b) such that
> $$\frac{f(b) - f(a)}{g(b) - g(a)} = \frac{f'(w)}{g'(w)}.$$

PROOF We first note that $g(b) - g(a) \neq 0$, because otherwise $g(a) = g(b)$ and, by Rolle's theorem (3.10), there is a number c in (a, b) such that $g'(c) = 0$, contrary to our assumption about g'.

Let us introduce a new function h as follows:

$$h(x) = [f(b) - f(a)]g(x) - [g(b) - g(a)]f(x)$$

for every x in $[a, b]$. It follows that h is continuous on $[a, b]$ and differentiable on (a, b) and that $h(a) = h(b)$. By Rolle's theorem, there is a number w in (a, b) such that $h'(w) = 0$—that is,

$$[f(b) - f(a)]g'(w) - [g(b) - g(a)]f'(w) = 0.$$

This is equivalent to Cauchy's formula. ■

Cauchy's formula is a generalization of the mean value theorem (3.12) for if we let $g(x) = x$ in (6.50), we obtain

$$\frac{f(b) - f(a)}{b - a} = \frac{f'(w)}{1} = f'(w).$$

The next result is the main theorem on indeterminate forms.

l'Hôpital's Rule* 6.51

> Suppose that f and g are differentiable on an open interval (a, b) containing c, except possibly at c itself. If $f(x)/g(x)$ has the indeterminate form $0/0$ or ∞/∞ at $x = c$ and if $g'(x) \neq 0$ for $x \neq c$, then
> $$\lim_{x \to c} \frac{f(x)}{g(x)} = \lim_{x \to c} \frac{f'(x)}{g'(x)},$$
> provided either
> $$\lim_{x \to c} \frac{f'(x)}{g'(x)} \text{ exists} \quad \text{or} \quad \lim_{x \to c} \frac{f'(x)}{g'(x)} = \infty.$$

---

*G. l'Hôpital (1661–1704) was a French nobleman who published the first calculus book. The rule appeared in that book; however, it was actually discovered by John Bernoulli (1667–1748), who communicated the result to l'Hôpital in 1694. (See *Mathematicians and Their Times*, Chapter 5.)

PROOF Suppose that $f(x)/g(x)$ has the indeterminate form $0/0$ at $x = c$ and $\lim_{x \to c}[f'(x)/g'(x)] = L$ for some number L. We wish to prove that $\lim_{x \to c}[f(x)/g(x)] = L$. Let us introduce two functions F and G as follows:

$$F(x) = f(x) \quad \text{if } x \neq c \quad \text{and} \quad F(c) = 0$$
$$G(x) = g(x) \quad \text{if } x \neq c \quad \text{and} \quad G(c) = 0$$

Since
$$\lim_{x \to c} F(x) = \lim_{x \to c} f(x) = 0 = F(c),$$

the function F is continuous at c and hence is continuous *throughout* the interval (a, b). Similarly, G is continuous on (a, b). Moreover, at every $x \neq c$, we have $F'(x) = f'(x)$ and $G'(x) = g'(x)$. It follows from Cauchy's formula, applied either to the interval $[c, x]$ or to $[x, c]$, that there is a number w between c and x such that

$$\frac{F(x) - F(c)}{G(x) - G(c)} = \frac{F'(w)}{G'(w)} = \frac{f'(w)}{g'(w)}.$$

Using the fact that $F(x) = f(x)$, $G(x) = g(x)$, and $F(c) = G(c) = 0$ gives us

$$\frac{f(x)}{g(x)} = \frac{f'(w)}{g'(w)}.$$

Since w is always between c and x (see Figure 6.48), it follows that

$$\lim_{x \to c} \frac{f(x)}{g(x)} = \lim_{x \to c} \frac{f'(w)}{g'(w)} = \lim_{w \to c} \frac{f'(w)}{g'(w)} = L,$$

which is what we wished to prove.

A similar argument may be given if $\lim_{x \to c}[f'(x)/g'(x)] = \infty$. The proof for the indeterminate form ∞/∞ is more difficult and may be found in texts on advanced calculus. ■

Figure 6.48

$c < w < x$:

$\quad\quad\quad c \quad w \quad x$

$x < w < c$:

$\quad\quad\quad x \quad w \quad c$

L'Hôpital's rule is sometimes used incorrectly, by applying the quotient rule to $f(x)/g(x)$. Note that (6.51) states that the derivatives of $f(x)$ and $g(x)$ are taken *separately*, after which the limit of $f'(x)/g'(x)$ is investigated.

EXAMPLE ■ 1 Find $\lim\limits_{x \to 0} \dfrac{\cos x + 2x - 1}{3x}$.

SOLUTION Both the numerator and the denominator have the limit 0 as $x \to 0$. Hence the quotient has the indeterminate form $0/0$ at $x = 0$. By l'Hôpital's rule (6.51),

$$\lim_{x \to 0} \frac{\cos x + 2x - 1}{3x} = \lim_{x \to 0} \frac{-\sin x + 2}{3},$$

provided the limit on the right exists or equals ∞. Since

$$\lim_{x \to 0} \frac{-\sin x + 2}{3} = \frac{2}{3},$$

it follows that

$$\lim_{x \to 0} \frac{\cos x + 2x - 1}{3x} = \frac{2}{3}.$$

Sometimes it is necessary to use l'Hôpital's rule several times in the same problem, as illustrated in the next example.

EXAMPLE ▪ 2   Find $\displaystyle \lim_{x \to 0} \frac{e^x + e^{-x} - 2}{1 - \cos 2x}$.

SOLUTION   The given quotient has the indeterminate form $0/0$. By l'Hôpital's rule,

$$\lim_{x \to 0} \frac{e^x + e^{-x} - 2}{1 - \cos 2x} = \lim_{x \to 0} \frac{e^x - e^{-x}}{2 \sin 2x},$$

provided the second limit exists. Because the last quotient has the indeterminate form $0/0$, we apply l'Hôpital's rule a second time, obtaining

$$\lim_{x \to 0} \frac{e^x - e^{-x}}{2 \sin 2x} = \lim_{x \to 0} \frac{e^x + e^{-x}}{4 \cos 2x} = \frac{2}{4} = \frac{1}{2}.$$

It follows that the given limit exists and equals $\frac{1}{2}$.

L'Hôpital's rule is also valid for one-sided limits, as illustrated in the following example.

EXAMPLE ▪ 3   Find $\displaystyle \lim_{x \to (\pi/2)^-} \frac{4 \tan x}{1 + \sec x}$.

SOLUTION   The indeterminate form is ∞/∞. By l'Hôpital's rule,

$$\lim_{x \to (\pi/2)^-} \frac{4 \tan x}{1 + \sec x} = \lim_{x \to (\pi/2)^-} \frac{4 \sec^2 x}{\sec x \tan x} = \lim_{x \to (\pi/2)^-} \frac{4 \sec x}{\tan x}.$$

The last quotient again has the indeterminate form ∞/∞ at $x = \pi/2$; however, additional applications of l'Hôpital's rule always produce the form ∞/∞ (verify this fact). In this case, the limit may be found by using trigonometric identities to change the quotient as follows:

$$\frac{4 \sec x}{\tan x} = \frac{4/\cos x}{\sin x/\cos x} = \frac{4}{\sin x}$$

Consequently,

$$\lim_{x \to (\pi/2)^-} \frac{4 \tan x}{1 + \sec x} = \lim_{x \to (\pi/2)^-} \frac{4}{\sin x} = \frac{4}{1} = 4.$$

There is another form of l'Hôpital's rule that can be proved for $x \to \infty$ or $x \to -\infty$. Let us give a partial proof of this fact. Suppose that

$$\lim_{x \to \infty} f(x) = \lim_{x \to \infty} g(x) = 0.$$

If we let $u = 1/x$ and apply l'Hôpital's rule, then

$$\lim_{x \to \infty} \frac{f(x)}{g(x)} = \lim_{u \to 0^+} \frac{f(1/u)}{g(1/u)} = \lim_{u \to 0^+} \frac{(d/du)(f(1/u))}{(d/du)(g(1/u))}.$$

By the chain rule,

$$\frac{d}{du}(f(1/u)) = f'(1/u)(-1/u^2) \quad \text{and} \quad \frac{d}{du}(g(1/u)) = g'(1/u)(-1/u^2).$$

Substituting in the last limit and simplifying, we obtain

$$\lim_{x \to \infty} \frac{f(x)}{g(x)} = \lim_{u \to 0^+} \frac{f'(1/u)}{g'(1/u)} = \lim_{x \to \infty} \frac{f'(x)}{g'(x)}.$$

We shall also refer to this result as l'Hôpital's rule. The next two examples illustrate the application of the rule to the form ∞/∞.

EXAMPLE■4 Find $\displaystyle\lim_{x \to \infty} \frac{\ln x}{\sqrt{x}}$.

SOLUTION The indeterminate form is ∞/∞. By l'Hôpital's rule,

$$\lim_{x \to \infty} \frac{\ln x}{\sqrt{x}} = \lim_{x \to \infty} \frac{1/x}{1/(2\sqrt{x})}.$$

The last expression has the indeterminate form $0/0$. However, further applications of l'Hôpital's rule would again lead to $0/0$ (verify this fact). If, instead, we simplify the expression algebraically, we can find the limit as follows:

$$\lim_{x \to \infty} \frac{1/x}{1/(2\sqrt{x})} = \lim_{x \to \infty} \frac{2\sqrt{x}}{x} = \lim_{x \to \infty} \frac{2}{\sqrt{x}} = 0$$

EXAMPLE■5 Find $\displaystyle\lim_{x \to \infty} \frac{e^{3x}}{x^2}$, if it exists.

SOLUTION The indeterminate form is ∞/∞. We apply l'Hôpital's rule:

$$\lim_{x \to \infty} \frac{e^{3x}}{x^2} = \lim_{x \to \infty} \frac{3e^{3x}}{2x}$$

The last quotient has the indeterminate form ∞/∞, so we apply l'Hôpital's rule a second time, obtaining

$$\lim_{x \to \infty} \frac{3e^{3x}}{2x} = \lim_{x \to \infty} \frac{9e^{3x}}{2} = \infty.$$

Thus, e^{3x}/x^2 has no limit, increasing without bound as $x \to \infty$.

It is extremely important to verify that a given quotient has the indeterminate form $0/0$ *or* ∞/∞ *before using l'Hôpital's rule.* If we apply
the rule to a form that is not indeterminate, we may obtain an incorrect
conclusion, as illustrated in the next example.

EXAMPLE 6   Find $\lim\limits_{x\to 0}\dfrac{e^{x}+e^{-x}}{x^{2}}$, if it exists.

SOLUTION   The quotient does *not* have either of the indeterminate
forms, $0/0$ or ∞/∞, at $x = 0$. To investigate the limit, we write

$$\lim_{x\to 0}\frac{e^{x}+e^{-x}}{x^{2}} = \lim_{x\to 0}(e^{x}+e^{-x})\left(\frac{1}{x^{2}}\right).$$

Since
$$\lim_{x\to 0}(e^{x}+e^{-x}) = 2 \quad \text{and} \quad \lim_{x\to 0}\frac{1}{x^{2}} = \infty,$$

it follows that

$$\lim_{x\to 0}\frac{e^{x}+e^{-x}}{x^{2}} = \infty.$$

If we had overlooked the fact that the quotient does not have the indeterminate form $0/0$ or ∞/∞ at $x = 0$ and had (incorrectly) applied
l'Hôpital's rule, we would have obtained

$$\lim_{x\to 0}\frac{e^{x}-e^{-x}}{2x}.$$

Since the last quotient has the indeterminate form $0/0$, we might have
applied l'Hôpital's rule, obtaining

$$\lim_{x\to 0}\frac{e^{x}-e^{-x}}{2x} = \lim_{x\to 0}\frac{e^{x}+e^{-x}}{2} = \frac{1+1}{2} = 1.$$

This would have given us the (wrong) conclusion that the given limit exists
and equals 1.

The next example illustrates an application of an indeterminate form in
the analysis of an electrical circuit.

Figure 6.49

EXAMPLE 7   The schematic diagram in Figure 6.49 illustrates an
electrical circuit consisting of an electromotive force V, a resistor R, and
an inductor L. The current I at time t is given by

$$I = \frac{V}{R}(1 - e^{-Rt/L}).$$

When the voltage is first applied (at $t = 0$), the inductor opposes the rate
of increase of current and I is small; however, as t increases, I approaches
V/R.

(a) If L is the only independent variable, find $\lim_{L\to 0^{+}} I$.
(b) If R is the only independent variable, find $\lim_{R\to 0^{+}} I$.

SOLUTION

(a) If we consider V, R, and t as constants and L as a variable, then the expression for I is not indeterminate at $L = 0$. Using standard limit theorems, we obtain

$$\lim_{L \to 0^+} I = \lim_{L \to 0^+} \frac{V}{R} \left(1 - e^{-Rt/L} \right)$$

$$= \frac{V}{R} \left(1 - \lim_{L \to 0^+} e^{-Rt/L} \right)$$

$$= \frac{V}{R}(1 - 0) = \frac{V}{R}.$$

Thus, if $L \approx 0$, then the current can be approximated by Ohm's law $I = V/R$.

(b) If V, L, and t are constant and if R is a variable, then I has the indeterminate form $0/0$ at $R = 0$. Applying l'Hôpital's rule, we have

$$\lim_{R \to 0^+} I = V \lim_{R \to 0^+} \frac{1 - e^{-Rt/L}}{R}$$

$$= V \lim_{R \to 0^+} \frac{0 - e^{-Rt/L}(-t/L)}{1}$$

$$= V[0 - (1)(-t/L)] = \frac{V}{L}t.$$

This result may be interpreted as follows. As $R \to 0^+$, the current I is directly proportional to the time t, with the constant of proportionality V/L. Thus, at $t = 1$, the current is V/L; at $t = 2$, it is $(V/L)(2)$; at $t = 3$, it is $(V/L)(3)$; and so on.

THE FORMS $0 \cdot \infty$, 0^0, ∞^0, 1^∞, AND $\infty - \infty$

There are a number of other indeterminate forms whose limits can be found by rewriting the expressions as quotients and applying l'Hôpital's rule. We begin with products that may lead to the indeterminate form $0 \cdot \infty$, as defined in the following table.

Indeterminate form	Limit form: $\lim_{x \to c}[f(x)g(x)]$	
$0 \cdot \infty$	$\lim_{x \to c} f(x) = 0 \quad$ and	$\lim_{x \to c} g(x) = \infty$ or $-\infty$

In the exercises, we shall also consider the indeterminate form $0 \cdot \infty$ for the case $x \to \infty$ or $x \to -\infty$. The following guidelines may be used.

Guidelines for Investigating
$\lim_{x \to c} [f(x)g(x)]$ *for*
the Form $0 \cdot \infty$ **6.52**

1 Write $f(x)g(x)$ as

$$\frac{f(x)}{1/g(x)} \quad \text{or} \quad \frac{g(x)}{1/f(x)}.$$

2 Apply l'Hôpital's rule (6.51) to the resulting indeterminate form $0/0$ or ∞/∞.

The choice in guideline (1) is not arbitrary. The following example shows that using $f(x)/[1/g(x)]$ gives us the limit, whereas using $g(x)/[1/f(x)]$ leads to a more complicated expression.

EXAMPLE■8 Find $\lim_{x \to 0^+} x^2 \ln x$.

SOLUTION The indeterminate form is $0 \cdot \infty$. Applying guideline (1) of (6.52), we write

$$x^2 \ln x = \frac{\ln x}{1/x^2}.$$

Because the quotient on the right has the indeterminate form ∞/∞ at $x = 0$, we may apply l'Hôpital's rule:

$$\lim_{x \to 0^+} x^2 \ln x = \lim_{x \to 0^+} \frac{\ln x}{1/x^2} = \lim_{x \to 0^+} \frac{1/x}{-2/x^3}$$

The last quotient has the indeterminate form ∞/∞; however, further applications of l'Hôpital's rule would again lead to ∞/∞. In this case, we simplify the quotient algebraically and find the limit as follows:

$$\lim_{x \to 0^+} \frac{1/x}{-2/x^3} = \lim_{x \to 0^+} \frac{x^3}{-2x} = \lim_{x \to 0^+} \frac{x^2}{-2} = 0$$

If, in applying guideline (1), we had rewritten the given expression as

$$x^2 \ln x = \frac{x^2}{1/\ln x} = \frac{x^2}{(\ln x)^{-1}},$$

then the resulting indeterminate form would have been $0/0$. By l'Hôpital's rule,

$$\lim_{x \to 0^+} x^2 \ln x = \lim_{x \to 0^+} \frac{x^2}{(\ln x)^{-1}}$$

$$= \lim_{x \to 0^+} \frac{2x}{-(\ln x)^{-2}(1/x)}$$

$$= \lim_{x \to 0^+} [-2x^2 (\ln x)^2].$$

The expression $-2x^2 (\ln x)^2$ is more complicated than $x^2 \ln x$, so this choice in guideline (1) does *not* give us the limit.

EXAMPLE • 9 Find $\lim\limits_{x\to(\pi/2)^-}(2x-\pi)\sec x$.

SOLUTION The indeterminate form is $0\cdot\infty$. Using guideline (1) of (6.52), we begin by writing

$$(2x-\pi)\sec x = \frac{2x-\pi}{1/\sec x} = \frac{2x-\pi}{\cos x}.$$

Because the last expression has the indeterminate form $0/0$ at $x=\pi/2$, l'Hôpital's rule may be applied as follows:

$$\lim_{x\to(\pi/2)^-}\frac{2x-\pi}{\cos x} = \lim_{x\to(\pi/2)^-}\frac{2}{-\sin x} = \frac{2}{-1} = -2$$

The indeterminate forms defined in the next table may occur in investigating limits involving exponential expressions.

Indeterminate form	Limit form: $\quad\lim\limits_{x\to c} f(x)^{g(x)}$	
0^0	$\lim\limits_{x\to c} f(x)=0$ and	$\lim\limits_{x\to c} g(x)=0$
∞^0	$\lim\limits_{x\to c} f(x)=\infty$ or $-\infty$ and	$\lim\limits_{x\to c} g(x)=0$
1^∞	$\lim\limits_{x\to c} f(x)=1$ and	$\lim\limits_{x\to c} g(x)=\infty$ or $-\infty$

In exercises, we will also consider cases in which $x\to\infty$ or $x\to-\infty$. One method for investigating these forms is to consider

$$y = f(x)^{g(x)}$$

and take the natural logarithm of both sides, obtaining

$$\ln y = \ln f(x)^{g(x)} = g(x)\ln f(x).$$

If the indeterminate form for y is 0^0 or ∞^0, then the indeterminate form for $\ln y$ is $0\cdot\infty$, which may be handled using earlier methods. Similarly, if y has the form 1^∞, then the indeterminate form for $\ln y$ is $\infty\cdot 0$. It follows that

$$\text{if}\quad \lim_{x\to c}\ln y = \ln\left(\lim_{x\to c} y\right) = L,\quad\text{then}\quad \lim_{x\to c} y = e^L;$$

that is,

$$\lim_{x\to c} f(x)^{g(x)} = e^L.$$

This procedure may be summarized as follows.

Guidelines for Investigating $\lim_{x\to c} f(x)^{g(x)}$ ***for the Forms*** $0^0, 1^\infty,$ ***and*** ∞^0 **6.53**

1 Let $y = f(x)^{g(x)}$.

2 Take natural logarithms in guideline (1):

$$\ln y = \ln f(x)^{g(x)} = g(x)\ln f(x)$$

3 Investigate $\lim_{x \to c} \ln y = \lim_{x \to c} [g(x) \ln f(x)]$ and conclude the following:

(a) If $\lim_{x \to c} \ln y = L$, then $\lim_{x \to c} y = e^L$.

(b) If $\lim_{x \to c} \ln y = \infty$, then $\lim_{x \to c} y = \infty$.

(c) If $\lim_{x \to c} \ln y = -\infty$, then $\lim_{x \to c} y = 0$.

A common error is to stop after showing $\lim_{x \to c} \ln y = L$ and conclude that the given expression has the limit L. Remember that *we wish to find the limit of y*. Thus, if $\ln y$ has the limit L, then y has the limit e^L. The guidelines may also be used if $x \to \infty$ or if $x \to -\infty$ or for one-sided limits.

EXAMPLE ▪ 10 Find $\lim_{x \to 0^+} x^x$.

SOLUTION The indeterminate form is 0^0. (See the discussion and graph of the function x^x in Example 5 of Section 6.5.) Using Guidelines (6.53), we proceed as follows:

Guideline 1 $y = x^x$

Guideline 2 $\ln y = x \ln x$

Guideline 3 This expression has the indeterminate form $0 \cdot \infty$. We apply guideline (1) of (6.52) to write

$$x \ln x = \frac{\ln x}{1/x}.$$

Since the quotient on the right has the indeterminate form ∞/∞ at $x = 0$, we may apply l'Hôpital's rule:

$$\lim_{x \to 0^+} \ln y = \lim_{x \to 0^+} x \ln x = \lim_{x \to 0^+} \frac{\ln x}{1/x} = \lim_{x \to 0^+} \frac{1/x}{-1/x^2}$$

We can evaluate this last limit by an algebraic simplification:

$$\lim_{x \to 0^+} \frac{1/x}{-1/x^2} = \lim_{x \to 0^+} \left(-\frac{x^2}{x} \right) = \lim_{x \to 0^+} (-x) = 0$$

Since $\lim_{x \to 0^+} \ln y = 0$, by guideline (3a) of (6.53), we have

$$\lim_{x \to 0^+} x^x = \lim_{x \to 0^+} y = e^0 = 1.$$

Here we have a rigorous proof of a property of x^x that we observed from the graph in Figure 6.23.

The final indeterminate form we shall consider is defined in the following table.

Indeterminate form	Limit form: $\lim\limits_{x \to c}[f(x) - g(x)]$
$\infty - \infty$	$\lim\limits_{x \to c} f(x) = \infty$ and $\lim\limits_{x \to c} g(x) = \infty$

When investigating $\infty - \infty$, we try to change the form of $f(x) - g(x)$ to a quotient or a product and then apply l'Hôpital's rule or some other method of evaluation, as illustrated in the next example.

EXAMPLE ■ 11   Find $\lim\limits_{x \to 0^+} \left(\dfrac{1}{e^x - 1} - \dfrac{1}{x} \right)$.

SOLUTION   The form is $\infty - \infty$; however, if the difference is written as a single fraction, then

$$\lim_{x \to 0^+} \left(\frac{1}{e^x - 1} - \frac{1}{x} \right) = \lim_{x \to 0^+} \frac{x - e^x + 1}{xe^x - x}.$$

This gives us the indeterminate form $0/0$. It is necessary to apply l'Hôpital's rule twice, since the first application leads to the indeterminate form $0/0$. Thus,

$$\lim_{x \to 0^+} \frac{x - e^x + 1}{xe^x - x} = \lim_{x \to 0^+} \frac{1 - e^x}{xe^x + e^x - 1}$$

$$= \lim_{x \to 0^+} \frac{-e^x}{xe^x + 2e^x} = -\frac{1}{2}.$$

EXAMPLE ■ 12   The velocity v of an electrical impulse in an insulated cable is given by

$$v = -k \left(\frac{r}{R} \right)^2 \ln \left(\frac{r}{R} \right),$$

where k is a positive constant, r is the radius of the cable, and R is the distance from the center of the cable to the outside of the insulation, as shown in Figure 6.50. Find

(a) $\lim\limits_{R \to r^+} v$ **(b)** $\lim\limits_{r \to 0^+} v$

Figure 6.50   Insulation

Cable

SOLUTION

(a) The limit notation implies that r is fixed and R is a variable. In this case, the expression for v is not indeterminate, and

$$\lim_{R \to r^+} v = -k \lim_{R \to r^+} \left(\frac{r}{R}\right)^2 \ln \left(\frac{r}{R}\right) = -k(1)^2 \ln 1 = -k(0) = 0.$$

(b) If R is fixed and r is a variable, then the expression for v has the indeterminate form $0 \cdot \infty$ at $r = 0$, and we first change the form of the expression algebraically, as follows:

$$\lim_{r \to 0^+} v = -k \lim_{r \to 0^+} \frac{\ln(r/R)}{(r/R)^{-2}} = -\frac{k}{R^2} \lim_{r \to 0^+} \frac{\ln r - \ln R}{r^{-2}}$$

The last quotient has the indeterminate form ∞/∞ at $r = 0$, so we may apply l'Hôpital's rule, obtaining

$$\lim_{r \to 0^+} v = -\frac{k}{R^2} \lim_{r \to 0^+} \frac{(1/r) - 0}{-2r^{-3}}$$

$$= -\frac{k}{R^2} \lim_{r \to 0^+} \left(\frac{r^2}{-2}\right) = -\frac{k}{R^2}(0) = 0.$$

EXERCISES 6.9

Exer. 1–36: Find the limit, if it exists.

1. $\displaystyle \lim_{x \to 0} \frac{\sin x}{2x}$

2. $\displaystyle \lim_{x \to 0} \frac{5x}{\tan x}$

3. $\displaystyle \lim_{x \to 5} \frac{\sqrt{x-1}-2}{x^2-25}$

4. $\displaystyle \lim_{x \to -3} \frac{x^2+2x-3}{2x^2+3x-9}$

5. $\displaystyle \lim_{x \to 2} \frac{2x^2-5x+2}{5x^2-7x-6}$

6. $\displaystyle \lim_{x \to 1} \frac{x^3-3x+2}{x^2-2x-1}$

7. $\displaystyle \lim_{x \to 0} \frac{\sin x - x}{\tan x - x}$

8. $\displaystyle \lim_{x \to 0} \frac{x+1-e^x}{x^2}$

9. $\displaystyle \lim_{x \to 0} \frac{x - \sin x}{x^3}$

10. $\displaystyle \lim_{x \to \pi/2} \frac{1 - \sin x}{\cos x}$

11. $\displaystyle \lim_{x \to \pi/2} \frac{1 + \sin x}{\cos^2 x}$

12. $\displaystyle \lim_{x \to 0^+} \frac{\ln x}{\cot x}$

13. $\displaystyle \lim_{x \to (\pi/2)^-} \frac{2 + \sec x}{3 \tan x}$

14. $\displaystyle \lim_{x \to \infty} \frac{x^2}{\ln x}$

15. $\displaystyle \lim_{x \to 0^+} \frac{\ln \sin x}{\ln \sin 2x}$

16. $\displaystyle \lim_{x \to 0} \frac{e^x - e^{-x} - 2 \sin x}{x \sin x}$

17. $\displaystyle \lim_{x \to 0} \frac{x \cos x + e^{-x}}{x^2}$

18. $\displaystyle \lim_{x \to 0} \frac{2e^x - 3x - e^{-x}}{x^2}$

19. $\displaystyle \lim_{x \to \infty} \frac{2x^2+3x+1}{5x^2+x+4}$

20. $\displaystyle \lim_{x \to \infty} \frac{e^{3x}}{\ln x}$

21. $\displaystyle \lim_{x \to \infty} \frac{x^n}{e^x}, n > 0$

22. $\displaystyle \lim_{x \to \infty} \frac{e^x}{x^n}, n > 0$

23. $\displaystyle \lim_{x \to \infty} \frac{x \ln x}{x + \ln x}$

24. $\displaystyle \lim_{x \to 2^+} \frac{\ln(x-1)}{(x-2)^2}$

25. $\displaystyle \lim_{x \to 0} \frac{\sin^{-1} 2x}{\sin^{-1} x}$

26. $\displaystyle \lim_{x \to 0} \frac{\tan x - \sin x}{x^3 \tan x}$

27. $\displaystyle \lim_{x \to -\infty} \frac{3 - 3^x}{5 - 5^x}$

28. $\displaystyle \lim_{x \to 1} \frac{2x^3-5x^2+6x-3}{x^3-2x^2+x-1}$

29. $\displaystyle \lim_{x \to 1} \frac{x^4-x^3-3x^2+5x-2}{x^4-5x^3+9x^2-7x+2}$

30. $\displaystyle \lim_{x \to 1} \frac{x^4+x^3-3x^2-x+2}{x^4-5x^3+9x^2-7x+2}$

31. $\displaystyle \lim_{x \to 0} \frac{x - \tan^{-1} x}{x \sin x}$

32. $\displaystyle \lim_{x \to \infty} \frac{x^{3/2}+5x-4}{x \ln x}$

33 $\lim\limits_{x\to\infty} \dfrac{\sqrt{x^2+1}}{\tan^{-1}x}$ **34** $\lim\limits_{x\to\infty} \dfrac{2e^{3x}+\ln x}{e^{3x}+x^2}$

35 $\lim\limits_{x\to\infty} \dfrac{x-\cos x}{x}$ **36** $\lim\limits_{x\to 0^+} \dfrac{e^{-1/x}}{x}$

[c] **Exer. 37–38: Predict the limit after substituting the indicated values of x for $k = 1, 2, 3,$ and 4.**

37 $\lim\limits_{x\to 0^+} \dfrac{\ln(\tan x + \cos x)}{\sqrt{\ln(x^2+1)}}$; $x = 10^{-k}$

38 $\lim\limits_{x\to 0} \dfrac{\tan^2(\sin^{-1}x)}{1-\cos[\ln(1+x)]}$; $x = \pm 10^{-k}$

39 An object of mass m is released from a hot-air balloon. If the force of resistance due to air is directly proportional to the velocity $v(t)$ of the object at time t, then it can be shown that

$$v(t) = (mg/k)(1 - e^{-(k/m)t}),$$

where $k > 0$ and g is a gravitational constant. Find $\lim_{k\to 0^+} v(t)$.

40 If a steel ball of mass m is released into water and the force of resistance is directly proportional to the square of the velocity, then the distance $s(t)$ that the ball travels in time t is given by

$$s(t) = (m/k)\ln\cosh(\sqrt{gk/mt}),$$

where $k > 0$ and g is a gravitational constant. Find $\lim_{k\to 0^+} s(t)$.

41 Refer to Definition (3.24) for simple harmonic motion. The following is an example of the phenomenon of resonance. A weight of mass m is attached to a spring suspended from a support. The weight is set in motion by moving the support up and down according to the formula $h = A\cos\omega t$, where A and ω are positive constants and t is time. If frictional forces are negligible, then the displacement s of the weight from its initial position at time t is given by

$$s = \dfrac{A\omega^2}{\omega_0^2 - \omega^2}(\cos\omega t - \cos\omega_0 t),$$

with $\omega_0 = \sqrt{k/m}$ for some constant k and with $\omega \neq \omega_0$. Find $\lim_{\omega\to\omega_0} s$, and show that the resulting oscillations increase in magnitude.

42 The logistic model for population growth predicts the size $y(t)$ of a population at time t by means of the formula $y(t) = K/(1 + ce^{-rt})$, where r and K are positive constants and $c = [K - y(0)]/y(0)$. Ecologists call K the *carrying capacity* and interpret it as the maximum number of individuals that the environment can sustain. Find $\lim_{t\to\infty} y(t)$ and $\lim_{K\to\infty} y(t)$, and discuss the graphical significance of these limits.

43 The *sine integral* $Si(x) = \int_0^x [(\sin u)/u]\,du$ is a special function in applied mathematics. Find

(a) $\lim\limits_{x\to 0} \dfrac{Si(x)}{x}$ **(b)** $\lim\limits_{x\to 0} \dfrac{Si(x) - x}{x^3}$

44 The *Fresnel cosine integral* $C(x) = \int_0^x \cos u^2\,du$ is used in the analysis of the diffraction of light. Find

(a) $\lim\limits_{x\to 0} \dfrac{C(x)}{x}$ **(b)** $\lim\limits_{x\to 0} \dfrac{C(x) - x}{x^5}$

[c] **45 (a)** Refer to Exercise 44. Use Simpson's rule, with $n = 2$, to approximate $C(x)$ for $x = \frac{1}{4}, \frac{1}{2}, \frac{3}{4},$ and 1.

(b) Graph C on $[0, 1]$ using the values found in part (a).

[c] **46** Refer to Exercise 45. Let R be the region under the graph of C from $x = 0$ to $x = 1$ and V the volume of the solid obtained by revolving R about the x-axis. Approximate V by using Simpson's rule, with $n = 2$.

47 Let $x > 0$. If $n \neq -1$, then $\int_1^x t^n\,dt = [t^{n+1}/(n+1)]_1^x$. Show that

$$\lim\limits_{n\to -1} \int_1^x t^n\,dt = \int_1^x t^{-1}\,dt.$$

48 Find $\lim_{x\to\infty} f(x)/g(x)$ if

$$f(x) = \int_0^x e^{(t^2)}\,dt \quad \text{and} \quad g(x) = e^{(x^2)}.$$

Exer. 49–76: Find the limit, if it exists.

49 $\lim\limits_{x\to 0^+} x\ln x$ **50** $\lim\limits_{x\to(\pi/2)^-} \tan x\ln\sin x$

51 $\lim\limits_{x\to\infty} (x^2-1)e^{-x^2}$ **52** $\lim\limits_{x\to-\infty} x\tan^{-1}x$

53 $\lim\limits_{x\to 0} e^{-x}\sin x$ **54** $\lim\limits_{x\to 0^+} \sin x\ln\sin x$

55 $\lim\limits_{x\to\infty} x\sin\dfrac{1}{x}$ **56** $\lim\limits_{x\to 0} x\sec^2 x$

57 $\lim\limits_{x\to\infty} \left(1+\dfrac{1}{x}\right)^{5x}$ **58** $\lim\limits_{x\to 0^+} (e^x+3x)^{1/x}$

59 $\lim\limits_{x\to 0^+} (e^x-1)^x$ **60** $\lim\limits_{x\to\infty} x^{1/x}$

61 $\lim\limits_{x\to(\pi/2)^-} (\tan x)^x$ **62** $\lim\limits_{x\to 0^+} (1+3x)^{\csc x}$

63 $\lim\limits_{x\to 0^+} (2x+1)^{\cot x}$ **64** $\lim\limits_{x\to\infty} \left(\dfrac{x^2}{x-1} - \dfrac{x^2}{x+1}\right)$

65 $\lim\limits_{x\to 0^-} \left(\dfrac{1}{x} - \dfrac{1}{\sin x}\right)$ **66** $\lim\limits_{x\to 1^-} (1-x)^{\ln x}$

67 $\lim\limits_{x\to 0^+} \left(\dfrac{1}{\sqrt{x^2+1}} - \dfrac{1}{x}\right)$ **68** $\lim\limits_{x\to 0} (\cot^2 x - \csc^2 x)$

69 $\lim\limits_{x\to 0^+} \cot 2x\tan^{-1}x$ **70** $\lim\limits_{x\to 0^+} (1+ax)^{b/x}$

71 $\lim\limits_{x\to(\pi/2)^-} (1+\cos x)^{\tan x}$

72 $\lim\limits_{x\to -3^-} \left(\dfrac{x}{x^2+2x-3} - \dfrac{4}{x+3}\right)$

73 $\lim_{x \to 0^+} (x + \cos 2x)^{\csc 3x}$

74 $\lim_{x \to \infty} (\sqrt{x^4 + 5x^2 + 3} - x^2)$

75 $\lim_{x \to \infty} (\sinh x - x)$

76 $\lim_{x \to \infty} [\ln(4x + 3) - \ln(3x + 4)]$

$\boxed{c}$ Exer. 77–78: Graph f on the given interval and use the graph to estimate $\lim_{x \to 0} f(x)$.

77 $f(x) = (x \tan x)^{(x^2)}$; $\quad [-1, 1]$

78 $f(x) = \left[\dfrac{\ln(x + 1)}{\tan x}\right]^{1/x}$; $\quad [-0.5, 0.5]$

Exer. 79–80: (a) Find the local extrema and discuss the behavior of $f(x)$ near $x = 0$. (b) Find horizontal asymptotes, if they exist. (c) Sketch the graph of f for $x > 0$.

79 $f(x) = x^{1/x}$

80 $f(x) = x^{\sqrt{x}}$

81 The *geometric mean* of two positive real numbers a and b is defined as $\sqrt{ab}$. Use l'Hôpital's rule to prove that

$$\sqrt{ab} = \lim_{x \to \infty} \left(\frac{a^{1/x} + b^{1/x}}{2}\right)^x.$$

82 If a sum of money P is invested at an interest rate of $100r$ percent per year, compounded m times per year, then the principal at the end of t years is given by $P(1 + rm^{-1})^{mt}$. If we regard m as a real number and let m increase without bound, then the interest is said to be *compounded continuously*. Use l'Hôpital's rule to show that, in this case, the principal after t years is Pe^{rt}.

83 Refer to Exercise 39. In the velocity formula

$$v(t) = (mg/k)(1 - e^{-(k/m)t}),$$

m represents the mass of the falling object. Find $\lim_{m \to \infty} v(t)$ and conclude that $v(t)$ is approximately proportional to time t if the mass is very large.

CHAPTER 6 REVIEW EXERCISES

Exer. 1–2: Find $f^{-1}(x)$.

1 $f(x) = 10 - 15x$

2 $f(x) = 9 - 2x^2$, $\quad x \le 0$

Exer. 3–4: Show that the function f has an inverse function, and find $[(d/dx)(f^{-1}(x))]_{x=a}$ for the given number a.

3 $f(x) = 2x^3 - 8x + 5$, $\quad -1 \le x \le 1$; $\quad a = 5$

4 $f(x) = e^{3x} + 2e^x - 5$, $\quad x \ge 0$; $\quad a = -2$

Exer. 5–26: Find $f'(x)$ if $f(x)$ is the given expression.

5 $\ln |4 - 5x^3|^5$

6 $(1 - 2x) \ln |1 - 2x|$

7 $\ln \dfrac{(3x + 2)^4 \sqrt{6x - 5}}{8x - 7}$

8 $\dfrac{\ln x}{e^{2x} + 1}$

9 $\dfrac{1}{\ln(2x^2 + 3)}$

10 $\dfrac{x}{\ln x}$

11 $e^{\ln(x^2 + 1)}$

12 $\ln(e^{4x} + 9)$

13 $10^x \log x$

14 $5^{3x} + (3x)^5$

15 $\sqrt{\ln \sqrt{x}}$

16 $(1 + \sqrt{x})^e$

17 $x^2 e^{-x^2}$

18 $\sqrt{e^{3x} + e^{-3x}}$

19 $10^{\ln x}$

20 $7^{\ln |x|}$

21 $x^{\ln x}$

22 $\ln |\tan x - \sec x|$

23 $\csc e^{-2x} \cot e^{-2x}$

24 $3^{\sin 3x}$

25 $\ln \cos^4 4x$

26 $(\sin x)^{\cos x}$

Exer. 27–28: Use implicit differentiation to find y'.

27 $1 + xy = e^{xy}$

28 $\ln(x + y) + x^2 - 2y^3 = 1$

Exer. 29–30: Use logarithmic differentiation to find dy/dx.

29 $y = (x + 2)^{4/3}(x - 3)^{3/2}$

30 $y = \sqrt[3]{(3x - 1)\sqrt{2x + 5}}$

Exer. 31–54: Evaluate the integral.

31 (a) $\displaystyle\int \dfrac{1}{\sqrt{x}e^{\sqrt{x}}} \, dx$ $\quad$ (b) $\displaystyle\int_1^4 \dfrac{1}{\sqrt{x}e^{\sqrt{x}}} \, dx$

32 (a) $\displaystyle\int x 4^{-x^2} \, dx$ $\quad$ (b) $\displaystyle\int_0^1 x 4^{-x^2} \, dx$

33 $\displaystyle\int x \tan x^2 \, dx$

34 $\displaystyle\int \cot \left(x + \dfrac{\pi}{6}\right) dx$

35 $\displaystyle\int x^e \, dx$

36 $\displaystyle\int \dfrac{1}{x - x \ln x} \, dx$

37 $\int \dfrac{(1+e^x)^2}{e^{2x}}\, dx$

38 $\int \dfrac{(e^{2x}+e^{3x})^2}{e^{5x}}\, dx$

39 $\int \dfrac{x^2}{x+2}\, dx$

40 $\int \dfrac{e^{1/x}}{x^2}\, dx$

41 $\int \dfrac{e^{4/x^2}}{x^3}\, dx$

42 $\int \dfrac{x}{x^4+2x^2+1}\, dx$

43 $\int \dfrac{e^x}{1+e^x}\, dx$

44 $\int (1+e^{-3x})^2\, dx$

45 $\int 5^x e^x\, dx$

46 $\int \dfrac{1}{x\sqrt{\log x}}\, dx$

47 $\int e^{-x}\sin e^{-x}\, dx$

48 $\int \tan x e^{\sec x}\sec x\, dx$

49 $\int \dfrac{\csc^2 x}{1+\cot x}\, dx$

50 $\int \dfrac{\cos 2x}{1-2\sin 2x}\, dx$

51 $\int e^x \tan e^x\, dx$

52 $\int \dfrac{\sec(1/x)}{x^2}\, dx$

53 $\int (\csc 3x + 1)^2\, dx$

54 $\int (\cot 9x + \csc 9x)\, dx$

55 Solve the differential equation $y'' = -e^{-3x}$ subject to the conditions $y = -1$ and $y' = 2$ if $x = 0$.

56 In *seasonal population growth*, the population $q(t)$ at time t (in years) increases during the spring and summer but decreases during the fall and winter. A differential equation that is sometimes used to describe this type of growth is $q'(t)/q(t) = k\sin 2\pi t$, where $k > 0$ and $t = 0$ corresponds to the first day of spring.

(a) Show that the population $q(t)$ is seasonal.

(b) If $q_0 = q(0)$, find a formula for $q(t)$.

57 A particle moves on a coordinate line with an acceleration at time t of $e^{t/2}$ cm/sec^2. At $t = 0$, the particle is at the origin and its velocity is 6 cm/sec. How far does it travel during the time interval $[0, 4]$?

58 Find the local extrema of $f(x) = x^2 \ln x$ for $x > 0$. Discuss concavity, find the points of inflection, and sketch the graph of f.

59 Find an equation of the tangent line to the graph of the equation $y = xe^{1/x^3} + \ln|2 - x^2|$ at the point $P(1, e)$.

60 Find the area of the region bounded by the graphs of the equations $y = e^{2x}$, $y = x/(x^2 + 1)$, $x = 0$, and $x = 1$.

61 The region bounded by the graphs of $y = e^{4x}$, $x = -2$, $x = -3$, and $y = 0$ is revolved about the x-axis. Find the volume of the resulting solid.

62 The 1993 population estimate for India was 907 million, and the population has been increasing at a rate of about 2% per year, with the rate of increase proportional to the number of people. If t denotes the time (in years) after 1993, find a formula for $N(t)$, the population (in millions) at time t. Assuming that this rapid growth rate continues, estimate the population and the rate of population growth in the year 2000.

63 A radioactive substance has a half-life of 5 days. How long will it take for an amount A to disintegrate to the extent that only 1% of A remains?

64 The carbon-14 dating equation $T = -8310 \ln x$ is used to predict the age T (in years) of a fossil in terms of the percentage $100x$ of carbon still present in the specimen (see Exercise 19, Section 6.6).

(a) If $x = 0.04$, estimate the age of the fossil to the nearest 1000 years.

(b) If the maximum error in estimating x in part (a) is ± 0.005, use differentials to approximate the maximum error in T.

65 The rate at which sugar dissolves in water is proportional to the amount that remains undissolved. Suppose that 10 lb of sugar is placed in a container of water at 1:00 P.M., and one half is dissolved at 4:00 P.M.

(a) How long will it take two more pounds to dissolve?

(b) How much of the 10 lb will be dissolved at 8:00 P.M.?

66 According to Newton's law of cooling, the rate at which an object cools is directly proportional to the difference in temperature between the object and its surrounding medium. If $f(t)$ denotes the temperature at time t, show that $f(t) = T + [f(0) - T]e^{-kt}$, where T is the temperature of the surrounding medium and k is a positive constant.

67 The bacterium *E. coli* undergoes cell division approximately every 20 min. Starting with 100,000 cells, determine the number of cells after 2 hr.

c **68** By letting $h = 0.1, 0.01$, and 0.001, predict which of the following expressions gives the best approximation of e for small values of h:

$$(1+h)^{1/h}, \quad (1+h+h^2)^{1/h}, \quad (1+h+\tfrac{1}{2}h^2)^{1/h}$$

Exer. 69–84: Find $f'(x)$ if $f(x)$ is the given expression.

69 $\arctan \sqrt{x-1}$

70 $\tan^{-1}(\ln 3x)$

71 $x^2 \operatorname{arcsec}(x^2)$

72 $2^{\arctan 2x}$

73 $\ln \tan^{-1}(x^2)$

74 $\dfrac{1-x^2}{\arccos x}$

75 $\sin^{-1}\sqrt{1-x^2}$

76 $(\tan x + \tan^{-1} x)^4$

77 $\tan^{-1}(\tan^{-1} x)$

78 $e^{4x}\sec^{-1} e^{4x}$

79 $\cosh e^{-5x}$

80 $e^{-x}\sinh e^{-x}$

81 $\dfrac{\sinh x}{\cosh x - \sinh x}$

82 $\ln \tanh(5x + 1)$

83 $\sinh^{-1}(x^2)$

84 $\tanh^{-1}(\tanh \sqrt[3]{x})$

Exer. 85–98: Evaluate the integral.

85 $\displaystyle\int \frac{1}{4+9x^2}\,dx$

86 $\displaystyle\int \frac{x}{4+9x^2}\,dx$

87 $\displaystyle\int \frac{e^{2x}}{\sqrt{1-e^{2x}}}\,dx$

88 $\displaystyle\int \frac{e^x}{\sqrt{1-e^{2x}}}\,dx$

89 $\displaystyle\int \frac{x}{\operatorname{sech}(x^2)}\,dx$

90 $\displaystyle\int \frac{\sinh(\ln x)}{x}\,dx$

91 $\displaystyle\int_{-1/2}^{1/2} \frac{1}{\sqrt{1-x^2}}\,dx$

92 $\displaystyle\int_{0}^{\pi/2} \frac{\cos x}{1+\sin^2 x}\,dx$

93 $\displaystyle\int \frac{1}{\sqrt{9-4x^2}}\,dx$

94 $\displaystyle\int \frac{x}{\sqrt{9-4x^2}}\,dx$

95 $\displaystyle\int \frac{1}{x\sqrt{9-4x^2}}\,dx$

96 $\displaystyle\int \frac{1}{x\sqrt{4x^2-9}}\,dx$

97 $\displaystyle\int \frac{x}{\sqrt{25x^2+36}}\,dx$

98 $\displaystyle\int \frac{1}{\sqrt{25x^2+36}}\,dx$

99 Find the points on the graph of $y = \sin^{-1} 3x$ at which the tangent line is parallel to the line through $A(2, -3)$ and $B(4, 7)$.

100 Find the points of inflection, and discuss the concavity of the graph of $y = x\sin^{-1} x$.

101 Find the local extrema of $f(x) = 8\sec x + \csc x$ on the interval $(0, \pi/2)$, and describe where $f(x)$ is increasing or is decreasing on that interval.

102 Find the area of the region bounded by the graphs of $y = x/(x^4 + 1)$, $x = 1$, and $y = 0$.

103 Damped oscillations are oscillations of decreasing magnitude that occur when frictional forces are considered. Shown in the figure is a graph of the damped oscillations given by $f(x) = e^{-x/2}\sin 2x$.

   (a) Find the x-coordinates of the extrema of f for $0 \le x \le 2\pi$.

   (b) Approximate the x-coordinates in part (a) to two decimal places.

Exercise 103

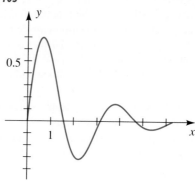

104 Find the arc length of the graph of $y = \ln \tanh \frac{1}{2}x$ from $x = 1$ to $x = 2$.

105 A balloon is released from level ground, 500 m away from a person who observes its vertical ascent. If the balloon rises at a constant rate of 2 m/sec, use inverse trigonometric functions to find the rate at which the angle of elevation of the observer's line of sight is changing at the instant the balloon is at a height of 100 m. (Disregard the observer's height.)

106 A square picture with sides 2 ft long is hung on a wall with the base 6 ft above the floor. A person whose eye level is 5 ft above the floor approaches the picture at a rate of 2 ft/sec. If θ is the angle between the line of sight and the top and bottom of the picture, find

   (a) the rate at which θ is changing when the person is 8 ft from the wall

   (b) the distance from the wall at which θ has its maximum value

107 A stunt man jumps from a hot-air balloon that is hovering at a constant altitude, 100 ft above a lake. A movie camera on shore, 200 ft from a point directly below the balloon, follows the stunt man's descent (see figure). At what rate is the angle of elevation θ of the camera changing 2 sec after the stunt man jumps? (Disregard the height of the camera.)

Exercise 107

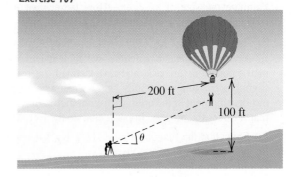

108 A person on a small island I, which is k miles from the closest point A on a straight shoreline, wishes to reach a camp that is d miles downshore from A by swimming to some point P on shore and then walking the rest of the way (see figure). Suppose the person burns c_1 calories per mile while swimming and c_2 calories per mile while walking, where $c_1 > c_2$.

    (a) Find a formula for the total number c of calories burned in completing the trip.

    (b) For what angle AIP does c have a minimum value?

Exercise 108

Exer. 109–120: Find the limit, if it exists.

109 $\displaystyle \lim_{x \to 0} \frac{\ln(2-x)}{1+e^{2x}}$

110 $\displaystyle \lim_{x \to 0} \frac{\sin 2x - \tan 2x}{x^2}$

111 $\displaystyle \lim_{x \to \infty} \frac{x^2 + 2x + 3}{\ln(x+1)}$

112 $\displaystyle \lim_{x \to 0} \frac{e^{2x} - e^{-2x} - 4x}{x^3}$

113 $\displaystyle \lim_{x \to \infty} \frac{x^e}{e^x}$

114 $\displaystyle \lim_{x \to (\pi/2)^-} \cos x \ln \cos x$

115 $\displaystyle \lim_{x \to \infty} (1 - 2e^{1/x})x$

116 $\displaystyle \lim_{x \to 0} (1 + 8x^2)^{1/x^2}$

117 $\displaystyle \lim_{x \to \infty} (e^x + 1)^{1/x}$

118 $\displaystyle \lim_{x \to 0^+} \left(\frac{1}{\tan x} - \frac{1}{x} \right)$

119 $\displaystyle \lim_{x \to \infty} \frac{\sqrt{x^2 + 1}}{x}$

120 $\displaystyle \lim_{x \to \infty} \frac{3^x + 2x}{x^3 + 1}$

121 Find $\lim_{x \to \infty} f(x)/g(x)$ if $f(x) = \int_1^x (\sin t)^{2/3} \, dt$ and $g(x) = x^2$.

122 Gauss's error integral erf $(x) = (2/\sqrt{\pi}) \int_0^x e^{-u^2} \, du$ is used in probability theory. It has the special property $\lim_{x \to \infty}$ erf $(x) = 1$. Find $\lim_{x \to \infty} e^{(x^2)}[1 - \text{erf}(x)]$.

EXTENDED PROBLEMS AND GROUP PROJECTS

1 For each positive integer n, let

$$s_n = \frac{1}{1} + \frac{1}{2} + \frac{1}{3} + \cdots + \frac{1}{n}.$$

By the use of circumscribed and inscribed rectangles, show that

$$\ln(n+1) < s_n < 1 + \ln n.$$

Estimate the size of $s_{1,000,000}$. How large must n be to guarantee that $s_n > 100$? What happens to s_n as $n \to \infty$?

2 In our development of calculus, we took the trigonometric functions as basic and *then* defined the inverse trigonometric functions. Show that we can give a rigorous development of the ordinary trigonometric functions that reverses the process. In particular, do the following.

    (a) Show that the function $f(t) = 1/(1 + t^2)$ is continuous and positive for all $t > 0$.

    (b) Define the function

$$A(x) = \int_0^x \frac{1}{1+t^2} \, dt.$$

    Show that A is differentiable and increasing for all real numbers x. What can you say about the range of the function A?

    (c) Define T as the inverse of the function A of part (b). What properties does the function T have? What is $T'(x)$? What are the similarities between $T(x)$ and $\tan x$?

    (d) Define S as the inverse of the function B, where

$$B(x) = \int_0^x \frac{1}{\sqrt{1-t^2}} \, dt \quad \text{for} -1 < x < 1,$$

    and define C as the inverse of the function D, where

$$D(x) = \int_0^x \frac{-1}{\sqrt{1-t^2}} \, dt \quad \text{for} -1 < x < 1.$$

Use the fundamental theorem of calculus and other results to establish properties of the functions S and C. Show that $S'(x) = C(x)$ and $C'(x) = -S(x)$. Is it true that $S^2(x) + C^2(x) = 1$ for all x?

(e) Suppose we define the sine, cosine, and tangent functions by the formulas $\sin x = S(x)$, $\cos x = C(x)$, and $\tan x = T(x)$. Do these functions have all the same properties as the usual sine, cosine, and tangent functions? If so, what are the advantages and the disadvantages of defining the trigonometric functions in this way?

3 To what extent are exponential and logarithmic functions determined by their arithmetic properties? In particular, do the following.

(a) Show that if f is a function such that $f' = f$ and $f(x + y) = f(x)f(y)$ for all x and y, then f must be either the natural exponential function or the function that is identically zero.

(b) Prove that if f is continuous and if $f(x + y) = f(x)f(y)$, then either f is identically zero or $f(x) = [f(1)]^x$ for all x.

(c) Prove that if f is a continuous function defined on the positive real numbers such that $f(xy) = f(x) + f(y)$ for all positive numbers x and y, then f is identically zero or $f(x) = f(e) \ln x$ for all $x > 0$.

INTRODUCTION

S **KI JUMPING DEMANDS** a high level of skill and intense concentration. Jumpers slide down a track from a hill 70 to 90 meters high and then leap from a platform, flying through the air for some 90 meters before landing on the ground and gliding to a stop. The competitors require a well-designed ski jump, one that maximizes performance and minimizes danger. The design of a ski jump is a complex task that must address a number of concerns. The goal may be to move the jumpers from one point to another in the fastest time or to ensure that they reach the platform with the greatest velocity. To minimize time or maximize speed may require selecting the appropriate curve for the shape of different sections of the track. The curve may be the graph of a function $y = f(x)$. From physical assumptions, we may find a differential equation whose solution is the desired function f. However a differential equation occurs, it usually involves the derivative f' or the second derivative f'' in an explicit or implicit manner. Solving a differential equation requires recovering functions from their derivatives—that is, finding indefinite integrals.

In this chapter, we consider additional ways to simplify integrals. Foremost among these is *integration by parts,* which we discuss in Section 7.1. This powerful device allows us to obtain indefinite integrals of $\ln x$, $\tan^{-1} x$, and other important transcendental functions. In Sections 7.2–7.5, we develop techniques for simplifying integrals that contain powers of trigonometric functions, radicals, and rational expressions. In Section 7.6, we examine the use of tables of integrals and computer algebra systems. Such tables and systems are always incomplete, and we must frequently use the skills introduced in the preceding sections *before* trying these approaches. Finally, we extend the definition of definite integrals $\int_a^b f(x)\,dx$ in Section 7.7 to handle certain cases where the function f has an infinite discontinuity on the interval $[a, b]$ or where the interval becomes infinitely long.

The techniques we investigate in this chapter extend the range of functions for which we can find antiderivatives explicitly. Sometimes it is impossible to obtain an antiderivative in the form of an expression involving a finite number of sums, products, quotients, or compositions of rational functions, trigonometric functions, or the exponential and logarithmic functions. In such cases, the trapezoidal rule or Simpson's rule can be used to obtain numerical approximations. Then, either a computer or a programmable calculator is invaluable, since it can usually arrive at an accurate approximation in a matter of seconds.

*Designing a complicated structure
such as a ski jump often involves
differential equations whose solutions
require evaluating indefinite integrals.*

Techniques of Integration

7.1 INTEGRATION BY PARTS

Up to this stage of our work, we have been unable to evaluate integrals such as the following:

$$\int \ln x \, dx, \qquad \int xe^x \, dx, \qquad \int x^2 \sin x \, dx, \qquad \int \tan^{-1} x \, dx$$

The next formula will enable us to evaluate not only these, but also many other types of integrals.

Integration by Parts Formula 7.1

If $u = f(x)$ and $v = g(x)$ and if f' and g' are continuous, then

$$\int u \, dv = uv - \int v \, du.$$

PROOF By the product rule,

$$\frac{d}{dx}(f(x)g(x)) = f(x)g'(x) + g(x)f'(x),$$

or, equivalently,

$$f(x)g'(x) = \frac{d}{dx}(f(x)g(x)) - g(x)f'(x).$$

Integrating both sides of the preceding equation gives us

$$\int f(x)g'(x) \, dx = \int \frac{d}{dx}(f(x)g(x)) \, dx - \int g(x)f'(x) \, dx.$$

By Theorem (4.5)(i), the first integral on the right side is $f(x)g(x) + C$. Because another constant of integration is obtained from the second integral, we may omit C in the formula—that is,

$$\int f(x)g'(x) \, dx = f(x)g(x) - \int g(x)f'(x) \, dx \,.$$

Since $dv = g'(x) \, dx$ and $du = f'(x) \, dx$, we may write the preceding formula as in (7.1). ■

When applying Formula (7.1) to an integral, we begin by letting one part of the integrand correspond to dv. The expression we choose for dv must include the differential dx. After selecting dv, we designate the remaining part of the integrand by u and then find du. Since this process involves splitting the integrand into two parts, the use of (7.1) is referred to as **integrating by parts**. A proper choice for dv is crucial. We generally *let dv equal the most complicated part of the integrand that can be readily integrated.* The following examples illustrate this method of integration.

EXAMPLE ▪ 1 Evaluate $\int xe^{2x}\,dx$.

SOLUTION The following list contains all possible choices for dv:

$$dx, \quad x\,dx, \quad e^{2x}\,dx, \quad xe^{2x}\,dx$$

The most complicated of these expressions that can be readily integrated is $e^{2x}\,dx$. Thus, we let

$$dv = e^{2x}\,dx.$$

The remaining part of the integrand is u—that is, $u = x$. To find v, we integrate dv, obtaining $v = \frac{1}{2}e^{2x}$. Note that a constant of integration is not added at this stage of the solution. (In Exercise 51, you are asked to prove that if a constant *is* added to v, the same final result is obtained.) If $u = x$, then $du = dx$. For ease of reference, let us display these expressions as follows:

$$dv = e^{2x}\,dx \qquad u = x$$
$$v = \tfrac{1}{2}e^{2x} \qquad du = dx$$

Substituting these expressions in Formula (7.1)—that is, *integrating by parts*—we obtain

$$\int xe^{2x}\,dx = x(\tfrac{1}{2}e^{2x}) - \int \tfrac{1}{2}e^{2x}\,dx.$$

We may find the integral on the right side as in Section 6.4. This gives us

$$\int xe^{2x}\,dx = \tfrac{1}{2}xe^{2x} - \tfrac{1}{4}e^{2x} + C.$$

It takes considerable practice to become proficient in making a suitable choice for dv. To illustrate, if we had chosen $dv = x\,dx$ in Example 1, then it would have been necessary to let $u = e^{2x}$, giving us

$$dv = x\,dx \qquad u = e^{2x}$$
$$v = \tfrac{1}{2}x^2 \qquad du = 2e^{2x}\,dx.$$

Integrating by parts, we obtain

$$\int xe^{2x}\,dx = \tfrac{1}{2}x^2e^{2x} - \int x^2e^{2x}\,dx.$$

Since the exponent associated with x has increased, the integral on the right is more complicated than the given integral. This indicates that we have made an incorrect choice for dv.

EXAMPLE ▪ 2 Evaluate

(a) $\int x\sec^2 x\,dx$ **(b)** $\int_0^{\pi/3} x\sec^2 x\,dx$

SOLUTION

(a) The possible choices for dv are

$$dx, \quad x\,dx, \quad \sec x\,dx, \quad x\sec x\,dx, \quad \sec^2 x\,dx, \quad x\sec^2 x\,dx.$$

The most complicated of these expressions that can be readily integrated is $\sec^2 x\,dx$. Thus, we let

$$dv = \sec^2 x\,dx \qquad u = x$$
$$v = \tan x \qquad du = dx.$$

Integrating by parts gives us

$$\int x\sec^2 x\,dx = x\tan x - \int \tan x\,dx$$

$$= x\tan x + \ln|\cos x| + C.$$

(b) The indefinite integral obtained in part (a) is an antiderivative of $x\sec^2 x$. Using the fundamental theorem of calculus (and dropping the constant of integration C), we obtain

$$\int_0^{\pi/3} x\sec^2 x\,dx = \Big[x\tan x + \ln|\cos x|\Big]_0^{\pi/3}$$

$$= \left(\frac{\pi}{3}\tan\frac{\pi}{3} + \ln\left|\cos\frac{\pi}{3}\right|\right) - (0 + \ln 1)$$

$$= \left(\frac{\pi}{3}\sqrt{3} + \ln\frac{1}{2}\right) - (0 + 0)$$

$$= \frac{\pi}{3}\sqrt{3} - \ln 2 \approx 1.12.$$

If, in Example 2, we had chosen $dv = x\,dx$ and $u = \sec^2 x$, then the integration by parts formula (7.1) would have led to a more complicated integral. (Verify this fact.)

In the next example, we use integration by parts to find an antiderivative of the natural logarithmic function.

EXAMPLE ■ 3   Evaluate $\displaystyle\int \ln x\,dx$.

SOLUTION   Let

$$dv = dx \qquad u = \ln x$$
$$v = x \qquad du = \frac{1}{x}\,dx$$

and integrate by parts as follows:

$$\int \ln x\,dx = (\ln x)x - \int (x)\frac{1}{x}\,dx$$

$$= x\ln x - \int dx$$

$$= x\ln x - x + C$$

Sometimes it is necessary to use integration by parts more than once in the same problem, as illustrated in the next example.

EXAMPLE ■ 4   Evaluate $\int x^2 e^{2x}\, dx$.

SOLUTION   Let

$$dv = e^{2x}\, dx \qquad u = x^2$$
$$v = \tfrac{1}{2}e^{2x} \qquad du = 2x\, dx$$

and integrate by parts as follows:

$$\int x^2 e^{2x}\, dx = x^2(\tfrac{1}{2}e^{2x}) - \int (\tfrac{1}{2}e^{2x})2x\, dx$$
$$= \tfrac{1}{2}x^2 e^{2x} - \int x e^{2x}\, dx$$

To evaluate the integral on the right side of the last equation, we must again integrate by parts. Proceeding exactly as in Example 1 leads to

$$\int x^2 e^{2x}\, dx = \tfrac{1}{2}x^2 e^{2x} - \tfrac{1}{2}x e^{2x} + \tfrac{1}{4}e^{2x} + C.$$

The following example illustrates another device for evaluating an integral by means of two applications of the integration by parts formula.

EXAMPLE ■ 5   Evaluate $\int e^x \cos x\, dx$.

SOLUTION   We could either let $dv = \cos x\, dx$ or let $dv = e^x\, dx$, since each of these expressions is readily integrable. Let us choose

$$dv = \cos x\, dx \qquad u = e^x$$
$$v = \sin x \qquad du = e^x\, dx$$

and integrate by parts as follows:

$$\int e^x \cos x\, dx = e^x \sin x - \int (\sin x)e^x\, dx$$

(1)
$$\int e^x \cos x\, dx = e^x \sin x - \int e^x \sin x\, dx$$

We next apply integration by parts to the integral on the right side of equation (1). Since we chose a trigonometric form for dv in the first integration by parts, we shall also choose a trigonometric form for the second. Letting

$$dv = \sin x\, dx \qquad u = e^x$$
$$v = -\cos x \qquad du = e^x\, dx$$

and integrating by parts, we have

$$\int e^x \sin x\, dx = e^x(-\cos x) - \int (-\cos x)e^x\, dx$$

(2)
$$\int e^x \sin x\, dx = -e^x \cos x + \int e^x \cos x\, dx.$$

If we now use equation (2) to substitute on the right side of equation (1), we obtain

$$\int e^x \cos x \, dx = e^x \sin x - \left[-e^x \cos x + \int e^x \cos x \, dx \right],$$

or $\qquad \int e^x \cos x \, dx = e^x \sin x + e^x \cos x - \int e^x \cos x \, dx.$

Adding $\int e^x \cos x \, dx$ to both sides of the last equation gives us

$$2 \int e^x \cos x \, dx = e^x (\sin x + \cos x).$$

Finally, dividing both sides by 2 and adding the constant of integration yields

$$\int e^x \cos x \, dx = \tfrac{1}{2} e^x (\sin x + \cos x) + C.$$

We could have evaluated the given integral by using $dv = e^x \, dx$ for both the first and second applications of the integration by parts formula.

We must choose substitutions carefully when evaluating an integral of the type given in Example 5. To illustrate, suppose that in the evaluation of the integral on the right in equation (1) of the solution we had used

$$dv = e^x \, dx \qquad u = \sin x$$
$$v = e^x \qquad du = \cos x \, dx.$$

Integration by parts then leads to

$$\int e^x \sin x \, dx = (\sin x) e^x - \int e^x \cos x \, dx$$
$$= e^x \sin x - \int e^x \cos x \, dx.$$

If we now substitute in equation (1), we obtain

$$\int e^x \cos x \, dx = e^x \sin x - \left[e^x \sin x - \int e^x \cos x \, dx \right],$$

which reduces to

$$\int e^x \cos x \, dx = \int e^x \cos x \, dx.$$

Although this is a true statement, it is not an evaluation of the given integral.

EXAMPLE • 6 Evaluate $\int \sec^3 x \, dx$.

SOLUTION The possible choices for dv are

$$dx, \quad \sec x \, dx, \quad \sec^2 x \, dx, \quad \sec^3 x \, dx.$$

The most complicated of these expressions that can be readily integrated is $\sec^2 x\,dx$. Thus, we let

$$dv = \sec^2 x\,dx \qquad u = \sec x$$
$$v = \tan x \qquad du = \sec x \tan x\,dx$$

and integrate by parts as follows:

$$\int \sec^3 x\,dx = \sec x \tan x - \int \sec x \tan^2 x\,dx$$

Instead of applying another integration by parts, let us change the form of the integral on the right by using the identity $1 + \tan^2 x = \sec^2 x$. This gives us

$$\int \sec^3 x\,dx = \sec x \tan x - \int \sec x (\sec^2 x - 1)\,dx,$$

or

$$\int \sec^3 x\,dx = \sec x \tan x - \int \sec^3 x\,dx + \int \sec x\,dx.$$

Adding $\int \sec^3 x\,dx$ to both sides of the last equation gives us

$$2\int \sec^3 x\,dx = \sec x \tan x + \int \sec x\,dx.$$

If we now evaluate $\int \sec x\,dx$ and divide both sides of the resulting equation by 2 (and then add the constant of integration), we obtain

$$\int \sec^3 x\,dx = \tfrac{1}{2} \sec x \tan x + \tfrac{1}{2} \ln |\sec x + \tan x| + C.$$

Integration by parts may sometimes be employed to obtain **reduction formulas** for integrals. We can use such formulas to write an integral involving powers of an expression in terms of integrals that involve lower powers of the expression.

EXAMPLE ■ 7   Find a reduction formula for $\displaystyle\int \sin^n x\,dx$.

SOLUTION   Let

$$dv = \sin x\,dx \qquad u = \sin^{n-1} x$$
$$v = -\cos x \qquad du = (n-1)\sin^{n-2} x \cos x\,dx$$

and integrate by parts as follows:

$$\int \sin^n x\,dx = -\cos x \sin^{n-1} x + (n-1)\int \sin^{n-2} x \cos^2 x\,dx$$

Since $\cos^2 x = 1 - \sin^2 x$, we may write

$$\int \sin^n x\,dx$$
$$= -\cos x \sin^{n-1} x + (n-1)\int \sin^{n-2} x\,dx - (n-1)\int \sin^n x\,dx.$$

Consequently,

$$\int \sin^n x \, dx + (n-1) \int \sin^n x \, dx$$

$$= -\cos x \sin^{n-1} x + (n-1) \int \sin^{n-2} x \, dx.$$

The left side of the last equation reduces to $n \int \sin^n x \, dx$. Dividing both sides by n, we obtain

$$\int \sin^n x \, dx = -\frac{1}{n} \cos x \sin^{n-1} x + \frac{n-1}{n} \int \sin^{n-2} x \, dx.$$

EXAMPLE ■ 8 Use the reduction formula in Example 7 to evaluate $\int \sin^4 x \, dx$.

SOLUTION Using the formula with $n = 4$ gives us

$$\int \sin^4 x \, dx = -\tfrac{1}{4} \cos x \sin^3 x + \tfrac{3}{4} \int \sin^2 x \, dx.$$

Applying the reduction formula, with $n = 2$, to the integral on the right, we have

$$\int \sin^2 x \, dx = -\tfrac{1}{2} \cos x \sin x + \tfrac{1}{2} \int dx = -\tfrac{1}{2} \cos x \sin x + \tfrac{1}{2} x + C.$$

Consequently,

$$\int \sin^4 x \, dx = -\tfrac{1}{4} \cos x \sin^3 x - \tfrac{3}{8} \cos x \sin x + \tfrac{3}{8} x + D$$

with $D = \tfrac{3}{4} C$.

It should be evident that by repeated applications of the formula in Example 7 we can find $\int \sin^n x \, dx$ for any positive integer n, because these reductions end with either $\int \sin x \, dx$ or $\int dx$, and each of these can be evaluated easily.

EXERCISES 7.1

Exer. 1–38: Evaluate the integral.

1 $\int xe^{-x} \, dx$

2 $\int x \sin x \, dx$

7 $\int x \sec x \tan x \, dx$

8 $\int x \csc^2 3x \, dx$

3 $\int x^2 e^{3x} \, dx$

4 $\int x^2 \sin 4x \, dx$

9 $\int x^2 \cos x \, dx$

10 $\int x^3 e^{-x} \, dx$

5 $\int x \cos 5x \, dx$

6 $\int xe^{-2x} \, dx$

11 $\int \tan^{-1} x \, dx$

12 $\int \sin^{-1} x \, dx$

13 $\int \sqrt{x}\ln x\, dx$

14 $\int x^2 \ln x\, dx$

15 $\int x\csc^2 x\, dx$

16 $\int x\tan^{-1} x\, dx$

17 $\int e^{-x}\sin x\, dx$

18 $\int e^{3x}\cos 2x\, dx$

19 $\int \sin x\ln\cos x\, dx$

20 $\int_0^1 x^3 e^{-x^2}\, dx$

21 $\int \csc^3 x\, dx$

22 $\int \sec^5 x\, dx$

23 $\int_0^1 \dfrac{x^3}{\sqrt{x^2+1}}\, dx$

24 $\int \sin\ln x\, dx$

25 $\int_0^{\pi/2} x\sin 2x\, dx$

26 $\int x\sec^2 5x\, dx$

27 $\int x(2x+3)^{99}\, dx$

28 $\int \dfrac{x^5}{\sqrt{1-x^3}}\, dx$

29 $\int e^{4x}\sin 5x\, dx$

30 $\int x^3\cos(x^2)\, dx$

31 $\int (\ln x)^2\, dx$

32 $\int x\,2^x\, dx$

33 $\int x^3\sinh x\, dx$

34 $\int (x+4)\cosh 4x\, dx$

35 $\int \cos\sqrt{x}\, dx$

36 $\int \tan^{-1} 3x\, dx$

37 $\int \cos^{-1} x\, dx$

38 $\int (x+1)^{10}(x+2)\, dx$

Exer. 39–42: Use integration by parts to derive the reduction formula.

39 $\int x^m e^x\, dx = x^m e^x - m\int x^{m-1} e^x\, dx$

40 $\int x^m\sin x\, dx = -x^m\cos x + m\int x^{m-1}\cos x\, dx$

41 $\int (\ln x)^m\, dx = x(\ln x)^m - m\int (\ln x)^{m-1}\, dx$

42 $\int \sec^m x\, dx = \dfrac{\sec^{m-2} x\tan x}{m-1} + \dfrac{m-2}{m-1}\int \sec^{m-2} x\, dx$
 for $m \neq 1$.

43 Use Exercise 39 to evaluate $\int x^5 e^x\, dx$.

44 Use Exercise 41 to evaluate $\int (\ln x)^4\, dx$.

45 If $f(x) = \sin\sqrt{x}$, find the area of the region under the graph of f from $x = 0$ to $x = \pi^2$.

46 The region between the graph of $y = x\sqrt{\sin x}$ and the x-axis from $x = 0$ to $x = \pi/2$ is revolved about the x-axis. Find the volume of the resulting solid.

47 The region bounded by the graphs of $y = \ln x$, $y = 0$, and $x = e$ is revolved about the y-axis. Find the volume of the resulting solid.

48 Suppose that the force $f(x)$ acting at the point with coordinate x on a coordinate line l is given by $f(x) = x^5\sqrt{x^3+1}$. Find the work done in moving an object from $x = 0$ to $x = 1$.

49 Find the centroid of the region bounded by the graphs of the equations $y = e^x$, $y = 0$, $x = 0$, and $x = \ln 3$.

50 The velocity (at time t) of a point moving along a coordinate line is t/e^{2t} ft/sec. If the point is at the origin at $t = 0$, find its position at time t.

51 When applying the integration by parts formula (7.1), show that if, after choosing dv, we use $v + C$ in place of v, the same result is obtained.

52 In Section 5.3, the discussion of finding volumes by means of cylindrical shells was incomplete because we did not show that the same result is obtained if the disk method is also applicable. Use integration by parts to prove that if f is differentiable and either $f'(x) > 0$ on $[a, b]$ or $f'(x) < 0$ on $[a, b]$, and if V is the volume of the solid obtained by revolving the region bounded by the graphs of f, $x = a$, and $x = b$ about the x-axis, then the same value of V is obtained using either the disk method or the shell method. (*Hint:* Let g be the inverse function of f, and use integration by parts on $\int_a^b \pi[f(x)]^2\, dx$.)

53 Discuss the following use of Formula (7.1): Given $\int (1/x)\, dx$, let $dv = dx$ and $u = 1/x$ so that $v = x$ and $du = (-1/x^2)\, dx$. Hence

$$\int \frac{1}{x}\, dx = \left(\frac{1}{x}\right)x - \int x\left(-\frac{1}{x^2}\right)dx,$$

or $\quad \displaystyle\int \frac{1}{x}\, dx = 1 + \int \frac{1}{x}\, dx.$

Consequently, $0 = 1$.

54 If $u = f(x)$ and $v = g(x)$, prove that the analogue of Formula (7.1) for definite integrals is

$$\int_a^b u\, dv = \big[uv\big]_a^b - \int_a^b v\, du$$

for values a and b of x.

Mathematicians and Their Times

JOSEPH-LOUIS LAGRANGE

HAILED BY NAPOLEON as "the lofty pyramid of the mathematical sciences," Joseph-Louis Lagrange (1736–1813) was the greatest mathematician of his time. He made fundamental contributions in mechanics, sound, astronomy, and almost every branch of pure mathematics: analysis, calculus of variations, probability, number theory, algebra, differential equations, analytical geometry, and, of course, calculus.

Lagrange was of mixed French and Italian background. His grandfather and father both served in the government of Sardinia. In an era of high infant mortality, only one of Lagrange's ten siblings survived with him beyond early childhood. As a schoolboy, Lagrange was originally attracted to the classics. An essay by Edmund Halley ex-

tolling the virtues of calculus captivated Lagrange. He thus turned his attention to mathematics, where he made rapid progress. At the age of sixteen, Lagrange became professor of geometry at the Royal Artillery School in Turin.

In 1764, Lagrange won the Grand Prize of the French Academy of Sciences for his solution to the problem of the "libration" of the moon: Why does the moon always present the same face toward the earth? Soon afterward, Lagrange accepted appointment as court mathematician to Frederick the Great and as director of the physics and mathematics division of the Berlin Academy, serving there for twenty years.

King Louis XVI of France invited Lagrange to return to Paris to continue his work in mathematics as a member of the French Academy, where he remained during the French Revolution. Although he was the beneficiary of royal support for most of his career, he was not sympathetic to the royalists. However, he did not support the revolutionists either because he was indignant at the excesses of terror in the movement, particularly the guillotining of his friend, the chemist Lavosier.

When the École Normale opened in 1795, Lagrange accepted the post of professor of mathematics and turned his considerable talents to teaching. In an effort to lead his students more clearly through calculus, Lagrange wrote two books developing the subject: *Theory of Analytic*

7.2 TRIGONOMETRIC INTEGRALS

In this section, we examine integrals of functions that are products of powers of the trigonometric functions. In particular, we consider integrals of the form

$$\int \sin^m x \cos^n x \, dx,$$

where m and n are integers. In Example 7 of Section 7.1, we obtained a reduction formula for $\int \sin^n x \, dx$. Integrals of this type may also be found without using integration by parts. If n is an odd positive integer, we begin by writing

$$\int \sin^n x \, dx = \int \sin^{n-1} x \sin x \, dx.$$

Since the integer $n - 1$ is even, we may then use the trigonometric identity $\sin^2 x = 1 - \cos^2 x$ to obtain a form that is easy to integrate, as illustrated in the next example.

EXAMPLE ■ 1   Evaluate $\int \sin^5 x \, dx$.

SOLUTION   As in the preceding discussion, we write

$$\int \sin^5 x \, dx = \int \sin^4 x \sin x \, dx$$

$$= \int (\sin^2 x)^2 \sin x \, dx$$

$$= \int (1 - \cos^2 x)^2 \sin x \, dx$$

$$= \int (1 - 2\cos^2 x + \cos^4 x) \sin x \, dx.$$

If we substitute

$$u = \cos x, \qquad du = -\sin x\, dx,$$

we obtain

$$\int \sin^5 x\, dx = -\int (1 - 2\cos^2 x + \cos^4 x)(-\sin x)\, dx$$

$$= -\int (1 - 2u^2 + u^4)\, du$$

$$= -u + \tfrac{2}{3}u^3 - \tfrac{1}{5}u^5 + C$$

$$= -\cos x + \tfrac{2}{3}\cos^3 x - \tfrac{1}{5}\cos^5 x + C.$$

Similarly, for odd powers of $\cos x$, we write

$$\int \cos^n x\, dx = \int \cos^{n-1} x \cos x\, dx$$

and use the fact that $\cos^2 x = 1 - \sin^2 x$ to obtain an integrable form.

If the integrand is $\sin^n x$ or $\cos^n x$ and n is *even*, then the half-angle formula

$$\sin^2 x = \frac{1 - \cos 2x}{2} \quad \text{or} \quad \cos^2 x = \frac{1 + \cos 2x}{2}$$

may be used to simplify the integrand.

EXAMPLE ▪ 2   Evaluate $\int \cos^2 x\, dx$.

SOLUTION   Using a half-angle formula, we have

$$\int \cos^2 x\, dx = \tfrac{1}{2}\int (1 + \cos 2x)\, dx$$

$$= \tfrac{1}{2}x + \tfrac{1}{4}\sin 2x + C.$$

EXAMPLE ▪ 3   Evaluate $\int \sin^4 x\, dx$.

SOLUTION

$$\int \sin^4 x\, dx = \int (\sin^2 x)^2\, dx$$

$$= \int \left(\frac{1 - \cos 2x}{2} \right)^2 dx$$

$$= \tfrac{1}{4}\int (1 - 2\cos 2x + \cos^2 2x)\, dx$$

We apply a half-angle formula again and write

$$\cos^2 2x = \tfrac{1}{2}(1 + \cos 4x) = \tfrac{1}{2} + \tfrac{1}{2}\cos 4x.$$

Substituting in the last integral and simplifying gives us

$$\int \sin^4 x \, dx = \tfrac{1}{4} \int (\tfrac{3}{2} - 2 \cos 2x + \tfrac{1}{2} \cos 4x) \, dx$$

$$= \tfrac{3}{8}x - \tfrac{1}{4} \sin 2x + \tfrac{1}{32} \sin 4x + C.$$

Integrals involving only products of $\sin x$ and $\cos x$ may be evaluated using the following guidelines.

Guidelines for Evaluating
$\int \sin^m x \cos^n x \, dx$ **7.2**

1 **If m is an odd integer:** Write the integral as

$$\int \sin^m x \cos^n x \, dx = \int \sin^{m-1} x \cos^n x \sin x \, dx$$

and express $\sin^{m-1} x$ in terms of $\cos x$ by using the trigonometric identity $\sin^2 x = 1 - \cos^2 x$. Make the substitution

$$u = \cos x, \quad du = -\sin x \, dx$$

and evaluate the resulting integral.

2 **If n is an odd integer:** Write the integral as

$$\int \sin^m x \cos^n x \, dx = \int \sin^m x \cos^{n-1} x \cos x \, dx$$

and express $\cos^{n-1} x$ in terms of $\sin x$ by using the trigonometric identity $\cos^2 x = 1 - \sin^2 x$. Make the substitution

$$u = \sin x, \quad du = \cos x \, dx$$

and evaluate the resulting integral.

3 **If m and n are even:** Use half-angle formulas for $\sin^2 x$ and $\cos^2 x$ to reduce the exponents by one-half.

EXAMPLE ■ 4 Evaluate $\int \cos^3 x \sin^4 x \, dx$.

SOLUTION By guideline (2) of (7.2),

$$\int \cos^3 x \sin^4 x \, dx = \int \cos^2 x \sin^4 x \cos x \, dx$$

$$= \int (1 - \sin^2 x) \sin^4 x \cos x \, dx.$$

If we let $u = \sin x$, then $du = \cos x \, dx$, and the integral may be written

$$\int \cos^3 x \sin^4 x \, dx = \int (1 - u^2) u^4 \, du = \int (u^4 - u^6) \, du$$

$$= \tfrac{1}{5} u^5 - \tfrac{1}{7} u^7 + C$$

$$= \tfrac{1}{5} \sin^5 x - \tfrac{1}{7} \sin^7 x + C.$$

The following guidelines are analogous to those in (7.2) for integrands of the form $\tan^m x \sec^n x$.

Guidelines for Evaluating
$\int \tan^m x \sec^n x \, dx$ **7.3**

1 **If *m* is an odd integer:** Write the integral as

$$\int \tan^m x \sec^n x \, dx = \int \tan^{m-1} x \sec^{n-1} x \sec x \tan x \, dx$$

and express $\tan^{m-1} x$ in terms of $\sec x$ by using the trigonometric identity $\tan^2 = \sec^2 x - 1$. Make the substitution

$$u = \sec x, \qquad du = \sec x \tan x \, dx$$

and evaluate the resulting integral.

2 **If *n* is an even integer:** Write the integral as

$$\int \tan^m x \sec^n x \, dx = \int \tan^m x \sec^{n-2} x \sec^2 x \, dx$$

and express $\sec^{n-2} x$ in terms of $\tan x$ by using the trigonometric identity $\sec^2 x = 1 + \tan^2 x$. Make the substitution

$$u = \tan x, \qquad du = \sec^2 x \, dx$$

and evaluate the resulting integral.

3 **If *m* is even and *n* is odd:** There is no standard method of evaluation. Possibly use integration by parts.

EXAMPLE 5 Evaluate $\displaystyle\int \tan^3 x \sec^5 x \, dx$.

SOLUTION By guideline (1) of (7.3),

$$\int \tan^3 x \sec^5 x \, dx = \int \tan^2 x \sec^4 x (\sec x \tan x) \, dx$$

$$= \int (\sec^2 x - 1) \sec^4 x (\sec x \tan x) \, dx.$$

Substituting $u = \sec x$ and $du = \sec x \tan x \, dx$, we obtain

$$\int \tan^3 x \sec^5 x \, dx = \int (u^2 - 1) u^4 \, du$$

$$= \int (u^6 - u^4) \, du$$

$$= \tfrac{1}{7} u^7 - \tfrac{1}{5} u^5 + C$$

$$= \tfrac{1}{7} \sec^7 x - \tfrac{1}{5} \sec^5 x + C.$$

EXAMPLE ■ 6 Evaluate $\int \tan^2 x \sec^4 x \, dx$.

SOLUTION By guideline (2) of (7.3),

$$\int \tan^2 x \sec^4 x \, dx = \int \tan^2 x \sec^2 x \sec^2 x \, dx$$

$$= \int \tan^2 x (\tan^2 x + 1) \sec^2 x \, dx.$$

If we let $u = \tan x$, then $du = \sec^2 x \, dx$, and

$$\int \tan^2 x \sec^4 x \, dx = \int u^2(u^2 + 1) \, du$$

$$= \int (u^4 + u^2) \, du$$

$$= \tfrac{1}{5}u^5 + \tfrac{1}{3}u^3 + C$$

$$= \tfrac{1}{5}\tan^2 x + \tfrac{1}{3}\tan^3 x + C.$$

Integrals of the form $\int \cot^m x \csc^n x \, dx$ may be evaluated in similar fashion.

Finally, if an integrand has one of the forms $\cos mx \cos nx$, $\sin mx \sin nx$, or $\sin mx \cos nx$, we use a product-to-sum formula to help evaluate the integral, as illustrated in the next example.

EXAMPLE ■ 7 Evaluate $\int \cos 5x \cos 3x \, dx$.

SOLUTION Using the product-to-sum formula for $\cos u \cos v$, we obtain

$$\int \cos 5x \cos 3x \, dx = \int \tfrac{1}{2}(\cos 8x + \cos 2x) \, dx$$

$$= \tfrac{1}{16} \sin 8x + \tfrac{1}{4} \sin 2x + C.$$

EXERCISES 7.2

Exer. 1–30: Evaluate the integral.

1 $\int \cos^3 x \, dx$

2 $\int \sin^2 2x \, dx$

5 $\int \sin^3 x \cos^2 x \, dx$

6 $\int \sin^5 x \cos^3 x \, dx$

3 $\int \sin^2 x \cos^2 x \, dx$

4 $\int \cos^7 x \, dx$

7 $\int \sin^6 x \, dx$

8 $\int \sin^4 x \cos^2 x \, dx$

9 $\int \tan^3 x \sec^4 x \, dx$

10 $\int \sec^6 x \, dx$

11 $\int \tan^3 x \sec^3 x \, dx$

12 $\int \tan^5 x \sec x \, dx$

13 $\int \tan^6 x \, dx$

14 $\int \cot^4 x \, dx$

15 $\int \sqrt{\sin x} \, \cos^3 x \, dx$

16 $\int \dfrac{\cos^3 x}{\sqrt{\sin x}} \, dx$

17 $\int (\tan x + \cot x)^2 \, dx$

18 $\int \cot^3 x \csc^3 x \, dx$

19 $\int_0^{\pi/4} \sin^3 x \, dx$

20 $\int_0^1 \tan^2(\tfrac{1}{4}\pi x) \, dx$

21 $\int \sin 5x \sin 3x \, dx$

22 $\int_0^{\pi/4} \cos x \cos 5x \, dx$

23 $\int_0^{\pi/2} \sin 3x \cos 2x \, dx$

24 $\int \sin 4x \cos 3x \, dx$

25 $\int \csc^4 x \cot^4 x \, dx$

26 $\int (1 + \sqrt{\cos x})^2 \sin x \, dx$

27 $\int \dfrac{\cos x}{2 - \sin x} \, dx$

28 $\int \dfrac{\tan^2 x - 1}{\sec^2 x} \, dx$

29 $\int \dfrac{\sec^2 x}{(1 + \tan x)^2} \, dx$

30 $\int \dfrac{\sec x}{\cot^5 x} \, dx$

31 The region bounded by the x-axis and the graph of $y = \cos^2 x$ from $x = 0$ to $x = 2\pi$ is revolved about the x-axis. Find the volume of the resulting solid.

32 The region between the graphs of $y = \tan^2 x$ and $y = 0$ from $x = 0$ to $x = \pi/4$ is revolved about the x-axis. Find the volume of the resulting solid.

33 The velocity (at time t) of a point moving on a coordinate line is $\cos^2 \pi t$ ft/sec. How far does the point travel in 5 sec?

34 The acceleration (at time t) of a point moving on a coordinate line is $\sin^2 t \cos t$ ft/sec^2. At $t = 0$, the point is at the origin and its velocity is 10 ft/sec. Find its position at time t.

35 (a) Prove that if m and n are positive integers,

$$\int \sin mx \sin nx \, dx$$

$$= \begin{cases} \dfrac{\sin(m-n)x}{2(m-n)} - \dfrac{\sin(m+n)x}{2(m+n)} + C & \text{if } m \neq n \\[2mm] \dfrac{x}{2} - \dfrac{\sin 2mx}{4m} + C & \text{if } m = n \end{cases}$$

(b) Obtain formulas similar to that in part (a) for

$$\int \sin mx \cos nx \, dx$$

and

$$\int \cos mx \cos nx \, dx.$$

36 (a) Use part (a) of Exercise 35 to prove that

$$\int_{-\pi}^{\pi} \sin mx \sin nx \, dx = \begin{cases} 0 & \text{if } m \neq n \\ \pi & \text{if } m = n \end{cases}$$

(b) Find

(i) $\int_{-\pi}^{\pi} \sin mx \cos nx \, dx$

(ii) $\int_{-\pi}^{\pi} \cos mx \cos nx \, dx$

7.3 TRIGONOMETRIC SUBSTITUTIONS

In Example 4 on p. 42, we saw how to change an expression of the form $\sqrt{a^2 - x^2}$, with $a > 0$, into a trigonometric expression without radicals, by using the *trigonometric substitution* $x = a \sin \theta$. We can use a similar procedure for $\sqrt{a^2 + x^2}$ and $\sqrt{x^2 - a^2}$. This technique is useful for eliminating radicals from certain types of integrands. The substitutions are listed in the following table.

Trigonometric Substitutions 7.4

Expression in integrand	Trigonometric substitution
$\sqrt{a^2 - x^2}$	$x = a \sin \theta$
$\sqrt{a^2 + x^2}$	$x = a \tan \theta$
$\sqrt{x^2 - a^2}$	$x = a \sec \theta$

When making a trigonometric substitution, we shall assume that θ is in the range of the corresponding inverse trigonometric function. Thus, for the substitution $x = a \sin \theta$, we have $-\pi/2 \le \theta \le \pi/2$. In this case, $\cos \theta \ge 0$ and

$$\sqrt{a^2 - x^2} = \sqrt{a^2 - a^2 \sin^2 \theta}$$
$$= \sqrt{a^2(1 - \sin^2 \theta)}$$
$$= \sqrt{a^2 \cos^2 \theta}$$
$$= a \cos \theta.$$

If $\sqrt{a^2 - x^2}$ occurs in a denominator, we add the restriction $|x| \ne a$, or, equivalently, $-\pi/2 < \theta < \pi/2$.

EXAMPLE ▪ 1   Evaluate $\displaystyle\int \frac{1}{x^2 \sqrt{16 - x^2}}\, dx$.

SOLUTION   The integrand contains $\sqrt{16 - x^2}$, which is of the form $\sqrt{a^2 - x^2}$ with $a = 4$. Hence, by (7.4), we let

$$x = 4 \sin \theta \quad \text{for} \quad -\pi/2 < \theta < \pi/2.$$

It follows that

$$\sqrt{16 - x^2} = \sqrt{16 - 16 \sin^2 \theta}$$
$$= 4\sqrt{1 - \sin^2 \theta} = 4\sqrt{\cos^2 \theta} = 4 \cos \theta.$$

Since $x = 4 \sin \theta$, we have $dx = 4 \cos \theta \, d\theta$. Substituting in the given integral yields

$$\int \frac{1}{x^2 \sqrt{16 - x^2}}\, dx = \int \frac{1}{(16 \sin^2 \theta) \, 4 \cos \theta}\, 4 \cos \theta \, d\theta$$

$$= \frac{1}{16} \int \frac{1}{\sin^2 \theta}\, d\theta$$

$$= \tfrac{1}{16} \int \csc^2 \theta \, d\theta$$

$$= -\tfrac{1}{16} \cot \theta + C.$$

We must now return to the original variable of integration, x. Since $\theta = \arcsin(x/4)$, we could write $-\frac{1}{16} \cot \theta$ as $-\frac{1}{16} \cot \arcsin(x/4)$, but this is a cumbersome expression. Since the integrand contains $\sqrt{16 - x^2}$, it is preferable that the evaluated form also contain this radical. There is a simple geometric method for ensuring that it does. If $0 < \theta < \pi/2$ and $\sin \theta = x/4$, we may interpret θ as an acute angle of a right triangle having opposite side and hypotenuse of lengths x and 4, respectively (see Figure 7.1). By the Pythagorean theorem, the length of the adjacent side is $\sqrt{16 - x^2}$. Referring to the triangle, we find

$$\cot \theta = \frac{\sqrt{16 - x^2}}{x}.$$

Figure 7.1

$\sin \theta = \dfrac{x}{4}$

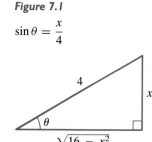

It can be shown that the last formula is also true if $-\pi/2 < \theta < 0$. Thus, Figure 7.1 may be used if θ is either positive or negative.

Substituting $\sqrt{16 - x^2}/x$ for $\cot\theta$ in our integral evaluation gives us

$$\int \frac{1}{x^2\sqrt{16 - x^2}}\, dx = -\frac{1}{16} \cdot \frac{\sqrt{16 - x^2}}{x} + C$$

$$= -\frac{\sqrt{16 - x^2}}{16x} + C.$$

If an integrand contains $\sqrt{a^2 + x^2}$ for $a > 0$, then, by (7.4), we use the substitution $x = a\tan\theta$ to eliminate the radical. When using this substitution, we assume that θ is in the range of the inverse tangent function—that is, $-\pi/2 < \theta < \pi/2$. In this case, $\sec\theta > 0$ and

$$\sqrt{a^2 + x^2} = \sqrt{a^2 + a^2\tan^2\theta}$$

$$= \sqrt{a^2(1 + \tan^2\theta)}$$

$$= \sqrt{a^2\sec^2\theta}$$

$$= a\sec\theta.$$

After substituting and evaluating the resulting trigonometric integral, it is necessary to return to the variable x. We can do so by using the formula $\tan\theta = x/a$ and referring to the right triangle shown in Figure 7.2.

Figure 7.2

$\tan\theta = \dfrac{x}{a}$

EXAMPLE ■ 2 Evaluate $\displaystyle\int \frac{1}{\sqrt{4 + x^2}}\, dx$.

SOLUTION The denominator of the integrand has the form $\sqrt{a^2 + x^2}$ with $a = 2$. Hence, by (7.4), we make the substitution

$$x = 2\tan\theta, \qquad dx = 2\sec^2\theta\, d\theta.$$

Consequently,

$$\sqrt{4 + x^2} = \sqrt{4 + 4\tan^2\theta}$$

$$= 2\sqrt{1 + \tan^2\theta} = 2\sqrt{\sec^2\theta} = 2\sec\theta$$

and

$$\int \frac{1}{\sqrt{4 + x^2}}\, dx = \int \frac{1}{2\sec\theta} 2\sec^2\theta\, d\theta$$

$$= \int \sec\theta\, d\theta$$

$$= \ln|\sec\theta + \tan\theta| + C.$$

Using $\tan\theta = x/2$, we sketch the triangle in Figure 7.3, from which we obtain

$$\sec\theta = \frac{\sqrt{4 + x^2}}{2}.$$

Figure 7.3

$\tan\theta = \dfrac{x}{2}$

Hence,

$$\int \frac{1}{\sqrt{4+x^2}}\, dx = \ln \left| \frac{\sqrt{4+x^2}}{2} + \frac{x}{2} \right| + C.$$

The expression on the right may be written

$$\ln \left| \frac{\sqrt{4+x^2}+x}{2} \right| + C = \ln \left| \sqrt{4+x^2}+x \right| - \ln 2 + C.$$

Since $\sqrt{4+x^2}+x > 0$ for every x, the absolute value sign is unnecessary. If we also let $D = -\ln 2 + C$, then

$$\int \frac{1}{\sqrt{4+x^2}}\, dx = \ln \left(\sqrt{4+x^2}+x \right) + D.$$

If an integrand contains $\sqrt{x^2-a^2}$, then using (7.4) we substitute $x = a\sec\theta$, where θ is chosen in the range of the inverse secant function—that is, either $0 \le \theta < \pi/2$ or $\pi \le \theta < 3\pi/2$. In this case, $\tan\theta \ge 0$ and

$$\begin{aligned}
\sqrt{x^2-a^2} &= \sqrt{a^2\sec^2\theta - a^2} \\
&= \sqrt{a^2(\sec^2\theta - 1)} \\
&= \sqrt{a^2\tan^2\theta} \\
&= a\tan\theta.
\end{aligned}$$

Since

$$\sec\theta = \frac{x}{a},$$

we may refer to the triangle in Figure 7.4 when changing from the variable θ to the variable x.

Figure 7.4

$$\sec\theta = \frac{x}{a}$$

EXAMPLE ▪ 3   Evaluate $\displaystyle\int \frac{\sqrt{x^2-9}}{x}\, dx$.

SOLUTION   The integrand contains $\sqrt{x^2-9}$, which is of the form $\sqrt{x^2-a^2}$ with $a = 3$. Referring to (7.4), we substitute as follows:

$$x = 3\sec\theta, \qquad dx = 3\sec\theta\tan\theta\, d\theta$$

Consequently,

$$\begin{aligned}
\sqrt{x^2-9} &= \sqrt{9\sec^2\theta - 9} \\
&= 3\sqrt{\sec^2\theta - 1} = 3\sqrt{\tan^2\theta} = 3\tan\theta
\end{aligned}$$

and

$$\int \frac{\sqrt{x^2 - 9}}{x} \, dx = \int \frac{3 \tan \theta}{3 \sec \theta} 3 \sec \theta \tan \theta \, d\theta$$

$$= 3 \int \tan^2 \theta \, d\theta$$

$$= 3 \int (\sec^2 \theta - 1) \, d\theta = 3 \int \sec^2 \theta \, d\theta - 3 \int d\theta$$

$$= 3 \tan \theta - 3\theta + C.$$

Since $\sec \theta = x/3$, we may refer to the right triangle in Figure 7.5. Using $\tan \theta = \sqrt{x^2 - 9}/3$ and $\theta = \sec^{-1}(x/3)$, we obtain

$$\int \frac{\sqrt{x^2 - 9}}{x} \, dx = 3 \frac{\sqrt{x^2 - 9}}{3} - 3 \sec^{-1} \left(\frac{x}{3} \right) + C$$

$$= \sqrt{x^2 - 9} - 3 \sec^{-1} \left(\frac{x}{3} \right) + C.$$

Figure 7.5

$\sec \theta = \dfrac{x}{3}$

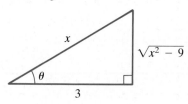

As shown in the next example, we can use trigonometric substitutions to evaluate certain integrals that involve $(a^2 - x^2)^r$, $(a^2 + x^2)^r$, or $(x^2 - a^2)^r$, in cases other than $r = \frac{1}{2}$.

EXAMPLE ■ 4   Evaluate $\displaystyle \int \frac{(1 - x^2)^{3/2}}{x^6} \, dx$.

SOLUTION   The integrand contains the expression $1 - x^2$, which is of the form $a^2 - x^2$ with $a = 1$. Using (7.4), we substitute

$$x = \sin \theta, \qquad dx = \cos \theta \, d\theta.$$

Thus, $1 - x^2 = 1 - \sin^2 \theta = \cos^2 \theta$, and

$$\int \frac{(1 - x^2)^{3/2}}{x^6} \, dx = \int \frac{(\cos^2 \theta)^{3/2}}{\sin^6 \theta} \cos \theta \, d\theta$$

$$= \int \frac{\cos^4 \theta}{\sin^6 \theta} \, d\theta = \int \frac{\cos^4 \theta}{\sin^4 \theta} \cdot \frac{1}{\sin^2 \theta} \, d\theta$$

$$= \int \cot^4 \theta \csc^2 \theta \, d\theta$$

$$= -\frac{1}{5} \cot^5 \theta + C.$$

Figure 7.6

$\sin \theta = x$

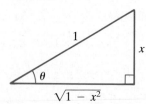

To return to the variable x, we note that $\sin \theta = x = x/1$ and refer to the right triangle in Figure 7.6, obtaining $\cot \theta = \sqrt{1 - x^2}/x$. Hence,

$$\int \frac{(1 - x^2)^{3/2}}{x^6} \, dx = -\frac{1}{5} \left(\frac{\sqrt{1 - x^2}}{x} \right)^5 + C = -\frac{(1 - x^2)^{5/2}}{5x^5} + C.$$

Trigonometric substitutions may also be used with trigonometric identities in the evaluation of definite integrals. In the next example, we use the substitution $x = a \sin \theta$ and the identity $2 \cos^2 \theta = 1 + \cos 2\theta$ to find the area bounded by an ellipse.

EXAMPLE ▪ 5   Find the area of the region bounded by an ellipse whose major and minor axes have lengths $2a$ and $2b$, respectively.

SOLUTION   By Theorem 34 (page 70), we see that an equation for the ellipse is $(x^2/a^2) + (y^2/b^2) = 1$. Solving for y gives us

$$y = \pm \frac{b}{a} \sqrt{a^2 - x^2}.$$

The graph of the ellipse has the general shape shown in Figure 64 (page 69) and hence, by symmetry, it is sufficient to find the area of the region in the first quadrant and multiply the result by 4. By Theorem (4.19),

$$A = 4 \int_0^a y \, dx = 4 \frac{b}{a} \int_0^a \sqrt{a^2 - x^2} \, dx.$$

If we make the trigonometric substitution $x = a \sin \theta$, then

$$\sqrt{a^2 - x^2} = a \cos \theta \quad \text{and} \quad dx = a \cos \theta \, d\theta.$$

Since the values of θ that correspond to $x = 0$ and $x = a$ are $\theta = 0$ and $\theta = \pi/2$, respectively, we obtain

$$A = 4 \frac{b}{a} \int_0^{\pi/2} a^2 \cos^2 \theta \, d\theta = 4ab \int_0^{\pi/2} \frac{1 + \cos 2\theta}{2} \, d\theta$$

$$= 2ab \left[\theta + \tfrac{1}{2} \sin 2\theta \right]_0^{\pi/2}$$

$$= 2ab \left[\frac{\pi}{2} \right] = \pi ab.$$

Thus, the area of an ellipse with axes of lengths $2a$ and $2b$ is πab. As a special case, if $b = a$, the ellipse is a circle and $A = \pi a^2$.

Although we now have additional integration techniques available, it is a good idea to keep earlier methods in mind. For example, the integral $\int (x/\sqrt{9 + x^2}) \, dx$ could be evaluated by means of the trigonometric substitution $x = 3 \tan \theta$. However, it is simpler to use the algebraic substitution $u = 9 + x^2$ and $du = 2x \, dx$, for in this event the integral takes on the form $\frac{1}{2} \int u^{-1/2} \, du$, which is readily integrated by means of the power rule. The following exercises include integrals that can be evaluated using simpler techniques than trigonometric substitutions.

EXERCISES 7.3

Exer. 1–22: Evaluate the integral.

1 $\int \dfrac{1}{x\sqrt{4-x^2}}\,dx$

2 $\int \dfrac{\sqrt{4-x^2}}{x^2}\,dx$

3 $\int \dfrac{1}{x\sqrt{9+x^2}}\,dx$

4 $\int \dfrac{1}{x^2\sqrt{x^2+9}}\,dx$

5 $\int \dfrac{1}{x^2\sqrt{x^2-25}}\,dx$

6 $\int \dfrac{1}{x^3\sqrt{x^2-25}}\,dx$

7 $\int \dfrac{x}{\sqrt{4-x^2}}\,dx$

8 $\int \dfrac{x}{x^2+9}\,dx$

9 $\int \dfrac{1}{(x^2-1)^{3/2}}\,dx$

10 $\int \dfrac{1}{\sqrt{4x^2-25}}\,dx$

11 $\int \dfrac{1}{(36+x^2)^2}\,dx$

12 $\int \dfrac{1}{(16-x^2)^{5/2}}\,dx$

13 $\int \dfrac{1}{\sqrt{9-x^2}}\,dx$

14 $\int \dfrac{1}{49+x^2}\,dx$

15 $\int \dfrac{x}{(16-x^2)^2}\,dx$

16 $\int x\sqrt{x^2-9}\,dx$

17 $\int \dfrac{x^3}{\sqrt{9x^2+49}}\,dx$

18 $\int \dfrac{1}{x\sqrt{25x^2+16}}\,dx$

19 $\int \dfrac{1}{x^4\sqrt{x^2-3}}\,dx$

20 $\int \dfrac{x^2}{(1-9x^2)^{3/2}}\,dx$

21 $\int \dfrac{(4+x^2)^2}{x^3}\,dx$

22 $\int \dfrac{3x-5}{\sqrt{1-x^2}}\,dx$

23 The region bounded by the graphs of $y=0$, $x=5$, and $y=x(x^2+25)^{-1/2}$ is revolved about the y-axis. Find the volume of the resulting solid.

24 Find the area of the region bounded by the graph of $y=x^3(10-x^2)^{-1/2}$, the x-axis, and the line $x=1$.

25 The shape of the earth's surface can be approximated by revolving the ellipse $(x^2/a^2)+(y^2/b^2)=1$, with $a=6378$ km and $b=6356$ km, about the x-axis. Approx-

imate the surface area of the earth to the nearest 10^6 km². (*Hint:* Use (5.19) with $f(x)=\sqrt{b^2-(b^2/a^2)x^2}$, and make the substitution $u=(b/a)x$.)

26 Let R be the region bounded by the right branch of the hyperbola $x^2-y^2=8$ and the vertical line through the focus. Find the area of the curved surface of the solid obtained by revolving R about the x-axis.

Exer. 27–28: Solve the differential equation subject to the given initial condition.

27 $x\,dy=\sqrt{x^2-16}\,dx;\quad y=0$ if $x=4$

28 $\sqrt{1-x^2}\,dy=x^3\,dx;\quad y=0$ if $x=0$

Exer. 29–34: Use a trigonometric substitution to derive the formula. (See Formulas 21, 27, 31, 36, 41, and 44 in Appendix II.)

29 $\int \sqrt{a^2+u^2}\,du$

$$=\frac{u}{2}\sqrt{a^2+u^2}+\frac{a^2}{2}\ln\left|u+\sqrt{a^2+u^2}\right|+C$$

30 $\int \dfrac{1}{u\sqrt{a^2+u^2}}\,du=-\dfrac{1}{a}\ln\left|\dfrac{\sqrt{a^2+u^2}+a}{u}\right|+C$

31 $\int u^2\sqrt{a^2-u^2}\,du$

$$=\frac{u}{8}(2u^2-a^2)\sqrt{a^2-u^2}+\frac{a^4}{8}\sin^{-1}\frac{u}{a}+C$$

32 $\int \dfrac{1}{u^2\sqrt{a^2-u^2}}\,du=-\dfrac{1}{a^2u}\sqrt{a^2-u^2}+C$

33 $\int \dfrac{\sqrt{u^2-a^2}}{u}\,du=\sqrt{u^2-a^2}-a\sec^{-1}\dfrac{u}{a}+C$

34 $\int \dfrac{u^2}{\sqrt{u^2-a^2}}\,du$

$$=\frac{u}{2}\sqrt{u^2-a^2}+\frac{a^2}{2}\ln\left|u+\sqrt{u^2-a^2}\right|+C$$

7.4 INTEGRALS OF RATIONAL FUNCTIONS

Recall that if q is a rational function, then $q(x)=f(x)/g(x)$, where $f(x)$ and $g(x)$ are polynomials. In this section, we examine the rules for evaluating $\int q(x)\,dx$.

Let us consider the specific case $q(x) = 2/(x^2 - 1)$. It is easy to verify that

$$\frac{1}{x-1} + \frac{-1}{x+1} = \frac{2}{x^2 - 1}.$$

The expression on the left side of the equation is called the *partial fraction decomposition* of $2/(x^2 - 1)$. To find $\int q(x)\, dx$, we integrate each of the fractions that make up the decomposition, obtaining

$$\int \frac{2}{x^2 - 1}\, dx = \int \frac{1}{x-1}\, dx + \int \frac{-1}{x+1}\, dx$$

$$= \ln|x-1| - \ln|x+1| + C$$

$$= \ln\left|\frac{x-1}{x+1}\right| + C.$$

It is theoretically possible to write *any* rational expression $f(x)/g(x)$ as a sum of rational expressions whose denominators involve powers of polynomials of degree not greater than two. Specifically, if $f(x)$ and $g(x)$ are polynomials *and the degree of $f(x)$ is less than the degree of $g(x)$*, then it can be proved that

$$\frac{f(x)}{g(x)} = F_1 + F_2 + \cdots + F_r$$

such that each term F_k of the sum has one of the forms

$$\frac{A}{(ax+b)^n} \quad \text{or} \quad \frac{Ax+B}{(ax^2 + bx + c)^n}$$

for real numbers A and B and a nonnegative integer n, where $ax^2 + bx + c$ is **irreducible** in the sense that this quadratic polynomial has no real zeros (that is, $b^2 - 4ac < 0$). In this case, $ax^2 + bx + c$ cannot be expressed as a product of two first-degree polynomials with real coefficients.

The sum $F_1 + F_2 + \cdots + F_r$ is the **partial fraction decomposition** of $f(x)/g(x)$, and each F_k is a **partial fraction**. We shall not prove this algebraic result but shall, instead, state guidelines for obtaining the decomposition.

The guidelines for finding the partial fraction decomposition of $f(x)/g(x)$ should be used only if $f(x)$ has lower degree than $g(x)$. If this is not the case, then we may use long division to arrive at the proper form. For example, given

$$\frac{x^3 - 6x^2 + 5x - 3}{x^2 - 1},$$

we obtain, by long division,

$$\frac{x^3 - 6x^2 + 5x - 3}{x^2 - 1} = x - 6 + \frac{6x - 9}{x^2 - 1}.$$

We then find the partial fraction decomposition for $(6x - 9)/(x^2 - 1)$.

1 If the degree of $f(x)$ is not lower than the degree of $g(x)$, use long division to obtain the proper form.

2 Express $g(x)$ as a product of linear factors $ax + b$ or irreducible quadratic factors $ax^2 + bx + c$, and collect repeated factors so that $g(x)$ is a product of *different* factors of the form $(ax + b)^n$ or $(ax^2 + bx + c)^n$ for a nonnegative integer n.

3 Apply the following rules.

Rule a For each factor $(ax + b)^n$ with $n \geq 1$, the partial fraction decomposition contains a sum of n partial fractions of the form

$$\frac{A_1}{ax + b} + \frac{A_2}{(ax + b)^2} + \cdots + \frac{A_n}{(ax + b)^n},$$

where each numerator A_k is a real number.

Rule b For each factor $(ax^2 + bx + c)^n$ with $n \geq 1$ and with $ax^2 + bx + c$ irreducible, the partial fraction decomposition contains a sum of n partial fractions of the form

$$\frac{A_1 x + B_1}{ax^2 + bx + c} + \frac{A_2 x + B_2}{(ax^2 + bx + c)^2} + \cdots + \frac{A_n x + B_n}{(ax^2 + bx + c)^n},$$

where each A_k and B_k is a real number.

EXAMPLE ▪ 1 Evaluate $\displaystyle\int \frac{4x^2 + 13x - 9}{x^3 + 2x^2 - 3x} \, dx$.

SOLUTION We may factor the denominator of the integrand as follows:

$$x^3 + 2x^2 - 3x = x(x^2 + 2x - 3) = x(x + 3)(x - 1)$$

Each factor has the form stated in rule (a) of (7.5), with $n = 1$. Thus, to the factor x there corresponds a partial fraction of the form A/x. Similarly, to the factors $x + 3$ and $x - 1$ there correspond partial fractions $B/(x + 3)$ and $C/(x - 1)$, respectively. Therefore, the partial fraction decomposition has the form

$$\frac{4x^2 + 13x - 9}{x(x + 3)(x - 1)} = \frac{A}{x} + \frac{B}{x + 3} + \frac{C}{x - 1}.$$

Multiplying by the lowest common denominator gives us

(∗) $4x^2 + 13x - 9 = A(x + 3)(x - 1) + Bx(x - 1) + Cx(x + 3)$.

In a case such as this, in which the factors are all linear and nonrepeated, the values for A, B, and C can be found by substituting values for x that make the various factors zero. If we let $x = 0$ in (∗), then

$$-9 = -3A, \quad \text{or} \quad A = 3.$$

Letting $x = 1$ in (∗) gives us

$$8 = 4C, \quad \text{or} \quad C = 2.$$

Finally, if $x = -3$ in $(*)$, then

$$-12 = 12B, \quad \text{or} \quad B = -1.$$

The partial fraction decomposition is, therefore,

$$\frac{4x^2 + 13x - 9}{x(x + 3)(x - 1)} = \frac{3}{x} + \frac{-1}{x + 3} + \frac{2}{x - 1}.$$

Integrating and letting K denote the sum of the constants of integration, we have

$$\int \frac{4x^2 + 13x - 9}{x(x + 3)(x - 1)} \, dx = \int \frac{3}{x} \, dx + \int \frac{-1}{x + 3} \, dx + \int \frac{2}{x - 1} \, dx$$

$$= 3 \ln |x| - \ln |x + 3| + 2 \ln |x - 1| + K$$

$$= \ln \left| x^3 \right| - \ln |x + 3| + \ln |x - 1|^2 + K$$

$$= \ln \left| \frac{x^3(x - 1)^2}{x + 3} \right| + K.$$

Another technique for finding A, B, and C is to expand the right-hand side of $(*)$ and collect like powers of x as follows:

$$4x^2 + 13x - 9 = (A + B + C)x^2 + (2A - B + 3C)x - 3A$$

We now use the fact that if two polynomials are equal, then coefficients of like powers of x are the same. It is convenient to arrange our work in the following way, which we call **comparing coefficients of x**:

$$\begin{aligned}
\textit{coefficients of } x^2: \quad & A + B + C = 4 \\
\textit{coefficients of } x: \quad & 2A - B + 3C = 13 \\
\textit{constant terms}: \quad & -3A = -9
\end{aligned}$$

We may show that the solution of this system of equations is $A = 3$, $B = -1$, and $C = 2$.

EXAMPLE ▪ 2 Evaluate $\displaystyle\int \frac{3x^3 - 18x^2 + 29x - 4}{(x + 1)(x - 2)^3} \, dx$.

SOLUTION By rule (a) of (7.5), there is a partial fraction of the form $A/(x + 1)$ corresponding to the factor $x + 1$ in the denominator of the integrand. For the factor $(x - 2)^3$, we apply rule (a), with $n = 3$, obtaining a sum of three partial fractions $B/(x - 2)$, $C/(x - 2)^2$, and $D/(x - 2)^3$. Consequently, the partial fraction decomposition has the form

$$\frac{3x^3 - 18x^2 + 29x - 4}{(x + 1)(x - 2)^3} = \frac{A}{x + 1} + \frac{B}{x - 2} + \frac{C}{(x - 2)^2} + \frac{D}{(x - 2)^3}.$$

Multiplying both sides by $(x + 1)(x - 2)^3$ gives us

$$\begin{aligned}
(*) \qquad 3x^3 - 18x^2 + 29x - 4 = {} & A(x - 2)^3 + B(x + 1)(x - 2)^2 \\
& + C(x + 1)(x - 2) + D(x + 1).
\end{aligned}$$

Two of the unknown constants may be determined easily. If we let $x = 2$ in (∗), we obtain

$$6 = 3D, \quad \text{or} \quad D = 2.$$

Similarly, letting $x = -1$ in (∗) yields

$$-54 = -27A, \quad \text{or} \quad A = 2.$$

The remaining constants may be found by comparing coefficients. Examining the right-hand side of (∗), we see that the coefficient of x^3 is $A + B$. This must equal the coefficient of x^3 on the left. Thus, by comparison,

$$\textit{coefficients of } x^3: \quad 3 = A + B.$$

Since $A = 2$, it follows that $B = 1$.

Finally, we compare the constant terms in (∗) by letting $x = 0$. This gives us the following:

$$\textit{constant terms:} \quad -4 = -8A + 4B - 2C + D$$

Substituting the values we have found for A, B, and D into the preceding equation yields

$$-4 = -16 + 4 - 2C + 2,$$

which has the solution $C = -3$. The partial fraction decomposition is, therefore,

$$\frac{3x^3 - 18x^2 + 29x - 4}{(x + 1)(x - 2)^3} = \frac{2}{x + 1} + \frac{1}{x - 2} + \frac{-3}{(x - 2)^2} + \frac{2}{(x - 2)^3}.$$

To find the given integral, we integrate each of the partial fractions on the right side of the last equation, obtaining

$$2 \ln |x + 1| + \ln |x - 2| + \frac{3}{x - 2} - \frac{1}{(x - 2)^2} + K$$

with K the sum of the four constants of integration. This may be written in the form

$$\ln \left[(x + 1)^2 \, |x - 2| \right] + \frac{3}{x - 2} - \frac{1}{(x - 2)^2} + K.$$

EXAMPLE ▪ 3 Evaluate $\displaystyle\int \frac{x^2 - x - 21}{2x^3 - x^2 + 8x - 4} \, dx$.

SOLUTION The denominator may be factored by grouping as follows:

$$2x^3 - x^2 + 8x - 4 = x^2(2x - 1) + 4(2x - 1) = (x^2 + 4)(2x - 1)$$

Applying rule (b) of (7.5) to the irreducible quadratic factor $x^2 + 4$, we see that one of the partial fractions has the form $(Ax + B)/(x^2 + 4)$. By rule (a), there is also a partial fraction $C/(2x - 1)$ corresponding to the factor $2x - 1$. Consequently,

$$\frac{x^2 - x - 21}{2x^3 - x^2 + 8x - 4} = \frac{Ax + B}{x^2 + 4} + \frac{C}{2x - 1}.$$

As in previous examples, this result leads to

$$(*) \qquad x^2 - x - 21 = (Ax + B)(2x - 1) + C(x^2 + 4).$$

We can find one constant easily. Substituting $x = \frac{1}{2}$ in $(*)$ gives us

$$-\tfrac{85}{4} = \tfrac{17}{4}C, \quad \text{or} \quad C = -5.$$

The remaining constants may be found by comparing coefficients of x in $(*)$:

$$
\begin{aligned}
\textit{coefficients of } x^2: & \quad 1 = 2A + C \\
\textit{coefficients of } x: & \quad -1 = -A + 2B \\
\textit{constant terms}: & \quad -21 = -B + 4C
\end{aligned}
$$

Since $C = -5$, it follows from $1 = 2A + C$ that $A = 3$. Similarly, using the coefficients of x with $A = 3$ gives us $-1 = -3 + 2B$, or $B = 1$. Thus, the partial fraction decomposition of the integrand is

$$
\begin{aligned}
\frac{x^2 - x - 21}{2x^3 - x^2 + 8x - 4} &= \frac{3x + 1}{x^2 + 4} + \frac{-5}{2x - 1} \\
&= \frac{3x}{x^2 + 4} + \frac{1}{x^2 + 4} - \frac{5}{2x - 1}.
\end{aligned}
$$

The given integral may now be found by integrating the right side of the last equation. This gives us

$$\frac{3}{2}\ln(x^2 + 4) + \frac{1}{2}\tan^{-1}\frac{x}{2} - \frac{5}{2}\ln|2x - 1| + K.$$

EXAMPLE ■ 4 Evaluate $\displaystyle\int \frac{5x^3 - 3x^2 + 7x - 3}{(x^2 + 1)^2}\,dx$.

SOLUTION Applying rule (b) of (7.5), with $n = 2$, yields

$$\frac{5x^3 - 3x^2 + 7x - 3}{(x^2 + 1)^2} = \frac{Ax + B}{x^2 + 1} + \frac{Cx + D}{(x^2 + 1)^2}.$$

Multiplying by the lowest common denominator $(x^2 + 1)^2$ gives us

$$
\begin{aligned}
5x^3 - 3x^2 + 7x - 3 &= (Ax + B)(x^2 + 1) + Cx + D \\
5x^3 - 3x^2 + 7x - 3 &= Ax^3 + Bx^2 + (A + C)x + (B + D).
\end{aligned}
$$

We next compare coefficients as follows:

$$
\begin{aligned}
\textit{coefficients of } x^3: & \quad 5 = A \\
\textit{coefficients of } x^2: & \quad -3 = B \\
\textit{coefficients of } x: & \quad 7 = A + C \\
\textit{constant terms}: & \quad -3 = B + D
\end{aligned}
$$

We now have $A = 5$, $B = -3$, $C = 7 - A = 2$, and $D = -3 - B = 0$; therefore,

$$\frac{5x^3 - 3x^2 + 7x - 3}{(x^2 + 1)^2} = \frac{5x - 3}{x^2 + 1} + \frac{2x}{(x^2 + 1)^2}$$

$$= \frac{5x}{x^2 + 1} - \frac{3}{x^2 + 1} + \frac{2x}{(x^2 + 1)^2}.$$

Integrating yields

$$\int \frac{5x^3 - 3x^2 + 7x - 3}{(x^2 + 1)^2} \, dx = \frac{5}{2} \ln(x^2 + 1) - 3 \tan^{-1} x - \frac{1}{x^2 + 1} + K.$$

Many of the steps in the partial fraction decomposition of $f(x)/g(x)$ are straightforward algebraic manipulations that may be tedious to perform if the degree of g is large. Computer algebra systems provide a useful tool for automating some of these steps in such cases. The next example illustrates a few of the capabilities of these systems.

EXAMPLE ■ 5   Use a computer algebra system to evaluate

$$\int \frac{264x^3 - 553x^2 - 310x - 543}{24x^4 - 142x^3 - 59x^2 + 267x + 90} \, dx.$$

SOLUTION   We first use a CAS to help factor the denominator. The exact commands and the rules for using them depend on the particular CAS being used. In this example, we illustrate some commands available in *Theorist®*. We *enter* the rational function and then *select* the denominator. The command to *factor* yields

$$\frac{264x^3 - 553x^2 - 310x - 543}{24x^4 - 142x^3 - 59x^2 + 267x + 90}$$

$$= \frac{264x^3 - 553x^2 - 310x - 543}{24(x - 6)(x + \frac{5}{4})(x - \frac{3}{2})(x + \frac{1}{3})}.$$

We see that the denominator is the product of distinct linear factors. Next, we *select* the entire expression on the right side of the equation and give the command to *expand*. The CAS responds with the decomposition:

$$\frac{264x^3 - 553x^2 - 310x - 543}{24(x - 6)(x + \frac{5}{4})(x - \frac{3}{2})(x + \frac{1}{3})}$$

$$= 7\frac{1}{x - 6} + \frac{7}{2}\frac{1}{x + \frac{5}{4}} + \frac{5}{2}\frac{1}{x - \frac{3}{2}} - 2\frac{1}{x + \frac{1}{3}}$$

We may now integrate the original fraction by integrating each of the four terms on the right:

$$\int \frac{264x^3 - 553x^2 - 310x - 543}{24x^4 - 142x^3 - 59x^2 + 267x + 90}\,dx$$

$$= 7 \int \frac{dx}{x-6} + \frac{7}{2}\int \frac{dx}{x+\frac{5}{4}} + \frac{5}{2}\int \frac{dx}{x-\frac{3}{2}} - 2\int \frac{dx}{x+\frac{1}{3}}$$

$$= 7\ln|x-6| + \frac{7}{2}\ln\left|x+\frac{5}{4}\right| + \frac{5}{2}\ln\left|x-\frac{3}{2}\right| - 2\ln\left|x+\frac{1}{3}\right| + K$$

The next example is an application in which a partial fraction decomposition is used to solve a differential equation.

EXAMPLE ■ 6   If x represents the number of people in a population of constant size N who have certain information, then a model of social diffusion for the rate by which x changes is

$$\frac{dx}{dt} = kx(N-x)$$

for some positive constant k (see Section 3.8).

(a) Find the number of people $x(t)$ who have the information at time t as an explicit function of t.

(b) Find $\lim_{t\to\infty} x(t)$ and interpret the result.

SOLUTION

(a) Beginning with the differential equation

$$\frac{dx}{dt} = kx(N-x),$$

we separate the variables and integrate to obtain

$$\int dt = \int \frac{1}{kx(N-x)}\,dx.$$

To integrate the expression on the right side of this equation, we make the partial fraction decomposition

$$\frac{1}{kx(N-x)} = \frac{1/N}{kx} + \frac{1/kN}{N-x},$$

so that

$$\int N\,dt = \int \frac{N}{kx(N-x)}\,dx$$

$$= \int \left(\frac{1}{kx} + \frac{1/k}{N-x}\right)dx$$

$$= \int \frac{1}{kx}\,dx + \int \frac{1/k}{N-x}\,dx.$$

Integrating, we have

$$Nt = \frac{1}{k} \ln |x| - \frac{1}{k} \ln |N - x| + D,$$

or

$$kNt = \ln |x| - \ln |N - x| + kD$$

for some constant D. Since x and $N - x$ represent numbers of people, they are positive, so $\ln |x| = \ln x$ and $\ln |N - x| = \ln(N - x)$. Thus,

$$kNt = \ln x - \ln(N - x) + kD.$$

Using properties of the logarithm,

$$\ln A + \ln B = \ln(AB) \quad \text{and} \quad \ln A - \ln B = \ln(A/B),$$

we have

$$kNt = \ln \frac{x}{N - x} + \ln e^{kD} = \ln \frac{Cx}{N - x},$$

where the constant C represents e^{kD}. Since $y = \ln x$ is equivalent to $x = e^y$, we can write

$$kNt = \ln \frac{Cx}{N - x} \quad \text{as} \quad \frac{Cx}{N - x} = e^{kNt}.$$

We now solve this equation for x:

$$Cx = e^{kNt}(N - x)$$
$$Cx + e^{kNt}x = Ne^{kNt}$$
$$x(C + e^{kNt}) = Ne^{kNt}$$
$$x = \frac{Ne^{kNt}}{C + e^{kNt}}$$

Thus, the solution of the differential equation gives $x(t)$, the number of people x who have the information at time t, as

$$x(t) = \frac{Ne^{kNt}}{C + e^{kNt}}.$$

If the number of people who have the information at time $t = 0$ is x_0, then we can determine the value of C since

$$x_0 = \frac{Ne^0}{C + e^0} = \frac{N}{C + 1}.$$

Thus,

$$C = \frac{N}{x_0} - 1 = \frac{N - x_0}{x_0},$$

and we can write $x(t)$ as

$$x(t) = \frac{Nx_0 e^{kNt}}{N - x_0 + x_0 e^{kNt}}.$$

(b) To determine

$$\lim_{t \to \infty} x(t) = \lim_{t \to \infty} \frac{Ne^{kNt}}{C + e^{kNt}},$$

we first divide the numerator and the denominator of the fraction by e^{kNt} to obtain

$$x = \frac{N}{Ce^{-kNt} + 1},$$

so that

$$\lim_{t \to \infty} x(t) = \lim_{t \to \infty} \frac{N}{Ce^{-kNt} + 1}.$$

Since k and N are positive,

$$\lim_{t \to \infty} e^{-kNt} = 0$$

and hence

$$\lim_{t \to \infty} x(t) = \lim_{t \to \infty} \frac{N}{Ce^{-kNt} + 1} = \frac{N}{C \cdot 0 + 1} = N.$$

We conclude that the model predicts that eventually everyone in the population will have the information.

NOTE   The differential equation $dx/dt = kx(N - x)$ in Example 6 is called the *logistic model*. It occurs in many applications in which the growth of a population is under consideration.

EXERCISES 7.4

Exer. 1–28: Evaluate the integral.

1 $\displaystyle\int \frac{5x - 12}{x(x - 4)}\, dx$

2 $\displaystyle\int \frac{x + 34}{(x - 6)(x + 2)}\, dx$

3 $\displaystyle\int \frac{37 - 11x}{(x + 1)(x - 2)(x - 3)}\, dx$

4 $\displaystyle\int \frac{4x^2 + 54x + 134}{(x - 1)(x + 5)(x + 3)}\, dx$

5 $\displaystyle\int \frac{6x - 11}{(x - 1)^2}\, dx$

6 $\displaystyle\int \frac{-19x^2 + 50x - 25}{x^2(3x - 5)}\, dx$

7 $\displaystyle\int \frac{x + 16}{x^2 + 2x - 8}\, dx$

8 $\displaystyle\int \frac{11x + 2}{2x^2 - 5x - 3}\, dx$

9 $\displaystyle\int \frac{5x^2 - 10x - 8}{x^3 - 4x}\, dx$

10 $\displaystyle\int \frac{4x^2 - 5x - 15}{x^3 - 4x^2 - 5x}\, dx$

11 $\displaystyle\int \frac{2x^2 - 25x - 33}{(x + 1)^2(x - 5)}\, dx$

12 $\displaystyle\int \frac{2x^2 - 12x + 4}{x^3 - 4x^2}\, dx$

13 $\displaystyle\int \frac{9x^4 + 17x^3 + 3x^2 - 8x + 3}{x^5 + 3x^4}\, dx$

14 $\displaystyle\int \frac{5x^2 + 30x + 43}{(x + 3)^3}\, dx$

15 $\displaystyle\int \frac{x^3 + 6x^2 + 3x + 16}{x^3 + 4x}\, dx$

16 $\displaystyle\int \frac{2x^2 + 7x}{x^2 + 6x + 9}\, dx$

17 $\displaystyle\int \frac{5x^2 + 11x + 17}{x^3 + 5x^2 + 4x + 20}\, dx$

18 $\displaystyle\int \frac{4x^3 - 3x^2 + 6x - 27}{x^4 + 9x^2}\, dx$

19 $\displaystyle\int \frac{x^2 + 3x + 1}{x^4 + 5x^2 + 4}\, dx$

20 $\displaystyle\int \frac{4x}{(x^2 + 1)^3}\, dx$

21 $\displaystyle\int \frac{2x^3 + 10x}{(x^2 + 1)^2}\, dx$

22 $\displaystyle\int \frac{x^4 + 2x^2 + 4x + 1}{(x^2 + 1)^3}\, dx$

23 $\displaystyle\int \frac{x^3 + 3x - 2}{x^2 - x}\, dx$

24 $\displaystyle\int \frac{x^4 + 2x^2 + 3}{x^3 - 4x}\, dx$

25 $\displaystyle\int \frac{x^6 - x^3 + 1}{x^4 + 9x^2}\, dx$

26 $\displaystyle\int \frac{x^5}{(x^2 + 4)^2}\, dx$

27 $\displaystyle\int \frac{2x^3 - 5x^2 + 46x + 98}{(x^2 + x - 12)^2}\, dx$

28 $\displaystyle\int \frac{-2x^4 - 3x^3 - 3x^2 + 3x + 1}{x^2(x + 1)^3}\, dx$

c **Exer. 29–32:** Using a computer algebra system, determine the partial fraction decomposition of the integrand and then evaluate the integral.

29 $\displaystyle\int \frac{282x^3 + 1021x^2 - 509x - 398}{36x^4 + 96x^3 - 131x^2 - 71x + 70}\, dx$

30 $\displaystyle\int \frac{-1302x^3 + 4075x^2 + 1742x - 13}{1980x^4 - 1641x^3 - 2684x^2 - 849x - 70}\, dx$

31 $\displaystyle\int \frac{-329x^2 - 440x + 4570}{30x^3 + 187x^2 - 555x - 250}\, dx$

32 $\displaystyle\int \frac{3244x^2 + 437x - 57}{6188x^3 + 1574x^2 - 420x + 18}\, dx$

Exer. 33–36: Use partial fractions to evaluate the integral (see Formulas 19, 49, 50, and 52 of the table of integrals in Appendix II).

33 $\displaystyle\int \frac{1}{a^2 - u^2}\, du$

34 $\displaystyle\int \frac{1}{u(a + bu)}\, du$

35 $\displaystyle\int \frac{1}{u^2(a + bu)}\, du$

36 $\displaystyle\int \frac{1}{u(a + bu)^2}\, du$

37 If $f(x) = x/(x^2 - 2x - 3)$, find the area of the region under the graph of f from $x = 0$ to $x = 2$.

38 The region bounded by the graphs of $y = 0$, $x = 2$, $x = 3$, and $y = 1/(x - 1)(4 - x)$ is revolved about the y-axis. Find the volume of the resulting solid.

39 If the region described in Exercise 38 is revolved about the x-axis, find the volume of the resulting solid.

40 Suppose $g(x) = (x - c_1)(x - c_2)\cdots(x - c_n)$ for a positive integer n and distinct real numbers $c_1, c_2, \ldots, c_n$. If $f(x)$ is a polynomial of degree less than n, show that

$$\frac{f(x)}{g(x)} = \frac{A_1}{x - c_1} + \frac{A_2}{x - c_2} + \cdots + \frac{A_n}{x - c_n}$$

with $A_k = f(c_k)/g'(c_k)$ for $k = 1, 2, \ldots, n$. (This is a method for finding the partial fraction decomposition if the denominator can be factored into distinct linear factors.)

41 Use Exercise 40 to find the partial fraction decomposition of

$$\frac{2x^4 - x^3 - 3x^2 + 5x + 7}{x^5 - 5x^3 + 4x}.$$

7.5 **QUADRATIC EXPRESSIONS AND MISCELLANEOUS SUBSTITUTIONS**

In this section, we study some additional techniques for finding antiderivatives. We first examine integrands that involve quadratic expressions, and then we consider a variety of integrals that can be handled by substitutions.

INTEGRALS INVOLVING QUADRATIC EXPRESSIONS

Partial fraction decompositions may lead to integrands containing an irreducible quadratic expression $ax^2 + bx + c$. If $b \neq 0$, it is sometimes necessary to complete the square as follows:

$$ax^2 + bx + c = a\left(x^2 + \frac{b}{a}x\right) + c$$

$$= a\left(x + \frac{b}{2a}\right)^2 + c - \frac{b^2}{4a}$$

The substitution $u = x + b/(2a)$ may then lead to an integrable form.

EXAMPLE ∎ 1   Evaluate $\int \dfrac{2x-1}{x^2-6x+13}\,dx$.

SOLUTION   Note that the quadratic expression $x^2-6x+13$ is irreducible, since $b^2-4ac=-16<0$. We complete the square as follows:

$$x^2-6x+13 = (x^2-6x \qquad)+13$$
$$= (x^2-6x+9)+13-9 = (x-3)^2+4$$

Thus,

$$\int \frac{2x-1}{x^2-6x+13}\,dx = \int \frac{2x-1}{(x-3)^2+4}\,dx.$$

We now make the substitution

$$u = x-3, \quad x = u+3, \quad dx = du.$$

Thus,

$$\int \frac{2x-1}{x^2-6x+13}\,dx = \int \frac{2(u+3)-1}{u^2+4}\,du$$

$$= \int \frac{2u+5}{u^2+4}\,du$$

$$= \int \frac{2u}{u^2+4}\,du + 5\int \frac{1}{u^2+4}\,du$$

$$= \ln(u^2+4) + \frac{5}{2}\tan^{-1}\frac{u}{2} + C$$

$$= \ln(x^2-6x+13) + \frac{5}{2}\tan^{-1}\frac{x-3}{2} + C.$$

We may also use the technique of completing the square if a quadratic expression appears under a radical sign. In the next example, we make a trigonometric substitution after completing the square.

EXAMPLE ∎ 2   Evaluate $\int \dfrac{1}{\sqrt{x^2+8x+25}}\,dx$.

SOLUTION   We complete the square for the quadratic expression as follows:

$$x^2+8x+25 = (x^2+8x \qquad)+25$$
$$= (x^2+8x+16)+25-16$$
$$= (x+4)^2+9$$

Thus, $\displaystyle \int \frac{1}{\sqrt{x^2+8x+25}}\,dx = \int \frac{1}{\sqrt{(x+4)^2+9}}\,dx.$

If we make the trigonometric substitution

$$x+4 = 3\tan\theta, \quad dx = 3\sec^2\theta\,d\theta,$$

then

$$\sqrt{(x+4)^2 + 9} = \sqrt{9\tan^2\theta + 9} = 3\sqrt{\tan^2\theta + 1} = 3\sec\theta$$

and

$$\int \frac{1}{\sqrt{x^2 + 8x + 25}}\, dx = \int \frac{1}{3\sec\theta} 3\sec^2\theta\, d\theta$$

$$= \int \sec\theta\, d\theta$$

$$= \ln|\sec\theta + \tan\theta| + C.$$

To return to the variable x, we use the triangle in Figure 7.7, obtaining

$$\int \frac{1}{\sqrt{x^2 + 8x + 25}}\, dx = \ln\left|\frac{\sqrt{x^2 + 8x + 25}}{3} + \frac{x+4}{3}\right| + C$$

$$= \ln\left|\sqrt{x^2 + 8x + 25} + x + 4\right| - \ln|3| + C$$

$$= \ln\left|\sqrt{x^2 + 8x + 25} + x + 4\right| + K$$

with $K = C - \ln 3$.

Figure 7.7

$$\tan\theta = \frac{x+4}{3}$$

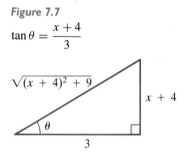

MISCELLANEOUS SUBSTITUTIONS

We now consider substitutions that are useful for evaluating certain types of integrals. The first example illustrates that if an integral contains an expression of the form $\sqrt[n]{f(x)}$, then one of the substitutions $u = \sqrt[n]{f(x)}$ or $u = f(x)$ may simplify the evaluation.

EXAMPLE 3   Evaluate $\displaystyle\int \frac{x^3}{\sqrt[3]{x^2 + 4}}\, dx$.

SOLUTION I   The substitution $u = \sqrt[3]{x^2 + 4}$ leads to the following equivalent equations:

$$u = \sqrt[3]{x^2 + 4}, \qquad u^3 = x^2 + 4, \qquad x^2 = u^3 - 4$$

Taking the differential of each side of the last equation, we obtain

$$2x\, dx = 3u^2\, du, \quad \text{or} \quad x\, dx = \tfrac{3}{2}u^2\, du.$$

We now substitute as follows:

$$\int \frac{x^3}{\sqrt[3]{x^2 + 4}}\, dx = \int \frac{x^2}{\sqrt[3]{x^2 + 4}} \cdot x\, dx$$

$$= \int \frac{u^3 - 4}{u} \cdot \frac{3}{2}u^2\, du = \frac{3}{2}\int (u^4 - 4u)\, du$$

$$= \tfrac{3}{2}\left(\tfrac{1}{5}u^5 - 2u^2\right) + C = \tfrac{3}{10}u^2(u^3 - 10) + C$$

$$= \tfrac{3}{10}(x^2 + 4)^{2/3}(x^2 - 6) + C$$

SOLUTION 2   If we substitute u for the expression *underneath* the radical, then

$$u = x^2 + 4, \quad \text{or} \quad x^2 = u - 4$$

and $\qquad 2x\,dx = du, \quad \text{or} \quad x\,dx = \tfrac{1}{2}\,du.$

In this case, we may write

$$\int \frac{x^3}{\sqrt[3]{x^2+4}}\,dx = \int \frac{x^2}{\sqrt[3]{x^2+4}} \cdot x\,dx$$

$$= \int \frac{u-4}{u^{1/3}} \cdot \frac{1}{2}\,du = \frac{1}{2}\int (u^{2/3} - 4u^{-1/3})\,du$$

$$= \tfrac{1}{2}(\tfrac{3}{5}u^{5/3} - 6u^{2/3}) + C = \tfrac{3}{10}u^{2/3}(u-10) + C$$

$$= \tfrac{3}{10}(x^2+4)^{2/3}(x^2 - 6) + C.$$

EXAMPLE ■ 4   Evaluate $\displaystyle\int \frac{1}{\sqrt{x} + \sqrt[3]{x}}\,dx.$

SOLUTION   To obtain a substitution that will eliminate the two radicals $\sqrt{x} = x^{1/2}$ and $\sqrt[3]{x} = x^{1/3}$, we use $u = x^{1/n}$, where n is the least common denominator of $\tfrac{1}{2}$ and $\tfrac{1}{3}$. Thus, we let

$$u = x^{1/6}, \quad \text{or, equivalently,} \quad x = u^6.$$

Hence,

$$dx = 6u^5\,du, \qquad x^{1/2} = (u^6)^{1/2} = u^3, \qquad x^{1/3} = (u^6)^{1/3} = u^2$$

and, therefore,

$$\int \frac{1}{\sqrt{x} + \sqrt[3]{x}}\,dx = \int \frac{1}{u^3 + u^2}6u^5\,du = 6\int \frac{u^3}{u+1}\,du.$$

By long division,

$$\frac{u^3}{u+1} = u^2 - u + 1 - \frac{1}{u+1}.$$

Consequently,

$$\int \frac{1}{\sqrt{x} + \sqrt[3]{x}}\,dx = 6\int \left(u^2 - u + 1 - \frac{1}{u+1}\right) du$$

$$= 6(\tfrac{1}{3}u^3 - \tfrac{1}{2}u^2 + u - \ln|u+1|) + C$$

$$= 2\sqrt{x} - 3\sqrt[3]{x} + 6\sqrt[6]{x} - 6\ln(\sqrt[6]{x}+1) + C.$$

If an integrand is a rational expression in $\sin x$ and $\cos x$, then the substitution

$$u = \tan \frac{x}{2} \quad \text{for} \quad -\pi < x < \pi$$

will transform the integrand into a rational (algebraic) expression in u. To prove this, first note that

$$\cos \frac{x}{2} = \frac{1}{\sec(x/2)} = \frac{1}{\sqrt{1 + \tan^2(x/2)}} = \frac{1}{\sqrt{1 + u^2}}$$

$$\sin \frac{x}{2} = \tan \frac{x}{2} \cos \frac{x}{2} = u \frac{1}{\sqrt{1 + u^2}}.$$

Consequently,

$$\sin x = 2 \sin \frac{x}{2} \cos \frac{x}{2} = \frac{2u}{1 + u^2}$$

$$\cos x = 1 - 2 \sin^2 \frac{x}{2} = 1 - \frac{2u^2}{1 + u^2} = \frac{1 - u^2}{1 + u^2}.$$

Moreover, since $x/2 = \tan^{-1} u$, we have $x = 2 \tan^{-1} u$, and, therefore,

$$dx = \frac{2}{1 + u^2} \, du.$$

The following theorem summarizes this discussion.

Theorem 7.6

> If an integrand is a rational expression in $\sin x$ and $\cos x$, the following substitutions will produce a rational expression in u:
>
> $$\sin x = \frac{2u}{1 + u^2}, \qquad \cos x = \frac{1 - u^2}{1 + u^2}, \qquad dx = \frac{2}{1 + u^2} \, du,$$
>
> where $u = \tan \dfrac{x}{2}$ for $-\pi < x < \pi$.

EXAMPLE ▪ 5 Evaluate $\displaystyle \int \frac{1}{4 \sin x - 3 \cos x} \, dx$.

SOLUTION Applying Theorem (7.6) and simplifying the integrand yields

$$\int \frac{1}{4 \sin x - 3 \cos x} \, dx = \int \frac{1}{4 \left(\dfrac{2u}{1 + u^2} \right) - 3 \left(\dfrac{1 - u^2}{1 + u^2} \right)} \cdot \frac{2}{1 + u^2} \, du$$

$$= \int \frac{2}{8u - 3(1 - u^2)} \, du$$

$$= 2 \int \frac{1}{3u^2 + 8u - 3} \, du.$$

Using partial fractions, we have

$$\frac{1}{3u^2 + 8u - 3} = \frac{1}{10} \left(\frac{3}{3u - 1} - \frac{1}{u + 3} \right)$$

and hence

$$\int \frac{1}{4 \sin x - 3 \cos x} \, dx = \frac{1}{5} \int \left(\frac{3}{3u - 1} - \frac{1}{u + 3} \right) du$$

$$= \frac{1}{5} (\ln |3u - 1| - \ln |u + 3|) + C$$

$$= \frac{1}{5} \ln \left| \frac{3u - 1}{u + 3} \right| + C$$

$$= \frac{1}{5} \ln \left| \frac{3 \tan(x/2) - 1}{\tan(x/2) + 3} \right| + C.$$

Theorem (7.6) may be used for any integrand that is a rational expression in $\sin x$ and $\cos x$. However, it is also important to consider simpler substitutions, as illustrated in the next example.

EXAMPLE • 6 Evaluate $\int \dfrac{\cos x}{1 + \sin^2 x} \, dx$.

SOLUTION We could use the formulas in Theorem (7.6) to change the integrand into a rational expression in u. The following substitution is simpler:

$$u = \sin x, \qquad du = \cos x \, dx$$

Thus,

$$\int \frac{\cos x}{1 + \sin^2 x} \, dx = \int \frac{1}{1 + u^2} \, du$$

$$= \arctan u + C$$

$$= \arctan \sin x + C.$$

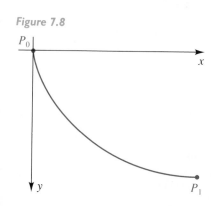

Figure 7.8

The evaluation of an integral may well involve the application of several techniques in succession. John Bernoulli's solution of the *brachistochrone problem*, for example, requires an algebraic substitution, a trigonometric substitution, and the use of several trigonometric identities. In 1696, Bernoulli published this problem as a challenge to other mathematicians: Find among all smooth curves lying in a vertical plane and connecting a given higher point P_0 to a given lower point P_1, the curve along which a particle will slide in the shortest possible time.

In Figure 7.8, we have set up a coordinate system with the higher point P_0 at the origin and with the positive direction of the y-axis downward. Figure 7.8 also shows a curve joining the points. Such a curve occurs in planning the design of a ski jump, for example. Bernoulli wished to find explicitly an expression for the function $y = f(x)$ whose graph is the curve of fastest descent from P_0 to P_1. Assuming that gravity is the only force acting on the particle, Bernoulli used ideas from optics, mechanics, and calculus to discover that the function f must satisfy the differential equation

$$y \left[1 + \left(\frac{dy}{dx} \right)^2 \right] = c$$

for some constant c.

We may rewrite this equation as

$$\left[1 + \left(\frac{dy}{dx} \right)^2 \right] = \frac{c}{y}$$

or, equivalently,

$$\left(\frac{dy}{dx} \right)^2 = \frac{c}{y} - 1 = \frac{c - y}{y},$$

so that

$$\frac{dy}{dx} = \sqrt{\frac{c - y}{y}},$$

where we have chosen the positive square root, since in our coordinate system y increases as x increases, making $dy/dx > 0$. Since the right side of this differential equation is a function only of the variable y, we separate the variables and obtain

$$\int dx = \int \sqrt{\frac{y}{c - y}} \, dy,$$

so that

$$x = \int \sqrt{\frac{y}{c - y}} \, dy.$$

To carry out the integration of the right side, we first make the algebraic substitution $u = \sqrt{y/(c - y)}$, so that $u^2 = y/(c - y)$, which gives $y = cu^2/(1 + u^2)$ and, with differentiation, $dy = [2cu/(1 + u^2)^2] \, du$. Thus,

$$x = \int \sqrt{\frac{y}{c - y}} \, dy = \int u \cdot \frac{2cu}{(1 + u^2)^2} \, du = \int \frac{2cu^2}{(1 + u^2)^2} \, du.$$

To evaluate the last integral, we make the trigonometric substitution $u = \tan \phi$, $du = \sec^2 \phi \, d\phi$. Thus, the integrand becomes

$$\frac{2cu^2}{(1 + u^2)^2} = \frac{2c \tan^2 \phi}{(1 + \tan^2 \phi)^2} = \frac{2c \tan^2 \phi}{(\sec^2 \phi)^2}$$

so that

$$x = \int \frac{2cu^2}{(1 + u^2)^2} \, du = \int \frac{2c \tan^2 \phi}{(\sec^2 \phi)^2} \sec^2 \phi \, d\phi = \int 2c \sin^2 \phi \, d\phi.$$

Our final substitution is to use a variation of the double-angle formula for the cosine, $2 \sin^2 \phi = 1 - \cos 2\phi$, obtaining

$$x = c \int (1 - \cos 2\phi) \, d\phi.$$

Integration yields

$$x = \frac{c}{2}(2\phi - \sin 2\phi) + D$$

for some constant D. Substituting $\phi = 0$ into this equation, we see that D is equal to the value of x when $\phi = 0$. At $\phi = 0$, $u = \tan 0 = 0$ and hence $y = cu^2/(1 + u^2) = 0$. But when $y = 0$, we also have $x = 0$ because the point $P_0(0, 0)$ lies on the curve. Thus, $D = 0$.

Using $u = \tan \phi$ gives us a formula for y:

$$y = \frac{cu^2}{1 + u^2} = \frac{c \tan^2 \phi}{1 + \tan^2 \phi} = \frac{c \tan^2 \phi}{\sec^2 \phi} = c \sin^2 \phi = \frac{c}{2}(1 - \cos 2\phi)$$

As a final simplification, we let $a = c/2$ and $\theta = 2\phi$. The curve of fastest descent is then described by the set of points $P(x, y)$, where x and y are given by a pair of equations:

$$x = a(\theta - \sin \theta), \qquad y = a(1 - \cos \theta)$$

Here we have described the curve by giving separate equations for the x- and y-coordinates in terms of a third variable, θ. This description is called a *parametric representation* of a curve; we will study such representations in more detail in Chapter 9. In this case, we may regard θ as representing time. At $\theta = 0$, $x = 0$ and $y = 0$, and the particle is at the point P_0. As θ increases, the particle moves down the curve toward the point P_1.

EXERCISES 7.5

Exer. 1–18: Evaluate the integral.

1 $\displaystyle\int \frac{1}{(x+1)^2 + 4}\, dx$

2 $\displaystyle\int \frac{1}{\sqrt{16 - (x-3)^2}}\, dx$

3 $\displaystyle\int \frac{1}{x^2 - 4x + 8}\, dx$

4 $\displaystyle\int \frac{1}{x^2 - 2x + 2}\, dx$

5 $\displaystyle\int \frac{1}{\sqrt{4x - x^2}}\, dx$

6 $\displaystyle\int \frac{1}{\sqrt{7 + 6x - x^2}}\, dx$

7 $\displaystyle\int \frac{2x + 3}{\sqrt{9 - 8x - x^2}}\, dx$

8 $\displaystyle\int \frac{x + 5}{9x^2 + 6x + 17}\, dx$

9 $\displaystyle\int \frac{1}{(x^2 + 4x + 5)^2}\, dx$

10 $\displaystyle\int \frac{1}{(x^2 - 6x + 34)^{3/2}}\, dx$

11 $\displaystyle\int \frac{1}{(x^2 + 6x + 13)^{3/2}}\, dx$

12 $\displaystyle\int \sqrt{x(6 - x)}\, dx$

13 $\displaystyle\int \frac{1}{2x^2 - 3x + 9}\, dx$

14 $\displaystyle\int \frac{2x}{(x^2 + 2x + 5)^2}\, dx$

15 $\displaystyle\int \frac{e^x}{e^{2x} + 3e^x + 2}\, dx$

16 $\displaystyle\int \sqrt{x^2 + 10x}\, dx$

17 $\displaystyle\int_2^3 \frac{x^2 - 4x + 6}{x^2 - 4x + 5}\, dx$

18 $\displaystyle\int_0^1 \frac{x - 1}{x^2 + x + 1}\, dx$

19 Find the area of the region bounded by the graphs of $y = 1/(x^2 + 4x + 29)$, $y = 0$, $x = -2$, and $x = 3$.

20 The region bounded by the graph of

$$y = 1/(x^2 + 2x + 10),$$

the coordinate axes, and the line $x = 2$ is revolved about the x-axis. Find the volume of the resulting solid.

Exer. 21–46: Evaluate the integral.

21 $\displaystyle\int x\sqrt[3]{x + 9}\, dx$

22 $\displaystyle\int x^2\sqrt{2x + 1}\, dx$

23 $\displaystyle\int \frac{x}{\sqrt[5]{3x + 2}}\, dx$

24 $\displaystyle\int \frac{5x}{(x + 3)^{2/3}}\, dx$

25 $\displaystyle\int_4^9 \frac{1}{\sqrt{x} + 4}\, dx$

26 $\displaystyle\int_0^{25} \frac{1}{\sqrt{4 + \sqrt{x}}}\, dx$

27 $\displaystyle\int \frac{\sqrt{x}}{1+\sqrt[3]{x}}\, dx$

28 $\displaystyle\int \frac{1}{\sqrt[4]{x}+\sqrt[3]{x}}\, dx$

29 $\displaystyle\int \frac{1}{(x+1)\sqrt{x-2}}\, dx$

30 $\displaystyle\int_0^4 \frac{2x+3}{\sqrt{1+2x}}\, dx$

31 $\displaystyle\int \frac{x+1}{(x+4)^{1/3}}\, dx$

32 $\displaystyle\int \frac{x^{1/3}+1}{x^{1/3}-1}\, dx$

33 $\displaystyle\int e^{3x}\sqrt{1+e^x}\, dx$

34 $\displaystyle\int \frac{e^{2x}}{\sqrt[3]{1+e^x}}\, dx$

35 $\displaystyle\int \frac{e^{2x}}{e^x+4}\, dx$

36 $\displaystyle\int \frac{\sin 2x}{\sqrt{1+\sin x}}\, dx$

37 $\displaystyle\int \sin\sqrt{x+4}\, dx$

38 $\displaystyle\int \sqrt{x}\, e^{\sqrt{x}}\, dx$

39 $\displaystyle\int_2^3 \frac{x}{(x-1)^6}\, dx$

40 $\displaystyle\int \frac{x^2}{(3x+4)^{10}}\, dx$

41 $\displaystyle\int \frac{\sin x}{\cos x(\cos x-1)}\, dx$

42 $\displaystyle\int \frac{\cos x}{\sin^2 x-\sin x-2}\, dx$

43 $\displaystyle\int \frac{e^x}{e^{2x}-1}\, dx$

44 $\displaystyle\int \frac{1}{e^x+e^{-x}}\, dx$

45 $\displaystyle\int \frac{\sin 2x}{\sin^2 x-2\sin x-8}\, dx$

46 $\displaystyle\int \frac{\sin x}{5\cos x+\cos^2 x}\, dx$

Exer. 47–52: Use Theorem (7.6) to evaluate the integral.

47 $\displaystyle\int \frac{1}{2+\sin x}\, dx$

48 $\displaystyle\int \frac{1}{3+2\cos x}\, dx$

49 $\displaystyle\int \frac{1}{1+\sin x+\cos x}\, dx$

50 $\displaystyle\int \frac{1}{\tan x+\sin x}\, dx$

51 $\displaystyle\int \frac{\sec x}{4-3\tan x}\, dx$

52 $\displaystyle\int \frac{1}{\sin x-\sqrt{3}\cos x}\, dx$

Exer. 53–54: Use Theorem (7.6) to derive the formula.

53 $\displaystyle\int \sec x\, dx = \ln\left|\frac{1+\tan\frac{1}{2}x}{1-\tan\frac{1}{2}x}\right| + C$

54 $\displaystyle\int \csc x\, dx = \frac{1}{2}\ln\left(\frac{1-\cos x}{1+\cos x}\right) + C$

7.6　TABLES OF INTEGRALS AND COMPUTER ALGEBRA SYSTEMS

Mathematicians and scientists who use integrals in their work may refer to a table of integrals or make use of a computer algebra system. We explore these approaches in this section. Many of the formulas contained in tables of integrals can be obtained by the methods that we have studied. A CAS can correctly apply the techniques that we have learned and also use more advanced methods. You should use a table of integrals or a CAS only after gaining some experience with the standard methods of integration.

A CAS may not always be able to perform an integration. However, making a substitution or implementing one of the other techniques may transform the original integral into a new integral that can be found in a table or can be successfully integrated by the CAS. To guard against errors, including data entry errors, when working with a CAS, always check the proposed answers by differentiation. Bear in mind that two answers can look quite different when found using different methods, even though they may be the same (or differ by only a constant).

TABLES OF INTEGRALS

Appendix II contains a brief table of integrals. We shall examine several examples illustrating the use of some of the formulas in this table.

EXAMPLE ▪ 1   Evaluate $\displaystyle\int x^3 \cos x \, dx$.

SOLUTION   We first use reduction Formula 85 in the table of integrals with $n = 3$ and $u = x$, obtaining

$$\int x^3 \cos x \, dx = x^3 \sin x - 3 \int x^2 \sin x \, dx.$$

Next we apply Formula 84 with $n = 2$, and then Formula 83, obtaining

$$\int x^2 \sin x \, dx = -x^2 \cos x + 2 \int x \cos x \, dx$$
$$= -x^2 \cos x + 2(\cos x + x \sin x) + C.$$

Substitution in the first expression gives us

$$\int x^3 \cos x \, dx = x^3 \sin x + 3x^2 \cos x - 6 \cos x - 6x \sin x + C.$$

EXAMPLE ▪ 2   Evaluate $\displaystyle\int \frac{1}{x^2\sqrt{3 + 5x^2}} \, dx$ for $x > 0$.

SOLUTION   The integrand suggests that we use that part of the table dealing with the form $\sqrt{a^2 + u^2}$. Specifically, Formula 28 states that

$$\int \frac{du}{u^2\sqrt{a^2 + u^2}} = -\frac{\sqrt{a^2 + u^2}}{a^2 u} + C.$$

(In tables, the differential du is placed in the numerator instead of to the right of the integrand.) To use this formula, we must adjust the given integral so that it matches *exactly* with the formula. If we let

$$a^2 = 3 \quad \text{and} \quad u^2 = 5x^2,$$

then the expression underneath the radical is taken care of; however, we also need

   (i) u^2 to the left of the radical
   (ii) du in the numerator

We can obtain (i) by writing the integral as

$$5 \int \frac{1}{5x^2\sqrt{3 + 5x^2}} \, dx.$$

For (ii), we note that

$$u = \sqrt{5}x \quad \text{and} \quad du = \sqrt{5} \, dx$$

and write the preceding integral as

$$5 \cdot \frac{1}{\sqrt{5}} \int \frac{1}{5x^2\sqrt{3 + 5x^2}} \sqrt{5} \, dx.$$

The last integral matches exactly with that in Formula 28, and hence

$$\int \frac{1}{x^2\sqrt{3+5x^2}}\, dx = \sqrt{5}\left[-\frac{\sqrt{3+5x^2}}{3(\sqrt{5}x)}\right] + C$$

$$= -\frac{\sqrt{3+5x^2}}{3x} + C.$$

As illustrated in the next example, it may be necessary to make a substitution of some type before a table can be used to help evaluate an integral.

EXAMPLE ■ 3   Evaluate $\displaystyle\int \frac{\sin 2x}{\sqrt{3-5\cos x}}\, dx$.

SOLUTION   Let us begin by rewriting the integral:

$$\int \frac{\sin 2x}{\sqrt{3-5\cos x}}\, dx = \int \frac{2\sin x \cos x}{\sqrt{3-5\cos x}}\, dx$$

Since no formulas in the table have this form, we consider making the substitution $u = \cos x$. In this case, $du = -\sin x\, dx$ and the integral may be written

$$2\int \frac{\sin x \cos x}{\sqrt{3-5\cos x}}\, dx = -2\int \frac{\cos x}{\sqrt{3-5\cos x}}(-\sin x)\, dx$$

$$= -2\int \frac{u}{\sqrt{3-5u}}\, du.$$

Referring to the table of integrals, we see that Formula 55 is

$$\int \frac{u\, du}{\sqrt{a+bu}} = \frac{2}{3b^2}(bu - 2a)\sqrt{a+bu} + C.$$

Using this result with $a = 3$ and $b = -5$ gives us

$$-2\int \frac{u}{\sqrt{3-5u}}\, du = -2\left(\frac{2}{75}\right)(-5u - 6)\sqrt{3-5u} + C.$$

Finally, since $u = \cos x$, we obtain

$$\int \frac{\sin 2x}{\sqrt{3-5\cos x}}\, dx = \frac{4}{75}(5\cos x + 6)\sqrt{3-5\cos x} + C.$$

COMPUTER ALGEBRA SYSTEMS

The next examples illustrate what can be expected from a computer algebra system. In using a CAS to find an antiderivative, the user typically specifies a command to do an indefinite integration, then enters the integrand, and finally gives the variable of integration.

EXAMPLE ▪ 4   Evaluate $\displaystyle\int \frac{1}{3 + \sqrt{x}}\, dx$ using a CAS.

SOLUTION   We use the CAS called *Maple*, which is available for many computer systems. *Maple* displays the symbol · as a prompt to the user that it is ready for a command. The user then types

```
int(1/(3+sqrt(x)),x);
```

where int indicates a request for the indefinite integral of the function, which will appear within the parentheses. The term ,x specifies the variable, and the semicolon at the end of the line denotes the end of the request. The user then presses the ENTER key, and *Maple* responds with an antiderivative, which is displayed on the screen as follows. Note that *Maple* does not include a constant of integration in its answer.

```
·  int(1/(3+sqrt(x)),x);

                                                           1/2
                          1/2         - 9 - x + 6 x
   - 3 ln(- 9 + x) + 2 x      + 3 ln(------------------)
                                          - 9 + x
```

Note that *Maple* and some other computer algebra systems give $\ln u$ as the antiderivative of $1/u$ rather than the more precise result

$$\int \frac{1}{u}\, du = \ln|u| + C.$$

Thus, we may write the answer as

$$\int \frac{1}{3 + \sqrt{x}}\, dx = -3\ln|-9 + x| + 2\sqrt{x} + 3\ln\left|\frac{-9 - x + 6\sqrt{x}}{-9 + x}\right| + C.$$

EXAMPLE ▪ 5   Evaluate $\displaystyle\int x e^{-\sqrt{x}}\, dx$ using a CAS.

SOLUTION   We use the CAS called *Mathematica*, which is also available for many computer systems. Note that *Mathematica* also does not include a constant of integration. The bold-faced characters on the screen display show what the user enters; *Mathematica* prints the rest:

```
In[6]:=
        Integrate[x*Exp[-1*Sqrt[x]],x]
Output[6]=

         -12 - 12 Sqrt[x] - 6 x - 2 x^3/2
         ─────────────────────────────────
                   E^Sqrt[x]
```

Mathematica's prompt to the user in this case has the form *In[6]:=*. *Mathematica* uses Sqrt[x] to denote the function $\sqrt{x}$ and Exp or E for the exponential function. Thus, *Mathematica* writes the function $e^{\sqrt{x}}$ as E $^{\text{Sqrt [x]}}$. We may write the answer as

$$\int xe^{-\sqrt{x}}\, dx = \frac{-12 - 12\sqrt{x} - 6x - 2x^{3/2}}{e^{\sqrt{x}}} + C.$$

EXAMPLE ■ 6   Evaluate $\displaystyle\int \frac{1}{3 + 2\sin x + \cos x}\, dx$ using a CAS.

SOLUTION   We now use a CAS for MS-DOS machines called *Derive*. *Derive* also does not include a constant of integration.

1: $\dfrac{1}{3 + 2\ \text{SIN}(x) + \text{COS}(x)}$

2: $\displaystyle\int \dfrac{1}{3 + 2\ \text{SIN}(x) + \text{COS}(x)}\ dx$

3: $\text{ATAN}\left[\dfrac{\text{COS}(x) + \text{SIN}(x) + 1}{\text{COS}(x) + 1}\right] - \text{ATAN}\left[\dfrac{\text{SIN}(x)}{\text{COS}(x) + 1}\right] + \dfrac{x}{2}$

With *Derive*, the user selects commands from a set of on-screen menus by either typing in the first letter of the command or using a "mouse" or "trackball." The screen then shows the result of executing the command. In this example, the user selects an <u>A</u>uthor command before entering the function on line 1 and then selects <u>C</u>alculus to request a new menu from which <u>I</u>ntegrate is chosen. There are also commands for selecting x as the variable of integration and for requesting a simplification of the resulting antiderivative.

Derive uses ATAN to represent the inverse tangent function that we have written as $\tan^{-1}$. We may write the answer as

$$\int \frac{1}{3 + 2\sin x + \cos x}\, dx$$
$$= \tan^{-1}\left(\frac{\cos x + \sin x + 1}{\cos x + 1}\right) - \tan^{-1}\left(\frac{\sin x}{\cos x + 1}\right) + \frac{x}{2} + C.$$

Each CAS uses its own set of rules for simplification. Thus, the same integrand may give different-looking antiderivatives for different computer algebra systems. It may take some effort to see that these results are either the same or differ only by a constant. For example, the three computer algebra systems discussed so far give the following results for $\int \sec x \, dx$:

Maple: `ln(sec(x) + tan(x))`

Mathematica: $-\text{Log}[\text{Cos}\left[\frac{x}{2}\right] - \text{Sin}\left[\frac{x}{2}\right]] + \text{Log}[\text{Cos}\left[\frac{x}{2}\right] + \text{Sin}\left[\frac{x}{2}\right]]$

Derive: $\text{LN}\left[\dfrac{\text{SIN}(x)+1}{\text{COS}(x)}\right]$

Note that each of these three computer algebra systems uses a different notation for the natural logarithm: ln in *Maple*, Log in *Mathematica*, and LN in *Derive*. Although these may appear to be three different functions, they are all equivalent. For example,

$$\frac{\sin x + 1}{\cos x} = \frac{\sin x}{\cos x} + \frac{1}{\cos x} = \tan x + \sec x.$$

As an exercise in trigonometric identities and elementary properties of logarithms, you may want to show that

$$\text{LN}\left[\frac{\text{SIN}(x)+1}{\text{COS}(x)}\right]$$

$$= -\text{Log}[\text{Cos}\left[\frac{x}{2}\right] - \text{Sin}\left[\frac{x}{2}\right]] + \text{Log}[\text{Cos}\left[\frac{x}{2}\right] + \text{Sin}\left[\frac{x}{2}\right]].$$

As in Example 2, we observe that computer algebra systems often display solutions to integration problems in the form $\ln u$ when the more precise answer should be $\ln |u|$. The results for $\int \sec x \, dx$ should be written as

$$\ln\left|\frac{1 + \sin x}{\cos x}\right| + C, \quad \ln|\sec x + \tan x| + C,$$

and $-\ln\left|\cos\dfrac{x}{2} - \sin\dfrac{x}{2}\right| + \ln\left|\cos\dfrac{x}{2} + \sin\dfrac{x}{2}\right| + C.$

We have discussed various methods for evaluating indefinite integrals; however, the types of integrals we have considered constitute only a small percentage of those that occur in applications. The following are examples of indefinite integrals for which antiderivatives of the integrands cannot be expressed in terms of a finite number of algebraic or transcendental functions:

$$\int \sqrt[3]{x^2 + 4x - 1} \, dx, \qquad \int \sqrt{3\cos^2 x + 1} \, dx, \qquad \int e^{-x^2} \, dx$$

In Chapter 8, we shall consider methods involving *infinite* sums that are sometimes useful in evaluating such integrals.

EXERCISES 7.6

Exer. 1–30: Use the table of integrals in Appendix II to evaluate the integral.

1. $\displaystyle\int \frac{\sqrt{4+9x^2}}{x}\,dx$

2. $\displaystyle\int \frac{1}{x\sqrt{2+3x^2}}\,dx$

3. $\displaystyle\int (16-x^2)^{3/2}\,dx$

4. $\displaystyle\int x^2\sqrt{4x^2-16}\,dx$

5. $\displaystyle\int x\sqrt{2-3x}\,dx$

6. $\displaystyle\int x^2\sqrt{5+2x}\,dx$

7. $\displaystyle\int \sin^6 3x\,dx$

8. $\displaystyle\int x\cos^5(x^2)\,dx$

9. $\displaystyle\int \csc^4 x\,dx$

10. $\displaystyle\int \sin 5x\cos 3x\,dx$

11. $\displaystyle\int x\sin^{-1} x\,dx$

12. $\displaystyle\int x^2\tan^{-1} x\,dx$

13. $\displaystyle\int e^{-3x}\sin 2x\,dx$

14. $\displaystyle\int x^5\ln x\,dx$

15. $\displaystyle\int \frac{\sqrt{5x-9x^2}}{x}\,dx$

16. $\displaystyle\int \frac{1}{x\sqrt{3x-2x^2}}\,dx$

17. $\displaystyle\int \frac{x}{5x^4-3}\,dx$

18. $\displaystyle\int \cos x\sqrt{\sin^2 x-\tfrac{1}{4}}\,dx$

19. $\displaystyle\int e^{2x}\cos^{-1} e^x\,dx$

20. $\displaystyle\int \sin^2 x\cos^3 x\,dx$

21. $\displaystyle\int x^3\sqrt{2+x}\,dx$

22. $\displaystyle\int \frac{7x^3}{\sqrt{2-x}}\,dx$

23. $\displaystyle\int \frac{\sin 2x}{4+9\sin x}\,dx$

24. $\displaystyle\int \frac{\tan x}{\sqrt{4+3\sec x}}\,dx$

25. $\displaystyle\int \frac{\sqrt{9+2x}}{x}\,dx$

26. $\displaystyle\int \sqrt{8x^3-3x^2}\,dx$

27. $\displaystyle\int \frac{1}{x(4+\sqrt[3]{x})}\,dx$

28. $\displaystyle\int \frac{1}{2x^{3/2}+5x^2}\,dx$

29. $\displaystyle\int \sqrt{16-\sec^2 x}\,\tan x\,dx$

30. $\displaystyle\int \frac{\cot x}{\sqrt{4-\csc^2 x}}\,dx$

c Exer. 31–40: Use a CAS to evaluate the integral, if possible.

31. $\displaystyle\int \frac{dx}{3+2\cos x+3\sin x}$

32. $\displaystyle\int \frac{dx}{2+2\sin x+\cos x}$

33. $\displaystyle\int x^3 e^{4x}\sin(2x)\,dx$

34. $\displaystyle\int x^2 e^{-x}\sin(5x)\,dx$

35. $\displaystyle\int \frac{\sqrt{x}}{3+x+\sqrt{x}}\,dx$

36. $\displaystyle\int \frac{\sin x}{\sin x+\cos x}\,dx$

37. $\displaystyle\int \csc x\,dx$

38. $\displaystyle\int \sqrt{\tan x}\,dx$

39. $\displaystyle\int \frac{dx}{1+\sqrt{x}}$

40. $\displaystyle\int \frac{dx}{\sqrt{1+\sqrt[3]{x}}}$

7.7 IMPROPER INTEGRALS

In our work with definite integrals of the form $\int_a^b f(x)\,dx$, we have considered almost exclusively *proper integrals*—that is, situations in which the function f is continuous on a closed interval $[a, b]$ of finite length. In this section, we extend the definite integral to cases where the interval may be of infinite length or where the function f has isolated discontinuities on the interval.

INTEGERS WITH INFINITE LIMITS OF INTEGRATION

Suppose that a function f is continuous and nonnegative on an infinite interval $[a, \infty)$ and $\lim_{x\to\infty} f(x) = 0$. If $t > a$, then the area $A(t)$ under the graph of f from a to t, as illustrated in Figure 7.9, is

$$A(t) = \int_a^t f(x)\,dx.$$

If $\lim_{t\to\infty} A(t)$ exists, then the limit may be interpreted as the area of the region that lies under the graph of f, over the x-axis, and to the right of $x = a$, as illustrated in Figure 7.10. The symbol $\int_a^\infty f(x)\,dx$ is used to denote this number. If $\lim_{t\to\infty} A(t) = \infty$, we cannot assign an area to this (unbounded) region.

Figure 7.9 $\int_a^t f(x)\,dx$

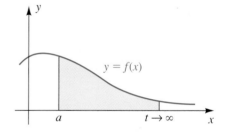

Figure 7.10 $\int_a^\infty f(x)\,dx$

Part (i) of the next definition generalizes the preceding remarks to the case where $f(x)$ may be negative for some x in $[a, \infty)$.

Definition 7.7

(i) If f is continuous on $[a, \infty)$, then

$$\int_a^\infty f(x)\,dx = \lim_{t\to\infty} \int_a^t f(x)\,dx,$$

provided the limit exists.

(ii) If f is continuous on $(-\infty, a]$, then

$$\int_{-\infty}^a f(x)\,dx = \lim_{t\to-\infty} \int_t^a f(x)\,dx,$$

provided the limit exists.

If $f(x) \geq 0$ for every x, then the limit in Definition (7.7)(ii) may be regarded as the area under the graph of f, over the x-axis, and to the *left* of $x = a$ (see Figure 7.11).

The expressions in Definition (7.7) are **improper integrals**. They differ from definite integrals in that one of the limits of integration is not a real number. An improper integral is said to **converge** if the limit exists, and the limit is the **value** of the improper integral. If the limit does not exist, the improper integral **diverges**.

Definition (7.7) is useful in many applications. In Example 4, we shall use an improper integral to calculate the work required to project an object

Figure 7.11

$\int_{-\infty}^a f(x)\,dx$

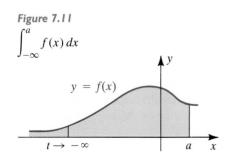

from the surface of the earth to a point outside of the earth's gravitational field. Another important application occurs in the investigation of infinite series.

EXAMPLE ▪ 1   Determine whether the integral converges or diverges, and if it converges, find its value.

(a) $\displaystyle\int_2^{\infty} \frac{1}{(x-1)^2}\, dx$ **(b)** $\displaystyle\int_2^{\infty} \frac{1}{x-1}\, dx$

SOLUTION

(a) By Definition (7.7)(i),

$$\int_2^{\infty} \frac{1}{(x-1)^2}\, dx = \lim_{t\to\infty} \int_2^t \frac{1}{(x-1)^2}\, dx = \lim_{t\to\infty} \left[\frac{-1}{x-1}\right]_2^t$$

$$= \lim_{t\to\infty} \left(\frac{-1}{t-1} + \frac{1}{2-1}\right) = 0 + 1 = 1.$$

Thus, the integral converges and has the value 1.

(b) By Definition (7.7)(i),

$$\int_2^{\infty} \frac{1}{x-1}\, dx = \lim_{t\to\infty} \int_2^t \frac{1}{x-1}\, dx$$

$$= \lim_{t\to\infty} \left[\ln(x-1)\right]_2^t$$

$$= \lim_{t\to\infty} \left[\ln(t-1) - \ln(2-1)\right]$$

$$= \lim_{t\to\infty} \ln(t-1) = \infty.$$

Since the limit does not exist, the improper integral diverges.

Figure 7.12

$y = \dfrac{1}{(x-1)^2}$

Figure 7.13

$y = \dfrac{1}{x-1}$

The graphs of the two functions given by the integrands in Example 1, together with the (unbounded) regions that lie under the graphs for $x \geq 2$, are sketched in Figures 7.12 and 7.13. Note that although the graphs have the same general shape for $x \geq 2$, we may assign an area to the region under the graph shown in Figure 7.12, but not to that shown in Figure 7.13.

The graph in Figure 7.13 has an interesting property. If we revolve the region under the graph of $y = 1/(x-1)$ about the x-axis, we obtain an unbounded solid of revolution. We may regard the improper integral

$$\int_2^{\infty} \pi \frac{1}{(x-1)^2}\, dx$$

as the volume of this solid. By Example 1(a), the value of this improper integral is π. This gives us the curious fact that although we cannot assign an area to the region in Figure 7.13, the volume of the solid of revolution generated by the region is finite.

Figure 7.14

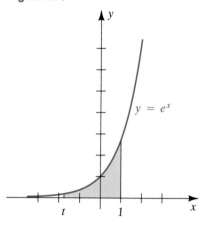

EXAMPLE 2 Assign an area to the region that lies under the graph of $y = e^x$, over the x-axis, and to the left of $x = 1$.

SOLUTION The region bounded by the graphs of $y = e^x$, $y = 0$, $x = 1$, and $x = t$, for $t < 1$, is sketched in Figure 7.14. The area of the *unbounded* region to the left of $x = 1$ is

$$\int_{-\infty}^{1} e^x \, dx = \lim_{t \to -\infty} \int_{t}^{1} e^x \, dx = \lim_{t \to -\infty} \left[e^x \right]_{t}^{1}$$

$$= \lim_{t \to -\infty} (e - e^t) = e - 0 = e.$$

An improper integral may have *two* infinite limits of integration, as in the following definition.

Definition 7.8

Let f be continuous for every x. If a is any real number, then

$$\int_{-\infty}^{\infty} f(x) \, dx = \int_{-\infty}^{a} f(x) \, dx + \int_{a}^{\infty} f(x) \, dx,$$

provided both of the improper integrals on the right converge.

If either of the integrals on the right in Definition (7.8) diverges, then $\int_{-\infty}^{\infty} f(x) \, dx$ is said to **diverge**. It can be shown that (7.8) does not depend on the choice of the real number a. It can also be shown that $\int_{-\infty}^{\infty} f(x) \, dx$ is not necessarily the same as $\lim_{t \to \infty} \int_{-t}^{t} f(x) \, dx$ (consider $f(x) = x$).

EXAMPLE 3

(a) Evaluate $\displaystyle\int_{-\infty}^{\infty} \frac{1}{1 + x^2} \, dx$.

(b) Sketch the graph of $f(x) = 1/(1 + x^2)$ and interpret the integral in part (a) as an area.

SOLUTION
(a) Using Definition (7.8), with $a = 0$, yields

$$\int_{-\infty}^{\infty} \frac{1}{1 + x^2} \, dx = \int_{-\infty}^{0} \frac{1}{1 + x^2} \, dx + \int_{0}^{\infty} \frac{1}{1 + x^2} \, dx.$$

Next, applying Definition (7.7)(i), we have

$$\int_{0}^{\infty} \frac{1}{1 + x^2} \, dx = \lim_{t \to \infty} \int_{0}^{t} \frac{1}{1 + x^2} \, dx = \lim_{t \to \infty} \left[\arctan x \right]_{0}^{t}$$

$$= \lim_{t \to \infty} (\arctan t - \arctan 0) = \frac{\pi}{2} - 0 = \frac{\pi}{2}.$$

Figure 7.15

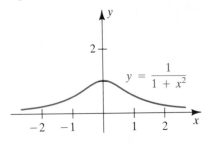

Similarly, we may show, by using Definition (7.7)(ii), that

$$\int_{-\infty}^{0} \frac{1}{1+x^2}\, dx = \frac{\pi}{2}.$$

Consequently, the given improper integral converges and has the value $(\pi/2) + (\pi/2) = \pi$.

(b) The graph of $y = 1/(1+x^2)$ is sketched in Figure 7.15. As in our previous discussion, the unbounded region that lies under the graph and above the x-axis may be assigned an area of π square units.

Figure 7.16

We now consider a physical application of an improper integral. If a and b are the coordinates of two points A and B on a coordinate line l (see Figure 7.16) and if $f(x)$ is the force acting at the point P with coordinate x, then, by Definition (5.21), the work done as P moves from A to B is given by

$$W = \int_{a}^{b} f(x)\, dx.$$

In similar fashion, the improper integral $\int_{a}^{\infty} f(x)\, dx$ may be used to define the work done as P moves indefinitely to the right (in applications, we use the terminology *P moves to infinity*). For example, if $f(x)$ is the force of attraction between a particle fixed at point A and a (movable) particle at P and if $c > a$, then $\int_{c}^{\infty} f(x)\, dx$ represents the work required to move P from the point with coordinate c to infinity.

Figure 7.17

EXAMPLE ▪ 4   Let l be a coordinate line with origin O at the center of the earth, as shown in Figure 7.17. The gravitational force exerted at a point on l that is a distance x from O is given by $f(x) = k/x^2$, for some constant k. Using 4000 mi as the radius of the earth, find the work required to project an object weighing 100 lb along l, from the surface to a point outside of the earth's gravitational field.

SOLUTION   Theoretically, there is *always* a gravitational force $f(x)$ acting on the object; however, we may think of projecting the object from the surface to infinity. From the preceding discussion, we wish to find

$$W = \int_{4000}^{\infty} f(x)\, dx.$$

By definition, $f(x) = k/x^2$ is the weight of an object that is a distance x from O, and hence

$$100 = f(4000) = \frac{k}{(4000)^2},$$

or, equivalently,

$$k = 100(4000)^2 = 10^2 \cdot 16 \cdot 10^6 = 16 \cdot 10^8.$$

Thus,

$$f(x) = (16 \cdot 10^8)\frac{1}{x^2}$$

and the required work is

$$W = \int_{4000}^{\infty} (16 \cdot 10^8)\frac{1}{x^2}\,dx = 16 \cdot 10^8 \lim_{t \to \infty} \int_{4000}^{t} \frac{1}{x^2}\,dx$$

$$= 16 \cdot 10^8 \lim_{t \to \infty} \left[-\frac{1}{x}\right]_{4000}^{t} = 16 \cdot 10^8 \lim_{t \to \infty} \left(-\frac{1}{t} + \frac{1}{4000}\right)$$

$$= \frac{16 \cdot 10^8}{4000} = 4 \cdot 10^5 \text{ mi-lb.}$$

In terms of foot-pounds,

$$W = 5280 \cdot 4 \cdot 10^5 \approx (2.1)10^9 \text{ ft-lb,}$$

or approximately 2 billion ft-lb.

In applications, we frequently encounter integrals for which there exist no antiderivatives that can be expressed in simple terms involving standard functions we have studied. The indefinite integral $\int e^{-x^2}\,dx$ is an example. If one of these integrals occurs as a definite integral, then we must use numerical integration techniques for evaluation. The next example illustrates how we may use such techniques for an improper integral.

EXAMPLE ▪ 5 The improper integral $\int_{-\infty}^{\infty} e^{-x^2}\,dx$ occurs frequently in the study of probability and statistics. Estimate the value of this integral using Simpson's rule.

SOLUTION The function $f(x) = e^{-x^2}$ has the property that $f(-x) = f(x)$. Hence, the graph of $y = f(x)$ is symmetric about the y-axis (see Figure 7.18). Thus,

Figure 7.18
$f(x) = e^{-x^2}$

$$\int_{-\infty}^{\infty} e^{-x^2}\,dx = 2\int_{0}^{\infty} e^{-x^2}\,dx.$$

Since

$$\int_{0}^{\infty} e^{-x^2}\,dx = \lim_{t \to \infty} \int_{0}^{t} e^{-x^2}\,dx,$$

we estimate the improper integral by numerical integration of $\int_{0}^{t} e^{-x^2}\,dx$ for successively larger values of t. Using Simpson's rule with $\Delta x = 0.01$, we obtain the values listed in the following table:

t	$\int_0^t e^{-x^2}\,dx$
1	0.746824132818
2	0.882081390760
3	0.886207348259
4	0.886226911790
5	0.886226925451
6	0.886226925453
7	0.886226925453

The numerical values in the table appear to converge rather rapidly for relatively small values of t. This result is consistent with Figure 7.18, which shows the graph of $f(x) = e^{-x^2}$ quickly approaching 0 as x moves away from 0.

It appears then that $\int_0^\infty e^{-x^2}\,dx \approx 0.886226925453$. Thus, our final estimate is $\int_{-\infty}^\infty e^{-x^2}\,dx \approx 2(0.886226925453) = 1.772453850906$. In Chapter 9, we shall determine that the exact value of the improper integral is $\sqrt{\pi} \approx 1.772453850906$.

In Example 5, we estimated $\int_0^\infty e^{-x^2}\,dx$ by a numerical approximation of $\int_0^7 e^{-x^2}\,dx$. With this approach, we ignored the contribution to the improper integral due to $\int_7^\infty e^{-x^2}\,dx$. By comparing our integral to one whose antiderivative we *can* find, we can estimate the error made by using this approach.

EXAMPLE ▪ 6   Obtain an upper bound for $\int_7^\infty e^{-x^2}\,dx$ by comparing this improper integral to $\int_7^\infty xe^{-x^2}\,dx$.

SOLUTION   For $x > 1$, we have $0 < e^{-x^2} < xe^{-x^2}$. Hence,

$$\int_t^\infty e^{-x^2}\,dx < \int_t^\infty xe^{-x^2}\,dx = \lim_{s \to \infty} \int_t^s xe^{-x^2}\,dx$$

$$= \lim_{s \to \infty} \left(-\tfrac{1}{2}e^{-x^2}\right)\Big|_t^s = \tfrac{1}{2}e^{-t^2}.$$

Thus, the error made by ignoring $\int_7^\infty e^{-x^2}\,dx$ is less than $\tfrac{1}{2}e^{-49}$, or about 2.621E−22.

In economics, improper integrals often occur when considering the entire *future* amount of a quantity whose rate of flow is known as a function of time. For example, if the revenue flow from sales of a particular item is estimated to be $R(t)$ dollars per time unit at time t, with $t = 0$ corresponding to the present, then the entire future revenue from sales is given by $\int_0^\infty R(t)\,dt$. Since t is the variable of integration, we can modify Definition (7.7) as follows: $\int_0^\infty R(t)\,dt = \lim_{N\to\infty} \int_0^N R(t)\,dt$. In the next example, we consider another application from economics.

EXAMPLE ■ 7 In assessing the potential revenue or profit from a mineral or energy source, economists must estimate the total amount of the resource that can be recovered from the site. Mining engineers determine that t years from now, a newly opened natural gas well will produce gas at a rate of

$$W(t) = 750e^{-0.1t} - 450e^{-0.3t}$$

thousand cubic feet per year. Estimate the total amount of gas that this well could produce.

SOLUTION We wish to estimate the entire future production of the well if it continues to pump indefinitely. This amount is given by

$$\int_0^\infty W(t)\,dt$$

$$= \int_0^\infty (750e^{-0.1t} - 450e^{-0.3t})\,dt$$

$$= \lim_{N\to\infty} \int_0^N (750e^{-0.1t} - 450e^{-0.3t})\,dt$$

$$= \lim_{N\to\infty} \left[\frac{750}{-0.1}e^{-0.1t} - \frac{450}{-0.3}e^{-0.3t} \right]_{t=0}^{t=N}$$

$$= \lim_{N\to\infty} \left[\frac{750}{-0.1}e^{-0.1N} - \frac{450}{-0.3}e^{-0.3N} - \left(\frac{750}{-0.1}e^0 - \frac{450}{-0.3}e^0 \right) \right]$$

$$= \lim_{N\to\infty} \left[\frac{750}{(-0.1)e^{0.1N}} - \frac{450}{(-0.3)e^{0.3N}} - (-7500 + 1500) \right]$$

$$= 0 - 0 - (-6000) = 6000.$$

Thus, we estimate that this well will produce $(6000)(1000) = 6$ million cubic feet of natural gas.

INTEGRALS WITH DISCONTINUOUS INTEGRANDS

If a function f is continuous on a closed interval $[a, b]$, then, by Theorem (4.20), the definite integral $\int_a^b f(x)\,dx$ exists. If f has an infinite discontinuity at some number in the interval, it may still be possible to assign

Figure 7.19 $\displaystyle\int_a^t f(x)\,dx$ **Figure 7.20** $\displaystyle\int_t^b f(x)\,dx$

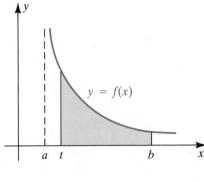

a value to the integral. Suppose, for example, that f is continuous and nonnegative on the half-open interval $[a, b)$ and $\lim_{x \to b^-} f(x) = \infty$. If $a < t < b$, then the area $A(t)$ under the graph of f from a to t (see Figure 7.19) is

$$A(t) = \int_a^t f(x)\,dx.$$

If $\lim_{t \to b^-} A(t)$ exists, then the limit may be interpreted as the area of the unbounded region that lies under the graph of f, over the x-axis, and between $x = a$ and $x = b$. We shall denote this number by $\int_a^b f(x)\,dx$.

For the situation illustrated in Figure 7.20, $\lim_{x \to a^+} f(x) = \infty$, and we define $\int_a^b f(x)\,dx$ as the limit of $\int_t^b f(x)\,dx$ as $t \to a^+$.

These remarks are the motivation for the following definition.

Definition 7.9

(i) If f is continuous on $[a, b)$ and discontinuous at b, then

$$\int_a^b f(x)\,dx = \lim_{t \to b^-} \int_a^t f(x)\,dx,$$

provided the limit exists.

(ii) If f is continuous on $(a, b]$ and discontinuous at a, then

$$\int_a^b f(x)\,dx = \lim_{t \to a^+} \int_t^b f(x)\,dx,$$

provided the limit exists.

As in the preceding section, the integrals defined in (7.9) are referred to as *improper integrals* and they *converge* if the limits exist. The limits are called the *values* of the improper integrals. If the limits do not exist, the improper integrals *diverge*.

Another type of improper integral is defined as follows.

Definition 7.10

If f has a discontinuity at a number c in the open interval (a, b) but is continuous elsewhere on $[a, b]$, then

$$\int_a^b f(x)\, dx = \int_a^c f(x)\, dx + \int_c^b f(x)\, dx,$$

provided *both* of the improper integrals on the right converge. If both converge, then the value of the improper integral $\int_a^b f(x)\, dx$ is the sum of the two values.

Figure 7.21

Figure 7.22

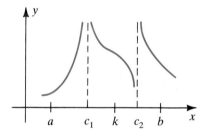

The graph of a function satisfying the conditions of Definition (7.10) is sketched in Figure 7.21.

A definition similar to (7.10) is used if f has any finite number of discontinuities in (a, b). For example, suppose f has discontinuities at c_1 and c_2, with $c_1 < c_2$, but is continuous elsewhere on $[a, b]$. One possibility is illustrated in Figure 7.22. In this case, we choose a number k between c_1 and c_2 and express $\int_a^b f(x)\, dx$ as a sum of four improper integrals over the intervals $[a, c_1]$, $[c_1, k]$, $[k, c_2]$, and $[c_2, b]$, respectively. By definition, $\int_a^b f(x)\, dx$ converges if and only if each of the four improper integrals in the sum converges. We can show that this definition is independent of the number k.

Finally, if f is continuous on (a, b) but has infinite discontinuities at a and b, then we again define $\int_a^b f(x)\, dx$ by means of (7.10).

EXAMPLE ■ 8 Evaluate $\displaystyle\int_0^3 \frac{1}{\sqrt{3 - x}}\, dx.$

SOLUTION Since the integrand has an infinite discontinuity at the number $x = 3$, we apply Definition (7.9)(i) as follows:

$$\int_0^3 \frac{1}{\sqrt{3 - x}}\, dx = \lim_{t \to 3^-} \int_0^t \frac{1}{\sqrt{3 - x}}\, dx$$

$$= \lim_{t \to 3^-} \left[-2\sqrt{3 - x} \right]_0^t$$

$$= \lim_{t \to 3^-} (-2\sqrt{3 - t} + 2\sqrt{3})$$

$$= 0 + 2\sqrt{3} = 2\sqrt{3}$$

EXAMPLE ■ 9   Determine whether the improper integral $\int_0^1 \frac{1}{x} \, dx$ converges or diverges.

SOLUTION   The integrand is undefined at $x = 0$. Applying (7.9)(ii) gives us

$$\int_0^1 \frac{1}{x} \, dx = \lim_{t \to 0^+} \int_t^1 \frac{1}{x} \, dx = \lim_{t \to 0^+} [\ln x]_t^1 = \lim_{t \to 0^+} (0 - \ln t) = \infty.$$

Since the limit does not exist, the improper integral diverges.

EXAMPLE ■ 10   Determine whether the improper integral

$$\int_0^4 \frac{1}{(x - 3)^2} \, dx$$

converges or diverges.

SOLUTION   The integrand is undefined at $x = 3$. Since this number is in the interval $(0, 4)$, we use Definition (7.10) with $c = 3$:

$$\int_0^4 \frac{1}{(x - 3)^2} \, dx = \int_0^3 \frac{1}{(x - 3)^2} \, dx + \int_3^4 \frac{1}{(x - 3)^2} \, dx$$

For the integral on the left to converge, *both* integrals on the right must converge. Equivalently, the integral on the left diverges if either of the integrals on the right diverges. Applying Definition (7.9)(i) to the first integral on the right gives us

$$\int_0^3 \frac{1}{(x - 3)^2} \, dx = \lim_{t \to 3^-} \int_0^t \frac{1}{(x - 3)^2} \, dx = \lim_{t \to 3^-} \left[\frac{-1}{x - 3} \right]_0^t$$

$$= \lim_{t \to 3^-} \left(\frac{-1}{t - 3} - \frac{1}{3} \right) = \infty.$$

Thus, the given improper integral diverges.

It is important to note that the fundamental theorem of calculus cannot be applied to the integral in Example 10, since the function given by the integrand is not continuous on $[0, 4]$. If we had (incorrectly) applied the fundamental theorem, we would have obtained

$$\left[\frac{-1}{x - 3} \right]_0^4 = -1 - \frac{1}{3} = -\frac{4}{3}.$$

This result is obviously incorrect, since the integrand is never negative.

An improper integral may have both a discontinuity in the integrand and an infinite limit of integration. Integrals of this type may be investigated by expressing them as sums of improper integrals, each of which has *one* of the forms previously defined. As an illustration, since the integrand of $\int_0^\infty (1/\sqrt{x})\,dx$ is discontinuous at $x = 0$, we choose any number greater than 0—say, 1—and write

$$\int_0^\infty \frac{1}{\sqrt{x}}\,dx = \int_0^1 \frac{1}{\sqrt{x}}\,dx + \int_1^\infty \frac{1}{\sqrt{x}}\,dx.$$

We can show that the first integral on the right side of the equation converges and the second diverges. Hence (by definition) the given integral diverges.

Improper integrals of the types considered in this section occur in physical applications. Figure 7.23 is a schematic drawing of a spring with an attached weight that is oscillating between points with coordinates $-c$ and c on a coordinate line y (the y-axis has been positioned at the right for clarity). The **period** T is the time required for one complete oscillation—that is, *twice* the time required for the weight to cover the interval $[-c, c]$. The next example illustrates how an improper integral results when we derive a formula for T.

Figure 7.23

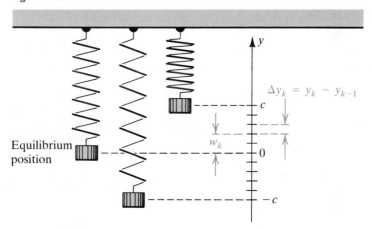

EXAMPLE ▪ 11   Let $v(y)$ denote the velocity of the weight in Figure 7.23 when it is at the point with coordinate y in $[-c, c]$. Show that the period T is given by

$$T = 2 \int_{-c}^c \frac{1}{v(y)}\,dy\,.$$

SOLUTION   Let us partition $[-c, c]$ in the usual way, and let $\Delta y_k = y_k - y_{k-1}$ denote the distance that the weight travels during the time interval Δt_k. If w_k is any number in the subinterval $[y_{k-1}, y_k]$, then $v(w_k)$ is the velocity of the weight when it is at the point with coordinate w_k. If the norm of the partition is small and if we assume that v is a continuous function, then the distance Δy_k may be approximated by the product $v(w_k)\Delta t_k$; that is,

$$\Delta y_k \approx v(w_k)\Delta t_k.$$

Hence, the time required for the weight to cover the distance Δy_k may be approximated by

$$\Delta t_k \approx \frac{1}{v(w_k)} \Delta y_k$$

and, therefore,

$$T = 2 \sum_k \Delta t_k \approx 2 \sum_k \frac{1}{v(w_k)} \Delta y_k.$$

By considering the limit of the sums on the right and using the definition of definite integral, we conclude that

$$T = 2 \int_{-c}^{c} \frac{1}{v(y)} \, dy.$$

Note that $v(c) = 0$ and $v(-c) = 0$, so the integral is improper.

EXERCISES 7.7

Exer. 1–20: Determine whether the integral converges or diverges, and if it converges, find its value.

1 $\displaystyle\int_1^\infty \frac{1}{x^{4/3}} \, dx$

2 $\displaystyle\int_{-\infty}^0 \frac{1}{(x-1)^3} \, dx$

3 $\displaystyle\int_1^\infty \frac{1}{x^{3/4}} \, dx$

4 $\displaystyle\int_0^\infty \frac{x}{1+x^2} \, dx$

5 $\displaystyle\int_{-\infty}^2 \frac{1}{5-2x} \, dx$

6 $\displaystyle\int_{-\infty}^\infty \frac{x}{x^4+9} \, dx$

7 $\displaystyle\int_0^\infty e^{-2x} \, dx$

8 $\displaystyle\int_{-\infty}^0 e^x \, dx$

9 $\displaystyle\int_{-\infty}^{-1} \frac{1}{x^3} \, dx$

10 $\displaystyle\int_0^\infty \frac{1}{\sqrt[3]{x+1}} \, dx$

11 $\displaystyle\int_{-\infty}^0 \frac{1}{(x-8)^{2/3}} \, dx$

12 $\displaystyle\int_1^\infty \frac{x}{(1+x^2)^2} \, dx$

13 $\displaystyle\int_0^\infty \frac{\cos x}{1+\sin^2 x} \, dx$

14 $\displaystyle\int_{-\infty}^2 \frac{1}{x^2+4} \, dx$

15 $\displaystyle\int_{-\infty}^\infty x e^{-x^2} \, dx$

16 $\displaystyle\int_{-\infty}^\infty \cos^2 x \, dx$

17 $\displaystyle\int_1^\infty \frac{\ln x}{x} \, dx$

18 $\displaystyle\int_3^\infty \frac{1}{x^2-1} \, dx$

19 $\displaystyle\int_{-\infty}^{\pi/2} \sin 2x \, dx$

20 $\displaystyle\int_0^\infty x e^{-x} \, dx$

Exer. 21–24: If f and g are continuous functions and $0 \le f(x) \le g(x)$ for every x in $[a, \infty)$, then the following *comparison tests for improper integrals* are true:

(i) If $\int_a^\infty g(x)\,dx$ converges, then $\int_a^\infty f(x)\,dx$ converges.

(ii) If $\int_a^\infty f(x)\,dx$ diverges, then $\int_a^\infty g(x)\,dx$ diverges.

Determine whether the first integral converges by comparing it with the second integral.

21 $\displaystyle\int_1^\infty \frac{1}{1+x^4} \, dx;$ $\displaystyle\int_1^\infty \frac{1}{x^4} \, dx$

22 $\displaystyle\int_2^\infty \frac{1}{\sqrt[3]{x^2-1}} \, dx;$ $\displaystyle\int_2^\infty \frac{1}{\sqrt[3]{x^2}} \, dx$

23 $\displaystyle\int_2^\infty \frac{1}{\ln x} \, dx;$ $\displaystyle\int_2^\infty \frac{1}{x} \, dx$

24 $\displaystyle\int_1^\infty e^{-x^2} \, dx;$ $\displaystyle\int_1^\infty e^{-x} \, dx$

Exer. 25–26: Assign, if possible, a value to **(a)** the area of the region R and **(b)** the volume of the solid obtained by revolving R about the x-axis.

25 $R = \{(x,\, y): x \ge 1,\, 0 \le y \le 1/x\}$

26 $R = \{(x,\, y): x \ge 1,\, 0 \le y \le 1/\sqrt{x}\}$

27 The unbounded region to the right of the y-axis and between the graphs of $y = e^{-x^2}$ and $y = 0$ is revolved about the y-axis. Show that a volume can be assigned to the resulting unbounded solid, and find the volume.

28 The graph of $y = e^{-x}$ for $x \geq 0$ is revolved about the x-axis. Show that an area can be assigned to the resulting unbounded surface, and find the area.

29 The solid of revolution known as *Gabriel's horn* is generated by rotating the region under the graph of $y = 1/x$ for $x \geq 1$ about the x-axis (see figure).

(a) Show that Gabriel's horn has a finite volume of π cubic units.

(b) Is a finite volume obtained if the graph is rotated about the y-axis?

(c) Show that the surface area of Gabriel's horn is given by $\int_1^{\infty} 2\pi (1/x)\sqrt{1 + (1/x^4)}\, dx$. Use a comparison test (see Exercises 21–24) with $f(x) = 2\pi/x$ to establish that this integral diverges.

(d) Comment on the following: "If Gabriel's Horn has finite volume but infinite surface area, then we can fill it with a finite amount of paint but we would never be able to paint its surface. On the other hand, if we fill it with paint, then the entire inside surface area is also covered with paint. Thus, we *can* paint the inside surface area. But the outside and inside surface areas are equal, so we can paint it with only finitely much paint!"

Exercise 29

$y = \dfrac{1}{x}, \quad x \geqslant 1$

30 A spacecraft carries a fuel supply of mass m. As a conservation measure, the captain decides to burn fuel at a rate of $R(t) = mke^{-kt}$ g/sec, for some positive constant k.

(a) What does the improper integral $\int_0^{\infty} R(t)\, dt$ represent?

(b) When will the spacecraft run out of fuel?

31 The force (in joules) with which two electrons repel one another is inversely proportional to the square of the distance (in meters) between them. If, in Figure 7.16, one electron is fixed at A, find the work done if another electron is repelled along l from a point B, which is 1 meter from A, to infinity.

32 An electric dipole consists of opposite charges separated by a small distance d. Suppose that charges of $+q$ and $-q$ units are located on a coordinate line l at $\frac{1}{2}d$ and

$-\frac{1}{2}d$, respectively (see figure). By Coulomb's law, the net force acting on a unit charge of -1 unit at $x > \frac{1}{2}d$ is given by

$$f(x) = \frac{-kq}{(x - \frac{1}{2}d)^2} + \frac{kq}{(x + \frac{1}{2}d)^2}$$

for some positive constant k. If $a > \frac{1}{2}d$, find the work done in moving the unit charge along l from a to infinity.

Exercise 32

$$\begin{array}{ccccc} -q & & +q & -1 & \\ \bullet & & \bullet & \bullet & \longrightarrow \\ -\frac{1}{2}d & 0 & \frac{1}{2}d & x & l \end{array}$$

33 The reliability $R(t)$ of a product is the probability that it will not require repair for at least t years. To design a warranty guarantee, a manufacturer must know the average time of service before first repair of a product. This is given by the improper integral $\int_0^{\infty} (-t)R'(t)\, dt$.

(a) For many high-quality products, $R(t)$ has the form e^{-kt} for some positive constant k. Find an expression in terms of k for the average time of service before repair.

(b) Is it possible to manufacture a product for which $R(t) = 1/(t + 1)$?

34 A sum of money is deposited into an account that pays interest at 8% per year, compounded continuously. Starting T years from now, money will be withdrawn at the *capital flow rate* of $f(t)$ dollars per year, continuing indefinitely. For future income to be generated at this rate, the minimum amount A that must be deposited, or the *present value of the capital flow*, is given by the improper integral $A = \int_T^{\infty} f(t)e^{-0.08t}\, dt$. Find A if the income desired 20 years from now is

(a) 12,000 dollars per year

(b) $12{,}000e^{0.04t}$ dollars per year

35 (a) Use integration by parts to establish the formula

$$\int_0^{\infty} x^2 e^{-ax^2}\, dx = \frac{1}{2a^{3/2}} \int_0^{\infty} e^{-u^2}\, du.$$

It can be shown that the value of this integral is $\sqrt{\pi}/2$.

(b) The relative number of gas molecules in a container that travel at a speed of v cm/sec can be found by using the *Maxwell–Boltzmann speed distribution F*:

$$F(v) = cv^2 e^{-mv^2/(2kT)},$$

where T is the temperature (in $^{\circ}$K), m is the mass of a molecule, and c and k are positive constants. The constant c must be selected so that $\int_0^{\infty} F(v)\, dv = 1$. Use part (a) to express c in terms of k, T, and m.

36 The *Fourier transform* is useful for solving certain differential equations. The *Fourier cosine transform* of a function f is defined by

$$F_c[f(x)] = \int_0^\infty f(x) \cos sx\, dx$$

for every real number s for which the improper integral converges. Find $F_c[e^{-ax}]$ for $a > 0$.

Exer. 37–42: In the theory of differential equations, if f is a function, then the *Laplace transform* L of $f(x)$ is defined by

$$L[f(x)] = \int_0^\infty e^{-sx} f(x)\, dx$$

for every real number s for which the improper integral converges. Find $L[f(x)]$ if $f(x)$ is the given expression.

37 1 **38** x **39** $\cos x$

40 $\sin x$ **41** e^{ax} **42** $\sin ax$

43 The *gamma function* Γ is defined by $\Gamma(n) = \int_0^\infty x^{n-1} e^{-x}\, dx$ for every positive real number n.

   (a) Find $\Gamma(1)$, $\Gamma(2)$, and $\Gamma(3)$.

   (b) Prove that $\Gamma(n+1) = n\Gamma(n)$.

   (c) Use mathematical induction to prove that if n is any positive integer, then $\Gamma(n+1) = n!$. (This shows that factorials are special values of the gamma function.)

44 Refer to Exercise 43. Functions given by $f(x) = cx^k e^{-ax}$ with $x > 0$ are called *gamma distributions* and play an important role in probability theory. The constant c must be selected so that $\int_0^\infty f(x)\, dx = 1$. Express c in terms of the positive constants k and a and the gamma function Γ.

⊡ **Exer. 45–46:** Approximate the improper integral by making the substitution $u = 1/x$ and then using Simpson's rule with $n = 2$.

45 $\displaystyle\int_2^\infty \frac{1}{\sqrt{x^4 + x}}\, dx$ **46** $\displaystyle\int_{-\infty}^{-10} \frac{\sqrt{|x|}}{x^3 + 1}\, dx$

Exer. 47–70: Determine whether the integral converges or diverges, and if it converges, find its value.

47 $\displaystyle\int_0^8 \frac{1}{\sqrt[3]{x}}\, dx$ **48** $\displaystyle\int_0^9 \frac{1}{\sqrt{x}}\, dx$

49 $\displaystyle\int_{-3}^1 \frac{1}{x^2}\, dx$ **50** $\displaystyle\int_{-2}^{-1} \frac{1}{(x+2)^{5/4}}\, dx$

51 $\displaystyle\int_0^{\pi/2} \sec^2 x\, dx$ **52** $\displaystyle\int_0^1 \frac{e^{\sqrt{x}}}{\sqrt{x}}\, dx$

53 $\displaystyle\int_0^4 \frac{1}{(4-x)^{3/2}}\, dx$ **54** $\displaystyle\int_0^{-1} \frac{1}{\sqrt[3]{x+1}}\, dx$

55 $\displaystyle\int_0^4 \frac{1}{(4-x)^{2/3}}\, dx$ **56** $\displaystyle\int_1^2 \frac{x}{x^2 - 1}\, dx$

57 $\displaystyle\int_{-2}^2 \frac{-1}{(x+1)^3}\, dx$ **58** $\displaystyle\int_{-1}^1 x^{-4/3}\, dx$

59 $\displaystyle\int_{-2}^0 \frac{1}{\sqrt{4-x^2}}\, dx$ **60** $\displaystyle\int_{-2}^0 \frac{x}{\sqrt{4-x^2}}\, dx$

61 $\displaystyle\int_{-1}^2 \frac{1}{x}\, dx$ **62** $\displaystyle\int_0^4 \frac{1}{x^2 - x - 2}\, dx$

63 $\displaystyle\int_0^1 x \ln x\, dx$ **64** $\displaystyle\int_0^{\pi/2} \tan^2 x\, dx$

65 $\displaystyle\int_0^{\pi/2} \tan x\, dx$ **66** $\displaystyle\int_0^{\pi/2} \frac{1}{1 - \cos x}\, dx$

67 $\displaystyle\int_2^4 \frac{x-2}{x^2 - 5x + 4}\, dx$ **68** $\displaystyle\int_{1/e}^e \frac{1}{x(\ln x)^2}\, dx$

69 $\displaystyle\int_{-1}^2 \frac{1}{x^2} \cos \frac{1}{x}\, dx$ **70** $\displaystyle\int_0^\pi \sec x\, dx$

Exer. 71–74: Suppose that f and g are continuous and $0 \le f(x) \le g(x)$ for every x in (a, b). If f and g are discontinuous at $x = a$, then the following *comparison tests* can be proved:

   (i) If $\int_a^b g(x)\, dx$ converges, then $\int_a^b f(x)\, dx$ converges.

   (ii) If $\int_a^b f(x)\, dx$ diverges, then $\int_a^b g(x)\, dx$ diverges.

Analogous tests may be stated for continuity on $[a, b)$ with a discontinuity at $x = b$. Determine whether the first integral converges or diverges by comparing it with the second integral.

71 $\displaystyle\int_0^\pi \frac{\sin x}{\sqrt{x}}\, dx;$ $\displaystyle\int_0^\pi \frac{1}{\sqrt{x}}\, dx$

72 $\displaystyle\int_0^{\pi/4} \frac{\sec x}{x^3}\, dx;$ $\displaystyle\int_0^{\pi/4} \frac{1}{x^3}\, dx$

73 $\displaystyle\int_0^2 \frac{\cosh x}{(x-2)^2}\, dx;$ $\displaystyle\int_0^2 \frac{1}{(x-2)^2}\, dx$

74 $\displaystyle\int_0^1 \frac{e^{-x}}{x^{2/3}}\, dx;$ $\displaystyle\int_0^1 \frac{1}{x^{2/3}}\, dx$

Exer. 75–76: Find all values of n for which the integral converges.

75 $\displaystyle\int_0^1 x^n\, dx$ **76** $\displaystyle\int_0^1 x^n \ln x\, dx$

Exer. 77–78: Assign, if possible, a value to **(a)** the area of the region R and **(b)** the volume of the solid obtained by revolving R about the x-axis.

77 $R = \{(x, y): 0 \le x \le 1, 0 \le y \le 1/\sqrt{x}\}$

78 $R = \{(x, y): 0 \le x \le 1, 0 \le y \le 1/\sqrt[3]{x}\}$

c 79 Approximate $\int_0^1 \dfrac{\cos x}{\sqrt{x}} \, dx$ by making the substitution $u = \sqrt{x}$ and then using the trapezoidal rule with $n = 4$.

c 80 Approximate $\int_0^1 \dfrac{\sin x}{x} \, dx$ by removing the discontinuity at $x = 0$ and then using Simpson's rule, with $n = 2$.

81 Refer to Example 11. If the weight in Figure 7.23 has mass m and if the spring obeys Hooke's law (with spring constant $k > 0$), then, in the absence of frictional forces, the velocity v of the weight is a solution of the differential equation

$$mv\frac{dv}{dy} + ky = 0.$$

(a) Use separation of variables (see Section 6.6) to show that $v^2 = (k/m)(c^2 - y^2)$. (*Hint:* Recall from Example 11 that $v(c) = v(-c) = 0$.)

(b) Find the period T of the oscillation.

82 A simple pendulum consists of a bob of mass m attached to a string of length L (see figure). If we assume that the string is weightless and that no other frictional forces are present, then the angular velocity $v = d\theta/dt$ is a solution of the differential equation

$$v\frac{dv}{d\theta} + \frac{g}{L}\sin\theta = 0,$$

where g is a gravitational constant.

(a) If $v = 0$ at $\theta = \pm\theta_0$, use separation of variables to show that

$$v^2 = \frac{2g}{L}(\cos\theta - \cos\theta_0).$$

(b) The period T of the pendulum is twice the amount of time needed for θ to change from $-\theta_0$ to θ_0. Show that T is given by the improper integral

$$T = 2\sqrt{\frac{2L}{g}} \int_0^{\theta_0} \frac{1}{\sqrt{\cos\theta - \cos\theta_0}} \, d\theta.$$

Exercise 82

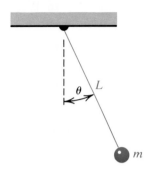

83 When a dose of y_0 milligrams of a drug is injected directly into the bloodstream, the average length of time T that a molecule remains in the bloodstream is given by the formula $T = (1/y_0)\int_0^{y_0} t \, dy$ for the time t at which exactly y milligrams is still present.

(a) If $y = y_0 e^{-kt}$ for some positive constant k, explain why the integral for T is improper.

(b) If τ is the half-life of the drug in the bloodstream, show that $T = \tau/\ln 2$.

84 In fishery science, the collection of fish that results from one annual reproduction is referred to as a *cohort*. The number N of fish still alive after t years is usually given by an exponential function. For North Sea haddock with initial size of a cohort N_0, $N = N_0 e^{-0.2t}$. The average life expectancy T (in years) of a fish in a cohort is given by $T = (1/N_0)\int_0^{N_0} t \, dN$ for the time t when precisely N fish are still alive.

(a) Find the value of T for North Sea haddock.

(b) Is it possible to have a species such that $N = N_0/(1 + kN_0 t)$ for some positive constant k? If so, compute T for such a species.

CHAPTER 7 REVIEW EXERCISES

Exer. 1–100: Evaluate the integral.

1 $\displaystyle\int x\sin^{-1} x \, dx$

2 $\displaystyle\int \sec^3(3x) \, dx$

3 $\displaystyle\int_0^1 \ln(1 + x) \, dx$

4 $\displaystyle\int_0^1 e^{\sqrt{x}} \, dx$

5 $\displaystyle\int \cos^3 2x \sin^2 2x \, dx$

6 $\displaystyle\int \cos^4 x \, dx$

7 $\displaystyle\int \tan x \sec^5 x \, dx$

8 $\displaystyle\int \tan x \sec^6 x \, dx$

9. $\displaystyle\int \frac{1}{(x^2+25)^{3/2}}\,dx$

10. $\displaystyle\int \frac{1}{x^2\sqrt{16-x^2}}\,dx$

11. $\displaystyle\int \frac{\sqrt{4-x^2}}{x}\,dx$

12. $\displaystyle\int \frac{x}{(x^2+1)^2}\,dx$

13. $\displaystyle\int \frac{x^3+1}{x(x-1)^3}\,dx$

14. $\displaystyle\int \frac{1}{x+x^3}\,dx$

15. $\displaystyle\int \frac{x^3-20x^2-63x-198}{x^4-81}\,dx$

16. $\displaystyle\int \frac{x-1}{(x+2)^5}\,dx$

17. $\displaystyle\int \frac{x}{\sqrt{4+4x-x^2}}\,dx$

18. $\displaystyle\int \frac{x}{x^2+6x+13}\,dx$

19. $\displaystyle\int \frac{\sqrt[3]{x+8}}{x}\,dx$

20. $\displaystyle\int \frac{\sin x}{2\cos x+3}\,dx$

21. $\displaystyle\int e^{2x}\sin 3x\,dx$

22. $\displaystyle\int \cos(\ln x)\,dx$

23. $\displaystyle\int \sin^3 x\cos^3 x\,dx$

24. $\displaystyle\int \cot^2 3x\,dx$

25. $\displaystyle\int \frac{x}{\sqrt{4-x^2}}\,dx$

26. $\displaystyle\int \frac{1}{x\sqrt{9x^2+4}}\,dx$

27. $\displaystyle\int \frac{x^5-x^3+1}{x^3+2x^2}\,dx$

28. $\displaystyle\int \frac{x^3}{x^3-3x^2+9x-27}\,dx$

29. $\displaystyle\int \frac{1}{x^{3/2}+x^{1/2}}\,dx$

30. $\displaystyle\int \frac{2x+1}{(x+5)^{100}}\,dx$

31. $\displaystyle\int e^x\sec e^x\,dx$

32. $\displaystyle\int x\tan x^2\,dx$

33. $\displaystyle\int x^2\sin 5x\,dx$

34. $\displaystyle\int \sin 2x\cos x\,dx$

35. $\displaystyle\int \sin^3 x\cos^{1/2} x\,dx$

36. $\displaystyle\int \sin 3x\cot 3x\,dx$

37. $\displaystyle\int e^x\sqrt{1+e^x}\,dx$

38. $\displaystyle\int x(4x^2+25)^{-1/2}\,dx$

39. $\displaystyle\int \frac{x^2}{\sqrt{4x^2+25}}\,dx$

40. $\displaystyle\int \frac{3x+2}{x^2+8x+25}\,dx$

41. $\displaystyle\int \sec^2 x\tan^2 x\,dx$

42. $\displaystyle\int \sin^2 x\cos^5 x\,dx$

43. $\displaystyle\int x\cot x\csc x\,dx$

44. $\displaystyle\int (1+\csc 2x)^2\,dx$

45. $\displaystyle\int x^2(8-x^3)^{1/3}\,dx$

46. $\displaystyle\int x(\ln x)^2\,dx$

47. $\displaystyle\int \sqrt{x}\sin\sqrt{x}\,dx$

48. $\displaystyle\int x\sqrt{5-3x}\,dx$

49. $\displaystyle\int \frac{e^{3x}}{1+e^x}\,dx$

50. $\displaystyle\int \frac{e^{2x}}{4+e^{4x}}\,dx$

51. $\displaystyle\int \frac{x^2-4x+3}{\sqrt{x}}\,dx$

52. $\displaystyle\int \frac{\cos^3 x}{\sqrt{1+\sin x}}\,dx$

53. $\displaystyle\int \frac{x^3}{\sqrt{16-x^2}}\,dx$

54. $\displaystyle\int \frac{x}{25-9x^2}\,dx$

55. $\displaystyle\int \frac{1-2x}{x^2+12x+35}\,dx$

56. $\displaystyle\int \frac{7}{x^2-6x+18}\,dx$

57. $\displaystyle\int \tan^{-1}5x\,dx$

58. $\displaystyle\int \sin^4 3x\,dx$

59. $\displaystyle\int \frac{e^{\tan x}}{\cos^2 x}\,dx$

60. $\displaystyle\int \frac{x}{\csc 5x^2}\,dx$

61. $\displaystyle\int \frac{1}{\sqrt{7+5x^2}}\,dx$

62. $\displaystyle\int \frac{2x+3}{x^2+4}\,dx$

63. $\displaystyle\int \cot^6 x\,dx$

64. $\displaystyle\int \cot^5 x\csc x\,dx$

65. $\displaystyle\int x^3\sqrt{x^2-25}\,dx$

66. $\displaystyle\int (\sin x)10^{\cos x}\,dx$

67. $\displaystyle\int (x^2-\operatorname{sech}^2 4x)\,dx$

68. $\displaystyle\int x\cosh x\,dx$

69. $\displaystyle\int x^2 e^{-4x}\,dx$

70. $\displaystyle\int x^5\sqrt{x^3+1}\,dx$

71. $\displaystyle\int \frac{3}{\sqrt{11-10x-x^2}}\,dx$

72. $\displaystyle\int \frac{12x^3+7x}{x^4}\,dx$

73. $\displaystyle\int \tan 7x\cos 7x\,dx$

74. $\displaystyle\int e^{1+\ln 5x}\,dx$

75. $\displaystyle\int \frac{4x^2-12x^2-10}{(x-2)(x^2-4x+3)}\,dx$

76. $\displaystyle\int \frac{1}{x^4\sqrt{16-x^2}}\,dx$

77. $\displaystyle\int (x^3+1)\cos x\,dx$

78. $\displaystyle\int (x-3)^2(x+1)\,dx$

79. $\displaystyle\int \frac{\sqrt{9-4x^2}}{x^2}\,dx$

80. $\displaystyle\int \frac{4x^3-15x^2-6x+81}{x^4-18x^2+81}\,dx$

81. $\displaystyle\int (5-\cot 3x)^2\,dx$

82. $\displaystyle\int x(x^2+5)^{3/4}\,dx$

83. $\displaystyle\int \frac{1}{x(\sqrt{x}+\sqrt[4]{x})}\,dx$

84. $\displaystyle\int \frac{x}{\cos^2 4x}\,dx$

85. $\displaystyle\int \frac{\sin x}{\sqrt{1+\cos x}}\,dx$

86. $\displaystyle\int \frac{4x^2-6x+4}{(x^2+4)(x-2)}\,dx$

87. $\displaystyle\int \frac{x^2}{(25+x^2)^2}\,dx$

88. $\displaystyle\int \sin^4 x\cos^3 x\,dx$

89. $\displaystyle\int \tan^3 x\sec x\,dx$

90. $\displaystyle\int \frac{x}{\sqrt{4+9x^2}}\,dx$

91 $\displaystyle\int \frac{2x^3 + 4x^2 + 10x + 13}{x^4 + 9x^2 + 20}\,dx$

92 $\displaystyle\int \frac{\sin x}{(1 + \cos x)^3}\,dx$

93 $\displaystyle\int \frac{(x^2 - 2)^2}{x}\,dx$

94 $\displaystyle\int \cot^2 x \csc x\,dx$

95 $\displaystyle\int x^{3/2} \ln x\,dx$

96 $\displaystyle\int \frac{x}{\sqrt[3]{x} - 1}\,dx$

97 $\displaystyle\int \frac{x^2}{\sqrt[3]{2x + 3}}\,dx$

98 $\displaystyle\int \frac{1 - \sin x}{\cot x}\,dx$

99 $\displaystyle\int x^3 e^{(x^2)}\,dx$

100 $\displaystyle\int (x + 2)^2 (x + 1)^{10}\,dx$

Exer. 101–112: Determine whether the integral converges or diverges, and if it converges, find its value.

101 $\displaystyle\int_4^\infty \frac{1}{\sqrt{x}}\,dx$

102 $\displaystyle\int_4^\infty \frac{1}{x\sqrt{x}}\,dx$

103 $\displaystyle\int_{-\infty}^0 \frac{1}{x + 2}\,dx$

104 $\displaystyle\int_0^\infty \sin x\,dx$

105 $\displaystyle\int_{-8}^1 \frac{1}{\sqrt[3]{x}}\,dx$

106 $\displaystyle\int_{-4}^0 \frac{1}{x + 4}\,dx$

107 $\displaystyle\int_0^2 \frac{x}{(x^2 - 1)^2}\,dx$

108 $\displaystyle\int_1^2 \frac{1}{x\sqrt{x^2 - 1}}\,dx$

109 $\displaystyle\int_{-\infty}^\infty \frac{1}{e^x + e^{-x}}\,dx$

110 $\displaystyle\int_{-\infty}^0 xe^x\,dx$

111 $\displaystyle\int_0^1 \frac{\ln x}{x}\,dx$

112 $\displaystyle\int_0^{\pi/2} \csc x\,dx$

$\boxed{c}$ **Exer. 113–114:** Approximate the improper integral by making the substitution $u = 1/x$ and then using Simpson's rule, with $n = 2$.

113 $\displaystyle\int_1^\infty e^{-x^2}\,dx$

114 $\displaystyle\int_1^\infty e^{-x} \sin \sqrt{x}\,dx$

Exer. 115–118: Assign, if possible, a value to **(a)** the area of the region R and **(b)** the volume of the solid obtained by revolving R about the x-axis.

115 $R = \{(x, y) : x \geq 4, 0 \leq y \leq x^{-3/2}\}$

116 $R = \{(x, y) : x \geq 8, 0 \leq y \leq x^{-2/3}\}$

117 $R = \{(x, y) : -4 \leq x \leq 4, 0 \leq y \leq 1/(x + 4)\}$

118 $R = \{(x, y) : 1 \leq x \leq 2, 0 \leq y \leq 1/(x - 1)\}$

EXTENDED PROBLEMS AND GROUP PROJECTS

1 (a) As an alternative to partial fractions, show that an integral of the form

$$\int \frac{1}{ax^2 + bx}\,dx$$

may be evaluated by writing it as

$$\int \frac{1/x^2}{a + (b/x)}\,dx$$

and using the substitution $u = a + (b/x)$.

(b) Generalize part (a) to integrals of the form

$$\int \frac{1}{ax^n + bx}\,dx.$$

2 (a) Use integration by parts on $\int f(x)\,dx$ with $u = f(x)$ and $dv = dx$ to find

(i) $\displaystyle\int \ln x\,dx$

(ii) $\displaystyle\int \tan^{-1} x\,dx$

(iii) $\displaystyle\int \sin^{-1} x\,dx$

(iv) $\displaystyle\int \cos^{-1} x\,dx$

(v) $\displaystyle\int \sqrt{x}\,dx$

(b) Use integration by parts on $\int f^{-1}(x)\,dx$ with $u = f^{-1}(x)$ and $dv = dx$ to show that

$$\int f^{-1}(x)\,dx = xf^{-1}(x) - F(f^{-1}(x)),$$

where F is any antiderivative of f.

(c) Verify that the formula for $\int f^{-1}(x)\,dx$ given in part (b) is valid for the functions appearing in part (a).

(d) In what sense is the statement "If we can integrate f, then we can integrate f^{-1}" true?

3 If $f(x)$ and $g(x)$ are polynomials with f having a smaller degree than g, then we claimed that the rational function $f(x)/g(x)$ can be decomposed as a finite sum, where each term has the form

$$\frac{A}{(ax + b)^n} \quad \text{or} \quad \frac{Ax + B}{(ax^2 + bx + c)^n},$$

where A and B are real numbers, n is a nonnegative integer, and $(ax^2 + bx + c)$ is an irreducible quadratic. Prove this claim. Are the terms in the partial fraction decomposition unique? (For a set of exercises outlining an approach to this problem, see Nathan Jacobson, *Basic Algebra I*, New York: Freeman, 1985, p. 150.)

INTRODUCTION

MATHEMATICAL MODELS of population growth—whether populations of people, bacteria in a petri dish, or radioactive atoms—all reflect the assumption that at least at some stage of growth or decay, the rate of change of population is proportional to the size of the population. When such assumptions are written in terms of differential equations, the solution invariably involves the natural exponential and logarithmic functions. For example, the simplest model $dP/dt = aP$ has the solution $P(t) = P_o e^{at}$. Predicting the population at various times from this solution requires the evaluation of the exponential function at specific values of the variables.

Models of other important applications also frequently involve the transcendental functions. If x is a real number, we generally find $\arcsin x$, e^x, $\ln x$, $\cosh x$, and other values of transcendental functions by using a calculator or a table. A more fundamental problem is determining *how* calculators compute these numbers or *how* a table is constructed. A principal goal of this chapter is to demonstrate how *infinite series* can be used to find function values.

We begin with a careful study of *sequences* in Section 8.1. These are basic to the definition (Section 8.2) of convergence or divergence of a *series*. We then develop various tests for the convergence of a series of positive constants in Sections 8.3 and 8.4. In Section 8.5, we again consider series of constants, but without restrictions on their signs.

We see in Section 8.6 how to use infinite series to find function values. Specifically, if a function f satisfies certain conditions, we develop techniques for *representing* $f(x)$ as an infinite series whose terms contain powers of x. Substituting a number c for x and then finding (or approximating) the resulting infinite sum gives us the value (or an approximation) of $f(c)$. This method is essentially the same as that which a calculator uses when it approximates function values. We explore these techniques further in Sections 8.7 and 8.8.

This new way of representing functions is the most important reason for developing the theory in the first five sections of the chapter. Infinite series representations for $\sin x$, e^x, and other expressions allow us to consider problems that cannot be solved by finite methods. For example, if x is suitably restricted, we can evaluate integrals such as $\int \sin \sqrt{x} \, dx$ and $\int e^{-x^2} \, dx$, something we could not do in Chapter 7. As another application, in Chapter 15 we use infinite series to extend the definitions of $\sin x$, e^x, and other expressions to the case where x is a *complex number* $a + bi$ with a and b real and $i^2 = -1$.

Using calculus to predict future behavior or population often requires using infinite series to estimate the numerical value of transcendental functions.

Infinite Series

8.1   SEQUENCES

An arbitrary *infinite sequence* (or simply a *sequence*) is often denoted as follows.

Sequence Notation 8.1

$$a_1, a_2, a_3, \ldots, a_n, \ldots$$

We may regard (8.1) as a collection of real numbers that is in one-to-one correspondence with the positive integers. Each number a_k is a **term** of the sequence. The sequence is *ordered* in the sense that there is a **first term** a_1, a **second term** a_2, and, if n denotes an arbitrary positive integer, an **nth term** a_n.

We may also define a sequence as a function. Recall that a function f is a correspondence that associates with each number x in the domain exactly one number $f(x)$ in the range. If we restrict the domain to the positive integers $1, 2, 3, \ldots$, we obtain a sequence.

Definition 8.2

A **sequence** is a function f whose domain is the set of positive integers.

In this text, the range of a sequence will be a set of real numbers. If a function f is a sequence, then to each positive integer k there corresponds a real number $f(k)$. The numbers in the range of f may be denoted by

$$f(1), f(2), f(3), \ldots, f(n), \ldots$$

The three dots at the end indicate that the sequence does not terminate.

Note that Definition (8.2) leads to the subscript form (8.1) if we let $a_k = f(k)$ for each positive integer k. Conversely, given (8.1), we can obtain the function f in (8.2) by letting $f(k) = a_k$ for each k.

If we regard a sequence as a function f, then we may consider its graph in an xy-plane. Since the domain of f is the set of positive integers, the only points on the graph are

$$(1, a_1), (2, a_2), (3, a_3), \ldots, (n, a_n), \ldots,$$

where a_n is the nth term of the sequence (see Figure 8.1). We sometimes use the graph of a sequence to illustrate the behavior of the nth term a_n as n increases without bound.

Another notation for a sequence with nth term a_n is $\{a_n\}$. For example, the sequence $\{2^n\}$ has nth term 2^n. Using the notation in (8.1), we write this sequence as follows:

$$2^1, 2^2, 2^3, \ldots, 2^n, \ldots$$

Figure 8.1 Graph of a sequence

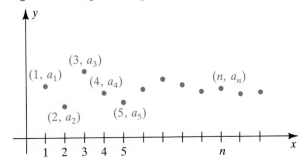

By Definition (8.2), the sequence $\{2^n\}$ is the function f with $f(n) = 2^n$ for every positive integer n.

EXAMPLE ■ 1 List the first four terms and the tenth term of each sequence.

(a) $\left\{ \dfrac{n}{n+1} \right\}$ **(b)** $\{2 + (0.1)^n\}$ **(c)** $\left\{ (-1)^{n+1} \dfrac{n^2}{3n-1} \right\}$ **(d)** $\{4\}$

SOLUTION To find the first four terms, we substitute, successively, $n = 1, 2, 3,$ and 4 in the formula for a_n. The tenth term is found by substituting 10 for n. Doing this and simplifying gives us the following:

	Sequence	nth term a_n	First four terms	Tenth term
(a)	$\left\{ \dfrac{n}{n+1} \right\}$	$\dfrac{n}{n+1}$	$\dfrac{1}{2}, \dfrac{2}{3}, \dfrac{3}{4}, \dfrac{4}{5}$	$\dfrac{10}{11}$
(b)	$\{2 + (0.1)^n\}$	$2 + (0.1)^n$	$2.1, 2.01, 2.001, 2.0001$	2.0000000001
(c)	$\left\{ (-1)^{n+1} \dfrac{n^2}{3n-1} \right\}$	$(-1)^{n+1} \dfrac{n^2}{3n-1}$	$\dfrac{1}{2}, -\dfrac{4}{5}, \dfrac{9}{8}, -\dfrac{16}{11}$	$-\dfrac{100}{29}$
(d)	$\{4\}$	4	$4, 4, 4, 4$	4

For some sequences, we state the first term a_1, together with a rule for obtaining any term a_{k+1} from the preceding term a_k whenever $k \geq 1$. We call this a **recursive definition**, and the sequence is said to be defined **recursively**.

EXAMPLE ■ 2 Find the first four terms and the nth term of the sequence defined recursively as follows:

$$a_1 = 3 \quad \text{and} \quad a_{k+1} = 2a_k \quad \text{for } k \geq 1$$

SOLUTION The sequence is defined recursively, since the first term is given, as well as a rule for finding a_{k+1} whenever a_k is known. Thus, the first four terms of the sequence are

$$a_1 = 3$$

$$a_2 = 2a_1 = 2 \cdot 3 = 6$$

$$a_3 = 2a_2 = 2 \cdot 2 \cdot 3 = 2^2 \cdot 3 = 12$$

$$a_4 = 2a_3 = 2 \cdot 2 \cdot 2 \cdot 3 = 2^3 \cdot 3 = 24.$$

We have written the terms as products to gain insight into the nature of the nth term. Continuing, we obtain $a_5 = 2^4 \cdot 3$ and $a_6 = 2^5 \cdot 3$; it appears that $a_n = 2^{n-1} \cdot 3$. We can use mathematical induction to prove that this guess is correct. Using the notation in (8.1), we write the sequence as

$$3, 2 \cdot 3, 2^2 \cdot 3, 2^3 \cdot 3, \ldots, 2^{n-1} \cdot 3, \ldots .$$

A sequence $\{a_n\}$ may have the property that as n increases, a_n gets very close to some real number L—that is, $|a_n - L| \approx 0$ if n is large. As an illustration, suppose that

$$a_n = 2 + \left(-\tfrac{1}{2}\right)^n .$$

The first few terms of the sequence $\{a_n\}$ are

$$2 - \tfrac{1}{2}, 2 + \tfrac{1}{4}, 2 - \tfrac{1}{8}, 2 + \tfrac{1}{16}, 2 - \tfrac{1}{32}, 2 + \tfrac{1}{64}, \ldots,$$

or, equivalently,

$$1.5, 2.25, 1.875, 2.0625, 1.96875, 2.015625, \ldots .$$

It appears that the terms get closer to 2 as n increases. Note that for every positive integer n,

$$|a_n - 2| = \left| 2 + \left(-\frac{1}{2}\right)^n - 2 \right| = \left| \left(-\frac{1}{2}\right)^n \right| = \left(\frac{1}{2} \right)^n = \frac{1}{2^n}.$$

The number $1/2^n$, and hence $|a_n - 2|$, *can be made arbitrarily close to* 0 *by choosing n sufficiently large.* According to the next definition, the sequence *has the limit* 2, or *converges to* 2, and we write

$$\lim_{n \to \infty} \left[2 + \left(-\tfrac{1}{2}\right)^n \right] = 2.$$

This type of limit is almost the same as $\lim_{x \to \infty} f(x) = L$, given in Chapter 1. The only difference is that if $f(n) = a_n$, the domain of f is the set of positive integers and not an infinite interval of real numbers. As in Definition (1.16), but using a_n instead of $f(x)$, we state the following.

Definition 8.3

A sequence $\{a_n\}$ **has the limit** L, or **converges to** L, denoted by either

$$\lim_{n \to \infty} a_n = L \quad \text{or} \quad a_n \to L \text{ as } n \to \infty,$$

if for every $\epsilon > 0$ there exists a positive number N such that

$$|a_n - L| < \epsilon \quad \text{whenever} \quad n > N.$$

If such a number L does not exist, the sequence **has no limit**, or **diverges**.

A graphical interpretation similar to that shown for the limit of a function in Figure 1.34 can be given for the limit of a sequence. The only difference is that the x-coordinate of each point on the graph is a positive integer. Figure 8.2 is the graph of a sequence $\{a_n\}$ for a specific case in which $\lim_{n \to \infty} a_n = L$. Note that for any $\epsilon > 0$, the points (n, a_n) lie between the lines $y = L \pm \epsilon$, provided n is sufficiently large. Of course, the approach to L may vary from that illustrated in the figure (see, for example, Figures 8.3 and 8.6).

Figure 8.2

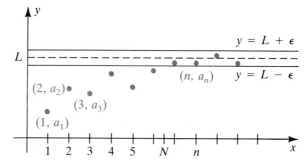

If we can make a_n as large as desired by choosing n sufficiently large, then the sequence $\{a_n\}$ diverges, but we still use the limit notation and write $\lim_{n \to \infty} a_n = \infty$. A more precise definition follows.

Definition 8.4

The notation

$$\lim_{n \to \infty} a_n = \infty$$

means that for every positive real number P there exists a number N such that $a_n > P$ whenever $n > N$.

As was the case for functions in Section 1.4, $\lim_{n \to \infty} a_n = \infty$ does *not* mean that the limit exists, but rather that the number a_n increases without bound as n increases. Similarly, $\lim_{n \to \infty} a_n = -\infty$ means that a_n decreases without bound as n increases.

The next theorem is important because it allows us to use results from Chapter 1 to investigate convergence or divergence of sequences. The proof follows from Definitions (8.3) and (1.16).

Theorem 8.5

Let $\{a_n\}$ be a sequence, let $f(n) = a_n$, and suppose that $f(x)$ exists for every real number $x \geq 1$.

(i) If $\lim\limits_{x\to\infty} f(x) = L$, then $\lim\limits_{n\to\infty} f(n) = L$.

(ii) If $\lim\limits_{x\to\infty} f(x) = \infty$ (or $-\infty$), then $\lim\limits_{n\to\infty} f(n) = \infty$ (or $-\infty$).

The following example illustrates the use of Theorem (8.5).

EXAMPLE ■ 3   If $a_n = 1 + (1/n)$, determine whether $\{a_n\}$ converges or diverges.

SOLUTION   We let $f(n) = 1 + (1/n)$ and consider

$$f(x) = 1 + \frac{1}{x} \quad \text{for every real number } x \geq 1.$$

From our work in Section 1.4,

$$\lim_{x\to\infty} f(x) = \lim_{x\to\infty} \left(1 + \frac{1}{x}\right) = \lim_{x\to\infty} 1 + \lim_{x\to\infty} \frac{1}{x} = 1 + 0 = 1.$$

Hence, by Theorem (8.5),

$$\lim_{n\to\infty} \left(1 + \frac{1}{n}\right) = 1.$$

Thus, the sequence $\{a_n\}$ converges to 1.

The difference between

$$\lim_{x\to\infty} \left(1 + \frac{1}{x}\right) = 1 \quad \text{and} \quad \lim_{n\to\infty} \left(1 + \frac{1}{n}\right) = 1$$

is illustrated in Figure 8.3. Note that for $1 + (1/x)$, the function f is continuous if $x \geq 1$, and the graph has a horizontal asymptote $y = 1$. For

Figure 8.3

$$\lim_{x\to\infty} \left(1 + \frac{1}{x}\right) = 1 \qquad\qquad \lim_{n\to\infty} \left(1 + \frac{1}{n}\right) = 1$$

$1 + (1/n)$, we consider only the points whose x-coordinates are positive integers.

Figure 8.4

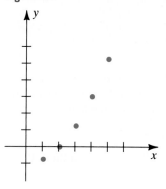

EXAMPLE ▪ 4 Determine whether the sequence converges or diverges.

(a) $\{\frac{1}{4}n^2 - 1\}$ **(b)** $\{(-1)^{n-1}\}$

SOLUTION

(a) If we let $f(x) = \frac{1}{4}x^2 - 1$, then $f(x)$ exists for every $x \geq 1$ and

$$\lim_{x \to \infty} \left(\tfrac{1}{4}x^2 - 1\right) = \infty.$$

Hence, by Theorem (8.5),

$$\lim_{n \to \infty} \left(\tfrac{1}{4}n^2 - 1\right) = \infty.$$

Since the limit does not exist, the sequence diverges. The graph in Figure 8.4 illustrates the manner in which the sequence diverges.

Figure 8.5

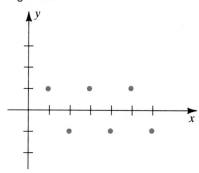

(b) Letting $n = 1, 2, 3, \ldots$, we see that the terms of $(-1)^{n-1}$ oscillate between 1 and -1 as follows:

$$1, -1, 1, -1, 1, -1, \ldots$$

This result is illustrated graphically in Figure 8.5. Thus, since

$$\lim_{n \to \infty} (-1)^{n-1}$$

does not exist, the sequence diverges.

The next example shows how we may use l'Hôpital's rule (6.51) to find limits of certain sequences.

EXAMPLE ▪ 5 Determine whether the sequence $\{5n/e^{2n}\}$ converges or diverges.

SOLUTION Let $f(x) = 5x/e^{2x}$ for every real number x. Since f takes on the indeterminate form ∞/∞ as $x \to \infty$, we may use l'Hôpital's rule, obtaining

$$\lim_{x \to \infty} \frac{5x}{e^{2x}} = \lim_{x \to \infty} \frac{5}{2e^{2x}} = 0.$$

Hence, by Theorem (8.5), $\lim_{n \to \infty}(5n/e^{2n}) = 0$. Thus, the sequence converges to 0.

The proof of the next theorem illustrates the use of Definition (8.3).

Theorem 8.6

> **(i)** $\lim\limits_{n \to \infty} r^n = 0$ if $|r| < 1$
>
> **(ii)** $\lim\limits_{n \to \infty} |r^n| = \infty$ if $|r| > 1$

PROOF If $r = 0$, it follows trivially that the limit is 0. Let us assume that $0 < |r| < 1$. To prove (i) by means of Definition (8.3), we must show that for every $\epsilon > 0$, there exists a positive number N such that

$$\text{if} \quad n > N, \quad \text{then} \quad |r^n - 0| < \epsilon.$$

The inequality $|r^n - 0| < \epsilon$ is equivalent to each inequality in the following list:

$$|r|^n < \epsilon, \qquad \ln |r|^n < \ln \epsilon, \qquad n \ln |r| < \ln \epsilon, \qquad n > \frac{\ln \epsilon}{\ln |r|}$$

The final inequality sign is reversed because $\ln |r|$ is negative if $0 < |r| < 1$. The last inequality in the list provides a clue to the choice of N. Let us consider the two cases $\epsilon < 1$ and $\epsilon \geq 1$ separately. If $\epsilon < 1$, then $\ln \epsilon < 0$ and we let $N = \ln \epsilon / \ln |r| > 0$. In this event, if $n > N$, then the last inequality in the list is true and hence so is the first, which is what we wished to prove. If $\epsilon \geq 1$, then $\ln \epsilon \geq 0$ and hence $\ln \epsilon / \ln |r| \leq 0$. In this case, if N is *any* positive number, then whenever $n > N$, the last inequality in the list is again true.

To prove (ii), let $|r| > 1$ and consider any positive real number P. The following inequalities are equivalent:

$$|r|^n > P, \qquad \ln |r|^n > \ln P, \qquad n \ln |r| > \ln P, \qquad n > \frac{\ln P}{\ln |r|}$$

If we choose $N = \ln P / \ln |r|$, then whenever $n > N$, the last inequality is true and hence so is the first—that is, $|r|^n > P$. By Definition (8.4), this means that $\lim_{n \to \infty} |r|^n = \infty$. ∎

EXAMPLE 6 List the first four terms of the sequence, and determine whether the sequence converges or diverges.

(a) $\left\{ \left(-\frac{2}{3} \right)^n \right\}$ **(b)** $\{ (1.01)^n \}$

SOLUTION

(a) The first four terms of $\left\{ \left(-\frac{2}{3} \right)^n \right\}$ are

$$-\frac{2}{3}, \frac{4}{9}, -\frac{8}{27}, \frac{16}{81}.$$

If we let $r = -\frac{2}{3}$, then, by Theorem (8.6)(i), with $|r| = \frac{2}{3} < 1$,

$$\lim_{n \to \infty} \left(-\frac{2}{3} \right)^n = 0.$$

Hence, the sequence converges to 0.

(b) The first four terms of $\{(1.01)^n\}$ are

$$1.01, 1.0201, 1.030301, 1.04060401.$$

If we let $r = 1.01$, then, by Theorem (8.6)(ii),

$$\lim_{n \to \infty} (1.01)^n = \infty.$$

Since the limit does not exist, the sequence diverges.

Limit theorems that are analogous to those stated in Chapter 1 for sums, differences, products, and quotients of functions can be established for sequences. For example, if $\{a_n\}$ and $\{b_n\}$ are convergent sequences, then

$$\lim_{n \to \infty} (a_n + b_n) = \lim_{n \to \infty} a_n + \lim_{n \to \infty} b_n,$$

$$\lim_{n \to \infty} (a_n b_n) = \left(\lim_{n \to \infty} a_n \right) \left(\lim_{n \to \infty} b_n \right),$$

and so on.

If $a_n = c$ for every n, the sequence $\{a_n\}$ is $c, c, \ldots, c, \ldots$ and

$$\lim_{n \to \infty} c = c.$$

Similarly, if c is a real number and k is a positive rational number, then, as in Theorem (1.18),

$$\lim_{n \to \infty} \frac{c}{n^k} = 0.$$

EXAMPLE ▪ 7 Find the limit of the sequence $\left\{ \dfrac{2n^2}{5n^2 - 3} \right\}$.

SOLUTION To find $\lim_{n \to \infty} a_n$, where $a_n = 2n^2/(5n^2 - 3)$, we divide both the numerator and the denominator of a_n by n^2 and apply limit theorems to obtain

$$\lim_{n \to \infty} \frac{2n^2}{5n^2 - 3} = \lim_{n \to \infty} \frac{2}{5 - (3/n^2)} = \frac{\lim\limits_{n \to \infty} 2}{\lim\limits_{n \to \infty} [5 - (3/n^2)]}$$

$$= \frac{2}{\lim\limits_{n \to \infty} 5 - \lim\limits_{n \to \infty} (3/n^2)} = \frac{2}{5 - 0} = \frac{2}{5}.$$

Hence, the sequence has the limit $\frac{2}{5}$. We can also prove this by applying l'Hôpital's rule to $2x^2/(5x^2 - 3)$.

The next theorem, which is similar to Theorem (1.15), states that if the terms of a sequence are always sandwiched between corresponding terms of two sequences that have the same limit L, then the given sequence also has the limit L.

**Sandwich Theorem
for Sequences 8.7**

If $\{a_n\}$, $\{b_n\}$, and $\{c_n\}$ are sequences and $a_n \le b_n \le c_n$ for every n and if

$$\lim_{n \to \infty} a_n = L = \lim_{n \to \infty} c_n,$$

then

$$\lim_{n \to \infty} b_n = L.$$

EXAMPLE 8 Find the limit of the sequence $\left\{ \dfrac{\cos^2 n}{3^n} \right\}$.

SOLUTION Since $0 < \cos^2 n < 1$ for every positive integer n,

$$0 < \frac{\cos^2 n}{3^n} < \frac{1}{3^n}.$$

Applying Theorem (8.6)(i) with $r = \frac{1}{3}$, we have

$$\lim_{n \to \infty} \frac{1}{3^n} = \lim_{n \to \infty} \left(\frac{1}{3} \right)^n = 0.$$

Moreover, $\lim_{n \to \infty} 0 = 0$. It follows from the sandwich theorem (8.7), with $a_n = 0$, $b_n = (\cos^2 n)/3^n$, and $c_n = \left(\frac{1}{3} \right)^n$, that

$$\lim_{n \to \infty} \frac{\cos^2 n}{3^n} = 0.$$

Hence, the limit of the sequence is 0.

The next theorem can be proved using Definition (8.3).

Theorem 8.8

Let $\{a_n\}$ be a sequence. If $\lim_{n \to \infty} |a_n| = 0$, then $\lim_{n \to \infty} a_n = 0$.

EXAMPLE 9 Suppose the nth term of a sequence is

$$a_n = (-1)^{n+1} \frac{1}{n}.$$

Prove that $\lim_{n \to \infty} a_n = 0$.

SOLUTION The terms of the sequence are alternately positive and negative. For example, the first seven terms are

$$1, -\tfrac{1}{2}, \tfrac{1}{3}, -\tfrac{1}{4}, \tfrac{1}{5}, -\tfrac{1}{6}, \tfrac{1}{7}.$$

Since

$$\lim_{n \to \infty} |a_n| = \lim_{n \to \infty} \frac{1}{n} = 0,$$

it follows from Theorem (8.8) that $\lim_{n \to \infty} a_n = 0$.

A sequence is **monotonic** if successive terms are nondecreasing:

$$a_1 \leq a_2 \leq \cdots \leq a_n \leq \cdots;$$

or if they are nonincreasing:

$$a_1 \geq a_2 \geq \cdots \geq a_n \geq \cdots.$$

A sequence is **bounded** if there is a positive real number M such that $|a_k| \leq M$ for every k. To illustrate, the sequence

$$\frac{1}{2}, \frac{2}{3}, \frac{3}{4}, \frac{4}{5}, \ldots, \frac{n}{n+1}, \cdots$$

Figure 8.6
$$\left\{ \frac{n}{n+1} \right\}$$

is both monotonic (the terms are increasing) and bounded (since we have $k/(k+1) < 1$ for every k). The graph of the sequence is illustrated in Figure 8.6. Note that any number $M \geq 1$ is a bound for the sequence; however, if $K < 1$, then K is not a bound, since $K < k/(k+1)$ when k is sufficiently large.

The next theorem is fundamental for later developments.

Theorem 8.9

A bounded, monotonic sequence has a limit.

To prove Theorem (8.9), it is necessary to use an important property of real numbers. Let us first state several definitions. If S is a nonempty set of real numbers, then a real number u is an **upper bound** of S if $x \leq u$ for every x in S. A number v is a **least upper bound** of S if v is an upper bound and no number less than v is an upper bound of S. Thus, *the least upper bound is the smallest real number that is greater than or equal to every number in S*. To illustrate, if S is the open interval (a, b), then any number greater than b is an upper bound of S; however, the least upper bound of S is unique and equals b. The monotonic sequence $\{n/(n+1)\}$ illustrated in Figure 8.6 has the least upper bound (and limit) 1.

The following statement is an axiom for the real number system.

Completeness Property 8.10

If a nonempty set S of real numbers has an upper bound, then S has a least upper bound.

PROOF OF THEOREM (8.9) Let $\{a_n\}$ be a bounded, monotonic sequence with nondecreasing terms. Thus,

$$a_1 \leq a_2 \leq \cdots \leq a_n \leq \cdots,$$

and there is a number M such that $a_k \le M$ for every positive integer k. Since M is an upper bound for the set S of all numbers in the sequence, it follows from the completeness property (8.10) that S has a least upper bound L such that $L \le M$ (see Figure 8.7).

Figure 8.7

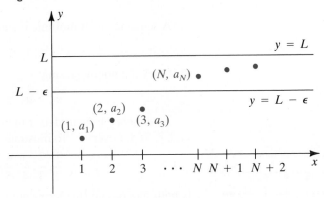

If $\epsilon > 0$, then $L - \epsilon$ is not an upper bound of S, and hence at least one term of $\{a_n\}$ is greater than $L - \epsilon$; that is,

$$L - \epsilon < a_N \quad \text{for some positive integer } N,$$

as shown in Figure 8.7. Since the terms of $\{a_n\}$ are nondecreasing,

$$a_N \le a_{N+1} \le a_{N+2} \le \cdots$$

and, therefore,

$$L - \epsilon < a_n \quad \text{for every } n \ge N.$$

It follows that if $n > N$, then

$$0 \le L - a_n < \epsilon \quad \text{or} \quad |L - a_n| < \epsilon.$$

By Definition (8.3), this result means that

$$\lim_{n \to \infty} a_n = L \le M.$$

That is, $\{a_n\}$ has a limit.

We may obtain the proof for a sequence $\{a_n\}$ of nonincreasing terms in a similar fashion or by considering the sequence $\{-a_n\}$. ∎

Programmable calculators and computers have features that allow us to easily investigate sequences that are defined recursively.

EXAMPLE ▪ 10   If a sequence is defined recursively by $a_1 = 5$, $a_{k+1} = f(a_k)$, where $f(x) = e^{x/4} - 2$, find the first five terms and discuss what happens to the terms of the sequence as k increases.

SOLUTION   On a programmable or graphing calculator that permits storage of variables, we can easily compute many terms of this sequence.

For example, on most graphing calculators, the first term can be stored in the variable memory by the command.

$$5 \rightarrow X.$$

The command line

$$\boxed{e^x}(X/4) \ - \ 2 \ \rightarrow \ X \ \boxed{\text{ENTER}}$$

calculates the second term. Repeatedly pressing $\boxed{\text{ENTER}}$ causes the previous command to execute again, using the most recently stored value in X. This repetition gives the successive terms in the sequence, which are approximately

$$5, \quad 1.490343, \quad -0.548517, \quad -1.128142, \quad -1.245753$$

Since the range of the function $f(x) = e^{x/4} - 2$ is $(-2, \infty)$, this sequence is bounded below by the number -2. So long as the sequence continues to be monotone decreasing, it must converge. With this assurance, we look at more terms to approximate the limit:

$$a_{10} \approx -1.27248018104, \quad a_{15} \approx -1.27248552218,$$

$$a_{20} \approx -1.27248552324$$

The sequence appears to converge to a number that is approximately equal to -1.27248552.

---

*The notation $S \rightarrow I \rightarrow S$ is an abbreviation for *Susceptible $\rightarrow$ Infected $\rightarrow$ Susceptible* and signifies that an infected person who becomes cured is not immune to the disease, but may contract it again. Examples of such diseases are gonorrhea and strep throat. Recall the discussions of epidemics in Section 3.8.

There are many applications of sequences. In particular, sequences may be applied to the investigation of the time course of an $S \rightarrow I \rightarrow S$ epidemic.* Suppose that physicians issue daily reports indicating the number of persons who have become infected with a particular disease and those who have been cured. We shall label the reporting days as $1, 2, \ldots, n, \ldots$ and let N denote the total population. In addition, let

$$I_n = \text{number of persons who have the disease on day } n$$
$$F_n = \text{number of newly infected persons on day } n$$
$$C_n = \text{number of persons cured on day } n.$$

It follows that for every $n \geq 1$,

$$I_{n+1} = I_n + F_{n+1} - C_{n+1}.$$

Suppose health officials decide that the number of new cases on a given day is directly proportional to the product of the number ill and the number not infected on the previous day. (This is known as the *law of mass action* and is typical of a population of students on a college campus.) Moreover, suppose that the number cured each day is directly proportional to the number ill the previous day. Hence,

$$F_{n+1} = aI_n(N - I_n) \quad \text{and} \quad C_{n+1} = bI_n,$$

where a and b are positive constants that can be approximated from early data. Substituting in the preceding formula for I_{n+1}, we have

$$I_{n+1} = I_n + aI_n(N - I_n) - bI_n.$$

In the early stages of an epidemic, I_n will be very small compared to N, and from the point of view of public health, it is better to *overestimate* the number ill than to underestimate and be unprepared for the spread of the disease. With this in mind, we drop the term $-a_n I_n^2$ in the formula for I_{n+1} and investigate the early dynamics of the epidemic by examining the equation

$$I_{n+1} = I_n + aN I_n - bI_n = (1 + aN - b)I_n.$$

If we let $r = 1 + aN - b$, then $I_{n+1} = rI_n$ and, therefore,

$$I_2 = rI_1, \quad I_3 = rI_2 = r^2 I_1, \quad I_4 = rI_3 = r^3 I_1, \quad \ldots, \quad I_n = r^{n-1} I_1, \quad \ldots.$$

This gives us the following sequence of numbers of infected individuals:

$$I_1, rI_1, r^2 I_1, \ldots, r^{n-1} I_1, \ldots$$

The number $r = 1 + aN - b$ is of critical import. If $r > 1$, then, by Theorem (8.6)(ii), $\lim_{n \to \infty} I_n = \infty$ and an epidemic is in progress. In this case, when n is large, I_n is no longer small compared to N, and the formula for I_{n+1} becomes invalid. If $r < 1$, then, by Theorem (8.6)(i), $\lim_{n \to \infty} I_n = 0$ and health officials need not be concerned. The case $r = 1$ results in the constant sequence $I_1, I_1, \ldots, I_1, \ldots$

EXERCISES 8.1

Exer. 1–16: The expression is the *n*th term a_n of a sequence $\{a_n\}$. Find the first four terms and $\lim_{n \to \infty} a_n$, if it exists.

1 $\dfrac{n}{3n + 2}$

2 $\dfrac{6n - 5}{5n + 1}$

3 $\dfrac{7 - 4n^2}{3 + 2n^2}$

4 $\dfrac{4}{8 - 7n}$

5 -5

6 $\sqrt{2}$

7 $\dfrac{(2n - 1)(3n + 1)}{n^3 + 1}$

8 $8n + 1$

9 $\dfrac{2}{\sqrt{n^2 + 9}}$

10 $\dfrac{100n}{n^{3/2} + 4}$

11 $(-1)^{n+1} \dfrac{3n}{n^2 + 4n + 5}$

12 $(-1)^{n+1} \dfrac{\sqrt{n}}{n + 1}$

13 $1 + (0.1)^n$

14 $1 - \dfrac{1}{2^n}$

15 $1 + (-1)^{n+1}$

16 $\dfrac{n + 1}{\sqrt{n}}$

Exer. 17–42: Determine whether the sequence converges or diverges, and if it converges, find the limit.

17 $\left\{ 6\left(-\dfrac{5}{6}\right)^n \right\}$

18 $\left\{ 8 - \left(\dfrac{7}{8}\right)^n \right\}$

19 $\{\arctan n\}$

20 $\left\{ \dfrac{\tan^{-1} n}{n} \right\}$

21 $\{1000 - n\}$

22 $\left\{ \dfrac{(1.0001)^n}{1000} \right\}$

23 $\left\{ (-1)^n \dfrac{\ln n}{n} \right\}$

24 $\left\{ \dfrac{n^2}{\ln(n + 1)} \right\}$

25 $\left\{ \dfrac{4n^4 + 1}{2n^2 - 1} \right\}$

26 $\left\{ \dfrac{\cos n}{n} \right\}$

27 $\left\{ \dfrac{e^n}{n^4} \right\}$

28 $\{e^{-n} \ln n\}$

29 $\left\{ \left(1 + \dfrac{1}{n}\right)^n \right\}$

30 $\{(-1)^n n^3 3^{-n}\}$

31 $\{2^{-n} \sin n\}$

32 $\left\{ \dfrac{4n^3 + 5n + 1}{2n^3 - n^2 + 5} \right\}$

33 $\left\{ \dfrac{n^2}{2n - 1} - \dfrac{n^2}{2n + 1} \right\}$

34 $\left\{ n \sin \dfrac{1}{n} \right\}$

35 $\{\cos \pi n\}$

36 $\left\{ 4 + \sin \dfrac{1}{2} \pi n \right\}$

37 $\{n^{1/n}\}$

38 $\left\{ \dfrac{n^2}{2^n} \right\}$

39 $\left\{\dfrac{n^{-10}}{\sec n}\right\}$ **40** $\left\{(-1)^n\,\dfrac{n^2}{1+n^2}\right\}$

41 $\{\sqrt{n+1}-\sqrt{n}\}$ **42** $\{\sqrt{n^2+n}-n\}$

43 A stable population of 35,000 birds lives on three islands. Each year, 10% of the population on island A migrates to island B, 20% of the population on island B migrates to island C, and 5% of the population on island C migrates to island A. Let A_n, B_n, and C_n denote the numbers of birds on islands A, B, and C, respectively, in year n before migration takes place.

(a) Show that

$$A_{n+1} = 0.9A_n + 0.05C_n$$
$$B_{n+1} = 0.1A_n + 0.80B_n$$

and

$$C_{n+1} = 0.95C_n + 0.20B_n.$$

(b) Assuming that $\lim_{n\to\infty} A_n$, $\lim_{n\to\infty} B_n$, and $\lim_{n\to\infty} C_n$ exist, approximate the number of birds on each island after many years.

44 A bobcat population is classified by age as kittens (less than one year old) and adults (at least one year old). All adult females, including those born the preceding year, have a litter each June, with an average litter size of three kittens. The survival rate of kittens is 50%, whereas that of adults is $66\frac{2}{3}\%$ per year. Let K_n be the number of newborn kittens in June of the nth year, let A_n be the number of adults, and assume that the ratio of males to females is always 1.

(a) Show that

$$K_{n+1} = \tfrac{3}{2}A_{n+1} \quad \text{and} \quad A_{n+1} = \tfrac{2}{3}A_n + \tfrac{1}{2}K_n.$$

(b) Conclude that $A_{n+1} = \frac{17}{12}A_n$ and $K_{n+1} = \frac{17}{12}K_n$, and that $A_n = \left(\frac{17}{12}\right)^{n-1}A_1$ and $K_n = \left(\frac{17}{12}\right)^{n-1}K_1$. What can you conclude about the population?

c 45 Terms of the sequence defined recursively by $a_1 = 5$ and $a_{k+1} = \sqrt{a_k}$ may be generated on a calculator. We enter 5 and then either repeatedly press $\boxed{\sqrt{x}}$ for a scientific calculator or repeatedly use $\boxed{\sqrt{}}\ \boxed{\text{ANS}}$ on a graphing calculator.

(a) Describe what happens to the terms of the sequence as k increases.

(b) Show that $a_n = 5^{1/2^n}$, and find $\lim_{n\to\infty} a_n$.

c 46 If a sequence is generated by entering a number and repeatedly performing the operation of $\boxed{1/x}$, under what conditions does the sequence have a limit?

c 47 Terms of the sequence defined recursively by $a_1 = 1$ and $a_{k+1} = \cos a_k$ may be generated on a calculator. On most graphing calculators, we enter $1 \to A$ and then $\boxed{\cos}\,A \to A\,\boxed{\text{ENTER}}$. Repeatedly pressing $\boxed{\text{ENTER}}$ will produce successive terms in the sequence.

(a) Describe what happens to the terms of the sequence as k increases.

(b) Assuming that $\lim_{n\to\infty} a_n = L$, prove that $L = \cos L$. (*Hint*: $\lim_{n\to\infty} a_{n+1} = L$.)

c 48 A sequence $\{x_n\}$ is defined recursively by the formula $x_{k+1} = x_k - \tan x_k$.

(a) If $x_1 = 3$, approximate the first five terms of the sequence. Predict $\lim_{n\to\infty} x_n$.

(b) If $x_1 = 6$, approximate the first five terms of the sequence. Predict $\lim_{n\to\infty} x_n$.

(c) Assuming that $\lim_{n\to\infty} x_n = L$, prove that $L = \pi n$ for some integer n.

c 49 Approximations to $\sqrt{N}$ may be generated from the sequence defined recursively by

$$x_1 = \frac{N}{2}, \qquad x_{k+1} = \frac{1}{2}\left(x_k + \frac{N}{x_k}\right).$$

(a) Approximate x_2, x_3, x_4, x_5, x_6 if $N = 10$.

(b) Assuming that $\lim_{n\to\infty} x_n = L$, prove that $L = \sqrt{N}$.

c 50 The famous *Fibonacci sequence* is defined recursively by $a_{k+1} = a_k + a_{k-1}$ with $a_1 = a_2 = 1$.

(a) Find the first ten terms of the sequence.

(b) The terms of the sequence $r_k = a_{k+1}/a_k$ give approximations to τ, the *golden ratio*. Approximate the first ten terms of this sequence.

(c) Assuming that $\lim_{n\to\infty} r_n = \tau$, prove that

$$\tau = \tfrac{1}{2}(1 + \sqrt{5}).$$

c Exer. 51–52: If f is differentiable, then a sequence $\{a_n\}$ defined recursively by $a_{k+1} = f(a_k)$, for $k \geq 1$, will converge for any a_1 if the derivative f' is continuous and $|f'(x)| \leq B < 1$ for some positive constant B. (a) For the given f, verify that the sequence $\{a_n\}$ converges for any a_1 by finding a suitable B. (b) Approximate, to two decimal places, $\lim_{n\to\infty} a_n$, if $a_1 = 1$ and also if $a_1 = -100$.

51 $f(a_k) = \tfrac{1}{4}\sin a_k \cos a_k + 1$

52 $f(a_k) = \dfrac{a_k^2}{a_k^2 + 1} + 2$

Mathematicians and Their Times

CARL FRIEDRICH GAUSS

BORN IN A HUMBLE COTTAGE in Germany, Carl Friedrich Gauss (1777–1855), arguably the greatest mathematician who ever lived, seemed at first destined for a life of poverty and hard physical labor. His father, Gerhard, worked as a gardener and bricklayer, and he expected his son to do likewise. Although he was scrupulously honest, Gerhard's harsh ways came close to brutality as he tried to prevent Carl from acquiring a suitable education. Fortunately, the boy's mother, Dorothea, recognized and encouraged Carl's talents.

Gauss's mental prowess was evident at an extremely young age. Before his third birthday, he found an error in his father's calculation of a weekly payroll. His schoolmaster confessed that at age 10, Gauss had mastered arithmetic so well that "I can teach him nothing more." Eventually, the Duke of Brunswick learned of Gauss's abilities and took responsibility for financing his education. At age 15, Gauss mastered infinite series (the subject of this chapter) and gave the first rigorous proof of the general binomial theorem, a result that had been conjectured and used by Newton.

Intellectual historians see Gauss as a transition figure. Felix Klein describes Gauss as "the point where historical epochs separate: he is the highest development of the past, which he closes, and the foundation of the new . . . Gauss is like the highest peak among our Bavarian mountains . . . the gradually ascending foothills culminate in the one gigantic Colossus, which falls away steeply into the lowlands of a new formation, into which its spurs reach out for many miles and in which the waters gushing from it begets new life." * Gauss saw the essence of analysis, the branch of mathematics including calculus, as the rigorous use of infinite processes. Newton, Leibniz, Euler, and Lagrange all manipulated infi-

---

*Felix Klein, *Vorlesungen über die Entwicklung der Mathematik*, Teil I. Berlin: J. Springer, 1926, p. 62.

nite series masterfully but failed to prove that the results obtained were correct. Gauss's insistence on rigor fundamentally changed mathematics.

Gauss made profound discoveries in nearly all areas of pure and applied mathematics. His work established new directions in number theory, algebra, non-Euclidean geometry, statistics, differential geometry, analytical dynamics, potential theory, magnetism, and optics. His doctoral thesis, for example, gave the first proof of the fundamental theorem of algebra: Every polynomial with complex coefficients has at least one complex root. Not only did this work of a 22-year-old establish an important theorem, it also saw the introduction of a coherent account of complex numbers and their geometric representation, a subject of central importance in mathematics.

8.2 CONVERGENT OR DIVERGENT SERIES

We may use sequences to define expressions of the form

$$0.6 + 0.06 + 0.006 + 0.0006 + 0.00006 + \cdots,$$

where the three dots indicate that the sum continues indefinitely. In Definition (8.11), we call such an expression an *infinite series*. Since only finite sums may be added algebraically, we must *define* what is meant by this "infinite sum." As we shall see, the key to the definition is to consider the *sequence of partial sums* $\{S_n\}$, where S_k *is the sum of the first k numbers of the infinite series.* For the preceding illustration,

$$S_1 = 0.6$$

$$S_2 = 0.6 + 0.06 = 0.66$$

$$S_3 = 0.6 + 0.06 + 0.006 = 0.666$$

$$S_4 = 0.6 + 0.06 + 0.006 + 0.0006 = 0.6666$$

and so on. Thus, the sequence of partial sums $\{S_n\}$ may be written

$$0.6, 0.66, 0.666, 0.6666, 0.66666, \ldots .$$

It will follow from Theorem (8.15) that

$$S_n \to \tfrac{2}{3} \quad \text{as} \quad n \to \infty.$$

From an intuitive point of view, the more numbers of the infinite series that we add, the closer the sum gets to $\frac{2}{3}$. Thus, we write

$$\tfrac{2}{3} = 0.6 + 0.06 + 0.006 + 0.0006 + \cdots$$

and call $\frac{2}{3}$ the *sum* of the infinite series.

With this special case in mind, let us introduce terminology that will be used throughout the remainder of this chapter. In the following definition, we assume that $a_1, a_2, \ldots, a_n, \ldots$ are the terms of some sequence.

Definition 8.11

An **infinite series** (or simply a **series**) is an expression of the form

$$a_1 + a_2 + \cdots + a_n + \cdots,$$

or, in summation notation,

$$\sum_{n=1}^{\infty} a_n, \quad \text{or} \quad \sum a_n.$$

Each number a_k is a **term** of the series, and a_n is the **nth term**.

Sometimes there is confusion between the concept of a series and that of a sequence. Remember that a series is an expression that represents an *infinite sum* of numbers. A sequence is a collection of numbers that are in one-to-one correspondence with the positive integers. The sequence of partial sums in the next definition is a special type of sequence that we obtain by using the terms of a series.

As in the special case introduced at the beginning of this section, we define the *sequence of partial sums* of a series as follows.

Definition 8.12

(i) The **kth partial sum** S_k of the series $\sum a_n$ is

$$S_k = a_1 + a_2 + \cdots + a_k.$$

(ii) The **sequence of partial sums** of the series $\sum a_n$ is

$$S_1, S_2, S_3, \ldots, S_n, \ldots.$$

By Definition (8.12)(i),

$$S_1 = a_1$$
$$S_2 = a_1 + a_2$$
$$S_3 = a_1 + a_2 + a_3$$
$$S_4 = a_1 + a_2 + a_3 + a_4.$$

To calculate S_5, S_6, S_7, and so on, we add more terms of the series. Thus, S_{1000} is the sum of the first one thousand terms of $\sum a_n$. If the sequence $\{S_n\}$ has a limit S, we call S the *sum* of the series $\sum a_n$, as in the next definition.

Definition 8.13

A series $\sum a_n$ is **convergent** (or **converges**) if its sequence of partial sums $\{S_n\}$ converges—that is, if

$$\lim_{n \to \infty} S_n = S \quad \text{for some real number } S.$$

The limit S is the **sum** of the series $\sum a_n$, and we write

$$S = a_1 + a_2 + \cdots + a_n + \cdots.$$

The series $\sum a_n$ is **divergent** (or **diverges**) if $\{S_n\}$ diverges. A divergent series has no sum.

For most series, it is very difficult to find a formula for S_n. However, as we shall see in later sections, it may be possible to establish the convergence or divergence of a series using other methods. In the remainder of this section, we consider several important series for which we *can* find a formula for S_n.

EXAMPLE■1 Given the series

$$\frac{1}{1 \cdot 2} + \frac{1}{2 \cdot 3} + \frac{1}{3 \cdot 4} + \cdots + \frac{1}{n(n+1)} + \cdots,$$

(a) find S_1, S_2, S_3, S_4, S_5, and S_6
(b) find S_n
(c) show that the series converges and find its sum

SOLUTION

(a) By Definition (8.12), the first six partial sums are as follows:

$$S_1 = \frac{1}{1 \cdot 2} = \frac{1}{2}$$

$$S_2 = \frac{1}{1 \cdot 2} + \frac{1}{2 \cdot 3} = \frac{2}{3}$$

$$S_3 = \frac{1}{1 \cdot 2} + \frac{1}{2 \cdot 3} + \frac{1}{3 \cdot 4} = \frac{3}{4}$$

$$S_4 = \frac{1}{1 \cdot 2} + \frac{1}{2 \cdot 3} + \frac{1}{3 \cdot 4} + \frac{1}{4 \cdot 5} = \frac{4}{5}$$

$$S_5 = S_4 + a_5 = \frac{4}{5} + \frac{1}{5 \cdot 6} = \frac{5}{6}$$

$$S_6 = S_5 + a_6 = \frac{5}{6} + \frac{1}{6 \cdot 7} = \frac{6}{7}$$

(b) To find S_n, we shall write the terms of the series in a different way. Using partial fractions, we can show that

$$a_n = \frac{1}{n(n+1)} = \frac{1}{n} - \frac{1}{n+1}.$$

Consequently, the nth partial sum of the series may be written

$$S_n = a_1 + a_2 + a_3 + \cdots + a_n$$

$$= \left(1 - \frac{1}{2}\right) + \left(\frac{1}{2} - \frac{1}{3}\right) + \left(\frac{1}{3} - \frac{1}{4}\right) + \cdots + \left(\frac{1}{n} - \frac{1}{n+1}\right).$$

Regrouping, we see that all numbers except the first and last cancel, and hence

$$S_n = 1 - \frac{1}{n+1} = \frac{n}{n+1}.$$

(c) Using the formula for S_n obtained in part (b), we obtain

$$\lim_{n \to \infty} S_n = \lim_{n \to \infty} \frac{n}{n+1} = 1.$$

Thus, the series converges and has the sum 1. As in Definition (8.13), we may write

$$1 = \frac{1}{1 \cdot 2} + \frac{1}{2 \cdot 3} + \frac{1}{3 \cdot 4} + \cdots + \frac{1}{n(n+1)} + \cdots.$$

The series $\sum 1/[n(n+1)]$ of Example 1 is called a **telescoping series**, since writing S_n as shown in part (b) of the solution causes the terms to *telescope* to $1 - [1/(n+1)]$.

EXAMPLE ■ 2 Given the series

$$\sum_{n=1}^{\infty} (-1)^{n-1} = 1 + (-1) + 1 + (-1) + \cdots + (-1)^{n-1} + \cdots,$$

(a) find S_1, S_2, S_3, S_4, S_5, and S_6

(b) find S_n

(c) show that the series diverges

SOLUTION

(a) By Definition (8.12),

$$S_1 = 1, \quad S_2 = 0, \quad S_3 = 1, \quad S_4 = 0, \quad S_5 = 1, \quad \text{and} \quad S_6 = 0.$$

(b) We can write S_n as follows:

$$S_n = \begin{cases} 1 & \text{if } n \text{ is odd} \\ 0 & \text{if } n \text{ is even} \end{cases}$$

(c) Since the sequence of partial sums $\{S_n\}$ oscillates between 1 and 0, it follows that $\lim_{n \to \infty} S_n$ does not exist. Hence, the series diverges.

EXAMPLE ■ 3 Prove that the following series is divergent:

$$1 + \frac{1}{2} + \frac{1}{3} + \frac{1}{4} + \cdots + \frac{1}{n} + \cdots$$

SOLUTION   Let us group the terms of the series as follows:

$$1 + \tfrac{1}{2} + (\tfrac{1}{3} + \tfrac{1}{4}) + (\tfrac{1}{5} + \tfrac{1}{6} + \tfrac{1}{7} + \tfrac{1}{8})$$

$$+ (\tfrac{1}{9} + \cdots + \tfrac{1}{16}) + (\tfrac{1}{17} + \cdots + \tfrac{1}{32}) + \cdots$$

Note that each group contains twice the number of terms as the preceding group. Moreover, since increasing the denominator *decreases* the value of a fraction, we have the following:

$$\tfrac{1}{3} + \tfrac{1}{4} > \tfrac{1}{4} + \tfrac{1}{4} = \tfrac{1}{2}$$

$$\tfrac{1}{5} + \tfrac{1}{6} + \tfrac{1}{7} + \tfrac{1}{8} > \tfrac{1}{8} + \tfrac{1}{8} + \tfrac{1}{8} + \tfrac{1}{8} = \tfrac{1}{2}$$

$$\tfrac{1}{9} + \tfrac{1}{10} + \cdots + \tfrac{1}{16} > \tfrac{1}{16} + \tfrac{1}{16} + \cdots + \tfrac{1}{16} = \tfrac{1}{2}$$

$$\tfrac{1}{17} + \tfrac{1}{18} + \cdots + \tfrac{1}{32} > \tfrac{1}{32} + \tfrac{1}{32} + \cdots + \tfrac{1}{32} = \tfrac{1}{2}$$

Since the sum of the terms within each set of parentheses is greater than $\tfrac{1}{2}$, we obtain the following inequalities:

$$S_4 > 1 + \tfrac{1}{2} + \tfrac{1}{2} > 3(\tfrac{1}{2})$$

$$S_8 > 1 + \tfrac{1}{2} + \tfrac{1}{2} + \tfrac{1}{2} > 4(\tfrac{1}{2})$$

$$S_{16} > 1 + \tfrac{1}{2} + \tfrac{1}{2} + \tfrac{1}{2} + \tfrac{1}{2} > 5(\tfrac{1}{2})$$

$$S_{32} > 1 + \tfrac{1}{2} + \tfrac{1}{2} + \tfrac{1}{2} + \tfrac{1}{2} + \tfrac{1}{2} > 6(\tfrac{1}{2})$$

It can be shown, by mathematical induction, that

$$S_{2^k} > (k+1)(\tfrac{1}{2}) \quad \text{for every positive integer } k.$$

It follows that S_n can be made as large as desired by taking n sufficiently large—that is, $\lim_{n \to \infty} S_n = \infty$. Since $\{S_n\}$ diverges, the given series diverges.

We can add numerical support to the inequalities in Example 3 by computing a few of the partial sums on a calculator or a computer. Using $k = 9, 10, 11,$ and 12 gives

$$S_{512} \approx 6.81652 \qquad S_{1024} \approx 7.50918$$

$$S_{2048} \approx 8.20208 \qquad S_{4096} \approx 8.89510.$$

Although a calculator or a computer will not provide a proof of convergence or divergence, we frequently have reason to compute partial sums.

   The series in Example 3 will be useful in later developments. It is given the following special name.

Definition 8.14

The **harmonic series** is the divergent series

$$1 + \frac{1}{2} + \frac{1}{3} + \cdots + \frac{1}{n} + \cdots.$$

In the next section, we shall give another proof of the divergence of the harmonic series.

Certain types of series occur frequently in solutions of applied problems. One of the most important is the **geometric series**

$$a + ar + ar^2 + \cdots + ar^{n-1} + \cdots,$$

where a and r are real numbers, with $a \neq 0$.

Theorem 8.15

Let $a \neq 0$. The geometric series

$$a + ar + ar^2 + \cdots + ar^{n-1} + \cdots$$

(i) converges and has the sum $S = \dfrac{a}{1 - r}$ if $|r| < 1$

(ii) diverges if $|r| \geq 1$

PROOF   If $r = 1$, then $S_n = a + a + \cdots + a = na$ and the series diverges, since $\lim_{n \to \infty} S_n$ does not exist.

If $r = -1$, then $S_k = a$ if k is odd and $S_k = 0$ if k is even. Since the sequence of partial sums oscillates between a and 0, the series diverges.

If $r \neq 1$, then

$$S_n = a + ar + ar^2 + \cdots + ar^{n-1}$$

and

$$rS_n = ar + ar^2 + ar^3 + \cdots + ar^n.$$

Subtracting corresponding sides of these equations, we obtain

$$(1 - r)S_n = a - ar^n.$$

Dividing both sides by $1 - r$ gives us

$$S_n = \frac{a}{1 - r} - \frac{ar^n}{1 - r}.$$

Consequently,

$$\lim_{n \to \infty} S_n = \lim_{n \to \infty} \left(\frac{a}{1 - r} - \frac{ar^n}{1 - r} \right)$$

$$= \lim_{n \to \infty} \frac{a}{1 - r} - \lim_{n \to \infty} \frac{ar^n}{1 - r}$$

$$= \frac{a}{1 - r} - \frac{a}{1 - r} \lim_{n \to \infty} r^n.$$

If $|r| < 1$, then $\lim_{n \to \infty} r^n = 0$, by Theorem (8.6)(i), and hence

$$\lim_{n \to \infty} S_n = \frac{a}{1 - r} = S.$$

If $|r| > 1$, then $\lim_{n \to \infty} r^n$ does not exist, by Theorem (8.6)(ii), and hence $\lim_{n \to \infty} S_n$ does not exist. In this case, the series diverges. ∎

EXAMPLE ▪ 4   Prove that the following series converges, and find its sum:

$$0.6 + 0.06 + 0.006 + \cdots + \frac{6}{10^n} + \cdots$$

SOLUTION This is the series considered at the beginning of this section. It is geometric with $a = 0.6$ and $r = 0.1$. Since $|r| < 1$, we conclude from Theorem (8.15)(i) that the series converges and has the sum

$$S = \frac{a}{1 - r} = \frac{0.6}{1 - 0.1} = \frac{0.6}{0.9} = \frac{2}{3}.$$

Thus,

$$\frac{2}{3} = 0.6 + 0.06 + 0.006 + \cdots + \frac{6}{10^n} + \cdots.$$

This justifies the nonterminating decimal notation $\frac{2}{3} = 0.66666\ldots.$

EXAMPLE 5 Prove that the following series converges, and find its sum:

$$2 + \frac{2}{3} + \frac{2}{3^2} + \cdots + \frac{2}{3^{n-1}} + \cdots$$

SOLUTION The series converges, since it is geometric with $r = \frac{1}{3} < 1$. By Theorem (8.15)(i), the sum is

$$S = \frac{a}{1 - r} = \frac{2}{1 - \frac{1}{3}} = \frac{2}{\frac{2}{3}} = 3.$$

Theorem 8.16

If a series $\sum a_n$ is convergent, then $\lim_{n \to \infty} a_n = 0$.

PROOF The nth term a_n of the series can be expressed as

$$a_n = S_n - S_{n-1}.$$

If S is the sum of the series $\sum a_n$, then we know $\lim_{n \to \infty} S_n = S$ and also $\lim_{n \to \infty} S_{n-1} = S$. Hence,

$$\lim_{n \to \infty} a_n = \lim_{n \to \infty} (S_n - S_{n-1}) = \lim_{n \to \infty} S_n - \lim_{n \to \infty} S_{n-1} = S - S = 0. \quad \blacksquare$$

CAUTION The preceding theorem states that *if* a series converges, *then* the limit of its nth term a_n as $n \to \infty$ is 0. The converse is false—that is, *if* $\lim_{n \to \infty} a_n = 0$, *it does not necessarily follow* that the series $\sum a_n$ is convergent. The harmonic series (8.14) is an illustration of a divergent series $\sum a_n$ for which $\lim_{n \to \infty} a_n = 0$. Consequently, to establish convergence of a series, *it is not enough* to prove that $\lim_{n \to \infty} a_n = 0$, since that may be true for divergent as well as for convergent series.

The next result is a corollary of Theorem (8.16) and the preceding remarks.

nth-Term Test 8.17

(i) If $\lim\limits_{n \to \infty} a_n \ne 0$, then the series $\sum a_n$ is divergent.

(ii) If $\lim\limits_{n \to \infty} a_n = 0$, then further investigation is necessary to determine whether the series $\sum a_n$ is convergent or divergent.

The next illustration shows how to apply the *n*th-term test to a series.

ILLUSTRATION

Series	nth-term test	Conclusion
$\displaystyle\sum_{n=1}^{\infty} \frac{n}{2n+1}$	$\lim\limits_{n\to\infty} \dfrac{n}{2n+1} = \dfrac{1}{2} \ne 0$	Diverges, by (8.17)(i)
$\displaystyle\sum_{n=1}^{\infty} \frac{1}{n^2}$	$\lim\limits_{n\to\infty} \dfrac{1}{n^2} = 0$	Further investigation is necessary, by (8.17)(ii)
$\displaystyle\sum_{n=1}^{\infty} \frac{1}{\sqrt{n}}$	$\lim\limits_{n\to\infty} \dfrac{1}{\sqrt{n}} = 0$	Further investigation is necessary, by (8.17)(ii)
$\displaystyle\sum_{n=1}^{\infty} \frac{e^n}{n}$	$\lim\limits_{n\to\infty} \dfrac{e^n}{n} = \infty$	Diverges, by (8.17)(i)

We shall see in the next section that the second series in the illustration converges and that the third series diverges.

The next theorem states that if corresponding terms of two series are identical after a certain term, then both series converge or both series diverge.

Theorem 8.18

If $\sum a_n$ and $\sum b_n$ are series such that $a_j = b_j$ for every $j > k$, where k is a positive integer, then both series converge or both series diverge.

PROOF   By hypothesis, we may write the following:

$$\sum a_n = a_1 + a_2 + \cdots + a_k + a_{k+1} + \cdots + a_n + \cdots$$

$$\sum b_n = b_1 + b_2 + \cdots + b_k + a_{k+1} + \cdots + a_n + \cdots$$

Let S_n and T_n denote the *n*th partial sums of $\sum a_n$ and $\sum b_n$, respectively. It follows that if $n \ge k$, then

$$S_n - S_k = T_n - T_k,$$

or

$$S_n = T_n + (S_k - T_k).$$

Consequently,

$$\lim_{n\to\infty} S_n = \lim_{n\to\infty} T_n + (S_k - T_k),$$

and hence either both of the limits exist or both do not exist. This gives us the desired conclusion. If both series converge, then their sums differ by $S_k - T_k$. ∎

Theorem (8.18) implies that changing a finite number of terms of a series has no effect on its convergence or divergence (although it does change the sum of a convergent series). In particular, if we replace the first k terms of $\sum a_n$ by 0, convergence is unaffected. It follows that the series

$$a_{k+1} + a_{k+2} + \cdots + a_n + \cdots$$

converges or diverges if $\sum a_n$ converges or diverges, respectively. The series $a_{k+1} + a_{k+2} + \cdots$ is obtained from $\sum a_n$ by **deleting the first k terms**.

Let us state this result for reference as follows.

Theorem 8.19

For any positive integer k, the series

$$\sum_{n=1}^{\infty} a_n = a_1 + a_2 + \cdots \quad \text{and} \quad \sum_{n=k+1}^{\infty} a_n = a_{k+1} + a_{k+2} + \cdots$$

either both converge or both diverge.

EXAMPLE • 6 Show that the following series converges:

$$\frac{1}{3 \cdot 4} + \frac{1}{4 \cdot 5} + \cdots + \frac{1}{(n+2)(n+3)} + \cdots$$

SOLUTION The series can be obtained by deleting the first two terms of the convergent telescoping series of Example 1. Hence, by Theorem (8.19), the given series converges.

The proof of the next theorem follows directly from Definition (8.13).

Theorem 8.20

If $\sum a_n$ and $\sum b_n$ are convergent series with sums A and B, respectively, then

(i) $\sum (a_n + b_n)$ converges and has sum $A + B$

(ii) $\sum c a_n$ converges and has sum cA for every real number c

(iii) $\sum (a_n - b_n)$ converges and has sum $A - B$

It is also easy to show that if $\sum a_n$ diverges, then so does $\sum c a_n$ for every $c \neq 0$.

EXAMPLE ▪ 7 Prove that the following series converges, and find its sum:

$$\sum_{n=1}^{\infty} \left[\frac{7}{n(n+1)} + \frac{2}{3^{n-1}} \right]$$

SOLUTION The telescoping series $\sum 1/[n(n+1)]$ was considered in Example 1, where we found that it converges and has the sum 1. Using Theorem (8.20)(ii) with $c = 7$ and $a_n = 1/[n(n+1)]$, we see that the series $\sum 7/[n(n+1)]$ converges and has the sum $7(1) = 7$.

The geometric series $\sum 2/3^{n-1}$ converges and has the sum 3 (see Example 5). Hence, by Theorem (8.20)(i), the given series converges and has the sum $7 + 3 = 10$.

Theorem 8.21

If $\sum a_n$ is a convergent series and $\sum b_n$ is divergent, then the series $\sum (a_n + b_n)$ is divergent.

PROOF As in the statement of the theorem, let $\sum a_n$ be convergent and $\sum b_n$ be divergent. We shall give an indirect proof—that is, we shall assume that the conclusion of the theorem is *false* and arrive at a contradiction. Thus, *suppose* that $\sum (a_n + b_n)$ is convergent. Applying Theorem (8.20)(iii), we find that the series

$$\sum [(a_n + b_n) - a_n] = \sum b_n$$

is convergent. This result contradicts the fact that $\sum b_n$ is divergent, and hence our supposition is false—that is, $\sum (a_n + b_n)$ is divergent. ▪

EXAMPLE ▪ 8 Determine the convergence or divergence of the series

$$\sum_{n=1}^{\infty} \left(\frac{1}{5^n} + \frac{1}{n} \right).$$

SOLUTION Since $\sum (1/5^n)$ is a convergent geometric series and $\sum (1/n)$ is the divergent harmonic series, then by Theorem (8.21), the given series diverges.

Infinite series often occur in applications in which we want to estimate the long-term behavior of a process that changes at regularly spaced intervals. The next example illustrates such a situation.

EXAMPLE ▪ 9 A chemical plant produces pesticide that contains a molecule potentially harmful to people if the concentration is too high. The plant flushes out the tanks containing the pesticide once a week, and the discharge flows into a river that feeds the water reservoir of a nearby town.

The dangerous molecule breaks down gradually in water so that 90% of the amount remaining each week is dissipated by the end of the next week. Suppose that D units of the molecule are discharged each week.

(a) Find the number of units of the molecule in the river after n weeks.

(b) Estimate the amount of the molecule in the water supply after a very long time.

(c) If the toxic level of the molecule is T units, how large an amount of the molecule can the plant discharge each week?

SOLUTION

(a) Let A_n denote the amount of the molecule in the river immediately after the nth weekly discharge. The amount A_n is equal to the amount of the current discharge plus the amount remaining from previous discharges. Since 90% of the molecule that was in the river the week before is now gone, only 10% remains, so we have

$$A_n = D + \tfrac{1}{10}A_{n-1}.$$

Hence,

$$A_1 = D, \qquad A_2 = D + \tfrac{1}{10}A_1 = D + \tfrac{1}{10}D = D\left(1 + \tfrac{1}{10}\right)$$

and

$$A_3 = D + \tfrac{1}{10}A_2 = D + \tfrac{1}{10}\left[D\left(1 + \tfrac{1}{10}\right)\right] = D\left[1 + \left(\tfrac{1}{10}\right) + \left(\tfrac{1}{10}\right)^2\right].$$

Similarly, we obtain

$$A_4 = D + \tfrac{1}{10}A_3 = D\left[1 + \left(\tfrac{1}{10}\right) + \left(\tfrac{1}{10}\right)^2 + \left(\tfrac{1}{10}\right)^3\right].$$

We can show, by mathematical induction, that the amount of the molecule in the reservoir after n weeks is

$$A_n = D\left[1 + \left(\tfrac{1}{10}\right) + \left(\tfrac{1}{10}\right)^2 + \cdots + \left(\tfrac{1}{10}\right)^{n-1}\right].$$

(b) As n increases, the amount of the molecule in the water supply approaches

$$D\left[1 + \left(\tfrac{1}{10}\right) + \left(\tfrac{1}{10}\right)^2 + \cdots + \left(\tfrac{1}{10}\right)^{n-1} + \cdots\right],$$

which is a geometric series with $a = D$ and $r = \tfrac{1}{10}$. By Theorem (8.15)(i), the series converges to

$$S = \frac{D}{1 - \tfrac{1}{10}} = \frac{10D}{9}.$$

Hence, in the long run, the water supply will contain about $\tfrac{10}{9}D$ units.

(c) To keep the long-term level below T units, we must have

$$\frac{10D}{9} < T$$

or

$$D < \tfrac{9}{10}T,$$

so the plant can discharge up to 90% of the toxic level each week.

We may apply infinite series to the $S \to I \to S$ epidemic discussed at the end of Section 8.1. Suppose that instead of I_n (the number ill on day n), we are interested in the *total number* S_n of individuals who have been ill at some time between the first and nth days. As in our earlier discussion, let us overestimate S_n by approximating the number F_{n+1} of new cases on day $n + 1$ by aNI_n. Thus,

$$S_n = I_1 + F_2 + F_3 + F_4 + \cdots + F_n$$
$$= I_1 + aNI_1 + aNI_2 + aNI_3 + \cdots + aNI_{n-1}.$$

Recalling that $I_n = r^{n-1}I_1$, with $r = 1 + aN - b$, we obtain

$$S_n = I_1 + aNI_1 + aNrI_1 + aNr^2I_1 + \cdots + aNr^{n-2}I_1$$
$$= I_1 + aNI_1(1 + r + r^2 + \cdots + r^{n-2}).$$

As in the proof of Theorem (8.15), this may be written

$$S_n = I_1 + aNI_1 \left(\frac{1}{1-r} - \frac{r^{n-1}}{1-r} \right).$$

If $r < 1$, then

$$\lim_{n \to \infty} S_n = I_1 + aNI_1 \left(\frac{1}{1-r} \right)$$
$$= I_1 \left(1 + \frac{aN}{1-r} \right)$$
$$= I_1 \left(1 + \frac{aN}{b - aN} \right)$$
$$= I_1 \left(\frac{b}{b - aN} \right).$$

If a and b are approximated from early data, this result enables health officials to determine an upper bound for the total number of individuals who will be ill at some stage of the epidemic.

EXERCISES 8.2

Exer. 1–6: Use the method of Example 1 to find (a) S_1, S_2, and S_3; (b) S_n; and (c) the sum of the series, if it converges.

1 $\displaystyle\sum_{n=1}^{\infty} \frac{-2}{(2n+5)(2n+3)}$

2 $\displaystyle\sum_{n=1}^{\infty} \frac{5}{(5n+2)(5n+7)}$

3 $\displaystyle\sum_{n=1}^{\infty} \frac{1}{4n^2 - 1}$

4 $\displaystyle\sum_{n=1}^{\infty} \frac{-1}{9n^2 + 3n - 2}$

5 $\displaystyle\sum_{n=1}^{\infty} \ln \frac{n}{n+1}$

6 $\displaystyle\sum_{n=1}^{\infty} \frac{1}{\sqrt{n+1} + \sqrt{n}}$ (*Hint:* Rationalize the denominator.)

Exer. 7–16: Use Theorem (8.15) to determine whether the geometric series converges or diverges; if it converges, find its sum.

7 $3 + \dfrac{3}{4} + \cdots + \dfrac{3}{4^{n-1}} + \cdots$

8 $3 + \dfrac{3}{(-4)} + \cdots + \dfrac{3}{(-4)^{n-1}} + \cdots$

9 $1 + \left(\dfrac{-1}{\sqrt{5}}\right) + \cdots + \left(\dfrac{-1}{\sqrt{5}}\right)^{n-1} + \cdots$

10 $1 + \left(\dfrac{e}{3}\right) + \cdots + \left(\dfrac{e}{3}\right)^{n-1} + \cdots$

11 $0.37 + 0.0037 + \cdots + \dfrac{37}{(100)^n} + \cdots$

12 $0.628 + 0.000628 + \cdots + \dfrac{628}{(1000)^n} + \cdots$

13 $\displaystyle\sum_{n=1}^{\infty} 2^{-n} 3^{n-1}$ 　　　 **14** $\displaystyle\sum_{n=1}^{\infty} (-5)^{n-1} 4^{-n}$

15 $\displaystyle\sum_{n=1}^{\infty} (-1)^{n-1}$ 　　　 **16** $\displaystyle\sum_{n=1}^{\infty} (\sqrt{2})^{n-1}$

Exer. 17–20: Use Theorem (8.15) to find all values of x for which the series converges, and find the sum of the series.

17 $1 - x + x^2 - x^3 + \cdots + (-1)^n x^n + \cdots$

18 $1 + x^2 + x^4 + \cdots + x^{2n} + \cdots$

19 $\dfrac{1}{2} + \dfrac{(x-3)}{4} + \dfrac{(x-3)^2}{8} + \cdots + \dfrac{(x-3)^n}{2^{n+1}} + \cdots$

20 $3 + (x-1) + \dfrac{(x-1)^2}{3} + \cdots + \dfrac{(x-1)^n}{3^{n-1}} + \cdots$

Exer. 21–24: The overbar indicates that the digits underneath repeat indefinitely. Express the repeating decimal as a series, and find the rational number it represents.

21 $0.\overline{23}$ 　　　　　 **22** $5.1\overline{46}$

23 $3.2\overline{394}$ 　　　　 **24** $2.7\overline{1828}$

Exer. 25–32: Use Example 1 or 3 and Theorem (8.19) or (8.20) to determine whether the series converges or diverges.

25 $\dfrac{1}{4 \cdot 5} + \dfrac{1}{5 \cdot 6} + \cdots + \dfrac{1}{(n+3)(n+4)} + \cdots$

26 $\dfrac{1}{10 \cdot 11} + \dfrac{1}{11 \cdot 12} + \cdots + \dfrac{1}{(n+9)(n+10)} + \cdots$

27 $\dfrac{5}{1 \cdot 2} + \dfrac{5}{2 \cdot 3} + \cdots + \dfrac{5}{n(n+1)} + \cdots$

28 $\dfrac{-1}{1 \cdot 2} + \dfrac{-1}{2 \cdot 3} + \cdots + \dfrac{-1}{n(n+1)} + \cdots$

29 $\dfrac{1}{4} + \dfrac{1}{5} + \cdots + \dfrac{1}{n+3} + \cdots$

30 $6^{-1} + 7^{-1} + \cdots + (n+5)^{-1} + \cdots$

31 $3 + \dfrac{3}{2} + \cdots + \dfrac{3}{n} + \cdots$

32 $-4 - 2 - \dfrac{4}{3} - \cdots - \dfrac{4}{n} - \cdots$

Exer. 33–40: Use the nth-term test (8.17) to determine whether the series diverges or needs further investigation.

33 $\displaystyle\sum_{n=1}^{\infty} \dfrac{3n}{5n-1}$ 　　 **34** $\displaystyle\sum_{n=1}^{\infty} \dfrac{1}{1+(0.3)^n}$

35 $\displaystyle\sum_{n=1}^{\infty} \dfrac{1}{n^2+3}$ 　　 **36** $\displaystyle\sum_{n=1}^{\infty} \dfrac{1}{e^n+1}$

37 $\displaystyle\sum_{n=1}^{\infty} \dfrac{1}{\sqrt[n]{e}}$ 　　 **38** $\displaystyle\sum_{n=1}^{\infty} n \sin \dfrac{1}{n}$

39 $\displaystyle\sum_{n=1}^{\infty} \dfrac{n}{\ln(n+1)}$ 　 **40** $\displaystyle\sum_{n=1}^{\infty} \ln\left(\dfrac{2n}{7n-5}\right)$

Exer. 41–48: Use known convergent or divergent series, together with Theorem (8.20) or (8.21), to determine whether the series is convergent or divergent; if it converges, find its sum.

41 $\displaystyle\sum_{n=3}^{\infty} \left[\left(\dfrac{1}{4}\right)^n + \left(\dfrac{3}{4}\right)^n\right]$ 　 **42** $\displaystyle\sum_{n=1}^{\infty} \left[\left(\dfrac{3}{2}\right)^n + \left(\dfrac{2}{3}\right)^n\right]$

43 $\displaystyle\sum_{n=1}^{\infty} (2^{-n} - 2^{-3n})$ 　 **44** $\displaystyle\sum_{n=1}^{\infty} \left(\dfrac{1}{3^n} - \dfrac{1}{4^n}\right)$

45 $\displaystyle\sum_{n=1}^{\infty} \left[\dfrac{1}{8^n} + \dfrac{1}{n(n+1)}\right]$

46 $\displaystyle\sum_{n=1}^{\infty} \left[\dfrac{1}{n(n+1)} - \dfrac{4}{n}\right]$

47 $\displaystyle\sum_{n=1}^{\infty} \left(\dfrac{5}{n+2} - \dfrac{5}{n+3}\right)$

48 $\displaystyle\sum_{n=1}^{\infty} \left(\dfrac{1}{n+1} - \dfrac{1}{n}\right)$

c **Exer. 49–50:** For the given convergent series, **(a)** approximate S_1, S_2, and S_3 to five decimal places and **(b)** approximate the sum of the series to three decimal places.

49 $\displaystyle\sum_{n=1}^{\infty} \dfrac{\sin n}{4^n}$ 　　 **50** $\displaystyle\sum_{n=1}^{\infty} \dfrac{\sqrt{n}}{e^{(n^2)}}$

c **51** Let S_n be the nth partial sum of the harmonic series. If $M = 3$, use the method of Example 3 to find a positive integer m such that $S_m \geq M$, and approximate S_m to two decimal places.

c **52** Work Exercise 51 if $M = 8$.

c Exer. 53–56: Compute partial sums S_n for the series, using $n = 4, 8, 12, 16,$ and $20.$

53 $\displaystyle\sum_{n=1}^{\infty} \frac{1}{n^2}$

54 $\displaystyle\sum_{n=1}^{\infty} \frac{1}{\sqrt{n}}$

55 $\displaystyle\sum_{n=1}^{\infty} \frac{n}{n^2 + n + 1}$

56 $\displaystyle\sum_{n=1}^{\infty} \frac{n}{e^n}$

57 Prove or disprove: If $\sum a_n$ and $\sum b_n$ both diverge, then $\sum(a_n + b_n)$ diverges.

58 What is wrong with the following "proof" that the divergent geometric series $\sum_{n=1}^{\infty}(-1)^{n+1}$ has the sum 0? (See Example 2.)

$$\sum_{n=1}^{\infty}(-1)^{n+1}$$
$$= [1 + (-1)] + [1 + (-1)] + [1 + (-1)] + \cdots$$
$$= 0 + 0 + 0 + \cdots = 0$$

59 A rubber ball is dropped from a height of 10 m. If it rebounds approximately one-half the distance after each fall, use a geometric series to approximate the total distance that the ball travels before coming to rest.

60 The bob of a pendulum swings through an arc 24 cm long on its first swing. If each successive swing is approximately five-sixths the length of the preceding swing, use a geometric series to approximate the total distance that the bob travels before coming to rest.

61 If a dosage of Q units of a certain drug is administered to an individual, then the amount remaining in the bloodstream at the end of t minutes is given by Qe^{-ct}, where $c > 0$. Suppose this same dosage is given at successive T-minute intervals.

(a) Show that the amount $A(k)$ of the drug in the bloodstream immediately after the kth dose is given by $A(k) = \sum_{n=0}^{k-1} Qe^{-ncT}$.

(b) Find an upper bound for the amount of the drug in the bloodstream after any number of doses.

(c) Find the smallest time between doses that will ensure that $A(k)$ does not exceed a certain level M for $M > Q$.

62 Suppose that each dollar introduced into the economy recirculates as follows: 85% of the original dollar is spent, then 85% of that $0.85 is spent, and so on. Find the economic impact (the total amount spent) if $1,000,000 is introduced into the economy.

63 In a pest eradication program, N sterilized male flies are released into the general population each day, and 90% of these flies will survive a given day.

(a) Show that the number of sterilized flies in the population after n days is

$$N + (0.9)N + \cdots + (0.9)^{n-1}N.$$

(b) If the *long-range* goal of the program is to keep 20,000 sterilized males in the population, how many such flies should be released each day?

64 A certain drug has a half-life in the bloodstream of about 2 hr. Doses of K milligrams will be administered every 4 hr, with K still to be determined.

(a) Show that the number of milligrams of drug in the bloodstream after the nth dose has been administered is

$$K + \tfrac{1}{4}K + \cdots + (\tfrac{1}{4})^{n-1}K,$$

and that this sum is approximately $\tfrac{4}{3}K$ for large values of n.

(b) If more than 500 mg of the drug in the bloodstream is considered to be a dangerous level, find the largest possible dose that can be given repeatedly over a long period of time.

(c) Refer to Exercise 61. If the dose K is 50 mg, how frequently can the drug be safely administered?

65 The first figure shows some terms of a sequence of squares $S_1, S_2, \ldots, S_k, \ldots$. Let $a_k, A_k,$ and P_k denote the side, area, and perimeter, respectively, of the square S_k. The square S_{k+1} is constructed from S_k by connecting four points on S_k, with each point a distance of $\tfrac{1}{4}a_k$ from a vertex, as shown in the second figure.

(a) Find a relationship between a_{k+1} and a_k.

(b) Find $a_n, A_n,$ and P_n.

(c) Calculate $\displaystyle\sum_{n=1}^{\infty} P_n$ and $\displaystyle\sum_{n=1}^{\infty} A_n$.

Exercise 65

66 The figure shows several terms of a sequence consisting of alternating circles and squares. Each circle is inscribed in a square, and each square (excluding the largest) is inscribed in a circle. Let S_n denote the area of the nth square and C_n the area of the nth circle.

(a) Find relationships between S_n and C_n and between C_n and S_{n+1}.

(b) What portion of the largest square is shaded in the figure?

Exercise 66

8.3 POSITIVE-TERM SERIES

In the preceding section, we established the convergence or divergence of several series by finding a formula for the nth partial sum S_n and then determining whether or not $\lim_{n \to \infty} S_n$ exists. Unfortunately, except in special cases such as a geometric series or a telescoping series, it is often impossible to find an explicit formula for S_n. However, we can develop tests for convergence or divergence of a series $\sum a_n$ that use the nth term a_n. *These tests will not give us the sum S of the series*, but instead will tell us only whether the sum *exists*. This result is sufficient in most applications, because knowing that the sum exists, we can usually approximate it to any degree of accuracy by adding a sufficient number of terms of the series.

In this section, we consider only **positive-term series**—that is, series $\sum a_n$ such that $a_n > 0$ for every n. Although this approach may appear to be very specialized, positive-term series are the foundation for all of our future work with series. As we shall see later, the convergence or divergence of an *arbitrary* series can often be determined from that of a related positive-term series.

The next theorem shows that to establish convergence or divergence of a positive-term series, it is sufficient to determine whether the sequence of partial sums $\{S_n\}$ is bounded.

Theorem **8.22**

If $\sum a_n$ is a positive-term series and if there exists a number M such that

$$S_n = a_1 + a_2 + \cdots + a_n < M$$

for every n, then the series converges and has a sum $S \le M$. If no such M exists, the series diverges.

PROOF   If $\{S_n\}$ is the sequence of partial sums of the positive-term series $\sum a_n$, then

$$S_1 < S_2 < \cdots < S_n < \cdots$$

and therefore $\{S_n\}$ is monotonic. If there exists a number M such that $S_n < M$ for every n, then $\{S_n\}$ is bounded monotonic. As in the proof of Theorem (8.9),

$$\lim_{n \to \infty} S_n = S \le M$$

for some S, and hence the series converges. If no such M exists, then $\lim_{n \to \infty} S_n = \infty$ and the series diverges.   ■

We may use the nth term a_n of a series $\sum a_n$ to define a function f such that $f(n) = a_n$ for every positive integer n. In some cases, if we replace n with x, we obtain a function that is defined for every *real* number $x \ge 1$. For example,

$$\text{given} \quad \sum_{n=1}^{\infty} \frac{1}{n^2}, \quad \text{let} \quad f(n) = \frac{1}{n^2}.$$

Replacing n with x, we obtain $f(x) = 1/x^2$, which gives us the desired function f. Note that

$$\sum_{n=1}^{\infty} \frac{1}{n^2} = \sum_{n=1}^{\infty} f(n) = f(1) + f(2) + \cdots + f(n) + \cdots.$$

The next result shows that if a function f obtained in this way satisfies certain conditions, then we may use the improper integral $\int_1^{\infty} f(x)\,dx$ to test the series $\sum_{n=1}^{\infty} f(n)$ for convergence or divergence.

Integral Test   8.23

> If $\sum a_n$ is a series, let $f(n) = a_n$ and let f be the function obtained by replacing n with x. If f is positive-valued, continuous, and decreasing for every real number $x \ge 1$, then the series $\sum a_n$
>
> (i) converges if $\int_1^{\infty} f(x)\,dx$ converges
>
> (ii) diverges if $\int_1^{\infty} f(x)\,dx$ diverges

Figure 8.8

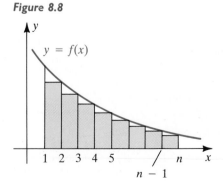

PROOF   As in the hypotheses, we let $f(n) = a_n$ and consider $f(x)$ for every real number $x \ge 1$. A typical graph of this positive-valued, continuous, decreasing function is sketched in Figure 8.8. If n is a positive integer greater than 1, the area of the inscribed rectangular polygon illustrated in Figure 8.8 is

$$\sum_{k=2}^{n} f(k) = f(2) + f(3) + \cdots + f(n).$$

Figure 8.9

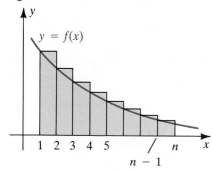

Similarly, the area of the circumscribed rectangular polygon illustrated in Figure 8.9 is

$$\sum_{k=1}^{n-1} f(k) = f(1) + f(2) + \cdots + f(n-1).$$

Since $\int_1^n f(x)\, dx$ is the area under the graph of f from 1 to n,

$$\sum_{k=2}^{n} f(k) \leq \int_1^n f(x)\, dx \leq \sum_{k=1}^{n-1} f(k).$$

Let S_n be the nth partial sum of the series $f(1) + f(2) + \cdots + f(n) + \cdots$, then this inequality may be written

$$S_n - f(1) \leq \int_1^n f(x)\, dx \leq S_{n-1}.$$

The preceding inequality implies that if the integral $\int_1^\infty f(x)\, dx$ converges and equals $K > 0$, then

$$S_n - f(1) \leq K, \quad \text{or} \quad S_n \leq K + f(1)$$

for every positive integer n. Hence, by Theorem (8.22), the series $\sum f(n)$ converges.

If the improper integral diverges, then

$$\lim_{n \to \infty} \int_1^n f(x)\, dx = \infty,$$

and since $\int_1^n f(x)\, dx \leq S_{n-1}$, we also have $\lim_{n \to \infty} S_{n-1} = \infty$—that is, the series $\sum f(n)$ diverges. ■

In using the integral test (8.23), it is necessary to consider

$$\int_1^\infty f(x)\, dx = \lim_{t \to \infty} \int_1^t f(x)\, dx.$$

Thus, we must integrate $f(x)$ and then take a limit. If $f(x)$ is not readily integrable, a different test for convergence or divergence should be used.

EXAMPLE ■ I   Use the integral test (8.23) to prove that the harmonic series

$$1 + \frac{1}{2} + \frac{1}{3} + \cdots + \frac{1}{n} + \cdots$$

diverges (see Example 3 of Section 8.2).

SOLUTION   Since $a_n = 1/n$, we let $f(n) = 1/n$. Replacing n by x gives us $f(x) = 1/x$. Because f is positive-valued, continuous, and decreasing for $x \geq 1$, we can apply the integral test (8.23):

$$\int_1^\infty \frac{1}{x}\, dx = \lim_{t \to \infty} \int_1^t \frac{1}{x}\, dx = \lim_{t \to \infty} \left[\ln x\right]_1^t$$

$$= \lim_{t \to \infty} \left[\ln t - \ln 1\right] = \infty$$

The series diverges, by (8.23)(ii).

EXAMPLE ▪ 2   Determine whether the infinite series $\sum ne^{-n^2}$ converges or diverges.

SOLUTION   Since $a_n = ne^{-n^2}$, we let $f(n) = ne^{-n^2}$ and consider $f(x) = xe^{-x^2}$. If $x \geq 1$, then f is positive-valued and continuous. The first derivative may be used to determine whether f is decreasing. Since

$$f'(x) = e^{-x^2} - 2x^2 e^{-x^2} = e^{-x^2}(1 - 2x^2) < 0,$$

f is decreasing on $[1, \infty)$. We may therefore apply the integral test as follows:

$$\int_1^\infty xe^{-x^2}\, dx = \lim_{t \to \infty} \int_1^t xe^{-x^2}\, dx = \lim_{t \to \infty} \left[\left(-\tfrac{1}{2}\right) e^{-x^2}\right]_1^t$$

$$= \left(-\frac{1}{2}\right) \lim_{t \to \infty} \left[\frac{1}{e^{(t^2)}} - \frac{1}{e}\right] = \frac{1}{2e}$$

Hence the series converges, by (8.23)(i).

In Example 2, we proved that the series $\sum ne^{-n^2}$ converges and therefore has a sum S. However, *we have not found the numerical value of S.* The number $1/(2e)$ in the solution is the value of an improper integral, not the sum of the series. If desired, we could *approximate* S by using a partial sum S_n, with n sufficiently large. (See Exercise 59.)

An integral test may also be used if the function f satisfies the conditions of (8.23) for every $x \geq k$ for some positive integer k. In this case, we merely replace the integral in (8.23) by $\int_k^\infty f(x)\, dx$. This corresponds to deleting the first $k - 1$ terms of the series.

The following series, which is a generalization of the harmonic series (8.14), will be useful when we apply comparision tests later in this section.

Definition 8.24

A *p*-series, or a **hyperharmonic series**, is a series of the form

$$\sum_{n=1}^\infty \frac{1}{n^p} = 1 + \frac{1}{2^p} + \frac{1}{3^p} + \cdots + \frac{1}{n^p} + \cdots,$$

where p is a positive real number.

Note that if $p = 1$ in (8.24), we obtain the harmonic series. The following theorem provides information about convergence or divergence of *p*-series.

Theorem 8.25

The *p*-series $\displaystyle\sum_{n=1}^{\infty} \frac{1}{n^p}$

(i) converges if $p > 1$

(ii) diverges if $p \le 1$

PROOF The special case $p = 1$ is the divergent harmonic series. Suppose that p is a positive real number and $p \ne 1$. We use the integral test (8.23), letting $f(n) = 1/n^p$ and considering $f(x) = 1/x^p = x^{-p}$. The function f is positive-valued and continuous for $x \ge 1$. Moveover, for these values of x we see that $f'(x) = -px^{-p-1} < 0$, and hence f is decreasing. Thus, f satisfies the conditions stated in the integral test (8.23), and we consider

$$\int_1^\infty \frac{1}{x^p}\, dx = \lim_{t \to \infty} \int_1^t x^{-p}\, dx = \lim_{t \to \infty} \left[\frac{x^{1-p}}{1-p} \right]_1^t$$

$$= \frac{1}{1-p} \lim_{t \to \infty} (t^{1-p} - 1).$$

If $p > 1$, then $p - 1 > 0$ and the last expression may be written

$$\frac{1}{1-p} \lim_{t \to \infty} \left(\frac{1}{t^{p-1}} - 1 \right) = \frac{1}{1-p}(0 - 1) = \frac{1}{p-1}.$$

Thus, by (8.23)(i), the *p*-series converges if $p > 1$.

If $0 < p < 1$, then $1 - p > 0$ and

$$\frac{1}{1-p} \lim_{t \to \infty} (t^{1-p} - 1) = \infty.$$

Hence, by (8.23)(ii), the *p*-series diverges.

If $p \le 0$, then $\lim_{n \to \infty}(1/n^p) \ne 0$ and, by the nth-term test (8.17)(i), the series diverges. ■

The following illustration contains some specific *p*-series.

ILLUSTRATION

	p-Series	Value of *p*	Conclusion
▪	$\displaystyle\sum_{n=1}^{\infty} \frac{1}{n^2} = 1 + \frac{1}{2^2} + \frac{1}{3^2} + \cdots$	$p = 2$	Converges, by (8.25)(i), since $2 > 1$
▪	$\displaystyle\sum_{n=1}^{\infty} \frac{1}{\sqrt{n}} = 1 + \frac{1}{\sqrt{2}} + \frac{1}{\sqrt{3}} + \cdots$	$p = \frac{1}{2}$	Diverges, by (8.25)(ii), since $\frac{1}{2} < 1$
▪	$\displaystyle\sum_{n=1}^{\infty} \frac{1}{n^{3/2}} = 1 + \frac{1}{2^{3/2}} + \frac{1}{3^{3/2}} + \cdots$	$p = \frac{3}{2}$	Converges, by (8.25)(i), since $\frac{3}{2} > 1$
▪	$\displaystyle\sum_{n=1}^{\infty} \frac{1}{\sqrt[3]{n}} = 1 + \frac{1}{\sqrt[3]{2}} + \frac{1}{\sqrt[3]{3}} + \cdots$	$p = \frac{1}{3}$	Diverges, by (8.25)(ii), since $\frac{1}{3} < 1$

The next theorem allows us to use known convergent (divergent) series to establish the convergence (divergence) of other series.

Basic Comparision Tests 8.26

Let $\sum a_n$ and $\sum b_n$ be positive-term series.

(i) If $\sum b_n$ converges and $a_n \le b_n$ for every positive integer n, then $\sum a_n$ converges.

(ii) If $\sum b_n$ diverges and $a_n \ge b_n$ for every positive integer n, then $\sum a_n$ diverges.

PROOF   Let S_n and T_n denote the nth partial sums of $\sum a_n$ and $\sum b_n$, respectively. Suppose $\sum b_n$ converges and has the sum T. If $a_n \le b_n$ for every n, then $S_n \le T_n < T$ and hence, by Theorem (8.22), $\sum a_n$ converges. This proves part (i).

To prove (ii), suppose $\sum b_n$ diverges and $a_n \ge b_n$ for every n. Then $S_n \ge T_n$, and since T_n increases without bound as n becomes infinite, so does S_n. Consequently, $\sum a_n$ diverges.   ∎

The convergence or divergence of a series is not affected by deleting a finite number of terms, so the condition $a_n \le b_n$ or $a_n \ge b_n$ of (8.26) is only required from the kth term on, for some positive integer k.

A series $\sum d_n$ is said to **dominate** a series $\sum c_n$ if $d_n \ge c_n$ for every positive integer n. In this terminology, (8.26)(i) states that *a positive-term series that is dominated by a convergent series is also convergent.* Part (ii) states that *a series that dominates a divergent positive-term series is also divergent.*

EXAMPLE ▪ 3   Determine whether the series converges or diverges:

(a) $\displaystyle\sum_{n=1}^{\infty} \frac{1}{2 + 5^n}$ **(b)** $\displaystyle\sum_{n=2}^{\infty} \frac{3}{\sqrt{n} - 1}$

SOLUTION

(a) For every $n \ge 1$,

$$\frac{1}{2 + 5^n} < \frac{1}{5^n} = \left(\frac{1}{5}\right)^n.$$

Since $\sum (1/5)^n$ is a convergent geometric series, the given series converges, by the basic comparison test (8.26)(i).

(b) The p-series $\sum 1/\sqrt{n}$ diverges, and hence so does the series obtained by disregarding the first term $1/\sqrt{1}$. If $n \ge 2$, then

$$\frac{1}{\sqrt{n} - 1} > \frac{1}{\sqrt{n}} \quad \text{and hence} \quad \frac{3}{\sqrt{n} - 1} > \frac{1}{\sqrt{n}}.$$

It follows from the basic comparison test (8.26)(ii) that the given series diverges.

When we use a basic comparison test, we must first decide on a suitable series $\sum b_n$ and then prove that either $a_n \leq b_n$ or $a_n \geq b_n$ for every n greater than some positive integer k. This proof can be very difficult if a_n is a complicated expression. The following comparison test is often easier to apply, because after deciding on $\sum b_n$, we need only take a limit of the quotient a_n/b_n as $n \to \infty$.

Limit Comparison Test 8.27

Let $\sum a_n$ and $\sum b_n$ be positive-term series. If there is a positive real number c such that

$$\lim_{n \to \infty} \frac{a_n}{b_n} = c > 0,$$

then either both series converge or both series diverge.

PROOF If $\lim_{n \to \infty}(a_n/b_n) = c > 0$, then a_n/b_n is close to c if n is large. Hence, there exists a number N such that

$$\frac{c}{2} < \frac{a_n}{b_n} < \frac{3c}{2} \quad \text{whenever} \quad n > N$$

Figure 8.10

(see Figure 8.10). This is equivalent to

$$\frac{c}{2}b_n < a_n < \frac{3c}{2}b_n \quad \text{whenever} \quad n > N.$$

If the series $\sum a_n$ converges, then $\sum (c/2)b_n$ also converges, because it is dominated by $\sum a_n$. Applying (8.20)(ii), we find that the series

$$\sum b_n = \sum \left(\frac{2}{c}\right)\left(\frac{c}{2}\right)b_n$$

converges.

Conversely, if $\sum b_n$ converges, then so does $\sum a_n$, since it is dominated by the convergent series $\sum (3c/2)b_n$. We have proved that $\sum a_n$ converges if and only if $\sum b_n$ converges. Consequently, $\sum a_n$ diverges if and only if $\sum b_n$ diverges. ∎

If, in (8.27), the limit equals 0 or ∞, it may be possible to determine whether the series $\sum a_n$ converges or diverges by using the comparison test stated in Exercise 51 or 52, respectively.

To find a suitable series $\sum b_n$ to use in the limit comparison test (8.27) when a_n is a quotient, a good procedure is to *delete all terms in the numerator and the denominator of a_n except those that have the greatest effect on the magnitude*. We may also replace any constant factor c by 1, since $\sum b_n$ and $\sum cb_n$ either both converge or both diverge (see Theorem 8.20). The next illustration demonstrates this procedure for several series $\sum a_n$.

ILLUSTRATION

a_n	Deleting terms of least magnitude	Choice of b_n in (8.27)
$\dfrac{3n+1}{4n^3+n^2-2}$	$\dfrac{3n}{4n^3} = \dfrac{3}{4n^2}$	$\dfrac{1}{n^2}$
$\dfrac{5}{\sqrt{n^2+2n+7}}$	$\dfrac{5}{\sqrt{n^2}} = \dfrac{5}{n}$	$\dfrac{1}{n}$
$\dfrac{\sqrt[3]{n^2+4}}{6n^2-n-1}$	$\dfrac{\sqrt[3]{n^2}}{6n^2} = \dfrac{n^{2/3}}{6n^2} = \dfrac{1}{6n^{4/3}}$	$\dfrac{1}{n^{4/3}}$

EXAMPLE ■ 4 Determine whether the series converges or diverges.

(a) $\displaystyle\sum_{n=1}^{\infty} \frac{1}{\sqrt[3]{n^2+1}}$ **(b)** $\displaystyle\sum_{n=1}^{\infty} \frac{3n^2+5n}{2^n(n^2+1)}$

SOLUTION

(a) The nth term of the series is

$$a_n = \frac{1}{\sqrt[3]{n^2+1}}.$$

If we delete the number 1 in the radicand, we obtain $b_n = 1/\sqrt[3]{n^2}$, which is the nth term of a divergent p-series, with $p = \frac{2}{3}$. Applying the limit comparison test (8.27) gives us the following:

$$\lim_{n\to\infty} \frac{a_n}{b_n} = \lim_{n\to\infty} \frac{\sqrt[3]{n^2}}{\sqrt[3]{n^2+1}} = \lim_{n\to\infty} \sqrt[3]{\frac{n^2}{n^2+1}} = 1 > 0$$

Since $\sum b_n$ diverges, so does $\sum a_n$.

It is important to note that we cannot use $b_n = 1/\sqrt[3]{n^2}$ with the basic comparison test (8.26), because $a_n < b_n$ instead of $a_n \geq b_n$.

(b) The nth term of the series is

$$a_n = \frac{3n^2+5n}{2^n n^2+2^n}.$$

Deleting the terms of least magnitude in the numerator and the denominator, we obtain

$$\frac{3n^2}{2^n n^2} = \frac{3}{2^n},$$

and hence we choose $b_n = 1/2^n$. Applying the limit comparison test (8.27) gives us

$$\lim_{n\to\infty} \frac{a_n}{b_n} = \lim_{x\to\infty} \frac{3n^2+5n}{2^n(n^2+1)} \cdot \frac{2^n}{1} = \lim_{x\to\infty} \frac{3n^2+5n}{n^2+1} = 3 > 0.$$

Since, by Theorem (8.15)(i), $\sum b_n$ is a convergent geometric series (with $r = \frac{1}{2} < 1$), the series $\sum a_n$ is also convergent.

EXAMPLE ▪ 5   Let $a_n = \dfrac{8n + \sqrt{n}}{5 + n^2 + n^{7/2}}$. Determine whether $\sum a_n$ converges or diverges.

SOLUTION   To find a suitable comparison series $\sum b_n$, we delete all but the highest powers of n in the numerator and the denominator, obtaining

$$\frac{8n}{n^{7/2}} = \frac{8}{n^{5/2}}.$$

Applying the limit comparison test (8.27), with $b_n = 1/n^{5/2}$, we find

$$\lim_{n \to \infty} \frac{a_n}{b_n} = \lim_{n \to \infty} \frac{8n + n^{1/2}}{5 + n^2 + n^{7/2}} \cdot \frac{n^{5/2}}{1}$$

$$= \lim_{n \to \infty} \frac{8n^{7/2} + n^3}{5 + n^2 + n^{7/2}} = 8 > 0.$$

Since $\sum b_n$ is a convergent p-series with $p = \frac{5}{2} > 1$, it follows from (8.27) that $\sum a_n$ is also convergent.

With a programmable calculator or a computer, it is relatively easy to find the nth partial sum S_n for a given infinite series. If the infinite series converges to the sum S, then given an $\epsilon > 0$, we can find an N such that S_N is within ϵ of S. In some cases, we can determine a value for N without explicitly knowing S. Whenever this is possible, we can obtain good approximations to the sum of the infinite series. One such case occurs when the terms of $\{a_n\}$ form a positive, decreasing sequence.

As in the integral test (8.23), if $\sum a_n$ is a series, let $f(n) = a_n$ and let f be the function obtained by replacing n with x. If f is continuous and decreasing for $x > N$ for some integer N, then it can be shown that the error in approximating the sum of the given series by $\sum_{n=1}^{N} a_n$ is less than $\int_{N}^{\infty} f(x)\,dx$ (see Exercise 53).

 EXAMPLE ▪ 6   Approximate the sum of the series $\sum_{n=1}^{\infty} (1/n^3)$ with an error smaller than 10^{-5}.

SOLUTION   The given series converges by Theorem (8.25) with $p = 3$. The function $f(x) = 1/x^3$ is positive, continuous, and decreasing for all $x > 0$—the last condition is true since $f'(x) = -3/x^4$ is negative. We also have

$$\int_{N}^{\infty} f(x)\,dx = \int_{N}^{\infty} \frac{1}{x^3}\,dx = \lim_{t \to \infty} \int_{N}^{t} \frac{1}{x^3}\,dx$$

$$= \lim_{t \to \infty} \left[\frac{-1}{2x^2} \right]_{N}^{t} = \lim_{t \to \infty} \left[\frac{-1}{2t^2} + \frac{1}{2N^2} \right] = \frac{1}{2N^2}.$$

From the result of Exercise 53, we see that $\sum_{n=1}^{N}(1/n^3)$ is within $1/(2N^2)$ of the sum of the infinite series $\sum_{n=1}^{\infty}(1/n^3)$. If we wish the error in our approximation to be less than 10^{-5}, then we must choose N such that

$$\frac{1}{2N^2} < 10^{-5},$$

which is equivalent to $N^2 > 10^5/2 = 50{,}000$, or $N > \sqrt{50{,}000} \approx 223.6$.

Thus, by summing the first 224 terms, we can approximate $\sum_{n=1}^{\infty}(1/n^3)$ with an error smaller than 10^{-5}. A simple computer program yields the result $\sum_{n=1}^{224}(1/n^3) \approx 1.20204698$, which is our approximation to $\sum_{n=1}^{\infty}(1/n^3)$. By our error estimate, we have

$$1.20204698 < \sum_{n=1}^{\infty}\frac{1}{n^3} < 1.20204698 + 0.000005.$$

Thus, the true value of $\sum_{n=1}^{\infty}(1/n^3)$ lies in the interval

$$[1.20204698,\ 1.20205198].$$

We conclude this section with several general remarks about positive-term series. Suppose that $\sum a_n$ is a positive-term series and the terms are grouped in some manner, such as

$$(a_1 + a_2) + a_3 + (a_4 + a_5 + a_6 + a_7) + \cdots.$$

If we denote the last series by $\sum b_n$, so that

$$b_1 = a_1 + a_2, \quad b_2 = a_3, \quad b_3 = a_4 + a_5 + a_6 + a_7, \quad \ldots,$$

then any partial sum of the series $\sum b_n$ is also a partial sum of $\sum a_n$. It follows that if $\sum a_n$ converges, then $\sum b_n$ converges and has the same sum. A similar argument may be used for any grouping of the terms of $\sum a_n$. Thus, *if a positive-term series converges, then the series obtained by grouping the terms in any manner also converges and has the same sum.* We cannot make a similar statement about arbitrary divergent series. For example, the terms of the divergent series $\sum(-1)^n$ may be grouped to produce a convergent series (see Exercise 58 of Section 8.2).

Next, suppose that a convergent positive-term series $\sum a_n$ has the sum S and that a new series $\sum b_n$ is formed by rearranging the terms in some way. For example, $\sum b_n$ could be the series

$$a_2 + a_8 + a_1 + a_5 + a_7 + a_3 + \cdots.$$

If T_n is the nth partial sum of $\sum b_n$, then it is a sum of terms of $\sum a_n$. If m is the largest of the subscripts associated with the terms a_k in T_n, then $T_n \le S_m < S$. Consequently, $T_n < S$ for every n. Applying Theorem (8.22), we find that $\sum b_n$ converges and has a sum $T \le S$. The preceding proof is independent of the particular rearrangement of terms. We may also regard the series $\sum a_n$ as having been obtained by rearranging the terms of $\sum b_n$ and hence, by the same argument, $S \le T$. We have proved that *if the terms of a convergent positive-term series $\sum a_n$ are rearranged in any manner, then the resulting series converges and has the same sum.*

EXERCISES 8.3

Exer. 1–12: (a) Show that the function f determined by the nth term of the series satisfies the hypotheses of the integral test. (b) Use the integral test to determine whether the series converges or diverges.

1 $\displaystyle\sum_{n=1}^{\infty} \frac{1}{(3+2n)^2}$

2 $\displaystyle\sum_{n=1}^{\infty} \frac{1}{(4+n)^{3/2}}$

3 $\displaystyle\sum_{n=1}^{\infty} \frac{1}{4n+7}$

4 $\displaystyle\sum_{n=1}^{\infty} \frac{n}{n^2+1}$

5 $\displaystyle\sum_{n=1}^{\infty} n^2 e^{-n^3}$

6 $\displaystyle\sum_{n=3}^{\infty} \frac{1}{n(2n-5)}$

7 $\displaystyle\sum_{n=3}^{\infty} \frac{\ln n}{n}$

8 $\displaystyle\sum_{n=2}^{\infty} \frac{1}{n(\ln n)^2}$

9 $\displaystyle\sum_{n=2}^{\infty} \frac{1}{n\sqrt{n^2-1}}$

10 $\displaystyle\sum_{n=4}^{\infty} \left(\frac{1}{n-3} - \frac{1}{n} \right)$

11 $\displaystyle\sum_{n=1}^{\infty} \frac{\arctan n}{1+n^2}$

12 $\displaystyle\sum_{n=1}^{\infty} \frac{1}{1+16n^2}$

Exer. 13–20: Use a basic comparison test to determine whether the series converges or diverges.

13 $\displaystyle\sum_{n=1}^{\infty} \frac{1}{n^4+n^2+1}$

14 $\displaystyle\sum_{n=1}^{\infty} \frac{\sqrt{n}}{n^2+1}$

15 $\displaystyle\sum_{n=1}^{\infty} \frac{1}{n3^n}$

16 $\displaystyle\sum_{n=1}^{\infty} \frac{2+\cos n}{n^2}$

17 $\displaystyle\sum_{n=1}^{\infty} \frac{\arctan n}{n}$

18 $\displaystyle\sum_{n=1}^{\infty} \frac{\operatorname{arcsec} n}{(0.5)^n}$

19 $\displaystyle\sum_{n=1}^{\infty} \frac{1}{n^n}$

20 $\displaystyle\sum_{n=1}^{\infty} \frac{1}{n!}$

Exer. 21–28: Use the limit comparison test to determine whether the series converges or diverges.

21 $\displaystyle\sum_{n=1}^{\infty} \frac{\sqrt{n}}{n+4}$

22 $\displaystyle\sum_{n=1}^{\infty} \frac{2}{3+\sqrt{n}}$

23 $\displaystyle\sum_{n=2}^{\infty} \frac{1}{\sqrt{4n^3-5n}}$

24 $\displaystyle\sum_{n=1}^{\infty} \frac{1}{\sqrt{n(n+1)(n+2)}}$

25 $\displaystyle\sum_{n=1}^{\infty} \frac{8n^2-7}{e^n(n+1)^2}$

26 $\displaystyle\sum_{n=1}^{\infty} \frac{3n+5}{n2^n}$

27 $\displaystyle\sum_{n=1}^{\infty} \frac{1}{\sqrt{n}+9}$

28 $\displaystyle\sum_{n=1}^{\infty} \frac{n^2}{n^3+1}$

Exer. 29–46: Determine whether the series converges or diverges.

29 $\displaystyle\sum_{n=1}^{\infty} \frac{2n+n^2}{n^3+1}$

30 $\displaystyle\sum_{n=1}^{\infty} \frac{n^5+4n^3+1}{2n^8+n^4+2}$

31 $\displaystyle\sum_{n=1}^{\infty} \frac{1+2^n}{1+3^n}$

32 $\displaystyle\sum_{n=4}^{\infty} \frac{3n}{2n^2-7}$

33 $\displaystyle\sum_{n=1}^{\infty} \frac{1}{\sqrt[3]{5n^2+1}}$

34 $\displaystyle\sum_{n=1}^{\infty} \frac{\ln n}{n^4}$

35 $\displaystyle\sum_{n=1}^{\infty} \frac{1}{\sqrt[3]{2n+1}}$

36 $\displaystyle\sum_{n=1}^{\infty} \frac{n+\ln n}{n^3+n+1}$

37 $\displaystyle\sum_{n=1}^{\infty} ne^{-n}$

38 $\displaystyle\sum_{n=1}^{\infty} \frac{1}{n(n+1)(n+2)}$

39 $\displaystyle\sum_{n=1}^{\infty} \sin \frac{1}{n^2}$

40 $\displaystyle\sum_{n=1}^{\infty} \tan \frac{1}{n}$

41 $\displaystyle\sum_{n=1}^{\infty} \frac{(2n+1)^3}{(n^3+1)^2}$

42 $\displaystyle\sum_{n=1}^{\infty} \frac{n+\ln n}{n^2+1}$

43 $\displaystyle\sum_{n=1}^{\infty} \frac{n^2+2^n}{n+3^n}$

44 $\displaystyle\sum_{n=1}^{\infty} \ln\left(1+\frac{1}{2^n}\right)$

45 $\displaystyle\sum_{n=1}^{\infty} \frac{\ln n}{n^3}$

46 $\displaystyle\sum_{n=1}^{\infty} \frac{\sin n + 2^n}{n+5^n}$

Exer. 47–48: Find every real number k for which the series converges.

47 $\displaystyle\sum_{n=2}^{\infty} \frac{1}{n^k \ln n}$

48 $\displaystyle\sum_{n=2}^{\infty} \frac{1}{n(\ln n)^k}$

49 (a) Use the proof of the integral test (8.23) to show that, for every positive integer $n > 1$,

$$\ln(n+1) < 1 + \frac{1}{2} + \frac{1}{3} + \cdots + \frac{1}{n} < 1 + \ln n.$$

(b) Estimate the number of terms of the harmonic series that should be added so that $S_n > 100$.

50 Consider the hypothetical problem illustrated in the figure on the following page: Starting with a ball of radius 1 ft, a person stacks balls vertically such that if r_k is the radius of the kth ball, then $r_{n+1} = r_n \sqrt{n/(n+1)}$ for each positive integer n.

(a) Show that the height of the stack can be made arbitrarily large.

(b) If the balls are made of a material that weighs 1 lb/ft^3, show that the total weight of the stack is always less than 4π pounds.

Exercise 50

1 ft

51 Suppose that $\sum a_n$ and $\sum b_n$ are positive-term series. Prove that if $\lim_{n\to\infty}(a_n/b_n) = 0$ and $\sum b_n$ converges, then $\sum a_n$ converges. (This is not necessarily true for series that contain negative terms.)

52 Prove that if $\lim_{n\to\infty}(a_n/b_n) = \infty$ and $\sum b_n$ diverges, then $\sum a_n$ diverges.

53 Let $\sum a_n$ be a convergent, positive-term series. Let $f(n) = a_n$, and suppose f is continuous and decreasing for $x \geq N$ for some integer N. Prove that the error in approximating the sum of the given series by $\sum_{n=1}^{N} a_n$ is less than $\int_N^\infty f(x)\,dx$.

Exer. 54–56: Use Exercise 53 to estimate the smallest number of terms that can be added to approximate the sum of the series with an error less than E.

54 $\displaystyle\sum_{n=1}^{\infty} \frac{1}{n^2}$; $E = 0.001$ **55** $\displaystyle\sum_{n=1}^{\infty} \frac{1}{n^4}$; $E = 0.01$

56 $\displaystyle\sum_{n=2}^{\infty} \frac{1}{n(\ln n)^2}$; $E = 0.05$

57 Prove that if a positive-term series $\sum a_n$ converges, then $\sum (1/a_n)$ diverges.

58 Prove that if a positive-term series $\sum a_n$ converges, then $\sum \sqrt{a_n a_{n+1}}$ converges. (*Hint:* First show that the following is true: $\sqrt{a_n a_{n+1}} \leq (a_n + a_{n+1})/2$.)

c **Exer. 59–64:** Approximate the sum of the given series to three decimal places. (Use Exercise 53 to justify the accuracy of your answer.)

59 $\displaystyle\sum_{n=1}^{\infty} ne^{-n^2}$ **60** $\displaystyle\sum_{n=1}^{\infty} n^2 e^{-n^3}$

61 $\displaystyle\sum_{n=1}^{\infty} \frac{1}{n^4}$ **62** $\displaystyle\sum_{n=1}^{\infty} \frac{e^{1/n}}{n^2}$

63 $\displaystyle\sum_{n=1}^{\infty} \frac{\ln n}{n^2}$ **64** $\displaystyle\sum_{n=1}^{\infty} \frac{e^{-\sqrt{n}}}{\sqrt{n}}$

c **65** Graph, on the same coordinate axes, $y = x$ and $y = \ln(x^k)$ for $k = 1, 2, 3$ and $1 \leq x \leq 20$. Then use the graphs to predict whether the series $\sum_{n=2}^{\infty}(1/\ln(n^k))$ converges or diverges for $k = 1, 2,$ and 3.

c **66** Graph, on the same coordinate axes, $y = x$ and $y = (\ln x)^k$ for $k = 1, 2, 3$ and $1 \leq x \leq 200$. Then use the graphs to predict whether the series $\sum_{n=2}^{\infty}(1/(\ln n)^k)$ converges or diverges for $k = 1, 2,$ and 3.

8.4   THE RATIO AND ROOT TESTS

For the integral test to be applied to a positive-term series $\sum a_n$ with $a_n = f(n)$, the terms must be decreasing and we must be able to integrate $f(x)$. These conditions often rule out series that involve factorials and other complicated expressions. In this section, we examine two tests that can be used to help determine convergence or divergence when other tests are not applicable. Unfortunately, as indicated by part (iii) of both tests, they are inconclusive for certain series.

Ratio Test 8.28

Let $\sum a_n$ be a positive-term series, and suppose that

$$\lim_{n \to \infty} \frac{a_{n+1}}{a_n} = L.$$

(i) If $L < 1$, the series is convergent.

(ii) If $L > 1$ or $\lim_{n \to \infty} \frac{a_{n+1}}{a_n} = \infty$, the series is divergent.

(iii) If $L = 1$, apply a different test; the series may be convergent or divergent.

PROOF

(i) Suppose that $\lim_{n \to \infty}(a_{n+1}/a_n) = L < 1$. Let r be any number such that $0 \leq L < r < 1$. Since a_{n+1}/a_n is close to L if n is large, there exists an integer N such that whenever $n \geq N$,

$$\frac{a_{n+1}}{a_n} < r \quad \text{or} \quad a_{n+1} < a_n r.$$

Substituting $N, N + 1, N + 2, \ldots$ for n, we obtain

$$a_{N+1} < a_N r$$
$$a_{N+2} < a_{N+1} r < a_N r^2$$
$$a_{N+3} < a_{N+2} r < a_N r^3$$

and, in general,

$$a_{N+m} < a_N r^m \quad \text{whenever} \quad m > 0.$$

It follows from the basic comparison test (8.26)(i) that the series

$$a_{N+1} + a_{N+2} + \cdots + a_{N+m} + \cdots$$

converges, since its terms are less than the corresponding terms of the convergent geometric series

$$a_N r + a_N r^2 + \cdots + a_N r^n + \cdots.$$

Since convergence or divergence is unaffected by discarding a finite number of terms (see Theorem (8.19)), the series $\sum_{n=1}^{\infty} a_n$ also converges.

(ii) Suppose that $\lim_{n \to \infty}(a_{n+1}/a_n) = L > 1$. If r is a real number such that $L > r > 1$, then there exists an integer N such that

$$\frac{a_{n+1}}{a_n} > r > 1 \quad \text{whenever} \quad n \geq N.$$

Consequently, $a_{n+1} > a_n$ if $n \geq N$. Thus, $\lim_{n \to \infty} a_n \neq 0$ and, by the nth-term test (8.17)(i), the series $\sum a_n$ diverges.

The proof for $\lim_{n \to \infty}(a_{n+1}/a_n) = \infty$ is similar and is left as an exercise.

(iii) The ratio test is inconclusive if

$$\lim_{n\to\infty} \frac{a_{n+1}}{a_n} = 1,$$

for it is easy to verify that the limit is 1 for both the convergent series $\sum(1/n^2)$ and the divergent series $\sum(1/n)$. Consequently, *if the limit is 1, then a different test must be used.* ■

EXAMPLE ■ 1   Determine whether the series is convergent or divergent.

(a) $\displaystyle\sum_{n=1}^{\infty} \frac{3^n}{n!}$ **(b)** $\displaystyle\sum_{n=1}^{\infty} \frac{3^n}{n^2}$

SOLUTION

(a) Applying the ratio test (8.28), we have

$$\lim_{n\to\infty} \frac{a_{n+1}}{a_n} = \lim_{n\to\infty} \left(a_{n+1} \cdot \frac{1}{a_n}\right)$$

$$= \lim_{n\to\infty} \frac{3^{n+1}}{(n+1)!} \cdot \frac{n!}{3^n}$$

$$= \lim_{n\to\infty} \frac{3}{n+1} = 0.$$

Since $0 < 1$, the series is convergent.

(b) Applying the ratio test (8.28), we obtain

$$\lim_{n\to\infty} \frac{a_{n+1}}{a_n} = \lim_{n\to\infty} \frac{3^{n+1}}{(n+1)^2} \cdot \frac{n^2}{3^n}$$

$$= \lim_{n\to\infty} \frac{3n^2}{n^2 + 2n + 1} = 3.$$

Since $3 > 1$, the series diverges, by (8.28)(ii).

EXAMPLE ■ 2   Determine the convergence or divergence of $\displaystyle\sum_{n=1}^{\infty} \frac{n^n}{n!}$.

SOLUTION   Applying the ratio test gives us

$$\lim_{n\to\infty} \frac{a_{n+1}}{a_n} = \lim_{n\to\infty} \frac{(n+1)^{n+1}}{(n+1)!} \cdot \frac{n!}{n^n}$$

$$= \lim_{n\to\infty} \frac{(n+1)^{n+1}}{(n+1)} \cdot \frac{1}{n^n}$$

$$= \lim_{n\to\infty} \frac{(n+1)^n}{n^n} = \lim_{n\to\infty} \left(\frac{n+1}{n}\right)^n$$

$$= \lim_{n\to\infty} \left(1 + \frac{1}{n}\right)^n = e.$$

The last equality is a consequence of Theorem (6.32)(ii). Since $e > 1$, the series diverges.

If $\sum a_n$ is a series such that $\lim_{n \to \infty}(a_{n+1}/a_n) = 1$, we must use a different test (see (iii) of (8.28)). The next illustration contains several series of this type and suggestions on how to show convergence or divergence.

ILLUSTRATION

Series $\sum a_n$	$\displaystyle\lim_{n\to\infty} \frac{a_{n+1}}{a_n}$	Suggestion
$\displaystyle\sum_{n=1}^{\infty} \frac{2n^2 + 3n + 4}{5n^5 - 7n^3 + n}$	1	Show convergence by using the limit comparison test (8.27) with $b_n = 1/n^3$.
$\displaystyle\sum_{n=1}^{\infty} \frac{2n + 1}{\sqrt{n^3 + 5n + 3}}$	1	Show divergence by using the limit comparison test (8.27) with $b_n = 1/\sqrt{n}$.
$\displaystyle\sum_{n=1}^{\infty} \frac{\ln n}{n}$	1	Show divergence by using the integral test (8.23).

The following test is often useful if a_n contains powers of n.

Root Test 8.29

Let $\sum a_n$ be a positive-term series, and suppose that

$$\lim_{n\to\infty} \sqrt[n]{a_n} = L.$$

 (i) If $L < 1$, the series is convergent.
 (ii) If $L > 1$ or $\lim_{n\to\infty} \sqrt[n]{a_n} = \infty$, the series is divergent.
 (iii) If $L = 1$, apply a different test; the series may be convergent or divergent.

PROOF If $L < 1$, consider any number r such that $0 \le L < r < 1$. By the definition of limit, there exists a positive integer N such that if $n \ge N$, then

$$\sqrt[n]{a_n} < r \quad \text{or} \quad a_n < r^n.$$

Since $0 < r < 1$, $\sum_{n=N}^{\infty} r^n$ is a convergent geometric series, and hence, by the basic comparison test (8.26), $\sum_{n=N}^{\infty} a_n$ converges. Consequently, $\sum_{n=1}^{\infty} a_n$ converges. This proves (i). The remainder of the proof is similar to that used for the ratio test. ∎

EXAMPLE 3 Determine the convergence or divergence of

$$\sum_{n=1}^{\infty} \frac{2^{3n+1}}{n^n}.$$

SOLUTION Applying the root test (8.29) yields

$$\lim_{n \to \infty} \sqrt[n]{\frac{2^{3n+1}}{n^n}} = \lim_{n \to \infty} \left(\frac{2^{3n+1}}{n^n} \right)^{1/n}$$

$$= \lim_{n \to \infty} \frac{2^{3+(1/n)}}{n} = 0.$$

Since $0 < 1$, the series converges. We could have applied the ratio test (8.28); however, the process of evaluating the limit would have been more complicated.

EXERCISES 8.4

Exer. 1–10: Find $\lim_{n \to \infty} (a_{n+1}/a_n)$, and use the ratio test (8.28) to determine if the series converges or diverges or if the test is inconclusive.

1 $\displaystyle\sum_{n=1}^{\infty} \frac{3n+1}{2^n}$
2 $\displaystyle\sum_{n=1}^{\infty} \frac{3^n}{n^2+4}$

3 $\displaystyle\sum_{n=1}^{\infty} \frac{5^n}{n(3^{n+1})}$
4 $\displaystyle\sum_{n=1}^{\infty} \frac{2^{n-1}}{5^n(n+1)}$

5 $\displaystyle\sum_{n=1}^{\infty} \frac{100^n}{n!}$
6 $\displaystyle\sum_{n=1}^{\infty} \frac{n^{10}+10}{n!}$

7 $\displaystyle\sum_{n=1}^{\infty} \frac{n+3}{n^2+2n+5}$
8 $\displaystyle\sum_{n=1}^{\infty} \frac{3n}{\sqrt{n^3+1}}$

9 $\displaystyle\sum_{n=1}^{\infty} \frac{n!}{e^n}$
10 $\displaystyle\sum_{n=1}^{\infty} \frac{n!}{(n+1)^5}$

Exer. 11–18: Find $\lim_{n \to \infty} \sqrt[n]{a_n}$, and use the root test (8.29) to determine if the series converges or diverges or if the test is inconclusive.

11 $\displaystyle\sum_{n=1}^{\infty} \frac{1}{n^n}$
12 $\displaystyle\sum_{n=1}^{\infty} \frac{(\ln n)^n}{n^{n/2}}$
13 $\displaystyle\sum_{n=1}^{\infty} \frac{2^n}{n^2}$

14 $\displaystyle\sum_{n=2}^{\infty} \frac{5^{n+1}}{(\ln n)^n}$
15 $\displaystyle\sum_{n=1}^{\infty} \frac{n}{3^n}$
16 $\displaystyle\sum_{n=1}^{\infty} \frac{n^{10}}{10^n}$

17 $\displaystyle\sum_{n=1}^{\infty} \left(\frac{n}{2n+1} \right)^n$
18 $\displaystyle\sum_{n=2}^{\infty} \left(\frac{n}{\ln n} \right)^n$

Exer. 19–40: Determine whether the series converges or diverges.

19 $\displaystyle\sum_{n=1}^{\infty} \frac{\sqrt{n}}{n^2+1}$
20 $\displaystyle\sum_{n=1}^{\infty} \frac{\sqrt{n}}{3n+4}$

21 $\displaystyle\sum_{n=1}^{\infty} \frac{99^n(n^5+2)}{n^2 10^{2n}}$
22 $\displaystyle\sum_{n=1}^{\infty} \frac{n3^{2n}}{5^{n-1}}$

23 $\displaystyle\sum_{n=1}^{\infty} \frac{2}{n^3+e^n}$
24 $\displaystyle\sum_{n=1}^{\infty} \frac{n+1}{n^3+1}$

25 $\displaystyle\sum_{n=1}^{\infty} \left(\frac{2}{n} \right)^n n!$
26 $\displaystyle\sum_{n=1}^{\infty} \frac{n!}{n^n}$
27 $\displaystyle\sum_{n=1}^{\infty} \frac{n^n}{10^{n+1}}$

28 $\displaystyle\sum_{n=1}^{\infty} \frac{10+2^n}{n!}$
29 $\displaystyle\sum_{n=1}^{\infty} \frac{(n!)^2}{(2n)!}$
30 $\displaystyle\sum_{n=1}^{\infty} \frac{(2n)!}{2^n}$

31 $\displaystyle\sum_{n=2}^{\infty} \frac{1}{n\sqrt[3]{\ln n}}$
32 $\displaystyle\sum_{n=1}^{\infty} \frac{(2n)^n}{(5n+3n^{-1})^n}$

33 $\displaystyle\sum_{n=1}^{\infty} \frac{\ln n}{(1.01)^n}$
34 $\displaystyle\sum_{n=1}^{\infty} 3^{1/n}$
35 $\displaystyle\sum_{n=1}^{\infty} n \tan \frac{1}{n}$

36 $\displaystyle\sum_{n=1}^{\infty} \frac{\arctan n}{n^2}$
37 $\displaystyle\sum_{n=1}^{\infty} \left(1 + \frac{1}{n} \right)^n$
38 $\displaystyle\sum_{n=2}^{\infty} \frac{1}{(\ln n)^n}$

39 $1 + \dfrac{1 \cdot 3}{2!} + \dfrac{1 \cdot 3 \cdot 5}{3!} + \cdots$

$$+ \frac{1 \cdot 3 \cdot 5 \cdot \cdots \cdot (2n-1)}{n!} + \cdots$$

40 $\dfrac{1}{2} + \dfrac{1 \cdot 4}{2 \cdot 4} + \dfrac{1 \cdot 4 \cdot 7}{2 \cdot 4 \cdot 6} + \cdots$

$$+ \frac{1 \cdot 4 \cdot 7 \cdot \cdots \cdot (3n-2)}{2 \cdot 4 \cdot 6 \cdot \cdots \cdot (2n)} + \cdots$$

ALTERNATING SERIES AND ABSOLUTE CONVERGENCE

The tests for convergence that we have discussed thus far can be applied only to positive-term series. We now consider infinite series that contain both positive and negative terms. One of the simplest, and most useful, series of this type is an **alternating series**, in which the terms are alternately positive and negative. It is customary to express an alternating series in one of the forms

$$a_1 - a_2 + a_3 - a_4 + \cdots + (-1)^{n-1} a_n + \cdots$$

or $\qquad -a_1 + a_2 - a_3 + a_4 - \cdots + (-1)^n a_n + \cdots$

with $a_k > 0$ for every k. The next theorem provides the main test for convergence of these series. For convenience, we consider $\sum_{n=1}^{\infty} (-1)^{n-1} a_n$. A similar proof holds for $\sum_{n=1}^{\infty} (-1)^n a_n$.

Alternating Series Test 8.30

The alternating series

$$\sum_{n=1}^{\infty} (-1)^{n-1} a_n = a_1 - a_2 + a_3 - a_4 + \cdots + (-1)^{n-1} a_n + \cdots$$

is convergent if the following two conditions are satisfied:

(i) $a_k \geq a_{k+1} > 0$ for every k

(ii) $\lim_{n \to \infty} a_n = 0$

PROOF   By condition (i), we may write

$$a_1 \geq a_2 \geq a_3 \geq a_4 \geq a_5 \geq \cdots \geq a_k \geq a_{k+1} \geq \cdots.$$

Let us consider the partial sums

$$S_2, S_4, S_6, \ldots, S_{2n}, \ldots,$$

which contain an even number of terms of the series. Since

$$S_{2n} = (a_1 - a_2) + (a_3 - a_4) + \cdots + (a_{2n-1} - a_{2n})$$

and $a_k - a_{k+1} \geq 0$ for every k, we see that

$$0 \leq S_2 \leq S_4 \leq \cdots \leq S_{2n} \leq \cdots;$$

that is, $\{S_{2n}\}$ is a monotonic sequence. This fact is also evident from Figure 8.11, where we have used a coordinate line l to represent the following four partial sums of the series:

$$S_1 = a_1, \quad S_2 = a_1 - a_2, \quad S_3 = a_1 - a_2 + a_3, \quad S_4 = a_1 - a_2 + a_3 - a_4$$

You may find it instructive to locate the points on l that correspond to S_5 and S_6.

Figure 8.11

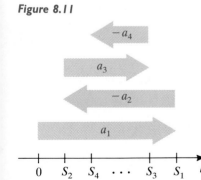

Referring to Figure 8.11, we see that $S_{2n} \leq a_1$ for every positive integer n. This may also be proved algebraically by observing that

$$S_{2n} = a_1 - (a_2 - a_3) - (a_4 - a_5) - \cdots - (a_{2n-2} - a_{2n-1}) - a_{2n} \leq a_1.$$

Thus, $\{S_{2n}\}$ is a *bounded* monotonic sequence. As in the proof of Theorem (8.9),

$$\lim_{n \to \infty} S_{2n} = S \leq a_1$$

for some number S.

If we next consider a partial sum S_{2n+1} having an *odd* number of terms of the series, then $S_{2n+1} = S_{2n} + a_{2n+1}$ and, since $\lim_{n \to \infty} a_{2n+1} = 0$,

$$\lim_{n \to \infty} S_{2n+1} = \lim_{n \to \infty} S_{2n} = S.$$

Because both the sequence of even partial sums and the sequence of odd partial sums have the same limit S, it follows that

$$\lim_{n \to \infty} S_n = S \leq a_1.$$

That is, the series converges.   ■

EXAMPLE ■ I   Determine whether the alternating series converges or diverges.

(a) $\displaystyle\sum_{n=1}^{\infty} (-1)^{n-1} \frac{2n}{4n^2 - 3}$ **(b)** $\displaystyle\sum_{n=1}^{\infty} (-1)^{n-1} \frac{2n}{4n - 3}$

SOLUTION

(a) Let $$a_n = f(n) = \frac{2n}{4n^2 - 3}.$$

To apply the alternating series test (8.30), we must show that

(i) $a_k \geq a_{k+1}$ for every positive integer k

(ii) $\lim_{n \to \infty} a_n = 0$

There are several ways to prove (i). One method is to show that $f(x) = 2x/(4x^2 - 3)$ is decreasing for $x \geq 1$. By the quotient rule,

$$f'(x) = \frac{(4x^2 - 3)(2) - (2x)(8x)}{(4x^2 - 3)^2} = \frac{-8x^2 - 6}{(4x^2 - 3)^2} < 0.$$

By Theorem (3.15), $f(x)$ is decreasing and, therefore, $f(k) \geq f(k + 1)$; that is, $a_k \geq a_{k+1}$ for every positive integer k.

We can also prove (i) directly, by proving that $a_k - a_{k+1} \geq 0$. Thus, if $a_n = 2n/(4n^2 - 3)$, then for every positive integer k,

$$a_k - a_{k+1} = \frac{2k}{4k^2 - 3} - \frac{2(k+1)}{4(k+1)^2 - 3} = \frac{8k^2 + 8k + 6}{(4k^2 - 3)(4k^2 + 8k + 1)} \geq 0.$$

Still another technique for proving that $a_k \geq a_{k+1}$ is to show that $a_{k+1}/a_k \leq 1$.

To prove (ii), we see that

$$\lim_{n \to \infty} a_n = \lim_{n \to \infty} \frac{2n}{4n^2 - 3} = 0.$$

Thus, the alternating series converges.

(b) We can show that $a_k \geq a_{k+1}$ for every k; however,

$$\lim_{n \to \infty} a_n = \lim_{n \to \infty} \frac{2n}{4n - 3} = \frac{1}{2} \neq 0,$$

and hence the series diverges, by the nth-term test (8.17)(i).

The alternating series test (8.30) may be used if condition (i) holds for $k > m$ for some positive integer m, because this corresponds to deleting the first m terms of the series.

If a series converges, then the nth partial sum S_n can be used to approximate the sum S of the series. In many cases, it is difficult to determine the accuracy of the approximation. However, for an *alternating series*, the next theorem provides a simple way of estimating the error that is involved.

Theorem 8.31

Let $\sum_{n=1}^{\infty} (-1)^{n-1} a_n$ be an alternating series that satisfies conditions (i) and (ii) of the alternating series test. If S is the sum of the series and S_n is a partial sum, then

$$|S - S_n| \leq a_{n+1};$$

that is, the error involved in approximating S by S_n is less than or equal to a_{n+1}.

PROOF The series obtained by deleting the first n terms of $\sum (-1)^{n-1} a_n$, namely,

$$(-1)^n a_{n+1} + (-1)^{n+1} a_{n+2} + (-1)^{n+2} a_{n+3} + \cdots,$$

also satisfies the conditions of (8.30) and therefore has a sum R_n. Thus,

$$S - S_n = R_n = (-1)^n (a_{n+1} - a_{n+2} + a_{n+3} - \cdots)$$

and

$$|R_n| = a_{n+1} - a_{n+2} + a_{n+3} - \cdots.$$

Employing the same argument used in the proof of the alternating series test, we see that $|R_n| \leq a_{n+1}$. Consequently,

$$E = |S - S_n| = |R_n| \leq a_{n+1},$$

which is what we wished to prove. ■

In the next example, we use Theorem (8.31) to approximate the sum of an alternating series. In order to discuss the accuracy of an approximation, we must first agree on what is meant by one-decimal-place accuracy, two-decimal-place accuracy, and so on. Let us adopt the following convention. If E is the error in an approximation, then the approximation will be considered accurate to k decimal places if $|E| < 0.5 \times 10^{-k}$. For example, we have

$$1\text{-decimal-place accuracy if } |E| < 0.5 \times 10^{-1} = 0.05$$

$$2\text{-decimal-place accuracy if } |E| < 0.5 \times 10^{-2} = 0.005$$

$$3\text{-decimal-place accuracy if } |E| < 0.5 \times 10^{-3} = 0.0005.$$

EXAMPLE■2   Prove that the series

$$1 - \frac{1}{3!} + \frac{1}{5!} - \cdots + (-1)^{n-1}\frac{1}{(2n-1)!} + \cdots$$

is convergent, and approximate its sum S to five decimal places.

SOLUTION   The nth term $a_n = 1/(2n-1)!$ has limit 0 as $n \to \infty$, and $a_k > a_{k+1}$ for every positive integer k. Hence the series converges, by the alternating series test. If we use S_n to approximate S, then, by Theorem (8.31), the error involved is less than or equal to $a_{n+1} = 1/(2n+1)!$. Calculating several values of a_{n+1}, we find that for $n = 4$,

$$a_5 = \frac{1}{9!} \approx 0.0000028 < 0.000005.$$

Hence, the partial sum S_4 approximates S to five decimal places. Since

$$S_4 = 1 - \frac{1}{3!} + \frac{1}{5!} - \frac{1}{7!}$$

$$= 1 - \tfrac{1}{6} + \tfrac{1}{120} - \tfrac{1}{5040} \approx 0.841468,$$

we have $S \approx 0.84147$.

It will follow from (8.48)(a) that the sum of the series is $\sin 1$, and hence $\sin 1 \approx 0.84147$.

The following concept is useful in investigating a series that contains both positive and negative terms but is not alternating. It allows us to use tests for positive-term series to establish convergence for other types of series (see Theorem 8.34).

Definition 8.32

A series $\sum a_n$ is **absolutely convergent** if the series

$$\sum |a_n| = |a_1| + |a_2| + \cdots + |a_n| + \cdots$$

is convergent.

Note that if $\sum a_n$ is a positive-term series, then $|a_n| = a_n$, and in this case, absolute convergence is the same as convergence.

EXAMPLE■3   Prove that the following alternating series is absolutely convergent:

$$1 - \frac{1}{2^2} + \frac{1}{3^2} - \frac{1}{4^2} + \cdots + (-1)^n\frac{1}{n^2} + \cdots$$

SOLUTION   Taking the absolute value of each term gives us

$$1 + \frac{1}{2^2} + \frac{1}{3^2} + \frac{1}{4^2} + \cdots + \frac{1}{n^2} + \cdots,$$

which is a convergent p-series. Hence, by Definition (8.32), the alternating series is absolutely convergent.

EXAMPLE ■ 4   The **alternating harmonic series** is

$$\sum_{n=1}^{\infty} (-1)^{n-1} \frac{1}{n} = 1 - \frac{1}{2} + \frac{1}{3} - \frac{1}{4} + \cdots + (-1)^{n-1} \frac{1}{n} + \cdots.$$

Show that this series is

(a) convergent   **(b)** not absolutely convergent

SOLUTION

(a) Conditions (i) and (ii) of the alternating series test (8.30) are satisfied, because

$$\frac{1}{k} > \frac{1}{k+1} \quad \text{for every } k \quad \text{and} \quad \lim_{n \to \infty} \frac{1}{n} = 0.$$

Hence, the alternating harmonic series is convergent.

(b) To examine the series for absolute convergence, we apply Definition (8.32) and consider

$$\sum_{n=1}^{\infty} \left| (-1)^{n-1} \frac{1}{n} \right| = 1 + \frac{1}{2} + \frac{1}{3} + \frac{1}{4} + \cdots + \frac{1}{n} + \cdots.$$

This series is the divergent harmonic series (see Example 3 of Section 8.2). Hence, by Definition (8.32), the alternating harmonic series is not absolutely convergent.

Series that are convergent but not absolutely convergent, such as the alternating harmonic series in Example 4, are given a special name, as indicated in the next definition.

Definition 8.33

A series $\sum a_n$ is **conditionally convergent** if $\sum a_n$ is convergent and $\sum |a_n|$ is divergent.

The following theorem tells us that absolute convergence implies convergence.

Theorem 8.34

If a series $\sum a_n$ is absolutely convergent, then $\sum a_n$ is convergent.

PROOF   If we let $b_n = a_n + |a_n|$ and we make use of the property that $-|a_n| \le a_n \le |a_n|$, then

$$0 \le a_n + |a_n| \le 2|a_n|, \quad \text{or} \quad 0 \le b_n \le 2|a_n|.$$

If $\sum a_n$ is absolutely convergent, then $\sum |a_n|$ is convergent and hence, by Theorem (8.20)(ii), $\sum 2|a_n|$ is convergent. If we apply the basic comparison test (8.26), it follows that $\sum b_n$ is convergent. By (8.20)(iii), $\sum (b_n - |a_n|)$ is convergent. Since $b_n - |a_n| = a_n$, the proof is complete.   ∎

EXAMPLE = 5   Let $\sum a_n$ be the series

$$\frac{1}{2} + \frac{1}{2^2} - \frac{1}{2^3} - \frac{1}{2^4} + \frac{1}{2^5} + \frac{1}{2^6} - \frac{1}{2^7} - \frac{1}{2^8} + \cdots,$$

where the signs of the terms vary in pairs as indicated and where $|a_n| = 1/2^n$. Determine whether $\sum a_n$ converges or diverges.

SOLUTION   The series is neither alternating nor geometric nor positive-term, so none of the earlier tests can be applied. Let us consider the series of absolute values:

$$\sum |a_n| = \frac{1}{2} + \frac{1}{2^2} + \frac{1}{2^3} + \frac{1}{2^4} + \cdots + \frac{1}{2^n} + \cdots$$

This series is geometric, with $r = \frac{1}{2}$, and since $\frac{1}{2} < 1$, it is convergent, by Theorem (8.15)(i). Thus the given series is absolutely convergent and hence, by Theorem (8.34), it is convergent.

EXAMPLE = 6   Determine whether the following series is convergent or divergent:

$$\sin 1 + \frac{\sin 2}{2^2} + \frac{\sin 3}{3^2} + \cdots + \frac{\sin n}{n^2} + \cdots$$

SOLUTION   The series contains both positive and negative terms, but it is not an alternating series, because, for example, the first three terms are positive and the next three are negative. The series of absolute values is

$$\sum_{n=1}^{\infty} \left| \frac{\sin n}{n^2} \right| = \sum_{n=1}^{\infty} \frac{|\sin n|}{n^2}.$$

Since

$$\frac{|\sin n|}{n^2} < \frac{1}{n^2},$$

the series of absolute values $\sum |(\sin n)/n^2|$ is dominated by the convergent p-series $\sum (1/n^2)$ and hence is convergent. Thus, the given series is absolutely convergent and therefore is convergent, by Theorem (8.34).

We see from the preceding discussion that an arbitrary series may be classified in exactly *one* of the following ways:

(i) absolutely convergent   **(ii)** conditionally convergent   **(iii)** divergent

Of course, for positive-term series, we need only determine convergence or divergence.

The following form of the ratio test may be used to investigate absolute convergence.

Ratio Test for Absolute Convergence   8.35

Let $\sum a_n$ be a series of nonzero terms, and suppose

$$\lim_{n \to \infty} \left| \frac{a_{n+1}}{a_n} \right| = L.$$

(i) If $L < 1$, the series is absolutely convergent.

(ii) If $L > 1$ or $\lim_{n \to \infty} \left| \frac{a_{n+1}}{a_n} \right| = \infty$, the series is divergent.

(iii) If $L = 1$, apply a different test; the series may be absolutely convergent, conditionally convergent, or divergent.

The proof is similar to that of (8.28). Note that for positive-term series the two ratio tests are identical.

We can also state a root test for absolute convergence. The statement is the same as that of (8.29), except that we replace $\sqrt{a_n}$ with $\sqrt{|a_n|}$.

EXAMPLE ■ 7   Determine whether the following series is absolutely convergent, conditionally convergent, or divergent:

$$\sum_{n=1}^{\infty} (-1)^n \frac{n^2 + 4}{2^n}$$

SOLUTION   Using the ratio test (8.35), we obtain

$$\lim_{n \to \infty} \left| \frac{a_{n+1}}{a_n} \right| = \lim_{n \to \infty} \left| \frac{(n+1)^2 + 4}{2^{n+1}} \cdot \frac{2^n}{n^2 + 4} \right|$$

$$= \lim_{n \to \infty} \frac{1}{2} \left(\frac{n^2 + 2n + 5}{n^2 + 4} \right) = \frac{1}{2}(1) = \frac{1}{2} < 1.$$

Hence, by (8.35)(i), the series is absolutely convergent.

It can be proved that if a series $\sum a_n$ is absolutely convergent and if the terms are rearranged in any manner, then the resulting series converges and has the same sum as the given series, which is not true for conditionally convergent series. If $\sum a_n$ is conditionally convergent, then by suitably rearranging terms, we can obtain either a divergent series or a series that converges and has any desired sum S. (See an advanced calculus text for details.)

We now have a variety of tests that can be used to investigate a series for convergence or divergence. Considerable skill is needed to determine which test is best suited for a particular series. This skill can be obtained by working many exercises involving different types of series. The following summary may be helpful in deciding which test to apply; however, some series cannot be investigated by any of these tests. In those cases, it may be necessary to use results from advanced mathematics courses.

Summary of Convergence and Divergence Tests for Series

Test	Series	Convergence or divergence	Comments
nth-term	$\sum a_n$	Diverges if $\lim_{n \to \infty} a_n \neq 0$	Inconclusive if $\lim_{n \to \infty} a_n = 0$
Geometric series	$\displaystyle\sum_{n=1}^{\infty} ar^{n-1}$	(i) Converges with sum $S = \dfrac{a}{1-r}$ if $\lvert r \rvert < 1$ (ii) Diverges if $\lvert r \rvert \geq 1$	Useful for comparison tests if the nth term a_n of a series is *similar* to ar^{n-1}
p-series	$\displaystyle\sum_{n=1}^{\infty} \dfrac{1}{n^p}$	(i) Converges if $p > 1$ (ii) Diverges if $p \leq 1$	Useful for comparison tests if the nth term a_n of a series is *similar* to $1/n^p$
Integral	$\displaystyle\sum_{n=1}^{\infty} a_n$ $a_n = f(n)$	(i) Converges if $\displaystyle\int_1^{\infty} f(x)\,dx$ converges (ii) Diverges if $\displaystyle\int_1^{\infty} f(x)\,dx$ diverges	The function f obtained from $a_n = f(n)$ must be continuous, positive, decreasing, and readily integrable.
Comparison	$\sum a_n,\ \sum b_n$ $a_n > 0,\ b_n > 0$	(i) If $\sum b_n$ converges and $a_n \leq b_n$ for every n, then $\sum a_n$ converges. (ii) If $\sum b_n$ diverges and $a_n \geq b_n$ for every n, then $\sum a_n$ diverges. (iii) If $\lim_{n \to \infty}(a_n/b_n) = c$ for some positive real number c, then both series converge or both diverge.	The comparison series $\sum b_n$ is often a geometric series or a p-series. To find b_n in (iii), consider only the terms of a_n that have the greatest effect on the magnitude.
Ratio	$\sum a_n$	If $\displaystyle\lim_{n \to \infty} \left\lvert \dfrac{a_{n+1}}{a_n} \right\rvert = L$ (or ∞), the series (i) converges (absolutely) if $L < 1$ (ii) diverges if $L > 1$ (or ∞)	Inconclusive if $L = 1$ Useful if a_n involves factorials or nth powers If $a_n > 0$ for every n, the absolute value sign may be disregarded.
Root	$\sum a_n$	If $\lim_{n \to \infty} \sqrt[n]{\lvert a_n \rvert} = L$ (or ∞), the series (i) converges (absolutely) if $L < 1$ (ii) diverges if $L > 1$ (or ∞)	Inconclusive if $L = 1$ Useful if a_n involves nth powers If $a_n > 0$ for every n, the absolute value sign may be disregarded.
Alternating series	$\sum (-1)^n a_n$ $a_n > 0$	Converges if $a_k \geq a_{k+1}$ for every k and $\lim_{n \to \infty} a_n = 0$	Applicable only to an alternating series
$\sum \lvert a_n \rvert$	$\sum a_n$	If $\sum \lvert a_n \rvert$ converges, then $\sum a_n$ converges.	Useful for series that contain both positive and negative terms

EXERCISES 8.5

Exer. 1–4: Determine whether the series (a) satisfies conditions (i) and (ii) of the alternating series test (8.30) and (b) converges or diverges.

1. $\displaystyle\sum_{n=1}^{\infty}(-1)^{n-1}\frac{1}{n^2+7}$

2. $\displaystyle\sum_{n=1}^{\infty}(-1)^{n-1}n5^{-n}$

3. $\displaystyle\sum_{n=1}^{\infty}(-1)^n(1+e^{-n})$

4. $\displaystyle\sum_{n=1}^{\infty}(-1)^n\frac{e^{2n}+1}{e^{2n}-1}$

Exer. 5–32: Determine whether the series is absolutely convergent, conditionally convergent, or divergent.

5. $\displaystyle\sum_{n=1}^{\infty}(-1)^{n-1}\frac{1}{\sqrt{2n+1}}$

6. $\displaystyle\sum_{n=1}^{\infty}(-1)^{n-1}\frac{1}{n^{2/3}}$

7. $\displaystyle\sum_{n=1}^{\infty}(-1)^{n+1}\frac{1}{\ln(n+1)}$

8. $\displaystyle\sum_{n=1}^{\infty}(-1)^{n+1}\frac{n}{n^2+4}$

9. $\displaystyle\sum_{n=2}^{\infty}(-1)^n\frac{n}{\ln n}$

10. $\displaystyle\sum_{n=1}^{\infty}(-1)^n\frac{\ln n}{n}$

11. $\displaystyle\sum_{n=1}^{\infty}(-1)^n\frac{5}{n^3+1}$

12. $\displaystyle\sum_{n=1}^{\infty}(-1)^n e^{-n}$

13. $\displaystyle\sum_{n=1}^{\infty}\frac{(-10)^n}{n!}$

14. $\displaystyle\sum_{n=1}^{\infty}\frac{n!}{(-5)^n}$

15. $\displaystyle\sum_{n=1}^{\infty}(-1)^n\frac{n^2+3}{(2n-5)^2}$

16. $\displaystyle\sum_{n=1}^{\infty}\frac{\sin\sqrt{n}}{\sqrt{n^3+4}}$

17. $\displaystyle\sum_{n=1}^{\infty}(-1)^{n-1}\frac{\sqrt[3]{n}}{n+1}$

18. $\displaystyle\sum_{n=1}^{\infty}(-1)^n\frac{(n+1)^2}{n^5+1}$

19. $\displaystyle\sum_{n=1}^{\infty}\frac{\cos\frac{1}{6}\pi n}{n^2}$

20. $\displaystyle\sum_{n=1}^{\infty}(-1)^n\frac{\ln n}{(1.5)^n}$

21. $\displaystyle\sum_{n=1}^{\infty}(-1)^n n\sin\frac{1}{n}$

22. $\displaystyle\sum_{n=1}^{\infty}(-1)^n\frac{\arctan n}{n^2}$

23. $\displaystyle\sum_{n=2}^{\infty}(-1)^n\frac{1}{n\sqrt{\ln n}}$

24. $\displaystyle\sum_{n=1}^{\infty}(-1)^n\frac{2^{1/n}}{n!}$

25. $\displaystyle\sum_{n=1}^{\infty}\frac{n^n}{(-5)^n}$

26. $\displaystyle\sum_{n=1}^{\infty}\frac{(n^2+1)^n}{(-n)^n}$

27. $\displaystyle\sum_{n=1}^{\infty}(-1)^n\frac{1+4^n}{1+3^n}$

28. $\displaystyle\sum_{n=1}^{\infty}(-1)^n\frac{n^4}{e^n}$

29. $\displaystyle\sum_{n=1}^{\infty}(-1)^n\frac{\cos\pi n}{n}$

30. $\displaystyle\sum_{n=1}^{\infty}\frac{1}{n}\sin\frac{(2n-1)\pi}{2}$

31. $\displaystyle\sum_{n=1}^{\infty}(-1)^n\frac{1}{(n-4)^2+5}$

32. $\displaystyle\sum_{n=1}^{\infty}(-1)^n\frac{\ln n}{\sqrt[3]{n}}$

[c] Exer. 33–38: Approximate the sum of each series to three decimal places.

33. $\displaystyle\sum_{n=0}^{\infty}(-1)^n\frac{1}{n!}$

34. $\displaystyle\sum_{n=0}^{\infty}(-1)^{n+1}\frac{1}{(2n)!}$

35. $\displaystyle\sum_{n=1}^{\infty}(-1)^{n-1}\frac{1}{n^3}$

36. $\displaystyle\sum_{n=1}^{\infty}(-1)^{n-1}\frac{1}{n^5}$

37. $\displaystyle\sum_{n=1}^{\infty}(-1)^{n-1}\frac{n+1}{5^n}$

38. $\displaystyle\sum_{n=1}^{\infty}(-1)^n\frac{1}{n}\left(\frac{1}{2}\right)^n$

[c] Exer. 39–42: Use Theorem (8.31) to find a positive integer n such that S_n approximates the sum of the series to four decimal places.

39. $\displaystyle\sum_{n=1}^{\infty}(-1)^n\frac{1}{n^2}$

40. $\displaystyle\sum_{n=1}^{\infty}(-1)^n\frac{1}{\sqrt{n}}$

41. $\displaystyle\sum_{n=1}^{\infty}(-1)^n\frac{1}{n^n}$

42. $\displaystyle\sum_{n=1}^{\infty}(-1)^n\frac{1}{n^3+1}$

Exer. 43–44: Show that the alternating series converges for every positive integer k.

43. $\displaystyle\sum_{n=1}^{\infty}(-1)^n\frac{(\ln n)^k}{n}$

44. $\displaystyle\sum_{n=1}^{\infty}(-1)^n\frac{1}{\sqrt[k]{n}}$

45. If $\sum a_n$ and $\sum b_n$ are both convergent series, is $\sum a_n b_n$ convergent? Explain.

46. If $\sum a_n$ and $\sum b_n$ are both divergent series, is $\sum a_n b_n$ divergent? Explain.

8.6 POWER SERIES

The most important reason for developing the theory in the previous sections is to represent functions as *power series*—that is, as series whose terms contain powers of a variable x. To illustrate, if we use the formula

$S = a/(1 - r)$ for the sum of a geometric series (see Theorem (8.15)(i)), we obtain

$$1 + x + x^2 + \cdots + x^n + \cdots = \frac{1}{1 - x},$$

provided $|x| < 1$. If we let $f(x) = 1/(1 - x)$ with $|x| < 1$, then

$$f(x) = 1 + x + x^2 + \cdots + x^n + \cdots.$$

We say that $f(x)$ is *represented* by this power series. To find a function value $f(c)$, we can let $x = c$ and find the sum of a series. For example,

$$f\left(\frac{1}{2}\right) = 1 + \frac{1}{2} + \left(\frac{1}{2}\right)^2 + \cdots + \left(\frac{1}{2}\right)^n + \cdots = \frac{1}{1 - \frac{1}{2}} = 2.$$

Later we shall apply other techniques to express many different types of functions as series.

The following definition may be considered as a generalization of the notion of a polynomial to an infinite series.

Definition 8.36

Let x be a variable. A **power series in x** is a series of the form

$$\sum_{n=0}^{\infty} a_n x^n = a_0 + a_1 x + a_2 x^2 + \cdots + a_n x^n + \cdots,$$

where each a_k is a real number.

If a number c is substituted for x in the power series $\sum_{n=0}^{\infty} a_n x^n$, we obtain

$$\sum_{n=0}^{\infty} a_n c^n = a_0 + a_1 c + a_2 c^2 + \cdots + a_n c^n + \cdots.$$

This series of constant terms may then be tested for convergence or divergence. To simplify the nth term, we assume that $x^0 = 1$, even if $x = 0$. The main objective of this section is to determine all values of x for which a power series converges. Every power series in x converges if $x = 0$, since

$$a_0 + a_1(0) + a_2(0)^2 + \cdots + a_n(0)^n + \cdots = a_0.$$

To find other values of x that produce convergent series, we often use the ratio test for absolute convergence (8.35), as illustrated in the following examples.

EXAMPLE ■ I   Find all values of x for which the following power series is absolutely convergent:

$$1 + \frac{1}{5}x + \frac{2}{5^2}x^2 + \cdots + \frac{n}{5^n}x^n + \cdots$$

SOLUTION   If we let

$$u_n = \frac{n}{5^n}x^n = \frac{nx^n}{5^n},$$

then

$$\lim_{n \to \infty} \left| \frac{u_{n+1}}{u_n} \right| = \lim_{n \to \infty} \left| \frac{(n+1)x^{n+1}}{5^{n+1}} \cdot \frac{5^n}{nx^n} \right|$$

$$= \lim_{n \to \infty} \left| \frac{(n+1)x}{5n} \right| = \lim_{n \to \infty} \left(\frac{n+1}{5n} \right) |x| = \frac{1}{5} |x|.$$

By the ratio test (8.35), with $L = \frac{1}{5}|x|$, the series is absolutely convergent if the following equivalent inequalities are true:

$$\tfrac{1}{5}|x| < 1, \qquad |x| < 5, \qquad -5 < x < 5$$

The series diverges if $\frac{1}{5}|x| > 1$—that is, if $x > 5$ or $x < -5$.

If $\frac{1}{5}|x| = 1$, the ratio test is inconclusive, and hence the numbers 5 and -5 require special consideration. Substituting 5 for x in the power series, we obtain

$$1 + 1 + 2 + 3 + \cdots + n + \cdots,$$

which is divergent, by the nth-term test (8.17), because $\lim_{n \to \infty} a_n \neq 0$. If we let $x = -5$, we obtain

$$1 - 1 + 2 - 3 + \cdots + (-1)^n n + \cdots,$$

which is also divergent, by the nth-term test. Consequently, the power series is absolutely convergent for every x in the open interval $(-5, 5)$ and diverges elsewhere.

EXAMPLE ■ 2 Find all values of x for which the following power series is absolutely convergent:

$$1 + \frac{1}{1!}x + \frac{1}{2!}x^2 + \cdots + \frac{1}{n!}x^n + \cdots$$

SOLUTION We shall employ the same technique as was used in Example 1. If we let

$$u_n = \frac{1}{n!}x^n = \frac{x^n}{n!},$$

then

$$\lim_{n \to \infty} \left| \frac{u_{n+1}}{u_n} \right| = \lim_{n \to \infty} \left| \frac{x^{n+1}}{(n+1)!} \cdot \frac{n!}{x^n} \right|$$

$$= \lim_{n \to \infty} \left| \frac{x}{n+1} \right| = \lim_{n \to \infty} \frac{1}{n+1} |x| = 0.$$

The limit 0 is less than 1 for every value of x, and hence, from the ratio test (8.35), the power series is absolutely convergent for *every* real number x.

EXAMPLE ■ 3   Find all values of x for which $\sum n! \, x^n$ is convergent.

SOLUTION   Let $u_n = n! \, x^n$. If $x \neq 0$, then

$$\lim_{n \to \infty} \left| \frac{u_{n+1}}{u_n} \right| = \lim_{n \to \infty} \left| \frac{(n+1)! \, x^{n+1}}{n! \, x^n} \right|$$

$$= \lim_{n \to \infty} |(n+1)x| = \lim_{n \to \infty} (n+1) |x| = \infty$$

and, by the ratio test (8.35), the series diverges. Hence, the power series is convergent only if $x = 0$.

Theorem (8.38) will show that the solutions of the preceding examples are typical in the sense that if a power series converges for nonzero values of x, then either it is absolutely convergent for every real number or it is absolutely convergent throughout some open interval $(-r, r)$ and diverges outside of the closed interval $[-r, r]$. The proof of this fact depends on the next theorem.

Theorem 8.37

> **(i)** If a power series $\sum a_n x^n$ converges for a nonzero number c, then it is absolutely convergent whenever $|x| < |c|$.
>
> **(ii)** If a power series $\sum a_n x^n$ diverges for a nonzero number d, then it diverges whenever $|x| > |d|$.

PROOF   If $\sum a_n c^n$ converges and $c \neq 0$, then, by Theorem (8.16), $\lim_{n \to \infty} a_n c^n = 0$. Using Definition (8.3) with $\epsilon = 1$, we know that there is a positive integer N such that

$$|a_n c^n| < 1 \quad \text{whenever} \quad n \geq N.$$

Consequently,

$$|a_n x^n| = \left| \frac{a_n c^n x^n}{c^n} \right| = |a_n c^n| \left| \frac{x}{c} \right|^n < \left| \frac{x}{c} \right|^n,$$

provided $n \geq N$. If $|x| < |c|$, then $|x/c| < 1$ and $\sum |x/c|^n$ is a convergent geometric series. Hence, by the basic comparison test (8.26), the series obtained by deleting the first N terms of $\sum |a_n x^n|$ is convergent. It follows that the series $\sum |a_n x^n|$ is also convergent, which proves (i).

To prove (ii), suppose the series diverges for $x = d \neq 0$. If the series converges for some number c_1 with $|c_1| > |d|$, then, by (i), it converges whenever $|x| < |c_1|$. In particular, the series converges for $x = d$, contrary to our supposition. Hence, the series diverges whenever $|x| > |d|$. ■

We may now prove the following.

Theorem 8.38

If $\sum a_n x^n$ is a power series, then exactly one of the following is true:

(i) The series converges only if $x = 0$.

(ii) The series is absolutely convergent for every x.

(iii) There is a number $r > 0$ such that the series is absolutely convergent if x is in the open interval $(-r, r)$ and divergent if $x < -r$ or $x > r$.

PROOF   If neither (i) nor (ii) is true, then there exist nonzero numbers c and d such that the series converges if $x = c$ and diverges if $x = d$. Let S denote the set of all real numbers for which the series is absolutely convergent. By Theorem (8.37), the series diverges if $|x| > |d|$, and hence every number in S is less than $|d|$. By the completeness property (8.10), S has a least upper bound r. It follows that the series is absolutely convergent if $|x| < r$ and diverges if $|x| > r$. ■

Figure 8.12
$\sum a_n x^n$ with radius of convergence r

Case (iii) of Theorem (8.38) is illustrated graphically in Figure 8.12. The number r is called the **radius of convergence** of the series. Either convergence or divergence may occur at $-r$ or r, depending on the nature of the series.

The totality of numbers for which a power series converges is called its **interval of convergence**. If the radius of convergence r is positive, then the interval of convergence is one of the following (see Figure 8.13):

$$(-r, r), \quad (-r, r], \quad [-r, r), \quad [-r, r]$$

To determine which of these intervals occurs, we must conduct separate investigations for the numbers $x = r$ and $x = -r$.

In (i) or (ii) of Theorem (8.38), the radius of convergence is denoted by 0 or ∞, respectively. In Example 1, the interval of the convergence is $(-5, 5)$ and the radius of convergence is 5. In Example 2, the interval of convergence is $(-\infty, \infty)$ and we write $r = \infty$. In Example 3, $r = 0$. The next example illustrates the case of a half-open interval of convergence.

Figure 8.13
Intervals of convergence

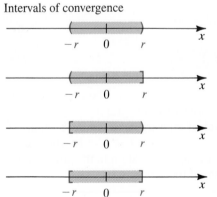

EXAMPLE ▪4

(a) Find the interval of convergence of the power series

$$\sum_{n=1}^{\infty} \frac{1}{\sqrt{n}} x^n.$$

(b) Use a graphing utility to plot the polynomials

$$p_k(x) = \sum_{n=1}^{k} \frac{1}{\sqrt{n}} x^n$$

for $k = 3, 4, 5,$ and 6.

SOLUTION

(a) Note that the coefficient of x^0 is 0, and the summation begins with $n = 1$. We let $u_n = x^n/\sqrt{n}$ and consider

$$\lim_{n \to \infty} \left| \frac{u_{n+1}}{u_n} \right| = \lim_{n \to \infty} \left| \frac{x^{n+1}}{\sqrt{n+1}} \cdot \frac{\sqrt{n}}{x^n} \right| = \lim_{n \to \infty} \left| \frac{\sqrt{n}}{\sqrt{n+1}} x \right|$$

$$= \lim_{n \to \infty} \sqrt{\frac{n}{n+1}} \, |x| = (1)|x| = |x|.$$

It follows from the ratio test (8.35) that the power series is absolutely convergent if $|x| < 1$—that is, if x is in the open interval $(-1, 1)$. The series diverges if $x > 1$ or $x < -1$. The numbers 1 and -1 must be investigated separately by substitution in the power series.

If we substitute $x = 1$, we obtain

$$\sum_{n=1}^{\infty} \frac{1}{\sqrt{n}} (1)^n = 1 + \frac{1}{\sqrt{2}} + \frac{1}{\sqrt{3}} + \cdots + \frac{1}{\sqrt{n}} + \cdots,$$

which is a divergent p-series, with $p = \frac{1}{2}$. If we substitute $x = -1$, we obtain

$$\sum_{n=1}^{\infty} \frac{1}{\sqrt{n}} (-1)^n = -1 + \frac{1}{\sqrt{2}} - \frac{1}{\sqrt{3}} + \cdots + \frac{(-1)^n}{\sqrt{n}} + \cdots,$$

which converges, by the alternating series test. Thus, the power series converges if $-1 \le x < 1$.

(b) Using a computer, we plot the polynomials

$$P_k(x) = \sum_{n=1}^{k} \frac{1}{\sqrt{n}} x^n$$

for $k = 3, 4, 5,$ and 6 on the same coordinate axes. These polynomials are defined for all real numbers, and we expect the graphs to approximate the power series for x in the interval of convergence. Figure 8.14 shows the convergence of the graphs on the interval $-1 \le x < 1$, where each successive polynomial provides a better approximation to the power series because it contains an additional term of the series.

Figure 8.14

$-1.5 \le x \le 1.5, \quad -3 \le y \le 16$

We next consider the following more general types of power series.

Definition 8.39

Let c be a real number and x a variable. A **power series in** $x - c$ is a series of the form

$$\sum_{n=0}^{\infty} a_n (x - c)^n$$

$$= a_0 + a_1(x - c) + a_2(x - c)^2 + \cdots + a_n(x - c)^n + \cdots,$$

where each a_k is a real number.

To simplify the nth term in (8.39), we assume that $(x - c)^0 = 1$ even if $x = c$. As in the proof of Theorem (8.38), but with x replaced by $x - c$, exactly one of the following cases is true:

(i) The series converges only if $x - c = 0$—that is, if $x = c$.

(ii) The series is absolutely convergent for every x.

(iii) There is a number $r > 0$ such that the series is absolutely convergent if x is in the open interval $(c - r, c + r)$ and divergent if $x < c - r$ or $x > c + r$.

Figure 8.15

$\sum a_n (x - c)^n$ with radius of convergence r

Divergent   Absolutely convergent   Divergent

$c - r \quad c \quad c + r \qquad x$

Case (iii) is illustrated in Figure 8.15. The endpoints $c - r$ and $c + r$ of the interval must be investigated separately. As before, the totality of numbers for which the series converges is called the *interval of convergence*, and r is the *radius of convergence*.

EXAMPLE ▪ 5   Find the interval of convergence of the series

$$1 - \frac{1}{2}(x - 3) + \frac{1}{3}(x - 3)^2 + \cdots + (-1)^n \frac{1}{n+1}(x - 3)^n + \cdots.$$

SOLUTION   If we let

$$u_n = (-1)^n \frac{(x - 3)^n}{n + 1},$$

then

$$\lim_{n \to \infty} \left| \frac{u_{n+1}}{u_n} \right| = \lim_{n \to \infty} \left| \frac{(x - 3)^{n+1}}{n + 2} \cdot \frac{n + 1}{(x - 3)^n} \right|$$

$$= \lim_{n \to \infty} \left| \frac{n + 1}{n + 2}(x - 3) \right|$$

$$= \lim_{n \to \infty} \left(\frac{n + 1}{n + 2} \right) |x - 3|$$

$$= (1)|x - 3| = |x - 3|.$$

By the ratio test (8.35), the series is absolutely convergent if $|x - 3| < 1$; that is, if,

$$-1 < x - 3 < 1, \quad \text{or} \quad 2 < x < 4.$$

Thus, the series is absolutely convergent for every x in the open interval $(2, 4)$. The series diverges if $x < 2$ or $x > 4$. The numbers 2 and 4 each require separate investigation.

If we substitute $x = 4$ in the series, we obtain

$$1 - \frac{1}{2} + \frac{1}{3} - \cdots + (-1)^n \frac{1}{n+1} + \cdots,$$

Figure 8.16

Interval of convergence of

$$\sum (-1)^n \frac{1}{n+1}(x - 3)^n$$

$$0 \quad 1 \quad 2 \quad 3 \quad 4 \quad 5 \qquad x$$

which converges, by the alternating series test (8.30). Substituting $x = 2$ gives us

$$1 + \frac{1}{2} + \frac{1}{3} + \cdots + \frac{1}{n+1} + \cdots,$$

which is the divergent harmonic series. Hence, the interval of convergence is $(2, 4]$, as illustrated in Figure 8.16.

EXERCISES 8.6

Exer. 1–30: Find the interval of convergence of the power series.

1 $\displaystyle\sum_{n=0}^{\infty} \frac{1}{n+4} x^n$

2 $\displaystyle\sum_{n=0}^{\infty} \frac{1}{n^2+4} x^n$

3 $\displaystyle\sum_{n=0}^{\infty} \frac{n^2}{2^n} x^n$

4 $\displaystyle\sum_{n=1}^{\infty} \frac{(-3)^n}{n} x^{n+1}$

5 $\displaystyle\sum_{n=1}^{\infty} (-1)^{n-1} \frac{1}{\sqrt{n}} x^n$

6 $\displaystyle\sum_{n=1}^{\infty} \frac{1}{\ln(n+1)} x^n$

7 $\displaystyle\sum_{n=2}^{\infty} \frac{n}{n^2+1} x^n$

8 $\displaystyle\sum_{n=1}^{\infty} \frac{1}{4^n \sqrt{n}} x^n$

9 $\displaystyle\sum_{n=2}^{\infty} \frac{\ln n}{n^3} x^n$

10 $\displaystyle\sum_{n=0}^{\infty} \frac{10^{n+1}}{3^{2n}} x^n$

11 $\displaystyle\sum_{n=0}^{\infty} \frac{n+1}{10^n} (x-4)^n$

12 $\displaystyle\sum_{n=1}^{\infty} \frac{1}{n(n+1)} (x-2)^n$

13 $\displaystyle\sum_{n=0}^{\infty} \frac{n!}{100^n} x^n$

14 $\displaystyle\sum_{n=0}^{\infty} \frac{(3n)!}{(2n)!} x^n$

15 $\displaystyle\sum_{n=0}^{\infty} \frac{1}{(-4)^n} x^{2n+1}$

16 $\displaystyle\sum_{n=1}^{\infty} (-1)^{n-1} \frac{1}{\sqrt[3]{n}\, 3^n} x^n$

17 $\displaystyle\sum_{n=0}^{\infty} \frac{2^n}{(2n)!} x^{2n}$

18 $\displaystyle\sum_{n=0}^{\infty} \frac{10^n}{n!} x^n$

19 $\displaystyle\sum_{n=0}^{\infty} \frac{3^{2n}}{n+1} (x-2)^n$

20 $\displaystyle\sum_{n=1}^{\infty} \frac{1}{n5^n} (x-5)^n$

21 $\displaystyle\sum_{n=0}^{\infty} \frac{n^2}{2^{3n}} (x+4)^n$

22 $\displaystyle\sum_{n=0}^{\infty} \frac{1}{2n+1} (x+3)^n$

23 $\displaystyle\sum_{n=1}^{\infty} (-1)^n \frac{n^n}{n+1} (x-3)^n$

24 $\displaystyle\sum_{n=1}^{\infty} (-1)^n \frac{n!}{n^3} (x+2)^n$

25 $\displaystyle\sum_{n=1}^{\infty} \frac{\ln n}{e^n} (x-e)^n$

26 $\displaystyle\sum_{n=0}^{\infty} \frac{n}{3^{2n-1}} (x-1)^{2n}$

27 $\displaystyle\sum_{n=1}^{\infty} (-1)^n \frac{1}{n6^n} (2x-1)^n$

28 $\displaystyle\sum_{n=0}^{\infty} \frac{1}{\sqrt{3n+4}} (3x+4)^n$

29 $\displaystyle\sum_{n=1}^{\infty} (-1)^n \frac{3^n}{n!} (x-4)^n$

30 $\displaystyle\sum_{n=1}^{\infty} (-1)^n \frac{e^{n+1}}{n^n} (x-1)^n$

Exer. 31–34: (a) Find the radius of convergence of the power series. (b) Graph, on the same coordinate axes, the polynomials

$$p_k(x) = \sum_{n=1}^{k} a_n x^n$$

associated with the power series for $k = 3, 4$, and 5.

31 $\displaystyle\sum_{n=1}^{\infty} (-1)^n \frac{1 \cdot 3 \cdot 5 \cdot \,\cdots\, \cdot (2n-1)}{3 \cdot 6 \cdot 9 \cdot \,\cdots\, \cdot (3n)} x^n$

32 $\displaystyle\sum_{n=1}^{\infty} \frac{2 \cdot 4 \cdot 6 \cdot \,\cdots\, \cdot (2n)}{4 \cdot 7 \cdot 10 \cdot \,\cdots\, \cdot (3n+1)} x^n$

33 $\displaystyle\sum_{n=1}^{\infty} \frac{n^n}{n!} x^n$

34 $\displaystyle\sum_{n=0}^{\infty} \frac{(n+1)!}{10^n} (x-5)^n$

Exer. 35–36: Find the radius of convergence of the power series for positive integers c and d.

35 $\displaystyle\sum_{n=0}^{\infty} \frac{(n+c)!}{n!\,(n+d)!} x^n$

36 $\displaystyle\sum_{n=0}^{\infty} \frac{(cn)!}{(n!)^c} x^n$

37 *Bessel functions* are useful in the analysis of problems that involve oscillations. If α is a positive integer, the Bessel function $J_\alpha(x)$ *of the first kind of order* α is defined by the power series

$$J_\alpha(x) = \sum_{n=0}^{\infty} \frac{(-1)^n}{n!\,(n+\alpha)!} \left(\frac{x}{2}\right)^{2n+\alpha}$$

Show that this power series is convergent for every real number.

38 Refer to Exercise 37. The sixth-degree polynomial

$$1 - \frac{x^2}{4} + \frac{x^4}{64} - \frac{x^6}{2304}$$

is sometimes used to approximate the Bessel function $J_0(x)$ of the first kind of order zero for $0 \le x \le 1$. Show that the error E involved in this approximation is less than 0.00001.

Exer. 39–40: Refer to Exercise 37. For the given α, find the first four terms of the series for $J_\alpha(x)$ and graph J_α on the given interval.

39 $\alpha = 0$; $[0, 2]$

40 $\alpha = 1$; $[0, 4]$

41 If $\lim_{n\to\infty} |a_{n+1}/a_n| = k$ and $k \ne 0$, prove that the radius of convergence of $\sum a_n x^n$ is $1/k$.

42 If $\lim_{n \to \infty} \sqrt[n]{|a_n|} = k$ and $k \neq 0$, prove that the radius of convergence of $\sum a_n x^n$ is $1/k$.

43 If $\sum a_n x^n$ has radius of convergence r, prove that $\sum a_n x^{2n}$ has radius of convergence $\sqrt{r}$.

44 If $\sum a_n$ is absolutely convergent, prove that $\sum a_n x^n$ is absolutely convergent for every x in the interval $[-1, 1]$.

45 If the interval of convergence of $\sum a_n x^n$ is $(-r, r]$, prove that the series is conditionally convergent at r.

46 If $\sum a_n x^n$ is absolutely convergent at one endpoint of its interval of convergence, prove that it is also absolutely convergent at the other endpoint.

8.7 POWER SERIES REPRESENTATIONS OF FUNCTIONS

A power series $\sum a_n x^n$ determines a function f whose domain is the interval of convergence of the series. Specifically, for each x in this interval, we let $f(x)$ equal the sum of the series—that is,

$$f(x) = a_0 + a_1 x + a_2 x^2 + \cdots + a_n x^n + \cdots.$$

If a function f is defined in this way, we say that $\sum a_n x^n$ is a **power series representation for $f(x)$** (or *of* $f(x)$). We also use the phrase **f is represented by the power series.**

Numerical computations using power series provide the basis for the design of calculators and the construction of mathematical tables. In addition to this use, power series representations for functions have far-reaching consequences in advanced mathematics and applications. The proof of Theorem (8.41) will show that e^x may be represented as follows:

$$e^x = 1 + x + \frac{x^2}{2!} + \frac{x^3}{3!} + \cdots + \frac{x^n}{n!} + \cdots$$

We can thus consider e^x as a *series* instead of as the inverse of the natural logarithmic function. As we shall see, algebraic manipulations, differentiation, and integration can be performed by using the series for e^x, instead of previous methods. The same will be true for trigonometric, inverse trigonometric, logarithmic, and hyperbolic functions. In the next example, we consider a power series representation for a simple algebraic function.

EXAMPLE ■ 1 Find a function f that is represented by the power series

$$1 - x + x^2 - x^3 + \cdots + (-1)^n x^n + \cdots.$$

SOLUTION If $|x| < 1$, then, by Theorem (8.15)(i), the given geometric series converges and has the sum

$$S = \frac{a}{1 - r} = \frac{1}{1 - (-x)} = \frac{1}{1 + x}.$$

Hence, we may write

$$\frac{1}{1+x} = 1 - x + x^2 - x^3 + \cdots + (-1)^n x^n + \cdots.$$

This result is a power series representation for $f(x) = 1/(1+x)$ on the interval $(-1, 1)$.

If a function f is represented by a power series in x, then

$$f(x) = \sum_{n=0}^{\infty} a_n x^n$$

for every x in the interval of convergence of the series. Since a polynomial in x is a *finite* sum of terms of the form $a_n x^n$, it may not be surprising that f has properties similar to those for polynomial functions. In particular, in the next theorem (stated without proof), we see that f has a derivative f' whose power series representation can be found by differentiating each term of the series for f. Similarly, definite integrals of $f(x)$ may be obtained by integrating each term of the series $\sum a_n x^n$. In the statement of the theorem, note that for the nth term $a_n x^n$ of the series, we have

$$\frac{d}{dx}(a_n x^n) = na_n x^{n-1} \quad \text{and} \quad \int_0^x a_n t^n \, dt = \left[a_n \frac{t^{n+1}}{n+1} \right]_0^x = a_n \frac{x^{n+1}}{n+1}.$$

Theorem 8.40

Suppose that a power series $\sum a_n x^n$ has a radius of convergence $r > 0$, and let f be defined by

$$f(x) = \sum_{n=0}^{\infty} a_n x^n = a_0 + a_1 x + a_2 x^2 + a_3 x^3 + \cdots + a_n x^n + \cdots$$

for every x in the interval of convergence. If $-r < x < r$, then

(i) $f'(x) = a_1 + 2a_2 x + 3a_3 x^2 + \cdots + na_n x^{n-1} + \cdots$

$$= \sum_{n=1}^{\infty} na_n x^{n-1}$$

(ii) $\int_0^x f(t) \, dt = a_0 x + a_1 \frac{x^2}{2} + a_2 \frac{x^3}{3} + \cdots + a_n \frac{x^{n+1}}{n+1} + \cdots$

$$= \sum_{n=0}^{\infty} \frac{a_n}{n+1} x^{n+1}$$

The series obtained by differentiation in (i) or integration in (ii) of Theorem (8.40) has the same radius of convergence as $\sum a_n x^n$. However, convergence at the endpoints $x = r$ and $x = -r$ of the interval may change. As usual, these numbers require separate investigation.

As a corollary of Theorem (8.40)(i), *a function that is represented by a power series in an interval* $(-r, r)$ *is continuous throughout* $(-r, r)$ (see Theorem (2.12)). Similar results are true for functions represented by power series of the form $\sum a_n (x - c)^n$.

EXAMPLE ▪ 2   Use a power series representation for $1/(1 + x)$ to obtain a power series representation for

$$\frac{1}{(1 + x)^2} \quad \text{if} \quad |x| < 1.$$

SOLUTION   From Example 1,

$$\frac{1}{1 + x} = 1 - x + x^2 - x^3 + \cdots + (-1)^n x^n + \cdots \quad \text{if} \quad |x| < 1.$$

If we differentiate each term of this series, then, by Theorem (8.40)(i),

$$-\frac{1}{(1 + x)^2} = -1 + 2x - 3x^2 - \cdots + (-1)^n n x^{n-1} + \cdots.$$

By Theorem (8.20)(ii), we may multiply both sides by -1, obtaining

$$\frac{1}{(1 + x)^2} = 1 - 2x + 3x^2 + \cdots + (-1)^{n+1} n x^{n-1} + \cdots$$

if $|x| < 1$.

EXAMPLE ▪ 3   Find a power series representation for $\ln(1 + x)$ if $|x| < 1$.

SOLUTION   If $|x| < 1$, then

$$\ln(1 + x) = \int_0^x \frac{1}{1 + t} \, dt$$

$$= \int_0^x [1 - t + t^2 - \cdots + (-1)^n t^n + \cdots] \, dt,$$

where the last equality follows from Example 1. By Theorem (8.40)(ii), we may integrate each term of the series as follows:

$$\ln(1 + x) = \int_0^x 1 \, dt - \int_0^x t \, dt + \int_0^x t^2 \, dt - \cdots + (-1)^n \int_0^x t^n \, dt + \cdots$$

$$= \left[t \right]_0^x - \left[\frac{t^2}{2} \right]_0^x + \left[\frac{t^3}{3} \right]_0^x - \cdots + (-1)^n \left[\frac{t^{n+1}}{n + 1} \right]_0^x + \cdots$$

Hence,

$$\ln(1 + x) = x - \frac{x^2}{2} + \frac{x^3}{3} - \cdots + (-1)^n \frac{x^{n+1}}{n + 1} + \cdots$$

if $|x| < 1$.

EXAMPLE ▪ 4   Use the results of Example 3 to calculate $\ln(1.1)$ to five decimal places.

SOLUTION   In Example 3, we found a series representation for $\ln(1 + x)$ if $|x| < 1$. Substituting 0.1 for x in that series gives us the alternating series

$$\ln(1.1) = 0.1 - \frac{(0.1)^2}{2} + \frac{(0.1)^3}{3} - \frac{(0.1)^4}{4} + \frac{(0.1)^5}{5} - \cdots$$

$$\approx 0.1 - 0.005 + 0.000333 - 0.000025 + 0.000002 - \cdots .$$

If we sum the first four terms on the right and round off to five decimal places, we obtain

$$\ln(1.1) \approx 0.09531.$$

By Theorem (8.31), the error E is less than or equal to the absolute value 0.000002 of the fifth term of the series, and therefore the number 0.09531 is accurate to five decimal places.

EXAMPLE ▪ 5   Find a power series representation for $\arctan x$.

SOLUTION   We first observe that

$$\arctan x = \int_0^x \frac{1}{1 + t^2}\, dt.$$

Next, we note that if $|t| < 1$, then, by Theorem (8.15)(i) with $a = 1$ and $r = -t^2$,

$$\frac{1}{1 + t^2} = 1 - t^2 + t^4 - \cdots + (-1)^n t^{2n} + \cdots .$$

By Theorem (8.40)(ii), we may integrate each term of the series from 0 to x to obtain

$$\arctan x = x - \frac{x^3}{3} + \frac{x^5}{5} - \cdots + (-1)^n \frac{x^{2n+1}}{2n + 1} + \cdots ,$$

provided $|x| < 1$. It can be proved that this series representation is also valid if $|x| = 1$.

In the next theorem, we find a power series representation for e^x.

Theorem 8.41

If x is any real number,

$$e^x = 1 + x + \frac{x^2}{2!} + \frac{x^3}{3!} + \cdots + \frac{x^n}{n!} + \cdots .$$

P R O O F   We considered the indicated power series in Example 2 of the preceding section and found that it is absolutely convergent for every real number x. If we let f denote the function represented by the series, then

$$f(x) = \sum_{n=0}^{\infty} \frac{x^n}{n!}.$$

Applying Theorem (8.40)(i) gives us

$$f'(x) = \sum_{n=1}^{\infty} \frac{nx^{n-1}}{n!} = \sum_{n=1}^{\infty} \frac{x^{n-1}}{(n-1)!}$$

$$= 1 + x + \frac{x^2}{2!} + \frac{x^3}{3!} + \cdots + \frac{x^n}{n!} + \cdots.$$

That is,

$$f'(x) = f(x) \quad \text{for every } x.$$

If, in Theorem (6.33), we let $y = f(t), t = x$, and $c = 1$, we obtain

$$f(x) = f(0)e^x.$$

However,

$$f(0) = 1 + 0 + \frac{0^2}{2!} + \cdots + \frac{0^n}{n!} + \cdots = 1$$

and hence

$$f(x) = e^x,$$

which is what we wished to prove.   ■

Note that Theorem (8.41) allows us to express the number e as the sum of a convergent positive-term series, namely,

$$e = 1 + 1 + \frac{1}{2!} + \frac{1}{3!} + \cdots + \frac{1}{n!} + \cdots.$$

We can use a power series representation for a function to obtain representations for other related functions by making algebraic substitutions. Thus, by Theorem (8.41), if x is any real number,

$$e^x = 1 + x + \frac{x^2}{2!} + \frac{x^3}{3!} + \cdots + \frac{x^n}{n!} + \cdots.$$

To obtain a power series representation for e^{-x}, we need only substitute $-x$ for x:

$$e^{-x} = 1 + (-x) + \frac{(-x)^2}{2!} + \frac{(-x)^3}{3!} + \cdots + \frac{(-x)^n}{n!} + \cdots,$$

or

$$e^{-x} = 1 - x + \frac{x^2}{2!} - \frac{x^3}{3!} + \cdots + (-1)^n \frac{x^n}{n!} + \cdots$$

By Theorem (8.20)(i), we may add corresponding terms of the series for e^x and e^{-x}, obtaining

$$e^x + e^{-x} = 2 + 2 \cdot \frac{x^2}{2!} + 2 \cdot \frac{x^4}{4!} + \cdots + 2 \cdot \frac{x^{2n}}{(2n)!} + \cdots.$$

(Note that odd powers of x cancel.) We can now find a power series for $\cosh x = \frac{1}{2}(e^x + e^{-x})$ by multiplying each term of the last series by $\frac{1}{2}$ (see Theorem (8.20)(ii)). Thus,

$$\cosh x = 1 + \frac{x^2}{2!} + \frac{x^4}{4!} + \cdots + \frac{x^{2n}}{(2n)!} + \cdots.$$

We could find a power series representation for $\sinh x$ either by using $\frac{1}{2}(e^x - e^{-x})$ or by differentiating each term of the series for $\cosh x$. It is left as an exercise to show that

$$\sinh x = x + \frac{x^3}{3!} + \frac{x^5}{5!} + \cdots + \frac{x^{2n+1}}{(2n+1)!} + \cdots.$$

EXAMPLE 6 Find a power series representation for xe^{-2x}.

SOLUTION First we substitute $-2x$ for x in Theorem (8.41):

$$e^{-2x} = 1 + (-2x) + \frac{(-2x)^2}{2!} + \frac{(-2x)^3}{3!} + \cdots + \frac{(-2x)^n}{n!} + \cdots,$$

or $\quad e^{-2x} = 1 - 2x + (2^2)\frac{x^2}{2!} - (2^3)\frac{x^3}{3!} + \cdots + (-2)^n\frac{x^n}{n!} + \cdots$

Multiplying both sides by x gives us

$$xe^{-2x} = x - 2x^2 + (2^2)\frac{x^3}{2!} - (2^3)\frac{x^4}{3!} + \cdots + (-2)^n\frac{x^{n+1}}{n!} + \cdots,$$

which may be written as

$$xe^{-2x} = \sum_{n=0}^{\infty}(-2)^n\frac{x^{n+1}}{n!}.$$

EXAMPLE 7

(a) If g is the function defined by

$$g(t) = \begin{cases} \dfrac{e^t - 1}{t} & \text{if } t \neq 0 \\[2mm] 1 & \text{if } t = 0 \end{cases}$$

show that g is continuous at 0.

(b) Find a power series representation $\sum_{n=1}^{\infty} a_n x^n$ for the function represented by $\int_0^x g(t)\, dt$.

(c) Plot the graphs of the polynomials $p_k(x) = \sum_{n=1}^{k} a_n x^n$ associated with the power series in part (a) for $k = 3, 4$, and 5.

SOLUTION

(a) To show that g is continuous at 0, we need to show that $\lim_{t \to 0} g(t) = g(0) = 1$. Since $\lim_{t \to 0}(e^t - 1) = \lim_{t \to 0} t = 0$, we have the indeterminate form $0/0$. By l'Hôpital's rule,

$$\lim_{t \to 0} \frac{e^t - 1}{t} = \lim_{t \to 0} \frac{(e^t - 1)'}{(t)'} = \lim_{t \to 0} \frac{e^t}{1} = \frac{1}{1} = 1.$$

(b) From Theorem (8.41), we have

$$e^t - 1 = t + \frac{t^2}{2!} + \frac{t^3}{3!} + \frac{t^4}{4!} + \cdots + \frac{t^n}{n!} + \cdots.$$

Dividing through by t gives

$$\frac{e^t - 1}{t} = 1 + \frac{t}{2!} + \frac{t^2}{3!} + \frac{t^3}{4!} + \cdots + \frac{t^{n-1}}{n!} + \cdots.$$

Since the power series in this equation has value 1 when $t = 0$ and $g(0) = 1$, we have

$$g(t) = 1 + \frac{t}{2!} + \frac{t^2}{3!} + \frac{t^3}{4!} + \cdots + \frac{t^{n-1}}{n!} + \cdots$$

for all t.

Applying Theorem (8.40)(ii) yields

$$\int_0^x g(t)\, dt = \left[t \right]_0^x + \left[\frac{t^2}{2 \cdot 2!} \right]_0^x + \left[\frac{t^3}{3 \cdot 3!} \right]_0^x + \cdots + \left[\frac{t^n}{n \cdot n!} \right]_0^x + \cdots$$

$$= x + \frac{x^2}{2 \cdot 2!} + \frac{x^3}{3 \cdot 3!} + \cdots + \frac{x^n}{n \cdot n!} + \cdots.$$

Thus, the power series representation of $\int_0^x g(t)\, dt$ is $\sum_{n=1}^{\infty} a_n x^n$, where $a_n = 1/(n \cdot n!)$.

(c) The polynomials are

$$p_3(x) = x + \frac{x^2}{2 \cdot 2!} + \frac{x^3}{3 \cdot 3!}$$

$$p_4(x) = x + \frac{x^2}{2 \cdot 2!} + \frac{x^3}{3 \cdot 3!} + \frac{x^4}{4 \cdot 4!}$$

$$p_5(x) = x + \frac{x^2}{2 \cdot 2!} + \frac{x^3}{3 \cdot 3!} + \frac{x^4}{4 \cdot 4!} + \frac{x^5}{5 \cdot 5!}.$$

Using a graphing utility and an x-range restricted to the interval $[-4, 4]$, we plot the graphs of each polynomial on the same coordinate axes to

Figure 8.17

$-4 \leq x \leq 4, -5 \leq y \leq 18$

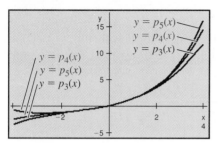

obtain the results shown in Figure 8.17. For $-2 < x < 2$, we see that the graphs converge to approximate the function represented by the polynomials in the power series.

EXAMPLE ■ 8 Approximate $\int_0^{0.1} e^{-x^2} \, dx$.

SOLUTION We cannot use the fundamental theorem of calculus to evaluate the integral, because we do not know of an antiderivative for e^{-x^2}. Although we could use the trapezoidal rule (4.37) or Simpson's rule (4.38), the following method is simpler and, in addition, produces a high degree of accuracy using a sum of only two terms. Letting $x = -t^2$ in Theorem (8.41), we obtain

$$e^{-t^2} = 1 - t^2 + \frac{t^4}{2!} - \cdots + \frac{(-1)^n t^{2n}}{n!} + \cdots$$

for every t. Applying Theorem (8.40)(ii) yields

$$\int_0^{0.1} e^{-x^2} \, dx = \int_0^{0.1} e^{-t^2} \, dt$$

$$= \left[t \right]_0^{0.1} - \left[\frac{t^3}{3} \right]_0^{0.1} + \left[\frac{t^5}{10} \right]_0^{0.1} - \cdots$$

$$= 0.1 - \frac{(0.1)^3}{3} + \frac{(0.1)^5}{10} - \cdots .$$

If we use the first two terms to approximate the sum of this convergent alternating series, then, by Theorem (8.31), the error is less than the third term $(0.1)^5/10 = 0.000001$. Hence,

$$\int_0^{0.1} e^{-x^2} \, dx \approx 0.1 - \frac{0.001}{3}$$

$$\approx 0.09967,$$

which is accurate to five decimal places.

The method used in Example 8 is accurate because the numbers in the interval $[0, 0.1]$ are close to 0. The method would be much less accurate (for the same number of terms of the series) if, for example, the limits of integration were 3 and 3.1. Recall also Example 5 of Section 7.7.

Thus far, the methods we have used to obtain power series representations of functions are *indirect* in the sense that we started with known series and then differentiated or integrated. In the next section, we shall discuss a *direct* method that can be used to find power series representations for a large variety of functions.

EXERCISES 8.7

Exer. 1–4: (a) Find a power series representation for $f(x)$. (b) Use Theorem (8.40) to find power series representations for $f'(x)$ and $\int_0^x f(t)\,dt$.

1 $f(x) = \dfrac{1}{1-3x}$; $|x| < \frac{1}{3}$

2 $f(x) = \dfrac{1}{1+5x}$; $|x| < \frac{1}{5}$

3 $f(x) = \dfrac{1}{2+7x}$; $|x| < \frac{2}{7}$

4 $f(x) = \dfrac{1}{3-2x}$; $|x| < \frac{3}{2}$

Exer. 5–10: Find a power series in x that has the given sum, and specify the radius of convergence. (Hint: Use (8.15), (8.40), or long division, as necessary.)

5 $\dfrac{x^2}{1-x^2}$

6 $\dfrac{x}{1-x^4}$

7 $\dfrac{x}{2-3x}$

8 $\dfrac{x^3}{4-x^3}$

9 $\dfrac{x^2+1}{x-1}$

10 $\dfrac{x^2-3}{x-2}$

11 (a) Prove that

$$\ln(1-x) = -\sum_{n=1}^{\infty} \frac{x^n}{n} \quad \text{if} \quad |x| < 1.$$

(b) Use the series in part (a) to approximate $\ln(1.2)$ to three decimal places, and compare the approximation with that obtained using a calculator.

12 Use the first three terms of the series in Exercise 11(a) to approximate $\ln(0.9)$, and compare the approximation with that obtained using a calculator.

13 Use Example 5 to prove that

$$\frac{\pi}{6} = \frac{1}{\sqrt{3}} \sum_{n=0}^{\infty} (-1)^n \frac{1}{3^n(2n+1)}.$$

14 (a) Use the first five terms of the series in Example 5 to approximate $\pi/4$.

(b) Estimate the error in the approximation obtained in part (a).

Exer. 15–26: Use a power series representation obtained in this section to find a power series representation for $f(x)$.

15 $f(x) = xe^{3x}$

16 $f(x) = x^2 e^{(x^2)}$

17 $f(x) = x^3 e^{-x}$

18 $f(x) = xe^{-3x}$

19 $f(x) = x^2 \ln(1+x^2)$; $|x| < 1$

20 $f(x) = x \ln(1-x)$; $|x| < 1$

21 $f(x) = \arctan\sqrt{x}$; $|x| < 1$

22 $f(x) = x^4 \arctan(x^4)$; $|x| < 1$

23 $f(x) = \sinh(-5x)$

24 $f(x) = \sinh(x^2)$

25 $f(x) = x^2 \cosh(x^3)$

26 $f(x) = \cosh(-2x)$

Exer. 27–32: Use an infinite series to approximate the integral to four decimal places.

27 $\displaystyle\int_0^{1/3} \frac{1}{1+x^6}\,dx$

28 $\displaystyle\int_0^{1/2} \arctan x^2\,dx$

29 $\displaystyle\int_{0.1}^{0.2} \frac{\arctan x}{x}\,dx$

30 $\displaystyle\int_0^{0.2} \frac{x^3}{1+x^5}\,dx$

31 $\displaystyle\int_0^1 e^{-x^2/10}\,dx$

32 $\displaystyle\int_0^{0.5} e^{-x^3}\,dx$

33 Use the power series representation for $(1-x^2)^{-1}$ to find a power series representation for $2x(1-x^2)^{-2}$.

34 Use the method of Example 3 to find a power series representation for $\ln(3+2x)$.

35 Refer to Exercise 37 of Section 8.6. Use Theorem (8.40) to prove the following.

(a) If $J_0(x)$ and $J_1(x)$ are Bessel functions of the first kind of orders 0 and 1, respectively, then

$$\frac{d}{dx}(J_0(x)) = -J_1(x).$$

(b) If $J_2(x)$ and $J_3(x)$ are Bessel functions of the first kind of orders 2 and 3, respectively, then

$$\int x^3 J_2(x)\,dx = x^3 J_3(x) + C.$$

36 Light is absorbed by rods and cones in the retina of the eye. The number of photons absorbed by a photoreceptor during a given flash of light is governed by the *Poisson distribution*. More precisely, the probability p_n that a photoreceptor absorbs exactly n photons is given by the formula $p_n = e^{-\lambda} \lambda^n / n!$ for some $\lambda > 0$.

(a) Show that $\sum_{n=0}^{\infty} p_n = 1$.

(b) Sight usually occurs when two or more photons are absorbed by a photoreceptor. Show that the probability that this will occur is $1 - e^{-\lambda}(\lambda + 1)$.

Exer. 37–38: Find a power series representation for $f(x)$. (If the integrand is denoted by $g(t)$, assume that the value of $g(0)$ is $\lim_{t \to 0} g(t)$.)

37 $f(x) = \int_0^x \dfrac{\ln(1-t)}{t} \, dt$ **38** $f(x) = \int_0^x \dfrac{\sin t}{t} \, dt$

c Exer. 39–42: (a) Find a power series to represent the function. (b) Plot the graphs of the polynomials

$p_k(x) = \sum_{n=1}^k a_n x^n$ associated with the power series in part (a) for $k = 3$, 4, and 5.

39 $\int_0^x \dfrac{1}{1+t^4} \, dt$

40 $\int_0^x \dfrac{t}{(1+t)^3} \, dt$

41 $\int_0^x \dfrac{1-e^{-t}}{t} \, dt$

42 $\int_0^x e^{-t^2/4} \, dt$

8.8   MACLAURIN AND TAYLOR SERIES

In the preceding section, we considered power series representations for several special functions, including those where $f(x)$ has the form

$$\frac{1}{1+x}, \quad \ln(1+x), \quad \arctan x, \quad e^x, \quad \text{or} \quad \cosh x,$$

provided x is suitably restricted. We now wish to consider the following two general questions.

Question 1: If a function f has a power series representation

$$f(x) = \sum_{n=0}^{\infty} a_n x^n \quad \text{or} \quad f(x) = \sum_{n=0}^{\infty} a_n (x - c)^n,$$

what is the form of a_n?

Question 2: What conditions are sufficient for a function f to have a power series representation?

Let us begin with question 1. Suppose that

$$f(x) = \sum_{n=0}^{\infty} a_n x^n = a_0 + a_1 x + a_2 x^2 + a_3 x^3 + a_4 x^4 + \cdots$$

and the radius of convergence of the series is $r > 0$. By Theorem (8.40)(i), a power series representation for $f'(x)$ may be obtained by differentiating each term of the series for $f(x)$. We may then find a series for $f''(x)$ by differentiating the terms of the series for $f'(x)$. Series for $f'''(x)$, $f^{(4)}(x)$, and so on, can be found in similar fashion. Thus,

$$f'(x) = a_1 + 2a_2 x + 3a_3 x^2 + 4a_4 x^3 + \cdots = \sum_{n=1}^{\infty} n a_n x^{n-1}$$

$$f''(x) = 2a_2 + (3 \cdot 2)a_3 x + (4 \cdot 3)a_4 x^2 + \cdots = \sum_{n=2}^{\infty} n(n-1) a_n x^{n-2}$$

$$f'''(x) = (3 \cdot 2)a_3 + (4 \cdot 3 \cdot 2)a_4 x + \cdots = \sum_{n=3}^{\infty} n(n-1)(n-2) a_n x^{n-3},$$

and for every positive integer k,

$$f^{(k)}(x) = \sum_{n=k}^{\infty} n(n-1)(n-2)\cdots(n-k+1)a_n x^{n-k}.$$

Each series obtained by differentiation has the same radius of convergence r as the series for $f(x)$. Substituting 0 for x in each of these series representations, we obtain

$$f(0) = a_0, \quad f'(0) = a_1, \quad f''(0) = 2a_2, \quad f'''(0) = (3 \cdot 2)a_3,$$

and for every positive integer k,

$$f^{(k)}(0) = k(k-1)(k-2)\cdots(1)a_k.$$

If we let $k = n$, then

$$f^{(n)}(0) = n!\, a_n.$$

Solving the preceding equations for $a_0, a_1, a_2, \ldots$, we see that

$$a_0 = f(0), \quad a_1 = f'(0), \quad a_2 = \frac{f''(0)}{2}, \quad a_3 = \frac{f'''(0)}{3 \cdot 2},$$

and, in general,

$$a_n = \frac{f^{(n)}(0)}{n!}.$$

We have proved that the power series for $f(x)$ has the form stated in the next theorem. It is called a *Maclaurin series for* $f(x)$—named after the Scottish mathematician Colin Maclaurin (1698–1746).

Maclaurin Series for $f(x)$ 8.42

If a function f has a power series representation

$$f(x) = \sum_{n=0}^{\infty} a_n x^n$$

with radius of convergence $r > 0$, then $f^{(k)}(0)$ exists for every positive integer k and $a_n = f^{(n)}(0)/n!$. Thus,

$$f(x) = f(0) + f'(0)x + \frac{f''(0)}{2!}x^2 + \cdots + \frac{f^{(n)}(0)}{n!}x^n + \cdots.$$

Employing the type of proof used for (8.42) gives us the next theorem. If $c \neq 0$, we call the series a *Taylor series for* $f(x)$ *at* c—named after the English mathematician Brook Taylor (1685–1731).

Taylor Series for f(x)   **8.43**

If a function f has a power series representation

$$f(x) = \sum_{n=0}^{\infty} a_n (x - c)^n$$

with radius of convergence $r > 0$, then $f^{(k)}(c)$ exists for every positive integer k and $a_n = f^{(n)}(c)/n!$. Thus,

$$f(x) = f(c) + f'(c)(x - c) + \frac{f''(c)}{2!}(x - c)^2 + \cdots$$
$$+ \frac{f^{(n)}(c)}{n!}(x - c)^n + \cdots.$$

Note that the special Taylor series with $c = 0$ is the Maclaurin series (8.42). If we use the convention $f^{(0)}(c) = f(c)$, then the Maclaurin and Taylor series for f may be written in the following summation forms:

$$f(x) = \sum_{n=0}^{\infty} \frac{f^{(n)}(0)}{n!} x^n \quad \text{and} \quad f(x) = \sum_{n=0}^{\infty} \frac{f^{(n)}(c)}{n!}(x - c)^n$$

EXAMPLE ■ I   By Theorem (8.41), e^x has the following power series representation:

$$e^x = 1 + x + \frac{1}{2!}x^2 + \frac{1}{3!}x^3 + \cdots + \frac{1}{n!}x^n + \cdots$$

Verify that this is a Maclaurin series.

SOLUTION   If $f(x) = e^x$, then the nth derivative of f is $f^{(n)}(x) = e^x$ and

$$f^{(n)}(0) = e^0 = 1 \quad \text{for } n = 0, 1, 2, \ldots.$$

Hence, the Maclaurin series (8.42) is

$$\sum_{n=0}^{\infty} \frac{f^{(n)}(0)}{n!} x^n = \sum_{n=0}^{\infty} \frac{1}{n!} x^n = 1 + x + \frac{1}{2!}x^2 + \cdots + \frac{1}{n!}x^n + \cdots,$$

which is the same as the given series.

Theorems (8.42) and (8.43) imply that *if* a function f is represented by a power series in x or in $x - c$, then the series *must* be a Maclaurin or Taylor series, respectively. However, the theorems do *not* answer question 2 posed at the beginning of this section: What conditions on a function guarantee that a power series representation *exists*? We shall next obtain such conditions for any series in $x - c$ (including $c = 0$). Let us begin with the following definition.

Definition 8.44

Let c be a real number, and let f be a function that has n derivatives at c: $f'(c), f''(c), \ldots, f^{(n)}(c)$. The **nth-degree Taylor polynomial** $P_n(x)$ of f at c is

$$P_n(x) = f(c) + f'(c)(x - c) + \frac{f''(c)}{2!}(x - c)^2$$

$$+ \cdots + \frac{f^{(n)}(c)}{n!}(x - c)^n.$$

In summation notation,

$$P_n(x) = \sum_{k=0}^{n} \frac{f^{(k)}(c)}{k!}(x - c)^k.$$

If $c = 0$ in (8.44), we call $P_n(x)$ the **nth-degree Maclaurin polynomial of f**. Note that $P_n(x)$ in (8.44) is the $(n + 1)$st partial sum of the Taylor series (8.43). If we let $c = 0$, then $P_n(x)$ is the $(n + 1)$st partial sum of the Maclaurin series (8.42). The next result will lead to an answer to question 2.

Taylor's Formula with Remainder 8.45

Let f have $n + 1$ derivatives throughout an interval containing c. If x is any number in the interval that is different from c, then there is a number z between c and x such that

$$f(x) = P_n(x) + R_n(x), \quad \text{where} \quad R_n(x) = \frac{f^{(n+1)}(z)}{(n + 1)!}(x - c)^{n+1}.$$

PROOF If x is any number in the interval that is different from c, let us *define* $R_n(x)$ as follows:

$$R_n(x) = f(x) - P_n(x)$$

This equation may be rewritten as

$$f(x) = P_n(x) + R_n(x).$$

All we need to show is that for a suitable number z, $R_n(x)$ has the form stated in the conclusion of the theorem.

If t is any number in the interval, let g be the function defined by

$$g(t) = f(x) - \left[f(t) + f'(t)(x - t) + \frac{f''(t)}{2!}(x - t)^2 + \cdots + \frac{f^{(n)}(t)}{n!}(x - t)^n \right] - R_n(x)\frac{(x - t)^{n+1}}{(x - c)^{n+1}}.$$

If we differentiate each side of the equation *with respect to t* (regarding x as a constant), then many terms on the right-hand side cancel. You may verify that

$$g'(t) = -\frac{f^{(n+1)}(t)}{n!}(x - t)^n + R_n(x) \cdot (n + 1)\frac{(x - t)^n}{(x - c)^{n+1}}.$$

By referring to the formula for $g(t)$, we can verify that $g(x) = 0$. We also see that

$$g(c) = f(x) - [P_n(x)] - R_n(x)\frac{(x-c)^{n+1}}{(x-c)^{n+1}}$$
$$= f(x) - P_n(x) - R_n(x)$$
$$= f(x) - [P_n(x) + R_n(x)]$$
$$= f(x) - f(x) = 0.$$

Hence, by Rolle's theorem (3.10), there is a number z between c and x such that $g'(z) = 0$—that is,

$$-\frac{f^{(n+1)}(z)}{n!}(x-z)^n + R_n(x) \cdot (n+1)\frac{(x-z)^n}{(x-c)^{n+1}} = 0.$$

Solving for $R_n(x)$, we obtain

$$R_n(x) = \frac{f^{(n+1)}(z)}{(n+1)!}(x-c)^{n+1},$$

which is what we wished to prove. ∎

If $c = 0$, we refer to (8.45) as **Maclaurin's formula with remainder**. The expression $R_n(x)$ obtained in Theorem (8.45) is called the **Taylor remainder of f at c**. If $c = 0$, $R_n(x)$ is the **Maclaurin remainder of f**. In the next theorem, we use the Taylor remainder to obtain sufficient conditions for the existence of power series representations for a function f.

Theorem 8.46

Let f have derivatives of all orders throughout an interval containing c, and let $R_n(x)$ be the Taylor remainder of f at c. If

$$\lim_{n\to\infty} R_n(x) = 0$$

for every x in the interval, then $f(x)$ is represented by the Taylor series for $f(x)$ at c.

PROOF The Taylor polynomial $P_n(x)$ is the $(n+1)$st term for the sequence of partial sums of the Taylor series for $f(x)$ at c. By Theorem (8.45), $P_n(x) = f(x) - R_n(x)$, and hence

$$\lim_{n\to\infty} P_n(x) = \lim_{n\to\infty} f(x) - \lim_{n\to\infty} R_n(x) = f(x) - 0 = f(x).$$

Thus, the sequence of partial sums converges to $f(x)$, which proves the theorem. ∎

In Example 2 of Section 8.6, we proved that the power series $\sum x^n/n!$ is absolutely convergent for every real number x. Since the nth term of a convergent series must approach 0 as $n \to \infty$ (see Theorem (8.16)), we obtain the following result.

Theorem 8.47

If x is any real number,

$$\lim_{n \to \infty} \frac{|x|^n}{n!} = 0.$$

We shall use Theorem (8.47) in the solution of the following example.

EXAMPLE ■ 2   Find the Maclaurin series for $\sin x$, and prove that it represents $\sin x$ for every real number x.

SOLUTION   Let us arrange our work as follows:

$$
\begin{aligned}
f(x) &= \sin x & f(0) &= 0 \\
f'(x) &= \cos x & f'(0) &= 1 \\
f''(x) &= -\sin x & f''(0) &= 0 \\
f'''(x) &= -\cos x & f'''(0) &= -1
\end{aligned}
$$

Successive derivatives follow this pattern. Substitution in (8.42) gives us the following Maclaurin series:

$$\sin x = x - \frac{x^3}{3!} + \frac{x^5}{5!} - \frac{x^7}{7!} + \cdots + (-1)^n \frac{x^{2n+1}}{(2n+1)!} + \cdots$$

At this stage, all we know is that *if* $\sin x$ is represented by a power series in x, then it is given by the preceding series. To prove that $\sin x$ *is* actually represented by this Maclaurin series, let us use Theorem (8.46) with $c = 0$. If n is a positive integer, then either

$$\left| f^{(n+1)}(x) \right| = \left| \cos x \right| \quad \text{or} \quad \left| f^{(n+1)}(x) \right| = \left| \sin x \right|.$$

Hence, $\left| f^{(n+1)}(z) \right| \le 1$ for every number z. Using the formula for $R_n(x)$ in Theorem (8.45), with $c = 0$, we obtain

$$\left| R_n(x) \right| = \frac{\left| f^{(n+1)}(z) \right|}{(n+1)!} |x|^{n+1} \le \frac{|x|^{n+1}}{(n+1)!}.$$

It follows from Theorem (8.47) and the sandwich theorem (8.7) that $\lim_{n \to \infty} \left| R_n(x) \right| = 0$. Consequently, $\lim_{n \to \infty} R_n(x) = 0$, and the Maclaurin series representation for $\sin x$ is true for every x.

EXAMPLE ■ 3

(a) Find the Maclaurin series for $\cos x$.

(b) Plot the graphs of several polynomial approximations to the Maclaurin series of part (a).

Figure 8.18

$-6 \leq x \leq 6, -2 \leq y \leq 2$

(a) $k = 3$

(b) $k = 4$

(c) $k = 5$

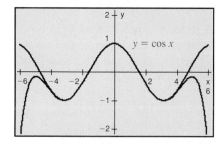

SOLUTION

(a) We could proceed directly, as in Example 2; however, let us obtain the series for $\cos x$ by differentiating the series for $\sin x$ obtained in Example 2:

$$\cos x = 1 - \frac{x^2}{2!} + \frac{x^4}{4!} - \frac{x^6}{6!} + \cdots + (-1)^n \frac{x^{2n}}{(2n)!} + \cdots$$

(b) We use the polynomials

$$P_{2k}(x) = \sum_{n=0}^{k} (-1)^n \frac{x^{2n}}{(2n)!}$$

for $k = 3, 4$, and 5. Although we can plot the graphs of $\cos x$ and the polynomials on the same coordinate axes, we see the graphs more distinctly if we plot $\cos x$ and each of the polynomials separately, as in Figure 8.18. From the three views shown in the figure, we note that each successive polynomial approximates $\cos x$ over a larger interval of x, and we easily see how rapidly the Maclaurin series approaches the cosine function that it represents.

The Maclaurin series for e^x was obtained in Theorem (8.41) by an indirect technique (see also Example 1 of this section). We next give a direct derivation of this important formula.

EXAMPLE ■ 4 Find a Maclaurin series that represents e^x for every real number x.

SOLUTION If $f(x) = e^x$, then $f^{(k)}(x) = e^x$ for every positive integer k. Hence, $f^{(k)}(0) = 1$, and substitution in (8.42) gives us

$$e^x = 1 + x + \frac{x^2}{2!} + \frac{x^3}{3!} + \cdots + \frac{x^n}{n!} + \cdots.$$

As in the solution of Example 2, we now use Theorem (8.46) to prove that this power series representation of e^x is true for every real number x. Using the formula for $R_n(x)$ with $c = 0$, we obtain

$$R_n(x) = \frac{f^{(n+1)}(z)}{(n+1)!} x^{n+1} = \frac{e^z}{(n+1)!} x^{n+1},$$

where z is a number between 0 and x. If $0 < x$, then $e^z < e^x$, since the natural exponential function is increasing, and hence for every positive integer n,

$$0 < R_n(x) < \frac{e^x}{(n+1)!} x^{n+1}.$$

By Theorem (8.47),

$$\lim_{n \to \infty} \frac{e^x}{(n+1)!} x^{n+1} = e^x \lim_{n \to \infty} \frac{x^{n+1}}{(n+1)!} = 0,$$

and by the sandwich theorem (8.7),

$$\lim_{n \to \infty} R_n(x) = 0.$$

If $x < 0$, then $z < 0$, and hence $e^z < e^0 = 1$. Consequently,

$$0 < |R_n(x)| < \left| \frac{x^{n+1}}{(n+1)!} \right|,$$

and we again see that $R_n(x)$ has the limit 0 as $n \to \infty$. It follows from Theorem (8.46) that the power series representation for e^x is valid for all nonzero x. Finally, note that if $x = 0$, then the series reduces to $e^0 = 1$.

EXAMPLE ■ 5 Find the Taylor series for the function $f(x) = \sin x$ in powers of $x - (\pi/6)$.

SOLUTION The derivatives of $f(x) = \sin x$ are listed in Example 2. If we evaluate them at $c = \pi/6$, we obtain

$$f\left(\frac{\pi}{6}\right) = \frac{1}{2}, \quad f'\left(\frac{\pi}{6}\right) = \frac{\sqrt{3}}{2}, \quad f''\left(\frac{\pi}{6}\right) = -\frac{1}{2}, \quad f'''\left(\frac{\pi}{6}\right) = -\frac{\sqrt{3}}{2},$$

and this pattern of four numbers repeats itself indefinitely. Substitution in (8.43) gives us

$$\sin x = \frac{1}{2} + \frac{\sqrt{3}}{2}\left(x - \frac{\pi}{6}\right) - \frac{1}{2(2!)}\left(x - \frac{\pi}{6}\right)^2 - \frac{\sqrt{3}}{2(3!)}\left(x - \frac{\pi}{6}\right)^3 + \cdots.$$

The nth term u_n of this series is given by

$$u_n = \begin{cases} (-1)^{n/2} \dfrac{1}{2(n!)} \left(x - \dfrac{\pi}{6}\right)^n & \text{if } n = 0, 2, 4, 6, \ldots \\[2mm] (-1)^{(n-1)/2} \dfrac{\sqrt{3}}{2(n!)} \left(x - \dfrac{\pi}{6}\right)^n & \text{if } n = 1, 3, 5, 7, \ldots \end{cases}$$

The proof that the series represents $\sin x$ for every x is similar to that given in Example 2 and is therefore omitted.

The next example brings out the fact that a function f may have derivatives of all orders at some number c, but may not have a Taylor series representation at that number. This shows that an additional condition, such as $\lim_{n \to \infty} R_n(x) = 0$, is required to guarantee the existence of a Taylor series.

EXAMPLE ■ 6 Let f be the function defined by

$$f(x) = \begin{cases} e^{-1/x^2} & \text{if } x \neq 0 \\ 0 & \text{if } x = 0 \end{cases}$$

Show that $f(x)$ cannot be represented by a Maclaurin series.

SOLUTION   By Definition (2.6), the derivative of f at 0 is

$$f'(0) = \lim_{x \to 0} \frac{f(x) - f(0)}{x - 0} = \lim_{x \to 0} \frac{e^{-1/x^2}}{x} = \lim_{x \to 0} \frac{(1/x)}{e^{1/x^2}}.$$

The last expression has the indeterminate form ∞/∞. Applying l'Hôpital's rule (6.51), we see that

$$f'(0) = \lim_{x \to 0} \frac{(-1/x^2)}{(-2/x^3)e^{1/x^2}} = \lim_{x \to 0} \frac{x}{2e^{1/x^2}} = 0.$$

It can be proved that $f''(0) = 0$, $f'''(0) = 0$, and, in general, $f^{(n)}(0) = 0$ for every positive integer n. According to Theorem (8.42), if $f(x)$ has a Maclaurin series representation, then it is given by

$$f(x) = 0 + 0x + \frac{0}{2!}x^2 + \cdots + \frac{0}{n!}x^n + \cdots,$$

which implies that $f(x) = 0$ throughout an interval containing 0. However, this contradicts the definition of f. Consequently, $f(x)$ does not have a Maclaurin series representation.

As a by-product of Example 6, it follows from Theorem (8.46) that for the given function f, $\lim_{n \to \infty} R_n(x) \neq 0$ at $c = 0$.

We next list, for reference, Maclaurin series that have been obtained in examples in this section and Section 8.7. These series are important because of their uses in advanced mathematics and applications.

Important Maclaurin Series 8.48

Maclaurin series	Interval of convergence
(a) $\sin x = x - \dfrac{x^3}{3!} + \dfrac{x^5}{5!} - \dfrac{x^7}{7!} + \cdots + (-1)^n \dfrac{x^{2n+1}}{(2n+1)!} + \cdots$	$(-\infty, \infty)$
(b) $\cos x = 1 - \dfrac{x^2}{2!} + \dfrac{x^4}{4!} - \dfrac{x^6}{6!} + \cdots + (-1)^n \dfrac{x^{2n}}{(2n)!} + \cdots$	$(-\infty, \infty)$
(c) $e^x = 1 + x + \dfrac{x^2}{2!} + \dfrac{x^3}{3!} + \cdots + \dfrac{x^n}{n!} + \cdots$	$(-\infty, \infty)$
(d) $\ln(1 + x) = x - \dfrac{x^2}{2} + \dfrac{x^3}{3} - \dfrac{x^4}{4} + \cdots + (-1)^n \dfrac{x^{n+1}}{n+1} + \cdots$	$(-1, 1]$
(e) $\tan^{-1} x = x - \dfrac{x^3}{3} + \dfrac{x^5}{5} - \dfrac{x^7}{7} + \cdots + (-1)^n \dfrac{x^{2n+1}}{2n+1} + \cdots$	$[-1, 1]$
(f) $\sinh x = x + \dfrac{x^3}{3!} + \dfrac{x^5}{5!} + \dfrac{x^7}{7!} + \cdots + \dfrac{x^{2n+1}}{(2n+1)!} + \cdots$	$(-\infty, \infty)$
(g) $\cosh x = 1 + \dfrac{x^2}{2!} + \dfrac{x^4}{4!} + \dfrac{x^6}{6!} + \cdots + \dfrac{x^{2n}}{(2n)!} + \cdots$	$(-\infty, \infty)$

We can use Maclaurin or Taylor series to approximate values of functions and definite integrals, as illustrated in the next two examples.

EXAMPLE ■ 7   Use the first two nonzero terms of a Maclaurin series to approximate the following, and estimate the error in the approximation.

(a) $\sin(0.1)$ **(b)** $\sin x$ for any nonzero real number x in $[-1, 1]$

SOLUTION

(a) Letting $x = 0.1$ in the Maclaurin series for $\sin x$ (see (8.48)(a)) yields

$$\sin(0.1) = 0.1 - \frac{0.001}{6} + \frac{0.00001}{120} - \cdots.$$

By Theorem (8.31), the error involved in approximating $\sin(0.1)$ by using the first two terms of this alternating series is less than the third term, $0.00001/120 \approx 0.00000008$. To six decimal places,

$$\sin(0.1) \approx 0.1 - \frac{0.001}{6} \approx 0.099833.$$

(b) Using the first two terms of (8.48)(a) gives us the approximation formula

$$\sin x \approx x - \frac{x^3}{6}.$$

By Theorem (8.31), the error involved in using this formula for a real number x in $[-1, 1]$ is less than $|x^5|/5!$.

EXAMPLE ■ 8   Approximate $\displaystyle\int_0^1 \sin(x^2)\, dx$ to four decimal places.

SOLUTION   Substituting x^2 for x in (8.48)(a) gives us

$$\sin(x^2) = x^2 - \frac{x^6}{3!} + \frac{x^{10}}{5!} - \frac{x^{14}}{7!} + \cdots.$$

Integrating each term of this series, we obtain

$$\int_0^1 \sin(x^2)\, dx = \frac{1}{3} - \frac{1}{42} + \frac{1}{1320} - \frac{1}{75,600} + \cdots.$$

Summing the first three terms yields

$$\int_0^1 \sin(x^2)\, dx \approx 0.31028.$$

By Theorem (8.31), the error is less than $\frac{1}{75,600} \approx 0.00001$.

Note that in the preceding example we achieved accuracy to four decimal places by summing only *three* terms of the integrated series for

$\sin(x^2)$. To obtain this degree of accuracy by means of the trapezoidal rule or Simpson's rule, it would be necessary to use a large value of n for the interval $[0, 1]$. However, if the interval were $[10, 11]$, the efficiency of each method would be quite different. An important point for numerical applications is that in addition to analyzing a given problem, we should also strive to find the most efficient method for computing the answer.

To obtain a Taylor or Maclaurin series representation for a function f, it is necessary to find a general formula for $f^{(n)}(x)$ and, in addition, to investigate $\lim_{n \to \infty} R_n(x)$. For this reason, our examples have been restricted to expressions such as $\sin x$, $\cos x$, and e^x. The method cannot be used if, for example, $f(x)$ equals $\tan x$ or $\sin^{-1} x$, because $f^{(n)}(x)$ becomes very complicated as n increases. Most of the exercises that follow are based on functions whose nth derivatives can be determined easily or on series representations that we have already established. In more complicated cases, we shall restrict our attention to only the first few terms of a Taylor or Maclaurin series representation.

EXERCISES 8.8

Exer. 1–6: If $f(x) = \sum_{n=0}^{\infty} a_n x^n$, find a_n by using the formula for a_n in (8.42).

1 $f(x) = e^{3x}$

2 $f(x) = e^{-2x}$

3 $f(x) = \sin 2x$

4 $f(x) = \cos 3x$

5 $f(x) = \dfrac{1}{1 + 3x}$

6 $f(x) = \dfrac{1}{1 - 2x}$

7 Let $f(x) = \cos x$.

   (a) Use the method of Example 2 to prove that $\lim_{n \to \infty} R_n(x) = 0$.

   (b) Use (8.42) to find a Maclaurin series for $f(x)$.

8 Let $f(x) = e^{-x}$.

   (a) Use the method of Example 4 to prove that $\lim_{n \to \infty} R_n(x) = 0$.

   (b) Use (8.42) to find a Maclaurin series for $f(x)$.

Exer. 9–14: Use a Maclaurin series obtained in this section to obtain a Maclaurin series for $f(x)$.

9 $f(x) = x \sin 3x$

10 $f(x) = x^2 \sin x$

11 $f(x) = \cos(-2x)$

12 $f(x) = \cos(x^2)$

13 $f(x) = \cos^2 x$ (*Hint:* Use a half-angle formula.)

14 $f(x) = \sin^2 x$

Exer. 15–16: Find a Maclaurin series for $f(x)$. (Do not verify that $\lim_{n \to \infty} R_n(x) = 0$.)

15 $f(x) = 10^x$

16 $f(x) = \ln(3 + x)$

Exer. 17–20: Find a Taylor series for $f(x)$ at c. (Do not verify that $\lim_{n \to \infty} R_n(x) = 0$.)

17 $f(x) = \sin x$; $c = \pi/4$ **18** $f(x) = \cos x$; $c = \pi/3$

19 $f(x) = 1/x$; $c = 2$ **20** $f(x) = e^x$; $c = -3$

21 Find a series representation for e^{2x} in powers of $x + 1$.

22 Find a series representation of $\ln x$ in powers of $x - 1$.

Exer. 23–28: Find the first three terms of the Taylor series for $f(x)$ at c.

23 $f(x) = \sec x$; $c = \pi/3$

24 $f(x) = \tan x$; $c = \pi/4$

25 $f(x) = \sin^{-1} x$; $c = \frac{1}{2}$

26 $f(x) = \tan^{-1} x$; $c = 1$

27 $f(x) = xe^x$; $c = -1$

28 $f(x) = \csc x$; $c = 2\pi/3$

Exer. 29–38: Use the first two nonzero terms of a Maclaurin series to approximate the number, and estimate the error in the approximation.

29 $\dfrac{1}{\sqrt{e}}$ **30** $\dfrac{1}{e}$ **31** $\cos 3°$

32 $\sin 1°$ **33** $\tan^{-1} 0.1$ **34** $\ln 1.5$

35 $\displaystyle\int_0^1 e^{-x^2}\, dx$ **36** $\displaystyle\int_0^{1/2} x \cos(x^3)\, dx$

37 $\int_0^{0.5} \cos(x^2)\,dx$ **38** $\int_0^{0.1} \tan^{-1}(x^2)\,dx$

Exer. 39–42: Approximate the improper integral to four decimal places. (Assume that if the integrand is $f(x)$, then $f(0) = \lim_{x \to 0} f(x)$.)

39 $\int_0^1 \dfrac{1 - \cos x}{x^2}\,dx$ **40** $\int_0^1 \dfrac{\sin x}{x}\,dx$

41 $\int_0^{1/2} \dfrac{\ln(1 + x)}{x}\,dx$ **42** $\int_0^1 \dfrac{1 - e^{-x}}{x}\,dx$

Exer. 43–44: **(a)** Let $g(x)$ be the sum of the first two nonzero terms of the Maclaurin series for $f(x)$. Use $g(x)$ to approximate $\int_0^1 f(x)\,dx$ and $\int_1^2 f(x)\,dx$. **(b)** First sketch the graphs, on the same coordinate axes, of f and g for $0 \le x \le 2$, and then use the graphs to compare the accuracy of the approximations in part (a).

43 $f(x) = \sin(x^2)$ **44** $f(x) = \sinh x$

45 Use (8.48)(d) to find the Maclaurin series for

$$f(x) = \ln \frac{1 + x}{1 - x}.$$

46 Use the first five terms of the series in Exercise 45 to calculate $\ln 2$, and compare your answer to the value obtained using a calculator.

47 **(a)** Use (8.48)(e) with $x = 1$ to represent π as the sum of an infinite series.

(b) What accuracy is obtained by using the first five terms of the series to approximate π?

(c) Approximately how many terms of the series are required to obtain four-decimal-place accuracy for π?

48 **(a)** Use the identity

$$\tan^{-1} \frac{1}{2} + \tan^{-1} \frac{1}{3} = \frac{\pi}{4}$$

to express π as the sum of two infinite series.

(b) Use the first five terms of each series in part (a) to approximate π, and compare the result with that obtained in Exercise 47.

49 In planning a highway across a desert, a surveyor must make compensations for the curvature of the earth when measuring differences in elevation (see figure).

(a) If s is the length of the highway and R is the radius of the earth, show that the correction C is given by $C = R[\sec(s/R) - 1]$.

(b) Use the Maclaurin series for $\sec x$ to show that C is approximately $s^2/(2R) + (5s^4)/(24R^3)$.

(c) The average radius of the earth is 3959 mi. Estimate the correction, to the nearest 0.1 ft, for a stretch of highway 5 mi long.

Exercise 49   Line of sight

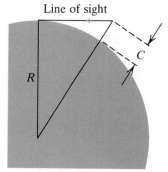

50 The velocity v of a water wave is related to its length L and the depth h of the water by

$$v^2 = \frac{gL}{2\pi} \tanh \frac{2\pi h}{L},$$

where g is a gravitational constant.

(a) Show that $\tanh x \approx x - \frac{1}{3}x^3$ if $x \approx 0$.

(b) Use the approximation $\tanh x \approx x$ to show that $v^2 \approx gh$ if h/L is small.

(c) Use part (a) and the fact that the Maclaurin series for $\tanh x$ is an alternating series to show that if $L > 20h$, then the error involved in using $v^2 \approx gh$ is less than $0.002gL$.

51 If too large a downward force of P pounds is applied to a cantilever column of length L at a point x units to the right of center (see figure), the column will buckle. The horizontal deflection δ can be expressed as

$$\delta = x(\sec kL - 1) \quad \text{with} \quad k = \sqrt{P/R},$$

where R is a constant called the *flexural rigidity* of the material and $0 \le kL < \pi/2$. Use Exercise 49(b) to show that $\delta \approx \frac{1}{2}PxL^2/R$ if PL^2 is small compared to $2R$.

Exercise 51

52 Show that $\cos x \approx 1 - \frac{1}{2}x^2 + \frac{1}{24}x^4 - \frac{1}{720}x^6$ is accurate to five decimal places if $0 \le x \le \pi/4$.

c **Exer. 53–60:** (a) **Find a Maclaurin series for the function.** (b) **Plot the graphs of the polynomials $p_k(x) = \sum_{n=0}^{k} a_n x^n$ associated with the series in part (a) for $k = 3$, 4, and 5.**

53 $\sin(0.6x)$

54 $\cosh 2x$

55 $\cos(x^2)$

56 $\tan^{-1}(0.4x)$

57 $\displaystyle\int_0^x \frac{\sin t}{t}\,dt$

58 $\displaystyle\int_0^x \frac{1 - \cos t}{t^2}\,dt$

59 $\displaystyle\int_0^x \sin(t^3)\,dt$

60 $\displaystyle\int_0^x e^{-t^4/81}\,dt$

8.9 APPLICATIONS OF TAYLOR POLYNOMIALS

In this section, we consider how to use Taylor polynomials to approximate transcendental functions. In particular, we investigate how accurately a Taylor polynomial of particular degree is in estimating values such as $\sin x$, $\cos x$, e^x, or $\ln x$.

In Example 7(b) of Section 8.8, we used the first two nonzero terms of the Maclaurin series for $\sin x$ to obtain the approximation formula

$$\sin x \approx x - \frac{x^3}{6}.$$

By (8.44), the expression on the right-hand side of this formula is the third-degree Taylor polynomial $P_3(x)$ of $\sin x$ at $c = 0$. Thus, we could write

$$\sin x \approx P_3(x).$$

Using additional terms of the Maclaurin series for $\sin x$ would give us other approximation formulas. To illustrate,

$$\sin x \approx x - \frac{x^3}{3!} + \frac{x^5}{5!} = P_5(x).$$

By Theorem (8.31), the error involved in using this formula is less than $|x^7|/7!$. Thus, the approximation is very accurate if x is close to 0.

We can use this procedure for any function f that has a sufficient number of derivatives. Specifically, if f satisfies the hypotheses of Taylor's formula with remainder (8.45), then

$$f(x) = P_n(x) + R_n(x),$$

where $P_n(x)$ is the nth-degree Taylor polynomial of f at c and $R_n(x)$ is the Taylor remainder. If $\lim_{n\to\infty} R_n(x) = 0$, then, as n increases, we have $P_n(x) \to f(x)$; hence the approximation formula $f(x) \approx P_n(x)$ improves as n gets larger. Thus, we can approximate values of many different transcendental functions by using *polynomial* functions. This is a very important fact, because polynomial functions are the simplest functions to use for calculations—their values can be found by employing only additions and multiplications of real numbers.

As another illustration, consider the exponential function given by $f(x) = e^x$. From (8.48)(c), the Maclaurin series for e^x is

$$e^x = 1 + x + \frac{x^2}{2!} + \frac{x^3}{3!} + \cdots + \frac{x^n}{n!} + \cdots.$$

Figure 8.19

Figure 8.20

Figure 8.21

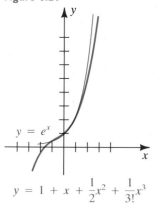

If we approximate e^x by means of Taylor polynomials (with $c = 0$), we obtain

$$e^x \approx P_1(x) = 1 + x$$

$$e^x \approx P_2(x) = 1 + x + \frac{x^2}{2}$$

$$e^x \approx P_3(x) = 1 + x + \frac{x^2}{2} + \frac{x^3}{6}$$

and so on. These approximation formulas are accurate only if x is close to 0. To approximate e^x for larger values of x, we may use Taylor polynomials with $c \neq 0$.

The accuracy of the preceding three approximation formulas for e^x is illustrated by the graphs of the functions P_1, P_2, and P_3 in Figures 8.19–8.21. It is of interest to note that the graph of $y = P_1(x) = 1 + x$ in Figure 8.19 is the tangent line to the graph of $y = e^x$ at the point $(0, 1)$. You may verify that the parabola $y = P_2(x) = 1 + x + \frac{1}{2}x^2$ in Figure 8.20 has the same tangent line and the same concavity as the graph of $y = e^x$ at $(0, 1)$. The graph of $y = P_3(x)$ in Figure 8.21 has the same tangent line and concavity at $(0, 1)$ and also the same *rate of change of concavity*, since $P_3'''(0) = (d^3/dx) |e^x|_{x=0}$. In general, we can show that for any positive integer n, $P_n^{(n)}(0) = (d^n/dx) |e^x|_{x=0}$. As n increases without bound, the graph of the equation $y = P_n(x)$ more closely resembles the graph of $y = e^x$.

The following table indicates the accuracy of the approximation formula $e^x \approx P_3(x)$ for several values of x, approximated to the nearest hundredth.

x	-1.5	-1.0	-0.5	0	0.5	1.0	1.5
e^x	0.22	0.37	0.61	1	1.65	2.72	4.48
$P_3(x)$	0.06	0.33	0.60	1	1.65	2.67	4.19

If more accuracy is desired, we can use any larger positive integer n to obtain

$$e^x \approx P_n(x) = 1 + x + \frac{x^2}{2!} + \frac{x^3}{3!} + \cdots + \frac{x^n}{n!}.$$

We can use this remarkably simple formula to approximate e^x to any degree of accuracy.

In the remainder of this section, we shall use Taylor polynomials to approximate values of functions that satisfy the hypotheses of Taylor's formula with remainder (8.45). Using the conclusion $f(x) = P_n(x) + R_n(x)$ of that theorem, we see that the error involved in approximating $f(x)$ by $P_n(x)$ is

$$|f(x) - P_n(x)| = |R_n(x)|.$$

The complete statement of this result is given in the next theorem.

Theorem 8.49

Let f have $n + 1$ derivatives throughout an interval containing c. If x is any number in the interval and $x \neq c$, then the error in approximating $f(x)$ by the nth-degree Taylor polynomial of f at c,

$$P_n(x) = f(c) + f'(c)(x - c) + \frac{f''(c)}{2!}(x - c)^2$$
$$+ \cdots + \frac{f^{(n)}(c)}{n!}(x - c)^n,$$

is equal to $|R_n(x)|$, where

$$R_n(x) = \frac{f^{(n+1)}(z)}{(n + 1)!}(x - c)^{n+1}$$

and z is the number between c and x given by (8.45).

In the next two examples, Taylor polynomials are used to approximate function values. If we are interested in k-decimal-place accuracy in the approximation of a sum, we often approximate each term of the sum to $k + 1$ decimal places and then round off the final result to k decimal places. In certain cases, this may fail to produce the required degree of accuracy; however, it is customary to proceed in this way for elementary approximations. More precise techniques may be found in texts on *numerical analysis*.

EXAMPLE 1   Let $f(x) = \ln x$.

(a) Find $P_3(x)$ and $R_3(x)$ at $c = 1$.

(b) Approximate $\ln 1.1$ to four decimal places by means of $P_3(1.1)$, and use $R_3(1.1)$ to estimate the error in this approximation.

SOLUTION

(a) As in Theorem (8.49), the general Taylor polynomial $P_3(x)$ and Taylor remainder $R_3(x)$ at $c = 1$ are

$$P_3(x) = f(1) + f'(1)(x - 1) + \frac{f''(1)}{2!}(x - 1)^2 + \frac{f'''(1)}{3!}(x - 1)^3$$

and
$$R_3(x) = \frac{f^{(4)}(z)}{4!}(x - 1)^4,$$

where z is a number between 1 and x. Thus, we need the first four derivatives of f. It is convenient to arrange our work as follows:

$$\begin{array}{ll}
f(x) = \ln x & f(1) = 0 \\
f'(x) = x^{-1} & f'(1) = 1 \\
f''(x) = -x^{-2} & f''(1) = -1 \\
f'''(x) = 2x^{-3} & f'''(1) = 2 \\
f^{(4)}(x) = -6x^{-4} & f^{(4)}(z) = -6z^{-4}
\end{array}$$

Substituting in $P_3(x)$ and $R_3(x)$, we obtain

$$P_3(x) = 0 + 1(x - 1) - \frac{1}{2!}(x - 1)^2 + \frac{2}{3!}(x - 1)^3$$

and
$$R_3(x) = \frac{-6z^{-4}}{4!}(x - 1)^4 = -\frac{1}{4z^4}(x - 1)^4,$$

where z is between 1 and x.

(b) From part (a),

$$\ln 1.1 \approx P_3(1.1) = 0.1 - \tfrac{1}{2}(0.1)^2 + \tfrac{1}{3}(0.1)^3,$$

or
$$\ln 1.1 \approx 0.0953.$$

To estimate the error in this approximation, we consider

$$|R_3(1.1)| = \left| -\frac{(0.1)^4}{4z^4} \right|, \qquad \text{where} \quad 1 < z < 1.1.$$

Since $z > 1$, $1/z < 1$ and therefore $1/z^4 < 1$. Consequently,

$$|R_3(1.1)| = \left| -\frac{(0.1)^4}{4z^4} \right| < \left| -\frac{0.0001}{4} \right| = 0.000025.$$

Because $0.000025 < 0.00005$, it follows from Theorem (8.49) that the approximation $\ln 1.1 \approx 0.0953$ is accurate to four decimal places.

If we wish to approximate a function value $f(x)$ for some x, it is desirable to choose the number c in Theorem (8.49) such that the remainder $R_n(x)$ is very close to 0 when n is relatively small (say, $n = 3$ or $n = 4$). We obtain this result if we choose c close to x. In addition, we should choose c so that values of the first $n + 1$ derivatives of f at c are easy to calculate, as was done in Example 1, where to approximate $\ln x$ for $x = 1.1$ we selected $c = 1$. The next example provides another illustration of a suitable choice of c.

EXAMPLE ▪ 2 Use a Taylor polynomial to approximate $\cos 61°$, and estimate the accuracy of the approximation.

SOLUTION We wish to approximate $f(x) = \cos x$ if $x = 61°$. Let us begin by observing that $61°$ is close to $60°$, or $\pi/3$ radians, and that it is easy to calculate values of trigonometric functions at $\pi/3$. This suggests that we choose $c = \pi/3$ in (8.49). The choice of n will depend on the degree of accuracy we wish to attain. Let us try $n = 2$. In this case, the first three derivatives of f are required and we arrange our work as follows:

$$
\begin{array}{ll}
f(x) = \cos x & f\left(\pi/3\right) = 1/2 \\
f'(x) = -\sin x & f'\left(\pi/3\right) = -\sqrt{3}/2 \\
f''(x) = -\cos x & f''\left(\pi/3\right) = -1/2 \\
f'''(x) = \sin x & f'''(z) = \sin z
\end{array}
$$

As in (8.49), the second-degree Taylor polynomial of f at $c = \pi/3$ is

$$P_2(x) = \frac{1}{2} - \frac{\sqrt{3}}{2}\left(x - \frac{\pi}{3}\right) - \frac{1/2}{2!}\left(x - \frac{\pi}{3}\right)^2.$$

Since x represents a real number, we must convert $61°$ to radian measure before substituting into $P_2(x)$. Writing

$$61° = 60° + 1° = \frac{\pi}{3} + \frac{\pi}{180}$$

and substituting, we obtain

$$P_2\left(\frac{\pi}{3} + \frac{\pi}{180}\right) = \frac{1}{2} - \left(\frac{\sqrt{3}}{2}\right)\left(\frac{\pi}{180}\right) - \frac{1}{4}\left(\frac{\pi}{180}\right)^2 \approx 0.48481,$$

and hence $\qquad\qquad \cos 61° \approx 0.48481.$

To estimate the accuracy of this approximation, we consider

$$|R_2(x)| = \left|\frac{f'''(z)}{3!}\left(x - \frac{\pi}{3}\right)^3\right| = \left|\frac{\sin z}{3!}\left(x - \frac{\pi}{3}\right)^3\right|,$$

where z is between $\pi/3$ and x. Substituting $x = (\pi/3) + (\pi/180)$ and using the fact that $|\sin z| \le 1$, we obtain

$$\left|R_2\left(\frac{\pi}{3} + \frac{\pi}{180}\right)\right| = \left|\frac{\sin z}{3!}\left(\frac{\pi}{180}\right)^3\right| \le \left|\frac{1}{3!}\left(\frac{\pi}{180}\right)^3\right| \le 0.000001.$$

Thus, by (8.49), the approximation $\cos 61° \approx 0.48481$ is accurate to five decimal places. For greater accuracy, we must find a value of n such that the maximum value of $|R_n[(\pi/3) + (\pi/180)]|$ is within the desired range.

EXAMPLE ▪ 3   If $f(x) = e^x$, use the Taylor polynomial $P_9(x)$ of f at $c = 0$ to approximate e, and estimate the error in the approximation.

SOLUTION   For every positive integer k, $f^{(k)}(x) = e^x$, and hence $f^{(k)}(0) = e^0 = 1$. Thus, using $n = 9$ and $c = 0$ in Theorem (8.49) yields

$$P_9(x) = 1 + x + \frac{x^2}{2!} + \frac{x^3}{3!} + \cdots + \frac{x^9}{9!},$$

and therefore $\quad e \approx P_9(1) = 1 + 1 + \frac{1}{2!} + \frac{1}{3!} + \cdots + \frac{1}{9!}.$

This result gives us $e \approx 2.71828153$.

To estimate the error, we consider

$$R_9(x) = \frac{e^z}{10!}x^{10}.$$

If $x = 1$, then $0 < z < 1$. Using results about e^x from Chapter 6, we have $e^z < e^1 < 3$, and

$$|R_9(1)| = \left|\frac{e^z}{10!}(1)\right| < \frac{3}{10!} < 0.000001.$$

Hence, the approximation $e \approx 2.71828$ is accurate to five decimal places.

EXERCISES 8.9

Exer. 1–4: (a) Find the Maclaurin polynomials $P_1(x)$, $P_2(x)$, and $P_3(x)$ for $f(x)$. (b) Sketch, on the same coordinate axes, the graphs of P_1, P_2, P_3, and f. (c) Approximate $f(a)$ to four decimal places by means of $P_3(a)$, and use $R_3(a)$ to estimate the error in this approximation.

1 $f(x) = \sin x;$ $a = 0.05$

2 $f(x) = \cos x;$ $a = 0.2$

3 $f(x) = \ln(x + 1);$ $a = 0.9$

4 $f(x) = \tan^{-1} x;$ $a = 0.1$

[c] Exer. 5–6: Graph, on the same coordinate axes, f, P_1, P_3, and P_5 for $-3 \le x \le 3$.

5 $f(x) = \sinh x$ 6 $f(x) = \cosh x$

Exer. 7–18: Find Taylor's formula with remainder (8.45) for the given $f(x)$, c, and n.

7 $f(x) = \sin x;$ $c = \pi/2,$ $n = 3$

8 $f(x) = \cos x;$ $c = \pi/4,$ $n = 3$

9 $f(x) = \sqrt{x};$ $c = 4,$ $n = 3$

10 $f(x) = e^{-x};$ $c = 1,$ $n = 3$

11 $f(x) = \tan x;$ $c = \pi/4,$ $n = 2$

12 $f(x) = 1/(x - 1)^2;$ $c = 2,$ $n = 5$

13 $f(x) = 1/x;$ $c = -2,$ $n = 5$

14 $f(x) = \sqrt[3]{x};$ $c = -8,$ $n = 3$

15 $f(x) = \tan^{-1} x;$ $c = 1,$ $n = 2$

16 $f(x) = \ln \sin x;$ $c = \pi/6,$ $n = 3$

17 $f(x) = xe^x;$ $c = -1,$ $n = 4$

18 $f(x) = \log x;$ $c = 10,$ $n = 2$

Exer. 19–30: Find Maclaurin's formula with remainder for the given $f(x)$ and n.

19 $f(x) = \ln(x + 1);$ $n = 4$

20 $f(x) = \sin x;$ $n = 7$

21 $f(x) = \cos x;$ $n = 8$

22 $f(x) = \tan^{-1} x;$ $n = 3$

23 $f(x) = e^{2x};$ $n = 5$

24 $f(x) = \sec x;$ $n = 3$

25 $f(x) = 1/(x - 1)^2;$ $n = 5$

26 $f(x) = \sqrt{4 - x};$ $n = 3$

27 $f(x) = \arcsin x;$ $n = 2$

28 $f(x) = e^{-x^2};$ $n = 3$

29 $f(x) = 2x^4 - 5x^3;$ $n = 4$ and $n = 5$

30 $f(x) = \cosh x;$ $n = 4$ and $n = 5$

[c] Exer. 31–34: Approximate the number to four decimal places by using the indicated exercise and the fact that $\pi/180 \approx 0.0175$. Prove that your answer is correct by showing that $\left|R_n(x)\right| < 0.5 \times 10^{-4}$.

31 $\sin 89°$ (Exercise 7)

32 $\cos 47°$ (Exercise 8)

33 $\sqrt{4.03}$ (Exercise 9)

34 $e^{-1.02}$ (Exercise 10)

[c] Exer. 35–40: Approximate the number by using the indicated exercise, and estimate the error in the approximation by means of $R_n(x)$.

35 $-1/(2.2)$ (Exercise 13)

36 $\sqrt[3]{-8.5}$ (Exercise 14)

37 $\ln 1.25$ (Exercise 19)

38 $\sin 0.1$ (Exercise 20)

39 $\cos 30°$ (Exercise 21)

40 $\log 10.01$ (Exercise 18)

Exer. 41–46: Use Maclaurin's formula with remainder to establish the approximation formula, and state, in terms of decimal places, the accuracy of the approximation if $|x| \le 0.1$.

41 $\cos x \approx 1 - \dfrac{x^2}{2}$ 42 $\sqrt[3]{1 + x} \approx 1 + \dfrac{1}{3}x$

43 $e^x \approx 1 + x + \dfrac{x^2}{2}$ 44 $\sin x \approx x - \dfrac{x^3}{6}$

45 $\ln(1 + x) \approx x - \dfrac{x^2}{2} + \dfrac{x^3}{3}$

46 $\cosh x \approx 1 + \dfrac{x^2}{2}$

47 Let $P_n(x)$ be the nth-degree Maclaurin polynomial. If $f(x)$ is a polynomial of degree n, prove that $f(x) = P_n(x)$.

CHAPTER 8 REVIEW EXERCISES

Exer. 1–6: Determine whether the sequence converges or diverges; if it converges, find the limit.

1 $\left\{ \dfrac{\ln(n^2 + 1)}{n} \right\}$ **2** $\{100(0.99)^n\}$

3 $\left\{ \dfrac{10^n}{n^{10}} \right\}$ **4** $\left\{ \dfrac{1}{n} + (-2)^n \right\}$

5 $\left\{ \dfrac{n}{\sqrt{n + 4}} - \dfrac{n}{\sqrt{n + 9}} \right\}$ **6** $\left\{ \left(1 + \dfrac{2}{n}\right)^{2n} \right\}$

c **Exer. 7–8:** For the recursively defined sequence, determine what happens to terms of the sequence as k increases.

7 $a_1 = 1$ and $a_{k+1} = 0.5 \cosh a_k$

8 $a_1 = 2$ and $a_{k+1} = \cosh^{-1} a_k + 1$

Exer. 9–34: If the series is positive-term, determine whether it is convergent or divergent; if the series contains negative terms, determine whether it is absolutely convergent, conditionally convergent, or divergent.

9 $\displaystyle\sum_{n=1}^{\infty} \dfrac{1}{\sqrt[3]{n(n + 1)(n + 2)}}$ **10** $\displaystyle\sum_{n=0}^{\infty} \dfrac{(2n + 3)^2}{(n + 1)^3}$

11 $\displaystyle\sum_{n=1}^{\infty} \left(-\dfrac{2}{3}\right)^{n-1}$ **12** $\displaystyle\sum_{n=0}^{\infty} \dfrac{1}{2 + (\frac{1}{2})^n}$

13 $\displaystyle\sum_{n=1}^{\infty} \dfrac{3^{2n+1}}{n 5^{n-1}}$ **14** $\displaystyle\sum_{n=1}^{\infty} \dfrac{1}{3^n + 2}$

15 $\displaystyle\sum_{n=1}^{\infty} \dfrac{n!}{\ln(n + 1)}$ **16** $\displaystyle\sum_{n=1}^{\infty} \dfrac{n^2 - 1}{n^2 + 1}$

17 $\displaystyle\sum_{n=1}^{\infty} (n^2 + 9)(-2)^{1-n}$ **18** $\displaystyle\sum_{n=1}^{\infty} \dfrac{n + \cos n}{n^3 + 1}$

19 $\displaystyle\sum_{n=1}^{\infty} \dfrac{e^n}{n^e}$ **20** $\displaystyle\sum_{n=1}^{\infty} (-1)^{n-1} \dfrac{n}{n^2 + 1}$

21 $\displaystyle\sum_{n=1}^{\infty} (-1)^n \dfrac{1}{\sqrt[n]{n}}$ **22** $\displaystyle\sum_{n=2}^{\infty} (-1)^n \dfrac{(0.9)^n}{\ln n}$

23 $\displaystyle\sum_{n=1}^{\infty} \dfrac{\sin \frac{5\pi}{3} n}{n^{5\pi/3}}$ **24** $\displaystyle\sum_{n=2}^{\infty} (-1)^n \dfrac{\sqrt[3]{n} - 1}{n^2 - 1}$

25 $\displaystyle\sum_{n=1}^{\infty} (-1)^{n-1} \dfrac{\sqrt{n}}{n + 1}$ **26** $\displaystyle\sum_{n=1}^{\infty} (-1)^n \dfrac{2n + 3}{n!}$

27 $\displaystyle\sum_{n=1}^{\infty} \dfrac{1 - \cos n}{n^2}$

28 $\dfrac{2}{1!} - \dfrac{2 \cdot 4}{2!} + \cdots + (-1)^{n-1} \dfrac{2 \cdot 4 \cdot \, \cdots \, \cdot (2n)}{n!} + \cdots$

29 $\displaystyle\sum_{n=1}^{\infty} \dfrac{(2n)^n}{n^{2n}}$ **30** $\displaystyle\sum_{n=1}^{\infty} \dfrac{3^{n-1}}{n^2 + 9}$

31 $\displaystyle\sum_{n=1}^{\infty} \dfrac{e^{2n}}{(2n - 1)!}$ **32** $\displaystyle\sum_{n=1}^{\infty} \left(\dfrac{1}{3^n} - \dfrac{5}{\sqrt{n}}\right)$

33 $\displaystyle\sum_{n=2}^{\infty} (-1)^n \dfrac{\sqrt{\ln n}}{n}$ **34** $\displaystyle\sum_{n=1}^{\infty} \dfrac{\tan^{-1} n}{\sqrt{1 + n^2}}$

Exer. 35–40: Use the integral test (8.23) to determine the convergence or divergence of the series.

35 $\displaystyle\sum_{n=1}^{\infty} \dfrac{1}{(3n + 2)^3}$ **36** $\displaystyle\sum_{n=2}^{\infty} \dfrac{n}{\sqrt{n^2 - 1}}$

37 $\displaystyle\sum_{n=1}^{\infty} n^{-2} e^{1/n}$ **38** $\displaystyle\sum_{n=2}^{\infty} \dfrac{1}{n(\ln n)^3}$

39 $\displaystyle\sum_{n=1}^{\infty} \dfrac{10}{\sqrt[3]{n + 8}}$ **40** $\displaystyle\sum_{n=5}^{\infty} \dfrac{1}{n^2 - 4n}$

c **Exer. 41–42:** Approximate the sum of the series to three decimal places.

41 $\displaystyle\sum_{n=1}^{\infty} (-1)^{n-1} \dfrac{1}{(2n + 1)!}$ **42** $\displaystyle\sum_{n=1}^{\infty} (-1)^{n-1} \dfrac{1}{n^2(n^2 + 1)}$

c **Exer. 43–44:** Approximate the sum of the given series to four decimal places by using an integral of the form $\int_N^{\infty} f(x)\,dx$.

43 $\displaystyle\sum_{n=1}^{\infty} n 2^{-n^2}$ **44** $\displaystyle\sum_{n=1}^{\infty} \dfrac{2n + 3}{(n^2 + 3n - 1)^2}$

Exer. 45–48: Find the interval of convergence of the series.

45 $\displaystyle\sum_{n=0}^{\infty} \dfrac{n + 1}{(-3)^n} x^n$ **46** $\displaystyle\sum_{n=0}^{\infty} (-1)^n \dfrac{4^{2n}}{\sqrt{n + 1}} x^n$

47 $\displaystyle\sum_{n=1}^{\infty} \dfrac{1}{n \, 2^n} (x + 10)^n$ **48** $\displaystyle\sum_{n=2}^{\infty} \dfrac{1}{n(\ln n)^2} (x - 1)^n$

Exer. 49–50: Find the radius of convergence of the series.

49 $\displaystyle\sum_{n=0}^{\infty} \dfrac{(2n)!}{(n!)^2} x^n$ **50** $\displaystyle\sum_{n=0}^{\infty} \dfrac{1}{(n + 5)!} (x + 5)^n$

Exer. 51–54: Find the Maclaurin series for $f(x)$, and state the radius of convergence.

51 $f(x) = \begin{cases} \dfrac{1 - \cos x}{x} & \text{if } x \neq 0 \\ 0 & \text{if } x = 0 \end{cases}$

52 $f(x) = xe^{-2x}$

53 $f(x) = \sin x \cos x$

54 $f(x) = \ln(2 + x)$

Exer. 55–58: (a) Find a Maclaurin series for the function. (b) Plot the graphs of the polynomials $P_k(x) = \sum_{n=0}^{k} a_n x^n$ associated with the power series in part (a) for $k = 3, 4$, and 5.

55 $\sinh(0.6x)$

56 $e^{0.5x}$

57 $\displaystyle\int_0^x \cosh t^2 \, dt$

58 $\displaystyle\int_0^x e^{-0.3t^2} \, dt$

59 Find a series representation for e^{-x} in powers of $x + 2$.

60 Find a series representation for $\cos x$ in powers of $x - (\pi/2)$.

Exer. 61–64: Use an infinite series to approximate the number to three decimal places.

61 $\displaystyle\int_0^1 x^2 e^{-x^2} \, dx$

62 $1/\sqrt[3]{e}$

63 $\displaystyle\int_0^1 f(x) \, dx$ with $f(x) = (\sin x)/\sqrt{x}$ if $x \neq 0$ and $f(0) = 0$

64 $e^{-0.25}$

Exer. 65–66: Find Taylor's formula with remainder for the given $f(x)$, c, and n.

65 $f(x) = \ln \cos x$, $\quad c = \pi/6$, $\quad n = 3$

66 $f(x) = \sqrt{x - 1}$, $\quad c = 2$, $\quad n = 4$

Exer. 67–68: Find Maclaurin's formula with remainder for the given $f(x)$ and n.

67 $f(x) = e^{-x^2}$, $\quad n = 3$

68 $f(x) = \dfrac{1}{1 - x}$, $\quad n = 6$

69 Use Taylor's formula with remainder to approximate $\cos 43°$ to four decimal places.

70 Use Taylor's formula with remainder to show that the approximation formula $\sin x \approx x - \frac{1}{6}x^3 + \frac{1}{120}x^5$ is accurate to four decimal places for $0 \leq x \leq \pi/4$.

EXTENDED PROBLEMS AND GROUP PROJECTS

1 (a) Define *lower bound* and *greatest lower bound* analogous to *upper bound* and *least upper bound*. Show that the completeness property is equivalent to the following statement: Every set of real numbers that has a lower bound has a greatest lower bound.

(b) A sequence $I_1, I_2, \ldots$ of sets is **nested** if I_{n+1} is a subset of I_n for each $n = 1, 2, \ldots$. Show that the completeness property is equivalent to the **nested interval property:** Every sequence of nested closed intervals has at least one real number that belongs to every interval.

(c) Does the nested interval property remain true if it is not required that each interval be closed?

2 Investigate the representation of the real numbers as decimals.

(a) Assuming that there is a one-to-one correspondence between points on the coordinate line and real numbers, show that every real number x can be written in the form $x = N.d_1 d_2 d_3 \ldots d_k \ldots$, where N is an integer and each d_k is an integer between 0 and 9.

(b) Use the nested interval property to show that every decimal expression of the form $N.d_1 d_2 d_3 \ldots d_k \ldots$ corresponds to a real number.

(c) Show that 0.5 and 0.499999... are both decimal representations of the number $\frac{1}{2}$ and that 0.9999... represents the same real number as 1.000....

(d) From part (c), we see that some real numbers have two different decimal representations. Characterize those numbers. Show that every other real number has a unique decimal representation.

(e) Show that a real number is rational if and only if it has decimal representation in which some block of digits repeats indefinitely; for example,

$$\frac{665}{3333} = 0.1995199519951995\ldots.$$

3 There is a theory of **infinite products** analogous to the theory of infinite series. Given a sequence of nonzero numbers $\{a_n\}$, we can form the *kth partial product*:

$$P_k = \prod_{n=1}^{k} a_n = (a_1)(a_2) \cdots (a_k).$$

Formulate a definition for the convergence of an infinite product

$$\prod_{n=1}^{\infty} a_n$$

in terms of limits of partial products. Prove that if an infinite product converges, then a_n approaches 1. Find examples of convergent and divergent infinite products. Discover other tests for convergence of infinite products that may be analogous to tests for convergence of infinite series.

INTRODUCTION

THE NATURAL WORLD is filled with curves that excite both the imagination of the artist and the curiosity of the scientist: the outline of the moon against the evening sky, the delicate folds of a flower, the sinuous curve of a river, the graceful silhouette of a bird in flight, the curls and spirals of a cresting wave. In this chapter, we examine several different ways of representing curves in mathematical forms that will enable us both to understand the curves better and to gain new appreciation of their beauty.

For an equation of the form $y = f(x)$, where f is a function, the graph is a *curve* in the xy-plane. The concept of curve is more general, however, than that of the graph of a function, since a curve may cross itself in figure-eight style, be closed (as are circles and ellipses), or spiral around a fixed point. In fact, some curves studied in advanced mathematics pass through every point in a coordinate plane!

The curves discussed in this chapter lie in an xy-plane, and each has the property that the coordinates x and y of an arbitrary point P on the curve can be expressed as functions of a variable t, called a *parameter*. We choose the letter t because in many applications this variable denotes time and P represents a moving object that has position (x, y) at time t. In later chapters, we will use such representations to define velocity, acceleration, and other concepts associated with motion. In Section 9.1, we consider the definitions of a curve and parametric equations, and we discuss a number of examples and applications. The determination of tangent lines and arc length from a parametric representation of a curve is the topic of Section 9.2; we also discuss how to determine the area of a surface of revolution of a curve given its parametric equations.

In Sections 9.3 and 9.4, we discuss polar (or circular) coordinates and use definite integrals to find areas enclosed by graphs of polar equations. Our methods are analogous to those developed in Chapter 5. The principal difference is that we consider limits of sums of circular sectors instead of vertical or horizontal rectangles. Switching from an xy-coordinate system to a polar coordinate system often yields a much simpler equation for a plane curve. The circles and spirals evident in such natural phenomena as the curl of an ocean wave have much simpler representations in polar coordinates than in rectangular xy-coordinates. Thus, we may be able to describe and understand some facets of nature more easily by using polar coordinates. Equations for a curve in the xy-coordinate system can also be simplified by adopting a rectangular coordinate system obtained from the standard xy-coordinates by a *translation* or *rotation* of axes, which we discuss in Section 9.5.

Parametric equations and polar coordinates provide useful frameworks for the analysis of many natural phenomena such as those exhibited by ocean waves.

Parametric Equations and Polar Coordinates

9.1   PARAMETRIC EQUATIONS

In this section, we introduce a new way to describe curves in the plane by using parametric equations. If f is a continuous function, the graph of the equation $y = f(x)$ is often called a *plane curve*. However, this definition is restrictive because it excludes many useful graphs. The following definition is more general.

Definition 9.1

> A **plane curve** is a set C of ordered pairs $(f(t), g(t))$, where f and g are continuous functions on an interval I.

For simplicity, we often refer to a plane curve as a **curve.** The **graph** of C in Definition (9.1) consists of all points $P(t) = (f(t), g(t))$ in an xy-plane, for t in I. We shall use the term *curve* interchangeably with *graph of a curve*. We sometimes regard the point $P(t)$ as tracing the curve C as t varies through the values of the interval I.

The graphs of several curves are sketched in Figure 9.1, where I is a closed interval $[a, b]$. In Figure 9.1(a), $P(a) \neq P(b)$, and $P(a)$ and $P(b)$ are called the **endpoints** of C. The curve in (a) intersects itself; that is, two different values of t produce the same point. If $P(a) = P(b)$, as in Figure 9.1(b), then C is a **closed curve.** If $P(a) = P(b)$ and C does not intersect itself at any other point, as in Figure 9.1(c), then C is a **simple closed curve.**

Figure 9.1

(a) Curve

(b) Closed curve

(c) Simple closed curve

A convenient way to represent curves is given in the next definition.

Definition 9.2

Let C be the curve consisting of all ordered pairs $(f(t), g(t))$, where f and g are continuous on an interval I. The equations

$$x = f(t), \qquad y = g(t),$$

for t in I, are **parametric equations** for C with **parameter** t.

The curve C in this definition is referred to as a **parametrized curve**, and the parametric equations are a **parametrization** for C. We often use the notation

$$x = f(t), \quad y = g(t); \quad t \text{ in } I$$

to indicate the domain I of f and g. Sometimes it may be possible to eliminate the parameter and obtain a familiar equation in x and y for C. In simple cases, we may sketch a graph of a parametrized curve by plotting points and connecting them in the order of increasing t, as illustrated in the next example.

Figure 9.2
(a) $x = 2t, y = t^2 - 1; -1 \le t \le 2$

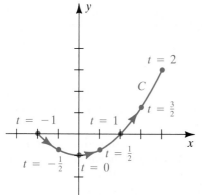

EXAMPLE ■ 1 Let C be the curve that has parametrization

$$x = 2t, \quad y = t^2 - 1; \quad -1 \le t \le 2.$$

(a) Sketch the graph of C by hand by plotting several points and joining them with a smooth curve.

(b) Obtain an equation for the curve in the form $y = f(x)$ for some function f.

(c) Use a graphing utility to plot a graph of C. Set the viewing window so that it contains the entire graph.

SOLUTION

(a) We use the parametric equations to tabulate coordinates of points $P(x, y)$ on C as follows.

(b) $-1 \le t \le 2, -4.7 \le x \le 4.8,$
$-2.1 \le y \le 4.2$

t	-1	$-\frac{1}{2}$	0	$\frac{1}{2}$	1	$\frac{3}{2}$	2
x	-2	-1	0	1	2	3	4
y	0	$-\frac{3}{4}$	-1	$-\frac{3}{4}$	0	$\frac{5}{4}$	3

Plotting points leads to the sketch in Figure 9.2(a). The arrowheads on the graph indicate the direction in which $P(x, y)$ traces the curve as t *increases* from -1 to 2.

(b) We may obtain a clearer description of the graph by eliminating the parameter. Solving the first parametric equation for t, we obtain $t = \frac{1}{2}x$. Substituting this expression for t in the second equation gives us

$$y = (\tfrac{1}{2}x)^2 - 1.$$

The graph of this equation in x and y is a parabola symmetric with respect to the y-axis with vertex $(0, -1)$. However, since $x = 2t$ and it satisfies $-1 \le t \le 2$, we see that $-2 \le x \le 4$ for points (x, y) on C, and hence C is that part of the parabola between the points $(-2, 0)$ and $(4, 3)$ shown in Figure 9.2(a).

(c) We set the graphing utility to parametric mode and enter the parametric equations. We also specify the interval for the parameter t as $[-1, 2]$. To select a viewing window that will contain the entire graph, we first note that since $x = 2t$, x will range from -2 to 4 as t ranges from -1 to 2. Similarly, since $y = t^2 - 1$, y has a minimum value of -1 at $t = 0$ and a maximum value of 3 at $t = 2$. Thus, the smallest viewing window that will accommodate the entire graph is $-2 \le x \le 4$, $-1 \le y \le 3$. We will use the slightly larger viewing window shown in Figure 9.2(b).

The graphing utility may show the curve tracing out its path in the direction indicated by the arrowheads in Figure 9.2(a). If not, we can use the trace operation to verify that the graph begins at $(-2, 0)$, moves downward through the third quadrant to $(0, -1)$, and then moves upward through quadrants IV and I until it reaches $(4, 3)$.

As indicated by the arrowheads in Figure 9.2(a), the point $P(x, y)$ traces the curve C from *left to right* as t increases. The parametric equations

$$x = -2t, \quad y = t^2 - 1; \quad -2 \le t \le 1$$

give us the same graph; however, as t increases, $P(x, y)$ traces the curve from *right to left*. For other parametrizations, the point $P(x, y)$ may oscillate back and forth as t increases.

The **orientation** of a parametrized curve C is the direction determined by *increasing* values of the parameter. We often indicate an orientation by placing arrowheads on C as in Figure 9.2(a). If $P(x, y)$ moves back and forth as t increases, we may place arrows *alongside* of C. As we have observed, a curve may have different orientations, depending on the parametrization.

The next example demonstrates that it is sometimes useful to eliminate the parameter *before* plotting points.

EXAMPLE■2   A point moves in a plane such that its position $P(x, y)$ at time t is given by

$$x = a \cos t, \quad y = a \sin t; \quad t \ge 0,$$

where $a > 0$. Describe the motion of the point.

Figure 9.3
$x = a \cos t, y = a \sin t; t \geq 0$

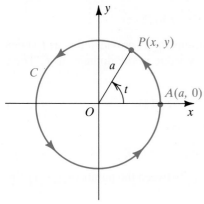

SOLUTION We may eliminate the parameter by rewriting the parametric equations as

$$\frac{x}{a} = \cos t, \qquad \frac{y}{a} = \sin t$$

and using the identity $\cos^2 t + \sin^2 t = 1$ to obtain

$$\left(\frac{x}{a}\right)^2 + \left(\frac{y}{a}\right)^2 = 1,$$

or

$$x^2 + y^2 = a^2.$$

This result shows that the point $P(x, y)$ moves on the circle C of radius a with center at the origin (see Figure 9.3). The point is at $A(a, 0)$ when $t = 0$, at $(0, a)$ when $t = \pi/2$, at $(-a, 0)$ when $t = \pi$, at $(0, -a)$ when $t = 3\pi/2$, and back at $A(a, 0)$ when $t = 2\pi$. Thus, P moves around C in a counterclockwise direction, making one revolution every 2π units of time. The orientation of C is indicated by the arrowheads in the figure.

Note that in this example we may interpret t geometrically as the radian measure of the angle generated by the line segment OP.

Figure 9.4
(a)

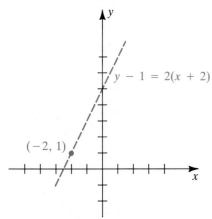

EXAMPLE ■ 3 Sketch the graph of the curve C that has the parametrization

$$x = -2 + t^2, \quad y = 1 + 2t^2; \quad t \text{ in } \mathbb{R}$$

and indicate the orientation.

SOLUTION To eliminate the parameter, we use the first equation to obtain $t^2 = x + 2$ and then substitute for t^2 in the second equation. Thus,

$$y = 1 + 2(x + 2).$$

This result is an equation of the line of slope 2 through the point $(-2, 1)$, as indicated by the dashes in Figure 9.4(a). However, since $t^2 \geq 0$, we see from the parametric equations for C that

$$x = -2 + t^2 \geq -2 \quad \text{and} \quad y = 1 + 2t^2 \geq 1.$$

(b)

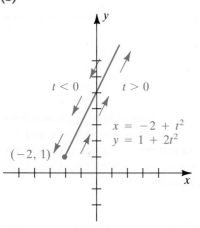

Thus, the graph of C is that part of the line to the right of $(-2, 1)$ (the point corresponding to $t = 0$), as shown in Figure 9.4(b). The orientation is indicated by the arrows alongside of C. As t increases in the interval $(-\infty, 0]$, $P(x, y)$ moves down the curve toward the point $(-2, 1)$. As t increases in $[0, \infty)$, $P(x, y)$ moves up the curve away from $(-2, 1)$.

If a curve C is described by an equation $y = f(x)$ for a continuous function f, then an easy way to obtain parametric equations for C is to let

$$x = t, \qquad y = f(t),$$

where t is in the domain of f. For example, if $y = x^3$, then parametric equations are

$$x = t, \quad y = t^3; \quad t \text{ in } \mathbb{R}.$$

We can use many different substitutions for x, provided that as t varies through some interval, x takes on every value in the domain of f. Thus, the graph of $y = x^3$ is also given by

$$x = t^{1/3}, \quad y = t; \quad t \text{ in } \mathbb{R}.$$

Note, however, that the parametric equations

$$x = \sin t, \quad y = \sin^3 t; \quad t \text{ in } \mathbb{R}$$

give only that part of the graph of $y = x^3$ between the points $(-1, -1)$ and $(1, 1)$.

E X A M P L E ▪ 4 Find three parametrizations for the line of slope m through the point (x_1, y_1).

S O L U T I O N By the point–slope form, an equation for the line is

$$y - y_1 = m(x - x_1).$$

If we let $x = t$, then $y - y_1 = m(t - x_1)$ and we obtain the parametrization

$$x = t, \quad y = y_1 + m(t - x_1); \quad t \text{ in } \mathbb{R}.$$

We obtain another parametrization for the line if we let $x - x_1 = t$. In this case, $y - y_1 = mt$, and we have

$$x = x_1 + t, \quad y = y_1 + mt; \quad t \text{ in } \mathbb{R}.$$

As a third illustration, if we let $x - x_1 = \tan t$, then

$$x = x_1 + \tan t, \quad y = y_1 + m \tan t; \quad -\frac{\pi}{2} < t < \frac{\pi}{2}.$$

There are many other parametrizations for the line.

Parametric equations of the form

$$x = a \sin \omega_1 t, \quad y = b \cos \omega_2 t; \quad t \geq 0,$$

where a, b, ω_1, and ω_2 are constants, occur in electrical theory. The variables x and y usually represent voltages or currents at time t. The resulting curve is often difficult to sketch; however, using an oscilloscope and imposing voltages or currents on the input terminals, we can represent the graph, a **Lissajous figure**,[*] on the screen of the oscilloscope. Computers are also useful in obtaining these complicated graphs.

---

[*]Jules Antoine Lissajous (1822–1890) was a French physicist known for his research in acoustics and optics. He invented a system of optical telegraphy used during the 1871 siege of Paris.

Figure 9.5
$x = \sin 2t, y = \cos t; 0 \le t \le 2\pi$

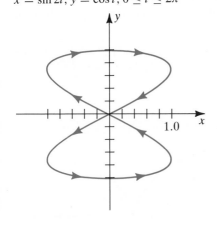

EXAMPLE ▪ 5 A graph of the Lissajous figure

$$x = \sin 2t, \quad y = \cos t; \quad 0 \le t \le 2\pi$$

is shown in Figure 9.5, with the arrowheads indicating the orientation. Verify the orientation and find an equation in x and y for the curve.

SOLUTION Referring to the parametric equations, we see that as t increases from 0 to $\pi/2$, the point $P(x, y)$ starts at $(0, 1)$ and traces the part of the curve in quadrant I (in a generally clockwise direction). As t increases from $\pi/2$ to π, $P(x, y)$ traces the part in quadrant III (in a counterclockwise direction). For $\pi \le t \le 3\pi/2$, we obtain the part in quadrant IV; and $3\pi/2 \le t \le 2\pi$ gives us the part in quadrant II.

We may find an equation in x and y for the curve by using trigonometric identities and algebraic manipulations. Writing $x = 2 \sin t \cos t$ and squaring, we have

$$x^2 = 4 \sin^2 t \cos^2 t,$$

or $$x^2 = 4(1 - \cos^2 t) \cos^2 t.$$

Using $y = \cos t$ gives us

$$x^2 = 4(1 - y^2)y^2.$$

To express y in terms of x, let us rewrite the last equation as

$$4y^4 - 4y^2 + x^2 = 0$$

and use the quadratic formula to solve for y^2 as follows:

$$y^2 = \frac{4 \pm \sqrt{16 - 16x^2}}{8} = \frac{1 \pm \sqrt{1 - x^2}}{2}$$

Taking square roots, we obtain

$$y = \pm \sqrt{\frac{1 \pm \sqrt{1 - x^2}}{2}}.$$

These complicated equations should indicate the advantage of expressing the curve in parametric form.

A curve C is **smooth** if it has a parametrization $x = f(t), y = g(t)$ on an interval I such that the derivatives f' and g' are continuous and not simultaneously zero, except possibly at endpoints of I. A curve C is **piecewise smooth** if the interval I can be partitioned into closed subintervals with C smooth on each subinterval. The graph of a smooth curve has no corners or cusps. The curves given in Examples 1–5 are smooth. The curve in the next example is piecewise smooth.

EXAMPLE ▪ 6 The curve traced by a fixed point P on the circumference of a circle as the circle rolls along a line in a plane is called a **cycloid**. Find parametric equations for a cycloid and determine the intervals on which it is smooth.

SOLUTION   Suppose the circle has radius a and that it rolls along (and above) the x-axis in the positive direction. If one position of P is the origin, then Figure 9.6 displays part of the curve and a possible position of the circle.

Figure 9.6

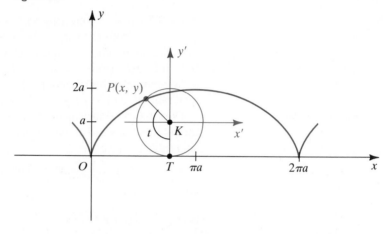

Let K denote the center of the circle and T the point of tangency with the x-axis. We introduce, as a parameter t, the radian measure of angle TKP. The distance that the circle has rolled is $d(O, T) = at$. Consequently, the coordinates of K are (at, a). We set up a new rectangular coordinate system centered at $K(at, a)$ with the horizontal and vertical axes designated by x' and y', respectively. In this $x'y'$-coordinate system, if $P(x', y')$ denotes the point P, then we have

$$x = at + x', \qquad y = a + y'.$$

Figure 9.7

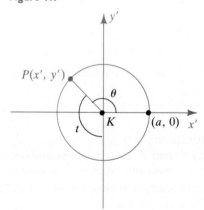

If, as in Figure 9.7, θ denotes an angle in standard position on the $x'y'$-plane, then $\theta = (3\pi/2) - t$. Hence,

$$x' = a \cos \theta = a \cos \left(\frac{3\pi}{2} - t \right) = -a \sin t$$

$$y' = a \sin \theta = a \sin \left(\frac{3\pi}{2} - t \right) = -a \cos t,$$

and substitution in $x = at + x'$, $y = a + y'$ gives us parametric equations for the cycloid:

$$x = a(t - \sin t), \quad y = a(1 - \cos t); \quad t \text{ in } \mathbb{R}$$

Differentiating the parametric equations of the cycloid yields

$$\frac{dx}{dt} = a(1 - \cos t), \qquad \frac{dy}{dt} = a \sin t.$$

These derivatives are continuous for every t, but are simultaneously 0 at $t = 2\pi n$ for every integer n. The points corresponding to $t = 2\pi n$ are the x-intercepts of the graph, and the cycloid has a cusp at each such point (see Figure 9.6). The graph is piecewise smooth, since it is smooth on the t-interval $[2\pi n, 2\pi (n + 1)]$ for every integer n.

Figure 9.8

If $a < 0$, then the graph of $x = a(t - \sin t)$, $y = a(1 - \cos t)$ is the inverted cycloid that results if the circle of Example 6 rolls *below* the x-axis. This curve has a number of important physical properties. To illustrate, suppose that a thin wire passes through two fixed points A and B, as shown in Figure 9.8, and that the shape of the wire can be changed by bending it in any manner. Suppose further that a bead is allowed to slide along the wire and the only force acting on the bead is gravity. We now ask which of all the possible paths will allow the bead to slide from A to B in the least amount of time. This question is the brachistochrone problem discussed in Section 7.5. It is natural to believe that the desired path is the straight line segment from A to B; however, this is not the correct answer. In Section 7.5, we saw that the path requiring the least amount of time coincides with the graph of an inverted cycloid with A at the origin. Because the velocity of the bead increases more rapidly along the cycloid than along the line through A and B, the bead reaches B more rapidly, even though the distance is greater.

There is another interesting property of this **curve of least descent.** Suppose that A is the origin and B is the point with x-coordinate $\pi |a|$—that is, the lowest point on the cycloid in the first arc to the right of A. If the bead is released at *any* point between A and B, it can be shown that the time required for the bead to reach B is always the same.

Variations of the cycloid occur in applications. For example, if a motorcycle wheel rolls along a straight road, then the curve traced by a fixed point on one of the spokes is a cycloidlike curve. In this case, the curve does not have corners or cusps, nor does it intersect the road (the x-axis) as does the graph of a cycloid. If the wheel of a train rolls along a railroad track, then the curve traced by a fixed point on the circumference of the wheel (which extends below the track) contains loops at regular intervals. Other cycloids are defined in Exercises 33 and 34.

As another application of parametric equations, we consider how they can be used to study projectile motion. Suppose that an object is projected into the air with an initial horizontal velocity of h ft/sec and an initial vertical velocity of v ft/sec. If an xy-coordinate system is set up with the origin at the object's initial position at time $t = 0$, then we can find parametric equations that describe the position of the object at subsequent times if we assume that the only force acting on the object is the gravitational attraction of the earth. (We ignore air resistance, for example.) Then since there is no force to accelerate or decelerate the horizontal motion, the horizontal velocity remains constant, and we have $x(t) = ht$. For the vertical motion, we have, from our discussion of free fall in Section 3.7, $y(t) = -16t^2 + vt$.

In Chapter 11, we will see that if the object is projected into the air at an angle of θ with an initial speed of s ft/sec, then the initial horizontal and vertical velocities are given by

$$h = s \cos \theta \quad \text{and} \quad v = s \sin \theta.$$

We summarize our discussion with a slightly generalized statement of parametric equations for the motion of a projectile acted on only by a constant gravitational force.

*Parametric Equations
for Projectile Motion*   **9.3**

The equations of motion of a projectile in a plane launched from an initial position (x_0, y_0) at time $t = 0$ with an initial horizontal velocity h_0 and an initial vertical velocity v_0 are

(i) $x(t) = x_0 + h_0 t, \quad y(t) = -\frac{1}{2}gt^2 + v_0 t + y_0,$

where g is the magnitude of the assumed constant acceleration of gravity. If the projectile is launched at an angle of elevation θ with an initial speed s_0, then

(ii) $h_0 = s_0 \cos\theta, \quad v_0 = s_0 \sin\theta.$

Figure 9.9

90 ft

30 ft

EXAMPLE ▪ 7   A pitcher on a baseball team throws a ball to a friend who is standing on the roof of a building 90 ft high. The pitcher stands 30 ft from the base of the building and releases the ball from a height of 8 ft with an initial horizontal velocity of 23.5 ft/sec and an initial vertical velocity of 84.8 ft/sec. (See Figure 9.9.)

(a) Determine whether the ball will reach the top of the building.

(b) Estimate the initial speed of the ball and the angle of release.

SOLUTION

(a) The parametric equations for the motion of the ball are

$$x(t) = 23.5t \quad \text{and} \quad y(t) = -16t^2 + 84.8t + 8.$$

When the ball reaches the wall, $x = 30$ and the corresponding time T satisfies $23.5T = 30$, so

$$T = \frac{30}{23.5} \approx 1.2765957 \, \text{sec}.$$

At this time T, the y-coordinate is given by

$$y(T) \approx -16(1.2765957)^2 + 84.8(1.2765957) + 8 \approx 90.18 \, \text{ft}.$$

Thus, since the ball will be 90.18 ft high when it reaches the building, it will just clear the top of the building.

(b) If s is the initial speed and the ball is released at an angle of θ, then

$$s \sin\theta = 84.8 \quad \text{and} \quad s \cos\theta = 23.5.$$

If we square each equation and then add them, we obtain

$$s^2 \sin^2\theta + s^2 \cos^2\theta = (84.8)^2 + (23.5)^2$$
$$s^2(\sin^2\theta + \cos^2\theta) = 7191.04 + 552.25$$
$$s^2 = 7743.29$$
$$s = \sqrt{7743.29} \approx 87.99596582.$$

Thus, the original speed was approximately 88 ft/sec (about 60 mi/hr). To determine the angle θ of release, we note that

$$\frac{s \sin \theta}{s \cos \theta} = \frac{84.8}{23.5} \quad \text{or, equivalently,} \quad \tan \theta = \frac{84.8}{23.5}.$$

Thus, the angle of release is

$$\theta = \tan^{-1}\left(\frac{84.8}{23.5}\right) \approx 1.3 \text{ radians} \quad \text{or about } 74.5°.$$

We next examine an important application of parametric curves that was first introduced by the French scientist Pierre Bézier.* **Bézier curves** are special parametric curves commonly used in computer-aided design, microcomputer drawing applications, and the mathematical representation of different fonts for laser printers. Bézier was trying to solve a problem plaguing the designers of stamped parts such as car-body panels: The curves they created at the drawing board did not coincide exactly with what was produced. He wanted to devise a method to represent in a computer "an accurate, complete and indisputable definition of freeform shapes." He discovered that this could be done by piecing together particular types of cubic polynomials.

A cubic Bézier curve is specified by four control points in the plane, $P_0(p_0, q_0)$, $P_1(p_1, q_1)$, $P_2(p_2, q_2)$, and $P_3(p_3, q_3)$. The curve starts at the first point for the parameter $t = 0$, ends at the last point for $t = 1$, and roughly "heads toward" the middle points for parameter values between 0 and 1. Artists and engineering designers can move the control points to adjust the end locations and the shape of the parametric curve until what appears on the computer screen is the shape they want. The cubic Bézier curve for the four control points has the following parametric equations:

$$x(t) = p_0(1-t)^3 + 3p_1(1-t)^2 t + 3p_2(1-t)t^2 + p_3 t^3,$$

$$y(t) = q_0(1-t)^3 + 3q_1(1-t)^2 t + 3q_2(1-t)t^2 + q_3 t^3; \quad 0 \leq t \leq 1.$$

Note that

$$x(0) = p_0(1-0)^3 + 3p_1(1-0)^2 \cdot 0 + 3p_2(1-0) \cdot 0^2 + p_3 \cdot 0^3 = p_0.$$

Similarly, we have $y(0) = q_0$, so the curve passes through the control point P_0 at $t = 0$. We may also compute that $x(1) = p_3$ and $y(1) = q_3$ so the curve passes through P_3 at $t = 1$. For values of t between 0 and 1, $x(t)$ is a "weighted average" of the x-coordinates of all four control points and $y(t)$ is a weighted average of their y-coordinates. For example, at $t = \frac{1}{2}$, we have

$$x\left(\tfrac{1}{2}\right) = \frac{p_0 + 3p_1 + 3p_2 + p_3}{8} \quad \text{and} \quad y\left(\tfrac{1}{2}\right) = \frac{q_0 + 3q_1 + 3q_2 + q_3}{8}.$$

---

*Pierre E. Bézier is a contemporary French scientist whose mathematical, engineering, and design work for the Rénault automobile company beginning in the mid-1960s led to the development of the field of computer-aided geometric design.

In general, the cubic Bézier curve will *not* pass through the control points P_1 and P_2. The relative locations of these points, however, will determine the shape of the Bézier curve.

EXAMPLE ▪ 8   Use a graphing utility to obtain the graph of the cubic Bézier curve with the control points $P_0(32, 6)$, $P_1(85, 30)$, $P_2(6, 35)$, and $P_3(45, 8)$. Set the viewing window so that the control points appear within it and plot the control points.

SOLUTION   Using the general form for the parametric equations for the cubic Bézier curve with

$$p_0 = 32 \quad p_1 = 85 \quad p_2 = 6 \quad p_3 = 45$$
$$q_0 = 6 \quad q_1 = 30 \quad q_2 = 35 \quad q_3 = 8,$$

Figure 9.10
$-10 \le x \le 90, -10 \le y \le 50$

we obtain the parametric equations for this Bézier curve:

$$x(t) = 32(1 - t)^3 + 255(1 - t)^2 t + 18(1 - t)t^2 + 45t^3,$$
$$y(t) = 6(1 - t)^3 + 90(1 - t)^2 t + 105(1 - t)t^2 + 8t^3; \quad 0 \le t \le 1$$

The x-coordinates of the control points range from 6 to 85, and the y-coordinates of the control points range from 6 to 35. To ensure that all four control points and the coordinate axes will appear on the screen, we set the viewing window to $-10 \le x \le 90, -10 \le y \le 50$, and plot the equations and the control points to obtain the curve shown in Figure 9.10.

Several Bézier curves can be pieced together continuously by making the last control point on one curve the first control point on the next curve. Piecewise parametric equations can be constructed in a similar manner. For simplicity, we treat each piece as a separate parametric curve so that we may use the same form (repeatedly) in a graphing utility. If we use the four control points P_0, P_1, P_2, and P_3 to determine the first piece of the curve with parametric equations for $0 \le t \le 1$, then control points P_3, P_4, P_5, and P_6 are used for the next piece, again with $0 \le t \le 1$. Because the fourth control point for the first piece is the first control point for the next piece, the two pieces fit together continuously. The equations for the second piece of the curve are

$$x(t) = p_3(1 - t)^3 + 3p_4(1 - t)^2 t + 3p_5(1 - t)t^2 + p_6 t^3,$$
$$y(t) = q_3(1 - t)^3 + 3q_4(1 - t)^2 t + 3q_5(1 - t)t^2 + q_6 t^3; \quad 0 \le t \le 1.$$

EXAMPLE ▪ 9   Use a graphing utility to obtain the graph of the continuous piecewise Bézier curve with the control points $P_0(10, 15)$, $P_1(16, 14)$, $P_2(25, 38)$, $P_3(30, 40)$ (repeated), $P_4(18, 5)$, $P_5(50, 20)$, and $P_6(16, 30)$. Set the viewing window so that the control points appear within it and plot the control points.

SOLUTION Since seven control points have been specified, the curve will have two pieces. We use the first four control points (P_0, P_1, P_2, P_3) for the first piece and the last four (P_3, P_4, P_5, P_6) for the second piece. Using the general form for the parametric equations of a cubic Bézier curve, we obtain the parametric equations:

$$x_1(t) = 10(1-t)^3 + 48(1-t)^2 t + 75(1-t)t^2 + 30t^3,$$
$$y_1(t) = 15(1-t)^3 + 42(1-t)^2 t + 114(1-t)t^2 + 40t^3,$$

and

$$x_2(t) = 30(1-t)^3 + 54(1-t)^2 t + 150(1-t)t^2 + 16t^3,$$
$$y_2(t) = 40(1-t)^3 + 15(1-t)^2 t + 60(1-t)t^2 + 30t^3,$$

both for $0 \le t \le 1$.

Figure 9.11 shows a plot of the equations and the control points.

Figure 9.11
$-10 \le x \le 85, -10 \le y \le 53$

Cubic Bézier curves are often used in the design of characters (letters, numerals, and punctuation marks) that will be printed in different font sizes on a laser printer. Stored in the memory of the laser printer are the control points for each character. When it needs to print a letter "A," for example, the printer recalls the control points for "A" and then directs a graphing utility to plot the cubic Bézier curve with those points. Since there are a relatively small number of control points for each character, the memory requirements for storing the information about all the characters on the keyboard is not large. Another advantage is that only one size for the font need be stored in the memory of the laser printer. A simple scaling or rotation of the control points will yield characters of different size or rotational orientation. The general form of the cubic equations that give the parametrization of the Bézier curves and the specific control points for a particular character form a *mathematical representation* of that character.

The mathematical representation of other designs besides characters is also given by specifying particular collections of control points. To obtain such a mathematical representation for a given figure, we begin by sketching the figure on a sheet of rectangularly ruled graph paper, mark some points on the figure, and then use these points as "end" control points for a piecewise continuous sequence of Bézier curves. Other control points needed between these "ends" will not lie on the final curve, but will "pull" the curve in certain directions.

EXAMPLE ▪ 10 Use several Bézier curves to obtain a parametrization of a curve whose graph is the numeral "2."

SOLUTION We begin with a crude drawing of the numeral "2" on a large sheet of graph paper. We draw it in a cursive style with no

Figure 9.12
$-3 \le x \le 92, \ -3 \le y \le 60$
(a)

(b)

line segments, and then subdivide the character into pieces that appear to be possible to obtain by a single Bézier curve. For the segments we have chosen, we try control points $P_0(40, 40)$, $P_1(48, 55)$, $P_2(70, 45)$, $P_3(50, 25)$ (repeated), $P_4(43, 18)$, $P_5(40, 13)$, $P_6(40, 10)$ (repeated), $P_7(46, 14)$, $P_8(53, 9)$, and $P_9(58, 10)$. Plotting the associated equations and the control points gives us the graph shown in Figure 9.12(a).

The use of a graphing utility makes it quite easy to experiment with modifications in the locations of the control points until we are satisfied that the Bézier curve produced represents the figure we have designed. If the result of the first attempt is not satisfactory, we move some control points and try again. Software for computer-aided geometric design allows the designer to use a mouse or trackball to drag a control point from one part of the computer screen to another. The software then recalculates the equations for the Bézier curves on the basis of the new coordinates of the control points and immediately plots the graph of the new curve.

EXERCISES 9.1

Exer. 1–24: **(a)** Find an equation in x and y whose graph contains the points on the curve C. **(b)** Sketch the graph of C and indicate the orientation.

1 $x = t - 2$, $y = 2t + 3$; $0 \le t \le 5$

2 $x = 1 - 2t$, $y = 1 + t$; $-1 \le t \le 4$

3 $x = t^2 + 1$, $y = t^2 - 1$; $-2 \le t \le 2$

4 $x = t^3 + 1$, $y = t^3 - 1$; $-2 \le t \le 2$

5 $x = 4t^2 - 5$, $y = 2t + 3$; t in $\mathbb{R}$

6 $x = t^3$, $y = t^2$; t in $\mathbb{R}$

7 $x = e^t$, $y = e^{-2t}$; t in $\mathbb{R}$

8 $x = \sqrt{t}$, $y = 3t + 4$; $t \ge 0$

9 $x = 2 \sin t$, $y = 3 \cos t$; $0 \le t \le 2\pi$

10 $x = \cos t - 2$, $y = \sin t + 3$; $0 \le t \le 2\pi$

11 $x = \sec t$, $y = \tan t$; $-\pi/2 < t < \pi/2$

12 $x = \cos 2t$, $y = \sin t$; $-\pi \le t \le \pi$

13 $x = t^2$, $y = 2 \ln t$; $t > 0$

14 $x = \cos^3 t$, $y = \sin^3 t$; $0 \le t \le 2\pi$

15 $x = \sin t$, $y = \csc t$; $0 < t \le \pi/2$

16 $x = e^t$, $y = e^{-t}$; t in $\mathbb{R}$

17 $x = \cosh t$, $y = \sinh t$; t in $\mathbb{R}$

18 $x = 3 \cosh t$, $y = 2 \sinh t$; t in $\mathbb{R}$

19 $x = t$, $y = \sqrt{t^2 - 1}$; $|t| \ge 1$

20 $x = -2\sqrt{1 - t^2}$, $y = t$; $|t| \le 1$

21 $x = t$, $y = \sqrt{t^2 - 2t + 1}$; $0 \le t \le 4$

22 $x = 2t$, $y = 8t^3$; $-1 \le t \le 1$

23 $x = (t + 1)^3$, $y = (t + 2)^2$; $0 \le t \le 2$

24 $x = \tan t$, $y = 1$; $-\pi/2 < t < \pi/2$

Exer. 25–26: **Curves C_1, C_2, C_3, and C_4 are given parametrically, for t in $\mathbb{R}$. Sketch their graphs and indicate orientations.**

25 C_1: $x = t^2$, $y = t$
    C_2: $x = t^4$, $y = t^2$
    C_3: $x = \sin^2 t$, $y = \sin t$
    C_4: $x = e^{2t}$, $y = -e^t$

26 C_1: $x = t$, $y = 1 - t$
    C_2: $x = 1 - t^2$, $y = t^2$
    C_3: $x = \cos^2 t$, $y = \sin^2 t$
    C_4: $x = \ln t - t$, $y = 1 + t - \ln t$; $t > 0$

Exer. 27–28: The parametric equations specify the position of a moving point $P(x, y)$ at time t. Sketch the graph and indicate the motion of P as t increases.

27 (a) $x = \cos t,$ $\quad y = \sin t;$ $\qquad 0 \le t \le \pi$

 (b) $x = \sin t,$ $\quad y = \cos t;$ $\qquad 0 \le t \le \pi$

 (c) $x = t,$ $\qquad y = \sqrt{1 - t^2};$ $\quad -1 \le t \le 1$

28 (a) $x = t^2,$ $\qquad y = 1 - t^2;$ $\qquad 0 \le t \le 1$

 (b) $x = 1 - \ln t,$ $\quad y = \ln t;$ $\qquad 1 \le t \le e$

 (c) $x = \cos^2 t,$ $\quad y = \sin^2 t;$ $\qquad 0 \le t \le 2\pi$

29 Show that

$$x = a \cos t + h, \quad y = b \sin t + k; \quad 0 \le t \le 2\pi$$

are parametric equations of an ellipse with center (h, k) and axes of lengths $2a$ and $2b$.

30 Show that

$$x = a \sec t + h, \quad y = b \tan t + k;$$

$$-\frac{\pi}{2} < t < \frac{3\pi}{2} \quad \text{and} \quad t \ne \frac{\pi}{2}$$

are parametric equations of a hyperbola with center (h, k), transverse axis of length $2a$, and conjugate axis of length $2b$. Determine the values of t for each branch.

31 If $P_1(x_1, y_1)$ and $P_2(x_2, y_2)$ are distinct points, show that

$$x = (x_2 - x_1)t + x_1, \quad y = (y_2 - y_1)t + y_1; \quad t \text{ in } \mathbb{R}$$

are parametric equations for the line l through P_1 and P_2.

32 Describe the difference between the graph of the hyperbola $(x^2/a^2) - (y^2/b^2) = 1$ and the graph of

$$x = a \cosh t, \quad y = b \sinh t; \quad t \text{ in } \mathbb{R}.$$

(*Hint:* Use Theorem (6.42).)

33 A circle C of radius b rolls on the outside of the circle $x^2 + y^2 = a^2$, and $b < a$. Let P be a fixed point on C, and let the initial position of P be $A(a, 0)$, as shown in the figure. If the parameter t is the angle from the positive x-axis to the line segment from O to the center of C, show that parametric equations for the curve traced by P (an *epicycloid*) are

$$x = (a + b) \cos t - b \cos \left(\frac{a + b}{b} t \right),$$

$$y = (a + b) \sin t - b \sin \left(\frac{a + b}{b} t \right); \quad 0 \le t \le 2\pi.$$

Exercise 33

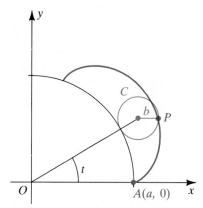

34 If the circle C of Exercise 33 rolls on the inside of the second circle (see figure), then the curve traced by P is a *hypocycloid*.

 (a) Show that parametric equations for this curve are

$$x = (a - b) \cos t + b \cos \left(\frac{a - b}{b} t \right),$$

$$y = (a - b) \sin t - b \sin \left(\frac{a - b}{b} t \right); \quad 0 \le t \le 2\pi.$$

 (b) If $b = \frac{1}{4}a$, show that $x = a \cos^3 t$, $y = a \sin^3 t$ and sketch the graph.

Exercise 34

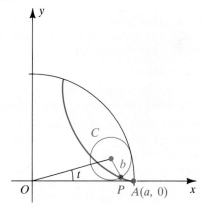

35 If $b = \frac{1}{3}a$ in Exercise 33, find the parametric equations for the epicycloid and sketch the graph.

36 The radius of circle B is one third that of circle A. How many revolutions will circle B make as it rolls around circle A until it reaches its starting point? (*Hint:* Use Exercise 35.)

37 If a string is unwound from around a circle of radius a and is kept tight in the plane of the circle, then a fixed point P on the string will trace a curve called the *involute of the circle.* Let the circle be chosen as in the figure. If the parameter t is the measure of the indicated angle and the initial position of P is $A(a, 0)$, show that parametric equations for the involute are

$$x = a(\cos t + t \sin t), \quad y = a(\sin t - t \cos t).$$

Exercise 37

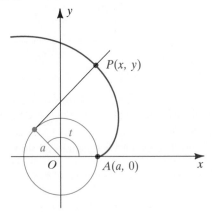

38 Generalize the cycloid of Example 6 to the case where P is any point on a fixed line through the center C of the circle. If $b = d(C, P)$, show that

$$x = at - b \sin t, \quad y = a - b \cos t.$$

Sketch a typical graph if $b < a$ (a *curtate cycloid*) and if $b > a$ (a *prolate cycloid*). The term *trochoid* is sometimes used for either of these curves.

39 Refer to Example 5.

(a) Describe the Lissajous figure given by $f(t) = a \sin \omega t$ and $g(t) = b \cos \omega t$ for $t \geq 0$ and $a \neq b$.

(b) Suppose $f(t) = a \sin \omega_1 t$ and $g(t) = b \sin \omega_2 t$, where ω_1 and ω_2 are positive rational numbers, and write ω_2/ω_1 as m/n for positive integers m and n. Show that if $p = 2\pi n/\omega_1$, then $f(t + p) = f(t)$ and $g(t + p) = g(t)$. Conclude that the curve retraces itself every p units of time.

40 Shown in the figure is the Lissajous figure given by

$$x = 2 \sin 3t, \quad y = 3 \sin 1.5t; \quad t \geq 0.$$

(a) Find the period of the figure—that is, the length of the smallest t-interval that traces the curve.

(b) Find the maximum distance from the origin to a point on the graph.

Exercise 40

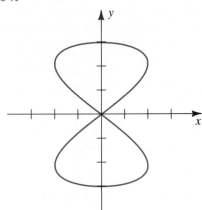

c **Exer. 41–44: Graph the curve.**

41 $x = 3 \sin^5 t$, $\qquad y = 3 \cos^5 t$;
$\qquad\qquad\qquad\qquad 0 \leq t \leq 2\pi$

42 $x = 8 \cos t - 2 \cos 4t$, $\quad y = 8 \sin t - 2 \sin 4t$;
$\qquad\qquad\qquad\qquad\qquad\qquad 0 \leq t \leq 2\pi$

43 $x = 3t - 2 \sin t$, $\qquad y = 3 - 2 \cos t$;
$\qquad\qquad\qquad\qquad -8 \leq t \leq 8$

44 $x = 2t - 3 \sin t$, $\qquad y = 2 - 3 \cos t$;
$\qquad\qquad\qquad\qquad -8 \leq t \leq 8$

Exer. 45–48: Graph the given curves on the same coordinate axes and describe the shape of the resulting figure.

c **45** $C_1: x = 2 \sin 3t, \ y = 3 \cos 2t$; $\qquad \dfrac{-\pi}{2} \leq t \leq \dfrac{\pi}{2}$

$C_2: x = \dfrac{1}{4} \cos t + \dfrac{3}{4}, \ y = \dfrac{1}{4} \sin t + \dfrac{3}{2}$; $\ \ 0 \leq t \leq 2\pi$

$C_3: x = \dfrac{1}{4} \cos t - \dfrac{3}{4}, \ y = \dfrac{1}{4} \sin t + \dfrac{3}{2}$; $\ \ 0 \leq t \leq 2\pi$

$C_4: x = \dfrac{3}{4} \cos t, \ y = \dfrac{1}{4} \sin t$; $\qquad\qquad 0 \leq t \leq 2\pi$

$C_5: x = \dfrac{1}{4} \cos t, \ y = \dfrac{1}{8} \sin t + \dfrac{3}{4}$; $\qquad \pi \leq t \leq 2\pi$

46 $C_1: x = \dfrac{3}{2} \cos t + 1, \ y = \sin t - 1$; $\quad \dfrac{-\pi}{2} \leq t \leq \dfrac{\pi}{2}$

$C_2: x = \dfrac{3}{2} \cos t + 1, \ y = \sin t + 1$; $\quad \dfrac{-\pi}{2} \leq t \leq \dfrac{\pi}{2}$

$C_3: x = 1, \ y = 2 \tan t$; $\qquad\qquad \dfrac{-\pi}{4} \leq t \leq \dfrac{\pi}{4}$

47 $C_1: x = \tan t, \ y = 3 \tan t$; $\qquad\qquad 0 \leq t \leq \dfrac{\pi}{4}$

$C_2: x = 1 + \tan t, \ y = 3 - 3 \tan t$; $\quad 0 \leq t \leq \dfrac{\pi}{4}$

$C_3: x = \dfrac{1}{2} + \tan t, \ y = \dfrac{3}{2}$; $\qquad\qquad 0 \leq t \leq \dfrac{\pi}{4}$

48 C_1: $x = 1 + \cos t$, $y = 1 + \sin t$; $\pi/3 \le t \le 2\pi$

C_2: $x = 1 + \tan t$, $y = 1$; $0 \le t \le \dfrac{\pi}{4}$

49 If a rock is thrown from a point 3 ft above the ground with a horizontal velocity of 90 ft/sec and a vertical velocity of 47 ft/sec, how far away will it land if nothing is obstructing its path? If there is a 7-ft high fence 9 ft in front of you, will the rock sail over the fence?

50 A basketball player shoots a ball with a speed of 25 ft/sec from a point 15 ft horizontally away from the center of the basket. The basket is 10 ft above the floor and the player releases the ball from a height of 8 ft. At what angle should the player shoot the ball?

51 If a projectile is launched at an angle θ to the horizontal, show that its horizontal range is $4M/(\tan \theta)$, where M is the maximum height reached by the projectile.

52 Anne kicks a soccer ball toward her brother Sasha with an initial velocity of 48 ft/sec at an angle of elevation of $\pi/6$. At the moment of the kick, he is 90 ft away and starts running to meet the ball. Sasha's top speed is 20 ft/sec.

(a) Can Sasha reach the ball before it hits the ground?

(b) If Sasha is 5 ft 6 in. tall, can he reach the ball in time to bounce it off the top of his head?

(c) For what range of angles can Anne kick the ball so that Sasha has time to hit the ball with his head?

[c] **Exer. 53–58:** Plot the graph of the continuous piecewise Bézier curve for the given control points. Set the viewing window so that all control points appear within the window. If possible, use equally scaled axes and also plot the control points. For the first piece of each curve, use control points P_0, P_1, P_2, and P_3. If more points are given, use P_3, P_4, P_5, and P_6 for the second piece and P_6, P_7, P_8, and P_9 for the third piece.

53 $P_0(10, 2)$, $P_1(2, 60)$, $P_2(100, 56)$, and $P_3(110, 10)$

54 $P_0(1, 32)$, $P_1(25, 85)$, $P_2(30, 1)$, and $P_3(3, 40)$

55 $P_0(5, 10)$, $P_1(4, 16)$, $P_2(28, 25)$, $P_3(30, 30)$, $P_4(1, 18)$, $P_5(18, 40)$, and $P_6(20, 16)$

56 $P_0(60, 40)$, $P_1(50, 30)$, $P_2(43, 10)$, $P_3(55, 10)$, $P_4(65, 10)$, $P_5(68, 25)$, and $P_6(50, 22)$

57 $P_0(30, 30)$, $P_1(58, 10)$, $P_2(12, 12)$, $P_3(45, 10)$, $P_4(40, 5)$, $P_5(66, 31)$, and $P_6(25, 30)$

58 $P_0(48, 20)$, $P_1(20, 15)$, $P_2(20, 50)$, $P_3(48, 45)$, $P_4(28, 47)$, $P_5(28, 18)$, $P_6(48, 20)$, $P_7(48, 36)$, $P_8(52, 32)$, and $P_9(40, 32)$

[c] **Exer. 59–60:** Experiment with the locations for control points to obtain piecewise Bézier curves approximating the given shape or object. Give the final control points chosen, and sketch the resulting parametric curve.

59 Find a piecewise Bézier curve with two components that approximates the letter "S" in a simple font. Improve the sketch by using a piecewise Bézier curve with three components.

60 Find a piecewise Bézier curve that approximates the Gateway Arch to the West. (See Exercise 35 of Section 6.8.)

Mathematicians and Their Times

AUGUSTIN-LOUIS CAUCHY

THE PEOPLE OF PARIS rose on July 14, 1789, to storm the Bastille and begin the struggle and violence of the French Revolution. The Revolution promised a new era of democracy, liberty, and equality to replace the supreme power of the king. It was tarnished by the Reign of Terror begun in 1793 by the Committee of Public Safety, which swept hundreds to the guillotine in its zeal to protect France's internal security.

Among those in gravest danger were government officials suspected of loyalty to the king. Many fled to small villages to find safety. One such man was the father of Augustin-Louis Cauchy. A child of the Revolution, Cauchy was born on August 21, 1789. His earliest years were spent in fear and exile. Tumultuous political events continued to dominate France for most of his life: revolution in 1789, the rise and fall of Napoleon, the restoration of the Bourbon kings in 1814, more revolutions in 1830 and 1848, the overthrow of the second republic and establishment of the second empire by Napoleon III.

Cauchy's conservative political and religious attitudes put him at odds with many of his fellow scientists. His detractors called him self-righteous and arrogant, a narrow-minded bigot, and a smug hypocrite. His defenders regarded him as a pious believer in traditional religion, highly principled, and sincere but naive in his politics. All agree now, however, that Cauchy was one of the most influential mathematicians of the nineteenth century.

Cauchy's specific contributions to pure and applied mathematics were both deep and broad. He essentially created, for example, both the theory of functions of a complex variable and finite group theory. Even more important was his lead in raising the standards of rigor. Both Gauss and Cauchy were the "apostles of rigor" who transformed mathematics. "It is difficult to find an adequate simile for the magnitude of this advance," wrote E. T. Bell. "Suppose that for centuries an entire people had been worshipping false gods and that suddenly their error is revealed to them." Although Gauss preceded him, Cauchy had greater impact because of his gift for effective teaching, and his many textbooks and research papers that showed how calculus could be developed in a rigorous manner.

Cauchy died unexpectedly on May 23, 1857. Although estranged from many colleagues who charged him with judging scientists more on the basis of their political and religious views than on their scientific achievements, Cauchy was actively planning new charitable works. His last words were addressed to the Archbishop of Paris: "Men pass away, but their deeds abide."

9.2 ARC LENGTH AND SURFACE AREA

If a curve is described by an equation of the form $y = f(x)$, where f is a differentiable function, we know from earlier chapters how to find the slope of the tangent line at a point on the curve, the length of a segment of the curve, and the area of the surface of revolution obtained by revolving the curve about an axis. In this section, we discuss how to find these quantities when the curve is described by parametric equations.

The curve C given parametrically by

$$x = 2t, \quad y = t^2 - 1; \quad -1 \le t \le 2$$

can also be represented by an equation of the form $y = k(x)$, where k is a function defined on a suitable interval. In Example 1 of Section 9.1, we eliminated the parameter t, obtaining

$$y = k(x) = \tfrac{1}{4}x^2 - 1 \quad \text{for} \quad -2 \le x \le 4.$$

The slope of the tangent line at any point $P(x, y)$ on C is

$$k'(x) = \tfrac{1}{2}x, \quad \text{or} \quad k'(x) = \tfrac{1}{2}(2t) = t.$$

Since it is often difficult to eliminate a parameter, we next derive a formula that can be used to find the slope directly from the parametric equations.

Theorem 9.4

If a smooth curve C is given parametrically by $x = f(t)$, $y = g(t)$, then the slope dy/dx of the tangent line to C at $P(x, y)$ is

$$\frac{dy}{dx} = \frac{dy/dt}{dx/dt}, \quad \text{provided} \quad \frac{dx}{dt} \ne 0.$$

PROOF   If $dx/dt \neq 0$ at $x = c$, then, since f is continuous at c, it follows from the intermediate value theorem (1.26) that $dx/dt > 0$ or $dx/dt < 0$ throughout an interval $[a, b]$, with $a < c < b$. Applying Theorem (6.6) or the analogous result for decreasing functions, we know that f has an inverse function f^{-1}, and we may consider $t = f^{-1}(x)$ for x in $[f(a), f(b)]$. Applying the chain rule to $y = g(t)$ and $t = f^{-1}(x)$, we obtain

$$\frac{dy}{dx} = \frac{dy}{dt}\frac{dt}{dx} = \frac{dy/dt}{dx/dt},$$

where the last equality follows from Corollary (6.8).   ■

EXAMPLE ▪ I   Let C be the curve with parametrization

$$x = 2t, \quad y = t^2 - 1; \quad -1 \leq t \leq 2.$$

Find the slopes of the tangent line and normal line to C at $P(x, y)$.

SOLUTION   The curve C was considered in Example 1 of Section 9.1 (see Figure 9.2). Using Theorem (9.4) with $x = 2t$ and $y = t^2 - 1$, we find that the slope of the tangent line at $P(x, y)$ is

$$\frac{dy}{dx} = \frac{dy/dt}{dx/dt} = \frac{2t}{2} = t.$$

This result agrees with that of the discussion at the beginning of this section, where we used the form $y = k(x)$ to show that $m = \frac{1}{2}x = t$.

The slope of the normal line is the negative reciprocal $-1/t$, provided $t \neq 0$.

EXAMPLE ▪ 2   Let C be the curve with parametrization

$$x = t^3 - 3t, \quad y = t^2 - 5t - 1; \quad t \text{ in } \mathbb{R}.$$

(a) Find an equation of the tangent line to C at the point corresponding to $t = 2$.

(b) For what values of t is the tangent line horizontal or vertical?

Figure 9.13
$x = t^3 - 3t, y = t^2 - 5t - 1; t \text{ in } \mathbb{R}$

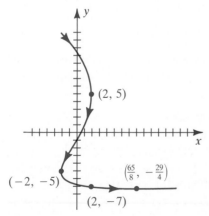

SOLUTION

(a) A portion of the graph of C is sketched in Figure 9.13, where we have also plotted several points and indicated the orientation. Using the parametric equations for C, we find that the point corresponding to $t = 2$ is $(2, -7)$. By Theorem (9.4),

$$\frac{dy}{dx} = \frac{dy/dt}{dx/dt} = \frac{2t - 5}{3t^2 - 3}.$$

The slope m of the tangent line at $(2, -7)$ is

$$m = \frac{dy}{dx}\bigg]_{t=2} = \frac{2(2) - 5}{3(2^2) - 3} = -\frac{1}{9}.$$

Applying the point–slope form, we obtain an equation of the tangent line:

$$y + 7 = -\tfrac{1}{9}(x - 2), \quad \text{or} \quad x + 9y = -61$$

(b) The tangent line is horizontal if $dy/dx = 0$—that is, if $2t - 5 = 0$, or $t = \tfrac{5}{2}$. The corresponding point on C is $(\tfrac{65}{8}, -\tfrac{29}{4})$, as shown in Figure 9.13.

The tangent line is vertical if $3t^2 - 3 = 0$. Thus, there are vertical tangent lines at the points corresponding to $t = 1$ and $t = -1$—that is, at $(-2, 5)$ and $(2, 5)$.

If a curve C is parametrized by $x = f(t)$, $y = g(t)$ and if y' is a differentiable function of t, we can find d^2y/dx^2 by applying Theorem (9.4) to y' as follows.

**Second Derivative
in Parametric Form 9.5**

$$\frac{d^2y}{dx^2} = \frac{d}{dx}(y') = \frac{dy'/dt}{dx/dt}$$

It is important to observe that

$$\frac{d^2y}{dx^2} \neq \frac{d^2y/dt^2}{d^2x/dt^2}.$$

EXAMPLE ■ 3 Let C be the curve with parametrization

$$x = e^{-t}, \quad y = e^{2t}; \quad t \text{ in } \mathbb{R}.$$

(a) Sketch the graph of C and indicate the orientation.

(b) Use (9.4) and (9.5) to find dy/dx and d^2y/dx^2.

(c) Find a function k that has the same graph as C, and use $k'(x)$ and $k''(x)$ to check the answers to part (b).

(d) Discuss the concavity of C.

Figure 9.14
$x = e^{-t}, y = e^{2t}; t$ in $\mathbb{R}$

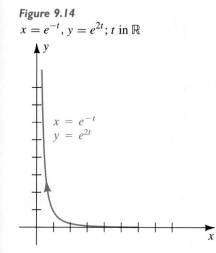

$x = e^{-t}$
$y = e^{2t}$

SOLUTION

(a) To help us sketch the graph, let us first eliminate the parameter. Using $x = e^{-t} = 1/e^t$, we see that $e^t = 1/x$. Substituting in $y = e^{2t} = (e^t)^2$ gives us

$$y = \left(\frac{1}{x}\right)^2 = \frac{1}{x^2}.$$

Remembering that $x = e^{-t} > 0$ leads to the graph in Figure 9.14. Note that the point $(1, 1)$ corresponds to $t = 0$. If t increases in $(-\infty, 0]$, the point $P(x, y)$ approaches $(1, 1)$ from the right, as indicated by the arrowhead. If t increases in $[0, \infty)$, $P(x, y)$ moves up the curve, approaching the y-axis.

(b) By (9.4) and (9.5),

$$y' = \frac{dy}{dx} = \frac{dy/dt}{dx/dt} = \frac{2e^{2t}}{-e^{-t}} = -2e^{3t}$$

$$\frac{d^2y}{dx^2} = \frac{dy'}{dx} = \frac{dy'/dt}{dx/dt} = \frac{-6e^{3t}}{-e^{-t}} = 6e^{4t}.$$

(c) From part (a), a function k that has the same graph as C is given by

$$k(x) = \frac{1}{x^2} = x^{-2} \quad \text{for} \quad x > 0.$$

Differentiating twice yields

$$k'(x) = -2x^{-3} = -2(e^{-t})^{-3} = -2e^{3t}$$

$$k''(x) = 6x^{-4} = 6(e^{-t})^{-4} = 6e^{4t},$$

which is in agreement with part (b).

(d) Since $d^2y/dx^2 = 6e^{4t} > 0$ for every t, the curve C is concave upward at every point.

Figure 9.15

If a curve C is the graph of $y = f(x)$ and the function f is smooth on $[a, b]$, then the length of C is given by $\int_a^b \sqrt{1 + [f'(x)]^2} \, dx$ (see Definition (5.14)). We shall next obtain a formula for finding lengths of parametrized curves.

Suppose a smooth curve C is given parametrically by

$$x = f(t), \quad y = g(t); \quad a \le t \le b.$$

Furthermore, suppose C does not intersect itself—that is, different values of t between a and b determine different points on C. Consider a partition P of $[a, b]$ given by $a = t_0 < t_1 < t_2 < \cdots < t_n = b$. Let $\Delta t_k = t_k - t_{k-1}$ and let $P_k = (f(t_k), g(t_k))$ be the point on C that corresponds to t_k. If $d(P_{k-1}, P_k)$ is the length of the line segment $P_{k-1}P_k$, then the length L_P of the broken line in Figure 9.15 is

$$L_P = \sum_{k=1}^{n} d(P_{k-1}, P_k).$$

As in Section 5.5, we define

$$L = \lim_{\|P\| \to 0} L_P$$

and call L the **length of C** from P_0 to P_n if for every $\epsilon > 0$ there exists a $\delta > 0$ such that $|L_P - L| < \epsilon$ for every partition P with $\|P\| < \delta$.

By the distance formula,

$$d(P_{k-1}, P_k) = \sqrt{[f(t_k) - f(t_{k-1})]^2 + [g(t_k) - g(t_{k-1})]^2}.$$

By the mean value theorem (3.12), there exist numbers w_k and z_k in the open interval (t_{k-1}, t_k) such that

$$f(t_k) - f(t_{k-1}) = f'(w_k)\Delta t_k$$

$$g(t_k) - g(t_{k-1}) = g'(z_k)\Delta t_k.$$

Substituting these in the formula for $d(P_{k-1}, P_k)$ and removing the common factor $(\Delta t_k)^2$ from the radicand gives us

$$d(P_{k-1}, P_k) = \sqrt{[f'(w_k)]^2 + [g'(z_k)]^2}\,\Delta t_k.$$

Consequently,

$$L = \lim_{\|P\|\to 0} L_P = \lim_{\|P\|\to 0} \sum_{k=1}^{n} \sqrt{[f'(w_k)]^2 + [g'(z_k)]^2}\,\Delta t_k,$$

provided the limit exists. If $w_k = z_k$ for every k, then the sums are Riemann sums for the function defined by $\sqrt{[f'(t)]^2 + [g'(t)]^2}$. The limit of these sums is

$$L = \int_a^b \sqrt{[f'(t)]^2 + [g'(t)]^2}\,dt.$$

The limit exists even if $w_k \neq z_k$; however, the proof requires advanced methods and is omitted. The next theorem summarizes this discussion.

Theorem 9.6

If a smooth curve C is given parametrically by $x = f(t)$, $y = g(t)$; $a \le t \le b$, and if C does not intersect itself, except possibly for $t = a$ and $t = b$, then the length L of C is

$$L = \int_a^b \sqrt{[f'(t)]^2 + [g'(t)]^2}\,dt = \int_a^b \sqrt{\left(\frac{dx}{dt}\right)^2 + \left(\frac{dy}{dt}\right)^2}\,dt.$$

The integral formula in Theorem (9.6) is not necessarily true if C intersects itself. For example, if C has the parametrization $x = \cos t$, $y = \sin t$; $0 \le t \le 4\pi$, then the graph is a unit circle with center at the origin. If t varies from 0 to 4π, the circle is traced twice and hence intersects itself infinitely many times. If we use Theorem (9.6) with $a = 0$ and $b = 4\pi$, we obtain the incorrect value 4π for the length of C. The correct value 2π can be obtained by using the t-interval $[0, 2\pi]$. Note that in this case the curve intersects itself only at the points corresponding to $t = 0$ and $t = 2\pi$, which is allowable by the theorem.

If a curve C is given by $y = k(x)$, with k' continuous on $[a, b]$, then parametric equations for C are

$$x = t, \quad y = k(t); \quad a \le t \le b.$$

In this case,

$$\frac{dx}{dt} = 1, \quad \frac{dy}{dt} = k'(t) = k'(x), \quad dt = dx,$$

and from Theorem (9.6),

$$L = \int_a^b \sqrt{1 + [k'(x)]^2}\,dx.$$

This result agrees with the arc length formula given in Definition (5.14).

EXAMPLE ■ 4   Find the length of one arch of the cycloid that has the parametrization

$$x = t - \sin t, \quad y = 1 - \cos t; \quad t \text{ in } \mathbb{R}.$$

SOLUTION   The graph has the shape illustrated in Figure 9.6. The radius a of the circle is 1. One arch is obtained if t varies from 0 to 2π. Applying Theorem (9.6) yields

$$L = \int_0^{2\pi} \sqrt{(1 - \cos t)^2 + (\sin t)^2} \, dt$$

$$= \int_0^{2\pi} \sqrt{1 - 2\cos t + \cos^2 t + \sin^2 t} \, dt.$$

Since $\cos^2 t + \sin^2 t = 1$, the integrand reduces to

$$\sqrt{2 - 2\cos t} = \sqrt{2}\sqrt{1 - \cos t}.$$

Thus, $$L = \int_0^{2\pi} \sqrt{2}\sqrt{1 - \cos t} \, dt.$$

By a half-angle formula, $\sin^2 \frac{1}{2}t = \frac{1}{2}(1 - \cos t)$, or, equivalently,

$$1 - \cos t = 2\sin^2 \tfrac{1}{2}t.$$

Hence, $\sqrt{1 - \cos t} = \sqrt{2\sin^2 \tfrac{1}{2}t} = \sqrt{2} \left| \sin \tfrac{1}{2}t \right|.$

The absolute value sign may be deleted, since if $0 \le t \le 2\pi$, then we have $0 \le \frac{1}{2}t \le \pi$ and hence $\sin \frac{1}{2}t \ge 0$. Consequently,

$$L = \int_0^{2\pi} \sqrt{2}\sqrt{2} \sin \tfrac{1}{2}t \, dt = 2\int_0^{2\pi} \sin \tfrac{1}{2}t \, dt$$

$$= -4 \left[\cos \tfrac{1}{2}t \right]_0^{2\pi} = -4(-1 - 1) = 8.$$

To remember Theorem (9.6), recall that if ds is the differential of arc length, then, by Theorem (5.17),

$$(ds)^2 = (dx)^2 + (dy)^2.$$

Assuming that ds and dt are positive, we have the following.

Parametric Differential
of Arc Length   9.7

$$ds = \sqrt{(dx)^2 + (dy)^2} = \sqrt{\left(\frac{dx}{dt}\right)^2 + \left(\frac{dy}{dt}\right)^2} \, dt$$

Using (9.7), we can rewrite the formula for arc length in Theorem (9.6) as

$$L = \int_{t=a}^{t=b} ds.$$

The limits of integration specify that the independent variable is t, not s

Figure 9.16

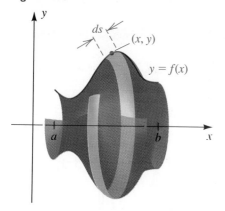

If a function f is smooth and nonnegative for $a \le x \le b$, then, by Definition (5.19), the area S of the surface that is generated by revolving the graph of $y = f(x)$ about the x-axis (see Figure 9.16) is given by

$$S = \int_{x=a}^{x=b} 2\pi y \, ds,$$

where $ds = \sqrt{1 + [f'(x)]^2} \, dx$. We can regard $2\pi y \, ds$ as the surface area of a frustum of a cone of slant height ds and average radius y (see (5.18)).

If a curve C is given parametrically by $x = f(t)$, $y = g(t)$; $a \le t \le b$ and if $g(t) \ge 0$ throughout $[a, b]$, we can use an argument similar to that given in Section 5.5 to show that the area of the surface generated by revolving C about the y-axis is $S = \int_{t=a}^{t=b} 2\pi y \, ds$, where ds is the parametric differential of arc length (9.7). Let us state this for reference as follows.

Theorem 9.8

Let a smooth curve C be given by $x = f(t)$, $y = g(t)$; $a \le t \le b$, and suppose C does not intersect itself, except possibly at the point corresponding to $t = a$ and $t = b$. If $g(t) \ge 0$ throughout $[a, b]$, then the area S of the surface of revolution obtained by revolving C about the x-axis is

$$S = \int_{t=a}^{t=b} 2\pi y \, ds = \int_a^b 2\pi g(t) \sqrt{\left(\frac{dx}{dt}\right)^2 + \left(\frac{dy}{dt}\right)^2} \, dt.$$

Figure 9.17

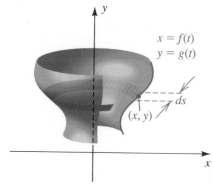

The formula for S in Theorem (9.8) can be extended to the case in which $y = g(t)$ is negative for some t in $[a, b]$ by replacing the variable y that precedes ds by $|y|$.

If the curve C in Theorem (9.8) is revolved about the y-axis and if $x = f(t) \ge 0$ for $a \le t \le b$ (see Figure 9.17), then

$$S = \int_{t=a}^{t=b} 2\pi x \, ds = \int_a^b 2\pi f(t) \sqrt{\left(\frac{dx}{dt}\right)^2 + \left(\frac{dy}{dt}\right)^2} \, dt.$$

In this case, we may regard $2\pi x \, ds$ as the surface area of a frustum of a cone of slant height ds and average radius x.

EXAMPLE ▪ 5 Verify that the surface area of a sphere of radius a is $4\pi a^2$.

SOLUTION If C is the upper half of the circle $x^2 + y^2 = a^2$, then the spherical surface may be obtained by revolving C about the x-axis. Parametric equations for C are

$$x = a \cos t, \quad y = a \sin t; \quad 0 \le t \le \pi.$$

Applying Theorem (9.8) and using the identity $\sin^2 t + \cos^2 t = 1$, we have

$$S = \int_0^\pi 2\pi a \sin t \sqrt{a^2 \sin^2 t + a^2 \cos^2 t} \, dt = 2\pi a^2 \int_0^\pi \sin t \, dt$$
$$= -2\pi a^2 [\cos t]_0^\pi = -2\pi a^2 [-1 - 1] = 4\pi a^2.$$

EXERCISES 9.2

Exer. 1–8: Find the slopes of the tangent line and the normal line at the point on the curve that corresponds to $t = 1$.

1   $x = t^2 + 1,$ $y = t^2 - 1;$ $-2 \le t \le 2$

2   $x = t^3 + 1,$ $y = t^3 - 1;$ $-2 \le t \le 2$

3   $x = 4t^2 - 5,$ $y = 2t + 3;$ t in $\mathbb{R}$

4   $x = t^3,$ $y = t^2;$ t in $\mathbb{R}$

5   $x = e^t,$ $y = e^{-2t};$ t in $\mathbb{R}$

6   $x = \sqrt{t},$ $y = 3t + 4;$ $t \ge 0$

7   $x = 2 \sin t,$ $y = 3 \cos t;$ $0 \le t \le 2\pi$

8   $x = \cos t - 2,$ $y = \sin t + 3;$ $0 \le t \le 2\pi$

Exer. 9–10: Let C be the curve with the given parametrization, for t in $\mathbb{R}$. Find the points on C at which the slope of the tangent line is m.

9   $x = -t^3,$ $y = -6t^2 - 18t;$ $m = 2$

10   $x = t^2 + t,$ $y = 5t^2 - 3;$ $m = 4$

Exer. 11–18: **(a)** Find the points on the curve C at which the tangent line is either horizontal or vertical. **(b)** Find d^2y/dx^2. **(c)** Sketch the graph of C.

11   $x = 4t^2,$ $y = t^3 - 12t;$ t in $\mathbb{R}$

12   $x = t^3 - 4t,$ $y = t^2 - 4;$ t in $\mathbb{R}$

13   $x = t^3 + 1,$ $y = t^2 - 2t;$ t in $\mathbb{R}$

14   $x = 12t - t^3,$ $y = t^2 - 5t;$ t in $\mathbb{R}$

15   $x = 3t^2 - 6t,$ $y = \sqrt{t};$ $t \ge 0$

16   $x = \sqrt[3]{t},$ $y = \sqrt[3]{t} - t;$ t in $\mathbb{R}$

17   $x = \cos^3 t,$ $y = \sin^3 t;$ $0 \le t \le 2\pi$

18   $x = \cosh t,$ $y = \sinh t;$ t in $\mathbb{R}$

Exer. 19–20: Shown is a Lissajous figure (see Example 5 of Section 9.1). Determine where the tangent line is horizontal or vertical.

19   $x = 4 \sin 2t,$
     $y = 2 \cos 3t$

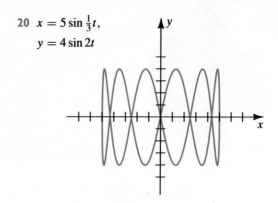

20   $x = 5 \sin \frac{1}{3}t,$
     $y = 4 \sin 2t$

Exer. 21–26: Find the length of the curve.

21   $x = 5t^2,$ $y = 2t^3;$ $0 \le t \le 1$

22   $x = 3t,$ $y = 2t^{3/2};$ $0 \le t \le 4$

23   $x = e^t \cos t,$ $y = e^t \sin t;$ $0 \le t \le \pi/2$

24   $x = \cos 2t,$ $y = \sin^2 t;$ $0 \le t \le \pi$

25   $x = t \cos t - \sin t,$ $y = t \sin t + \cos t;$ $0 \le t \le \pi/2$

26   $x = \cos^3 t,$ $y = \sin^3 t;$ $0 \le t \le \pi/2$

$\boxed{\text{c}}$ **Exer. 27–28:** Use Simpson's rule, with $n = 3$, to approximate the length of the curve.

27   $x = 2 \cos t,$ $y = 3 \sin t;$ $0 \le t \le 2\pi$

28   $x = 4t^3 - t,$ $y = 2t^2;$ $0 \le t \le 1$

Exer. 29–34: Find the area of the surface generated by revolving the curve about the x-axis.

29   $x = t^2,$ $y = 2t;$ $0 \le t \le 4$

30   $x = 4t,$ $y = t^3;$ $1 \le t \le 2$

31   $x = t^2,$ $y = t - \frac{1}{3}t^3;$ $0 \le t \le 1$

32   $x = 4t^2 + 1,$ $y = 3 - 2t;$ $-2 \le t \le 0$

33   $x = t - \sin t,$ $y = 1 - \cos t;$ $0 \le t \le 2\pi$

34   $x = t,$ $y = \frac{1}{3}t^3 + \frac{1}{4}t^{-1};$ $1 \le t \le 2$

Exer. 35–38: Find the area of the surface generated by revolving the curve about the y-axis.

35   $x = 4t^{1/2},$ $y = \frac{1}{2}t^2 + t^{-1};$ $1 \le t \le 4$

36   $x = 3t,$ $y = t + 1;$ $0 \le t \le 5$

37 $x = e^t \sin t$, $y = e^t \cos t$; $0 \le t \le \pi/2$

38 $x = 3t^2$, $y = 2t^3$; $0 \le t \le 1$

[c] **Exer. 39–40:** Use Simpson's rule, with $n = 2$, to approximate the area of the surface generated by revolving the curve about the given axis.

39 $x = \cos(t^2)$, $y = \sin^2 t$; $0 \le t \le 1$; the x-axis

40 $x = t^2 + 2t$, $y = t^4$; $0 \le t \le 1$; the y-axis

41 Prove that the tangent line at the initial control point P_0 on any Bézier curve will pass through the second control point P_1.

42 Prove that the tangent line at the final control point P_3 on any Bézier curve will pass through the third control point P_2.

[c] **43** Approximate the arc length for the Bézier curve in Exercise 53 of Section 9.1. Approximate the length of the piecewise *linear* curve made up of the line segments from P_0 to P_1, from P_1 to P_2, and from P_2 to P_3. Compare the two lengths.

[c] **44** Work Exercise 43 for the Bézier curve in Exercise 54 of Section 9.1. Make a conjecture about a general result concerning these two lengths.

9.3 POLAR COORDINATES

The polar coordinate system, which we discuss in this section, provides a useful alternative to the rectangular coordinate system in investigating plane curves, particularly circles, ellipses, spirals, and other curves with similar symmetries.

In a rectangular coordinate system, the ordered pair (a, b) denotes the point whose directed distances from the x- and y-axes are b and a, respectively. Another method for representing points is to use *polar coordinates*. We begin with a fixed point O (the **origin**, or **pole**) and a directed half-line (the **polar axis**) with endpoint O. Next we consider any point P in the plane different from O. If, as illustrated in Figure 9.18, $r = d(O, P)$ and θ denotes the measure of any angle determined by the polar axis and OP, then r and θ are **polar coordinates** of P, and the symbols (r, θ) or $P(r, \theta)$ are used to denote P. As usual, θ is considered positive if the angle is generated by a counterclockwise rotation of the polar axis and negative if the rotation is clockwise. Either radian or degree measure may be used for θ.

The polar coordinates of a point are not unique. For example, $(3, \pi/4)$, $(3, 9\pi/4)$, and $(3, -7\pi/4)$ all represent the same point (see Figure 9.19).

Figure 9.18

Figure 9.19

Figure 9.20

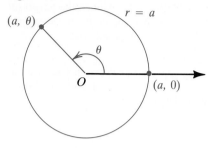

We shall also allow r to be negative. In this case, instead of measuring $|r|$ units along the terminal side of the angle θ, we measure along the half-line with endpoint O that has direction *opposite* that of the terminal side. The points corresponding to the pairs $(-3, 5\pi/4)$ and $(-3, -3\pi/4)$ are also plotted in Figure 9.19.

We agree that the pole O has polar coordinates $(0, \theta)$ for *any* θ. An assignment of ordered pairs of the form (r, θ) to points in a plane is a **polar coordinate system**, and the plane is an $r\theta$**-plane**.

A **polar equation** is an equation in r and θ. A **solution** of a polar equation is an ordered pair (a, b) that leads to equality if a is substituted for r and b for θ. The **graph** of a polar equation is the set of all points (in an $r\theta$-plane) that correspond to the solutions. The simplest polar equations are $r = a$ and $\theta = a$, where a is a nonzero real number. Since the solutions of the polar equation $r = a$ are of the form (a, θ) for *any* angle θ, it follows that the graph is a circle of radius $|a|$ with center at the pole. A graph for $a > 0$ is sketched in Figure 9.20. The same graph is obtained for $r = -a$.

Figure 9.21

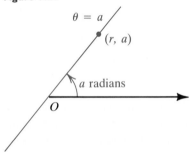

The advantages of using polar coordinates to represent naturally occurring curves is already becoming apparent. In the xy-coordinate system, the equation for the circle of radius a with center at the origin is a quadratic expression in both variables, $x^2 + y^2 = a^2$. In polar coordinates, we have one of the simplest possible equations, a variable equals a constant, for the same circle.

The solutions of the polar equation $\theta = a$ are of the form (r, a) for *any* real number r. Since the (angle) coordinate a is constant, the graph is a line through the origin, as illustrated in Figure 9.21 for the case $0 < a < \pi/2$.

In the following examples, we obtain the graphs of polar equations by plotting points. As you proceed through this section, you should try to recognize forms of polar equations so that you will be able to sketch their graphs by plotting few, if any, points.

EXAMPLE ■ I Sketch the graph of the polar equation $r = 4\sin\theta$.

SOLUTION The following table displays some solutions of the equation. We have included a third row in the table that contains one-decimal-place approximations to r.

Figure 9.22

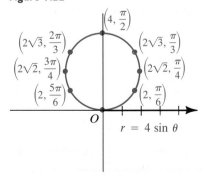

θ	0	$\dfrac{\pi}{6}$	$\dfrac{\pi}{4}$	$\dfrac{\pi}{3}$	$\dfrac{\pi}{2}$	$\dfrac{2\pi}{3}$	$\dfrac{3\pi}{4}$	$\dfrac{5\pi}{6}$	π
r	0	2	$2\sqrt{2}$	$2\sqrt{3}$	4	$2\sqrt{3}$	$2\sqrt{2}$	2	0
r **(approx.)**	0	2	2.8	3.4	4	3.4	2.8	2	0

The points in an $r\theta$-plane that correspond to the pairs in the table appear to lie on a circle of radius 2, and we draw the graph accordingly (see Figure 9.22). As an aid to plotting points, we have extended the polar axis in the negative direction and introduced a vertical line through the pole.

The proof that the graph of $r = 4 \sin \theta$ is a circle is given in Example 6. Additional points obtained by letting θ vary from π to 2π lie on the same circle. For example, the solution $(-2, 7\pi/6)$ gives us the same point as $(2, \pi/6)$; the point corresponding to $(-2\sqrt{2}, 5\pi/4)$ is the same as that obtained from $(2\sqrt{2}, \pi/4)$; and so on. If we let θ increase through all real numbers, we obtain the same points again and again because of the periodicity of the sine function.

EXAMPLE ▪ 2 Sketch the graph of the polar equation $r = 2 + 2\cos\theta$.

SOLUTION Since the cosine function decreases from 1 to -1 as θ varies from 0 to π, it follows that r decreases from 4 to 0 in this θ-interval. The following table exhibits some solutions of $r = 2 + 2\cos\theta$, together with one-decimal-place approximations to r.

θ	0	$\dfrac{\pi}{6}$	$\dfrac{\pi}{4}$	$\dfrac{\pi}{3}$	$\dfrac{\pi}{2}$	$\dfrac{2\pi}{3}$	$\dfrac{3\pi}{4}$	$\dfrac{5\pi}{6}$	π
r	4	$2+\sqrt{3}$	$2+\sqrt{2}$	3	2	1	$2-\sqrt{2}$	$2-\sqrt{3}$	0
r (approx.)	4	3.7	3.4	3	2	1	0.6	0.3	0

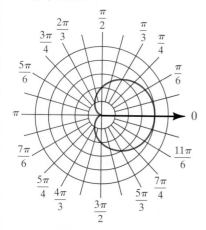

Figure 9.23
$r = 2 + 2\cos\theta$

Plotting points in an $r\theta$-plane leads to the upper half of the graph sketched in Figure 9.23. (We have used polar coordinate graph paper, which displays lines through O at various angles and concentric circles with centers at the pole.)

If θ increases from π to 2π, then $\cos\theta$ increases from -1 to 1 and, consequently, r increases from 0 to 4. Plotting points for $\pi \le \theta \le 2\pi$ gives us the lower half of the graph.

The same graph may be obtained by taking other intervals of length 2π for θ.

The heart-shaped graph in Example 2 is a **cardioid**. In general, the graph of any of the following polar equations, with $a \ne 0$, is a cardioid:

$$r = a(1 + \cos\theta), \qquad r = a(1 + \sin\theta),$$
$$r = a(1 - \cos\theta), \qquad r = a(1 - \sin\theta)$$

If a and b are not zero, then the graphs of the following polar equations are **limaçons**:

$$r = a + b\cos\theta, \qquad r = a + b\sin\theta$$

Note that the special limaçons in which $|a| = |b|$ are cardioids. Some limaçons contain a loop, as shown in the next example.

EXAMPLE ■ 3   Sketch the graph of the polar equation $r = 2 + 4\cos\theta$.

SOLUTION   Coordinates of some points in an $r\theta$-plane that correspond to $0 \le \theta \le \pi$ are listed in the following table.

θ	0	$\dfrac{\pi}{6}$	$\dfrac{\pi}{4}$	$\dfrac{\pi}{3}$	$\dfrac{\pi}{2}$	$\dfrac{2\pi}{3}$	$\dfrac{3\pi}{4}$	$\dfrac{5\pi}{6}$	π
r	6	$2+2\sqrt{3}$	$2+2\sqrt{2}$	4	2	0	$2-2\sqrt{2}$	$2-2\sqrt{3}$	-2
r (approx.)	6	5.4	4.8	4	2	0	-0.8	-1.4	-2

Note that $r = 0$ at $\theta = 2\pi/3$. The values of r are negative if $2\pi/3 < \theta \le \pi$, and this leads to the lower half of the small loop in Figure 9.24. Letting θ range from π to 2π gives us the upper half of the small loop and the lower half of the large loop.

Figure 9.24

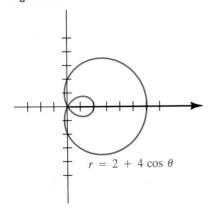

$r = 2 + 4\cos\theta$

EXAMPLE ■ 4   Sketch the graph of the polar equation $r = a\sin 2\theta$ for $a > 0$.

SOLUTION   Instead of tabulating solutions, let us reason as follows. If θ increases from 0 to $\pi/4$, then 2θ varies from 0 to $\pi/2$ and hence $\sin 2\theta$ increases from 0 to 1. It follows that r increases from 0 to a in the θ-interval $[0, \pi/4]$. If we next let θ increase from $\pi/4$ to $\pi/2$, then 2θ changes from $\pi/2$ to π and hence $\sin 2\theta$ decreases from 1 to 0. Thus, r decreases from a to 0 in the θ-interval $[\pi/4, \pi/2]$. The corresponding points on the graph constitute the first-quadrant loop illustrated in Figure 9.25. Note that the point $P(r, \theta)$ traces the loop in a *counterclockwise* direction (indicated by the arrows) as θ increases from 0 to $\pi/2$.

If $\pi/2 \le \theta \le \pi$, then $\pi \le 2\theta \le 2\pi$ and, therefore, $r = a\sin 2\theta \le 0$. Thus, *if $\pi/2 < \theta < \pi$, then r is negative and the points $P(r, \theta)$ are in the fourth quadrant.* If θ increases from $\pi/2$ to π, then we can show, by plotting points, that $P(r, \theta)$ traces (in a counterclockwise direction) the loop shown in the fourth quadrant.

Similarly, for $\pi \le \theta \le 3\pi/2$ we get the loop in the third quadrant, and for $3\pi/2 \le \theta \le 2\pi$ we get the loop in the second quadrant. Both loops are traced in a counterclockwise direction as θ increases. You should verify these facts by plotting some points with, say, $a = 1$. In Figure 9.25, we have plotted only those points on the graph that correspond to the largest numerical values of r.

Figure 9.25

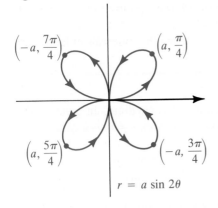

$\left(-a, \dfrac{7\pi}{4}\right)$ $\left(a, \dfrac{\pi}{4}\right)$

$\left(a, \dfrac{5\pi}{4}\right)$ $\left(-a, \dfrac{3\pi}{4}\right)$

$r = a\sin 2\theta$

The graph in Example 4 is a **four-leafed rose**. In general, a polar equation of the form

$$r = a\sin n\theta \quad \text{or} \quad r = a\cos n\theta$$

for any positive integer n greater than 1 and any nonzero real number a has a graph that consists of a number of loops through the origin. If n is even, there are $2n$ loops, and if n is odd, there are n loops (see Exercises 15–18).

The graph of the polar equation $r = a\theta$ for any nonzero real number a is a **spiral of Archimedes**. The case $a = 1$ is considered in the next example.

EXAMPLE • 5 Sketch the graph of the polar equation $r = \theta$ for $\theta \geq 0$.

SOLUTION The graph consists of all points that have polar coordinates of the form (c, c) for every real number $c \geq 0$. Thus, the graph contains the points $(0, 0)$, $(\pi/2, \pi/2)$, (π, π), and so on. As θ increases, r increases at the same rate, and the spiral winds around the origin in a counterclockwise direction, intersecting the polar axis at $0, 2\pi, 4\pi, \dots,$ as illustrated in Figure 9.26.

If θ is allowed to be negative, then as θ decreases through negative values, the resulting spiral winds around the origin and is the symmetric image, with respect to the vertical axis, of the curve sketched in Figure 9.26.

Figure 9.26

Spirals seen in nature, such as those apparent in the curl of an ocean wave, may have quite complicated equations in the rectangular xy-coordinate system, but much simpler and more elegant representations in polar coordinates. For the spiral of Archimedes in Example 5, the corresponding equation in xy-coordinates is a quite complex expression, $\sqrt{x^2 + y^2} = \tan^{-1}(y/x)$. The graphs of polar coordinates illustrating other spirals are given in Exercises 21, 24, and 66.

Let us next superimpose an xy-plane on an $r\theta$-plane so that the positive x-axis coincides with the polar axis. Any point P in the plane may then be assigned rectangular coordinates (x, y) or polar coordinates (r, θ). If $r > 0$, we have a situation similar to that illustrated in Figure 9.27(a). If $r < 0$, we have that shown in Figure 9.27(b), where, for later purposes,

Figure 9.27

(a) $r > 0$ (b) $r < 0$

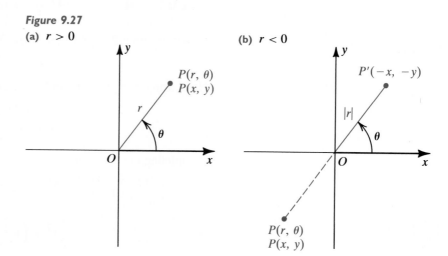

we have also plotted the point P' having polar coordinates $(|r|, \theta)$ and rectangular coordinates $(-x, -y)$.

The following result specifies relationships between (x, y) and (r, θ), where it is assumed that the positive x-axis coincides with the polar axis.

Relationships between Rectangular and Polar Coordinates **9.9**

> The rectangular coordinates (x, y) and polar coordinates (r, θ) of a point P are related as follows:
>
> **(i)** $x = r\cos\theta, \quad y = r\sin\theta$
>
> **(ii)** $r^2 = x^2 + y^2, \quad \tan\theta = y/x \quad \text{if} \quad x \neq 0$

PROOF Although we have pictured θ as an acute angle in Figure 9.27, the discussion that follows is valid for all angles. If $r > 0$ as in Figure 9.27(a), then $\cos\theta = x/r$, $\sin\theta = y/r$, and hence

$$x = r\cos\theta, \quad y = r\sin\theta.$$

If $r < 0$, then $|r| = -r$, and from Figure 9.27(b) we see that

$$\cos\theta = \frac{-x}{|r|} = \frac{-x}{-r} = \frac{x}{r}, \quad \sin\theta = \frac{-y}{|r|} = \frac{-y}{-r} = \frac{y}{r}.$$

Multiplication by r gives us relationship (i), and therefore these formulas hold if r is either positive or negative. If $r = 0$, then the point is the pole and we again see that the formulas in (i) are true.

The formulas in (ii) follow readily from Figure 9.27. ∎

We may use this result to change from one system of coordinates to the other. A more important use is for transforming a polar equation to an equation in x and y, and vice versa, illustrated in Examples 6–8.

EXAMPLE 6 Find an equation in x and y that has the same graph as the polar equation $r = a\sin\theta$, with $a \neq 0$. Sketch the graph.

SOLUTION From (9.9)(i), a relationship between $\sin\theta$ and y is given by $y = r\sin\theta$. To introduce this expression into the equation $r = a\sin\theta$, we multiply both sides by r, obtaining

$$r^2 = ar\sin\theta.$$

Next, using $r^2 = x^2 + y^2$ and $y = r\sin\theta$, we have

$$x^2 + y^2 = ay,$$

or

$$x^2 + y^2 - ay = 0.$$

Completing the square in y gives us

$$x^2 + y^2 - ay + \left(\frac{a}{2}\right)^2 = \left(\frac{a}{2}\right)^2,$$

or

$$x^2 + \left(y - \frac{a}{2}\right)^2 = \left(\frac{a}{2}\right)^2.$$

Figure 9.28

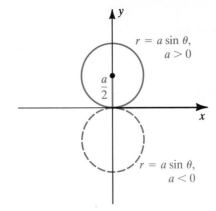

In the xy-plane, the graph of the last equation is a circle with center $(0, a/2)$ and radius $|a|/2$, as illustrated in Figure 9.28 for the case $a > 0$ (the solid circle) and $a < 0$ (the dashed circle).

Figure 9.29

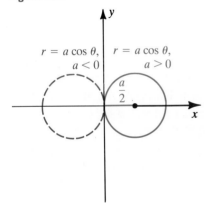

Using the same method as in the preceding example, we can show that the graph of $r = a \cos \theta$, with $a \neq 0$, is a circle of radius $a/2$ of the type illustrated in Figure 9.29.

E X A M P L E ■ 7   Find a polar equation for the hyperbola given by $x^2 - y^2 = 16$.

S O L U T I O N   Using the formulas $x = r \cos \theta$ and $y = r \sin \theta$, we obtain the following polar equations:

$$(r \cos \theta)^2 - (r \sin \theta)^2 = 16$$
$$r^2 \cos^2 \theta - r^2 \sin^2 \theta = 16$$
$$r^2(\cos^2 \theta - \sin^2 \theta) = 16$$
$$r^2 \cos 2\theta = 16$$
$$r^2 = \frac{16}{\cos 2\theta} \quad \text{or} \quad r^2 = 16 \sec 2\theta$$

The division by $\cos 2\theta$ is allowable because $\cos 2\theta \neq 0$. (Note that if $\cos 2\theta = 0$, then $r^2 \cos 2\theta \neq 16$.)

E X A M P L E ■ 8   Find a polar equation of an arbitrary line.

S O L U T I O N   Every line in an xy-coordinate plane is the graph of a linear equation $ax + by = c$. Using the formulas $x = r \cos \theta$ and $y = r \sin \theta$ gives us the following equivalent polar equations:

$$ar \cos \theta + br \sin \theta = c$$
$$r(a \cos \theta + b \sin \theta) = c$$
$$r = \frac{c}{a \cos \theta + b \sin \theta}$$

Figure 9.30

Symmetries of graphs of polar equations

(a) Polar axis

(b) Line $\theta = \pi/2$

(c) Pole

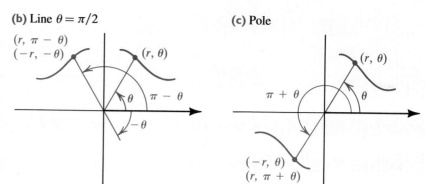

If we superimpose an xy-plane on an $r\theta$-plane, then the graph of a polar equation may be symmetric with respect to the x-axis (the polar axis), the y-axis (the line $\theta = \pi/2$), or the origin (the pole). Some typical symmetries are illustrated in Figure 9.30. The next result states tests for symmetry using polar coordinates.

Tests for Symmetry   9.10

> **(i)** The graph of $r = f(\theta)$ is symmetric with respect to the polar axis if substitution of $-\theta$ for θ leads to an equivalent equation.
>
> **(ii)** The graph of $r = f(\theta)$ is symmetric with respect to the vertical line $\theta = \pi/2$ if substitution of either (a) $\pi - \theta$ for θ or (b) $-r$ for r and $-\theta$ for θ leads to an equivalent equation.
>
> **(iii)** The graph of $r = f(\theta)$ is symmetric with respect to the pole if substitution of either (a) $-r$ for r or (b) $\pi + \theta$ for θ leads to an equivalent equation.

To illustrate, since $\cos(-\theta) = \cos\theta$, the graph of the polar equation $r = 2 + 4\cos\theta$ in Example 3 is symmetric with respect to the polar axis, by test (i). Since $\sin(\pi - \theta) = \sin\theta$, the graph in Example 1 is symmetric with respect to the line $\theta = \pi/2$, by test (ii). The graph in Example 4 is symmetric to the polar axis, the line $\theta = \pi/2$, and the pole. Other tests for symmetry may be stated; however, those we have listed are among the easiest to apply.

Unlike the graph of an equation in x and y, the graph of a polar equation $r = f(\theta)$ can be symmetric with respect to the polar axis, the line $\theta = \pi/2$, or the pole *without* satisfying one of the preceding tests for symmetry. This is true because of the many different ways of specifying a point in polar coordinates.

Another difference between rectangular and polar coordinate systems is that the points of intersection of two graphs cannot always be found by solving the polar equations simultaneously. To illustrate, from Example 1, the graph of $r = 4\sin\theta$ is a circle of diameter 4 with center at $(2, \pi/2)$ (see Figure 9.31). Similarly, the graph of $r = 4\cos\theta$ is a circle of diameter

Figure 9.31

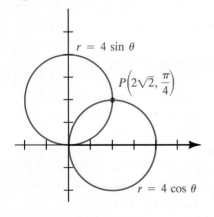

4, with center at $(2, 0)$ on the polar axis. Referring to Figure 9.31, we see that the coordinates of the point of intersection $P(2\sqrt{2}, \pi/4)$ in quadrant I satisfy both equations; however, the origin O, which is on each circle, *cannot* be found by solving the equations simultaneously. Thus, in searching for points of intersection of polar graphs, it is sometimes necessary to refer to the graphs themselves, *in addition* to solving the two equations simultaneously. An alternative method is to use different (equivalent) equations for the graphs.

Tangent lines to graphs of polar equations may be found by means of the next theorem.

Theorem 9.11

The slope m of the tangent line to the graph of $r = f(\theta)$ at the point $P(r, \theta)$ is

$$m = \frac{\dfrac{dr}{d\theta}\sin\theta + r\cos\theta}{\dfrac{dr}{d\theta}\cos\theta - r\sin\theta}.$$

PROOF   If (x, y) are the rectangular coordinates of $P(r, \theta)$, then, by Theorem (9.9),

$$x = r\cos\theta = f(\theta)\cos\theta$$

$$y = r\sin\theta = f(\theta)\sin\theta.$$

These may be considered as parametric equations for the graph with parameter θ. Applying Theorem (9.4), we find that the slope of the tangent line at (x, y) is

$$\frac{dy}{dx} = \frac{dy/d\theta}{dx/d\theta} = \frac{f(\theta)\cos\theta + f'(\theta)\sin\theta}{f(\theta)(-\sin\theta) + f'(\theta)\cos\theta}$$

$$= \frac{f'(\theta)\sin\theta + f(\theta)\cos\theta}{f'(\theta)\cos\theta - f(\theta)\sin\theta},$$

which is equivalent to the formula for m in Theorem (9.11). ∎

Horizontal tangent lines occur if the numerator in the formula for m is 0 and the denominator is not 0. Vertical tangent lines occur if the denominator is 0 and the numerator is not 0. The case $0/0$ requires further investigation.

To find the slopes of the tangent lines at the pole, we must determine the values of θ for which $r = f(\theta) = 0$. For such values (and with $r = 0$ and $dr/d\theta \neq 0$), the formula in Theorem (9.11) reduces to $m = \tan\theta$. These remarks are illustrated in the next example.

EXAMPLE ▪ 9   For the cardioid $r = 2 + 2\cos\theta$ with $0 \le \theta < 2\pi$, find

(a) the slope of the tangent line at $\theta = \pi/6$

(b) the points at which the tangent line is horizontal

(c) the points at which the tangent line is vertical

SOLUTION

(a) The graph of $r = 2 + 2\cos\theta$ was considered in Example 2 and is re-sketched in Figure 9.32. Applying Theorem (9.11), we find that the slope m of the tangent line is

$$m = \frac{(-2\sin\theta)\sin\theta + (2 + 2\cos\theta)\cos\theta}{(-2\sin\theta)\cos\theta - (2 + 2\cos\theta)\sin\theta}$$

$$= \frac{2(\cos^2\theta - \sin^2\theta) + 2\cos\theta}{-2(2\sin\theta\cos\theta) - 2\sin\theta}$$

$$= -\frac{\cos 2\theta + \cos\theta}{\sin 2\theta + \sin\theta}.$$

At $\theta = \pi/6$ (that is, at the point $(2 + \sqrt{3}, \pi/6)$),

$$m = -\frac{\cos(\pi/3) + \cos(\pi/6)}{\sin(\pi/3) + \sin(\pi/6)} = -\frac{(1/2) + (\sqrt{3}/2)}{(\sqrt{3}/2) + (1/2)} = -1.$$

(b) To find horizontal tangents, we let

$$\cos 2\theta + \cos\theta = 0.$$

This equation may be written as

$$2\cos^2\theta - 1 + \cos\theta = 0,$$

or $$(2\cos\theta - 1)(\cos\theta + 1) = 0.$$

From $\cos\theta = \frac{1}{2}$, we obtain $\theta = \pi/3$ and $\theta = 5\pi/3$. The corresponding points are $(3, \pi/3)$ and $(3, 5\pi/3)$.

Using $\cos\theta = -1$ gives us $\theta = \pi$. The denominator in the formula for m is 0 at $\theta = \pi$, and hence further investigation is required. If $\theta = \pi$, then $r = 0$ and the formula for m in Theorem (9.11) reduces to $m = \tan\theta$. Thus, the slope at $(0, \pi)$ is $m = \tan\pi = 0$, and therefore the tangent line is horizontal at the pole.

(c) To find vertical tangent lines, we let

$$\sin 2\theta + \sin\theta = 0.$$

Equivalent equations are

$$2\sin\theta\cos\theta + \sin\theta = 0$$

and $$\sin\theta(2\cos\theta + 1) = 0.$$

Letting $\sin\theta = 0$ and $\cos\theta = -\frac{1}{2}$ leads to the following values of θ: 0, π, $2\pi/3$, and $4\pi/3$. We found, in part (b), that π gives us a horizontal tangent. The remaining values result in the points $(4, 0)$, $(1, 2\pi/3)$, and $(1, 4\pi/3)$, at which the graph has vertical tangent lines.

Figure 9.32
$r = 2 + 2\cos\theta$

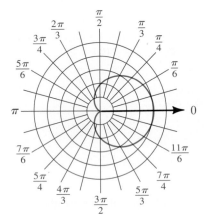

EXERCISES 9.3

Exer. 1–26: Sketch the graph of the polar equation.

1 $r = 5$

2 $r = -2$

3 $\theta = -\pi/6$

4 $\theta = \pi/4$

5 $r = 3\cos\theta$

6 $r = -2\sin\theta$

7 $r = 4 - 4\sin\theta$

8 $r = -6(1 + \cos\theta)$

9 $r = 2 + 4\sin\theta$

10 $r = 1 + 2\cos\theta$

11 $r = 2 - \cos\theta$

12 $r = 5 + 3\sin\theta$

13 $r = 4\csc\theta$

14 $r = -3\sec\theta$

15 $r = 8\cos 3\theta$

16 $r = 2\sin 4\theta$

17 $r = 3\sin 2\theta$

18 $r = 8\cos 5\theta$

19 $r^2 = 4\cos 2\theta$ (lemniscate)

20 $r^2 = -16\sin 2\theta$

21 $r = e^\theta$, $\theta \geq 0$ (logarithmic spiral)

22 $r = 6\sin^2(\theta/2)$

23 $r = 2\theta$, $\theta \geq 0$

24 $r\theta = 1$, $\theta > 0$ (spiral)

25 $r = 2 + 2\sec\theta$ (conchoid)

26 $r = 1 - \csc\theta$

Exer. 27–36: Find a polar equation that has the same graph as the equation in x and y.

27 $x = -3$

28 $y = 2$

29 $x^2 + y^2 = 16$

30 $x^2 = 8y$

31 $2y = -x$

32 $y = 6x$

33 $y^2 - x^2 = 4$

34 $xy = 8$

35 $(x^2 + y^2)\tan^{-1}(y/x) = ay$, $a > 0$ (cochleoid, or Ouija board curve)

36 $x^3 + y^3 - 3axy = 0$ (Folium of Descartes)

Exer. 37–50: Find an equation in x and y that has the same graph as the polar equation and use it to help sketch the graph in an $r\theta$-plane.

37 $r\cos\theta = 5$

38 $r\sin\theta = -2$

39 $r = -3\csc\theta$

40 $r = 4\sec\theta$

41 $r^2\cos 2\theta = 1$

42 $r^2\sin 2\theta = 4$

43 $r(\sin\theta - 2\cos\theta) = 6$

44 $r(3\cos\theta - 4\sin\theta) = 12$

45 $r(\sin\theta + r\cos^2\theta) = 1$

46 $r(r\sin^2\theta - \cos\theta) = 3$

47 $r = 8\sin\theta - 2\cos\theta$

48 $r = 2\cos\theta - 4\sin\theta$

49 $r = \tan\theta$

50 $r = 6\cot\theta$

Exer. 51–60: Find the slope of the tangent line to the graph of the polar equation at the point corresponding to the given value of θ.

51 $r = 2\cos\theta$; $\qquad \theta = \pi/3$

52 $r = -2\sin\theta$; $\qquad \theta = \pi/6$

53 $r = 4(1 - \sin\theta)$; $\quad \theta = 0$

54 $r = 1 + 2\cos\theta$; $\quad \theta = \pi/2$

55 $r = 8\cos 3\theta$; $\qquad \theta = \pi/4$

56 $r = 2\sin 4\theta$; $\qquad \theta = \pi/4$

57 $r^2 = 4\cos 2\theta$; $\qquad \theta = \pi/6$

58 $r^2 = -2\sin 2\theta$; $\quad \theta = 3\pi/4$

59 $r = 2^\theta$; $\qquad\qquad \theta = \pi$

60 $r\theta = 1$; $\qquad\qquad \theta = 2\pi$

61 If $P_1(r_1, \theta_1)$ and $P_2(r_2, \theta_2)$ are points in an $r\theta$-plane, use the law of cosines to prove that

$$[d(P_1, P_2)]^2 = r_1^2 + r_2^2 - 2r_1r_2\cos(\theta_2 - \theta_1).$$

62 If a and b are nonzero real numbers, prove that the graph of $r = a\sin\theta + b\cos\theta$ is a circle, and find its center and radius.

63 If the graphs of the polar equations $r = f(\theta)$ and $r = g(\theta)$ intersect at $P(r, \theta)$, prove that the tangent lines at P are perpendicular if and only if

$$f'(\theta)g'(\theta) + f(\theta)g(\theta) = 0.$$

(The graphs are said to be *orthogonal* at P.)

64 Use Exercise 63 to prove that the graphs of each pair of equations are orthogonal at their point of intersection:

(a) $r = a\sin\theta$, $\quad r = a\cos\theta$

(b) $r = a\theta$, $\quad r\theta = a$

65 If $\cos\theta \neq 0$, show that the slope of the tangent line to the graph of $r = f(\theta)$ is

$$m = \frac{(dr/d\theta)\tan\theta + r}{(dr/d\theta) - r\tan\theta}.$$

66 A logarithmic spiral has a polar equation of the form $r = ae^{b\theta}$ for nonzero constants a and b (see Exercise 21). A famous *four bugs problem* illustrates such a curve. Four bugs A, B, C, and D are placed at the four corners of a square. The center of the square corresponds to the pole. The bugs begin to crawl simultaneously—bug A crawls toward B, B toward C, C toward D, and D toward A, as shown in the figure. Assume that all bugs crawl at the same rate, that they move directly toward the next bug at all times, and that they approach one another but never meet. (The bugs are infinitely small!) At any instant, the positions of the bugs are the vertices of a square, which shrinks and rotates toward the center of the original square as the bugs continue to crawl. If the position of bug A has polar coordinates (r, θ), then the position of bug B has coordinates $(r, \theta + \pi/2)$.

(a) Show that the line through A and B has slope

$$\frac{\sin\theta - \cos\theta}{\sin\theta + \cos\theta}.$$

(b) The line through A and B is tangent to the path of bug A. Use the formula in Exercise 65 to conclude that $dr/d\theta = -r$.

(c) Prove that the path of bug A is a logarithmic spiral. (*Hint:* Solve the differential equation in part (b) by separating variables.)

Exercise 66

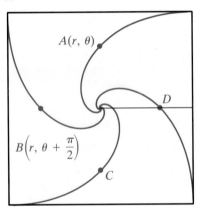

$\boxed{\text{c}}$ **Exer. 67–68:** Graph the polar equation for the given values of θ, and use the graph to determine symmetries.

67 $r = 2\sin^2\theta\tan^2\theta; \quad -\pi/3 \le \theta \le \pi/3$

68 $r = \dfrac{4}{1 + \sin^2\theta}; \quad 0 \le \theta \le 2\pi$

$\boxed{\text{c}}$ **Exer. 69–70:** Graph the polar equations on the same coordinate plane, and estimate the points of intersection of the graphs.

69 $r = 8\cos 3\theta, \quad r = 4 - 2.5\cos\theta$

70 $r = 2\sin^2\theta, \quad r = \tfrac{3}{4}(\theta + \cos^2\theta)$

9.4 INTEGRALS IN POLAR COORDINATES

Figure 9.33

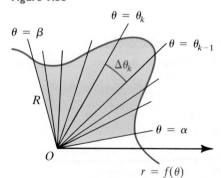

The areas of certain regions bounded by graphs of polar equations can be found by using limits of sums of areas of circular sectors. We shall call a region R in the $r\theta$-plane an R_θ **region** (for integration with respect to θ) if R is bounded by lines $\theta = \alpha$ and $\theta = \beta$ for $0 \le \alpha < \beta \le 2\pi$ and by the graph of a polar equation $r = f(\theta)$, where f is continuous and $f(\theta) \ge 0$ on $[\alpha, \beta]$. An R_θ region is illustrated in Figure 9.33.

Let P denote a partition of $[\alpha, \beta]$ determined by

$$\alpha = \theta_0 < \theta_1 < \theta_2 < \cdots < \theta_n = \beta$$

and let $\Delta\theta_k = \theta_k - \theta_{k-1}$ for $k = 1, 2, \ldots, n$. The lines $\theta = \theta_k$ divide R into wedge-shaped subregions. If $f(u_k)$ is the minimum value and $f(v_k)$ is the maximum value of f on $[\theta_{k-1}, \theta_k]$, then, as illustrated in Figure 9.34, the area ΔA_k of the kth subregion is between the areas of the inscribed and

Figure 9.34

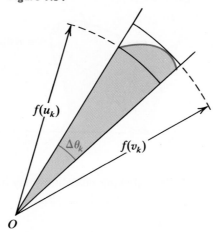

circumscribed circular sectors having central angle $\Delta\theta_k$ and radii $f(u_k)$ and $f(v_k)$, respectively. Hence, by the formula for finding the area of a circular sector (page 38),

$$\tfrac{1}{2}[f(u_k)]^2\Delta\theta_k \le \Delta A_k \le \tfrac{1}{2}[f(v_k)]^2\Delta\theta_k.$$

Summing from $k = 1$ to $k = n$ and using the fact that the sum of the ΔA_k is the area A of R, we obtain

$$\sum_{k=1}^{n}\tfrac{1}{2}[f(u_k)]^2\Delta\theta_k \le A \le \sum_{k=1}^{n}\tfrac{1}{2}[f(v_k)]^2\Delta\theta_k.$$

The limits of the sums, as the norm $\|P\|$ of the subdivision approaches zero, both equal the integral $\int_\alpha^\beta \tfrac{1}{2}[f(\theta)]^2\,d\theta$. This gives us the following result.

Theorem 9.12

If f is continuous and $f(\theta) \ge 0$ on $[\alpha, \beta]$, where $0 \le \alpha < \beta \le 2\pi$, then the area A of the region bounded by the graphs of $r = f(\theta)$, $\theta = \alpha$, and $\theta = \beta$ is

$$A = \int_\alpha^\beta \tfrac{1}{2}[f(\theta)]^2\,d\theta = \int_\alpha^\beta \tfrac{1}{2}r^2\,d\theta.$$

Figure 9.35

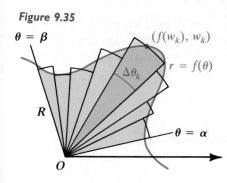

The integral in Theorem (9.12) may be interpreted as a limit of sums by writing

$$A = \int_\alpha^\beta \tfrac{1}{2}[f(\theta)]^2\,d\theta = \lim_{\|P\|\to 0}\sum_{k=1}^{n}\tfrac{1}{2}[f(w_k)]^2\Delta\theta_k$$

for *any* number w_k in the subinterval $[\theta_{k-1}, \theta_k]$ of $[\alpha, \beta]$. Figure 9.35 is a geometric illustration of a typical Riemann sum.

Figure 9.36

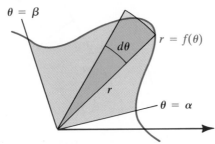

The following guidelines may be useful for remembering this limit of sums formula (see Figure 9.36).

Guidelines for Finding the Area of an R_θ Region   9.13

1   Sketch the region, labeling the graph of $r = f(\theta)$. Find the smallest value $\theta = \alpha$ and the largest value $\theta = \beta$ for points (r, θ) in the region.

2   Sketch a typical circular sector and label its central angle $d\theta$.

3   Express the area of the sector in guideline (2) as $\frac{1}{2}r^2 \, d\theta$.

4   Apply the limit of sums operator $\int_\alpha^\beta$ to the expression in guideline (3) and evaluate the integral.

EXAMPLE■1   Find the area of the region bounded by the cardioid $r = 2 + 2\cos\theta$.

Figure 9.37

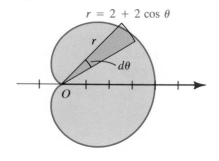

$r = 2 + 2 \cos \theta$

SOLUTION   Following guideline (1), we first sketch the region as in Figure 9.37. The cardioid is obtained by letting θ vary from 0 to 2π; however, using symmetry we may find the area of the top half and multiply by 2. Thus, we use $\alpha = 0$ and $\beta = \pi$ for the smallest and largest values of θ. As in guideline (2), we sketch a typical circular sector and label its central angle $d\theta$. To apply guideline (3), we refer to the figure, obtaining the following:

radius of circular sector: $r = 2 + 2\cos\theta$

area of sector: $\frac{1}{2}r^2 \, d\theta = \frac{1}{2}(2 + 2\cos\theta)^2 \, d\theta$

We next use guideline (4), with $\alpha = 0$ and $\beta = \pi$, remembering that applying $\int_0^\pi$ to the expression $\frac{1}{2}(2 + 2\cos\theta)^2 \, d\theta$ represents taking a limit of sums of areas of circular sectors, *sweeping out* the region by letting θ vary from 0 to π. Thus,

$$A = 2\int_0^\pi \frac{1}{2}(2 + 2\cos\theta)^2 \, d\theta$$

$$= \int_0^\pi (4 + 8\cos\theta + 4\cos^2\theta) \, d\theta.$$

Using the fact that $\cos^2 \theta = \frac{1}{2}(1 + \cos 2\theta)$ yields

$$A = \int_0^\pi (6 + 8\cos\theta + 2\cos 2\theta)\, d\theta$$

$$= \left[6\theta + 8\sin\theta + \sin 2\theta\right]_0^\pi = 6\pi.$$

We could also have found the area by using $\alpha = 0$ and $\beta = 2\pi$.

Figure 9.38

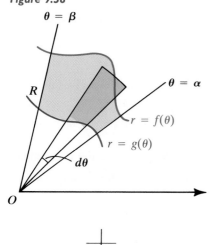

$\theta = \beta$

$\theta = \alpha$

R

$r = f(\theta)$

$r = g(\theta)$

$d\theta$

O

A region R between the graphs of two polar equations $r = f(\theta)$ and $r = g(\theta)$ and the lines $\theta = \alpha$ and $\theta = \beta$ is sketched in Figure 9.38. We may find the area A of R by subtracting the area of the *inner* region bounded by $r = g(\theta)$ from the area of the *outer* region bounded by $r = f(\theta)$ as follows:

$$A = \int_\alpha^\beta \tfrac{1}{2}[f(\theta)]^2\, d\theta - \int_\alpha^\beta \tfrac{1}{2}[g(\theta)]^2\, d\theta$$

We use this technique in the next example.

EXAMPLE 2   Find the area A of the region R that is inside the cardioid $r = 2 + 2\cos\theta$ and outside the circle $r = 3$.

Figure 9.39

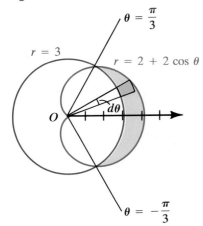

$\theta = \dfrac{\pi}{3}$

$r = 3$

$r = 2 + 2\cos\theta$

O

$d\theta$

$\theta = -\dfrac{\pi}{3}$

SOLUTION   Figure 9.39 shows the region R and circular sectors that extend from the pole to the graphs of the two polar equations. The points of intersection $(3, -\pi/3)$ and $(3, \pi/3)$ can be found by solving the equations simultaneously. Since the angles α and β in Guidelines (9.13) are nonnegative, we shall find the area of the top half of R (using $\alpha = 0$ and $\beta = \pi/3$) and then double the result. Subtracting the area of the inner region (bounded by $r = 3$) from the area of the outer region (bounded by $r = 2 + 2\cos\theta$), we obtain

$$A = 2\left[\int_0^{\pi/3} \tfrac{1}{2}(2 + 2\cos\theta)^2\, d\theta - \int_0^{\pi/3} \tfrac{1}{2}(3)^2\, d\theta\right]$$

$$= \int_0^{\pi/3} (4\cos^2\theta + 8\cos\theta - 5)\, d\theta.$$

As in Example 1, the integral may be evaluated by using the substitution $\cos^2\theta = \frac{1}{2}(1 + \cos 2\theta)$. It can be shown that

$$A = \tfrac{9}{2}\sqrt{3} - \pi \approx 4.65.$$

If a curve C is the graph of a polar equation $r = f(\theta)$ from $\theta = \alpha$ to $\theta = \beta$, we can find its length L by using parametric equations. Thus, as in the proof of Theorem (9.11), a parametrization for C is

$$x = f(\theta)\cos\theta, \quad y = f(\theta)\sin\theta; \quad \alpha \le \theta \le \beta.$$

Differentiating with respect to θ, we obtain

$$\frac{dx}{d\theta} = -f(\theta)\sin\theta + f'(\theta)\cos\theta$$

$$\frac{dy}{d\theta} = f(\theta)\cos\theta + f'(\theta)\sin\theta.$$

Using the trigonometric identity $\sin^2\theta + \cos^2\theta = 1$, we can show that

$$\left(\frac{dx}{d\theta}\right)^2 + \left(\frac{dy}{d\theta}\right)^2 = [f(\theta)]^2 + [f'(\theta)]^2.$$

Substitution in Theorem (9.6) with $t = \theta$, $a = \alpha$, and $b = \beta$ gives us

$$L = \int_\alpha^\beta \sqrt{[f(\theta)]^2 + [f'(\theta)]^2}\, d\theta = \int_\alpha^\beta \sqrt{r^2 + \left(\frac{dr}{d\theta}\right)^2}\, d\theta.$$

As an aid to remembering this formula, we may use the differential of arc length $ds = \sqrt{(dx)^2 + (dy)^2}$ in (9.7). The preceding manipulations give us the following.

Differential of Arc Length in Polar Coordinates   9.14

$$ds = \sqrt{r^2 + \left(\frac{dr}{d\theta}\right)^2}\, d\theta$$

We may now write the formula for L as

$$L = \int_{\theta=\alpha}^{\theta=\beta} ds.$$

The limits of integration specify that the independent variable is θ, not s.

EXAMPLE 3   Find the length of the cardioid $r = 1 + \cos\theta$.

SOLUTION   The cardioid is sketched in Figure 9.40. Making use of symmetry, we shall find the length of the upper half and double the result. Applying (9.14), we have

$$ds = \sqrt{(1 + \cos\theta)^2 + (-\sin\theta)^2}\, d\theta$$

$$= \sqrt{1 + 2\cos\theta + \cos^2\theta + \sin^2\theta}\, d\theta$$

$$= \sqrt{2 + 2\cos\theta}\, d\theta$$

$$= \sqrt{2}\sqrt{1 + \cos\theta}\, d\theta.$$

Hence,

$$L = 2\int_{\theta=0}^{\theta=\pi} ds = 2\int_0^\pi \sqrt{2}\sqrt{1 + \cos\theta}\, d\theta.$$

Figure 9.40

$r = 1 + \cos\theta$

The last integral may be evaluated by employing the trigonometric identity $\cos^2 \frac{1}{2}\theta = \frac{1}{2}(1 + \cos\theta)$, or, equivalently, $1 + \cos\theta = 2\cos^2 \frac{1}{2}\theta$. Thus,

$$L = 2\sqrt{2} \int_0^\pi \sqrt{2\cos^2 \tfrac{1}{2}\theta} \, d\theta$$

$$= 4 \int_0^\pi \cos \tfrac{1}{2}\theta \, d\theta$$

$$= 8 \left[\sin \tfrac{1}{2}\theta \right]_0^\pi = 8.$$

Figure 9.41

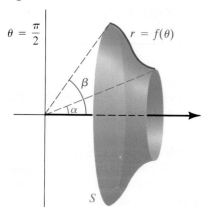

$\theta = \dfrac{\pi}{2}$

$r = f(\theta)$

β

α

S

In the solution to Example 3, it was legitimate to replace $\sqrt{\cos^2 \frac{1}{2}\theta}$ by $\cos \frac{1}{2}\theta$, because if $0 \le \theta \le \pi$, then $0 \le \frac{1}{2}\theta \le \pi/2$, and hence $\cos \frac{1}{2}\theta$ is *positive* on $[0, \pi]$. If we had *not* used symmetry, but had written L as $\int_0^{2\pi} \sqrt{r^2 + (dr/d\theta)^2} \, d\theta$, this simplification would not have been valid. Generally, in determining areas or arc lengths that involve polar coordinates, it is a good idea to use any symmetries that exist.

Let C be the graph of a polar equation $r = f(\theta)$ for $\alpha \le \theta \le \beta$. Let us obtain a formula for the area S of the surface generated by revolving C about the polar axis, as illustrated in Figure 9.41. Since parametric equations for C are

$$x = f(\theta)\cos\theta, \quad y = f(\theta)\sin\theta; \quad \alpha \le \theta \le \beta,$$

we may find S by using Theorem (9.8) with $\theta = t$. This gives us the following result, where the arc length differential ds is given by (9.14).

Surfaces of Revolution in Polar Coordinates 9.15

$$\text{About the polar axis:} \quad S = \int_{\theta=\alpha}^{\theta=\beta} 2\pi y \, ds = \int_{\theta=\alpha}^{\theta=\beta} 2\pi r \sin\theta \, ds$$

$$\text{About the line } \theta = \pi/2: \quad S = \int_{\theta=\alpha}^{\theta=\beta} 2\pi x \, ds = \int_{\theta=\alpha}^{\theta=\beta} 2\pi r \cos\theta \, ds$$

Figure 9.42

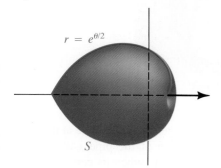

$r = e^{\theta/2}$

S

When applying (9.15), *we must choose α and β so that the surface does not retrace itself when C is revolved*, as would be the case if the circle $r = \cos\theta$, with $0 \le \theta \le \pi$, were revolved about the polar axis.

EXAMPLE ■ 4 The part of the spiral $r = e^{\theta/2}$ from $\theta = 0$ to $\theta = \pi$ is revolved about the polar axis. Find the area of the resulting surface.

SOLUTION The surface is illustrated in Figure 9.42. By (9.14), the polar differential of arc length in polar coordinates is

$$ds = \sqrt{(e^{\theta/2})^2 + (\tfrac{1}{2}e^{\theta/2})^2} \, d\theta = \sqrt{\tfrac{5}{4}e^\theta} \, d\theta = \frac{\sqrt{5}}{2}e^{\theta/2} \, d\theta.$$

Hence, by (9.15),

$$S = \int_{\theta=0}^{\theta=\pi} 2\pi y \, ds = \int_{\theta=0}^{\theta=\pi} 2\pi r \sin\theta \, ds$$

$$= \int_0^\pi 2\pi e^{\theta/2} \sin\theta \left(\frac{\sqrt{5}}{2} e^{\theta/2}\right) d\theta$$

$$= \sqrt{5}\pi \int_0^\pi e^\theta \sin\theta \, d\theta.$$

Using integration by parts or Formula 98 in the table of integrals (see Appendix II), we have

$$S = \frac{\sqrt{5}\pi}{2} \left[e^\theta(\sin\theta - \cos\theta)\right]_0^\pi = \frac{\sqrt{5}\pi}{2}(e^\pi + 1) \approx 84.8.$$

We have already seen many applications of calculus that involve the calculation of area. The next example illustrates an application in which the area is most naturally represented as an integral using polar coordinates.

EXAMPLE ▪ 5   An industrial plant releases its discharge through a circular pipe of diameter 10 cm. The flow is controlled by a valve consisting of a circular disk of the same diameter. Moving the valve back and forth across the pipe increases or decreases the rate of discharge. (See Figure 9.43.) If the flow of discharge is proportional to the area of the opening, what percentage of the maximum flow occurs when the center of the valve disk is 5 cm from the center of the pipe?

Figure 9.43

SOLUTION   We set up a coordinate system with the pole at the center of the pipe. The polar coordinate equation of the pipe will then be $r = 5$. Since the radius of the valve disk is 5 cm, when its center is 5 cm from the center of the pipe, the center is at the point with rectangular coordinates $(5, 0)$. The polar coordinate equation for the valve disk is then

Figure 9.44

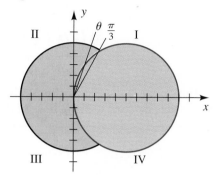

$r = 10 \cos \theta$. The maximum flow is proportional to the area of the circular opening of the pipe, which is $\pi (5)^2$. With the valve blocking the flow, the actual flow is proportional to the shaded area shown in Figure 9.44.

To find this area A, we begin by noting that it is made up of the left semicircle of the pipe plus additional regions in quadrants I and IV. By symmetry, the regions in quadrants I and quadrant IV have the same area. Thus,

$A = $ (area of semicircle of radius 5) $ + 2$(area of region in quadrant I).

To find the area in quadrant I, we first determine the intersection point of the two circles. They intersect when $5 = 10 \cos \theta$, or $\cos \theta = 1/2$. Thus, $\theta = \pi/3$ or $5\pi/3$. The first value, $\theta = \pi/3$, is the one in quadrant I.

The area of the region in quadrant I is the shaded area between the pipe and the valve as θ varies from $\pi/3$ to $\pi/2$. This area is

$$\int_{\pi/3}^{\pi/2} \tfrac{1}{2}[5^2 - (10 \cos \theta)^2] \, d\theta$$

$$= \tfrac{1}{2} \int_{\pi/3}^{\pi/2} [25 - 100 \cos^2 \theta] \, d\theta$$

$$= \tfrac{1}{2} \int_{\pi/3}^{\pi/2} [25 - 50(1 + \cos 2\theta)] \, d\theta$$

$$= \tfrac{1}{2} \int_{\pi/3}^{\pi/2} [-25 - 50 \cos 2\theta] \, d\theta = \tfrac{1}{2} \left[-25\theta - 25 \sin 2\theta \right]_{\pi/3}^{\pi/2}$$

$$= \frac{1}{2} \left[\left(-\frac{25\pi}{2} - 0 \right) - \left(-\frac{25\pi}{3} - \frac{25\sqrt{3}}{2} \right) \right]$$

$$= \frac{1}{2} \left(\frac{25\sqrt{3}}{2} - \frac{25\pi}{6} \right).$$

Thus, the total shaded area is

$$\frac{25\pi}{2} + 2 \left[\frac{1}{2} \left(\frac{25\sqrt{3}}{2} - \frac{25\pi}{6} \right) \right] = 25 \left(\frac{\pi}{3} + \frac{\sqrt{3}}{2} \right).$$

Since the entire area of the pipe is 25π, the fraction that is left uncovered is

$$\frac{25\left(\pi/3 + \sqrt{3}/2\right)}{25\pi} = \frac{1}{3} + \frac{\sqrt{3}}{2\pi} \approx 0.61.$$

Hence, when the center of the valve is 5 cm from the center of the pipe, it restricts the flow to about 61% of the maximum possible flow.

EXERCISES 9.4

Exer. 1–6: Find the area of the region bounded by the graph of the polar equation.

1 $r = 2\cos\theta$ **2** $r = 5\sin\theta$

3 $r = 1 - \cos\theta$ **4** $r = 6 - 6\sin\theta$

5 $r = \sin 2\theta$ **6** $r^2 = 9\cos 2\theta$

Exer. 7–8: Find the area of region R.

7 $R = \{(r, \theta): 0 \le \theta \le \pi/2, 0 \le r \le e^{\theta}\}$

8 $R = \{(r, \theta): 0 \le \theta \le \pi, 0 \le r \le 2\theta\}$

Exer. 9–12: Find the area of the region bounded by one loop of the graph of the polar equation.

9 $r^2 = 4\cos 2\theta$ **10** $r = 2\cos 3\theta$

11 $r = 3\cos 5\theta$ **12** $r = \sin 6\theta$

Exer. 13–16: Set up integrals in polar coordinates that can be used to find the area of the region shown in the figure.

13

14

15

16

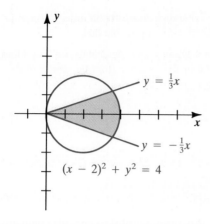

$y = \frac{1}{3}x$

$y = -\frac{1}{3}x$

$(x - 2)^2 + y^2 = 4$

Exer. 17–18: Set up integrals in polar coordinates that can be used to find the area of (a) the blue region and (b) the green region.

17

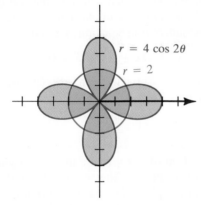

$r = 4 \cos 2\theta$

$r = 2$

18

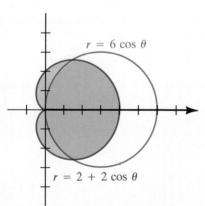

$r = 6 \cos \theta$

$r = 2 + 2 \cos \theta$

Exer. 19–22: Find the area of the region that is outside the graph of the first equation and inside the graph of the second equation.

19 $r = 2 + 2 \cos \theta$, $\quad r = 3$

20 $r = 2$, $\qquad\qquad r = 4 \cos \theta$

21 $r = 2$, $\qquad\qquad r^2 = 8 \sin 2\theta$

22 $r = 1 - \sin \theta$, $\quad r = 3 \sin \theta$

Exer. 23–26: Find the area of the region that is inside the graphs of *both* equations.

23 $r = \sin \theta$, $\qquad r = \sqrt{3} \cos \theta$

24 $r = 2(1 + \sin \theta)$, $\quad r = 1$

25 $r = 1 + \sin \theta$, $\qquad r = 5 \sin \theta$

26 $r^2 = 4 \cos 2\theta$, $\qquad r = 1$

Exer. 27–32: Find the length of the curve.

27 $r = e^{-\theta}$ from $\theta = 0$ to $\theta = 2\pi$

28 $r = \theta$ from $\theta = 0$ to $\theta = 4\pi$

29 $r = \cos^2 \frac{1}{2}\theta$ from $\theta = 0$ to $\theta = \pi$

30 $r = 2^{\theta}$ from $\theta = 0$ to $\theta = \pi$

31 $r = \sin^3 \frac{1}{3}\theta$

32 $r = 2 - 2 \cos \theta$

c **Exer. 33–34:** Use Simpson's rule, with $n = 2$, to approximate the length of the curve.

33 $r = \theta + \cos \theta$ from $\theta = 0$ to $\theta = \pi/2$

34 $r = \sin \theta + \cos^2 \theta$ from $\theta = 0$ to $\theta = \pi$

Exer. 35–38: Find the area of the surface generated by revolving the graph of the equation about the polar axis.

35 $r = 2 + 2 \cos \theta$ $\qquad$ **36** $r^2 = 4 \cos 2\theta$

37 $r = 2a \sin \theta$ $\qquad\qquad$ **38** $r = 2a \cos \theta$

c **Exer. 39–40:** Use the trapezoidal rule, with $n = 4$, to approximate the area of the surface generated by revolving the graph of the polar equation about the line $\theta = \pi/2$. (Use symmetry when setting up the integral.)

39 $r = \sin^2 \theta$ $\qquad\qquad$ **40** $r = \cos^2 \theta$

41 A *torus* is the surface generated by revolving a circle about a nonintersecting line in its plane. Use polar coordinates to find the surface area of the torus generated by revolving the circle $x^2 + y^2 = a^2$ about the line $x = b$, where $0 < a < b$.

42 Let OP be the ray from the pole to the point $P(r, \theta)$ on the spiral $r = a\theta$, where $a > 0$. If the ray makes two revolutions (starting from $\theta = 0$), find the area of the region swept out in the second revolution that was not swept out in the first revolution (see figure).

Exercise 42 $\qquad r = a\theta$

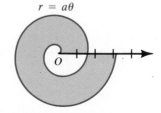

O

43 The part of the spiral $r = e^{-\theta}$ from $\theta = 0$ to $\theta = \pi/2$ is revolved about the line $\theta = \pi/2$. Find the area of the resulting surface.

$\boxed{c}$ **Exer. 44–45: Refer to Example 5.**

44 Determine how far from the center of the pipe the center

of the valve disk should be in order to limit the flow to half the maximum possible flow.

45 Obtain a graph of the percentage of flow blocked as a function of the distance from the center of the valve to the center of the pipe.

9.5   TRANSLATION AND ROTATION OF AXES

In the study of plane curves, it is often helpful to consider new coordinate systems obtained by translating or rotating the original coordinate axes. In these new systems, the equation of a curve may be much simpler than it was in the original system. We define and illustrate the use of these translations and rotations in this section. One of our main goals is to show that the graph of the general second-degree equation in x and y

$$Ax^2 + Bxy + Cy^2 + Dx + Ey + F = 0$$

is either a conic or a degenerate conic. You may wish to review the material on conic sections in the Precalculus Review.

TRANSLATION OF AXES

Figure 9.45 illustrates a **translation of axes**, where the x- and y-axes are shifted to positions—denoted by x' and y'—that are parallel to their original positions. Every point P in the plane then has two different ordered-pair representations: $P(x, y)$ in the xy-system and $P(x', y')$ in the $x'y'$-system. If the origin of the new $x'y'$-system has coordinates (h, k) in the xy-plane, as shown in Figure 9.45, we see that

$$x = x' + h \quad \text{and} \quad y = y' + k.$$

These formulas are true for all values of h and k. Equivalent formulas are

$$x' = x - h \quad \text{and} \quad y' = y - k.$$

Figure 9.45

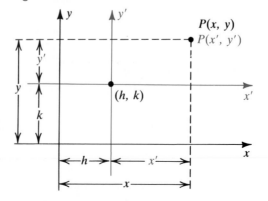

We may summarize this discussion by the following.

Translation of Axes Formulas 9.16

If (x, y) are the coordinates of a point P in an xy-plane and if (x', y') are the coordinates of P in an $x'y'$-plane with origin at the point (h, k) of the xy-plane, then

  (i) $x = x' + h, \quad y = y' + k$
  (ii) $x' = x - h, \quad y' = y - k$

If a certain collection of points in the xy-plane is the graph of an equation in x and y, then to find an equation in x' and y' that has the same graph in the $x'y'$-plane, we substitute $x' + h$ for x and $y' + k$ for y. Conversely, if a set of points in the $x'y'$-plane is the graph of an equation in x' and y', then to find the corresponding equation in x and y, we substitute $x - h$ for x' and $y - k$ for y'.

To illustrate the use of (9.16), we begin by noting that

$$(x')^2 + (y')^2 = 1$$

is an equation of a circle of radius 1 with center at the origin O' of the $x'y'$-plane. Using translation of axes formulas (9.16)(ii), we see that

$$(x - h)^2 + (y - k)^2 = 1$$

is an equation of the same circle in the xy-plane with center (h, k).

The next example shows another application of translation of axes to simplify a second-degree equation.

EXAMPLE ▪ 1 Discuss and sketch the graph of

$$25x^2 + 250x - 16y^2 + 32y + 109 = 0.$$

SOLUTION We first complete the square in x and y by rewriting the equation as

$$25(x^2 + 10x) - 16(y^2 - 2y) = -109.$$

Then we add constant terms to obtain squares of binomials in x and y:

$$25(x^2 + 10x + 25) - 16(y^2 - 2y + 1) = -109 + 25(25) - 16(1),$$

which we can write as

$$25(x + 5)^2 - 16(y - 1)^2 = 500.$$

We now use the translation of axes formulas (9.16)(ii), with $x' = x + 5$ and $y' = y - 1$, in order to write the equation as

$$25(x')^2 - 16(y')^2 = 500$$

or, equivalently,

$$\frac{(x')^2}{20} - \frac{(y')^2}{\frac{125}{4}} = 1,$$

Figure 9.46

$25x^2 + 250x - 16y^2 + 32y + 109 = 0,$
$[(x')^2/20] - [4(y')^2/125] = 1$

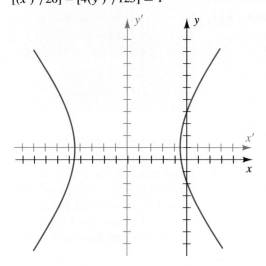

which has the form

$$\frac{(x')^2}{a^2} - \frac{(y')^2}{b^2} = 1$$

with $a^2 = 20$ and $b^2 = \frac{125}{4}$. In this form, we recognize that the curve represented by the equation is a hyperbola with vertices on the x'-axis and center at the origin of the $x'y'$-coordinate system. From Theorem 38 in the Precalculus Review (page 75), the hyperbola has vertices with $x'y'$-coordinates $(\pm a, 0) = (\pm 2\sqrt{5}, 0)$. The foci have $x'y'$-coordinates $(\pm c, 0)$, where $c = \sqrt{a^2 + b^2} = \sqrt{20 + (125/4)} = \sqrt{205}/2$, and the asymptotes have equations $y' = \pm(b/a)x' = \pm(5/4)x'$. Figure 9.46 shows a sketch of the curve.

Since $x = x' - 5$ and $y = y' + 1$, we can translate the information about the center, the vertices, and the foci to xy-coordinates. The vertices, for example, have xy-coordinates $(-5 \pm 2\sqrt{5}, 1)$, and the asymptotes are the lines $(y - 1) = \pm\frac{5}{4}(x + 5)$.

In a similar manner, by completing the square and translating axes, we can replace

$$Ax^2 + Bxy + Cy^2 + Dx + Ey + F = 0$$

by an equivalent equation of the form

$$A'(x')^2 + B'x'y' + C'(y')^2 + F' = 0.$$

That is, we can eliminate the linear terms. Translation of axes, however, will still retain the second-degree term $x'y'$. To remove such a term, we must introduce rotation of axes.

Figure 9.47

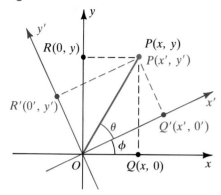

As we have seen, translation of axes gives a new coordinate system that may result in simpler representations for the equations of curves. Rotation of axes, which we now investigate, provides an additional way to build new coordinate systems in which we may obtain even further simplifications of such equations. We obtained the $x'y'$-plane used in a translation of axes by moving the origin O of the xy-plane to a new position $C(h, k)$ without changing the positive directions of the axes or the units of length. We now consider a new coordinate plane obtained by keeping the origin O fixed and rotating the x- and y-axes about O to another position, denoted by x' and y'. A transformation of this type is a **rotation of axes**.

Consider the rotation of axes in Figure 9.47, and let ϕ denote the acute angle through which the positive x-axis must be rotated in order to coincide with the positive x'-axis. If (x, y) are the coordinates of a point P relative to the xy-plane, then (x', y') will denote its coordinates relative to the new $x'y'$-plane.

Let the projection of P on the various axes be denoted as in Figure 9.47, and let θ denote angle POQ'. If $p = d(O, P)$, then

$$x' = p \cos \theta, \qquad y' = p \sin \theta$$
$$x = p \cos(\theta + \phi), \qquad y = p \sin(\theta + \phi).$$

Applying the addition formulas for the sine and cosine, we see that

$$x = p \cos \theta \cos \phi - p \sin \theta \sin \phi$$
$$y = p \sin \theta \cos \phi + p \cos \theta \sin \phi.$$

Using the fact that $x' = p \cos \theta$ and $y' = p \sin \theta$ gives us (i) of the next theorem. The formulas in (ii) may be obtained from (i) by solving for x' and y'.

Rotation of Axes Formulas   9.17

If the x- and y-axes are rotated about the origin O, through an acute angle ϕ, then the coordinates (x, y) and (x', y') of a point P in the xy- and $x'y'$-planes are related as follows:

(i) $x = x' \cos \phi - y' \sin \phi$, $y = x' \sin \phi + y' \cos \phi$

(ii) $x' = x \cos \phi + y \sin \phi$, $y' = -x \sin \phi + y \cos \phi$

Figure 9.48
$y = 1/x$

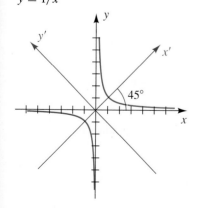

EXAMPLE ■ 2   The graph of $xy = 1$, or, equivalently, $y = 1/x$, is sketched in Figure 9.48. If the coordinate axes are rotated through an angle of $45°$, find an equation of the graph relative to the new $x'y'$-plane.

SOLUTION   We let $\phi = 45°$ in rotation of axes formulas (9.17)(i):

$$x = x'\left(\frac{\sqrt{2}}{2}\right) - y'\left(\frac{\sqrt{2}}{2}\right) = \frac{\sqrt{2}}{2}(x' - y')$$

$$y = x'\left(\frac{\sqrt{2}}{2}\right) + y'\left(\frac{\sqrt{2}}{2}\right) = \frac{\sqrt{2}}{2}(x' + y')$$

Substituting for x and y in the equation $xy = 1$ gives us

$$\frac{\sqrt{2}}{2}(x' - y') \cdot \frac{\sqrt{2}}{2}(x' + y') = 1.$$

This equation reduces to

$$\frac{(x')^2}{2} - \frac{(y')^2}{2} = 1,$$

which is an equation of a hyperbola with vertices $(\pm\sqrt{2}, 0)$ on the x'-axis. Note that the asymptotes for the hyperbola have equations $y' = \pm x'$ in the new system. These correspond to the original x- and y-axes.

Example 2 illustrates a method for eliminating a term of an equation that contains the product xy. This method can be used to transform any equation of the form

$$Ax^2 + Bxy + Cy^2 + Dx + Ey + F = 0,$$

where $B \neq 0$, into an equation in x' and y' that contains no $x'y'$-term. Let us prove that this may always be done. If we rotate the axes through an angle ϕ, then using rotation of axes formulas (9.17)(i) to substitute for x and y gives us

$$A\,(x' \cos \phi - y' \sin \phi)^2 + B(x' \cos \phi - y' \sin \phi)(x' \sin \phi + y' \cos \phi)$$
$$+ C(x' \sin \phi + y' \cos \phi)^2 + D(x' \cos \phi - y' \sin \phi)$$
$$+ E(x' \sin \phi + y' \cos \phi) + F$$
$$= 0.$$

By performing the multiplications and rearranging terms, we may write this equation in the form

$$A'(x')^2 + B'x'y' + C'(y')^2 + D'x' + E'y' + F' = 0$$

with

$$A' = A \cos^2 \phi + B \cos \phi \sin \phi + C \sin^2 \phi$$
$$B' = 2(C - A) \sin \phi \cos \phi + B(\cos^2 \phi - \sin^2 \phi)$$
$$C' = A \sin^2 \phi - B \sin \phi \cos \phi + C \cos^2 \phi$$
$$D' = D \cos \phi + E \sin \phi$$
$$E' = -D \sin \phi + E \cos \phi$$
$$F' = F.$$

To eliminate the $x'y'$-term, we must select ϕ such that $B' = 0$—that is,

$$2(C - A) \sin \phi \cos \phi + B(\cos^2 \phi - \sin^2 \phi) = 0.$$

Using double-angle formulas, we may write this equation as

$$(C - A) \sin 2\phi + B \cos 2\phi = 0,$$

which is equivalent to

$$\cot 2\phi = \frac{A - C}{B}.$$

This formulation proves the next result.

Theorem 9.18

To eliminate the xy-term from the equation

$$Ax^2 + Bxy + Cy^2 + Dx + Ey + F = 0,$$

where $B \neq 0$, choose an angle ϕ such that

$$\cot 2\phi = \frac{A - C}{B} \quad \text{with} \quad 0 < 2\phi < \pi$$

and use the rotation of axes formulas.

The graph of any equation in x and y of the type displayed in the preceding theorem is a conic, except for certain degenerate cases.

In using Theorem (9.18), note that $\sin 2\phi > 0$, since $0 < 2\phi < \pi$. Moreover, because $\cot 2\phi = \cos 2\phi / \sin 2\phi$, the signs of $\cot 2\phi$ and $\cos 2\phi$ are always the same.

EXAMPLE ■ 3   Discuss and sketch the graph of the equation

$$41x^2 - 24xy + 34y^2 - 25 = 0.$$

SOLUTION   Use the notation of Theorem (9.18):

$$A = 41, \qquad B = -24, \qquad C = 34$$

$$\cot 2\phi = \frac{41 - 34}{-24} = -\frac{7}{24}$$

Since $\cot 2\phi$ is negative, we choose 2ϕ such that $\pi/2 < 2\phi < \pi$, and consequently, $\cos 2\phi = -\frac{7}{25}$. We now use the half-angle formulas to obtain

$$\sin \phi = \sqrt{\frac{1 - \cos 2\phi}{2}} = \sqrt{\frac{1 - (-\frac{7}{25})}{2}} = \frac{4}{5}$$

Figure 9.49

$$\cos \phi = \sqrt{\frac{1 + \cos 2\phi}{2}} = \sqrt{\frac{1 + (-\frac{7}{25})}{2}} = \frac{3}{5}.$$

Thus, the desired rotation of axes formulas are

$$x = \tfrac{3}{5}x' - \tfrac{4}{5}y' \quad \text{and} \quad y = \tfrac{4}{5}x' + \tfrac{3}{5}y'.$$

After substituting for x and y in the given equation and simplifying, we obtain the equation

$$(x')^2 + 2(y')^2 = 1.$$

The graph therefore is an ellipse with vertices at $(\pm 1, 0)$ on the x'-axis. Since $\tan \phi = \sin \phi / \cos \phi = (\frac{4}{5})/(\frac{3}{5}) = \frac{4}{3}$, we obtain $\phi = \tan^{-1}(\frac{4}{3})$. The angle ϕ is approximately 0.927 radian; to the nearest minute, $\phi \approx 53°8'$. The graph is sketched in Figure 9.49.

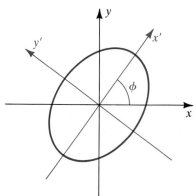

The next theorem states rules that we can apply to identify the type of conic *before* rotating the axes.

Identification Theorem 9.19

The graph of the equation

$$Ax^2 + Bxy + Cy^2 + Dx + Ey + F = 0$$

is either a conic or a degenerate conic. If the graph is a conic, then it is

   (i)   a parabola if $B^2 - 4AC = 0$

   (ii)   an ellipse if $B^2 - 4AC < 0$

   (iii)   a hyperbola if $B^2 - 4AC > 0$

PROOF   If the x- and y-axes are rotated through an angle ϕ, using the rotation of axes formulas gives us

$$A'(x')^2 + B'x'y' + C'(y')^2 + D'x' + E'y' + F' = 0.$$

Using the formulas for A', B', and C' on page 836, we can show that

$$(B')^2 - 4A'C' = B^2 - 4AC.$$

For a suitable rotation of axes, we obtain $B' = 0$ and

$$A'(x')^2 + C'(y')^2 + D'x' + E'y' + F' = 0.$$

Except for degenerate cases, the graph of this equation is an ellipse if $A'C' > 0$ (A' and C' have the same sign), a hyperbola if $A'C' < 0$ (A' and C' have opposite signs), or a parabola if $A'C' = 0$ (either $A' = 0$ or $C' = 0$). However, if $B' = 0$, then $B^2 - 4AC = -4A'C'$, and hence the graph is an ellipse if $B^2 - 4AC < 0$, a hyperbola if $B^2 - 4AC > 0$, or a parabola if $B^2 - 4AC = 0$.   ∎

The expression $B^2 - 4AC$ is called the **discriminant** of the equation in the identification theorem (9.19). We say that this discriminant is **invariant** under a rotation of axes, because it is unchanged by any such rotation.

EXAMPLE ■ 4   Use the identification theorem (9.19) to determine if the graph of the equation

$$41x^2 - 24xy + 34y^2 - 25 = 0$$

is a parabola, an ellipse, or a hyperbola.

SOLUTION   We considered this equation in Example 3, where we performed a rotation of axes. Since $A = 41$, $B = -24$, and $C = 34$, the discriminant is

$$B^2 - 4AC = 576 - 4(41)(34) = -5000 < 0.$$

Hence, by the identification theorem, the graph is an ellipse.

In some cases, after eliminating the xy-term, it may be necessary to translate the axes of the $x'y'$-coordinate system to obtain the graph, as illustrated in the next example.

EXAMPLE • 5 Discuss and sketch the graph of the equation

$$x^2 + 2\sqrt{3}xy + 3y^2 + 8\sqrt{3}x - 8y + 32 = 0.$$

SOLUTION Using $A = 1$, $B = 2\sqrt{3}$, and $C = 3$, we see that

$$B^2 - 4AC = 12 - 12 = 0.$$

By the identification theorem (9.19), the graph is a parabola.

To apply a rotation of axes, we calculate

$$\cot 2\phi = \frac{A - C}{B} = \frac{1 - 3}{2\sqrt{3}} = -\frac{1}{\sqrt{3}}.$$

Hence $2\phi = 2\pi/3$, $\phi = \pi/3$, and

$$\sin \phi = \frac{\sqrt{3}}{2}, \qquad \cos \phi = \frac{1}{2}.$$

Figure 9.50

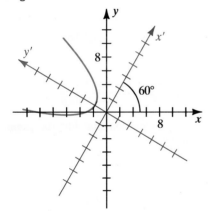

The rotation of axes formulas (9.17)(i) are as follows:

$$x = \frac{1}{2}x' - \frac{\sqrt{3}}{2}y' = \frac{1}{2}(x' - \sqrt{3}y')$$

$$y = \frac{\sqrt{3}}{2}x' + \frac{1}{2}y' = \frac{1}{2}(\sqrt{3}x' + y')$$

Substituting for x and y in the given equation and simplifying leads to

$$4(x')^2 - 16y' + 32 = 0,$$

or, equivalently, $\qquad (x')^2 = 4(y' - 2).$

The parabola is sketched in Figure 9.50, where each tic represents two units. Note that the vertex is at the point $(0, 2)$ in the $x'y'$-plane, and the graph is symmetric with respect to the y'-axis.

EXERCISES 9.5

Exer. 1–6: Find the vertices and the foci of the conic, and use the translation of axes formulas to sketch its graph.

1 $y^2 - 8x + 8y + 32 = 0$

2 $x = 2y^2 + 8y + 3$

3 $4x^2 + 9y^2 + 24x - 36y + 36 = 0$

4 $3x^2 + 4y^2 - 18x + 8y + 19 = 0$

5 $x^2 - 9y^2 + 8x + 90y - 210 = 0$

6 $4x^2 - y^2 - 40x - 8y + 88 = 0$

Exer. 7–19: (a) Use the identification theorem (9.19) to determine whether the graph of the equation is a parabola, an ellipse, or a hyperbola. (b) Use a suitable rotation of axes to find an equation for the graph in an $x'y'$-plane, and sketch the graph, labeling vertices.

7 $x^2 - 2xy + y^2 - 2\sqrt{2}x - 2\sqrt{2}y = 0$

8 $x^2 - 2xy + y^2 + 4x + 4y = 0$

9 $5x^2 - 8xy + 5y^2 = 9$

10 $x^2 - xy + y^2 = 3$

11 $11x^2 + 10\sqrt{3}xy + y^2 = 4$

12 $7x^2 - 48xy - 7y^2 = 225$

13 $16x^2 - 24xy + 9y^2 - 60x - 80y + 100 = 0$

14 $x^2 + 4xy + 4y^2 + 6\sqrt{5}x - 18\sqrt{5}y + 45 = 0$

15 $40x^2 - 36xy + 25y^2 - 8\sqrt{13}x - 12\sqrt{13}y = 0$

16 $18x^2 - 48xy + 82y^2 + 6\sqrt{10}x + 2\sqrt{10}y - 80 = 0$

17 $5x^2 + 6\sqrt{3}xy - y^2 + 8x - 8\sqrt{3}y - 12 = 0$

18 $15x^2 + 20xy - 4\sqrt{5}x + 8\sqrt{5}y - 100 = 0$

19 $32x^2 - 72xy + 53y^2 = 80$

[c] Exer. 20–22: Graph the equation.

20 $1.1x^2 - 1.3xy + y^2 - 2.9x - 1.9y = 0$

21 $2.1x^2 - 4xy + 1.5y^2 - 4x + y - 1 = 0$

22 $3.2x^2 - 4\sqrt{2}xy + 2.5y^2 + 2.1y + 3x - 2.1 = 0$

CHAPTER 9 REVIEW EXERCISES

Exer. 1–4: (a) Find an equation in x and y whose graph contains the points on the curve C. (b) Sketch the graph of C and indicate the orientation.

1 $x = \dfrac{1}{t} + 1, \qquad y = \dfrac{2}{t} - t; \qquad 0 < t \le 4$

2 $x = \cos^2 t - 2, \quad y = \sin t + 1; \qquad 0 \le t \le 2\pi$

3 $x = \sqrt{t}, \qquad y = 2^{-t}; \qquad t \ge 0$

4 $x = 3\cos t + 2, \quad y = -3\sin t - 1; \quad 0 \le t \le 2\pi$

Exer. 5–6: Sketch the graphs of C_1, C_2, C_3, and C_4, and indicate their orientations.

5 $C_1: x = t, \qquad\qquad y = \sqrt{16 - t^2}; \qquad -4 \le t \le 4$
$C_2: x = -\sqrt{16 - t}, \quad y = -\sqrt{t}; \qquad 0 \le t \le 16$
$C_3: x = 4\cos t, \qquad y = 4\sin t; \qquad 0 \le t \le 2\pi$
$C_4: x = e^t, \qquad\qquad y = -\sqrt{16 - e^{2t}}; \quad t \le \ln 4$

6 $C_1: x = t^2, \qquad y = t^3; \qquad t \text{ in } \mathbb{R}$
$C_2: x = t^4, \qquad y = t^6; \qquad t \text{ in } \mathbb{R}$
$C_3: x = e^{2t}, \qquad y = e^{3t}; \qquad t \text{ in } \mathbb{R}$
$C_4: x = 1 - \sin^2 t, \quad y = \cos^3 t; \qquad t \text{ in } \mathbb{R}$

Exer. 7–8: Let C be the given parametrized curve. (a) Express dy/dx in terms of t. (b) Find the values of t that correspond to horizontal or vertical tangent lines to the graph of C. (c) Express d^2y/dx^2 in terms of t.

7 $x = t^2, \qquad\qquad y = 2t^3 + 4t - 1; \quad t \text{ in } \mathbb{R}$

8 $x = t - 2\sin t, \quad y = 1 - 2\cos t; \qquad t \text{ in } \mathbb{R}$

Exer. 9–26: Sketch the graph of the polar equation.

9 $r = -4\sin\theta$

10 $r = 10\cos\theta$

11 $r = 6 - 3\cos\theta$

12 $r = 3 + 2\cos\theta$

13 $r^2 = 9\sin 2\theta$

14 $r^2 = -4\sin 2\theta$

15 $r = 3\sin 5\theta$

16 $r = 2\sin 3\theta$

17 $2r = \theta$

18 $r = e^{-\theta}, \quad \theta \ge 0$

19 $r = 8\sec\theta$

20 $r(3\cos\theta - 2\sin\theta) = 6$

21 $r = 4 - 4\cos\theta$

22 $r = 4\cos^2 \tfrac{1}{2}\theta$

23 $r = 6 - r\cos\theta$

24 $r = 6\cos 2\theta$

25 $r = \dfrac{8}{3 + \cos\theta}$

26 $r = \dfrac{8}{1 - 3\sin\theta}$

Exer. 27–32: Find a polar equation that has the same graph as the given equation.

27 $y^2 = 4x$

28 $x^2 + y^2 - 3x + 4y = 0$

29 $2x - 3y = 8$

30 $x^2 + y^2 = 2xy$

31 $y^2 = x^2 - 2x$

32 $x^2 = y^2 + 3y$

Exer. 33–38: Find an equation in x and y that has the same graph as the polar equation.

33 $r^2 = \tan\theta$

34 $r = 2\cos\theta + 3\sin\theta$

35 $r^2 = 4\sin 2\theta$

36 $r^2 = \sec 2\theta$

37 $\theta = \sqrt{3}$

38 $r = -6$

Exer. 39–40: Find the slope of the tangent line to the graph of the polar equation at the point corresponding to the given value of θ.

39 $r = \dfrac{3}{2 + 2\cos\theta}; \quad \theta = \pi/2$

40 $r = e^{3\theta}; \qquad \theta = \pi/4$

41 Find the area of the region bounded by one loop of $r^2 = 4\sin 2\theta$.

42 Find the area of the region that is inside the graph of $r = 3 + 2\sin\theta$ and outside the graph of $r = 4$.

43 The position (x, y) of a moving point at time t is given by $x = 2\sin t$, $y = \sin^2 t$. Find the distance that the point travels from $t = 0$ to $t = \pi/2$.

44 Find the length of the spiral $r = 1/\theta$ from $\theta = 1$ to $\theta = 2$.

45 The curve with parametrization $x = 2t^2 + 1$, $y = 4t - 3$; $0 \le t \le 1$ is revolved about the y-axis. Find the area of the resulting surface.

46 The arc of the spiral $r = e^\theta$ from $\theta = 0$ to $\theta = 1$ is revolved about the line $\theta = \pi/2$. Find the area of the resulting surface.

47 Find the area of the surface generated by revolving the lemniscate $r^2 = a^2 \cos 2\theta$ about the polar axis.

48 A line segment of fixed length has endpoints A and B on the y-axis and x-axis, respectively. A fixed point P on AB is selected with $d(A, P) = a$ and $d(B, P) = b$ (see figure). If A and B may slide freely along their

Exercise 48

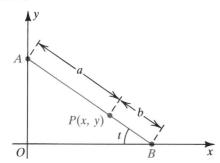

respective axes, a curve C is traced by P. If t is the radian measure of angle ABO, find parametric equations for C with parameter t and describe C.

Exer. 49–51: Find the vertices and the foci of the conic, and use the translation of axes formulas to sketch its graph.

49 $y^2 - 2y + 4x - 7 = 0$

50 $4x^2 + y^2 - 24x + 4y + 36 = 0$

51 $x^2 - 9y^2 + 8x + 7 = 0$

52 Use the discriminant to identify the graph of each equation. (Do not sketch the graph.)

(a) $2x^2 - 3xy + 4y^2 + 6x - 2y - 6 = 0$

(b) $3x^2 + 2xy - y^2 - 2x + y + 4 = 0$

(c) $x^2 - 6xy + 9y^2 + x - 3y + 5 = 0$

Exer. 53–54: Use a suitable rotation of axes to find an equation for the graph in an $x'y'$-plane, and sketch the graph, labeling vertices.

53 $x^2 - 8xy + 16y^2 - 12\sqrt{17}x - 3\sqrt{17}y = 0$

54 $8x^2 + 12xy + 17y^2 - 16\sqrt{5}x - 12\sqrt{5}y = 0$

EXTENDED PROBLEMS AND GROUP PROJECTS

I Investigate the representation of conic sections in polar coordinates.

(a) Let F be a fixed point and l a fixed line in a plane. Prove that the set of all points P in the plane, such that the ratio $d(P, F)/d(P, Q)$ is a positive constant e with $d(P, Q)$ the distance from P to l, is a conic section. Show that the conic is a parabola if $e = 1$, an ellipse if $0 < e < 1$, and a hyperbola if $e > 1$.

(b) Prove the following theorem: A polar equation that has one of the four forms

$$r = \frac{de}{1 \pm e\cos\theta}, \qquad r = \frac{de}{1 \pm e\sin\theta}$$

is a conic section. Show that the conic is a parabola if $e = 1$, an ellipse if $0 < e < 1$, or a hyperbola if $e > 1$.

(c) Describe and sketch the graph of each of the following polar equations:

(i) $r = 10/(3 + 2\cos\theta)$

(ii) $r = 10/(2 + 3\sin\theta)$

(iii) $r = 15/(4 - 4\cos\theta)$

(d) Find a polar equation of the conic with a focus at the pole, eccentricity $e = 1/2$, and directrix $r = -3\sec\theta$.

2 Develop formulas in polar coordinates for the center of mass using the following as an outline for a possible approach.

(a) Show that the center of mass of a triangle is located on each median, two thirds of the way from a vertex to the opposite side.

(b) Consider a thin triangle as shown in the figure. Discuss why it is reasonable to assume that its center of mass has polar coordinates that are approximately $(\frac{2}{3}r, \theta)$.

Problem 2

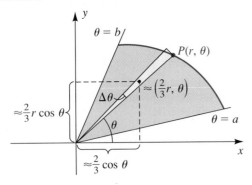

(c) Consider a region in the plane bounded by the lines $\theta = a$, $\theta = b$, and the curve $r = f(\theta)$. Slice the region into triangular wedges. Show that the moment of the region about the x-axis is approximately $\sum \frac{1}{3}r^3 \sin\theta \, \Delta\theta$.

(d) Show that the sum in part (c) is a Riemann sum and find the form for the definite integral that is the limit as $\Delta\theta \to 0$.

(e) Show that the coordinates of the center of mass are given by

$$\bar{x} = \frac{\int_a^b \frac{2}{3}r^3 \cos\theta \, d\theta}{\int_a^b r^2 \, d\theta}, \qquad \bar{y} = \frac{\int_a^b \frac{2}{3}r^3 \sin\theta \, d\theta}{\int_a^b r^2 \, d\theta}.$$

(f) Show that the formulas in part (e) give the correct answer if the region is a circle centered at the origin.

(g) Find the center of mass of the region enclosed by a semicircle of radius a.

(h) Find the center of mass of the region enclosed by the cardioid $r = a(1 + \sin\theta)$.

3 Investigate *space-filling curves*, which are curves whose graphs fill a solid region of the plane. The discovery of such curves in the late nineteenth century dramatically revealed that our intuitive understanding of the "one dimensionality" of curves was deeply flawed. This

Problem 3

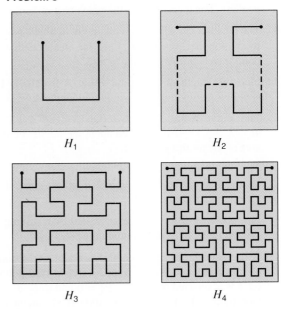

H_1 H_2

H_3 H_4

revelation led to an intensive research into a precise definition of *dimension* and the properties of curves that has actively continued to the present.

The first such curve was discovered in 1890 by Giuseppe Peano (1858–1932).* In this problem, we consider an example published in 1891 by David Hilbert (1862–1943). The Hilbert curve is the limit of a sequence of curves $H_1, H_2, H_3, \ldots$, each of which is contained in the unit square S of side 1. The first four curves in this sequence are shown in the figure.

(a) Examine the first stage, H_1, which is made up of three line segments. Show that the curve H_1 can be given a simple parametric representation, $(f_1(t), g_1(t))$, for $0 \le t \le 1$.

(b) Show that if the unit square is divided into four congruent squares, then H_1 passes through the center of each of these squares.

(c) Show that every point in the unit square is within a distance of $\sqrt{2}/4$ units of a point on the curve H_1.

To obtain H_2, a smaller copy of H_1 is placed in each of the four small squares together with extending segments, connecting the copies of H_1 to form a curve made up of line segments.

---

*For a description of Peano's curve and other attempts to create space-filling curves, see Heinz-Otto Peitgen, Hartmut Jurgens, and Dietmar Saupe, *Fractals for the Classroom, Part One: Introduction to Fractals and Chaos*, New York: Springer-Verlag, 1991.

(d) Determine the number of line segments in H_2. Show that it is possible to give a parametric representation of H_2. Show that if the unit square is divided into 16 congruent smaller squares, then H_2 passes through the center of each of these squares, and that every point in the unit square is within $\sqrt{2}/8$ units of a point on H_2.

At each stage in the construction, the unit square is divided into four congruent subsquares and a shrunken copy of H_n is placed inside each of the four subsquares; these are joined to form H_{n+1}.

(e) Show that if the unit square is divided into 4^n congruent subsquares, then the curve H_n passes through the center of each of these subsquares. Show that each point of the unit square is within $\sqrt{2}/2^{n+1}$ units of a point on H_n. Show that each H_n has a parametric representation $(f_n(t), g_n(t))$ for $0 \le t \le 1$.

(f) Let $H = \lim_{n \to \infty} H_n$. Show that H has a parametric representation $(f(t), g(t))$, where $f(t) = \lim_{n \to \infty} f_n(t)$ and $g(t) = \lim_{n \to \infty} g_n(t)$. Prove that $f(t)$ and $g(t)$ are continuous functions and that the graph of H passes through every point in the unit square.

INTRODUCTION

IN OUR STUDY OF CALCULUS SO FAR, the focus has been on functions whose domains and codomains (or ranges) are sets of real numbers. The emphasis has been a relationship that can be described by an expression of the form $y = f(x)$ or $r = f(\theta)$, where f is a function and x, y, r, and θ are real numbers. The graph of such a relationship is a curve in the plane. The solution to many real-world problems, however, requires us to leave the plane and examine solids and surfaces in three-dimensional space. The design of a radio telescope, for example, calls for finding a surface that maximizes the likelihood of clearly receiving faint signals from deep space.

We need to generalize the main ideas of calculus—limits, continuity, derivatives, integrals—to relationships whose graphs may be curves or surfaces in three-dimensional space. Such relationships are expressed through functions whose domains or codomains may be *multivariable*. We will develop multivariable concepts in the next five chapters. In this chapter, we begin our study of *vectors* as a tool for examining points in the plane and in space. A vector has two different natures, one geometric and the other algebraic, both of which are useful in applications. Generally, geometry helps us to develop an intuitive understanding of vector concepts; however, rigorous proofs require algebraic techniques.

There are different ways to approach our study of vectors. One way is to define vectors in three dimensions and then *specialize* to vectors in a coordinate plane. The other is to define vectors in a coordinate plane and then *generalize* to three dimensions. We shall use the latter method, beginning with the familiar xy-plane in Section 10.1 and then proceeding to an xyz-coordinate system in Section 10.2. An advantage of this approach is that we can plot (and visualize) initial and terminal points of vectors more easily in the plane than in three dimensions. Moreover, proofs of properties of vector addition and scalar multiplication are shorter, because there is one fewer vector component to consider. Since the corresponding proofs in three dimensions follow exactly the same steps, it is unnecessary to reprove these properties in Section 10.2.

Two important operations involving vectors in the plane and space, the dot product and the cross product, are introduced in Sections 10.3 and 10.4, respectively. We then apply the language of vectors in Section 10.5 to discuss representations of lines and planes. In Section 10.6, we examine surfaces in space.

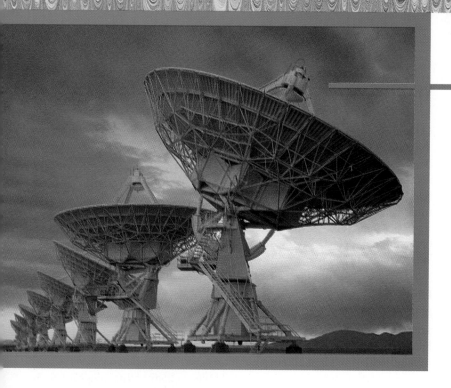

The optimal design of radio telescopes and other electronic technologies requires a sophisticated understanding and use of vectors and surfaces.

Vectors and Surfaces

10.1 VECTORS IN TWO DIMENSIONS

Quantities such as area, volume, arc length, temperature, and time have magnitude only and can be completely characterized by a single real number (with an appropriate unit of measurement such as in^2, ft^3, cm, deg, or sec). A quantity of this type is a **scalar quantity,** and the corresponding real number is a **scalar.** A concept such as velocity or force has both magnitude and direction and is often represented by a **directed line segment**— that is, a line segment to which a direction has been assigned. Another name for a directed line segment is a **vector.** In this section, we consider vectors in the plane.

As shown in Figure 10.1, we use $\overrightarrow{PQ}$ to denote the vector with **initial point** P and **terminal point** Q and indicate the direction of the vector by placing an arrowhead at Q. The **magnitude** of $\overrightarrow{PQ}$ is the length of the segment PQ and is denoted by $\| \overrightarrow{PQ} \|$. As in the figure, we use boldface letters such as **u** and **v** to denote vectors whose endpoints are not specified. In handwritten work, a notation such as $\vec{u}$ or $\vec{v}$ is often used.

Vectors that have the same magnitude and direction are said to be **equivalent.** In calculus, a vector is determined only by its magnitude and direction, not by its location. Thus, we regard equivalent vectors, such as those in Figure 10.1, as **equal** and write

$$\mathbf{u} = \overrightarrow{PQ}, \qquad \mathbf{v} = \overrightarrow{PQ}, \quad \text{and} \quad \mathbf{u} = \mathbf{v}.$$

Thus, *a vector may be translated from one location to another, provided neither the magnitude nor the direction is changed.*

We can represent many physical concepts by vectors. To illustrate, suppose an airplane is descending at a constant speed of 100 mi/hr and the line of flight makes an angle of 20° with the horizontal. Both of these facts are represented by the vector **v** of magnitude 100 in Figure 10.2. The vector **v** is a **velocity vector.**

A vector that represents a pull or push of some type is a **force vector.** The force exerted when a person holds a 5-lb weight is illustrated by the vector **F** of magnitude 5 in Figure 10.3. This force has the same magnitude as the force exerted on the weight by gravity, but acts in the opposite direction. As a result, there is no movement upward or downward.

We sometimes use $\overrightarrow{AB}$ to represent the path of a point (or particle) as it moves along the line segment from A to B. We then refer to $\overrightarrow{AB}$ as a **displacement** of the point (or particle). As in Figure 10.4, a displacement $\overrightarrow{AB}$ followed by a displacement $\overrightarrow{BC}$ leads to the same point as the single displacement $\overrightarrow{AC}$. By definition, the vector $\overrightarrow{AC}$ is the **sum** of $\overrightarrow{AB}$ and $\overrightarrow{BC}$, and we write

$$\overrightarrow{AC} = \overrightarrow{AB} + \overrightarrow{BC}.$$

Since vectors may be translated from one location to another, *any* two vectors may be added by placing the initial point of the second vector on

Figure 10.1

Figure 10.2

Figure 10.3

Figure 10.4
$\overrightarrow{AC} = \overrightarrow{AB} + \overrightarrow{BC}$

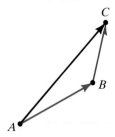

Figure 10.5
$$\vec{PS} = \vec{PQ} + \vec{PR}$$

Figure 10.6

Figure 10.7

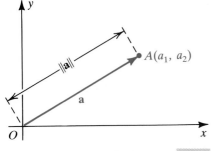

Figure 10.8

the terminal point of the first and then drawing the line segment from the initial point of the first to the terminal point of the second, as in Figure 10.4.

Another way to find the sum is to choose vectors $\vec{PQ}$ and $\vec{PR}$ that are equal to $\vec{AB}$ and $\vec{BC}$, respectively, and have the same initial point P, as shown in Figure 10.5. If we construct parallelogram $RPQS$, then since $\vec{PR} = \vec{QS}$, it follows that $\vec{PS} = \vec{PQ} + \vec{PR}$. If $\vec{PQ}$ and $\vec{PR}$ are two forces acting at P, then $\vec{PS}$ is the **resultant force**—that is, the single force that produces the same effect as the two combined forces.

If c is a scalar and $\mathbf{v}$ is a vector, then $c\mathbf{v}$ is defined as a vector whose magnitude is $|c|$ times the magnitude $\|\mathbf{v}\|$ of $\mathbf{v}$ and whose direction is either the same as that of $\mathbf{v}$ (if $c > 0$) or opposite that of $\mathbf{v}$ (if $c < 0$). Illustrations are given in Figure 10.6. We refer to $c\mathbf{v}$ as a **scalar multiple** of $\mathbf{v}$.

Throughout the remainder of this section, we shall restrict our discussion to the special case in which all vectors lie in an xy-plane. If $\vec{PQ}$ is such a vector, then, as indicated in Figure 10.7, there are many vectors that are equivalent to $\vec{PQ}$; however, there is precisely *one* equivalent vector $\mathbf{a} = \vec{OA}$ with initial point at the origin. This vector $\vec{OA}$ is designated as the **position vector** for $\vec{PQ}$. Thus, *each vector determines a unique ordered pair of real numbers*, the coordinates (a_1, a_2) of the terminal point of its position vector. Conversely, every ordered pair (a_1, a_2) determines the vector $\vec{OA}$, where A has coordinates (a_1, a_2). Thus, *there is a one-to-one correspondence between vectors in an xy-plane and ordered pairs of real numbers*. This correspondence allows us to interpret a vector as both a directed line segment *and* an ordered pair of real numbers. To avoid confusion with the notation for open intervals or points, we use the symbol $\langle a_1, a_2 \rangle$ for an ordered pair that represents a vector, and we denote it by a boldface letter, such as $\mathbf{a} = \langle a_1, a_2 \rangle$. The numbers a_1 and a_2 are the **components** of the vector $\langle a_1, a_2 \rangle$. Two vectors $\langle a_1, a_2 \rangle$ and $\langle b_1, b_2 \rangle$ are **equal** if and only if $a_1 = b_1$ and $a_2 = b_2$. If A is the point (a_1, a_2), as in Figure 10.7, we call $\vec{OA}$ the **position vector** for $\langle a_1, a_2 \rangle$, or for the *point* A.

The preceding discussion shows that vectors have two different natures, one geometric and the other algebraic. Often we do not distinguish between the two. It should always be clear from our discussion whether we are referring to ordered pairs or directed line segments. We shall use the symbol V_2 for the set of all vectors $\langle x, y \rangle$, where x and y are real numbers.

The **magnitude** $\|\mathbf{a}\|$ of the vector $\mathbf{a} = \langle a_1, a_2 \rangle$ is, by definition, the length of its position vector $\vec{OA}$, as illustrated in Figure 10.8. Another way of stating this result is given in the following definition.

Definition 10.1

> The **magnitude** $\|\mathbf{a}\|$ of the vector $\mathbf{a} = \langle a_1, a_2 \rangle$ is
> $$\|\mathbf{a}\| = \|\langle a_1, a_2 \rangle\| = \sqrt{a_1^2 + a_2^2}.$$

Figure 10.9

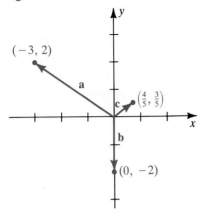

EXAMPLE ▪ I   Sketch the position vectors for

$$\mathbf{a} = \langle -3, 2 \rangle, \qquad \mathbf{b} = \langle 0, -2 \rangle, \qquad \mathbf{c} = \langle \tfrac{4}{5}, \tfrac{3}{5} \rangle$$

and find the magnitude of each vector.

SOLUTION   The position vectors are sketched in Figure 10.9. By Definition (10.1),

$$\|\mathbf{a}\| = \|\langle -3, 2 \rangle\| = \sqrt{(-3)^2 + 2^2} = \sqrt{13}$$

$$\|\mathbf{b}\| = \|\langle 0, -2 \rangle\| = \sqrt{0^2 + (-2)^2} = \sqrt{4} = 2$$

$$\|\mathbf{c}\| = \|\langle \tfrac{4}{5}, \tfrac{3}{5} \rangle\| = \sqrt{(\tfrac{4}{5})^2 + (\tfrac{3}{5})^2} = \sqrt{\tfrac{25}{25}} = 1.$$

Figure 10.10

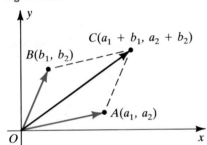

Let $\mathbf{a} = \langle a_1, a_2 \rangle$ and $\mathbf{b} = \langle b_1, b_2 \rangle$ be vectors in V_2, and let $\overrightarrow{OA}$ and $\overrightarrow{OB}$ be their position vectors, as illustrated in Figure 10.10. If $\overrightarrow{OC}$ is the position vector for $\mathbf{c} = \langle a_1 + b_1, a_2 + b_2 \rangle$, then it can be shown, using slopes, that O, A, C, and B are vertices of a parallelogram—that is,

$$\overrightarrow{OA} + \overrightarrow{OB} = \overrightarrow{OC}.$$

This result motivates the following definition of addition in V_2.

Addition of Vectors   10.2

$$\langle a_1, a_2 \rangle + \langle b_1, b_2 \rangle = \langle a_1 + b_1, a_2 + b_2 \rangle$$

Figure 10.11

Note that to add two vectors of V_2, we add corresponding components.

It can also be shown that if c is a scalar, then the ordered pair determined by $c\overrightarrow{OA}$ is (ca_1, ca_2), as illustrated in Figure 10.11 for $c > 1$, which suggests the following definition of scalar multiples in V_2.

Scalar Multiple of a Vector   10.3

$$c\langle a_1, a_2 \rangle = \langle ca_1, ca_2 \rangle$$

Thus, to find a scalar multiple of a vector in V_2, we multiply each component by the scalar.

EXAMPLE ▪ 2   Let $\mathbf{a} = \langle -1, 3 \rangle$ and $\mathbf{b} = \langle 4, 2 \rangle$.

(a) Find $\mathbf{a} + \mathbf{b}$, $-\tfrac{3}{2}\mathbf{b}$, and $2\mathbf{a} + 3\mathbf{b}$.

(b) Sketch position vectors for $\mathbf{a}$, $\mathbf{b}$, $\mathbf{a} + \mathbf{b}$, and $-\tfrac{3}{2}\mathbf{b}$.

Figure 10.12

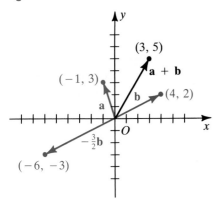

SOLUTION

(a) Applying Definitions (10.2) and (10.3), we have the following:

$$\mathbf{a} + \mathbf{b} = \langle -1, 3 \rangle + \langle 4, 2 \rangle = \langle -1 + 4, 3 + 2 \rangle = \langle 3, 5 \rangle$$

$$-\tfrac{3}{2}\mathbf{b} = -\tfrac{3}{2}\langle 4, 2 \rangle = \langle -\tfrac{3}{2}(4), -\tfrac{3}{2}(2) \rangle = \langle -6, -3 \rangle$$

$$2\mathbf{a} + 3\mathbf{b} = 2\langle -1, 3 \rangle + 3\langle 4, 2 \rangle = \langle -2, 6 \rangle + \langle 12, 6 \rangle = \langle 10, 12 \rangle$$

(b) Position vectors for $\mathbf{a}$, $\mathbf{b}$, $\mathbf{a} + \mathbf{b}$, and $-\tfrac{3}{2}\mathbf{b}$ are sketched in Figure 10.12.

The **zero vector 0** and the **negative** $-\mathbf{a}$ of a vector $\mathbf{a} = \langle a_1, a_2 \rangle$ are defined as follows.

Definition 10.4

$$\mathbf{0} = \langle 0, 0 \rangle, \qquad -\mathbf{a} = -\langle a_1, a_2 \rangle = \langle -a_1, -a_2 \rangle$$

In the next theorem, we state some important properties of addition and scalar multiples of vectors for any vectors $\mathbf{a}$, $\mathbf{b}$, and $\mathbf{c}$ in V_2 and scalars c and d. There should be no difficulty in remembering these properties, since they resemble familiar laws of real numbers.

Theorem 10.5

(i) $\mathbf{a} + \mathbf{b} = \mathbf{b} + \mathbf{a}$

(ii) $\mathbf{a} + (\mathbf{b} + \mathbf{c}) = (\mathbf{a} + \mathbf{b}) + \mathbf{c}$

(iii) $\mathbf{a} + \mathbf{0} = \mathbf{a}$

(iv) $\mathbf{a} + (-\mathbf{a}) = \mathbf{0}$

(v) $c(\mathbf{a} + \mathbf{b}) = c\mathbf{a} + c\mathbf{b}$

(vi) $(c + d)\mathbf{a} = c\mathbf{a} + d\mathbf{a}$

(vii) $(cd)\mathbf{a} = c(d\mathbf{a}) = d(c\mathbf{a})$

(viii) $1\mathbf{a} = \mathbf{a}$

(ix) $0\mathbf{a} = \mathbf{0} = c\mathbf{0}$

PROOF Let $\mathbf{a} = \langle a_1, a_2 \rangle$ and $\mathbf{b} = \langle b_1, b_2 \rangle$. To prove (i), we note that

$$\mathbf{a} + \mathbf{b} = \langle a_1 + b_1, a_2 + b_2 \rangle = \langle b_1 + a_1, b_2 + a_2 \rangle = \mathbf{b} + \mathbf{a}.$$

The proof of (v) is as follows:

$$
\begin{aligned}
c(\mathbf{a} + \mathbf{b}) &= c\langle a_1 + b_1, a_2 + b_2 \rangle \\
&= \langle c(a_1 + b_1), c(a_2 + b_2) \rangle \\
&= \langle ca_1 + cb_1, ca_2 + cb_2 \rangle \\
&= \langle ca_1, ca_2 \rangle + \langle cb_1, cb_2 \rangle \\
&= c\mathbf{a} + c\mathbf{b}
\end{aligned}
$$

Proofs of the remaining properties are similar and are left as exercises. ∎

The operation of **subtraction** of vectors in V_2, denoted by $-$, is defined as follows.

Definition 10.6

> Let $\mathbf{a} = \langle a_1, a_2 \rangle$ and $\mathbf{b} = \langle b_1, b_2 \rangle$. The **difference** $\mathbf{a} - \mathbf{b}$ is
>
> $$\mathbf{a} - \mathbf{b} = \mathbf{a} + (-\mathbf{b}).$$

Using Definitions (10.6), (10.4), and (10.2), we see that

$$\mathbf{a} - \mathbf{b} = \mathbf{a} + (-\mathbf{b})$$
$$= \langle a_1, a_2 \rangle + \langle -b_1, -b_2 \rangle$$
$$= \langle a_1 - b_1, a_2 - b_2 \rangle.$$

Thus, to find $\mathbf{a} - \mathbf{b}$, we merely subtract the components of $\mathbf{b}$ from the corresponding components of $\mathbf{a}$.

EXAMPLE ■ 3 If $\mathbf{a} = \langle 5, -4 \rangle$ and $\mathbf{b} = \langle -3, 2 \rangle$, find the vectors $\mathbf{a} - \mathbf{b}$ and $2\mathbf{a} - 3\mathbf{b}$.

SOLUTION

$$\mathbf{a} - \mathbf{b} = \langle 5, -4 \rangle - \langle -3, 2 \rangle = \langle 5 - (-3), -4 - 2 \rangle = \langle 8, -6 \rangle$$
$$2\mathbf{a} - 3\mathbf{b} = 2\langle 5, -4 \rangle - 3\langle -3, 2 \rangle = \langle 10, -8 \rangle - \langle -9, 6 \rangle = \langle 19, -14 \rangle$$

Figure 10.13

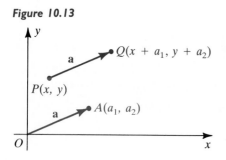

We sometimes represent a nonzero vector $\mathbf{a} = \langle a_1, a_2 \rangle$ of V_2 by a directed line segment $\overrightarrow{PQ}$ with *any* initial point $P(x, y)$ and terminal point $Q(x + a_1, y + a_2)$, as illustrated in Figure 10.13. Note that the position vectors for $\overrightarrow{PQ}$ and $\mathbf{a}$ are both $\overrightarrow{OA}$, and that $\| \overrightarrow{PQ} \| = \| \mathbf{a} \|$. We say that $\overrightarrow{PQ}$ *corresponds to* $\mathbf{a}$, or that $\mathbf{a}$ corresponds to $\overrightarrow{PQ}$. Strictly speaking, $\overrightarrow{PQ}$ *represents* $\mathbf{a}$; however, we often refer to $\overrightarrow{PQ}$ as the *vector* $\mathbf{a}$. The zero vector will be represented by any point in the plane.

We shall make frequent use of the next theorem.

Theorem 10.7

> If $P_1(x_1, y_1)$ and $P_2(x_2, y_2)$ are any points, the vector $\mathbf{a}$ in V_2 that corresponds to $\overrightarrow{P_1 P_2}$ is
>
> $$\mathbf{a} = \langle x_2 - x_1, y_2 - y_1 \rangle.$$

Figure 10.14

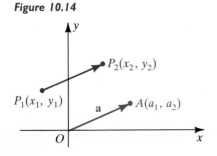

PROOF The vector $\overrightarrow{P_1 P_2}$ and the position vector for $\mathbf{a} = \langle a_1, a_2 \rangle$ are illustrated in Figure 10.14. Since $\overrightarrow{P_1 P_2}$ corresponds to $\mathbf{a}$, we see that

$$x_2 = x_1 + a_1 \quad \text{and} \quad y_2 = y_1 + a_2.$$

Solving for a_1 and a_2 gives us $\mathbf{a} = \langle x_2 - x_1, y_2 - y_1 \rangle$. ■

EXAMPLE • 4 Given the points $P(-2, 3)$ and $Q(4, 5)$,

(a) sketch $\overrightarrow{PQ}$ and $\overrightarrow{QP}$

(b) find vectors **a** and **b** in V_2 that correspond to $\overrightarrow{PQ}$ and $\overrightarrow{QP}$

(c) sketch the position vectors for **a** and **b**

SOLUTION

(a) The vectors $\overrightarrow{PQ}$ and $\overrightarrow{QP}$ are sketched in Figure 10.15.

(b) By Theorem (10.7),

$$\mathbf{a} = \langle 4 - (-2), 5 - 3 \rangle = \langle 6, 2 \rangle$$
$$\mathbf{b} = \langle -2 - 4, 3 - 5 \rangle = \langle -6, -2 \rangle = -\langle 6, 2 \rangle.$$

(c) The position vectors $\overrightarrow{OA}$ and $\overrightarrow{OB}$ for **a** and **b** are sketched in Figure 10.15.

Figure 10.15

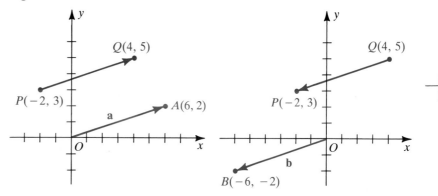

If **a** and **b** are arbitrary vectors, then

$$\mathbf{b} + (\mathbf{a} - \mathbf{b}) = \mathbf{a};$$

that is, $\mathbf{a} - \mathbf{b}$ *is the vector that, when added to* **b***, gives us* **a***.* If we represent **a** and **b** by vectors $\overrightarrow{PQ}$ and $\overrightarrow{PR}$ *with the same initial point,* as in Figure 10.16, then $\overrightarrow{RQ}$ represents $\mathbf{a} - \mathbf{b}$.

At the beginning of this section, we used the symbol $c\overrightarrow{AB}$ to denote a vector whose direction is the same as that of $\overrightarrow{AB}$ if $c > 0$ or opposite that of $\overrightarrow{AB}$ if $c < 0$. The next definition is the analogous concept for vectors in V_2.

Figure 10.16

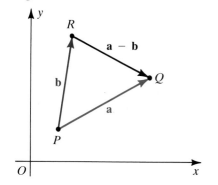

Definition 10.8

Nonzero vectors **a** and **b** in V_2 have

    (i) the **same direction** if $\mathbf{b} = c\mathbf{a}$ for some scalar $c > 0$

    (ii) the **opposite direction** if $\mathbf{b} = c\mathbf{a}$ for some scalar $c < 0$

Figure 10.17

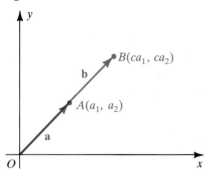

Figure 10.17

Let $\mathbf{a} = \langle a_1, a_2 \rangle$ and $\mathbf{b} = \langle b_1, b_2 \rangle$. If $\mathbf{b} = c\mathbf{a}$, then $b_1 = ca_1$ and $b_2 = ca_2$, and the position vectors $\overrightarrow{OA}$ and $\overrightarrow{OB}$ for $\mathbf{a}$ and $\mathbf{b}$ lie on the same line through the origin, as illustrated in Figure 10.17. Moreover, if $c > 0$, then A and B are in the same quadrant (or on the same positive or negative coordinate axis). If $c < 0$ and A is not on a coordinate axis, then A and B are in diagonally opposite quadrants.

We shall assume that the zero vector $\mathbf{0}$ has *every* direction. Do not think of this peculiar property as something that can be proved. It is merely a definition of convenience, which will help us state certain results about a vector $\mathbf{a}$ without adding the condition $\mathbf{a} \neq \mathbf{0}$. Remember that $\mathbf{0}$ is also special in another way. It is the only vector that has zero magnitude!

By definition, two vectors $\mathbf{a}$ and $\mathbf{b}$ in V_2 are **parallel** if and only if $\mathbf{b} = c\mathbf{a}$ for some scalar c. Thus, nonzero vectors $\mathbf{a}$ and $\mathbf{b}$ are parallel if they have the same or opposite direction. Note that the zero vector is parallel to *every* vector $\mathbf{a}$, since we can write $\mathbf{0} = 0\mathbf{a}$.

The next theorem specifies the relationship between the magnitudes of $\mathbf{a}$ and $c\mathbf{a}$.

Theorem 10.9

If $\mathbf{a}$ is a vector and c is a scalar, then $\|c\mathbf{a}\| = |c| \, \|\mathbf{a}\|$.

PROOF If $\mathbf{a} = \langle a_1, a_2 \rangle$, then $c\mathbf{a} = \langle ca_1, ca_2 \rangle$. Using Definition (10.1) and properties of real numbers gives us

$$\|c\mathbf{a}\| = \sqrt{(ca_1)^2 + (ca_2)^2} = \sqrt{c^2(a_1^2 + a_2^2)}$$
$$= \sqrt{c^2}\sqrt{a_1^2 + a_2^2} = |c| \, \|\mathbf{a}\|. \quad \blacksquare$$

Theorem (10.9) implies that the length of a directed line segment that represents $c\mathbf{a}$ is $|c|$ times the length of a directed line segment that represents $\mathbf{a}$. This agrees with the geometric interpretation discussed at the beginning of this section and illustrated in Figure 10.6.

The vectors $\mathbf{i}$ and $\mathbf{j}$ defined next will be useful in future developments.

Definition 10.10

$$\mathbf{i} = \langle 1, 0 \rangle, \qquad \mathbf{j} = \langle 0, 1 \rangle$$

Both $\mathbf{i}$ and $\mathbf{j}$ have magnitude 1 since, by Definition (10.1),

$$\|\mathbf{i}\| = \sqrt{1^2 + 0^2} = 1 \quad \text{and} \quad \|\mathbf{j}\| = \sqrt{0^2 + 1^2} = 1.$$

We may use $\mathbf{i}$ and $\mathbf{j}$ to obtain another way of denoting vectors in V_2, as indicated by the next result.

i, j Form for Vectors 10.11

If $\mathbf{a} = \langle a_1, a_2 \rangle$, then $\mathbf{a} = a_1\mathbf{i} + a_2\mathbf{j}$.

PROOF   Using Definitions (10.1) and (10.2), we have

$$\langle a_1, a_2 \rangle = \langle a_1, 0 \rangle + \langle 0, a_2 \rangle = a_1 \langle 1, 0 \rangle + a_2 \langle 0, 1 \rangle = a_1 \mathbf{i} + a_2 \mathbf{j}. \quad \blacksquare$$

The formula $\mathbf{a} = a_1 \mathbf{i} + a_2 \mathbf{j}$ for the vector $\mathbf{a} = \langle a_1, a_2 \rangle$ has an interesting geometric interpretation. Position vectors for $\mathbf{i}$, $\mathbf{j}$, and $\mathbf{a}$ are sketched in Figure 10.18(a). Since $\mathbf{i}$ and $\mathbf{j}$ both have magnitude 1, we see from Theorem (10.9) that

$$\|a_1 \mathbf{i}\| = |a_1| \|\mathbf{i}\| = |a_1| \quad \text{and} \quad \|a_2 \mathbf{j}\| = |a_2| \|\mathbf{j}\| = |a_2|.$$

Hence, as in Figure 10.18(b), $a_1 \mathbf{i}$ and $a_2 \mathbf{j}$ may be represented by horizontal and vertical vectors of magnitudes $|a_1|$ and $|a_2|$, respectively. The position vector for $\mathbf{a}$ may be regarded as the sum of these vectors. For this reason, a_1 is called the **horizontal component** and a_2 the **vertical component** of the vector $\mathbf{a}$.

Figure 10.18

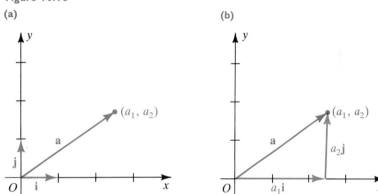

(a)   (b)

The vector sum $a_1 \mathbf{i} + a_2 \mathbf{j}$ is called a **linear combination** of $\mathbf{i}$ and $\mathbf{j}$. Note that if $\mathbf{b} = \langle b_1, b_2 \rangle = b_1 \mathbf{i} + b_2 \mathbf{j}$ and if c is a scalar, then

$$(a_1 \mathbf{i} + a_2 \mathbf{j}) + (b_1 \mathbf{i} + b_2 \mathbf{j}) = (a_1 + b_1) \mathbf{i} + (a_2 + b_2) \mathbf{j}$$
$$(a_1 \mathbf{i} + a_2 \mathbf{j}) - (b_1 \mathbf{i} + b_2 \mathbf{j}) = (a_1 - b_1) \mathbf{i} + (a_2 - b_2) \mathbf{j}$$
$$c(a_1 \mathbf{i} + a_2 \mathbf{j}) = (ca_1) \mathbf{i} + (ca_2) \mathbf{j}.$$

The preceding formulas show that we may regard linear combinations of $\mathbf{i}$ and $\mathbf{j}$ as ordinary algebraic sums.

EXAMPLE ∎ 5   If $\mathbf{a} = 5\mathbf{i} + \mathbf{j}$ and $\mathbf{b} = 4\mathbf{i} - 7\mathbf{j}$, express $3\mathbf{a} - 2\mathbf{b}$ as a linear combination of $\mathbf{i}$ and $\mathbf{j}$.

SOLUTION

$$\begin{aligned} 3\mathbf{a} - 2\mathbf{b} &= 3(5\mathbf{i} + \mathbf{j}) - 2(4\mathbf{i} - 7\mathbf{j}) \\ &= (15\mathbf{i} + 3\mathbf{j}) - (8\mathbf{i} - 14\mathbf{j}) \\ &= 7\mathbf{i} + 17\mathbf{j} \end{aligned}$$

A **unit vector** is a vector of magnitude 1. The vectors **i** and **j** are unit vectors.

Theorem 10.12

If $\mathbf{a} \neq \mathbf{0}$, then the unit vector $\mathbf{u}$ that has the same direction as $\mathbf{a}$ is

$$\mathbf{u} = \frac{1}{\|\mathbf{a}\|}\mathbf{a}.$$

PROOF   If we let $c = 1/\|\mathbf{a}\|$, then $c > 0$ and, by Definition (10.8), the vector $\mathbf{u} = c\mathbf{a}$ has the same direction as the vector $\mathbf{a}$. Moreover, by Theorem (10.9),

$$\|\mathbf{u}\| = \|c\mathbf{a}\| = |c|\,\|\mathbf{a}\| = \frac{1}{\|\mathbf{a}\|}\,\|\mathbf{a}\| = 1. \quad \blacksquare$$

EXAMPLE ▪ 6   Find the unit vector $\mathbf{u}$ that has the same direction as the vector $3\mathbf{i} - 4\mathbf{j}$.

SOLUTION   By Definition (10.1), $\|\mathbf{a}\| = \sqrt{9 + 16} = 5$. Hence, from Theorem (10.12),

$$\mathbf{u} = \tfrac{1}{5}(3\mathbf{i} - 4\mathbf{j}) = \tfrac{3}{5}\mathbf{i} - \tfrac{4}{5}\mathbf{j}.$$

Vectors have many applications in physics and engineering. We conclude this section with an example from the field of aerodynamics.

Figure 10.19

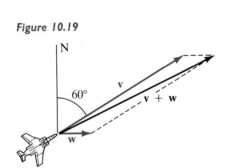

EXAMPLE ▪ 7   In air navigation, directions are specified by measuring from the north in a *clockwise* direction. Suppose an airplane with an airspeed of 200 mi/hr is flying in the direction 60°, and a wind is blowing directly from the west at 40 mi/hr. As in Figure 10.19, these velocities may be represented by vectors **v** and **w** of magnitudes 200 and 40, respectively. The direction of the resultant **v** + **w** is the **true course** of the airplane relative to the ground, and the magnitude $\|\mathbf{v} + \mathbf{w}\|$ is the **ground speed** of the airplane. Approximate the ground speed to the nearest mile per hour and the true course to the nearest degree.

SOLUTION   As in Figure 10.20, let us consider an xy-plane with the airplane at the origin, the y-axis on the north–south line, and $\mathbf{v} = \langle v_1, v_2 \rangle$. Since $d(O, A) = \|\mathbf{v}\| = 200$, it follows that

Figure 10.20

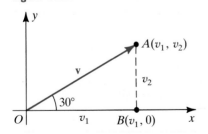

$$v_1 = 200 \cos 30° = 100\sqrt{3} \quad \text{and} \quad v_2 = 200 \sin 30° = 100.$$

Hence $\mathbf{v} = \langle 100\sqrt{3}, 100 \rangle \quad \text{and} \quad \mathbf{w} = \langle 40, 0 \rangle.$

The resultant is

$$\mathbf{v} + \mathbf{w} = \langle 100\sqrt{3} + 40, 100 \rangle,$$

and the ground speed is

$$\|\mathbf{v} + \mathbf{w}\| = \sqrt{(100\sqrt{3} + 40)^2 + (100)^2} \approx 235 \text{ mi/hr.}$$

If θ is the angle between $\mathbf{v} + \mathbf{w}$ and the positive x-axis, then

$$\tan \theta = \frac{100}{100\sqrt{3} + 40} \approx 0.469 \quad \text{and} \quad \theta \approx \tan^{-1}(0.469) \approx 25°.$$

Hence, the true course of the airplane is approximately $90° - 25° = 65°$.

EXERCISES 10.1

Exer. 1–6: Sketch the position vector of a and find $\|\mathbf{a}\|$.

1 $\mathbf{a} = \langle 2, 5 \rangle$ **2** $\mathbf{a} = \langle -4, -7 \rangle$

3 $\mathbf{a} = \langle -5, 0 \rangle$ **4** $\mathbf{a} = -18\mathbf{j}$

5 $\mathbf{a} = -4\mathbf{i} + 5\mathbf{j}$ **6** $\mathbf{a} = 2\mathbf{i} - 3\mathbf{j}$

Exer. 7–12: Find $\mathbf{a} + \mathbf{b}$, $\mathbf{a} - \mathbf{b}$, $2\mathbf{a}$, $-3\mathbf{b}$, and $4\mathbf{a} - 5\mathbf{b}$.

7 $\mathbf{a} = \langle 2, -3 \rangle$, $\mathbf{b} = \langle 1, 4 \rangle$

8 $\mathbf{a} = \langle -2, -5 \rangle$, $\mathbf{b} = \langle -3, 4 \rangle$

9 $\mathbf{a} = -\langle 7, -2 \rangle$, $\mathbf{b} = 4\langle -2, 1 \rangle$

10 $\mathbf{a} = 2\langle 1, 5 \rangle$, $\mathbf{b} = -3\langle -1, -4 \rangle$

11 $\mathbf{a} = 3\mathbf{i} + 2\mathbf{j}$, $\mathbf{b} = -\mathbf{i} + 5\mathbf{j}$

12 $\mathbf{a} = -5\mathbf{i} + 2\mathbf{j}$, $\mathbf{b} = \mathbf{i} - 3\mathbf{j}$

Exer. 13–18: Use components to express the sum or difference as a scalar multiple of one of the vectors a, b, c, d, e, or f shown in the figure.

Exercises 13–18

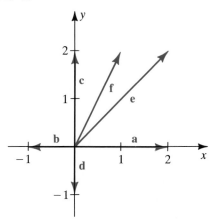

13 $\mathbf{a} + \mathbf{b}$ **14** $\mathbf{c} - \mathbf{d}$ **15** $\mathbf{b} + \mathbf{e}$

16 $\mathbf{f} - \mathbf{b}$ **17** $\mathbf{b} + \mathbf{d}$ **18** $\mathbf{e} + \mathbf{c}$

Exer. 19–24: Find the vector a in V_2 that corresponds to $\overrightarrow{PQ}$. Sketch $\overrightarrow{PQ}$ and the position vector for a.

19 $P(1, -4)$, $Q(5, 3)$

20 $P(7, -3)$, $Q(-2, 4)$

21 $P(2, 5)$, $Q(-4, 5)$

22 $P(-4, 6)$, $Q(-4, -2)$

23 $P(-3, -1)$, $Q(6, -4)$

24 $P(2, 3)$, $Q(-6, 0)$

Exer. 25–28: Find a unit vector that has (a) the same direction as a and (b) the opposite direction of a.

25 $\mathbf{a} = -8\mathbf{i} + 15\mathbf{j}$ **26** $\mathbf{a} = 5\mathbf{i} - 3\mathbf{j}$

27 $\mathbf{a} = \langle 2, -5 \rangle$ **28** $\mathbf{a} = \langle 0, 6 \rangle$

29 Find a vector that has the same direction as $\langle -6, 3 \rangle$ and

     (a) twice the magnitude

     (b) one-half the magnitude

30 Find a vector that has the opposite direction of $8\mathbf{i} - 5\mathbf{j}$ and

     (a) three times the magnitude

     (b) one-third the magnitude

31 Find a vector of magnitude 6 that has the same direction as $\mathbf{a} = 4\mathbf{i} - 7\mathbf{j}$.

32 Find a vector of magnitude 4 that has the opposite direction of $\mathbf{a} = \langle 2, -5 \rangle$.

Exer. 33–34: Find all real numbers c such that (a) $\|c\mathbf{a}\| = 3$, (b) $\|c\mathbf{a}\| = -3$, and (c) $\|c\mathbf{a}\| = 0$.

33 $\mathbf{a} = 3\mathbf{i} - 4\mathbf{j}$ **34** $\mathbf{a} = \langle -5, 12 \rangle$

[c] **Exer. 35–38: Approximate the horizontal and vertical components of the vector that is described.**

35 A quarterback releases a football with a velocity of 50 ft/sec at an angle of elevation of $35°$.

36 A child pulls a sled through the snow by exerting a force of 20 lb at an angle of $40°$ with the horizontal.

37 The biceps muscle, in supporting the forearm and a weight held in the hand, exerts a force of 20 lb. As shown in the figure, the muscle makes an angle of $108°$ with the forearm.

Exercise 37

38 A jet airplane approaches a runway at an angle of $7.5°$ with the horizontal, traveling at a velocity of 160 mi/hr.

39 An airplane is flying in the direction $150°$ with an airspeed of 300 mi/hr, and the wind is blowing in the direction $60°$ at 30 mi/hr. Approximate the true course and the ground speed of the airplane.

40 An airplane pilot wishes to maintain a true course in the direction $240°$ with a ground speed of 400 mi/hr when the wind is blowing directly north at 50 mi/hr. Find the required airspeed and compass heading.

41 Two tugboats are towing a large ship into port, as shown in the figure. The larger tug exerts a force of 4000 lb on its cable, and the smaller tug exerts a force of 3200 lb on its cable. If the ship is to travel in a straight line from A to B, find the angle θ that the larger tug must make with the line segment AB.

Exercise 41

[c] **42** Shown in the figure is an apparatus used to simulate gravity conditions on other planets. A rope is attached to an astronaut who maneuvers on an inclined plane that makes an angle of θ degrees with the horizontal.

(a) If the astronaut weighs 160 lb, find the x- and y-components of this downward force (see figure for axes).

(b) The y-component in part (a) is the weight of the astronaut relative to the inclined plane. The astronaut would weigh 27 lb on the moon and 60 lb on Mars. Approximate each angle θ (to the nearest $0.01°$) so that the inclined-plane apparatus will simulate walking on these surfaces.

Exercise 42

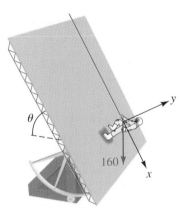

43 Find scalars p and q such that

$$p\langle 3, -1\rangle + q\langle 4, 3\rangle = \langle -6, -11\rangle.$$

44 If $\mathbf{a} = \langle a_1, a_2\rangle$ and $\mathbf{b} = \langle b_1, b_2\rangle$ are any nonzero, nonparallel vectors and if $\mathbf{c}$ is any other vector, prove that there exist scalars p and q such that $\mathbf{c} = p\mathbf{a} + q\mathbf{b}$. Interpret this fact geometrically.

45 Under what conditions is $\|\mathbf{a} + \mathbf{b}\| = \|\mathbf{a}\| + \|\mathbf{b}\|$?

46 (a) Let $\mathbf{a} = \langle a_1, a_2\rangle$ be any nonzero vector, and let $\overrightarrow{OA}$ be the position vector for $\mathbf{a}$. If θ is the smallest nonnegative angle from the positive x-axis to $\overrightarrow{OA}$, show that $\mathbf{a} = \|\mathbf{a}\| (\cos\theta \mathbf{i} + \sin\theta \mathbf{j})$.

(b) Show that every unit vector in V_2 can be expressed in the form $\cos\theta \mathbf{i} + \sin\theta \mathbf{j}$ for some θ.

47 Let $\mathbf{r}_0 = \langle x_0, y_0\rangle, \mathbf{r} = \langle x, y\rangle$, and $c > 0$. Describe the set of all points $P(x, y)$ such that $\|\mathbf{r} - \mathbf{r}_0\| = c$.

48 Let $\mathbf{r}_0 = \langle x_0, y_0\rangle, \mathbf{r} = \langle x, y\rangle$, and $\mathbf{a} = \langle a_1, a_2\rangle \neq \mathbf{0}$. Describe the set of all points $P(x, y)$ such that the difference $\mathbf{r} - \mathbf{r}_0$ is a scalar multiple of $\mathbf{a}$.

Exer. 49–56: Prove the given property if $\mathbf{a} = \langle a_1, a_2 \rangle$, $\mathbf{b} = \langle b_1, b_2 \rangle$, $\mathbf{c} = \langle c_1, c_2 \rangle$, and p and q are real numbers.

49 $\mathbf{a} + (\mathbf{b} + \mathbf{c}) = (\mathbf{a} + \mathbf{b}) + \mathbf{c}$

50 $\mathbf{a} + \mathbf{0} = \mathbf{a}$

51 $0\mathbf{a} = \mathbf{0} = p\mathbf{0}$

52 $1\mathbf{a} = \mathbf{a}$

53 $(p + q)\mathbf{a} = p\mathbf{a} + q\mathbf{a}$

54 $p(\mathbf{a} - \mathbf{b}) = p\mathbf{a} - p\mathbf{b}$

55 If $p\mathbf{a} = \mathbf{0}$ and $p \neq 0$, then $\mathbf{a} = \mathbf{0}$.

56 If $p\mathbf{a} = \mathbf{0}$ and $\mathbf{a} \neq \mathbf{0}$, then $p = 0$.

10.2 VECTORS IN THREE DIMENSIONS

The concepts discussed for vectors in the plane can be generalized for vectors in three-dimensional space. We investigate how to carry out this generalization in this section.

To discuss vectors in space, we use *ordered triples* and a *three-dimensional coordinate system*. An **ordered triple** (a, b, c) is a set $\{a, b, c\}$ of three numbers in which a is considered the first number of the set, b the second, and c the third. Two ordered triples (a_1, a_2, a_3) and (b_1, b_2, b_3) are **equal**, and we write $(a_1, a_2, a_3) = (b_1, b_2, b_3)$, if and only if $a_1 = b_1$, $a_2 = b_2$, and $a_3 = b_3$. The totality of ordered triples is denoted by $\mathbb{R}^3$.

To obtain a rectangular coordinate system in three dimensions, we choose a fixed point O (the **origin**) and consider three mutually perpendicular coordinate lines (the x-, y-, and z-axes) with common origin O, as illustrated in Figure 10.21. We may regard the y- and z-axes as lying in the plane of the paper and the x-axis as projecting out from the paper. The three coordinate lines determine three coordinate planes: the **xy-plane**, the **yz-plane**, and the **xz-plane**, shown in Figure 10.22.

The three axes determine a **right-handed** coordinate system. (If the x- and y-axes are interchanged, the coordinate system is **left-handed**.) The terminology *right-handed* may be justified as follows. If, as in Figure 10.23, the fingers of the right hand are curled in the direction of a 90°

Figure 10.21

Figure 10.22 Coordinate planes

Figure 10.23 Right-handed system

Figure 10.24

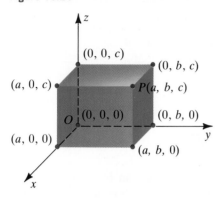

rotation of axes in the xy-plane (so that the positive x-axis is rotated toward the positive y-axis), then the extended thumb points in the direction of the positive z-axis. In this text, we generally use right-handed coordinate systems.

If P is a point, then the (perpendicular) projection of P on the x-axis has coordinate a, the **x-coordinate** of P. Similarly, the coordinates b and c of the projections of P on the y- and z-axes, respectively, are the **y-coordinate** and **z-coordinate** of P. We use $P(a, b, c)$ or (a, b, c) to denote the point P with coordinates a, b, and c. If P is not on a coordinate plane, then the three planes through P that are parallel to the coordinate planes, together with the coordinate planes, determine a rectangular box of the type illustrated in Figure 10.24.

The concept of plotting points in three dimensions is similar to that in two dimensions. Several points are plotted in Figure 10.25. To plot $(3, 4, -2)$, we first locate $(3, 4, 0)$ on the xy-plane and then move *down* 2 units. To plot $(-2, 3, 4)$, we first locate $(-2, 3, 0)$ on the xy-plane and then move *up* 4 units.

Figure 10.25

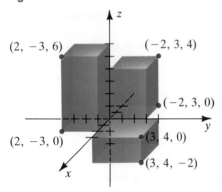

The one-to-one correspondence between points in space and ordered triples of real numbers is a **rectangular coordinate system in three dimensions,** or an ***xyz*-coordinate system**. The three coordinate planes partition space into eight parts, called **octants**. The part consisting of all points $P(a, b, c)$ with a, b, and c positive is the **first octant**. The remaining octants are usually not numbered.

Let us derive a formula for the distance $d(P_1, P_2)$ between two points P_1 and P_2. If P_1 and P_2 are on a line parallel to the z-axis, as in Figure 10.26, and if their projections on the z-axis are $A_1(0, 0, z_1)$ and $A_2(0, 0, z_2)$, respectively, then $d(P_1, P_2) = d(A_1, A_2) = |z_2 - z_1|$. Similar formulas hold if the line through P_1 and P_2 is parallel to the x- or y-axis.

To find the distance between two points P_1 and P_2 that are not parallel to an axis, we consider a situation of the type illustrated in Figure 10.27.

Figure 10.26

Figure 10.27

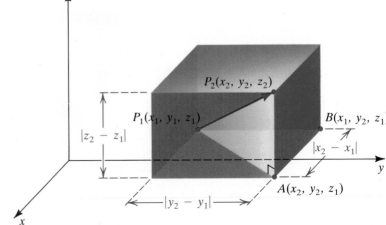

Triangle $P_1 A P_2$ is a right triangle, and hence, by the Pythagorean theorem,

$$d(P_1, P_2) = \sqrt{[d(P_1, A)]^2 + [d(A, P_2)]^2}.$$

Since P_1 and A are in a plane parallel to the xy-plane, the two-dimensional distance formula gives us $[d(P_1, A)]^2 = (x_2 - x_1)^2 + (y_2 - y_1)^2$. From our previous remarks, $[d(A, P_2)]^2 = (z_2 - z_1)^2$. Substituting in the preceding formula for $d(P_1, P_2)$, we obtain the following.

Distance Formula 10.13

The distance between $P_1(x_1, y_1, z_1)$ and $P_2(x_2, y_2, z_2)$ is

$$d(P_1, P_2) = \sqrt{(x_2 - x_1)^2 + (y_2 - y_1)^2 + (z_2 - z_1)^2}.$$

If P_1 and P_2 are on the xy-plane so that $z_1 = z_2 = 0$, then (10.13) reduces to the two-dimensional distance formula.

EXAMPLE 1   Find the distance between the points $A(-1, -3, 1)$ and $B(3, 4, -2)$.

SOLUTION   Using the distance formula (10.13), we have

$$d(A, B) = \sqrt{(3 + 1)^2 + (4 + 3)^2 + (-2 - 1)^2}$$
$$= \sqrt{16 + 49 + 9} = \sqrt{74}.$$

To discuss vectors in three dimensions, we include a third component and use symbols of the form

$$\mathbf{a} = \langle a_1, a_2, a_3 \rangle, \qquad \mathbf{b} = \langle b_1, b_2, b_3 \rangle.$$

The set of all such vectors is denoted by V_3. Each concept discussed in the preceding section for V_2 readily extends to V_3. For example, if A is the point (a_1, a_2, a_3), we refer to the position vector $\overrightarrow{OA}$ for $\mathbf{a}$, illustrated in Figure 10.28, which also shows another vector $\overrightarrow{PQ}$ that corresponds to $\mathbf{a}$. As in Section 10.1, we seldom distinguish between the geometric and algebraic natures of vectors. Magnitude of a vector, vector addition, and scalar multiples of vectors are defined as before by simply including an additional component as follows:

Figure 10.28

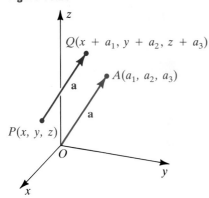

$$\|\mathbf{a}\| = \sqrt{a_1^2 + a_2^2 + a_3^2}$$
$$\mathbf{a} + \mathbf{b} = \langle a_1 + b_1, a_2 + b_2, a_3 + b_3 \rangle$$
$$c\mathbf{a} = \langle ca_1, ca_2, ca_3 \rangle$$

The zero vector in V_3 is $\mathbf{0} = \langle 0, 0, 0 \rangle$, and the negative of $\mathbf{a}$ is $-\mathbf{a} = \langle -a_1, -a_2, -a_3 \rangle$. The theorems and definitions stated in (10.5)–(10.9) extend to V_3, the only change being the inclusion of a third component for each vector. The following examples contain specific illustrations of these concepts.

EXAMPLE ■ 2   Given $\mathbf{a} = \langle 2, 5, -3 \rangle$ and $\mathbf{b} = \langle -4, 1, 7 \rangle$, find $\mathbf{a} + \mathbf{b}$, $2\mathbf{a} - 3\mathbf{b}$, and $\|\mathbf{a}\|$.

SOLUTION

$$\mathbf{a} + \mathbf{b} = \langle 2, 5, -3 \rangle + \langle -4, 1, 7 \rangle = \langle -2, 6, 4 \rangle$$
$$2\mathbf{a} - 3\mathbf{b} = 2\langle 2, 5, -3 \rangle - 3\langle -4, 1, 7 \rangle = \langle 4, 10, -6 \rangle - \langle -12, 3, 21 \rangle$$
$$= \langle 16, 7, -27 \rangle$$
$$\|\mathbf{a}\| = \sqrt{(2)^2 + (5)^2 + (-3)^2} = \sqrt{38}$$

EXAMPLE ■ 3   Given the vectors $\mathbf{a} = \langle 15, -6, 24 \rangle$, $\mathbf{b} = \langle 5, -2, 8 \rangle$, and $\mathbf{c} = \langle -\frac{15}{2}, 3, -12 \rangle$, show that $\mathbf{a}$ and $\mathbf{b}$ have the same direction and that $\mathbf{a}$ and $\mathbf{c}$ have opposite directions.

SOLUTION   We see, by inspection, that

$$\mathbf{a} = 3\mathbf{b}, \quad \text{or} \quad \mathbf{b} = \tfrac{1}{3}\mathbf{a}.$$

Since the scalar 3 (or $\frac{1}{3}$) is positive, $\mathbf{a}$ and $\mathbf{b}$ have the same direction. Again, by inspection,

$$\mathbf{a} = -2\mathbf{c}, \quad \text{or} \quad \mathbf{c} = -\tfrac{1}{2}\mathbf{a}.$$

Since the scalar -2 (or $-\frac{1}{2}$) is negative, $\mathbf{a}$ and $\mathbf{c}$ have opposite directions.

Figure 10.29

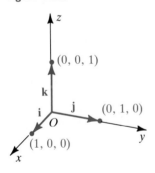

EXAMPLE ■ 4   Given the points $P_1(5, 6, -2)$ and $P_2(-3, 8, 7)$, find the vector $\mathbf{a}$ in V_3 that corresponds to $\overrightarrow{P_1 P_2}$.

SOLUTION   By the analogue of Theorem (10.7) in V_3,

$$\mathbf{a} = \langle -3 - 5, 8 - 6, 7 + 2 \rangle = \langle -8, 2, 9 \rangle.$$

A vector $\mathbf{a}$ in V_3 is a **unit vector** if $\|\mathbf{a}\| = 1$. The special unit vectors

$$\mathbf{i} = \langle 1, 0, 0 \rangle, \qquad \mathbf{j} = \langle 0, 1, 0 \rangle, \qquad \mathbf{k} = \langle 0, 0, 1 \rangle$$

are important, since any vector $\mathbf{a} = \langle a_1, a_2, a_3 \rangle$ can be expressed as a linear combination of $\mathbf{i}$, $\mathbf{j}$, and $\mathbf{k}$. Specifically,

$$\mathbf{a} = \langle a_1, a_2, a_3 \rangle = a_1\mathbf{i} + a_2\mathbf{j} + a_3\mathbf{k}.$$

Figure 10.30

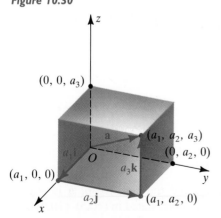

Position vectors for $\mathbf{i}$, $\mathbf{j}$, and $\mathbf{k}$ are sketched in Figure 10.29. Figure 10.30 illustrates how the position vector for $\mathbf{a} = \langle a_1, a_2, a_3 \rangle$ may be regarded as the sum of three vectors corresponding to $a_1\mathbf{i}$, $a_2\mathbf{j}$, and $a_3\mathbf{k}$. It is often convenient to regard V_2 as a subset of V_3 by identifying the vector $\langle a_1, a_2 \rangle$ with $\langle a_1, a_2, 0 \rangle$. Thus, the unit vectors $\mathbf{i} = \langle 1, 0 \rangle$ and $\mathbf{j} = \langle 0, 1 \rangle$ are essentially the same as $\mathbf{i} = \langle 1, 0, 0 \rangle$ and $\mathbf{j} = \langle 0, 1, 0 \rangle$.

E X A M P L E ■ 5   Express $\mathbf{a} = \langle 3, -4, 2 \rangle$ in terms of $\mathbf{i}$, $\mathbf{j}$, and $\mathbf{k}$, and find the unit vector $\mathbf{u}$ that has the same direction as $\mathbf{a}$.

S O L U T I O N   We may write

$$\mathbf{a} = \langle 3, -4, 2 \rangle = 3\mathbf{i} - 4\mathbf{j} + 2\mathbf{k}.$$

The magnitude of $\mathbf{a}$ is

$$\|\mathbf{a}\| = \sqrt{3^2 + (-4)^2 + 2^2} = \sqrt{29}.$$

As in Theorem (10.12), a unit vector $\mathbf{u}$ having the same direction as $\mathbf{a}$ is

$$\mathbf{u} = \frac{1}{\|\mathbf{a}\|}\mathbf{a} = \frac{1}{\sqrt{29}}(3\mathbf{i} - 4\mathbf{j} + 2\mathbf{k}) = \frac{3}{\sqrt{29}}\mathbf{i} - \frac{4}{\sqrt{29}}\mathbf{j} + \frac{2}{\sqrt{29}}\mathbf{k}.$$

We shall conclude this section by using vectors to derive several useful formulas. The first may be stated as follows.

Midpoint Formula 10.14

The midpoint of the line segment from $P_1(x_1, y_1, z_1)$ to $P_2(x_2, y_2, z_2)$ is

$$\left(\frac{x_1 + x_2}{2}, \frac{y_1 + y_2}{2}, \frac{z_1 + z_2}{2} \right).$$

Figure 10.31

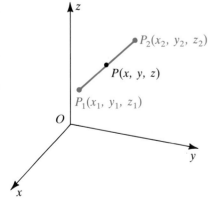

P R O O F   Let $P(x, y, z)$ be the midpoint of the line segment $P_1 P_2$ (see Figure 10.31). The fact that P is halfway from P_1 to P_2 may be expressed in vector form as follows:

$$\overrightarrow{P_1 P} = \tfrac{1}{2}\overrightarrow{P_1 P_2},$$

or

$$\langle x - x_1, y - y_1, z - z_1 \rangle = \tfrac{1}{2}\langle x_2 - x_1, y_2 - y_1, z_2 - z_1 \rangle$$

Equating components, we have

$$x - x_1 = \tfrac{1}{2}(x_2 - x_1), \qquad y - y_1 = \tfrac{1}{2}(y_2 - y_1), \qquad z - z_1 = \tfrac{1}{2}(z_2 - z_1).$$

Solving these three equations for x, y, and z gives us the following coordinates of the midpoint P:

$$x = \tfrac{1}{2}(x_1 + x_2), \qquad y = \tfrac{1}{2}(y_1 + y_2), \qquad z = \tfrac{1}{2}(z_1 + z_2) \quad ■$$

E X A M P L E ■ 6   Find the midpoint of the line segment from $(2, -3, 6)$ to $(3, 4, -2)$.

SOLUTION The points are plotted in Figure 10.25. By Theorem (10.14), the midpoint is

$$\left(\frac{2+3}{2}, \frac{-3+4}{2}, \frac{6+(-2)}{2}\right) = \left(\frac{5}{2}, \frac{1}{2}, 2\right).$$

Figure 10.32

The **graph of an equation** in three variables x, y, and z is the set of all points $P(a, b, c)$ in a rectangular coordinate system such that the ordered triple (a, b, c) is a solution of the equation—that is, equality is obtained when a, b, and c are substituted for x, y, and z, respectively. The graph of such an equation is a **surface**. We can easily derive an equation for a sphere of radius r with center at $P_0(x_0, y_0, z_0)$. As illustrated in Figure 10.32, a point $P(x, y, z)$ is on the sphere if and only if $\| \overrightarrow{P_0 P} \| = r$. An equivalent statement is

$$\sqrt{(x - x_0)^2 + (y - y_0)^2 + (z - z_0)^2} = r.$$

Squaring both sides of the last equation gives us the following theorem.

Theorem 10.15

An equation of a sphere of radius r and center $P_0(x_0, y_0, z_0)$ is

$$(x - x_0)^2 + (y - y_0)^2 + (z - z_0)^2 = r^2.$$

If we square the expressions in Theorem (10.15) and simplify, then an equation of a sphere may be written in the form

$$x^2 + y^2 + z^2 + ax + by + cz + d = 0,$$

where a, b, c, and d are real numbers. Conversely, if we begin with an equation of this form and if the graph exists, then by completing squares we can obtain the form in Theorem (10.15), and hence the graph is a sphere (or a point if $r = 0$).

EXAMPLE • 7 Discuss the graph of the equation

$$x^2 + y^2 + z^2 - 6x + 8y + 4z + 4 = 0.$$

SOLUTION We complete squares as follows:

$$(x^2 - 6x) + (y^2 + 8y) + (z^2 + 4z) = -4$$
$$(x^2 - 6x + 9) + (y^2 + 8y + 16) + (z^2 + 4z + 4) = -4 + 9 + 16 + 4$$
$$(x - 3)^2 + (y + 4)^2 + (z + 2)^2 = 25$$

Comparing the last equation with Theorem (10.15), we see that the graph is a sphere of radius 5 with center $(3, -4, -2)$.

Some calculators and many computer mathematical utilities provide for vector representation and vector operations. Consult the appropriate manual to learn how to enter a two- or three-dimensional vector. Check which operations are provided. It should be easy to add two vectors, find a scalar multiple of a vector, and determine the norm or the magnitude of a vector. The operations introduced in this section and the next several sections can often be done by a simple command.

The command for the *norm* of the vector $\mathbf{a} = \langle 3, -4, 2 \rangle$ in Example 5 gives the numerical value 5.38516481, and the command for a *unit vector* that has the same direction as $\mathbf{a}$ gives the numerical components $\langle 0.55708601, -0.74278135, 0.37139068 \rangle$.

EXERCISES 10.2

Exer. 1–6: Plot A and B and find (a) $d(A, B)$, (b) the midpoint of AB, and (c) the vector in V_3 that corresponds to $\overrightarrow{AB}$.

1 $A(2, 4, -5)$, $B(4, -2, 3)$

2 $A(1, -2, 7)$, $B(2, 4, -1)$

3 $A(-4, 0, 1)$, $B(3, -2, 1)$

4 $A(0, 5, -4)$, $B(1, 1, 0)$

5 $A(1, 0, 0)$, $B(0, 1, 1)$

6 $A(0, 0, 0)$, $B(-8, -1, 4)$

Exer. 7–12: Find (a) $\mathbf{a} + \mathbf{b}$, (b) $\mathbf{a} - \mathbf{b}$, (c) $5\mathbf{a} - 4\mathbf{b}$, (d) $\|\mathbf{a}\|$, and (e) $\|-3\mathbf{a}\|$.

7 $\mathbf{a} = \langle -2, 6, 1 \rangle$, $\mathbf{b} = \langle 3, -3, -1 \rangle$

8 $\mathbf{a} = \langle 1, 2, -3 \rangle$, $\mathbf{b} = \langle -4, 0, 1 \rangle$

9 $\mathbf{a} = 3\mathbf{i} - 4\mathbf{j} + 2\mathbf{k}$, $\mathbf{b} = \mathbf{i} + 2\mathbf{j} - 5\mathbf{k}$

10 $\mathbf{a} = 2\mathbf{i} - \mathbf{j} + 4\mathbf{k}$, $\mathbf{b} = \mathbf{i} - \mathbf{k}$

11 $\mathbf{a} = \mathbf{i} + \mathbf{j}$, $\mathbf{b} = -\mathbf{j} + \mathbf{k}$

12 $\mathbf{a} = 2\mathbf{i}$, $\mathbf{b} = 3\mathbf{k}$

Exer. 13–14: Sketch position vectors for a, b, 2a, $-3\mathbf{b}$, $\mathbf{a} + \mathbf{b}$, and $\mathbf{a} - \mathbf{b}$.

13 $\mathbf{a} = \langle 2, 3, 4 \rangle$, $\mathbf{b} = \langle 1, -2, 2 \rangle$

14 $\mathbf{a} = -\mathbf{i} + 2\mathbf{j} + 3\mathbf{k}$, $\mathbf{b} = -2\mathbf{j} + \mathbf{k}$

Exer. 15–16: Find the unit vector that has the same direction as a.

15 $\mathbf{a} = 2\langle -2, 5, -1 \rangle$ 16 $\mathbf{a} = 3\mathbf{i} - 7\mathbf{j} + 2\mathbf{k}$

Exer. 17–18: Find the vector that has (a) the same direction as a and twice the magnitude of a, (b) the opposite direction of a and one-third the magnitude of a, and (c) the same direction as a and magnitude 2.

17 $\mathbf{a} = 14\mathbf{i} - 15\mathbf{j} + 6\mathbf{k}$

18 $\mathbf{a} = \langle -6, -3, 6 \rangle$

Exer. 19–22: Find an equation of the sphere with center C and radius r.

19 $C(3, -1, 2)$; $r = 3$

20 $C(4, -5, 1)$; $r = 5$

21 $C(-5, 0, 1)$; $r = \frac{1}{2}$

22 $C(0, -3, -6)$; $r = \sqrt{3}$

Exer. 23–24: Find an equation of the sphere with center C that is tangent to (a) the xy-plane, (b) the xz-plane, and (c) the yz-plane.

23 $C(-2, 4, -6)$

24 $C(3, -1, 2)$

Exer. 25–26: Find an equation of the sphere that has endpoints of a diameter at A and B.

25 $A(1, 4, -2)$, $B(-7, 1, 2)$

26 $A(4, -3, 4)$, $B(0, 3, -10)$

27 Find an equation of the sphere that has radius 1 and center in the first octant and is tangent to each coordinate plane.

28 Find an equation of the sphere that has center $(2, 3, -1)$ and contains the point $(1, 7, -9)$.

Exer. 29–34: Find the center and the radius of the sphere.

29 $x^2 + y^2 + z^2 + 4x - 2y + 2z + 2 = 0$

30 $x^2 + y^2 + z^2 - 6x - 10y + 6z + 34 = 0$

31 $x^2 + y^2 + z^2 - 8x + 8z + 16 = 0$

32 $4x^2 + 4y^2 + 4z^2 - 4x + 8y - 3 = 0$

33 $x^2 + y^2 + z^2 + 4y = 0$

34 $x^2 + y^2 + z^2 - z = 0$

Exer. 35–42: Describe the region R in a three-dimensional coordinate system.

35 $R = \{(x, y, z): x^2 + y^2 + z^2 \leq 1\}$

36 $R = \{(x, y, z): x^2 + y^2 + z^2 > 1\}$

37 $R = \{(x, y, z): |x| \leq 1, |y| \leq 2, |z| \leq 3\}$

38 $R = \{(x, y, z): |x| \geq 1, |y| \geq 2, |z| \geq 3\}$

39 $R = \{(x, y, z): x^2 + y^2 \leq 25, |z| \leq 3\}$

40 $R = \{(x, y, z): (x^2/4) + (y^2/9) \geq 1\}$

41 $R = \{(x, y, z): xyz \neq 0\}$

42 $R = \{(x, y, z): 4 < x^2 + y^2 + z^2 < 9\}$

43 Show that the line segments $P_1 P_1'$, $P_2 P_2'$, and $P_3 P_3'$ (see figure) that join the midpoints of the opposite edges of a tetrahedron intersect at a point P, and each is bisected by that point. (*Hint:* Place vertex D on the z-axis and triangle ABC in the xy-plane.)

Exercise 43

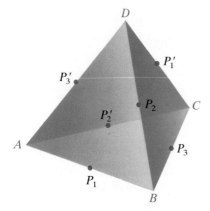

44 A cube has edges of length k. The centers of the six faces of the cube are the vertices of an octahedron, as shown in the figure.

(a) Find the coordinates of each vertex of the octahedron.

(b) Find the length of each edge of the octahedron in terms of k.

Exercise 44

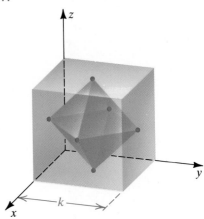

45 *Coulomb's law* states that the magnitude of the force of attraction between two oppositely charged particles is directly proportional to the product of the magnitudes q_1 and q_2 of the charges and inversely proportional to the square of the distance d between them. Show that if a particle with charge $+q$ is fixed at point A and a particle of charge -1 is placed at B (see figure), then the force of attraction $\mathbf{F}$ at A is given by

$$\mathbf{F} = \frac{kq}{\| \overrightarrow{BA} \|^3} \overrightarrow{BA} \quad \text{for some positive constant } k.$$

Exercise 45

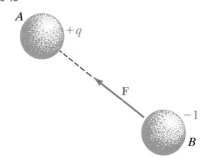

46 Refer to Exercise 45. Particles of charge $+q$ are placed and held fixed at points $(1, 0, 0)$, $(0, 1, 0)$, and $(0, 0, 1)$. A charge of -1 is then placed at $P(x, y, z)$.

(a) If $\mathbf{v} = \overrightarrow{OP}$, show that the net force $\mathbf{F}$ on the negatively charged particle is given by

$$\mathbf{F} = kq \left(\frac{\mathbf{v} - \mathbf{i}}{\| \mathbf{v} - \mathbf{i} \|^3} + \frac{\mathbf{v} - \mathbf{j}}{\| \mathbf{v} - \mathbf{j} \|^3} + \frac{\mathbf{v} - \mathbf{k}}{\| \mathbf{v} - \mathbf{k} \|^3} \right).$$

(b) The negatively charged particle is to be placed at a point $P(x, y, z)$, equidistant from the three positive charges, such that the net force acting on the particle is $\mathbf{0}$. Find the coordinates of P.

Mathematicians and Their Times

KARL WEIERSTRASS

A TALENTED YOUNG MAN achieves a distinguished secondary-school record, receiving several graduation prizes. He shows industriousness, securing a bookkeeper's job at age 15. His family makes financial sacrifices so that he can follow his father's wishes to study public finance and administration at a noted university.

On campus, he soon faces a conflict between duty and inclination. His family expects him to acquire the background needed for a high administrative government post. He is instead drawn to mathematics, which he begins to study on his own. He skips classes. Under increasing mental and physical strain, he seeks relief by plunging himself into student social and athletic life. He earns a reputation as a champion beer drinker and fencer. He also earns poor grades and leaves the university after eight semesters with no degree.

His family is appalled; their hopes for him seem dashed. They are angry at the young man and berate him for his failures. He becomes in his own words "sick of body and soul." It is at last decided that he will prepare for a career in school teaching. After completing this program, he begins a 15-year period of "infinite dreariness and boredom" teaching mathematics, physics, German, botany, geography, history, gymnastics, and even penmanship to the children of obscure towns.

Yet during this entire time, Karl Theodore Wilhelm Weierstrass (1815–1897) was secretly doing highly original mathematics even though he had no access to a mathematical library. He spoke to no one about this work and exchanged no scientific letters. He toiled late into the night, sometimes working until dawn. His health broke down; he suffered painful attacks of vertigo. Finally, at age 40, Weierstrass submitted a masterpiece of mathematics to a leading mathematics journal. Its value, as well as the quality of the author, were quickly realized. He was awarded an honorary doctor's degree, granted a year's leave to continue his studies, and eventually offered a life appointment at the University of Berlin.

For a quarter century, Weierstrass was the greatest analyst in Europe and Berlin's foremost teacher of mathematics. Although most of his achievements were in advanced mathematics, Weierstrass developed rigorous arguments for the foundations of calculus. He was the first, for example, to give rigorous proofs of the intermediate value (1.26) and extreme value (3.3) theorems. His discovery of a function continuous at every point but differentiable at no point startled the mathematical community and showed that intuitive understanding of functions required fundamental reassessment.

10.3   THE DOT PRODUCT

Two useful concepts that involve vectors **a** and **b** are the *dot product*, which is a scalar, and the *vector product*, which is a vector. In this section, we define the dot product. The vector product will be discussed in the next section.

The **dot product a · b** of two vectors may be defined either geometrically or algebraically. We use the algebraic approach for two reasons:

1. It can be readily generalized for use in advanced mathematics courses such as linear algebra, where vectors in four or more dimensions are studied.

2. It allows us to give easier proofs of properties of dot products.

We shall consider geometric and physical interpretations of **a · b** later in the section.

Definition 10.16

> The **dot product a · b** of $\mathbf{a} = \langle a_1, a_2, a_3 \rangle$ and $\mathbf{b} = \langle b_1, b_2, b_3 \rangle$ is
> $$\mathbf{a} \cdot \mathbf{b} = a_1 b_1 + a_2 b_2 + a_3 b_3.$$

The symbol **a · b** is read **a** *dot* **b**. The dot product is also referred to as the **scalar product,** or **inner product.** It is important to note that **a · b** is a scalar, not a vector. If $\mathbf{a} = \langle a_1, a_2 \rangle$ and $\mathbf{b} = \langle b_1, b_2 \rangle$ are vectors in V_2, then we define their dot product as $\mathbf{a} \cdot \mathbf{b} = a_1 b_1 + a_2 b_2$.

EXAMPLE ■ 1   Find **a · b** if

(a) $\mathbf{a} = \langle 2, 4, -3 \rangle, \quad \mathbf{b} = \langle -1, 5, 2 \rangle$

(b) $\mathbf{a} = 3\mathbf{i} - 2\mathbf{j} + \mathbf{k}, \quad \mathbf{b} = 4\mathbf{i} + 5\mathbf{j} - 2\mathbf{k}$

SOLUTION

(a) $\langle 2, 4, -3 \rangle \cdot \langle -1, 5, 2 \rangle = (2)(-1) + (4)(5) + (-3)(2) = 12$

(b) $(3\mathbf{i} - 2\mathbf{j} + \mathbf{k}) \cdot (4\mathbf{i} + 5\mathbf{j} - 2\mathbf{k}) = (3)(4) + (-2)(5) + (1)(-2) = 0$

Some properties of the dot product are listed next for any vectors **a, b, c** and a scalar m.

Properties of the Dot Product 10.17

(i) $\mathbf{a} \cdot \mathbf{a} = \|\mathbf{a}\|^2$

(ii) $\mathbf{a} \cdot \mathbf{b} = \mathbf{b} \cdot \mathbf{a}$

(iii) $\mathbf{a} \cdot (\mathbf{b} + \mathbf{c}) = \mathbf{a} \cdot \mathbf{b} + \mathbf{a} \cdot \mathbf{c}$

(iv) $(m\mathbf{a}) \cdot \mathbf{b} = m(\mathbf{a} \cdot \mathbf{b}) = \mathbf{a} \cdot (m\mathbf{b})$

(v) $\mathbf{0} \cdot \mathbf{a} = 0$

PROOF   We shall prove (iii) and leave proofs of the remaining properties as exercises. If $\mathbf{a} = \langle a_1, a_2, a_3 \rangle$, $\mathbf{b} = \langle b_1, b_2, b_3 \rangle$, and $\mathbf{c} = \langle c_1, c_2, c_3 \rangle$, then

$$\begin{aligned}
\mathbf{a} \cdot (\mathbf{b} + \mathbf{c}) &= \langle a_1, a_2, a_3 \rangle \cdot \langle b_1 + c_1, b_2 + c_2, b_3 + c_3 \rangle \\
&= a_1(b_1 + c_1) + a_2(b_2 + c_2) + a_3(b_3 + c_3) \\
&= a_1 b_1 + a_1 c_1 + a_2 b_2 + a_2 c_2 + a_3 b_3 + a_3 c_3 \\
&= (a_1 b_1 + a_2 b_2 + a_3 b_3) + (a_1 c_1 + a_2 c_2 + a_3 c_3) \\
&= \mathbf{a} \cdot \mathbf{b} + \mathbf{a} \cdot \mathbf{c}.
\end{aligned}$$

We next show that there is a close relationship between dot products and the angle between two vectors.

Angle Between a and b 10.18

Let **a** and **b** be nonzero vectors.

(i) If **b** is not a scalar multiple of **a** and if $\overrightarrow{OA}$ and $\overrightarrow{OB}$ are the position vectors for **a** and **b**, respectively, then **the angle θ between a and b** (or between $\overrightarrow{OA}$ and $\overrightarrow{OB}$) is angle AOB of the triangle determined by points A, O, and B (see Figure 10.33).

(ii) If $\mathbf{b} = c\mathbf{a}$ for some scalar c (that is, if **a** and **b** are parallel), then $\theta = 0$ if $c > 0$ (**a** and **b** have the same direction) and $\theta = \pi$ if $c < 0$ (**a** and **b** have opposite direction).

Figure 10.33

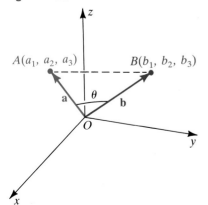

Note that if θ is the angle between two nonzero vectors, then $0 \leq \theta \leq \pi$.

The vectors **a** and **b** are **orthogonal**, or **perpendicular**, if $\theta = \pi/2$. We shall assume that the zero vector **0** is parallel and orthogonal to every vector **a**. (Recall that **0** has *every* direction.)

Theorem 10.19

If θ is the angle between nonzero vectors **a** and **b**, then

$$\mathbf{a} \cdot \mathbf{b} = \|\mathbf{a}\| \, \|\mathbf{b}\| \cos\theta.$$

PROOF If $\mathbf{b} \neq c\mathbf{a}$, we have the situation illustrated in Figure 10.33. Applying the law of cosines to triangle AOB yields

$$\| \overrightarrow{AB} \|^2 = \|\mathbf{a}\|^2 + \|\mathbf{b}\|^2 - 2\|\mathbf{a}\| \, \|\mathbf{b}\| \cos\theta.$$

Consequently,

$$(b_1 - a_1)^2 + (b_2 - a_2)^2 + (b_3 - a_3)^2$$
$$= (a_1^2 + a_2^2 + a_3^2) + (b_1^2 + b_2^2 + b_3^2) - 2\|\mathbf{a}\| \, \|\mathbf{b}\| \cos\theta,$$

which simplifies to

$$-2a_1 b_1 - 2a_2 b_2 - 2a_3 b_3 = -2\|\mathbf{a}\| \, \|\mathbf{b}\| \cos\theta.$$

Dividing both sides of the last equation by -2 gives us the desired conclusion.

If $\mathbf{b} = c\mathbf{a}$, then, by properties (iv) and (i) of Theorem (10.17),

$$\mathbf{a} \cdot \mathbf{b} = \mathbf{a} \cdot (c\mathbf{a}) = c(\mathbf{a} \cdot \mathbf{a}) = c\|\mathbf{a}\|^2.$$

Also,

$$\|\mathbf{a}\| \, \|\mathbf{b}\| \cos\theta = \|\mathbf{a}\| \, \|c\mathbf{a}\| \cos\theta = |c| \, \|\mathbf{a}\|^2 \cos\theta.$$

If $c > 0$, then $|c| = c$, $\theta = 0$, and $|c| \, \|\mathbf{a}\|^2 \cos\theta$ reduces to $c\|\mathbf{a}\|^2$. Thus, $\mathbf{a} \cdot \mathbf{b} = \|\mathbf{a}\| \, \|\mathbf{b}\| \cos\theta$. If $c < 0$, then $|c| = -c$, $\theta = \pi$, and again $|c| \, \|\mathbf{a}\|^2 \cos\theta$ reduces to $c\|\mathbf{a}\|^2$. This completes the proof of the theorem. ■

In words, Theorem (10.19) may be stated as follows: *The dot product of two vectors is equal to the product of their magnitudes and the cosine of the angle between the vectors.* In some texts, this statement is used as the definition of the dot product, because it has an interesting geometric interpretation that is useful in applications. Two disadvantages of making it the definition are that it is difficult to visualize vectors in four or more dimensions and that it is cumbersome to use it to prove some of the properties of the dot product.

Dividing both sides of the formula in Theorem (10.19) by $\|\mathbf{a}\| \, \|\mathbf{b}\|$ gives us the following.

Corollary 10.20

If θ is the angle between nonzero vectors **a** and **b**, then

$$\cos\theta = \frac{\mathbf{a} \cdot \mathbf{b}}{\|\mathbf{a}\| \, \|\mathbf{b}\|}.$$

E X A M P L E ▪ 2   Find the angle between the vectors $\mathbf{a} = \langle 4, -3, 1 \rangle$ and $\mathbf{b} = \langle -1, -2, 2 \rangle$.

S O L U T I O N   Applying Corollary (10.20), we have

$$\cos \theta = \frac{\mathbf{a} \cdot \mathbf{b}}{\|\mathbf{a}\| \, \|\mathbf{b}\|} = \frac{(4)(-1) + (-3)(-2) + (1)(2)}{\sqrt{16 + 9 + 1}\sqrt{1 + 4 + 4}} = \frac{4}{3\sqrt{26}}.$$

Hence, $\theta = \arccos(4/(3\sqrt{26})) \approx 1.31$ radians $\approx 74.84°$.

The next theorem follows from Theorem (10.19).

Theorem 10.21

> Two vectors $\mathbf{a}$ and $\mathbf{b}$ are orthogonal if and only if $\mathbf{a} \cdot \mathbf{b} = 0$.

E X A M P L E ▪ 3   Prove that the vectors are orthogonal:

(a) i, j   **(b)** $3\mathbf{i} - 7\mathbf{j} + 2\mathbf{k}$, $10\mathbf{i} + 4\mathbf{j} - \mathbf{k}$

S O L U T I O N   The proof follows from Theorem (10.21).

(a) $\mathbf{i} \cdot \mathbf{j} = \langle 1, 0, 0 \rangle \cdot \langle 0, 1, 0 \rangle = (1)(0) + (0)(1) + (0)(0) = 0$
(b) $(3\mathbf{i} - 7\mathbf{j} + 2\mathbf{k}) \cdot (10\mathbf{i} + 4\mathbf{j} - \mathbf{k}) = 30 - 28 - 2 = 0$

Each of the next two results about magnitudes is true for all vectors $\mathbf{a}$ and $\mathbf{b}$.

Cauchy–Schwarz Inequality 10.22

> $$|\mathbf{a} \cdot \mathbf{b}| \leq \|\mathbf{a}\| \, \|\mathbf{b}\|$$

P R O O F   The result is trivial if either $\mathbf{a}$ or $\mathbf{b}$ is $\mathbf{0}$. If $\mathbf{a}$ and $\mathbf{b}$ are nonzero vectors, then, by Theorem (10.19), $|\mathbf{a} \cdot \mathbf{b}| = \|\mathbf{a}\| \, \|\mathbf{b}\| \, |\cos \theta|$, where θ is the angle between $\mathbf{a}$ and $\mathbf{b}$. Since $|\cos \theta| \leq 1$, $|\mathbf{a} \cdot \mathbf{b}| \leq \|\mathbf{a}\| \, \|\mathbf{b}\|$. ▪

Figure 10.34

If we represent $\mathbf{a}$, $\mathbf{b}$, and $\mathbf{a} + \mathbf{b}$ geometrically, as in Figure 10.34, and use the fact that the length $\|\mathbf{a} + \mathbf{b}\|$ of one side of the triangle cannot exceed the sum of the lengths of the other two sides, we obtain the following result.

Triangle Inequality 10.23

> $$\|\mathbf{a} + \mathbf{b}\| \leq \|\mathbf{a}\| + \|\mathbf{b}\|$$

Let us give an alternative *algebraic* proof of the triangle inequality, using the dot product.

ALGEBRAIC PROOF OF (10.23)   Using properties of the dot product, we obtain

$$\|\mathbf{a} + \mathbf{b}\|^2 = (\mathbf{a} + \mathbf{b}) \cdot (\mathbf{a} + \mathbf{b})$$
$$= \mathbf{a} \cdot \mathbf{a} + 2(\mathbf{a} \cdot \mathbf{b}) + \mathbf{b} \cdot \mathbf{b}$$
$$= \|\mathbf{a}\|^2 + 2(\mathbf{a} \cdot \mathbf{b}) + \|\mathbf{b}\|^2 .$$

Since $\mathbf{a} \cdot \mathbf{b} \le |\mathbf{a} \cdot \mathbf{b}| \le \|\mathbf{a}\| \, \|\mathbf{b}\|$ (see (10.22)),

$$\|\mathbf{a} + \mathbf{b}\|^2 \le \|\mathbf{a}\|^2 + 2 \, \|\mathbf{a}\| \, \|\mathbf{b}\| + \|\mathbf{b}\|^2$$
$$= (\|\mathbf{a}\| + \|\mathbf{b}\|)^2.$$

Taking square roots, we have

$$\|\mathbf{a} + \mathbf{b}\| \le \|\mathbf{a}\| + \|\mathbf{b}\| . \quad \blacksquare$$

Figure 10.35

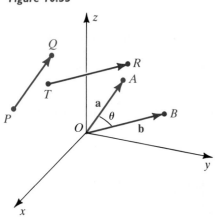

The angle between vectors in V_3 was defined in (10.18). We define the angle between two vectors (directed line segments) $\overrightarrow{PQ}$ and $\overrightarrow{TR}$ as the angle θ between their corresponding vectors $\mathbf{a}$ and $\mathbf{b}$ in V_3. A typical angle is illustrated in Figure 10.35, where $\overrightarrow{OA}$ and $\overrightarrow{OB}$ are position vectors for $\mathbf{a}$ and $\mathbf{b}$, respectively. If $\theta = \pi/2$, then $\overrightarrow{PQ}$ and $\overrightarrow{TR}$ are **orthogonal.**

The **dot product** of $\overrightarrow{PQ}$ and $\overrightarrow{TR}$ in Figure 10.35 is defined by

$$\overrightarrow{PQ} \cdot \overrightarrow{TR} = \mathbf{a} \cdot \mathbf{b} = \|\mathbf{a}\| \, \|\mathbf{b}\| \cos \theta.$$

Since $\| \overrightarrow{PQ} \| = \|\mathbf{a}\|$ and $\| \overrightarrow{TR} \| = \|\mathbf{b}\|$,

$$\overrightarrow{PQ} \cdot \overrightarrow{TR} = \| \overrightarrow{PQ} \| \| \overrightarrow{TR} \| \cos \theta.$$

If $\overrightarrow{PQ}$ and $\overrightarrow{PR}$ have the same initial point and if S is the projection of Q on the line through P and R (see Figure 10.36), then we call the scalar $\| \overrightarrow{PQ} \| \cos \theta$ the **component of $\overrightarrow{PQ}$ along $\overrightarrow{PR}$**, abbreviated $\mathbf{comp}_{\overrightarrow{PR}} \, \overrightarrow{PQ}$.

Figure 10.36 $\mathrm{comp}_{\overrightarrow{PR}} \, \overrightarrow{PQ}$

$\|\overrightarrow{PQ}\| \cos \theta > 0$ $\|\overrightarrow{PQ}\| \cos \theta < 0$

Note that the scalar $\| \overrightarrow{PQ} \| \cos \theta$ is positive if $0 \leq \theta < \pi/2$ or negative if $\pi/2 < \theta \leq \pi$. If $\theta = \pi/2$, the component is 0. It follows that

$$\text{comp}_{\overrightarrow{PR}} \overrightarrow{PQ} = \| \overrightarrow{PQ} \| \cos \theta$$

$$= \| \overrightarrow{PQ} \| \, \frac{\overrightarrow{PQ} \cdot \overrightarrow{PR}}{\| \overrightarrow{PQ} \| \| \overrightarrow{PR} \|}$$

$$= \overrightarrow{PQ} \cdot \frac{1}{\| \overrightarrow{PR} \|} \overrightarrow{PR}.$$

One way to remember this fact is to use the following statement.

Theorem 10.24

> The component of $\overrightarrow{PQ}$ along $\overrightarrow{PR}$ equals the dot product of $\overrightarrow{PQ}$ with the unit vector that has the same direction as $\overrightarrow{PR}$.

We use Theorem (10.24) to define the component of a vector $\mathbf{a}$ along a vector $\mathbf{b}$ in V_3 as follows.

Definition 10.25

> Let $\mathbf{a}$ and $\mathbf{b}$ be vectors in V_3 with $\mathbf{b} \neq \mathbf{0}$. The **component of a along b,** denoted by $\text{comp}_{\mathbf{b}} \, \mathbf{a}$, is
>
> $$\text{comp}_{\mathbf{b}} \, \mathbf{a} = \mathbf{a} \cdot \frac{1}{\| \mathbf{b} \|} \mathbf{b}.$$

If $\mathbf{a} = a_1 \mathbf{i} + a_2 \mathbf{j} + a_3 \mathbf{k}$, then, by Definition (10.25),

$$\text{comp}_{\mathbf{i}} \, \mathbf{a} = \mathbf{a} \cdot \mathbf{i} = a_1, \quad \text{comp}_{\mathbf{j}} \, \mathbf{a} = \mathbf{a} \cdot \mathbf{j} = a_2, \quad \text{and} \quad \text{comp}_{\mathbf{k}} \, \mathbf{a} = \mathbf{a} \cdot \mathbf{k} = a_3.$$

Thus, the components of $\mathbf{a}$ along $\mathbf{i}, \mathbf{j},$ and $\mathbf{k}$ are the same as the components $a_1, a_2,$ and a_3 of $\mathbf{a}$.

EXAMPLE ■ 4   If $\mathbf{a} = 4\mathbf{i} - \mathbf{j} + 5\mathbf{k}$ and $\mathbf{b} = 6\mathbf{i} + 3\mathbf{j} - 2\mathbf{k}$, find
(a) $\text{comp}_{\mathbf{b}} \, \mathbf{a}$ **(b)** $\text{comp}_{\mathbf{a}} \, \mathbf{b}$

SOLUTION
(a) Using Definition (10.25), we have

$$\text{comp}_{\mathbf{b}} \, \mathbf{a} = \mathbf{a} \cdot \frac{1}{\| \mathbf{b} \|} \mathbf{b} = (4\mathbf{i} - \mathbf{j} + 5\mathbf{k}) \cdot \frac{1}{7}(6\mathbf{i} + 3\mathbf{j} - 2\mathbf{k})$$

$$= \frac{24 - 3 - 10}{7} = \frac{11}{7} \approx 1.6.$$

(b) Interchanging the roles of **a** and **b** in (10.25) yields

$$\text{comp}_{\mathbf{a}}\,\mathbf{b} = \mathbf{b} \cdot \frac{1}{\|\mathbf{a}\|}\mathbf{a} = (6\mathbf{i} + 3\mathbf{j} - 2\mathbf{k}) \cdot \frac{1}{\sqrt{42}}(4\mathbf{i} - \mathbf{j} + 5\mathbf{k})$$

$$= \frac{24 - 3 - 10}{\sqrt{42}} = \frac{11}{\sqrt{42}} \approx 1.7.$$

Figure 10.37

We conclude this section with an important physical interpretation for the dot product. Recall from Definition (5.20) that the work done if a constant force F is exerted through a distance d is given by $W = Fd$. This formula is very restrictive, since it can be used only if the force is applied along the line of motion. More generally, suppose a vector $\overrightarrow{PQ}$ represents a force, and its point of application moves along a vector $\overrightarrow{PR}$. This is illustrated in Figure 10.37, where a force $\overrightarrow{PQ}$ is used to pull an object along a level path from P to R. Since

$$\overrightarrow{PQ} = \overrightarrow{PS} + \overrightarrow{SQ},$$

$\overrightarrow{SQ}$ does not contribute to the horizontal movement, and hence the motion from P to R is caused by $\overrightarrow{PS}$ alone. Applying (5.20) with $F = \|\overrightarrow{PS}\|$ and $d = \|\overrightarrow{PR}\|$, we find that the work W is

$$W = \left(\|\overrightarrow{PS}\|\right)\|\overrightarrow{PR}\|$$

$$= \left(\|\overrightarrow{PR}\|\cos\theta\right)\|\overrightarrow{PR}\| = \overrightarrow{PQ} \cdot \overrightarrow{PR}.$$

This is the motivation for the following definition.

Definition   10.26

> The **work done by a constant force** $\overrightarrow{PQ}$ as its point of application moves along the vector $\overrightarrow{PR}$ is $\overrightarrow{PQ} \cdot \overrightarrow{PR}$.

EXAMPLE ▪ 5   The magnitude and direction of a constant force are given by $\mathbf{a} = 5\mathbf{i} + 2\mathbf{j} + 6\mathbf{k}$. Find the work done if the point of application of the force moves from $P(1, -1, 2)$ to $R(4, 3, -1)$.

SOLUTION   The vector in V_3 that corresponds to $\overrightarrow{PR}$ is

$$\mathbf{b} = \langle 4 - 1, 3 - (-1), -1 - 2 \rangle = \langle 3, 4, -3 \rangle.$$

If $\overrightarrow{PQ}$ corresponds to **a**, then, by Definition (10.26), the work done is

$$\overrightarrow{PQ} \cdot \overrightarrow{PR} = \mathbf{a} \cdot \mathbf{b} = \langle 5, 2, 6 \rangle \cdot \langle 3, 4, -3 \rangle = 15 + 8 - 18 = 5.$$

If, for example, distance is in feet and the magnitude of the force is in pounds, then the work done is 5 ft-lb. If distance is in meters and force is in Newtons, then the work done is 5 Newton-meters, or 5 joules.

EXAMPLE ▪ 6   A small cart weighing 100 lb is pushed up an incline that makes an angle of 30° with the horizontal, as shown in Figure 10.38(a). Find the work done against gravity in pushing the cart a distance of 80 ft.

SOLUTION   We shall treat this as a two-dimensional problem and introduce an xy-coordinate system, as shown in Figure 10.38(b). The vector $\overrightarrow{PQ}$ represents the force of gravity acting vertically downward with a magnitude of 100 lb. The corresponding vector $\mathbf{F}$ in V_2 is $0\mathbf{i} - 100\mathbf{j}$. The point of application of this force moves along the vector $\overrightarrow{PR}$ of magnitude 80. If $\overrightarrow{PR}$ corresponds to $\mathbf{a} = a_1\mathbf{i} + a_2\mathbf{j}$, then referring to triangle PTR, we see that

$$a_1 = 80\cos 30° = 40\sqrt{3}, \qquad a_2 = 80\sin 30° = 40,$$

and hence

$$\mathbf{a} = 40\sqrt{3}\mathbf{i} + 40\mathbf{j}.$$

The work done *by* gravity is $\overrightarrow{PQ} \cdot \overrightarrow{PR}$, or $\mathbf{F} \cdot \mathbf{a}$. Applying the definition of dot product yields

$$\mathbf{F} \cdot \mathbf{a} = (0\mathbf{i} - 100\mathbf{j}) \cdot (40\sqrt{3}\mathbf{i} + 40\mathbf{j}) = 0 - 4000 = -4000 \text{ ft-lb.}$$

Thus, the work done *against* gravity is

$$-\mathbf{F} \cdot \mathbf{a} = 4000 \text{ ft-lb.}$$

Figure 10.38

(a)

(b)

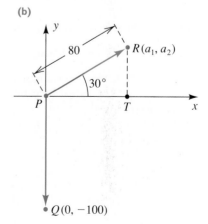

EXERCISES 10.3

Exer. 1–10: Given $\mathbf{a} = \langle -2, 3, 1 \rangle$, $\mathbf{b} = \langle 7, 4, 5 \rangle$, and $\mathbf{c} = \langle 1, -5, 2 \rangle$, find the number.

1 $\mathbf{a} \cdot \mathbf{b}$ 2 $\mathbf{b} \cdot \mathbf{c}$

3 (a) $\mathbf{a} \cdot (\mathbf{b} + \mathbf{c})$ (b) $\mathbf{a} \cdot \mathbf{b} + \mathbf{a} \cdot \mathbf{c}$

4 (a) $(\mathbf{a} - \mathbf{c}) \cdot \mathbf{b}$ (b) $\mathbf{a} \cdot \mathbf{b} - \mathbf{c} \cdot \mathbf{b}$

5 $(2\mathbf{a} + \mathbf{b}) \cdot 3\mathbf{c}$ 6 $(\mathbf{a} - \mathbf{b}) \cdot (\mathbf{b} + \mathbf{c})$

7 $\text{comp}_{\mathbf{c}} \mathbf{b}$ 8 $\text{comp}_{\mathbf{b}} \mathbf{c}$

9 $\text{comp}_{\mathbf{b}} (\mathbf{a} + \mathbf{c})$ 10 $\text{comp}_{\mathbf{c}} \mathbf{c}$

Exer. 11–14: Find the angle between a and b.

11 $\mathbf{a} = -4\mathbf{i} + 8\mathbf{j} - 3\mathbf{k}$, $\mathbf{b} = 2\mathbf{i} + \mathbf{j} + \mathbf{k}$

12 $\mathbf{a} = \mathbf{i} - 7\mathbf{j} + 4\mathbf{k}$, $\mathbf{b} = 5\mathbf{i} - \mathbf{k}$

13 $\mathbf{a} = \langle -2, -3, 0 \rangle$, $\mathbf{b} = \langle -6, 0, 4 \rangle$

14 $\mathbf{a} = \langle 3, -5, -1 \rangle$, $\mathbf{b} = \langle 2, 1, -3 \rangle$

Exer. 15–16: Show that a and b are orthogonal.

15 $\mathbf{a} = 3\mathbf{i} - 2\mathbf{j} + \mathbf{k}$, $\mathbf{b} = 4\mathbf{i} + 5\mathbf{j} - 2\mathbf{k}$

16 $\mathbf{a} = \langle 4, -1, -2 \rangle$, $\mathbf{b} = \langle 2, -2, 5 \rangle$

Exer. 17–18: Find all values of c such that a and b are orthogonal.

17 $\mathbf{a} = \langle c, -2, 3 \rangle$, $\mathbf{b} = \langle c, c, -5 \rangle$

18 $\mathbf{a} = 4\mathbf{i} + 2\mathbf{j} + c\mathbf{k}$, $\mathbf{b} = \mathbf{i} + 22\mathbf{j} - 3c\mathbf{k}$

Exer. 19–24: Given points $P(3, -2, -1)$, $Q(1, 5, 4)$, $R(2, 0, -6)$, and $S(-4, 1, 5)$, find the indicated quantity.

19 $\overrightarrow{PQ} \cdot \overrightarrow{RS}$ 20 $\overrightarrow{QS} \cdot \overrightarrow{RP}$

21 The angle between $\overrightarrow{PQ}$ and $\overrightarrow{RS}$

22 The angle between $\overrightarrow{QS}$ and $\overrightarrow{RP}$

23 The component of $\overrightarrow{PS}$ along $\overrightarrow{QR}$

24 The component of $\overrightarrow{QR}$ along $\overrightarrow{PS}$

Exer. 25–26: If the vector a represents a constant force, find the work done when its point of application moves along the line segment from P to Q.

25 $\mathbf{a} = -\mathbf{i} + 5\mathbf{j} - 3\mathbf{k}$; $P(4, 0, -7)$, $Q(2, 4, 0)$

26 $\mathbf{a} = \langle 8, 0, -4 \rangle$; $P(-1, 2, 5)$, $Q(4, 1, 0)$

27 A constant force of magnitude 4 lb has the same direction as the vector $\mathbf{a} = \mathbf{i} + \mathbf{j} + \mathbf{k}$. If distance is measured in feet, find the work done if the point of

application moves along the y-axis from $(0, 2, 0)$ to $(0, -1, 0)$.

28 A constant force of magnitude 5 N (Newtons) has the same direction as the positive z-axis. If distance is measured in meters, find the work done if the point of application moves along a line from the origin to the point $P(1, 2, 3)$.

29 A child pulls a wagon along level ground by exerting a force of 20 lb on a handle that makes an angle of $30°$ with the horizontal (see figure). Find the work done in pulling the wagon 100 ft.

Exercise 29

30 Refer to Exercise 29. Find the work done if the wagon is pulled, with the same force, 100 ft up an incline that makes an angle of $30°$ with the horizontal (see figure).

Exercise 30

31 If AB is a diameter of a sphere with center O and radius r and if P is a third point on the sphere, use vectors to show that APB is a right triangle. (*Hint:* Let $\mathbf{v}_1 = \overrightarrow{OA}$ and $\mathbf{v}_2 = \overrightarrow{OP}$, and write $\overrightarrow{PA}$ and $\overrightarrow{PB}$ in terms of $\mathbf{v}_1$ and $\mathbf{v}_2$.)

32 A rectangular box has length a, width b, and height c (see figure). If P is the center of the box, use vectors to find an expression for angle APB in terms of $a, b,$ and c.

Exercise 32

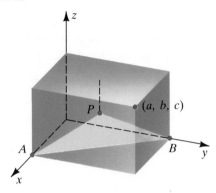

33 Refer to Exercise 32. In the mineral sphalerite, each zinc atom is surrounded by four sulphur atoms, which form a tetrahedron with the zinc atom at its center (see figure). The *bond angle* θ is the angle formed by the S–Zn–S combination. Use vectors to show that the *tetrahedral angle* θ is approximately $109.5°$.

Exercise 33

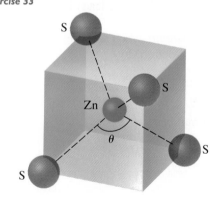

34 Given a sequence A–B–C–D of four bonded atoms, the angle between the plane formed by A, B, and C and the plane formed by $B, C,$ and D is called the *torsion angle* θ of the bond. This torsion angle is used to explain the stability of molecular structures. If segment BC is placed along the z-axis (see figure), how can θ be computed in terms of the components of vectors $\overrightarrow{BA}$ and $\overrightarrow{CD}$?

Exercise 34

35 The *direction angles* of a nonzero vector $\mathbf{a} = \langle a_1, a_2, a_3 \rangle$ are defined as the angles α, β, and γ between the vectors $\mathbf{i}$, $\mathbf{j}$, and $\mathbf{k}$, respectively, and the vector $\mathbf{a}$. The *direction cosines* of $\mathbf{a}$ are $\cos \alpha$, $\cos \beta$, and $\cos \gamma$. Prove the following:

(a) $\cos \alpha = \dfrac{a_1}{\|\mathbf{a}\|}, \quad \cos \beta = \dfrac{a_2}{\|\mathbf{a}\|}, \quad \cos \gamma = \dfrac{a_3}{\|\mathbf{a}\|}$

(b) $\cos^2 \alpha + \cos^2 \beta + \cos^2 \gamma = 1$

36 Refer to Exercise 35.

(a) Find the direction cosines of $\mathbf{a} = \langle -2, 1, 5 \rangle$.

(b) Find the direction angles and the direction cosines of $\mathbf{i}, \mathbf{j},$ and $\mathbf{k}$.

(c) Find two unit vectors that satisfy the condition

$$\cos \alpha = \cos \beta = \cos \gamma.$$

37 Three nonzero numbers l, m, and n are *direction numbers* of a nonzero vector $\mathbf{a}$ if they are proportional to the direction cosines—that is, if there exists a positive number k such that

$$l = k \cos \alpha, \qquad m = k \cos \beta, \qquad n = k \cos \gamma.$$

If $d = (l^2 + m^2 + n^2)^{1/2}$, prove that

$$\cos \alpha = l/d, \qquad \cos \beta = m/d, \qquad \cos \gamma = n/d.$$

38 Refer to Exercise 37. If l_1, m_1, n_1 and l_2, m_2, n_2 are direction numbers of $\mathbf{a}$ and $\mathbf{b}$, respectively, prove that

(a) $\mathbf{a}$ and $\mathbf{b}$ are orthogonal if and only if

$$l_1 l_2 + m_1 m_2 + n_1 n_2 = 0$$

(b) $\mathbf{a}$ and $\mathbf{b}$ are parallel if and only if there is a number k such that $l_1 = kl_2, m_1 = km_2,$ and $n_1 = kn_2$

Exer. 39–42: Prove the given property of vectors if $\mathbf{a} = \langle a_1, a_2, a_3 \rangle$, $\mathbf{b} = \langle b_1, b_2, b_3 \rangle$, and c is a scalar.

39 $\mathbf{a} \cdot \mathbf{a} = \|\mathbf{a}\|^2$ **40** $\mathbf{a} \cdot \mathbf{b} = \mathbf{b} \cdot \mathbf{a}$

41 $(c\mathbf{a}) \cdot \mathbf{b} = c(\mathbf{a} \cdot \mathbf{b}) = \mathbf{a} \cdot (c\mathbf{b})$

42 $\mathbf{0} \cdot \mathbf{a} = 0$

Exer. 43–52: Work *without using components* for the vectors.

43 Let $\mathbf{a}$ and $\mathbf{b}$ be nonzero, nonparallel vectors such that $c\mathbf{a} + d\mathbf{b} = p\mathbf{a} + q\mathbf{b}$, where c, d, p, and q are scalars. Prove that $c = p$ and $d = q$.

44 Prove that if $\mathbf{c}$ is orthogonal to both $\mathbf{a}$ and $\mathbf{b}$, then $\mathbf{c}$ is orthogonal to $p\mathbf{a} + q\mathbf{b}$ for all scalars p and q.

45 Prove that $\|\mathbf{a} + \mathbf{b}\|^2 = \|\mathbf{a}\|^2 + 2(\mathbf{a} \cdot \mathbf{b}) + \|\mathbf{b}\|^2$.

46 Prove that $\|\mathbf{a} + \mathbf{b}\|^2 + \|\mathbf{a} - \mathbf{b}\|^2 = 2(\|\mathbf{a}\|^2 + \|\mathbf{b}\|^2)$.

47 In the Cauchy–Schwarz inequality, under what conditions is $|\mathbf{a} \cdot \mathbf{b}| = \|\mathbf{a}\| \, \|\mathbf{b}\|$?

48 In the triangle inequality, under what conditions is $\|\mathbf{a} + \mathbf{b}\| = \|\mathbf{a}\| + \|\mathbf{b}\|$?

49 Prove that $\|\mathbf{a} - \mathbf{b}\| \geq \|\mathbf{a}\| - \|\mathbf{b}\|$. (*Hint:* Consider $\mathbf{a} = \mathbf{b} + (\mathbf{a} - \mathbf{b})$ and use the triangle inequality.)

50 Prove that $(\mathbf{a} + \mathbf{b}) \cdot (\mathbf{a} - \mathbf{b}) = \mathbf{a} \cdot \mathbf{a} - \mathbf{b} \cdot \mathbf{b}$.

51 Prove that $\mathbf{a} \cdot \mathbf{b} = \frac{1}{4}(\|\mathbf{a} + \mathbf{b}\|^2 - \|\mathbf{a} - \mathbf{b}\|^2)$.

52 Prove that $\operatorname{comp}_{\mathbf{c}}(\mathbf{a} + \mathbf{b}) = \operatorname{comp}_{\mathbf{c}} \mathbf{a} + \operatorname{comp}_{\mathbf{c}} \mathbf{b}$.

10.4   THE VECTOR PRODUCT

In this section, we consider the *vector product* (or *cross product*) $\mathbf{a} \times \mathbf{b}$ of two vectors $\mathbf{a}$ and $\mathbf{b}$ in V_3. Unlike the dot product $\mathbf{a} \cdot \mathbf{b}$, which is a scalar, the cross product is a vector. As we shall see, the vector $\mathbf{a} \times \mathbf{b}$ is orthogonal to both $\mathbf{a}$ and $\mathbf{b}$. It is used in physics and engineering to describe rotational effects produced by forces. We could define $\mathbf{a} \times \mathbf{b}$ geometrically and then obtain an $\mathbf{i}$, $\mathbf{j}$, $\mathbf{k}$-form by introducing a rectangular coordinate system; however, we begin with an algebraic definition. This approach disguises the geometric nature of $\mathbf{a} \times \mathbf{b}$, but it leads to simpler proofs of properties.

It is convenient to use *determinants* when working with vector products. A **determinant of order 2** is defined by

$$\begin{vmatrix} a_1 & a_2 \\ b_1 & b_2 \end{vmatrix} = a_1 b_2 - a_2 b_1,$$

where all letters represent real numbers.

ILLUSTRATION

$$\begin{vmatrix} 2 & -3 \\ 4 & 5 \end{vmatrix} = (2)(5) - (-3)(4) = 10 + 12 = 22$$

$$\begin{vmatrix} 4 & 5 \\ 2 & -3 \end{vmatrix} = (4)(-3) - (5)(2) = -12 - 10 = -22$$

The preceding illustration demonstrates a property that is true for determinants of all orders: *interchanging two rows changes the sign of a determinant*.

A **determinant of order 3** is given by

$$
\begin{vmatrix} c_1 & c_2 & c_3 \\ a_1 & a_2 & a_3 \\ b_1 & b_2 & b_3 \end{vmatrix} = \begin{vmatrix} a_2 & a_3 \\ b_2 & b_3 \end{vmatrix} c_1 - \begin{vmatrix} a_1 & a_3 \\ b_1 & b_3 \end{vmatrix} c_2 + \begin{vmatrix} a_1 & a_2 \\ b_1 & b_2 \end{vmatrix} c_3.
$$

This equation is sometimes called the *expansion of the determinant by the first row*. The value can be found by evaluating the second-order determinants on the right side of the equation.

ILLUSTRATION

$$
\begin{vmatrix} 2 & -1 & 3 \\ -2 & 5 & 1 \\ 1 & 2 & -4 \end{vmatrix} = \begin{vmatrix} 5 & 1 \\ 2 & -4 \end{vmatrix} (2) - \begin{vmatrix} -2 & 1 \\ 1 & -4 \end{vmatrix} (-1) + \begin{vmatrix} -2 & 5 \\ 1 & 2 \end{vmatrix} (3)
$$

$$
= (-20 - 2)(2) - (8 - 1)(-1) + (-4 - 5)(3)
$$

$$
= -44 + 7 - 27 = -64
$$

Definition 10.27

> The **vector product** (or **cross product**) $\mathbf{a} \times \mathbf{b}$ of $\mathbf{a} = \langle a_1, a_2, a_3 \rangle$ and $\mathbf{b} = \langle b_1, b_2, b_3 \rangle$ is
>
> $$
> \mathbf{a} \times \mathbf{b} = \begin{vmatrix} a_2 & a_3 \\ b_2 & b_3 \end{vmatrix} \mathbf{i} - \begin{vmatrix} a_1 & a_3 \\ b_1 & b_3 \end{vmatrix} \mathbf{j} + \begin{vmatrix} a_1 & a_2 \\ b_1 & b_2 \end{vmatrix} \mathbf{k}.
> $$

The symbol $\mathbf{a} \times \mathbf{b}$ is read $\mathbf{a}$ *cross* $\mathbf{b}$. Note that the formula for $\mathbf{a} \times \mathbf{b}$ can be obtained by replacing c_1, c_2, c_3 in our definition of a determinant of order 3 by the unit vectors $\mathbf{i}, \mathbf{j}, \mathbf{k}$. This suggests the following notation for the formula in Definition (10.27).

Notation for Vector Products 10.28

> $$
> \mathbf{a} \times \mathbf{b} = \begin{vmatrix} \mathbf{i} & \mathbf{j} & \mathbf{k} \\ a_1 & a_2 & a_3 \\ b_1 & b_2 & b_3 \end{vmatrix}
> $$

The symbol on the right side in (10.28) is *not* a determinant, since the first row contains vectors instead of scalars. However, determinant notation is useful for remembering the more cumbersome formula in Definition (10.27). With this distinction in mind, we shall use (10.28) to find vector products, as in the following example.

EXAMPLE 1   Find $\mathbf{a} \times \mathbf{b}$ if $\mathbf{a} = \langle 2, -1, 6 \rangle$ and $\mathbf{b} = \langle -3, 5, 1 \rangle$.

SOLUTION   We proceed as follows:

$$\mathbf{a} \times \mathbf{b} = \begin{vmatrix} \mathbf{i} & \mathbf{j} & \mathbf{k} \\ 2 & -1 & 6 \\ -3 & 5 & 1 \end{vmatrix}$$

$$= \begin{vmatrix} -1 & 6 \\ 5 & 1 \end{vmatrix} \mathbf{i} - \begin{vmatrix} 2 & 6 \\ -3 & 1 \end{vmatrix} \mathbf{j} + \begin{vmatrix} 2 & -1 \\ -3 & 5 \end{vmatrix} \mathbf{k}$$

$$= (-1 - 30)\mathbf{i} - (2 + 18)\mathbf{j} + (10 - 3)\mathbf{k}$$

$$= -31\mathbf{i} - 20\mathbf{j} + 7\mathbf{k}$$

If $\mathbf{a}$ is any vector in V_3, then

$$\mathbf{a} \times \mathbf{0} = \mathbf{0} = \mathbf{0} \times \mathbf{a},$$

for if one of the vectors in Definition (10.27) is $\mathbf{0}$, then each determinant has a row of zeros, and hence, its value is 0 (verify this fact). It is also easy to show that $\mathbf{a} \times \mathbf{a} = \mathbf{0}$ for every $\mathbf{a}$.

The next theorem brings out an important property of vector products.

Theorem 10.29

> The vector $\mathbf{a} \times \mathbf{b}$ is orthogonal to both $\mathbf{a}$ and $\mathbf{b}$.

PROOF   By Theorem (10.21), it is sufficient to show that

$$(\mathbf{a} \times \mathbf{b}) \cdot \mathbf{a} = 0 \quad \text{and} \quad (\mathbf{a} \times \mathbf{b}) \cdot \mathbf{b} = 0.$$

Taking the dot product of $\mathbf{a} \times \mathbf{b}$ in (10.27) with $\mathbf{a} = \langle a_1, a_2, a_3 \rangle$ gives us

$$(\mathbf{a} \times \mathbf{b}) \cdot \mathbf{a} = \begin{vmatrix} a_2 & a_3 \\ b_2 & b_3 \end{vmatrix} a_1 - \begin{vmatrix} a_1 & a_3 \\ b_1 & b_3 \end{vmatrix} a_2 + \begin{vmatrix} a_1 & a_2 \\ b_1 & b_2 \end{vmatrix} a_3$$

$$= (a_2 b_3 - a_3 b_2)a_1 - (a_1 b_3 - a_3 b_1)a_2 + (a_1 b_2 - a_2 b_1)a_3$$

$$= a_2 b_3 a_1 - a_3 b_2 a_1 - a_1 b_3 a_2 + a_3 b_1 a_2 + a_1 b_2 a_3 - a_2 b_1 a_3$$

$$= 0.$$

Hence, $\mathbf{a} \times \mathbf{b}$ is orthogonal to $\mathbf{a}$. The proof that $(\mathbf{a} \times \mathbf{b}) \cdot \mathbf{b} = 0$ is similar. ∎

Let us interpret Theorem (10.29) geometrically. As illustrated in Figure 10.39(a), if nonzero vectors $\mathbf{a}$ and $\mathbf{b}$ correspond to vectors $\overrightarrow{PQ}$ and $\overrightarrow{PR}$ with the same initial point P, then $\mathbf{a} \times \mathbf{b}$ corresponds to a vector $\overrightarrow{PS}$ that is orthogonal to both $\overrightarrow{PQ}$ and $\overrightarrow{PR}$. Hence, $\overrightarrow{PS}$ is perpendicular to the plane

Figure 10.39

(a)

(b)

determined by P, Q, and R. We shall write

$$\overrightarrow{PS} = \overrightarrow{PQ} \times \overrightarrow{PR}.$$

The direction of $\overrightarrow{PS}$ may be obtained using the right-hand rule illustrated in Figure 10.39(b). Thus, if θ is the angle between $\overrightarrow{PQ}$ and $\overrightarrow{PR}$ and if the fingers of the right hand are curled such that a rotation through θ will transform $\overrightarrow{PQ}$ into a vector that has the same direction as $\overrightarrow{PR}$, then the extended thumb points in the direction of $\overrightarrow{PQ} \times \overrightarrow{PR}$.

The following result describes the magnitude of $\mathbf{a} \times \mathbf{b}$.

Theorem 10.30

If θ is the angle between nonzero vectors $\mathbf{a}$ and $\mathbf{b}$, then

$$\|\mathbf{a} \times \mathbf{b}\| = \|\mathbf{a}\| \, \|\mathbf{b}\| \sin \theta.$$

PROOF By Theorem (10.17)(i) and Definition (10.27),

$$\|\mathbf{a} \times \mathbf{b}\|^2 = (\mathbf{a} \times \mathbf{b}) \cdot (\mathbf{a} \times \mathbf{b})$$

$$= \begin{vmatrix} a_2 & a_3 \\ b_2 & b_3 \end{vmatrix}^2 + \begin{vmatrix} a_1 & a_3 \\ b_1 & b_3 \end{vmatrix}^2 + \begin{vmatrix} a_1 & a_2 \\ b_1 & b_2 \end{vmatrix}^2$$

$$= (a_2 b_3 - a_3 b_2)^2 + (a_1 b_3 - a_3 b_1)^2 + (a_1 b_2 - a_2 b_1)^2$$

$$= a_2^2 b_3^2 - 2a_2 a_3 b_2 b_3 + a_3^2 b_2^2 + a_1^2 b_3^2 - 2a_1 a_3 b_1 b_3$$

$$\quad + a_3^2 b_1^2 + a_1^2 b_2^2 - 2a_1 a_2 b_1 b_2 + a_2^2 b_1^2$$

$$= (a_1^2 + a_2^2 + a_3^2)(b_1^2 + b_2^2 + b_3^2) - (a_1 b_1 + a_2 b_2 + a_3 b_3)^2.$$

The last equality may be verified by multiplying the indicated expressions. In vector form, the preceding identity is

$$\|\mathbf{a} \times \mathbf{b}\|^2 = (\|\mathbf{a}\| \|\mathbf{b}\|)^2 - (\mathbf{a} \cdot \mathbf{b})^2,$$

or, since $\mathbf{a} \cdot \mathbf{b} = \|\mathbf{a}\| \|\mathbf{b}\| \cos \theta$,

$$\|\mathbf{a} \times \mathbf{b}\|^2 = (\|\mathbf{a}\| \|\mathbf{b}\|)^2 - (\|\mathbf{a}\| \|\mathbf{b}\|)^2 \cos^2 \theta$$
$$= (\|\mathbf{a}\| \|\mathbf{b}\|)^2 (1 - \cos^2 \theta).$$

Hence, $$\|\mathbf{a} \times \mathbf{b}\|^2 = (\|\mathbf{a}\| \|\mathbf{b}\|)^2 \sin^2 \theta.$$

Taking square roots and observing that $\sin \theta \geq 0$ (since $0 \leq \theta \leq \pi$), we obtain

$$\|\mathbf{a} \times \mathbf{b}\| = \|\mathbf{a}\| \|\mathbf{b}\| \sin \theta. \quad \blacksquare$$

Corollary 10.31

Two vectors $\mathbf{a}$ and $\mathbf{b}$ are parallel if and only if $\mathbf{a} \times \mathbf{b} = \mathbf{0}$.

PROOF   Suppose that $\mathbf{a}$ and $\mathbf{b}$ are nonzero vectors. If θ is the angle between $\mathbf{a}$ and $\mathbf{b}$, then the vectors are parallel if and only if $\theta = 0$ or $\theta = \pi$, or, equivalently, $\sin \theta = 0$. By Theorem (10.30), the last statement is equivalent to $\mathbf{a} \times \mathbf{b} = \mathbf{0}$. If either $\mathbf{a}$ or $\mathbf{b}$ is the zero vector, the proof is trivial, since $\mathbf{0}$ is parallel to every vector. $\blacksquare$

Figure 10.40

To interpret $\|\mathbf{a} \times \mathbf{b}\|$ geometrically, let us represent $\mathbf{a}$ and $\mathbf{b}$ by vectors $\overrightarrow{PQ}$ and $\overrightarrow{PR}$ having the same initial point P. As in Figure 10.40, let S be the point such that segments PQ and PR are adjacent sides of a parallelogram with vertices P, Q, R, and S. An altitude of the parallelogram is $\|\mathbf{b}\| \sin \theta$, and hence its area is $\|\mathbf{a}\| \|\mathbf{b}\| \sin \theta$. Comparing this with Theorem (10.30), we see that *the magnitude of the vector product $\mathbf{a} \times \mathbf{b}$ equals the area of the parallelogram determined by $\mathbf{a}$ and $\mathbf{b}$.*

EXAMPLE ▪ 2   Find the area of the triangle determined by $P(4, -3, 1)$, $Q(6, -4, 7)$, and $R(1, 2, 2)$.

SOLUTION   The vectors in V_3 corresponding to $\overrightarrow{PQ}$ and $\overrightarrow{PR}$ are given by

$$\mathbf{a} = \langle 6 - 4, -4 - (-3), 7 - 1 \rangle = \langle 2, -1, 6 \rangle$$
$$\mathbf{b} = \langle 1 - 4, 2 - (-3), 2 - 1 \rangle = \langle -3, 5, 1 \rangle.$$

From Example 1,

$$\mathbf{a} \times \mathbf{b} = -31\mathbf{i} - 20\mathbf{j} + 7\mathbf{k}.$$

Hence, the area of the parallelogram with adjacent sides PQ and PR is

$$\|\mathbf{a} \times \mathbf{b}\| = \sqrt{(-31)^2 + (-20)^2 + 7^2} = \sqrt{1410}.$$

It follows that the area of the triangle is $\frac{1}{2}\sqrt{1410} \approx 18.8$.

The vector products of the special unit vectors $\mathbf{i}$, $\mathbf{j}$, and $\mathbf{k}$ are of interest. For example, using Definition (10.27) with $\mathbf{a} = \mathbf{i} = \langle 1, 0, 0 \rangle$ and $\mathbf{b} = \mathbf{j} = \langle 0, 1, 0 \rangle$, we have

$$\mathbf{i} \times \mathbf{j} = \begin{vmatrix} 0 & 0 \\ 1 & 0 \end{vmatrix} \mathbf{i} - \begin{vmatrix} 1 & 0 \\ 0 & 0 \end{vmatrix} \mathbf{j} + \begin{vmatrix} 1 & 0 \\ 0 & 1 \end{vmatrix} \mathbf{k} = \mathbf{k}.$$

We can prove the following in similar fashion.

Properties of $\mathbf{i}$, $\mathbf{j}$, $\mathbf{k}$ 10.32

$\mathbf{i} \times \mathbf{j} = \mathbf{k}$	$\mathbf{j} \times \mathbf{k} = \mathbf{i}$	$\mathbf{k} \times \mathbf{i} = \mathbf{j}$
$\mathbf{j} \times \mathbf{i} = -\mathbf{k}$	$\mathbf{k} \times \mathbf{j} = -\mathbf{i}$	$\mathbf{i} \times \mathbf{k} = -\mathbf{j}$
$\mathbf{i} \times \mathbf{i} = \mathbf{j} \times \mathbf{j} = \mathbf{k} \times \mathbf{k} = \mathbf{0}$		

Figure 10.41

Figure 10.41 is a simple mnemonic device for remembering six of the formulas in (10.32). If we place the symbols $\mathbf{i}$, $\mathbf{j}$, $\mathbf{k}$ around a circle as shown and proceed in the *clockwise* direction (as denoted by the arrows), then the vector product of any two successive vectors is the next vector. To illustrate, $\mathbf{i} \times \mathbf{j} = \mathbf{k}$. For the *counterclockwise* direction, take the *negative* of the next vector, as in $\mathbf{j} \times \mathbf{i} = -\mathbf{k}$.

The fact that $\mathbf{i} \times \mathbf{j} \neq \mathbf{j} \times \mathbf{i}$ shows that the *vector product is not commutative.* Moreover, *the associative law does not hold*, since, for example,

$$\mathbf{i} \times (\mathbf{j} \times \mathbf{j}) = \mathbf{i} \times \mathbf{0} = \mathbf{0},$$

but

$$(\mathbf{i} \times \mathbf{j}) \times \mathbf{j} = \mathbf{k} \times \mathbf{j} = -\mathbf{i}.$$

The following properties are true for any vectors $\mathbf{a}$, $\mathbf{b}$, and $\mathbf{c}$ and a scalar m.

Properties of the Vector Product   10.33

(i) $\mathbf{b} \times \mathbf{a} = -(\mathbf{a} \times \mathbf{b})$

(ii) $(m\mathbf{a}) \times \mathbf{b} = m(\mathbf{a} \times \mathbf{b}) = \mathbf{a} \times (m\mathbf{b})$

(iii) $\mathbf{a} \times (\mathbf{b} + \mathbf{c}) = (\mathbf{a} \times \mathbf{b}) + (\mathbf{a} \times \mathbf{c})$

(iv) $(\mathbf{a} + \mathbf{b}) \times \mathbf{c} = (\mathbf{a} \times \mathbf{c}) + (\mathbf{b} \times \mathbf{c})$

(v) $(\mathbf{a} \times \mathbf{b}) \cdot \mathbf{c} = \mathbf{a} \cdot (\mathbf{b} \times \mathbf{c})$

(vi) $\mathbf{a} \times (\mathbf{b} \times \mathbf{c}) = (\mathbf{a} \cdot \mathbf{c})\mathbf{b} - (\mathbf{a} \cdot \mathbf{b})\mathbf{c}$

PROOF   Each property may be established by straightforward (but sometimes lengthy) applications of Definition (10.27). To prove (i), we let $\mathbf{a} = \langle a_1, a_2, a_3 \rangle$ and $\mathbf{b} = \langle b_1, b_2, b_3 \rangle$ and find

$$\mathbf{b} \times \mathbf{a} = \begin{vmatrix} b_2 & b_3 \\ a_2 & a_3 \end{vmatrix} \mathbf{i} - \begin{vmatrix} b_1 & b_3 \\ a_1 & a_3 \end{vmatrix} \mathbf{j} + \begin{vmatrix} b_1 & b_2 \\ a_1 & a_2 \end{vmatrix} \mathbf{k}.$$

Since interchanging two rows of a determinant changes its sign, we obtain

$$\mathbf{b} \times \mathbf{a} = -\begin{vmatrix} a_2 & a_3 \\ b_2 & b_3 \end{vmatrix} \mathbf{i} + \begin{vmatrix} a_1 & a_3 \\ b_1 & b_3 \end{vmatrix} \mathbf{j} - \begin{vmatrix} a_1 & a_2 \\ b_1 & b_2 \end{vmatrix} \mathbf{k} = -(\mathbf{a} \times \mathbf{b}).$$

To prove (iii), consider $\mathbf{c} = \langle c_1, c_2, c_3 \rangle$. The $\mathbf{i}$ component of $\mathbf{a} \times (\mathbf{b} + \mathbf{c})$ is

$$\begin{vmatrix} a_2 & a_3 \\ b_2 + c_2 & b_3 + c_3 \end{vmatrix} = a_2(b_3 + c_3) - a_3(b_2 + c_2)$$

$$= (a_2 b_3 - a_3 b_2) + (a_2 c_3 - a_3 c_2)$$

$$= \begin{vmatrix} a_2 & a_3 \\ b_2 & b_3 \end{vmatrix} + \begin{vmatrix} a_2 & a_3 \\ c_2 & c_3 \end{vmatrix},$$

which is equal to the $\mathbf{i}$ component of $(\mathbf{a} \times \mathbf{b}) + (\mathbf{a} \times \mathbf{c})$. A similar calculation can be used to prove that the $\mathbf{j}$ and $\mathbf{k}$ components of $\mathbf{a} \times (\mathbf{b} + \mathbf{c})$ are the same as those of $(\mathbf{a} \times \mathbf{b}) + (\mathbf{a} \times \mathbf{c})$. This establishes (iii). The proofs of the remaining properties are left as exercises. ■

Formula (vi) of Theorem (10.33) is called the **triple vector product** of $\mathbf{a}$, $\mathbf{b}$, and $\mathbf{c}$.

EXAMPLE ■ 3   Find a formula for the distance d from a point R to a line l.

SOLUTION   Let P and Q be points on l, as shown in Figure 10.42, and let θ be the angle between $\overrightarrow{PQ}$ and $\overrightarrow{PR}$. Since $d = \| \overrightarrow{PR} \| \sin \theta$, we obtain

$$\| \overrightarrow{PQ} \times \overrightarrow{PR} \| = \| \overrightarrow{PQ} \| \| \overrightarrow{PR} \| \sin \theta = \| \overrightarrow{PQ} \| d.$$

Hence,

$$d = \frac{\| \overrightarrow{PQ} \times \overrightarrow{PR} \|}{\| \overrightarrow{PQ} \|}.$$

Figure 10.42

R

d

θ

P   Q l

Figure 10.43

$\mathbf{a} \times \mathbf{b}$

h

θ

$\mathbf{c}$

$\mathbf{b}$

$\mathbf{a}$

Area of base
$\| \mathbf{a} \times \mathbf{b} \|$

EXAMPLE ■ 4   Suppose that $\mathbf{a}$, $\mathbf{b}$, and $\mathbf{c}$ represent adjacent sides of the oblique box in Figure 10.43. Show that $|(\mathbf{a} \times \mathbf{b}) \cdot \mathbf{c}|$ is the volume of the box.

SOLUTION   The area of the base of the box is $\| \mathbf{a} \times \mathbf{b} \|$. Let θ be the angle between $\mathbf{c}$ and $\mathbf{a} \times \mathbf{b}$. Since $\mathbf{a} \times \mathbf{b}$ is perpendicular to the base, the altitude h of the box is given by $h = \| \mathbf{c} \| | \cos \theta |$. We must use the absolute value $|\cos \theta|$, since θ may be an obtuse angle. Thus, the volume V of the box is

$$V = \text{(area of base)(altitude)}$$

$$= \| \mathbf{a} \times \mathbf{b} \| \| \mathbf{c} \| | \cos \theta |$$

$$= |(\mathbf{a} \times \mathbf{b}) \cdot \mathbf{c}|,$$

where the last equality follows from Theorem (10.19).

The number $(\mathbf{a} \times \mathbf{b}) \cdot \mathbf{c}$ considered in Example 4 is called the **triple scalar product** of $\mathbf{a}$, $\mathbf{b}$, and $\mathbf{c}$. This number can be expressed as a determinant of order 3 that involves the components of the three vectors (see Exercise 21).

Vector products are used to study rotational effects produced by forces. Suppose a force $\overrightarrow{PQ}$ acts at a point P in an object, as illustrated in Figure 10.44, where we have superimposed an xyz-coordinate system. The force tends to make the object rotate about a line through O that is perpendicular to the plane determined by $\overrightarrow{PQ}$ and the position vector $\overrightarrow{OP}$ of P. The **torque vector** $\overrightarrow{OT}$ for the force (also called **the moment of $\overrightarrow{PQ}$ about** O) is defined by

$$\overrightarrow{OT} = \overrightarrow{OP} \times \overrightarrow{PQ}.$$

Figure 10.44

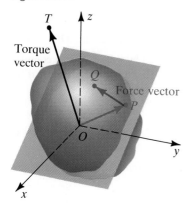

T

z

Torque
vector

Q

Force vector

P

O

y

x

In physics or mechanics, $\overrightarrow{OT}$ specifies both the magnitude and the direction of the rotational effect induced by the force $\overrightarrow{PQ}$. If $\overrightarrow{OP}$ represents the handle of a wrench and $\overrightarrow{PQ}$ is the force applied to the end of the handle, then $\overrightarrow{OT}$ points in the direction that a (right-handed) bolt moves when the wrench is turned.

Vector products are also useful in electromagnetic field theory. To illustrate, let us consider an **electric dipole**—that is, two point charges $+q$ and $-q$ that are always a constant distance d apart. If $\mathbf{d}$ is the vector shown in Figure 10.45(a), then $\|\mathbf{d}\| = d$ and the **electric dipole moment $\mathbf{p}$** is defined as $\mathbf{p} = q\mathbf{d}$.

Figure 10.45

(a)   (b)

Suppose, as illustrated in Figure 10.45(b), the electric dipole is placed in an electromagnetic field where every force vector $\mathbf{E}$ has the same magnitude and direction. The electric dipole will tend to rotate until it is aligned with the direction of the field—that is, the angle θ between $\mathbf{d}$ and $\mathbf{E}$ will approach 0. The torque $\mathbf{T}$ caused by $\mathbf{E}$ at P is given by $\mathbf{T} = \mathbf{p} \times \mathbf{E}$, and the work done by $\mathbf{E}$ in the alignment process is $W = \mathbf{p} \cdot \mathbf{E}$.

We shall consider other interpretations of the vector product in Chapter 11.

EXERCISES 10.4

Exer. 1–10: Find $\mathbf{a} \times \mathbf{b}$.

1 $\mathbf{a} = \langle 1, -2, 3 \rangle$, $\mathbf{b} = \langle 2, 1, -4 \rangle$

2 $\mathbf{a} = \langle -5, 1, -1 \rangle$, $\mathbf{b} = \langle 3, 6, -2 \rangle$

3 $\mathbf{a} = \langle 0, 1, 2 \rangle$, $\mathbf{b} = \langle 1, 2, 0 \rangle$

4 $\mathbf{a} = \langle 0, 0, 4 \rangle$, $\mathbf{b} = \langle -7, 1, 0 \rangle$

5 $\mathbf{a} = 5\mathbf{i} - 6\mathbf{j} - \mathbf{k}$, $\mathbf{b} = 3\mathbf{i} + \mathbf{k}$

6 $\mathbf{a} = 2\mathbf{i} + \mathbf{j}$, $\mathbf{b} = -5\mathbf{j} + 2\mathbf{k}$

7 $\mathbf{a} = -3\mathbf{i} + \mathbf{j} + 2\mathbf{k}$, $\mathbf{b} = 9\mathbf{i} - 3\mathbf{j} - 6\mathbf{k}$

8 $\mathbf{a} = 3\mathbf{i} - \mathbf{j} + 8\mathbf{k}$, $\mathbf{b} = 5\mathbf{j}$

9 $\mathbf{a} = 4\mathbf{i} - 6\mathbf{j} + 2\mathbf{k}$, $\mathbf{b} = -2\mathbf{i} + 3\mathbf{j} - \mathbf{k}$

10 $\mathbf{a} = 3\mathbf{i}$, $\mathbf{b} = 4\mathbf{k}$

Exer. 11–12: Use the vector product to show that a and b are parallel.

11 $\mathbf{a} = \langle -6, -10, 4 \rangle$, $\mathbf{b} = \langle 3, 5, -2 \rangle$

12 $\mathbf{a} = 2\mathbf{i} - \mathbf{j} + 4\mathbf{k}$, $\mathbf{b} = -6\mathbf{i} + 3\mathbf{j} - 12\mathbf{k}$

Exer. 13–14: Let $\mathbf{a} = \langle 2, 0, -1 \rangle, \mathbf{b} = \langle -3, 1, 0 \rangle$, and $\mathbf{c} = \langle 1, -2, 4 \rangle$.

13 Find $\mathbf{a} \times (\mathbf{b} \times \mathbf{c})$ and $(\mathbf{a} \times \mathbf{b}) \times \mathbf{c}$.

14 Find $\mathbf{a} \times (\mathbf{b} - \mathbf{c})$ and $(\mathbf{a} \times \mathbf{b}) - (\mathbf{a} \times \mathbf{c})$.

Exer. 15–18: (a) Find a vector perpendicular to the plane determined by P, Q, and R. (b) Find the area of the triangle PQR.

15 $P(1, -1, 2)$, $Q(0, 3, -1)$, $R(3, -4, 1)$

16 $P(-3, 0, 5)$, $Q(2, -1, -3)$, $R(4, 1, -1)$

17 $P(4, 0, 0)$, $Q(0, 5, 0)$, $R(0, 0, 2)$

18 $P(-1, 2, 0)$, $Q(0, 2, -3)$, $R(5, 0, 1)$

Exer. 19–20: Refer to Example 3. Find the distance from P to the line through Q and R.

19 $P(3, 1, -2)$, $Q(2, 5, 1)$, $R(-1, 4, 2)$

20 $P(-2, 5, 1)$, $Q(3, -1, 4)$, $R(1, 6, -3)$

21 If $\mathbf{a} = \langle a_1, a_2, a_3 \rangle, \mathbf{b} = \langle b_1, b_2, b_3 \rangle, \mathbf{c} = \langle c_1, c_2, c_3 \rangle$, prove that

$$\mathbf{a} \cdot (\mathbf{b} \times \mathbf{c}) = (\mathbf{a} \times \mathbf{b}) \cdot \mathbf{c} = \begin{vmatrix} a_1 & a_2 & a_3 \\ b_1 & b_2 & b_3 \\ c_1 & c_2 & c_3 \end{vmatrix}.$$

Exer. 22–23: Use Example 4 and Exercise 21 to find the volume of the box having adjacent sides AB, AC, and AD.

22 $A(0, 0, 0)$, $B(1, -1, 2)$, $C(0, 3, -1)$, $D(3, -4, 1)$

23 $A(2, 1, -1)$, $B(3, 0, 2)$, $C(4, -2, 1)$, $D(5, -3, 0)$

24 If $\mathbf{a}$, $\mathbf{b}$, and $\mathbf{c}$ are represented by vectors with a common initial point, show that $\mathbf{a} \cdot (\mathbf{b} \times \mathbf{c}) = 0$ if and only if the vectors are coplanar.

25 Prove that $(\mathbf{a} \times \mathbf{b}) \cdot \mathbf{b} = 0$ for all vectors $\mathbf{a}$ and $\mathbf{b}$.

26 If $\mathbf{a} \times \mathbf{b} = \mathbf{a} \times \mathbf{c}$ and $\mathbf{a} \neq \mathbf{0}$, does it follow that $\mathbf{b} = \mathbf{c}$? Explain.

27 Let $\mathbf{a} \neq \mathbf{0}$. If $\mathbf{a} \times \mathbf{b} = \mathbf{a} \times \mathbf{c}$ *and* $\mathbf{a} \cdot \mathbf{b} = \mathbf{a} \cdot \mathbf{c}$, prove that $\mathbf{b} = \mathbf{c}$.

Exer. 28–31: Prove the given property if $\mathbf{a} = \langle a_1, a_2, a_3 \rangle$, $\mathbf{b} = \langle b_1, b_2, b_3 \rangle$, $\mathbf{c} = \langle c_1, c_2, c_3 \rangle$, and m is a scalar.

28 $(m\mathbf{a}) \times \mathbf{b} = m(\mathbf{a} \times \mathbf{b}) = \mathbf{a} \times (m\mathbf{b})$

29 $(\mathbf{a} + \mathbf{b}) \times \mathbf{c} = (\mathbf{a} \times \mathbf{c}) + (\mathbf{b} \times \mathbf{c})$

30 $(\mathbf{a} \times \mathbf{b}) \cdot \mathbf{c} = \mathbf{a} \cdot (\mathbf{b} \times \mathbf{c})$

31 $\mathbf{a} \times (\mathbf{b} \times \mathbf{c}) = (\mathbf{a} \cdot \mathbf{c})\mathbf{b} - (\mathbf{a} \cdot \mathbf{b})\mathbf{c}$

Exer. 32–37: Verify *without using components* for the vectors.

32 $(\mathbf{a} + \mathbf{b}) \times (\mathbf{a} - \mathbf{b}) = 2(\mathbf{b} \times \mathbf{a})$

33 $\mathbf{a} \times (\mathbf{b} \times \mathbf{c}) + \mathbf{b} \times (\mathbf{c} \times \mathbf{a}) + \mathbf{c} \times (\mathbf{a} \times \mathbf{b}) = \mathbf{0}$
    (*Hint:* Use (vi) of Theorem (10.33).)

34 $(\mathbf{a} \times \mathbf{b}) \times \mathbf{c} = (\mathbf{a} \cdot \mathbf{c})\mathbf{b} - (\mathbf{b} \cdot \mathbf{c})\mathbf{a}$

35 $(\mathbf{a} \times \mathbf{b}) \cdot (\mathbf{c} \times \mathbf{d}) = \begin{vmatrix} \mathbf{a} \cdot \mathbf{c} & \mathbf{b} \cdot \mathbf{c} \\ \mathbf{a} \cdot \mathbf{d} & \mathbf{b} \cdot \mathbf{d} \end{vmatrix}$

36 $(\mathbf{a} \times \mathbf{b}) \times (\mathbf{c} \times \mathbf{d}) = (\mathbf{a} \times \mathbf{b} \cdot \mathbf{d})\mathbf{c} - (\mathbf{a} \times \mathbf{b} \cdot \mathbf{c})\mathbf{d}$

37 $(\mathbf{a} \times \mathbf{b}) \cdot (\mathbf{b} \times \mathbf{c}) \times (\mathbf{c} \times \mathbf{a}) = (\mathbf{a} \cdot \mathbf{b} \times \mathbf{c})^2$

10.5 LINES AND PLANES

In this section, we describe lines and planes by using the vector concepts of *parallel* and *orthogonal*, respectively. We will assume that all lines and planes lie in an xyz-coordinate system.

Let $\mathbf{a} = \langle a_1, a_2, a_3 \rangle$ be a nonzero vector in V_3, let $P_1(x_1, y_1, z_1)$ be any point, and let $\overrightarrow{OA}$ be the position vector for $\mathbf{a}$. As illustrated in Figure 10.46, we describe **the line l through $P_1(x_1, y_1, z_1)$ parallel to $\mathbf{a}$** as the set of all points $P(x, y, z)$ such that $\overrightarrow{P_1P}$ is parallel to $\overrightarrow{OA}$—that is,

$$\overrightarrow{P_1P} = t\,\overrightarrow{OA} \quad \text{for some scalar } t.$$

In terms of vectors in V_3,

$$\langle x - x_1, y - y_1, z - z_1 \rangle = t \langle a_1, a_2, a_3 \rangle = \langle a_1 t, a_2 t, a_3 t \rangle.$$

Equating components and solving for x, y, and z gives us

$$x = x_1 + a_1 t, \quad y = y_1 + a_2 t, \quad z = z_1 + a_3 t,$$

where t is a real number. These are parametric equations for the line l, with parameter t. The points $P(x, y, z)$ on l are obtained by letting t take on all real values. We have proved the following.

Figure 10.46

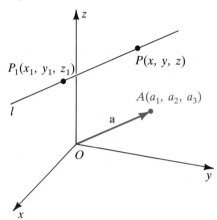

Theorem 10.34

> Parametric equations for the line through $P_1(x_1, y_1, z_1)$ parallel to $\mathbf{a} = \langle a_1, a_2, a_3 \rangle$ are
>
> $$x = x_1 + a_1 t, \quad y = y_1 + a_2 t, \quad z = z_1 + a_3 t; \quad t \text{ in } \mathbb{R}.$$

Note that the coefficients of t in (10.34) *are the components of the vector* $\mathbf{a}$. We may obtain the same line by using any nonzero vector that is parallel to $\mathbf{a}$, for in this case $\mathbf{b} = c\mathbf{a} = \langle ca_1, ca_2, ca_3 \rangle$ and parametric equations for the line are

$$x = x_1 + (ca_1)v, \quad y = y_1 + (ca_2)v, \quad z = z_1 + (ca_3)v; \quad v \text{ in } \mathbb{R}$$

These equations determine the same line, since the point given by t can be obtained by letting $v = t/c$.

EXAMPLE ■ 1

(a) Find the parametric equations for the line l through $P(5, -2, 4)$ that is parallel to $\mathbf{a} = \langle \frac{1}{2}, 2, -\frac{2}{3} \rangle$.

(b) Where does l intersect the xy-plane?

(c) Sketch the position vector for $\mathbf{a}$ and the line l.

SOLUTION

(a) To avoid fractions, we shall use the vector $\mathbf{b} = 6\mathbf{a} = \langle 3, 12, -4 \rangle$ instead of $\mathbf{a}$. Applying Theorem (10.34), we obtain the following parametric equations for l:

$$x = 5 + 3t, \quad y = -2 + 12t, \quad z = 4 - 4t; \quad t \text{ in } \mathbb{R}$$

(b) The line intersects the xy-plane at the point $R(x, y, z)$ if $z = 4 - 4t = 0$ (that is, if $t = 1$). Letting $t = 1$ in the parametric equations from part (a), we find that the x- and y-coordinates of R are

$$x = 5 + 3(1) = 8 \quad \text{and} \quad y = -2 + 12(1) = 10.$$

Hence, R is the point with coordinates $(8, 10, 0)$.

(c) The position vector $\overrightarrow{OA}$ for $\mathbf{a}$ is shown in Figure 10.47. The line l may be sketched by using the given point $P(5, -2, 4)$ and the point $R(8, 10, 0)$ found in part (b).

Figure 10.47

Figure 10.48

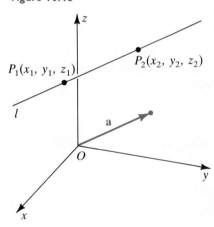

To find parametric equations for the line through two arbitrary points $P_1(x_1, y_1, z_1)$ and $P_2(x_2, y_2, z_2)$, as illustrated in Figure 10.48, we use the vector $\mathbf{a}$ that corresponds to $\overrightarrow{P_1 P_2}$—that is,

$$\mathbf{a} = \langle x_2 - x_1, y_2 - y_1, z_2 - z_1 \rangle.$$

Substituting in Theorem (10.34), we obtain the parametric equations

$$x = x_1 + (x_2 - x_1)t, \quad y = y_1 + (y_2 - y_1)t, \quad z = z_1 + (z_2 - z_1)t; \quad t \text{ in } \mathbb{R}$$

Note that $t = 0$ gives us the point P_1, $t = \frac{1}{2}$ the midpoint of $P_1 P_2$, and $t = 1$ the point P_2. Generally, as t varies from 0 to 1, the point $P(x, y, z)$ traces the line segment $P_1 P_2$. Other values of t give different points on l.

EXAMPLE ■ 2   Find parametric equations for the line l through $P_1(3, 1, -2)$ and $P_2(-2, 7, -4)$.

SOLUTION   The vector $\mathbf{a}$ in V_3 corresponding to $\overrightarrow{P_1 P_2}$ is

$$\mathbf{a} = \langle -2 - 3, 7 - 1, -4 + 2 \rangle = \langle -5, 6, -2 \rangle.$$

Applying Theorem (10.34), we find that parametric equations for l are

$$x = 3 - 5t, \quad y = 1 + 6t, \quad z = -2 - 2t; \quad t \text{ in } \mathbb{R}.$$

EXAMPLE ■ 3   Let lines l_1 and l_2 have the respective parametrizations

$$x = -2 + 3t, \quad y = 5 - 4t, \quad z = 1 + 2t; \quad t \text{ in } \mathbb{R}$$
$$x = 1 - v, \quad y = 3 + 2v, \quad z = -4 - 3v; \quad v \text{ in } \mathbb{R}.$$

Determine whether l_1 and l_2 intersect, and if so, find the point of intersection.

SOLUTION The lines intersect if there are values of t (for l_1) and v (for l_2) that give us the same point $P(x, y, z)$. Hence, we consider the following system of three equations in two variables:

$$\begin{cases} -2 + 3t = \quad 1 - \quad v \\ 5 - 4t = \quad 3 + 2v \\ 1 + 2t = -4 - 3v \end{cases} \quad \text{or, equivalently,} \quad \begin{cases} 3t + \quad v = \quad 3 \\ -4t - 2v = -2 \\ 2t + 3v = -5 \end{cases}$$

Solving the first two equations of the system simultaneously, we obtain $t = 2$ and $v = -3$. Substituting into the third equation, $2t + 3v = -5$, we obtain $2(2) + 3(-3) = -5$, or $-5 = -5$. Since $t = 2$ and $v = -3$ is a solution of each of the three equations, the lines intersect. The point of intersection may be found by letting $t = 2$ in the parametrization for l_1 or by letting $v = -3$ in the parametrization for l_2. In either case, we obtain the point $P(4, -3, 5)$.

It is important to note that if the solution $t = 2$ and $v = -3$ of the first two equations in the system had *not* satisfied the third equation, then the lines would have no point of intersection.

In the next definition, we use the angle between two vectors (see (10.18)) to define the angle between two lines. We also use vectors to define parallel lines and orthogonal lines.

Definition 10.35

Let θ be the angle between nonzero vectors **a** and **b** and let l_1 and l_2 be lines that are parallel to the position vectors of **a** and **b** (see Figure 10.49).

    (i) The **angles between l_1 and l_2** are θ and $\pi - \theta$.

    (ii) The lines l_1 and l_2 are **parallel** if and only if **a** and **b** are parallel—that is, $\mathbf{b} = c\mathbf{a}$ for some scalar c.

    (iii) The lines l_1 and l_2 are **orthogonal** if and only if **a** and **b** are orthogonal—that is, $\mathbf{a} \cdot \mathbf{b} = 0$.

Figure 10.49

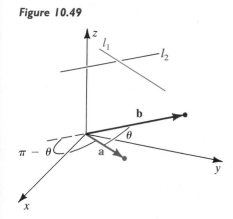

Note that the angle between l_1 and l_2 is defined for either intersecting or nonintersecting lines.

EXAMPLE 4 Let lines l_1 and l_2 have the respective parametrizations for t in $\mathbb{R}$:

$$x = 1 + 2t, \quad y = 3 - 4t, \quad z = -2 + t$$
$$x = -5 - t, \quad y = 2 - 3t, \quad z = 4 + 3t$$

Find the angles between l_1 and l_2.

SOLUTION Referring to the coefficients of t, we see that l_1 and l_2 are parallel, respectively, to

$$\mathbf{a} = \langle 2, -4, 1 \rangle \quad \text{and} \quad \mathbf{b} = \langle -1, -3, 3 \rangle.$$

If θ is the angle between the position vectors of **a** and **b**, then, by (10.20),

$$\cos \theta = \frac{\mathbf{a} \cdot \mathbf{b}}{\|\mathbf{a}\|\,\|\mathbf{b}\|} = \frac{(2)(-1) + (-4)(-3) + (1)(3)}{\sqrt{4 + 16 + 1}\sqrt{1 + 9 + 9}} = \frac{13}{\sqrt{399}}.$$

Hence, by Definition (10.35), the angles between l_1 and l_2 are given by

$$\theta = \cos^{-1} \frac{13}{\sqrt{399}} \approx 0.86 \text{ radian} \approx 49°$$

$$\pi - \theta \approx 2.28 \text{ radians} \approx 131°.$$

As illustrated in Figure 10.50(a), if $P_1(x_1, y_1, z_1)$ is a point on a line l, then **the plane through P_1 with normal line l** is the set of points on all lines l' that are orthogonal to l at P_1. To obtain an equivalent definition using vectors, choose another point P_2 on l and consider the vector $\overrightarrow{P_1 P_2}$ as in Figure 10.50(b). **The plane through P_1 with normal vector $\overrightarrow{P_1 P_2}$** is defined as the set of all points $P(x, y, z)$ such that $\overrightarrow{P_1 P}$ is orthogonal to $\overrightarrow{P_1 P_2}$.

Figure 10.50

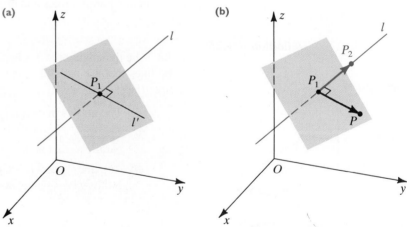

(a) (b)

By Theorem (10.21), P is on the plane if and only if $\overrightarrow{P_1 P_2} \cdot \overrightarrow{P_1 P} = 0$. If $\mathbf{a} = \langle a_1, a_2, a_3 \rangle$ corresponds to $\overrightarrow{P_1 P_2}$, this is equivalent to

$$\langle a_1, a_2, a_3 \rangle \cdot \langle x - x_1, y - y_1, z - z_1 \rangle = 0.$$

Applying the definition of dot product (10.16) gives us the next result.

Theorem 10.36

An equation of the plane through $P_1(x_1, y_1, z_1)$ with normal vector $\mathbf{a} = \langle a_1, a_2, a_3 \rangle$ is

$$a_1(x - x_1) + a_2(y - y_1) + a_3(z - z_1) = 0.$$

EXAMPLE ▪ 5 Find an equation of the plane through the point $(5, -2, 4)$ with normal vector $\mathbf{a} = \langle 1, 2, 3 \rangle$.

SOLUTION Using Theorem (10.36), we obtain

$$1(x - 5) + 2(y + 2) + 3(z - 4) = 0,$$

or

$$x + 2y + 3z - 13 = 0.$$

EXAMPLE ▪ 6 Find an equation of the plane determined by the points $P(4, -3, 1)$, $Q(6, -4, 7)$, and $R(1, 2, 2)$.

SOLUTION The points P, Q, and R determine a plane that contains the triangle shown in Figure 10.51. Vectors $\mathbf{a}$ and $\mathbf{b}$ corresponding to $\overrightarrow{PQ}$ and $\overrightarrow{PR}$ are

$$\mathbf{a} = \langle 2, -1, 6 \rangle \quad \text{and} \quad \mathbf{b} = \langle -3, 5, 1 \rangle.$$

The vector $\mathbf{a} \times \mathbf{b}$ is normal to the plane determined by P, Q, and R. From Example 1 of Section 10.4,

$$\mathbf{a} \times \mathbf{b} = -31\mathbf{i} - 20\mathbf{j} + 7\mathbf{k}.$$

Figure 10.51

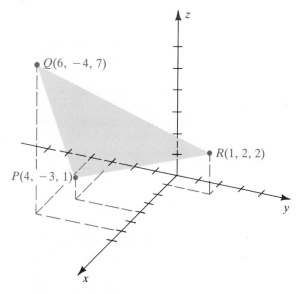

Using Theorem (10.36) with $P_1 = P(4, -3, 1)$ and normal vector $\mathbf{a} \times \mathbf{b}$ gives us the equation

$$-31(x - 4) - 20(y + 3) + 7(z - 1) = 0,$$

or

$$-31x - 20y + 7z + 57 = 0.$$

The equation of the plane in Theorem (10.36) may be written in the form

$$ax + by + cz + d = 0,$$

where $a = a_1, b = a_2, c = a_3$, and $d = -a_1 x_1 - a_2 y_1 - a_3 z_1$. Conversely, given $ax + by + cz + d = 0$, with a, b, and c not all zero, we may choose numbers x_1, y_1, and z_1 such that $ax_1 + by_1 + cz_1 + d = 0$. It follows that $d = -ax_1 - by_1 - cz_1$, and hence

$$ax + by + cz - ax_1 - by_1 - cz_1 = 0,$$

or $$a(x - x_1) + b(y - y_1) + c(z - z_1) = 0.$$

According to Theorem (10.36), the graph of the last equation is a plane through $P(x_1, y_1, z_1)$ with normal vector $\langle a, b, c \rangle$. An equation of the form $ax + by + cz + d = 0$, with a, b, and c not all zero, is called a **linear equation in three variables** x, y, and z. We have proved the following theorem.

Theorem 10.37

> The graph of every linear equation $ax + by + cz + d = 0$ is a plane with normal vector $\langle a, b, c \rangle$.

For simplicity, we use the phrase *the plane $ax + by + cz + d = 0$* instead of *the plane that has equation $ax + by + cz + d = 0$* which is more accurate.

To sketch the graph of a linear equation, we find, if possible, the **trace** of the graph in each coordinate plane—that is, the line in which the graph intersects the coordinate plane. To find the trace in the xy-plane, we substitute 0 for z, since this will lead to all points of the graph that lie on the xy-plane. Similarly, to find the trace in the yz-plane or the xz-plane, we let $x = 0$ or $y = 0$, respectively, in the equation $ax + by + cz + d = 0$.

Figure 10.52

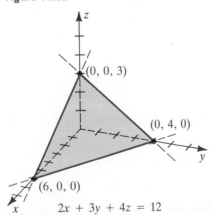

$2x + 3y + 4z = 12$

EXAMPLE ▪ 7   Sketch the graph of the equation $2x + 3y + 4z = 12$.

SOLUTION   We can easily find three points on the plane—namely, the points of intersection of the plane with the coordinate axes. Substituting 0 for both y and z in the equation, we obtain $2x = 12$, or $x = 6$. Thus, the point $(6, 0, 0)$ is on the graph. As in two dimensions, 6 is called the x-*intercept* of the graph. Similarly, substitution of 0 for x and z gives us the y-*intercept* 4, and hence the point $(0, 4, 0)$ is on the graph. The point $(0, 0, 3)$ (or z-*intercept* 3) is obtained by substituting 0 for x and y. The trace in the xy-plane is found by substituting 0 for z in the given equation. This leads to $2x + 3y = 12$, which has as its graph in the xy-plane a line with x-intercept 6 and y-intercept 4. This trace and the traces of the graph in the xz- and yz-planes are illustrated in Figure 10.52.

Definition 10.38

Two planes with normal vectors **a** and **b** are

(i) **parallel** if **a** and **b** are parallel

(ii) **orthogonal** if **a** and **b** are orthogonal

EXAMPLE ■ 8   Prove that the planes $2x - 3y - z - 5 = 0$ and $-6x + 9y + 3z + 2 = 0$ are parallel.

SOLUTION   By Theorem (10.37), the planes have normal vectors $\mathbf{a} = \langle 2, -3, -1 \rangle$ and $\mathbf{b} = \langle -6, 9, 3 \rangle$. Since $\mathbf{b} = -3\mathbf{a}$, the vectors **a** and **b** are parallel and hence, by Definition (10.38), so are the planes.

EXAMPLE ■ 9   Find an equation of the plane through $P(5, -2, 4)$ that is parallel to the plane $3x + y - 6z + 8 = 0$.

SOLUTION   By Theorem (10.37), the plane $3x + y - 6z + 8 = 0$ has a normal vector $\mathbf{a} = \langle 3, 1, -6 \rangle$. Hence, an equation for a parallel plane has the form

$$3x + y - 6z + d = 0$$

for some real number d. If $P(5, -2, 4)$ is on this plane, then its coordinates must satisfy the equation—that is,

$$3(5) + 1(-2) - 6(4) + d = 0, \quad \text{or} \quad d = 11.$$

This gives us the equation $3x + y - 6z + 11 = 0$.

The vector $\mathbf{i} = \langle 1, 0, 0 \rangle$ is a normal vector for the yz-plane. A plane that has an equation of the form $x - x_1 = 0$ (or $x = x_1$) also has normal vector **i** and hence is parallel to the yz-plane (and orthogonal to both the xy- and xz-planes). A portion of the graph $x = a$ that lies in the first octant is sketched in Figure 10.53(a). Similarly, the graph of $y = b$ is a

Figure 10.53

(a) $x = a$

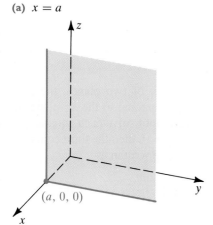

$(a, 0, 0)$

(b) $y = b$

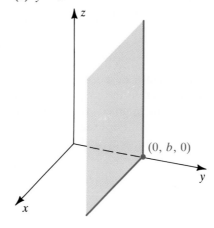

$(0, b, 0)$

(c) $z = c$

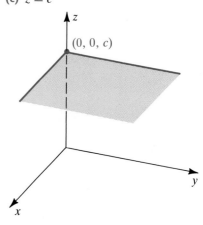

$(0, 0, c)$

plane parallel to the xz-plane with y-intercept b, and the graph of $z = c$ is a plane parallel to the xy-plane with z-intercept c (see (b) and (c) of Figure 10.53).

A plane with an equation of the form $by + cz + d = 0$ has normal vector $\mathbf{a} = \langle 0, b, c \rangle$ and is orthogonal to the yz-plane, since $\mathbf{a} \cdot \mathbf{i} = 0$. Similarly, graphs of $ax + by + d = 0$ and $ax + cz + d = 0$ are planes that are orthogonal to the xy-plane and xz-plane, respectively.

EXAMPLE ■ 10   Sketch the plane $3x + 5z = 10$.

SOLUTION   The plane is orthogonal to the xz-plane and has x-intercept $\frac{10}{3}$ and z-intercept 2. Note that the trace in the yz-plane has equation $5z = 10$ and hence is a line parallel to the y-axis with z-intercept 2. Similarly, the trace in the xy-plane has equation $3x = 10$ and is a line parallel to the y-axis with x-intercept $\frac{10}{3}$. A portion of the graph showing traces in the three coordinate planes is sketched in Figure 10.54.

Figure 10.54

$3x + 5z = 10$

Lines may be described as intersections of planes. If a line l is given parametrically as in Theorem (10.34) and if a_1, a_2, a_3 are different from zero, we may solve each equation for t, obtaining

$$t = \frac{x - x_1}{a_1}, \qquad t = \frac{y - y_1}{a_2}, \qquad t = \frac{z - z_1}{a_3}.$$

It follows that a point $P(x, y, z)$ is on l if and only if the following equations, called *the symmetric form for l*, are satisfied.

Symmetric Form for a Line   10.39

$$\frac{x - x_1}{a_1} = \frac{y - y_1}{a_2} = \frac{z - z_1}{a_3}$$

The symmetric form for a line is not unique, since in (10.39) we may use any three numbers b_1, b_2, b_3 that are proportional to a_1, a_2, a_3 or any point on l other than (x_1, y_1, z_1).

If, in (10.39), we take the indicated expressions in pairs, say,

$$\frac{x - x_1}{a_1} = \frac{y - y_1}{a_2} \quad \text{and} \quad \frac{x - x_1}{a_1} = \frac{z - z_1}{a_3},$$

we obtain a description of l as an intersection of two planes, the first orthogonal to the xy-plane and the second orthogonal to the xz-plane. If one of the numbers $a_1, a_2,$ or a_3 is zero, we cannot solve each equation in (10.34) for t. For example, if $a_3 = 0$ and $a_1 a_2 \neq 0$, then the third equation reduces to $z = z_1$, and a symmetric form may be written as

$$\frac{x - x_1}{a_1} = \frac{y - y_1}{a_2}, \qquad z = z_1,$$

which again expresses l as an intersection of two planes, the first orthogonal to the xy-plane and the second, $z = z_1$, parallel to the xy-plane. A similar situation exists if $a_1 = 0$ or $a_2 = 0$.

EXAMPLE ■ 11   Find a symmetric form for the line through the points $P_1(3, 1, -2)$ and $P_2(-2, 7, -4)$.

SOLUTION   As in Example 2, a parametric representation for the line is

$$x = 3 - 5t, \quad y = 1 + 6t, \quad z = -2 - 2t; \quad t \text{ in } \mathbb{R}.$$

Solving each equation for t and equating the results, we obtain the symmetric form

$$\frac{x - 3}{-5} = \frac{y - 1}{6} = \frac{z + 2}{-2}.$$

EXAMPLE ■ 12   Let l be the line of intersection of the two planes

$$2x - y + 4z = 4 \quad \text{and} \quad x + 3y - 2z = 1.$$

Find parametric equations for l.

SOLUTION   A point $P(x, y, z)$ is on l if and only if P is on each of the planes—that is, if and only if (x, y, z) is a solution of the system of equations

$$\begin{cases} 2x - y + 4z = 4 \\ x + 3y - 2z = 1 \end{cases}$$

or, equivalently,

$$\begin{cases} 2x - y = 4 - 4z \\ x + 3y = 1 + 2z \end{cases}$$

Let us regard these results as a system of two equations in x and y. By eliminating y from the system, we can express x in terms of z. Similarly, eliminating x from the system leads to a formula for y in terms of z. This procedure gives us the equivalent system

$$\begin{cases} x = \frac{13}{7} - \frac{10}{7}z \\ y = -\frac{2}{7} + \frac{8}{7}z \end{cases}$$

By letting z take on every real value t, we obtain the solutions of the system—that is, the points on l. Thus, parametric equations for l are

$$x = \frac{13}{7} - \frac{10}{7}t, \quad y = -\frac{2}{7} + \frac{8}{7}t, \quad z = t; \quad t \text{ in } \mathbb{R}.$$

In the next example, we use vector methods to find the distance from a point to a plane.

EXAMPLE ▪ 13   Find a formula for the distance h from a point $P_0(x_0, y_0, z_0)$ to the plane $ax + by + cz + d = 0$.

SOLUTION   As in Figure 10.55, let $P_1(x_1, y_1, z_1)$ be any point in the plane, and let $\mathbf{n}$ be a normal vector to the plane. The vector $\mathbf{p}$ corresponding to $\overrightarrow{P_1 P_0}$ is

$$\mathbf{p} = \langle x_0 - x_1, y_0 - y_1, z_0 - z_1 \rangle.$$

Referring to the figure, we see that

$$h = |\text{comp}_{\mathbf{n}}\, \mathbf{p}| = \left| \mathbf{p} \cdot \frac{1}{\|\mathbf{n}\|} \mathbf{n} \right|.$$

Since $\langle a, b, c \rangle$ is a normal vector to the plane, we may let

$$\frac{1}{\|\mathbf{n}\|} \mathbf{n} = \frac{1}{\sqrt{a^2 + b^2 + c^2}} \langle a, b, c \rangle.$$

Consequently,

$$h = \left| \mathbf{p} \cdot \frac{1}{\|\mathbf{n}\|} \mathbf{n} \right| = \left| \frac{a(x_0 - x_1) + b(y_0 - y_1) + c(z_0 - z_1)}{\sqrt{a^2 + b^2 + c^2}} \right|$$

$$= \frac{|(ax_0 + by_0 + cz_0) + (-ax_1 - by_1 - cz_1)|}{\sqrt{a^2 + b^2 + c^2}}.$$

Since P_1 is on the plane, (x_1, y_1, z_1) is a solution of $ax + by + cz + d = 0$, and hence $ax_1 + by_1 + cz_1 + d = 0$, or $d = -ax_1 - by_1 - cz_1$. Thus, the preceding formula for h may be written

$$h = \frac{|ax_0 + by_0 + cz_0 + d|}{\sqrt{a^2 + b^2 + c^2}}.$$

Two lines are **skew** if they are not parallel and do not intersect. In the next example, we find a formula for the shortest distance between skew lines.

EXAMPLE ▪ 14   Find a formula for the shortest distance d between two skew lines l_1 and l_2.

SOLUTION   Choose points P_1, Q_1 on l_1 and P_2, Q_2 on l_2, as illustrated in Figure 10.56. By Theorem (10.29), $\overrightarrow{P_1 Q_1} \times \overrightarrow{P_2 Q_2}$ is orthogonal to both $\overrightarrow{P_1 Q_1}$ and $\overrightarrow{P_2 Q_2}$, and hence a *unit* vector $\mathbf{n}$ orthogonal to both $\overrightarrow{P_1 Q_1}$ and $\overrightarrow{P_2 Q_2}$ is

$$\mathbf{n} = \frac{1}{\|\overrightarrow{P_1 Q_1} \times \overrightarrow{P_2 Q_2}\|} (\overrightarrow{P_1 Q_1} \times \overrightarrow{P_2 Q_2}).$$

Let us consider planes through P_1 and P_2, respectively, each having normal vector $\mathbf{n}$. These planes are parallel and contain l_1 and l_2, respec-

Figure 10.55

Figure 10.56

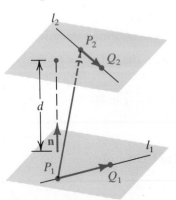

tively. The distance d between the planes is measured along a line parallel to the common normal $\mathbf{n}$, as shown in the figure. It follows that d is the shortest distance between l_1 and l_2. Moreover,

$$d = |\text{comp}_\mathbf{n} \overrightarrow{P_1 P_2}| = |\mathbf{n} \cdot \overrightarrow{P_1 P_2}|$$

$$= \frac{1}{\| \overrightarrow{P_1 Q_1} \times \overrightarrow{P_2 Q_2} \|} |(\overrightarrow{P_1 Q_1} \times \overrightarrow{P_2 Q_2}) \cdot \overrightarrow{P_1 P_2}|.$$

We can combine the vector representation of lines, calculus methods for finding slopes of tangent lines, and a graphing calculator to approximate the point of intersection of two curves and the angles between tangent lines to the curves, as demonstrated in the next example.

EXAMPLE ▪ 15 Let $f(x) = \cos x^2$ and $g(x) = (2 + x^2)/4$.

(a) Plot the graphs of f and g on the same coordinate plane for $0 \le x \le 1$.

(b) Estimate the coordinates of the point P of intersection of the curves.

(c) Find parametric representations for the tangent lines to f and g at P.

(d) Approximate the angles between the tangent lines to the graphs at P.

SOLUTION

Figure 10.57
$0 \le x \le 1, 0 \le y \le 1$

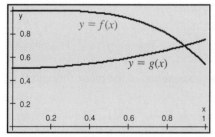

(a) The graphs of f and g are shown in Figure 10.57.

(b) Using the trace operation or Newton's method, we find that the point P of intersection is approximately $(0.8925, 0.6991)$.

(c) From $f(x) = \cos x^2$, we find that $f'(x) = -2x \sin x^2$, so the slope of the tangent line to the graph of f at P is approximately $f'(0.8925) \approx -1.276$. From Example 4 of Section 9.1, we can write parametric equations for the tangent line as

$$x = 0.8925 + 1t, \quad y = 0.6991 + (-1.276)t.$$

This line is parallel to the vector $\mathbf{a} = \langle 1, -1.276 \rangle$.

Similarly, from $g(x) = (2 + x^2)/4$, we have $g'(x) = x/2$, so the slope of the tangent line to the graph of g at P is approximately $g'(0.8925) \approx 0.4463$. We can write parametric equations for this tangent line as

$$x = 0.8925 + 1t, \quad y = 0.6991 + 0.4463t.$$

This line is parallel to the vector $\mathbf{b} = \langle 1, 0.4463 \rangle$.

(d) The acute angle between the tangent lines is equal to the acute angle between the vectors $\mathbf{a}$ and $\mathbf{b}$. Since $\mathbf{a} \cdot \mathbf{b} = \|\mathbf{a}\| \|\mathbf{b}\| \cos \theta$, where θ is the acute angle between the tangent lines, we have

$$\theta = \cos^{-1}\left(\frac{\mathbf{a} \cdot \mathbf{b}}{\|\mathbf{a}\| \|\mathbf{b}\|}\right) \approx \cos^{-1}\left[\frac{0.4305}{(1.621)(1.095)}\right] \approx \cos^{-1}(0.2425) \approx 76°.$$

Thus the angles between the tangent lines are approximately $76°$ and $104°$.

EXERCISES 10.5

Exer. 1–4: Find parametric equations for the line through P parallel to a.

1 $P(4, 2, -3)$; $\mathbf{a} = \langle \frac{1}{3}, 2, \frac{1}{2} \rangle$

2 $P(5, 0, -2)$; $\mathbf{a} = \langle -1, -4, 1 \rangle$

3 $P(0, 0, 0)$; $\mathbf{a} = \mathbf{j}$

4 $P(1, 2, 3)$; $\mathbf{a} = \mathbf{i} + 2\mathbf{j} + 3\mathbf{k}$

Exer. 5–8: Find parametric equations for the line through P_1 and P_2. Determine (if possible) the points at which the line intersects each of the coordinate planes.

5 $P_1(5, -2, 4)$, $P_2(2, 6, 1)$

6 $P_1(-3, 1, -1)$, $P_2(7, 11, -8)$

7 $P_1(2, 0, 5)$, $P_2(-6, 0, 3)$

8 $P_1(2, -2, 4)$, $P_2(2, -2, -3)$

9 If l has parametric equations $x = 5 - 3t$, $y = -2 + t$, $z = 1 + 9t$, find parametric equations for the line through $P(-6, 4, -3)$ that is parallel to l.

10 Find parametric equations for the line through the point $P(4, -1, 0)$ that is parallel to the line through the points $P_1(-3, 9, -2)$ and $P_2(5, 7, -3)$.

Exer. 11–14: Determine whether the two lines intersect, and if so, find the point of intersection.

11 $x = 1 + 2t$, $y = 1 - 4t$, $z = 5 - t$
    $x = 4 - v$, $y = -1 + 6v$, $z = 4 + v$

12 $x = 1 - 6t$, $y = 3 + 2t$, $z = 1 - 2t$
    $x = 2 + 2v$, $y = 6 + v$, $z = 2 + v$

13 $x = 3 + t$, $y = 2 - 4t$, $z = t$
    $x = 4 - v$, $y = 3 + v$, $z = -2 + 3v$

14 $x = 2 - 5t$, $y = 6 + 2t$, $z = -3 - 2t$
    $x = 4 - 3v$, $y = 7 + 5v$, $z = 1 + 4v$

Exer. 15–18: Equations for two lines l_1 and l_2 are given. Find the angles between l_1 and l_2.

15 $x = 7 - 2t$, $y = 4 + 3t$, $z = 5t$
    $x = -1 + 4t$, $y = 3 + 4t$, $z = 1 + t$

16 $x = 5 + 3t$, $y = 4 - t$, $z = 3 + 2t$
    $x = -t$, $y = 1 - 2t$, $z = 3 + t$

17 $\dfrac{x - 1}{-3} = \dfrac{y + 2}{8} = \dfrac{z}{-3}$; $\dfrac{x + 2}{10} = \dfrac{y}{10} = \dfrac{z - 4}{-7}$

18 $\dfrac{x}{3} = \dfrac{y - 2}{3} = z - 1$; $\dfrac{x + 5}{4} = \dfrac{y - 1}{-3} = \dfrac{z + 7}{-9}$

Exer. 19–26: Find an equation of the plane that satisfies the stated conditions.

19 Through $P(6, -7, 4)$ and parallel to
   (a) the xy-plane   (b) the yz-plane   (c) the xz-plane

20 Through $P(-2, 5, -8)$ with normal vector
   (a) $\mathbf{i}$ (b) $\mathbf{j}$ (c) $\mathbf{k}$

21 Through $P(-11, 4, -2)$ with normal vector $\mathbf{a} = 6\mathbf{i} - 5\mathbf{j} - \mathbf{k}$

22 Through $P(4, 2, -9)$ with normal vector $\overrightarrow{OP}$

23 Through $P(2, 5, -6)$ and parallel to the plane $3x - y + 2z = 10$

24 Through the origin and parallel to the plane $x - 6y + 4z = 7$

25 Through $P(-4, 1, 6)$ and having the same trace in the xz-plane as the plane $x + 4y - 5z = 8$

26 Through the origin and the points $P(0, 2, 5)$ and $Q(1, 4, 0)$

Exer. 27–28: Find an equation of the plane through P, Q, and R.

27 $P(1, 1, 3)$, $Q(-1, 3, 2)$, $R(1, -1, 2)$

28 $P(3, 2, 1)$, $Q(-1, 1, -2)$, $R(3, -4, 1)$

Exer. 29–36: Sketch the graph of the equation in an xyz-coordinate system.

29 (a) $x = 3$ (b) $y = -2$ (c) $z = 5$

30 (a) $x = -4$ (b) $y = 0$ (c) $z = -\frac{2}{3}$

31 $2x + y - 6 = 0$ 32 $3x - 2z - 24 = 0$

33 $2y - 3z - 9 = 0$ 34 $5x + y - 4z + 20 = 0$

35 $2x - y + 5z + 10 = 0$ 36 $x + y + z = 0$

Exer. 37–40: Find an equation of the plane.

37

38

39

40

Exer. 41–42: Find an equation of the plane through P that is parallel to the given plane.

41 $P(1, 2, -3)$; $\quad 4x - y + 3z - 7 = 0$

42 $P(3, -2, 4)$; $\quad -2x + 3y - z + 5 = 0$

Exer. 43–46: Find a symmetric form for the line through P_1 and P_2.

43 $P_1(5, -2, 4)$, $\quad P_2(2, 6, 1)$

44 $P_1(-3, 1, -1)$, $\quad P_2(7, 11, -8)$

45 $P_1(4, 2, -3)$, $\quad P_2(-3, 2, 5)$

46 $P_1(5, -7, 4)$, $\quad P_2(-2, -1, 4)$

Exer. 47–50: Find parametric equations for the line of intersection of the two planes.

47 $x + 2y - 9z = 7, \qquad 2x - 3y + 17z = 0$

48 $2x + 5y + 16z = 13, \quad -x - 2y - 6z = -5$

49 $-2x + 3y + 9z = 12, \quad x - 2y - 5z = -8$

50 $5x - y - 12z = 15, \qquad 2x + 3y + 2z = 6$

Exer. 51–52: Refer to Example 13. Find the distance from P to the plane.

51 $P(1, -1, 2)$; $\quad 3x - 7y + z - 5 = 0$

52 $P(3, 1, -2)$; $\quad 2x + 4y - 5z + 1 = 0$

Exer. 53–54: Show that the two planes are parallel and find the distance between the planes.

53 $4x - 2y + 6z = 3, \qquad -6x + 3y - 9z = 4$

54 $3x + 12y - 6z = -2, \quad 5x + 20y - 10z = 7$

Exer. 55–56: Refer to Example 14. Let l_1 be the line through A and B, and let l_2 be the line through C and D. Find the shortest distance between l_1 and l_2.

55 $A(1, -2, 3)$, $B(2, 0, 5)$; $C(4, 1, -1)$, $\quad D(-2, 3, 4)$

56 $A(1, 3, 0)$, $\quad B(0, 4, 5)$; $C(-2, -1, 2)$, $D(5, 1, 0)$

Exer. 57–58: Find an equation of the plane that contains the point P and the line.

57 $P(5, 0, 2)$; $\quad x = 3t + 1, \ y = -2t + 4, \ z = t - 3$

58 $P(4, -3, 0)$; $x = t + 5, \quad y = 2t - 1, \quad z = -t + 7$

Exer. 59–60: Use a dot product to find the distance from A to the line through B and C.

59 $A(2, -6, 1)$; $B(3, 4, -2)$, $\quad C(7, -1, 5)$

60 $A(1, 5, 0)$; $\quad B(-2, 1, -4)$, $\quad C(0, -3, 2)$

Exer. 61–62: Find the distance from the point P to the line.

61 $P(2, 1, -2)$; $x = 3 - 2t$, $y = -4 + 3t$, $z = 1 + 2t$

62 $P(3, 1, -1)$; $x = 1 + 4t$, $y = 3 - t$, $z = 3t$

Exer. 63–64: If a plane has nonzero x-, y-, and z-intercepts a, b, and c, respectively, then its *intercept form* is

$$\frac{x}{a} + \frac{y}{b} + \frac{z}{c} = 1.$$

Find the intercept form for the given plane.

63 $10x - 15y + 6z = 30$ **64** $12x + 15y - 20z = 60$

Exer. 65–66: Find an equation for the plane of the form $Ax + By + Cz = D$.

65

66

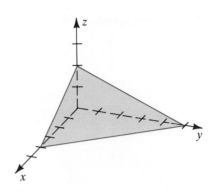

Exer. 67–68: Graph f and g on the same coordinate plane for $-2 \le x \le 2$. (a) Estimate the coordinates of their point P of intersection. (b) Approximate the angles between the tangent lines to the graphs at P.

67 $f(x) = \sin(x^2)$, $g(x) = \cos x - x$

68 $f(x) = 1 - 3x + x^3$, $g(x) = x^5 + \frac{1}{2}$

10.6 SURFACES

In this section, we examine some techniques for obtaining accurate sketches of surfaces in space that are described by equations in x, y, and z. To sketch a surface with pencil and paper, we usually choose the coordinate axes as in Figure 10.21, regarding the y- and z-axes as lying in the plane of the paper and the x-axis as projecting out from the paper. This technique is illustrated in Figures 10.58–10.68 and the charts in this section. A disadvantage of choosing coordinate axes in this way is that when a specific equation is graphed, the shape of the resulting surface may seem distorted. For example, circular cross-sections may appear to be elliptical and vice versa. For this reason, graphs in three dimensions that are illustrated later in the text are computer-generated, with axes and units of distance chosen to provide an undistorted view of a surface. In such an *axonometric system*, none of the coordinate planes coincides with the plane of the paper. After you have studied this section carefully and learned how to identify a surface from its equation, you should strive to become proficient in using any choice of coordinate axes.

In the preceding section, we defined the *trace* of the graph of a linear equation in a coordinate plane. More generally, if S is any surface (that is, the graph of an equation in x, y, and z), then the **trace** of S in a plane is the intersection of S and the plane. To sketch a surface, we make considerable use of traces. Of special importance are traces in the coordinate planes. These three traces are the **xy-trace**, the **yz-trace**, and the **xz-trace**. It may also be convenient to find traces in planes that are *parallel* to the coordinate planes. The following charts show how to find equations of traces from an equation of a surface. The indicated sketch of the trace is only one possibility. In typical problems, the trace may be a conic section, a line, or some other curve. Sometimes there is *no* trace—that is, the surface may not intersect a given plane.

Trace	To find equation of trace	Sketch of trace
xy-trace	Let $z = 0$	
yz-trace	Let $x = 0$	
xz-trace	Let $y = 0$	

Trace	To find equation of trace	Sketch of trace
On $z = k$	Let $z = k$	![z = k trace]
On $x = k$	Let $x = k$	![x = k trace]
On $y = k$	Let $y = k$	![y = k trace]

Figure 10.58
$z = x^2 + y^2$

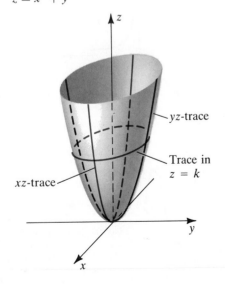

EXAMPLE ■ 1   Sketch the graph of $z = x^2 + y^2$.

SOLUTION   The graph is a surface in an xyz-coordinate system. The chart on the following page indicates pertinent traces on planes.

We next sketch these traces using only *one* set of coordinate axes. Noting that the radius $\sqrt{k}$ of the circular trace on $z = k$ increases as z increases leads to the sketch in Figure 10.58.

Traces on the planes $x = k$ and $y = k$ are parabolas. To illustrate, the trace on the plane $y = 1$ is the parabola $z = x^2 + 1$, and the trace on the plane $x = 2$ is the parabola $z = y^2 + 4$. We shall not sketch these, since the circular traces on $z = k$ are sufficient for indicating the graph. The surface in this example may be regarded as having been generated by revolving the graph of the parabola $z = y^2$ in the yz-plane about the z-axis. This surface is called a **circular paraboloid**, or **paraboloid of revolution**.

Trace	Equation of trace	Description of trace	Sketch of trace
xy-trace	$0 = x^2 + y^2$	Origin	
yz-trace	$z = y^2$	Parabola	
xz-trace	$z = x^2$	Parabola	
On $z = k$	$k = x^2 + y^2$, or $x^2 + y^2 = k$	Circle, point, or no graph	

You may have previously regarded a cylinder as an object that has the shape of a tin can with circular cross sections. The next definition indicates that cylinders may have a variety of shapes.

Definition 10.40

> Let C be a curve in a plane, and let l be a line that is not in a parallel plane. The set of points on all lines that are parallel to l and intersect C is a **cylinder**.

Two examples of cylinders are sketched in Figure 10.59. The curve C in the plane is called a **directrix** for the cylinder, and each line through C parallel to l is a **ruling** of the cylinder. The cylinder in Figure 10.59(a) is a **right circular cylinder**, obtained if C is a circle in a plane and l is a line perpendicular to the plane. Although we have cut off the cylinder in the figure, the rulings extend indefinitely. As illustrated in Figure 10.59(b), the *directrix C is not necessarily a closed curve.*

Figure 10.59

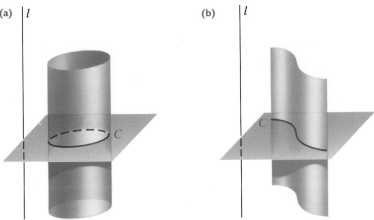

(a)   (b)

Figure 10.60
$y = f(x)$

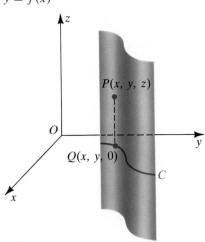

Let us consider a cylinder in which the directrix C is on the xy-plane and has equation $y = f(x)$ for some function f. Suppose also that the rulings are parallel to the z-axis. As in Figure 10.60, a point $P(x, y, z)$ is on the cylinder if and only if $Q(x, y, 0)$ is on C—that is, if and only if the first two coordinates x and y of P satisfy the equation $y = f(x)$. Thus, an equation of the cylinder is $y = f(x)$, which is the same as the equation of the directrix in the xy-plane.

EXAMPLE • 2   Sketch the graph of $(x^2/4) + (y^2/9) = 1$ in three dimensions.

SOLUTION   From our previous remarks, the graph is a cylinder with rulings parallel to the z-axis. We begin by sketching the graph of the

Figure 10.61
$$\frac{x^2}{4} + \frac{y^2}{9} = 1$$

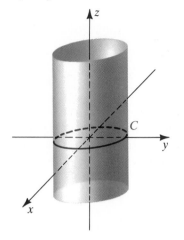

Figure 10.61
$$\frac{x^2}{4} + \frac{y^2}{9} = 1$$

Figure 10.62
$$y^2 = 9 - z$$

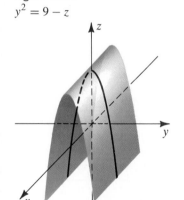

Figure 10.63
$$z = \sin x$$

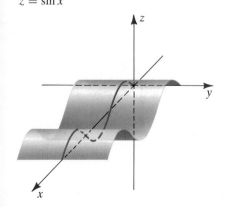

equation $(x^2/4) + (y^2/9) = 1$ in the xy-plane. This ellipse is a directrix C for the cylinder. All traces in planes parallel to the xy-plane are ellipses congruent to this directrix. When sketching a cylinder, we usually draw at least two such traces, as indicated in Figure 10.61. Since the directrix is an ellipse, we call this surface an **elliptic cylinder**.

The graph of an equation that contains only the variables y and z is a cylinder with rulings parallel to the x-axis and whose trace (directrix) in the yz-plane is the graph of the given equation. Similarly, the graph of an equation that does not contain the variable y is a cylinder with rulings parallel to the y-axis and whose directrix is the graph of the given equation in the xz-plane.

EXAMPLE 3 Sketch the graphs of the following equations in three dimensions:

(a) $y^2 = 9 - z$ **(b)** $z = \sin x$

SOLUTION

(a) The graph of $y^2 = 9 - z$ is a cylinder with rulings parallel to the x-axis. A directrix in the yz-plane is the parabola $y^2 = 9 - z$. Part of the graph (a **parabolic cylinder**) is sketched in Figure 10.62.

(b) The graph of $z = \sin x$ is a cylinder with rulings parallel to the y-axis and whose directrix in the xz-plane is the sine curve $z = \sin x$. A portion of the graph is sketched in Figure 10.63.

From our earlier work, we know that, in two dimensions, the graph of any second-degree equation in x and y,

$$Ax^2 + Bxy + Cy^2 + Dx + Ey + F = 0,$$

is a conic section (except for degenerate cases). In three dimensions, the graph of a second-degree equation in x, y, and z,

$$Ax^2 + By^2 + Cz^2 + Dxy + Exz + Fyx + Gx + Hy + Iz + J = 0,$$

is a **quadric surface** (except for degenerate cases). For simplicity, we shall limit our discussion to the case in which the coefficients D, E, F, G, H, and I are all zero. More general equations can be reduced to this case by using translations and possibly rotations of axes.

There are three types of quadric surfaces: *ellipsoids*, *hyperboloids*, and *paraboloids*. The names are derived from the fact that traces in planes parallel to coordinate planes are usually ellipses, hyperbolas, and parabolas, respectively. In the remainder of this section, unless noted otherwise, a, b, and c denote positive real numbers. The graph of the following equation is an ellipsoid. The chart that follows indicates traces of this surface in the coordinate planes.

Ellipsoid 10.41

$$\frac{x^2}{a^2} + \frac{y^2}{b^2} + \frac{z^2}{c^2} = 1$$

Trace	Equation of trace	Description of trace	Sketch of trace
xy-trace	$\dfrac{x^2}{a^2} + \dfrac{y^2}{b^2} = 1$	Ellipse	
yz-trace	$\dfrac{y^2}{b^2} + \dfrac{z^2}{c^2} = 1$	Ellipse	
xz-trace	$\dfrac{x^2}{a^2} + \dfrac{z^2}{c^2} = 1$	Ellipse	

Figure 10.64

$$\frac{x^2}{a^2} + \frac{y^2}{b^2} + \frac{z^2}{c^2} = 1$$

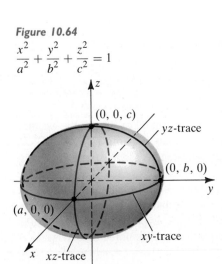

We next sketch these traces using only *one* set of coordinate axes, as illustrated in Figure 10.64. After you become familiar with quadric surfaces, these three traces will suffice to sketch an accurate graph. However, to complete our discussion, let us consider traces on planes parallel to the coordinate planes.

To find the trace on a plane $z = k$ parallel to the xy-plane, we let $z = k$ in the equation of the ellipsoid, obtaining

$$\frac{x^2}{a^2} + \frac{y^2}{b^2} = 1 - \frac{k^2}{c^2}.$$

If $|k| > c$, then $1 - (k^2/c^2) < 0$ and there is no graph. Thus, the graph of (10.41) lies between the planes $z = -k$ and $z = k$. If $|k| < c$, then $1 - (k^2/c^2) > 0$ and hence the trace in the plane $z = k$ is an ellipse, as illustrated in Figure 10.65. Viewable portions of other traces in planes parallel to the xy-plane are also shown in the figure.

Figure 10.65

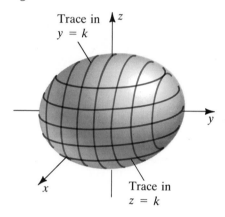

Trace in $y = k$

Trace in $z = k$

The trace on a plane $y = k$ parallel to the xz-plane is

$$\frac{x^2}{a^2} + \frac{z^2}{c^2} = 1 - \frac{k^2}{b^2}.$$

Viewable portions of several of these traces (ellipses) for $-b < k < b$ are shown in Figure 10.65.

Finally, letting $x = k$ for $-a < k < a$, we see that the trace in a plane parallel to the yz-plane is an ellipse. (These ellipses are not shown in Figure 10.65.)

Note that if $a = b = c$, then the graph of (10.41) is a sphere of radius a with center at the origin.

The graph of the following equation is a *hyperboloid of one sheet*. The chart that follows indicates traces in the coordinate planes.

Hyperboloid of One Sheet   10.42

$$\frac{x^2}{a^2} + \frac{y^2}{b^2} - \frac{z^2}{c^2} = 1$$

Trace	Equation of trace	Description of trace	Sketch of trace
xy-trace	$\dfrac{x^2}{a^2} + \dfrac{y^2}{b^2} = 1$	Ellipse	
yz-trace	$\dfrac{y^2}{b^2} - \dfrac{z^2}{c^2} = 1$	Hyperbola	
xz-trace	$\dfrac{x^2}{a^2} - \dfrac{z^2}{c^2} = 1$	Hyperbola	

Figure 10.66

$$\frac{x^2}{a^2} + \frac{y^2}{b^2} - \frac{z^2}{c^2} = 1$$

Trace in $z = k$

Trace in $z = -k$

These traces are sketched on one set of coordinate axes in Figure 10.66. The z-axis is the **axis of the hyperboloid**.

The trace in a plane $z = k$ parallel to the xy-plane is given by

$$\frac{x^2}{a^2} + \frac{y^2}{b^2} = 1 + \frac{k^2}{c^2}$$

and hence is an ellipse. As k increases through positive values, the lengths of the axes of the ellipse increase.

Traces in planes $x = k$ or $y = k$ (that is, in planes parallel to the yz-plane or xz-plane, respectively) are hyperbolas (verify this fact).

The graphs of

$$\frac{x^2}{a^2} - \frac{y^2}{b^2} + \frac{z^2}{c^2} = 1 \quad \text{and} \quad -\frac{x^2}{a^2} + \frac{y^2}{b^2} + \frac{z^2}{c^2} = 1$$

are also hyperboloids of one sheet; however, in the first case the axis of the hyperboloid is the y-axis, and in the second case the axis of the hyperboloid coincides with the x-axis. Thus, the term that is negative in these equations indicates the axis of the hyperboloid.

The graph of the following equation is a different type of hyperboloid, a *hyperboloid of two sheets*.

Hyperboloid of Two Sheets   10.43

$$-\frac{x^2}{a^2} - \frac{y^2}{b^2} + \frac{z^2}{c^2} = 1$$

Trace	Equation of trace	Description of trace	Sketch of trace
xy-trace	$-\dfrac{x^2}{a^2} - \dfrac{y^2}{b^2} = 1$	None	No graph
yz-trace	$-\dfrac{y^2}{b^2} + \dfrac{z^2}{c^2} = 1$	Hyperbola	
xz-trace	$-\dfrac{x^2}{a^2} + \dfrac{z^2}{c^2} = 1$	Hyperbola	

Figure 10.67

$$-\frac{x^2}{a^2} - \frac{y^2}{b^2} + \frac{z^2}{c^2} = 1$$

Trace in $z = k$

Trace in $z = -k$

These traces are sketched on one set of coordinate axes in Figure 10.67, where we have also shown traces on the planes $z = k$ and $z = -k$ for $k > c$. (Verify that these traces are ellipses if $a \neq b$.) Traces in planes parallel to the yz-plane or xz-plane are hyperbolas. The z-axis is the axis of the hyperboloid.

By using minus signs on different terms, we can obtain a hyperboloid of two sheets whose axis is the x-axis or the y-axis. (Which terms should be negative?)

A (double-napped) *cone* may be thought of as the degenerate hyperboloid obtained by replacing the number 1 in (10.42) or (10.43) by 0. We thus obtain the following equation.

Cone   10.44

$$\frac{x^2}{a^2} + \frac{y^2}{b^2} - \frac{z^2}{c^2} = 0$$

As indicated in the following chart, the yz-trace and the xz-trace are the lines $z = \pm(c/b)y$ and $z = \pm(c/a)x$, respectively.

Trace	Equation of trace	Description of trace	Sketch of trace
xy-trace	$\dfrac{x^2}{a^2} + \dfrac{y^2}{b^2} = 0$	Origin	
yz-trace	$\dfrac{y^2}{b^2} - \dfrac{z^2}{c^2} = 0$	Two intersecting lines	
xz-trace	$\dfrac{x^2}{a^2} - \dfrac{z^2}{c^2} = 0$	Two intersecting lines	

909

The trace in a plane $z = k$ parallel to the xy-plane has the equation

$$\frac{x^2}{a^2} + \frac{y^2}{b^2} = \frac{k^2}{c^2}$$

and hence is an ellipse. Traces in planes parallel to the other coordinate axes are hyperbolas. The graph is sketched in Figure 10.68. The z-axis is the axis of the cone.

Figure 10.68

$$\frac{x^2}{a^2} + \frac{y^2}{b^2} - \frac{z^2}{c^2} = 0$$

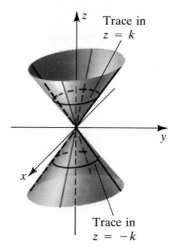

Trace in $z = k$

Trace in $z = -k$

The graph of the next equation, for c either positive or negative, is a *paraboloid*.

Paraboloid 10.45

$$\frac{x^2}{a^2} + \frac{y^2}{b^2} = cz$$

Example 1 is a special case of (10.45) with $a = b = c = 1$. If $c > 0$, then the graph of (10.45) is similar to that shown in Figure 10.58, except that if $a \neq b$, then traces in planes $z = k$ parallel to the xy-plane are ellipses instead of circles. If $c < 0$, then the paraboloid *opens downward*. The z-axis is the **axis of the paraboloid**. The graphs of the equations

$$\frac{x^2}{a^2} + \frac{z^2}{b^2} = cy \quad \text{and} \quad \frac{y^2}{a^2} + \frac{z^2}{b^2} = cx$$

are paraboloids whose axes are the y-axis and x-axis, respectively. If we change the sign of one of the terms on the left in (10.45), we obtain a *hyperbolic paraboloid*.

Hyperbolic Paraboloid **10.46**

$$\frac{y^2}{a^2} - \frac{x^2}{b^2} = cz$$

Figure 10.69

$\dfrac{y^2}{a^2} - \dfrac{x^2}{b^2} = cz$

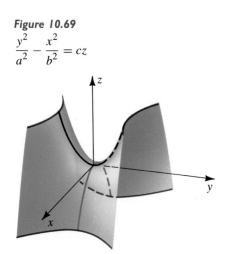

A typical sketch of this saddle-shaped surface for the case $c > 0$ is shown in Figure 10.69. Variations are obtained by interchanging x, y, and z in (10.46). A hyperbolic paraboloid is the most difficult quadric surface to visualize, and it takes considerable practice to become proficient at sketching the graph. Traces of the type shown in Figure 10.69 can be helpful. Note that the trace in the xy-plane has equation

$$\frac{y^2}{a^2} - \frac{x^2}{b^2} = 0, \quad \text{or} \quad y = \pm\frac{a}{b}x,$$

and hence is a pair of intersecting lines through the origin. This trace is not shown in Figure 10.69.

The next two examples are special cases of quadric surfaces.

EXAMPLE 4 Sketch the graph of $16x^2 - 9y^2 + 36z^2 = 144$, and identify the surface.

SOLUTION Dividing both sides of the equation by 144 leads to

$$\frac{x^2}{9} - \frac{y^2}{16} + \frac{z^2}{4} = 1,$$

where the negative y-term indicates that the axis of this hyperboloid of one sheet is the y-axis. Traces in coordinate planes are as follows.

Figure 10.70

$16x^2 - 9y^2 + 36z^2 = 144$

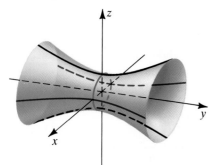

Trace	Equation of trace	Description of trace
xy-plane	$\dfrac{x^2}{9} - \dfrac{y^2}{16} = 1$	Hyperbola
yz-plane	$\dfrac{z^2}{4} - \dfrac{y^2}{16} = 1$	Hyperbola
xz-plane	$\dfrac{x^2}{9} + \dfrac{z^2}{4} = 1$	Ellipse

The graph of the equation is sketched in Figure 10.70. Traces in planes parallel to the xz-plane are ellipses, and traces in planes parallel to the xy- or yz-planes are hyperbolas.

Figure 10.71
$y^2 + 4z^2 = x$

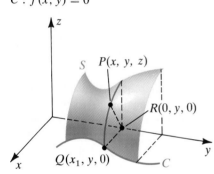

EXAMPLE ■ 5   Sketch the graph of $y^2 + 4z^2 = x$, and identify the surface.

SOLUTION   Traces are as follows.

Trace	Equation of trace	Description of trace
xy-plane	$y^2 = x$	Parabola
yz-plane	$y^2 + 4z^2 = 0$	Origin
xz-plane	$4z^2 = x$	Parabola

The trace in a plane $x = k$ parallel to the yz-plane has an equation of the form $y^2 + 4z^2 = k$, which is an ellipse if $k > 0$. Traces in planes parallel to the xz- or xy-planes are parabolas. The surface, a paraboloid having the x-axis as its axis, is sketched in Figure 10.71.

Figure 10.72
$C : f(x, y) = 0$

S $P(x, y, z)$

$R(0, y, 0)$

$Q(x_1, y, 0)$ C

A *surface of revolution* is obtained by revolving a plane curve C about a line (the axis of revolution) in the plane. In the following discussion, C will always lie in a coordinate plane and the axis of revolution will be one of the coordinate axes. We use the symbol $f(x, y)$ for an expression in the variables x and y. In this case, $f(a, b)$ denotes the number obtained by substituting a for x and b for y. This notation will be discussed further in Chapter 12.

The graph of the equation $f(x, y) = 0$ in the xy-plane is a curve C. (We are interested here only in the graph in the xy-plane and not in the three-dimensional graph, which is a cylinder.) Suppose, for simplicity, that x and y are nonnegative for all points (x, y) on C, and let S denote the surface obtained by revolving C about the y-axis, as illustrated in Figure 10.72. A point $P(x, y, z)$ is on S if and only if $Q(x_1, y, 0)$ is on C and $x_1 = \sqrt{x^2 + z^2}$. Consequently, $P(x, y, z)$ is on S if and only if $f(\sqrt{x^2 + z^2}, y) = 0$. Thus, to find an equation for S, we replace the variable x in the equation for C by $\sqrt{x^2 + z^2}$. Similarly, if the graph of $f(x, y) = 0$ is revolved about the x-axis, then an equation for the resulting surface may be found by replacing y by $\sqrt{y^2 + z^2}$. For some curves that contain points (x, y) with x or y negative, the preceding discussion can be extended by substituting $\pm\sqrt{x^2 + z^2}$ for x or $\pm\sqrt{y^2 + z^2}$ for y. If, as in the next example, x or y occurs only in even powers, then this distinction is unnecessary, since the radical disappears when the equation is simplified.

EXAMPLE ■ 6   The graph of $9x^2 + 4y^2 = 36$ is revolved about the y-axis. Find an equation for the resulting surface.

SOLUTION   We find an equation for the surface by substituting $x^2 + z^2$ for x^2. This gives us

$$9(x^2 + z^2) + 4y^2 = 36.$$

The surface is an *ellipsoid of revolution*. If we divide both sides by 36 and rearrange terms, we obtain

$$\frac{x^2}{4} + \frac{y^2}{9} + \frac{z^2}{4} = 1,$$

which is of the form in (10.41).

A similar discussion can be given for curves that lie in the yz-plane or the xz-plane. For example, if a suitable curve C in the xz-plane is revolved about the z-axis, then we may find an equation for the resulting surface by replacing x by $\sqrt{x^2 + y^2}$. If C is revolved about the x-axis, we replace z by $\sqrt{y^2 + z^2}$.

Finally, note that equations for surfaces of revolution are characterized by the fact that two of the variables occur in combinations such as $x^2 + y^2$, $y^2 + z^2$, or $x^2 + z^2$.

Computer plotting of three-dimensional surfaces is extremely helpful for surfaces more complex than those related to conic sections. Fortunately, there are many three-dimensional plotting routines available for computers. Such routines require a very large number of numerical computations. Since hand-held calculators generally operate at much slower speeds and have a cruder screen resolution than computers, reasonably good surface plots may be more difficult to obtain on them. Recently developed "palm-top" computers, which are close to calculators in size and price, often can quickly generate a high-resolution plot of a surface.

In three dimensions, a graphing utility may require that all or part of a **viewing box** be specified. A *viewing box* is analogous to the *viewing window* we considered earlier. A frequent option is to outline the viewing box on the screen to help give the illusion of the third dimension. A three-dimensional graphing utility provides a variety of ways to display the surface (wire frame mesh, hidden lines, patches, shading, coloring, and simulation of lighting). The graphing utility may also provide a choice for the viewer's location, or perspective, and allow rotation of the surface to view it from different angles.

EXAMPLE • 7   Plot several views of the surface $z = 2 - 3x^2 - y^2$.

Figure 10.73
$-1 \le x \le 1, -1 \le y \le 1, -2 \le z \le 2$

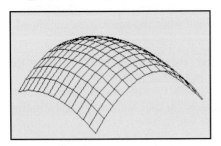

SOLUTION   Since $z = 2 - 3x^2 - y^2 = 2 - (3x^2 + y^2) \le 2$, the maximum value of z is 2. A good viewing box will include the point $(0, 0, 2)$. We select $-1 \le x \le 1$, $-1 \le y \le 1$, and $-2 \le z \le 2$ to form the viewing box. We plot $z = 2 - 3x^2 - y^2$ and direct the graphing utility to display the surface by drawing a wire frame representation for the surface, with hidden lines removed to give the illusion of three dimensions (see Figure 10.73). The graphing utility has plotted the surface in the available space on the computer screen without scaling the three axes equally. To keep the representation as simple as possible, the coordinate axes have not been drawn.

Next, we direct the same graphing utility to add shading to simulate the effects of lighting, as in Figure 10.74. Color might be used instead of shading to make the image look more three-dimensional.

Figure 10.74

Finally, in Figure 10.75, we use a different graphing utility to display the same surface. Here the three axes have been scaled equally and the axes have been displayed. This particular computer application happens to draw the lines in the wire frame and the shaded panels from the back forward, with more forward panels covering up some of the back ones, which are hidden in the final figure. Thus more information can be gained by *watching* the creation of some plots rather than by just looking at the final drawing. Most graphing utilities also make it very easy to turn the surface and obtain a different view point.

Figure 10.75

Figure 10.76

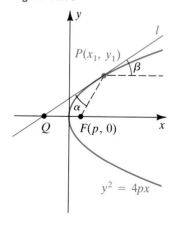

As an application of surfaces, we consider the use of paraboloids of revolution for collectors and reflectors for light, sound, radio, and television signals. Radio telescopes, satellite TV "dishes," and field microphones used at football games, for example, have the shape of a paraboloid. The reason for selecting this shape is to take advantage of an important property of tangent lines to parabolas. Suppose that l is the tangent line at a point $P(x_1, y_1)$ on the parabola $y^2 = 4px$ and that F is the focus. As in Figure 10.76, we let α denote the angle between l and the line segment FP, and β denote the angle between l and the horizontal half-line with endpoint P. It can be shown that $\alpha = \beta$ (see Exercises 58 and 59).

Figure 10.77

(a) Searchlight mirror

(b) Telescope mirror

(c) Radio telescope

The equality of the angles α and β is called the *reflective property* of the parabola. If a light source is placed at F in Figure 10.76, then, by a law of physics (*the angle of reflection equals the angle of incidence*), a beam of light will be reflected along a line parallel to the axis (see Figure 10.77a).

The same principle is used in the construction of mirrors for optical telescopes and bowls for radio telescopes. In an optical telescope, a beam of light coming toward the parabola, parallel to the axis, will be reflected into the focus (see Figure 10.77b). In a radio telescope, the receiver is placed at the focus, enabling the clearer reception of a faint signal as it is concentrated at the focus (see Figure 10.77c).

For a parabola with equation of the form $y^2 = 4px$, the constant p is the distance between the focus and the vertex of the parabola (see the discussion of parabolas in the Precalculus Review). If the parabola is rotated around its axis to form a paraboloid of revolution, then p is called the *focal length* of the paraboloid. Since every signal that strikes the surface of the bowl of a radio telescope is reflected to the receiver at the focus, signals from very distant stars can be amplified by making the surface area of the bowl very large.

Figure 10.78

$f(x) = 2\sqrt{p}\sqrt{x}$

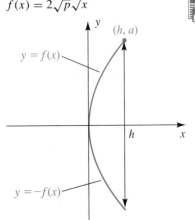

EXAMPLE ▪ 8   A radio telescope has the shape of a paraboloid of revolution with focal length p. The opening has diameter $2a$ and the telescope has altitude h.

(a) Find the surface area S available for collecting radio waves.

(b) The largest radio telescope in the world is located in Arecibo, Puerto Rico, and has a diameter of 305 m. If the focal length is 100 m, approximate its surface area to the nearest square foot.

SOLUTION

(a) The surface of the telescope is obtained by revolving the upper branch of the parabola $y^2 = 4px$ from $x = 0$ to $x = h$ about the x-axis. The upper branch of the parabola is given by the function $y = f(x) = 2\sqrt{px} = 2\sqrt{p}\sqrt{x}$ (see Figure 10.78).

By Theorem (5.19), the surface area is given by

$$S = \int_0^h 2\pi f(x)\sqrt{1 + [f'(x)]^2}\, dx.$$

Now $f'(x) = 2\sqrt{p}(\tfrac{1}{2})x^{-1/2} = \sqrt{p}/\sqrt{x}$, and so

$$\sqrt{1 + [f'(x)]^2} = \sqrt{1 + \frac{p}{x}} = \frac{\sqrt{x+p}}{\sqrt{x}},$$

which gives

$$f(x)\sqrt{1 + [f'(x)]^2} = 2\sqrt{p}\sqrt{x}\frac{\sqrt{x+p}}{\sqrt{x}} = 2\sqrt{p}\sqrt{x+p}.$$

Thus the surface area is

$$S = 2\pi 2\sqrt{p}\int_0^h \sqrt{x+p}\, dx = 4\pi\sqrt{p}\left[\frac{2}{3}(x+p)^{3/2}\right]_0^h$$

$$= \frac{8\pi\sqrt{p}}{3}\left[(h+p)^{3/2} - p^{3/2}\right] = \frac{8\pi\sqrt{p}}{3}\left\{p^{3/2}\left[\left(\frac{h+p}{p}\right)^{3/2} - 1\right]\right\}$$

$$= \frac{8\pi p^2}{3}\left[\left(1 + \frac{h}{p}\right)^{3/2} - 1\right].$$

Since the point $P(h, a)$ lies on the parabola $y^2 = 4px$, we have $a^2 = 4ph$, and we can express the altitude h in terms of the focal length p and the radius of the base a:

$$h = \frac{a^2}{4p}$$

Thus, $h/p = a^2/4p^2$, and the surface area may be written as

$$S = \frac{8\pi p^2}{3}\left[\left(1 + \frac{a^2}{4p^2}\right)^{3/2} - 1\right].$$

(b) Using the final formula of part (a) with $a = \frac{305}{2}$ and $p = 100$, we have

$$\text{the surface area} = \frac{8\pi(100^2)}{3}\left[\left(1 + \frac{152.5^2}{4(100^2)}\right)^{3/2} - 1\right] \approx 82{,}828 \text{ m}^2.$$

EXERCISES 10.6

Exer. 1–8: Sketch the graph of the cylinder in an *xyz*-coordinate system.

1 $x^2 + y^2 = 9$

2 $y^2 + z^2 = 16$

3 $4y^2 + 9z^2 = 36$

4 $x^2 + 5z^2 = 25$

5 $x^2 = 9z$

6 $x^2 - 4y = 0$

7 $y^2 - x^2 = 16$

8 $xz = 1$

Exer. 9–20: Match each graph with one of the equations.

A. $\dfrac{x^2}{25} + \dfrac{y^2}{9} + \dfrac{z^2}{4} = 1$

B. $x = z^2 + \dfrac{y^2}{4}$

C. $y^2 + z^2 - x^2 = 1$

D. $\dfrac{x^2}{4} + \dfrac{y^2}{9} - \dfrac{z^2}{4} = 0$

E. $z = \dfrac{x^2}{9} - \dfrac{y^2}{4}$

F. $z^2 - \dfrac{x^2}{4} - y^2 = 1$

G. $\dfrac{z^2}{9} + \dfrac{y^2}{4} - \dfrac{x^2}{4} = 0$

H. $\dfrac{x^2}{4} - y^2 - z^2 = 1$

I. $y = \dfrac{x^2}{4} - \dfrac{z^2}{9}$

J. $x^2 + \dfrac{y^2}{4} + \dfrac{z^2}{16} = 1$

K. $z = \dfrac{x^2}{9} + y^2$

L. $\dfrac{x^2}{4} + \dfrac{y^2}{16} + \dfrac{z^2}{9} = 1$

M. $y = \dfrac{z^2}{9} - \dfrac{x^2}{4}$

N. $y = \dfrac{x^2}{4} + \dfrac{z^2}{4}$

O. $z^2 + \dfrac{x^2}{4} - y^2 = 1$

P. $\dfrac{x^2}{4} + \dfrac{z^2}{9} - \dfrac{y^2}{4} = 0$

Q. $y^2 - \dfrac{x^2}{4} - z^2 = 1$

R. $x^2 + \dfrac{y^2}{4} - z^2 = 1$

9

10

11

12

13

14

15

16

17

18

19

20

Exer. 21–32: Sketch the graph of the quadric surface.

Ellipsoids

21 $\dfrac{x^2}{4} + \dfrac{y^2}{9} + \dfrac{z^2}{16} = 1$
22 $x^2 + \dfrac{y^2}{9} + \dfrac{z^2}{4} = 1$

Hyperboloids of one sheet

23 (a) $\dfrac{x^2}{4} + y^2 - z^2 = 1$ (b) $x^2 + \dfrac{z^2}{4} - y^2 = 1$

24 (a) $z^2 + x^2 - y^2 = 1$ (b) $y^2 + \dfrac{z^2}{4} - x^2 = 1$

Hyperboloids of two sheets

25 (a) $x^2 - \dfrac{y^2}{4} - z^2 = 1$ (b) $\dfrac{z^2}{4} - y^2 - x^2 = 1$

26 (a) $z^2 - \dfrac{x^2}{4} - \dfrac{y^2}{4} = 1$ (b) $\dfrac{y^2}{4} - x^2 - \dfrac{z^2}{9} = 1$

Cones

27 (a) $\dfrac{x^2}{9} + \dfrac{y^2}{4} = \dfrac{z^2}{4}$ (b) $\dfrac{x^2}{4} - y^2 + \dfrac{z^2}{9} = 0$

28 (a) $\dfrac{x^2}{25} + \dfrac{y^2}{9} - z^2 = 0$ (b) $x^2 = 4y^2 + z^2$

Paraboloids

29 (a) $y = \dfrac{x^2}{4} + \dfrac{z^2}{9}$ (b) $x^2 + \dfrac{y^2}{4} - z = 0$

30 (a) $z = x^2 + \dfrac{y^2}{9}$ (b) $\dfrac{z^2}{25} + \dfrac{y^2}{9} - x = 0$

Hyperbolic paraboloids

31 (a) $z = x^2 - y^2$ (b) $z = y^2 - x^2$

32 (a) $z = \dfrac{y^2}{9} - \dfrac{x^2}{4}$ (b) $z = \dfrac{x^2}{4} - \dfrac{y^2}{9}$

Exer. 33–46: Sketch the graph of the equation in an *xyz*-coordinate system, and identify the surface.

33 $16x^2 - 4y^2 - z^2 + 1 = 0$

34 $8x^2 + 4y^2 + z^2 = 16$ 35 $36x = 9y^2 + z^2$

36 $16x^2 + 100y^2 - 25z^2 = 400$

37 $x^2 - 16y^2 = 4z^2$ 38 $3x^2 - 4y^2 - z^2 = 12$

39 $9x^2 + 4y^2 + z^2 = 36$ 40 $16y = x^2 + 4z^2$

41 $z = e^y$ 42 $x^2 + (y - 2)^2 = 1$

43 $4x - 3y = 12$ 44 $2x + 4y + 3z = 12$

45 $y^2 - 9x^2 - z^2 - 9 = 0$ 46 $36x^2 - 16y^2 + 9z^2 = 0$

c **Exer. 47–50: Graph the surface.**

47 $z = \frac{1}{5}y^2 - 3\,|x|$ 48 $z = x^2 + 3xy + 4y^2$

49 $z = xy + x^2$ 50 $z = \dfrac{y^2}{16} - \dfrac{xy}{15} - \dfrac{x^2}{9}$

Exer. 51–56: Find an equation of the surface obtained by revolving the graph of the equation about the indicated axis.

51 $x^2 + 4y^2 = 16$; *y*-axis   52 $y^2 = 4x$; *x*-axis

53 $z = 4 - y^2$;   *z*-axis   54 $z = e^{-y^2}$; *y*-axis

55 $z^2 - x^2 = 1$;   *x*-axis   56 $xz = 1$;   *z*-axis

57 Although we often use a sphere as a model of the earth, a more precise relationship is needed for surveying the earth's surface. The *Clarke ellipsoid* (1866), with equation $(x^2/a^2) + (y^2/b^2) + (z^2/c^2) = 1$ for $a = b = 6378.2064$ km and $c = 6356.5838$ km, is used to establish the geographic positions of control points in the U.S. national geodetic network.

(a) Explain briefly the difference between the Clarke ellipsoid and the usual spherical representation of the earth's surface.

(b) Curves of equal latitude are traces in a plane of the form $z = k$. Describe these curves.

(c) Curves of equal longitude (or *meridians*) are traces in a plane of the form $y = mx$. Describe these curves.

58 Let *m* and *b* be nonzero real numbers.

(a) If the line $y = mx + b$ intersects the parabola $y^2 = 4px$ in only one point, show that $p = mb$.

(b) Show that the slope of the tangent line to $y^2 = 4px$ at $P(x_1, y_1)$ is $y_1/(2x_1)$.

59 Establish the reflective property of the parabola. (*Hint:* Show that $d(Q, F) = d(F, P)$ in Figure 10.76 and use the result in Exercise 58.)

CHAPTER 10 REVIEW EXERCISES

Exer. 1–20: If $a = 3i - j - 4k$, $b = 2i + 5j - 2k$, and $c = -i + 6k$, find the vector or scalar.

1 $3a - 2b$ **2** $a \cdot (b - c)$ **3** $\|-3b\|$

4 $\|b + c\|$ **5** $\|a\|^2$ **6** $\|a \times a\|$

7 The angle between a and c

8 The angles between a and the unit vectors i, j, and k

9 A unit vector having the same direction as a

10 A vector having the opposite direction of a and twice the magnitude of b

11 $a \times b$ **12** $(b + c) \times a$

13 $\text{comp}_b \, a$ **14** $\text{comp}_a \, (b \times c)$

15 $(2a) \cdot (3a)$ **16** $(2a) \times (3a)$

17 $(a \times c) + (c \times a)$ **18** $(a \times c) \times (c \times a)$

19 The volume of the box determined by a, b, and c

20 The area of the triangle determined by a and b

21 Show that $a = \langle 4, 3, -5 \rangle$ and $b = \langle -1, 3, 1 \rangle$ are orthogonal.

22 A constant force of magnitude 18 lb has the same direction as the vector $a = 4i + 7j + 4k$. If the distance is measured in feet, find the work done if the point of application of the force moves along the line segment from $P(1, 1, 1)$ to $Q(3, 5, 4)$.

23 Given the points $A(5, -3, 2)$ and $B(-1, -4, 3)$, find the following:

  (a) $d(A, B)$

  (b) the midpoint of AB

  (c) an equation of the sphere with center B and tangent to the xz-plane

  (d) an equation of the plane through B parallel to the xz-plane

  (e) parametric equations for the line through A and B

  (f) an equation of the plane through A with normal vector $\overrightarrow{AB}$

24 Find an equation of the plane through $A(0, 4, 9)$ and $B(0, -3, 7)$ that is perpendicular to the yz-plane.

25 Find an equation of the plane that has x-intercept 5, y-intercept -2, and z-intercept 6.

26 Find a symmetric form for the line $x = 3 + 4t$, $y = 2 - t$, $z = 2t$.

27 Find parametric equations for the line of intersection of the planes $2x + y + 4z = 8$ and $x + 3y - z = -1$.

28 Find an equation of the plane through $P(1, 3, -2)$ that is parallel to the plane $5x - 4y + 3z = 7$.

29 Find an equation of the plane through $P(4, 1, 2)$ that has the same yz-trace as the plane $2x + 3y - 4z = 11$.

30 Find an equation of the cylinder that has rulings parallel to the z-axis and that has, for its directrix, the circle in the xy-plane with center $C(4, -3, 0)$ and radius 5.

31 Find an equation of the ellipsoid with center O, x-intercept 8, y-intercept 3, and z-intercept 1.

32 Find an equation of the surface obtained by revolving the graph of the equation $z = x$ about the z-axis.

Given the points $P(2, -1, 1)$, $Q(-3, 2, 0)$, and $R(4, -5, 3)$, find the following:

  (a) a unit normal vector to the plane determined by P, Q, and R

  (b) an equation of the plane through P, Q, and R

  (c) parametric equations for a line through P that is parallel to the line through Q and R

  (d) $\overrightarrow{QP} \cdot \overrightarrow{QR}$

  (e) the angle between $\overrightarrow{QP}$ and $\overrightarrow{QR}$

  (f) the area of triangle PQR

  (g) the distance from R to the line through P and Q

34 Find the angles between the two lines

$$\frac{x - 3}{2} = \frac{y + 1}{-4} = \frac{z - 5}{8}$$

and

$$\frac{x + 1}{7} = \frac{6 - y}{2} = \frac{2z + 7}{-4}.$$

35 Find parametric equations for each of the lines in Exercise 34.

36 Determine whether the following lines intersect, and if so, find the point of intersection:

$$\begin{array}{lll} x = 2 + t, & y = 1 + t, & z = 4 + 7t \\ x = -4 + 5v, & y = 2 - 2v, & z = 1 - 4v \end{array}$$

37 Find the angles between the lines in Exercise 36.

38 Suppose a line l has parametric equations $x = 2t + 1$, $y = -t + 3$, $z = 5t$. Find the distance from $A(3, 1, -1)$ to l.

Exer. 39–50: Sketch the graph of the equation.

39 $x^2 + y^2 + z^2 - 14x + 6y - 8z + 10 = 0$

40 $4y - 3z - 15 = 0$

41 $3x - 5y + 2z = 10$ **42** $y = z^2 + 1$

43 $9x^2 + 4z^2 = 36$

44 $x^2 + 4y + 9z^2 = 0$

45 $z^2 - 4x^2 = 9 - 4y^2$

46 $2x^2 + 4z^2 - y^2 = 0$

47 $z^2 - 4x^2 - y^2 = 4$

48 $x^2 + 2y^2 + 4z^2 = 16$

49 $x^2 - 4y^2 = 4z$

50 $z = 9 - x^2 - y^2$

EXTENDED PROBLEMS AND GROUP PROJECTS

1 Let $\mathbf{v} = \langle a, b \rangle$ and $\mathbf{w} = \langle c, d \rangle$ be vectors in $\mathbb{R}^2$. Define vector addition by $\mathbf{v} + \mathbf{w} = \langle a + c, b + d \rangle$, scalar multiplication by $c\mathbf{v} = \langle ca, cb \rangle$, and a new vector multiplication by $\mathbf{v} \otimes \mathbf{w} = \langle ac - bd, ad + bc \rangle$.

(a) Show that $\otimes$ is a commutative operation—that is, for all $\mathbf{v}$ and $\mathbf{w}$,

$$\mathbf{v} \otimes \mathbf{w} = \mathbf{w} \otimes \mathbf{v}.$$

(b) Show that $\otimes$ is an associative operation—that is, for all $\mathbf{u}$, $\mathbf{v}$, and $\mathbf{w}$,

$$\mathbf{u} \otimes (\mathbf{v} \otimes \mathbf{w}) = (\mathbf{u} \otimes \mathbf{v}) \otimes \mathbf{w}.$$

(c) Show that for every vector $\mathbf{v}$, $\mathbf{v} \otimes \langle 1, 0 \rangle = \mathbf{v}$. In what way does this result make the statement "$\langle 1, 0 \rangle$ behaves like the number 1 for multiplication" true?

(d) Show that if $\mathbf{v} = \langle a, b \rangle$ is not the zero vector, then there is a unique vector $\mathbf{u}$ such that $\mathbf{v} \otimes \mathbf{u} = \langle 1, 0 \rangle$.

(*Hint*: Try $\mathbf{u} = [1/(a^2 + b^2)]\langle a, -b \rangle$.)

(e) Show that a distributive law holds:

$$\mathbf{u} \otimes (\mathbf{v} + \mathbf{w}) = (\mathbf{u} \otimes \mathbf{v}) + (\mathbf{u} \otimes \mathbf{w}).$$

(f) A vector of the form $\mathbf{v} = \langle a, 0 \rangle$ is called a *pure real*. Show that if $\mathbf{v}$ and $\mathbf{w}$ are pure reals, then so are $\mathbf{v} + \mathbf{w}$ and $\mathbf{v} \otimes \mathbf{w}$. What are the first components of $\mathbf{v} + \mathbf{w}$ and $\mathbf{v} \otimes \mathbf{w}$?

(g) A vector of the form $\langle 0, b \rangle$ is called a *pure imaginary*. If $\mathbf{v}$ is a pure imaginary, describe $\mathbf{v} \otimes \mathbf{v}$.

(h) For any vector $\mathbf{v}$, show that $\|\mathbf{v} \otimes \mathbf{w}\| = \|\mathbf{v}\|^2$.

(i) Denote a pure real $\mathbf{v} = \langle a, 0 \rangle$ by a and a pure imaginary vector $\mathbf{w} = \langle 0, b \rangle$ by bi. Show that any vector $\langle a, b \rangle$ can then be represented in the form $a + bi$.

(j) Show that the arithmetic of sums and products that we have defined satisfies the following description: Treat the numbers of the form $a + bi$ as real numbers for the purposes of adding and multiplying except that whenever i^2 appears, replace it by -1. The arithmetic of sums and products introduced in this problem is called the **complex number system**.

2 Is there a way to define a vector multiplication # between three-dimensional vectors that will be commutative and associative and will satisfy the condition that for all $\mathbf{v}$, we have $\mathbf{v}$ # $\langle 1, 0, 0 \rangle = \mathbf{v}$? What properties does the vector multiplication show if the product of $\mathbf{v} = \langle a, b, c \rangle$ and $\mathbf{w} = \langle d, e, f \rangle$ is $\langle ad, be, cf \rangle$?

3 The terms *vector* and *scalar* and the notation $\mathbf{i}$, $\mathbf{j}$, and $\mathbf{k}$ are all credited to the Irish mathematician William Rowan Hamilton (1805–1865), who introduced the treatment of complex numbers as two-dimensional vectors. While trying to answer the question posed in Problem 2, Hamilton discovered a method of multiplying four-dimensional vectors that he called *quaternions*. A quaternion is an expression of the form

$$\mathbf{q} = q_1 + q_2\mathbf{i} + q_3\mathbf{j} + q_4\mathbf{k},$$

where each q_i is a real number. Quaternions are added and multiplied using all the standard rules for real numbers with the additional stipulations that

$$\mathbf{i}^2 = \mathbf{j}^2 = \mathbf{k}^2 = -1,$$

$$\mathbf{ij} = \mathbf{k}, \quad \mathbf{jk} = \mathbf{i}, \quad \mathbf{ki} = \mathbf{j}, \quad \mathbf{ji} = -\mathbf{k}, \quad \mathbf{kj} = -\mathbf{i}, \quad \mathbf{ik} = -\mathbf{j}.$$

Show that if

$$\mathbf{q} = q_1 + q_2\mathbf{i} + q_3\mathbf{j} + q_4\mathbf{k} \quad \text{and} \quad \mathbf{r} = r_1 + r_2\mathbf{i} + r_3\mathbf{j} + r_4\mathbf{k}$$

are quaternions, then

$$\mathbf{q} + \mathbf{r} = (q_1 + r_1) + (q_2 + r_2)\mathbf{i} + (q_3 + r_3)\mathbf{j} + (q_4 + r_4)\mathbf{k}.$$

Hamilton called $q_2\mathbf{i} + q_3\mathbf{j} + q_4\mathbf{k}$ the *vector part* of the quaternion $\mathbf{q}$ and q_1 the *scalar part* of the quaternion $\mathbf{q}$. A *pure vector* is a quaternion $\mathbf{p}$ with scalar part equal to zero. Show that the quaternion product of two pure vectors $\mathbf{p}$ and $\mathbf{r}$ is given by

$$\text{quaternion}(\mathbf{p}, \mathbf{r}) = \mathbf{p} \times \mathbf{r} - (\mathbf{p} \cdot \mathbf{r}).$$

Is quaternion multiplication commutative? Is it associative? In what ways can quaternions be fruitfully viewed as four-dimensional vectors?

INTRODUCTION

THE CULMINATION OF THIS CHAPTER'S DISCUSSION is a presentation of an achievement that launched the modern age of science and fundamentally changed the course of human history and thought: Newton's use of vector methods to derive Kepler's laws of planetary motion.

The motion of the sun, moon, stars, and planets has fascinated humans since the dawn of history. For centuries, most people believed the earth to be the center of the universe, with the sun and planets moving in circular orbits about us. In the early seventeenth century, after years of prolonged calculation, Johannes Kepler found beautiful patterns in the masses of data on planetary positions. He stated these as three empirical laws but had no theory as to why these laws might be true or how they related to each other.

In the 1660s, Sir Isaac Newton used the newly forged tools of calculus to show the essential equivalence between Kepler's laws and an inverse square law of gravitational attraction. Newton's spectacular success in applying new mathematical methods to explain the physical world was an epoch-changing moment in history. To understand Newton's accomplishment, we need some new concepts.

In this chapter, we begin to examine functions in a multivariable setting. We introduce functions whose values are vectors in Section 11.1 and discuss how to represent such functions by triples of real-valued functions. From this representation, we can apply earlier methods to define limits, derivatives, and integrals of vector-valued functions. This approach, presented in Section 11.2, enables us to readily obtain properties that are analogous to those obtained for real-valued functions in Chapters 2 and 4.

If a vector-valued function is continuous, then the terminal points of the vectors in its range determine a curve in three dimensions, and, conversely, to each curve there corresponds a continuous vector-valued function. We use this one-to-one correspondence in applications involving a point moving along a curve in Section 11.3, where we study velocity and acceleration and apply Newton's Second Law to analyze the motion of a projectile.

In Sections 11.4 and 11.5, we use vector-valued functions to define *curvature*, which is a concept for determining how much a curve bends or changes its shape. The final section (11.6) states and derives Kepler's laws. The proofs are challenging, but a careful reading should give you an appreciation of the power of vector methods.

The calculus of vector-valued functions provides the means to derive many important physical laws, including those of planetary motion.

Vector-Valued Functions

11.1   VECTOR-VALUED FUNCTIONS AND SPACE CURVES

By definition, a *function* is a correspondence that assigns to each element of its domain D exactly one element of its range E. If the function values are real numbers, we sometimes refer to the function as a **real-valued function**, or as a **scalar function**. In this section, we consider *vector-valued functions*, denoted by **r** (or by $\vec{r}$), for which the function values are *vectors*.

Definition 11.1

Let D be a set of real numbers. A **vector-valued function r** with domain D is a correspondence that assigns to each number t in D exactly one vector $\mathbf{r}(t)$ in V_3.

Figure 11.1

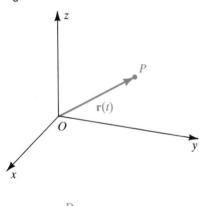

The range of **r** in Definition (11.1) consists of all possible vectors $\mathbf{r}(t)$, for t in D. In Figure 11.1, the domain D is a closed interval; however, D could be *any* set of real numbers. We also have sketched the position vector for $\mathbf{r}(t)$. In the future, we shall not make a distinction between the vector $\mathbf{r}(t)$ in V_3 and its position vector $\overrightarrow{OP}$. Thus, to *sketch* $\mathbf{r}(t)$ will mean to sketch $\overrightarrow{OP}$. The terminal point P of $\overrightarrow{OP}$ will be referred to as the **endpoint** of $\mathbf{r}(t)$.

Since for each t in D the three components of $\mathbf{r}(t)$ are uniquely determined, we may write

$$\mathbf{r}(t) = f(t)\mathbf{i} + g(t)\mathbf{j} + h(t)\mathbf{k},$$

where f, g, and h are *scalar* functions with domain D. Conversely, a formula of this type for $\mathbf{r}(t)$ determines a vector-valued function **r**. This gives us the following.

Theorem 11.2

If D is a set of real numbers, then **r** is a vector-valued function with domain D if and only if there are scalar functions f, g, and h such that

$$\mathbf{r}(t) = f(t)\mathbf{i} + g(t)\mathbf{j} + h(t)\mathbf{k}$$

for every t in D.

Figure 11.2

EXAMPLE 1   Let $\mathbf{r}(t) = (t + 1)\mathbf{i} + (t^2 - 4)\mathbf{j} + t^2\mathbf{k}$, for t in $\mathbb{R}$.

(a) Sketch $\mathbf{r}(1)$ and $\mathbf{r}(3)$.

(b) Find the vectors $\mathbf{r}(t)$ that are in a coordinate plane.

SOLUTION

(a) Letting $t = 1$ and $t = 3$ in the formula for $\mathbf{r}(t)$, we obtain

$$\mathbf{r}(1) = 2\mathbf{i} - 3\mathbf{j} + \mathbf{k} \quad \text{and} \quad \mathbf{r}(3) = 4\mathbf{i} + 5\mathbf{j} + 9\mathbf{k}.$$

These vectors are sketched in Figure 11.2.

(b) The following chart displays the vectors $\mathbf{r}(t)$ that are in a coordinate plane.

Plane containing $\mathbf{r}(t)$	Zero component of $\mathbf{r}(t)$	Value of t	$\mathbf{r}(t)$
xy-plane	$\mathbf{k}$-component	$t = 0$	$\mathbf{r}(0) = \mathbf{i} - 4\mathbf{j}$
yz-plane	$\mathbf{i}$-component	$t = -1$	$\mathbf{r}(-1) = -3\mathbf{j} + \mathbf{k}$
xz-plane	$\mathbf{j}$-component	$t = \pm 2$	$\mathbf{r}(2) = 3\mathbf{i} + 4\mathbf{k}$ $\mathbf{r}(-2) = -\mathbf{i} + 4\mathbf{k}$

EXAMPLE ■ 2 Let $\mathbf{r}(t) = (9 - 4t)\mathbf{i} + (-4 + 6t)\mathbf{j} + (3 + 3t)\mathbf{k}$, for t in $\mathbb{R}$.

(a) Sketch $\mathbf{r}(0)$, $\mathbf{r}(1)$, and $\mathbf{r}(2)$.

(b) Give a graphical description of the set of all endpoints of $\mathbf{r}(t)$.

SOLUTION

(a) Let $t = 0$, 1, and 2 in the formula for $\mathbf{r}(t)$, as follows:

$$\mathbf{r}(0) = 9\mathbf{i} - 4\mathbf{j} + 3\mathbf{k}$$
$$\mathbf{r}(1) = 5\mathbf{i} + 2\mathbf{j} + 6\mathbf{k}$$
$$\mathbf{r}(2) = \mathbf{i} + 8\mathbf{j} + 9\mathbf{k}$$

These are the vectors $\overrightarrow{OA}$, $\overrightarrow{OB}$, and $\overrightarrow{OC}$ sketched in Figure 11.3.

Figure 11.3

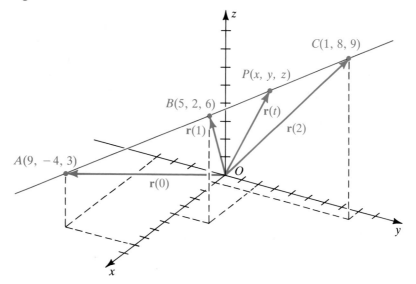

(b) If $P(x, y, z)$ is the endpoint of $\mathbf{r}(t)$, as indicated in Figure 11.3, then

$$x = 9 - 4t, \quad y = -4 + 6t, \quad z = 3 + 3t; \quad t \text{ in } \mathbb{R}.$$

By Theorem (10.34), these are parametric equations for the line l through $A(9, -4, 3)$ parallel to the vector $\langle -4, 6, 3 \rangle$. Since $\langle -4, 6, 3 \rangle$ corresponds to $\overrightarrow{AB}$, we see that the endpoints of $\mathbf{r}(t)$ coincide with the line l.

The endpoints of $\mathbf{r}(t)$ in Example 2 determine a line l in an xyz-coordinate system. In general, the endpoints of

$$\mathbf{r}(t) = f(t)\mathbf{i} + g(t)\mathbf{j} + h(t)\mathbf{k},$$

where the scalar functions f, g, and h are continuous on an interval I, determine a **space curve** (or simply a **curve**)—that is, a set C of ordered triples $(f(t), g(t), h(t))$. The **graph** of C consists of all points $P(f(t), g(t), h(t))$ in an xyz-coordinate system that correspond to the ordered triples. (As in two dimensions, the terms *curve* and *graph of a curve* are used interchangeably.) The equations

$$x = f(t), \quad y = g(t), \quad z = h(t); \quad t \text{ in } I$$

are **parametric equations** for C. We call C a **parametrized curve** and the equations a **parametrization** for C. The **orientation** of C is the direction determined by increasing values of t.

The notions of **endpoints, closed curve,** and **simple closed curve** are defined exactly as for plane curves (see Section 9.1). A curve C is **smooth** if it has a **smooth parametrization:** $x = f(t), y = g(t), z = h(t)$, with f', g', and h' continuous and not simultaneously zero (that is, not all zero for some t), except possibly at endpoints. A smooth curve has no sharp corners, cusps, or breaks. A **piecewise-smooth curve** is defined as in Section 9.1.

If $\mathbf{r}(t) = f(t)\mathbf{i} + g(t)\mathbf{j} + h(t)\mathbf{k}$, with f, g, and h continuous on an interval I, then, as illustrated in Figure 11.4, as t varies through I, the endpoint of $\mathbf{r}(t)$ traces the curve C with parametrization

$$x = f(t), \quad y = g(t), \quad z = h(t); \quad t \text{ in } I.$$

We refer to C as **the curve determined by $\mathbf{r}(t)$.** In Section 11.3, we shall use this vector representation of C, with t as time, to describe the motion of a point P as it moves in space.

A **twisted cubic** is a curve that has a parametrization

$$x = at, \quad y = bt^2, \quad z = ct^3,$$

where a, b, and c are nonzero constants. The special case $a = 1, b = 1$, and $c = 1$ is the curve determined by $\mathbf{r}(t)$ in the next example.

Figure 11.4

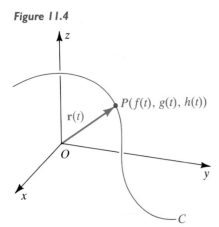

EXAMPLE ■ 3 Let $\mathbf{r}(t) = t\mathbf{i} + t^2\mathbf{j} + t^3\mathbf{k}$ for $t \geq 0$. Sketch the curve C determined by $\mathbf{r}(t)$, and indicate the orientation.

SOLUTION Parametric equations for C are

$$x = t, \quad y = t^2, \quad z = t^3; \quad t \geq 0.$$

Since x, y, and z are nonnegative, C lies in the first octant.

To help us visualize the graph, let us eliminate the parameter from the first two equations, obtaining $y = x^2$. This implies that every point $P(x, y, z)$ on C is also on the parabolic cylinder $y = x^2$.

If we eliminate the parameter from $x = t$ and $z = t^3$, we obtain $z = x^3$, which is an equation of a cylinder with rulings parallel to the y-axis. The curve C is the intersection of the two cylinders $y = x^2$ and $z = x^3$, as shown in Figure 11.5.

At $t = 0$, the endpoint $P(x, y, z)$ of the position vector for $\mathbf{r}(t)$ is $(0, 0, 0)$. As t increases, the endpoint traces the curve C, passing through the points $(1, 1, 1)$, $(2, 4, 8)$, $(3, 9, 27)$, and so on. The arrowheads in the figure indicate the orientation of C. Note that y increases more rapidly than x, and z more rapidly than y. Thus, at $t = 10$, the endpoint of the position vector is $(10, 100, 1000)$.

Figure 11.5

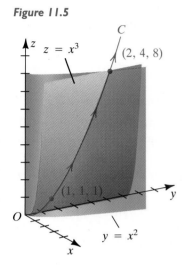

EXAMPLE ■ 4 Let $\mathbf{r}(t) = a \cos t\, \mathbf{i} + a \sin t\, \mathbf{j} + bt\mathbf{k}$ for $t \geq 0$ and positive constants a and b. Sketch the curve C determined by $\mathbf{r}(t)$, and indicate the orientation.

SOLUTION Parametric equations for C are

$$x = a \cos t, \quad y = a \sin t, \quad z = bt; \quad t \geq 0.$$

If we eliminate the parameter in the first two equations (see Example 2 of Section 9.1), we obtain

$$x^2 + y^2 = a^2.$$

Thus, every point $P(x, y, z)$ on C is on the circular cylinder $x^2 + y^2 = a^2$. If t varies from 0 to 2π, the point P starts at $(a, 0, 0)$ and moves upward, while making one revolution around the cylinder, as illustrated in Figure 11.6. Additional t-intervals of length 2π result in similar portions of the curve. The orientation of C is indicated by arrowheads.

Figure 11.6

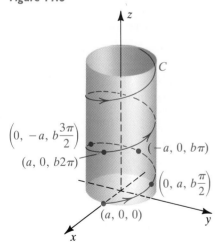

The curve C in Figure 11.6 is called a **circular helix**.

If we regard V_2 as the subset of V_3 consisting of all vectors whose third component is 0, then $\mathbf{r}(t)$ in Theorem (11.2) may be written

$$\mathbf{r}(t) = f(t)\mathbf{i} + g(t)\mathbf{j}.$$

In this case, the curve determined by $\mathbf{r}(t)$ is in the xy-plane.

EXAMPLE ■ 5   Let $\mathbf{r}(t) = 2t\mathbf{i} + (8 - 2t^2)\mathbf{j}$ for $-2 \le t \le 2$.

(a) Sketch the curve C determined by $\mathbf{r}(t)$, and indicate the orientation.

(b) Sketch $\mathbf{r}(t)$ for $t = -2, -1, 0, \frac{3}{2}$, and 2.

SOLUTION

(a) The curve C is in the xy-plane and has the parametrization

$$x = 2t, \quad y = 8 - 2t^2; \quad -2 \le t \le 2.$$

Figure 11.7

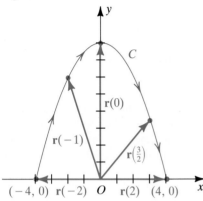

Eliminating the parameter, we see that C is on the parabola

$$y = 8 - 2(\tfrac{1}{2}x)^2, \quad \text{or} \quad y = 8 - \tfrac{1}{2}x^2.$$

The values $-2 \le t \le 2$ give us the part of the parabola between $(-4, 0)$ and $(4, 0)$, sketched in Figure 11.7, where the arrowheads on C indicate the orientation.

(b) Letting $t = -2, -1, 0, \frac{3}{2}$, and 2 in the formula for $\mathbf{r}(t)$, we obtain the following:

$$\mathbf{r}(-2) = -4\mathbf{i} \qquad \mathbf{r}(-1) = -2\mathbf{i} + 6\mathbf{j} \qquad \mathbf{r}(0) = 8\mathbf{j}$$

$$\mathbf{r}(\tfrac{3}{2}) = 3\mathbf{i} + \tfrac{7}{2}\mathbf{j} \qquad \mathbf{r}(2) = 4\mathbf{i}$$

These vectors are sketched in Figure 11.7.

EXAMPLE ■ 6   Let $\mathbf{r}(t) = (\tfrac{1}{2}t \cos t)\mathbf{i} + (\tfrac{1}{2}t \sin t)\mathbf{j}$ for $0 \le t \le 3\pi$.

(a) Sketch the curve determined by $\mathbf{r}(t)$, and indicate the orientation.

(b) Sketch $\mathbf{r}(6)$.

SOLUTION

(a) The curve C lies in the xy-plane and has parametric equations

$$x = \tfrac{1}{2}t \cos t, \quad y = \tfrac{1}{2}t \sin t; \quad 0 \le t \le 3\pi.$$

We may assume that t is the radian measure of an angle θ and write

$$x = \tfrac{1}{2}\theta \cos \theta, \quad y = \tfrac{1}{2}\theta \sin \theta; \quad 0 \le \theta \le 3\pi.$$

Moreover, since $\qquad \dfrac{y}{x} = \dfrac{\tfrac{1}{2}\theta \sin \theta}{\tfrac{1}{2}\theta \cos \theta} = \tan \theta,$

it follows from Theorem (9.9) that we can use θ to express (x, y) in polar coordinates. We next note that

$$x^2 + y^2 = \tfrac{1}{4}\theta^2 \cos^2 \theta + \tfrac{1}{4}\theta^2 \sin^2 \theta$$
$$= \tfrac{1}{4}\theta^2(\cos^2 \theta + \sin^2 \theta) = \tfrac{1}{4}\theta^2.$$

Changing $x^2 + y^2$ to polar coordinates gives us

$$r^2 = \tfrac{1}{4}\theta^2, \quad \text{or} \quad r = \tfrac{1}{2}\theta.$$

(Note that $r \ne -\tfrac{1}{2}\theta$, because $r = \|\mathbf{r}(t)\| \ge 0$.)

Figure 11.8

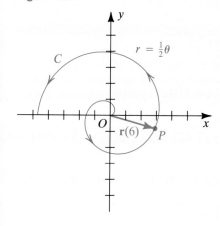

The graph of the polar equation $r = \frac{1}{2}\theta$ for $0 \leq \theta \leq 3\pi$ is the portion of the spiral of Archimedes illustrated in Figure 11.8 (see Example 5 of Section 9.3). The orientation is in the counterclockwise direction, as indicated by the arrowheads.

(b) Substituting 6 for t gives us

$$\mathbf{r}(6) = (3 \cos 6)\mathbf{i} + (3 \sin 6)\mathbf{j} \approx 2.9\mathbf{i} - 0.8\mathbf{j},$$

and we sketch $\mathbf{r}(6)$ as in Figure 11.8.

We may define the **length of a curve** C in three dimensions exactly as we did for plane curves in Section 9.1.

The next result is analogous to Theorem (9.6).

Theorem 11.3

If a curve C has a smooth parametrization

$$x = f(t), \quad y = g(t), \quad z = h(t); \quad a \leq t \leq b$$

and if C does not intersect itself, except possibly for $t = a$ and $t = b$, then the length L of C is

$$L = \int_a^b \sqrt{[f'(t)]^2 + [g'(t)]^2 + [h'(t)]^2} \, dt$$

$$= \int_a^b \sqrt{\left(\frac{dx}{dt}\right)^2 + \left(\frac{dy}{dt}\right)^2 + \left(\frac{dz}{dt}\right)^2} \, dt.$$

EXAMPLE 7 Find the length L of the circular helix

$$x = a \cos t, \quad y = a \sin t, \quad z = bt; \quad 0 \leq t \leq 2\pi.$$

SOLUTION The helix is that part of the curve in Figure 11.6 between the points $(a, 0, 0)$ and $(a, 0, 2\pi b)$. Using Theorem (11.3), we obtain

$$L = \int_0^{2\pi} \sqrt{(-a \sin t)^2 + (a \cos t)^2 + (b)^2} \, dt$$

$$= \int_0^{2\pi} \sqrt{a^2 \sin^2 t + a^2 \cos^2 t + b^2} \, dt$$

$$= \int_0^{2\pi} \sqrt{a^2 + b^2} \, dt$$

$$= \left[\sqrt{a^2 + b^2} \, t \right]_0^{2\pi} = 2\pi \sqrt{a^2 + b^2}.$$

EXERCISES 11.1

Exer. 1–8: **(a)** Sketch the two vectors listed after the formula for r(t). **(b)** Sketch, on the same coordinate system, the curve C determined by r(t), and indicate the orientation for the given values of t.

1 $\mathbf{r}(t) = 3t\mathbf{i} + (1 - 9t^2)\mathbf{j}$,
$\mathbf{r}(0)$, $\mathbf{r}(1)$; t in $\mathbb{R}$

2 $\mathbf{r}(t) = (1 - t^3)\mathbf{i} + t\mathbf{j}$,
$\mathbf{r}(1)$, $\mathbf{r}(2)$; $t \geq 0$

3 $\mathbf{r}(t) = (t^3 - 1)\mathbf{i} + (t^2 + 2)\mathbf{j}$,
$\mathbf{r}(1)$, $\mathbf{r}(2)$; $-2 \leq t \leq 2$

4 $\mathbf{r}(t) = (2 + \cos t)\mathbf{i} - (3 - \sin t)\mathbf{j}$,
$\mathbf{r}(\pi/2)$, $\mathbf{r}(\pi)$; $0 \leq t \leq 2\pi$

5 $\mathbf{r}(t) = (3 + t)\mathbf{i} + (2 - t)\mathbf{j} + (1 + 2t)\mathbf{k}$,
$\mathbf{r}(-1)$, $\mathbf{r}(0)$; $t \geq -1$

6 $\mathbf{r}(t) = t\mathbf{i} - 3\sin t\mathbf{j} + 3\cos t\mathbf{k}$,
$\mathbf{r}(0)$, $\mathbf{r}(\pi/2)$; $t \geq 0$

7 $\mathbf{r}(t) = t\mathbf{i} + 4\cos t\mathbf{j} + 9\sin t\mathbf{k}$,
$\mathbf{r}(0)$, $\mathbf{r}(\pi/2)$; $t \geq 0$

8 $\mathbf{r}(t) = \tan t\mathbf{i} + \sec t\mathbf{j} + 2\mathbf{k}$,
$\mathbf{r}(0)$, $\mathbf{r}(\pi/4)$; $-\pi/2 < t < \pi/2$

Exer. 9–16: Sketch the curve C determined by r(t), and indicate the orientation.

9 $\mathbf{r}(t) = e^t \cos t\mathbf{i} + e^t \sin t\mathbf{j}$; $0 \leq t \leq \pi$

10 $\mathbf{r}(t) = 2\cosh t\mathbf{i} + 3\sinh t\mathbf{j}$; t in $\mathbb{R}$

11 $\mathbf{r}(t) = t\mathbf{i} + 2t^2\mathbf{j} + 3t^3\mathbf{k}$; t in $\mathbb{R}$

12 $\mathbf{r}(t) = t^3\mathbf{i} + t^2\mathbf{j} + t\mathbf{k}$; $0 \leq t \leq 4$

13 $\mathbf{r}(t) = (t^2 + 1)\mathbf{i} + t\mathbf{j} + 3\mathbf{k}$; t in $\mathbb{R}$

14 $\mathbf{r}(t) = 6\sin t\mathbf{i} + 4\mathbf{j} + 25\cos t\mathbf{k}$; $-2\pi \leq t \leq 2\pi$

15 $\mathbf{r}(t) = t\mathbf{i} + t\mathbf{j} + \sin t\mathbf{k}$; t in $\mathbb{R}$

16 $\mathbf{r}(t) = t\mathbf{i} + 2t\mathbf{j} + e^t\mathbf{k}$; t in $\mathbb{R}$

[c] **Exer. 17–20:** Plot the curve C determined by r(t), and indicate the orientation.

17 $\mathbf{r}(t) = 3\sin(t^2)\mathbf{i} + (4 - t^{3/2})\mathbf{j}$; $0 \leq t \leq 5$

18 $\mathbf{r}(t) = e^{\sin 3t}\mathbf{i} + e^{-\cos t}\mathbf{j}$; $0 \leq t \leq 2\pi$

19 $\mathbf{r}(t) = (4 + \sin 2t)\mathbf{i} + (1 - 3\cos 3t)\mathbf{j}$; $-\pi \leq t \leq \pi$

20 $\mathbf{r}(t) = (1 + 3\sin 2t)\mathbf{i} + (2\cos 3t)\mathbf{j}$; $0 \leq t \leq \pi$

Exer. 21–26: Find the arc length of the parametrized curve. Estimate with numerical integration if needed, and express answers to four decimal places of accuracy.

21 $x = 5t$, $y = 4t^2$, $z = 3t^2$; $0 \leq t \leq 2$

22 $x = t^2$, $y = t\sin t$, $z = t\cos t$; $0 \leq t \leq 1$

23 $x = e^t \cos t$, $y = e^t$, $z = e^t \sin t$; $0 \leq t \leq 2\pi$

24 $x = 2t$, $y = 4\sin 3t$, $z = 4\cos 3t$; $0 \leq t \leq 2\pi$

[c] **25** $x = 3\cos t$, $y = 2\sin t$, $z = 2 - t^2$; $-\sqrt{2} \leq t \leq \sqrt{2}$

[c] **26** $x = (1 - t)^2$, $y = t^2$, $z = 2(1 - t)t$; $0 \leq t \leq 1$

27 A *concho-spiral* is a curve C that has a parametrization $x = ae^{\mu t}\cos t$, $y = ae^{\mu t}\sin t$, $z = be^{\mu t}$; $t \geq 0$, where a, b, and μ are constants.

(a) Show that C lies on the cone $a^2 z^2 = b^2(x^2 + y^2)$.

(b) Sketch C for $a = b = 4$ and $\mu = -1$.

(c) Find the length of C corresponding to the t-interval $[0, \infty)$.

28 A curve C has the parametrization

$$x = a\sin t\sin\alpha, \quad y = b\sin t\cos\alpha, \quad z = c\cos t; \quad t \geq 0,$$

where a, b, c, and α are positive constants.

(a) Show that C lies on the ellipsoid

$$\frac{x^2}{a^2} + \frac{y^2}{b^2} + \frac{z^2}{c^2} = 1.$$

(b) Show that C also lies on a plane that contains the z-axis.

(c) Describe the curve C.

29 (a) Show that a twisted cubic having parametrization $x = at$, $y = bt^2$, $z = ct^3$; $t \geq 0$ intersects a given plane in at most three points.

(b) Determine the length of the twisted cubic $x = 6t$, $y = 3t^2$, $z = t^3$ between the points corresponding to $t = 0$ and $t = 1$.

30 A rectangle can be made into a cylinder by joining together two opposite and parallel edges. Shown in the figure on the following page is a rectangle $ABCD$ of width 2π. Edge AD is joined to edge BC, and point A is then positioned at $(1, 0, 0)$ to form part of the cylinder $x^2 + y^2 = 1$, sketched on the right of the rectangle.

Exercise 30

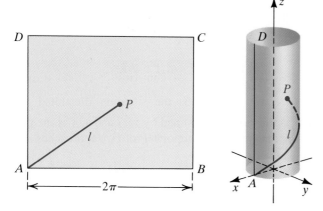

(a) If l is the line segment from A to another point P in the rectangle and m is the slope of l, show that, as a curve on the cylinder, l has a parametrization $x = \cos t,\ y = \sin t,\ z = mt$.

(b) Use part (a) to show that the curve on the cylinder with shortest length from $(1,\ 0,\ 0)$ to another point P is a helix (see Example 4).

11.2 LIMITS, DERIVATIVES, AND INTEGRALS

Limits, derivatives, and integrals for vector-valued functions are natural generalizations of the corresponding concepts for real-valued functions of a single real variable. We consider these generalizations in this section. We could define the limit of a vector-valued function $\mathbf{r}$ by using an ϵ–δ approach similar to that used in Definition (1.4) (see Exercise 45); however, since $\mathbf{r}(t)$ may be expressed in terms of $\mathbf{i}$, $\mathbf{j}$, and $\mathbf{k}$, where the components are scalar functions f, g, and h, it is simpler to use the following definition.

Definition 11.4

Let $\mathbf{r}(t) = f(t)\mathbf{i} + g(t)\mathbf{j} + h(t)\mathbf{k}$. The **limit of** $\mathbf{r}(t)$ as t approaches a is

$$\lim_{t \to a} \mathbf{r}(t) = \left[\lim_{t \to a} f(t) \right]\mathbf{i} + \left[\lim_{t \to a} g(t) \right]\mathbf{j} + \left[\lim_{t \to a} h(t) \right]\mathbf{k},$$

provided f, g, and h have limits as t approaches a.

Thus, to find $\lim_{t \to a} \mathbf{r}(t)$, we take the limit of each component of $\mathbf{r}(t)$. We may state a definition similar to (11.4) for one-sided limits.

ILLUSTRATION

$$\lim_{t \to 2}(t^2\mathbf{i} + 3t\mathbf{j} + 5\mathbf{k}) = \left[\lim_{t \to 2} t^2 \right]\mathbf{i} + \left[\lim_{t \to 2} 3t \right]\mathbf{j} + \left[\lim_{t \to 2} 5 \right]\mathbf{k}$$
$$= 4\mathbf{i} + 6\mathbf{j} + 5\mathbf{k}$$

Figure 11.9

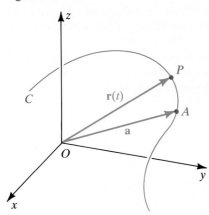

If, in Definition (11.4),

$$\lim_{t \to a} f(t) = a_1, \qquad \lim_{t \to a} g(t) = a_2, \quad \text{and} \quad \lim_{t \to a} h(t) = a_3,$$

then

$$\lim_{t \to a} \mathbf{r}(t) = a_1\mathbf{i} + a_2\mathbf{j} + a_3\mathbf{k}.$$

Letting $\mathbf{a} = a_1\mathbf{i} + a_2\mathbf{j} + a_3\mathbf{k}$ gives us the situation illustrated in Figure 11.9, where C is the curve determined by $\mathbf{r}(t)$ and $\overrightarrow{OA}$ is the position vector for $\mathbf{a}$. As t approaches a, the endpoint P of $\mathbf{r}(t)$ approaches A—that is, $\mathbf{r}(t)$ approaches $\overrightarrow{OA}$.

Continuity of a vector function $\mathbf{r}$ is defined in the same manner as for a scalar function.

Definition 11.5

A vector-valued function $\mathbf{r}$ is **continuous** at a if

$$\lim_{t \to a} \mathbf{r}(t) = \mathbf{r}(a).$$

It follows that if $\mathbf{r}(t) = f(t)\mathbf{i} + g(t)\mathbf{j} + h(t)\mathbf{k}$, then *$\mathbf{r}$ is continuous at a if and only if f, g, and h are continuous at a*. Continuity on an interval is defined in the usual way.

Definition 11.6

Let $\mathbf{r}$ be a vector-valued function. The **derivative** of $\mathbf{r}$ is the vector-valued function $\mathbf{r}'$ defined by

$$\mathbf{r}'(t) = \lim_{\Delta t \to 0} \frac{1}{\Delta t}[\mathbf{r}(t + \Delta t) - \mathbf{r}(t)]$$

for every t such that the limit exists.

If we write

$$\frac{1}{\Delta t}[\mathbf{r}(t + \Delta t) - \mathbf{r}(t)] = \frac{\mathbf{r}(t + \Delta t) - \mathbf{r}(t)}{\Delta t},$$

then Definition (11.6) takes on the familiar form for derivatives of scalar functions, introduced in Chapter 2.

The next theorem states that to find $\mathbf{r}'(t)$, we differentiate each component of $\mathbf{r}(t)$.

Theorem 11.7

If $\mathbf{r}(t) = f(t)\mathbf{i} + g(t)\mathbf{j} + h(t)\mathbf{k}$ and f, g, and h are differentiable, then

$$\mathbf{r}'(t) = f'(t)\mathbf{i} + g'(t)\mathbf{j} + h'(t)\mathbf{k}.$$

PROOF By Definition (11.6),

$$
\begin{aligned}
\mathbf{r}'(t) &= \lim_{\Delta t \to 0} \frac{\mathbf{r}(t + \Delta t) - \mathbf{r}(t)}{\Delta t} \\
&= \lim_{\Delta t \to 0} \frac{[f(t + \Delta t)\mathbf{i} + g(t + \Delta t)\mathbf{j} + h(t + \Delta t)\mathbf{k}] - [f(t)\mathbf{i} + g(t)\mathbf{j} + h(t)\mathbf{k}]}{\Delta t} \\
&= \lim_{\Delta t \to 0} \left[\frac{f(t + \Delta t) - f(t)}{\Delta t}\mathbf{i} + \frac{g(t + \Delta t) - g(t)}{\Delta t}\mathbf{j} + \frac{h(t + \Delta t) - h(t)}{\Delta t}\mathbf{k} \right].
\end{aligned}
$$

Taking the limit of each component gives us the conclusion of the theorem. ■

If $\mathbf{r}'(t)$ exists, we say that $\mathbf{r}$ is **differentiable** at t. We also denote derivatives as follows:

$$
\mathbf{r}'(t) = \frac{d}{dt}\mathbf{r}(t)
$$

Higher derivatives may be obtained in similar fashion. For example, if f, g, and h have second derivatives, then

$$
\mathbf{r}''(t) = f''(t)\mathbf{i} + g''(t)\mathbf{j} + h''(t)\mathbf{k}.
$$

EXAMPLE ■ 1 Let $\mathbf{r}(t) = (\ln t)\mathbf{i} + e^{-3t}\mathbf{j} + t^2\mathbf{k}$.

(a) Find the domain of $\mathbf{r}$, and determine where $\mathbf{r}$ is continuous.

(b) Find $\mathbf{r}'(t)$ and $\mathbf{r}''(t)$.

SOLUTION

(a) Since $\ln t$ is undefined if $t \le 0$, the domain of $\mathbf{r}$ is the set of positive real numbers. Moreover, $\mathbf{r}$ is continuous throughout its domain, since each component determines a continuous function.

(b) Applying Theorem (11.7), we have

$$
\mathbf{r}'(t) = \frac{1}{t}\mathbf{i} - 3e^{-3t}\mathbf{j} + 2t\mathbf{k}
$$

$$
\mathbf{r}''(t) = -\frac{1}{t^2}\mathbf{i} + 9e^{-3t}\mathbf{j} + 2\mathbf{k}.
$$

The next theorem lists differentiation formulas for vector-valued functions. Note the similarity to those for scalar functions.

Theorem 11.8

If $\mathbf{u}$ and $\mathbf{v}$ are differentiable vector-valued functions and c is a scalar, then

(i) $[\mathbf{u}(t) + \mathbf{v}(t)]' = \mathbf{u}'(t) + \mathbf{v}'(t)$

(ii) $[c\mathbf{u}(t)]' = c\mathbf{u}'(t)$

(iii) $[\mathbf{u}(t) \cdot \mathbf{v}(t)]' = \mathbf{u}(t) \cdot \mathbf{v}'(t) + \mathbf{u}'(t) \cdot \mathbf{v}(t)$

(iv) $[\mathbf{u}(t) \times \mathbf{v}(t)]' = \mathbf{u}(t) \times \mathbf{v}'(t) + \mathbf{u}'(t) \times \mathbf{v}(t)$

PROOF We shall prove (iii) and leave the other parts as exercises. Let

$$\mathbf{u}(t) = f_1(t)\mathbf{i} + f_2(t)\mathbf{j} + f_3(t)\mathbf{k}$$
$$\mathbf{v}(t) = g_1(t)\mathbf{i} + g_2(t)\mathbf{j} + g_3(t)\mathbf{k},$$

where each scalar function f_k and g_k is a differentiable function of t. By the definition of dot product,

$$\mathbf{u}(t) \cdot \mathbf{v}(t) = f_1(t)g_1(t) + f_2(t)g_2(t) + f_3(t)g_3(t) = \sum_{k=1}^{3} f_k(t)g_k(t).$$

Figure 11.10

(a)

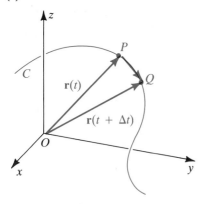

Consequently,

$$[\mathbf{u}(t) \cdot \mathbf{v}(t)]' = \left[\sum_{k=1}^{3} f_k(t)g_k(t) \right]' = \sum_{k=1}^{3} [f_k(t)g_k(t)]'$$

$$= \sum_{k=1}^{3} [f_k(t)g_k'(t) + f_k'(t)g_k(t)]$$

$$= \sum_{k=1}^{3} f_k(t)g_k'(t) + \sum_{k=1}^{3} f_k'(t)g_k(t)$$

$$= \mathbf{u}(t) \cdot \mathbf{v}'(t) + \mathbf{u}'(t) \cdot \mathbf{v}(t). \quad \blacksquare$$

To obtain a geometric interpretation for the derivative $\mathbf{r}'(t)$ in Definition (11.6), consider

$$\mathbf{r}(t) = f(t)\mathbf{i} + g(t)\mathbf{j} + h(t)\mathbf{k},$$

where f, g, and h are differentiable (scalar) functions, and let C be the curve determined by $\mathbf{r}(t)$. As shown in Figure 11.10(a), if $\overrightarrow{OP}$ and $\overrightarrow{OQ}$ correspond to $\mathbf{r}(t)$ and $\mathbf{r}(t + \Delta t)$, respectively, then

(b)

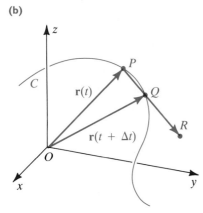

$$\overrightarrow{PQ} = \overrightarrow{OQ} - \overrightarrow{OP} \quad \text{corresponds to} \quad \mathbf{r}(t + \Delta t) - \mathbf{r}(t).$$

Hence, for any nonzero real number Δt,

$$\frac{1}{\Delta t}\overrightarrow{PQ} \quad \text{corresponds to} \quad \frac{1}{\Delta t}[\mathbf{r}(t + \Delta t) - \mathbf{r}(t)].$$

If $\Delta t > 0$, then $(1/\Delta t)\overrightarrow{PQ}$ is a vector $\overrightarrow{PR}$ having the same direction as $\overrightarrow{PQ}$. Moreover, if $0 < \Delta t < 1$, then $(1/\Delta t) > 1$ and $\| \overrightarrow{PR} \| > \| \overrightarrow{PQ} \|$, as illustrated in Figure 11.10(b).

(c)

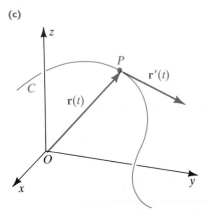

If we let $\Delta t \to 0^+$, then Q approaches P along C, and since $\overrightarrow{PR}$ lies on the secant line through P and Q, the vector PR should approach a vector that lies on the tangent line to C at P. A similar discussion may be given if $\Delta t < 0$. For this reason, we refer to $\mathbf{r}'(t)$, or any of its geometric representations, as a **tangent vector** to C at P. *We shall always represent $\mathbf{r}'(t)$ by a vector with initial point P on the curve C,* as shown in Figure 11.10(c). *The vector $\mathbf{r}'(t)$ points in the direction determined by the orientation of C.* By definition, the **tangent line** to C at P is the line through P that is parallel to $\mathbf{r}'(t)$.

Figure 11.11

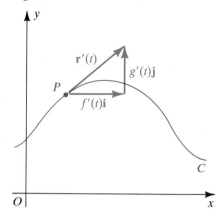

For the two-dimensional case $\mathbf{r}(t) = f(t)\mathbf{i} + g(t)\mathbf{j}$, the tangent vector is

$$\mathbf{r}'(t) = f'(t)\mathbf{i} + g'(t)\mathbf{j}.$$

If $f'(t) \neq 0$, then the slope of the line through P parallel to $\mathbf{r}'(t)$ is $g'(t)/f'(t)$ (see Figure 11.11). This is in agreement with the formula in Theorem (9.4) obtained for tangent lines to plane curves.

EXAMPLE ■ 2   Let $\mathbf{r}(t) = 2t\mathbf{i} + (t^2 - 4)\mathbf{j}$ for $-2 \leq t \leq 3$.

(a) Sketch the curve C determined by $\mathbf{r}(t)$, and indicate the orientation.
(b) Find $\mathbf{r}'(t)$ and sketch $\mathbf{r}(1)$ and $\mathbf{r}'(1)$.

SOLUTION

(a) Since $\mathbf{r}(t)$ is in V_2, we shall use an xy-plane. Parametric equations for C are

$$x = 2t, \quad y = t^2 - 4; \quad -2 \leq t \leq 3.$$

Eliminating the parameter, we see that C is on the parabola

$$y = (\tfrac{1}{2}x)^2 - 4 = \tfrac{1}{4}x^2 - 4.$$

The values $-2 \leq t \leq 3$ give us the part of the parabola between $(-4, 0)$ and $(6, 5)$ sketched in Figure 11.12, where the orientation is from left to right.

Figure 11.12

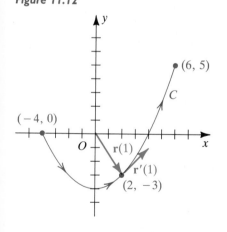

(b) Differentiating $\mathbf{r}(t) = 2t\mathbf{i} + (t^2 - 4)\mathbf{j}$ yields

$$\mathbf{r}'(t) = 2\mathbf{i} + 2t\mathbf{j}.$$

Letting $t = 1$ in the formulas for $\mathbf{r}(t)$ and $\mathbf{r}'(t)$, we obtain

$$\mathbf{r}(1) = 2\mathbf{i} - 3\mathbf{j}, \qquad \mathbf{r}'(1) = 2\mathbf{i} + 2\mathbf{j}.$$

As mentioned in our discussion, we always represent $\mathbf{r}'(t)$ graphically by taking the initial point as the point on C that corresponds to t. Thus, in Figure 11.12, the initial point of $\mathbf{r}'(1)$ is the point $P(2, -3)$ that corresponds to $t = 1$. Since $\mathbf{r}'(1) = 2\mathbf{i} + 2\mathbf{j}$, the terminal point of $\mathbf{r}'(1)$ is $(2 + 2, (-3) + 2)$, or $(4, -1)$. Note that $\mathbf{r}'(1)$ points in the direction determined by the orientation of C.

EXAMPLE ■ 3   Let C be the curve with parametric equations

$$x = t, \quad y = t^2, \quad z = t^3; \quad t \geq 0.$$

Find parametric equations for the tangent line to C at the point corresponding to $t = 2$.

SOLUTION   The curve C is determined by

$$\mathbf{r}(t) = t\mathbf{i} + t^2\mathbf{j} + t^3\mathbf{k} \quad \text{for } t \geq 0$$

and is the twisted cubic discussed in Example 3 of Section 11.1 (see Figure 11.5). From our previous discussion, a tangent vector to C at the point corresponding to t is

$$\mathbf{r}'(t) = \mathbf{i} + 2t\mathbf{j} + 3t^2\mathbf{k}.$$

In particular, the point on C corresponding to $t = 2$ is $(2, 4, 8)$, and a tangent vector is

$$\mathbf{r}'(2) = \mathbf{i} + 4\mathbf{j} + 12\mathbf{k}.$$

By Theorem (10.34), parametric equations for the corresponding tangent lines are

$$x = 2 + t, \quad y = 4 + 4t, \quad z = 8 + 12t; \quad t \text{ in } \mathbb{R}.$$

Figure 11.13

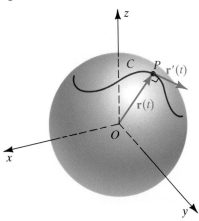

If $\mathbf{r}$ is differentiable and if the magnitude of $\mathbf{r}(t)$ is constant—say, $\|\mathbf{r}(t)\| = c$ for some c—then the curve C determined by $\mathbf{r}(t)$ lies on a sphere of radius c with center at O, as illustrated in Figure 11.13. Our geometric intuition tells us that the tangent vector $\mathbf{r}'(t)$ at any point P on C is orthogonal to the vector $\mathbf{r}(t)$ from O to P. We shall prove this fact next.

Theorem 11.9

> If $\mathbf{r}$ is differentiable and $\|\mathbf{r}(t)\|$ is constant, then $\mathbf{r}'(t)$ is orthogonal to $\mathbf{r}(t)$ for every t in the domain of $\mathbf{r}'$.

PROOF   If $\|\mathbf{r}(t)\| = c$ for some constant c, then

$$\mathbf{r}(t) \cdot \mathbf{r}(t) = \|\mathbf{r}(t)\|^2 = c^2$$

and, therefore,

$$[\mathbf{r}(t) \cdot \mathbf{r}(t)]' = \frac{d}{dt}(c^2) = 0.$$

Using Theorem (11.8)(iii), we obtain

$$\mathbf{r}(t) \cdot \mathbf{r}'(t) + \mathbf{r}'(t) \cdot \mathbf{r}(t) = 0,$$

or $$2[\mathbf{r}(t) \cdot \mathbf{r}'(t)] = 0.$$

Hence, by Theorem (10.21), $\mathbf{r}(t)$ and $\mathbf{r}'(t)$ are orthogonal.   ■

Definite integrals of vector-valued functions may be defined as follows.

Definition 11.10

> Let $\mathbf{r}(t) = f(t)\mathbf{i} + g(t)\mathbf{j} + h(t)\mathbf{k}$, where f, g, and h are integrable on $[a, b]$. The **definite integral of r from a to b** is
>
> $$\int_a^b \mathbf{r}(t)\, dt = \left[\int_a^b f(t)\, dt\right]\mathbf{i} + \left[\int_a^b g(t)\, dt\right]\mathbf{j} + \left[\int_a^b h(t)\, dt\right]\mathbf{k}.$$

If $\mathbf{R}'(t) = \mathbf{r}(t)$, then $\mathbf{R}(t)$ is an **antiderivative** of $\mathbf{r}(t)$. The next result is analogous to the fundamental theorem of calculus.

Theorem 11.11

If $\mathbf{R}(t)$ is an antiderivative of $\mathbf{r}(t)$ on $[a, b]$, then

$$\int_a^b \mathbf{r}(t)\, dt = \mathbf{R}(t)\Big]_a^b = \mathbf{R}(b) - \mathbf{R}(a).$$

EXAMPLE 4   Find $\int_0^2 \mathbf{r}(t)\, dt$ if $\mathbf{r}(t) = 12t^3\mathbf{i} + 4e^{2t}\mathbf{j} + (t+1)^{-1}\mathbf{k}$.

SOLUTION   Finding an antiderivative for each component of $\mathbf{r}(t)$, we obtain

$$\mathbf{R}(t) = 3t^4\mathbf{i} + 2e^{2t}\mathbf{j} + \ln(t+1)\mathbf{k}.$$

Since $\mathbf{R}'(t) = \mathbf{r}(t)$, it follows from Theorem (11.11) that

$$\int_0^2 \mathbf{r}(t)\, dt = \mathbf{R}(2) - \mathbf{R}(0)$$
$$= (48\mathbf{i} + 2e^4\mathbf{j} + \ln 3\mathbf{k}) - (0\mathbf{i} + 2\mathbf{j} + 0\mathbf{k})$$
$$= 48\mathbf{i} + 2(e^4 - 1)\mathbf{j} + \ln 3\mathbf{k}.$$

The theory of indefinite integrals of vector-valued functions is similar to that developed for real-valued functions in Chapter 4. The proofs of theorems require only minor modifications of those given earlier and are thus omitted. If $\mathbf{R}(t)$ is an antiderivative of $\mathbf{r}(t)$, then every antiderivative has the form $\mathbf{R}(t) + \mathbf{c}$ for some (constant) vector $\mathbf{c}$, and we write

$$\int \mathbf{r}(t)\, dt = \mathbf{R}(t) + \mathbf{c}, \quad \text{where} \quad \mathbf{R}'(t) = \mathbf{r}(t).$$

EXERCISES 11.2

Exer. 1–8: (a) Find the domain of r. (b) Find r′(t) and r″(t).

1 $\mathbf{r}(t) = \sqrt{t-1}\,\mathbf{i} + \sqrt{2-t}\,\mathbf{j}$

2 $\mathbf{r}(t) = \dfrac{1}{t}\mathbf{i} + \sin 3t\,\mathbf{j}$ 3 $\mathbf{r}(t) = \tan t\,\mathbf{i} + (t^2 + 8t)\mathbf{j}$

4 $\mathbf{r}(t) = e^{(t^2)}\mathbf{i} + \sin^{-1} t\,\mathbf{j}$ 5 $\mathbf{r}(t) = t^2\mathbf{i} + \tan t\,\mathbf{j} + 3\mathbf{k}$

6 $\mathbf{r}(t) = \sqrt[3]{t}\,\mathbf{i} + \dfrac{1}{t}\mathbf{j} + e^{-t}\mathbf{k}$ 7 $\mathbf{r}(t) = \sqrt{t}\,\mathbf{i} + e^{2t}\mathbf{j} + t\mathbf{k}$

8 $\mathbf{r}(t) = \ln(1-t)\mathbf{i} + \sin t\,\mathbf{j} + t^2\mathbf{k}$

Exer. 9–16: (a) Sketch the curve (in the xy-plane) determined by r(t) and indicate the orientation. (b) Find r′(t) and sketch r(t) and r′(t) for the indicated value of t.

9 $\mathbf{r}(t) = -\tfrac{1}{4}t^4\mathbf{i} + t^2\mathbf{j};$ $t = 2$

10 $\mathbf{r}(t) = e^{2t}\mathbf{i} + e^{-4t}\mathbf{j};$ $t = 0$

11 $\mathbf{r}(t) = 4\cos t\,\mathbf{i} + 2\sin t\,\mathbf{j};$ $t = 3\pi/4$

12 $\mathbf{r}(t) = 2\sec t\,\mathbf{i} + 3\tan t\,\mathbf{j};$ $t = \pi/4,\ |t| < \pi/2$

13 $\mathbf{r}(t) = t^3\mathbf{i} + t^{-3}\mathbf{j};$ $t = 1,$ $t > 0$

14 $\mathbf{r}(t) = t^2\mathbf{i} + t^3\mathbf{j};$ $t = -1$

15 $\mathbf{r}(t) = (2t - 1)\mathbf{i} + (4 - t)\mathbf{j};$ $t = 3$

16 $\mathbf{r}(t) = 5\mathbf{i} + t^3\mathbf{j};$ $t = 2$

Exer. 17–20: A curve C is given parametrically. Find parametric equations for the tangent line to C at P.

17 $x = 2t^3 - 1,$ $y = -5t^2 + 3,$ $z = 8t + 2;$ $P(1, -2, 10)$

18 $x = 4\sqrt{t},$ $y = t^2 - 10,$ $z = 4/t;$ $P(8, 6, 1)$

19 $x = e^t,$ $y = te^t,$ $z = t^2 + 4;$ $P(1, 0, 4)$

20 $x = t \sin t,$ $y = t \cos t,$ $z = t;$ $P(\pi/2, 0, \pi/2)$

Exer. 21–22: A curve C is given parametrically. Find two unit tangent vectors to C at P.

21 $x = e^{2t},$ $y = e^{-t},$ $z = t^2 + 4;$ $P(1, 1, 4)$

22 $x = \sin t + 2,$ $y = \cos t,$ $z = t;$ $P(2, 1, 0)$

23 Refer to Exercise 27 of Section 11.1. Show that the concho-spiral has the special property that the angle between $\mathbf{k}$ and the tangent vector $\mathbf{r}'(t)$ is a constant.

24 The *general helix* is a curve whose tangent vector makes a constant angle with a fixed unit vector $\mathbf{u}$. Show that the curve with parametrization $x = 3t - t^3,$ $y = 3t^2, z = 3t + t^3; t$ in $\mathbb{R}$ is a general helix by finding an appropriate vector $\mathbf{u}$.

25 A point P moves along a curve C in such a way that the position vector $\mathbf{r}(t)$ of P is equal to the tangent vector $\mathbf{r}'(t)$ for every t. Find parametric equations for C, and describe the graph.

26 A point moves along a curve C in such a way that the position vector $\mathbf{r}(t)$ and the tangent vector $\mathbf{r}'(t)$ are always orthogonal. Prove that C lies on a sphere with center at the origin. (*Hint:* Show that $(d/dt)\,\|\mathbf{r}(t)\|^2 = 0$.)

Exer. 27–30: Evaluate the integral.

27 $\displaystyle\int_0^2 (6t^2\mathbf{i} - 4t\mathbf{j} + 3\mathbf{k})\,dt$

28 $\displaystyle\int_{-1}^1 (-5t\mathbf{i} + 8t^3\mathbf{j} - 3t^2\mathbf{k})\,dt$

29 $\displaystyle\int_0^{\pi/4} (\sin t\mathbf{i} - \cos t\mathbf{j} + \tan t\mathbf{k})\,dt$

30 $\displaystyle\int_0^1 [te^{(t^2)}\mathbf{i} + \sqrt{t}\mathbf{j} + (t^2 + 1)^{-1}\mathbf{k}]\,dt$

Exer. 31–34: Find $\mathbf{r}(t)$ subject to the given conditions.

31 $\mathbf{r}'(t) = t^2\mathbf{i} + (6t + 1)\mathbf{j} + 8t^3\mathbf{k},$
$\mathbf{r}(0) = 2\mathbf{i} - 3\mathbf{j} + \mathbf{k}$

32 $\mathbf{r}'(t) = 2\mathbf{i} - 4t^3\mathbf{j} + 6\sqrt{t}\mathbf{k},$
$\mathbf{r}(0) = \mathbf{i} + 5\mathbf{j} + 3\mathbf{k}$

33 $\mathbf{r}''(t) = 6t\mathbf{i} - 12t^2\mathbf{j} + \mathbf{k},$
$\mathbf{r}'(0) = \mathbf{i} + 2\mathbf{j} - 3\mathbf{k},$ $\mathbf{r}(0) = 7\mathbf{i} + \mathbf{k}$

34 $\mathbf{r}''(t) = 6t\mathbf{i} + 3\mathbf{j},$
$\mathbf{r}'(0) = 4\mathbf{i} - \mathbf{j} + \mathbf{k},$ $\mathbf{r}(0) = 5\mathbf{j}$

Exer. 35–36: If a curve C has a tangent vector **a** at a point P, then the *normal plane* to C at P is the plane through P with normal vector **a**. Find an equation of the normal plane to the given curve at P.

35 $x = e^t,$ $y = te^t,$ $z = t^2 + 4;$ $P(1, 0, 4)$

36 $x = t \sin t,$ $y = t \cos t,$ $z = t;$ $P(\pi/2, 0, \pi/2)$

Exer. 37–38: Find $[\mathbf{u}(t) \cdot \mathbf{v}(t)]'$ and $[\mathbf{u}(t) \times \mathbf{v}(t)]'$.

37 $\mathbf{u}(t) = t\mathbf{i} + t^2\mathbf{j} + t^3\mathbf{k},$ $\mathbf{v}(t) = \sin t\mathbf{i} + \cos t\mathbf{j} + 2\sin t\mathbf{k}$

38 $\mathbf{u}(t) = 2t\mathbf{i} + 6t\mathbf{j} + t^2\mathbf{k},$ $\mathbf{v}(t) = e^{-t}\mathbf{i} - e^{-t}\mathbf{j} + \mathbf{k}$

39 If **u** and **v** are vector-valued functions that have limits as $t \to a$, prove the following:

(a) $\displaystyle\lim_{t \to a}[\mathbf{u}(t) + \mathbf{v}(t)] = \lim_{t \to a}\mathbf{u}(t) + \lim_{t \to a}\mathbf{v}(t)$

(b) $\displaystyle\lim_{t \to a}[\mathbf{u}(t) \cdot \mathbf{v}(t)] = \lim_{t \to a}\mathbf{u}(t) \cdot \lim_{t \to a}\mathbf{v}(t)$

(c) $\displaystyle\lim_{t \to a} c\mathbf{u}(t) = c\lim_{t \to a}\mathbf{u}(t),$ where c is a scalar

40 If a scalar function f and a vector-valued function **u** have limits as $t \to a$, prove that

$$\lim_{t \to a} f(t)\mathbf{u}(t) = \left[\lim_{t \to a} f(t)\right]\left[\lim_{t \to a}\mathbf{u}(t)\right].$$

41 Prove that $\lim_{t \to a}\mathbf{u}(t) = \mathbf{b}$ if and only if for every $\epsilon > 0$ there is a $\delta > 0$ such that $\|\mathbf{u}(t) - \mathbf{b}\| < \epsilon$ whenever $0 < |t - a| < \delta$. Give a graphical description of this result.

42 If **u** and **v** have limits as $t \to a$, prove that

$$\lim_{t \to a}[\mathbf{u}(t) \times \mathbf{v}(t)] = \left[\lim_{t \to a}\mathbf{u}(t)\right] \times \left[\lim_{t \to a}\mathbf{v}(t)\right].$$

Exer. 43–44: If **u** and **v** are differentiable, prove the stated rule for derivatives.

43 $[\mathbf{u}(t) + \mathbf{v}(t)]' = \mathbf{u}'(t) + \mathbf{v}'(t)$

44 $[\mathbf{u}(t) \times \mathbf{v}(t)]' = \mathbf{u}(t) \times \mathbf{v}'(t) + \mathbf{u}'(t) \times \mathbf{v}(t)$

45 If f and **u** are differentiable, prove that

$$[f(t)\mathbf{u}(t)]' = f(t)\mathbf{u}'(t) + f'(t)\mathbf{u}(t).$$

46 If f and **u** are differentiable with suitably restricted domains, prove the chain rule

$$[\mathbf{u}(f(t))]' = f'(t)\mathbf{u}'(f(t)).$$

47 If **u**, **v**, and **w** are differentiable, prove that

$$[\mathbf{u}(t) \cdot \mathbf{v}(t) \times \mathbf{w}(t)]' = [\mathbf{u}'(t) \cdot \mathbf{v}(t) \times \mathbf{w}(t)]$$
$$+ [\mathbf{u}(t) \cdot \mathbf{v}'(t) \times \mathbf{w}(t)]$$
$$+ [\mathbf{u}(t) \cdot \mathbf{v}(t) \times \mathbf{w}'(t)].$$

48 If $\mathbf{u}'(t)$ and $\mathbf{u}''(t)$ exist, prove that

$$[\mathbf{u}(t) \times \mathbf{u}'(t)]' = \mathbf{u}(t) \times \mathbf{u}''(t).$$

49 If **u** and **v** are integrable on $[a, b]$ and if c is a scalar,

prove the following:

(a) $\displaystyle \int_a^b [\mathbf{u}(t) + \mathbf{v}(t)] \, dt = \int_a^b \mathbf{u}(t) \, dt + \int_a^b \mathbf{v}(t) \, dt$

(b) $\displaystyle \int_a^b c\mathbf{u}(t) \, dt = c \int_a^b \mathbf{u}(t) \, dt$

50 If **u** is integrable on $[a, b]$ and **c** is in V_3, prove that

$$\int_a^b \mathbf{c} \cdot \mathbf{u}(t) \, dt = \mathbf{c} \cdot \int_a^b \mathbf{u}(t) \, dt.$$

11.3   CURVILINEAR MOTION

In Section 3.7, we analyzed the motion of an object moving along a straight line using a function f whose value $f(t)$ was the position of the object at time t. In this section, we investigate the motion of an object along a curve in the plane or in space. Motion often takes place in a plane. For example, although the earth moves through space, its orbit lies in a plane. (This will be proved in Section 11.6.) To study the motion of a point P in a coordinate plane, it is essential to know its position (x, y) at every instant. As usual, for *objects* in motion we assume that the mass is concentrated at P. Suppose the coordinates of P are given by the parametric equations

$$x = f(t), \quad y = g(t),$$

where t is in some interval I. If we let

$$\mathbf{r}(t) = f(t)\mathbf{i} + g(t)\mathbf{j},$$

then as t varies through I, the endpoint of $\mathbf{r}(t)$ traces the path C of the point. We refer to $\mathbf{r}(t)$ as the **position vector of P**. As in Figure 11.14, we represent

$$\mathbf{r}'(t) = f'(t)\mathbf{i} + g'(t)\mathbf{j}$$

as a tangent vector to C with initial point P. The vector $\mathbf{r}'(t)$ points in the direction of increasing values of t and has magnitude

$$\|\mathbf{r}'(t)\| = \sqrt{[f'(t)]^2 + [g'(t)]^2}.$$

Let t_0 be any number in I, and let P_0 be the point on C that corresponds to t_0 (see Figure 11.14). If C is smooth, then, by Theorem (9.6), the arc length $s(t)$ of C from P_0 to P is

$$s(t) = \int_{t_0}^t \sqrt{[f'(t)]^2 + [g'(t)]^2} \, dt = \int_{t_0}^t \|\mathbf{r}'(t)\| \, dt.$$

Applying Theorem (4.35), we obtain

$$\frac{d}{dt}[s(t)] = \frac{d}{dt} \int_{t_0}^t \|\mathbf{r}'(t)\| \, dt = \|\mathbf{r}'(t)\|.$$

Figure 11.14

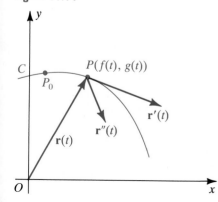

Thus, $\|\mathbf{r}'(t)\|$ *is the rate of change of arc length with respect to time.* For this reason, we refer to $\|\mathbf{r}'(t)\|$ as the *speed* of the point. The tangent vector $\mathbf{r}'(t)$ is defined as the *velocity* of the point P at time t, and the vector $\mathbf{r}''(t)$ is the *acceleration* of P. As with $\mathbf{r}'(t)$, we shall represent $\mathbf{r}''(t)$ graphically by a vector with initial point P. In most cases, $\mathbf{r}''(t)$ is directed toward the concave side of C, as illustrated in Figure 11.14.

The next definition summarizes this discussion and introduces the symbols $\mathbf{v}$ and $\mathbf{a}$ for velocity and acceleration.

Definition 11.12

Let the position vector for a point $P(x, y)$ moving in an xy-plane be

$$\mathbf{r}(t) = x\mathbf{i} + y\mathbf{j} = f(t)\mathbf{i} + g(t)\mathbf{j},$$

where t is time and f and g have first and second derivatives. The velocity, speed, and acceleration of P at time t are as follows:

Velocity: $\quad \mathbf{v}(t) = \mathbf{r}'(t) = \dfrac{dx}{dt}\mathbf{i} + \dfrac{dy}{dt}\mathbf{j}$

Speed: $\quad v(t) = \|\mathbf{v}(t)\| = \|\mathbf{r}'(t)\| = \sqrt{\left(\dfrac{dx}{dt}\right)^2 + \left(\dfrac{dy}{dt}\right)^2}$

Acceleration: $\quad \mathbf{a}(t) = \mathbf{v}'(t) = \mathbf{r}''(t) = \dfrac{d^2x}{dt^2}\mathbf{i} + \dfrac{d^2y}{dt^2}\mathbf{j}$

Most graphing utilities offer a *dot mode* or a *point style* for plotting as an alternative to the default connected mode in which a line segment (or a curve segment) is drawn between calculated points. Plotting parametric equations representing motion along the curve in the plane or in space using the dot mode can help us visualize speed and velocity. When plotting

$$x = f(t), \quad y = g(t); \quad a \leq t \leq b,$$

most graphing utilities choose a regular partition $a = t_0 < t_1 < \cdots < t_n = b$ of the parameter interval with $t_{i+1} - t_i = \Delta t$. The dot mode will display only the points $P_i(f(t_i), g(t_i))$, for $i = 0, 1, \ldots, n$. Visually, we can relate a greater distance between plotted points to greater speed:

$$\text{dist}(P_i, P_{i+1}) = \sqrt{\left[f(t_{i+1}) - f(t_i)\right]^2 + \left[g(t_{i+1}) - g(t_i)\right]^2}$$

$$= \sqrt{\left[\frac{f(t_{i+1}) - f(t_i)}{\Delta t}\right]^2 + \left[\frac{g(t_{i+1}) - g(t_i)}{\Delta t}\right]^2}\,\Delta t$$

$$\approx v(t_i)\Delta t.$$

We can also visualize the direction of the velocity vector $\mathbf{v}(t)$ as being in about the same direction as the vector between plotted points:

$$\overrightarrow{P_i P_{i+1}} = \left[f(t_{i+1}) - f(t_i)\right]\mathbf{i} + \left[g(t_{i+1}) - g(t_i)\right]\mathbf{j}$$

$$= \left\{\left[\frac{f(t_{i+1}) - f(t_i)}{\Delta t}\right]\mathbf{i} + \left[\frac{g(t_{i+1}) - g(t_i)}{\Delta t}\right]\mathbf{j}\right\}\Delta t$$

$$\approx \mathbf{v}(t_i)\Delta t.$$

EXAMPLE ▪ 1   The position vector of a point P moving in an xy-plane is

$$\mathbf{r}(t) = (t^2 + t + 0.5)\mathbf{i} + (t^3 + 0.5)\mathbf{j}, \quad \text{for} \quad 0 \le t \le 2.$$

(a) Determine the velocity $\mathbf{v}(t)$ and the acceleration $\mathbf{a}(t)$ of P at time t.

(b) Sketch the path C of the point, together with $\mathbf{v}(1)$ and $\mathbf{a}(1)$.

(c) Use a graphing utility in dot mode to graph the path C.

SOLUTION

Figure 11.15

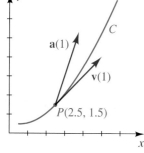

(a) The velocity and the acceleration are

$$\mathbf{v}(t) = \mathbf{r}'(t) = (2t + 1)\mathbf{i} + 3t^2\mathbf{j} \quad \text{and} \quad \mathbf{a}(t) = \mathbf{r}''(t) = 2\mathbf{i} + 6t\mathbf{j}.$$

(b) Parametric equations for C obtained from $\mathbf{r}(t)$ are

$$x = t^2 + t + 0.5, \quad y = t^3 + 0.5; \quad 0 \le t \le 2,$$

and coordinates of several points (x, y) on C are listed in the following table.

t	0	0.5	1	1.5	2
x	0.5	1.25	2.5	4.25	6.5
y	0.5	0.625	1.5	3.875	8.5

Figure 11.16
$-2.7 \le x \le 11.55, -0.6 \le y \le 8.85,$
$0 \le t \le 2, \Delta t = 0.1$

Plotting points and using the continuity of the scalar components of $\mathbf{r}(t)$ give us the sketch of C in Figure 11.15.

At $t = 1$, the point is at $P(2.5, 1.5)$ with

$$\mathbf{v}(1) = \mathbf{r}'(1) = 3\mathbf{i} + 3\mathbf{j} \quad \text{and} \quad \mathbf{a}(1) = \mathbf{r}''(1) = 2\mathbf{i} + 6\mathbf{j}.$$

(c) We set the graphing utility in dot mode and parametric mode, and enter the parametric equations in part (b). We next let t vary from 0 to 2 in increments of 0.1 to obtain the points plotted in Figure 11.16. Observe that as t increases, the gaps between successive plotted points grow larger. Thus, we conclude that the point is speeding up as t increases.

Figure 11.17

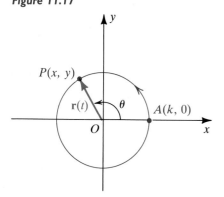

EXAMPLE ▪ 2   Show that if a point P moves around a circle of radius k at a constant speed v, then the acceleration vector has constant magnitude v^2/k and is directed from P toward the center of the circle.

SOLUTION   Let us assume that the center of the circle is the origin O in an xy-plane, and the orientation is counterclockwise. Suppose that at time $t = 0$ the point P is at $A(k, 0)$ and that θ is the angle generated by $\overrightarrow{OP}$ after t units of time (see Figure 11.17). Since P moves around the circle at a constant speed, the rate of change of θ with respect to t (the **angular speed**) is a constant ω. Thus,

$$\frac{d\theta}{dt} = \omega, \quad \text{or} \quad d\theta = \omega \, dt.$$

Integration gives us $\int d\theta = \int \omega\,dt$, or

$$\theta = \omega t + c$$

for some constant c. Since $\theta = 0$ when $t = 0$, we see that $c = 0$—that is, $\theta = \omega t$. Thus, the coordinates (x, y) of P are

$$x = k \cos \omega t, \qquad y = k \sin \omega t,$$

and the position vector of P is

$$\mathbf{r}(t) = k \cos \omega t\,\mathbf{i} + k \sin \omega t\,\mathbf{j}.$$

Consequently,

$$\mathbf{v}(t) = \mathbf{r}'(t) = -\omega k \sin \omega t\,\mathbf{i} + \omega k \cos \omega t\,\mathbf{j},$$

$$\mathbf{a}(t) = \mathbf{r}''(t) = -\omega^2 k \cos \omega t\,\mathbf{i} - \omega^2 k \sin \omega t\,\mathbf{j},$$

and hence

$$\mathbf{a}(t) = -\omega^2 (k \cos \omega t\,\mathbf{i} + k \sin \omega t\,\mathbf{j}) = -\omega^2 \mathbf{r}(t).$$

The negative sign shows that the direction of the acceleration vector $\mathbf{a}(t)$ is opposite that of $\mathbf{r}(t)$, and hence $\mathbf{a}(t)$ is directed from P toward O, as illustrated in Figure 11.18, which also shows the velocity vector $\mathbf{v}(t)$.

The magnitude of $\mathbf{a}(t)$ is

$$\|\mathbf{a}(t)\| = \left\|-\omega^2 \mathbf{r}(t)\right\| = \left|-\omega^2\right|\,\|\mathbf{r}(t)\| = \omega^2 k,$$

which is a constant. Moreover,

$$v = \|\mathbf{r}'(t)\| = \sqrt{(-\omega k)^2 \sin^2 \omega t + (\omega k)^2 \cos^2 \omega t} = \sqrt{\omega^2 k^2} = \omega k,$$

and hence $\omega = v/k$. Substitution in the formula $\|\mathbf{a}(t)\| = \omega^2 k$ gives us the constant magnitude

$$\|\mathbf{a}(t)\| = \frac{v^2}{k}.$$

Figure 11.18

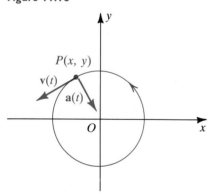

The acceleration vector $\mathbf{a}(t)$ in Example 2 is called a **centripetal acceleration vector**, and the force that produces $\mathbf{a}(t)$ is a **centripetal force**. Note that the magnitude $\|\mathbf{a}(t)\| = v^2/k$ will increase if we either increase v or decrease k, a fact that is evident to anyone who has attached an object to a string or rope and twirled it about in a circular path.

Our discussion of motion may be extended to a point P moving in a three-dimensional coordinate system. Suppose the coordinates of P at time t are $(f(t), g(t), h(t))$, where f, g, and h are defined on an interval I. The position vector of P is

$$\mathbf{r}(t) = f(t)\mathbf{i} + g(t)\mathbf{j} + h(t)\mathbf{k}.$$

As t varies, the endpoint P of $\mathbf{r}(t)$ traces the path. As in Definition (11.12), the derivative $\mathbf{r}'(t)$, if it exists, is the **velocity** $\mathbf{v}(t)$ at time t and is a tangent vector to C at P. The vector $\mathbf{a}(t) = \mathbf{r}''(t)$ is the **acceleration** at time t. The **speed** $v(t)$ is the magnitude $\|\mathbf{v}(t)\|$ of the velocity and, as in two dimensions, is equal to the rate of change of arc length with respect to time.

Figure 11.19

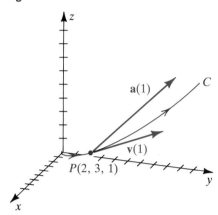

EXAMPLE ▪3 The position vector of a point P is given by $\mathbf{r}(t) = 2t\mathbf{i} + 3t^2\mathbf{j} + t^3\mathbf{k}$ for $0 \le t \le 2$. Find the velocity and the acceleration of P at time t. Sketch the path of P and show $\mathbf{v}(1)$ and $\mathbf{a}(1)$.

SOLUTION The velocity and the acceleration are

$$\mathbf{v}(t) = \mathbf{r}'(t) = 2\mathbf{i} + 6t\mathbf{j} + 3t^2\mathbf{k} \quad \text{and} \quad \mathbf{a}(t) = \mathbf{r}''(t) = 6\mathbf{j} + 6t\mathbf{k}.$$

At $t = 1$, the point is at $P(2, 3, 1)$ and

$$\mathbf{v}(1) = 2\mathbf{i} + 6\mathbf{j} + 3\mathbf{k} \quad \text{and} \quad \mathbf{a}(1) = 6\mathbf{j} + 6\mathbf{k}.$$

These vectors and the path of the point are sketched in Figure 11.19. The path is a twisted cubic (see Example 3 and Exercise 29 of Section 11.1).

In the discussion of torque at the end of Section 10.4, we noted that the cross product of vectors is used in the investigation of rotational motion. The following example is another illustration of this fact.

Figure 11.20

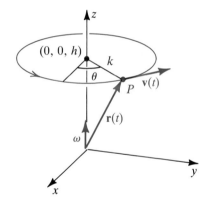

EXAMPLE ▪4 A point P is rotating about the z-axis on a circle of radius k that lies in the plane $z = h$, as illustrated in Figure 11.20. The angular speed $d\theta/dt$ is a constant ω. The vector $\boldsymbol{\omega} = \omega\mathbf{k}$ directed along the z-axis and having magnitude ω is the **angular velocity** of P. Show that the velocity $\mathbf{v}(t)$ of P is the cross product of $\boldsymbol{\omega}$ and the position vector $\mathbf{r}(t)$ for P.

SOLUTION Extending Example 2 to three dimensions, we find that the motion of P is given by

$$\mathbf{r}(t) = k \cos \omega t\,\mathbf{i} + k \sin \omega t\,\mathbf{j} + h\mathbf{k}.$$

Using the definition of cross product, we have

$$\boldsymbol{\omega} \times \mathbf{r}(t) = \begin{vmatrix} \mathbf{i} & \mathbf{j} & \mathbf{k} \\ 0 & 0 & \omega \\ k \cos \omega t & k \sin \omega t & h \end{vmatrix}$$

$$= -\omega k \sin \omega t\,\mathbf{i} + \omega k \cos \omega t\,\mathbf{j} = \mathbf{r}'(t) = \mathbf{v}(t).$$

The following law is used frequently in the investigation of objects in motion.

Newton's Second Law of Motion 11.13

The force $\mathbf{F}$ acting on an object of constant mass m is related to the acceleration $\mathbf{a}$ of the object as follows:

$$\mathbf{F} = m\mathbf{a}$$

In the next example, we shall use (11.13) to determine the position of an object that is moving near the surface of the earth under the influence of gravity alone—that is, air resistance and other forces that could affect acceleration are negligible. We shall also assume that the ground is level and the curvature of the earth is not a factor in determining the path of the object.

EXAMPLE ▪ 5   A projectile is fired, with an initial velocity $\mathbf{v}_0$, from a point h_0 feet above ground. If the only force acting on the projectile is caused by the gravitational acceleration $\mathbf{g}$, determine its position after t seconds.

Figure 11.21

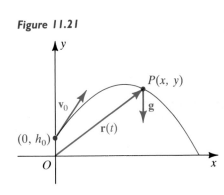

SOLUTION   Let us introduce an xy-plane, as in Figure 11.21, where $(0, h_0)$ is the point from which the projectile is fired. Let $P(x, y)$ denote the position of the projectile after t seconds, and let $\mathbf{r}(t)$ be the position vector for P. Since $\mathbf{g}$ acts in the downward direction,

$$\mathbf{g} = -g\mathbf{j} \quad \text{and} \quad \|\mathbf{g}\| = g \approx 32 \text{ ft/sec}^2.$$

By Newton's second law (11.13), $\mathbf{F} = m\mathbf{a}$, where $\mathbf{a}$ is the acceleration of the projectile. In the present situation, $m\mathbf{a} = m\mathbf{g}$, or $\mathbf{a} = \mathbf{g}$, which leads to the vector differential equation

$$\mathbf{r}''(t) = \mathbf{g}.$$

Indefinite integration gives us

$$\mathbf{r}'(t) = t\mathbf{g} + \mathbf{c}$$

for some constant vector $\mathbf{c}$. Since $\mathbf{r}'(t)$ is the velocity at time t,

$$\mathbf{v}_0 = \mathbf{r}'(0) = \mathbf{c}$$

and hence

$$\mathbf{r}'(t) = t\mathbf{g} + \mathbf{v}_0.$$

By indefinite integration,

$$\mathbf{r}(t) = \tfrac{1}{2}t^2\mathbf{g} + t\mathbf{v}_0 + \mathbf{d}$$

for some constant vector $\mathbf{d}$. Since $\mathbf{r}(0) = h_0\mathbf{j}$ (see Figure 11.21), it follows that $\mathbf{d} = h_0\mathbf{j}$. Consequently,

$$\mathbf{r}(t) = \tfrac{1}{2}t^2\mathbf{g} + t\mathbf{v}_0 + h_0\mathbf{j}.$$

Since $\mathbf{g} = -g\mathbf{j}$, this may also be written

$$\mathbf{r}(t) = (-\tfrac{1}{2}t^2g + h_0)\mathbf{j} + t\mathbf{v}_0.$$

EXAMPLE ▪ 6   Suppose, in Example 5, that the projectile is fired from O and the initial velocity vector $\mathbf{v}_0$ makes an angle α with the horizontal.

(a) Find parametric equations and an equation in x and y for the path of the projectile.

(b) Find the range of the projectile—that is, the horizontal distance it travels before hitting the ground. What is the maximum range?

(c) Find the maximum altitude of the projectile.

Figure 11.22

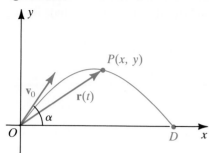

SOLUTION

(a) This special case of Example 5 is illustrated in Figure 11.22. A unit vector $\mathbf{u}$ having the same direction as $\mathbf{v}_0$ is

$$\mathbf{u} = \cos \alpha \, \mathbf{i} + \sin \alpha \, \mathbf{j}.$$

Thus, if $\|\mathbf{v}_0\| = v_0$, then

$$\mathbf{v}_0 = v_0 \mathbf{u} = (v_0 \cos \alpha) \mathbf{i} + (v_0 \sin \alpha) \mathbf{j}.$$

Using the formula for $\mathbf{r}(t)$ obtained in Example 5 with $h_0 = 0$, we obtain

$$\mathbf{r}(t) = (-\tfrac{1}{2} g t^2) \mathbf{j} + t[(v_0 \cos \alpha) \mathbf{i} + (v_0 \sin \alpha) \mathbf{j}]$$

$$= t(v_0 \cos \alpha) \mathbf{i} + (-\tfrac{1}{2} g t^2 + t v_0 \sin \alpha) \mathbf{j}.$$

Hence, parametric equations for the path of the projectile are

$$x = (v_0 \cos \alpha) t, \qquad y = -\tfrac{1}{2} g t^2 + (v_0 \sin \alpha) t.$$

Eliminating the parameter gives us the following equation in rectangular coordinates:

$$y = \frac{-g}{2 v_0^2 \cos^2 \alpha} x^2 + (\tan \alpha) x$$

This shows that the path of the projectile is parabolic.

(b) To find the range of the projectile, we must find the point D in Figure 11.22 at which the projectile hits the ground. Since the y-coordinate of D is 0, we let $y = 0$ in the second parametric equation of part (a) and factor, obtaining

$$t(-\tfrac{1}{2} g t + v_0 \sin \alpha) = 0.$$

Thus, either

$$t = 0 \quad \text{or} \quad -\tfrac{1}{2} g t + v_0 \sin \alpha = 0.$$

It follows that the projectile is at O if $t = 0$ and at D if

$$\tfrac{1}{2} g t = v_0 \sin \alpha, \quad \text{or} \quad t = \frac{2 v_0 \sin \alpha}{g}.$$

Using the second parametric equation of part (a), we can find the distance $d(O, D)$ from O to D:

$$d(O, D) = (v_0 \cos \alpha) \left(\frac{2 v_0 \sin \alpha}{g} \right) = \frac{v_0^2 \sin 2\alpha}{g}$$

In particular, the range will have its maximum value v_0^2/g if $\sin 2\alpha = 1$, or $\alpha = 45°$.

(c) Differentiating $\mathbf{r}(t)$ (see part (a)), we obtain the velocity vector

$$\mathbf{v}(t) = \mathbf{r}'(t) = (v_0 \cos \alpha) \mathbf{i} + (-g t + v_0 \sin \alpha) \mathbf{j}.$$

At the maximum altitude, this tangent vector to the path of the projectile is horizontal—that is, the $\mathbf{j}$-component is zero. Solving for t gives us $t = (v_0 \sin \alpha)/g$. We can also obtain this by taking one-half the time required for the projectile to reach D. Substituting this value of t into the parametric equation for y found in part (a), we find that the maximum altitude is

$$-\frac{1}{2}g\left(\frac{v_0\sin\alpha}{g}\right)^2 + (v_0\sin\alpha)\left(\frac{v_0\sin\alpha}{g}\right) = \frac{v_0^2\sin^2\alpha}{2g}.$$

EXERCISES 11.3

Exer. 1–8: Let $r(t)$ be the position vector of a moving point P. Sketch the path C of P together with $v(t)$ and $a(t)$ for the given value of t.

1 $r(t) = 2t\mathbf{i} + (4t^2 + 1)\mathbf{j}$; $t = 1$

2 $r(t) = (4 - 9t^2)\mathbf{i} + 3t\mathbf{j}$; $t = 1$

3 $r(t) = \sin t\mathbf{i} + 4\cos 2t\mathbf{j}$; $t = \pi/6$

4 $r(t) = \cos^2 t\mathbf{i} + 2\sin t\mathbf{j}$; $t = 3\pi/4$

5 $r(t) = \cos t\mathbf{i} + \sin t\mathbf{j} + t\mathbf{k}$; $t = \pi/2$

6 $r(t) = 4\sin t\mathbf{i} + 2t\mathbf{j} + 9\cos t\mathbf{k}$; $t = 3\pi/4$

7 $r(t) = t^2\mathbf{i} + t\mathbf{j} + 2t\mathbf{k}$; $t = 1$

8 $r(t) = t^2\mathbf{i} + t^3\mathbf{j} + t\mathbf{k}$; $t = 2$

Exer. 9–16: If $r(t)$ is the position vector of a moving point P, find its velocity, acceleration, and speed at the given time t.

9 $r(t) = \dfrac{2}{t}\mathbf{i} + \dfrac{3}{t+1}\mathbf{j}$; $t = 2$

10 $r(t) = \sqrt{t}\mathbf{i} + (1 + \sqrt{t})\mathbf{j}$; $t = 4$

11 $r(t) = e^{2t}\mathbf{i} + e^{-t}\mathbf{j}$; $t = 0$

12 $r(t) = 2t\mathbf{i} + e^{-t^2}\mathbf{j}$; $t = 1$

13 $r(t) = e^t(\cos t\mathbf{i} + \sin t\mathbf{j} + \mathbf{k})$; $t = \pi/2$

14 $r(t) = t(\cos t\mathbf{i} + \sin t\mathbf{j} + t\mathbf{k})$; $t = \pi/2$

15 $r(t) = (1 + t)\mathbf{i} + 2t\mathbf{j} + (2 + 3t)\mathbf{k}$; $t = 2$

16 $r(t) = 2t\mathbf{i} + \mathbf{j} + 9t^2\mathbf{k}$; $t = 2$

17 If a point moves at a constant speed, prove that the velocity and acceleration vectors are orthogonal. (*Hint:* See Theorem (11.9).)

18 If the acceleration of a moving point is always $\mathbf{0}$, prove that the motion is along a line.

c **Exer. 19–20:** Let $r(t)$ be the position vector of a moving point. Plot the path C traced out by the point (preferably in a dot or a point mode). From the graph, estimate the value of t for which the velocity is largest.

19 $r(t) = 28\cos t\mathbf{i} + 10\sin(0.8t)\mathbf{j}$; $0 \le t \le 2\pi$

20 $r(t) = (5 + \cos 3t)\mathbf{i} + 1.6\sin 2t\mathbf{j}$; $0 \le t \le \pi$

Exer. 21–22: Solve, using the results of Example 2 and 4000 mi for the radius of the earth.

21 A space shuttle is in a circular orbit 150 mi above the surface of the earth. Approximate

(a) its speed

(b) the time required for one revolution

22 An earth satellite is in a circular orbit. If the time required for one revolution is 88 min, approximate the satellite's altitude.

Exer. 23–26: Solve, using the results of Example 5.

23 A projectile is fired from level ground with an initial speed of 1500 ft/sec and angle of elevation 30°. Find

(a) the velocity at time t

(b) the maximum altitude

(c) the range

(d) the speed at which the projectile strikes the ground

24 Work Exercise 23 if the angle of elevation is 60°.

25 A baseball player throws a ball a distance of 250 ft. If the ball is released at an angle of 45° with the horizontal, find its initial speed.

26 A projectile is fired horizontally with a velocity of 1800 ft/sec from an altitude of 1000 ft above level ground. When and where does it strike the ground?

27 To test ability to withstand G-forces, an astronaut is placed at the end of a centrifuge device (see figure) that

rotates at an angular velocity ω. If the arm is 30 ft in length, find the number of revolutions per second that will result in an acceleration that is eight times that of gravity $\mathbf{g}$. (Use $\|\mathbf{g}\| = 32$ ft/sec^2.)

Exercise 27

28 The orbits of Earth, Venus, and Neptune are nearly circular. Given the information in the table, estimate the (average) speed of each planet to the nearest 0.1 km/sec.

Planet	Period (days)	Distance from sun (10^6 km)
Earth	365.3	149.6
Venus	224.7	108.2
Neptune	60,188	4498

29 A satellite moves in a circular orbit about the earth at a distance of d miles from the earth's surface. The magnitude of the force of attraction $\mathbf{F}$ between the satellite and the earth is

$$\|\mathbf{F}\| = G\frac{mM}{(R+d)^2},$$

where m is the mass of the satellite, M is the mass of the earth, R is the radius of the earth, and G is a gravitational constant. Use the results of Example 2 to establish *Kepler's third law* for circular orbits:

$$T^2 = \frac{4\pi^2}{GM}(R+d)^3,$$

where T is the period of the satellite. (*Hint:* $\|\mathbf{F}\|$ is also given by $m\|\mathbf{r}''(t)\|$.)

[c] 30 Refer to Exercise 29. If the period of a satellite is measured in days, Kepler's third law for circular orbits may be written $T^2 = 0.00346[1 + (d/R)]^3$.

(a) A satellite is moving in a circular orbit that is 1000 mi above the earth's surface. Assuming that the radius of the earth is 3959 mi, estimate the period of the satellite's orbit to the nearest 0.01 hr.

(b) In a *geosynchronous orbit*, a satellite is always located in the same position relative to the earth— that is, the period of the satellite is one day. Estimate, to the nearest mile, the distance of such a satellite from the earth's surface. (Information of this type is needed for positioning communication satellites.)

Exer. 31–32: Solve, using the results of Example 6.

31 A major-league pitcher releases a ball at a point 6 ft above the ground and 58 ft from home plate at a speed of 100 mi/hr. If gravity had no effect, the ball would travel along a line and cross home plate 4 ft off the ground, as shown in the figure. Find the drop d caused by gravity.

Exercise 31

32 A quarterback on a football team throws a pass, releasing the ball at an angle of 30° with the horizontal. Approximate the velocity at which the football must be released to reach a receiver 150 ft downfield. (Neglect air resistance.)

33 Vertical wind shear in the lowest 300 ft of the atmosphere is of great importance to aircraft during takeoffs and landings. Vertical wind shear is defined as $d\mathbf{v}/dh$, where $\mathbf{v}$ is the wind velocity and h is the height above the ground. During strong wind gusts at a certain airport, the wind velocity (in miles per hour) for altitudes h between 0 and 200 ft is estimated to be

$$\mathbf{v} = (12 + 0.006h^{3/2})\mathbf{i} + (10 + 0.005h^{3/2})\mathbf{j}.$$

Calculate the magnitude of the vertical wind shear 150 ft above ground.

Mathematicians and Their Times

BERNHARD RIEMANN

MANY GREAT MATHEMATICIANS AND SCIENTISTS had long and productive careers and were honored for their achievements during their lifetimes. Some rose from humble origins or broken families to achieve prominence and authority. Many, however, were far less fortunate. Évariste Galois (1811–1832) was expelled from college for his political activities, his path-breaking papers on groups and fields were lost or ignored, and he was killed in a meaningless duel at the age of 20. Niels Henrik Abel (1802–1829), worn down by poverty and consumed by tuberculosis, died just as his work was beginning to draw recognition for its brilliance. Georg Friedrich Bernhard Riemann (1826–1866) was another nineteenth-century figure stalked by tragedy and premature death.

The son of an impoverished Lutheran minister, Riemann suffered undernourishment and poor health as a child. He entered Göttingen University as a theology student but soon switched to mathematics. Riemann transferred to Berlin University, which despite the presence of Gauss at Göttingen, had a stronger mathematics faculty overall. In Berlin, Riemann formed ideas on the subject of complex variables that led to most of his great contributions. He eventually returned to Göttingen, where he also studied physics and completed his doctoral degree. He secured a job there as an unpaid lecturer, supporting himself by collecting fees from his students. He was jubilant when eight students enrolled in his lectures on partial differential equations.

At many points in his life, misfortune overtook Riemann soon after any minor success. He eventually obtained a salaried position at the university. Although the pay was a "miserable pittance," it marked an important step forward. He had little time to celebrate, however, as his father and elder sister soon died. Not long after another promotion, a brother and a sister died. Riemann finally obtained a professorship and married in 1862. A month later, he had an attack of pleurisy, which began a long bout with tuberculosis that ended with his death before the age of 40.

Although he wrote relatively few papers, each has had a profound impact on twentieth-century mathematics and physics. His masterpiece on the foundations of geometry, for example, created precisely the right structure that Einstein needed for his general theory of relativity. In Chapter 4, we saw some of Riemann's contribution to calculus with a careful development of integration. Riemann posed what is the most important unsolved problem in mathematics today: For which complex numbers s does the infinite series

$$\frac{1}{1^s} + \frac{1}{2^s} + \cdots + \frac{1}{n^s} + \cdots$$

have value 0?

11.4 CURVATURE

As a point moves along a curve C, it may change direction rapidly or slowly, depending on whether C bends sharply or gradually. To measure the rate at which C bends, or changes shape, we use the notion of *curvature*. In this section, we begin by considering certain unit tangent and normal vectors to curves that will be useful in our discussion of this concept. If $\mathbf{r}$ is a vector-valued function and the curve C determined by $\mathbf{r}(t)$ is smooth, then we know (from our previous work) that $\mathbf{r}'(t)$ is a tangent vector to C that points in a direction determined by the orientation of C. If $\mathbf{r}'(t) \neq \mathbf{0}$, then the *unit tangent vector* $\mathbf{T}(t)$ to C is defined as follows.

Unit Tangent Vector 11.14

$$\mathbf{T}(t) = \frac{1}{\|\mathbf{r}'(t)\|}\mathbf{r}'(t)$$

Since $\|\mathbf{T}(t)\| = 1$ for every t, we see from Theorem (11.9) that if $\mathbf{T}$ is differentiable, then $\mathbf{T}'(t)$ is orthogonal to $\mathbf{T}(t)$. The *principal unit normal vector* $\mathbf{N}(t)$ to C is defined as the *unit* vector having the same direction as $\mathbf{T}'(t)$, as follows.

Principal Unit Normal Vector 11.15

$$\mathbf{N}(t) = \frac{1}{\|\mathbf{T}'(t)\|}\mathbf{T}'(t)$$

Figure 11.23

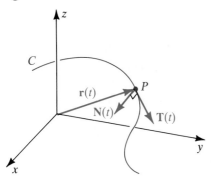

Formulas (11.14) and (11.15) may be applied to either plane curves or space curves.

When we represent $\mathbf{T}(t)$ or $\mathbf{N}(t)$ by directed line segments, we shall take the initial point at the point P on C corresponding to t, as illustrated in Figure 11.23, where we have used the right angle symbol $\lnot$ to emphasize that $\mathbf{N}(t)$ and $\mathbf{T}(t)$ are orthogonal. Since the tangent vector $\mathbf{r}'(t)$ points in the direction of increasing values of t, so does $\mathbf{T}(t)$.

EXAMPLE ■ 1 Let C be the plane curve determined by the function $\mathbf{r}(t) = t^2\mathbf{i} + t\mathbf{j}$.

(a) Find the unit tangent and normal vectors $\mathbf{T}(t)$ and $\mathbf{N}(t)$.

(b) Sketch C, $\mathbf{T}(1)$, and $\mathbf{N}(1)$.

SOLUTION

(a) By (11.14) with $\mathbf{r}'(t) = 2t\mathbf{i} + \mathbf{j}$,

$$\mathbf{T}(t) = \frac{1}{(4t^2 + 1)^{1/2}}(2t\mathbf{i} + \mathbf{j}) = \frac{2t}{(4t^2 + 1)^{1/2}}\mathbf{i} + \frac{1}{(4t^2 + 1)^{1/2}}\mathbf{j}.$$

Figure 11.24

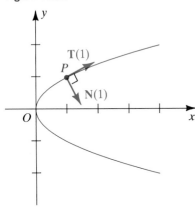

Differentiating the components of $\mathbf{T}(t)$ gives us

$$\mathbf{T}'(t) = \frac{2}{(4t^2 + 1)^{3/2}}\mathbf{i} - \frac{4t}{(4t^2 + 1)^{3/2}}\mathbf{j} = \frac{2}{(4t^2 + 1)^{3/2}}(\mathbf{i} - 2t\mathbf{j}).$$

It is easy to verify that

$$\|\mathbf{T}'(t)\| = \frac{2}{4t^2 + 1}.$$

Applying (11.15) and simplifying, we obtain

$$\mathbf{N}(t) = \frac{1}{(4t^2 + 1)^{1/2}}(\mathbf{i} - 2t\mathbf{j}).$$

(b) The point P corresponding to $t = 1$ is $(1, 1)$. By substitution,

$$\mathbf{T}(1) = \frac{1}{\sqrt{5}}(2\mathbf{i} + \mathbf{j}) \quad \text{and} \quad \mathbf{N}(1) = \frac{1}{\sqrt{5}}(\mathbf{i} - 2\mathbf{j}).$$

To check the orthogonality of $\mathbf{T}(1)$ and $\mathbf{N}(1)$, we can verify the fact that $\mathbf{T}(1) \cdot \mathbf{N}(1) = 0$. These vectors and the curve (a parabola) are sketched in Figure 11.24.

Figure 11.25

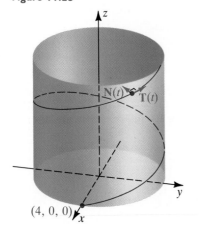

EXAMPLE ■ 2 Let C be determined by

$$\mathbf{r}(t) = 4\cos t\,\mathbf{i} + 4\sin t\,\mathbf{j} + 3t\mathbf{k} \quad \text{for} \quad t \geq 0.$$

Sketch C and find $\mathbf{T}(t)$ and $\mathbf{N}(t)$.

SOLUTION The curve is the circular helix of radius 4 shown in Figure 11.25 (see Example 4 of Section 11.1). Differentiating $\mathbf{r}(t)$ and applying Definition (11.14) gives us the following:

$$\mathbf{r}'(t) = -4\sin t\mathbf{i} + 4\cos t\mathbf{j} + 3\mathbf{k}$$

$$\|\mathbf{r}'(t)\| = \sqrt{16\sin^2 t + 16\cos^2 t + 9} = \sqrt{16 + 9} = \sqrt{25} = 5$$

$$\mathbf{T}(t) = \frac{1}{\|\mathbf{r}'(t)\|}\mathbf{r}'(t) = -\frac{4}{5}\sin t\mathbf{i} + \frac{4}{5}\cos t\mathbf{j} + \frac{3}{5}\mathbf{k}$$

Differentiating $\mathbf{T}(t)$ and using Definition (11.15), we obtain

$$\mathbf{T}'(t) = -\tfrac{4}{5}\cos t\mathbf{i} - \tfrac{4}{5}\sin t\mathbf{j}$$

$$\|\mathbf{T}'(t)\| = \sqrt{\tfrac{16}{25}\cos^2 t + \tfrac{16}{25}\sin^2 t} = \sqrt{\tfrac{16}{25}} = \tfrac{4}{5}$$

$$\mathbf{N}(t) = \frac{1}{\|\mathbf{T}'(t)\|}\mathbf{T}'(t) = -\cos t\mathbf{i} - \sin t\mathbf{j}.$$

To check the orthogonality of $\mathbf{T}(t)$ and $\mathbf{N}(t)$, we can verify the fact that $\mathbf{T}(t) \cdot \mathbf{N}(t) = 0$.

Typical unit vectors $\mathbf{T}(t)$ and $\mathbf{N}(t)$ are shown in Figure 11.25. Note that the principal unit normal $\mathbf{N}(t)$ to the circular helix is always parallel to the xy-plane and points toward the z-axis.

To define curvature, we shall begin with plane curves. Space curves will be considered later in this section.

A smooth plane curve C has many different parametrizations. Sometimes it is convenient to use arc length measured along C as a parameter. As illustrated in Figure 11.26, let A be a fixed point on C and let s denote the length of arc AP from A to an arbitrary point $P(x, y)$ on C. Suppose that C is given parametrically by

$$x = f(s), \quad y = g(s); \quad s = \text{length of } \overparen{AP}.$$

Thus, to each value of s there corresponds a point $P(f(s), g(s))$ on C that is s units from A, measured along C. The orientation of C is determined by increasing values of s. We shall call s an *arc length parameter* for the curve C.

The next definition summarizes these remarks.

Figure 11.26

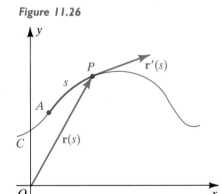

Definition 11.16

A variable s is an **arc length parameter** for a plane curve C if there is a parametrization

$$x = f(s), \qquad y = g(s)$$

such that s is the length of C from a fixed point A on the curve to the point $P(x, y)$. The parametric equations are called an **arc length parametrization** for C.

Figure 11.27

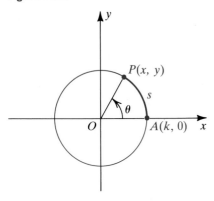

EXAMPLE ■ 3 Find an arc length parametrization for the circle with equation $x^2 + y^2 = k^2$, where $k > 0$.

SOLUTION Let $A(k, 0)$ be the fixed point in Definition (11.16), and let $P(x, y)$ be any point on the circle. Let s denote the length of the circular arc AP, measured in the counterclockwise direction from A to P (see Figure 11.27). If θ is the radian measure of angle AOP, then parametric equations for the circle are

$$x = k \cos\theta, \quad y = k \sin\theta; \quad 0 \le \theta \le 2\pi.$$

By formula (13) for the length of a circular arc (see page 38),

$$s = k\theta, \quad \text{or} \quad \theta = \frac{s}{k}.$$

Since $0 \le s \le 2\pi k$ corresponds to $0 \le \theta \le 2\pi$, an arc length parametrization for the circle is

$$x = k \cos\frac{s}{k}, \quad y = k \sin\frac{s}{k}; \quad 0 \le s \le 2\pi k.$$

By using a different fixed point, we could obtain other arc length parametrizations.

Figure 11.28

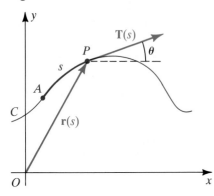

A general method for finding arc length parametrizations for curves is outlined in the instructions for Exercises 53–56. It is a very tedious process to say the least. However, as we shall see, there is great value in working with such parametrizations.

If a curve C has an arc length parametrization as in Definition (11.16), then the position vector for the point $P(x, y)$ on C is

$$\mathbf{r}(s) = x\mathbf{i} + y\mathbf{j} = f(s)\mathbf{i} + g(s)\mathbf{j}.$$

Differentiating with respect to the arc length parameter s gives us a tangent vector to C,

$$\mathbf{r}'(s) = \frac{dx}{ds}\mathbf{i} + \frac{dy}{ds}\mathbf{j} = f'(s)\mathbf{i} + g'(s)\mathbf{j}.$$

The magnitude of $\mathbf{r}'(s)$ is

Figure 11.29

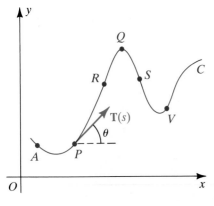

$$\|\mathbf{r}'(s)\| = \sqrt{\left(\frac{dx}{ds}\right)^2 + \left(\frac{dy}{ds}\right)^2} = \sqrt{\left(\frac{ds}{ds}\right)^2} = 1,$$

where we have used (9.7). Thus, *for an arc length parametrization, $\mathbf{r}'(s)$ is a unit tangent vector* to C at P. As in (11.14), with $t = s$, we shall denote this unit tangent vector by $\mathbf{T}(s)$.

For each value of s, let θ denote the angle between $\mathbf{T}(s)$ and $\mathbf{i}$, as illustrated in Figure 11.28. Note that θ *is a function of s,* since $\mathbf{T}(s)$ is a function of s. We may use the rate of change $d\theta/ds$ of θ with respect to s to measure how much the curve C bends at various points. To illustrate, suppose a point P moves along the curve C in Figure 11.29. At point R the curve bends gradually, and the angle θ between $\mathbf{T}(s)$ and $\mathbf{i}$ changes slowly—that is, $|d\theta/ds|$ is small. At point Q the curve bends sharply, and $|d\theta/ds|$ is large. We use these observations to motivate the next definition.

Definition 11.17

Let a smooth plane curve C have an arc length parametrization $x = f(s)$, $y = g(s)$, and let θ be the angle between the unit tangent vector $\mathbf{T}(s)$ and $\mathbf{i}$. The **curvature** K of C at the point $P(x, y)$ is

$$K = \left| \frac{d\theta}{ds} \right|.$$

For the curve in Figure 11.29, the curvature at points R and S is relatively small compared to the curvature at Q and V. The next two examples are illustrations of Definition (11.17).

Figure 11.30

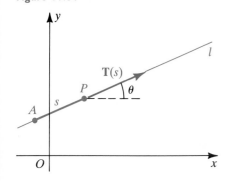

EXAMPLE • 4 Prove that the curvature of a line l is 0 at every point on l.

SOLUTION As in Figure 11.30, let A be a fixed point on l, and let $s = d(A, P)$. Since the unit tangent vector $\mathbf{T}(s)$ to l always lies on l, the angle θ is the same for every s—that is, θ is constant. Hence,

$$K = \left| \frac{d\theta}{ds} \right| = |0| = 0.$$

EXAMPLE • 5 Prove that the curvature at every point on a circle of radius k is $1/k$.

SOLUTION As in Figure 11.31, we assume that the circle has its center at O, that $A(k, 0)$ is the fixed point, and that the point P is in the first quadrant. Let s be the length of the arc AP. If α is the radian measure of angle POA, then

$$s = k\alpha, \quad \text{or} \quad \alpha = \frac{s}{k}.$$

Figure 11.31

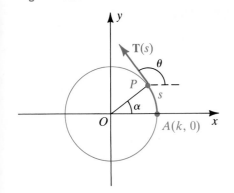

Referring to Figure 11.31 and using the fact that $\mathbf{T}(s)$ is orthogonal to $\overrightarrow{OP}$, we see that

$$\theta = \alpha + \frac{\pi}{2} = \frac{s}{k} + \frac{\pi}{2}.$$

Differentiating yields

$$\frac{d\theta}{ds} = \frac{d}{ds}\left(\frac{s}{k} + \frac{\pi}{2} \right) = \frac{1}{k} + 0 = \frac{1}{k}.$$

Hence,

$$K = \left| \frac{d\theta}{ds} \right| = \frac{1}{k}.$$

Figure 11.32

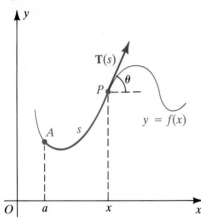

Note from Example 5 that as the radius k of a circle increases, its curvature $K = 1/k$ decreases. If we let k increase without bound, K approaches 0, which is the curvature of a line. As the radius k approaches 0, the curvature K increases without bound.

Let us next derive a formula that can be used to find the curvature K at a point (x, y) on the graph of $y = f(x)$, where f' is continuous on some interval. Let $\mathbf{T}(s)$ and θ be defined as in (11.17) (see Figure 11.32). Since y' is the slope of the tangent line at P, we see that

(1) $$\tan \theta = y', \quad \text{or} \quad \theta = \tan^{-1} y'.$$

From Definition (5.16), the arc length function s may be defined by

(2) $$s(x) = \int_a^x \sqrt{1 + (y')^2}\, dx,$$

where a is the x-coordinate of a fixed point A on C. If y'' exists, then, by the chain rule,

$$\frac{d\theta}{dx} = \frac{d\theta}{ds}\frac{ds}{dx}, \quad \text{or} \quad \frac{d\theta}{ds} = \frac{d\theta/dx}{ds/dx}.$$

Hence, by Definition (11.17),

(3) $$K = \left| \frac{d\theta}{ds} \right| = \left| \frac{d\theta/dx}{ds/dx} \right|.$$

Referring to equations (1) and (2), we have

$$\frac{d\theta}{dx} = \frac{d}{dx} \tan^{-1} y' = \frac{y''}{1 + (y')^2},$$

$$\frac{ds}{dx} = \frac{d}{dx} \int_a^x \sqrt{1 + (y')^2}\, dx = \sqrt{1 + (y')^2}.$$

Substitution in equation (3) gives us the following theorem.

Theorem 11.18

If a smooth curve C is the graph of $y = f(x)$, then the curvature K at $P(x, y)$ is

$$K = \frac{|y''|}{[1 + (y')^2]^{3/2}}.$$

Figure 11.33

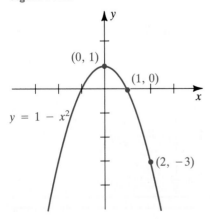

$(0, 1)$

$(1, 0)$

$y = 1 - x^2$

$(2, -3)$

EXAMPLE 6 Sketch the graph of $y = 1 - x^2$, and find the curvature at the points (x, y), $(0, 1)$, $(1, 0)$, and $(2, -3)$.

SOLUTION The graph (a parabola) is sketched in Figure 11.33. Since $y' = -2x$ and $y'' = -2$, we have, from Theorem (11.18),

$$K = \frac{2}{(1 + 4x^2)^{3/2}}.$$

Letting $x = 0, 1$, and 2, we obtain the following table.

Point on C	$(0, 1)$	$(1, 0)$	$(2, -3)$
Curvature K	2	$\dfrac{2}{5^{3/2}} \approx 0.18$	$\dfrac{2}{17^{3/2}} \approx 0.03$

Note that the maximum curvature occurs at $(0, 1)$ and as x increases (or decreases) without bound, the curvature at $P(x, y)$ approaches 0.

Let us next derive a formula for finding K if C is described in terms of *any* parameter t. Suppose C is given parametrically by

$$x = f(t), \qquad y = g(t)$$

and that f'' and g'' exist for every t in some interval I. As in the discussion preceding Theorem (11.18),

$$K = \left| \frac{d\theta}{ds} \right| = \left| \frac{d\theta/dt}{ds/dt} \right|.$$

Since the slope of the tangent line at $P(x, y)$ is $g'(t)/f'(t)$ (see Theorem (9.4)),

$$\tan \theta = \frac{g'(t)}{f'(t)}, \quad \text{or} \quad \theta = \tan^{-1} \frac{g'(t)}{f'(t)},$$

provided $f'(t) \neq 0$. Hence,

$$\frac{d\theta}{dt} = \frac{1}{1 + [g'(t)/f'(t)]^2} \frac{f'(t)g''(t) - g'(t)f''(t)}{[f'(t)]^2}$$

$$= \frac{f'(t)g''(t) - g'(t)f''(t)}{[f'(t)]^2 + [g'(t)]^2}.$$

Moreover, by (9.7),

$$\left| \frac{ds}{dt} \right| = \sqrt{[f'(t)]^2 + [g'(t)]^2}.$$

Substituting in the formula $K = |(d\theta/dt)/(ds/dt)|$ gives us the following.

Theorem 11.19

> If a plane curve C has a parametrization $x = f(t)$, $y = g(t)$ and if f'' and g'' exist, then the curvature K at $P(x, y)$ is
>
> $$K = \frac{|f'(t)g''(t) - g'(t)f''(t)|}{[(f'(t))^2 + (g'(t))^2]^{3/2}}.$$

In the next example, we use Theorem (11.19) to rework Example 5.

EXAMPLE ■ 7   Prove that the curvature at every point on a circle of radius k is $1/k$.

SOLUTION   Parametric equations for a circle of radius k with center at the origin are

$$x = k \cos t = f(t), \quad y = k \sin t = g(t); \quad 0 \le t \le 2\pi.$$

Differentiating gives us

$$f'(t) = -k \sin t \quad f''(t) = -k \cos t$$
$$g'(t) = k \cos t \quad g''(t) = -k \sin t.$$

Substituting in Theorem (11.19), we obtain

$$K = \frac{|(-k \sin t)(-k \sin t) - (k \cos t)(-k \cos t)|}{[(-k \sin t)^2 + (k \cos t)^2]^{3/2}}$$

$$= \frac{k^2 \sin^2 t + k^2 \cos^2 t}{(k^2 \sin^2 t + k^2 \cos^2 t)^{3/2}}$$

$$= \frac{k^2}{(k^2)^{3/2}} = \frac{k^2}{k^3} = \frac{1}{k}.$$

Figure 11.34

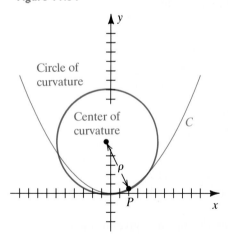

If the curvature K at a point P on a curve C is not 0, then the circle of radius $\rho = 1/K$ whose center lies on the concave side of C and that has the same tangent line at P as C is the **circle of curvature** of the curve C at the point P. Its radius ρ and center are the **radius of curvature** and **center of curvature**, respectively, for P. According to Examples 5 and 7, the curvature of the circle of curvature is $1/\rho$, or K, and hence is the same as the curvature of C. For this reason, *the circle of curvature may be thought of as the circle that best coincides with C at P.* A circle of curvature is shown in Figure 11.34.

EXAMPLE ■ 8   A curve C has parametrization $x = t^2$, $y = t^3$ for all t in $\mathbb{R}$. Find the curvature at the point P corresponding to $t = 0.5$. Sketch the graph of C and the circle of curvature for C at P.

SOLUTION   Letting $f(t) = t^2$, $g(t) = t^3$ and differentiating, we have

$$f'(t) = 2t, \quad f''(t) = 2, \quad g'(t) = 3t^2, \quad g''(t) = 6t.$$

Substituting into the formula for K in Theorem (11.19) gives us

$$K = \frac{|(2t)(6t) - (3t^2)(2)|}{[(2t)^2 + (3t^2)^2]^{3/2}} = \frac{6t^2}{[4t^2 + 9t^4]^{3/2}}.$$

If $t = 0.5$, then $K = \frac{96}{125} = 0.768$. The radius of curvature ρ at the point P is $1/K$, or $\rho = \frac{125}{96} \approx 1.302$.

The point P corresponding to $t = 0.5$ has coordinates $(0.25, 0.125)$. The vector $\mathbf{a} = f'(0.5)\mathbf{i} + g'(0.5)\mathbf{j} = \mathbf{i} + 0.75\mathbf{j}$ gives the direction of the tangent vector. Locating $\mathbf{a}$ at P will position it along the tangent line. Let $\mathbf{b} = b_1\mathbf{i} + b_2\mathbf{j}$ be the vector from P to the center of the circle of curvature. Since

Figure 11.35

$$\mathbf{a} \cdot \mathbf{b} = 0 \quad \text{and} \quad \|\mathbf{b}\| = \rho,$$

we must have $b_1 + 0.75b_2 = 0$ and $b_1^2 + b_2^2 = \rho^2$. These equations have two solutions, which result in two choices for $\mathbf{b}$:

$$\mathbf{b} = \langle -0.6\rho, 0.8\rho \rangle \quad \text{and} \quad \mathbf{b} = \langle 0.6\rho, -0.8\rho \rangle$$

Looking at Figure 11.35, we choose the first vector so that the center will be above and to the left of the point P. The coordinates of the center of the circle of curvature are

$$(0.25 - 0.6\rho, 0.125 + 0.8\rho) = (-\tfrac{17}{32}, \tfrac{7}{6})$$

$$\approx (-0.531, 1.167).$$

The definition of curvature $K = |d\theta/ds|$ for plane curves has no immediate analogue in three dimensions, because the unit tangent vector $\mathbf{T}(s)$ cannot be specified in terms of a single angle θ. Thus, it is necessary to use a different approach for space curves. Of course, we then must show that if we specialize to vectors in a plane, the new definition of curvature agrees with (11.17).

To find a clue to a suitable definition, we first observe that in two dimensions the unit tangent vector $\mathbf{T}(s)$ can be written

$$\mathbf{T}(s) = \cos\theta\mathbf{i} + \sin\theta\mathbf{j},$$

where θ is the angle between $\mathbf{T}(s)$ and $\mathbf{i}$ (see Figure 11.29). Regarding θ as a function of s and differentiating yields

$$\mathbf{T}'(s) = \left(-\sin\theta\frac{d\theta}{ds}\right)\mathbf{i} + \left(\cos\theta\frac{d\theta}{ds}\right)\mathbf{j} = \frac{d\theta}{ds}(-\sin\theta\mathbf{i} + \cos\theta\mathbf{j}).$$

Hence,

$$\|\mathbf{T}'(s)\| = \left|\frac{d\theta}{ds}\right| \|-\sin\theta\mathbf{i} + \cos\theta\mathbf{j}\| = \left|\frac{d\theta}{ds}\right| = K.$$

We shall use this fact to define curvature in three dimensions. Our plan is to describe the unit tangent vector $\mathbf{T}(s)$ *without* referring to the angle θ and then *define* K as $\|\mathbf{T}'(s)\|$.

Suppose that a curve C has an arc length parametrization

$$x = f(s), \quad y = g(s), \quad z = h(s),$$

where f'', g'', and h'' exist. As illustrated in Figure 11.36, s is the arc length along C from a fixed point A to the point $P(f(s), g(s), h(s))$. As in two dimensions, if $\mathbf{r}(s) = f(s)\mathbf{i} + g(s)\mathbf{j} + h(s)\mathbf{k}$ is the position vector for P, then $\mathbf{r}'(s)$ is a unit tangent vector to C at P, denoted by $\mathbf{T}(s)$. Since $\|\mathbf{T}(s)\|$ is a constant, it follows that $\mathbf{T}'(s)$ is orthogonal to $\mathbf{T}(s)$ (see Theorem (11.9)). If $\mathbf{T}'(s) \neq \mathbf{0}$, we let

$$\mathbf{N}(s) = \frac{1}{\|\mathbf{T}'(s)\|}\mathbf{T}'(s).$$

The vector $\mathbf{N}(s)$ is a unit vector orthogonal to $\mathbf{T}(s)$ and is referred to as the **principal unit normal vector** to C at the point P. These concepts are illustrated in Figure 11.36.

Figure 11.36

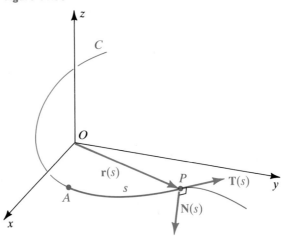

Having obtained a suitable unit tangent vector $\mathbf{T}(s)$ and assuming that $\mathbf{T}'(s)$ exists, we follow our plan and define curvature in three dimensions, as follows.

Definition 11.20

Let a smooth space curve C have an arc length parametrization

$$x = f(s), \quad y = g(s), \quad z = h(s).$$

Let $\mathbf{r}(s) = f(s)\mathbf{i} + g(s)\mathbf{j} + h(s)\mathbf{k}$ and let $\mathbf{T}(s) = \mathbf{r}'(s)$. The **curvature** K of C at the point $P(x, y, z)$ is

$$K = \|\mathbf{T}'(s)\|.$$

As we have already proved, the preceding definition reduces to (11.17) if C is a plane curve. Note that

$$\mathbf{N}(s) = \frac{1}{K}\mathbf{T}'(s), \quad \text{or} \quad \mathbf{T}'(s) = K\mathbf{N}(s).$$

The formula for K in Definition (11.20) is usually cumbersome to apply to specific problems. In the next section, we shall derive a more practical formula that can be used to find curvature.

EXERCISES 11.4

Exer. 1–6: (a) Find the unit tangent and normal vectors $T(t)$ and $N(t)$ for the curve C determined by $r(t)$. (b) Sketch the graph of C, and show $T(t)$ and $N(t)$ for the given value of t.

1 $r(t) = t\mathbf{i} - \frac{1}{2}t^2\mathbf{j}$; $t = 1$

2 $r(t) = -t^2\mathbf{i} + 2t\mathbf{j}$; $t = 1$

3 $r(t) = t^3\mathbf{i} + 3t\mathbf{j}$; $t = 1$

4 $r(t) = (4 + \cos t)\mathbf{i} - (3 - \sin t)\mathbf{j}$; $t = \pi/6$

5 $r(t) = 2\sin t\mathbf{i} + 3\mathbf{j} + 2\cos t\mathbf{k}$; $t = \pi/4$

6 $r(t) = t\mathbf{i} + \frac{1}{2}t^2\mathbf{j} + t^2\mathbf{k}$; $t = 1$

Exer. 7–18: Find the curvature of the curve at P.

7 $y = 2 - x^3$; $P(1, 1)$

8 $y = x^4$; $P(1, 1)$

9 $y = e^{(x^2)}$; $P(0, 1)$

10 $y = \ln(x - 1)$; $P(2, 0)$

11 $y = \cos 2x$; $P(0, 1)$

12 $y = \sec x$; $P(\pi/3, 2)$

13 $x = t - 1$, $y = \sqrt{t}$; $P(3, 2)$

14 $x = t + 1$, $y = t^2 + 4t + 3$; $P(1, 3)$

15 $x = t - t^2$, $y = 1 - t^3$; $P(0, 1)$

16 $x = t - \sin t$, $y = 1 - \cos t$; $P(\pi/2 - 1, 1)$

17 $x = 2\sin t$, $y = 3\cos t$; $P(1, \frac{3}{2}\sqrt{3})$

18 $x = \cos^3 t$, $y = \sin^3 t$; $P(\frac{1}{4}\sqrt{2}, \frac{1}{4}\sqrt{2})$

Exer. 19–22: For the given curve and point P, (a) find the radius of curvature, (b) find the center of curvature, and (c) sketch the graph and the circle of curvature for P.

19 $y = \sin x$; $P(\pi/2, 1)$

20 $y = \sec x$; $P(0, 1)$

21 $y = e^x$; $P(0, 1)$

22 $xy = 1$; $P(1, 1)$

$\boxed{\text{c}}$ Exer. 23–26: For the given curve C and point P, approximate the curvature of C at P and plot the curve C along with the circle of curvature for C at P.

23 $x = \cos t$, $y = \sin(0.8t)$; $P\left(\cos\left(\frac{\pi}{3}\right), \sin\left(\frac{4\pi}{15}\right)\right)$

24 $x = 1 + \cos t$, $y = \sin(1.2t)$; $P(1 + \cos 1, \sin 1.2)$

25 $x = 5t^2$, $y = 8t - 7t^2$; $P(\frac{5}{4}, \frac{9}{4})$

26 $x = 9t^2(3 - 2t)$, $y = 12t(3 - 3t + 2t^2)$; $P(0.936, 5.952)$, $t = 0.2$

Exer. 27–32: Find the points on the given curve at which the curvature is a maximum.

27 $y = e^{-x}$ 28 $y = \cosh x$

29 $9x^2 + 4y^2 = 36$ 30 $9x^2 - 4y^2 = 36$

31 $y = \ln x$ 32 $y = \sin x$

Exer. 33–36: Find the points on the graph of the equation at which the curvature is 0.

33 $y = x^4 - 12x^2$ 34 $y = \tan x$

35 $y = \sinh x$ 36 $y = e^{-x^2}$

37 Suppose that a curve C is the graph of a polar equation $r = f(\theta)$. If $r' = dr/d\theta$ and $r'' = d^2r/d\theta^2$, show that the curvature K at $P(r, \theta)$ is

$$K = \frac{\left|2(r')^2 - rr'' + r^2\right|}{[(r')^2 + r^2]^{3/2}}.$$

(*Hint:* Use $x = r\cos\theta$ and $y = r\sin\theta$ to express C in parametric form.)

Exer. 38–40: Use the formula in Exercise 37 to find the curvature of the polar curve at $P(r, \theta)$.

38 $r = a(1 - \cos\theta)$; $0 < \theta < 2\pi$

39 $r = \sin 2\theta$; $0 < \theta < 2\pi$

40 $r = e^{a\theta}$

41 Let $P(x, y)$ be a point on the graph of $y = f(x)$ at which $K \neq 0$. If (h, k) is the center of curvature for P, show that

$$h = x - \frac{y'[1 + (y')^2]}{y''}, \qquad k = y + \frac{[1 + (y')^2]}{y''}.$$

Exer. 42–46: Use the formulas in Exercise 41 to find the center of curvature for the point P on the graph of the equation. (Refer to Exercises 7–11.)

42 $y = 2 - x^3$; $P(1, 1)$

43 $y = x^4$; $P(1, 1)$

44 $y = e^{(x^2)}$; $P(0, 1)$

45 $y = \ln(x - 1)$; $P(2, 0)$

46 $y = \cos 2x$; $P(0, 1)$

47 The path of a highway and exit ramp are superimposed on a rectangular coordinate system such that the highway coincides with the x-axis. The exit ramp begins at the origin O. After following the graph of $y = -\frac{1}{27}x^3$ from O to the point $P(3, -1)$, the path follows along the arc of a circle, as shown in the figure. If $K(x)$ is the curvature of the exit ramp at (x, y), find the center of the circular arc that makes the curvature at $P(3, -1)$ continuous at $x = 3$.

Exercise 47

48 Use the equation $y = mx + b$ to prove that the curvature at every point on a line is 0 (see Example 4).

49 Prove that the maximum curvature of a parabola is at the vertex.

50 Prove that the maximum and minimum curvatures of an ellipse are at the ends of the major and minor axes, respectively.

51 Prove that the maximum curvature of a hyperbola is at the ends of the transverse axis.

52 Prove that lines and circles are the only *plane* curves that have a constant curvature. (Exercise 17 of Section 11.5 shows that this result is not true for space curves.)

Exer. 53–56: If the curve C in Figure 11.23 has a smooth parametrization $x = f(t)$, $y = g(t)$, then, by Theorem (9.6), the relationship between t and the arc length parameter s is given by

$$s = \int_a^t \sqrt{[f'(u)]^2 + [g'(u)]^2} \, du,$$

where a is the value of t corresponding to the fixed point A. Use this relationship to express the given curve in terms of the arc length parameter s if the fixed point A corresponds to $t = 0$. (*Hint:* First evaluate the integral to find the relationship between t and s, and then substitute for t in the parametric equations.)

53 $x = 4t - 3$, $y = 3t + 5$; $t \geq 0$

54 $x = 3t^2$, $y = 2t^3$; $t \geq 0$

55 $x = 4\cos t$, $y = 4\sin t$; $0 \leq t \leq 2\pi$

56 $x = e^t \cos t$, $y = e^t \sin t$; $t \geq 0$

57 Prove that if k is a nonnegative function that is nonzero and continuous on an interval $[0, a]$, then there is a plane curve C such that k represents the curvature of C as a function of arc length. (*Hint:* If s is in $[0, a]$, define

$$h(s) = \int_0^s k(t) \, dt, \qquad x = f(s) = \int_0^s \cos h(t) \, dt,$$

$$\text{and} \quad y = g(s) = \int_0^s \sin h(t) \, dt.)$$

58 Use Exercise 57 to work Exercise 52.

11.5 TANGENTIAL AND NORMAL COMPONENTS OF ACCELERATION

In this section, we use the concept of curvature to help analyze the motion of objects.

Theorem 11.21

Let $\mathbf{r}(t)$ be the position vector of a moving point P at time t, and let s be an arc length parameter for the curve C determined by $\mathbf{r}(t)$. If $\mathbf{T}(s)$ and $\mathbf{N}(s)$ are the unit tangent and principal unit normal vectors and if K is the curvature of C at the point corresponding to s, then the velocity and the acceleration of P may be written as follows:

$$\mathbf{v}(t) = \frac{ds}{dt}\mathbf{T}(s)$$

$$\mathbf{a}(t) = \frac{d^2s}{dt^2}\mathbf{T}(s) + K\left(\frac{ds}{dt}\right)^2\mathbf{N}(s)$$

Figure 11.37

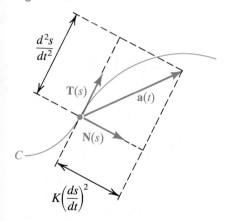

PROOF The sketch in Figure 11.37 is a geometric interpretation of the formula for $\mathbf{a}(t)$ in the statement of the theorem. Recall that the unit tangent vector $\mathbf{T}(s)$ to C is given by

$$\mathbf{T}(s) = \frac{1}{\|\mathbf{r}'(t)\|}\mathbf{r}'(t),$$

and hence

$$\mathbf{r}'(t) = \|\mathbf{r}'(t)\|\,\mathbf{T}(s).$$

Since $\mathbf{r}'(t) = \mathbf{v}(t)$ and $\|\mathbf{r}'(t)\| = \|\mathbf{v}(t)\| = ds/dt$, this gives us the velocity formula

$$\mathbf{v}(t) = \frac{ds}{dt}\mathbf{T}(s).$$

Differentiating the preceding equation with respect to t and using the chain rules in Exercises 45 and 46 of Section 11.2, we obtain

$$\mathbf{a}(t) = \mathbf{v}'(t) = \frac{d^2s}{dt^2}\mathbf{T}(s) + \frac{ds}{dt}\frac{d}{dt}\mathbf{T}(s)$$

$$= \frac{d^2s}{dt^2}\mathbf{T}(s) + \frac{ds}{dt}\frac{ds}{dt}\mathbf{T}'(s).$$

From the remark following Definition (11.20), $\mathbf{T}'(s) = K\mathbf{N}(s)$. Thus, the last formula may be written

$$\mathbf{a}(t) = \frac{d^2s}{dt^2}\mathbf{T}(s) + K\left(\frac{ds}{dt}\right)^2\mathbf{N}(s). \quad \blacksquare$$

If we denote the speed $\|\mathbf{v}(t)\| = ds/dt$ of P by v, then the formulas in Theorem (11.21) are

$$\mathbf{v}(t) = v\mathbf{T}(s) \quad \text{and} \quad \mathbf{a}(t) = \frac{dv}{dt}\mathbf{T}(s) + Kv^2\mathbf{N}(s).$$

The second formula expresses the acceleration $\mathbf{a}(t)$ in terms of a *tangential component* $a_{\mathbf{T}}$ and a *normal component* $a_{\mathbf{N}}$—that is,

$$\mathbf{a}(t) = a_{\mathbf{T}}\mathbf{T}(s) + a_{\mathbf{N}}\mathbf{N}(s),$$

where $a_{\mathbf{T}} = \dfrac{dv}{dt} = \dfrac{d^2s}{dt^2}$ and $a_{\mathbf{N}} = Kv^2 = K\left(\dfrac{ds}{dt}\right)^2.$

Tangential and normal components of acceleration are useful in the investigation of the forces that act on a moving object as it travels along a curve C. Note that the normal component $a_{\mathbf{N}} = Kv^2$ depends only on the speed of P and the curvature of C. If the speed or curvature is large, then the normal component of acceleration is large. This result, obtained theoretically, proves the well-known fact that an automobile driver should slow down when attempting to negotiate a sharp turn.

We can also derive formulas for the tangential and normal components of acceleration that depend only on $\mathbf{r}(t)$. Taking the dot product of $\mathbf{v}(t)$ and $\mathbf{a}(t)$ in Theorem (11.21), we obtain

$$\mathbf{v}(t) \cdot \mathbf{a}(t) = \left[\frac{ds}{dt}\mathbf{T}(s)\right] \cdot \left[\frac{d^2s}{dt^2}\mathbf{T}(s) + K\left(\frac{ds}{dt}\right)^2\mathbf{N}(s)\right]$$

$$= \left(\frac{ds}{dt}\right)\left(\frac{d^2s}{dt^2}\right)[\mathbf{T}(s) \cdot \mathbf{T}(s)] + K\left(\frac{ds}{dt}\right)^3[\mathbf{T}(s) \cdot \mathbf{N}(s)].$$

Since $\mathbf{T}(s)$ has magnitude 1 and is orthogonal to $\mathbf{N}(s)$, this formula reduces to

$$\mathbf{v}(t) \cdot \mathbf{a}(t) = \left(\frac{ds}{dt}\right)\left(\frac{d^2s}{dt^2}\right),$$

or, in terms of $\mathbf{r}$,

$$\mathbf{r}'(t) \cdot \mathbf{r}''(t) = \|\mathbf{r}'(t)\|\frac{d^2s}{dt^2} = \|\mathbf{r}'(t)\| a_{\mathbf{T}}.$$

If $\mathbf{r}'(t) \neq 0$, this gives us the following formula.

Tangential Component of Acceleration 11.22

$$a_{\mathbf{T}} = \frac{d^2s}{dt^2} = \frac{\mathbf{r}'(t) \cdot \mathbf{r}''(t)}{\|\mathbf{r}'(t)\|}$$

If, instead of the dot product, we take the *cross* product of $\mathbf{a}(t)$ and $\mathbf{v}(t)$, we obtain

$$\mathbf{v}(t) \times \mathbf{a}(t) = \left(\frac{ds}{dt}\right)\left(\frac{d^2s}{dt^2}\right)[\mathbf{T}(s) \times \mathbf{T}(s)] + K\left(\frac{ds}{dt}\right)^3[\mathbf{T}(s) \times \mathbf{N}(s)].$$

Since the cross product of a vector with itself is $\mathbf{0}$, $\mathbf{T}(s) \times \mathbf{T}(s) = \mathbf{0}$. Since $\mathbf{T}(s)$ and $\mathbf{N}(s)$ are orthogonal unit vectors, $\|\mathbf{T}(s) \times \mathbf{N}(s)\| = 1$ (see Theorem (10.30)). Hence, the formula for $\mathbf{v}(t) \times \mathbf{a}(t)$ gives us

$$\|\mathbf{v}(t) \times \mathbf{a}(t)\| = K \left(\frac{ds}{dt}\right)^3$$

and, therefore,

$$a_{\mathbf{N}} = K \left(\frac{ds}{dt}\right)^2 = \frac{\|\mathbf{v}(t) \times \mathbf{a}(t)\|}{ds/dt}.$$

Rewriting this in terms of $\mathbf{r}$, we obtain the following.

Normal Component of Acceleration 11.23

$$a_{\mathbf{N}} = K \left(\frac{ds}{dt}\right)^2 = \frac{\|\mathbf{r}'(t) \times \mathbf{r}''(t)\|}{\|\mathbf{r}'(t)\|}$$

EXAMPLE ▪ 1   The position vector of a moving point at time t is $\mathbf{r}(t) = t\mathbf{i} + t^2\mathbf{j} + t^3\mathbf{k}$ for $1 \le t \le 4$.

(a) Find the tangential and normal components of acceleration at time t.

(b) Find two-decimal-place approximations for $a_{\mathbf{N}}$, $a_{\mathbf{T}}$, and $\|\mathbf{v}(t)\|$ at times $t = 1, 2, 3$, and 4, and describe the motion of the point.

SOLUTION

(a) Differentiating yields

$$\mathbf{v}(t) = \mathbf{r}'(t) = \mathbf{i} + 2t\mathbf{j} + 3t^2\mathbf{k}$$
$$\mathbf{a}(t) = \mathbf{r}''(t) = 2\mathbf{j} + 6t\mathbf{k}$$
$$\|\mathbf{v}(t)\| = \|\mathbf{r}'(t)\| = (1 + 4t^2 + 9t^4)^{1/2}.$$

By (11.22),

$$a_{\mathbf{T}} = \frac{\mathbf{r}'(t) \cdot \mathbf{r}''(t)}{\|\mathbf{r}'(t)\|} = \frac{4t + 18t^3}{(1 + 4t^2 + 9t^4)^{1/2}}.$$

To find $a_{\mathbf{N}}$, we first calculate

$$\mathbf{r}'(t) \times \mathbf{r}''(t) = \begin{vmatrix} \mathbf{i} & \mathbf{j} & \mathbf{k} \\ 1 & 2t & 3t^2 \\ 0 & 2 & 6t \end{vmatrix} = 6t^2\mathbf{i} - 6t\mathbf{j} + 2\mathbf{k}.$$

Applying (11.23) gives us

$$a_{\mathbf{N}} = \frac{\|\mathbf{r}'(t) \times \mathbf{r}''(t)\|}{\|\mathbf{r}'(t)\|} = \frac{(36t^4 + 36t^2 + 4)^{1/2}}{(1 + 4t^2 + 9t^4)^{1/2}} = 2\left(\frac{9t^4 + 9t^2 + 1}{9t^4 + 4t^2 + 1}\right)^{1/2}.$$

(b) The path of the point—that is, the curve C determined by $\mathbf{r}(t)$—is the portion of the twisted cubic discussed in Example 3 of Section 11.1 that lies between $(1, 1, 1)$ and $(4, 16, 64)$ (see Figure 11.5). Substitution into the formulas obtained in part (a) gives us the following table of approximations.

t	1	2	3	4
Position of P	$(1, 1, 1)$	$(2, 4, 8)$	$(3, 9, 27)$	$(4, 16, 64)$
a_N	2.33	2.12	2.06	2.03
a_T	5.88	11.98	17.99	24.00
$\|\mathbf{v}(t)\|$	3.74	12.69	27.68	48.67

Thus, as t increases from 1 to 4, the point moves along C from $(1, 1, 1)$ to $(4, 16, 64)$, gaining speed rapidly. The normal component of acceleration a_N approaches 2 (note that $\lim_{t \to \infty} a_N = 2$). We see from the table that the tangential component a_T increases at approximately 6 units per unit change in time.

We can obtain an alternative formula for the normal component of acceleration. To simplify the notation, let us denote $\mathbf{T}(s)$, $\mathbf{N}(s)$, and $\mathbf{a}(t)$ in Theorem (11.21) by $\mathbf{T}$, $\mathbf{N}$, and $\mathbf{a}$, respectively. Thus,

$$\mathbf{a} = a_T \mathbf{T} + a_N \mathbf{N}.$$

Since $\mathbf{T}$ and $\mathbf{N}$ are mutually orthogonal unit vectors, $\mathbf{T} \cdot \mathbf{N} = 0$, $\mathbf{T} \cdot \mathbf{T} = 1$, and $\mathbf{N} \cdot \mathbf{N} = 1$. Consequently,

$$\|\mathbf{a}\|^2 = \mathbf{a} \cdot \mathbf{a} = (a_T \mathbf{T} + a_N \mathbf{N}) \cdot (a_T \mathbf{T} + a_N \mathbf{N})$$
$$= a_T^2 (\mathbf{T} \cdot \mathbf{T}) + 2 a_N a_T (\mathbf{N} \cdot \mathbf{T}) + a_N^2 (\mathbf{N} \cdot \mathbf{N})$$
$$= a_T^2 + a_N^2.$$

This relationship can also be obtained by applying the Pythagorean theorem to one of the right triangles with hypotenuse $\|\mathbf{a}(t)\|$ in Figure 11.37. Solving for a_N gives us the following.

Theorem 11.24

$$a_N = \sqrt{\|\mathbf{a}\|^2 - a_T^2}$$

Theorem (11.24) should be used if the normal component is difficult to find by means of (11.23).

EXAMPLE ■ 2   Find a_N in Example 1 by using Theorem (11.24).

SOLUTION   As in the solution to Example 1, we obtain

$$a_T = \frac{4t + 18t^3}{(1 + 4t^2 + 9t^4)^{1/2}} \quad \text{and} \quad \mathbf{a} = \mathbf{r}''(t) = 2\mathbf{j} + 6t\mathbf{k}.$$

Applying (11.24) gives us

$$a_N = \sqrt{\|\mathbf{a}\|^2 - a_T^2}$$

$$= \sqrt{(4 + 36t^2) - \frac{(4t + 18t^3)^2}{1 + 4t^2 + 9t^4}}.$$

This simplifies to the expression for a_N obtained in Example 1.

EXAMPLE ■ 3   If a point P moves around a circle of radius k with a constant speed v, find the tangential and normal components of acceleration.

SOLUTION   Since $v = ds/dt$ is a constant, the tangential component $a_T = d^2s/dt^2$ is 0.

By Example 5 of Section 11.4, the curvature of the circle is $1/k$. Hence, the normal component of acceleration (see (11.23)) is $(1/k)v^2$. This shows that the acceleration is a vector of constant magnitude v^2/k directed from P toward the center of the circle, as was shown in Example 2 of Section 11.3.

We may use (11.23) to obtain a formula for curvature of a space curve C. If C has the parametrization

$$x = f(t), \quad y = g(t), \quad z = h(t),$$

then C is determined by the position vector

$$\mathbf{r}(t) = f(t)\mathbf{i} + g(t)\mathbf{j} + h(t)\mathbf{k}.$$

If we solve the equation in (11.23) for K and use $ds/dt = \|\mathbf{r}'(t)\|$, we obtain the next theorem.

Theorem 11.25

Let a space curve C have the parametrization $x = f(t), y = g(t), z = h(t)$, where f'', g'', and h'' exist. The curvature K at the point $P(x, y, z)$ on C is

$$K = \frac{\|\mathbf{r}'(t) \times \mathbf{r}''(t)\|}{\|\mathbf{r}'(t)\|^3} = a_N \frac{1}{\|\mathbf{r}'(t)\|^2}.$$

This formula may also be used for plane curves (see Exercise 16).

EXAMPLE ■ 4

(a) Find the curvature K of the twisted cubic $x = t, y = t^2, z = t^3$ at the point (x, y, z).

(b) Find four-decimal-place approximations for K at the points corresponding to $t = 1, 2, 3,$ and 4.

SOLUTION

(a) If we let

$$\mathbf{r}(t) = t\mathbf{i} + t^2\mathbf{j} + t^3\mathbf{k},$$

then the curve is the same as that considered in Example 1. Substituting the expressions obtained there for $\mathbf{r}'(t)$ and $\mathbf{r}'(t) \times \mathbf{r}''(t)$ into the formula for K in Theorem (11.25) yields

$$K = \frac{2(9t^4 + 9t^2 + 1)^{1/2}}{(9t^4 + 4t^2 + 1)^{3/2}}.$$

We could also find K by substituting for $a_\mathbf{N}$ and $\|\mathbf{r}'(t)\|$ in Theorem (11.25).

(b) Substituting $t = 1, 2, 3,$ and 4 into the formula for K obtained in part (a), we obtain the following approximations for K (compare with the table on page 962).

t	1	2	3	4
(x, y, z)	$(1, 1, 1)$	$(2, 4, 8)$	$(3, 9, 27)$	$(4, 16, 64)$
K	0.1664	0.0132	0.0027	0.0009

Using limit theorems, we can show that $\lim_{t \to \infty} K = 0$—that is, the curvature of the curve approaches that of a line as t increases.

EXERCISES 11.5

Exer. 1–8: Find general formulas for the tangential and normal components of acceleration and for the curvature of the curve C determined by $\mathbf{r}(t)$.

1 $\mathbf{r}(t) = t^2\mathbf{i} + (3t + 2)\mathbf{j}$ **2** $\mathbf{r}(t) = (2t^2 - 1)\mathbf{i} + 5t\mathbf{j}$

3 $\mathbf{r}(t) = 3t\mathbf{i} + t^3\mathbf{j} + 3t^2\mathbf{k}$ **4** $\mathbf{r}(t) = 4t\mathbf{i} + t^2\mathbf{j} + 2t^2\mathbf{k}$

5 $\mathbf{r}(t) = t(\cos t\mathbf{i} + \sin t\mathbf{j})$ **6** $\mathbf{r}(t) = \cosh t\mathbf{i} + \sinh t\mathbf{j}$

7 $\mathbf{r}(t) = 4\cos t\mathbf{i} + 9\sin t\mathbf{j} + t\mathbf{k}$

8 $\mathbf{r}(t) = e^t(\sin t\mathbf{i} + \cos t\mathbf{j} + \mathbf{k})$

9 A point moves along the parabola $y = x^2$ such that the horizontal component of velocity is always 3. Find the tangential and normal components of acceleration at $P(1, 1)$.

10 Work Exercise 9 if the point moves along the graph of $y = 2x^3 - x$.

11 Prove that if a point moves along a curve C with a constant speed, then the acceleration is always normal to C.

12 Use Theorem (11.25) to prove that if a point moves through space with an acceleration that is always $\mathbf{0}$, then the motion is on a line.

13 If a point P moves along a curve C with a constant speed, show that the magnitude of the acceleration is directly proportional to the curvature of the curve.

14 If, in Exercise 13, a second point Q moves along C with a speed twice that of P, show that the magnitude of the acceleration of Q is four times greater than that of P.

15 Show that if a point moves along the graph of $y = f(x)$ for $a \le x \le b$, then the normal component of acceleration is 0 at a point of inflection.

16 If a plane curve is given parametrically by $x = f(t)$, $y = g(t)$ and if f'' and g'' exist, use Theorem (11.25) to prove that the curvature at the point $P(x, y)$ is given by Theorem (11.19).

17 Show that the curvature at every point on the circular helix $x = a \cos t$, $y = a \sin t$, $z = bt$, where $a > 0$, is given by $K = a/(a^2 + b^2)$.

18 An *elliptic helix* has parametric equations $x = a \cos t$, $y = b \sin t$, $z = ct$, where a, b, and c are positive real numbers and $a \neq b$. Find the curvature at (x, y, z).

11.6 KEPLER'S LAWS

It is fitting to conclude this chapter with a display of the power and beauty of vector methods when applied to the derivation of three classical physical laws. The discussion in this section is not simple, for it is not a simple problem that we intend to consider. There is no exercise set at the end of this section. The reason is that we are not interested in numerical calculations involving the laws to be developed. Your objective should be to carefully read and understand each step of the discussion. Proceed slowly. You will gain considerable insight into vector methods by studying the material that follows.

After many years of analyzing an enormous amount of empirical data, the German astronomer Johannes Kepler (1571–1630) formulated three laws that describe the motion of planets about the sun. These laws may be stated as follows.

Kepler's Laws 11.26

First Law:	The orbit of each planet is an ellipse with the sun at one focus.
Second Law:	The vector from the sun to a moving planet sweeps out area at a constant rate.
Third Law:	If the time required for a planet to travel once around its elliptical orbit is T and if the major axis of the ellipse is $2a$, then $T^2 = ka^3$ for some constant k.

Approximately 50 years later, Sir Isaac Newton (see *Mathematicians and Their Times*, Chapter 3) proved that Kepler's laws were consequences of Newton's law of universal gravitation and second law of motion. The achievements of both men were monumental, because these laws clarified all astronomical observations that had been made up to that time.

In this section, we shall prove Kepler's laws through the use of vectors. Since the force of gravity that the sun exerts on a planet far exceeds that exerted by other celestial bodies, we shall neglect all other forces acting on a planet. From this point of view, we have only two objects to consider: the sun and a planet revolving around it.

Figure 11.38

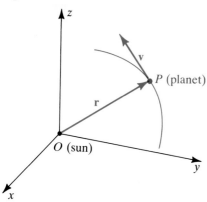

It is convenient to introduce a coordinate system with the center of mass of the sun at the origin, O, as illustrated in Figure 11.38. The point P represents the center of mass of the planet. To simplify the notation, we shall denote the position vector of P by $\mathbf{r}$ instead of $\mathbf{r}(t)$ and use $\mathbf{v}$ and $\mathbf{a}$ to denote the velocity $\mathbf{r}'(t)$ and acceleration $\mathbf{r}''(t)$, respectively.

Before proving Kepler's laws, let us show that the motion of the planet takes place in one plane. If we let $r = \|\mathbf{r}\|$, then $\mathbf{u} = (1/r)\mathbf{r}$ is a unit vector having the same direction as $\mathbf{r}$. According to Newton's law of gravitation, the force $\mathbf{F}$ of gravitational attraction on the planet is given by

$$\mathbf{F} = -G\frac{Mm}{r^2}\mathbf{u},$$

where M is the mass of the sun, m is the mass of the planet, and G is a gravitational constant. Newton's second law of motion (11.13) states that

$$\mathbf{F} = m\mathbf{a}.$$

If we equate these two expressions for $\mathbf{F}$ and solve for $\mathbf{a}$, we obtain

(1)
$$\mathbf{a} = -\frac{GM}{r^2}\mathbf{u}.$$

This shows that $\mathbf{a}$ is parallel to $\mathbf{r} = r\mathbf{u}$ and hence $\mathbf{r} \times \mathbf{a} = \mathbf{0}$. In addition, since $\mathbf{v} \times \mathbf{v} = \mathbf{0}$, we see that

$$\frac{d}{dt}(\mathbf{r} \times \mathbf{v}) = \mathbf{r} \times \frac{d\mathbf{v}}{dt} + \frac{d\mathbf{r}}{dt} \times \mathbf{v}$$

$$= \mathbf{r} \times \mathbf{a} + \mathbf{v} \times \mathbf{v} = \mathbf{0}.$$

It follows that

(2)
$$\mathbf{r} \times \mathbf{v} = \mathbf{c}$$

for a constant vector $\mathbf{c}$. The vector $\mathbf{c}$ will play an important role in the proofs of Kepler's laws.

Since $\mathbf{r} \times \mathbf{v} = \mathbf{c}$, the vector $\mathbf{r}$ is orthogonal to $\mathbf{c}$ for every value of t. This implies that the curve traced by P lies in one plane—that is, *the orbit of the planet is a plane curve*, as illustrated in Figure 11.39.

Let us now prove Kepler's first law. We may assume that the motion of the planet takes place in the xy-plane. In this case, the vector $\mathbf{c}$ is perpendicular to the xy-plane, and we may assume that $\mathbf{c}$ has the same direction as the positive z-axis, as illustrated in Figure 11.39.

Since $\mathbf{r} = r\mathbf{u}$, we see that

$$\mathbf{v} = \frac{d\mathbf{r}}{dt} = r\frac{d\mathbf{u}}{dt} + \frac{dr}{dt}\mathbf{u}.$$

Substitution in $\mathbf{c} = \mathbf{r} \times \mathbf{v}$ and use of properties of vector products gives us

$$\mathbf{c} = r\mathbf{u} \times \left(r\frac{d\mathbf{u}}{dt} + \frac{dr}{dt}\mathbf{u} \right)$$

$$= r^2\left(\mathbf{u} \times \frac{d\mathbf{u}}{dt} \right) + r\frac{dr}{dt}(\mathbf{u} \times \mathbf{u}).$$

Figure 11.39

Since $\mathbf{u} \times \mathbf{u} = \mathbf{0}$, this reduces to

(3)
$$\mathbf{c} = r^2 \left(\mathbf{u} \times \frac{d\mathbf{u}}{dt} \right).$$

Using equations (3) and (1) together with (ii) and (vi) of Theorem (10.33), we see that

$$\mathbf{a} \times \mathbf{c} = \left(-\frac{GM}{r^2}\mathbf{u} \right) \times \left[r^2 \left(\mathbf{u} \times \frac{d\mathbf{u}}{dt} \right) \right]$$

$$= -GM \left[\mathbf{u} \times \left(\mathbf{u} \times \frac{d\mathbf{u}}{dt} \right) \right]$$

$$= -GM \left[\left(\mathbf{u} \cdot \frac{d\mathbf{u}}{dt} \right) \mathbf{u} - (\mathbf{u} \cdot \mathbf{u})\frac{d\mathbf{u}}{dt} \right].$$

Since $\|\mathbf{u}\| = 1$, it follows from Theorem (11.9) that $\mathbf{u} \cdot (d\mathbf{u}/dt) = 0$. In addition, $\mathbf{u} \cdot \mathbf{u} = \|\mathbf{u}\|^2 = 1$, and hence the last formula for $\mathbf{a} \times \mathbf{c}$ reduces to

$$\mathbf{a} \times \mathbf{c} = GM\frac{d\mathbf{u}}{dt} = \frac{d}{dt}(GM\mathbf{u}).$$

We may also write

$$\mathbf{a} \times \mathbf{c} = \frac{d\mathbf{v}}{dt} \times \mathbf{c} = \frac{d}{dt}(\mathbf{v} \times \mathbf{c})$$

and, consequently,

$$\frac{d}{dt}(\mathbf{v} \times \mathbf{c}) = \frac{d}{dt}(GM\mathbf{u}).$$

Integrating both sides of this equation gives us

(4)
$$\mathbf{v} \times \mathbf{c} = GM\mathbf{u} + \mathbf{b},$$

where $\mathbf{b}$ is a constant vector.

The vector $\mathbf{v} \times \mathbf{c}$ is orthogonal to $\mathbf{c}$ and therefore is in the xy-plane. Since $\mathbf{u}$ is also in the xy-plane, it follows from (4) that $\mathbf{b}$ is in the xy-plane.

Up to this point, our proof has been independent of the positions of the x- and y-axes. Let us now choose a coordinate system such that the positive x-axis has the direction of the constant vector $\mathbf{b}$, as illustrated in Figure 11.40.

Let (r, θ) be polar coordinates for the point P, with $r = \|\mathbf{r}\|$. It follows that

$$\mathbf{u} \cdot \mathbf{b} = \|\mathbf{u}\| \|\mathbf{b}\| \cos \theta = b \cos \theta,$$

where $b = \|\mathbf{b}\|$. If we let $c = \|\mathbf{c}\|$, then using equation (2), together with properties of the dot and vector products, and also equation (4), yields

$$c^2 = \mathbf{c} \cdot \mathbf{c} = (\mathbf{r} \times \mathbf{v}) \cdot \mathbf{c} = \mathbf{r} \cdot (\mathbf{v} \times \mathbf{c})$$

$$= (r\mathbf{u}) \cdot (GM\mathbf{u} + \mathbf{b})$$

$$= rGM(\mathbf{u} \cdot \mathbf{u}) + r(\mathbf{u} \cdot \mathbf{b})$$

$$= rGM + rb \cos \theta.$$

Figure 11.40

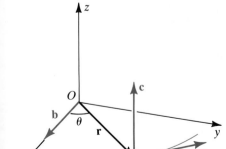

Solving the last equation for r gives us

$$r = \frac{c^2}{GM + b\cos\theta}.$$

Dividing the numerator and the denominator of this fraction by GM, we obtain

(5)
$$r = \frac{p}{1 + e\cos\theta},$$

with $p = c^2/(GM)$ and $e = b/(GM)$. It can be shown that the graph of this polar equation is a conic with eccentricity e and focus at the origin. We can convert the polar equation to x, y-coordinates using the relationships $x = r\cos\theta$ and $r = \sqrt{x^2 + y^2}$. Algebraic manipulation produces an equation in the form

(6)
$$\frac{[x - (-pe/(1 - e^2))]^2}{p^2/(1 - e^2)^2} + \frac{y^2}{p^2/(1 - e^2)} = 1,$$

which is a standard form of a conic (see Section E of the Precalculus Review). Since the orbit is a closed curve, it follows that $0 < e < 1$ and that the conic is an ellipse. This completes the proof of Kepler's first law.

Note that for some artificial satellites or "space probes," the path of motion is not closed, but is a hyperbola carrying the spacecraft out of the solar system.

Let us next prove Kepler's second law. We may assume that the orbit of the planet is an ellipse in the xy-plane. Let $r = f(\theta)$ be a polar equation of the orbit, with the center of the sun at the focus O. Let P_0 denote the position of the planet at time t_0 and P its position at any time $t \ge t_0$. As illustrated in Figure 11.41, θ_0 and θ will denote the angles measured from the positive x-axis to $\overrightarrow{OP_0}$ and $\overrightarrow{OP}$, respectively.

By Theorem (9.12), the area A swept out by $\overrightarrow{OP}$ in the time interval $[t_0, t]$ is

$$A = \int_{\theta_0}^{\theta} \frac{1}{2} r^2 d\theta,$$

and hence

$$\frac{dA}{d\theta} = \frac{d}{d\theta} \int_{\theta_0}^{\theta} \frac{1}{2} r^2 d\theta = \frac{1}{2} r^2.$$

Using this fact and the chain rule gives us

(7)
$$\frac{dA}{dt} = \frac{dA}{d\theta} \frac{d\theta}{dt} = \frac{1}{2} r^2 \frac{d\theta}{dt}.$$

Next we observe that since $\mathbf{r} = r\cos\theta\mathbf{i} + r\sin\theta\mathbf{j} + 0\mathbf{k}$, the unit vector $\mathbf{u} = (1/r)\mathbf{r}$ may be expressed in the form

$$\mathbf{u} = \cos\theta\mathbf{i} + \sin\theta\mathbf{j} + 0\mathbf{k}.$$

Consequently,

$$\frac{d\mathbf{u}}{dt} = -\sin\theta\frac{d\theta}{dt}\mathbf{i} + \cos\theta\frac{d\theta}{dt}\mathbf{j} + 0\mathbf{k}.$$

Figure 11.41

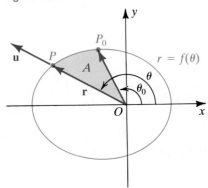

A direct calculation may be used to show that

$$\mathbf{u} \times \frac{d\mathbf{u}}{dt} = \frac{d\theta}{dt}\mathbf{k}.$$

If $\mathbf{c}$ is the vector obtained in the proof of Kepler's first law, then, by (3) and the last equation,

$$\mathbf{c} = r^2 \left(\mathbf{u} \times \frac{d\mathbf{u}}{dt} \right) = r^2 \frac{d\theta}{dt}\mathbf{k}$$

and hence

(8)
$$c = |\mathbf{c}| = r^2 \frac{d\theta}{dt}.$$

Combining (7) and (8), we see that

(9)
$$\frac{dA}{dt} = \frac{1}{2}c;$$

that is, the rate at which A is swept out by $\overrightarrow{OP}$ is a constant. This establishes Kepler's second law.

To prove Kepler's third law, we shall retain the notation used in the proofs of the first two laws. In particular, we assume that a polar equation of the planetary orbit is given by

$$r = \frac{p}{1 + e\cos\theta}$$

with $p = c^2/(GM)$ and $e = b/(GM)$.

Let T denote the time required for the planet to make one complete revolution about the sun. By equation (9), the area swept out in the time interval $[0, T]$ is given by

$$A = \int_0^T \frac{dA}{dt}\,dt = \int_0^T \frac{1}{2}c\,dt = \frac{1}{2}cT.$$

This also equals the area of the plane region bounded by the ellipse. However, in Example 5 of Section 7.3, we found that $A = \pi ab$ for an ellipse whose major and minor axes have lengths $2a$ and $2b$, respectively. Consequently,

$$\frac{1}{2}cT = \pi ab, \quad \text{or} \quad T = \frac{2\pi ab}{c}.$$

By the definition of the eccentricity of an ellipse (see page 73),

$$e = \frac{\sqrt{a^2 - b^2}}{a}$$

and hence $\qquad a^2 e^2 = a^2 - b^2, \quad \text{or} \quad b^2 = a^2(1 - e^2).$

Thus, $\qquad\qquad T^2 = \frac{4\pi^2 a^2 b^2}{c^2} = \frac{4\pi^2 a^4(1 - e^2)}{c^2}.$

From equation (6), we know that

$$a^2 = \frac{p^2}{(1 - e^2)^2}, \quad \text{or} \quad a = \frac{p}{1 - e^2},$$

and hence

$$T^2 = \frac{4\pi^2 a^4}{c^2}\left(\frac{p}{a}\right).$$

Since $p = c^2/(GM)$, this reduces to

$$T^2 = \frac{4\pi^2}{GM}a^3 = ka^3$$

with $k = 4\pi^2/(GM)$. This completes the proof.

Remember that in our proofs of Kepler's laws we assumed that the only gravitational force acting on a planet was that of the sun. If forces exerted by other planets are taken into account, then irregularities in the elliptical orbits may occur. Observed irregularities in the motion of Uranus led both the British astronomer J. Adams (1819–1892) and the French astronomer U. Leverrier (1811–1877) to predict the presence of an unknown planet that was causing the irregularities. On the basis of their predictions, this planet, later named Neptune, was first observed with a telescope by the German astronomer J. Galle in 1846.

CHAPTER 11 REVIEW EXERCISES

Exer. 1–2: Let $r(t)$ be the position vector of a moving point P. Sketch the path C of P together with $v(t)$ and $a(t)$ for the given value of t.

1 $\mathbf{r}(t) = \tan t\,\mathbf{i} - \sec t\,\mathbf{j}$; $t = \pi/4$, $|t| < \pi/2$

2 $\mathbf{r}(t) = t^2\mathbf{i} + 3\cos t\,\mathbf{j} + 5\sin t\,\mathbf{k}$; $t = \pi$

Exer. 3–4: The position of a point moving in a coordinate plane is given by $\mathbf{r}(t)$. Find its velocity, acceleration, and speed at time t.

3 $\mathbf{r}(t) = t^2\mathbf{i} + (4t^2 - t^4)\mathbf{j}$

4 $\mathbf{r}(t) = (2 - \sin t)\mathbf{i} + (1 - \cos t)\mathbf{j}$

5 Let C be the curve determined by

$$\mathbf{r}(t) = (e^t \sin t)\mathbf{i} + (e^t \cos t)\mathbf{j} + e^t\mathbf{k} \quad \text{for} \quad 0 \le t \le 1.$$

  (a) Find a unit tangent vector to C at the point corresponding to $t = 0$.

  (b) Find an equation of the tangent line to C at the point corresponding to $t = 0$.

6 Find the length of the curve C in Exercise 5.

Exer. 7–8: Let

$$\mathbf{u}(t) = t^2\mathbf{i} + 6t\mathbf{j} + t\mathbf{k} \quad \text{and} \quad \mathbf{v}(t) = t\mathbf{i} - 5t\mathbf{j} + 4t^2\mathbf{k}.$$

7 Find the values of t for which $\mathbf{u}(t)$ and $\mathbf{v}'(t)$ are orthogonal.

8 Find $[\mathbf{u}(t) \times \mathbf{v}(t)]'$ and $[\mathbf{u}(t) \cdot \mathbf{v}(t)]'$.

9 The position vector of a point at time t is

$$\mathbf{r}(t) = 3t\mathbf{i} + t^3\mathbf{j} + t^4\mathbf{k} \quad \text{for} \quad t \text{ in } \mathbb{R}.$$

Find the velocity, acceleration, and speed at t and at $t = 1$.

Exer. 10–11: Evaluate the integral.

10 $\displaystyle\int (\sin 3t\mathbf{i} + e^{-2t}\mathbf{j} + \cos t\mathbf{k})\,dt$

11 $\displaystyle\int_0^1 (4t\mathbf{i} + t^3\mathbf{j} - \mathbf{k})\,dt$

12 Find $\mathbf{u}(t)$ if

$$\mathbf{u}'(t) = e^{-t}\mathbf{i} - 4\sin 2t\mathbf{j} + 3\sqrt{t}\mathbf{k} \quad \text{and} \quad \mathbf{u}(0) = -\mathbf{i} + 2\mathbf{j}.$$

13 A point P is moving with an acceleration $12t\mathbf{j} + 5\mathbf{k}$ at time t. If, at $t = 1$, the coordinates of P are $(-1, 3, \frac{3}{2})$ and its velocity is $3\mathbf{i} - 2\mathbf{j} + 4\mathbf{k}$, find its position vector $\mathbf{r}(t)$.

Exer. 14–15: Verify the identity *without using components*.

14 $\dfrac{d}{dt}\|\mathbf{u}(t)\|^2 = 2\mathbf{u}(t) \cdot \mathbf{u}'(t)$

15 $\dfrac{d}{dt}[\mathbf{u}(t) \cdot \mathbf{u}'(t) \times \mathbf{u}''(t)] = \mathbf{u}(t) \cdot \mathbf{u}'(t) \times \mathbf{u}'''(t)$

Exer. 16–18: Find the curvature of the curve at the point P.

16 $y = xe^x$; $P(0, 0)$

17 $x = \dfrac{1}{1+t}$, $y = \dfrac{1}{1-t}$; $P(\tfrac{2}{3}, 2)$

18 $x = 2t^2$, $y = t^4$, $z = 4t$; $P(x, y, z)$

19 Find the x-coordinates of the points on the graph of $y = x^3 - 3x$ at which the curvature is a maximum.

20 If C is the graph of $y = \cosh x$, find an equation of the circle of curvature for the point $P(0, 1)$.

21 Find the radius of curvature for the point $P(2, \pi)$ on the graph of the polar equation $r = 2 + \sin\theta$.

22 Let C be the curve determined by $\mathbf{r}(t) = (t^2 + 1)\mathbf{i} + 4t\mathbf{j}$, for t in $\mathbb{R}$.

   (a) Find the unit tangent and normal vectors $\mathbf{T}(t)$ and $\mathbf{N}(t)$.

   (b) Sketch the graph of C and show $\mathbf{T}(1)$ and $\mathbf{N}(1)$.

Exer. 23–24: If $\mathbf{r}(t)$ is the position vector for a point, find the tangential and normal components of acceleration at time t.

23 $\mathbf{r}(t) = \sin 2t\,\mathbf{i} + \cos t\,\mathbf{j}$ **24** $\mathbf{r}(t) = 3t\mathbf{i} + t^3\mathbf{j} + t\mathbf{k}$

c Exer. 25–26: Let $\mathbf{r}(t)$ be the position vector of a moving point. Plot the path C traced out by the point (preferably in a dot or a point mode). From the graph, estimate the value of t for which the velocity is largest.

25 $\mathbf{r}(t) = (1 + 18t - 15t^2)\mathbf{i} + (2 + 12t - 20t^2)\mathbf{j}$;
$$0 \le t \le 2\pi$$

26 $\mathbf{r}(t) = t\cos 3t\,\mathbf{i} + 5\sin 3t\,\mathbf{j}$; $1.5 \le t \le 2.5$

EXTENDED PROBLEMS AND GROUP PROJECTS

1 Investigate what is known about the Riemann zeta function $\zeta(s)$, where

$$\zeta(s) = \frac{1}{1^s} + \frac{1}{2^s} + \cdots + \frac{1}{n^s} + \cdots.$$

Some of its properties are relatively easy to prove, whereas others are still major unsolved problems. All are fascinating.

Show, for example, that $\zeta(s)$ is equal to the product of the numbers $(1 - p^{-s})^{-1}$ for all primes p. From this fact, show that the divergence of the harmonic series implies that there are infinitely many prime numbers.

Euler was able to find a formula for $\zeta(s)$ whenever s is a positive even integer. In particular, he showed

$$\sum_{n=1}^{\infty} \frac{1}{n^2} = \frac{\pi^2}{6}, \quad \sum_{n=1}^{\infty} \frac{1}{n^4} = \frac{\pi^4}{90}, \quad \text{and} \quad \sum_{n=1}^{\infty} \frac{1}{n^6} = \frac{\pi^6}{945}.$$

What is the general pattern? What can be said about $\zeta(s)$ when s is a positive odd integer? In particular, is the value of $\sum_{n=1}^{\infty}(1/n^3)$ known?

What is the Riemann hypothesis and why is it considered such an important problem?

2 The **binormal** is the cross product of the unit tangent vector and the principal unit normal vector—that is, $\mathbf{B} = \mathbf{T} \times \mathbf{N}$. Investigate how twists in a curve correspond to changes in $\mathbf{B}$. In particular, do the following.

   (a) Show that $\mathbf{B}$ is a unit vector that is orthogonal to both $\mathbf{T}$ and $\mathbf{N}$.

   (b) Prove that for each t at which $\mathbf{T}'(t) \ne 0$, there is a number $\tau(t)$ such that $\mathbf{B}'(t) = \tau(t)\mathbf{N}(t)$. The number $\tau(t)$ is called the **torsion** of the curve at t.

   (c) Show that $\|\mathbf{B}'\| = |\tau|$ and that a positive value for τ corresponds to the curve twisting toward $\mathbf{N}$.

   (d) Verify the validity of the following *Frenet–Serret formulas:*

   $$\mathbf{T}' = K\mathbf{N}$$
   $$\mathbf{B}' = \tau\mathbf{N}$$
   $$\mathbf{N}' = -K\mathbf{T} - \tau\mathbf{B}$$

   (e) Show that if the curvature is identically zero, then the curve is a straight line.

   (f) Prove that if the torsion is identically zero, then the curve lies in a plane.

   (g) If the torsion is identically zero and the curvature is a nonzero constant, then show that the curve is a circle.

3 Let $\mathbf{r}$ be a differentiable vector-valued function, where $\mathbf{r}(t)$ is the position at time t of an object whose mass $m(t)$ may vary with time. The vector-valued function $\mathbf{M}(t) = m(t)\mathbf{v}(t)$ is the **linear momentum** of the particle. The vector-valued function $\mathbf{F}(t) = [m(t)\mathbf{v}(t)]'$ is the **force vector**. The **angular momentum** about the origin is defined as $\mathbf{L}(t) = \mathbf{r}(t) \cdot \mathbf{M}(t)$ and the **torque** about the origin is $\mathbf{T}(t) = \mathbf{r}(t) \cdot \mathbf{F}(t)$.

   (a) Prove the *law of conservation of linear momentum:* If the force vector is identically zero, then the linear momentum is constant.

   (b) Show that torque is the derivative of angular momentum.

   (c) Prove the *law of conservation of angular momentum:* If the torque is identically zero, then the angular momentum is constant.

INTRODUCTION

I **N THE CASE OF VECTOR-VALUED FUNCTIONS** of a real variable (the subject of Chapter 11), each element of the domain of a function is a real number, whereas each element of the range is a vector. In this chapter, we begin the study of functions in which the situation is reversed: Each element of the domain is a vector whereas each element of the range is a single real number. These functions are called *real-valued functions of several variables*. In many applications, the geometric properties of a vector such as $\langle x, y, z \rangle$—its length and direction, for example—are not considered, and we regard the vector simply as an ordered triple of numbers.

Consider a manufacturer analyzing the production cost C of a commodity. The cost depends on the quality of the material used, the hourly wage of employees, the type of machinery needed, maintenance charges, and overhead. We say, then, that C is a function of five variables, because it depends on five different quantities. In this chapter, we begin with the case of two independent variables and proceed to any number of variables.

First, we examine how to graphically represent the function of two or three variables by the use of *level curves* and *level surfaces* in Section 12.1. Next, we generalize the single-variable concepts of limits and continuity to functions of several variables in Section 12.2. We consider the concept of *partial differentiation* in Section 12.3, where we study how a function of several variables changes if only one of the variables changes. We then discuss in Section 12.4 how to generalize our earlier work with increments and differentials. Chain rules for differentiating compositions involving functions of several variables form the subject of Section 12.5.

Among the applications considered in examples and exercises are finding the rate at which the temperature of a solid changes, examining the rate at which a pollutant emitted from a smokestack dissipates, measuring the intensity of winds inside a hurricane, designing computer chips, and investigating relationships between the market price of a commodity and consumer demand.

In Section 12.6, we introduce an important tool for analyzing values of a function of several variables: the *gradient*. We shall use gradients to find the rate at which a scalar quantity such as temperature or pressure changes as a point moves in any specified direction. We present in Section 12.7 a geometric view of the gradient, which emphasizes tangent planes and normal lines. Another important use is the determination of extreme values of functions, which we study in Sections 12.8 and 12.9.

The calculus of functions of several variables helps in the analysis of the production cost of manufactured goods, such as these computer chips.

Partial Differentiation

973

12.1 FUNCTIONS OF SEVERAL VARIABLES

The functions discussed in previous chapters involved only one independent variable. Such functions have many applications; however, in some problems, *several* independent variables occur.

ILLUSTRATION

- The area of a rectangle depends on *two* quantities, length and width.
- If an object is located in space, the temperature at a point P in the object may depend on *three* rectangular coordinates, x, y, and z of P.
- If the temperature of an object in space changes with time t, then *four* variables, x, y, z, and t, are involved.

In this section, we define functions of several variables and discuss some of their properties. Recall that a function f is a correspondence that assigns to each element in its domain D exactly one element in its range E. If D is a subset of $\mathbb{R}^2$ (the set of all ordered pairs (x, y) of real numbers) and E is a subset of $\mathbb{R}$, then f is a *function of two (real) variables*. Another way of stating this is as follows.

Definition 12.1

> Let D be a set of ordered pairs of real numbers. **A function f of two variables** is a correspondence that assigns to each pair (x, y) in D exactly one real number, denoted by $f(x, y)$. The set D is the **domain** of f. The **range** of f consists of all real numbers $f(x, y)$, where (x, y) is in D.

Figure 12.1

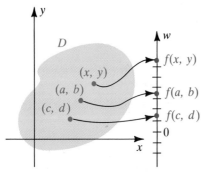

We may represent the domain D in Definition (12.1) by points in an xy-plane and the range by points on a real line—say, a w-axis—as illustrated in Figure 12.1. Several curved arrows are drawn from ordered pairs in D to the corresponding numbers in the range. As an application, suppose a flat metal plate has the shape of D. To each point (x, y) on the plate there corresponds a temperature $f(x, y)$ that can be recorded on a thermometer, represented by the w-axis. We could also regard D as the surface of a lake and let $f(x, y)$ denote the depth of the water under the point (x, y). There are many other physical interpretations for Definition (12.1).

We often use an expression in x and y to specify $f(x, y)$ and assume that the domain is the set of all pairs (x, y) for which the given expression is meaningful. We then call f a **function of x and y**. From our discussions in Chapters 10 and 11, we may regard the point $P(x, y)$ as the terminal point of the position vector OP, where O is the origin. We can represent the ordered pair (x, y) as the vector $\langle x, y \rangle = x\mathbf{i} + y\mathbf{j}$. Thus, the domain of a function of two variables may also be conveniently regarded as a set of vectors.

EXAMPLE ■ 1 Let $f(x, y) = \dfrac{xy - 5}{2\sqrt{y - x^2}}$.

(a) Sketch the domain D of f.

(b) Represent the numbers $f(2, 5)$, $f(1, 2)$, and $f(-1, 2)$ on a w-axis, as in Figure 12.1.

SOLUTION

(a) Referring to the radical $\sqrt{y - x^2}$ in the denominator of $f(x, y)$, we see that the domain D is the set of all pairs (x, y) such that $y - x^2$ is positive—that is, $y > x^2$. The graph of D is the set of all points that lie above the parabola $y = x^2$, as shown in Figure 12.2.

Figure 12.2

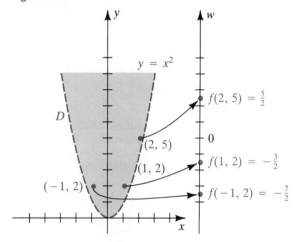

(b) By substitution,

$$f(2, 5) = \frac{(2)(5) - 5}{2\sqrt{5 - 4}} = \frac{5}{2}.$$

Similarly, $f(1, 2) = -\frac{3}{2}$ and $f(-1, 2) = -\frac{7}{2}$. These function values are indicated on the w-axis in Figure 12.2.

Figure 12.3

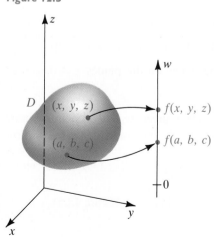

Formulas may be used to define functions of two variables. For example, the formula $V = \pi r^2 h$ expresses the volume V of a right circular cylinder as a function of the altitude h and base radius r. The symbols r and h are referred to as **independent variables**, and V is the **dependent variable**.

A function f of three (real) variables is defined as in (12.1), except that the domain D is a subset of $\mathbb{R}^3$. In this case, to each ordered triple (x, y, z) in D there is assigned exactly one real number $f(x, y, z)$. If we represent D by a region in three dimensions, as illustrated in Figure 12.3, then to each point (x, y, z) in D there corresponds a unique point on the w-axis with coordinate $f(x, y, z)$. As an application, we could regard D as a solid object and let $f(x, y, z)$ be the temperature at (x, y, z).

Functions of three variables are often defined by expressions. For example,

$$f(x, y, z) = \frac{xe^y + \sqrt{z^2 + 1}}{xy \sin z}$$

determines a function of x, y, and z. Formulas, such as $V = lwh$ for the volume of a rectangular box of dimensions l, w, and h, may also be used to define functions of three variables. By representing the ordered triple (x, y, z) as the vector $\langle x, y, z \rangle = x\mathbf{i} + y\mathbf{j} + z\mathbf{k}$, we can regard the domain of a function of three variables as a set of vectors.

Let us return to a function f of two variables x and y. The **graph** of f is, by definition, the graph of the equation $z = f(x, y)$ in an xyz-coordinate system and, hence, is usually a surface S of some type. If we represent the domain D by a region in the xy-plane, then the pair (x, y) in D is represented by the point $(x, y, 0)$. Function values $f(x, y)$ are the (signed) distances from the xy-plane to S, as illustrated in Figure 12.4.

Figure 12.4

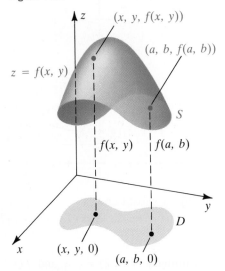

Figure 12.5
Circular traces on the planes $z = k$

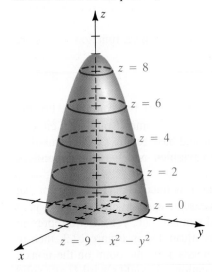

EXAMPLE 2 Let f be the function with domain D given by

$$f(x, y) = 9 - x^2 - y^2 \quad \text{and} \quad D = \{(x, y): x^2 + y^2 \leq 9\}.$$

Sketch the graph of f and show the traces on the planes $z = 0$, $z = 2$, $z = 4$, $z = 6$, and $z = 8$.

SOLUTION The domain D may be represented by all points within and on the circle $x^2 + y^2 = 9$ in the xy-plane. The graph of f is the portion of the graph of

$$z = 9 - x^2 - y^2$$

(a paraboloid) that lies above or on the xy-plane (see Figure 12.5).

Figure 12.6

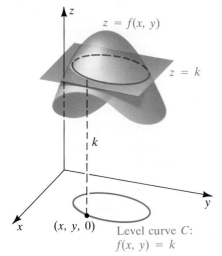

Level curve C:
$f(x, y) = k$

To find the trace on the plane $z = k$, we consider

$$k = 9 - x^2 - y^2, \quad \text{or} \quad x^2 + y^2 = 9 - k.$$

Letting $k = 0, 2, 4, 6$, and 8, we obtain the circles of radii $3, \sqrt{7}, \sqrt{5}, \sqrt{3}$, and 1, respectively, shown in Figure 12.5.

Let f be a function of two variables, and consider the trace of the graph of f on the plane $z = k$, as illustrated in Figure 12.6. If we project this trace onto the xy-plane, we obtain a curve C with equation $f(x, y) = k$. Note that if a point $(x, y, 0)$ moves along C, the corresponding function values $f(x, y)$ always equal k. We call C a **level curve** of f.

EXAMPLE ▪ 3 Sketch some level curves of the function f in Example 2.

SOLUTION The level curves are graphs, in the xy-plane, of equations of the form $f(x, y) = k$—that is,

$$9 - x^2 - y^2 = k, \quad \text{or} \quad x^2 + y^2 = 9 - k.$$

These are circles, provided $0 \le k < 9$. Figure 12.7 shows level curves corresponding to $k = 0, 2, 4, 6$, and 8. These level curves are the projections, in the xy-plane, of the circular traces shown in Figure 12.5.

Figure 12.7
Level curves: $9 - x^2 - y^2 = k$

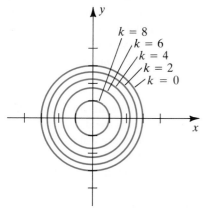

If f is a function of two variables and we sketch the level curves $f(x, y) = k$ for equispaced values of k, such as $k = 0, 2, 4, 6$, and 8 in Example 3, then the nearness of successive curves gives us information about the steepness of the graph of f. In Figure 12.7, the level curves corresponding to $k = 0$ and $k = 2$ are closer than those corresponding to $k = 6$ and $k = 8$, which indicates that the surface shown in Figure 12.5 is steeper at points near the xy-plane than at points farther away.

EXAMPLE ▪ 4 If $f(x, y) = y^2 - x^2$, sketch some level curves of f.

SOLUTION The graph of f is the hyperbolic paraboloid with equation $z = y^2 - x^2$ obtained by letting $a = b = c = 1$ in (10.46). The graph is sketched in Figure 12.8.

The level curves in the xy-plane are the graphs of the equations $f(x, y) = k$—that is, of

$$y^2 - x^2 = k, \quad \text{for } k \text{ in } \mathbb{R}.$$

Figure 12.8

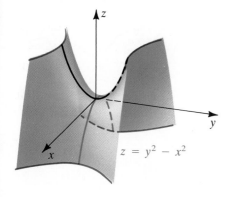

$z = y^2 - x^2$

Figure 12.9

Level curves: $y^2 - x^2 = k$

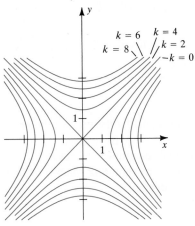

These curves are described in the following table.

Value of k	Equation of level curve C	Description of C in xy-plane
$k > 0$	$y^2 - x^2 = k$	Hyperbola with vertices $(0, \pm\sqrt{k})$
$k = 0$	$y^2 - x^2 = 0$, or $y = \pm x$	Two lines of slopes ± 1 through the origin
$k < 0$	$x^2 - y^2 = k$, or $y^2 - x^2 = -k$	Hyperbola with vertices $(\pm\sqrt{-k}, 0)$

The level curves corresponding to $k = 0, \pm 2, \pm 4, \pm 6$, and ± 8 are sketched in Figure 12.9.

Figure 12.10

Mountain elevations

Level curves are often used in making **topographic**, or **contour**, **maps**. For example, suppose $f(x, y)$ denotes the elevation (in feet) at a point (x, y) of latitude x and longitude y. On the hill pictured in Figure 12.10, we have sketched curves (in three dimensions) corresponding to elevations of 0, 100, 200, 300, and 400 ft. We may regard these curves as having been obtained by slicing the hill with planes parallel to the base. A person walking along one side of these curves would always remain at the same elevation. The (two-dimensional) level curves corresponding to the same elevations are shown in Figure 12.11. They represent the view obtained by looking down on the hill from an airplane.

A similar map is used to indicate the depth of the water in a lake. One example is sketched in Figure 12.12, where $f(x, y)$ is the depth under the point (x, y). What parts of the lake should be avoided by water skiers?

As another illustration of level curves, Figure 12.13 shows a weather map of the United States, where $f(x, y)$ denotes the high temperature at (x, y) during a certain day. On the level curves, called **isothermal curves** or **isotherms**, temperature is constant. A different weather map could be

Figure 12.11

Topographic map of mountain

Figure 12.12 Depth of water in a lake

Figure 12.13 Isothermal curves

drawn in which $f(x, y)$ represented the barometric pressure at (x, y). In this case, the level curves would be called **isobars**.

In economics, level curves are often called **isoquants**. In a simple production application, for example, the cost of production may be treated as a function of two variables, capital and labor. The set of all combinations of capital and labor that yield the same production cost is then called an **isocost**. If profit is modeled as a function $f(x, y)$ of the number x of units sold and the price y per unit, then an **isoprofit** is a level curve of the function f.

If f is a function of three variables x, y, and z, then by definition the **level surfaces** of f are the graphs of $f(x, y, z) = k$ for suitable values of k. If we let $k = w_0$, w_1, and w_2, the resulting graphs are surfaces S_0, S_1, and S_2, as illustrated in Figure 12.14. As a point (x, y, z) moves along one of these surfaces, $f(x, y, z)$ does not change. If $f(x, y, z)$ is the temperature at (x, y, z), the level surfaces are **isothermal surfaces**, and the temperature is constant on each surface. If $f(x, y, z)$ represents electrical potential, the level surfaces are **equipotential surfaces**, and the voltage does not change if (x, y, z) remains on such a surface.

Figure 12.14 Level surfaces of $f(x, y, z)$

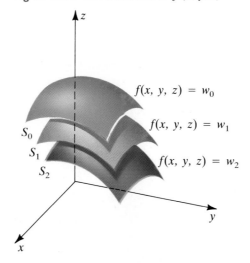

$f(x, y, z) = w_0$

$f(x, y, z) = w_1$

$f(x, y, z) = w_2$

S_0
S_1
S_2

Figure 12.15

Level surfaces: $z - \sqrt{x^2 + y^2} = k$

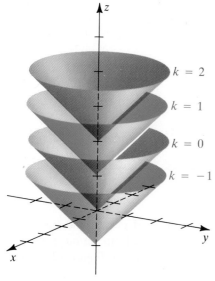

$k = 2$

$k = 1$

$k = 0$

$k = -1$

EXAMPLE = 5 If $f(x, y, z) = z - \sqrt{x^2 + y^2}$, sketch some level surfaces of f.

SOLUTION The level surfaces are graphs of $f(x, y, z) = k$—that is, of

$$z - \sqrt{x^2 + y^2} = k, \quad \text{or} \quad z = \sqrt{x^2 + y^2} + k,$$

where k is any real number. For each k, we obtain a right circular cone whose axis is along the z-axis. The cases $k = -1, 0, 1$, and 2 are sketched in Figure 12.15.

EXAMPLE ■ 6 If $f(x, y, z) = x^2 - y^2 + z^2$, describe the level surfaces of f.

SOLUTION The level surfaces are graphs of equations of the form $f(x, y, z) = k$—that is, of

$$x^2 - y^2 + z^2 = k,$$

where k is a real number. These surfaces are described in the following table.

Value of k	Equation of level surface	Description of surface
$k > 0$	$x^2 - y^2 + z^2 = k$	Hyperboloid of one sheet with axis along the y-axis
$k = 0$	$x^2 - y^2 + z^2 = 0$	Cone with axis along the y-axis
$k < 0$	$-x^2 + y^2 - z^2 = -k$	Hyperboloid of two sheets with axis along the y-axis

A level surface of each type is sketched in Figure 12.16.

Figure 12.16

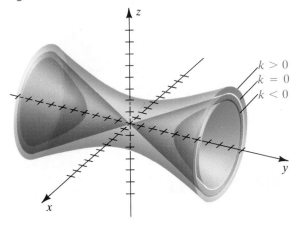

Computers can be programmed to show level curves and traces of surfaces on various planes and to represent surfaces from various perspectives. Several *computer-generated graphs* are shown in Figure 12.17 and in Exercises 39–44 on the following pages.

Figure 12.17

(a) $z = -(x^2 + y^{2/3})$

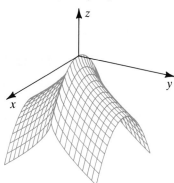

(b) Level curves: $k = -(x^2 + y^{2/3})$

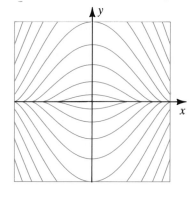

(c) $z = \dfrac{1}{9x^2 + y^2}$

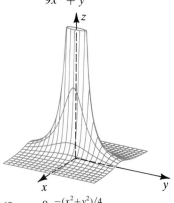

(d) $z = x^3 - 2xy^2 - x^2 - y^2$

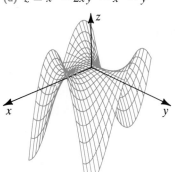

(e) Level curves: $k = x^3 - 2xy^2 - x^2 - y^2$

(f) $z = 8e^{-(x^2+y^2)/4}$

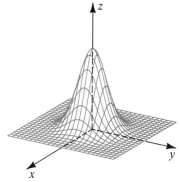

(g) $z = \cos x + \cos y$

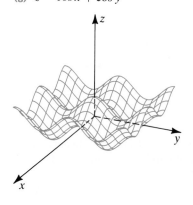

(h) $z = \dfrac{\frac{1}{2}\cos(2x^2 + y^2)}{1 + 2x^2 + y^2}$

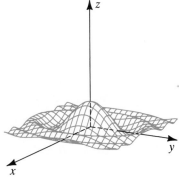

(i) $z = \dfrac{1 - 2\cos(x^2 + y^2)}{x^2 + y^2}$

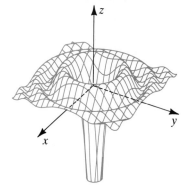

(j) $z = \cos x + 3\cos(3x + y)$

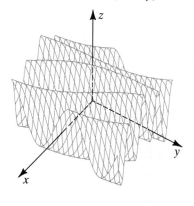

(k) $z = \ln(2x^2 + y^2)$

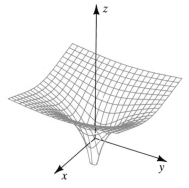

(l) $z = \begin{cases} \sqrt{1 - y^2} & \text{if } |x| \geq 1 \text{ and } |y| \leq 1, \text{ or} \\ & \text{if } |x| \leq 1 \text{ and } |y| \leq |x| \\ \sqrt{1 - x^2} & \text{if } |y| \geq 1 \text{ and } |x| \leq 1, \text{ or} \\ & \text{if } |y| \leq 1 \text{ and } |x| \leq |y| \\ 0 & \text{if } |x| \geq 1 \text{ and } |y| \geq 1 \end{cases}$

EXERCISES 12.1

Exer. 1–6: Describe the domain of f, and find the indicated function values.

1 $f(x, y) = 2x - y^2;$
$$f(-2, 5), \quad f(5, -2), \quad f(0, -2)$$

2 $f(x, y) = \dfrac{y + 2}{x};$
$$f(3, 1), \quad f(1, 3), \quad f(2, 0)$$

3 $f(u, v) = \dfrac{uv}{u - 2v};$
$$f(2, 3), \quad f(-1, 4), \quad f(0, 1)$$

4 $f(r, s) = \sqrt{1 - r} - e^{r/s};$
$$f(1, 1), \quad f(0, 4), \quad f(-3, 3)$$

5 $f(x, y, z) = \sqrt{25 - x^2 - y^2 - z^2};$
$$f(1, -2, 2), \qquad f(-3, 0, 2)$$

6 $f(x, y, z) = 2 + \tan x + y \sin z;$
$$f(\pi/4, 4, \pi/6), \qquad f(0, 0, 0)$$

Exer. 7–14: Sketch the graph of f.

7 $f(x, y) = \sqrt{1 - x^2 - y^2}$

8 $f(x, y) = 4 - x^2 - 4y^2$

9 $f(x, y) = x^2 + y^2 - 1$

10 $f(x, y) = \frac{1}{6}\sqrt{9x^2 + 4y^2}$

11 $f(x, y) = 6 - 2x - 3y$

12 $f(x, y) = \sqrt{72 + 4x^2 - 9y^2}$

13 $f(x, y) = \sqrt{y^2 - 4x^2 - 16}$

14 $f(x, y) = \sqrt{x^2 + 4y^2 + 25}$

c **Exer. 15–20:** Plot the graph of f using the indicated viewing box. Rotate the graph to select a "good" viewpoint to save and/or print.

15 $f(x, y) = \dfrac{x^3}{12} + \dfrac{x(y^2 - 1)}{4} + \dfrac{2 - y}{8};$
$$-3 \le x \le 3, \, -3 \le y \le 3, \, -9 \le z \le 9$$

16 $f(x, y) = \dfrac{5x^2 + 2y^2 - 2xy - 4x - 4y - 25}{5};$
$$-2 \le x \le 2, \, -2 \le y \le 2, \, -7 \le z \le 5$$

17 $f(x, y) = \dfrac{-8x}{x^2 + y^2 + 1};$
$$-4 \le x \le 4, \, -3 \le y \le 3, \, -4 \le z \le 4$$

18 $f(x, y) = e^{(2 - x^2 - y^2)};$
$$-3 \le x \le 3, \, -3 \le y \le 3, \, 0 \le z \le 10$$

19 $f(x, y) = \cos\left(\dfrac{2x^2 + 3y^2}{6}\right);$
$$-4 \le x \le 4, \, -3 \le y \le 3, \, -1 \le z \le 1$$

20 $f(x, y) = \cos x \, e^{-\sqrt{(x^2 + 3y^2)/7}};$
$$-5 \le x \le 5, \, -3 \le y \le 3, \, -1 \le z \le 1$$

Exer. 21–26: Sketch the level curves of f for the given values of k.

21 $f(x, y) = y^2 - x^2;$ $\qquad\qquad$ $k = -4, 0, 9$

22 $f(x, y) = 3x - 2y;$ $\qquad\qquad$ $k = -4, 0, 6$

23 $f(x, y) = x^2 - y;$ $\qquad\qquad$ $k = -2, 0, 3$

24 $f(x, y) = xy;$ $\qquad\qquad\qquad$ $k = -4, 1, 4$

25 $f(x, y) = (x - 2)^2 + (y + 3)^2;$ $\quad$ $k = 1, 4, 9$

26 $f(x, y) = 4x^2 + y^2;$ $\qquad\qquad$ $k = 4, 9, 16$

Exer. 27–28: Find an equation of the level curve of f that contains the point P.

27 $f(x, y) = y \arctan x;$ $\qquad$ $P(1, 4)$

28 $f(x, y) = (2x + y^2)e^{xy};$ $\quad$ $P(0, 2)$

Exer. 29–30: Find an equation of the level surface of f that contains the point P.

29 $f(x, y, z) = x^2 + 4y^2 - z^2;$ $\quad$ $P(2, -1, 3)$

30 $f(x, y, z) = z^2y + x;$ $\qquad\qquad$ $P(1, 4, -2)$

Exer. 31–34: Describe the level curves, whenever they exist, for the surface in the indicated figure if **(a)** $k > 0$, **(b)** $k = 0$, and **(c)** $k < 0$.

31 Figure 12.17(c)

32 Figure 12.17(k)

33 Figure 12.17(f)

34 Figure 12.17(l)

c **Exer. 35–38:** Plot some level curves of f.

35 $f(x, y) = \dfrac{x^2 - xy}{x^2 + y^2 + 1};$ $\quad$ $-3 \le x \le 3, -3 \le y \le 3$

36 $f(x, y) = e^{(1 - 0.6x^2 - 0.4y^2)};$ $\quad$ $-2 \le x \le 2, -3 \le y \le 3$

37 $f(x, y) = \sec\left(\dfrac{x^2}{x^2 + 3y^2}\right);$ $-4 \le x \le 4, -3 \le y \le 3$

38 $f(x, y) = \sqrt{|xy|};$ $\qquad\qquad$ $-5 \le x \le 5, -5 \le y \le 5$

Exercises 12.1

Exer. 39 – 41: Correctly match one of the graphs shown in parts (a) – (c) to the system of level curves of the function given by $z = f(x, y)$.

39 $z = e^{-x^2} + e^{-4y^2}$;
$-2 \le x \le 2, -2 \le y \le 2$

(a)

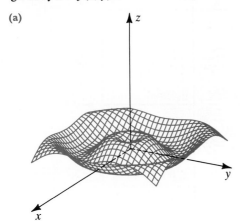

40 $z = \sin \sqrt{x^2 + y^2}$;
$-7 \le x \le 7, -7 \le y \le 7$

(b)

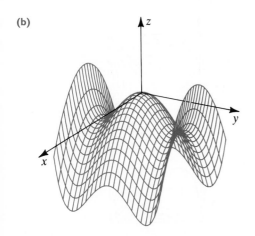

41 $z = y^4 - 8y^2 - 4x^2$;
$-3 \le x \le 3, -3 \le y \le 3$

(c)

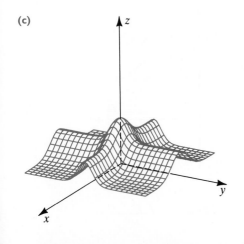

Exer. 42–44: Correctly match one of the graphs shown in parts (a)–(c) to the system of level curves of the function given by $z = f(x, y)$.

(a)

(b)

(c)

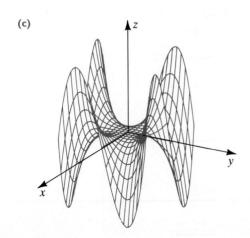

42 $z = \dfrac{15x^2 y^2 e^{-x^2 - y^2}}{x^2 + y^2}$

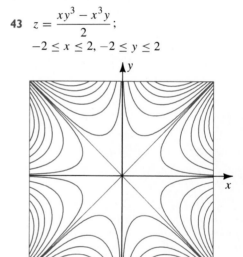

43 $z = \dfrac{xy^3 - x^3 y}{2};$
$-2 \le x \le 2, \ -2 \le y \le 2$

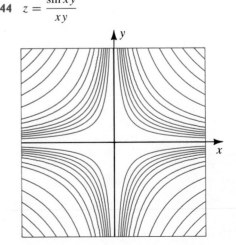

44 $z = \dfrac{\sin xy}{xy}$

Exer. 45–46: Graph, on the same coordinate plane, the level curves for $k = 1$, 16, and 81, using polar coordinates.

45 $f(x, y) = x^4 + 8x^2y^2 + y^4$

46 $f(x, y) = 3x^4 + x^2y^2 + y^4$

Exer. 47–52: Describe the level surfaces of f for the given values of k.

47 $f(x, y, z) = x^2 + y^2 + z^2;$ $k = -1, 0, 4$

48 $f(x, y, z) = z + x^2 + 4y^2;$ $k = -1, 0, 2$

49 $f(x, y, z) = x + 2y + 3z;$ $k = -6, 6, 12$

50 $f(x, y, z) = x^2 + y^2 - z^2;$ $k = -1, 0, 1$

51 $f(x, y, z) = x^2 + y^2;$ $k = -1, 0, 4$

52 $f(x, y, z) = z;$ $k = -2, 0, 3$

53 A flat metal plate is situated in an xy-plane such that the temperature T (in $°C$) at the point (x, y) is inversely proportional to the distance from the origin.

(a) Describe the isotherms.

(b) If the temperature at the point $P(4, 3)$ is $40°C$, find an equation of the isotherm for a temperature of $20°C$.

54 If the voltage V at the point $P(x, y, z)$ is given by $V = 6/(x^2 + 4y^2 + 9z^2)^{1/2}$,

(a) describe the equipotential surfaces

(b) find an equation of the equipotential surface $V = 120$

55 According to Newton's law of universal gravitation, if a particle of mass m_0 is at the origin of an xyz-coordinate system, then the magnitude F of the force exerted on a particle of mass m located at the point (x, y, z) is given by

$$F = \frac{Gm_0m}{x^2 + y^2 + z^2},$$

where G is the universal gravitational constant. How many independent variables are present? If m_0 and m are constant, describe the level surfaces of the resulting function of x, y, and z. What is the physical significance of these level surfaces?

56 According to the *ideal gas law*, the pressure P, volume V and temperature T of a confined gas are related by the formula $PV = kT$ for a constant k. Express P as a function of V and T, and describe the level curves associated with this function. What is the physical significance of these level curves?

57 The power P generated by a wind rotor is proportional to the product of the area A swept out by the blades and the third power of the wind velocity v.

(a) Express P as a function of A and v.

(b) Describe the level curves of P and explain their physical significance.

(c) When the diameter of the circular area swept out by the blades is 10 ft and the wind velocity is 20 mi/hr, then $P = 3000$ watts. Find an equation of the level curve $P = 4000$.

58 If x is wind velocity (in m/sec) and y is temperature (in $°C$), then the *windchill factor F* (in (kcal/m^2)/hr) is given by $F = (33 - y)(10\sqrt{x} - x + 10.5)$.

(a) Find the velocities and temperatures for which the windchill factor is 0. (Assume that $0 \le x \le 50$ and $-50 \le y \le 50$.)

(b) If $F \ge 1400$, frostbite will occur on exposed human skin. Sketch the graph of the level curve $F = 1400$.

59 Shown in the figure are level curves for the surface area of the human body (in tenths of a square meter) as a function of weight x (in kilograms) and height y (in centimeters). Use this system of level curves to estimate your own body surface area. Compare this estimate to the more exact estimate given by the *DuBois and DuBois formula*,

$$S = 0.007184x^{0.425}y^{0.725}.$$

Exercise 59

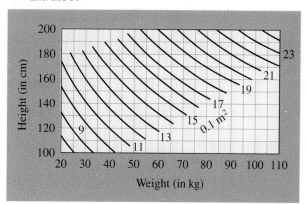

60 The upper boundary of the semicircular region shown in the figure is kept at a temperature of $10°C$, while the lower boundary is maintained at $0°C$. The long-range,

Exercise 60

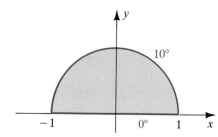

or *steady state*, temperature T at a point (x, y) inside the region is given by

$$T(x, y) = \frac{20}{\pi} \tan^{-1} \left(\frac{2y}{1 - x^2 - y^2} \right).$$

Show that the isotherms are arcs of circles that have centers on the negative y-axis and pass through the points $(-1, 0)$ and $(1, 0)$. Sketch the isotherm corresponding to a temperature of $5\,^\circ\mathrm{C}$.

61 A garbage incineration plant will be built to service two cities. Each city would like to maximize its distance from the plant, but for economic reasons the sum of the distances from each city to the plant cannot exceed M miles. Show that the level curves for the location of the plant are ellipses.

62 The atmospheric pressure near ground level in a certain region is given by

$$p(x, y) = ax^2 + by^2 + c,$$

where a, b, and c are positive constants.

(a) Describe the isobars in this region for pressures greater than c.

(b) Is this a region of high or low pressure?

12.2   LIMITS AND CONTINUITY

Limits and continuity are fundamental concepts in calculus. In this section, we discuss limits and continuity for real-valued functions of two or three variables. If f is a function of two variables, we may wish to consider changes in the function values $f(x, y)$ as (x, y) varies through the domain D of f. As a physical illustration, suppose a flat metal plate has the shape of the region D in Figure 12.18. To each point (x, y) on the plate there corresponds a temperature $f(x, y)$, which is recorded on a thermometer represented by the w-axis. As the point (x, y) moves on the plate, the temperature may increase, decrease, or remain constant; therefore, the point on the w-axis that corresponds to $f(x, y)$ will move in a positive direction, move in a negative direction, or remain fixed, respectively. If the temperature $f(x, y)$ gets closer to a fixed value L as (x, y) gets closer to a fixed point (a, b), we use the following notation.

Figure 12.18

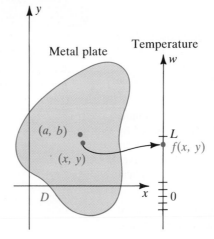

Limit Notation 12.2

$$\lim_{(x,y)\to(a,b)} f(x, y) = L, \quad \text{or} \quad f(x, y) \to L \text{ as } (x, y) \to (a, b)$$

Figure 12.19

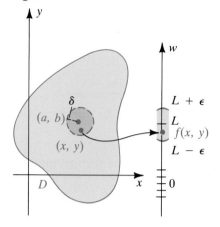

This notation may be read *the limit of $f(x, y)$ as (x, y) approaches (a, b) is L.*

To make (12.2) mathematically precise, let us proceed as follows. For any $\epsilon > 0$, consider the open interval $(L - \epsilon, L + \epsilon)$ on the w-axis, as illustrated in Figure 12.19. If (12.2) is true, then there is a $\delta > 0$ such that for every point (x, y) inside the circle of radius δ with center (a, b), except possibly (a, b) itself, the function value $f(x, y)$ is in the open interval $(L - \epsilon, L + \epsilon)$. This is equivalent to the following statement:

$$\text{If} \quad 0 < \sqrt{(x - a)^2 + (y - b)^2} < \delta, \quad \text{then} \quad |f(x, y) - L| < \epsilon.$$

Thus, we have the following.

Definition of Limit 12.3

Let a function f of two variables be defined throughout the interior of a circle with center (a, b), except possibly at (a, b) itself. The statement

$$\lim_{(x,y)\to(a,b)} f(x, y) = L$$

means that for every $\epsilon > 0$ there is a $\delta > 0$ such that

$$\text{if} \quad 0 < \sqrt{(x - a)^2 + (y - b)^2} < \delta, \quad \text{then} \quad |f(x, y) - L| < \epsilon.$$

Let us consider the graph S of f illustrated in Figure 12.20. Intuitively, we know that Definition (12.3) means that as the point $(x, y, 0)$ approaches $(a, b, 0)$ in the xy-plane, the corresponding point $(x, y, f(x, y))$ on S approaches (a, b, L) (which may or may not be on S). We can show that *if the limit L exists, it is unique.*

If f and g are functions of two variables, then $f + g$, $f - g$, fg, and f/g are defined in the usual way and Theorem (1.8) concerning limits of sums, products, and quotients can be extended. For example, if f and g have limits as (x, y) approaches (a, b), then

Figure 12.20

$$\lim_{(x,y)\to(a,b)} [f(x, y) + g(x, y)] = \lim_{(x,y)\to(a,b)} f(x, y) + \lim_{(x,y)\to(a,b)} g(x, y),$$

$$\lim_{(x,y)\to(a,b)} \frac{f(x, y)}{g(x, y)} = \frac{\lim_{(x,y)\to(a,b)} f(x, y)}{\lim_{(x,y)\to(a,b)} g(x, y)} \quad \text{if} \quad \lim_{(x,y)\to(a,b)} g(x, y) \neq 0,$$

$$\lim_{(x,y)\to(a,b)} \sqrt{f(x, y)} = \sqrt{\lim_{(x,y)\to(a,b)} f(x, y)} \quad \text{if} \quad \lim_{(x,y)\to(a,b)} f(x, y) > 0,$$

and so on.

A function f of two variables is a **polynomial function** if $f(x, y)$ can be expressed as a sum of terms of the form $cx^m y^n$, where c is a real number and m and n are nonnegative integers. A **rational function** is a quotient of two polynomial functions. As for single-variable functions, limits of polynomial and rational functions in two variables may be found by substituting for x and y.

EXAMPLE■1 Find

(a) $\displaystyle\lim_{(x,y)\to(2,-3)} (x^3 - 4xy^2 + 5y - 7)$ **(b)** $\displaystyle\lim_{(x,y)\to(3,4)} \frac{x^2 - y^2}{\sqrt{x^2 + y^2}}$

SOLUTION

(a) Since $x^3 - 4xy^2 + 5y - 7$ is a polynomial, we may find the limit by substituting 2 for x and -3 for y. Thus,

$$\lim_{(x,y)\to(2,-3)} (x^3 - 4xy^2 + 5y - 7) = 2^3 - 4(2)(-3)^2 + 5(-3) - 7$$
$$= 8 - 72 - 15 - 7 = -86.$$

(b) We may proceed as follows:

$$\textbf{(b)} \quad \lim_{(x,y)\to(3,4)} \frac{x^2 - y^2}{\sqrt{x^2 + y^2}} = \frac{\displaystyle\lim_{(x,y)\to(3,4)} (x^2 + y^2)}{\displaystyle\lim_{(x,y)\to(3,4)} \sqrt{x^2 + y^2}}$$

$$= \frac{9 - 16}{\sqrt{\displaystyle\lim_{(x,y)\to(3,4)} (x^2 + y^2)}} = \frac{-7}{\sqrt{9 + 16}} = -\frac{7}{5}$$

EXAMPLE■2 Show that $\displaystyle\lim_{(x,y)\to(0,0)} \frac{x^2 - y^2}{x^2 + y^2}$ does not exist.

SOLUTION Let $f(x, y) = \dfrac{x^2 - y^2}{x^2 + y^2}$.

The function f is rational; however, we cannot substitute 0 for both x and y, since that would lead to a zero denominator.

If we consider any point $(x, 0)$ on the x-axis, then

$$f(x, 0) = \frac{x^2 - 0}{x^2 + 0} = 1, \quad \text{provided} \quad x \neq 0.$$

For any point $(0, y)$ on the y-axis, we have

$$f(0, y) = \frac{0 - y^2}{0 + y^2} = -1, \quad \text{provided} \quad y \neq 0.$$

Consequently, as illustrated in Figure 12.21, *every* circle with center $(0, 0)$ contains points at which the value of f is 1 and points at which the value of f is -1. It follows that the limit does not exist, for if we take $\epsilon = 1$

Figure 12.21

Figure 12.22

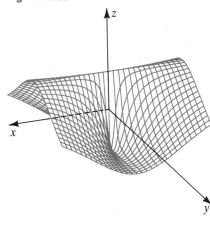

in Definition (12.3), there is no open interval $(L - \epsilon, L + \epsilon)$ on the w-axis containing both 1 and -1. Hence it is impossible to find a $\delta > 0$ that satisfies the conditions of the definition.

Figure 12.22 is a graph of f. Note that for $x \neq 0$ and $y \neq 0$, the graph contains all points $(x, 0, 1)$ and $(0, y, -1)$, as we have already observed. We can show that the entire graph lies between the planes $z = -1$ and $z = 1$.

In Chapter 1, for a function of one variable with a jump discontinuity at $x = a$, we proved that $\lim_{x \to a} f(x)$ does not exist by showing that $\lim_{x \to a^-} f(x)$ and $\lim_{x \to a^+} f(x)$ are not equal. When considering such one-sided limits, we may regard the point on the x-axis with coordinate x as approaching the point with coordinate a either from the left or from the right, respectively. The analogous situation for functions of two variables is more complicated, since in a coordinate plane there are an infinite number of different curves, or **paths**, along which (x, y) can approach (a, b), as illustrated in Figure 12.23. However, if the limit in Definition (12.3) exists, then $f(x, y)$ must have the limit L, regardless of the path taken. This illustrates the following rule for investigating limits.

Figure 12.23

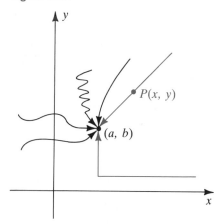

Two-Path Rule | **12.4**

If two different paths to a point $P(a, b)$ produce two different limiting values for f, then $\lim_{(x,y) \to (a,b)} f(x, y)$ does not exist.

We next rework Example 2 using (12.4).

EXAMPLE ■ 3 Show that $\lim\limits_{(x,y) \to (0,0)} \dfrac{x^2 - y^2}{x^2 + y^2}$ does not exist.

SOLUTION   If $P(x, y)$ approaches $(0, 0)$ along the x-axis (see Figure 12.24a), the y-coordinate of P is always zero and the expression $(x^2 - y^2)/(x^2 + y^2)$ reduces to x^2/x^2, or 1. Hence, the limiting value along this path is 1.

If $P(x, y)$ approaches $(0, 0)$ along the y-axis (see Figure 12.24b), the x-coordinate of P is 0 and $(x^2 - y^2)/(x^2 + y^2)$ reduces to $-y^2/y^2$, or -1. Since two different values are obtained, the limit does not exist, by the two-path rule (12.4).

We could, of course, have chosen other paths to the origin $(0, 0)$. For example, if we let $P(x, y)$ approach $(0, 0)$ along the line $y = 2x$ (see Figure 12.24c), then

$$\frac{x^2 - y^2}{x^2 + y^2} = \frac{x^2 - (2x)^2}{x^2 + (2x)^2} = \frac{-3x^2}{5x^2} = -\frac{3}{5}$$

and the limiting value along the line $y = 2x$ is $-\frac{3}{5}$.

Figure 12.24

(a)

(b)

(c)

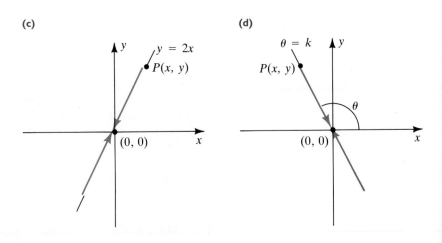

(d)

As an alternative solution, we could change the expression in x and y to polar coordinates, as follows:

$$\frac{x^2 - y^2}{x^2 + y^2} = \frac{r^2 \cos^2\theta - r^2 \sin^2\theta}{r^2} = \cos^2\theta - \sin^2\theta = \cos 2\theta$$

As $P(x, y) \to (0, 0)$ along any ray $\theta = k$ (see Figure 12.24d), the function value is always $\cos 2k$, and hence $f(x, y)$ would have this limiting value. By choosing suitable values of k, we could make $f(x, y)$ approach any value between -1 and 1. Hence, by (12.4), the limit does not exist.

EXAMPLE ▪ 4 Show that $\displaystyle\lim_{(x,y)\to(0,0)} \frac{x^2 y}{x^4 + y^2}$ does not exist.

SOLUTION If we let $P(x, y)$ approach $(0, 0)$ along any line $y = mx$ that passes through the origin (see Figure 12.25a), we see that if $m \neq 0$,

$$\lim_{(x,y)\to(0,0)} \frac{x^2 y}{x^4 + y^2} = \lim_{(x,y)\to(0,0)} \frac{x^2(mx)}{x^4 + (mx)^2} = \lim_{(x,y)\to(0,0)} \frac{mx^3}{x^4 + m^2 x^2}$$

$$= \lim_{(x,y)\to(0,0)} \frac{mx}{x^2 + m^2} = \frac{0}{0 + m^2} = 0.$$

It is tempting to conclude that the limit is 0 as (x, y) approaches $(0, 0)$; however, if we let $P(x, y)$ approach $(0, 0)$ along the parabola $y = x^2$ (see Figure 12.25b), then

$$\lim_{(x,y)\to(0,0)} \frac{x^2 y}{x^4 + y^2} = \lim_{(x,y)\to(0,0)} \frac{x^2(x^2)}{x^4 + (x^2)^2}$$

$$= \lim_{(x,y)\to(0,0)} \frac{x^4}{2x^4} = \lim_{(x,y)\to(0,0)} \frac{1}{2} = \frac{1}{2}.$$

Therefore, not every path to $(0, 0)$ leads to the same limiting value, and hence, by the two-path rule (12.4), the limit does not exist.

If, in the preceding solution, we had obtained the limit 0 using the path $y = x^2$, we *still* could not have concluded that the limit is 0 as the point $P(x, y) \to (0, 0)$, because there could be some *other* path that yields a nonzero number. Remember that the two-path rule cannot be used to prove that a limit exists—only that a limit does *not* exist.

When investigating properties of a function of two variables, we often restrict the domain to points (x, y) that lie in certain types of regions of the xy-plane. In particular, an **open disk** consists of all points that lie *inside* a circle. A **closed disk** contains both the points inside and the points *on* the circle. A point (a, b) is an **interior point** of a region R if there is an open disk with center (a, b) that lies completely within R. An interior point is

Figure 12.25

(a)

(b)

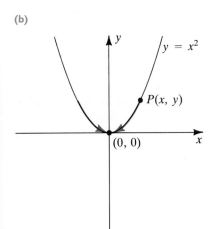

illustrated in Figure 12.26(a). A point (a, b) is a **boundary point** of R if every disk with center (a, b) contains points that are in R and points that are not in R, as illustrated in Figure 12.26(b). The concept of an interior (or boundary) point (a, b, c) in a *three*-dimensional region R is defined in similar fashion using spherical regions instead of disks.

Figure 12.26

(a) Interior point of R

(b) Boundary point of R

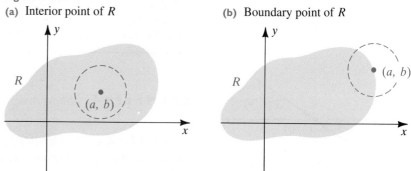

A region is **closed** if it contains all of its boundary points. A region is **open** if it contains *none* of its boundary points—that is, every point of the region is an interior point. A region that contains some, but not all, of its boundary points is neither open nor closed. These concepts are analogous to those of closed, open, and half-open intervals of real numbers.

Suppose a function f of two variables is defined for every (x, y) in a region R, except possibly at (a, b). If (a, b) is an interior point, then Definition (12.3) can be used to investigate the limit of f as (x, y) approaches (a, b). *If (a, b) is a boundary point, we shall use Definition (12.3) with the added restriction that (x, y) must be both in R and inside the circle of radius δ.* Theorems on limits may then be extended to boundary points.

The definition of continuity of a function f of two variables is analogous to that for a function of one variable (see Definition (1.20)).

Definition 12.5

A function f of two variables is **continuous** at an interior point (a, b) of its domain if

$$\lim_{(x,y)\to(a,b)} f(x, y) = f(a, b).$$

In order to define continuity at a boundary point (a, b), we add the restriction that (x, y) is in the domain D of f. Intuitively, we know that $f(x, y) \to f(a, b)$ if $(x, y) \to (a, b)$ along every path C that lies entirely within D. The function f is **continuous on its domain D** if it is continuous at every pair (a, b) in D. If f is continuous on D, then a small change in (x, y) produces a small change in $f(x, y)$. Referring to the graph S of f (see Figure 12.20), we note that if (x, y) is close to (a, b), then the point

$(x, y, f(x, y))$ on S is close to $(a, b, f(a, b))$. Thus, there are no holes or vertical steps in the graph of a continuous function of two variables.

We may prove theorems on continuity for functions of two variables that are analogous to those for functions of one variable. Thus, sums, products, and quotients of continuous functions are continuous, provided no zero denominators occur for the case of quotients. In particular, polynomial functions are continuous everywhere, and rational functions are continuous everywhere except at points where the denominator is zero.

The preceding discussion on limits and continuity can be extended to functions of three or more variables. For example, if f is a function of three variables, we have the following definition.

Definition 12.6

Let a function f of three variables be defined throughout the interior of a sphere with center (a, b, c), except possibly at (a, b, c) itself. The statement

$$\lim_{(x,y,z) \to (a,b,c)} f(x, y, z) = L$$

means that for every $\epsilon > 0$ there is a $\delta > 0$ such that if

$$0 < \sqrt{(x - a)^2 + (y - b)^2 + (z - c)^2} < \delta,$$

then

$$|f(x, y, z) - L| < \epsilon.$$

Figure 12.27

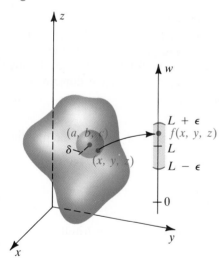

We may give a graphical interpretation for Definition (12.6) that is similar to the one illustrated in Figure 12.19. The difference is that in place of the *circle* in the xy-plane, we use a *sphere* of radius δ in an xyz-coordinate system. Specifically, for any $\epsilon > 0$, consider the open interval $(L - \epsilon, L + \epsilon)$ on the w-axis, as in Figure 12.27. If (12.6) is true, then there is a $\delta > 0$ such that for every point (x, y, z) within the sphere of radius δ with center (a, b, c), except possibly for (a, b, c) itself, the function value $f(x, y, z)$ is in the interval $(L - \epsilon, L + \epsilon)$.

The limit definition (12.6) may be extended to a boundary point (a, b, c) of a region R by adding the restriction that (x, y, z) is both in R and inside the sphere of radius δ.

The two-path rule (12.4) can be extended to functions of three variables. In the next example, we use this rule to show that a limit does not exist.

EXAMPLE 5 Show that

$$\lim_{(x,y,z) \to (1,2,-1)} \frac{(x + y + 3z)^2}{(x - 1)^2 + (y - 2)^2 + (z + 1)^2}$$

does not exist.

SOLUTION The given expression is defined at every point except $(1, 2, -1)$. Suppose we let $P(x, y, z)$ approach $(1, 2, -1)$ along the line l that is parallel to the vector $\langle a, b, c \rangle$, where a, b, and c are real numbers.

By (10.34), parametric equations for l are

$$x = 1 + at, \quad y = 2 + bt, \quad z = -1 + ct; \quad t \text{ in } \mathbb{R}.$$

Hence, for any $P(x, y, z)$ on l different from $(1, 2, -1)$,

$$\frac{(x + y + 3z)^2}{(x - 1)^2 + (y - 2)^2 + (z + 1)^2} = \frac{(at + bt + 3ct)^2}{(at)^2 + (bt)^2 + (ct)^2}$$

$$= \frac{(a + b + 3c)^2 t^2}{(a^2 + b^2 + c^2) t^2} = \frac{(a + b + 3c)^2}{(a^2 + b^2 + c^2)}.$$

Since the last expression may have different values, depending on the choice of a, b, and c, the limit does not exist by the two-path rule.

A function f of three variables is **continuous** at an interior point (a, b, c) of a region if

$$\lim_{(x,y,z) \to (a,b,c)} f(x, y, z) = f(a, b, c).$$

Continuity at a boundary point is defined by using the extension of Definition (12.6) to boundary points.

We may extend the definition of limits to functions of four or more variables; however, in these cases, we do not give geometric interpretations.

In Section 12.5, we shall discuss composite functions of several variables. As a simple illustration, suppose that f is a function of two variables x and y and that g is a function of one variable t. A function h of two variables may be obtained by substituting $f(x, y)$ for t—that is, by letting $h(x, y) = g(f(x, y))$—provided the range of f is in the domain of g. This is illustrated in the next example.

E X A M P L E ▪ 6　Express $g(f(x, y))$ in terms of x and y, and find the domain of the resulting composite function.

(a) $f(x, y) = xe^y, \quad g(t) = 3t^2 + t + 1$

(b) $f(x, y) = y - 4x^2, \quad g(t) = \sin \sqrt{t}$

S O L U T I O N　In each case, we substitute $f(x, y)$ for t in the expression for $g(t)$:

(a) $g(f(x, y)) = g(xe^y) = 3x^2 e^{2y} + xe^y + 1$

(b) $g(f(x, y)) = g(y - 4x^2) = \sin \sqrt{y - 4x^2}$

The domain in part (a) is $\mathbb{R}^2$, and the domain in part (b) consists of every ordered pair (x, y) such that $y \geq 4x^2$.

The following analogue of Theorem (1.24) may be proved for functions of the type illustrated in Example 6.

Theorem 12.7

> If a function f of two variables is continuous at (a, b) and a function g of one variable is continuous at $f(a, b)$, then the function h defined by $h(x, y) = g(f(x, y))$ is continuous at (a, b).

Theorem (12.7) allows us to establish the continuity of composite functions of several variables, as illustrated in the next example.

EXAMPLE ■ 7 If $h(x, y) = e^{x^2 + 5xy + y^3}$, show that h is continuous at every pair (a, b).

SOLUTION If we let $f(x, y) = x^2 + 5xy + y^3$ and $g(t) = e^t$, it follows that $h(x, y) = g(f(x, y))$. Since f is a polynomial function, it is continuous at every pair (a, b). Moreover, g is continuous at every $t = f(a, b)$. Thus, by Theorem (12.7), h is continuous at (a, b).

EXERCISES 12.2

Exer. 1–10: Find the limit.

1. $\displaystyle\lim_{(x,y)\to(0,0)} \frac{x^2 - 2}{3 + xy}$

2. $\displaystyle\lim_{(x,y)\to(2,1)} \frac{4 + x}{2 - y}$

3. $\displaystyle\lim_{(x,y)\to(\pi/2,1)} \frac{y + 1}{2 - \cos x}$

4. $\displaystyle\lim_{(x,y)\to(-1,3)} \frac{y^2 + x}{(x - 1)(y + 2)}$

5. $\displaystyle\lim_{(x,y)\to(0,0)} \frac{x^4 - y^4}{x^2 + y^2}$

6. $\displaystyle\lim_{(x,y)\to(1,0)} \frac{x^4 - 2x^3 + x^2 + x^2y^2}{(x - 1)^2 + y^2}$

7. $\displaystyle\lim_{(x,y)\to(0,0)} \frac{3x^3 - 2x^2y + 3y^2x - 2y^3}{x^2 + y^2}$

8. $\displaystyle\lim_{(x,y)\to(0,0)} \frac{x^3 - x^2y + xy^2 - y^3}{x^2 + y^2}$

9. $\displaystyle\lim_{(x,y,z)\to(0,2,0)} \frac{x^2y^2 + y^4 - 4y^3 + 4y^2 + y^2z^2}{x^2 + (y - 2)^2 + z^2}$

10. $\displaystyle\lim_{(x,y,z)\to(0,2,-3)} \frac{x^4 - (z + 3)^4}{x^2 + (z + 3)^2}$

Exer. 11–20: Show that the limit does not exist.

11. $\displaystyle\lim_{(x,y)\to(0,0)} \frac{2x^2 - y^2}{x^2 + 2y^2}$

12. $\displaystyle\lim_{(x,y)\to(0,0)} \frac{x^2 - 2xy + 5y^2}{3x^2 + 4y^2}$

13. $\displaystyle\lim_{(x,y)\to(1,2)} \frac{xy - 2x - y + 2}{x^2 + y^2 - 2x - 4y + 5}$

14. $\displaystyle\lim_{(x,y)\to(-2,1)} \frac{xy - x + 2y - 2}{(x + 2)^2 + (y - 1)^2}$

15. $\displaystyle\lim_{(x,y)\to(0,0)} \frac{4x^3 y}{2x^4 + 3y^4}$

16. $\displaystyle\lim_{(x,y)\to(0,0)} \frac{3xy}{5x^4 + 2y^4}$

17. $\displaystyle\lim_{(x,y,z)\to(0,0,0)} \frac{xy + yz + xz}{x^2 + y^2 + z^2}$

18. $\displaystyle\lim_{(x,y,z)\to(0,0,0)} \frac{2x^2 + 3y^2 + z^2}{x^2 + y^2 + z^2}$

19. $\displaystyle\lim_{(x,y,z)\to(-3,0,0)} \frac{(x - z + 3)^2}{(x + 3)^2 + y^2 + z^2}$

20. $\displaystyle\lim_{(x,y,z)\to(2,0,0)} \frac{(x - 2)yz^2}{(x - 2)^4 + y^4}$

Exer. 21–24: Use polar coordinates to find the limit, if it exists.

21 $\lim\limits_{(x,y)\to(0,0)} \dfrac{xy^2}{x^2+y^2}$

22 $\lim\limits_{(x,y)\to(0,0)} \dfrac{x^3-y^3}{x^2+y^2}$

23 $\lim\limits_{(x,y)\to(0,0)} \dfrac{x^2+y^2}{\sin(x^2+y^2)}$

24 $\lim\limits_{(x,y)\to(0,0)} \dfrac{\sinh(x^2+y^2)}{x^2+y^2}$

Exer. 25–28: Describe the set of all points in the *xy*-plane at which f is continuous.

25 $f(x,\,y) = \ln(x+y-1)$

26 $f(x,\,y) = \dfrac{xy}{x^2-y^2}$

27 $f(x,\,y) = \sqrt{x}\,e^{\sqrt{1-y^2}}$

28 $f(x,\,y) = \sqrt{25-x^2-y^2}$

Exer. 29–32: Describe the set of all points in an *xyz*-coordinate system at which f is continuous.

29 $f(x,\,y,\,z) = \dfrac{1}{x^2+y^2-z^2}$

30 $f(x,\,y,\,z) = \sqrt{xy}\,\tan z$

31 $f(x,\,y,\,z) = \sqrt{x-2}\,\ln(yz)$

32 $f(x,\,y,\,z) = \sqrt{4-x^2-y^2-z^2}$

Exer. 33–34: The graph of f is shown in the figure. Use polar coordinates to investigate $\lim_{(x,y)\to(0,0)} f(x,\,y)$.

33 $f(x,\,y) = \dfrac{x^2+y^2}{\ln(x^2+y^2)}$

34 $f(x,\,y) = \dfrac{\sin(2x^2+y^2)}{x^2+y^2}$

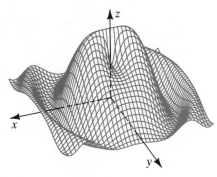

Exer. 35–38: Find $h(x,\,y) = g(f(x,\,y))$ and use Theorem (12.7) to determine where h is continuous.

35 $f(x,\,y) = x^2-y^2$; $g(t) = (t^2-4)/t$

36 $f(x,\,y) = 3x+2y-4$; $g(t) = \ln(t+5)$

37 $f(x,\,y) = x+\tan y$; $g(z) = z^2+1$

38 $f(x,\,y) = y\ln x$; $g(w) = e^w$

39 If $f(x,\,y) = x^2+2y$, $g(t) = e^t$, and $h(t) = t^2-3t$, find $g(f(x,\,y))$, $h(f(x,\,y))$, and $f(g(t),\,h(t))$.

40 If $f(x,\,y,\,z) = 2x+ye^z$ and $g(t) = t^2$, find $g(f(x,\,y,\,z))$.

41 If $f(u,\,v) = uv-3u+v$, $g(x,\,y) = x-2y$, and $k(x,\,y) = 2x+y$, find $f(g(x,\,y),\,k(x,\,y))$.

42 If $f(x,\,y) = 2x+y$, find $f(f(x,\,y),\,f(x,\,y))$.

43 Extend Definition (12.3) to functions of four variables.

44 Prove that if f is a continuous function of two variables and $f(a,\,b) > 0$, then there is a circle C in the *xy*-plane with center $(a,\,b)$ such that $f(x,\,y) > 0$ for every pair $(x,\,y)$ that is in the domain of f and within C.

45 Prove, directly from Definition (12.3), that

(a) $\lim\limits_{(x,\,y)\to(a,\,b)} x = a$ (b) $\lim\limits_{(x,\,y)\to(a,\,b)} y = b$

46 If $\lim_{(x,\,y)\to(a,\,b)} f(x,\,y) = L$ and c is any real number, prove, directly from Definition (12.3), that

$$\lim\limits_{(x,\,y)\to(a,\,b)} cf(x,\,y) = cL.$$

Mathematicians and Their Times

SOFIYA VASILIEVNA KOVALEVSKAYA

THE GREATEST WOMAN MATHEMATICIAN before the twentieth century, Sofiya Vasilievna (Sonya) Kovalevskaya was actively engaged in the scientific, political, and literary issues of her times. Her dramatic life was the subject of a commercial film, *A Hill on the Dark Side of the Moon*.

Kovalevskaya was born in Moscow in January 1850, the daughter of well-educated Russian nobility. In her au-

tobiography, she described with vividness the life at the family's country estate, her father's unhappiness with the "new" ideas of his daughters, and the family's social circle, which included the Russian novelist Fyodor Dostoevsky. Sonya had an unusual introduction to calculus. When she was 14, temporary wallpaper was put up in one of the children's rooms: pages of an old calculus book. At age 17, she studied with a professor at the St. Petersburg naval academy who immediately recognized her mathematical potential.

In the mid-nineteenth century, European women were very restricted in their efforts to gain higher education. Serious young mathematicians studied at one of the leading French or German universities. But single women could not travel abroad alone. Kovalevskaya was part of a movement to promote the emancipation of women. One tactic of this movement was to arrange "marriages of convenience" that would permit unchaperoned travel outside the country. Sonya contracted such a nominal marriage with a young paleontologist.

The couple went to Heidelberg, where Kovalevskaya took courses with several prominent mathematicians. Two years later, she moved to Berlin to study with Karl Weierstrass. Because women could not attend university lectures, she had to work privately with Weierstrass. To overcome any possible objection to granting a mathematics doctorate to a woman, Kovalevskaya submitted three independent research papers, each worthy of a degree. Göttingen University awarded her the doctorate *in absentia*, as women were not permitted to participate in the ceremonies.

Although she had proven her mathematical abilities, prejudice against women temporarily prevented Kovalevskaya from obtaining an academic position anywhere in Europe. Eventually, she secured a lectureship at the University of Stockholm, where she completed research on many topics in analysis. Her paper *On the Rotation of a Solid Body about a Fixed Point* won a major prize from the French Academy of Sciences. Subsequent research was recognized by honors from the Swedish and Russian scientific academies.

In addition to mathematical works, she also wrote a number of novels and plays. In 1891, at the height of her career, Sofiya Kovalevskaya died of influenza complicated by pneumonia.

12.3   PARTIAL DERIVATIVES

Derivatives have been a central concern in our development of calculus. We begin, in this section, to investigate how differentiation can be defined for real-valued functions of two or three variables. In Chapter 2, we defined the derivative $f'(x)$ of a function of *one* variable as

$$f'(x) = \lim_{h \to 0} \frac{f(x + h) - f(x)}{h}.$$

We may interpret this formula as follows. First change the independent variable x by an amount h; then divide the corresponding change in f, $f(x + h) - f(x)$, by h; and finally, let h approach 0. An analogous procedure can be applied to functions of several variables. Given $f(x, y)$, first change *one* of the variables, say x, by an amount h; then divide the corresponding change in f, $f(x + h, y) - f(x, y)$, by h; and finally let h approach 0. This leads to the concept of the *partial derivative* $f_x(x, y)$ *of f with respect to x*, defined in (12.8). A similar process is used for the variable y.

Definition   12.8

Let f be a function of two variables. The **first partial derivatives of f with respect to x and y** are the functions f_x and f_y such that

$$f_x(x, y) = \lim_{h \to 0} \frac{f(x + h, y) - f(x, y)}{h}$$

$$f_y(x, y) = \lim_{h \to 0} \frac{f(x, y + h) - f(x, y)}{h}.$$

In Definition (12.8), x and y are fixed (but arbitrary) and h is the only variable; hence we use the notation for limits of functions of one variable instead of the $(x, y) \to (a, b)$ notation introduced in Section 12.2. We can find partial derivatives without using limits, as follows. If we let $y = b$ and define a function g of one variable by $g(x) = f(x, b)$, then $g'(x) = f_x(x, b) = f_x(x, y)$. Thus, *to find $f_x(x, y)$, we regard y as a constant and differentiate $f(x, y)$ with respect to x.* Similarly, *to find $f_y(x, y)$, we regard the variable x as a constant and differentiate $f(x, y)$ with respect to y.* This technique is illustrated in Example 1.

Other common notations for partial derivatives are listed in the next box.

Notations for Partial Derivatives 12.9

If $w = f(x, y)$, then

$$f_x = \frac{\partial f}{\partial x}, \qquad f_y = \frac{\partial f}{\partial y}$$

$$f_x(x, y) = \frac{\partial}{\partial x} f(x, y) = \frac{\partial w}{\partial x} = w_x$$

$$f_y(x, y) = \frac{\partial}{\partial y} f(x, y) = \frac{\partial w}{\partial y} = w_y.$$

For brevity, we often speak of $\partial f / \partial x$ or $\partial f / \partial y$ as *the partial of f with respect to x or to y,* respectively.

EXAMPLE 1   If $f(x, y) = x^3 y^2 - 2x^2 y + 3x$, find

(a) $f_x(x, y)$ and $f_y(x, y)$ **(b)** $f_x(2, -1)$ and $f_y(2, -1)$

SOLUTION

(a) We regard y as a constant and differentiate with respect to x:

$$f_x(x, y) = 3x^2 y^2 - 4xy + 3$$

We regard x as a constant and differentiate with respect to y:

$$f_y(x, y) = 2x^3 y - 2x^2$$

(b) Substituting $x = 2$ and $y = -1$ in part (a), we have

$$f_x(2, -1) = 3(2)^2(-1)^2 - 4(2)(-1) + 3 = 23,$$
$$f_y(2, -1) = 2(2)^3(-1) - 2(2)^2 = -24.$$

Formulas similar to those for functions of one variable are true for partial derivatives. For example, if $u = f(x, y)$ and $v = g(x, y)$, then a product rule and quotient rule for partial derivatives are

$$\frac{\partial}{\partial x}(uv) = u \frac{\partial v}{\partial x} + v \frac{\partial u}{\partial x}, \qquad \frac{\partial}{\partial x}\left(\frac{u}{v}\right) = \frac{v \dfrac{\partial u}{\partial x} - u \dfrac{\partial v}{\partial x}}{v^2}$$

or, in subscript notation,

$$(fg)_x = fg_x + gf_x, \qquad \left(\frac{f}{g}\right)_x = \frac{gf_x - fg_x}{g^2}.$$

A power rule for partial differentiation is

$$\frac{\partial}{\partial x}(u^n) = nu^{n-1}\frac{\partial u}{\partial x},$$

where n is a real number. Similarly,

$$\frac{\partial}{\partial x}\cos u = -\sin u\frac{\partial u}{\partial x}, \qquad \frac{\partial}{\partial x}e^u = e^u\frac{\partial u}{\partial x},$$

and so on.

EXAMPLE ■ 2   Find $\dfrac{\partial w}{\partial y}$ if $w = xy^2 e^{xy}$.

SOLUTION   We write $w = (xy^2)(e^{xy})$ and use a product rule:

$$\frac{\partial w}{\partial y} = (xy^2)\frac{\partial}{\partial y}(e^{xy}) + (e^{xy})\frac{\partial}{\partial y}(xy^2)$$

$$= xy^2(xe^{xy}) + e^{xy}(2xy) = xy(xy + 2)e^{xy}$$

Figure 12.28

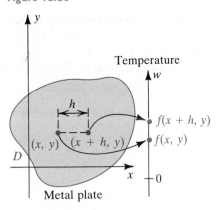

Temperature

Metal plate

Figure 12.29

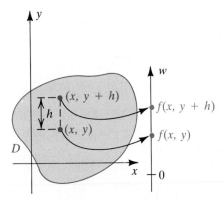

Let us apply Definition (12.8) to the physical illustration given at the beginning of Section 12.2: $f(x, y)$ is the temperature at the point (x, y) of a flat metal plate in an xy-plane. Note that the points (x, y) and $(x + h, y)$ are on a horizontal line, as indicated in Figure 12.28. If a point moves horizontally from (x, y) to $(x + h, y)$, then $f(x + h, y) - f(x, y)$ is the net change in temperature, and we have

average change in temperature: $\dfrac{f(x + h, y) - f(x, y)}{h}$

For example, if the temperature change is 2 degrees and the distance h is 4, then the average change in temperature from (x, y) to $(x + h, y)$ is $\frac{2}{4}$, or $\frac{1}{2}$. Thus, *on the average*, the temperature changes at a rate of $\frac{1}{2}$ degree per unit change in distance. Taking the limit of the average change as h approaches 0, we see that $f_x(x, y)$ *is the (instantaneous) rate of change of temperature with respect to distance as the point* (x, y) *moves in a horizontal direction*.

Similarly, if a point moves on a vertical line from (x, y) to $(x, y + h)$, as shown in Figure 12.29, we obtain

average change in temperature: $\dfrac{f(x, y + h) - f(x, y)}{h}$

Letting h approach 0, we see that $f_y(x, y)$ *is the (instantaneous) rate of change of temperature as the point* (x, y) *moves in a vertical direction*.

As another application, suppose D represents the surface of a lake and $f(x, y)$ is the depth of the water under the point (x, y) on the surface. In this case, $f_x(x, y)$ is the rate at which the depth changes as a point moves

Figure 12.30

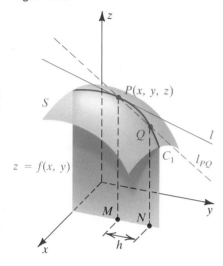

away from (x, y) parallel to the x-axis. Similarly, $f_y(x, y)$ is the rate of change of the depth in the direction of the y-axis.

We shall next use the graph S of f to obtain a geometric interpretation of Definition (12.8). To illustrate $f_y(x, y)$, consider the points $M(x, y, 0)$ and $N(x, y + h, 0)$. The plane parallel to the yz-plane and passing through M and N intersects S in a curve C_1 (see Figure 12.30). Let P and Q be the points on C_1 that have projections M and N on the xy-plane.

If we introduce a rectangular coordinate system in the plane containing M, N, P, and Q, then the slope m_{PQ} of the secant line l_{PQ} through P and Q is

$$m_{PQ} = \frac{f(x, y + h) - f(x, y)}{h}.$$

The limit of m_{PQ} as h approaches 0—that is, the partial derivative $f_y(x, y)$—is the slope of the tangent line l to C_1 at P.

Similarly, if C_2 is the trace of S on the plane parallel to the xz-plane and passing through M, then $f_x(x, y)$ is the slope of the tangent line to C_2 at P.

The following theorem summarizes this discussion.

Theorem 12.10

Let S be the graph of $z = f(x, y)$, and let $P(a, b, f(a, b))$ be a point on S at which f_x and f_y exist. Let C_1 and C_2 be the traces of S on the planes $x = a$ and $y = b$, respectively, and let l_1 and l_2 be the tangent lines to C_1 and C_2 at P (see Figure 12.31).

(i) The slope of l_1 in the plane $x = a$ is $f_y(a, b)$.

(ii) The slope of l_2 in the plane $y = b$ is $f_x(a, b)$.

Figure 12.31

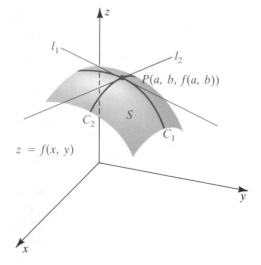

First partial derivatives of functions of three or more variables are defined in the same manner as in Definition (12.8). Specifically, all variables except one are regarded as constant, and we differentiate with respect to the remaining variable. Given $f(x, y, z)$, we may find f_x, f_y, and f_z (or, equivalently, $\partial f/\partial x$, $\partial f/\partial y$, and $\partial f/\partial z$). For example,

$$f_z(x, y, z) = \lim_{h \to 0} \frac{f(x, y, z+h) - f(x, y, z)}{h}.$$

As an application, let $f(x, y, z)$ be the temperature at the point $P(x, y, z)$. The partial derivative $f_z(x, y, z)$ is the instantaneous rate of change of temperature with respect to distance along a line through P that is parallel to the z-axis. The partial derivatives $f_x(x, y, z)$ and $f_y(x, y, z)$ give the rates of change in the directions of the x- and y-axes, respectively.

EXAMPLE ■ 3 If $w = x^2 y^3 \sin z + e^{xz}$, find $\partial w/\partial x$, $\partial w/\partial y$, and $\partial w/\partial z$.

SOLUTION   We proceed as follows:

Regard y and z as constant: $\dfrac{\partial w}{\partial x} = 2xy^3 \sin z + z e^{xz}$

Regard x and z as constant: $\dfrac{\partial w}{\partial y} = 3x^2 y^2 \sin z$

Regard x and y as constant: $\dfrac{\partial w}{\partial z} = x^2 y^3 \cos z + x e^{xz}$

If f is a function of two variables x and y, then f_x and f_y are also functions of two variables, and we may consider *their* first partial derivatives. These are called the **second partial derivatives of f** and are denoted as follows.

Second Partial Derivatives 12.11

$$\frac{\partial}{\partial x} f_x = (f_x)_x = f_{xx} = \frac{\partial}{\partial x}\left(\frac{\partial f}{\partial x}\right) = \frac{\partial^2 f}{\partial x^2}$$

$$\frac{\partial}{\partial y} f_x = (f_x)_y = f_{xy} = \frac{\partial}{\partial y}\left(\frac{\partial f}{\partial x}\right) = \frac{\partial^2 f}{\partial y\, \partial x}$$

$$\frac{\partial}{\partial x} f_y = (f_y)_x = f_{yx} = \frac{\partial}{\partial x}\left(\frac{\partial f}{\partial y}\right) = \frac{\partial^2 f}{\partial x\, \partial y}$$

$$\frac{\partial}{\partial y} f_y = (f_y)_y = f_{yy} = \frac{\partial}{\partial y}\left(\frac{\partial f}{\partial y}\right) = \frac{\partial^2 f}{\partial y^2}$$

If $w = f(x, y)$, we write

$$\frac{\partial^2}{\partial x^2} f(x, y) = f_{xx}(x, y) = \frac{\partial^2 w}{\partial x^2} = w_{xx},$$

$$\frac{\partial^2}{\partial y\, \partial x} f(x, y) = f_{xy}(x, y) = \frac{\partial^2 w}{\partial y\, \partial x} = w_{xy},$$

and so on.

Note that f_{xy} means that we differentiate first with respect to x and then with respect to y, and f_{yx} means that we differentiate first with respect to y and then with respect to x. For the "∂" notation, the reverse order is used—that is, for $\partial^2 f / \partial x\, \partial y$, we differentiate first with respect to y and then with respect to x.

We shall refer to f_{xy} and f_{yx} as the **mixed second partial derivatives of f** (or simply *mixed partials of f*). The next theorem states that, under suitable conditions, these mixed partials are equal—that is, the order of differentiation is immaterial. The proof may be found in texts on advanced calculus.

Theorem 12.12

Let f be a function of two variables x and y. If f, f_x, f_y, f_{xy}, and f_{yx} are continuous on an open region R, then $f_{xy} = f_{yx}$ throughout R.

The hypothesis of Theorem (12.12) is satisfied for most functions encountered in calculus and its applications. Similarly, if $w = f(x, y, z)$ and f has continuous first and second partial derivatives, then the following equalities hold for mixed partials:

$$\frac{\partial^2 w}{\partial y\, \partial x} = \frac{\partial^2 w}{\partial x\, \partial y}, \qquad \frac{\partial^2 w}{\partial z\, \partial x} = \frac{\partial^2 w}{\partial x\, \partial z}, \qquad \frac{\partial^2 w}{\partial z\, \partial y} = \frac{\partial^2 w}{\partial y\, \partial z}$$

EXAMPLE ■ 4 Find the second partial derivatives of f if

$$f(x, y) = x^3 y^2 - 2x^2 y + 3x.$$

SOLUTION This function was considered in Example 1, where we obtained

$$f_x(x, y) = 3x^2 y^2 - 4xy + 3 \quad \text{and} \quad f_y(x, y) = 2x^3 y - 2x^2.$$

Hence, the second partial derivatives are as follows:

$$f_{xx}(x, y) = \frac{\partial}{\partial x} f_x(x, y) = \frac{\partial}{\partial x}(3x^2 y^2 - 4xy + 3) = 6xy^2 - 4y$$

$$f_{xy}(x, y) = \frac{\partial}{\partial y} f_x(x, y) = \frac{\partial}{\partial y}(3x^2 y^2 - 4xy + 3) = 6x^2 y - 4x$$

$$f_{yx}(x, y) = \frac{\partial}{\partial x} f_y(x, y) = \frac{\partial}{\partial x}(2x^3 y - 2x^2) = 6x^2 y - 4x$$

$$f_{yy}(x, y) = \frac{\partial}{\partial y} f_y(x, y) = \frac{\partial}{\partial y}(2x^3 y - 2x^2) = 2x^3$$

Third and higher partial derivatives are defined in similar fashion. For example,

$$\frac{\partial}{\partial x} f_{xx} = f_{xxx} = \frac{\partial}{\partial x}\left(\frac{\partial^2 f}{\partial x^2}\right) = \frac{\partial^3 f}{\partial x^3},$$

$$\frac{\partial}{\partial x} f_{xy} = f_{xyx} = \frac{\partial}{\partial x}\left(\frac{\partial^2 f}{\partial y\,\partial x}\right) = \frac{\partial^3 f}{\partial x\,\partial y\,\partial x},$$

and so on. If first, second, and third partial derivatives are continuous, then the order of differentiation is immaterial—that is,

$$f_{xyx} = f_{yxx} = f_{xxy} \quad \text{and} \quad f_{yxy} = f_{xyy} = f_{yyx}.$$

Of course, letters other than x and y may be used. If f is a function of r and s, then symbols such as

$$f_r(r, s), \qquad f_s(r, s), \qquad f_{rs}, \qquad \frac{\partial f}{\partial r}, \qquad \frac{\partial^2 f}{\partial r^2}$$

are used for partial derivatives.

Similar notations and results apply to partials of functions of more than two variables.

EXERCISES 12.3

Exer. 1–18: Find the first partial derivatives of f.

1 $f(x, y) = 2x^4y^3 - xy^2 + 3y + 1$

2 $f(x, y) = (x^3 - y^2)^5$

3 $f(r, s) = \sqrt{r^2 + s^2}$

4 $f(s, t) = \dfrac{t}{s} - \dfrac{s}{t}$

5 $f(x, y) = xe^y + y\sin x$

6 $f(x, y) = e^x \ln xy$

7 $f(t, v) = \ln\sqrt{\dfrac{t + v}{t - v}}$

8 $f(u, w) = \arctan\dfrac{u}{w}$

9 $f(x, y) = x\cos\dfrac{x}{y}$

10 $f(x, y) = \sqrt{4x^2 - y^2}\sec x$

11 $f(x, y, z) = 3x^2z + xy^2$

12 $f(x, y, z) = x^2y^3z^4 + 2x - 5yz$

13 $f(r, s, t) = r^2e^{2s}\cos t$

14 $f(x, y, t) = \dfrac{x^2 - t^2}{1 + \sin 3y}$

15 $f(x, y, z) = xe^z - ye^x + ze^{-y}$

16 $f(r, s, v, p) = r^3\tan s + \sqrt{s}e^{(v^2)} - v\cos 2p$

17 $f(q, v, w) = \sin^{-1}\sqrt{qv} + \sin vw$

18 $f(x, y, z) = xyze^{xyz}$

Exer. 19–24: Verify that $w_{xy} = w_{yx}$.

19 $w = xy^4 - 2x^2y^3 + 4x^2 - 3y$

20 $w = \dfrac{x^2}{x + y}$

21 $w = x^3e^{-2y} + y^{-2}\cos x$

22 $w = y^2e^{(x^2)} + \dfrac{1}{x^2y^3}$

23 $w = x^2\cosh\dfrac{z}{y}$

24 $w = \sqrt{x^2 + y^2 + z^2}$

25 If $w = 3x^2y^3z + 2xy^4z^2 - yz$, find w_{xyz}.

26 If $w = u^4vt^2 - 3uv^2t^3$, find w_{tut}.

27 If $u = v \sec rt$, find u_{rvr}.

28 If $v = y \ln(x^2 + z^4)$, find v_{zzy}.

29 If $w = \sin xyz$, find $\dfrac{\partial^3 w}{\partial z\, \partial y\, \partial x}$.

30 If $w = \dfrac{x^2}{y^2 + z^2}$, find $\dfrac{\partial^3 w}{\partial z\, \partial y^2}$.

31 If $w = r^4s^3t - 3s^2e^{rt}$, verify that $w_{rrs} = w_{rsr} = w_{srr}$.

32 If $w = \tan uv + 2\ln(u + v)$, verify that $w_{uvv} = w_{vuv} = w_{vvu}$.

Exer. 33 – 36: A function f of x and y is *harmonic* if

$$\frac{\partial^2 f}{\partial x^2} + \frac{\partial^2 f}{\partial y^2} = 0$$

throughout the domain of f. Prove that the given function is harmonic.

33 $f(x, y) = \ln\sqrt{x^2 + y^2}$

34 $f(x, y) = \arctan\dfrac{y}{x}$

35 $f(x, y) = \cos x \sinh y + \sin x \cosh y$

36 $f(x, y) = e^{-x}\cos y + e^{-y}\cos x$

37 If $w = \cos(x - y) + \ln(x + y)$, show that

$$\frac{\partial^2 w}{\partial x^2} - \frac{\partial^2 w}{\partial y^2} = 0$$

38 If $w = (y - 2x)^3 - \sqrt{y - 2x}$, show that $w_{xx} - 4w_{yy} = 0$.

39 If $w = e^{-c^2 t}\sin cx$, show that $w_{xx} = w_t$ for every real number c.

40 The ideal gas law may be stated as $PV = knT$, where n is the number of moles of gas, V is the volume, T is the temperature, P is the pressure, and k is a constant. Show that

$$\frac{\partial V}{\partial T}\frac{\partial T}{\partial P}\frac{\partial P}{\partial V} = -1.$$

Exer. 41– 42: Show that v satisfies the *wave equation*

$$\frac{\partial^2 v}{\partial t^2} = a^2\frac{\partial^2 v}{\partial x^2}.$$

41 $v = (\sin akt)(\sin kx)$

42 $v = (x - at)^4 + \cos(x + at)$

Exer. 43 – 46: Show that the functions u and v satisfy the *Cauchy–Riemann equations* $u_x = v_y$ and $u_y = -v_x$.

43 $u(x, y) = x^2 - y^2$; $v(x, y) = 2xy$

44 $u(x, y) = \dfrac{y}{x^2 + y^2}$; $v(x, y) = \dfrac{x}{x^2 + y^2}$

45 $u(x, y) = e^x\cos y$; $v(x, y) = e^x\sin y$

46 $u(x, y) = \cos x \cosh y + \sin x \sinh y$;
$v(x, y) = \cos x \cosh y - \sin x \sinh y$

47 List all possible second partial derivatives of $w = f(x, y, z)$.

48 If $w = f(x, y, z, t, v)$, define w_t as a limit.

49 A flat metal plate lies on an xy-plane such that the temperature T at (x, y) is given by $T = 10(x^2 + y^2)^2$, where T is in degrees and x and y are in centimeters. Find the instantaneous rate of change of T with respect to distance at $(1, 2)$ in the direction of

  (a) the x-axis

  (b) the y-axis

50 The surface of a certain lake is represented by a region D in an xy-plane such that the depth under the point corresponding to (x, y) is given by the function $f(x, y) = 300 - 2x^2 - 3y^2$, where x, y, and $f(x, y)$ are in feet. If a water skier is in the water at the point $(4, 9)$, find the instantaneous rate at which the depth changes in the direction of

  (a) the x-axis

  (b) the y-axis

51 Suppose the electrical potential V at the point (x, y, z) is given by $V = 100/(x^2 + y^2 + z^2)$, where V is in volts and x, y, and z are in inches. Find the instantaneous rate of change of V with respect to distance at $(2, -1, 1)$ in the direction of

  (a) the x-axis

  (b) the y-axis

  (c) the z-axis

52 An object is situated in a rectangular coordinate system such that the temperature T at the point $P(x, y, z)$ is given by $T = 4x^2 - y^2 + 16z^2$, where T is in degrees and x, y, and z are in centimeters. Find the instantaneous rate of change of T with respect to distance at the point $P(4, -2, 1)$ in the direction of

  (a) the x-axis

  (b) the y-axis

  (c) the z-axis

53 When a pollutant such as nitric oxide is emitted from a smokestack of height h meters, the long-range concentration $C(x, y)$ (in $\mu g/m^3$) of the pollutant at a point x kilometers from the smokestack and at a height of y meters (see figure) can often be represented by

$$C(x, y) = \frac{a}{x^2}[e^{-b(y-h)^2/x^2} + e^{-b(y+h)^2/x^2}],$$

where a and b are positive constants that depend on atmospheric conditions and the pollution emission rate. Suppose that

$$C(x, y) = \frac{200}{x^2}[e^{-0.02(y-10)^2/x^2} + e^{-0.02(y+10)^2/x^2}].$$

Compute and interpret $\partial C/\partial x$ and $\partial C/\partial y$ at the point $(2, 5)$.

Exercise 53

54 The analysis of certain electrical circuits involves the formula $I = V/\sqrt{R^2 + L^2\omega^2}$, where I is the current, V the voltage, R the resistance, L the inductance, and ω a positive constant. Find and interpret $\partial I/\partial R$ and $\partial I/\partial L$.

55 Most computers have only one processor that can be used for computations. Modern supercomputers, however, have anywhere from two to several thousand processors. A multiprocessor supercomputer is compared with a uniprocessor computer in terms of speedup. The *speedup S* is the number of times faster that a given computation can be accomplished with a multiprocessor than with a uniprocessor. A formula used to determine S is *Amdahl's law*,

$$S(p, q) = \frac{p}{q + p(1 - q)},$$

where p is the number of processors and q is the fraction of the computation that can be performed using all available processors in parallel—that is, using them in such a way that data are processed concurrently by separate units. The ideal situation, *complete parallelism*, occurs when $q = 1$.

(a) If $q = 0.8$, find the speedup when $p = 10$, 100, and 1000. Show that the speedup S cannot exceed 5, regardless of the number of processors available.

(b) Find the instantaneous rate of change of S with respect to q.

(c) What is the rate of change in part (b) if there is complete parallelism, and how does the number of processors affect this rate of change?

56 Refer to Exercise 55. The efficiency E of a multiprocessor computation can be calculated using the equation

$$E = \frac{S(p, q)}{p} = \frac{1}{q + p(1 - q)}.$$

Show that if $0 \leq q < 1$, E is a decreasing function of p and therefore, without complete parallelism, increasing the number of processors does not increase the efficiency of the computation.

57 In the study of frost penetration in highway engineering, the temperature T at time t hours and depth x feet can be approximated by

$$T = T_0 e^{-\lambda x} \sin(\omega t - \lambda x),$$

where T_0, ω, and λ are constants. Assume that the period of $\sin(\omega t - \lambda x)$ is 24 hr.

(a) Find and interpret $\partial T/\partial t$ and $\partial T/\partial x$.

(b) Show that T satisfies the one-dimensional heat equation

$$\frac{\partial T}{\partial t} = k \frac{\partial^2 T}{\partial x^2},$$

where k is a constant.

58 Show that any function given by

$$w = (\sin ax)(\cos by)e^{-\sqrt{a^2+b^2}\,z}$$

satisfies Laplace's equation in three dimensions:

$$\frac{\partial^2 w}{\partial x^2} + \frac{\partial^2 w}{\partial y^2} + \frac{\partial^2 w}{\partial z^2} = 0$$

59 The vital capacity V of the lungs is the largest volume (in milliliters) that can be exhaled after a maximum inhalation of air. For a typical male x years old and y centimeters tall, V may be approximated by the formula $V = 27.63y - 0.112xy$. Compute and interpret

(a) $\partial V/\partial x$ **(b)** $\partial V/\partial y$

60 On a clear day, the intensity of sunlight $I(x, t)$ (in foot-candles) at t hours after sunrise and at ocean depth x (in meters) can be approximated by

$$I(x, t) = I_0 e^{-kx} \sin^3(\pi t/D),$$

where I_0 is the intensity at midday, D is the length of the day (in hours), and k is a positive constant. If $I_0 = 1000$, $D = 12$, and $k = 0.10$, compute and interpret $\partial I / \partial t$ and $\partial I / \partial x$ when $t = 6$ and $x = 5$.

61 In economics, the *price elasticity of demand* for a commodity indicates the responsiveness of consumers to a change in the market price of the commodity. Suppose n commodities $C_1, C_2, \ldots, C_n$ have prices $p_1, p_2, \ldots, p_n$, respectively, and consumer demand for C_k is a function q_k of $p_1, p_2, \ldots, p_n$. The price elasticity of C_k is the function e_k defined by

$$e_k = \frac{p_k}{q_k} \frac{\partial q_k}{\partial p_k}.$$

If, for each k,

$$q_k = b_k p_1^{-a_{k1}} p_2^{-a_{k2}} \cdots p_n^{-a_{kn}},$$

where $a_{k1}, a_{k2}, \ldots, a_{kn}$ and b_k are nonnegative constants, show that e_k is a constant function.

62 Refer to Exercise 61. Commodity C_k is said to be *independent* of commodity C_j if a change in price p_k does not affect demand q_j. This is equivalent to the condition $\partial q_j / \partial p_k = 0$. If q_k has the form in Exercise 61, show that C_k is independent of C_j if and only if $a_{jk} = 0$.

Exer. 63–64: Use Theorem (12.10).

63 Let C be the trace of the paraboloid $z = 9 - x^2 - y^2$ on the plane $x = 1$. Find parametric equations of the tangent line l to C at the point $P(1, 2, 4)$. Sketch the paraboloid, C, and l.

64 Let C be the trace of the graph of the equation $z = \sqrt{36 - 9x^2 - 4y^2}$ on the plane $y = 2$. Find parametric equations of the tangent line l to C at the point $P(1, 2, \sqrt{11})$. Sketch the surface, C, and l.

[c] **Exer. 65–68:** Use the following formulas with $h = 0.01$ to approximate $f_x(0.5, 0.2)$ and $f_y(0.5, 0.2)$, and compare the results with the values obtained from $f_x(x, y)$ and $f_y(x, y)$.

$$f_x(x, y) \approx \frac{f(x + h, y) - f(x - h, y)}{2h}$$

$$f_y(x, y) \approx \frac{f(x, y + h) - f(x, y - h)}{2h}$$

65 $f(x, y) = y^2 \sin(xy)$

66 $f(x, y) = xy^3 + 4x^3 y^2$

67 $f(x, y) = \dfrac{x^2 y^2}{3x^2 + y^2}$

68 $f(x, y) = e^{(1 - 2x^2 - y^2)}$

[c] **Exer. 69–70:** Use the following formulas with $h = 0.01$ to approximate $f_{xx}(0.6, 0.8)$ and $f_{yy}(0.6, 0.8)$.

$$f_{xx}(x, y) \approx \frac{f(x + h, y) - 2f(x, y) + f(x - h, y)}{h^2}$$

$$f_{yy}(x, y) \approx \frac{f(x, y + h) - 2f(x, y) + f(x, y - h)}{h^2}$$

69 $f(x, y) = \sec^2[\tan(xy^2)] \sin(xy)$

70 $f(x, y) = \dfrac{x^3 + xy^2}{\tan(xy) + 4x^2 y^3}$

12.4 INCREMENTS AND DIFFERENTIALS

We continue, in this section, to consider differentiation of real-valued functions of two or three variables. We introduce *increments* and *differentials* and develop a definition of *differentiability* for functions of two variables.

If f is a function of two variables x and y, then the symbols Δx and Δy denote increments of x and y. Note that Δy is an increment of the *independent* variable y and is not the same as that defined in (2.30) for a *dependent* variable y. In terms of this increment notation, Definition (12.8) may be written

$$f_x(x, y) = \lim_{\Delta x \to 0} \frac{f(x + \Delta x, y) - f(x, y)}{\Delta x}$$

$$f_y(x, y) = \lim_{\Delta y \to 0} \frac{f(x, y + \Delta y) - f(x, y)}{\Delta y}.$$

The increment of the *dependent* variable $w = f(x, y)$ is defined as follows.

Definition 12.13

Let $w = f(x, y)$, and let Δx and Δy be increments of x and y, respectively. The **increment** Δw of $w = f(x, y)$ is

$$\Delta w = f(x + \Delta x, y + \Delta y) - f(x, y).$$

The *increment Δw represents the change in the function value if (x, y) changes to $(x + \Delta x, y + \Delta y)$.* If we regard $f(x, y)$ as the temperature at the point (x, y) in a metal plate D, as shown in Figure 12.32, then Δw is the net change in temperature in going from (x, y) to $(x + \Delta x, y + \Delta y)$.

Figure 12.32

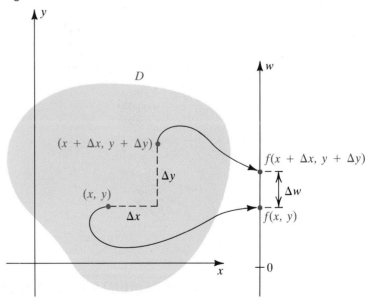

EXAMPLE 1 Let $w = f(x, y) = 3x^2 - xy$.

(a) If Δx and Δy are increments of x and y, find Δw.

(b) Use Δw to calculate the change in $f(x, y)$ if (x, y) changes from $(1, 2)$ to $(1.01, 1.98)$.

SOLUTION

(a) Using Definition (12.13), we have

$$\Delta w = f(x + \Delta x, y + \Delta y) - f(x, y)$$
$$= [3(x + \Delta x)^2 - (x + \Delta x)(y + \Delta y)] - (3x^2 - xy)$$
$$= [3x^2 + 6x\,\Delta x + 3(\Delta x)^2 - (xy + x\,\Delta y + y\,\Delta x + \Delta x\,\Delta y)]$$
$$-3x^2 + xy$$
$$= 6x\,\Delta x + 3(\Delta x)^2 - x\,\Delta y - y\,\Delta x - \Delta x\,\Delta y.$$

Figure 12.33

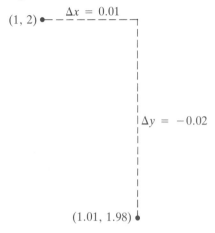

$(1, 2)$ •
$\Delta x = 0.01$

$\Delta y = -0.02$

$(1.01, 1.98)$ •

(b) If (x, y) changes from $(1, 2)$ to $(1.01, 1.98)$, we have the situation shown in Figure 12.33. Substituting

$$x = 1, \quad y = 2, \quad \Delta x = 0.01, \quad \text{and} \quad \Delta y = -0.02$$

into the formula for Δw gives us

$$\Delta w = 6(1)(0.01) + 3(0.01)^2 - (1)(-0.02) - 2(0.01) - (0.01)(-0.02)$$
$$= 0.0605.$$

This number can also be found by calculating $f(1.01, 1.98) - f(1, 2)$.

The formula for Δw in Definition (12.13) is used only for finding differences in function values between two points. It is not well suited for establishing other results about the variation of $f(x, y)$. The following theorem provides a formula that is more useful for applications. To simplify the statement, we shall use the abbreviations ϵ_1 and ϵ_2 for certain functions of Δx and Δy, as well as for their *values* $\epsilon_1(\Delta x, \Delta y)$ and $\epsilon_2(\Delta x, \Delta y)$ at $(\Delta x, \Delta y)$.

Theorem 12.14

Let $w = f(x, y)$, where the function f is defined on a rectangular region $R = \{(x, y): a < x < b, \ c < y < d\}$. Suppose f_x and f_y exist throughout R and are continuous at the point (x_0, y_0) in R. If $(x_0 + \Delta x, y_0 + \Delta y)$ is in R and

$$\Delta w = f(x_0 + \Delta x, y_0 + \Delta y) - f(x_0, y_0),$$

then

$$\Delta w = f_x(x_0, y_0)\,\Delta x + f_y(x_0, y_0)\,\Delta y + \epsilon_1\,\Delta x + \epsilon_2\,\Delta y,$$

where ϵ_1 and ϵ_2 are functions of Δx and Δy that have the limit 0 as $(\Delta x, \Delta y) \to (0, 0)$.

P R O O F The graph of R and the points we wish to consider are illustrated in Figure 12.34.

Let us rewrite the increment Δw by subtracting and adding the term $f(x_0, y_0 + \Delta y)$ as follows:

(1)
$$\Delta w = f(x_0 + \Delta x, y_0 + \Delta y) - f(x_0, y_0)$$
$$= [f(x_0 + \Delta x, y_0 + \Delta y) - f(x_0, y_0 + \Delta y)]$$
$$+ [f(x_0, y_0 + \Delta y) - f(x_0, y_0)]$$

Figure 12.34

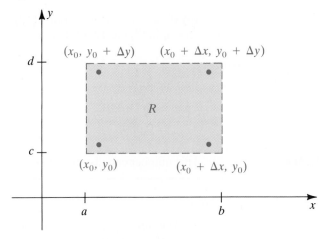

We next consider Δy as a constant and define a function g of *one* variable x by letting

$$g(x) = f(x, y_0 + \Delta y) \quad \text{for} \quad a < x < b.$$

It follows that $g'(x) = f_x(x, y_0 + \Delta y)$ for $a < x < b$. Applying the mean value theorem (3.12) to g on the interval $[x_0, x_0 + \Delta x]$, we obtain

$$g(x_0 + \Delta x) - g(x_0) = g'(u)\,\Delta x,$$

where u is between x_0 and $x_0 + \Delta x$. Since $g(x) = f(x, y_0 + \Delta y)$ and $g'(x) = f_x(x, y_0 + \Delta y)$, the last equation can be written

(2) $\quad f(x_0 + \Delta x, y_0 + \Delta y) - f(x_0, y_0 + \Delta y) = f_x(u, y_0 + \Delta y)\,\Delta x.$

Similarly, if we let $h(y) = f(x_0, y)$ for $c < y < d$, then $h'(y) = f_y(x_0, y)$. Applying the mean value theorem to the function h on the interval $[y_0, y_0 + \Delta y]$, we obtain

$$h(y_0 + \Delta y) - h(y_0) = h'(v)\,\Delta y,$$

where v is between y_0 and $y_0 + \Delta y$, or, equivalently,

(3) $\quad f(x_0, y_0 + \Delta y) - f(x_0, y_0) = f_y(x_0, v)\,\Delta y.$

Figure 12.35

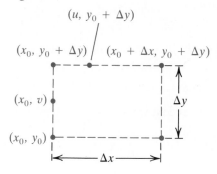

Substituting (2) and (3) into the expression for Δw in (1) gives us

(4) $$\Delta w = f_x(u, y_0 + \Delta y)\, \Delta x + f_y(x_0, v)\, \Delta y.$$

Typical points corresponding to $(u, y_0 + \Delta y)$ and (x_0, v) are shown in Figure 12.35.

Let us define ϵ_1 and ϵ_2 as follows:

$$\epsilon_1 = f_x(u, y_0 + \Delta y) - f_x(x_0, y_0)$$

$$\epsilon_2 = f_y(x_0, v) - f_y(x_0, y_0)$$

Using the fact that f_x and f_y are continuous and noting from Figure 12.35 that $u \to x_0$ as $\Delta x \to 0$ and $v \to y_0$ as $\Delta y \to 0$, we can show that ϵ_1 and ϵ_2 have the limit 0 as $(\Delta x, \Delta y) \to (0, 0)$. If we write the preceding equations for ϵ_1 and ϵ_2 in the form

$$f_x(u, y_0 + \Delta y) = f_x(x_0, y_0) + \epsilon_1$$

$$f_y(x_0, v) = f_y(x_0, y_0) + \epsilon_2$$

and then substitute into (4), the expression for Δw, we see that

$$\Delta w = [f_x(x_0, y_0) + \epsilon_1]\, \Delta x + [f_y(x_0, y_0) + \epsilon_2]\, \Delta y,$$

which leads to the conclusion of the theorem. ■

EXAMPLE ■ 2 If $w = 3x^2 - xy$, find expressions for ϵ_1 and ϵ_2 that satisfy the conclusion of Theorem (12.14) with $(x_0, y_0) = (x, y)$.

SOLUTION From the solution of Example 1,

$$\Delta w = 6x\, \Delta x + 3(\Delta x)^2 - x\, \Delta y - y\, \Delta x - \Delta x\, \Delta y,$$

or, equivalently,

$$\Delta w = (6x - y)\, \Delta x + (-x)\, \Delta y + (3\, \Delta x)(\Delta x) + (-\Delta x)\, \Delta y.$$

This is of the form given in Theorem (12.14) with

$$f_x(x, y) = 6x - y, \quad f_y(x, y) = -x, \quad \epsilon_1 = 3\, \Delta x, \quad \epsilon_2 = -\Delta x.$$

Note that ϵ_1 and ϵ_2 are not unique, since we may also write

$$\Delta w = (6x - y)\, \Delta x + (-x)\, \Delta y + (3\, \Delta x - \Delta y)\, \Delta x + (0)\, \Delta y,$$

in which case $\epsilon_1 = 3\, \Delta x - \Delta y$ and $\epsilon_2 = 0$.

In Chapter 2, we defined differentials for functions of one variable. Recall from (2.34) that if $u = f(x)$ and Δx is an increment of x, then $du = f'(x)\, \Delta x$. As we noted in Chapter 4, we consider the differential of an independent variable to be its increment. The next definition extends these concepts to functions of two variables.

Definition 12.15

Let $w = f(x, y)$, and let Δx and Δy be increments of x and y, respectively.

(i) The **differentials dx and dy** of the independent variables x and y are

$$dx = \Delta x \quad \text{and} \quad dy = \Delta y.$$

(ii) The **differential dw** of the dependent variable w is

$$dw = f_x(x, y)\, dx + f_y(x, y)\, dy = \frac{\partial w}{\partial x}\, dx + \frac{\partial w}{\partial y}\, dy.$$

If f satisfies the hypotheses of Theorem (12.14), then using the conclusion of that theorem, with the pair (x_0, y_0) replaced by (x, y), we see that

$$\Delta w = dw + \epsilon_1\, \Delta x + \epsilon_2\, \Delta y.$$

Hence, $\qquad\qquad\qquad \Delta w - dw = \epsilon_1\, \Delta x + \epsilon_2\, \Delta y,$

where both ϵ_1 and ϵ_2 approach 0 as $(\Delta x, \Delta y) \to (0, 0)$. It follows that if Δx and Δy are small, then $\Delta w - dw \approx 0$—that is, $dw \approx \Delta w$. We may use this fact to approximate the change in w corresponding to small changes in x and y.

EXAMPLE ■ 3 If $w = 3x^2 - xy$, find dw and use it to approximate the change in w if (x, y) changes from $(1, 2)$ to $(1.01, 1.98)$. How does this compare with the exact change in w?

SOLUTION This is the same function considered in Examples 1 and 2. Applying Definition (12.15) yields

$$dw = \frac{\partial w}{\partial x}\, dx + \frac{\partial w}{\partial y}\, dy$$
$$= (6x - y)\, dx + (-x)\, dy.$$

Substituting $x = 1$, $y = 2$, $dx = \Delta x = 0.01$, and $dy = \Delta y = -0.02$, we obtain

$$dw = (6 - 2)(0.01) + (-1)(-0.02) = 0.06.$$

In Example 1, we showed that $\Delta w = 0.0605$. Hence, the error involved in using dw is 0.0005.

EXAMPLE ■ 4 The radius and the altitude of a right circular cylinder are measured as 3 in. and 8 in., respectively, with a possible error in measurement of ± 0.05 in. Use differentials to approximate the maximum error in the calculated volume of the cylinder.

SOLUTION The method is analogous to that used for functions of one variable. Thus, we begin by considering the *general* formula $V =$

Figure 12.36

$\pi r^2 h$ for the volume V of a cylinder of radius r and altitude h. We regard r and h as measured values, with maximum errors in measurement dr and dh, respectively. The case where both dr and dh are positive is illustrated in Figure 12.36. Of course, the error in one or both measurements could be negative. The error in the calculated volume is the change in V that corresponds to dr and dh. Using differentials, we have, by our previous remarks and Definition (12.15)(ii),

$$\Delta V \approx dV = \frac{\partial V}{\partial r}\, dr + \frac{\partial V}{\partial h}\, dh = 2\pi r h\, dr + \pi r^2\, dh.$$

Finally, we substitute specific values for the variables. Letting $r = 3$, $h = 8$, and $dr = dh = \pm 0.05$ gives us the following approximation to the maximum error:

$$dV = 48\pi(\pm 0.05) + 9\pi(\pm 0.05) = \pm 2.85\pi \approx \pm 8.95 \text{ in}^3$$

Recall from Definition (2.36) that if a measurement w has an error (approximated by) dw, then the *relative error* is approximately dw/w. The *percentage error* is the average error multiplied by 100%. In the next example, we estimate the percentage error caused by errors in two related measurements.

EXAMPLE ■ 5   A weather balloon is released at sea level, a distance R from the eye of a hurricane, and its altitude increases as it moves toward the eye. The altitude h that the balloon will attain at the eye may be approximated by

$$h = \pi^2 g \frac{R^4}{C^2},$$

where g is a gravitational constant and C is a meteorological measurement called the *circulation* of the wind velocity. (Circulation is closely related to the intensity and direction of the winds inside the hurricane.) If the maximum percentage errors in the measurements of R and h are $\pm 2\%$ and $\pm 5\%$, respectively, estimate the maximum percentage error in the calculation of C.

SOLUTION   Solving the given formula for C, we obtain

$$C = \pi g^{1/2} \frac{R^2}{h^{1/2}}.$$

If the errors in measurement of R and h are dR and dh, then the error in C may be approximated by

$$dC = \frac{\partial C}{\partial R}\, dR + \frac{\partial C}{\partial h}\, dh$$

$$= \pi g^{1/2} \frac{2R}{h^{1/2}}\, dR - \pi g^{1/2} \frac{1}{2} \frac{R^2}{h^{3/2}}\, dh.$$

We are interested in percentage errors, so we divide both sides of this equation by $C = \pi g^{1/2} R^2/h^{1/2}$, obtaining

$$\frac{dC}{C} = 2\frac{dR}{R} - \frac{1}{2}\frac{dh}{h}.$$

The maximum percentage error will occur when dR and dh have opposite signs. Thus, if $dR/R = 2\%$ and $dh/h = -5\%$,

$$\frac{dC}{C} = 2(2\%) - \frac{1}{2}(-5\%) = 6.5\%.$$

If $dR/R = -2\%$ and $dh/h = 5\%$,

$$\frac{dC}{C} = 2(-2\%) - \frac{1}{2}(5\%) = -6.5\%.$$

For a function of one variable, the term *differentiable* means that the derivative exists. For functions of two variables, we use the condition stated in the next definition, which is based on the formula for Δw in Theorem (12.14).

Definition 12.16

Let $w = f(x, y)$. The function f is **differentiable** at (x_0, y_0) if Δw can be expressed in the form

$$\Delta w = f_x(x_0, y_0)\,\Delta x + f_y(x_0, y_0)\,\Delta y + \epsilon_1\,\Delta x + \epsilon_2\,\Delta y,$$

where ϵ_1 and ϵ_2 are functions of Δx and Δy that have the limit 0 as $(\Delta x, \Delta y) \to (0, 0)$.

A function f of two variables is said to be **differentiable on a region R** if it is differentiable at every point of R. The next theorem follows directly from Theorem (12.14) and Definition (12.16). The term *rectangular region* means a region of the type described in (12.14).

Theorem 12.17

If $w = f(x, y)$ and if f_x and f_y are continuous on a rectangular region R, then f is differentiable on R.

The following result shows that a differentiable function is continuous, and it is one of the reasons for using Definition (12.16) as the definition of differentiability.

Theorem 12.18

If a function f of two variables is differentiable at (x_0, y_0), then f is continuous at (x_0, y_0).

PROOF Let Δx and Δy be increments. We can write Δw in the following two ways, using (12.13) and (12.16):

$$\Delta w = f(x_0 + \Delta x, y_0 + \Delta y) - f(x_0, y_0)$$

$$\Delta w = [f_x(x_0, y_0) + \epsilon_1]\,\Delta x + [f_y(x_0, y_0) + \epsilon_2]\,\Delta y$$

If we equate these expressions and let $x = x_0 + \Delta x$, $y = y_0 + \Delta x$, then

$$f(x, y) - f(x_0, y_0) = [f_x(x_0, y_0) + \epsilon_1](x - x_0)$$
$$+ [f_y(x_0, y_0) + \epsilon_2](y - y_0).$$

It follows that

$$\lim_{(x,y)\to(x_0,y_0)} [f(x, y) - f(x_0, y_0)] = 0,$$

or, equivalently,

$$\lim_{(x,y)\to(x_0,y_0)} f(x, y) = f(x_0, y_0).$$

Hence, f is continuous at (x_0, y_0). ▪

If f_x and f_y are continuous, then, by Theorem (12.17), f is differentiable. This gives us the following corollary to Theorem (12.18).

Corollary 12.19

> If f is a function of two variables and if f_x and f_y are continuous on a rectangular region R, then f is continuous on R.

We can show by means of examples that the mere *existence* of f_x and f_y is not enough to ensure continuity of f (see Exercise 41). This is different from the single-variable case, where the existence of f' implies continuity of f.

The preceding discussion can be extended to functions of more than two variables. For example, suppose $w = f(x, y, z)$, where f is defined on a suitable region R (such as a rectangular box) and f_x, f_y, f_z exist in R and are continuous at (x, y, z). If x, y, and z are given increments $\Delta x, \Delta y$, and Δz, respectively, then the corresponding increment

$$\Delta w = f(x + \Delta x, y + \Delta y, z + \Delta z) - f(x, y, z)$$

can be written in the form

$$\Delta w = f_x(x, y, z)\,\Delta x + f_y(x, y, z)\,\Delta y + f_z(x, y, z)\,\Delta z$$
$$+ \epsilon_1\,\Delta x + \epsilon_2\,\Delta y + \epsilon_3\,\Delta z,$$

where ϵ_1, ϵ_2, and ϵ_3 are functions of Δx, Δy, and Δz that have the limit 0 as $(\Delta x, \Delta y, \Delta z) \to (0, 0, 0)$. Results that are analogous to (12.17)–(12.19) may be established for functions of three variables.

The following definition is an extension of (12.15) to functions of three variables.

Definition 12.20

Let $w = f(x, y, z)$, and let Δx, Δy, and Δz be increments of x, y, and z, respectively.

(i) The **differentials dx, dy, and dz** of the independent variables x, y, and z are

$$dx = \Delta x, \qquad dy = \Delta y, \qquad dz = \Delta z.$$

(ii) The **differential dw** of the dependent variable w is

$$dw = \frac{\partial w}{\partial x}\,dx + \frac{\partial w}{\partial y}\,dy + \frac{\partial w}{\partial z}\,dz.$$

We can use dw to approximate Δw if the increments of x, y, and z are small. Differentiability is defined as in (12.16). The extension to four or more variables is made in similar fashion.

EXAMPLE 6   Suppose the dimensions (in inches) of a rectangular box change from 9, 6, and 4 to 9.02, 5.97, and 4.01, respectively.

(a) Use differentials to approximate the change in volume.

(b) Find the exact change in volume.

SOLUTION

(a) The method of solution is similar to that used for the right circular cylinder in Example 4. Thus, we begin with the general formula $V = xyz$ for the volume of a rectangular box of measured dimensions x, y, and z. We next regard dx, dy, and dz as errors in measurement. The error in the calculated volume is then

$$\Delta V \approx dV = \frac{\partial V}{\partial x}\,dx + \frac{\partial V}{\partial y}\,dy + \frac{\partial V}{\partial z}\,dz$$

$$= yz\,dx + xz\,dy + xy\,dz.$$

Finally, we substitute the given values $x = 9$, $y = 6$, $z = 4$, $dx = 0.02$, $dy = -0.03$, and $dz = 0.01$, obtaining

$$dV = 24(0.02) + 36(-0.03) + 54(0.01)$$

$$= 0.48 - 1.08 + 0.54 = -0.06.$$

Thus, the volume decreases by approximately 0.06 in^3.

(b) The exact change in volume is

$$\Delta V = (9.02)(5.97)(4.01) - (9)(6)(4) = -0.063906 \text{ in}^3.$$

Figure 12.37

Figure 12.38

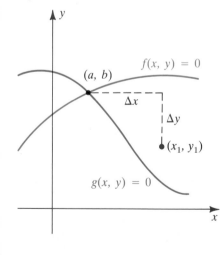

In precalculus, we study methods of solving systems of linear equations, such as

$$\begin{cases} 2x - 3y = 4 \\ 5x + 2y = 9 \end{cases} \quad \text{or} \quad \begin{cases} x + 3y - z = 4 \\ 2x - y + 3z = 1 \\ x + 2y - z = -5 \end{cases}$$

In applications, it is often necessary to solve a system of *nonlinear* equations, such as

$$\begin{cases} x^2 + y^2 - 1 = 0 \\ y - \sin x = 0 \end{cases}$$

Since we cannot find exact solutions of this system using algebraic techniques, we must develop methods for *approximating* solutions. We could graph both equations on the same coordinate axes, as in Figure 12.37, and estimate the points of intersection (a, b) and (c, d); however, for most applications, we need more accuracy. We shall next discuss a method based on differentials that can be used to approximate solutions to any degree of accuracy.

Let us consider a general system of two equations in x and y:

$$\begin{cases} f(x, y) = 0 \\ g(x, y) = 0 \end{cases}$$

where f and g are functions of two variables. (In the previous discussion, $f(x, y) = x^2 + y^2 - 1$ and $g(x, y) = y - \sin x$.) We wish to approximate a point of intersection (a, b) of the graphs of the two equations (see Figure 12.38). Let us begin with a *first approximation* (x_1, y_1), where, as indicated in the figure, there are increments Δx and Δy such that

$$a = x_1 + \Delta x, \qquad b = y_1 + \Delta y.$$

Let Δf and Δg be the increments of f and g corresponding to the change from (x_1, y_1) to (a, b). Since $f(a, b) = 0$ and $g(a, b) = 0$, we see that

$$\Delta f = f(a, b) - f(x_1, y_1) = -f(x_1, y_1)$$

$$\Delta g = g(a, b) - g(x_1, y_1) = -g(x_1, y_1).$$

Thus, if df and dg are the differentials of f and g evaluated at (x_1, y_1), then

$$df \approx -f(x_1, y_1) \quad \text{and} \quad dg \approx -g(x_1, y_1).$$

Using the form for df and dg given by Definition (12.15)(ii) gives us the following system of *approximate equalities* in the unknowns Δx and Δy:

(1)
$$\begin{cases} f_x(x_1, y_1)\, \Delta x + f_y(x_1, y_1)\, \Delta y \approx -f(x_1, y_1) \\ g_x(x_1, y_1)\, \Delta x + g_y(x_1, y_1)\, \Delta y \approx -g(x_1, y_1) \end{cases}$$

In practice, this system is often written in *matrix form*, as follows, where all functions are evaluated at (x_1, y_1):

(2)
$$\begin{bmatrix} f_x & f_y \\ g_x & g_y \end{bmatrix} \begin{bmatrix} \Delta x \\ \Delta y \end{bmatrix} \approx \begin{bmatrix} -f \\ -g \end{bmatrix}$$

We solve the approximate equalities in (1) for Δx and Δy by treating $\approx$ as $=$, which gives us *approximations* for Δx and Δy. Although $(x_2, y_2) = (x_1 + \Delta x, y_1 + \Delta y)$ will not usually be the *exact* solution (a, b), it is often a better approximation than (x_1, y_1). The process can be repeated using the approximation (x_2, y_2) instead of (x_1, y_1) in (1) to obtain a third approximation, (x_3, y_3). We continue the process until $|\Delta x| \le \epsilon$ and $|\Delta y| \le \epsilon$ for some positive error tolerance $\epsilon > 0$. The convergence of x_k and y_k to a and b, respectively, is usually rapid, and the number of correct digits approximately doubles with each step. This method of successive approximations is known as **Newton's method** for a system of two equations in two unknowns. It can be extended to systems of n equations in n unknowns. This is one of the most widely used methods for approximating solutions of systems of nonlinear equations.

EXAMPLE ■ 7   Use Newton's method to approximate the solutions of the following system:

$$\begin{cases} x^2 + y^2 - 1 = 0 \\ y - \sin x = 0 \end{cases}$$

SOLUTION   The graphs of the two equations are sketched in Figure 12.37. We shall take $(x_1, y_1) = (0.5, 0.5)$ as a first approximation for the solution corresponding to the point of intersection in the first quadrant. Letting $f(x, y) = x^2 + y^2 - 1$ and $g(x, y) = y - \sin x$, we obtain

$$f_x(x, y) = 2x, \quad f_y(x, y) = 2y, \quad g_x(x, y) = -\cos x, \quad g_y(x, y) = 1.$$

We now let $x_1 = 0.5$ and $y_1 = 0.5$ in (1) and solve for (the approximate values) Δx and Δy. We then take, as a second approximation,

$$(x_2, y_2) = (x_1 + \Delta x, y_1 + \Delta y).$$

Using (x_2, y_2) in place of (x_1, y_1) in (1), we find (approximate values for) Δx_1 and Δy_1 and take, as a third approximation,

$$(x_3, y_3) = (x_2 + \Delta x_1, y_2 + \Delta y_1).$$

The process continues until we reach the desired accuracy. The following table lists eight-decimal-place approximations to the values used in equations (1). Note that if $k = 5$, we have $|\Delta x| \le \epsilon$ and $|\Delta y| \le \epsilon$ with $\epsilon = 0.000000005$. Hence, the solution (x_5, y_5) is accurate to eight decimal places.

k	x_k	y_k	$f(x_k, y_k)$	$g(x_k, y_k)$
1	0.5	0.5	-0.5	0.02057446
2	0.77725783	0.72274217	0.12648598	0.02141484
3	0.74030063	0.67498279	0.00364679	0.00047290
4	0.73908622	0.67361333	0.00000335	0.00000050
5	0.73908513	0.67361203	0.00000000	0.00000000

$f_x(x_k, y_k)$	$f_y(x_k, y_k)$	$g_x(x_k, y_k)$	$g_y(x_k, y_k)$	Δx_k	Δy_k
1	1	-0.87758256	1	0.27725783	0.22274217
1.55451565	1.44548435	-0.71283938	1	-0.03695719	-0.04775939
1.48060127	1.34996558	-0.73826581	1	-0.00121442	-0.00136946
1.47817243	1.34722665	-0.73908440	1	-0.00000108	-0.00000130
1.47817027	1.34722406	-0.73908513	1	0.00000000	0.00000000

From the table,

$$(x_5, y_5) \approx (0.73908513, 0.67361203).$$

An approximation to the solution corresponding to the point in the third quadrant in Figure 12.37 is, by symmetry, $(-x_5, -y_5)$.

EXERCISES 12.4

Exer. 1–2: If Δx and Δy are increments of x and y, find (a) Δw, (b) dw, and (c) $dw - \Delta w$.

1 $w = 5y^2 - xy$ 2 $w = xy - y^2 + 3x$

Exer. 3–6: Find expressions for ϵ_1 and ϵ_2 that satisfy Definition (12.16).

3 $f(x, y) = 4y^2 - 3xy + 2x$

4 $f(x, y) = (2x - y)^2$ 5 $f(x, y) = x^3 + y^3$

6 $f(x, y) = 2x^2 - xy^2 + 3y$

Exer. 7–18: Find dw.

7 $w = x^3 - x^2y + 3y^2$ 8 $w = 5x^2 + 4y - 3xy^3$

9 $w = x^2 \sin y + 2y^{3/2}$ 10 $w = ye^{-2x} - 3x^4$

11 $w = x^2e^{xy} + (1/y^2)$

12 $w = \ln(x^2 + y^2) + x \tan^{-1} y$

13 $w = x^2 \ln(y^2 + z^2)$ 14 $w = x^2y^3z + e^{-2z}$

15 $w = \dfrac{xyz}{x + y + z}$ 16 $w = x^2e^{yz} + y \ln z$

17 $w = x^2z + 4yt^3 - xz^2t$ 18 $w = x^2y^3zt^{-1}v^4$

Exer. 19–22: Use differentials to approximate the change in f if the independent variables change as indicated.

19 $f(x, y) = x^2 - 3x^3y^2 + 4x - 2y^3 + 6$;

$(-2, 3)$ to $(-2.02, 3.01)$

20 $f(x, y) = x^2 - 2xy + 3y$; $(1, 2)$ to $(1.03, 1.99)$

21 $f(x, y, z) = x^2z^3 - 3yz^2 + x^{-3} + 2y^{1/2}z$;

$(1, 4, 2)$ to $(1.02, 3.97, 1.96)$

22 $f(x, y, z) = xy + xz + yz$;

$(-1, 2, 3)$ to $(-0.98, 1.99, 3.03)$

23 The dimensions of a closed rectangular box are measured as 3 ft, 4 ft, and 5 ft, with a possible error of $\pm\frac{1}{16}$ in. for each measurement. Use differentials to approximate the maximum error in the calculated value of

(a) the surface area   (b) the volume

24 The two shortest sides of a right triangle are measured as 3 cm and 4 cm, respectively, with a possible error of ±0.02 cm in each measurement. Use differentials to approximate the maximum error in the calculated value of

(a) the hypotenuse   (b) the area of the triangle

25 The *withdrawal resistance* of a nail indicates its holding strength in wood. An empirical formula used for bright, common nails is $P = 15{,}700S^{5/2}RD$, where P is the maximum withdrawal resistance (in pounds), S is the specific gravity of the wood at 12% moisture content, R is the radius of the nail (in inches), and D is the depth (in inches) to which the nail has penetrated the wood. A 6d bright, common nail of length 2 in. and diameter 0.113 in. is driven completely into a piece of Douglas fir that has a specific gravity of 0.54.

(a) Approximate the maximum withdrawal resistance. (In applications, only one sixth of this maximum is considered safe for extended periods of time.)

(b) When nails are manufactured, R and D can vary by $\pm 2\%$, and the specific gravity of different samples of Douglas fir can vary by $\pm 3\%$. Approximate the maximum percentage error in the calculated value of P.

26 The total resistance R of three resistances R_1, R_2, and R_3 connected in parallel is given by

$$\frac{1}{R} = \frac{1}{R_1} + \frac{1}{R_2} + \frac{1}{R_3}.$$

If measurements of R_1, R_2, and R_3 are 100, 200, and 400 ohms, respectively, with a maximum error of $\pm 1\%$ in each measurement, approximate the maximum error in the calculated value of R.

27 The specific gravity of an object more dense than water is given by $s = A/(A - W)$, where A and W are the weights (in pounds) of the object in air and water, respectively. If measurements are $A = 12$ lb and $W = 5$ lb, with maximum errors of $\pm\frac{1}{2}$ oz in air and ± 1 oz in water, what is the maximum error in the calculated value of s?

28 The pressure P, volume V, and temperature T (in $^\circ$K) of a confined gas are related by the ideal gas law $PV = kT$, where k is a constant. If $P = 0.5$ lb/in^2 when $V = 64$ in^3 and $T = 350\,^\circ$K, approximate the change in P if V and T change to 70 in^3 and 345 $^\circ$K, respectively.

29 Suppose that when the specific gravity formula $s = A/(A - W)$ is used (see Exercise 27), there are percentage errors of $\pm 2\%$ and $\pm 4\%$ in the measurements of A and W, respectively. Express the maximum percentage error in the calculated value of s as a function of A and W.

30 Suppose that when the ideal gas law $PV = kT$ is used (see Exercise 28), there are percentage errors of $\pm 0.8\%$ and $\pm 0.5\%$ in the measurements of T and P, respectively. Approximate the maximum percentage error in the calculated value of V.

31 The electrical resistance R of a wire is directly proportional to its length and inversely proportional to the square of its diameter. If the length is measured with a possible error of $\pm 1\%$ and the diameter is measured with a possible error of $\pm 3\%$, what is the maximum percentage error in the calculated value of R?

32 The flow of blood through an arteriole is given by $F = \pi P R^4 / (8 v l)$, where l is the length of the arteriole, R is the radius, P is the pressure difference between the two ends, and v is the viscosity of the blood (see Section 5.8). Suppose that v and l are constant. Use differentials to approximate the percentage change in the blood flow if the radius decreases by 2% and the pressure increases by 3%.

33 The temperature T at the point $P(x,\ y,\ z)$ in an xyz-coordinate system is given by

$$T = 8(2x^2 + 4y^2 + 9z^2)^{1/2}.$$

Use differentials to approximate the temperature difference between the points $(6,\ 3,\ 2)$ and $(6.1,\ 3.3,\ 1.98)$.

34 Approximate the change in area of an isosceles triangle if each of the two equal sides increases from 100 to 101 and the angle between them decreases from 120° to 119°.

35 If a mountaintop is viewed from the point P shown in the figure, the angle of elevation is α. From a point Q that is a distance x units closer to the mountain, the angle of elevation is β. From trigonometry, the height h of the mountain is given by

$$h = \frac{x}{\cot \alpha - \cot \beta}.$$

A surveyor measures α and β to an accuracy of $30''$ (approximately 0.000145 radian). Suppose that $\alpha = 15^\circ$, $\beta = 20^\circ$, and $x = 2000$ ft. Use differentials to estimate, to the nearest 0.1 ft, how accurate the length measurement must be so that the maximum error in the calculated value of h is no greater than ± 10 ft.

Exercise 35

36 If a drug is taken orally, the time T at which the largest amount of drug is in the bloodstream can be calculated using the half-life x of the drug in the stomach and the half-life y of the drug in the bloodstream. For many common drugs (such as penicillin), T is given by

$$T = \frac{xy(\ln x - \ln y)}{(x - y) \ln 2}.$$

For a certain drug, $x = 30$ min and $y = 1$ hr. If the maximum error in estimating each half-life is $\pm 10\%$, find the maximum error in the calculated value of T.

37 Assume that the cylinder described in Example 4 has a closed top and bottom. Use differentials to approximate the maximum error in the calculated total surface area.

38 Use differentials to approximate the change in surface area of the box described in Example 6. What is the exact change?

Exer. 39–40: Prove that f is differentiable throughout its domain.

39 $f(x, y) = \dfrac{x^2 - y^2}{x^2 + y^2}$

40 $f(x, y, z) = \dfrac{x + y + z}{x^2 + y^2 + z^2}$

41 Let

$$f(x, y) = \begin{cases} \dfrac{xy}{x^2 + y^2} & \text{if } (x, y) \neq (0, 0) \\ 0 & \text{if } (x, y) = (0, 0) \end{cases}$$

(a) Prove that $f_x(0, 0)$ and $f_y(0, 0)$ exist. (*Hint:* Use Definition (12.8).)

(b) Prove that f is not continuous at $(0, 0)$.

(c) Prove that f is not differentiable at $(0, 0)$.

42 Let

$$f(x, y, z) = \begin{cases} \dfrac{xyz}{x^3 + y^3 + z^3} & \text{if } (x, y, z) \neq (0, 0, 0) \\ 0 & \text{if } (x, y, z) = (0, 0, 0) \end{cases}$$

(a) Prove that f_x, f_y, and f_z exist at $(0, 0, 0)$.

(b) Prove that f is not differentiable at $(0, 0, 0)$.

[c] **Exer. 43–44: Refer to Example 7. Use Newton's method to approximate a solution of the system of equations to four decimal places using the given first approximation.**

43
$$\begin{cases} \dfrac{x^2}{4} + \dfrac{y^2}{9} = 1 \\ \dfrac{(x - 1)^2}{10} + \dfrac{(y + 1)^2}{5} = 1 \end{cases} \qquad (x_1, y_1) = (2, 1)$$

44
$$\begin{cases} \dfrac{x^2}{9} + \dfrac{y^2}{4} = 1 \\ \dfrac{(x - 1)^2}{2} - \dfrac{(y - 1)^2}{3} = 1 \end{cases} \qquad (x_1, y_1) = (-1.5, -1.5)$$

45 Derive Newton's method for a system of three equations,

$$\begin{cases} f(x, y, z) = 0 \\ g(x, y, z) = 0 \\ h(x, y, z) = 0 \end{cases}$$

where f, g, and h are functions of three variables. Express the solution in a matrix form similar to (2) on page 1017.

12.5 CHAIN RULES

If f and g are functions of *one* variable such that

$$w = f(u) \quad \text{and} \quad u = g(x),$$

then the composite function of f and g is given by

$$w = f(g(x)).$$

Applying the chain rule (2.26), we may find the derivative of w with respect to x as follows:

$$\frac{dw}{dx} = \frac{dw}{du} \frac{du}{dx}$$

In this section, we shall extend this formula to functions of several variables.

Let f, g, and h be functions of two variables such that

$$w = f(u, v), \quad \text{with} \quad u = g(x, y), \quad v = h(x, y).$$

If, for each pair (x, y) in a subset D of $\mathbb{R}^2$, the corresponding pair (u, v) is in the domain of f, then

$$w = f(g(x, y), h(x, y))$$

defines w as a (composite) function of x and y with domain D. For example, if

$$w = u^2 + u \sin v, \quad \text{with} \quad u = xe^{2y}, v = xy,$$

then

$$w = x^2 e^{4y} + xe^{2y} \sin xy.$$

The next theorem provides formulas for expressing $\partial w / \partial x$ and $\partial w / \partial y$ in terms of the first partial derivatives of the functions g, h, and f. In the statement of the theorem, we have assumed that domains are chosen so that the composite function is defined on a suitable domain D. Each of the formulas stated in Theorem (12.21) is referred to as a *chain rule*.

Chain Rules 12.21

If $w = f(u, v)$, with $u = g(x, y)$, $v = h(x, y)$, and if f, g, and h are differentiable, then

$$\frac{\partial w}{\partial x} = \frac{\partial w}{\partial u} \frac{\partial u}{\partial x} + \frac{\partial w}{\partial v} \frac{\partial v}{\partial x}$$

$$\frac{\partial w}{\partial y} = \frac{\partial w}{\partial u} \frac{\partial u}{\partial y} + \frac{\partial w}{\partial v} \frac{\partial v}{\partial y}.$$

PROOF If x is given an increment Δx and y is held constant (that is, $\Delta y = 0$), we obtain the following increments of u and v:

(1)
$$\Delta u = g(x + \Delta x, y) - g(x, y)$$
$$\Delta v = h(x + \Delta x, y) - h(x, y)$$

These in turn lead to the following increment of w:

$$\Delta w = f(u + \Delta u, v + \Delta v) - f(u, v)$$

Since f is differentiable, we have, by Definition (12.16),

(2)
$$\Delta w = \frac{\partial w}{\partial u} \Delta u + \frac{\partial w}{\partial v} \Delta v + \epsilon_1 \Delta u + \epsilon_2 \Delta v,$$

where ϵ_1 and ϵ_2 are functions of Δu and Δv that have the limit 0 as $(\Delta u, \Delta v) \to (0, 0)$. Moreover, we may assume that both ϵ_1 and ϵ_2 equal 0 if $(\Delta u, \Delta v) = (0, 0)$, because if they do not, they can be replaced by functions μ_1 and μ_2 that have this property and are equal to ϵ_1 and ϵ_2 elsewhere. With this agreement, the functions ϵ_1 and ϵ_2 in equation (2) are continuous at $(0, 0)$. Dividing both sides of (2) by Δx gives us

(3)
$$\frac{\Delta w}{\Delta x} = \frac{\partial w}{\partial u} \frac{\Delta u}{\Delta x} + \frac{\partial w}{\partial v} \frac{\Delta v}{\Delta x} + \epsilon_1 \frac{\Delta u}{\Delta x} + \epsilon_2 \frac{\Delta v}{\Delta x}.$$

If we regard w as a function of x and y, then

$$\lim_{\Delta x \to 0} \frac{\Delta w}{\Delta x} = \frac{\partial w}{\partial x}.$$

Also, from equations (1),

$$\lim_{\Delta x \to 0} \frac{\Delta u}{\Delta x} = \frac{\partial u}{\partial x} \quad \text{and} \quad \lim_{\Delta x \to 0} \frac{\Delta v}{\Delta x} = \frac{\partial v}{\partial x}.$$

If Δx approaches 0, we see from (1) that Δu and Δv also approach 0 and hence so do ϵ_1 and ϵ_2. Consequently, if we take the limit in equation (3) as $\Delta x \to 0$, we obtain

$$\frac{\partial w}{\partial x} = \frac{\partial w}{\partial u}\frac{\partial u}{\partial x} + \frac{\partial w}{\partial v}\frac{\partial v}{\partial x}.$$

The second formula in the statement of the theorem is established in similar fashion. ▬

Figure 12.39

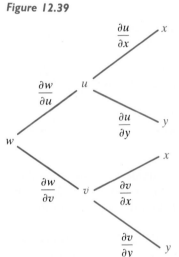

The *tree diagram* in Figure 12.39 is a device for remembering chain rules (12.21). We first draw *branches* (line segments) from w to u and v, to indicate that w is a function of those two variables. Since u is a function of x and y, we next draw branches from u to x and y. Similarly, branches are drawn from v to x and y. In this diagram, we have also indicated the various partial derivatives that are involved for the given variables.

To find $\partial w / \partial x$ by means of the diagram, we consider all pairs of branches (paths) that lead from w to x. There are two possibilities:

$$w \xrightarrow[\frac{\partial w}{\partial u}]{} u \xrightarrow[\frac{\partial u}{\partial x}]{} x \quad \text{and} \quad w \xrightarrow[\frac{\partial w}{\partial v}]{} v \xrightarrow[\frac{\partial v}{\partial x}]{} x$$

We next add the products of the corresponding pairs of partial derivatives, to obtain

$$\frac{\partial w}{\partial x} = \frac{\partial w}{\partial u}\frac{\partial u}{\partial x} + \frac{\partial w}{\partial v}\frac{\partial v}{\partial x}.$$

The formula for $\partial w/\partial y$ may be found in a like manner, using the pairs of branches that lead from w to y.

EXAMPLE ▪ I Use a chain rule to find $\partial w/\partial p$ and $\partial w/\partial q$ if

$$w = r^3 + s^2, \quad \text{with} \quad r = pq^2, \quad s = p^2 \sin q.$$

SOLUTION Note that w is a (composite) function of p and q. We could substitute for r and s, obtaining

$$w = (pq^2)^3 + (p^2 \sin q)^2,$$

and then find $\partial w/\partial p$ and $\partial w/\partial q$ directly; however, we wish to illustrate chain rules. Moreover, in some cases, such substitutions lead to very complicated expressions. The procedure here is analogous to procedures used in computer programming languages to express a complex computation as a succession of simpler ones.

Figure 12.40

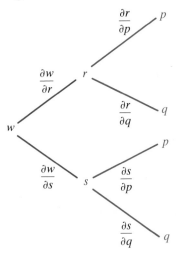

Since w is a function of r and s and both r and s are functions of p and q, we begin by constructing the tree diagram in Figure 12.40. Referring to the diagram, we take the products of all pairs of partial derivatives that lead from w to p, obtaining

$$\frac{\partial w}{\partial p} = \frac{\partial w}{\partial r}\frac{\partial r}{\partial p} + \frac{\partial w}{\partial s}\frac{\partial s}{\partial p}$$

$$= (3r^2)(q^2) + (2s)(2p\sin q).$$

We now substitute $r = pq^2$ and $s = p^2\sin q$:

$$\frac{\partial w}{\partial p} = 3(pq^2)^2(q^2) + (2p^2\sin q)(2p\sin q)$$

$$= 3p^2q^6 + 4p^3\sin^2 q$$

To find $\partial w/\partial q$, we again refer to the diagram in Figure 12.40:

$$\frac{\partial w}{\partial q} = \frac{\partial w}{\partial r}\frac{\partial r}{\partial q} + \frac{\partial w}{\partial s}\frac{\partial s}{\partial q}$$

$$= (3r^2)(2pq) + (2s)(p^2\cos q)$$

Substituting for r and s, we obtain

$$\frac{\partial w}{\partial q} = 3(pq^2)^2(2pq) + 2(p^2\sin q)(p^2\cos q)$$

$$= 6p^3q^5 + 2p^4\sin q\cos q.$$

Figure 12.41

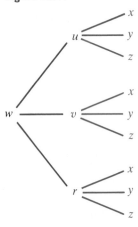

Note that after applying a chain rule in Example 1, we substituted for r and s, thereby expressing $\partial w/\partial p$ and $\partial w/\partial q$ in terms of p and q. This was done to emphasize the fact that w is a (composite) function of the *two* variables p and q.

We can apply chain rules to composite functions of any number of variables and construct tree diagrams to help formulate these rules. Hereafter, we shall not attach partial derivative symbols to the branches. It will be understood that if a branch leads from a variable w to another variable r, as in Figure 12.40, then the partial derivative is $\partial w/\partial r$. However, if w is a function of only *one* variable r, then we write dw/dr instead of $\partial w/\partial r$.

As an illustration, suppose that w is a function of u, v, and r and that u, v, and r are each functions of x, y, and z. This is indicated by the diagram in Figure 12.41. If we wish to find $\partial w/\partial y$, we take the pairs of products of the partials that lead from w to y and add, obtaining

$$\frac{\partial w}{\partial y} = \frac{\partial w}{\partial u}\frac{\partial u}{\partial y} + \frac{\partial w}{\partial v}\frac{\partial v}{\partial y} + \frac{\partial w}{\partial r}\frac{\partial r}{\partial y}.$$

EXAMPLE ■ 2 Use a chain rule to find $\partial w/\partial z$ if

$$w = r^2 + sv + t^3,$$

with $r = x^2 + y^2 + z^2, \quad s = xyz, \quad v = xe^y, \quad t = yz^2.$

Figure 12.42

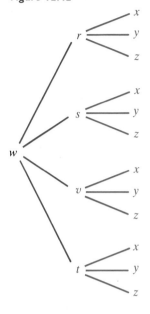

SOLUTION Note that w is a function of r, s, v, and t and that each of these four variables is a function of x, y, and z. An appropriate tree diagram is shown in Figure 12.42. We wish to find $\partial w/\partial z$, so we trace all branches that lead from w to z. This gives us

$$\frac{\partial w}{\partial z} = \frac{\partial w}{\partial r}\frac{\partial r}{\partial z} + \frac{\partial w}{\partial s}\frac{\partial s}{\partial z} + \frac{\partial w}{\partial v}\frac{\partial v}{\partial z} + \frac{\partial w}{\partial t}\frac{\partial t}{\partial z}$$

$$= (2r)(2z) + v(xy) + s(0) + (3t^2)(2yz)$$

$$= 4z(x^2 + y^2 + z^2) + xe^y(xy) + 0 + 3(yz^2)^2(2yz)$$

$$= 4z(x^2 + y^2 + z^2) + x^2ye^y + 6y^3z^5.$$

If w is a function of several variables, each of which is a function of only *one* variable, say t, then w is a function of the one variable t and we may consider dw/dt. A chain rule may be applied, as in the next example.

EXAMPLE ■ 3 Use a chain rule to find dw/dt if

$$w = x^2 + yz, \quad \text{with} \quad x = 3t^2 + 1, \quad y = 2t - 4, \quad z = t^3.$$

Figure 12.43

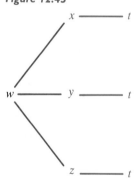

SOLUTION To apply a chain rule, we construct the tree diagram in Figure 12.43. The pairs of branches that lead from w to t give us the following, where we use d/dt for differentiation with respect to the *single* variable t:

$$\frac{dw}{dt} = \frac{\partial w}{\partial x}\frac{dx}{dt} + \frac{\partial w}{\partial y}\frac{dy}{dt} + \frac{\partial w}{\partial z}\frac{dz}{dt}$$

$$= (2x)(6t) + z(2) + y(3t^2)$$

$$= 2(3t^2 + 1)6t + t^3(2) + (2t - 4)3t^2$$

$$= 44t^3 - 12t^2 + 12t$$

The problem could also be solved without a chain rule by writing

$$w = (3t^2 + 1)^2 + (2t - 4)t^3$$

and then finding dw/dt by single-variable methods.

Chain rules are useful for solving related rate problems, as illustrated in the next example.

EXAMPLE ■ 4 A simple electrical circuit consists of a resistor R and an electromotive force V. At a certain instant, V is 80 volts and is increasing at a rate of 5 volts/min, while R is 40 ohms and is decreasing at a rate of 2 ohms/min. Use Ohm's law, $I = V/R$, and a chain rule to find the rate at which the current I (in amperes) is changing.

Figure 12.44

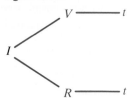

SOLUTION   Since I is a function of V and R and both V and R are functions of time t (in minutes), we consider the tree diagram in Figure 12.44. Applying a chain rule gives us

$$\frac{dI}{dt} = \frac{\partial I}{\partial V}\frac{dV}{dt} + \frac{\partial I}{\partial R}\frac{dR}{dt}$$

$$= \left(\frac{1}{R}\right)\frac{dV}{dt} + \left(-\frac{V}{R^2}\right)\frac{dR}{dt}.$$

Substituting

$$V = 80, \qquad \frac{dV}{dt} = 5, \qquad R = 40, \quad \text{and} \quad \frac{dR}{dt} = -2,$$

we obtain

$$\frac{dI}{dt} = \left(\frac{1}{40}\right)(5) + \left(-\frac{80}{1600}\right)(-2) = \frac{9}{40} = 0.225 \text{ amp/min.}$$

Partial derivatives can be used to find derivatives of functions that are determined implicitly. Suppose, as in Section 2.6, an equation $F(x, y) = 0$ determines a differentiable function f such that $y = f(x)$—that is, $F(x, f(x)) = 0$ for every x in the domain D of f. Let us introduce the following composite function F:

$$w = F(u, y), \quad \text{where} \quad u = x \quad \text{and} \quad y = f(x)$$

Figure 12.45

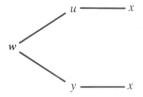

This leads to the tree diagram in Figure 12.45. Using a chain rule and the fact that u and y are functions of *one* variable x yields

$$\frac{dw}{dx} = \frac{\partial w}{\partial u}\frac{du}{dx} + \frac{\partial w}{\partial y}\frac{dy}{dx}.$$

Since $w = F(x, f(x)) = 0$ for every x, it follows that $dw/dx = 0$. Moreover, since $u = x$ and $y = f(x)$,

$$\frac{du}{dx} = 1 \quad \text{and} \quad \frac{dy}{dx} = f'(x).$$

Substituting in the preceding chain rule formula for dw/dx, we obtain

$$0 = \frac{\partial w}{\partial u}(1) + \frac{\partial w}{\partial y}f'(x).$$

If $\partial w/\partial y \neq 0$, then (since $u = x$)

$$f'(x) = -\frac{\partial w/\partial u}{\partial w/\partial y} = -\frac{\partial w/\partial x}{\partial w/\partial y} = -\frac{F_x(x, y)}{F_y(x, y)}.$$

We may summarize the preceding discussion as follows.

Theorem 12.22

If an equation $F(x, y) = 0$ determines, implicitly, a differentiable function f of one variable x such that $y = f(x)$, then

$$\frac{dy}{dx} = -\frac{F_x(x, y)}{F_y(x, y)}.$$

EXAMPLE 5 Find dy/dx if $y = f(x)$ is determined implicitly by

$$y^4 + 3y - 4x^3 - 5x - 1 = 0.$$

SOLUTION If $F(x, y)$ is the expression on the left side of the equation, then, by Theorem (12.22),

$$\frac{dy}{dx} = -\frac{-12x^2 - 5}{4y^3 + 3} = \frac{12x^2 + 5}{4y^3 + 3}.$$

Compare this solution with that of Example 2 in Section 2.6, which was obtained using single-variable methods.

Given the equation

$$x^2 - 4y^3 + 2z - 7 = 0,$$

we can solve for z, obtaining

$$z = \tfrac{1}{2}(-x^2 + 4y^3 + 7),$$

which is of the form

$$z = f(x, y).$$

In analogy with the single-variable case, we say that the function f of two variables x and y is determined *implicitly* by the given equation. The next theorem gives us formulas for finding f_x and f_y, or, equivalently, $\partial z/\partial x$ and $\partial z/\partial y$, without actually solving the equation for z.

Theorem 12.23

If an equation $F(x, y, z) = 0$ determines an implicit differentiable function f of two variables x and y such that $z = f(x, y)$ for every (x, y) in the domain of f, then

$$\frac{\partial z}{\partial x} = -\frac{F_x(x, y, z)}{F_z(x, y, z)}, \qquad \frac{\partial z}{\partial y} = -\frac{F_y(x, y, z)}{F_z(x, y, z)}.$$

PROOF The statement $F(x, y, z) = 0$ *determines a function f such that* $z = f(x, y)$ means that $F(x, y, f(x, y)) = 0$ for every (x, y) in the domain of f. Consider the composite function F of x and y defined as follows:

$$w = F(u, v, z), \quad \text{where} \quad u = x, \quad v = y, \quad z = f(x, y).$$

Note that u and v are functions of x and y, since we can write u and v as $u = x + (0 \cdot y)$ and $v = y + (0 \cdot x)$. Referring to the diagram in Figure 12.46 and considering the branches that lead from w to x, we obtain

$$\frac{\partial w}{\partial x} = \frac{\partial w}{\partial u}\frac{\partial u}{\partial x} + \frac{\partial w}{\partial v}\frac{\partial v}{\partial x} + \frac{\partial w}{\partial z}\frac{\partial z}{\partial x}.$$

Since $w = F(x, y, f(x, y)) = 0$ for every x and every y, it follows that $\partial w/\partial x = 0$. Moreover, since $\partial u/\partial x = 1$ and $\partial v/\partial x = 0$, our chain rule formula for $\partial w/\partial x$ may be written

$$0 = \frac{\partial w}{\partial x}(1) + \frac{\partial w}{\partial y}(0) + \frac{\partial w}{\partial z}\frac{\partial z}{\partial x},$$

and if $\partial w/\partial z \neq 0$,

$$\frac{\partial z}{\partial x} = -\frac{\partial w/\partial x}{\partial w/\partial z} = -\frac{F_x(x, y, z)}{F_z(x, y, z)}.$$

The formula for $\partial z/\partial y$ may be obtained in similar fashion. ∎

Figure 12.46

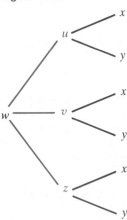

EXAMPLE 6 Find $\partial z/\partial x$ and $\partial z/\partial y$ if $z = f(x, y)$ is determined implicitly by

$$x^2 z^2 + xy^2 - z^3 + 4yz - 5 = 0.$$

SOLUTION If we let $F(x, y, z)$ denote the expression on the left side of the given equation, then, by Theorem (12.23),

$$\frac{\partial z}{\partial x} = -\frac{2xz^2 + y^2}{2x^2 z - 3z^2 + 4y}$$

$$\frac{\partial z}{\partial y} = -\frac{2xy + 4z}{2x^2 z - 3z^2 + 4y}.$$

EXERCISES 12.5

Use a chain rule in Exercises 1–14.

Exer. 1–2: Find $\partial w/\partial x$ and $\partial w/\partial y$.

1 $w = u \sin v; \quad u = x^2 + y^2, \quad v = xy$

2 $w = uv + v^2; \quad u = x \sin y, \quad v = y \sin x$

Exer. 3–4: Find $\partial w/\partial r$ and $\partial w/\partial s$.

3 $w = u^2 + 2uv; \quad u = r \ln s, \quad v = 2r + s$

4 $w = e^{tv}; \qquad t = r + s, \quad v = rs$

Exer. 5–6: Find $\partial z/\partial x$ and $\partial z/\partial y$.

5 $z = r^3 + s + v^2$;
   $r = xe^y$, $s = ye^x$, $v = x^2y$

6 $z = pq + qw$;
   $p = 2x - y$, $q = x - 2y$, $w = -2x + 2y$

Exer. 7–8: Find $\partial r/\partial u$, $\partial r/\partial v$, and $\partial r/\partial t$.

7 $r = x \ln y$; $x = 3u + vt$, $y = uvt$

8 $r = w^2 \cos z$; $w = u^2vt$, $z = ut^2$

9 If $p = u^2 + 3v^2 - 4w^2$, where $u = x - 3y + 2r - s$,
   $v = 2x + y - r + 2s$, and $w = -x + 2y + r + s$, find
   $\partial p/\partial r$.

10 Find $\partial s/\partial y$ if $s = tr + ue^v$, where $t = xy^2z$, $r = x^2yz$,
    $u = xyz^2$, and $v = xyz$.

Exer. 11–14: Find dw/dt.

11 $w = x^3 - y^3$; $x = \dfrac{1}{t+1}$, $y = \dfrac{t}{t+1}$

12 $w = \ln(u + v)$; $u = e^{-2t}$, $v = t^3 - t^2$

13 $w = r^2 - s \tan v$; $r = \sin^2 t$, $s = \cos t$, $v = 4t$

14 $w = x^2y^3z^4$; $x = 2t + 1$, $y = 3t - 2$,
                          $z = 5t + 4$

Exer. 15–18: Use partial derivatives to find dy/dx if $y = f(x)$ is determined implicitly by the given equation.

15 $2x^3 + x^2y + y^3 = 1$

16 $x^4 + 2x^2y^2 - 3xy^3 + 2x = 0$

17 $6x + \sqrt{xy} = 3y - 4$

18 $x^{2/3} + y^{2/3} = 4$

Exer. 19–22: Find $\partial z/\partial x$ and $\partial z/\partial y$ if $z = f(x, y)$ is determined implicitly by the given equation.

19 $2xz^3 - 3yz^2 + x^2y^2 + 4z = 0$

20 $xz^2 + 2x^2y - 4y^2z + 3y - 2 = 0$

21 $xe^{yz} - 2ye^{xz} + 3ze^{xy} = 1$

22 $yx^2 + z^2 + \cos xyz = 4$

Exer. 23–32: Use a chain rule.

23 The radius r and altitude h of a right circular cylinder
    are increasing at rates of 0.01 in./min and 0.02 in./min,
    respectively.

   (a) Find the rate at which the volume is increasing at the
       time when $r = 4$ in. and $h = 7$ in.

   (b) At what rate is the curved surface area changing at
       this time?

24 The equal sides and the included angle of an isosceles
    triangle are increasing at rates of 0.1 ft/hr and 2°/hr,
    respectively. Find the rate at which the area of the
    triangle is increasing at the time when the length of each
    of the equal sides is 20 ft and the included angle is 60°.

25 The pressure P, volume V, and temperature T of a
    confined gas are related by the ideal gas law $PV = kT$,
    where k is a constant. If P and V are changing at the
    rates dP/dt and dV/dt, respectively, find a formula for
    dT/dt.

26 If the base radius r and altitude h of a right circular
    cylinder are changing at the rates dr/dt and dh/dt,
    respectively, find a formula for dV/dt, where V is the
    volume of the cylinder.

27 A certain gas obeys the ideal gas law $PV = 8T$.
    Suppose that the gas is being heated at a rate of 2°/min
    and the pressure is increasing at a rate of $\frac{1}{2}$ (lb/in^2)/min.
    If, at a certain instant, the temperature is $200°$ and the
    pressure is 10 lb/in^2, find the rate at which the volume is
    changing.

28 Sand is leaking from a hole in a container at a rate of
    6 in^3/min. As it leaks out, it forms a pile in the shape of
    a right circular cone whose base radius is increasing at a
    rate of $\frac{1}{4}$ in./min. If, at the instant that 40 in^3 has leaked
    out, the radius is 5 in., find the rate at which the height
    of the pile is increasing.

29 At age 2 yr, a typical boy is 86 cm tall, weighs 13 kg,
    and is growing at the rate of 9 cm/year and 2 kg/year.
    Use the DuBois and DuBois surface area formula,
    $S = 0.007184x^{0.425}y^{0.725}$ for weight x and height y,
    to estimate the rate at which the body surface area is
    growing (see Exercise 59 of Section 12.1).

30 When the size of the molecules and their forces of
    attraction are taken into account, the pressure P, volume
    V, and temperature T of a mole of confined gas are
    related by the *van der Waals equation*,

$$\left(P + \frac{a}{V^2}\right)(V - b) = kT,$$

   where a, b, and k are positive constants. If t is time,
   find a formula for dT/dt in terms of dP/dt, dV/dt,
   P, and V.

31 If n resistances R_1, R_2, ..., R_n are connected in
    parallel, then the total resistance R is given by

$$\frac{1}{R} = \sum_{k=1}^{n} \frac{1}{R_k}.$$

Prove that for $k = 1, 2, \ldots, n$,

$$\frac{\partial R}{\partial R_k} = \left(\frac{R}{R_k}\right)^2.$$

32 A function f of two variables is *homogeneous of degree n* if $f(tx, ty) = t^n f(x, y)$ for every t such that (tx, ty) is in the domain of f. Show that, for such functions, $xf_x(x, y) + yf_y(x, y) = nf(x, y)$. (*Hint:* Differentiate $f(tx, ty)$ with respect to t.)

Exer. 33–36: Refer to Exercise 32. Find the degree n of the homogeneous function f and verify the formula

$$xf_x(x, y) + yf_y(x, y) = nf(x, y).$$

33 $f(x, y) = 2x^3 + 3x^2 y + y^3$

34 $f(x, y) = \dfrac{x^3 y}{x^2 + y^2}$

35 $f(x, y) = \arctan \dfrac{y}{x}$

36 $f(x, y) = xye^{y/x}$

37 If $w = f(x, y)$, where $x = r\cos\theta$ and $y = r\sin\theta$, show that

$$\left(\frac{\partial w}{\partial x}\right)^2 + \left(\frac{\partial w}{\partial y}\right)^2 = \left(\frac{\partial w}{\partial r}\right)^2 + \frac{1}{r^2}\left(\frac{\partial w}{\partial \theta}\right)^2.$$

38 If $w = f(x, y)$, where $x = e^r \cos\theta$ and $y = e^r \sin\theta$, show that

$$\frac{\partial^2 w}{\partial x^2} + \frac{\partial^2 w}{\partial y^2} = e^{-2r}\left(\frac{\partial^2 w}{\partial r^2} + \frac{\partial^2 w}{\partial \theta^2}\right).$$

39 If $w = f(x, y)$, where $x = r\cos\theta$ and $y = r\sin\theta$, show that

$$\frac{\partial^2 w}{\partial x^2} + \frac{\partial^2 w}{\partial y^2} = \frac{\partial^2 w}{\partial r^2} + \frac{1}{r^2}\frac{\partial^2 w}{\partial \theta^2} + \frac{1}{r}\frac{\partial w}{\partial r}.$$

40 If $v = f(x - at) + g(x + at)$ and f and g have second partial derivatives, show that v satisfies the wave equation

$$\frac{\partial^2 v}{\partial t^2} = a^2\frac{\partial^2 v}{\partial x^2}.$$

(Compare with Exercise 42 of Section 12.3.)

41 If $w = \cos(x + y) + \cos(x - y)$, show, without using addition formulas, that $w_{xx} - w_{yy} = 0$.

42 If $w = f(x^2 + y^2)$, show that $y(\partial w/\partial x) - x(\partial w/\partial y) = 0$. (*Hint:* Let $u = x^2 + y^2$.)

43 If $w = f(u, v)$, where $u = g(x, y)$ and $v = k(x, y)$, show that

$$\frac{\partial^2 w}{\partial x^2} = \frac{\partial^2 w}{\partial u^2}\left(\frac{\partial u}{\partial x}\right)^2 + \left(\frac{\partial^2 w}{\partial v\,\partial u} + \frac{\partial^2 w}{\partial u\,\partial v}\right)\frac{\partial u}{\partial x}\frac{\partial v}{\partial x}$$
$$+ \frac{\partial^2 w}{\partial v^2}\left(\frac{\partial v}{\partial x}\right)^2 + \frac{\partial w}{\partial u}\frac{\partial^2 u}{\partial x^2} + \frac{\partial w}{\partial v}\frac{\partial^2 v}{\partial x^2}.$$

44 For $w, u,$ and v as given in Exercise 43, show that

$$\frac{\partial^2 w}{\partial y\,\partial x} = \frac{\partial^2 w}{\partial u^2}\frac{\partial u}{\partial x}\frac{\partial u}{\partial y} + \frac{\partial^2 w}{\partial v\,\partial u}\frac{\partial u}{\partial x}\frac{\partial v}{\partial y} + \frac{\partial^2 w}{\partial u\,\partial v}\frac{\partial u}{\partial y}\frac{\partial v}{\partial x}$$
$$+ \frac{\partial^2 w}{\partial v^2}\frac{\partial v}{\partial x}\frac{\partial v}{\partial y} + \frac{\partial w}{\partial u}\frac{\partial^2 u}{\partial y\,\partial x} + \frac{\partial w}{\partial v}\frac{\partial^2 v}{\partial y\,\partial x}.$$

45 Suppose that $u = f(x, y)$ and $v = g(x, y)$ satisfy the Cauchy–Riemann equations $u_x = v_y$ and $u_y = -v_x$. If $x = r\cos\theta$ and $y = r\sin\theta$, show that

$$\frac{\partial u}{\partial r} = \frac{1}{r}\frac{\partial v}{\partial \theta} \quad \text{and} \quad \frac{\partial v}{\partial r} = -\frac{1}{r}\frac{\partial u}{\partial \theta}.$$

46 If $u = y/(x^2 + y^2)$ and $v = x/(x^2 + y^2)$, verify the formulas for $\partial u/\partial r$ and $\partial v/\partial r$ in Exercise 45 directly, by substituting $r\cos\theta$ for x and $r\sin\theta$ for y and then differentiating.

12.6   DIRECTIONAL DERIVATIVES

In Section 12.3, we discussed the fact that if $f(x, y)$ is the temperature of a flat metal plate at the point $P(x, y)$ in an xy-plane, then the partial derivatives $f_x(x, y)$ and $f_y(x, y)$ give us the instantaneous rates of change of temperature with respect to distance in the horizontal and vertical directions, respectively (see Figures 12.28 and 12.29). In this section, we generalize this fact to the rate of change of $f(x, y)$ in *any* direction.

Let $\mathbf{u} = u_1\mathbf{i} + u_2\mathbf{j}$ be a unit vector. If we represent $\mathbf{u}$ by a vector with initial point $P(x, y)$, as in Figure 12.47(a), then the terminal point has coordinates $(x + u_1, y + u_2)$. We wish to define the rate of change of $f(x, y)$ with respect to distance in the direction determined by $\mathbf{u}$.

Consider the line l through P parallel to $\mathbf{u}$, and let Q be any point on l, as in Figure 12.47(b). The vector $\overrightarrow{PQ}$ corresponds to a scalar multiple $s\mathbf{u} = (su_1)\mathbf{i} + (su_2)\mathbf{j}$ for some s. Hence, the coordinates of the point Q are $(x + su_1, y + su_2)$. Since $\mathbf{u}$ is a unit vector,

$$\|\overrightarrow{PQ}\| = \|s\mathbf{u}\| = |s|\,\|\mathbf{u}\| = |s|.$$

Thus, s is the (signed) distance from P, measured along l. If $s > 0$, then $s\mathbf{u}$ has the same direction as $\mathbf{u}$. If $s < 0$, then $s\mathbf{u}$ has the opposite direction.

Figure 12.47

(a)

(b)

If the position of a point changes from P to Q, then the increment Δw of $w = f(x, y)$ is

$$\Delta w = f(x + su_1, y + su_2) - f(x, y).$$

The *average rate of change* in $f(x, y)$ is

$$\frac{\Delta w}{s} = \frac{f(x + su_1, y + su_2) - f(x, y)}{s}.$$

To illustrate, suppose $w = f(x, y)$ is the temperature at (x, y). If the temperature at P is $50°$ and the temperature at Q is $51.5°$, then $\Delta w = 1.5°$. If $\|\overrightarrow{PQ}\| = 3$ cm, then the average rate of change in temperature as a point moves from P to Q is

$$\frac{\Delta w}{s} = 0.5° \text{ per centimeter.}$$

As in our previous work with derivatives, to find the *instantaneous* rate of change of $f(x, y)$ at P in the direction determined by $\mathbf{u}$, we consider the limit of the average rate of change $\Delta w / s$ as $s \to 0$. This is the motivation for the following definition.

Definition   12.24

Let $w = f(x, y)$, and let $\mathbf{u} = u_1\mathbf{i} + u_2\mathbf{j}$ be a unit vector. The **directional derivative of f at $P(x, y)$ in the direction of u**, denoted by $D_{\mathbf{u}} f(x, y)$, is

$$D_{\mathbf{u}} f(x, y) = \lim_{s \to 0} \frac{f(x + su_1, y + su_2) - f(x, y)}{s}.$$

If **a** is any vector that has the same direction as **u**, we shall also refer to $D_{\mathbf{u}} f(x, y)$ as the **directional derivative of f in the direction of a.**

EXAMPLE • 1   Estimate the directional derivative of

$$f(x, y) = x^2 y \ln(xy^2 + 5x + 2y)$$

at $P(1, 2)$ in the direction $\mathbf{a} = 1.2\mathbf{i} + 3.7\mathbf{j}$ using the quotient in Definition (12.24) with $s = 0.02, 0.01$, and 0.005.

SOLUTION   The magnitude of the vector **a** is

$$\|\mathbf{a}\| = \sqrt{(1.2)^2 + (3.7)^2} = \sqrt{15.13} \approx 3.8897.$$

We find a unit vector **u** in the same direction as **a** by using Theorem (10.12):

$$\mathbf{u} = \frac{1}{\|\mathbf{a}\|}\mathbf{a} \approx \frac{1.2}{3.8897}\mathbf{i} + \frac{3.7}{3.8897}\mathbf{j} \approx 0.3085\mathbf{i} + 0.9512\mathbf{j} \approx u_1\mathbf{i} + u_2\mathbf{j}$$

Using Definition (12.24) with $x = 1$, $y = 2$, and $s = 0.02$, we have

$$D_{\mathbf{u}} f(1, 2) \approx \frac{f(1 + 0.02u_1, 2 + 0.02u_2) - f(1, 2)}{0.02}.$$

With $u_1 = 0.3085$ and $u_2 = 0.9512$, this quotient is

$$\frac{f(1.00617, 2.019024) - f(1, 2)}{0.02} \approx \frac{5.26943 - 5.12990}{0.02} \approx 6.977.$$

Similar computations give the quotient approximate values of 6.943 and 6.923 when $s = 0.01$ and $s = 0.005$, respectively. Thus, we estimate the value of the directional derivative to be approximately 6.9.

The following theorem provides a formula for finding directional derivatives.

Theorem   12.25

If f is a differentiable function of two variables and $\mathbf{u} = u_1\mathbf{i} + u_2\mathbf{j}$ is a unit vector, then

$$D_{\mathbf{u}} f(x, y) = f_x(x, y)u_1 + f_y(x, y)u_2.$$

P R O O F Consider x, y, u_1, and u_2 as fixed (but arbitrary), and let g be the function of *one* variable s defined by

$$g(s) = f(x + su_1, y + su_2).$$

From Definitions (2.6) and (12.24),

$$g'(0) = \lim_{s \to 0} \frac{g(s) - g(0)}{s - 0}$$

$$= \lim_{s \to 0} \frac{f(x + su_1, y + su_2) - f(x, y)}{s}$$

$$= D_{\mathbf{u}} f(x, y).$$

We next consider g as a composite function, with

$$w = g(s) = f(r, v) \quad \text{and} \quad r = x + su_1, \quad v = y + su_2.$$

As indicated by the tree diagram in Figure 12.48, w is a function of r and v, and r and v are functions of *one* variable s. Applying a chain rule, we obtain

$$\frac{dw}{ds} = \frac{\partial w}{\partial r} \frac{dr}{ds} + \frac{\partial w}{\partial v} \frac{dv}{ds}.$$

Using the definitions of $w = g(s)$, r, and v, we may rewrite this as

$$g'(s) = f_r(r, v)u_1 + f_v(r, v)u_2.$$

If we let $s = 0$, then $r = x$, $v = y$, and, by the first part of the proof, $g'(0) = D_{\mathbf{u}} f(x, y)$. Hence,

$$D_{\mathbf{u}} f(x, y) = f_x(x, y)u_1 + f_y(x, y)u_2. \quad \blacksquare$$

The first partial derivatives of f are special cases of the directional derivative $D_{\mathbf{u}} f(x, y)$. Thus, if we let $\mathbf{u} = \mathbf{i}$, then $u_1 = 1$, $u_2 = 0$, and the formula in (12.25) becomes

$$D_{\mathbf{i}} f(x, y) = f_x(x, y).$$

If $\mathbf{u} = \mathbf{j}$, then $u_1 = 0$, $u_2 = 1$, and we obtain

$$D_{\mathbf{j}} f(x, y) = f_y(x, y).$$

Figure 12.48

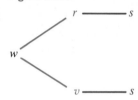

EXAMPLE ▪ 2 Let $f(x, y) = x^3 y^2$.

(a) Find the directional derivative of f at the point $P(-1, 2)$ in the direction of the vector $\mathbf{a} = 4\mathbf{i} - 3\mathbf{j}$.

(b) Discuss the significance of part (a) if $f(x, y)$ is the temperature at (x, y).

SOLUTION

(a) The vector $\mathbf{a} = 4\mathbf{i} - 3\mathbf{j}$ with initial point $P(-1, 2)$ is represented in Figure 12.49. We wish to find $D_{\mathbf{u}} f(-1, 2)$ for the *unit* vector $\mathbf{u}$ that has

Figure 12.49

the direction of **a**. By Theorem (10.12),

$$\mathbf{u} = \frac{1}{\|\mathbf{a}\|}\mathbf{a} = \frac{1}{5}(4\mathbf{i} - 3\mathbf{j}) = \frac{4}{5}\mathbf{i} - \frac{3}{5}\mathbf{j}.$$

Since $\qquad f_x(x, y) = 3x^2y^2 \quad$ and $\quad f_y(x, y) = 2x^3y,$

it follows from (12.25) that

$$D_{\mathbf{u}}f(x, y) = 3x^2y^2\left(\tfrac{4}{5}\right) + 2x^3y\left(-\tfrac{3}{5}\right).$$

Hence, at $P(-1, 2)$,

$$D_{\mathbf{u}}f(-1, 2) = 3(-1)^2(2)^2\left(\tfrac{4}{5}\right) + 2(-1)^3(2)\left(-\tfrac{3}{5}\right) = 12.$$

(b) If $f(x, y)$ is the temperature (in degrees) at (x, y), then $D_{\mathbf{u}}f(-1, 2) = 12$ tells us that if a point moves in the direction of **u**, the temperature at P is increasing at the rate of $12°$ per unit change in distance. It is of interest to compare this value with $f_x(-1, 2) = 12$ and $f_y(-1, 2) = -4$, which are the rates of change in the horizontal and vertical directions, respectively.

We may use Theorem (12.25) to express a directional derivative as a dot product of two vectors, as follows:

$$D_{\mathbf{u}}f(x, y) = [f_x(x, y)\mathbf{i} + f_y(x, y)\mathbf{j}] \cdot [u_1\mathbf{i} + u_2\mathbf{j}]$$

The vector in the first bracket, whose components are the first partial derivatives of $f(x, y)$, is very important. It is denoted by $\nabla f(x, y)$ (read *del $f(x, y)$*) and is given the following special name.

Definition 12.26

> Let f be a function of two variables. The **gradient** of f (or of $f(x, y)$) is the vector function given by
>
> $$\nabla f(x, y) = f_x(x, y)\mathbf{i} + f_y(x, y)\mathbf{j}.$$

The gradient $\nabla f(x, y)$ is sometimes denoted by **grad** $f(x, y)$. From the previous discussion, we have the following.

Directional Derivative
(Gradient Form) 12.27

> $$D_{\mathbf{u}}f(x, y) = \nabla f(x, y) \cdot \mathbf{u}$$

Thus, *to find the directional derivative of f in the direction of the unit vector* **u**, *we may dot the gradient of f with* **u**. Henceforth, we shall use this formula instead of Theorem (12.25) to find $D_{\mathbf{u}}f(x, y)$.

The symbol ∇ (called **del**) is a *vector differential operator* and is defined by

$$\nabla = \mathbf{i}\frac{\partial}{\partial x} + \mathbf{j}\frac{\partial}{\partial y}.$$

Figure 12.50

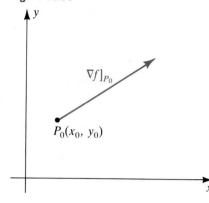

Its properties are similar to those of the operator d/dx (see Exercises 39–44). Operating on $f(x, y)$, it produces the gradient vector $\nabla f(x, y)$.

If $P_0(x_0, y_0)$ is a specific point in the xy-plane, we sometimes denote the gradient vector at P_0 by $\nabla f]_{P_0}$. Thus,

$$\nabla f]_{P_0} = \nabla f(x_0, y_0)$$
$$= f_x(x_0, y_0)\mathbf{i} + f_y(x_0, y_0)\mathbf{j}.$$

When we represent the vector $\nabla f]_P$ graphically, we always take the initial point at P_0, as illustrated in Figure 12.50.

EXAMPLE • 3 Let $f(x, y) = x^2 - 4xy$.

(a) Find the gradient of f at the point $P(1, 2)$, and sketch the vector $\nabla f]_P$.
(b) Use the gradient to find the directional derivative of f at $P(1, 2)$ in the direction from $P(1, 2)$ to $Q(2, 5)$.

SOLUTION
(a) By Definition (12.26),

$$\nabla f(x, y) = (2x - 4y)\mathbf{i} - 4x\mathbf{j}.$$

At $P(1, 2)$,

$$\nabla f]_P = \nabla f(1, 2)$$
$$= (2 - 8)\mathbf{i} - 4\mathbf{j} = -6\mathbf{i} - 4\mathbf{j}.$$

This vector is sketched in Figure 12.51, with initial point $P(1, 2)$.

(b) If we let $\mathbf{a} = \overrightarrow{PQ}$, then

$$\mathbf{a} = \langle 2 - 1, \ 5 - 2 \rangle = \langle 1, 3 \rangle = \mathbf{i} + 3\mathbf{j}$$

Figure 12.51

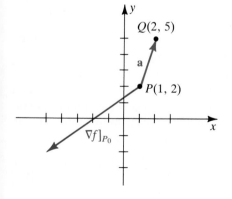

(see Figure 12.51). A unit vector having the direction of $\overrightarrow{PQ}$ is

$$\mathbf{u} = \frac{1}{\|\mathbf{a}\|}\mathbf{a} = \frac{1}{\sqrt{10}}(\mathbf{i} + 3\mathbf{j}).$$

Applying (12.27), we have

$$D_{\mathbf{u}} f(1, 2) = \nabla f(1, 2) \cdot \mathbf{u}$$

$$= (-6\mathbf{i} - 4\mathbf{j}) \cdot \frac{1}{\sqrt{10}}(\mathbf{i} + 3\mathbf{j})$$

$$= \frac{1}{\sqrt{10}}(-6 - 12) = -\frac{18}{\sqrt{10}} \approx -5.7.$$

Let $P(x, y)$ be a fixed point, and consider the directional derivative $D_{\mathbf{u}} f(x, y)$ as $\mathbf{u} = \langle u_1, u_2 \rangle$ varies. For a given unit vector $\mathbf{u}$, the directional derivative may be positive (that is, $f(x, y)$ may increase) or negative

($f(x, y)$ may decrease) or 0. In many applications, it is important to find the direction in which $f(x, y)$ increases most rapidly and the maximum rate of change. The next theorem provides this information.

Gradient Theorem　12.28

Let f be a function of two variables that is differentiable at the point $P(x, y)$.

   (i)　The maximum value of $D_{\mathbf{u}} f(x, y)$ at $P(x, y)$ is $\|\nabla f(x, y)\|$.

  (ii)　The maximum rate of increase of $f(x, y)$ at $P(x, y)$ occurs in the direction of $\nabla f(x, y)$.

Figure 12.52

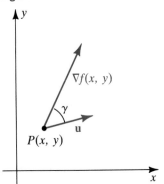

PROOF

   (i) We shall consider the point $P(x, y)$ and the vector $\nabla f(x, y)$ as fixed (but arbitrary) and the unit vector $\mathbf{u}$ as variable (see Figure 12.52). Let γ be the angle between $\mathbf{u}$ and $\nabla f(x, y)$. By (12.27) and Theorem (10.22),

$$D_{\mathbf{u}} f(x, y) = \nabla f(x, y) \cdot \mathbf{u}$$

$$= \|\nabla f(x, y)\| \, \|\mathbf{u}\| \cos \gamma$$

$$= \|\nabla f(x, y)\| \cos \gamma.$$

Since $-1 \le \cos \gamma \le 1$, the maximum value of $D_{\mathbf{u}} f(x, y)$ occurs if $\cos \gamma = 1$; in this case, $D_{\mathbf{u}} f(x, y) = \|\nabla f(x, y)\|$.

  (ii) The directional derivative $D_{\mathbf{u}} f(x, y)$ is the rate of change of $f(x, y)$ with respect to distance at $P(x, y)$ in the direction determined by $\mathbf{u}$. As in part (i), this rate of change has its maximum value if $\cos \gamma = 1$—that is, if $\gamma = 0$. However, if $\gamma = 0$, then $\mathbf{u}$ has the same direction as $\nabla f(x, y)$. ∎

Figure 12.53

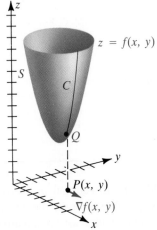

   The gradient theorem (12.28) has an interesting and useful geometric interpretation. Let S denote the graph of $z = f(x, y)$ in an xyz-coordinate system. As illustrated in Figure 12.53, the point $P(x, y)$ and the vector $\nabla f(x, y)$ are in the xy-plane, and $Q(x, y, f(x, y))$ is the point on S corresponding to P. If a (variable) point in the xy-plane moves through P in the direction of $\nabla f(x, y)$, then, by (12.28)(ii), the corresponding point on S moves up a curve C of *steepest ascent* at the point Q on the graph of f. In Section 12.7, we discuss another geometric aspect of $\nabla f(x, y)$, which involves the level curves of f.

   Referring to the proof of the gradient theorem (12.28), we see that the *minimum* value of $D_{\mathbf{u}} f(x, y)$ occurs if $\cos \gamma = -1$. In this case, $\gamma = \pi$ (or $\gamma = 180°$) and $D_{\mathbf{u}} f(x, y) = - \|\nabla f(x, y)\|$, which is the maximum rate of *decrease* of $f(x, y)$. This gives us the following corollary to Theorem (12.28).

Corollary **12.29**

Let f be a function of two variables that is differentiable at the point $P(x, y)$.

(i) The minimum value of $D_{\mathbf{u}} f(x, y)$ at the point $P(x, y)$ is $-\|\nabla f(x, y)\|$.

(ii) The minimum rate of increase (or maximum rate of decrease) of $f(x, y)$ at the point $P(x, y)$ occurs in the direction of $-\nabla f(x, y)$.

Figure 12.54

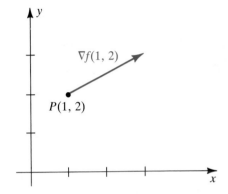

EXAMPLE 4 Let $f(x, y) = 2 + x^2 + \frac{1}{4} y^2$.

(a) Find the direction in which $f(x, y)$ increases most rapidly at the point $P(1, 2)$, and find the maximum rate of increase of f at P.

(b) Interpret part (a) using the graph of f.

SOLUTION

(a) The gradient of f is

$$\nabla f(x, y) = 2x\mathbf{i} + \tfrac{1}{2} y\mathbf{j}.$$

At $P(1, 2)$,

$$\nabla f(1, 2) = 2\mathbf{i} + \mathbf{j}.$$

This vector, sketched in Figure 12.54, determines the direction in which $f(x, y)$ increases most rapidly at $P(1, 2)$.

The maximum rate of increase of f at $P(1, 2)$ is

$$\|\nabla f(1, 2)\| = \|2\mathbf{i} + \mathbf{j}\| = \sqrt{2^2 + 1^2} = \sqrt{5} \approx 2.2.$$

Figure 12.55

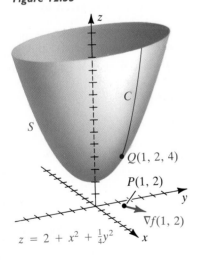

(b) The graph S of f, an elliptic paraboloid, is sketched in Figure 12.55. The point $P(1, 2)$ and the vector $\nabla f(1, 2)$ are shown in the xy-plane. The point on S corresponding to P is $Q(1, 2, 4)$. If a (variable) point in the xy-plane moves through P in the direction of $\nabla f(1, 2)$, the corresponding point on the graph moves along a curve C of steepest ascent on the paraboloid.

The directional derivative of a function f of three variables is defined in the same manner as in Definition (12.24). Specifically, if the vector $\mathbf{u} = u_1\mathbf{i} + u_2\mathbf{j} + u_3\mathbf{k}$ is a unit vector, we have the following.

Directional Derivative of $f(x, y, z)$ **12.30**

$$D_{\mathbf{u}} f(x, y, z) = \lim_{s \to 0} \frac{f(x + su_1, y + su_2, z + su_3) - f(x, y, z)}{s}$$

As in the two-variable case, $D_{\mathbf{u}} f(x, y, z)$ is the rate of change of f with respect to distance at $P(x, y, z)$ in the direction of $\mathbf{u}$. The **gradient** of f (or of $f(x, y, z)$), denoted by $\nabla f(x, y, z)$ or **grad** $f(x, y, z)$, is defined as follows.

Gradient of $f(x, y, z)$ 12.31

$$\nabla f(x, y, z) = f_x(x, y, z)\mathbf{i} + f_y(x, y, z)\mathbf{j} + f_z(x, y, z)\mathbf{k}$$

As in two dimensions, the gradient at a specific point $P_0(x_0, y_0, z_0)$ is denoted by $\nabla f]_{P_0}$.

The next result is the three-dimensional version of Theorem (12.25).

Theorem   12.32

If f is a differentiable function of three variables and the vector $\mathbf{u} = u_1\mathbf{i} + u_2\mathbf{j} + u_3\mathbf{k}$ is a unit vector, then

$$D_{\mathbf{u}} f(x, y, z) = \nabla f(x, y, z) \cdot \mathbf{u}$$
$$= f_x(x, y, z)u_1 + f_y(x, y, z)u_2 + f_z(x, y, z)u_3.$$

As in Theorem (12.28), of all possible directional derivatives $D_{\mathbf{u}} f(x, y, z)$ at the point $P(x, y, z)$, the one in the direction of $\nabla f(x, y, z)$ has the largest value, and that maximum value is $\| \nabla f(x, y, z) \|$.

EXAMPLE ■ 5   Suppose an xyz-coordinate system is located in space such that the temperature T at the point (x, y, z) is given by the formula $T = 100/(x^2 + y^2 + z^2)$.

(a) Find the rate of change of T with respect to distance at the point $P(1, 3, -2)$ in the direction of the vector $\mathbf{a} = \mathbf{i} - \mathbf{j} + \mathbf{k}$.

(b) In what direction from P does T increase most rapidly? What is the maximum rate of change of T at P?

SOLUTION

(a) By Definition (12.31), the gradient of $T = 100/(x^2 + y^2 + z^2)$ is

$$\nabla T = \frac{\partial T}{\partial x}\mathbf{i} + \frac{\partial T}{\partial y}\mathbf{j} + \frac{\partial T}{\partial z}\mathbf{k}.$$

Since

$$\frac{\partial T}{\partial x} = \frac{-200x}{(x^2 + y^2 + z^2)^2}, \quad \frac{\partial T}{\partial y} = \frac{-200y}{(x^2 + y^2 + z^2)^2}, \quad \frac{\partial T}{\partial z} = \frac{-200z}{(x^2 + y^2 + z^2)^2},$$

we obtain

$$\nabla T = \frac{-200}{(x^2 + y^2 + z^2)^2}(x\mathbf{i} + y\mathbf{j} + z\mathbf{k}).$$

Hence,

$$\nabla T]_P = \frac{-200}{196}(\mathbf{i} + 3\mathbf{j} - 2\mathbf{k}).$$

A unit vector **u** having the same direction as $\mathbf{a} = \mathbf{i} - \mathbf{j} + \mathbf{k}$ is

$$\mathbf{u} = \frac{1}{\sqrt{3}}(\mathbf{i} - \mathbf{j} + \mathbf{k}).$$

By Theorem (12.32), the rate of change of T at P in the direction of **a** is

$$D_{\mathbf{u}} T]_P = \nabla T]_P \cdot \mathbf{u} = \frac{-200}{196} \frac{(1 - 3 - 2)}{\sqrt{3}} = \frac{200}{49\sqrt{3}} \approx 2.4.$$

If, for example, T is measured in degrees and distance in inches, then T is increasing at a rate of 2.4 degrees per inch at P in the direction of **a**.

(b) The maximum rate of change of T at P occurs in the direction of the gradient—that is, in the direction of the vector $-\mathbf{i} - 3\mathbf{j} + 2\mathbf{k}$. The maximum rate of change equals the magnitude of the gradient—that is,

$$\|\nabla T]_P\| = \frac{200}{196}\sqrt{1 + 9 + 4} \approx 3.8.$$

In this section, we used the gradient of a function f primarily for finding directional derivatives. Later in the text, other important aspects will be considered. We shall conclude this section with an application to thermodynamics.

Let $T = f(x, y)$ be the steady state temperature at the point (x, y) of a two-dimensional region R. (*Steady state* means that T is independent of time.) *Fourier's law of heat conduction* asserts that the rate q at which heat flows past a point $P(x, y)$ on the boundary of R is directly proportional to the component of the temperature gradient ∇T along the unit (outer) normal vector **n** to the boundary at P (see Figure 12.56). Thus,

Figure 12.56

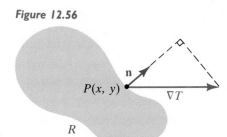

$$q = k \operatorname{comp}_{\mathbf{n}} \nabla T = k\nabla T \cdot \mathbf{n}$$

for some scalar k (see Definition (10.25)). If $\nabla T \cdot \mathbf{n} = 0$, the component in the direction of the normal will be 0 (∇T and **n** are orthogonal), and hence no heat will escape from the region at P. The boundary is then said to be *insulated* at the point P. The region is *insulated along part of a boundary* if it is insulated at every point on that part. Analogous statements can be made for three-dimensional regions.

EXAMPLE ▪ 6 Let R be the rectangular region in the xy-plane shown in Figure 12.57, and let $T = f(x, y)$ be the steady state temperature at (x, y). Find conditions on T that make the following insulated:

Figure 12.57

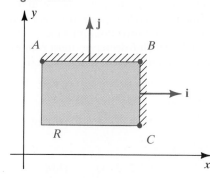

(a) edge BC **(b)** edge AB

SOLUTION

(a) We may use the vector **i** for the unit normal **n** to edge BC. Thus, BC is insulated if and only if, at each point P on BC,

$$\nabla T \cdot \mathbf{i} = 0, \quad \text{or} \quad \left(\frac{\partial T}{\partial x}\mathbf{i} + \frac{\partial T}{\partial y}\mathbf{j}\right) \cdot \mathbf{i} = \frac{\partial T}{\partial x} = 0.$$

This means that the rate of change of T in the horizontal direction is 0.

(b) A normal to edge AB is $\mathbf{j}$. In this case, AB is insulated if and only if

$$\nabla T \cdot \mathbf{j} = 0 \quad \text{or, equivalently,} \quad \frac{\partial T}{\partial y} = 0;$$

that is, the rate of change of T in the vertical direction is 0.

EXERCISES 12.6

Exer. 1–6: Find the gradient of f at P.

1   $f(x, y) = \sqrt{x^2 + y^2}$; $P(-4, 3)$

2   $f(x, y) = 7y - 5x$; $P(2, 6)$

3   $f(x, y) = e^{3x} \tan y$; $P(0, \pi/4)$

4   $f(x, y) = x \ln(x - y)$; $P(5, 4)$

5   $f(x, y, z) = yz^3 - 2x^2$; $P(2, -3, 1)$

6   $f(x, y, z) = xy^2 e^z$; $P(2, -1, 0)$

Exer. 7–10: Use (12.24) or (12.30) to estimate the directional derivative of f at P in the indicated direction with $s = 0.02$, 0.01, and 0.005.

7   $f(x, y) = \dfrac{x^3 \tanh(x + y)}{4 + x^2 + y^2}$;

     $P(2, 5)$, $\mathbf{a} = \sqrt{2}\mathbf{i} + \sqrt{3}\mathbf{j}$

8   $f(x, y) = x \ln(5x^2 + 4xy + y^2)$;

     $P(\sqrt{5}, 3)$, $\mathbf{a} = -0.89\mathbf{i} + 1.75\mathbf{j}$

9   $f(x, y, z) = y^2 e^{(z^3 + 5xy)} + 6x^2 yz$;

     $P(0, 1.2, -2.5)$, $\mathbf{a} = 3.7\mathbf{i} + 1.9\mathbf{j} - 2.1\mathbf{k}$

10   $f(x, y, z) = \dfrac{x^3 \sinh(xyz)}{1 + y^2 + z^4}$;

     $P(-1, 2, 1)$, $\mathbf{a} = 2\mathbf{i} - \mathbf{j} + 4\mathbf{k}$

Exer. 11–24: Find the directional derivative of f at the point P in the indicated direction.

11   $f(x, y) = x^2 - 5xy + 3y^2$;

     $P(3, -1)$, $\mathbf{u} = (\sqrt{2}/2)(\mathbf{i} + \mathbf{j})$

12   $f(x, y) = x^3 - 3x^2 y - y^3$;

     $P(1, -2)$, $\mathbf{u} = \frac{1}{2}(-\mathbf{i} + \sqrt{3}\mathbf{j})$

13   $f(x, y) = \arctan \dfrac{y}{x}$;

     $P(4, -4)$, $\mathbf{a} = 2\mathbf{i} - 3\mathbf{j}$

14   $f(x, y) = x^2 \ln y$;

     $P(5, 1)$, $\mathbf{a} = -\mathbf{i} + 4\mathbf{j}$

15   $f(x, y) = \sqrt{9x^2 - 4y^2 - 1}$;

     $P(3, -2)$, $\mathbf{a} = \mathbf{i} + 5\mathbf{j}$

16   $f(x, y) = \dfrac{x - y}{x + y}$;

     $P(2, -1)$, $\mathbf{a} = 3\mathbf{i} + 4\mathbf{j}$

17   $f(x, y) = x \cos^2 y$;

     $P(2, \pi/4)$, $\mathbf{a} = \langle 5, 1 \rangle$

18   $f(x, y) = xe^{3y}$;

     $P(4, 0)$, $\mathbf{a} = \langle -1, 3 \rangle$

19   $f(x, y, z) = xy^3 z^2$;

     $P(2, -1, 4)$, $\mathbf{a} = \mathbf{i} + 2\mathbf{j} - 3\mathbf{k}$

20   $f(x, y, z) = x^2 + 3yz + 4xy$;

     $P(1, 0, -5)$, $\mathbf{a} = 2\mathbf{i} - 3\mathbf{j} + \mathbf{k}$

21   $f(x, y, z) = z^2 e^{xy}$;

     $P(-1, 2, 3)$, $\mathbf{a} = 3\mathbf{i} + \mathbf{j} - 5\mathbf{k}$

22   $f(x, y, z) = \sqrt{xy} \sin z$;

     $P(4, 9, \pi/4)$, $\mathbf{a} = 2\mathbf{i} + 3\mathbf{j} - 2\mathbf{k}$

23   $f(x, y, z) = (x + y)(y + z)$;

     $P(5, 7, 1)$, $\mathbf{a} = \langle -3, 0, 1 \rangle$

24   $f(x, y, z) = z^2 \tan^{-1}(x + y)$;

     $P(0, 0, 4)$, $\mathbf{a} = \langle 6, 0, 1 \rangle$

Exer. 25–28: (a) Find the directional derivative of f at P in the direction from P to Q. (b) Find a unit vector in the direction in which f increases most rapidly at P, and find the rate of change of f in that direction. (c) Find a unit vector in the direction in which f decreases most rapidly at P, and find the rate of change of f in that direction.

25   $f(x, y) = x^2 e^{-2y}$; $P(2, 0)$, $Q(-3, 1)$

26   $f(x, y) = \sin(2x - y)$;

     $P(-\pi/3, \pi/6)$, $Q(0, 0)$

27   $f(x, y, z) = \sqrt{x^2 + y^2 + z^2}$;

     $P(-2, 3, 1)$, $Q(0, -5, 4)$

28 $f(x, y, z) = \dfrac{x}{y} - \dfrac{y}{z}$; $P(0, -1, 2)$, $Q(3, 1, -4)$

29 A metal plate is located in an xy-plane such that the temperature T at (x, y) is inversely proportional to the distance from the origin, and the temperature at $P(3, 4)$ is $100\,°\text{F}$.

  (a) Find the rate of change of T at P in the direction of $\mathbf{i} + \mathbf{j}$.

  (b) In what direction does T increase most rapidly at P?

  (c) In what direction does T decrease most rapidly at P?

  (d) In what direction is the rate of change 0?

30 The surface of a lake is represented by a region D in the xy-plane such that the depth (in feet) under the point (x, y) is $f(x, y) = 300 - 2x^2 - 3y^2$.

  (a) In what direction should a boat at $P(4, 9)$ sail in order for the depth of the water to decrease most rapidly?

  (b) In what direction does the depth remain the same?

31 The electrical potential V at (x, y, z) is

$$V = x^2 + 4y^2 + 9z^2.$$

  (a) Find the rate of change of V at $P(2, -1, 3)$ in the direction from P to the origin.

  (b) Find the direction that produces the maximum rate of change of V at P.

  (c) What is the maximum rate of change at P?

32 The temperature T at (x, y, z) is given by

$$T = 4x^2 - y^2 + 16z^2.$$

  (a) Find the rate of change of T at $P(4, -2, 1)$ in the direction of $2\mathbf{i} + 6\mathbf{j} - 3\mathbf{k}$.

  (b) In what direction does T increase most rapidly at P?

  (c) What is this maximum rate of change?

  (d) In what direction does T decrease most rapidly at P?

  (e) What is this rate of change?

Exer. 33–34: Refer to the discussion that precedes Example 6. In each case, T is the temperature at (x, y).

33 Shown in the figure is a semicircular region R.

  (a) Use polar coordinates to show that the upper boundary $\overset{\frown}{AB}$ is insulated if and only if $\partial T / \partial r = 0$. (*Hint:* Show that if $T = f(x, y)$, with $x = r\cos\theta$ and $y = r\sin\theta$, then $\partial T / \partial r = (\partial T / \partial x)\cos\theta + (\partial T / \partial y)\sin\theta$.)

  (b) Interpret $\partial T / \partial r$ as a rate of change of T.

Exercise 33

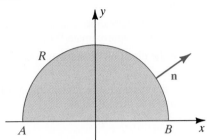

34 Shown in the figure is a circular sector whose boundary AB is insulated.

  (a) Use polar coordinates to show that the insulation condition is equivalent to $\partial T / \partial\theta = 0$ for every point on the segment AB.

  (b) Interpret $\partial T / \partial\theta$ as a rate of change of T.

Exercise 34

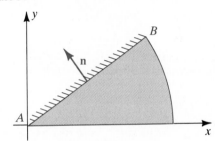

c **35** In some applications, it may be difficult to directly calculate the gradient $\nabla f(x, y)$. The following approximations are sometimes used for the components, where $h \approx 0$:

$$f_x(x, y) \approx \frac{f(x + h, y) - f(x - h, y)}{2h}$$

$$f_y(x, y) \approx \frac{f(x, y + h) - f(x, y - h)}{2h}$$

  (a) Show that these approximations improve as $h \to 0$.

  (b) If $f(x, y) = x^3 / (1 + y)$, approximate $\nabla f(1, 2)$ using $h = 0.01$ and compare the approximation with the exact result.

36 An analysis of the temperature of each component is crucial to the design of a computer chip. Suppose that for a chip to operate properly, the temperature of each component must not exceed $78\,°\text{F}$. If a component is likely to become too warm, engineers will usually place it in a cool portion of the chip. The designing of chips is aided by computer simulation in which temperature gradients are analyzed. A computer simulation for a new chip has resulted in the temperature grid (in °F) shown in the figure.

(a) If $T(x, y)$ is the temperature at (x, y), use Exercise 35 with $h = 1$ to approximate $\nabla T(3, 3)$.

(b) Estimate the direction of maximum heat transfer at $(3, 3)$.

(c) Estimate the instantaneous rate of change of T in the direction of $\mathbf{a} = -\mathbf{i} + 2\mathbf{j}$ at $(3, 3)$.

Exercise 36

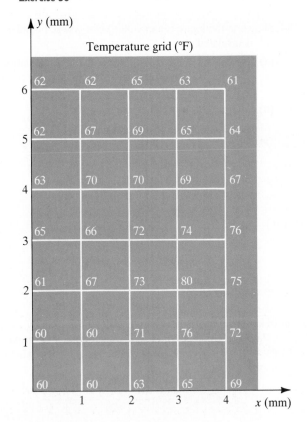

Exer. 37–38: Extend the approximation formulas in Exercise 35 to the first partial derivatives of $f(x, y, z)$, and then use $h = 0.01$ to approximate the directional derivative of f at $P_0(1, 1, 1)$ in the direction of **u**.

37 $f(x, y, z) = \dfrac{x^2 \sin y \tan z}{x^3 + y^2 z^4}$; $\quad \mathbf{u} = \dfrac{1}{\sqrt{3}}(\mathbf{i} + \mathbf{j} + \mathbf{k})$

38 $f(x, y, z) = \dfrac{x^4 + 3x^2 z^2}{4x^2 y^2 + \cosh(yz)}$; $\quad \mathbf{u} = \dfrac{1}{\sqrt{6}}(2\mathbf{i} - \mathbf{j} + \mathbf{k})$

Exer. 39–44: If $u = f(x, y)$, $v = g(x, y)$, and f and g are differentiable, prove the identity.

39 $\nabla(cu) = c \nabla u \quad$ for a constant c

40 $\nabla(u + v) = \nabla u + \nabla v$

41 $\nabla(uv) = u \nabla v + v \nabla u$

42 $\nabla\left(\dfrac{u}{v}\right) = \dfrac{v \nabla u - u \nabla v}{v^2} \quad$ with $v \neq 0$

43 $\nabla u^n = n u^{n-1} \nabla u \quad$ for every real number n

44 If $w = h(u)$, then $\nabla w = \dfrac{dw}{du} \nabla u$.

45 Let **u** be a unit vector and let θ be the angle, measured in the counterclockwise direction, from the positive x-axis to the position vector corresponding to **u**.

(a) Show that
$$D_{\mathbf{u}} f(x, y) = f_x(x, y) \cos \theta + f_y(x, y) \sin \theta.$$

(b) If $f(x, y) = x^2 + 2xy - y^2$ and $\theta = 5\pi/6$, find $D_{\mathbf{u}} f(2, -3)$.

46 Refer to Exercise 45. If $f(x, y) = (xy + y^2)^4$ and $\theta = \pi/3$, find $D_{\mathbf{u}}(f(2, -1))$.

47 Suppose that $w = f(x, y)$, $x = g(t)$, $y = h(t)$, and all functions are differentiable. If $\mathbf{r}(t) = x\mathbf{i} + y\mathbf{j}$, prove that
$$\frac{dw}{dt} = \nabla w \cdot \mathbf{r}'(t).$$

12.7 TANGENT PLANES AND NORMAL LINES

In this section, we use the gradient of a function to study tangent planes to surfaces in three-dimensional space and their associated normal lines. Suppose that a surface S is the graph of an equation $F(x, y, z) = 0$ and that F has continuous first partial derivatives. Let $P_0(x_0, y_0, z_0)$ be a point on S at which F_x, F_y, and F_z are not all zero. A **tangent line** to S at P_0 is,

Figure 12.58

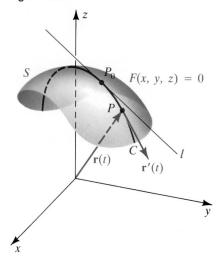

by definition, a tangent line l to any curve C that lies on S and contains P_0 (see Figure 12.58). If C has the parametrization

$$x = f(t), \quad y = g(t), \quad z = h(t)$$

for t in some interval I and if $\mathbf{r}(t)$ is the position vector of $P(x, y, z)$, then

$$\mathbf{r}(t) = f(t)\mathbf{i} + g(t)\mathbf{j} + h(t)\mathbf{k}.$$

Hence,

$$\mathbf{r}'(t) = f'(t)\mathbf{i} + g'(t)\mathbf{j} + h'(t)\mathbf{k}$$

is a tangent vector to C at $P(x, y, z)$, as indicated in Figure 12.58.

For each t, the point $(f(t), g(t), h(t))$ on C is also on S, and therefore

$$F(f(t), g(t), h(t)) = 0.$$

If we let

$$w = F(x, y, z), \quad \text{with} \quad x = f(t), \quad y = g(t), \quad z = h(t),$$

then using a chain rule and the fact that $w = 0$ for every t, we have

$$\frac{dw}{dt} = \frac{\partial w}{\partial x}\frac{dx}{dt} + \frac{\partial w}{\partial y}\frac{dy}{dt} + \frac{\partial w}{\partial z}\frac{dz}{dt} = 0.$$

Figure 12.59

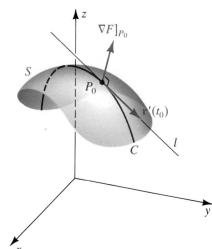

Thus, for every point $P(x, y, z)$ on C,

$$F_x(x, y, z)f'(t) + F_y(x, y, z)g'(t) + F_z(x, y, z)h'(t) = 0,$$

or, equivalently,

$$\nabla F(x, y, z) \cdot \mathbf{r}'(t) = 0.$$

In particular, if $P_0(x_0, y_0, z_0)$ corresponds to $t = t_0$, then

$$\nabla F]_{P_0} \cdot \mathbf{r}'(t_0) = \nabla F(x_0, y_0, z_0) \cdot \mathbf{r}'(t_0) = 0.$$

Since $\mathbf{r}'(t_0)$ is a tangent vector to C at P_0, this implies that *the vector* $\nabla F]_{P_0}$ *is orthogonal to every tangent line l to S at P_0* (see Figure 12.59).

The plane through P_0 with normal vector $\nabla F]_{P_0}$ is the **tangent plane to S at P_0.** We have shown that *every tangent line l to S at P_0 lies in the tangent plane at P_0.* The following theorem summarizes our discussion.

Theorem 12.33

Suppose that $F(x, y, z)$ has continuous first partial derivatives and that S is the graph of $F(x, y, z) = 0$. If P_0 is a point on S and if F_x, F_y, and F_z are not all 0 at P_0, then the vector $\nabla F]_{P_0}$ is normal to the tangent plane to S at P_0.

We shall refer to $\nabla F]_{P_0}$ in Theorem (12.33) as a vector that is *normal to the surface S at P_0.* Applying Theorem (10.36) gives us the following corollary, where we assume that F_x, F_y, and F_z are not all 0 at P_0.

Corollary 12.34

An equation for the tangent plane to the graph of $F(x, y, z) = 0$ at the point $P_0(x_0, y_0, z_0)$ is

$$F_x(x_0, y_0, z_0)(x - x_0) + F_y(x_0, y_0, z_0)(y - y_0)$$
$$+ F_z(x_0, y_0, z_0)(z - z_0) = 0.$$

EXAMPLE■1 Find an equation for the tangent plane to the ellipsoid $\frac{3}{4}x^2 + 3y^2 + z^2 = 12$ at the point $P_0(2, 1, \sqrt{6})$, and illustrate graphically.

SOLUTION To use Corollary (12.34), we first express the equation of the surface in the form $F(x, y, z) = 0$ by letting

$$F(x, y, z) = \tfrac{3}{4}x^2 + 3y^2 + z^2 - 12 = 0.$$

The partial derivatives of F are

$$F_x(x, y, z) = \tfrac{3}{2}x, \quad F_y(x, y, z) = 6y, \quad F_z(x, y, z) = 2z,$$

and hence, at $P_0(2, 1, \sqrt{6})$,

$$F_x(2, 1, \sqrt{6}) = 3, \quad F_y(2, 1, \sqrt{6}) = 6, \quad F_z(2, 1, \sqrt{6}) = 2\sqrt{6}.$$

We now apply Corollary (12.34), obtaining the equation

$$3(x - 2) + 6(y - 1) + 2\sqrt{6}(z - \sqrt{6}) = 0,$$

or, equivalently,

$$3x + 6y + 2\sqrt{6}z = 24.$$

The ellipsoid has x-, y-, and z-intercepts ± 4, ± 2, and $\pm\sqrt{12}$, respectively, and is sketched in Figure 12.60. A normal vector to the tangent plane at the point $P_0(2, 1, \sqrt{6})$ is

$$\nabla F]_{P_0} = \nabla F(2, 1, \sqrt{6}) = 3\mathbf{i} + 6\mathbf{j} + 2\sqrt{6}\mathbf{k},$$

as shown in the figure.

Figure 12.60

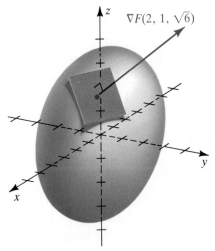

If $z = f(x, y)$ is an equation for a surface S and we let $F(x, y, z) = f(x, y) - z$, then the equation in Corollary (12.34) takes on the form

$$f_x(x_0, y_0)(x - x_0) + f_y(x_0, y_0)(y - y_0) + (-1)(z - z_0) = 0.$$

This gives us the following.

Theorem 12.35

An equation for the tangent plane to the graph of $z = f(x, y)$ at the point (x_0, y_0, z_0) is

$$z - z_0 = f_x(x_0, y_0)(x - x_0) + f_y(x_0, y_0)(y - y_0).$$

The line perpendicular to the tangent plane at a point $P_0(x_0, y_0, z_0)$ on a surface S is a **normal line** to S at P_0. If S is the graph of $F(x, y, z) = 0$, then the normal line is parallel to the vector $\nabla F(x_0, y_0, z_0)$.

EXAMPLE ▪ 2 Find an equation of the normal line to the ellipsoid $\frac{3}{4}x^2 + 3y^2 + z^2 = 12$ at the point $P_0(2, 1, \sqrt{6})$.

SOLUTION This surface is the same as that considered in Example 1. Consequently, the vector

$$\nabla F(2, 1, \sqrt{6}) = 3\mathbf{i} + 6\mathbf{j} + 2\sqrt{6}\mathbf{k},$$

shown in Figure 12.60, is parallel to the normal line. Using Theorem (10.34), we obtain the following parametric equations for the normal line:

$$x = 2 + 3t, \quad y = 1 + 6t, \quad z = \sqrt{6} + 2\sqrt{6}t; \quad t \text{ in } \mathbb{R}$$

EXAMPLE ▪ 3 Let P_0 be the point $(3, -4, 2)$ on the hyperboloid $16x^2 - 9y^2 + 36z^2 = 144$. Find equations for the tangent plane and the normal line at P_0.

SOLUTION The hyperboloid sketched in Figure 12.61 is the same as that sketched in Figure 10.70. Letting

$$F(x, y, z) = 16x^2 - 9y^2 + 36z^2 - 144,$$

we obtain

$$\nabla F(x, y, z) = 32x\mathbf{i} - 18y\mathbf{j} + 72z\mathbf{k}.$$

Figure 12.61 $16x^2 - 9y^2 + 36z^2 = 144$

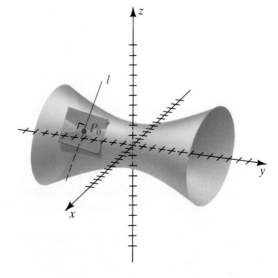

Hence, a normal vector to the hyperboloid at $P_0(3, -4, 2)$ is

$$\nabla F]_{P_0} = 96\mathbf{i} + 72\mathbf{j} + 144\mathbf{k}.$$

Any scalar multiple of $\nabla F]_{P_0}$ is also a normal vector. In particular, since

$$\nabla F]_{P_0} = 24(4\mathbf{i} + 3\mathbf{j} + 6\mathbf{k}),$$

we may use $\mathbf{a} = 4\mathbf{i} + 3\mathbf{j} + 6\mathbf{k}$. Thus, an equation for the tangent plane at $P_0(3, -4, 2)$ is

$$4(x - 3) + 3(y + 4) + 6(z - 2) = 0,$$

or, equivalently,

$$4x + 3y + 6z = 12.$$

This plane has x-, y-, and z-intercepts 3, 4, and 2, respectively (see Figure 12.61). Note that unlike the one in Example 1, the tangent plane at P_0 intersects the surface at many points.

Using $\mathbf{a}$ as a normal vector leads to the following parametric equations for the normal line l in the figure:

$$x = 3 + 4t, \quad y = -4 + 3t, \quad z = 2 + 6t; \quad t \text{ in } \mathbb{R}$$

Suppose that S is the graph of the equation $z = f(x, y)$. The tangent plane to S at $P_0(x_0, y_0, z_0)$ may be used to obtain a geometric interpretation for the differential

$$dz = f_x(x_0, y_0)\, \Delta x + f_y(x_0, y_0)\, \Delta y,$$

which was defined in (12.15). Since

$$\Delta z = f(x_0 + \Delta x, y_0 + \Delta y) - f(x_0, y_0),$$

Figure 12.62

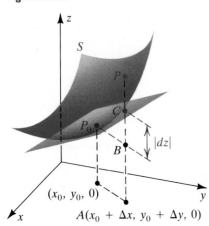

the point $P(x_0 + \Delta x, y_0 + \Delta y, z_0 + \Delta z)$ is on S (see Figure 12.62).

Let $C(x_0 + \Delta x, y_0 + \Delta y, z)$ be on the tangent plane, and consider the points $A(x_0 + \Delta x, y_0 + \Delta y, 0)$ and $B(x_0 + \Delta x, y_0 + \Delta y, z_0)$ shown in Figure 12.62. Since C is on the tangent plane, its coordinates satisfy the equation in Theorem (12.35)—that is,

$$z - z_0 = f_x(x_0, y_0)(x_0 + \Delta x - x_0) + f_y(x_0, y_0)(y_0 + \Delta y - y_0)$$
$$= f_x(x_0, y_0)\Delta x + f_y(x_0, y_0)\Delta y = dz.$$

In terms of distances, we have shown that $|dz| = \|\overrightarrow{BC}\|$; that is, $|dz|$ *is the distance from B to the point on the tangent plane that lies above (or below) B.* This is analogous to the single-variable case $y = f(x)$ discussed in Section 2.8, where we interpreted dy as a distance to a point on a tangent line to the graph of f (see Figure 2.41 and Definition (2.34)). It follows that when dz is used as an approximation to the change in z, we assume that the surface S almost coincides with its tangent plane at points close to P_0.

As an application of the discussion in this section, suppose a solid, liquid, or gas is situated in a three-dimensional coordinate system such that the temperature at the point $P(x, y, z)$ is $w = F(x, y, z)$. If $P_0(x_0, y_0, z_0)$

Figure 12.63

Isothermal surfaces

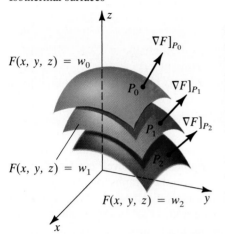

is a fixed point, then the graph of the equation

$$F(x, y, z) = F(x_0, y_0, z_0)$$

is the level (isothermal) surface that passes through P_0. For every point on this surface, the temperature is $w_0 = F(x_0, y_0, z_0)$. It is convenient to visualize all possible level surfaces that can be obtained by choosing different points. Several surfaces through P_0, P_1, and P_2 corresponding to temperatures w_0, w_1, and w_2 are sketched in Figure 12.63. By Theorem (12.33), $\nabla F]_{P_0}$, $\nabla F]_{P_1}$, and $\nabla F]_{P_2}$ are normal vectors to their corresponding surfaces at the points P_0, P_1, and P_2, respectively.

In Section 12.6, we saw that the maximum rate of change of $F(x, y, z)$ at $P_0(x_0, y_0, z_0)$ is in the direction of $\nabla F]_{P_0}$. From the preceding remarks, this maximum rate of change takes place in a direction that is normal to the level surface of F that contains P_0.

The following theorem summarizes these remarks.

Theorem 12.36

> Let a function F of three variables be differentiable at $P_0(x_0, y_0, z_0)$, and let S be the level surface of F containing P_0. If the gradient vector $\nabla F(x_0, y_0, z_0) \neq \mathbf{0}$, then it is normal to S at P_0. Thus, the direction for the maximum rate of change of $F(x, y, z)$ at P_0 is normal to S.

The preceding theorem is illustrated in Figure 12.63, where the maximum rate of change of temperature takes place in directions normal to the isothermal surfaces. Another illustration is found in electrical theory, where $F(x, y, z)$ is the potential at (x, y, z). If a particle moves on one of the level (equipotential) surfaces, the potential remains constant. The potential changes most rapidly when a particle moves in a direction that is normal to an equipotential surface.

EXAMPLE ■ 4   If $F(x, y, z) = x^2 + y^2 + z$, sketch the level surface of F that passes through $P(1, 2, 4)$ and sketch $\nabla F(1, 2, 4)$.

Figure 12.64

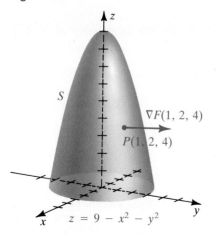

SOLUTION   The level surfaces of F are graphs of equations of the form $F(x, y, z) = k$, (k is constant). Since $F(1, 2, 4) = 1^2 + 2^2 + 4 = 9$, the level surface S that passes through $P(1, 2, 4)$ is the graph of the equation $F(x, y, z) = 9$—that is, of

$$x^2 + y^2 + z = 9, \quad \text{or} \quad z = 9 - x^2 - y^2.$$

The graph of this level surface, a paraboloid of revolution, is sketched in Figure 12.64.

The gradient of F is

$$\nabla F(x, y, z) = 2x\mathbf{i} + 2y\mathbf{j} + \mathbf{k}.$$

At $P(1, 2, 4)$,

$$\nabla F(1, 2, 4) = 2(1)\mathbf{i} + 2(2)\mathbf{j} + \mathbf{k} = 2\mathbf{i} + 4\mathbf{j} + \mathbf{k}.$$

The vector $\nabla F(1, 2, 4)$ is sketched in Figure 12.64. By Theorem (12.36), $\nabla F(1, 2, 4)$ is orthogonal to the level surface S at P.

The following result for functions of *two* variables is analogous to Theorem (12.36).

Theorem 12.37

Let a function f of two variables be differentiable at $P_0(x_0, y_0)$, and let C be the level curve of f that contains P_0. If $\nabla f(x_0, y_0) \neq \mathbf{0}$, then this gradient vector is orthogonal to C at P_0. Thus, the direction for the maximum rate of change of $f(x, y)$ at P_0 is orthogonal to C.

Figure 12.65

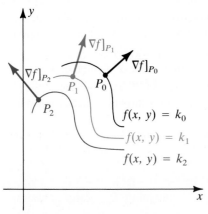

Theorem (12.37) is illustrated graphically in Figure 12.65 for an arbitrary function f of two variables. The level curves through $P_0(x_0, y_0)$, $P_1(x_1, y_1)$, and $P_2(x_2, y_2)$ are the graphs of $f(x, y) = k_0$, $f(x, y) = k_1$, and $f(x, y) = k_2$, respectively, where $k_0 = f(x_0, y_0)$, $k_1 = f(x_1, y_1)$, and $k_2 = f(x_2, y_2)$. By Theorem (12.37), the vectors $\nabla f]_{P_0}$, $\nabla f]_{P_1}$, and $\nabla f]_{P_2}$ are orthogonal to the level curves, as indicated in the figure. If each level curve C is an isobar on a weather map, then the barometric pressure does not change as a point moves along C. The vector $\nabla f]_P$ at a point P on an isobar points in the direction in which the barometric pressure changes most rapidly.

As an illustration of Theorem (12.37), if $f(x, y) = 9 - x^2 - y^2$, then the level curves are given by $9 - x^2 - y^2 = k$, where k is a real number (see Example 3 of Section 12.1). Some of these level curves (circles) are shown in Figure 12.66. By Theorem (12.37), the maximum (or minimum) rate of change of $f(x, y)$ occurs if (x, y) moves in a direction orthogonal to these circles and hence along lines through the origin. This movement corresponds to movement of the point $(x, y, f(x, y))$ up (or down) the steepest part of the graph of f sketched in Figure 12.64.

Figure 12.66

Level curves: $9 - x^2 - y^2 = k$

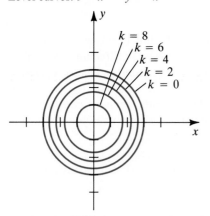

EXAMPLE • 5 Let $f(x, y) = x^2 + 2y^2$.

(a) Sketch the level curve C of f that passes through the point $P(3, 1)$, and sketch $\nabla f(3, 1)$.

(b) Discuss the significance of part (a) in terms of the graph of f.

SOLUTION

(a) Since $f(3, 1) = 3^2 + 2(1)^2 = 9 + 2 = 11$, the level curve of f that passes through the point $P(3, 1)$ is given by $f(x, y) = 11$—that is, by

$$x^2 + 2y^2 = 11.$$

The graph is the ellipse sketched in Figure 12.67. The gradient of f is

$$\nabla f(x, y) = 2x\mathbf{i} + 4y\mathbf{j},$$

and hence,

$$\nabla f(3, 1) = 2(3)\mathbf{i} + 4(1)\mathbf{j} = 6\mathbf{i} + 4\mathbf{j}.$$

The vector $\nabla f(3, 1)$ is shown in Figure 12.67. By Theorem (12.37), this vector is orthogonal to the level curve C.

Figure 12.67

Figure 12.68

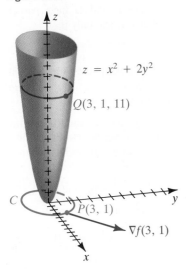

$z = x^2 + 2y^2$

$Q(3, 1, 11)$

C

$P(3, 1)$

$\nabla f(3, 1)$

(b) The graph of f (that is, of $z = x^2 + 2y^2$) is an elliptic paraboloid (see Figure 12.68). The level curve $x^2 + 2y^2 = 11$ in the xy-plane corresponds to the trace of the paraboloid on the plane $z = 11$. The maximum rate of change of $f(x, y)$ occurs if the point (x, y) in the xy-plane moves in the direction of $\nabla f(3, 1)$ at $P(3, 1)$. This movement corresponds to movement of the point $(x, y, f(x, y))$ up the steepest part of the paraboloid at $Q(3, 1, 11)$.

Tangent planes can be used to obtain successive approximations to a solution of a system of two nonlinear equations, $f(x, y) = 0$, $g(x, y) = 0$, from a first approximation (x_1, y_1) by performing the following steps.

Step 1 Use Theorem (12.35) to find equations of the tangent planes to the graphs of f and g at the points $(x_1, y_1, f(x_1, y_1))$ and $(x_1, y_1, g(x_1, y_1))$.

Step 2 Find the trace in the xy-plane of each tangent plane in step (1).

Step 3 Find the point of intersection (x_2, y_2) of the traces found in step (2).

Step 4 Take (x_2, y_2) as a second approximation and repeat steps (1)–(3).

This process is a geometric description of Newton's method, discussed at the end of Section 12.4.

EXERCISES 12.7

Exer. 1–10: Find equations for the tangent plane and the normal line to the graph of the equation at the point P.

1 $4x^2 - y^2 + 3z^2 = 10$; $P(2, -3, 1)$

2 $9x^2 - 4y^2 - 25z^2 = 40$; $P(4, 1, -2)$

3 $z = 4x^2 + 9y^2$; $P(-2, -1, 25)$

4 $z = 4x^2 - y^2$; $P(5, -8, 36)$

5 $xy + 2yz - xz^2 + 10 = 0$; $P(-5, 5, 1)$

6 $x^3 - 2xy + z^3 + 7y + 6 = 0$; $P(1, 4, -3)$

7 $z = 2e^{-x} \cos y$; $P(0, \pi/3, 1)$

8 $z = \ln xy$; $P(\frac{1}{2}, 2, 0)$

9 $x = \ln \dfrac{y}{2z}$; $P(0, 2, 1)$

10 $xyz - 4xz^3 + y^3 = 10$; $P(-1, 2, 1)$

Exer. 11–14: Sketch both the level curve C of f that contains P and $\nabla f]_P$.

11 $f(x, y) = y^2 - x^2$; $P(2, 1)$

12 $f(x, y) = 3x - 2y$; $P(-2, 1)$

13 $f(x, y) = x^2 - y$; $P(-3, 5)$

14 $f(x, y) = xy$; $P(3, 2)$

Exer. 15–20: Sketch both the level surface S of F that contains P and $\nabla F]_P$.

15 $F(x, y, z) = x^2 + y^2 + z^2$; $P(1, 5, 2)$

16 $F(x, y, z) = z - x^2 - y^2$; $P(2, -2, 1)$

17 $F(x, y, z) = x + 2y + 3z$; $P(3, 4, 1)$

18 $F(x, y, z) = x^2 + y^2 - z^2$; $P(3, -1, 1)$

19 $F(x, y, z) = x^2 + y^2$; $P(2, 0, 3)$

20 $F(x, y, z) = z$; $P(2, 3, 4)$

Exer. 21–24: Prove that an equation of the tangent plane to the given quadric surface at the point $P_0(x_0, y_0, z_0)$ may be written in the indicated form.

21 $\dfrac{x^2}{a^2} + \dfrac{y^2}{b^2} + \dfrac{z^2}{c^2} = 1$; $\dfrac{xx_0}{a^2} + \dfrac{yy_0}{b^2} + \dfrac{zz_0}{c^2} = 1$

22 $\dfrac{x^2}{a^2} - \dfrac{y^2}{b^2} + \dfrac{z^2}{c^2} = 1$; $\dfrac{xx_0}{a^2} - \dfrac{yy_0}{b^2} + \dfrac{zz_0}{c^2} = 1$

23 $\dfrac{x^2}{a^2} - \dfrac{y^2}{b^2} - \dfrac{z^2}{c^2} = 1;$ $\dfrac{xx_0}{a^2} - \dfrac{yy_0}{b^2} - \dfrac{zz_0}{c^2} = 1$

24 $\dfrac{x^2}{a^2} + \dfrac{y^2}{b^2} = cz;$ $\dfrac{2xx_0}{a^2} + \dfrac{2yy_0}{b^2} = c(z + z_0)$

25 Find the points on the hyperboloid of two sheets with equation $x^2 - 2y^2 - 4z^2 = 16$ at which the tangent plane is parallel to the plane $4x - 2y + 4z = 5$.

26 Show that the sum of the squares of the x-, y-, and z-intercepts of every tangent plane to the graph of the equation $x^{2/3} + y^{2/3} + z^{2/3} = a^{2/3}$ is the constant a^2.

27 Prove that every normal line to a sphere passes through the center of the sphere.

28 Find the points on the paraboloid $z = 4x^2 + 9y^2$ at which the normal line is parallel to the line through $P(-2, 4, 3)$ and $Q(5, -1, 2)$.

29 Two surfaces are said to be orthogonal at a point of intersection $P(x, y, z)$ if their normal lines at P

are orthogonal. Show that the graphs of $F(x, y, z) = 0$ and $G(x, y, z) = 0$ (where F and G have partial derivatives) are orthogonal at P if and only if

$$F_x G_x + F_y G_y + F_z G_z = 0.$$

30 Refer to Exercise 29. Prove that the sphere with equation $x^2 + y^2 + z^2 = a^2$ and the cone $x^2 + y^2 - z^2 = 0$ are orthogonal at every point of intersection.

|c| **Exer. 31–34: For the given system of equations and first approximate solution (x_1, y_1), use Newton's method as outlined at the end of the section to find two more approximations.**

31 $x^2 - y^3 = 0,$ $x^2 + y^2 - 3 = 0;$ $(1.3, 1.1)$

32 $\sin x - \cos y = 0,$ $x^2 - y^2 - 1.3 = 0;$ $(1.2, 0.3)$

33 $2e^{(-x^2 - y^2)} - 1 = 0,$ $4xy - x^4 - y^4 = 0;$ $(0.8, 0.2)$

34 $2 \sin x \sin y - 1 = 0,$ $y^2 - x^2 + 1 = 0;$ $(1.1, 0.65)$

12.8 EXTREMA OF FUNCTIONS OF SEVERAL VARIABLES

In Chapter 3, we discussed local and absolute extrema for functions of one variable. In this section, we extend these concepts to functions of several variables.

A function f of two variables has a **local maximum** at (a, b) if there is an open disk R containing (a, b) such that $f(x, y) \le f(a, b)$ for every (x, y) in R. The local maxima correspond to the high points on the graph S of f, as illustrated in Figure 12.69. Similarly, the function f has a **local minimum** at (c, d) if there is an open disk R containing (c, d) such that $f(x, y) \ge f(c, d)$ for every (x, y) in R. The local minima correspond to the low points on the graph of f, as illustrated in Figure 12.70.

Figure 12.69 Local maximum $f(a, b)$

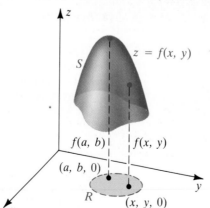

Figure 12.70 Local minimum $f(c, d)$

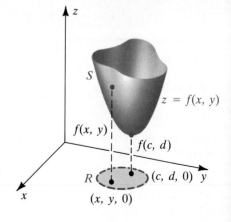

A region in the xy-plane is **bounded** if it is a subregion of a closed disk. If f is continuous on a closed and bounded region R, then f has a **maximum** $f(a, b)$ and **minimum** $f(c, d)$ for some (a, b) and (c, d) in R—that is,

$$f(c, d) \le f(x, y) \le f(a, b)$$

for every (x, y) in R. The proof may be found in texts on advanced calculus.

The local maxima and minima are the **local extrema** of f. The **extrema** also include the maximum and minimum (if they exist). If f has continuous first partial derivatives at (x_0, y_0) and if $f(x_0, y_0)$ is a local extremum of f, then the tangent plane to the graph of $z = f(x, y)$ at (x_0, y_0, z_0) is parallel to the xy-plane, and hence its equation is $z = z_0$. It follows from Theorem (12.35) that $f_x(x_0, y_0) = 0$ and $f_y(x_0, y_0) = 0$. Therefore, the pairs that yield local extrema must be solutions of *both*

$$f_x(x, y) = 0 \quad \text{and} \quad f_y(x, y) = 0.$$

As was the case for functions of one variable, local extrema can also occur at (x, y) if either $f_x(x, y)$ or $f_y(x, y)$ does not exist. All of these pairs are crucial for finding local extrema, so we shall give them a special name.

Definition 12.38

Let f be a function of two variables. A pair (a, b) is a **critical point** of f if either

(i) $f_x(a, b) = 0$ and $f_y(a, b) = 0$, or

(ii) $f_x(a, b)$ or $f_y(a, b)$ does not exist.

When searching for local extrema of a function, we usually begin by finding the critical points. We then test each pair in some way to determine if it yields a local maximum or minimum.

A maximum or minimum of a function of two variables may occur at a boundary point of its domain R. The investigation of such **boundary extrema** usually requires a separate procedure, as was the case with endpoint extrema for functions of one variable. If there are no boundary points—for example, when R is the entire xy-plane or an open disk—then there can be no boundary extrema.

Figure 12.71

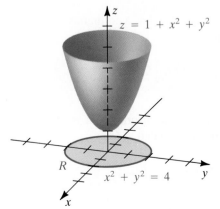

EXAMPLE 1 Let $f(x, y) = 1 + x^2 + y^2$, with $x^2 + y^2 \le 4$. Find the extrema of f.

SOLUTION The restriction $x^2 + y^2 \le 4$ corresponds to the closed disk R of radius 2 and center O in the xy-plane (see Figure 12.71). By Definition (12.38)(i), the critical points are solutions of the following system of two equations:

$$\begin{cases} f_x(x, y) = 0 \\ f_y(x, y) = 0 \end{cases} \quad \text{or} \quad \begin{cases} 2x = 0 \\ 2y = 0 \end{cases}$$

The only pair that satisfies both equations is $(0, 0)$; hence, $f(0, 0) = 1$ is the only possible local extremum. Moreover, because

$$f(x, y) = 1 + x^2 + y^2 > 1 \quad \text{if} \quad (x, y) \neq (0, 0),$$

f has the local minimum 1 at $(0, 0)$. This fact may also be seen from the graph of f, sketched in Figure 12.71. The function value $f(0, 0) = 1$ is also the minimum of f on R.

To find possible boundary extrema, we investigate points (a, b) that lie on the boundary of R. Referring to the figure, we see that any such point (for example, $(0, 2)$) leads to the maximum $f(0, 2) = 5$.

As illustrated in the next example, not every critical point may lead to an extremum.

EXAMPLE■2   If $f(x, y) = y^2 - x^2$ and the domain of f is $\mathbb{R}^2$, find the extrema of f.

Figure 12.72

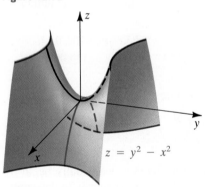

$z = y^2 - x^2$

SOLUTION   By Definition (12.38)(i), the critical points are solutions of the following system of two equations:

$$\begin{cases} f_x(x, y) = 0 \\ f_y(x, y) = 0 \end{cases} \quad \text{or} \quad \begin{cases} -2x = 0 \\ 2y = 0 \end{cases}$$

As in Example 1, the only possible local extremum is $f(0, 0) = 0$. However, if $y \neq 0$, then $f(0, y) = y^2 > 0$, and if $x \neq 0$, then $f(x, 0) = -x^2 < 0$. Thus, every open disk in the xy-plane containing $(0, 0)$ contains pairs at which function values are greater than $f(0, 0)$ and also pairs at which values are less than $f(0, 0)$. Consequently, f has no local extrema. This fact is also evident from the graph of f, a hyperbolic paraboloid, sketched in Figure 12.72.

To determine the extrema of a more complicated function f of two variables, it is convenient to use the following function D, called the *discriminant* of f.

Definition   12.39

Let f be a function of two variables that has continuous second partial derivatives. The **discriminant** D of f is given by

$$D(x, y) = f_{xx}(x, y) f_{yy}(x, y) - [f_{xy}(x, y)]^2.$$

One way to remember the formula for the discriminant is to use the following determinant, with all functions evaluated at (x, y):

$$D = \begin{vmatrix} f_{xx} & f_{xy} \\ f_{yx} & f_{yy} \end{vmatrix} = f_{xx} f_{yy} - (f_{xy})^2$$

Note that we have used the fact that $f_{yx} = f_{xy}$.

The following test is stated without proof. It is the counterpart of the second derivative test for functions of one variable.

Test for Local Extrema 12.40

Let f be a function of two variables that has continuous second partial derivatives throughout an open disk R containing (a, b). If $f_x(a, b) = f_y(a, b) = 0$ and $D(a, b) > 0$, then $f(a, b)$ is

    (i) a local maximum of f if $f_{xx}(a, b) < 0$
    (ii) a local minimum of f if $f_{xx}(a, b) > 0$

Note that if $D(a, b) > 0$, then $f_{xx}(a, b) f_{yy}(a, b) > [f_{xy}(a, b)]^2$ and hence $f_{xx}(a, b) f_{yy}(a, b)$ is positive—that is, $f_{xx}(a, b)$ and $f_{yy}(a, b)$ have the same sign. Thus, we may replace $f_{xx}(a, b)$ in (i) and (ii) of the preceding test by $f_{yy}(a, b)$.

A point $P(a, b, f(a, b))$ on the graph of f is a **saddle point**, or a *saddle point for* f, if $f_x(a, b) = f_y(a, b) = 0$ and if there is an open disk R containing (a, b) such that $f(x, y) > f(a, b)$ for some points (x, y) in R and $f(x, y) < f(a, b)$ for other points. Note that the graph of the hyperbolic paraboloid in Figure 12.72 has the saddle point $(0, 0, 0)$. The following theorem is proved in advanced calculus texts.

Theorem 12.41

Let f have continuous second partial derivatives throughout an open disk R containing (a, b). If $f_x(a, b) = f_y(a, b) = 0$ and $D(a, b)$ is negative, then $P(a, b, f(a, b))$ is a saddle point on the graph of f.

We *cannot* use (12.40) to determine whether $f(a, b)$ is a local extremum if $D(a, b) = 0$ or if either $f_x(a, b)$ or $f_y(a, b)$ does not exist. In these cases, we must analyze the variation of $f(x, y)$ near (a, b) by referring to the definition of $f(x, y)$ or the graph of f. To illustrate, if $f(x, y) = -(x^2 + y^{2/3})$, then the partial derivative $f_y(x, y) = -\frac{2}{3} y^{-1/3}$ does not exist if $y = 0$. Hence, *every* point $(a, 0)$ on the x-axis is a critical point of f. From the graph of f in Figure 12.17(a), it is evident that $f(0, 0) = 0$ is both a local maximum and a maximum of f in $\mathbb{R}^2$. We can show this algebraically by observing that if $(x, y) \neq (0, 0)$, then $f(x, y) < 0$. Since many such exceptions to (12.40) require complicated analyses, we will not consider examples or exercises of this type.

EXAMPLE ▪ 3 If $f(x, y) = x^2 - 4xy + y^3 + 4y$, find the local extrema and saddle points of f.

SOLUTION The first partial derivatives of f are

$$f_x(x, y) = 2x - 4y \quad \text{and} \quad f_y(x, y) = -4x + 3y^2 + 4.$$

Since f_x and f_y exist for every (x, y), the only critical points are the solutions of the following system of two equations in two unknowns:

$$\begin{cases} 2x - 4y = 0 \\ -4x + 3y^2 + 4 = 0 \end{cases}$$

Solving this system, we find that f has the two critical points $(4, 2)$ and $(\frac{4}{3}, \frac{2}{3})$. The second partial derivatives of f are

$$f_{xx}(x, y) = 2, \quad f_{xy}(x, y) = -4, \quad f_{yy}(x, y) = 6y.$$

Hence, the discriminant D is given by

$$D(x, y) = (2)(6y) - (-4)^2 = 12y - 16.$$

The remainder of the solution is arranged in the following table.

Critical point	Value of the discriminant	Value of f_{xx}	Conclusion
$(4, 2)$	$D(4, 2) = 8 > 0$	$f_{xx}(4, 2) = 2 > 0$	$f(4, 2) = 0$ is a local minimum by (12.40)(ii)
$(\frac{4}{3}, \frac{2}{3})$	$D(\frac{4}{3}, \frac{2}{3}) = -8 < 0$	irrelevant	$(\frac{4}{3}, \frac{2}{3}, f(\frac{4}{3}, \frac{2}{3}))$ is a saddle point by (12.41)

EXAMPLE ■ 4   If $f(x, y) = x^2 - 4xy + y^3 + 4y$, find the extrema of f on the triangular region R that has vertices $(-1, -1)$, $(7, -1)$, and $(7, 7)$.

SOLUTION   The boundary of R consists of the line segments C_1, C_2, and C_3, as shown in Figure 12.73. In Example 3, we found that f has a local minimum 0 at the point $(4, 2)$, which is within R. Thus, we need only check for boundary extrema.

On C_1 we have $y = -1$, so values of f are given by

$$f(x, -1) = x^2 + 4x - 1 - 4 = x^2 + 4x - 5.$$

This equation determines a function of one variable whose domain is the interval $[-1, 7]$. The first derivative is $2x + 4$, which equals 0 at $x = -2$, a number outside the interval $[-1, 7]$. Thus, there is no local extremum of $f(x, -1)$ on $[-1, 7]$. Since $f(x, -1)$ is increasing throughout this interval, we obtain the endpoint extrema

$$f(-1, -1) = -8 \quad \text{and} \quad f(7, -1) = 72.$$

On C_2 we have $x = 7$, and values of f are given by

$$f(7, y) = 49 - 28y + y^3 + 4y = y^3 - 24y + 49$$

for $-1 \le y \le 7$. The first derivative of this function of y is $3y^2 - 24$, and hence there is a critical number if $3y^2 = 24$, or $y = \sqrt{8} = 2\sqrt{2}$. Using the second derivative $6y$ of $f(7, y)$, we see that $f(7, 2\sqrt{2}) = 49 - 32\sqrt{2} \approx$

Figure 12.73

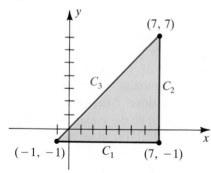

3.7 is a local minimum for f *on the segment* C_2. However, it is not a minimum for f on R, since $f(-1, -1) = -8 < 3.7$. The values of f at the endpoints of C_2 are

$$f(7, -1) = 72 \quad \text{and} \quad f(7, 7) = 224.$$

Finally, on C_3 we have $y = x$, and values of f are given by

$$f(x, x) = x^2 - 4x^2 + x^3 + 4x = x^3 - 3x^2 + 4x.$$

The first derivative $3x^2 - 6x + 4$ has no real roots, and hence there are no critical numbers for $f(x, x)$. We have already calculated the endpoint values $f(-1, -1) = -8$ and $f(7, 7) = 224$. These, therefore, are the minimum and maximum values for f on the triangular region R.

EXAMPLE ▪ 5 If $f(x, y) = \frac{1}{3}x^3 + \frac{4}{3}y^3 - x^2 - 3x - 4y - 3$, find the local extrema and saddle points of f.

SOLUTION The first partial derivatives of f are

$$f_x(x, y) = x^2 - 2x - 3 \quad \text{and} \quad f_y(x, y) = 4y^2 - 4.$$

Since f_x and f_y exist for every (x, y), the only critical points are the solutions of the following system:

$$\begin{cases} x^2 - 2x - 3 = 0 \\ 4y^2 - 4 = 0 \end{cases}$$

Solving this system, we obtain the four critical points $(3, 1)$, $(3, -1)$, $(-1, 1)$, and $(-1, -1)$. The second partial derivatives of f are

$$f_{xx}(x, y) = 2x - 2, \quad f_{xy}(x, y) = 0, \quad f_{yy}(x, y) = 8y.$$

Hence, the discriminant D is given by

$$D(x, y) = (2x - 2)(8y) - 0^2 = 16y(x - 1).$$

Let us arrange our work in the following table.

Critical point	Value of the discriminant	Value of f_{xx}	Conclusion
$(3, 1)$	$D(3, 1) = 32 > 0$	$f_{xx}(3, 1) = 4 > 0$	$f(3, 1) = -\frac{44}{3}$ is a local minimum by (12.40)(ii)
$(3, -1)$	$D(3, -1) = -32 < 0$	irrelevant	$(3, -1, f(3, -1))$ is a saddle point by (12.41)
$(-1, 1)$	$D(-1, 1) = -32 < 0$	irrelevant	$(-1, 1, f(-1, 1))$ is a saddle point by (12.41)
$(-1, -1)$	$D(-1, -1) = 32 > 0$	$f_{xx}(-1, -1) = -4 < 0$	$f(-1, -1) = \frac{4}{3}$ is a local maximum by (12.40)(i)

EXAMPLE ▪ 6   A rectangular box with no top is to be constructed to have a volume $V = 12$ ft^3. The cost per square foot of the material to be used is \$4 for the bottom, \$3 for two of the opposite sides, and \$2 for the remaining pair of opposite sides. Find the dimensions of the box that will minimize the cost.

Figure 12.74

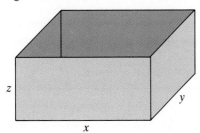

SOLUTION   As in Figure 12.74, let x and y denote the dimensions (in feet) of the base and z the altitude (in feet). Since the area of the base is xy and there are two sides of area xz and two sides of area yz, the cost C (in dollars) of the material is

$$C = 4xy + 3(2xz) + 2(2yz),$$

where x, y, and z are positive. Since $V = xyz = 12$, it follows that $z = 12/(xy)$. Substituting for z in the formula for C and simplifying, we obtain

$$C = 4xy + \frac{72}{y} + \frac{48}{x}.$$

There are no boundary points, since $x > 0$ and $y > 0$ for every (x, y). Hence there are no boundary extrema. To determine possible local extrema, we solve the following system of two equations:

$$C_x = 4y - \frac{48}{x^2} = 0, \qquad C_y = 4x - \frac{72}{y^2} = 0.$$

Equivalent equations are

$$y = \frac{12}{x^2} \quad \text{and} \quad xy^2 = 18.$$

Substituting $y = 12/x^2$ into the second equation give us

$$x\left(\frac{144}{x^4}\right) = 18, \quad \text{or} \quad 144 = 18x^3.$$

Hence $x^3 = 8$, or $x = 2$. Since $y = 12/x^2$, the corresponding value of y is $\frac{12}{4}$, or 3. Using Theorem (12.40), we can show that these values of x and y determine a minimum value of C. Finally, using $z = 12/(xy)$, we obtain $z = 12/(2 \cdot 3) = 2$. Thus, the minimum cost occurs if the dimensions of the base are 3 ft by 2 ft and the altitude is 2 ft. The longer sides should be made from the \$2 material and the shorter sides from the \$3 material.

EXAMPLE ▪ 7   In the design of modern computers, it is necessary to consider the speed of light (approximately 186,000 mi/sec) when positioning electrical components, so that they can communicate with one another as rapidly as possible. Suppose that three components are located in a coordinate plane at $P_1(x_1, y_1)$, $P_2(x_2, y_2)$, and $P_3(x_3, y_3)$ and that each angle of triangle $P_1 P_2 P_3$ is less than $120°$. A fourth component, which will communicate equally with the other three, is to be placed at a point $P(x, y)$ that will minimize the total delay time between the transmission and

Figure 12.75

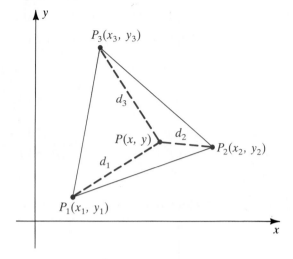

reception of signals (see Figure 12.75). If P is positioned properly, find the angles between the following pairs of vectors:

$$\overrightarrow{P_1P}, \overrightarrow{P_2P}; \qquad \overrightarrow{P_1P}, \overrightarrow{P_3P}; \qquad \overrightarrow{P_2P}, \overrightarrow{P_3P}$$

S O L U T I O N Let d_1, d_2, and d_3 be the distances between P and P_1, P and P_2, and P and P_3, respectively. Thus,

$$d_k = \sqrt{(x - x_k)^2 + (y - y_k)^2} \quad \text{for} \quad k = 1, 2, 3.$$

The minimal delay time between the transmission and the reception of signals will occur if (x, y) gives the minimal value of

$$f(x, y) = d_1 + d_2 + d_3.$$

To find the critical points of f, we consider the following system of equations:

$$\begin{cases} f_x(x, y) = \dfrac{x - x_1}{d_1} + \dfrac{x - x_2}{d_2} + \dfrac{x - x_3}{d_3} = 0 \\[2mm] f_y(x, y) = \dfrac{y - y_1}{d_1} + \dfrac{y - y_2}{d_2} + \dfrac{y - y_3}{d_3} = 0 \end{cases}$$

If we transpose the last term in each equation to the right-hand side, square each side, and add, we obtain

$$\left(\frac{x - x_1}{d_1} + \frac{x - x_2}{d_2}\right)^2 + \left(\frac{y - y_1}{d_1} + \frac{y - y_2}{d_2}\right)^2 = \frac{(x - x_3)^2 + (y - y_3)^2}{d_3^2} = 1.$$

You should verify that this simplifies to

$$(*) \qquad \frac{(x - x_1)(x - x_2) + (y - y_1)(y - y_2)}{d_1 d_2} = -\frac{1}{2}.$$

Let θ be the angle between the vectors

$$\overrightarrow{P_1 P} = (x - x_1)\mathbf{i} + (y - y_1)\mathbf{j} \quad \text{and} \quad \overrightarrow{P_2 P} = (x - x_2)\mathbf{i} + (y - y_2)\mathbf{j}.$$

Applying Corollary (10.20) and using (∗), we have

$$\cos \theta = \frac{\overrightarrow{P_1 P} \cdot \overrightarrow{P_2 P}}{\| \overrightarrow{P_1 P} \| \| \overrightarrow{P_2 P} \|} = -\frac{1}{2}.$$

Hence, $\theta = \arccos(-\frac{1}{2}) = 120°$. By symmetry, the angles between $\overrightarrow{P_1 P}$ and $\overrightarrow{P_3 P}$ and between $\overrightarrow{P_2 P}$ and $\overrightarrow{P_3 P}$ must also equal $120°$, as illustrated in Figure 12.76.

Figure 12.76

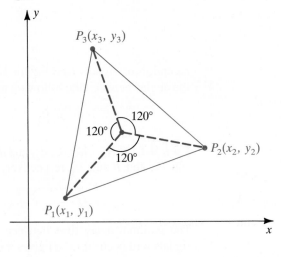

The discussion in this section can be generalized to functions of more than two variables. For example, given $f(x, y, z)$, we define local maxima and minima in a manner analogous to that used for the two-variable case. If f has first partial derivatives, then a local extremum can occur only at a point where f_x, f_y, and f_z are simultaneously 0. It is difficult to obtain tests for determining whether such a point corresponds to a maximum, a minimum, or neither. However, in applications we can often determine this by analyzing the physical nature of the problem.

EXERCISES 12.8

Exer. 1–20: Find the extrema and saddle points of f.

1 $f(x, y) = -x^2 - 4x - y^2 + 2y - 1$

2 $f(x, y) = x^2 - 2x + y^2 - 6y + 12$

3 $f(x, y) = x^2 + 4y^2 - x + 2y$

4 $f(x, y) = 5 + 4x - 2x^2 + 3y - y^2$

5 $f(x, y) = x^2 + 2xy + 3y^2$

6 $f(x, y) = x^2 - 3xy - y^2 + 2y - 6x$

7 $f(x, y) = x^3 + 3xy - y^3$

8 $f(x, y) = x^2 + xy$

9 $f(x, y) = \frac{1}{2}x^2 + 2xy - \frac{1}{2}y^2 + x - 8y$

10 $f(x, y) = -2x^2 - 2xy - \frac{3}{2}y^2 - 14x - 5y$

11 $f(x, y) = \frac{1}{3}x^3 - \frac{2}{3}y^3 + \frac{1}{2}x^2 - 6x + 32y + 4$

12 $f(x, y) = \frac{1}{3}x^3 + \frac{1}{3}y^3 - \frac{3}{2}x^2 - 4y$

13 $f(x, y) = \frac{1}{2}x^4 - 2x^3 + 4xy + y^2$

14 $f(x, y) = \frac{1}{3}x^3 + 4xy - 9x - y^2$

15 $f(x, y) = x^4 + y^3 + 32x - 9y$

16 $f(x, y) = -\frac{1}{3}x^3 + xy + \frac{1}{2}y^2 - 12y$

17 $f(x, y) = e^x \sin y$

18 $f(x, y) = x \sin y$

19 $f(x, y) = \dfrac{4y + x^2y^2 + 8x}{xy}$

20 $f(x, y) = \dfrac{x}{x + y}$

21 Shown in the figure is a graph of

$$f(x, y) = (x^2 + 3y^2)e^{-(x^2+y^2)}.$$

Show that there are five critical points, and find the extrema of f.

Exercise 21

22 Shown in the figure is a graph of

$$f(x, y) = xy^2 e^{-(x^2+y^2)/4}.$$

(a) Show that there is an infinite number of critical points.

(b) Find the coordinates of the four critical points shown in the figure.

Exercise 22

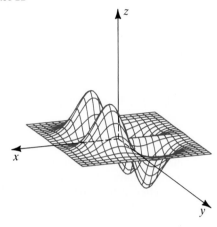

Exer. 23–28: Find the maximum and minimum values of f on R. (Refer to Exercises 3–8 for local extrema.)

23 $f(x, y) = x^2 + 4y^2 - x + 2y$;
the region R bounded by the ellipse $x^2 + 4y^2 = 1$

24 $f(x, y) = 5 + 4x - 2x^2 + 3y - y^2$;
the triangular region R bounded by the lines $y = x$, $y = -x$, and $y = 2$

25 $f(x, y) = x^2 + 2xy + 3y^2$;
$R = \{(x, y): -2 \le x \le 4, -1 \le y \le 3\}$

26 $f(x, y) = x^2 - 3xy - y^2 + 2y - 6x$;
$R = \{(x, y): |x| \le 3, |y| \le 2\}$

27 $f(x, y) = x^3 + 3xy - y^3$;
the triangular region R with vertices $(1, 2), (1, -2)$, and $(-1, -2)$

28 $f(x, y) = x^2 + xy$;
the region R bounded by the graphs of $y = x^2$ and $y = 9$

29 Find the shortest distance from the point $P(2, 1, -1)$ to the plane $4x - 3y + z = 5$.

30 Find the shortest distance between the parallel planes $2x + 3y - z = 2$ and $2x + 3y - z = 4$.

31 Find the points on the graph of $xy^3z^2 = 16$ that are closest to the origin.

32 Find three positive real numbers whose sum is 1000 and whose product is a maximum.

33 If an open rectangular box is to have a fixed volume V, what relative dimensions will make the surface area a minimum?

34 If an open rectangular box is to have a fixed surface area A, what relative dimensions will make the volume a maximum?

35 Find the dimensions of the rectangular box of maximum volume with faces parallel to the coordinate planes that can be inscribed in the ellipsoid

$$16x^2 + 4y^2 + 9z^2 = 144.$$

36 Generalize Exercise 35 to any ellipsoid

$$\frac{x^2}{a^2} + \frac{y^2}{b^2} + \frac{z^2}{c^2} = 1.$$

37 Find the dimensions of the rectangular box of maximum volume that has three of its faces in the coordinate planes, one vertex at the origin, and another vertex in the first octant on the plane $4x + 3y + z = 12$.

38 Generalize Exercise 37 to any plane

$$\frac{x}{a} + \frac{y}{b} + \frac{z}{c} = 1,$$

where a, b, and c are positive real numbers.

39 A company plans to manufacture closed rectangular boxes that have a volume of 8 ft^3. Find the dimensions that will minimize the cost if the material for the top and bottom costs twice as much as the material for the sides.

40 A window has the shape of a rectangle surmounted by an isosceles triangle, as illustrated in the figure. If the perimeter of the window is 12 ft, what values of x, y, and θ will maximize the total area?

Exercise 40

41 The U.S. Postal Service will not accept a rectangular box if the sum of its length and girth (the perimeter of a cross section that is perpendicular to the length) is more than 108 in. Find the dimensions of the box of maximum volume that can be mailed.

42 Find a vector in three dimensions having magnitude 8 such that the sum of its components is as large as possible.

Exer. 43–44: Refer to Example 7.

43 Three electrical components of a computer are located at $P_1(0, 0)$, $P_2(4, 0)$, and $P_3(0, 4)$. Locate the position of a fourth component so that the signal delay time is minimal.

44 Show that if triangle $P_1 P_2 P_3$ contains an angle that is greater than or equal to $120°$, then P cannot be located as in Figure 12.76.

45 In scientific experiments, corresponding values of two quantities x and y are often tabulated as follows:

x-values	x_1	x_2	$\cdots$	x_n
y-values	y_1	y_2	$\cdots$	y_n

Plotting the points (x_k, y_k) may lead the investigator to conjecture that x and y are related *linearly*—that is, $y = mx + b$ for some m and b. Thus, it is desirable to find a line l having that equation which best fits the data, as illustrated in the figure. Statisticians call l a *linear regression line*.

One technique for finding l is to use the *method of least squares*. To do so, consider, for each k, the *vertical deviation* $d_k = y_k - (mx_k + b)$ of the point (x_k, y_k) from the line $y = mx + b$ (see figure). Values of m and b are then determined that minimize the sum of the squares $\sum_{k=1}^{n} d_k^2$ (the squares d_k^2 are used because some

Exercise 45

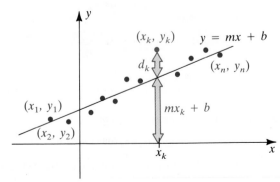

of the d_k may be negative). Substituting for d_k produces the following function f of m and b:

$$f(m, b) = \sum_{k=1}^{n} (y_k - mx_k - b)^2$$

Show that the line $y = mx + b$ of best fit occurs if

$$\left(\sum_{k=1}^{n} x_k\right) m + nb = \sum_{k=1}^{n} y_k$$

and

$$\left(\sum_{k=1}^{n} x_k^2\right) m + \left(\sum_{k=1}^{n} x_k\right) b = \sum_{k=1}^{n} x_k y_k.$$

Thus, the line can be found by solving this system of two equations for the two unknowns m and b.

46 Given the equations in Exercise 45, show that the sum $\sum_{k=1}^{n} d_k$ of the deviations is 0. (This means that the positive and negative deviations cancel one another, and it is one reason for using $\sum_{k=1}^{n} d_k^2$ in the method of least squares.)

Exer. 47–48: Use the method of least squares (see Exercise 45) to find a line $y = mx + b$ that best fits the given data.

47

x-values	1	4	7
y-values	3	5	6

48

x-values	1	4	6	8
y-values	1	3	2	4

c 49 The following table lists the relationship between semester averages and scores on the final examination for ten students in a mathematics class.

Semester average	40 55 62 68 72 76 80 86 90 94
Final examination	30 45 65 72 60 82 76 92 88 98

Fit these data to a line, and use the line to estimate the final examination grade of a student with an average of 70.

c 50 In studying the stress–strain diagram of an elastic material, an engineer finds that part of the curve appears to be linear. Experimental values are listed in the following table.

Stress (lb)	2	2.2	2.4	2.6	2.8	3.0
Strain (in.)	0.10	0.30	0.40	0.60	0.70	0.90

Fit these data to a straight line, and estimate the strain when the stress is 2.5 lb.

51 Shown in the figure are the relative positions of three towns, A, B, and C. City planners want to use the least-squares criterion to decide where to construct a new high school that will serve all three communities. They will construct the school about a point $P(x, y)$ at which the sum of the squares of the distances from towns A, B, and C is a minimum. Find the relative position of the construction site.

Exercise 51

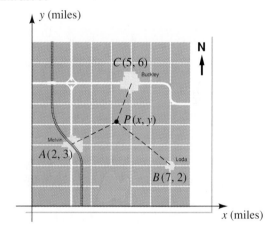

52 Generalize Exercise 51 to the case of n towns at positions $Q_1(x_1, y_1), Q_2(x_2, y_2), \ldots, Q_n(x_n, y_n)$.

53 Exercise 45 may be generalized to solve the problem of finding the plane $z = ax + by + c$ that best fits data $(x_1, y_1, z_1), (x_2, y_2, z_2), \ldots, (x_n, y_n, z_n)$. The method of least squares attempts to determine values a, b, and c such that

$$f(a, b, c) = \sum_{k=1}^{n} (z_k - ax_k - by_k - c)^2$$

is minimized.

(a) Find a system of three equations for the three unknowns a, b, and c.

(b) Find the plane that best fits the data $(0, 0, 0)$, $(0, 1, 0), (0, 0, 1)$, and $(1, 1, 2)$. (*Hint:* Solve the system of three equations $f_a = 0$, $f_b = 0$, $f_c = 0$.)

54 Three alleles (alternative forms of a gene) A, B, and O determine the four human blood types: A (AA or AO), B (BB or BO), O (OO), and AB. The *Hardy–Weinberg law* asserts that the proportion of individuals in a population who carry two different alleles is given by the formula

$$P = 2pq + 2pr + 2rq,$$

where $p, q,$ and r are the proportions of alleles A, B, and O, respectively, in the population. Show that P must be less than or equal to $\frac{2}{3}$. (*Hint:* $p \geq 0, q \geq 0, r \geq 0,$ and $p + q + r = 1$.)

c **Exer. 55–56: Estimate the critical points of f on $R = \{(x, y): |x| \leq 1.5 \text{ and } |y| \leq 1.5\}$ by graphing $f_x(x, y) = 0$ and $f_y(x, y) = 0$ on the same coordinate plane.**

55 $f(x, y) = x^3 \sin x - xy + 4y^2 + y$

56 $f(x, y) = xy - \arctan x - y^{5/4}$

12.9   LAGRANGE MULTIPLIERS

In many applications, we must find the extrema of a function f of several variables when the variables are restricted in some manner. In this section, we introduce the technique of *Lagrange multipliers** to locate such extrema. As an illustration, suppose we wish to find the volume of the largest rectangular box with faces parallel to the coordinate planes that can be inscribed in the ellipsoid $16x^2 + 4y^2 + 9z^2 = 144$. Note that, by symmetry, it is sufficient to examine the part in the first octant illustrated in Figure 12.77. If $P(x, y, z)$ is the vertex shown in the figure, then the volume V of the entire box is $V = 8xyz$. We must find the maximum value of V subject to the **constraint** (or **side condition**)

$$16x^2 + 4y^2 + 9z^2 - 144 = 0.$$

Figure 12.77

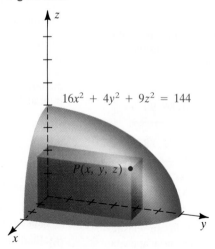

*This method was invented by the French mathematician Joseph-Louis Lagrange (see *Mathematicians and Their Times*, Chapter 7).

Figure 12.78

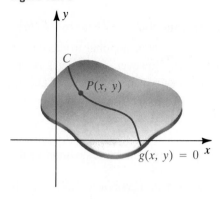

Solving this equation for z and substituting in the formula for V, we obtain

$$V = (8xy)\tfrac{1}{3}\sqrt{144 - 16x^2 - 4y^2}.$$

We may now find the extrema by (12.40); however, this method is cumbersome because of the manipulations involved in finding partial derivatives and critical points. Another disadvantage of this technique is that for similar problems it may be impossible to solve for z. For these reasons, it is sometimes simpler to employ the method of Lagrange multipliers discussed in this section.

As another illustration, let $f(x, y)$ be the temperature at the point $P(x, y)$ on the flat metal plate illustrated in Figure 12.78, and let C be a curve that has the equation $g(x, y) = 0$. Suppose we wish to find the points on C at which the temperature attains its largest or smallest value. This problem amounts to finding the extrema of $f(x, y)$ *subject to the constraint* $g(x, y) = 0$. One technique for solving this problem is given by the following definition.

Lagrange's Theorem   12.42

Suppose that f and g are functions of two variables having continuous first partial derivatives and that $\nabla g \neq \mathbf{0}$ throughout a region of the xy-plane. If f has an extremum $f(x_0, y_0)$ subject to the constraint $g(x, y) = 0$, then there is a real number λ such that

$$\nabla f(x_0, y_0) = \lambda \nabla g(x_0, y_0).$$

PROOF   The graph of $g(x, y) = 0$ is a curve C in the xy-plane. It can be shown that, under the given conditions, C has a smooth parametrization

$$x = h(t), \quad y = k(t)$$

for t in some interval I. Let

$$\mathbf{r}(t) = x\mathbf{i} + y\mathbf{j} = h(t)\mathbf{i} + k(t)\mathbf{j}$$

be the position vector of the point $P(x, y)$ on C (see Figure 12.79), and let the point $P_0(x_0, y_0)$ at which f has an extremum correspond to $t = t_0$. That is,

$$\mathbf{r}(t_0) = x_0\mathbf{i} + y_0\mathbf{j} = h(t_0)\mathbf{i} + k(t_0)\mathbf{j}.$$

If we define a function F of one variable t by

$$F(t) = f(h(t), k(t)),$$

Figure 12.79

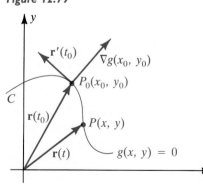

then, as t varies, we obtain function values $f(x, y)$ that correspond to (x, y) on C—that is, $f(x, y)$ is subject to the constraint $g(x, y) = 0$. Since $f(x_0, y_0)$ is an extremum of f under these conditions, it follows that $F(t_0) = f(h(t_0), k(t_0))$ is an extremum of $F(t)$. Thus, $F'(t_0) = 0$. If we regard F as a composite function, then, by a chain rule,

$$F'(t) = f_x(x, y)\frac{dx}{dt} + f_y(x, y)\frac{dy}{dt} = f_x(x, y)h'(t) + f_y(x, y)k'(t).$$

Letting $t = t_0$, we have

$$0 = F'(t_0) = f_x(x_0, y_0)h'(t_0) + f_y(x_0, y_0)k'(t_0) = \nabla f(x_0, y_0) \cdot \mathbf{r}'(t_0).$$

This result shows that the vector $\nabla f(x_0, y_0)$ is orthogonal to the tangent vector $\mathbf{r}'(t_0)$ to C (see Theorem (12.37)). However, $\nabla g(x_0, y_0)$ is also orthogonal to $\mathbf{r}'(t_0)$, because C is a level curve for g. Since $\nabla f(x_0, y_0)$ and $\nabla g(x_0, y_0)$ are orthogonal to the same vector, they are parallel—that is, $\nabla f(x_0, y_0) = \lambda \nabla g(x_0, y_0)$ for some λ. ∎

The number λ in the preceding theorem is called a **Lagrange multiplier**.

The equation $\nabla f(x_0, y_0) = \lambda \nabla g(x_0, y_0)$ in Theorem (12.42) may be written

$$f_x(x_0, y_0)\mathbf{i} + f_y(x_0, y_0)\mathbf{j} = \lambda g_x(x_0, y_0)\mathbf{i} + \lambda g_y(x_0, y_0)\mathbf{j}.$$

Equating the **i**- and **j**-components and using the constraint $g(x, y) = 0$ leads to the following non-vector restatement of Lagrange's theorem, in which it is assumed that f and g have continuous first partial derivatives and $g_x^2 + g_y^2 \neq 0$—that is, g_x and g_y are not simultaneously zero.

Corollary 12.43

The points at which a function f of two variables has relative extrema subject to the constraint $g(x, y) = 0$ are included among the points (x, y) determined by the first two coordinates of the solutions (x, y, λ) of the system of equations

$$\begin{cases} f_x(x, y) = \lambda g_x(x, y) \\ f_y(x, y) = \lambda g_y(x, y) \\ g(x, y) = 0 \end{cases}$$

When using Corollary (12.43), we first find all solutions

$$(x_1, y_1, \lambda_1), \quad (x_2, y_2, \lambda_2), \quad (x_3, y_3, \lambda_3), \quad \ldots$$

of the indicated system of equations. (Sometimes there is only *one* solution.) The points at which the relative extrema of f occur are among

$$(x_1, y_1), \quad (x_2, y_2), \quad (x_3, y_3), \quad \ldots.$$

Thus, the Lagrange multipliers $\lambda_1, \lambda_2, \lambda_3, \ldots$ are discarded after the solutions (x_k, y_k, λ_k) are found. Each point (x_k, y_k) is then examined to determine whether $f(x_k, y_k)$ is a relative maximum, a relative minimum, or neither. There is a complicated test (proved in advanced mathematics) for distinguishing between maxima and minima; however, it is usually easier to use graphical or physical considerations to make the distinction. In solutions to problems, we do not usually give rigorous proofs that our results yield extrema.

EXAMPLE ■ 1   Find the extrema of $f(x, y) = xy$ if (x, y) is restricted to the ellipse $4x^2 + y^2 = 4$.

SOLUTION   In this example, the constraint is given by $g(x, y) = 4x^2 + y^2 - 4 = 0$. Setting $\nabla f(x, y) = \lambda \nabla g(x, y)$ (see (12.42)), we obtain

$$y\mathbf{i} + x\mathbf{j} = \lambda(8x\mathbf{i} + 2y\mathbf{j}).$$

Thus, the system of equations in Corollary (12.43) is

$$\begin{cases} y = 8x\lambda \\ x = 2y\lambda \\ 4x^2 + y^2 - 4 = 0 \end{cases}$$

There are many ways to find the solutions of this system. One method is to let $y = 8x\lambda$ in $x = 2y\lambda$, obtaining

$$x = 2(8x\lambda)\lambda = 16x\lambda^2.$$

An equivalent equation is

$$x - 16x\lambda^2 = 0, \quad \text{or} \quad x(1 - 16\lambda^2) = 0.$$

Therefore, either $x = 0$ or $\lambda = \pm\frac{1}{4}$.

If $\lambda = \pm\frac{1}{4}$, then

$$y = 8x\lambda = 8x(\pm\tfrac{1}{4}), \quad \text{or} \quad y = \pm 2x.$$

Substituting for y in the equation $4x^2 + y^2 - 4 = 0$, we obtain

$$4x^2 + (\pm 2x)^2 - 4 = 0, \qquad 8x^2 = 4, \quad \text{or} \quad x = \pm\frac{1}{\sqrt{2}} = \pm\frac{\sqrt{2}}{2}.$$

The corresponding values of y are

$$y = \pm 2x = \pm 2\left(\pm\frac{\sqrt{2}}{2}\right) = \pm\sqrt{2}.$$

Discarding the Lagrange multipliers gives us the four points $(\sqrt{2}/2, \pm\sqrt{2})$ and $(-\sqrt{2}/2, \pm\sqrt{2})$.

If $x = 0$, then $y = 8x\lambda$ is also 0 but the point $(0, 0)$ does not lie on the ellipse $4x^2 + y^2 = 4$, so there is no solution to the system of equations in Corollary (12.43) when $x = 0$.

The value of f at each of the points we have found is listed in the following table.

(x, y)	$(\sqrt{2}/2, \sqrt{2})$	$(\sqrt{2}/2, -\sqrt{2})$	$(-\sqrt{2}/2, \sqrt{2})$	$(-\sqrt{2}/2, -\sqrt{2})$
$f(x, y)$	1	-1	-1	1

Thus, $f(x, y)$ takes on a maximum value of 1 at either $(\sqrt{2}/2, \sqrt{2})$ or $(-\sqrt{2}/2, -\sqrt{2})$ and a minimum value of -1 at $(\sqrt{2}/2, -\sqrt{2})$ or

$(-\sqrt{2}/2, \sqrt{2})$. These facts are indicated on the ellipse shown in Figure 12.80.

Figure 12.80

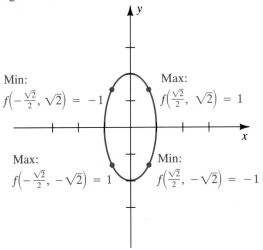

Min:
$f\left(-\frac{\sqrt{2}}{2}, \sqrt{2}\right) = -1$

Max:
$f\left(\frac{\sqrt{2}}{2}, \sqrt{2}\right) = 1$

Max:
$f\left(-\frac{\sqrt{2}}{2}, -\sqrt{2}\right) = 1$

Min:
$f\left(\frac{\sqrt{2}}{2}, -\sqrt{2}\right) = -1$

Lagrange's theorem can be extended to functions of more than two variables. The next result is one of many such extensions. We can state similar theorems for functions of any number of variables.

Extension of Lagrange's Theorem 12.44

Suppose that f and g are functions of three variables having continuous first partial derivatives and $\nabla g \neq \mathbf{0}$ throughout a region in an xyz-coordinate system. If f has an extremum $f(x_0, y_0, z_0)$ subject to the constraint $g(x, y, z) = 0$, then there is a real number λ such that

$$\nabla f(x_0, y_0, z_0) = \lambda \nabla g(x_0, y_0, z_0).$$

The following is an analogue of Corollary (12.43).

Corollary 12.45

The points at which a function f of three variables has relative extrema subject to the constraint $g(x, y, z) = 0$ are included among the points (x, y, z) determined by the first three coordinates of the solutions (x, y, z, λ) of the system of equations

$$\begin{cases} f_x(x, y, z) = \lambda g_x(x, y, z) \\ f_y(x, y, z) = \lambda g_y(x, y, z) \\ f_z(x, y, z) = \lambda g_z(x, y, z) \\ g(x, y, z) = 0 \end{cases}$$

The following example was discussed briefly at the beginning of this section.

EXAMPLE ∎ 2 Find the volume of the largest rectangular box with faces parallel to the coordinate planes that can be inscribed in the ellipsoid $16x^2 + 4y^2 + 9z^2 = 144$.

SOLUTION A typical position for the box is illustrated in Figure 12.77. We wish to maximize the volume

$$V = f(x, y, z)$$
$$= 8xyz$$

subject to the constraint

$$g(x, y, z) = 16x^2 + 4y^2 + 9z^2 - 144$$
$$= 0.$$

Setting $\nabla f(x, y, z) = \lambda \nabla g(x, y, z)$ (see (12.44)), we obtain

$$8yz\mathbf{i} + 8xz\mathbf{j} + 8xy\mathbf{k} = \lambda(32x\mathbf{i} + 8y\mathbf{j} + 18z\mathbf{k}).$$

This equation, together with the constraint $g(x, y, z) = 0$, gives us the following system of four equations (see Corollary (12.45)):

$$\begin{cases} 8yz = 32x\lambda \\ 8xz = 8y\lambda \\ 8xy = 18z\lambda \\ 16x^2 + 4y^2 + 9z^2 - 144 = 0 \end{cases}$$

Multiplying the first equation by x, the second by y, and the third by z and adding gives us

$$24xyz = 32x^2\lambda + 8y^2\lambda + 18z^2\lambda$$
$$= 2\lambda(16x^2 + 4y^2 + 9z^2).$$

This result, coupled with the constraint $16x^2 + 4y^2 + 9z^2 = 144$, implies that

$$24xyz = 2\lambda(144), \quad \text{or} \quad xyz = 12\lambda.$$

The last equation may be used to find x, y, and z. For example, multiplying both sides of the equation $8yz = 32x\lambda$ by x and using $xyz = 12\lambda$, we obtain

$$8xyz = 32x^2\lambda$$
$$8(12\lambda) = 32x^2\lambda$$
$$96\lambda - 32x^2\lambda = 0$$
$$32\lambda(3 - x^2) = 0.$$

Consequently, either $\lambda = 0$ or $x = \sqrt{3}$. We may reject $\lambda = 0$, since this leads to $xyz = 0$ and, hence, $V = 8xyz = 0$. Thus, the only possibility is $x = \sqrt{3}$.

Similarly, multiplying both sides of the equation $8xz = 8y\lambda$ by y leads to

$$8xyz = 8y^2\lambda$$
$$8(12\lambda) = 8y^2\lambda$$
$$96\lambda - 8y^2\lambda = 0$$
$$8\lambda(12 - y^2) = 0$$

and thus to $y = \sqrt{12} = 2\sqrt{3}$.

Finally, multiplying the equation $8xy = 18z\lambda$ by z and using the same technique, we obtain $z = 4/\sqrt{3}$. It follows that the desired volume is

$$V = 8xyz = 8(\sqrt{3})(2\sqrt{3})(4/\sqrt{3})$$
$$= 64\sqrt{3} \approx 111.$$

EXAMPLE = 3 If $f(x, y, z) = 4x^2 + y^2 + 5z^2$, find the point on the plane $2x + 3y + 4z = 12$ at which $f(x, y, z)$ has its least value.

SOLUTION We shall examine $f(x, y, z) = 4x^2 + y^2 + 5z^2$ subject to the constraint $g(x, y, z) = 2x + 3y + 4z - 12 = 0$. As in Example 2, we consider $\nabla f(x, y, z) = \lambda \nabla g(x, y, z)$ —that is,

$$\nabla(4x^2 + y^2 + 5z^2) = \lambda\nabla(2x + 3y + 4z - 12),$$

from which we obtain

$$8x\mathbf{i} + 2y\mathbf{j} + 10z\mathbf{k} = \lambda(2\mathbf{i} + 3\mathbf{j} + 4\mathbf{k}).$$

Equating components and using $g(x, y, z) = 0$ leads to the following system of equations (see Corollary (12.45)):

$$\begin{cases} 8x = 2\lambda \\ 2y = 3\lambda \\ 10z = 4\lambda \\ 2x + 3y + 4z - 12 = 0 \end{cases}$$

Solving the first three equations for λ gives us

$$\lambda = 4x = \tfrac{2}{3}y = \tfrac{5}{2}z.$$

These conditions imply that

$$y = 6x \quad \text{and} \quad z = \tfrac{8}{5}x.$$

Substituting into the constraint equation, $2x + 3y + 4z - 12 = 0$, we obtain

$$2x + 18x + \tfrac{32}{5}x - 12 = 0,$$

or $x = \tfrac{5}{11}$. Hence, $y = 6(\tfrac{5}{11}) = \tfrac{30}{11}$ and $z = (\tfrac{8}{5})(\tfrac{5}{11}) = \tfrac{8}{11}$. Since there is only one critical point, it follows that the minimum value occurs at that point, $(\tfrac{5}{11}, \tfrac{30}{11}, \tfrac{8}{11})$.

Some applications may involve more than one constraint. In particular, consider the problem of finding the extrema of $f(x, y, z)$ subject to the *two* constraints

$$g(x, y, z) = 0 \quad \text{and} \quad h(x, y, z) = 0.$$

If f has an extremum subject to these constraints, then the following condition must be satisfied for some real numbers λ and μ:

$$\nabla f(x, y, z) = \lambda \nabla g(x, y, z) + \mu \nabla h(x, y, z).$$

By equating components and using the constraints, we obtain a system of five equations in the five unknowns x, y, z, λ, and μ. A specific illustration of this situation is given in the next example. This method can also be extended to functions of more than three variables and to more than two constraints.

EXAMPLE ■ 4   Let C denote the first-octant arc of the curve in which the paraboloid $2z = 16 - x^2 - y^2$ and the plane $x + y = 4$ intersect. Find the points on C that are closest to and farthest from the origin. Find the minimum and maximum distances from the origin to C.

Figure 12.81

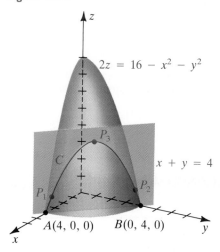

SOLUTION   The curve C is sketched in Figure 12.81. If $P(x, y, z)$ is an arbitrary point on C, then we wish to find the largest and smallest values of $d(O, P) = \sqrt{x^2 + y^2 + z^2}$. These values may be found by determining the triples (x, y, z) that give the extrema of the radicand

$$f(x, y, z) = x^2 + y^2 + z^2$$

subject to the two constraints

$$g(x, y, z) = x^2 + y^2 + 2z - 16 = 0$$
$$h(x, y, z) = x + y - 4 = 0.$$

As in the discussion preceding this example, consider

$$\nabla(x^2 + y^2 + z^2) = \lambda \nabla(x^2 + y^2 + 2z - 16) + \mu \nabla(x + y - 4).$$

Thus,

$$2x\mathbf{i} + 2y\mathbf{j} + 2z\mathbf{k} = \lambda(2x\mathbf{i} + 2y\mathbf{j} + 2\mathbf{k}) + \mu(\mathbf{i} + \mathbf{j})$$
$$= (2x\lambda + \mu)\mathbf{i} + (2y\lambda + \mu)\mathbf{j} + (2\lambda)\mathbf{k}.$$

Equating components and using the two constraints gives us the following system of five equations:

$$\begin{cases} 2x = 2x\lambda + \mu \\ 2y = 2y\lambda + \mu \\ 2z = 2\lambda \\ x^2 + y^2 + 2z - 16 = 0 \\ x + y - 4 = 0 \end{cases}$$

Subtracting the second equation from the first, we obtain the following equivalent equations:

$$2x - 2y = (2x\lambda + \mu) - (2y\lambda + \mu) = 2x\lambda - 2y\lambda$$
$$2(x - y) - 2\lambda(x - y) = 0$$
$$2(x - y)(1 - \lambda) = 0$$

Consequently, either $\lambda = 1$ or $x = y$.

If $\lambda = 1$, we have $2z = 2\lambda = 2(1)$, or $z = 1$. The first constraint—that is, $x^2 + y^2 + 2z - 16 = 0$—then gives us $x^2 + y^2 - 14 = 0$. Solving this equation simultaneously with $x + y - 4 = 0$, we find that either

$$x = 2 + \sqrt{3}, \quad y = 2 - \sqrt{3} \quad \text{or} \quad x = 2 - \sqrt{3}, \quad y = 2 + \sqrt{3}.$$

Thus, points on C that may lead to extrema are

$$P_1(2 + \sqrt{3}, 2 - \sqrt{3}, 1) \quad \text{and} \quad P_2(2 - \sqrt{3}, 2 + \sqrt{3}, 1).$$

The corresponding distances from O are

$$d(O, P_1) = \sqrt{15} = d(O, P_2).$$

If $x = y$, then, using the constraint $x + y - 4 = 0$, we obtain the equivalent equations $x + x - 4 = 0$, $2x = 4$, or $x = 2$. This gives us $P_3(2, 2, 4)$ and $d(O, P_3) = 2\sqrt{6}$.

Referring to Figure 12.81, we may now make the following observations. As a point moves continuously along C from $A(4, 0, 0)$ to $B(0, 4, 0)$, its distance from the origin starts at $d(O, A) = 4$, decreases to a minimum value $\sqrt{15}$ at P_1, and then increases to a maximum value $2\sqrt{6}$ at P_3. The distance then decreases to $\sqrt{15}$ at P_2 and again increases to 4 at B.

As a check on the solution, note that parametric equations for C are

$$x = 4 - t, \quad y = t, \quad z = 4t - t^2; \quad 0 \le t \le 4.$$

In this case,

$$f(x, y, z) = (4 - t)^2 + t^2 + (4t - t^2)^2,$$

and the extrema of f may be found using single-variable methods. It can be verified that the same points are obtained.

EXERCISES 12.9

Exer. 1–10: Use Lagrange multipliers to find the extrema of f subject to the stated constraints.

1 $f(x, y) = y^2 - 4xy + 4x^2$;
$$x^2 + y^2 = 1$$

2 $f(x, y) = 2x^2 + xy - y^2 + y$;
$$2x + 3y = 1$$

3 $f(x, y, z) = x + y + z$;
$$x^2 + y^2 + z^2 = 25$$

4 $f(x, y, z) = x^2 + y^2 + z^2$;
$$x + y + z = 25$$

5 $f(x, y, z) = x^2 + y^2 + z^2$;
$$x - y + z = 1$$

6 $f(x, y, z) = x + 2y - 3z;$
$$z = 4x^2 + y^2$$

7 $f(x, y, z) = x^2 + y^2 + z^2;$
$$x - y = 1, \qquad y^2 - z^2 = 1$$

8 $f(x, y, z) = z - x^2 - y^2;$
$$x + y + z = 1, \quad x^2 + y^2 = 4$$

9 $f(x, y, z, t) = xyzt;$
$$x - z = 2, \qquad y^2 + t = 4$$

10 $f(x, y, z, t) = x^2 + y^2 + z^2 + t^2;$
$$3x + 4y = 5, \qquad z + t = 2$$

11 Find the point on the sphere $x^2 + y^2 + z^2 = 9$ that is closest to the point $(2, 3, 4)$.

12 Find the point on the line of intersection of the planes $x + 3y - 2z = 11$ and $2x - y + z = 3$ that is closest to the origin.

13 A closed rectangular box having a volume of 2 ft^3 is to be constructed. If the cost per square foot of the material for the sides, bottom, and top is $1.00, $2.00, and $1.50, respectively, find the dimensions that will minimize the cost.

14 Prove that a closed rectangular box of fixed volume and minimal surface area is a cube.

15 Find the volume of the largest rectangular box that has three of its vertices on the positive x-, y-, and z-axes, respectively, and a fourth vertex on the plane $2x + 3y + 4z = 12$.

16 Find the dimensions of the rectangular box of maximum volume that has three of its faces in the coordinate planes, one vertex at the origin, and another vertex in the first octant on the plane $2x + 3y + 5z = 90$.

17 A container with a closed top and fixed surface area is to be constructed in the shape of a right circular cylinder. Find the relative dimensions that maximize the volume.

18 Find the dimensions of the rectangular box of maximum volume, with faces parallel to the coordinate planes, that can be inscribed in the ellipsoid $4x^2 + 4y^2 + z^2 = 36$.

19 Prove that the triangle of maximum area and fixed perimeter p is equilateral. (*Hint:* If the sides are x, y, z and if $s = \frac{1}{2}p$, then the area A is given by Heron's formula, $A = \sqrt{s(s-x)(s-y)(s-z)}$.)

20 Prove that the product of the sines of the angles of a triangle is greatest when the triangle is equilateral.

21 The strength of a rectangular beam varies as the product of its width and the square of its depth. Find the dimensions of the strongest rectangular beam that can be cut from a cylindrical log whose cross sections are elliptical with major and minor axes of lengths 24 in. and 16 in., respectively.

22 If x units of capital and y units of labor are required to manufacture $f(x, y)$ units of a certain commodity, the *Cobb–Douglas production function* is defined by $f(x, y) = kx^a y^b$, where k is a constant and a and b are positive numbers such that $a + b = 1$. Suppose that $f(x, y) = x^{1/5} y^{4/5}$ and that each unit of capital costs C dollars and each unit of labor costs L dollars. If the total amount available for these costs is M dollars, so $xC + yL = M$, how many units of capital and labor will maximize production?

c **Exer. 23–24: Use Lagrange multipliers and graphs to estimate the extrema of $f(x, y)$ subject to the constraint $g(x, y) = 0$.**

23 $f(x, y) = y - \cos x + 2x; \quad g(x, y) = x^2 + 2y^2 - 1$

24 $f(x, y) = \frac{1}{5}x^5 + \frac{1}{3}y^3; \qquad g(x, y) = x^2 + y^2 - 1$

CHAPTER 12 REVIEW EXERCISES

Exer. 1–4: Describe the domain of f and the level curve or surface through P.

1 $f(x, y) = \sqrt{36 - 4x^2 + 9y^2};$ $P(3, 4)$

2 $f(x, y) = \ln xy;$ $P(2, 3)$

3 $f(x, y, z) = (z^2 - x^2 - y^2)^{-3/2}; \quad P(0, 0, 1)$

4 $f(x, y, z) = \dfrac{\sec z}{x - y}; \quad P(5, 3, 0)$

Exer. 5–8: Find the limit or show that it does not exist.

5 $\lim\limits_{(x,\,y)\to(0,\,0)} \dfrac{3xy+5}{y^2+4}$

6 $\lim\limits_{(x,\,y,\,z)\to(0,\,0,\,3)} \dfrac{2x^2z+2y^2z}{x^2+y^2}$

7 $\lim\limits_{(x,\,y)\to(0,\,0)} \left(\dfrac{x^2-y^2}{x^2+y^2}\right)^2$

8 $\lim\limits_{(x,\,y)\to(0,\,0)} \dfrac{x^2y^2}{x^4+2y^4}$

9 Use polar coordinates to help describe the level curves of

$$f(x,\,y) = \dfrac{xy}{(x^2+y^2)^{3/2}},$$

and investigate $\lim\limits_{(x,\,y)\to(0,\,0)} f(x,\,y)$.

10 The graph of $f(x,\,y) = x^2y^2/(x^2+y^2)^2$ shown in the figure indicates that f has first partial derivatives at $(0,\,0)$ but is not continuous there. Assume $f(0,\,0)=0$ and verify each of these facts.

Exercise 10

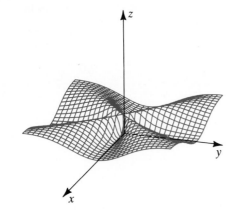

Exer. 11–16: Find the first partial derivatives of f.

11 $f(x,\,y) = x^3\cos y - y^2 + 4x$

12 $f(r,\,s) = r^2 e^{rs}$

13 $f(x,\,y,\,z) = \dfrac{x^2+y^2}{y^2+z^2}$

14 $f(u,\,v,\,t) = u\ln\dfrac{v}{t}$

15 $f(x,\,y,\,z,\,t) = x^2z\sqrt{2y+t}$

16 $f(v,\,w) = v^2\cos w + w^2\cos v$

Exer. 17–18: Find the second partial derivatives of f.

17 $f(x,\,y) = x^3y^2 - 3xy^3 + x^4 - 3y + 2$

18 $f(x,\,y,\,z) = x^2 e^{y^2-z^2}$

19 If $u = (x^2+y^2+z^2)^{-1/2}$, prove that

$$\dfrac{\partial^2 u}{\partial x^2} + \dfrac{\partial^2 u}{\partial y^2} + \dfrac{\partial^2 u}{\partial z^2} = 0.$$

20 Find dw if

(a) $w = y^3\tan^{-1}x^2 + 2x - y$

(b) $w = x^2\sin yz$

21 (a) Find Δw and dw if $w = x^2 + 3xy - y^2$.

(b) If $(x,\,y)$ changes from $(-1,\,2)$ to $(-1.1,\,2.1)$, use Δw to find the exact change in w, and use dw to find an approximate change in w.

22 Suppose that when Ohm's law, $R = V/I$, is used, there are percentage errors of $\pm3\%$ and $\pm2\%$ in the measurements of V and I, respectively. Use differentials to estimate the maximum percentage error in the calculated value of R.

Exer. 23–24: Prove that f is differentiable throughout its domain.

23 $f(x,\,y) = \dfrac{xy}{y^2-x^2}$ 24 $f(x,\,y,\,z) = \dfrac{xyz}{x-y}$

Exer. 25–27: Use a chain rule.

25 Find $\partial s/\partial x$ and $\partial s/\partial y$ if $s = uv + vw - uw, u = 2x + 3y, v = 4x - y$, and $w = -x + 2y$.

26 Find $\partial z/\partial r, \partial z/\partial s$, and $\partial z/\partial t$ if $z = ye^x, x = r + st$, and $y = 2r + 3s - t$.

27 Find dw/dt if $w = x\sin yz, x = 3e^{-t}, y = t^2$, and $z = 3t$.

28 Use partial derivatives to find dy/dx if $y = f(x)$ is determined implicitly by

$$x^3 - 4xy^3 - 3y + x - 2 = 0.$$

29 Find $\partial z/\partial x$ and $\partial z/\partial y$ if $z = f(x,\,y)$ is determined implicitly by $x^2y + z\cos y - xz^3 = 0$.

30 (a) Find the directional derivative of the function $f(x,\,y) = 3x^2 - y^2 + 5xy$ at $P(2,\,-1)$ in the direction of $\mathbf{a} = -3\mathbf{i} - 4\mathbf{j}$.

(b) Find the maximum rate of increase of $f(x,\,y)$ at P.

31 Suppose that the temperature at the point $(x,\,y,\,z)$ is given by $T(x,\,y,\,z) = 3x^2 + 2y^2 - 4z$.

(a) Find the instantaneous rate of change of T at the point $P(-1, -3, 2)$ in the direction from P to the point $Q(-4, 1, -2)$.

(b) Find the maximum rate of change of T at P.

32 An *urban density model* is a formula that relates the population density (in number of people per square mile) to the distance from the center of the city. Suppose a rectangular coordinate system is introduced with the center of the city at the origin, and let $Q(x, y)$ denote the population density near the point $P(x, y)$. Decide the direction in which the population density increases most rapidly at $P(x, y)$ for the given density function Q, where a, b, and c are positive constants.

(a) $Q(x, y) = ae^{-b\sqrt{x^2+y^2}}$

(b) $Q(x, y) = ae^{-b\sqrt{x^2+y^2}+c(x^2+y^2)}$

33 Find equations of the tangent plane and the normal line to the graph of $7z = 4x^2 - 2y^2$ at $P(-2, -1, 2)$.

34 A curve C has parametrization $x = t$, $y = t^2$, $z = t^3$; t in $\mathbb{R}$. If $f(x, y, z) = y^2 + xz$ and $\mathbf{u}$ is a unit tangent vector to C at $P(2, 4, 8)$, find $D_{\mathbf{u}} f(2, 4, 8)$.

35 Show that every plane tangent to the following cone passes through the origin:

$$\frac{x^2}{a^2} - \frac{y^2}{b^2} + \frac{z^2}{c^2} = 0$$

36 If $f(x, y) = \frac{1}{4}x^2 + \frac{1}{25}y^2$, sketch the level curve of f that contains $P(0, 5)$, and sketch $\nabla f]_P$.

37 If $F(x, y, z) = z + 4x^2 + 9y^2$, sketch the level surface of F that contains $P(1, 0, 0)$, and sketch $\nabla F]_P$.

Exer. 38–39: Find the extrema of f.

38 $f(x, y) = \dfrac{4}{x} + \dfrac{2}{y} - xy$

39 $f(x, y) = x^2 + 3y - y^3$

40 The material for the bottom of a rectangular box costs twice as much per square inch as the material for the sides and top. If the volume is fixed, find the relative dimensions that minimize the cost.

Exer. 41–42: Use Lagrange multipliers to find the extrema of f subject to the stated constraints.

41 $f(x, y, z) = xyz$;
$$x^2 + 4y^2 + 2z^2 = 8$$

42 $f(x, y, z) = 4x^2 + y^2 + z^2$;
$$2x - y + z = 4, \quad x + 2y - z = 1$$

43 Find the point on the paraboloid

$$z = \frac{x^2}{25} + \frac{y^2}{4}$$

that is closest to the point $(0, 5, 0)$.

44 A grain elevator hopper has the shape of a right circular cone of radius 2 ft surmounted by a right circular cylinder, as shown in the figure. If the total volume is 100 ft^3, find the altitudes h and k of the cylinder and cone, respectively, that will minimize the curved surface area.

Exercise 44

2 ft

h

k

45 A *pipeline processor* in a supercomputer is like an assembly line designed to compute a large number of similar arithmetic operations in stages. Once the pipeline is full, it will compute one arithmetic operation in each clock period. Pipelines usually have less than 10 stages. The *performance–cost ratio* P for a pipeline processor is given by

$$P = \frac{s}{(T + sk_0)(C + sk_1)},$$

where s is the number of stages, T is the time required to process the operation on a nonpipeline computer, C is the total cost of the pipeline, and k_0 and k_1 are constants that depend on the hardware used. For the optimal design $\partial P / \partial s = 0$. Determine the number of stages in the optimal design.

46 A thin slab of metal having the shape of a square lamina with vertices on a coordinate plane at points $(0, 0)$, $(1, 0)$, $(1, 1)$, and $(0, 1)$ has each of its four edges inserted in ice at $0\,°\text{C}$. The initial temperature T at $P(x, y)$ is given by

$$T = 20 \sin(\pi x) \sin(\pi y).$$

(a) Find the point at which the initial temperature is greatest.

(b) If $U(x, y, t)$ is the temperature at $P(x, y)$ at time t, then it can be shown that U satisfies the two-dimensional heat equation

$$\frac{\partial U}{\partial t} - k\left(\frac{\partial^2 U}{\partial x^2} + \frac{\partial^2 U}{\partial y^2}\right) = 0,$$

where k is a constant that depends on the thermal conductivity and specific heat of the slab. Show that the following is a solution of this heat equation:

$$U(x, y, t) = 20e^{-2k\pi^2 t} \sin(\pi x) \sin(\pi y)$$

(c) Describe the temperature of the slab over a long period of time.

$\boxed{c}$ Exer. 47–50: Plot the graph of f using the indicated viewing box. Rotate the graph to select a "good" viewpoint to save and/or print.

47 $f(x, y) = \dfrac{1}{\sqrt{1 + 3x^2 + 5y^2}}$;

$-4 \le x \le 4, \quad -3 \le y \le 3, \quad 0 \le z \le 1$

48 $f(x, y) = |2x|\,|y|$;

$-2 \le x \le 2, \quad -3 \le y \le 3, \quad 0 \le z \le 12$

49 $f(x, y) = \cos(x - y) - \cos(x + y)$;

$-3 \le x \le 3, \quad -3 \le y \le 3, \quad -2 \le z \le 2$

50 $f(x, y) = \cos(xy)$;

$-\pi \le x \le \pi, \quad -\pi \le y \le \pi, \quad -1 \le z \le 1$

$\boxed{c}$ Exer. 51–52: Plot level curves of f.

51 $f(x, y) = \dfrac{2x}{y^2 - x}$; $\quad -3 \le x \le 3, \quad -3 \le y \le 3$

52 $f(x, y) = x + \sqrt{x + y}$; $\quad 0 \le x \le 5, \quad 0 \le y \le 5$

$\boxed{c}$ Exer. 53–54: (a) Use (12.24) to estimate the directional derivative of f at P in the indicated direction with $s = 0.02, 0.01,$ and 0.005. (b) Find the directional derivative of f at P in the indicated direction.

53 $f(x, y) = xy^2 \sin(x + y)$;

$$P(2, 1), \qquad \mathbf{u} = \frac{2}{\sqrt{5}}\mathbf{i} + \frac{1}{\sqrt{5}}\mathbf{j}$$

54 $f(x, y) = e^{(1+x^2-y^2)}$;

$$P(1.5, 2.7), \quad \mathbf{a} = -0.53\mathbf{i} + 0.41\mathbf{j}$$

$\boxed{c}$ Exer. 55–56: For the given system of equations and first approximate solution (x_1, y_1), use Newton's method to find two more approximations.

55 $x^2 + 3y^3 - 70 = 0, \quad \ln(x + y) - 2 = 0; \quad (4, 4)$

56 $e^{(-2x+y^2)} - 5 = 0, \quad x^2 - y^2 - 2 = 0; \quad (3.1, 2.8)$

57 For spacecraft rotation calculations, there are occasions when space scientists must work with the following functions:

$$F(\psi, \theta, \varphi) = \cos\psi \sin\theta \cos\varphi + \sin\psi \cos\theta \sin\varphi$$
$$G(\psi, \theta, \varphi) = \cos\psi \cos\theta \sin\varphi + \sin\psi \sin\theta \cos\varphi$$
$$H(\psi, \theta, \varphi) = \cos\psi \sin\theta \sin\varphi + \sin\psi \cos\theta \cos\varphi$$
$$K(\psi, \theta, \varphi) = \cos\psi \cos\theta \cos\varphi - \sin\psi \sin\theta \sin\varphi$$

for very small angles $\psi, \theta,$ and φ. Use differentials to find formulas that approximate the values of each of these functions near $\psi = 0, \theta = 0, \varphi = 0$.

EXTENDED PROBLEMS AND GROUP PROJECTS

1 A region R of the plane is **bounded** if there is some open disk in the plane that contains all the points of R. A region R is **compact** if it is both closed and bounded.

(a) Show that the x-axis is closed but not bounded.

(b) Show that the set of all points in the plane less than 1 unit from the origin is bounded but not closed.

(c) Find an example of a region that is neither bounded nor closed.

(d) Give some examples of regions that are both closed and bounded.

(e) Investigate the validity of the following analogue of the extreme value theorem (3.3): If a function $f(x, y)$ is continuous on a compact region R, then f takes on a minimum value and a maximum value at least once in R.

2 Let R_1 be the region made up of all the points that are inside the circle of radius 1 centered at the origin, and let R_2 be the region of all the points that are outside the circle of radius 2 centered at the origin. Let R be the region made up of R_1 and R_2. Which points of

the plane are not included in R? Define a function f with domain R by the rules $f(x, y) = 1$ if (x, y) is in R_1 and $f(x, y) = 2$ if (x, y) is in R_2. Show that f is differentiable at all points of R and that $\nabla f(x, y) = 0$ for every point in R, but that f is not a constant function. Thus, the direct generalization of Corollary (3.13) to functions of two variables fails.

A **polygonally connected set** R has the property that any two points in it can be joined by a finite sequence of line segments, each of which lies entirely in R. Show that the set R described above is not polygonally connected. Draw some sketches of polygonally connected regions in the plane. Prove the following analogue of Corollary (3.13): If f is differentiable on a polygonally connected open region R and $\nabla f(x, y) = 0$ for every (x, y) in R, then f is constant on R.

3 In what ways is the test for local extrema (12.40) analogous to the second derivative test? Show that the discriminant of a function $f(x, y)$ of two variables is the determinant of the **Hessian matrix**

$$\begin{bmatrix} f_{xx} & f_{xy} \\ f_{yx} & f_{yy} \end{bmatrix}$$

first studied by Ludwig Otto Hesse (1811–1874). What would the Hessian matrix of a function of three variables $f(x, y, z)$ look like? Could its determinant be used to find a test for local extrema of a function of three variables? Develop a complete and careful proof of (12.40).

INTRODUCTION

W **E NOW CONTINUE OUR STUDY** of the calculus of real-valued functions of several variables by examining the concept of integration. We will define *multiple integrals* using functions of several variables and limits of sums analogous to those used to define the definite integral $\int_a^b f(x)\,dx$ in Chapter 4. The principal difference is that rather than begin with a partition of the interval $[a, b]$, we subdivide a region R of the xy-plane or an xyz-coordinate system. Evaluation theorems are based on equations of curves and surfaces and hence are more complicated than the fundamental theorem of calculus for definite integrals.

We begin with functions of two variables in Section 13.1 and see how the idea of Riemann sums can be extended and used to define a double integral. We then discuss how to evaluate double integrals by iterated definite integrals. In Sections 13.2 and 13.4, we apply these ideas to find areas in the plane and volumes and surface areas in three-dimensional space. Since the points in the plane can be described in terms of rectangular (x, y) coordinates or polar (r, θ) coordinates, double integrals can also be developed in polar coordinates, which we discuss in Section 13.3. We consider the generalization of double integrals to triple integrals for real-valued functions of three variables in Section 13.5.

Multiple integrals enable us to consider applications involving *nonhomogeneous* solids and surfaces that are more general than those obtained by revolving a plane region or curve about a line. In particular, in Section 13.6, we examine ways in which we may find moments and centers of gravity of irregularly shaped solids in which the density is not constant—something we were unable to do earlier in the text.

In Sections 13.7 and 13.8, we define cylindrical and spherical coordinate systems in three dimensions. These systems are useful for investigating properties of solids determined by quadric surfaces, especially of cylinders, spheres, and cones. The chapter closes by introducing the concept of the *Jacobian* in Section 13.9 and applying it to the problem of changing variables in multiple integrals.

Multiple integrals help measure such quantities as the surface area of billowing sails or the volume and average depth of a lake.

Multiple Integrals

DOUBLE INTEGRALS

In Chapter 4, we defined the definite integral $\int_a^b f(x)\,dx$ of a function f of one variable. We can also consider integrals of functions of several variables, called *double integrals, triple integrals, surface integrals,* and *line integrals.* Each integral is defined in a similar manner. The principal difference is the domain of the integrand. In this section, we consider double integrals.

Recall that $\int_a^b f(x)\,dx$ may be defined by applying the following four steps. (Steps 1 and 2 are illustrated in Figure 13.1.)

Figure 13.1

Step 1 Partition $[a, b]$ by choosing $a = x_0 < x_1 < x_2 < \cdots < x_n = b$.

Step 2 For each k, select any number w_k in the subinterval $[x_{k-1}, x_k]$.

Step 3 Form the Riemann sum $\sum_k f(w_k)\Delta x_k$, where $\Delta x_k = x_k - x_{k-1}$.

Step 4 If $\|P\|$ is the norm of the partition (the largest Δx_k), then

$$\int_a^b f(x)\,dx = \lim_{\|P\|\to 0} \sum_k f(w_k)\Delta x_k.$$

If f is nonnegative on $[a, b]$, then the Riemann sum $\sum_k f(w_k)\Delta x_k$ in Step (3) is a sum of areas of rectangles of the type illustrated in Figure 13.2. This sum approximates the area of the region that lies under the graph of f from $x = a$ to $x = b$. If the limit of Riemann sums exists as $\|P\| \to 0$ (Step 4), we obtain the definite integral $\int_a^b f(x)\,dx$, whose value is the (exact) area under the graph.

Let us now consider a function f of *two* variables such that $f(x, y)$ exists throughout a region R of the xy-plane. We shall define the double integral $\iint_R f(x, y)\,dA$ by using a four-step process similar to that used for $\int_a^b f(x)\,dx$. Throughout this chapter, R will denote a region that can be subdivided into a finite number of R_x and R_y regions, as defined in Section 5.1. Every such region R is contained in a closed rectangular region W. If W is divided into smaller rectangles by means of a grid of horizontal and vertical lines, then the collection of all closed rectangular subregions that lie completely *within* R is an **inner partition** P of R. (This partitioning corresponds to Step 1 of the four-step process.)

The blue rectangular regions in Figure 13.3 illustrate an inner partition of R. If we label them $R_1, R_2, \ldots, R_n$, then this inner partition is denoted by $\{R_k\}$. The length of the longest diagonal of all the R_k is the **norm** $\|P\|$ **of the partition** P. The symbol ΔA_k denotes the area of R_k. For each k,

Figure 13.2

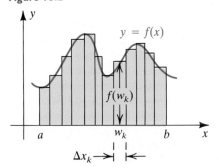

Figure 13.3
Inner partition of R

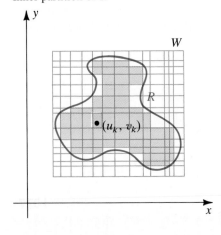

we choose any point (u_k, v_k) in R_k, as shown in Figure 13.3. (This point selection corresponds to Step 2.) Riemann sums (Step 3) are defined as follows.

Definition 13.1

Let f be a function of two variables that is defined on a region R, and let $P = \{R_k\}$ be an inner partition of R. A **Riemann sum** of f for P is any sum of the form

$$\sum_k f(u_k, v_k) \Delta A_k,$$

where (u_k, v_k) is a point in R_k and ΔA_k is the area of R_k. The summation extends over all the subregions $R_1, R_2, \ldots, R_n$ of P.

Finally, we consider a limit of Riemann sums as $\|P\| \to 0$ (Step 4). It can be shown that if f is continuous on R, the Riemann sums in Definition (13.1) approach a real number L as $\|P\| \to 0$, regardless of which points (u_k, v_k) are chosen in the subregions R_k. The number L is the *double integral* $\iint_R f(x, y) \, dA$. The mathematically precise definition of a limit of Riemann sums may be stated as follows.

Definition 13.2

Let f be a function of two variables that is defined on a region R, and let L be a real number. The statement

$$\lim_{\|P\| \to 0} \sum_k f(u_k, v_k) \Delta A_k = L$$

means that for every $\epsilon > 0$ there is a $\delta > 0$ such that if $P = \{R_k\}$ is an inner partition of R with $\|P\| < \delta$, then

$$\left| \sum_k f(u_k, v_k) \Delta A_k - L \right| < \epsilon$$

for every choice of (u_k, v_k) in R_k.

Definition (13.2) states that *we can make every Riemann sum of f as close as desired to L by choosing an inner partition of sufficiently small norm* $\|P\|$.

The double integral of f may be defined as follows.

Definition 13.3

Let f be a function of two variables that is defined on a region R. The **double integral of f over R**, denoted by $\iint_R f(x, y) \, dA$, is

$$\iint_R f(x, y) \, dA = \lim_{\|P\| \to 0} \sum_k f(u_k, v_k) \Delta A_k,$$

provided the limit exists.

If the double integral of f over R exists, then f is said to be **integrable over R**. It is proved in more advanced texts that if f is continuous on R, then f is integrable over R.

EXAMPLE ▪ 1   Approximate $\iint_R \cos(x^2 y + 1) \, dA$ for the rectangle R with $-1 \le x \le 1, 0 \le y \le \pi/2$, using Riemann sums with regular partitions of each side of the rectangle and evaluation points where each coordinate is as small as possible.

SOLUTION   For positive integers n and m, let the regular partitions be

$$-1 = x_0 < x_1 < \cdots < x_n = 1, \qquad \Delta x = \frac{2}{n},$$

$$0 = y_0 < y_1 < \cdots < y_m = \frac{\pi}{2}, \qquad \Delta y = \frac{\pi}{2m}.$$

These regular partitions divide the rectangle R into mn subrectangles, each of area $\Delta x \Delta y$. The vertices of each subrectangle are of the form $(x_{i-1}, y_{j-1}), (x_{i-1}, y_j), (x_i, y_{j-1}), (x_i, y_j)$ for $1 \le i \le n, 1 \le j \le m$, so the point with the smallest coordinates is (x_{i-1}, y_{j-1}). To add up the terms of the Riemann sum in a systematic way, we first fix a value of i and sum over the rectangles that lie above the subinterval $[x_{i-1}, x_i]$. This gives the sum

$$\sum_{j=1}^{m} f(x_{i-1}, y_{j-1}) \Delta x \Delta y = \sum_{j=1}^{m} \cos(x_{i-1}^2 y_{j-1} + 1) \frac{2}{n} \frac{\pi}{2m}.$$

Next, we add up these sums as i varies from 1 to n, which yields a Riemann sum I given by

$$I = \sum_{i=1}^{n} \left(\sum_{j=1}^{m} \cos(x_{i-1}^2 y_{j-1} + 1) \frac{2}{n} \frac{\pi}{2m} \right)$$

$$= \frac{2}{n} \frac{\pi}{2m} \sum_{i=1}^{n} \left(\sum_{j=1}^{m} \cos(x_{i-1}^2 y_{j-1} + 1) \right).$$

For even moderate values of n and m, the number of terms mn in this sum can be quite large. Thus, we use a programmable calculator or a computer to carry out the computation. The following table shows values of I for several choices of n and m.

n	m	Number of subrectangles	I
10	10	100	0.996659
30	20	600	0.974834
50	30	1500	0.959850

Figure 13.4

Figure 13.4

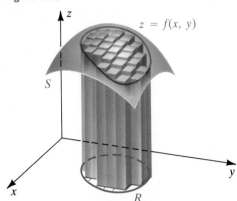

Figure 13.5

There is a useful geometric interpretation for Riemann sums and double integrals if f is continuous and $f(x, y) \geq 0$ throughout R. Let S denote the graph of f, and let Q denote the solid that lies under S and over R, as illustrated in Figure 13.4. If $P_k(u_k, v_k, 0)$ is a point in the subregion R_k of an inner partition P of R, then $f(u_k, v_k)$ is the distance from the xy-plane to the point B_k on S, directly above P_k. The product $f(u_k, v_k) \Delta A_k$ is the volume of the prism with a rectangular base of area ΔA_k illustrated in Figure 13.4. The sum of all such volumes of prisms (see Figure 13.5) is an approximation to the volume V of Q. Since this approximation to the volume improves as $\| P \|$ approaches zero, we define V as the limit of sums of the numbers $f(u_k, v_k) \Delta A_k$. Applying (13.3) gives us the following.

Definition 13.4

Let f be a continuous function of two variables such that $f(x, y)$ is nonnegative for every (x, y) in a region R. The **volume** V of the solid that lies under the graph of $z = f(x, y)$ and over R is

$$V = \iint\limits_{R} f(x, y) \, dA.$$

If the region R describes the base of a mountain and $f(x, y)$ is the height at the point (x, y), then the double integral of f over R is the volume of the mountain. If the region R describes the surface of a lake and $f(x, y)$ is the depth of the water at (x, y), we consider the z-axis pointing toward the bottom of the lake and conclude that the volume of water in the lake is given by the double integral of f over R.

If $f(x, y) \leq 0$ throughout R, the double integral of f over R is the *negative* of the volume of the solid that lies *over* the graph of f and *under* the region R.

In the next theorem, we list, without proof, some properties of double integrals that correspond to those given for definite integrals in Chapter 4. It is assumed that all regions and functions are suitably restricted so that the indicated integrals exist. The term *nonoverlapping*, in (iii) of the theorem, means that the regions R_1 and R_2 have at most only boundary points in common, as illustrated in Figure 13.6.

Theorem 13.5

Figure 13.6
$R = R_1 \cup R_2$

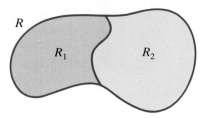

(i) $\displaystyle\iint\limits_R cf(x, y)\, dA = c \iint\limits_R f(x, y)\, dA$ for every real number c

(ii) $\displaystyle\iint\limits_R [f(x, y) + g(x, y)]\, dA = \iint\limits_R f(x, y)\, dA + \iint\limits_R g(x, y)\, dA$

(iii) If R is the union of two nonoverlapping regions R_1 and R_2,

$$\iint\limits_R f(x, y)\, dA = \iint\limits_{R_1} f(x, y)\, dA + \iint\limits_{R_2} f(x, y)\, dA.$$

(iv) If $f(x, y) \geq 0$ throughout R, then $\displaystyle\iint\limits_R f(x, y)\, dA \geq 0$.

Except for elementary cases, it is virtually impossible to find the value of a double integral $\iint_R f(x, y)\, dA$ directly from Definition (13.3). If R is an R_x or R_y region, however, the double integral can be evaluated by using two successive integrals, each involving only one independent variable.

Let us begin with a simple case, a function f that is continuous on a closed rectangular region R of the type illustrated in Figure 13.7. It is shown in advanced calculus that the double integral $\iint_R f(x, y)\, dA$ can be evaluated by using an **iterated integral** of the following type:

Figure 13.7

$$\int_a^b \left[\int_c^d f(x, y)\, dy \right] dx$$

As indicated by the brackets, we first perform a **partial integration** *with respect to y*, regarding x as a constant. Substituting the limits of integration c and d for y in the usual way, we obtain an expression in x, which is then integrated from a to b. We can also use the following iterated integral:

$$\int_c^d \left[\int_a^b f(x, y)\, dx \right] dy$$

In this case, we first perform a partial integration *with respect to x*, regarding y as a constant. After substituting the limits of integration a and b for x, we integrate the resulting expression in y from c to d.

The notation for iterated integrals is usually shortened by omitting the brackets, as in the following definition.

Definition 13.6

(i) $\displaystyle\int_a^b \int_c^d f(x, y)\, dy\, dx = \int_a^b \left[\int_c^d f(x, y)\, dy \right] dx$

(ii) $\displaystyle\int_c^d \int_a^b f(x, y)\, dx\, dy = \int_c^d \left[\int_a^b f(x, y)\, dx \right] dy$

Note that the *first* differential to the *right* of the integrand $f(x, y)$ in Definition (13.6) specifies the variables in the *first* partial integration. The first integral sign to the *left* of $f(x, y)$ specifies the limits of integration for that variable. Thus, when evaluating an iterated integral, work with the *innermost* integral first.

EXAMPLE = 2   Evaluate $\displaystyle\int_1^4 \int_{-1}^2 (2x + 6x^2 y) \, dy \, dx$.

SOLUTION   By Definition (13.6)(i), the integral equals

$$\int_1^4 \left[\int_{-1}^2 (2x + 6x^2 y) \, dy \right] dx = \int_1^4 \left[2xy + 6x^2 \left(\frac{y^2}{2} \right) \right]_{-1}^2 dx$$

$$= \int_1^4 [(4x + 12x^2) - (-2x + 3x^2)] \, dx$$

$$= \int_1^4 (6x + 9x^2) \, dx$$

$$= \left[3x^2 + 3x^3 \right]_1^4 = 234.$$

EXAMPLE = 3   Evaluate $\displaystyle\int_{-1}^2 \int_1^4 (2x + 6x^2 y) \, dx \, dy$.

SOLUTION   By Definition (13.6)(ii), the integral equals

$$\int_{-1}^2 \left[\int_1^4 (2x + 6x^2 y) \, dx \right] dy = \int_{-1}^2 \left[2 \left(\frac{x^2}{2} \right) + 6 \left(\frac{x^3}{3} \right) y \right]_1^4 dy$$

$$= \int_{-1}^2 [(16 + 128y) - (1 + 2y)] \, dy$$

$$= \int_{-1}^2 (126y + 15) \, dy$$

$$= \left[63y^2 + 15y \right]_{-1}^2 = 234.$$

The fact that the iterated integrals in Examples 2 and 3 are equal is no accident. If f is continuous, then the two iterated integrals in (13.6) are *always* equal. We say that the *order of integration* is immaterial.

An iterated double integral may be defined over an R_x or R_y region of the type shown in Figure 13.8, as follows.

Figure 13.8
(a) R_x region

(b) R_y region

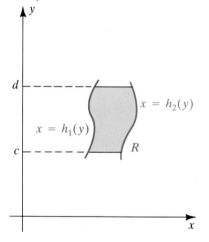

Definition of
Iterated Integrals   13.7

(i) $\displaystyle\int_a^b \int_{g_1(x)}^{g_2(x)} f(x, y) \, dy \, dx = \int_a^b \left[\int_{g_1(x)}^{g_2(x)} f(x, y) \, dy \right] dx$

(ii) $\displaystyle\int_c^d \int_{h_1(y)}^{h_2(y)} f(x, y) \, dx \, dy = \int_c^d \left[\int_{h_1(y)}^{h_2(y)} f(x, y) \, dx \right] dy$

In Definition (13.7)(i), we first perform a partial integration with respect to y and then substitute $g_2(x)$ and $g_1(x)$ for y in the usual way. The resulting expression in x is then integrated from a to b. In (ii) of the definition, we first integrate with respect to x and then, after substituting $h_2(y)$ and $h_1(y)$ for x, integrate the resulting expression with respect to y from c to d. Thus, as in Definition (13.6) for rectangular regions, we work with the *innermost* integral first.

EXAMPLE ▪ 4   Evaluate $\displaystyle\int_0^2 \int_{x^2}^{2x} (x^3 + 4y)\, dy\, dx$.

SOLUTION   By Definition (13.7)(i), the integral equals

$$\int_0^2 \left[\int_{x^2}^{2x} (x^3 + 4y)\, dy \right] dx = \int_0^2 \left[x^3 y + 4\left(\frac{y^2}{2}\right) \right]_{x^2}^{2x} dx$$

$$= \int_0^2 [(2x^4 + 8x^2) - (x^5 + 2x^4)]\, dx$$

$$= \left[\tfrac{8}{3}x^3 - \tfrac{1}{6}x^6 \right]_0^2 = \tfrac{32}{3}.$$

EXAMPLE ▪ 5   Evaluate $\displaystyle\int_1^3 \int_{\pi/6}^{y^2} 2y \cos x \, dx\, dy$.

SOLUTION   By Definition (13.7)(ii), the integral equals

$$\int_1^3 \left[\int_{\pi/6}^{y^2} 2y \cos x \, dx \right] dy = \int_1^3 2y [\sin x]_{\pi/6}^{y^2} \, dy$$

$$= \int_1^3 2y(\sin y^2 - \tfrac{1}{2})\, dy$$

$$= \int_1^3 (2y \sin y^2 - y)\, dy$$

$$= \left[-\cos y^2 - \tfrac{1}{2}y^2 \right]_1^3$$

$$= (-\cos 9 - \tfrac{9}{2}) - (-\cos 1 - \tfrac{1}{2}) \approx -2.55.$$

The next theorem states that if the region R is an R_x or R_y region, then $\iint_R f(x, y)\, dA$ may be evaluated by means of an iterated integral.

Evaluation Theorem for Double Integrals   13.8

(i) Let R be the R_x region shown in Figure 13.8(a). If f is continuous on R, then

$$\iint_R f(x, y)\, dA = \int_a^b \int_{g_1(x)}^{g_2(x)} f(x, y)\, dy\, dx.$$

(ii) Let R be the R_y region shown in Figure 13.8(b). If f is continuous on R, then

$$\iint\limits_{R} f(x, y)\, dA = \int_{c}^{d} \int_{h_1(y)}^{h_2(y)} f(x, y)\, dx\, dy.$$

For more complicated regions, we divide R into R_x or R_y subregions, apply (13.8) to each, and add the values of the resulting integrals.

A proof of Theorem (13.8) may be found in texts on advanced calculus. The following intuitive discussion makes the result plausible, at least for nonnegative-valued functions. Suppose $f(x, y) \geq 0$ throughout the R_x region R illustrated in Figure 13.8(a). Let S denote the graph of f, Q the solid that lies under S and over R, and V the volume of Q. Consider the plane that is parallel to the yz-plane and intersects the x-axis at $(x, 0, 0)$, with $a \leq x \leq b$, and let C be the trace of S on this plane (see Figure 13.9). This plane intersects the boundaries of the region—$y = g_1(x)$ and $y = g_2(x)$—at $P_1(x, g_1(x), 0)$ and $P_2(x, g_2(x), 0)$.

Figure 13.9

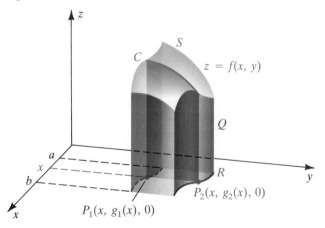

From Chapter 5, we know that the area $A(x)$ of this portion of the plane is

$$A(x) = \int_{g_1(x)}^{g_2(x)} f(x, y)\, dy.$$

Since $A(x)$ is the area of a typical cross section of Q, it follows from (5.13) that

$$V = \int_{a}^{b} A(x)\, dx = \int_{a}^{b} \int_{g_1(x)}^{g_2(x)} f(x, y)\, dy\, dx.$$

Using the fact that V is also given by Definition (13.4), we obtain the evaluation formula in part (i) of Theorem (13.8). A similar discussion can be given for part (ii).

Figure 13.10

Figure 13.11

Figure 13.12

Figure 13.13

Before using Theorem (13.8) to evaluate a double integral, *it is important to sketch the region R* and determine its boundaries. The following examples illustrate the use of Theorem (13.8).

EXAMPLE ▪ 6   Let R be the region in the xy-plane bounded by the graphs of $y = x^2$ and $y = 2x$. Evaluate $\iint_R (x^3 + 4y) \, dA$ using

(a) Theorem (13.8)(i)   **(b)** Theorem (13.8)(ii)

SOLUTION

(a) The region R is sketched in Figure 13.10. Note that R is both an R_x region and an R_y region. Let us regard R as an R_x region having lower boundary $y = x^2$ and upper boundary $y = 2x$, with $0 \le x \le 2$. We have drawn a vertical line segment between these boundaries to indicate that the first integration is with respect to y (from the lower boundary to the upper boundary). By Theorem (13.8)(i),

$$\iint\limits_R (x^3 + 4y) \, dA = \int_0^2 \int_{x^2}^{2x} (x^3 + 4y) \, dy \, dx.$$

From Example 4, we know that the last integral equals $\frac{32}{3}$.

(b) To use Theorem (13.8)(ii), we regard R as an R_y region and solve each of the given equations for x in terms of y, obtaining

$$x = \tfrac{1}{2}y \quad \text{and} \quad x = \sqrt{y}, \quad \text{with} \quad 0 \le y \le 4.$$

The horizontal line segment in Figure 13.11 extends from the left boundary to the right boundary, indicating that the first integration is with respect to x. By Theorem (13.8)(ii),

$$\iint\limits_R f(x, y) \, dA = \int_0^4 \int_{y/2}^{\sqrt{y}} (x^3 + 4y) \, dx \, dy$$

$$= \int_0^4 \left[\tfrac{1}{4}x^4 + 4yx \right]_{y/2}^{\sqrt{y}} dy$$

$$= \int_0^4 \left[(\tfrac{1}{4}y^2 + 4y^{3/2}) - (\tfrac{1}{64}y^4 + 2y^2) \right] dy = \tfrac{32}{3}.$$

EXAMPLE ▪ 7   Let R be the region bounded by the graphs of the equations $y = \sqrt{x}$, $y = \sqrt{3x - 18}$, and $y = 0$. If f is an arbitrary continuous function on R, express the double integral $\iint_R f(x, y) \, dA$ in terms of iterated integrals using only

(a) Theorem (13.8)(i)   **(b)** Theorem (13.8)(ii)

SOLUTION   The graphs of $y = \sqrt{x}$ and $y = \sqrt{3x - 18}$ are the top halves of the parabolas $y^2 = x$ and $y^2 = 3x - 18$. The region R is sketched in Figures 13.12 and 13.13.

(a) If we wish to use only Theorem 13.8(i), then it is necessary to employ two iterated integrals, because if $0 \le x \le 6$, the lower boundary of the region is the graph of $y = 0$, and if $6 \le x \le 9$, the lower boundary is the graph of $y = \sqrt{3x - 18}$. (The vertical line segments in Figure 13.12 extend from the lower boundary to the upper boundary of R.)

If R_1 denotes the part of the region R that lies between $x = 0$ and $x = 6$ and if R_2 denotes the part between $x = 6$ and $x = 9$, then both R_1 and R_2 are R_x regions. Hence,

$$\iint\limits_{R} f(x, y) \, dA = \iint\limits_{R_1} f(x, y) \, dA + \iint\limits_{R_2} f(x, y) \, dA$$

$$= \int_0^6 \int_0^{\sqrt{x}} f(x, y) \, dy \, dx + \int_6^9 \int_{\sqrt{3x-18}}^{\sqrt{x}} f(x, y) \, dy \, dx.$$

(b) To use Theorem (13.8)(ii), we must solve each of the given equations for x in terms of y, obtaining

$$x = y^2 \quad \text{and} \quad x = \frac{y^2 + 18}{3} = \frac{1}{3} y^2 + 6, \quad \text{with} \quad 0 \le y \le 3.$$

The horizontal line segment in Figure 13.13 extends from the left boundary to the right boundary. Only one iterated integral is required in this case, since R is an R_y region. Thus,

$$\iint\limits_{R} f(x, y) \, dA = \int_0^3 \int_{y^2}^{(y^2/3)+6} f(x, y) \, dx \, dy.$$

Examples 6 and 7 indicate how certain double integrals may be evaluated by using either (i) or (ii) of Theorem (13.8). Generally, the choice of the *order of integration dy dx* or *dx dy* depends on the form of $f(x, y)$ and the region R. Sometimes it is extremely difficult, or even impossible, to evaluate a given iterated double integral. However, by *reversing the order of integration* from *dy dx* to *dx dy*, or vice versa, it may be possible to find an equivalent iterated double integral that can be easily evaluated. This technique is illustrated in the next example.

Figure 13.14

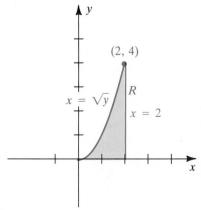

EXAMPLE ■ 8   Given $\int_0^4 \int_{\sqrt{y}}^2 y \cos x^5 \, dx \, dy$, reverse the order of integration and evaluate the resulting integral.

SOLUTION   The given order of integration, *dx dy*, indicates that the region is an R_y region. As illustrated in Figure 13.14, the left and right boundaries are the graphs of the equations $x = \sqrt{y}$ and $x = 2$, respectively, with $0 \le y \le 4$.

Note that R is also an R_x region whose lower and upper boundaries are given by $y = 0$ and $y = x^2$, respectively, with $0 \le x \le 2$. Hence, by

Theorem (13.8)(i), the integral can be evaluated as follows:

$$\int_0^4 \int_{\sqrt{y}}^2 y\cos x^5 \, dx\, dy = \iint\limits_R y\cos x^5 \, dA = \int_0^2 \int_0^{x^2} y\cos x^5 \, dy\, dx$$

$$= \int_0^2 \left[\frac{y^2}{2}\cos x^5\right]_0^{x^2} dx = \int_0^2 \frac{x^4}{2}\cos x^5 \, dx$$

$$= \tfrac{1}{10}\int_0^2 (\cos x^5)(5x^4)\, dx$$

$$= \tfrac{1}{10}\left[\sin x^5\right]_0^2 = \tfrac{1}{10}\sin 32 \approx 0.055$$

EXERCISES 13.1

Exer. 1–10: Determine whether the region is R_x, R_y, or neither.

1

2

3

4

5

6

7

8

9

10

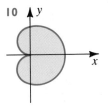

11 Suppose $f(x, y) = 4x + 2y + 1$ and R is the region in the first quadrant bounded by the coordinate axes and the graph of $y = 4 - \frac{1}{4}x^2$. Let $\{R_k\}$, with $k = 1, 2, \ldots, 7$, be the inner partition of R shown in the figure. Calculate the Riemann sum $\sum_k f(u_k, v_k)\Delta A_k$ if, for each k, (u_k, v_k) is the point

(a) in the lower left corner of R_k

(b) in the upper right corner of R_k

(c) at the center of R_k

Exercise 11

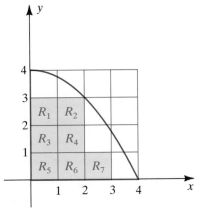

12 Suppose $f(x, y) = xy$ and R is the region bounded by the trapezoid with vertices $(0, 0)$, $(4, 4)$, $(8, 4)$, and $(12, 0)$. Let P be the inner partition of R determined by vertical lines with x-intercepts 0, 2, 4, 6, 8, 10, and 12 and by horizontal lines with y-intercepts 0, 2, and 4. Calculate the Riemann sums as in parts (a)–(c) of Exercise 11.

Exer. 13–20: Evaluate the iterated integral.

13 $\int_{1}^{2} \int_{-1}^{2} (12xy^2 - 8x^3) \, dy \, dx$

14 $\int_{0}^{3} \int_{-2}^{-1} (4xy^3 + y) \, dx \, dy$

15 $\int_{1}^{2} \int_{1-x}^{\sqrt{x}} x^2 y \, dy \, dx$

16 $\int_{-1}^{1} \int_{x^3}^{x+1} (3x + 2y) \, dy \, dx$

17 $\int_{0}^{2} \int_{y^2}^{2y} (4x - y) \, dx \, dy$

18 $\int_{0}^{1} \int_{-y-1}^{y-1} (x^2 + y^2) \, dx \, dy$

19 $\int_{1}^{2} \int_{x^3}^{x} e^{y/x} \, dy \, dx$

20 $\int_{0}^{\pi/6} \int_{0}^{\pi/2} (x \cos y - y \cos x) \, dy \, dx$

Exer. 21–26: Sketch the region R bounded by the graphs of the given equations. If $f(x, y)$ is an arbitrary continuous function, express $\iint_R f(x, y) \, dA$ as an iterated integral by using (a) only Theorem (13.8)(i) and (b) only Theorem (13.8)(ii).

21 $y = \sqrt{x}$, $\quad x = 4$, $\quad y = 0$

22 $y = \sqrt{x}$, $\quad x = 0$, $\quad y = 2$

23 $y = x^3$, $\quad x = 0$, $\quad y = 8$

24 $y = x^3$, $\quad x = 2$, $\quad y = 0$

25 $y = \sqrt{x}$, $\quad y = x^3$

26 $y = \sqrt{1 - x^2}$, $\quad y = 0$

Exer. 27–32: Express the double integral over the indicated region R as an iterated integral, and find its value.

27 $\iint_R (y + 2x) \, dA$;

the rectangular region with vertices $(-1, -1)$, $(2, -1)$, $(2, 4)$, $(-1, 4)$

28 $\iint_R (x - y) \, dA$;

the triangular region with vertices $(2, 9)$, $(2, 1)$, $(-2, 1)$

29 $\iint_R xy^2 \, dA$;

the triangular region with vertices $(0, 0)$, $(3, 1)$, $(-2, 1)$

30 $\iint_R (y + 1) \, dA$;

the region between the graphs of $y = \sin x$ and $y = \cos x$ from $x = 0$ to $x = \pi/4$

31 $\iint_R x^3 \cos xy \, dA$;

the region bounded by the graphs of $y = x^2$, $y = 0$, and $x = 2$

32 $\iint_R e^{x/y} \, dA$;

the region bounded by the graphs of $y = 2x$, $y = -x$, and $y = 4$

Exer. 33–38: Sketch the region R bounded by the graphs of the given equations. If f is an arbitrary continuous function on R, express $\iint_R f(x, y) \, dA$ as a sum of two iterated integrals of the type used in (a) Theorem (13.8)(i) and (b) Theorem (13.8)(ii).

33 $8y = x^3$, $\qquad y - x = 4$, $\quad 4x + y = 9$

34 $x = 2\sqrt{y}$, $\quad \sqrt{3}x = \sqrt{y}$, $\quad y = 2x + 5$

35 $x = \sqrt{3 - y}$, $\quad y = 2x$, $\qquad x + y + 3 = 0$

36 $x + 2y = 5$, $\quad x - y = 2$, $\quad 2x + y = -2$

37 $y = e^x$, $\qquad y = \ln x$, $\quad x + y = 1$, $\quad x + y = 1 + e$

38 $y = \sin x$, $\qquad \pi y = 2x$

Exer. 39–44: Sketch the region of integration for the iterated integral.

39 $\int_{-1}^{2} \int_{-\sqrt{4-x^2}}^{4-x^2} f(x, y) \, dy \, dx$

40 $\int_{-1}^{2} \int_{x^2-4}^{x-2} f(x, y) \, dy \, dx$

41 $\int_{0}^{1} \int_{\sqrt{y}}^{\sqrt[3]{y}} f(x, y) \, dx \, dy$

42 $\int_{-2}^{-1} \int_{3y}^{2y} f(x, y) \, dx \, dy$

43 $\int_{-3}^{1} \int_{\arctan x}^{e^x} f(x, y) \, dy \, dx$

44 $\int_{\pi}^{2\pi} \int_{\sin y}^{\ln y} f(x, y) \, dx \, dy$

Exer. 45–50: Reverse the order of integration, and evaluate the resulting integral.

45 $\int_0^1 \int_{2x}^2 e^{(y^2)} \, dy \, dx$

46 $\int_0^9 \int_{\sqrt{y}}^3 \sin x^3 \, dx \, dy$

47 $\int_0^2 \int_{y^2}^4 y \cos x^2 \, dx \, dy$

48 $\int_1^e \int_0^{\ln x} y \, dy \, dx$

49 $\int_0^8 \int_{\sqrt[3]{y}}^2 \frac{y}{\sqrt{16 + x^7}} \, dx \, dy$

50 $\int_0^2 \int_x^2 y^4 \cos(xy^2) \, dy \, dx$

---

c Exer. 51–52: Approximate the double integral over the given rectangle using Riemann sums with regular partitions of each side of the rectangle that have *n* partition points on the *x*-axis and *m* partition points on the *y*-axis. Choose evaluation points where each coordinate is as small as possible.

51 $\int_0^1 \int_0^1 e^{x^2 y^2} \, dy \, dx;$ $\qquad n = 10, \quad m = 15$

52 $\int_0^1 \int_0^1 \sin(x^2 + y^3) \, dy \, dx;$ $\quad n = 20, \quad m = 15$

---

c Exer. 53–54: Consider double integrals over the triangular region *R*, where $0 \le x \le 2, 0 \le y \le x$. For any positive integer *n*, let

$$0 = x_0 < x_1 < \cdots < x_n = 2, \quad \Delta x = \frac{2}{n},$$

$$0 = y_0 < y_1 < \cdots < y_n = 2, \quad \Delta y = \frac{2}{n},$$

be regular partitions of [0, 2]. Approximate the double integral by the Riemann sum

$$\sum_{i=1}^n \sum_{j=1}^i f(x_i, y_j) \, \Delta y \, \Delta x.$$

53 $\int_0^2 \int_0^x \sqrt{\cos(xy) + 1} \, dy \, dx; \quad n = 10$

54 $\int_0^2 \int_0^x \ln(x + y + 1) \, dy \, dx; \quad n = 12$

---

c Exer. 55–56: Approximate the double integral by evaluating the innermost integral exactly and then approximating the outermost integral.

55 $\int_{-1}^1 \int_0^{x^2} \cos y \, dy \, dx$

56 $\int_1^4 \int_1^{\sqrt{x}} \frac{1}{x + y} \, dy \, dx$

13.2 AREA AND VOLUME

In Section 13.1, we saw that if $f(x, y) \ge 0$ and f is continuous, then the volume V of the solid that lies under the graph of $z = f(x, y)$ and over an R_x region in the xy-plane is given by

$$V = \int_a^b A(x) \, dx = \int_a^b \int_{g_1(x)}^{g_2(x)} f(x, y) \, dy \, dx,$$

where $A(x)$ is the area of a typical cross section of the solid (see Figure 13.9). In this section, we discuss a method for interpreting this formula for V as a limit of (double) sums. By the volume-by-cross-section formula (5.13),

$$V = \int_a^b A(x) \, dx = \lim_{\|P'\| \to 0} \sum_k A(u_k) \Delta x_k,$$

where P' is a partition of the interval $[a, b]$, u_k is any number in the kth subinterval $[x_{k-1}, x_k]$ of P', and $\Delta x_k = x_k - x_{k-1}$. As illustrated in Figure 13.15, $A(u_k) \Delta x_k$ is the volume of a lamina L_k with face parallel to the yz-plane and rectangular base of width Δx_k in the xy-plane. Thus, *the volume V is a limit of a sum of volumes of such laminas.*

Figure 13.15

Figure 13.16

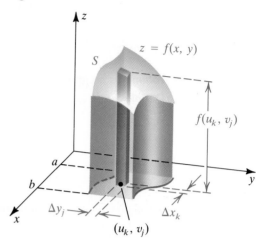

The iterated integral $V = \int_a^b \int_{g_1(x)}^{g_2(x)} f(x, y)\, dy\, dx$ may also be expressed in terms of limits of sums. First we apply the definition of the definite integral as follows. If x is in $[a, b]$, then

$$A(x) = \int_{g_1(x)}^{g_2(x)} f(x, y)\, dy = \lim_{\|P''\| \to 0} \sum_j f(x, v_j)\Delta y_j,$$

where P'' is a partition of the y-interval $[g_1(x), g_2(x)]$, v_j is any number in the jth subinterval $[y_{j-1}, y_j]$ of P'', and $\Delta y_j = y_j - y_{j-1}$. Hence, for each u_k in $[a, b]$,

$$A(u_k) = \lim_{\|P''\| \to 0} \sum_j f(u_k, v_j)\Delta y_j.$$

Consequently,

$$V = \int_a^b A(x)\, dx = \lim_{\|P'\| \to 0} \sum_k A(u_k)\Delta x_k$$

$$= \lim_{\|P'\| \to 0} \sum_k \left[\lim_{\|P''\| \to 0} \sum_j f(u_k, v_j)\Delta y_j \right] \Delta x_k.$$

If we disregard the limits in this formula, we obtain a **double sum**:

$$\sum_k \sum_j f(u_k, v_j)\Delta y_j \Delta x_k$$

Referring to Figure 13.16, we see that $f(u_k, v_j)\Delta y_j \Delta x_k$ is the volume of a prism having base area $\Delta y_j \Delta x_k$ and altitude $f(u_k, v_j)$. The double sum may be interpreted as follows. *Holding x fixed*, we sum the volumes of the prisms in the direction of the y-axis, thereby obtaining the volume of the lamina illustrated in Figure 13.15. The volumes of all such laminas are then added by summing in the direction of the x-axis. In a certain sense, the iterated integral does this and, at the same time, takes limits as the bases of the prisms shrink to points. This interpretation is often useful in visualizing applications of the double integral. To emphasize this point of view, we sometimes express the volume as follows, where $\|P\|$ is the length of the longest diagonal of the bases of the prisms.

Volume as a Limit of Double Sums   13.9

$$V = \lim_{\|P\| \to 0} \sum_k \sum_j f(u_k, v_j) \Delta y_j \, \Delta x_k = \int_a^b \int_{g_1(x)}^{g_2(x)} f(x, y) \, dy \, dx$$

Areas can also be considered as limits of double sums. We begin by denoting the double integral $\iint_R 1 \, dA$ by $\iint_R dA$. If R is the R_x region in Figure 13.17, then, by Theorem (13.8)(i),

Figure 13.17

$$\iint_R dA = \int_a^b \int_{g_1(x)}^{g_2(x)} 1 \, dy \, dx = \int_a^b \left[y \right]_{g_1(x)}^{g_2(x)} dx$$

$$= \int_a^b [g_2(x) - g_1(x)] \, dx$$

which, according to Theorem (5.1), equals the area A of R. A similar formula holds if R is an R_y region. These facts are also evident from the definition of the double integral, for if $f(x, y) = 1$ throughout R, then the Riemann sum in Definition (13.3) is a sum of areas of rectangles in an inner partition P of R. As the norm $\|P\|$ approaches zero, these rectangles cover more of R and the limit equals the area of R.

If an iterated integral is used to find an area, it may be regarded as a limit of double sums in a manner similar to that used for volumes. Specifically, as in (13.9), with $f(x, y) = 1$, we have the following result.

Area as a Limit of Double Sums   13.10

$$A = \iint_R dA = \int_a^b \int_{g_1(x)}^{g_2(x)} dy \, dx = \lim_{\|P\| \to 0} \sum_k \sum_j \Delta y_j \, \Delta x_k$$

As for functions of one variable (see Guidelines (5.3)), we may employ the following *intuitive* method for remembering the limit of sums formula (13.10).

Guidelines for Finding the Area of an R_x Region by Means of a Double Integral   13.11

1 Sketch the region, as in Figure 13.18(a), and show a typical rectangle of dimensions dx and dy.

2 *Holding x fixed*, regard $\int_{g_1(x)}^{g_2(x)}$ as an operator that sums the area elements $dy \, dx$ in the same direction as the y-axis, from the lower

Figure 13.18

(a)

(b)

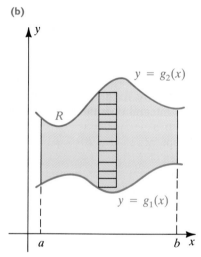

boundary to the upper boundary. The expression $[\int_{g_1(x)}^{g_2(x)} dy]\, dx$ represents the area of the vertical rectangle in Figure 13.18(b).

3 Apply the operator $\int_a^b$ to $[\int_{g_1(x)}^{g_2(x)} dy]\, dx$, thereby taking a limit of sums of areas of the vertical rectangles in guideline (2), from $x = a$ to $x = b$.

It is important to remember that when Guidelines (13.11) are applied to an R_x region, the first integration is with respect to y and the second with respect to x. We can, of course, obtain the area of the region in Figure 13.18 by using the methods of Section 5.1. The reason for using a double integral is to gain practice in working with limits of double sums. This same technique will be used later for problems that *cannot* be solved using single-variable methods.

EXAMPLE ■ 1   An artificially created lake is bordered on one side by a straight dam. The shape of the lake's surface is that of a region in the xy-plane bounded by the graphs of $2y = 16 - x^2$ and $x + 2y = 4$. Find the area A of the surface of the lake.

SOLUTION   Following guideline (1), we sketch the region and a typical rectangle of area $dy\,dx$, as in Figure 13.19(a). We solve the equations for y in terms of x, and we label the boundaries of the region $y = 2 - \frac{1}{2}x$ and $y = 8 - \frac{1}{2}x^2$.

As in guideline (2), the partial integration

$$\int_{2-(x/2)}^{8-(x^2/2)} dy\, dx = \left[\int_{2-(x/2)}^{8-(x^2/2)} dy\right] dx$$

Figure 13.19

(a)

(b)

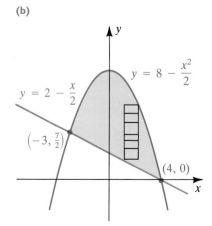

represents the sum of the area elements $dy\,dx$ from the lower boundary to the upper boundary. This sum gives us the area of the vertical rectangle in Figure 13.19(b).

Finally, using guideline (3), we apply the operator $\int_{-3}^{4}$, thereby taking a limit of sums of these vertical rectangles from $x = -3$ to $x = 4$:

$$A = \int_{-3}^{4} \int_{2-(x/2)}^{8-(x^2/2)} dy\,dx = \int_{-3}^{4} \left[y\right]_{2-(x/2)}^{8-(x^2/2)} dx$$

$$= \int_{-3}^{4} \left[\left(8 - \frac{x^2}{2}\right) - \left(2 - \frac{x}{2}\right)\right] dx$$

$$= \left[6x - \frac{x^3}{6} + \frac{x^2}{4}\right]_{-3}^{4} = \frac{343}{12} \approx 28.6$$

EXAMPLE ■ 2   If the depth of the lake described in Example 1 at the point (x, y) is given by $f(x, y) = x^2 + y$, find the volume of water in the lake.

SOLUTION   Since the depth is nonnegative for all points in the region, the volume V of water is the double integral over the region of the depth function. Thus we have

$$V = \int_{-3}^{4} \int_{2-(x/2)}^{8-(x^2/2)} f(x, y)\,dy\,dx$$

$$= \int_{-3}^{4} \int_{2-(x/2)}^{8-(x^2/2)} (x^2 + y)\,dy\,dx$$

$$= \int_{-3}^{4} \left[x^2 y + \frac{y^2}{2}\right]_{2-(x/2)}^{8-(x^2/2)} dx$$

$$= \int_{-3}^{4} \left\{x^2(8 - \tfrac{1}{2}x^2) + \tfrac{1}{2}(8 - \tfrac{1}{2}x^2)^2 - \left[x^2(2 - \tfrac{1}{2}x) + \tfrac{1}{2}(2 - \tfrac{1}{2}x)^2\right]\right\} dx$$

$$= \int_{-3}^{4} \left(-\tfrac{3}{8}x^4 + \tfrac{1}{2}x^3 + \tfrac{15}{8}x^2 + x + 30\right) dx$$

$$= \left[\tfrac{-3}{40}x^5 + \tfrac{1}{8}x^4 + \tfrac{5}{8}x^3 + \tfrac{1}{2}x^2 + 30x\right]_{-3}^{4}$$

$$= \tfrac{616}{5} - \tfrac{-2961}{40} = \tfrac{7889}{40} = 197.225 \text{ cubic units.}$$

We next state guidelines similar to (13.11) for R_y regions.

**Guidelines for Finding the Area of
an R_y Region by Means of a
Double Integral 13.12**

1 Sketch the region, as in Figure 13.20(a), and show a typical rectangle of dimensions dx and dy.

2 *Holding y fixed*, regard $\int_{h_1(y)}^{h_2(y)}$ as an operator that sums the area elements $dx\,dy$ in the same direction as the x-axis, from the left boundary to the right boundary. The expression $[\int_{h_1(y)}^{h_2(y)} dx]\,dy$ represents the area of the horizontal rectangle in Figure 13.20(b).

3 Apply the operator $\int_c^d$ to $[\int_{h_1(y)}^{h_2(y)} dx]\,dy$, thereby taking a limit of sums of areas of the horizontal rectangles in guideline (2), from $y = c$ to $y = d$.

Figure 13.20

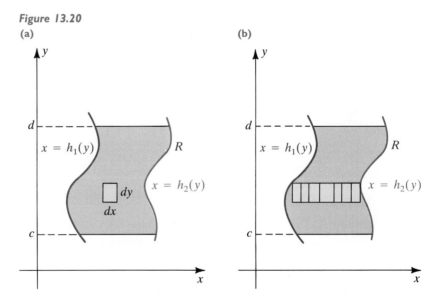

(a)   (b)

Figure 13.21

(a)

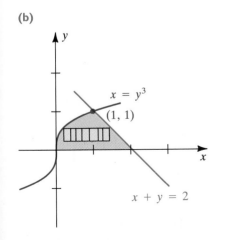

(b)

Note that when Guidelines (13.12) are applied to an R_y region, the first integration is with respect to x and the second with respect to y.

EXAMPLE 3   Find the area A of the region in the xy-plane bounded by the graphs of $x = y^3$, $x + y = 2$, and $y = 0$.

SOLUTION   The region is sketched in Figure 13.21(a), together with a rectangle of area $dx\,dy$. As in guideline (2) of (13.12), the partial integration

$$\int_{y^3}^{2-y} dx\,dy = \left[\int_{y^3}^{2-y} dx\right] dy$$

represents the sum of the area elements $dx\,dy$ from the left boundary to the right boundary. This gives us the area of the horizontal rectangle in Figure 13.21(b).

Following guideline (3), we apply the operator $\int_0^1$, thereby taking a limit of sums of these horizontal rectangles from $y = 0$ to $y = 1$:

$$A = \iint\limits_R dA = \int_0^1 \int_{y^3}^{2-y} dx\, dy = \int_0^1 \big[x\big]_{y^3}^{2-y}\, dy$$

$$= \int_0^1 (2 - y - y^3)\, dy = \left[2y - \frac{y^2}{2} - \frac{y^4}{4} \right]_0^1 = \frac{5}{4}$$

The area can also be found by using Theorem (13.8)(i); however, in this case, it is necessary to divide R into two parts by means of a vertical line through $(1, 1)$. We then have

$$A = \int_0^1 \int_0^{\sqrt[3]{x}} dy\, dx + \int_1^2 \int_0^{2-x} dy\, dx.$$

Figure 13.22

(a)

(b)

In the next two examples, we find volumes by means of (13.9).

EXAMPLE ■ 4   Find the volume V of the solid in the first octant bounded by the coordinate planes, the paraboloid $z = x^2 + y^2 + 1$, and the plane $2x + y = 2$.

SOLUTION   As illustrated in Figure 13.22(a), the solid lies under the paraboloid and over the triangular region R in the xy-plane bounded by the coordinate axes and the line $y = 2 - 2x$.
By Definition (13.4), with $f(x, y) = x^2 + y^2 + 1$,

$$V = \iint\limits_R (x^2 + y^2 + 1)\, dA.$$

We may interpret this integral as a limit of double sums, as in (13.9), and evaluate it using Guidelines (13.11). If we represent dA by $dx\, dy$, then

$$(x^2 + y^2 + 1)\, dy\, dx$$

represents the volume of the prism illustrated in Figure 13.22(a). Figure 13.22(b) shows the region R in the xy-plane and a rectangle corresponding to the following partial integration with respect to y (holding x fixed):

$$\int_0^{2-2x} (x^2 + y^2 + 1)\, dy\, dx = \left[\int_0^{2-2x} (x^2 + y^2 + 1)\, dy \right] dx$$

This formula represents the sum of the prism volumes $(x^2 + y^2 + 1)\, dy\, dx$ *in the direction parallel to the y-axis* and gives us the volume of a lamina whose face is parallel to the yz-plane. Finally, we apply the operator $\int_0^1$; that is, we take a limit of sums of these laminar volumes from $x = 0$ to

$x = 1$. Thus,

$$V = \int_0^1 \left[\int_0^{2-2x} (x^2 + y^2 + 1)\, dy \right] dx = \int_0^1 \left[x^2 y + \tfrac{1}{3}y^3 + y \right]_0^{2-2x} dx$$

$$= \int_0^1 \left(-\tfrac{14}{3}x^3 + 10x^2 - 10x + \tfrac{14}{3} \right) dx$$

$$= \left[-\tfrac{7}{6}x^4 + \tfrac{10}{3}x^3 - 5x^2 + \tfrac{14}{3}x \right]_0^1 = \tfrac{11}{6}.$$

We may also find V by integrating first with respect to x, in which case the iterated integral has the form.

$$V = \int_0^2 \int_0^{(2-y)/2} (x^2 + y^2 + 1)\, dx\, dy.$$

EXAMPLE ▪ 5 Find the volume V of the solid that lies in the first octant and is bounded by the three coordinate planes and the cylinders $x^2 + y^2 = 9$ and $y^2 + z^2 = 9$.

SOLUTION The cylinder $y^2 + z^2 = 9$ has radius 3 and axis on the x-axis, with the part in the first octant above the xy-plane, as illustrated in Figure 13.23(a). The cylinder $x^2 + y^2 = 9$ has radius 3 and axis on the z-axis, with the part in the first octant intersecting the xy-plane in a quarter-circle. Thus, the solid lies *under* the graph of $z = \sqrt{9 - y^2}$ and *over* the region R shown in Figure 13.23(b). By Definition (13.4),

$$V = \iint_R \sqrt{9 - y^2}\, dA.$$

Figure 13.23

(a)

(b)

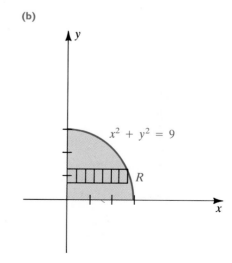

We may regard the integral as a limit of double sums of volumes of prisms of the type shown in Figure 13.23(a). We first sum in the direction of the x-axis from $x = 0$ to $x = \sqrt{9 - y^2}$. (Figure 13.23b shows a strip in the xy-plane corresponding to this first summation.) This sum gives us the volume of a lamina whose face is parallel to the xz-plane. We then sum these laminar volumes in the direction of the y-axis from $y = 0$ to $y = 3$. Using an iterated integral, we have

$$
\begin{aligned}
V &= \int_0^3 \int_0^{\sqrt{9-y^2}} \sqrt{9 - y^2} \, dx \, dy \\
&= \int_0^3 \sqrt{9 - y^2} \, [x]_0^{\sqrt{9-y^2}} \, dy \\
&= \int_0^3 (9 - y^2) \, dy \\
&= \left[9y - \tfrac{1}{3}y^3 \right]_0^3 = 18.
\end{aligned}
$$

The double integral could also be evaluated by integrating first with respect to y; however, this order of integration is much more complicated.

We may apply any numerical integration technique to both the innermost and the outermost integrals of an iterated double integral. For simplicity, suppose that the region R is the rectangle $a \le x \le b$, $c \le y \le d$. Applying the trapezoidal rule with n subintervals for the outermost x-interval and m subintervals for the innermost y-integral yields the **iterated trapezoidal rule:**

$$
\int_a^b \int_c^d f(x, y) \, dy \, dx \approx \frac{\Delta x}{2} \left[G(a) + 2G(x_1) + \cdots + 2G(x_{n-1}) + G(b) \right],
$$

where

$$
G(x_i) = \frac{\Delta y}{2} \left[f(x_i, c) + 2f(x_i, y_1) + \cdots + 2f(x_i, y_{m-1}) + f(x_i, d) \right]
$$

and

$$
\Delta x = \frac{b - a}{n}, \qquad \Delta y = \frac{d - c}{m}.
$$

E X A M P L E ▪ 6   Approximate the volume V of the solid that lies in the first octant above the rectangle $0 \le x \le 2$, $0 \le y \le 3$ and under the graph of $z = \tfrac{1}{5}\sqrt{225 - 25x^2 - 9y^2}$ using the iterated trapezoidal rule.

S O L U T I O N   The volume V is given by the iterated integral

$$
V = \int_0^2 \int_0^3 \tfrac{1}{5}\sqrt{225 - 25x^2 - 9y^2} \, dy \, dx.
$$

To approximate the iterated integral with a regular partition using n subintervals for the x-interval and m subintervals for the y-interval, we have $\Delta x = (2 - 0)/n = 2/n$ and $\Delta y = (3 - 0)/m$. We also have $x_i = 2i/n$

for $0 \le i \le n$ and $y_j = 3j/m$ for $0 \le j \le m$. Thus,

$$G(x_i) = \frac{3}{2m}\left[f(x_i, 0) + 2f\left(x_i, \frac{3}{m}\right) + \cdots + 2f\left(x_i, \frac{3j}{m}\right) + \cdots \right.$$
$$\left. + 2f\left(x_i, \frac{3(m-1)}{m}\right) + f(x_i, 3)\right]$$

where

$$f\left(x_i, \frac{3j}{m}\right) = \frac{1}{5}\sqrt{225 - 25x_i^2 - 9\left(\frac{3j}{m}\right)^2},$$

so

$$V \approx \frac{\Delta x}{2}[G(a) + 2G(x_1) + \cdots + 2G(x_i) + \cdots + 2G(x_{n-1}) + G(b)]$$
$$= \frac{1}{n}\left[G(0) + 2G\left(\frac{2}{n}\right) + \cdots + 2G\left(\frac{2i}{n}\right) + \cdots \right.$$
$$\left. + 2G\left(\frac{2(n-1)}{n}\right) + G(2)\right].$$

The following table shows the approximating sums obtained with the iterated trapezoidal rule for several different choices of n and m.

n	m	V
10	15	15.2683
20	30	15.2787
30	45	15.2807

The volume is approximately 15.28 cubic units.

EXERCISES 13.2

Exer. 1–4: Set up an iterated double integral that can be used to find the area of the region.

1

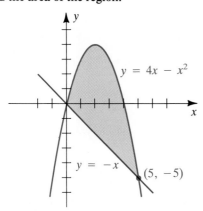

$y = 4x - x^2$

$y = -x$

$(5, -5)$

2

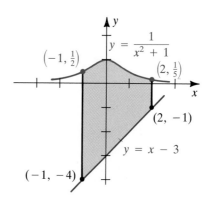

$y = \dfrac{1}{x^2 + 1}$

$\left(-1, \frac{1}{2}\right)$

$\left(2, \frac{1}{5}\right)$

$(2, -1)$

$y = x - 3$

$(-1, -4)$

3

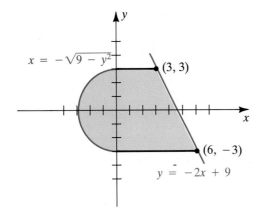

$x = -\sqrt{9 - y^2}$

$(3, 3)$

x

$(6, -3)$

$y = -2x + 9$

4

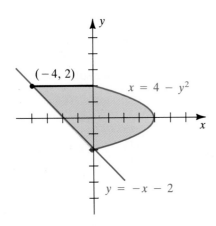

$(-4, 2)$

$x = 4 - y^2$

x

$y = -x - 2$

Exer. 5–12: Sketch the region bounded by the graphs of the equations, and find its area by using one or more double integrals.

5 $y = 1/x^2,$ $\quad y = -x^2,$ $\quad x = 1,$ $\quad\quad x = 2$

6 $y = \sqrt{x},$ $\quad y = -x,$ $\quad x = 1,$ $\quad\quad x = 4$

7 $y^2 = -x,$ $\quad x - y = 4,$ $\quad y = -1,$ $\quad y = 2$

8 $x = y^2,$ $\quad y - x = 2,$ $\quad y = -2,$ $\quad y = 3$

9 $y = x,$ $\quad\quad y = 3x,$ $\quad x + y = 4$

10 $x - y = -1,$ $\; 7x - y = 17,$ $\; 2x + y = -2$

11 $y = e^x,$ $\quad y = \sin x,$ $\quad x = -\pi,$ $\quad x = \pi$

12 $y = x^2,$ $\quad\quad y = 1/(x^2 + 1)$

Exer. 13–16: Set up an iterated double integral that can be used to find the volume of the solid.

13

14

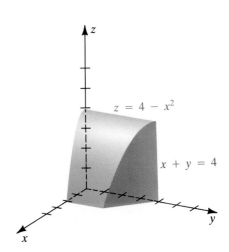

$z = 4 - x^2$

$x + y = 4$

15

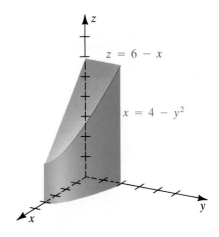

$z = 6 - x$

$x = 4 - y^2$

16

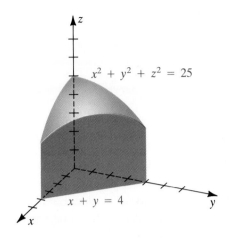

$x^2 + y^2 + z^2 = 25$

$x + y = 4$

Exer. 17–20: The iterated double integral represents the volume of a solid under a surface S and over a region R in the xy-plane. Describe S and sketch R.

17 $\displaystyle\int_0^4 \int_{-1}^2 3 \, dy \, dx$

18 $\displaystyle\int_0^1 \int_{3-x}^{3-x^2} \sqrt{25 - x^2 - y^2} \, dy \, dx$

19 $\displaystyle\int_{-2}^1 \int_{x-1}^{1-x^2} (x^2 + y^2) \, dy \, dx$

20 $\displaystyle\int_0^4 \int_0^{\sqrt{y}} \sqrt{x^2 + y^2} \, dx \, dy$

Exer. 21–22: Find the volume of the solid that lies under the graph of the equation and over the region in the xy-plane bounded by the polygon with the given vertices.

21 $z = 4x^2 + y^2$; $(0, 0), (0, 1), (2, 0), (2, 1)$

22 $z = x^2 + 4y^2$; $(0, 0), (1, 0), (1, 2)$

Exer. 23–30: Sketch the solid in the first octant bounded by the graphs of the equations, and find its volume.

23 $x^2 + z^2 = 9$, $y = 2x$, $y = 0$, $z = 0$

24 $z = 4 - x^2$, $x + y = 2$, $x = 0$, $y = 0$, $z = 0$

25 $2x + y + z = 4$, $x = 0$, $y = 0$, $z = 0$

26 $y^2 = z$, $y = x$, $x = 4$, $z = 0$

27 $z = x^2 + y^2$, $y = 4 - x^2$, $x = 0$, $y = 0$, $z = 0$

28 $z = y^3$, $y = x^3$, $x = 0$, $z = 0$, $y = 1$

29 $z = x^3$, $x = 4y^2$, $16y = x^2$, $z = 0$

30 $x^2 + y^2 = 16$, $x = z$, $y = 0$, $z = 0$

Exer. 31–32: Find the volume of the solid bounded by the graphs of the equations.

31 $z = x^2 + 4$, $y = 4 - x^2$, $x + y = 2$, $z = 0$

32 $y = x^3$, $y = x^4$, $z - x - y = 4$, $z = 0$

C **Exer. 33–36:** Approximate the volume

$$V = \int_0^1 \int_0^{1/2} f(x, y) \, dy \, dx$$

by using the iterated trapezoidal rule for a partition of n subintervals of the x-interval $[0, 1]$ and m subintervals of the y-interval $[0, \frac{1}{2}]$.

33 $f(x, y) = \sin[\cos(xy)]$; $n = 10$, $m = 5$

34 $f(x, y) = \sqrt{x^4 + y^4}$; $n = 15$, $m = 10$

35 $f(x, y) = e^{-x^2 - y^2}$; $n = 10$, $m = 10$

36 $f(x, y) = \cos(e^{-xy})$; $n = 20$, $m = 10$

C **Exer. 37–40:** Approximate the volume

$$V = \int_0^2 \int_0^3 f(x, y) \, dy \, dx$$

to three decimal places.

37 $f(x, y) = e^{-x^3 - y^2}$

38 $f(x, y) = \sqrt{x^3 + y^3}$

39 $f(x, y) = \cos(xy + 1)$

40 $f(x, y) = \sin(\sin(xy))$

Mathematicians and Their Times

SRINIVASA RAMANUJAN

IN JANUARY 1913, the eminent British mathematician G. H. Hardy (1877–1947) received the most remarkable letter in the history of mathematics. "I beg to introduce myself to you as a clerk in the Accounts Department of the Port Trust Office at Madras," the letter, stuffed in a large envelope covered with stamps from India, began. "I am now about 23 years of age. I have had no university education but I have undergone the ordinary school course. After leaving school, I have been employing the spare time at my disposal to work at Mathematics. I have not trodden through the conventional regular course which is followed in a University course, but I am striking out a new path for myself."

This letter from Srinivasa Ramanujan (1887–1920) was crammed with cryptic statements of 120 mathematical theorems. Some were familiar; others Hardy himself had recently proved or thought he could prove. But then there appeared some extraordinary assertions that, Hardy wrote, "defeated me completely; I had never seen anything in the least like them before. A single look at them is enough to show that they could only be written down by a mathematician of the highest class. They must be true because, if they were not true, no one would have the imagination to invent them."

Ramanujan came from a poor Brahman family. He showed extraordinary mental powers as a youth and began doing mathematics intensely at age 15, working his way through a borrowed book from a local library. Unfortunately, he neglected his other studies and twice failed university entrance exams. Although he was forced to work at low-paying jobs, he continued to create mathematics. After writing to two well-known English mathematicians who rebuffed his requests for advice and assistance, he tried Hardy.

Convinced at once of Ramanujan's genius, Hardy bent his full efforts to assist him. He invited Ramanujan to England, where they spent five years working together. Known as the "Apostle of Proof" for his insistence on rigor, Hardy had keen respect for Ramanujan, "the Prince of Intuition," who had the deepest insight into mathematical truth, but no training in what constituted proof. The harsh English climate and declining health led Ramanujan to return to India, however, where he tragically died of tuberculosis at age 32.

Ramanujan left a rich legacy of discoveries that mathematicians are still trying to understand and prove. Primarily of interest in pure mathematics, Ramanujan's ideas are now appearing in many applications including super string theory in physics, statistical mechanics, computer science, cryptology, space travel, and cancer studies.

13.3 DOUBLE INTEGRALS IN POLAR COORDINATES

Figure 13.53

$\Delta A = \bar{r} \, \Delta r \, \Delta \theta$

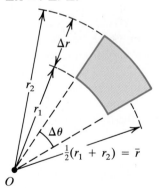

In this section, we discuss how to evaluate a double integral over a region R that is bounded by graphs of polar equations. Let us first consider the *elementary polar region* in Figure 13.24, bounded by arcs of circles of radii r_1 and r_2, with centers at the origin, and by two rays from the origin. If $\Delta\theta$ is the radian measure of the angle between the rays and if $\Delta r = r_2 - r_1$, then the area ΔA of the region is

$$\Delta A = \tfrac{1}{2} r_2^2 \Delta\theta - \tfrac{1}{2} r_1^2 \Delta\theta.$$

This formula may also be written

$$\Delta A = \tfrac{1}{2}(r_2^2 - r_1^2)\Delta\theta = \tfrac{1}{2}(r_2 + r_1)(r_2 - r_1)\Delta\theta.$$

If we denote the **average radius** $\tfrac{1}{2}(r_2 + r_1)$ by $\bar{r}$, then

$$\Delta A = \bar{r} \, \Delta r \, \Delta\theta.$$

Next, consider the region R of the type illustrated in Figure 13.25(a), bounded by two rays that make positive angles α and β with the polar axis

Figure 13.25   **(a)**   **(b)**

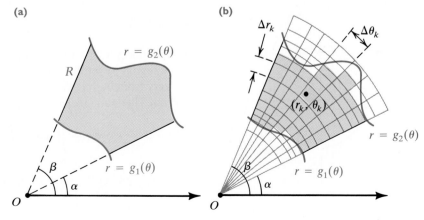

and by the graphs of two polar equations $r = g_1(\theta)$ and $r = g_2(\theta)$, where g_1 and g_2 are continuous functions and $g_1(\theta) \le g_2(\theta)$ for $\alpha \le \theta \le \beta$. If R is subdivided by means of circular arcs and rays as shown in Figure 13.25(b), then the collection of elementary polar regions $R_1, R_2, \ldots, R_n$ that lie completely within R is called an **inner polar partition** P of R. The norm $\|P\|$ of the partition is the length of the longest diagonal of the R_k. If we choose a point (r_k, θ_k) in R_k such that r_k is the average radius, then

$$\Delta A_k = r_k \, \Delta r_k \, \Delta \theta_k.$$

If f is a continuous function of the polar variables r and θ, then the following can be proved.

Evaluation Theorem 13.13

$$\lim_{\|P\| \to 0} \sum_k f(r_k, \theta_k) \, r_k \, \Delta r_k \, \Delta \theta_k = \iint\limits_R f(r, \theta) \, dA$$

$$= \int_\alpha^\beta \int_{g_1(\theta)}^{g_2(\theta)} f(r, \theta) \, r \, dr \, d\theta.$$

Figure 13.26

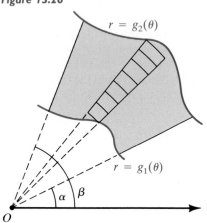

Note that the integrand on the right in (13.13) is the product of $f(r, \theta)$ and r. This is due to the fact that ΔA_k equals $r_k \Delta r_k \Delta \theta_k$.

We may regard the iterated integral in (13.13) as a limit of double sums. We *first hold θ fixed* and sum along the wedge-shaped region shown in Figure 13.26, from the graph of g_1 to the graph of g_2. For the second summation, we *sweep out* the region by letting θ vary from α to β. The iterated integral denotes the limit of these sums as $\|P\| \to 0$.

If $f(r, \theta) = 1$ throughout R, then the integral in (13.13) equals the area of R. This also follows from our work in Chapter 9, since after performing the partial integration with respect to r, we obtain the formula in Theorem (9.12).

Figure 13.27

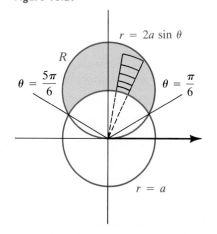

EXAMPLE ▪ 1 Find the area of the region R that lies outside the circle $r = a$ and inside the circle $r = 2a \sin \theta$.

SOLUTION The region is sketched in Figure 13.27, together with a typical wedge of elementary polar regions obtained by summing from the inner boundary $r = a$ to the outer boundary $r = 2a \sin \theta$. We then sweep out the region by letting θ vary from $\pi/6$ to $5\pi/6$. (A common error is to let θ vary from 0 to π. Why is this incorrect?)

From Theorem (13.13) with $f(r, \theta) = 1$,

$$A = \iint\limits_R dA = \int_{\pi/6}^{5\pi/6} \int_a^{2a \sin \theta} r \, dr \, d\theta.$$

To simplify the evaluation, we may use the symmetry of R with respect to the y-axis. In this case, we let θ vary from $\pi/6$ to $\pi/2$ and double the integral. Thus,

$$A = \iint\limits_{R} dA = 2 \int_{\pi/6}^{\pi/2} \int_{a}^{2a\sin\theta} r \, dr \, d\theta$$

$$= 2 \int_{\pi/6}^{\pi/2} \left[\tfrac{1}{2} r^2 \right]_{a}^{2a\sin\theta} d\theta = \int_{\pi/6}^{\pi/2} (4a^2 \sin^2\theta - a^2) \, d\theta$$

$$= a^2 \int_{\pi/6}^{\pi/2} \left(4 \frac{1 - \cos 2\theta}{2} - 1 \right) d\theta = a^2 \int_{\pi/6}^{\pi/2} (1 - 2\cos 2\theta) \, d\theta$$

$$= a^2 \left[\theta - \sin 2\theta \right]_{\pi/6}^{\pi/2} = a^2 \left[\left(\frac{\pi}{2} - 0 \right) - \left(\frac{\pi}{6} - \frac{\sqrt{3}}{2} \right) \right]$$

$$= a^2 \left(\frac{\pi}{3} + \frac{\sqrt{3}}{2} \right) \approx 1.9a^2.$$

EXAMPLE ■ 2 Find the area of the region R bounded by one loop of the lemniscate $r^2 = a^2 \sin 2\theta$, where $a > 0$.

SOLUTION The lemniscate is sketched in Figure 13.28, together with a wedge of elementary polar regions obtained by summing from the origin ($r = 0$) to the boundary $r^2 = a^2 \sin 2\theta$ of R in the first quadrant. We then sweep out the loop by letting θ vary from 0 to $\pi/2$. (Note that r is *undefined* if $\pi/2 < \theta < \pi$ or $3\pi/2 < \theta < 2\pi$.)

By Theorem (13.13),

$$A = \iint\limits_{R} dA = \int_{0}^{\pi/2} \int_{0}^{a\sqrt{\sin 2\theta}} r \, dr \, d\theta$$

$$= \int_{0}^{\pi/2} \left[\tfrac{1}{2} r^2 \right]_{0}^{a\sqrt{\sin 2\theta}} d\theta = \tfrac{1}{2} \int_{0}^{\pi/2} a^2 \sin 2\theta \, d\theta$$

$$= -\tfrac{1}{4} a^2 \left[\cos 2\theta \right]_{0}^{\pi/2} = -\tfrac{1}{4} a^2 (-1 - 1) = \tfrac{1}{2} a^2.$$

Figure 13.28

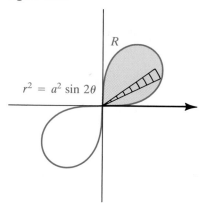

$r^2 = a^2 \sin 2\theta$

Under suitable conditions, an iterated double integral in rectangular coordinates can be transformed into a double integral in polar coordinates. First, the variables x and y in the integrand are replaced by $r \cos\theta$ and $r \sin\theta$. Next, in the iterated integral, $dy \, dx$ or $dx \, dy$ is replaced by $r \, dr \, d\theta$ or $r \, d\theta \, dr$. The following formula indicates the form of a typical integrand. The double integral sign $\iint_R$ should be replaced by suitable *iterated* integral signs and $dA = dy \, dx$ by $r \, dr \, d\theta$.

Change of Variables
Formula 13.14

$$\iint\limits_{R} f(x, y)\, dy\, dx = \iint\limits_{R} f(r\cos\theta, r\sin\theta) r\, dr\, d\theta$$

If the integrand $f(x, y)$ of a double integral $\iint_{R} f(x, y)\, dA$ contains the expression $x^2 + y^2$ or if the region R involves circular arcs with centers at the origin, then the introduction of polar coordinates often leads to a simple evaluation, because $x^2 + y^2 = r^2$ and the circular arcs have equations of the form $r = k$.

EXAMPLE 3   Use polar coordinates to evaluate

$$\int_{-a}^{a} \int_{0}^{\sqrt{a^2-x^2}} (x^2 + y^2)^{3/2}\, dy\, dx.$$

SOLUTION   The region of integration is bounded by the graphs of $y = 0$ (the x-axis) and $y = \sqrt{a^2 - x^2}$ (a semicircle), as shown in Figure 13.29. We first replace $x^2 + y^2$ in the integrand by r^2 and $dy\, dx$ by $r\, dr\, d\theta$. Next we change the limits to polar coordinates. Referring to Figure 13.29, we find that

$$\int_{-a}^{a} \int_{0}^{\sqrt{a^2-x^2}} (x^2 + y^2)^{3/2}\, dy\, dx = \int_{0}^{\pi} \int_{0}^{a} r^3 r\, dr\, d\theta$$

$$= \int_{0}^{\pi} \left[\tfrac{1}{5} r^5 \right]_{0}^{a} d\theta = \tfrac{1}{5} a^5 \int_{0}^{\pi} d\theta$$

$$= \tfrac{1}{5} a^5 \left[\theta \right]_{0}^{\pi} = \tfrac{1}{5}\pi a^5.$$

Figure 13.29

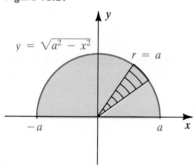

Polar coordinates can also be used for double integrals if R is a region of the type illustrated in Figure 13.30(a). In this case, R is bounded by the arcs of two circles of radii a and b and by the graphs of two polar equations

Figure 13.30

(a)

$\theta = h_2(r)$

R

$\theta = h_1(r)$

O a b

(b)

$\theta = h_2(r)$

$\theta = h_1(r)$

a b

$\theta = h_1(r)$ and $\theta = h_2(r)$, where h_1 and h_2 are continuous functions and $h_1(r) \le h_2(r)$ for $a \le r \le b$. If f is a function of r and θ and if f is continuous on R, then the limit in (13.13) exists and the evaluation formula is given in the next theorem.

Evaluation Theorem 13.15

$$\iint\limits_{R} f(r, \theta)\, dA = \int_a^b \int_{h_1(r)}^{h_2(r)} f(r, \theta)r\, d\theta\, dr$$

The iterated integral in Theorem (13.15) may be interpreted in terms of limits of double sums. We first hold r fixed and sum along a circular arc, illustrated by the pink region in Figure 13.30(b). We next sweep out R by summing terms that correspond to such ring-shaped regions from $r = a$ to $r = b$.

E X A M P L E ■ 4   Find the area of the smaller of the regions bounded by the polar axis, the graphs of $r = 1$ and $r = 2$, and the part of the spiral $r\theta = 1$ from $\theta = \frac{1}{2}$ to $\theta = 1$.

Figure 13.31

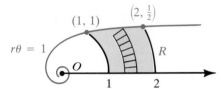

S O L U T I O N   The region R, together with a ring of elementary polar regions, is sketched in Figure 13.31. We apply Theorem (13.15) with $f(r, \theta) = 1$:

$$A = \iint\limits_{R} dA = \int_1^2 \int_0^{1/r} r\, d\theta\, dr$$

$$= \int_1^2 r\left[\theta\right]_0^{1/r} dr = \int_1^2 r\left(\frac{1}{r}\right) dr$$

$$= \int_1^2 dr = \left[r\right]_1^2 = 2 - 1 = 1$$

As in the preceding section, if $f(x, y) \ge 0$, then the volume V of the solid that lies under the graph of $z = f(x, y)$ and over a region R in the xy-plane is given by

$$V = \iint\limits_{R} f(x, y)\, dA.$$

Sometimes it is convenient to evaluate this double integral by using polar coordinates, as in the next example.

E X A M P L E ■ 5   Find the volume V of the solid that is bounded by the paraboloid $z = 4 - x^2 - y^2$ and the xy-plane.

SOLUTION   The first-octant portion of the solid is sketched in Figure 13.32(a). By symmetry, we may find the volume of this portion and multiply the result by 4.

The region R in the xy-plane is bounded by the coordinate axes and one fourth of the circle with polar equation $r = 2$, sketched in Figure 13.32(b), where we have also shown a wedge of elementary polar regions. The base of the prism in Figure 13.32(a) corresponds to *one* elementary polar region. We can obtain the volume of the solid by using $r^2 = x^2 + y^2$ and the change of variables formula (13.14):

$$V = 4 \iint\limits_R (4 - x^2 - y^2)\, dA$$

$$= 4 \int_0^{\pi/2} \int_0^2 (4 - r^2) r \, dr \, d\theta$$

$$= 4 \int_0^{\pi/2} \left[2r^2 - \tfrac{1}{4}r^4 \right]_0^2 d\theta = 4 \int_0^{\pi/2} 4 \, d\theta = 16 \left[\theta \right]_0^{\pi/2} = 8\pi$$

Figure 13.32

(a)

(b)

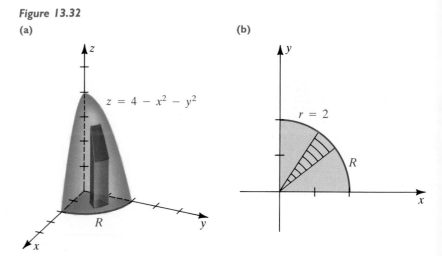

The same problem, stated in terms of rectangular coordinates, leads to the following double integral:

$$V = 4 \iint\limits_R (4 - x^2 - y^2)\, dA = 4 \int_0^2 \int_0^{\sqrt{4-x^2}} (4 - x^2 - y^2) \, dy \, dx$$

Evaluating the last integral should thoroughly convince anyone of the advantage of using polar coordinates in certain problems.

EXERCISES 13.3

Exer. 1–6: Express the area of the region as an iterated double integral in polar coordinates, using symmetry whenever possible.

1

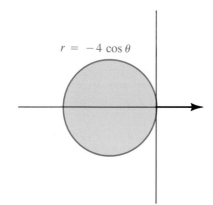

$r = 4 \sin \theta$

2

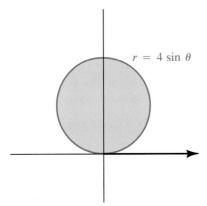

$r = -4 \cos \theta$

3

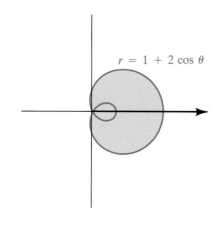

$r = 1 + 2 \cos \theta$

4

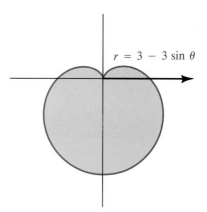

$r = 3 - 3 \sin \theta$

5

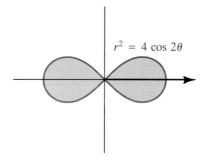

$r^2 = 4 \cos 2\theta$

6

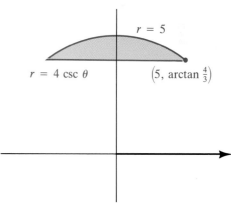

$r = 5$

$r = 4 \csc \theta$

$\left(5, \arctan \frac{4}{3}\right)$

Exer. 7–12: Use a double integral to find the area of the region that has the indicated shape.

7 One loop of $r = 4 \sin 3\theta$

8 One loop of $r = 2 \cos 4\theta$

9 Inside $r = 2 - 2 \cos \theta$ and outside $r = 3$

10 Bounded by $r = 3 + 2 \sin \theta$

11 One loop of $r^2 = 9\cos 2\theta$

12 Inside $r = 3\sin\theta$ and outside $r = 1 + \sin\theta$

Exer. 13–24: Use polar coordinates to evaluate the integral.

13 $\displaystyle\iint\limits_{R} (x^2 + y^2)^{3/2}\, dA$;

R is bounded by the circle $x^2 + y^2 = 4$.

14 $\displaystyle\iint\limits_{R} x^2(x^2 + y^2)^3\, dA$;

R is bounded by the semicircle $y = \sqrt{1 - x^2}$ and the x-axis.

15 $\displaystyle\iint\limits_{R} \frac{x^2}{x^2 + y^2}\, dA$;

R is the annular region bounded by $x^2 + y^2 = a^2$ and $x^2 + y^2 = b^2$ with $0 < a < b$.

16 $\displaystyle\iint\limits_{R} (x + y)\, dA$;

R is bounded by the circle $x^2 + y^2 = 2y$.

17 $\displaystyle\iint\limits_{R} \sqrt{x^2 + y^2}\, dA$;

R is bounded by the triangle with vertices $(0, 0)$, $(3, 0)$, $(3, 3)$.

18 $\displaystyle\iint\limits_{R} \sqrt{x^2 + y^2}\, dA$;

R is bounded by the semicircle $y = \sqrt{2x - x^2}$ and the line $y = x$.

19 $\displaystyle\int_{-a}^{a}\int_{0}^{\sqrt{a^2 - x^2}} e^{-(x^2 + y^2)}\, dy\, dx$

20 $\displaystyle\int_{0}^{a}\int_{0}^{\sqrt{a^2 - x^2}} (x^2 + y^2)^{3/2}\, dy\, dx$

21 $\displaystyle\int_{1}^{2}\int_{0}^{x} \frac{1}{\sqrt{x^2 + y^2}}\, dy\, dx$

22 $\displaystyle\int_{0}^{1}\int_{0}^{\sqrt{1 - x^2}} e^{\sqrt{x^2 + y^2}}\, dy\, dx$

23 $\displaystyle\int_{0}^{2}\int_{0}^{\sqrt{4 - y^2}} \cos(x^2 + y^2)\, dx\, dy$

24 $\displaystyle\int_{0}^{2}\int_{-\sqrt{2y - y^2}}^{\sqrt{2y - y^2}} x\, dx\, dy$

Exer. 25–30: Use polar coordinates to find the volume of the solid that has the shape of Q.

25 Q is the region inside the sphere $x^2 + y^2 + z^2 = 25$ and outside the cylinder $x^2 + y^2 = 9$.

26 Q is cut out of the ellipsoid $4x^2 + 4y^2 + z^2 = 16$ by the cylinder $x^2 + y^2 = 1$.

27 Q is bounded by the cone $z^2 = x^2 + y^2$ and the cylinder $x^2 + y^2 = 2x$.

28 Q is bounded by the paraboloid $z = 4x^2 + 4y^2$, the cylinder $x^2 + y^2 = 3y$, and the plane $z = 0$.

29 Q is the largest region that is inside both the sphere $x^2 + y^2 + z^2 = 16$ and the cylinder $x^2 + y^2 = 4y$.

30 Q is bounded by the paraboloid $z = 9 - x^2 - y^2$ and the plane $z = 5$.

31 Let R_a be the region bounded by the circle with equation $x^2 + y^2 = a^2$. If we define

$$\int_{-\infty}^{\infty}\int_{-\infty}^{\infty} e^{-(x^2 + y^2)}\, dx\, dy = \lim_{a \to \infty} \iint\limits_{R_a} e^{-(x^2 + y^2)}\, dA,$$

evaluate the improper double integral.

32 (a) It can be proved, using advanced methods, that

$$\int_{-\infty}^{\infty}\int_{-\infty}^{\infty} e^{-(x^2 + y^2)}\, dx\, dy = \int_{-\infty}^{\infty} e^{-x^2}\, dx \int_{-\infty}^{\infty} e^{-y^2}\, dy.$$

Use this fact and Exercise 31 to show that $\int_{-\infty}^{\infty} e^{-x^2}\, dx = \sqrt{\pi}$. Interpret this result geometrically.

(b) Use part (a) to prove the following result, which is important in the field of statistics:

$$\frac{1}{\sqrt{2\pi}} \int_{-\infty}^{\infty} e^{-x^2/2}\, dx = 1$$

[c] **Exer. 33–38:** We can sometimes evaluate a double integral by changing it to a single integral. Use polar coordinates to change the given double integral to a single integral involving only the variable r, and approximate the single integral to three decimal places.

33 $\displaystyle\iint\limits_{R} \sqrt{1 + (x^2 + y^2)^2}\, dA$;

R is the region in the first quadrant bounded by the circle $x^2 + y^2 = 4$ and the coordinate axes.

34 $\displaystyle\iint\limits_{R} \sin\sqrt[3]{x^2 + y^2}\, dA$;

R is the region bounded by the semicircle $x = \sqrt{1 - y^2}$ and the y-axis.

35 $\displaystyle\int_0^3 \int_0^{\sqrt{9-x^2}} \cos\sqrt{x^2+y^2}\,dy\,dx$

36 $\displaystyle\int_0^2 \int_{-\sqrt{4-y^2}}^{\sqrt{4-y^2}} e^{-(x^2+y^2)^2}\,dx\,dy$

37 $\displaystyle\int_{-1}^1 \int_{-\sqrt{1-x^2}}^{\sqrt{1-x^2}} \sqrt{(x^2+y^2)^3+1}\,dy\,dx$

38 $\displaystyle\int_{-5}^5 \int_0^{\sqrt{25-x^2}} \sin\left[(x^2+y^2)^{3/2}-3(x^2+y^2)+5\right]\,dy\,dx$

13.4 SURFACE AREA

Figure 13.33

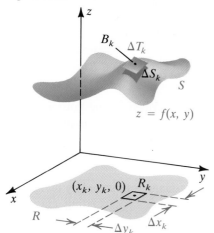

In Section 5.5, we obtained formulas for the area of a surface of revolution. We now consider a method for finding areas of more general surfaces. Suppose that $f(x, y) \geq 0$ throughout a region R in the xy-plane and that f has continuous first partial derivatives on R. Let S denote the portion of the graph of f whose projection on the xy-plane is R, as illustrated in Figure 13.33. To simplify the discussion, we assume that no normal vector to S is parallel to the xy-plane. We wish to define the area A of S and find a formula that can be used to calculate A.

Let $P = \{R_k\}$ be an inner partition of R, and let the dimensions of the rectangle R_k be Δx_k and Δy_k. For each k, choose any point $(x_k, y_k, 0)$ in R_k, and let $B_k(x_k, y_k, f(x_k, y_k))$ be the corresponding point on S. Next consider the tangent plane to S at B_k (see Figure 13.33), and let ΔT_k and ΔS_k be the areas of the regions on the tangent plane and on S, respectively, obtained by projecting R_k vertically upward. If the norm $\|P\|$ of the partition is small, then ΔT_k is an approximation to ΔS_k, and $\sum_k \Delta T_k$ is an approximation to the area A of S. This approximation should improve as $\|P\|$ approaches 0, so we arrive at the following.

Definition 13.16

$$A = \lim_{\|P\|\to 0} \sum_k \Delta T_k$$

Figure 13.34

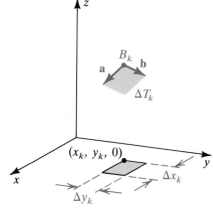

To find an integral formula for A, let us choose $(x_k, y_k, 0)$ as the corner of R_k closest to the origin. As in the isolated view of ΔT_k shown in Figure 13.34, consider the vectors $\mathbf{a}$ and $\mathbf{b}$ with initial point $B_k(x_k, y_k, f(x_k, y_k))$, which are tangent to the traces of S on the planes $y = y_k$ and $x = x_k$, respectively. By Theorem (12.10), the slopes of the lines determined by $\mathbf{a}$ and $\mathbf{b}$ in these planes are $f_x(x_k, y_k)$ and $f_y(x_k, y_k)$, respectively. It follows that

$$\mathbf{a} = \Delta x_k \mathbf{i} + f_x(x_k, y_k)\Delta x_k \mathbf{k},$$

$$\mathbf{b} = \Delta y_k \mathbf{j} + f_y(x_k, y_k)\Delta y_k \mathbf{k}.$$

From Section 10.4, the area ΔT_k of the parallelogram determined by $\mathbf{a}$ and $\mathbf{b}$ is $\|\mathbf{a} \times \mathbf{b}\|$. Using

$$\mathbf{a} \times \mathbf{b} = \begin{vmatrix} \mathbf{i} & \mathbf{j} & \mathbf{k} \\ \Delta x_k & 0 & f_x(x_k, y_k)\Delta x_k \\ 0 & \Delta y_k & f_y(x_k, y_k)\Delta y_k \end{vmatrix}$$

gives us

$$\mathbf{a} \times \mathbf{b} = -f_x(x_k, y_k)\Delta x_k \Delta y_k \mathbf{i} - f_y(x_k, y_k)\Delta x_k \Delta y_k \mathbf{j} + \Delta x_k \Delta y_k \mathbf{k}.$$

Consequently,

$$\Delta T_k = \|\mathbf{a} \times \mathbf{b}\| = \sqrt{[f_x(x_k, y_k)]^2 + [f_y(x_k, y_k)]^2 + 1}\ \Delta x_k \Delta y_k,$$

where $\Delta x_k \Delta y_k = \Delta A_k$. If, as in (13.16), we take the limit of sums of the ΔT_k and apply the definition of the double integral, we obtain the following.

Integral for Surface Area | **13.17**

$$A = \iint_R \sqrt{[f_x(x, y)]^2 + [f_y(x, y)]^2 + 1}\ dA$$

The formula (13.17) may also be used if $f(x, y) \le 0$ on R.

EXAMPLE ■ 1 A billowing sail is described as the portion of the graph of $z = 3x + y^2$ that lies over the triangular region R in the xy-plane with vertices $(0, 0, 0)$, $(0, 1, 0)$, and $(1, 1, 0)$. Find the surface area A of the sail.

Figure 13.35

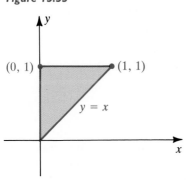

SOLUTION The region R in the xy-plane is bounded by the graphs of $y = x$, $x = 0$, and $y = 1$, as shown in Figure 13.35. Letting $f(x, y) = 3x + y^2$ and applying (13.17) gives us

$$A = \iint_R \sqrt{3^2 + (2y)^2 + 1}\ dA = \int_0^1 \int_0^y (10 + 4y^2)^{1/2}\ dx\, dy$$

$$= \int_0^1 (10 + 4y^2)^{1/2} [x]_0^y\, dy = \int_0^1 (10 + 4y^2)^{1/2} y\, dy$$

$$= \left[\frac{1}{12}(10 + 4y^2)^{3/2}\right]_0^1 = \frac{14^{3/2} - 10^{3/2}}{12} \approx 1.7 \text{ square units.}$$

EXAMPLE ■ 2 Find the surface area of the paraboloid given by $z = 4 - x^2 - y^2$ for $z \ge 0$.

SOLUTION The paraboloid is sketched in Figure 13.36(a). Applying (13.17) with $f(x, y) = 4 - x^2 - y^2$, we have

$$A = \iint_R \sqrt{(-2x)^2 + (-2y)^2 + 1}\ dA$$

$$= \iint_R \sqrt{4x^2 + 4y^2 + 1}\ dA,$$

Figure 13.36

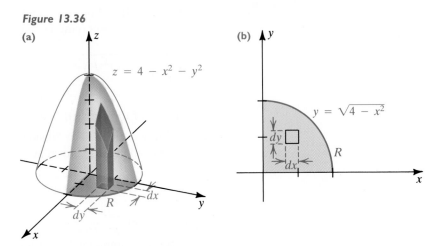

(a) $z = 4 - x^2 - y^2$

(b) $y = \sqrt{4 - x^2}$ R

where R is the region in the xy-plane bounded by the circle $x^2 + y^2 = 4$. By symmetry, we can find the surface area corresponding to the first-quadrant portion of R shown in Figure 13.36(b) and then multiply this result by 4. Using an iterated integral yields

$$A = 4 \int_0^2 \int_0^{\sqrt{4-x^2}} \sqrt{4x^2 + 4y^2 + 1} \, dy \, dx.$$

It would be very tedious to evaluate this integral, so we change to polar coordinates and proceed as follows:

$$A = 4 \int_0^{\pi/2} \int_0^2 \sqrt{4r^2 + 1} \, r \, dr \, d\theta$$

$$= 4 \int_0^{\pi/2} \left[\tfrac{1}{12}(4r^2 + 1)^{3/2} \right]_0^2 d\theta$$

$$= 4 \int_0^{\pi/2} \tfrac{1}{12}(17^{3/2} - 1) \, d\theta$$

$$= \frac{1}{3}(17^{3/2} - 1) \left[\theta \right]_0^{\pi/2} = \frac{\pi}{6}(17^{3/2} - 1) \approx 36.177$$

Formulas similar to (13.17) may be stated if the surface S has suitable projections on the yz- or xz-planes. Thus, if S is the graph of an equation $y = g(x, z)$ and if the projection on the xz-plane is R_1, then

$$A = \iint_{R_1} \sqrt{[g_x(x, z)]^2 + [g_z(x, z)]^2 + 1} \, dA.$$

A formula of this type may also be stated if S is given by $x = h(y, z)$.

EXAMPLE = 3   Approximate the surface area of the portion of the graph of $f(x, y) = 5 - 3x^2 - y^2$ that lies above the square R in the xy-plane with vertices $(-1, -1)$, $(1, -1)$, $(1, 1)$, and $(-1, 1)$.

SOLUTION For the given function f, we have $f_x(x, y) = -6x$ and $f_y(x, y) = -2y$. Applying Definition (13.17), we obtain

$$A = \iint\limits_R \sqrt{(-6x)^2 + (-2y)^2 + 1}\, dA = \int_{-1}^{1} \int_{-1}^{1} \sqrt{36x^2 + 4y^2 + 1}\, dy\, dx.$$

Evaluation of this iterated integral in closed form by antidifferentiation is exceedingly difficult. The "inside integral," $\int_{-1}^{1} \sqrt{36x^2 + 4y^2 + 1}\, dy$, can be determined in closed form as

$$g(x) = \sqrt{36x^2 + 5} + \frac{36x^2 + 1}{4} \ln\left(\frac{2 + \sqrt{36x^2 + 5}}{-2 + \sqrt{36x^2 + 5}}\right)$$

so that $A = \int_{-1}^{1} g(x)\, dx$, but we cannot easily find an antiderivative for g. Thus, we use a numerical integration method to find an approximate value for the surface area, which is 13.9508.

EXERCISES 13.4

Exer. 1–4: Set up an iterated double integral that can be used to find the surface area of the portion of the graph of the equation that lies over the region R in the xy-plane that has the given boundary. Use symmetry whenever possible.

1 $x^2 + y^2 + z^2 = 4$;
the square with vertices $(1, 1)$, $(1, -1)$, $(-1, 1)$, $(-1, -1)$

2 $x^2 - y^2 + z^2 = 1$;
the square with vertices $(0, 1)$, $(1, 0)$, $(-1, 0)$, $(0, -1)$

3 $36z^2 = 16x^2 + 9y^2 + 144$;
the circle with center at the origin and radius 3

4 $9y^2 = 225z^2 + 25x^2$;
the triangle with vertices $(0, 0)$, $(3, 5)$, $(-3, 5)$

Exer. 5–6: Find the surface area of the portion of the graph of the equation that lies over the region R in the xy-plane that has the given boundary.

5 $z = y + \frac{1}{2}x^2$;
the square with vertices $(0, 0)$, $(1, 0)$, $(1, 1)$, $(0, 1)$

6 $z = y^2$;
the triangle with vertices $(0, 0)$, $(0, 2)$, $(2, 2)$

7 A portion of the plane $(x/a) + (y/b) + (z/c) = 1$ is cut out by the cylinder $x^2 + y^2 = k^2$, where a, b, c, and k are positive. Find the area of that portion.

8 Find the surface area of the first-octant portion of the cylinder $y^2 + z^2 = 9$ that lies inside the cylinder $x^2 + y^2 = 9$ (see Figure 13.23).

Exer. 9–12: Find the area of the surface S.

9 S is the part of the paraboloid $z = x^2 + y^2$ cut off by the plane $z = 1$.

10 S is the part of the plane $z = y + 1$ that is inside the cylinder $x^2 + y^2 = 1$.

11 S is the part of the sphere $x^2 + y^2 + z^2 = a^2$ that is inside the cylinder $x^2 + y^2 = ay$.

12 S is the first-octant part of the hyperbolic paraboloid $z = x^2 - y^2$ that is inside the cylinder $x^2 + y^2 = 1$.

13 A dome-shaped camping tent is designed to have a circular floor of radius 5 ft and a roof described by the graph of $z = 7 - \frac{7}{25}(x^2 + y^2)$ for $z \geq 0$. Approximate the number of square feet of material needed to construct the tent. (Do not consider wasted or overlapping material.)

14 A Van de Graaff generator can be used to produce millions of volts of static electricity and large sparks. Part of the generator consists of a spherical metal shell with a circular hole cut out of the bottom. The total charge C (in coulombs) generated is given by $C = \delta A$, where A is the area of the metal surface and δ is the

area charge density on the surface. Suppose the sphere can be described by $x^2 + y^2 + z^2 = 1$, and the hole is obtained by intersecting the cylinder $x^2 + y^2 = \frac{1}{16}$ with the sphere for $z < 0$, where x, y, and z are in meters. Approximate C if $\delta = 1.2 \times 10^{-5}$ coulomb/m². (This value of C corresponds to a voltage potential of approximately 1.35 million volts.)

Exer. 15 – 20: Approximate the surface area of the portion of the graph of f that lies over the rectangular region R with the given vertices in the *xy*-plane to three decimal places.

15 $f(x, y) = \frac{1}{3}x^3 + \cos y$; (0, 0), (0, 1), (1, 1), (1, 0)

16 $f(x, y) = x^3 y$; $(\frac{1}{4}, \frac{1}{4})$, $(\frac{1}{4}, \frac{3}{4})$, $(\frac{3}{4}, \frac{1}{4})$, $(\frac{3}{4}, \frac{3}{4})$

17 $f(x, y) = x^4 y^2 + xy + 2$;

(0, 0), (1, 0), (1, 2), (0, 2)

18 $f(x, y) = e^{-(x^2+y^2)}$;

(−1, −1), (−1, 1), (1, 1), (1, −1)

19 $f(x, y) = (x + y)^3$;

(0, 0), (0, 0.5), (1, 0.5), (1, 0)

20 $f(x, y) = \sin(x + y)$;

(0, 0), (0, 0.3), (0.2, 0.3), (0.2, 0)

13.5 TRIPLE INTEGRALS

We consider in this section how to generalize the development of double integrals to *triple integrals*. *Triple integrals* for a function f of *three* variables x, y, and z can be defined using a four-step process similar to that given in Section 13.1 for functions of two variables. The simplest case occurs if f is continuous throughout a box-shaped region Q in three dimensions of the following type:

$$Q = \{(x, y, z) : a \le x \le b, c \le y \le d, m \le z \le n\}$$

A possible position for Q is illustrated in Figure 13.37(a). If Q is divided into subregions $Q_1, Q_2, \ldots, Q_r$ by means of planes parallel to the three coordinate planes (see Figure 13.37b), then the collection $\{Q_k\}$ is a **partition** P of Q. The **norm** $\|P\|$ of the partition is the length of the longest

Figure 13.37

(a)   (b)   (c)

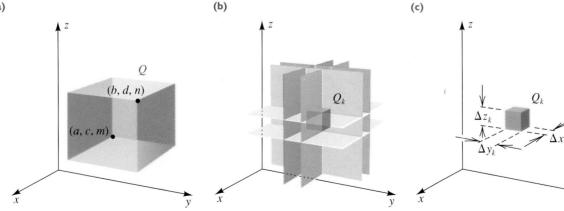

diagonal of all the Q_k. As shown in Figure 13.37(c), if Δx_k, Δy_k, and Δz_k are the dimensions of Q_k, then its volume ΔV_k is

$$\Delta V_k = \Delta x_k \Delta y_k \Delta z_k.$$

A **Riemann sum** of f for P is

$$\sum_k f(u_k, v_k, w_k) \Delta V_k,$$

where (u_k, v_k, w_k) is any point in Q_k. A limit of Riemann sums, as $\|P\| \to 0$, is defined as it was for functions of two variables. If the limit exists, it is called the **triple integral of f over Q** and is denoted by $\iiint_Q f(x, y, z)\, dV$. Using limit notation, we have the following.

Triple Integral of f over Q 13.18

$$\iiint_Q f(x, y, z)\, dV = \lim_{\|P\| \to 0} \sum_k f(u_k, v_k, w_k) \Delta V_k$$

It can be shown that if Q is the region in Figure 13.37(a), then

$$\iiint_Q f(x, y, z)\, dV = \int_m^n \int_c^d \int_a^b f(x, y, z)\, dx\, dy\, dz.$$

The iterated integral on the right is evaluated by beginning with the *innermost* integral and working outward. Thus, the first integration is with respect to x (with y and z fixed), the second is with respect to y (with z fixed), and the third is with respect to z. There are five other iterated integrals that equal the triple integral of f over Q. For example, if we integrate in the order y, z, x, then

$$\iiint_Q f(x, y, z)\, dV = \int_a^b \int_m^n \int_c^d f(x, y, z)\, dy\, dz\, dx.$$

It can be shown that for the box-shaped region Q in Figure 13.37(a), the order of integration is immaterial.

EXAMPLE ▪ 1   Evaluate $\displaystyle\iiint_Q (xy^2 + yz^3)\, dV$ if

$$Q = \{(x, y, z): -1 \le x \le 1, 3 \le y \le 4, 0 \le z \le 2\}.$$

SOLUTION   Of the six possible iterated integrals, we shall use the following:

$$\int_3^4 \int_{-1}^1 \int_0^2 (xy^2 + yz^3)\, dz\, dx\, dy = \int_3^4 \int_{-1}^1 \left[\int_0^2 (xy^2 + yz^3)\, dz \right] dx\, dy$$

$$= \int_3^4 \int_{-1}^1 \left[xy^2 z + y \left(\frac{z^4}{4} \right) \right]_0^2 dx\, dy$$

$$= \int_3^4 \int_{-1}^1 (2xy^2 + 4y)\, dx\, dy$$

$$= \int_3^4 \left[2 \left(\frac{x^2}{2} \right) y^2 + 4yx \right]_{-1}^1 dy$$

$$= \int_3^4 [(y^2 + 4y) - (y^2 - 4y)]\, dy$$

$$= \int_3^4 8y\, dy = 8 \left[\frac{y^2}{2} \right]_3^4 = 28$$

Triple integrals may be defined over a region more complicated than a rectangular box. Suppose, for example, that R is a region in the xy-plane that can be divided into R_x and R_y regions and that Q is the region in three dimensions defined by

$$Q = \{(x, y, z): (x, y) \text{ is in } R \text{ and } k_1(x, y) \le z \le k_2(x, y)\},$$

where k_1 and k_2 are functions that have continuous first partial derivatives throughout R. The region Q lies between the graphs of $z = k_1(x, y)$ and $z = k_2(x, y)$ and over or under the region R (see Figure 13.38). If Q is subdivided by means of planes parallel to the three coordinate planes, then the resulting (small) box-shaped regions $Q_1, Q_2, \ldots, Q_n$ that lie completely within Q form an **inner partition** P of Q. A typical element Q_k of an inner partition of Q is shown in Figure 13.38.

A **Riemann sum** of f for P is a sum of the form $\sum_k f(u_k, v_k, w_k)\Delta V_k$, where (u_k, v_k, w_k) is an arbitrary point in Q_k and ΔV_k is the volume of Q_k. The triple integral of f over Q is again defined as the limit (13.18). If f is continuous throughout Q, then the following can be proved.

Figure 13.38

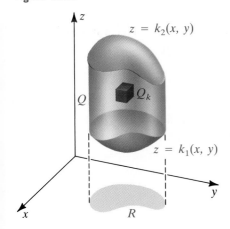

z = k_2(x, y)

Q   *Q_k*

z = k_1(x, y)

R

Evaluation Theorem   13.19

$$\iiint\limits_Q f(x, y, z)\, dV = \iint\limits_R \left[\int_{k_1(x, y)}^{k_2(x, y)} f(x, y, z)\, dz \right] dA$$

The notation on the right in (13.19) means that after first integrating with respect to z, we evaluate the resulting double integral over the region

R in the xy-plane using the methods of Section 13.1. Thus, for an R_x region such as that in Figure 13.8(a),

$$\iiint_Q f(x, y, z)\, dV = \int_a^b \int_{g_1(x)}^{g_2(x)} \int_{k_1(x,y)}^{k_2(x,y)} f(x, y, z)\, dz\, dy\, dx.$$

The symbol on the right-hand side of this equation is an **iterated triple integral**. It is evaluated by means of partial integrations of $f(x, y, z)$ in the order z, y, x, with the indicated limits substituted in the usual way. Similarly, for an R_y region such as that in Figure 13.8(b),

$$\iiint_Q f(x, y, z)\, dV = \int_c^d \int_{h_1(y)}^{h_2(y)} \int_{k_1(x,y)}^{k_2(x,y)} f(x, y, z)\, dz\, dx\, dy.$$

It is convenient to regard either of the last two integrals as a limit of a *triple sum*,

$$\sum_i \sum_j \sum_k f(u_i, v_j, w_k)\Delta z_k \Delta y_j \Delta x_i,$$

where the first summation (with respect to k) corresponds to a column of (small) boxes, in the direction parallel to the z-axis, from the lower surface (with equation $z = k_1(x, y)$) to the upper surface (with equation $z = k_2(x, y)$), as illustrated in Figure 13.39. The remaining two summations take place over the region R in the xy-plane, as in our work with double integrals in Section 13.1.

Figure 13.39

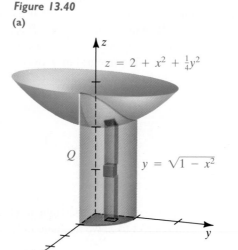

EXAMPLE ■ 2   Express $\iiint_Q f(x, y, z)\, dV$ as an iterated integral if Q is the region in the first octant bounded by the coordinate planes, the paraboloid $z = 2 + x^2 + \frac{1}{4}y^2$, and the cylinder $x^2 + y^2 = 1$.

SOLUTION   As illustrated in Figure 13.40(a), Q lies under the paraboloid and over the xy-plane. The region R in the xy-plane is bounded by the coordinate axes and the graph of $y = \sqrt{1 - x^2}$. Also

Figure 13.40

(a)

(b)

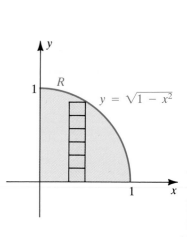

shown in the figure is a column that corresponds to a first summation over the boxes of volumes $\Delta z_k \Delta y_j \Delta x_i$ in the direction of the z-axis. Since the column extends from the xy-plane to the paraboloid, the lower limit for the integration with respect to z is $z = 0$ and the upper limit is $z = 2 + x^2 + \frac{1}{4}y^2$. The second and third integrations are taken over the region R in the xy-plane (see Figure 13.40b). Thus, the integral in (13.19) has the form

$$\int_0^1 \int_0^{\sqrt{1-x^2}} \int_0^{2+x^2+(1/4)y^2} f(x, y, z)\, dz\, dy\, dx.$$

If $f(x, y, z) = 1$ throughout a region Q, then the triple integral of f over Q is written $\iiint_Q dV$ and its value is the volume of Q. In Example 2, the volume V of the region Q in Figure 13.40(a) is given by

$$V = \int_0^1 \int_0^{\sqrt{1-x^2}} \int_0^{2+x^2+(1/4)y^2} dz\, dy\, dx.$$

We could, of course, obtain the volume using a double integral, as in Section 13.2. The reason for using a triple integral here and in the following examples is to gain practice in working with limits of triple sums. This same technique will be used later for problems that *cannot* be solved using double integrals.

EXAMPLE ■ 3   Find the volume V of the solid that is bounded by the cylinder $y = x^2$ and by the planes $y + z = 4$ and $z = 0$.

SOLUTION   The solid is sketched in Figure 13.41(a), together with a column corresponding to a first summation in the direction of the z-axis. Note that the column extends from $z = 0$ to $z = 4 - y$. The region R in the xy-plane is shown in Figure 13.41(b), together with a rectangle corre-

Figure 13.41

(a)

(b)

sponding to the first (of a double) integration with respect to y. Applying (13.19) with $f(x, y, z) = 1$, we have

$$V = \int_{-2}^{2} \int_{x^2}^{4} \int_{0}^{4-y} dz\, dy\, dx = \int_{-2}^{2} \int_{x^2}^{4} [z]_0^{4-y}\, dy\, dx$$

$$= \int_{-2}^{2} \int_{x^2}^{4} (4 - y)\, dy\, dx = \int_{-2}^{2} \left[4y - \tfrac{1}{2}y^2\right]_{x^2}^{4} dx$$

$$= \int_{-2}^{2} (8 - 4x^2 + \tfrac{1}{2}x^4)\, dx$$

$$= \left[8x - \tfrac{4}{3}x^3 + \tfrac{1}{10}x^5\right]_{-2}^{2}$$

$$= \tfrac{256}{15} \approx 17.07.$$

If we use the order $dx\, dy$ for the double integral, then the rectangle in Figure 13.41(b) will be horizontal and the formula for V will become

$$V = \int_0^4 \int_{-\sqrt{y}}^{\sqrt{y}} \int_0^{4-y} dz\, dx\, dy.$$

Some triple integrals may be evaluated by means of an iterated triple integral in which the first integration is with respect to y. Thus, consider

$$Q = \{(x, y, z): a \le x \le b, h_1(x) \le z \le h_2(x), k_1(x, z) \le y \le k_2(x, z)\},$$

where h_1 and h_2 are functions that are continuous on $[a, b]$ and k_1 and k_2 are functions that have continuous first partial derivatives on the region R in the xz-plane shown in Figure 13.42. Note that Q lies between the graphs of $y = k_1(x, z)$ and $y = k_2(x, z)$. The projection R of Q onto the xz-plane is an R_x region. In this case, we have the following.

Figure 13.42

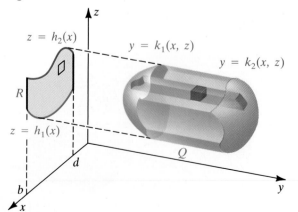

Evaluation Theorem **13.20**

$$\iiint\limits_{Q} f(x, y, z) \, dV = \int_a^b \int_{h_1(x)}^{h_2(x)} \int_{k_1(x,z)}^{k_2(x,z)} f(x, y, z) \, dy \, dz \, dx$$

The iterated integral in (13.20) may be interpreted as a limit of triple sums obtained by first summing over a row of small boxes in the direction parallel to the y-axis from the left surface (with equation $y = k_1(x, z)$) to the right surface (with equation $y = k_2(x, z)$), as indicated in Figure 13.42. The resulting double integral is then evaluated over the region R in the xz-plane, as illustrated in the next example.

EXAMPLE 4   Find the volume of the region Q bounded by the graphs of $z = 3x^2$, $z = 4 - x^2$, $y = 0$, and $z + y = 6$.

SOLUTION   As shown in Figure 13.43(a), Q lies under the cylinder $z = 4 - x^2$, over the cylinder $z = 3x^2$, to the right of the xz-plane, and to the left of the plane $z + y = 6$. Hence Q is a region of the type illustrated in Figure 13.42 with $k_1(x, z) = 0$ and $k_2(x, z) = 6 - z$. The region R in the xz-plane is sketched in Figure 13.43(b). Applying (13.20), we have

$$V = \iiint\limits_{Q} dV = \int_{-1}^{1} \int_{3x^2}^{4-x^2} \int_{0}^{6-z} dy \, dz \, dx = \int_{-1}^{1} \int_{3x^2}^{4-x^2} \left[y \right]_0^{6-z} dz \, dx$$

$$= \int_{-1}^{1} \int_{3x^2}^{4-x^2} (6 - z) \, dz \, dx = \int_{-1}^{1} \left[6z - \tfrac{1}{2} z^2 \right]_{3x^2}^{4-x^2} dx$$

$$= \int_{-1}^{1} (16 - 20x^2 + 4x^4) \, dx = \tfrac{304}{15} \approx 20.3.$$

If we used a different order of integration, it would be necessary to use several triple integrals. (Do you see why this is true?)

Figure 13.43

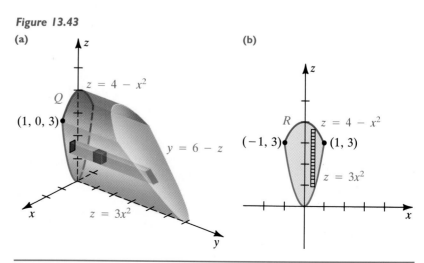

(a)

$z = 4 - x^2$

Q

$(1, 0, 3)$

$y = 6 - z$

$z = 3x^2$

x

y

(b)

z

R

$z = 4 - x^2$

$(-1, 3)$ $(1, 3)$

$z = 3x^2$

x

Figure 13.44

Figure 13.45

(a)

(b)

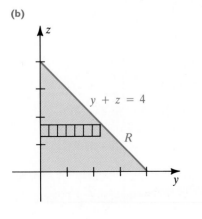

Finally, if Q is a region of the type illustrated in Figure 13.44, where the functions p_1 and p_2 have continuous first partial derivatives on a suitable region R in the yz-plane, then

$$\iiint\limits_{Q} f(x, y, z)\, dV = \iint\limits_{R} \left[\int_{p_1(y,z)}^{p_2(y,z)} f(x, y, z)\, dx \right] dA.$$

In the final iterated double integral, dA will be replaced by $dz\, dy$ or $dy\, dz$.

EXAMPLE ■ 5   In Example 3, we considered the solid bounded by the cylinder $y = x^2$ and the planes $y + z = 4$ and $z = 0$. Set up a triple integral for the volume of this solid in which the first integration is with respect to x.

SOLUTION   The solid is resketched in Figure 13.45(a). A first integration with respect to x corresponds to a summation along a column of small boxes in the direction parallel to the x-axis, extending from the graph of $x = -\sqrt{y}$ to the graph of $x = \sqrt{y}$. The region R in the yz-plane lies between these graphs and is bounded by the y- and z-axes and the line $y + z = 4$, as shown in Figure 13.45(b).

If the second integration is with respect to y, as indicated by the row of rectangles in Figure 13.45(b), then

$$V = \int_0^4 \int_0^{4-z} \int_{-\sqrt{y}}^{\sqrt{y}} dx\, dy\, dz.$$

If we take the second integration with respect to z instead of y, then the iterated integral is

$$V = \int_0^4 \int_0^{4-y} \int_{-\sqrt{y}}^{\sqrt{y}} dx\, dz\, dy.$$

In the next section, we shall use triple integrals to find the *center of mass* of a solid. We conclude this section by discussing the preliminary application of finding the mass of a solid.

If a solid has mass m and volume V and if the mass is uniformly distributed throughout the solid, we say that the solid is **homogeneous**. The **mass density** δ is defined by

$$\delta = \frac{m}{V}, \quad \text{or} \quad m = \delta V.$$

Thus, δ is mass per unit volume. If, for example, m is in kilograms and V in cubic meters, then δ is in kilograms per cubic meter.

Let us next consider a **nonhomogeneous solid**, in which the mass density is not the same throughout. For instance, an object might consist of different metals, such as copper, iron, and gold, formed into one solid. We begin by introducing a coordinate system as in Figure 13.46, where the solid has the shape of the region Q. To define the mass density $\delta(x, y, z)$ at a point $P(x, y, z)$, consider a box-shaped subregion Q_k that contains P and has volume ΔV_k and mass Δm_k (see Figure 13.46). If the longest diagonal $\|\Delta V_k\|$ of the box is small, we expect $\Delta m_k / \Delta V_k$ to be an approximation to $\delta(x, y, z)$. This motivates the following definition.

Figure 13.46

Definition of Mass Density 13.21

$$\delta(x, y, z) = \lim_{\|\Delta V_k\| \to 0} \frac{\Delta m_k}{\Delta V_k}$$

If the limit in (13.21) exists and $\|\Delta V_k\| \approx 0$, then

$$\Delta m_k \approx \delta(x, y, z)\Delta V_k.$$

In some cases, we may know the density $\delta(x, y, z)$ and wish to find the mass. If δ is a continuous function and Q is a suitable region, then we consider an inner partition $\{Q_k\}$ of Q, choose a point (x_k, y_k, z_k) in each Q_k, and form the Riemann sum $\sum_k \delta(x_k, y_k, z_k)\Delta V_k$. We may define the mass m of Q as a limit of such sums. This gives us the following.

Mass of a Solid 13.22

$$m = \iiint_Q \delta(x, y, z)\, dV$$

EXAMPLE 6 A solid has the shape of a right circular cylinder of base radius a and height h. Find the mass if the density at a point P is directly proportional to the distance from one of the bases to P.

Figure 13.47

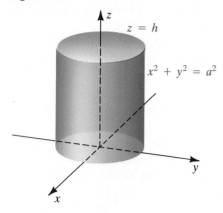

SOLUTION If we introduce a coordinate system as in Figure 13.47, the solid is bounded by the graphs of $x^2 + y^2 = a^2$, $z = 0$, and $z = h$. By hypothesis, the density at (x, y, z) is $\delta(x, y, z) = kz$ for some constant k. From the form of δ and the symmetry of the solid, we may calculate m for the first-octant portion and multiply by 4. Using (13.22), we have

$$m = 4 \int_0^a \int_0^{\sqrt{a^2-x^2}} \int_0^h kz \, dz \, dy \, dx$$

$$= 4k \int_0^a \int_0^{\sqrt{a^2-x^2}} \left[\tfrac{1}{2}z^2\right]_0^h dy \, dx$$

$$= 4k \int_0^a \int_0^{\sqrt{a^2-x^2}} \tfrac{1}{2}h^2 \, dy \, dx$$

$$= 4k \cdot \tfrac{1}{2}h^2 \int_0^a \left[y\right]_0^{\sqrt{a^2-x^2}} dx$$

$$= 2kh^2 \int_0^a \sqrt{a^2 - x^2} \, dx$$

$$= 2kh^2 \cdot \tfrac{1}{4}\pi a^2 = \tfrac{1}{2}\pi kh^2 a^2.$$

Figure 13.48

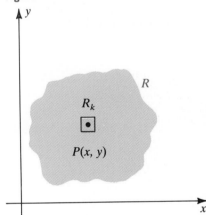

A discussion similar to the preceding one can be given for a lamina that has the shape of a region R in the xy-plane. Consider a rectangular subregion R_k that contains the point P and has area ΔA_k (see Figure 13.48). In this case, the **area mass density** at $P(x, y)$ is defined by

$$\delta(x, y) = \lim_{\|\Delta A_k\| \to 0} \frac{\Delta m_k}{\Delta A_k}.$$

The units for $\delta(x, y)$ are *mass per unit area*, such as kg/m². We can state a formula analogous to (13.22) for a lamina, as follows.

Mass of a Lamina 13.23

$$m = \iint_R \delta(x, y) \, dA$$

Figure 13.49

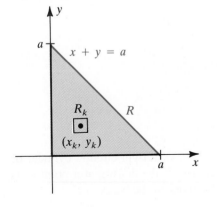

EXAMPLE 7 A lamina has the shape of an isosceles right triangle with equal sides of length a. Find the mass if the area mass density at a point P is directly proportional to the square of the distance from P to the vertex that is opposite the hypotenuse.

SOLUTION It is convenient to introduce a coordinate system, as in Figure 13.49, with the vertex at the origin and the hypotenuse of the triangle along the line $x + y = a$. Also shown in the figure is a typical rectangle R_k (of area ΔA_k) of an inner partition with a point (x_k, y_k) in the rectangle.

By hypothesis, the area mass density at (x, y) is $\delta(x, y) = k(x^2 + y^2)$ for some constant k. Applying Definition (13.23) and Theorem (13.8), we find that the mass is

$$m = \iint\limits_{R} k(x^2 + y^2)\, dA = \int_0^a \int_0^{a-x} k(x^2 + y^2)\, dy\, dx$$

$$= k \int_0^a \left[x^2 y + \tfrac{1}{3}y^3\right]_0^{a-x} dx = k \int_0^a \left[x^2(a - x) + \tfrac{1}{3}(a - x)^3\right] dx.$$

The last integral equals $\frac{1}{6}ka^4$.

EXERCISES 13.5

Exer. 1–6: Evaluate the iterated integral.

1. $\displaystyle\int_0^3 \int_{-1}^0 \int_1^2 (x + 2y + 4z)\, dx\, dy\, dz$

2. $\displaystyle\int_0^1 \int_{-1}^2 \int_1^3 (6x^2 z + 5xy^2)\, dz\, dx\, dy$

3. $\displaystyle\int_0^1 \int_{x+1}^{2x} \int_z^{x+z} x\, dy\, dz\, dx$

4. $\displaystyle\int_1^2 \int_0^{z^2} \int_{x+z}^{x-z} z\, dy\, dx\, dz$

5. $\displaystyle\int_{-1}^2 \int_1^{x^2} \int_0^{x+y} 2x^2 y\, dz\, dy\, dx$

6. $\displaystyle\int_2^3 \int_0^{3y} \int_1^{yz} (2x + y + z)\, dx\, dz\, dy$

Exer. 7–10: If f is an arbitrary continuous function of three variables and Q is the region shown in the figure, express $\iiint_Q f(x, y, z)\, dV$ as an iterated triple integral in six different ways.

8

9

7

10

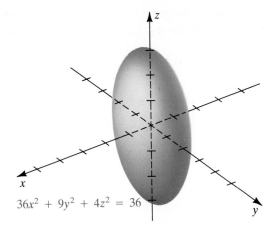

$$36x^2 + 9y^2 + 4z^2 = 36$$

Exer. 11–20: Sketch the region bounded by the graphs of the equations, and use a triple integral to find its volume.

11 $z + x^2 = 4$, $y + z = 4$, $y = 0$, $z = 0$

12 $x^2 + z^2 = 4$, $y^2 + z^2 = 4$

13 $y = 2 - z^2$, $y = z^2$, $x + z = 4$, $x = 0$

14 $z = 4y^2$, $z = 2$, $x = 2$, $x = 0$

15 $y^2 + z^2 = 1$, $x + y + z = 2$, $x = 0$

16 $z = x^2 + y^2$, $y + z = 2$

17 $z = 9 - x^2$, $z = 0$, $y = -1$, $y = 2$

18 $z = e^{x+y}$, $y = 3x$, $x = 2$, $y = 0$, $z = 0$

19 $z = x^2$, $z = x^3$, $y = z^2$, $y = 0$

20 $y = x^2 + z^2$, $z = x^2$, $z = 4$, $y = 0$

Exer. 21–22: Let a, b, and c be positive numbers.

21 Find the volume of the tetrahedron bounded by the coordinate planes and the plane

$$\frac{x}{a} + \frac{y}{b} + \frac{z}{c} = 1.$$

22 Find the volume of the solid bounded by the ellipsoid

$$\frac{x^2}{a^2} + \frac{y^2}{b^2} + \frac{z^2}{c^2} = 1.$$

Exer. 23–28: The iterated integral represents the volume of a region Q in an xyz-coordinate system. Describe Q.

23 $\int_0^1 \int_{\sqrt{1-z}}^{\sqrt{4-z}} \int_2^3 dx\, dy\, dz$

24 $\int_0^1 \int_{z^3}^{\sqrt{z}} \int_0^{4-x} dy\, dx\, dz$

25 $\int_0^2 \int_{x^2}^{2x} \int_0^{x+y} dz\, dy\, dx$

26 $\int_0^1 \int_x^{3x} \int_0^{xy} dz\, dy\, dx$

27 $\int_1^2 \int_{-\sqrt{z}}^{\sqrt{z}} \int_{-\sqrt{z-x^2}}^{\sqrt{z-x^2}} dy\, dx\, dz$

28 $\int_1^4 \int_{-z}^{z} \int_{-\sqrt{z^2-y^2}}^{\sqrt{z^2-y^2}} dx\, dy\, dz$

Exer. 29–30: A lamina having area mass density $\delta(x, y)$ has the shape of the region bounded by the graphs of the equations. Set up an iterated double integral that can be used to find the mass of the lamina.

29 $\delta(x, y) = y^2$; $y = e^{-x}$, $x = 0$, $x = 1$, $y = 0$

30 $\delta(x, y) = x^2 + y^2$; $xy^2 = 1$, $x = 0$, $y = 1$, $y = 2$

Exer. 31–32: A solid having density $\delta(x, y, z)$ has the shape of the region bounded by the graphs of the equations. Set up an iterated triple integral that can be used to find the mass of the solid.

31 $\delta(x, y, z) = x^2 + y^2$;
      $x + 2y + z = 4$, $x = 0$, $y = 0$, $z = 0$

32 $\delta(x, y, z) = z + 1$; $z = 4 - x^2 - y^2$, $z = 0$

33 Near sea level, the density δ of the earth's atmosphere at a height of z meters can be approximated by $\delta = 1.225 - 0.000113z$ kg/m^3. Approximate the mass of a region of the atmosphere that has the shape of a cube with edge of length 1 km and one face on the surface of the earth.

34 For an altitude z up to 30,000 m, the density δ (in kilograms per cubic meter) of the earth's atmosphere can be approximated by

$$\delta = 1.2 - (1.096 \times 10^{-4})z + (3.42 \times 10^{-9})z^2$$
$$-(3.6 \times 10^{-14})z^3.$$

The average pressure P (in newtons) on a region of the earth's surface of area A caused by a volume Q of the atmosphere can be estimated from

$$P = \frac{1}{A} \iiint_Q \delta g\, dV,$$

where $g = 9.80$ m/sec^2 is a gravitational constant. Estimate the pressure exerted by a column of air 30,000 m high on one square meter of the earth's surface. (Standard atmospheric pressure on one square meter is 101,325 newtons.)

Exer. 35–36: A cubical solid bounded by the coordinate planes and the planes $x = 1$, $y = 1$, and $z = 1$ has the given density $\delta(x, y, z)$. Approximate its mass by evaluating

$$\sum_{i=1}^{n} \sum_{j=1}^{n} \sum_{k=1}^{n} \delta(u_i, v_j, w_k) \, \Delta z_k \, \Delta y_j \, \Delta x_i$$

for the indicated value of n, where each increment equals $1/n$ and $u_i = (i - \frac{1}{2})/n$, $v_j = (j - \frac{1}{2})/n$, and $w_k = (k - \frac{1}{2})/n$.

35   $\delta(x, y, z) = \sqrt{x^3 + y^3 + z^3}; \quad n = 2$

36   $\delta(x, y, z) = \tan(xy \cos z); \quad n = 4$

13.6   MOMENTS AND CENTER OF MASS

In Section 5.7, we considered moments and the center of mass of a homogeneous lamina. In this section, we use double and triple integrals to extend these concepts to *nonhomogeneous* regions and solids.

We can use double integrals to find moments and the center of mass of a nonhomogeneous lamina L that has the shape of a region R in the xy-plane. If the area mass density at the point (x, y) is $\delta(x, y)$ and δ is a continuous function on R, then, by (13.23), the mass m of L is given by

$$m = \iint_R \delta(x, y) \, dA.$$

Figure 13.50

Consider an inner partition $P = \{R_k\}$ of R, and for each k, let (x_k, y_k) be any point in R_k (see Figure 13.50). Since δ is continuous, a small change in (x, y) produces a small change in the density $\delta(x, y)$—that is, δ is *almost constant* on R_k. Hence, if $\|P\| \approx 0$, the mass Δm_k that corresponds to R_k may be approximated by $\delta(x_k, y_k) \Delta A_k$, where ΔA_k is the area of R_k. If we assume that the mass Δm_k is concentrated at (x_k, y_k), then the moment with respect to the x-axis of this element of L is the product $y_k \delta(x_k, y_k) \Delta A_k$. Since the sum of the moments $\sum_k y_k \delta(x_k, y_k) \Delta A_k$ should approximate the moment of the *lamina*, we define the moment M_x of L with respect to the x-axis as follows:

$$M_x = \lim_{\|P\| \to 0} \sum_k y_k \delta(x_k, y_k) \Delta A_k = \iint_R y \delta(x, y) \, dA$$

Similarly, the moment M_y of L with respect to the y-axis is

$$M_y = \lim_{\|P\| \to 0} \sum_k x_k \delta(x_k, y_k) \Delta A_k = \iint_R x \delta(x, y) \, dA.$$

As in Definition (5.24), we define the center of mass of the lamina as the point $(\overline{x}, \overline{y})$ such that $\overline{x} = M_y/m$, $\overline{y} = M_x/m$.

This derivation gives us the following.

Definition 13.24

Let L be a lamina that has the shape of a region R in the xy-plane. If the area mass density at (x, y) is $\delta(x, y)$ and if δ is continuous on R, then the **mass** m, the **moments** M_x and M_y, and the **center of mass** $(\overline{x}, \overline{y})$ are

(i) $m = \iint\limits_{R} \delta(x, y) \, dA$

(ii) $M_x = \iint\limits_{R} y\delta(x, y) \, dA, \quad M_y = \iint\limits_{R} x\delta(x, y) \, dA$

(iii) $\overline{x} = \dfrac{M_y}{m} = \dfrac{\iint_{R} x\delta(x, y) \, dA}{\iint_{R} \delta(x, y) \, dA}, \quad \overline{y} = \dfrac{M_x}{m} = \dfrac{\iint_{R} y\delta(x, y) \, dA}{\iint_{R} \delta(x, y) \, dA}$

If L is homogeneous, then the area mass density $\delta(x, y)$ is a constant and can be canceled in (13.24)(iii). Thus, as in Section 5.7, the center of mass of a homogeneous lamina depends only on its shape, and we refer to $(\overline{x}, \overline{y})$ as the **centroid** of the region R.

EXAMPLE ■ 1   A lamina has the shape of an isosceles right triangle with equal sides of length a. The area mass density at a point P is directly proportional to the square of the distance from P to the vertex that is opposite the hypotenuse. Find the center of mass.

SOLUTION   The lamina is the same as that considered in Example 7 of Section 13.5, where we placed the triangle as in Figure 13.51. Using the area mass density $\delta(x, y) = k(x^2 + y^2)$, we found that $m = \frac{1}{6}ka^4$. By Definition (13.24)(ii),

$$M_y = \iint\limits_{R} xk(x^2 + y^2) \, dA = \int_0^a \int_0^{a-x} xk(x^2 + y^2) \, dy \, dx,$$

which is equal to $\frac{1}{15}ka^5$. From Definition (13.24)(iii),

$$\overline{x} = \frac{\frac{1}{15}ka^5}{\frac{1}{6}ka^4} = \frac{2}{5}a.$$

Figure 13.51

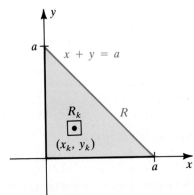

Similarly, $M_x = \frac{1}{15}ka^5$ and $\bar{y} = \frac{2}{5}a$. Thus, the center of mass (the *balance point*) of the lamina is $(\frac{2}{5}a, \frac{2}{5}a)$.

EXAMPLE = 2 A lamina has the shape of the region R in the xy-plane bounded by the parabola $x = y^2$ and the line $x = 4$. The area mass density at the point $P(x, y)$ is directly proportional to the distance from the y-axis to P. Find the center of mass.

SOLUTION The region is sketched in Figure 13.52. By hypothesis, the area mass density at (x, y) is $\delta(x, y) = kx$ for some constant k. It follows from the form of δ and the symmetry of the region that the center of mass is on the x-axis—that is, $\bar{y} = 0$.

Using Definition (13.24)(i) and then integrating first with respect to x, as indicated by the row of rectangles in Figure 13.52, we have

$$m = \iint_R kx \, dA = k \int_{-2}^{2} \int_{y^2}^{4} x \, dx \, dy$$

$$= k \int_{-2}^{2} \left[\tfrac{1}{2}x^2\right]_{y^2}^{4} dy = \tfrac{1}{2}k \int_{-2}^{2} (16 - y^4) \, dy = \tfrac{128}{5}k.$$

By Definition (13.24)(ii),

$$M_y = \iint_R x(kx) \, dA = k \int_{-2}^{2} \int_{y^2}^{4} x^2 \, dx \, dy$$

$$= k \int_{-2}^{2} \left[\tfrac{1}{3}x^3\right]_{y^2}^{4} dy = \tfrac{1}{3}k \int_{-2}^{2} (64 - y^6) \, dy = \tfrac{512}{7}k.$$

Consequently,

$$\bar{x} = \frac{M_y}{m} = \frac{512k}{7} \cdot \frac{5}{128k} = \frac{20}{7} \approx 2.86.$$

Hence, the center of mass is $(\frac{20}{7}, 0)$.

Figure 13.52

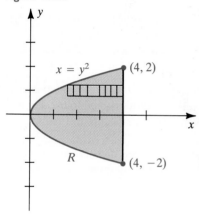

The moments M_x and M_y in Definition (13.24) are also called the *first moments* of L with respect to the coordinate axes. If we use the *squares* of the distances from the coordinate axes, we obtain the *second moments*, or **moments of inertia**, I_x and I_y with respect to the x-axis and y-axis, respectively. The sum $I_O = I_x + I_y$ is the **polar moment of inertia**, or the **moment of inertia with respect to the origin**. In the following formula, I_O can be obtained as a limit of a sum, using the *square* of the distance from the origin to (x_k, y_k) in Figure 13.50.

**Moments of Inertia
of a Lamina 13.25**

$$I_x = \lim_{\|P\| \to 0} \sum_k y_k^2 \delta(x_k, y_k) \Delta A_k = \iint\limits_R y^2 \delta(x, y)\, dA$$

$$I_y = \lim_{\|P\| \to 0} \sum_k x_k^2 \delta(x_k, y_k) \Delta A_k = \iint\limits_R x^2 \delta(x, y)\, dA$$

$$I_O = \lim_{\|P\| \to 0} \sum_k (x_k^2 + y_k^2) \delta(x_k, y_k) \Delta A_k$$

$$= \iint\limits_R (x^2 + y^2) \delta(x, y)\, dA$$

Figure 13.53

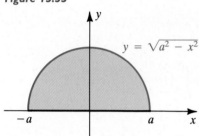

$y = \sqrt{a^2 - x^2}$

EXAMPLE 3 A lamina has the semicircular shape illustrated in Figure 13.53. The area mass density is directly proportional to the distance from the x-axis. Find the moment of inertia with respect to the x-axis.

SOLUTION By hypothesis, the area mass density at (x, y) is $\delta(x, y) = ky$. Applying (13.25), we find that the moment of inertia with respect to the x-axis is

$$I_x = \int_{-a}^{a} \int_0^{\sqrt{a^2 - x^2}} y^2 (ky)\, dy\, dx = k \int_{-a}^{a} \left[\tfrac{1}{4} y^4 \right]_0^{\sqrt{a^2 - x^2}} dx$$

$$= \tfrac{1}{4} k \int_{-a}^{a} (a^4 - 2a^2 x^2 + x^4)\, dx = \tfrac{4}{15} ka^5.$$

Figure 13.54

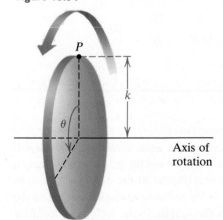

Axis of
rotation

Moments of inertia are useful in problems that involve rotation of an object about a fixed axis, such as rotation of a wheel (or disk) about an axle (see Figure 13.54). If a particle P on the wheel has mass m and is a distance k from the axis of rotation, then the moment of inertia I of P with respect to the axis is mk^2. If the angular speed $d\theta/dt$ is a constant ω, then the speed v of the particle is $k\omega$. From physics, the **kinetic energy** K.E. of P is

$$\text{K.E.} = \tfrac{1}{2} mv^2.$$

Since $v = k\omega$, $\quad \text{K.E.} = \tfrac{1}{2} mk^2 \omega^2 = \tfrac{1}{2} I \omega^2.$

If we represent the wheel by a disk and introduce a limit of sums, then the same formula may be derived for the kinetic energy of the wheel. The formula can also be extended to laminas having noncircular shapes. The kinetic energy of a rotating object tells an engineer or physicist the amount of work necessary to bring the object to rest. Since K.E. $= \tfrac{1}{2} I \omega^2$, *the kinetic energy is directly proportional to the moment of inertia*. For a fixed ω, the larger the moment of inertia, the larger the amount of work required to stop the rotation.

Figure 13.55

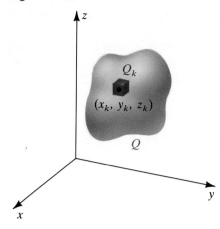

Moments and Center of Mass in Three Dimensions   13.26

Let us next consider a solid that has the shape of a three-dimensional region Q. Suppose that the mass density (mass per unit volume) at (x, y, z) is $\delta(x, y, z)$ and that δ is continuous throughout Q. Let $\{Q_k\}$ be an inner partition of Q, and let ΔV_k be the volume of Q_k. If (x_k, y_k, z_k) is a point in Q_k (see Figure 13.55), then the corresponding mass is approximately $\delta(x_k, y_k, z_k) \Delta V_k$. By (13.22), the mass of the solid is the limit of such sums—that is, $\iiint_Q \delta(x, y, z)\, dV$.

The *moment with respect to the xy-plane* of the portion of the solid that corresponds to Q_k is approximated by $z_k \delta(x_k, y_k, z_k) \Delta V_k$ (see Figure 13.55). Summing and taking a limit gives us the **moment M_{xy} of the solid with respect to the xy-plane** (see (13.26)(ii)). The **moments M_{xz} and M_{yz}** with respect to the xz-plane and yz-plane, respectively, are obtained in similar fashion. The **center of mass** $(\overline{x}, \overline{y}, \overline{z})$ of the solid is defined in (13.26)(iii).

(i) $\displaystyle m = \iiint_Q \delta(x, y, z)\, dV$

(ii) $\displaystyle M_{xy} = \iiint_Q z\delta(x, y, z)\, dV$

$\displaystyle M_{xz} = \iiint_Q y\delta(x, y, z)\, dV$

$\displaystyle M_{yz} = \iiint_Q x\delta(x, y, z)\, dV$

(iii) $\displaystyle \overline{x} = \frac{M_{yz}}{m}; \qquad \overline{y} = \frac{M_{xz}}{m}; \qquad \overline{z} = \frac{M_{xy}}{m}$

From (iii) of (13.26),

$$m\overline{x} = M_{yz}, \qquad m\overline{y} = M_{xz}, \qquad m\overline{z} = M_{xy}.$$

Since $m\overline{x}, m\overline{y}$, and $m\overline{z}$ are the moments with respect to the yz-plane, the xz-plane, and the xy-plane, respectively, of a single point mass located at $(\overline{x}, \overline{y}, \overline{z})$, we may interpret the center of mass of a solid as the point at which the total mass can be concentrated to obtain the moments M_{yz}, M_{xz}, and M_{xy} of the solid.

If the solid is homogeneous, the mass density δ is a constant and hence δ can be canceled after substitution in (13.26)(iii). Consequently, the center of mass of a homogeneous solid depends only on the shape of Q. As in two dimensions, the corresponding point for geometric solids is called the **centroid** of the solid. To find the centroid, we let $\delta(x, y, z) = 1$ in (13.26).

E X A M P L E ■ 4   A solid has the shape of a right circular cylinder of base radius a and height h. The density at a point P is directly proportional to the distance from one of the bases to P. Find the center of mass.

Figure 13.56

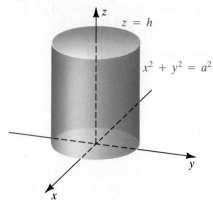

SOLUTION   In Example 6 of Section 13.5, we positioned the solid as shown in Figure 13.56. Using $\delta(x, y, z) = kz$, we found that its mass m is $m = \frac{1}{2}\pi k h^2 a^2$.

The center of mass is on the z-axis, so it is sufficient to find only $\bar{z} = M_{xy}/m$. Moreover, because of the form of δ and the symmetry of the solid, we may calculate M_{xy} for the first-octant portion and multiply by 4. Using (13.26), we have

$$M_{xy} = 4\int_0^a \int_0^{\sqrt{a^2-x^2}} \int_0^h z(kz)\, dz\, dy\, dx$$

$$= 4k\int_0^a \int_0^{\sqrt{a^2-x^2}} \left[\frac{z^3}{3}\right]_0^h dy\, dx$$

$$= 4k\int_0^a \int_0^{\sqrt{a^2-x^2}} \frac{h^3}{3}\, dy\, dx$$

$$= \tfrac{4}{3}kh^3 \int_0^a [y]_0^{\sqrt{a^2-x^2}}\, dx = \tfrac{4}{3}kh^3 \int_0^a \sqrt{a^2-x^2}\, dx$$

$$= \tfrac{4}{3}kh^3 \cdot \tfrac{1}{4}\pi a^2 = \tfrac{1}{3}\pi k h^3 a^2.$$

Finally, we calculate the center of mass:

$$\bar{z} = \frac{M_{xy}}{m} = \frac{\frac{1}{3}\pi k h^3 a^2}{\frac{1}{2}\pi k h^2 a^2} = \frac{2}{3}h$$

Hence, the center of mass is on the axis of the cylinder, two thirds of the way from the lower base.

EXAMPLE ■ 5   A solid has the shape of the region in the first octant bounded by the paraboloid $z = 4 - 9x^2 - y^2$ and planes $y = 4x$, $z = 0$, and $y = 0$. The density at $P(x, y, z)$ is proportional to the distance from the origin to P. Set up iterated integrals that can be used to find $\bar{x}$.

Figure 13.57

SOLUTION   The region is sketched in Figure 13.57. The density at (x, y, z) is $\delta(x, y, z) = k(x^2 + y^2 + z^2)^{1/2}$ for some k. Using (13.26) gives us

$$m = \int_0^{8/5} \int_{y/4}^{\sqrt{4-y^2}/3} \int_0^{4-9x^2-y^2} k(x^2 + y^2 + z^2)^{1/2}\, dz\, dx\, dy$$

$$M_{yz} = \int_0^{8/5} \int_{y/4}^{\sqrt{4-y^2}/3} \int_0^{4-9x^2-y^2} xk(x^2 + y^2 + z^2)^{1/2}\, dz\, dx\, dy$$

$$\bar{x} = \frac{M_{yz}}{m}.$$

If a particle of mass m is at the point (x, y, z), then its distance to the z-axis is $(x^2 + y^2)^{1/2}$ and its **moment of inertia I_z with respect to the z-axis** is defined as $(x^2 + y^2)m$. Similarly, **moments of inertia I_x and I_y with respect to the x-axis and the y-axis** are $(y^2 + z^2)m$ and $(x^2 + z^2)m$, respectively. For solids having the shape of Q in Figure 13.55, we use limits of sums in the usual way to obtain the following.

Moments of Inertia of Solids 13.27

$$I_z = \iiint\limits_Q (x^2 + y^2)\delta(x, y, z)\, dV$$

$$I_x = \iiint\limits_Q (y^2 + z^2)\delta(x, y, z)\, dV$$

$$I_y = \iiint\limits_Q (x^2 + z^2)\delta(x, y, z)\, dV$$

EXAMPLE ■ 6   Find the moment of inertia, with respect to the axis of symmetry, of the cylindrical solid described in Example 4.

SOLUTION   The solid is sketched in Figure 13.56. Employing (13.27) with $\delta(x, y, z) = kz$ and making use of symmetry and the fact that δ is a function of z alone, we have

$$I_z = 4 \int_0^a \int_0^{\sqrt{a^2 - x^2}} \int_0^h (x^2 + y^2)kz\, dz\, dy\, dx$$

$$= 4k \int_0^a \int_0^{\sqrt{a^2 - x^2}} (x^2 + y^2) \left[\frac{z^2}{2}\right]_0^h dy\, dx$$

$$= 2kh^2 \int_0^a \int_0^{\sqrt{a^2 - x^2}} (x^2 + y^2)\, dy\, dx$$

$$= 2kh^2 \int_0^a \left[x^2 y + \tfrac{1}{3}y^3\right]_0^{\sqrt{a^2 - x^2}} dx$$

$$= 2kh^2 \int_0^a \left[x^2\sqrt{a^2 - x^2} + \tfrac{1}{3}\sqrt{(a^2 - x^2)^3}\right] dx.$$

The last integral may be evaluated by using a trigonometric substitution or a table of integrals. This results in $I_z = \tfrac{1}{4}\pi kh^2 a^4$.

EXAMPLE ■ 7   A homogeneous solid has the shape of the region Q bounded by the cone $x^2 - y^2 + z^2 = 0$ and the plane $y = 3$. Set up a triple integral that can be used to find its moment of inertia with respect to the y-axis.

Figure 13.58

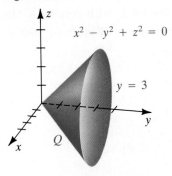

SOLUTION   The region is sketched in Figure 13.58. Note that the trace of the cone on the xy-plane is the pair of lines $x = \pm y$. If we denote the constant density by k, then applying (13.27) yields

$$I_y = \int_0^3 \int_{-y}^y \int_{-\sqrt{y^2-x^2}}^{\sqrt{y^2-x^2}} (x^2 + z^2)k \, dz \, dx \, dy.$$

We could also find I_y by multiplying the moment of inertia of that part of the solid that lies in the first octant by 4. Thus,

$$I_y = 4 \int_0^3 \int_0^y \int_0^{\sqrt{y^2-x^2}} (x^2 + z^2)k \, dz \, dx \, dy.$$

We conclude this section with a proof of the following theorem, which was stated without proof in (5.26).

Theorem of Pappus   **13.28**

Let R be a region in a plane that lies entirely on one side of a line l in the plane. If R is revolved once about l, the volume of the resulting solid is the product of the area of R and the distance traveled by the centroid of R.

Figure 13.59

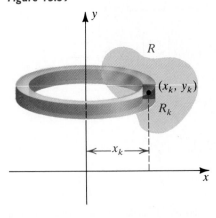

PROOF   We may assume that l is the y-axis and R is a region in the first quadrant, as shown in Figure 13.59. Let $\{R_k\}$ be an inner partition P of R, and let ΔA_k denote the area of R_k. For each k, choose (x_k, y_k) as the center of the rectangle R_k.

If R is revolved about the y-axis, then, as shown in Figure 13.59, R_k generates a cylindrical shell having volume $2\pi x_k \Delta A_k$. Thus, the volume V of the solid generated by R is

$$V = \lim_{\|P\| \to 0} \sum_k 2\pi x_k \Delta A_k = \iint_R 2\pi x \, dA.$$

If we apply Definition (13.24) to centroids, by taking $\delta(x, y) = 1$ and $m = A$, then

$$V = 2\pi \iint_R x \, dA = 2\pi M_y = 2\pi \overline{x} A.$$

Thus, we may find the volume V by multiplying the area A of R by the distance $2\pi \overline{x}$ traveled by the centroid when R is revolved once about the y-axis. ∎

EXERCISES 13.6

Exer. 1–8: Find the mass and the center of mass of the lamina that has the shape of the region bounded by the graphs of the given equations and has the indicated area mass density.

1 $y = \sqrt{x}$, $x = 9$, $y = 0$; $\delta(x, y) = x + y$

2 $y = \sqrt[3]{x}$, $x = 8$, $y = 0$; $\delta(x, y) = y^2$

3 $y = x^2$, $y = 4$;
   density at $P(x, y)$ is directly proportional to the distance from the y-axis to P

4 $y = x^3$, $y = 2x$;
   density at $P(x, y)$ is directly proportional to the distance from the x-axis to P

5 $y = e^{-x^2}$, $y = 0$, $x = -1$, $x = 1$; $\delta(x, y) = |xy|$

6 $y = \sin x$, $y = 0$, $x = 0$, $x = \pi$; $\delta(x, y) = y$

7 $y = \sec x$, $x = -\dfrac{\pi}{4}$, $x = \dfrac{\pi}{4}$, $y = \dfrac{1}{2}$; $\delta(x, y) = 4$

8 $y = \ln x$, $y = 0$, $x = 2$; $\delta(x, y) = \dfrac{1}{x}$

9 Find I_x, I_y, and I_O for the lamina of Exercise 1.

10 Find I_x, I_y, and I_O for the lamina of Exercise 2.

11 Find I_x, I_y, and I_O for the lamina of Exercise 3.

12 Find I_x, I_y, and I_O for the lamina of Exercise 4.

13 A homogeneous lamina has the shape of a square of side a. Find the moment of inertia with respect to

   (a) a side **(b)** a diagonal **(c)** the center of mass

14 A homogeneous lamina has the shape of an equilateral triangle of side a. Find the moment of inertia with respect to

   (a) an altitude **(b)** a side **(c)** a vertex

15 If a solid of mass m has moment of inertia I with respect to a line, then the *radius of gyration* is, by definition, the number d such that $I = md^2$. This formula implies that the radius of gyration is the distance from the line at which all the mass could be concentrated without changing the moment of inertia of the solid. Find the radius of gyration in part (a) of Exercise 13.

16 Refer to Exercise 15. Find the radius of gyration in part (a) of Exercise 14.

Exer. 17–18: Find the center of mass of Q.

17 The density at a point P in a cubical solid Q of edge a is directly proportional to the square of the distance from P to a fixed corner of the cube.

18 Let Q be the tetrahedron bounded by the coordinate planes and the plane $2x + 5y + z = 10$. The density at the point $P(x, y, z)$ is directly proportional to the distance from the xz-plane to P.

Exer. 19–20: Set up iterated integrals that can be used to find the center of mass of the solid that has the shape of the region Q and density $\delta(x, y, z)$.

19 Q is bounded by the paraboloid $x = y^2 + 4z^2$ and the plane $x = 4$; $\delta(x, y, z) = x^2 + z^2$.

20 Q is bounded by the hyperboloid $y^2 - x^2 - z^2 = 1$ and the plane $y = 2$; $\delta(x, y, z) = x^2 y^2 z^2$.

Exer. 21–22: Set up iterated integrals that can be used to find the centroid of the solid shown in the figure.

21

$(a, 0, 0)$

22

23 Let Q be the solid in the first octant bounded by the coordinate planes and the graphs of $z = 9 - x^2$ and $2x + y = 6$.

    (a) Set up iterated integrals that can be used to find the centroid.

    (b) Find the centroid.

24 Set up iterated integrals that can be used to find the centroid of the solid bounded by the graphs of the equations $z = x^2$, $y = x^2$, $y = x^3$, and $z = 0$.

Exer. 25–28: Set up an iterated integral that can be used to find the moment of inertia with respect to the z-axis of the indicated solid.

25 The sphere of radius a with center at the origin; $\delta(x, y, z) = x^2 + y^2 + z^2$

26 The cone bounded by the graphs of $x^2 + 9y^2 - z^2 = 0$ and $z = 36$; $\delta(x, y, z) = x^2 + y^2$

27 The homogeneous tetrahedron bounded by the coordinate planes and the plane $(x/a) + (y/b) + (z/c) = 1$ for positive numbers a, b, and c

28 The homogeneous solid bounded by the ellipsoid with equation $(x^2/a^2) + (y^2/b^2) + (z^2/c^2) = 1$

13.7 CYLINDRICAL COORDINATES

It is easy to extend the system of polar coordinates to three dimensions. We merely represent a point P by an ordered triple (r, θ, z), where z is the usual (third) rectangular coordinate of P and r and θ are polar coordinates for the projection P' of P onto the xy-plane, as illustrated in Figure 13.60. We call (r, θ, z) **cylindrical coordinates** for P. Our principal use for cylindrical coordinates will be to simplify certain types of multiple integrals.

Since polar coordinates are used in the xy-plane, the formulas in Theorem (9.9) give relationships between rectangular coordinates and cylindrical coordinates. Let us restate these formulas as follows.

Theorem   13.29

The rectangular coordinates (x, y, z) and the cylindrical coordinates (r, θ, z) of a point P are related as follows:

$$x = r\cos\theta, \qquad y = r\sin\theta, \qquad \tan\theta = \frac{y}{x},$$

$$r^2 = x^2 + y^2, \quad z = z$$

Figure 13.60

If $r_0 > 0$, then the graph of the equation $r = r_0$, or, equivalently, $x^2 + y^2 = r_0^2$, is a circular cylinder of radius r_0 with axis along the z-axis. If θ_0 and z_0 are numbers, then the graph of $\theta = \theta_0$ is a plane containing the z-axis and the graph of $z = z_0$ is a plane perpendicular to the z-axis. Typical graphs are sketched in Figure 13.61 on the following page.

Figure 13.61

Figure 13.62

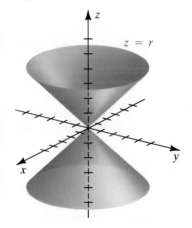

EXAMPLE ■ 1 Change the equation $z^2 = x^2 + y^2$ to cylindrical coordinates, and sketch its graph.

SOLUTION Using Theorem (13.29), we obtain

$$z^2 = r^2, \quad \text{or} \quad z = \pm r.$$

Since r can be positive or negative, $z = \pm r$ is equivalent to $z = r$. The graph is a circular cone with axis along the z-axis, as illustrated in Figure 13.62.

EXAMPLE ■ 2 Change the equation to rectangular coordinates, and sketch its graph in an xyz-coordinate system.

(a) $z = 4r^2$
(b) $r = 4 \sin \theta$

Figure 13.63

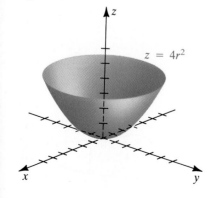

SOLUTION
(a) Applying Theorem (13.29), we have

$$z = 4(x^2 + y^2), \quad \text{or} \quad z = 4x^2 + 4y^2.$$

The graph is the paraboloid of revolution sketched in Figure 13.63.
(b) Multiplying both sides of $r = 4 \sin \theta$ by r gives us $r^2 = 4r \sin \theta$. Using (13.29), we obtain $x^2 + y^2 = 4y$, or, equivalently,

$$x^2 + (y - 2)^2 = 4.$$

Figure 13.64

$r = 4 \sin \theta$

The graph is a cylinder with rulings parallel to the z-axis, as shown in Figure 13.64. A directrix of the cylinder is a circle of radius 2 in the xy-plane whose center in rectangular coordinates is $(0, 2, 0)$.

Sometimes it is convenient to use cylindrical coordinates to evaluate triple integrals. The simplest case occurs when a function f of r, θ, and z is continuous throughout a region of the form

$$Q = \{(r, \theta, z) : a \le r \le b, c \le \theta \le d, m \le z \le n\}.$$

We divide Q into subregions $Q_1, Q_2, \ldots, Q_r$ that have the same shape as Q by using graphs of equations of the form $r = a_k$, $\theta = c_k$, and $z = m_k$, where a_k, c_k, and m_k are in the intervals $[a, b]$, $[c, d]$, and $[m, n]$, respectively. If $a_k \ne 0$, these graphs are circular cylinders, planes containing the z-axis, and planes parallel to the xy-plane, respectively (see Figure 13.61).

A typical subregion Q_k is sketched in Figure 13.65(a). If $\overline{r_k}$ is the *average radius* of the base of Q_k, as shown in Figure 13.65(b), then the volume ΔV_k of Q_k is the area of the base $\overline{r_k}\,\Delta r_k\,\Delta\theta_k$ (see Section 13.3) times the altitude Δz_k:

$$\Delta V_k = \overline{r_k}\,\Delta r_k\,\Delta\theta_k\,\Delta z_k$$

If (r_k, θ_k, z_k) is any point in Q_k and $\|P\|$ is the length of the longest diagonal of the Q_k, then the triple integral of f over Q is defined as

$$\iiint\limits_{Q} f(r, \theta, z)\, dV = \lim_{\|P\| \to 0} \sum_k f(r_k, \theta_k, z_k)\,\Delta V_k.$$

It can be proved that

$$\iiint\limits_{Q} f(r, \theta, z)\, dV = \int_m^n \int_c^d \int_a^b f(r, \theta, z)\, r\, dr\, d\theta\, dz.$$

There are five other possible orders of integration for this iterated integral.

Figure 13.65

(a)   (b)

Figure 13.66

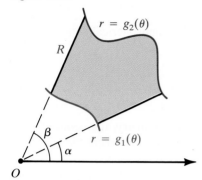

Triple integrals in cylindrical coordinates may be defined over regions more complicated than the one just discussed by using inner partitions. Thus, if R is a polar region of the type discussed in Section 13.5 and if

$$Q = \{(r, \theta, z): (r, \theta) \text{ is in } R \text{ and } k_1(r, \theta) \le z \le k_2(r, \theta)\},$$

where k_1 and k_2 are functions that have continuous first partial derivatives throughout R, then it can be shown that

$$\iiint\limits_Q f(r, \theta, z) \, dV = \iint\limits_R \left[\int_{k_1(r,\theta)}^{k_2(r,\theta)} f(r, \theta, z) \, dz \right] dA.$$

In particular, if R is a region of the type illustrated in Figure 13.66, we may evaluate the triple integral of f over Q as follows.

**Evaluation Theorem
(Cylindrical Coordinates) 13.30**

$$\iiint\limits_Q f(r, \theta, z) \, dV = \int_\alpha^\beta \int_{g_1(\theta)}^{g_2(\theta)} \int_{k_1(r,\theta)}^{k_2(r,\theta)} f(r, \theta, z) \, r \, dz \, dr \, d\theta$$

The iterated integral in (13.30) may be interpreted as a limit of triple sums. The first summation takes place along a column of the subregions Q_k from the lower surface (with equation $z = k_1(r, \theta)$) to the upper surface (with equation $z = k_2(r, \theta)$) (see Figure 13.67). The second summation extends over a wedge of these columns, with θ fixed and r allowed to vary. At this stage, we have summed over a slice of Q, as illustrated in Figure 13.67. Finally, we sweep out Q by letting θ vary from α to β.

Figure 13.67

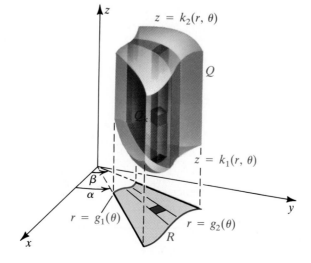

The following examples illustrate the use of (13.30).

EXAMPLE ▪ 3   Find the centroid of a hemispherical solid Q of radius a.

Figure 13.68

$z = \sqrt{a^2 - r^2}$

$(a, 0, 0)$

SOLUTION   If we introduce an xyz-coordinate system, as in Figure 13.68, then an equation of the hemisphere is $z = \sqrt{a^2 - x^2 - y^2}$, or, in cylindrical coordinates, $z = \sqrt{a^2 - r^2}$. By symmetry, the centroid is on the z-axis, and hence we need only find $\bar{z} = M_{xy}/V$, where the volume V is given by $V = \frac{1}{2} \cdot \frac{4}{3}\pi a^3 = \frac{2}{3}\pi a^3$.

We use (13.26)(ii) with $\delta(x, y, z) = 1$, obtaining

$$M_{xy} = \iiint\limits_{Q} z \, dV.$$

Applying the evaluation theorem (13.30), we note that the first integration (with respect to z) is from $z = 0$ (the xy-plane) to the hemisphere with equation $z = \sqrt{a^2 - r^2}$. The remaining integrations with respect to r and θ are over the region bounded by the circle $r = a$. Thus,

$$M_{xy} = \int_0^{2\pi} \int_0^a \int_0^{\sqrt{a^2-r^2}} zr \, dz \, dr \, d\theta$$

$$= \int_0^{2\pi} \int_0^a \left[\frac{z^2}{2}\right]_0^{\sqrt{a^2-r^2}} r \, dr \, d\theta$$

$$= \frac{1}{2} \int_0^{2\pi} \int_0^a (a^2 - r^2) r \, dr \, d\theta$$

$$= \frac{1}{2} \int_0^{2\pi} \int_0^a (a^2 r - r^3) \, dr \, d\theta$$

$$= \frac{1}{2} \int_0^{2\pi} \left[a^2 \left(\frac{r^2}{2}\right) - \frac{r^4}{4}\right]_0^a d\theta$$

$$= \frac{a^4}{8} \int_0^{2\pi} d\theta$$

$$= \frac{a^4}{8} \left[\theta\right]_0^{2\pi} = \frac{1}{4}\pi a^4.$$

Hence,

$$\bar{z} = \frac{M_{xy}}{V} = \frac{\frac{1}{4}\pi a^4}{\frac{2}{3}\pi a^3} = \frac{3}{8}a.$$

EXAMPLE 4 A solid Q is bounded by the cone $z = \sqrt{x^2 + y^2}$ and the plane $z = 2$. The density at $P(x, y, z)$ is directly proportional to the square of the distance from the origin to P. Find its mass.

SOLUTION The solid is sketched in Figure 13.69. An equation of the cone in cylindrical coordinates is $z = r$, with $r \geq 0$. Note that if $z = 2$, then $2 = \sqrt{x^2 + y^2}$, or $4 = x^2 + y^2$. Therefore, the top of the cone projects onto the circle $x^2 + y^2 = 4$, or $r = 2$, in the xy-plane. The density at (x, y, z) is given by

$$\delta = k(x^2 + y^2 + z^2), \quad \text{or} \quad \delta = k(r^2 + z^2).$$

By (13.26) and (13.30),

$$m = \iiint_Q \delta \, dV = \int_0^{2\pi} \int_0^2 \int_r^2 k(r^2 + z^2) \, r \, dz \, dr \, d\theta.$$

Note that the first integration (with respect to z) is from the cone $z = r$ (the lower surface) to the plane $z = 2$ (the upper surface). The remaining integrations with respect to r and θ are over the circular region bounded by $r = 2$. Integrating, we obtain

$$m = k \int_0^{2\pi} \int_0^2 \left[r^2 z + \frac{z^3}{3} \right]_r^2 r \, dr \, d\theta$$

$$= k \int_0^{2\pi} \int_0^2 \left[\left(2r^2 + \frac{8}{3} \right) - \left(r^3 + \frac{r^3}{3} \right) \right] r \, dr \, d\theta$$

$$= k \int_0^{2\pi} \int_0^2 \left(2r^3 + \tfrac{8}{3} r - \tfrac{4}{3} r^4 \right) dr \, d\theta$$

$$= k \int_0^{2\pi} \left[2 \frac{r^4}{4} + \frac{8}{3} \frac{r^2}{2} - \frac{4}{3} \frac{r^5}{5} \right]_0^2 d\theta$$

$$= k \int_0^{2\pi} \tfrac{24}{5} \, d\theta = \tfrac{24}{5} k \Big[\theta \Big]_0^{2\pi} = \tfrac{48}{5} \pi k.$$

EXAMPLE 5 A solid has the shape of the region Q that lies inside the cylinder $r = a$, within the sphere $r^2 + z^2 = 4a^2$, and above the xy-plane. The density at a point P is directly proportional to the distance from the xy-plane to P. Find the mass and the moment of inertia I_z of the solid.

SOLUTION The region Q is sketched in Figure 13.70, together with a column that corresponds to a first summation in the direction parallel to the z-axis. Since the density at the point P is given by kz for some constant k, we may find the mass m by applying (13.26)(i), with $\delta(x, y, z) = kz$ and

Figure 13.69

Figure 13.70

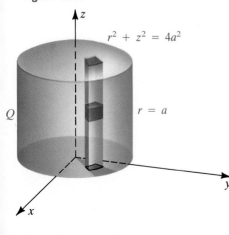

$dV = r \, dz \, dr \, d\theta$. Thus,

$$m = \int_0^{2\pi} \int_0^a \int_0^{\sqrt{4a^2-r^2}} (kz) \, r \, dz \, dr \, d\theta = k \int_0^{2\pi} \int_0^a \left[\frac{z^2}{2}\right]_0^{\sqrt{4a^2-r^2}} r \, dr \, d\theta$$

$$= \frac{1}{2}k \int_0^{2\pi} \int_0^a (4a^2 r - r^3) \, dr \, d\theta = \frac{1}{2}k \int_0^{2\pi} \left[2a^2 r^2 - \frac{r^4}{4}\right]_0^a d\theta$$

$$= \frac{7}{8}a^4 k \int_0^{2\pi} d\theta = \frac{7}{8}a^4 k \left[\theta\right]_0^{2\pi} = \frac{7}{4}a^4 \pi k.$$

To find I_z, we employ (13.27) and use cylindrical coordinates, obtaining

$$I_z = \int_0^{2\pi} \int_0^a \int_0^{\sqrt{4a^2-r^2}} r^2(kz) \, r \, dz \, dr \, d\theta = k \int_0^{2\pi} \int_0^a \left[\frac{z^2}{2}\right]_0^{\sqrt{4a^2-r^2}} r^3 \, dr \, d\theta$$

$$= \frac{1}{2}k \int_0^{2\pi} \int_0^a (4a^2 r^3 - r^5) \, dr \, d\theta = \frac{5}{6}a^6 \pi k.$$

EXAMPLE ■ 6 A glass chamber is to be used in experiments to test the effects of exposure to radon gas. The chamber has the shape of the paraboloid $z = 25 - x^2 - y^2$, with $z \geq 0$ and units in centimeters. If the density of radioactive energy is to be 6×10^{-11} joule/cm^3, calculate the total amount of radioactive energy that is required in the chamber.

SOLUTION The shape of the first-quadrant part of the chamber is similar to that of the surface in Figure 13.36. In this example, we let $\delta(x, y, z)$ denote the *energy density* at (x, y, z) instead of the mass density. If dV is the volume of a subregion dQ within the chamber and (x, y, z) is a point in dQ, then the radioactive energy within dQ may be approximated by $\delta(x, y, z) \, dV$. Taking a limit of triple sums with energy density $\delta(x, y, z) = \delta = 6 \times 10^{-11}$, we find that the total radioactive energy E is

$$E = \iiint_V \delta \, dV.$$

Using cylindrical coordinates yields

$$E = \delta \int_0^{2\pi} \int_0^5 \int_0^{25-r^2} r \, dz \, dr \, d\theta$$

$$= \delta \int_0^{2\pi} \int_0^5 r\left[z\right]_0^{25-r^2} dr \, d\theta$$

$$= \delta \int_0^{2\pi} \int_0^5 (25r - r^3) \, dr \, d\theta$$

$$= \frac{625}{2}\pi\delta \approx 5.89 \times 10^{-8} \text{ joule.}$$

EXERCISES 13.7

Exer. 1–14: Describe the graph of the equation in three dimensions.

1 (a) $r = 4$ (b) $\theta = -\pi/2$ (c) $z = 1$

2 (a) $r = -3$ (b) $\theta = \pi/4$ (c) $z = -2$

3 $r = -3 \sec \theta$ 4 $r = -\csc \theta$

5 $z = 4r^2$ 6 $z = 4 - r^2$

7 $r = 6 \sin \theta$ 8 $r \sec \theta = 4$

9 $z = 2r$ 10 $3z = r$

11 $r^2 = 9 - z^2$ 12 $r^2 + z^2 = 16$

13 $r = 2 \csc \theta \cot \theta$ 14 $r = \tan \theta \sec \theta$

Exer. 15–24: Change the equation to cylindrical coordinates.

15 $x^2 + y^2 + z^2 = 4$ 16 $x^2 + y^2 = 4z$

17 $3x + y - 4z = 12$ 18 $y = x$

19 $x^2 = 4 - y^2$

20 $x^2 + (y - 2)^2 = 4$

21 $x^2 - 4z^2 + y^2 = 0$

22 $x^2 - y^2 - z^2 = 1$

23 $y^2 + z^2 = 9$

24 $x^2 + z^2 = 9$

Exer. 25–28: Let f be an arbitrary continuous function of r, θ, and z, and let Q be the region shown in the figure. Set up an iterated triple integral in cylindrical coordinates for $\iiint_Q f(r, \theta, z)\, dV$.

25

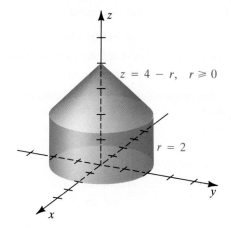

26

$z = 4 - r, \quad r \geqslant 0$

$r = 2$

27

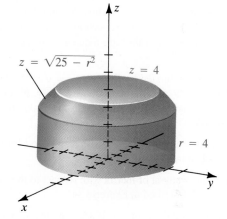

$z = \sqrt{25 - r^2}$

$z = 4$

$r = 4$

$z = 16 - r^2$

28

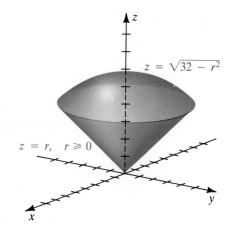

$z = \sqrt{32 - r^2}$

$z = r, \quad r \geqslant 0$

Exer. 29–38: Use cylindrical coordinates.

29 A solid is bounded by the paraboloid $z = x^2 + y^2$, the cylinder $x^2 + y^2 = 4$, and the xy-plane. Find

(a) its volume   (b) its centroid

30 A solid is bounded by the cone $z = \sqrt{x^2 + y^2}$, the cylinder $x^2 + y^2 = 4$, and the xy-plane. Find

(a) its volume   (b) its centroid

31 A homogeneous right circular cylinder of density δ has altitude h and radius of base a. Find its moment of inertia with respect to

(a) the axis of the cylinder

(b) a diameter of the base

32 A homogeneous solid of density δ is bounded by the cone $z = \sqrt{x^2 + y^2}$ and the paraboloid $z = x^2 + y^2$. Find

(a) its center of mass

(b) its moment of inertia with respect to the z-axis

33 A spherical solid has radius a, and the density at $P(x, y, z)$ is directly proportional to the distance from P to a fixed line l through the center of the solid. Find its mass.

34 A solid is bounded by the cone $z = \sqrt{x^2 + y^2}$ and the plane $z = 4$, and the density at $P(x, y, z)$ is directly proportional to the distance from the z-axis to P. Find its mass.

35 Find the moment of inertia with respect to l for the solid described in Exercise 33.

36 Find the moment of inertia with respect to the z-axis for the solid described in Exercise 34.

c **37** For an altitude z up to 10,000 m, the density δ (in kilograms per cubic meter) of the earth's atmosphere can be approximated by

$$\delta = 1.2 - (1.05 \times 10^{-4})z + (2.6 \times 10^{-9})z^2.$$

Estimate the mass of a column of air 10 km high that has a circular base of radius 3 m.

38 A solid is cut out of the sphere $x^2 + y^2 + z^2 = 4$ by the cylinder $x^2 + y^2 = 2y$. The density at $P(x, y, z)$ is directly proportional to the distance from the xy-plane to P. Find

(a) its mass   (b) its center of mass

Exer. 39–40: Use cylindrical coordinates to evaluate the integral.

39 $\displaystyle\int_0^1 \int_0^{\sqrt{1-y^2}} \int_0^{\sqrt{4-x^2-y^2}} z \, dz \, dx \, dy$

40 $\displaystyle\int_0^2 \int_{-\sqrt{2x-x^2}}^{\sqrt{2x-x^2}} \int_0^{x^2+y^2} \sqrt{x^2 + y^2} \, dz \, dy \, dx$

c **Exer. 41–44: (a) Set up an iterated triple integral in cylindrical coordinates for the given integral. (b) Approximate the integral in part (a) to three decimal places.**

41 $\displaystyle\iiint_Q z e^{-(x^2+y^2+5xy)} \, dV$; Q is bounded by the cylinder $x^2 + y^2 = 4$ and the planes $z = 0$ and $z = 4$

42 $\displaystyle\iiint_Q z^2 \sqrt{x^2 + y^2 + z^2} \, dV$; Q is bounded by the cylinder $x^2 + y^2 = 5$ and the planes $z = 1$ and $z = 5$

43 $\displaystyle\iiint_Q \sqrt{x^2 + y^2 + 2xy + z^3} \, dV$; Q is in the first octant bounded by the cylinder $x^2 + y^2 = 4$ and the planes $z = 0$ and $z = 4$

44 $\displaystyle\iiint_Q xy e^{-(x^2+y^2)} \, dV$; Q is bounded by the cylinder $x^2 + y^2 = 3$, and the planes $z = 0$ and $z = 2$, and with $y \geq -\sqrt{3}x$ and $y \geq (\sqrt{3}/3)x$

13.8 SPHERICAL COORDINATES

In this section, we discuss the **spherical coordinate system** in three dimensions. The *spherical coordinates* of a point P are given by an ordered triple (ρ, ϕ, θ), where, as shown in Figure 13.71 on the following page, $\rho = \|\overrightarrow{OP}\|$, ϕ is the angle between $\overrightarrow{OP}$ and the vector $\mathbf{k}$, and θ is a polar

Figure 13.71

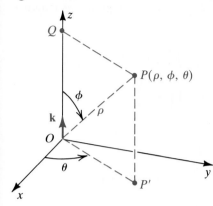

angle associated with the projection P' of P onto the xy-plane. The point Q is the projection of P onto the z-axis. Note that $\rho \geq 0$ and $0 \leq \phi \leq \pi$.

If $\rho_0 > 0$, the graph of $\rho = \rho_0$ is a sphere of radius ρ_0 with center O (see Figure 13.72a). The graph of $\rho = 0$ consists of one point, the origin.

If $0 < \phi_0 < \pi$, the graph of $\phi = \phi_0$ is a half-cone with vertex at O (see Figure 13.72b). The graph of $\phi = 0$ is the nonnegative part of the z-axis, and the graph of $\phi = \pi$ is the nonpositive part.

The graph of $\theta = \theta_0$ is a half-plane containing the z-axis (see Figure 13.72c).

The relationship between spherical coordinates (ρ, ϕ, θ) and rectangular coordinates (x, y, z) of a point P can be found by referring to Figure 13.71. Using the fact that

$$x = \| \overrightarrow{OP'} \| \cos \theta \quad \text{and} \quad y = \| \overrightarrow{OP'} \| \sin \theta$$

together with

$$\| \overrightarrow{OP'} \| = \| \overrightarrow{QP} \| = \rho \sin \phi \quad \text{and} \quad \| \overrightarrow{OQ} \| = \rho \cos \phi$$

gives us (i) of the next theorem. Part (ii) is a consequence of the distance formula (10.13).

Figure 13.72

(a)

(b)

(c)

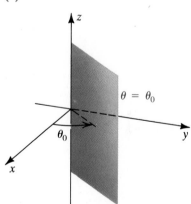

Theorem 13.31

The rectangular coordinates (x, y, z) and the spherical coordinates (ρ, ϕ, θ) of a point P are related as follows:

(i) $x = \rho \sin \phi \cos \theta, \quad y = \rho \sin \phi \sin \theta, \quad z = \rho \cos \phi$

(ii) $\rho^2 = x^2 + y^2 + z^2$

EXAMPLE ■ I If a point P has spherical coordinates $(4, \pi/6, \pi/3)$, find rectangular and cylindrical coordinates for P.

Figure 13.73

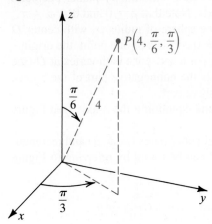

SOLUTION   The point P is plotted in Figure 13.73. Using Theorem (13.31)(i), with $\rho = 4$, $\phi = \pi/6$, and $\theta = \pi/3$, we obtain the rectangular coordinates:

$$x = 4 \sin \frac{\pi}{6} \cos \frac{\pi}{3} = 4 \left(\frac{1}{2} \right) \left(\frac{1}{2} \right) = 1$$

$$y = 4 \sin \frac{\pi}{6} \sin \frac{\pi}{3} = 4 \left(\frac{1}{2} \right) \left(\frac{\sqrt{3}}{2} \right) = \sqrt{3}$$

$$z = 4 \cos \frac{\pi}{6} = 4 \left(\frac{\sqrt{3}}{2} \right) = 2\sqrt{3}$$

To find cylindrical coordinates for P, we first note that

$$r^2 = x^2 + y^2 = 1 + 3 = 4,$$

and hence $r = \pm 2$. Thus, cylindrical coordinates (r, θ, z) for P are $(2, \pi/3, 2\sqrt{3})$.

EXAMPLE ■ 2   Find an equation in spherical coordinates whose graph is the paraboloid $z = x^2 + y^2$.

SOLUTION   Substituting from (13.31)(i), we obtain

$$\rho \cos \phi = \rho^2 \sin^2 \phi \cos^2 \theta + \rho^2 \sin^2 \phi \sin^2 \theta$$
$$= \rho^2 \sin^2 \phi (\cos^2 \theta + \sin^2 \theta)$$
$$= \rho^2 \sin^2 \phi,$$

or, equivalently,

$$\rho (\cos \phi - \rho \sin^2 \phi) = 0.$$

Hence either

$$\rho = 0 \quad \text{or} \quad \rho \sin^2 \phi = \cos \phi.$$

The graph of $\rho = 0$ is the origin, which is also on the graph of the equation $\rho \sin^2 \phi = \cos \phi$. Hence, $z = x^2 + y^2$ is equivalent to $\rho \sin^2 \phi = \cos \phi$.

We cannot have $\sin \phi = 0$ in the equation $\rho \sin^2 \phi = \cos \phi$, because in this case $\phi = 0$ or $\phi = \pi$, and hence $\cos \phi = 1$ or $\cos \phi = -1$, which gives us the absurdity $0 = 1$ or $0 = -1$. Therefore, we may divide both sides of the equation $\rho \sin^2 \phi = \cos \phi$ by $\sin^2 \phi$, obtaining

$$\rho = \frac{\cos \phi}{\sin^2 \phi} = \frac{\cos \phi}{\sin \phi} \frac{1}{\sin \phi},$$

or, equivalently,

$$\rho = \cot \phi \csc \phi.$$

EXAMPLE ■ 3   Change the equation $\rho = 2 \sin \phi \cos \theta$ to rectangular coordinates, and describe its graph.

SOLUTION   Since we plan to use the first formula in (13.31)(i), we multiply both sides of the equation by ρ, obtaining

$$\rho^2 = 2\rho \sin \phi \cos \theta.$$

Applying (13.31) gives us the following equivalent equation in x, y, and z:

$$x^2 + y^2 + z^2 = 2x$$
$$(x - 1)^2 + y^2 + z^2 = 1$$

By Theorem (10.15), the graph is a sphere with radius 1 whose center, in rectangular coordinates, is $(1, 0, 0)$.

Sometimes it is convenient to use spherical coordinates to evaluate triple integrals. The simplest case occurs when a function f of ρ, ϕ, and θ is continuous throughout a region of the form

$$Q = \{(\rho, \phi, \theta): a \le \rho \le b, c \le \phi \le d, m \le \theta \le n\}.$$

We divide Q into subregions $Q_1, Q_2, \ldots, Q_r$ that have the same shape as Q by using graphs of equations $\rho = a_k$, $\phi = c_k$, and $\theta = m_k$, where a_k, c_k, and m_k are numbers in the intervals $[a, b]$, $[c, d]$, and $[m, n]$, respectively. These graphs are spheres with centers at the origin, cones with axes along the z-axis, and half-planes containing the z-axis, respectively (see Figure 13.72).

A typical subregion Q_k is sketched in Figure 13.74(a). The increment $\Delta\rho_k$ is the distance between two successive spheres, $\Delta\phi_k$ is the change in the vertex angle between successive cones, and $\Delta\theta_k$ is the angle between two successive half-planes. If, as shown in Figure 13.74(b), P has spherical coordinates $(\rho_k, \phi_k, \theta_k)$ and P' is the projection of P onto the z-axis, then, from trigonometry, the length of segment $P'P$ is $\rho_k \sin \phi_k$ and hence, by Theorem (13) on page 38, the length of arc PS is $\rho_k \sin \phi_k \Delta\theta_k$. The

Figure 13.74
(a)   **(b)**

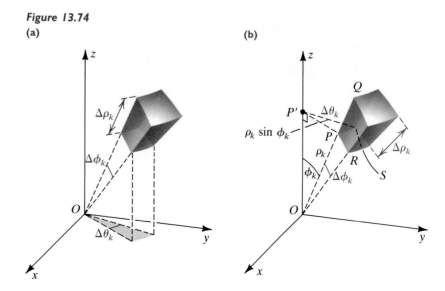

distance from O to P is ρ_k, so the length of arc PR is $\rho_k \Delta \phi_k$. The length of segment PQ is $\Delta \rho_k$. We shall approximate the volume ΔV_k of Q_k by regarding it as a rectangular box with these three dimensions. Thus,

$$\Delta V_k \approx (\rho_k \sin \phi_k \Delta \theta_k)(\rho_k \Delta \phi_k) \Delta \rho_k,$$

or

$$\Delta V_k \approx \rho_k^2 \sin \phi_k \Delta \rho_k \Delta \phi_k \Delta \theta_k.$$

We define the triple integral of f over Q as a limit of Riemann sums $\sum_k f(\rho_k, \phi_k, \theta_k) \Delta V_k$. The following theorem can be proved.

Evaluation Theorem
(Spherical Coordinates)   13.32

$$\iiint\limits_Q f(\rho, \phi, \theta) \, dV = \int_m^n \int_c^d \int_a^b f(\rho, \phi, \theta) \, \rho^2 \sin \phi \, d\rho \, d\phi \, d\theta$$

There are five other possible orders of integration for this iterated integral.

Spherical coordinates may also be employed over regions more complicated than the one just discussed by using inner partitions. In such cases, the limits of integration in the iterated integral must be suitably adjusted.

In the following two examples, we reconsider Examples 3 and 4 of the preceding section, which were solved using cylindrical coordinates.

EXAMPLE ■ 4   Find the centroid of a hemispherical solid Q of radius a.

Figure 13.75

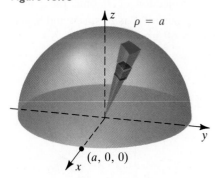

SOLUTION   The solid is resketched in Figure 13.75, together with a spherical coordinates element of volume and a column indicating a first (partial) integration with respect to ρ. As in Example 3 of Section 13.7, it is sufficient to find $\bar{z}$. Using spherical coordinates, we have

$$M_{xy} = \iiint\limits_Q z \, dV$$

$$= \int_0^{2\pi} \int_0^{\pi/2} \int_0^a (\rho \cos \phi) \, \rho^2 \sin \phi \, d\rho \, d\phi \, d\theta$$

$$= \int_0^{2\pi} \int_0^{\pi/2} \left[\frac{\rho^4}{4} \right]_0^a \sin \phi \cos \phi \, d\phi \, d\theta$$

$$= \tfrac{1}{4} a^4 \int_0^{2\pi} \int_0^{\pi/2} \sin \phi \cos \phi \, d\phi \, d\theta$$

$$= \tfrac{1}{4} a^4 \int_0^{2\pi} \left[\tfrac{1}{2} \sin^2 \phi \right]_0^{\pi/2} d\theta$$

$$= \tfrac{1}{8} a^4 \left[\theta \right]_0^{2\pi} = \tfrac{1}{4} \pi a^4.$$

Hence,

$$\bar{z} = \frac{M_{xy}}{V} = \frac{\tfrac{1}{4} \pi a^4}{\tfrac{2}{3} \pi a^3} = \frac{3}{8} a.$$

Figure 13.76

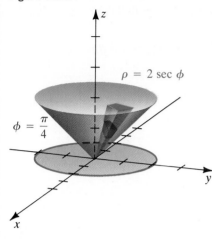

$\rho = 2 \sec \phi$

$\phi = \dfrac{\pi}{4}$

EXAMPLE ▪ 5   A solid Q is bounded by the cone $z = \sqrt{x^2 + y^2}$ and the plane $z = 2$. The density at $P(x, y, z)$ is directly proportional to the square of the distance from the origin to P. Find its mass.

SOLUTION   The solid is resketched in Figure 13.76, together with a spherical coordinates element of volume and a column indicating a first (partial) integration with respect to ρ. An equation of the plane $z = 2$ in spherical coordinates is $\rho \cos \phi = 2$, or $\rho = 2 \sec \phi$. The density at (x, y, z) is given by

$$\delta = k(x^2 + y^2 + z^2) = k\rho^2$$

for some constant k. Hence,

$$m = \iiint\limits_{Q} \delta \, dV$$

$$= k \int_0^{2\pi} \int_0^{\pi/4} \int_0^{2\sec\phi} \rho^2 \cdot \rho^2 \sin\phi \, d\rho \, d\phi \, d\theta$$

$$= k \int_0^{2\pi} \int_0^{\pi/4} \left[\frac{\rho^5}{5}\right]_0^{2\sec\phi} \sin\phi \, d\phi \, d\theta$$

$$= k \int_0^{2\pi} \int_0^{\pi/4} \tfrac{32}{5} \sec^5\phi \sin\phi \, d\phi \, d\theta$$

$$= \tfrac{32}{5}k \int_0^{2\pi} \int_0^{\pi/4} \sec^3\phi \sec\phi \tan\phi \, d\phi \, d\theta$$

$$= \tfrac{32}{5}k \int_0^{2\pi} \left[\tfrac{1}{4} \sec^4\phi\right]_0^{\pi/4} d\theta = \tfrac{8}{5}k \int_0^{2\pi} (4 - 1) \, d\theta = \tfrac{48}{5}\pi k.$$

Figure 13.77

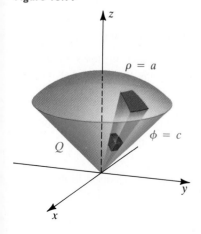

$\rho = a$

$\phi = c$

Q

EXAMPLE ▪ 6   Find the volume and the centroid of the region Q that is bounded above by the sphere $\rho = a$ and below by the cone $\phi = c$, where $0 < c < \pi/2$.

SOLUTION   The region Q is sketched in Figure 13.77, together with a spherical coordinates element of volume and a column indicating a first (partial) integration with respect to ρ. The volume V is

$$V = \iiint\limits_{Q} dV = \int_0^{2\pi} \int_0^c \int_0^a \rho^2 \sin\phi \, d\rho \, d\phi \, d\theta = \int_0^{2\pi} \int_0^c \left[\frac{\rho^3}{3}\right]_0^a \sin\phi \, d\phi \, d\theta$$

$$= \tfrac{1}{3}a^3 \int_0^{2\pi} \int_0^c \sin\phi \, d\phi \, d\theta = \tfrac{1}{3}a^3 \int_0^{2\pi} \left[-\cos\phi\right]_0^c d\theta$$

$$= \tfrac{1}{3}a^3 \int_0^{2\pi} (1 - \cos c) \, d\theta = \tfrac{2}{3}\pi a^3 (1 - \cos c).$$

By symmetry, the centroid is on the z-axis, so we need find only $\bar{z} = M_{xy}/V$. Using (13.26)(ii) with $\delta(x, y, z) = 1$, we find that

$$M_{xy} = \iiint_Q z \, dV = \int_0^{2\pi} \int_0^c \int_0^a (\rho \cos\phi)\rho^2 \sin\phi \, d\rho \, d\phi \, d\theta$$

$$= \int_0^{2\pi} \int_0^c \left[\frac{\rho^4}{4}\right]_0^a \sin\phi \cos\phi \, d\phi \, d\theta$$

$$= \tfrac{1}{4}a^4 \int_0^{2\pi} \int_0^c \sin\phi \cos\phi \, d\phi \, d\theta$$

$$= \tfrac{1}{4}a^4 \int_0^{2\pi} \left[\tfrac{1}{2}\sin^2\phi\right]_0^c d\theta$$

$$= \tfrac{1}{8}a^4 \sin^2 c \int_0^{2\pi} d\theta = \tfrac{1}{4}\pi a^4 \sin^2 c.$$

The centroid, in rectangular coordinates, is $(0, 0, \bar{z})$ with

$$\bar{z} = \frac{M_{xy}}{V} = \frac{3}{8}a(1 + \cos c).$$

EXERCISES 13.8

Exer. 1–2: Change the spherical coordinates to (a) rectangular coordinates and (b) cylindrical coordinates.

1 $(4, \pi/6, \pi/2)$

2 $(1, 3\pi/4, 2\pi/3)$

Exer. 3–4: Change the rectangular coordinates to (a) spherical coordinates and (b) cylindrical coordinates.

3 $(1, 1, -2\sqrt{2})$

4 $(1, \sqrt{3}, 0)$

Exer. 5–20: Describe the graph of the equation in three dimensions.

5 (a) $\rho = 3$ **(b)** $\phi = \pi/6$ **(c)** $\theta = \pi/3$

6 (a) $\rho = 5$ **(b)** $\phi = 2\pi/3$ **(c)** $\theta = \pi/4$

7 $\rho = 4\cos\phi$ **8** $\rho\sec\phi = 6$

9 $\rho\cos\phi = 3$ **10** $\rho = 4\sec\phi$

11 $\rho = 6\sin\phi\cos\theta$ **12** $\rho = 8\sin\phi\sin\theta$

13 $\rho = 5\csc\phi\csc\theta$ **14** $\rho = 2\csc\phi\sec\theta$

15 $\rho = 5\csc\phi$ **16** $\rho\sin\phi = 3$

17 $\tan\phi = 2$ **18** $\tan\theta = 4$

19 $\rho = 6\cot\phi\csc\phi$ **20** $\rho^2 - 3\rho + 2 = 0$

Exer. 21–30: Change the equation to spherical coordinates.

21 $x^2 + y^2 + z^2 = 4$ **22** $x^2 + y^2 = 4z$

23 $3x + y - 4z = 12$ **24** $y = x$

25 $x^2 = 4 - y^2$ **26** $x^2 + (y - 2)^2 = 4$

27 $x^2 - 4z^2 + y^2 = 0$ **28** $x^2 - y^2 - z^2 = 1$

29 $y^2 + z^2 = 9$ **30** $x^2 + z^2 = 9$

Exer. 31–38: Use spherical coordinates.

31 Find the mass and the center of mass of a solid hemisphere of radius a if the density at a point P is directly proportional to the distance from the center of the base to P.

32 Find the volume and the centroid of the solid bounded by the cone $z = \sqrt{x^2 + y^2}$, the cylinder $x^2 + y^2 = 4$, and the xy-plane.

33 Find the moment of inertia with respect to the axis of symmetry for the hemisphere in Exercise 31.

34 Find the moment of inertia with respect to a diameter of the base of a homogeneous solid hemisphere of radius a.

35 The points of intersection of the cone $z^2 = x^2 + y^2$ and the sphere $x^2 + y^2 + z^2 = 4z$ form a circle (in addition to the origin). Find the volume of the solid that lies above the plane containing this circle and inside the sphere.

36 Find the volume of the solid that lies outside the cone $z^2 = x^2 + y^2$ and inside the sphere $x^2 + y^2 + z^2 = 1$.

37 Find the mass of the solid that lies outside the sphere $x^2 + y^2 + z^2 = 1$ and inside the sphere $x^2 + y^2 + z^2 = 4$ if the density at a point P is directly proportional to the square of the distance from the center of the spheres to P.

38 If the earth is assumed to be spherical with a radius of 6370 km, the density δ (in kilograms per cubic meter) of the atmosphere a distance ρ meters from the center of the earth can be approximated by

$$\delta = 619.09 - (9.7 \times 10^{-5})\rho$$

$$6{,}370{,}000 \le \rho \le 6{,}373{,}000.$$

(a) Estimate the mass of the atmosphere between ground level and an altitude of 3 km.

(b) The atmosphere extends beyond an altitude of 100 km and has a total mass of approximately 5.1×10^{18} kg. What percentage of the mass is in the lowest 3 km of the atmosphere?

Exer. 39–40: Evaluate the integral by changing to spherical coordinates.

39 $\displaystyle\int_{-2}^{2} \int_{-\sqrt{4-x^2}}^{\sqrt{4-x^2}} \int_{\sqrt{x^2+y^2}}^{\sqrt{8-x^2-y^2}} (x^2 + y^2 + z^2)\, dz\, dy\, dx$

40 $\displaystyle\int_{0}^{\sqrt{2}} \int_{y}^{\sqrt{4-y^2}} \int_{0}^{\sqrt{4-x^2-y^2}} \sqrt{x^2 + y^2 + z^2}\, dz\, dx\, dy$

41 The relationship between spherical and rectangular coordinates is important in the design of joints for robot arms. The movement caused by three joints is illustrated in the figure: A hinge joint rotates the arm about the z-axis through an angle θ in the xy-plane, a sliding joint extends and contracts the length L of the arm, and another hinge joint tilts the arm up and down through the (spherical coordinate) angle ϕ.

(a) If the joints are currently set at $\theta = 120°$, $\phi = 135°$, and $L = 12$ in., find the rectangular coordinates of the joint P in the hand of the arm.

(b) Describe the changes in θ, ϕ, and L that are necessary to pick up an object located at the point with rectangular coordinates $(-8, -8, 8)$.

Exercise 41

42 Small movements in the joints of robot arms, called *differential motions*, are used to describe the change in position of the hand when joint settings are not exact. Suppose that the joint settings θ, ϕ, and L for the arm in Exercise 41 have errors $\Delta\theta$, $\Delta\phi$, and ΔL.

(a) Use differentials to find formulas for the corresponding errors Δx, Δy, and Δz in the rectangular coordinates of P.

(b) Use the formulas found in part (a) to estimate Δx, Δy, and Δz if $\theta = 135°$, $\phi = 45°$, $L = 12$ in., $\Delta\theta = -1°$, $\Delta\phi = 0.5°$, and $\Delta L = 0.1$.

c **Exer. 43–44:** **(a)** Set up an iterated triple integral in spherical coordinates for the given integral in a fashion that allows at least one of the innermost integrals to be evaluated. **(b)** Approximate the integral in part (a) to three decimal places.

43 $\displaystyle\int_{-3}^{3} \int_{0}^{\sqrt{9-x^2}} \int_{0}^{\sqrt{9-x^2-y^2}} \sqrt{x^2 + y^2 + z^2}\, dz\, dy\, dx$

44 $\displaystyle\int_{0}^{4} \int_{-\sqrt{16-x^2}}^{\sqrt{16-x^2}} \int_{-\sqrt{16-x^2-y^2}}^{\sqrt{16-x^2-y^2}} e^{-(x^2+y^2+z^2)^3}\, dz\, dy\, dz$

13.9   CHANGE OF VARIABLES AND JACOBIANS

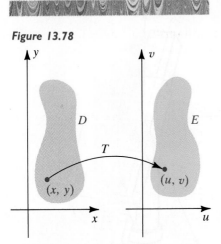

In Section 13.3, we discussed how to change a double integral in rectangular coordinates to a double integral in polar coordinates. In the preceding two sections, we considered triple integrals in cylindrical and spherical coordinates. We now introduce a more general method for changing variables in multiple integrals. This method is closely connected to *transformations* (or *correspondences*) from one rectangular coordinate system to another.

Let us begin by considering a function T whose domain D is a region in the xy-plane and whose range E is a region in a uv-plane. As illustrated in Figure 13.78, to each point (x, y) in D there corresponds exactly one point (u, v) in E such that $T(x, y) = (u, v)$. We call T a **transformation of coordinates** from the xy-plane to the uv-plane. Since each pair (u, v) is uniquely determined by (x, y), both u and v are functions of x and y. Thus, we have the following *transformation formulas*, where f and g are functions that have the same domain D as T.

Transformation of Coordinates Formulas   13.33

$$u = f(x, y), \quad v = g(x, y); \quad (x, y) \text{ in } D, \ (u, v) \text{ in } E$$

Given the transformation of coordinates (13.33), let us partition a region in the uv-plane by means of vertical lines $u = c_1, u = c_2, u = c_3, \ldots$ and horizontal lines $v = d_1, v = d_2, v = d_3, \ldots$, as illustrated in Figure 13.79(a). The corresponding level curves for f and g in the xy-plane are graphs of

$$u = f(x, y) = c_j \quad \text{and} \quad v = g(x, y) = d_k$$

for $j = 1, 2, 3, \ldots$ and $k = 1, 2, 3, \ldots$ These curves determine a *curvilinear partition* of a region in the xy-plane, as illustrated in Figure 13.79(b).

Figure 13.79

(a)

(b)

Figure 13.80

(a)

(b)

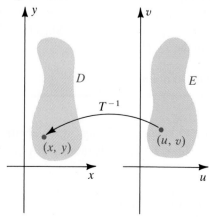

We refer to the curves $u = f(x, y) = c_j$ and $v = g(x, y) = d_k$ in the xy-plane as **u-curves** and **v-curves**, respectively. Of course, the types of curves obtained depend on the nature of the functions f and g. The next example illustrates a case for which each u-curve and v-curve is a line.

EXAMPLE■1 Let T be the transformation of coordinates from the xy-plane to the uv-plane determined by

$$u = x + 2y, \quad v = x - 2y.$$

Sketch, in the uv-plane, the vertical lines $u = 2, u = 4, u = 6, u = 8$ and the horizontal lines $v = -1, v = 1, v = 3, v = 5$. Sketch the corresponding u-curves and v-curves in the xy-plane.

SOLUTION The vertical and horizontal lines in the uv-plane are sketched in Figure 13.80(a).

The u-curves in the xy-plane are the lines

$$x + 2y = 2, \quad x + 2y = 4, \quad x + 2y = 6, \quad x + 2y = 8,$$

and the v-curves are the lines

$$x - 2y = -1, \quad x - 2y = 1, \quad x - 2y = 3, \quad x - 2y = 5.$$

These lines are sketched in Figure 13.80(b).

The transformation T in Example 1 is one-to-one; that is, if $(x_1, y_1) \neq (x_2, y_2)$ in the xy-plane, then $T(x_1, y_1) \neq T(x_2, y_2)$ in the uv-plane. In general, if T is a one-to-one transformation of coordinates, then by reversing the correspondence, we obtain a transformation T^{-1} from the uv-plane to the xy-plane called the **inverse** of T. We may specify T^{-1} by means of equations of the form

$$x = F(u, v), \quad y = G(u, v)$$

for certain functions F and G. If a one-to-one transformation T is illustrated by an arrow, as in Figure 13.78, then the inverse T^{-1} can be illustrated by *reversing* the arrow, as in Figure 13.81. Note that $T^{-1}(T(x, y)) = (x, y)$ and that $T(T^{-1}(u, v)) = (u, v)$ for every (x, y) in D and every (u, v) in E.

Figure 13.81

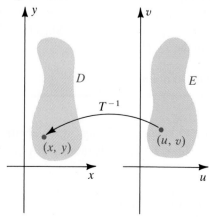

EXAMPLE■2

(a) Find the inverse of the transformation T in Example 1.

(b) Find the curve in the uv-plane that T^{-1} transforms onto the ellipse $x^2 + 4y^2 = 1$.

SOLUTION

(a) The transformation T is given by

$$u = x + 2y, \quad v = x - 2y.$$

If we add corresponding sides of these equations, we obtain $u + v = 2x$.

Figure 13.82

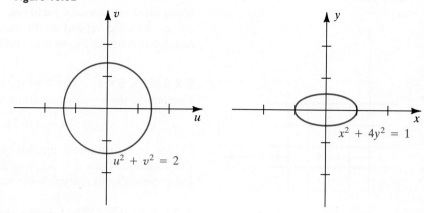

$$u^2 + v^2 = 2$$

$$x^2 + 4y^2 = 1$$

Subtracting corresponding sides leads to $u - v = 4y$. Thus, T^{-1} is given by

$$x = \tfrac{1}{2}(u + v), \quad y = \tfrac{1}{4}(u - v).$$

(b) Since x and y are related by means of the last two equations in part (a), the points (u, v) corresponding to $x^2 + 4y^2 = 1$ must satisfy the equation

$$[\tfrac{1}{2}(u + v)]^2 + 4[\tfrac{1}{4}(u - v)]^2 = 1.$$

This simplifies to

$$u^2 + v^2 = 2.$$

It follows that the circle of radius $\sqrt{2}$ with center at the origin in the uv-plane is transformed onto the ellipse $x^2 + 4y^2 = 1$ by T^{-1} (see Figure 13.82).

Figure 13.83

(a)

(b)

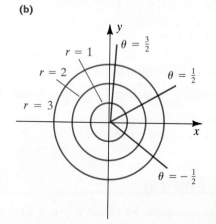

EXAMPLE ▪ 3   If $P(x, y)$ is any point with $x \neq 0$ in the xy-plane, let

$$r = \sqrt{x^2 + y^2}, \qquad \theta = \arctan \frac{y}{x}$$

and let $P(0, b)$ correspond to $r = b, \theta = \pi/2$. Thus, r and θ are polar coordinates for P, and the preceding equalities determine a transformation of coordinates T from the xy-plane to the $r\theta$-plane. Sketch, in the $r\theta$-plane, the graphs of $r = 1$, $r = 2$, $r = 3$ and $\theta = -\frac{1}{2}$, $\theta = \frac{1}{2}$, $\theta = \frac{3}{2}$. Sketch the corresponding r-curves and θ-curves in the xy-plane.

SOLUTION   The graphs are sketched in Figure 13.83.

In Section 4.6, we discussed how to change variables in a definite integral $\int_a^b f(x)\, dx$. Under suitable conditions, if we substitute $x = g(u)$, then $dx = g'(u)\, du$ and

$$\int_a^b f(x)\, dx = \int_c^d f(g(u))g'(u)\, du,$$

where $a = g(c)$ and $b = g(d)$ (see Theorem (4.33)). We shall next obtain a formula for changing variables in a double integral.

Consider $\iint_R F(x, y)\, dA$, where R is a region in the xy-plane, and suppose we make the substitution

$$x = f(u, v), \quad y = g(u, v),$$

where f and g are functions that have continuous second partial derivatives. These equations define a transformation of coordinates W from the uv-plane to the xy-plane. After the substitution for x and y, the integrand becomes a function of u and v. One of our objectives is to find a region S in the uv-plane that is transformed onto R by W, as illustrated in Figure 13.84, such that

$$\iint_R F(x, y)\, dA = \iint_S F(f(u, v), g(u, v))\, dA.$$

Figure 13.84

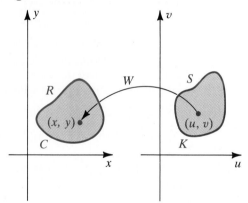

For a suitable region S to exist, it is necessary to place restrictions on the region R and on the integrand $F(x, y)$. We shall assume that R consists of all points that are either inside or on a piecewise-smooth simple closed curve C, and that F has continuous first partial derivatives throughout an open region containing R. The **positive direction along C** is such that as a point P traces C, the region R is always on the left. For the **negative direction**, R is on the right as P traces C. We shall also require that W transform a region S of the uv-plane in a one-to-one manner onto R, and that S be bounded by a piecewise-smooth simple closed curve K that W transforms onto C. Our final requirement is that, as (u, v) traces K once in the positive direction, the corresponding point (x, y) trace C once in either the positive or the negative direction. It is possible to weaken these conditions; however, that is beyond the scope of our work.

The function of u and v introduced in the next definition will be used in the change of variable process. It is named after the German mathematician C. G. Jacobi (1804–1851).

Definition 13.34

If $x = f(u, v)$ and $y = g(u, v)$, then the **Jacobian** of x and y with respect to u and v, denoted by $\partial(x, y)/\partial(u, v)$, is

$$\frac{\partial(x, y)}{\partial(u, v)} = \begin{vmatrix} \dfrac{\partial x}{\partial u} & \dfrac{\partial x}{\partial v} \\[2mm] \dfrac{\partial y}{\partial u} & \dfrac{\partial y}{\partial v} \end{vmatrix} = \frac{\partial x}{\partial u}\frac{\partial y}{\partial v} - \frac{\partial y}{\partial u}\frac{\partial x}{\partial v}.$$

In the next theorem, *all symbols have the same meanings as before, and we assume that the regions and functions satisfy the conditions we have discussed.* In the statement of the theorem, the notations $dx\,dy$ and $du\,dv$ are used in place of dA so that there is no misunderstanding about the region of integration.

Theorem 13.35

If $x = f(u, v)$, $y = g(u, v)$ is a transformation of coordinates, then

$$\iint\limits_{R} F(x, y)\,dx\,dy = \pm \iint\limits_{S} F(f(u, v), g(u, v))\frac{\partial(x, y)}{\partial(u, v)}\,du\,dv.$$

As (u, v) traces the boundary K of S once in the positive direction, the corresponding point (x, y) traces the boundary C of R once in either the positive direction, in which case the plus sign is chosen, or the negative direction, in which case the minus sign is chosen.

A proof of Theorem (13.35), based on Green's theorem (14.19), is given in Appendix I.

If we let $F(x, y) = 1$ for every (x, y) in R, then the integral on the left in (13.35) is the area A of R and we obtain

$$A = \pm \iint\limits_{S} \frac{\partial(x, y)}{\partial(u, v)}\,du\,dv.$$

If $\partial(x, y)/\partial(u, v) > 0$ throughout S, then the double integral on the right side of this equation is positive, and hence the $+$ sign should be used. This shows that if the Jacobian is always positive, then as the point (u, v) traces the boundary K of S once in the positive direction, the point (x, y) traces the boundary C of R once in the positive direction. If the Jacobian is always negative, then (x, y) traces C once in the negative direction. This gives us the following corollary to Theorem (13.35).

Corollary 13.36

If $x = f(u, v)$, $y = g(u, v)$ is a transformation of coordinates and if the Jacobian $\partial(x, y)/\partial(u, v)$ does not change sign in S, then

$$\iint\limits_{R} F(x, y)\,dx\,dy = \iint\limits_{S} F(f(u, v), g(u, v))\left|\frac{\partial(x, y)}{\partial(u, v)}\right|\,du\,dv.$$

EXAMPLE ■ 4   Evaluate

$$\iint\limits_{R} e^{(y-x)/(y+x)} \, dx \, dy,$$

where R is the region in the xy-plane bounded by the trapezoid with vertices $(0, 1)$, $(0, 2)$, $(2, 0)$, and $(1, 0)$.

SOLUTION   The fact that $y - x$ and $y + x$ appear in the integrand suggests the following substitution:

$$u = y - x, \quad v = y + x$$

These equations define a transformation T from the xy-plane to the uv-plane. To apply Theorem (13.35), we must use the inverse transformation T^{-1}. To find T^{-1}, we solve the preceding equations for x and y in terms of u and v, obtaining

$$x = \tfrac{1}{2}(-u + v), \quad y = \tfrac{1}{2}(u + v).$$

By (13.34), the Jacobian is

$$\frac{\partial(x, y)}{\partial(u, v)} = \begin{vmatrix} -\tfrac{1}{2} & \tfrac{1}{2} \\ \tfrac{1}{2} & \tfrac{1}{2} \end{vmatrix} = -\frac{1}{4} - \frac{1}{4} = -\frac{1}{2}.$$

The region R is sketched in Figure 13.85(a). To determine the region S in the uv-plane that transforms onto R, we note that the sides of R lie on the lines

$$x = 0, \quad x + y = 2, \quad y = 0, \quad x + y = 1.$$

Substituting $x = \tfrac{1}{2}(-u + v)$ and $y = \tfrac{1}{2}(u + v)$, we see that the corresponding curves in the uv-plane are, respectively,

$$v = u, \quad v = 2, \quad v = -u, \quad v = 1.$$

These lines form the boundary of the trapezoidal region S shown in Figure 13.85(b).

If the point (u, v) moves along the line $v = u$ from $(1, 1)$ to $(2, 2)$, then the corresponding point (x, y) moves along the line $x = 0$ from $(0, 1)$ to $(0, 2)$. Similarly, if (u, v) moves along the line $v = 2$ from $(2, 2)$ to $(-2, 2)$, then (x, y) moves along the line $x + y = 2$ from $(0, 2)$ to $(2, 0)$. Proceeding in this manner, we see that if (u, v) traces the boundary of S once in the *positive* direction, then (x, y) traces the boundary of R once in the *negative* direction. It can also be shown, using inequalities, that interior points of S correspond to interior points of R. Applying Theorem (13.35) yields

$$\iint\limits_{R} e^{(y-x)/(y+x)} \, dx \, dy = -\iint\limits_{S} e^{u/v} \left(-\tfrac{1}{2}\right) du \, dv = \tfrac{1}{2} \int_{1}^{2} \int_{-v}^{v} e^{u/v} \, du \, dv$$

$$= \tfrac{1}{2} \int_{1}^{2} \left[v e^{u/v} \right]_{-v}^{v} dv = \tfrac{1}{2} \int_{1}^{2} v(e - e^{-1}) \, dv$$

$$= \tfrac{1}{2}(e - e^{-1}) \left[\tfrac{1}{2} v^2 \right]_{1}^{2} = \tfrac{3}{4}(e - e^{-1}) \approx 1.763.$$

Figure 13.85

(a)

(b)

Figure 13.86

Figure 13.87

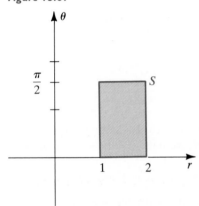

EXAMPLE 5 Evaluate $\iint_R e^{-(x^2+y^2)} \, dx \, dy$ for the first-quadrant region R in the xy-plane between the circular arcs shown in Figure 13.86.

SOLUTION This integral could be evaluated by changing to polar coordinates as we did in Section 13.3; however, our objective is to demonstrate the use of Theorem (13.35). The polar substitution

$$x = r \cos \theta, \quad y = r \sin \theta$$

determines a transformation from the $r\theta$-plane to the xy-plane, whose Jacobian is

$$\frac{\partial(x, y)}{\partial(r, \theta)} = \begin{vmatrix} \cos \theta & -r \sin \theta \\ \sin \theta & r \cos \theta \end{vmatrix} = r.$$

We can verify that the rectangular region S in the $r\theta$-plane bounded by $r = 1$, $r = 2$, $\theta = 0$, and $\theta = \pi/2$ (see Figure 13.87) corresponds to R under the transformation. Moreover, as (r, θ) traces the boundary of S once in the positive direction, the corresponding point (x, y) traces R once in the positive direction. Hence, from Theorem (13.35),

$$\iint_R e^{-(x^2+y^2)} \, dx \, dy = \iint_S e^{-r^2} r \, dr \, d\theta = \int_0^{\pi/2} \int_1^2 e^{-r^2} r \, dr \, d\theta$$

$$= \int_0^{\pi/2} \left[-\tfrac{1}{2} e^{-r^2} \right]_1^2 d\theta = -\tfrac{1}{2}(e^{-4} - e^{-1}) \int_0^{\pi/2} d\theta$$

$$= -\tfrac{1}{4}\pi(e^{-4} - e^{-1}) \approx 0.275.$$

Definition (13.34) can be extended to functions of three variables, as follows.

Definition 13.37

If $x = f(u, v, w)$, $y = g(u, v, w)$, and $z = h(u, v, w)$, then the **Jacobian** of x, y, and z with respect to u, v, and w, denoted by $\partial(x, y, z)/\partial(u, v, w)$, is defined by

$$\frac{\partial(x, y, z)}{\partial(u, v, w)} = \begin{vmatrix} \dfrac{\partial x}{\partial u} & \dfrac{\partial x}{\partial v} & \dfrac{\partial x}{\partial w} \\[2mm] \dfrac{\partial y}{\partial u} & \dfrac{\partial y}{\partial v} & \dfrac{\partial y}{\partial w} \\[2mm] \dfrac{\partial z}{\partial u} & \dfrac{\partial z}{\partial v} & \dfrac{\partial z}{\partial w} \end{vmatrix}$$

The following theorem is a three-dimensional analogue of Corollary (13.36).

Theorem 13.38

If $x = f(u, v, w)$, $y = g(u, v, w)$, $z = h(u, v, w)$ is a transformation of coordinates from a region S in a uvw-coordinate system onto a region R in an xyz-coordinate system and if the Jacobian $\partial(x, y, z)/\partial(u, v, w)$ does not change sign in S, then

$$\iiint\limits_{R} F(x, y, z)\, dx\, dy\, dz = \iiint\limits_{S} G(u, v, w) \left| \frac{\partial(x, y, z)}{\partial(u, v, w)} \right| du\, dv\, dw,$$

where $G(u, v, w)$ is the expression obtained by substituting for x, y, and z in $F(x, y, z)$.

To illustrate Definition (13.37), if we use the spherical coordinate formulas (13.31), then

$$x = \rho \sin \phi \cos \theta, \quad y = \rho \sin \phi \sin \theta, \quad z = \rho \cos \phi.$$

In Exercise 33, you are asked to show that

$$\frac{\partial(x, y, z)}{\partial(\rho, \phi, \theta)} = \rho^2 \sin \phi.$$

Since $\rho^2 \sin \phi \geq 0$, the conclusion of Theorem (13.38) takes on the form

$$\iiint\limits_{R} F(x, y, z)\, dx\, dy\, dz = \iiint\limits_{S} G(\rho, \phi, \theta) \rho^2 \sin \phi \, d\rho\, d\phi\, d\theta.$$

This is in agreement with formula (13.32), which was obtained in an intuitive manner.

We conclude this section with an applied problem that involves a change of variables in a double integral.

Figure 13.88

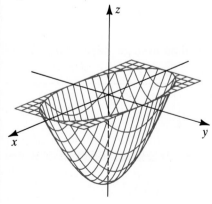

EXAMPLE ▪ 6 *Elliptic sinusoids* are good approximations to the shapes of many lake beds. Suppose that the boundary of a lake has the shape of the ellipse $(x^2/a^2) + (y^2/b^2) = 1$, where a and b are positive. If the maximum depth of the water is h_M, then an elliptic sinusoid is given by

$$f(x, y) = -h_M \cos \left(\frac{\pi}{2} \sqrt{\frac{x^2}{a^2} + \frac{y^2}{b^2}} \right),$$

where $(x^2/a^2) + (y^2/b^2) \leq 1$ (see Figure 13.88). Find the volume V and the average depth h_{av} of the water in the lake.

SOLUTION If R is the elliptical region that corresponds to the surface of the lake (see Figure 13.89a on the following page), then since the water depth at (x, y) is $\left| f(x, y) \right|$,

$$V = \iint\limits_{R} |f(x, y)|\, dA = h_M \iint\limits_{R} \cos \left(\frac{\pi}{2} \sqrt{\frac{x^2}{a^2} + \frac{y^2}{b^2}} \right) dA.$$

Figure 13.89

(a)

(b)

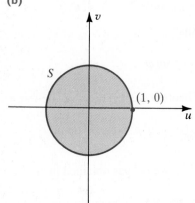

The form of the integrand suggests the substitution

$$x = au, \quad y = bv.$$

In this case, the Jacobian is $\partial(x, y)/\partial(u, v) = ab > 0$, and the region S that corresponds to R is the circular disk $u^2 + v^2 \le 1$ (see Figure 13.89b). Applying Corollary (13.36) gives us

$$V = abh_M \iint\limits_S \cos\left(\frac{\pi}{2}\sqrt{u^2 + v^2}\right) du\, dv.$$

We may evaluate the last integral by means of the polar coordinate substitution $u = r\cos\theta, \; v = r\sin\theta$. This gives us

$$V = abh_M \int_0^{2\pi} \int_0^1 \cos\left(\frac{\pi}{2}r\right) r\, dr\, d\theta.$$

Using integration by parts, or Formula 83 of the table of integrals, we have

$$V = abh_M \int_0^{2\pi} \left[\frac{4}{\pi^2}\cos\left(\frac{\pi}{2}r\right) + \frac{2}{\pi}r\sin\left(\frac{\pi}{2}r\right)\right]_0^1 d\theta$$

$$= abh_M \int_0^{2\pi} \left(\frac{2}{\pi} - \frac{4}{\pi^2}\right) d\theta$$

$$= abh_M \left(\frac{2}{\pi} - \frac{4}{\pi^2}\right) 2\pi \approx 1.4535 abh_M.$$

This formula can be used to estimate lake volume from the three measurements a, b, and h_M.

The area of the elliptical region R is πab. Applying an analogue of Definition (4.29) to double integrals, we obtain the average depth:

$$h_{av} = \frac{1}{\pi ab} \iint\limits_R |f(x, y)|\, dA.$$

Hence, $\quad h_{av} \approx \dfrac{1}{\pi ab}(1.4535)abh_M \approx 0.4627 h_M.$

A study of 107 lakes worldwide yielded an average value of h_{av}/h_M of 0.467.

EXERCISES 13.9

Exer. 1–8: Let T be the transformation from the xy-plane to the uv-plane determined by the given formulas. **(a)** Describe the u-curves and the v-curves. **(b)** Find formulas $x = F(u, v)$, $y = G(u, v)$ that specify T^{-1}.

1   $u = 3x, \qquad\qquad v = 5y$

2   $u = \frac{1}{2}y, \qquad\qquad v = \frac{1}{3}x$

3   $u = x - y, \qquad\quad v = 2x + 3y$

4   $u = -5x + 4y, \quad v = 2x - 3y$

5   $u = x^3, \qquad\qquad v = x + y$

6   $u = x + 1, \qquad\quad v = 2 - y^3$

7   $u = e^x, \qquad\qquad v = e^y$

8   $u = e^{2y}, \qquad\qquad v = e^{-3x}$

Exer. 9–12: Let T be the indicated transformation of coordinates (see Exercises 1–4). Find the curve in the uv-plane that corresponds to the curve in the xy-plane specified in parts (a) and (b).

9 $u = 3x, \quad v = 5y$

(a) the rectangle with vertices $(0, 0), (0, 1), (2, 1), (2, 0)$

(b) the circle $x^2 + y^2 = 1$

10 $u = \frac{1}{2}y, \quad v = \frac{1}{3}x$

(a) the triangle with vertices $(0, 0), (3, 6), (9, 4)$

(b) the line $3x - 2y = 4$

11 $u = x - y, \quad v = 2x + 3y$

(a) the triangle with vertices $(0, 0), (0, 1), (2, 0)$

(b) the line $x + 2y = 1$

12 $u = -5x + 4y, \quad v = 2x - 3y$

(a) the square with vertices $(0, 0), (1, -1), (2, 0), (1, 1)$

(b) the circle $x^2 + y^2 = 1$

Exer. 13–16: Find the Jacobian $\dfrac{\partial(x, y)}{\partial(u, v)}$.

13 $x = u^2 - v^2, \quad y = 2uv$

14 $x = e^u \sin v, \quad y = e^u \cos v$

15 $x = ve^{-2u}, \quad y = u^2 e^{-v}$

16 $x = \dfrac{u}{u^2 + v^2}, \quad y = \dfrac{v}{u^2 + v^2}$

Exer. 17–18: Find the Jacobian $\dfrac{\partial(x, y, z)}{\partial(u, v, w)}$.

17 $x = 2u + 3v - w, \quad y = v - 5w, \quad z = u + 4w$

18 $x = u^2 + vw, \quad y = 2v + u^2 w, \quad z = uvw$

Exer. 19–22: Express the integral as an iterated double integral over a region S in the uv-plane by making the indicated change of variables.

19 $\displaystyle\iint\limits_R (y - x)\, dx\, dy$

boundary of R: $y = 2x, \ y = 0, \ x = 2$
change of variables: $x = u + v, \ y = 2v$

20 $\displaystyle\iint\limits_R (3x - 4y)\, dx\, dy$

boundary of R: $y = 3x, \ y = \frac{1}{2}x, \ x = 4$
change of variables: $x = u - 2v, \ y = 3u - v$

21 $\displaystyle\iint\limits_R (\frac{1}{4}x^2 + \frac{1}{9}y^2)\, dx\, dy$

boundary of R: $\frac{1}{4}x^2 + \frac{1}{9}y^2 = 1$
change of variables: $x = 2u, \ y = 3v$

22 $\displaystyle\iint\limits_R xy\, dx\, dy$

boundary of R: $y = x, \ y = 2 - x, \ y = 0$
change of variables: $x = u + v, \ y = u - v$

Exer. 23–28: Evaluate the integral by making the indicated change of variables.

23 $\displaystyle\iint\limits_R (x - y)^2 \cos^2(x + y)\, dx\, dy$

boundary of R: the square with vertices $(0, 1), (1, 2), (2, 1), (1, 0)$
change of variables: $u = x - y, \ v = x + y$

24 $\displaystyle\iint\limits_R \sin\frac{y - x}{y + x}\, dx\, dy$

boundary of R: the trapezoid with vertices $(1, 1), (2, 2), (4, 0), (2, 0)$
change of variables: $u = y - x, \ v = y + x$

25 $\displaystyle\iint\limits_R (x^2 + 2y^2)\, dx\, dy$

boundary of R: $xy = 1, \ xy = 2, \ y = |x|, \ y = 2x$
change of variables: $x = u/v, \ y = v$

26 $\displaystyle\iint\limits_R \frac{1}{(4x + 4y + 1)^2}\, dx\, dy$

boundary of R: $\sqrt[3]{x} + \sqrt[3]{y} = 1, \ x = 0, \ y = 0$
change of variables: $x = u^3, \ y = v^3$

27 $\displaystyle\iint\limits_R \frac{2y + x}{y - 2x}\, dx\, dy$

boundary of R: the trapezoid with vertices $(-1, 0), (-2, 0), (0, 4), (0, 2)$
change of variables: $u = y - 2x, \quad v = 2y + x$

28 $\displaystyle\iint\limits_R (\sqrt{x - 2y} + \frac{1}{4}y^2)\, dx\, dy$

boundary of R: the triangle with vertices $(0, 0), (4, 0), (4, 2)$
change of variables: $u = \frac{1}{2}y, \quad v = x - 2y$

29 Because of rotation, the earth is not perfectly spherical but is slightly flattened at the poles, with a polar radius of 6356 km and equatorial radius of 6378 km. As a result, the shape of the earth's surface can be approximated by the ellipsoid

$$\frac{x^2}{a^2} + \frac{y^2}{b^2} + \frac{z^2}{c^2} = 1,$$

with $a = b = 6378$ and $c = 6356$. Estimate the volume of the earth. (*Hint:* Let $x = au, \ y = bv$, and $z = cw$.)

30 A balloon has the shape of the sphere $x^2 + y^2 + z^2 = 9$, where x, y, and z are in meters. The balloon is slowly heated at a constant pressure of one atmosphere—that is, 1.01×10^5 N/m^2—and expands until it has the shape of the ellipsoid $(x^2/10) + (y^2/16) + (z^2/12) = 1$. For an ideal gas, the work W done by the gas is given by

$$W = \iiint\limits_{V_1}^{V_2} P \, dV,$$

where P is the pressure and V_1 and V_2 are the initial and final volumes. Use Exercise 29 to find W.

31 Refer to Exercise 41 of Section 13.8. It is very difficult to find formulas that describe the final position of the hand if five or six joints of a robotics arm are moved. The process can be simplified by using the three *Euler angles* ϕ, θ, and ψ for rotations of coordinate axes. Referring to the figure, first rotate the xy-plane about the z-axis through the angle ϕ, obtaining the new x''- and y''-axes. Second, rotate the $x''z$-plane about the y''-axis through the angle θ, obtaining the new x'''- and z'-axes. Third, rotate the $x'''y''$-plane about the z'-axis through the angle ψ, obtaining the new x'- and y'-axes. If the coordinates of point P in the hand are (x, y, z) in the xyz-system, then its coordinates (x', y', z') in the transformed system are given by

$$x' = (\cos \phi \cos \theta \cos \psi - \sin \phi \sin \psi)x$$
$$+ (-\cos \phi \cos \theta \sin \psi - \sin \phi \cos \psi)y$$
$$+ (\cos \phi \sin \theta)z$$
$$y' = (\sin \phi \cos \theta \cos \psi + \cos \phi \sin \psi)x$$
$$+ (-\sin \phi \cos \theta \sin \psi + \cos \phi \cos \psi)y$$
$$+ (\sin \phi \sin \theta)z$$
$$z' = (-\sin \theta \cos \psi)x + (\sin \theta \sin \psi)y + (\cos \theta)z.$$

(a) A robot's hand is positioned at $P(\sqrt{2}, 0, 1)$ and then rotated through $\phi = 45°$. Find the new location of the hand.

Exercise 31

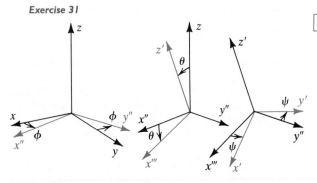

(b) The hand is next rotated through $\theta = 90°$. Find the new location of the hand.

(c) Finally, the hand is rotated through $\psi = 90°$. Find the final location of the hand.

32 If a robot arm has several joints, it may be possible to position its hand in a specific location in more than one way by rotating the joints differently. There will be exactly one choice for the rotation if and only if the Jacobian of the transformation of coordinates is nonzero.

(a) Find the Jacobian $\partial(x', y', z')/\partial(x, y, z)$ in Exercise 31 when $\phi = 45°$, $\theta = 90°$, and $\psi = 90°$.

(b) Can the hand be positioned in the same location by using different values of ϕ, θ, and ψ?

33 Verify that, for the spherical coordinate transformation,

$$\frac{\partial(x, y, z)}{\partial(\rho, \phi, \theta)} = \rho^2 \sin \phi.$$

34 Use Theorem (13.38) to derive a formula for

$$\iiint\limits_{R} F(x, y, z) \, dx \, dy \, dz$$

for a transformation from rectangular to cylindrical coordinates.

35 If the transformation of coordinates $x = f(u, v)$, $y = g(u, v)$ is one-to-one, show that

$$\frac{\partial(x, y)}{\partial(u, v)} \frac{\partial(u, v)}{\partial(x, y)} = 1.$$

Hint: Use the following property of determinants:

$$\begin{vmatrix} a & b \\ c & d \end{vmatrix} \begin{vmatrix} p & q \\ r & s \end{vmatrix} = \begin{vmatrix} ap + br & aq + bs \\ cp + dr & cq + ds \end{vmatrix}$$

36 Given the transformation $x = f(u, v)$, $y = g(u, v)$ and the transformation $u = h(r, s)$, $v = k(r, s)$, show that

$$\frac{\partial(x, y)}{\partial(u, v)} \frac{\partial(u, v)}{\partial(r, s)} = \frac{\partial(x, y)}{\partial(r, s)}.$$

(Hint: Use the hint given in Exercise 35.)

⚬ **Exer. 37–38: Let T be the indicated transformation from the uv-plane to the xy-plane. Graph the points (u, v) such that $\partial(x, y)/\partial(u, v) = 0$ for the indicated values of u. (This graph shows where T is not one-to-one.)**

37 $x = \frac{1}{4}v^4 + u^2,$

$$y = u - \cos v; \quad -\pi \le u \le \pi$$

38 $x = \frac{1}{20}u^4 - 10\cos u + \frac{1}{2}v^2,$

$$y = u + v; \quad -5 \le u \le 5$$

CHAPTER 13 REVIEW EXERCISES

Exer. 1–6: Evaluate the iterated integral.

1 $\displaystyle\int_{-1}^{0}\int_{x+1}^{x^3}(x^2-2y)\,dy\,dx$

2 $\displaystyle\int_{1}^{2}\int_{1}^{y^2}\frac{1}{y}\,dx\,dy$

3 $\displaystyle\int_{0}^{3}\int_{r}^{r^2+1}r\,d\theta\,dr$

4 $\displaystyle\int_{2}^{0}\int_{0}^{z^2}\int_{x}^{z}(x+z)\,dy\,dx\,dz$

5 $\displaystyle\int_{0}^{2}\int_{\sqrt{y}}^{1}\int_{z^2}^{y}xy^2z^3\,dx\,dz\,dy$

6 $\displaystyle\int_{0}^{\pi/2}\int_{0}^{\pi/4}\int_{0}^{a\cos\phi}\rho^2\sin\phi\,d\rho\,d\phi\,d\theta$

Exer. 7–10: If $f(x,y)$ is an arbitrary continuous function, express $\iint_R f(x,y)\,dA$ as an iterated double integral, where R is the region bounded by the graphs of the equations.

7 $x^2-y^2=4,\quad x=4$

8 $x^2-y^2=4,\quad y=4,\qquad y=0$

9 $y^2=4+x,\qquad y^2=4-x$

10 $y=-x^2+4,\quad y=3x^2$

Exer. 11–14: Sketch the region of integration for the iterated integral.

11 $\displaystyle\int_{-1}^{1}\int_{e^y}^{y^3}dx\,dy$

12 $\displaystyle\int_{-1}^{0}\int_{x}^{-x^2}dy\,dx$

13 $\displaystyle\int_{0}^{\pi}\int_{0}^{2+4\cos\theta}r\,dr\,d\theta$

14 $\displaystyle\int_{0}^{2}\int_{-\sqrt{y^2+1}}^{\sqrt{y^2+1}}\int_{0}^{\sqrt{1-x^2+y^2}}dz\,dx\,dy$

Exer. 15–16: Reverse the order of integration and evaluate the resulting integral.

15 $\displaystyle\int_{0}^{3}\int_{y^2}^{9}ye^{-x^2}\,dx\,dy$

16 $\displaystyle\int_{0}^{1}\int_{x}^{\sqrt{x}}e^{x/y}\,dy\,dx$

Exer. 17–18: Find the volume of the solid that is under the graph of the equation and over the region R in the xy-plane.

17 $z=xy^2$; R is the rectangle with vertices $(1,1),(2,1),(1,3),(2,3)$.

18 $z=7-x^2-y^2$; R is the circle $x^2+y^2=4$.

19 Find the surface area of the part of the cone $z=\sqrt{x^2+y^2}$ that is inside the cylinder $x^2+y^2=4x$.

c 20 Find the surface area of the part of the hyperboloid $x^2-y^2+z^2=1$ that lies over the region $x^2+y^2\le 1$.

21 Use a double integral to find the area of the region bounded by the polar axis and the graphs of $r=e^\theta$ and $r=2$ from $\theta=0$ to $\theta=\ln 2$.

22 Use polar coordinates to evaluate
$$\int_{-a}^{0}\int_{-\sqrt{a^2-x^2}}^{0}\sqrt{x^2+y^2}\,dy\,dx.$$

23 Use spherical coordinates to evaluate
$$\int_{0}^{4}\int_{0}^{\sqrt{16-x^2}}\int_{\sqrt{x^2+y^2}}^{\sqrt{32-x^2-y^2}}\sqrt{x^2+y^2+z^2}\,dz\,dy\,dx.$$

24 Let f be an arbitrary continuous function of x, y, and z, and let Q be the region bounded by the paraboloid $y=x^2+4z^2$ and the plane $y=4$. Express the triple integral $\iiint_Q f(x,y,z)\,dV$ as an iterated triple integral in six different ways.

Exer. 25–26: Find the mass and the center of mass of the lamina that has the shape of the region bounded by the graphs of the equations and has the indicated density.

25 $y=x,y=2x,x=3$; area mass density at $P(x,y)$ is directly proportional to the distance from the y-axis to P.

26 $y^2=x,x=4$; area mass density at $P(x,y)$ is directly proportional to the distance from the line $x=-1$ to P.

27 A lamina has the shape of the region that is inside the limaçon $r=2+\sin\theta$ and outside the circle $r=1$. The area mass density at $P(r,\theta)$ is inversely proportional to the distance from the pole to P. Find the mass.

28 A lamina has the shape of the region bounded by the graphs of $y=x^2$ and $y=x^3$, and the area mass density

at $P(x, y)$ is directly proportional to the distance from the y-axis to P. Find I_x, I_y, and I_O.

29 A lamina has the shape of a right triangle with sides of lengths a, b, and $\sqrt{a^2 + b^2}$. The area mass density is directly proportional to the distance from the side of length a. Find the moment of inertia with respect to a line containing the side of length a.

30 A lamina has the shape of the region between concentric circles of radii a and b, with $a < b$. The area mass density at a point P is directly proportional to the distance from the center of the circles to P. Use polar coordinates to find the moment of inertia with respect to a line through the center.

31 Use triple integrals to find the volume and centroid of the solid bounded by the graphs of the cylinder $z = x^2$ and the planes $y = 0$ and $y + z = 4$.

32 A solid is bounded by the paraboloid $z = 9x^2 + y^2$ and the plane $z = 9$. The density at $P(x, y, z)$ is inversely proportional to the square of the distance from $(0, 0, -1)$ to P. Set up an iterated triple integral that can be used to find the moment of inertia with respect to the z-axis.

33 A solid is bounded by the hyperboloid of one sheet $x^2 - y^2 + z^2 = 1$ and the planes $y = 0$ and $y = 4$. The density at $P(x, y, z)$ is directly proportional to the distance from the y-axis to P. Set up an iterated triple integral that can be used to find the moment of inertia with respect to the y-axis.

34 A homogeneous solid of density δ is bounded by the paraboloid $z = 9 - x^2 - y^2$, the interior of the cylinder $x^2 + y^2 = 4$, and the xy-plane. Use cylindrical coordinates to find

(a) the mass   (b) the center of mass

(c) the moment of inertia with respect to the z-axis

35 A spherical solid has radius a, and the density at any point is directly proportional to its distance from the center of the sphere. Use spherical coordinates to find the mass.

Exer. 36–37: Complete the table.

Rectangular coordinates	Cylindrical coordinates	Spherical coordinates
36 $(2, -2, 1)$	_____	_____
37 _____	_____	$(12, \pi/6, 3\pi/4)$

Exer. 38–43: Change the equation to rectangular coordinates, and describe its graph in three dimensions.

38 $r = -\csc \theta$

39 $z + 3r^2 = 9$

40 $z = \frac{1}{4} r$

41 $\rho \sin \phi = 4$

42 $\rho \sin \phi \cos \theta = 1$

43 $\rho^2 - 3\rho = 0$

Exer. 44–47: Change the equation to (a) cylindrical coordinates and (b) spherical coordinates.

44 $x^2 + y^2 = 1$

45 $z = x^2 - y^2$

46 $x^2 + y^2 + z^2 - 2z = 0$

47 $2x + y - 3z = 4$

Exer. 48–49: Express the area of the region in terms of one or more iterated double integrals that involve only the order of integration (a) $dy\,dx$, (b) $dx\,dy$, and (c) $dr\,d\theta$.

48

49

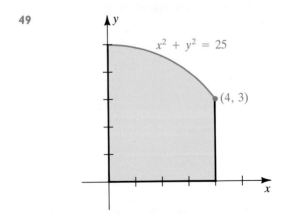

Exer. 50–51: Express the volume of the region in terms of one or more iterated triple integrals in **(a)** rectangular coordinates, **(b)** cylindrical coordinates, and **(c)** spherical coordinates.

50

51

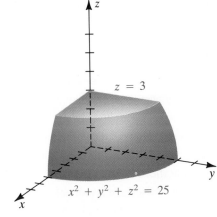

52 If a transformation T of the xy-plane to the uv-plane is given by $u = 2x + 5y$, $v = 3x - 4y$, find

  (a) the u-curves   **(b)** the v-curves

  (c) T^{-1} **(d)** $\dfrac{\partial(x,\ y)}{\partial(u,\ v)}$

  (e) the curve in the uv-plane corresponding to the line $ax + by + c = 0$

  (f) the curve in the uv-plane corresponding to the circle $x^2 + y^2 = a^2$

53 Evaluate the integral

$$\int_0^1 \int_y^{2-y} e^{(x-y)/(x+y)}\, dx\, dy$$

by means of the change of variables given by $u = x - y$, $v = x + y$.

     EXTENDED PROBLEMS AND GROUP PROJECTS

---

1 Develop careful and complete proofs of two theorems.

  (a) If f is a real-valued function of one variable that is continuous on an interval I, then f is integrable on I.

  (b) If f is a real-valued function of two variables that is continuous on a region R of the plane, then f is integrable over R.

  Discuss the analogous result for a real-valued function of three variables.

2 An **outer partition** P of a region R is a collection of closed rectangular subregions that collectively completely contain R. Develop a definition of Riemann sum and double integral over R that is based on outer partitions. Will both approaches yield the same value for the double integral?

3 Discuss the assertion that "differentiation and antidifferentiation (or indefinite integration) for functions of one variable are inverse operations of each other." In what sense does the fundamental theorem of calculus reinforce this view? Investigate possible extensions of this assertion for real-valued functions of two or three variables, using the results of Chapters 12 and 13. If the gradient of f plays a role similar to the derivative for functions of one variable, what would antidifferentiation mean for functions of two or three variables? Is there a reasonable notion of an *indefinite* double or triple integral? Is there a version of the fundamental theorem of calculus for real-valued functions of several variables?

INTRODUCTION

ONE OF NATURE'S MOST POWERFUL and violent creations, a tornado causes widespread death and destruction, uprooting large trees, overturning railroad cars, and carrying such heavy objects as automobiles hundreds of feet. A tornado is a twisting wind storm that may last several hours, travel up to 200 miles, and have winds that whirl around the center at 300 miles per hour. A greater understanding of why tornadoes develop can provide the information for more accurate predictions of their occurrence and location. This in turn might lead to new ways to control or prevent tornadoes and save more lives. Since tornadoes are three-dimensional objects with velocity and force vectors at each point, a mathematical analysis of their behavior requires the use of vector-valued functions of several variables.

To understand such functions, we extend previous methods to new concepts in this chapter and obtain results that have many uses throughout the sciences. We first discuss *vector fields* in Section 14.1. Principal applications that we examine involve *velocity fields* and *force fields* of either solids, liquids, or gases. Of primary importance are *conservative* vector fields; these include the gravitational and electromagnetic fields that occur throughout the universe.

Line integrals, which are generalizations of the definite integral, enable us to find the work done as a particle moves through a force field. We examine the definition and applications of line integrals in Section 14.2. In a conservative vector field, the work done is independent of the path taken; in Section 14.3, we see how this result leads to an important law of conservation of energy. Green's theorem, the subject of Section 14.4, relates the line integral along a simple closed curve bounding a region to a double integral over that region.

We consider *surface integrals* in Section 14.5. They are extensions of double integrals that we can apply to problems about *flux:* the rate of flow of energy, fluids, and gases across a surface. We use line and surface integrals to interpret the *divergence* and the *curl* of a vector field. Divergence measures the rate at which a physical entity, such as heat, is absorbed or generated near a point, and curl is used to describe rotational aspects of a vector field. The final sections contain discussions of the *divergence theorem* (Section 14.6) and *Stokes's theorem* (Section 14.7), which relate divergence to flux and work to curl, respectively. These two outstanding results have far-reaching consequences in science and engineering, such as studying the direction and the magnitude of wind circulation inside a tornado.

Vector calculus provides new concepts and techniques that help us model and understand complex natural phenomena such as tornadoes.

Vector Calculus

14.1 VECTOR FIELDS

Figure 14.1

Figure 14.2

In our study of multivariable calculus, we have examined vector-valued functions of a single variable and real-valued functions of several variables. In the former case, elements of the domain are real numbers and elements of the range are vectors; in the latter situation, domain elements are vectors and range elements are real numbers. We now turn to a generalized setting in which domain and range elements are both vectors—that is, we will be considering vector-valued functions of several variables.

If to each point K in a region there is assigned exactly one vector having initial point K, then the collection of all such vectors is a **vector field**. Figure 14.1 illustrates a vector field determined by a wheel rotating about an axle. To each point on the wheel there corresponds a velocity vector. A vector field of this type is a **velocity field**. A velocity field may also be determined by the flow of water or of wind. Figure 14.2 displays velocity vectors associated with fluid particles moving in a stream. Sketches of this type show only a few vectors of the vector field. Remember that a vector is assigned to *every* point in the region.

A **force field** is a vector field in which a force vector is assigned to each point. Force fields are common in the study of mechanics and electricity. In our illustrations, we have assumed that the vectors are independent of time. Such **steady vector fields** are the only types that we shall consider in this chapter.

Given a vector field, let us introduce an xyz-coordinate system and denote the vector assigned to the point $K(x, y, z)$ by $\mathbf{F}(x, y, z)$ (see Figure 14.3). Since the components of $\mathbf{F}(x, y, z)$ depend on the coordinates x, y, and z of K, we may write

$$\mathbf{F}(x, y, z) = M(x, y, z)\mathbf{i} + N(x, y, z)\mathbf{j} + P(x, y, z)\mathbf{k},$$

where M, N, and P are scalar functions. Conversely, every such equation determines a vector field. Thus, a vector field may be regarded as a function $\mathbf{F}$ of the type stated in the following definition.

Figure 14.3

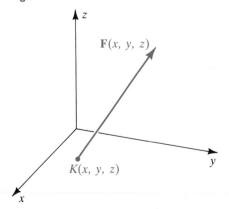

Definition 14.1

A **vector field in three dimensions** is a function **F** whose domain D is a subset of $\mathbb{R}^3$ and whose range is a subset of V_3. If (x, y, z) is in D, then

$$\mathbf{F}(x, y, z) = M(x, y, z)\mathbf{i} + N(x, y, z)\mathbf{j} + P(x, y, z)\mathbf{k},$$

where M, N, and P are scalar functions.

Figure 14.4

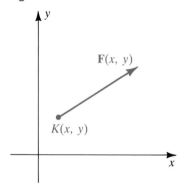

As a special case of (14.1), if the domain D corresponds to a region in the xy-plane and if the range is a subset of V_2 (see Figure 14.4), then a **vector field F in two dimensions** is given by

$$\mathbf{F}(x, y) = M(x, y)\mathbf{i} + N(x, y)\mathbf{j},$$

where M and N are scalar functions. We sometimes describe a vector field by sketching vectors that correspond to $\mathbf{F}(x, y, z)$ or to $\mathbf{F}(x, y)$ for typical points (x, y, z) or (x, y), as illustrated in the following example.

Figure 14.5

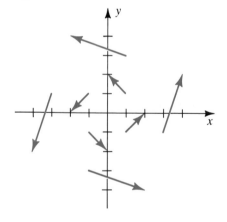

EXAMPLE ■ 1 Describe the vector field **F** if $\mathbf{F}(x, y) = -y\mathbf{i} + x\mathbf{j}$.

SOLUTION The vectors $\mathbf{F}(x, y)$ associated with several points (x, y) are listed in the following table and are sketched in Figure 14.5.

(x, y)	$\mathbf{F}(x, y)$		(x, y)	$\mathbf{F}(x, y)$
$(1, 1)$	$-\mathbf{i} + \mathbf{j}$		$(1, 3)$	$-3\mathbf{i} + \mathbf{j}$
$(-1, 1)$	$-\mathbf{i} - \mathbf{j}$		$(-3, 1)$	$-\mathbf{i} - 3\mathbf{j}$
$(-1, -1)$	$\mathbf{i} - \mathbf{j}$		$(-1, -3)$	$3\mathbf{i} - \mathbf{j}$
$(1, -1)$	$\mathbf{i} + \mathbf{j}$		$(3, -1)$	$\mathbf{i} + 3\mathbf{j}$

The vectors are similar to those associated with the rotating wheel in Figure 14.1.

To arrive at a general description of this vector field **F**, consider an arbitrary point $K(x, y)$ and let $\mathbf{r} = x\mathbf{i} + y\mathbf{j}$ be the position vector for $K(x, y)$ (see Figure 14.6). It appears that $\mathbf{F}(x, y)$ is orthogonal to $\mathbf{r}$ and is therefore tangent to the circle of radius $\|\mathbf{r}\|$ with center at the origin. We may prove this fact by showing that $\mathbf{r} \cdot \mathbf{F}(x, y) = 0$, as follows:

$$\mathbf{r} \cdot \mathbf{F}(x, y) = (x\mathbf{i} + y\mathbf{j}) \cdot (-y\mathbf{i} + x\mathbf{j})$$
$$= -xy + yx = 0$$

Figure 14.6

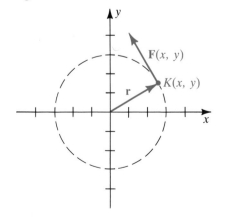

We also note that

$$\|\mathbf{F}(x, y)\| = \sqrt{y^2 + x^2} = \|\mathbf{r}\|.$$

Hence, the magnitude of $\mathbf{F}(x, y)$ equals the radius of the circle. This implies that as the point $K(x, y)$ moves away from the origin, the magnitude of $\mathbf{F}(x, y)$ increases, as is the case for the rotating wheel in Figure 14.1.

The next definition introduces one of the most important vector fields that occurs in the physical sciences.

Definition 14.2

Let $\mathbf{r} = x\mathbf{i} + y\mathbf{j} + z\mathbf{k}$ be the position vector for (x, y, z), and let the vector $\mathbf{u} = (1/\|\mathbf{r}\|)\mathbf{r}$ denote the unit vector that has the same direction as $\mathbf{r}$. A vector field $\mathbf{F}$ is an **inverse square field** if

$$\mathbf{F}(x, y, z) = \frac{c}{\|\mathbf{r}\|^2}\mathbf{u} = \frac{c}{\|\mathbf{r}\|^3}\mathbf{r},$$

where c is a scalar.

Figure 14.7
Inverse square vector fields
(a) $c < 0$

(b) $c > 0$

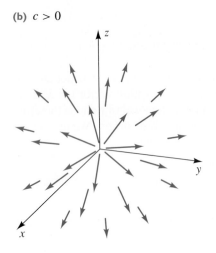

EXAMPLE 2   Describe the inverse square field $\mathbf{F}(x, y, z)$ of Definition (14.2).

SOLUTION   Since $\mathbf{r} = x\mathbf{i} + y\mathbf{j} + z\mathbf{k}$,

$$\mathbf{F}(x, y, z) = \frac{c}{\|\mathbf{r}\|^3}\mathbf{r} = \frac{c}{(x^2 + y^2 + z^2)^{3/2}}(x\mathbf{i} + y\mathbf{j} + z\mathbf{k}).$$

We may write the last expression as in Definition (14.1); however, it is simpler to analyze the vectors in the field by using $\mathbf{r}$. If $c < 0$, then $\mathbf{F}(x, y, z)$ is a negative scalar multiple of $\mathbf{r}$, and hence the direction of $\mathbf{F}(x, y, z)$ is toward the origin O. Moreover, since

$$\|\mathbf{F}(x, y, z)\| = \frac{|c|}{\|\mathbf{r}\|^3}\|\mathbf{r}\| = \frac{|c|}{\|\mathbf{r}\|^2},$$

the magnitude of $\mathbf{F}(x, y, z)$ is inversely proportional to the square of the distance from O to the point (x, y, z). This means that as the point $K(x, y, z)$ moves away from the origin, the length of the vector $\mathbf{F}(x, y, z)$ decreases. Some typical vectors in $\mathbf{F}$ are sketched in Figure 14.7(a) for $c < 0$. If $c > 0$, the vectors point *away* from the origin, as in Figure 14.7(b).

The force of gravity determines inverse square vector fields. According to **Newton's law of universal gravitation**, if a particle of mass m_0 is located at the origin of an xyz-coordinate system, then the force exerted on a particle of mass m located at $K(x, y, z)$ is

$$\mathbf{F}(x, y, z) = -G\frac{m_0 m}{\|\mathbf{r}\|^2}\mathbf{u},$$

where G is a gravitational constant, $\mathbf{r}$ is the position vector for the point K, and $\mathbf{u} = (1/\|\mathbf{r}\|)\mathbf{r}$ is a unit vector.

Inverse square fields also occur in electrical theory. **Coulomb's law** states that if a point charge of Q coulombs is at the origin, then the force $\mathbf{F}(x, y, z)$ that it exerts on a point charge of q coulombs located at $K(x, y, z)$ is

$$\mathbf{F}(x, y, z) = c\frac{Qq}{\|\mathbf{r}\|^2}\mathbf{u},$$

where c is a constant and $\mathbf{r}$ and $\mathbf{u}$ are as in Definition (14.2). Note that Coulomb's law has the same form as Newton's law of universal gravitation.

If f is a function of three variables, then, as in (12.31), the *gradient* of $f(x, y, z)$ is the vector field defined as follows:

$$\nabla f(x, y, z) = f_x(x, y, z)\mathbf{i} + f_y(x, y, z)\mathbf{j} + f_z(x, y, z)\mathbf{k}$$

Vector fields that are gradients of scalar functions are given the following special name.

Definition 14.3

A vector field $\mathbf{F}$ is **conservative** if

$$\mathbf{F}(x, y, z) = \nabla f(x, y, z)$$

for some scalar function f.

Figure 14.8

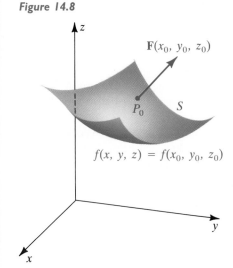

$\mathbf{F}(x_0, y_0, z_0)$

P_0 S

$f(x, y, z) = f(x_0, y_0, z_0)$

If $\mathbf{F}$ is conservative, then the function f in (14.3) is called a **potential function** for $\mathbf{F}$ and $f(x, y, z)$ is the **potential** at the point (x, y, z). Recall from Theorem (12.36) that if (x_0, y_0, z_0) is in the domain of $\mathbf{F} = \nabla f$, then the gradient vector $\nabla f(x_0, y_0, z_0) = \mathbf{F}(x_0, y_0, z_0)$ is normal to the level surface S of f that contains $P_0(x_0, y_0, z_0)$. The surface S is the graph of the equation $f(x, y, z) = f(x_0, y_0, z_0)$. Thus, *every vector $\mathbf{F}(x_0, y_0, z_0)$ in a conservative vector field is normal to the level surface of a potential function f for $\mathbf{F}$ that contains $P_0(x_0, y_0, z_0)$.* A typical case is illustrated in Figure 14.8. In Section 12.6, we proved that $\|\mathbf{F}(x_0, y_0, z_0)\|$ is the maximum rate of change of the potential at P_0. In applications where $f(x, y, z)$ is the temperature (or potential) at (x, y, z), this maximum rate of change of f is normal to the isothermal (or equipotential) surface containing P_0.

The next theorem is important in applications.

Theorem 14.4

Every inverse square vector field is conservative.

PROOF If $\mathbf{F}$ is an inverse square field, then, as in the solution to Example 2,

$$\mathbf{F}(x, y, z) =$$

$$\frac{cx}{(x^2 + y^2 + z^2)^{3/2}}\mathbf{i} + \frac{cy}{(x^2 + y^2 + z^2)^{3/2}}\mathbf{j} + \frac{cz}{(x^2 + y^2 + z^2)^{3/2}}\mathbf{k},$$

where c is a constant. By Definition (14.3), if $\mathbf{F}$ is conservative, then $\mathbf{F}(x, y, z) = \nabla f(x, y, z)$ for some scalar function f, and the components of $\mathbf{F}$ are $f_x(x, y, z)$, $f_y(x, y, z)$, and $f_z(x, y, z)$, respectively. Partial integrations of these components with respect to x, y, and z, respectively, *suggest* that

$$f(x, y, z) = \frac{-c}{(x^2 + y^2 + z^2)^{1/2}}.$$

Calculation of the partial derivatives of f proves that this guess is correct. This gives us the following:

$$\mathbf{F}(x, y, z) = \nabla f(x, y, z) = \nabla\left(\frac{-c}{r}\right),$$

where $r = \|\mathbf{r}\| = (x^2 + y^2 + z^2)^{1/2}$. ■

The level surfaces for the potential function f of the inverse square vector field $\mathbf{F}$ in the proof of Theorem (14.4) are graphs of equations of the form

$$\frac{-c}{(x^2 + y^2 + z^2)^{1/2}} = k,$$

where $k < 0$. Squaring both sides and taking reciprocals gives us

$$x^2 + y^2 + z^2 = \frac{c^2}{k^2}.$$

Thus, the level surfaces are spheres with centers at the origin O. The vectors $\mathbf{F}(x, y, z)$ are orthogonal to these spheres and hence are directed toward O or away from O, as in Figure 14.7.

In physics, the potential function of a conservative vector field $\mathbf{F}$ is defined as the function p such that $\mathbf{F}(x, y, z) = -\nabla p(x, y, z)$. In this case, letting $p = -f$ in the proof of Theorem (14.4), we obtain $p(x, y, z) = c/r$. We shall say more about conservative vector fields later in this chapter. In particular, in Section 14.3, we discuss a method for finding $f(x, y, z)$ when given $\mathbf{F}(x, y, z) = \nabla f(x, y, z)$.

The vector differential operator ∇ in three dimensions is

$$\nabla = \mathbf{i}\frac{\partial}{\partial x} + \mathbf{j}\frac{\partial}{\partial y} + \mathbf{k}\frac{\partial}{\partial z}.$$

From (12.31), if ∇ operates on a *scalar* function f, it produces the gradient of f,

$$\text{grad } f = \nabla f = \frac{\partial f}{\partial x}\mathbf{i} + \frac{\partial f}{\partial y}\mathbf{j} + \frac{\partial f}{\partial z}\mathbf{k}.$$

We next use ∇ as an operator on a *vector* field to define another vector field called the *curl* of $\mathbf{F}$, which we denote by curl $\mathbf{F}$ or $\nabla \times \mathbf{F}$.

Definition 14.5

Let $\mathbf{F}(x, y, z) = M(x, y, z)\mathbf{i} + N(x, y, z)\mathbf{j} + P(x, y, z)\mathbf{k}$, where M, N, and P have partial derivatives in some region. The **curl of F**, denoted by curl $\mathbf{F}$ or $\nabla \times \mathbf{F}$, is given by

$$\text{curl } \mathbf{F} = \nabla \times \mathbf{F}$$

$$= \left(\frac{\partial P}{\partial y} - \frac{\partial N}{\partial z}\right)\mathbf{i} + \left(\frac{\partial M}{\partial z} - \frac{\partial P}{\partial x}\right)\mathbf{j} + \left(\frac{\partial N}{\partial x} - \frac{\partial M}{\partial y}\right)\mathbf{k}.$$

We shall also use the symbol curl $\mathbf{F}$ or $\nabla \times \mathbf{F}$ to denote the vector curl $\mathbf{F}(x, y, z)$ or $\nabla \times \mathbf{F}(x, y, z)$ associated with (x, y, z). The formula for curl $\mathbf{F}$ can be written in the following *determinant form*.

Determinant Notation for curl F 14.6

$$\text{curl } \mathbf{F} = \nabla \times \mathbf{F} = \begin{vmatrix} \mathbf{i} & \mathbf{j} & \mathbf{k} \\ \dfrac{\partial}{\partial x} & \dfrac{\partial}{\partial y} & \dfrac{\partial}{\partial z} \\ M & N & P \end{vmatrix}$$

The expression on the right in (14.6) is not actually a determinant, since the first row is made up of vectors, the second row of partial derivative operators, and the third row of scalar functions; however, it is an extremely useful device for remembering the cumbersome formula given in Definition (14.5).

EXAMPLE 3 If $\mathbf{F}(x, y, z) = xy^2z^4\mathbf{i} + (2x^2y + z)\mathbf{j} + y^3z^2\mathbf{k}$, find $\nabla \times \mathbf{F}$.

SOLUTION Applying (14.6), we have

$$\nabla \times \mathbf{F} = \begin{vmatrix} \mathbf{i} & \mathbf{j} & \mathbf{k} \\ \dfrac{\partial}{\partial x} & \dfrac{\partial}{\partial y} & \dfrac{\partial}{\partial z} \\ xy^2z^4 & (2x^2y + z) & y^3z^2 \end{vmatrix}$$

$$= (3y^2z^2 - 1)\mathbf{i} + 4xy^2z^3\mathbf{j} + (4xy - 2xyz^4)\mathbf{k}.$$

Figure 14.9

In Section 14.7, we shall see that if $\mathbf{F}$ is the velocity field of a fluid or gas that is moving through an xyz-coordinate system, then curl $\mathbf{F}$ provides information about rotational aspects of the motion. If we consider a point $K(x, y, z)$ at which the fluid or gas is rotating, or swirling (or *curl*ing), then curl $\mathbf{F}$ lies on the axis of rotation (see Figure 14.9) and can be used to describe rotational properties of the field.

The operator ∇ also can be used to obtain a *scalar* function from a vector field $\mathbf{F}$, as follows.

Definition 14.7

Let $\mathbf{F}(x, y, z) = M(x, y, z)\mathbf{i} + N(x, y, z)\mathbf{j} + P(x, y, z)\mathbf{k}$, where M, N, and P have partial derivatives in some region. The **divergence** of $\mathbf{F}$, denoted by div $\mathbf{F}$ or $\nabla \cdot \mathbf{F}$, is given by

$$\text{div } \mathbf{F} = \nabla \cdot \mathbf{F} = \frac{\partial M}{\partial x} + \frac{\partial N}{\partial y} + \frac{\partial P}{\partial z}.$$

We use the symbol $\nabla \cdot \mathbf{F}$ for divergence because the formula may be obtained by taking what *appears* to be the dot product of ∇ and $\mathbf{F}$, as follows:

$$\nabla \cdot \mathbf{F} = \left(\frac{\partial}{\partial x}\mathbf{i} + \frac{\partial}{\partial y}\mathbf{j} + \frac{\partial}{\partial z}\mathbf{k}\right) \cdot (M\mathbf{i} + N\mathbf{j} + P\mathbf{k})$$

$$= \frac{\partial}{\partial x}(M) + \frac{\partial}{\partial y}(N) + \frac{\partial}{\partial z}(P)$$

As in the next example, we also use the symbol $\nabla \cdot \mathbf{F}$ to denote the function value $(\nabla \cdot \mathbf{F})(x, y, z)$. The value of $\nabla \cdot \mathbf{F}$, or div $\mathbf{F}$, at a particular point K is denoted by $[\nabla \cdot \mathbf{F}]_K$ or $[\text{div } \mathbf{F}]_K$.

EXAMPLE ▪ 4   If $\mathbf{F}(x, y, z) = xy^2z^4\mathbf{i} + (2x^2y + z)\mathbf{j} + y^3z^2\mathbf{k}$, find $\nabla \cdot \mathbf{F}$.

SOLUTION   By Definition (14.7),

$$\nabla \cdot \mathbf{F} = \frac{\partial}{\partial x}(xy^2z^4) + \frac{\partial}{\partial y}(2x^2y + z) + \frac{\partial}{\partial z}(y^3z^2)$$

$$= y^2z^4 + 2x^2 + 2y^3z.$$

In Section 14.6, we shall see that if $\mathbf{F}$ is the velocity field of a fluid or gas, then div $\mathbf{F}$ provides information about the flow of mass. If K is a point and $[\text{div } \mathbf{F}]_K < 0$, there is more mass flowing toward the point than away from it, and we say that there is a **sink** at K. If $[\text{div } \mathbf{F}]_K > 0$, more mass is flowing away from K than toward K and there is a **source** at K. If $[\text{div } \mathbf{F}]_K = 0$, as is true for incompressible fluids, there is neither a sink nor a source at K. Similarly, if $\mathbf{F}$ represents the flow of heat and if $[\nabla \cdot \mathbf{F}]_K > 0$, then there is a source of heat at K (or heat is *leaving* K and therefore the temperature at K is decreasing). If $[\nabla \cdot \mathbf{F}]_K < 0$, then heat is being *absorbed* at K (or the temperature at K is increasing).

Various algebraic properties can be established for *div* and *curl*. A property that is analogous to the product rule for derivatives is considered in the next example. Other properties are stated in Exercises 25–28.

EXAMPLE ▪ 5   Let f be a scalar function and $\mathbf{F}$ a vector function. If partial derivatives exist, show that

$$\nabla \cdot (f\mathbf{F}) = f(\nabla \cdot \mathbf{F}) + (\nabla f) \cdot \mathbf{F}.$$

SOLUTION   If we write $\mathbf{F} = M\mathbf{i} + N\mathbf{j} + P\mathbf{k}$, where M, N, and P are functions of x, y, and z, then

$$f\mathbf{F} = fM\mathbf{i} + fN\mathbf{j} + fP\mathbf{k}.$$

Applying Definition (14.7) yields

$$\nabla \cdot (f\mathbf{F}) = \frac{\partial}{\partial x}(fM) + \frac{\partial}{\partial y}(fN) + \frac{\partial}{\partial z}(fP)$$

$$= f\frac{\partial M}{\partial x} + \frac{\partial f}{\partial x}M + f\frac{\partial N}{\partial y} + \frac{\partial f}{\partial y}N + f\frac{\partial P}{\partial z} + \frac{\partial f}{\partial z}P.$$

Rearranging terms gives us

$$\nabla \cdot (f\mathbf{F}) = f\left(\frac{\partial M}{\partial x} + \frac{\partial N}{\partial y} + \frac{\partial P}{\partial z}\right) + \left(\frac{\partial f}{\partial x}M + \frac{\partial f}{\partial y}N + \frac{\partial f}{\partial z}P\right)$$
$$= f(\nabla \cdot \mathbf{F}) + (\nabla f) \cdot \mathbf{F}.$$

Finally, if a vector field $\mathbf{F}$ is described as in (14.1), then we may define limits, continuity, partial derivatives, and multiple integrals by using the components of $\mathbf{F}(x, y, z)$ as we did for vector-valued functions in Chapter 11. For example, if we wish to differentiate or integrate $\mathbf{F}(x, y, z)$, we differentiate or integrate each component. The usual theorems may be established: $\mathbf{F}$ is continuous if and only if the component functions M, N, and P are continuous; $\mathbf{F}$ has partial derivatives if and only if the same is true for M, N, and P; and so on.

EXERCISES 14.1

Exer. 1–10: Sketch a sufficient number of vectors to illustrate the pattern of the vectors in the field F.

1 $\mathbf{F}(x, y) = x\mathbf{i} - y\mathbf{j}$

2 $\mathbf{F}(x, y) = -x\mathbf{i} + y\mathbf{j}$

3 $\mathbf{F}(x, y) = 2x\mathbf{i} + 3y\mathbf{j}$

4 $\mathbf{F}(x, y) = 3\mathbf{i} + x\mathbf{j}$

5 $\mathbf{F}(x, y) = (x^2 + y^2)^{-1/2}(x\mathbf{i} + y\mathbf{j})$

6 $\mathbf{F}(x, y, z) = x\mathbf{i} + z\mathbf{k}$

7 $\mathbf{F}(x, y, z) = -x\mathbf{i} - y\mathbf{j} - z\mathbf{k}$

8 $\mathbf{F}(x, y, z) = x\mathbf{i} + y\mathbf{j} + z\mathbf{k}$

9 $\mathbf{F}(x, y, z) = \mathbf{i} + \mathbf{j} + \mathbf{k}$

10 $\mathbf{F}(x, y, z) = 2\mathbf{k}$

c **Exer. 11–12:** Plot some of the vectors in F, as in Figure 14.5.

11 $\mathbf{F}(x, y) = -\dfrac{x\mathbf{i} + y\mathbf{j}}{(x^2 + y^2)^2}$

12 $\mathbf{F}(x, y) = \dfrac{x\mathbf{i} + y\mathbf{j}}{e^{(x^2+y^2)}}$

Exer. 13–16: Find a conservative vector field that has the given potential.

13 $f(x, y, z) = x^2 - 3y^2 + 4z^2$

14 $f(x, y, z) = \sin(x^2 + y^2 + z^2)$

15 $f(x, y) = \arctan(xy)$

16 $f(x, y) = y^2 e^{-3x}$

Exer. 17–20: Find $\nabla \times \mathbf{F}$ and $\nabla \cdot \mathbf{F}$.

17 $\mathbf{F}(x, y, z) = x^2 z\mathbf{i} + y^2 x\mathbf{j} + (y + 2z)\mathbf{k}$

18 $\mathbf{F}(x, y, z) = (3x + y)\mathbf{i} + xy^2 z\mathbf{j} + xz^2\mathbf{k}$

19 $\mathbf{F}(x, y, z) = 3xyz^2\mathbf{i} + y^2 \sin z\mathbf{j} + xe^{2z}\mathbf{k}$

20 $\mathbf{F}(x, y, z) = x^3 \ln z\mathbf{i} + xe^{-y}\mathbf{j} - (y^2 + 2z)\mathbf{k}$

21 If $\mathbf{r} = x\mathbf{i} + y\mathbf{j} + z\mathbf{k}$, prove that

(a) $\nabla \cdot \mathbf{r} = 3$ (b) $\nabla \times \mathbf{r} = \mathbf{0}$ (c) $\nabla \|\mathbf{r}\| = \mathbf{r}/\|\mathbf{r}\|$

22 If $\mathbf{r} = x\mathbf{i} + y\mathbf{j} + z\mathbf{k}$ and $\mathbf{a}$ is a constant vector, prove that

(a) curl $(\mathbf{a} \times \mathbf{r}) = 2\mathbf{a}$ (b) div $(\mathbf{a} \times \mathbf{r}) = 0$

23 Prove that the curl and the divergence of an inverse square vector field are $\mathbf{0}$ and 0, respectively.

24 If a vector field $\mathbf{F}$ is conservative and has continuous partial derivatives, prove that curl $\mathbf{F} = \mathbf{0}$.

Exer. 25–28: Verify the identity.

25 $\nabla \times (\mathbf{F} + \mathbf{G}) = \nabla \times \mathbf{F} + \nabla \times \mathbf{G}$

26 $\nabla \cdot (\mathbf{F} + \mathbf{G}) = \nabla \cdot \mathbf{F} + \nabla \cdot \mathbf{G}$

27 $\nabla \times (f\mathbf{F}) = f(\nabla \times \mathbf{F}) + (\nabla f) \times \mathbf{F}$

28 $\nabla \cdot (\mathbf{F} \times \mathbf{G}) = (\nabla \times \mathbf{F}) \cdot \mathbf{G} - (\nabla \times \mathbf{G}) \cdot \mathbf{F}$

Exer. 29–32: If f and $\mathbf{F}$ have continuous second partial derivatives and $\mathbf{a}$ is a constant vector, verify the identity.

29 curl grad $f = \mathbf{0}$

30 div curl $\mathbf{F} = 0$

31 curl (grad f + curl $\mathbf{F}$) = curl curl $\mathbf{F}$

32 curl $\mathbf{a} = \mathbf{0}$

33 Let $\mathbf{r} = x\mathbf{i} + y\mathbf{j} + z\mathbf{k}$ and $r = \|\mathbf{r}\|$. If $\mathbf{F}(x, y, z) = (c/r^k)\mathbf{r}$, where c and k are constants with $k > 0$, prove that curl $\mathbf{F} = \mathbf{0}$. (*Hint:* Use Exercise 27.)

34 The differential operator $\nabla^2 = \nabla \cdot \nabla$ is defined by

$$\nabla^2 = \frac{\partial^2}{\partial x^2} + \frac{\partial^2}{\partial y^2} + \frac{\partial^2}{\partial z^2}.$$

If ∇^2 operates on $f(x, y, z)$, it produces a scalar function called the *Laplacian* of f, given by

$$\nabla^2 f = \frac{\partial^2 f}{\partial x^2} + \frac{\partial^2 f}{\partial y^2} + \frac{\partial^2 f}{\partial z^2}.$$

If f and g are scalar functions that have second partial derivatives, prove that

(a) $\nabla \cdot (\nabla f) = \nabla^2 f$

(b) $\nabla^2 (fg) = f\nabla^2 g + g\nabla^2 f + 2\nabla f \cdot \nabla g$

Exer. 35–36: Using the notation of Exercise 34, prove that the function satisfies *Laplace's equation* $\nabla^2 f = 0$. (Functions of this type are *harmonic* and are important in physical applications.)

35 $f(x, y, z) = (x^2 + y^2 + z^2)^{-1/2}$

36 $f(x, y, z) = ax^2 + by^2 + cz^2$ for $a + b + c = 0$

37 If $\mathbf{F}$ is given by (14.1), define

$$\lim_{(x, y, z) \to (x_0, y_0, z_0)} \mathbf{F}(x, y, z) = \mathbf{a}$$

in a manner analogous to that of Definition (11.4). Give an ϵ–δ definition for this limit. What is the geometric significance of the limit?

38 If $\mathbf{F}$ is given by (14.1), define the notion of *continuity* at (x_0, y_0, z_0) in a manner similar to that of Definition (11.5). What is the geometric significance of a continuous vector function?

$\boxed{c}$ **Exer. 39–42:** Use a centered difference quotient such as

$$\frac{\partial^2 f}{\partial x^2} \approx \frac{f(x + h, y, z) - 2f(x, y, z) + f(x - h, y, z)}{h^2}$$

for each variable to approximate $\nabla^2 f(0.3, 0.5, 0.2)$ using $h = 0.05$ as the change in each variable.

39 $f(x, y, z) = \dfrac{\tan^2 \sqrt{xyz}}{\cos z \sec^2(x^2 y^2)}$

40 $f(x, y, z) = \dfrac{x^2 + xy^2 z + z^3}{4x^3 + 2y^2 z + 4xz^2}$

41 $f(x, y, z) = \dfrac{\sqrt{xyz + 1}}{(x^2 y^2 + 5xy^3 z + 1)^2}$

42 $f(x, y, z) = \sinh\left(\dfrac{x^2 - yz}{1 + x + 2y + 5z + z^2}\right)$

14.2 LINE INTEGRALS

Line integrals, introduced in this section, are an important generalization of the definite integral $\int_a^b f(x)\, dx$. To define $\int_a^b f(x)\, dx$ in Chapter 4, we began by dividing the interval $[a, b]$ into subintervals of lengths $\Delta x_1, \Delta x_2, \ldots, \Delta x_n$. We then chose any number w_k in each subinterval and considered the limit of Riemann sums $\sum_k f(w_k)\Delta x_k$ as $\Delta x_k \to 0$. We may use a similar process to define *line integrals* of functions of several variables along (or over) curves in two or three dimensions.

Recall that a plane curve C is *smooth* if it has a parametrization

$$x = g(t), \quad y = h(t); \quad a \le t \le b$$

such that g' and h' are continuous and not simultaneously 0 on $[a, b]$. For space curves, we also consider a third function k of the same type with $z = k(t)$. The **positive direction**, or *orientation*, of C is the direction determined by increasing values of t. The curve C is *piecewise smooth* if $[a, b]$ can be partitioned into closed subintervals such that C is smooth for each subinterval.

Suppose f is a function of two variables x and y that is continuous on a region D containing a smooth curve C, with parametrization $x = g(t), y = h(t); a \le t \le b$. We shall define three different integrals of f along C. Let us begin by partitioning the parameter interval $[a, b]$, choosing

$$a = t_0 < t_1 < t_2 < \cdots < t_n = b.$$

The norm of this partition (that is, the length of the largest subinterval $[t_{k-1}, t_k]$) will be denoted by $\|P\|$. If $P_k(x_k, y_k)$ is the point on C corresponding to t_k, then the points $P_0, P_1, P_2, \ldots, P_n$ divide C into n parts $\overset{\frown}{P_{k-1}P_k}$ as illustrated in Figure 14.10. Let

$$\Delta x_k = x_k - x_{k-1}, \quad \Delta y_k = y_k - y_{k-1}, \quad \Delta s_k = \text{length of } \overset{\frown}{P_{k-1}P_k}.$$

For each k, let $Q_k(u_k, v_k)$ be a point in $\overset{\frown}{P_{k-1}P_k}$ obtained by choosing some number in $[t_{k-1}, t_k]$(see Figure 14.10). We now consider the three sums

$$\sum_k f(u_k, v_k)\Delta s_k, \qquad \sum_k f(u_k, v_k)\Delta x_k, \qquad \sum_k f(u_k, v_k)\Delta y_k.$$

Figure 14.10

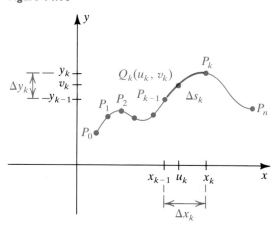

If the limits of these sums exist as $\|P\| \to 0$, they are the **line integrals of f along** (or **over**) C with respect to $s, x,$ and y, respectively, and are denoted as follows.

Line Integrals in Two Dimensions 14.8

$$\int_C f(x, y)\, ds = \lim_{\|P\| \to 0} \sum_k f(u_k, v_k)\Delta s_k$$

$$\int_C f(x, y)\, dx = \lim_{\|P\| \to 0} \sum_k f(u_k, v_k)\Delta x_k$$

$$\int_C f(x, y)\, dy = \lim_{\|P\| \to 0} \sum_k f(u_k, v_k)\Delta y_k$$

The term *line integral* is a misnomer, since C is not necessarily a line. It would be more descriptive to use the term *curve integral*.

If f is continuous on D, then the limits in (14.8) exist and are the same for every parametrization of C (provided the same orientation is used). Moreover, the integrals can be evaluated by substituting $x = g(t)$ and $y = h(t)$ from the parametrization of C and replacing the differentials by

$$ds = \sqrt{(dx)^2 + (dy)^2} = \sqrt{[g'(t)]^2 + [h'(t)]^2}\, dt$$

$$dx = g'(t)\, dt, \qquad dy = h'(t)\, dt.$$

Note that the formula for ds is the differential of arc length from (9.7). Let us state these facts for reference as follows.

Evaluation Theorem for Line Integrals   14.9

If a smooth curve C is given by $x = g(t)$, $y = h(t)$; $a \le t \le b$ and if $f(x, y)$ is continuous on a region D containing C, then

(i) $\displaystyle\int_C f(x, y)\, ds = \int_a^b f(g(t), h(t))\sqrt{[g'(t)]^2 + [h'(t)]^2}\, dt$

(ii) $\displaystyle\int_C f(x, y)\, dx = \int_a^b f(g(t), h(t))g'(t)\, dt$

(iii) $\displaystyle\int_C f(x, y)\, dy = \int_a^b f(g(t), h(t))h'(t)\, dt$

The preceding discussion can be extended to more complicated curves. In particular, suppose C is a piecewise-smooth curve that can be expressed as the union of a finite number of smooth curves $C_1, C_2, \ldots, C_n$, where the terminal point of C_k is the initial point of C_{k+1} for $k = 1, 2, \ldots, n - 1$. In this case, the line integral of f along C is defined as the sum of the line integrals along the individual curves.

We may establish properties of line integrals that are similar to those obtained for definite integrals in Chapter 4. For example, reversing the direction of integration changes the sign of the integral; the integral of a sum of two functions is the sum of the integrals of the individual functions; and so on.

Figure 14.11

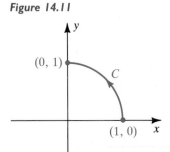

EXAMPLE ■ 1   Evaluate $\displaystyle\int_C xy^2\, ds$ if C has the parametrization

$$x = \cos t, \quad y = \sin t; \quad 0 \le t \le \pi/2.$$

SOLUTION   The curve C is the first-quadrant part of the unit circle with center at the origin, as shown in Figure 14.11, where the arrowhead indicates the orientation of C and the direction of integration. Applying

Theorem (14.9)(i) gives us

$$\int_C xy^2\, ds = \int_0^{\pi/2} \cos t \sin^2 t \sqrt{\sin^2 t + \cos^2 t}\; dt$$

$$= \int_0^{\pi/2} \sin^2 t \cos t\; dt = \left[\tfrac{1}{3}\sin^3 t\right]_0^{\pi/2} = \tfrac{1}{3}.$$

EXAMPLE ■ 2   Evaluate $\int_C xy^2\, dx$ and $\int_C xy^2\, dy$ if C is the portion of the parabola $y = x^2$ from $(0, 0)$ to $(2, 4)$.

SOLUTION   The graph of $y = x^2$ from $(0, 0)$ to $(2, 4)$ is sketched in Figure 14.12. Parametric equations for C are

$$x = t, \quad y = t^2; \quad 0 \le t \le 2.$$

The differentials are

$$dx = dt, \quad dy = 2t\, dt.$$

Applying Theorem (14.9)(ii) and (iii), we have

$$\int_C xy^2\, dx = \int_0^2 t(t^2)^2\, dt = \int_0^2 t^5\, dt = \left[\tfrac{1}{6}t^6\right]_0^2 = \tfrac{32}{3}$$

$$\int_C xy^2\, dy = \int_0^2 t(t^2)^2\, 2t\, dt = \int_0^2 2t^6\, dt = \left[\tfrac{2}{7}t^7\right]_0^2 = \tfrac{256}{7}.$$

Figure 14.12

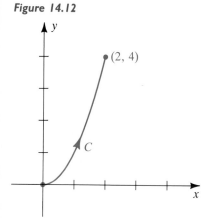

If, as in Example 2, C is the graph of an equation $y = g(x)$ such that $a \le x \le b$, then parametric equations for C are

$$x = t, \quad y = g(t); \quad a \le t \le b.$$

The line integrals (ii) and (iii) of Theorem (14.9) may then be evaluated as follows:

$$\int_C f(x, y)\, dx = \int_a^b f(t, g(t))\, dt = \int_a^b f(x, g(x))\, dx$$

$$\int_C f(x, y)\, dy = \int_a^b f(t, g(t))g'(t)\, dt = \int_a^b f(x, g(x))g'(x)\, dx$$

This result shows that *for curves given in the form $y = g(x)$ with $a \le x \le b$, we may bypass the parametric equations by substituting $y = g(x)$, $dy = g'(x)\, dx$ and then using a and b for the limits of integration.* To illustrate, in Example 2, where $y = x^2$, we could have written

$$\int_C xy^2\, dx = \int_0^2 x(x^4)\, dx = \int_0^2 x^5\, dx = \tfrac{32}{3}$$

$$\int_C xy^2\, dy = \int_0^2 x(x^4)2x\, dx = \int_0^2 2x^6\, dx = \tfrac{256}{7}.$$

A physical application of the line integral $\int_C f(x, y)\, ds$ may be obtained by regarding the curve as a thin wire of variable density. If the wire is represented by the curve C in Figure 14.10 and the **linear mass density** (the mass per unit length) at the point (x, y) is given by $\delta(x, y)$, then $\delta(u_k, v_k)\Delta s_k$ is an approximation to the mass Δm_k of the part of the wire between P_{k-1} and P_k. The sum

$$\sum_k \Delta m_k = \sum_k \delta(u_k, v_k)\Delta s_k$$

is an approximation to the total mass m of the wire. To define m, we take the limit of these sums, obtaining the following:

Mass of a Wire   14.10

$$m = \int_C \delta(x, y)\, ds$$

EXAMPLE ▪ 3   A thin wire is bent into the shape of a semicircle of radius a. If the linear mass density at a point P is directly proportional to its distance from the line through the endpoints, find the mass of the wire.

SOLUTION   Let us introduce a coordinate system such that the shape of the wire coincides with the upper half of a circle of radius a with center at the origin (see Figure 14.13). Parametric equations for C are

$$x = a\cos t, \quad y = a\sin t; \quad 0 \le t \le \pi.$$

By hypothesis, the linear mass density at $P(x, y)$ is given by $\delta(x, y) = ky$, where k is a constant. By (14.10), the mass of the wire is

$$m = \int_C (ky)\, ds = \int_0^\pi (ka\sin t)\sqrt{a^2\sin^2 t + a^2\cos^2 t}\, dt$$

$$= \int_0^\pi (ka\sin t)a\, dt$$

$$= ka^2 \int_0^\pi \sin t\, dt = ka^2\big[-\cos t\big]_0^\pi = 2ka^2.$$

Figure 14.13

In applications involving work, line integrals often occur in the combination

$$\int_C M(x, y)\, dx + \int_C N(x, y)\, dy,$$

where the functions M and N are continuous on a domain D containing C. This sum is usually abbreviated as

$$\int_C M(x, y)\, dx + N(x, y)\, dy.$$

Figure 14.14

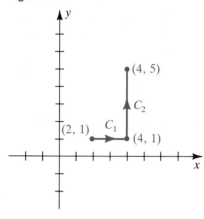

EXAMPLE ■ 4 Evaluate $\displaystyle\int_C xy\,dx + x^2 dy$ if

(a) C consists of line segments from $(2, 1)$ to $(4, 1)$ and from $(4, 1)$ to $(4, 5)$

(b) C is the line segment from $(2, 1)$ to $(4, 5)$

(c) parametric equations for C are $x = 3t - 1$, $y = 3t^2 - 2t$; $1 \le t \le \frac{5}{3}$

SOLUTION

(a) If C is subdivided into two parts C_1 and C_2, as shown in Figure 14.14, then parametric equations for these curves are

$$C_1: \quad x = t, \quad y = 1; \quad 2 \le t \le 4$$
$$C_2: \quad x = 4, \quad y = t; \quad 1 \le t \le 5.$$

The line integral along C may be expressed as a sum of two line integrals, the first along C_1 and the second along C_2. On C_1 we have $dy = 0$, $dx = dt$, and hence

$$\int_{C_1} xy\,dx + x^2\,dy = \int_2^4 t(1)\,dt + t^2(0) = \left[\tfrac{1}{2}t^2\right]_2^4 = 6.$$

Figure 14.15

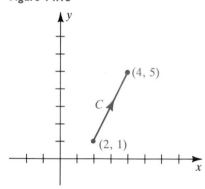

On C_2 we have $dx = 0$, $dy = dt$, and therefore

$$\int_{C_2} xy\,dx + x^2\,dy = \int_1^5 4t(0) + 16\,dt = 16\big[t\big]_1^5 = 64.$$

Consequently, the line integral along C equals $6 + 64$, or 70.

(b) The graph of C is sketched in Figure 14.15. An equation for C is $y = 2x - 3$ with $2 \le x \le 4$. In this case, $dy = 2\,dx$ and

$$\int_C xy\,dx + x^2\,dy = \int_2^4 x(2x - 3)\,dx + x^2(2)\,dx$$
$$= \int_2^4 (4x^2 - 3x)\,dx = \tfrac{170}{3}.$$

Figure 14.16

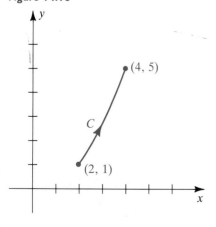

(c) The graph of C is part of a parabola (see Figure 14.16). In this case, using the parametric equations $x = 3t - 1$, $y = 3t^2 - 2t$; $1 \le t \le \frac{5}{3}$, we obtain $dx = 3\,dt$, $dy = (6t - 2)\,dt$, and the line integral equals

$$\int_1^{5/3} (3t - 1)(3t^2 - 2t)3\,dt + (3t - 1)^2(6t - 2)\,dt.$$

We can show that the value is 58.

Another method of solution is to use the equation $y = \tfrac{1}{3}(x^2 - 1)$ for the parabola, with $2 \le x \le 4$. We then obtain the integral

$$\int_C xy\,dx + x^2\,dy = \int_2^4 x \cdot \tfrac{1}{3}(x^2 - 1)\,dx + x^2(\tfrac{2}{3}x)\,dx$$
$$= \int_2^4 (x^3 - \tfrac{1}{3}x)\,dx = \left[\tfrac{1}{4}x^4 - \tfrac{1}{6}x^2\right]_2^4 = 58.$$

In the preceding example, we obtained three different values for the line integral along three different paths from (2, 1) to (4, 5). In Section 14.3, we consider line integrals that have the same value along every curve joining two points A and B. For such integrals, we use the phrase *independent of path*.

If a smooth curve C in three dimensions has the parametrization

$$x = g(t), \quad y = h(t), \quad z = k(t); \quad a \le t \le b,$$

then line integrals of a function f of *three* variables are defined in a manner similar to that used for integrals of a function of two variables. In this case, instead of using (x_k, y_k) and (u_k, v_k) as the coordinates of P_k and Q_k on C as in Figure 14.10, we use (x_k, y_k, z_k) and (u_k, v_k, w_k), respectively (see Figure 14.17). We now have

$$\int_C f(x, y, z) \, ds = \lim_{\|P\| \to 0} \sum_k f(u_k, v_k, w_k) \Delta s_k.$$

This integral may be evaluated by using the formula

$$\int_a^b f(g(t), h(t), k(t)) \sqrt{[g'(t)]^2 + [h'(t)]^2 + [k'(t)]^2} \, dt.$$

If a wire has the shape of C and if $f(x, y, z)$ is the linear mass density at (x, y, z), then the value of this integral is the total mass of the wire.

Figure 14.17

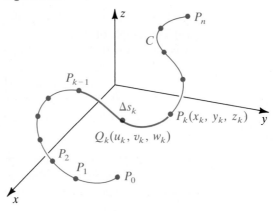

In addition to line integrals $\int_C f(x, y, z) \, dx$ and $\int_C f(x, y, z) \, dy$, a function of three variables also has a line integral with respect to z, given by

$$\int_C f(x, y, z) \, dz = \lim_{\|P\| \to 0} \sum_k f(u_k, v_k, w_k) \Delta z_k,$$

where $\Delta z_k = z_k - z_{k-1}$. As in two dimensions, these line integrals often occur as sums, which we abbreviate as

$$\int_C M(x, y, z) \, dx + N(x, y, z) \, dy + P(x, y, z) \, dz,$$

Figure 14.18

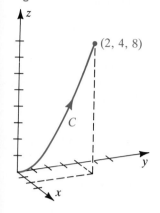

where the functions M, N, and P are continuous throughout a region containing C. If C is given parametrically, then this line integral may be evaluated by substituting for x, y, and z in the same manner as in the two-variable case.

EXAMPLE ■ 5 Evaluate $\int_C yz\,dx + xz\,dy + xy\,dz$ if C is given by

$$x = t, \quad y = t^2, \quad z = t^3; \quad 0 \le t \le 2.$$

SOLUTION The curve C (a twisted cubic) is sketched in Figure 14.18. We substitute for x, y, and z and use $dx = dt$, $dy = 2t\,dt$, and $dz = 3t^2\,dt$:

$$\int_0^2 t^5\,dt + 2t^5\,dt + 3t^5\,dt = \int_0^2 6t^5\,dt = \left[t^6\right]_0^2 = 64$$

Figure 14.19

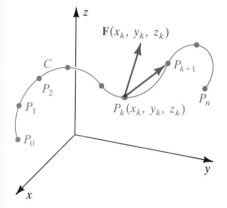

One of the most important physical applications of line integrals involves force fields. Suppose that the force acting at the point (x, y, z) is

$$\mathbf{F}(x, y, z) = M(x, y, z)\mathbf{i} + N(x, y, z)\mathbf{j} + P(x, y, z)\mathbf{k},$$

where M, N, and P are continuous functions. We shall formulate a definition for the work done as the point of application of $\mathbf{F}(x, y, z)$ moves along a smooth curve C that has a parametrization

$$x = g(t), \quad y = h(t), \quad z = k(t); \quad a \le t \le b.$$

We assume that the motion is in the direction determined by *increasing* values of t.

Let us subdivide C by means of points $P_0, P_1, P_2, \ldots, P_n$, where P_k has coordinates (x_k, y_k, z_k) (see Figure 14.19). If the norm $\|P\|$ is sufficiently small, then P_k is close to P_{k+1} for each k. Hence the work done by $\mathbf{F}(x, y, z)$ from P_k to P_{k+1} can be approximated by the work ΔW_k done by the *constant* force $\mathbf{F}(x_k, y_k, z_k)$ as its point of application moves along $\overrightarrow{P_k P_{k+1}}$.

Figure 14.20

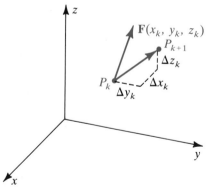

If $\Delta x_k = x_{k+1} - x_k$, $\Delta y_k = y_{k+1} - y_k$, and $\Delta z_k = z_{k+1} - z_k$, then $\overrightarrow{P_k P_{k+1}}$ corresponds to the vector $\Delta x_k\mathbf{i} + \Delta y_k\mathbf{j} + \Delta z_k\mathbf{k}$ (see Figure 14.20). By (10.26), the work ΔW_k done by $\mathbf{F}(x_k, y_k, z_k)$ along $\overrightarrow{P_k P_{k+1}}$ is

$$\Delta W_k = \mathbf{F}(x_k, y_k, z_k) \cdot (\Delta x_k\mathbf{i} + \Delta y_k\mathbf{j} + \Delta z_k\mathbf{k})$$
$$= M(x_k, y_k, z_k)\Delta x_k + N(x_k, y_k, z_k)\Delta y_k + P(x_k, y_k, z_k)\Delta z_k.$$

The **work W done by $\mathbf{F}$ along C** is defined as follows.

Definition of Work 14.11

$$W = \lim_{\|P\| \to 0} \sum_k \Delta W_k$$

$$= \int_C M(x, y, z)\,dx + N(x, y, z)\,dy + P(x, y, z)\,dz$$

For applications, the integral in (14.11) is usually expressed in vector form. If we let

$$\mathbf{r}(t) = x\mathbf{i} + y\mathbf{j} + z\mathbf{k},$$

where $x = g(t)$, $y = h(t)$, and $z = k(t)$, then $\mathbf{r}(t)$ is the position vector for the point $Q(x, y, z)$ on C. If s is an arc length parameter for C, then, as in Section 11.4, a unit tangent vector $\mathbf{T}(s)$ to C at Q is

$$\mathbf{T}(s) = \frac{d}{ds}\mathbf{r}(t) = \frac{dx}{ds}\mathbf{i} + \frac{dy}{ds}\mathbf{j} + \frac{dz}{ds}\mathbf{k}.$$

Figure 14.21

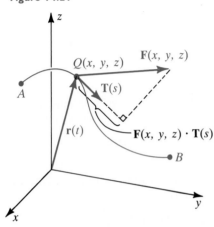

The vectors $\mathbf{r}(t)$, $\mathbf{F}(x, y, z)$, and $\mathbf{T}(s)$ are illustrated in Figure 14.21.

The **tangential component of F at Q** (see Figure 14.21) is

$$\mathbf{F}(x, y, z) \cdot \mathbf{T}(s) = M(x, y, z)\frac{dx}{ds} + N(x, y, z)\frac{dy}{ds} + P(x, y, z)\frac{dz}{ds}.$$

Note that this component is positive if the angle between $\mathbf{F}(x, y, z)$ and $\mathbf{T}(s)$ is acute and negative if this angle is obtuse. Formula (14.11) may now be rewritten as follows:

$$W = \int_C \mathbf{F}(x, y, z) \cdot \mathbf{T}(s)\, ds$$

Thus, *the work done as the point of application of $\mathbf{F}(x, y, z)$ moves along C equals the line integral with respect to s of the tangential component of $\mathbf{F}$ along C.* To simplify the notation, we shall denote $\mathbf{F}(x, y, z)$ by $\mathbf{F}$ and $\mathbf{T}(s)$ by $\mathbf{T}$ and let

$$d\mathbf{r} = dx\mathbf{i} + dy\mathbf{j} + dz\mathbf{k} = \mathbf{T}\, ds.$$

Our discussion may be summarized as follows.

Definition 14.12

Let C be a smooth space curve, let $\mathbf{T}$ be a unit tangent vector to C at (x, y, z), and let $\mathbf{F}$ be the force acting at (x, y, z). The **work W done by $\mathbf{F}$ along C** is

$$W = \int_C \mathbf{F} \cdot \mathbf{T}\, ds = \int_C \mathbf{F} \cdot d\mathbf{r},$$

where $\mathbf{r} = x\mathbf{i} + y\mathbf{j} + z\mathbf{k}$.

Intuitively, we may regard $\mathbf{F} \cdot d\mathbf{r}$ in Definition (14.12) as the work done as the point of application of $\mathbf{F}$ moves along the tangent vector $d\mathbf{r}$ to C. The integral sign represents the limit of sums of the elements $\mathbf{F} \cdot d\mathbf{r}$ of work.

EXAMPLE ■ 6   If an inverse force field $\mathbf{F}$ is given by

$$\mathbf{F}(x, y, z) = \frac{k}{\|\mathbf{r}\|^3}\mathbf{r},$$

where k is a constant, find the work done by $\mathbf{F}$ as its point of application moves along the x-axis from $A(1, 0, 0)$ to $B(2, 0, 0)$.

SOLUTION   Let C denote the line segment from A to B. Parametric equations for C are $x = t, y = 0, z = 0; 1 \le t \le 2$. As in Example 2 of Section 14.1,

$$\mathbf{F}(x, y, z) = \frac{k}{(x^2 + y^2 + z^2)^{3/2}} (x\mathbf{i} + y\mathbf{j} + z\mathbf{k}).$$

Applying Definition (14.12), we have

$$W = \int_C \mathbf{F} \cdot d\mathbf{r} = \int_C \frac{k}{(x^2 + y^2 + z^2)^{3/2}} (x\,dx + y\,dy + z\,dz).$$

Substituting for x, y, and z from the parametric equations for C gives us

$$W = \int_1^2 \frac{k}{(t^2)^{3/2}} t\,dt = \int_1^2 \frac{k}{t^2}\,dt = \left[-\frac{k}{t} \right]_1^2 = \frac{k}{2}.$$

The units for W depend on those for distance and $\|\mathbf{F}(x, y, z)\|$.

If the preceding discussion is restricted to two dimensions, then a force field may be expressed in the form

$$\mathbf{F}(x, y) = M(x, y)\mathbf{i} + N(x, y)\mathbf{j}.$$

If C is a finite piecewise-smooth plane curve, then the work done as the point of application of $\mathbf{F}(x, y)$ moves along C is

$$W = \int_C M(x, y)\,dx + N(x, y)\,dy = \int_C \mathbf{F} \cdot d\mathbf{r},$$

where $\mathbf{r} = x\mathbf{i} + y\mathbf{j}$.

EXAMPLE ■ 7   Let C be the part of the parabola $y = x^2$ between $(0, 0)$ and $(3, 9)$. If $\mathbf{F}(x, y) = -y\mathbf{i} + x\mathbf{j}$ is a force acting at (x, y), find the work done by $\mathbf{F}$ along C from

(a) $(0, 0)$ to $(3, 9)$ **(b)** $(3, 9)$ to $(0, 0)$

SOLUTION   The vector field $\mathbf{F}$ was discussed in Example 1 of Section 14.1. Some typical force vectors acting along C are shown in Figure 14.22. For example,

$$\mathbf{F}(1, 1) = -\mathbf{i} + \mathbf{j}, \qquad \mathbf{F}(2, 4) = -4\mathbf{i} + 2\mathbf{j}, \qquad \mathbf{F}(3, 9) = -9\mathbf{i} + 3\mathbf{j}.$$

Note the change in direction and magnitude of $\mathbf{F}(x, y)$ between $(0, 0)$ and $(3, 9)$.

(a) Using the parametrization $x = t, y = t^2; 0 \le t \le 3$, we obtain

$$W = \int_C \mathbf{F} \cdot d\mathbf{r} = \int_C -y\,dx + x\,dy$$

$$= \int_0^3 -t^2\,dt + t(2t)\,dt = \int_0^3 t^2\,dt = \left[\frac{t^3}{3} \right]_0^3 = 9.$$

If, for example, $\|\mathbf{F}\|$ is in pounds and s is in feet, then $W = 9$ ft-lb.

Figure 14.22

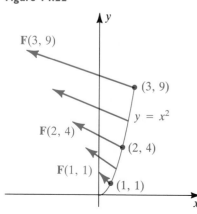

(b) We could use the parametrization $x = -t$, $y = t^2$; $-3 \le t \le 0$ for C and evaluate $\int_C \mathbf{F} \cdot d\mathbf{r}$; however, it is simpler to reverse the limits of integration in part (a). This gives us $W = -9$.

The significance of the sign difference between 9 and -9 in solutions (a) and (b) of Example 7 may be interpreted as follows. In (a), imagine a particle moving through the force field $\mathbf{F}$ along the path $y = x^2$ from $(0, 0)$ to $(3, 9)$. The particle moves in a northeasterly direction, and the force vectors act in a northwesterly direction. The angle θ between $\mathbf{F}(x, y)$ and a tangent vector to the path (in the direction of the motion) is always acute, and hence $\mathbf{F}$ *assists* the movement of the particle. Thus, the work done by $\mathbf{F}$ is positive.

In (b), the particle moves from $(3, 9)$ to $(0, 0)$ in a southwesterly direction. The angle θ is always obtuse, and hence $\mathbf{F}$ *hinders* the movement of the particle. Thus, the work done by $\mathbf{F}$ is negative.

EXERCISES 14.2

Exer. 1–2: Evaluate the line integrals $\int_C f(x, y)\,ds$, $\int_C f(x, y)\,dx$, and $\int_C f(x, y)\,dy$ if C has the given parametrization.

1 $f(x, y) = x^3 + y$, $x = 3t$, $y = t^3$; $0 \le t \le 1$

2 $f(x, y) = xy^{2/5}$, $x = \frac{1}{2}t$, $y = t^{5/2}$; $0 \le t \le 1$

c Exer. 3–4: Approximate, to four decimal places, the line integrals $\int_C f(x, y)\,ds$, $\int_C f(x, y)\,dx$, and $\int_C f(x, y)\,dy$ if C has the given parametrization.

3 $f(x, y) = \sqrt{x^2 + y^2}$, $x = 2t^2$, $y = 4t - 3t^2$;
$$0 \le t \le 1$$

4 $f(x, y) = \ln(x + y^2)$, $x = 2\cos t$, $y = \sin t$;
$$0 \le t \le \pi/2$$

Exer. 5–8: Evaluate the line integral along C.

5 $\displaystyle\int_C 6x^2 y\,dx + xy\,dy$;

C is the graph of $y = x^3 + 1$ from $(-1, 0)$ to $(1, 2)$.

6 $\displaystyle\int_C y\,dx + (x + y)\,dy$;

C is the graph of $y = x^2 + 2x$ from $(0, 0)$ to $(2, 8)$.

7 $\displaystyle\int_C (x - y)\,dx + x\,dy$;

C is the graph of $y^2 = x$ from $(4, -2)$ to $(4, 2)$.

8 $\displaystyle\int_C xy\,dx + x^2 y^3\,dy$;

C is the graph of $x = y^3$ from $(0, 0)$ to $(1, 1)$.

c Exer. 9–10: Approximate the line integral to three decimal places if C has the given parametrization.

9 $\displaystyle\int_C \cos(yz)\,dx + \cos(xz)\,dy + \cos(xy)\,dz$,

$x = 1 + 2t - 3t^2$, $y = 1 + 4t - 5t^2$, $z = 1 + 4t - 2t^2$;
$$0 \le t \le 1$$

10 $\displaystyle\int_C e^y\,dx + e^x\,dy$, $x = \cos 2t$, $y = \sin 3t$;
$$0 \le t \le \pi$$

11 Evaluate $\int_C xy\,dx + (x + y)\,dy$ along each curve C from $(0, 0)$ to $(1, 3)$.

(a)   (b)

(c)

(d)

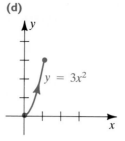

$y = 3x^2$

12 Evaluate $\int_C (x^2 + y^2)\, dx + 2x\, dy$ along each curve C from $(1, 2)$ to $(-2, 8)$.

(a)

(b)

(c)

(d)

$y = 2x^2$

13 Evaluate $\int_C xz\, dx + (y + z)\, dy + x\, dz$ if C is the graph of $x = e^t, y = e^{-t}, z = e^{2t}; 0 \le t \le 1$.

14 Evaluate $\int_C y\, dx + z\, dy + x\, dz$ if C is the graph of $x = \sin t, y = 2\sin t, z = \sin^2 t; 0 \le t \le \pi/2$.

15 Evaluate

$$\int_C (x + y + z)\, dx + (x - 2y + 3z)\, dy + (2x + y - z)\, dz,$$

where C is the curve from $(0, 0, 0)$ to $(2, 3, 4)$, if

(a) C consists of three line segments, the first parallel to the x-axis, the second parallel to the y-axis, and third parallel to the z-axis

(b) C consists of three line segments, the first parallel to the z-axis, the second parallel to the x-axis, and the third parallel to the y-axis

(c) C is a line segment

16 Evaluate $\int_C (x - y)\, dx + (y - z)\, dy + x\, dz$ if C is the curve from $(1, -2, 3)$ to $(-4, 5, 2)$ of the type described in parts (a)–(c) of Exercise 15.

17 Evaluate $\int_C xyz\, ds$ if C is the line segment from $(0, 0, 0)$ to $(1, 2, 3)$.

18 Evaluate $\int_C (xy + z)\, ds$ if C is the helix $x = a\cos t, y = a\sin t, z = bt; 0 \le t \le 2\pi$.

19 If the force at (x, y) is $\mathbf{F}(x, y) = xy^2\mathbf{i} + x^2 y\mathbf{j}$, find the work done by $\mathbf{F}$ along the curves in parts (a)–(d) of Exercise 11.

20 If the force at (x, y) is

$$\mathbf{F}(x, y) = (2x + y)\mathbf{i} + (x + 2y)\mathbf{j},$$

find the work done by $\mathbf{F}$ along the curves in parts (a)–(d) of Exercise 12.

21 The force acting at a point (x, y) in a coordinate plane is $\mathbf{F}(x, y) = (4/\|\mathbf{r}\|^3)\mathbf{r}$, where $\mathbf{r} = x\mathbf{i} + y\mathbf{j}$. Find the work done by $\mathbf{F}$ along the upper half of the circle $x^2 + y^2 = a^2$ from $(-a, 0)$ to $(a, 0)$.

22 The force at a point (x, y) in a coordinate plane is given by $\mathbf{F}(x, y) = (x^2 + y^2)\mathbf{i} + xy\mathbf{j}$. Find the work done by $\mathbf{F}(x, y)$ along the graph of $y = x^3$ from $(0, 0)$ to $(2, 8)$.

23 The force at a point (x, y, z) in three dimensions is given by $\mathbf{F}(x, y, z) = y\mathbf{i} + z\mathbf{j} + x\mathbf{k}$. Find the work done by $\mathbf{F}(x, y, z)$ along the twisted cubic $x = t, y = t^2, z = t^3$ from $(0, 0, 0)$ to $(2, 4, 8)$.

24 Work Exercise 23 if $\mathbf{F}(x, y, z) = e^x\mathbf{i} + e^y\mathbf{j} + e^z\mathbf{k}$.

25 If an object moves through a force field $\mathbf{F}$ such that at each point (x, y, z) its velocity vector is orthogonal to $\mathbf{F}(x, y, z)$, show that the work done by $\mathbf{F}$ on the object is 0.

26 If a constant force $\mathbf{c}$ acts on a moving object as it travels once around a circle, show that the work done by $\mathbf{c}$ on the object is 0.

27 If a thin wire has the shape of a plane curve C and if the linear mass density at (x, y) is $\delta(x, y)$, then the *moments with respect to the x- and y-axes* are defined by

$$M_x = \int_C y\delta(x, y)\, ds \quad \text{and} \quad M_y = \int_C x\delta(x, y)\, ds.$$

Use limits of sums to show that these are natural definitions for M_x and M_y. The center of mass $(\bar{x}, \bar{y})$ of the wire is defined by $\bar{x} = M_y/m$ and $\bar{y} = M_x/m$, where m is the mass of the wire. Find the center of mass of the wire in Example 3.

28 Refer to Exercise 27. A thin wire is situated in a coordinate plane such that its shape coincides with the part of the parabola $y = 4 - x^2$ between $(-2, 0)$ and $(2, 0)$. The density at the point (x, y) is directly proportional to its distance from the y-axis. Find the mass and the center of mass.

29 Extend the definitions of moments and center of mass of a wire to three dimensions (see Exercise 27).

30 A wire of constant density is bent into the shape of the helix $x = a\cos t$, $y = a\sin t$, $z = bt$; $0 \le t \le 3\pi$. Find the mass and the center of mass of the wire (see Exercise 29).

31 If a thin wire of variable density is represented by a curve C in the xy-plane, define the moments of inertia I_x and I_y with respect to the x- and y-axes, respectively. Find I_x and I_y for the wire in Example 3.

32 Find the moments of inertia I_x and I_y for the wire in Exercise 28.

33 If a thin wire of variable density is represented by a curve in three dimensions, define the moments of inertia with respect to the x-, y-, and z-axes.

34 Given the wire in Exercise 30, find the moment of inertia with respect to the z-axis (see also Exercise 33).

[c] Exer. 35–37: Let $a = t_0 < t_1 < \cdots < t_n = b$ be a regular partition of the parameter interval, and let (x_k, y_k) be the point on the curve C associated with the parameter value t_k. Approximate the line integral by the indicated sum.

35 $\displaystyle\int_C \sin(xy)\, dx, \quad x = \cos t, \quad y = \sin(0.8t);$

$$0 \le t \le 2$$

by $\displaystyle\sum_{k=1}^{20} \sin(x_k y_k)\Delta x_k$

36 $\displaystyle\int_C \ln\sqrt{xy}\, dy, \quad x = t^2, \quad y = 1 + 3t - t^2;$

$$0 \le t \le 1$$

by $\displaystyle\sum_{k=1}^{10} \ln\sqrt{x_k y_k}\,\Delta y_k$

37 $\displaystyle\int_C \sqrt{x^2 + y^2}\, ds, \quad x = 2 + 4t^2, \quad y = 2t - 3t^2;$

$$0 \le t \le 1$$

by $\displaystyle\sum_{k=1}^{10} \sqrt{x_k^2 + y_k^2}\sqrt{(\Delta x_k)^2 + (\Delta y_k)^2}$

14.3   INDEPENDENCE OF PATH

A piecewise-smooth curve with endpoints A and B is sometimes called a **path** from A to B. In this section, we obtain conditions under which a line integral is **independent of path** in a region in the sense that if A and B are arbitrary points, then the same value is obtained for *every* path in that region from A to B. The results will be established for line integrals in two dimensions. Proofs for the three-dimensional case are similar and are omitted.

If the line integral $\int_C f(x, y)\, ds$ is independent of path, we may denote it by $\int_A^B f(x, y)\, ds$, since the value of the integral depends only on the endpoints A and B of the curve C. A similar notation is used for $\int_C f(x, y)\, dx$, $\int_C f(x, y)\, dy$, and line integrals in three dimensions.

Throughout this section, we assume that all regions are **connected**—that is, any two points in a region can be joined by a piecewise-smooth curve that lies in the region. The next theorem is a fundamental result about independence of path.

Theorem 14.13

> If $\mathbf{F}(x, y) = M(x, y)\mathbf{i} + N(x, y)\mathbf{j}$ is continuous on an open connected region D, then the line integral $\int_C \mathbf{F} \cdot d\mathbf{r}$ is independent of path if and only if $\mathbf{F}$ is conservative—that is, $\mathbf{F}(x, y) = \nabla f(x, y)$ for some scalar function f.

P R O O F Suppose the integral is independent of path in D. If (x_0, y_0) is a fixed point in D, let f be defined by

$$f(x, y) = \int_{(x_0, y_0)}^{(x, y)} \mathbf{F} \cdot d\mathbf{r}$$

for every point (x, y) in D. We shall show that $\mathbf{F}(x, y) = \nabla f(x, y)$.

Since the integral is independent of path, f depends only on x and y, not on the path C from (x_0, y_0) to (x, y). Choose a circle in D with center (x, y) and let (x_1, y) be a point within the circle such that $x_1 \neq x$. As illustrated in Figure 14.23, if C_1 is any path from (x_0, y_0) to (x_1, y) and C_2 is the horizontal segment from (x_1, y) to (x, y), then

Figure 14.23

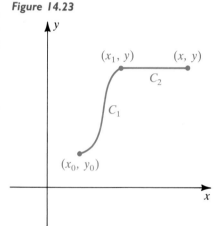

$$f(x, y) = \int_{C_1} \mathbf{F} \cdot d\mathbf{r} + \int_{C_2} \mathbf{F} \cdot d\mathbf{r} = \int_{(x_0, y_0)}^{(x_1, y)} \mathbf{F} \cdot d\mathbf{r} + \int_{(x_1, y)}^{(x, y)} \mathbf{F} \cdot d\mathbf{r}.$$

Since the first integral does not depend on x,

$$\frac{\partial}{\partial x} f(x, y) = 0 + \frac{\partial}{\partial x} \int_{(x_1, y)}^{(x, y)} \mathbf{F} \cdot d\mathbf{r}.$$

If we write $\mathbf{F} \cdot d\mathbf{r} = M(x, y)\, dx + N(x, y)\, dy$ and use the fact that $dy = 0$ on C_2 (see Figure 14.23), we obtain

$$\frac{\partial}{\partial x} f(x, y) = \frac{\partial}{\partial x} \int_{(x_1, y)}^{(x, y)} M(x, y)\, dx.$$

Since y is fixed in this partial differentiation, we may regard the integrand as a function of one variable x. Applying Theorem (4.35) gives us

$$\frac{\partial}{\partial x} f(x, y) = M(x, y).$$

Figure 14.24

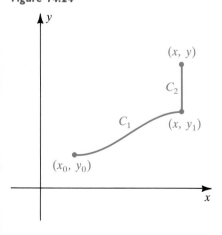

Similarly, if we choose the path shown in Figure 14.24 and differentiate with respect to y, we obtain

$$\frac{\partial}{\partial y} f(x, y) = N(x, y).$$

This proves that $\nabla f(x, y) = M(x, y)\mathbf{i} + N(x, y)\mathbf{j} = \mathbf{F}(x, y)$, which is what we wished to show.

Conversely, if there exists a function f such that $\mathbf{F}(x, y) = \nabla f(x, y)$, then

$$M(x, y)\mathbf{i} + N(x, y)\mathbf{j} = f_x(x, y)\mathbf{i} + f_y(x, y)\mathbf{j}$$

Hence, $M(x, y) = f_x(x, y)$ and $N(x, y) = f_y(x, y)$.

Figure 14.25

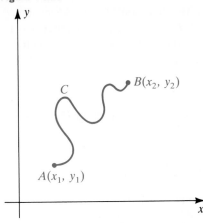

If $A(x_1, y_1)$ and $B(x_2, y_2)$ are points in the region D and if C is any piecewise-smooth curve with endpoints A and B (see Figure 14.25), then

$$\int_C \mathbf{F} \cdot d\mathbf{r} = \int_C M(x, y)\, dx + N(x, y)\, dy$$

$$= \int_C f_x(x, y)\, dx + f_y(x, y)\, dy.$$

If C has the smooth parametrization $x = g(t)$, $y = h(t)$; $t_1 \le t \le t_2$, then, by Theorem (14.9),

$$\int_C \mathbf{F} \cdot d\mathbf{r} = \int_{t_1}^{t_2} \Big[f_x(g(t), h(t)) g'(t) + f_y(g(t), h(t)) h'(t) \Big]\, dt.$$

Applying a chain rule and the fundamental theorem of calculus, we have

$$\int_C \mathbf{F} \cdot d\mathbf{r} = \int_{t_1}^{t_2} \frac{d}{dt} [f(g(t), h(t))]\, dt$$

$$= f(g(t_2), h(t_2)) - f(g(t_1), h(t_1))$$

$$= f(x_2, y_2) - f(x_1, y_1)$$

$$= \big[f(x, y) \big]_{(x_1, y_1)}^{(x_2, y_2)},$$

where the last bracket symbol is analogous to that used for functions of one variable. Thus, the line integral depends only on the coordinates of A and B, not on the path C—that is, $\int_C \mathbf{F} \cdot d\mathbf{r}$ is independent of path. The proof can be extended to piecewise-smooth curves by subdividing C into a finite number of smooth curves. ■

The proof of Theorem (14.13) contains a method for evaluating line integrals that are independent of path. We may state this result, which is analogous to the fundamental theorem of calculus, as follows.

Theorem   14.14

Let $\mathbf{F}(x, y) = M(x, y)\mathbf{i} + N(x, y)\mathbf{j}$ be continuous on an open connected region D, and let C be a piecewise-smooth curve in D with endpoints $A(x_1, y_1)$ and $B(x_2, y_2)$. If $\mathbf{F}(x, y) = \nabla f(x, y)$, then

$$\int_C M(x, y)\, dx + N(x, y)\, dy = \int_{(x_1, y_1)}^{(x_2, y_2)} \mathbf{F} \cdot d\mathbf{r} = \big[f(x, y) \big]_{(x_1, y_1)}^{(x_2, y_2)}.$$

If a line integral $\int_C \mathbf{F} \cdot d\mathbf{r}$ is independent of path, then using Theorem (14.14) with $(x_1, y_1) = (x_2, y_2)$, we see that $\int_C \mathbf{F} \cdot d\mathbf{r} = 0$ *for every simple closed curve C.*

EXAMPLE ■ 1   Let $\mathbf{F}(x, y) = (2x + y^3)\mathbf{i} + (3xy^2 + 4)\mathbf{j}$.

(a) Show that $\displaystyle\int_C \mathbf{F} \cdot d\mathbf{r}$ is independent of path.

(b) Evaluate $\displaystyle\int_{(0,1)}^{(2,3)} \mathbf{F} \cdot d\mathbf{r}$.

SOLUTION

(a) By Theorem (14.13), the line integral is independent of path if and only if there exists a differentiable function f of x and y such that $\nabla f(x, y) = \mathbf{F}(x, y)$—that is,

$$f_x(x, y)\mathbf{i} + f_y(x, y)\mathbf{j} = (2x + y^3)\mathbf{i} + (3xy^2 + 4)\mathbf{j},$$

or, equivalently,

(∗) $\qquad\qquad f_x(x, y) = 2x + y^3 \quad \text{and} \quad f_y(x, y) = 3xy^2 + 4.$

If we (partially) integrate $f_x(x, y) = 2x + y^3$ with respect to x, we obtain

$$f(x, y) = x^2 + xy^3 + g(y),$$

where g is a function of y alone. (We must use $g(y)$ instead of a constant of integration in order to obtain the most *general* expression for $f(x, y)$ such that $f_x(x, y) = 2x + y^3$.)

We next differentiate $f(x, y) = x^2 + xy^3 + g(y)$ with respect to y and compare the result with the expression $f_y(x, y) = 3xy^2 + 4$ in (∗). Thus,

$$f_y(x, y) = 0 + 3xy^2 + g'(y) = 3xy^2 + 4$$

and hence $\qquad\qquad\qquad g'(y) = 4.$

Integrating with respect to y, we obtain

$$g(y) = 4y + c,$$

where c is a constant of integration. Thus,

$$f(x, y) = x^2 + xy^3 + 4y + c$$

gives us a function f such that $\nabla f(x, y) = \mathbf{F}(x, y)$.

(b) Since any potential function may be used to evaluate the integral, we apply Theorem (14.14) with $f(x, y) = x^2 + xy^3 + 4y$, obtaining

$$\int_{(0,1)}^{(2,3)} \mathbf{F} \cdot d\mathbf{r} = \left[f(x, y) \right]_{(0,1)}^{(2,3)} = \left[x^2 + xy^3 + 4y \right]_{(0,1)}^{(2,3)}$$

$$= (4 + 54 + 12) - 4 = 66.$$

The following result is an application of the preceding discussion to conservative vector fields.

Theorem 14.15

If $\mathbf{F}$ is a conservative force field in two dimensions, then the work done by $\mathbf{F}$ along any path C from $A(x_1, y_1)$ to $B(x_2, y_2)$ is equal to the difference in potentials between A and B.

PROOF If $\mathbf{F}$ is conservative, then $\mathbf{F}(x, y) = \nabla f(x, y)$ for some potential function f. By Definition (14.12) and Theorem (14.14), the work

W done along any path from $A(x_1, y_1)$ to $B(x_2, y_2)$ is

$$W = \int_{(x_1, y_1)}^{(x_2, y_2)} \mathbf{F} \cdot d\mathbf{r} = f(x_2, y_2) - f(x_1, y_1),$$

which is the difference in the potentials between A and B.   ■

Figure 14.26

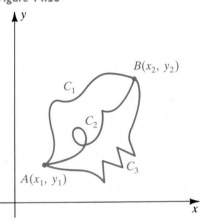

The preceding theorem is illustrated in Figure 14.26 for piecewise-smooth curves C_1, C_2, and C_3. The same amount of work is done no matter which path is taken from A to B. If C is a closed curve—that is, if $A = B$—then the difference in potentials is 0, and hence the work done in traversing C is 0. A converse of this result is also true—namely, if $\int_C \mathbf{F} \cdot d\mathbf{r} = 0$ for every simple closed curve C, then the line integral is independent of path and hence the field is conservative. These facts are important in applications, since many of the vector fields that occur in nature are inverse square fields and hence are conservative (see Theorem (14.4)). In physical terms, if a unit particle moves completely around a closed curve in a conservative force field, then the work done is 0. We will discuss conservative fields in more detail at the end of this section.

Theorems (14.14) and (14.15) can be extended to a vector field in three dimensions. The next example is an illustration of these results.

EXAMPLE ■ 2   Let $\mathbf{F}$ be the gravitational field of a particle of mass m_0 located at the origin of an xyz-coordinate system. Find the work done by $\mathbf{F}$ as a particle of mass m moves from $A(2, 3, 4)$ to $B(1, 0, 0)$.

SOLUTION   From page 1170, the force exerted on a particle of mass m located at $K(x, y, z)$ is

$$\mathbf{F}(x, y, z) = -G \frac{m_0 m}{\|\mathbf{r}\|^3} \mathbf{r},$$

where $\mathbf{r} = x\mathbf{i} + y\mathbf{j} + z\mathbf{k}$. As in the proof of Theorem (14.4), $\mathbf{F}(x, y, z) = \nabla f(x, y, z)$, where

$$f(x, y, z) = \frac{Gm_0 m}{(x^2 + y^2 + z^2)^{1/2}}.$$

Hence, by the extension of Theorems (14.14) and (14.15) to three dimensions,

$$W = \int_A^B \mathbf{F} \cdot d\mathbf{r} = \left[\frac{Gm_0 m}{(x^2 + y^2 + z^2)^{1/2}} \right]_{(2,3,4)}^{(1,0,0)} = Gm_0 m \left(1 - \frac{1}{\sqrt{29}} \right).$$

If the integral $\int_C M(x, y)\, dx + N(x, y)\, dy$ is independent of path, then, by Theorem (14.13), there is a function f such that

$$M = \frac{\partial f}{\partial x} \quad \text{and} \quad N = \frac{\partial f}{\partial y}.$$

Consequently, $\dfrac{\partial M}{\partial y} = \dfrac{\partial^2 f}{\partial y\, \partial x} \quad \text{and} \quad \dfrac{\partial N}{\partial x} = \dfrac{\partial^2 f}{\partial x\, \partial y}.$

If M and N have continuous first partial derivatives, then f has continuous *second* partial derivatives, and hence the order of differentiation is immaterial—that is,

$$\frac{\partial M}{\partial y} = \frac{\partial N}{\partial x}.$$

The converse of this result is false, unless we place additional restrictions on the domain D of $\mathbf{F}(x, y)$. In particular, if D is a **simply connected region** in the sense that every simple closed curve C in D encloses only points in D (there are no *holes* in the region), then the condition $\partial M/\partial y = \partial N/\partial x$ implies that the line integral $\int_C M(x, y)\,dx + N(x, y)\,dy$ is independent of path. (A proof may be found in books on advanced calculus.) Our discussion can be summarized as follows.

Theorem 14.16

If $M(x, y)$ and $N(x, y)$ have continuous first partial derivatives on a simply connected region D, then the line integral

$$\int_C M(x, y)\,dx + N(x, y)\,dy$$

is independent of path in D if and only if

$$\frac{\partial M}{\partial y} = \frac{\partial N}{\partial x}.$$

EXAMPLE ■ 3 Show that the line integral

$$\int_C (e^{3y} - y^2 \sin x)\,dx + (3xe^{3y} + 2y \cos x)\,dy$$

is independent of path in a simply connected region.

SOLUTION Letting

$$M = e^{3y} - y^2 \sin x \quad \text{and} \quad N = 3xe^{3y} + 2y \cos x,$$

we see that $\quad \dfrac{\partial M}{\partial y} = 3e^{3y} - 2y \sin x = \dfrac{\partial N}{\partial x}.$

Hence, by Theorem (14.16), the line integral is independent of path.

EXAMPLE ■ 4 Determine whether $\int_C x^2 y\,dx + 3xy^2\,dy$ is independent of path.

SOLUTION If we let $M = x^2 y$ and $N = 3xy^2$, then

$$\frac{\partial M}{\partial y} = x^2 \quad \text{and} \quad \frac{\partial N}{\partial x} = 3y^2.$$

Since $\partial M/\partial y \neq \partial N/\partial x$, the integral is not independent of path, by Theorem (14.16).

In Example 1, we determined a potential function f for a two-dimensional conservative vector field. The solution of the next example illustrates a similar technique that can be used in three dimensions.

EXAMPLE • 5   Let

$$\mathbf{F}(x, y, z) = y^2 \cos x \mathbf{i} + (2y \sin x + e^{2z})\mathbf{j} + 2ye^{2z}\mathbf{k}.$$

(a) Show that $\int_C \mathbf{F} \cdot d\mathbf{r}$ is independent of path, and find a potential function f for $\mathbf{F}$.

(b) If $\mathbf{F}$ is a force field, find the work done by $\mathbf{F}$ along any curve C from $(0, 1, \frac{1}{2})$ to $(\pi/2, 3, 2)$.

SOLUTION

(a) The integral is independent of path if there exists a differentiable function f of x, y, and z such that $\nabla f(x, y, z) = \mathbf{F}(x, y, z)$—that is,

$$f_x(x, y, z) = y^2 \cos x$$

(1)
$$f_y(x, y, z) = 2y \sin x + e^{2z}$$

$$f_z(x, y, z) = 2ye^{2z}.$$

If we (partially) integrate $f_x(x, y, z)$ with respect to x, we see that

(2)
$$f(x, y, z) = y^2 \sin x + g(y, z)$$

for some function g of y and z. (We must use $g(y, z)$ in order to obtain the *most general* expression $f(x, y, z)$ such that $f_x(x, y, z) = y^2 \cos x$.)

Differentiating $f(x, y, z)$ in (2) with respect to y and then comparing the result with the formula for $f_y(x, y, z)$ in (1), we obtain

$$f_y(x, y, z) = 2y \sin x + g_y(y, z) = 2y \sin x + e^{2z}$$

and hence
$$g_y(y, z) = e^{2z}.$$

A partial integration of $g_y(y, z)$ with respect to y gives us

$$g(y, z) = ye^{2z} + h(z),$$

where h is a function of the one variable z. Using formula (2) yields

$$f(x, y, z) = y^2 \sin x + ye^{2z} + h(z).$$

Differentiating the preceding expression with respect to z and then comparing the result with the formula for $f_z(x, y, z)$ in (1), we have

$$f_z(x, y, z) = 2ye^{2z} + h'(z) = 2ye^{2z}.$$

It follows that $h'(z) = 0$ or $h(z) = c$ for some constant c. Thus,

$$f(x, y, z) = y^2 \sin x + ye^{2z} + c$$

gives us a potential function f for $\mathbf{F}$.

(b) Using the extension of Theorem (14.15) to vector fields in three dimensions, we find that the work done by **F** along any curve C from $(0, 1, \frac{1}{2})$ to $(\pi/2, 3, 2)$ is

$$\int_C \mathbf{F} \cdot d\mathbf{r} = \left[y^2 \sin x + y e^{2z} \right]_{(0,1,1/2)}^{(\pi/2,3,2)}$$

$$= (9 + 3e^4) - (0 + e) = 9 + 3e^4 - e \approx 170.08.$$

We shall conclude this section by considering some applications of independence of path. The next definition is used in physics.

Definition 14.17

Let $\mathbf{F}(x, y, z)$ be a conservative vector force field with potential function f. The **potential energy** $p(x, y, z)$ of a particle at the point (x, y, z) is

$$p(x, y, z) = -f(x, y, z).$$

Since the potential energy is the *negative* of the potential $f(x, y, z)$,

$$\mathbf{F}(x, y, z) = -\nabla p(x, y, z).$$

If A and B are two points, then, as in Theorem (14.15), the work done by **F** along a smooth curve C with endpoints A and B is

$$W = \int_A^B \mathbf{F} \cdot d\mathbf{r} = \left[-p(x, y, z) \right]_A^B = p(A) - p(B),$$

where $p(A)$ and $p(B)$ denote the potential energy at A and B, respectively. Thus, the work W is the difference in the potential energies at A and B. In particular, if B is a point at which the potential is 0, then $W = p(A)$. This is in agreement with the classical physical description of potential energy as the type of energy that a body has by virtue of its position.

If a particle has mass m and speed v, then its *kinetic energy* is $\frac{1}{2}mv^2$. Thus, kinetic energy is the energy that a body has by virtue of its speed and mass.

Let us establish the following fundamental physical law, which is the principal reason for referring to certain vector fields as *conservative*.

Law of Conservation of Energy 14.18

If a particle moves from one point to another in a conservative vector force field, then the sum of the potential and kinetic energies remains constant—that is, the total energy does not change.

PROOF Let **F** be a conservative vector field, and suppose that a particle moves from A to B along a piecewise-smooth curve C such that its position at time t is given by $x = g(t), y = h(t), z = k(t); a \leq t \leq b$. If

Figure 14.27

$\mathbf{r} = x\mathbf{i} + y\mathbf{j} + z\mathbf{k}$ is the position vector for the particle (see Figure 14.27), then the velocity and the acceleration at time t are

$$\mathbf{v} = \frac{d\mathbf{r}}{dt} \quad \text{and} \quad \mathbf{a} = \frac{d\mathbf{v}}{dt},$$

where we have used the abbreviations $\mathbf{v}$ and $\mathbf{a}$ for $\mathbf{v}(t)$ and $\mathbf{a}(t)$, respectively. The speed of the particle is $v(t) = v = \|\mathbf{v}\|$. The work done by $\mathbf{F}(x, y, z)$ along C is

$$W = \int_A^B \mathbf{F} \cdot d\mathbf{r} = \int_a^b \left(\mathbf{F} \cdot \frac{d\mathbf{r}}{dt} \right) dt = \int_a^b (\mathbf{F} \cdot \mathbf{v}) \, dt.$$

By Newton's second law of motion (11.13),

$$\mathbf{F} = m\mathbf{a} = m\frac{d\mathbf{v}}{dt}$$

and hence

$$\begin{aligned} W &= \int_a^b \left(m\frac{d\mathbf{v}}{dt} \cdot \mathbf{v} \right) dt \\ &= m \int_a^b \left(\frac{d\mathbf{v}}{dt} \cdot \mathbf{v} \right) dt = m \int_a^b \frac{1}{2}\frac{d}{dt} (\mathbf{v} \cdot \mathbf{v}) \, dt \\ &= \frac{1}{2}m \int_a^b \frac{d}{dt} (v^2) \, dt = \frac{1}{2}m \left[v^2 \right]_a^b \\ &= \frac{1}{2}m \, [v(b)]^2 - \frac{1}{2}m \, [v(a)]^2. \end{aligned}$$

Since the kinetic energy K of a particle of mass m and velocity v is $\frac{1}{2}mv^2$, the last equality may be written

$$W = K(B) - K(A),$$

where $K(A)$ and $K(B)$ denote the kinetic energy of the particle at A and B, respectively. From the discussion following Definition (14.17), $W = p(A) - p(B)$, and hence

$$p(A) - p(B) = K(B) - K(A),$$

or

$$p(A) + K(A) = p(B) + K(B).$$

The last formula states that if a particle moves from one point to another in a conservative vector field, then the sum of the potential and kinetic energies is constant. ■

EXERCISES 14.3

Exer. 1–10: Show that $\int_C \mathbf{F} \cdot d\mathbf{r}$ is independent of path by finding a potential function f for **F**.

1 $\mathbf{F}(x, y) = (3x^2y + 2)\mathbf{i} + (x^3 + 4y^3)\mathbf{j}$

2 $\mathbf{F}(x, y) = (6xy^2 + 2y)\mathbf{i} + (6x^2y + 2x)\mathbf{j}$

3 $\mathbf{F}(x, y) = (2x \sin y + 4e^x)\mathbf{i} + x^2 \cos y\mathbf{j}$

4 $\mathbf{F}(x, y) = (2xe^{2y} + 4y^3)\mathbf{i} + (2x^2e^{2y} + 12xy^2)\mathbf{j}$

5 $\mathbf{F}(x, y) = -2y^3 \sin x\mathbf{i} + (6y^2 \cos x + 5)\mathbf{j}$

6 $\mathbf{F}(x, y) = (5y^3 + 4y^3 \sec^2 x)\mathbf{i} + (15xy^2 + 12y^2 \tan x)\mathbf{j}$

7 $F(x, y, z) = 8xz\mathbf{i} + (1 - 6yz^3)\mathbf{j} + (4x^2 - 9y^2z^2)\mathbf{k}$

8 $F(x, y, z) = (y + z)\mathbf{i} + (x + z)\mathbf{j} + (x + y)\mathbf{k}$

9 $F(x, y, z) = (y \sec^2 x - ze^x)\mathbf{i} + \tan x\mathbf{j} - e^x\mathbf{k}$

10 $F(x, y, z) =$
$$2x \sin z\mathbf{i} + 2y \cos z\mathbf{j} + (x^2 \cos z - y^2 \sin z)\mathbf{k}$$

Exer. 11–14: Show that the line integral is independent of path, and find its value.

11 $\displaystyle\int_{(-1, 2)}^{(3, 1)} (y^2 + 2xy)\,dx + (x^2 + 2xy)\,dy$

12 $\displaystyle\int_{(0, 0)}^{(1, \pi/2)} e^x \sin y\,dx + e^x \cos y\,dy$

13 $\displaystyle\int_{(1, 0, 2)}^{(-2, 1, 3)} (6xy^3 + 2z^2)\,dx + 9x^2y^2\,dy + (4xz + 1)\,dz$

14 $\displaystyle\int_{(4, 0, 3)}^{(-1, 1, 2)} (yz + 1)\,dx + (xz + 1)\,dy + (xy + 1)\,dz$

Exer. 15–18: Use Theorem (14.16) to show that $\int_C F \cdot d\mathbf{r}$ is not independent of path.

15 $F(x, y) = 4xy^3\mathbf{i} + 2xy^3\mathbf{j}$

16 $F(x, y) = (6x^2 - 2xy^2)\mathbf{i} + (2x^2y + 5)\mathbf{j}$

17 $F(x, y) = e^x\mathbf{i} + (3 - e^x \sin y)\mathbf{j}$

18 $F(x, y) = y^3 \cos x\mathbf{i} - 3y^2 \sin x\mathbf{j}$

19 If $\int_C M(x, y, z)\,dx + N(x, y, z)\,dy + P(x, y, z)\,dz$ is independent of path and M, N, and P have continuous first partial derivatives, prove that
$$\frac{\partial M}{\partial y} = \frac{\partial N}{\partial x}, \qquad \frac{\partial M}{\partial z} = \frac{\partial P}{\partial x}, \qquad \frac{\partial N}{\partial z} = \frac{\partial P}{\partial y}.$$

Exer. 20–22: Use Exercise 19 to show that the line integral is not independent of path.

20 $\displaystyle\int_C 5y\,dx + 5x\,dy + yz^2\,dz$

21 $\displaystyle\int_C 2xy\,dx + (x^2 + z^2)\,dy + yz\,dz$

22 $\displaystyle\int_C e^y \cos z\,dx + xe^y \cos z\,dy + xe^y \sin z\,dz$

23 Suppose a force $F(x, y, z)$ is directed toward the origin with a magnitude that is inversely proportional to the distance from the origin. Prove that F is conservative by finding a potential function f for F.

24 Suppose a force $F(x, y, z)$ is directed away from the origin with a magnitude that is directly proportional to the distance from the origin. Prove that F is conservative by finding a potential function f for F.

25 If F is a constant force, prove that the work done along any curve with endpoints P and Q is $F \cdot \overrightarrow{PQ}$.

26 If $F(x, y, z) = g(x)\mathbf{i} + h(y)\mathbf{j} + k(z)\mathbf{k}$, where g, h, and k are continuous functions, show that F is conservative.

27 Let $F(x, y) = M(x, y)\mathbf{i} + N(x, y)\mathbf{j}$, where
$$M(x, y) = \frac{-y}{x^2 + y^2} \quad \text{and} \quad N(x, y) = \frac{x}{x^2 + y^2}.$$
Show that $\partial M/\partial y = \partial N/\partial x$ for every (x, y) in the domain D of F but that $\int_C F \cdot d\mathbf{r}$ is not independent of path in D. (*Hint:* Consider two different semicircles with endpoints $(-1, 0)$ and $(1, 0)$.) Why doesn't this contradict Theorem (14.16)?

28 Suppose that a satellite of mass m is orbiting the earth at a constant altitude of h miles and that it completes one revolution every k minutes. If the radius of the earth is taken as 4000 mi, find the work done by the gravitational field of the earth during any time interval.

29 Let F be an inverse square field such that $F(x, y, z) = (c/\|\mathbf{r}\|^3)\mathbf{r}$ for $\mathbf{r} = x\mathbf{i} + y\mathbf{j} + z\mathbf{k}$, where c is a constant. Let P_1 and P_2 be any points whose distances from the origin are d_1 and d_2, respectively. Express, in terms of d_1 and d_2, the work done by F along any piecewise-smooth curve joining P_1 to P_2.

30 Suppose that as a particle moves through a conservative vector field, its potential energy is decreasing at a rate of k units per second. At what rate is its kinetic energy changing?

Mathematicians and Their Times

DOROTHY BERNSTEIN

THE MATHEMATICS COMMUNITY has suffered a discrimination that has often afflicted society: prejudice against women. As noted earlier, Hypatia was murdered and Kovalevskaya was denied university admission and professorships. In our own century, Dorothy Bernstein overcame both sexism and anti-Semitism to achieve distinction in mathematics.

Born in Chicago to Russian immigrants, Dorothy Lewis Bernstein (1914–1988) grew up in Milwaukee. At the University of Wisconsin's experimental college, she spent her junior and senior years entirely in independent study. She graduated first in her class, earning both B.A. and M.A. degrees simultaneously. Her master's thesis concerned extending Newton's method to find complex roots

of polynomials. Although her professors urged her to pursue graduate work, they suggested she should broaden her training at another university.

"So I came to Brown in 1935," Bernstein later wrote, "and here became aware for the first time that some people made a distinction between men and women in mathematics." She and a male graduate student, Hugh Hamilton, were assigned to teach the same course. Bernstein was given all the women students (3) and Hamilton all the men (45). Hamilton and Bernstein suggested making two sections of 24 students each, but the department chair refused, saying that "Brown boys would not stand being taught by a woman instructor."

When Bernstein went to see Brown's graduate school dean, who was also secretary of the American Mathematical Society, for advice on getting her first full-time teaching job, he took out a map of the United States, covered the region west of the Mississippi, and said: "You can't get a job there, because you are a woman." Then he covered the part south of the Ohio River and said: "And you can't get a job there, because you are Jewish." Bernstein finally secured a position at Mt. Holyoke College, where she taught for three years, completing her Ph.D. thesis on multiple Laplace transformations in 1939. She then

14.4 GREEN'S THEOREM

Green's theorem, which we investigate in this section, enables us to evaluate certain line integrals around the boundary of a plane region by evaluating a double integral over the region. Let C be a smooth plane curve with parametrization $x = g(t)$, $y = h(t)$; $a \le t \le b$. Recall that if $A = (g(a), h(a)) = (g(b), h(b)) = B$, then C is a smooth *closed* curve. If C does not cross itself between A and B, it is *simple*. Some common examples of smooth simple closed curves are circles and ellipses. A *piecewise-smooth simple closed curve* is a finite union of smooth curves C_k such that as t varies from a to b, the point $P(t)$ obtained from parametrizations of the C_k traces C exactly once, with the exception that $P(a) = P(b)$. A curve of this type forms the boundary of a region R in the xy-plane, and, by definition, the **orientation**, or **positive direction**, along C is such that R is on the left as $P(t)$ traces C, as illustrated by the arrowheads in Figure 14.28. The notation

Figure 14.28

$$\oint_C M(x, y)\, dx + N(x, y)\, dy$$

denotes a line integral along a simple closed curve C once in the positive direction.

The next theorem, named after the English mathematical physicist George Green (1793–1841), specifies a relationship between the line integral around C and a double integral over R. To simplify the statement, the symbols M, N, $\partial N/\partial x$, and $\partial M/\partial y$ are used in the integrands to denote the values of these functions at (x, y).

Green's Theorem   14.19

Let C be a piecewise-smooth simple closed curve, and let R be the region consisting of C and its interior. If M and N are continuous functions that have continuous first partial derivatives throughout an open region D containing R, then

$$\oint_C M\,dx + N\,dy = \iint_R \left(\frac{\partial N}{\partial x} - \frac{\partial M}{\partial y} \right) dA.$$

Figure 14.29

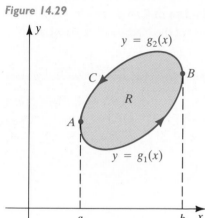

PROOF FOR A SPECIAL CASE   We shall prove the theorem if R is *both* an R_x region and an R_y region of the following types:

$$R = \{(x, y): a \le x \le b, g_1(x) \le y \le g_2(x)\}$$
$$R = \{(x, y): c \le y \le d, h_1(y) \le x \le h_2(y)\},$$

where g_1, g_2, h_1, and h_2 are smooth functions. It is sufficient to show that each of the following is true, since addition of these integrals produces the desired conclusion:

(1)
$$\oint_C M\,dx = -\iint_R \frac{\partial M}{\partial y}\,dA$$

(2)
$$\oint_C N\,dy = \iint_R \frac{\partial N}{\partial x}\,dA$$

To prove (1), we refer to Figure 14.29 and note that C consists of two curves C_1 and C_2 that have equations $y = g_1(x)$ and $y = g_2(x)$, respectively. The line integral $\oint_C M\,dx$ may be written

$$\oint_C M\,dx = \int_{C_1} M(x, y)\,dx + \int_{C_2} M(x, y)\,dx$$

$$= \int_a^b M(x, g_1(x))\,dx + \int_b^a M(x, g_2(x))\,dx$$

$$= \int_a^b M(x, g_1(x))\,dx - \int_a^b M(x, g_2(x))\,dx.$$

Figure 14.30

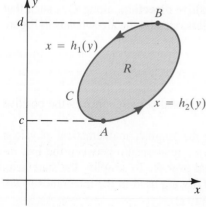

We next apply Theorem (13.8)(i) to the double integral $\iint_R (\partial M/\partial y)\,dA$, obtaining

$$\iint_R \frac{\partial M}{\partial y}\,dA = \int_a^b \int_{g_1(x)}^{g_2(x)} \frac{\partial M}{\partial y}\,dy\,dx$$

$$= \int_a^b \Big[M(x, y) \Big]_{g_1(x)}^{g_2(x)}\,dx$$

$$= \int_a^b \Big[M(x, g_2(x)) - M(x, g_1(x)) \Big]\,dx.$$

Comparing the last expression with that obtained for $\oint_C M\,dx$ gives us (1). The formula in (2) may be established in similar fashion by referring to Figure 14.30. ∎

Figure 14.31

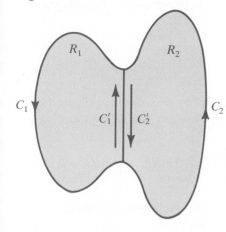

Although we shall not prove it, Green's theorem may be extended to a region R where part of the boundary consists of horizontal or vertical line segments. The theorem may then be further extended to the case where R is a finite union of such regions. For example, if $R = R_1 \cup R_2$ and if the boundary of R_1 is $C_1 \cup C_1'$ and the boundary of R_2 is $C_2 \cup C_2'$, as illustrated in Figure 14.31, then

$$\iint_{R_1} \left(\frac{\partial N}{\partial x} - \frac{\partial M}{\partial y} \right) dy\, dx = \oint_{C_1 \cup C_1'} M\, dx + N\, dy$$

$$\iint_{R_2} \left(\frac{\partial N}{\partial x} - \frac{\partial M}{\partial y} \right) dy\, dx = \oint_{C_2 \cup C_2'} M\, dx + N\, dy.$$

The sum of the two double integrals equals a double integral over R. We next observe that a line integral along C_1' in the direction indicated in Figure 14.31 is the negative of that along C_2', since the curve is the same but the direction is opposite. Hence the sum of the two line integrals reduces to a line integral along $C_1 \cup C_2$, which is the boundary C of R. Thus,

$$\iint_R \left(\frac{\partial N}{\partial x} - \frac{\partial M}{\partial y} \right) dy\, dx = \oint_C M\, dx + N\, dy.$$

We may now proceed, by mathematical induction, to any finite union. The proof of Green's theorem for the most general case is beyond the scope of this book.

EXAMPLE ■ 1   Use Green's theorem to evaluate $\oint_C 5xy\, dx + x^3\, dy$, where C is the closed curve consisting of the graphs of $y = x^2$ and $y = 2x$ between the points $(0, 0)$ and $(2, 4)$.

Figure 14.32

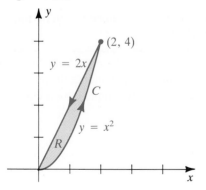

SOLUTION   The region R bounded by C is shown in Figure 14.32. Applying Green's theorem, with $M(x, y) = 5xy$ and $N(x, y) = x^3$, gives us

$$\oint_C 5xy\, dx + x^3\, dy = \iint_R \left[\frac{\partial}{\partial x}(x^3) - \frac{\partial}{\partial y}(5xy) \right] dA$$

$$= \int_0^2 \int_{x^2}^{2x} (3x^2 - 5x)\, dy\, dx$$

$$= \int_0^2 \left[3x^2 y - 5xy \right]_{x^2}^{2x} dx$$

$$= \int_0^2 (11x^3 - 10x^2 - 3x^4)\, dx = -\tfrac{28}{15}.$$

The line integral can also be evaluated directly.

Figure 14.33

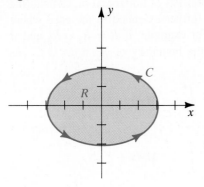

Figure 14.33

EXAMPLE■2 Use Green's theorem to evaluate the line integral

$$\oint_C 2xy\,dx + (x^2 + y^2)\,dy$$

if C is the ellipse $4x^2 + 9y^2 = 36$.

SOLUTION The region R bounded by C is shown in Figure 14.33. Applying Green's theorem, with $M(x, y) = 2xy$ and $N(x, y) = x^2 + y^2$, we have

$$\oint_C 2xy\,dx + (x^2 + y^2)\,dy = \iint_R (2x - 2x)\,dA = \iint_R 0\,dA = 0.$$

Note that we did not use the curve C in the evaluation. We can show that the line integral is independent of path and hence is zero for *every* simple closed curve C (see the remark following Theorem (14.14)).

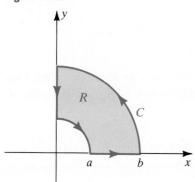

Figure 14.34

EXAMPLE■3 Evaluate $\oint_C (4 + e^{\cos x})\,dx + (\sin y + 3x^2)\,dy$ if C is the boundary of the region R between quarter-circles of radii a and b and segments on the x- and y-axes, as shown in Figure 14.34.

SOLUTION If we apply Green's theorem and then change the double integral to polar coordinates, the line integral equals

$$\iint_R (6x - 0)\,dA = \int_0^{\pi/2} \int_a^b (6r\cos\theta)r\,dr\,d\theta$$

$$= 6\int_0^{\pi/2} \cos\theta \left[\tfrac{1}{3}r^3\right]_a^b d\theta = 2(b^3 - a^3)\int_0^{\pi/2} \cos\theta\,d\theta$$

$$= 2(b^3 - a^3)\left[\sin\theta\right]_0^{\pi/2} = 2(b^3 - a^3).$$

A deeper appreciation of Green's theorem can be gained by attempting to evaluate the line integral in this example directly.

Green's theorem may be used to derive a formula for finding the area A of a region R that is bounded by a piecewise-smooth simple closed curve C. If we let $M = 0$ and $N = x$ in (14.19), we obtain

$$A = \iint_R dA = \oint_C x\,dy.$$

However, if we let $M = -y$ and $N = 0$, the result is

$$A = \iint_R dA = -\oint_C y\,dx.$$

We may combine these two formulas for A by adding both sides of the equations and dividing by 2. This gives us the third formula in the next theorem.

Theorem 14.20

If a region R in the xy-plane is bounded by a piecewise-smooth simple closed curve C, then the area A of R is

$$A = \oint_C x\,dy = -\oint_C y\,dx = \tfrac{1}{2}\oint_C x\,dy - y\,dx.$$

Although the first two formulas in (14.20) appear to be easier to apply, for certain curves the third formula leads to a simpler integration. If you doubt this, try to find the area in Exercise 20 by using either $\oint_C x\,dy$ or $-\oint_C y\,dx$.

EXAMPLE ■ 4 Use Theorem (14.20) to find the area of the ellipse $(x^2/a^2) + (y^2/b^2) = 1$.

SOLUTION Parametric equations for the ellipse C are $x = a\cos t$, $y = b\sin t$; $0 \le t \le 2\pi$. Applying the third formula in Theorem (14.20) yields

$$
\begin{aligned}
A &= \tfrac{1}{2}\oint_C x\,dy - y\,dx \\
&= \tfrac{1}{2}\int_0^{2\pi} (a\cos t)(b\cos t)\,dt - (b\sin t)(-a\sin t)\,dt \\
&= \tfrac{1}{2}\int_0^{2\pi} ab(\cos^2 t + \sin^2 t)\,dt \\
&= \tfrac{1}{2}ab\int_0^{2\pi} dt = \tfrac{1}{2}ab(2\pi) = \pi ab.
\end{aligned}
$$

Figure 14.35

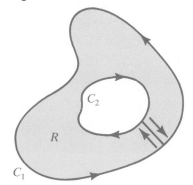

Green's theorem can be extended to a region R that contains holes, provided we integrate over the *entire* boundary and always keep the region R to the left of C. This is illustrated by the region in Figure 14.35, where the double integral over R equals the sum of the line integrals along C_1 and C_2 in the indicated directions. The proof consists of making a slit in R as illustrated in the figure, then noting that the sum of two line integrals in opposite directions along the same curve is zero. A similar argument can be used if the region has several holes. We shall use this observation in the solution of the next example.

EXAMPLE ■ 5 Let C_1 and C_2 be two nonintersecting piecewise-smooth simple closed curves, each having the origin O as an interior point. If

$$M = \frac{-y}{x^2 + y^2}, \qquad N = \frac{x}{x^2 + y^2},$$

prove that

$$\oint_{C_1} M\,dx + N\,dy = \oint_{C_2} M\,dx + N\,dy.$$

SOLUTION If R denotes the region between C_1 and C_2, then we have a situation similar to that illustrated in Figure 14.35, with O inside

C_2. By the remarks preceding this example, Green's theorem gives us

$$\oint_{C_1} M\,dx + N\,dy + \oint_{C_2} M\,dx + N\,dy = \iint_R \left(\frac{\partial N}{\partial x} - \frac{\partial M}{\partial y}\right)dA,$$

where we have used the symbol $\circlearrowright$ to indicate the positive direction along C_2 *with respect to* R. Since

$$\frac{\partial N}{\partial x} = \frac{(x^2 + y^2)(1) - x(2x)}{(x^2 + y^2)^2} = \frac{y^2 - x^2}{(x^2 + y^2)^2} = \frac{\partial M}{\partial y},$$

the double integral over R is zero. Consequently,

$$\oint_{C_1} M\,dx + N\,dy = -\oint_{C_2} M\,dx + N\,dy.$$

Since the positive direction along C_2 *with respect to the region in the interior of* C_2 is opposite to that indicated in the preceding integral, we obtain

$$\oint_{C_1} M\,dx + N\,dy = \oint_{C_2} M\,dx + N\,dy,$$

which is what we wished to prove.

EXAMPLE • 6 If **F** is defined by

$$\mathbf{F}(x, y) = \frac{1}{x^2 + y^2}(-y\mathbf{i} + x\mathbf{j}),$$

prove that $\oint_C \mathbf{F} \cdot d\mathbf{r} = 2\pi$ for every piecewise-smooth simple closed curve C having the origin in its interior.

Figure 14.36

SOLUTION If we let $\mathbf{F}(x, y) = M\mathbf{i} + N\mathbf{j}$, then M and N are the same as in Example 5. Hence, if we choose a circle C_1 of radius a with center at the origin that lies entirely within C (see Figure 14.36), then, by Example 5,

$$\oint_C \mathbf{F} \cdot d\mathbf{r} = \oint_{C_1} \mathbf{F} \cdot d\mathbf{r}.$$

Since parametric equations for C_1 are

$$x = a\cos t, \quad y = a\sin t; \quad 0 \le t \le 2\pi,$$

we obtain

$$\oint_C \mathbf{F} \cdot d\mathbf{r} = \oint_{C_1} \frac{-y}{x^2 + y^2}\,dx + \frac{x}{x^2 + y^2}\,dy$$

$$= \int_0^{2\pi} \frac{-a\sin t}{a^2}(-a\sin t)\,dt + \frac{a\cos t}{a^2}(a\cos t)\,dt$$

$$= \int_0^{2\pi} (\sin^2 t + \cos^2 t)\,dt = \int_0^{2\pi} dt = 2\pi.$$

The conclusion of Green's theorem (14.19) may be expressed in vector form. First note that if $\mathbf{F}$ is a vector field in two dimensions, then we may write

$$\mathbf{F}(x, y) = M\mathbf{i} + N\mathbf{j} + 0\mathbf{k},$$

where $M = M(x, y)$ and $N = N(x, y)$. The curl of $\mathbf{F}$ is

$$\nabla \times \mathbf{F} = \begin{vmatrix} \mathbf{i} & \mathbf{j} & \mathbf{k} \\ \dfrac{\partial}{\partial x} & \dfrac{\partial}{\partial y} & \dfrac{\partial}{\partial z} \\ M & N & 0 \end{vmatrix} = 0\mathbf{i} + 0\mathbf{j} + \left(\dfrac{\partial N}{\partial x} - \dfrac{\partial M}{\partial y} \right) \mathbf{k}.$$

The coefficient of $\mathbf{k}$ has the same form as the integrand of the double integral in Green's theorem (14.19). Let s be an arc length parameter for C, and consider the unit tangent vector

$$\mathbf{T} = \frac{dx}{ds}\mathbf{i} + \frac{dy}{ds}\mathbf{j} + \frac{dz}{ds}\mathbf{k}.$$

Using this notation, we may write Green's theorem as follows.

Vector Form of Green's Theorem 14.21

$$\oint_C \mathbf{F} \cdot \mathbf{T}\, ds = \iint_R (\nabla \times \mathbf{F}) \cdot \mathbf{k}\, dA$$

Since $(\nabla \times \mathbf{F}) \cdot \mathbf{k}$ is the component of curl $\mathbf{F}$ in the direction parallel to the z-axis, we refer to it as the **normal component** (to R) of curl $\mathbf{F}$. In words, (14.21) may be phrased: *The line integral of the tangential component of* $\mathbf{F}$ *taken along C once in the positive direction is equal to the double integral over R of the normal component of* curl $\mathbf{F}$. The three-dimensional analogue of this result is *Stokes's theorem*, which will be discussed in Section 14.7. We will also discuss physical interpretations of curl $\mathbf{F}$ at that time.

Green's theorem is important in the area of mathematics known as *complex variables*, a subject that is fundamental for many applications in the physical sciences and engineering.

EXERCISES 14.4

Exer. 1–14: Use Green's theorem to evaluate the line integral.

1 $\oint_C (x^2 + y)\, dx + (xy^2)\, dy$;

  C is the closed curve determined by $y^2 = x$ and $y = -x$ with $0 \leq x \leq 1$.

2 $\oint_C (x + y^2)\, dx + (1 + x^2)\, dy$;

  C is the closed curve determined by $y = x^3$ and $y = x^2$ with $0 \leq x \leq 1$.

3 $\oint_C x^2 y^2\, dx + (x^2 - y^2)\, dy$;

  C is the square with vertices $(0, 0), (1, 0), (1, 1), (0, 1)$.

4 $\oint_C \sqrt{y}\, dx + \sqrt{x}\, dy$;

  C is the triangle with vertices $(1, 1), (3, 1), (2, 2)$.

5 $\oint_C xy\, dx + (y + x)\, dy$;

  C is the circle $x^2 + y^2 = 1$.

6 $\oint_C y^2\, dx + x^2\, dy$;

C is the boundary of the region bounded by the semicircle $y = \sqrt{4 - x^2}$ and the x-axis.

7 $\oint_C xy\, dx + \sin y\, dy$;

C is the triangle with vertices $(1, 1), (2, 2), (3, 0)$.

8 $\oint_C \tan^{-1} x\, dx + 3x\, dy$;

C is the rectangle with vertices $(1, 0), (0, 1),$ $(2, 3), (3, 2)$.

9 $\oint_C y^2(1 + x^2)^{-1}\, dx + 2y \tan^{-1} x\, dy$;

C is the hypocycloid $x^{2/3} + y^{2/3} = 1$.

10 $\oint_C (x^2 + y^2)\, dx + 2xy\, dy$;

C is the boundary of the region bounded by the graphs of $y = \sqrt{x}, y = 0,$ and $x = 4$.

11 $\oint_C (x^4 + 4)\, dx + xy\, dy$;

C is the cardioid $r = 1 + \cos\theta$.

12 $\oint_C xy\, dx + (x^2 + y^2)\, dy$;

C is the first-quadrant loop of $r = \sin 2\theta$.

13 $\oint_C (x + y)\, dx + (y + x^2)\, dy$;

C is the boundary of the region between the circles $x^2 + y^2 = 1$ and $x^2 + y^2 = 4$.

14 $\oint_C (1 - x^2 y)\, dx + \sin y\, dy$;

C is the boundary of the region that lies inside the square with vertices $(\pm 2, \pm 2)$ and outside the square with vertices $(\pm 1, \pm 1)$.

Exer. 15–18: Use Theorem (14.20) to find the area of the region bounded by the graphs of the equations.

15 $y = 4x^2$, $\qquad y = 16x$ **16** $y = x^3$, $\;\; y^2 = x$

17 $y^2 - x^2 = 5$, $\;\; y = 3$ **18** $xy = 2$, $\;\; x + y = 3$

Exer. 19–20: Use Theorem (14.20) to find the area of the region bounded by the curve C.

19 C is the hypocycloid $x = a \cos^3 t, y = a \sin^3 t$;
$$0 \le t \le 2\pi.$$

20 C is the piecewise-smooth curve consisting of the folium of Descartes
$$x = \frac{3t}{t^3 + 1}, \quad y = \frac{3t^2}{t^3 + 1}; \quad 0 \le t \le 1$$
and the line segment from $(\tfrac{3}{2}, \tfrac{3}{2})$ to $(0, 0)$.

21 If $\mathbf{F}(x, y)$ is a two-dimensional vector field and $\int_A^B \mathbf{F} \cdot d\mathbf{r}$ is independent of path in a region D, use Green's theorem to prove that $\oint_C \mathbf{F} \cdot d\mathbf{r} = 0$ for every piecewise-smooth simple closed curve in D.

22 If f and g are differentiable functions of one variable, prove that $\oint_C f(x)\, dx + g(y)\, dy = 0$ for every piecewise-smooth simple closed curve C.

23 Let $M = y/(x^2 + y^2)$ and $N = -x/(x^2 + y^2)$. If R is the region bounded by the unit circle C with center at the origin, show that
$$\oint_C M\, dx + N\, dy \ne \iint\limits_R \left(\frac{\partial N}{\partial x} - \frac{\partial M}{\partial y} \right) dA.$$

Explain why Green's theorem is not true in this instance.

24 Suppose $\mathbf{F}(x, y)$ has continuous first partial derivatives in a simply connected region R. If curl $\mathbf{F} = \mathbf{0}$ in R, prove that $\oint_C \mathbf{F} \cdot d\mathbf{r} = 0$ for every piecewise-smooth simple closed curve C in R.

25 Let R be the region bounded by a piecewise-smooth simple closed curve C in the xy-plane. If the area of R is A, use Green's theorem to prove that the centroid $(\overline{x}, \overline{y})$ is
$$\overline{x} = \frac{1}{2A} \oint_C x^2\, dy, \qquad \overline{y} = -\frac{1}{2A} \oint_C y^2\, dx.$$

26 Suppose a homogeneous lamina of density k has the shape of a region in the xy-plane that is bounded by a piecewise-smooth simple closed curve C. Prove that the moments of inertia with respect to the x- and y-axes are
$$I_x = -\frac{k}{3} \oint_C y^3\, dx, \qquad I_y = \frac{k}{3} \oint_C x^3\, dy.$$

27 Use Exercise 25 to find the centroid of a semicircular region of radius a.

28 Use Exercise 26 to find the moment of inertia of a homogeneous circular disk of radius a with respect to a diameter.

14.5   SURFACE INTEGRALS

Line integrals are evaluated along curves. Double and triple integrals are defined on regions in two and three dimensions, respectively. In this section, we consider integrals of functions over surfaces. For simplicity, we re-

Figure 14.37

(a)

(b)

(c)

Figure 14.38

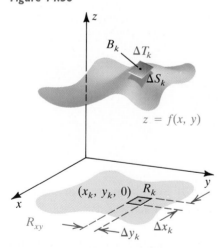

strict our discussion to surfaces that are graphs of rather simple equations, and our demonstrations will be at an intuitive level. Rigorous treatments may be found in texts on advanced calculus.

If the projection of a surface S on a coordinate plane is a region of a type considered for double integrals, we say that S has a **regular projection** on the coordinate plane. If S has a regular projection on the xy-plane, the xz-plane, or the yz-plane, we sometimes denote the region by R_{xy}, R_{xz}, or R_{yz}, respectively. For such projections, we assume that S is the graph of an equation of the form $z = f(x, y)$, $y = h(x, z)$, or $x = k(y, z)$, respectively, as shown in Figure 14.37. Moreover, we assume that the function f, h, or k has continuous first partial derivatives on its respective region.

Let us first consider the case of the graph S of $z = f(x, y)$ (see Figure 14.37a). In (13.16), we defined the area A of S. A similar technique may be used to define an integral of $g(x, y, z)$ over the surface S, if the function g is continuous throughout a region containing S. We use the notation of Section 13.2. Thus, as illustrated in Figure 14.38, ΔS_k and ΔT_k will denote areas of portions of the surface S and the tangent plane at $B_k(x_k, y_k, z_k)$, respectively, that project onto the rectangle R_k of an inner partition of R_{xy}. We evaluate g at B_k for each k and form the sum $\sum_k g(x_k, y_k, z_k)\Delta T_k$. As in the following definition, the **surface integral** $\iint_S g(x, y, z)\, dS$ **of g over** S is the limit of such sums as the norms of the partitions approach 0.

Surface Integral 14.22

$$\iint_S g(x, y, z)\, dS = \lim_{\|P\| \to 0} \sum_k g(x_k, y_k, z_k)\Delta T_k$$

If S is the union of several surfaces of the proper types, then the surface integral is defined as the sum of the individual surface integrals.

If $g(x, y, z) = 1$ for every (x, y, z), then (14.22) reduces to (13.17) and hence the surface integral gives us the area of the surface S.

A physical application of surface integrals may be obtained by considering a thin metal sheet, such as tinfoil, that has the shape of S. If the area mass density at (x, y, z) is $g(x, y, z)$, then (14.22) is the mass of the sheet. The center of mass and the moments of inertia may be obtained by employing the methods used for solids in Chapter 13 (see Exercise 21).

In a manner similar to the way in which formula (13.17) for surface area was developed, the surface integral (14.22) may be evaluated by means of formula (i) of the next theorem. Formulas (ii) and (iii) are used for surfaces of the type illustrated in Figure 14.37(b) and (c).

Evaluation Theorem for Surface Integrals 14.23

(i) $\displaystyle\iint_S g(x, y, z)\, dS =$

$$\iint_{R_{xy}} g(x, y, f(x, y))\sqrt{[f_x(x, y)]^2 + [f_y(x, y)]^2 + 1}\, dA$$

(ii) $\displaystyle\iint_S g(x, y, z)\, dS =$

$$\iint_{R_{xz}} g(x, h(x, z), z)\sqrt{[h_x(x, z)]^2 + [h_z(x, z)]^2 + 1}\, dA$$

(iii) $\displaystyle\iint_S g(x, y, z)\, dS =$

$$\iint_{R_{yz}} g(k(y, z), y, z)\sqrt{[k_y(y, z)]^2 + [k_z(y, z)]^2 + 1}\, dA$$

Figure 14.39

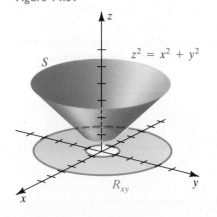

EXAMPLE■1 Evaluate $\iint_S x^2 z\, dS$ if S is the portion of the cone $z^2 = x^2 + y^2$ that lies between the planes $z = 1$ and $z = 4$.

SOLUTION As shown in Figure 14.39, the projection R_{xy} of S onto the xy-plane is the annular region bounded by circles of radii 1 and 4 with centers at the origin. If we write the equation for S in the form

$$z = (x^2 + y^2)^{1/2} = f(x, y),$$

then

$$f_x(x, y) = \frac{x}{(x^2 + y^2)^{1/2}} \quad \text{and} \quad f_y(x, y) = \frac{y}{(x^2 + y^2)^{1/2}}.$$

Applying (14.23)(i) and noting that the radical reduces to $\sqrt{2}$, we obtain

$$\iint_S x^2 z \, dS = \iint_{R_{xy}} x^2 (x^2 + y^2)^{1/2} \sqrt{2} \, dA \, .$$

We use polar coordinates to evaluate the double integral as follows:

$$\iint_{R_{xy}} x^2 (x^2 + y^2)^{1/2} \sqrt{2} \, dA = \int_0^{2\pi} \int_1^4 (r^2 \cos^2 \theta) r \sqrt{2} \, r \, dr \, d\theta$$

$$= \sqrt{2} \int_0^{2\pi} \cos^2 \theta \left[\frac{r^5}{5} \right]_1^4 d\theta$$

$$= \frac{1023\sqrt{2}}{5} \int_0^{2\pi} \frac{1 + \cos 2\theta}{2} \, d\theta$$

$$= \frac{1023\sqrt{2}}{10} \left[\theta + \frac{1}{2} \sin 2\theta \right]_0^{2\pi}$$

$$= \frac{1023\sqrt{2}\pi}{5} \approx 909$$

E X A M P L E ■ 2   Evaluate $\iint_S (xz/y) \, dS$ if S is the portion of the cylinder $x = y^2$ that lies in the first octant between the planes $z = 0, z = 5$, $y = 1$, and $y = 4$.

S O L U T I O N   The surface S is sketched in Figure 14.40. The projection R_{yz} of S on the yz-plane is the rectangle with vertices $(0, 1, 0), (0, 4, 0), (0, 4, 5)$, and $(0, 1, 5)$. Thus, by Theorem (14.23)(iii) with $k(y, z) = y^2$,

$$\iint_S \frac{xz}{y} \, dS = \int_1^4 \int_0^5 \frac{y^2 z}{y} \sqrt{(2y)^2 + 0^2 + 1} \, dz \, dy$$

$$= \int_1^4 \int_0^5 y \sqrt{4y^2 + 1} \, z \, dz \, dy = \int_1^4 y \sqrt{4y^2 + 1} \left[\frac{z^2}{2} \right]_0^5 dy$$

$$= \frac{25}{2} \int_1^4 y \sqrt{4y^2 + 1} \, dy = \frac{25}{24} \left[(4y^2 + 1)^{3/2} \right]_1^4$$

$$= \frac{25}{24} (65^{3/2} - 5^{3/2}) \approx 534.2.$$

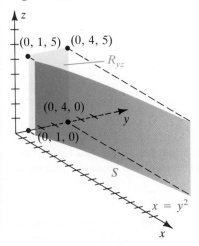

Figure 14.40

The formulas in Theorem (14.23) are stated under the assumption that the functions f, h, and k have continuous first partial derivatives on R_{xy}, R_{xz}, and R_{yz}, respectively. In certain cases, we can remove these restrictions by using an improper integral, as in the next example.

EXAMPLE ▪ 3   Evaluate $\iint_S (z + y) \, dS$ if S is the part of the graph of $z = \sqrt{1 - x^2}$ in the first octant between the xz-plane and the plane $y = 3$.

SOLUTION   The surface S is the first-octant portion of the cylinder $x^2 + z^2 = 1$ from $y = 0$ to $y = 3$. The graph is sketched in Figure 14.41. The projection R_{xy} of S onto the xy-plane is the rectangular region with vertices $(0, 0, 0)$, $(1, 0, 0)$, $(1, 3, 0)$, and $(0, 3, 0)$. If we wish to use Theorem (14.23)(i) with $f(x, y) = \sqrt{1 - x^2}$, then

$$\sqrt{[f_x(x, y)]^2 + [f_y(x, y)]^2 + 1} = \sqrt{\left(\frac{-x}{\sqrt{1 - x^2}}\right)^2 + (0)^2 + 1} = \frac{1}{\sqrt{1 - x^2}}.$$

Note that $f_x(x, y)$ is undefined at $x = 1$ and hence is not continuous throughout R_{xy}. However, we may apply part (i) of Theorem (14.23) with $0 \leq x < 1$ and then use an improper integral, as follows:

$$\iint_S (z + y) \, dS = \iint_{R_{xy}} (\sqrt{1 - x^2} + y) \frac{1}{\sqrt{1 - x^2}} \, dA$$

$$= \int_0^1 \int_0^3 \left(1 + \frac{y}{\sqrt{1 - x^2}}\right) dy \, dx$$

$$= \int_0^1 \left[y + \frac{1}{2} \frac{y^2}{\sqrt{1 - x^2}} \right]_0^3 dx = \int_0^1 \left[3 + \frac{1}{2} \frac{9}{\sqrt{1 - x^2}} \right] dx$$

$$= \lim_{t \to 1^-} \int_0^t \left[3 + \frac{9}{2} \frac{1}{\sqrt{1 - x^2}} \right] dx$$

$$= \lim_{t \to 1^-} (3t + \tfrac{9}{2} \arcsin t) = 3 + \tfrac{9}{4}\pi \approx 10.1$$

Figure 14.41

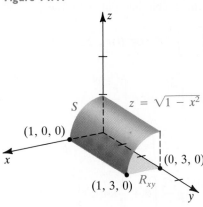

In Section 5.8, we discussed how to find the force exerted on one face of a lamina when it is submerged in a fluid (see Definition (5.27)). We can use surface integrals to solve similar problems for non-laminar surfaces, as illustrated in the next example.

EXAMPLE ▪ 4   A paper cup has the shape of a right circular cone of altitude 6 in. and radius of base 3 in. If the cup is full of water weighing 62.5 lb/ft³, find the total force exerted by the water on the inside surface of the cup.

SOLUTION   If we position the cup in an xyz-coordinate system as illustrated in Figure 14.42 (where the units are inches), then an equation of the cone has the form $z = k\sqrt{x^2 + y^2}$ for some constant k. Since the point $(0, 3, 6)$ is on the cone, $6 = k\sqrt{0^2 + 3^2}$, or $k = 2$. Hence $z = 2\sqrt{x^2 + y^2}$.

Figure 14.42

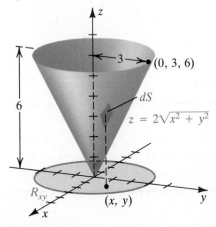

Let dS represent an element of surface area of the cone that is (approximately) a distance z from the xy-plane. The corresponding depth of the water in the cup is $6 - z$. Changing the density of the water to lb/in³, we obtain $(62.5/12^3)$ lb/in³. Thus, the pressure at a depth of $6 - z$ inches is $(62.5/12^3)(6 - z)$ lb/in², and the force exerted by the fluid on this portion of the cup is approximately

$$\frac{62.5}{12^3}(6 - z)\,dS \text{ lb.}$$

Taking a limit of sums, we find that the total force exerted by the water on the inside surface S of the cone is given by

$$F = \frac{62.5}{12^3}\iint\limits_{S}(6 - z)\,dS.$$

Note that the region R_{xy} that lies under the cone is bounded by the circle $x^2 + y^2 = 9$. Applying (14.23)(i) with $f(x, y) = 2\sqrt{x^2 + y^2} = z$, we can verify that

$$F = \frac{62.5}{12^3}\iint\limits_{R_{xy}}(6 - z)\sqrt{5}\,dA.$$

Using polar coordinates, we have

$$F = \frac{62.5}{12^3}\int_0^{2\pi}\int_0^3 (6 - 2r)\sqrt{5}\,r\,dr\,d\theta$$

$$= \frac{125\sqrt{5}}{192}\pi \text{ lb} \approx 4.57 \text{ lb.}$$

Figure 14.43

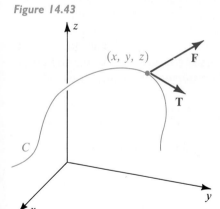

The most important applications of surface integrals occur when working with vector fields. Let $\mathbf{F}$ be a vector field such that

$$\mathbf{F}(x, y, z) = M(x, y, z)\mathbf{i} + N(x, y, z)\mathbf{j} + P(x, y, z)\mathbf{k},$$

where M, N, and P are continuous (scalar) functions. In Section 14.2, we considered the following line integral for a piecewise-smooth curve C:

$$\int_C \mathbf{F}\cdot\mathbf{T}\,ds,$$

where s is an arc length parameter for C. As in Figure 14.43, $\mathbf{T} = d\mathbf{r}/ds$ is a unit tangent vector to C at the point (x, y, z), where $\mathbf{r} = x\mathbf{i} + y\mathbf{j} + z\mathbf{k}$ is the position vector of (x, y, z). If $\mathbf{F}$ is a force field, then, from (14.12), the value of this line integral is the work done by $\mathbf{F}$ along C.

Let us next consider a surface S and a unit normal vector $\mathbf{n}$ to S at the point (x, y, z), as illustrated in Figure 14.44. If the components of $\mathbf{n}$ are continuous functions of x, y, and z, then $\mathbf{F}\cdot\mathbf{n}$ is a continuous (scalar) function and we may consider the following surface integral over S, which (for reasons we will discuss at the end of this section) is called the **flux integral** of $\mathbf{F}$ over S.

Figure 14.44

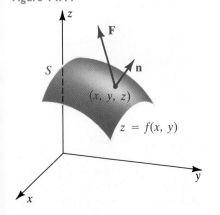

Flux Integral of F Over S   14.24

$$\iint\limits_{S} \mathbf{F} \cdot \mathbf{n} \, dS$$

To evaluate (14.24), we must be more precise about properties of the unit vector **n**. If the surface S is the graph of an equation $z = f(x, y)$ and if we let $g(x, y, z) = z - f(x, y)$, then S is also the graph of the equation $g(x, y, z) = 0$. Since the gradient $\nabla g(x, y, z)$ is a normal vector to the graph of $g(x, y, z) = 0$ at the point (x, y, z), a *unit* normal vector **n** can be obtained as follows:

$$\mathbf{n} = \frac{\nabla g(x, y, z)}{\|\nabla g(x, y, z)\|} = \frac{-f_x(x, y)\mathbf{i} - f_y(x, y)\mathbf{j} + \mathbf{k}}{\sqrt{[f_x(x, y)]^2 + [f_y(x, y)]^2 + 1}}$$

Similar formulas can be stated if S is given by $y = h(x, z)$ or by $x = k(y, z)$. Note that for the case $z = f(x, y)$, the unit vector **n** is an **upper normal** to S, since the **k** component is positive. A **lower normal** can be obtained by using $-\mathbf{n}$.

In the remainder of this chapter, we shall assume that every surface S is **oriented** (or **orientable**) in the sense that a unit normal vector **n** exists at every nonboundary point (x, y, z), and that the components of **n** are continuous functions of $x, y,$ and z. We say that **n** *varies continuously* over the surface S. We shall also assume that S has two sides, such as the *top side* and *bottom side* of the graph of $z = f(x, y)$ in Figure 14.44. For a *closed* surface S such as a sphere, we may consider the *outside* and *inside* of S. We then refer to **n** as the **unit outer normal** or the **unit inner normal** (see Figure 14.45).

Figure 14.45

(a) Unit outer normals

(b) Unit inner normals

Figure 14.46
Möbius strip

Our restrictions on S rule out *one-sided surfaces*, such as a *Möbius strip*, named after the German mathematician A. F. Möbius (1790–1868). A Möbius strip can be constructed from a long rectangular strip of paper by making a half twist in the paper and taping the short edges together, producing the surface illustrated in Figure 14.46. Imagine a painter who

Figure 14.47

starts at some point and attempts to paint only one side of the original strip. The painter will eventually return to the same point, having painted the entire strip without crossing any edge. This one-sided surface is not orientable.

To interpret the physical significance of the flux integral (14.24), let us regard the surface S as a thin membrane through which a fluid can pass. Suppose S is submerged in a fluid having a velocity field $\mathbf{F}(x, y, z)$. Some typical vectors of the field are shown in the lower part of Figure 14.47. Each vector of the field is the velocity of a fluid particle (or molecule).

Let dS represent a small element of area of S. If $\mathbf{F}$ is continuous, then it is almost constant on dS, and the amount of fluid crossing dS per unit of time may be approximated by the volume of a prism of base area dS and altitude $\mathbf{F} \cdot \mathbf{n}$, as illustrated in Figure 14.47. If dV denotes the volume of the prism in the figure, then $dV = \mathbf{F} \cdot \mathbf{n}\, dS$. Since dV represents the amount of fluid crossing dS per unit time, the flux integral is a limit of sums of elements of volume, and hence $\iint_S \mathbf{F} \cdot \mathbf{n}\, dS$ *is the volume of fluid crossing S per unit time*. This quantity is called the *flux* of $\mathbf{F}$ *through* (or *over*) S. If $\mathbf{F}$ and S satisfy the conditions we have imposed, then we have the following.

Definition 14.25

The **flux** of a vector field $\mathbf{F}$ through (or over) a surface S is

$$\iint_S \mathbf{F} \cdot \mathbf{n}\, dS.$$

If the fluid in the previous discussion has density $\delta = \delta(x, y, z)$, then the value of the flux integral $\iint_S \delta \mathbf{F} \cdot \mathbf{n}\, dS$ is the mass of the fluid crossing S.

The vector field $\mathbf{F}$ in Definition (14.25) may also represent steady state heat flow, a uniformly expanding gas, or fields encountered in the theory of electricity. In these cases, flux is a measure of the amount of heat crossing S, the flow of gas through S, or magnetic or electric flux through S, respectively.

If, for a closed surface such as a sphere, we choose $\mathbf{n}$ as the unit *outer normal*, then the flux measures the *net outward flow* per unit time. If the flux integral is positive, then the flow out of S exceeds the flow into S, and we say there is a **source** of $\mathbf{F}$ within S. If the integral is negative, then the flow into S exceeds the flow out of S, and there is a **sink** within S. If the integral is 0, then the flow into and the flow out of S are equal—that is, the sources and sinks balance one another.

Figure 14.48

$z = 9 - x^2 - y^2$

EXAMPLE 5 Let S be the part of the graph of $z = 9 - x^2 - y^2$ with $z \geq 0$. If $\mathbf{F}(x, y, z) = 3x\mathbf{i} + 3y\mathbf{j} + z\mathbf{k}$, find the flux of $\mathbf{F}$ through S.

SOLUTION The graph is sketched in Figure 14.48, together with a typical unit upper normal $\mathbf{n}$ and the vector $\mathbf{F}(x, y, z)$. As in our discussion following (14.24), to find a formula for $\mathbf{n}$ we consider

$$g(x, y, z) = z - (9 - x^2 - y^2) = z - 9 + x^2 + y^2$$

and let
$$\mathbf{n} = \frac{\nabla g(x, y, z)}{\|\nabla g(x, y, z)\|} = \frac{2x\mathbf{i} + 2y\mathbf{j} + \mathbf{k}}{\sqrt{4x^2 + 4y^2 + 1}}.$$

By (14.25), the flux of $\mathbf{F}$ through S is

$$\iint_S \mathbf{F} \cdot \mathbf{n}\, dS = \iint_S \frac{6x^2 + 6y^2 + z}{\sqrt{4x^2 + 4y^2 + 1}}\, dS.$$

Applying Theorem (14.23)(i) gives us

$$\iint_S \mathbf{F} \cdot \mathbf{n}\, dS = \iint_{R_{xy}} \frac{6x^2 + 6y^2 + 9 - x^2 - y^2}{\sqrt{4x^2 + 4y^2 + 1}}\sqrt{4x^2 + 4y^2 + 1}\, dA$$

$$= \iint_{R_{xy}} (5x^2 + 5y^2 + 9)\, dA,$$

where R_{xy} is the circular region in the xy-plane bounded by the graph of $x^2 + y^2 = 9$. Changing to polar coordinates, we have

$$\iint_S \mathbf{F} \cdot \mathbf{n}\, dS = \int_0^{2\pi} \int_0^3 (5r^2 + 9)r\, dr\, d\theta = \frac{567\pi}{2} \approx 890.6.$$

If, for example, $\mathbf{F}$ is the velocity field of an expanding gas and $\|\mathbf{F}\|$ is measured in meters per second, then the unit of flux is cubic meters per second. Hence the flow of gas across the surface S in Figure 14.48 is approximately 890.6 m^3/sec.

In the next section, we discuss a result, called the *divergence theorem*, that will enable us to calculate the flux over a closed surface by using a triple integral.

EXERCISES 14.5

Exer. 1–4: Evaluate $\iint_S g(x, y, z)\, dS$.

1 $g(x, y, z) = x^2$;
  S is the upper half of the sphere $x^2 + y^2 + z^2 = a^2$.

2 $g(x, y, z) = x^2 + y^2 + z^2$;
  S is the part of the plane $z = y + 4$ that is inside the cylinder $x^2 + y^2 = 4$.

3 $g(x, y, z) = x + y$;
  S is the first-octant portion of the plane with equation $2x + 3y + z = 6$.

4 $g(x, y, z) = (x^2 + y^2 + 1)^{1/2}$;
  S is the portion of the paraboloid $2z = x^2 + y^2$ that lies inside the cylinder $x^2 + y^2 = 2y$.

Exer. 5–8: Express the surface integral as an iterated double integral by using a projection of S on **(a)** the yz-plane and **(b)** the xz-plane.

5 $\iint_S xy^2z^3\, dS$;
  S is the first-octant portion of the plane with equation $2x + 3y + 4z = 12$.

6 $\iint\limits_{S} (xz + 2y)\, dS$;

   S is the portion of the graph of $y = x^3$ between the planes $y = 0$, $y = 8$, $z = 2$, and $z = 0$.

7 $\iint\limits_{S} (x^2 - 2y + z)\, dS$;

   S is the portion of the graph of $4x + y = 8$ bounded by the coordinate planes and the plane $z = 6$.

8 $\iint\limits_{S} (x^2 + y^2 + z^2)\, dS$;

   S is the first-octant portion of the graph of $x^2 + y^2 = 4$ bounded by the coordinate planes and the plane $x + z = 2$.

9 Interpret $\iint_S g(x, y, z)\, dS$ geometrically if f is the constant function $g(x, y, z) = c$ with $c > 0$ and S has a regular projection on the xy-plane.

10 Show that a double integral $\iint_R f(x, y)\, dA$ is a special case of a surface integral.

Exer. 11–14: Find $\iint_S \mathbf{F} \cdot \mathbf{n}\, dS$ if $\mathbf{n}$ is a unit upper normal to S.

11 $\mathbf{F} = x\mathbf{i} + y\mathbf{j} + z\mathbf{k}$;

   S is the upper half of the sphere $x^2 + y^2 + z^2 = a^2$.

12 $\mathbf{F} = x\mathbf{i} - y\mathbf{j}$;

   S is the first-octant portion of the sphere $x^2 + y^2 + z^2 = a^2$.

13 $\mathbf{F} = 2\mathbf{i} + 5\mathbf{j} + 3\mathbf{k}$;

   S is the portion of the cone $z = (x^2 + y^2)^{1/2}$ that is inside the cylinder $x^2 + y^2 = 1$.

14 $\mathbf{F} = x\mathbf{i} + y\mathbf{j} + z\mathbf{k}$;

   S is the portion of the plane $3x + 2y + z = 12$ cut out by the planes $x = 0$, $y = 0$, $x = 1$, and $y = 2$.

Exer. 15–16: Find the flux of $\mathbf{F}$ through S.

15 $\mathbf{F}(x, y, z) = x\mathbf{i} + y\mathbf{j} + z\mathbf{k}$;

   S is the first-octant portion of the plane with equation $2x + 3y + z = 6$.

16 $\mathbf{F}(x, y, z) = (x^2 + z)\mathbf{i} + y^2 z\mathbf{j} + (x^2 + y^2 + z)\mathbf{k}$;

   S is the first-octant portion of the paraboloid $z = x^2 + y^2$ that is cut off by the plane $z = 4$.

Exer. 17–18: Find the flux of $\mathbf{F}$ over the closed surface S. (Use the *outer* normal to S.)

17 $\mathbf{F}(x, y, z) = (x + y)\mathbf{i} + z\mathbf{j} + xz\mathbf{k}$;

   S is the surface of the cube having vertices $(\pm 1, \pm 1, \pm 1)$.

18 $\mathbf{F}(x, y, z) = x\mathbf{i} - y\mathbf{j} + z\mathbf{k}$;

   S is the surface of the solid bounded by the graphs of $z = x^2 + y^2$ and $z = 4$.

19 If S is given by $x = k(y, z)$ and $\mathbf{F} = M\mathbf{i} + N\mathbf{j} + P\mathbf{k}$, show that

$$\iint\limits_{S} \mathbf{F} \cdot \mathbf{n}\, dS = \iint\limits_{R_{yz}} [M - Nk_y(y, z) - Pk_z(y, z)]\, dy\, dz\,.$$

20 By Coulomb's law, if a point charge of q coulombs is located at the origin, then the force exerted on a unit charge at (x, y, z) is given by $\mathbf{F}(x, y, z) = (cq/\|\mathbf{r}\|^3)\mathbf{r}$, where $\mathbf{r} = x\mathbf{i} + y\mathbf{j} + z\mathbf{k}$. If S is any sphere with center at the origin, show that the flux of $\mathbf{F}$ through S is $4\pi cq$.

21 If a thin metal sheet T has the shape of a surface S and has area mass density $\delta(x, y, z)$, then the *moments* of T with respect to the coordinate planes are defined by

$$M_{xy} = \iint\limits_{S} z\delta(x, y, z)\, dS$$

$$M_{xz} = \iint\limits_{S} y\delta(x, y, z)\, dS$$

$$M_{yz} = \iint\limits_{S} x\delta(x, y, z)\, dS.$$

Use limits of sums to show that these are natural definitions. The *center of mass* $(\bar{x}, \bar{y}, \bar{z})$ is defined by

$$\bar{x}m = M_{yz}, \qquad \bar{y}m = M_{xz}, \qquad \bar{z}m = M_{xy},$$

where m is the mass of T. Suppose a metal funnel has the shape of the surface S described in Example 1 (see Figure 14.39). If the unit of length is the centimeter and the area mass density at the point (x, y, z) is z^2 gm/cm^2, find

(a) the mass and the center of mass of the funnel

(b) the moment of inertia of the funnel with respect to the z-axis

22 Refer to Exercise 21. A thin metal sheet of constant density k has the shape of the hemisphere

$$z = \sqrt{a^2 - x^2 - y^2}.$$

Find its center of mass.

14.6   THE DIVERGENCE THEOREM

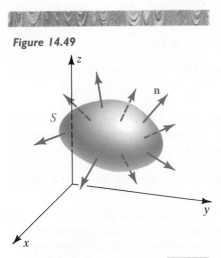

Figure 14.49

One of the most important theorems of vector calculus is the *divergence theorem*. It is often called *Gauss's theorem* in honor of Carl Friedrich Gauss (see Mathematicians and Their Times, Chapter 8). The theorem specifies the flux of a vector field over a **closed surface** S that is the boundary of a region Q in three dimensions. For example, S could be a sphere, an ellipsoid, a cube, or a tetrahedron. The divergence theorem can be proved for very complicated regions; however, the proof would take us into the field of advanced calculus. Instead, *throughout this section we shall assume that Q is a region over which triple integrals can be evaluated* using methods developed in Chapter 13. We will also assume that surface integrals can be evaluated over S. In this section, $\mathbf{n}$ will denote a unit *outer* normal to S, as illustrated by some typical vectors in Figure 14.49.

Under the restrictions we have placed on Q and S, the divergence theorem may be stated as follows.

Divergence Theorem   14.26

Let Q be a region in three dimensions bounded by a closed surface S, and let $\mathbf{n}$ denote the unit outer normal to S at (x, y, z). If $\mathbf{F}$ is a vector function that has continuous partial derivatives on Q, then

$$\iint\limits_{S} \mathbf{F} \cdot \mathbf{n} \, dS = \iiint\limits_{Q} \nabla \cdot \mathbf{F} \, dV;$$

that is, *the flux of $\mathbf{F}$ over S equals the triple integral of the divergence of $\mathbf{F}$ over Q.*

PROOF   If $\mathbf{F}(x, y, z) = M(x, y, z)\mathbf{i} + N(x, y, z)\mathbf{j} + P(x, y, z)\mathbf{k}$, let us simplify the notation by writing $\mathbf{F} = M\mathbf{i} + N\mathbf{j} + P\mathbf{k}$. Using properties of the dot product and the definition of $\mathbf{F}$, we may write the conclusion of the theorem as

$$\iint\limits_{S} (M\mathbf{i} \cdot \mathbf{n} + N\mathbf{j} \cdot \mathbf{n} + P\mathbf{k} \cdot \mathbf{n}) \, dS = \iiint\limits_{Q} \left(\frac{\partial M}{\partial x} + \frac{\partial N}{\partial y} + \frac{\partial P}{\partial z} \right) dV.$$

To prove this equality, it is sufficient to show that

$$\iint\limits_{S} M\mathbf{i} \cdot \mathbf{n} \, dS = \iiint\limits_{Q} \frac{\partial M}{\partial x} \, dV$$

$$\iint\limits_{S} N\mathbf{j} \cdot \mathbf{n} \, dS = \iiint\limits_{Q} \frac{\partial N}{\partial y} \, dV$$

$$\iint\limits_{S} P\mathbf{k} \cdot \mathbf{n} \, dS = \iiint\limits_{Q} \frac{\partial P}{\partial z} \, dV.$$

Figure 14.50

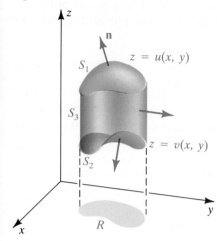

Since the proofs of all three formulas are similar, we shall consider only the third. Moreover, the proof will be restricted to the case, illustrated in Figure 14.50, where S is the surface of a region Q that lies between the graphs of $z = v(x, y)$ and $z = u(x, y)$ and over a suitable region R in the xy-plane. We shall denote the upper surface by S_1, the lower surface by S_2, and the lateral surface by S_3. Some typical unit outer normal vectors are shown in the figure.

On S_3, the $\mathbf{k}$ component of $\mathbf{n}$ is 0, and hence $\mathbf{k} \cdot \mathbf{n} = 0$. Thus, the flux integral over S_3 is 0, and we may write

$$\iint_S P\mathbf{k} \cdot \mathbf{n}\, dS = \iint_{S_1} P\mathbf{k} \cdot \mathbf{n}\, dS + \iint_{S_2} P\mathbf{k} \cdot \mathbf{n}\, dS.$$

As in Section 14.5, to find a unit (upper) normal to S_1, we let

$$g_1(x, y, z) = z - u(x, y)$$

and calculate

$$\mathbf{n} = \frac{\nabla g_1(x, y, z)}{\| \nabla g_1(x, y, z) \|} = \frac{-u_x(x, y)\mathbf{i} - u_y(x, y)\mathbf{j} + \mathbf{k}}{\sqrt{[u_x(x, y)]^2 + [u_x(x, y)]^2 + 1}}.$$

Hence, $\mathbf{k} \cdot \mathbf{n} = 1 / \sqrt{[u_x(x, y)]^2 + [u_y(x, y)]^2 + 1}$. If we apply Theorem (14.23)(i) with $R = R_{xy}$ and $u(x, y) = f(x, y)$, the radicals cancel and we obtain

$$\iint_{S_1} P\mathbf{k} \cdot \mathbf{n}\, dS = \iint_R P(x, y, u(x, y))\, dA.$$

On S_2, we let

$$g_2(x, y, z) = z - v(x, y)$$

and use the *lower* unit normal $\mathbf{n}$ given by

$$\mathbf{n} = -\frac{\nabla g_2(x, y, z)}{\| \nabla g_2(x, y, z) \|} = \frac{v_x(x, y)\mathbf{i} + v_y(x, y)\mathbf{j} - \mathbf{k}}{\sqrt{[v_x(x, y)]^2 + [v_y(x, y)]^2 + 1}}.$$

Again applying (14.23)(i) gives us the flux integral over S_2,

$$\iint_{S_2} P\mathbf{k} \cdot \mathbf{n}\, dS = -\iint_R P(x, y, v(x, y))\, dA.$$

Adding the flux integrals over S_1 and S_2 yields

$$\iint_S P\mathbf{k} \cdot \mathbf{n}\, dS = \iint_R [P(x, y, u(x, y)) - P(x, y, v(x, y))]\, dA$$

$$= \iint_R \left[\int_{v(x,y)}^{u(x,y)} \frac{\partial P}{\partial z}\, dz \right] dA = \iiint_Q \frac{\partial P}{\partial z}\, dV,$$

which is what we wished to prove. Our proof can be extended to finite unions of regions of the type we have considered. The proofs of the formulas for $\iint_S M\mathbf{i} \cdot \mathbf{n}\, dS$ and $\iint_S N\mathbf{j} \cdot \mathbf{n}\, dS$ are similar, provided Q and S are suitably restricted. ■

EXAMPLE ■ I   Let Q be the region bounded by the circular cylinder $x^2 + y^2 = 4$ and the planes $z = 0$ and $z = 3$, and let S denote the surface of Q. If $\mathbf{F}(x, y, z) = x^3\mathbf{i} + y^3\mathbf{j} + z^3\mathbf{k}$, use the divergence theorem to find $\iint_S \mathbf{F} \cdot \mathbf{n}\, dS$.

Figure 14.51

SOLUTION   The surface S and some typical positions of $\mathbf{n}$ are sketched in Figure 14.51. Since

$$\nabla \cdot \mathbf{F} = 3x^2 + 3y^2 + 3z^2 = 3(x^2 + y^2 + z^2),$$

we have, by the divergence theorem (14.26),

$$\iint_S \mathbf{F} \cdot \mathbf{n}\, dS = 3 \iiint_Q (x^2 + y^2 + z^2)\, dV.$$

If we use cylindrical coordinates to evaluate the triple integral, then

$$\iint_S \mathbf{F} \cdot \mathbf{n}\, dS = 3 \int_0^{2\pi} \int_0^2 \int_0^3 (r^2 + z^2)\, r\, dz\, dr\, d\theta$$

$$= 3 \int_0^{2\pi} \int_0^2 \left[r^2 z + \tfrac{1}{3}z^3 \right]_0^3 r\, dr\, d\theta = 3 \int_0^{2\pi} \int_0^2 (3r^2 + 9)r\, dr\, d\theta$$

$$= 3 \int_0^{2\pi} \left[\tfrac{3}{4}r^4 + \tfrac{9}{2}r^2 \right]_0^2 d\theta = 3 \int_0^{2\pi} 30\, d\theta = 90 \left[\theta \right]_0^{2\pi} = 180\pi.$$

Figure 14.52

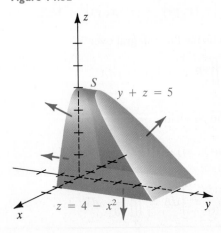

EXAMPLE ■ 2   Let Q be the region bounded by the cylinder $z = 4 - x^2$, the plane $y + z = 5$, and the xy- and xz-planes, and let S be the surface of Q. If $\mathbf{F}(x, y, z) = (x^3 + \sin z)\mathbf{i} + (x^2 y + \cos z)\mathbf{j} + e^{x^2+y^2}\mathbf{k}$, find $\iint_S \mathbf{F} \cdot \mathbf{n}\, dS$.

SOLUTION   The region Q is sketched in Figure 14.52 with several unit outer normal vectors. It would be an incredibly difficult job to evaluate the surface integral directly; however, using the divergence theorem, we can obtain the value from the triple integral

$$\iiint_Q (3x^2 + x^2)\, dV = \iiint_Q 4x^2\, dV.$$

Referring to Figure 14.52 and using (13.20), we see that this integral equals

$$\int_{-2}^{2} \int_{0}^{4-x^2} \int_{0}^{5-z} 4x^2 \, dy \, dz \, dx = \int_{-2}^{2} \int_{0}^{4-x^2} 4x^2 (5-z) \, dz \, dx$$

$$= \int_{-2}^{2} \left[20x^2 z - 2x^2 z^2 \right]_{0}^{4-x^2} dx$$

$$= \int_{-2}^{2} (48x^2 - 4x^4 - 2x^6) \, dx$$

$$= \frac{4608}{35} \approx 131.7.$$

We may use the divergence theorem to obtain a physical interpretation for the divergence of a vector field. Let us begin by recalling from the mean value theorem for definite integrals (4.28) that if a function f of one variable is continuous on a closed interval $[a, b]$, then there is a number c in (a, b) such that

$$\int_{a}^{b} f(x) \, dx = f(c)L,$$

where $L = b - a$ is the length of $[a, b]$. Analogous results may be proved for double and triple integrals. Thus, if a function f of three variables is continuous throughout a spherical region Q, then there is a point $A(x_1, y_1, z_1)$ in the interior of Q such that

$$\iiint_{Q} f(x, y, z) \, dV = [f(x, y, z)]_A V,$$

where V is the volume of Q and $[f(x, y, z)]_A = f(x_1, y_1, z_1)$. Let us call this result the **mean value theorem for triple integrals**. It follows that if $\mathbf{F}$ is a continuous vector function, then

$$\iiint_{Q} \nabla \cdot \mathbf{F} \, dV = [\nabla \cdot \mathbf{F}]_A V,$$

or, equivalently,

$$[\nabla \cdot \mathbf{F}]_A = \frac{1}{V} \iiint_{Q} \nabla \cdot \mathbf{F} \, dV.$$

Applying the divergence theorem gives us

$$[\nabla \cdot \mathbf{F}]_A = \frac{1}{V} \iint_{S} \mathbf{F} \cdot \mathbf{n} \, dS,$$

where S is the surface of Q. The quantity on the right may be regarded as *the flux of* $\mathbf{F}$ *per unit volume over the sphere*.

Next, let P be an arbitrary point, and suppose $\mathbf{F}$ is continuous throughout a region containing P in its interior. Let S_k be the surface of a sphere of radius k with center at P. From the previous discussion, for each k there is a point P_k within S_k such that

$$[\nabla \cdot \mathbf{F}]_{P_k} = \frac{1}{V_k} \iint\limits_{S_k} \mathbf{F} \cdot \mathbf{n} \, dS,$$

where V_k is the volume of the sphere (see Figure 14.53).

If we let $k \to 0$, then $P_k \to P$ and $[\nabla \cdot \mathbf{F}]_{P_k} \to [\nabla \cdot \mathbf{F}]_P = [\text{div } \mathbf{F}]_P$. This gives us the following.

Divergence as a Limit   14.27

$$[\text{div } \mathbf{F}]_P = \lim_{k \to 0} \frac{1}{V_k} \iint\limits_{S_k} \mathbf{F} \cdot \mathbf{n} \, dS$$

Figure 14.53

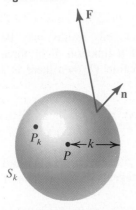

Thus, the *divergence of* $\mathbf{F}$ *at* P *is the limiting value of the flux per unit volume over a sphere with center* P, *as the radius of the sphere approaches* 0. In particular, if $\mathbf{F}$ represents the velocity of a fluid, then $[\text{div } \mathbf{F}]_P$ can be interpreted as the rate of loss or gain of fluid per unit volume at P per unit time. If $\delta = \delta(x, y, z)$ is the density at (x, y, z), then $[\text{div } \delta \mathbf{F}]_P$ is the change in *mass* per unit volume per unit time. It follows from the remarks at the end of Section 14.5 that there is a source or sink at P if $[\text{div } \mathbf{F}]_P > 0$ or $[\text{div } \mathbf{F}]_P < 0$, respectively. If the fluid is incompressible and no source or sink is present, then there can be no gain or loss within the volume element V_k and hence, at every point P, div $\mathbf{F} = 0$. The limit form for divergence in (14.27) may also be applied to physical concepts such as magnetic or electric flux, since these entities have many characteristics that are similar to those of fluids.

EXAMPLE 3   Suppose a closed surface S forms the boundary of a region Q and the origin O is an interior point of Q. If an inverse square field is given by $\mathbf{F} = (q/r^3)\mathbf{r}$, where q is a constant, $\mathbf{r} = x\mathbf{i} + y\mathbf{j} + z\mathbf{k}$, and $r = \|\mathbf{r}\|$, prove that the flux of $\mathbf{F}$ over S is $4\pi q$ regardless of the shape of Q.

Figure 14.54

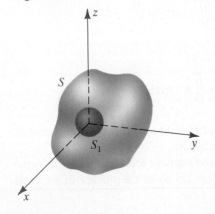

SOLUTION   Since $\mathbf{F}$ is not continuous at O, the divergence theorem cannot be applied directly; however, we may proceed as follows. Let S_1 be a sphere of radius a and center O that lies completely within S (see Figure 14.54), and let Q_1 denote the region that lies *outside* S_1 and *inside* S. Since $\mathbf{F}$ is continuous throughout Q_1, we may apply the divergence theorem, obtaining

$$\iiint\limits_{Q_1} \nabla \cdot \mathbf{F} \, dV = \iint\limits_{S} \mathbf{F} \cdot \mathbf{n} \, dS + \iint\limits_{S_1} \mathbf{F} \cdot \mathbf{n} \, dS.$$

We can show that if $\mathbf{F}$ is an inverse square field, then $\nabla \cdot \mathbf{F} = 0$ (see Exercise 21 of Section 14.1) and therefore,

$$\iint_S \mathbf{F} \cdot \mathbf{n}\, dS = -\iint_{S_1} \mathbf{F} \cdot \mathbf{n}\, dS.$$

Since the unit normal $\mathbf{n}$ on S_1 is an *outer* normal, it points *away* from the region Q_1 that has S_1 as part of its boundary. Hence on S_1, the unit normal $\mathbf{n}$ points *toward* the origin. Thus, $\mathbf{n} = (-1/r)\mathbf{r}$ with $r = a$, and

$$\iint_S \mathbf{F} \cdot \mathbf{n}\, dS = -\iint_{S_1} \left(\frac{q}{r^3}\right)\mathbf{r} \cdot \left(-\frac{1}{r}\mathbf{r}\right) ds$$

$$= \iint_{S_1} \frac{q}{r^4}(\mathbf{r} \cdot \mathbf{r})\, dS = \iint_{S_1} \frac{q}{r^2}\, dS$$

$$= \frac{q}{a^2} \iint_{S_1} dS = \frac{q}{a^2}(4\pi a^2) = 4\pi q.$$

Example 3 has an important application in the theory of electricity. By Coulomb's law, if a point charge of q coulombs is located at the origin, then the force exerted on a unit charge at (x, y, z) is $\mathbf{F}(x, y, z) = (cq/\|\mathbf{r}\|^3)\mathbf{r}$, where $\mathbf{r} = x\mathbf{i} + y\mathbf{j} + z\mathbf{k}$ and c is a constant. It follows that the *electric flux* of $\mathbf{F}$ over any closed surface containing the origin is $4\pi cq$. Thus, the electric flux is independent of the size or shape of S and depends only on q. It is not difficult to extend this to the case of more than one point charge within S. This result, called *Gauss's law*, has far-reaching consequences in the study of electric fields.

EXERCISES 14.6

Exer. 1–4: Use the divergence theorem (14.26) to find $\iint_S \mathbf{F} \cdot \mathbf{n}\, dS$.

1 $\mathbf{F} = y \sin x\mathbf{i} + y^2 z\mathbf{j} + (x + 3z)\mathbf{k}$;
S is the surface of the region bounded by the planes $x = \pm 1$, $y = \pm 1$, $z = \pm 1$.

2 $\mathbf{F} = y^3 e^z\mathbf{i} - xy\mathbf{j} + x \arctan y\mathbf{k}$;
S is the surface of the region bounded by the coordinate planes and the plane $x + y + z = 1$.

3 $\mathbf{F} = (x^2 + \sin yz)\mathbf{i} + (y - xe^{-z})\mathbf{j} + z^2\mathbf{k}$;
S is the surface of the region bounded by the cylinder $x^2 + y^2 = 4$ and the planes $x + z = 2$ and $z = 0$.

4 $\mathbf{F} = 2xy\mathbf{i} + z \cosh x\mathbf{j} + (z^2 + y \sin^{-1} x)\mathbf{k}$;
S is the surface of the region bounded by the paraboloid $z = x^2 + y^2$ and the plane $z = 1$.

Exer. 5–10: Use the divergence theorem (14.26) to find the flux of $\mathbf{F}$ through S.

5 $\mathbf{F}(x, y, z) = yz\mathbf{i} + xz\mathbf{j} + xy\mathbf{k}$;
S is the graph of $x^{2/3} + y^{2/3} + z^{2/3} = 1$.

6 $\mathbf{F}(x, y, z) = (x^2 + z^2)\mathbf{i} + (y^2 - 2xy)\mathbf{j} + (4z - 2yz)\mathbf{k}$;
S is the surface of the region bounded by the cone $x = \sqrt{y^2 + z^2}$ and the plane $x = 9$.

7 $\mathbf{F}(x, y, z) = 3x\mathbf{i} + xz\mathbf{j} + z^2\mathbf{k}$;
S is the surface of the region bounded by the paraboloid $z = 4 - x^2 - y^2$ and the xy-plane.

8 $\mathbf{F}(x, y, z) = xy^2\mathbf{i} + yz^2\mathbf{j} + zx^2\mathbf{k}$;
S is the surface of the region that lies between the cylinders $x^2 + y^2 = 4$ and $x^2 + y^2 = 9$ and between the planes $z = -1$ and $z = 2$.

9 $F(x, y, z) = 2xz\mathbf{i} + xyz\mathbf{j} + yz\mathbf{k}$;
S is the surface of the region bounded by the coordinate planes and the planes $x + 2z = 4$ and $y = 2$.

10 $F(x, y, z) = x^3\mathbf{i} + y^3\mathbf{j} + z^3\mathbf{k}$;
S is the surface of the region that is inside both the cone $z = \sqrt{x^2 + y^2}$ and the sphere $x^2 + y^2 + z^2 = 25$.

Exer. 11–14: Verify the divergence theorem (14.26) by evaluating both the surface integral and the triple integral.

11 $F = x\mathbf{i} + y\mathbf{j} + z\mathbf{k}$;
S is the sphere $x^2 + y^2 + z^2 = a^2$.

12 $F = x^2\mathbf{i} + y^2\mathbf{j} + z^2\mathbf{k}$;
S is the surface of the cube bounded by the coordinate planes and the planes $x = a$, $y = a$, $z = a$ with $a > 0$.

13 $F = (x + z)\mathbf{i} + (y + z)\mathbf{j} + (x + y)\mathbf{k}$;
S is the surface of the region
$$Q = \left\{ (x, y, z): 0 \le y^2 + z^2 \le 1, \ 0 \le x \le 2 \right\}.$$

14 $F = \|\mathbf{r}\|^2 \mathbf{r}$ for $\mathbf{r} = x\mathbf{i} + y\mathbf{j} + z\mathbf{k}$;
S is the sphere $x^2 + y^2 + z^2 = a^2$.

15 If $\iint_S F \cdot \mathbf{n} \, dS = 0$ for every closed surface of the type considered in the divergence theorem, prove that div $F = 0$.

16 Use the divergence theorem to prove that if a scalar function f has continuous second partial derivatives, then
$$\iiint_Q \nabla^2 f \, dV = \iint_S D_\mathbf{n} f \, dS,$$
where $D_\mathbf{n} f$ is the directional derivative of f in the direction of an outer unit normal $\mathbf{n}$ to S.

Exer. 17–18: Assume that S and Q satisfy the conditions of the divergence theorem and that f and g are scalar functions that have continuous second partial derivatives. Prove the identity.

17 $\iiint_Q (f\nabla^2 g + \nabla f \cdot \nabla g) \, dV = \iint_S (f\nabla g) \cdot \mathbf{n} \, dS$

(*Hint:* Let $F = f\nabla g$ in (14.26).)

18 $\iiint_Q (f\nabla^2 g - g\nabla^2 f) \, dV = \iint_S (f\nabla g - g\nabla f) \cdot \mathbf{n} \, dS$

(*Hint:* Use the identity in Exercise 17 together with that obtained by interchanging f and g.)

Exer. 19–22: Assume that S and Q satisfy the conditions of the divergence theorem.

19 If F is a conservative vector field with potential function f and if div $F = 0$, prove that
$$\iiint_Q F \cdot F \, dV = \iint_S fF \cdot \mathbf{n} \, dS.$$

20 If $\mathbf{r} = x\mathbf{i} + y\mathbf{j} + z\mathbf{k}$ and V is the volume of Q, prove that $V = \frac{1}{3} \iint_S \mathbf{r} \cdot \mathbf{n} \, dS$.

21 If F has continuous second partial derivatives, prove that $\iint_S$ curl $F \cdot \mathbf{n} \, dS = 0$.

22 If $\mathbf{a}$ is a constant vector, prove that $\iint_S \mathbf{a} \cdot \mathbf{n} \, dS = 0$.

Exer. 23–24: A surface (or triple) integral of a vector function is defined as the sum of the surface (or triple) integrals of each component of the function. Using this fact and assuming that S and Q satisfy the conditions of the divergence theorem, establish the result.

23 $\iint_S F \times \mathbf{n} \, dS = -\iiint_Q \nabla \times F \, dV$

(*Hint:* Apply the divergence theorem and Exercise 28 of Section 14.1 to $\mathbf{c} \times F$, where $\mathbf{c}$ is an arbitrary constant vector.)

24 $\iint_S f\mathbf{n} \, dS = \iiint_Q \nabla f \, dV$

(*Hint:* Apply the divergence theorem to $f\mathbf{c}$, where $\mathbf{c}$ is an arbitrary constant vector.)

25 If F is orthogonal to S at each point (x, y, z), prove that $\iiint_Q$ curl $F \, dV = \mathbf{0}$.

26 If $\mathbf{r}$ is the position vector for (x, y, z) and $r = \|\mathbf{r}\|$, prove that $\iiint_Q \mathbf{r} \, dV = \frac{1}{2} \iint_S r^2\mathbf{n} \, dS$.

27 An empty cylindrical tank of diameter 2 ft and length 10 ft is immersed in water having a density of 62.5 lb/ft^3. The *buoyancy force* on the tank is given by $-\iint_S p\mathbf{n} \, dS$, where $\mathbf{n}$ is the unit outer normal to the surface S of the tank and p is the pressure exerted by the water on S. Use the formula
$$\iint_S p\mathbf{n} \, dS = \iiint_Q \nabla p \, dV$$
(see Exercise 24) to calculate the buoyancy force. (*Hint:* If the surface of the water coincides with the xy-plane, then $p = -62.5z$, where z is the depth of the water.)

14.7 STOKES'S THEOREM

Stokes's theorem, the topic of this section, is a generalization of Green's theorem to curves and surfaces in three-dimensional space. In (14.21), we stated the following vector form for Green's theorem:

$$\oint_C \mathbf{F} \cdot \mathbf{T} \, ds = \iint_R (\operatorname{curl} \mathbf{F}) \cdot \mathbf{k} \, dA,$$

Figure 14.55

where the plane curve C is the boundary of R. This result may be extended to a piecewise-smooth simple closed curve C in three dimensions that forms the boundary of a surface S. The sketch in Figure 14.55 illustrates a case in which S is the graph of $z = f(x, y)$, f has continuous first partial derivatives, and the projection C_1 of C onto the xy-plane is a curve that bounds a region R of the type considered in Green's theorem. In the figure, $\mathbf{n}$ is a unit *upper* normal to S. We take the positive direction along C as that which corresponds to the positive direction along C_1. The vector $\mathbf{T}$ is a unit tangent vector to C that points in the positive direction. If $\mathbf{F}$ is a vector field whose components have continuous partial derivatives in a region containing S, then we have the following theorem, named after the English mathematical physicist George G. Stokes (1819–1903).

Stokes's Theorem 14.28

$$\oint_C \mathbf{F} \cdot \mathbf{T} \, ds = \iint_S (\operatorname{curl} \mathbf{F}) \cdot \mathbf{n} \, dS$$

Stokes's theorem may be stated: *The line integral of the tangential component of* $\mathbf{F}$ *taken along* C *once in the positive direction equals the surface integral of the normal component of curl* $\mathbf{F}$ *over* S. If $\mathbf{F}$ is a force field, the theorem states that the work done by $\mathbf{F}$ along C equals the flux of curl $\mathbf{F}$ over S. The line integral in (14.28) may also be written $\oint_C \mathbf{F} \cdot d\mathbf{r}$, where $\mathbf{r}$ is the position vector of the point (x, y, z) on C. To consider situations more general than the one pictured in Figure 14.55, we must consider an *oriented* surface S and define a positive direction along C in a suitable manner. (A proof of Stokes's theorem may be found in more advanced texts.)

EXAMPLE ■ 1   Let S be the part of the paraboloid $z = 9 - x^2 - y^2$ with $z \geq 0$, and let C be the trace of S on the xy-plane. Verify Stokes's theorem (14.28) for the vector field $\mathbf{F} = 3z\mathbf{i} + 4x\mathbf{j} + 2y\mathbf{k}$.

SOLUTION   We wish to show that the two integrals in Theorem (14.28) have the same value.

The surface is the same as that considered in Example 5 of Section 14.5 (see Figure 14.48), where we found that

$$\mathbf{n} = \frac{2x\mathbf{i} + 2y\mathbf{j} + \mathbf{k}}{\sqrt{4x^2 + 4y^2 + 1}}.$$

By (14.6),

$$\text{curl } \mathbf{F} = \begin{vmatrix} \mathbf{i} & \mathbf{j} & \mathbf{k} \\ \dfrac{\partial}{\partial x} & \dfrac{\partial}{\partial y} & \dfrac{\partial}{\partial z} \\ 3z & 4x & 2y \end{vmatrix} = 2\mathbf{i} + 3\mathbf{j} + 4\mathbf{k}.$$

Consequently,

$$\iint\limits_S (\text{curl } \mathbf{F}) \cdot \mathbf{n}\, dS = \iint\limits_S \frac{4x + 6y + 4}{\sqrt{4x^2 + 4y^2 + 1}}\, dS.$$

Using (14.23)(i) to evaluate this surface integral, we have

$$\iint\limits_S (\text{curl } \mathbf{F}) \cdot \mathbf{n}\, dS = \iint\limits_R (4x + 6y + 4)\, dA,$$

where R is the region in the xy-plane bounded by the circle of radius 3 with center at the origin. Changing to polar coordinates, we obtain

$$\iint\limits_S (\text{curl } \mathbf{F}) \cdot \mathbf{n}\, dS = \int_0^{2\pi} \int_0^3 (4r\cos\theta + 6r\sin\theta + 4)r\, dr\, d\theta$$

$$= \int_0^{2\pi} \int_0^3 \left[(4\cos\theta + 6\sin\theta)r^2 + 4r\right] dr\, d\theta$$

$$= \int_0^{2\pi} \left[(4\cos\theta + 6\sin\theta)\tfrac{1}{3}r^3 + 2r^2\right]_0^3 d\theta$$

$$= \int_0^{2\pi} (36\cos\theta + 54\sin\theta + 18)\, d\theta$$

$$= \left[36\sin\theta - 54\cos\theta + 18\theta\right]_0^{2\pi} = 36\pi.$$

The line integral in Stokes's theorem (14.28) may be written

$$\oint_C \mathbf{F} \cdot \mathbf{T}\, ds = \oint_C \mathbf{F} \cdot d\mathbf{r} = \oint_C 3z\, dx + 4x\, dy + 2y\, dz,$$

where C is the circle $x^2 + y^2 = 9$ in the xy-plane. Since $z = 0$ on C, this line integral reduces to

$$\oint_C \mathbf{F} \cdot d\mathbf{r} = \oint_C 4x\, dy = 4\oint_C x\, dy.$$

By Theorem (14.20), $\oint_C x\, dy$ is the area of the region (a circle of radius 3) bounded by C, and hence

$$\oint \mathbf{F} \cdot d\mathbf{r} = 4(9\pi) = 36\pi,$$

which is the same as the value of the surface integral.

Figure 14.56

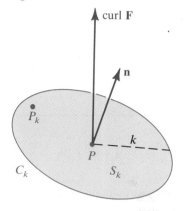

We may use Stokes's theorem to obtain a physical interpretation for curl **F**. If P is any point, let S_k be a circular disk of radius k with center at P, and let C_k denote the boundary of S_k (see Figure 14.56). Applying Stokes's theorem and a mean value theorem for double integrals leads to

$$\oint_{C_k} \mathbf{F} \cdot \mathbf{T} \, ds = \iint_{S_k} (\text{curl } \mathbf{F}) \cdot \mathbf{n} \, dS = [(\text{curl } \mathbf{F}) \cdot \mathbf{n}]_{P_k} (\pi k^2),$$

where P_k is some interior point of S_k and πk^2 is the area of S_k. Thus,

$$[(\text{curl } \mathbf{F}) \cdot \mathbf{n}]_{P_k} = \frac{1}{\pi k^2} \oint_{C_k} \mathbf{F} \cdot \mathbf{T} \, ds.$$

If we let $k \to 0$, then $P_k \to P$ and we obtain the following.

(curl F) · n as a Limit | **14.29**

$$[(\text{curl } \mathbf{F}) \cdot \mathbf{n}]_P = \lim_{k \to 0} \frac{1}{\pi k^2} \oint_{C_k} \mathbf{F} \cdot \mathbf{T} \, ds$$

Figure 14.57
Curl meter

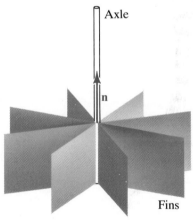

If **F** is the velocity field of a fluid, then the line integral $\oint_C \mathbf{F} \cdot \mathbf{T} \, ds$ is the **circulation around** C. It measures the average tendency of the fluid to move, or *circulate*, around the curve. We see from (14.29) that $[(\text{curl } \mathbf{F}) \cdot \mathbf{n}]_P$ provides information about the motion of the fluid around the circumference of a circular disk that is perpendicular to **n**, as the disk shrinks to a point. Since $[(\text{curl } \mathbf{F}) \cdot \mathbf{n}]_P$ has its maximum value when **n** is parallel to curl **F**, the direction of curl **F** at P is that for which the circulation around the boundary of a disk perpendicular to curl **F** will have its maximum value as the disk shrinks to a point.

To visualize these ideas, consider a miniature (imaginary) paddlewheel, or a **curl meter**, of the type shown in Figure 14.57. A velocity field acting on the fins may cause the wheel to turn on its axle. If **n** is a unit vector directed along the axle, then the curl meter will rotate most rapidly when **n** is parallel to curl **F**.

The scalar $(\text{curl } \mathbf{F}) \cdot \mathbf{n}$ is sometimes called the **rotation** of **F** about **n** and is denoted by rot **F**, since it measures the tendency of the vector field to *rotate* about P. If rot **F** > 0 throughout a region containing the curl meter in Figure 14.57, the fins rotate in a counterclockwise direction (a direction determined by the right-hand rule with thumb along **n**). If rot **F** < 0, the fins rotate in a clockwise direction. If rot **F** $= 0$, as happens in the case when curl **F** $= \mathbf{0}$, then the circulation around every closed curve C is 0 and the fins do not rotate. For this reason, we define an **irrotational vector field F** as one for which curl **F** $= \mathbf{0}$.

Figure 14.58

EXAMPLE = 2 If $\mathbf{F}(x, y) = a\mathbf{i} + b\mathbf{j} + c\mathbf{k}$ and a, b, and c are constants, discuss the rotational properties of the field **F**.

SOLUTION Every vector of the field **F** has the same length and direction, as illustrated in Figure 14.58, where $a, b,$ and c are positive. If we inserted a curl meter into this vector field at any point, the fins would not rotate. Note that

$$\text{curl } \mathbf{F} = \nabla \times \mathbf{F} = \begin{vmatrix} \mathbf{i} & \mathbf{j} & \mathbf{k} \\ \dfrac{\partial}{\partial x} & \dfrac{\partial}{\partial y} & \dfrac{\partial}{\partial z} \\ a & b & c \end{vmatrix} = \mathbf{0}.$$

Hence $\mathbf{F}$ is irrotational.

EXAMPLE • 4   Let $\mathbf{F}(x, y, z) = 3(y^2 + 1)^{-1}\mathbf{i} + 0\mathbf{j} + 0\mathbf{k}$ with $|y| \le 4$.

(a) Describe the vector field $\mathbf{F}$ and discuss the circulation of $\mathbf{F}$ around several circles C_k that lie in the xy-plane.

(b) Where does $(\text{curl } \mathbf{F}) \cdot \mathbf{n}$ have its maximum value?

SOLUTION

(a) If we restrict our attention to the xy-plane, the vectors in $\mathbf{F}$ form the pattern shown in Figure 14.59. (Verify this fact.) The same pattern is repeated in any plane parallel to the xy-plane.

Figure 14.59

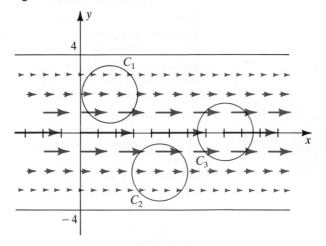

Let us regard $\mathbf{F}$ as the velocity field of a fluid that is flowing in a culvert. Note that the speed of the fluid is greatest along the x-axis and decreases as the point (x, y) approaches either of the lines $y = \pm 4$. If we consider the circle C_1, then evidently $\oint_{C_1} \mathbf{F} \cdot \mathbf{T} \, ds > 0$, since as C_1 is traversed in the positive direction, the tangential component $\mathbf{F} \cdot \mathbf{T}$ is large and positive on the lower half of C_1 and negative (but close to 0) on the upper half. If a curl meter were inserted at the center of C_1, with its axle perpendicular to the xy-plane, the fins would rotate in the counterclockwise direction.

Again considering values of $\mathbf{F} \cdot \mathbf{T}$, we see that $\oint_{C_2} \mathbf{F} \cdot \mathbf{T} \, ds < 0$ and $\oint_{C_3} \mathbf{F} \cdot \mathbf{T} \, ds = 0$. A curl meter for C_2 would rotate in a clockwise direction, and that for C_3 would not rotate at all.

(b) Applying (14.6) gives us

$$\operatorname{curl} \mathbf{F} = \frac{6y}{(y^2+1)^2}\mathbf{k}.$$

Since the unit normal $\mathbf{n}$ is $\mathbf{k}$,

$$(\operatorname{curl} \mathbf{F}) \cdot \mathbf{n} = (\operatorname{curl} \mathbf{F}) \cdot \mathbf{k} = \frac{6y}{(y^2+1)^2}.$$

Observe that $(\operatorname{curl} \mathbf{F}) \cdot \mathbf{n}$ is positive for points above the x-axis, negative for points below the x-axis, and 0 for points on the x-axis. Differentiating with respect to y yields

$$\frac{d}{dy}[(\operatorname{curl} \mathbf{F}) \cdot \mathbf{n}] = \frac{6(1-3y^2)}{(y^2+1)^3},$$

and hence critical numbers for $(\operatorname{curl} \mathbf{F}) \cdot \mathbf{n}$ are $y = \pm 1/\sqrt{3} \approx \pm 0.577$. We see from the first derivative test that these lead to maximum values. Thus, the maximum values for $(\operatorname{curl} \mathbf{F}) \cdot \mathbf{n}$ occur at any point on either of the horizontal lines $y = \pm 1/\sqrt{3}$.

To demonstrate another connection between curl $\mathbf{F}$ and rotational aspects of motion, consider a fluid rotating uniformly about the z-axis, as if it were a rigid body. If we concentrate on a single fluid particle at the point $P(x, y, z)$ and if x, y, and z are functions of time t, then we have a situation similar to that illustrated in Figure 14.60, where $\mathbf{r}(t) = x\mathbf{i} + y\mathbf{j} + z\mathbf{k}$ is the position vector of P and $\boldsymbol{\omega} = \omega_1\mathbf{i} + \omega_2\mathbf{j} + \omega_3\mathbf{k}$ is the (constant) angular velocity. Referring to Figure 14.60, we see that $d\theta/dt = \|\boldsymbol{\omega}\|$. If we denote the velocity $\mathbf{r}'(t)$ of P by $\mathbf{F}$, then, as in Example 4 of Section 11.3,

$$\mathbf{F} = \boldsymbol{\omega} \times \mathbf{r} = (\omega_2 z - \omega_3 y)\mathbf{i} + (\omega_3 x - \omega_1 z)\mathbf{j} + (\omega_1 y - \omega_2 x)\mathbf{k}.$$

By (14.6),

$$\operatorname{curl} \mathbf{F} = \begin{vmatrix} \mathbf{i} & \mathbf{j} & \mathbf{k} \\ \dfrac{\partial}{\partial x} & \dfrac{\partial}{\partial y} & \dfrac{\partial}{\partial z} \\ \omega_2 z - \omega_3 y & \omega_3 x - \omega_1 z & \omega_1 y - \omega_2 x \end{vmatrix}$$

Using the fact that $\boldsymbol{\omega}$ is a constant vector, we may verify that

$$\operatorname{curl} \mathbf{F} = 2\omega_1\mathbf{i} + 2\omega_2\mathbf{j} + 2\omega_3\mathbf{k} = 2\boldsymbol{\omega}.$$

This shows that the magnitude of curl $\mathbf{F}$ is twice the angular velocity, and the direction of curl $\mathbf{F}$ is along the axis of rotation.

We can apply the preceding discussion to the velocity field $\mathbf{F}$ of the wind inside the core of a tornado. (We often call $\mathbf{F}$ the *tangential wind velocity*.) The study of the circulation around a curve C inside the core is important in the analysis of a tornado. In the next example, we calculate the circulation if C is a circle.

Figure 14.60

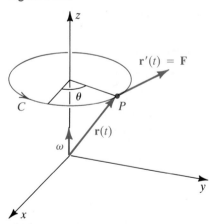

EXAMPLE ▪ 4   The core of a tornado rotates as if it were a rigid body, in the manner illustrated in Figure 14.60, where the angular wind speed $\omega = d\theta/dt$ is constant. If C is a circle of radius a in a cross section of the core, as shown in the figure, the magnitude of the tangential wind velocity $\mathbf{v}$ is $\|\mathbf{v}\| = \omega a$. Find the circulation of $\mathbf{v}$ around C by

(a) using Stokes's theorem   **(b)** evaluating $\int_C \mathbf{v} \cdot \mathbf{T}\, ds$

SOLUTION

(a) Applying Stokes's theorem, we find that the circulation is

$$\int_C \mathbf{v} \cdot \mathbf{T}\, ds = \iint_S (\text{curl } \mathbf{v}) \cdot \mathbf{n}\, dS,$$

where S is the disk bounded by C and where $\mathbf{n} = \mathbf{k}$. From the discussion preceding this example,

$$\text{curl } \mathbf{v} = 2\boldsymbol{\omega} = 2\omega\mathbf{k}.$$

Hence,

$$\int_C \mathbf{v} \cdot \mathbf{T}\, ds = \iint_S (2\omega\mathbf{k}) \cdot \mathbf{k}\, dS$$

$$= 2\omega \iint_S dS$$

$$= 2\omega\pi a^2.$$

This shows that as the radius a of C increases, the tendency of the winds to move around C also increases.

(b) Since $\mathbf{T}$ is a unit tangent vector to C having the same direction as $\mathbf{v}$,

$$\mathbf{v} \cdot \mathbf{T} = \|\mathbf{v}\|\, \|\mathbf{T}\| = \|\mathbf{v}\|$$

(see Theorem (10.19)). Hence,

$$\int_C \mathbf{v} \cdot \mathbf{T}\, ds = \int_C \|\mathbf{v}\|\, ds = \omega a \int_C ds$$

$$= \omega a (2\pi a) = 2\omega\pi a^2.$$

EXAMPLE ▪ 5   For points outside the core, a tornado does not rotate as if it were a rigid body. Let r_0 be the radius of the core and ω the (constant) angular wind speed inside the core, as in Example 4. If C is a cross-sectional circle of radius $r > r_0$, as illustrated in Figure 14.61, then the magnitude of the tangential wind velocity $\mathbf{v}$ on C is (approximately) $\|\mathbf{v}\| = \omega r_0^2/r$.

(a) Show that the circulation around C is the same for every $r > r_0$.

(b) Show that curl $\mathbf{v} = \mathbf{0}$ outside the core.

SOLUTION

(a) The circulation around every circle C of radius $r > r_0$ is

$$\int_C \mathbf{v} \cdot \mathbf{T}\, ds = \int_C \|\mathbf{v}\|\, ds$$

$$= \frac{\omega r_0^2}{r} \int_C ds$$

$$= \frac{\omega r_0^2}{r} 2\pi r = 2\pi \omega r_0^2.$$

(b) A cross-sectional view of the tornado is shown in Figure 14.61, where $\mathbf{r} = x\mathbf{i} + y\mathbf{j}$ is the position vector of a point $P(x, y)$ on C and $\|\mathbf{r}\| = r$. In Example 1 of Section 14.1, we proved that $\mathbf{F}(x, y) = -y\mathbf{i} + x\mathbf{j}$ is orthogonal to $\mathbf{r}$ at every point (x, y), as illustrated in the figure. Hence a *unit* vector $\mathbf{u}$ in the direction of $\mathbf{F}$ is

$$\mathbf{u} = \frac{1}{\sqrt{x^2 + y^2}}(-y\mathbf{i} + x\mathbf{j}).$$

Since the magnitude of $\mathbf{v}$ is $\omega r_0^2/r$ and $r = \sqrt{x^2 + y^2}$,

$$\mathbf{v} = \frac{\omega r_0^2}{r}\mathbf{u} = \frac{\omega r_0^2}{x^2 + y^2}(-y\mathbf{i} + x\mathbf{j}).$$

Applying Definition (14.6), we obtain

$$\text{curl } \mathbf{v} = \nabla \times \mathbf{v} = \mathbf{0}.$$

Figure 14.61

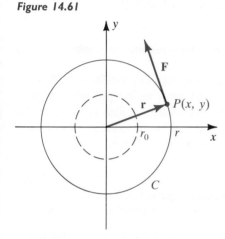

In our discussion of line integrals that are independent of path, we introduced the notion of a simply connected region in the plane. This concept can be extended to three dimensions; however, the usual definition requires properties of curves and surfaces studied in more advanced courses. For our purposes, a region D in three dimensions will be **simply connected** if every simple closed curve C in D is the boundary of a surface S in D that satisfies the conditions of Stokes's theorem. In regions of this type, any simple closed curve can be continuously shrunk to a point in D without crossing the boundary of D. For example, the region inside of a sphere or a rectangular box is simply connected. The region inside of a torus (a doughnut-shaped surface) is not simply connected. Using this restriction, we can establish the following result.

Theorem 14.30

If $\mathbf{F}(x, y, z)$ has continuous partial derivatives throughout a simply connected region D, then curl $\mathbf{F} = \mathbf{0}$ in D if and only if $\oint_C \mathbf{F} \cdot d\mathbf{r} = 0$ for every simple closed curve C in D.

PROOF   If curl $\mathbf{F} = \mathbf{0}$, then, by Stokes's theorem (14.28),

$$\oint_C \mathbf{F} \cdot d\mathbf{r} = \iint_S (\text{curl } \mathbf{F}) \cdot \mathbf{n}\, dS = 0.$$

Conversely, suppose $\oint_C \mathbf{F} \cdot d\mathbf{r} = 0$ for every simple closed curve C. If, at some point P, curl $\mathbf{F} \neq \mathbf{0}$, then, by continuity, there is a subregion of D containing P throughout which curl $\mathbf{F} \neq \mathbf{0}$. If, in this subregion, we choose a circular disk as illustrated in Figure 14.56 with $\mathbf{n}$ parallel to curl $\mathbf{F}$, then

$$\oint_{C_k} \mathbf{F} \cdot d\mathbf{r} = \iint_{S_k} (\text{curl } \mathbf{F}) \cdot \mathbf{n}\, dS > 0,$$

a contradiction. Consequently, curl $\mathbf{F} = \mathbf{0}$ throughout D.   ■

If $\mathbf{F}$ is continuous on a suitable region, then the condition $\oint_C \mathbf{F} \cdot d\mathbf{r} = 0$ for every simple closed curve is equivalent to independence of path. Moreover, as in Theorem (14.13), independence of path is equivalent to $\mathbf{F} = \nabla f$ for some scalar function f (that is, $\mathbf{F}$ is conservative). Combining these facts with Theorem (14.30), we obtain the following.

Theorem 14.31

If $\mathbf{F}(x, y, z)$ has continuous first partial derivatives throughout a simply connected region D, then the following conditions are equivalent:

  (i) $\mathbf{F}$ is conservative; that is, $\mathbf{F} = \nabla f$ for some scalar function f.
  (ii) $\int_C \mathbf{F} \cdot d\mathbf{r}$ is independent of path in D.
  (iii) $\oint_C \mathbf{F} \cdot d\mathbf{r} = 0$ for every simple closed curve C in D.
  (iv) $\mathbf{F}$ is irrotational—that is, curl $\mathbf{F} = \mathbf{0}$.

EXAMPLE ■ 6   Prove that if

$$\mathbf{F}(x, y, z) = (3x^2 + y^2)\mathbf{i} + 2xy\mathbf{j} - 3z^2\mathbf{k},$$

then $\mathbf{F}$ is conservative.

SOLUTION   The function $\mathbf{F}$ has continuous first partial derivatives for every (x, y, z). Since

$$\text{curl } \mathbf{F} = \begin{vmatrix} \mathbf{i} & \mathbf{j} & \mathbf{k} \\ \dfrac{\partial}{\partial x} & \dfrac{\partial}{\partial y} & \dfrac{\partial}{\partial z} \\ 3x^2 + y^2 & 2xy & -3z^2 \end{vmatrix}$$
$$= (0 - 0)\mathbf{i} + (0 - 0)\mathbf{j} + (2y - 2y)\mathbf{k} = \mathbf{0},$$

$\mathbf{F}$ is conservative, by Theorem (14.31)(iv).

EXERCISES 14.7

Exer. 1–4: Verify Stokes's theorem (14.28) for F and S.

1 $F = y^2\mathbf{i} + z^2\mathbf{j} + x^2\mathbf{k}$;
S is the first-octant portion of the plane $x + y + z = 1$.

2 $F = 2y\mathbf{i} - z\mathbf{j} + 3\mathbf{k}$;
S is the portion of the paraboloid $z = 4 - x^2 - y^2$ that lies inside the cylinder $x^2 + y^2 = 1$.

3 $F = z\mathbf{i} + x\mathbf{j} + y\mathbf{k}$;
S is the hemisphere $z = (a^2 - x^2 - y^2)^{1/2}$.

4 $F = x^2\mathbf{i} + y^2\mathbf{j} + z^2\mathbf{k}$;
S is the part of the cone $z = (x^2 + y^2)^{1/2}$ cut off by the plane $z = 1$.

5 If $F = (3z - \sin x)\mathbf{i} + (x^2 + e^y)\mathbf{j} + (y^3 - \cos z)\mathbf{k}$, use Stokes's theorem to evaluate $\oint_C F \cdot d\mathbf{r}$, where C is the curve given by $x = \cos t, y = \sin t, z = 1; 0 \le t \le 2\pi$.

6 If $F = yz\mathbf{i} + xy\mathbf{j} + xz\mathbf{k}$ and C is the square with vertices $(0, 0, 2), (1, 0, 2), (1, 1, 2)$, and $(0, 1, 2)$, use Stokes's theorem to evaluate $\oint_C F \cdot d\mathbf{r}$.

7 If $F = 2y\mathbf{i} + e^z\mathbf{j} - \arctan x\mathbf{k}$ and S is the portion of the paraboloid $z = 4 - x^2 - y^2$ cut off by the xy-plane, use Stokes's theorem to evaluate $\iint_S \text{curl } F \cdot \mathbf{n}\, dS$.

8 If $F = xy^2\mathbf{i} + yz^2\mathbf{j} + zx^2\mathbf{k}$ and S is the region bounded by the triangle with vertices $(1, 0, 0), (0, 2, 0)$, and $(0, 0, 3)$, use Stokes's theorem to evaluate the integral $\iint_S \text{curl } F \cdot \mathbf{n}\, dS$.

Exer. 9–12: Sketch some of the field vectors F(x, y, z), and discuss the rotational properties of F by using both a curl meter and curl F. Where does (curl F) · n have its maximum value?

9 $F(x, y, z) = (y^2 - 2y)\mathbf{i} + 0\mathbf{j} + 0\mathbf{k}$; $0 \le y \le 2$

10 $F(x, y, z) = 0\mathbf{i} + \sin x\mathbf{j} + 0\mathbf{k}$; $0 \le x \le \pi$

11 $F(x, y, z) = -y\mathbf{i} + x\mathbf{j} + 0\mathbf{k}$ (see Example 1 of Section 14.1 and the discussion following Example 3 on p. 1227).

12 The inverse square field of Definition (14.2)

13 If $F(x, y, z) = y\mathbf{i} + (x + e^z)\mathbf{j} + (1 + ye^z)\mathbf{k}$, prove that F is irrotational.

14 A vector field F is a *central force field* if $F = f(r)\mathbf{r}$, where $\mathbf{r} = x\mathbf{i} + y\mathbf{j} + z\mathbf{k}, r = \|\mathbf{r}\|$, and f is a scalar function. If f is differentiable, show that F is irrotational.

15 Let $\mathbf{n}$ denote the unit outer normal at any point P on the surface of a sphere S. If F has continuous first partial derivatives within and on S, prove that $\iint_S \text{curl } F \cdot \mathbf{n}\, dS = 0$ by using

(a) the divergence theorem

(b) Stokes's theorem

16 If S and C satisfy the conditions of Stokes's theorem and if F is a constant vector function, use (14.28) to prove that $\oint_C F \cdot \mathbf{T}\, ds = 0$.

Exer. 17–19: Establish the identity under the assumption that C and S satisfy the conditions of Stokes's theorem.

17 $\oint_C f\nabla g \cdot d\mathbf{r} = \iint_S (\nabla f \times \nabla g) \cdot \mathbf{n}\, dS$, where f and g are scalar functions

18 $\oint_C \mathbf{a} \times \mathbf{r} \cdot d\mathbf{r} = 2\mathbf{a} \cdot \iint_S \mathbf{n}\, dS$, where $\mathbf{a}$ is a constant vector

19 $\oint_C f\, d\mathbf{r} = \iint_S \mathbf{n} \times \nabla f\, dS$, where f is a scalar function

20 If M, N, and P are functions of x, y, and z that have continuous first partial derivatives on a simply connected region, prove that

$$\int_C M(x, y, z)\, dx + N(x, y, z)\, dy + P(x, y, z)\, dz$$

is independent of path if and only if

$$\frac{\partial M}{\partial y} = \frac{\partial N}{\partial x}, \quad \frac{\partial M}{\partial z} = \frac{\partial P}{\partial x}, \quad \frac{\partial N}{\partial z} = \frac{\partial P}{\partial y}.$$

CHAPTER 14 REVIEW EXERCISES

Exer. 1–4: Sketch some typical vectors in the vector field F.

1 $\mathbf{F}(x, y) = 2x\mathbf{i} + y\mathbf{j}$

2 $\mathbf{F}(x, y, z) = x\mathbf{i} + y\mathbf{j} + \mathbf{k}$

3 $\mathbf{F}(x, y, z) = -\mathbf{k}$

4 $\mathbf{F}(x, y, z) = \nabla(x^2 + y^2 + z^2)^{-1/2}$

5 Find a conservative vector field in two dimensions that has the potential function $f(x, y) = y^2 \tan x$.

6 Find a conservative vector field in three dimensions that has the potential function $f(x, y, z) = \ln(x + y + z)$.

c **Exer. 7–8: Plot some of the vectors in F, as in Figure 14.5.**

7 $\mathbf{F}(x, y) = \dfrac{x^3\mathbf{i} + y^3\mathbf{j}}{x^2 + y^2}$ 8 $\mathbf{F}(x, y) = \dfrac{y\mathbf{i} + x\mathbf{j}}{e^{x^2+y^2}}$

Exer. 9–12: Evaluate $\int_C y^2\, dx + xy\, dy$, where C is the curve from $(1, 0)$ to $(-1, 4)$ shown in the figure.

9

10

11

$x = 1 - t$
$y = t^2$

12

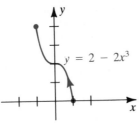

$y = 2 - 2x^3$

13 Evaluate $\int_C xy\, ds$, where C is the graph of $y = x^4$ from $(-1, 1)$ to $(2, 16)$.

Exer. 14–16: Evaluate

$$\int_C x\, dx + (x + y)\, dy + (x + y + z)\, dz$$

for the given curve C from $A(0, 0, 0)$ to $B(2, 4, 8)$.

14 C consists of three line segments, the first parallel to the z-axis, the second parallel to the x-axis, and the third parallel to the y-axis.

15 C consists of the line segment from A to B.

16 C has the parametrization $x = t$, $y = t^2$, $z = t^3$.

17 If $\mathbf{F}(x, y) = (x + y)\mathbf{i} + (x - y)\mathbf{j}$ and C is given by $x = \cos t$, $y = \sin t$; $-\pi \le t \le 0$, evaluate $\int_C \mathbf{F} \cdot d\mathbf{r}$.

18 The force $\mathbf{F}$ at a point (x, y, z) in three dimensions is $\mathbf{F}(x, y, z) = xy\mathbf{i} + y^2z\mathbf{j} + xz^2\mathbf{k}$. If C is the square with vertices $A_1(1, 1, 1)$, $A_2(-1, 1, 1)$, $A_3(-1, -1, 1)$, and $A_4(1, -1, 1)$, find the work done by $\mathbf{F}$ as its point of application moves around C once in the direction determined by increasing subscripts on the A_k.

c **Exer. 19–20: Approximate, to three decimal places, the line integrals $\int_C f(x, y)\, ds, \int_C f(x, y)\, dx$, and $\int_C f(x, y)\, dy$ if C has the given parametrization.**

19 $f(x, y) = \sqrt{x^3 + y^3}$, $x = 2t^2$, $y = 4t + t^2$;
$$0 \le t \le 1$$

20 $f(x, y) = \cos(x + y^2)$, $x = 1 + \cos t$, $y = 2 - 2\sin t$;
$$0 \le t \le 2\pi$$

c **Exer. 21–22: Approximate the line integral to three decimal places if C has the given parametrization.**

21 $\int_C \cos(y + z)\, dx + \cos(x + z)\, dy + \cos(x + y)\, dz$,
$x = 2t - 3t^2$, $y = 4t - 5t^2$, $z = 4t - 2t^2$;
$$0 \le t \le 1$$

22 $\int_C e^y\, dx + \sin x\, dy$, $x = \cos(0.5t)$, $y = \sin(0.7t)$;
$$0 \le t \le 2$$

Exer. 23–24: Prove that the line integral is independent of path, and find its value.

23 $\displaystyle\int_{(1, -1)}^{(2, 3)} (x + y)\, dx + (x + y)\, dy$

24 $\displaystyle\int_{(0, 0, 0)}^{(2, 1, 3)} (8x^3 + z^2)\, dx - 3z\, dy + (2xz - 3y)\, dz$

25 If

$$\mathbf{F}(x,\ y,\ z) = 2xe^{2y}\mathbf{i} + 2(x^2e^{2y} + y\cot z)\mathbf{j} - y^2\csc^2 z\mathbf{k},$$

prove that $\int_C \mathbf{F} \cdot d\mathbf{r}$ is independent of path, and find a potential function f for $\mathbf{F}$.

Exer. 26–28: Use Green's theorem to evaluate the line integral

$$\oint_C xy\,dx + (x^2 + y^2)\,dy$$

if C is the given curve.

26 C is the closed curve determined by $y = x^2$ and $y - x = 2$ between $(-1,\ 1)$ and $(2,\ 4)$.

27 C is the triangle with vertices $(0,\ 0)$, $(1,\ 0)$, $(0,\ 1)$.

28 C is the circle $x^2 + y^2 - 2x = 0$.

29 If $\mathbf{F}(x,\ y,\ z) = x^3z^4\mathbf{i} + xyz^2\mathbf{j} + x^2y^2\mathbf{k}$, find div $\mathbf{F}$ and curl $\mathbf{F}$.

30 If f and g are scalar functions of three variables, prove that $\nabla \cdot (f\nabla g) = f\nabla^2 g + \nabla f \cdot \nabla g$.

31 Evaluate $\iint_S xyz\,dS$ if S is the part of the plane $z = x + y$ that lies over the triangular region in the xy-plane having vertices $(0,\ 0,\ 0)$, $(1,\ 0,\ 0)$, and $(0,\ 2,\ 0)$.

32 Evaluate $\iint_S x^2z^2\,dS$ if S is the top half of the cylinder $y^2 + z^2 = 4$ between $x = 0$ and $x = 1$.

33 Let Q be the region bounded by the cylinder $x^2 + y^2 = 1$ and the planes $z = 0$ and $z = 1$. If the vector field $\mathbf{F} = x^3\mathbf{i} + y^3\mathbf{j} + z^3\mathbf{k}$, use the divergence theorem to find $\iint_S \mathbf{F} \cdot \mathbf{n}\,dS$, where S is the surface of Q and $\mathbf{n}$ is the unit outer normal to S.

34 Verify the divergence theorem if $\mathbf{F} = 2x\mathbf{i} + y\mathbf{j} - z\mathbf{k}$ and S is the surface of the rectangular box bounded by the planes $x = \pm 1$, $y = \pm 2$, $z = \pm 3$.

Exer. 35–36: Verify Stokes's theorem for $\mathbf{F}$ and S.

35 $\mathbf{F} = y^2\mathbf{i} + 2x\mathbf{j} + 5y\mathbf{k}$;
S is the hemisphere $z = (4 - x^2 - y^2)^{1/2}$.

36 $\mathbf{F} = (x + y)\mathbf{i} + (y + z)\mathbf{j} + (x + z)\mathbf{k}$;
S is the region bounded by the triangle with vertices $(1,\ 0,\ 0)$, $(0,\ 1,\ 0)$, $(0,\ 0,\ 1)$.

EXTENDED PROBLEMS AND GROUP PROJECTS

1 In what ways does the potential function f for a vector field $\mathbf{F}$ play the role of an antiderivative? Is it reasonable to think of the vector differential operator ∇ as analogous to the differentiation operation for a real-valued function of one variable? Since ∇ also operates on a vector field to produce curl and divergence, starting with a scalar function f and operating twice with ∇ might be analogous to taking a second derivative. Discuss this analogy. Note that curl $\mathbf{F} = \nabla \times \mathbf{F} = \nabla \times (\nabla f)$ and div $\mathbf{F} = \nabla \cdot (\nabla f)$, so there are essentially two different ways of applying ∇ twice.

2 Let $\gamma(t) = (g(t), h(t))$ for $a \le t \le b$ be a parametrization of a smooth curve C. Show that the conclusion of Theorem (14.14) may be represented as

$$\int_C \nabla f = f(\gamma(b)) - f(\gamma(a)).$$

Discuss the claim that "This formula is a sort of fundamental theorem of calculus for line integrals, expressing the integral of the derivative of f as the difference of the values of f at the endpoints of the curve of integration." In what ways can Green's theorem, Stokes's theorem, and the divergence theorem be considered "fundamental theorems of calculus"?

3 Our proof of Green's theorem assumed that the region R is both an R_x region and an R_y region. Extend the proof to consider first the case in which part of the boundary of R consists of horizontal or vertical segments and then the more general case in which R is a finite union of such regions. What is the most general case in which Green's theorem remains true?

INTRODUCTION

THE GOLDEN GATE BRIDGE is one of the largest and most spectacular suspension bridges in the world, spanning a channel about four miles long and one mile wide. Because the channel is relatively narrow, strong winds blowing against the bridge pose potential problems of vibration that could threaten the safety and stability of the bridge. The Tacoma Narrows Bridge, built about the same time as the Golden Gate Bridge, collapsed only a few months after completion when its deck began to oscillate wildly. Civil engineers who design and build large structures such as bridges and skyscrapers need to understand the behavior of vibrating systems. Differential equations provide the mathematical language to help gain this understanding.

Earlier in the text, we solved certain types of differential equations by separating variables and then using indefinite integration. Section 15.1 contains additional examples of *separable* equations, and also presents basic definitions about *solutions* of differential equations. In later sections, we discuss several other varieties that can be solved systematically using standard techniques and formulas. Section 15.2 discusses first-order linear differential equations, which can also be solved in principle by indefinite integration. In Section 15.3, we consider some of the theory underlying the solution of second-order linear differential equations and present the general solution for the homogeneous version of these equations with constant coefficients. We consider the solution of the nonhomogeneous version of such equations in Section 15.4. Although the latter differential equations are more general than separable equations, they are still very specialized. In modern applications that involve placing satellites into orbit or sending a spacecraft to distant parts of the solar system, the differential equations are very complicated and supercomputers are used to obtain approximations to solutions. We discuss Euler's method (Section 15.1) and the Runge–Kutta method (Section 15.2) as examples of two numerical approximation methods.

The major application we study in this chapter is vibrations in mechanical systems that are caused by external forces. In Section 15.5, we see how to use second-order linear differential equations to study both damped and undamped vibrations. We also discuss the phenomenon of *resonance*, which can cause even large and apparently stable structures, such as airplane wings, bridges, and buildings, to undergo destructive oscillations.

Differential equations help model the
responses of large structures to
external vibrations so that planners
can create designs that ensure safety.

Differential Equations

SEPARABLE DIFFERENTIAL EQUATIONS

If y is a function of x and if n is a positive integer, then an equation that involves x, y, y', y'', $\ldots$, $y^{(n)}$ is called an **ordinary differential equation of order n**.

ILLUSTRATION

Differential equation	Order
$y' = 2x$	1
$\dfrac{d^2 y}{dx^2} + x^2 \left(\dfrac{dy}{dx}\right)^3 - 15y = 0$	2
$(y''')^4 - x^2(y'')^5 + 4xy = xe^x$	3
$\left(\dfrac{d^4 y}{dx^4}\right)^2 - 1 = x^3 \dfrac{dy}{dx}$	4

In this section, we discuss what it means to *solve* a differential equation, examine separable differential equations, and discuss an application to the problem of orthogonal trajectories. We also introduce the first of several methods for obtaining numerical approximations to solutions of differential equations.

We considered differential equations briefly in Sections 4.1 and 6.6. Recall that a function f (or $f(x)$) is a **solution** of a differential equation if substitution of $f(x)$ for y results in an identity for every x in some interval. For example, the differential equation

$$y' = 6x^2 - 5$$

has the solution

$$f(x) = 2x^3 - 5x + C$$

for every real number C, since substitution of $f(x)$ for y leads to the identity $6x^2 - 5 = 6x^2 - 5$. We call $f(x) = 2x^3 - 5x + C$ the **general solution** of $y' = 6x^2 - 5$, since every solution is of this form. A **particular solution** is obtained by assigning to C a specific value. To illustrate, if $C = 0$, we obtain the particular solution $y = 2x^3 - 5x$. Sometimes **initial conditions** are stated that determine a particular solution, as illustrated in the following example.

EXAMPLE ■ 1 Given the differential equation $y' = 2x$,

(a) find the general solution and illustrate it graphically

(b) find the particular solution that satisfies the condition $y = 3$ if $x = 0$

Figure 15.1

$y = x^2 + C$

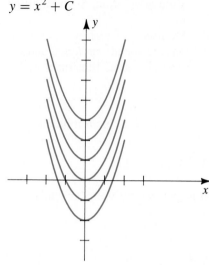

SOLUTION

(a) We use indefinite integration as follows:

$$y' = 2x$$

$$\int y'\,dx = \int 2x\,dx$$

$$y = x^2 + C$$

This is the general solution of the differential equation. Particular solutions are obtained by assigning values to C. This leads to the family of parabolas illustrated in Figure 15.1.

(b) If $y = 3$ when $x = 0$, then substitution in the general solution $y = x^2 + C$ gives us $3 = 0 + C$, or $C = 3$. Hence the particular solution is $y = x^2 + 3$. The graph is the parabola in Figure 15.1 with y-intercept 3.

Differential equations of the type given in the next example occur in the analysis of vibrations. We will consider these in detail in Sections 15.3–15.5.

EXAMPLE ■ 2 Show that the differential equation $y'' - 25y = 0$ has the solution

$$y = C_1 e^{5x} + C_2 e^{-5x}$$

for all real numbers C_1 and C_2.

SOLUTION Differentiating y twice, we have

$$y' = 5C_1 e^{5x} - 5C_2 e^{-5x}$$

$$y'' = 25C_1 e^{5x} + 25C_2 e^{-5x}.$$

Substituting for y and y'' in the differential equation $y'' - 25y = 0$ gives us

$$(25C_1 e^{5x} + 25C_2 e^{-5x}) - 25(C_1 e^{5x} + C_2 e^{-5x}) = 0,$$

or $\quad (25C_1 e^{5x} - 25C_1 e^{5x}) + (25C_2 e^{-5x} - 25C_2 e^{-5x}) = 0.$

Since the left side is 0 for every x, it follows that $y = C_1 e^{5x} + C_2 e^{-5x}$ is a solution of $y'' - 25y = 0$.

The solution $y = C_1 e^{5x} + C_2 e^{-5x}$ in Example 2 is the general solution of $y'' - 25y = 0$. Observe that the differential equation is of order 2 and the general solution contains two arbitrary constants (called **parameters**) C_1 and C_2. The precise definition of a general solution involves the concept of **independent parameters** and is left for more advanced courses. General solutions of nth-order differential equations contain n independent parameters $C_1, C_2, \ldots, C_n$. A particular solution is obtained by assigning a specific value to each parameter. Some differential equations have solutions that are not special cases of the general solution. Such **singular solutions** will not be discussed in this text.

The solutions in Examples 1 and 2 express y *explicitly* in terms of x. Solutions of certain differential equations are stated *implicitly*, and we use implicit differentiation to check such solutions, as illustrated in the following example.

EXAMPLE ■ 3   Show that $x^3 + x^2 y - 2y^3 = C$ is an implicit solution of

$$(x^2 - 6y^2)y' + 3x^2 + 2xy = 0.$$

SOLUTION   If the first equation is satisfied by $y = f(x)$, then differentiating implicitly yields

$$3x^2 + 2xy + x^2 y' - 6y^2 y' = 0,$$

or
$$(x^2 - 6y^2)y' + 3x^2 + 2xy = 0.$$

Thus, $f(x)$ is a solution of the differential equation.

We examine several types of first-order differential equations that can be written in the form

$$y' = f(x, y), \quad \text{or} \quad \frac{dy}{dx} = f(x, y),$$

where f is a specified function of two variables, x and y.

The simplest type of such a first-order differential equation has the form

$$y' = f(x), \quad \text{or} \quad \frac{dy}{dx} = f(x),$$

where f is a specified function only of the variable x. We examined such differential equations in Section 4.1, where we saw that the solution of the differential equation is any antiderivative of f—that is, $y = \int f(x)\,dx$.

Another simple type of differential equation is

$$M(x) + N(y)y' = 0, \quad \text{or} \quad M(x) + N(y)\frac{dy}{dx} = 0,$$

where M and N are continuous functions. If $y = f(x)$ is a solution, then

$$M(x) + N(f(x))f'(x) = 0.$$

If $f'(x)$ is continuous, then indefinite integration leads to

$$\int M(x)\,dx + \int N(f(x))f'(x)\,dx = C,$$

or
$$\int M(x)\,dx + \int N(y)\,dy = C.$$

The last equation is an (implicit) solution of the differential equation. The differential equation $M(x) + N(y)y' = 0$ is **separable**, since the variables x and y may be separated as we have indicated.

An easy way to remember the method of separating the variables is to change the equation

$$M(x) + N(y)\frac{dy}{dx} = 0$$

to the *differential form*

$$M(x)\,dx + N(y)\,dy = 0$$

and then integrate each term.

An equivalent form for a separable differential equation is

$$y' = f(x)g(y), \quad \text{or} \quad \frac{dy}{dx} = f(x)g(y),$$

where the derivative of the unknown function can be written as the product of a function of x and a function of y. If we divide the equation by $g(y)$, we have

$$f(x) = \frac{1}{g(y)}\frac{dy}{dx}, \quad \text{or, equivalently,} \quad f(x) - \frac{1}{g(y)}\frac{dy}{dx} = 0,$$

which has the required form

$$M(x) + N(y)\frac{dy}{dx} = 0$$

with $M(x) = f(x)$ and $N(y) = 1/g(y)$.

EXAMPLE ■ 4   Solve the differential equation $y^4 e^{2x} + \dfrac{dy}{dx} = 0$.

SOLUTION   We first express the equation in differential form,

$$y^4 e^{2x}\,dx + dy = 0.$$

If $y \neq 0$, we may separate the variables by dividing by y^4, as follows:

$$e^{2x}\,dx + \frac{1}{y^4}\,dy = 0$$

Integrating each term, we obtain the (implicit) solution

$$\frac{1}{2}e^{2x} - \frac{1}{3y^3} = C.$$

If we multiply both sides by 6, then another form of the solution is

$$3e^{2x} - \frac{2}{y^3} = K$$

with $K = 6C$. If an explicit form is desired, we may solve for y, obtaining

$$y = \left(\frac{2}{3e^{2x} - K}\right)^{1/3}.$$

We have assumed that $y \neq 0$; however, $y = 0$ *is* a solution of the differential equation. In future examples, we will not point out such singular solutions.

EXAMPLE ▪ 5 Solve the differential equation

$$2y + (xy + 3x)\frac{dy}{dx} = 0,$$

where $x \neq 0$.

SOLUTION We first express the equation in differential form as follows:

$$2y\,dx + (xy + 3x)\,dy = 0$$

$$2y\,dx + x(y + 3)\,dy = 0$$

The variables may be separated by dividing both sides of the last equation by xy. This gives us

$$\frac{2}{x}\,dx + \left(\frac{y + 3}{y}\right)dy = 0,$$

or

$$\frac{2}{x}\,dx + \left(1 + \frac{3}{y}\right)dy = 0.$$

Integrating each term leads to the following equivalent equations:

$$2\ln|x| + y + 3\ln|y| = c$$

$$\ln|x|^2 + \ln|y|^3 = c - y$$

$$\ln|x|^2\,|y|^3 = c - y$$

Changing to exponential form, we have

$$|x|^2\,|y|^3 = e^{c-y} = e^c e^{-y} = ke^{-y},$$

where $k = e^c$. We can show, by differentiation, that

$$x^2 y^3 = ke^{-y}, \quad \text{or} \quad x^2 y^3 e^y = k,$$

is an implicit solution of the differential equation.

EXAMPLE ▪ 6 Solve the differential equation

$$(1 + y^2) + (1 + x^2)\frac{dy}{dx} = 0.$$

SOLUTION We first express the equation in differential form and then separate the variables, obtaining

$$(1 + y^2)\,dx + (1 + x^2)\,dy = 0$$

$$\frac{1}{1 + x^2}\,dx + \frac{1}{1 + y^2}\,dy = 0.$$

Integrating each term, we obtain the (implicit) solution

$$\tan^{-1} x + \tan^{-1} y = C.$$

To find an explicit solution, we may solve for y as follows:

$$\tan^{-1} y = C - \tan^{-1} x$$
$$y = \tan(C - \tan^{-1} x)$$

The form of this solution may be changed by using the subtraction formula for the tangent function. Thus,

$$y = \frac{\tan C - \tan(\tan^{-1} x)}{1 + \tan C \tan(\tan^{-1} x)}.$$

If we let $k = \tan C$ and use the fact that $\tan(\tan^{-1} x) = x$, this may be written

$$y = \frac{k - x}{1 + kx}.$$

Figure 15.2
$y = 2x + b; \ y = -\frac{1}{2}x + c$

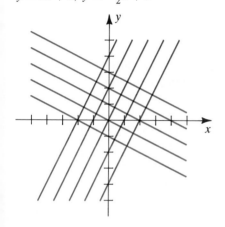

An **orthogonal trajectory** of a family of curves is a curve that intersects each curve of the family orthogonally. We shall restrict our discussion to curves in a coordinate plane. To illustrate, for the family $y = 2x + b$ of all lines of slope 2, every line $y = -\frac{1}{2}x + c$ of slope $-\frac{1}{2}$ is an orthogonal trajectory. Several curves from each family are sketched in Figure 15.2. We say that two such families of curves are **mutually orthogonal**. As another example, the family $y = mx$ of all lines through the origin and the family $x^2 + y^2 = a^2$ of all concentric circles with centers at the origin are mutually orthogonal (see Figure 15.3).

Pairs of mutually orthogonal families of curves occur frequently in applications. In the theory of electricity and magnetism, the lines of force associated with a given field are orthogonal trajectories of the corresponding equipotential curves. Similarly, the stream lines studied in aerodynamics and hydrodynamics are orthogonal trajectories of the *velocity-equipotential* curves. As a final illustration, in the study of thermodynamics, the flow of heat across a plane surface is orthogonal to the isothermal curves.

The next example illustrates a technique for finding the orthogonal trajectories of a family of curves.

Figure 15.3
$y = mx; \ x^2 + y^2 = a^2$

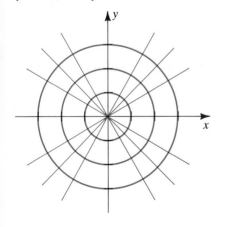

EXAMPLE ■ 7 Find the orthogonal trajectories of the family of ellipses $x^2 + 3y^2 = c$, and sketch several members of each family.

SOLUTION Differentiating the given equation implicitly, we obtain

$$2x + 6yy' = 0, \quad \text{or} \quad y' = -\frac{x}{3y}.$$

Hence the slope of the tangent line at any point (x, y) on each of the ellipses is $y' = -x/(3y)$. If dy/dx is the slope of the tangent line on a corresponding orthogonal trajectory, then it must equal the negative

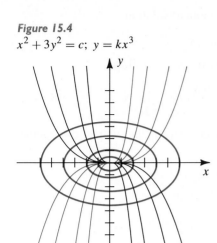

Figure 15.4
$x^2 + 3y^2 = c$; $y = kx^3$

reciprocal of y'. This gives us the following differential equation for the family of orthogonal trajectories:

$$\frac{dy}{dx} = \frac{3y}{x}$$

Separating the variables, we have

$$\frac{dy}{y} = 3\frac{dx}{x}.$$

Integrating and writing the constant of integration as $\ln |k|$ gives us

$$\ln |y| = 3 \ln |x| + \ln |k| = \ln \left|kx^3\right|.$$

It follows that $y = kx^3$ is an equation for the family of orthogonal trajectories. We have sketched several members of the family of ellipses (in blue) and corresponding orthogonal trajectories (in red or in black) in Figure 15.4.

In this chapter, we study some special types of differential equations for which it is possible to find a general solution in the form of an explicit or implicit function that does not involve derivatives. Most differential equations that occur in the study of complex phenomena, however, do not belong to one of these special types. In such cases, the best we can do is find numerical approximations to the values of the solution.

To be more precise, suppose we have a differential equation of the form $y' = f(x, y)$ for a specified function f. Let $y = g(x)$ be the (unknown) solution of this differential equation, which satisfies an *initial condition* $g(a) = A$. Next, suppose we want to know the value $g(b)$ for some number $b > a$. If we could find an explicit formula for $g(x)$, we could obtain the answer by evaluating the function g at $x = b$. If we cannot find such an explicit formula for g, then we may be able to approximate $g(b)$ numerically. The simple methods that we explore here and in the exercises allow us to compute approximate values for the function g by using the differential equation directly.

The first technique for numerical solution that we will examine is called *Euler's method*, introduced in 1768 by Leonhard Euler (see *Mathematicians and Their Times*, Chapter 6). Euler's method uses the idea of linear approximation (Section 2.8) in which the tangent line to a curve at a point is used to approximate the values of the function near the point of tangency. From the linear approximation formula (2.33), we obtain

$$g(x_1) \approx g(x_0) + g'(x_0)(x_1 - x_0).$$

From the differential equation, we have

$$g'(x_0) = y'(x_0) = f(x_0, y_0),$$

which gives the approximation

$$g(x_1) \approx g(x_0) + f(x_0, y_0)(x_1 - x_0).$$

If we let $y_1 = g(x_0) + f(x_0, y_0)(x_1 - x_0)$, then we obtain the point (x_1, y_1) as an approximation to the point $(x_1, g(x_1))$ on the solution curve.

If we partition the interval $[a, b]$ into n equal subintervals, each of width $h = (b - a)/n$ with $a = x_0$, $x_1 = x_0 + h$, $x_2 = x_1 + h$, ..., $x_n = x_{n-1} + h$, then the approximation formula for $g(x_1)$ becomes

$$g(x_1) \approx g(x_0) + hf(x_0, y_0) = y_1.$$

We then use the point (x_1, y_1) as the starting point to generate the next approximation,

$$g(x_2) \approx g(x_1) + hf(x_1, y_1) = y_2.$$

We repeat this procedure until we find y_n as an approximation for $g(x_n) = g(b)$.

We may summarize Euler's method as follows.

Euler's Method 15.1

To approximate the solution of the differential equation

$$y' = f(x, y) \text{ for } a \leq x \leq b \quad \text{with } y(a) = A:$$

Step 1 Select a regular partition $a = x_0 < x_1 < \cdots < x_n = b$, where n is a positive integer and $x_{k+1} - x_k = (b - a)/n = h$ for $k = 0, 1, \ldots, n - 1$.

Step 2 Let $y_0 = A$.

Step 3 Repeatedly let $y_k = y_{k-1} + hf(x_{k-1}, y_{k-1})$ until $k = n$. The computed y_k is approximately the value of the solution at $x = x_k$ for $k = 1, 2, \ldots, n$.

As with numerical integration, making n larger requires more computational effort. Larger values for n mean smaller values for the increment h of x, since $h = (b - a)/n$. Thus, larger values for n give more accurate results in general because tangent-line approximations are more accurate for smaller increments of x. Other methods that give more accuracy for the same computational effort will be presented in the exercises and in Section 15.2.

EXAMPLE ▪ 8 Use Euler's method to obtain numerical solutions to the differential equation

$$y' = 1 + (y - x)^2 \quad \text{for } 1 \leq x \leq 2.4$$

with initial condition $y_0 = 1.59$ at $x = 1$.

(a) For $n = 10$, list the computed approximations for the solution in a table.

(b) For $n = 25$, obtain a plot of the pairs (x_k, y_k) for $k = 0, 1, 2, \ldots, 25$.

SOLUTION

(a) For $n = 10$, the regular partition has $h = (2.4 - 1)/10 = 0.14$. The first two stages in Euler's method yield

$$y_1 = y_0 + hf(x_0, y_0) = 1.59 + (0.14)[1 + (1.59 - 1)^2] = 1.778734$$

and

$$y_2 = y_1 + hf(x_1, y_1)$$
$$= 1.778734 + (0.14)[1 + (1.778734 - 1.14)^2] \approx 1.975851.$$

The results of the computations for Step 3 are presented in the following table for the remaining stages.

k	x_k	y_k	k	x_k	y_k
1	1.14	1.778734	6	1.84	2.910417
2	1.28	1.975851	7	1.98	3.210828
3	1.42	2.183641	8	2.12	3.562919
4	1.56	2.405281	9	2.26	3.994402
5	1.70	2.645311	10	2.40	4.555427

Figure 15.5
$1 \le x \le 2.4, 0 \le y \le 6$

(b) For $n = 25, h = (2.4 - 1)/25 = 0.056$. The computational procedure is the same. Instead of displaying the results in a table, we graph the pairs (x_k, y_k) for $k = 0, 1, \ldots, 25$ to get a better sense of how the graph of the solution might appear (see Figure 15.5).

For comparison, using $n = 25$ in Euler's method yields a value of $y_{25} \approx 5.057507$ as an approximation for the solution at $x = 2.4$, whereas using $n = 10$ gives $y_{10} \approx 4.555427$.

EXERCISES 15.1

Exer. 1–4: (a) Find the general solution of the differential equation, and illustrate it graphically. **(b)** Find the particular solution that satisfies the condition $y = 2$ when $x = 0$.

1 $y' = 3x^2$

2 $y' = x - 1$

3 $y' = \dfrac{-x}{\sqrt{4 - x^2}}$

4 $y' = 3$

Exer. 5–10: Prove that y is a solution of the differential equation.

5 $y'' - 3y' + 2y = 0$; $y = C_1 e^x + C_2 e^{2x}$

6 $y' + 3y = 0$; $y = Ce^{-3x}$

7 $2xy^3 + 3x^2 y^2 \dfrac{dy}{dx} = 0$; $y = Cx^{-2/3}$

8 $x^3 y''' + x^2 y'' - 3xy' - 3y = 0$; $y = Cx^3$

9 $(x - 2y)\dfrac{dy}{dx} + 2x + y = 0$; $y^2 - x^2 - xy = C$

10 $y\dfrac{dy}{dx} = x$; $x^2 - y^2 = C$

Exer. 11–26: Solve the differential equation.

11 $\sec x \, dy - 2y \, dx = 0$

12 $x^2 dy - \csc 2y \, dx = 0$

13 $x \, dy - y \, dx = 0$

14 $(4 + y^2) dx + (9 + x^2) dy = 0$

15 $3y \, dx + (xy + 5x) dy = 0$

16 $(xy - 4x) dx + (x^2 y + y) dy = 0$

17 $y' = x - 1 + xy - y$

18 $(y + yx^2) dy + (x + xy^2) dx = 0$

19 $e^{x+2y} dx - e^{2x-y} dy = 0$

20 $\cos x \, dy - y \, dx = 0$

21 $y(1 + x^3)y' + x^2(1 + y^2) = 0$

22 $x^2 y' - yx^2 = y$

23 $x \tan y - y' \sec x = 0$

24 $xy + y' e^{(x^2)} \ln y = 0$

25 $e^y \sin x \, dx - \cos^2 x \, dy = 0$

26 $\sin y \cos x \, dx + (1 + \sin^2 x) \, dy = 0$

Exer. 27–34: Find the particular solution of the differential equation that satisfies the given condition.

27 $2y^2 y' = 3y - y'$; $y = 1$ when $x = 3$

28 $\sqrt{x}\, y' - \sqrt{y} = x\sqrt{y}$; $y = 4$ when $x = 9$

29 $x \, dy - (2x + 1)e^{-y} \, dx = 0$; $y = 2$ when $x = 1$

30 $\sec 2y \, dx - \cos^2 x \, dy = 0$; $y = \dfrac{\pi}{6}$ when $x = \dfrac{\pi}{4}$

31 $(xy + x) \, dx + \sqrt{4 + x^2} \, dy = 0$; $y = 1$ when $x = 0$

32 $x \, dy - \sqrt{1 - y^2} \, dx = 0$; $y = \frac{1}{2}$ when $x = 1$

33 $\cot x \, dy - (1 + y^2) \, dx = 0$; $y = 1$ when $x = 0$

34 $\csc y \, dx - e^x \, dy = 0$; $y = 0$ when $x = 0$

Exer. 35–40: Find the orthogonal trajectories of the family of curves. Describe the graphs.

35 $x^2 - y^2 = c$ **36** $xy = c$

37 $y^2 = cx$ **38** $y = cx^2$

39 $y^2 = cx^3$ **40** $y = ce^{-x}$

---

[c] **Exer. 41–44:** Use Euler's method to obtain a numerical solution to the differential equation on the indicated interval. (a) For $n = 10$, list the computed approximations for the solution in a table. (b) For $n = 25$, obtain a plot of the pairs (x_k, y_k) for $k = 0, 1, 2, \ldots, 25$.

41 $y' = -\dfrac{xy}{x^2 + y^2}$; $y_0 = 0.56$ at $x = 1$, $1 \le x \le 3$

42 $y' = 1.4 + (y - x)^{1.5}$; $y_0 = 1.2$ at $x = 1$, $1 \le x \le 2$

43 $y' = 0.3 + x \sin(xy)$; $y_0 = 0.12$ at $x = 0$, $0 \le x \le 2$

44 $y' = x + y^2$; $y_0 = 0.67$ at $x = 0$, $0 \le x \le 0.9$

[c] **Exer. 45–48:** A more accurate method than Euler's is the *improved Euler's method* (or *Heun's method*) in which the formula in Step 3 is replaced by

$$y_k = y_{k-1} + \tfrac{1}{2}h[f(x_{k-1}, y_{k-1})$$
$$+ f(x_{k-1} + h, y_{k-1} + hf(x_{k-1}, y_{k-1}))].$$

Work the following exercises using this method.

45 Exercise 41   **46** Exercise 42

47 Exercise 43   **48** Exercise 44

15.2 FIRST-ORDER LINEAR DIFFERENTIAL EQUATIONS

In this section, we investigate first-order differential equations

$$y' = f(x, y)$$

in which the function f can be written in the form $Q(x) - P(x)y$, where P and Q are continuous functions of x. Such equations occur frequently in applications of calculus, particularly in the study of physical phenomena.

Definition 15.2

A **first-order linear differential equation** is an equation of the form

$$y' + P(x)y = Q(x),$$

where P and Q are continuous functions.

If, in Definition (15.2), $Q(x) = 0$ for every x, we may separate the variables and then integrate as follows (provided $y \neq 0$):

$$\frac{dy}{dx} + P(x)y = 0$$

$$\frac{1}{y}\frac{dy}{dx} = -P(x)$$

$$\frac{1}{y}\,dy = -P(x)\,dx$$

$$\ln|y| = -\int P(x)\,dx + \ln|C|$$

We have expressed the constant of integration as $\ln|C|$ in order to change the form of the last equation as follows:

$$\ln|y| - \ln|C| = -\int P(x)\,dx$$

$$\ln\left|\frac{y}{C}\right| = -\int P(x)\,dx$$

$$\frac{y}{C} = e^{-\int P(x)\,dx}$$

$$ye^{\int P(x)\,dx} = C$$

We next observe that, by the product rule,

$$\frac{d}{dx}\left(ye^{\int P(x)\,dx}\right) = \left(\frac{dy}{dx}\right)e^{\int P(x)\,dx} + y\left(\frac{d}{dx}e^{\int P(x)\,dx}\right)$$

$$= y'e^{\int P(x)\,dx} + yP(x)e^{\int P(x)\,dx}$$

$$= (y' + P(x)y)e^{\int P(x)\,dx}.$$

Consequently, if we multiply both sides of $y' + P(x)y = Q(x)$ by $e^{\int P(x)\,dx}$, then the resulting equation may be written

$$\frac{d}{dx}(ye^{\int P(x)\,dx}) = Q(x)e^{\int P(x)\,dx}.$$

Integrating both sides gives us the following implicit solution of the first-order differential equation in Definition (15.2):

$$ye^{\int P(x)\,dx} = \int Q(x)e^{\int P(x)\,dx}\,dx + K$$

for a constant K. Solving this equation for y leads to an explicit solution. The expression $e^{\int P(x)\,dx}$ is an **integrating factor** of the differential equation. We have proved the following result.

Theorem 15.3

The first-order linear differential equation $y' + P(x)y = Q(x)$ may be transformed into a separable differential equation by multiplying both sides by the integrating factor $e^{\int P(x)\,dx}$.

EXAMPLE ▪ 1   Solve the differential equation $\dfrac{dy}{dx} - 3x^2 y = x^2$.

SOLUTION   The differential equation has the form in Definition (15.2) with $P(x) = -3x^2$ and $Q(x) = x^2$. By Theorem (15.3), an integrating factor is

$$e^{\int -3x^2\, dx} = e^{-x^3}.$$

There is no need to introduce a constant of integration, since $e^{-x^3+c} = e^c e^{-x^3}$, which differs from e^{-x^3} by a constant factor e^c. Multiplying both sides of the given differential equation by the integrating factor e^{-x^3}, we obtain

$$e^{-x^3}\frac{dy}{dx} - 3x^2 e^{-x^3} y = x^2 e^{-x^3},$$

or, equivalently, $\dfrac{d}{dx}(e^{-x^3} y) = x^2 e^{-x^3}.$

Integrating both sides of the last equation gives us

$$e^{-x^3} y = \int x^2 e^{-x^3}\, dx = -\tfrac{1}{3} e^{-x^3} + C.$$

Finally, multiplying by e^{x^3} gives us the explicit solution

$$y = -\tfrac{1}{3} + Ce^{(x^3)}.$$

EXAMPLE ▪ 2   Solve the differential equation $x^2 y' + 5xy + 3x^5 = 0$, where $x \neq 0$.

SOLUTION   To find an integrating factor, we begin by expressing the differential equation in the standardized form (15.3), where y' has the coefficient 1. Thus, dividing both sides by x^2, we obtain

$$y' + \frac{5}{x} y = -3x^3,$$

which has the form given in (15.2) with $P(x) = 5/x$. By Theorem (15.3), an integrating factor is

$$e^{\int P(x)\, dx} = e^{5\ln|x|} = e^{\ln|x|^5} = |x|^5.$$

If $x > 0$, then $|x|^5 = x^5$. If $x < 0$, then $|x|^5 = -x^5$. In either case, multiplying both sides of the standardized form by $|x|^5$ yields

$$x^5 y' + 5x^4 y = -3x^8,$$

or, equivalently, $\dfrac{d}{dx}(x^5 y) = -3x^8.$

Integrating both sides of the last equation gives us the solution

$$x^5 y = \int -3x^8 \, dx = -\frac{x^9}{3} + C,$$

or

$$y = -\frac{x^4}{3} + \frac{C}{x^5}.$$

EXAMPLE ▪ 3   Solve the differential equation

$$y' + y \tan x = \sec x + 2x \cos x.$$

SOLUTION   The equation is a first-order linear differential equation. By Theorem (15.3), with $P(x) = \tan x$, an integrating factor is

$$e^{\int \tan x \, dx} = e^{\ln|\sec x|} = |\sec x|.$$

Multiplying both sides of the differential equation by $|\sec x|$ and discarding the absolute value sign gives us

$$y' \sec x + y \sec x \tan x = \sec^2 x + 2x \cos x \sec x,$$

or

$$\frac{d}{dx}(y \sec x) = \sec^2 x + 2x.$$

Integrating both sides yields the (implicit) solution

$$y \sec x = \tan x + x^2 + C.$$

Finally, multiplying both sides of the last equation by $1/\sec x = \cos x$, we obtain the explicit solution

$$y = \sin x + (x^2 + C) \cos x.$$

We have previously used indefinite integration to derive laws of motion for falling bodies, assuming that air resistance could be disregarded (see Section 3.7). This is a valid assumption for small, slowly moving objects; however, in many cases, air resistance must be taken into account. This frictional force often increases as the speed of the object increases. In the following example, we shall derive the law of motion for a falling body under the assumption that the resistance due to the air is directly proportional to the speed of the body.

Figure 15.6

EXAMPLE ▪ 4   An object of mass m is released from a hot-air balloon. Find the distance that it falls in t seconds if the force of resistance due to the air is directly proportional to the speed of the object.

SOLUTION   Let us introduce a vertical axis with positive direction downward and origin at the point of release, as illustrated in Figure 15.6. We wish to find the distance $s(t)$ from the origin to the object at time t. The speed of the object is $v = s'(t)$, and the magnitude of the acceleration is $a = dv/dt = s''(t)$. If g is the gravitational constant, then the object is attracted toward the earth with a force of magnitude mg. By hypothesis, the force of resistance due to the air is kv for some constant k, and this force is directed opposite to the motion. It follows that the downward force F on the object is $mg - kv$. Since Newton's second law of motion states that $F = ma = m(dv/dt)$, we arrive at the following differential equation:

$$m \frac{dv}{dt} = mg - kv,$$

or, equivalently,

$$\frac{dv}{dt} + \frac{k}{m} v = g.$$

If we denote the constant k/m by c, this equation may be written

$$\frac{dv}{dt} + cv = g,$$

which is a first-order differential equation with t as the independent variable. By (15.3), an integrating factor is

$$e^{\int c \, dt} = e^{ct}.$$

Multiplying both sides of the last differential equation by e^{ct} gives us

$$e^{ct} \frac{dv}{dt} + ce^{ct} v = ge^{ct},$$

or, equivalently,

$$\frac{d}{dt}(ve^{ct}) = ge^{ct}.$$

Integrating both sides, we have

$$ve^{ct} = \frac{g}{c} e^{ct} + K, \quad \text{or} \quad v = \frac{g}{c} + Ke^{-ct},$$

where K is a constant.

If we let $t = 0$, then $v = 0$ and hence

$$0 = \frac{g}{c} + K, \quad \text{or} \quad K = -\frac{g}{c}.$$

Consequently,

$$v = \frac{g}{c} - \frac{g}{c} e^{-ct}.$$

Integrating both sides of this equation with respect to t and using the fact that $v = s'(t)$, we see that

$$s(t) = \frac{g}{c} t + \frac{g}{c^2} e^{-ct} + E.$$

We may find the constant of integration E by letting $t = 0$. Since $s(0) = 0$,

$$0 = 0 + \frac{g}{c^2} + E, \quad \text{or} \quad E = -\frac{g}{c^2}.$$

Thus, the distance that the object falls in t seconds is

$$
\begin{aligned}
s(t) &= \frac{g}{c} t + \frac{g}{c^2} e^{-ct} - \frac{g}{c^2} \\
&= \frac{g}{c^2}(ct + e^{-ct} - 1) \\
&= \frac{gm^2}{k^2} \left(\frac{k}{m} t + e^{-(k/m)t} - 1 \right).
\end{aligned}
$$

It is interesting to compare this formula for $s(t)$ with that obtained when the air resistance is disregarded. In the latter case, the differential equation $m(dv/dt) = mg - kv$ reduces to $dv/dt = g$, and hence $s'(t) = v = gt$. Integrating both sides leads to the much simpler formula $s(t) = \frac{1}{2}gt^2$.

Figure 15.7

EXAMPLE ■ 5   A simple electrical circuit consists of a resistance R and an inductance L connected in series, as illustrated schematically in Figure 15.7, with a constant electromotive force V. If the switch S is closed at $t = 0$, then it follows from one of Kirchhoff's rules for electrical circuits that if $t > 0$, the current I satisfies the differential equation

$$L\frac{dI}{dt} + RI = V.$$

Express I as a function of t.

SOLUTION   The differential equation may be written

$$\frac{dI}{dt} + \frac{R}{L}I = \frac{V}{L},$$

which is a first-order linear differential equation. Applying (15.3), we multiply both sides by the integrating factor $e^{\int (R/L)\, dt} = e^{(R/L)t}$, obtaining

$$e^{(R/L)t}\frac{dI}{dt} + \frac{R}{L}e^{(R/L)t}I = \frac{V}{L}e^{(R/L)t},$$

or

$$\frac{d}{dt}(Ie^{(R/L)t}) = \frac{V}{L}e^{(R/L)t}.$$

Integrating with respect to t yields

$$Ie^{(R/L)t} = \int \frac{V}{L}e^{(R/L)t}\, dt = \frac{V}{R}e^{(R/L)t} + C.$$

Since $I = 0$ when $t = 0$, it follows that $C = -V/R$. Substituting for C leads to

$$Ie^{(R/L)t} = \frac{V}{R}e^{(R/L)t} - \frac{V}{R}.$$

Finally, multiplying both sides by $e^{-(R/L)t}$ gives us

$$I = \frac{V}{R}[1 - e^{-(R/L)t}].$$

Observe that as t increases without bound, I approaches V/R, which is the current predicted by Ohm's law when no inductance is present.

Although Euler's method and the improved Euler's method are easy to understand and implement, neither is very efficient. Each requires a fairly substantial value for n to make the increment h sufficiently small to produce highly accurate approximations. We consider now a better method, which is analogous to replacing the midpoint and the trapezoidal rules with Simpson's rule for numerical integration. Consider the following first-order problem (which may or may not be linear):

$$y' = f(x, y) \quad \text{for} \quad a \le x \le b$$

with initial condition $y_0 = A$ at $x = a$

The following technique for numerical solution is called the *Runge–Kutta method*. It was devised by Carl D. T. Runge (1856–1927), the first person to hold a professorship in a German university specifically designated for applied mathematics, and introduced in 1895. The method was extended in 1901 by another German applied mathematician, Martin Wilhelm Kutta (1867–1944). The following states the method in an algorithmic format.

Runge–Kutta Method 15.4

Step 1 Select a regular partition $a = x_0 < x_1 < \cdots < x_n = b$, where n is a positive integer and

$$x_{k+1} - x_k = \frac{b - a}{n} = h$$

for $k = 0, 1, \ldots, n - 1$.

Step 2 Let $y_0 = A$.

Step 3 Repeatedly, until $k = n$, let

$$y_k = y_{k-1} + \tfrac{1}{6}(K_1 + 2K_2 + 2K_3 + K_4),$$

where

$$\begin{aligned}
K_1 &= hf(x_{k-1}, y_{k-1}) \\
K_2 &= hf(x_{k-1} + \tfrac{1}{2}h, y_{k-1} + \tfrac{1}{2}K_1) \\
K_3 &= hf(x_{k-1} + \tfrac{1}{2}h, y_{k-1} + \tfrac{1}{2}K_2) \\
K_4 &= hf(x_{k-1} + h, y_{k-1} + K_3).
\end{aligned}$$

The computed y_k is approximately the value of the solution at $x = x_k$, for $k = 1, 2, \ldots, n$.

Compared to Euler's method, each execution of Step 3 in the Runge–Kutta method takes about four times the computational effort. The reward for this extra computation is a much more accurate approximation. The next example shows the results of applying the Runge–Kutta method to the same differential equation we considered in Example 8 of Section 15.1.

 EXAMPLE ▪ 6 Use the Runge–Kutta method **(a)** with $n = 10$ and **(b)** with $n = 25$ to obtain numerical solutions for the differential equation

$$y' = 1 + (y - x)^2 \quad \text{for} \quad 1 \le x \le 2.4$$

with initial condition $y_0 = 1.59$ at $x = 1$.

SOLUTION

(a) For $n = 10$, the regular partition has $h = (2.4 - 1)/10 = 0.14$. We present the details of calculating y_1. First we determine K_1, K_2, K_3, and K_4 using (15.4):

$$K_1 = hf(x_0, y_0) = (0.14)[1 + (1.59 - 1)^2] = 0.188734,$$

$$\begin{aligned}
K_2 &= hf[x_0 + (h/2), y_0 + (K_1/2)] \\
&= (0.14)f(1 + 0.07, 1.59 + 0.094367) \\
&= (0.14)f(1.07, 1.684367) = (0.14)[1 + (1.684367 - 1.07)^2] \\
&\approx 0.1928426,
\end{aligned}$$

$$\begin{aligned}
K_3 &= hf[x_0 + (h/2), y_0 + (K_2/2)] \\
&= (0.14)f(1 + 0.07, 1.59 + 0.0964213) \\
&= (0.14)f(1.07, 1.6864213) = (0.14)[1 + (1.6864213 - 1.07)^2] \\
&\approx 0.1931965,
\end{aligned}$$

$$\begin{aligned}
K_4 &= hf(x_0 + h, y_0 + K_3) = (0.14)f(1 + 0.14, 1.59 + 0.1931965) \\
&= (0.14)f(1.14, 1.7831965) = (0.14)[1 + (1.7831965 - 1.14)^2] \\
&\approx 0.1979182.
\end{aligned}$$

Now we calculate y_1:

$$\begin{aligned}
y_1 &= y_0 + \tfrac{1}{6}(K_1 + 2K_2 + 2K_3 + K_4) \\
&= 1.59 + \tfrac{1}{6}(0.188734 + 0.3856852 + 0.386393 + 0.1979182) \\
&= 1.59 + \tfrac{1}{6}(1.1587304) \approx 1.7831217
\end{aligned}$$

We have carried out the computations to determine y_1 by hand in order to demonstrate the differences between the Runge–Kutta method and Euler's method. In practice, each of these methods and other numerical methods for approximating solutions of differential equations would be programmed on a calculator or a computer. The results of the computations for Step 3 of the Runge–Kutta method for $n = 1, \ldots, 10$ are listed in the following table.

k	x_k	y_k	k	x_k	y_k
1	1.14	1.783122	6	1.84	3.009697
2	1.28	1.986756	7	1.98	3.378742
3	1.42	2.204365	8	2.12	3.859311
4	1.56	2.441121	9	2.26	4.558999
5	1.70	2.705107	10	2.40	5.788904

(b) For $n = 25$, $h = (2.4 - 1)/25 = 0.056$. The computational steps are the same, but we record only the last value $y_{25} = 5.790747$ as an approximation for the solution at $x = 2.4$.

There is a close connection between our methods for the numerical solution of first-order differential equations and the methods we considered previously for numerical integration. Consider the special differential equation of the form

$$y' = f(x) \quad \text{for} \quad a \leq x \leq b$$
$$\text{with initial condition } y_0 = 0 \text{ at } x = a.$$

We concentrate on obtaining the value of the function y at $x = b$. The solution is exactly $y = \int_a^b f(x)\, dx$. Applying Euler's method to this special differential equation with initial value 0 is the same as approximating the definite integral using the same partition with evaluations at left endpoints. Using the improved Euler's method is the same as approximating the definite integral using the trapezoidal rule. In this special case, using the Runge–Kutta method corresponds to applying Simpson's rule. Error estimates for these numerical methods also resemble the error estimates for numerical integration. They are given here without proof.

Theorem 15.5

The error made in numerically approximating the value Y of the function y at $x = b$ for the first-order differential equation

$$y' = f(x, y) \quad \text{for} \quad a \leq x \leq b$$
$$\text{with initial condition } y_0 = A \text{ at } x = a$$

can be estimated by

(a) $|Y - y_n| \leq M_1 h$ for Euler's method,

(b) $|Y - y_n| \leq M_2 h^2$ for the improved Euler's method,

(c) $|Y - y_n| \leq M_4 h^4$ for the Runge–Kutta method.

The constants M_1, M_2, and M_4 are dependent on the length of the interval $[a, b]$, the function f in the differential equation, and the initial condition $y_0 = A$ at $x = a$.

Doubling the value of n (where n describes the size of the partition) halves the value of the increment h. From Theorem (15.5), we see that doubling n divides the error bound in Euler's method by 2; in the improved Euler's method by 4; and in the Runge–Kutta method by 16. In the next example, we see how quickly each of these methods converges to the exact solution of a differential equation.

EXAMPLE ■ 7 For the differential equation

$$y' = 1 + (y - x)^2 \quad \text{for} \quad 1 \leq x \leq 2.4$$
$$\text{with initial condition } y_0 = 1.59 \text{ at } x = 1$$

(a) verify that $y = x - \dfrac{1}{x + C}$ is a solution

(b) find the value of C for which $y = 1.59$ at $x = 1$

(c) compute the exact value of y at $x = 2.4$

(d) compare the exact solution with the approximate solutions obtained by Euler's method, the improved Euler's method, and the Runge–Kutta method for $n = 10, 20, 40, 80$, and 160

SOLUTION

(a) For
$$y = x - \frac{1}{x + C} = x - (x + C)^{-1},$$

we have $y' = 1 + (x + C)^{-2}$, whereas

$$1 + (y - x)^2 = 1 + \left(x - \frac{1}{x + C} - x\right)^2$$
$$= 1 + \left(\frac{-1}{x + C}\right)^2 = 1 + \frac{1}{(x + C)^2} = 1 + (x + C)^{-2}.$$

Thus, the proposed function is a solution to the differential equation.

(b) When $x = 1$, $y = x - [1/(x + C)]$ has the value $1 - [1/(1 + C)]$, which is equal to 1.59 when $1 - 1.59 = 1/(1 + C)$ or $1 + C = -1/0.59$. Thus, $C = -1.59/0.59 \approx -2.694915$.

(c) Substituting the value of C found in part (b), we obtain the solution

$$y = x - \frac{1}{x - 1.59/0.59}.$$

Thus, the exact value of the solution at $x = 2.4$ is

$$2.4 - \frac{1}{2.4 - 1.59/0.59} \approx 5.7908046.$$

For Euler's method, the improved Euler's method, and the Runge–Kutta method, we begin with $n = 10$ and proceed to double n in the following tables. From Theorem (15.5), we expect the ratio of successive errors for each method to be about 2, 4, and 16, respectively. Note that the errors listed are not equal to $Y - y_n$ due to rounding of results.

n	Euler's y_n	Exact Y	Error E_n	Ratio $E_{n/2}/E_n$
10	4.555542	5.790805	1.235262	—
20	4.944401	5.790805	0.846403	1.46
40	5.265825	5.790805	0.524979	1.61
80	5.490908	5.790805	0.299897	1.75
160	5.629037	5.790805	0.161767	1.85
320	5.706550	5.790805	0.084255	1.91

n	Improved y_n	Exact Y	Error E_n	Ratio $E_{n/2}/E_n$
10	5.581352	5.790805	0.209452	—
20	5.725189	5.790805	0.065615	3.19
40	5.772649	5.790805	0.018156	3.61
80	5.786059	5.790805	0.004745	3.82
160	5.789594	5.790805	0.001210	3.92

n	Runge–Kutta y_n	Exact Y	Error E_n	Ratio $E_{n/2}/E_n$
10	5.788903938	5.790804598	0.001901	—
20	5.790665965	5.790804598	0.000139	13.71
40	5.790795488	5.790804598	9.1 E -6	15.22
80	5.790804020	5.790804598	5.7 E -7	15.77
160	5.790804562	5.790804598	3.6 E -8	15.83

Note that for a fixed n, the Runge–Kutta method is much more accurate than either the Euler or the improved Euler method. For $n = 20$, for example, Euler's method has an error of 0.846403 and the improved Euler's method has an error of 0.065615, but the Runge–Kutta method has an error of only 0.000139. Although the Runge–Kutta method requires four times as many computations as the Euler method, in this case, the error for the Euler method is more than 6000 times greater than that for the Runge–Kutta method.

EXERCISES 15.2

Exer. 1–22: Solve the differential equation.

1 $y' + 2y = e^{2x}$

2 $y' - 3y = 2$

3 $xy' - 3y = x^5$

4 $y' + y \cot x = \csc x$

5 $xy' + y + x = e^x$

6 $xy' + (1 + x)y = 5$

7 $x^2 \, dy + (2xy - e^x) \, dx = 0$

8 $x^2 \, dy + (x - 3xy + 1) \, dx = 0$

9 $y' + y \cot x = 4x^2 \csc x$

10 $y' + y \tan x = \sin x$

11 $(y \sin x - 2) \, dx + \cos x \, dy = 0$

12 $(x^2 y - 1) \, dx + x^3 \, dy = 0$

13 $(x^2 \cos x + y) \, dx - x \, dy = 0$

14 $y' + y = \sin x$

15 $xy' + (2 + 3x)y = xe^{-3x}$

16 $(x + 4)y' + 5y = x^2 + 8x + 16$

17 $x^{-1}y' + 2y = 3$

18 $y' - 5y = e^{5x}$

19 $\tan x \, dy + (y - \sin x) \, dx = 0$

20 $\cos x \, dy - (y \sin x + e^{-x}) \, dx = 0$

21 $y' + 3x^2 y = x^2 + e^{-x^3}$

22 $y' + y \tan x = \cos^3 x$

Exer. 23–26: Find the particular solution of the differential equation that satisfies the given condition.

23 $xy' - y = x^2 + x;$ $y = 2$ when $x = 1$

24 $y' + 2y = e^{-3x};$ $y = 2$ when $x = 0$

25 $xy' + y + xy = e^{-x};$ $y = 0$ when $x = 1$

26 $y' + 2xy - e^{-x^2} = x;$ $y = 1$ when $x = 0$

Exer. 27–28: Solve the differential equation by (a) using an integrating factor and (b) separating the variables.

27 The differential equation

$$R\frac{dQ}{dt} + \frac{Q}{C} = V$$

describes the charge Q on a capacitor with capacitance C during a charging process involving a resistance R and electromotive force V. If the charge is 0 when $t = 0$, express Q as a function of t.

28 The differential equation

$$R\frac{dI}{dt} + \frac{I}{C} = \frac{dV}{dt}$$

describes an electrical circuit consisting of an electromotive force V with a resistance R and capacitance C connected in series. If V is constant and if $I = I_0$ when $t = 0$, express I as a function of t.

29 At time $t = 0$, a tank contains K pounds of salt dissolved in 80 gal of water. Suppose that water containing $\frac{1}{3}$ lb of salt per gallon is being added to the tank at a rate of 6 gal/min, and that the well-stirred solution is being drained from the tank at the same rate. Find a formula for the amount $f(t)$ of salt in the tank at time t.

30 An object of mass m is moving on a coordinate line subject to a force given by $F(t) = e^{-t}$, where t is time. The motion is resisted by a frictional force that is numerically equal to twice the speed of the object. If $v = 0$ at $t = 0$, find a formula for v at any time $t > 0$.

31 A *learning curve* is used to describe the rate at which a skill is acquired. Suppose a manufacturer estimates that a new employee will produce A items the first day on the job, and that as the employee's proficiency increases, items will be produced more rapidly until the employee produces a maximum of M items per day. Let $f(t)$ denote the number produced on day t, where $t \geq 1$. Suppose that the rate of production $f'(t)$ is proportional to $M - f(t)$.

(a) Find a formula for $f(t)$.

(b) If $M = 30$, $f(1) = 5$, and $f(2) = 8$, estimate the number of items produced on day 20.

32 A room having dimensions 10 ft $\times$ 15 ft $\times$ 8 ft originally contains 0.001% carbon monoxide (CO). Suppose that at time $t = 0$, fumes containing 5% CO begin entering the room at a rate of 0.12 ft^3/min, and that the well-circulated mixture is eliminated from the room at the same rate.

(a) Find a formula for the volume $f(t)$ of CO in the room at time t.

(b) The minimum level of CO considered to be hazardous to health is 0.015%. After approximately how many minutes will the room contain this level of CO?

33 The von Bertalanffy model for the growth of an animal assumes that there is an upper bound of y_L centimeters on length. If y is the length at age t years, the rate of growth is assumed to be proportional to the length yet to be achieved. Find the general solution of the resulting differential equation.

34 A cell is in a liquid containing a solute, such as potassium, of constant concentration c_0. If $c(t)$ is the concentration of the solute inside the cell at time t, then *Fick's principle* for passive diffusion across the cell membrane asserts that the rate at which the concentration changes is proportional to the *concentration gradient* $c_0 - c(t)$. Find $c(t)$ if $c(0) = 0$.

35 Many common drugs (such as forms of penicillin) are eliminated from the bloodstream at a rate that is proportional to the amount y still present.

(a) If y_0 milligrams are injected directly into the bloodstream, show that $y = y_0 e^{-kt}$ for some $k > 0$.

(b) If the drug is fed intravenously into the bloodstream at a rate of I milligrams per minute, then $y' = -ky + I$. If $y = 0$ at $t = 0$, express y as a function of t and find $\lim_{t \to \infty} y$.

(c) If the half-life of the drug in the bloodstream is 2 hr, estimate the infusion rate that will result in a long-term amount of 100 mg in the bloodstream.

36 Shown in the figure is a two-tank system in which water flows in and out of the tanks at the rate of 5 gal/min. Each tank contains 50 gal of water.

(a) If 1 lb of dye is thoroughly mixed into the water in tank 1 and if $x(t)$ denotes the amount of dye in the tank after t minutes, show that $x'(t) = -0.1x(t)$ and that $x(t) = e^{-0.1t}$.

(b) From part (a), the dye is entering tank 2 at the rate of $5 \cdot \frac{1}{50}x(t) = 0.1e^{-0.1t}$ lb/min. Show that if $y(t)$ is the amount of dye in tank 2 after t minutes, then $y'(t) = -0.1y(t) + 0.1e^{-0.1t}$. Assuming that $y(0) = 0$, find $y(t)$.

(c) What is the maximum amount of dye in tank 2?

(d) Rework part (b) if the volume of tank 2 is 40 gal.

Exercise 36

Tank 1   Tank 2

5 gal/min   5 gal/min   5 gal/min

[c] **Exer. 37–38:** Refer to Exercises 41–44 of Section 15.1. Show that numerical methods can fail to provide accurate approximations by first verifying that the differential equation $y' = 10y - 11e^{-x}$ with initial condition $y = 1$ at $x = 0$ has the solution $y = e^{-x}$ for every x and then approximating y at $x = 1$ using the indicated method with $n = 8$.

37 Euler's method

38 Improved Euler's method

[c] **Exer. 39–42:** Use the Runge–Kutta method to obtain a numerical solution to the differential equation on the indicated interval. (a) For $n = 5$, list the computed approximations for the solution in a table. (b) For $n = 25$, obtain a plot of the pairs (x_k, y_k) for $k = 0, 1, 2, \ldots, 25$.

39 $y' = -\dfrac{xy}{x^2 + y^2}$; $y_0 = 0.56$ at $x = 1$, $1 \leq x \leq 3$

40 $y' = 1.4 + (y - x)^{1.5}$; $y_0 = 1.2$ at $x = 1$, $1 \leq x \leq 2$

41 $y' = 0.3 + x \sin(xy)$; $y_0 = 0.12$ at $x = 0$, $0 \leq x \leq 2$

42 $y' = x + y^2$; $y_0 = 0.67$ at $x = 0$, $0 \leq x \leq 0.9$

c | **Exer. 43–44:** Use the Runge–Kutta method with $n = 4$, 8, 16, and 32 to obtain numerical approximations for y at $x = b$, where y solves the differential equation on the indicated interval. Compare these approximations to the given exact solution by finding the error E_n and the ratio $E_{n/2}/E_n$.

43 $y' = 1 + \dfrac{y}{x}$; $y_0 = 3$ at $x = 1$, $1 \leq x \leq 2$

with exact solution $y = x \ln x + 3x$

44 $y' = -(y + 1)(y + 3)$; $y_0 = -1.5$ at $x = 0$, $0 \leq x \leq 2$

with exact solution $y = -3 + 6(3 + e^{-2x})^{-1}$

15.3 SECOND-ORDER LINEAR DIFFERENTIAL EQUATIONS

The following definition is a generalization of (15.2).

Definition 15.6

A **linear differential equation of order n** is an equation of the form

$$y^{(n)} + f_1(x)y^{(n-1)} + \cdots + f_{n-1}(x)y' + f_n(x)y = k(x),$$

where $f_1, f_2, \ldots, f_n$ and k are functions of one variable that have the same domain. If $k(x) = 0$ for every x, the equation is **homogeneous**. If $k(x) \neq 0$ for some x, the equation is **nonhomogeneous**.

A thorough analysis of (15.6) may be found in textbooks on differential equations. We shall restrict our work to second-order equations in which f_1 and f_2 are constant functions. In this section, we shall consider the homogeneous case. Nonhomogeneous equations will be discussed in the next section. Applications are considered in Section 15.5.

The general second-order homogeneous linear differential equation with constant coefficients has the form

$$y'' + by' + cy = 0,$$

where b and c are constants. Before attempting to find particular solutions, let us establish the following result.

Theorem 15.7

If $y = f(x)$ and $y = g(x)$ are solutions of $y'' + by' + cy = 0$, then

$$y = C_1 f(x) + C_2 g(x)$$

is a solution for all real numbers C_1 and C_2.

PROOF Since $f(x)$ and $g(x)$ are solutions of $y'' + by' + cy = 0$,

$$f''(x) + bf'(x) + cf(x) = 0$$

and

$$g''(x) + bg'(x) + cg(x) = 0.$$

If we multiply the first of these equations by C_1 and the second by C_2 and add, the result is

$$[C_1 f''(x) + C_2 g''(x)] + b[C_1 f'(x) + C_2 g'(x)]$$

$$+c[C_1 f(x) + C_2 g(x)] = 0.$$

Thus, $C_1 f(x) + C_2 g(x)$ is a solution.   ■

We can show that if the solutions f and g in Theorem (15.7) have the property that $f(x) \neq Cg(x)$ for every real number C and if $g(x)$ is not identically 0, then $y = C_1 f(x) + C_2 g(x)$ is the general solution of the differential equation $y'' + by' + cy = 0$. Thus, to determine the general solution, it is sufficient to find two such functions f and g and use (15.7).

In our search for a solution of $y'' + by' + cy = 0$, we shall use $y = e^{mx}$ as a trial solution. Since $y' = me^{mx}$ and $y'' = m^2 e^{mx}$, it follows that $y = e^{mx}$ is a solution if and only if

$$m^2 e^{mx} + bm e^{mx} + ce^{mx} = 0$$

or, since $e^{mx} \neq 0$, if and only if

$$m^2 + bm + c = 0.$$

The last equation is very important in finding solutions of the equation $y'' + by' + cy = 0$ and is given the following special name.

Definition 15.8

> The **auxiliary equation** of the differential equation $y'' + by' + cy = 0$ is $m^2 + bm + c = 0$.

Note that the auxiliary equation can be obtained from the differential equation by replacing y'' by m^2, y' by m, and y by 1. Since the auxiliary equation is quadratic, it has either unequal roots m_1 and m_2, a double real root m, or two complex conjugate roots. The next theorem is a consequence of the remark following the proof of Theorem (15.7).

Theorem 15.9

> If the roots m_1, m_2 of the auxiliary equation are real and unequal, then the general solution of $y'' + by' + cy = 0$ is
>
> $$y = C_1 e^{m_1 x} + C_2 e^{m_2 x}.$$

EXAMPLE ■ I   Solve the differential equation $y'' - 3y' - 10y = 0$.

SOLUTION   The auxiliary equation is $m^2 - 3m - 10 = 0$, or, equivalently, $(m - 5)(m + 2) = 0$. Since the roots $m_1 = 5$ and $m_2 = -2$ are real and unequal, it follows from Theorem (15.9) that the general solution is

$$y = C_1 e^{5x} + C_2 e^{-2x}.$$

Theorem 15.10

> If the auxiliary equation has a double root m, then the general solution of $y'' + by' + cy = 0$ is
>
> $$y = C_1 e^{mx} + C_2 x e^{mx}.$$

PROOF   By the quadratic formula, the roots of $m^2 + bm + c = 0$ are $m = (-b \pm \sqrt{b^2 - 4c})/2$. If $b^2 - 4c = 0$, then m is a double root and we obtain $m = -b/2$, or $2m + b = 0$. Since m satisfies the auxiliary equation, $y = e^{mx}$ is a solution of the differential equation. According to the remark following the proof of Theorem (15.7), it is sufficient to show that $y = x e^{mx}$ is also a solution. Substitution of $x e^{mx}$ for y in the equation $y'' + by' + cy = 0$ gives us

$$(2m e^{mx} + m^2 x e^{mx}) + b(m x e^{mx} + e^{mx}) + c x e^{mx}$$
$$= (m^2 + bm + c) x e^{mx} + (2m + b) e^{mx}$$
$$= 0 x e^{mx} + 0 e^{mx} = 0,$$

which is what we wished to show.   ■

EXAMPLE ■ 2   Solve the differential equation $y'' - 6y' + 9y = 0$.

SOLUTION   The auxiliary equation $m^2 - 6m + 9 = 0$, or, equivalently, $(m - 3)^2 = 0$, has a double root 3. Hence, by Theorem (15.10), the general solution is

$$y = C_1 e^{3x} + C_2 x e^{3x} = e^{3x}(C_1 + C_2 x).$$

We may also consider second-order differential equations of the form

$$a y'' + b y' + c y = 0,$$

with $a \neq 1$. It is possible to obtain the form stated in Theorems (15.9) and (15.10) by dividing both sides by a; however, it is generally simpler to use the auxiliary equation

$$a m^2 + b m + c = 0,$$

as illustrated in the next example.

EXAMPLE ■ 3   Solve the differential equation $6y'' - 7y' + 2y = 0$.

SOLUTION   The auxiliary equation is $6m^2 - 7m + 2 = 0$, or, equivalently, $(2m - 1)(3m - 2) = 0$. Hence, the roots are $m_1 = \frac{1}{2}$ and $m_2 = \frac{2}{3}$. By Theorem (15.9), the general solution of the given equation is

$$y = C_1 e^{x/2} + C_2 e^{2x/3}.$$

The final case to consider is that in which the roots of the auxiliary equation $m^2 + bm + c = 0$ of $y'' + by' + cy = 0$ are conjugate complex numbers of the form

$$z_1 = s + ti \quad \text{and} \quad z_2 = s - ti,$$

where s and t are real numbers and $i^2 = -1$. We may anticipate, from Theorem (15.9), that the general solution of the differential equation is

$$y = C_1 e^{z_1 x} + C_2 e^{z_2 x} = C_1 e^{(s+ti)x} + C_2 e^{(s-ti)x}.$$

To define complex exponents, we must extend some of the concepts of calculus to include functions whose domains include complex numbers. Since a complete development requires advanced methods, we shall merely outline the main ideas.

In Section 8.8, we discussed how certain functions can be represented by power series. We can easily extend the definitions and theorems of Chapter 8 to infinite series that involve complex numbers. Since this is true, we shall use the power series representations (8.48) to *define* e^z, $\sin z$, and $\cos z$ for every complex number z as follows.

Functions of $z = a + bi$ 15.11

$$\text{(i)} \quad e^z = 1 + z + \frac{z^2}{2!} + \cdots + \frac{z^n}{n!} + \cdots$$

$$\text{(ii)} \quad \sin z = z - \frac{z^3}{3!} + \frac{z^5}{5!} - \cdots + (-1)^n \frac{z^{2n+1}}{(2n+1)!} + \cdots$$

$$\text{(iii)} \quad \cos z = 1 - \frac{z^2}{2!} + \frac{z^4}{4!} - \cdots + (-1)^n \frac{z^{2n}}{(2n)!} + \cdots$$

Using (15.11)(i) gives us

$$e^{iz} = 1 + (iz) + \frac{(iz)^2}{2!} + \frac{(iz)^3}{3!} + \frac{(iz)^4}{4!} + \frac{(iz)^5}{5!} + \cdots$$

$$= 1 + iz + i^2 \frac{z^2}{2!} + i^3 \frac{z^3}{3!} + i^4 \frac{z^4}{4!} + i^5 \frac{z^5}{5!} + \cdots.$$

Since $i^2 = -1$, $i^3 = -i$, $i^4 = 1$, $i^5 = i$, and so on, we see that

$$e^{iz} = 1 + iz - \frac{z^2}{2!} - i \frac{z^3}{3!} + \frac{z^4}{4!} + i \frac{z^5}{5!} - \cdots,$$

which may also be written in the form

$$e^{iz} = \left(1 - \frac{z^2}{2!} + \frac{z^4}{4!} - \cdots \right) + i \left(z - \frac{z^3}{3!} + \frac{z^5}{5!} - \cdots \right).$$

If we now use formulas (iii) and (ii) of (15.11), we obtain the following result, also named after Leonhard Euler.

Euler's Formula 15.12

If z is a complex number,

$$e^{iz} = \cos z + i \sin z.$$

The laws of exponents are true for complex numbers. In addition, formulas for derivatives may be extended to functions of a *complex* variable z. For example, $(d/dz)(e^{kz}) = ke^{kz}$, where k is a complex number. If the auxiliary equation of $y'' + by' + cy = 0$ has complex roots $s \pm ti$, then the general solution of this differential equation may be written in the following equivalent forms:

$$y = C_1 e^{(s+ti)x} + C_2 e^{(s-ti)x}$$
$$y = C_1 e^{sx+txi} + C_2 e^{sx-txi}$$
$$y = C_1 e^{sx} e^{txi} + C_2 e^{sx} e^{-txi}$$
$$y = e^{sx}(C_1 e^{itx} + C_2 e^{-itx})$$

This can be further simplified by using Euler's formula. Specifically, from (15.12),

$$e^{itx} = \cos tx + i \sin tx$$

and

$$e^{-itx} = \cos tx - i \sin tx,$$

from which it follows that

$$\cos tx = \frac{e^{itx} + e^{-itx}}{2} \quad \text{and} \quad \sin tx = \frac{e^{itx} - e^{-itx}}{2i}.$$

If we let $C_1 = C_2 = \frac{1}{2}$ in the preceding discussion and then use the formula for $\cos tx$, we obtain

$$y = \tfrac{1}{2} e^{sx}(e^{itx} + e^{-itx})$$
$$= \tfrac{1}{2} e^{sx}(2 \cos tx) = e^{sx} \cos tx.$$

Thus, $y = e^{sx} \cos tx$ is a particular solution of $y'' + by' + cy = 0$. Letting $C_1 = -C_2 = i/2$ leads to the particular solution $y = e^{sx} \sin tx$. This is a partial proof of the next theorem.

Theorem 15.13

If the auxiliary equation $m^2 + bm + c = 0$ has distinct complex roots $s \pm ti$, then the general solution of $y'' + by' + cy = 0$ is

$$y = e^{sx}(C_1 \cos tx + C_2 \sin tx).$$

EXAMPLE ■ 4 Solve the differential equation $y'' - 10y' + 41y = 0$.

SOLUTION The roots of the auxiliary equation $m^2 - 10m + 41 = 0$ are

$$m = \frac{10 \pm \sqrt{100 - 164}}{2} = \frac{10 \pm 8i}{2} = 5 \pm 4i.$$

Applying Theorem (15.13) yields the general solution of the differential equation:

$$y = e^{5x}(C_1 \cos 4x + C_2 \sin 4x)$$

EXERCISES 15.3

Exer. 1–22: Solve the differential equation.

1 $y'' - 5y' + 6y = 0$

2 $y'' - y' - 2y = 0$

3 $y'' - 3y' = 0$

4 $y'' + 6y' + 8y = 0$

5 $y'' + 4y' + 4y = 0$

6 $y'' - 4y' + 4y = 0$

7 $y'' - 4y' + y = 0$

8 $6y'' - 7y' - 3y = 0$

9 $y'' + 2\sqrt{2}y' + 2y = 0$

10 $4y'' + 20y' + 25y = 0$

11 $8y'' + 2y' - 15y = 0$

12 $y'' + 4y' + y = 0$

13 $9y'' - 24y' + 16y = 0$

14 $4y'' - 8y' + 7y = 0$

15 $2y'' - 4y' + y = 0$

16 $2y'' + 7y' = 0$

17 $y'' - 2y' + 2y = 0$

18 $y'' - 2y' + 5y = 0$

19 $y'' - 4y' + 13y = 0$

20 $y'' + 4 = 0$

21 $\dfrac{d^2y}{dx^2} + 6\dfrac{dy}{dx} + 2y = 0$

22 $\dfrac{d^2y}{dx^2} + 2\dfrac{dy}{dx} + 6y = 0$

Exer. 23–30: Find the particular solution of the differential equation that satisfies the stated conditions.

23 $y'' - 3y' + 2y = 0$; $y = 0$ and $y' = 2$ when $x = 0$

24 $y'' - 2y' + y = 0$; $y = 1$ and $y' = 2$ when $x = 0$

25 $y'' + y = 0$; $y = 1$ and $y' = 2$ when $x = 0$

26 $y'' - y' - 6y = 0$; $y = 0$ and $y' = 1$ when $x = 0$

27 $y'' + 8y' + 16y = 0$; $y = 2$ and $y' = 1$ when $x = 0$

28 $y'' + 5y = 0$; $y = 4$ and $y' = 2$ when $x = 0$

29 $\dfrac{d^2y}{dx^2} - 2\dfrac{dy}{dx} + 5y = 0$; $y = 0$ and $\dfrac{dy}{dx} = 1$ when $x = 0$

30 $\dfrac{d^2y}{dx^2} - 6\dfrac{dy}{dx} + 13y = 0$;

$y = 2$ and $\dfrac{dy}{dx} = 3$ when $x = 0$

15.4 NONHOMOGENEOUS LINEAR DIFFERENTIAL EQUATIONS

In this section, we consider second-order nonhomogeneous linear differential equations with constant coefficients—that is, equations of the form

$$y'' + by' + cy = k(x),$$

where b and c are constants and k is a continuous function.

It is convenient to use the differential operator symbols D and D^2 such that if $y = f(x)$, then

$$Dy = y' = f'(x) \quad \text{and} \quad D^2y = y'' = f''(x).$$

Definition 15.14

If $y = f(x)$ and if f'' exists, then the **linear differential operator** $L = D^2 + bD + c$ is defined by

$$L(y) = (D^2 + bD + c)y = D^2y + bDy + cy = y'' + by' + cy.$$

Using L, we may write the differential equation $y'' + by' + cy = k(x)$ in the compact form $L(y) = k(x)$. In Exercises 19 and 20, you are asked to verify that for every real number C,

$$L(Cy) = CL(y),$$

and that if $y_1 = f_1(x)$ and $y_2 = f_2(x)$, then

$$L(y_1 \pm y_2) = L(y_1) \pm L(y_2).$$

For the differential equation $y'' + by' + cy = k(x)$ (that is, $L(y) = k(x)$), the corresponding homogeneous equation $L(y) = 0$ is the **complementary equation**.

Theorem 15.15

Let $y'' + by' + cy = k(x)$ be a second-order nonhomogeneous linear differential equation. If y_p is a particular solution of $L(y) = k(x)$ and if y_c is the general solution of the complementary equation $L(y) = 0$, then the general solution of $L(y) = k(x)$ is $y = y_p + y_c$.

PROOF Since $L(y_p) = k(x)$ and $L(y_c) = 0$,

$$L(y_p + y_c) = L(y_p) + L(y_c) = k(x) + 0 = k(x),$$

which means that $y_p + y_c$ is a solution of $L(y) = k(x)$. Moreover, if $y = f(x)$ is any other solution, then

$$L(y - y_p) = L(y) - L(y_p) = k(x) - k(x) = 0.$$

Consequently, $y - y_p$ is a solution of the complementary equation. Thus, $y - y_p = y_c$ for some y_c, which is what we wished to prove. ∎

If we use the results of Section 15.3 to find the general solution y_c of $L(y) = 0$, then according to Theorem (15.15), all that is needed to determine the general solution of $L(y) = k(x)$ is *one* particular solution y_p.

EXAMPLE ▪ 1 Solve the differential equation $y'' - 4y = 6x - 4x^3$ using the particular solution $y_p = x^3$.

SOLUTION The complementary equation is $y'' - 4y = 0$, which, by Theorem (15.9), has the general solution

$$y_c = C_1 e^{2x} + C_2 e^{-2x}.$$

Applying Theorem (15.15), we find that the general solution of the given nonhomogeneous equation is

$$y = C_1 e^{2x} + C_2 e^{-2x} + x^3.$$

If a particular solution of $y'' + by' + cy = k(x)$ cannot be found by inspection, then the following technique, called **variation of parameters**, may be used. Given the differential equation $L(y) = k(x)$, let y_1 and y_2 be the expressions that appear in the general solution $y = C_1 y_1 + C_2 y_2$ of the complementary equation $L(y) = 0$. For example, as in Theorem (15.9), we might have $y_1 = e^{m_1 x}$ and $y_2 = e^{m_2 x}$. Let us now attempt to find a particular solution of $L(y) = k(y)$ that has the form

$$y_p = uy_1 + vy_2,$$

where $u = g(x)$ and $v = h(x)$ for some functions g and h. The first and second derivatives of y_p are

$$y'_p = (uy'_1 + vy'_2) + (u'y_1 + v'y_2)$$
$$y''_p = (uy''_1 + vy''_2) + (u'y'_1 + v'y'_2) + (u'y_1 + v'y_2)'.$$

Substituting these into $L(y_p) = y''_p + by'_p + cy_p$ and rearranging terms, we obtain

$$L(y_p) = u(y''_1 + by'_1 + cy_1) + v(y''_2 + by'_2 + cy_2)$$
$$+ b(u'y_1 + v'y_2) + (u'y_1 + v'y_2)' + (u'y'_1 + v'y'_2).$$

Since y_1 and y_2 are solutions of $y'' + by' + cy = 0$, the first two terms on the right side are 0. Hence, to obtain $L(y_p) = k(x)$, it is sufficient to choose u and v such that

$$\begin{cases} u'y_1 + v'y_2 = 0 \\ u'y'_1 + v'y'_2 = k(x) \end{cases}$$

We can show that this system of equations always has a unique solution u' and v'. We may then determine u and v by integration and use the fact that $y_p = uy_1 + vy_2$ to find a particular solution of the differential equation. Our discussion may be summarized as follows.

Variation of Parameters 15.16

If $y = C_1 y_1 + C_2 y_2$ is the general solution of the complementary equation $L(y) = 0$ of $y'' + by' + cy = k(x)$, then a particular solution of $L(y) = k(x)$ is

$$y_p = uy_1 + vy_2,$$

where $u = g(x)$ and $v = h(x)$ satisfy the following system of equations:

$$\begin{cases} u'y_1 + v'y_2 = 0 \\ u'y'_1 + v'y'_2 = k(x) \end{cases}$$

EXAMPLE ▪ 2   Solve the differential equation $y'' + y = \cot x$.

SOLUTION   The complementary equation is $y'' + y = 0$. Since the auxiliary equation $m^2 + 1 = 0$ has roots $\pm i$, we see from Theorem (15.13) that the general solution of the second-order homogeneous differential equation $y'' + y = 0$ is

$$y = C_1 \cos x + C_2 \sin x.$$

Thus, we seek a particular solution of the differential equation that has the form $y_p = uy_1 + vy_2$, where $y_1 = \cos x$ and $y_2 = \sin x$. The system of equations in Theorem (15.16) is, therefore,

$$\begin{cases} u' \cos x + v' \sin x = 0 \\ -u' \sin x + v' \cos x = \cot x \end{cases}$$

Solving for u' and v' gives us

$$u' = -\cos x, \qquad v' = \csc x - \sin x.$$

If we integrate each of these expressions (and discard the constants of integration), we obtain

$$u = -\sin x, \qquad v = \ln |\csc x - \cot x| + \cos x.$$

Applying Theorem (15.16), we find that a particular solution of the given equation is

$$y_p = -\sin x \cos x + \sin x \ln |\csc x - \cot x| + \sin x \cos x,$$

or, equivalently,

$$y_p = \sin x \ln |\csc x - \cot x|.$$

Finally, by Theorem (15.15), the general solution of $y'' + y = \cot x$ is

$$y = C_1 \cos x + C_2 \sin x + \sin x \ln |\csc x - \cot x|.$$

Given the differential equation

$$L(y) = y'' + by' + cy = e^{nx},$$

where e^{nx} is not a solution of $L(y) = 0$, we can reasonably expect that there exists a particular solution of the form $y_p = Ae^{nx}$, since a constant multiple of e^{nx} is the result of finding $L(Ae^{nx})$. This suggests that we use Ae^{nx} as a trial solution in the given equation and attempt to find the value of the coefficient A. This procedure is called the **method of undetermined coefficients** and is illustrated in the next example.

EXAMPLE ▪ 3   Solve the differential equation $y'' + 2y' - 8y = e^{3x}$.

SOLUTION   The auxiliary equation $m^2 + 2m - 8 = 0$ of the differential equation $y'' + 2y' - 8y = 0$ has roots 2 and -4. By Theorem (15.9), the general solution of the complementary equation is

$$y_c = C_1 e^{2x} + C_2 e^{-4x}.$$

From the preceding remarks, we seek a particular solution of the form $y_p = Ae^{3x}$. Since $y_p' = 3Ae^{3x}$ and $y_p'' = 9Ae^{3x}$, substitution in the given equation leads to

$$9Ae^{3x} + 6Ae^{3x} - 8Ae^{3x} = e^{3x}.$$

Dividing both sides by e^{3x}, we obtain

$$9A + 6A - 8A = 1, \quad \text{or} \quad A = \tfrac{1}{7}.$$

Thus, $y_p = \tfrac{1}{7}e^{3x}$, and, by Theorem (15.15), the general solution is

$$y = C_1 e^{2x} + C_2 e^{-4x} + \tfrac{1}{7}e^{3x}.$$

Three rules for arriving at trial solutions of second-order nonhomogeneous differential equations with constant coefficients are stated without proof in the next theorem. (See texts on differential equations for a more extensive treatment of this topic.)

Theorem on Particular Solutions 15.17

(i) If $y'' + by' + cy = e^{nx}$ and n is not a root of the auxiliary equation $m^2 + bm + c = 0$, then there is a particular solution of the form $y_p = Ae^{nx}$.

(ii) If $y'' + by' + cy = xe^{nx}$ and n is not a solution of the auxiliary equation $m^2 + bm + c = 0$, then there is a particular solution of the form $y_p = (A + Bx)e^{nx}$.

(iii) If either
$$y'' + by' + cy = e^{sx} \sin tx$$

or
$$y'' + by' + cy = e^{sx} \cos tx$$

and the complex number $s + ti$ is not a solution of the auxiliary equation $m^2 + bm + c = 0$, then there is a particular solution of the form

$$y_p = Ae^{sx} \cos tx + Be^{sx} \sin tx.$$

Part (i) of Theorem (15.17) was used in the solution of Example 3. Illustrations of parts (ii) and (iii) are given in the following examples.

EXAMPLE ■ 4 Solve $y'' - 3y' - 18y = xe^{4x}$.

SOLUTION Since the auxiliary equation $m^2 - 3m - 18 = 0$ has roots 6 and -3, it follows from the preceding section that the general solution of $y'' - 3y' - 18y = 0$ is

$$y = C_1 e^{6x} + C_2 e^{-3x}.$$

Since 4 is not a root of the auxiliary equation, we see from Theorem (15.17)(ii) that there is a particular solution of the form

$$y_p = (A + Bx)e^{4x}.$$

Differentiating twice, we have

$$y_p' = (4A + 4Bx + B)e^{4x}$$
$$y_p'' = (16A + 16Bx + 8B)e^{4x}.$$

Substitution in the given differential equation produces

$$(16A + 16Bx + 8B)e^{4x} - 3(4A + 4Bx + B)e^{4x} - 18(A + Bx)e^{4x}$$
$$= xe^{4x},$$

which reduces to

$$-14A + 5B - 14Bx = x.$$

Thus, y_p is a solution, provided

$$-14A + 5B = 0 \quad \text{and} \quad -14B = 1.$$

This gives us $B = -\frac{1}{14}$ and $A = -\frac{5}{196}$. Consequently,

$$y_p = (-\tfrac{5}{196} - \tfrac{1}{14}x)e^{4x} = -\tfrac{1}{196}(5 + 14x)e^{4x}.$$

Applying Theorem (15.15), we find that the general solution is

$$y = C_1 e^{6x} + C_2 e^{-3x} - \tfrac{1}{196}(5 + 14x)e^{4x}.$$

EXAMPLE ▪ 5   Solve $y'' - 10y' + 41y = \sin x$.

SOLUTION   In Example 4 of Section 15.3, we found the general solution

$$y_c = e^{5x}(C_1 \cos 4x + C_2 \sin 4x).$$

of the complementary equation. Referring to Theorem (15.17)(iii) with $s = 0$ and $t = 1$, we seek a particular solution of the form

$$y_p = A \cos x + B \sin x.$$

Since

$$y_p' = -A \sin x + B \cos x \quad \text{and} \quad y_p'' = -A \cos x - B \sin x,$$

substitution in the given equation produces

$$-A \cos x - B \sin x + 10A \sin x - 10B \cos x + 41A \cos x + 41B \sin x$$
$$= \sin x,$$

which can be written

$$(40A - 10B) \cos x + (10A + 40B) \sin x = \sin x.$$

Consequently, y_p is a solution, provided

$$40A - 10B = 0 \quad \text{and} \quad 10A + 40B = 1.$$

The solution of this system of equations is $A = \frac{1}{170}$ and $B = \frac{4}{170}$, and hence

$$y_p = \tfrac{1}{170} \cos x + \tfrac{4}{170} \sin x = \tfrac{1}{170}(\cos x + 4 \sin x).$$

Thus, the general solution is

$$y = e^{5x}(C_1 \cos 4x + C_2 \sin 4x) + \tfrac{1}{170}(\cos x + 4 \sin x).$$

EXERCISES 15.4

Exer. 1–10: Solve the differential equation by using the method of variation of parameters.

1 $y'' + y = \tan x$

2 $y'' + y = \sec x$

3 $y'' - 6y' + 9y = x^2 e^{3x}$

4 $y'' + 3y' = e^{-3x}$

5 $y'' - y = e^x \cos x$

6 $y'' - 4y' + 4y = x^{-2} e^{2x}$

7 $y'' - 9y = e^{3x}$

8 $y'' + y = \sin x$

9 $\dfrac{d^2 y}{dx^2} - 3\dfrac{dy}{dx} - 4y = 2$

10 $\dfrac{d^2 y}{dx^2} - \dfrac{dy}{dx} = x + 1$

Exer. 11–18: Solve the differential equation by using undetermined coefficients.

11 $y'' - 3y' + 2y = 4e^{-x}$

12 $y'' + 6y' + 9y = e^{2x}$

13 $y'' + 2y' = \cos 2x$

14 $y'' + y = \sin 5x$

15 $y'' - y = xe^{2x}$

16 $y'' + 3y' - 4y = xe^{-x}$

17 $\dfrac{d^2 y}{dx^2} - 6\dfrac{dy}{dx} + 13y = e^x \cos x$

18 $\dfrac{d^2 y}{dx^2} - 2\dfrac{dy}{dx} + 2y = e^{-x} \sin 2x$

Exer. 19–20: Prove the identity if L is the linear differential operator (15.14) and y, y_1, and y_2 are functions of x.

19 $L(Cy) = CL(y)$

20 $L(y_1 \pm y_2) = L(y_1) \pm L(y_2)$

c Exer. 21–22: A solution of a differential equation of the form $y'' = f(x, y)$ with initial conditions $y = y_0$ at $x = a$ and $y = y_{-1}$ at $x = a - h$ is sometimes approximated using the formula $y_{k+1} = 2y_k - y_{k-1} + h^2 f(x_k, y_k)$ for $k = 0, 1, 2, \ldots, n - 1$, where $h = (b - a)/n$ and where $x_k = a + kh$. If $h \approx 0$, then y_k is an approximation to y at $x = x_k$. Use this formula, with $n = 4$ and $a = 0$, to approximate y at $x = \frac{1}{2}$ for the given differential equation and initial conditions.

21 $y'' = (4x^2 - 2)y$; $\quad y_0 = 1$, $\quad y_{-1} = 0.984496$

22 $y'' = y - x$; $\qquad y_0 = 1$, $\quad y_{-1} = 0.882823$

15.5   VIBRATIONS

Vibrations in mechanical systems are caused by external forces. A violin string vibrates if bowed, a steel beam vibrates if struck by a hammer, and a bridge vibrates if a marching band crosses it in cadence. In this section, we use differential equations to analyze vibrations that may occur in a spring.

According to Hooke's law, the force required to stretch a spring y units beyond its natural length is ky, where k is a positive real number called the *spring constant*. The **restoring force** of the spring is $-ky$. Suppose that a weight W is attached to the spring and, at the equilibrium position, the

Figure 15.8

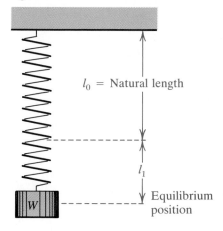

l_0 = Natural length

l_1

Equilibrium position

spring is stretched a distance l_1 beyond its natural length l_0, as illustrated in Figure 15.8. If g is a gravitational constant and m is the mass of the weight, then $W = mg$, and at the equilibrium position

$$mg = kl_1, \quad \text{or} \quad mg - kl_1 = 0,$$

provided that the mass of the spring is negligible compared to m.

Suppose that the weight is pulled down and released. Consider the co-ordinate line with origin at the equilibrium position and positive direction downward, as shown in Figure 15.9, where y is the coordinate of the center of mass of the weight after t seconds. The force F acting on the weight when its acceleration is a is given by Newton's second law of motion $F = ma$. If we assume that the motion is **undamped**—that is, there is no external retarding force—and the weight moves in a frictionless medium, we see that

$$F = mg - k(l_1 + y) = mg - kl_1 - ky = -ky.$$

Since $F = ma$ and $a = d^2y/dt^2$, this implies that

$$m\frac{d^2y}{dt^2} = -ky.$$

Dividing both sides of this equation by m leads to the following differential equation.

Free, Undamped Vibration | **15.18**

$$\frac{d^2y}{dt^2} + \frac{k}{m}y = 0$$

Figure 15.9

l_0

l_1

0

y

EXAMPLE■1 If a weight of mass m is in free, undamped vibration, show that the motion is simple harmonic. Find the displacement if the weight in Figure 15.9 is pulled down a distance l_2 and released with zero velocity.

SOLUTION If we denote k/m by ω^2, then (15.18) may be written

$$\frac{d^2y}{dt^2} + \omega^2 y = 0.$$

The solutions of the auxiliary equation $m^2 + \omega^2 = 0$ are $\pm\omega i$. Hence, from Theorem (15.13), the general solution is

$$y = C_1 \cos \omega t + C_2 \sin \omega t.$$

It follows that the weight is in simple harmonic motion (see Exercise 26 of Section 3.7).

If the weight is pulled down a distance l_2 and is released with zero velocity, then at $t = 0$

$$l_2 = C_1(1) + C_2(0), \quad \text{or} \quad C_1 = l_2.$$

Figure 15.10
Simple harmonic motion

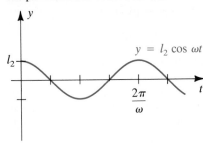

Since
$$\frac{dy}{dt} = -\omega C_1 \sin \omega t + \omega C_2 \cos \omega t,$$

we also have (at $t = 0$)

$$0 = -\omega C_1(0) + \omega C_2(1), \quad \text{or} \quad C_2 = 0.$$

Hence, the displacement y of the weight at time t is

$$y = l_2 \cos \omega t.$$

This type of motion was discussed earlier in the text (see Example 3 of Section 3.7). The *amplitude* (the maximum displacement) is l_2, and the *period* (the time for one complete vibration) is $2\pi/\omega = 2\pi\sqrt{m/k}$. A typical graph indicating this type of motion is illustrated in Figure 15.10.

EXAMPLE ▪ 2 An 8-lb weight stretches a spring 2 ft beyond its natural length. The weight is then pulled down another $\frac{1}{2}$ ft and released with an initial upward velocity of 6 ft/sec. Find a formula for the displacement of the weight at any time t.

SOLUTION From Hooke's law, $8 = k(2)$, or $k = 4$. If y is the displacement of the weight from the equilibrium position at time t, then, by (15.18),

$$\frac{d^2y}{dt^2} + \frac{4}{m}y = 0.$$

Since $W = mg$, it follows that $m = W/g = \frac{8}{32} = \frac{1}{4}$. Hence,

$$\frac{d^2y}{dt^2} + 16y = 0.$$

As in the solution to Example 1, the general solution is

$$y = C_1 \cos 4t + C_2 \sin 4t.$$

At $t = 0$ we have $y = \frac{1}{2}$, and therefore

$$\tfrac{1}{2} = C_1(1) + C_2(0), \quad \text{or} \quad C_1 = \tfrac{1}{2}.$$

Figure 15.11
Damping force

Since
$$\frac{dy}{dt} = -4C_1 \sin 4t + 4C_2 \cos 4t$$

and $dy/dt = -6$ at $t = 0$, we obtain

$$-6 = -4C_1(0) + 4C_2(1), \quad \text{or} \quad C_2 = -\tfrac{3}{2}.$$

Hence, the displacement at time t is given by

$$y = \tfrac{1}{2} \cos 4t - \tfrac{3}{2} \sin 4t.$$

Let us next consider the motion of the spring when a *damping* (or frictional) force is present, as is the case when the weight moves through a fluid (see Figure 15.11). The situation of a shock absorber in an automobile is a good illustration. We shall assume that the direction of the damping force is opposite that of the motion and that the force is directly proportional to the velocity of the weight. Thus, the damping force is given by $-c(dy/dt)$ for some positive constant c. According to Newton's second law, the differential equation that describes the motion is

$$m\frac{d^2y}{dt^2} = -ky - c\frac{dy}{dt}.$$

Dividing by m and rearranging terms gives us the following differential equation.

Free, Damped Vibration 15.19

$$\frac{d^2y}{dt^2} + \frac{c}{m}\frac{dy}{dt} + \frac{k}{m}y = 0$$

EXAMPLE ■ 3 Discuss the motion of a weight of mass m that is in free, damped vibration.

SOLUTION As in Example 1, let $k/m = \omega^2$. To simplify the roots of the auxiliary equation, we also let $c/m = 2p$. Using this notation, we may write (15.19) as

$$\frac{d^2y}{dt^2} + 2p\frac{dy}{dt} + \omega^2 y = 0.$$

The roots of the auxiliary equation $m^2 + 2pm + \omega^2 = 0$ are

$$\frac{-2p \pm \sqrt{4p^2 - 4\omega^2}}{2} = -p \pm \sqrt{p^2 - \omega^2}.$$

The following three possibilities for roots correspond to three possible types of motion of the weight:

$$p^2 - \omega^2 > 0, \quad p^2 - \omega^2 = 0, \quad \text{or} \quad p^2 - \omega^2 < 0.$$

Let us classify the three types of motion as follows.

Case I $p^2 - \omega^2 > 0$ **(overdamped vibration)**
The roots of the auxiliary equation are real and unequal, and, by Theorem (15.9), the general solution of the differential equation is

$$y = e^{-pt}(C_1 e^{\sqrt{p^2-\omega^2}\,t} + C_2 e^{-\sqrt{p^2-\omega^2}\,t}).$$

In the case under consideration,

$$p^2 - \omega^2 = \frac{c^2}{4m^2} - \frac{k}{m} = \frac{c^2 - 4mk}{4m^2} > 0,$$

so $c^2 > 4mk$. This shows that c is usually large compared to k—that is, the damping force dominates the restoring force of the spring, and the weight returns to the equilibrium position relatively quickly. In particular, this happens if the fluid has a high viscosity, as is true for heavy oil or grease.

Figure 15.12
Overdamped vibration

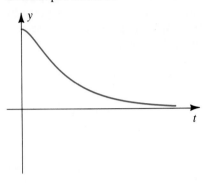

The manner in which y approaches 0 depends on the constants C_1 and C_2 in the general solution. The graph of a particular solution is sketched in Figure 15.12. (See also Figure 15.15.)

Case 2 $p^2 - \omega^2 = 0$ **(critically damped vibration)**
In this case, the auxiliary equation has a double root $-p$, and, by Theorem (15.10), the general solution of the differential equation is

$$y = e^{-pt}(C_1 + C_2 t).$$

The graph of a special case of this equation is sketched in Figure 15.13. The graph is similar to that in Figure 15.12; however, it decreases more rapidly for values of t near 0. Any decrease in the damping force leads to the oscillatory motion discussed in the following case.

Case 3 $p^2 - \omega^2 < 0$ **(underdamped vibration)**
The roots of the auxiliary equation are complex conjugates $a \pm bi$, and, by Theorem (15.13), the general solution of the differential equation is

$$y = e^{-at}(C_1 \sin bt + C_2 \cos bt).$$

In this case, c is usually smaller than k, and the spring oscillates while returning to the equilibrium position, as illustrated in Figure 15.14. Under-damped vibration could occur in a worn shock absorber in an automobile.

Figure 15.13
Critically damped vibration

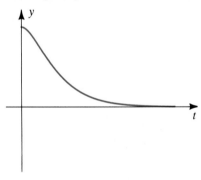

EXAMPLE ■ 4   A 24-lb weight stretches a spring 1 ft beyond its natural length. If the weight is pushed downward from its equilibrium position with an initial velocity of 2 ft/sec and if the damping force is $-9(dy/dt)$, find a formula for the displacement y of the weight at any time t.

SOLUTION   By Hooke's law, $24 = k(1)$, or $k = 24$. The mass of the weight is $m = W/g = \frac{24}{32} = \frac{3}{4}$. Using the notation introduced in Example 3, we have

$$2p = \frac{c}{m} = \frac{9}{\frac{3}{4}} = 12 \quad \text{and} \quad \omega^2 = \frac{k}{m} = \frac{24}{\frac{3}{4}} = 32.$$

Equation (15.19) is $\dfrac{d^2y}{dt^2} + 12\dfrac{dy}{dt} + 32y = 0.$

We can verify that the roots of the auxiliary equation are -8 and -4. Since the roots are real and unequal, the motion is overdamped, as in Case 1 of Example 3, and the general solution of the differential equation is

$$y = C_1 e^{-4t} + C_2 e^{-8t}.$$

Letting $t = 0$ gives us

$$0 = C_1 + C_2, \quad \text{or} \quad C_2 = -C_1.$$

Figure 15.14
Underdamped vibration

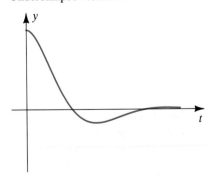

Since $\dfrac{dy}{dt} = -4C_1 e^{-4t} - 8C_2 e^{-8t},$

we have, at $t = 0$,

$$2 = -4C_1 - 8C_2 = -4C_1 - 8(-C_1) = 4C_1.$$

Figure 15.15

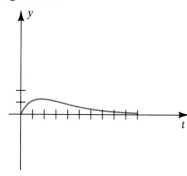

Consequently, $C_1 = \frac{1}{2}$ and $C_2 = -C_1 = -\frac{1}{2}$. Thus, the displacement y of the weight at time t is

$$y = \tfrac{1}{2}e^{-4t} - \tfrac{1}{2}e^{-8t} = \tfrac{1}{2}e^{-4t}(1 - e^{-4t}).$$

The graph is sketched in Figure 15.15. In Exercise 9, you are asked to find the damping force that will produce critical damping.

The final situation we consider is that of a mass in undamped vibration that is subjected to a continuing external force, called a *forcing function*. A violinist may continually bow a string, a construction worker may repeatedly hammer a steel beam, or a strong wind may blow against the side of the deck of a suspension bridge. The displacement y of the object is then described by a second-order nonhomogeneous differential equation of the form

$$y'' + \omega^2 y = x(t),$$

where $x(t)$ is the forcing function.

In many applications, the forcing function is a simple periodic function such as sine or cosine or it can be represented as a sum of sines and cosines. As we saw in Example 1, the general solution of the complementary homogeneous equation $y'' + \omega^2 y = 0$ is

$$y = C_1 \cos \omega t + C_2 \sin \omega t.$$

Thus, in the absence of a forcing function, the system oscillates with a *natural frequency* of $\omega/2\pi$.

Suppose that the forcing function has the form $x(t) = B\omega^2 \sin \omega t$ for some constant B—that is, the frequency of the forcing function agrees with the natural frequency of the oscillating system. This case is called the *resonance case*. In the next example, we will verify that a particular solution to the nonhomogeneous equation

$$y'' + \omega^2 y = B\omega^2 \sin \omega t$$

is given by

$$y = -\frac{B}{2}\omega t \cos \omega t,$$

so that the general solution is

$$y = C_1 \cos \omega t + C_2 \sin \omega t - \frac{B}{2}\omega t \cos \omega t.$$

In the case of resonance, the amplitude of the oscillations *increases without bound* over time because of the factor t in the third term of the solution.

EXAMPLE ▪ 5 For the forced resonant system

$$y'' + \omega^2 y = B\omega^2 \sin \omega t,$$

(a) Verify that $y = (-B/2)\omega t \cos \omega t$ is a particular solution.

(b) Show that $y = \cos 3t + 2 \sin 3t - 6t \cos 3t$ is a solution to the forced resonant system with $y(0) = 1$ and $y'(0) = 0$. Plot the graph of this solution on the interval $0 \le t \le 10$.

SOLUTION

(a) Differentiation of $y = (-B/2)\omega t \cos \omega t$ yields

$$y' = -\frac{B}{2}\omega(\cos \omega t - \omega t \sin \omega t)$$

and

$$y'' = B\omega^2 \sin \omega t + \frac{B}{2}\omega^3 t \cos \omega t.$$

Hence,

$$y'' + \omega^2 y = B\omega^2 \sin \omega t + \frac{B}{2}\omega^3 t \cos \omega t + \omega^2 \left(-\frac{B}{2}\omega t \cos \omega t\right)$$

$$= B\omega^2 \sin \omega t.$$

(b) The equation $y(t) = \cos 3t + 2 \sin 3t - 6t \cos 3t$ is a specific case of the general solution with $C_1 = 1$, $C_2 = 2$, $B = 4$, and $\omega = 3$. Evaluating y at $t = 0$ yields

$$y(0) = \cos 0 + 2 \sin 0 - (6)(0) \cos 0 = 1 + 0 - 0 = 1.$$

Differentiation gives

$$y' = -3 \sin 3t + 6 \cos 3t - 6(\cos 3t - 3t \sin 3t),$$

so that

$$y'(0) = -3 \sin 0 + 6 \cos 0 - 6[\cos 0 - 3(0) \sin 0] = 0 + 6 - 6(1 - 0) = 0.$$

We use a graphing utility to plot the function on the t-interval $[0, 10]$. The result is shown in Figure 15.16, where we see how rapidly the amplitude is increasing as t increases.

Figure 15.16

$y = \cos 3t + 2 \sin 3t - 6t \cos 3t = f(t)$,
$0 \le t \le 10, -60 \le y \le 60$

For a resonant system, the values of the displacement for large values of t are hardly affected by the initial conditions and rapidly grow in amplitude. Since no physical material can withstand unbounded oscillation, it will eventually collapse. Resonance is therefore of great importance in the physical sciences and engineering. The failure of airplane wings and propeller shafts has been attributed to resonance, as was the collapse of the Broughton suspension bridge in 1831. A column of soldiers marched in cadence over this bridge, located near Manchester, England. The rhythm of their march set up a periodic force of great amplitude with a frequency equal to the natural frequency of the bridge and induced very large oscillations in the bridge, thus causing its collapse. For this reason, troops are instructed to break cadence when crossing bridges. For similar reasons, audiences at indoor rock concerts are sometimes asked to avoid rhythmic clapping because of fears that resonance may occur and collapse the walls of the building.

The collapse of the Tacoma Narrows Bridge at Puget Sound in the state of Washington has also been attributed to resonance. Opened to the public on July 1, 1940, the bridge almost immediately displayed unusual vertical oscillations. On the morning of November 7, 1940, the span started swaying up and down, at first a few feet and eventually by as much as 28 feet. At 10:30 A.M., the bridge began to crack and at 11:10, the entire structure crashed into the water. Wind blowing horizontally against the bridge's deck set up wind swirls above and below the deck, creating periodic vertical forces. One explanation for the collapse of the Tacoma Narrows Bridge was that the frequency of these forces matched the natural frequency of the deck and resonance resulted. Recent studies have proposed alternative reasons, but the failure of the Tacoma Narrows Bridge led engineers to rethink bridge design and to consider new ways of "stiffening" the decks of suspension bridges.

EXERCISES 15.5

1 A 5-lb weight stretches a spring 6 in. beyond its natural length. If the weight is raised 4 in. and then released with zero velocity, find a formula for the displacement of the weight at any time t.

2 A 10-lb weight stretches a spring 8 in. beyond its natural length. If the weight is pulled down another 3 in. and then released with a downward velocity of 6 in./sec, find a formula for the displacement at any time t.

3 A 10-lb weight stretches a spring 1 ft beyond its natural length. If the weight is released from its equilibrium position with a downward velocity of 2 ft/sec and if the frictional force is $-5(dy/dt)$, find a formula for the displacement of the weight.

4 A 6-lb weight is suspended from a spring whose spring constant is 48 lb/ft. Initially the weight has a downward velocity of 4 ft/sec from a position 5 in. below the equilibrium point. Find a formula for the displacement.

5 A 4-lb weight stretches a spring 3 in. beyond its natural length. If the weight is pulled down another 4 in. and released with zero velocity in a medium where the damping force is $-2(dy/dt)$, find the displacement y at any time t.

6 A spring stretches $\frac{1}{2}$ ft when an 8-lb weight is attached. If the weight is released from the equilibrium position with an upward velocity of 1 ft/sec and if the damping force is $-4(dy/dt)$, find a formula for the displacement y at any time t.

7 Describe a spring-weight system that leads to the following differential equation and initial values:

$$\frac{1}{4}\frac{d^2y}{dt^2} + \frac{dy}{dt} + 6y = 0$$

at $t = 0$, $y = -2$ and $\frac{dy}{dt} = -1$

8 A 4-lb weight stretches a spring 1 ft. If the weight moves in a medium whose damping force is $-c(dy/dt)$, where $c > 0$, determine the values of c for which the motion is

(a) overdamped   (b) critically damped

(c) underdamped

9 In Example 4 of this section, find the damping force that will produce critical damping.

10 Suppose an undamped weight of mass m is attached to a spring having spring constant k, and an external periodic force given by $F \sin \alpha t$ is applied to the weight.

(a) Show that the differential equation that describes the motion of the weight is

$$\frac{d^2y}{dt^2} + \omega^2 y = \frac{F}{m} \sin \alpha t,$$

where $\omega^2 = k/m$.

(b) Prove that the displacement y of the weight is given by

$$y = C_1 \cos \omega t + C_2 \sin \omega t + C \sin \alpha t,$$

where $C = F/[m(\omega^2 - \alpha^2)]$. This type of motion is called *forced vibration*.

[c] Exer. 11–12: (a) Determine which parameters yield a damped system that is overdamped, critically damped, and underdamped. (b) Plot the solutions in each category on the same screen given that the spring is released from rest ($y'(0) = 0$) from a point 1 ft above the equilibrium position ($y(0) = -1$).

11 $m = 0.3$ slug, $k = 4.2$ lb/ft, $c = 1.7, 1.8, 1.9, 2.0, 2.1, 2.2, 2.3, 2.4$ lb-s/ft

12 $m = 0.3$ slug, $c = 2.2$ lb-s/ft, $k = 3.9, 4.0, 4.1, 4.2, 4.3, 4.4, 4.5, 4.6$ lb/ft

[c] Exer. 13–14: Plot the solutions on the same screen given that the underdamped spring has $m = 0.3$ slug, $k = 4.2$ lb/ft, and $c = 1.9$ lb-s/ft.

13 $y(0) = -1.1$ ft, $y'(0) = -0.5, -0.4, -0.3, -0.2, -0.1, 0, 0.1, 0.2, 0.3, 0.4$ ft/s

14 $y'(0) = -0.1$ ft/s, $y(0) = -0.7, -0.8, -0.9, -1.0, -1.1, -1.2, -1.3, -1.4$ ft

Mathematicians and Their Times

DAVID BLACKWELL

THE LIVES OF DOROTHY BERNSTEIN and David Blackwell reflect both the growing importance of the United States in science and the racism and sexism that have blocked many from scientific careers.

David Harold Blackwell, an African-American, was born on April 24, 1919, in Centralia, Illinois. Although he rarely encountered discrimination as a youth, Blackwell judges that "Southern Illinois was probably fairly racist even when I was growing up there." At age 16, Blackwell enrolled at the University of Illinois, intending to obtain an undergraduate degree and become an elementary school teacher. He went on to earn his Ph.D. degree in mathematics.

Blackwell faced racial discrimination when he joined the Institute for Advanced Study in 1941. Institute members were customarily appointed honorary members of the Princeton University faculty. But, Blackwell notes, "Princeton objected to appointing a black man as an honorary member of the faculty. The director of the institute just insisted and threatened" the university until it withdrew its objections.

When Blackwell sought a permanent position, he applied only to black colleges and universities, believing that no other institution would hire someone of his race. He taught at the Southern Agricultural and Mechanical College and at Clark University before moving to Howard University, where he served for a decade.

Racial prejudice stopped Blackwell's appointment in 1942 to the faculty of the University of California at Berkeley. Blackwell's thesis adviser recommended him for the job, noting "If not for his color, this student would be just the man you want, but I don't suppose you could use him. The democracy we are to fight for abroad could stand some extension at home in some directions."

Although Blackwell's hiring had been approved by the university president, the offer was not made. The wife of a mathematics professor stated that she would not invite any African-American to her house or attend a departmental function at which one was present. Fear of splitting the department blocked his appointment.

A dozen years later, however, Blackwell did join the faculty of Berkeley, where he still teaches. His achievements have brought many honors, including membership in the National Academy of Sciences. He has made contributions to statistics, probability, game theory, set theory, dynamic programming, and information theory.

Blackwell's view on mathematics is refreshing: "Basically, I'm not interested in doing research and I never have been. I'm interested in *understanding*, which is quite a different thing. And often to understand something you have to work it out yourself because no one else has done it."

CHAPTER 15 REVIEW EXERCISES

Exer. 1–34: Solve the differential equation.

1 $xe^y \, dx - \csc x \, dy = 0$

2 $x \, dy - (x + 1)y \, dx = 0$

3 $x(1 + y^2) \, dx + \sqrt{1 - x^2} \, dy = 0$

4 $y' + 4y = e^{-x}$

5 $y' + (\sec x)y = 2 \cos x$

6 $(x^2 y + x^2) \, dy + y \, dx = 0$

7 $y \tan x + y' = 2 \sec x$

8 $y' + (\tan x)y = 3$

9 $y\sqrt{1 - x^2}\, y' = \sqrt{1 - y^2}$

10 $(2y + x^3) \, dx - x \, dy = 0$

11 $e^x \cos y \, dy - \sin^2 y \, dx = 0$

12 $(\cos x)y' + (\sin x)y = 3\cos^2 x$

13 $y' + (2\cos x)y = \cos x$

14 $y'' + y' - 6y = 0$

15 $y'' - 8y' + 16y = 0$

16 $y'' - 6y' + 25y = 0$

17 $\dfrac{d^2 y}{dx^2} - 2\dfrac{dy}{dx} = 0$

18 $\dfrac{d^2 y}{dx^2} = y + \sin x$

19 $y'' - y = e^x \sin x$

20 $y'' - y' - 6y = e^{2x}$

21 $y' + y = e^{4x}$

22 $y'' + 2y' = 0$

23 $y'' - 3y' + 2y = e^{5x}$

24 $xe^y \, dx - (x + 1)y \, dy = 0$

25 $xy' + y = (x - 2)^2$

26 $\sec^2 y \, dx = \sqrt{1 - x^2} \, dy - x\sec^2 y \, dx$

27 $\dfrac{d^2 y}{dx^2} + 5\dfrac{dy}{dx} + 7y = 0$

28 $\dfrac{d^2 y}{dx^2} + y = \csc x$

29 $e^{x+y} \, dx - \csc x \, dy = 0$

30 $y'' + 10y' + 25y = 0$

31 $\cot x \, dy = (y - \cos x) \, dx$

32 $y'' + y' + y = e^x \cos x$

33 $y' + y\csc x = \tan x$

34 $y'' - y' - 20y = xe^{-x}$

35 The bird population of an island experiences seasonal growth described by $dy/dt = (3 \sin 2\pi t)y$, where t is the time in years and $t = 0$ corresponds to the beginning of spring. Migration to and from the island is also seasonal. The rate of migration is given by

$M(t) = 2000 \sin 2\pi t$ birds per year. Hence the complete differential equation for the population is

$$dy/dt = (3 \sin 2\pi t)y + 2000 \sin 2\pi t.$$

Solve for y if the population is 500 at $t = 0$. Determine the maximum population.

36 As a jar containing 10 moles of gas A is heated, the velocity of the gas molecules increases and a second gas B is formed. When two molecules of gas A collide, two molecules of gas B are formed. The rate dy/dt at which gas B is formed is proportional to $(10 - y)^2$, the number of pairs of molecules of gas A. Find a formula for y if $y = 2$ moles after 30 sec.

37 In chemistry, the notation $A + B \rightarrow Y$ is used to denote the production of a substance Y due to the interaction of two substances A and B. Let a and b be the initial amounts of A and B, respectively. If, at time t, the concentration of Y is $y = f(t)$, then the concentrations of A and B are $a - f(t)$ and $b - f(t)$, respectively. If the rate at which the production of Y takes place is given by $dy/dt = k(a - y)(b - y)$ for some positive constant k and if $f(0) = 0$, find $f(t)$.

38 Let $y = f(t)$ be the population, at time t, of some collection, such as insects or bacteria. If the rate of growth dy/dt is proportional to y—that is, if $dy/dt = cy$ for some positive constant c—then $f(t) = f(0)e^{ct}$ (see Theorem (6.33)). In most cases, the rate of growth is dependent on available resources such as food supplies, and as t becomes large, $f'(t)$ begins to decrease. An equation that is often used to describe this type of variation in population is the *law of logistic growth* $dy/dt = y(c - by)$ for some positive constants c and b.

(a) If $f(0) = k$, find $f(t)$.

(b) Find $\lim_{t \to \infty} f(t)$.

(c) Show that $f'(t)$ is increasing if $f(t) < c/(2b)$ and is decreasing if $f(t) > c/(2b)$.

(d) Sketch a typical graph of f. (A graph of this type is a *logistic curve*.)

39 Use a differential equation to describe all functions such that the tangent line at any point $P(x, y)$ on the graph is perpendicular to the line segment joining P to the origin. What is the graph of a typical solution of the differential equation?

40 The following differential equation occurs in the study of electrostatic potentials in spherical regions:

$$\frac{d}{d\theta}\left(\sin\theta \frac{dV}{d\theta}\right) = 0$$

Find a solution $V(\theta)$ that satisfies the conditions $V(\pi/2) = 0$ and $V(\pi/4) = V_0$.

[c] **Exer. 41–42: Use Euler's method to obtain a numerical solution to the differential equation on the indicated interval. (a) For $n = 10$, list the computed approximations for the solution in a table. (b) For $n = 25$, obtain a plot of the pairs (x_k, y_k) for $k = 0, 1, 2, \ldots, 25$.**

41 $y' = \sin x^2 + \cos y^2$; $y_0 = 1.02$ at $x = 0$, $0 \le x \le 2$

42 $y' = \dfrac{y - x}{y^2 + x^2}$; $y_0 = 1.4$ at $x = 1$, $1 \le x \le 2$

[c] **Exer. 43–44: Use the Runge–Kutta method to work the following exercises.**

43 Exercise 41   **44** Exercise 42

[c] **Exer. 45–46: For the given differential equation with initial condition $y_0 = 0.1$ at $x = 0$, approximate y at $x = 0.5$ with $n = 4$, 8, and 16 using (a) Euler's method, (b) the improved Euler's method, and (c) the Runge–Kutta method.**

45 $y' = x^2 + y^2$ **46** $y' = \sin(x + y^2)$

EXTENDED PROBLEMS AND GROUP PROJECTS

1 Investigate *homogeneous differential equations*—that is, differential equations of the form

$$\frac{dy}{dx} = g(x, y),$$

where $g(x, y)$ can be written in the form $f(y/x)$, or

$$\frac{dy}{dx} = f\left(\frac{y}{x}\right).$$

(a) Show that the differential equation $dy/dx = \cos(y/x)$ is of this form.

(b) Show that the differential equation

$$\frac{dy}{dx} = \frac{x + y}{x - y}$$

is homogeneous. *Hint*: Write the fraction as

$$\frac{1 + (y/x)}{1 - (y/x)}.$$

(c) Show that the differential equation

$$\frac{dy}{dx} = \frac{2xy}{x^2 + y^2}$$

is homogeneous but the differential equation

$$\frac{dy}{dx} = \frac{2x^2 y}{x + y^2}$$

is not.

(d) For a homogeneous differential equation, the right side can be considered to be a function of the single variable $v = y/x$. Show that the substitution $v = y/x$ or $y = xv$ changes the homogeneous differential equation into the equivalent equation

$$x\frac{dv}{dx} + v = f(v),$$

which is separable.

(e) Use the technique in part (d) to solve the homogeneous differential equation

$$\frac{dy}{dx} = 2\left(\frac{y}{x}\right) + \left(\frac{y}{x}\right)^2$$

and to solve the homogeneous differential equations of parts (b) and (c).

(f) Show that a differential equation $dy/dx = f(x, y)$ is homogeneous if there is a constant n such that $f(tx, ty) = t^n f(x, y)$ for every real number t.

2 (a) Develop a proof for the claim made after Theorem (15.7) that if the solutions f and g have the property that $f(x) \neq Cg(x)$ for every real number C and if $g(x)$ is not identically zero, then the solution $y = C_1 f(x) + C_2 g(x)$ is the general solution of the differential equation $y'' + by' + cy = 0$.

(b) A pair of functions $\{f(x), g(x)\}$ is called *linearly independent* if the only constants a and b that make the combination $af(x) + bg(x)$ identically zero are $a = b = 0$. Show that the pair $\{e^x, e^{2x}\}$ is linearly independent and the pair $\{\sin x, \cos x\}$ is linearly independent, but that the pair $\{x, 3x\}$ is not linearly independent. Show that $\{f(x), g(x)\}$ is linearly independent if and only if f and g satisfy the property stated in part (a).

(c) A triple of functions $\{f(x), g(x), h(x)\}$ is *linearly independent* if the only constants a, b, and c

that make the combination $af(x) + bg(x) + ch(x)$ identically zero are $a = b = c = 0$. Show that the triple $\{e^x, e^{2x}, e^{3x}\}$ is linearly independent, as is the triple $\{e^x, xe^x, x^2 e^x\}$.

3 Extend the theory developed in Section 15.3 to homogeneous third-order linear differential equations with constant coefficients,

$$y''' + by'' + cy' + d = 0.$$

In particular, do the following:

(a) Prove that if $y = f(x)$, $y = g(x)$, and $y = h(x)$ are each solutions of $y''' + by'' + cy' + d = 0$, then $y = C_1 f(x) + C_2 g(x) + C_3 h(x)$ is a solution for all real numbers C_1, C_2, and C_3.

(b) Define the **auxiliary equation** of the differential equation $y''' + by'' + cy' + d = 0$ in a fashion analogous to Definition (15.6) in such a way that if the roots m_1, m_2, and m_3 of the auxiliary equation are real and unequal, then the general solution of the differential equation is

$$y = C_1 e^{m_1} + C_2 e^{m_2} + C_3 e^{m_3}.$$

(You may assume the theorem that if the triple $\{f(x), g(x), h(x)\}$ is a linearly independent set of solutions of the differential equation, then the general solution is of the form

$$C_1 f(x) + C_2 g(x) + C_3 h(x);$$

see Problem 2 for the definition of linear independence.)

(c) What form does the general solution to

$$y''' + by'' + cy' + d = 0$$

have if the auxiliary equation has a triple root—that is, if it is of the form $(m - m_1)^3 = 0$ for some real number m_1? What can you say about the nature of the general solution if the auxiliary equation can be written as $(m - m_1)^2(m - m_2) = 0$ for unequal real numbers m_1 and m_2? What if the auxiliary equation has one real and two complex roots?

APPENDICES

<table>
<tr><td>**I**</td><td>## THEOREMS ON LIMITS, DERIVATIVES, AND INTEGRALS</td></tr>
</table>

This appendix contains proofs for some theorems stated in the text. The numbering system corresponds to that given in previous chapters.

Uniqueness Theorem for Limits

> If $f(x)$ has a limit as x approaches a, then the limit is unique.

PROOF Suppose $\lim_{x \to a} f(x) = L_1$ and $\lim_{x \to a} f(x) = L_2$ with $L_1 \neq L_2$. We may assume that $L_1 < L_2$. Choose $\epsilon > 0$ such that $\epsilon < \frac{1}{2}(L_2 - L_1)$ and consider the open intervals $(L_1 - \epsilon, L_1 + \epsilon)$ and $(L_2 - \epsilon, L_2 + \epsilon)$ on the coordinate line l' (see Figure 1). Since $\epsilon < \frac{1}{2}(L_2 - L_1)$, these two intervals do not intersect. By Definition (1.5), there is a $\delta_1 > 0$ such that whenever x is in $(a - \delta_1, a + \delta_1)$ and $x \neq a$, then $f(x)$ is in $(L_1 - \epsilon, L_1 + \epsilon)$. Similarly, there is a $\delta_2 > 0$ such that whenever x is in $(a - \delta_2, a + \delta_2)$ and $x \neq a$, then $f(x)$ is in $(L_2 - \epsilon, L_2 + \epsilon)$. This is illustrated in Figure 1, with $\delta_1 < \delta_2$. If an x is selected that is in *both* $(a - \delta_1, a + \delta_1)$ and $(a - \delta_2, a + \delta_2)$, then $f(x)$ is in $(L_1 - \epsilon, L_1 + \epsilon)$ and also in $(L_2 - \epsilon, L_2 + \epsilon)$, contrary to the fact that these two intervals do not intersect. Hence our original supposition is false, and consequently $L_1 = L_2$. ■

Figure 1

Theorem 1.8

If $\lim\limits_{x \to a} f(x)$ and $\lim\limits_{x \to a} g(x)$ both exist, then

(i) $\lim\limits_{x \to a} [f(x) + g(x)] = \lim\limits_{x \to a} f(x) + \lim\limits_{x \to a} g(x)$

(ii) $\lim\limits_{x \to a} [f(x) \cdot g(x)] = \lim\limits_{x \to a} f(x) \cdot \lim\limits_{x \to a} g(x)$

(iii) $\lim\limits_{x \to a} \left[\dfrac{f(x)}{g(x)} \right] = \dfrac{\lim\limits_{x \to a} f(x)}{\lim\limits_{x \to a} g(x)}$, provided $\lim\limits_{x \to a} g(x) \neq 0$

PROOF   Suppose that $\lim_{x \to a} f(x) = L$ and $\lim_{x \to a} g(x) = M$.

(i) According to Definition (1.4), we must show that for every $\epsilon > 0$ there is a $\delta > 0$ such that

(1)   if $0 < |x - a| < \delta$,   then $|f(x) + g(x) - (L + M)| < \epsilon$.

We begin by writing

(2) $|f(x) + g(x) - (L + M)| = |(f(x) - L) + (g(x) - M)|$.

Using the *triangle inequality*

$$|b + c| \leq |b| + |c|$$

for any real numbers b and c, we obtain

$$|(f(x) - L) + (g(x) - M)| \leq |f(x) - L| + |g(x) - M|.$$

Combining the last inequality with (2) gives us

(3) $|f(x) + g(x) - (L + M)| \leq |f(x) - L| + |g(x) - M|$.

Since $\lim_{x \to a} f(x) = L$ and $\lim_{x \to a} g(x) = M$, the numbers $|f(x) - L|$ and $|g(x) - M|$ can be made arbitrarily small by choosing x sufficiently close to a. In particular, they can be made less than $\epsilon/2$. Thus, there exist $\delta_1 > 0$ and $\delta_2 > 0$ such that

(4)   if $0 < |x - a| < \delta_1$,   then $|f(x) - L| < \epsilon/2$,   and
         if $0 < |x - a| < \delta_2$,   then $|g(x) - M| < \epsilon/2$.

If δ denotes the *smaller* of δ_1 and δ_2, then whenever $0 < |x - a| < \delta$, the inequalities in (4) involving $f(x)$ and $g(x)$ are both true. Consequently, if $0 < |x - a| < \delta$, then, from (4) and (3),

$$|f(x) + g(x) - (L + M)| < \epsilon/2 + \epsilon/2 = \epsilon,$$

which is the desired statement (1).

(ii) We first show that if k is a function and

(5)   if $\lim\limits_{x \to a} k(x) = 0$,   then $\lim\limits_{x \to a} f(x)k(x) = 0$.

Since $\lim_{x \to a} f(x) = L$, it follows from Definition (1.4) (with $\epsilon = 1$) that there is a $\delta_1 > 0$ such that if $0 < |x - a| < \delta_1$, then $|f(x) - L| < 1$ and hence also

$$|f(x)| = |f(x) - L + L| \leq |f(x) - L| + |L| < 1 + |L|.$$

Consequently,

(6) if $0 < |x - a| < \delta_1$, then $|f(x)k(x)| < (1 + |L|)\,|k(x)|$.

Since $\lim_{x \to a} k(x) = 0$, for every $\epsilon > 0$ there is a $\delta_2 > 0$ such that

(7) if $0 < |x - a| < \delta_2$, then $|k(x) - 0| < \dfrac{\epsilon}{1 + |L|}$.

If δ denotes the smaller of δ_1 and δ_2, then whenever $0 < |x - a| < \delta$, both inequalities (6) and (7) are true and, consequently,

$$|f(x)k(x)| < (1 + |L|) \cdot \frac{\epsilon}{1 + |L|}.$$

Therefore,

$$\text{if} \quad 0 < |x - a| < \delta, \quad \text{then} \quad |f(x)k(x) - 0| < \epsilon,$$

which proves (5).

Next consider the identity

(8) $f(x)g(x) - LM = f(x)[g(x) - M] + M[f(x) - L].$

Since $\lim_{x \to a}[g(x) - M] = 0$, it follows from (5), with $k(x) = g(x) - M$, that $\lim_{x \to a} f(x)[g(x) - M] = 0$. In addition, $\lim_{x \to a} M[f(x) - L] = 0$ and hence, from (8), $\lim_{x \to a}[f(x)g(x) - LM] = 0$. The last statement is equivalent to $\lim_{x \to a} f(x)g(x) = LM$.

(iii) It is sufficient to show that $\lim_{x \to a} 1/g(x) = 1/M$, for once this is done, the desired result may be obtained by applying (ii) to the product $f(x) \cdot 1/g(x)$. Consider

(9) $\left| \dfrac{1}{g(x)} - \dfrac{1}{M} \right| = \left| \dfrac{M - g(x)}{g(x)M} \right| = \dfrac{1}{|M|\,|g(x)|} |g(x) - M|.$

Since $\lim_{x \to a} g(x) = M$, there is a $\delta_1 > 0$ such that if $0 < |x - a| < \delta_1$, then $|g(x) - M| < |M|/2$. Consequently, for every such x,

$$|M| = |g(x) + (M - g(x))|$$
$$\leq |g(x)| + |M - g(x)|$$
$$< |g(x)| + |M|/2$$

and therefore,

$$\frac{|M|}{2} < |g(x)|, \quad \text{or} \quad \frac{1}{|g(x)|} < \frac{2}{|M|}.$$

Substitution in (9) leads to

(10) $\left| \dfrac{1}{g(x)} - \dfrac{1}{M} \right| < \dfrac{2}{|M|^2} |g(x) - M|, \quad \text{provided } 0 < |x - a| < \delta_1.$

Again, since $\lim_{x \to a} g(x) = M$, it follows that for every $\epsilon > 0$ there is a $\delta_2 > 0$ such that

(11) if $0 < |x - a| < \delta_2$, then $|g(x) - M| < \dfrac{|M|^2}{2}\epsilon.$

If δ denotes the smaller of δ_1 and δ_2, then both inequalities (10) and (11) are true. Thus,

$$\text{if} \quad 0 < |x - a| < \delta, \quad \text{then} \quad \left| \frac{1}{g(x)} - \frac{1}{M} \right| < \epsilon,$$

which means that $\lim_{x \to a} 1/g(x) = 1/M$. ■

Theorem 1.13

> If $a > 0$ and n is a positive integer, or if $a \leq 0$ and n is an odd positive integer, then
>
> $$\lim_{x \to a} \sqrt[n]{x} = \sqrt[n]{a}.$$

PROOF Suppose $a > 0$ and n is any positive integer. We must show that for every $\epsilon > 0$ there is a $\delta > 0$ such that

$$\text{if} \quad 0 < |x - a| < \delta, \quad \text{then} \quad \left| \sqrt[n]{x} - \sqrt[n]{a} \right| < \epsilon,$$

or, equivalently,

(1) if $\quad -\delta < x - a < \delta \quad$ and $\quad x \neq a, \quad$ then $\quad -\epsilon < \sqrt[n]{x} - \sqrt[n]{a} < \epsilon.$

It is sufficient to prove (1) if $\epsilon < \sqrt[n]{a}$, for if a δ exists under this condition, then the same δ can be used for any *larger* value of ϵ. Thus, in the remainder of the proof, $\sqrt[n]{a} - \epsilon$ is considered to be a positive number less than $\sqrt[n]{a}$. The inequalities in the following list are all equivalent:

$$-\epsilon < \sqrt[n]{x} - \sqrt[n]{a} < \epsilon$$

$$\sqrt[n]{a} - \epsilon < \sqrt[n]{x} < \sqrt[n]{a} + \epsilon$$

$$(\sqrt[n]{a} - \epsilon)^n < x < (\sqrt[n]{a} + \epsilon)^n$$

$$(\sqrt[n]{a} - \epsilon)^n - a < x - a < (\sqrt[n]{a} + \epsilon)^n - a$$

$$-[a - (\sqrt[n]{a} - \epsilon)^n] < x - a < (\sqrt[n]{a} + \epsilon)^n - a$$

If δ denotes the smaller of the two positive numbers $a - (\sqrt[n]{a} - \epsilon)^n$ and $(\sqrt[n]{a} + \epsilon)^n - a$, then whenever $-\delta < x - a < \delta$, the last inequality in the list is true and hence so is the first. This gives us (1).

Next suppose $a < 0$ and n is an odd positive integer. In this case, $-a$ and $\sqrt[n]{-a}$ are positive and, by the first part of the proof, we may write

$$\lim_{-x \to -a} \sqrt[n]{-..} = \sqrt[n]{-a}.$$

Thus, for every $\epsilon > 0$ there is a $\delta > 0$ such that

$$\text{if} \quad 0 < |-x - (-a)| < \delta, \quad \text{then} \quad \left| \sqrt[n]{-x} - \sqrt[n]{-a} \right| < \epsilon,$$

or equivalently,

$$\text{if} \quad 0 < |x - a| < \delta, \quad \text{then} \quad \left| \sqrt[n]{x} - \sqrt[n]{a} \right| < \epsilon.$$

The last inequalities imply that $\lim_{x \to a} \sqrt[n]{x} = \sqrt[n]{a}$. ■

Sandwich Theorem 1.15

Suppose $f(x) \le h(x) \le g(x)$ for every x in an open interval containing a, except possibly at a.

If $\lim_{x \to a} f(x) = L = \lim_{x \to a} g(x)$, then $\lim_{x \to a} h(x) = L$.

P R O O F For every $\epsilon > 0$, there is a $\delta_1 > 0$ and a $\delta_2 > 0$ such that

(1)
$$\text{if} \quad 0 < |x - a| < \delta_1, \quad \text{then} \quad |f(x) - L| < \epsilon, \quad \text{and}$$
$$\text{if} \quad 0 < |x - a| < \delta_2, \quad \text{then} \quad |g(x) - L| < \epsilon.$$

If δ denotes the smaller of δ_1 and δ_2, then whenever $0 < |x - a| < \delta$, both inequalities in (1) that involve ϵ are true—that is,

$$-\epsilon < f(x) - L < \epsilon \quad \text{and} \quad -\epsilon < g(x) - L < \epsilon.$$

Thus, if $0 < |x - a| < \delta$, then $L - \epsilon < f(x)$ and $g(x) < L + \epsilon$. Since $f(x) \le h(x) \le g(x)$, if $0 < |x - a| < \delta$, then $L - \epsilon < h(x) < L + \epsilon$, or, equivalently, $|h(x) - L| < \epsilon$, which is what we wished to prove. ∎

Theorem 1.18

If k is a positive rational number and c is any real number, then

$$\lim_{x \to \infty} \frac{c}{x^k} = 0 \quad \text{and} \quad \lim_{x \to -\infty} \frac{c}{x^k} = 0,$$

provided x^k is always defined.

P R O O F To use Definition (1.16) to prove that $\lim_{x \to \infty} (c/x^k) = 0$, we must show that for every $\epsilon > 0$ there is a positive number N such that

$$\left| \frac{c}{x^k} - 0 \right| < \epsilon \quad \text{whenever} \quad x > N.$$

If $c = 0$, any $N > 0$ will suffice. If $c \ne 0$, the following four inequalities are equivalent for $x > 0$:

$$\left| \frac{c}{x^k} - 0 \right| < \epsilon, \qquad \frac{|x|^k}{|c|} > \frac{1}{\epsilon}, \qquad |x|^k > \frac{|c|}{\epsilon}, \qquad x > \left(\frac{|c|}{\epsilon} \right)^{1/k}$$

The last inequality gives us a clue to a choice for N. Letting $N = (|c|/\epsilon)^{1/k}$, we see that whenever $x > N$, the fourth, and hence the first, inequality is true, which is what we wished to show. The second part of the theorem may be proved in similar fashion. ∎

Theorem 1.24

If $\lim_{x \to c} g(x) = b$ and if f is continuous at b, then

$$\lim_{x \to c} f(g(x)) = f(b) = f \left(\lim_{x \to c} g(x) \right).$$

Figure 2

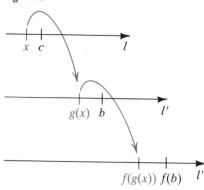

PROOF   The composite function $f(g(x))$ may be represented geometrically by means of three real lines l, l', and l'', as shown in Figure 2. To each coordinate x on l, there corresponds the coordinate $g(x)$ on l' and then, in turn, $f(g(x))$ on l''. We wish to prove that $f(g(x))$ has the limit $f(b)$ as x approaches c. In terms of Definition (1.4), we must show that for every $\epsilon > 0$ there exists a $\delta > 0$ such that

(1)   if $\quad 0 < |x - c| < \delta, \quad$ then $\quad |f(g(x)) - f(b)| < \epsilon.$

Let us begin by considering the interval $(f(b) - \epsilon, f(b) + \epsilon)$ on l'', shown in Figure 3. Since f is continuous at b, $\lim_{z \to b} f(z) = f(b)$ and hence, as illustrated in the figure, there exists a number $\delta_1 > 0$ such that

(2)   if $\quad |z - b| < \delta_1, \quad$ then $\quad |f(z) - f(b)| < \epsilon.$

In particular, if we let $z = g(x)$ in (2), it follows that

(3)   if $\quad |g(x) - b| < \delta_1, \quad$ then $\quad |f(g(x)) - f(b)| < \epsilon.$

Next, turning our attention to the interval $(b - \delta_1, b + \delta_1)$ on l' and using the definition of $\lim_{x \to c} g(x) = b$, we obtain the fact illustrated in Figure 4 — that there exists a $\delta > 0$ such that

(4)   if $\quad 0 < |x - c| < \delta, \quad$ then $\quad |g(x) - b| < \delta_1.$

Figure 3

Figure 4

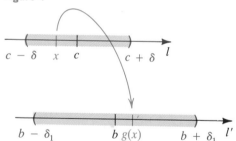

Finally, combining (4) and (3), we see that

$$\text{if} \quad 0 < |x - c| < \delta, \quad \text{then} \quad |f(g(x)) - f(b)| < \epsilon,$$

which is the desired conclusion (1).   ■

Theorem **2.10a**

> If n is a positive integer and $f(x) = x^{1/n}$, then
>
> $$f'(x) = \frac{1}{n} x^{(1/n)-1}.$$

PROOF   By Definition 2.5,

$$f'(x) = \lim_{h \to 0} \frac{(x + h)^{1/n} - x^{1/n}}{h}.$$

Consider the identity

$$u^n - v^n = (u - v)(u^{n-1} + u^{n-2}v + \cdots + uv^{n-2} + v^{n-1}).$$

If $u \neq v$, then

$$\frac{u - v}{u^n - v^n} = \frac{1}{u^{n-1} + u^{n-2}v + \cdots + uv^{n-2} + v^{n-1}}.$$

Substituting $u = (x + h)^{1/n}$ and $v = x^{1/n}$, we obtain

$$\frac{(x + h)^{1/n} - x^{1/n}}{(x + h) - x} =$$

$$\frac{1}{(x + h)^{(n-1)/n} + (x + h)^{(n-2)/n}x^{1/n} + \cdots + (x + h)^{1/n}x^{(n-2)/n} + x^{(n-1)/n}}$$

Letting $h \to 0$, we have

$$f'(x) = \frac{1}{x^{(n-1)/n} + x^{(n-1)/n} + \cdots + x^{(n-1)/n} + x^{(n-1)/n}}$$

$$= \frac{1}{nx^{1-(1/n)}} = \frac{1}{n}x^{(1/n)-1}. \quad \blacksquare$$

Chain Rule 2.26

If $y = f(u)$, $u = g(x)$, and the derivatives $\dfrac{dy}{du}$ and $\dfrac{du}{dx}$ both exist, then the composite function defined by $y = f(g(x))$ has a derivative given by

$$\frac{dy}{dx} = \frac{dy}{du}\frac{du}{dx} = f'(u)g'(x) = f'(g(x))g'(x).$$

PROOF Since f is differentiable at $g(x)$, we have

$$\lim_{k \to 0} \frac{f(g(x) + k) - f(g(x))}{k} = f'(g(x))$$

so that if $\epsilon > 0$, there is a $\delta_1 > 0$ such that

(1) if $0 < |k| < \delta_1$, then $\left| \dfrac{f(g(x) + k) - f(g(x))}{k} - f'(g(x)) \right| < \epsilon$.

Since g is differentiable at x, g is also continuous at x, and hence there is a $\delta > 0$ such that

(2) if $|h| < \delta$, then $|g(x + h) - g(x)| < \delta_1$.

We define a function T by the rule

$$T(h) = \begin{cases} \dfrac{f(g(x + h)) - f(g(x))}{g(x + h) - g(x)} & \text{if } g(x + h) \neq g(x) \\ f'(g(x)) & \text{if } g(x + h) = g(x) \end{cases}$$

Note that

$$\frac{f(g(x + h)) - f(g(x))}{h} = T(h)\frac{g(x + h) - g(x)}{h}$$

for *all* nonzero h (both sides are 0 if $g(x + h) = g(x)$).

Let h be any real number with $|h| < \delta$ and let $k = g(x + h) - g(x)$. Thus, by (2), $|k| < \delta_1$. If $k \neq 0$, then

$$T(h) = \frac{f(g(x + h)) - f(g(x))}{g(x + h) - g(x)} = \frac{f(g(x) + k) - f(g(x))}{k}.$$

Since $|k| < \delta_1$, by (1) we also have $|T(h) - f'(g(x))| < \epsilon$. If $k = 0$, then $T(h) = f'(g(x))$ so that $|T(h) - f'(g(x))| = 0 < \epsilon$. Thus, given an $\epsilon > 0$, there is a $\delta > 0$ such that if $|h| < \delta$, then $|T(h) - f'(g(x))| < \epsilon$. Hence, T is continuous at 0 with $\lim_{h \to 0} T(h) = f'(g(x))$.

Finally, we note that

$$f'(g(x))g'(x) = \lim_{h \to 0} T(h) \lim_{h \to 0} \frac{g(x + h) - g(x)}{h}$$

$$= \lim_{h \to 0} T(h) \frac{g(x + h) - g(x)}{h}$$

$$= \lim_{h \to 0} \frac{f(g(x + h)) - f(g(x))}{h} = (f \circ g)'(x). \quad \blacksquare$$

Theorem   4.22

> If f is integrable on $[a, b]$ and c is any number, then cf is integrable on $[a, b]$ and
>
> $$\int_a^b cf(x)\, dx = c \int_a^b f(x)\, dx.$$

PROOF   If $c = 0$, the result follows from Theorem (4.21). Assume, therefore, that $c \neq 0$. Since f is integrable, $\int_a^b f(x)\, dx = I$ for some number I. If P is a partition of $[a, b]$, then each Riemann sum R_P for the function cf has the form $\sum_k cf(w_k)\Delta x_k$ such that for every k, w_k is in the kth subinterval $[x_{k-1}, x_k]$ of P. We wish to show that for every $\epsilon > 0$ there is a $\delta > 0$ such that whenever $\|P\| < \delta$,

(1)
$$\left| \sum_k cf(w_k)\Delta x_k - cI \right| < \epsilon$$

for every w_k in $[x_{k-1}, x_k]$. If we let $\epsilon' = \epsilon/|c|$, then, since f is integrable, there exists a $\delta > 0$ such that whenever $\|P\| < \delta$,

$$\left| \sum_k f(w_k)\Delta x_k - I \right| < \epsilon' = \frac{\epsilon}{|c|}.$$

Multiplying both sides of this inequality by $|c|$ leads to (1). Hence,

$$\lim_{\|P\| \to 0} \sum_k cf(w_k)\Delta x_k = cI$$

$$= c \int_a^b f(x)\, dx. \quad \blacksquare$$

Theorem 4.23

If f and g are integrable on $[a, b]$, then $f + g$ and $f - g$ are integrable on $[a, b]$ and

(i) $\displaystyle\int_a^b [f(x) + g(x)]\, dx = \int_a^b f(x)\, dx + \int_a^b g(x)\, dx$

(ii) $\displaystyle\int_a^b [f(x) - g(x)]\, dx = \int_a^b f(x)\, dx - \int_a^b g(x)\, dx$

PROOF **(i)** By hypothesis, there exist real numbers I_1 and I_2 such that

$$\int_a^b f(x)\, dx = I_1 \quad \text{and} \quad \int_a^b g(x)\, dx = I_2.$$

Let P denote a partition of $[a, b]$ and let R_P denote an arbitrary Riemann sum for $f + g$ associated with P—that is,

(1) $$R_P = \sum_k \left[f(w_k) + g(w_k) \right] \Delta x_k$$

such that w_k is in $\left[x_{k-1}, x_k \right]$ for every k. We wish to show that for every $\epsilon > 0$ there is a $\delta > 0$ such that whenever $\|P\| < \delta$, $|R_P - (I_1 + I_2)| < \epsilon$. Using Theorem (4.11)(i), we may write (1) in the form

$$R_P = \sum_k f(w_k) \Delta x_k + \sum_k g(w_k) \Delta x_k.$$

Rearranging terms and using the triangle inequality, we obtain

(2) $$|R_P - (I_1 + I_2)| = \left| \left(\sum_k f(w_k) \Delta x_k - I_1 \right) + \left(\sum_k g(w_k) \Delta x_k - I_2 \right) \right|$$

$$\leq \left| \sum_k f(w_k) \Delta x_k - I_1 \right| + \left| \sum_k g(w_k) \Delta x_k - I_2 \right|.$$

By the integrability of f and g, if $\epsilon' = \epsilon/2$, then there exist $\delta_1 > 0$ and $\delta_2 > 0$ such that whenever $\|P\| < \delta_1$ and $\|P\| < \delta_2$,

(3) $$\left| \sum_k f(w_k) \Delta x_k - I_1 \right| < \epsilon' = \epsilon/2 \quad \text{and}$$

$$\left| \sum_k g(w_k) \Delta x_k - I_2 \right| < \epsilon' = \epsilon/2$$

for every w_k in $\left[x_{k-1}, x_k \right]$. If δ denotes the smaller of δ_1 and δ_2, then whenever $\|P\| < \delta$, both inequalities in (3) are true and hence, from (2),

$$|R_P - (I_1 + I_2)| < (\epsilon/2) + (\epsilon/2) = \epsilon,$$

which is what we wished to prove.

(ii) By Theorem (4.22) with $c = -1$, we know that $-g$ is integrable on $[a, b]$ and $\int_a^b -g(x)\, dx = -1 \int_a^b g(x)\, dx$. Thus, by part (i), $f - g =$

$f + (-g)$ is integrable with

$$\int_a^b [f(x) - g(x)]\, dx = \int_a^b [f(x) + (-g(x))]\, dx$$

$$= \int_a^b f(x)\, dx + \int_a^b -g(x)\, dx$$

$$= \int_a^b f(x)\, dx - \int_a^b g(x)\, dx. \quad \blacksquare$$

Theorem 4.24

If $a < c < b$ and if f is integrable on both $[a, c]$ and $[c, b]$, then f is integrable on $[a, b]$ and

$$\int_a^b f(x)\, dx = \int_a^c f(x)\, dx + \int_c^b f(x)\, dx.$$

PROOF   By hypothesis, there exist real numbers I_1 and I_2 such that

(1)
$$\int_a^c f(x)\, dx = I_1 \quad \text{and} \quad \int_c^b f(x)\, dx = I_2.$$

Let us denote a partition of $[a, c]$ by P_1, of $[c, b]$ by P_2, and of $[a, b]$ by P. Arbitrary Riemann sums associated with P_1, P_2, and P will be denoted by R_{P_1}, R_{P_2}, and R_P, respectively. We must show that for every $\epsilon > 0$ there is a $\delta > 0$ such that if $\|P\| < \delta$, then $|R_P - (I_1 + I_2)| < \epsilon$.

If we let $\epsilon' = \epsilon/4$, then, by (1), there exist positive numbers δ_1 and δ_2 such that if $\|P_1\| < \delta_1$ and $\|P_2\| < \delta_2$, then

(2)
$$\left| R_{P_1} - I_1 \right| < \epsilon' = \epsilon/4 \quad \text{and} \quad \left| R_{P_2} - I_2 \right| < \epsilon' = \epsilon/4.$$

If δ denotes the smaller of δ_1 and δ_2, then both inequalities in (2) are true whenever $\|P\| < \delta$. Moreover, since f is integrable on $[a, c]$ and $[c, b]$, it is bounded on both intervals and hence there exists a number M such that $|f(x)| \leq M$ for every x in $[a, b]$. We shall now assume that δ has been chosen so that, in addition to the previous requirement, we also have $\delta < \epsilon/(4M)$.

Let P be a partition of $[a, b]$ such that $\|P\| < \delta$. If the numbers that determine P are

$$a = x_0, x_1, x_2, \ldots, x_n = b,$$

then there is a unique half-open interval of the form $(x_{d-1}, x_d]$ that contains c. If $R_P = \sum_{k=1}^n f(w_k)\Delta x_k$, we may write

(3)
$$R_P = \sum_{k=1}^{d-1} f(w_k)\Delta x_k + f(w_d)\Delta x_d + \sum_{k=d+1}^n f(w_k)\Delta x_k.$$

Let P_1 denote the partition of $[a, c]$ determined by $\{a, x_1, \ldots, x_{d-1}, c\}$, let P_2 denote the partition of $[c, b]$ determined by $\{c, x_d, \ldots, x_{n-1}, b\}$, and consider the Riemann sums

$$R_{P_1} = \sum_{k=1}^{d-1} f(w_k)\Delta x_k + f(c)(c - x_{d-1}) \quad \text{and}$$

(4)

$$R_{P_2} = f(c)(x_d - c) + \sum_{k=d+1}^{n} f(w_k)\Delta x_k.$$

Using the triangle inequality and (2), we obtain

$$\left|(R_{P_1} + R_{P_2}) - (I_1 + I_2)\right| = \left|(R_{P_1} - I_1) + (R_{P_2} - I_2)\right|$$

(5)
$$\leq \left|R_{P_1} - I_1\right| + \left|R_{P_2} - I_2\right|$$

$$< \frac{\epsilon}{4} + \frac{\epsilon}{4} = \frac{\epsilon}{2}.$$

It follows from (3) and (4) that

$$\left|R_P - (R_{P_1} + R_{P_2})\right| = \left|f(w_d) - f(c)\right|\Delta x_d.$$

Employing the triangle inequality and the choice of δ gives us

$$\left|R_P - (R_{P_1} + R_{P_2})\right| \leq (|f(w_d)| + |f(c)|)\Delta x_d$$

(6)
$$\leq (M + M)[\epsilon/(4M)] = \epsilon/2,$$

provided $\|P\| < \delta$. If we write

$$\left|R_P - (I_1 + I_2)\right| = \left|R_P - (R_{P_1} + R_{P_2}) + (R_{P_1} + R_{P_2}) - (I_1 + I_2)\right|$$

$$\leq \left|R_P - (R_{P_1} + R_{P_2})\right| + \left|(R_{P_1} + R_{P_2}) - (I_1 + I_2)\right|,$$

then it follows from (6) and (5) that whenever $\|P\| < \delta$,

$$\left|R_P - (I_1 + I_2)\right| < (\epsilon/2) + (\epsilon/2) = \epsilon$$

for every Riemann sum R_P. This completes the proof. ■

Theorem 4.26

If f is integrable on $[a, b]$ and $f(x) \geq 0$ for every x in $[a, b]$, then

$$\int_a^b f(x)\, dx \geq 0.$$

PROOF We shall give an indirect proof. Let $\int_a^b f(x)\, dx = I$, and *suppose* that $I < 0$. Consider any partition P of $[a, b]$, and let $R_P = \sum_k f(w_k)\Delta x_k$ be an arbitrary Riemann sum associated with P. Since $f(w_k) \geq 0$ for every w_k in $[x_{k-1}, x_k]$, it follows that $R_P \geq 0$. If we let

$\epsilon = -(I/2)$, then, according to Definition (4.15), whenever $\| P \|$ is sufficiently small,

$$|R_P - I| < \epsilon = -\frac{I}{2}.$$

It follows that $R_P < I - (I/2) = I/2 < 0$, a contradiction. Therefore, the supposition $I < 0$ is false and hence $I \geq 0$. ∎

Theorem 6.6

> If f is continuous and increasing on $[a, b]$, then f has an inverse function f^{-1} that is continuous and increasing on $[f(a), f(b)]$.

PROOF If f is increasing, then f is one-to-one and so f^{-1} exists. To prove that f^{-1} is increasing, we must show that if $w_1 < w_2$ in $[f(a), f(b)]$, then $f^{-1}(w_1) < f^{-1}(w_2)$ in $[a, b]$. Let us give an indirect proof of this fact. *Suppose* $f^{-1}(w_2) \leq f^{-1}(w_1)$. Since f is increasing, it follows that $f(f^{-1}(w_2)) \leq f(f^{-1}(w_1))$ and hence $w_2 \leq w_1$, which is a contradiction. Consequently, $f^{-1}(w_1) < f^{-1}(w_2)$.

We next prove that f^{-1} is continuous on $[f(a), f(b)]$. Recall that $y = f(x)$ if and only if $x = f^{-1}(y)$. In particular, if y_0 is in an open interval $(f(a), f(b))$, let x_0 denote the number in the interval (a, b) such that $y_0 = f(x_0)$, or, equivalently, $x_0 = f^{-1}(y_0)$. We wish to show that

(1) $$\lim_{y \to y_0} f^{-1}(y) = f^{-1}(y_0) = x_0.$$

A geometric representation of f and its inverse f^{-1} is shown in Figure 5. The domain $[a, b]$ of f is represented by points on an x-axis and the domain $[f(a), f(b)]$ of f^{-1} by points on a y-axis. Arrows are drawn from one axis to the other to represent function values. To prove (1), consider any interval $(x_0 - \epsilon, x_0 + \epsilon)$ for $\epsilon > 0$. It is sufficient to find an interval $(y_0 - \delta, y_0 + \delta)$, of the type sketched in Figure 6 on the following page, such that whenever y is in $(y_0 - \delta, y_0 + \delta)$, $f^{-1}(y)$ is in $(x_0 - \epsilon, x_0 + \epsilon)$. We may assume that $x_0 - \epsilon$ and $x_0 + \epsilon$ are in $[a, b]$. As in Figure 7 on the following page, let $\delta_1 = y_0 - f(x_0 - \epsilon)$ and $\delta_2 = f(x_0 + \epsilon) - y_0$.

Figure 5

Figure 6

Figure 7

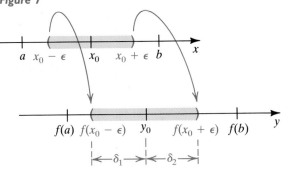

Since f determines a one-to-one correspondence between the numbers in the intervals $(x_0 - \epsilon, x_0 + \epsilon)$ and $(y_0 - \delta_1, y_0 + \delta_2)$, the function values of f^{-1} that correspond to numbers in $(y_0 - \delta_1, y_0 + \delta_2)$ must lie in $(x_0 - \epsilon, x_0 + \epsilon)$. Let δ denote the smaller of δ_1 and δ_2. It follows that if y is in $(y_0 - \delta, y_0 + \delta)$, then $f^{-1}(y)$ is in $(x_0 - \epsilon, x_0 + \epsilon)$, which is what we wished to prove.

The continuity at the endpoints $f(a)$ and $f(b)$ of the domain of f^{-1} may be proved in á similar manner using one-sided limits. ■

Theorem 13.35

If $x = f(u, v)$, $y = g(u, v)$ is a transformation of coordinates, then

$$\iint\limits_{R} F(x, y)\, dx\, dy = \pm \iint\limits_{S} F(f(u, v), g(u, v)) \frac{\partial(x, y)}{\partial(u, v)}\, du\, dv.$$

As (u, v) traces the boundary K of S once in the positive direction, the corresponding point (x, y) traces the boundary C of R once in either the positive direction, in which case the plus sign is chosen, or the negative direction, in which case the minus sign is chosen.

P R O O F   Let us begin by choosing $G(x, y)$ such that $\partial G/\partial x = F$. Applying Green's theorem (14.19) with $G = N$ gives us

(1) $$\iint\limits_{R} F(x, y)\, dx\, dy = \iint\limits_{R} \frac{\partial}{\partial x}[G(x, y)]\, dx\, dy = \oint_{C} G(x, y)\, dy.$$

Suppose the curve K in the uv-plane has a parametrization

$$u = \phi(t), v = \psi(t); \quad a \le t \le b.$$

From our assumptions on the transformation, parametric equations for the curve C in the xy-plane are

(2) $$x = f(u, v) = f(\phi(t), \psi(t))$$
$$y = g(u, v) = g(\phi(t), \psi(t))$$

for $a \le t \le b$. We may therefore evaluate the line integral $\oint_{C} G(x, y)\, dy$ in (1) through formal substitutions for x and y. To simplify the notation, let

$$H(t) = G[f(\phi(t), \psi(t)), g(\phi(t), \psi(t))].$$

Applying a chain rule to y in (2) gives us

$$\frac{dy}{dt} = \frac{\partial y}{\partial u}\frac{du}{dt} + \frac{\partial y}{\partial v}\frac{dv}{dt} = \frac{\partial y}{\partial u}\phi'(t) + \frac{\partial y}{\partial v}\psi'(t).$$

Consequently,

$$\oint_C G(x, y)\, dy = \oint_C H(t)\frac{dy}{dt}\, dt$$

$$= \int_a^b H(t)\left[\frac{\partial y}{\partial u}\phi'(t) + \frac{\partial y}{\partial v}\psi'(t)\right] dt.$$

Since $du = \phi'(t)\, dt$ and $dv = \psi'(t)\, dt$, we may regard the last line integral as a line integral around the curve K in the uv-plane. Thus,

(3)
$$\oint_C G(x, y)\, dy = \pm\oint_K G\frac{\partial y}{\partial u}\, du + G\frac{\partial y}{\partial v}\, dv.$$

For simplicity, we have used G as an abbreviation for $G(f(u, v), g(u, v))$. The choice of the $+$ sign or the $-$ sign is made by letting t vary from a to b and noting whether (x, y) traces C in the same direction or the opposite direction, respectively, as (u, v) traces K.

The line integral on the right in (3) has the form

$$\oint_K M\, du + N\, dv$$

with
$$M = G\frac{\partial y}{\partial u} \quad\text{and}\quad N = G\frac{\partial y}{\partial v}.$$

Applying Green's theorem, we obtain

$$\oint_K M\, du + N\, dv$$

$$= \iint_S \left(\frac{\partial N}{\partial u} - \frac{\partial M}{\partial v}\right) du\, dv$$

$$= \iint_S \left(G\frac{\partial^2 y}{\partial u\, \partial v} + \frac{\partial G}{\partial y}\frac{\partial y}{\partial v} - G\frac{\partial^2 y}{\partial v\, \partial u} - \frac{\partial G}{\partial v}\frac{\partial y}{\partial u}\right) du\, dv$$

$$= \iint_S \left[\left(\frac{\partial G}{\partial x}\frac{\partial x}{\partial u} + \frac{\partial G}{\partial y}\frac{\partial y}{\partial u}\right)\frac{\partial y}{\partial v} - \left(\frac{\partial G}{\partial x}\frac{\partial x}{\partial v} + \frac{\partial G}{\partial y}\frac{\partial y}{\partial v}\right)\frac{\partial y}{\partial u}\right] du\, dv$$

$$= \iint_S \frac{\partial G}{\partial x}\left(\frac{\partial x}{\partial u}\frac{\partial y}{\partial v} - \frac{\partial y}{\partial u}\frac{\partial v}{\partial x}\right) du\, dv.$$

Using the fact that $\partial G/\partial x = F(x, y)$, together with the definition of Jacobian (13.34), gives us

$$\oint_K M\, du + N\, dv = \iint_S F(f(u, v), g(u, v))\frac{\partial(x, y)}{\partial(u, v)}\, du\, dv.$$

Combining this formula with (1) and (3) leads to the desired result. ∎

II TABLE OF INTEGRALS

Basic forms

$$1 \quad \int u\, dv = uv - \int v\, du$$

$$2 \quad \int u^n\, du = \frac{1}{n+1}u^{n+1} + C, \; n \neq -1$$

$$3 \quad \int \frac{du}{u} = \ln|u| + C$$

$$4 \quad \int e^u\, du = e^u + C$$

$$5 \quad \int a^u\, du = \frac{1}{\ln a}a^u + C$$

$$6 \quad \int \sin u\, du = -\cos u + C$$

$$7 \quad \int \cos u\, du = \sin u + C$$

$$8 \quad \int \sec^2 u\, du = \tan u + C$$

$$9 \quad \int \csc^2 u\, du = -\cot u + C$$

$$10 \quad \int \sec u \tan u\, du = \sec u + C$$

$$11 \quad \int \csc u \cot u\, du = -\csc u + C$$

$$12 \quad \int \tan u\, du = -\ln|\cos u| + C$$

$$13 \quad \int \cot u\, du = \ln|\sin u| + C$$

$$14 \quad \int \sec u\, du = \ln|\sec u + \tan u| + C$$

$$15 \quad \int \csc u\, du = \ln|\csc u - \cot u| + C$$

$$16 \quad \int \frac{du}{\sqrt{a^2 - u^2}} = \sin^{-1}\frac{u}{a} + C$$

$$17 \quad \int \frac{du}{a^2 + u^2} = \frac{1}{a}\tan^{-1}\frac{u}{a} + C$$

$$18 \quad \int \frac{du}{u\sqrt{u^2 - a^2}} = \frac{1}{a}\sec^{-1}\frac{u}{a} + C$$

$$19 \quad \int \frac{du}{a^2 - u^2} = \frac{1}{2a}\ln\left|\frac{u + a}{u - a}\right| + C$$

$$20 \quad \int \frac{du}{\sqrt{u^2 - a^2}} = \ln\left|u + \sqrt{u^2 - a^2}\right| + C$$

Forms involving $\sqrt{a^2 + u^2}$

$$21 \quad \int \sqrt{a^2 + u^2}\, du = \frac{u}{2}\sqrt{a^2 + u^2} + \frac{a^2}{2}\ln\left|u + \sqrt{a^2 + u^2}\right| + C$$

$$22 \quad \int u^2\sqrt{a^2 + u^2}\, du = \frac{u}{8}(a^2 + 2u^2)\sqrt{a^2 + u^2} - \frac{a^4}{8}\ln\left|u + \sqrt{a^2 + u^2}\right| + C$$

$$23 \quad \int \frac{\sqrt{a^2 + u^2}}{u}\, du = \sqrt{a^2 + u^2} - a\ln\left|\frac{a + \sqrt{a^2 + u^2}}{u}\right| + C$$

$$24 \quad \int \frac{\sqrt{a^2 + u^2}}{u^2}\, du = -\frac{\sqrt{a^2 + u^2}}{u} + \ln\left|u + \sqrt{a^2 + u^2}\right| + C$$

$$25 \quad \int \frac{du}{\sqrt{a^2 + u^2}} = \ln\left|u + \sqrt{a^2 + u^2}\right| + C$$

26 $\displaystyle\int \frac{u^2\,du}{\sqrt{a^2+u^2}} = \frac{u}{2}\sqrt{a^2+u^2} - \frac{a^2}{2}\ln\left|u + \sqrt{a^2+u^2}\right| + C$

27 $\displaystyle\int \frac{du}{u\sqrt{a^2+u^2}} = -\frac{1}{a}\ln\left|\frac{\sqrt{a^2+u^2}+a}{u}\right| + C$

28 $\displaystyle\int \frac{du}{u^2\sqrt{a^2+u^2}} = -\frac{\sqrt{a^2+u^2}}{a^2u} + C$

29 $\displaystyle\int \frac{du}{(a^2+u^2)^{3/2}} = \frac{u}{a^2\sqrt{a^2+u^2}} + C$

Forms involving $\sqrt{a^2-u^2}$

30 $\displaystyle\int \sqrt{a^2-u^2}\,du = \frac{u}{2}\sqrt{a^2-u^2} + \frac{a^2}{2}\sin^{-1}\frac{u}{a} + C$

31 $\displaystyle\int u^2\sqrt{a^2-u^2}\,du = \frac{u}{8}(2u^2-a^2)\sqrt{a^2-u^2} + \frac{a^4}{8}\sin^{-1}\frac{u}{a} + C$

32 $\displaystyle\int \frac{\sqrt{a^2-u^2}}{u}\,du = \sqrt{a^2-u^2} - a\ln\left|\frac{a+\sqrt{a^2-u^2}}{u}\right| + C$

33 $\displaystyle\int \frac{\sqrt{a^2-u^2}}{u^2}\,du = -\frac{1}{u}\sqrt{a^2-u^2} - \sin^{-1}\frac{u}{a} + C$

34 $\displaystyle\int \frac{u^2\,du}{\sqrt{a^2-u^2}} = -\frac{u}{2}\sqrt{a^2-u^2} + \frac{a^2}{2}\sin^{-1}\frac{u}{a} + C$

35 $\displaystyle\int \frac{du}{u\sqrt{a^2-u^2}} = -\frac{1}{a}\ln\left|\frac{a+\sqrt{a^2-u^2}}{u}\right| + C$

36 $\displaystyle\int \frac{du}{u^2\sqrt{a^2-u^2}} = -\frac{1}{a^2u}\sqrt{a^2-u^2} + C$

37 $\displaystyle\int (a^2-u^2)^{3/2}\,du = -\frac{u}{8}(2u^2-5a^2)\sqrt{a^2-u^2} + \frac{3a^4}{8}\sin^{-1}\frac{u}{a} + C$

38 $\displaystyle\int \frac{du}{(a^2-u^2)^{3/2}} = \frac{u}{a^2\sqrt{a^2-u^2}} + C$

Forms involving $\sqrt{u^2-a^2}$

39 $\displaystyle\int \sqrt{u^2-a^2}\,du = \frac{u}{2}\sqrt{u^2-a^2} - \frac{a^2}{2}\ln\left|u + \sqrt{u^2-a^2}\right| + C$

40 $\displaystyle\int u^2\sqrt{u^2-a^2}\,du = \frac{u}{8}(2u^2-a^2)\sqrt{u^2-a^2} - \frac{a^4}{8}\ln\left|u + \sqrt{u^2-a^2}\right| + C$

41 $\displaystyle\int \frac{\sqrt{u^2-a^2}}{u}\,du = \sqrt{u^2-a^2} - a\sec^{-1}\frac{u}{a} + C$

42 $\displaystyle\int \frac{\sqrt{u^2-a^2}}{u^2}\,du = -\frac{\sqrt{u^2-a^2}}{u} + \ln\left|u + \sqrt{u^2-a^2}\right| + C$

43 $\displaystyle\int \frac{u^2\,du}{\sqrt{u^2-a^2}} = \frac{u}{2}\sqrt{u^2-a^2} + \frac{a^2}{2}\ln\left|u + \sqrt{u^2-a^2}\right| + C$

44 $\displaystyle\int \frac{du}{u^2\sqrt{u^2-a^2}} = \frac{\sqrt{u^2-a^2}}{a^2u} + C$

45 $\displaystyle\int \frac{du}{(u^2-a^2)^{3/2}} = -\frac{u}{a^2\sqrt{u^2-a^2}} + C$

46 $\displaystyle\int \frac{u^2\,du}{(u^2-a^2)^{3/2}} = \frac{-u}{\sqrt{u^2-a^2}} + \ln\left|u + \sqrt{u^2-a^2}\right| + C$

Forms involving $a + bu$

47 $\displaystyle\int \frac{u\,du}{a+bu} = \frac{1}{b^2}(a+bu-a\ln|a+bu|)+C$

48 $\displaystyle\int \frac{u^2\,du}{a+bu} = \frac{1}{2b^3}[(a+bu)^2 - 4a(a+bu) + 2a^2\ln|a+bu|]+C$

49 $\displaystyle\int \frac{du}{u(a+bu)} = \frac{1}{a}\ln\left|\frac{u}{a+bu}\right|+C$

50 $\displaystyle\int \frac{du}{u^2(a+bu)} = -\frac{1}{au} + \frac{b}{a^2}\ln\left|\frac{a+bu}{u}\right|+C$

51 $\displaystyle\int \frac{u\,du}{(a+bu)^2} = \frac{a}{b^2(a+bu)} + \frac{1}{b^2}\ln|a+bu|+C$

52 $\displaystyle\int \frac{du}{u(a+bu)^2} = \frac{1}{a(a+bu)} - \frac{1}{a^2}\ln\left|\frac{a+bu}{u}\right|+C$

53 $\displaystyle\int \frac{u^2\,du}{(a+bu)^2} = \frac{1}{b^3}\left(a+bu - \frac{a^2}{a+bu} - 2a\ln|a+bu|\right)+C$

54 $\displaystyle\int u\sqrt{a+bu}\,du = \frac{2}{15b^2}(3bu-2a)(a+bu)^{3/2}+C$

55 $\displaystyle\int \frac{u\,du}{\sqrt{a+bu}} = \frac{2}{3b^2}(bu-2a)\sqrt{a+bu}+C$

56 $\displaystyle\int \frac{u^2\,du}{\sqrt{a+bu}} = \frac{2}{15b^3}(8a^2+3b^2u^2-4abu)\sqrt{a+bu}+C$

57 $\displaystyle\int \frac{du}{u\sqrt{a+bu}} = \frac{1}{\sqrt{a}}\ln\left|\frac{\sqrt{a+bu}-\sqrt{a}}{\sqrt{a+bu}+\sqrt{a}}\right|+C, \quad \text{if } a>0$

$\displaystyle\qquad\qquad = \frac{2}{\sqrt{-a}}\tan^{-1}\sqrt{\frac{a+bu}{-a}}+C, \qquad \text{if } a<0$

58 $\displaystyle\int \frac{\sqrt{a+bu}}{u}\,du = 2\sqrt{a+bu} + a\int \frac{du}{u\sqrt{a+bu}}$

59 $\displaystyle\int \frac{\sqrt{a+bu}}{u^2}\,du = -\frac{\sqrt{a+bu}}{u} + \frac{b}{2}\int \frac{du}{u\sqrt{a+bu}}$

60 $\displaystyle\int u^n\sqrt{a+bu}\,du = \frac{2}{b(2n+3)}\left[u^n(a+bu)^{3/2} - na\int u^{n-1}\sqrt{a+bu}\,du\right]$

61 $\displaystyle\int \frac{u^n\,du}{\sqrt{a+bu}} = \frac{2u^n\sqrt{a+bu}}{b(2n+1)} - \frac{2na}{b(2n+1)}\int \frac{u^{n-1}\,du}{\sqrt{a+bu}}$

62 $\displaystyle\int \frac{du}{u^n\sqrt{a+bu}} = -\frac{\sqrt{a+bu}}{a(n-1)u^{n-1}} - \frac{b(2n-3)}{2a(n-1)}\int \frac{du}{u^{n-1}\sqrt{a+bu}}$

Trigonometric forms

63 $\displaystyle\int \sin^2 u\,du = \tfrac{1}{2}u - \tfrac{1}{4}\sin 2u + C$

64 $\displaystyle\int \cos^2 u\,du = \tfrac{1}{2}u + \tfrac{1}{4}\sin 2u + C$

65 $\displaystyle\int \tan^2 u\,du = \tan u - u + C$

66 $\displaystyle\int \cot^2 u\,du = -\cot u - u + C$

67 $\displaystyle\int \sin^3 u\,du = -\tfrac{1}{3}(2+\sin^2 u)\cos u + C$

68 $\displaystyle\int \cos^3 u\,du = \tfrac{1}{3}(2+\cos^2 u)\sin u + C$

69 $\displaystyle\int \tan^3 u\,du = \tfrac{1}{2}\tan^2 u + \ln|\cos u| + C$

70 $\displaystyle\int \cot^3 u\,du = -\tfrac{1}{2}\cot^2 u = -\ln|\sin u| + C$

71 $\int \sec^3 u \, du = \frac{1}{2} \sec u \tan u + \frac{1}{2} \ln |\sec u + \tan u| + C$

72 $\int \csc^3 u \, du = -\frac{1}{2} \csc u \cot u + \frac{1}{2} \ln |\csc u - \cot u| + C$

73 $\int \sin^n u \, du = -\frac{1}{n} \sin^{n-1} u \cos u + \frac{n-1}{n} \int \sin^{n-2} u \, du$

74 $\int \cos^n u \, du = \frac{1}{n} \cos^{n-1} u \sin u + \frac{n-1}{n} \int \cos^{n-2} u \, du$

75 $\int \tan^n u \, du = \frac{1}{n-1} \tan^{n-1} u - \int \tan^{n-2} u \, du$

76 $\int \cot^n u \, du = \frac{-1}{n-1} \cot^{n-1} u - \int \cot^{n-2} u \, du$

77 $\int \sec^n u \, du = \frac{1}{n-1} \tan u \sec^{n-2} u + \frac{n-2}{n-1} \int \sec^{n-2} u \, du$

78 $\int \csc^n u \, du = \frac{-1}{n-1} \cot u \csc^{n-2} u + \frac{n-2}{n-1} \int \csc^{n-2} u \, du$

79 $\int \sin au \sin bu \, du = \frac{\sin(a-b)u}{2(a-b)} - \frac{\sin(a+b)u}{2(a+b)} + C$

80 $\int \cos au \cos bu \, du = \frac{\sin(a-b)u}{2(a-b)} + \frac{\sin(a+b)u}{2(a+b)} + C$

81 $\int \sin au \cos bu \, du = -\frac{\cos(a-b)u}{2(a-b)} - \frac{\cos(a+b)u}{2(a+b)} + C$

82 $\int u \sin u \, du = \sin u - u \cos u + C$

83 $\int u \cos u \, du = \cos u + u \sin u + C$

84 $\int u^n \sin u \, du = -u^n \cos u + n \int u^{n-1} \cos u \, du$

85 $\int u^n \cos u \, du = u^n \sin u - n \int u^{n-1} \sin u \, du$

86 $\int \sin^n u \cos^m u \, du = -\frac{\sin^{n-1} u \cos^{m+1} u}{n+m} + \frac{n-1}{n+m} \int \sin^{n-2} u \cos^m u \, du$

$\qquad = \frac{\sin^{n+1} u \cos^{m-1} u}{n+m} + \frac{m-1}{n+m} \int \sin^n u \cos^{m-2} u \, du$

Inverse trigonometric forms

87 $\int \sin^{-1} u \, du = u \sin^{-1} u + \sqrt{1-u^2} + C$

88 $\int \cos^{-1} u \, du = u \cos^{-1} u - \sqrt{1-u^2} + C$

89 $\int \tan^{-1} u \, du = u \tan^{-1} u - \frac{1}{2} \ln(1+u^2) + C$

90 $\int u \sin^{-1} u \, du = \frac{2u^2-1}{4} \sin^{-1} u + \frac{u\sqrt{1-u^2}}{4} + C$

91 $\int u \cos^{-1} u \, du = \frac{2u^2-1}{4} \cos^{-1} u - \frac{u\sqrt{1-u^2}}{4} + C$

92 $\int u \tan^{-1} u \, du = \frac{u^2+1}{2} \tan^{-1} u - \frac{u}{2} + C$

93 $\int u^n \sin^{-1} u \, du = \frac{1}{n+1} \left[u^{n+1} \sin^{-1} u - \int \frac{u^{n+1} \, du}{\sqrt{1-u^2}} \right], \quad n \neq -1$

94 $\int u^n \cos^{-1} u \, du = \frac{1}{n+1} \left[u^{n+1} \cos^{-1} u + \int \frac{u^{n+1} \, du}{\sqrt{1-u^2}} \right], \quad n \neq -1$

95 $\int u^n \tan^{-1} u \, du = \frac{1}{n+1} \left[u^{n+1} \tan^{-1} u - \int \frac{u^{n+1} \, du}{1+u^2} \right], \quad n \neq -1$

Exponential and logarithmic forms

96 $\displaystyle\int ue^{au}\,du = \frac{1}{a^2}(au-1)e^{au} + C$

97 $\displaystyle\int u^n e^{au}\,du = \frac{1}{a}u^n e^{au} - \frac{n}{a}\int u^{n-1}e^{au}\,du$

98 $\displaystyle\int e^{au}\sin bu\,du = \frac{e^{au}}{a^2+b^2}(a\sin bu - b\cos bu) + C$

99 $\displaystyle\int e^{au}\cos bu\,du = \frac{e^{au}}{a^2+b^2}(a\cos bu + b\sin bu) + C$

100 $\displaystyle\int \ln u\,du = u\ln u - u + C$

101 $\displaystyle\int u^n \ln u\,du = \frac{u^{n+1}}{(n+1)^2}[(n+1)\ln u - 1] + C$

102 $\displaystyle\int \frac{1}{u\ln u}\,du = \ln|\ln u| + C$

Hyperbolic forms

103 $\displaystyle\int \sinh u\,du = \cosh u + C$

104 $\displaystyle\int \cosh u\,du = \sinh u + C$

105 $\displaystyle\int \tanh u\,du = \ln\cosh u + C$

106 $\displaystyle\int \coth u\,du = \ln|\sinh u| + C$

107 $\displaystyle\int \operatorname{sech} u\,du = \tan^{-1}\sinh u + C$

108 $\displaystyle\int \operatorname{csch} u\,du = \ln\left|\tanh\tfrac{1}{2}u\right| + C$

109 $\displaystyle\int \operatorname{sech}^2 u\,du = \tanh u + C$

110 $\displaystyle\int \operatorname{csch}^2 u\,du = -\coth u + C$

111 $\displaystyle\int \operatorname{sech} u\tanh u\,du = -\operatorname{sech} u + C$

112 $\displaystyle\int \operatorname{csch} u\coth u\,du = -\operatorname{csch} u + C$

Forms involving $\sqrt{2au-u^2}$

113 $\displaystyle\int \sqrt{2au-u^2}\,du = \frac{u-a}{2}\sqrt{2au-u^2} + \frac{a^2}{2}\cos^{-1}\left(\frac{a-u}{a}\right) + C$

114 $\displaystyle\int u\sqrt{2au-u^2}\,du = \frac{2u^2-au-3a^2}{6}\sqrt{2au-u^2} + \frac{a^3}{2}\cos^{-1}\left(\frac{a-u}{a}\right) + C$

115 $\displaystyle\int \frac{\sqrt{2au-u^2}}{u}\,du = \sqrt{2au-u^2} + a\cos^{-1}\left(\frac{a-u}{a}\right) + C$

116 $\displaystyle\int \frac{\sqrt{2au-u^2}}{u^2}\,du = -\frac{2\sqrt{2au-u^2}}{u} - \cos^{-1}\left(\frac{a-u}{a}\right) + C$

117 $\displaystyle\int \frac{du}{\sqrt{2au-u^2}} = \cos^{-1}\left(\frac{a-u}{a}\right) + C$

118 $\displaystyle\int \frac{u\,du}{\sqrt{2au-u^2}} = -\sqrt{2au-u^2} + a\cos^{-1}\left(\frac{a-u}{a}\right) + C$

119 $\displaystyle\int \frac{u^2\,du}{\sqrt{2au-u^2}} = -\frac{u+3a}{2}\sqrt{2au-u^2} + \frac{3a^2}{2}\cos^{-1}\left(\frac{a-u}{a}\right) + C$

120 $\displaystyle\int \frac{du}{u\sqrt{2au-u^2}} = -\frac{\sqrt{2au-u^2}}{au} + C$

III   THE BINOMIAL SERIES

The binomial theorem states that if k is a positive integer, then for all numbers a and b,

$$(a + b)^k = a^k + ka^{k-1}b + \frac{k(k-1)}{2!}a^{k-2}b^2 + \cdots$$

$$+ \frac{k(k-1)\cdots(k-n+1)}{n!}a^{k-n}b^n + \cdots + b^k.$$

If we let $a = 1$ and $b = x$, then

$$(1+x)^k = 1 + kx + \frac{k(k-1)}{2!}x^2 + \cdots$$

$$+ \frac{k(k-1)\cdots(k-n+1)}{n!}x^n + \cdots + x^k.$$

If k is not a positive integer (or 0), it is useful to study the power series $\sum a_n x^n$ with $a_0 = 1$ and $a_n = k(k-1)\cdots(k-n+1)/n!$ for $n \geq 1$. This infinite series has the form

$$1 + kx + \frac{k(k-1)}{2!}x^2 + \cdots + \frac{k(k-1)\cdots(k-n+1)}{n!}x^n + \cdots$$

and is called the **binomial series**. If k is a nonnegative integer, the series reduces to the finite sum given in the binomial theorem. Otherwise, the series does not terminate. Using the formula for a_n, we can show that

$$\lim_{n \to \infty} \left| \frac{a_{n+1}x^{n+1}}{a_n x^n} \right| = \lim_{n \to \infty} \left| \frac{k-n}{n+1} \right| |x| = |x|.$$

Hence, by the ratio test (8.35), the series is absolutely convergent if $|x| < 1$ and is divergent if $|x| > 1$. Thus, the binomial series represents a function f such that

$$f(x) = 1 + \sum_{n=1}^{\infty} \frac{k(k-1)\cdots(k-n+1)}{n!}x^n \quad \text{if} \quad |x| < 1.$$

We have already noted that if k is a nonnegative integer, then $f(x) = (1+x)^k$. We shall now prove that the same is true for *every* real number k. Differentiating each term of the binomial series gives us

$$f'(x) = k + k(k-1)x + \cdots + \frac{nk(k-1)\cdots(k-n+1)}{n!}x^{n-1} + \cdots$$

and therefore,

$$xf'(x) = kx + k(k-1)x^2 + \cdots + \frac{nk(k-1)\cdots(k-n+1)}{n!}x^n + \cdots.$$

If we add corresponding terms of the preceding two power series, then the coefficient of x^n is

$$\frac{(n+1)k(k-1)\cdots(k-n)}{(n+1)!} + \frac{nk(k-1)\cdots(k-n+1)}{n!},$$

which simplifies to

$$[(k-n)+n]\frac{k(k-1)\cdots(k-n+1)}{n!} = ka_n.$$

Consequently,

$$f'(x) + xf'(x) = \sum_{n=0}^{\infty} ka_n x^n = kf(x),$$

or, equivalently,

$$f'(x)(1+x) - kf(x) = 0.$$

If we define the function g by $g(x) = f(x)/(1+x)^k$, then

$$g'(x) = \frac{(1+x)^k f'(x) - f(x)k(1+x)^{k-1}}{(1+x)^{2k}}$$

$$= \frac{(1+x)f'(x) - kf(x)}{(1+x)^{k+1}} = 0.$$

It follows that $g(x) = c$ for some constant c—that is,

$$\frac{f(x)}{(1+x)^k} = c.$$

Since $f(0) = 1$, we see that $c = 1$ and hence $f(x) = (1+x)^k$, which is what we wished to prove. The next statement summarizes this discussion.

Binomial Series

If $|x| < 1$, then for every real number k,

$$(1+x)^k = 1 + kx + \frac{k(k-1)}{2!}x^2 + \cdots$$

$$+ \frac{k(k-1)\cdots(k-n+1)}{n!}x^n + \cdots.$$

EXAMPLE ■ 1

(a) Find a power series representation for $f(x) = \sqrt[3]{1+x}$.

(b) Plot the graphs of f and $g(x) = 1 + \frac{1}{3}x$.

(c) Use the graphs to estimate the largest closed interval of $[-1, 1]$ on which $|f(x) - g(x)| < \frac{1}{10}$.

SOLUTION

(a) Using the binomial series with $k = \frac{1}{3}$, we obtain

$$\sqrt[3]{1+x} = 1 + \frac{1}{3}x + \frac{\frac{1}{3}(\frac{1}{3}-1)}{2!}x^2 + \frac{\frac{1}{3}(\frac{1}{3}-1)(\frac{1}{3}-2)}{3!}x^3 + \cdots$$

$$+ \frac{\frac{1}{3}(\frac{1}{3}-1)\cdots(\frac{1}{3}-n+1)}{n!}x^n + \cdots,$$

which may be written as

$$\sqrt[3]{1+x} = 1 + \frac{1}{3}x - \frac{2}{3^2 \cdot 2!}x^2 + \frac{1 \cdot 2 \cdot 5}{3^3 \cdot 3!}x^3 + \cdots$$

$$+ (-1)^{n+1}\frac{1 \cdot 2 \cdots (3n-4)}{3^n \cdot n!}x^n + \cdots$$

for $|x| < 1$. The formula for the nth term of this series is valid provided $n \geq 2$.

(b) The function $g(x) = 1 + \frac{1}{3}x$ consists of the first two terms of the power series representation of f. By graphing these functions, we can gain a sense of how closely the first-degree polynomial g approximates f. We use a graphing utility to plot f and g for $-1 \leq x \leq 1$ and $0 \leq y \leq 1.4$ on the same axes, as shown in Figure 1. We see that $g(x)$ appears to be at least as large as $f(x)$ over the entire interval. We also note that the values of f and g are relatively close to each other for $-0.5 < x < 1$. The closer x is to -1, the farther apart are the values $f(x)$ and $g(x)$.

(c) To find the x-values where f and g are within $\frac{1}{10}$ unit of each other, we graph the constant function $\frac{1}{10}$ and the function $h(x) = |f(x) - g(x)|$. Figure 2 shows the graphs. By using the trace operation or Newton's method, we find that the graphs cross at approximately -0.70664905. Thus, we conclude that $|f(x) - g(x)| \leq \frac{1}{10}$ on the interval $[-0.70664905, 1]$. If we use $1 + (x/3)$ to approximate $\sqrt[3]{1+x}$ for any x in this interval, the error will be less than $\frac{1}{10}$.

Figure 1

$f(x) = \sqrt[3]{1+x}, g(x) = 1 + \frac{1}{3}x$

$-1 \leq x \leq 1, 0 \leq y \leq 1.4$

Figure 2

$h(x) = |f(x) - g(x)|$

$-1 \leq x \leq 1, 0 \leq y \leq 0.25$

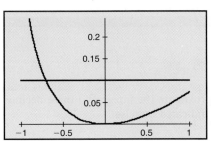

EXAMPLE ▪ 2   Find a power series representation for $\sqrt[3]{1+x^4}$.

SOLUTION   The power series can be obtained by substituting x^4 for x in the series of Example 1. Hence, if $|x| < 1$, then

$$\sqrt[3]{1+x^4} = 1 + \frac{1}{3}x^4 - \frac{2}{3^2 \cdot 2!}x^8 + \cdots$$

$$+ (-1)^{n+1}\frac{1 \cdot 2 \cdots (3n-4)}{3^n \cdot n!}x^{4n} + \cdots.$$

EXAMPLE ▪ 3 Approximate $\int_0^{0.3} \sqrt[3]{1+x^4}\,dx$.

SOLUTION Integrating the terms of the series obtained in Example 2 gives us

$$\int_0^{0.3} \sqrt[3]{1+x^4}\,dx = 0.3 + 0.000162 - 0.000000243 + \cdots.$$

Consequently, the integral may be approximated by 0.300162, which is accurate to six decimal places, since the error is less than 0.000000243. (Why?)

The binomial series can be used to obtain polynomial approximation formulas for $(1+x)^k$. To illustrate, if $|x| < 1$, then from Example 1,

$$\sqrt[3]{1+x} \approx 1 + \tfrac{1}{3}x.$$

Since the series is alternating and satisfies (8.30) from the second term onward, the error involved in this approximation is less than the third term, $\tfrac{1}{9}x^2$.

EXERCISES

Exer. 1–12: Find a power series representation for the expression, and state the radius of convergence.

1 (a) $\sqrt{1+x}$ (b) $\sqrt{1-x^3}$

2 (a) $\dfrac{1}{\sqrt[3]{1+x}}$ (b) $\dfrac{1}{\sqrt[3]{1-x^2}}$

3 $(1+x)^{-2/3}$ 4 $(1+x)^{1/4}$

5 $(1-x)^{3/5}$ 6 $(1-x)^{2/3}$

7 $(1+x)^{-2}$ 8 $(1+x)^{-4}$

9 $(1+x)^{-3}$ 10 $x(1+2x)^{-2}$

11 $\sqrt[3]{8+x}$ (*Hint:* Consider $2\sqrt[3]{1+\tfrac{1}{8}x}$.)

12 $(4+x)^{3/2}$

Exer. 13–14: (a) Obtain a power series representation for $f(x)$ by using the given relationship. (b) Find the radius of convergence.

13 $f(x) = \sin^{-1}x$; $\sin^{-1}x = \displaystyle\int_0^x (1/\sqrt{1-t^2})\,dt$

14 $f(x) = \sinh^{-1}x$; $\sinh^{-1}x = \displaystyle\int_0^x (1/\sqrt{1+t^2})\,dt$

c **Exer. 15–18:** Approximate the integral to three decimal places, using the indicated exercise.

15 $\displaystyle\int_0^{1/2} \sqrt{1+x^3}\,dx$ (Exercise 1)

16 $\displaystyle\int_0^{1/2} \frac{1}{\sqrt[3]{1+x^2}}\,dx$ (Exercise 2)

17 $\displaystyle\int_0^{0.3} \frac{1}{(1+x^3)^2}\,dx$ (Exercise 7)

18 $\displaystyle\int_0^{0.1} \frac{1}{(1+5x^2)^4}\,dx$ (Exercise 8)

c **Exer. 19–20:** For the given k, graph $f(x) = (1+x)^k$ and $g(x) = 1 + kx$ on the same xy-plane for $-1 \le x \le 1$. Use the graphs to estimate the largest closed interval of $[-1, 1]$ on which $|f(x) - g(x)| \le \tfrac{1}{10}$.

19 $k = \tfrac{1}{2}$ 20 $k = \tfrac{5}{2}$

21 Refer to Exercise 82 of Section 7.7. The formula for the period T of a pendulum of length L, initially displaced from equilibrium through an angle of θ_0 radians, is given by the improper integral

$$T = 2\sqrt{\frac{2L}{g}} \int_0^{\theta_0} \frac{1}{\sqrt{\cos\theta - \cos\theta_0}} \, d\theta.$$

By making the substitution $\sin u = (1/k)\sin\frac{1}{2}\theta$, with $k = \sin\frac{1}{2}\theta_0$, it can be shown that

$$T = 4\sqrt{\frac{L}{g}} \int_0^{\pi/2} \frac{1}{\sqrt{1 - k^2 \sin^2 u}} \, du.$$

(a) Use the binomial series for $(1 - x)^{-1/2}$ to show that

$$T \approx 2\pi \sqrt{\frac{L}{g}} \left(1 + \frac{1}{4}k^2 \right).$$

(b) Approximate T if $\theta_0 = \pi/6$.

A N S W E R S

to Selected Exercises

A Student's Solutions Manual to accompany this text is available from your college bookstore. The guide, by Jeffery A. Cole and Gary K. Rockswold, contains detailed solutions to approximately one-third of the exercises as well as strategies for solving other exercises in the text.

Answers are usually not provided for exercises that require lengthy proofs.

PREALGEBRA REVIEW

Exercises A

1 (a) -15 **(b)** -3 **(c)** 11

3 (a) $4 - \pi$ **(b)** $4 - \pi$ **(c)** $1.5 - \sqrt{2}$ **5** $-x - 3$

7 $2 - x$ **9** $-\dfrac{6}{5}, \dfrac{2}{3}$ **11** $-\dfrac{9}{2}, \dfrac{3}{4}$ **13** $-2 \pm \sqrt{2}$

15 $\dfrac{3}{4} \pm \dfrac{1}{4}\sqrt{41}$ **17** $(12, \infty)$ **19** $[9, 19)$ **21** $(-2, 3)$

23 $(-\infty, -2) \cup (4, \infty)$ **25** $\left(-\infty, -\dfrac{5}{2}\right] \cup [1, \infty)$

27 $\left(\dfrac{3}{2}, \dfrac{7}{3}\right)$ **29** $(-\infty, -1) \cup \left(2, \dfrac{7}{2}\right]$ **31** $(-3.01, -2.99)$

33 $(-\infty, -2.001] \cup [-1.999, \infty)$

35 $\left(-\dfrac{9}{2}, -\dfrac{1}{2}\right)$ **37** $\left[\dfrac{3}{5}, \dfrac{9}{5}\right]$

39 (a) The line parallel to the y-axis that intersects the x-axis at $(-2, 0)$
(b) The line parallel to the x-axis that intersects the y-axis at $(0, 3)$
(c) All points to the right of and on the y-axis
(d) All points in quadrants I and III
(e) All points below the x-axis
(f) All points within the rectangle such that $-2 \le x \le 2$ and $-1 \le y \le 1$

41 (a) $\sqrt{29}$ **(b)** $\left(5, -\dfrac{1}{2}\right)$

43 $d(A, C)^2 = d(A, B)^2 + d(B, C)^2$; area $= 28$

45

47

49

51

53

55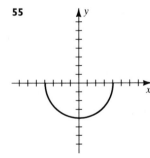

57 $(x - 2)^2 + (y + 3)^2 = 25$
59 $(x + 4)^2 + (y - 4)^2 = 16$
61 $4x + y = 17$ **63** $3x - 4y = 12$ **65** $5x - 2y = 18$
67 $-4.04, -0.53$ **69** $0.05, 2.40$ **71** $200 \le m \le 600$
73 $4 \le p < 6$

75 (a) 1 cm   **(b)** Capsule: $\dfrac{11\pi}{96}$ cm³; tablet: $\dfrac{\pi}{8}$ cm³

77 $0 \le v < 30$ **79 (a)** $0\,°C$ **(b)** $\dfrac{1}{273}$ **(c)** $163.8\,°C$

Exercises B

1 $-12;\ -22;\ -36$

3 (a) $5a - 2$ **(b)** $-5a - 2$ **(c)** $-5a + 2$
    (d) $5a + 5h - 2$ **(e)** $5a + 5h - 4$ **(f)** 5

5 (a) $a^2 - a + 3$
    (b) $a^2 + a + 3$
    (c) $-a^2 + a - 3$
    (d) $a^2 + 2ah + h^2 - a - h + 3$
    (e) $a^2 + h^2 - a - h + 6$
    (f) $2a + h - 1$

7 All real numbers except -2, 0, and 2

9 $\left[\dfrac{3}{2}, 4\right) \cup (4, \infty)$

11 (a) Odd   **(b)** Even   **(c)** Neither

13

15

17 (a)

(b)

(c)

(d)

19

21

23

25

27

29 (a)

(b)

(c)

(d)

(e)

(f)

(g)

(h)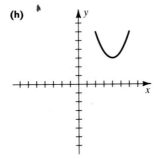

41 (a) $2\sqrt{x+5}$; 0; $x+5$; 1 **(b)** $[-5, \infty)$; $(-5, \infty)$

43 (a) $\dfrac{3x^2+6x}{(x-4)(x+5)}$; $\dfrac{x^2+14x}{(x-4)(x+5)}$;

$\dfrac{2x^2}{(x-4)(x+5)}$; $\dfrac{2x+10}{x-4}$

(b) All real numbers except -5 and 4; all real numbers except -5, 0, and 4

45 (a) $x+2-3\sqrt{x+2}$; $[-2, \infty)$

(b) $\sqrt{x^2-3x+2}$; $(-\infty, 1] \cup [2, \infty)$

47 (a) $\sqrt{\sqrt{x+5}-2}$; $[-1, \infty)$

(b) $\sqrt{\sqrt{x-2}+5}$; $[2, \infty)$

49 (a) $\sqrt{28-x}$; $[3, 28]$

(b) $\sqrt{\sqrt{25-x^2}-3}$; $[-4, 4]$

51 (a) $\dfrac{1}{x+3}$; all real numbers except -3 and 0

(b) $\dfrac{6x+4}{x}$; all real numbers except $-\dfrac{2}{3}$ and 0

Exer. 53–60: Answers are not unique.

53 $u = x^2 + 3x$, $y = u^{1/3}$ **55** $u = x-3$, $y = \dfrac{1}{u^4}$

57 $u = x^4 - 2x^2 + 5$, $y = u^5$

59 $u = \sqrt{x+4}$, $y = \dfrac{u-2}{u+2}$

61 7.91; 5.05 **63** $V = 4x^3 - 100x^2 + 600x$

65 $d = 2\sqrt{t^2 + 2500}$

67 (a) $y = \sqrt{h^2 + 2hr}$ **(b)** 1280.6 mi

69 $d = \sqrt{90{,}400 + x^2}$

71 (a) $y = \dfrac{bh}{a-b}$ **(b)** $V = \dfrac{\pi}{3}h(a^2 + ab + b^2)$

(c) $\dfrac{200}{7\pi}$ ft

Exercises C

1 (a) $\dfrac{5\pi}{6}$ **(b)** $\dfrac{2\pi}{3}$ **(c)** $\dfrac{5\pi}{2}$ **(d)** $-\dfrac{\pi}{3}$

3 (a) $120°$ **(b)** $150°$ **(c)** $135°$ **(d)** $-630°$

5 $\dfrac{20\pi}{9}$; $\dfrac{80\pi}{9}$ **7** $x = 8$, $y = 4\sqrt{3}$

Exer. 9–16: Answers are in the order sin, cos, tan, cot, sec, csc.

9 $\dfrac{3}{5}, \dfrac{4}{5}, \dfrac{3}{4}, \dfrac{4}{3}, \dfrac{5}{4}, \dfrac{5}{3}$ **11** $\dfrac{5}{13}, \dfrac{12}{13}, \dfrac{5}{12}, \dfrac{12}{5}, \dfrac{13}{12}, \dfrac{13}{5}$

13 $-\dfrac{3}{5}, \dfrac{4}{5}, -\dfrac{3}{4}, -\dfrac{4}{3}, \dfrac{5}{4}, -\dfrac{5}{3}$

Exer. 15–20: Answers are not unique.

15 $4\cos\theta$ **17** $\sin\theta$ **19** $\sin\theta$

21 (a) $\dfrac{\sqrt{3}}{2}$ **(b)** $\dfrac{\sqrt{2}}{2}$ **23 (a)** $-\dfrac{\sqrt{3}}{3}$ **(b)** $-\sqrt{3}$

25 (a) -2 **(b)** $\dfrac{2}{\sqrt{3}}$

27 (a) 0.9205 **(b)** 2.3662

29 (a) 0.9781 **(b)** 1.2868

31 (a) -0.8560 **(b)** -0.2958

33 (a) **(b)**

35 (a) **(b)**

37 (a)   **(b)**

          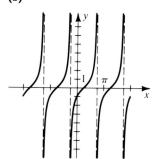

Exer. 39–42: Answers are not unique.

39 $u = \tan^2 x + 4,\ y = \sqrt{u}$ **41** $u = x + \dfrac{\pi}{4},\ y = \sec u$

43 $\dfrac{f(x+h) - f(x)}{h} = \dfrac{\cos(x+h) - \cos x}{h}$

$\qquad = \dfrac{\cos x \cos h - \sin x \sin h - \cos x}{h}$

$\qquad = \dfrac{\cos x \cos h - \cos x}{h} - \dfrac{\sin x \sin h}{h}$

$\qquad = \cos x \left(\dfrac{\cos h - 1}{h} \right) - \sin x \left(\dfrac{\sin h}{h} \right)$

Exer. 45–54: Typical verifications are given.

45 $(1 - \sin^2 t)(1 + \tan^2 t) = (\cos^2 t)(\sec^2 t)$

$\qquad\qquad = (\cos^2 t)(1/\cos^2 t) = 1$

47 $\dfrac{\csc^2 \theta}{1 + \tan^2 \theta} = \dfrac{\csc^2 \theta}{\sec^2 \theta} = \dfrac{1/\sin^2 \theta}{1/\cos^2 \theta} = \dfrac{\cos^2 \theta}{\sin^2 \theta}$

$\qquad\qquad = \left(\dfrac{\cos \theta}{\sin \theta} \right)^2 = \cot^2 \theta$

49 $\dfrac{1 + \csc \beta}{\sec \beta} - \cot \beta = \dfrac{1}{\sec \beta} + \dfrac{\csc \beta}{\sec \beta} - \cot \beta$

$\qquad\qquad = \cos \beta + \dfrac{\cos \beta}{\sin \beta} - \cot \beta = \cos \beta$

51 $\sin 3u = \sin(2u + u) = \sin 2u \cos u + \cos 2u \sin u$

$\qquad = (2 \sin u \cos u) \cos u + (1 - 2 \sin^2 u) \sin u$

$\qquad = 2 \sin u \cos^2 u + \sin u - 2 \sin^3 u$

$\qquad = 2 \sin u(1 - \sin^2 u) + \sin u - 2 \sin^3 u$

$\qquad = 2 \sin u - 2 \sin^3 u + \sin u - 2 \sin^3 u$

$\qquad = 3 \sin u - 4 \sin^3 u = \sin u(3 - 4 \sin^2 u)$

53 $\cos^4 \dfrac{\theta}{2} = \left(\cos^2 \dfrac{\theta}{2} \right)^2 = \left(\dfrac{1 + \cos \theta}{2} \right)^2$

$\qquad = \dfrac{1 + 2 \cos \theta + \cos^2 \theta}{4}$

$\qquad = \dfrac{1}{4} + \dfrac{1}{2} \cos \theta + \dfrac{1}{4} \left(\dfrac{1 + \cos 2\theta}{2} \right)$

$\qquad = \dfrac{1}{4} + \dfrac{1}{2} \cos \theta + \dfrac{1}{8} + \dfrac{1}{8} \cos 2\theta$

$\qquad = \dfrac{3}{8} + \dfrac{1}{2} \cos \theta + \dfrac{1}{8} \cos 2\theta$

55 $\dfrac{\pi}{12} + \pi n,\ \dfrac{11\pi}{12} + \pi n$, where n denotes any integer

57 $\dfrac{\pi}{6}, \dfrac{5\pi}{6}, \dfrac{3\pi}{2}$ **59** $\dfrac{\pi}{4}, \dfrac{5\pi}{4}$ **61** $0, \pi, \dfrac{2\pi}{3}, \dfrac{4\pi}{3}$ **63** $0, \pi$

65 3.7408, 5.6840   **67** 1.2275, 4.3691   **69** 2.6816, 3.6016

71 The graph of f appears to pass through the point $(\pi, -1)$.

73 $-0.7, 0.4$

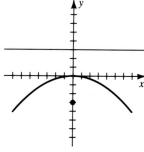

75 $h = \dfrac{10}{\cot 0.17 - \cot 1.2}$

$\qquad \approx 1.84$ km

Exercises D

7 5   **9** $-1, 3$ **11** 6   **13** $\dfrac{18}{5}$

15 (a) 900   **(b)** 590   **(c)** 349

17 (a) 1.15 mg   **(b)** 30%   **19** $3 = \log_5 125$

21 $x = \log_3(7 + t)$ **23** $t = \log_{0.7}(2/3)$ **25** $2^5 = 32$

27 $10^3 = 1000$ **29** $7^{5x+3} = m$ **31** $t = 5 \log_a(5/2)$

33 $t = \dfrac{1}{C} \log_a \dfrac{A - D}{B}$ **35** 0   **37** Not possible   **39** 8

41 4   **43** $-2, -1$ **51 (a)** 10   **(b)** 30

53 (a) 253 million; 271.36 million   **(b)** 2089

Exercises E

1 $V(0, 0); F(0, -3);$ **3** $V(0, 0); F\left(-\dfrac{3}{8}, 0\right); x = \dfrac{3}{8}$
$y = 3$

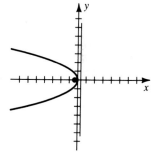

5 $V(0, 0); F\left(0, \dfrac{1}{32}\right);$ **7** $V(2, -2); F\left(2, -\dfrac{7}{4}\right)$
$y = -\dfrac{1}{32}$

9 $V(-1, 0)$; $F(2, 0)$

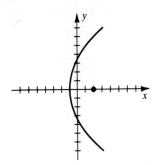

11 $V(-4, 2)$; $F\left(-\dfrac{7}{2}, 2\right)$

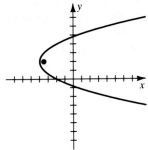

33 $V(4 \pm 3, 2)$; $F(4 \pm \sqrt{5}, 2)$

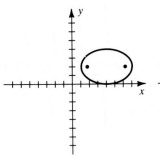

35 $V(-3 \pm 4, 1)$;
$F(-3 \pm \sqrt{7}, 1)$

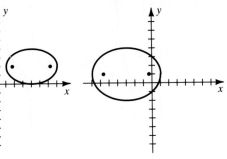

13 $V(-5, -6)$; $F\left(-5, -\dfrac{97}{16}\right)$ **15** $V\left(0, \dfrac{1}{2}\right)$; $F\left(0, -\dfrac{9}{2}\right)$

37 $V(5, 2 \pm 5)$; $F(5, 2 \pm \sqrt{21})$ **39** $\dfrac{x^2}{64} + \dfrac{y^2}{39} = 1$

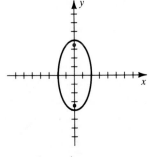

41 $\dfrac{4x^2}{9} + \dfrac{y^2}{25} = 1$

43 $\dfrac{8x^2}{81} + \dfrac{y^2}{36} = 1$

45 $\dfrac{x^2}{7} + \dfrac{y^2}{16} = 1$

47 $\dfrac{x^2}{4} + 9y^2 = 1$

17 $y^2 = 8x$ **19** $(y + 5)^2 = 4(x - 3)$
21 $y^2 = -12(x + 1)$ **23** $3x^2 = -4y$
25 $V(\pm 3, 0)$; $F(\pm\sqrt{5}, 0)$ **27** $V(0, \pm 4)$; $F(0, \pm 2\sqrt{3})$

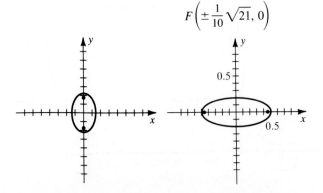

49 $V(\pm 3, 0)$; $F(\pm\sqrt{13}, 0)$ **51** $V(0, \pm 3)$; $F(0, \pm\sqrt{13})$

 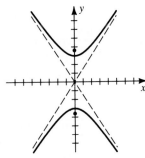

29 $V(0, \pm\sqrt{5})$; $F(0, \pm\sqrt{3})$ **31** $V\left(\pm\dfrac{1}{2}, 0\right)$;
$F\left(\pm\dfrac{1}{10}\sqrt{21}, 0\right)$

53 $V(0, \pm 4)$; $F(0, \pm 2\sqrt{5})$ **55** $V(\pm 1, 0)$; $F(\pm\sqrt{2}, 0)$

57 $V(\pm 5, 0); F(\pm\sqrt{30}, 0)$ **59** $V(0, \pm\sqrt{3}); F(0, \pm 2)$

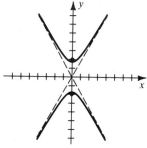

61 $V(-5 \pm 2\sqrt{5}, 1);$
$\quad F\left(-5 \pm \dfrac{1}{2}\sqrt{205}, 1\right)$

63 $V(-2, -5 \pm 3);$
$\quad F(-2, -5 \pm 3\sqrt{5})$

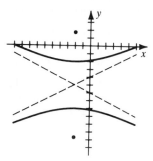

65 $V(6, 2 \pm 2);$
$\quad F(6, 2 \pm 2\sqrt{10})$

67 $y^2 - \dfrac{x^2}{15} = 1$

69 $\dfrac{x^2}{9} - \dfrac{y^2}{16} = 1$

71 $\dfrac{y^2}{21} - \dfrac{x^2}{4} = 1$

73 $\dfrac{x^2}{9} - \dfrac{y^2}{36} = 1$

75 $\dfrac{x^2}{25} - \dfrac{y^2}{100} = 1$

77 The graphs have the same asymptotes.

79 $\sqrt{84} \approx 9.165$ ft
81 $x = \sqrt{9 + 4y^2}$

CHAPTER ▪ 1

Exercises 1.1

DNE denotes Does Not Exist.

1 -7 **3** 4 **5** 7 **7** π **9** -3 **11** $\dfrac{7}{2}$

13 4 **15** $\dfrac{1}{9}$ **17** 32 **19** $2x$ **21** 12 **23** DNE

25 (a) -1 (b) 1 (c) DNE
27 (a) DNE   (b) -6 (c) DNE
29 (a) DNE   (b) DNE   (c) DNE
31 (a) 3 (b) 1 (c) DNE   (d) 2 (e) 2 (f) 2
33 (a) 1 (b) 1 (c) 1 (d) 3 (e) 3 (f) 3
35 (a) 1 (b) 0 (c) DNE   (d) 1 (e) 0 (f) DNE
37 (a) DNE   (b) DNE   (c) DNE   (d) DNE   (e) 0
   (f) DNE
39 (a) -1 (b) -1 (c) -1 (d) DNE   (e) 1
   (f) DNE
41   (a) 0 (b) 3 (c) DNE

43   (a) 2 (b) 2 (c) 2

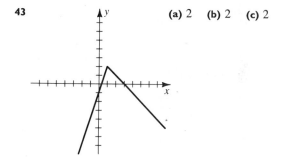

45   (a) 2 (b) 2 (c) 2

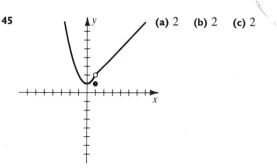

47 (a) $T(x) = \begin{cases} 0.15x & \text{if } x \le 20{,}000 \\ 0.20x - 1000 & \text{if } x > 20{,}000 \end{cases}$
   (b) \$3000; \$3000

49 (a) $S(x) = \begin{cases} 4 & \text{if } 0 < x \le 10 \\ 4 + 0.4[\![x - 9]\!] & \text{if } x > [\![x]\!] \text{ and } x > 10 \\ 4 + 0.4(x - 10) & \text{if } x = [\![x]\!] \text{ and } x > 10 \end{cases}$

(b) $0.4a$; $0.4(a + 1)$

51 (a) $2g$'s, the g-force at liftoff

(b) Left-hand limit of 8—the g-force just before the second booster is released; right-hand limit of 1—the g-force just after the second booster is released

(c) Left-hand limit of 3—the g-force just before the engines are shut down; right-hand limit of 0—the g-force just after the engines are shut down

Exer. 53–60: A calculator cannot *prove* results on limits. It can only suggest that certain limits exist.

53 (a)

x	$(1 + x)^{1/x}$
-0.02	2.745973
-0.0002	2.718554
-0.000002	2.718285
0.000002	2.718279
0.0002	2.718010
0.02	2.691588

55 (a)

x	$(3^x - 9)/(x - 2)$
1.99	9.833396
1.9999	9.886967
1.999999	9.887505
2.000001	9.887516
2.0001	9.888054
2.01	9.942023

57 (a)

x	$\left(\dfrac{4^{|x|} + 9^{|x|}}{2}\right)^{1/|x|}$
-0.1	6.049510
-0.01	6.004934
-0.001	6.000493
0.001	6.000493
0.01	6.004934
0.1	6.049510

59 (a)

x	$\dfrac{\sin x - 7x}{x \cos x}$
-0.3	-6.296140
-0.03	-6.002851
-0.003	-6.000029
0.004	-6.000051
0.04	-6.005070
0.4	-6.542948

61 (a) Approximate values: 1.0000, 1.0000, 1.0000; -1.2802, 0.6290, -0.8913

(b) The limit does not exist.

Exercises 1.2

1 (a) $\lim\limits_{t \to c} v(t) = K$ means that for every $\epsilon > 0$, there is a $\delta > 0$ such that if $0 < |t - c| < \delta$, then $|v(t) - K| < \epsilon$.

(b) $\lim\limits_{t \to c} v(t) = K$ means that for every $\epsilon > 0$, there is a $\delta > 0$ such that if t is in the open interval $(c - \delta, c + \delta)$ and $t \ne c$, then $v(t)$ is in the open interval $(K - \epsilon, K + \epsilon)$.

3 (a) $\lim\limits_{x \to p^-} g(x) = C$ means that for every $\epsilon > 0$, there is a $\delta > 0$ such that if $p - \delta < x < p$, then $|g(x) - C| < \epsilon$.

(b) $\lim\limits_{x \to p^-} g(x) = C$ means that for every $\epsilon > 0$, there is a $\delta > 0$ such that if x is in the open interval $(p - \delta, p)$, then $g(x)$ is in the open interval $(C - \epsilon, C + \epsilon)$.

5 (a) $\lim\limits_{z \to t^+} f(z) = N$ means that for every $\epsilon > 0$, there is a $\delta > 0$ such that if $t < z < t + \delta$, then $|f(z) - N| < \epsilon$.

(b) $\lim\limits_{z \to t^+} f(z) = N$ means that for every $\epsilon > 0$, there is a $\delta > 0$ such that if z is in the open interval $(t, t + \delta)$, then $f(z)$ is in the open interval $(N - \epsilon, N + \epsilon)$.

7 0.005   **9** $\sqrt{16.1} - 4$ **11** $|(3.9)^2 - 16| = 0.79$

13 Approximately 0.02396

15 Given any ϵ, choose $\delta \le \epsilon/5$.

17 Given any ϵ, choose $\delta \le \epsilon/2$.

19 Given any ϵ, choose $\delta \le \epsilon/9$.

21 Given any ϵ, let δ be any positive number.

23 Given any ϵ, let δ be any positive number.

31 Every interval $(3 - \delta, 3 + \delta)$ contains numbers for which the quotient equals 1 and other numbers for which the quotient equals -1.

33 Every interval $(-1 - \delta, -1 + \delta)$ contains numbers for which the quotient equals 3 and other numbers for which the quotient equals -3.

35 $1/x^2$ can be made as large as desired by choosing x sufficiently close to 0.

37 $1/(x + 5)$ can be made as large (positively or negatively) as desired by choosing x sufficiently close to -5.

39 *Hint:* Use Theorem (1.3).

41 There are many examples; one is $f(x) = (x^2 - 1)/(x - 1)$ if $x \ne 1$ and $f(1) = 3$.

43 Every interval $(a - \delta, a + \delta)$ contains numbers such that $f(x) = 0$ and other numbers such that $f(x) = 1$.

Exercises 1.3

1 15 **3** −2 **5** 8 **7** $\dfrac{7}{5}$ **9** 81 **11** 0

13 −13 **15** $5\sqrt{2} - 20$ **17** $\pi - 3.1416$ **19** −23

21 −7 **23** DNE **25** $-\dfrac{3}{8}$ **27** $-\dfrac{1}{4}$ **29** 2

31 $\dfrac{72}{7}$ **33** −2 **35** −2 **37** $-\dfrac{1}{8}$ **39** $\dfrac{3}{5}$

41 −810 **43** 3 **45** 1 **47** $\dfrac{1}{8}$

49 (a) 0 **(b)** DNE **(c)** DNE

51 (a) 0 **(b)** 0 **(c)** 0

53 $(-1)^{n-1}$; $(-1)^{n}$ **55** 0; 0

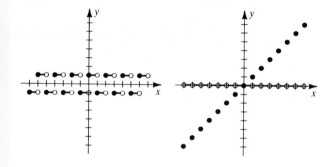

57 (a) $n - 1$ **(b)** n **59 (a)** n **(b)** $n + 1$
65 *Hint:* Let $g(x) = cx^2$.
67 Because Theorem (1.8) is applicable only when the individual limits exist, and $\lim\limits_{x \to 0} \sin \dfrac{1}{x}$ does not exist
69 (a) 0
   (b) If $T < -273\,°C$, the volume V is negative, an absurdity.
71 (a) DNE **(b)** The image is moving farther to the right.

Exercises 1.4

1 (a) $-\infty$ **(b)** ∞ **(c)** DNE
3 (a) $-\infty$ **(b)** ∞ **(c)** DNE
5 (a) $-\infty$ **(b)** $-\infty$ **(c)** $-\infty$
7 (a) ∞ **(b)** $-\infty$ **(c)** DNE
9 (a) ∞ **(b)** ∞ **(c)** ∞ **11** $\dfrac{5}{2}$ **13** $-\dfrac{7}{3}$
15 0 **17** $-\infty$ **19** ∞ **21** 1 **23** DNE
25 0.996664442, 0.999966666, 0.999999666, 0.999999996;
   the limit appears to be 1.
27 $x = -2$, $x = 2$; $y = 0$ **29** None; $y = 2$
31 $x = -3$, $x = 0$, $x = 2$; $y = 0$
33 $x = -3$, $x = 1$; $y = 1$ **35** $x = 4$; $y = 0$

Exer. 37–40: Answers are not unique.

37 **39**

41 (a) $V(t) = 50 + 5t$; $A(t) = 0.5t$ **(b)** $c(t) = t/(10t + 100)$
   (c) $c(t)$ approaches 0.1.

Exercises 1.5

1 Jump **3** Removable **5** Jump **7** Infinite **9** Removable
11 Jump **13** Removable **15** Removable **17** Removable
19 $\lim\limits_{x \to 4} f(x) = 12 + \sqrt{3} = f(4)$

21 $\lim\limits_{x \to -2} f(x) = 19 - \dfrac{1}{\sqrt{2}} = f(-2)$

23 f is not defined at -2. **25** $\lim\limits_{x \to 3} f(x) = 6 \neq 4 = f(3)$
27 $\lim\limits_{x \to 3} f(x) = 1 \neq 0 = f(3)$ **29** $\lim\limits_{x \to 0} f(x) = 1 \neq 0 = f(0)$
31 -3, 2 **33** -2, 1
35 If $4 < c < 8$, $\lim\limits_{x \to c} f(x) = \sqrt{c - 4} = f(c)$.

   $\lim\limits_{x \to 4^+} f(x) = 0 = f(4)$ and $\lim\limits_{x \to 8^-} f(x) = 2 = f(8)$

37 If $c > 0$, $\lim\limits_{x \to c} f(x) = \dfrac{1}{c^2} = f(c)$. **39** $\left\{x: x \neq -1, \dfrac{3}{2}\right\}$

41 $\left[\dfrac{3}{2}, \infty\right)$ **43** $(-\infty, -1) \cup (1, \infty)$ **45** $\{x: x \neq -9\}$
47 $\{x: x \neq 0, 1\}$ **49** $[-5, -3] \cup [3, 4) \cup (4, 5]$
51 $\left\{x: x \neq \dfrac{\pi}{4} + \dfrac{\pi}{2}n\right\}$ **53** $\{x: x \neq 2\pi n\}$

55 $\dfrac{5}{2}$ **57** $c = d = 8$ **59** $c = \sqrt[3]{w - 1}$

61 $c = \dfrac{1}{2} + \dfrac{1}{2}\sqrt{4w + 1}$

63 $f(0) = -9 < 100$ and $f(10) = 561 > 100$. Since f is
   continuous on $[0, 10]$, there is at least one number a in
   $[0, 10]$ such that $f(a) = 100$.
65 $h(3) = -12 < 0$ and $h(4) = 58 > 0$. Since h is continuous on $[3, 4]$, there is at least one number a in $[3, 4]$ such that $h(a) = 0$.
67 $g(35°) \approx 9.79745 < 9.8$ and $g(40°) \approx 9.80180 > 9.8$.
   Since g is continuous on $[35°, 40°]$, there is at least one
   latitude θ between $35°$ and $40°$ such that $g(\theta) = 9.8$.
69 -1.341 **71** -0.921, -0.154, 0.936, 1.888
73 ±12.141 **75** $x \approx 5.586$, 8.414 **77** $x \approx -1.521$

Chapter I Review Exercises

1 13   **3** $-4 - \sqrt{14}$ **5** $\dfrac{7}{8}$ **7** $\dfrac{32}{3}$ **9** ∞

11 3   **13** -1 **15** $4a^3$ **17** $\dfrac{1}{3}$ **19** $\dfrac{3}{2}$

21 0   **23** $-\infty$ **25** $-\infty$

27   **(a)** 6   **(b)** 4   **(c)** DNE

29   **(a)** $\dfrac{1}{11}$ **(b)** -1 **(c)** DNE

31   **(a)** 1   **(b)** 3   **(c)** DNE

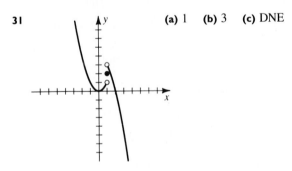

33 Given any ϵ, choose $\delta \le \epsilon/5$.   **35** ± 4 **37** 0, 2

39 $\mathbb{R}$ **41** $[-3, -2) \cup (-2, 2) \cup (2, 3]$

43 $\displaystyle\lim_{x \to 8} f(x) = 7 = f(8)$

45 (a)

x	$\dfrac{x^3 + 2x^2 - 9x - 18}{x - 3}$
2.99	29.8901
2.999	29.989001
2.9999	29.99890001
3.0001	30.00110001
3.001	30.011001
3.01	30.1101

47 (a)

x	$\dfrac{\cos(\pi x)}{x - (3/2)}$
1.45	3.1286893
1.495	3.1414635
1.4995	3.1415914
1.5005	3.1415914
1.505	3.1414635
1.55	3.1286893

49 $-0.874, 1.941$ **51** $x \approx -1.618, 0.618$

CHAPTER ▪ 2

Exercises 2.1

1 (a) $10a - 4$ **(b)** $y = 16x - 20$

3 (a) $3a^2$ **(b)** $y = 12x - 16$

5 (a) 3   **(b)** $y = 3x + 2$

7 (a) $\dfrac{1}{2\sqrt{a}}$

   (b) $y = \dfrac{1}{4}x + 1$

   (c)

9 (a) $-\dfrac{1}{a^2}$

   (b) $y = -\dfrac{1}{4}x + 1$

   (c)

11 (a)   **(b)** $(3, 9)$

13 In cm/sec: **(a)** 11.8; 11.4; 11.04   **(b)** 11

15 In ft/sec: **(a)** -32 **(b)** $-32\sqrt{10}$

17 (a) Creature at $x = 3$ **(b)** No hit

19 (a) 6.5   **(b)** 6   **21 (a)** $p_v = -\dfrac{200}{v^2}$ **(b)** -2

23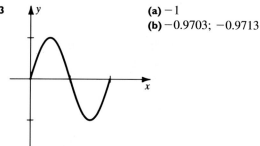

(a) -1
(b) $-0.9703;\ -0.9713$

25 In ft/sec: $-0.06864;\ -0.06426;\ -0.06382$
27 (a) $-1.851,\ -2.986,\ -0.966$
29 (a) $1.129,\ -0.253,\ -0.500$
31 (a) $0.322;\ 0.341;\ 0.360$
   (b) $-0.222;\ -0.239;\ -0.255$

Exercises 2.2

1 (a) $-10x + 8$ **(b)** $\mathbb{R}$ **(c)** $y = 18x + 7$
  (d) $\left(\dfrac{4}{5}, \dfrac{26}{5}\right)$

3 (a) $3x^2 + 1$ **(b)** $\mathbb{R}$ **(c)** $y = 4x - 2$ **(d)** None
5 (a) 9 **(b)** $\mathbb{R}$ **(c)** $y = 9x - 2$ **(d)** None
7 (a) 0 **(b)** $\mathbb{R}$ **(c)** $y = 37$ **(d)** All
9 (a) $\dfrac{-3}{x^4}$ **(b)** $(-\infty, 0) \cup (0, \infty)$ **(c)** $y = -\dfrac{3}{16}x + \dfrac{1}{2}$
  (d) None
11 (a) $\dfrac{1}{x^{3/4}}$ **(b)** $(0, \infty)$ **(c)** $y = \dfrac{1}{27}x + 9$ **(d)** None

13 $18x^5;\ 90x^4;\ 360x^3$ **15** $6x^{-1/3};\ -2x^{-4/3};\ \dfrac{8}{3}x^{-7/3}$

17 $36t^{-1/5}$ **19** 0
21 (a) No, because f is not differentiable at $x = 0$
  (b) Yes, because f' exists for every number in $[1, 3)$
23 (a) No   **(b)** Yes   **25 (a)** Yes   **(b)** No
27 (a) Yes   **(b)** Yes   **29 (a)** No   **(b)** No
31 $f'(-1) = 1, f'(1) = 0, f'(2)$ is undefined, $f'(3) = -1$
33 The right-hand and left-hand derivatives are unequal at $a = 5$.
35 The right-hand and left-hand derivatives are unequal at $a = 2$.
37 $\{x : x \neq 0\}$ **39** $\{x : x \neq -1\}$

   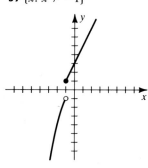

41 f is not differentiable at $\pm 1, \pm 2$.

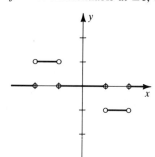

43 (a) $f'(x) = \begin{cases} 6x - 6 & \text{if } 1 \le x < a \\ -14x + 54 & \text{if } a \le x < b \\ 16x - 96 & \text{if } b \le x \le 6 \end{cases}$

$f''(x) = \begin{cases} 6 & \text{if } 1 \le x < a \\ -14 & \text{if } a \le x < b \\ 16 & \text{if } b \le x \le 6 \end{cases}$

  (b) f' exists at $x = a$ and at $x = b$.
45 $\dfrac{1}{8}$ **47** $F_C = \dfrac{9}{5}$ **49** $V_r = 4\pi r^2$
51 (a) $A_r = 2\pi r$ **(b)** 1000π ft²/ft
53 (a) The formula gives an approximation of the slope of
  the tangent line at $(a, f(a))$ by using the slope of the
  secant line through $P(a - h, f(a - h))$ and
  $Q(a + h, f(a + h))$.
  (b) Subtract and add $f(a)$ in the numerator and consider
  two limits.
  (c) $-2.0406;\ -2.0004;\ -2.0000$ **(d)** -2
55 (a) 53.2 ft/sec   **(b)** 88.3 ft/sec
57 $x \approx -0.7$

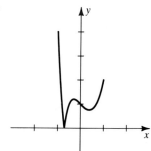

59 (b) Horizontal: $x = 0$
61 (b) Horizontal: $x = 0, x \approx \pm 2.029$
63 (b) Not differentiable: $x = 0, \pm 1, 2$;
  Horizontal: $x = 0.5, x \approx -0.618, 1.618$

Exercises 2.3

1 $10t^{2/3}$ **3** $-20s^3 + 8s - 1$ **5** $6x + \dfrac{4}{3}x^{1/3}$

7 $10x^4 + 9x^2 - 28x$ **9** $\dfrac{5}{2}x^{3/2} + \dfrac{3}{2}x^{1/2} - 2x^{-1/2}$

11 $18r^5 - 21r^2 + 4r$ **13** $416x^3 - 195x^2 + 64x - 20$

15 $\dfrac{23}{(3x+2)^2}$ **17** $\dfrac{-27z^2+12z+70}{(2-9z)^2}$ **19** $\dfrac{6v^2}{(v^3+1)^2}$

21 $-\dfrac{3t+10}{3\sqrt[3]{t}(3t-5)^2}$ **23** $-\dfrac{1+2x+3x^2}{(1+x+x^2+x^3)^2}$ **25** $\dfrac{-14x}{(x^2+5)^2}$

27 $2t-\dfrac{2}{t^3}$ **29** $-\dfrac{4}{81}s^{-5}$ **31** $10(5x-4)$ **33** $\dfrac{-10}{(5r-4)^3}$

35 (a) $3x^2-10x+8$ **37** (a) $\dfrac{24x^4+8x+3}{x^4}$

39 (a) $\dfrac{12x^2+16x-13}{(3x+2)^2}$ **41** $-2,3$ **43** $0,4$ **45** $-5,3$

47 $\dfrac{-3x+2}{x^3}$ **49** $\dfrac{4x-3}{3\sqrt[3]{x^2}}$ **51** $\dfrac{2}{(x+1)^3}$ **53** $y=\dfrac{4}{5}x+\dfrac{13}{5}$

55 (a) $-2,\dfrac{2}{3}$ (b) $-\dfrac{4}{3},0$ **57** $(1,0)$

59 $x=0$

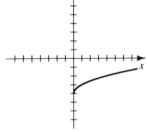

61 In ft/sec: (a) 4, 10, 18   (b) $6\sqrt{5}$

63 $-\dfrac{1}{2}$ **65** $y=2x-1,\;y=18x-81$

67 (a) 1   (b) -3 (c) -4 (d) 11   (e) $-\dfrac{1}{25}$ (f) $\dfrac{1}{9}$

69 (a) -4 (b) 1   (c) -20 (d) $-\dfrac{1}{4}$

73 $(8x-1)(x^2+4x+7)(3x^2)+$
   $(8x-1)(2x+4)(x^3-5)+8(x^2+4x+7)(x^3-5)$
75 $x(2x^3-5x-1)(12x)+x(6x^2-5)(6x^2+7)+$
   $(2x^3-5x-1)(6x^2+7)$

77 (a) $\dfrac{1}{4}$ cm/min   (b) 36π cm³/min   (c) 12π cm²/min

79 In cm²/sec: (a) 3200π (b) 6400π (c) 9600π

81 (a) $R=40$ when $x=0$ and R decreases to 6 as $x\to\infty$,
   since $dR/dx<0$ for $x>0$.
   (b) $\dfrac{dR}{dx}=\dfrac{-537.2x^3}{(1+3.95x^4)^2}$ (c) $x\approx0.624$

83 (a) 1.01   (b)

(c) 1; l_2 is nearly parallel to the tangent line, but l_1 is not.

Exercises 2.4

1 1   **3** $\dfrac{1}{8}$ **5** $\dfrac{2}{3}$ **7** 0   **9** $-\dfrac{3}{4}$ **11** 0   **13** 7

15 1   **17** 0   **19** 2   **21** 1   **23** 1   **25** -1

31 $-4\sin x$ **33** $5\csc v(1-v\cot v)$

35 $t^2\sin t-2t\cos t+1$ **37** $\dfrac{\theta\cos\theta-\sin\theta}{\theta^2}$

39 $t^2(t\cos t+3\sin t)$

41 $-2x\csc^2 x+2\cot x+x^2\sec^2 x+2x\tan x$

43 $\dfrac{2\sin z}{(1+\cos z)^2}$ **45** $-\csc x(1+2\cot^2 x)$

47 $-x\csc^2 x-\csc^3 x+\cot x-\csc x\cot^2 x$

49 $-\sin x$ **51** $\dfrac{\sec^2 x+x^2\sec^2 x-2x\tan x}{(1+x^2)^2}$ **53** $-\csc^2 v$

55 $-\cos x-\sin x$ **57** $\sin\phi+\sec\phi\tan\phi$

59 $y-\sqrt{2}=\sqrt{2}\left(x-\dfrac{\pi}{4}\right);\;y-\sqrt{2}=-\dfrac{1}{\sqrt{2}}\left(x-\dfrac{\pi}{4}\right)$

61 $\left(\dfrac{\pi}{4},\sqrt{2}\right),\left(\dfrac{5\pi}{4},-\sqrt{2}\right)$ **63** $\left(\dfrac{\pi}{4},2\sqrt{2}\right)$

65 (a) $\dfrac{\pi}{6}+2\pi n,\dfrac{5\pi}{6}+2\pi n$ (b) $y=x+2$

67 (a) $\dfrac{\pi}{4}+2\pi n,\dfrac{7\pi}{4}+2\pi n$ (b) $y-4=\sqrt{3}\left(x-\dfrac{\pi}{6}\right)$

69 $x\approx0.9,2.4,3.7$ **71** $\dfrac{\pi}{6}+2\pi n,\dfrac{5\pi}{6}+2\pi n$

73 $(16,96)$

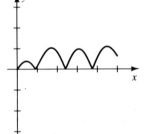

75 (a) $-\sin x;\;-\cos x;\;\sin x;\;\cos x$ (b) $\sin x$
77 $2\sec^2 x(3\tan^2 x+1)$

79 $D_x\cot x=D_x\left(\dfrac{\cos x}{\sin x}\right)=\dfrac{(\sin x)(-\sin x)-(\cos x)(\cos x)}{\sin^2 x}$
   $=\dfrac{-1(\sin^2 x+\cos^2 x)}{\sin^2 x}=-\dfrac{1}{\sin^2 x}=-\csc^2 x$

81 $D_x\sin 2x=D_x(2\sin x\cos x)$
   $=2[\sin x(-\sin x)+\cos x\cos x]$
   $=2(\cos^2 x-\sin^2 x)=2\cos 2x$

Exercises 2.5

1 $6x^2(x^3-4)$ **3** $\dfrac{-3}{2(3x-2)^{3/2}}$ **5** $6x\sec^2(3x^2)$

7 $3(x^2-3x+8)^2(2x-3)$ **9** $-40(8x-7)^{-6}$

11 $-\dfrac{7x^2+1}{(x^2-1)^5}$

13 $5(8x^3 - 2x^2 + x - 7)^4(24x^2 - 4x + 1)$

15 $17,000(17v - 5)^{999}$

17 $2(6x - 7)^2(8x^2 + 9)(168x^2 - 112x + 81)$

19 $12\left(z^2 - \dfrac{1}{z^2}\right)^5\left(z + \dfrac{1}{z^3}\right)$ **21** $8r^2(8r^3 + 27)^{-2/3}$

23 $-5v^4(v^5 - 32)^{-6/5}$ **25** $\dfrac{w^2 + 4w - 9}{2w^{5/2}}$ **27** $\dfrac{6(3 - 2x)}{(4x^2 + 9)^{3/2}}$

29 $2x \cos(x^2 + 2)$ **31** $-15 \cos^4 3\theta \sin 3\theta$

33 $4(2z + 1) \sec(2z + 1)^2 \tan(2z + 1)^2$

35 $(2 - 3s^2) \csc^2(s^3 - 2s)$

37 $-6x \sin(3x^2) - 6 \cos 3x \sin 3x$

39 $-4 \csc^2 2\phi \cot 2\phi$ **41** $2z \cot 5z - 5z^2 \csc^2 5z$

43 $2 \tan \theta \sec^5 \theta + 3 \tan^3 \theta \sec^3 \theta$

45 $25(\sin 5x - \cos 5x)^4(\cos 5x + \sin 5x)$

47 $-9 \cot^2(3w + 1) \csc^2(3w + 1)$ **49** $\dfrac{4}{1 - \sin 4w}$

51 $6 \tan 2x \sec^2 2x (\tan 2x - \sec 2x)$

53 $\dfrac{\cos\sqrt{x}}{2\sqrt{x}} + \dfrac{\cos x}{2\sqrt{\sin x}}$

55 $\dfrac{8 \cos\sqrt{3 - 8\theta} \sin\sqrt{3 - 8\theta}}{\sqrt{3 - 8\theta}}$

57 $x \sec^2\sqrt{x^2 + 1} + \dfrac{x \tan\sqrt{x^2 + 1}}{\sqrt{x^2 + 1}}$

59 $\dfrac{2 \sec\sqrt{4x + 1} \tan\sqrt{4x + 1}}{\sqrt{4x + 1}}$ **61** $-\dfrac{3 \csc^2 3x \cot 3x}{\sqrt{4 + \csc^2 3x}}$

63 (a) $y - 81 = 864(x - 2)$; $y - 81 = -\dfrac{1}{864}(x - 2)$

   (b) $\dfrac{1}{2}, 1, \dfrac{3}{2}$

65 (a) $y = 32$; $x = 1$ **(b)** ± 1

67 (a) $y = 6x$; $y = -\dfrac{1}{6}x$ **(b)** $\dfrac{\pi}{3} + \dfrac{2\pi}{3}n$

69 $\dfrac{3}{2(3z + 1)^{1/2}}$; $-\dfrac{9}{4(3z + 1)^{3/2}}$

71 $20(4r + 7)^4$; $320(4r + 7)^3$

73 $3 \sin^2 x \cos x$; $6 \sin x \cos^2 x - 3 \sin^3 x$

75 $\dfrac{dK}{dt} = mv\dfrac{dv}{dt}$ **77** -0.1819 lb/sec   **79** -4; 15

81 $-\dfrac{2}{5}$ **83** 4.91

85 (a) *Hint:* Differentiate both sides of $f(-x) = f(x)$ using the chain rule.

   (b) *Hint:* Differentiate both sides of $f(-x) = -f(x)$ using the chain rule.

87 (a) $\dfrac{dW}{dt} = (1.644 \times 10^{-4})L^{1.74}\dfrac{dL}{dt}$

   (b) 7.876 cm/month

89 $\dfrac{dd}{ds} = 60cs - 3cs^2$

91 (b) $\dfrac{dL}{d\theta} = \dfrac{b}{8}\sec^2\left(\dfrac{\theta}{4}\right)$

Exercises 2.6

1 $-\dfrac{8x}{y}$ **3** $-\dfrac{6x^2 + 2xy}{x^2 + 3y^2}$ **5** $\dfrac{10x - y}{x + 8y}$ **7** $-\sqrt{\dfrac{y}{x}}$

9 $\dfrac{-4x\sqrt{xy} - y}{x}$ **11** $\dfrac{1}{6 \sin 3y \cos 3y - 1} = \dfrac{1}{3 \sin 6y - 1}$

13 $\dfrac{-y \cot(xy) \csc(xy)}{1 + x \cot(xy) \csc(xy)}$ **15** $\dfrac{\cos y}{x \sin y + 2y}$

17 $\dfrac{4x\sqrt{\sin y}}{4y\sqrt{\sin y} - \cos y}$ **19** $-\dfrac{\sqrt{2}}{5}$ **21** -1 **23** 4

25 $-\dfrac{36}{23}$ **27** -2π **29** $-\dfrac{3}{4y^3}$ **31** $-\dfrac{2x}{y^5}$

33 $\dfrac{\sin y}{(1 + \cos y)^3}$ **35** An infinite number   **37** None

39 Let $f_c(x) = \begin{cases} \sqrt{x} & \text{if } 0 \le x \le c \\ -\sqrt{x} & \text{if } x > c \end{cases}$ for any $c > 0$

41 $\left(\pm\sqrt{3}, \dfrac{3}{2}\right)$ **43** $y - 3 = \dfrac{5}{6}(x + 2)$

45 $5x - 6y = -28$; $6x + 5y = 3$

47 $4x + 5y = -3$; $5x - 4y = -14$

49 (a) $y' = -\dfrac{x_1 b^2}{y_1 a^2}$

   (b) Horizontal: $(0, \pm b)$; vertical: $(\pm a, 0)$

Exercises 2.7

1 60 **3** $\dfrac{4}{15}$ **5** 3 **7** $-\dfrac{24}{17}$

9 $0.15\pi \approx 0.471$ cm^2/min   **11** $\dfrac{20}{9\pi} \approx 0.707$ ft/min

13 $-\dfrac{3}{8}\sqrt{336} \approx -6.9$ ft/sec   **15** $\dfrac{64}{11}$ ft/sec; $\dfrac{20}{11}$ ft/sec

17 $-7442\pi \approx -23,380$ in^3/hr   **19** $\dfrac{10}{3}$ ft/sec

21 5 in^3/min (increasing)   **23** $\dfrac{15}{32}\sqrt{3} \approx 0.81$ ft/min

25 $-\dfrac{\sqrt{2}}{5\sqrt[4]{3}} \approx -0.2149$ cm/min   **27** π m/sec

29 $\dfrac{11}{1600} = 0.006875$ ohm/sec   **31** $\dfrac{13.37}{112\pi} \approx 0.038$ ft/min

33 64 ft/sec   **35** $\dfrac{180(6 + \sqrt{2})}{\sqrt{10 + 3\sqrt{2}}} \approx 353.6$ mi/hr

37 $-\dfrac{27}{25\pi} \approx -0.3438$ in./hr   **39** $\dfrac{10,000\pi}{135} \approx 232.7$ ft/sec

41 $\dfrac{\pi}{10}\sqrt{3} \approx 0.54$ in^2/min

43 $\dfrac{2,640,000}{\sqrt{180,400}} \approx 6215.6$ ft/min ≈ 70.63 mi/hr

45 $\dfrac{1000\pi}{3}$ ft/sec ≈ 714.0 mi/hr

47 Ground speed is $\dfrac{175}{88}\dfrac{d\theta}{dt}$ mi/hr.

49 (a) $2v \dfrac{dv}{dt} = gr(1 + \sec^2 \theta)\dfrac{d\theta}{dt}$

(b) $2v \dfrac{dv}{dt} = g \tan \theta \, (1 + \sec^2 \theta)\dfrac{dr}{dt}$ **51** 19.25 mi/hr

Exercises 2.8

1 -3.94 **3** 0.92 **5** 1.80 **7** 2.12

9 (a) $(4x - 4)\Delta x + 2(\Delta x)^2$; $(4x - 4)dx$
(b) -0.72; -0.8

11 (a) $\dfrac{-(2x + \Delta x)\Delta x}{x^2(x + \Delta x)^2}$; $-\dfrac{2}{x^3}\,dx$

(b) $-\dfrac{7}{363} \approx -0.01928$; $-\dfrac{1}{45} = -0.0\overline{2}$

13 (a) $-9\,\Delta x$ **(b)** $-9\,dx$ **(c)** 0
15 (a) $(6x + 5)\,\Delta x + 3(\Delta x)^2$ **(b)** $(6x + 5)\,dx$
(c) $-3(\Delta x)^2$

17 (a) $\dfrac{-\Delta x}{x(x + \Delta x)}$ **(b)** $-\dfrac{1}{x^2}\,dx$ **(c)** $\dfrac{-(\Delta x)^2}{x^2(x + \Delta x)}$

19 (a) With $h = 0.001$, $y \approx -0.98451 - 0.27315(x - 2.5)$.
(b) -1.011825 **(c)** -1.011825
(d) They are equal because the tangent line approximation is equivalent to using (2.35).

21 (a) 4.0208 **(b)** 4.0207 **23 (a)** 3.666 **(b)** 3.659
25 (a) 0.51511 **(b)** 0.51504
27 ± 0.02; $\pm 2\%$ **29** ± 0.04; $\pm 4\%$ **31** 1.1
33 $\pm 45\%$ **35** ± 0.06
37 $\pm 1.92\pi$ in$^2 \approx \pm 6.03$ in^2; ± 0.0075; $\pm 0.75\%$
39 30 in^3; 30.301 in^3
41 3301.661 ft^2; ± 11.464 ft^2; ± 0.00347; $\pm 0.347\%$

43 $\dfrac{1}{50\pi}$ cm ≈ 0.00637 cm   **45** -1 cm   **47** 40% increase

49 $\dfrac{5\sqrt{2}\pi}{81}$ lb ≈ 0.274 lb   **51** $\pm \dfrac{\pi}{9}$ ft $\approx \pm 0.35$ ft   **53** $\pm 0.19°$

55 *Hint:* Show that $v\,dp = -\dfrac{c}{v}\,dv$.

57 dA is the shaded region.

$\Delta A - dA$

59 (a) 60π cm^2; ± 1.508 cm^2 **(b)** $\pm 0.8\%$
61 0.09 **63 (a)** 1.28 **(b)** $c \approx 1.25$

Exercises 2.9

1 (a) 3.3166 **(b)** 3.3166 **3** 1.2599 **5** 1.3315
7 -1.7321 **9** 4.6458 **11** 0.56 **13** 1.50

15 ± 3.34 **17** $-1, 1.35$ **19** $-1.88, 0.35, 1.53$
21 2.71 **23** $-1.16, 1.45$ **25** ± 2.99
27 (a) $3, 3.1425465, 3.1415927, 3.1415926, 3.1415926$
(b) They approach 2π.

29 $f'\left(\dfrac{1}{2}\right) = 0$ and hence the expression for x_2 would be undefined.

31 (a) f: $x_1 = 1.1$, $x_2 = 1.066485$, $x_3 = 1.044237$,
$x_4 = 1.029451$
g: $x_1 = 1.1$, $x_2 = 0.9983437$, $x_3 = 0.9999995$,
$x_4 = 1.000000$
(b) Because $f'(1) = 0$

33 $x_5 = 0.525$
35 (a) 1.5 **(b)** 1.34

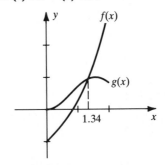

37 (a) $x_1 = 2$, $x_{12} \approx x_{13} \approx 1.93456$
(b) $x_1 = 0.5$, $x_8 \approx x_9 \approx 0.45018$

Chapter 2 Review Exercises

1 $\dfrac{-24x}{(3x^2 + 2)^2}$ **3** $6x^2 - 7$ **5** $\dfrac{3}{\sqrt{6t + 5}}$

7 $\dfrac{2(7z - 2)}{3(7z^2 - 4z + 3)^{2/3}}$ **9** $-\dfrac{144x}{(3x^2 - 1)^5}$ **11** $-\dfrac{4(r + r^{-3})}{(r^2 - r^{-2})^3}$

13 $\dfrac{12}{5(3x + 2)^{1/5}}$ **15** $\dfrac{1024s(2s^2 - 1)^3(18s^3 - 27s + 4)}{(1 - 9s^3)^5}$

17 $3(x^6 + 1)^4(3x + 2)^2(33x^6 + 20x^5 + 3)$

19 $(9s - 1)^3(108s^2 - 139s + 39)$ **21** $12x + \dfrac{5}{x^2} - \dfrac{4}{3x^{5/3}}$

23 $\dfrac{-53}{2\sqrt{(2w + 5)(7w - 9)^3}}$ **25** 0 **27** $\dfrac{2}{3}$ **29** $\dfrac{3}{5}$ **31** 2

33 $-\dfrac{\sin 2r}{\sqrt{1 + \cos 2r}}$

35 $12x^2 \sin 8x^3$ **37** $5 \sec x(\sec x + \tan x)^5$

39 $2x(\cot 2x - x \csc^2 2x)$ **41** $\dfrac{2}{1 + \cos 2\theta}$

43 $-\dfrac{(\cos \sqrt[3]{x} - \sin \sqrt[3]{x})^2(\cos \sqrt[3]{x} + \sin \sqrt[3]{x})}{\sqrt[3]{x^2}}$

45 $\dfrac{\csc u(1 - \cot u + \csc u)}{(\cot u + 1)^2}$ **47** $10 \tan 5x \sec^2 5x$

49 $\dfrac{\tan^3 (\sqrt[4]{\theta}) \sec^2 (\sqrt[4]{\theta})}{\sqrt[4]{\theta^3}}$ **51** $\dfrac{4xy^2 - 15x^2}{12y^2 - 4x^2y}$

53 $\dfrac{1}{\sqrt{x}(3\sqrt{y}+2)}$

55 $\dfrac{\cos(x+2y)-y^2}{2xy-2\cos(x+2y)}$ **57** $y=\dfrac{9}{4}x-3;\ y=-\dfrac{4}{9}x+\dfrac{70}{9}$

59 $\dfrac{7\pi}{12}+\pi n,\ \dfrac{11\pi}{12}+\pi n$

61 $15x^2+\dfrac{2}{\sqrt{x}};\ 30x-\dfrac{1}{\sqrt{x^3}};\ 30+\dfrac{3}{2\sqrt{x^5}}$

63 $\dfrac{5(y^2-4xy-x^2)}{(y-2x)^3}=-\dfrac{40}{(y-2x)^3}$

65 (a) $6x\,\Delta x+3(\Delta x)^2$ (b) $6x\,dx$ (c) $-3(\Delta x)^2$
67 $\pm 0.06\sqrt{3}\approx\pm 0.104\ \text{in}^2;\ \pm 1.5\%$ **69** -0.57

71 (a) 2 (b) -7 (c) -14 (d) 21 (e) $-\dfrac{10}{9}$

(f) $-\dfrac{19}{27}$

73 (a) Vertical tangent line at $(-1,-4)$
(b) Cusp at $(8,-1)$

75 2% **77** $\dfrac{68\pi}{5}$ ft²/ft **79** $\dfrac{5}{6}$ ft³/min **81** $\dfrac{dp}{dv}=-\dfrac{p}{v}$

83 (a) $h(t)=60-50\cos\dfrac{\pi}{15}t$ (b) 10.4 ft/sec

85 4.493

CHAPTER ▪ 3

Exercises 3.1

1 Maximum of 4 at 2; minimum of 0 at 4; local maximum at $x=2$, $6\le x\le 8$; local minimum at $x=4$, $6<x<8$, $x=10$
3 (a) Min: $f(-3)=-6$; max: none

(b) Min: none; max: $f(-1)=\dfrac{2}{3}$

(c) Min: none; max: $f(-1)=\dfrac{2}{3}$

(d) Min: $f(1)=-\dfrac{2}{3}$; max: $f(3)=6$

5
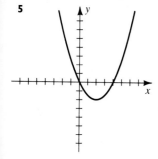
(a) Min: $f(2)=-2$; max: none
(b) Min: none; max: none
(c) Min: $f(2)=-2$; max: none
(d) Min: $f(2)=-2$; max: $f(5)=\dfrac{5}{2}$

7 Min: $f(-2)=f(1)=-3$; max: $f(-3)=f(0)=5$

9 Min: $f(8)=-3$; max: $f(0)=1$ **11** $\dfrac{3}{8}$ **13** $-2,\ \dfrac{5}{3}$

15 2 **17** ± 4 **19** $\dfrac{5+\sqrt{153}}{8},\ \pm 2$ **21** $0,\ \dfrac{15}{7},\ \dfrac{5}{2}$

23 None **25** $\pi n,\ \dfrac{2\pi}{3}+2\pi n,\ \dfrac{4\pi}{3}+2\pi n$

27 $\dfrac{\pi}{6}+2\pi n,\ \dfrac{5\pi}{6}+2\pi n,\ \dfrac{3\pi}{2}+2\pi n$ **29** $\dfrac{3\pi}{2}+2\pi n$

31 πn **33** None

35 $0,\ \pm\sqrt{k\pi-1}$ for $k=1,2,3,\ldots$

37 (a) Since $f'(x)=\dfrac{1}{3}x^{-2/3}$, $f'(0)$ does not exist. If $a\ne 0$, then $f'(a)\ne 0$. Hence, 0 is the only critical number of f. The number $f(0)=0$ is not a local extremum, since $f(x)<0$ if $x<0$ and $f(x)>0$ if $x>0$.
(b) The only critical number is 0, for the same reasons given in part (a). The number $f(0)=0$ is a local minimum, since $f(x)>0$ if $x\ne 0$.
39 (a) There is a critical number, 0, but $f(0)$ is not a local extremum, since $f(x)<f(0)$ if $x<0$ and $f(x)>f(0)$ if $x>0$.
(b)

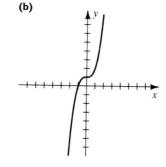

(c) The function is continuous at every number a, since $\lim_{x\to a}f(x)=f(a)$. If $0<x_1<x_2<1$, then $f(x_1)<f(x_2)$ and hence there is neither a maximum nor a minimum on $(0,1)$.
(d) This does not contradict Theorem (3.3) because the interval $(0,1)$ is open.
41 (a) If $f(x)=cx+d$ and $c\ne 0$, then $f'(x)=c\ne 0$. Hence, there are no critical numbers.
(b) On $[a,b]$, the function has absolute extrema at a and b.
43 If $x=n$ is an integer, then $f'(n)$ does not exist. Otherwise, $f'(x)=0$ for every $x\ne n$.
45 If $f(x)=ax^2+bx+c$ and $a\ne 0$, then $f'(x)=2ax+b$. Hence, $-b/(2a)$ is the only critical number of f.
47 Since $f'(x)=nx^{n-1}$, the only possible critical number is $x=0$, and $f(0)=0$. If n is even, then $f(x)>0$ if $x\ne 0$ and hence 0 is a local minimum. If n is odd, then 0 is

not an extremum, since $f(x) < 0$ if $x < 0$ and $f(x) > 0$ if $x > 0$.

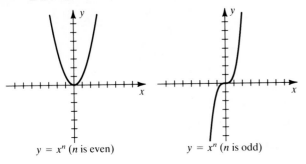

$y = x^n$ (n is even) $y = x^n$ (n is odd)

49 Min: $f(0.48) \approx 0.36$;
max: $f(-1) = f(1) = 2$

51 $-2.41, -0.92, 0.41, 1.41$

53 $-1.662, 0, 2.175$
55 $0.131, 2.535, 3, 4$
57 $-0.222, 0, 0.818, 15.404$

Exercises 3.2

1 $3, 7$ **3** 2 **5** 0 **7** $\dfrac{\pi}{4}, \dfrac{3\pi}{4}$ **9** 2

11 f is not continuous on $[0, 2]$.
13 f is not differentiable on $(-8, 8)$.   **15** 2

17 $\dfrac{1}{3}(2 - \sqrt{7}) \approx -0.22$ **19** 2 **21** 2

23 The number c such that $\cos c = 2/\pi$ ($c \approx 0.88$)
25 $-0.371, 1.307$
27 -0.5
29 4.6926
31 f is not differentiable on $(1, 4)$.
33 $f(-1) = f(1) = 1$. $f'(x) = 1$ if $x > 0$, $f'(x) = -1$ if $x < 0$, and $f'(0)$ does not exist. This does not contradict Rolle's theorem, because f is not differentiable throughout the open interval $(-1, 1)$.
35 *Hint:* Show that $c^2 = -4$.
37 *Hint:* Let $f(x) = px + q$.
39 *Hint:* If f has degree 3, then $f'(x)$ is a polynomial of degree 2.

41 Let x be any number in (a, b). Applying the mean value theorem to the interval $[a, x]$ yields
$f(x) - f(a) = f'(c)(x - a) = 0(x - a) = 0$. Thus, $f(x) = f(a)$, and hence f is a constant function.
43 *Hint:* Use the method of Example 4.
45 *Hint:* Show that $dW/dt < -44$ lb/mo.
49 *Hint:* Show that $dI/dt > 3500$ cases/mo.
51 $-1/x + C$
53 $\sin x + C$
55 (a) $f' = g' = 2 \sin x \cos x$
   (b) No, f and g differ by a constant.
57 0.64

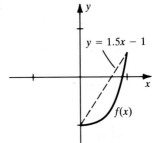

$y = 1.5x - 1$

$f(x)$

Exercises 3.3

1 Max: $f\left(-\dfrac{7}{8}\right) = \dfrac{129}{16}$;
increasing on $\left(-\infty, -\dfrac{7}{8}\right]$;
decreasing on $\left[-\dfrac{7}{8}, \infty\right)$

3 Max: $f(-2) = 29$; min: $f\left(\dfrac{5}{3}\right) = -\dfrac{548}{27}$; increasing on $(-\infty, -2]$ and $\left[\dfrac{5}{3}, \infty\right)$; decreasing on $\left[-2, \dfrac{5}{3}\right]$

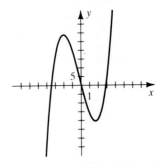

5 Max: $f(0) = 1$; min: $f(-2) = f(2) = -15$; increasing on $[-2, 0]$ and $[2, \infty)$; decreasing on $(-\infty, -2]$ and $[0, 2]$

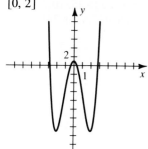

7 Max: $f\left(\dfrac{3}{5}\right) = \dfrac{216}{625} \approx 0.35$; min: $f(1) = 0$; increasing on $\left(-\infty, \dfrac{3}{5}\right]$ and $[1, \infty)$; decreasing on $\left[\dfrac{3}{5}, 1\right]$

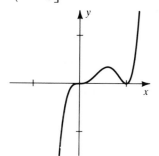

9 Min: $f(-1) = -3$; increasing on $[-1, \infty)$; decreasing on $(-\infty, -1]$

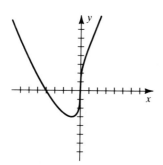

11 Max: $f\left(\dfrac{7}{4}\right) = \dfrac{441}{16}\sqrt[3]{\dfrac{49}{16}} + 2 \approx 42.03$;

min: $f(0) = f(7) = 2$;

increasing on $\left[0, \dfrac{7}{4}\right]$

and $[7, \infty)$;

decreasing on $(-\infty, 0]$

and $\left[\dfrac{7}{4}, 7\right]$

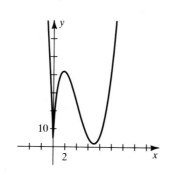

13 Max: $f(0) = 0$; min: $f(-\sqrt{3}) = f(\sqrt{3}) = -3$; increasing on $[-\sqrt{3}, 0]$ and $[\sqrt{3}, \infty)$; decreasing on $(-\infty, -\sqrt{3}]$ and $[0, \sqrt{3}]$

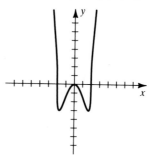

15 No extrema; increasing on $(-\infty, -3]$ and $[3, \infty)$

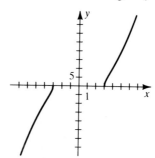

17 Max: $f\left(\dfrac{\pi}{4}\right) = \sqrt{2}$; min: $f\left(\dfrac{5\pi}{4}\right) = -\sqrt{2}$;

increasing on $\left[0, \dfrac{\pi}{4}\right]$ and $\left[\dfrac{5\pi}{4}, 2\pi\right]$;

decreasing on $\left[\dfrac{\pi}{4}, \dfrac{5\pi}{4}\right]$

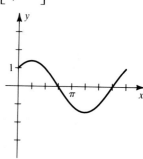

19 Max: $f\left(\dfrac{5\pi}{3}\right) = \dfrac{5\pi}{6} + \dfrac{\sqrt{3}}{2}$;

min: $f\left(\dfrac{\pi}{3}\right) = \dfrac{\pi}{6} - \dfrac{\sqrt{3}}{2}$;

increasing on $\left[\dfrac{\pi}{3}, \dfrac{5\pi}{3}\right]$;

decreasing on $\left[0, \dfrac{\pi}{3}\right]$

and $\left[\dfrac{5\pi}{3}, 2\pi\right]$

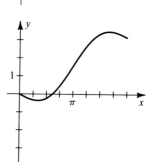

21 Max: $f\left(\dfrac{\pi}{6}\right) = \dfrac{3\sqrt{3}}{2}$; min: $f\left(\dfrac{5\pi}{6}\right) = -\dfrac{3\sqrt{3}}{2}$;

increasing on $\left[0, \dfrac{\pi}{6}\right]$

and $\left[\dfrac{5\pi}{6}, 2\pi\right]$;

decreasing on $\left[\dfrac{\pi}{6}, \dfrac{5\pi}{6}\right]$

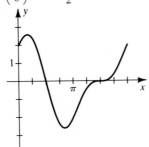

23 Max: $f(-\sqrt{3}) = \sqrt[3]{6\sqrt{3}} \approx 2.18$;

min: $f(\sqrt{3}) = -\sqrt[3]{6\sqrt{3}}$

25 Max: $f(-1) = 0$; min: $f\left(\dfrac{5}{7}\right) = -\dfrac{9^3 12^4}{7^7} \approx -18.36$

27 Max: $f(4) = \dfrac{1}{16}$ **29** Min: $f(0) = 1$

31 Max: $f\left(\dfrac{\pi}{4}\right) = 1$ **33** $-\dfrac{11\pi}{6}, -\dfrac{7\pi}{6}, \dfrac{\pi}{6}, \dfrac{5\pi}{6}$

35

37

39

41

(a) Max: $f(-1.31) \approx 10.13$
(b) increasing on $[-2, -1.31]$; decreasing on $[-1.31, 2]$

43 (a) Max: $f(2.55) \approx 105.63$; min: $f(0.78) \approx 102.89$, $f(6.35) \approx 12.69$
(b) Increasing on $[0.78, 2.55]$ and $[6.35, 10]$; decreasing on $[-4, 0.78]$ and $[2.55, 6.35]$

45 Max at $x \approx -0.51$; min at $x \approx 0.49$

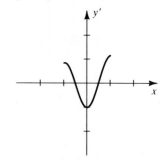

47 Max at $x \approx 0.46, 1.78, 5.97$; min at $x \approx 1.03, 5.22$

Exercises 3.4

1 Since $f''\left(\dfrac{1}{3}\right) = -2 < 0, f\left(\dfrac{1}{3}\right) = \dfrac{31}{27}$ is a maximum; since $f''(1) = 2 > 0, f(1) = 1$ is a minimum; CU on $\left(\dfrac{2}{3}, \infty\right)$; CD on $\left(-\infty; \dfrac{2}{3}\right)$; x-coordinate of PI is $\dfrac{2}{3}$.

3 Since $f''(1) = 12 > 0, f(1) = 5$ is a minimum; CU on $(-\infty, 0)$ and $\left(\dfrac{2}{3}, \infty\right)$; CD on $\left(0, \dfrac{2}{3}\right)$; x-coordinates of PI are 0 and $\dfrac{2}{3}$.

5 Since $f''(0) = 0$, use the first derivative test to show that $f(0) = 0$ is a maximum; since $f''(\pm\sqrt{2}) = 96 > 0$, $f(\pm\sqrt{2}) = -8$ are minima; CU on $\left(-\infty, -\sqrt{\dfrac{6}{5}}\right)$ and $\left(\sqrt{\dfrac{6}{5}}, \infty\right)$; CD on $\left(-\sqrt{\dfrac{6}{5}}, \sqrt{\dfrac{6}{5}}\right)$; x-coordinates of PI are $\pm\sqrt{\dfrac{6}{5}}$.

7 Since $f''(0) = -4 < 0$, $f(0) = 1$ is a maximum; since $f''(\pm 1) = 8 > 0$, $f(\pm 1) = 0$ are minima; CU on $\left(-\infty, -\sqrt{\dfrac{1}{3}}\right)$ and $\left(\sqrt{\dfrac{1}{3}}, \infty\right)$; CD on $\left(-\sqrt{\dfrac{1}{3}}, \sqrt{\dfrac{1}{3}}\right)$; x-coordinates of PI are $\pm\sqrt{\dfrac{1}{3}}$.

9 No local extrema; CU on $(-\infty, 0)$; CD on $(0, \infty)$; x-coordinate of PI is 0.

11 Since $f''\left(-\dfrac{4}{3}\right) < 0$, $f\left(-\dfrac{4}{3}\right) \approx 7.27$ is a maximum; since $f''(0)$ is undefined, use the first derivative test to show that $f(0) = 0$ is a minimum; CU on $\left(\dfrac{2}{3}, \infty\right)$; CD on $(-\infty, 0)$ and $\left(0, \dfrac{2}{3}\right)$; x-coordinate of PI is $\dfrac{2}{3}$.

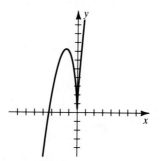

13 Since $f''(0) < 0$, $f(0) = 0$ is a maximum; since $f''\left(\dfrac{10}{7}\right) > 0$, $f\left(\dfrac{10}{7}\right) \approx -1.82$ is a minimum. Let $a = \dfrac{20 - 5\sqrt{2}}{14} \approx 0.92$ and $b = \dfrac{20 + 5\sqrt{2}}{14} \approx 1.93$. CU on $\left(a, \dfrac{5}{3}\right)$ and (b, ∞); CD on $(-\infty, a)$ and $\left(\dfrac{5}{3}, b\right)$; x-coordinates of PI are a, $\dfrac{5}{3}$, and b.

15 Since $f''(-2) > 0$, $f(-2) \approx -7.55$ is a minimum; CU on $(-\infty, 0)$ and $(4, \infty)$; CD on $(0, 4)$; x-coordinates of PI are 0 and 4.

since $f''(\pm\sqrt{6}) < 0, f(\pm\sqrt{6}) \approx 10.4$ are maxima;
since $f''(0) > 0, f(0) = 0$ is a minimum. Let
$a = -\dfrac{1}{2}\sqrt{27 - 3\sqrt{33}} \approx -1.56$ and $b = -a$. CU on
(a, b); CD on $(-3, a)$ and $(b, 3)$; x-coordinates of PI
are a and b.

19 Since $f''\left(\dfrac{\pi}{4}\right) = -\sqrt{2} < 0, f\left(\dfrac{\pi}{4}\right) = \sqrt{2}$ is a maximum;

since $f''\left(\dfrac{5\pi}{4}\right) = \sqrt{2} > 0, f\left(\dfrac{5\pi}{4}\right) = -\sqrt{2}$ is a minimum.

21 Since $f''\left(\dfrac{5\pi}{3}\right) = -\dfrac{\sqrt{3}}{2} < 0, f\left(\dfrac{5\pi}{3}\right) = \dfrac{5\pi}{6} + \dfrac{\sqrt{3}}{2}$ is a

maximum; since $f''\left(\dfrac{\pi}{3}\right) = \dfrac{\sqrt{3}}{2} > 0, f\left(\dfrac{\pi}{3}\right) = \dfrac{\pi}{6} - \dfrac{\sqrt{3}}{2}$ is
a minimum.

23 Since $f''\left(\dfrac{\pi}{6}\right) = -3\sqrt{3} < 0, f\left(\dfrac{\pi}{6}\right) = \dfrac{3\sqrt{3}}{2}$ is a maxi-

mum; since $f''\left(\dfrac{5\pi}{6}\right) = 3\sqrt{3} > 0, f\left(\dfrac{5\pi}{6}\right) = -\dfrac{3\sqrt{3}}{2}$ is a
minimum.

25 Since $f''(0) = \dfrac{1}{4} > 0, f(0) = 1$ is a minimum.

27 Since $f''\left(\dfrac{\pi}{4}\right) = -8 < 0, f\left(\dfrac{\pi}{4}\right) = 1$ is a maximum.

29 Since $f''\left(-\dfrac{11\pi}{6}\right) = f''\left(\dfrac{\pi}{6}\right) = -\dfrac{\sqrt{3}}{2} < 0,$

$f\left(-\dfrac{11\pi}{6}\right) \approx -2.01$ and $f\left(\dfrac{\pi}{6}\right) \approx 1.13$ are local maxima.

Since $f''\left(-\dfrac{7\pi}{6}\right) = f''\left(\dfrac{5\pi}{6}\right) = \dfrac{\sqrt{3}}{2} > 0,$

$f\left(-\dfrac{7\pi}{6}\right) \approx -2.70$ and $f\left(\dfrac{5\pi}{6}\right) \approx 0.44$ are local minima.

31 **33**

35 **37**

39 If $f(x) = ax^2 + bx + c$, then $f''(x) = 2a$, which does
not change sign. Thus, there is no point of inflection.
(a) CU if $a > 0$.   **(b)** CD if $a < 0$.

41
(a) CU on $(-0.48, 1)$;
   CD on $(-1, -0.48)$
(b) -0.48

43
(a) CU on $(0, 3)$
(b) No PI on $(0, 3)$

45 (a) Min: $f(2.42) \approx -0.90$
(b) PI: $(0, 17)$, $\left(\dfrac{3}{2}, \dfrac{95}{16}\right)$; CU on $[-10, 0)$ and $\left(\dfrac{3}{2}, 10\right]$;
   CD on $\left(0, \dfrac{3}{2}\right)$

47 (a) Max: $f(0.21) \approx 1, f(1.37) \approx 1, f(3.50) \approx -0.17$,
$f(5.63) \approx 1$;
min: $f(0.73) \approx -1, f(2.32) \approx -1, f(4.68) \approx -1$
(b) PI: $(0.45, 0.05), (1.01, -0.08), (1.73, 0.16)$,
$(2.75, -0.67), (4.25, -0.67), (5.27, 0.16)$,
$(5.99, -0.08)$;
CU on $(0.45, 1.01), (1.73, 2.75), (4.25, 5.27)$,
$(5.99, 6]$;
CD on $[0, 0.45), (1.01, 1.73), (2.75, 4.25)$,
$(5.27, 5.99)$
49 (a) Max: $f(-1.48) \approx 15.48, f(0.67) \approx 1.32$;
min: $f(0) = 0, f(3.21) \approx -93.73$
(b) PI: $(-0.96, 9.26), (0.35, 0.67), (2.41, -57.38)$;
CU on $(-0.96, 0.35)$ and $(2.41, 6]$;
CD on $[-4, -0.96)$ and $(0.35, 2.41)$

Exercises 3.5

1 No extrema

3 Max: $f(5 + 2\sqrt{6}) \approx 1.05$;
min: $f(5 - 2\sqrt{6}) \approx 5.95$

5 Max: $f(12 - 2\sqrt{30}) \approx 0.25$;
min: $f(12 + 2\sqrt{30}) \approx 2.93$

7 Min: $f(0) = 0$

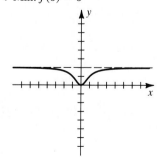

9 Min: $f(4) = 4$

11 No extrema

13 No extrema

15 Max: $f(-2) = -4$;
min: $f(0) = 0$

17 Max: $f(-3 + \sqrt{5}) \approx 1.53$; min: $f(-3 - \sqrt{5}) \approx 10.47$

19 Max: $f(0.36) \approx 3.63, f(2.54) \approx 1.42$;
min: $f(1.46) \approx -1.42, f(3.64) \approx -3.63$
21 Max: $f(\pm 0.44) \approx -4.49$;
min: $f(0) = -4.8, f(\pm 1.25) \approx -11.37$
23 Max: $f(3.86) \approx 17.14$; min: 0 for $x \geq 6$ or $x < 1$
25 Max: $f(-0.10) \approx -65.10, f(6.77) \approx 96.58$;
min: $f(-2.37) \approx 0, f(2.89) \approx 0, f(9.49) \approx 0$
27 Max: $f(8) = \dfrac{3}{32}$;
PI: $\left(16, \dfrac{1}{12}\right)$

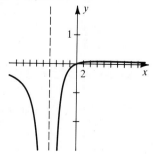

29 Max: $f(1) = \frac{3}{2}$;

min: $f(-1) = -\frac{3}{2}$;

PI: $\left(\pm \sqrt{3}, \pm \frac{3}{4} \sqrt{3} \right)$,

$(0, 0)$

31 No extrema;
PI: $(\pm 3, 6)$

45

47 (b)

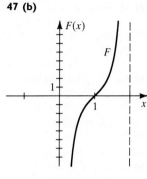

(a) CU on $(-0.43, 2)$;
CD on $(-2, -0.43)$

(b) -0.43

49

33

$(-2, 7)$

35

$\left(1, -\frac{1}{2}\right)$

53 (b) $m + 3n - 2$ (before simplification)

Exercises 3.6

1 225   **3** 71   **5** 800

7 Side of base = 2 ft; height = 1 ft

9 Radius of base = height = $\dfrac{1}{\sqrt[3]{\pi}}$

11 $x = 166\frac{2}{3}$ ft; $y = 125$ ft

13 Approximately 2:23:05 P.M.   **15** $5\sqrt{5} \approx 11.18$ ft

17 Length = $2\sqrt[3]{300} \approx 13.38$ ft;

width = $\dfrac{3}{2}\sqrt[3]{300} \approx 10.04$ ft;

height = $\sqrt[3]{300} \approx 6.69$ ft

21 55   **23** Radius = $\dfrac{1}{2}\sqrt[3]{15}$; length of cylinder = $2\sqrt[3]{15}$

25 Length of base = $\sqrt{2}\,a$; height = $\dfrac{1}{2}\sqrt{2}\,a$ **27** $\dfrac{32}{81}\pi a^3$

29 $(1, 2)$ **31** Width = $\dfrac{2}{\sqrt{3}}a$; depth = $\dfrac{2\sqrt{2}}{\sqrt{3}}a$ **33** 500

35 (a) Use $\dfrac{36\sqrt{3}}{2 + \sqrt{3}} \approx 16.71$ cm for the rectangle.

(b) Use all the wire for the rectangle.

37 Width = $\dfrac{12}{6 - \sqrt{3}} \approx 2.81$ ft;

height = $\dfrac{18 - 6\sqrt{3}}{6 - \sqrt{3}} \approx 1.78$ ft

37

39

41

43

41 37   **43** 18 in., 18 in., 36 in.

45 $\dfrac{4}{1 + \sqrt[4]{\dfrac{1}{2}}} \approx 2.17$ mi from A

49 (c) $4\sqrt{30} \approx 21.9$ mi/hr

51 60°   **53** $2\pi\left(1 - \dfrac{1}{3}\sqrt{6}\right)$ radians $\approx 66.06°$

55 $\tan\theta = \dfrac{\sqrt{2}}{2}$; $\theta \approx 35.3°$

59 $\tan\theta = \sqrt[3]{\dfrac{4}{3}}$; $\theta \approx 47.74°$; $L = \dfrac{4}{\sin\theta} + \dfrac{3}{\cos\theta} \approx 9.87$ ft

61 (b) $\cos\theta = \dfrac{2}{3}$; $\theta \approx 48.2°$

Exercises 3.7

1 $v(t) = 6(t-2)$; $a(t) = 6$; left in $[0, 2)$; right in $(2, 5]$

$t = 5$
$t = 2$
$t = 0$
-10 0 10 l

3 $v(t) = 3(t^2 - 3)$; $a(t) = 6t$; right in $[-3, -\sqrt{3})$; left in $(-\sqrt{3}, \sqrt{3})$; right in $(\sqrt{3}, 3]$

$t = \sqrt{3}$
$t = 3$
$t = -\sqrt{3}$
$t = 0$
$t = -3$
-8 -4 0 4 8 l

5 $v(t) = -6(t-1)(t-4)$; $a(t) = -6(2t - 5)$; left in $[0, 1)$; right in $(1, 4)$; left in $(4, 5]$

$t = 5$
$t = 2.5$ $t = 4$
$t = 1$
$t = 0$
-20 -10 0 10 l

7 $v(t) = 4t(2t^2 - 3)$; $a(t) = 12(2t^2 - 1)$; left in $\left[-2, -\sqrt{\dfrac{3}{2}}\right)$; right in $\left(-\sqrt{\dfrac{3}{2}}, 0\right)$; left in $\left(0, \sqrt{\dfrac{3}{2}}\right)$; right in $\left(\sqrt{\dfrac{3}{2}}, 2\right]$

$t = 2$
$t = \sqrt{3/2}$
$t = 0$
$t = -\sqrt{3/2}$ $t = -2$
-8 -4 0 4 8 l

9 (a) 30 ft/sec   **(b)** 2.8 sec

11 (a) $v(t) = 16(9 - 2t)$; $a(t) = -32$ **(b)** 324 ft
(c) 9 sec

13 $5; 8; \dfrac{1}{8}$ **15** $6; 3; \dfrac{1}{3}$ **17** $79{,}200\pi$; $3600\sqrt{2}\pi$

19 (a) $y = 4.5 \sin\left[\dfrac{\pi}{6}(t - 10)\right] + 7.5$

$= 4.5 \sin\left(\dfrac{\pi}{6}t - \dfrac{5\pi}{3}\right) + 7.5$

(b) 1.178 ft/hr

21 (a) In in./sec: $0, -\pi, 0, \pi, 0$
(b) $(n, n + 1)$, where n is an odd positive integer

27

t	0	1	2	3	4	5
s	0	4.21	1.82	0.14	-0.45	-0.37
v	10	-1.51	-2.29	-1.07	-0.18	0.25
a	0	-5.40	1.11	1.12	0.66	0.20

29 (a) 1600 ft   **(b)** No, speed is about 218 mi/hr on impact.   **(c)** 1089 ft

31 (a) 3 sec   **(b)** 40 ft/sec
33 For $a > 0$ and measured in ft/sec², $v_0 = v_1 + \dfrac{15}{22}\sqrt{2as}$.

Exercises 3.8

1 (a) 806
(b) $c(x) = \dfrac{800}{x} + 0.04 + 0.0002x$;
$C'(x) = 0.04 + 0.0004x$; $c(100) = 8.06$;
$C'(100) = 0.08$

3 (a) 11,250
(b) $c(x) = \dfrac{250}{x} + 100 + 0.001x^2$;
$C'(x) = 100 + 0.003x^2$; $c(100) = 112.50$;
$C'(100) = 130$

5 $C'(5) = \$46$; $C(6) - C(5) \approx \$46.67$
7 (a) -0.1 **(b)** $50x - 0.1x^2$ **(c)** $48x - 0.1x^2 - 10$
(d) $48 - 0.2x$ **(e)** 5750   **(f)** 2
9 (a) $1800x - 2x^2$ **(b)** $1799x - 2.01x^2 - 1000$
(c) 100   **(d)** \$158,800

11 (a) 3990 mills   **(b)** $15,420.10

13 The stable point occurs at $\left(\dfrac{m}{n}, \dfrac{a}{b}\right)$.

Chapter 3 Review Exercises

1 Max: $f(3) = 1$; min: $f(6) = -8$ **3** $-2, -1, \dfrac{1}{3}$

5 Max: $f(2) = 28$; min: $f\left(-\dfrac{1}{2}\right) = -\dfrac{13}{4}$; increasing on

$\left[-\dfrac{1}{2}, 2\right]$; decreasing on $\left(-\infty, -\dfrac{1}{2}\right]$ and $[2, \infty)$

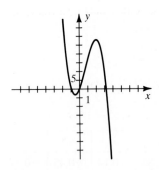

7 Max: $f(1) = 3$;
increasing on $(-\infty, 1]$;
decreasing on $[1, \infty)$

9 Since $f''(0) = 0$ and $f''(2)$ is undefined, use the first derivative test to show that there are no extrema; CU on $(-\infty, 0)$ and $(2, \infty)$; CD on $(0, 2)$; x-coordinates of PI are 0 and 2.

11 Since $f''(0) = -2 < 0$, $f(0) = 1$ is a maximum;
CU on $\left(-\infty, -\dfrac{1}{3}\sqrt{3}\right)$ and $\left(\dfrac{1}{3}\sqrt{3}, \infty\right)$;
CD on $\left(-\dfrac{1}{3}\sqrt{3}, \dfrac{1}{3}\sqrt{3}\right)$;
x-coordinates of
PI are $\pm\dfrac{1}{3}\sqrt{3}$.

13 Max: $f\left(\dfrac{\pi}{2}\right) = 3$ and $f\left(\dfrac{3\pi}{2}\right) = -1$;
min: $f\left(\dfrac{7\pi}{6}\right) = f\left(\dfrac{11\pi}{6}\right) = -\dfrac{3}{2}$

15

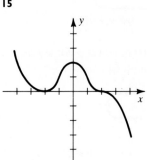

17 Max: $f(0) = 0$

19 No extrema

$y = x - 1$

21 Max: $f(3 + \sqrt{7}) \approx 0.08$;
min: $f(3 - \sqrt{7}) \approx 0.37$

23 $\dfrac{\sqrt{61}-1}{3}$ **25** 125 yd by 250 yd   **27** $\dfrac{\pi}{2}$

29 Radius of semicircle is $\dfrac{1}{8\pi}$ mi, length of rectangle is $\dfrac{1}{8}$ mi.

31 (a) Use all the wire for the circle.

(b) Use length $\dfrac{5\pi}{4+\pi} \approx 2.2$ ft for the circle and the

remainder for the square.

33 $v(t) = \dfrac{3(1-t^2)}{(t^2+1)^2}$; $a(t) = \dfrac{6t(t^2-3)}{(t^2+1)^3}$; left in $[-2,-1)$;

right in $(-1,1)$; left in $(1,2]$

35 $C'(100) = 116$; $C(101) - C(100) = 116.11$

37 (a) $18x$ **(b)** $-0.02x^2 + 12x - 500$ **(c)** 300

(d) $1300

39 98 ft/sec^2 **41** 2.27   **43** ± 0.79

45 Min: $f(1.5345) \approx -10.2624$; PI: none

47 Max: $f(0.3666) \approx 0.3240$; min: $f(0.4780) \approx 0$,
$f(0.2527) \approx 0$; PI: $(0.4780, 0)$ and $(0.2527, 0)$

49 Max: $f(1.0810) \approx 2.2948$; min: $f(0.5643) \approx 2.1902$;
PI: $(-0.8281, 5.5559)$ and $(0.8281, 2.2434)$

CHAPTER ▪ 4

Exercises 4.1

1 $2x^2 + 3x + C$ **3** $3t^3 - 2t^2 + 3t + C$

5 $-\dfrac{1}{2z^2} + \dfrac{3}{z} + C$ **7** $2u^{3/2} + 2u^{1/2} + C$

9 $\dfrac{8}{9}v^{9/4} + \dfrac{24}{5}v^{5/4} - v^{-3} + C$ **11** $3x^3 - 3x^2 + x + C$

13 $\dfrac{2}{3}x^3 + \dfrac{3}{2}x^2 + C$ **15** $\dfrac{24}{5}x^{5/3} - \dfrac{15}{2}x^{2/3} + C$

17 $\dfrac{1}{3}x^3 + \dfrac{1}{2}x^2 + x + C$ **19** $-t^{-1} - 2t^{-3} - \dfrac{9}{5}t^{-5} + C$

21 $\dfrac{3}{4}\sin u + C$ **23** $-7\cos x + C$

25 $\dfrac{2}{3}t^{3/2} + \sin t + C$ **27** $\tan t + C$ **29** $-\cot v + C$

31 $\sec w + C$ **33** $-\csc z + C$ **35** $\sqrt{x^2+4} + C$

37 $\sin \sqrt[3]{x} + C$ **39** $x^3\sqrt{x-4}$

43 $a^2 x + C$ **45** $\dfrac{1}{2}at^2 + bt + C$ **47** $(a+b)u + C$

49 $f(x) = 4x^3 - 3x^2 + x + 3$ **51** $y = \dfrac{8}{3}x^{3/2} - \dfrac{1}{3}$

53 $f(x) = \dfrac{2}{3}x^3 - \dfrac{1}{2}x^2 - 8x + \dfrac{65}{6}$

55 $y = -3\sin x + 4\cos x + 5x + 3$ **57** $t^2 - t^3 - 5t + 4$

59 (a) $s(t) = -16t^2 + 1600t$ **(b)** $s(50) = 40,000$ ft

61 (a) $s(t) = -16t^2 - 16t + 96$ **(b)** $t = 2$ sec

(c) -80 ft/sec

63 Solve the differential equation $s''(t) = -g$ for $s(t)$.

65 10 ft/sec^2 **67** 19.62

69 $C(x) = 20x - 0.0075x^2 + 5.0075$; $C(50) \approx 986.26$

71 $10x^4 + 4x^3 + 27x^2 - 10x + 4$;

$\dfrac{1}{3}x^6 + \dfrac{1}{5}x^5 + \dfrac{9}{4}x^4 - \dfrac{5}{3}x^3 + 2x^2 + 10x + C$

73 $e^{3x}[(3x^2 + 2x)\cos(4x) - 4x^2 \sin(4x)]$;

$\dfrac{1}{15,625}e^{3x}[(1875x^2 + 350x - 234)\cos(4x)$

$\qquad + 4(625x^2 - 300x + 22)\sin(4x)] + C$

75 $\dfrac{-(4t^3 - 27t^2 - 30t + 31)}{(2t^3 + 3t^2 - 5t - 6)^2}$;

$-\ln(2t-3)^{1/5} - \ln(t+2) + \dfrac{6}{5}\ln(t+1) + C$

77 (b) Each pair of functions differs only by a constant.

Exercises 4.2

1 $\dfrac{1}{44}(2x^2 + 3)^{11} + C$ **3** $\dfrac{1}{12}(3x^3 + 7)^{4/3} + C$

5 $\dfrac{1}{2}(1 + \sqrt{x})^4 + C$ **7** $\dfrac{2}{3}\sin\sqrt{x^3} + C$

9 $\dfrac{2}{9}(3x - 2)^{3/2} + C$ **11** $\dfrac{3}{32}(8t + 5)^{4/3} + C$

13 $\dfrac{1}{15}(3z + 1)^5 + C$ **15** $\dfrac{2}{9}(v^3 - 1)^{3/2} + C$

17 $-\dfrac{3}{8}(1 - 2x^2)^{2/3} + C$ **19** $\dfrac{1}{5}s^5 + \dfrac{2}{3}s^3 + s + C$

21 $\dfrac{2}{5}(\sqrt{x} + 3)^5 + C$ **23** $-\dfrac{1}{4(t^2 - 4t + 3)^2} + C$

25 $-\dfrac{3}{4}\cos 4x + C$ **27** $\dfrac{1}{4}\sin(4x - 3) + C$

29 $-\dfrac{1}{2}\cos(v^2) + C$ **31** $\dfrac{1}{4}(\sin 3x)^{4/3} + C$

33 $x - \dfrac{1}{2}\cos 2x + C$ **35** $-\cos x - \cos^2 x - \dfrac{1}{3}\cos^3 x + C$

37 $\dfrac{1}{3\cos^3 x} + C$ **39** $\dfrac{1}{1 - \sin t} + C$ **41** $\dfrac{1}{3}\tan(3x - 4) + C$

43 $\dfrac{1}{6}\sec^2 3x + C$ **45** $-\dfrac{1}{5}\cot 5x + C$

47 $-\dfrac{1}{2}\csc(x^2) + C$ **49** $f(x) = \dfrac{1}{4}(3x + 2)^{4/3} + 5$

51 $f(x) = 3\sin x - 4\cos 2x + x + 2$

53 (a) $\dfrac{1}{3}(x + 4)^3 + C_1$

(b) $\dfrac{1}{3}x^3 + 4x^2 + 16x + C_2$; $C_2 = C_1 + \dfrac{64}{3}$

55 (a) $\dfrac{2}{3}(\sqrt{x} + 3)^3 + C_1$

(b) $\dfrac{2}{3}x^{3/2} + 6x + 18x^{1/2} + C_2$; $C_2 = C_1 + 18$

59 474,592 ft^3 **61 (a)** $\dfrac{dV}{dt} = 0.6 \sin\left(\dfrac{2\pi}{5}t\right)$ **(b)** $\dfrac{3}{\pi} \approx 0.95$ L

63 *Hint:* (i) Let $u = \sin x$.   (ii) Let $u = \cos x$.
(iii) Use the double angle formula for the sine. The three answers differ by constants.

Exercises 4.3

1 34   **3** 40   **5** 10   **7** 500

9 $\dfrac{1}{3}n(n^2 + 6n + 20)$ **11** $\dfrac{1}{12}n(3n^3 + 14n^2 + 9n + 46)$

Exer. 13–18: Answers are not unique.

13 $\displaystyle\sum_{k=1}^{5}(4k - 3)$ **15** $\displaystyle\sum_{k=1}^{4}\dfrac{k}{3k - 1}$ **17** $1 + \displaystyle\sum_{k=1}^{n}(-1)^k\dfrac{x^{2k}}{2k}$

19 111,142.3744   **21** 7.4855

23 0.9441   **25** 21,781,332

27 (a) 10   **(b)** 14

29 (a) $\dfrac{35}{4}$ **(b)** $\dfrac{51}{4}$ **31 (a)** 1.04   **(b)** 1.19

Exer. 33–38: Answers for (a) and (b) are the same.

33 28   **35** 18   **37** 6   **39 (a)** 20   **(b)** $\dfrac{1}{4}(b^4 - a^4)$

Exercises 4.4

1 (a) 1.1, 1.5, 1.1, 0.4, 0.9   **(b)** 1.5
3 (a) 0.3, 1.7, 1.4, 0.5, 0.1   **(b)** 1.7
5 (a) 30   **(b)** 42   **(c)** 36
7 (a) 15.127   **(b)** 15.283   **(c)** 15.3975
9 (a) 141   **(b)** 551   **(c)** 307
11 (a) 292.5   **(b)** 348.5   **(c)** 319.75
13 (a) 0.2668   **(b)** 0.2962   **(c)** 0.2813

15 $\displaystyle\int_{-1}^{2}(3x^2 - 2x + 5)\,dx$ **17** $\displaystyle\int_{0}^{4} 2\pi x(1 + x^3)\,dx$

19 $-\dfrac{14}{3}$ **21** $\dfrac{14}{3}$ **23** $-\dfrac{14}{3}$

25 $\displaystyle\int_{0}^{4}\left(-\dfrac{5}{4}x + 5\right)dx$ **27** $\displaystyle\int_{-1}^{5}\sqrt{9 - (x - 2)^2}\,dx$

29 36   **31** 25   **33** 2.5   **35** $\dfrac{9\pi}{4}$ **37** $12 + 2\pi$

Exercises 4.5

1 30   **3** -12 **5** 2   **7** 78   **9** $-\dfrac{291}{2}$

11 Use Corollary (4.27).   **13** Use Theorem (4.26).

15 Use Theorem (4.26).   **17** $\displaystyle\int_{-3}^{1} f(x)\,dx$

19 $\displaystyle\int_{e}^{d} f(x)\,dx$ **21** $\displaystyle\int_{h}^{c+h} f(x)\,dx$ **23 (a)** $\sqrt{3}$ **(b)** 9

25 (a) -1 **(b)** 2   **27 (a)** 3   **(b)** 6

29 (a) $\sqrt[3]{\dfrac{15}{4}}$ **(b)** 14

31 1.426   **33** Use (4.22) and (4.23)(i).

Exercises 4.6

1 -18 **3** $\dfrac{265}{2}$ **5** 5   **7** $\dfrac{31}{32}$ **9** $\dfrac{20}{3}$ **11** $\dfrac{352}{5}$

13 $\dfrac{13}{3}$ **15** $-\dfrac{7}{2}$ **17** 0   **19** $\dfrac{10}{3}$ **21** $\dfrac{53}{2}$ **23** $\dfrac{14}{3}$

25 0   **27** $\dfrac{1}{3}$ **29** $\dfrac{5}{36}$ **31** $\dfrac{3}{2}(\sqrt{3} - 1) \approx 1.10$

33 $1 - \sqrt{2} \approx -0.41$ **35** 0

37 No, $\sec^2 x$ is not continuous on $[0, \pi]$.

39 Yes, since $\displaystyle\int_{-1}^{0} f(x)\,dx + \int_{0}^{1} f(x)\,dx = \int_{-1}^{1} f(x)\,dx$.

41 (a) $\sqrt{3}$ **(b)** $\dfrac{1}{2}$ **43 (a)** $\dfrac{544}{225}$ **(b)** $\dfrac{38}{15}$

45 0   **47** $\dfrac{1}{x + 1}$ **51 (a)** $\dfrac{6}{7}cd^{1/6}$

55 *Hint:* Use Part I of the fundamental theorem of calculus (4.30) and the chain rule.

57 $\dfrac{4x^7}{\sqrt{x^{12} + 2}}$ **59** $3x^2(x^9 + 1)^{10} - 3(27x^3 + 1)^{10}$

Exercises 4.7

1 $L_6 = 10.95$; $R_6 = 11.95$; $M_3 = 11.1$; $T_6 = 11.45$; $S_3 = 11\frac{1}{3}$
3 $L_8 = 12.33375$; $R_8 = 13.60875$; $M_4 = 12.6975$; $T_8 = 12.97125$; $S_4 = 12.88$
5 (a) $L_8 = 1.1501$; $R_8 = 1.2597$ **(b)** 1.2049
7 (a) $L_3 = 0.84$; $L_6 = 0.9$; $L_{12} = 0.93$
  (b) 0.96; $E_3 = 0.12$; $E_6 = 0.06$; $E_{12} = 0.03$
  (c) The error is reduced by $\frac{1}{2}$ when n doubles.
9 (a) $M_2 = 144$; $M_4 = 153$; $M_8 = 155.25$
  (b) 156; $E_2 = 12$; $E_4 = 3$; $E_8 = 0.75$
  (c) The error is reduced by $\frac{1}{4}$ when n doubles.
11 (a) $T_2 = 180$; $T_4 = 162$; $T_8 = 157.5$
  (b) 156; $E_2 = -24$; $E_4 = -6$; $E_8 = -1.5$
  (c) The error is reduced by $\frac{1}{4}$ when n doubles.
13 (a) $S_2 = S_4 = S_8 = 156$
  (b) 156; $E_2 = E_4 = E_8 = 0$
  (c) Simpson's rule is exact for all n.
15 (a) $T_5 \approx 6.249806$; $T_{10} \approx 6.234926$; $T_{20} \approx 6.231201$; $T_{40} \approx 6.230270$
  (b) At least two decimal places
17 (a) $S_2 \approx 2.3987529621$; $S_6 \approx S_{18} \approx S_{54} \approx 2.4039394306$
  (b) At least ten decimal places
19 (a) 0.26   **(b)** 4.2×10^{-5}
21 (a) 0.125   **(b)** 6.5×10^{-4}
23 (a) 3,386,880   **(b)** 642   **(c)** 10

25 (a) 25   **(b)** 3   **(c)** 1   **29 (a)** 127.5   **(b)** 131.7
31 0.174 m/sec   **33** 0.28   **35** 1.48

Chapter 4 Review Exercises

1 $-\dfrac{8}{x} + \dfrac{2}{x^2} - \dfrac{5}{3x^3} + C$ **3** $100x + C$ **5** $\dfrac{1}{16}(2x+1)^8 + C$

7 $-\dfrac{1}{16}(1-2x^2)^4 + C$ **9** $-\dfrac{2}{1+\sqrt{x}} + C$

11 $3x - x^2 - \dfrac{5}{4}x^4 + C$ **13** $\dfrac{1}{6}(4x^2 + 2x - 7)^3 + C$

15 $-\dfrac{1}{x^2} - x^3 + C$ **17** $\dfrac{3}{5}$ **19** $\dfrac{1}{6}$ **21** $\sqrt{8} - \sqrt{3} \approx 1.10$

23 $\dfrac{52}{9}$ **25** $-\dfrac{37}{6}$ **27** $8\sqrt{3} + 16 \approx 29.86$

29 $\dfrac{1}{5}\cos(3-5x) + C$ **31** $\dfrac{1}{15}\sin^5 3x + C$

33 $-\dfrac{1}{6\sin^2 3x} + C$ **35** $\dfrac{2}{15}(16\sqrt{2} - 3\sqrt{3}) \approx 2.32$ **37** $\dfrac{1}{6}$

39 $\sqrt[5]{x^4 + 2x^2 + 1} + C$ **41** 0   **43** $y = x^3 - 2x^2 + x + 2$

45 $\dfrac{135}{4}$ **47** Use Corollary (4.27).   **49** $\displaystyle\int_a^e f(x)\,dx$

51 (a) $-16t^2 - 30t + 900$ **(b)** -190 ft/sec

   (c) $\dfrac{15}{16}(-1 + \sqrt{65}) \approx 6.6$ sec

53 $\displaystyle\int_{-2}^{3} \sqrt{1 + 3x^2}\,dx$ **55** $M_5 \approx 0.824279$; $M_{10} \approx 0.8092539$

57 $S_4 \approx 11.105304$; $S_8 \approx 11.105302$ **59** $81.625\,°F$

CHAPTER ▪ 5

Exercises 5.1

Exer. 1–4: Answers are not unique.

1 $\displaystyle\int_{-2}^{2} [(x^2 + 1) - (x - 2)]\,dx$

3 $\displaystyle\int_{-2}^{1} [(-3y^2 + 4) - y^3]\,dy$

5 $\displaystyle\int_0^4 (4x - x^2)\,dx = \dfrac{32}{3}$ **7** $2\displaystyle\int_0^2 [5 - (x^2 + 1)]\,dx = \dfrac{32}{3}$

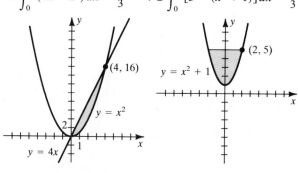

9 $\displaystyle\int_1^2 \left[\dfrac{1}{x^2} - (-x^2)\right] dx = \dfrac{17}{6}$

$y = \dfrac{1}{x^2}$

$y = -x^2$

11 $\displaystyle\int_{-1}^{2} [(4 + y) - (-y^2)]\,dy = \dfrac{33}{2}$

$x = -y^2$

$y = x - 4$

13 $2\displaystyle\int_0^{\sqrt{3}} [(2 - y^2) - (y^2 - 4)]\,dy = 8\sqrt{3}$

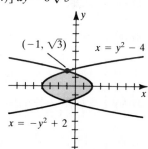

$(-1, \sqrt{3})$ $x = y^2 - 4$

$x = -y^2 + 2$

15 $2\displaystyle\int_0^2 [(4y - y^3) - 0]\,dy = 8$

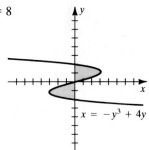

$x = -y^3 + 4y$

17 $2\displaystyle\int_0^1 [0 - (x^3 - x)]\,dx = \dfrac{1}{2}$

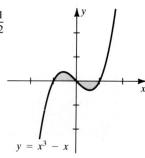

$y = x^3 - x$

19 $\displaystyle\int_{-3}^{0} [(y^3 + 2y^2 - 3y) - 0]\, dy +$

$\displaystyle\int_{0}^{1} [0 - (y^3 + 2y^2 - 3y)]\, dy = \frac{71}{6}$

$x = y^3 + 2y^2 - 3y$

21 $\displaystyle\int_{0}^{2} x\sqrt{4 - x^2}\, dx = \frac{16}{3}$

$y = x\sqrt{4 - x^2}$

23 $3 + \dfrac{3}{2}\sqrt{3} \approx 5.74$

25 (a) $\displaystyle\int_{0}^{1}(3x - x)\, dx + \int_{1}^{2}[(4 - x) - x]\, dx$

(b) $\displaystyle\int_{0}^{2}\left(y - \frac{1}{3}y\right) dy + \int_{2}^{3}\left[(4 - y) - \frac{1}{3}y\right] dy$

27 (a) $\displaystyle\int_{1}^{4}[\sqrt{x} - (-x)]\, dx$

(b) $\displaystyle\int_{-4}^{-1}[4 - (-y)]\, dy + \int_{-1}^{1}(4 - 1)\, dy + \int_{1}^{2}(4 - y^2)\, dy$

29 (a) $\displaystyle\int_{-6}^{-1}[(x + 3) - (-\sqrt{3 - x})]\, dx + 2\int_{-1}^{3}\sqrt{3 - x}\, dx$

(b) $\displaystyle\int_{-3}^{2}[(3 - y^2) - (y - 3)]\, dy$

31 9 **33** 12 **35** $4\sqrt{2}$

37 $\displaystyle\int_{0}^{1}(x^2 - 6x + 5)\, dx + \int_{1}^{5}-(x^2 - 6x + 5)\, dx +$

$\displaystyle\int_{5}^{7}(x^2 - 6x + 5)\, dx$

41 $\displaystyle\int_{-1.5}^{-1.1}-(x^3 - 0.7x^2 - 0.8x + 1.3)\, dx +$

$\displaystyle\int_{-1.1}^{1.5}(x^3 - 0.7x^2 - 0.8x + 1.3)\, dx$

43 (a) $(a, 0.9052)$, $(b, 5.3623)$, $a \approx 0.0819$, $b \approx 2.8754$

(b) $\displaystyle\int_{a}^{b}[\sqrt{10x} - (x^3 - 2x^2 - x + 1)]\, dx$ **(c)** 10.3259

45 (a) $(\pm a, -8.0061)$, $a \approx 3.4632$

(b) $\displaystyle\int_{-a}^{a}[50\cos(0.5x) - (x^2 - 20)]\, dx$ **(c)** 308.2566

47 (a) $\displaystyle\int_{-5}^{5}[\sqrt{25 - x^2} - (\sqrt{29 - x^2} - 2)]\, dx$ **(b)** 14.7515

49 (a) $\displaystyle\int_{0}^{\pi}[\sin x - \sin(\sin x)]\, dx$ **(b)** 0.2135

51 (a) $[0, 1]$ **(b)** $\dfrac{1}{6}$ **53 (a)** $[0, 1]$ **(b)** 2

55 (a) $(\pm 1.540, 0.618)$

(b) $\displaystyle 2\int_{0}^{1.54}\left[\sqrt{\frac{1}{2.9}(6.09 - 2.1x^2)} - \left(2.1 - \sqrt{\frac{1}{4.3}(21.07 - 4.9x^2)}\right)\right] dx$

57 (a) $(0.741, 2.206)$

(b) $\displaystyle\int_{0}^{0.74}\left\{0.5 + \sqrt{\frac{1}{5.3}[14.31 + 2.7(x - 0.1)^2]} - [0.1 + \sqrt{1.6 + 3.2(x + 0.2)^2}]\right\} dx$

Exercises 5.2

1 $\displaystyle\pi\int_{-1}^{2}\left(\frac{1}{2}x^2 + 2\right)^2 dx$ **3** $\displaystyle 2 \cdot \pi\int_{0}^{4}[(\sqrt{25 - y^2})^2 - 3^2]\, dy$

5 $\displaystyle\pi\int_{1}^{3}\left(\frac{1}{x}\right)^2 dx = \frac{2\pi}{3}$

$y = \dfrac{1}{x}$

7 $\pi \int_0^4 (x^2 - 4x)^2 \, dx = \dfrac{512\pi}{15}$

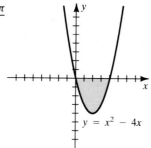

$y = x^2 - 4x$

9 $\pi \int_0^2 (\sqrt{y})^2 \, dy = 2\pi$ **11** $\pi \int_0^4 (4y - y^2)^2 \, dy = \dfrac{512\pi}{15}$

$y = x^2$ $x = 4y - y^2$

13 $2 \cdot \pi \int_0^{\sqrt{2}} [(4 - x^2)^2 - (x^2)^2] \, dx = \dfrac{64\pi\sqrt{2}}{3}$

15 $\pi \int_0^2 [(4 - x)^2 - (x)^2] \, dx = 16\pi$

13

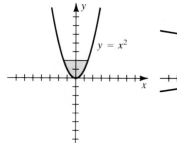

$y = x^2$ $y = 4 - x^2$

15

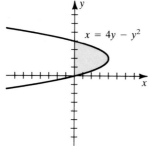

$y = x$ $y = -x + 4$

17 $\pi \int_0^2 [(2y)^2 - (y^2)^2] \, dy = \dfrac{64\pi}{15}$

19 $\pi \int_{-1}^2 [(y + 2)^2 - (y^2)^2] \, dy = \dfrac{72\pi}{5}$

17

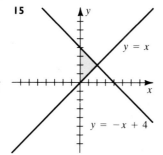

$y = \frac{1}{2}x$ $(4, 2)$ $x = y^2$

19

$(4, 2)$ $x = y^2$ $y = x - 2$ $(1, -1)$

21 $\pi \int_0^\pi (\sin 2x)^2 \, dx = \dfrac{1}{2}\pi^2$

23 $\pi \int_0^{\pi/4} [(\cos x)^2 - (\sin x)^2] \, dx = \dfrac{\pi}{2}$

21

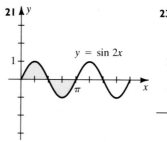

$y = \sin 2x$

23

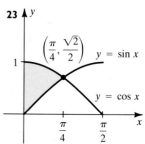

$\left(\dfrac{\pi}{4}, \dfrac{\sqrt{2}}{2}\right)$ $y = \sin x$ $y = \cos x$

25

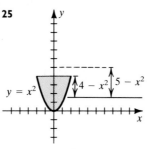

$y = x^2$ $4 - x^2$ $5 - x^2$

$3 - (-\sqrt{y})$ $2 - (-\sqrt{y})$ $3 - \sqrt{y}$ $2 - \sqrt{y}$ $x = -\sqrt{y}$ $x = \sqrt{y}$

(a) $2 \cdot \pi \int_0^2 (4 - x^2)^2 \, dx = \dfrac{512\pi}{15}$

(b) $2 \cdot \pi \int_0^2 [(5 - x^2)^2 - (5 - 4)^2] \, dx = \dfrac{832\pi}{15}$

(c) $\pi \int_0^4 \{[2 - (-\sqrt{y})]^2 - [2 - \sqrt{y}]^2\} \, dy = \dfrac{128\pi}{3}$

(d) $\pi \int_0^4 \{[3 - (-\sqrt{y})]^2 - [3 - \sqrt{y}]^2\} \, dy = 64\pi$

27 (a) $\pi \int_0^4 \left\{ \left[\left(-\dfrac{1}{2}x + 2\right) - (-2) \right]^2 - [0 - (-2)]^2 \right\} \, dx$

(b) $\pi \int_0^4 \left\{ (5 - 0)^2 - \left[5 - \left(-\dfrac{1}{2}x + 2\right) \right]^2 \right\} \, dx$

(c) $\pi \int_0^2 \{(7 - 0)^2 - [7 - (-2y + 4)]^2\} \, dy$

(d) $\pi \int_0^2 \{[(-2y + 4) - (-4)]^2 - [0 - (-4)]^2\} \, dy$

29 $\pi \int_{-2}^0 [(8 - 4x)^2 - (8 - x^3)^2] \, dx +$
$\pi \int_0^2 [(8 - x^3)^2 - (8 - 4x)^2] \, dx$

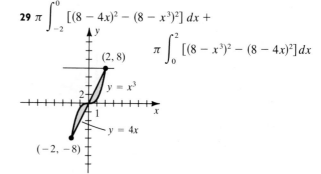

$(2, 8)$ $y = x^3$ $y = 4x$ $(-2, -8)$

31 $\pi \int_2^3 \{[2-(3-y)]^2 - [2-\sqrt{3-y}]^2\} \, dy$

$x = 3 - y$

$(1, 2)$

$x = \sqrt{3-y}$

33 $2 \cdot \pi \int_0^1 \{[5-(-\sqrt{1-y^2})]^2 - [5-\sqrt{1-y^2}]^2\} \, dy$

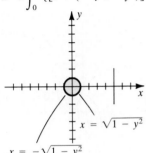

$x = \sqrt{1-y^2}$

$x = -\sqrt{1-y^2}$

35 $\pi \int_0^h r^2 \, dy = \pi r^2 h$ **37** $\pi \int_0^h \left(\frac{r}{h}x\right)^2 dx = \frac{1}{3}\pi r^2 h$

39 $\pi \int_0^h \left(\frac{R-r}{h}x + r\right)^2 dx = \frac{1}{3}\pi h(R^2 + Rr + r^2)$ **41** $\frac{63\pi}{2}$

43

$f(x)$ $g(x)$

0.45 2.01

(a) $0.45, 2.01$ **(b)** 0.28

45 $\frac{4}{3}\pi ab^2$ **47 (a)** $p = \frac{r^2}{4h}$

(b) $\frac{1}{2}\pi r^2 h$

Exercises 5.3

1 $2\pi \int_2^{11} x\sqrt{x-2} \, dx$ **3** $2\pi \int_0^6 y\left(-\frac{1}{2}y + 3\right) dy$

5 $2\pi \int_0^4 x\sqrt{x} \, dx = \frac{128\pi}{5}$

$y = \sqrt{x}$

$(4, 2)$

7 $2\pi \int_0^2 x(\sqrt{8x} - x^2) \, dx = \frac{24\pi}{5}$

$y = x^2$

$(2, 4)$

$x = \frac{1}{8}y^2$

9 $2\pi \int_4^7 x\left[\left(\frac{1}{2}x - \frac{3}{2}\right) - (2x-12)\right] dx = \frac{135\pi}{2}$

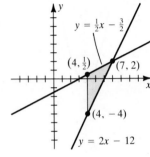

$y = \frac{1}{2}x - \frac{3}{2}$

$(4, \frac{1}{2})$ $(7, 2)$

$(4, -4)$

$y = 2x - 12$

11 $2\pi \int_0^2 x[0 - (2x-4)] \, dx = \frac{16\pi}{3}$

$y = 2x - 4$

13 $2 \cdot 2\pi \int_0^4 y\sqrt{4y} \, dy = \frac{512\pi}{5}$

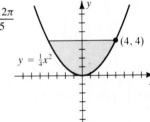

$(4, 4)$

$y = \frac{1}{4}x^2$

15 $2\pi \int_0^6 y\left(\frac{1}{2}y\right) dy = 72\pi$

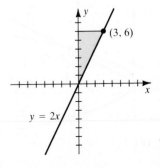

$(3, 6)$

$y = 2x$

17 $2\pi \int_0^2 y[0 - (y^2 - 4)]\,dy = 8\pi$

$y = \sqrt{x} + 4$

19 (a) $2\pi \int_0^2 (3 - x)(x^2 + 1)\,dx$

(b) $2\pi \int_0^2 [x - (-1)](x^2 + 1)\,dx$

21 (a) $2 \cdot 2\pi \int_0^4 (4 - y)\sqrt{y}\,dy$

(b) $2 \cdot 2\pi \int_0^4 (5 - y)\sqrt{y}\,dy$

(c) $2\pi \int_{-2}^2 (2 - x)(4 - x^2)\,dx$

(d) $2\pi \int_{-2}^2 [x - (-3)](4 - x^2)\,dx$

23 $2\pi \int_0^1 (2 - x)[(3 - x^2) - (3 - x)]\,dx$

25 $2 \cdot 2\pi \int_{-1}^1 (5 - x)\sqrt{1 - x^2}\,dx$

27 (a) $2\pi \int_0^{1/2} y(4 - 1)\,dy + 2\pi \int_{1/2}^1 y[(1/y^2) - 1]\,dy$

(b) $\pi \int_1^4 \left(\dfrac{1}{\sqrt{x}}\right)^2 dx$

29 (a) $2\pi \int_0^1 x(x^2 + 2)\,dx$

(b) $\pi \int_0^2 (1)^2\,dy + \pi \int_2^3 [(1)^2 - (\sqrt{y} - 2)^2]\,dy$

31 76π

33

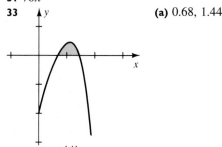

(a) $0.68, 1.44$

(b) $2\pi \int_{0.68}^{1.44} x(-x^4 + 2.21x^3 - 3.21x^2 + 4.42x - 2)\,dx$

35 (a) $\dfrac{8}{3}$ **(b)** 2π **(c)** $\dfrac{16\pi}{5}$

Exercises 5.4

Exer. 1–26: The first integral represents a general formula for the volume. In Exercises 1–8, the vertical distance between the graphs of $y = \sqrt{x}$ and $y = -\sqrt{x}$ is $[\sqrt{x} - (-\sqrt{x})]$, denoted by $2\sqrt{x}$.

1 $\displaystyle\int_c^d s^2\,dx = \int_0^9 (2\sqrt{x})^2\,dx = 162$

3 $\displaystyle\int_c^d \frac{1}{2}\pi r^2\,dx = \int_0^9 \frac{1}{2}\pi\,(\sqrt{x})^2\,dx = \frac{81\pi}{4}$

5 $\displaystyle\int_c^d \frac{\sqrt{3}}{4}s^2\,dx = \int_0^9 \frac{\sqrt{3}}{4}(2\sqrt{x})^2\,dx = \frac{81\sqrt{3}}{2}$

7 $\displaystyle\int_c^d \frac{1}{2}(B + b)h\,dx$

$= \displaystyle\int_0^9 \frac{1}{2}\left[2\sqrt{x} + \frac{1}{2}(2\sqrt{x})\right]\left[\frac{1}{4}(2\sqrt{x})\right]dx = \frac{243}{8}$

9 $\displaystyle\int_c^d s^2\,dx$

$= 2\displaystyle\int_0^a [\sqrt{a^2 - x^2} - (-\sqrt{a^2 - x^2})]^2\,dx = \frac{16}{3}a^3$

11 $\displaystyle\int_c^d \frac{1}{2}bh\,dx$

$= 2\displaystyle\int_0^2 \frac{1}{2}\left[\frac{1}{\sqrt{2}}(4 - x^2)\right]\left[\frac{1}{\sqrt{2}}(4 - x^2)\right]dx = \frac{128}{15}$

13 $\displaystyle\int_c^d lw\,dx = \int_0^h \left(\frac{2ax}{h}\right)\left(\frac{ax}{h}\right)dx = \frac{2}{3}a^2 h$

15 $\displaystyle\int_c^d \frac{1}{2}\pi r^2\,dy = 2\int_0^4 \frac{1}{2}\pi\left[\frac{1}{2}\left(4 - \frac{1}{4}y^2\right)\right]^2 dy = \frac{128\pi}{15}$

17 $\displaystyle\int_c^d lw\,dy = \int_0^a [\sqrt{a^2 - y^2} - (-\sqrt{a^2 - y^2})]y\,dy$

$= \dfrac{2}{3}a^3$

19 $\displaystyle\int_c^d \frac{1}{2}bh\,dx = \int_{-a}^a \frac{1}{2}[\sqrt{a^2 - x^2} - (-\sqrt{a^2 - x^2})]h\,dx$

$= \dfrac{1}{2}\pi a^2 h$

21 $\displaystyle\int_c^d \frac{1}{2}bh\,dx = \int_0^4 \frac{1}{2}\left(\frac{2}{4}x\right)\left(\frac{3}{4}x\right)dx = 4 \text{ cm}^3$

23 $\displaystyle\int_c^d \frac{1}{2}\pi r^2\,dy = \int_0^a \frac{1}{2}\pi\left[\frac{1}{2}(a - y)\right]^2 dy = \frac{\pi}{24}a^3$

25 The areas of cross sections of typical disks and washers are $\pi[f(x)]^2$ and $\pi\{[f(x)]^2 - [g(x)]^2\}$, respectively. In each case, the integrand represents $A(x)$ in (5.13).

27 864

Exercises 5.5

1 (a) $\displaystyle\int_1^3 \sqrt{1 + (3x^2)^2}\,dx$

(b) $\displaystyle\int_2^{28} \sqrt{1 + \left[\frac{1}{3}(y - 1)^{-2/3}\right]^2}\,dy$

3 (a) $\displaystyle\int_{-3}^{-1} \sqrt{1 + (-2x)^2}\, dx$

(b) $\displaystyle\int_{-5}^{3} \sqrt{1 + \left[\frac{1}{2}(4-y)^{-1/2}\right]^2}\, dy$

5 $\displaystyle\int_{1}^{8} \sqrt{1 + \left(\frac{4}{9}x^{-1/3}\right)^2}\, dx = \left(4 + \frac{16}{81}\right)^{3/2} - \left(1 + \frac{16}{81}\right)^{3/2}$
≈ 7.29

7 $\displaystyle\int_{1}^{4} \sqrt{1 + \left(-\frac{3}{2}x^{1/2}\right)^2}\, dx = \frac{8}{27}\left[10^{3/2} - \left(\frac{13}{4}\right)^{3/2}\right] \approx 7.63$

9 $\displaystyle\int_{1}^{2} \sqrt{1 + \left(\frac{1}{4}x^2 - \frac{1}{x^2}\right)^2}\, dx = \frac{13}{12}$

11 $\displaystyle\int_{1}^{2} \sqrt{1 + \left(\frac{3}{2}y^{-4} + \frac{1}{6}y^4\right)^2}\, dy = \frac{353}{240}$

13 $\displaystyle\int_{0}^{2} \sqrt{1 + \left(\frac{7}{2} - 3y^2\right)^2}\, dy$

15 $\displaystyle 8\int_{a}^{1} \sqrt{1 + [(-x^{-1/3})(1 - x^{2/3})^{1/2}]^2}\, dx = 6,$
where $a = \left(\frac{1}{2}\right)^{3/2}$

17 (a) $\displaystyle\int_{1}^{1.1} \sqrt{1 + \frac{4}{9}x^{-2/3}}\, dx \approx 0.119599$

(b) $\sqrt{13}/30 \approx 0.120185$　**(c)** 0.119598

19 (a) $\displaystyle\int_{2}^{2.1} \sqrt{1 + 4x^2}\, dx \approx 0.422021$

(b) $\sqrt{17}(0.1) \approx 0.412311$　**(c)** $\sqrt{0.1781} \approx 0.422019$

21 (a) $\displaystyle\int_{\pi/6}^{31\pi/180} \sqrt{1 + \sin^2 x}\, dx \approx 0.0195733$

(b) $\pi\sqrt{5}/360 \approx 0.0195134$　**(c)** 0.0195725

23 9.778303　**25** 1.849432

27 (a) 3.7900; 3.8125; it is smaller

(b) $\displaystyle\int_{0}^{\pi} \sqrt{1 + \cos^2 x}\, dx$; 3.8199; 3.8202

29 $\displaystyle 2\pi\int_{0}^{1} \sqrt{4x}\sqrt{1 + (x^{-1/2})^2}\, dx = \frac{8\pi}{3}(2^{3/2} - 1) \approx 15.32$

31 $\displaystyle 2\pi\int_{1}^{2} \left(\frac{1}{4}x^4 + \frac{1}{8}x^{-2}\right)\sqrt{1 + \left(x^3 - \frac{1}{4}x^{-3}\right)^2}\, dx$
$= \frac{16{,}911\pi}{1024} \approx 51.88$

33 $\displaystyle 2\pi\int_{2}^{4} \frac{1}{8}y^3 \sqrt{1 + \left(\frac{3}{8}y^2\right)^2}\, dy$
$= \frac{\pi}{27}[8(37)^{3/2} - 13^{3/2}] \approx 204.04$

35 $\displaystyle 2\pi\int_{4}^{5} \sqrt{25 - y^2}\sqrt{1 + [(-y)(25 - y^2)^{-1/2}]^2}\, dy = 10\pi$

37 $\displaystyle 2\pi\int_{0}^{h} \left(\frac{r}{h}x\right)\sqrt{1 + \left(\frac{r}{h}\right)^2}\, dx = \pi r\sqrt{h^2 + r^2}$

39 $\displaystyle 2 \cdot 2\pi\int_{0}^{r} \sqrt{r^2 - x^2}\sqrt{1 + [(-x)(r^2 - x^2)^{-1/2}]^2}\, dx$
$= 4\pi r^2$

41 *Hint*: Regard ds as the slant height of the frustum of a cone that has average radius x.

43 (a) 13.6862; 14.2384; it is smaller

(b) $\displaystyle 2\pi\int_{0}^{\pi} \sin x\sqrt{1 + \cos^2 x}\, dx$; 13.4821; 14.1937

45 201 in^2

47 (a) $x^2 = 500(y - 10)$　**(b)** $\displaystyle\int_{-200}^{200} \sqrt{1 + \left(\frac{1}{250}x\right)^2}\, dx$

(c) 282 ft

49 (a) *Hint*: $\displaystyle S = \int_{0}^{a} 2\pi x\sqrt{1 + \left(\frac{1}{2p}x\right)^2}\, dx$　**(b)** $64{,}968$ ft^2

Exercises 5.6

1 (a) and **(b)** 6000 ft-lb　**3 (a)** $\frac{128}{3}$ in.-lb　**(b)** $\frac{64}{3}$ in.-lb

5 $W_2 = 3W_1$　**7** $27{,}945$ ft-lb　**9** 276 ft-lb　**11** 2250 ft-lb

13 (a) $\frac{81\pi}{2}(62.5) \approx 7952$ ft-lb　**(b)** $\frac{189\pi}{2}(62.5) \approx 18{,}555$ ft-lb

15 500 ft-lb　**17** $575\left(\frac{1}{2} - 40^{-1/5}\right) \approx 12.55$ in.-lb

19 $W = \dfrac{Gm_1m_2h}{(4000)(4000 + h)}$　**21** 36.85 ft-lb

23 (a) $\frac{3}{10}k$ J (k a constant)　**(b)** $\frac{9}{40}k$ J

Exercises 5.7

1 250; 140; 0.56　**3** 14; -27; -46; $\left(-\frac{23}{7}, -\frac{27}{14}\right)$

5 $\frac{1}{4}$; $\frac{1}{14}$; $\frac{1}{5}$; $\left(\frac{4}{5}, \frac{2}{7}\right)$　**7** $\frac{32}{3}$; $\frac{256}{15}$; 0; $\left(0, \frac{8}{5}\right)$

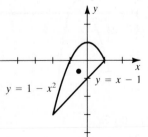

9 $\frac{4}{3}$; $\frac{4}{3}$; $\frac{32}{15}$; $\left(\frac{8}{5}, 1\right)$　**11** $\frac{9}{2}$; $-\frac{27}{10}$; $-\frac{9}{4}$; $\left(-\frac{1}{2}, -\frac{3}{5}\right)$

13 $\dfrac{9}{2}; \dfrac{9}{4}; \dfrac{36}{5}; \left(\dfrac{8}{5}, \dfrac{1}{2}\right)$

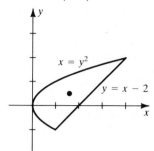

15 $\left(\dfrac{4a}{3\pi}, \dfrac{4a}{3\pi}\right)$

17 With the center of the circle at the origin, the centroid is $\left(0, -\dfrac{20a}{3(8 + \pi)}\right)$.

19 Show that the centroid is $\left(\dfrac{1}{3}a, \dfrac{1}{3}(b + c)\right)$.

21 $(2\pi \cdot 3)(\sqrt{2}\sqrt{18}) = 36\pi$ **23** $\left(\dfrac{4a}{3\pi}, \dfrac{4a}{3\pi}\right)$

25

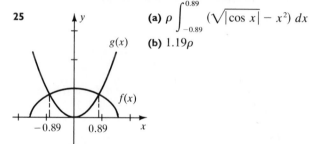

(a) $\rho \displaystyle\int_{-0.89}^{0.89} (\sqrt{|\cos x|} - x^2)\, dx$

(b) 1.19ρ

Exercises 5.8

1 (a) $\dfrac{1}{2}(62.5)$ lb   (b) $\dfrac{3}{2}(62.5)$ lb

3 (a) $\dfrac{\sqrt{3}}{3}(62.5)$ lb   (b) $\dfrac{\sqrt{3}}{24}(62.5)$ lb

5 $\dfrac{16}{3}(60)$ lb   **7** $\dfrac{592}{3}(62.5)$ lb

9 (a) $90(50)$ lb   (b) $54(50)$ lb; $36(50)$ lb   **11** 1.56 L/min

13 In min: (a) 20   (b) 66   (c) 115   (d) 197

15 $10\sqrt{11} - 10 \approx 23.17$ min

17 666   **19** 11   **21** (a) and (b) 150 J

23 $9 - \dfrac{5\sqrt{5}}{3} \approx 5.27$ gal   **25** 1.45 coulombs

27 (a) $\displaystyle\int_0^{1/30} 12,450\pi \sin(30\pi t)\, dt = 830$ cm^3

(b) It is not safe, since approximately 0.027 joule is inhaled.

29 32   **31** $x_c = 320$; 2560   **33** $x_c = 800$; 120,000

Chapter 5 Review Exercises

1 (a) $2 \displaystyle\int_0^2 [(-x^2) - (x^2 - 8)]\, dx = \dfrac{64}{3}$

(b) $4 \displaystyle\int_{-4}^0 \sqrt{-y}\, dy = \dfrac{64}{3}$

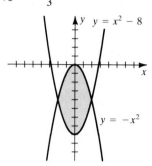

3 $\displaystyle\int_a^b [(1 - y) - y^2]\, dy = \dfrac{5\sqrt{5}}{6}$, where $a = \dfrac{1}{2}(-1 - \sqrt{5})$ and $b = \dfrac{1}{2}(-1 + \sqrt{5})$

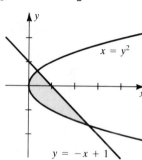

5 $\displaystyle\int_{\pi/3}^{\pi} \left(\sin x - \cos \dfrac{1}{2}x\right) dx = \dfrac{1}{2}$

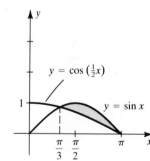

7 $\pi \displaystyle\int_0^2 (\sqrt{4x + 1})^2\, dx = 10\pi$

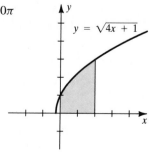

9 $2\pi \displaystyle\int_0^1 x[2 - (x^3 + 1)]\,dx = \dfrac{3\pi}{5}$

11 $2\pi \displaystyle\int_0^{\sqrt{\pi/2}} x(\cos x^2)\,dx = \pi$

13 (a) $\pi \displaystyle\int_{-2}^1 [(-4x + 8)^2 - (4x^2)^2]\,dx = \dfrac{1152\pi}{5}$

(b) $2\pi \displaystyle\int_{-2}^1 (1 - x)[(-4x + 8) - 4x^2]\,dx = 54\pi$

(c) $\pi \displaystyle\int_{-2}^1 \{(16 - 4x^2)^2 - [16 - (-4x + 8)]^2\}\,dx$
$$= \dfrac{1728\pi}{5}$$

15 $\displaystyle\int_{-2}^5 \sqrt{1 + \left[\dfrac{1}{3}(x + 3)^{-1/3}\right]^2}\,dx$
$$= \dfrac{1}{27}(37^{3/2} - 10^{3/2}) \approx 7.16$$

17 $\displaystyle\int_0^4 (5 - y)(62.5)\pi(6)^2\,dy = 432\pi(62.5)$ ft-lb

19 $\rho \displaystyle\int_0^{\sqrt{8}} (6 - y)2(\sqrt{8} - y)\,dy +$
$$\rho \int_{-\sqrt{8}}^0 (6 - y)2(y + \sqrt{8})\,dy = 96(62.5) \text{ lb}$$

21 $4;\ -\dfrac{4}{21};\ \dfrac{16}{15};\ \left(\dfrac{4}{15}, -\dfrac{1}{21}\right)$

23 $2\pi \displaystyle\int_1^2 \left(\dfrac{1}{3}x^3 + \dfrac{1}{4}x^{-1}\right)\sqrt{1 + \left(x^2 - \dfrac{1}{4}x^{-2}\right)^2}\,dx = \dfrac{515\pi}{64}$
$$\approx 25.3$$

25 900 ft

27 (a) The area under the graph of $y = 2\pi x^4$

(b) (i) The volume obtained by revolving $y = \sqrt{2}x^2$ about the x-axis

(ii) The volume obtained by revolving $y = x^3$ about the y-axis

(c) The work done by a force of magnitude $y = 2\pi x^4$ as it moves from $x = 0$ to $x = 1$.

29 (a) $(a, 0.67), (b, 1.91), a \approx -0.82, b \approx 1.38$

(b) $\displaystyle\int_a^b (\sqrt{1 + x^3} - x^2)\,dx \approx 1.43$

31 (a) $(a, 2.40), (b, 9.53), a \approx 0.29, b \approx 4.54$

(b) $\displaystyle\int_a^b [\sqrt{20x} - (x^3 - 4x^2 - x + 3)]\,dx \approx 44.42$

CHAPTER ▪ 6

Exercises 6.1

1 $\dfrac{x - 5}{3}$ **3** $\dfrac{2x + 1}{3x}$ **5** $\dfrac{5x + 2}{2x - 3}$ **7** $-\dfrac{1}{3}\sqrt{6 - 3x}$

9 $3 - x^2, x \geq 0$ **11** $(x - 1)^3$

13 (a) The graph of f is a line of slope $a \neq 0$ and hence is one-to-one. $f^{-1}(x) = \dfrac{x - b}{a}$

(b) No (not one-to-one)

15 (a)

(b) $[-1, 2];\ \left[\dfrac{1}{2}, 4\right]$

(c) $\left[\dfrac{1}{2}, 4\right];\ [-1, 2]$

17 (a)

(b) $[-3, 3];\ [-2, 2]$
(c) $[-2, 2];\ [-3, 3]$

19 (a) $[-0.27, 1.22]$
  (b) $[-0.20, 3.31]$; $[-0.27, 1.22]$
21 (a) $[-1.43, 1.43]$ **(b)** $[-0.84, 0.84]$; $[-1.43, 1.43]$
23 (a) $[-2.14, 1]$ **(b)** $[0.5, 2]$; $[-2.14, 1]$
25 (a) f is increasing on $\left[-\dfrac{3}{2}, \infty\right)$ and hence is one-to-
  one.   **(b)** $[0, \infty)$ **(c)** x
27 (a) f is decreasing on $[0, \infty)$ and hence is one-to-one.
  (b) $(-\infty, 4]$ **(c)** $-\dfrac{1}{2\sqrt{4-x}}$
29 (a) f is decreasing on $(-\infty, 0)$ and $(0, \infty)$ and hence is
  one-to-one.
  (b) All real numbers except zero   **(c)** $-\dfrac{1}{x^2}$
31 (a) f is increasing, since $f'(x) > 0$ for every x **(b)** $\dfrac{1}{16}$
33 (a) f is decreasing, since $f'(x) < 0$ for $x > 0$ **(b)** $-\dfrac{2}{7}$
35 (a) f is increasing, since $f'(x) > 0$ for every x **(b)** $\dfrac{1}{16}$

Exercises 6.2

1 $\dfrac{9}{9x+4}$ **3** $\dfrac{2(3x-1)}{3x^2-2x+1}$ **5** $\dfrac{2}{2x-3}$ **7** $\dfrac{15}{3x-2}$
9 $\dfrac{3x^2}{2x^3-7}$ **11** $1 + \ln x$ **13** $\dfrac{1}{2x}\left(1 + \dfrac{1}{\sqrt{\ln x}}\right)$
15 $-\dfrac{1}{x}\left[\dfrac{1}{(\ln x)^2} + 1\right]$ **17** $\dfrac{20}{5x-7} + \dfrac{6}{2x+3}$
19 $\dfrac{x}{x^2+1} - \dfrac{18}{9x-4}$ **21** $\dfrac{x}{x^2-1} - \dfrac{x}{x^2+1}$ **23** $\dfrac{1}{\sqrt{x^2-1}}$
25 $-2 \tan 2x$ **27** $9 \csc 3x \sec 3x$ **29** $\dfrac{2 \tan 2x}{\ln \sec 2x}$
31 $\tan x$ **33** $\sec x$ **35** $\dfrac{y(2x^2-1)}{x(3y+1)}$ **37** $\dfrac{y(y-x \ln y)}{x(x-y \ln x)}$
39 $(5x+2)^2(6x+1)(150x+39)$ **41** $\dfrac{(14x+11)(x-5)^2}{\sqrt{4x+7}}$
43 $\dfrac{(19x^2+20x-3)(x^2+3)^4}{2(x+1)^{3/2}}$ **45** $y = 8x - 15$
47 $(10, 5 \ln 10 - 5) \approx (10, 6.51)$; $y'' = -(5/x^2) < 0$ implies that the graph is CD for $x > 0$.   **49** ± 0.73 yr
51 (a) $s'(0) = 0$ m/sec; $s''(0) = \dfrac{bc}{m_1+m_2}$ m/sec^2
  (b) $s'\left(\dfrac{m_2}{b}\right) = c \ln\left(\dfrac{m_1+m_2}{m_1}\right)$; $s''\left(\dfrac{m_2}{b}\right) = \dfrac{bc}{m_1}$
53 The graphs coincide if $x > 0$; however, the graph of $y = \ln(x^2)$ contains points with negative x-coordinates.
55 (a) $-3.18 \le y \le 0$
  (b) x-int.: $\pi/2 \approx 1.57$; max: $f(\pi/2) = 0$
57 (a) $1.33 \le y \le 2.18$
  (b) y-int.: 2

59 (a) $-1.97 \le y \le 3.79$
  (b) x-int.: 0.55; max: $f(2.47) \approx 1.56$, $f(8.14) \approx 2.91$,
    $f(14.30) \approx 3.49$; min: $f(4.65) \approx 0.34$,
    $f(10.97) \approx 1.19$, $f(17.26) \approx 1.65$
61 0.5671   **63** -3.2088, 2.0435   **65** 1.7477   **67** 1.8929
69 12.0536   **71** 9.3392

Exercises 6.3

1 $-5e^{-5x}$ **3** $6xe^{3x^2}$ **5** $\dfrac{e^{2x}}{\sqrt{1+e^{2x}}}$ **7** $\dfrac{e^{\sqrt{x+1}}}{2\sqrt{x+1}}$
9 $2xe^{-2x}(1-x)$ **11** $\dfrac{e^x(x-1)^2}{(x^2+1)^2}$ **13** $12e^{4x}(e^{4x}-5)^2$
15 $-\dfrac{e^{1/x}}{x^2} - e^{-x}$ **17** $\dfrac{4}{(e^x+e^{-x})^2}$ **19** $e^{-2x}\left(\dfrac{1}{x} - 2\ln x\right)$
21 $5e^{5x} \cos e^{5x}$ **23** $e^{-x} \tan e^{-x}$
25 $e^{3x}\left(\dfrac{\sec^2 \sqrt{x}}{2\sqrt{x}} + 3 \tan \sqrt{x}\right)$
27 $-8e^{-4x} \sec^2(e^{-4x}) \tan(e^{-4x})$ **29** $e^{\cot x}(1 - x \csc^2 x)$
31 $\dfrac{3x^2 - ye^{xy}}{xe^{xy} + 6y}$ **33** $\dfrac{e^x \cot y - e^{2y}}{2xe^{2y} + e^x \csc^2 y}$
35 $y = (e+3)x - (e+1)$
37 Min: $f(-1) = -e^{-1} \approx -0.368$; increasing on $[-1, \infty)$;
  decreasing on $(-\infty, -1]$; CU on $(-2, \infty)$; CD on
  $(-\infty, -2)$; PI: $(-2, -2e^{-2}) \approx (-2, -0.271)$

39 Decreasing on $(-\infty, 0)$ and $(0, \infty)$; CU on $\left(-\dfrac{1}{2}, 0\right)$
  and $(0, \infty)$; CD on $\left(-\infty, -\dfrac{1}{2}\right)$; PI: $\left(-\dfrac{1}{2}, e^{-2}\right)$

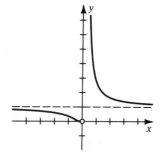

41 Min: $f(e^{-1}) = -e^{-1}$; increasing on $[e^{-1}, \infty)$; decreasing on $(0, e^{-1}]$; CU on $(0, \infty)$; no PI

43 $q'(t) = -cq(t)$ **45 (a)** $\dfrac{\ln(a/b)}{a-b}$ **(b)** $\lim\limits_{t \to \infty} C(t) = 0$

47 (a) 75.8 cm; 15.98 cm/yr   **(b)** 3 mo; 6 yr

49 (a) $f\left(\dfrac{n}{a}\right)$ **(b)** At $x = \dfrac{2}{a}$

51

55 Max: $f(\mu) = \dfrac{1}{\sigma\sqrt{2\pi}}$; increasing on $(-\infty, \mu]$;
decreasing on $[\mu, \infty)$; CU on $(-\infty, \mu - \sigma)$ and $(\mu + \sigma, \infty)$; CD on $(\mu - \sigma, \mu + \sigma)$;
PI: $\left(\mu \pm \sigma, \dfrac{1}{\sigma\sqrt{2\pi e}}\right)$; both limits equal 0

57 (a) $[0.054, 1]$ **(b)** y-int.: 1
59 (a) $[-3.18, 6.13]$
  (b) x-int.: ± 0.84, 2.52, 4.20, 5.88, 7.56; y-int.: 6; max: $f(-0.11) \approx 6.13$, $f(3.25) \approx 1.65$, $f(6.61) \approx 0.45$; min: $f(1.57) \approx -3.18$, $f(4.93) \approx -0.86$
61 0.5671   **63** 1.2022   **65** $e^{-1/2} \approx 0.607$

Exercises 6.4

1 (a) $\dfrac{1}{2}\ln|2x + 7| + C$ **(b)** $\ln\sqrt{3}$

3 (a) $2\ln|x^2 - 9| + C$ **(b)** $\ln\dfrac{25}{64}$

5 (a) $-\dfrac{1}{4}e^{-4x} + C$ **(b)** $-\dfrac{1}{4}(e^{-12} - e^{-4})$

7 (a) $-\dfrac{1}{2}\ln|\cos 2x| + C$ **(b)** $\dfrac{1}{4}\ln 2$

9 (a) $2\ln\left|\csc\dfrac{1}{2}x - \cot\dfrac{1}{2}x\right| + C$ **(b)** $2\ln(2 + \sqrt{3})$

11 $\dfrac{1}{2}\ln|x^2 - 4x + 9| + C$

13 $\dfrac{1}{2}x^2 + 4x + 4\ln|x| + C$

15 $\dfrac{1}{2}(\ln x)^2 + C$ **17** $\dfrac{1}{2}x^2 + \dfrac{1}{5}e^{5x} + C$

19 $-\dfrac{3}{2}\ln|1 + 2\cos x| + C$ **21** $e^x + 2x - e^{-x} + C$

23 $\ln(e^x + e^{-x}) + C$ **25** $3\ln|\sin\sqrt[3]{x}| + C$

27 $\dfrac{1}{2}\ln|\sec 2x + \tan 2x| + C$ **29** $-\dfrac{1}{3}\ln|\sec e^{-3x}| + C$

31 $\ln|\csc x - \cot x| + \cos x + C$ **33** $\ln|\csc x| + C$
35 $x + 2\ln|\sec x + \tan x| + \tan x + C$ **37** 4

39 $\pi(1 - e^{-1})$ **41** $y = 2e^{2x} - \dfrac{3}{2}e^{-2x} + \dfrac{7}{2}$

43 $y = 3e^{-x} + 4x - 4$ **45** $\dfrac{2}{\ln(13/4)} \approx 1.697$

47 (a) 25   **(b)** 205   **(c)** 12   **49** $\Delta S = c\ln\dfrac{T_2}{T_1}$

51 (a) $\dfrac{5}{2}(1 - e^{-4t})$ **(b)** $\lim\limits_{t \to \infty} Q(t) = \dfrac{5}{2}$ coulombs
53 (a) $s(t) = kv_0(1 - e^{-t/k})$ **(b)** $\lim\limits_{t \to \infty} s(t) = kv_0$
55 0.7468   **57** 127.2930   **59** 6.43   **61** 9.34

Exercises 6.5

1 $7^x \ln 7$ **3** $8^{x^2+1}(2x \ln 8)$ **5** $\dfrac{4x^3 + 6x}{(x^4 + 3x^2 + 1)\ln 10}$

7 $5^{3x-4}(3 \ln 5)$ **9** $\dfrac{-(x^2 + 1)10^{1/x}(\ln 10)}{x^2} + (2x)10^{1/x}$

11 $\dfrac{30x}{(3x^2 + 2)\ln 10}$ **13** $\left(\dfrac{6}{6x + 4} - \dfrac{2}{2x - 3}\right)\dfrac{1}{\ln 5}$

15 $\dfrac{1}{x \ln x \ln 10}$ **17** $ex^{e-1} + e^x$

19 $(x + 1)^x\left[\dfrac{x}{x + 1} + \ln(x + 1)\right]$ **21** $2^{\sin^2 x}(\sin 2x)\ln 2$

23 (a) 0   **(b)** $5x^4$ **(c)** $\sqrt{5}x^{\sqrt{5}-1}$ **(d)** $(\sqrt{5})^x \ln\sqrt{5}$
  (e) $x^{1+x^2}(1 + 2\ln x)$

29 (a) $\dfrac{7^x}{\ln 7} + C$ **(b)** $\dfrac{342}{49\ln 7} \approx 3.59$

31 (a) $\dfrac{-5^{-2x}}{2 \ln 5} + C$ **(b)** $\dfrac{12}{625 \ln 5} \approx 0.012$ **33** $\dfrac{10^{3x}}{3 \ln 10} + C$

35 $\dfrac{-3^{-x^2}}{2 \ln 3} + C$ **37** $\dfrac{\ln(2^x + 1)}{\ln 2} + C$

39 $(\ln 10) \ln \left| \log x \right| + C$ **41** $-\dfrac{3^{\cos x}}{\ln 3} + C$

43 (a) $\pi^\pi x + C$ **(b)** $\dfrac{1}{5}x^5 + C$ **(c)** $\dfrac{x^{\pi + 1}}{\pi + 1} + C$

    (d) $\dfrac{\pi^x}{\ln \pi} + C$ **45** $\dfrac{1}{\ln 2} - \dfrac{1}{2} \approx 0.94$

47 (a) \$0.05/yr   **(b)** \$0.95

49 (a) In trout/yr: 95; 62; 53   **(b)** 9.36

51 pH ≈ 2.201; $\pm 0.1\%$

53 (b) $S = \dfrac{k}{x}$, where $k = \dfrac{a}{\ln 10}$;

    $S(x) = 2\,S(2x)$ (twice as sensitive)

55 (a) With $n = r/h$,

$$\ln A = \ln[P(1 + h)^{rt/h}] = \ln P + rt \ln(1 + h)^{1/h}.$$

    (b) Since $h = r/n$, $n \to \infty$ if and only if $h \to 0^+$. Thus,

$$\ln A = \lim_{h \to 0^+}[\ln P + rt \ln(1 + h)^{1/h}]$$
$$= \ln P + rt \ln e = \ln(Pe^{rt})$$

    and $A = Pe^{rt}$.

57 Let $h = x/n$. Then

$$\lim_{n \to \infty}\left(1 + \frac{x}{n}\right)^n = \lim_{h \to 0^+}(1 + h)^{x/h} = \lim_{h \to 0^+}[(1 + h)^{1/h}]^x = e^x.$$

Exercises 6.6

1 $q(t) = 5000(3)^{t/10}$; 45,000; $\dfrac{10 \ln 10}{\ln 3} \approx 20.96$ hr

3 $30\left(\dfrac{29}{30}\right)^5 \approx 25.32$ in.

5 $\dfrac{\ln(40/5.5)}{0.02} \approx 99.21$ yr after Jan. 1, 1993

    (March 17, 2092)

7 $\dfrac{5 \ln(1/6)}{\ln(1/3)} \approx 8.15$ min

9 Proceed as in the solution to Example 1.

11 $P(z) = \left(\dfrac{288 - 0.01z}{288}\right)^{3.42}$ **13** $\dfrac{29 \ln(2/5)}{\ln(1/2)} \approx 38.34$ yr

15 $600\left(\dfrac{1}{2}\right)^{-3/16} \approx 683.27$ mg

17 $v(y) = \sqrt{2k\left(\dfrac{1}{y} - \dfrac{1}{y_0}\right) + v_0^2}$

19 $\dfrac{5700 \ln(0.2)}{\ln(1/2)} \approx 13{,}235$ yr   **21** Use Theorem (4.35).

23 $V(t) = \dfrac{1}{27}(kt + C)^3$

25 (c)

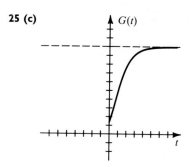

Exercises 6.7

1 (a) $-\dfrac{\pi}{4}$ **(b)** $\dfrac{2\pi}{3}$ **(c)** $-\dfrac{\pi}{3}$

3 (a) Not defined   **(b)** Not defined   **(c)** $\dfrac{\pi}{4}$

5 (a) $\dfrac{\pi}{3}$ **(b)** $\dfrac{5\pi}{6}$ **(c)** $-\dfrac{\pi}{6}$ **7 (a)** $\dfrac{\pi}{3}$ **(b)** $\dfrac{2\pi}{3}$ **(c)** $\dfrac{\pi}{6}$

9 (a) $\dfrac{\sqrt{3}}{2}$ **(b)** 0   **(c)** Not defined

11 (a) $-\dfrac{\sqrt{21}}{2}$ **(b)** $\dfrac{\sqrt{65}}{4}$ **(c)** $\dfrac{5}{\sqrt{24}}$

13 (a) -1.1971 **(b)** 0.2712 **15 (a)** 1.0556 **(b)** 0.6183

17 $\dfrac{x}{\sqrt{x^2 + 1}}$ **19** $\dfrac{3}{\sqrt{9 - x^2}}$

21   **23**

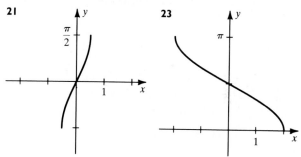

27 (a) $y = \cot^{-1} x$ if and only if $x = \cot y$ for $0 < y < \pi$.

    (b)

29 (a) $\alpha = \theta - \sin^{-1}\dfrac{d}{k}$ **(b)** $40°$ **31** $\dfrac{1}{2\sqrt{x}\sqrt{1 - x}}$

33 $\dfrac{3}{9x^2 - 30x + 26}$ **35** $\dfrac{-e^{-x}}{\sqrt{e^{-2x} - 1}} - e^{-x} \operatorname{arcsec} e^{-x}$

37 $\dfrac{2x}{(1 + x^4) \arctan (x^2)}$ **39** $-\dfrac{9(1 + \cos^{-1} 3x)^2}{\sqrt{1 - 9x^2}}$

41 $\left(\dfrac{1}{x^2}\right) \sin \left(\dfrac{1}{x}\right) + \sec x \tan x - \dfrac{1}{\sqrt{1 - x^2}}$

43 $3^{\arcsin (x^3)} \dfrac{(3 \ln 3)x^2}{\sqrt{1 - x^6}}$

45 $\dfrac{1 - 2x \arctan x}{(x^2 + 1)^2}$ **47** $\dfrac{1}{2\sqrt{x}}\left(\dfrac{1}{\sqrt{x - 1}} + \sec^{-1}\sqrt{x}\right)$

49 $\dfrac{ye^x - 2x - \sin^{-1} y}{\dfrac{x}{\sqrt{1 - y^2}} - e^x}$

51 (a) $\dfrac{1}{4} \tan^{-1}\left(\dfrac{x}{4}\right) + C$ (b) $\dfrac{\pi}{16}$

53 (a) $\dfrac{1}{2} \sin^{-1} (x^2) + C$ (b) $\dfrac{\pi}{12}$

55 $2 \tan^{-1}\sqrt{x} + C$ **57** $\sin^{-1}\left(\dfrac{e^x}{4}\right) + C$

59 $\dfrac{1}{2} \ln (x^2 + 9) + C$ **61** $\dfrac{1}{5} \sec^{-1}\left(\dfrac{e^x}{5}\right) + C$

63 $\pm\dfrac{7}{3576}$ rad **65** $-\dfrac{25}{1044}$ rad/sec **67** $\sqrt{4800} \approx 69.3$ ft

69 $\dfrac{2\pi}{27} \approx 0.233$ mi/sec **75** $x \sin^{-1} (2x) + \dfrac{1}{2}\sqrt{1 - 4x^2} + C$

77 $\dfrac{1}{2}x^2 \tan^{-1} (x^2) - \dfrac{1}{4} \ln (x^4 + 1) + C$

79 0.7241 **81** 2.0570 **83** 31.9285

Exercises 6.8

1 (a) 27.2899 (b) 2.1250 (c) -0.9951
 (d) 1.0000 (e) 0.2658 (f) -0.8509
3 $5 \cosh 5x$ **5** $3x^2 \sinh (x^3)$

7 $\dfrac{1}{2\sqrt{x}}(\sqrt{x} \operatorname{sech}^2 \sqrt{x} + \tanh \sqrt{x})$

9 $\left(\dfrac{1}{x^2}\right) \operatorname{csch}^2 \left(\dfrac{1}{x}\right)$

11 $\dfrac{-2x \operatorname{sech} (x^2)[(x^2 + 1) \tanh (x^2) + 1]}{(x^2 + 1)^2}$

13 $-12 \operatorname{csch}^2 6x \coth 6x$

15 (a) $\mathbb{R}$ (b) $\dfrac{4x \sinh \sqrt{4x^2 + 3}}{\sqrt{4x^2 + 3}}$

17 (a) $\mathbb{R}$ (b) $-\dfrac{\operatorname{sech}^2 x}{(\tanh x + 1)^2}$

19 $\dfrac{1}{3} \sinh (x^3) + C$ **21** $2 \cosh \sqrt{x} + C$

23 $\dfrac{1}{3} \tanh 3x + C$ **25** $-2 \coth \left(\dfrac{1}{2}x\right) + C$

27 $-\dfrac{1}{3} \operatorname{sech} 3x + C$ **29** $-\operatorname{csch} x + C$

31 $(\ln (2 \pm \sqrt{3}), \pm\sqrt{3})$

33 Show that $A = \dfrac{1}{2}(\cosh t)(\sinh t) - \displaystyle\int_1^{\cosh t} \sqrt{x^2 - 1} \, dx$

and that $\dfrac{dA}{dt} = \dfrac{1}{2}$.

35 (a) $286{,}574$ ft^2 (b) 1494 ft **37** 34.94 ft
39 $10.5 \sinh^{-1} \frac{4}{3} \approx 11.54$ ft

41 (b) $y = \dfrac{1}{\alpha} \ln [\cosh (\sqrt{g\alpha}\, t + v_0)] + h_0$

43 (a) $\displaystyle\lim_{h\to\infty} v^2 = \dfrac{gL}{2\pi}$ (b) *Hint*: Let $f(h) = v^2$.

45 (a) 0.7 (b) 0.722

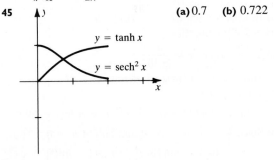

47 $\cosh x + \sinh x = \dfrac{e^x + e^{-x}}{2} + \dfrac{e^x - e^{-x}}{2} = e^x$

49 $\sinh x \cosh y + \cosh x \sinh y$

$= \dfrac{(e^x - e^{-x})(e^y + e^{-y})}{4} + \dfrac{(e^x + e^{-x})(e^y - e^{-y})}{4}$

$= \dfrac{(e^{x+y} + e^{x-y} - e^{-x+y} - e^{-x-y}) + (e^{x+y} - e^{x-y} + e^{-x+y} - e^{-x-y})}{4}$

$= \dfrac{2e^{x+y} - 2e^{-x-y}}{4} = \dfrac{e^{x+y} - e^{-(x+y)}}{2} = \sinh (x + y)$

51 $\sinh(x - y)$
$= \sinh(x + (-y))$
$= \sinh x \cosh(-y) + \cosh x \sinh(-y)$ (Exer. 49)
$= \sinh x \cosh y - \cosh x \sinh y$ (Exer. 48)

53 Let $y = x$ in Exercise 49.
55 From Exercise 54,

$$\cosh 2y = \cosh^2 y + \sinh^2 y$$
$$= (1 + \sinh^2 y) + \sinh^2 y$$
$$= 1 + 2 \sinh^2 y,$$

and hence

$$\sinh^2 y = \dfrac{\cosh 2y - 1}{2}.$$

Let $y = \dfrac{x}{2}$ to obtain the identity.

57 $\cosh nx + \sinh nx = \dfrac{e^{nx} + e^{-nx}}{2} + \dfrac{e^{nx} - e^{-nx}}{2}$

$= e^{nx} = (e^x)^n = (\cosh x + \sinh x)^n$

59 (a) 0.8814 (b) 1.3170 (c) -0.5493 (d) 1.3170

61 $\dfrac{5}{\sqrt{25x^2 + 1}}$ **63** $\dfrac{1}{2\sqrt{x}\sqrt{x - 1}}$ **65** $\dfrac{4}{16x^2 - 1}$

67 $-\dfrac{2}{x\sqrt{1-x^4}}$

69 (a) $(\tfrac{1}{4}, \infty)$ **(b)** $\dfrac{4}{\sqrt{16x^2-1}\ \cosh^{-1}(4x)}$

7I (a) $(-2, 0)$ **(b)** $-\dfrac{1}{x(x+2)}$

73 $\dfrac{1}{4}\sinh^{-1}\left(\dfrac{4}{9}x\right)+C$ **75** $\dfrac{1}{14}\tanh^{-1}\left(\dfrac{2}{7}x\right)+C$

77 $\cosh^{-1}\left(\dfrac{e^x}{4}\right)+C$ **79** $-\dfrac{1}{6}\operatorname{sech}^{-1}\left(\dfrac{x^2}{3}\right)+C$

8I $y=\sinh 3t$

83

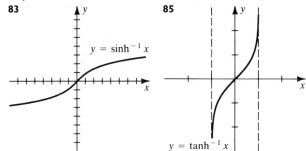
$y=\sinh^{-1}x$

85

$y=\tanh^{-1}x$

Exer. 87–91: (a) Use a procedure similar to that given in the text for $\sinh^{-1}x$. **(b)** Let $u=x$ in Theorem (6.48) and differentiate $\cosh^{-1}u$. **(c)** Differentiate the right-hand side.

Exercises 6.9

I $\dfrac{1}{2}$ **3** $\dfrac{1}{40}$ **5** $\dfrac{3}{13}$ **7** $-\dfrac{1}{2}$ **9** $\dfrac{1}{6}$ **II** ∞ **13** $\dfrac{1}{3}$ **15** 1

17 ∞ **19** $\dfrac{2}{5}$ **2I** 0   **23** ∞ **25** 2   **27** $\dfrac{3}{5}$ **29** -3

3I 0   **33** ∞ **35** 1

37 0.9129, 0.9901, 0.9990, 0.9999; predict limit of 1

39 gt **4I** $\dfrac{1}{2}A\omega_0 t \sin \omega_0 t$ **43 (a)** 1   **(b)** $-\dfrac{1}{18}$

45 (a) 0.2499, 0.4969, 0.7266, 0.9045

(b)

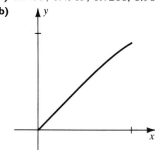

49 0   **5I** 0   **53** 0   **55** 1   **57** e^5 **59** 1   **6I** ∞

63 e^2 **65** 0   **67** $-\infty$ **69** $\dfrac{1}{2}$ **7I** e **73** $e^{1/3}$ **75** ∞

77 1

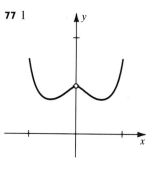

79 (a) Max: $f(e)=e^{1/e}$

≈ 1.44;

$\displaystyle\lim_{x\to 0^+} x^{1/x}=0$

(b) $y=1$

83 gt

Chapter 6 Review Exercises

I $\dfrac{10-x}{15}$

3 f is decreasing, since $f'(x)<0$ for $-1\le x\le 1$; $-\dfrac{1}{8}$

5 $\dfrac{75x^2}{5x^3-4}$ **7** $\dfrac{12}{3x+2}+\dfrac{3}{6x-5}-\dfrac{8}{8x-7}$

9 $\dfrac{-4x}{(2x^2+3)[\ln(2x^2+3)]^2}$ **II** $2x$

13 $\dfrac{10^x}{x\ln 10}+10^x(\ln 10)\log x$ **15** $\dfrac{1}{4x\sqrt{\ln\sqrt{x}}}$

17 $2xe^{-x^2}(1-x^2)$ **19** $\dfrac{10^{\ln x}\ln 10}{x}$ **2I** $\dfrac{2\ln x\,(x^{\ln x})}{x}$

23 $2e^{-2x}\csc e^{-2x}(\csc^2 e^{-2x}+\cot^2 e^{-2x})$

25 $-16\tan 4x$ **27** $-\dfrac{y}{x}$

29 $\left[\dfrac{4}{3(x+2)}+\dfrac{3}{2(x-3)}\right](x+2)^{4/3}(x-3)^{3/2}$

3I (a) $-2e^{-\sqrt{x}}+C$ **(b)** $2(e^{-1}-e^{-2})\approx 0.465$

33 $-\dfrac{1}{2}\ln|\cos x^2|+C$ **35** $\dfrac{x^{e+1}}{e+1}+C$

37 $-\dfrac{1}{2}e^{-2x}-2e^{-x}+x+C$

39 $\dfrac{1}{2}x^2-2x+4\ln|x+2|+C$ **4I** $-\dfrac{1}{8}e^{4/x^2}+C$

43 $\ln(1+e^x)+C$ **45** $\dfrac{(5e)^x}{\ln(5e)}+C$

47 $\cos e^{-x}+C$ **49** $-\ln|1+\cot x|+C$

5I $-\ln|\cos e^x|+C$

53 $-\dfrac{1}{3}\cot 3x+\dfrac{2}{3}\ln|\csc 3x-\cot 3x|+x+C$

55 $y=-\dfrac{1}{9}e^{-3x}+\dfrac{5}{3}x-\dfrac{8}{9}$ **57** $4e^2+12\approx 41.56$ cm

59 $y-e=-2(1+e)(x-1)$

6I $\dfrac{\pi}{8}(e^{-16}-e^{-24})\approx 4.42\times 10^{-8}$

63 $\dfrac{5 \ln(1/100)}{\ln(1/2)} \approx 33.2$ days

65 (a) $\dfrac{3 \ln(3/10)}{\ln(1/2)} \approx 5.2$ hr or 2.2 additional hr

(b) $10\left[1 - \left(\dfrac{1}{2}\right)^{7/3}\right] \approx 8.016$ lb

67 $100{,}000(2)^6 = 6{,}400{,}000$

69 $\dfrac{1}{2x\sqrt{x-1}}$ **71** $\dfrac{2x}{\sqrt{x^4-1}} + 2x \operatorname{arcsec}(x^2)$

73 $\dfrac{2x}{(1+x^4)\tan^{-1}(x^2)}$ **75** $\dfrac{-x}{\sqrt{x^2(1-x^2)}}$

77 $\dfrac{1}{(1+x^2)[1+(\tan^{-1}x)^2]}$ **79** $-5e^{-5x}\sinh e^{-5x}$

81 $(\cosh x - \sinh x)^{-2}$, or e^{2x} **83** $\dfrac{2x}{\sqrt{x^4+1}}$

85 $\dfrac{1}{6}\tan^{-1}\left(\dfrac{3}{2}x\right) + C$ **87** $-\sqrt{1-e^{2x}} + C$

89 $\dfrac{1}{2}\sinh(x^2) + C$ **91** $\dfrac{\pi}{3}$

93 $\dfrac{1}{2}\sin^{-1}\left(\dfrac{2}{3}x\right) + C$ **95** $-\dfrac{1}{3}\operatorname{sech}^{-1}\left(\dfrac{2}{3}|x|\right) + C$

97 $\dfrac{1}{25}\sqrt{25x^2 + 36} + C$ **99** $\left(\pm\dfrac{4}{15}, \sin^{-1}\left(\pm\dfrac{4}{5}\right)\right)$

101 Let $c = \tan^{-1}\dfrac{1}{2}$. Min: $f(c) = 5\sqrt{5}$; increasing on $\left[c, \dfrac{\pi}{2}\right)$; decreasing on $(0, c]$.

103 (a) $\dfrac{1}{2}\tan^{-1}4 + \dfrac{\pi}{2}n$ for $n = 0, 1, 2, 3$

(b) 0.66, 2.23, 3.80, 5.38

105 $\dfrac{1}{260}$ rad/sec $\approx 0.22°/$sec

107 $-\dfrac{800}{2581} \approx -0.31$ rad/sec **109** $\dfrac{1}{2}\ln 2$ **111** ∞

113 0 **115** $-\infty$ **117** e **119** 1 **121** 0

CHAPTER ■ 7

Exercises 7.1

1 $-(x+1)e^{-x} + C$ **3** $\dfrac{1}{27}e^{3x}(9x^2 - 6x + 2) + C$

5 $\dfrac{1}{5}x \sin 5x + \dfrac{1}{25}\cos 5x + C$

7 $x \sec x - \ln|\sec x + \tan x| + C$

9 $x^2 \sin x + 2x \cos x - 2 \sin x + C$

11 $x \tan^{-1} x - \dfrac{1}{2}\ln(1 + x^2) + C$

13 $\dfrac{2}{9}x^{3/2}(3 \ln x - 2) + C$ **15** $-x \cot x + \ln|\sin x| + C$

17 $-\dfrac{1}{2}e^{-x}(\sin x + \cos x) + C$

19 $\cos x(1 - \ln \cos x) + C$

21 $-\dfrac{1}{2}\csc x \cot x + \dfrac{1}{2}\ln|\csc x - \cot x| + C$

23 $\dfrac{1}{3}(2 - \sqrt{2}) \approx 0.20$ **25** $\dfrac{\pi}{4}$

27 $\dfrac{1}{40{,}400}(2x + 3)^{100}(200x - 3) + C$

29 $\dfrac{1}{41}e^{4x}(4 \sin 5x - 5 \cos 5x) + C$

31 $x(\ln x)^2 - 2x \ln x + 2x + C$

33 $x^3 \cosh x - 3x^2 \sinh x + 6x \cosh x - 6 \sinh x + C$

35 $2\sqrt{x}\sin\sqrt{x} + 2\cos\sqrt{x} + C$

37 $x \cos^{-1} x - \sqrt{1 - x^2} + C$ **39** Let $u = x^m$.

41 Let $u = (\ln x)^m$.

43 $e^x(x^5 - 5x^4 + 20x^3 - 60x^2 + 120x - 120) + C$

45 2π **47** $\dfrac{\pi}{2}(e^2 + 1) \approx 13.18$ **49** $\left(\dfrac{3 \ln 3 - 2}{2}, 1\right)$

Exercises 7.2

1 $\sin x - \dfrac{1}{3}\sin^3 x + C$ **3** $\dfrac{1}{8}x - \dfrac{1}{32}\sin 4x + C$

5 $-\dfrac{1}{3}\cos^3 x + \dfrac{1}{5}\cos^5 x + C$

7 $\dfrac{1}{8}\left(\dfrac{5}{2}x - 2 \sin 2x + \dfrac{3}{8}\sin 4x + \dfrac{1}{6}\sin^3 2x\right) + C$

9 $\dfrac{1}{4}\tan^4 x + \dfrac{1}{6}\tan^6 x + C$ **11** $\dfrac{1}{5}\sec^5 x - \dfrac{1}{3}\sec^3 x + C$

13 $\dfrac{1}{5}\tan^5 x - \dfrac{1}{3}\tan^3 x + \tan x - x + C$

15 $\dfrac{2}{3}\sin^{3/2} x - \dfrac{2}{7}\sin^{7/2} x + C$ **17** $\tan x - \cot x + C$

19 $\dfrac{2}{3} - \dfrac{5}{6\sqrt{2}} \approx 0.08$ **21** $\dfrac{1}{2}\left(\dfrac{1}{2}\sin 2x - \dfrac{1}{8}\sin 8x\right) + C$

23 $\dfrac{3}{5}$ **25** $-\dfrac{1}{5}\cot^5 x - \dfrac{1}{7}\cot^7 x + C$

27 $-\ln(2 - \sin x) + C$ **29** $-\dfrac{1}{1 + \tan x} + C$

31 $\dfrac{3}{4}\pi^2 \approx 7.40$ **33** $\dfrac{5}{2}$

35 (a) Use the trigonometric product-to-sum formulas.

(b) $\displaystyle\int \sin mx \cos nx \, dx$

$= \begin{cases} -\dfrac{\cos(m+n)x}{2(m+n)} - \dfrac{\cos(m-n)x}{2(m-n)} + C & \text{if } m \neq n \\[2mm] -\dfrac{\cos 2mx}{4m} + C & \text{if } m = n \end{cases}$

$\displaystyle\int \cos mx \cos nx \, dx$

$= \begin{cases} \dfrac{\sin(m+n)x}{2(m+n)} + \dfrac{\sin(m-n)x}{2(m-n)} + C & \text{if } m \neq n \\[2mm] \dfrac{x}{2} + \dfrac{\sin 2mx}{4m} + C & \text{if } m = n \end{cases}$

Exercises 7.3

1. $\dfrac{1}{2}\ln\left|\dfrac{2}{x} - \dfrac{\sqrt{4-x^2}}{x}\right| + C$ 3. $\dfrac{1}{3}\ln\left|\dfrac{\sqrt{x^2+9}}{x} - \dfrac{3}{x}\right| + C$

5. $\dfrac{\sqrt{x^2-25}}{25x} + C$ 7. $-\sqrt{4-x^2} + C$

9. $-\dfrac{x}{\sqrt{x^2-1}} + C$ 11. $\dfrac{1}{432}\left[\tan^{-1}\left(\dfrac{x}{6}\right) + \dfrac{6x}{x^2+36}\right] + C$

13. $\sin^{-1}\left(\dfrac{x}{3}\right) + C$ 15. $\dfrac{1}{2(16-x^2)} + C$

17. $\dfrac{1}{243}(9x^2+49)^{3/2} - \dfrac{49}{81}\sqrt{9x^2+49} + C$

19. $\dfrac{(3+2x^2)\sqrt{x^2-3}}{27x^3} + C$ 21. $-\dfrac{8}{x^2} + 8\ln|x| + \dfrac{1}{2}x^2 + C$

23. $25\pi[\sqrt{2} - \ln(\sqrt{2}+1)] \approx 41.85$ 25. 509×10^6 km²

27. $y = \sqrt{x^2-16} - 4\sec^{-1}\dfrac{x}{4}$

29. Let $u = a\tan\theta$.   31. Let $u = a\sin\theta$.
33. Let $u = a\sec\theta$.

Exercises 7.4

Answers are expressed as sums that correspond to partial fraction decompositions. Logarithms can be combined. Thus, an equivalent answer for Exercise 1 is
$\ln|x|^3(x-4)^2 + C$.

1. $3\ln|x| + 2\ln|x-4| + C$
3. $4\ln|x+1| - 5\ln|x-2| + \ln|x-3| + C$
5. $6\ln|x-1| + \dfrac{5}{x-1} + C$
7. $3\ln|x-2| - 2\ln|x+4| + C$
9. $2\ln|x| - \ln|x-2| + 4\ln|x+2| + C$
11. $5\ln|x+1| - \dfrac{1}{x+1} - 3\ln|x-5| + C$
13. $5\ln|x| - \dfrac{2}{x} + \dfrac{3}{2x^2} - \dfrac{1}{3x^3} + 4\ln|x+3| + C$
15. $x + 4\ln|x| + \ln(x^2+4) - \dfrac{1}{2}\tan^{-1}\left(\dfrac{x}{2}\right) + C$
17. $\ln(x^2+4) + \dfrac{1}{2}\tan^{-1}\left(\dfrac{x}{2}\right) + 3\ln|x+5| + C$
19. $-\dfrac{1}{2}\ln(x^2+4) + \dfrac{1}{2}\tan^{-1}\left(\dfrac{x}{2}\right) + \dfrac{1}{2}\ln(x^2+1) + C$
21. $\ln(x^2+1) - \dfrac{4}{x^2+1} + C$
23. $\dfrac{1}{2}x^2 + x + 2\ln|x| + 2\ln|x-1| + C$
25. $\dfrac{1}{3}x^3 - 9x - \dfrac{1}{9x} - \dfrac{1}{2}\ln(x^2+9) + \dfrac{728}{27}\tan^{-1}\left(\dfrac{x}{3}\right) + C$
27. $2\ln|x+4| + \dfrac{6}{x+4} - \dfrac{5}{x-3} + C$
29. $\dfrac{13}{6}\ln(6x+5) + \dfrac{8}{3}\ln(3x-2) - \ln(2x+7) +$

$4\ln(x-1) + C$

31. $-\dfrac{34}{5}\ln(5x+2) - \dfrac{17}{3}\ln(3x+25) + \dfrac{3}{2}\ln(2x-5) + C$

33. $\dfrac{1}{2a}(\ln|a+u| - \ln|a-u|) + C = \dfrac{1}{2a}\ln\left|\dfrac{a+u}{a-u}\right| + C$

35. $-\dfrac{b}{a^2}\ln|u| - \dfrac{1}{au} + \dfrac{b}{a^2}\ln|a+bu| + C =$

$-\dfrac{1}{au} + \dfrac{b}{a^2}\ln\left|\dfrac{a+bu}{u}\right| + C$

37. $\dfrac{1}{2}\ln 3 \approx 0.55$ 39. $\dfrac{\pi}{27}(4\ln 2 + 3) \approx 0.67$

41. $\dfrac{\frac{7}{4}}{x} + \dfrac{-\frac{5}{3}}{x-1} + \dfrac{-\frac{1}{3}}{x+1} + \dfrac{\frac{29}{24}}{x-2} + \dfrac{\frac{25}{24}}{x+2}$

Exercises 7.5

1. $\dfrac{1}{2}\tan^{-1}\dfrac{x+1}{2} + C$ 3. $\dfrac{1}{2}\tan^{-1}\dfrac{x-2}{2} + C$

5. $\sin^{-1}\dfrac{x-2}{2} + C$

7. $-2\sqrt{9-8x-x^2} - 5\sin^{-1}\dfrac{x+4}{5} + C$

9. $\dfrac{1}{2}\left[\tan^{-1}(x+2) + \dfrac{x+2}{x^2+4x+5}\right] + C$

11. $\dfrac{x+3}{4\sqrt{x^2+6x+13}} + C$ 13. $\dfrac{2}{3\sqrt{7}}\tan^{-1}\dfrac{4x-3}{3\sqrt{7}} + C$

15. $\ln\left(\dfrac{e^x+1}{e^x+2}\right) + C$ 17. $1 + \dfrac{\pi}{4} \approx 1.79$ 19. $\dfrac{\pi}{20} \approx 0.16$

21. $\dfrac{3}{7}(x+9)^{7/3} - \dfrac{27}{4}(x+9)^{4/3} + C$

23. $\dfrac{5}{81}(3x+2)^{9/5} - \dfrac{5}{18}(3x+2)^{4/5} + C$

25. $2 + 8\ln\dfrac{6}{7} \approx 0.767$

27. $\dfrac{6}{7}x^{7/6} - \dfrac{6}{5}x^{5/6} + 2x^{1/2} - 6x^{1/6} + 6\tan^{-1}(x^{1/6}) + C$

29. $\dfrac{2}{\sqrt{3}}\tan^{-1}\sqrt{\dfrac{x-2}{3}} + C$

31. $\dfrac{3}{5}(x+4)^{5/3} - \dfrac{9}{2}(x+4)^{2/3} + C$

33. $\dfrac{2}{7}(1+e^x)^{7/2} - \dfrac{4}{5}(1+e^x)^{5/2} + \dfrac{2}{3}(1+e^x)^{3/2} + C$

35. $e^x - 4\ln(e^x+4) + C$

37. $2\sin\sqrt{x+4} - 2\sqrt{x+4}\cos\sqrt{x+4} + C$ 39. $\dfrac{137}{320}$

41. $\ln|\cos x| - \ln(1-\cos x) + C$

43. $\dfrac{1}{2}\ln|e^x-1| - \dfrac{1}{2}\ln(e^x+1) + C$

45. $\dfrac{4}{3}\ln(4-\sin x) + \dfrac{2}{3}\ln(\sin x + 2) + C$

47. $\dfrac{2}{\sqrt{3}}\tan^{-1}\dfrac{2\tan(x/2)+1}{\sqrt{3}} + C$ 49. $\ln\left|\tan\dfrac{x}{2} + 1\right| + C$

51. $-\dfrac{1}{5}\ln\left|2\tan\dfrac{x}{2} - 1\right| + \dfrac{1}{5}\ln\left|\tan\dfrac{x}{2} + 2\right| + C$

Exercises 7.6

1 $\sqrt{4 + 9x^2} - 2 \ln\left|\dfrac{2 + \sqrt{4 + 9x^2}}{3x}\right| + C$

3 $-\dfrac{x}{8}(2x^2 - 80)\sqrt{16 - x^2} + 96 \sin^{-1}\dfrac{x}{4} + C$

5 $-\dfrac{2}{135}(9x + 4)(2 - 3x)^{3/2} + C$

7 $-\dfrac{1}{18}\sin^5 3x \cos 3x - \dfrac{5}{72}\sin^3 3x \cos 3x -$

$$\dfrac{5}{48}\sin 3x \cos 3x + \dfrac{5}{16}x + C$$

9 $-\dfrac{1}{3}\cot x \csc^2 x - \dfrac{2}{3}\cot x + C$

11 $\dfrac{2x^2 - 1}{4}\sin^{-1} x + \dfrac{x\sqrt{1 - x^2}}{4} + C$

13 $\dfrac{1}{13}e^{-3x}(-3 \sin 2x - 2 \cos 2x) + C$

15 $\sqrt{5x - 9x^2} + \dfrac{5}{6}\cos^{-1}\dfrac{5 - 18x}{5} + C$

17 $\dfrac{1}{4\sqrt{15}}\ln\left|\dfrac{\sqrt{5}x^2 - \sqrt{3}}{\sqrt{5}x^2 + \sqrt{3}}\right| + C$

19 $\dfrac{1}{4}(2e^{2x} - 1)\cos^{-1} e^x - \dfrac{1}{4}e^x\sqrt{1 - e^{2x}} + C$

21 $\dfrac{2}{315}(35x^3 - 60x^2 + 96x - 128)(2 + x)^{3/2} + C$

23 $\dfrac{2}{81}(4 + 9 \sin x - 4 \ln|4 + 9 \sin x|) + C$

25 $2\sqrt{9 + 2x} + 3 \ln\left|\dfrac{\sqrt{9 + 2x} - 3}{\sqrt{9 + 2x} + 3}\right| + C$

27 $\dfrac{3}{4}\ln\left|\dfrac{\sqrt[3]{x}}{4 + \sqrt[3]{x}}\right| + C$

29 $\sqrt{16 - \sec^2 x} - 4 \ln\left|\dfrac{4 + \sqrt{16 - \sec^2 x}}{\sec x}\right| + C$

31 $\dfrac{1}{2}\ln(\cos x + \sin x + 1) - \dfrac{1}{2}\ln(5 \cos x + \sin x + 5) + C$

33 $e^{4x}\left[\dfrac{1}{5000}(1000x^3 - 450x^2 + 60x + 21) \sin 2x -\right.$

$$\left.\dfrac{1}{2500}(250x^3 - 300x^2 + 165x - 36) \cos 2x\right] + C$$

35 $2\sqrt{x} - \ln(x + \sqrt{x} + 3) -$

$$\dfrac{10}{11}\sqrt{11} \tan^{-1}\left(\dfrac{\sqrt{11}(2\sqrt{x} + 1)}{11}\right) + C$$

37 $\ln\left(\dfrac{1 - \cos x}{\sin x}\right) + C$ **39** $2\sqrt{x} - 2 \ln(\sqrt{x} + 1) + C$

Exercises 7.7

C denotes that the integral converges; D denotes that it diverges.

1 C; 3 **3** D **5** D **7** C; $\dfrac{1}{2}$ **9** C; $-\dfrac{1}{2}$

11 D **13** D **15** C; 0 **17** D **19** D **21** C **23** D

25 (a) Not possible **(b)** π **27** π

29 (b) No **31** If $F(x) = \dfrac{k}{x^2}$, then $W = k$.

33 (a) $\dfrac{1}{k}$ **(b)** No, the improper integral diverges.

35 (b) $c = \dfrac{4}{\sqrt{\pi}}\left(\dfrac{m}{2kT}\right)^{3/2}$ **37** $\dfrac{1}{s}, s > 0$ **39** $\dfrac{s}{s^2 + 1}, s > 0$

41 $\dfrac{1}{s - a}, s > a$

43 (a) 1; 1; 2 **(b)** *Hint:* Let $u = x^n$ and integrate by parts.

45 0.49

47 C; 6 **49** D **51** D **53** D **55** C; $3\sqrt[3]{4}$ **57** D

59 C; $\dfrac{\pi}{2}$ **61** D **63** C; $-\dfrac{1}{4}$ **65** D **67** D

69 D **71** C **73** D

75 $n > -1$ **77 (a)** 2 **(b)** Not possible **79** 1.79

81 (b) $T = 2\pi\sqrt{\dfrac{m}{k}}$ **83 (a)** t is undefined at $y = 0$.

Chapter 7 Review Exercises

1 $\dfrac{1}{2}x^2 \sin^{-1} x - \dfrac{1}{4}\sin^{-1}x + \dfrac{1}{4}x\sqrt{1 - x^2} + C$

3 $2 \ln 2 - 1 \approx 0.39$ **5** $\dfrac{1}{6}\sin^3 2x - \dfrac{1}{10}\sin^5 2x + C$

7 $\dfrac{1}{5}\sec^5 x + C$ **9** $\dfrac{x}{25\sqrt{x^2 + 25}} + C$

11 $2 \ln\left|\dfrac{2 - \sqrt{4 - x^2}}{x}\right| + \sqrt{4 - x^2} + C$

13 $2 \ln|x - 1| - \ln|x| - \dfrac{x}{(x - 1)^2} + C$

15 $-5 \ln|x - 3| + 2 \ln|x + 3| + 2 \ln(x^2 + 9) +$

$$\dfrac{1}{3}\tan^{-1}\dfrac{x}{3} + C$$

17 $-\sqrt{4 + 4x - x^2} + 2 \sin^{-1}\dfrac{x - 2}{\sqrt{8}} + C$

19 $3(x + 8)^{1/3} + \ln[(x + 8)^{1/3} - 2]^2 -$

$$\ln|(x + 8)^{2/3} + 2(x + 8)^{1/3} + 4| -$$

$$\dfrac{6}{\sqrt{3}}\tan^{-1}\dfrac{(x + 8)^{1/3} + 1}{\sqrt{3}} + C$$

21 $\dfrac{1}{13}e^{2x}(2 \sin 3x - 3 \cos 3x) + C$

23 $\dfrac{1}{4}\sin^4 x - \dfrac{1}{6}\sin^6 x + C$ **25** $-\sqrt{4 - x^2} + C$

27 $\dfrac{1}{3}x^3 - x^2 + 3x - \dfrac{1}{4}\ln|x| - \dfrac{1}{2x} - \dfrac{23}{4}\ln|x + 2| + C$

29 $2 \tan^{-1}\sqrt{x} + C$ **31** $\ln|\sec e^x + \tan e^x| + C$

33 $\dfrac{1}{125}[10x \sin 5x - (25x^2 - 2) \cos 5x] + C$

35 $\dfrac{2}{7}\cos^{7/2} x - \dfrac{2}{3}\cos^{3/2} x + C$ **37** $\dfrac{2}{3}(1 + e^x)^{3/2} + C$

39 $\dfrac{1}{16}[2x\sqrt{4x^2+25}-25\ln(\sqrt{4x^2+25}+2x)]+C$

41 $\dfrac{1}{3}\tan^3 x+C$ **43** $-x\csc x+\ln|\csc x-\cot x|+C$

45 $-\dfrac{1}{4}(8-x^3)^{4/3}+C$

47 $-2x\cos\sqrt{x}+4\sqrt{x}\sin\sqrt{x}+4\cos\sqrt{x}+C$

49 $\dfrac{1}{2}e^{2x}-e^x+\ln(1+e^x)+C$

51 $\dfrac{2}{5}x^{5/2}-\dfrac{8}{3}x^{3/2}+6x^{1/2}+C$

53 $\dfrac{1}{3}(16-x^2)^{3/2}-16(16-x^2)^{1/2}+C$

55 $\dfrac{11}{2}\ln|x+5|-\dfrac{15}{2}\ln|x+7|+C$

57 $x\tan^{-1}5x-\dfrac{1}{10}\ln(1+25x^2)+C$ **59** $e^{\tan x}+C$

61 $\dfrac{1}{\sqrt{5}}\ln|\sqrt{7+5x^2}+\sqrt{5}x|+C$

63 $-\dfrac{1}{5}\cot^5 x+\dfrac{1}{3}\cot^3 x-\cot x-x+C$

65 $\dfrac{1}{5}(x^2-25)^{5/2}+\dfrac{25}{3}(x^2-25)^{3/2}+C$

67 $\dfrac{1}{3}x^3-\dfrac{1}{4}\tanh 4x+C$

69 $-\dfrac{1}{4}x^2e^{-4x}-\dfrac{1}{8}xe^{-4x}-\dfrac{1}{32}e^{-4x}+C$

71 $3\sin^{-1}\dfrac{x+5}{6}+C$ **73** $-\dfrac{1}{7}\cos 7x+C$

75 $-9\ln|x-1|+18\ln|x-2|-5\ln|x-3|+C$

77 $x^3\sin x+3x^2\cos x-6x\sin x-6\cos x+\sin x+C$

79 $-\dfrac{\sqrt{9-4x^2}}{x}-2\sin^{-1}\left(\dfrac{2}{3}x\right)+C$

81 $24x-\dfrac{10}{3}\ln|\sin 3x|-\dfrac{1}{3}\cot 3x+C$

83 $-\ln x-\dfrac{4}{\sqrt[4]{x}}+4\ln(\sqrt[4]{x}+1)+C$

85 $-2\sqrt{1+\cos x}+C$

87 $-\dfrac{x}{2(25+x^2)}+\dfrac{1}{10}\tan^{-1}\dfrac{x}{5}+C$

89 $\dfrac{1}{3}\sec^3 x-\sec x+C$

91 $\dfrac{7}{\sqrt{5}}\tan^{-1}\left(\dfrac{x}{\sqrt{5}}\right)-\dfrac{3}{2}\tan^{-1}\left(\dfrac{x}{2}\right)+\ln(x^2+4)+C$

93 $\dfrac{1}{4}x^4-2x^2+4\ln|x|+C$

95 $\dfrac{2}{5}x^{5/2}\ln x-\dfrac{4}{25}x^{5/2}+C$

97 $\dfrac{3}{64}(2x+3)^{8/3}-\dfrac{9}{20}(2x+3)^{5/3}+\dfrac{27}{16}(2x+3)^{2/3}+C$

99 $\dfrac{1}{2}e^{(x^2)}(x^2-1)+C$

101 D **103** D **105** C; $-\dfrac{9}{2}$ **107** D

109 C; $\dfrac{\pi}{2}$ **111** D **113** 0.14

115 (a) 1 (b) $\dfrac{\pi}{32}$

117 (a) Not possible (b) Not possible

CHAPTER ▪ 8

Exercises 8.1

1 $\dfrac{1}{5},\dfrac{1}{4},\dfrac{3}{11},\dfrac{2}{7};\dfrac{1}{3}$ **3** $\dfrac{3}{5},-\dfrac{9}{11},-\dfrac{29}{21},-\dfrac{57}{35};-2$

5 $-5,-5,-5,-5;-5$ **7** $2,\dfrac{7}{3},\dfrac{25}{14},\dfrac{7}{5};0$

9 $\dfrac{2}{\sqrt{10}},\dfrac{2}{\sqrt{13}},\dfrac{2}{\sqrt{18}},\dfrac{2}{5};0$ **11** $\dfrac{3}{10},-\dfrac{6}{17},\dfrac{9}{26},-\dfrac{12}{37};0$

13 $1.1, 1.01, 1.001, 1.0001; 1$ **15** $2, 0, 2, 0;$ DNE

17 C; 0 **19** C; $\dfrac{\pi}{2}$ **21** D **23** C; 0 **25** D **27** D

29 C; e **31** C; 0 **33** C; $\dfrac{1}{2}$ **35** D **37** C; 1

39 C; 0 **41** C; 0

43 (b) 10,000 on A; 5000 on B; 20,000 on C

45 (a) The sequence appears to converge to 1.
(b) Use mathematical induction; 1

47 (a) The sequence appears to converge to approximately 0.739.

49 (a) $x_2=3.5$, $x_3=3.178571429$, $x_4=3.162319422$, $x_5=3.162277660$, $x_6=3.162277660$

51 (a) $B=\dfrac{1}{4}$ (b) 1.10

Exercises 8.2

1 (a) $-\dfrac{2}{35},-\dfrac{4}{45},-\dfrac{6}{55}$ (b) $-\dfrac{2n}{5(2n+5)}$ (c) C; $-\dfrac{1}{5}$

3 (a) $\dfrac{1}{3},\dfrac{2}{5},\dfrac{3}{7}$ (b) $\dfrac{n}{2n+1}$ (c) C; $\dfrac{1}{2}$

5 (a) $-\ln 2, -\ln 3, -\ln 4$ (b) $-\ln(n+1)$ (c) D

7 C; 4 **9** C; $\dfrac{\sqrt{5}}{\sqrt{5}+1}$ **11** C; $\dfrac{37}{99}$ **13** D **15** D

17 $-1<x<1$; $\dfrac{1}{1+x}$ **19** $1<x<5$; $\dfrac{1}{5-x}$ **21** $\dfrac{23}{99}$

23 $\dfrac{16,181}{4995}$ **25** C **27** C **29** D **31** D **33** D

35 Needs further investigation **37** D **39** D

41 C; $\dfrac{41}{24}$ **43** C; $\dfrac{6}{7}$ **45** C; $\dfrac{8}{7}$ **47** C; $\dfrac{5}{3}$

49 (a) 0.21037; 0.26720; 0.26940 (b) 0.265

51 $S_{32}\approx 4.06$

53 1.423611; 1.527422; 1.564977; 1.584347; 1.596163
55 1.040293; 1.573514; 1.921645; 2.179883; 2.385110
57 Disprove; let $a_n = 1$ and $b_n = -1$ **59** 30 m
61 (b) $\dfrac{Q}{1 - e^{-cT}}$ **(c)** $-\dfrac{1}{c}\ln\dfrac{M - Q}{M}$ **63 (b)** 2000

65 (a) $a_{k+1} = \dfrac{1}{4}\sqrt{10}\,a_k$

(b) $a_n = \left(\dfrac{1}{4}\sqrt{10}\right)^{n-1} a_1$; $A_n = \left(\dfrac{5}{8}\right)^{n-1} A_1$;

$P_n = \left(\dfrac{1}{4}\sqrt{10}\right)^{n-1} P_1$

(c) $\dfrac{16}{4 - \sqrt{10}}a_1$; $\dfrac{8}{3}a_1^2$

Exercises 8.3

Exer. 1–12: (a) Each function f is positive-valued and continuous on the interval of integration. Since $f'(x)$ is negative, f is decreasing. (b) The value of the improper integral is given, if it exists.

1 (a) $f'(x) = \dfrac{-4}{(2x + 3)^3} < 0$ if $x \geq 1$

(b) $\displaystyle\int_1^\infty f(x)\,dx = \dfrac{1}{10}$; C

3 (a) $f'(x) = \dfrac{-4}{(4x + 7)^2} < 0$ if $x \geq 1$

(b) $\displaystyle\int_1^\infty f(x)\,dx = \infty$; D

5 (a) $f'(x) = x(2 - 3x^3)e^{-x^3} < 0$ if $x \geq 1$

(b) $\displaystyle\int_1^\infty f(x)\,dx = \dfrac{1}{3e}$; C

7 (a) $f'(x) = \dfrac{1 - \ln x}{x^2} < 0$ if $x \geq 3$

(b) $\displaystyle\int_3^\infty f(x)\,dx = \infty$; D

9 (a) $f'(x) = \dfrac{1 - 2x^2}{x^2(x^2 - 1)^{3/2}} < 0$ if $x \geq 2$

(b) $\displaystyle\int_2^\infty f(x)\,dx = \dfrac{\pi}{6}$; C

11 (a) $f'(x) = \dfrac{1 - 2x \arctan x}{(1 + x^2)^2} < 0$ if $x \geq 1$

(b) $\displaystyle\int_1^\infty f(x)\,dx = \dfrac{3\pi^2}{32}$; C

Exer. 13–28: A typical b_n is listed; however, there are many other possible choices.

13 $b_n = \dfrac{1}{n^4}$; C   **15** $b_n = \dfrac{1}{3^n}$; C   **17** $b_n = \dfrac{\pi/4}{n}$; D

19 $b_n = \dfrac{1}{n^2}$; C   **21** $b_n = \dfrac{1}{\sqrt{n}}$; D   **23** $b_n = \dfrac{1}{n^{3/2}}$; C

25 $b_n = \dfrac{1}{e^n}$; C   **27** $b_n = \dfrac{1}{\sqrt{n}}$; D   **29** D   **31** C

33 D   **35** D   **37** C   **39** C   **41** C   **43** C

45 C   **47** $k > 1$ **49 (b)** $n > e^{100} - 1 \approx 2.688 \times 10^{43}$

51 Since $\displaystyle\lim_{n\to\infty} \dfrac{a_n}{b_n} = 0$, there is an M such that if $K > M$,

then $\dfrac{a_k}{b_k} < 1$, or $a_k < b_k$. Since Σb_n converges and $a_n < b_n$ for all but at most a finite number of terms, Σa_n must also converge.

53 $\displaystyle\sum_{k=1}^\infty a_k = \sum_{k=1}^n a_k + \sum_{k=n+1}^\infty a_k$, where the error

$E = \displaystyle\sum_{k=n+1}^\infty a_k < \int_n^\infty f(x)\,dx$. (See Figure 8.8.)

55 4

57 Since Σa_n converges, $\displaystyle\lim_{n\to\infty} a_n = 0$ and $\displaystyle\lim_{n\to\infty} \dfrac{1}{a_n} = \infty$. By

(8.17), $\Sigma\dfrac{1}{a_n}$ diverges.

59 $S_3 \approx 0.40488$ **61** $S_9 \approx 1.08194$ **63** $S_{21,998} \approx 0.93705$

65 The series diverges for $k = 1, 2$, and 3.

Exercises 8.4

1 $\dfrac{1}{2}$; C   **3** $\dfrac{5}{3}$; D   **5** 0; C   **7** 1; inconclusive

9 ∞; D   **11** 0; C   **13** 2; D   **15** $\dfrac{1}{3}$; C   **17** $\dfrac{1}{2}$; C

19 C   **21** C   **23** C   **25** C   **27** D   **29** C
31 D   **33** C   **35** D   **37** D   **39** D

Exercises 8.5

1 (a) Conditions (i) and (ii) are satisfied.
(b) Converges, by (8.30)
3 (a) Condition (i) is satisfied, but (ii) is not.
(b) Diverges, by (8.17)
5 CC   **7** CC   **9** D   **11** AC   **13** AC   **15** D   **17** CC
19 AC   **21** D   **23** CC   **25** D   **27** D   **29** D   **31** AC
33 0.368   **35** 0.901   **37** 0.306   **39** 141   **41** 5

45 No. If $a_n = b_n = \dfrac{(-1)^n}{\sqrt{n}}$, then both Σa_n and Σb_n

converge by the alternating series test. However,

$\Sigma a_n b_n = \Sigma\dfrac{1}{n}$, which diverges.

Exercises 8.6

1 $[-1, 1)$ **3** $(-2, 2)$ **5** $(-1, 1]$ **7** $[-1, 1)$
9 $[-1, 1]$ **11** $(-6, 14)$ **13** Converges only for $x = 0$
15 $(-2, 2)$ **7** $(-\infty, \infty)$ **19** $\left[\dfrac{17}{9}, \dfrac{19}{9}\right)$ **21** $(-12, 4)$

23 Converges only for $x = 3$ **25** $(0, 2e)$ **27** $\left(-\dfrac{5}{2}, \dfrac{7}{2}\right]$

29 $(-\infty, \infty)$

31 (a) $\dfrac{3}{2}$ **33 (a)** $\dfrac{1}{e}$ **35** ∞ **37** Use (8.35).

39 $J_0(x) \approx 1 - \dfrac{x^2}{4} + \dfrac{x^4}{64} - \dfrac{x^6}{2304}$ **41** Use (8.35).
43 Use (8.37).

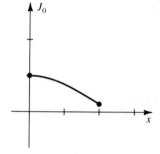

45 Assume that $\Sigma a_n x^n$ is absolutely convergent at $x = r$.
Let $x = -r$. Then $\Sigma |a_n(-r)^n| = \Sigma |a_n r^n|$ is absolutely
convergent, which implies that $\Sigma a_n(-r)^n$ is convergent.
This is a contradiction.

Exercises 8.7

1 (a) $\displaystyle\sum_{n=0}^{\infty} 3^n x^n$ **(b)** $\displaystyle\sum_{n=1}^{\infty} n 3^n x^{n-1}$; $\displaystyle\sum_{n=0}^{\infty} \dfrac{3^n}{n+1} x^{n+1}$

3 (a) $\dfrac{1}{2}\displaystyle\sum_{n=0}^{\infty} (-1)^n \left(\dfrac{7}{2}\right)^n x^n$

(b) $\dfrac{1}{2}\displaystyle\sum_{n=1}^{\infty} (-1)^n \dfrac{n7^n}{2^n} x^{n-1}$; $\dfrac{1}{2}\displaystyle\sum_{n=0}^{\infty} (-1)^n \dfrac{7^n}{(n+1)2^n} x^{n+1}$

5 $\displaystyle\sum_{n=0}^{\infty} x^{2n+2}$; $r = 1$ **7** $\displaystyle\sum_{n=0}^{\infty} \dfrac{3^n}{2^{n+1}} x^{n+1}$; $r = \dfrac{2}{3}$

9 $-1 - x - 2\displaystyle\sum_{n=2}^{\infty} x^n$; $r = 1$ **11 (b)** 0.183; 0.182321557

15 $\displaystyle\sum_{n=0}^{\infty} \dfrac{3^n}{n!} x^{n+1}$ **17** $\displaystyle\sum_{n=0}^{\infty} (-1)^n \dfrac{1}{n!} x^{n+3}$

19 $\displaystyle\sum_{n=0}^{\infty} (-1)^n \dfrac{1}{n+1} x^{2n+4}$ **21** $\displaystyle\sum_{n=0}^{\infty} (-1)^n \dfrac{1}{2n+1} x^{(2n+1)/2}$

23 $\displaystyle\sum_{n=0}^{\infty} \dfrac{-5^{2n+1}}{(2n+1)!} x^{2n+1}$ **25** $\displaystyle\sum_{n=0}^{\infty} \dfrac{1}{(2n)!} x^{6n+2}$ **27** 0.3333

29 0.0992 **31** 0.9677 **33** $\displaystyle\sum_{n=1}^{\infty} (2n) x^{2n-1}$ **37** $-\displaystyle\sum_{n=1}^{\infty} \dfrac{1}{n^2} x^n$

39 (a) $\displaystyle\sum_{n=0}^{\infty} (-1)^n \dfrac{x^{4n+1}}{4n+1}$ **41 (a)** $\displaystyle\sum_{n=0}^{\infty} (-1)^n \dfrac{x^{n+1}}{(n+1)(n+1)!}$

Exercises 8.8

1 $\dfrac{3^n}{n!}$

3 $a_n = 0$ if $n = 2k$, and $a_n = (-1)^k \dfrac{2^{2k+1}}{(2k+1)!}$ if $n = 2k+1$

5 $(-1)^n 3^n$ **7 (b)** $\displaystyle\sum_{n=0}^{\infty} (-1)^n \dfrac{1}{(2n)!} x^{2n}$

9 $\displaystyle\sum_{n=0}^{\infty} (-1)^n \dfrac{3^{2n+1}}{(2n+1)!} x^{2n+2}$ **11** $\displaystyle\sum_{n=0}^{\infty} (-1)^n \dfrac{2^{2n}}{(2n)!} x^{2n}$

13 $1 + \displaystyle\sum_{n=1}^{\infty} (-1)^n \dfrac{2^{2n-1}}{(2n)!} x^{2n}$ **15** $\displaystyle\sum_{n=0}^{\infty} \dfrac{(\ln 10)^n}{n!} x^n$

17 $\displaystyle\sum_{n=0}^{\infty} (-1)^n \dfrac{1}{\sqrt{2}(2n+1)!} \left(x - \dfrac{\pi}{4}\right)^{2n+1} +$

$\displaystyle\sum_{n=0}^{\infty} (-1)^n \dfrac{1}{\sqrt{2}(2n)!} \left(x - \dfrac{\pi}{4}\right)^{2n}$

19 $\displaystyle\sum_{n=0}^{\infty} (-1)^n \dfrac{1}{2^{n+1}} (x - 2)^n$ **21** $\displaystyle\sum_{n=0}^{\infty} \dfrac{2^n}{e^2 n!} (x + 1)^n$

23 $2 + 2\sqrt{3}\left(x - \dfrac{\pi}{3}\right) + 7\left(x - \dfrac{\pi}{3}\right)^2$

25 $\dfrac{\pi}{6} + \dfrac{2}{\sqrt{3}}\left(x - \dfrac{1}{2}\right) + \dfrac{2}{3\sqrt{3}}\left(x - \dfrac{1}{2}\right)^2$

27 $-\dfrac{1}{e} + \dfrac{1}{2e}(x + 1)^2 + \dfrac{1}{3e}(x + 1)^3$ **29** 0.5; 0.125

31 0.9986; 3.13×10^{-7} **33** 0.0997; 2×10^{-6}
35 0.6667; 0.1 **37** 0.4969; 9.04×10^{-6} **39** 0.4864
41 0.4484
43 (a) 0.309524, -0.690476

(b)

The first approximation
is more accurate.

45 $2\displaystyle\sum_{n=0}^{\infty} \dfrac{1}{2n+1} x^{2n+1}$

47 (a) $\pi = 4\left[1 - \dfrac{1}{3} + \dfrac{1}{5} - \dfrac{1}{7} + \cdots + (-1)^n \dfrac{1}{2n+1} + \cdots\right]$

(b) 3.34 with an error of less than $\dfrac{4}{11}$ **(c)** 40,000

49 (c) 16.7 ft

53 (a) $\displaystyle\sum_{n=1}^{\infty} (-1)^{n+1} \dfrac{(3x/5)^{2n-1}}{(2n-1)!}$ **55 (a)** $\displaystyle\sum_{n=0}^{\infty} (-1)^n \dfrac{x^{4n}}{(2n)!}$

57 (a) $\displaystyle\sum_{n=0}^{\infty} (-1)^n \dfrac{x^{2n+1}}{(2n+1)(2n+1)!}$

59 (a) $\displaystyle\sum_{n=1}^{\infty} (-1)^{n+1} \dfrac{x^{6n-2}}{(6n-2)(2n-1)!}$

Exercises 8.9

1 (a) x; x; $x - \dfrac{1}{6}x^3$

(b)

$P_1 = P_2$ P_3

(c) 0.0500; 2.6×10^{-7}

3 (a) x; $x - \dfrac{1}{2}x^2$; $x - \dfrac{1}{2}x^2 + \dfrac{1}{3}x^3$

(b)

(c) 0.7380; 0.164

5

$P_1(x) = x$;

$P_3(x) = x + \dfrac{1}{6}x^3$;

$P_5(x) = x + \dfrac{1}{6}x^3 + \dfrac{1}{120}x^5$

7 $\sin x = 1 - \dfrac{1}{2}\left(x - \dfrac{\pi}{2}\right)^2 + \dfrac{1}{24}\sin z \left(x - \dfrac{\pi}{2}\right)^4$,

z is between x and $\dfrac{\pi}{2}$.

9 $\sqrt{x} = 2 + \dfrac{1}{4}(x - 4) - \dfrac{1}{64}(x - 4)^2 + \dfrac{1}{512}(x - 4)^3 -$

$\dfrac{5}{128}z^{-7/2}(x - 4)^4$, z is between x and 4.

11 $\tan x = 1 + 2\left(x - \dfrac{\pi}{4}\right) + 2\left(x - \dfrac{\pi}{4}\right)^2 +$

$\dfrac{1}{3}(3\tan^4 z + 4\tan^2 z + 1)\left(x - \dfrac{\pi}{4}\right)^3$, z is between x and $\dfrac{\pi}{4}$.

13 $\dfrac{1}{x} = -\dfrac{1}{2} - \dfrac{1}{4}(x + 2) - \dfrac{1}{8}(x + 2)^2 - \dfrac{1}{16}(x + 2)^3$

$- \dfrac{1}{32}(x + 2)^4 - \dfrac{1}{64}(x + 2)^5 + z^{-7}(x + 2)^6$,

z is between x and -2.

15 $\tan^{-1} x = \dfrac{\pi}{4} + \dfrac{1}{2}(x - 1) - \dfrac{1}{4}(x - 1)^2 + \dfrac{3z^2 - 1}{3(1 + z^2)^3}(x - 1)^3$,

z is between x and 1.

17 $xe^x = -\dfrac{1}{e} + \dfrac{1}{2e}(x + 1)^2 + \dfrac{1}{3e}(x + 1)^3 + \dfrac{1}{8e}(x + 1)^4 +$

$\dfrac{ze^z + 5e^z}{120}(x + 1)^5$, z is between x and -1.

Exer. 19–30: Since $c = 0$, z is between x and 0.

19 $\ln(x + 1) = x - \dfrac{1}{2}x^2 + \dfrac{1}{3}x^3 - \dfrac{1}{4}x^4 + \dfrac{x^5}{5(z + 1)^5}$

21 $\cos x = 1 - \dfrac{x^2}{2!} + \dfrac{x^4}{4!} - \dfrac{x^6}{6!} + \dfrac{x^8}{8!} - \dfrac{\sin z}{9!}x^9$

23 $e^{2x} = 1 + 2x + 2x^2 + \dfrac{4}{3}x^3 + \dfrac{2}{3}x^4 + \dfrac{4}{15}x^5 + \dfrac{4}{45}e^{2z}x^6$

25 $\dfrac{1}{(x - 1)^2} = 1 + 2x + 3x^2 + 4x^3 + 5x^4 + 6x^5 +$

$7x^6(z - 1)^{-8}$

27 $\arcsin x = x + \dfrac{1 + 2z^2}{6(1 - z^2)^{5/2}}x^3$ **29** $f(x) = -5x^3 + 2x^4$

31 0.9998; $|R_3(x)| < 4 \times 10^{-9}$

33 2.0075; $|R_3(x)| < 3 \times 10^{-10}$

35 -0.454545; $|R_5(x)| \leq 5 \times 10^{-7}$

37 0.223; $|R_4(x)| < 2 \times 10^{-4}$

39 0.8660254; $|R_8(x)| < 8.2 \times 10^{-9}$

41 Five decimal places, since

$|R_3(x)| \leq 4.2 \times 10^{-6} < 0.5 \times 10^{-5}$

43 Three decimal places, since

$|R_2(x)| \leq 1.85 \times 10^{-4} < 0.5 \times 10^{-3}$

45 Four decimal places, since

$|R_3(x)| \leq 3.82 \times 10^{-5} < 0.5 \times 10^{-4}$

47 If f is a polynomial of degree n, then the Taylor remainder $R_n(x) = 0$, since $f^{(n+1)}(x) = 0$. By (8.45), we have $f(x) = P_n(x)$.

Chapter 8 Review Exercises

1 C; 0   **3** D   **5** C; 5   **7** The terms approach 0.589388.

9 D   **11** AC   **13** D   **15** D   **17** AC   **19** D

21 D   **23** AC   **25** CC   **27** C   **29** C   **31** C

33 CC   **35** C   **37** C   **39** D   **41** 0.158

43 $S_4 \approx 0.63092$ **45** $(-3, 3)$ **47** $[-12, -8)$ **49** $\dfrac{1}{4}$

51 $\displaystyle\sum_{n=1}^{\infty}(-1)^{n+1}\dfrac{1}{(2n)!}x^{2n-1}$; ∞

53 $\displaystyle\sum_{n=0}^{\infty}(-1)^n\dfrac{2^{2n}}{(2n + 1)!}x^{2n+1}$; ∞ **55 (a)** $\displaystyle\sum_{n=1}^{\infty}\dfrac{(3x/5)^{2n-1}}{(2n - 1)!}$

57 (a) $\displaystyle\sum_{n=0}^{\infty} \frac{x^{4n+1}}{(4n+1)(2n)!}$ **59** $e^{-x} = e^2 \displaystyle\sum_{n=0}^{\infty} (-1)^n \frac{1}{n!}(x+2)^n$

61 0.189   **63** 0.621

65 $\ln \cos x = \ln\left(\frac{1}{2}\sqrt{3}\right) - \frac{1}{3}\sqrt{3}\left(x - \frac{\pi}{6}\right) - \frac{2}{3}\left(x - \frac{\pi}{6}\right)^2 -$

$\frac{4}{27}\sqrt{3}\left(x - \frac{\pi}{6}\right)^3 - \frac{1}{12}(\sec^4 z + 2\sec^2 z \tan^2 z)\left(x - \frac{\pi}{6}\right)^4,$

z is between x and $\frac{\pi}{6}$.

67 $e^{-x^2} = 1 - x^2 + \frac{1}{6}(4z^4 - 12z^2 + 3)e^{-z^2}x^4,$

z is between x and 0.

69 0.7314

CHAPTER ▪ 9

Exercises 9.1

1 $y = 2x + 7$

3 $y = x - 2$

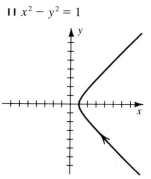

5 $(y - 3)^2 = x + 5$

7 $y = \dfrac{1}{x^2}$

9 $\dfrac{x^2}{4} + \dfrac{y^2}{9} = 1$ **11** $x^2 - y^2 = 1$

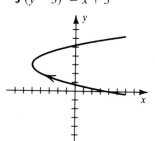

13 $y = \ln x$ **15** $y = \dfrac{1}{x}$

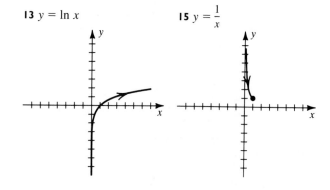

17 $x^2 - y^2 = 1$ **19** $y = \sqrt{x^2 - 1}$

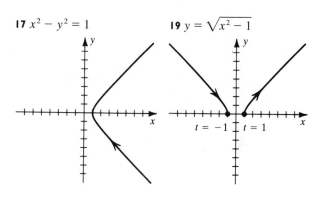

21 $y = |x - 1|$ **23** $y = (x^{1/3} + 1)^2$

23 C_1 C_2

23 C_3 C_4

41   **43**

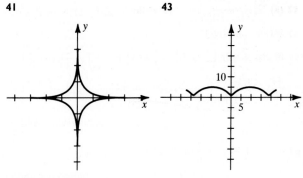

27 (a)   **(b)**

45 A mask with a mouth, nose, and eyes   **47** The letter A

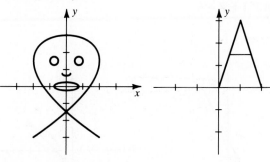

(c)

49 270 ft; yes

59 Try using: $P_0(50, 35)$, $P_1(55, \ 49)$, $P_2(15, 49)$, $P_3(35, 25)$ (repeated), $P_4(60, -5)$, $P_5(15, -5)$, $P_6(20, 15)$

Exercises 9.2

1 $1; -1$ **3** $\dfrac{1}{4}; -4$ **5** $-\dfrac{2}{e^3}; \dfrac{1}{2}e^3$

7 $-\dfrac{3}{2}\tan 1 \approx -2.34; \dfrac{2}{3}\cot 1 \approx 0.43$

9 $(-27, -108)$, $(1, 12)$

11 (a) Horizontal: $(16, \pm16)$; vertical: $(0, 0)$

     (b) $\dfrac{3t^2 + 12}{64t^3}$

     (c)

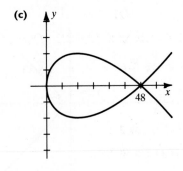

35 $x = 4b\cos t - b\cos 4t$, $y = 4b\sin t - b\sin 4t$

39 (a) The figure is an ellipse with center $(\overline{0}, 0)$ and axes of lengths $2a$ and $2b$.

13 (a) Horizontal; $(2, -1)$; vertical: $(1, 0)$ **(b)** $\dfrac{-2t + 4}{9t^5}$

(c)

15 (a) Horizontal: none; vertical: $(0, 0)$, $(-3, 1)$

(b) $\dfrac{1 - 3t}{144t^{3/2}(t - 1)^3}$

(c)

17 (a) Horizontal: $(\pm 1, 0)$; vertical: $(0, \pm 1)$

(b) $\dfrac{1}{3} \sec^4 t \csc t$

(c)

19 Horizontal: $(0, \pm 2)$, $(2\sqrt{3}, \pm 2)$, $(-2\sqrt{3}, \pm 2)$;
vertical: $(4, \pm\sqrt{2})$, $(-4, \pm\sqrt{2})$

21 $\dfrac{2}{27}(34^{3/2} - 125) \approx 5.43$ **23** $\sqrt{2}(e^{\pi/2} - 1) \approx 5.39$

25 $\dfrac{1}{8}\pi^2 \approx 1.23$ **27** 15.9 **29** $\dfrac{8\pi}{3}(17^{3/2} - 1) \approx 578.83$

31 $\dfrac{11\pi}{9} \approx 3.84$ **33** $\dfrac{64\pi}{3} \approx 67.02$ **35** $\dfrac{536\pi}{5} \approx 336.78$

37 $\dfrac{2}{5}\sqrt{2}\pi(2e^\pi + 1) \approx 84.03$ **39** 2.2

43 Arc length: 142.29; segments: 203.7

Exercises 9.3

I

3

5

7

9

II

13

15

17

19

21

23

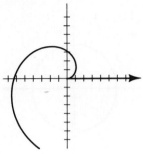

45 $y = -x^2 + 1$

47 $(x + 1)^2 + (y - 4)^2 = 17$

25

27 $r = -3 \sec \theta$ **29** $r = 4$

31 $\theta = \tan^{-1}\left(-\dfrac{1}{2}\right)$

33 $r^2 = -4 \sec 2\theta$

35 $r\theta = a \sin \theta$

49 $y^2 = \dfrac{x^4}{1 - x^2}$

51 $\sqrt{3}/3$ **53** -1

55 2   **57** 0   **59** $\dfrac{1}{\ln 2}$

37 $x = 5$

39 $y = -3$

61 Let $P_1(r_1, \theta_1)$ and $P_2(r_2, \theta_2)$ be points in an $r\theta$-plane. Let $a = r_1$, $b = r_2$, $c = d(P_1, P_2)$, and $\gamma = \theta_2 - \theta_1$. Substituting into the law of cosines, $c^2 = a^2 + b^2 - 2ab \cos \gamma$, gives us the formula.

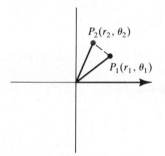

65 Use (9.11).

67 Symmetric with respect to the polar axis

41 $x^2 - y^2 = 1$

43 $y - 2x = 6$

69 The approximate polar coordinates are $(1.75, \pm 0.45)$, $(4.49, \pm 1.77)$, and $(5.76, \pm 2.35)$.

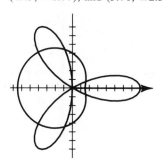

Exercises 9.4

1 π **3** $\dfrac{3\pi}{2}$ **5** $\dfrac{\pi}{2}$ **7** $\dfrac{1}{4}(e^{\pi} - 1) \approx 5.54$ **9** 2

11 $\dfrac{9\pi}{20}$ **13** $\displaystyle\int_{0}^{\arctan(3/4)} \dfrac{1}{2}(4 \sec \theta)^2 \, d\theta + \int_{\arctan(3/4)}^{\pi/2} \dfrac{1}{2}(5)^2 \, d\theta$

15 $\displaystyle\int_{\pi/4}^{\arctan 3} \dfrac{1}{2}[(4 \csc \theta)^2 - (2)^2] \, d\theta$

17 (a) $8 \displaystyle\int_{0}^{\pi/6} \dfrac{1}{2}[(4 \cos 2\theta)^2 - (2)^2] \, d\theta$

(b) $8 \left[\displaystyle\int_{0}^{\pi/6} \dfrac{1}{2}(2)^2 \, d\theta + \int_{\pi/6}^{\pi/4} \dfrac{1}{2}(4 \cos 2\theta)^2 \, d\theta \right]$

19 $2\pi + \dfrac{9}{2}\sqrt{3} \approx 14.08$ **21** $4\sqrt{3} - \dfrac{4\pi}{3} \approx 2.74$

23 $\dfrac{5\pi}{24} - \dfrac{1}{4}\sqrt{3} \approx 0.22$

25 $\dfrac{3\pi}{4} + 11 \arcsin \dfrac{1}{4} - \dfrac{1}{4}\sqrt{15} \approx 4.17$

27 $\sqrt{2}(1 - e^{-2\pi}) \approx 1.41$ **29** 2   **31** $\dfrac{3\pi}{2}$ **33** 2.4

35 $\dfrac{128\pi}{5} \approx 80.42$ **37** $4\pi^2 a^2$ **39** 4.2   **41** $4\pi^2 ab$

43 $\dfrac{2}{5}\pi\sqrt{2}\,(2 + e^{-\pi}) \approx 3.63$

Exercises 9.5

1 $V(2, -4)$; $F(4, -4)$ **3** $V(-3 \pm 3, 2)$; $F(-3 \pm \sqrt{5}, 2)$

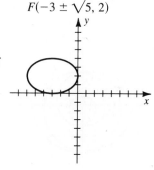

5 $V(-4 \pm 1, 5)$; $F\left(-4 \pm \dfrac{1}{3}\sqrt{10}, 5\right)$

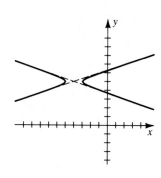

Exer. 7–19: The answer in part (a) gives the value of $B^2 - 4AC$ in the identification theorem.

7

(a) 0, parabola
(b) $(y')^2 = 2(x')$

9 (a) -36, ellipse
(b) $\dfrac{(x')^2}{9} + (y')^2 = 1$

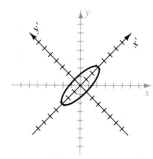

11 (a) 256, hyperbola
(b) $\dfrac{(x')^2}{1/4} - (y')^2 = 1$

13 (a) 0, parabola
(b) $(y')^2 = 4(x' - 1)$

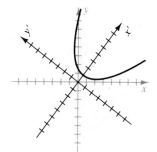

15 (a) -2704, ellipse
(b) $\dfrac{(x' - 2)^2}{4} + (y')^2 = 1$

17 (a) 128, hyperbola

(b) $(y' + 2)^2 - \dfrac{(x')^2}{1/2} = 1$

19 (a) −1600, ellipse

(b) $\dfrac{(x')^2}{16} + (y')^2 = 1$

C_3 C_4

21

7 (a) $\dfrac{3t^2 + 2}{t}$ **(b)** Horizontal: none; vertical: 0

(c) $\dfrac{3t^2 - 2}{2t^3}$

9   **11**

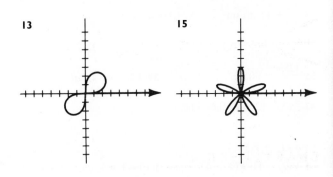

Chapter 9 Review Exercises

1 (a) $y = \dfrac{2x^2 - 4x + 1}{x - 1}$

(b)

$\left(\dfrac{5}{4}, -\dfrac{7}{2}\right)$

3 (a) $y = 2^{-x^2}$

(b)

13   **15**

5 C_1 C_2

17   **19**

21

23

5 $\sqrt{41}$

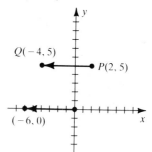
$\langle -4, 5 \rangle$

25

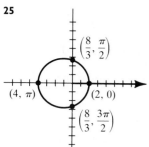
$\left(\frac{8}{3}, \frac{\pi}{2} \right)$
$(4, \pi)$ $(2, 0)$
$\left(\frac{8}{3}, \frac{3\pi}{2} \right)$

27 $r = 4 \cot \theta \csc \theta$
29 $r(2 \cos \theta - 3 \sin \theta) = 8$
31 $r = 2 \cos \theta \sec 2\theta$
33 $x^3 + xy^2 = y$
35 $(x^2 + y^2)^2 = 8xy$
37 $y = (\tan \sqrt{3})x$
39 -1
41 2

7 $\langle 3, 1 \rangle, \langle 1, -7 \rangle, \langle 4, -6 \rangle, \langle -3, -12 \rangle, \langle 3, -32 \rangle$
9 $\langle -15, 6 \rangle, \langle 1, -2 \rangle, \langle -14, 4 \rangle, \langle 24, -12 \rangle, \langle 12, -12 \rangle$
11 $2\mathbf{i} + 7\mathbf{j}, 4\mathbf{i} - 3\mathbf{j}, 6\mathbf{i} + 4\mathbf{j}, 3\mathbf{i} - 15\mathbf{j}, 17\mathbf{i} - 17\mathbf{j}$
13 $-\mathbf{b}$ **15** $\mathbf{f}$ **17** $-\frac{1}{2}\mathbf{e}$

19 $\langle 4, 7 \rangle$

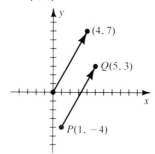
$(4, 7)$
$Q(5, 3)$
$P(1, -4)$

21 $\langle -6, 0 \rangle$

$Q(-4, 5)$ $P(2, 5)$
$(-6, 0)$

43 $\sqrt{2} + \ln(1 + \sqrt{2}) \approx 2.30$
45 $2\pi[5\sqrt{2} + \ln(1 + \sqrt{2})] \approx 49.97$
47 $2\pi a^2(2 - \sqrt{2}) \approx 3.68a^2$ **49** $V(2, 1); F(1, 1)$
51 $V(-4 \pm 3, 0);$ **53** $(y')^2 = 3x'$
   $F(-4 \pm \sqrt{10}, 0)$

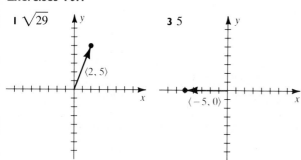

23 $\langle 9, -3 \rangle$

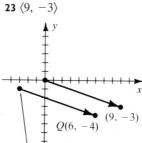
$Q(6, -4)$ $(9, -3)$
$P(-3, -1)$

25 (a) $-\frac{8}{17}\mathbf{i} + \frac{15}{17}\mathbf{j}$

(b) $\frac{8}{17}\mathbf{i} - \frac{15}{17}\mathbf{j}$

27 (a) $\left\langle \frac{2}{\sqrt{29}}, -\frac{5}{\sqrt{29}} \right\rangle$

(b) $\left\langle -\frac{2}{\sqrt{29}}, \frac{5}{\sqrt{29}} \right\rangle$

CHAPTER ■ 10

Exercises 10.1

1 $\sqrt{29}$

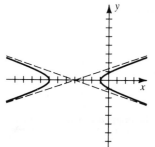
$\langle 2, 5 \rangle$

3 5

$\langle -5, 0 \rangle$

29 (a) $\langle -12, 6 \rangle$ (b) $\left\langle -3, \frac{3}{2} \right\rangle$ **31** $\frac{24}{\sqrt{65}}\mathbf{i} - \frac{42}{\sqrt{65}}\mathbf{j}$

33 (a) $\pm\frac{3}{5}$ (b) None (c) 0 **35** $40.96; 28.68$

37 $-6.180; 19.021$ **39** 301.5 mi/hr; $144.3°$
41 $\sin^{-1}(0.4) \approx 23.6°$ **43** $p = 2; q = -3$
45 If $\mathbf{a}$ and $\mathbf{b}$ have the same direction
47 The circle with center (x_0, y_0) and radius c

Exercises 10.2

1 (a) $\sqrt{104}$ (b) $(3, 1, -1)$ (c) $\langle 2, -6, 8 \rangle$

3 (a) $\sqrt{53}$ (b) $\left(-\frac{1}{2}, -1, 1 \right)$ (c) $\langle 7, -2, 0 \rangle$

5 (a) $\sqrt{3}$ **(b)** $\left(\frac{1}{2}, \frac{1}{2}, \frac{1}{2}\right)$ **(c)** $\langle -1, 1, 1\rangle$

7 (a) $\langle 1, 3, 0\rangle$ **(b)** $\langle -5, 9, 2\rangle$ **(c)** $\langle -22, 42, 9\rangle$
   (d) $\sqrt{41}$ **(e)** $3\sqrt{41}$

9 (a) $4\mathbf{i} - 2\mathbf{j} - 3\mathbf{k}$ **(b)** $2\mathbf{i} - 6\mathbf{j} + 7\mathbf{k}$
   (c) $11\mathbf{i} - 28\mathbf{j} + 30\mathbf{k}$ **(d)** $\sqrt{29}$ **(e)** $3\sqrt{29}$

11 (a) $\mathbf{i} + \mathbf{k}$ **(b)** $\mathbf{i} + 2\mathbf{j} - \mathbf{k}$ **(c)** $5\mathbf{i} + 9\mathbf{j} - 4\mathbf{k}$ **(d)** $\sqrt{2}$
   (e) $3\sqrt{2}$

13

15 $\dfrac{1}{\sqrt{30}} \langle -2, 5, -1\rangle$

17 (a) $28\mathbf{i} - 30\mathbf{j} + 12\mathbf{k}$ **(b)** $-\dfrac{14}{3}\mathbf{i} + 5\mathbf{j} - 2\mathbf{k}$

   (c) $\dfrac{2}{\sqrt{457}}(14\mathbf{i} - 15\mathbf{j} + 6\mathbf{k})$

19 $(x - 3)^2 + (y + 1)^2 + (z - 2)^2 = 9$

21 $(x + 5)^2 + y^2 + (z - 1)^2 = \dfrac{1}{4}$

23 (a) $(x + 2)^2 + (y - 4)^2 + (z + 6)^2 = 36$
   (b) $(x + 2)^2 + (y - 4)^2 + (z + 6)^2 = 16$
   (c) $(x + 2)^2 + (y - 4)^2 + (z + 6)^2 = 4$

25 $(x + 3)^2 + \left(y - \dfrac{5}{2}\right)^2 + z^2 = \dfrac{89}{4}$

27 $(x - 1)^2 + (y - 1)^2 + (z - 1)^2 = 1$

29 $(-2, 1, -1); 2$ **31** $(4, 0, -4); 4$ **33** $(0, -2, 0); 2$

35 All points inside or on the sphere of radius 1 with center at the origin

37 All points inside or on a rectangular box with center at the origin and having edges of lengths 2, 4, and 6 in the $x, y,$ and z directions, respectively

39 All points inside or on a cylindrical region of radius 5 and altitude 6 with center at the origin and axis along the z-axis

41 All points not on a coordinate plane

43 *Hint*: P is $\left(\dfrac{x_1 + x_2 + x_3}{4}, \dfrac{y_1 + y_2 + y_3}{4}, \dfrac{z_4}{4}\right)$.

Exercises 10.3

1 3   **3 (a)** -12 **(b)** -12 **5** -99 **7** $-\dfrac{3}{\sqrt{30}}$

9 0   **11** $\arccos\dfrac{-3}{\sqrt{534}} \approx 97.5°$ **13** $\arccos\dfrac{6}{13} \approx 62.5°$

15 *Hint*: Use Theorem (10.21).   **17** $-3, 5$ **19** 74

21 $\arccos\dfrac{37}{\sqrt{3081}} \approx 48.2°$ **23** $\dfrac{-82}{\sqrt{126}}$ **25** 1

27 $-4\sqrt{3} \approx 6.93$ ft-lb   **29** $1000\sqrt{3} \approx 1732$ ft-lb

35 (a) *Hint*: $a_1 = \mathbf{a} \cdot \mathbf{i} = \|\mathbf{a}\| \|\mathbf{i}\| \cos \alpha$.

47 When $\mathbf{a}$ and $\mathbf{b}$ have the same or opposite direction

Exercises 10.4

1 $\langle 5, 10, 5\rangle$ **3** $\langle -4, 2, -1\rangle$ **5** $-6\mathbf{i} - 8\mathbf{j} + 18\mathbf{k}$

7 $0\mathbf{i} + 0\mathbf{j} + 0\mathbf{k} = \mathbf{0}$ **9** $0\mathbf{i} + 0\mathbf{j} + 0\mathbf{k} = \mathbf{0}$

11 *Hint*: Use Corollary (10.31).

13 $\langle 12, -14, 24\rangle; \langle 16, -2, -5\rangle$

Exer. 15–18: c is a nonzero scalar.

15 (a) $c\langle 13, 7, 5\rangle$ **(b)** $\dfrac{9}{2}\sqrt{3}$

17 (a) $c\langle -10, -8, -20\rangle$ **(b)** $\sqrt{141}$

19 $\sqrt{\dfrac{282}{11}} \approx 5.06$

23 4

Exercises 10.5

In answers it is assumed that the domain of each parameter is $\mathbb{R}$.

1 $x = 4 + \dfrac{1}{3}t, y = 2 + 2t, z = -3 + \dfrac{1}{2}t$

3 $x = 0, y = t, z = 0$

5 $x = 5 - 3t, y = -2 + 8t, z = 4 - 3t;$
   $\left(1, \dfrac{26}{3}, 0\right), \left(\dfrac{17}{4}, 0, \dfrac{13}{4}\right), \left(0, \dfrac{34}{3}, -1\right)$

7 $x = 2 - 8t, y = 0, z = 5 - 2t;$
   $(-18, 0, 0),$ lies in xz-plane, $\left(0, 0, \dfrac{9}{2}\right)$

9 $x = -6 - 3s, y = 4 + s, z = -3 + 9s$ **11** $(5, -7, 3)$

13 Do not intersect

15 $\theta = \cos^{-1}\left(\dfrac{9}{\sqrt{38}\sqrt{33}}\right) \approx 75°$ and $180° - \theta$

17 $\theta = \cos^{-1}\left(\dfrac{71}{\sqrt{82}\sqrt{249}}\right) \approx 60°$ and $180° - \theta$

19 (a) $z = 4$ **(b)** $x = 6$ **(c)** $y = -7$

21 $6x - 5y - z = -84$ **23** $3x - y + 2z = -11$
25 $x + 42y - 5z = 8$ **27** $2x + y - 2z = -3$
29 (a)

(b)

(c)

31

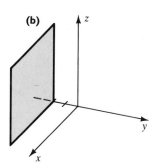

$(3, 0, 0)$ $(0, 6, 0)$

33

$(0, \frac{9}{2}, 0)$
$(0, 0, -3)$

35

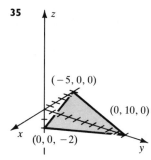

$(-5, 0, 0)$
$(0, 10, 0)$
$(0, 0, -2)$

37 $x + z = 5$
39 $3x + 2y = 6$
41 $4x - y + 3z + 7 = 0$
43 $\dfrac{x - 5}{-3} = \dfrac{y + 2}{8} = \dfrac{z - 4}{-3}$
45 $\dfrac{x - 4}{-7} = \dfrac{z + 3}{8}; y = 2$

47 $x = 3 - t, y = 2 + 5t, z = t$
49 $x = 3t, y = 4 - t, z = t$
51 $\dfrac{7}{\sqrt{59}} \approx 0.91$ **53** $\dfrac{17}{6\sqrt{14}} \approx 0.76$ **55** $\dfrac{89}{\sqrt{521}} \approx 3.90$
57 $6x + 11y + 4z = 38$ **59** $\sqrt{\dfrac{5411}{90}} \approx 7.75$
61 $\sqrt{\dfrac{474}{17}} \approx 5.28$ **63** $\dfrac{x}{3} + \dfrac{y}{-2} + \dfrac{z}{5} = 1$
65 $6x + 4y + 3z = 12$

67

(a) $(0.55, 0.30)$
(b) $103°; 77°$

Exercises 10.6

1

3

5

7

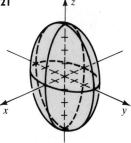

9 K   **11** C   **13** Q
15 P   **17** A   **19** E

21

23 (a)

(b)

25 (a)   **(b)**

33 Hyperboloid of one sheet   **35** Paraboloid

27 (a)   **(b)**

37 Cone   **39** Ellipsoid

29 (a)   **(b)**

41 Exponential cylinder   **43** Plane

31 (a)   **(b)**

45 Hyperboloid of two sheets   **47**

49

51 $x^2 + z^2 + 4y^2 = 16$ **53** $z = 4 - x^2 - y^2$
55 $y^2 + z^2 - x^2 = 1$
57 (a) The Clarke ellipsoid is flatter at the north and south poles.
   (b) Ellipses   **(c)** Ellipses

Chapter 10 Review Exercises

1 $5\mathbf{i} - 13\mathbf{j} - 8\mathbf{k}$ **3** $3\sqrt{33}$ **5** 26

7 $\arccos \dfrac{-27}{\sqrt{962}} \approx 150.52°$ **9** $\dfrac{1}{\sqrt{26}}(3\mathbf{i} - \mathbf{j} - 4\mathbf{k})$

11 $22\mathbf{i} - 2\mathbf{j} + 17\mathbf{k}$ **13** $\dfrac{9}{\sqrt{33}} \approx 1.57$ **15** 156 **17** 0

19 80 **21** *Hint*: Use Theorem (10.21).

23 (a) $\sqrt{38}$

   (b) $\left(2, -\dfrac{7}{2}, \dfrac{5}{2}\right)$

   (c) $(x + 1)^2 + (y + 4)^2 + (z - 3)^2 = 16$
   (d) $y = -4$
   (e) $x = 5 + 6t, y = -3 + t, z = 2 - t$
   (f) $6x + y - z = 25$
25 $6x - 15y + 5z = 30$
27 $x = -13t + 5, y = 6t - 2, z = 5t$

29 $4x + 3y - 4z = 11$ **31** $\dfrac{x^2}{64} + \dfrac{y^2}{9} + z^2 = 1$

33 (a) $\dfrac{1}{\sqrt{66}}\langle 1, 4, 7 \rangle$ **(b)** $x + 4y + 7z = 5$

   (c) $x = 2 + 7t, y = -1 - 7t, z = 1 + 3t$ **(d)** 59

   (e) $\arccos \dfrac{59}{\sqrt{3745}} \approx 15.40°$

   (f) $\sqrt{66} \approx 8.12$ **(g)** $\sqrt{\dfrac{264}{35}} \approx 2.75$

35 $x = 3 + 2t, y = -1 - 4t, z = 5 + 8t$;

   $x = -1 + 7t, y = 6 - 2t, z = -\dfrac{7}{2} - 2t$

37 $\theta = \arccos \dfrac{-25}{\sqrt{2295}} \approx 121.46°$ and $180° - \theta$

39

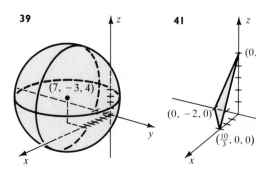

41

$(0, 0, 5)$
$(0, -2, 0)$
$\left(\dfrac{10}{3}, 0, 0\right)$

43

45

47

49

CHAPTER ▪ 11

Exercises 11.1

1

$\mathbf{r}(0)$
$\mathbf{r}(1)$

3

$(-9, 6)$ $(7, 6)$
$\mathbf{r}(1)$ $\mathbf{r}(2)$

5

7

$r(\pi/2)$

$r(0)$

9

11

13

15

17

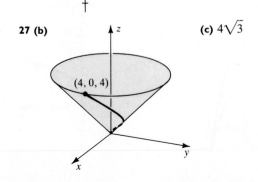

$21 \frac{5}{4}[4\sqrt{17} + \ln(4 + \sqrt{17})]$

≈ 23.2339

$23 \sqrt{3}(e^{2\pi} - 1) \approx 925.7667$

$25\ 8.2182$

27 (b)

$(4, 0, 4)$

(c) $4\sqrt{3}$

29 (a) *Hint*: Let $Ax + By + Cz = D$ be the equation of an arbitrary plane.   **(b)** 7

Exercises 11.2

1 (a) $[1, 2]$

 (b) $\frac{1}{2}(t - 1)^{-1/2}\mathbf{i} - \frac{1}{2}(2 - t)^{-1/2}\mathbf{j};$

 $-\frac{1}{4}(t - 1)^{-3/2}\mathbf{i} - \frac{1}{4}(2 - t)^{-3/2}\mathbf{j}$

3 (a) $\left\{t: t \neq \frac{\pi}{2} + \pi n\right\}$

 (b) $\sec^2 t\,\mathbf{i} + (2t + 8)\mathbf{j};\ 2\sec^2 t\tan t\,\mathbf{i} + 2\mathbf{j}$

5 (a) $\left\{t: t \neq \frac{\pi}{2} + \pi n\right\}$

 (b) $2t\,\mathbf{i} + \sec^2 t\,\mathbf{j};\ 2\mathbf{i} + 2\sec^2 t\tan t\,\mathbf{j}$

7 (a) $\{t: t \geq 0\}$ **(b)** $\frac{1}{2\sqrt{t}}\mathbf{i} + 2e^{2t}\mathbf{j} + \mathbf{k};\ -\frac{1}{4t\sqrt{t}}\mathbf{i} + 4e^{2t}\mathbf{j}$

9 $-t^3\mathbf{i} + 2t\mathbf{j}$

11 $-4\sin t\,\mathbf{i} + 2\cos t\,\mathbf{j}$

13 $3t^2\mathbf{i} - 3t^{-4}\mathbf{j}$

15 $2\mathbf{i} - \mathbf{j}$

17 $x = 1 + 6t,\ y = -2 - 10t,\ z = 10 + 8t$

19 $x = 1 + t,\ y = t,\ z = 4$ **21** $\pm\frac{1}{\sqrt{5}}(2\mathbf{i} - \mathbf{j})$

25 $x = ae^t,\ y = be^t,$ and $z = ce^t$ for constants a, b, and c; the graph is a half-line with endpoint O deleted.

27 $16\mathbf{i} - 8\mathbf{j} + 6\mathbf{k}$ **29** $\left(1 - \frac{1}{\sqrt{2}}\right)\mathbf{i} - \frac{1}{\sqrt{2}}\mathbf{j} + (\ln\sqrt{2})\mathbf{k}$

31 $\left(\frac{1}{3}t^3 + 2\right)\mathbf{i} + (3t^2 + t - 3)\mathbf{j} + (2t^4 + 1)\mathbf{k}$

33 $(t^3 + t + 7)\mathbf{i} + (2t - t^4)\mathbf{j} + \left(\frac{1}{2}t^2 - 3t + 1\right)\mathbf{k}$

35 $x + y = 1$

37 $(1 + 5t^2) \sin t + (2t^3 + 3t) \cos t$;
$[(t^3 + 4t) \sin t - t^2 \cos t]\mathbf{i} +$
$\quad [(3t^2 - 2) \sin t + (t^3 - 2t) \cos t]\mathbf{j} +$
$\quad\quad [-3t \sin t + (1 - t^2) \cos t]\mathbf{k}$

Exercises 11.3

1

3

5

7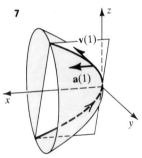

9 $-\dfrac{1}{2}\mathbf{i} - \dfrac{1}{3}\mathbf{j}; \dfrac{1}{2}\mathbf{i} + \dfrac{2}{9}\mathbf{j}; \dfrac{1}{6}\sqrt{13}$ **11** $2\mathbf{i} - \mathbf{j}; 4\mathbf{i} + \mathbf{j}; \sqrt{5}$

13 $e^{\pi/2}(-\mathbf{i} + \mathbf{j} + \mathbf{k}); e^{\pi/2}(-2\mathbf{i} + \mathbf{k}); \sqrt{3}e^{\pi/2}$

15 $\mathbf{i} + 2\mathbf{j} + 3\mathbf{k}; \mathbf{0}; \sqrt{14}$ **19** 4.7

21 (a) 18,054 mi/hr   **(b)** 86.7 min

23 (a) $750\sqrt{3}\mathbf{i} + (-gt + 750)\mathbf{j}$ **(b)** $\dfrac{(1500)^2}{8g} \approx 8789$ ft

(c) $\dfrac{(1500)^2\sqrt{3}}{2g} \approx 60{,}892$ ft   **(d)** 1500 ft/sec

25 $\sqrt{250g} \approx 89.4$ ft/sec   **27** 0.46 rev/sec

31 2.51 ft   **33** 0.14 (mi/hr)/ft

Exercises 11.4

1 (a) $\dfrac{1}{(1 + t^2)^{1/2}}\mathbf{i} - \dfrac{t}{(1 + t^2)^{1/2}}\mathbf{j}; -\dfrac{t}{(1 + t^2)^{1/2}}\mathbf{i} - \dfrac{1}{(1 + t^2)^{1/2}}\mathbf{j}$

(b)

3 (a) $\dfrac{t^2}{(t^4 + 1)^{1/2}}\mathbf{i} + \dfrac{1}{(t^4 + 1)^{1/2}}\mathbf{j}; \dfrac{1}{(t^4 + 1)^{1/2}}\mathbf{i} - \dfrac{t^2}{(t^4 + 1)^{1/2}}\mathbf{j}$

(b)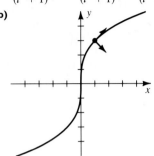

5 (a) $\cos t\,\mathbf{i} - \sin t\,\mathbf{k}; -\sin t\,\mathbf{i} - \cos t\,\mathbf{k}$

(b)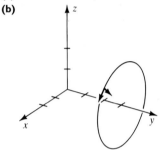

7 $\dfrac{6}{10^{3/2}} \approx 0.19$ **9** 2   **11** 4   **13** $\dfrac{2}{17^{3/2}} \approx 0.03$

15 0   **17** $\dfrac{48}{21^{3/2}} \approx 0.50$

19 (a) 1   **(b)** $\left(\dfrac{\pi}{2}, 0\right)$

(c)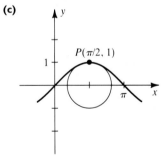

21 (a) $2\sqrt{2}$ **(b)** $(-2, 3)$

(c)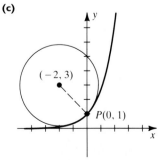

23 0.6439   **25** 0.6034

27 $\left(\ln\sqrt{2}, \dfrac{1}{\sqrt{2}}\right)$ **29** $(0, \pm3)$ **31** $\left(\dfrac{1}{\sqrt{2}}, -\dfrac{1}{2}\ln 2\right)$

33 $(\pm\sqrt{2}, -20)$ **35** $(0, 0)$ **39** $\dfrac{8 - 3\sin^2 2\theta}{(1 + 3\cos^2 2\theta)^{3/2}}$

43 $\left(-\dfrac{14}{3}, \dfrac{29}{12}\right)$ **45** $(4, -2)$ **47** $(0, -4)$

53 $x = \dfrac{4}{5}s - 3, \; y = \dfrac{3}{5}s + 5; \; s \geq 0$

55 $x = 4\cos\dfrac{1}{4}s, \; y = 4\sin\dfrac{1}{4}s; \; 0 \leq s \leq 8\pi$

Exercises 11.5

1 $\dfrac{4t}{(4t^2 + 9)^{1/2}}; \dfrac{6}{(4t^2 + 9)^{1/2}}; \dfrac{6}{(4t^2 + 9)^{3/2}}$

3 $\dfrac{6t(t^2 + 2)}{(t^4 + 4t^2 + 1)^{1/2}}; \dfrac{6(t^4 + t^2 + 1)^{1/2}}{(t^4 + 4t^2 + 1)^{1/2}}; \dfrac{2(t^4 + t^2 + 1)^{1/2}}{3(t^4 + 4t^2 + 1)^{3/2}}$

5 $\dfrac{t}{(1 + t^2)^{1/2}}; \dfrac{2 + t^2}{(1 + t^2)^{1/2}}; \dfrac{2 + t^2}{(1 + t^2)^{3/2}}$

7 $\dfrac{-65\sin t \cos t}{(16\sin^2 t + 81\cos^2 t + 1)^{1/2}};$

$\dfrac{(81\sin^2 t + 16\cos^2 t + 1296)^{1/2}}{(16\sin^2 t + 81\cos^2 t + 1)^{1/2}};$

$\dfrac{(81\sin^2 t + 16\cos^2 t + 1296)^{1/2}}{(16\sin^2 t + 81\cos^2 t + 1)^{3/2}}$

9 $\dfrac{36}{\sqrt{5}} \approx 16.10; \dfrac{18}{\sqrt{5}} \approx 8.05$

Chapter 11 Review Exercises

1

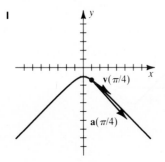

3 $2t\mathbf{i} + (8t - 4t^3)\mathbf{j}; \; 2\mathbf{i} + (8 - 12t^2)\mathbf{j};$
$2|t|\sqrt{17 - 16t^2 + 4t^4}$

5 (a) $\dfrac{1}{\sqrt{3}}(\mathbf{i} + \mathbf{j} + \mathbf{k})$ (b) $x = t, \; y = 1 + t, \; z = 1 + t$

7 $0, \dfrac{10}{3}$

9 $3\mathbf{i} + 3t^2\mathbf{j} + 4t^3\mathbf{k}, \; 6t\mathbf{j} + 12t^2\mathbf{k}, \; \sqrt{9 + 9t^4 + 16t^6};$
$3\mathbf{i} + 3\mathbf{j} + 4\mathbf{k}, \; 6\mathbf{j} + 12\mathbf{k}, \; \sqrt{34}$

11 $2\mathbf{i} + \dfrac{1}{4}\mathbf{j} - \mathbf{k}$

13 $\mathbf{r}(t) = (3t - 4)\mathbf{i} + (2t^3 - 8t + 9)\mathbf{j} + \left(\dfrac{5}{2}t^2 - t\right)\mathbf{k}$

17 $\dfrac{108}{82^{3/2}} \approx 0.15$

19 $\pm\sqrt{\dfrac{36 + \sqrt{3096}}{90}} \approx \pm1.009$ **21** $\dfrac{5}{6}\sqrt{5} \approx 1.86$

23 $\dfrac{-8\cos 2t \sin 2t + \sin t \cos t}{(4\cos^2 2t + \sin^2 t)^{1/2}}; \dfrac{2|\cos 2t \cos t + 2\sin 2t \sin t|}{(4\cos^2 2t + \sin^2 t)^{1/2}}$

25 2π

CHAPTER ▪ 12

Exercises 12.1

1 $\mathbb{R}^2; \; -29, 6, -4$ **3** $\{(u, v): u \neq 2v\}; -\dfrac{3}{2}, \dfrac{4}{9}, 0$

5 $\{(x, y, z): x^2 + y^2 + z^2 \leq 25\}; \; 4, 2\sqrt{3}$

7   **9**

11   **13**

21   **23**

25

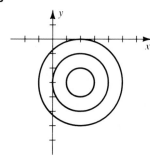

27 $y \arctan x = \pi$ **29** $x^2 + 4y^2 - z^2 = -1$
31 (a) Ellipses   **(b)** None   **(c)** None
33 (a) $k > 8$: none; $k = 8$: the point $(0, 0, 8)$; $0 < k < 8$:
   circles
   (b) None   **(c)** None
35 $k = \dfrac{1}{4}, \dfrac{1}{2}, \dfrac{3}{4}, 1$
37 $k = 1.1, 1.2, 1.4, 1.6, 1.8$
39 (c)   **41 (b)**   **43 (c)**
45

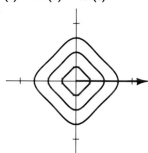

47 None; the origin; the sphere with center $(0, 0, 0)$ and
   radius 2
49 Planes with x-intercept k, y-intercept $\dfrac{k}{2}$, and z-intercept $\dfrac{k}{3}$
51 None; the z-axis; the right circular cylinder with the z-
   axis as its axis and radius 2
53 (a) Circles with center at the origin   **(b)** $x^2 + y^2 = 100$
55 Five; spheres with centers at the origin; the force F is
   constant if (x, y, z) moves along a level surface.
57 (a) $P = kAv^3$ for $k > 0$
   (b) A typical level curve (see figure) shows the
   combinations of areas and wind velocities that result
   in a fixed power $P = c$.

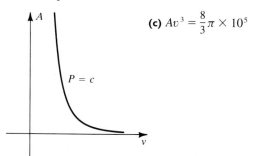

   (c) $Av^3 = \dfrac{8}{3}\pi \times 10^5$

59 *Example*: 5'11" and 175 lb are approximately 180 cm
   and 80 kg. From the graph, we have a surface area of
   approximately 2.0 m². Using the formula, we obtain
   $S \approx 1.996$ m².

Exercises 12.2

1 $-\dfrac{2}{3}$ **3** 1   **5** 0   **7** 0   **9** 4

Exer. 11–20: The answer gives equations of possible
paths, and their resulting values, to use in (12.4).

11 $x = 0, -\dfrac{1}{2}; y = 0, 2$ **13** $y - 2 = m(x - 1), \dfrac{m}{1 + m^2}$
15 $y = mx, \dfrac{4m}{2 + 3m^4}$ **17** $x = y = 0, 0; x = y = z, 1$
19 $x = -3 + at, y = bt, z = ct, \dfrac{(a - c)^2}{a^2 + b^2 + c^2}$ **21** 0   **23** 1
25 $\{(x, y): x + y > 1\}$ **27** $\{(x, y): x \geq 0 \text{ and } |y| \leq 1\}$
29 $\{(x, y, z): z^2 \neq x^2 + y^2\}$ **31** $\{(x, y, z): x \geq 2, yz > 0\}$
33 0   **35** $\dfrac{x^4 - 2x^2y^2 + y^4 - 4}{x^2 - y^2}; \{(x, y): x^2 \neq y^2\}$
37 $x^2 + 2x \tan y + \tan^2 y + 1; \left\{(x, y): y \neq \dfrac{\pi}{2} + \pi n\right\}$
39 $e^{x^2+2y}; (x^2 + 2y)(x^2 + 2y - 3); e^{2t} + 2(t^2 - 3t)$
41 $2x^2 - 3xy - 2y^2 - x + 7y$
43 The statement $\displaystyle\lim_{(x,y,z,w)\to(a,b,c,d)} f(x, y, z, w) = L$ means
   that for every $\epsilon > 0$ there is a $\delta > 0$ such that if
   $0 < \sqrt{(x - a)^2 + (y - b)^2 + (z - c)^2 + (w - d)^2} < \delta$,
   then $|f(x, y, z, w) - L| < \epsilon$.

Exercises 12.3

1 $f_x(x, y) = 8x^3y^3 - y^2; f_y(x, y) = 6x^4y^2 - 2xy + 3$
3 $f_r(r, s) = \dfrac{r}{(r^2 + s^2)^{1/2}}; f_s(r, s) = \dfrac{s}{(r^2 + s^2)^{1/2}}$
5 $f_x(x, y) = e^y + y \cos x; f_y(x, y) = xe^y + \sin x$
7 $f_t(t, v) = -\dfrac{v}{t^2 - v^2}; f_v(t, v) = \dfrac{t}{t^2 - v^2}$
9 $f_x(x, y) = \cos\dfrac{x}{y} - \dfrac{x}{y}\sin\dfrac{x}{y}; f_y(x, y) = \left(\dfrac{x}{y}\right)^2 \sin\dfrac{x}{y}$
11 $f_x(x, y, z) = 6xz + y^2; f_y(x, y, z) = 2xy;$
   $f_z(x, y, z) = 3x^2$
13 $f_r(r, s, t) = 2re^{2s} \cos t; f_s(r, s, t) = 2r^2e^{2s} \cos t;$
   $f_t(r, s, t) = -r^2e^{2s} \sin t$
15 $f_x(x, y, z) = e^z - ye^x; f_y(x, y, z) = -e^x - ze^{-y};$
   $f_z(x, y, z) = xe^z + e^{-y}$
17 $f_q(q, v, w) = \dfrac{v}{2\sqrt{qv}\sqrt{1 - qv}};$
   $f_v(q, v, w) = \dfrac{q}{2\sqrt{qv}\sqrt{1 - qv}} + w \cos vw;$
   $f_w(q, v, w) = v \cos vw$
19 $w_{xy} = w_{yx} = 4y^3 - 12xy^2$
21 $w_{xy} = w_{yx} = -6x^2e^{-2y} + 2y^{-3} \sin x$

23 $w_{xy} = w_{yx} = -\dfrac{2xz}{y^2}\sinh\dfrac{z}{y}$ **25** $18xy^2 + 16y^3z$

27 $t^2(\sec^2 rt)(\sec^2 rt + \tan^2 rt)$

29 $(1 - x^2y^2z^2)\cos xyz - 3xyz\sin xyz$

31 $w_{rrs} = w_{rsr} = w_{srr} = 36r^2s^2t - 6st^2e^{rt}$

33 Show that $\dfrac{\partial^2 f}{\partial x^2} = \dfrac{y^2 - x^2}{(x^2 + y^2)^2} = -\dfrac{\partial^2 f}{\partial y^2}.$

35 Show that $\dfrac{\partial^2 f}{\partial x^2} = -\cos x \sinh y - \sin x \cosh y = -\dfrac{\partial^2 f}{\partial y^2}.$

37 Show that $\dfrac{\partial^2 w}{\partial x^2} = -\cos(x - y) - \dfrac{1}{(x + y)^2} = \dfrac{\partial^2 w}{\partial y^2}.$

39 Show that $w_{xx} = -c^2e^{-c^2t}\sin cx = w_t.$

41 Show that $\dfrac{\partial^2 v}{\partial t^2} = a^2\left[-k^2(\sin akt)(\sin kx)\right] = a^2\dfrac{\partial^2 v}{\partial x^2}.$

43 Show that $u_x = 2x = v_y$ and $u_y = -2y = -v_x.$

45 Show that $u_x = e^x\cos y = v_y$ and $u_y = -e^x\sin y =$

47 $w_{xx},\ w_{xy},\ w_{xz},\ w_{yx},\ w_{yy},\ w_{yz},\ w_{zx},\ w_{zy},\ w_{zz}$

49 In deg/cm: **(a)** 200   **(b)** 400

51 In volts/in.:   **(a)** $-\dfrac{100}{9}$ **(b)** $\dfrac{50}{9}$ **(c)** $-\dfrac{50}{9}$

53 $\dfrac{\partial C}{\partial x} \approx -36.58\ (\mu\text{g/m}^3)/\text{m}$ is the rate at which the concentration changes in the horizontal direction at $(2, 5)$.

$\dfrac{\partial C}{\partial y} \approx -0.229\ (\mu\text{g/m}^3)/\text{m}$ is the rate at which the concentration changes in the vertical direction at $(2, 5)$.

55 (a) 3.57; 4.81; 4.98; $\displaystyle\lim_{p\to\infty}\dfrac{p}{0.8 + 0.2p} = 5$

(b) $\dfrac{p(p - 1)}{[q + p(1 - q)]^2}$

(c) $p(p - 1)$; as the number of processors increases, the rate of change of the speedup increases.

57 (a) $\partial T/\partial t = T_0\omega e^{-\lambda x}\cos(\omega t - \lambda x)$ is the rate of change of temperature with respect to time at depth x.
$\partial T/\partial x = -T_0\lambda e^{-\lambda x}[\cos(\omega t - \lambda x) + \sin(\omega t - \lambda x)]$ is the rate of change of temperature with respect to the depth at time t.   **(b)** Show that $k = \omega/(2\lambda^2)$

59 (a) $\partial V/\partial x = -0.112y$ mL/yr is the rate at which lung capacity decreases with age for an adult male.

(b) $\partial V/\partial y = 27.63 - 0.112x$ mL/cm is difficult to interpret because we usually think of adult height y as fixed instead of a function of age x.

61 $e_k = -a_{kk}$ for every k

63 $x = 1,\ y = t,\ z = -4t + 12$

65 0.007960028, 0.059882618; 0.007960033, 0.05983345

67 0.00256544, 0.12014571; 0.00256369, 0.12017305

69 1.8369; 4.1743

Exercises 12.4

1 (a) $10y\Delta y - x\Delta y - y\Delta x + 5(\Delta y)^2 - \Delta x\Delta y$

(b) $-y\,dx + (10y - x)\,dy$ **(c)** $\Delta x\Delta y - 5(\Delta y)^2$

Exer. 3–6: The expressions for ϵ_1 and ϵ_2 are not unique.

3 $\epsilon_1 = -3\Delta y;\ \epsilon_2 = 4\Delta y$

5 $\epsilon_1 = 3x\Delta x + (\Delta x)^2;\ \epsilon_2 = 3y\Delta y + (\Delta y)^2$

7 $(3x^2 - 2xy)\,dx + (-x^2 + 6y)\,dy$

9 $(2x\sin y)\,dx + (x^2\cos y + 3y^{1/2})\,dy$

11 $xe^{xy}(xy + 2)\,dx + (x^3e^{xy} - 2y^{-3})\,dy$

13 $[2x\ln(y^2 + z^2)]\,dx + \left(\dfrac{2x^2y}{y^2 + z^2}\right)dy + \left(\dfrac{2x^2z}{y^2 + z^2}\right)dz$

15 $\left[\dfrac{yz(y + z)}{(x + y + z)^2}\right]dx + \left[\dfrac{xz(x + z)}{(x + y + z)^2}\right]dy + \left[\dfrac{xy(x + y)}{(x + y + z)^2}\right]dz$

17 $(2xz - z^2t)\,dx + (4t^3)\,dy + (x^2 - 2xzt)\,dz + (12yt^2 - xz^2)\,dt$

19 7.38   **21** 1.87   **23 (a)** $\pm\dfrac{1}{4}\text{ft}^2$ **(b)** $\pm\dfrac{47}{192}\text{ft}^3$

25 (a) 380 lb   **(b)** $\pm11.5\%$ **27** ±0.0185

29 $\pm\dfrac{6W}{A - W}\%$ **31** $\pm7\%$ **33** 2.96

35 Maximum error in x must not exceed ±2.9 ft.

37 $\pm1.7\pi$ in^2 **39** Use Theorem (12.17).

43 $(x_4, y_4) \approx (1.8460, 1.1546)$

45 $\begin{bmatrix} f_x & f_y & f_z \\ g_x & g_y & g_z \\ h_x & h_y & h_z \end{bmatrix}\begin{bmatrix} \Delta x \\ \Delta y \\ \Delta z \end{bmatrix} = \begin{bmatrix} -f \\ -g \\ -h \end{bmatrix}$

Exercises 12.5

1 $2x\sin(xy) + y(x^2 + y^2)\cos(xy);$
$2y\sin(xy) + x(x^2 + y^2)\cos(xy)$

3 $2r(\ln s)^2 + 8r\ln s + 2s\ln s;$

$\dfrac{2r^2\ln s}{s} + \dfrac{4r^2}{s} + 2r + 2r\ln s$

5 $3x^2e^{3y} + ye^x + 4x^3y^2;\ 3x^3e^{3y} + e^x + 2x^4y$

7 $3\ln(uvt) + 3 + \dfrac{vt}{u};\ t\ln(uvt) + \dfrac{3u}{v} + t;$

$v\ln(uvt) + \dfrac{3u}{t} + v$

9 $-34y + 6r - 24s$ **11** $\dfrac{-3(1 + t^2)}{(t + 1)^4}$

13 $4\sin^3 t\cos t + \tan 4t\sin t - 4\cos t\sec^2 4t$

15 $-\dfrac{6x^2 + 2xy}{x^2 + 3y^2}$ **17** $\dfrac{12\sqrt{xy} + y}{6\sqrt{xy} - x}$

19 $-\dfrac{2z^3 + 2xy^2}{6xz^2 - 6yz + 4};\ -\dfrac{2x^2y - 3z^2}{6xz^2 - 6yz + 4}$

21 $-\dfrac{e^{yz} - 2yze^{xz} + 3yze^{xy}}{xye^{yz} - 2xye^{xz} + 3e^{xy}};\ -\dfrac{xze^{yz} - 2e^{xz} + 3xze^{xy}}{xye^{yz} - 2xye^{xz} + 3e^{xy}}$

23 (a) $0.88\pi \approx 2.76 \text{ in}^3/\text{min}$ **(b)** $0.3\pi \approx 0.94 \text{ in}^2/\text{min}$

25 $\dfrac{dT}{dt} = \dfrac{V}{k}\dfrac{dP}{dt} + \dfrac{P}{k}\dfrac{dV}{dt}$ **27** $-6.4 \text{ in}^3/\text{min}$

29 $762.6 \text{ cm}^2/\text{yr}$ **33** 3   **35** 0

Exercises 12.6

1 $-\dfrac{4}{5}\mathbf{i} + \dfrac{3}{5}\mathbf{j}$ **3** $3\mathbf{i} + 2\mathbf{j}$ **5** $-8\mathbf{i} + \mathbf{j} - 9\mathbf{k}$

7 0.154841, 0.154669, 0.154583

9 $-0.229378, -0.114097, -0.056901$

11 $-\dfrac{10}{\sqrt{2}} \approx -7.07$ **13** $-\dfrac{1}{8\sqrt{13}} \approx -0.03$

15 $\dfrac{67}{8\sqrt{26}} \approx 1.64$ **17** $\dfrac{1}{2\sqrt{26}} \approx 0.098$ **19** $16\sqrt{14} \approx 59.87$

21 $\dfrac{15e^{-2}}{\sqrt{35}} \approx 0.34$ **23** $-\dfrac{12}{\sqrt{10}} \approx -3.79$

25 (a) $-\dfrac{28}{\sqrt{26}}$ **(b)** $\dfrac{1}{\sqrt{5}}\mathbf{i} - \dfrac{2}{\sqrt{5}}\mathbf{j}; \sqrt{80}$

(c) $-\dfrac{1}{\sqrt{5}}\mathbf{i} + \dfrac{2}{\sqrt{5}}\mathbf{j}; -\sqrt{80}$

27 (a) $-\dfrac{25}{7\sqrt{22}}$ **(b)** $-\dfrac{2}{\sqrt{14}}\mathbf{i} + \dfrac{3}{\sqrt{14}}\mathbf{j} + \dfrac{1}{\sqrt{14}}\mathbf{k}; 1$

(c) $\dfrac{2}{\sqrt{14}}\mathbf{i} - \dfrac{3}{\sqrt{14}}\mathbf{j} - \dfrac{1}{\sqrt{14}}\mathbf{k}; -1$

29 (a) $-\dfrac{28}{\sqrt{2}}$ **(b)** The direction of $-12\mathbf{i} - 16\mathbf{j}$

(c) The direction of $12\mathbf{i} + 16\mathbf{j}$

(d) The direction of $4\mathbf{i} - 3\mathbf{j}$

31 (a) $-\dfrac{178}{\sqrt{14}}$ **(b)** The direction of $4\mathbf{i} - 8\mathbf{j} + 54\mathbf{k}$

(c) $\sqrt{2996} \approx 54.7$

33 (b) $\partial T/\partial r$ is the rate of change of temperature in the direction normal to the circular boundary.

35 (b) $\nabla f(1, 2) \approx 1.00003333\mathbf{i} - 0.11111235\mathbf{j}$;

$\nabla f(1, 2) = \mathbf{i} - \dfrac{1}{9}\mathbf{j}$

37 0.1294   **45 (b)** $5 + \sqrt{3}$

Exercises 12.7

1 $16(x - 2) + 6(y + 3) + 6(z - 1) = 0$;
$x = 2 + 16t, y = -3 + 6t, z = 1 + 6t$

3 $16(x + 2) + 18(y + 1) + (z - 25) = 0$;
$x = -2 + 16t, y = -1 + 18t, z = 25 + t$

5 $4(x + 5) - 3(y - 5) + 20(z - 1) = 0$;
$x = -5 + 4t, y = 5 - 3t, z = 1 + 20t$

7 $x + \sqrt{3}\left(y - \dfrac{\pi}{3}\right) + (z - 1) = 0; x = t, y = \dfrac{\pi}{3} + \sqrt{3}t,$
$z = 1 + t$

9 $-x + \dfrac{1}{2}(y - 2) - (z - 1) = 0; x = -t, y = 2 + \dfrac{1}{2}t,$
$z = 1 - t$

11

13

15

17

19
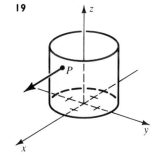

25 $\left(\dfrac{8\sqrt{2}}{\sqrt{5}}, \dfrac{2\sqrt{2}}{\sqrt{5}}, -\dfrac{2\sqrt{2}}{\sqrt{5}}\right),$
$\left(-\dfrac{8\sqrt{2}}{\sqrt{5}}, -\dfrac{2\sqrt{2}}{\sqrt{5}}, \dfrac{2\sqrt{2}}{\sqrt{5}}\right)$

31 (1.2718, 1.1787),
(1.2729, 1.1746)

33 (0.8239, 0.1372),
(0.8210, 0.1384)

Exercises 12.8

1 Max: $f(-2, 1) = 4$ **3** Min: $f\left(\dfrac{1}{2}, -\dfrac{1}{4}\right) = -\dfrac{1}{2}$

5 Min: $f(0, 0) = 0$

7 SP: $(0, 0, f(0, 0))$; min: $f(1, -1) = -1$

9 SP: $(3, -2, f(3, -2))$

11 SP: $(2, 4, f(2, 4)), (-3, -4, f(-3, -4))$;
min: $f(2, -4) = -\dfrac{266}{3}$; max: $f(-3, 4) = \dfrac{617}{6}$

13 SP: $(0, 0, f(0, 0))$; min: $f(4, -8) = -64$,
$f(-1, 2) = -\dfrac{3}{2}$

15 SP: $(-2, -\sqrt{3}, f(-2, -\sqrt{3}))$;
min: $f(-2, \sqrt{3}) = -48 - 6\sqrt{3}$

17 No extrema or saddle points

19 Min: $f(\sqrt[3]{2}, 2\sqrt[3]{2}) = \dfrac{12}{\sqrt[3]{2}}$

21 $(0, 0), (\pm 1, 0), (0, \pm 1)$; min: $f(0, 0) = 0$;
max: $f(0, \pm 1) = \dfrac{3}{e}$

23 Min: $f\left(\dfrac{1}{2}, -\dfrac{1}{4}\right) = -\dfrac{1}{2}$; max: $f\left(-\dfrac{1}{\sqrt{2}}, \dfrac{1}{2\sqrt{2}}\right) = 1 + \sqrt{2}$

25 Min: $f(0, 0) = 0$; max: $f(4, 3) = 67$

27 Min: $f(1, 2) = f(1, -1) = -1$; max: $f(-1, -2) = 13$

29 $\dfrac{1}{\sqrt{26}}$

31 $\left(\dfrac{2}{\sqrt[4]{12}}, \sqrt[4]{12}, \pm\dfrac{2\sqrt{2}}{\sqrt[4]{12}}\right), \left(-\dfrac{2}{\sqrt[4]{12}}, -\sqrt[4]{12}, \pm\dfrac{2\sqrt{2}}{\sqrt[4]{12}}\right)$

33 Square base, altitude $\dfrac{1}{2}$ the length of the side of the base

35 $\dfrac{8}{\sqrt{3}}, \dfrac{6}{\sqrt{3}}, \dfrac{12}{\sqrt{3}}$ **37** $1, \dfrac{4}{3}, 4$

39 Square base of side $\sqrt[3]{4}$ ft, height $2\sqrt[3]{4}$ ft

41 $(18 \text{ in.}) \times (18 \text{ in.}) \times (36 \text{ in.})$

43 $\left(2 - \dfrac{2}{3}\sqrt{3}, 2 - \dfrac{2}{3}\sqrt{3}\right)$ **47** $y = \dfrac{1}{2}x + \dfrac{8}{3}$

49 $y = mx + b$ with $m \approx 1.23$, $b \approx -18.09$; grade of 68

51 $\left(\dfrac{14}{3}, \dfrac{11}{3}\right)$

53 (b) $4x - y - 2z + 1 = 0$

55 $(-0.35, -0.17)$

Exercises 12.9

1 Min: $f\left(\dfrac{1}{\sqrt{5}}, \dfrac{2}{\sqrt{5}}\right) = f\left(-\dfrac{1}{\sqrt{5}}, -\dfrac{2}{\sqrt{5}}\right) = 0$;

max: $f\left(\dfrac{2}{\sqrt{5}}, -\dfrac{1}{\sqrt{5}}\right) = f\left(-\dfrac{2}{\sqrt{5}}, \dfrac{1}{\sqrt{5}}\right) = 5$

3 Min: $f\left(-\dfrac{5}{\sqrt{3}}, -\dfrac{5}{\sqrt{3}}, -\dfrac{5}{\sqrt{3}}\right) = -5\sqrt{3}$;

max: $f\left(\dfrac{5}{\sqrt{3}}, \dfrac{5}{\sqrt{3}}, \dfrac{5}{\sqrt{3}}\right) = 5\sqrt{3}$

5 Min: $f\left(\dfrac{1}{3}, -\dfrac{1}{3}, \dfrac{1}{3}\right) = \dfrac{1}{3}$

7 Min: $f(0, -1, 0) = 1$ and $f(2, 1, 0) = 5$

9 Min: $f\left(1, \dfrac{2}{\sqrt{3}}, -1, \dfrac{8}{3}\right) = -\dfrac{16}{3\sqrt{3}}$;

max: $f\left(1, -\dfrac{2}{\sqrt{3}}, -1, \dfrac{8}{3}\right) = \dfrac{16}{3\sqrt{3}}$

11 $\left(\dfrac{6}{\sqrt{29}}, \dfrac{9}{\sqrt{29}}, \dfrac{12}{\sqrt{29}}\right)$

13 Square base of side $\dfrac{2}{\sqrt[3]{7}}$, height $\dfrac{7}{2\sqrt[3]{7}}$

15 $\dfrac{8}{3}$ **17** Height is twice the radius.

21 Width $= 8\sqrt{3}$ in., depth $= \dfrac{16}{3}\sqrt{6}$ in.

23 Max: $f(0.97, 0.17) \approx 1.55$;
min: $f(-0.87, -0.35) \approx -2.73$

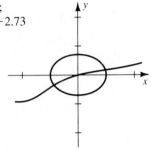

Chapter 12 Review Exercises

1 $\{(x, y): 4x^2 - 9y^2 \le 36\}$;
the hyperbola $9y^2 - 4x^2 = 108$

3 $\{(x, y, z): z^2 > x^2 + y^2\}$;
the hyperboloid of two sheets $z^2 - x^2 - y^2 = 1$

5 $\dfrac{5}{4}$ **7** DNE   **9** Halves of four-leafed roses; DNE

11 $f_x(x, y) = 3x^2 \cos y + 4$; $f_y(x, y) = -x^3 \sin y - 2y$

13 $f_x(x, y, z) = -\dfrac{2x}{y^2 + z^2}$; $f_y(x, y, z) = \dfrac{2y(z^2 - x^2)}{(y^2 + z^2)^2}$;

$f_z(x, y, z) = -\dfrac{2z(x^2 + y^2)}{(y^2 + z^2)^2}$

15 $f_x(x, y, z, t) = 2xz\sqrt{2y + t}$; $f_y(x, y, z, t) = \dfrac{x^2 z}{\sqrt{2y + t}}$;

$f_z(x, y, z, t) = x^2\sqrt{2y + t}$; $f_t(x, y, z, t) = \dfrac{x^2 z}{2\sqrt{2y + t}}$

17 $f_{xx}(x, y) = 6xy^2 + 12x^2$;
$f_{xy}(x, y) = f_{yx}(x, y) = 6x^2 y - 9y^2$; $f_{yy}(x, y) = 2x^3 - 18xy$

21 (a) $(2x + 3y) \Delta x + (3x - 2y) \Delta y + (\Delta x)^2 + 3\Delta x \Delta y - (\Delta y)^2$;
$(2x + 3y) dx + (3x - 2y) dy$ (b) -1.13; -1.1

25 $12x + 18y$; $18x - 22y$ **27** $3e^{-t}(9t^2 \cos 3t^3 - \sin 3t^3)$

29 $\dfrac{z^3 - 2xy}{\cos y - 3xz^2}$; $\dfrac{z \sin y - x^2}{\cos y - 3xz^2}$ **31** (a) $-\dfrac{14}{\sqrt{41}}$ (b) 14

33 $-16(x + 2) + 4(y + 1) - 7(z - 2) = 0$;
$x = -2 - 16t$, $y = -1 + 4t$, $z = 2 - 7t$

37

39 Min: $f(0, -1) = -2$

41 Min: $f\left(-\sqrt{\dfrac{8}{3}}, -\sqrt{\dfrac{2}{3}}, -\sqrt{\dfrac{4}{3}}\right)$

$$= f\left(-\sqrt{\dfrac{8}{3}}, \sqrt{\dfrac{2}{3}}, \sqrt{\dfrac{4}{3}}\right)$$

$$= f\left(\sqrt{\dfrac{8}{3}}, -\sqrt{\dfrac{2}{3}}, \sqrt{\dfrac{4}{3}}\right)$$

$$= f\left(\sqrt{\dfrac{8}{3}}, \sqrt{\dfrac{2}{3}}, -\sqrt{\dfrac{4}{3}}\right) = -\dfrac{8}{9}\sqrt{3};$$

max: $f\left(\sqrt{\dfrac{8}{3}}, \sqrt{\dfrac{2}{3}}, \sqrt{\dfrac{4}{3}}\right) = f\left(-\sqrt{\dfrac{8}{3}}, -\sqrt{\dfrac{2}{3}}, \sqrt{\dfrac{4}{3}}\right)$

$$= f\left(-\sqrt{\dfrac{8}{3}}, \sqrt{\dfrac{2}{3}}, -\sqrt{\dfrac{4}{3}}\right)$$

$$= f\left(\sqrt{\dfrac{8}{3}}, -\sqrt{\dfrac{2}{3}}, -\sqrt{\dfrac{4}{3}}\right)$$

$$= \dfrac{8}{9}\sqrt{3}$$

43 $(0, 4, 4)$ **45** $\sqrt{\dfrac{TC}{k_0 k_1}}$

51 $k = \pm\frac{1}{2}, \pm 1, \pm\frac{3}{2}$

53 (a) $-2.351176, -2.314361, -2.296035$
(b) -2.277764

55 $(4.3418, 3.0227), (4.8003, 2.5887)$

57 $F(\psi, \theta, \varphi) \approx \theta, G(\psi, \theta, \varphi) \approx \varphi, H(\psi, \theta, \varphi) \approx \psi,$
$K(\psi, \theta, \varphi) \approx 1$

CHAPTER ▪ 13

Exercises 13.1

1 R_x **3** R_y **5** Neither   **7** R_x and R_y **9** Neither

11 (a) 39   **(b)** 81   **(c)** 60   **13** -36 **15** $\dfrac{163}{120}$ **17** $\dfrac{36}{5}$

19 $\dfrac{1}{2}(4e - e^4) \approx -21.86$

21 (a) $\displaystyle\int_0^4 \int_0^{\sqrt{x}} f(x, y)\, dy\, dx$ **23 (a)** $\displaystyle\int_0^2 \int_{x^3}^8 f(x, y)\, dy\, dx$

(b) $\displaystyle\int_0^2 \int_{y^2}^4 f(x, y)\, dx\, dy$ **(b)** $\displaystyle\int_0^8 \int_0^{y^{1/3}} f(x, y)\, dx\, dy$

25

(a) $\displaystyle\int_0^1 \int_{x^3}^{\sqrt{x}} f(x, y)\, dy\, dx$

(b) $\displaystyle\int_0^1 \int_{y^2}^{y^{1/3}} f(x, y)\, dx\, dy$

Exer. 27–32: Answers are not unique.

27 $\displaystyle\int_{-1}^4 \int_{-1}^2 (y + 2x)\, dx\, dy = \dfrac{75}{2}$

29 $\displaystyle\int_0^1 \int_{-2y}^{3y} xy^2\, dx\, dy = \dfrac{1}{2}$

31 $\displaystyle\int_0^2 \int_0^{x^2} x^3\cos xy\, dy\, dx = \dfrac{1}{3}(1 - \cos 8) \approx 0.38$

33

(a) $\displaystyle\int_1^2 \int_{9-4x}^{x+4} f(x, y)\, dy\, dx + \int_2^4 \int_{x^3/8}^{x+4} f(x, y)\, dy\, dx$

(b) $\displaystyle\int_1^5 \int_{(9-y)/4}^{2y^{1/3}} f(x, y)\, dx\, dy + \int_5^8 \int_{y-4}^{2y^{1/3}} f(x, y)\, dx\, dy$

35

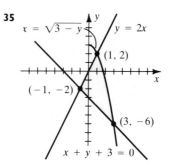

(a) $\displaystyle\int_{-1}^1 \int_{-x-3}^{2x} f(x, y)\, dy\, dx + \int_1^3 \int_{-x-3}^{3-x^2} f(x, y)\, dy\, dx$

(b) $\displaystyle\int_{-6}^{-2} \int_{-y-3}^{\sqrt{3-y}} f(x, y)\, dx\, dy + \int_{-2}^2 \int_{y/2}^{\sqrt{3-y}} f(x, y)\, dx\, dy$

37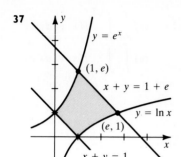

(a) $\displaystyle\int_0^1\int_{1-x}^{e^x}f(x,y)\,dy\,dx+\int_1^e\int_{\ln x}^{1+e-x}f(x,y)\,dy\,dx$

(b) $\displaystyle\int_0^1\int_{1-y}^{e^y}f(x,y)\,dx\,dy+\int_1^e\int_{\ln y}^{1+e-y}f(x,y)\,dx\,dy$

39   **41**

43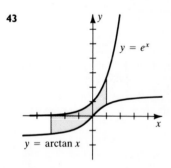

45 $\displaystyle\int_0^2\int_0^{y/2}e^{(y^2)}dx\,dy=\frac14(e^4-1)\approx13.40$

47 $\displaystyle\int_0^4\int_0^{\sqrt x}y\cos x^2\,dy\,dx=\frac14\sin16\approx-0.07$

49 $\displaystyle\int_0^2\int_0^{x^3}\frac{y}{\sqrt{16+x^7}}\,dy\,dx=\frac87$

51 1.10   **53** 2.35   **55** 0.62

Exercises 13.2

1 $\displaystyle\int_0^5\int_{-x}^{4x-x^2}dy\,dx$ **3** $\displaystyle\int_{-3}^3\int_{-\sqrt{9-y^2}}^{(9-y)/2}dx\,dy$

5 $\displaystyle\int_1^2\int_{-x^2}^{1/x^2}dy\,dx=\frac{17}{6}$ **7** $\displaystyle\int_{-1}^2\int_{-y^2}^{y+4}dx\,dy=\frac{33}{2}$

9 $\displaystyle\int_0^1\int_x^{3x}dy\,dx+\int_1^2\int_x^{4-x}dy\,dx=2$

11 $\displaystyle\int_{-\pi}^{\pi}\int_{\sin x}^{e^x}dy\,dx=e^{\pi}-e^{-\pi}\approx23.10$

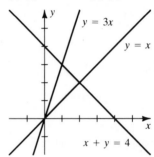

13 $\displaystyle\int_0^3\int_0^{(12-4x)/3}\frac1{12}(60-20x-15y)\,dy\,dx$

15 $\displaystyle\int_0^2\int_0^{4-y^2}(6-x)\,dx\,dy$

17 The plane $z=3$

19 The paraboloid $z = x^2 + y^2$

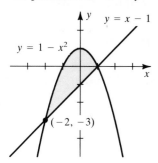

21 $\dfrac{34}{3}$

23 $\displaystyle\int_0^3 \int_0^{2x} (9 - x^2)^{1/2} \, dy \, dx = 18$

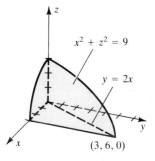

25 $\displaystyle\int_0^2 \int_0^{4-2x} (4 - 2x - y) \, dy \, dx = \dfrac{16}{3}$

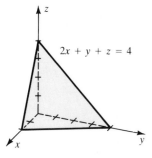

27 $\displaystyle\int_0^2 \int_0^{4-x^2} (x^2 + y^2) \, dy \, dx = \dfrac{832}{35}$

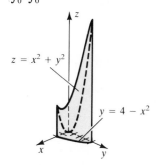

29 $\displaystyle\int_0^4 \int_{x^2/16}^{\sqrt{x}/2} x^3 \, dy \, dx = \dfrac{128}{9}$

31 $\displaystyle\int_{-1}^2 \int_{2-x}^{4-x^2} (x^2 + 4) \, dy \, dx = \dfrac{423}{20}$

33 0.417   **35** 0.344

37 0.791   **39** -1.281

Exercises 13.3

1 $\displaystyle 2\int_0^{\pi/2} \int_0^{4\sin\theta} r \, dr \, d\theta$ **3** $\displaystyle 2\int_0^{2\pi/3} \int_0^{1+2\cos\theta} r \, dr \, d\theta$

5 $\displaystyle 4\int_0^{\pi/4} \int_0^{2\sqrt{\cos 2\theta}} r \, dr \, d\theta$ **7** $\displaystyle \int_0^{\pi/3} \int_0^{4\sin 3\theta} r \, dr \, d\theta = \dfrac{4\pi}{3}$

9 $\displaystyle 2\int_{2\pi/3}^{\pi} \int_3^{2-2\cos\theta} r \, dr \, d\theta = \dfrac{9}{2}\sqrt{3} - \pi \approx 4.65$

11 $\displaystyle 2\int_0^{\pi/4} \int_0^{3\sqrt{\cos 2\theta}} r \, dr \, d\theta = \dfrac{9}{2}$

13 $\displaystyle \int_0^{2\pi} \int_0^2 (r^3) r \, dr \, d\theta = \dfrac{64\pi}{5}$

15 $\displaystyle \int_0^{2\pi} \int_a^b (\cos^2\theta) r \, dr \, d\theta = \dfrac{\pi}{2}(b^2 - a^2)$

17 $\displaystyle \int_0^{\pi/4} \int_0^{3\sec\theta} (r) r \, dr \, d\theta = \dfrac{9}{2}[\sqrt{2} + \ln(\sqrt{2} + 1)] \approx 10.33$

19 $\displaystyle \int_0^{\pi} \int_0^a e^{-r^2} r \, dr \, d\theta = \dfrac{\pi}{2}(1 - e^{-a^2})$

21 $\displaystyle \int_0^{\pi/4} \int_{\sec\theta}^{2\sec\theta} \left(\dfrac{1}{r}\right) r \, dr \, d\theta = \ln(\sqrt{2} + 1) \approx 0.88$

23 $\displaystyle \int_0^{\pi/2} \int_0^2 \cos(r^2) r \, dr \, d\theta = \dfrac{\pi}{4}\sin 4 \approx -0.59$

25 $\displaystyle 8\int_0^{\pi/2} \int_3^5 (25 - r^2)^{1/2} r \, dr \, d\theta = \dfrac{256\pi}{3}$

27 $\displaystyle 2\int_{-\pi/2}^{\pi/2} \int_0^{2\cos\theta} (r) r \, dr \, d\theta = \dfrac{64}{9}$

29 $4\displaystyle\int_0^{\pi/2}\int_0^{4\sin\theta}(16-r^2)^{1/2}r\,dr\,d\theta=\dfrac{128}{9}(3\pi-4)\approx77.15$

31 π

33 $\displaystyle\int_0^2\int_0^{\pi/2}\sqrt{1+r^4}\,r\,d\theta\,dr\approx7.299$

35 $\displaystyle\int_0^3\int_0^{\pi/2}(\cos r)\,r\,d\theta\,dr\approx-2.461$

37 $\displaystyle\int_0^1\int_0^{2\pi}\sqrt{r^6+1}\,r\,d\theta\,dr\approx3.492$

Exercises 13.4

1 $4\displaystyle\int_0^1\int_0^1\sqrt{\left(\dfrac{-x}{\sqrt{4-x^2-y^2}}\right)^2+\left(\dfrac{-y}{\sqrt{4-x^2-y^2}}\right)^2+1}\,dy\,dx$

3 $4\displaystyle\int_0^3\int_0^{\sqrt{9-x^2}}\sqrt{\left(\dfrac{8x}{3\sqrt{16x^2+9y^2+144}}\right)^2+\left(\dfrac{3y}{2\sqrt{16x^2+9y^2+144}}\right)^2+1}\,dy\,dx$

5 $\displaystyle\int_0^1\int_0^1\sqrt{(x)^2+(1)^2+1}\,dy\,dx$

$\quad=\dfrac{1}{2}[\sqrt{3}+2\ln(1+\sqrt{3})-\ln2]\approx1.52$

7 $\pi c k^2\sqrt{\left(\dfrac{1}{a}\right)^2+\left(\dfrac{1}{b}\right)^2+\left(\dfrac{1}{c}\right)^2}$

9 $\dfrac{\pi}{6}(5^{3/2}-1)\approx5.33$

11 $2a^2(\pi-2)$ **13** 247.4 ft^2

15 1.205 **17** 5.800 **19** 1.559

Exercises 13.5

1 $\dfrac{39}{2}$ **3** $-\dfrac{1}{12}$ **5** $\dfrac{513}{8}$

7 $\displaystyle\int_0^6\int_0^{(6-x)/2}\int_0^{(6-x-2y)/3}f(x,y,z)\,dz\,dy\,dx;$

$\displaystyle\int_0^3\int_0^{6-2y}\int_0^{(6-x-2y)/3}f(x,y,z)\,dz\,dx\,dy;$

$\displaystyle\int_0^3\int_0^{(6-2y)/3}\int_0^{6-2y-3z}f(x,y,z)\,dx\,dz\,dy;$

$\displaystyle\int_0^2\int_0^{(6-3z)/2}\int_0^{6-2y-3z}f(x,y,z)\,dx\,dy\,dz;$

$\displaystyle\int_0^6\int_0^{(6-x)/3}\int_0^{(6-x-3z)/2}f(x,y,z)\,dy\,dz\,dx;$

$\displaystyle\int_0^2\int_0^{6-3z}\int_0^{(6-x-3z)/2}f(x,y,z)\,dy\,dx\,dz$

9 $\displaystyle\int_{-3/2}^{3/2}\int_{-\sqrt{9-4x^2}}^{\sqrt{9-4x^2}}\int_0^{9-4x^2-y^2}f(x,y,z)\,dz\,dy\,dx;$

$\displaystyle\int_{-3}^{3}\int_{-\sqrt{9-y^2}/2}^{\sqrt{9-y^2}/2}\int_0^{9-4x^2-y^2}f(x,y,z)\,dz\,dx\,dy;$

$\displaystyle\int_{-3}^{3}\int_0^{9-y^2}\int_{-\sqrt{9-z-y^2}/2}^{\sqrt{9-z-y^2}/2}f(x,y,z)\,dx\,dz\,dy;$

$\displaystyle\int_0^9\int_{-\sqrt{9-z}}^{\sqrt{9-z}}\int_{-\sqrt{9-z-y^2}/2}^{\sqrt{9-z-y^2}/2}f(x,y,z)\,dx\,dy\,dz;$

$\displaystyle\int_{-3/2}^{3/2}\int_0^{9-4x^2}\int_{-\sqrt{9-4x^2-z}}^{\sqrt{9-4x^2-z}}f(x,y,z)\,dy\,dz\,dx;$

$\displaystyle\int_0^9\int_{-\sqrt{9-z}/2}^{\sqrt{9-z}/2}\int_{-\sqrt{9-4x^2-z}}^{\sqrt{9-4x^2-z}}f(x,y,z)\,dy\,dx\,dz$

11 $2\displaystyle\int_0^4\int_0^{4-y}\int_0^{\sqrt{4-z}}dx\,dz\,dy=\dfrac{128}{5}$

13 $\displaystyle\int_{-1}^1\int_{z^2}^{2-z^2}\int_0^{4-z}dx\,dy\,dz=\dfrac{32}{3}$

15 $\displaystyle\int_{-1}^1\int_{-\sqrt{1-y^2}}^{\sqrt{1-y^2}}\int_0^{2-y-z}dx\,dy\,dz=2\pi$

17 $\int_{-3}^{3} \int_{-1}^{2} \int_{0}^{9-x^2} dz\, dy\, dx = 108$

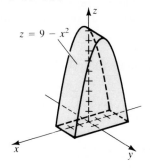

$z = 9 - x^2$

19 $\int_{0}^{1} \int_{x^3}^{x^2} \int_{0}^{z^2} dy\, dz\, dx = \dfrac{1}{70}$

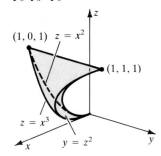

$(1, 0, 1)$ $z = x^2$

$(1, 1, 1)$

$z = x^3$

$y = z^2$

21 $\dfrac{1}{6} abc$

23 The region bounded by the planes $z = 0$, $z = 1$, $x = 2$, $x = 3$ and the cylinders $y = \sqrt{1 - z}$ and $y = \sqrt{4 - z}$

25 The region under the plane $z = x + y$ and over the region in the xy-plane bounded by the parabola $y = x^2$ and the line $y = 2x$

27 The region bounded by the paraboloid $z = x^2 + y^2$ and the planes $z = 1$ and $z = 2$

29 $\int_{0}^{1} \int_{0}^{e^{-x}} y^2\, dy\, dx$

31 $\int_{0}^{2} \int_{0}^{4-2y} \int_{0}^{4-x-2y} (x^2 + y^2)\, dz\, dx\, dy$

33 1.1685×10^9 kg   **35** 0.77

Exercises 13.6

1 $m = \dfrac{2349}{20}$; $\bar{x} = \dfrac{1290}{203}$, $\bar{y} = \dfrac{38}{29}$

3 $m = 8k$ (k a proportionality constant); $\bar{x} = 0$, $\bar{y} = \dfrac{8}{3}$

5 $m = \dfrac{1}{4}(1 - e^{-2}) \approx 0.22$; $\bar{x} = 0$, $\bar{y} = \dfrac{4(1 - e^{-3})}{9(1 - e^{-2})} \approx 0.49$

7 $m = 4 \ln(\sqrt{2} + 1) - 4 \ln(\sqrt{2} - 1) - \pi \approx 3.91$;
$\bar{x} = 0$, $\bar{y} = \dfrac{16 - \pi}{4m} \approx 0.82$

9 $3^5\left(\dfrac{31}{28}\right)$; $3^7\left(\dfrac{19}{8}\right)$; $3^5\left(\dfrac{1259}{56}\right)$ **11** $64k$; $\dfrac{32}{3}k$; $\dfrac{224}{3}k$

13 ($\delta = $ density)   **(a)** $\dfrac{1}{3}a^4\delta$ **(b)** $\dfrac{1}{12}a^4\delta$ **(c)** $\dfrac{1}{6}a^4\delta$ **15** $\dfrac{a}{\sqrt{3}}$

17 $\bar{x} = \bar{y} = \bar{z} = \dfrac{7}{12}a$ (with fixed corner at O)

19 $m = \int_{0}^{4} \int_{-\sqrt{x}/2}^{\sqrt{x}/2} \int_{-\sqrt{x-4z^2}}^{\sqrt{x-4z^2}} (x^2 + z^2)\, dy\, dz\, dx$; the integrals for M_{yz}, M_{xz}, and M_{xy} have the same limits, but the integrands are $x(x^2 + z^2)$, $y(x^2 + z^2)$, and $z(x^2 + z^2)$, respectively.

21 $m = \int_{-a}^{a} \int_{-\sqrt{a^2-x^2}}^{\sqrt{a^2-x^2}} \int_{0}^{\sqrt{a^2-x^2-y^2}} dz\, dy\, dx$; the integral for M_{xy} has the same limits, but the integrand is z. By symmetry, $\bar{x} = \bar{y} = 0$.

23 (a) $m = \int_{0}^{3} \int_{0}^{9-x^2} \int_{0}^{6-2x} dy\, dz\, dx$; the integrals for M_{yz}, M_{xz}, and M_{xy} have the same limits, but the integrands are x, y, and z, respectively.

(b) $m = \dfrac{135}{2}$; $M_{yz} = \dfrac{567}{10}$; $M_{xz} = \dfrac{729}{5}$; $M_{xy} = \dfrac{2673}{10}$; $\bar{x} = \dfrac{21}{25}$, $\bar{y} = \dfrac{54}{25}$, $\bar{z} = \dfrac{99}{25}$

25 $I_z = \int_{-a}^{a} \int_{-\sqrt{a^2-x^2}}^{\sqrt{a^2-x^2}} \int_{-\sqrt{a^2-x^2-y^2}}^{\sqrt{a^2-x^2-y^2}} (x^2 + y^2)(x^2 + y^2 + z^2)\, dz\, dy\, dx$

27 $I_z = \int_{0}^{a} \int_{0}^{b(1-x/a)} \int_{0}^{c(1-x/a-y/b)} (x^2 + y^2)\,\delta\, dz\, dy\, dx$

Exercises 13.7

1 (a) The right circular cylinder of radius 4 with axis along the z-axis   **(b)** The yz-plane
(c) The plane parallel to the xy-plane with z-intercept 1
3 The plane parallel to the yz-plane with x-intercept -3
5 A paraboloid with vertex $(0, 0, 0)$ and opening upward
7 The right circular cylinder with trace $x^2 + (y - 3)^2 = 9$ in the xy-plane
9 The cone $z^2 = 4x^2 + 4y^2$
11 The sphere with center at the origin and radius 3
13 The cylinder with trace $y^2 = 2x$ in the xy-plane and rulings parallel to the z-axis
15 $r^2 + z^2 = 4$ **17** $3r\cos\theta + r\sin\theta - 4z = 12$
19 $r = 2$ **21** $r = 2z$ **23** $r^2\sin^2\theta + z^2 = 9$

25 $\int_{0}^{2\pi} \int_{0}^{4} \int_{0}^{16-r^2} f(r, \theta, z)\, r\, dz\, dr\, d\theta$

27 $\int_{0}^{2\pi} \int_{0}^{3} \int_{0}^{4} f(r, \theta, z)\, r\, dz\, dr\, d\theta\ +$
$\int_{0}^{2\pi} \int_{3}^{4} \int_{0}^{\sqrt{25-r^2}} f(r, \theta, z)\, r\, dz\, dr\, d\theta$

29 (a) 8π **(b)** $\left(0, 0, \dfrac{4}{3}\right)$

31 (a) $\dfrac{1}{2}\pi h a^4\delta$ **(b)** $\pi h a^2\left(\dfrac{1}{4}a^2 + \dfrac{1}{3}h^2\right)\delta$ **33** $\dfrac{1}{4}k\pi^2 a^4$

35 $\frac{1}{8}k\pi^2 a^6$ **37** 215,360 kg   **39** $\frac{7\pi}{16}$

41 (a) $\displaystyle\int_0^{2\pi}\int_0^2\int_0^4 ze^{-(r^2+5r^2\cos\theta\sin\theta)} r\,dz\,dr\,d\theta$ **(b)** 973.947

43 (a) $\displaystyle\int_0^{\pi/2}\int_0^2\int_0^4 \sqrt{r^2+2r^2\cos\theta\sin\theta+z^3}\,r\,dz\,dr\,d\theta$
   (b) 48.848

Exercises 13.8

1 (a) $(0, 2, 2\sqrt{3})$ **(b)** $\left(2, \dfrac{\pi}{2}, 2\sqrt{3}\right)$

3 (a) $\left(\sqrt{10}, \cos^{-1}\!\left(\dfrac{-2}{\sqrt{5}}\right), \dfrac{\pi}{4}\right)$ **(b)** $\left(\sqrt{2}, \dfrac{\pi}{4}, -2\sqrt{2}\right)$

5 (a) The sphere of radius 3 and center O
(b) A half-cone with vertex O and vertex angle $\pi/3$
(c) A half-plane with edge on the z-axis and making an angle of $\pi/3$ with the xz-plane
7 The sphere of radius 2 and center $(0, 0, 2)$
9 The plane $z = 3$
11 The sphere of radius 3 and center $(3, 0, 0)$
13 The plane $y = 5$
15 The right circular cylinder of radius 5 with axis along the z-axis
17 The cone $x^2 + y^2 = 4z^2$
19 The paraboloid $6z = x^2 + y^2$ **21** $\rho = 2$
23 $\rho(3\sin\phi\cos\theta + \sin\phi\sin\theta - 4\cos\phi) = 12$
25 $\rho = 2\csc\phi$ **27** $\tan^2\phi = 4$
29 $\rho^2(\sin^2\phi\sin^2\theta + \cos^2\phi) = 9$

31 $\frac{1}{2}k\pi a^4$ (k a proportionality constant); center of mass is $\frac{2}{5}a$ from base along the axis of symmetry.

33 $\frac{2}{9}k\pi a^6$ **35** $\frac{16\pi}{3}$ **37** $\frac{124}{5}k\pi$

39 $\frac{256\pi}{5}(\sqrt{2}-1) \approx 66.63$

41 (a) $(-3\sqrt{2}, 3\sqrt{6}, -6\sqrt{2})$

   (b) θ should be increased by $105°$, ϕ decreased by $\left(45 + \arctan\dfrac{8}{\sqrt{128}}\right) \approx 80.26°$, and L increased by $\sqrt{192} - 12 \approx 1.86$ in.

43 (a) $\displaystyle\int_0^\pi\int_0^{\pi/2}\int_0^3 \sqrt{\rho^2}\,\rho^2\sin\phi\,d\rho\,d\phi\,d\theta$ **(b)** 63.617

Exercises 13.9

1 (a) Vertical lines; horizontal lines   **(b)** $x = \dfrac{1}{3}u, \; y = \dfrac{1}{5}v$

3 (a) Lines with slopes 1 and $-\dfrac{2}{3}$

   (b) $x = \dfrac{3}{5}u + \dfrac{1}{5}v, \; y = -\dfrac{2}{5}u + \dfrac{1}{5}v$

5 (a) Vertical lines; lines with slope -1
   (b) $x = u^{1/3}, \; y = v - u^{1/3}$
7 (a) Vertical lines; horizontal lines   **(b)** $x = \ln u, \; y = \ln v$
9 (a) The rectangle with vertices $(0, 0)$, $(6, 0)$, $(6, 5)$, $(0, 5)$
   (b) The ellipse $\dfrac{u^2}{9} + \dfrac{v^2}{25} = 1$
11 (a) The triangle with vertices $(0, 0)$, $(-1, 3)$, $(2, 4)$
   (b) The line $-u + 3v = 5$
13 $4u^2 + 4v^2$ **15** $2u(uv - 1)e^{-(2u+v)}$ **17** -6

19 $\displaystyle\int_0^2\int_0^{2-u}(v - u)2\,dv\,du$ **21** $\displaystyle\int_{-1}^1\int_{-\sqrt{1-v^2}}^{\sqrt{1-v^2}}(u^2 + v^2)6\,du\,dv$

23 $\displaystyle\int_1^3\int_{-1}^1 (u^2\cos^2 v)\frac{1}{2}\,du\,dv = \frac{1}{3} + \frac{1}{12}\sin 6 - \frac{1}{12}\sin 2 \approx 0.23$

25 $\displaystyle\int_1^2\int_{\sqrt{u}}^{\sqrt{2u}}\left(\frac{u^2}{v^2} + 2v^2\right)\frac{1}{v}\,dv\,du = \frac{15}{8}$

27 $\displaystyle -\int_2^4\int_{-u/2}^{2u}\left(\frac{v}{u}\right)\left(-\frac{1}{5}\right)dv\,du = \frac{9}{4}$ **29** 1.08×10^{12} km^3

31 (a) $(1, 1, 1)$ **(b)** $\left(\dfrac{1}{\sqrt{2}}, \dfrac{1}{\sqrt{2}}, -\sqrt{2}\right)$

   (c) $\left(\dfrac{1}{\sqrt{2}} - 1, \dfrac{1}{\sqrt{2}} + 1, 0\right)$

37

Chapter 13 Review Exercises

1 $-\dfrac{5}{84}$ **3** $\dfrac{63}{4}$ **5** $-\dfrac{107}{210}$

7 $\displaystyle\int_2^4\int_{-\sqrt{x^2-4}}^{\sqrt{x^2-4}} f(x, y)\,dy\,dx$ **9** $\displaystyle\int_{-2}^2\int_{y^2-4}^{4-y^2} f(x, y)\,dx\,dy$

11

$x = y^3$ $x = e^y$

13

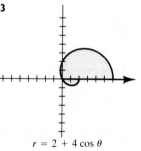

$r = 2 + 4\cos\theta$

15 $\dfrac{1}{4}(1 - e^{-81}) \approx 0.25$ **17** 13   **19** $4\sqrt{2\pi} \approx 17.77$

21 $2\ln 2 - \dfrac{3}{4} \approx 0.64$

23 $\displaystyle\int_0^{\pi/2}\int_0^{\pi/4}\int_0^{4\sqrt{2}}(\rho\cdot\rho^2\sin\phi)\,d\rho\,d\phi\,d\theta = 64(2-\sqrt{2})\pi$
≈ 117.78

25 $9k$ (k a proportionality constant); $\left(\dfrac{9}{4},\dfrac{27}{8}\right)$

27 $2\pi k$ **29** $\dfrac{1}{20}ab^4k$ **31** $\dfrac{256}{15}$; $\left(0,\dfrac{8}{7},\dfrac{12}{7}\right)$

33 $I_y = \displaystyle\int_0^4\int_{-\sqrt{1+y^2}}^{\sqrt{1+y^2}}\int_{-\sqrt{1-x^2+y^2}}^{\sqrt{1-x^2+y^2}}k(x^2+z^2)\sqrt{x^2+z^2}\,dz\,dx\,dy$

35 $\pi a^4 k$

37 Rectangular: $(-3\sqrt{2}, 3\sqrt{2}, 6\sqrt{3})$;
cylindrical: $\left(6, \dfrac{3\pi}{4}, 6\sqrt{3}\right)$

39 $z = 9 - 3x^2 - 3y^2$; a paraboloid with vertex $(0, 0, 9)$ and opening downward

41 $x^2 + y^2 = 16$; a right circular cylinder of radius 4 with axis along the z-axis

43 $\sqrt{x^2 + y^2 + z^2}(\sqrt{x^2 + y^2 + z^2} - 3) = 0$; the sphere of radius 3 with center at the origin, together with its center

45 (a) $z = r^2\cos 2\theta$ (b) $\cos\phi = \rho\sin^2\phi\cos 2\theta$

47 (a) $2r\cos\theta + r\sin\theta - 3z = 4$
(b) $2\rho\sin\phi\cos\theta + \rho\sin\phi\sin\theta - 3\rho\cos\phi = 4$

49 (a) $\displaystyle\int_0^4\int_0^{\sqrt{25-x^2}}dy\,dx$ (b) $\displaystyle\int_0^3\int_0^4 dx\,dy + \int_3^5\int_0^{\sqrt{25-y^2}}dx\,dy$

(c) $\displaystyle\int_0^{\arctan(3/4)}\int_0^{4\sec\theta}r\,dr\,d\theta + \int_{\arctan(3/4)}^{\pi/2}\int_0^5 r\,dr\,d\theta$

51 (a) $\displaystyle\int_0^5\int_0^{\sqrt{25-x^2}}\int_0^{\sqrt{25-x^2-y^2}}dz\,dy\,dx -$
$\displaystyle\int_0^4\int_0^{\sqrt{16-x^2}}\int_3^{\sqrt{25-x^2-y^2}}dz\,dy\,dx$

(b) $\displaystyle\int_0^{\pi/2}\int_0^4\int_0^3 r\,dz\,dr\,d\theta + \int_0^{\pi/2}\int_4^5\int_0^{\sqrt{25-r^2}}r\,dz\,dr\,d\theta$

(c) $\displaystyle\int_0^{\pi/2}\int_0^{\arctan(4/3)}\int_0^{3\sec\phi}\rho^2\sin\phi\,d\rho\,d\phi\,d\theta +$
$\displaystyle\int_0^{\pi/2}\int_{\arctan(4/3)}^{\pi/2}\int_0^5\rho^2\sin\phi\,d\rho\,d\phi\,d\theta$

53 $e - 1 \approx 1.72$

CHAPTER ▪ 14

Exercises 14.1

1

3

5

7

9
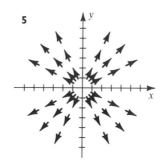

13 $\mathbf{F}(x, y, z) = 2x\mathbf{i} - 6y\mathbf{j} + 8z\mathbf{k}$

15 $\mathbf{F}(x, y) = \dfrac{1}{1 + x^2y^2}(y\mathbf{i} + x\mathbf{j})$

17 $\mathbf{i} + x^2\mathbf{j} + y^2\mathbf{k}$; $2xz + 2xy + 2$

19 $-y^2\cos z\mathbf{i} + (6xyz - e^{2z})\mathbf{j} - 3xz^2\mathbf{k}$;
$3yz^2 + 2y\sin z + 2xe^{2z}$

39 0.145 **41** -1.807

Exercises 14.2

1 $14(2^{3/2} - 1) \approx 25.60$; 21; 14   **3** 3.8185; 3.1918; 0.2550

5 $\dfrac{34}{7}$ **7** $-\dfrac{16}{3}$ **9** -0.060

11 (a) $\dfrac{15}{2}$ (b) 6   (c) 7   (d) $\dfrac{29}{4}$

13 $\dfrac{1}{12}(3e^4 + 6e^{-2} - 12e + 8e^3 - 5) \approx 23.97$

15 (a) 19   (b) 35   (c) 27   **17** $\dfrac{3}{2}\sqrt{14}$

19 $\dfrac{9}{2}$ (for all paths)   **21** 0   **23** $\dfrac{412}{15}$ **27** $\bar{x} = 0, \bar{y} = \dfrac{1}{4}\pi a$

31 $I_x = \dfrac{4}{3}ka^4$; $I_y = \dfrac{2}{3}ka^4$

33 If the density at (x, y, z) is $\delta(x, y, z)$, then
$I_x = \int_C(y^2 + z^2)\,\delta(x, y, z)\,ds$,
$I_y = \int_C(x^2 + z^2)\,\delta(x, y, z)\,ds$, and
$I_z = \int_C(x^2 + y^2)\,\delta(x, y, z)\,ds$.

35 -0.1584 **37** 18.8815

Exercises 14.3

1 $f(x, y) = x^3y + 2x + y^4 + c$
3 $f(x, y) = x^2\sin y + 4e^x + c$

5 $f(x, y) = 2y^3 \cos x + 5y + c$
7 $f(x, y, z) = 4x^2z + y - 3y^2z^3 + c$
9 $f(x, y, z) = y \tan x - ze^x + c$

Exer. 11–14: A potential function f is given along with the value of the integral.

11 $f(x, y) = xy^2 + x^2y$; 14
13 $f(x, y, z) = 3x^2y^3 + 2xz^2 + z$; -31
15 $\dfrac{\partial}{\partial y}(4xy^3) \ne \dfrac{\partial}{\partial x}(2xy^3)$ **17** $\dfrac{\partial}{\partial y}(e^x) \ne \dfrac{\partial}{\partial x}(3 - e^x \sin y)$
21 $\dfrac{\partial N}{\partial z} \ne \dfrac{\partial P}{\partial y}$
23 $f(x, y, z) = -\dfrac{1}{2}c \ln(x^2 + y^2 + z^2) + d$, where $c > 0$ and d is a constant
27 This does not violate Theorem (14.16) since D is not simply connected. M and N are not continuous at $(0, 0)$.
29 $W = c\,\dfrac{d_2 - d_1}{d_1 d_2}$

Exercises 14.4

1 $-\dfrac{7}{60}$ **3** $\dfrac{2}{3}$ **5** π **7** -3 **9** 0   **11** 0
13 -3π **15** $\dfrac{128}{3}$ **17** $6 - \dfrac{5}{2}\ln 5 \approx 1.98$ **19** $\dfrac{3}{8}\pi a^2$
23 Green's theorem does not apply since M and N are undefined at $(0, 0)$ and hence are not continuous everywhere inside the unit circle.
27 $\bar{x} = 0$, $\bar{y} = \dfrac{4a}{3\pi}$

Exercises 14.5

1 $\dfrac{2}{3}\pi a^4$ **3** $5\sqrt{14}$
5 (a) $\displaystyle\int_0^4 \int_0^{(12-3y)/4} \dfrac{1}{2}(12 - 3y - 4z)y^2z^3\left(\dfrac{1}{2}\sqrt{29}\right) dz\, dy$
(b) $\displaystyle\int_0^3 \int_0^{6-2z} x\left[\dfrac{1}{3}(12 - 2x - 4z)\right]^2 z^3\left(\dfrac{1}{3}\sqrt{29}\right) dx\, dz$
7 (a) $\displaystyle\int_0^8 \int_0^6 \left(4 - 3y + \dfrac{1}{16}y^2 + z\right)\left(\dfrac{1}{4}\sqrt{17}\right) dz\, dy$
(b) $\displaystyle\int_0^2 \int_0^6 [x^2 - 2(8 - 4x) + z]\sqrt{17}\, dz\, dx$
9 Since $\iint_S g(x, y, z)\, dS = c\iint_R dA$, the value of the integral equals the volume of a cylinder of altitude c, with rulings parallel to the z-axis, whose base is the projection of S on the xy-plane.
11 $2\pi a^3$ **13** 3π **15** 18   **17** 8
21 (a) $\dfrac{255}{2}\pi\sqrt{2}$; $\bar{x} = \bar{y} = 0$, $\bar{z} = \dfrac{1364}{425}$ **(b)** $1365\pi\sqrt{2}$

Exercises 14.6

1 24   **3** 20π **5** 0   **7** $136\pi/3$ **9** 24
11 Both integrals equal $4\pi a^3$.   **13** Both integrals equal 4π.
27 625π lb upward

Exercises 14.7

1 Both integrals equal -1.   **3** Both integrals equal πa^2.
5 0   **7** -8π
9 The curl meter rotates counterclockwise for $0 < y < 1$ and clockwise for $1 < y < 2$. There is no rotation if $y = 1$. curl $\mathbf{F} = 2(1 - y)\mathbf{k}$; $|(\text{curl }\mathbf{F}) \cdot \mathbf{k}| = |2(1 - y)|$ has a maximum value 2 at $y = 0$ and $y = 2$ and a minimum value 0 at $y = 1$.

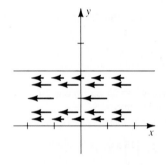

11 Typical field vectors are shown in Figure 14.5. A curl meter rotates counterclockwise for every $(x, y) \ne (0, 0)$. curl $\mathbf{F} = 2\mathbf{k}$; $|(\text{curl }\mathbf{F}) \cdot \mathbf{k}| = 2$ for every (x, y).

Chapter 14 Review Exercises

1   **3**

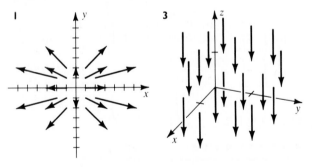

5 $\mathbf{F}(x, y) = (y^2 \sec^2 x)\mathbf{i} + (2y \tan x)\mathbf{j}$
9 -8 **11** $-\dfrac{56}{5}$ **13** $\dfrac{1}{144}(1025^{3/2} - 17^{3/2}) \approx 227.40$
15 70   **17** 0   **19** 25.905; 12.168; 22.744   **21** 2.361
23 $\dfrac{25}{2}$ **25** $f(x, y, z) = x^2e^{2y} + y^2 \cot z + c$ **27** $\dfrac{1}{6}$
29 $3x^2z^4 + xz^2$;
$(2x^2y - 2xyz)\mathbf{i} + (4x^3z^3 - 2xy^2)\mathbf{j} + (yz^2)\mathbf{k}$
31 $\dfrac{1}{5}\sqrt{3}$ **33** $\dfrac{5\pi}{2}$ **35** Both integrals equal 8π.

CHAPTER ■ 15

Exercises 15.1

1 (a) $y = x^3 + C$

3 (a) $y = \sqrt{4 - x^2} + C$

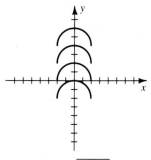

(b) $y = x^3 + 2$ **(b)** $y = \sqrt{4 - x^2}$

11 $y = Ce^{2\sin x}$ **13** $y = Cx$ **15** $x^3 e^y y^5 = C$

17 $y = -1 + Ce^{(x^2/2)-x}$ **19** $y = -\dfrac{1}{3}\ln(3C + 3e^{-x})$

21 $y^2 = C(1 + x^3)^{-2/3} - 1$

23 $x \sin x + \cos x - \ln|\sin y| = C$

25 $\sec x + e^{-y} = C$ **27** $y^2 + \ln y = 3x - 8$

29 $y = \ln(2x + \ln x + e^2 - 2)$ **31** $y = 2e^{2-\sqrt{4+x^2}} - 1$

33 $\tan^{-1} y - \ln|\sec x| = \dfrac{\pi}{4}$ **35** $xy = k$; hyperbolas

37 $2x^2 + y^2 = k$; ellipses   **39** $2x^2 + 3y^2 = k$; ellipses

41 (a)

x	y
1.0	0.5600
1.2	0.4747
1.4	0.4063
1.6	0.3528
1.8	0.3107
2.0	0.2772
2.2	0.2500
2.4	0.2276
2.6	0.2088
2.8	0.1928
3.0	0.1791

43 (a)

x	y
0.0	0.1200
0.2	0.1800
0.4	0.2414
0.6	0.3092
0.8	0.3913
1.0	0.5006
1.2	0.6565
1.4	0.8867
1.6	1.2116
1.8	1.5702
2.0	1.7418

45 (a)

x	y
1.0	0.5600
1.2	0.4832
1.4	0.4213
1.6	0.3720
1.8	0.3323
2.0	0.2999
2.2	0.2732
2.4	0.2507
2.6	0.2316
2.8	0.2151
3.0	0.2009

47 (a)

x	y
0.0	0.1200
0.2	0.1807
0.4	0.2453
0.6	0.3204
0.8	0.4173
1.0	0.5540
1.2	0.7578
1.4	1.0507
1.6	1.3772
1.8	1.5824
2.0	1.6256

Exercises 15.2

1 $y = \dfrac{1}{4}e^{2x} + Ce^{-2x}$ **3** $y = \dfrac{1}{2}x^5 + Cx^3$

5 $y = \dfrac{e^x}{x} - \dfrac{1}{2}x + \dfrac{C}{x}$ **7** $y = \dfrac{e^x + C}{x^2}$

9 $y = \dfrac{4}{3}x^3 \csc x + C \csc x$ **11** $y = 2 \sin x + C \cos x$

13 $y = x \sin x + Cx$ **15** $y = \left(\dfrac{1}{3}x + \dfrac{C}{x^2}\right)e^{-3x}$

17 $y = \dfrac{3}{2} + Ce^{-x^2}$ **19** $y = \dfrac{1}{2}\sin x + \dfrac{C}{\sin x}$

21 $y = \dfrac{1}{3} + (x + C)e^{-x^3}$ **23** $y = x(x + \ln x + 1)$

25 $y = e^{-x}(1 - x^{-1})$ **27** $Q = CV(1 - e^{-t/RC})$

29 $f(t) = \dfrac{80}{3}(1 - e^{-0.075t}) + Ke^{-0.075t}$

31 (a) $f(t) = M + (A - M)e^{k(1-t)}$ (k a constant)
(b) 28 items

33 $y = y_L(1 - ce^{-kt})$, $k > 0$

35 (b) $y = \dfrac{I}{k}(1 - e^{-kt})$; $\dfrac{I}{k}$ **(c)** 0.58 mg/min

37 -3.23103

39 (a)

x	y
1.0	0.5600
1.4	0.4209
1.8	0.3318
2.2	0.2728
2.6	0.2312
3.0	0.2006

41 (a)

x	y
0.0	0.1200
0.4	0.2445
0.8	0.4148
1.2	0.7547
1.6	1.3881
2.0	1.6540

43

n	y	E_n	$E_{n/2}/E_n$
4	7.38624259	5.2×10^{-5}	—
8	7.38629082	3.5×10^{-6}	15
16	7.38629413	2.3×10^{-7}	15
32	7.38629435	1.5×10^{-8}	15

Exercises 15.3

1 $y = C_1 e^{2x} + C_2 e^{3x}$ **3** $y = C_1 + C_2 e^{3x}$

5 $y = C_1 e^{-2x} + C_2 x e^{-2x}$ **7** $y = C_1 e^{(2+\sqrt{3})x} + C_2 e^{(2-\sqrt{3})x}$

9 $y = C_1 e^{-\sqrt{2}x} + C_2 x e^{-\sqrt{2}x}$ **11** $y = C_1 e^{-3x/2} + C_2 e^{5x/4}$

13 $y = C_1 e^{4x/3} + C_2 x e^{4x/3}$

15 $y = C_1 e^{(2+\sqrt{2})x/2} + C_2 e^{(2-\sqrt{2})x/2}$

17 $y = e^x(C_1 \cos x + C_2 \sin x)$

19 $y = e^{2x}(C_1 \cos 3x + C_2 \sin 3x)$

21 $y = C_1 e^{(-3+\sqrt{7})x} + C_2 e^{(-3-\sqrt{7})x}$ **23** $y = -2e^x + 2e^{2x}$

25 $y = \cos x + 2 \sin x$ **27** $y = e^{-4x}(2 + 9x)$

29 $y = \dfrac{1}{2}e^x \sin 2x$

Exercises 15.4

1 $y = C_1 \cos x + C_2 \sin x - \cos x \ln|\sec x + \tan x|$

3 $y = \left(C_1 + C_2 x + \dfrac{1}{12}x^4\right)e^{3x}$

5 $y = C_1 e^x + C_2 e^{-x} + \dfrac{2}{5}e^x \sin x - \dfrac{1}{5}e^x \cos x$

7 $y = \left(C_3 + \dfrac{1}{6}x\right)e^{3x} + C_2 e^{-3x}$, where $C_3 = C_1 - \dfrac{1}{36}$

9 $y = C_1 e^{-x} + C_2 e^{4x} - \dfrac{1}{2}$ **11** $y = C_1 e^x + C_2 e^{2x} + \dfrac{2}{3}e^{-x}$

13 $y = C_1 + C_2 e^{-2x} + \dfrac{1}{8}\sin 2x - \dfrac{1}{8}\cos 2x$

15 $y = C_1 e^x + C_2 e^{-x} + \dfrac{1}{9}(-4 + 3x)e^{2x}$

17 $y = e^{3x}(C_1 \cos 2x + C_2 \sin 2x) +$

$$\dfrac{1}{65}e^x(7 \cos x - 4 \sin x)$$

21 0.776805

Exercises 15.5

1 $y = -\dfrac{1}{3}\cos 8t$ **3** $y = \dfrac{1}{8}\sqrt{2}e^{-8t}(e^{4\sqrt{2}t} - e^{-4\sqrt{2}t})$

5 $y = \dfrac{1}{3}e^{-8t}(\sin 8t + \cos 8t)$

7 If m is the mass of the weight, then the spring constant is $24m$ and the damping force is $-4m\dfrac{dy}{dt}$. The motion is begun by releasing the weight from 2 ft above the equilibrium position with an initial velocity of 1 ft/sec in the upward direction.

9 $-6\sqrt{2}\dfrac{dy}{dt}$

11 (a) Overdamped: 2.3, 2.4; underdamped: 1.7, 1.8, 1.9, 2.0, 2.1, 2.2

Chapter 15 Review Exercises

1 $\sin x - x \cos x + e^{-y} = C$ **3** $y = \tan(\sqrt{1 - x^2} + C)$

5 $y = \dfrac{2x - 2 \cos x + C}{\sec x + \tan x}$ **7** $y = 2 \sin x + C \cos x$

9 $\sqrt{1 - y^2} + \sin^{-1} x = C$ **11** $\csc y = e^{-x} + C$

13 $y = \dfrac{1}{2} + Ce^{-2\sin x}$ **15** $y = (C_1 + C_2 x)e^{4x}$

17 $y = C_1 + C_2 e^{2x}$

19 $y = C_1 e^{-x} + C_2 e^x - \dfrac{1}{5}e^x(2 \cos x + \sin x)$

21 $y = \dfrac{1}{5}e^{4x} + Ce^{-x}$ **23** $y = C_1 e^x + C_2 e^{2x} + \dfrac{1}{12}e^{5x}$

25 $y = \dfrac{(x - 2)^3}{3x} + \dfrac{C}{x}$

27 $y = e^{-5x/2}\left[C_1 \cos\left(\dfrac{1}{2}\sqrt{3}x\right) + C_2 \sin\left(\dfrac{1}{2}\sqrt{3}x\right)\right]$

29 $\dfrac{1}{2}e^x(\sin x - \cos x) + e^{-y} = C$

31 $y = \dfrac{1}{2}\cos x + C \sec x$

33 $y = \dfrac{\ln|\sec x + \tan x| - x + C}{\csc x - \cot x}$

35 $y = \dfrac{3500}{3}e^{(3/(2\pi))(1-\cos 2\pi t)} - \dfrac{2000}{3}$; 2366

37 $f(t) = \dfrac{ab[e^{k(b-a)t} - 1]}{be^{k(b-a)t} - a}$

39 $y\,dy + x\,dx = 0$; a circle with center at the origin

41 (a)

x	y
0.0	1.0200
0.2	1.1212
0.4	1.1909
0.6	1.2532
0.8	1.3237
1.0	1.4071
1.2	1.4958
1.4	1.5704
1.6	1.5993
1.8	1.5423
2.0	1.3781

43 (a)

x	y
0.0	1.0200
0.2	1.1068
0.4	1.1776
0.6	1.2478
0.8	1.3256
1.0	1.4110
1.2	1.4939
1.4	1.5526
1.6	1.5537
1.8	1.4680
2.0	1.3077

45 (a) 0.132956, 0.140085, 0.144016
(b) 0.149392, 0.148477, 0.148247
(c) 0.148170, 0.148170, 0.148170

ALGEBRA

EXPONENTS AND RADICALS

$$a^m a^n = a^{m+n} \qquad a^{m/n} = \sqrt[n]{a^m} = (\sqrt[n]{a})^m$$

$$(a^m)^n = a^{mn} \qquad \sqrt[n]{ab} = \sqrt[n]{a}\,\sqrt[n]{b}$$

$$(ab)^n = a^n b^n \qquad \sqrt[n]{\frac{a}{b}} = \frac{\sqrt[n]{a}}{\sqrt[n]{b}}$$

$$\left(\frac{a}{b}\right)^n = \frac{a^n}{b^n} \qquad \sqrt[m]{\sqrt[n]{a}} = \sqrt[mn]{a}$$

$$\frac{a^m}{a^n} = a^{m-n} \qquad a^{-n} = \frac{1}{a^n}$$

ABSOLUTE VALUE $(d > 0)$

$|x| < d$ if and only if $-d < x < d$

$|x| > d$ if and only if either $x > d$ or $x < -d$

$|a + b| \le |a| + |b|$ (Triangle inequality)

$-|a| \le a \le |a|$

INEQUALITIES

If $a > b$ and $b > c$, then $a > c$

If $a > b$, then $a + c > b + c$

If $a > b$ and $c > 0$, then $ac > bc$

If $a > b$ and $c < 0$, then $ac < bc$

QUADRATIC FORMULA

If $a \ne 0$, the roots of $ax^2 + bx + c = 0$ are

$$x = \frac{-b \pm \sqrt{b^2 - 4ac}}{2a}$$

LOGARITHMS

$y = \log_a x$ means $a^y = x \qquad \log_a 1 = 0$

$\log_a xy = \log_a x + \log_a y \qquad \log_a a = 1$

$\log_a \dfrac{x}{y} = \log_a x - \log_a y \qquad \log x = \log_{10} x$

$\log_a x^r = r \log_a x \qquad\qquad \ln x = \log_e x$

BINOMIAL THEOREM

$$(x + y)^n = x^n + \binom{n}{1}x^{n-1}y + \binom{n}{2}x^{n-2}y^2 +$$

$$\cdots + \binom{n}{k}x^{n-k}y^k + \cdots + y^n,$$

where $\dbinom{n}{k} = \dfrac{n!}{k!(n-k)!}$

ANALYTIC GEOMETRY

DISTANCE FORMULA

$$d(P_1, P_2) = \sqrt{(x_2 - x_1)^2 + (y_2 - y_1)^2}$$

EQUATION OF A CIRCLE

$$(x - h)^2 + (y - k)^2 = r^2$$

SLOPE m OF A LINE

$$m = \frac{y_2 - y_1}{x_2 - x_1}$$

POINT–SLOPE FORM

$$y - y_1 = m(x - x_1)$$

SLOPE–INTERCEPT FORM

$$y = mx + b$$

GRAPH OF A QUADRATIC FUNCTION

$$y = ax^2, \, a > 0 \qquad\qquad y = ax^2 + bx + c, \, a > 0$$